KU-688-027

Organometallics in Synthesis
A Manual

Organometallics in Synthesis
A Manual

Edited by

Manfred Schlosser
Université de Lausanne, Switzerland

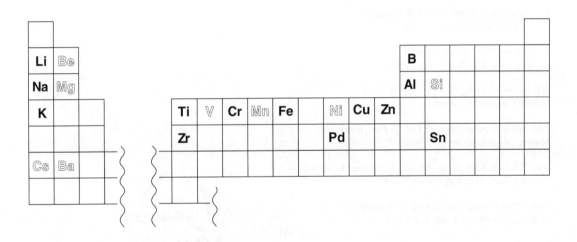

JOHN WILEY & SONS, LTD

Copyright © 2002 John Wiley & Sons Ltd.,
Baffins Lane, Chichester,
West Sussex PO19 1UD, England

Telephone (+44) 1243 779777

Email (for orders and customer service enquiries): cs-books@wiley.co.uk
Visit our Home Page on www.wiley.co.uk or www.wiley.com

Reprinted July 2002

All Rights Reserved. No part of this publication may be reproduced, stored in a retrieval system, or
transmitted, in any form or by any means, electronic, mechanical, photocopying, recording, scanning or
otherwise, except under the terms of the Copyright, Designs and Patents Act 1988 or under the terms of
a licence issued by the Copyright Licensing Agency Ltd, 90 Tottenham Court Road, London W1P
0LP, UK, without the permission in writing of the Publisher. Requests to the Publisher should be
addressed to the Permissions Department, John Wiley & Sons Ltd, Baffins Lane, Chichester, West
Sussex, PO19 1UD, England, or emailed to permreq@wiley.co.uk, or faxed to (+44) 1243 770571.

This publication is designed to provide accurate and authoritative information in regard to the subject
matter covered. It is sold on the understanding that the Publisher is not engaged in rendering
professional services. If professional advice or other expert assistance is required, the services of a
competent professional should be sought.

Other Wiley Editorial Offices

John Wiley & Sons Inc., 605 Third Avenue,
New York, NY 10158-0012, USA

Jossey-Bass, 989 Market Street,
San Francisco, CA 94103-1741, USA

WILEY-VCH Verlag GmbH, Pappelallee 3,
D-69469 Weinheim, Germany

John Wiley & Sons Australia Ltd, 33 Park Road,
Milton, Queensland 4064, Australia

John Wiley & Sons (Asia) Pte Ltd, 2 Clementi Loop #02-01,
Jin Xing Distripark, Singapore 129809

John Wiley & Sons Canada Ltd, 22 Worcester Road,
Etobicoke, Ontario, Canada M9W 1L1

Library of Congress Cataloging-in-Publication Data

British Library Cataloguing in Publication Data

A catalogue record for this book is available from the British Library

ISBN 0 471 98416 7

Typeset in 10/12pt Times by Laserwords Private Limited, Chennai, India
Printed and bound in Great Britain by Antony Rowe Limited, Chippenham, UK
This book is printed on acid-free paper responsibly manufactured from sustainable forestry
in which at least two trees are planted for each one used for paper production.

Contents

List of Contributors

LOUIS S. HEGEDUS — Colorado State University, Fort Collins, Colorado, USA

BRUCE H. LIPSHUTZ — Department of Chemistry, University of California, Santa Barbara, California, USA

JAMES A. MARSHALL — Department of Chemistry, University of Virginia, Charlottesville, Virginia, USA

EIICHI NAKAMURA — Department of Chemistry, The University of Tokyo, Tokyo, Japan

EI-ICHI NEGISHI — Department of Chemistry, Purdue University, West Lafayette, Indiana, USA

MANFRED T. REETZ — Max-Planck-Institue für Kohlenforschung, Mülheim/Ruhr, Germany

MANFRED SCHLOSSER — Scetion de Chimie, Université Lausanne, Switzerland

MARTIN F. SEMMELHACK — Department of Chemistry, Princeton University, Princeton, USA

KEITH SMITH — Department of Chemistry, University of Wales, Swansea, UK

HISASHI YAMAMOTO — School of Engineering, Nagoya University, Japan

Preface to the First Edition

The origin of this Manual, was a series of three post-graduate workshops which I had the privilege to host in the late eighties. These five-day seminars provided ample opportunity for the participants to talk to the lecturers, solve exercises and, in 1987 at least, to practise their new skills in the laboratory. One common feature became apparent in all such interactions between tutors and novices: the psychological and practical barrier for newcomers to enter the field of organometallic chemistry is still very high.

Therefore, this Manual is meant primarily for those researchers who have not yet had a chance to familiarize themselves with the basic concepts and techniques of organometallic reactions. It is a nuts-and-bolts text: full of useful hints, rules of thumb and, last but not least, carefully selected working procedures. Thus, it summarizes what organometallic reagents can do for modern organic synthesis and at the same time explains how they are conveniently employed.

However, as Ludwig Boltzmann recognized, 'nothing is more practical than a good theory'. The message applied to our case is that sound mechanistic insight is required if we wish not only to document the course of known reactions but also predict the outcome of future experiments. Therefore, despite its down-to-earth approach, this Manual puts great emphasis on the presentation of first principles that can provide a rational basis for understanding organometallic reactivity, a fascinating and at the same time still widely mysterious subject.

In order not to discourage the reader by too voluminous a book, we had to restrict the coverage to a selection of the most popular metals and methods. Hence, many important reagents and topics had to be neglected for the moment. They may, however, be included in a future, more comprehensive edition.

It is now my pleasure to express by profound gratitude to my co-authors who, notwithstanding their numerous other commitments, have agreed to share their expertise with the reader and thus to contribute to the propagation of organometallics in synthesis. I wish also to acknowledge the help of my coworkers who have checked many working procedures. Finally, I am indebted to my wife Elsbeth, who has made the Chem-Art china ink drawings (contributions Nos 1, 2, 3 and 8) with artistic skill and aesthetic taste.

Lausanne, Spring 1994   Manfred Schlosser

Preface to the Second Edition

Compared with the first edition of the *Manual*, the second has almost doubled in size. Thus, this monograph on organometallics is far from just a remake or update of a successful earlier version but, as the Publisher notes, is essentially a new book. Four additional chapters have been included featuring the chemistry of tin, zinc, zirconium, and chromium and iron. Such an extension of useful organometallic procedures reflects the continuing evolution of organic synthesis, which increasingly relies on a greater percentage of the Periodic Table. Previous chapters have been extensively reworked and updated. For example, the number of Tables to be found in the chapter covering alkali and alkali-earth derived reagents has increased from 12 to 166.

Nonetheless, maintained throughout is the original, underlying concept behind creation of the Manual. That is, it strives to offer guidance in the form of countless "cookbook recipes," along with recommendations on associated equipment needs as well as practical advice at critical stages. At the same time, indispensable mechanistic insight is provided which is crucial to those who wish to apply these existing tools rationally and to contribute to the further development of novel reagents and methodology.

Although not always easy, the role of the coordinating Editor can be quite rewarding. I have enjoyed and appreciated the numerous comments and suggestions colleagues and reviewers have shared with me. A good deal of the recent accomplishments have to be credited to them. To name at least one person, I wish to mention the invaluable input made by my friend Bruce Lipshutz who was always available when help or advice was sought.

Lausanne, Summer 2001   Manfred Schlosser

I

Organoalkali Chemistry

MANFRED SCHLOSSER

Section de Chimie (BCh), Université, CH-1015 Lausanne, Switzerland

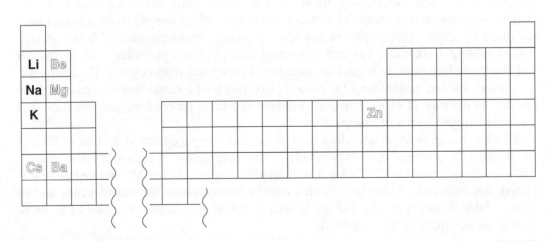

Organometallics in Synthesis: A Manual. Edited by Manfred Schlosser.
© 2002 John Wiley & Sons Ltd

Hints and Road Maps

Nomenclature and Formulas

It was not intended to apply systematic IUPAC rules strictly and at all events. On the other hand, we feel the time has come to abolish some anachronisms. The Old-German '*n*-alkyl' having never been part of an authorized nomenclature, should be definitively abrogated and $LiCH_2CH_2CH_2CH_3$ be unequivocally named butyllithium.

The formula drawings shown try to reflect the structure of the various organometallic species as faithfully as possible. There is one major exception. For easy comparison with their precursors, *N*-deprotonated carboxamides and carbamates (see, *e.g.*, Tables 121 and 124) are frequently represented as lithium *N*-acylamides rather than *O*-lithio azaenolates, as the true position of the metallomeric equilibrium would suggest:

Coverage of this Chapter

This Chapter deals with all kinds of organometallic species containing a carbon–alkali or a carbon–alkaline earth bond. However, weakly basic species ($pK_a \leq 30$) such as acetylides, cyclopentadienides, indenides, fluorenides and d-orbital resonance stabilized compounds (such as α-metalated dithioacetals) are only marginally treated, if at all. Enolates and *N*-deprotonated pyrroles, indoles *etc.* or α-deprotonated Schiff bases and picolines, sulfones and phosphonates as any other derivative carrying the metal at a hetero rather than a carbon atom are almost totally omitted from the present review.

Generation of Organometallic Species: Ordering Principles and Tables

The most voluminous part of this Chapter is Section 4, which summarizes the various methods available for the generation of reactive organometallic intermediates. The contents within each subcategory always follow in the same order: we start with sp^3-centers, first considering saturated aliphatic structures before turning to unsaturated ones stabilized by allylic, propargylic or benzylic resonance, and terminate with sp^2-centers, first examining 2-alkenylmetals and ferrocenylmetals before continuing with arylmetals and finally ending with five- and six-membered metalated heterocycles. If substituents are present, the one maintaining the closest proximity to the metal-bearing center has the priority. In the case of equal distances, nitrogen functions preceed oxygen functions and the latter halogen atoms or groups.

All three 'navigational charts' displayed on the opposite page refer to Section 4. Table 1 summarizes the principal options available for the preparation of organometallic species. Table 2 surveys over the 52 Tables that exemplify the metalation of saturated or unsaturated, unsubstituted or heterosubstituted acyclic hydrocarbons, of metallocenes and of arenes. Table 3 covers the 13 Tables, in which typical metalation reactions of 5- or 6-membered heterocycles are compiled.

Graphical Guideposts to Section I/4

Table 1. Principal Methods of M−R Generation

With elemental metal (M): reductive insertion		Electrofugal group	With organomet. reagent: permutat. interconversion	
⊕⊕⊕	pp. 86−101	X (halogen)	pp. 101−137	⊕⊕⊕
⊕⊕	pp. 138−148	Y (chalcogen)	pp. 148−155	⊕
⊕⊕	pp. 155−159	Q (metalloid)	pp. 159−171	⊕⊕
⊕	pp. 171−177	C (carbon)	pp. 177−181	⊕
⊕	pp. 181−185	H (hydrogen)	pp. 185−284	⊕⊕⊕

Table 2. Metalation of hydrocarbons and metallocenes

X =	−H −R −SiR$_3$	−NR$_2$ −N(M)R $\diagup$C−NR$_2$ $\diagup$C−N(M)R	−OR $\diagup$C−OM	−F −Cl −Br −I −CF$_3$
M−C−X	Table 80 (p. 186)	Tables 81−86 (pp. 187−192)	Tables 87−89 (pp. 93−195)	Table 90 (p. 196)
M−C−C=C X	Tables 91−99 (pp. 187−192)	Table 100 (p. 205)	Tables 101−102 (pp. 206−207)	Table 103 (p. 207)
M−C−⟨X⟩	Table 104 (p. 208)	Table 105 (p. 209)	Tables 106−107 (pp. 210−211)	Table 108 (p. 212)
M−C=C− X	Table 109 (p. 213)	Tables 110−113 (pp. 214−217)	Tables 114−115 (pp. 218−219)	Tables 116 and 117 (pp. 220−221)
M−⟨X⟩M″	Table 118 (p. 222)	Table 118 (p. 222)	—	—
M−⟨X⟩	Table 119 (p. 223)	Tables 120−124 (pp. 224−228)	Tables 125−128 (pp. 229−232)	Tables 129−131 (pp. 233−235)

Table 3. Metalation of five- and six-membered heterocycles

M−⟨N−R⟩	Table 132 (p. 236)
M−⟨indolyl N−R⟩	Table 133 (p. 237)
M−⟨N−N−R⟩	Table 134 (p. 238)
M−⟨N−N−R⟩	Table 135 (p. 239)
M−⟨O⟩	Table 136 (p. 240)
M−⟨S⟩	Table 137 (p. 241)
M−⟨O−N⟩	Table 138 (p. 242)
M−⟨N−O⟩	Table 139 (p. 243)
M−⟨S−N⟩	Table 140 (p. 243)
M−⟨N−S⟩	Table 244 (p. 243)
M−⟨N⟩	Tables 142−144 (p. 245−247)

Contents

1   Background

All chapters in this book focus on reactions and reagents. However, in order to understand what drives organometallic reactivity, we have to know about structures and mechanisms. Moreover, a look back shows us how things started and how much effort, combined with imagination, was needed to achieve the position organometallic methods nowadays occupy in organic synthesis.

1.1   Glimpses on History

The cradle of polar organometallic chemistry stood some 150 years ago in Northern Hesse[1-3]. There, in 1831, Friedrich Wöhler became a teacher at the newly founded Higher Industrial School (*Höhere Gewerbeschule*) in Cassel. Previously, he had held an appointment at a similar institution in Berlin where he had succeeded in the preparation of the 'natural product' urea from the indisputably inorganic precursors ammonia and silver or lead cyanate[4,5]. Thus, he had defeated the dogma of the indispensable *vis vitalis* and had fired the starting shot in the race towards organic synthesis. An outburst of cholera prompted him to leave the Prussian capital and to return to the proximity of his native town. Among his numerous noteworthy accomplishments in Cassel was the preparation of diethyltellurium[6] as the first main group organometallic compound. This discovery, however, remained without aftermath.

In 1836, Wöhler occupied the vacant chair of chemistry at the university of Göttingen while the young Privatdozent Robert Wilhelm Bunsen, became his successor in Cassel. Scientifically ambitious, Bunsen decided to tackle one of the greatest experimental challenges of his time. He was determined to solve the mysteries of the dreadful Cadet's liquor, an atrociously smelling, poisonous distillate that forms when arsenic and potassium acetate are exposed to red heat. At the risk of his health, he managed to isolate the main component and to identify its elementary composition as $C_4H_{12}As_2O$[7]. Although he initially called the new substance alcarsine oxyde, he soon adopted the more revealing name of cacodyl oxyde ($\kappa\alpha\kappa\acute{o}\varsigma$ = stinking, $\acute{o}\delta\eta\varsigma$ = odor).

Nowadays, we can unambiguously assign the structure of bis(dimethylarsanyl) oxide to this compound. However, in the middle of last century, the notions of chemical bonding, valency and connectivity had not yet become established, although a forerunner concept had just emerged. The great Swedish chemist Jons Jacob Berzelius, a dominant authority of the time, conceived organic molecules to consist of two electrostatically matched parts, for example ether of ethyl and oxide subunits, exactly as mineral salts are made up from cations and anions. If the organic world really were nothing but a reproduction of the inorganic, should one not be able to identify highly reactive molecular fragments in just the same way as metallic sodium and elementary chlorine had been obtained from sodium chloride? Indeed, there was a general consensus that the isolation of organic radicals in the free state would constitute compelling evidence for the correctness of Berzelius' electrostatic concepts which postulated a dualistic constitution of *all* matter and which was, at the beginning at least, cognate with the 'radical theory' promoted by Dumas, Liebig and Wöhler. However, all experiments undertaken in this respect so far, had failed.

Bunsen was the man to create the sensation. He treated cacodyl oxyde with concentrated hydrochloric acid to obtain cacodyl chloride (dimethylarsanyl chloride). This substance was painstakingly purified and filled into glass ampoules which contained activated zinc sheets and which had previously been purged with carbon dioxide before they were flame sealed and heated on a water bath. After three hours, the vessel was opened, the zinc chloride formed extracted and the remaining oil collected. When exposed to air, the compound spontaneously ignited. It was, as we now know, tetramethyldiarsane. Bunsen, however, believed he had isolated the free cacodyl radical, and many of his contemporaries shared this conviction.

$$
\begin{array}{c}
H_3C \\
As-O-As \\
H_3C CH_3 \\
CH_3
\end{array}
\xrightarrow{HCl}
\begin{array}{c}
H_3C \\
As-Cl \\
H_3C
\end{array}
\xrightarrow{Zn}
\left(
\begin{array}{c}
H_3C \\
As\cdot \\
H_3C
\end{array}
\right)
\longrightarrow
\begin{array}{c}
H_3C CH_3 \\
As-As \\
H_3C CH_3
\end{array}
$$

Otherwise feared as a merciless critic, Berzelius wholeheartedly praised Bunsen. The latter moved as an associate (1839) and later as a full professor (1841) to Marburg before, after a short interlude in Breslau, he ascended in 1852 to the then most prestigious position, the chemistry chair in Heidelberg. His rising renown attracted scholars from all over the world. One of them, Edward Frankland, hoped to obtain the free ethyl radical by applying his mentor's method, *i.e.*, the reduction of a halide with zinc, to ethyl iodide. He prepared the first organozinc compound instead[8]. A few years later, John A. Wanklyn, another English research fellow in Bunsen's laboratory, treated dialkylzincs with sodium and thus produced the first, though ill-defined, organosodium species (which actually combined with unconsumed precursor material to afford a sodium triethylzincate complex[9]).

$$
H_5C_2\text{-}I \xrightarrow{Zn} H_5C_2\text{-}ZnI \dashrightarrow (H_5C_2)_2Zn
$$

$$
(H_5C_2)_2Zn \xrightarrow{Na} H_5C_2\text{-}Na \dashrightarrow (H_5C_2)_3Zn^{\ominus}Na^{\oplus}
$$

The work of Wanklyn was followed up by Paul Schorigin (from 1906) and Wilhelm Schlenk (from 1922), while Frankland inspired Sergei Nikolaijevitch Reformatzky (from 1887), whose α-bromo ester-derived zinc enolates became the first synthetically useful organometallic reagents. Philippe Barbier (from 1899) and Victor Grignard (from 1900) switched from organozinc to organomagnesium compounds. The latter, being universally accessible and at the same time more reactive, soon became favorite tools for preparative chemistry. However, it still needed the ingenuity and perseverance of pioneers such as Morris S. Kharasch, Avery A. Morton, Henry Gilman, Georg Wittig and Karl Ziegler, who systematically developed, complemented and refined the methods that have become the foundations of the organometallic approach to synthesis. The seeds planted by such outstanding researchers in the 1930s and 1940s began to grow in the years after the second World War. Within a few decades, organic synthesis had undergone a complete metamorphosis and polar organometallic chemistry had played a crucial role in this evolutionary process (Fig. 1).

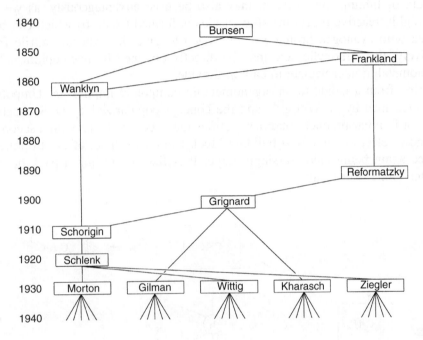

Figure 1. Genealogy of polar organometallic chemistry.

1.2   The Organometallic Approach to Synthesis

What is so special and novel about the organometallic approach to organic synthesis? To gain this perspective, first consider the most classical and typical organic reaction, the S_N2 process (route a_2, in the scheme overleaf): a standard nucleophile $M-Nu$ (iodide, cyanide, enolate, *etc.*) encounters in a bimolecular collision a carboelectrophile $\geqslant C-X$ (mostly derived from the corresponding hydrocarbon; route a_1) and, attacking from the rear, promotes the replacement of the nucleofugal leaving group X (halide, sulfonate, *etc.*) under a Walden inversion of configuration. In general, such nucleophilic substitution proceeds with great ease if the carboelectrophilic substrate has a methyl or primary alkyl group as the organic entity. With a secondary alkyl carbon as the reaction center, generally only a poor yield is obtained. Worse, tertiary alkyl, 1-alkenyl, 1-alkynyl, aryl, hetaryl and cyclopropyl halides or sulfonates are completely S_N2 inactive. In other words, the concerted bimolecular nucleophilic substitution has a fairly narrow scope of applicability.

Organometallic chemistry now offers a way out of this dilemma. Suppose we wish to convert an aryl bromide into the corresponding benzoic acid. It would be more than naive to treat it with potassium cyanide in the hope of obtaining in this way the aryl nitrile, which subsequently could be hydrolyzed. However, when it is allowed to react with magnesium or lithium (butyllithium may also be used advantageously, as we shall see later), a highly reactive organometallic species is formed (route **b**) which can be readily condensed with cyanogen bromide or cyanogen to give the expected nitrile (route **c₂**). Alternatively, the carboxylic acid may be directly obtained by nucleophilic addition of the organometallic intermediate to carbon dioxide.

Switching from a halide to an organometallic compound implies an 'Umpolung', the term being coined by G. Wittig[10] and the concept popularized by D. Seebach[11]. The inversion of familiar product polarities gratifies the experimentalist with a second option: he may now select his preferred building block from a fresh set of electrophiles *El*–X, the choice again being embarrassingly large. It is like reshuffling a pack of cards and starting the game over again.

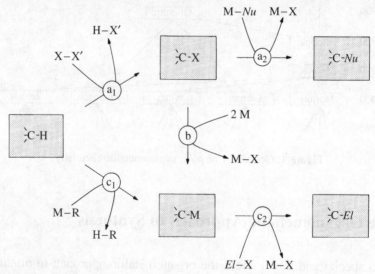

The reductive or permutational replacement of halogen by a metal M (route **b**) can hardly fail, and subsequent electrophilic replacement of the metal (route **c₂**) rarely causes trouble. However, the selective access to the required starting material (route **a₁**) may give us a true headache. In general, neither such a halide nor the corresponding alcohol exists as such in nature. Ultimately, most have to be prepared from a suitable, saturated or unsaturated hydrocarbon. The site-selective introduction of the heteroatom is far from being a trivial task. There is a great temptation to try a shortcut and to convert the hydrocarbon directly into an organometallic intermediate (route **c₁**).

This is easier said than done. Simple hydrocarbons are only minimally acidic and hence extremely reluctant to undergo deprotonation. The second obstacle is that several or even many different hydrogen atoms, not just one, are present in ordinary substrates (omitting the few high-symmetry structures known, such as benzene, neopentane and cyclohexane). We have to learn how to exploit subtle differences in their environment in order to discriminate between them chemically. This means that we have to conceive metalating reagents that unite two seemingly incompatible reactivity profiles: maximum reactivity and maximum selectivity. A substantial part of this Chapter is devoted to this fascinating problem. It will be shown how the two antagonistic features can be reconciled and what impressive practical results can be achieved in this way.

In summary, this Chapter assigns the highest priority to the *generation* of organometallic reagents, although their transformation, their reactivity or other properties will not be completely neglected. When we consider the various possibilities that exist for the preparation of key organometallic intermediates, emphasis is put on the most direct method, the permutational hydrogen/metal exchange.

This Chapter is restricted to the so-called *polar organometallics*[12] (Fig. 2). This class of compounds includes the derivatives of the most common alkali and alkaline earth metals and extends into the area of the organozinc compounds. However, the latter reagents deserve to be treated in a separate review (see Chapter V).

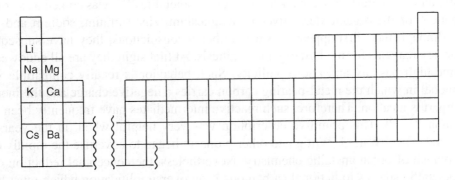

Figure 2. A periodic table featuring the major elements used as constituents of 'polar' organometallic reagents.

Although all these species are characterized by a polar carbon–metal bond, the degree and pattern of polarity vary significantly with the element. Whereas nucleophilicity and basicity are the absolutely dominant features of organic derivatives of potassium, cesium and barium, the reactions of lithium, magnesium and zinc compounds are, in increasing order, triggered by the electrophilicity (Lewis acidity) of the metal. As will be stressed over and over again, the individuality of the metal involved is the most critical parameter for designing tailor-made organometallic reactions.

1.3   Metal Tuning

The present contribution avoids dealing with organic derivatives of transition elements, no matter whether produced in a stoichiometric reaction or as a transient species in a catalytic cycle. However, one cannot ignore the interdependence or the complementarity of main group and transition block chemistry in modern organic synthesis. For example, organotitanium and organocopper compounds are most conveniently prepared through organolithium or organomagnesium reagents. Organocopper and organopalladium compounds allow us to escape definitively from the S_N2 restrictions even in such cases when, despite *Umpolung*, no suitable electrophile can be matched with a polar organometallic reagent. The area of transition element chemistry (like that of boron, aluminum and tin) is treated in the following Chapters by some of the most competent workers in the field.

2   Structure and Mobility

A quarter of a century ago, the term 'polar organometallics'[12] was coined as a common designation for the organic derivatives of magnesium, zinc, lithium, sodium and potassium. Although their first appearance was rather inconspicuous, they have subsequently become key reagents for modern organic synthesis. At first sight they are all endowed with the same high basicity and nucleophilicity. Such behavior is readily compatible with a bond model in which the metal-bearing carbon carries a negative charge and the inorganic counterpart is a cation. Therefore, such reactive intermediates have frequently been called 'carbanions'[13,14]. This primitive description was very helpful when, in the years after the Second World War, G. Wittig and other pioneers began to advertise the rapidly developing branch of organometallic chemistry. Nevertheless, the conceptual reduction of real organometallic species to fictional carbanions is an oversimplification which must lead to misjudgments. In fact, no difference in organometallic reactivity patterns can be rationalized unless the metal and its specific interactions with the accompanying carbon backbone, the surrounding solvent, and the substrate of the reaction, are explicitly taken into account.

In other words, in order to understand *reactivity* we need a detailed knowledge of the *structures* involved. Only if we have a realistic idea about the nature of an organometallic bond can we dare to predict what changes it may undergo under the influence of a suitable reaction partner until a thermodynamically more stable entity will emerge from such a molecular reorganization. For this reason, this Chapter aims to provide a suitable descriptions of the specific interactions that exist between metals and carbon, of the unique architecture of polar organometallic compounds, and of fluxionalitities that are characteristic of such species.

2.1   The Nature of the Organometallic Bond

Many of us may remember to have heard in highschool that metals want to get rid of their valence electrons in order to become cations having the same electron shell as noble gases. Although common, this belief is entirely wrong. We just have to look up the ionization potentials of monoatomic metals in order to convince ourselves of the contrary. Stripping off an electron from even the most 'electropositive' metals requires energies of 124 (lithium), 118 (sodium), 100 (potassium), 98 (rubidium) and 90 (cesium) kcal/mol[15]. Why should a metal want to become a cation at all? Actually it does not. As do other elements, it merely seeks to share electrons with other atoms by forming a chemical *compound*. However, in the case of metals this is not a trivial matter.

Let us consider a metal atom, say lithium, in the gas phase. When it is allowed to combine with another radical, binding energy is produced. If a lithium chloride molecule is formed, this gain amounts to 112 kcal/mol. In order to assess on this basis its heat of formation from the elements, one has to deduct half of the dissociation energy of elemental chlorine (58 kcal/mol) and, in addition, the heat of fusion, vaporization and atomization of bulk lithium (altogether some 70 kcal/mol). In other words, the formation of lithium chloride from the elements is an endothermic process as long as monomeric species are generated in the gas phase. This conclusion holds also for other alkali metal halides. They become thermodynamically strongly favored only in the solid state. What structural features make the crystal lattice so advantageous?

The answer can be found in any textbook of general chemistry. Due to the dense packing of the ball-like ions, each is surrounded by several, mostly six, counterions. In this way the attraction between oppositely charged particles increases steeply, whereas the repulsion of ions having the same sign remains moderate because of the longer inter-nuclear distances. The electrostatic model can also be used to describe smaller molecular packages such as aggregates or clusters. As an extension, solvation may be portrayed as a charge–dipole interaction.

Although nothing is wrong with these ideas, we wish to adopt a different point of view or, more properly, to use a different vocabulary. We prefer to conceive of the forces holding together a sodium chloride crystal as 'partial bonds' (electron-deficient bonds) rather than 'ionic bonds'. This approach has the merit of universality; it creates a continuum of binding interactions spanning from perfectly covalent to totally polar bonds as the extremes.

The reader may immediately wish to object. How can somebody dare to deny the ionic nature of the sodium chloride crystal? Has it not been proved that within the radii of 0.95 and 1.81 Å around sodium and chloride nuclei precisely 10 and 18 electrons, respectively, can be found and that, most revealingly, the electron density between such spheres drops to zero[16-22]? Yet electron population analyses may be fallicious[23-27]. A lithium atom in the gas phase would have to be visualized as an 'electride' if treated in the same way as solid lithium fluoride: at 0.65 Å from the nucleus the electron density is zero; inside of this 'ion radius' confinement we find two 1s, but outside the entire 2s valence electron[28,29]!

To feel reassured, the reader should not forget how imperfect any description of the physical reality is when man-made equations, images and semantics are used. Hence the liberty, if not necessity, to choose in a given case the model which appears to be the most appropriate, is required. For example, the τ ('banana bond') and the $\sigma-\pi$ depiction of ethylene are equivalent. Depending on the situation, one or the other representation may be more suitable. In the same way, it is a matter of convenience whether to deal with metal derivatives in terms of ionic and partial bonds. As all molecular ensembles are tied together by the electrostatic attraction between protons and electrons, ultimately both models must converge and lead to the same conclusions.

With all these precautions in mind, we now attempt to rationalize the most characteristic features of metal binding, in particular aggregation and coordination. The selection of examples is focused entirely on carbon–metal compounds, although metal amides, metal alkoxides or aroxides and metal halides exhibit the same structural phenomena.

2.2   Aggregation and Coordination

Subunits can assemble to afford larger chemical entities when the attractive forces overcompensate the repulsive forces. In order to minimize repulsion, electrons must be able to spread out over a maximum of space. In order to maximize attraction, the electrons have to approach the atomic nuclei as closely as possible. These seemingly contradictory requirements can best be reconciled if the electrons surround the atomic nuclei spherically. Within a diatomic molecule (1) such as lithium chloride, electrons will inevitably concentrate between the two poles of attraction, the metal and halogen nuclei, and thus create an energetically unfavorable asymmetric charge distribution. As we have already seen, a naked metal ion (2) is worse, being extremely energetic if kept in empty space. However, in a crystal lattice (3) it is surrounded by four, six or eight heteroatomic neighbors which enwrap it spherically and share valence electrons with the metal. The interaction between a metal atom and solvent molecules (Sv) in a solvate (4) can be treated in a similar manner.

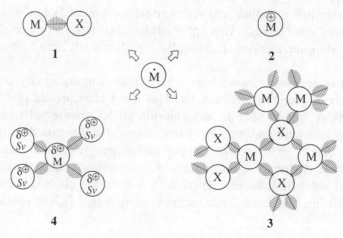

The typical coordination number of lithium[30], magnesium[31] and zinc[32] is four, and of larger cations such as sodium[33] or potassium[34] six. Special effects, in particular steric hindrance, may deplete the coordination sphere and reduce it to two ligands [*e.g.* lithium perchlorate dietherate (**5**)[35] or bis[tris(trimethylsilyl)methyl]magnesium[31]] whereas crown ether- or cryptand-like complexands achieve exceptionally high coordination numbers of six [lithium tetrafluorborate bis(*cis,cis*-1,2,3,4,5,6-triepoxycyclohexane) (**6**)[36], magnesium dibromide tetrakis(tetrahydrofuran)dihydrate[37]], seven [sodium perchlorate benzo-15-crown-5[38] or potassium 2-phenylethenolate 18-crown-6[39]], eight [diethylmagnesium 18-crown-6[40], diethylzinc 18-crown-6[40], sodium and potassium thiocyanate nonactine[41] or potassium ethyl acetoacetate 18-crown-6[42]] and ten [bis[1,1,1,5,5,5-hexafluoro-1,3-pentanedionato-*O,O'*)barium] 18-crown-6 (**7**)[43]].

5　　　　　　**6**　　　　　　**7**

The fascinating structures of organometallic compounds become immediately intelligible if we keep in mind that metals in organic, as in inorganic, derivatives always strive for maximum coordination numbers. Partial or electron-deficient bonds are the instruments with which this goal is materialized. This trick allows organometallic species to cluster into oligomeric or polymeric superstructures. These so-called *aggregates* exist in the vapor phase, in solution and in the solid state.

2.2.1　Structures in the Gas Phase

The prototype of all electron-deficient bonds is the double hydrogen bridge that keeps the diborane molecule together. Monomeric borane, being isoelectronic with the methyl cation, has only six electrons in the valence sphere. In order to attain tetracoordination, the metalloid atom abandons one ordinary boron–hydrogen bond in favor of two electron-deficient bonds which then allow two BH_3 units to be connected. Although only a total of four electrons is available to construct the four bridging B–H bonds, the resulting dimer (**8**) is thermodynamically very stable, the aggregation enthalpy amounting to 35 kcal/mol[44]. Boron–hydrogen, boron–boron and, where appropriate, boron–carbon electron-deficient bonds are also at the origin of the stability of polyboranes[45–47] (such as pentaborane, **9**) and carboranes[45–47] (such as the 1,2-carborane **10**; the open and filled circles representing BH and CH groups, respectively).

Unlike dimethylborane (**11**), trimethylborane is no longer capable of dimerization[48,49]. This is to some extent a consequence of the inferior strength of boron–carbon compared with boron–hydrogen electron-deficient bonds. The increased steric hindrance is another crucial, even determining factor. Despite their substantially longer metal–carbon bonds, triisopropylaluminum and trineopentylaluminum are monomeric too[50]. Unbranched trialkylaluminums can better accommodate two bridging carbon atoms and therefore dimerize. The binding forces being relatively moderate $(\Delta H^\circ 20 \text{ kcal/mol})$[51,52], monomers and dimers of trimethylaluminum (**12**) are concomitantly observed in the gas phase[53], whereas benzene solutions contain only the dimer[53]. An X-ray diffraction crystal analysis[54] shows the electron-deficient Al–C bonds to be markedly longer (2.24 Å) than the ordinary ones (2.00 Å). The exocyclic C–Al–C angle is found a little widened (124°) with respect to the monomer (120°) whereas the endocyclic one within the bridge approximates the tetrahedral geometry (110°). The Al–C–Al angle is compressed to 70°.

Dimethylberyllium vapor consists of an equilibrium mixture of monomers, dimers (**13**) and trimers (**14**)[55,56]. Upon slow condensation colorless needles form. The X-ray analysis of the latter reveals a polymer (**15**) ordered in Be–C–Be–C rhombus, alternating in two perpendicular planes. The bond angles around the metal atoms are close to tetrahedral (114°) with carbon-centered Be–C–Be angles of 66°[57]. Obviously, beryllium has to contribute both of its valencies to the construction of electron-deficient bonds in order to attain tetracoordination (as in **15**). If only one classic metal–carbon bond is sacrificed and converted into two half-bonds, the metal will be placed in a trigonal environment (as in **13**).

$$H_3C - Be - CH_3$$

13   14

15

For monovalent methyllithium it should be particularly difficult, if not impossible, to reach tetracoordination. However, its driving force for oligomer formation is remarkably high. The lightest alkali metal allows fairly reliable *ab initio* calculations to be performed if large basis sets are used and electron correlation is included. Thus, the aggregation enthalpy was found to approximate to 20, 25 and 30 kcal/mol per $LiCH_3$ unit in the methyllithium dimer, trimer and tetramer, respectively[58,59]. Actually, no monomers, only oligomers, can be observed in the gas phase[60–63]. The monomer can be generated, but it must be trapped in an argon matrix in order to be preserved[64]. In contrast, bis(trimethylsilyl)methyllithium was found to be purely monomeric in the gas phase although it is polymeric in the crystalline state[65].

Unfortunately, gas-phase studies of organometallic species are often hampered by practical problems such as insufficient volatility of the sample. Even the most powerful techniques for the structure elucidation of vapors, infrared spectroscopy and electron diffraction[66,67], do not always provide unambiguous and conclusive answers. Therefore, the available information is scarce when compared to crystal and solution data.

2.2.2   Crystal Structures

The first single crystal X-ray analysis of a solvent-free organometallic compound was that of ethyllithium reported in 1963 and refined two decades later[68]. Its lattice, like that of methyllithium (see below), consists of tetrameric subunits. Four lithium atoms and four methyl-bearing methylene groups alternate at the corners of a distorted cube. Thus, every α-carbon atom is hexacoordinated while every lithium site has three α-carbon nearest neighbors. In addition, however, short Li, H contacts and elongated α-C–H bonds suggest one of the proximal hydrogen atoms to act as a fourth pseudoligand of the metal. In the meantime, coordination of C−H or B−H bonds to lithium has been observed frequently as, for example in the polymeric lithium tetramethylborate (**16**)[69], the monomeric lithium tetrakis(di-*tert*-butylmethyleneimidoaluminate) (**17**)[70] and the dimeric lithium boronate− *N,N,N',N'*-tetramethylethylenediamine adduct (**18**)[71]. The 'agostic' C−H−Li interaction[72,73] has become a popular, though somehow nebulous and controversial[74] concept.

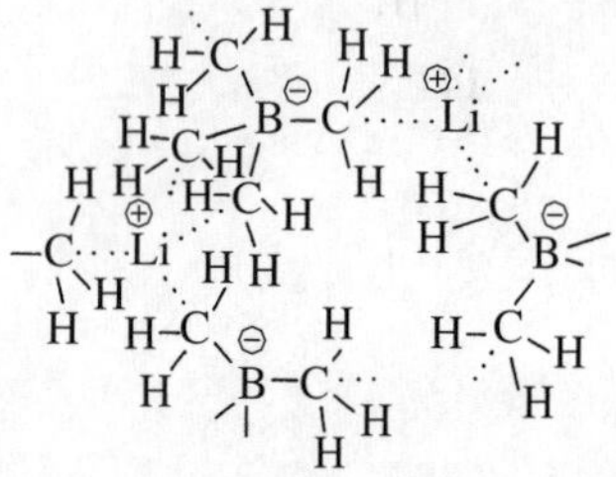

16 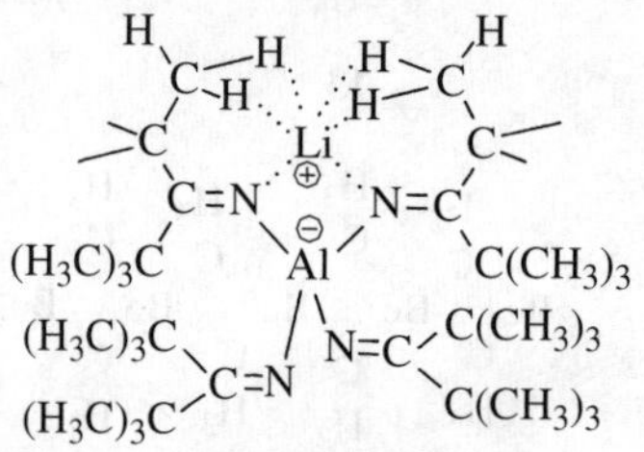   **17** 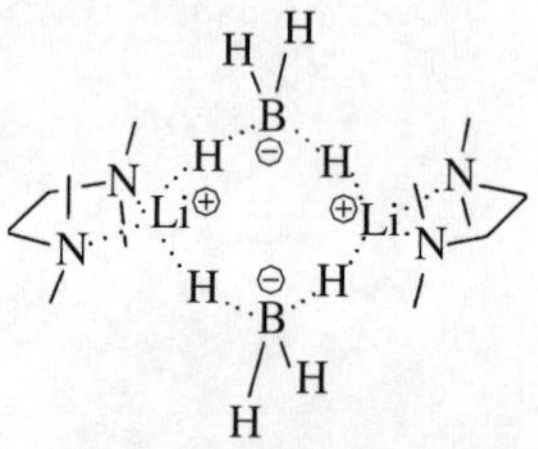   **18**

tert-Butyllithium also crystallizes as a tetramer[75]. In contrast, butyllithium[75], trimethylsilylmethyllithium (**19**)[76] and cyclohexyllithium[77] crystals are composed of hexameric units exhibiting a chair-like arrangement of the metal atoms. The structure may be visualized as a distorted octahedron. Its corners are occupied by the metal atoms and it has two small and four medium triangular faces which are each bridged by an organic group. Thus, every lithium atom is again tricoordinate.

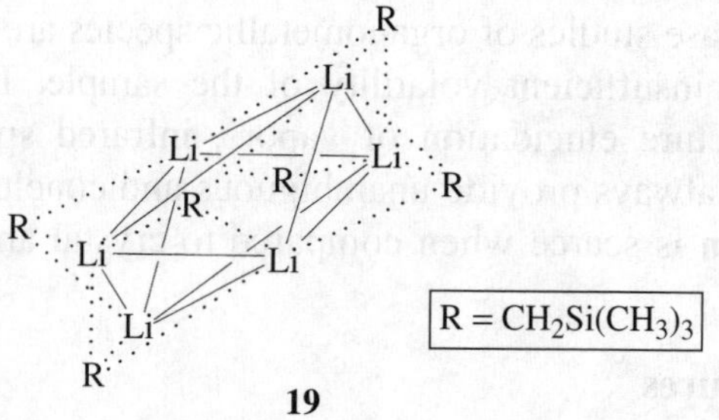

$R = CH_2Si(CH_3)_3$

19

Methylalkali compounds, the smallest organometallics, can only be isolated as powders. Nevertheless it was possible to collect accurate and meaningful data by submitting the perdeuterated species to a combination of neutron and synchroton radiation diffraction techniques at 1.5 and 290 K. Methyllithium[78,79] was found to contain tetrameric units. A carbon and a lithium tetrahedron penetrate each other to form a Li_4C_4 cube (**20**) which is distorted, as the Li−Li distances (2.6 Å) are considerably shorter than the C−C distances (3.7 Å). The methyl groups adopt a staggered position relative to the adjacent lithium triplet (Fig. 3).

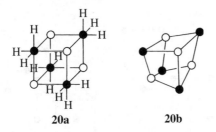

20a   **20b**

Figure 3. Tetrameric units of methyllithium as the constituents of its crystal lattice: idealized, regular cube (**20a**) and real, distorted cube (**20b**).

Methylpotassium[80,81] shows a repeating orthorhombic unit cell (**21**) containing four methylpotassium species in a trigonal-prismatic array (Fig. 4). The methyl groups have alternating orientations, each one facing with their electron-pair three potassiums at a close distance (3.0 Å) while turning their hydrogen covered reverse side (H−C−H = 105°) to the three others, kept at a longer distance of 3.3 Å. The structures of methylrubidium[82] and methylcesium[82] are quite similar and again resemble an ion lattice of the NiAs type.

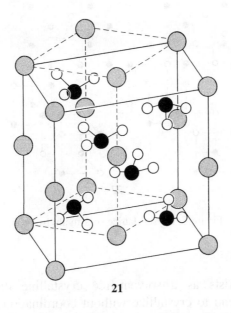

21

Figure 4. Crystal lattice of methylpotassium.

Methylsodium (**22**)[83,84] presents features characteristic of both the methyllithium and methylpotassium structures. The space group is orthorhombic and contains 16 molecules. Half of them are clustered together to give tetramers which are linked by the remaining 8 molecules arranged in Na zig-zag chains (Fig. 5).

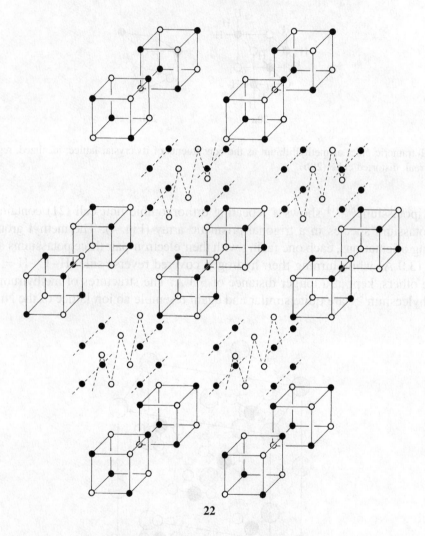

22

Figure 5. Crystal lattice of a methylsodium.

Diphenylzinc (**23**) exists as a solvent-free crystalline dimer[85]. Only sterically congested aryllithiums tend to crystallize without coordination to donor molecules: the polymeric 4-butylphenyllithium and 4-*tert*-butylphenyllithium, the hexameric 3,5-di(*tert*-

butyl)phenyllithium, the tetrameric 2,4,6-tri(isopropyl)phenyllithium and the dimeric 2,6-bismesitylphenyllithium and 2,6-bis(2,6-diisopropylphenyl)phenyllithium[86]. The sole ether- and amine-free monomeric structure, 2,4,6-tris(2,4,6-triisopropylphenyl)phenyl-lithium (**24**), still contains a benzene molecule providing intralattice solvation[87].

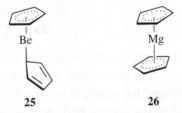

23   **24**

The tetrameric crystal structure of vinyllithium[87] resembles that of methyllithium although it is considerably distorted. (2-Anisyldimethylsilyl)ethynyllithium[88] was found to give a hexamer and lithium *tert*-butylacetylide (3,3-dimethyl-1-butynyllithium)[89] a dodecamer (see also p. 20), whereas ethynylsodium and 1-propynylsodium[90] exhibit a 'salt-like' layered lattice.

The dimeric 9-fluorenyllithium[91] is the only benzyl-type organoalkali compound of which solvent-free crystals have been grown so far. The disordered (η^1/η^6-bound, 'slipped sandwich') dicyclopentadienylberyllium[92,93] ('beryllocene', **25**) as well as dicyclopentadienylmagnesium[94–96] ('magnesiocene', **26**) show no sign of aggregation.

Be   Mg

25   **26**

As we have seen above, solvent- or ligand-free crystals of organometallic compounds are relatively rare. In general, ether or amine molecules have to be incorporated in the structure to saturate vacant coordination sites.

Butyllithium combines with tetrahydrofuran (THF) in such a way that each lithium corner of the tetrameric pseudocube is occupied by one solvent molecule. In addition, hexane molecules are accommodated to make up a $[(H_9C_4Li \cdot THF)_4 \cdot C_6H_{14}]$ stoichiometry[97]. *tert*-Butyllithium forms with diethyl ether (DEE) a $[(H_3C)_3CLi \cdot DEE]_2$ dimer having short (64°) Li−C−Li angles[75]. A non-exhaustive collection of amine-complexed alkylmetals contains the dimeric ethylmagnesium bromide trimethyl-amine[98] $[H_5C_2MgBr \cdot N(CH_3)_3]_2$, the polymeric tetrakis(methyllithium) *N,N,N',N'*-tetramethylethylenediamine (TMEDA) chains[99] $[(H_3CLi)_4 \cdot TMEDA]_\infty$ (**27**), the butyllithium/TMEDA dimer[97,100] [(H$_9$C$_4$Li · TMEDA]$_2$ (**28**) and the 1-bicyclo[1.1.0]-butyllithium/TMEDA dimer[101].

27   **28**

In the aryl series we find the monomeric alkali-earth derivatives phenylmagnesium bromide dietherate[102] H$_5$C$_6$MgBr · (DEE)$_2$ (**29**) and bis(pentafluorophenyl)zinc bis(tetrahydrofuranate)[103]. Depending on the complexand, phenyllithium is tetrameric ([H$_5$C$_6$Li · DEE]$_4$)[104], dimeric ([H$_5$C$_6$Li · TMEDA]$_2$)[105] or monomeric H$_5$C$_6$Li · PMDTA (PMDTA = *N,N,N′,N″,N″*-pentamethyldiethylenetriamine)[106]. The mesityllithium bis(tetrahydrofuranate)[107] (**30**) [2,4,6-(H$_3$C)$_3$C$_6$H$_2$Li · (THF)$_2$]$_2$, the 2,4,6-tri(isopropyl)phenyllithium diethyl etherate[108] [2,4,6-(^{i}H$_7$C$_3$)$_3$C$_6$H$_2$Li · DEE]$_2$ and the phenylsodium/PMDTA adduct[109] [H$_5$C$_6$Na · PMDTA]$_2$ are all dimeric.

29   **30**   R = CH$_3$; Sv = ⬡

The tetrameric 3,3-dimethyl-1-butynyllithium tetrahydrofuranate[89] (**31**) may be conceived as a core structure. By adding unsolvated acetylide dimers it can be continuously extended towards higher oligomers, for example to afford the dodecamer **32** carrying just four tetrahydrofuran molecules at the cluster periphery, and ultimately polymers.

31   **32**   ○ = Li; ●≡● = C≡C- C(CH$_3$)$_3$;   Sv = ⬡

The allyllithium/PMDTA adduct (**33**) shows a monomeric π-complex structure, the metal occupying an out-of-center position[110]. The benzyl group entertains an

η^1-interaction with the metal in both the monomeric benzyllithium/TMEDA/THF[111], the tetrameric benzylsodium/TMEDA[112] (**34**) and in the polymeric benzyllithium diethyl etherate[107], whereas the coordination is of the η^3-type in the polymeric benzyllithium/(DABCO)$_2$[113] (DABCO = 1,4-diazabicyclo[2.2.2]octane). A pure η^2-hapticity is found in the α-(trimethylsilyl)benzyllithium/TMEDA adduct[114]. The triphenylmethyl backbone[115] displays an unsurpassed diversity of bonding to the metal: $\eta^6(+\eta^3)$ in triphenylmethylcesium/TMDTA[116] and triphenylmethylpotassium/PMDTA/THF[116], $\eta^5+\eta^1$ in triphenylmethylsodium/PMEDA[117], η^4 in triphenylmethyllithium/PMEDA[118], η^3 in triphenylmethyllithium bis(diethyl etherate)[108] (**35**) and η^0 in the ion pair triphenylmethyllithium/12-crown-4[119] (12-crown-4 = 1,4,7,10-tetraoxacyclodecane).

33 **34** **35**

The crystal structures of solvated organometallics represent instructive information. They allow one to anticipate what may happen in solution.

2.2.3 Structures in Solution

As F. Hein and H. Schramm recognized in 1930, organolithium compounds tend to cluster also in solution[120]. Since then numerous systematic investigations have been carried out. The principle techniques of investigation are cryoscopy[120–122], ebullioscopy[123] or vapor pressure measurements[124,125] and ^{1}H, ^{6}Li, ^{7}Li or ^{13}C nuclear magnetic resonance spectroscopy[126–132]. Rather than provide direct information about details of the three-dimensional structures, these techniques basically monitor the aggregation numbers. A few pertinent trends become apparent.

- *The aggregation number may, though must not always, be higher in paraffinic or aromatic solvents when compared to ethereal ones.* Ethyllithium[60,133,134] and butyllithium[60,126,133–136] are hexameric in hexane, but tetrameric in diethyl ether or tetrahydrofuran; dilute solutions of trimethylsilylmethyllithium in hexane are hexameric but only tetrameric in benzene[135,137,138]. 2-Phenylisobutyllithium (2-methyl-2-phenylpropyl-lithium, 'neophyllithium') is tetrameric in pentane and dimeric in diethyl ether[139]. However, methyllithium is tetrameric in both the solid state (suspension)[78,79] and in ethereal solution[140].

- *Steric bulk hinders aggregation.* In contrast to primary alkyllithiums, isopropyl-lithiums[135], *sec*-butyllithium[141,142] and *tert*-butyllithium[135,143] are in hexanes not hexameric, but rather tetrameric. *tert*-Butyllithium appears to be dimeric in diethyl ether and monomeric in tetrahydrofuran, at least at low temperatures[122].

- *Hybridization towards increasing s character disfavors aggregation.* Whereas for example isopropyllithium[135] is always tetrameric in ether, phenyllithium[123,144–147] and 3,3-dimethyl-1-butynyllithium[99,120,125,148] exist as equilibrium mixtures in which, depending on the solvent, temperature and concentrations, monomers, dimers or tetramers dominate.

- *Charge delocalization also discourages aggregation.* Polyisoprenyllithium is dimeric in petroleum ether and monomeric in diethyl ether or tetrahydrofuran[149]; allyl type species such as 2-methylallyllithium[150] and 2-methylallylpotassium[151] in tetrahydro-furan are apparently monomeric.

- *Low temperatures cause deaggregation.* Below −100 °C in tetrahydrofuran, the butyllithium tetramer disintegrates to a large extent to form a dimer whereas the phenyllithium dimer establishes an equilibrium with its monomer[146,147].

Aggregate statics and dynamics are of prime importance for the reactivity[12] of organometallic reagents. Because, in general, the transformation promoting species is the monomer[12], and only occasionally a dimer[152–154] or higher aggregate fragment, a total or partial deaggregation has to take place before a reaction can occur. Organolithium reagents that combine to give particularly tight oligomers are handicapped in this respect. As a monomer, methyllithium should be more basic than phenyllithium by more than 10 pK units (see below). At high dilution in tetrahydrofuran methyllithium is, however, only little more reactive than phenyllithium and at 0.5 M concentration it is even of inferior reactivity[144].

Alkyllithium reagents have a high tendency to combine with other organo-lithium[155–157], organomagnesium or organozinc[158–160] and with organosodium compounds[161]. The mixed aggregates (heterooligomers) thus obtained are in general ther-modynamically more stable than the corresponding 'pure blood' homooligomers. Mixed aggregate formation can promote rapid intracomplex hydrogen–metal exchange processes and thus explain the ease with which certain dimetalation reactions can be brought about[12].

Delocalized species such as allyllithium[136], 2-methallylpotassium[151] and benzyl-lithium[124,136] are mainly, if not exclusively, monomeric in dilute solutions in tetrahydrofuran, whereas some of those may preferentially form dimers in diethyl ether[150]. ^{13}C spectral perturbation by alkyl substituents or deuterium labeling allow one to differentiate between a permanent time averaged symmetry of allyl species. Allylmagnesium bromide[162] (**36a**) proved to entertain an organometallic σ-bond (η^1-type interaction between the metal and the carbon backbone) as previously suspected[163–165].

Allylpotassium (**36c**) and allylcesium exist as perfectly symmetrical π-complexes (η^1-type interaction). Allyllithium[162] (**36b**) and allylsodium[162,166] adopt again a π-bound structure although this time the metal, too small to bridge easily the C_3 unit, positions itself nearer to one and more distantly to the other terminal carbon atom.

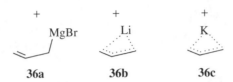

    36a   **36b**   **36c**

The more intensively an organometallic compound is resonance-stabilized, the weaker is the metal–carbon bond strength. It may even be small enough to be overcompensated by the solvation energy provided by an extra solvent molecule (Sv). If this is the case, the latter will replace the organic ligand from the coordination sphere of the metal and expel it as an anion. Nevertheless, this carbanion remains intimately associated with the metal counterion, both staying tied together by electrostatic forces. Thus, the organometallic *contact species* (sometimes incorrectly apostrophed as an 'intimate ion pair') metamorphoses to an *ion pair* (or 'solvent-separated ion pair'). An exemplary investigation has focused on the ionization behavior of fluorenylmetals. The contact species (**37a**) and ion pairs (**37b**) can be unambiguously differentiated by their electron excitation (UV/VIS) spectra (Table 4)[167].

Table 4. Absorption maxima (λ_{max}) of 9-fluorenylmetals in tetrahydrofuran (THF) at 25 °C as a function of the metal (M)[167].

M	C(9)–M distance[a] (Å)	λ_{max} (nm)	Species type
H[b]	1.0	305	hydrocarbon
Li	2.3	349	contact species
Na	2.7	356	contact species
K	3.2	362	contact species
Cs	3.6	364	contact species
$N(C_4H_9)_4$	4.5	368	contact species
$M(Sv)_n$	5.5	373	ion pair
—[c]	∞	374	free anion

[a] Average numbers taken from related crystallographic structures. [b] Fluorene (UV/VIS spectrum in hexane: d'Oliveira, J.M.R.; Pereira, V.R.; Martinho, J.M.G.; *J. Phys. Chem.* **1990**, *94*, 7090. [c] J. Smid, in *Ions and Ion Pairs in Organic Reactions*, vol. *1* (M. Swarc, ed.), Wiley, New York, **1972**, 85–151, spec. 98.

Small metals benefit from particularly strong solvation enthalpies. Only 9-fluorenyllithium is hence prodominantly ionized in tetrahydrofuran (see Table 5)[167].

In (mono)ethylene glycol dimethyl ether (MEGME, 'glyme') 9-fluorenylsodium also exists predominantly as an ion pair, as the bidentate ether is capable of securing entropy economic solvation (Table 5)[167].

37a   **37b**

Table 5. 9-Fluorenylmetals: proportion of ion pairs **37b** in equilibrium with contact species **37a** at 25 °C as a function of the solvent (THF, MEGME) and the metal (M)[167].

M	Sv = THF (%)	Sv = MEGME (%)
Li	80	100
Na	5	95
K	0	10
Cs	0	0

Nuclear magnetic resonance represents another technique with which to probe structural reorganizations of organometallic species. When the latter are 'titrated' with up to 10 equivalents of hexamethylphosphoric triamide (HMPA), one can differentiate three model cases. Tight aggregates such as methyllithium tetramer do not disintegrate but get peripherally solvated by the polar additive[168]. 9-Fluorenyllithium[168,169], triphenylmethyllithium[168] and lithium triphenylmercurate[168], which exist exclusively as ion pairs at low temperatures (−120 to −130 °C) and in THF-rich media, progressively replace all four THF molecules in the first coordination sphere of the lithium cation by HMPA. This is revealed by the ^{7}Li signal which, while experiencing a stepwise downfield shift over a total range of nearly 2 ppm, indicates by a gradual change of its multiplicity (s → d → t → q → quintet) how many phosphorus nuclei (0 → 1 → 2 → 3 → 4) are located in coupling distance to the metal. 2-Dimethylphenylsilyl-2-(1,3-dithianyl)lithium[169] (**38**) exists as a contact species in THF and other ethereal solvents, as testified by a 1 : 1 : 1 : 1 quartet in the ^{13}C spectrum. However, upon addition of HMPA (≥3 equivalents) at −130 °C, an ion pair forms as shown by the C(2) signal becoming a singlet (no ^{7}Li, ^{13}C coupling any more) and the ^{7}Li signal, a singlet in THF, presenting the quintet pattern characteristic for the [Li(HMPA)$_4$] entity[169]. If the metal-bearing carbon is a center of chirality, diastereotopic groups (marked by asterisks in **39**) can be used to monitor the rates for configurational changes. The coalescence temperatures determined for the methyls (at silicon) and the methylenes (in the pyrrolidine ring) in the amino-chelated bissilylmethyllithium **39** correspond to activation barriers of 11 and 10 kcal/mol[170].

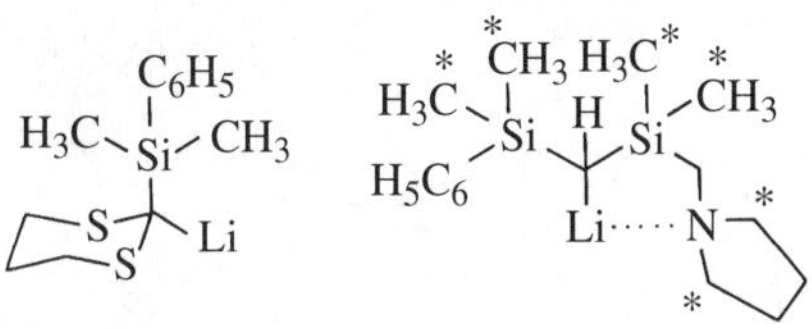

38   **39**

The dissociation of an ion pair into independently moving *free ions* essentially depends on the dielectric constant of the solvent. Only polar solvents such as *N,N*-dimethylformamide (DMF), dimethylsulfoxide (DMSO), hexamethylphosphoric triamide (HMPA), formic acid or water dispose of the required high ε (or E_T) values[171] to attenuate the electrostatic attraction between ions of opposite charge effectively. On the contrary, ethers, which have the advantage of being fairly inert towards most organometallic compounds, generally do not sustain ionic dissociation. Representative dissociation constants K (relative to the contact compound) in tetrahydrofuran fall in the range of 10^{-7} for sodium derivatives (9-fluorenyl sodium[167], polystyrylsodium[172], naphthalene–sodium 'radical anion'[173,174]), with sodium tetraphenylborate[172,175] ($K \approx 10^{-4}$) and sodium cyclopentadienide (K presumably[176,177] 10^{-9} to 10^{-10}) marking extreme deviations in both directions. The dissociation constants of the corresponding lithium derivatives are tenfold greater, whereas those of the potassium and cesium analogs are smaller by one and two powers of ten, respectively.

When assessing dissociation constants, one complication has to be taken into account. The equivalence conductivities abruptly diminish with increasing concentrations of the electrolyte, as one would predict on the basis of the ion product equilibrium. However, after having passed through a minimum (which typically falls in the 10^{-3} M range), the conductivities often rise again. The reason is the formation of 'triple ions'[178] (and higher ion 'swarms'), as first postulated for ethereal solutions of triphenylmethylsodium[179,180]:

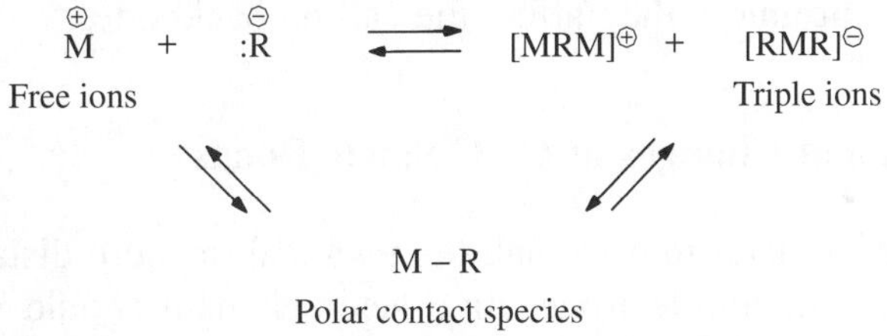

Related is the ionization–dimerization of organolithium species such as the bishomoaromatic[181] 3,6-bicyclo[3,2,1]octadien-2-yllithium[182] (**40**) and the aromatic cyclopentadienyllithium[183] (**41**) in tetrahydrofuran to afford lithium (persolvate)

diorganolithiate complexes. The di(cyclopentadienyl)lithiate and di(cyclopentadienyl)so-diate anions can be precipitated and crystallized when combined with tetraphenylphos-phonium as the counter-ion[184].

2.3   Stereomutations and Stereopreferences

Any description of the fascinating structures of transition element complexes would remain incomplete without full coverage of their numerous modes of internal mobility. Let us mention just the most prominent of such processes: the pseudorotation occurring at pentavalent centers[185–187], bridging–nonbridging or bonding–nonbonding tautomerisms[188–191], the rotation around a metal–π-ligand bond[192–198], the 'surfing' of the metal on fluorene, naphthalene or anthracene planes[196,199–201] and the face fidelity or face switching of η^3- or η^4-coordinated metals[202–205]. Polar organometallics lack this diversity of mechanisms to mediate structural metamorphosis. If dicyclopentadienylmagnesium (magnesiocene) displays in the crystal[96], but not in the gas phase[94,95], a staggered arrangement of its ten C−H bonds, this may simply reflect better packing forces. Moreover, organoalkaline and organoalkaline earth compounds are rarely requested in their own right but are rather prepared to be exploited as synthetic tools. For such practical reasons, metal-promoted stereomutations generally do not need to be taken into account unless they affect the carbon backbone.

2.3.1   Conformational Changes at C−C Single Bonds

The introduction of a metal in a vicinal, homovicinal or more distant position relative to either another metal or an electronegative heteroelement should give rise to dipolar interactions. No matter, whether attractive or repulsive in nature, these forces should cause conformational changes. Unfortunately, pertinent experimental or theoretical studies are still scarce.

In-depth investigations have been dedicated only to 1,4-dilithio derivatives[206], in particular *o,o′*-dilithiobiphenyls. Quantum chemical calculations had predicted an extra stabilization of the achiral, coplanar conformation to the extent of 9–17 kcal/mol, varying with the degree of solvation, due to intramolecular aggregation (see 'dimer' **42**) upon introduction of two lithium atoms at the *ortho* and *ortho'* positions[207,208]. The computational findings were at variance with work carried out with 2,2'-dilithio-1,1'-binaphthyl which was found to retain its atropisomerism in ethereal solution at temperatures up to $-45\,^{\circ}$C[209–211] much the same as 1,1'-binaphthyl itself. In other words, the torsional barrier was little affected by the absence or presence of metals capable of intermolecular bridging. However, the comparison between the hydrocarbon and its 2,2'-dilithio derivative may not be relevant, because the metal atoms may not face each other at the coplanar transition state of the racemization process, but rather collide with a hydrogen atom located at the 8-position of the neighboring naphthyl ring (**43**). This ambiguity can be obviated if one uses a rotationally restricted model compound. The 1,11-dilithio-5,5,7,7-tetramethyl-5,7-dihydrodibenz[*c,e*]oxepine (**44**) has no other choice than to orient the metal atoms towards each other when attaining the coplanar transition state. Nevertheless, the torsional barrier was determined to be only 2 kcal/mol smaller than that of the hydrocarbon[211]. Obviously, a favorable geometry is required to make aggregation profitable. This fits well into the concept of spatially oriented rather than ionic C−Li bonds[212].

2.3.2 Torsional Isomerization of Allyl and Benzyl Species

As all benzylmetal compounds, terminally substituted allylmetals are at least momentarily dissymmetric. Allylic derivatives of transition elements are capable of keeping this chirality at ordinary temperatures[203,204,213,214]. However, alkali or alkali-earth metals are too mobile to observe face or corner fidelity. They rapidly swing from one η^3-triangle across the $C^\alpha - C^i$ axis to the other side or slip from the upper to the lower π-face and back. According to all available evidence[215,216] such intrafacial and interfacial metallotropies occur almost effortlessly ($E_a < 8$ kcal/mol) within delocalized alkali(earth) metal compounds:

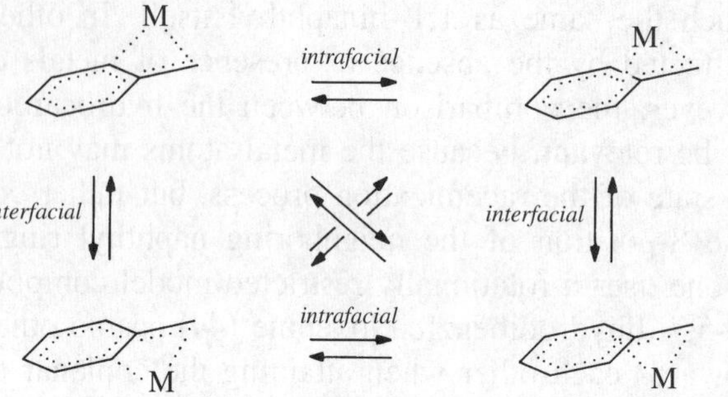

Metal mobility alone, however, does not suffice to remove dissymmetry if the benzyl-metal compound is isotopically labeled at both the α and one *ortho* position. The transition between the *syn* and *anti* isotopomer (*syn-* and *anti-*45) requires a 180° rotation of the benzylic center relative to the aromatic ring. At the transition state the C−M and π-orbitals of the two subunits occupy perpendicular planes. Consequently, all electronic delocalization is turned off. The resonance energy lost in this way has to be compensated by the torsional activation energy: some 35 kcal/mol[217] in the case of the naked carbanion, >20 kcal/mol[178] for benzylpotassium, 12 kcal/mol[218,219] for benzyllithium and an unspecified, though discrete, amount for benzylmercury chloride[220].

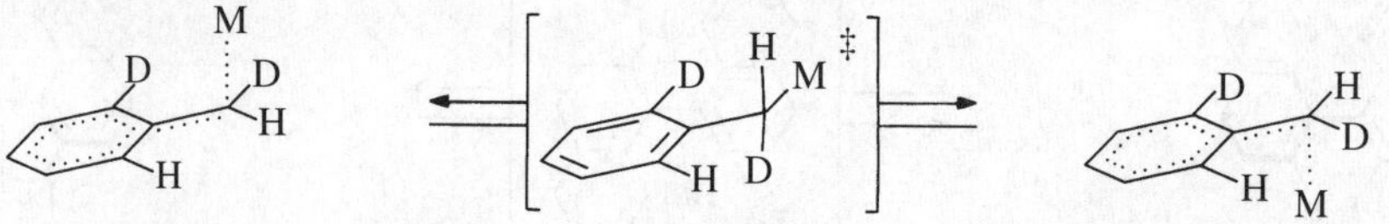

*syn-*45   *anti-*45

Torsional isomerism is a key feature of allylmetal chemistry. The rotation around one of the C−C axes is slow again in the free allyl anion ($E_a \approx 25$ kcal/mol[217]), or the potassium derivative ($E_a \approx 20$ kcal/mol[221]) and fast in the lithium ($E_a \approx 11$ kcal/mol[222,223]) and magnesium compound $E_a \approx 5$ kcal/mol[221]). It appears to be thermodynamically more advantageous to have the metal pointing inward (**46b**) rather than outward (**46a**) when it lies in the allyl plane at the transition state[224].

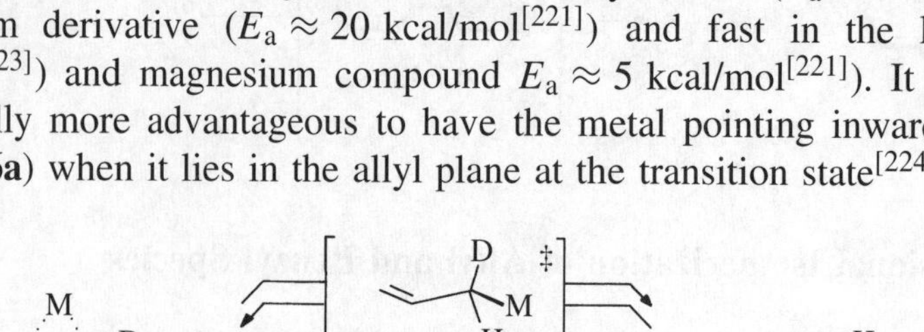

When terminally alkyl-substituted allylmetals were allowed to undergo torsional iso-merization, an amazing stereopreference was recognized. The sterically hindered (*Z*) or *endo* conformers (**endo-47**) proved to be preferred over the (*E*) or *exo* isomers (**exo-47**) despite their handicap of intramolecular steric hindrance. The *endo–exo* equilibrium ratios vary within wide limits, depending on the nature of the alkali metal attached to the organic backbone and the alkyl substituent R (see Table 6)[221,225,226]. The stereo-selectivities achieved with unbranched 2-alkenylpotassium intermediates are synthetically useful[227,228].

endo-**47**   *exo*-**47**

Table 6. Stereopreferences of polar 2-alkenylmetal compounds **47**: approxima-tive *endo/exo* ratios at equilibrium as a function of the metal (M) and the alkyl substituent R ($n = 1-\infty$)[225,226].

R	M = MgBr	Li	Na	K	Cs
H_3C	1 : 3	3 : 1	10 : 1	125 : 1	500 : 1
$H_3C(CH_2)_n$	—	—	—	15 : 1	—
$(H_3C)_2CH$	—	—	—	5 : 1	—
$(H_3C)_3C$	—	—	—	1 : 10	—
$(H_3C)_3Si$	—	—	—	1 : 20	—

An additional alkyl substituent R′ introduced at the 2-position, the nodal point, exerts little electronic effect but discriminates further against the *exo* conformer due to the steric repulsion caused. Typical *endo/exo* equilibrium ratios of 2-alkyl-2-alkenylpotassium intermediates (**48**) fall in the range 50–500[221,229].

endo-**48**   *exo*-**48**

Replacing the allyl-attached alkyl substituent by a vinyl group leads to 2,4-pentadienylmetal compounds. The extended area of delocalization and the added degrees of rotational freedom increases the number of approximately coplanar and hence privileged backbone structures. Again the metal and the substituent pattern, but sometimes also the solvent dictate the outcome of the torsional equilibration[221]. It was possible to "breed" several shapes with remarkably high stereoselectivities, for example the horseshoe-like pentadienylpotassium (***U*-49**, R = H)[230,231], 2,4-dimethylpentadienyllithium (***U*-50**, M = Li)[230] and -potassium (***U*-50**, M = K)[230] and 2,4-alkadienylpotassiums (***U*-49**, R = CH$_3$, C$_5$H$_{11}$, C$_6$H$_{13}$)[232] or the zigzag-like pentadienyllithium (***W*-51**, R = H)[230,231], 3-methyl-2,4-pentadienyllithium and -potassium (***W*-51**, R = CH$_3$, M = Li or K)[230] and *endo*- or *exo*-2,4-decadienyllithium and -undecadienyllithium (***endo*-*W*-52**, R = C$_5$H$_{11}$, C$_6$H$_{13}$; ***exo*-*W*-52**, R = C$_5$H$_{11}$, C$_6$H$_{13}$)[232]. Further extension of the area of delocalization favor the zigzag-band form. Heptatrienylpotassium (**53**, R = H, M = K), 3,7-dimethyl-2,4-6-octatrienyllithium or -potassium[233] (**53**, R = CH$_3$, M = Li or K), 3-methyl-5-(2,6,6-trimethyl-1-cyclohexenyl)-2,4-pentadienyllithium or -potassium[233] (**54**, M = Li or K) and retinyllithium or -potassium[233] (**55**, M = Li or K) all adopt preferentially or exclusively the most outstretched conformation.

u-49   *u*-50   *w*-51   *endo*-52   *exo*-52

53   54   55

Not only alkyl groups but also heteroatoms attempt to occupy *endo*-positions of allylmetal species. The metalation of allyl alkyl or allyl aryl ethers with *sec*-butyllithium immediately produces the alkoxy- or aroxyallyl species **56** (M = Li) as a pure *endo*-conformer[234–238]. The reductive cleavage (see pp. 138–148) of 1-methoxy-3-(phenylthio)propene with the naphthalene-potassium 1 : 1 adduct allows one to generate *exo*-methoxyallylpotassium (**56**, M = K) at −120 °C. At this temperature, isomerization to the thermodynamically more stable *endo*-form is slow with the potassium compound but instantaneous with the analogously prepared lithium species (**56**, M = Li)[239]. Alkoxy outperforms alkyl if in the competition for the more attractive *endo*-position as the *exo*-R-*endo*-OR′-structure of lithiated crotyl 2-tetrahydropyranyl ether (**57**) illustrates[240]. Upon α-lithiation of γ-branched allyl ethers (such as prenyl, neryl or geranyl methyl ether) the resulting organometallic intermediates **58** (R, R′ = CH$_3$ or 4-methyl-3-pentenyl) inevitably have to accommodate the methoxy group in the terminal *exo*-position[241].

endo-56   *exo*-56   57   58

2.3.3 Inversion of Metal-bearing Pyramidal Centers

Amines are known to undergo fast inversion of their pyramidal structure, the energy barrier associated with this process amounting to 6 kcal/mol in the case of ammonia[242–244]. Alkyl substituents, in particular if bulky, lower it moderately[242–244]. Resonance-active groups such as aryl, acyl or cyano reduce it considerably, even to the extent that the amine derivative (aniline, carboxamide or cyanamide) becomes nearly or perfectly flat[242–244]. However, ring strain obstructs planarization. Aziridine[245,246] and imines[247] (azomethines, Schiff bases) encounter barriers to inversion of about 18 and 28 kcal/mol (Fig. 6). Electronegative substituents can also diminish the conformational mobility of amines significantly. According to microwave spectroscopy, the inversion of fluoroamine requires an activation energy of 15 kcal/mol[248] (Fig. 6). Force field (MP-2) calculations set these threshold numbers at 18, 42 and 78 kcal/mol for fluoroamine, difluoroamine and trifluoroamine[249]. A combination of both effects, ring strain and electron-withdrawal, causes the conformations to 'freeze'. Thus, an inversion barrier of 77 kcal/mol was estimated for *N*-fluoromethaneimine[247] (Fig. 6).

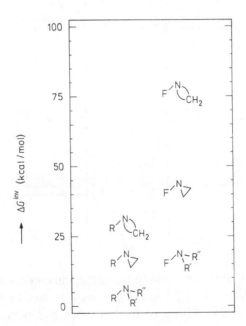

Figure 6. Approximate barriers to the pyramidal inversion of acyclic and cyclic amines (R, R′, R″ = hydrogen or alkyl).

Appropriately substituted oxaziridines[250–252] and *N*-chloroaziridines[253] are protected against inversion by barriers exceeding the critical threshold value of 23 kcal/mol and are hence configurationally stable at ambient temperature. Such diastereoisomers can be separated by chromatography.

As already briefly mentioned (see pp. 24–25, cpd. **39**), metal-bearing carbon atoms can also undergo pyramidal inversion. The isoelectronic relationship between amines and carbanions may guide us to understand and even predict the internal mobility of the latter species. As computational[254,255] and spectroscopic[256] investigations have revealed, the methanide anion adopts indeed a pyramidal shape (H−C−H $108 \pm 1.5°$ [256]) and its inversion requires an activation energy of only 1–2 kcal/mol[254–256]. Carbanions being elusive or often even fictuous species, what one would really like to know is how their organometallic derivatives behave.

The configurational inversion of such individuals must be more complicated because it does not make sense to turn the hydrocarbon part upside down if the metal does not move simultaneously from one face to the other. As long as the details of this transport problem have not yet been elucidated, one can only speculate whether an ion pair (**59**) is involved[257] or a S_E2-like rear-side attack of a lithium cation occurs[258] (transition state **60**) or a 'conveyer-belt' delivery mechanism[259,260] (**61**) operates through an 'open aggregate'[261].

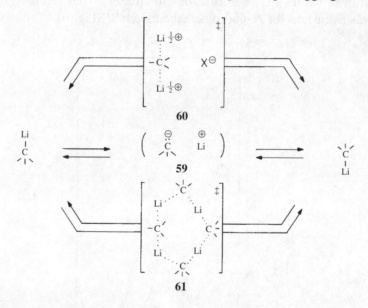

Optically active *sec*-alkyllithiums tend to racemize, but can conserve their configuration at least temporarily. When (−)-2-iodooctane was treated with an excess of *sec*-butyllithium in a 7 : 93 (v/v) mixture of diethyl ether and pentanes for 2 h at −70 °C, carboxylation afforded a small quantity of 2-methyloctanoic acid (5–10%) which had retained approximately 20% of its initial enantiomeric purity[262]. Unlike the halogen−metal exchange reaction, the metalloid−metal permutation between optically active bis(*sec*-butyl)mercury and 2-octyllithium can be brought about in a pure hydrocarbon medium. When the *sec*-butyllithium thus generated was carboxylated after 6 h at −40 °C or 0.5 h at −8 °C, 83% and 55% of the configuration were retained, respectively[263]. However, the addition of a little (6%) diethyl ether sufficed to cause complete racemization after a short while at −8 °C[263].

These trapping experiments have been corroborated and complemented by variable-temperature NMR studies. The barrier to pyramidal inversion at the metal-bearing methylene group was determined to be approximately 15 kcal/mol with 3,3-dimethylbutyllithium ('neohexyllithium') in diethyl ether[264] and with *sec*-butyllithium in pentanes[125]. Activation energies of 20 and 26 kcal/mol were estimated for the α-methylene inversion of dineohexylmagnesium[264] and dineohexylzinc[264], respectively. Trineopentylaluminium and dineopentylmercury were found to be configurationally stable at ambient and elevated temperatures (up to 160 °C)[264]. The epimerization of di-*sec*-butylmagnesium in diethyl ether at 45 °C follows a kinetic reaction order of 2.0 and is governed by activation parameters of $\Delta H^{\ddagger} = 18$ kcal/mol and $\Delta S^{\ddagger} = 5$ kcal/mol·K[265].

Aliphatic Grignard reagents appear in general to be configurationally less mobile than the corresponding lithium derivatives. Dicyclohexylmagnesium in diethyl ether at 25 °C rapidly interconverts axial and equatorial conformers. The 3 : 7 ratio corresponds to an *A* value of 0.5 kcal/mol[266]. The adduct **62** resulting from a Diels–Alder reaction between 1,2-dehydrobenzene ('benzyne') and cyclopentadienylmagnesium bromide conserves the configuration at the short bridge[267] despite the enhanced antiaromaticity[268,269] of this specific intermediate. *endo*-Norbornylmagnesium bromide (***endo*-63**) coexists with its *exo*-isomer in a 1 : 1 ratio at 25 °C, but can be obtained pure by crystallization at low temperatures[270]. It reacts with cuprous iodide or formaldehyde at −75 °C under complete retention of configuration[270]. In contrast, bornyllithium (**64**, the *endo*-form of which can again be isolated by crystallization from pentanes) isomerizes quickly to establish a 4 : 96 *endo/exo* equilibrium[271].

Similarly, *trans*-5-isopropyl-2-methyl-cyclohexyllithium (menthyllithium, **65**) rapidly equilibrates to a 11 : 89 *axial/equatorial* ratio[271,272]. The pyramidal inversion of *cis*-2,4-dimethyl-1,3-dithian-2-yllithium (**66**) favors the equatorial isomer again, this time even in an extreme way (*axial/equatorial* < 1 : 99)[273,274]. The axial 2-lithio derivative of *cis*-2,4-dimethyl-1,3-diselenane (**67**) is configurationally stable at −130 °C, but is rapidly converted into the equatorial epimer at −75 °C[275,276].

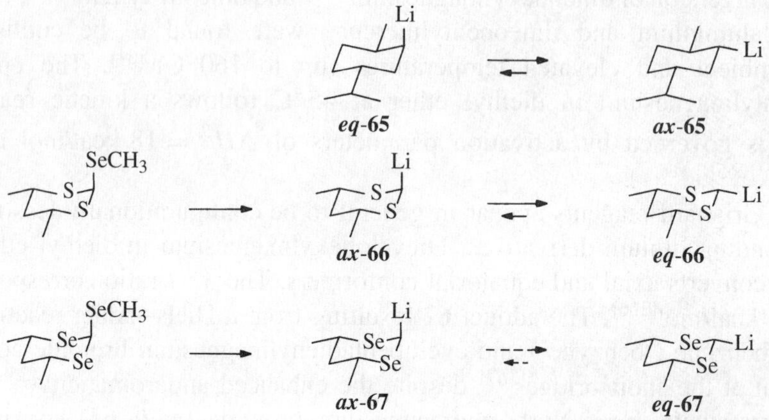

eq-65   *ax*-65

ax-66   *eq*-66

ax-67   *eq*-67

The cyclic intermediates **62–67** (see pp. 33–34) contain two or more asymmetric carbon atoms. All stereomutations were epimerizations rather than racemizations. The extreme positions of the diastereoisomeric equilibria monitored offer the possibility to secure stereochemical homogeneity. Such stereopreferences are generally less pronounced in the acyclic series. They can nevertheless reach useful levels and a few cases of practical exploitation are documented in the literature.

Metalloid/metal permutation (pages 159–171), by applying *sec*-butyllithium in THF at −75 °C to diastereomerically pure precursors, allows one to generate selectively the (*R**,*S**) and (*R**,*R**) forms of 1-phenylthio-2-(2-dioxolanyl)propyllithium (**68**), whereas metalation of the tin-free sulfide initially affords a mixture of both components. However, pyramidal inversion of the organometallic center leads rapidly to a 2 : 98 (*R**,*S**/*R**,*R**) composition[277]. Intramolecular coordination of lithium and a negative anomeric effect may be at the origin of this remarkable stereopreference.

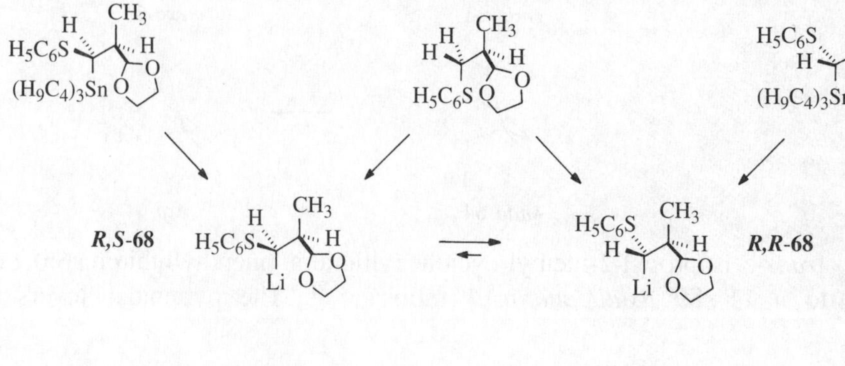

R,*S*-68   *R*,*R*-68

The *trans*-selective[278,279] and the 'three-dimensional' ('SCOOPY')[279,280] modification of the Wittig reaction rely again on such a low-temperature epimerization due to pyramidal inversion of a metal-bearing center. When a phosphorus ylid carrying an aliphatic side chain is allowed to combine with an aldehyde in ethereal, lithium bromide-containing medium, the *erythro*-betaine-LiBr (**erythro-69**) adduct is formed preferentially. Deprotonation of the α-position with phenyllithium creates a center of high configurational mobility. The resulting betaine-ylid **70** epimerizes instantaneously at −75 °C to afford a 1 : 200 *erythro/threo* mixture. Reprotonation or electrophilic substitution fixates the stereochemical uniformity thus acquired. Decomplexation followed by elimination of triphenylphosphine oxide leads to an alkene in which the hydrocarbon groups R and R′, provided by the aldehyde and the ylid, respectively, are positioned across the double bond.

In accordance with the amine analogy (see Fig. 6, p. 31), a series of cyclopropylmetal compounds was found to conserve their configuration under ordinary conditions (ethereal solvents, 25 °C). Examples are: 1-methyl-2,2-diphenylcyclopropyllithium[281–283] (**71**), 2-methylcyclopropyllithium[284], 2-phenylcyclopropyllithium[285], 2-([α-lithiumoxy-*N*-phenylimino]methyl)cyclopropyllithium[286] (**72**) and 3-(2-methyl-1-propenyl)-1,2-dimethylcyclopropylmagnesium bromide (**73**)[287].

   An indispensable condition is, however, to generate the organometallic intermediates
by a permutational process (*e.g.*, by halogen–metal or metalloid–metal exchange). If a
reductive method, in other words elemental metals or their 1 : 1 arene-adducts ('radical
anions') are employed, stereoequilibration becomes unpreventable. Thus, the treatment
of both *cis*- and *trans*-7-chloronorcarane with lithium/4,4'-di-*tert*-butylbiphenyl [4,4'-
di(*tert*-butyl)hydrobiphenylyllithium] afforded only *trans*-7-norcaranyl-lithium (***trans*-74**)
although the *cis*-isomer (***cis*-74**) of the latter intermediate, once formed, proved to be
perfectly stable[288].

   In the same way, β-alkyl- or β-aryl- substituted vinyllithiums (**75**, X = H) proved to be
configurationally stable[289–293]. However, partial or total stereoequilibration took place
again whenever elemental metals were involved in a reductive reaction mode[294–296].
Slow (Z/E)-isomerization was observed with 1-alkenyllithiums **75** carrying α substituents
capable of supporting p-orbital (phenyl[297], ethynyl[298]) or d-orbital (silyl[299,300],
bromo[301]) resonance. The conversion of a (Z)-1-cyano-1-alkenyllithium (**75**, X = CN)
into the (E) isomer even occurs rapidly[302]. However, α-chloro-[303], α-alkoxy-[304] and
α-arylthio-substituted[305] 1-alkenyllithiums (**75**, X = Cl, OC_2H_5, SC_6H_5) do not
isomerize at all under ordinary conditions.

| X = H: (*Z*)-**75** | | | | | | X = H: (*E*)-**75** |
| X ≠ H: (*E*)-**75** | R | Li | | R | X | X ≠ H: (*Z*)-**75** |

X	(Z/E)-isomerization
H, Cl, OC_2H_5, SC_6H_5	none
Br, $Si(CH_3)_3$, C≡CR, C_6H_5	slow
CN	fast

   α-Oxyalkyllithiums and α-aminoalkyllithiums are the most important in practical terms
among all chiral organometallic species. When considering their chemistry, one has to
distinguish the same three stages: generation, evolution and transformation.

Nonracemic α-oxyalkyllithiums can be obtained in two ways. Chromatographic separation of suitable diastereomeric precursors[306] or asymmetric reduction of acyltrialkylstannanes[307–310] affords homochiral (α-oxyalkyl)trialkylstannanes (**76**). Treatment of the latter with butyllithium or *sec*-butyllithium promotes a permutational metalloid–metal exchange. In accordance with all available evidence, this 'lithiodestanny-lation' can be assumed to proceed with retention in any case.

R-CH-OCH₂OCH₃
|
Cl

R-C=O
|
SnR′₃

R-CH-OCH₂OCH₃ ⟶ R-CH-OCH₂OCH₃
|
SnR′₃   **76**

R = R′ = alkyl

A second, less laborious access to nonracemic α-oxyalkyllithiums relies on the asymmetric α-deprotonation of *O*-carbamoyl-protected alcohols with *sec*-butyllithium in the presence of the natural alcaloid (−)sparteine, a superb achievement of D. Hoppe and his collaborators[311–315]. Beak *et al.*[316,317] discovered a major extension when they succeeded in the enantioselective α-lithiation of *N*-(*tert*-butoxycarbonyl)pyrrolidine. All such sparteine-assisted α-lithiations of resonance-inactive centers exhibit the same kind of stereoselection: (*S*)-configured intermediates **77** and **78** are produced with remarkably high enantiomeric excesses (e.e.) of 88–99%.

R-CH₂-R-C-N   O ⟶ R-C-O-C-N   O
             ‖ | ‖
             O   Li H O
                            SPA
R = alkyl, 1-alkenyl, aryl   **77**

N-C-OC(CH₃)₃ ⟶ N-C-OC(CH₃)₃
   ‖ | ‖
   O   Li H O
                   SPA
                        78

The survey of the configurational stability of α-oxy- and α-amino-substituted organolithium species given below (see Fig. 7) should be taken as only qualitatively valid. Despite the approximations made, distinct trends can be recognized. α-Alkoxy[306] and α-(carbamoyloxy)alkyllithiums[313] retain their configuration at any practically tolerable temperature. In contrast, the delocalized species derived from *O*-carbamoyl-protected allyl, propargyl or benzyl alcohols undergo rapid pyramidal inversion even at −75 °C[313]. Acyclic α-(dialkylamino)alkyllithiums rapidly racemize at temperatures around or above −50 °C[318]. Ring strain helps to preserve the chiral homogeneity: α-lithio-*N*-methylpiperidine maintains its configuration at −50 °C, whereas α-lithio-*N*-methylpyrrolidine and α-lithio-*N*-(4,5-dihydro-4-isopropyl-2-oxazolyl)piperidine show the same propensity even up to −25 °C[319]. Chelating appendices such as an *N*-2-methoxyethyl or

N-tert-butoxycarbonyl group tend to lower the barrier of pyramidal inversion, whereas a *N-tert*-butylcarbimino[318] substituent is only marginally detrimental and a *N*-(2-dihydrooxalyl) ring[319] appears to exert a stereostabilizing effect. Polar solvents (*e.g.*, glycol dimethyl ether) or complexands (*e.g.*, TMEDA, HMPA) may sometimes enhance the configurational mobility[319]. Benzylic *α*-aminoorganolithiums, such as 1-lithio-*N*-(4,5-dihydro-4-isopropyl-2-oxazolyl)-1,2,3,4-tetrahydroisoquinoline, rapidly racemize (or epimerize) even below −75 °C[320−322].

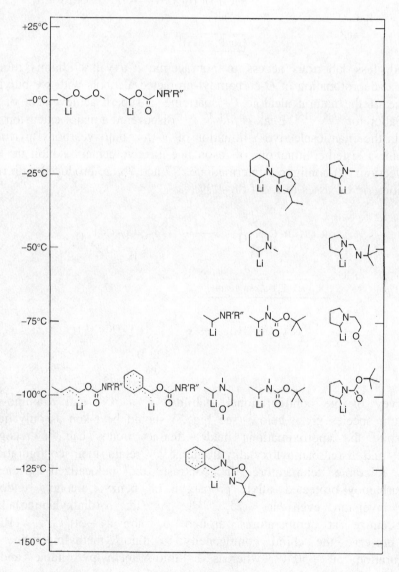

Figure 7. Configurational stability of *α*-oxy- and *α*-amino-substituted organolithium compounds: approximate threshold temperatures above which rapid racemization occurs.

Configurational mobility is not necessarily a misfortune. The equilibrium favors often one of the two diastereomeric sparteine adducts. Even if both are almost equally represented in solution, one of the diastereomers may precipitate and thus ascertain stereochemical uniformity. Such behavior was recognized first with *O*-α-lithiated crotyl *N,N*-diisopropylcarbamate[323,324] and benzyl *N,N*-diisopropylcarbamate[325]. In the latter case, ***R*-79**/(−)-sparteine *vs.* ***S*-79**/(−)-sparteine ratios of 43 : 57 and 31 : 69 were determined in diethyl ether and hexane solution, respectively, whereas the microcrystalline solid obtained upon addition of more hexane showed a 5 : 95 composition[325].

What complicates matters is the inability to predict the effect amine ligands and other reaction variables may have on the conformational mobility of individual chiral organo-lithium intermediates. Conclusive and, at the same time, confusing results in this respect have been reported with two benzyllithiums lacking an α-heterosubstituent but containing a secondary amide function at a more remote position[326].

A lithiated species ***R*-80** was obtained, when *N*-pivaloyl-2-ethylaniline was conse-cutively treated with *sec*-butyllithium (≥2 equiv.) in diethyl ether at −75 °C and (−)-sparteine ('SPA') at −25 °C (for 45 min). Quenching at −75 °C with silyl, stannyl and alkyl halides or aldehydes and ketones led to products **81** invariably under inversion of configuration. It was possible to regenerate from the stannane [**81**, *El* = Sn(CH₃)₃, 66% e.e.] with *sec*-butyllithium at −75 °C the intermediate **80**, although this time it had the opposite chirality (*S*). Upon trapping with chlorotrimethylsilane the antipodal silane ***ent*-81** [*El* = Si(CH₃)₃] was formed. However, the original (*R*) configuration of the silane was restored when, after the lithiodestannylation, the reaction mixture was kept for 45 min at −25 °C before chlorotrimethylsilane was added at −75 °C ('warm-and-cold protocol')[326].

The inference is obvious. At $-75\,°C$ and in the presence of sparteine, the benzyllithium species **80** maintains its configurational integrity whereas at $-25\,°C$ it undergoes stereo-equilibration which strongly favors the **R-80**/sparteine complex. A somewhat disturbing fact is that with *sec*-butyllithium alone or with *sec*-butyllithium in the presence of N,N,N',N'-tetramethylethylenediamine (TMEDA rather than SPA), the lithiodesilylation produced totally racemic material even at $-75\,°C^{[326]}$.

With *N*-methyl-3-phenylpropanoylamide exactly the opposite situation was encountered[327-329]. Regardless whether or not $(-)$-sparteine was present, the $(\pm)$-3-d_1 substrate was lithiated smoothly. The resulting racemic mixture of the 3-lithio species **82** equilibrated at $-75\,°C$ to afford instantaneously one pure enantiomer (supposedly **R-82**) after $(-)$-sparteine had been added[327]. The chiral alcaloid was undoubtedly involved in the equilibration step but apparently did not participate in the deprotonation process. Interception with chlorotrimethylsilane and chlorotrimethylstannane gave the silane **R-83a** (90% e.e.) and the stannane **R-83b** (60% e.e.)[328]. Consecutive reaction of the stannane with *sec*-butyllithium and chlorotrimethylsilane afforded the silane **R-83b** having the same sense of chirality (60% e.e.)[329].

One final mystery has to be mentioned. The metalation of the $(\pm)$-*N*-*tert*-butyloxy-carbonyl-*N*-methylbenzylamine-α-d_1 with *sec*-butyllithium/$(-)$-sparteine in hexane, diethyl ether or tetrahydrofuran at $-75\,°C$ occurred stereoselectively, only the *R*-enantiomer being consumed. In other words, stereoselection in conjunction with an extraordinarily high isotope effect brought about a kinetic racemate resolution.

Surprisingly, the resulting organometallic intermediate **84** nevertheless emerged as an antipodal mixture which only slowly, in the course of one or two hours, recovered its configurational preference in favor of one predominant **84** · SPA adduct (diastereomeric ratio presumably >9 : 1)[330].

Another unprecedented finding was the solvent dependence of the subsequent reaction with electrophiles. Deuteration (with phenylacetylene-ω-d_1), alkylation (with methyl iodide and dimethyl sulfate) and carboxylation (with carbon dioxide and carbon disulfide) invariably took place with retention in hexanes (55–90% e.e.) and diethyl ether (51–67% e.e.), but with inversion in tetrahydrofuran (68–85% e.e.)[330]. Two different explanations of the role of the solvent can be advanced. A plausible, though still tentative explanation of this solvent effect invokes differences in the tightness of carbon–metal interactions[330]. Structural extremes can be represented by the covalently bound species **85a** and by the 'false (hetero-connected) ion pair' **85b** in which one solvent molecule has pushed the carbon out from the first coordination sphere of the metal. The high electron density in the front lobe makes it attractive for the electrophile to attack under retention on the face occupied by the metal, despite severe steric congestion. However, when a more powerful donor solvent such as tetrahydrofuran secures sufficient stationary concentrations of ion pairs the inversion mode implying reaction at the unshielded opposite face, now of almost equal electron density, could prove more advantageous[330]. The sensitivity of the electrophile to steric repulsion, push-pull cooperative effects and long-range interactions ('softness') will ultimately dictate the stereochemical outcome.

Such model considerations may ultimately be extended to include true ion pairs (**85c**), a prerequisite for the formation of the latter being efficient charge delocalization and an inevitable consequence being racemization. 1-Methyl- and 1-butylindenyllithium represent such rare and very instructive examples[331]. In diethyl ether they combine with (−)-sparteine to form crystalline η^3-complexes which react with high regio- and stereoselectivity. In contrast, when tetrahydrofuran is added, the bisamine ligand has to give way to the latter solvent and a time-averaged achiral η^1-indenyllithium tri(tetrahydrofuranate) is obtained[331].

'Dynamic kinetic resolution' can serve as an alternative possibility to rationalize *retention/inversion* dichotomies. This concept has been invoked to account for strongly divergent enantiomeric composition (e.r. = enantiomeric ratio) of the products formed when α-lithio-2-ethyl-*N,N*-diisopropylaniline (**86**, generated by treatment of either the corresponding hydrogen- or stannyl-bearing precursor with *sec*-butyllithium and sparteine) was exposed to butyl chloride, bromide, iodide or *p*-toluenesulfonate[332]. The argument goes as follows. If the two diastereoisomeric **86** · SPA adducts equilibrate much faster than they are trapped by the electrophile, the structure of the latter and, in particular, the nature of its nucleofugal leaving group become crucial factors. Unequal matching between the reagent and either of the two competing diastereomeric substrates may cause differences in the substitution rates which, if pronounced enough, may override concentration effects. Thus, under special conditions, the major enantiomer formed may arise from the minor organolithium/(−)-sparteine diastereoisomer[332].

7-Phenyl-7-bicyclo[2.2.1]heptyllithium[333] and 1-phenylethyllithium[334] are configurationally very labile ($E_a^{inv} < 10$ kcal/mol). In view of this, the reduced pyramidal mobility of the benzylic organolithium intermediates **80** and **82** (see above) is astonishing unless one attributes a stabilizing effect to the intramolecular coordination of the lithium atom with the hetero atom of the imino (or, in other cases, carbonyl) group. It may be for the same reason that all electrophilic substitution reactions apparently occur with retention of configuration at the asymmetric center.

As emphasized above (taking the intermediates **85** and **86** as examples), the stereochemical course of an organometallic transformation depends on many parameters, including the nature of the electrophile and the solvent. The structure of the organometallic intermediate is of course a prime factor, controlling whether the retention or the inversion mode is favored. A few representative α-heterosubstituted organolithium compounds are examined in this respect to provide some crude orientation. The chirality of simple α-alkoxyalkyllithium appears to be delivered into products with retention under all conditions attempted[306–310]. The same holds for all undelocalized α-(carbamoyloxy)alkyllithiums[335,336] (see Table 7). On the other hand, the α-lithio- and α-titanio-drivatives of crotyl N,N-diisopropylcarbamate (**87a** and **87b**, respectively) combine with trimethylstannyl chloride exclusively under inversion of configuration regardless of whether an S_E or S'_E mode is employed[324,337,338]. The α-lithiated benzyl N,N-diisopropylcarbamate[339] favors inversion with most electrophiles but prefers retention when esters and anhydrides are applied (Table 7). However, the predominant (S) configuration of the lithiated intermediate **87a** · SPA must be considered to be the result of an equilibration of diastereoisomeric adducts ('dynamic thermodynamic resolution'), only species derived from *sec*-alkyl carbamates being configurationally stable in diethyl ether at $-70\,°C$[338].

N-Methyl-2-pyrrolidyllithium (see Table 7) and *N*-methyl-2-piperidyllithium retain their configuration when reacting with a variety of electrophiles. Exceptions are benzophenone, which causes racemization (presumably due to a single electron transfer mechanism), and primary alkyl iodides, which follow the inversion mode[319].

Table 7. Reaction between α-heterosubstituted alkyllithium and various electrophiles: retention (*ret*) or inversion (*inv*) of configuration.

Reaction type	Electrophile	ref.[335]	ref.[338]	ref.[319]
Protonation	HOOCCH$_3$	—	*inv*	—
Protonation	HC(C$_6$H$_5$)$_3$	—	*inv*	—
Protonation	HOCH$_3$	—	*ret*	—
Metalloidation	ClSi(CH$_3$)$_3$	*ret*	—	—
Metalloidation	ClSn(CH$_3$)$_3$	*ret*	*inv*	*ret*
Alkylation	ICH$_3$	*ret*	—	*inv*
Alkylation	IC$_4$H$_9$	*ret*	—	*inv*
Alkylation	BrCH$_2$CH=CH$_2$	*ret*	—	—
α-Hydroxyalkylation	O=CH−C$_6$H$_5$	—	—	*ret*
α-Hydroxyalkylation	O=C(CH$_2$)$_5$	—	—	—
Acylation	O=C(CH$_2$)$_2$	—	*ret*	*ret*
Acylation	H$_3$COCCOC$_6$H$_5$	—	*ret*	—
Acylation	O(COCH$_3$)$_2$	—	*ret*	—
Acylation	ClCOOCH$_3$	*ret*	*inv*	*ret*
Carboxylation	CO$_2$	*ret*	*inv*	—
Dithiocarboxylation	CS$_2$	—	*inv*	—
Carboxylation	OCNC$_6$H$_5$	—	*inv*	—

Whenever the trialkylstannylation occurs with inversion, the subsequent lithiodestannylation allows one to switch into the series of opposite chirality. This principle was first demonstrated with α-lithiated 1-phenyl *N,N*-diisopropylcarbamate[339]. A particular impressive example was elaborated starting with *N*-benzyl-*N*-BOC-*p*-anisidine (BOC = *tert*-butoxycarbonyl)[340]. Sparteine-assisted metalation with *sec*-butyllithium did not follow the precedents established with saturated carbamates (see p. 37), but produced the (*R*)-configured organometallic intermediate rather than its mirror image.

In fact, the abstraction of the *pro-(R)* hydrogen is not unfrequently observed with allylic or benzylic carbamates. The resulting species *R*-88 reacted with methyl trifluoromethane-sulfonate under retention[340]. A second metalation, this time performed with butyllithium in the presence of *N,N,N',N'*-tetramethylethylenediamine (TMEDA), produced a new intermediate *S*-89 which proved to be configurationally stable under the experimental conditions. Condensation with allyl trifluoromethanesulfonate and oxidation with cerium ammonium nitrate (CAN) to remove the 'protective' *p*-anisyl group gave (*S*)-*N*-BOC-1-methyl-1-phenyl-3-butenylamine as the final product[340].

OR = OC(CH$_3$)$_3$

The entire reaction sequence could be extended to the antipodal series. The key intermediate *S*-88 was generated by metalloid/metal interconversion. The precursor, (*S*)-*N*-BOC-*N*-[α-(trimethylstannyl)]-*p*-anisidine, was obtained by trapping the first lithiated intermediate *R*-88 with trimethylstannyl chloride[340].

3   Reactivity and Selectivity

The way a metal atom and a carbon radical establish a chemical bond between them-selves is the contrary of a Romeo and Julia story. The two elements *can* get together, but actually they do not want to. What is the reason for this lack of affinity? No matter whether we look at the sodium chloride crystal as a three-dimensional inorganic polymer kept together by electron-deficient bonds, or as a cubic lattice of electrostatically attracted cations or anions, we immediately realize how ideally the metal and the halide fit together due to their spherical symmetries and complementary electronegativities. The matching of an alkali or alkali-earth metal with an organic counterpart is energetically far less favorable. The two unlike binding partners have to enter into a compromise: the metal has to content itself with an imperfectly designed ligand and low coordination numbers, while the carbon binding partner has to tolerate hypervalency and a fractional negative charge imposed by the high polarity of the organometallic bonds.

Lack of thermodynamic stability means a high reactivity potential. The conversion of an organometallic compound into an essentially covalent hydrocarbon and a salt-like metal derivative is inevitably accompanied by a substantial gain in free reaction enthalpy and should thus provide an important driving force for the chemical transformation.

3.1   The Chemical Potential of Organometallic Reagents

A crude estimate of reaction enthalpies can be based on a comparison between the bond strengths of starting materials and reaction products. As a compilation of literature data[341] reveals, the homolytic dissociation enthalpies of metal species vary over a wide range (Fig. 8).

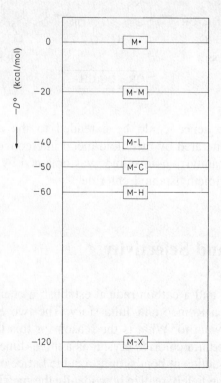

Figure 8. Approximate bond strengths-$D°$ (negative dissociation enthalpies) of elemental metals and archetypical derivatives (ligand L = CO, PH$_3$ etc.; X = F, Cl, Br, OCH$_3$ etc.).

Metal–metal bonds as present in small clusters[342–344] or in the bulk state are notoriously weak. On the other end of the scale we find molecular entities formed between metals and heteroatoms or heterofunctional groups. Metal halides of metal alkoxides, for example, are held together by exceptionally strong forces. Between these extremes are located metal hydrides, metal–ligand complexes and organometallic compounds. The respective bond strengths are moderate and may be compared to that of a carbon–iodine bond.

If we want to select the most suitable reagent for a given transformation, it is not enough to know that organometallics are globally reactive entities. What we need is kind of a yardstick that allows us to evaluate the relative chemical potential of any individual organometallic reagent.

Our attempt to solve this problem begins with the wrong assumption that naked carbanions were involved in our organometallic reactions. Then we would expect the reactivity to parallel roughly the proton affinity of those intermediates. As in reality we deal with organometallic contact species, we have to find out how and to what extent the metal attenuates the carbanion basicity. This is easily done if we just consider the polar organometallic bond as a chimera between a covalently bonded and an ionic nonbonded limiting structure. The former can be mimicked by the corresponding hydrocarbon, the latter by the free carbanion, both to be superposed in a phantom hybridization. With a highly electropositive metal such as potassium the charge-separated resonance form will become preponderant and the organometallic compound will conserve most of the carbanionic reactivity. At the other extreme, a weakly electropositive metal such as mercury will favor the homeopolar resonance form and make the reagent almost as inert as a pure hydrocarbon.

To quantify the chemical potential of all typical organometallic species, we may assign 'polarity coefficients' to various metals and metalloids: hydrogen 0%, mercury 5%, zinc 10%, magnesium 20%, lithium 40%, sodium 60%, potassium 80% and cesium 90%. Although these numbers[345] are more or less arbitrarily chosen, they help to illustrate the principle. The reactivity potential of a given carbanion can be most conveniently expressed by its gas phase basicity[217,346]. Such parameters are well documented nowadays (see Table 8). For example, the proton affinity of the methyl anion exceeds that of the phenyl anion by 15 kcal/mol. According to our set of polarity coefficients, organolithium compounds should exhibit only 40% of carbanion character. When we compare no-longer 'naked' anions but their lithium derivatives, the basicity difference between methyl and phenyl species should shrink to about 5 kcal/mol.

Table 8. Gas phase acidities: enthalpies ΔH_g° and free energies ΔG_g° of deprotonation (kcal/mol).

Anion	ΔH_g°	ΔG_g°	$\Delta\Delta G_g^{\circ\,[a]}$	Ref.[b]
tert-Butyl	413.1	405.7	−2.9	[346]
Isopropyl	419.4	411.4	+2.8	[346]
Ethyl	420.1	411.4	+2.8	[346]
Methyl	416.7	408.6	0.0	[347]
Cyclopropyl	412.0[c]	403.4[c]	−5.2	[346,348,349]
Ethenyl (vinyl)	409.4	401.0	−7.6	[350]
Amide	403.6	396.1	−12.4	[351]
Hydride	400.4	394.2	−14.4	[352]
Benzenide (phenyl)	401.7	392.9	−15.7	[353]
Hydroxy	390.7	384.1	−24.5	[354]
2-Propenyl (allyl)	390.7	384.1	−24.5	[355]
Trimethylsilylmethyl	388.2	379.0	−28.5	[356]
Phenylmethyl (benzyl)	380.8	373.7	−34.9	[355]
Ethynyl (acetylide)	378.0	369.8	−38.8	[350]
2,4-Pentadienyl	369.2	364.4	−44.2	[357]
Diphenylmethyl	363.6	358.2	−50.4	[355]
Triphenylmethyl	358.7	352.8	−55.8	[358]
Cyclopentadienyl	353.9	347.7	−60.9	[355]
9-Fluorenyl	351.7	344.0	−64.6	[358]
9-Phenyl-9-fluorenyl	343.2	335.5	−73.1	[358]
Iodide	314.3	309.2	−99.4	[359]

[a] Relative to methane ($\Delta\Delta G_g^\circ \equiv 0$). [b] Compilation from ref.[360]. [c] Average of three recent literature values.

The chemical potential of typical organometallic compounds can thus be represented as a function of the metal and the organic counterpart in a simple graph (see Fig. 9). Electropositive metals such as potassium and cesium retain a maximum of the inherent carbanion basicity. In contrast, when the metal is replaced by hydrogen, both polarity and chemical potential vanish.

It would of course be naive to believe that all organometallic reactivity could be described by merely multiplying two parameters, the polarity coefficient of the metal and the gas phase basicity of the carbanion. This approach must fail in many cases because it is based on oversimplifications. It postulates linearity in the ground state behavior and

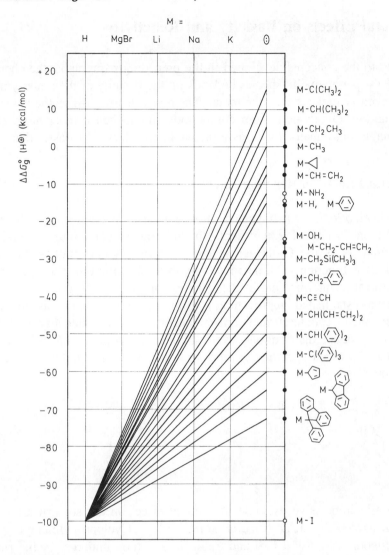

Figure 9. Gas phase basicities of free carbanions (listed on the right) and their attenuation as a function of the metal electropositivity.

ignores the nature of intermediates and transition states involved in the reaction under inspection. Moreover, for the sake of consistency, it attributes to secondary and tertiary sites in hydrocarbons lower C—H acidities than to ethane, and thus denies experimental reality (see Table 8). Despite such shortcomings, the model has the merit to reflect some essential features of organometallic species. On the whole, it should lead to reasonably reliable predictions about acid-base type processes, including hydrogen—metal interconversion, the arguably most fundamental organometallic reaction mode.

3.2   Metal Effects on Basicity and Reactivity

According to the concepts developed in the preceding paragraphs, the chemical potential of a polar organometallic species depends on the basicity of the underlying carbanion which, however, is attenuated by the metal. The more electropositive the latter, the weaker is the attenuation. Organopotassium compounds should hence react more exothermally than their magnesium analogs and, as a plausible corollary, also faster than the latter.

3.2.1   Metal Effects on Reaction Energies

The postulated role of the metal can be most conclusively demonstrated by equilibration processes. In one series of experiments the hydrogen–metal interconversion between toluene derivatives as the C–H acids, and benzylic organometallics **90** as the bases, was studied as a function of time in tetrahydrofuran solution. The appearance of the new organometallic intermediate at the expense of the initial one was monitored until equilibrium was established, approaching it from both sides. With potassium as the metal, the hydrogen–metal countercurrent transfer occurred much faster than with lithium and more extreme product distributions were attained. A slope of $pK^{M=K}/pK^{M=Li}$ of 1.4 resulted when the metal specific equilibrium constants were plotted against each other (see Fig. 10)[361].

A particularly impressive example of a metal effect on an isomerization equilibrium has been described by J. D. Roberts *et al.*[362,363]. Cyclopropylmethylmetals[364–366] as the analogous cations[367,368] and radicals[369] too, undergo facile ring opening affording the corresponding 3-butyl ('homoallyl') isomers. Exactly the same happens with (cyclopropyldiphenyl)methylmagnesium bromide (**91–MgBr**) which, as soon is generated, rearranges to give 4,4-diphenyl-3-butenylmagnesium bromide (**92–MgBr**)[362,363]. However, the other isomer is favored when potassium is involved. The colorless 4,4-diphenyl-3-butenylpotassium (**92–K**) disappears to reemerge instantaneously as the bright red (cyclopropyldiphenyl)methylpotassium (**91–K**)[362,363].

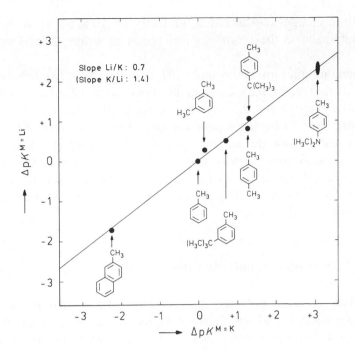

Figure 10. Countercurrent transfer of a proton and a metal between two benzyl entities: metal effect on the equilibrium constants.

To understand this metal-dependent reversal of relative thermodynamic stabilities it is helpful to consider the extreme cases, the hydrocarbons (**91** and **92**, M = H) and the carbanions (**91** and **92**, M = :). The olefinic hydrocarbon should be thermodynamically more stable by about 15 kcal/mol than its three-membered ring isomer. Contributions that favor the ring-opened form are relief of ring strain (cyclopropane: 27 kcal/mol; olefin: 22 kcal/mol, *i.e.* the difference in the C—C bond strength between two ethanes and one ethylene), relief of steric strain due to angle widening ($\ni$C—C—C of 112° and 122° in the cyclic and acyclic derivatives, respectively), double bond conjugation and free rotation around one extra carbon—carbon bond in the acyclic isomer. At the carbanion level, the same parameters are effective. In addition, however, delocalization of the negative charge into the two phenyl rings comes into play and proves to be a dominant factor. Resonance should indeed stabilize the cyclopropyldiphenylmethyl anion (**91**, M = :) relative to the isomeric primary carbanide (**92**, M = :) by 45, if not 50 kcal/mol.

The 'morphology bonus' of about 15 kcal/mol in favor of the open form is metal invariant and remains always the same. It is largely overcompensated for by the delocalization term, not only when the free carbanion is considered but also in the case of the potassium derivative (**91**, M = K). If we suppose, for sake of argument, that the latter species has still retained four-fifths of the full carbanion character, the resonance stabilization will still amount to some 35 kcal/mol. However, the magnesium analogue (**91**, M = MgBr) is carbanion-like only to the extent of about one-fifth, and consequently benefits little from charge delocalization, probably by 10 kcal/mol at best. This is not enough to counterbalance the morphological bonus of some 15 kcal/mol for the ring opening.

The corresponding lithium compounds (**91** and **92**, M = Li) behave as structural cameleons: in diethyl ether they exist in the open form (**92–Li**, colorless), whereas in tetrahydrofuran they prefer the ring closed shape (**91–Li**, red)[362,363]. Obviously, additional factors have to be taken into account. The open form should benefit from substantial extra stabilization due to the aggregation phenomenon which is so typical for alkyllithium compounds. Nevertheless, when diethyl ether is replaced by tetrahydrofuran, this superior donor solvent enhances the polarity of the organometallic bond in **91** or even breaks it up into ion pairs. In either case, the gain in resonance energy suffices to favor the delocalized cyclic isomer.

3.2.2   Metal Effects on Absolute Reaction Rates

Metals with low ionization potentials (see p. 11) boost the reactivity of organic entities to which they are bonded, and the elements themselves dispose of a powerful reducing capacity. The most 'electropositive' alkaline and alkaline earth metals are capable of transferring electrons to simple aromatic hydrocarbons. Magnesium reacts with anthracene in tetrahydrofuran solution to form the 'dianion' 9,10-magnesia-9,10-dihydroanthracene (**93**)[370,371]. When in contact with biaryls or naphthalene, lithium, sodium and potassium readily generate 1 : 1 adducts (radical anions, *e.g.* **94** or **95**)[372–379], but 1 : 2 adducts ('dianions') only under exceptional circumstances[380–382].

93   94   95

Lithium, sodium or calcium dissolved in ammonia or alkylamines can reversibly generate radical anions even with benzene. These species may be trapped *in situ* with proton sources such as water, methanol, ethanol or *tert*-butyl alcohol[383]. This is the basis for reductions with dissolved metals. Extending the work of W. Hückel[384] and C. B. Wooster[385], A. J. Birch[386,387] elaborated the method, named after him, which enables the conversion of benzene into 1,4-cyclohexadiene, naphthalene into 1,4,5,8-tetrahydronaphthalene and 3-methoxy-17-estrol into its 1,4-dihydro derivative[388,389]. In aprotic media, only cesium reacts with benzene[390]. The product, a black solid, was originally claimed to be phenylcesium[391], although soon doubts were expressed about this assignment[392]. It required a careful study at variable temperature to elucidate the truth[390]. The black 1 : 1 adduct **96** already forms at −75 °C. At −45 °C it dimerizes to afford the yellow 6,6′-bis(2,4-cyclohexadienylcesium) **97** which, upon warming to +35 °C, loses two equivalents of cesium hydride. The resulting biphenyl combines with still unconsumed metal to produce a black mixture containing the 'dianion' **99** as the main component[390] and presumably also some radical anion **98**.

While no cesium hydride is split off from the radical anion **96**, such an elimination does occur from the toluene–cesium 1 : 1 adduct. The transient benzyl radical thus generated is metal trapped to give benzylcesium[393,394].

The gain of rearomatization energy is one major factor favoring hydride elimination, the M−H bond strength[395] is another. Neither cyclohexadienyllithium (**100a**, M = Li)[396] nor cyclohexadienylsodium (**100b**, M = Na)[397] survive long enough to be intercepted by an electrophile. However, cyclohexadienylpotassium (**100c**, M = K)[398,399] and alkylated derivatives[400] thereof were found to be stable in tetrahydrofuran at −75 °C. An ethereal solution of 1,4-dihydro-1-naphthyllithium[401] can be even stored at +25 °C. The seemingly 'inverse-metal effect', that is the superior reactivity of the organolithium compared to the more polar organopotassium species, finds a plausible rationale. Evidently the metal hydride bond strength, rather than the relatively weak interaction of the metal with the pentadienyl-type organic backbone, dictates the course of the reaction this time[402].

The metal-promoted scission of arylated ethanes is doubtlessly initiated by single-electron transfer, generating transient radical anions and radicals as postulated above by the sequence of events leading to benzylcesium. Not unexpectedly, one sees the natural metal reactivity restored. Although the hexaphenylethane dimer[403] **101** is reductively cleaved by any alkali metal and 1,1,2,2-tetraphenylethane[404] by potassium, cesium is the only bulk metal to cope with 1,2-diphenylethane[405,406].

Both 'radical anions' and γ-aryl substituted alkylmetals are prone to carbanion elimination. 3,3-Diphenylpropyllithium (**102**) decomposes smoothly at ambient temperature to afford ethylene and the resonance stabilized diphenylmethyllithium (benzhydryllithium)[407]. The analogous magnesium compound is stable under such conditions. Surprizingly the opposite behavior was observed with 3-methyl-3-(4-pyridyl)butylmagnesium chloride (**103**) which rapidly decomposes under evolution of ethylene, whereas its lithium analog proved to be perfectly stable[408]. The inverse metal effect points at a mechanistic switch from a push to a push–pull mechanism.

Another push–pull mechanism appears to be operative with triphenylmethylmagnesium bromide. It readily reacts with tetrahydrofuran to form the adduct bromomagnesium

5,5,5-triphenylpentanolate **(104)**[409]. Triphenylmethyllithium, -sodium or -potassium are stable under the same conditions. Evidently, the ring opening of the cyclic ether requires electrophilic participation. The alkaline earth metal is, of course, a stronger Lewis acid than any alkali metal ion.

The most extreme incarnation of pull-driven reations are represented by carbocation mechanisms. They are treated in a subsequent subsection (see pp. 81–84).

Organometallics may be cleaved under carbanion expulsion even if the nucleofugal leaving group is only moderately resonance stabilized. Thus, 2,2,3-triphenylpropyllithium and -sodium (**105**, M = Li, Na) dissociate into 1,1-diphenylethylene and benzylmetal. However, neither of the two fragments can be directly detected as they recombine instantaneously to afford the extensively delocalized 1,1,3-triphenylpropylmetal. Crossover experiments using isotopically labelled material have provided unequivocal evidence for the two-step elimination–readdition sequence[410].

2,2,3-Triphenylpropylpotassium and -cesium (**105**, M = K, Cs) give mixtures of isomerization products by competing benzyl and phenyl 1,2-migrations. The latter reaction mode is known as the Grovenstein–Zimmerman rearrangement[411–421]. It is a strictly intramolecular process. Only aryl groups qualify for the migration as they can form at the transition state a bridged, pentadienylmetal-like structure. Like any spontaneous isomerization of an organometallic compound, the rearrangement aims at the conversion of a more basic into a less basic entity. Therefore, 2-arylethylmetals do not undergo degenerate 1,2-migration. Rearrangement does occur, although slowly (in diethyl ether at about +35 °C), with 2,2-diarylethylmetals such as 2,2-diphenylpropyllithium[417] and 9-[9-(2,2-dimethylpropyl)fluorenyl]methyllithium[421]. It is faster with 3,3,3-triphenylethylmetals. Moreover, the metal effects on the rearrangement rates are very pronounced. Bis(2,2,2-triphenylethyl)mercury is a perfectly stable compound which can serve as the precursor to the other metal derivatives **106** when submitted to metal–metal cleavage. Whereas the magnesium compound requires reflux temperatures to bring about slow 1,2-phenyl migration *via* the bridged transition state **107**, affording the 1,1,2-triphenylethylmetal isomer

108, rearrangement of the corresponding lithium, sodium, potassium and cesium species occur in tetrahydrofuran at about 0, −50, −75 and −100 °C, respectively[419].

The gain in resonance energy constitutes the driving force for the rearrangement. It increases with the electropositivity of the metal, as we have already seen. For the reaction kinetics, however, the extent of resonance stabilization at the pentadienylmetal-like 1,2-aryl bridged transition state becomes the crucial issue. The transient bridged species **107** can evolve into a stable intermediate **109** if its charge delocalization is further improved by replacing the migrating phenyl by a *p*-biphenyl group. At the same time, any resonance stabilization of the possible rearrangement product must be avoided (hence **109**, R = H, CH_3 or CD_3)[420].

3.2.3   Metal Effects on Relative Rates

A substrate and a reagent frequently have different reaction channels at their disposal, each one leading to different products. The latter may belong to different classes (types) of compounds or, if isomers, exhibit different connectivity or spatial orientation patterns. If different metals affect the competing reaction modes unequally, a smart choice of the element may suffice to secure typoselectivity (synonymous with the more diffuse expression 'chemoselectivity'), regioselectivity or stereoselectivity.

a) Typoselectivity

When an organometallic reagent encounters an aliphatic nitrile, it may act as a base or as a nucleophile. Depending on its intrinsic preference, it will promote either α-deprotonation or nucleophilic addition producing, respectively, an enolate-like metal ketene imide **110**, which can be readily alkylated at the α-position, or a metal ketimide **111** which gives a ketone upon hydrolysis. This fundamental dichotomy was first recognized in industry. Whereas organomagnesium reagents have long been known to undergo clean addition[422,423], lithium dialkylamides[424] and phenylsodium[425] were found to elicit exclusively the deprotonation mode. Thus, it needs nothing more than a primitive modification of the protocol to access selectively either of two functionally different products.

$$R\text{-}CH_2\text{-}C{\equiv}N \xrightarrow{M\text{-}R'}$$

$$R\text{-}CH{=}C{\equiv}N\text{-}M \qquad \mathbf{110} \xrightarrow{R''CH_2Br} \overset{R''CH_2}{\underset{}{R\text{-}CH\text{-}C{\equiv}N}}$$

$$R\text{-}CH_2\text{-}\underset{R'}{C{=}N}\text{-}M \qquad \mathbf{111} \xrightarrow{H_2O} R\text{-}CH_2\text{-}\underset{R'}{C{=}O}$$

Acetylides[426,427] are too weakly basic to deprotonate nitriles. Thus, phenylethynyl-magnesium bromide, -lithium, -sodium and -potassium react with both acetonitrile and benzonitrile under clean nucleophilic addition. Furthermore, they exhibit the ordinary metal effect: the more polar the organometallic bond, the faster the reaction[428,429]. As it is plausible to assume that alkyl- and arylmetals follow the same trend, polarity seems to affect deprotonation much more strongly than addition, isomerization or condensation reactions. As discussed later, this appears to be a general phenomenon.

The polarity of organometallic bonds can be mimicked by medium effects. Aliphatic nitriles may be deprotonated by organolithium reagents if the latter are employed in diethyl ether rather than in hexanes[430]. Grignard reagents add to diisopropyl ketone in ethereal solvents but generate the enolate by α-proton abstraction in the presence of hexamethylphosphoric triamide (HMPT)[431].

All aliphatic or aliphatic aromatic ketones are subject to the deprotonation/nucleophilic addition rivalry. The typoselective outcome of the reaction of acetophenone with phenylmetal was investigated by C. R. Hauser *et al.*[430]. The tertiary alkoxide **113** is formed quantitatively when the Grignard reagent is used. In contrast, phenylpotassium exclusively affords the enolate **112**, whereas phenylsodium and phenyllithium produce mixtures containing both intermediates (see Table 9)[430].

Table 9. Reaction between acetophenone and phenylmetal: yields of enolates **112** *vs. tert*-carbinolates **113**
as a function of the metal (M)[430].

M	Enolate **112**	Carbinolate **113**
MgBr	0%	97%
Li	75%	14%
Na	60%	4%
K	67%	0%

In general, organometallics considerably surpass alkoxides in their reactivity. This
does not necessarily hold for the weakly basic acetylides. The typoselective competition
between the two nucleophilic sites of potassium ω-potassiooxy-1-alkynides **114** (M = K)
upon treatment with alkyl bromides ends in favor of ether formation[432] since high
electron density is concentrated at the oxygen atom. The corresponding lithium derivatives (**114**, M = Li) afford carbon-chain-lengthened products[433]. Lithium alkoxides are
virtually inert towards electrophiles[432].

$$M\text{-}C{\equiv}C\text{-}(CH_2)_n\text{-}OM + BrCH_2R \quad \xrightarrow{M = K} \quad K\text{-}C{\equiv}C\text{-}(CH_2)_n\text{-}OCH_2R$$

114

$$\xrightarrow{M = Li} \quad RCH_2\text{-}C{\equiv}C\text{-}(CH_2)_n\text{-}OLi$$

n = 2–8

b) Regioselectivity

Regiochemical ambiguity can manifest itself in various ways. The arguably most
common one is the competition between two or more potential sites of deprotonation.
Terminal alkenes tend to afford directly 2-alkenylmetals (*i.e.* allylic species) when exposed
to alkylpotassiums, whereas alkylsodiums may generate predominantly 1-alkenylmetals,
at least in the initial stages of the reaction before extensive transmetalation sets in[434].
Such transmetalation processes can bring about the isomerization of 1-alkenes to 2-alkenes
if only catalytic amounts of the organometallic reagent are employed[434–437].

$$R\text{-}CH_2\text{-}CH{=}CH\text{-}M \quad \xrightarrow{R\text{-}CH_2\text{-}CH{=}CH_2} \quad R\text{-}\overset{M}{CH}\text{-}CH\text{-}CH_2$$

$$\Big\uparrow M\text{-}R' \quad -[H\text{-}R'] \qquad M\text{-}R' \quad -[H\text{-}R'] \qquad \Big\downarrow R\text{-}CH_2\text{-}CH{=}CH_2$$

$$R\text{-}CH_2\text{-}CH{=}CH_2 \qquad\qquad\qquad R\text{-}CH{=}CH\text{-}CH_3$$

The rate of isomerization increases steeply with the electropositivity of the metal[435]. The same metal dependence also holds for the conversion of 2-alkynes into metal 1-alkynides (acetylides)[438], a transformation which for obvious reasons requires stoichiometric amounts of a metalating reagent.

$$\left(\overset{M}{R\text{-}C\!\vdots\!C\text{-}CH_2} \right) \qquad \left(\overset{M}{R\text{-}CH\text{-}C\!\vdots\!CH} \right) \qquad R\text{-}CH_2\text{-}C\!\equiv\!C\text{-}M$$

M-R'↑   R-C≡C-CH₃⟍   R-C⋮C-CH₂↑   R-C≡C-CH₃⟍   R-C⋮C-CH₂↑

$$R\text{-}C\!\equiv\!C\text{-}CH_3 \qquad \left(R\text{-}CH\!=\!C\!=\!CH_2 \right) \qquad R\text{-}CH_2\text{-}C\!\equiv\!CH$$

A suspension of butylsodium in petroleum ether reacts with cyclopentene ten, if not up to hundred times faster than with cyclohexene, exclusively at the olefinic position (affording intermediate **115a**, M = Na)[439]. In contrast, butylpotassium discriminates little between the two cycloalkenes as far as the rates are concerned but, while attacking the smaller ring simultaneously at the olefinic and allylic positions (producing intermediates **115a** and **115b**, M = K), cyclohexene is solely converted into the allylic species (**116b**, M = K, no regioisomer **116a** being formed)[439].

With alkylarenes as substrates, the issue becomes aromatic *vs.* benzylic metalation. Cumene (isopropylbenzene) represents a model case. The selective deprotonation of the exocyclic α-position, which leads to the least basic cumylmetal isomer, can only be accomplished with trimethylsilylmethylpotassium[440], not even with the superbasic mixture of butyllithium and potassium *tert*-butoxide ('LIC-KOR'; see pp. 248–249)[441]. Ethylpotassium abstracts protons concomitantly from the *ortho*, *meta*, *para* and α positions of cumene in a ratio of 7 : 23 : 25 : 45[442,443]. Ethylbenzene and toluene can be cleanly metalated at their benzylic positions when trimethylsilylmethylpotassium or the LIC-KOR superbase are used[444], whereas the reaction is notoriously messy with alkylpotassiums[443–445], alkylsodiums[443] and butyllithium[446,447], irrespective of whether or not the latter reagent is complexed with *N,N,N',N'*-tetramethylethylenediamine (TMEDA).

Heterosubstituted toluenes offer further examples of competition between aromatic and benzylic sites. Butyllithium in the presence of TMEDA attacks 2-methylanisole simultaneously at the α- and 6-position[448], whereas pentylsodium gives pure 2-methoxybenzylsodium (**117**)[449]. The *para*-isomer 4-methylanisole behaves regioselectively towards butyllithium, just affording 2-methoxy-5-methylphenyllithium (**118**)[449]. In contrast, the LIC-KOR reagent abstracts protons randomly from the benzylic and the oxygen-adjacent position[444]. Finally, *N,N*,4-trimethylaniline undergoes metalation with TMEDA-activated butyllithium exclusively at the position next to the nitrogen atom[450], whereas the LIC-KOR reagent promotes the hydrogen–metal exchange solely at the benzylic position (89% and 65% of products derived from intermediates **119** and **120**, respectively)[451,452].

117   **118**   **119**   **120**

In summary, the more polar organometallic reagents (K > Na > Li) tend to produce the more polar and less basic allyl and benzyl species, whereas lithium and to some extent sodium reagents prefer the kinetically more efficient olefinic and aromatic positions. Attack at sp^2-centers occurs with particular ease if a nearby heterosubstituent can provide neighboring group assistance (see pp. 251–253).

Site-specific proton abstraction is also a prerequisite for the controlled generation of structurally uniform enolates. Asymmetric ketones such as 2-methylcyclohexanone (**121**, R = H$_3$C), 2-phenylcyclohexanone (**121**, R = H$_5$C$_6$) and 4-cholesten-3-one (**122**) can be irreversibly deprotonated at the readily accessible α-methylene group when strong and bulky bases such as triphenylmethylsodium[453–455], triphenylmethyllithium[456,457], triphenylmethylpotassium[458], lithium diisopropylamide (LIDA)[459–461] or lithium cyclo-hexyl(isopropyl)amide (LICA)[461] are employed. Alternatively, the branched enolate, the thermodynamically more stable isomer, can be selectively obtained when relatively weak bases are used under equilibrating conditions (*e.g.* sodium hydride in diethyl ether[462] or potassium *tert*-butoxide in refluxing *tert*-butyl alcohol[463]).

121

122

Lithium enolates can be trapped with chlorotrimethylsilane in a separate step under *in situ* conditions[464]. Treatment of the resulting *O*-trimethylsilyl enethers with methyl-lithium regenerates the original lithium enolates[464].

α,β-Unsaturated carbonyl compounds may add nucleophiles 'directly' at the functional group or 'vinologously' at the olefinic β carbon. This ambivalency of 1,2- *vs.*

1,4-reactivity represents another prominent case of regioselection. Famous examples are the clean 1,2 addition of Grignard reagents to cyclohexanones and the complete reversal of this intrinsic regioselectivity in favor of a 1,4-addition if catalytic amounts of a cuprous salt are present[465] (see Chapter 6). However, it does not always need a transition element to cause such a change in orientation. When benzylacetone (4-phenyl-3-butene-2-one) acts as the substrate, organomagnesium and -beryllium reagents spontaneously favor the 1,4-mode, whereas more polar alkali and alkali earth metal derivatives follow the 1,2-mode (Table 10) affording the enolate **123** and the carbinolate **124**, respectively[466].

Table 10. Reaction between benzylacetone and phenyl-metals: yields of 1,4-adducts **123** and 1,2-adducts **124** as a function of the metal (M).

M	1,4-adduct **123**	1,2-adduct **124**
CaI	45%	0%
K	52%	0%
Na	39%	4%
Li	69%	13%
MgI	0%	77%
BeC$_6$H$_5$	0%	90%

A word of warning is appropriate. Whenever the thermodynamically more stable 1,4-adduct is found to predominate, one has to refrain from any mechanistic interpretation until it has been checked whether this outcome is the result of an irreversible ('kinetically controlled') reaction or a subsequent equilibration (under 'thermodynamic control'). For example, the methyl isobutyrate-derived lithium enolate (**125**) combines with 2-cyclohexanone at $-75\,^\circ$C to give the 1,2-adduct exclusively, which quantitatively rearranges at ambient temperature to the 1,4-adduct[467].

So far, all regioselectivities considered have been linked to an ambipositional reactivity of the substrate. However, if allylic or otherwise delocalized, the organometallic reagent can display nucleophilicity at two or more sites. Most electrophiles attack a crotyl (2-butenyl) Grignard species at the inner electron-rich center. This entails an 'allyl shift' which has been extensively studied[468–470] and has found numerous practical applications (*e.g.* as a key step in the first total synthesis of vitamin D₃[471]). The 'vinologous' reactivity of allylmagnesiums is also mechanistically intriguing. The ring-opening addition of crotylmagnesium bromide to butene oxide occurs with inversion at one oxygen-bearing carbon atom and hence implies an open, rather than cyclic transition state (**126a** rather than **126b**)[472].

The intramolecular addition of allylmetals to built-in epoxides may selectively lead to three- and four-membered rings. The rational access to the cyclobutane derivative **127**, a precursor to the pheromone grandisol, illustrates this possibility[473].

The exact ratio in which the regioisomers are formed depends primarily on the metal involved, although the nature of the electrophile used also plays a role. As a comparison between prenyl-type organometallics **128** reveals, the magnesium derivatives react with all electrophiles except chlorotrimethylsilane preferentially at the inner allylic position, thus leading mainly to the branched products **129**. In contrast, the potassium analogs favor the chain-lengthened isomers **130** originating from electrophilic attack at the unsubstituted allylic terminus (see Table 11)[474,475].

Table 11. Reaction between 3-methyl-2-butenylmagnesium bromide (**128a**, R = H, M = MgBr) and -potassium (**128b**, R = H, M = K) and 3-ethyl-2-pentylmagnesium bromide (**128a**, R = H$_3$C, M = MgBr) and -potassium (**128b**, R = H$_3$C, M = K) with a variety of electrophiles X–*El*: ratio of branched *vs.* chain-elongated products (**129** and **130**, respectively)[474,475].

X–*El*	M = MgBr		M = K	
	R = H	R = H$_3$C	R = H	R = H$_3$C
ClSi(CH$_3$)$_3$	1 : 99	0 : 100	1 : 99	0 : 100
FB(OCH$_3$)$_2$	80 : 20	80 : 20	5 : 95	2 : 98
ICH$_3$	—	92 : 8	—	6 : 94
IC$_3$H$_2$	75 : 25	70 : 30	15 : 85	10 : 90
O(CH$_2$)$_2$	98 : 2	98 : 2	43 : 57	20 : 80
O=CH$_2$	99 : 1	99 : 1	28 : 72	30 : 70
O=C=O	99 : 1	99 : 1	10 : 90	4 : 96

Allylmetals can combine with allylic halides, sulfonates and phosphonates in four different, regioisomerically distinct ways. Whereas 2-alkenylmagnesium halides strongly favor the formation of the branched head-to-tail isomer **131**, analogous barium compounds show a very pronounced preference for the tail-to-tail isomer **132**[476].

An extension of the area of delocalization attenuates the metal dependence of the regioselectivity in electrophilic substitution reactions. Characteristic metal effects can nevertheless be observed with pentadienyl species, their oxy analogs (dienolates), their acyclic or cyclic aza analogs (deprotonated α,β-unsaturated Schiff bases and pyrrolides, respectively) and even benzyl derivatives. Deuterium chloride neutralizes α-cumylmagnesium and -mercury chloride by proton transfer to the *ortho*-position (producing essentially isotopomer **133a**)[477]. In contrast, ethereal α-cumylpotassium incorporates deuterium mainly in the benzylic position (providing 87% of the isotopomer **133b**) and only to a minor extent in the *ortho*- and *para*-positions (3% and 10%, respectively)[478]. No attack at aromatic positions is observed at all, if α-cumylpotassium is either dissolved in ethylene glycol dimethyl ether ('glyme') or suspended in pentanes[478].

Benzylmagnesium chloride is attacked by dicyan (cyanogen) at the *ortho*-position, affording 2-methylbenzonitrile[479], whereas benzylpotassium smoothly forms benzyl cyanide[480]. Similarly, 4-*tert*-butylmagnesium bromide attaches formaldehyde at the *ortho*-position, affording after tautomerization the rearomatized 5-*tert*-butyl-2-methylbenzyl alcoholate (**134**)[479,481,482]. Again, the potassium compound promotes the normal reaction mode leading to the 2-(4-*tert*-butylphenyl)ethoxide (**135**)[481].

Although the fascinating regioselectivity mystery of allyl- and benzylmetals has not yet been clarified in all its facets, the overall picture is unveiled. Grignard reagents having monohapto (η^1) structures, the metal is tightly bound to the terminal carbon atom. Enough charge density to interact efficiently with an electrophile can be mobilized neither in the front nor in the back lobe of the organometallic bond (*e.g.* transition state **136a**). If, however, the electrophile approaches the allyl entity at the vinologous position on the metal-remote face, it can draw from there a steady flow of electrons at the expense of the eventually disappearing organometallic bond (transition state **136b**).

Alkali metals, which establish with allylic and most benzylic backbones trihapto (η^3) π-bonds, behave differently. *A priori*, both electron-rich allyl termini are thus equally capable to attract an electrophile. Of course, the latter encounters still less steric hindrance when it approaches a primary rather than a secondary or tertiary carbon center. This explains the observed regioselectivity in favor of chain-lengthened products (transition state **136c**; R, R' ≠ H).

As mentioned above (Table 11, p. 62), allylic magnesium compounds condense with chlorotrimethylsilane regioselectively at the terminal position, but react with all other electrophiles preferentially, if not exclusively, at the secondary or tertiary carbon atom occupying the inner end of the allyl entity. Also in the potassium series one finds a striking exception. Ethylene reacts smoothly with 3-ethyl-2-pentylpotassium (**128b**, R = H_3C, M = K, Sv = tetrahydrofuran) while adding exclusively at the inner allylic position. The transient 3,3-diethyl-4-pentylpotassium (**137**) is unstable in an ethereal solvent such as tetrahydrofuran, from which it instantaneously abstracts a proton to afford a doubly branched, unsaturated hydrocarbon[475]. The complete reversal of regioselectivity points to a profound change of mechanism. The terminal η^1-allylpotassium species ('σ-complex') which is assumed to exist as a minute component in dynamic equilibrium with the dominant η^3 structure ('π-complex') presumably coordinates ethylene in a η^2 fashion and thus prepares the stage for an intramolecular nucleophilic addition.

c) Stereoselectivity

In general, a variation of the metal has little effect on the outcome of a diastereogenic organometallic reaction. If the diastereomeric ratios do differ significantly, organolithium and organozinc reagents often represent the extremes. Allylpotassium, -sodium and -lithium favor the *axial* over the *equatorial* approach to the extent of 2 : 1 when adding nucleophilically to 4-*tert*-butylcyclohexanone, the *cis*-1,4-disubstituted cyclohexyl alcoholate (*cis*-**138**) being the major product[483]. In contrast, the strongly coordinating allylzinc bromide gives predominantly the opposite stereoisomer (*trans*-**138**) resulting from an equatorial attack (Table 12)[483]. Allylmagnesium bromide affords a 1 : 1 mixture[483].

cis-**138**   +   trans-**138**

Table 12. Addition of allylmetals to 4-*tert*-butylcyclohexanone: face selectivity as a function of the metal (M)[483].

M	cis-**141** : trans-**141**
ZnBr	15 : 85
MgBr	55 : 45
Li	65 : 35
Na	65 : 35
K	63 : 37

The stereochemistry of reactions between α-hetero-substituted carbonyl compounds and organometallic reagents offers many examples of the struggle between steric and chelation control. When adding to the acrolein dimer 2-(3,4-dihydro-2*H*-pyranyl)carbaldehyde, ethyllithium favors the formation of the erythro diastereomer *erythro*-**139**, particularly so in the presence of the polar cosolvent hexamethylphosphorus triamide (HMPA)[484]. Conversely, the tightly coordinating diethylzinc and ethylmagnesium bromide afford preferentially the *threo* component *threo*-**139**[484] (Table 13).

erythro-**139**   threo-**139**

The technically important manufacture of polybutadiene and polyisoprene represents a noteworthy example of applied organometallic chemistry and marked element effects. The isoprene polymer is composed of repetitive C_5 modules which may adopt either of four possible shapes: the 1,2 and 3,4 structures having dangling vinyl or 1-methylvinyl tails from the carbon backbone and the *cis*- and *trans*-1,4 structures. The latter are most crucial for the physical properties of the material as Nature demonstrates. Latex rubber

Table 13. Reaction between the acrolein dimer and ethylmetals: diastereoselectivity of adduct formation as a function of the metal (M)[484].

M	*erythro*-**139** : *threo*-**139**
ZnC_2H_5	15 : 85
MgBr	30 : 70
Li	72 : 28
Li + HMPA	88 : 12

consists almost exclusively of *cis*-1,4 units which tend to coil up helically and thus confer elasticity. Guttapercha, another polyisoprene of plant origin is made up of zigzagging *trans*-1,4 units. As a consequence, it is hard and resistant to deformation and hence can be used as an insulator of transatlantic cables stretching over the ocean floor.

1,2 subunit   3,4 subunit   *cis*-1,4 subunit   *trans*-1,4 subunit

The *cis*-1,4 polymer, the only one suitable as an elastomer, can be made by treating a solution of isoprene in aliphatic hydrocarbons with butyllithium. In the absence of any better electron donor the lithium coordinates with the diene which, in order to act as an η^4-ligand, has to adopt the *cisoid* (*s-cis*) conformation. The subsequent nucleophilic addition within this complex generates an *endo*-allyllithium species *endo*-**140** which immediately combines with a second monomer, and so on. In fact, dienes and styrenes react with allylic and benzylic metal derivatives much faster than with the more basic alkyl counterparts[485]. Therefore, only catalytic amounts (typically 0.1%) of butyllithium (or another polymerization initiator) are required and high-molecular-weight products are obtained. Unless quenched by protolysis, the organometallic end groups remain active even after all monomer is consumed ('living polymer'[486]). Chain growth starts again, as soon as new monomer is added. Butyllithium in heptane produces 90–94% *cis*-1,4, 6–7% 3,4 and traces of *trans*-1,4 junctions[487,488]. Still higher *cis*-1,4 proportions can be achieved with Ziegler–Natta catalysts[489].

endo-**140**

Derivatives of the more electropositive alkali metals in whatever medium as well as butyllithium in ethereal solution do not give rise to *cis*-1,4 junctions at all. The final polymer results from the nucleophilic addition of an *exo*-allyl intermediate (***exo*-140**) to the *s-trans* form of isoprene and consists on an average of two times 45% of *trans*-1,4- and 3,4-connected subunits along with some 10% of 1,2 subunits[487,488]. Obviously, the poorly coordinating sodium and potassium are unable to bind butadiene in an η^4-fashion and lithium does so only if no better complexand is available.

***exo*-140**

3.3   Organometallic Reaction Mechanisms

Reactivity is not a quality label on which one can blindly rely. It is rather a potential which may or may not be exploited. It is the overall scenario that determines whether a chemical transformation can be brought about and, if so, how the bond breaking and making is accomplished. One has to know such details before one can hope to rationalize and even predict the course and the ease of a reaction. The following section summarizes the most common mechanistic patterns being at the disposal of organometallic reagents and which dictate the reactivity of the latter.

3.3.1   Carbanionic Processes

Polar organometallics are frequently called 'carbanions'[490,491]. However, such species rarely act as true reaction intermediates. Acid–base equilibria accomplished in the gas phase obviously represent such an exception. In the case of a degenerate process, for example the proton abstraction from toluene by the benzyl anion, the reaction passes through a symmetrical transition state **141** at which each benzylic carbon atom carries the same fraction of a negative charge. The estimate of 7 kcal/mol for the intrinsic barrier[491,492] means that a C–H–C hydrogen bond is weak at best.

141

A 'carbanionic transition state', which denies any active participation by the metal, can be realized in solution only if both the reacting and the newly formed organometallic entities are ion pairs already in their ground states, or if they can easily mutate from contact to solvent-separated structures. The conversion of fluorene into fluorenyllithium using triphenylmethyllithium in tetrahydrofuran as the base should follow a carbanion mechanism because both organometallics involved exist as ion pairs under such conditions[167,493–495]. The same can be said about the deprotonation of fluorene, 9-methylfluorene and 9-phenylfluorene by triphenylsilyllithium, methyldiphenylsilyllithium and dimethylphenyl-silyllithium in tetrahydrofuran at $-30\,°C$[495,496]. Whatever substrate–reagent combination studied, the activation parameters measured by means of stopped-flow techniques are identical within the limits of error[495,496]. The free energies of activation are quite small ($\Delta G^{\ddagger} \approx 13$ kcal/mol) and essentially reflect the considerable negative activation entropy ($\Delta S^{\ddagger} \approx -33$ e.u.), typical for reactions of organometallic ground state monomers. The almost negligible activation enthalpy ($\Delta H^{\ddagger} \approx 5$ kcal/mol) indicates that at the transition state (**142**) bond-breaking is widely compensated by simultaneous bond-making.

$$R = H, CH_3, C_6H_5; R', R'' = CH_3, C_6H_5$$

α-Ferrocenyllithium (**143**) sets the esthetically appealing stage for rapid reversible proton transfer[497] ($\Delta G^{\ddagger} \approx 10$ kcal/mol at $+25\,°C$[498]). Although in tetrahydrofuran the metal is contact-bound at the ground state and even occupies an *exo* position, it can easily switch into a permutation-mediating ion pair structure by coordinating an extra solvent molecule[499].

Whenever ion pairs act as the crucial intermediates, lithium should fare better than any of the heavier alkali metals because of the greater gain in heats of solvation (see pp. 23–24). All metal acetylides react very sluggishly when exposed in 0.5 M ethereal solution to a tenfold excess of methyl iodide and the 'normal' order of reactivity is obeyed, *i.e.* K > Na > Li > MgX[432]. However, in tetrahydrofuran as the solvent, the lithium acetylide outperforms all the others (MgBr $\ll$ Li $\gg$ K > Na). Its reaction half-life at

25 °C amounts to a few seconds, whereas those of the sodium and potassium derivative approximate 1 week and 25 min, respectively[499]. This reactivity jump can be plausibly rationalized by assuming that a carbanion mechanism (transition state **144a**) operates with the lithium species in tetrahydrofuran, and that a four- (or six)-center process (transition state **144b**) is effective in all other cases.

$$M^{\oplus} \quad {}^{\ominus}:C\!\equiv\!C\text{-}C(CH_3)_3 \quad \longrightarrow \quad \left[\begin{array}{c} M^{\oplus} \quad H \\ I \cdots C \cdots C\!\equiv\!C\text{-}C(CH_3)_3 \\ {}^{\ominus}\delta \quad H \quad H \quad {}^{\delta\ominus} \end{array} \right]^{\ddagger}$$

144a

$$H_3C\text{-}C\!\equiv\!C\text{-}C(CH_3)_3$$

$$M\text{-}C\!\equiv\!C\text{-}C(CH_3)_3 \quad \longrightarrow \quad \left[\begin{array}{c} X \cdots M \\ M \quad H \quad C\!\equiv\!C\text{-}C(CH_3)_3 \\ I \cdots C \cdot H \\ H \end{array} \right]^{\ddagger}$$

144b

3.3.2 Multicenter Processes

If the standard organometallic reaction mechanism is not carbanion-like, then what else is it? Several decades ago, R. Huisgen and J. Sauer postulated a four-center transition state as the mediator of the countercurrent transfer of proton metal to convert a stronger acid–stronger base into a weaker acid–weaker base couple[500]. R. Waak *et al.* have provided evidence in support of this hypothesis by measuring kinetic isotope effects[501].

$$\left[\begin{array}{c} M \\ {}_{-}C \cdots \cdots Bs \\ H \end{array} \right]^{\ddagger}$$

Bs (base) $= C\!\stackrel{\cdot}{\cdot}$, $N\!\stackrel{\cdot}{\cdot}$, O-

Other types of organometallic reactions also appear to be accomplished by connectivity changes simultaneously affecting more than three bonds. The conspicuously small Hammett response constants ($\rho = -0.02$[502], 0.01[502], 0.06[503], 0.17[504], 0.18[504], 0.18[505], 0.24[504], 0.27[506], 0.54[502], 0.59[502]; but also $\rho = 0.90$[502] and 1.45[503]) found for the nucleophilic addition of organomagnesium and -lithium reagents to aromatic aldehydes and ketones, suggest a four-center transition state, which may or may not be preceded by Pfeiffer–Swain precomplex formation[507]. There is some ambiguity with benzophenones as the substrates because they tend to act as single-electron acceptors thus generating transient ketyls (*i.e.* 'radical anions') in the rate determining step (see p. 79). As E. C. Ashby *et al.*[508] have demonstrated, single-electron transfer may be triggered by trace amounts of transition elements and can often be avoided when ultrapure crystalline magnesium is used for the preparation of the Grignard reagent.

Stereochemical arguments have been put forward to unveil the β elimination of hydrogen halides or alcohols by organometallics as another example of a multi-center process[12]. If methyl *trans*-2-phenyl cyclohexyl ether (***trans*-145**) reacts with butyllithium much faster than its *cis* isomer (***cis*-145**)[509] although its concerted elimination implies a boat-shaped transition state, the otherwise discriminated[510] *syn*-periplanar elimination must be much favored over the *anti*-periplanar one. The explanation is obvious: only the former can be accomplished in a six-center mechanism.

trans-145   *cis*-145

The same kind of analysis can be applied to the phenyllithium-promoted dehydrochlorination of 2-chloro-1-phenylpropene[12]. The (*E*) isomer (***E*-146**), which can easily enter a six-center type push-pull transition state, proves to be more reactive than the (*Z*) isomer (***Z*-146**).

E-146   **Z-146**

What had not be considered at that time[12], was whether a single organometallic molecule would be able to span the distance between the electrofugal and the nucleofugal leaving group in case of a *syn*-periplanar elimination. As in-depth investigations of the concentration dependence of the rates of *tert*-butoxide driven dehydrohalogenations have revealed[511–514], *syn* eliminations are in general brought about in arrays containing two equivalents of the base (transition state ***syn*-147**), whereas three up to four alcoxide molecules have to be assembled to construct the 'conveyer belt'[511–514] ensuring a push-pull controlled *anti* periplanar elimination (transition state ***anti*-147**). A similar situation is encountered when lithium diisopropylamide acts as the base[515].

***syn*-147**   ***anti*-147**

The concept of 'conveyer-belt' ('open-aggregate'[516])-mediated mechanisms was verified by quantum chemical calculations simulating the nucleophilic addition of a methyllithium dimer to formaldehyde[516], acetaldehyde[516] and oxirane[517]. Further computational work focused on hypothetical four-center transition states of the exothermal or endothermal proton transfer from acetylene to lithium hydride[518], methane to lithium hydride[518] and methane to lithium amide[519] as on the nucleophilic addition of magnesium hydride to formaldehyde[520], lithium hydride (or methyllithium) to carbon dioxide[521] and lithium hydride (or methyllithium) to ethylene or acetylene[522].

3.3.3   Ate Complex Processes

The term 'ate complex' was coined by G. Wittig for hypervalent structures organized around a central metal or nonmetal atom carrying a negative charge. He ascribed to them the role of ubiquitous and the most important intermediates in organometallic reactions[523–525], a prophecy which became true at least in part.

Currently, ate complexes receive much attention as potential and even probable turntables for halogen–metal and other exchange processes[526–528], although a four-center mechanism[529] would of course also be compatible with the observed retention mode. What put ate complexes fully into the limelight was the isolation and spectroscopic characterization of such species. The TMEDA-coordinated lithium bis(pentafluorophenyl)iodate (**148**) proved even to be momentarily stable at ordinary temperatures and crystallizable to obtain an X-ray structure[530].

148

The collision between an organic halide (say, an iodide) and an organolithium reagent (say, *tert*-butyllithium) may generate an ate complex or a pair of radicals (see Section 3.3.4)[531]. It seems to be tacitly understood that it has to be either this or that mechanism. However, even if the intermediacy of radicals is confirmed by all imaginable probes, this does not necessarily rule out the simultaneous existence of ate complexes. Still more, each kind of such species may convert into the other. Thus an ate complex (such as the intermediate **149**) may reversibly decompose to its organometallic and organohalide components, may undergo a concerted reductive elimination of metal halide after the example established by transient organocopper (II)[532] and other transition element compounds or, finally, break down to set free a radical pair. The latter mode, which implies a stepwise scission of both carbon–halogen bonds, should be a favorable process. The conceivable nonbond–bond resonance[533] between the ionic lithium iodate (**149a**) and the covalent lithioiodinane (**149b**) limiting structure should facilitate such a homolytic cleavage.

3.3.4   Radical Processes

Radicals frequently emerge from organometallic precursors like meteoric intermediates[534]. In general they are unwanted as they tend to pave the way for detrimental side reactions. However, they can also be a blessing when they open reaction channels closed otherwise. For example, the condensation of haloarenes with enolates can only occur according to the $S_{RN}1$ scheme[535–539], which implies radical anions and aryl radicals **150** as the key intermediates.

Radicals may be detected and identified spectroscopically. The electron paramagnetic resonance (EPR or electron spin resonance, ESR) is a powerful technique for the investigation of persistent radicals which can clarify many structural details (*e.g.* how lithiomethyl is embedded together with three methyllithium units in a tetrameric aggregate[540]). The phenomenon of chemically induced dynamic nuclear polarization (CIDNP)[541–545] enables the visualization of transient radicals by their memory effect in nuclear magnetic resonance (NMR). This method has revealed the intermediacy of radicals in virtually all reactions in which organometallic species are generated from alkyl or aryl halides by treatment with bulk metal (*e.g.* magnesium[546]) or metal–π-acceptor adducts (*e.g.* naphthalene-sodium[547]).

Radicals can be easily diagnosed by their characteristic reactivity. They abstract hydrogen from ethereal solvents and dimerize or co-dimerize with other radicals. Thus, 1-bromotrypticene, when exposed to magnesium in tetrahydrofuran, is quantitatively reduced to the halogen-free hydrocarbon, although 1-trypticylmagnesium bromide, prepared from 1-trypticyllithium by metal–metal exchange, is perfectly stable under such conditions[548]. Similarly, 1-chloro-, 1-bromo- and 1-iodoadamentane afford mainly adamantane when treated with magnesium in tetrahydrofuran[549]. In addition some 2-(1-adamantyl)tetrahydrofuran (151) is obtained as a by-product[549].

151

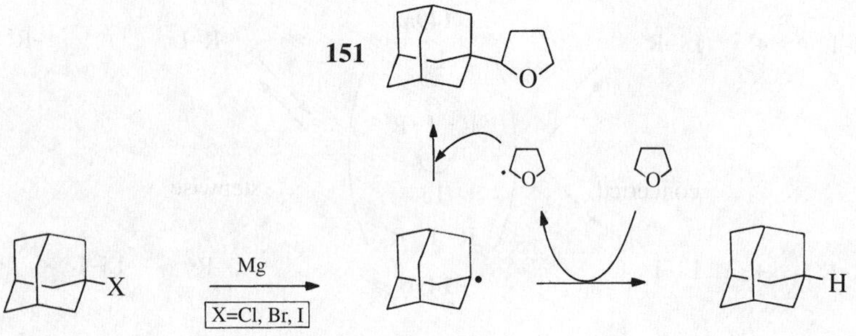

The dimerization of 'radical anions' has already been mentioned in the context of the adduct formation between benzene and cesium (see pp. 52–53). In the same way, the pyridine–sodium 1 : 1 adduct forms a N,N'-disodiobi-4,4'-(1,4-dihydropyridyl) at −75 °C and, after elimination of sodium hydride, bi-4,4'-pyridyl at 25 °C[550]. Particularly stable radical anions may not dimerize at all or, if they do, only establish a metal and concentration dependent equilibrium with the monomer. Although the lithium and sodium[551,552] 1 : 1 adducts of benzophenone are perfectly stable as such, the corresponding halomagnesium ketyls afford quantitatively tetraphenylpinacolates[553].

Transient radicals may further unmask themselves by isomerization. (1Z,5Z)-Bromo-1,5-cyclononadiene is quantitatively converted into cis,cis-1,5-cyclononadiene by sodium dissolved in liquid ammonia[554]. The initially generated (1E,5Z)-1,5-cyclononadienyl radical must have immediately sought and found relief of ring strain by changing the double-bond geometry.

The most popular probe for radical detection is the cycloalkylcarbinyl/(ω-1)alkenyl rearrangement[555,556]. Whereas the three-[557,558] and four-membered[559] cyclic radicals (e.g. 152[558]) undergo spontaneous ring opening, 5-hexenyl[560–562] and congeners[562–565] (e.g. 153[563]) rapidly form the cyclopentylmethyl radical, thus reflecting a profound decrease in ring strain and hence a change in relative thermodynamic stability. The absolute rates of ring-closure or ring-opening being precisely known[555–556] such rearrangments are often called 'radical clock reactions'. They allow one to probe the time scale on which simultaneously occurring, ultrafast processes take place.

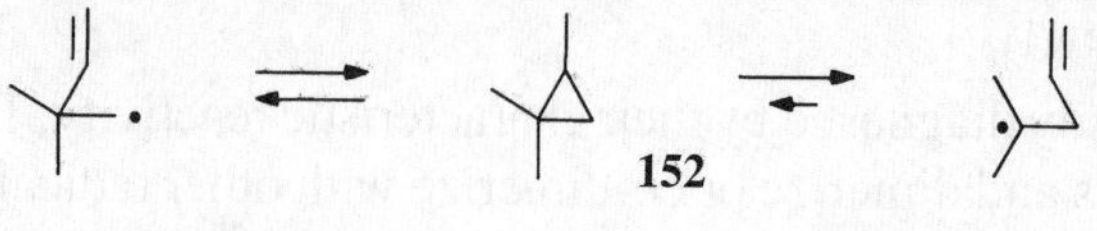

152

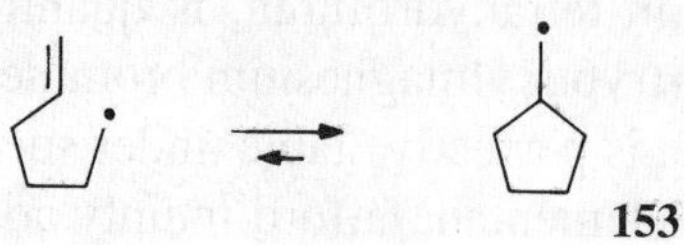

153

Alkynyl radicals may also cyclize. The first case of such a ring–chain transformation was the butyllithium-promoted reductive conversion of 6-bromo-1-phenyl-1-hexyne (**154**) into benzylidenecyclopentane[566]. An analogous cyclization had been reported previously but was erroneously interpreted as a polar organolithium reaction[567].

Another mechanistic revision deems warranted. When, upon reduction of 2,2,2-triphenylethyl chloride with sodium in liquid ammonia, 1,1,2-triphenylethane was obtained, the isomerization of a short-lived 2,2,2-triphenylethylsodium (**155**) was thought to be at the origin of this result[568]. However, as later systematic studies of the Grovenstein–Zimmerman rearrangement (see pp. 55–56) showed, this process would be too slow to compete with the presumably diffusion-controlled protonation of the alleged organometallic intermediate **155**. It is more plausible to envisage a different sequence of events. The transfer of the first electron inevitably generates the 2,2,2-triphenylethyl radical which can immediately reorganize itself by phenyl migration to the resonance stabilized 1,1,2-triphenylethyl isomer (**156**). The latter species eventually takes up the second electron to afford, after neutralization, the rearranged hydrocarbon.

Radicals may be generated either when organometallic reagents are prepared by reductive insertion of a metal into a carbon–halogen or another carbon–heteroelement bond or, alternatively, from an already available organometallic compound by a redox reaction of the latter with another radical or an unsaturated acceptor molecule simultaneously producing a 'radical anion'. If R stands for an organic entity, M for a metal, X for a heteroatom and A for an electron acceptor, this leads to three possible situations.

$$R-X + M \longrightarrow R\cdot + X-M$$

$$X\cdot + M-R \longrightarrow X-M + \cdot R$$

$$A: + M-R \longrightarrow [A^{\ominus}]M^{\oplus} + \cdot R$$

In the standard *de novo* preparation of organomagnesium and -lithium reagents two electrons are transferred one at a time from the bulk metal to the organic substrate from which a halide is detached as an anion. The scission of the carbon–halogen bond at the 'radical anion'-like adduct stage that is reached after the transfer of the first electron inevitably sets free a carboradical R. This may immediately become attached to the metal surface again to pick up the second electron and thus complete the formation of the organometallic species which will now leave the surface to enter into solution. Alternatively, the already nascent radical R may diffuse into the solution. From there it may eventually return to the metal surface unless consumed by alternative chemical processes such as reaction with an ethereal solvent or another radical.

Although there is general consensus about such principles, the quantitative aspects are discussed quite controversially. The 'free-diffusion' hypothesis, first advocated by F. Bickelhaupt, C. Blomberg et al.[546], assumes that most, if not all, radicals exist as temporarily surface-separated species[569–572]. This notion is supported by a wealth of experimental evidence. However, the 'surface-absorption model'[573] effectively explains memory effects such as the partial retention of configuration, when nonracemic 1-bromo-1-methyl-2,2-diphenylcyclopropane is converted into the corresponding Grignard reagent[574,575]. The persistence of radicals at metal surfaces should therefore not be rigorously ruled out, although such sequestered species may be relatively rare.

If soluble 'radical anions' such as naphthalene–sodium are used instead of the bulk metal, the rate of the second-electron transfer almost attains the diffusion control limit $(10^9–10^{10}\ L \cdot mol^{-1} \cdot s^{-1})$, as J. F. Garst et al.[576] have shown. Even alkyl fluorides undergo reduction under such conditions. The central role of metal halide formation as the thermodynamic driving force is confirmed by a strong inverse metal effect (Li $\gg$ Na $\gg$ K $\gg$ Cs)[576].

The cyclization of 5-hexenyl to cyclopentyl radicals being a relatively slow process $(k = 10^5\ s^{-1}$ at $25\ ^\circ C^{[555]})$, it can efficiently compete with the electron transfer in homogeneous media only if the reduction is performed under high dilution[576]. However, the inversion of cyclopropyl radicals is too fast ($k \approx 10^{11}$ s^{-1} at 25 °C[577–579]) not to outperform any bimolecular electron transfer. Therefore, *cis*- and *trans*-alkyl-substituted cyclopropyl halides give invariantly stereoconvergent product mixtures[580]. The net retention observed with aryl-substituted cyclopropyl halides (**157**, X = Cl, Br)[581] has been plausibly explained by G. Boche *et al.*[580] as the consequence of a change in mechanism. The first electron is no longer absorbed by the halogen but is temporarily 'parked' in the aryl ring. Although it is unknown what happens in detail, the arrival of the second electron triggers the intramolecular transfer of the first electron to the halogen-bearing center, thus generating instantaneously a configurationally stable organometallic species[580].

Most pertinent studies have been performed with molecular oxygen, a ground state triplet and hence a permanent biradical. Upon bubbling air into an ethereal solution of triphenylmethylsodium, the intense brick-red color fades to a brownish yellow, thus indicating the presence of the triphenylmethyl radical in equilibrium with its dimer as the largely predominant component[582]. The intermediacy of radicals can be demonstrated by characteristic isomerization reactions. Thus, the oxidation of 2-methyl-2-phenylpropyllithium produces a mixture of 2-methyl-2-phenyl-1-propanol and 2-methyl-3-phenyl-2-propanol, the formation of the latter evidently involving the rearrangement of the 'neophyl' to the 2-benzylisopropyl radical[583]. Treatment of *cis*- and *trans*-1-propenyllithium gives rise to a rapidly equilibrating mixture of geometrically isomeric radicals resulting in a mixture of lithium enolates **158** and, after derivatization with acetic anhydride, acetates[584]. However, if lithium *tert*-butylperoxide is employed as the oxygen source, a polar reaction takes place and the configuration is retained[584].

Other radical generators are the stable spin trap 2,2,6,6-tetramethylpiperidyl-N-oxyl ('TEMPO')[585] and salts of heavy metals such as lead chloride[586]. Even the 'Koelsch radical' [9-(9H-fluoren-9-ylidenephenylmethyl)-9H-fluoren-9-yl], a stable hydrocarbon having an unpaired electron, which when encountering butyllithium gives the ion pair **159** and butyl radical[587].

159

The reactions between a radical and an organometallic compound need not always follow the substitution pattern but can also result in an addition. Although this possibility has never been rigorously verified, it deserves to be considered as an attractive, if speculative hypothesis. The ethereal solution of phenyllithium prepared from bromobenzene and lithium chunks (see p. 100) invariably contains some 5–10% of biphenyl. A direct coupling reaction can be safely ruled out. The fundamental reductive insertion sequence consists without doubt of three steps: metal transfer to bromobenzene to afford a 'radical anion', which instantaneously eliminates lithium bromide to set free a phenyl radical which, by taking up the second metal, becomes stabilized as phenyllithium. The latter reagent should be equally capable of trapping the transient phenyl radical. The highly exothermal combination of the two species would generate the biphenyl–lithium 1 : 1 adduct **160**. This 'radical anion' could subsequently act as a soluble metal carrier and thus accelerate both electron-transfer steps.

At the moment, the 'radical anion'-forming interception of radicals by organometallics is nothing but an unproven idea. Notwithstanding this caution, aromatic hydrocarbons may play a still unsuspected role as vehicles transporting metals or electrons in many more instances.

The third category of radical-producing organometallic processes covers a wide variety of patterns. Nitroarenes having a particularly high electron affinity are capable of oxidizing all sorts of organometallics[588], including triphenylmethyllithium[589]. Benzophenone has long been known to strip an electron from triphenylmethylsodium, affording a sodium ketyl 'radical anion' and the dimerizing triphenylmethyl[590]. An analogous electron transfer initiates apparently the formation of 2- and 4-*tert*-butylphenyl phenyl ketone (52%, *ortho/para* ratio 1 : 9) along with 2,2-dimethyl-1,1-diphenyl-1-propanol from benzophenone and *tert*-butyllithium[591]. When facing a diaryl ketone substrate, simple Grignard reagents may get engaged in either a 'polar' (four-center) or a radical (single electron-transfer) mechanism[508,592–596]. Trace amounts of transition elements (in particular, iron and manganese) can open a third reaction channel which promotes the formation of symmetrical coupling products (notably pinacolates)[597–599] by reductive elimination[600].

Depending on the substituents, counterions, solvents and concentrations involved, metal ketyls can be transformed more or less completely to the corresponding pinacolates **161** (*e.g.* R = C_6H_5, M = MgBr). It is tempting to assume direct dimerization of the 'radical anion'. However, the nucleophilic addition of a ketone–metal 1 : 2 adduct ('dianion') to the unchanged carbonyl precursor cannot be ruled out either. This alternative is reminiscent of the famous controversy between K. Ziegler[601] and W. Schlenk[602] concerning the reductive duplication of 1,1-diphenylethylene to 1,1,4,4-tetraphenyl-1,4-butandiyldisodium. The observation is remarkable in this respect that the ketyl equally forms when benzophenone is added to benzhydrol (diphenylmethanol) which was beforehand treated with two equivalents of lithium diethylamide or lithium diisopropylamide, to deprotonate simultaneously the *O* and *α* sites[603].

161

Organic halides also qualify as electron acceptors from organometallics. 5-Hexenyl bromide[604] and 6-mono- or 6,6-disubstituted congeners[605,606] react with *tert*-butyllithium to form open-chain and cyclized products (hydrocarbons and primary organolithiums), the rearranged derivatives suggesting the intermediacy of radicals. The solvent may play a critical role for the mechanistic partitioning. 2,2-Dimethyl-5-hexenyl iodide gives, with 2-(1,3-dithianyl)lithium in tetrahydrofuran, the open chain and cyclized products **162** and **163** in an approximate 90 : 10 ratio, whereas in hexanes the rearranged skeleton is found exclusively[607]. The same organolithium reagent condenses with enantiomerically pure 2-butyl bromide with little (≤15%) or with complete racemization, depending on whether tetrahydrofuran or hexane is used as the solvent[607].

A remarkable case of radical-mediated isomerization has been recently reported by W. F. Bailey *et al.*[608]. When treated with phenyllithium in a 1 : 9 mixture of diethyl ether and pentanes at 25 °C, 3-methyl-5-hexenyl iodide is quantitatively converted into (2-methyl-cyclopentyl)methyl iodide. A rearrangement at the level of the corresponding lithium compounds by intramolecular nucleophilic addition to the double bond can be discarded. The implied halogen–metal exchange would proceed only slowly in an essentially hydrocarbon medium and the aryllithium component would be extremely favored at equilibrium ($K_\phi \approx 10^{4[609]}$). Moreover, the *cis/trans* ratio (0.15) of the cyclization product is incompatible with a polar (organolithium) mode of cyclization (*cis/trans* 0.086) but agrees perfectly well with the typical stereoisomer distribution resulting from a radical process. Thus, the phenyllithium clearly acts just as an initiator of a radical chain in which the (2-methylcyclopentyl)methyl radical produced by cyclization of its 3-methyl-5-hexenyl precursor is stabilized by iodine transfer from the educt, thus starting a new reaction cycle.

Even heteroatom-free unsaturated hydrocarbons will occasionally pick up an electron from an organometallic reagent. The first documented case of that type was the formation of 9,9′-bis(9,10-dihydro-10-phenanthrylpotassium) (**166**) and 2,3-dimethyl-2,

3-diphenyl-butane from phenanthrene and α-cumylpotassium through the respective monomeric radicals[610].

166

Since then numerous further examples of such single-electron-transfer processes have been described. Upon treatment of diphenylacetylene (tolane) with *tert*-butyllithium a mixture of *trans*-1,2-diphenyl-1,2-ethenediyldilithium (**168**) and of (*E,E*)-1,2,3,4-tetraphenyl-1,3-butadiene-1,4-diyldilithium is obtained[611]. The latter compound may be formed by dimerization of the transient tolane/lithium 1 : 1 adduct **167** or by nucleophilic addition of the 1 : 2 adduct **168** to tolane[612].

High concentrations of butyllithium in tetrahydrofuran and in the presence of hexamethylphosphorus triamide (HMPA) react with biphenyl to give the 'radical anion' **160**[613]. Ethereal triphenylmethylsodium consecutively transfers two electrons to cyclooctatetraene, thus converting the initially generated 'radical anion' instantaneously into the Hückel-aromatic dianion[614].

3.3.5 Cation Processes

Although at first sight appearing to be schizophrenic behavior, it is not unreasonable for polar organometallics to generate carbocations as reaction intermediates. Such reagents, being anything but carbanions, often exhibit a remarkably pronounced Lewis acid character. The latter increasing with the diminishing electropositivity of the metal, inverse metal effects are usually observed whenever a reaction pathway leads through a carbocationic intermediate.

Although *tert*-butyl chloride is virtually inert towards any Grignard reagent, it does react with butyllithium under β-elimination of hydrogen chloride. Organozinc compounds, which are neither particularly basic nor nucleophilic but strongly coordinating, can display a different kind of reactivity. By pulling off the chloride, they form initially a carbenium zincate intermediate **169** which stabilizes itself by anionotropic alkyl group transfer to the electrophilic carbon atom, thus providing a route to quaternary carbon structures[615,616].

$$
\begin{array}{c}
\xrightarrow{\text{LiC}_4\text{H}_9}
\left[
\begin{array}{c}
\text{H}_2\text{C}\cdots\text{H}\cdots\text{C}_4\text{H}_9 \\
\text{H}_3\text{C-C}\cdots\text{Cl}\cdots\text{Li} \\
\text{CH}_3
\end{array}
\right]^{\ddagger}
\longrightarrow (\text{H}_3\text{C})_2\text{C=CH}_2
\end{array}
$$

$$
\begin{array}{c}
\text{CH}_3 \\
\text{H}_3\text{C-C-Cl} \\
\text{CH}_3
\end{array}
\quad
\xrightarrow{\text{BrMgC}_4\text{H}_9}
\quad \text{||} \longrightarrow
$$

$$
\xrightarrow{\text{ClZnC}_4\text{H}_9}
\left(
\begin{array}{cc}
\overset{\text{CH}_3}{\underset{\text{CH}_3}{\text{H}_3\text{C-C}}}\!\oplus &
\text{Cl}_2\overset{\ominus}{\text{Zn}}\text{C}_4\text{H}_9
\end{array}
\right)
\longrightarrow (\text{H}_3\text{C})_3\text{C-C}_4\text{H}_9
$$

169

1-Bromoadamantane, another tertiary halide, condenses smoothly with the Reformatzky reagent derived from ethyl bromoacetate even at 25 °C in dichloromethane[617]. The *trans*-isomer of 8a-chlorodecahydronaphthalene reacts with both dimethylzinc and methyltitanium trichloride to afford a 1 : 1 mixture of the *cis*- and *trans*-8a-methyl substitution product. Two mechanisms appear to operate concomitantly when 3-phenyl-2-propenyllithium is treated with *tert*-butyl bromide. An ion-pair route leads mainly to 3,3-dimethyl-2-phenyl-1-pentene and a radical pair route preferentially to 4,4-dimethyl-1-phenyl-1-pentene[618].

An alkoxy substituent is more effective in stabilizing a positive charge than any alkyl group. Although unreactive towards organolithiums, orthoformates and other orthoesters undergo neat substitution with Grignard reagents. This offers convenient access to aldehydes, in particular aliphatic ones, *via* acetals[619–623]. Any rational explanation of the paradoxal metal behaviour has to allow for the intermediacy of a carbenium-dioxonium magnesiate **170** (R = alkyl, 1-alkynyl, aryl).

$$
\text{HC(OCH}_3)_3 \xrightarrow{\text{BrMg-R}}
\left(
\begin{array}{cc}
\overset{\text{OCH}_3}{\underset{\text{OCH}_3}{\text{HC}}}\!\oplus &
\overset{\text{Br}}{\underset{\text{OCH}_3}{\text{R-}\overset{\ominus}{\text{Mg}}}}
\end{array}
\right)
\longrightarrow (\text{H}_3\text{CO})_2\text{CH-R}
$$

170

As ion pairs like **170** are formed in an irreversible step, even the transition state has to be stabilized by charge delocalization in order to be energetically accessible. This is the case with *trans*-2-methoxy-*cis*-4,6-dimethyl-1,3-dioxane (***trans*-171**) which, when attacked by methylmagnesium bromide, rapidly undergoes a methoxy–methyl displacement with perfect retention of configuration[624]. The axially positioned leaving group fully benefits from the neighboring group assistance provided by both ring heteroatoms. In contrast, the *all-cis* stereoisomer (*cis*-**171**) proved to be unreactive, in tetrahydrofuran solution at least[624], because for steric reasons the triaxial conformer is practically unpopulated and the relevant orbitals occupy *gauche* positions in the dominant triequatorial form[625].

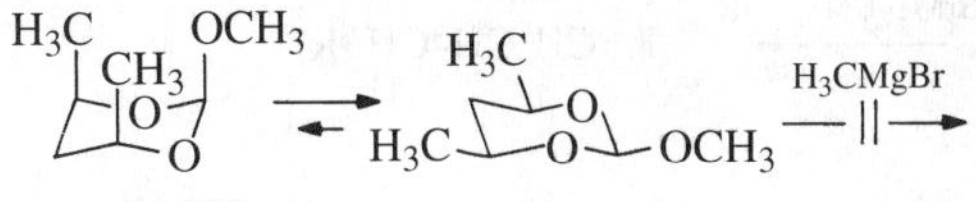

cis-**171**

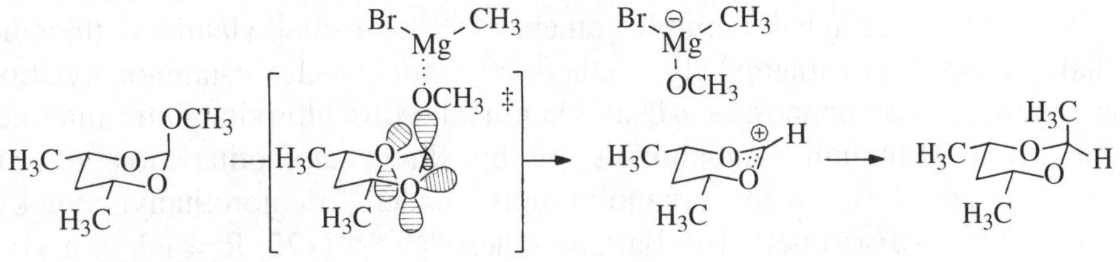

trans-**171**

Relief of ring strain facilitates heterolytic cleavages of the carbon–oxygen bond. Whereas most acyclic acetals are perfectly stable towards ordinary organometallics, dioxolanes undergo smooth ring-opening substitution with alkylmagnesiums[626] as with trimethylaluminum[627]. Therefore, Grignard reagents containing an acetal functionality should be prepared in tetrahydrofuran at low temperatures and using activated magnesium[628]. Ring-opening substitutions of acetonides (*e.g.* **172**) can be regioselectively conducted and have been applied to natural product synthesis[629].

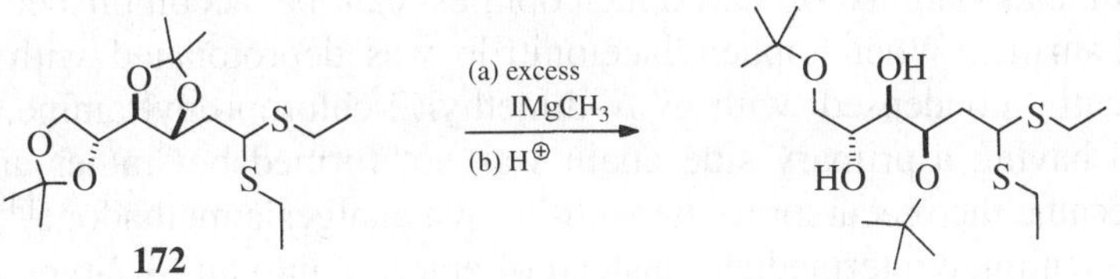

172

The Lewis acid properties of Grignard reagents are sufficiently strong to pull off even the poorly nucleofugal cyano group if the cation left behind is resonance stabilized by three alkoxy groups. In this way triethyl cyanoorthoformate (triethoxyacetonitrile, **173**) can be easily converted into orthoesters and, after acid hydrolysis, into carboxylic acids or their ethyl esters[630]. Organolithiums, however, add nucleophilically to the C≡N triple bond to afford, after hydrolysis, α-oxo-carboxylic acids or their ethyl esters or orthoesters[630].

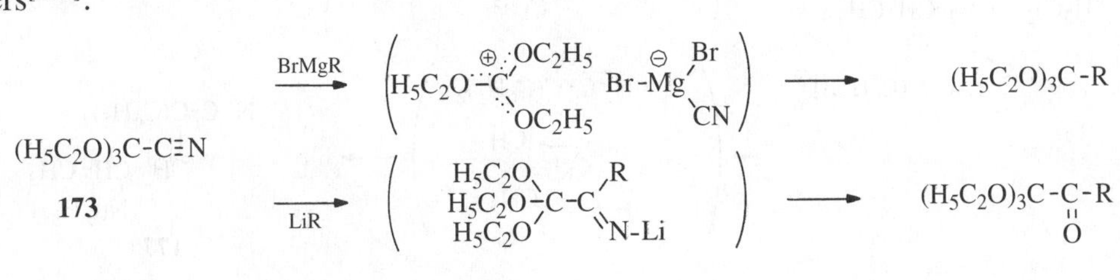

Dialkylamino groups are the most powerful electron donors. As a consequence, Strecker adducts (α-dialkylaminocarbonitriles, **174**) condense smoothly with Grignard reagents[631]. This so-called Bruylant reaction can, for example, be applied to the preparation of allylamines[632].

$$
\begin{array}{c}
\text{NR}''_2 \\
| \\
\text{R-CH} \xrightarrow{\text{BrMg-CH=CH-R'}} \\
| \\
\text{CN}
\end{array}
\qquad
\begin{array}{c}
\text{NR}''_2 \\
| \\
\text{R-CH-CH=CH-R'}
\end{array}
$$

174

It has become more customary to use hemiaminals such as α-dialkylamino ethers[633], α-bis(trimethylsilyl)amino ethers[634,635], α-dialkylamino thioethers[634] or α-dialkylamino *O*-trimethylsilyl ethers[635] for such α-aminoalkylations of organomagnesium compounds. (α-Dialkylaminoalkyl)trichlorotitanium intermediates, accessible from *O*-lithium hemiaminals or by direct aminotitanation of carbonyl compounds, react best with organolithiums under trichlorotitanyloxy/alkyl(aryl) displacement[636]. α-Bis(trimethylsilyl)amino ethers[637,638] (**175**, R = alkyl, aryl) can be used to prepare primary amines by condensation with a Grignard reagent (R' = alkyl, aryl) and and subsequent desilylation by treatment with an acid.

$$
\begin{array}{c}
\quad\ \text{Si(CH}_3)_3 \\
\quad\ \text{N} \\
\text{R-CH} \quad \text{Si(CH}_3)_3 \xrightarrow[\text{(b) H}^+]{\text{(a) BrMgR'}} \\
\quad\ \text{OCH}_3
\end{array}
\qquad
\begin{array}{c}
\text{NH}_2 \\
| \\
\text{R-CH-R'}
\end{array}
$$

175

Also, β-aminoalkylations of carbonucleophiles can be accomplished by means of cationic mechanisms. When diphenylacetonitrile was deprotonated with phenylsodium and subsequently condensed with *N,N*-dimethyl(2-chloropropyl)amine, the expected product (**176**) having a primary side chain was not formed but rather an isomer (**177**) which later became the precursor to the well-known analgesic methadone[425,639]. Clearly, the *N*-sodio ketenimide intermediate, instead of entering into an S_N2-type process, is still electrophilic enough to generate by chloride abstraction a transient aziridinium cation[640] which is opened by nucleophilic attack at the sterically least hindered 3-position.

$$
\begin{array}{c}
\text{N} \equiv \text{C-C(C}_6\text{H}_5)_2 \\[-2pt]
\delta\ominus \; \big\downarrow \\
(\text{H}_3\text{C})_2\text{N-CH}_2\text{-CH-CH}_3 \\[-2pt]
\big\downarrow \\
\text{Cl}^{\delta\ominus}
\end{array}
\Bigg]^{\ddagger}
\longrightarrow
\begin{array}{c}
\text{N} \equiv \text{C-C(C}_6\text{H}_5)_2 \\
\text{--}|\text{--} \\
(\text{H}_3\text{C})_2\text{N-CH}_2\text{-CH-CH}_3
\end{array}
$$

176

$$
(\text{H}_3\text{C})_2\text{N-CH}_2\text{-CH-CH}_3 \\
\qquad\qquad\quad \text{Cl} \\
+ \, \text{Na}^{\oplus} \, [\text{NC-C(C}_6\text{H}_5)_2]^{\ominus}
$$

$$
\left(
\begin{array}{c}
\qquad\quad \ominus \\
\text{N} \equiv \text{C-C(C}_6\text{H}_5)_2 \\
\qquad\qquad\qquad \text{CH}_3 \\
\quad \text{N}^{\oplus} \\
\text{H}_3\text{C} \quad\ \text{CH}_3
\end{array}
\right)
\longrightarrow
\begin{array}{c}
\text{N} \equiv \text{C-C(C}_6\text{H}_5)_2 \\
\text{--}|\text{--} \\
\text{CH}_2\text{-CH-CH}_3 \\
| \\
\text{N(CH}_3)_2
\end{array}
$$

177

4   Generation of Organometallic Reagents and Intermediates

A metal can be transferred to an organic backbone in either of two principally different ways[641]. One of these options involves of the use of another organometallic reagent as a carrier of the metal M to be introduced in exchange against a displaceable electrofugal group Z. Some of the most convenient and versatile methods for the preparation of organometallic intermediates follow such a *permutational interconversion* scheme.

$$\ce{>C-Z + M-R <=> >C-M + R-Z}$$

The use of a ready-made organometallic reagent of course only defers the problem of the *de novo* creation of carbon–metal bonds. The ultimate source of the metal can only be the metal itself. Therefore, the most straightforward and often also the most rational approach is the direct displacement of the leaving group Z by a *reductive insertion* of the metal.

$$\ce{>C-Z + 2 M. -> >C-M + M-Z}$$

Once the organometallic target compound has been identified, the most appropriate method of preparation will be selected taking into consideration the ease and repro-ducibility of execution, the cost, and the availability of the precursor material. Depending on the nature of the electrofugal leaving group and the reductive or permutational mode of metal introduction, one can differentiate two sets of five methods. As indicated by the number of plus signs appearing in the diagram below (Table 14), there are three favorite methods whereas four others have only restricted application.

Table 14. A survey of the standard methods applied to the preparation of organometallic compounds.

reductive insertion	Z (electrofugal group)	permutational interconversion
⊕⊕⊕	X (halogen)	⊕⊕⊕
⊕⊕	Y (chalcogen)	⊕
⊕⊕	Q (metalloid)	⊕⊕
⊕	C (carbon)	⊕
⊕	H (hydrogen)	⊕⊕⊕

4.1   Displacement of Halogens

Halogen scores high on both halves of the diagram (see Table 14). However, it should be pointed out that only chlorides, bromides and iodides are capable of incorporating reductively a metal atom under ordinary conditions and, in general, only bromides and iodides can efficaciously participate in a permutational interconversion process. In other words, fluorides do not qualify as precursors to organometallic compounds.

4.1.1   Reductive Metal Insertion in Carbon–Halogen Bonds

Halides outweigh all other potential starting materials in the *de novo* access to organometallics. In fact, tetraethyllead made from ethyl chloride and sodium–lead laminas at temperatures in the vicinity of $100\,^{\circ}C^{[642]}$ was by far the most important organometallic bulk material ever manufactured. Its worldwide production peaked in the 1970s at about 700 000 tons per year before declining steeply because of environmental concerns.

a) Representative Examples of Reductive Carbon-Halogen Scission

Countless organic halides have been converted into the corresponding zinc, magnesium, lithium, sodium and potassium compounds on a laboratory and sometimes even a technical scale. Anything but complete, the following compilation features the most common classes of organometal compounds: unbranched and branched alkyl (see Tables 15 and 16), 1-alkenyl (see Table 17), mononuclear and condensed aryl (see Tables 18, 19 and 20) and 5- and 6-membered heterocyclic derivatives (see Tables 21 and 22).

b) Scope and Limitations

Judging by this selection of examples (see Tables 15–23), one may feel tempted to consider the reductive insertion of metal in carbon–halogen bonds as an all-round and foolproof method that never fails. In reality there remains a constant menace that the organometallic compound, once generated, begins to 'cannibalize' its unconsumed precursor. Any metal halide-producing reaction mode will be highly exothermic and hence potentially competing with the reductive insertion, in particular β-elimination and carbon–carbon condensation. Depending on the precursor structure, transmetalation (*i.e.* a permutational hydrogen–metal interconversion, see p. 185) may also occur.

The most obvious solution to this problem is to minimize the duration of contact between the organometallic species and its halogenated precursor. This can be achieved by using surface-activated metals[734,756–761], amalgams and other alloys or metals of smaller particle size. Finest dispersions, clusters consisting of a handful metal atoms and even matrix-isolated individual atoms can be prepared by high-speed stirring, ultrasonification[762], intercalation in graphite, salt reduction ('Rieke metals'), vaporization and laser ablation[763–774]. Alternatively, the

Table 15. Unbranched primary alkylzincs, -magnesiums, -lithiums, -sodiums and -potassiums by metal insertion into halides R—X: trapping yields as a function of the solvent Sv, the metal M and the electrophile El—X'.

M—R	M	X	Sv[a]	Product, El—X'	Reference
	ZnX	Cl, Br, I	DMF	≈80%, [b]	[643]
M—CH$_3$	MgX	Cl, Br, I	DEE	≈80%, [b]	[644–645]
	Li	Cl, Br, I	DEE	≈80%, titration	[646–648]
	ZnX	Cl, Br, I	DMF	≈80%, [b]	[643]
M—C$_2$H$_5$	MgX	Br	DEE	≈80%, [b]	[644,649]
	Li	Cl, Br	PET[c,d]	50%, titration	[650–654]
	MgX	Br	DEE	≈80%, [b]	[644,655]
M—C$_3$H$_7$	Li	Cl	DEE	≈80%, titration	[650,656]
	Na	Cl	PET	26%[f], CO$_2$	[657]
	MgX	Br	DEE	≈80%, [b]	[658–661]
M—C$_4$H$_9$	Li	Cl, Br	PET[d]	~95%, [b]	[662–663]
	Na	Cl	PET	79%[g], CO$_2$	[657,664]
	MgX	Br	DEE	48%, HC(OC$_2$H$_5$)$_3$	[665]
M—C$_5$H$_{11}$	Na	Cl	PET	≈70%, [b]	[666]
	K	Cl	PET	35%, CO$_2$	[667]
	MgX	Br	DEE	≈80%, [b]	[644,655]
M—C$_6$H$_{13}$	Li	Cl	DEE	≥90%, titration	[668]
	Na	Cl	PET	[h]	[669]
	MgX	Br	DEE	≈80%, [b]	[644,655]
M—C$_8$H$_{17}$	Li[i]	Cl	THF	94%, H$_2$O	[378]
	Na	Cl	PET	64%, CO$_2$	[669]
M—C$_{10}$H$_{21}$	MgX	Br	DEE	80%, H$_3$CCOCl [j]	[670]
	Na	Cl	PET	28%, CO$_2$	[669]
	Li	Cl	DEE	77%, titration	[671]
M—C$_{12}$H$_{25}$	Na	Cl	PET	27%, CO$_2$	[671]
	K	Cl	PET	10%, CO$_2$	[671]

[a] PET = petroleum ether (pentanes, hexanes, heptanes), DEE = diethyl ether, THF = tetrahydrofuran, DMF = *N,N*-dimethylformamide; in the cases of mixtures (containing, for example, paraffinic cosolvents) only the most polar component is indicated. [b] Various electrophiles. [c] Or benzene. [d] Or DEE. [e] Or THF. [f] 10% Mono- + 16% diacid. [g] 58% Mono- + 21% diacid. [h] Not specified. [i] As the 1 : 1 adduct with 4,4'-di-*tert*-butylbiphenyl. [j] In the presence of catalytic Fe(acac)$_3$.

anthracene—magnesium 1 : 1 adduct ('dianion')[775,776] or 'radical anions' such as naphthalene—sodium[377], 1-dimethylamino)naphthalene/sodium[777] and, in particular, 4,4'-bis(*tert*-butyl)biphenyl/lithium (Freeman's reagent)[778,779] may be employed. Such arene—metal adducts immensely accelerate the generation of the organometallic compound because the critical electron transfer now occurs in a homogeneous phase

Table 16. Branched primary, secondary and tertiary alkylmagnesiums, -lithiums, and sodiums: yields of trapping products as a function the metal M, of the precursor halide X, the solvent Sv and the electrophile $El-X'$ used.

M–R	M	X	Sv[a]	Product, $El-X'$	Reference
M–CH$_2$Si(CH$_3$)$_3$	MgX	Cl	DEE	69%, (H$_5$C$_6$O)$_2$P(O)N$_3$	[672]
	Li	Cl	PET; DEE	90%, sublimation	[673]
M–CH$_2$C(CH$_3$)$_3$	MgX	Cl	DEE	55%, BF$_3$	[674]
	Li	Cl	PET; DEE	71%, (H$_3$C)$_2$CHCOCH(CH$_3$)$_2$	[675]
M–CH(CH$_3$)$_2$	MgX	Br	DEE	54%, H$_3$CCHO	[676]
	Li	Br	PET	45%, by titration	[650]
M–CH(CH$_3$)C$_2$H$_5$	MgX	Cl	DEE	81%, CO$_2$	[677]
	Li	Cl	PET	[b]	[678]
M–C(CH$_3$)$_3$	MgX	Cl	DEE	62%, CO$_2$	[679]
	Li	Cl	PET	89%, by titration	[680–682]
M–CH(CH$_2$)$_2$[c]	MgX	Cl	DEE	9%, CO$_2$	[683]
	Li	Br	DEE	88%, SnCl$_4$	[684]
M–CH(CH$_2$)$_5$[d]	MgX	Cl	DEE	66%, CH$_2$O	[685]
	Li	Cl	DEE	70%, GeCl$_4$	[686]
M–C$_7$H$_{11}$[e]	MgX	Cl	THF	65%, CO$_2$	[687]
	Li	Cl	PET	51%, CO$_2$	[688]
M–C$_{10}$H$_{15}$[f]	MgX	Br	DEE	58%, CO$_2$	[549]
	Li	Cl	PET	82%, D$_2$O	[689]

[a] Solvent Sv: PET = petroleum ether (pentanes, hexanes, heptanes), DEE = diethyl ether, THF = tetrahydrofuran, DMF = N,N-dimethylformamide. [b] Not specified.

rather than heterogeneously at a solid surface. Their main drawback is the loading of the reaction mixture with a stoichiometric amount of a nonvolatile hydrocarbon.

The preparation of Grignard reagents[780] does, in general, not require such precautions. In the case of a sluggish reaction, the 'entrainment method' may prove helpful. The trick is to add continuously small quantities of a very reactive halide, for example ethyl bromide, which then drags along the poorly reactive substrate. Initially, the 'pacemaker'

Table 17. 1-Alkenylmagnesiums, -lithiums, and sodiums: yields of trapping products as a function of the metal M, the precursor halide X, the solvent Sv, and the electrophile $El–X'$ (common temperature range $+25\,°C$ to $+50\,°C$).

M–R	M	X	Sv[a]	Product, $El–X'$	Reference
	MgX	Cl, Br	DEE[b]	94%, titration	[690–691]
	Li	Cl	THF	70–80%[c]	[681]
M–CH=CH$_2$	Na	Cl	PET	90%, CO$_2$	[692]
	K	Cl	PET	73%	[692–693]
M–CH=C(CH$_3$)$_2$	MgX	Br	THF	90%, Cl$_3$CCHO	[690–694]
	Li	Br	DEE	33%, H$_5$C$_6$CHO	[695]
M–C(CH$_3$)=CH–CH$_3$	MgX	Cl	THF	50%, by titration	[691]
	Li	Br	DEE	75%, H$_3$CCOOC$_2$H$_5$	[696]
M–CH(C$_6$H$_5$)=CH$_2$	Na	Cl	PET	21%, CO$_2$	[692]
M–CH=CH–C$_9$H$_{19}$	Na	Br	PET	32%, CO$_2$[d]	[697]
M–CH=CH–C$_{10}$H$_{13}$	Na	Br	PET	[e]CO$_2$[d]	[697]
M–C$_4$H$_5$[f]	Mg	Br	THF	62%, CO$_2$	[698]
M–C$_5$H$_7$[g]	Mg	Br	THF	63%, CO$_2$	[699]
	Mg	Br	THF	33%, CO$_2$	[699]
M–C$_6$H$_9$[h]	Li	Cl	DEE	40%, H$_5$C$_6$CHO	[695]
	Na	Cl	PET	60%, CO$_2$	[692]
M–C$_7$H$_{11}$[i]	Mg	Br	THF	20%, CO$_2$	[698]
M–C$_8$H$_{13}$[j]	Mg	Br	THF	27%, CO$_2$	[699]
M–C$_{12}$H$_{21}$[k]	Mg	Br	THF	42%, CO$_2$	[698]

[a] Solvent Sv: PET = petroleum ether (pentanes, hexanes, heptanes), DEE = diethyl ether, THF = tetrahydrofuran. [b] Or THF. [c] Various $El–X'$. [d] Contains isomers. [e] Not specified.

was believed to act merely as a surface sweeper, always restoring new areas of high reactivity by further corroding the metal when it becomes covered with oxide and halide deposits. Recent evidence points at a more specific role. The primary radical inevitably generated from the promotor halide, say ethyl bromide, will freely diffuse into the solution (see p. 76) where it encounters a relatively inert secondary or tertiary halide. Transfer of a bromine atom restitutes ethyl bromide and sets free a more complex radical which

Table 18. Phenylmagnesiums, -lithiums, and sodiums and congeners: yields of trapping products as a function of the metal M, the precursor halide X, the solvent Sv and the electrophile $El-X'$ used.

M−R	M	X	Sv[a]	Product, $El-X'$	Reference
M–⟨phenyl⟩	MgX	Cl, Br	DEE	95%[b]	[700–701]
	Li	Cl, Br	DEE[c]	≈85%[b]	[702–705]
	Na	Cl	PET[d]	87%, CO_2	[639,707,708]
M–⟨2-CH₃-phenyl (H₃C ortho)⟩	MgX	Cl	THF	99%, titration	[700,706]
	Li	Br	DEE	93%, titration	[706]
M–⟨3-CH₃-phenyl (CH₃ meta)⟩	MgX	Cl	THF	96%, titration	[700–706]
	Li	Br	DEE	86%, titration	[706]
M–⟨4-CH₃-phenyl⟩	MgX	Cl	THF	93%, titration	[700]
	Li	Br	DEE	95%, titration	[702–704,706]
	Na	Cl	PET	70%, CO_2	[707,708]
M–⟨2-C₂H₅-phenyl⟩	Li	I	DEE	50%, ⟨O-furanyl⟩–COOLi	[709]
M–⟨4-C₂H₅-phenyl⟩	MgX	Cl	THF	97%, titration	[700]
M–⟨2-C₆H₅-phenyl⟩	Li	Br, I	DEE	~75%[b]	[710,711]
M–⟨3-C₆H₅-phenyl⟩	Li	Br	DEE	72%, titration	[710]
M–⟨4-C₆H₅-phenyl⟩	Li	Br	DEE	87%, titration	[710,712]
M–⟨4-C₁₂H₂₅-phenyl⟩	Li	Br	DEE	75%, titration	[713]
M–⟨2,6-di-tert-butyl-phenyl⟩	MgX	Br	DEE[c]	50%, titration	[700]
M–⟨3,5-dimethyl-phenyl with 4-CH₃⟩	MgX	Br	DEE	85%, CO_2	[714,715]
	Li	Br	DEE	44%, titration	[716,717]
M–⟨2,3,5,6-tetramethyl-phenyl⟩	MgX	Br	DEE	70%, phthalide	[715]

[a] Solvent Sv: PET = petroleum ether (pentanes, hexanes, heptanes), DEE = diethyl ether, THF = tetrahydrofuran, DMF = N,N-dimethylformamide; in case of mixtures (containing for example paraffinic cosolvents) only the most polar component is indicated. [b] Various electrophiles. [c] Or THF. [d] Or benzene or toluene.

Table 19. 1- and 2-Naphthyl-, 9-anthracenyl- and 9-phenanthrenyl, 1-pyrenylmagnesiums, -lithiums and -sodiums: yields of trapping products as a function of the metal M, the precursor halide X, the solvent Sv and the electrophile $El-X'$ used.

M–R	M	X	Sv[a]	Product, $El-X'$	Reference
	MgX	Cl, Br	DEE	69%, CO_2	[700,718,719]
	Li	Br	DEE	80%, titration	[706]
	Na	Cl	PET	90%, CO_2	[720]
	MgX	Br	DEE	84%, titration	[721]
	Li	Br	DEE	81%, $SiCl_4$	[706,722]
	MgX	Br	DEE[b]	20%, $H_3CCOC_2H_5$	[723]
	Na	Cl	PET	55%, CO_2	[720]
	MgX	Br	DEE	41%, $HC(OC_2H_5)_3$ [c]	[724]
	Na	Cl	PET	35%, CO_2	[720]
	Na	Cl	PET	80%, CO_2	[720]

[a] PET = petroleum ether (pentanes, hexanes, heptanes), DEE = diethyl ether, THF = tetrahydrofuran; in case of mixtures (containing, for example, paraffinic cosolvents) only the most polar component is indicated. [b] Or THF. [c] Aldehyde isolated after acid hydrolysis.

merely needs to pick up from the metal surface a single electron to evolve to the desired organomagnesium[781].

Alkylmagnesium fluorides can be prepared either under entrainment conditions[782] or by means of atomized metals[783]. Metal activation is indispensable to make cyclopropylmethylmagnesium bromide[784], which undergoes ring-opening isomerization at temperatures above −75 °C, and 2-isoprenylmagnesium chloride[785–787], which exists in dynamic equilibrium with its 2,3-allenyl metallomer[788]. Also, ordinary allylic Grignard reagents, such as prenylmagnesium bromide[789], deserve care as they may rapidly condense with their precursor halides. Overall, reductive dimerization is an even bigger threat with α-halo ethers as the starting materials. Ethoxymethylmagnesium chloride can only be prepared under strictly controlled conditions[790,791]. The propensity of dioxolanes, and other acetals, to form transient carbenium-oxonium ions has already been emphasized (see pp. 82–83).

Table 20. Arylmagnesiums and aryllithiums carrying amino, alkoxy and halo substituents: yields of trapping products as a function of the metal M, the precursor halide X, the solvent Sv and the electrophile $El-X'$ used.

M–R	M	X	Sv[a]	Product, $El-X'$	Reference
M—⬡ N(CH₃)₂	MgX	Br	DEE	70%, $(H_3C)_2AsCl$	[725]
	Li	Br	DEE	50%, $(H_5C_6)_3PbCl$	[726]
M—⬡ N(CH₃)₂	Li	Br	DEE	58%, $(H_5C_6)_3SiCl$	[727]
M—⬡—N(CH₃)₂	MgX	Cl, Br	DEE[b]	88%, titration	[727]
	Li	Br	DEE	65%, CO_2	[680,728]
M—⬡—N(CH₂C₆H₅)₂	Li	Br	DEE	22%, $RCOOC_2H_5$ [c]	[729]
M—⬡ OCH₃	Li	Br	DEE	85%, titration	[707]
M—⬡—OCH₃	Mg	Cl	THF	77%, titration	[706]
	Li	Br	DEE	66%, titration	[707]
M—⬡ O O	MgX	Br	THF	90%, RCHO [d]	[730]
M—⬡ F	MgX	Cl	THF	50%, titration	[700]
M—⬡ CF₃	MgX	Cl	THF	62%, titration	[700]
M—⬡ Cl	MgX	Cl	THF	19%, titration	[700]
M—⬡—Cl	MgX	Cl	THF	96%, titration	[700]
Cl Cl M—⬡—Cl Cl Cl	MgX	Cl	THF	78%, titration	[700]

[a] PET = petroleum ether (pentanes, hexanes, heptanes), DEE = diethyl ether, THF = tetrahydrofuran, DMF = N,N-dimethylformamide; in case of mixtures (*e.g.* containing paraffinic cosolvents) the most polar component is indicated.
[b] Or THF. [c] R = $H_3CCH(OC_2H_5)$. [d] R = H_5C_6COO…O⎤, then oxidation with CrO_3.

Table 21. Reductive metal insertion in halogenated pyrazoles, furans and thiophenes: yields of trapping products as a function of the metal M, the precursor halide X, the solvent Sv and the electrophile $El-X'$ used.

$M-R^{[a]}$	M	X	$Sv^{[b]}$	Product, $El-X'$	Reference
(pyrazole, $N-C_6H_5$)	XMg	Cl	DEE	82%, CO_2	[731]
(furan)	XMg	I	HMPA	46%, [c]	[732]
(furan)	K	I	PET[d]	0.5%, CO_2	[733]
(methylfuran, CH_3)	XMg	Br	THF	40%, $(H_3C)_2C=CH-CH_2Br$	[734]
(thiophene)	XMg / Na	Br, I / Cl	DEE / BNZ	~100%, $(H_5C_6)_3SnCl$ / 84%, CO_2	[735–737] / [738]
(methylthiophene, CH_3)	XMg	I	DEE	50%, 3,6-dichlorophthalanhydride	[739]
(bromothiophene, Br)	XMg	Br	DEE	78%, S_8, then H_3CI	[740–742]
(bromothiophene, Br)	XMg	Br	DEE	57%, $(H_3CS)_2$	[741, 742]
(bromothiophene, Br)	XMg	Br	DEE	~65%, [e]	[741–743]
(chlorothiophene, Cl)	MgX	Br	DEE	87%, CO_2	[744]
(bromothiophene, Br)	MgX	Br, I	DEE	38%, CO_2	[741,742]
(dibromothiophene, Br, Br)	MgX	Br	DEE	50%, CO_2 [f]	[741,742]

[a] 3-Furylmethylmagnesium bromide [Tanis, S.P.; Head, D.B.; *Tetrahedron Lett.* **1982**, *23*, 5509; Araki, S.; Butsugan, Y.; *Bull. Chem. Soc. Jap.* **1983**, *56*, 1446] and other compounds carrying the metal in a side chain rather than at a ring position are not listed. [b] PET = petroleum ether (pentanes, hexanes, heptanes), BNZ = benzene, DEE = diethyl ether, THF = tetrahydrofuran, HMPA = hexamethylphosohoric triamide; in case of mixtures (containing, for example, paraffinic cosolvents) only the most polar component is indicated. [c] 1,4-Addition to methyl 5-methoxy-1,5-cyclohexadienylcarboxylate followed by interception with 2-bromoallyl bromide. [d] First neat; PET added just before carboxylation. [e] Various electrophiles. [f] The position of magnesium insertion and subsequent carboxylation was not unequivocally established.

Table 22. Reductive metal insertion in halogenated isoxazoles, thiazoles and benzothiazoles: yields of trapping products as a function of the metal M, the precursor halide X, the solvent Sv and the electrophile used.

M−R	M	X	Sv[a]	Product, $El-X'$	Reference
	XMg	Br	DEE	50%, titration[b]	[745]
	XMg	Br	DEE	90%, titration[b]	[745]
	XMg	Br	DEE	85%, titration[b]	[745]
	XMg	I	DEE	73%, titration[b]	[745]
	XMg	Br	DEE	31%, titration[b]	[745]
	XMg	Cl	THF	79%, titration[b]	[745,746]
	XMg	I	DEE	53%, CO_2	[747]

[a] DEE = diethyl ether, THF = tetrahydrofuran. [b] Assumed.

2-(2-Dioxolanyl)ethylmagnesium bromide[792] (**182a**, $n = 1$) is even less stable than its higher homologs[793–799] (**182b–e**, $n = 2$–4). It readily decomposes by ring-opening β-elimination.

The preparation of β-amino functionalized Grignard reagents, such as **184**[800,801] (R, R′ = alkyl, aryl) and **185**[800,802], again requires activated magnesium. Unsaturation, as present in **185**, enhances the stability of such organometallic intermediates. α-Amino[803,804] and γ-amino[805] analogs, such as **183** (R, R′ = alkyl, aryl, R″ = H, alkyl, aryl) and **186** (R, R′ = alkyl, aryl, R″ = H, CH_3) are remarkably robust species.

Table 23. Reductive metal insertion in halopyridines and quinolines: yields of trapping products as a function of the metal M, the precursor halide X, the solvent Sv and the electrophile used.

M–R	M	X	Sv [a]	Product, $El–X'$	Reference
M–(2-pyridyl)	XMg	Cl, Br	DEE, THF	58%, $HC(OC_2H_5)_3$	[748–750]
M–(3-pyridyl)	XMg	Cl, Br	DEE, THF	56%, $HC(OC_2H_5)_3$	[748,751]
M–(4-pyridyl)	XMg	Cl	DEE, THF	19%, H_5C_6CHO [b]	[748,752]
M–(2-quinolyl)	XMg	Cl	THF	62%, titration	[747]
M–(8-quinolyl)	XMg	Cl	THF	59%, titration	[747]
M–(tetrafluoropyridyl)	XMg	Br	THF	59%, CO_2	[753]
M–(tetrachloropyridyl)	XMg	Cl	THF	80%, $(H_3C)_3SiCl$	[754]
M–(tetrabromopyridyl)	XMg	Br	DEE	37%, CO_2	[755]

[a] DEE = diethyl ether, THF = tetrahydrofuran. [b] The main product is pyridine.

Perfluoroalkylmagnesium halides (such as **187**[806] and **188**[807]) are extremely labile, low-yielding reagents. Perfluoroalkylmagnesium compounds, such as **189**[808], carrying no halogen in either the α or β the positions are much easier to handle.

$$IMg-CF_2-CF_3 \qquad IMg-CF_2-CF_2-CF_3 \qquad BrMg-CH_2-CH_2-(CF_2)_5CF_3$$

187　　　　　**188**　　　　　**189**

Success or failure in the preparation of organometallic compounds does not exclusively depend on the appropriate activation of the metal. The solvent and the precursor halide are other crucial factors as a 1930 landmark study has revealed (see Table 24)[808]. Butyl iodide was found to destroy butyllithium in the course of a few minutes or a few hours, depending on whether diethylether or benzene was the solvent. The alkyllithium survived in the presence of butylbromide in benzene, but not in ether. Finally, virtually no reaction was observed between butyl chloride and butyllithium, no matter what solvent. The practical advice is obvious: the chloride is the best starting material with which to make butyllithium and the iodide is simply unsuitable.

Table 24. Butyllithium in the presence of butyl chloride, bromide and iodide (0.5 M initial concentrations): half-lives $\tau_{1/2}$ (in hours) as a function of the solvent benzene (BNZ) or diethyl ether (DEE)[808].

$H_9C_4-X + LiC_4H_9$	$\tau_{1/2}^{BNZ}$ [h]	$\tau_{1/2}^{DEE}$
X = I	3	<0.1
X = Br	40	0.5
X = Cl	>100	40

Secondary and tertiary alkyllithiums attack ethereal solvents even at temperatures around or below $-25\,^{\circ}C$ (see pp. 289–293). Therefore, low-boiling petroleum ether fractions are the medium of choice. Moreover, sodium-lithium alloy[680,809] or sodium–lithium codispersion[683,810] may have to be employed unless standard label lithium is activated by abrasion[682].

The use of 'radical anions' as metal sources will again often facilitate the envisaged transformations. For example, 7-norbornadienyllithium (**190**)[811,812] can be generated from its labile chloro precursor only by means of 4,4'-bis(*tert*-butyl)biphenyl/lithium. γ-*O*-functional alkyllithiums (such as **192**[813] and **193**[814,815]) and the corresponding *N*-analogs (*e.g.* **197**[816]) are readily accessible by the reductive insertion method. Double bonds (as in **191**[817] and **193**[814]) and the deprotonation of OH and NH sites (**194**[818], **195**[819], **196**[820,821] and **198**[822]) tend to stabilize that kind of organometallic intermediates.

190

$$Li-\overset{\overset{\displaystyle CH_2}{\|}}{C}-CH(OC_2H_5)_2$$

191

$$Li-CH_2-CH_2-CH_2-O-\overset{\overset{\displaystyle CH_3}{|}}{CH}-OC_2H_5$$

192

$$Li-\overset{\overset{\displaystyle H}{|}}{C}=\overset{\overset{\displaystyle }{}}{\underset{\underset{\displaystyle H}{|}}{C}}-CH(OC_2H_5)_2$$

193

$$Li-CH_2-\overset{\overset{\displaystyle R'}{|}}{\underset{\underset{\displaystyle R}{|}}{C}}-OLi$$

194

$$Li-CH_2-CH_2-\overset{\overset{\displaystyle O}{\|}}{C}-OLi$$

195

$$Li-CH_2-CH-N=C-OLi$$

with R above first CH and C$_6$H$_5$ below C

196

$$Li-CH_2-CH_2-CH_2-N(CH_3)_2$$

197

structure **198**

Although the preparation of organosodium compounds by reductive metal insertion has been frequently accomplished on a kilogram scale, this demands skills and equipment. Only chlorides are suitable precursors. Petroleum ether is mandatory as the reaction medium, not only to slow down subsequent elimination reactions but even to withdraw the organometallic compound as formed by precipitation from the solution. Alkylsodiums are obtained with fair to good yields only if ultrafine sodium dispersions (≤ 25 μm average particle diameter) are employed. Efficient mixing is ensured by high-speed stirring (5000–20 000 rpm) and vortical fluid motion in special glassware ('greased flasks') and the reaction temperature is kept around or below $-10\,°C$. The preparation of phenylsodium and other arylsodiums is less critical. For example, it suffices to use sodium 'sand'[708,823,824] rather than a metal dispersion.

c) Working Procedures

The protocols given below feature one organomagnesium, nine organolithium and two organosodium compounds. A few of these prominent reagents are commercially available but either their prices are prohibitive (vinylmagnesium bromide, vinyllithium) or their purity may be insufficient (*tert*-butyllithium) or their composition changes erratically (methyllithium containing 0.7–1.5 equivalents of lithium bromide). All reactions should be carried out under a blanket of an inert gas (*e.g.* under $\geq 99.99\%$ pure nitrogen) using appropriate equipment (see pp. 285–288), and anhydrous solvents.

Vinylmagnesium Chloride[691]

$$ClMg-CH=CH_2$$

Vinyl chloride (0.25 kg, 4.0 mol) is absorbed in tetrahydrofuran (1.0 L). A small portion of this solution (approx. 20 mL) is added together with some ethyl bromide (≈ 5 mL) to magnesium turnings (0.10 kg, 4.1 mol) placed in a round-bottomed flask equipped with a dry ice condenser. The mixture is vigorously stirred. As soon as the reaction starts, the remainder of the solution is added at a rate to keep the internal temperature below or around $50\,°C$. Stirring is continued for further 45 min at $50\,°C$. Yield: 90–98% by titration, 53% by carboxylation.

Vinylmagnesium chloride can also be prepared in a continuous process[825].

Salt-free Methyllithium[826]

$$Li-CH_3$$

Lithium shred (70 g, 10 mol), covered with diethyl ether (0.2 L) are placed in a 4 L three-necked round-bottomed flask, hermetically closed with a mercury seal. Under vigorous magnetic stirring, methyl chloride (0.28 mL at −25 °C, 0.25 kg, 5.0 mol) is bubbled into the mixture at a rate slow enough to prevent the gas from escaping through the mercury seal. As soon as the reaction starts, the suspension becomes turbid and the metal surface silvery shiny. At this moment, the flask is cooled with an ice-bath and a larger quantity of diethyl ether (2.0 L) is added portionwise. After roughly 1 h, all methyl chloride is absorbed and most of the lithium consumed. The mixture is heated briefly (some 15 min) under reflux to drive residual methyl chloride out. Finally, the solution (a total of about 2 L) is siphoned through a plug of glass-wool into storage burettes. According to double titration (see p. 295), the concentration of methyllithium approximates 2.0 M.

Lithium Bromide Containing Methyllithium[826]

$$Li-CH_3 \cdot Li-Br$$

Replacing methyl chloride by methyl bromide (0.14 mL, 0.47 kg, 5.0 mol), the same protocol as above was applied. The organometallic titer of the filtered solution (some 2 L) is again approximately 2.0 M.

Lithium Bromide Containing Butyllithium[827]

$$Li-C_4H_9 \cdot Li-Br$$

At 25 °C and under vigorous stirring, a small quantity of butyl bromide (about 25 g) is added to lithium shred (70 g, 10 mol) covered with diethyl ether (0.2 L). As soon as the exothermic reaction starts (silvery shiny metal!), the flask is placed in a salt/ice bath and the remaining butyl bromide (a total of 0.54 L, 0.69 kg, 5.0 mol) in diethyl ether (2.0 L) is added dropwise in the course of 1 h, thus keeping the interior temperature around −10 °C until the end of the reaction when it is allowed to rise to 0 °C. The solution (some 2 L) is siphoned through a plug of glass wool into burettes which are stored in a freezer (at −10 °C, better −25 °C) to avoid rapid decomposition (see p. 295). The organometallic titer determined by double titration approximates to 1.8 M.

The same protocol can be applied to the preparation of lithium bromide containing *cyclopropyllithium*[684] from cyclopropyl bromide (0.40 L, 0.60 kg, 5.0 mol).

Lithium Bromide Containing 1,4-Tetramethylenedilithium[828,829]

$$Li-CH_2-CH_2-CH_2-CH_2-Li \cdot Li-Br$$

The mineral oil is completely removed from a commercial 30% (by weight) lithium dispersion (approx. 60 mL, 58 g, 2.5 mol) by thorough washing with hexanes and pentanes. The powder is collected by filtration through a glass frit before being transferred in a three-necked 1 L flask and covered with diethyl ether (0.50 L). A solution of 1,4-dibromobutane (60 mL, 0.11 kg, 0.50 mol) in diethyl ether (0.50 L) is added dropwise in the course of 1 h under vigorous stirring. As soon as the reaction starts, the mixture is cooled from 25 °C to −15 °C. The solution is siphoned through a plug of glass wool or through a glass frit (porosity 3) into a burette to be stored around or below −10 °C. The yield approximates 65%. The organometallic titer can be reliably assessed by treating an aliquot of the solution with excess chlorotrimethylsilane (or another suitable electrophile, for example dry ice) and determining the amount of trapping product by gas chromatography (using an 'internal standard' and calibration factors). Double titration will give only a crude estimate as the solution also contains some butyllithium and hexyllithium along with hydrocarbons such as cyclobutane.

tert-*Butyllithium*[681]

$$Li-C(CH_3)_3$$

A suspension of 1% sodium containing lithium dispersion (6.9 g, 1.00 mol) in pentanes (1.0 L) is heated under reflux. Under vigorous stirring, *tert*-butyl chloride (47 mL, 0.40 kg, 0.43 mol) containing a small amount of *tert*-butyl alcohol (2 mL) is added dropwise. After a short while (some 5–10 mL of the halide being introduced) the reaction starts. The external heating is immediately removed and the remaining *tert*-butyl chloride added over 3 h at a rate sufficient to maintain a gentle reflux. If necessary, the flask is immersed in an ice-bath. Stirring is continued for 0.5 h as the mixture slowly cools. The resulting slurry is filtered through sintered glass (4–8 μm) in a storage burette, the flask and the filter cake being rinsed with pentanes (0.1 L). The pyrophoric filter cake is cautiously discarded into dry ice or mixed with an excess of dry sand. The organometallic titer, as determined by double titration (see p. 295) approximates to 0.3 M.

Vinyllithium[830]

$$Li-CH=CH_2$$

A 2 : 98 (w/w) sodium/lithium codispersion (preparation given below; 7.0 g, 1.0 mol) is covered with tetrahydrofuran (0.1 L). Vinyl chloride (27 mL, 25 g,

0.40 mol) is continuously condensed into the vigorously stirred suspension. As soon as the reaction starts, more tetrahydrofuran (0.4 L) is added and the mixture is cooled with an ice-bath. When after 2 h all vinyl chloride is absorbed, stirring continues for 1 h at 0 °C, before the solution is siphoned through a plug of glass wool into a storage burette. The organometallic concentration (about 0.5 M) can be assessed by double titration or, more accurately, by quenching an aliquot with benzaldehyde and gas chromatographic determination of the yield of 1-phenyl-2-propen-1-ol, the trapping product (using an 'internal standard' and calibration factors).

The 2 : 98 (w/w) sodium/lithium co-dispersion[831] is prepared in a three-necked 4 L flask by melting lithium metal lumps under argon in hot (200 °C) paraffin oil (3 L) containing a few drops of oleic acid while a high-speed stirrer (see p. 288) is operated. When the metal has become a fine gray suspension, the heating jacket is removed. After a few further seconds the stirrer is stopped. When, without external cooling, ambient temperature is reached, the metal dust is filtered using a large-pore frit and washed with hexanes before being dried and transferred into a Schlenk vessel.

Lithium Bromide Containing Phenyllithium[702,703]

$$\text{Li} - \langle \bigcirc \rangle \cdot \text{Li--Br}$$

Lithium shred or wire (35 g, 5.0 mol) is placed in a three-necked 4 L flask and covered with diethyl ether (0.2 L). The reaction is started by adding a small amount of bromobenzene (about 10 mL). The vessel is now placed in an ice-bath and the remaining bromobenzene(a total of 0.21 L, 0.31 kg, 2.0 mol) is added in the course of 1 h under vigorous stirring, the mixture being kept below its boiling point. The magenta-colored solution is siphoned through a glass wool plug into storage burettes. It is roughly 0.75 M in phenyllithium, as indicated by double titration (see p. 295). It also contains 6–8% of biphenyl and trace amounts of ter- and quaterphenyls, 'radical anions' of the latter being presumed to be at the origin of the coloration.

In the same way, phenyllithium can be prepared in tetrahydrofuran, either 'salt-free' from chlorobenzene[704] or from lithium bromide containing from bromo-benzene[704,810]. However, both types or reagents have to be conserved below −25 °C to avoid rapid decomposition by reaction with the solvent.

Methoxymethyllithium[832]

$$\text{Li--CH}_2\text{OCH}_3$$

The paraffin oil is removed from a commercial dispersion of lithium (6.9 g, 1.00 mol) containing 0.9% of sodium by repetitive washing with diethyl ether. After drying, it is transferred into a three-necked flask filled with formaldehyde dimethyl acetal (0.15 L). At 0 °C, chloromethyl methyl ether (30 mL, 31 g, 0.40 mol), dissolved in formaldehyde dimethyl acetal (0.10 L), is added dropwise under vigorous stirring. As soon as a temperature rise indicates the start of the reaction, the mixture is cooled to −25 °C and the remaining chloride added in the course of 3 h. After an additional 0.5 h of stirring at −25 °C, the solution is filtered through a glass wool plug into a storage burette. The organometallic titer is determined by double titration (see p. 295) or, better, by quenching of an aliquot with benzaldehyde and subsequent gas chromatographic determination of the yield of the trapping product. The concentration generally ranges from 0.32 to 0.35 M.

Pentylsodium[659,833]

$$Na-C_5H_{11}$$

At −10 °C and under vigorous high-speed stirring (5000–10 000 rpm, see p. 288), pentyl chloride (60 mL, 53 g, 0.50 mol) is added dropwise over 1 h to a sodium dispersion (23 g, 1.0 mol, obtained by filtering a commercial suspension in toluene through a glass frit and by repetitive washing with pentanes) in pentanes (0.25 L). After introduction of the entire chloride quantity, stirring is continued for another 0.5 h at −10 °C. The resulting bluish-gray suspension may be used as such. If only a small fraction thereof is needed or if an aliquot has to be taken for analytical purposes, the required volume can be easily withdrawn from the well-stirred suspension by means of a nitrogen-purged pipette.

Pentylsodium is the most prominent organosodium reagent just for historical reasons. Exactly the same protocol as above can be applied to make *butylsodium*[662].

4.1.2 Permutational Exchange of Halogen Against Metal

Independently discovered by G. Wittig[834] and H. Gilman[835] and collaborators, the permutational halogen/metal exchange is arguably the most straightforward method of access to a reactive organometallic intermediate. Next to hydrogen/metal interconversion ('metalation', Section 4.5.2) this is the most frequently applied reaction mode, enabling the selective generation of organolithium compounds.

a) Representative Examples of Halogen/Metal Permutation

Again typical examples have been compiled to illustrate the potential of the method. They are ordered starting with alkyl and cycloalkyl derivatives (see Tables 25–30), proceeding through 1-alkenyl and 1-cycloalkenyl derivatives (see Tables 31–40), aryl derivatives (see Tables 41–54) and metallocene derivatives (see Table 55) and

Table 25. Primary aliphatic organolithiums (M = Li) by halogen/metal permutation using reagents M–R' for exchange and electrophiles El–X' for trapping.

M–R	M–R'	X[a]	Sv[b]	T[°C]	Product, El–X'[c]	Reference
M–CH$_3$	LiC$_4$H$_9$	Br	DEE	−75	64%, by titration	[60, 836]
M–C$_4$H$_9$	LiC(CH$_3$)$_3$	I	DEE	−75	91%, H$_9$C$_4$CH=O	[837]
M–C$_8$H$_{17}$	LiC(CH$_3$)$_3$	I	DEE	−75	93%, H$_3$CCOCH$_3$	[837]
M–C$_{10}$H$_{15}$[d]	LiC(CH$_3$)$_3$	I	DEE	−45	66%, CO$_2$	[838]
M–CH$_2$-⟨cyclopentyl⟩	LiC(CH$_3$)$_3$	I	DEE	−75	91%, CO$_2$	[837]
M–(CH$_2$)$_2$C$_6$H$_5$	LiC(CH$_3$)$_3$	I	DEE	−75	91%, CO$_2$	[837–839]
M–(CH$_2$)$_2$-⟨1,3-dioxolan-2-yl⟩	LiC(CH$_3$)$_3$	I	DEE	−75	92%, BrCH$_2$CH$_2$Br	[837]
M–CH$_2$C(CH$_3$)$_3$	LiC(CH$_3$)$_3$	I	DEE	−75	89%, H$_7$C$_3$CH=O	[837]
M–(CH$_2$)$_2$CH=CH$_2$	LiC(CH$_3$)$_3$	I	DEE	−75	88%, ClSn(CH$_3$)$_3$	[839]
M–(CH$_2$)$_3$CH=CH$_2$	LiC(CH$_3$)$_3$	I	DEE	−75	87%, CO$_2$	[840]
M–(CH$_2$)$_2$–C≡C–C$_4$H$_9$	LiC(CH$_3$)$_3$	I	DEE	−75	88%, ClSn(CH$_3$)$_3$	[839]
M–(CH$_2$)$_4$–C≡C–C$_6$H$_5$	LiC(CH$_3$)$_3$	I	DEE	−75	87%, D$_2$O	[841]
M–(CH$_2$)$_4$–Cl	LiC(CH$_3$)$_3$	I	DEE	−75	55%, ClSi(CH$_3$)$_3$	[839]
M–(CH$_2$)$_4$–M	LiC(CH$_3$)$_3$	I	DEE	−75	70%, ClSi(CH$_3$)$_3$	[839]
M–(CH$_2$)$_5$–M	LiC(CH$_3$)$_3$	I	THF	−75	70%, ClSi(CH$_3$)$_3$	[839]
M–(CH$_2$)$_6$–M	LiC(CH$_3$)$_3$	I	[e]	−75	75%, ClSi(CH$_3$)$_3$	[839]

[a] X = halogen displaced by the metal. [b] Solvent (*Sv*): DEE = diethyl ether, THF = tetrahydrofuran; in case of mixtures (*e.g.* with paraffinic cosolvents) only the most polar component is indicated. [c] Electrophilic trapping reagent. [d] C$_{10}$H$_{15}$ = 2-adamantyl. [e] Not specified.

terminating with organometallic derivatives of five- and six-membered heterocycles (see Tables 56–60). As in the preceding Section, each structural group begins with the heteroatom-free or silylated hydrocarbons and turns consecutively to amino (or phosphino), oxy (or thio) and halo substituted congeners.

b) Scope and Limitations

Despite its universality and simplicity, the halogen/metal permutation cannot cope with all situations. To avoid pitfalls, one has to know the fundamental features and restrictions.

Table 26. α-Fluoroalkylmagnesiums and -lithiums by halogen/metal permutation using reagents M–R'.

M–R	M–R'	X[a]	Sv[b]	T[°C]	Product, El–X' [c]	Reference
M–C$_2$F$_5$	BrMgC$_2$H$_5$	I	DEE	−45	83%, (H$_3$C)$_3$SiCOC$_6$H$_5$	[842]
	LiCH$_3$[d]	I	DEE	−75	88%, H$_5$C$_6$COCH$_3$	[843]
M–C$_3$H$_7$	LiCH$_3$[d]	I	DEE	−40[e]	54%, H$_5$C$_6$CHO	[844]
M–C$_7$F$_{15}$	LiC$_4$H$_9$	I	DEE	−90	38%, CO$_2$	[845]
M–C$_8$F$_{17}$	BrMgC$_6$H$_5$	I	DEE	−70	90%, (H$_3$C)$_2$CO [d]	[846]
M–C$_{10}$F$_{21}$	BrMgC$_2$H$_5$	I	THF	−80	63%, ClSi(CH$_3$)$_3$	[847]
M–CF(CF$_3$)$_2$	LiCH$_3$[d]	I	DEE	−75	53%, H$_5$C$_2$CHO	[848]
M–(CF$_2$)$_2$–OCF(CF$_3$)$_2$	BrMgC$_2$H$_5$	I	THF	−75	65%, ClSi(CH$_3$)$_3$	[847]
M–(CF$_2$)$_4$–OCF(CF$_3$)$_2$	BrMgC$_2$H$_5$	I	THF	−75	58%, ClSn(CH$_3$)$_3$	[847]
M–(CF$_2$)$_6$–OCF(CF$_3$)$_2$	BrMgC$_2$H$_5$	I	THF	−75	60%, ClSn(CH$_3$)$_3$	[847]
M–(CF$_2$)$_8$–OCF(CF$_3$)$_2$	BrMgC$_2$H$_5$	I	THF	−75	62%, ClSn(CH$_3$)$_3$	[847]
M–(CF$_2$)$_4$–M	LiC$_4$H$_9$	I	DEE	−80	18%, H$_3$CCHO	[849]

[a] X = halogen displaced by the metal. [b] DEE = diethyl ether, THF = tetrahydrofuran. [c] El–X' = Electrophilic trapping agent. [d] Halogen/metal exchange performed in the presence (*in situ*) of the electrophile. [e] Or −75 °C.

- *Element Aptitudes*

Halogen/metal permutation is to a large extent synonymous with bromine/lithium or iodine/lithium interconversion. Only a handful of bromine/sodium interconversions[650,1089] and one single bromine/potassium interconversion (treatment of 3-bromothiophene with α-cumylpotassium for 24 h at 25 °C in diethyl ether, followed by carboxylation affording 62% of 2-thiophenecarboxylic acid and 81% of α,α'-dicumyl)[1090] have been reported. In contrast, halogen/metal interconversions involving aryl bromides or iodides and Grignard reagents are well documented and presently receive much attention (see below under the heading 'Group tolerance').

In general, chlorides do not participate in halogen/metal interconversions. Noteworthy exceptions are *gem*-dichlorocyclopropanes and *gem*-dichloroalkenes[860,861] on one side and chloroarenes in which the halogen is surrounded by more chlorine atoms or other electron-withdrawing substituents [hexachlorobenzene[1022], *N*,*N*-dimethylpentachlorobenzylamine[1091], 2-chloro-1,3,5-tris(trifluoromethyl)benzene[1092]] on the other side. Bromochloroarenes and iodochloroarenes exchange exclusively the heavier halogen against lithium.

The interconversion between an aryl bromide and an alkyllithium proceeds to completion in the course of a couple of minutes in diethyl ether and requires a few seconds in

Table 27. α-Alkoxy-, α-chloro-, α-bromo- and -α-iodoalkylmetal compounds (metal M = magnesium halide or lithium) by halogen/metal permutation using organometallic reagents M−R′.

M−R	M−R′	X[a]	Sv[b]	T[°C]	Product, El−X′[c]	Reference
M−CH$_2$Cl	LiC$_4$H$_9$	I	THF	−75	88% [d], H$_5$C$_6$CHO [d]	[850,851]
M−CH$_2$Br	ClMgCH(CH$_3$)$_2$	Br	THF	−75	50%, H$_5$C$_6$CHO [e]	[852,853]
M−CH$_2$I	ClMgCH(CH$_3$)$_2$	I	DEE[f]	−75	45%, H$_5$C$_6$CHO [e]	[853,854]
	LiC$_4$H$_9$	I	THF	−75	89%, [e]	[855]
M−CH(CH$_3$)Cl	LiCH(CH$_3$)C$_2$H$_5$	Br	THF	−115	63%, H$_5$C$_6$CHO	[856]
M−CH(CH$_3$)Br	LiCH(CH$_3$)C$_2$H$_5$	Br	THF	−115	59%, H$_5$C$_6$CHO	[856]
M−C(CH$_3$)$_2$Cl	LiCH(CH$_3$)C$_2$H$_5$	Br	THF	−115	45%, H$_5$C$_6$CHO	[856]
M−C(CH$_3$)$_2$Br	LiCH(CH$_3$)C$_2$H$_5$	Br	THF	−115	41%, H$_5$C$_6$CHO	[856]
M−C(C$_6$H$_5$)$_2$Cl	LiC$_4$H$_9$	Cl	THF	−100	75%, CO$_2$ [e]	[857]
M−CHCl$_2$	ClMgCH(CH$_3$)$_2$	Cl	THF	−65	25%, H$_5$C$_6$CHO	[853]
M−CHBr$_2$	ClMgCH(CH$_3$)$_2$	Br	THF	−75	71%, H$_5$C$_6$CHO	[853,858,859]
M−CHI$_2$	ClMgCH(CH$_3$)$_2$	I	THF	−95	72%, ClSn(CH$_3$)$_3$	[860]
M−CBr$_2$CH$_3$	LiC$_4$H$_9$	Br	THF	−100	by NMR	[861]
M−CCl$_3$	LiC$_4$H$_9$	Cl	THF	−100	76%, CO$_2$	[857,858,861]
M−CBr$_3$	LiC$_6$H$_5$[g]	Br	THF	−110	91%, CO$_2$	[861,862]

[a] X = halogen displaced by the metal. [b] DEE = diethyl ether, THF = tetrahydrofuran; in the case of mixtures only the most polar component is indicated. [c] $El-X'$ = electrophilic trapping reagent. [d] Trapped *in situ*. [e] The isolated product emanated from a subsequent reaction. [f] Or THF. [g] Or butyllithium.

tetrahydrofuran, both at −75 °C. However, the reaction occurs only at a moderate rate when an aryllithium is used as the exchange reagent[1093]. The halogen/metal permutation with butyllithium is so fast that it was even suspected to outperform the deprotonation of hydroxy or acylamino groups[1094,1095], erroneously as was demonstrated later[1096]. Presumably due to by-product formation, iodides sometimes give inferior yields compared to bromides.

● *Group Tolerance*

The extreme rapidity of halogen/metal permutations allows one to accomplish such processes in the presence of highly reactive functional groups. This holds for both the lithium and the magnesium series.

Table 28. Cyclopropyl- and other cycloalkyllithiums by halogen/metal permutation using reagents M−R′.

M−R	M−R′	X[a]	Sv[b]	T[°C]	Product, El−X′ [c]	Reference
M—(cyclopropyl)	LiCH(CH₃)₃	Br	THF	−75	91%, (H₅C₆)₂CO	[863,866]
M—(cyclopropyl, methyl)	LiCH(CH₃)₂	Br	DEE	0	15%, CO₂	[867]
M—(cyclopropyl, C₆H₅, C₆H₅)	LiC₄H₉	Br	DEE	+25	53%, CO₂	[868]
M—(cyclopropyl, dimethyl, C₆H₅, C₆H₅)	LiC(CH₃)₃	Br	DEE	−75	25%, CO₂[d]	[869]
M—(cyclopropyl, vinyl, methyl)	LiC₄H₉	Br	DEE	−60	[e], CO₂	[870]
M—(cyclopentyl)	LiC(CH₃)₃	I	DEE	−70	60–90%, H₅C₆CHO	[871]
M—(cyclohexyl)	LiC(CH₃)₃	I	DEE	−70	60–90%, H₅C₆CHO	[871]
M—(adamantyl)	LiC(CH₃)₃	I	DEE	−70	≈70%, H₅C₆CHO	[872]
M—(dibenzo tricyclic)	LiC₄H₉	Br	DEE	+25	41%, CO₂	[873]

[a] Halogen which is displaced by the metal. [b] DEE = diethyl ether, THF = tetrahydrofuran; in case of mixtures only the most polar component is indicated. [c] El−X′ = electrophilic trapping reagent. [d] The organolithium intermediate was treated with calcium dibromide before being carboxylated. [e] 'Main product'.

Perfluoroalkylmagnesium halides (X = Cl and I) can be prepared from perfluoroalkyl iodides by treatment with aliphatic or aromatic Grignard reagents. The chain length can vary from propyl (**199**, n = 2) up to decyl (**199**, n = 9)[1097,1099]. A *tert*-butyloxycarbonylmethylmagnesium halide **200** was prepared from the iodide with isopropylmagnesium chloride in tetrahydrofuran at −75 °C[1100]. Similarly, arylmagnesium halides were obtained carrying bromine, 1-piperidylcarbonyl, cyano and alkoxycarbonyl groups at *ortho*, *meta* or *para* positions[1101].

$$XMg - (CF_2)_nCF_3 \qquad XMg - CH_2\overset{\text{O}}{\underset{\|}{C}}C(CH_3)_3$$

199   **200**

Table 29. α-Alkoxy- and β-alkoxycyclopropyllithiums (M = Li) by halogen/metal permutation (to be trapped with electrophiles $El-X'$) using organometallic reagents M−R′.

M−R	M−R′	X[a]	Sv[b]	T[°C]	Product, El−X′	Reference
M−⊲ (cis), OCH₃	LiC(CH₃)₃	Br	DEE	−75	92%, (structure)	[874]
M−⊲−OCH₃ (trans)	LiC(CH₃)₃	Br	DEE	−75	85%, H₁₃C₆−CHO	[874]
M−⊲−OR [c], OR	LiC₄H₉	Br	DEE	−75	[d]	[875]
OC₂H₅, M−⊲	LiC(CH₃)₃	Br	DEE	−75	81%, (structure) [e]	[876]
OCH₃ OCH₃, M−⊲ OCH₃, OCH₃ OCH₃	LiC(CH₃)₃	Cl	DEE	−75	77%, CH₃I	[877]

[a] Halogen displaced by the metal. [b] DEE = diethyl ether. [c] M−⊲−OR (RO) = M−⊲(O-benzo-O)
[d] H₃CCOCH₂C(OCH₃)₃; yield not specified.
[e] Product isolated after subsequent transformation.

In addition to cyano[1102,1103], amide[1103] and ester functions[1103,1104], aryllithiums sustain chloromethyl[1104,1105], 2-bromoethyl [1104−1106], trimethylsilyloxy[1107], alkoxy-carbonyloxy[1107], carbamoyl[1107], azo[1108] and nitro groups[983,984]. Simple acetal entities are, of course, fairly inert toward organolithium reagents[1109]. But even unprotected carbonyl groups can coexist with organolithiums if sterically shielded as in the lithiated ketones **201**[1110], **202**[1111] and **203**[1112] (all generated from the corresponding bromides by halogen−metal permutation with butyllithium or *tert*-butyllithium).

201   **202**   **203**

Formyl groups are too exposed not to be attacked by organometallic reagents. D. L. Cumins *et al.*[1113,1114] found an elegant solution to this problem. The aromatic or aliphatic aldehyde is protected *in situ* by addition of a lithium amide such as *N*-lithio morpholine or lithium *N*-methyl-*N*-2-pyridylamide. The resulting *O*-lithiated hemiaminal **204** (see the scheme on p. 109) can now be submitted to a halogen−metal permutation followed by electrophilic trapping. Upon neutralization the aldehyde function is restored.

Table 30. α-Halocyclopropyllithiums (M = Li) by halogen/metal permutation using reagents M$-$R$'$.

M$-$R	M$-$R$'$	X[a]	S_v[b]	T[°C]	Product, $El-$X$'$ [c]	Reference
F, M (cyclopropane fused cyclohexane)	LiC$_4$H$_9$	Br	THF	-135	9%, H$_2$O	[878]
Cl, M (cyclopropane fused cyclohexane)	LiC$_4$H$_9$	Br	THF	-110	64%, Br$_2$[d]	[878,879]
Br, M (cyclopropane fused cyclohexane)	LiC$_4$H$_9$	Br	THF	-100	39%, (structure)	[880–882]
I, M (cyclopropane fused cyclohexane)	LiC$_4$H$_9$	I	THF	-80	[e]	[862]
Br, M, C$_4$H$_9$ (cyclopropane)	LiC$_4$H$_9$	Br	THF	-100	70%, (H$_3$C)$_2$CO	[883]
Br, M, C(CH$_3$)$_3$ (cyclopropane)	LiC$_4$H$_9$	Br	THF	-100	25%, (H$_3$C)$_2$CO	[883]
Br, M, C$_6$F$_5$ (cyclopropane)	LiC$_4$H$_9$	Br	THF	-100	47%, (H$_3$C)$_2$CO	[883,884]
Br, M, CH$_3$, CH$_3$ (cyclopropane)	LiC$_4$H$_9$	Br	THF	-100	60%, (H$_3$C)$_2$CO	[883,870]
Br, M, CH$_3$, CH$_3$, CH$_3$, CH$_3$ (cyclopropane)	LiC$_4$H$_9$	Br	THF	-100	80%, H$_9$C$_4$CHO	[883]
Br, M, CH$_2$OBn [f] (cyclopropane)	LiC$_4$H$_9$	Br	THF	-95	83%, HOC$_2$H$_5$	[884]
Br, M (cyclopropane with dioxolane)	LiC$_4$H$_9$	Br	THF	-100	77%, [g]	[885]
Br, M, OC$_2$H$_5$ (cyclopropane)	LiC$_4$H$_9$	Br	THF	-95	18%, (cyclopentanone)	[886]
Br, M (cyclopropane fused oxane)	LiCH$_3$	Br	DEE	-80	88%, CO$_2$	[887]
Br, M (cyclopropane fused oxane)	LiCH$_3$	Br	DEE	-80	65%, CO$_2$	[887]
Br, M, CH$_3$, CON(CH$_3$)$_2$ (cyclopropane)	LiC$_4$H$_9$	Br	DEE	-75	89%, H$_2$O[h]	[888]
Br, M, SC$_6$H$_5$ (cyclopropane)	LiC$_4$H$_9$	Br	DME	-105	18%, (cyclopentanone)	[886]

[a] X = halogen displaced by the metal. [b] DEE = diethyl ether, THF = tetrahydrofuran; DME = dimethyl ether. [c] $El-$X$'$ = electrophilic trapping reagent. [d] Only 26% with HgCl$_2$[878]. [e] Detected by NMR. [f] Bn = CH$_2$C$_6$H$_5$. [g] 3$'$,4$'$,5$'$-Trimethoxycinnamaldehyde. [h] (E) Isomer.

Table 31. 1-Alkenyllithiums (M = Li) and 1-alkenylmagnesium halides (M = XMg) by halogen/metal permutation using the organometallic reagents M–R'.

M–R	X[a]	M–R'	Sv[b]	T[°C]	Product, El–X'[c]	Reference
M–CH=CH2	Br	LiC(CH3)3	THF	−115	74%, (SC6H5)2	[889,890]
	I	BrMgCH(CH3)2	THF	+25	74%, p-H3CC6H4CN	[891]
M–CH=CH–C4H9	I	LiC(CH3)3	PET	+25	71%, H5C6CHO	[889,890]
M–CH=CH–C6H13	I	LiC4H9[d]	PET	+25	71%, H5C6CHO	[890,892]
M–CH=CH–C7H15	I	LiC4H9[e]	DEE	−50	91%, H3CCHO	[893]
M–CH=CH–C8H17	I	LiC(CH3)3	PET	+25	74%, H5C6CHO	[891]
M–CH=CH–C10H21	I	LiC4H9	PET	+25	77%, (cyclohexanone)	[892]
M–CH=CH–C10H21	I	LiC4H9	PET	+25	100%, (cycloheptanone)	[892]
M–CH=CH–C13H27	I	LiC(CH3)3	PET	+25	77%, H5C6CHO	[891]
M–CH=CH–C6H5	Br	LiC(CH3)3	DEE	−100	[f]	[894]
M–(cyclopropenyl)	Br	LiC(CH3)3	DEE	−80	24%, (H5C2)2O	[895]
M–C(CH3)=CH–C2H5	I	LiC4H9[e]	DEE	−50	91%, (H5C2)2CO	[893]
M–C(C4H9)=CH–C2H5	I	LiC4H9[e]	DEE	−50	88%, (C5H2)2CO	[893]
M–C(CH3)[g]=CH–R	I	LiC4H9[e]	DEE	−50	85%, (CH2O)∞	[893]
M–C(C6H5)=CH–C6H5	Br	LiC4H9	THF	−75	58%, CO2[h]	[896]
M–C(C6H5)=CH–CH=C(C6H5)–M	Br	LiC4H9	DEE	+25	39%, (H3C)2SnCl2	[897]

[a] X = halogen displaced by the metal. [b] PET = petroleum ether (pentanes, hexanes, heptanes), DEE = diethyl ether, THF = tetrahydrofuran. [c] *El–X'* = electrophilic trapping reagent. [d] Or LiC(CH3)3. [e] Or LiC2H5. [f] Intermediate not characterized. [g] R = CH2CH2CH=C(CH3)2. [h] Along with 4% of 2-bromo-3,3-diphenylpropenoic acid.

Table 32. α-Silyl-1-alkenyllithiums (M = Li) by halogen/metal permutation using reagents M–R′.

M–R	X[a]	M–R′	Sv[b]	T[c](°C)	Product, El–X′ [d]	Reference
CH₂Si(CH₃)₃ M	Br	LiC(CH₃)₃	DEE	0	77%, ZnCl₂[e]	[898]
Si(C₆H₅)₃ M	Br	LiC₄H₉	DEE	+25	82%, (H₃CCO)₂O	[899]
Si(C₆H₅)₃ M C₆H₅	Br	LiC₄H₉	DEE	+25	67%, (H₃CCO)₂O	[899]

[a] X = halogen displaced by the metal. [b] DEE = diethyl ether; in case of mixtures only the most polar component is indicated. [c] Only the highest temperature is given if in the course of a reaction the mixture was allowed to warm up. [d] El–X′ = electrophilic trapping reagent. [e] Followed by the addition of 3-ethyl-2-methyl-2-oxiranylcarbaldehyde.

● *Intramolecular Trapping of Halogen–Metal Exchange Intermediates*

The 2-(2-bromoethyl)phenyllithium generated by treatment of 1-bromo-2-(2-bromo-ethyl)-benzene is only stable at −100 °C[1104−1105]. At higher temperatures, intramolecular alkylation occurs affording 5,6-dihydrobenzocyclobutene. Analogously 7*H*-benzocyclopropene **205** can be prepared by elimination of lithium methoxide from a transient 2-(methoxymethyl)phenyllithium[1115].

K. B. Wiberg *et al.*[1116] have used this intramolecular alkylation approach to prepare the [2.2.2]propellane **206** which is only moderately stable even at low temperatures.

Table 33. 1-Cycloalkenyllithium compounds by halogen/metal permutation using reagents M−R'.

M−R	X[a]	M−R'	Sν[b]	T[°C]	Product, El−X'[c]	Reference
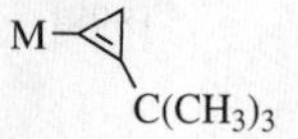 M—◁ C(CH₃)₃	Cl[d]	LiCH₃	DEE	+25	72%, CO_2	[900]
	Br[d]	LiCH₃	DEE	+25	78%, CO_2	[901]
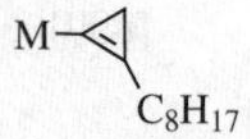 M—◁ C₈H₁₇	Br[d]	LiCH₃	DEE	+25	95%, D_2O	[900,901]
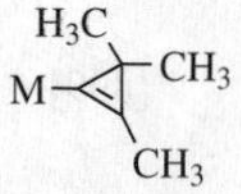 H₃C ─ CH₃ M—◁ CH₃	Cl[d]	LiCH₃	DEE	+25	63%, $(H_3C)_2CO$	[900]
	Br[d]	LiCH₃	DEE	+25	76%, CO_2	[900]
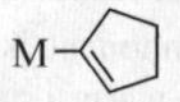 M—⬠	Br	LiC(CH₃)₃	THF	−75	73%, I_2	[889]
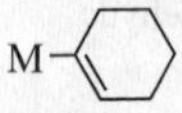 M—⬡	Br	LiC(CH₃)₃	THF	−75	52%, $(H_3C)_3SiOOSi(CH_3)_3$	[889]
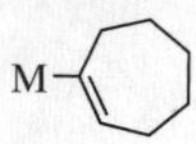 M—⬡(7)	Br	LiC(CH₃)₃	THF	−75	93%, $ClSi(CH_3)_3$	[889]
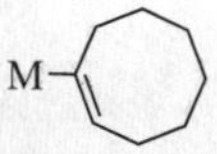 M—⬡(8)	Br	LiC(CH₃)₃	THF	−75	88%, $(H_5C_6)_2CO$	[902]
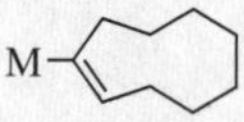 M—⬡(9)	Br	LiC(CH₃)₃	THF	−75	83%, $(SCH_3)_2$[e]	[902]
 M—⬡(diene)	Br	LiC₄H₉	THF	−60	7%, $FClO_3$[f]	[903,904]
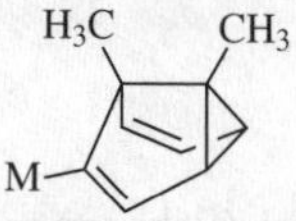 H₃C CH₃ M—	Br	LiC(CH₃)₃	DEE	−75	50%, $(CN)_2$	[905]

[a] X = halogen displaced by the metal. [b] DEE = diethyl ether, THF = tetrahydrofuran. [c] *El−X'* = electrophilic trapping reagent. [d] Prepared *in situ* from trihalocyclopropanes. [e] *Cis/trans* mixture? [f] 59% with CO_2.

Its lower homolog **207** is still more strained. As a consequence, the exchange reagent *tert*-butyllithium adds instantaneously to its central C−C bond forming 4-*tert*-butyl-1-norbornyllithium[1117].

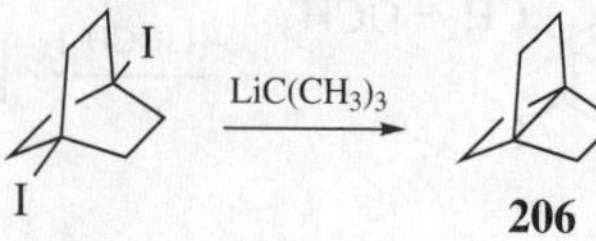

206

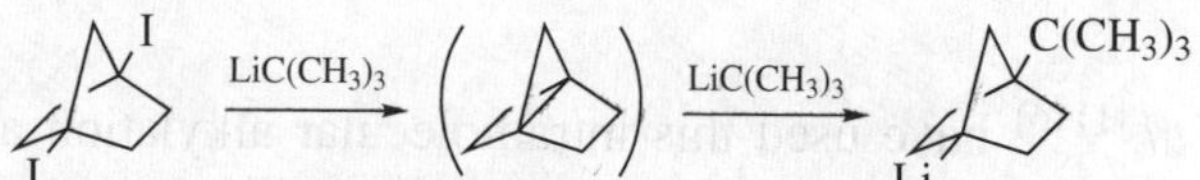

207

Table 34. β-Amino- and β′-amino-1-alkenyllithium compounds (M = Li) by halogen/metal permutation.

M–R	X[a]	M–R′[b]	Sv[c]	T[°C]	Product, El–X′[d]	Reference
CH_2NQ_2[e] M	Br	$LiC(CH_3)_3$	DEE	−80	82%, H_5C_6NCO	[906]
$CH_2NLiCH_2CH=CH_2$ M	Br	$LiC(CH_3)_3$	DEE	−80	88%, $(H_9C_4)_3SnCl$	[907]
$CH_2NLiCH_2C_6H_5$ M	Br	$LiC(CH_3)_3$	DEE	−80	90%, $(H_9C_4)_3SnCl$	[907]
$CH_2NLiSi(CH_3)_3$ M	Br	$LiC(CH_3)_3$	THF	+25	65%, $H_5C_6CH=NC_6H_5$	[906]
CH_3 M $N(C_2H_5)_2$	Br	$LiC(CH_3)_3$	THF	−70	100%, $(H_3C)_2NCHO$	[908,909]
C_2H_5 M N O	Br	$LiC(CH_3)_3$	THF	−70	95%, $(H_3C)_2NCHO$	[908,909]
$C(CH_3)_3$ M N O	Br	$LiC(CH_3)_3$	THF	−70	45%, $(H_3C)_2NCHO$	[908,909]

[a] X = halogen displaced by the metal. [b] M–R′ = organometallic exchange reagent. [c] DEE = diethyl ether, THF = tetrahydrofuran; in the case of mixtures only the most polar component is indicated. [d] El–X′ = electrophilic trapping reagent. [e]

$NQ_2 = N\begin{bmatrix} Si \\ Si \end{bmatrix}$

W. F. Bailey *et al.*[1117] have investigated the *tert*-butyllithium-promoted cyclization of 1,ω-diiodoalkanes. Whereas cyclopropanes, cyclobutanes and cyclopentanes are obtained almost quantitatively, the yield drops to approximately 5% with cyclohexanes and becomes zero from cycloheptanes on. However, minor stationary concentrations of a 3-, 4- or 5-iodoalkyllithium suffice to mediate its closure to a small ring as this is an extremely efficient process. Thus, phenyllithium smoothly converts *endo,endo*-2,3-di(iodomethyl)norbornene into *endo*-tricyclo[4.2.1.0^{2,5}]non-7-ene (**208**)[1118] although the halogen/metal exchange preceding the crucial C−C linking step is, energetically speaking an 'uphill process', an alkyllithium being more basic than any aryllithium species.

Table 35. β-, γ- and δ-Monoalkoxy or -dialkoxy-substituted 1-alkenyllithiums (M = Li) by halogen/metal permutation using alkyllithiums M−R′.

M−R	X[a]	M−R′	S_v[b]	T[°C]	Product, El−X′ [c]	Reference
M⟍ OC$_2$H$_5$	Br	LiC(CH$_3$)$_3$	THF	−75	84%, H$_5$C$_6$CHO	[910,911]
M⟍OC$_2$H$_5$	Br	LiC$_4$H$_9$	THF	−75	75%, (H$_5$C$_6$)$_2$CO	[911]
M⟍ (O,O-dioxolane)	Br	LiC$_4$H$_9$	THF	−75	72%, (H$_3$C)$_3$SiCl	[911,912]
M⟍ OCH$_3$ / OCH$_3$	Br	LiC$_4$H$_9$	THF	−75	67%, (H$_3$C)$_3$SiCl	[911,912]
M⟍ OC$_2$H$_5$ / OC$_2$H$_5$	Br	LiC$_4$H$_9$	THF	−75	64%, H$_5$C$_6$CH$_2$Br	[911,912]
(O,O-) M⟍ C$_5$H$_{11}$	I	LiC$_4$H$_9$	PET	−70	90%, D$_2$O	[913]
OC$_2$H$_5$ / OC$_2$H$_5$ M⟍	Br	LiC(CH$_3$)$_3$	THF	−120	93%, H$_5$C$_6$CHO	[914]
(bicyclic O,O) M⟍	Br	LiC(CH$_3$)$_3$	THF	−75	92%, H$_3$CI	[915]

[a] X = halogen displaced by the metal. [b] PET = petroleum ether (pentanes, hexanes, heptanes); DEE = diethyl ether, THF = tetrahydrofuran; in case of mixtures (containing for example paraffinic cosolvents) only the most polar component is indicated. [c] El−X′ = electrophile used to trap the organometallic intermediate.

Phenyllithium also proves to be the best choice to effect the reductive cyclization of 3,4″-bis(bromomethyl)-*o*-terphenyl. Obviously the simultaneous formation of the new C−C bond and a molecule of lithium bromide provides a sufficient driving force to overcompensate the unfavorable geometry of the heavily strained cyclophane **209**[1119].

Halogen/metal permutation may pave the way not only to intramolecular alkylation but also to S_N2'-allylations or acylations. This is exemplified by the synthesis of a carbocycline and an anthracyclinone precursor (**210**[1120] and **211**[1121], respectively). The sulfone **212** (OM = OH), a key intermediate on the way towards morphine, has been obtained by consecutive Michael addition and alkylation, both intramolecular, of an aryllithium generated by halogen−metal interconversion[1122].

Table 36. β-Alkoxy-substituted 1-alkenyllithiums by halogen/metal permutation (M−R′ = butyllithium).

M−R	X[a]	M−R′	Sv[b]	T[°C]	Product, El−X′[c]	Reference
CH$_2$OC(CH$_3$)$_3$ (M)	Br	LiC$_4$H$_9$	THF	−75	[d], CO$_2$	[800]
CH(OC$_2$H$_5$)$_2$ (M)	Br	LiC$_4$H$_9$	DEE[d]	−55	50%, CO$_2$	[916]
CH(OC$_2$H$_5$)$_2$, CH$_3$, CH$_3$ (M)	Br	LiC$_4$H$_9$	DEE[d]	−90	50%, CO$_2$	[916]
(spiro dioxolane cyclopentene, M)	Br	LiC$_4$H$_9$	THF	−75	71%, H$_{11}$C$_5$I	[917,918]
(spiro dioxolane cyclopentene, M, CH$_3$)	Br	LiC$_4$H$_9$	THF	+25	60%, H$_{11}$C$_5$I	[917,919]
H$_3$CO, OCH$_3$, M−, OCH$_3$, H$_3$CO	Br	LiC$_4$H$_9$	THF	−70	39%, (cyclohexanone structure)	[920]

[a] X = halogen displaced by the metal. [b] THF = tetrahydrofuran; DEE = diethyl ether. [c] El−X′ = electrophilic trapping reagent. [d] Not explicitly stated.

Table 37. ω-Alkoxy-1-dienyl, -trienyl, -tetraenyl and -pentadienyllithiums (M = Li) by halogen/metal permutation in diethyl ether (DEE) at −75 °C.

M−R [a]	X [b]	M−R′ [c]	Product, El−X′ [d]	Reference
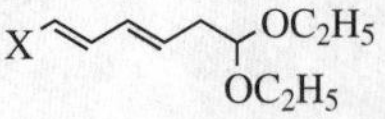 X⤳OC₂H₅ / OC₂H₅	Br	LiC(CH₃)₃	43%, β-ionone	[921]
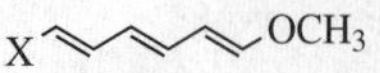 X⤳OCH₃	Br	LiC(CH₃)₃	68%, H₅C₆CHO	[922]
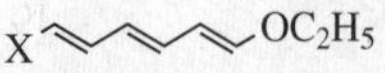 X⤳OC₂H₅	Br	LiC(CH₃)₃	81%, H₅C₆COCH₃	[921]
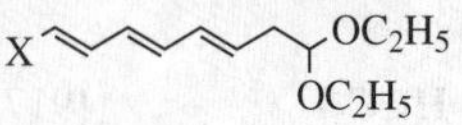 X⤳OC₂H₅ / OC₂H₅	Br	LiC(CH₃)₃	52%, H₃C)₂CHCHO	[921]
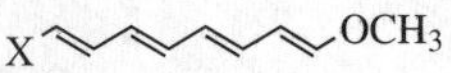 X⤳OCH₃	Br	LiC(CH₃)₃	56%,  [e]	[922]
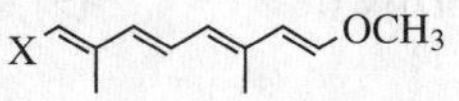 X⤳OCH₃	Br	LiC(CH₃)₃	68%, cyclocitral	[923]
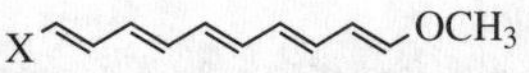 X⤳OCH₃	Br	LiC(CH₃)₃	60%,  [e]	[922]

[a] Stereoisomers, the alkoxy and the lithium-bearing double bonds having (Z) and (E) configurations.
[b] X = halogen in the precursor. [c] M−R′ = *tert*-butyllithium as the organometallic exchange reagent.
[d] El−X′ = trapping electrophile. [e] Product isolated after subsequent transformation.

● *Multiple Exchange*

There is *a priori* no reason why a substrate containing more than one bromine or iodine atom should not undergo a twofold or threefold halogen/metal interconversion. A series of di- and trilithio compounds have indeed been generated in that way: 2,2′-dilithiobiphenyl[1123], 3,3′-dilithiobiphenyl[1124], 4,4′-dilithiobiphenyl[951], 4,4′dilithiobiphenyl ether[951] (**213**), 9-ethyl-4,5-dilithiocarbazol[976] (**214**), 2,2′,2″-trilithiotriphenylamine[1125] (**215**) and 4,4′,4″-trilithiotriphenylamine[1126].

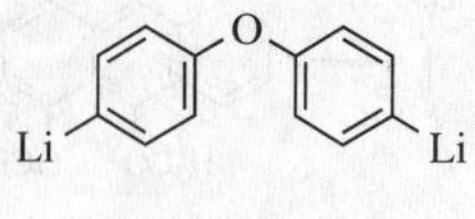

213 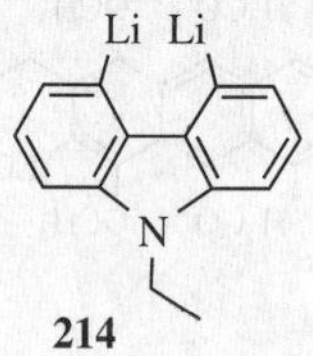   214 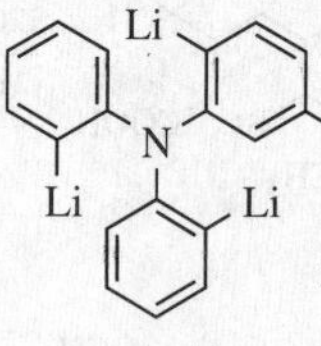   215

However, is it not always possible to impose a second or a third metal atom. 9,10-Dibromotriptycene can be readily converted with butyllithium into the monolithio **216**, but not into the dilithio compound[1127]. Depending on the reaction conditions, 1,3,5-tribromobenzene cleanly undergoes a single[951] or a double[951] halogen/lithium exchange. 1,3,5-Trilithiobenzene (**217**), still contaminated with mono- and dilithio

Table 38. 1-Alkenyllithiums (M = Li) carrying silyloxy substituents at the β or γ position: generation by halogen/metal permutation using butyllithium or *tert*-butyllithium as the exchange reagent M–R'.

M–R	X[a]	M–R'	Sv[b]	T[°C]	Product, El–X'[c]	Reference
(structure) OSi(CH₃)₃	Br	LiC(CH₃)₃	DEE	−70	97%, β-ionone	[924]
(structure) OSi(CH₃)₃	Br	LiC(CH₃)₃	DEE	−70	60%[d]	[924]
(structure) OSiR(CH₃)₃	Br, I	LiC(CH₃)₃	DEE	−75	63%, H₅C₆CHO	[925,926]
(structure) OSiR(CH₃)₃	Br, I	LiC(CH₃)₃	DEE	−75	86%, H₉C₄CHO	[925,926]
(structure) OSi(ⁱC₃H₇)₂	I	LiC(CH₃)₃	DEE	−75	70%, CO₂	[927]
(structure) OSi(CH₃)₂C(CH₃)₃	I	LiC(CH₃)₃	PET	[e]	[e] (structure) R [f]	[928]
(structure) OSi(CH₃)₃	I	LiC₄H₉	DEE	−75	19%, (structure) (CH₂)₆COOC₂H₅	[929]
(structure) OSi(CH₃)₂C(CH₃)₃	I	LiC₄H₉	DEE	−75	58%, (structure) BF₄⁻	[930]

[a] Halogen in the precursor. [b] PET = petroleum ether (pentanes, hexanes, heptanes), DEE = diethyl ether, THF = tetrahydrofuran. [c] El–X' = electrophilic trapping reagent. [d] β-Ionylidene–acetaldehyde ((Z)/(E) mixture). [e] Not specified. [f] R = *cis*–CH₂–CH=CH–(CH₂)₃–COOCH₃.

species, can only be made by reductive replacement of all bromines using 4,4'-di-*tert*-butylbiphenyl in tetrahydrofuran at −40°C as F. Bickelhaupt *et al.*[1128] have demonstrated. Further dilithio compounds have been generated by double halogen/metal permutation: *cis,cis*-1,4-dilithio-1,3-butadiene[1129] (**218**), 1,4-dilithiobenzene [951,976], 9,10-dilithioanthracene[1130] (**219**), 3,4-dichloro-2,5-dilithiothiophene[1131] and 2,6-dilithiopyridine [1073,1074] (**220**).

216 217 218 219 220

Table 39. 1-Alkenyllithiums (M = Li) carrying lithiooxy groups (at the β or γ position) or lithiooxycarbonyl groups (at the β position): generation by halogen/metal permutation using reagents M−R′.

M−R	X[a]	M−R′	Sν[b]	T[°C][c]	Product, El−X′ [d]	Reference
	Br[e]	LiC(CH₃)₃	THF	0	88%,	[931]
	Br[e]	LiC(CH₃)₃	THF	0	64%,	[931]
	Br[e]	LiC(CH₃)₃	THF	0	71%, (H₃C)₃CCl[f]	[932]
	Br[e]	LiC(CH₃)₃	THF	0	77%, H₅C₆CHO	[931]
	Br, I	LiC(CH₃)₃[h]	DEE	−75	55%, H₅C₆CHO	[933]
	Br, I	LiC(CH₃)₃	DEE	−75	57%, H₅C₆CHO	[933]
	Br, I	LiC(CH₃)₃[h]	DEE	−75	28%, H₅C₆CHO	[933,934]
	Br, I	LiC(CH₃)₃	DEE	−75	65%, (H₃C)₃CCHO	[933]
	Br	LiC₄H₉	THF	−100	38%, H₅C₆CHO[i]	[935]
	Br	LiC₄H₉	THF	−100	61%, H₅C₆CHO[i]	[935]
	Br	LiC₄H₉	THF	−100	47%, H₅C₆CHO[i]	[935]

[a] X = halogen displaced by the metal. [b] DEE = diethyl ether, THF = tetrahydrofuran. [c] If the temperature at which the organometallic intermediate was generated varied, only the highest one is given. [d] El−X′ = electrophile used to trap the organometallic intermediate. [e] Lithium β-bromoenolate generated from the α-bromoketone +LiCH₃ (2 equiv.). [f] C,O-Disilylated. [g] Similar results were obtained with O-*tert*-butyldimethylsilyl-protected 1-alkenyl bromides as the precursors. [h] Or *sec*-butyllithium. [i] Isolated as the lactone.

Table 40. Halogen-substituted 1-alkenyl metals by halogen/metal permutation using reagents M–R'.

M–R	X[a]	M–R'	Sv[b]	T[°C]	Product, El–X'[c]	Reference
F2 cyclobutene (M, Br)	Br	BrMgC2H5	DEE	35	59%, I_2	[936]
$M-CH=CH_2$	Br	LiC4H9	DEE	−75	25%, $(H_2C)_5CO$	[937]
$M-CF=CF-C_6H_{13}$	Br	BrMgC2H5	DEE	0	70%, H_5C_2CHO	[938]
$M-CF=CF-$⟨phenyl⟩	Cl	LiCH(CH3)C2H5	THF[d]	−110	70%, H_2O	[939]
$M-CF=CF-$⟨phenyl⟩$-OCH_3$	Cl	LiCH(CH3)C2H5	THF[d]	−110	90%, H_2O	[939]
$M-CF=CF-$⟨thienyl-S⟩$-Li$	Cl	LiCH(CH3)C2H5	THF[d]	−110	47%, [e]	[939]
$M-CF=CF-$⟨phenyl⟩	Cl	LiC4H9	DEE	−110	90%, H_2O	[940]
$M-CF=CF_2$	I	BrMgC2H5	DEE	−70	55%, $(F_3C)_2CO$	[941]
	Cl	LiC4H9	THF	−135	82%, $(H_2C)_5CO$)[f]	[942]
	Cl	LiC(CH3)3 [g]	DEE	−60	90%, $(H_2C)_5CO$ [f]	[943]
	Br	LiCH3	DEE	−75	66%, $(H_2C)_5CO$ [f]	[944]
$M-C(CF_3)=CH_2$	Br	LiC4H9	DEE	−85	51%, $H_5C_6COCH_3$	[945]
$M-CF=CF-Cl$	Cl	LiC4H9	THF	−115	85%, $(H_2C)_5CO$	[946]
$M-CCl=CF_2$	Cl	LiC4H9	THF	−90	82%, H_5C_6CHO	[947]
$M-CCl=CCl_2$	Cl	LiC4H9	THF	−110	20%, CO_2	[948]
	Br	LiC4H9	THF	−110	92%, CO_2	[948]
$M-CBr=C(C_6H_5)_2$	Br	LiC4H9	THF	−100	85%, CO_2	[860]
$M-CBr=C=C(C_6H_5)_2$	Br	LiC4H9	MTHF	−125	52%, $H_3COOCCl$	[949]
M cyclopentenyl (Br)	Br	LiC4H9	DEE	−70	65%, CO_2	[950]

[a] Halogen to be displaced from the precursor by the metal. [b] DEE = diethyl ether, MTHF = 2-methyltetrahydrofuran, THF = tetrahydrofuran; in case of mixtures (containing for example paraffinic cosolvents) only the most polar component is indicated. [c] $El-X'$ = electrophilic trapping agent. [d] Or DEE. [e] After LiF elimination, substitution of the terminal F by *tert*-butyl. [f] $(H_2C)_5CO$ = Cyclohexanone; product isolated after acid hydrolysis as an α,β-unsaturated carbonyl compound. [g] Or *sec*-butyllithium.

Table 41. Mononuclear heteroatom-free aryllithiums by halogen/metal permutation using butyllithium.

M−R [a]	X [b]	M−R′ [c]	Sv [d]	T[°C]	Product, El−X′ [e]	Reference
M—(phenyl)	Br, I	LiC_4H_9	PET[f]	[g]	65%, CO_2	[951–953]
M—(phenyl, 2-CH₃)	Br	LiC_4H_9	DEE	+50	84%, CO_2	[951]
M—(phenyl, 3-CH₃)	Br	LiC_4H_9	DEE	+50	65%, CO_2	[951]
M—(phenyl)—CH₃	Br	LiC_4H_9	PET	+25	86%, CO_2	[951]
M—(phenyl)—C(CH₃)₃	I	LiC_4H_9	DEE	−70	76%, CO_2	[954]
M—(phenyl)—(phenyl)	Br	LiC_4H_9	DEE	+25	62%, CO_2	[951,955]
M—(phenyl with C≡C-C₆H₅, CH₃)	Br	LiC_4H_9	DEE	−75	52%, CO_2	[956]
M—(phenyl, 2,6-di-C₂H₅)	Br	LiC_4H_9	THF	0	53%, $SnCl_2$	[957]
M—(phenyl, 2,6-di-isopropenyl)	Br	LiC_4H_9	DEE	+25	36%, [h]	[958]
M—(phenyl, 2,4,6-tri-CH₃)	Br	LiC_4H_9	BNZ	+25	15%, $SnCl_4$	[959]
M—(phenyl, 2,4,6-tri-C(CH₃)₃)	Br	LiC_4H_9	DEE	−75	60%, Cl_3SiBr	[960,961]
M—(phenyl, 2,4,6-tri-C₆H₅)	Br	LiC_4H_9	DEE	+25	87%, crystallized	[962]

[a] M = Li. [b] X = halogen displaced by the metal. [c] M−R′ = exchange reagent. [d] PET = petroleum ether (pentanes, hexanes, heptanes), DEE = diethyl ether. [e] El−X′ = electrophilic trapping reagent. [f] Or in benzene or diethyl ether. [g] In DEE in the range of −75 °C to +40 °C (reflux), in pure hydrocarbon media at +25 °C. [h] 9,9-Dimethylanthrone.

Table 42. Condensed heteroatom-free aryllithiums by halogen/metal permutation using reagents M−R′.

M−R [a]	X [b]	M−R′	Sv [c]	T[°C]	Product, El−X′ [d]	Reference
	Br	LiC_3H_7 [e]	DEE	+25	97%, CO_2	[650–963]
	Br	LiC_4H_9	DEE	+25	77%, CO_2	[963]
	Br	LiC_4H_9	DEE	+25	91%, CO_2	[951,963]
	Br	LiC_4H_9	DEE	+40	37%, CO_2	[964]
	Br	LiC_4H_9	DEE	+40	32%, CO_2	[964]
	Br	LiC_4H_9 [f]	DEE	+40	51%, CO_2	[964]
	Br	LiC_4H_9	PET	+25	72%, CO_2	[965]
	Br	LiC_6H_5	DEE	+25	82%, $(H_3C)_3SiCl$	[966]

[a] M = Li. [b] X = halogen displaced by the metal. [c] DEE = diethyl ether, PET = petroleum ether. [d] $El-X'$ = trapping electrophile. [e] LiC_2H_7 or LiC_4H_9. [f] In the presence of N,N,N',N'-tetramethylethylenediamine (TMEDA).

Table 43. Silylated aryllithiums (M = Li) by halogen/metal permutation using reagents M−R′.

M−R	X [a]	M−R′	Sv [b]	T[°C] [c]	Product, El−X′ [c]	Reference
$Si(CH_3)_3$	Br	LiC_4H_9	DEE	+40	37%, $(H_3C)_3CCHO$	[967]
$Si(CH_2C_6H_5)_3$	Br	LiC_4H_9	DEE	+25	37%, CO_2	[968,969]
CHQ_2 [d] CHQ_2 CHQ_2	Br	$Li(CH_3)_3$	[e]	[e]	39%, $(H_3C)_3SiCH_2N_3$	[970]

[a] X = halogen displaced by the metal. [b] DEE = diethyl ether, THF = tetrahydrofuran; in the case of mixtures (*e.g.*, containing alkanic cosolvents) only the most polar component is indicated. [c] $El-X'$ = electrophilic trapping reagent. [d] Q = $Si(CH_3)_3$. [e] Not specified.

Table 44. Amino- and aza-substituted aryllithiums by halogen/metal permutation.

M–R [a]	X [b]	M–R' [c]	Sv [d]	T[°C]	Product, El–X' [e]	Reference
M–C$_6$H$_4$–N(CH$_3$)$_2$ (ortho)	Br	LiC$_4$H$_9$	THF	−75	83%, (H$_5$C$_6$)$_2$CO	[971]
M–C$_6$H$_4$–N(CH$_3$)$_2$ (meta)	Br	LiC$_4$H$_9$	DEE	+40	32%, CO$_2$	[972]
M–C$_6$H$_4$–N(CH$_3$)$_2$ (para)	Br	LiC$_4$H$_9$	DEE	+40	56%, CO$_2$	[972,973]
M–C$_6$H$_4$–N(Si(CH$_3$)$_3$)$_2$	Br	LiC$_4$H$_9$	DEE	0	80%, (H$_5$C$_6$)$_2$CO	[974]
M–C$_6$H$_4$–(N-pyrrolyl)	Br	LiC$_4$H$_9$	THF	−80	75%, H$_5$C$_6$CHO	[975]
M–carbazolyl, N–C$_2$H$_5$	Br	LiC$_4$H$_9$	DEE	+40	71%, CO$_2$	[976]
M–quinolinyl	Br	LiC$_4$H$_9$	DEE	−70	51%, R$_4$As$^{\oplus}$I$^{\ominus}$ [f]	[977]

[a] M = Li. [b] X = halogen in the precursor. [c] Always butyllithium as the organometallic exchange reagent. [d] DEE = diethyl ether, THF = tetrahydrofuran. [e] El–X' = electrophilic trapping reagent. [f] R$_2$ = 4,4'-dimethyl-2,2'-biphenylylene.

Caution is advisable against all claims of double exchange at nearby positions. As A. Maercker *et al.* have conclusively demonstrated, the formation of 1-isopropyloxy-2-methyl-1-phenylpropene by consecutive treatment of β,β-diiodo-α-isopropyloxystyrene with excess *sec*-butyllithium and methyl iodide does not involve the postulated[1132] *gem*-dilithio intermediate **221**, but is brought about in two successive exchange/alkylation sequences[1133].

Table 45. Amino- and imino-substituted aryllithiums by halogen/metal permutation.

M−R [a]	X[b]	M−R′[c]	$S\nu$[d]	T[°C]	Product, El−X′ [e]	Reference
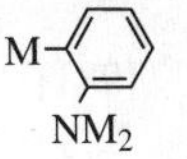 M−(aryl)NM₂ (ortho)	Br	LiC₄H₉	DEE	+40	40%, CO₂	[978]
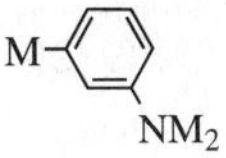 M−(aryl)NM₂ (meta)	Br	LiC₄H₉	DEE	+40	2%, (H₅C₆)₃SiCl	[978]
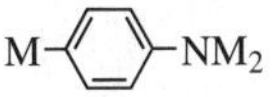 M−(aryl)−NM₂ (para)	Br	LiC₄H₉	DEE	−60	68%, CO₂	[979,980]
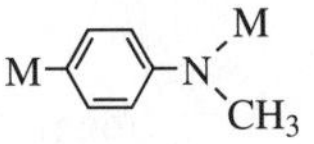 M−(aryl)−N(M)CH₃	Br	LiC₄H₉	DEE	+40	27%, CO₂	[978]
 M−(aryl) N=C(OM)CH₃	Br	LiC₄H₉	DEE	+40	52%, CO₂	[978]
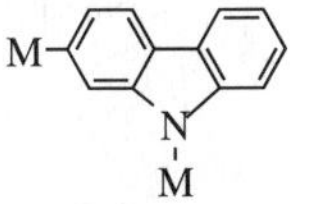 M−(carbazole)N−M	Br	LiC₄H₉	DEE	+40	58%, CO₂	[976]

[a] M = Li. [b] X = halogen displaced by the metal. [c] M−R′ = butyllithium as the exchange reagent. [d] DEE = diethyl ether, THF = tetrahydrofuran. [e] El−X′ = electrophilic trapping reagent.

● *Regioselectivity*

Unlike metalation reactions, halogen/metal permutations have the reputation of both site fidelity and site flexibility. All it needs is to position properly a bromine or iodine atom and one can rely on its later replacement by lithium and ultimately by a suitable electrophile. However, this is only true if he substrate does not contain two or more heavy halogen atoms.

2,5-Dibromotoluene inevitably produces a mixture of 4-bromo-2-methyl- and 4-bromo-3-methylphenyllithium[951]. In contrast, 2,4-dibromoanisole and 2,4,6-tribromoanisole exchange only one oxygen-adjacent halogen (to afford species **222**) when treated with isopropylmagnesium chloride[1121]. Similarly, 5,7-dibromo-2,3-dihydro-1-benzofuran reacts with butyllithium to give exclusively the 7-lithio intermediate[1134] (**223**). Amino or amido functions (as in the 2,4-dibromo-*N*-pivaloylaniline derivative **224**) display an analogous *ortho*-orienting effect as alkoxy groups[1109].

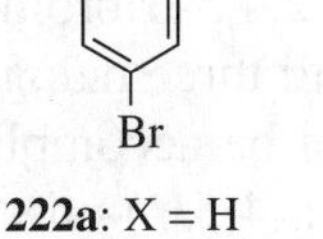

222a: X = H
222b: X = Br

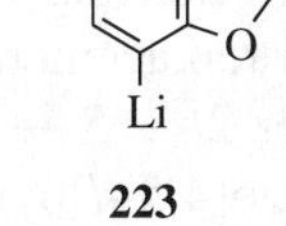

223

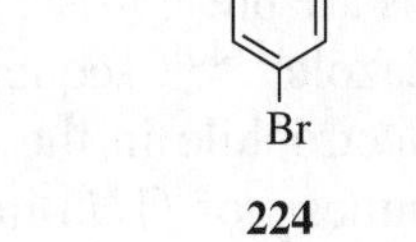

224

Table 46. Cyano- and nitro-substituted aryllithiums by halogen/metal permutation using reagents M–R'.

M–R [a]	X [b]	M–R'	Sv [c]	T [°C]	Product, El–X' [d]	Reference
M–(phenyl)–CN (ortho)	Br	LiC$_4$H$_9$	THF	−100	72%, H$_3$CN(structure)	[981]
M–(phenyl)–CN (para)	Br	LiC$_4$H$_9$	DEE	−70	17%, CO$_2$ [e]	[982]
M–(phenyl)–NO$_2$ (ortho)	Br	LiC$_6$H$_5$	THF	−100	97%, CO$_2$	[983]
M–(phenyl)–NO$_2$, NO$_2$	Br	LiC$_6$H$_5$	THF	−100	61%, CO$_2$	[983]
M–(phenyl)–CH$_3$, NO$_2$, NO$_2$	Br	LiC$_6$H$_5$	THF	−100	41%, CO$_2$	[983]
M–(phenyl)–NO$_2$, NO$_2$, CH$_3$	Br	LiC$_6$H$_5$	THF	−100	82%, CO$_2$	[983]

[a] M = Li. [b] X = halogen displaced by the metal. [c] DEE = diethyl ether, THF = tetrahydrofuran, in the case of mixtures, the most polar component is indicated. [d] El–X' = electrophilic trapping reagent. [e] Isolated as terephthalic acid.

2,4-Dibromo-1-nitrobenzene and 1,4-dibromo-1-nitrobenzene exchange only the nitro-adjacent halogen[1135]. H. Gerlach *et al.*[1135] have based their concise synthesis of 6,6'-dibromoindigo (**225**), the antique purple pigment, on this kind of regioselectivity.

225

Lithium has been introduced in all three vacant positions of imidazoles by halogen/metal permutation[1136–1140]. Exchange occurs first at the 2-position if several competing halogens are present[1136–1140]. Starting with a 2,4,5-tribromoimidazole[1139] or a 2,4,5-triiodoimidazole[1140], sequential replacement of all three halogens by lithium and subsequently an electrophile in the order 2 → 5 → 4 can be accomplished as illustrated by the ten-step synthesis of (1*H*)-imidazo[4,5-*d*]pyridazin-4(5*H*)-one[1140] (**226**).

Table 47. Phosphino-, arseno- and stibono-substituted aryllithiums by halogen/metal permutation.

M–R [a]	X[b]	M–R′[c]	Sv[d]	T[°C]	Product, El–X′ [e]	Reference
M–⟨⟩ P(CH₃)₂	Br	LiC₄H₉	DEE	+25	80%, $(H_3CTe)_2$	[984]
M–⟨⟩ P(C₆H₅)₂	Br	LiC₄H₉	THF	−75	91%, $(H_5C_6)_2PCl$	[985]
M–⟨⟩ P(C₆H₅)₂	Br	LiC₄H₉	DEE	+25	11%, CO_2	[986]
M–⟨⟩–P(C₆H₅)₂	Br	LiC₄H₉	DEE	+25	57%, CO_2	[986]
M–⟨⟩ As(CH₃)₂	Br	LiC₄H₉	PET	+70	68%, $(H_3CTe)_2$	[984]
M–⟨⟩ Sb(CH₃)₂	Br	LiC₄H₉	DEE	−65	15%, $(H_3CTe)_2$	[984]

[a] M = Li. [b] X = halogen displaced by the metal. [c] MR′ = butyllithium as the organometallic exchange reagent. [d] DEE = diethyl ether, in case of mixtures (containing, *e.g.* paraffinic cosolvents) only the most polar component is indicated. [e] El–X′ = electrophilic trapping reagent.

● *Stereoselectivity*

gem-Dibromoalkenes undergo halogen/metal permutations preferentially at the sterically least hindered position[884,1141]. However, subsequent equilibration[880] converts the (Z)-isomer (**Z-227**, see Scheme on p. 125) thus formed into the thermodynamically more stable (E)-isomer (**E-227**)[1141], which often becomes the only detectable species[1142].

Table 48. Alkoxy- and aroxy-substituted aryllithiums by halogen/metal permutation using reagents M−R′.

M−R [a]	X[b]	M−R′	Sv[c]	T[°C]	Product, El−X′ [d]	Reference
M—C₆H₄(OCH₃) (ortho)	Br, I	LiC₆H₅[e]	DEE	+25	90%, $(H_5C_6)_2CO$	[987,988]
M—C₆H₄(OCH₃) (meta)	Br	LiC₄H₉	THF	−75	52%, CO_2	[989]
M—C₆H₄—OCH₃ (para)	Br, I	LiC₄H₉	BNZ	+25	82%, CO_2	[951,952]
M—C₆H₄(OC₆H₅) (ortho)	Br, I	LiC₄H₉	DEE	+25	79%, CO_2	[990]
M—C₆H₄(OC₆H₅) (meta)	Br, I	LiC₄H₉	DEE	+25	75%, CO_2	[990]
M—C₆H₄—OC₆H₅ (para)	Br, I	LiC₄H₉	DEE	+25	65%, CO_2	[990]
H₃CO-naphthyl, M	Br	LiC₄H₉	PET	+25	[f], D_2O	[991]
M-naphthyl, H₃CO	Br	LiC₄H₉	PET	+25	[f], D_2O	[991]
M-dibenzofuran	Br	LiC₄H₉	DEE	+40	58%, CO_2	[992]
M-dibenzofuran	Br	LiC₄H₉	DEE	+40	27%, CO_2	[992]
M—C₆H₄(CH₂OCH₃)	Br	LiC₄H₉	DEE	+40	34%, CO_2	[993]
M—C₆H(CH₃)(CH₃)(CH₂OCH₃)	Br	LiC₄H₉	THF	−75	19%, $(CH_2)_2O$	[994]

[a] M = Li. [b] X = halogen displaced by the metal. [c] PET = paraffinic hydrocarbons; DEE = diethyl ether, THF = tetrahydrofuran, BNZ = benzene. [d] El−X′ = electrophilic trapping reagent. [e] Or LiC₄H₉ in PET at +25 °C or +40 °C. [f] Not specified.

Table 49. Dialkoxy- and acetal-substituted aryllithiums by halogen/metal permutation.

M–R [a]	X[b]	M–R′[c]	Sν[d]	T[°C]	Product, El–X′ [e]	Reference
(aryl) OCH₂OCH₃ [f]	Br	LiC₄H₉	THF	−75	83%, (1.) CO₂, (2.) CH₂N₂	[995]
(aryl-acetal)	Br	LiC₄H₉	THF	−75	73%, [g]	[996]
(aryl-O-acetal)	Br	LiC₄H₉	THF	−75	80%, [h]	[997]
(aryl) OCH₃ / OCH₃	Br	Li(CH₃)₃	THF	−75	72%, [i]	[998]
(aryl) OCH₂C₆H₅ / OCH₂C₆H₅	Br, I	LiC₄H₉	DEE	−75	42%, [j]	[999]
(aryl) OCH₃ / OCH₃ / OCH₃	Br	LiC₄H₉	THF	−75	52%, H₃C–□–O / (H₃C)₃CO–□=O	[1000]
(aryl) OCH₃ / OCH₃ / OCH₃	Br	LiC₄H₉	THF	−75	38%, HO– / H₃CO–(aryl)–CHO	[1001]
(aryl) O / O (acetal)	Br	LiC₄H₉	THF	−75	75%, ⟋∼N–CHO	[1002]

[a] M = Li. [b] X = halogen replaced by metal. [d] DEE = diethyl ether, THF = tetrahydrofuran, BNZ = benzene; in the case of mixtures (*e.g.* containing paraffinic cosolvents) only the most polar component is indicated. [e] El–X′ = electrophilic trapping reagent. [f] Ring A of *O,O′*-bis(methoxymethyl)-3,17-β-estradiol. [g] 1-*p*-Anisyl-3-phenyl-2-butanone. [h] An octaethylporphyrinone. [i] A 2-(2-oxazolinyl)naphthalene. [j] *N*-Benzyl-6-aza-2-bicyclo[2.2.2]octanone.

Although the exact equilibration mechanism is still unknown, the reversibility of exchange processes may be tentatively assumed to be at its origin.

R = H₁₁C₆, H₃COCH₂CH₂O–CH(CH₃), *etc.*
R′ = CH₃, C₄H₉, *etc.*

Table 50. Lithiooxy-substituted aryllithiums by halogen/metal permutation using reagents M−R′.

M−R [a]	X[b]	M−R′	Sν[c]	T[°C]	Product, El−X′ [d]	Reference
M—⟨benzene⟩ [e] MO	Br	LiC$_4$H$_9$	DEE	+25	67%, CO$_2$	[1003]
M—⟨benzene⟩ OM	Br	LiC(CH$_3$)$_3$	THF	0	61%, [f]	[1005]
M—⟨benzene⟩—OM	Br	LiC$_4$H$_9$	DEE,THF	+25	75%, CO$_2$	[1003]
C(CH$_3$)$_3$ [f] M—⟨benzene⟩—OM′ C(CH$_3$)$_3$	Br	LiC(CH$_3$)$_3$	THF	0	>61%, H$_5$C$_6$CN	[1006]
M—⟨benzene⟩ CH$_2$OM	Br	LiC$_4$H$_9$	DEE	+25	32%, CO$_2$	[982]
M—⟨benzene⟩—CH$_2$OM	Br	LiC$_4$H$_9$	DEE	+25	18%, CO$_2$	[982]
M—⟨benzene⟩—CH·OM CH$_3$	Br	LiC$_4$H$_9$	DEE	+25	45%, CO$_2$	[982]
M—⟨benzene⟩—CH$_2$CH$_2$OM	Br	LiC$_4$H$_9$	DEE	+25	52%, CO$_2$	[982]
M—⟨benzene⟩ COOM	Br	LiC$_4$H$_9$	DEE	−75	35%, CO$_2$	[981,1003]
M—⟨benzene⟩—COOM	Br	LiC$_4$H$_9$	DEE	−75	62%, CO$_2$	[976,1003]

[a] M = Li. [b] Halogen to be replaced by metal. [c] DEE = diethyl ether, THF = tetrahydrofuran. [d] El−X′ = trapping electrophile. [e] 2-(Trimethylsilyloxy)phenyllithium[1004]. [f] (1.) SO$_2$ (2.) HO$_3$SONH$_2$. [g] OM′ = OH → OLi.

Analogously the *exo*-positioned halogen of *gem*-dibromocyclopropanes is replaced under kinetic control until subsequent equilibration converts it completely into the *endo*-lithio species[884]. M. Neuenschwander *et al.*[1143] have designed a model compound (**228**) which passes with exceptional velocity from the intermediate to the final stage.

Table 51. Lithiothio- and sulfo-substituted aryllithiums by halogen/metal permutation using reagents M−R′.

M−R [a]	X[b]	M−R′	$S\nu$[c]	T[°C]	Product, El−X′[d]	Reference
M–⟨⟩–SCH₃	Br	LiC₄H₉	PET	+60	64%, (H₃C)₃SiCl	[1007]
M–⟨⟩–CH₂SCH₃	Br	LiC₄H₉	DEE	−75	49%, CuBr[e]	[1008]
M–⟨⟩–CH₂SC₂H₅	Br	LiC₄H₉	DEE	−75	54%, CuBr[e]	[1008]
M–⟨⟩–SO₂–⟨⟩	I	LiC₄H₉	BNZ	+25	51%, CO₂	[952]
M–⟨⟩–SO₂N(C₂H₅)₂	I	LiC₄H₉	DEE	−75	78%, CO₂	[1003]
M–⟨⟩–SO₂NM₂	Br	LiC₄H₉	DEE	+25	14%, CO₂	[982]
M–⟨⟩–SM	Br	LiC(CH₃)₃	THF[f]	−75	71%[g]	[1005]
M–⟨⟩–SM	Br	LiC(CH₃)₃	THF[f]	−75	76%[g]	[1005,1009]

[a] M = Li. [b] X = halogen displaced by the metal. [c] PET = petroleum ether (pentanes, hexanes, heptanes); DEE = diethyl ether, THF = tetrahydrofuran, BNZ = benzene. [d] El−X′ = electrophilic trapping reagent. [e] Gives the oxidative dimer. [f] Or DEE. [g] (1.) N-Methoxy-N-methylcyclohexanecarboxamide, (2.) acetyl chloride.

Unlike α-bromocyclopropyllithiums, the chloro counterparts appear to be configurationally stable. As evidence, the retention of the *exo*-lithio structure of 8-chloro-3,5-dioxybicyclo[5.1.0]octyllithium[1144] (**Z-229**) may be quoted.

R + R = CH₂−O−CH₂−O−CH₂   **Z-229**

Also, α-bromocyclopropylmagnesium halides do not equilibrate. In this way, the *exo*-trimethyltin derivative **230** can be prepared and subsequently submitted to a metalloid−metal permutation (see Section 4.3.2) generating 7-*endo*-bromo-7-*exo*-lithionorcarnane[1145].

R + R = −(CH₂)₄−   **230**

Table 52. Fluoro-substituted arylmetals by halogen/metal permutation.

M–R [a]	X [b]	M–R′ [c]	Sv [d]	T [°C]	Product, El–X′ [e]	Reference
M–C$_6$H$_4$F (ortho)	Br	LiC$_4$H$_9$	DEE	–70	84%, (H$_5$C$_6$)$_2$CO	[1010]
M–C$_6$H$_4$F (meta)	Br	LiC$_4$H$_9$	DEE	–40	65%, CO$_2$	[1011]
M–C$_6$H$_4$–F (para)	Br	LiC$_4$H$_9$	BNZ	+25	50%, CO$_2$	[952]
M–C$_6$H$_3$F$_2$	Br	LiC$_4$H$_9$	DEE	–75	83%, CO$_2$	[1012]
M–C$_6$H$_3$F$_2$	Br	LiC$_4$H$_9$	DEE	–75	77%, CO$_2$	[1013]
M–C$_6$F$_4$ (F,F / F,F,F)	Cl	BrMgC$_2$H$_5$	THF	+5	85%, aq. HCl	[1014]
	Br	LiC$_4$H$_9$	DEE	–75	51%, H$_2$O	[1015]
M–C$_6$H$_4$–CF$_3$	Br	LiC$_4$H$_9$	DEE	+25	61%, CO$_2$	[1016]
M–C$_6$H$_4$–CF$_3$ (para)	Br	LiC$_4$H$_9$	DEE	–50	64%, CO$_2$	[1017]
M–C$_6$H$_3$(CF$_3$)$_2$	Br	LiC$_4$H$_9$	DEE	0	48%, Cl$_2$C=CF$_2$	[1018]
M–C$_6$H$_3$(CF$_3$)$_2$	Br	LiC$_4$H$_9$	DEE	–75	94%, CO$_2$	[1019]
M–(fluoronaphthyl)	Br	LiC$_4$H$_9$	DEE	–70	78%, R$_4$As$^{\oplus}$I$^{\ominus}$ [f]	[1020]

[a] M = Li, XMg. [b] X = halogen in the precursor. [c] M–R′ = organometallic exchange reagent. [d] DEE = diethyl ether, THF = tetrahydrofuran, BNZ = benzene. [e] El–X′ = trapping electrophile. [f] R$_2$ = 4,4-dimethyl-2,2′-biphenyl-ylene.

c) Working Procedures

In view of the simplicity with which halogen/metal permutation can be executed, only a few protocols have been selected. They feature as substrates two bromoalkenes, two bromoarenes, one dibromoarene and 2-bromopyridine, the exchange reagent being either butyllithium or *tert*-butyllithium. Anhydrous solvents and an inert gas atmosphere should be used.

Table 53. Chloro- and iodo-substituted aryllithiums by halogen/metal permutation.

M−R [a]	X[b]	M−R′[c]	S_v[d]	T[°C]	Product, El−X′ [e]	Reference
M–⟨⟩–Cl	Br	LiC_4H_9	DEE	−90	93%, CO_2	[1021]
M–⟨⟩–Cl	Br, I	LiC_4H_9	DEE	+25	42%, CO_2	[951,976]
M–⟨⟩–Cl	Br, I	LiC_4H_9	BNZ, DEE	+25	90%, CO_2	[951,952]
M–(Cl,Cl,Cl,Cl,Cl)	Cl	LiC_4H_9	DEE	−10	71%, $(H_5C_6)_2CO$	[1022,1023]
M–⟨⟩, Cl–⟨⟩ (naphthalene)	Br	LiC_4H_9	DEE	−70	41%, $R_4P^{\oplus}I^{\ominus}$ [f]	[1020]
M–(anthracene)–Cl	Br	LiC_6H_5	DEE	+25	73%, CO_2	[1024]
M–⟨⟩–I	I	LiC_4H_9	DEE	+25	80%, $(H_5C_2)_3GeBr$	[1025]

[a] M = Li. [b] X = halogen in the precursor. [c] M−R′ = exchange reagent. [d] DEE = diethyl ether, THF = tetrahydrofuran, BNZ = benzene. [e] El−X′ = electrophilic trapping reagent. [f] R_2 = 4,4′-dimethyl-2,2′-biphenylylene.

(Z)-ω-Camphenyllithium[1146]

(Z)-2-(3,3-Dimethylbicyclo[2.2.1]hept-2-ylidene)methyl bromide[1146] (4.3 g, 20 mmol) and, 5 min later, *N,N*-dimethylformamide (2.0 mL, 1.9 g, 26 mmol) were added to a solution of *tert*-butyllithium (40 mmol) in neat tetrahydrofuran (50 mL) at −75 °C. Evaporation of the solvent, addition of water 25 mL, extraction with hexane (3 × 20 mL), washing with brine (20 mL) and distillation afforded the product as a colorless oil [m.p. −3 to −2 °C; b.p. 91–94 °C/2 mmHg; n_D^{20} 1.5230; yield 2.6 g (79%)].

Table 54. Bromosubstituted aryllithiums and arylmagnesium halides by halogen/metal permutation.

M−R [a]	X[b]	M−R′[c]	Sυ[d]	T[°C]	Product, El−X′ [e]	Reference
	Br	LiC$_4$H$_9$	DEE	−100	38%, CO$_2$	[951,1026]
	Br	LiC$_4$H$_9$	DEE	−35	44%, [f]	[1027,1028]
	Br	LiC$_4$H$_9$	DEE	+25	90%, CO$_2$	[952,1029]
	Br	ClMgCH(CH$_3$)$_2$	THF	+40	95%, H$_2$O	[1030]
	Br	ClMgCH(CH$_3$)$_2$	THF	+40	79%, H$_2$O	[1030]
	Br	LiC$_4$H$_9$	THF	−75	91%, CO$_2$	[1031]
	Br	LiC$_4$H$_9$	DEE	−75	93%, F$_7$C$_3$COOC$_2$H$_5$	[1032]
	Br	BrMgC$_6$H$_5$	THF	0	96%, H$_2$O	[1033,1034]
	Br	LiC$_4$H$_9$	DEE	−15	16%, CO$_2$	[755]
	Br	LiC$_4$H$_9$	DEE	+25	67%, H$_2$O	[951]
	Br	LiC$_4$H$_9$	DEE	−70	30%, R$_4$P$^{\oplus}$I$^{\ominus}$ [g]	[1020]
	Br	LiC$_4$H$_9$	DEE	+25	53%, H$_3$CI	[1024]

[a] M = Li or XMg. [b] X = halogen in the precursor. [c] M−R′ = organometallic exchange reagent. [d] DEE = diethyl ether, THF = tetrahydrofuran. [e] El−X′ = trapping electrophile. [f] Nucleophilic addition of 3-bromophenyllithium to the 4-position of pyrimidine followed by lithium hydride elimination. [g] R$_2$ = 4,4′-dimethyl-2,2′-biphenylylene.

Table 55. Ferrocenyllithiums by halogen/metal permutation.

M–R [a]	X[b]	M–R'[c]	$S\nu$[d]	T[°C]	Product, El–X' [e]	Reference
	Br	LiC$_4$H$_9$	DEE	0	~90%, Cl$_3$CCCl$_3$	[1035]
	Br	LiC$_4$H$_9$	THF	−25	60%, (H$_3$C)$_2$NCHO	[1036]
	Br	LiC$_4$H$_9$	THF	−25	67%, H$_5$C$_2$OOCCl	[1036]

[a] M = Li. [b] X = halogen displaced by the metal. [c] Using butyllithium as the exchange reagent M–R'. [d] DEE = diethyl ether, in the case of mixtures (containing, *e.g.* paraffinic cosolvents) only the most polar component is indicated. [e] El–X' = trapping electrophile.

(Z)-2-Ethoxyvinyllithium[1147,1148]

After having stripped off the commercial solvent (hexanes) from butyl-lithium (0.10 mol), the organolithium reagent is dissolved in precooled (−75 °C) tetrahydrofuran (50 mL). Still at −75 °C, (Z)-2-bromovinyl ethyl ether[911,1149] (11 mL, 15 g, 0.10 mol) and, 15 min later, β-ionone (20 mL, 19 g, 0.10 mol) are added. At 25 °C, the mixture is treated with 6 M hydrochloric acid (10 mL) and the homogeneous solution is kept for 2 h before it is concentrated to one-quarter of its initial volume. After addition of water (50 mL), the combined organic layers are washed with a saturated aqueous solution of sodium hydrogen carbonate (2 × 25 mL) and brine (25 mL) before being dried. Distillation affords β-ionylideneacetaldehyde as a colorless liquid [b.p. 95–98 °C/0.05 mmHg; n_D^{20} 1.5737; yield 18 g (72%)]. The 1 : 2 (Z/E) mixture (by gas chromatography) can be separated by elution with a 1 : 10 (v/v) mixture of ethyl acetate and hexanes from a column filled with silica gel (2 L).

Table 56. Pyrryl-, indolyl-, pyrazolyl- and imidazolyllithiums by halogen/metal permutation.

M−R [a]	X [b]	M−R′ [c]	Sv [d]	T [°C]	Product, El−X′ [e]	Reference
pyrrole, N−COOC(CH$_3$)$_3$	Br	LiC$_4$H$_9$	THF	−75	83%, (H$_3$C)$_3$SiCl	[1037]
pyrrole, N−Si(^{i}C$_3$H$_7$)$_3$	Br	LiC$_4$H$_9$	THF	−75	88%, CO$_2$	[1038,1039]
indole [f], N−SiR$_2$R′	Br	LiC(CH$_3$)$_3$	THF	−75	94%, (H$_3$C)$_3$SnCl	[1040–1042]
pyrazole, N−CH$_3$	Br	LiC(CH$_3$)$_3$	DEE	−100	77%, CO$_2$	[1043]
pyrazole, N−CH$_3$	Br	LiC$_4$H$_9$	DEE	−30	74%, (H$_5$C$_6$)$_2$CO	[1044]
H$_3$C, pyrazole, N−CH$_3$, H$_3$C	Br	LiC$_4$H$_9$	DEE	−30	81%, CO$_2$	[1044]
CH$_3$, pyrazole, N−CH$_3$, Cl	Br	LiC$_4$H$_9$	DEE	+25	97%, cyclohexyl-COCl	[1045]
Br, pyrazole, N−M, Br	Br	LiC$_4$H$_9$	DEE	−30	80%, CO$_2$	[1044]
imidazole, N−C(C$_6$H$_5$)$_3$	I	LiC$_4$H$_9$	THF	−75	51%, (H$_3$C)$_2$NCHO [g]	[1046]
CN, imidazole CN, N−CH$_3$	Br	LiC$_4$H$_9$	THF	−100	42%, CO$_2$	[1047]
Br, imidazole Br, N−CH$_2$C$_6$H$_5$	Br	LiC$_4$H$_9$	DEE	−75	54%, (H$_3$CS)$_2$ [h]	[1048–1050]

[a] M = Li. [b] X = halogen displaced by the metal. [c] M−R′ = exchange reagent. [d] DEE = diethyl ether, THF = tetrahydrofuran; in the case of mixtures (containing, *e.g.* paraffinic cosolvents) only the most polar component is indicated. [e] El−X′ = trapping electrophile. [f] SiR$_2$R′ = Si(CH$_3$)$_2$C(CH$_3$)$_3$. [g] Plus 7% 2-isomer. [h] Plus 14% 5-isomer.

Table 57. Furyl-, thienyl- and benzothienylmetals by halogen/metal permutation using reagents M–R'.

M–R [a]	X[b]	M–R'	$S\nu$[c]	T[°C]	Product, El–X' [d]	Reference
(2-furyl, M on ring)	Br	LiC$_4$H$_9$[e]	DEE[f]	−70	63%, (H$_9$C$_4$O)$_3$B/H$_2$O	[1051,1052]
(5-Si(CH$_3$)$_3$-2-furyl, M)	Br	LiCH(CH$_3$)C$_2$H$_5$	DEE	−70	78%, H$_5$C$_6$N(CH$_3$)CHO	[1053]
(3-Br-2-furyl, M)	Br	LiC$_4$H$_9$[e]	DEE	−45	49%, H$_3$CCON(CH$_3$)$_2$	[1052]
(3-Br-5-Si(CH$_3$)$_3$-2-furyl, M)	Br	LiC$_4$H$_9$	DEE	−80	65%, H$_2$O	[1053]
(3-Br-5-CH(OC$_2$H$_5$)$_2$-2-furyl, M)	Br	LiC$_4$H$_9$	DEE	−45	61%, (H$_9$C$_4$O)$_3$B/H$_2$O	[1052,1054]
(2-thienyl, M)	Br	ClMgCH(CH$_3$)$_2$	THF	−25	[g]	[1055]
	I	LiC$_6$H$_5$	DEE	+25	58%, CO$_2$	[1056]
(3-thienyl, M)	I	LiC$_4$H$_9$	DEE	−70	78%, CO$_2$	[1057,1058]
(4-H$_3$CO-benzothienyl, M)	Br	LiC$_4$H$_9$	THF	−75	51%, (H$_3$C)$_2$NCHO	[1059]
(3-Br-2-thienyl, M)	Br	LiC$_4$H$_9$	DEE	−70	90%, CO$_2$	[1058]
(3,4,5-Cl$_3$-2-thienyl, M)	Cl	LiC$_4$H$_9$[h]	DEE [f,h]	0	69%, CO$_2$	[1060,1061]
(2-selenophenyl, M)	Br	LiC$_6$H$_5$	DEE	+40	65%, CO$_2$	[1062]
(3,4-(CH$_3$)$_2$-2-selenophenyl, M)	I	LiC$_2$H$_5$	DEE	−100	51%, CO$_2$	[1063]

[a] M = Li or XMg. [b] X = halogen in the precursor. [c] DEE = diethyl ether, THF = tetrahydrofuran; in the case of mixtures (containing, *e.g.* paraffinic cosolvents) only the most polar component is indicated. [d] El–X' = electrophilic trapping reagent. [e] Or LiC$_2$H$_5$. [f] Or THF. [g] Product, trapping agent and yield not specified. [h] Or LiC(CH$_3$)$_3$.

Table 58. Pyridyl- and quinolyllithiums (M−R) by halogen/metal permutation.

M−R	M	$X^{[a]}$	$M-R'^{[b]}$	$Sv^{[c]}$	$T[°C]$	Product, $El-X'^{[d]}$	Reference
M—(3-pyridyl)	XMg	Br, I	$ClMgCH(CH_3)_2$	THF	+25	97%, I_2	[1064,1065]
	Li	Br	LiC_4H_9	DEE	−15	69%, H_5C_6CHO	[1066−1068]
M—(pyridyl)	XMg	Br, I	$ClMgCH(CH_3)_2$	THF	+25	78%, I_2	[1064,1065]
	Li	Br	LiC_4H_9	DEE	−35	62%, CO_2	[980,1066−1069]
M—(4-pyridyl)	XMg	I	$BrMgC_2H_5$	THF	+25	85%, H_5C_6CHO	[1064]
	Li	Br	LiC_4H_9	DEE	−75	55%, $(H_5C_6)_2CO$	[980,1070]
M—(5,6,7,8-tetrahydroisoquinolyl)	Li	Br	LiC_4H_9	DEE	−35	59%, H_5C_6CHO	[1071]
M—(2,4,6-triphenylpyridyl) H_5C_6, C_6H_5, C_6H_5	Li	Br	LiC_4H_9	DEE	−35	67%, CO_2	[982]
M—(pyridyl)—OCH_3	Li	Br	LiC_4H_9	DEE	−35	89%, H_5C_6CHO	[1072]
M—(quinolyl)	Li	Br, I	LiC_4H_9	DEE	−50	70%, $(H_5C_6)_2CO$	[830]
M—(isoquinolyl)	Li	Br	LiC_4H_9	DEE	−45	48%, CO_2	[976,1066−1070]

[a] X = halogen displaced by the metal. [b] Isopropylmagnesium chloride may be replaced by ethyl- or phenylmagnesium bromide as the organometallic exchange reagent M−R'. [c] DEE = diethyl ether, THF = tetrahydrofuran. [d] $El-X' =$ electrophilic trapping reagent.

1-Naphthyllithium[1149]

At 0 °C, 1-bromonaphthalene (18 mL, 41 g, 0.20 mol) and dry paraformalde- hyde (7.5 g, 0.25 mol) are consecutively added to butyllithium (0.20 mol) in diethyl ether (0.12 mL) and hexanes (0.13 mL). After 1 h of stirring at 25 °C, the mixture is neutralized with etheral hydrogen chloride (0.20 mol, 3.5 M), the solvents are stripped off and the residue is distilled [b.p. 163−164 °C/12 mmHg]. The 1-naphtha- lenemethanol crystallizes spontaneously [m.p. 61−62 °C; yield 28.8 g (91%)].

Table 59. Halogen-substituted pyridylmetals M–R by halogen/metal permutation using reagents M–R'.

M–R	M	X[a]	M–R'	Sv[b]	T[°C]	Product, (El–X')[c]	Reference
M–(pyridine), Br	XMg	Br	ClMgCH(CH$_3$)$_2$	THF	+25	90%, I$_2$	[1065]
	Li	Br	LiC$_4$H$_9$	DEE	−40	82%, H$_3$CCON(CH$_3$)$_2$	[1073–1075]
M–(pyridine), Cl	Li	Br	LiC$_4$H$_9$	THF	−60	80%, H$_3$CI	[1076]
M–(pyridine), Br	XMg	Br	ClMgCH(CH$_3$)$_2$	THF	+25	92%, H$_5$C$_6$CHO	[1065]
M–(pyridine), Br	XMg	Br	ClMgCH(CH$_3$)$_2$	THF	+25	76%, H$_5$C$_6$CHO	[1065]
M–(pyridine)–Br	XMg	Br	ClMgCH(CH$_3$)$_2$	THF	+25	86%, H$_5$C$_6$CHO	[1065]
	Li	Br	LiC$_4$H$_9$	THF	−100	85%, D$_2$SO$_4$	[1077]
M–(pyridine), Br	XMg	Br	ClMgCH(CH$_3$)$_2$	THF	+25	98%, H$_5$C$_6$CHO	[1065]
M–(pyridine), Cl Cl / Cl N(CH$_3$)$_2$	Li	Cl	LiC$_4$H$_9$	DEE	−35	84%, H$_2$O	[1078]
M–(pyridine), Cl Cl / Cl N(CH$_2$)$_4$	Li	Cl	LiC$_4$H$_9$	DEE	−35	77%, H$_2$O	[1078]
M–(pyridine), Cl Cl / Cl OCH$_3$	Li	Cl	LiC$_4$H$_9$	DEE	+25	88%, H$_2$O	[1078]
M–(pyridine), F F / F F	Li	I	LiC$_4$H$_9$	DEE	−35	65%, CO$_2$	[1079]
M–(pyridine), Cl Cl / Cl Cl	Li	Cl	LiC$_4$H$_9$	DEE	−75	29%, CO$_2$	[1080]
M–(pyridine), Br Br / Br Br	Li	Br	LiC$_4$H$_9$	DEE	−75	16%, CO$_2$	[1081]

[a] X = halogen in precursor. [b] Solvent (Sv): DEE = diethyl ether, THF = tetrahydrofuran. [c] El–X' = electrophilic trapping reagent.

Table 60. Halogen-substituted diazinyllithiums M−R by halogen/metal permutation.

M−R	M	X[a]	M−R'[b]	$S\nu$[c]	T[°C]	Product, $(El-X')$[d]	Reference
(diazinyl)	Li	I	LiC_4H_9	DEE	−65	30%, CO_2	[1082]
H_3C-(diazinyl)-CH_3	Li	I	LiC_4H_9	DEE	−50	45%, CO_2	[1083]
(diazinyl)	Li	Br	LiC_4H_9	THF	−100	59%, $HCOOC_2H_5$	[1084]
(diazinyl)-OCH_3, OCH_3	Li	Br	LiC_4H_9	THF	−65	40%, H_3CI	[1085]
(diazinyl)-$OCH_2C_6H_5$, $OCH_2C_6H_5$	Li	Br	LiC_4H_9	THF	−80	87%, $H_5C_6Se_2$	[1086]
(diazinyl)-SCH_3, $COOM$	Li	Br	LiC_4H_9	THF	−100	36%, H_5C_6COCl	[1087]
R[e], O, OM (diazinone)	Li	Br	LiC_4H_9	THF	−75	52%, H_3CI	[1088]

[a] X = halogen displaced by the metal. [b] Butyllithium as the exchange reagent M−R'. [c] DEE = diethyl ether, THF = tetrahydrofuran. [d] $El-X'$ = electrophilic trapping reagent. [e] R = O^3,O^5-bis(2'-tetrahydropyranyl)- or O^3,O^5-bis(*tert*-butyldimethylsilyl)-desoxy-1-ribosyl.

3-Methoxyphenyllithium[1150]

A solution of butyllithium (50 mmol) in hexanes (30 mL) is diluted with tetrahydrofuran (70 mL). At −75 °C, 3-bromoanisole (5.4 mL, 8.6 g, 50 mmol) and, 4 min later, when a white precipitate has formed, 6-methyl-5-hepten-2-one (7.4 mL, 6.4 g, 50 mmol) are added. The volatiles are evaporated and the residue is triturated with brine (35 mL). 2-(3-Methoxyphenyl)-6-methyl-5-hepten-2-ol is extracted with hexanes (3 × 20 mL) and isolated by distillation [b.p. 118−120 °C/0.5 mmHg; n_D^{20} 1.4219; yield 7.6 g (65%)].

2,2′-Biphenylidenedilithium[1021,1151,1152]

At 0 °C, 2,2′-dibromobiphenyl[1021,1151] (6.2 g, 20 mmol) is dissolved in a solution containing butyllithium (40 mmol) in diethyl ether (75 mL) and hexanes (25 mL). After 15 min at 0 °C, chlorodiphenylphosphine (7.2 mL, 8.8 g, 40 mmol) is added. Evaporation of the volatiles and crystallization from methanol affords 9-phenylphosphafluorene (5-phenyl-5*H*-benzo[*b*]phosphindole) as colorless platelets [m.p. 91–93 °C; yield 9.4 g (90%)].

2-Pyridyllithium[1067,1153]

At −75 °C, 2-bromopyridine (4.9 mL, 7.9 g, 50 mmol) and, 15 min later, 4-cyanopyridine (5.2 g, 50 mmol) are consecutively added to butyllithium (50 mmol) in diethyl ether (70 mL) and hexanes (30 mL). The mixture is allowed to stand for 2 h at +25 °C before being extracted with 2.0 M sulfuric acid (3 × 40 mL). The combined aqueous layers are heated for 15 min on a boiling water bath. After neutralization with solid sodium hydroxide, the aqueous phase is extracted with dichloromethane (3 × 50 mL). The organic solution is dried and evaporated. Crystallization of the residue from a 1 : 10 (v/v) mixture of acetone and hexane gives colorless prisms [m.p. 122–123 °C; yield 5.1 g (61%)].

4.2 Displacement of Chalcogens

This Subsection covers the preparation of organolithium and some organomagnesium compounds from derivatives of elements belonging to the fifth and sixth main groups, the so-called pnicogens and chalcogens. In practice, this boils down to the reductive insertion of metal into ethers or thioethers and the permutational exchange of selenium or tellurium, both metalloids rather than nonmetals, by lithium.

V	VI	VII	VIII
	N	O	
	P	S	
	As	Se	
	Sb	Te	

4.2.1   Reductive Insertion of Metals in Carbon–Chalcogen Bonds

The reductive cleavage of any bond between carbon and an electronegative element will produce a salt-like metal derivative and accordingly proceed with high exothermicity (see p. 46). To exploit this thermodynamic potential the precursor must be able to act as an electron sink (see pp. 73–77). Due to their moderate electron affinity[1154], derivatives of first-row elements such as trimethylamine, dimethyl ether and methyl fluoride are particularly disfavored in this respect. Unsaturation improves their electron-acceptor properties. *N,N*-Dimethylaniline can be cleaved by lithium in tetrahydrofuran at +25 °C to give phenyllithium and lithium dimethylamide, in other words, under conditions where the resulting aryllithium is inevitably converted to the arene by proton abstraction from the solvent. Ring strain, as present in *N*-phenylaziridine[1156] and *N*-phenylazetidine[1157], facilitates the cleavage process, particularly when radical anions such as 1 : 1 adducts of lithium with naphthalene or 4,4'-di-*tert*-butylbiphenyl are employed simultaneously. Benzylic amines[1158] such as *N*-phenyl-1,3-dehydroisoindole or *N*-phenyl-1,2,3,4-tetrahydroisoquinoline, just as aromatic nitriles[1159], react with particular ease under such conditions. However, all these methods are of little practical importance.

Triphenylphosphine is readily cleaved by lithium in tetrahydrofuran to diphenylphosphinyllithium **231** and phenyllithium, the latter being rapidly converted to benzene by attack of the solvent[1160]. This method offers an alternative, though not particularly advantageous route to diphenylphosphinyllithium, compared with its preparation from chlorodiphenylphosphine[1161].

231

a) Representative Examples of Ether and Thioether Cleavage

Ether cleavages[1162] aiming at synthetical applications center around three classes of substrates. Oxiranes afford β-lithiooxyalkyllithiums upon reductive ring opening (Table 61). Aryl ethers give in moderately polar solvents arylmetals and metal alkoxides

Table 61. β-Lithiooxyalkylmetals M−R by reductive cleavage of oxiranes.

M−R	M	Precursor	S_v[a]	T[°C]	Product, El−X′ [b]	Reference
M−CH₂−CH₂−OM	Li	(oxirane)	THF	−95	58%, H₁₁C₆CHO	[1164,1165]
M−CH₂−CH−OM [c] (CH₃)	Li	(methyloxirane, CH₃)	THF	−95	65%, *p*-anisaldehyde	[1164,1165]
M−CH₂−CH−OM (C₈H₁₇)	Li	(oxirane, C₈H₁₇)	THF	−75	~60%, H₃COH	[1165]
M−CH₂−CH−OM [c] (C₆H₁₁)	Li	(oxirane, C₆H₁₁)	THF	−75	<83%[d], H₃COH	[1165]
M−CH₂−CH−OM [c,e] (CH₂OR)	Li	(oxirane, CH₂OR)	THF	−75	60%, (structure)	[1166]
M−CH−CH−OM (H₅C₂ C₂H₅)	Li	(oxirane)	THF	−80	68%, H₂O	[1164]
M–(cyclohexyl) OM	Li	(cyclohexene oxide)	THF	−75	<75%[d], H₃COH	[1164,1165]
M–(cycloheptyl) OM	Li	(cycloheptene oxide)	THF	−75	<30%[d], H₃COH	[1165]
MO–(norbornyl) M	Li	(epoxide)	THF	+80	60%, H₂O	[1167]
M−CH₂−CH₂−OM (C₆H₅)	K	H₅C₆ (oxirane)	THF	−80	74%, H₃CI[f]	[1164]
M−CH–(cyclohexyl, OM) (SC₆H₅)	Li	H₅C₆S (oxirane)	THF	−75	63%, H₃C−I	[1165]

[a] THF = tetrahydrofuran. [b] El−X′ = electrophile. [c] Configuration of enantiopure oxirane retained. [d] Yield of isolated product not specified. [e] OR = OCH₂OCH₃. [f] Simultaneous *C*- and *O*-alkylation.

or phenolates (see Table 62). Allylic and benzylic ethers are readily cleaved to the corresponding 2-alkenyl or benzylmetals, an alkoxide or aroxide being again the by-product (see Table 63). The most useful modes of thioether cleavage[1163] promote the transformation of α-thioamines, α-thioethers or dithioacetals to α-heteroalkyllithiums (see Table 64) and of allylic or benzylic sulfides to allyl- or benzylmetals (see Table 65).

Table 62. Arylalkali metals (M−R) by reductive cleavage of alkyl aryl or diaryl ethers.

M−R	M	Precursor	$S\nu$[a]	T[°C]	Product, $El-X'$ [b]	Reference
M–C$_6$H$_5$	K[c]	[d]	PET	+25	60%, CO$_2$	[363,1168]
M–C$_6$H$_4$(CH$_3$) (ortho)	K	H$_3$CO–C$_6$H$_4$(CH$_3$)	THF	−100	[e]	[1169]
M–C$_6$H$_4$(OCH$_3$) (ortho)	K	H$_3$CO–C$_6$H$_4$(OCH$_3$)	THF	−100	[e]	[1169]
M–C$_6$H$_3$(OCH$_3$)(CH$_3$)	K	H$_3$CO–C$_6$H$_3$(OCH$_3$)(CH$_3$)	THF	−100	[e]	[1169]
M–C$_6$H(C(CH$_3$)$_3$)$_2$(CH$_3$)	K	H$_3$CO–C$_6$H(C(CH$_3$)$_3$)$_2$(CH$_3$)	MEGME	+25	100%[f]	[1170]
2-(2′-OM-phenyl)phenyl–M (biphenyl)	Li	dibenzofuran	DEE	+35	45%, (H$_5$C$_6$)$_2$PCl [g]	[1171,1172]
binaphthyl–M, –OM	Li	(binaphtho-furan)	DEE	+35	64%, (H$_5$C$_6$)$_2$PCl	[1171]

[a] PET = petroleum ether (pentanes, hexanes, heptanes); DEE = diethyl ether; THF = tetrahydrofuran; MEGME = (mono)-ethylene glycol dimethyl ether (glyme, 1,2′-dimethoxyethane). [b] $El-X'$ = electrophilic trapping reagent. [c] The cleavage can also be effected with lithium or sodium instead of potassium. [d] Anisol or diphenyl ether. [e] Cleavage occurs only partially in the indicated sense; estimated conversion ≤25%. [f] Protonation by the solvent. [g] Product isolated after subsequent transformations.

Table 63. Allylic and benzylic alkali and alkali-earth metal derivatives M−R by reductive cleavage of ethers R−OR′.

M−R	M	OR′	Sv[a]	T[°C]	Product, El−X′[b]	Reference
M−CH₂−CH=OM	XMg	OCH₃[c]	THF	+80	86%, CO_2	[1173–1175]
	Li	OC₆H₅	THF	−20	67%, R⌒⌒O [e]	[483,1176]
	Na	[d]	THF	−20	67%, R⌒⌒O [e]	[483,1177]
	K	[d]	PET	+30	57%, CO_2	[483,1178]
M−CH₂−C(CH₃)=CH₂	XMg	[f]	THF	+80	20%, CO_2	[1173]
M−CH₂−CH=CH−CH₃	Li	OC₆H₅	THF	+25	90%, [structure]	[1179,1180]
M−CH₂−CH=CH−C₆H₅	Li	OC₆H₅	THF	+25	76%, titration	[1179]
M−CH₂−CH=C(CH₃)₂	Li	OC₆H₅	THF	+25	85%, [structure]	[1180]
M−CH₂−CH=CH−C₆H₅	XMg	OC₆H₅	THF	+80	90%, CO_2	[1174]
M−CH₂−[phenyl]	XMg	OCH₃[g]	THF	+80	51%, D_2O	[1175,1181]
	Li	OCH₃, OC₂H₅	THF	−10	85%, $(H_5C_6)_2CO$	[1182]
M−C(CH₃)₂−[phenyl]	K	OCH₃	DEE	+25	>90%, titration	[1183,1184]
M−CH−[phenyl]₂	K	OC₆H₅	THF	[g]	[g]	[1170]
M−C(OCH₃)[phenyl]₂	Li	OCH₃	THF	+25	84%, H_3COH	[1185]
	Na	OCH₃	DOX	+25	82%, H_3COH	[1185]
	K	OCH₃	DOX	+25	92%, H_3COH	[1185]
M−C(C₆H₅)(C₆H₅)−[cyclopropyl]	K	OCH₃	DEE	+25	72%, CO_2	[363]
M−[fluorenyl], H₃CO	Na	OCH₃	DEE	+25	[g], CO_2	[1186]

[a] PET = petroleum ether (pentanes, hexanes, heptanes); DOX = 1,4-dioxane; DEE = diethyl ether; THF = tetrahydrofuran. [b] El−X′ = electrophilic trapping reagent. [c] OR′ = methoxy or 1- or 2-naphthyloxy. [d] OR′ = allyloxy or phenyloxy. [e] R = (H₃C)₃C. [f] OR′ = 1- or 2-naphthyloxy. [g] Not specified.

Table 64. α-Silyl-, α-stannyl-, α-amino-, α-oxy- and α-thio-substituted (cyclo)alkyllithiums (M—R) by reductive cleavage of thioethers R—SR'[a].

M—R	SR'	Sv[b]	T[°C]	Product, El—X'[c]	Reference
M—(cyclohexyl), Si(CH₃)₃	SC_6H_5	THF	−45	86%, $H_{11}C_5CHO$	[1187]
M—(NMCOC₆H₅), Sn(C₄H₉)₃	SC_6H_5	THF	−50	79%, H_2O	[1188]
$M-CH_2-N(C_2H_5)_2$	SC_6H_5	THF	−75	80%, $H_5C_6Si(CH_3)_2Cl$	[1189]
$M-CH_2-N$ (ring)	SC_6H_5	THF	−75	68%, $H_5C_6Si(CH_3)_2Cl$	[1189]
$M-CH-C_3H_7$, OCH_3	SC_6H_5	THF	−60	68%, $(H_3C)_3SiCl$	[1190]
M—(cyclopropyl), OCH_3	SC_6H_5	THF	−60	90%, (cyclohexanone)	[1190,1191]
M—(bicyclic), OCH_3	SC_6H_5	THF	−75	87%, $H_2C=CH-CHO$[d]	[1190]
M—(ring) OC_6H_4-p-OCH_3[e], $OSiRR'_2$	SC_6H_5	THF	−75	69%, (BnO, MOMO aldehyde)	[1192]
M—(O-ring)	SC_6H_5	THF	−60	65%, H_3CO—⟨⟩—CHO	[1190]
M—(O-bicyclic)	SC_6H_5	THF	−75	56%, H_5C_6CHO	[1193]
M—(ring) C_6H_{13}, O, O—$CH(CH_3)_2$	SC_6H_5	THF	−75	79%, $O_2S(OCH_3)_2$	[1194,1195]
M—(cyclopropyl), SC_6H_5	SC_6H_5	THF	−75	95%, $(H_3C)_3SiCl$	[1196]

[a] Using Li in the presence of 4,4'-di-*tert*-butylbiphenyl, 1-(dimethylamino)naphthalene *etc.* [b] THF = tetrahydrofuran. [c] *El*—X' = electrophilic trapping reagent. [d] Isolated as a cyclobutanone isomer after acidic treatment. [e] $SiRR'_2 = Si(CH_3)_2C(CH_3)_3$.

b) Scope and Limitations

Aliphatic ethers are perfectly inert towards elemental metals even at elevated temperatures. In contrast, C—O-bonds of alkyl esters, in particular sulfates[1204] and sulfonates[1205], readily undergo reductive cleavage with naphthalene/lithium.

Table 65. Allylmetals MR (M = XMg, Li) by reductive cleavage of thioethers R–SR′.

M–R	M[a]	SR′	Sv[b]	T[°C]	Product, El–X′[c]	Reference
M–CH₂–CH=CH₂	XMg	SC₆H₅	THF	+80	67%, CO₂	[1197]
M–CH₂–CH=CH–C₂H₅	Li	SC₆H₅	THF	−75	72%, (1.) CeCl₃ (2.) ∿O	[1198]
M–CH₂–CH=C(CH₃)₂	Li	SC₆H₅	THF	−75	54%, (1.) CeCl₃ (2.) ✓=O	[1199]
M–CH₂–C=CH₂ CH₃	XMg	SC₆H₅	THF	+80	88%, CO₂	[1197]
M–CH₂–⟨⟩	Li	SSi(CH₃)₃	THF	0	85%, (H₃C)₃SiCl[d]	[1201]
M–⟨⟩ 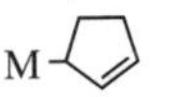	XMg	SC₆H₅	THF	+80	45%, CO₂	[1197]
M–⟨⟩ 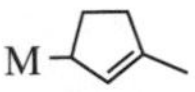	Li	SC₆H₅	THF	−60	71%, (1.) Ti(OR)₄[e] (2.) 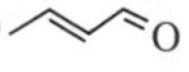 ∿∿O	[1200]
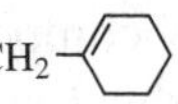 M–CH₂–⟨⟩	Li	SC₆H₅	THF	−60	92%, (1.) Ti(OR)₄[e] (2.) 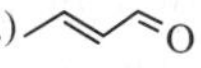 ∿∿O	[1200]
C(CH₃)₃ C(CH₃)₃ M–CH–CH=CH	Li	SC₆H₅	THF	−75	75%, (H₃C)SiCl	[1202]
H₃C CH₃ M–C–CH=C H₃C CH₃	XMg Li	SC₆H₅ SC₆H₅	THF THF	+80 −90	14%, CO₂ 88%, (H₃C)₃SnCl	[1197] [1203]

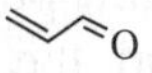

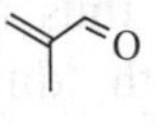

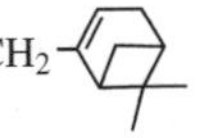

[a] XMg = H₅C₆SMg. [b] THF = tetrahydrofuran; in case of mixtures (*e.g.* containing paraffinic cosolvents) only the most polar component is indicated. [c] *El*–X′ = electrophilic trapping reagent. [d] *In situ* trapping. [e] OR = OCH(CH₃)₂.

Ring strain propels the fast reductive ring opening of oxiranes. However, depending on the given structures, the organolithium intermediates **232** are accompanied by more or less significant amounts of olefins resulting from lithium oxide β-elimination[1165,1167].

232

Oxetanes[1206] and β-lactones[1207] are equally prone to cleavage by lithium or potassium in the presence of carriers such as di-*tert*-butylbiphenyl or 1,4,7,10,13,16-hexaoxacyclooctadecane ('18-crown-6'), respectively. The tetrahydrofuran ring is reductively opened by naphthalene/lithium in the presence of stoichiometric or catalytic amounts of boron trifluoride[1208,1209] and by Lewis acid-activated magnesium (to afford 2-magnesiatetrahydropyran)[1210].

The splitting pattern of unsymmetrical aromatic or aliphatic ethers is particularly intriguing. With phenyl *p*-tolyl as the substrate, scission mainly occurs between the phenyl and the oxygen and only to a minor extent between the tolyl and the oxygen atom, thus apparently reflecting the relative electron affinities of the two aryl rings[1170]. Both, dihydrobenzofuran and *tert*-butyl phenyl ether give exclusively the respective phenolates when treated with lithium in tetrahydrofuran[1170]. Anisole is cleaved by lithium[1170] or potassium[1168] in paraffinic media between the phenyl-oxygen bond, but by lithium in tetrahydrofuran[1211,1212] or potassium in liquid ammonia between the methyl−oxygen bond.

One may feel tempted to rationalize the latter findings in terms of two competing pathways. The C−O bond scission would occur at the stage of either a 'radical anion' (ether/metal 1 : 1 adduct) or a 'dianion' (ether/metal 2 : 2 adduct). However, as no evidence for the intermediacy of such latter species has been found so far[1213], it seems safer to attribute differences in the reaction outcome to solvent effects on the thermodynamic stability of the two possible cleavage components, *i.e.* the phenyl radical/metal methoxide *vs.* the methyl radical/metal phenoxide pair.

The practical utility of allyl ether cleavage has been documented by several examples (listed in Table 63). Analogously, allylic diethoxy phosphates[1214] can be cleaved with metallic lithium and allylic triphenylsilyl ethers[1215] with di-*tert*-butylbiphenyl/lithium, both in tetrahydrofuran, to afford allyllithiums. The neryl and geranyl configurations are perfectly retained at the level of the intermediates (**Z-233** and **E-233**) and the products as long as the temperature is not raised above −75 °C.

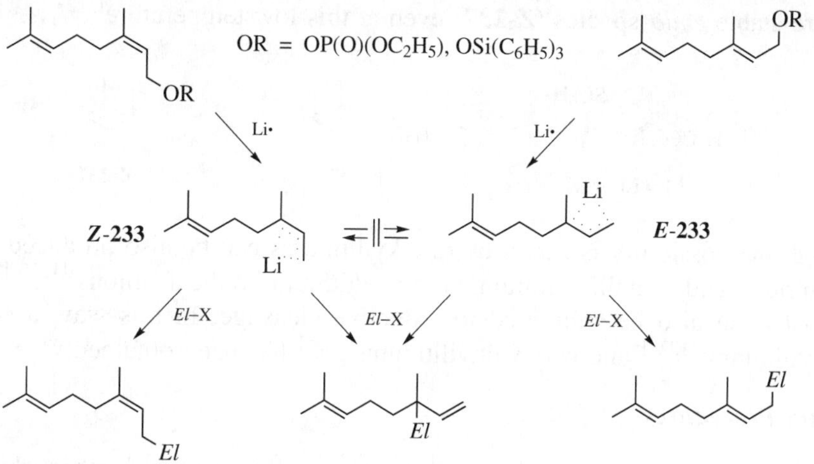

OR = OP(O)(OC$_2$H$_5$), OSi(C$_6$H$_5$)$_3$

A few examples of benzyl ether cleavages have already been presented (see Table 63). The same method applied to cyclic benzyl ethers[1216], such as dihydroisobenzofuran and tetrahydro-2-benzopyran (**234**) or to benzylic acetals[1217] (*e.g.* **235**), allows one to generate functionalized benzylmetals.

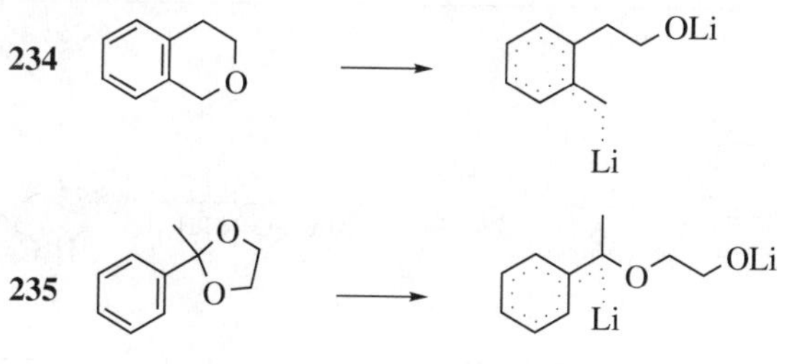

Earlier work on this topic remaining episodal[1218, 1222], the establishment of thioether cleavage as a practical method owes decisive impulses to the work of T. Cohen[1163] and D. R. Rychnovsky[1194, 1195] and their collaborators (see Tables 64–65). A synthesis of lavandulol (**236**), involving as a key step a reductive allyl sulfide cleavage, demonstrates the versatility of this approach.

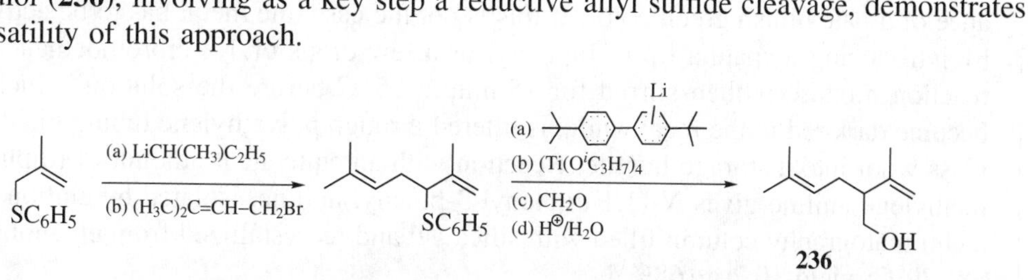

(a) LiCH(CH$_3$)C$_2$H$_5$
(b) (H$_3$C)$_2$C=CH–CH$_2$Br

(a) [structure with Li]
(b) (Ti(O^{i}C$_3$H$_7$)$_4$
(c) CH$_2$O
(d) H$^\oplus$/H$_2$O

236

Hardly any other method can rival the thioether cleavage as far as the mildness of reaction conditions is concerned. This allows one to generate extremely labile organometallic intermediates. Thus, the *exo*-3-methoxyallylpotassium (**E-237**) prepared from *trans*-3-methoxyallyl phenyl sulfide and naphthalene/potassium conserves its configuration at −120 °C, whereas the corresponding lithium compound isomerizes to the thermodynamically more stable *endo* species (**Z-237**) even at this low temperature[1223].

$$M = \text{Li or K} \qquad \textbf{E-237} \qquad \textbf{Z-237}$$

Although the possibility is rarely used, alkyllithiums can be also produced from alkyl phenyl sulfides and metallic lithium or arene/lithium 'radical anions'[1220,1221]. Alkyl aryl sulfones can also be submitted to reductive cleavage. In this way, a series of α-aminoalkyllithiums[1224] and α-oxyalkyllithiums[1225] has been obtained.

c) *Working Procedures*

The experimental details of three allyl ether and one benzyl ether cleavages are described. An additional example dealing with the reductive scission of a monothioacetal deserves attention because of its stereochemistry. When the organometallic intermediate is generated at −75 °C, the metal occupies the axial position almost exclusively, whereas epimerization occurs at −30 °C affording predominantly the equatorial conformer.

Allyllithium[1176]

Allyl phenyl ether (6.8 m, 6.7 g, 50 mmol) in diethyl ether (25 mL) is added over 45 min to a rapidly stirred suspension of freshly cut lithium wire (4.2 g, 0.61 mol) in tetrahydrofuran (0.10 mL) kept in a −15 °C cooling bath. Usually after the first few minutes a slightly exothermic reaction sets in, manifesting itself by the appearance of a pale bluish green color. If this is not the case, the metal has to be activated by introducing a spatula tip of biphenyl or a few drops of 1,2-dibromoethane. The reaction mixture is then stirred for 15 min at 25 °C before the solution, which has become dark red in the meanwhile, is filtered through polyethylene tubing filled with glass wool into a storage burette. Reaction with an equivalent amount of (diphenylmethylene)aniline gives *N*-(1,1-diphenyl-3-butenyl)aniline, isolated by elution from a chromatography column filled with silica gel and recrystallized from ethanol [m.p. 78–79°C; yield 10.2 g (68%)].

Allylsodium[1177]

A dispersion of sodium (5.8 g, 0.25 mol) is prepared in octane (see p. 101), which is then replaced by hexane (0.5 L). Diallyl ether (18 mL, 14 g, 0.15 mol) is added over a period of 1 h under high-speed stirring (5000 rpm) and at an average temperature of 35 °C. After a further 30 min of stirring the reaction mixture is siphoned onto dry-ice. After acidification, the 3-butenoic acid is extracted with diethyl ether and distilled [b.p. 61–62 °C/3 mmHg; yield 16.6 g (77%)].

2-Phenylisopropylpotassium (1-Methyl-1-phenylethylpotassium)[1183]

Potassium/sodium alloy[1226] (78 : 22 w/w; 18 mL, 15 g, 0.32 mol potassium) is transferred by means of a pipette into a solution of cumyl methyl ether[1183] (2-methoxy-2-phenylpropane; 15 g, 0.10 mol) in diethyl ether (0.50 L). The mixture is vigorously stirred or, if placed in a sealed tube, shaken. The bright red solution is carefully decanted through polyethylene tubing fitted with a glass-wool plug into a burette. Tolane (11.6 g, 65 mmol) was added in portions to an aliquot (0.3 L) of this roughly 0.2 M solution. An exothermal reaction caused the solvent to boil. The reaction mixture was poured on to dry-ice. Extraction with water (3 × 25 mL) and evaporation gave a colorless resinous material; yield 6.4 g (26%).

2-(β-Ionylidene)ethyllithium[1227]

At −75 °C, a 0.5 M solution of the lithium/biphenyl (1 : 1) adduct[1228] (20 mmol) in tetrahydrofuran (40 mL) is added to (*E*)-3-methoxy-3-methyl-1-(2,6,6-trimethyl-1-cyclohexenyl)-1,4-pentadiene (methyl vinyl β-ionyl ether; 4.7 g, 2.0 mmol) in tetrahydrofuran (10 mL). Immediately a bright red color develops. Still at $-75\,°C$, the mixture is consecutively treated with fluorodimethoxyboron etherate[1229,1230] (3.8 mL, 3.4 g, 20 mmol), 35% aqueous hydrogen peroxide (2.0 mL, 2.3 g, 23 mmol) and 4.0 M aqueous sodium hydroxide (20 mL). After 30 min of stirring, the aqueous phase is saturated with sodium chloride. The organic layer is decanted and evaporated. Chromatography of the residue on active alumina using a 1 : 1 (v/v) mixture of diethyl ether and hexane as the eluent gives analytically pure $(2E,4E)$-3-methyl-5-(2,6,6-trimethyl-1-cyclohexenyl)-2,4-pentadien-1-ol [(E)-2-(β-ionylidene)ethanol]; yield 3.3 g (75%).

(4-Methyl-1-piperazyl)methyllithium[1189]

At $-75\,°C$, lithium granules (0.69 g, 0.10 mol) are suspended in a solution of naphthalene (13 g, 0.10 mol) in tetrahydrofuran (0.15 L), which rapidly turns dark blue. After 2 h of vigorous stirring, still at $-75\,°C$, 4-methyl-1-(phenylthiomethyl)-piperazine (22 g, 0.10 mol) in precooled tetrahydrofuran (50 mL) is added. Some 15 min later, the mixture is treated with chloro(dimethyl)phenylsilane (17 g, 0.10 mol). After having reached $25\,°C$, it is concentrated by evaporation under reduced pressure to approximately one-third of its original volume, diluted with diethyl ether (0.15 L), washed with water (2 × 0.10 L) and brine (1 × 50 mL) and evaporated. The residue is purified by bulp-to-bulp ('Kugelrohr') distillation; b.p. $121\,°C/0.001$ mmHg (oven temperature) [yield 16.9 g (68%)].

4.2.2   Permutational Exchange of a Chalcogen Against Metal

As previously mentioned (see pp. 72–73), ate complexes are assumed to act as turntables for most permutational replacements of nonmetals or metalloids by metals. This possibility must, of course, be ruled out for first-row elements as they cannot expand their valence shell to 10 electrons. Organometallic reagents are therefore not capable of displacing alkyl or aryl groups from amines or ether.

Such an exchange can only be brought about indirectly, if at all. The Shapiro method[1231] of 1-alkenyllithium generation, the apolar–aprotic analog of the Bamford–Stevens reaction, can be regarded as the formal replacement of an enolate by a lithium lithiooxy function by a lithium atom. In reality, a doubly deprotonated (arenesulfonyl)hydrazone takes the role of the enolate. A β-elimination of sulfinate sets free the ephemeral *N*-lithiodiazene **238** which forms the final 1-alkenyllithium species by loss of elemental nitrogen.

Even though no immediate relationship to permutational processes is given, the Shapiro reaction is mentioned in this context because of its underestimated potential. The alkenyllithiums **239–241**, generated from the corresponding ketones (3-buten-2-one[1232], camphor[1233] and 2,2,6-trimethylcyclohexanone[1234]), have been selected to highlight the versatility of this method.

239   **240**   **241**

Sulfides undergo a permutational nonmetal/metal exchange only in exceptional cases. The clean conversion of cyclopropanone diphenyl dithioacetal into 1-(phenylthio)cyclopropyllithium belongs to the rare examples reported[1196]. In contrast, the substituent exchange between sulfoxides and Grignard reagents appears to be of wide applicability[1235–1237]. The reaction proceeds smoothly provided that it is accompanied by a substantial decrease in basicity. Chiral sulfoxides (such as **242**) retain the configurations at both the sulfur and the displaced carbon atom[1238].

Permutational interconversions can be carried out efficaciously with selenides[1239–1241]. Tellurides which form spectroscopically detectable ate complexes[1242] react even more rapidly[1243,1244].

a) Representative Examples of Selenium–Lithium Permutation

The exchange executed with simple alkyl and aryl selenides lacks practical merits. On the other hand, allylic and benzylic lithium compounds can be conveniently prepared in that way (see Table 66). From a synthetic point of view, the generation of α-heteroalkylli-thiums, in particular α-selenioalkyllithiums, is attractive (see Table 67).

Table 66. Benzylmetals (M−R) by permutational methylselenyl/lithium exchange.

M−R	M	Sv[a]	T[°C]	Product, El−X' [b]	Reference
M−CH$_3$	Li	THF	−75	32%, H$_5$C$_6$CHO[c]	[1245]
M−(phenyl)	Li	THF	−75	93%, H$_5$C$_6$CHO[d]	[1245]
M−CH$_2$−CH=CH$_2$	Li	THF	−75	95%, H$_5$C$_6$CHO	[1246]
M−CH$_2$−CH=CH−C$_6$H$_{13}$	Li	THF	−75	81%, H$_5$C$_6$CHO[e]	[1246]
M−CH$_2$−CH=C(CH$_3$)$_2$	Li	THF	−75	76%, H$_5$C$_6$CHO[e]	[1246]
M−CH$_2$−(phenyl)	Li	THF	−75	98%, H$_5$C$_6$CHO	[1247]
M−CH$_2$−(phenyl, *ortho*-Cl)	Li	THF	−75	94%, H$_5$C$_6$CHO	[1247]
M−CH$_2$−(phenyl, *para*-Cl)	Li	THF	−75	91%, H$_5$C$_6$CHO	[1247]
M−CH$_2$−(phenyl, *para*-F)	Li	THF	−75	83%, H$_5$C$_6$CHO	[1247]
M−CH$_2$−(phenyl, *para*-OCH$_3$)	Li	THF	−75	82%, H$_5$C$_6$CHO	[1247]
M−CH(CH$_3$)−(phenyl)	Li	THF	−75	83%, H$_5$C$_6$CHO	[1247]
M−C(CH$_3$)$_2$−(phenyl)	Li	THF	−75	72%, H$_5$C$_6$CHO	[1247]

[a] THF = tetrahydrofuran; in case of mixtures (containing for example paraffinic cosolvents) only the most polar component is indicated. [b] El−X' = electrophilic trapping reagent. [c] A major part of the butyllithium had remained unconsumed, as evidenced by the isolation of 44% of 1-phenyl-1-pentanol. [d] The yield was 72%, when diphenyl selenide rather than methyl phenyl selenide was used as the precursor. [e] Regioisomeric mixture of products.

Table 67. α-Methoxy, α-methylthio and α-alkylselenyl substituted alkyl- and benzyllithiums (M–R) by permutational SeR'/Li exchange.

M–R	M	Sv[a]	$T[°C]$	Product, El–X'[b]	Reference
M-CH(OCH$_3$)-C$_6$H$_5$	Li	THF	−75	95%, H$_{11}$C$_5$Br	[1248]
M-C(CH$_3$)(OCH$_3$)-C$_6$H$_5$	Li	THF	−75	85%, H$_{11}$C$_5$Br	[1248]
M-C(CH$_3$)$_2$-SCH$_3$	Li	THF	−75	84%, H$_5$C$_6$CHO	[1249]
M-CH$_2$-SeR'[c]	Li	DEE[d]	−75	88%, H$_5$C$_6$CHO	[1250]
M-CH(CH$_3$)-SeR'[c]	Li	THF	−75	81%, H$_5$C$_6$CHO	[1250]
M-CH(C$_6$H$_{13}$)-SeR'[c]	Li	THF	−75	89%, H$_5$C$_6$CHO	[1250]
M-CH(C(CH$_3$)$_3$)-SeR'[c]	Li	THF	−75	92%, H$_5$C$_6$CHO	[1250]
M-C(CH$_3$)$_2$-SCH$_3$[c]	Li	THF	−75	94%, H$_5$C$_6$CHO	[1250]
M-C[CH(CH$_3$)$_2$]$_2$-SeCH$_3$	Li	THF	−75	77%, H$_5$C$_6$CHO	[1250]
M-CH(C$_6$H$_5$)-SeR'[b]	Li	THF	−75	89%, H$_5$C$_6$CHO	[1250]
M-CH(C$_6$H$_4$·OCH$_3$)-SeCH$_3$	Li	THF	−75	77%, H$_5$C$_6$CHO	[1250]
M-CH(C$_6$H$_4$-OCH$_3$)-SeCH$_3$	Li	THF	−75	79%, H$_5$C$_6$CHO	[1250]
M-C(CH$_3$)(C$_6$H$_5$)-SeR'	Li	THF	−75	87%, H$_5$C$_6$CHO	[1250]
M-C(CH$_3$)(C$_6$H$_4$-CH$_3$)-SeCH$_3$	Li	THF	−75	90%, H$_5$C$_6$CHO	[1250]

[a] DEE = diethyl ether; THF = tetrahydrofuran; in case of mixtures (containing for example paraffinic cosolvents) only the most polar component is indicated. [b] El–X' = electrophilic trapping reagent. [c] SeR' = SeCH$_3$ or SeC$_6$H$_5$. [d] DEE or THF.

b) Scope and Limitations

Although a permutational exchange occurs with methyl selenides less rapidly than with phenyl selenides, it generally goes to completion and gives excellent yields (see Tables 66 and 67). Moreover, the phenyl is split off in preference to the alkyl group unless the anion of the latter is p- or d-orbital-resonance stabilized. Thus, both dimethylaminomethyl phenyl selenide and methoxymethyl phenylselenide react with *tert*-butyllithium to afford phenyllithium and the corresponding *tert*-butyl α-heteromethyl selenide[1241]. α-Alkoxy-alkyllithiums can, however, be generated by reductive cleavage of suitable selenide precursors[1251,1252] (such as the allyl ether **243**[1252]).

Allylic selenides can be obtained with particular ease by the condensation between the corresponding allyl halide and a metal selenolate[1253,1254]. The diselenide thus produced from 2-chloromethyl-3-chloro-1-propene can be readily converted with *sec*-butyllithium into dilithiotrimethylenemethane (**244**)[1253].

Diselenoacetals are extremely versatile components for synthesis. Both selenium functions may be consecutively replaced, each one first by lithium and then by the electrophile of choice. Alkenes are obtained from alkyl phenyl selenides by oxidation-triggered internal β-elimination of phenylselenol[1255] and from β-hydroxyalkyl selenides by Lewis acid promoted elimination of the hydroxy and a methylselenio group[1256]. Cyclopropanes can be obtained from 3-lithioalkyl selenide intermediates (such as **245**) by γ-elimination of lithium methaneselenoate[1257].

Selenide-derived organolithium species sometimes exhibit fascinating stereoselectivities. The cyclization of 1-methoxy-1-phenyl-5-hexenyllithium (**246**) produces one single stereoisomer, whatever the solvent and the reaction temperature[1257].

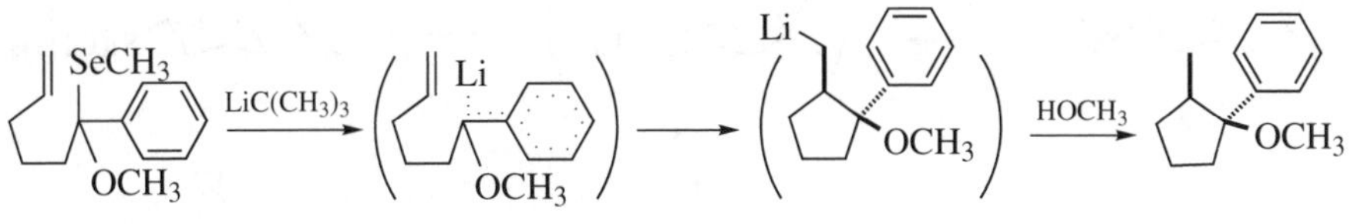

246

The heteroatom-free analogs 1-methyl-1-phenyl-5-hexenyllithium (**247a**) and 1-methyl-1-(2-methylpropenyl)-5-hexenyllithium (**247b**) favor either one of the two possible diastereomers depending on the reaction conditions. In tetrahydrofuran at −110 °C the (*E*)-isomers (methyl groups in *cis* position) are formed mainly if not exclusively. The same holds true for reactions performed in diethyl ether at low temperatures (≤ −75 °C). However, when the ethereal solution is warmed to −30 °C, the (*Z*) isomer becomes the predominant or even sole component[1258]. Obviously the stereoisomer originating from a kinetically controlled intramolecular nucleophilic addition can isomerize at higher temperatures and in the presence of a relatively apolar solvent to a thermodynamically more stable species.

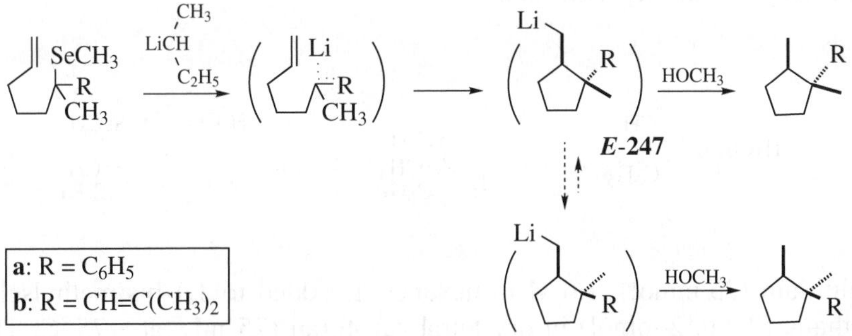

a: R = C₆H₅
b: R = CH=C(CH₃)₂

E-247

Z-247

Another case of equilibration was found when the cyclic diseleno acetal **248** was treated with *tert*-butyllithium in tetrahydrofuran and chlorotrimethylsilane was added after 10 min. The axial and the equatorial silanes were present in a ratio of 4 : 96. However, if the permutational Se/Li exchange was accomplished in the presence of chlorotrimethylsilane, *i.e.* under '*in situ* trapping conditions', the diastereomeric ratio changed to 88 : 12, thus revealing a kinetic preference for the equatorial and a thermodynamic one for the axial lithium species[1259].

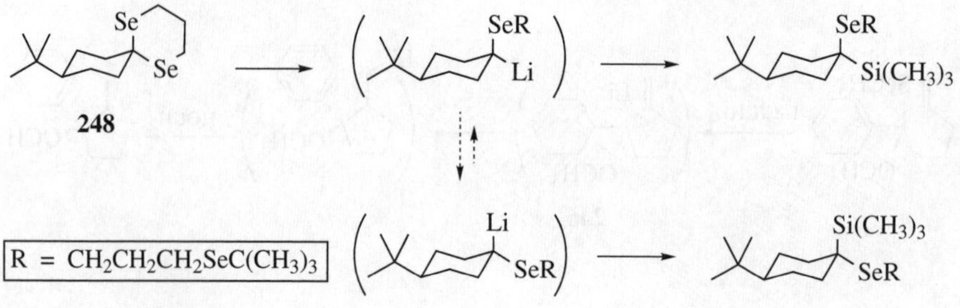

248

R = CH₂CH₂CH₂SeC(CH₃)₃ → displayed as $R = CH_2CH_2CH_2SeC(CH_3)_3$

c) Working Procedures

The two protocols given below are interconnected. The crucial intermediate of the entire reaction sequence is a tertiary selenide to which an allyl ether side chain has been attached. Its preparation starting with an acetophenone-derived diseleno acetal is described first. Then the remaining methylselenyl group is replaced by lithium and the new organometallic species cyclizes instantaneously by intramolecular nucleophilic addition to the olefinic double bond accompanied by simultaneous elimination of the vinylogous methoxy group. However, the intermediacy of the acyclic organolithium compound can be demonstrated by 'in situ trapping' with chlorosilane[1260].

1-Methylselenyl-1-phenylethyllithium[1260]

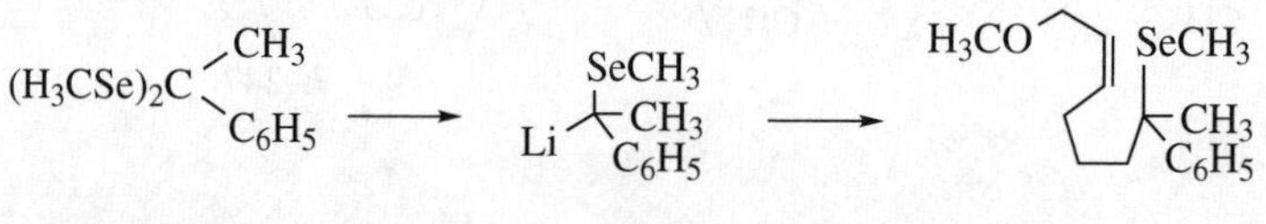

Butyllithium (25 mmol), 1.6 M in hexanes, is added to 1,1-bis(methylseleno)-1-phenylethane (7.7 g, 2 mmol) in dry tetrahydrofuran (25 mL) at −75 °C. The dark red mixture is kept 0.5 h at −75 °C, then a solution of the alkyl before being treated with (Z)-6-methoxy-4-hexenyl chloride (3.7 g, 25 mmol) in tetrahydrofuran (25 mL) is syringed into the reaction mixture at the same temperature. The volatile by-products are removed by rapid distillation (100 °C/0.1 mmHg). The resulting oil is purified by chromatography on silica gel. Upon elution with a 4 : 96 (v/v) mixture of diethyl ether and pentanes, (Z)-6-methoxy-1-methyl-1-phenyl methyl selenide is collected as a light yellow oil [b.p. 142−144 °C/0.5 mmHg; n_D^{20} 1.5541; yield 6.3 g (80%)].

(Z)-7-Methoxy-1-methyl-1-phenyl-5-heptenyllithium[1260]

Butyllithium (2.0 mmol), 1.6 M in hexanes, is added to (Z)-6-methoxy-1-methyl-1-phenyl methyl selenide (0.62 g, 2.0 mmol) in tetrahydrofuran (4 mL) at −75 °C. After 0.5 h at −75 °C, the reaction is quenched with methanol (1 mL). The solvents and the volatile by-products are removed *in vacuo* and the residue is absorbed on silica gel. Elution with pentanes affords 1-methyl-1-phenyl-2-vinylcyclopentane as a colorless oil which is composed of a minor and a major diastereomer (phenyl and vinyl *cis* and *trans* to each other, respectively) in ratios ranging from 5 : 95 to 9 : 91 [b.p. 52–53 °C/0.5 mmHg; n_D^{20} 1.5339; yield 0.34 g (90%)].

4.3 Displacement of Metalloids

'Metalloids' (or 'semimetals') are generally understood to be boron, silicon, germanium, arsenic, antimony, selenium and tellurium. In the given context we wish to extend this definition to tin and mercury, and even focus on them. No one can deny the metallic nature of these elements. However their organic compounds, for example dibutylmercury and tetrabutyltin, are perfectly nonpolar substances which show no propensity to aggregation and the volatility or solubility of which reminds one of paraffins. Moreover, such organometalloids participate in typical insertion and permutation reactions as if they were derivatives of nonmetals.

4.3.1 Reductive Insertion of Metals in Carbon–Metalloid Bonds

In principle any metals, not only metalloids, can be replaced by a more electropositive metal. For example, butyllithium exchanges its metal for the heavier one when in contact with elemental potassium[442].

In the early history of polar organometallic chemistry, the reductive cleavage of diorganomercurials had established itself as a favorite method of generation. Other organometalloid compounds (*e.g.* tribenzyltin chloride[1261]) were only rarely tested in such kinds of reactions.

a) Examples of Organomercurial Cleavage

The preparation of alkyl, cyclopropyl, vinyl, aryl and benzyl derivatives of lithium, sodium and potassium by the reductive cleavage of organomercury compounds has been widely reported (Table 68). Phenyllithium/phenylsodium ('phenyl-LiNa')[1262–1264] has also been prepared in this way. This early mixed-metal reagent is moderately stable in diethyl ether.

Table 68. Reductive cleavage of organomercurials affording alkali(earth) metal derivatives (M–R).

M–R	M	Sv[a]	Product, El–X'[b]	Reference
M–CH$_3$	Li	DEE	95%, titration	[1265]
	Na	without	~100%, isolated	[1266]
	K	without	~100%, isolated	[1266]
M–C$_2$H$_5$	Li	BNZ	[c], crystallized	[1267]
	Na	without	~100%, isolated	[1265–1271]
	K	BNZ	61%, CO$_2$	[1270]
M–C$_3$H$_7$	Li	PET	[c], isolated	[1267]
	Na	PET	79%, isolated	[1271]
M–C$_4$H$_9$	Li	PET	[c], isolated	[1267,1272]
	K	PET	[d]	[428,1273]
	Cs	PET	[d]	[1274]
M–CH$_2$Si(CH$_3$)$_3$	Li	DEE	83%, H$_5$C$_6$CH$_2$Br	[1275]
	K	PET	~100%, centrif., titration	[150,474,1276]
	Cs	PET	~100%, centrif., titration	[150]
M–◁	Li	PET	31%, titration	[156]
M–C(CH$_3$)$_3$	Li	PET	[c]	[833]
M–CH=CH$_2$	Li	PET	75%, H$_2$O	[1277,1278]
M–⬡	XMg	DEE[e]	65%, cryst. from DEE	[1279]
	Li	DEE	93%, titration	[1264]
	Na	PET	[c], isolated	[1265]
	K	PET	44%, CO$_2$	[1281]
M–⬡–CH$_3$	Li[f]	DEE	~100%, titration	[1263,1280]
	Na	DEE	75%, (H$_5$C$_6$)$_2$CO	[707,1263]
M–CH$_2$–⬡	Li	DEE	[c], crystallized	[123,1280]
M–CH$_2$–⬡–C(CH$_3$)$_2$–C$_2$H$_5$	Li	DEE[g]	72%, centrifug., titration	[1282]

[a] PET = petroleum ether (pentanes, hexanes, heptanes), BNZ = benzene; DEE = diethyl ether; THF = tetrahydrofuran; [b] El–X' = trapping electrophile. [c] Yield not specified. [d] *In situ* reaction with benzene. [e] Or tetrahydrofuran. [f] As a 1 : 1 complex with phenyllithium; when in PET at ≥100 °C, isomerization to benzylsodium[707]. [g] Or toluene.

b) Scope and Limitations

Uncomfortable in its execution, the reductive cleavage of organomercury compounds by alkali metals will never become a popular method. The mercury that is set free has to be bound in form of an amalgam. Otherwise the metal/metalloid displacement would not go to completion as it is principally a reversible process[808,1283]. Therefore a huge (5- to 15-fold) excess of the alkali metal is used and commonly the latter is shaken with an ethereal solution of the diorganomercurial in a sealed tube for several hours if not days.

Notwithstanding its inconveniences, the method offers distinct advantages. First, organometal solutions prepared this way do not contain salt-like cleavage by-products. Furthermore, the reductive cleavage of organomercurials is best suited to generate 1,2-phenylenedilithium[1284] and other dilithioarenes.

c) Working Procedures

Two protocols exemplify the smooth access to dilithioarenes. Two other procedures are devoted to the preparation of organopotassium reagents.

*1,2-Dilithiobenzene (*ortho-*phenylenedilithium)*[1284]

Lithium shred (6.0 g, 0.86 mol) is added to (hexameric) *o*-phenylenemercury[1285] (8.8 g, 32 mmol) in diethyl ether (0.10 L). The Schlenk vessel containing the mixture is flame-sealed and placed on a horizontal shaking machine for 4 days. The red solution is decanted into a Schlenk burette. According to double titration (see p. 295), the organometallic concentration approximates 0.5 M (corresponding to a 78% yield).

2,2′-Dilithiobiphenyl (2,2′-biphenylenedilithium)[1286]

A sealed tube filled with 2,2′-biphenylenemercury[1287] (1.0 g, 2.8 mmol), lithium shred (0.30 g, 43 mmol) and diethyl ether (45 mL) is shaken for 12 h at 25 °C. The yellow solution is decanted into a small Schlenk burette. Titration with acid reveals a 0.12 M organometallic concentration (*i.e.* 98% conversion).

Butylpotassium[1273]

$$Hg(C_4H_9)_2 \xrightarrow[-[K/Hg]]{K/Na} 2\ KC_4H_9$$

Liquid potassium/sodium alloy (see below; 6.2 mL, 5.0 g, 0.10 mol in potassium) is added to a solution of dibutylmercury[363,1288] (4.4 mL, 7.9 g, 25 mmol) in cyclopentane (50 mL). The mixture is vigorously stirred for 2 h at 25 °C. Then it is gently heated (<50 °C) until all the remaining metal clots together in one heavy clod from which the supernatant suspension is decanted under light shaking. The residue is washed with more cyclopentane (50 mL) which is transferred in the same storage vessel. An aliquot of the combined suspension is removed by means of a pipet and is treated with a slight excess of benzaldehyde. After neutralization, some 95% of 1-phenyl-1-pentanol are identified by gas chromatography.

Potassium/sodium alloy can be prepared in larger stock quantities by melting freshly cut pieces of potassium (78 g, 2.0 mol) and sodium (22 g, 0.96 mol) in gently boiling toluene (0.15 L) until one silvery shining lump has formed. At this stage the toluene may be siphoned off to be replaced by another solvent (*e.g.* cyclopentane) and this procedure may be repeated several times until the exchange is complete. The required amount of the alloy is withdrawn by pipet. To avoid any contact of the pyrophoric material with air, first a little bit of the solvent (≤0.1 mL) is sucked in, then the liquid alloy (d_4^0 0.8012) and finally the tip of the pipet is again filled with a little bit of the solvent.

Small *ad hoc* quantities of potassium/sodium alloy can be conveniently made in a test tube. After repetitive washing with cyclopentane (or another solvent, if toluene is unsuitable), the tube is held over an open neck of the reaction vessel, in which the alloy drops when the bottom of the tube is broken by using a sharply edged glass rod as a piston.

Trimethylsilylmethylpotassium[474,1276]

$$Hg[CH_2Si(CH_3)_3]_2 \xrightarrow[-[K/Hg]]{K/Na} 2\ KCH_2Si(CH_3)_3$$

Suspension in Pentanes

Potassium/sodium alloy[1289,1290] (see preceding paragraph; 78 : 22 w/w; 3.5 mL, 3.0 g, 64 mmol) is added to a solution of bis(trimethylsilylmethyl)-mercury[1291] (5.0 mL, 7.5 g, 20 mmol) in pentanes (50 mL) and the mixture vigorously stirred for 30 min. The organopotassium reagent is quantitatively formed. The resulting gray amalgam-containing suspension may be used as such.

Solution in Tetrahydrofuran

A suspension of trimethylsilylmethylpotassium (40 mmol) in pentane (50 mL) is prepared according to the procedure given above. The solvent is stripped off under reduced pressure and is replaced with precooled ($-75\,^{\circ}$C) tetrahydrofuran (80 mL). After 15 min of vigorous stirring at $-75\,^{\circ}$C, one waits until a sediment has deposited and transfers the colorless supernatant liquid by means of a pipet. The organometallic solution, approximately 0.5 M, is stored at temperatures around or below $-50\,^{\circ}$C.

4.3.2 Permutational Exchange of a Metalloid Against Metal

Metalloid/metal permutation is almost synonymous with tin/lithium interconversion. Although all other metalloids can get involved in such reactions too, pertinent examples are rare. They include the generation of vinyllithium[1292] from tetravinyllead with phenyllithium, of 1,1-dilithiohexane[1292] from 1,1-hexylidenebis(dicyclohexylborane) with butyllithium, of tris(trimethylsilyl)silyllithium[1293] from tetrakis(trimethylsilyl)silane with methyllithium, of trifluorovinyllithium[1294] from phenyltris(trifluorovinyl)silane with phenyllithium, of pentafluorophenyllithium[1295] from dimethyldi(pentafluorophenyl)silane with butyllithium, of 1-chloroallyllithium[1296], *gem*-dichloroallyllithium[1297] and dichloro(triphenylplumbyl)methyllithium[1298] from (3-chloroallyl)triphenylplumbane, (3,3-dichloroallyl)triphenylplumbane and dichloromethylenebis(triphenylplumbane), respectively, with butyllithium, of 4-bromophenyllithium[1299] from tri(*p*-bromophenyl)stibane and of benzyllithium[1299] from tribenzylstibane with ethyllithium. A few more cases dealing with

organomercurials as precursors are known. Dimethylmercury reacts with ethyllithium or butyllithium to form methyllithium[1265,1300], bis(7-norbornadienyl)mercury to generate 7-norbornyllithium[1301], diphenylmercury with butyllithium or butylsodium to give phenyllithium[1302] or phenylsodium and bis(2,4,6-tri-*tert*-butylphenyl)mercury to afford 2,4,6-tri-*tert*-butylphenyllithium[1303]. 2,5-Di-*tert*-butyl-1,4-phenylenebis(mercurio chloride) is cleanly converted into the corresponding bismagnesium compound when its solution in (mono)ethylene glycol dimethyl ether (1,2-dimethoxyethane) is heated under reflux for 18 h in the presence of 4 equivalents of methylmagnesium chloride[1304]. Tetralithiomethane (**249**), previously identified as one of the products formed in the reaction between tetrachloromethane and lithium vapor[1305], can be prepared selectively by treating tetrakis-(chloromercurio)methane, obtained from tetrachloromethane through tetrakis(dimethoxyboryl)methane, with an excess of *tert*-butyllithium[1306].

249

a) Representative Examples of Tin−Lithium Permutations

The permutational displacement of a triorganostannyl group by a lithium atom is a widely applicable method. It has been used to make a great variety of organometallic intermediates carrying heterofunctional or other substituents (see Tables 69–76). Organopotassium and -cesium compounds are also accessible in this way (see Table 70).

Table 69. Trimethylsilylmethylcyclopropyl-, 1-alkenyl- and aryllithiums (M−R) by permutational exchange between an organotin compound R−SnR''₃ and an organoalkali reagent M−R'.

M−R	M−R'	SnR''₃	Sv[a]	T[°C]	Product, El−X'[b]	Reference
M−CH₂Si(CH₃)₃	LiC₄H₉	Sn(C₄H₉)₃	THF	0	88%, ⌇COCl	[1307]
M−◁	LiC₄H₉	SnR₃[c]	PET	+25	79%, (H₃C)₃SnBr	[1308]
M−◁	LiC₄H₉	Sn(C₄H₉)₃	THF	0	76%, (structure)OTs, O	[1309]
M−CH=CH₂	LiC₆H₅	SnR₃[c]	DEE	+25	74%, (H₃C)₂CO	[156,1310]
M−C=C−CH₃ (H H)	LiC₄H₉	Sn(CH₃)₃[d]	DEE	+25	37%, CO₂	[1310]
M−C=C−CH₃ (H, H)	LiC₄H₉	Sn(CH₃)₃[d]	DEE	+25	28%, CO₂/CH₂N₂	[1310]
M−C=C−C₅H₁₁ (H H)	LiC₄H₉	Sn(C₄H₉)₃	THF	−40	95%, cuprate reaction	[1311]
M−C=C−CH₂−C=C−C₅H₁₁ (H H H H)	LiC₄H₉	Sn(C₄H₉)₃	THF	−40	89%, cuprate reaction	[1311]
M−C=C−Sn(C₄H₉)₃ (H, H)	LiC₄H₉	Sn(C₄H₉)₃	THF	−75	58%, (H₃C)₃SiCl	[1312]
(M−C=C)₂−C(C₆H₁₃)(OCH₃) (H H)	LiC₄H₉	Sn(C₄H₉)₂	PET	0	55%, (H₃C)₃CAsCl₂	[1313]
M⌇Sn(CH₃)₃	LiCH₃	Sn(CH₃)₃	DEE	−75	[e], H₂O	[1314]
M−(anthracenyl)	LiC₄H₉	Sn(C₂H₅)₃	PET	0	[e], isolated	[965]

[a] PET = petroleum ether (pentanes, hexanes, heptanes), DEE = diethyl ether, THF = tetrahydrofuran. [b] El−X' = electrophilic trapping reagent. [c] Precursors: tetracyclopropyltin, tetravinyltin. [d] Or Sn(C₄H₉)₃. [e] Not specified.

Table 70. Allylic and benzylic organolithium, -potassium and -cesium compounds (M−R) by permutational exchange between organotin precursors R−SnR″₃ and organoalkali reagents M−R′ (at −75 °C or higher temperatures).

M−R	M−R′	SnR''_3	Sv[a]	Product, $El-X'$[b]	Reference
$M-\overset{\displaystyle}{\diagup}$ (allyl)	LiC_4H_9	$Sn(C_4H_9)_3$	THF	61%, $H_{13}C_6CHO$	[1315]
$M-CH_2-CH=CH_2$	LiC_6H_5	$Sn(C_6H_5)_3$	THF	52%, $H_{11}C_5I$	[1316,1317]
$M-CH_2-\overset{H}{C}=\overset{H}{C}-CH_3$	LiC_4H_9 $KCH_2Si(CH_3)_3$	$Sn(CH_3)_3$ $Sn(CH_3)_3$[d]	DEE THF	95%, $(H_3C)_3SiCl$[c] 97%, $(H_3C)_3SiCl$	[1316–1318] [1317]
$M-CH_2-\underset{CH_3}{C}-CH_2$	$LiCH_3$ $KCH_2Si(CH_3)_3$	$Sn(CH_3)_3$ $Sn(CH_3)_3$	THF THF	53%, H_3CCHO 99%, $(CH_2)_2O$	[1316,1317] [1317]
$M-CH_2-\underset{C_6H_5}{C}=CH_2$	$LiCH_3$ $KCH_2Si(CH_3)_3$	$Sn(CH_3)_3$ $Sn(CH_3)_3$[d]	THF THF	64%, $(H_3C)_3SiCl$ 94%, $(H_3C)_3SiCl$	[1317] [1317]
$M-CH_2-\underset{o-C_6H_4-C_6H_5}{C}=CH_2$	$LiCH_3$ $KCH_2Si(CH_3)_3$	$Sn(CH_3)_3$ $Sn(CH_3)_3$[d]	THF THF	76%, $(H_3C)_3SiCl$ 63%, $(H_3C)_3SiCl$	[1317] [1317]
$M-CH_2-\underset{Si(CH_3)_3}{C}=CH_2$	$CsCH_2Si(CH_3)_3$	$Sn(CH_3)_3$	THF	95%, [e]	[1317]
$(M-CH_2-\overset{H}{C}=\overset{H}{C}-CH_2)_2$	$KCH_2Si(CH_3)_3$	$Sn(CH_3)_3$	THF	69%, H_3CI	[1317]
$M-CH_2-\overset{H_2C\ CH_2}{C}\diagup C-CH_2-M$	$LiCH_3$ $KCH_2Si(CH_3)_3$	$Sn(CH_3)_3$ $Sn(CH_3)_3$	THF THF	77%, $(H_3C)_3SiCl$ 82%, $(H_3C)_3SiCl$	[1317] [1317]
$M-CH_2-\bigcirc$	$LiCH_3$ $KCH_2Si(CH_3)_3$	$Sn(CH_3)_3$ $Sn(CH_3)_3$	DEE THF	93%, H_3CI 97%, [e]	[1316,1319] [1318]
$M-CH(CH_3)-\bigcirc$	LiC_4H_9	$Sn(C_4H_9)_3$	THF	92%, $\underset{H_5C_2}{\overset{H_3C}{>}}CHCHO$	[1280,1320]
$M-CH_2-$ / $M-CH_2-$ (perylene)	$LiCH_3$	$Sn(CH_3)_3$	THF	90%, H_3COCH_2Cl	[1321]

[a] DEE = diethyl ether, THF = tetrahydrofuran. [b] $El-X'$ = electrophilic trapping reagent. [c] $(Z/E) \approx 2 : 3$. [d] Or $Sn(C_4H_9)_3$. [e] By NMR integration.

Table 71. Aminoalkyllithiums (M–R) by trialkyltin–metal permutation (El–X' = electrophile).

M–R	M–R'[a]	SnR₃''[b]	Sv[c]	T[°C]	Product, El–X'[d]	Reference
M–CH₂N(CH₃)₂	LiC₄H₉	Sn(C₄H₉)₃	PET	0	73%, H₅C₆CHO	[1322]
M–CH₂N(CH₃)CH₂C₆H₅	LiC₄H₉	Sn(C₄H₉)₃	THF	−65	87%, (H₃C)₃SiCl	[1323]
M–CH(C₂H₅)–N(C₆H₅)–C(O)–N(CH₃)(CH₃)	LiC₄H₉	Sn(C₄H₉)₃	THF	−75	91%, (H₃C)₂CHCHO	[1324]
M–CH(CH₃)–N(CH₂C₆H₅)–C(O)O (oxazolidinone)	LiC₄H₉	Sn(C₄H₉)₃	THF	−75	69%, H₅C₆CHO	[1324]
M–CH(H₃C)–N–C(O)O (H₃C, C₆H₅)	LiC₄H₉	Sn(C₄H₉)₃	THF	−75	82%, H₅C₆CHO	[1324]
M–CH(CH₃)–CH=C(C₂H₅)–N~O (morpholine)	LiC₄H₉	Sn(C₄H₉)₃	THF	−75	73%, H₇C₃Cl	[1325]
M–(cyclopentenyl)–N~O	LiC₄H₉	Sn(C₄H₉)₃	THF	−75	63%, (H₃C)₃SiCl	[1325]
M–(cyclohexenyl)–N~O	LiC₄H₉	Sn(C₄H₉)₃	THF	−75	69%, (H₃C)₃CHCl	[1325]
M–CH₂–N=CH–C₃H₇	LiC₄H₉	Sn(CH₃)₃	THF	−75	62%, *trans*-stilbene	[1326]
M–CH₂–N=CH–C(CH₃)₃	LiCH₃	Sn(CH₃)₃	THF	−75	91%, [e]	[1326]
M–CH₂–N=CH–C₆H₅	LiCH₃	Sn(CH₃)₃	THF	−75	83%, *trans*-stilbene	[1326]
M–CH(CH(CH₃)₂)–N=CH–CH=CH–CH₃	LiC₄H₉	Sn(C₄H₉)₃	THF	−75	81%, H₅C₆SCH=CH₂	[1327]

[a] M–R' = organometallic exchange reagent. [b] SnR₃'' = organotin entity displaced by lithium. [c] PET = petroleum ether, THF = tetrahydrofuran. [d] El–X' = electrophilic trapping reagent. [e] (1.) (H₃C)₃SiCH=CH₂, (2.) ClCOOCH₃.

b) Scope and Limitations of Tin–Alkali Metal Permutations

Stannanes are attractive precursors to organolithium intermediates because of their accessibility by various complementary routes[306, 1333]. The standard method of preparation remains the reaction of a Grignard reagent with a trialkyltin chloride. Alternatively, a polar organometallic intermediate may be condensed with a

Table 72. Imino-substituted organolithiums (M−R) by tributyltin/metal permutation using butyllithium as the organometallic reagent M−R′.

M−R	M−R′	SnR″$_3$ [a]	Sv [b]	T[°C]	Product, El−X′ [c]	Reference
M−CH$_2$CH$_2$−C=NC$_6$H$_5$ | OM	LiC$_4$H$_9$	Sn(C$_4$H$_9$)$_3$	THF	−75	68%, H$_5$C$_6$COCH$_3$	[1328]
M−◁ C=NC$_6$H$_5$ | OM	LiC$_4$H$_9$	Sn(C$_4$H$_9$)$_3$	THF [d]	−75	43% [e], H$_5$C$_6$CHO	[1329]
SC$_6$H$_5$ | M−C=CH−C=N−M | OM	LiC$_4$H$_9$	Sn(C$_4$H$_9$)$_3$	THF	−10	63%, (H$_5$C$_6$)$_2$CO	[1330]
CH$_2$ // M−C \ CH$_2$−N−Si(CH$_3$)$_3$ | M	LiC$_4$H$_9$	Sn(C$_4$H$_9$)$_3$	THF	+25	65%, H$_5$C$_6$CH=NC$_6$H$_5$	[906]

[a] SnR″$_3$ = Organotin entity displaced by lithium. [b] THF = tetrahydrofuran; in the case of mixtures (containing for example paraffinic cosolvents) only the most polar component is indicated. [c] El−X′ = electrophilic trapping reagent. [d] Supposed, not explicitly stated. [e] At −75 °C exclusively the *trans* isomer, at 0 °C predominantly the *cis* isomer.

trialkylstannylmethyl halide. Finally, an alkyl halide or sulfonate may be combined with trimethyl-, tributyl- or triphenylstannyllithium (R″SnLi).

$$R''_3Sn-CH_2Br + Li-R$$
$$R''_3Sn-Cl + BrMg-CH_2R \longrightarrow R''_3Sn-CH_2R$$
$$R''_3Sn-Li + Br-CH_2R$$

A similar choice between old-established and more recent methods exists also for the practically important α-alkoxystannanes (R″ = methyl, butyl, phenyl). They can be obtained by treatment of an α-lithiated ether with a stannyl chloride or by reaction of a trialkylstannyllithium with either an α-chloro ether or by its addition to a carbonyl compound followed by alkylation of the resulting alcoxide.

$$R''_3Sn-Li + \begin{array}{l} 1.\ O=CH-R \\ 2.\ Br-R' \end{array}$$
$$R''_3Sn-Cl + Li-CH(OR')R \longrightarrow R''_3Sn-CH-R$$
$$ \underset{OR'}{|}$$
$$R''_3Sn-Li + Br-CH(OR')R$$

The metalloid/metal permutation occurs rapidly and under mildest conditions provided that it is accompanied by a sufficient decrease in basicity. It can be accomplished with phenyllithium, methyllithium or butyllithium but also with trimethylsilylmethyl-potassium[1318]. The method is well suited for the generation of sensitive intermediates such as 2-pyrimidyllithium[1367].

Table 73. α-Alkoxy(cyclo)alkyllithiums (M−R) by tin/metal permutation using butyllithium (M−R′).

M−R	M−R′	SnR₃″[a]	Sν[b]	T[°C]	Product, El−X′ [c]	Reference
M−CH₂OCH₃	LiC₄H₉	Sn(C₄H₉)₃[d]	DEE	−75	82%, H₅C₆CHO	[1331,1332]
M−CH₂OCH₂C₆H₅	LiC₄H₉	Sn(C₄H₉)₃	THF	−75	98%, (cyclohexyl-O structure)	[1333–1335]
M−CH₂OC(CH₃)₃	LiC₄H₉	Sn(C₄H₉)₃[d]	THF	−75	95%, H₅C₆CHO	[1331]
M−CH₂OC₆H₅	LiC₄H₉	Sn(C₄H₉)₃[d]	THF	−75	78%, H₅C₆CHO	[1331]
M−CH₂OCH₂CH₂OCH₃	LiC₄H₉	Sn(C₄H₉)₃[d]	THF	−75	82%, H₅C₆CHO	[1331]
M−CH₂OCH₂OCH₃	LiC₄H₉	Sn(C₄H₉)₃	THF	−70	93%, (cyclohexyl-O structure)	[1336–1338]
M−CH₂OCH(CH₃)OC₂H₅	LiC₄H₉	Sn(C₄H₉)₃	THF	−75	94%, (cyclohexyl-O structure)	[1333]
M−CHOCH₂OCH₂C₆H₅ | CH(CH₃)₂	LiC₄H₉	Sn(C₄H₉)₃	MEGE	−75	78%, (H₃C)₂NCHO	[1339]
M−CHOCH(CH₃)OC₂H₅ | CH₃	LiC₄H₉	Sn(C₄H₉)₃	THF	−75	83%, O=(cyclopentene structure)OCH₃	[1340]
M−CHOCH(CH₃)OC₂H₅ | C₆H₁₃	LiC₄H₉	Sn(C₄H₉)₃	THF	−75	82%, O=(cyclopentene structure)OCH₃	[1340]
M−CHOCH(CH₃)OC₂H₅ | C₆H₅	LiC₄H₉	Sn(C₄H₉)₃	THF	−75	80%, O=(cyclopentene structure)OCH₃	[1340]
M−CH−OCH(CH₃)OC₂H₅ (furyl structure)	LiC₄H₉	Sn(C₄H₉)₃	THF	−75	69%, geranyl-Cl	[1340]
M−(cyclobutyl)OCH₂OCH₃	LiC₄H₉	Sn(C₄H₉)₃	THF	−75	68%, (spiro structure)[e]	[1341,1342]
M−(cyclohexyl)OCH₂OCH₃	LiC₄H₉	Sn(C₄H₉)₃	THF	−75	85%, H₅C₂CHO	[1343]
M−CH(OC₆H₅)₂[f]	LiC₄H₉	Sn(C₄H₉)₃	THF	−80	65%, H₅C₆CHO	[1344]
M−(dioxolane with CH₃, CH₃)	LiC₄H₉	Sn(C₄H₉)₃	THF	−75	70%, H₅C₆CHO	[1345]

[a] SnR₃″ = Organotin entity replaced by lithium. [b] DEE = diethyl ether, THF = tetrahydrofuran. [c] El−X′ = electrophilic trapping agent. [d] In some of the work[1332] the organotin starting material was prepared *in situ* by treatment of the corresponding (alkoxymethyl)dibromochlorostannane with excess butyllithium. [e] Isolated after acid treatment as 2-spiro[3.5]nonanone. [f] Analogously, dimethoxymethyllithium.

Table 74. α-Lithiated cyclic ethers (M−R) by tin/metal permutation using organolithium reagents (M−R′).

M−R		M−R′	SnR₃″[a]	Sν[b]	T[°C]	Product, El−X′ [c]	Reference
M−CH₂O− (sugar structure)		LiC₆H₅	Sn(C₆H₅)₃	DEE	[d]	40%, (H₃C)₃SnCl	[1346]
M− (tetrahydrofuran ring)		LiC₄H₉	Sn(C₄H₉)₃	THF	−75	71%, H₅C₆CHO	[1343]
M− (tetrahydropyran ring)		LiC₄H₉	Sn(C₄H₉)₃	THF	−75	76%, H₅C₆CHO	[1343]
M− (pyran, OQ, OQ′)	[e,f]	LiC₄H₉	Sn(C₄H₉)₃	THF [g]	−75	71%, epoxide[i]	[1347]
M− (pyran, OR, OR, OR)	[i,j]	LiC₄H₉	Sn(C₄H₉)₃	THF	−75	76%, 80%, H₁₁C₅CHO	[1348]
M− (pyran with N, OM, OR, OR, OR)	[j,k]	LiC₄H₉	Sn(C₄H₉)₃	THF	−75	83%, CO₂	[1349]
M− (glycal, OR, OR, OR)	[j]	LiC₄H₉	Sn(C₄H₉)₃	THF	−75	70%, H₃CI	[1350]

[a] SnR₃″ = Organotin entity replaced by lithium. [b] THF = tetrahydrofuran. [c] El−X′ = electrophilic trapping reagent. [d] Not specified. [e] α epimer. [f] OQ = OSi(CH₃)₂C(CH₃)₃; OQ′ = OSi(C₆H₅)₂C(CH₃)₃. [g] Solvent not explicitly specified. [h] 4-Benzyloxy-1,2-epoxypentane. [i] Both α and β epimers, depending on the stannane configuration. [j] OR = OCH₂C₆H₅. [k] β epimer.

The high propensity of stannanes for ate complex formation can explain the ease with which tin/lithium permutations are brought about. 2-Bromo-5-(trimethylstannyl)pyridine exchanges cleanly first the metalloid group and only subsequently, after electrophilic interception of the first organolithium intermediate **250**, the halogen[1368].

250

Table 75. Oxygen-functional 1-alkenyllithiums (M—R) by tin/metal permutation.

M—R	M—R'[a]	SnR''$_3$[b]	Sv[c]	T[°C]	Product, El—X'[d]	Reference
M—(alkene)—OCH$_2$OCH$_3$	LiC$_4$H$_9$	Sn(C$_4$H$_9$)$_3$	THF	−75	98%, (H$_3$C)$_2$CHCHO	[1351,1352]
M—(alkene)—OCH$_2$OCH$_2$C$_6$H$_5$	LiC$_4$H$_9$	Sn(C$_4$H$_9$)$_3$	THF	−75	95%, (H$_3$C)$_2$CHCHO	[1351]
M—(alkene, H$_3$C)—OCH$_2$OCH$_3$	LiC$_4$H$_9$	Sn(C$_4$H$_9$)$_3$	THF	−75	85%, (H$_3$C)$_2$CHCHO	[1351]
M—(alkene, C$_4$H$_9$)—OCH$_2$OCH$_3$	LiC$_4$H$_9$	Sn(C$_4$H$_9$)$_3$	THF	−75	86%, D$_2$O	[1353]
M—(alkene, CH$_2$OCH$_2$C$_6$H$_5$)—OCH$_2$OCH$_3$	LiC$_4$H$_9$	Sn(C$_4$H$_9$)$_3$	THF	−75	98%, (H$_3$C)$_2$CHCHO	[1351]
M—CH=CH—OC$_2$H$_5$	LiC$_4$H$_9$	Sn(C$_4$H$_9$)$_3$	THF	−75	76%, H$_5$C$_6$CHO [e]	[1354,1355]
M—CH=CH—OCH$_2$OCH$_3$	LiC$_4$H$_9$	Sn(C$_4$H$_9$)$_3$	THF	−75	75%, (H$_3$C)$_2$CHCHO	[1352]
M—CH=CH—CH$_2$OR [f]	LiC$_4$H$_9$	Sn(C$_4$H$_9$)$_3$	THF	−75	85%, Br(CH$_2$)$_8$Br	[1356]
M—C(R''$_3$Sn)=CH—CH$_2$OR [g]	LiC$_4$H$_9$	Sn(C$_4$H$_9$)$_3$	THF	−75	89%, H$_5$C$_6$CHO [h]	[1357]
M—(cyclopentene with O-CH$_2$-O ring, OSi(CH$_3$)$_2$C(CH$_3$)$_3$)	LiC$_4$H$_9$	Sn(C$_4$H$_9$)$_3$	THF	−45	94%, H$_5$C$_6$OCH$_2$Cl	[1358]

[a] The exchange reagent M—R' is always butyllithium. [b] SnR''$_3$ = organotin entity replaced by lithium. [c] THF = tetrahydrofuran; in the case of mixtures (containing, *e.g.* paraffinic cosolvents) only the most polar component is indicated. [d] El—X' = electrophilic trapping reagent. [e] Cinnamaldehyde upon acid hydrolysis. [f] OR = *O*-2-tetrahydropyranyl. [g] OR = OCH$_2$OCH$_3$. [h] (*Z/E*) ≈ 1 : 4.

Two nearby stannyl groups cannot be displaced simultaneously. 3,4-Bis(tributylstannyl)-furan[1369] and 2,5-bis(trimethylstannyl)thiophene[1370] afford only the monolithio species even when treated with an excess of butyllithium. However, bis[(tributylstannyl)methyl]-sulfide generates the α,α'-dilithiodimethyl sulfide[1371] because the metal-bearing centers are separated by the electronegative sulfur atom.

Table 76. Halogenated organolithiums (M−R) by tin/metal permutation.

M−R	M−R'[a]	SnR''$_3$[b]	Sv[c]	T[°C]	Product, El−X'[d]	Reference
M−CH(C$_2$H$_5$)−I	LiC(CH$_3$)$_3$	Sn(C$_4$H$_9$)$_3$	DEE	−100	25%[e]	[1359]
M−(cyclohexenyl)−Br	LiC$_4$H$_9$	Sn(CH$_3$)$_3$	THF	−100	60%, CO$_2$	[1360]
M−CH$_2$−C(=CH$_2$)−CH$_2$Cl	LiCH$_3$	Sn(CH$_3$)$_3$	THF	−75	69%, (oxepanone)	[1361]
M−CF=CH$_2$	Pd(PR$_3$)$_4$[f]	Sn(C$_4$H$_9$)$_3$	THF	+65	66%, H$_3$CCO−C$_6$H$_4$−I	[1362]
M−CF=CF$_2$	LiC$_6$H$_5$	SnRR'$_2$[g]	DEE	−75	37%, CO$_2$	[1363]
M−CF$_2$−CH=CH$_2$	LiC$_4$H$_9$	Sn(CH$_3$)$_3$	THF	−95	52%, (H$_5$C$_6$)$_2$CO[h]	[1364]
M−CH$_2$−C$_6$H$_4$−F	LiC$_4$H$_9$	Sn(C$_4$H$_9$)$_3$	THF	−75	94%, CO$_2$	[1365]
M−CH$_2$−C$_6$H$_4$−Br	LiC$_4$H$_9$	Sn(CH$_3$)$_3$	THF	−70	92%, (H$_3$C)$_3$GeCl	[1366]

[a] M−R' = exchange reagent. [b] SnR''$_3$ = organotin entity displaced by the metal M. [c] THF = tetrahydrofuran; in case of mixtures (containing for example paraffinic cosolvents) only the most polar component is indicated. [d] El−X' = electrophilic trapping reagent. [e] (1.) CuBr, (2.) H$_5$C$_6$COCl. [f] R = C$_6$H$_5$. [g] R = C$_6$H$_5$, R' = CF=CF$_2$. [h] Small portions of butyllithium and benzophenone are added rapidly and alternately.

Both (methoxymethyl)tributylstannane[1332,1333] and (hydroxymethyl)tributylstannane[1372] undergo smooth permutational tin/lithium exchange with butyllithium. However, when the heteroatom is moved to the γ position with respect to the metalloid-bearing center, the differences in the stabilities of the equilibrating organolithiums become too small to sustain the exchange process any longer. The vanishing basicity gradient may be compensated by mixed aggregate formation. Whereas tributyl[3-(N-pyrrolidinyl)propyl]stannane does not react with butyllithium, tributyl[3-(N-methoxycarbonyl)aminopropyl]stannane (251) readily undergoes the tin/lithium permutation[1373]. Tributyl-3-(hydroxypropyl)stannane (252) forms the organolithium intermediate under the same conditions, whereas tributyl-3-(methoxypropyl)stannane proves to be totally inert[1373]. In the same way, butyllithium converts tributyl[cis-2-(hydroxymethyl)cyclopropyl]stannane smoothly into *cis*-2-(lithiooxymethyl)cyclopropyllithium[1311]. The crucial role of coordination at the exchange transition state is emphasized by a comparison between the *syn* and *anti* isomers of 5-tributylsilyl-2-bicyclo[2.2.1]hexanol (**253**). Only the former is capable of undergoing the permutational displacement[1374].

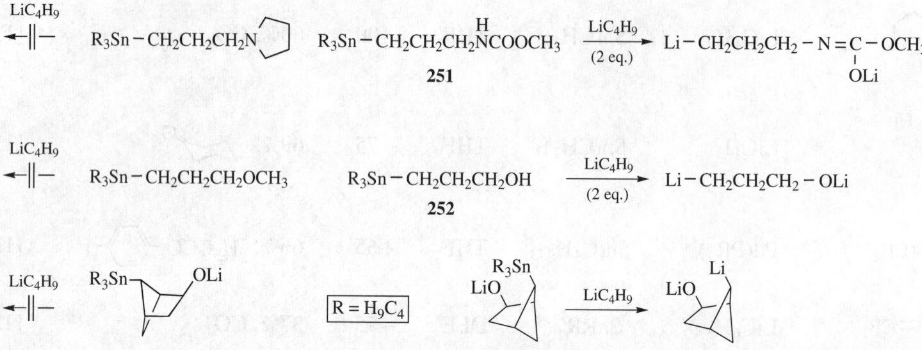

251 **252** ***anti*-253** ***syn*-253**

c) Working Procedures

The protocols selected testify to the great versatility of the tin/metal permutation method. They comprise the preparation of an α-amino-substituted alkyllithium, a 1-lithiated glucose derivative and three allylic potassium or cesium compounds.

N,N-Dimethylaminomethyllithium[1322]

$$(H_9C_4)_3Sn-CH_2N(CH_3)_2 \xrightarrow{LiC_4H_9} Li-CH_2N(CH_3)_2 \xrightarrow{H_5C_6CHO} HO-CH(C_6H_5)-CH_2N(CH_3)_2$$

At 0 °C, butyllithium (30 mmol) in hexanes (20 mL) is added to a solution of tributyl(*N,N*-dimethylaminomethyl)stannane (10.5 g, 30 mmol) in hexanes (10 mL). The conversion is instantaneous as evidenced by the quantitative formation of tetrabutylstannane (identified by gas chromatography) and a colorless precipitate an aliquot of which, when trapped with benzaldehyde, affords 2-dimethylamino-1-phenylethanol.

The starting material is prepared by condensation of *N,N*-dimethylmethyleneammonium chloride with tributylstannyllithium. The latter reagent is obtained by cleavage of hexabutyldisilane, performed at 0 °C with lithium wire containing 1% of sodium[1322,1375].

1-(2-Acetamido-3,4,6-tri-O-benzyl-2-deoxy-β-D-glucopyranosyl)lithium[1349]

At −75 °C, butyllithium (22 mmol) in hexanes (15 mL) is added to a solution of 1-(2-acetamido-3,4,6-tri-*O*-benzyl-2-deoxy-β-D-glucopyranosyl)tributylstannane[1349] (8.0 g, 10 mmol) in tetrahydrofuran (0.10 L). After 5 min at −65 °C, the mixture is poured on an excess of freshly crushed dry ice. The resulting 3-acetamido-2,6-anhydro-4,5,7-tri-*O*-benzyl-3-deoxy-β-D-*gulo*-heptonic acid is isolated by alkaline extraction and neutralization as a pale yellow solid [m.p. 230 °C, under decomposition; crystallized from ethyl acetate and acetic acid; yield 4.3 g (83%)].

The required stannane[1349] is obtained by treatment of a 0.1 M solution of 1-(2-acetamido-3,4,6-tri-*O*-benzyl-2-deoxy-β-D-glucopyranoxyl chloride[1349] in tetrahydrofuran for 90 min at −75 °C with an equivalent amount of tributylstannyllithium. The latter reagent is prepared by stirring (2 × 5 h) and ultrasound irradiating (2 × 30 min) a mixture of lithium pieces (1.7 g, 0.25 mol) and tributylstannyl chloride (27 mL, 33 g, 0.10 mol) in tetrahydrofuran (50 mL) and eventually decanting the dark green solution from excess metal[1376].

(Z)-2-Butenylpotassium[1318,1377]

At −75 °C, (Z)-2-butenyltrimethylstannane (2.2 g, 10 mmol) is added to a 0.10 M solution of trimethylsilylmethylpotassium (10 mmol) in tetrahydrofuran (see pp. 158–159). After 1 h, the temperature is raised and kept for 3 h at −50 °C before the mixture is treated with chlorotrimethylsilane (1.3 mL, 1.1 g, 10 mmol). Using authentic samples for comparison, trimethyl(1-methyl-2-propenyl)silane and (Z)-2-butenylsilane are identified by gas chromatography in a ratio of 25 : 75 (combined yield 97%).

(Z)-2-Butenyltrimethylstannane can be easily made by reaction of *cis*-butene (0.15 mol) with butyllithium and potassium *tert*-butoxide in tetrahydrofuran (see p. 268) and by pouring the resulting mixture portionwise into a solution of chlorotrimethylstannane (1.0 equiv.) in tetrahydrofuran. The product is isolated and purified by immediate distillation [b.p. 72–73 °C/46 mmHg; n_D^{20} 1.4853; yield 19 g (60%)].

(Z)-2-Phenylallylpotassium[1318,1377]

At −75 °C, trimethyl(2-phenyl-2-propenyl)stannane (2.8 g, 10 mmol) is introduced in a solution of trimethylsilylmethylpotassium[150,474,1276] (see pp. 158–159; 10 mmol) in tetrahydrofuran (20 mL) to which some diethyl ether (5 mL) and pentanes (5 mL) have been added. After 3 h at −100 °C and 1 h at −75 °C, the mixture is treated with chlorotrimethylsilane (1.3 mL, 1.1 g, 10 mmol). According to gas chromatographic analysis, 94% of trimethyl(2-phenyl-2-propenyl)silane are formed [b.p. 95–96 °C/ 10 mmHg; n_D^{20} 1.5079].

The stannane precursor is prepared by adding dropwise, in the course of 15 min, butyllithium (40 mmol) in neat tetrahydrofuran (25 mL; added precooled after the commercial paraffinic solvent has been stripped off) to a solution of α-methylstyrene (5.2 mL, 4.7 g, 40 mmol) and potassium *tert*-butoxide (4.5 g, 40 mmol) in tetrahydrofuran (125 mL) kept at −75 °C. After 1 h at −50 °C, the mixture is cooled again to −75 °C before being poured portionwise into a solution of chlorotrimethylstannane (8.0 g, 40 mmol) in tetrahydrofuran (50 mL). The product is isolated and purified by immediate distillation [b.p. 68–69 °C/0.1 mmHg; n_D^{20} 1.5542; yield 3.5 g (31%)].

2-(Trimethylsilyl)allylcesium[1318,1377]

By means of a syringe, trimethyl[1-(trimethylstannylmethyl)ethenyl]silane (0.14 g, 0.50 mmol) are introduced in the solution of trimethylsilylmethylcesium[150] in [²H₈]tetrahydrofuran (2.5 mL) kept at −100 °C. After 3 h at −100 °C and 1 h at −75 °C, the mixture is cooled again to −100 °C. Using a nitrogen-rinsed pipet, approximately half of the solution is transferred into a 5 mm wide NMR tube. The spectrum reveals the total absence of starting material and shows two characteristic singlets in a ratio of 4 : 9 (δ 2.52; −0.02). Neutralization of the solution with phenol produces the NMR pattern of trimethyl(1-methylethenyl)silane (δ 5.57, 5.26, 1.85, 0.11; relative intensities 1 : 1 : 3 : 9).

The silylated stannane is readily prepared by consecutive treatment of trimethyl-(1-methylethenyl)silane[1378] (2-trimethylsilylpropene; 2.3 g, 20 mmol) with butyllithium (20 mmol) and potassium *tert*-butoxide (2.2 g, 20 mmol) in neat tetrahydrofuran (40 mL) for 5 h at $-50\,°C$ and with chlorotrimethylstannane (4.0 g, 40 mmol) at $-75\,°C$. After evaporation of the mixture, the product is collected by immediate distillation [b.p. $33-34\,°C/1$ mmHg; n_D^{20} 1.4816; yield 2.5 g (45%)].

4.4 Displacement of Carbon

Hydrocarbon substituents are notoriously poor leaving groups. Stringent prerequisites have to be fulfilled in order to displace them in terms of a reductive insertion of metal or a permutational exchange with an organometallic reagent. The critical carbon–carbon bond must be either weakened by excessive internal strain or the emerging organometallic compound must be extraordinarily resonance stabilized.

$$\begin{array}{c} >C-C< \xrightarrow[\text{insertion}]{2\ M\cdot} >C-M\ +\ M-C< \\[2em] >C-C< \xrightarrow[\text{permutation}]{M-R} >C-R\ +\ M-C< \end{array}$$

4.4.1 Reductive Insertion of Metals in Carbon-Carbon Bonds

Carbon–carbon cleavage reactions are of practical use only if they do not produce mixtures of organometallic intermediates. Therefore, the hydrocarbon precursor should be either composed of two identical halves or be a cyclic entity which, upon reductive ring opening, can generate a well-defined dimetal species.

a) Representative Examples of Reductive Carbon-Carbon Scission

The electropositivity of the metal represents a third driving force besides the resonance stabilization of cleavage products and the destabilization of the hydrocarbon precursor by internal strain. The metal plays a particularly crucial role when resonance-stabilized products are formed. As already mentioned (see pp. 53–54), the triphenylmethyl dimer ('pseudo-hexaphenylethane') is readily cleaved by all alkali metals[403], 1,1,2,2-tetra-phenylethane by all of them except lithium[404] (unless in form of its 1 : 1 adduct with biphenyl[1379]) and finally 1,2-diphenylethane just by cesium[405] and possibly potassium mirrors[406].

The reductive scission of 1,2-diaryl-, 1,1,2,2-tetraaryl- or 1,1,1,2,2,2-hexa-arylethanes offers a very clean and convenient entry to benzylic organometallics, provided that the hydrocarbon precursor is readily available (see Table 77). For

Table 77. Reductive CC-cleavage of allylic, propargylic and benzylic hydrocarbons affording organoalkali compounds (M–R).

M–R	M	Sv[a]	T[°C]	Product, El–X' [b]	Reference
M–CH₂–CH=C(C₆H₅)₂	K	DOX	+25	[c], CO₂	[1380]
M–C(C≡C–C(CH₃)₃)₃	K	DEE	+25	70%[d], CO₂	[1381]
M–CH₂–C₆H₅	Cs	THF	−75	93%, CO₂	[405]
M–CH₂–C₆H₄–C₆H₅	K	DOX	+25	[c], HOC₂H₅	[1380]
M–CH(C₆H₅)₂	Li	THF	+25	93%, CO₂	[1379]
M–C(OC₆H₅)(C₆H₅)₂	Li	THF	+25	75%, H₂O	[1184,1382]
	Na	THF	+25	71%, H₂O	[1184,1382]
	K	THF	+25	58%, H₂O	[1184,1382]
M–(9-phenyl)fluorenyl	Na	DEE	+25	[c], isolated	[403]

[a] DOX = 1,4-dioxane, DEE = diethyl ether, THF = tetrahydrofuran; in the case of mixtures only the most polar component is indicated. [b] El–X' = trapping electrophile. [c] Yield not specified. [d] Crude.

example, 1,2-diphenoxy-1,1,2,2-tetraphenylethane is quantitatively formed by thermal rearrangement of bis(triphenylmethyl) peroxide, the autoxidation product of the triphenylmethyl radical.

b) Scope and Limitations

Strain as a driving force can manifest itself in many ways. Tricyclo[1.1.1.0^{1,3}]pentane, the smallest propellane, reacts with 4,4′-di-*tert*-butylbiphenyl/lithium in refluxing dimethyl ether (DME) to afford 1,3-dilithiobicyclo[1.1.1]pentane (**254**), which can be trapped as the diacid [1383]. Similarly, 1-phenyltricyclo[1.1.0]butane undergoes reductive ring opening to give 1,3-dilithio-1-phenylcyclobutane (**255**), this reaction occurring in tetrahydrofuran at −75 °C with elemental lithium[1384].

'Bent bonds are better'. L. Pauling's credo[1385] can serve as an intuitive starting point for the consistent rationalization of double bond behavior and reactivity. If the addition of two metal atoms can open a cyclopropane ring, the same should prove possible with the 'two-membered rings' of 'cycloethanes' (olefins) and 'bicyclo[0.0.0]ethanes' (acetylenes).

Ethylene does indeed readily combine with cesium to form a 1 : 2 adduct **256** which upon hydrolysis quantitatively sets free ethane[1386]. The analogous 1,2-dilithioethane (**257**) appears only as a transient species when a solution of ethylene in glycol dimethyl ether (MEGME, 'glyme') is stirred with a lithium dispersion in the presence of a carrier cocktail of biphenyl and naphthalene[1387]. Immediate β-elimination of lithium hydride affords a considerable amount of vinyllithium[1387] along with some butyllithium, 3-butenyllithium, 1,4-tetramethylenedilithium[1388], 1,6-hexamethylenedilithium[1387] and even acetylenedilithium (lithium carbide)[1389].

$$H_2C=CH_2 \xrightarrow{2\,Cs} Cs-CH_2-CH_2-Cs \xrightarrow{H_2O} H_3C-CH_3 + 2\,CsOH$$

$$\textbf{256}$$

$$H_2C=CH_2 \xrightarrow{2\,Li} Li-CH_2-CH_2-Li \xrightarrow[-[LiH]]{} H_2C=CH-Li$$

$$\downarrow H_2C=CH_2 \qquad \textbf{257} \qquad\qquad \downarrow$$

$$Li-(CH_2-CH_2)_2-Li \qquad LiCH_2CHLi_2 \longrightarrow LiCH=CH-Li \longrightarrow HC\equiv CLi$$

$$\downarrow H_2C=CH_2 \qquad\qquad\qquad\qquad\qquad\qquad\qquad \downarrow$$

$$Li-(CH_2-CH_2)_3-Li \qquad\qquad\qquad\qquad\qquad LiC\equiv CLi$$

2,5-Diphenyl-1,6aλ^4-trithiapentalene (**258**) in the presence of Lewis acids such as zinc(II) or iron(III) chloride acts as an efficient catalyst for the addition of lithium to 1-alkenes and the subsequent elimination of lithium hydride. In this way, not only ethylene can be converted into vinyllithium, but also propene into (*E*)-2-propenyllithium and 1-octene into (*E*)-2-octenyllithium with reasonable purities and yields (70–89%)[1390].

$$R-CH=CH_2 \xrightarrow[\{H_5C_6-\langle S\,S\,S\rangle-C_6H_5\}]{Li/THF\ \{ZnCl_2\}} \left(R-\underset{Li}{CH}-\underset{Li}{CH_2} \right) \longrightarrow R-\overset{H}{C}=\overset{}{\underset{H}{C}}-Li$$

$$\textbf{258}$$

Cyclopropylmethyllithium (see p. 50) and cyclobutylmethyllithium being prone to rapid ring-opening isomerization, metal addition to alkenic double bonds in the vicinity of small rings causes skeletal reorganizations. Methylenecyclopropane affords 2,4-dilithiobutene (**259**)[1391,1392], methylenespiro[2.2]pentane 2-lithio-3,3,4-trimethyl-1,4-pentadiene[1391] (**260**, after lithium hydride elimination), methylene(tetramethylcyclopropane) 2-lithio-3-(2-lithioethyl)-1,3-butadiene[1393] (**261**), methylenecyclobutane 2,5-dilithio-1-pentene (**262**)[1394] and tricyclo[6.4.0.0 2,7]dodeca-2,12-diene 2,2′-dilithiobi(2-cyclohexenyl) (**263**)[1395].

Homoconjugated bisallylic dimetal compounds result from the reductive addition of metal to *vic*-di-1-alkenylcyclopropanes. The C−C bond scission thus accomplished by treatment of semibullvalene (**264**)[1393,1394], barbaralane (**265**)[1395] and bullvalene (**266**)[1396] by treatment with lithium or potassium is reversible. Oxidation of the dimetal intermediate with air or iodine restores the original hydrocarbon structure.

Aliphatic alkynes react sluggishly with lithium powder. It takes 48 h to convert 3-octene, 4-octene and 5-octene into the corresponding (*E*)-*vic*-dilithioalkenes in 55–75% yield[1397,1398]. Cyclodecyne and cyclododecyne give the respective 1,2-dilithiocycloalkene in 95 : 5 and 73 : 27 (*Z/E*) ratios[1399]. Tolane (diphenylacetylene) affords (*Z*)-1,2-dilithiostilbene (**267**)[1400] or (*E,E*)-1,4-dilithio-1,2,3,4-tetraphenyl-1,3-butadiene (**268**)[1401,1402], depending on the reaction conditions.

Elemental lithium can also be added to allenes such as 1,2-cyclononadiene[1403] or tetraphenyl-1,2-propadiene[1404]. Conjugated double bonds capture electrons with particular ease. Stilbene[1405–1407] and 1,3-diphenyl-1,3-butadiene[1406,1408] form 1 : 2 adducts with all alkali metals. Butadiene[1409], isoprene[1410], 2,3-dimethyl-1,3-butadiene[1411], myrcene[1412,1413] and 1,2-dimethylenecyclohexane[1414] combine readily with activated magnesium producing 1-magnesia-3-cyclopentenes (such as **269**).

Anthracene reacts with magnesium to give a 1 : 1 adduct (see pp. 176–177) which, when crystallized from tetrahydrofuran, contains three solvent molecules[371,1415]. Benzene and simple congeners thereof are known to form 'radical anions' with alkali metals in very small equilibrium concentrations (see p. 52), but there is little evidence for the existence of 'dianions'. The reduction of benzene to 1,3-cyclohexadiene under Birch–Wooster–Hückel conditions or its reaction with lithium in the presence of chlorotrimethylsilane, leading to 1,4-bis(trimethylsilyl)-1,4-cyclohexadiene[1416], do not necessarily involve 'dianions' but can be explained on the basis of 'radical anion', radical and cyclohexadienylmetal intermediates. The same is true for the reductive opening of the four-membered ring in dihydrobenzocyclobutene[1417] and biphenylene[1418,1419] with lithium or potassium.

d) Working Procedures

Only a few applications of nonradical alkene–metal, alkyne–metal and arene–metal adducts to organic synthesis are known. 1,2,3,4-Tetraphenyl-1,3-butadiene-1,4-diyldilithium has been frequently employed for the preparation of metallacyclopentadienes and -spirononadienes[1409–1414,1420,1421]. The oligomeric or polymeric magnesium 1 : 1 adducts of butadiene and isoprene, and the monomeric dicyclopentadienylzirconium analogs[1422], can be used for the regio- and stereoselective synthesis of unsaturated hydrocarbons and functionalized derivatives thereof. Anthracene/magnesium can act as a metal carrier and, in addition, as a hydrogen storage device[1415,1423].

1,4-Dilithio-1,2,3,4-tetraphenyl-1,3-butadiene[1405,1420]

$$H_5C_6-C\equiv C-C_6H_5 \xrightarrow{\text{Li}} \quad \text{(diene)} \quad \xrightarrow{\text{H}_2\text{O}} \quad \text{(diene)}$$

Lithium shavings (1.0 g, 0.14 mol) are added to a solution of diphenylacetylene (8.9 g, 50 mmol) in diethyl ether (20 mL) and the mixture is stirred at 0 °C for 2 h. As indicated by the appearance of an intense red color, the reaction starts after 5–15 min. A yellow-orange suspension forms rapidly which eventually can be decanted from unconsumed lithium. The lithium is washed with more diethyl ether (2 × 15 mL). Neutralization of an aliquot affords (*E,E*)-1,2,3,4-tetraphenyl-1,3-butadiene [m.p. 182–183 °C; yield 70–85%]. It may be accompanied by small amounts of 1,2,3-triphenylnaphthalene (m.p. 151–152 °C) as a by-product.

1-Magnesia-3-cyclopentene polymer[1409]

$$\text{(butadiene)} \xrightarrow[\text{THF}]{\text{Mg}} \left(\text{(ring)}_{\text{Mg}} \right)_n \xrightarrow{\text{HN}} \text{(butene)}$$

Butadiene (5.4 g, 0.10 mol) is distilled into a suspension of magnesium turnings (4.8 g, 0.20 mol) in tetrahydrofuran (50 mL) kept at −75 °C. After addition of a small amount of iodobenzene (approx. 0.2 mL, 0.1 g, 0.5 mmol) as an initiator, the reaction vessel, a Schlenk flask, is sealed and the mixture stirred for 15 h at 40 °C. The tube is opened and the suspension formed decanted from the residual metal, which is washed with more tetrahydrofuran (2 × 25 mL). The suspension is filtered through a glass frit and the solid dried under reduced pressure (0.1 mmHg) at 70 °C for 3 h to afford a colorless product having the composition $C_4H_6Mg \cdot (THF)_2$ in about 50% yield. Aliquots are treated with excess piperidine and *tert*-butyl alcohol to give 2-butene and 1-butene, respectively.

9,10-Dihydro-9,10-anthracenediylmagnesium[1415]

$$\text{(anthracene)} \xrightarrow[\text{THF}]{\text{Mg}} \text{(Mg adduct)} \xrightarrow{\text{H}_2\text{O}} \text{(dihydroanthracene)}$$

Anthracene (0.20 kg, 1.1 mol) and magnesium powder (50 mesh particle size; 24 g, 1.0 mol) are placed in a three-necked 2 L flask equipped with a mechanical stirrer and a reflux condenser. After some 15 min of heating to 100–125 °C, tetrahydrofuran (1.0 L) and a few drops of 1,2-dibromoethane are added. The mixture is stirred at 60 °C for 24 h and at 25 °C for 12 h. The voluminous orange-red precipitate is filtered through a sintered glass frit. The solid is washed with tetrahydrofuran (2 × 0.25 L), dried *in vacuo* and collected (yield 0.37 kg, 89%). Having the composition $C_{14}H_{10}Mg \cdot (THF)_3$, an aliquot is quantitatively converted into 9,10-dihydroanthracene when treated with water.

4.4.2   Permutational Exchange of Carbon Against Metal

The most archetypical reaction falling in this category has already been mentioned: the addition of *tert*-butyllithium to the strained C—C bond of tricyclo-[2.2.1.0^{1,4}]heptane[1117] produced *in situ* (see p. 110). All permutational displacements of carbon by metal covered in this subsection involve the 'ring opening' (see p. 173) of double and triple bonds by the nucleophilic attack of organometallics at unsaturated substrates.

a) Representative Examples of Organometal Addition to Unsaturated Alcohols

The coordination-driven addition of organolithium[1424, 1425] and 'reactive' (allylic or benzylic) organomagnesium[1426–1428] reagents to allylic[1424–1426] and homoallylic[1427, 1428] or (homo)progargylic[1429] alcoholates is the only pertinent reaction type of sufficient generality and utility (see Table 78). It still has a fairly narrow profile of applicability as it does not tolerate any substituent at the 2-position and only aryl substituents at the 3-position. Lithium 2-cyclopentenoxide[1424] (**270**) and other 3-alkyl-bearing alcoholates undergo organolithium addition with inverse regioselectivity followed by instantaneous β-elimination of lithium oxide.

b) Scope and Limitations

Unlike butyllithium[1430, 1431] or any other primary alkyllithium, *sec*- and *tert*-butyllithium as well as cyclohexyllithium add smoothly to ethylene in ethereal solvents even at temperatures around or below −25 °C[1432–1435]. Propene proves to be inert under the same conditions. However, ω-metallo-1-alkenes[1436–1440], -1,2-allenes[1441] and -alkynes[1442]

Table 78. 3-(Metaloxy)alkylmetals (M−R) by reaction of allylic alcohols with alkyllithiums (in paraffinic solvents), other organolithiums or allylic and benzylic Grignard reagents (in diethyl ether) at +25 °C.

M−R	Substrates[a]	M−R′[b]	Product, El−X′[c]	Reference

M−CH₂−CH−CH₂OM
     |
    CH₂−CH=CH₂
	A	ClMgCH₂−CH=CH₂	50%, (H₃C)₂CO	[1426]

M−CH₂−CH−CH₂OM
     |
  H₃C−CH−CH=CH₂
	A	ClMgCH₂−CH=CH−CH₃	30%, H₂O	[1426]

M−CH₂−CH−CH₂OM
     |
    C₃H₇
	A	LiC₃H₇[d]	73%, H₂O	[1425]

M−CH₂−CH−CH₂OM
     |
    C₄H₉
	A	LiC₄H₉[d]	72%, H₂O	[1424]

M−CH₂−CH−CH₂OM
     |
  CH₂C₆H₅
	A	BrMgCH₂C₆H₅
		LiCH₂C₆H₅
			10%, H₂O
			60%, H₂O
				[1426]
				[1424]

M−CH₂−CH−CH₂OM
     |
  CH(CH₃)₂
	A	LiCH(CH₃)₂	48%, H₂O	[1424]

M−CH₂−CH−CH₂OM
     |
  C(CH₃)₃
	A	LiCH(CH₃)₃	22%, H₂O	[1424]

M−CH₂−CH−CH₂OM
     |
    C₆H₅
	A	LiC₆H₅	40%, H₂O	[1424]

CH₃
    |
M−CH₂−CH−CH−OM
       |
   CH₂CH=CH₂
	B
	B
		ClMgCH₂CH=CH₂
		LiCH₂CH=CH₂[d]
			16%, H₂O[e]
			[f], H₂O[g]
				[1426]
				[1425]

CH₃
    |
M−CH₂−CH−CH−OM
       |
    C₃H₇
	B	LiC₃H₇[d]	65%, H₂O[g]	[1425]

C₆H₅
    |
M−CH₂−CH−CH−OM
       |
    C₃H₇
	P	LiC₂H₅[d]	30%, H₂O[g]	[1425]

C₆H₅
  |
M−CH−CH−CH₂OM
     |
    C₂H₅
	C	LiC₃H₇[d]	70%, H₂O	[1425]

[a] A = Allyl alcohol (2-propen-1-ol), B = 3-buten-2-ol, C = cinnamyl alcohol (3-phenyl-2-propen-1-ol), P = 1-phenyl-2-propen-1-ol. [b] M−R′ = organometallic reagent undergoing addition onto the substrate. [c] El−X′ = electrophilic trapping reagent. [d] In the presence of *N,N,N′,N′*-tetramethylethylenediamine (TMEDA). [e] Mainly (89% d.s.) the diastereomer resulting from *anti*-periplanar addition. [f] Yield not specified. [g] Mainly or exclusively (d.s. >98%) the diastereoisomer resulting from *syn*-periplanar addition.

readily undergo intramolecular addition, in particular if 3-, 4- and 5-membered rings, even if strained[1439], are formed in this way.

Primary alkyllithiums and, to some extent, even organomagnesiums are capable of combining with strained olefinic double bonds as incorporated in cyclopropene[1443], norbornene[1444], norbornadiene[1445] and *trans*-cyclooctene. The same holds for strained triple bonds as present in cycloalkynes or dehydroarenes[1446].

Compared with relief of strain, resonance stabilization is even more powerful as a reaction promotor. Butyllithium adds smoothly to styrene. However, the resulting 1-phenylhexyllithium proves to be for more reactive towards the same substrate, thus unleashing a polymerization[1447]. In contrast, the adduct of butyllithium to 1,1-diphenylethylene is protected against any further transformation by steric and electronic stabilization[1448].

Resonance energies attain maximum values in cyclopentadienylmetals and 9-fluorenyl-metals. As a consequence, nucleophilic additions to pentafulvenes[1449-1451] (*e.g.* to **271**), 9-alkylidenefluorenes[1449] and azulenes[1452] proceed with particular ease.

Also d-orbital resonance can provide the driving force to promote nucleophilic addition to olefinic double bonds. Butyllithium, *tert*-butyllithium, phenyllithium and other organolithium species react readily with trimethylvinylsilane[1444,1453], triphenylvinylgermanium[1454], diphenylvinylphosphine[1455] and 2-methylene-1,3-di-thiane[1456] to give chain-elongated α-hetero-substituted alkyllithiums.

Strain and resonance also fuel the addition of organometallic reagents to alkynes. Both cyclooctyne[1457] and tolane (diphenylacetylene)[1458,1459] react with butyllithium. Both tolane and 1-phenyl-1-propyne combine with cumylpotassium (α,α-dimethylbenzyl-potassium)[1448].

Coordination, or more precisely mixed aggregate formation, has been recognized as instrumental for the reaction of organometallics with allylic alcoholates (see p. 177). The facile addition of allylic zinc derivatives to 1-alkenylmagnesium halices to afford δ,ε-unsaturated 1,1-dimetal species[1460–1464] (**272**) can be rationalized in the same manner.

$$BrMg-CH=CH-R + BrZn-CH_2-CH=CH-R' \xrightarrow[-50\,°C]{DEE} \begin{array}{c} BrZn \\ BrMg \end{array}\!\!\!\! CH-CH \begin{array}{c} CH-CH=CH_2 \\ R \end{array}$$

272

c) Working Procedures

The preparation of just two reagents is described below. 3,3-Dimethylbutyllithium displays virtually the same reactivity as butyllithium in metalation reactions but, unlike butane, 2,2-dimethylbutane (b.p. ~50 °C), the hydrocarbon by-product can be probed by gas chromatography and by nuclear magnetic resonance. The sterically hindered, hence nonnucleophilic, 1,1-diphenylhexyllithium may be employed as a moderately strong, self-indicating base to deprotonate ketones, esters and nitriles, or to check the water content in allegedly anhydrous solvents.

Salt-free 3,3-Dimethylbutyllithium[1465]

$$H_2C=CH_2 \xrightarrow[\text{DEE}]{\text{LiC(CH}_3)_3} Li-CH_2-CH_2-C(CH_3)_3$$

1,2-Dibromoethane (92 mL, 0.20 kg, 1.1 mol) is added to a suspension of magnesium turnings (40 g, 1.6 mol) in diethylene glycol dimethyl ether (0.15 L). The mixture is stirred slowly at the beginning, more rapidly later and vigorously at the end when the flask is heated to 75 °C. The evolving gas (ethylene) passes through glass tubing and a bubble counter filled with paraffin oil. It is condensed in a three-necked flask filled with a 2.0 M solution of *tert*-butyllithium (1.0 mol) in neat diethyl ether (0.50 L) cooled to −100 °C. (The initial solvent, pentanes, had been evaporated under reduced pressure before the residue was dissolved in diethyl ether precooled to −75 °C). As soon as the ethylene evolution ceases, the gas inlet is removed and the flask stopped. After 2 h at −100 °C, the solution is siphoned in a buret with graduated side arm which is stored in a deep-freezer at about $-25\,^{\circ}$C. The gas chromatographic analysis of an aliquot consecutively treated with dry ice, acid and diazomethane reveals the presence of methyl 4,4-dimethylpentanoate (95%) and trace amounts of methyl pivalate (<0.5%) and di-*tert*-butyl ketone ($\sim$1%).

Lithium Bromide Containing 3,3-Dimethylbutyllithium[1465]

$$H_2C{=}CH_2 \xrightarrow[\text{DEE}]{\text{LiC(CH}_3)_3 + 2\ \text{LiBr}} Li{-}CH_2{-}CH_2{-}C(CH_3)_3$$

A three-necked flask is filled with an approximately 1.5 M solution of *tert*-butyllithium (1.2 mol) in pentanes. The solvent is stripped off *in vacuo* and the residue taken up in precooled ($-75\,^{\circ}$C) diethyl ether (0.8 L). At $-100\,^{\circ}$C, 1,2-dibromoethane (34 mL, 74 g, 0.39 mol) is added dropwise, over the course of 5 min. After 2 h at $-100\,^{\circ}$C, the flask is placed in an ice-bath and stirred until homogeneous. The solution is siphoned in a Schlenk buret (see pp. 285–287). It may be stored in a refrigerator at $0\,^{\circ}$C for several days without significant loss of titer. The presence of 3,3-dimethylbutyllithium and the absence of *tert*-butyllithium is checked by carboxylation followed by gas chromatography of the esterified sample.

1,1-Diphenylhexyllithium[1426,1448]

Butyllithium (0.50 mol), from which the commercial solvent (hexanes) has been removed by evaporation *in vacuo*, is dissolved in precooled ($-75\,^{\circ}$C) tetrahydrofuran and 1,1-diphenylethylene (88 mL, 90 g, 0.50 mol) is added. After 1 h at $0\,^{\circ}$C, all butyllithium is consumed, as confirmed by the Gilman color test II (see p. 294) or the absence of characteristic smell of valeric acid after carboxylation and acidification of an aliquot. The solution can be stored at $25\,^{\circ}$C for several days, if not weeks.

4.5 Displacement of Hydrogen

Hydrogen being a poor nucleofugal leaving group; the reductive metal insertion into C$-$H bonds plays only a very marginal role in the generation of polar organometallic intermediates. In contrast, prototropic hydrogen transfer from a hydrocarbon to a meta-lating reagent in a permutational process is the method of choice whenever applicable.

4.5.1   Reductive Insertion of Metal in Carbon–Hydrogen Bonds

One single example of metal insertion in an unstrained and saturated hydrocarbon has been reported in the literature. Vapors of calcium and cyclohexane were claimed to generate cyclohexylcalcium hydride, albeit only in trace amounts (about 1%, as concluded after trapping with chlorotrimethylsilane)[1466]. If this reaction does occur in the described way, it cannot involve either single electron transfer (SET) or direct homolytical hydrogen abstraction during a molecule/atom collision, because the key steps would be highly endothermal (by about 40 and 60 kcal/mol when M = Li and K, respectively)[1467].

$$\text{C-H} + \cdot\text{M} \longrightarrow \left[\text{C}\cdots\text{H}\cdots\text{M} \right]^{\ddagger} \longrightarrow \text{C}\cdot + \text{H-M}$$

The same argument can also be put forward with respect to the metal hydride-producing reaction of alkali metals with elemental hydrogen[1468]. Again, a homolytic path can be ruled out on thermochemical grounds. A four-center type process between dihydrogen and a dimetal molecule remains as the most plausible possibility.

a) *Representative Examples of Reductive Carbon-Hydrogen Scission*

As previously mentioned (see p. 173), metals can displace hydrogen from unsaturated hydrocarbons through an addition–elimination process. Relative smooth transformations have been accomplished with substrates of the 1-alkene, 1-alkyne, 1,4-diene, phenyl-methane, (benzo)cyclopentadiene, furane and thiophene type (see Table 79).

b) *Scope and Limitations*

Metal insertion in C–H bonds suffers not only from its very restricted applicability but also from its lack of economy. The hydrogen to be displaced is eliminated as a metal hydride only in a minority of cases. More frequently, it is transferred to the substrate[1228].

Table 79. Direct (reductive) displacement of hydrogen by metals (M) affording organometallics (M–R).

M–R	M	Sv[a]	T[°C]	Product, El–X'[b]	Reference
M–CH=CH₂	Li[c]	THF	0	75%, H_2O	[1387,1390]
M–CH=CH–CH₃	Li[c]	THF	0	75%[d], H_2O	[1390]
M–CH=CH–C₆H₁₃	Li	THF	0	83%, $(H_3C)_3SiCl$	[1390]
M–C≡CH	Li	NH₃	−30	82%, ⟶⟍⟋⟍O	[1469]
	Na	NH₃	−30	~100%, isolated	[1470]
	K	NH₃	−30	73%, deh.androsterone	[1471]
M–C≡C–C₅H₁₁	Na	DEE	+40	~100%, isolated	[1472]
M–C≡C–C₆H₁₃	Na	NH₃	−30	[h]	[1473]
M–C≡C–C₆H₅	Na	DEE	+25	~100%, isolated	[1474]
M–CH₂–CH=CH–CH=CH₂	K	THF[e]	+25	97%, $(H_3C)_3SiCl$	[1475]
M–CH₂–CH=CH–CH=CH–CH₃	K	THF[e]	+25	90%, $(H_3C)_3SiCl$	[1475]
M–CH₂–⟨phenyl⟩	K[f]	TOL	+25	90%, CO_2	[1476]
M–C(⟨phenyl⟩)₃	K	MEGME	+25	37%, D_2O	[1477]
M–⟨fluorenyl⟩	Li	THF	+25	71%, CO_2	[1478]
	Na	[g]	+200	[h], isolated	[1479]
	K	DOX	+100	75%, D_2O	[1480]
M–⟨indenyl⟩	Na	[g]	+150	[h], isolated	[1481]
M–⟨cyclopentadienyl⟩	XMg	[g]	+550	>80%, isolated	[1482]
	Li	THF	+30%	30%, [i]	[1483]
	Na	THF	+30	~100%, isolated	[1484]
M–⟨methylcyclopentadienyl⟩	Na	PET[j]	+25	≤26%, CO_2	[1485]
M–⟨benzothienyl⟩	Na	DEE	+25	56%, CO_2	[1486]

[a] PET = petroleum ether (pentanes, hexanes, heptanes); TOL = toluene; DOX = 1,4-dioxane; THF = tetrahydrofuran; MEGME = (mono)ethyleneglycol dimethyl ether, in the case of mixtures (*e.g.* containing paraffinic cosolvents) only the most polar component is indicated. [b] El–X' = electrophilic trapping reagent. [c] In the presence of catalytic amounts of zinc dichloride and 2,5-diphenyl-1,6,6a,λ⁴-trithiapentalene. [d] Including small amounts (together <8%) of 1-methylvinyllithium and allyllithium. [e] In the presence of triethylamine. [f] Plus sodium oxide. [g] Neat, no solvent. [h] Not specified. [i] 2,4,6-Trimethylpyrylium perchlorate, affording 5,6,8-trimethylazulene. [j] Or DEE.

Thus, an important part of the hydrocarbon precursor is wasted and by-products accumulate. For example, the radical anion that is obtained by addition of a metal atom to cyclopentadiene attacks some of the unconsumed hydrocarbon under proton abstraction. The resulting cyclopentenyl radical combines with another metal atom and the resulting allylmetal species deprotonates another hydrocarbon molecule. Eventually, one cyclopentene per two cyclopentadienylmetals are formed.

c) Working Procedures

2,4-Pentadienylpotassium and 2,4-hexadienylpotassium have been employed to prepare open-sandwich complexes of transition elements[1486]. The required intermediates may be prepared from the respective conjugated or homoconjugated dienes either by reductive metal insertion or by hydrogen/metal permutation (see p. 203).

2,4-Pentadienylpotassium[1475]

At 0 °C and under vigorous stirring, 1,3-pentadiene (piperylene, *cis/trans* mixture; 20 mL, 14 g, 0.20 mol) is added dropwise, in the course of 1 h, to a dispersion of potassium (3.9 g, 0.10 mol) in tetrahydrofuran (40 mL) and triethylamine (21 mL, 15 g, 0.15 mol). Under continued stirring, the mixture is allowed to gradually attain 25 °C (in the course of some two further hours). More tetrahydrofuran (40 mL) is added. The mixture is briefly heated to 60 °C until a clear solution is obtained. Upon dilution with hexanes (50 mL) and cooling to 0 °C, a crystalline solid forms [yield 17 g, (96%)]. After recrystallization from a 2 : 1 (v/v) mixture of tetrahydrofuran and hexanes beautiful orange-colored needles are collected. The combined mother liquor contains $C_{10}H_{18}$ hydrocarbons, reductive dimers of 1,3-pentadiene (yield 6.8 g, 98%).

Treatment of a pentadienylpotassium solution in tetrahydrofuran at 0 °C with heavy water gives the *cis* isomer of 5-[²H]penta-1,3-diene (98%). However, the *trans*-isomer results (99%) if the crystalline material is deuterolyzed after having been heated for 1 h at 90 °C *in vacuo* (0.05 mmHg) and suspended in 1,2,3,4-tetrahydronaphthalene.

2,4-Pentadienylsodium, -rubidium and cesium have been prepared similarly (yields 96–99%). 2,4-Pentadienyllithium is accessible in better yields from 1,4- rather than from 1,3-pentadiene (90% *vs.* 52%).

4.5.2 Hydrogen/Metal Permutation ('Metalation')

The displacement of a specific hydrogen atom by metal is obviously the most straightforward way to selectively generate a reactive carbonucleophile. Two obstacles have to be surmounted to realize this goal. Strong bases are required to overcome the reluctance of hydrocarbons to participate in permutational exchange processes. Equally important is the operation of a subtle reaction mechanism which efficaciously discriminates between the various types of C−H bonds which are ordinarily present in the organic compounds considered as substrates.

a) Representative Examples of Metalation

This Subsection contains 66 tables featuring a wealth of organometallic intermediates generated by hydrogen/metal permutation. In view of this large number, the ordering principles have to be specified in detail. We move from sp^3 to sp^2 centers starting with resonance-inactive alkylmetals, then progressing to allylic, propargylic and benzylic species, before turning to olefinic, cyclopentadienic and aromatic species, and eventually terminating with five-ring and six-ring heteroaromatic organometals. Each of those categories is subdivided according to the same hierarchy: first the unsubstituted prototypes, next alkylated and silylated congeners and finally derivatives carrying, in this order, nitrogen, oxygen and halogen substituents. When two or more different substituents are present, the one which is closest to the metal-bearing center has precedence. A graphical survey of the various structural patterns covered is displayed in the preface to this Chapter (see p. 3).

Whereas unstrained alkanes and cycloalkanes are virtually inert towards polar organometallics (see pp. 212–248), cyclopropanes do react with organosodium reagents and bicyclobutanes even with organolithiums (see Table 80). Metalation of simple aliphatic amines such as trimethylamine or *N,N,N′,N′*-tetramethylethylenediamine (TMEDA) occurs slowly with alkyllithiums and rapidly with superbasic mixed-metal reagents (see Table 81). Formamidines (see Table 82), sterically hindered

Text continues on page 187

Table 80. Cyclopropyl- and 1-bicyclobutylmetals (M−R) by hydrogen/metal permutation between small-ring hydrocarbons and metalation reagents M−R'.

M−R	M−R'	Sv[a]	T[°C]	Product, El−X'[b]	Reference
M—◁	NaC_5H_{11}	PET[c]	+25	20%, CO_2	[1487]
M—(bicyclo structure)	$NaC_4H_9/KOC(CH_3)_3$	PET	+25	60%, $(H_3C)_3SiCl$	[440,1488]
M—(cyclopropyl-cyclopentane)	$NaC_4H_9/KOC(CH_3)_3$	PET	+25	30%, CO_2[d]	[1489]
M—(bicyclo[2.1.0])	NaC_5H_{11}	PET	+25	18%, CO_2[d,e]	[1489,1490]
M—(bicyclo)	NaC_5H_{11}	PET	+25	20%, CO_2[e]	[1489,1490]
M—(bicyclobutyl)	LiC_3H_7	PET	0	79%, $(H_3C)_3SiCl$	[1491,1492]
M—(bicyclo structure)	LiC_4H_9	DEE	[f]	~100%, D_2O	[1493]
M—◁ with CH₃	$LiCH_3$	DEE	[f]	~100%, H_3CI	[1494]
M—◁ with two CH₃	$LiCH_3$	DEE	[f]	~100%, H_3CI	[1494]
M—◁ CH-OM, H₃C	$LiCH(CH_3)_2$	PET	+25	45%, D_2O[g]	[1495]
M—(bicyclo) OM	$LiCH(CH_3)_2$	PET	+25	71%, D_2O	[1495]
M—(bicyclo) OM	$LiCH(CH_3)_2$	PET	+25	80%, D_2O	[1495]

[a] PET = petroleum ether (pentanes, hexanes, heptanes), DEE = diethyl ether. [b] El−X' = electrophilic trapping reagent. [c] Cyclopropane was employed as both the substrate and the solvent. [d] Metalation occurs exclusively at the *exo* position. [e] Bicyclo[2.1.0]pentane ('hausane') is attacked at both cyclopropanic positions in reagent-dependent ratios. [f] Not specified (+25 °C?). [g] The main product is accompanied by small amounts of its *cis* isomer (5%) and a regioisomer (11%) resulting from attack at the α-hydroxyalkyl-substituted position.

Table 81. α-Aminoalkyllithiums (M−R) by metalation of aliphatic amines using reagents M−R′.

M−R	M−R′	Sv[a]	T[°C]	Product, El−X′ [b]	Reference
M − CH$_2$N(CH$_3$)C$_6$H$_5$	LiC$_4$H$_9$	PET	+25	33%, H$_9$C$_4$I [c]	[1496]
M − CH$_2$N⁓	LiCH(CH$_3$)C$_2$H$_5$ [d]	PET	0	70%, H$_{17}$C$_8$Br	[1497]
M − CH$_2$N(CH$_3$)CH$_2$CH$_2$N(CH$_3$)$_2$	LiC(CH$_3$)$_3$	PET	+25	53%, (H$_3$C)$_3$SiCl	[1498−1501]
M−CH$_2$N⏜N⏜N−	LiCH(CH$_3$)C$_2$H$_5$	PET	0	50%, (H$_3$C)$_3$SnCl	[1501]
M−CH$_2$N⏜N⏜N−)$_2$	LiCH(CH$_3$)C$_2$H$_5$	PET	0	95%, H$_5$C$_6$CHO	[1502]
M−CH$_2$N⏜N⏜N⏜	LiCH(CH$_3$)C$_2$H$_5$	PET	0	31%, H$_3$COD	[1502]
M−CH$_2$−N⏜N⏜	LiCH(CH$_3$)C$_2$H$_5$	PET	−75	85%, (H$_3$C)$_3$SiCl	[1502]

[a] PET = petroleum ether (pentanes, hexanes, heptanes). [b] El−X′ = trapping electrophile. [c] Simultaneous addition of butyllithium and butyl iodide to the amine. [d] In the presence of potassium *tert*-butoxide.

carboxamides (see Table 83), *N-tert*-butoxycarbonyl (BOC)-protected amines (see Table 84), lithium carbamates and thiocarbamates (see Table 85) and *N*-nitrosoamines (see Table 86), all being dipole-activated substrates (see pp. 249 and 251), undergo a hydrogen/metal exchange at the nitrogen-adjacent α-carbon center with particular ease. Butylpotassium converts acyclic and cyclic aliphatic ethers efficaciously into the α-metalated derivatives (see Table 87). The ring strain facilitates the deprotonation of oxiranes. A trialkylsilyl or triarylsilyl substituent is nevertheless required to provide additional acidification and to stabilize sterically and electronically the resulting intermediate (see Table 88). The methoxy or primary alkoxy groups of esters and carbamates having a sterically congested carbonyl group are readily metalated by TMEDA-complexed *sec*-butyllithium, dipole activation being again a crucial attribute (see Table 89). Whereas monohalomethanes favor the substitution pattern when treated with a basic nucleophile, 1,1-dihaloalkanes react with organolithiums and trihalomethanes even with organomagnesiums preferentially or exclusively under deprotonation (see Table 90).

Text continues on page 191

Table 82. M−R′-Promoted metalation of *N*-(imidomethyl)- and *N*-(2-oxazolinyl)amines.

M−R [a]	M−R′	Sv [b]	T[°C]	Product, El−X$'$ [c]	Reference
$M{-}CH_2{-}N({CH_3}){-}CH{=}NR'$	$LiCH(CH_3)C_2H_5$	THF	−75	89%, $(H_3C)_3SiCl$	[1503,1504]
$M{-}CH_2{-}N({C_6H_5}){-}CH{=}NR'$	$LiCH(CH_3)C_2H_5$	THF	−75	90%, H_3CI	[1505]
	$LiCH(CH_3)C_2H_5$ [d]	THF	−25	85%, $H_3COOCCl$	[1506]
	$LiCH(CH_3)C_2H_5$	THF	−25	84%, H_9C_4I	[1505]
	$LiCH(CH_3)C_2H_5$	DEE	−25	30%, H_3CI	[1507]
	$LiC(CH_3)_3$	THF	−20	87%, $H_3COOCCl$	[1508]
	$LiC(CH_3)_3$	THF	−25	88%, H_3CI	[1509]
	$LiCH(CH_3)C_2H_5$	THF	−75	67%, $(H_2C)_2O$	[1505]
	LiC_4H_9	THF	−85	'high', H_3CI	[1510]
	LiC_4H_9	THF	−75	81%, H_5C_2I	[1510]
	$LiCH(CH_3)C_2H_5$	DEE	+10	81%, F_3CCH_2I	[1511]

[a] R′ = C(CH₃)₃. [b] THF = tetrahydrofuran; in the case of mixtures (containing for example paraffinic cosolvents) only the most polar component is indicated. [c] El−X$'$ = electrophilic trapping reagent. [d] Or *tert*- instead of *sec*-butyllithium. [e] X−C=NR″ = Δ³-2-(4-isopropyloxazolinyl). [f] R‴ = 2-(methoxymethyl)phenyl, this substrate being three powers of 10 more reactive than the analog with R‴ = C(CH₃)₃.

Table 83. α-Lithio-N-(thio)acylamines (M–R) by permutational hydrogen/metal exchange between tertiary carboxamides and metalation reagents M–R'.

M–R[a]	M–R'	Sν[b]	T[°C]	Product, El–X'[c]	Reference
M–CH$_2$–N(CH$_3$)–C(=O)–R	LiCH(CH$_3$)C$_2$H$_5$[d]	THF	−75	77%, H$_3$CI	[1512,1513]
M–CH$_2$–N(C$_2$H$_5$)–R	LiCH(CH$_3$)C$_2$H$_5$[d]	THF	−75	75%, H$_5$C$_6$CHO[e]	[1512,1513]
M–CH$_2$–N(C$_{10}$H$_{21}$)–C(=O)–R	LiCH(CH$_3$)C$_2$H$_5$[d]	THF	−75	85%, H$_{13}$C$_6$I	[1513]
M–CH$_2$–N(CH$_3$)–C(=O)–C(C$_6$H$_5$)$_3$	LiCH(CH$_3$)C$_2$H$_5$	THF	−40	88%, H$_{13}$C$_6$I	[1513]
M–CH$_2$–N[O CH$_2$C$_6$H$_5$][O CH$_2$C$_6$H$_5$]	LiCH(CH$_3$)C$_2$H$_5$[f]	THF	−100	95%, (cyclohexanone-CHO)	[1514]
M–CH(CH$_3$)–N(C$_2$H$_5$)–C(=O)–C(C$_2$H$_5$)$_3$	LiCH(CH$_3$)C$_2$H$_5$[d]	DEE	0	69%, H$_5$C$_6$CHO	[1515]
M–(azetidine)N–C(=O)–C(C$_6$H$_5$)$_3$	LiC(CH$_3$)$_3$	THF	−40	62%, H$_5$C$_6$CHO	[1516]
M–(pyrrolidine)N–C(=O)–C(C$_6$H$_5$)$_3$	LiC(CH$_3$)$_3$	THF	−40	52%, (H$_5$C$_6$)$_2$CO	[1516]
M–(4-C$_6$H$_5$-piperidine)N–C(=O)–C(C$_2$H$_5$)$_3$	LiCH(CH$_3$)C$_2$H$_5$[d]	DEE	0	78%, H$_5$C$_6$CHO	[1515]
M–(piperidine)N–C(=O)–R	LiCH(CH$_3$)C$_2$H$_5$[d]	THF	−75	30%, H$_5$C$_6$CHO	[1515]
M–CH$_2$–N(CH$_3$)–C(=S)–C(CH$_3$)$_3$	LiCH(CH$_3$)C$_2$H$_5$[d]	THF	−75	75%, H$_5$C$_6$CHO[e]	[1517]

[a] R = 2,4,6-triisopropylphenyl. [b] THF = tetrahydrofuran; in the case of mixtures (containing for example paraffinic cosolvents) only the most polar component is indicated. [c] El–X' = electrophilic trapping reagent. [d] In the presence of N,N,N',N'-tetramethylethylenediamine (TMEDA). [e] Raw compound consisting of two rotamers. [f] In the presence of hexamethylphosphoric triamide (HMPA).

Table 84. α-Lithio-*N*-carbamoylamines and α-lithio-*N*-(*tert*-butoxycarbonyl)amines (M–R) by permutational hydrogen/metal exchange between *N*-carbamoyl- or *N*-BOC-protected amines and metalation reagents M–R′.

M–R	M–R′	Sv [a]	T [°C]	Product, El–X′ [b]	Reference
	LiCH(CH$_3$)C$_2$H$_5$ [c]	THF	0	81%, (H$_5$C$_6$)$_2$CO	[1518]
	LiCH(CH$_3$)C$_2$H$_5$ [c]	THF	−75	91%, (H$_5$C$_6$)$_2$CO	[1515,1519]
	LiCH(CH$_3$)C$_2$H$_5$ [c]	DEE	0	48%, H$_5$C$_6$CHO	[1515,1519]
	LiCH(CH$_3$)C$_2$H$_5$ [c]	THF	−75	95%, H$_5$C$_6$Si(CH$_3$)$_2$Cl	[1520]
	LiCH(CH$_3$)C$_2$H$_5$ [c]	THF	−75	94%, H$_3$CI	[1520,1521]
	LiCH(CH$_3$)C$_2$H$_5$ [c]	THF	−75	58%, H$_2$C=CH−CH$_2$Br	[1520,1521]
	LiCH(CH$_3$)C$_2$H$_5$ [c]	THF	−75	80%, (H$_3$C)$_3$SiCl	[1521]
	LiCH(CH$_3$)C$_2$H$_5$ [c]	DEE	−75	94%, (H$_9$C$_4$)$_3$SnCl	[1522,1523]
	LiCH(CH$_3$)C$_2$H$_5$ [c]	DEE	−20	71%, (H$_3$CO)$_2$SO$_2$	[1523]
	LiC(CH$_3$)$_3$ [c]	DEE	−75	90%, O$_2$	[1524]
	LiCH(CH$_3$)C$_2$H$_5$ [c]	THF	−75	62%, H$_3$COOCCl	[1525]

[a] El–X′ = electrophilic trapping reagent. [b] DEE = diethyl ether; THF = tetrahydrofuran; in the case of mixtures (containing for example paraffinic cosolvents) only the most polar component is indicated. [c] In the presence of *N*,*N*,*N*′,*N*′-tetramethylethylenediamine (TMEDA).

Table 85. α,O-Dilithio-N-carboxyamines and α,S-dilithio-N-dithiocarboxyamines (M−R) by permutational hydrogen/metal exchange between N-CO$_2$- or N-CS$_2$-protected amines and metalation reagents M−R'.

M−R	M−R'	Sv[a]	T[°C]	Product, El−X'[b]	Reference
(indoline structure, M−N, MO−C=O)	LiC$_4$H$_9$/KOC(CH$_3$)$_3$	THF	−50	57%, (structure)	[1526]
(tetrahydroquinoline structure, M−N, MO−C=O)	LiC$_4$H$_9$/KOC(CH$_3$)$_3$	THF	−50	52%, (structure)	[1526]
M−CH$_2$−N(C$_4$H$_9$)(C=S)(MS)	LiCH(CH$_3$)C$_2$H$_5$	THF	−25	68%, (structure)	[1527]
M−CH$_2$−N(cyclohexyl)(C=S)(MS)	LiCH(CH$_3$)C$_2$H$_5$	THF	−30	90%, H$_5$C$_6$Si(CH$_3$)$_2$Cl	[1528]
M−CH$_2$−N(C$_6$H$_5$)(C=S)(MS)	LiCH(CH$_3$)C$_2$H$_5$	THF	−40	95%, (H$_3$C)$_3$SiCl	[1528]

[a] THF = tetrahydrofuran; in the case of mixtures (containing for example paraffinic cosolvents) only the most polar component is indicated. [b] El−X' = electrophilic trapping reagent.

Acyclic and cyclic alkenes (see Tables 91−95), alkenes (see Table 96), internal alkynes (see Table 97), conjugate or homoconjugated dienes (see Table 98) and allylic silanes (see Table 99) undergo smooth metalation with TMEDA-activated butyllithium or, better, superbasic mixed-metal reagents to afford allyl metal species which show attractive regioselective and stereoselective features. Both allyl amines and enamines can be converted into 1-aminoalkylmetals (see Table 100), whereas only allyl ethers are suitable precursors to 1-alkoxy(aroxy)allylmetals (see Table 101). The metalation of allylic and homoallylic metal alkenoates (see Table 102) can in general be accomplished without complications (*i.e.* 1,2- or 1,4-elimination of metal oxide). Allylic halides are easily deprotonated with bulky amide bases. However, the intermediates thus generated are too fragile to survive until transformed in a subsequent operation and hence can only be trapped *in situ* (see Table 103).

Despite some problems with the α *vs. ortho* selectivity (see pp. 253−257), alkylarenes can be cleanly metalated with superbases at the sterically least hindered benzylic site (see Table 104). Also N-pivaloyl- or N-BOC-protected o-toluidines (see Table 105), methoxy-substituted toluenes (see Table 106), 2-methylbenzylalcoholates (see Table 106), styrene,

Text continues on page 193

Table 86. Metalation of aliphatic N-nitrosoamines with reagents M–R'.

M–R	M–R'[a]	Sν[b]	T[°C]	Product, El–X'[c]	Reference
$M-CH_2-N-N=O$, CH_3	LIDA	THF	−75	90%, (cyclohexene-oxide structure)	[1529]
$M-CH_2-N-N=O$, C_2H_5	LIDA	THF	−75	75%, H_3CCHO	[1529]
$M-CH_2-N-N=O$, $CH(CH_3)_2$	LIDA	THF	−75	85%, H_3CCHO	[1529]
$M-CH_2-N-N=O$, $C(CH_3)_3$	LIDA	THF	−75	95%, H_3CCHO	[1529]
$M-CH_2-N-N=O$, C_6H_5	LIDA	THF	−75	58%, H_3CCHO	[1529]
$M-CH_2-N-N=O$, $CH(CH_3)OCH_3$	LIDA	THF	−80	78%, $(H_3C)_2CO$	[1530]
(pyrrolidine, M on C2, $N=O$)	LIDA	THF	−75	95%, $H_5C_6CH_2Br$	[1531]
(pyrrolidine, M on C2, C_2H_5, $N=O$)	LIDA	THF	−75	65%, $H_{15}C_7I$ [d]	[1531]
(piperidine, C_6H_5 at 4, M on C2, $N=O$)	LIDA	THF	−75	76%, CO_2	[1531]
(piperidine, C_6H_5 at 4, M on C2 with CH_3, $N=O$)	LIDA	THF	−75	62%, H_3CI	[1531]
H_3C, $M-C-N-N=O$, $H_3C \quad CH(CH_3)_2$	LIDA	THF	−75	40%, H_3CCHO	[1529]
C_6H_5, $M-CH-N-N=O$, $CH_2C_6H_5$	LIDA[e]	THF	−75	99%, CO_2	[1532]

[a] LIDA = lithium diisopropylamide. [b] THF = tetrahydrofuran. [c] El–X' = electrophilic trapping reagent. [d] *Cis/trans* mixture. [e] Or methyllithium instead of LIDA.

Table 87. α-Potassio ethers (M$-$R) by permutational hydrogen/metal exchange between acyclic and cyclic ether, aliphatic ethers and butylpotassium.

M$-$R	M$-$R'	Sν[a]	T[°C]	Product, $El-X'$ [b]	Reference
M$-$CH$_2$OCH$_3$	KC$_4$H$_9$	DME	-75	75%, H$_5$C$_6$CHO	[1533]
M$-$CH$_2$OC(CH$_3$)$_3$	KC$_4$H$_9$	THF	-75	73%, H$_5$C$_6$CHO	[1533,1534]
M$-$⟨O⟩	KC$_4$H$_9$	THF	-75	76%, (H$_3$C)$_3$SiCl	[1533]
M$-$⟨O⟩$-$CH$_3$	KC$_4$H$_9$	[c]	-75	21%, (H$_3$C)$_3$SiCl [d]	[1533]
M$-$⟨O⟩	KC$_4$H$_9$	[c]	-75	68%, (H$_3$C)$_3$SiCl	[1533]
M$-$⟨O⟩$-$C$_2$H$_5$	KC$_4$H$_9$	[c]	-75	52%, (H$_3$C)$_3$SiCl [e]	[1533]
M$-$⟨O⟩	KC$_4$H$_9$	[c]	-75	27%, (H$_3$C)$_3$SiCl	[1533]

[a] DME = dimethyl ether; THF = tetrahydrofuran. [b] $El-X'$ = electrophilic trapping reagent. [c] Employed in excess with respect to butylpotassium, the substrate acting as the solvent at the same time. [d] β-Elimination (affording potassium 4-pentenolate) is competing with the metalation; the 2-methyl-5-(trimethylsilyl)tetrahydrofuran is isolated as a 1 : 1 *cis/trans* mixture. [e] The 2-ethyl-6-(trimethylsilyl)tetrahydropyran is a pure *cis* isomer.

stilbene or indene oxides (see Table 107) and halogenated toluenes (see Table 108) can be successfully submitted to hydrogen/metal permutation at the lateral position.

Ethylene, 3,3-dimethyl-1-butene, camphene and other simple alkenes lacking effective allyl positions undergo metalation sluggishly (see Table 109). Not only ring strain, but also conjugation or homoconjugation of double bonds cause substantial acceleration (see Table 109).

Chelating enamines undergo metalation preferentially at the β-position, unlike 1-alkenyl-1- or 2-benzotriazoles which are attacked at the α position (see Table 110). Cyclic enethers of the di- and tetrahydropyridine type and numerous functionalized enethers also react at the α position (see Tables 111 and 112), whereas lithium allyl-, methallyl- or crotylamides are prone to hydrogen/lithium exchange at the γ-position (see Table 113). Acyclic or cyclic enethers (see Tables 114 and 115) and fluorinated or chlorinated alkenes (see Tables 116 and 117) react with organometallic bases at a heteroatom-bearing carbon whenever possible.

Ferrocene and congeners display high kinetic and thermodynamic acidity. Metalation occurs in the immediate neighborhood of a nitrogen-functional substituent, if one is present (see Table 118).

Lacking benzylic positions, benzene, naphthalene and *tert*-alkylarenes are far more reluctant to undergo metalation than, for example, toluene. Superbasic mixtures are

Table 88. α-Lithio-α-silyloxiranes (M$-$R) by permutational hydrogen/metal exchange between a 2-silyloxirane and the metalating reagent M$-$R'.

M$-$R	M$-$R'	Sν[a]	T[°C]	Product, $El-X'$ [b]	Reference
Si(CH$_3$)$_3$ · M—△—O	LiC$_4$H$_9$	PET	-85	80%, H$_3$CI	[1535–1537]
Si(C$_6$H$_5$)$_3$ · M—△—O	LiC$_4$H$_9$	THF	-75	73%, H$_3$CI	[1536]
(H$_3$C)$_3$Si C$_8$H$_{17}$ · M—△—O	LiCH(CH$_3$)C$_2$H$_5$ [c]	DEE	-100	83%, H$_7$C$_3$CHO	[1537]
Si(CH$_3$)$_3$ · M—△—O—C$_8$H$_{17}$	LiCH(CH$_3$)C$_2$H$_5$ [c]	DEE	-115	73%, H$_7$C$_3$CHO	[1537]
(H$_3$C)$_3$Si C(CH$_3$)$_3$ · M—△—O	LiCH(CH$_3$)C$_2$H$_5$ [c]	DEE	-115	21%, ⬡=O	[1537]
Si(CH$_3$)$_3$ · M—△—O—C(CH$_3$)$_3$	LiCH(CH$_3$)C$_2$H$_5$ [c]	DEE	-115	82%, H$_7$C$_3$CHO	[1537]
Si(CH$_3$)$_3$ [d] · M—△—O—CH$_2$OR	LiCH(CH$_3$)C$_2$H$_5$	THF	-115	77%, R'CHO [e]	[1538]
R$_2$Si—O—▭ [f] · M—△—O	LiC$_4$H$_9$	DEE	-95	38%, H$_2$C=C(CH$_3$)CHO	[1539]

[a] DEE = diethyl ether; PET = petroleum ether (pentanes, hexanes, heptanes); THF = tetrahydrofuran; in the case of mixtures (containing for example paraffinic cosolvents) only the most polar component is indicated. [b] $El-X'$ = electrophilic trapping reagent. [c] In the presence of N,N,N',N'-tetramethylethylenediamine (TMEDA). [d] OR = OCH(CH$_3$)OC$_2$H$_5$. [e] R'CHO = (E,E)-2,5-heptadienyl. [f] R = (H$_3$C)$_2$CH.

required to achieve satisfactory transformations (see Table 119). *N,N*-Dialkylanilines and *N*-alkylcarbazoles are only moderately more reactive (see Table 120). In contrast, *N*-lithiated *N*-BOC anilides (see Table 121), *N,N*-dimethylbenzylamines (see Table 122), *N,N*-diethyl or *N,N*-diisopropylbenzamides or other aromatic carboxamides (see Table 123) and aromatic *N*-lithio-*N*-methyl(thio)carboxamides (see Table 124) undergo *ortho* metalation quite readily. The selective metalation of aromatic ethers (see Tables 125 and 126), acetals (see Table 127) and benzylalcoholates (see Table 128) is abundantly

Table 89. Metalation of esters and carbamates at the α-position of the alkoxy group by reagents M–R'.

M–R [a]	M–R'	Sν [b]	T[°C]	Product, El–X' [c]	Reference
M-CH₂-O-C(=O)-C₆H₃(R')(R') (R' at ortho and para)	LiCH(CH₃)C₂H₅ [d]	THF	−75	84%, H₃CI	[1540–1542]
M-CH₂-O-C(=O)-C₆H₃(R'')(R'')	LiCH(CH₃)C₂H₅ [d]	THF	−75	37%, H₁₇C₈I	[1513]
M-CH(CH₃)-O-C(=O)-C₆H₃(R')(R')	LiCH(CH₃)C₂H₅ [d]	THF	−75	62%, H₂C=CHCH₂Br	[1542–1544]
M-CH(C₂H₅)-O-C(=O)-C₆H₃(R')(R')	LiCH(CH₃)C₂H₅ [d]	THF	−75	43%, (H₃C)₂CO	[1544]
M-CH(C₆H₅)-O-C(=O)-C₆H₃(R')(R')	LiCH(CH₃)C₂H₅ [d]	THF	−75	45%, H₃CI	[1544]
M-CH(CH₂N(CH₃)₂)-O-C(=O)-C₆H₃(R')(R')	LiCH(CH₃)C₂H₅ [d]	THF	−75	82%, D₂O	[1544]
M-CH(CH₃)-O-C(=O)-N(morpholine-type)	LiCH(CH₃)C₂H₅ [d]	DEE	−75	[e], D₂O	[1545,1546]
M-CH(CH₂NBn₂)-O-C(=O)-N [f]	LiCH(CH₃)C₂H₅ [d]	THF	−75	70%, (H₉C₄)₃SnCl	[1547]
M-CH(NBn-pyrrolidine)-O-C(=O)-N [f]	LiCH(CH₃)C₂H₅ [d]	THF	−75	85%, (H₃C)₂CHCHO	[1547]
M-CH(CHC₆H₅, N=C(C₆H₅)₂)-O-C(=O)-N	LiCH(CH₃)C₂H₅ [d]	THF	−75	80%, H₃CI	[1548]

[a] R' = CH(CH₃)₂; R'' = C(CH₃)₃. [b] DEE = diethyl ether; THF = tetrahydrofuran; in the case of mixtures (containing for example paraffinic cosolvents) only the most polar component is indicated. [c] *El* –X' = electrophilic trapping reagent. [d] In the presence of *N,N,N',N'*-tetramethylethylenediamine (TMEDA). [e] Not specified. [f] Bn = CH₂C₆H₅.

Table 90. Metalation of α-haloalkanes using reagents M−R′.

M−R	M−R′	Sν[a]	T[°C]	Product, El−X′[b]	Reference
(perfluorocubane structure, M−F)	LiCH$_3$	DEE	0	66%, (H$_3$C)$_3$SiCl	[1549]
M−CHCl$_2$	LiC$_4$H$_9$[c]	THF	−75	82%, HgCl$_2$	[1550]
Cl M−C−CH$_3$ Cl	LiC$_4$H$_9$[d]	THF	−100	88%, H$_{15}$C$_7$Br	[1551]
Cl　　OSi(CH$_3$)$_3$ M−C−CH Cl　　C$_6$H$_5$	LiC$_4$H$_9$[d]	THF	−110	65%, H$_5$C$_2$Br	[1551]
M−CCl$_3$	BrMgCH(CH$_3$)$_2$[e] LiC$_4$H$_9$[f]	THF THF	−65 −110	25%, H$_5$C$_6$CHO 96%, HgCl$_2$	[1552] [1553−1555]
M−CHBr$_2$	LiCHCl$_2$	THF	−100	72%, H$_5$C$_6$CHO	[1556]
Br M−C−C$_6$H$_5$ Br	LiCHCl$_2$	THF	−100	61%, CO$_2$[f]	[1556]
M−CBr$_3$	BrMgCH(CH$_3$)$_2$[e] LiCHCl$_2$[g]	DEE THF	−70 −100	20%, H$_5$C$_6$CHO 53%, H$_5$C$_6$CHO	[1552] [1554,1556]
M−CI$_3$	LiC$_4$H$_9$	THF	−100	[h]	[862]

[a] DEE = diethyl ether; THF = tetrahydrofuran; in the case of mixtures (containing for example paraffinic cosolvents) only the most polar component is indicated. [b] *El*−X′ = electrophilic trapping reagent. [c] Or at −100 °C rather than −75 °C if in the presence of hexamethylphosphoric triamide (HMPA). [d] In the presence of *N,N,N′,N′*-tetramethylethylenediamine (TMEDA). [e] In the presence of hexamethylphosphoric triamide (HMPA). [f] Isolated as phenylglyoxylic acid. [g] Or with lithium 2,2,6,6-tetramethylpiperidide or even butyllithium[1554]. [h] By NMR.

covered in the literature. The hydrogen/metal permutation has been applied to countless fluorinated and chlorinated arenes (see Tables 129 and 131) and several trifluoromethyl-substituted benzenes and naphthalines (see Table 130).

The metalation of five-membered heterocycles benefits simultaneously from ring strain and heteroelement effects. The reactions employing pyrroles (see Table 132), indoles (see Table 133), pyrazoles (see Table 134), imidazoles (see Table 135), furans (see Table 136), thiophenes (see Table 137), isoxazoles (see Table 138), oxazoles (see Table 139), isothiazoles (see Table 140) and thiazoles (see Table 141) as the substrates proceed rapidly and give mostly good yields.

Text continues on page 211

Table 91. Metalation of propene and 2-alkyl or 2-aryl substituted derivatives thereof at the terminal allylic position using reagents M–R′.

M–R	M–R′	Sv [a]	T[°C]	Product, El–X′ [b]	Reference
$M-CH_2-CH=CH_2$	NaC_5H_{11} [c]	PET	−25	41%, CO_2	[1557]
	$LiC_4H_9/KOC(CH_3)_3$	PET	−25	83%, $(H_3C)_3CSi(C_6H_5)_2Cl$	[236,1558]
	$LiC_4H_9/CsOC(CH_3)_3$	THF	−75	[d]	[150]
$M-CH_2-C=CH_2$ 　　　$\|$ 　　　CH_3	NaC_5H_{11} [c]	PET	+25	51%, CO_2 [e]	[1557]
	$LiC_4H_9/KOC(CH_3)_3$	THF	−50	100%, RCH_2I [f]	[1559,1560]
$M-CH_2-C=CH_2$ 　　　$\|$ 　　　C_2H_5	$LiC_4H_9/KOC(CH_3)_3$	THF	−50	63%, $(H_3CO)_2BF/H_2O_2$	[221,1560]
$M-CH_2-C=CH_2$ 　　　$\|$ 　　　$CH(CH_3)_2$	$LiC_4H_9/KOC(CH_3)_3$	THF	−50	98%, RCH_2I [f]	[221,1560]
$M-CH_2-C=CH_2$ 　　　$\|$ 　　$H_3CCHC_2H_5$	LiC_4H_9 [g]	THF	−50	95%, RCH_2I [f]	[1560]
$M-CH_2-C=CH_2$ 　　　$\|$ 　　　$C(CH_3)_3$	$LiC_4H_9/KOC(CH_3)_3$	THF	−50	88%, RCH_2I [f]	[221,1560]
$M-CH_2-C=CH_2$ 　　　$\|$ 　　$H_3CC(C_2H_5)_2$	LiC_4H_9 [g]	THF	−50	95%, RCH_2I [f]	[1560]
$M-CH_2-C=CH_2$ 　　　$\|$ 　　　$CH=CH_2$	Li-N⟨ ⟩ /KOC(CH_3)_3	THF	−70	82%, $(H_2C)_2O$	[1561]
$M-CH_2-C=CH_2$ 　　　$\|$ 　　$H_3C-C=CH_2$	Li-N⟨ ⟩ /KOC(CH_3)_3	THF	−75	93%, RCH_2I [f]	[1560]
$M-CH_2-C=CH_2$ [h]	LiC_4H_9 [g]	PET	+25	56%, $(H_3C)_3SnCl$	[1562,1563]
	$LiC_4H_9/KOC(CH_3)_3$	THF	−75	89%, $(H_3C)_2C=CHCH_2Br$ [i]	[233]
$M-CH_2-C=CH_2$ 　　　$\|$ 　　　C_6H_5	$LiC_4H_9/KOC(CH_3)_3$	THF	−50	31%, $(H_3C)_3SnCl$	[1318]
$M-CH_2-C=CH_2$ 　　　$\|$ 　$o\text{-}C_6H_4-C_6H_5$	$LiC_4H_9/KOC(CH_3)_3$	THF	−50	30%, $(H_3C)_3SnCl$	[1318]

[a] PET = petroleum ether (pentanes, hexanes, heptanes); THF = tetrahydrofuran; in the case of mixtures (containing for example paraffinic cosolvents) only the most polar component is indicated. [b] $El-X'$ = trapping electrophile. [c] In the presence of sodium isopropoxide ('Alfin complex'). [d] By NMR. [e] Mono- and diacid. [f] RCH_2I = (3β, 5α, 6β, 20S)-21-iodo-6-methoxy-20-methyl-3,5-cyclopregnane (see structure beneath Tables 92 and 93). [g] In the presence of N,N,N',N'-tetramethylethylenediamine (TMEDA). [h] Limonene as the precursor. [i] Linear and branched regioisomer (85 : 15).

Table 92. Metalation of 1-alkyl substituted propenes (2-alkenes) at the terminal allylic position using superbasic reagent mixtures M–R'.

M–R [a]	M–R'	Sν [b]	T[°C]	Product, El–X' [c]	Reference
M–CH$_2$–CH=CH–CH$_3$	LiC$_4$H$_9$/KOC(CH$_3$)$_3$	PET	+25	50%, (H$_3$CO)$_2$BF/H$_2$O$_2$	[225,1229]
M–CH$_2$–CH=CH–C$_2$H$_5$	LiC$_4$H$_9$/KOC(CH$_3$)$_3$	PET	+25	81%, (H$_3$C)$_3$SiCl	[1275]
M–CH$_2$–CH=CH–C$_3$H$_7$	LiC$_4$H$_9$/KOC(CH$_3$)$_3$	THP	0	60%, (H$_3$C)$_3$SiCl	[1558]
M–CH$_2$–CH=CH–C$_4$H$_9$	LiC$_4$H$_9$/KOC(CH$_3$)$_3$	THP	0	52%, (H$_3$C)$_3$SiCl	[1558]
M–CH$_2$–CH=CH–C$_5$H$_9$	LiC$_4$H$_9$/KOC(CH$_3$)$_3$	THP	0	70%, (H$_3$C)$_3$SiCl	[1558]
M–CH$_2$–CH=CH–C$_6$H$_{13}$	LiC$_4$H$_9$/KOC(CH$_3$)$_3$	THP	0	70%, (H$_3$C)$_3$SiCl	[1558]
M–CH$_2$–CH=CH–C$_7$H$_{15}$	LiC$_4$H$_9$/KOC(CH$_3$)$_3$	THP	0	72%, (H$_3$C)$_3$SiCl	[1558]
M–CH$_2$–CH=CH–C$_8$H$_{17}$	LiC$_4$H$_9$/KOC(CH$_3$)$_3$	THP	0	75%, (H$_3$C)$_3$SiCl	[1558]
M–CH$_2$–CH=CH–C$_9$H$_{19}$	LiC$_4$H$_9$/KOC(CH$_3$)$_3$	THP	0	67%, (H$_3$C)$_3$SiCl	[1558,1564]
M–CH$_2$–CH=CH–C(CH$_3$)$_3$	LiC$_4$H$_9$/KOC(CH$_3$)$_3$	THF	−50	79%, (H$_2$C)$_2$O [d]	[226,1276]
M–CH$_2$–CH=CH–C(C$_6$H$_5$)$_3$	LiC$_4$H$_9$/KOC(CH$_3$)$_3$	THF	−50	90%, CO$_2$	[1565]
M–CH$_2$–CH=C(CH$_3$)$_2$	LiC$_4$H$_9$/KOC(CH$_3$)$_3$	THF	−75	65%, (H$_2$C)$_2$O [e]	[474,475]
M–CH$_2$–CH=C(C$_2$H$_5$)$_2$	LiC$_4$H$_9$/KOC(CH$_3$)$_3$	THF	−50	42%, (H$_3$C)$_3$CCHO	[474,475]
M–CH$_2$–CH=C(CH$_3$)C$_2$H$_5$	LiC$_4$H$_9$/KOC(CH$_3$)$_3$	THF	−50	55%, RCH$_2$I [f]	[1560]

[a] The original alkene configuration is perfectly retained in the product if metalation and interception times are kept short. Conversely, under conditions of complete torsional equilibration[1558], (Z/E) ratios of >99 : 1 and ~94 : 6 are established for butenylpotassium and homologous 2-alkenylpotassiums, respectively. [b] PET = petroleum ether (pentanes, hexanes, heptanes); DEE = diethyl ether; THF = tetrahydrofuran; THP = tetrahydropyran; in the case of mixtures (containing for example paraffinic cosolvents) only the most polar component is indicated. [c] El–X' = electrophilic trapping reagent. Unbranched substrates afford mainly (>9 : 1) (Z) isomers. [d] The electrophile attacks the terminal and internal allylic positions in the ratio 9 : 1. [e] 2 : 1 mixture of regioisomers. [f] RCH$_2$I = (3β,5α,6β,20S)-21-iodo-6-methoxy-20-methyl-3,5-cyclopregnane:

Table 93. Metalation of 1,2-dialkyl-substituted propenes at the terminal allylic position using superbasic mixtures M−R′.

M−R [a]	M−R′	Sv [b]	T[°C]	Product, El−X′ [c]	Reference
M−CH₂−C=C−CH₃ with H₃C, H	LiC₄H₉/KOC(CH₃)₃	THF	−50	28%, RCH₂I [g]	[1560,1566]
M−CH₂−C=C−C₂H₅ with H₃C, H	LiC₄H₉/KOC(CH₃)₃	THF	−50	59%, RCH₂I [g]	[1560,1566]
M−CH₂−C=C−C₄H₉ with H₃C, H	LiC₄H₉/KOC(CH₃)₃	THF	−50	63%, (R′O)₂BF/H₂O₂ [d]	[221]
M−CH₂−C=C−CH₂C(CH₃)₃ with H₃C, H	LiC₄H₉/KOC(CH₃)₃	THF	−50	53%, (R′O)₂BF/H₂O₂	[221]
M−CH₂−C=C−(CH₂)₂ [e] with H₃C, H	LiC₄H₉/KOC(CH₃)₃	THF	−50	46%, (R′O)₂BF/H₂O₂	[1567]
M−CH₂−C=C−(CH₂)₂ [f] with H₃C, H	LiC₄H₉/KOC(CH₃)₃	THF	−50	46%, (R′O)₂BF/H₂O₂	[1568]
M−CH₂−C=C(CH₃)₂ with H₃C	LiC₄H₉/KOC(CH₃)₃	THF	−50	17%, RCH₂CH₂I [h]	[1560]
M−CH₂−C=C−C₄H₉ with H₅C₂, H	LiC₄H₉/KOC(CH₃)₃	THF	−50	60%, (H₃CO)₂BF/H₂O₂	[221]
M−CH₂−C=C−CH₃ with (H₃C)₂HC, H	LiC₄H₉/KOC(CH₃)₃	THF	−50	58%, RCH₂I [g]	[1560]
M−CH₂−C=C−C₂H₅ with (H₃C)₂HC, H	LiC₄H₉/KOC(CH₃)₃	THF	−50	59%, RCH₂I [g]	[1560]

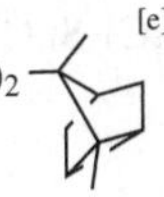

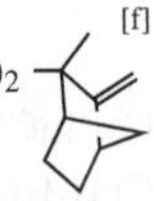

[a] Under conditions of complete torsional equilibration[1558], the (Z) configuration is strongly favored [(Z/E) ratios ≥500]. [b] THF = tetrahydrofuran; in the case of mixtures (containing for example paraffinic cosolvents) only the most polar component is indicated. [c] El−X′ = electrophilic trapping reagent. [d] R′O = H₃CO. [e] α-Santalene as the precursor. [f] Epi-β-santalene as the precursor. [g] RCH₂CH₂I = (3β,5α,6β)-3,5-cyclo-24-norcholane. [h] RCH₂I = (3β, 5α, 6β, 20S)-21-iodo-6-methoxy-20-methyl-3,5-cyclopregnane:

R =

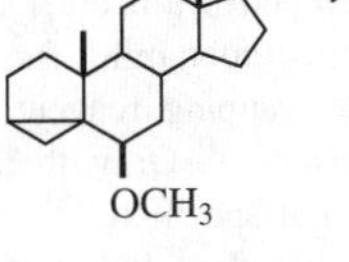

OCH₃

Table 94. Metalation of 1-methylcycloalkenes at the exocyclic allylic position using reagents M−R′ [a].

M−R	M−R′ [a]	Sv [b]	T[°C]	Product, El−X′ [c]	Reference
(structure: M–CH₂–cyclopropyl)	LiC₄H₉	THF	+10	71%, (oxepane-like structure with O)	[1569]
(structure: M–CH₂–cyclobutyl)	LiC₄H₉ [d]	PET	+25	73%, H₅C₆CH₂Br [e]	[1570]
(structure: M–CH₂–cyclopentenyl)	LiC₄H₉ [d,f]	PET	+25	[g], H₅C₆CHO	[1571]
(structure: M–CH₂–cyclohexenyl)	LiC₄H₉/KOC(CH₃)₃	PET	+25	65%, (H₃CO)₂BF/H₂O₂	[1572]
(structure: M–CH₂–cyclohexenyl with isopropyl)	LiC₄H₉ [d]	PET	+25	52%, (H₃C)₃SnCl	[1563]
(structure: M–CH₂–bicyclic gem-dimethyl)	LiC₄H₉ [d]	PET	+25	47%, (H₃C)₃SnCl	[1563]
(structure: M–CH₂–pinene-like bicyclic)	LiC₄H₉ [d]	PET	+25	59%, (H₃C)₃SnCl	[1563]
	LiC₄H₉/KOC(CH₃)₃	PET	+25	42%, (H₃CO)₂BF/H₂O₂	[1573]
(structure: M–CH₂–cycloheptenyl)	LiC₄H₉/KOC(CH₃)₃	THF	−75	61%, (H₃C)₃SiCl	[1574]
(structure: M–CH₂–cyclooctenyl)	LiC₄H₉/KOC(CH₃)₃	THF	−75	58%, (H₃C)₃SiCl	[1574]
(structure: M–CH₂– caryophyllene type) [h]	LiC₄H₉/KOC(CH₃)₃	THF	−75	10%, (H₃CO)₂BF/H₂O₂ [i]	[1575]
(structure: M–CH₂– isocaryophyllene type) [j]	LiC₄H₉/KOC(CH₃)₃	PET	+25	10%, (H₃CO)₂BF/H₂O₂ [i]	[1575]

[a] The same organometallic intermediate may be generated by metalation of the isomeric methylenecycloalkane at an allylic methylene group. [b] PET = petroleum ether (pentanes, hexanes, heptanes); THF = tetrahydrofuran; THP = tetrahydropyran; in the case of mixtures (containing for example paraffinic cosolvents) only the most polar component is indicated. [c] El−X′ = electrophilic trapping reagent. [d] In the presence of N,N,N′,N′-tetramethylethylenediamine (TMEDA). [e] Regioisomeric 1:1 mixture. [f] Or with butyllithium in the presence of potassium tert-butoxide rather than with LiC₄H₉/TMEDA. [g] Yield not specified. [h] Caryophyllene as the precursor. [i] Unaltered starting material is recovered to a large extent. [j] Isocaryophyllene [(Z)-isomer] as the precursor.

Table 95. Metalation of cycloalkenes at an allylic position using reagents M−R′.

M−R	M−R′	Sv [a]	T[°C]	Product, El−X′ [b]	Reference
M–⬠ [c]	KC$_4$H$_9$	PET	+25	8%, CO$_2$	[439]
M–⬡	LiCH(CH$_3$)C$_2$H$_5$ [d]	PET	+25	77%, (H$_2$C)$_2$O	[440]
M–⬡	LiC$_4$H$_9$/KOC(CH$_3$)$_3$	PET	+25	63%, (H$_3$C)$_3$SiCl	[233]
M–◯	LiC$_4$H$_9$/KOC(CH$_3$)$_3$	THP	−30	48%, (H$_3$C)$_3$SiCl	[1576]
M–◯	KCH$_2$Si(CH$_3$)$_3$	THF	−60	65%, (H$_3$C)$_3$SiCl [e]	[1576]
M–◯	LiC$_4$H$_9$/KOC(CH$_3$)$_3$	PET	+25	58%, (H$_5$CO)$_2$BF/H$_2$O$_2$	[233]

[a] PET = petroleum ether (pentanes, hexanes, heptanes); THP = tetrahydropyran; THF = tetrahydrofuran; in the case of mixtures (containing for example paraffinic cosolvents) only the most polar component is indicated. [b] El−X′ = electrophilic trapping reagent. [c] Minor product, the main reaction mode (36%) being the deprotonation of an olefinic position. [d] In the presence of KOC(CH$_3$)$_3$. [e] 2,5- and 2,4-Cyclooctadienyltrimethylsilane are formed in a ratio of 96 : 4.

Table 96. Metalation of aliphatic allenes using reagents M−R′.

M−R	M−R′	Sv [a]	T[°C]	Product, El−X′ [b,c]	Reference
M−CH=C=CH$_2$	LiC$_4$H$_9$	DEE	−15	89%, geranyl-Cl	[1577−1582]
M−CH=C=CH−C$_3$H$_7$	LiC$_4$H$_9$	THF	−75	66%, (H$_3$C)$_2$NCON(CH$_3$)$_2$	[1579,1580]
M−CH=C=CH−C$_5$H$_{11}$	LiC$_4$H$_9$	THF	−75	75%, CO$_2$	[1579,1580]
M−CH=C=CH−C$_6$H$_{13}$	LiC$_4$H$_9$	THF	−75	[d], H$_3$CI	[1579,1580]
M−CH=C=CH−C$_8$H$_{17}$	LiC$_4$H$_9$	THF	−75	62%, H$_3$CCON(CH$_3$)$_2$	[1579,1580]
M−CH=C=C(CH$_3$)$_2$	LiC$_4$H$_9$	THF	−75	85%, (H$_2$C)$_2$O	[1579,1580]
M−C=C=C(CH$_3$)$_2$ ‖ C$_4$H$_9$	LiC(CH$_3$)$_3$	THF	−40	88%, CO$_2$	[1579,1580]

[a] DEE = diethyl ether; THF = tetrahydrofuran; in the case of mixtures (containing for example paraffinic cosolvents) only the most polar component is indicated. [b] El−X′ = electrophilic trapping reagent. [c] The allenic derivatives are often contaminated by small amounts of acetylenic isomers resulting from electrophilic attack at the internal allylic position. [d] Yield not specified.

Table 97. Metalation of internal acetylenes at a propargylic position using reagents M−R′.

M−R	M−R′	Sv [a]	T[°C]	Product, El−X′	Reference
M−CH$_2$−C≡C−C$_4$H$_9$	LiC$_4$H$_9$ [d]	DEE	0	81%, (H$_3$C)$_3$SiCl	[1583]
M−CH$_2$−C≡C⁓⟋	LiC$_4$H$_9$ [d]	DEE	0	86%, (H$_3$C)$_3$SiCl	[1583]
M−CH$_2$−C≡C−C(CH$_3$)$_3$	LiC$_4$H$_9$ [d]	DEE	0	85%, (H$_3$C)$_3$SiCl	[1583]
M−CH$_2$−C≡C−◁	LiC$_4$H$_9$	THF	+25	90%, (H$_3$C)$_3$SiCl	[1584]
M−CH(CH$_3$)−C≡C−◁ [e]	LiC$_4$H$_9$	THF	+25	27%, (H$_3$C)$_3$SiCl	[1584]
M◁C≡C−C$_2$H$_5$ [e]	LiC$_4$H$_9$	THF	+25	48%, (H$_3$C)$_3$SiCl	[1584]
M◁C≡C−△	LiC$_4$H$_9$	THF	+25	73%, CO$_2$	[1585]
M◁C≡C−CH(CH$_3$)$_2$	LiC$_4$H$_9$	THF	+25	72%, H$_5$C$_2$Br	[1584]
M−CH(C$_5$H$_{11}$)−C≡C−Si(CH$_3$)$_3$	LiC(CH$_3$)$_3$	THF	−30	89%, (H$_3$C)$_3$SiCl	[1583]
M−CH(C$_6$H$_{11}$)−C≡C−Si(CH$_3$)$_3$	LiC(CH$_3$)$_3$	THF	−30	86%, (H$_3$C)$_3$SiCl	[1583]
M−CH(C$_6$H$_5$)−C≡C−Si(CH$_3$)$_3$	LiC(CH$_3$)$_3$	THF	−30	65%, (H$_3$C)$_3$SiCl	[1583]

[a] DEE = diethyl ether; THF = tetrahydrofuran; in the case of mixtures (containing for example paraffinic cosolvents) only the most polar component is indicated. [b] El−X′ = electrophilic trapping reagent. [c] Depending on the nature of the electrophile, more or less significant amounts of allenic isomers are formed along with the acetylenic main product. [d] In the presence of N,N,N′,N′-tetramethylethylenediamine (TMEDA). [e] 1-Cyclopropyl-1-butyne is concomitantly metalated at the propargylic methylene and the cyclopropylic methine group.

Table 98. Metalation of conjugated or homoconjugated dienes using reagents M−R′.

M−R [a]	M−R′[b]	Sv [c]	T[°C]	Product, El−X′[d]	Reference
$M-CH_2-\overset{\underset{\|}{H}}{C}=\overset{\underset{\|}{H}}{C}-CH=CH_2$	LiC$_4$H$_9$/KOR	THF	−75	95%, (H$_5$C$_2$)$_3$SiCl [e,f]	[230,231,233]
$M-CH_2-\overset{\overset{H}{\|}}{C}=C-CH=CH_2$	Li^{S}C$_4$H$_9$	THF	−75	93%, (H$_3$C)$_3$SiCl [e,g]	[230,396,1586]
	LiC$_4$H$_9$/KOR	PET	0	95%, (H$_3$C)$_3$SiCl [e,g]	[230,233]
$M-CH_2-\overset{\underset{\|}{H}}{C}=\overset{\underset{\|}{H}}{C}-\overset{\overset{H}{\|}}{C}=C-CH_3$ [h]	Li^{S}C$_4$H$_9$	THF	−50	65%, FB(OCH$_3$)$_2$/H$_2$O$_2$ [i,j]	[233]
	LiC$_4$H$_9$/KOR	PET	0	63%, FB(OCH$_3$)$_2$/H$_2$O$_2$ [i,k]	[233]
$M-CH_2-\overset{\overset{H}{\|}}{C}=C-\overset{\underset{\|}{H}}{C}=\overset{\underset{\|}{H}}{C}-CH_3$ [m]	Li^{S}C$_4$H$_9$	THF	−50	60%, FB(OCH$_3$)$_2$/H$_2$O$_2$ [i,m]	[233]
$M-CH_2-\overset{\underset{\|}{H}}{C}=\overset{\underset{\|}{H}}{C}-\overset{\overset{H}{\|}}{C}=C-C_5H_{11}$ [h]	Li^{S}C$_4$H$_9$/KOR	THF	0	52%, FB(OCH$_3$)$_2$/H$_2$O$_2$ [k]	[232]
$M-CH_2-\overset{\overset{H}{\|}}{C}=C-\overset{\underset{\|}{H}}{C}=\overset{\underset{\|}{H}}{C}-C_5H_{11}$ [l]	Li^{S}C$_4$H$_9$	THF	−75	80%, FB(OCH$_3$)$_2$/H$_2$O$_2$ [n]	[232]
$M-CH_2-\overset{\overset{H}{\|}}{C}=\overset{\overset{H}{\|}}{C}-\overset{\underset{\|}{H}}{C}=C-C_5H_{11}$	Li^{S}C$_4$H$_9$	THF	−75	73%, FB(OCH$_3$)$_2$/H$_2$O$_2$	[232]
$M-CH_2-\overset{\underset{\|}{H}}{C}=\overset{\underset{\|}{H}}{C}-\overset{\underset{\|}{CH_3}}{C}=CH_2$	LiC$_4$H$_9$/KOR	THF	−75	65%, FB(OCH$_3$)$_2$/H$_2$O$_2$ [o]	[230]
$M-CH_2-\overset{\overset{H}{\|}}{C}=\overset{\underset{\|}{H}}{C}-\overset{\underset{\|}{CH_3}}{C}=CH_2$	Li^{S}C$_4$H$_9$	THF	−75	77%, FB(OCH$_3$)$_2$/H$_2$O$_2$ [p]	[230]
$M-CH_2-\overset{\overset{H_3C}{\|}}{C}=\overset{\underset{\|}{H}}{C}-\overset{\overset{H}{\|}}{C}=CH_2$	Li^{S}C$_4$H$_9$	THF	−75	82%, FB(OCH$_3$)$_2$/H$_2$O$_2$ [f]	[1586]
$M-CH_2-\overset{\underset{\|}{H_3C}}{C}=\overset{\underset{\|}{H}}{C}-\overset{\underset{\|}{CH_3}}{C}=CH_2$	Li^{S}C$_4$H$_9$	THF	−50	57%, FB(OCH$_3$)$_2$/H$_2$O$_2$ [e]	[230]
	LiC$_4$H$_9$/KOR	PET	0	68%, FB(OCH$_3$)$_2$/H$_2$O$_2$ [e]	[230]
$M-CH_2-\overset{\underset{\|}{(H_3C)_3C}}{C}=\overset{\underset{\|}{H}}{C}-\overset{\underset{\|}{C(CH_3)_3}}{C}=CH_2$	LiC$_4$H$_9$/KOR	PET	+25	35%, solid, isolated	[1587]

[a] The corresponding 1,4-diene as the hydrocarbon precursor. [b] KOR = KOC(CH$_3$)$_3$; Li^{S}C$_4$H$_9$ = LiCH(CH$_3$)C$_2$H$_5$.
[c] Solvent (Sv) : PET = petroleum ether (pentanes, hexanes, heptanes); THF = tetrahydrofuran; in the case of mixtures only the most polar component is indicated. [d] El−X′ = trapping electrophile. [e] α/γ regioratios ranging from 5 : 1 to 30 : 1. [f] (Z/E) > 95 : 5. [g] (Z/E) ≤ 5 : 95. [h] (E)-isomer as the precursor. [i] B(OR)$_2$ = 4,4,5,5-tetramethyl-1,3-dioxa-2-borolan-2-yl. [j] (2E,4E/2Z,4E) = 82 : 18. [k] (2Z,4E/2E,4E) = 96 : 4. [l] (Z)-isomer as the precursor. [m] (2E,4Z/2Z,4Z) = 92 : 8. [n] (2E,4Z/2Z,4Z) = 97 : 3. [o] Head/tail attack ~1 : 2; (Z/E) = 97 : 3. [p] Head/tail attack ~2 : 3; (Z/E) = 28 : 72.

Table 99. Metalation of silylated alkenes at allylic positions with reagents M–R′.

M–R	M–R′	Sv [a]	T[°C]	Product, El–X′ [b]	Reference
M–CH$_2$–C=CH$_2$ $\mid$ Si(CH$_3$)$_3$	LiC$_4$H$_9$/KOC(CH$_3$)$_3$	THF	−50	45%, (H$_3$C)$_3$SnCl	[1318]
M–CH$_2$–CH=CH–Si(CH$_3$)$_3$	LiC$_4$H$_9$	THF	−75	73%, [c]	[1588,1589]
	LiC$_4$H$_9$/KOC(CH$_3$)$_3$	PET	+25	98%, H$_3$CI [d]	[226,1590]
M–CH$_2$–CH=CH–Si(C$_6$H$_5$)$_3$	LiC$_4$H$_9$/TMEDA	DEE	0	70%, H$_3$COCH$_2$Cl	[1591]
M–CH$_2$–CH=CH–Si(CH$_3$)$_2$–CH$_2$X [e]	LiCH(CH$_3$)C$_2$H$_5$	THF	−60	78%, H$_3$CI [f]	[1592]
M–CH$_2$–CH=CH–Si(CH$_3$)$_2$–CH$_2$X′ [g]	LiCH(CH$_3$)C$_2$H$_5$	THF	−60	94%, H$_3$CI [f]	[1592]
M–CH$_2$–C(CH$_3$)=CH–Si(CH$_3$)$_2$–X″ [h]	LiC$_4$H$_9$/KOC(CH$_3$)$_3$	THF	+25	80%, H$_{17}$C$_8$Br [c]	[1593]
M–CH(CH$_3$)–CH=CH–Si(CH$_3$)$_3$	LiC$_4$H$_9$/KOC(CH$_3$)$_3$	THF	−75	50%, H$_3$CI [f]	[1594]
M–CH(C$_2$H$_5$)–CH=CH–Si(CH$_3$)$_3$	LiC$_4$H$_9$/KOC(CH$_3$)$_3$	THF	−75	50%, (H$_3$CO)$_2$BF [c,i]	[1230]
M–CH(C$_3$H$_7$)–CH=CH–Si(CH$_3$)$_3$	LiC$_4$H$_9$/KOC(CH$_3$)$_3$	THF	−75	75%, (H$_3$C)$_3$SiCl [c,i]	[1230,1594]
M–CH(C$_5$H$_{11}$)–CH=CH–Si(CH$_3$)$_3$	LiC$_4$H$_9$/KOC(CH$_3$)$_3$	THF	−75	78%, (H$_3$C)$_3$SiCl [c]	[1230]
M–CH(C$_9$H$_{19}$)–CH=CH–Si(CH$_3$)$_3$	LiC$_4$H$_9$/KOC(CH$_3$)$_3$	THF	−75	77%, H$_2$CO [j]	[1230]
M–CH((CH$_2$)$_3$OM)–CH=CH–Si(CH$_3$)$_3$	LiC$_4$H$_9$/KOC(CH$_3$)$_3$	PET	+25	63%, (H$_3$CO)$_2$BF [c,i]	[1595]
M–CH((CH$_2$)$_5$OM)–CH=CH–Si(CH$_3$)$_3$	LiC$_4$H$_9$/KOC(CH$_3$)$_3$	THF	−75	44%, (H$_3$CO)$_2$BF [c,i]	[1595]

[a] PET = petroleum ether (pentanes, hexanes, heptanes); THF = tetrahydrofuran; in the case of mixtures (containing for example paraffinic cosolvents) only the most polar component is indicated. [b] El–X′ = electrophilic trapping reagent. [c] Electrophilic attack at the Si-remote γ position. [d] α : γ isomer = 1 : 5. [e] X = N(CH$_2$CH$_2$OCH$_3$)$_2$. [f] Electrophilic attack at the Si-adjacent α position. [g] X′ = 1-[2-(methoxymethyl)pyrrolidyl]. [h] X″ = N(^{i}C$_3$H$_7$)$_2$. [i] Subsequent conversion to the allylic alcohol by immediate oxidation with alkaline hydrogen peroxide. [j] Mixture of regioisomers.

Table 100. M–R' Promoted α metalation of allyl amines or γ-metalation of enamines.

M–R	M–R'	Sv[a]	T[°C]	Product, El–X'[b,c]	Reference
M–CH2–CH=CH–N⟨ring⟩	LiCH(CH3)C2H5	THF	−10	50%, H9C4Br	[1596]
M–CH2–CH=CH–N–C6H5 (CH3)	LiC4H9/KOC(CH3)3	PET	+25	70%, (structure)	[1597]
	LiC4H9	THF	0	75%, H5C6CH2Br[d]	[1598]
M–CH2–CH=CH–N⟨carbazol-9-yl⟩	LiC4H9[e]	DEE	−15	83%, (H3C)2CHCl	[1599]
M–CH2–CH=C–N⟨carbazol-9-yl⟩ (CH3)	LiC4H9[e]	DEE	−15	72%, (H3C)2CHCl	[1599]
M–CH2–C=CH–N⟨carbazol-9-yl⟩ (CH3)	LiC4H9[e]	DEE	−15	70%, (H3C)2CHCl	[1599]
M–CH2–CH=CH–N⟨benzimidazol-1-yl⟩	LiC4H9	THF	−75	85%, H3CI[f]	[1600]
M–CH2–CH=CH–N⟨benzotriazol-1-yl⟩	LiC4H9	THF	−75	91%, H9C4I[f]	[1600]
M–CH2⟨pyrrole ring⟩ N–CH3	LiCH(CH3)C2H5[g]	PET	−125	48%, (H3C)3SiCl	[1601]
M–CH2⟨pyrrole ring⟩ N–C(CH3)3	LiCH(CH3)C2H5[e]	PET	−75	~100%, H3CI	[1602]
M–CH2–CH=CH–N–BOC[h] (M)	LiCH(CH3)C2H5	THF	−75	70%, H5C6CHO[i]	[1603]

[a] DEE = diethyl ether; PET = petroleum ether (pentanes, hexanes, heptanes); THF = tetrahydrofuran; in the case of mixtures (containing for example paraffinic cosolvents) only the most polar component is indicated. [b] El–X' = electrophilic trapping reagent. [c] The electrophiles shown attack mainly at the nitrogen-remote γ-position giving (Z)-isomers, whereas other electrophiles may afford regioisomeric mixtures. [d] 4-Phenylbutanal isolated after acidic hydrolysis. [e] In the presence of N,N,N',N'-tetramethylethylenediamine (TMEDA). [f] Electrophilic attack exclusively at the nitrogen-adjacent α-position. [g] In the presence of 1,4-bis(dimethylamino)-2,3-(dimethoxy)butane. [h] BOC = COOC(CH3)3. [i] Clean attack at the nitrogen-adjacent α-position, affording mainly the *anti* diastereoisomer, only after preceding treatment with ZnCl2; otherwise regioisomeric mixtures are formed.

Table 101. Metalation of allylic ethers (including cyclic ones) using reagents M–R′.

M–R	M–R′	Sv[a]	T[°C]	Product, $El–X′$[b,c]	Reference
$M–CH_2–CH=CH–OCH_3$	$LiCH(CH_3)C_2H_5$	THF	–65	93%, (cyclohexyl–O structure)	[1604]
$M–CH_2–CH=CH–OC(CH_3)_3$	$LiCH(CH_3)C_2H_5$	THF	–65	80%, H_3CSSCH_3	[1604]
$M–CH_2–CH=CH–OC_6H_5$	LiC_4H_9 / $LiC_4H_9/KOC(CH_3)_3$	THF / PET	–75 / –30	95%, H_3CI / 57%, H_3CI	[1605] / [236]
$M–CH_2–\underset{\underset{CH_3}{\vert}}{C}=CH–OC_2H_5$	$LiCH(CH_3)C_2H_5$	THF	–65	79%, $H_{13}C_6I$	[1604]
$M–CH_2–CH=C(OCH_3)_2$	$LiCH(CH_3)C_2H_5$	THF	–90	72%, $(H_3C)_3SnCl$	[1606]
$M–CH_2–CH=C(OC_2H_5)_2$	$LiCH(CH_3)C_2H_5$	THF	–90	22%, $H_2C=CH–CH_2Br$	[1606]
(cyclic M–dihydropyran structure)	LiC_4H_9[d]	DEE	–65	73%, $(H_3C)_3SiCl$[e]	[1607]
(cyclic H_3C/M–dihydropyran structure)	$LiC(CH_3)_3$	THF	0	30%, $(H_5C_6)_2CO$	[1608]
(cyclic M–dioxine structure)	$LiCH(CH_3)C_2H_5$	THF	–75	70%, $H_3C–\overset{H}{\underset{H}{C}}=C–CHO$[e]	[1609]
$M–CH_2–$(dioxolane structure)	$LiCH(CH_3)C_2H_5$	THF	–75	100%, H_5C_6CHO[f]	[1610]
$M–CH_2–CH=$(epoxide-O structure)	$LiC(CH_3)_3$	THF	–110	45%, $(H_3C)_3SiCl$[e,g]	[1611]
(cyclooctene–O structure) M	$LiC(CH_3)_3$[d]	THF	–90	95%, $(H_3C)_3SiCl$[e,g]	[1611]

[a] DEE = diethyl ether; PET = petroleum ether (pentanes, hexanes, heptanes); THF = tetrahydrofuran; in the case of mixtures (containing for example paraffinic cosolvents) only the most polar component is indicated. [b] $El–X′$ = electrophilic trapping reagent. [c] Predominantly electrophilic attack at the oxygen-remote γ-position unless stated otherwise. [d] In the presence of $N,N,N′,N′$-tetramethylethylenediamine (TMEDA). [e] Electrophilic attack preferentially or exclusively at the oxygen-adjacent α-position only after preceding treatment with $ZnCl_2$; otherwise regioisomeric mixtures are formed. [g] The metalation is performed in the presence of chlorotrimethylsilane ('*in situ* trapping').

Table 102. M−R′ Promoted metalation of allylic and homoallylic alcoholates.

M−R	M−R′	Sv [a]	T[°C]	Product, El−X′	Reference
M−CH$_2$−C=CH−CH$_2$OM 　　　| 　　　CH$_3$	LiC$_4$H$_9$/KOC(CH$_3$)$_3$	PET	+25	7%, H$_3$CI [c]	[236]
M−CH$_2$−C=CH$_2$ 　　　| 　　　CH$_2$OM	LiC$_4$H$_9$/KOC(CH$_3$)$_3$	PET	+25	54%, epoxycyclohexane	[1612]
M−CH$_2$−C=CH−CH$_3$ 　　　| 　　　CH$_2$OM	LiC$_4$H$_9$/KOC(CH$_3$)$_3$	PET	−75	89%, [d]	1612
M−CH$_2$−C=CH$_2$ 　　　| 　　　CH$_2$CH$_2$OM	LiC$_4$H$_9$ [e] LiC$_4$H$_9$/KOC(CH$_3$)$_3$	PET PET	−75 +25	70%, (H$_3$C)$_2$C=CHCH$_2$Br 48%, H$_3$CI	[1613] [236]
M−CH$_2$−CH=C−CH=CH$_2$ 　　　　　　| 　　　　　　OM	LiC$_4$H$_9$/KOC(CH$_3$)$_3$	THF	−40	75%,	[1614]

[a] PET = petroleum ether (pentanes, hexanes, heptanes); THF = tetrahydrofuran; in the case of mixtures (containing for example paraffinic cosolvents) only the most polar component is indicated. [b] El−X′ = electrophilic trapping reagent. [c] Electrophilic attack the oxygen-remote γ position affording the (Z)-isomer. [d] Regioisomeric 2 : 1 mixture. [e] In the presence of N,N,N′,N′-tetramethylethylenediamine (TMEDA).

Table 103. Metalation of allylic halides using reagents M−R′.

M−R	M−R′	Sv [a]	T[°C]	Product, El−X′ [b,c]	Reference
M−CH$_2$−CH=CH−Cl	LiN$_2$	THF	−75	66%, (H$_3$C)$_3$SiCl	[1615]
M−CH$_2$−CH=CH−Br	LiN$_2$	THF	−75	55%, (H$_3$C)$_3$SiCl	[1615]
M−CH$_2$−CH=C−Cl 　　　　　| 　　　　　CH$_3$	LiN(^{i}C$_3$H$_7$)$_2$	THF	−75	74%, H$_9$C$_4$Br [d]	[1616]
M−CH$_2$−C=CH−Cl 　　　| 　　　CH$_3$	LiN$_2$	THF	−75	40%, (H$_3$C)$_3$SiCl	[1615]
M−CH−CH=CH−Cl 　| 　CH$_3$	LiN$_2$	THF	−75	32%, (H$_3$C)$_3$SiCl [e]	[1615]
M−CH$_2$−CH=CCl$_2$	LiN(^{i}C$_3$H$_7$)$_2$	THF	−95	81%, (H$_3$C)$_3$SiCl [d]	[1616−1618]

[a] THF = tetrahydrofuran; in the case of mixtures (containing for example paraffinic cosolvents) only the most polar component is indicated. [b] El−X′ = electrophilic trapping reagent. [c] The deprotonation is performed in the presence of the electrophile ('in situ trapping'); the halogen-adjacent α-position is preferentially or exclusively attacked unless stated otherwise. [d] Electrophilic attack occurs mainly at the halogen-remote γ-position (α/γ ratios 1 : 6−1 : 12). [e] (Z/E) ratio 15 : 85.

Table 104. Metalation of alkylarenes at benzylic positions using reagents M−R′.

M−R	M−R′	Sv [a]	T [°C]	Product, El−X′ [b]	Reference
M−CH₂−⬡	NaC₆H₅ [c]	TOL	−105	79%, CO₂	[706,1619]
M−CH₂−⬡(H₃C)	LiC₄H₉/KOC(CH₃)₃	PET	+25	[d], (H₃C)₃SiCl	[1620]
M−CH₂−⬡(CH₃)	LiC₄H₉ [e]	PET	+25	68%, (H₃C)₃SiCl	[1621]
M−CH₂−⬡−CH₃	LiC₄H₉/KOC(CH₃)₃	PET	+25	97%, (H₃CO)₂SO₂	[1622]
M−CH₂−⬡−C₂H₅	LiC₄H₉/KOC(CH₃)₃	PET	+25	87%, (H₃C)₂CHBr	[1622]
M−CH(CH₃)−⬡	RbCH₂Si(CH₃)₃	PET	+25	[d,f]	[1623]
M−CH(CH₃)−⬡−C₂H₅	LiC₄H₉/KOC(CH₃)₃	PET	+25	97%, CO₂	[1622]
M−CH(CH₃)−⬡−CH₂CH(CH₃)₂	LiC₄H₉/KOC(CH₃)₃	PET	+25	79%, CO₂	[1622]
M−CH(C₂H₅)−⬡	LiC₄H₉/KOC(CH₃)₃	THF	−50	85%, ⬡−CO−R [g]	[1624]
M−C(CH₃)₂−⬡	KCH₂Si(CH₃)₃	THF	−50	74%, H₃CI	[440,441]
M−(cyclopropyl)−⬡	KCH₂Si(CH₃)₃	THF	−50	33%, H₃CI	[1625]
M−(benzocyclobutene)	LiCH₂CH₂−⬡−Li	THF	+25	45%, H₂C=CH₂	1626

[a] PET = petroleum ether (pentanes, hexanes, heptanes); TOL = toluene; THF = tetrahydrofuran; in the case of mixtures (containing for example paraffinic cosolvents) only the most polar component is indicated. [b] El−X′ = electrophilic trapping reagent. [c] Phenylsodium is generated *in situ* from chlorobenzene and sodium. [d] Not specified. [e] In the presence of N,N,N',N',N'-pentamethyldiethylenetriamine (PMDTA). [f] By NMR. [g] R = p-C₆H₄OCH₂CH₂N(CH₃)₂, the product being isolated after dehydration of the adduct as the antimammary carcinoma drug tamoxifen.

Table 105. Lateral metalation of *N*-acylanilines, using reagents M−R′.

M−R	M−R′	Sv[a]	T[°C]	Product, El−X′[b]	Reference
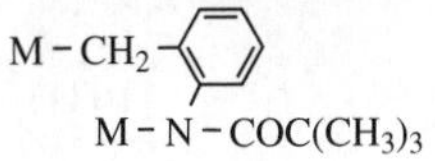 M−CH₂-⟨ring⟩ M−N−COC(CH₃)₃	LiC₄H₉	THF	0	71%, H₃CCHO	[1627]
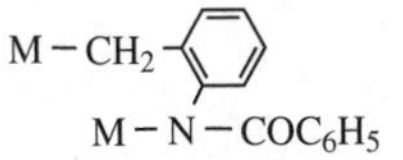 M−CH₂-⟨ring⟩ M−N−COC₆H₅	LiC₄H₉	THF	+25	90%, [c]	[1628]
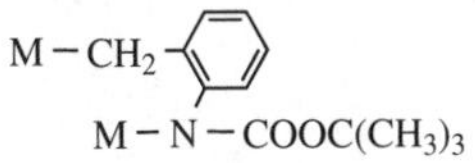 M−CH₂-⟨ring⟩ M−N−COOC(CH₃)₃	LiCH(CH₃)C₂H₅	THF	−20	98%, (CH₂)₃Cl₂	[1629]
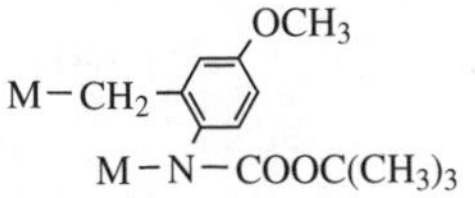 OCH₃ / M−CH₂-⟨ring⟩ M−N−COOC(CH₃)₃	LiCH(CH₃)C₂H₅	THF	−20	60%, (H₃C)₂NCHO [d]	[1630]
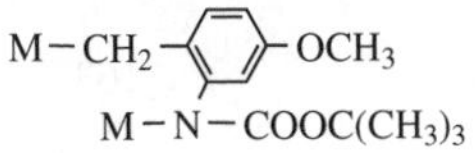 M−CH₂-⟨ring⟩-OCH₃ M−N−COOC(CH₃)₃	LiCH(CH₃)C₂H₅	THF	−15	75%, (H₃C)₂NCHO [d]	[1630]
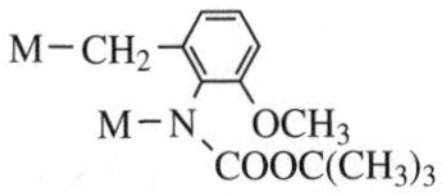 M−CH₂-⟨ring⟩ M−N(OCH₃)COOC(CH₃)₃	LiCH(CH₃)C₂H₅	THF	−20	64%, (H₃C)₂NCHO [d]	[1630]
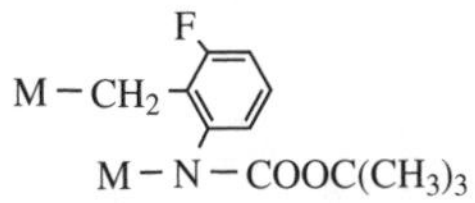 F / M−CH₂-⟨ring⟩ M−N−COOC(CH₃)₃	LiCH(CH₃)C₂H₅	THF	−20	73%, (H₃C)₂NCHO [d]	[1630]
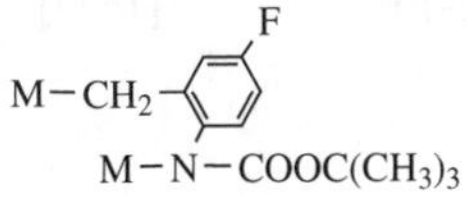 F / M−CH₂-⟨ring⟩ M−N−COOC(CH₃)₃	LiCH(CH₃)C₂H₅	THF	−20	86%, (H₃C)₂NCHO [d]	[1630]
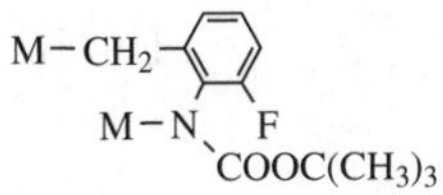 M−CH₂-⟨ring⟩ M−N(F)COOC(CH₃)₃	LiCH(CH₃)C₂H₅	THF	−20	66%, (H₃C)₂NCHO [d]	[1630]
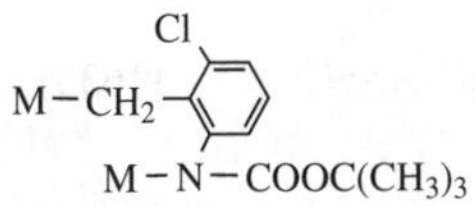 Cl / M−CH₂-⟨ring⟩ M−N−COOC(CH₃)₃	LiCH(CH₃)C₂H₅	THF	−20	70%, (H₃C)₂NCHO [d]	[1630]
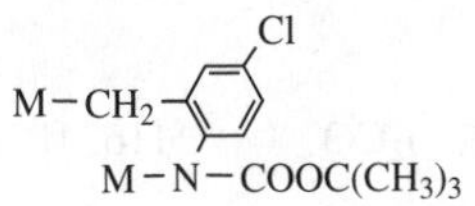 Cl / M−CH₂-⟨ring⟩ M−N−COOC(CH₃)₃	LiCH(CH₃)C₂H₅	THF	−20	83%, (H₃C)₂NCHO [d]	[1630]
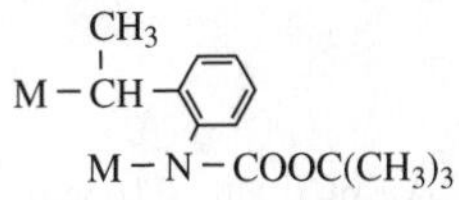 CH₃ / M−CH-⟨ring⟩ M−N−COOC(CH₃)₃	LiCH(CH₃)C₂H₅ [e]	THF	−20	80%, (H₃C)₂NCHO [d]	[1630]

[a] THF = tetrahydrofuran; in the case of mixtures (containing for example paraffinic cosolvents) only the most polar component is indicated. [b] El−X′ = electrophilic trapping reagent. [c] Intramolecular reaction affording 2-phenylindole after dehydration. [d] Indole formation by acid-catalyzed cyclization and dehydration. [e] In the presence of *N,N,N′,N′*-tetramethylethylenediamine (TMEDA).

Table 106. Lateral metalation of cresyl methyl ethers and in 2-methylbenzyl alcohols using reagents M–R′.

M–R	M–R′	Sv [a]	T[°C]	Product, El–X′[b]	Reference
$M-CH_2-$ (phenyl, OCH$_3$ ortho)	Li(N(^{i}C$_3$H$_7$)$_2$[c]	PET	+25	67%, CO$_2$	[1631]
$M-CH_2-$ (phenyl[c], OCH$_3$ meta)	LiC$_4$H$_9$/NaOC(CH$_3$)$_3$	PET	+25	71%, CO$_2$	[1631]
$M-CH_2-$ (phenyl)$-$OCH$_3$ (para)	LiN (pyrrolidide)[c]	THF	−50	43%, CO$_2$[d]	[1631]
$M-CH_2-$ (phenyl, H$_3$CO and CH$_3$)	LiC$_4$H$_9$	THF	0	48%, CO$_2$	[1632]
$M-CH_2-$ (phenyl, H$_3$C/OCH$_3$/H$_3$CO/CH$_3$)	LiC$_4$H$_9$	DEE	+25	79%, H$_3$CCHO	[1633]
$M-CH_2-$ (phenyl, MOCH$_2$)	LiC$_4$H$_9$	DEE	+25	92%, H$_5$C$_2$Br	[1634]
$M-CH_2-$ (phenyl)$-$CH$_3$, MOCH$_2$	LiC$_4$H$_9$	DEE	+25	53%, CO$_2$	[1634]
$M-CH_2-$ (phenyl, MO$-$CH$-$CH$_3$)	LiC$_4$H$_9$	DEE	+25	71%, (H$_9$C$_4$)$_3$SnCl	[1635]
$M-CH_2-$ (phenyl, H$_3$CO, MO$-$CH$-$CH$_3$)	LiC$_4$H$_9$	DEE	+25	59%, (H$_5$C$_6$)$_2$CO	[1634]
$M-CH_2-$ (phenyl, MO$-$CH$-$C$_6$H$_{13}$)	LiC$_4$H$_9$	DEE	+25	67%, (H$_9$C$_4$)$_3$SnCl	[1635]

[a] PET = petroleum ether (pentanes, hexanes, heptanes); DEE = diethyl ether; THF = tetrahydrofuran; in the case of mixtures (containing for example paraffinic cosolvents) only the most polar component is indicated. [b] El–X′ = electrophilic trapping reagent. [c] In the presence of potassium *tert*-butoxide. [d] Concomitant metalation at the oxygen-adjacent *ortho* position.

Table 107. M−R′ Promoted metalation of 2-aryloxiranes at the benzylic position.

M−R	M−R′	$S\nu$ [a]	$T[°C]$	Product, El−X′ [b]	Reference
	LiC(CH$_3$)$_3$ [c]	PET	−90	96%, (H$_3$C)$_3$SiCl [c]	[1611]
	LiC(CH$_3$)$_3$ [c]	THF	−95	91%, (H$_5$C$_2$)$_2$NCOC$_6$H$_5$	[1611]
	LiC(CH$_3$)$_3$ [c]	THF	+25	70%, (H$_3$C)$_3$SiCl [d]	[1611]
	LiC(CH$_3$)$_3$ [c]	THF	−80	75%, (H$_5$C$_2$)$_2$NCOC$_6$H$_5$	[1611]

[a] PET = petroleum ether (pentanes, hexanes, heptanes); THF = tetrahydrofuran; in the case of mixtures (containing for example paraffinic cosolvents) only the most polar component is indicated. [b] El−X′ = electrophilic trapping reagent. [c] In the presence of N,N,N',N'-tetramethylethylenediamine (TMEDA). [d] The deprotonation is performed in the presence of the electrophile ('*in situ* trapping').

Pyridines are less reactive and require activation by suitable substituents. Successful metalations have been reported of acylamino, dimethyloxazolinyl and N,N-diisopropyl-carbamoyl-substituted pyridines (see Table 142), methoxy- or carboxy-substituted pyridines (see Table 143) and halogenated pyridines (see Table 144).

b) Scope and Limitations

The preceding Subsection surveys organometallic intermediates generated by hydrogen/metal permutation, classified according to structural families. The present Subsection follows the same general theme but focuses on specific topics, all of which being somehow related to reactivity and selectivity issues.

• Metalation Thresholds and Superbases

The ease with which hydrogen/metal exchange can be accomplished depends on the basicity gradient between the reagent and the new organometallic intermediate formed and on the electropositivity of the metal involved (see Figure 10 on p. 248).

Table 108. M–R′ Promoted generation of halogenated benzylmetals.

M−R	M−R′	S_v [a]	T [°C]	Product, El−X′ [b]	Reference
M−CH₂−C₆H₄(F) (ortho)	LiN(ⁱC₃H₇)₂ [c]	THF	−75	37%, CO_2	[1636]
M−CH₂−C₆H₄(F) (meta)	LiN(ⁱC₃H₇)₂ [c]	THF	−75	62%, CO_2	[1636]
M−CH₂−C₆H₄(CF₃) (para)	LiC₄H₉/KOC(CH₃)₃	THF	−75	67%, CO_2	[1636]
M−CH₂−C₆H₄(Cl) (ortho)	LiN [d]	THF	−100	27%, CO_2	[1365]
M−CH₂−C₆H₄(Cl) (meta)	LiN [d]	THF	−100	16%, CO_2	[1365]
M−CH₂−C₆H₄−Cl (para)	LiN [d]	THF	−100	5%, CO_2	[1365]
M−CH₂−C₆H₄(Br) (ortho)	LiN [d]	THF	−100	23%, CO_2	[1365]
M−CH₂−C₆H₄(Br) (meta)	LiN [d]	THF	−100	20%, CO_2	[1365]
M−CH₂−C₆H₄−Br (para)	LiN [d]	THF	−100	13%, CO_2	[1365]

[a] THF = tetrahydrofuran; in the case of mixtures (containing for example paraffinic cosolvents) only the most polar component is indicated. [b] El−X′ = electrophilic trapping reagent. [c] In the presence of potassium *tert*-butoxide. [d] In the presence of potassium *tert*-butoxide and *N,N,N′,N″,N″*-pentamethyldiethylenetriamine (PMDTA).

Organomagnesiums react only with the most acidic hydrocarbons such as cyclopentadiene and terminal alkynes. Alkyllithiums are capable of deprotonating doubly activated substrates such as 1,4-pentadiene and diphenylmethane, whereas pentylsodium can even convert benzene in phenylsodium. The metalation of pentane, hexane and cyclohexane by butylpotassium has been claimed[1934] but should be considered with scepticism until conclusive experimental evidence is provided. For the time being, cyclopropanes are the least acidic hydrocarbons which can be submitted to a hydrogen/metal permutation under synthetically useful conditions.

Text continues on page 248

Table 109. Metalation of alkenic positions using reagents M–R'.

M–R	M–R'	Sv [a]	T[°C]	Product, El–X' [b]	Reference
M–CH=CH$_2$	NaC$_5$H$_{11}$ [c]	PET	+25	'good yield', CO$_2$	[664]
	LiC$_4$H$_9$/KOC(CH$_3$)$_3$ [d]	PET	−25	90%, CO$_2$	[1637]
M–CH=CH–C(CH$_3$)$_3$	NaC$_5$H$_{11}$ [c]	PET	+25	61%, CO$_2$	[664]
	NaC$_5$H$_{11}$ [e]	PET	+25	88%, CO$_2$ [f]	[1489]
(structure)	NaC$_5$H$_{11}$/KOC(CH$_3$)$_3$	PET	+25	71%, (H$_3$C)$_2$NCHO [e]	[440,1146]
(structure)	NaC$_5$H$_{11}$/KOC(CH$_3$)$_3$	PET	+25	70%, (H$_3$CO)$_2$BF [g]	[1489]
(structure)	LiC$_4$H$_9$/NaOC(CH$_3$)$_3$	PET	+25	92%, (H$_3$C)$_3$SiCl	[1638]
	NaC$_5$H$_{11}$	PET	+25	'good yield', CO$_2$	[1639]
(structure)	LiC$_4$H$_9$/KOC(CH$_3$)$_3$	PET	+25	80%, H$_5$C$_6$Si(CH$_3$)$_2$Cl [h]	[1640]
(structure)	NaC$_5$H$_9$/KOC(CH$_3$)$_3$	PET	+25	66%, (H$_3$C)$_3$SiCl	[1638,1641]
(structure)	NaNH$_2$	NH$_3$	−35	87%, (H$_5$C$_2$)$_2$CO	[1642]
(structure)	BrMgC$_2$H$_5$	DEE	+25	[i], CO$_2$	[1643]
	LiCH$_3$	DEE	+25	≤32%, H$_3$CCON(CH$_3$)$_2$	[1644]
(structure)	LiC(CH$_3$)$_3$ [c]	THF	−75	78%, (H$_3$C)$_3$SnCl [j]	[181,182]
(structure)	NaC$_5$H$_9$/KOC(CH$_3$)$_3$	PET	0	61%, BrCH$_2$CH$_2$Br [j]	[1645]
(structure)	LiC$_4$H$_9$/NaOC(CH$_3$)$_3$	THF	−50	88%, (H$_3$C)$_3$SiCl	[1638]
(structure)	NaC$_5$H$_{11}$	PET	+25	66%, (H$_3$C)$_3$SiCl	[1489]
(structure)	LiC$_4$H$_9$/NaOC(CH$_3$)$_3$	PET	+25	57%, (H$_3$C)$_3$SiCl	[1638]

[a] PET = petroleum ether (pentanes, hexanes, heptanes); DEE = diethyl ether; THF = tetrahydrofuran; in the case of mixtures (containing for example paraffinic cosolvents) only the most polar component is indicated. [b] El–X' = electrophilic trapping reagent. [c] In the presence of sodium isopropoxide. [d] In the presence of N,N,N',N'-tetramethylethylenediamine (TMEDA). [e] In the presence of disodium pinacolate. [f] Pure (E)-isomer. [g] 2-bicyclo[2.2.2]octanone isolated after oxidation with alkaline hydrogen peroxide. [h] Along with a minor regioisomer carrying the metal at the methyl-adjacent olefinic position in the ratio 7 : 1. [i] Yield not specified. [j] Along with a minor regioisomer carrying the metal at the neighboring olefinic positions in ratios 3 : 2–7 : 3.

Table 110. β- and α-Metalation of enamines using reagents M–R′.

M–R	M–R′	$S\nu$ [a]	T[°C]	Product, El–X′ [b]	Reference
(structure) C5H11; H3CN, N(CH3)2	LiC(CH3)3	PET	+25	87%, $H_{21}C_{10}I$ [c]	[1646]
(structure) H3CN, N(CH3)2	LiC(CH3)3	PET	+25	95%, H_3CI [c]	[1646]
(structure) H3CN, N(CH3)2	LiC(CH3)3	PET	+25	75%, $H_{21}C_{10}I$ [c]	[1646]
(structure) N=N, N	LiC4H9	THF	−75	48%, H_9C_4I	[1647]
(structure) N=N, N, CH3 [d]	LiC4H9	THF	−75	78%, H_3CI	[1647]
(structure) N=N, N, H3C	LiC4H9	THF	−75	78%, H_3CI	[1647]
(structure) N, N–N, CH3 [d]	LiC4H9	THF	−75	84%, H_3CI	[1647]
(structure) N, N–N, H3C	LiC4H9	THF	−75	82%, H_3CI	[1647]

[a] PET = petroleum ether (pentanes, hexanes, heptanes); THF = tetrahydrofuran; in the case of mixtures (containing for example paraffinic cosolvents) only the most polar component is indicated. [b] El–X′ = electrophilic trapping reagent. [c] Isolated as the ketone obtained after acid hydrolysis. [d] As the minor component in a mixture mainly containing the (E) isomer (see the line below).

Table 111. M−R′ Promoted metalation of dihydro- and tetrahydropyridines.

M−R	M−R′	Sv [a]	T[°C]	Product, El−X′ [b]	Reference
[pyridine structure, N−COOC(CH₃)₃, M−]	LiC(CH₃)C₂H₅	THF	−75	92%, (H₉C₄)₃SnCl	[1648,1649]
[pyridine structure, N−CH₃, M−]	LiC₄H₉/KOC(CH₃)₃	PET	+25	42%, (H₃C)₃SiCl	[1650]
[pyridine structure, N−C₆H₅, M−]	LiC₄H₉	THF	0	94%, D₂O [c]	[1651]
[pyridine structure, N−NC(CH₃)₃, M−]	LiC₄H₉ [d]	THF	−20	85%, H₇C₃I	[1652]
[pyridine structure, 4-C₄H₉, N−COOC(CH₃)₃, M−]	LiC(CH₃)C₂H₅	THF	−40	89%, (H₃CO)₂CO	[1653]
[pyridine structure, 4-C₆H₅, N−COOC(CH₃)₃, M−]	LiC(CH₃)C₂H₅	THF	−40	80%, H₃CI	[1653]
[pyridine structure, CH₃, CH(OM [e])NR₂, N−COOC(CH₃)₃, M−]	Li−[mesityl]	THF	−40	82%, H₃CI	[1654]
[pyridine structure, RO, CH₃ [f], N−COOC(CH₃)₃, M−]	LiC₆H₅	THF	−75	75%, H₃CI	[1654]
[pyridine structure, Br, CH₃, N−COOC(CH₃)₃, M−]	LiC₆H₅	THF	−40	87%, H₃CI	[1654]
[pyridine structure, Cl, N−C₁₁H₂₃, COOC(CH₃)₃, M−]	LiC₄H₉	THF	−40	83%, H₃CI	[1654]

[a] THF = tetrahydrofuran; in the case of mixtures (containing for example paraffinic cosolvents) only the most polar component is indicated. [b] El−X′ = electrophilic trapping reagent. [c] Progressive 2,6-dimetalation after prolonged exposure times. [d] Or LiC(CH₃)₃. [e] NR₂ = [pyrrolidine]NCH₃ [f] RO = (H₅C₂)₂NCOO (*N,N*-diethylcarbamoyloxy).

Table 112. M−R′ Promoted metalation of 'counter-polarized' enamines carrying powerful electron acceptor groups at the β-position.

M−R	M−R′	Sv[a]	T[°C]	Product, El−X′ [b]	Reference
NC, M, N (pyrrolidine enamine)	$LiN(^iC_3H_7)_2$	THF	−115	35%, H_5C_2I	[1655]
$OCN(C_2H_5)_2$, M, N (pyrrolidine enamine)	$LiC(CH_3)_3$	THF	−115	60%, H_5C_2I	[1655,1656]
$OCOC_2H_5$, M, N (pyrrolidine enamine)	$LiC(CH_3)_3$	THF	−115	41%, [c]	[1655]
NC, M, N−CH$_3$ (dihydropyridine)	$LiN(^iC_3H_7)_2$	THF	−80	37%, H_3COD	[1657]
NC, M, N, H_3C (dihydropyridine)	$LiN(^iC_3H_7)_2$	THF	−80	72%, H_3COSO_2F	[1657]
$OCN(C_2H_5)_2$, M, N, H_3C	$LiN(^iC_3H_7)_2$	THF	−80	82%, H_3COD	[1657]
NC, M, N, H_3C'	$LiN(^iC_3H_7)_2$	THF	−80	77%, H_3COSO_2F	[1657]
O, M, N, H_3C'	LiC_4H_9	THF	0	43%, H_3CI	[1658]
O, M, N, H_3C' (benzo)	$LiN(^iC_3H_7)_2$	THF	−75	71%, H_6C_6CHO	[1659]
F, O, M, N−CH$_3$, N, H_3C', O	$LiN(^iC_3H_7)_2$	THF	−75	40%, ICl	[1660]
O, M, N−COOM	$LiC(CH_3)_3$ [d]	THF	−70	60%, $(H_5C_6)_2CO$	[1661]

[a] THF = tetrahydrofuran; in the case of mixtures (containing for example paraffinic cosolvents) only the most polar component is indicated. [b] El−X′ = electrophilic trapping reagent. [c] 'Autoreaction', *i.e.* condensation of the metalated species with its precursor. [d] In the presence of hexamethylphosphoric triamide ('HMPA').

Table 113. M—R′ Promoted metalation of allylic metal enamides.

M—R	M—R′	Sv [a]	T[°C]	Product, El—X′ [b]	Reference
M—C=C—CH₂—N—CH₂—CH=CH₂ (H, H, M)	LiC(CH₃)₃	DEE	+25	87%, LiC(CH₃)₃[c]	[1662]
M—C=C—CH₂—N—C₆H₁₁ (H, H, M)	LiC(CH₃)₃	DEE	+25	68%, (H₅C₂O)₂CO	[1663]
M—C=C—CH₂—N—C(CH₃)₃ (H, H, M)	LiC₄H₉[d]	PET	+40	58%, (H₃C)₂SnCl₂	[1664,1665]
M—C=C—CH₂—N—C₆H₅ (H, H, M)	LiC(CH₃)₃	DEE	+25	71%, (H₅C₂O)₂CO	[1663]
M—C=C—CH₂—N—Si(CH₃)₃ (H, H, M)	LiC(CH₃)₃	PET	+70	32%, H₃CBBr₂	[1666]
M—C=C—CH₂—N—CH₂—CH=CH₂ (H, CH₃, M)	LiC(CH₃)₃	DEE	+25	86%, LiC(CH₃)₃[c]	[1662]
M—C=C—CH₂—N—CH(CH₃)₂ (H, CH₃, M)	LiC(CH₃)₃	DEE	+25	88%, (H₅C₂O)₂CO	[1663]
M—C=C—CH₂—N—C₆H₁₁ (H, CH₃, M)	LiC(CH₃)₃	DEE	+25	77%, (H₅C₂O)₂CO	[1663]
M—C=C—CH₂—N—C₆H₅ (H, CH₃, M)	LiC(CH₃)₃	DEE	+25	74%, (H₅C₂O)₂CO	[1663]
M—C=C—CH₂—N—Si(CH₃)₃ (H, CH₃, M)	LiC₄H₉	DEE	+25	50%, (H₃C)₃SiCl	[1665]
M—C=C—CH₂—N—C₆H₁₁ (H₃C, H, M)	LiC(CH₃)₃	DEE	+25	69%, (H₅C₂O)₂CO	[1663]
M—C=C—CH₂—N—C₆H₅ (H₃C, H, M)	LiC(CH₃)₃	DEE	+25	78%, (H₅C₂O)₂CO	[1663]

[a] DEE = diethyl ether; PET = petroleum ether (pentanes, hexanes, heptanes). [b] El—X′ = electrophilic trapping reagent. [c] Nucleophilic addition to the vinyl group. [d] Plus N,N,N',N'-tetramethylethylenediamine (TMEDA).

Table 114. α-Metalation of acyclic enethers using reagents M−R′.

M−R	M−R′	$S\nu$[a]	T[°C]	Product, El−X′[b]	Reference
M−C(=CH₂)OCH₃	LiC(CH₃)₃	THF	0	78%, H₅C₆CHO	[1667–1669]
M−C(=CH₂)OC₂H₅	LiC(CH₃)₃[c]	PET	−30	43%, H₅C₆CHO	[1670–1671]
M−C(=C, cyclic O/O-tetrahydropyranyl)	LiC(CH₃)₃/KOC(CH₃)₃	THF	−75	83%, H₃CI	[237]
M−CH(CH₃)(OCH₃)	LiC(CH₃)₃[c]	[d]	[d]	55%, H₅C₆CHO[e]	[1667]
M−CH(C₆H₅)(OCH₃)	LiC(CH₃)₃	TMEDA	−75	54%, (H₃C)₃SnCl	[1672]
M−C(=CH₂, OCH₃) allyl	LiC(CH₃)₃	THF	0	30%, H₅C₆CHO[e]	[1667]
M−C(=CH₂)CH₂R[f] (OCH₃)	LiC₄H₉/KOC(CH₃)₃	THF	−95	30%, H₃CI	[1673]
M−CH(OC₂H₅)(OC₂H₅)	LiC(CH₃)₃	THF	−75	75%, (H₃C)₃SiCl	[911]
M−C(=, OC₂H₅/OC₂H₅)	LiC(CH₃)₃	THF	−75	73%, (H₃C)₃SiCl	[911]
M−C(Cl)(OCH₃)(OCH₃)	LiC₄H₉	THF	−100	45%, CO₂	[1674]

[a] THF = tetrahydrofuran; in the case of mixtures (containing for example paraffinic cosolvents) only the most polar component is indicated. [b] El−X′ = electrophilic trapping reagent. [c] In the presence of N,N,N',N'-tetramethylethylene-diamine (TMEDA). [d] Not explicitly mentioned, presumably in THF at 0 °C. [e] Isolated after acid hydrolysis as 1-hydroxy-1-phenyl-2-butanone and 1-hydroxy-1-phenyl-3-buten-2-one, respectively. [f] R = prenyl (3-methyl-2-butenyl).

Table 115. α-Metalation of cyclic enethers using reagents M–R'.

M–R	M–R'	Sv [a]	T[°C]	Product, $El-X'$ [b]	Reference
(dihydrofuran, M at O)	LiC(CH₃)₃	THF	0	67%, (H₃C)₂C=CH–CH₂Br	[1675,1676]
(dimethyl dihydrofuran)	LiC(CH₃)₃	THF	+25	88%, H₃CCOC₆H₅	[1677]
(dihydropyran, M at O)	LiC(CH₃)₃	PET	0	60%, ⟨⟩–Br	[1607,1675]
	NaC₅H₁₁	PET	+25	49%, CO₂	[1678]
H₃C (methyl dihydropyran, M–O)	LiC(CH₃)₃	THF	0	0% (?)	[1675,1679]
CH₃ (4-methyl dihydropyran)	LiC(CH₃)₃	THF	0	60%, RCH₂I [c]	[1680]
(oxepine, M–O)	LiC₄H₉	THF	0	38%, (H₃C)₂CO	[1676]
CH₃ (methyl dihydropyran)	LiC₄H₉/KOC(CH₃)₃	PET	0	48%, (H₃C)₂SiCl	[1650]
(dioxine, M–O–O)	LiC(CH₃)₃	PET	–20	66%, ⌇O	[1681,1682]
OC₂H₅ / OC₂H₅ (diethoxy)	LiC₄H₉	PET	0	53%, (H₃C)₂CO	[1675]
H₃CO (methoxy dihydropyran)	LiC₄H₉	THF	+50	65%, H₅C₆CH₂O(CH₂)₃I	[1683]
OQ, OQ, OQ [d]	LiC(CH₃)₃/KOC(CH₃)₃	THF	–75	55%, (H₉C₄)₃SnCl	[1350,1684]
OR, OR, RO, OR [e]	LiC(CH₃)₃/KOC(CH₃)₃	THF	–100	47%, H₃CI	[1685]
Cl (chloro dihydrofuran)	LiC₄H₉	THF	–75	74%, H₃CI	[1686]
Cl (chloro dihydropyran)	LiC(CH₃)₃C₂H₅	THF	–75	78%, H₅C₂I	[1687]

[a] PET = petroleum ether (pentanes, hexanes, heptanes); THF = tetrahydrofuran; in the case of mixtures (containing for example paraffinic cosolvents) only the most polar component is indicated. [b] $El-X'$ = electrophilic trapping reagent. [c] R = (E)-(H₃C)₃CO(CH₂)₃C(CH₃)=CHCH₂. [d] Q = SiC(CH₃)₂C(CH₃)₃, Si(C₆H₅)₂ or Si(^{i}C₃H₇)₃. [e] OR = OBn (OCH₂C₆H₅).

Table 116. Metalation of fluoroalkenes using reagents M−R′.

M−R	M−R′	Sv [a]	T[°C]	Product, El−X′ [b]	Reference
$M-CH=CF_2$	LiC_4H_9 [c]	DEE	−75	52%, $(H_3C)_2CO$	[1688,1689]
$M-CF=CF_2$	LiC_4H_9	DEE	−75	63%, $(F_3C)_2CO$	[1688]
$\overset{F}{\underset{}{M}}-\overset{F}{C}=C-C_4H_9$	LiC_4H_9	THF	−110	72%, H_7C_3CHO	[943]
$M-\overset{F}{\underset{F}{C}}=C-CH(CH_3)C_2H_5$	LiC_4H_9	THF	−30	90%, CO_2	[943]
$M-\overset{F}{C}=\overset{F}{C}-C_7H_{15}$	LiC_4H_9	THF	−110	70%, CO_2	[943]
$M-\overset{F}{\underset{F}{C}}=C-C_7H_{15}$	LiC_4H_9	THF	−30	78%, CO_2	[943]
$M-\overset{F}{\underset{F}{C}}=C-C_6H_5$	LiC_4H_9	THF	−30	84%, CO_2	[943]
$M-\overset{F}{\underset{F}{C}}=C-Si(C_2H_5)_3$	LiC_4H_9	THF	−90	87%, I_2	[1690]
$M-C=C-CF_3$ with $F\ F$	LiC_4H_9	THF	−90	88%, I_2 [d]	[1691]
$M-C=CF_2$ with CF_3	$LiC(CH_3)_3$ [e]	DEE	−75	~100%, by ^{19}F NMR [d]	[1692]
$M-C=C-F$; F_2C-CF_2	$LiCH_3$	DEE	−70	23%, H_3CCHO	[1688–1693]
$M-C=C-F$; F_2C CF_2 / CF_2 (ring)	$LiCH_3$	DEE	−70	42%, H_3CCHO	[1693]
$M-C=C-F$; F_2C CF_2 / F_2C-CF_2 (ring)	$LiCH_3$	DEE	−70	63%, H_3CCHO	[1693]

[a] DEE = diethyl ether; THF = tetrahydrofuran; in the case of mixtures (containing for example paraffinic cosolvents) only the most polar component is indicated. [b] El−X′ = electrophilic trapping reagent. [c] Or *sec*-butyllithium. [d] After conversion into the corresponding organozinc species by treatment with zinc iodide. [e] Or $LiN(^iC_3H_7)_2$.

Table 117. Metalation of haloalkenes using reagents M−R′.

M−R	M−R′	Sv [a]	$T[°C]$	Product, El−X′ [b]	Reference
(anthracene-bridged structure, Cl, M)	LiC$_4$H$_9$	THF	−70	75%, CO$_2$	[1694]
(bicyclic structure, Cl, M)	LiC(CH$_3$)$_3$	THF	−75	44%, BrCH$_2$CH$_2$Cl	[1695]
M−CCl=CH$_2$	LiC$_4$H$_9$	THF	−110	100%, CO$_2$	[1696]
M−C(Cl)=C(H)−CH=CH$_2$	Li-N(structure)	THF	−90	68%, H$_5$C$_6$CHO	[1697]
M−C=C(Cl)(H)−CH=CH$_2$	Li-N(structure)	THF	−90	77%, H$_5$C$_6$CHO	[1697]
M−CCl=C(CH$_2$C$_6$H$_5$)$_2$	LiC$_4$H$_9$	THF	−100	97%, I$_2$	[1698]
M−C(Cl)=C(H)−C$_6$H$_5$	LiC$_4$H$_9$	THF	−115	13%, CO$_2$	[1699]
M−C=C(Cl)(H)−C$_6$H$_5$	LiC$_4$H$_9$	THF	−100	96%, CO$_2$	[1699]
M−CCl=C(C$_6$H$_5$)$_2$	LiC$_4$H$_9$	THF	−70	83%, CO$_2$	[1700]
M−C(Cl)=C(H)−OC$_2$H$_5$	LiC$_4$H$_9$	THF	−100	40%, CO$_2$	[303]
M−C=C(Cl)(H)−OC$_2$H$_5$	LiC$_4$H$_9$	THF	−100	100%, CO$_2$	[303]
M−C(Cl)=C(H)−Cl	LiC$_4$H$_9$	THF	−100	100%, CO$_2$	[1696]
M−C=C(Cl)(H)−Cl	LiC$_4$H$_9$	THF	−100	99%, CO$_2$	[1696]
M−CCl=CF$_2$	LiC$_4$H$_9$	DEE	−75	61%, H$_3$CCOCF$_3$	[1688]
M−CCl=CClF	LiC$_4$H$_9$	THF	−75	60%, (H$_3$C)$_2$CO	[1688]
M−CCl=CCl$_2$	LiC$_4$H$_9$	THF	−100	81%, CO$_2$	[1696]

[a] DEE = diethyl ether; THF = tetrahydrofuran; in the case of mixtures (containing for example paraffinic cosolvents) only the most polar component is indicated. [b] El−X′ = electrophilic trapping reagent.

Table 118. Metalation of ferrocene and nitrogen-bearing congeners using butyllithium (in the presence or absence of potassium *tert*-butoxide) as the metalating reagent M−R′.

M−R [a]	M−R′	$S\nu$ [b]	$T[°C]$	Product, El−X′ [c]	Reference
M—⬠ FeCp	LiC$_4$H$_9$/KOC(CH$_3$)$_3$	THF	−75	91%, (H$_3$C)$_2$NCHO	[1701–1703]
(H$_3$C)$_2$NCH$_2$ M—⬠ FeCp	LiC$_4$H$_9$	DEE	+25	71%, (H$_5$C$_6$)$_2$CO	[1704]
(H$_3$C)$_2$NCHCH$_3$ M—⬠ FeCp	LiC$_4$H$_9$	DEE	+25	58%, H$_2$CO	[1705]
(H$_3$C)$_2$NCHCH$_3$ M—⬠ [d] Q FeCp	LiC$_4$H$_9$	DEE	+25	45%, (H$_5$C$_6$)$_2$PCl	[1706]
N—CH$_2$ [d] M—⬠ Q FeCp	LiC$_4$H$_9$	DEE	+25	57%, (H$_3$C)$_2$SiCl	[1707]
N⟩ M—⬠ FeCp	LiC$_4$H$_9$	DEE	+25	82%, (H$_5$C$_6$)$_2$CO	[1708]
H$_5$C$_2$ ⟍N−C=O M⟋ M—⬠ FeCp	LiC$_4$H$_9$	THF	+25	30%, (H$_3$CO)$_2$SO$_2$	[1709]
(H$_3$C)$_2$NCH$_2$ CH$_2$ M—⬠ FeCp	LiC$_4$H$_9$	DEE	+25	68%, (H$_5$C$_6$)$_2$CO	[1710]

[a] Cp = C$_5$H$_5$ (cyclopentadienyl). [b] DEE = diethyl ether; THF = tetrahydrofuran; in the case of mixtures (containing for example paraffinic cosolvents) only the most polar component is indicated. [c] El–X′ = electrophilic trapping reagent. [d] Q = (H$_3$C)$_3$Si (trimethylsilyl).

Table 119. Superbase (M–R′) promoted metalations of benzene, naphthalene, *tert*-alkyl substituted benzenes, phenylacetylene and triisopropylsilylbenzene at aromatic positions.

M–R	M–R′	Sv [a]	T[°C]	Product, El–X′ [b]	Reference
M–⟨benzene⟩	$LiC_4H_9/KOC(CH_3)_3$	PET	+25	65%, CO_2[c]	[1711]
M–⟨naphthalene⟩	$LiC_4H_9/KOC(CH_3)_3$	THF	–50	23%, $(H_3CO)_2SO_2$[d]	[1711]
M–⟨phenylcyclopropyl⟩	$LiC_4H_9/KOC(CH_3)_3$	PET	+25	52%, CO_2[e]	[1711]
M–⟨$C(CH_3)_3$-benzene⟩	$LiC_4H_9/KOC(CH_3)_3$	PET	+25	53%, CO_2[e]	[1711]
M–⟨3,5-di-$C(CH_3)_3$-benzene⟩	$LiC_4H_9/KOC(CH_3)_3$	PET	+25	90%, CO_2	[1711,1712]
M–⟨1,4-di-$C(CH_3)_3$-benzene⟩	$LiC_4H_9/KOC(CH_3)_3$	PET	+75	24%, CO_2	[1712]
M–⟨1,1,3-trimethylindane⟩	$LiC_4H_9/KOC(CH_3)_3$	PET	+25	51%, CO_2	[1712]
M–⟨tetramethylindane⟩	$LiC_4H_9/KOC(CH_3)_3$	PET	+25	55%, CO_2	[1712]
M–⟨dimethylbenzofuran⟩	$LiC(CH_3)_3/KOC(CH_3)_3$	PET	+25	52%, CO_2	[1712]
M–⟨phenylacetylene, C≡C–M⟩	$LiC_4H_9/KOC(CH_3)_3$	PET	+5	88%, H_3CI	[1713]
M–⟨benzene⟩–$Si(^iC_3H_7)_3$	$LiC_4H_9/KOC(CH_3)_3$	PET	+25	42%, CO_2[f]	[1711]

[a] PET = petroleum ether (pentanes, hexanes, heptanes); THF = tetrahydrofuran. [b] El–X′ = electrophilic trapping reagent. [c] Along with 10% of dicarboxylic acids (*meta*- and *para*-isomers). [d] Roughly equal amounts of *α*- and *β*-isomers. [e] Approximately equal amounts of *meta*- and *para*-isomers. [f] Along with some 16% of the *meta*-isomer.

Table 120. M–R′ Promoted metalation of *N,N*-alkyl- or -aryl-substituted anilines.

M–R	M–R′	Sv [a]	T[°C]	Product, $El-X'$ [b]	Reference
M–(C6H4)–N(CH3)2 *(ortho)*	LiC$_4$H$_9$[c]	PET	+25	71%, (H$_5$C$_6$)$_2$CO	[1714–1716]
M–(C6H3), (H$_3$C)$_2$N, CH$_3$	LiC$_4$H$_9$	DEE	+25	38% [d], D$_2$O	[1716]
M–(C6H3), CH$_3$ *(top)*, N(CH$_3$)$_2$	LiC$_4$H$_9$[c]	PET	+25	80%, (H$_5$C$_6$)$_2$CO	[1716]
M–(C6H3)–CH$_3$, N(CH$_3$)$_2$	LiC$_4$H$_9$[c]	PET	+25	57%, (H$_5$C$_6$)$_2$CO	[1717]
M–(C6H3)–Si(CH$_3$)$_3$, N(CH$_3$)$_2$	LiC$_4$H$_9$[c]	PET	+70	81%, D$_2$O	[1718]
N(CH$_3$)$_2$ / M–(C6H3) / N(CH$_3$)$_2$	LiC$_4$H$_9$	PET	+70	70%, by X-ray and NMR	[1719]
M–(C6H3), N(CH$_3$)$_2$, N(CH$_3$)$_2$	LiC$_4$H$_9$	PET	+70	27%, morpholine–N-CHO	[1720]
M–(C6H4)–N(C$_6$H$_5$)$_2$	LiC$_4$H$_9$[e]	PET	+25	30%, CO$_2$	[1721]
M–(carbazole), H$_5$C$_2$–N	LiC$_4$H$_9$	THF	+25	21%, CO$_2$	[1722]
M–(phenothiazine), H$_5$C$_2$–N, S	LiC$_4$H$_9$	DEE	+25	13% [f], CO$_2$	[1723]

[a] DEE = diethyl ether; PET = petroleum ether (pentanes, hexanes, heptanes); THF = tetrahydrofuran; in the case of mixtures (containing for example paraffinic cosolvents) only the most polar component is indicated. [b] $El-X'$ = electrophilic trapping reagent. [c] In the presence of *N,N,N′,N′*-tetramethylethylenediamine (TMEDA). [d] Concomitant metalation at the benzylic methyl group. [e] Plus KOC(CH$_3$)$_3$. [f] Concomitant metalation at a sulfur-adjacent position.

Table 121. M—R′ Promoted metalation of *N*-BOC anilines[a].

M—R	M—R′	$S\nu$ [b]	T[°C]	Product, El–X′ [c]	Reference
M—⬡ MNCOOC(CH₃)₃	LiC(CH₃)₃	THF	−75	69%, H₅C₆NCS	[1724–1727]
CH₃—⬡ M— MNCOOC(CH₃)₃	LiC(CH₃)₃	THF	−20%	14%, [d]	[1728]
H₃CO—⬡ M— MNCOOC(CH₃)₃	LiC(CH₃)₃	THF	−20%	58%, [d]	[1727,1728]
OCH₃—⬡ M— MNCOOC(CH₃)₃	LiC(CH₃)₃	THF	−20%	39%, [d]	[1728]
H₃CO—⬡—F M— MN COOC(CH₃)₃	LiC(CH₃)₃	THF	−50%	86%, CO₂	[1729,1730]
M—⬡—F MNCOOC(CH₃)₃	LiCH(CH₃)C₂H₅	THF	−75[e]	72%, CO₂	[1731]
F—⬡ M— MNCOOC(CH₃)₃	LiC(CH₃)₃	THF	−50	80%, CO₂	[1728–1730,1732]
H₃CO—⬡—CF₃ M— MN COOC(CH₃)₃	LiC(CH₃)₃	THF	−50	97%, CO₂	[1730]
M—⬡—CF₃ MNCOOC(CH₃)₃	LiC(CH₃)₃	THF	−50	83%, CO₂	[1730,1733]
CF₃—⬡ M— MNCOOC(CH₃)₃	LiC(CH₃)₃	THF	−50	86%, CO₂	[1730]

[a] BOC = *tert*-butoxycarbonyl. [b] THF = tetrahydrofuran or mixtures rich in THF. [c] El–X′ = electrophilic trapping reagent. [d] Addition to 1,3-bis(dimethylamino)allyl perchlorate followed by cyclization to afford a quinoline. [e] In the presence of *N,N,N′,N″,N″*-pentamethyldiethylenetriamide (PMDTA); without this trident complexand 7-lithiated 2-*tert*-butoxybenzoxazole results from lithium fluoride elimination and subsequent cyclization (unless one works at −100 °C).

Table 122. Metalation of N,N-dialkyl benzylamines using reagents M−R′.

M−R	M−R′	Sv[a]	T[°C]	Product, El−X′[b]	Reference
M–(benzene)–CH$_2$N(CH$_3$)$_2$	LiC$_4$H$_9$	DEE	+25	78%, H$_5$C$_6$CHO	[1734,1735]
M–(benzene)–(H$_3$C)$_2$CN(CH$_3$)$_2$	LiC$_4$H$_9$	DEE	+25	57%, (H$_5$C$_6$)$_2$CO	[1734]
M–(benzene, CH$_3$)–CH$_2$N(CH$_3$)$_2$	LiC$_4$H$_9$	DEE	+25	82%, (H$_5$C$_6$)$_2$CO	[1735]
M–(benzene, OCH$_3$)–CH$_2$N(CH$_3$)$_2$	LiC$_4$H$_9$	DEE	+25	75%, (H$_5$C$_6$)$_2$CO	[1735]
M–(benzene, OCH$_3$)–CH$_2$N(CH$_3$)$_2$	LiC$_4$H$_9$	DEE	0	70%, (H$_5$C$_6$)$_2$CO	[1735–1737]
M–(benzene, H$_3$CO, OCH$_3$)–CH$_2$N(CH$_3$)$_2$	LiC$_4$H$_9$	DEE	0	92%, H$_2$CO	[1738]
M–(benzene, O–O methylenedioxy)–CH$_2$N(CH$_3$)$_2$	LiC$_4$H$_9$	THF	+25	90%, H$_2$CO	[1739]
M–(benzene, CF$_3$)–CH$_2$N(CH$_3$)$_2$	LiC$_4$H$_9$	DEE	+25	70%, (H$_5$C$_6$)$_2$CO	[1735]
M–(benzene)–H$_2$C, F, N(CH$_3$)$_2$	LiC$_4$H$_9$	DEE	+25	33%, (H$_5$C$_6$)$_2$CO	[1735]
M–(benzene)–H$_2$C, Cl, N(CH$_3$)$_2$	LiC$_4$H$_9$	DEE	+25	81%, (H$_5$C$_6$)$_2$CO	[1735]
M–(benzene, Cl)–CH$_2$N(CH$_3$)$_2$	LiC$_4$H$_9$	DEE	+25	82%, (H$_5$C$_6$)$_2$CO	[1735]

[a] DEE = diethyl ether. [b] El−X′ = electrophilic trapping reagent.

Table 123. Metalation of aromatic *N,N*-dialkyl carboxamides using reagents M–R'.

M–R	M–R'	$S\nu$[a]	T[°C]	Product, El–X'[b]	Reference
(phenyl) M– ring with O=C-N(C₂H₅)₂	LiCH(CH₃)C₂H₅[c]	THF	−75	75%, H₃CI	[1740,1741]
(phenyl) M– ring with –N(CH₃)₂; O=C-N(C₂H₅)₂	LiCH(CH₃)C₂H₅[c]	THF	−75	71%, H₅C₆CHO [d]	[1743,1744]
(phenyl) M– ring; O=C with OCH₃ and N(C₂H₅)₂	LiCH(CH₃)C₂H₅[c]	THF	−75	75%, (H₃C)₂NCHO	[1743,1744]
H₃CO (phenyl) M– ring; O=C-N(C₂H₅)₂	LiCH(CH₃)C₂H₅[c]	THF	−75	54%, CO₂	[1740,1744]
OCH₃ (phenyl) M– ring; O=C-N(C₂H₅)₂	LiCH(CH₃)C₂H₅[c]	THF	−75	>95%, (H₃C)₂SiCl	[1743–1745]
(phenyl) M– ring with –OCH₃; O=C with OCH₃ and N(C₂H₅)₂	LiCH(CH₃)C₂H₅[c]	THF	−75	72%, (furyl)–CHO	[1753,1746]
OCH₃ (phenyl) M– ring with –OCH₃; O=C with Si(CH₃)₃ and N(C₂H₅)₂	LiCH(CH₃)C₂H₅[c]	THF	−75	>95%, H₃CI	[1745]
OCH₃ (naphthyl) M–; O=C-N(ⁱC₃H₇)₂	LiC₄H₉	THF	−75	95%, H₅C₂I	[1747]
H₃CO OCH₃ (phenanthrene/biphenyl) M– ring with –OCH₃; O=C-N(C₂H₅)₂	LiCH(CH₃)C₂H₅[c]	THF	−75	72%, (pyridyl)–CHO [d]	[1748]

[a] THF = tetrahydrofuran; in the case of mixtures (containing for example paraffinic cosolvents) only the most polar component is indicated. [b] El–X' = electrophilic trapping reagent. [c] In the presence of *N,N,N',N'*-tetramethylethylene-diamine (TMEDA). [d] Followed by acid-catalyzed ring closure to afford the lactone.

Table 124. M—R′ Promoted metalation of aromatic *N*-alkyl(thio)carboxamides.

M—R	M—R′	Sv [a]	T[°C]	Product, $El-X'$ [b]	Reference
M—⟨aryl⟩, O=C-NCH₃, M	LiC₄H₉	THF	0	40%, ⟨structure⟩	[1749,1450]
M—⟨aryl⟩, O=C OCH₃, MNCH₃	LiC₄H₉	THF	~+45	50%, (H₂C)₂O	[1736]
H₃CO, M—⟨aryl⟩, O=C-NCH₃, M	LiC₄H₉	THF	~+45	91%, (H₅C₆)₂CO	[1736]
OCH₃, M—⟨aryl⟩, O=C-NCH₃, M	LiC₄H₉	THF	~+45	50%, (H₂C)₂O	[1736]
M—⟨naphthyl⟩, O=C, MNCH₃	LiC₄H₉	THF	+25	28%, (H₂C)₂O	[1751]
M—⟨naphthyl⟩, O=C, MNCH₃	LiC₄H₉	THF	+25	25%, (H₂C)₂O	[1751]
M—⟨aryl⟩, S=C-NCH₃, M	LiC₄H₉	THF	−10	68%, H₃CCHO	[1752]
M—⟨aryl⟩, S=C-NCH₃, M	LiC₄H₉	THF	0	49%, ClSi(CH₃)₃	[1752]
Cl, M—⟨aryl⟩, S=C-NCH₃, M	LiC₄H₉	THF	+25	78%, H₃CSSCH₃	[1752]

[a] THF = tetrahydrofuran; in the case of mixtures (containing for example paraffinic cosolvents) only the most polar component is indicated. [b] $El-X'$ = electrophilic trapping reagent.

Table 125. M—R′ Promoted metalation of alkyl aryl ethers, 1-alkenyl aryl ethers and diaryl ethers.

M–R	M–R′	$S\nu$ [a]	T[°C]	Product, El–X′ [b]	Reference
M–C6H4–OCH3	LiC4H9	THF	+25	40%, $H_{25}C_{12}Br$	[1753,1754]
	NaC6H5	BNZ	+25	64%, CO_2	[1239]
M–C6H4–OC2H5	LiC4H9	THF	+25	42%, CO_2	[1753,1755]
M–C6H4–OCH(CH3)2	LiC4H9	THF	+25	17%, CO_2	[1753,1755]
M–C6H4–OC(CH3)3	LiC(CH3)3	PET	+80	82%, CO_2	[1756]
M–C6H3(C6H5)–OCH3	LiC6H5	DEE	+25	10%, $(H_5C_6)_2CO$	[1757]
M–C6H4–OCH=CH2	LiCH(CH3)3C2H5 [c]	THF	0	20%, H_3CI	[1758]
M–C6H4–OC6H5	LiC4H9	DEE	+35	67%, $(H_3C)_3SiCl$	[1759] [1760,1761]
M–dibenzofuran	LiC4H9	THF	0	86%, CO_2	[1762,1763]
	NaC4H9	PET		36%, CO_2 [d]	[650,1764]
M–naphthyl–OCH3	LiC4H9	THF	+25	90%, H_5C_6CN [e]	[1765–1767]
M–phenanthryl–OC6H5	LiC6H5	DEE	+25	59%, $(H_5C_6)_2CO$	[1768]

[a] DEE = diethyl ether; PET = petroleum ether (pentanes, hexanes, heptanes); BNZ = benzene; THF = tetrahydrofuran; in the case of mixtures (containing for example paraffinic cosolvents) only the most polar component is indicated. [b] El–X′ = electrophilic trapping reagent. [c] α,*ortho*-dimethylation occurs when the LiC4H9/KOC(CH3)3 superbase is used. [d] Accompanied by varying amount of 1,8-dicarboxylic acid resulting from dimetalation. [e] Contaminated by small amounts (≤5%) of the α-regioisomer.

Table 126. Metalation of di-, tri- and tetraalkoxyarenes using reagents M−R′.

M−R	M−R′	Sv [a]	T[°C]	Product, El−X′ [b]	Reference
M—(arene), H_3CO OCH_3	LiC_4H_9	DEE	+25	60%, (cyclohexyl)–CHO	[1769]
H_3CO / M—(arene) / H_3CO	LiC_4H_9	THF	−20	70%, $H_{25}C_{12}Br$	[1754,1770]
M—(arene), OCH_3 / OCH_3	LiC_4H_9	DEE	0	68%, (furyl)–CH_2CHO	[1771]
M—(arene)—OCH_3, H_3CO OCH_3	LiC_4H_9	THF	+25	56%, H_5C_2Br	[1772]
H_3CO / M—(arene) / H_3CO OCH_3	$LiC_4H_9/KOC(CH_3)_3$	THF	−75	76%, CO_2	[1773]
H_3CO / M—(arene)—OCH_3 / H_3CO	LiC_4H_9	DEE	+25	65%, CuBr	[1774]
H_3CO / M—(arene)—OCH_3 / H_3CO OCH_3	LiC_4H_9	THF	+25	61%, CO_2	[1775]
H_3CO OCH_3 / M—(arene) / H_3CO OCH_3	LiC_4H_9	THF	+25	55%, $(H_3C)_2NCH_2Cl$	[1776]
M—(arene), O O / H_3C CH_3	LiC_4H_9	THF	+25	54%, D_2O	[1777]
M—(arene), O O (spirocyclohexane)	LiC_4H_9	THF	0	76%[c], $Br(CH_2)_{10}Br$	[1777]

[a] DEE = diethyl ether; THF = tetrahydrofuran; in the case of mixtures (containing for example paraffinic cosolvents) only the most polar component is indicated. [b] El−X′ = electrophilic trapping reagent. [c] 4,4″-(1,10-Decanediyl)biss-piro[1,3-benzodioxole-2,1′-cyclohexane].

Table 127. M−R′ Promoted metalation of methoxymethoxy (MOM) and 2-tetrahydropyranyl (THP) aryl ethers.

M−R [a]	M−R′	Sv [b]	T[°C]	Product, El−X′ [c]	Reference
(phenyl)M, OCH₂CH₂OCH₃	LiC₄H₉	DEE	+25	54%[d], D₂O	[1778]
(phenyl)M, OCH₂OCH₃	LiC₄H₉[e]	THF	+25	73%[f], (H₅C₆O)₃P	[1779,1780]
(phenyl)M, O O (THP)	LiC₄H₉	DEE	+35	52%, CO₂	[1781,1782]
(phenyl)M, RO OR	LiC₄H₉	DEE	+25	72%, (H₂C)₂O	[1782]
(phenyl)M, R′O OR′	LiC₄H₉	DEE	+35	48%, CO₂	[1781]
RO (phenyl)M, RO	LiC₄H₉	DEE	+35	63%, (H₃C)₂NCHO	[1783]
R′O (phenyl)M, R′O	LiC₄H₉	DEE	+35	60%, CO₂	[1781]
OR (phenyl)M, RO	LiC₄H₉	DEE	−25	92%, (H₂C)₂O	[1782]
OR′ (phenyl)M, R′O	LiC₄H₉	DEE	+35	65%, CO₂	[1781]
H₃CO— —OCH₃ anthracene M, RO	LiC₄H₉/KOC(CH₃)₃	THF	−75	86%, (H₉C₄)₃SnCl	[1784]

[a] RO = H₃COCH₂O (MOMO), R′O = ⌐O⌐O (THPO). [b] DEE = diethyl ether; THF = tetrahydrofuran; in the case of mixtures (containing for example paraffinic cosolvents) only the most polar component is indicated. [c] El−X′ = electrophilic trapping reagent. [d] Concomitant β-elimination affording styrene (8%) and phenolate (12%). [e] In the presence of $N,N,N′,N′$-tetramethylethylenediamine (TMEDA). [f] Tris[2-(methoxymethoxy)phenyl]phosphine.

Table 128. M−R′ Promoted metalation of benzyl alcoholates.

M−R	M−R′	Sv [a]	T[°C]	Product, El−X′ [b]	Reference
M–(C6H4), MO-CH2	LiC4H9 [c]	PET	+40	95%, H5C6CHO	[1785]
M–(C6H4), MO-CHCH3	LiC4H9 [c]	PET	+40	88%, H5C6CHO	[1785]
M–(C6H4), MO-C(CH3)2	LiC4H9 [c]	PET	+40	86%, H5C6CHO	[1785]
OCH3, M–(ring), MO-CH2	LiC4H9 [c]	PET	+60	53%, CO2	[1786]
OCH3, M–(ring), MO-CHCH3	LiC4H9 [c]	PET	+60	62%, CO2	[1786]
OCH3, M–(ring), MO-C(CH3)2	LiC4H9 [c]	PET	+60	46%, CO2	[1786]
OCH3, M–(ring), MO-CH2-O (chroman)	LiC4H9 [c]	PET	+60	80%, CO2	[1786]
OCH3, M–(ring), MO– (fluorene-type)	LiC4H9/NaOC(CH3)3	PET	+25	88%, CO2	[1787]
C6H5, M–(ring)–F, MO-CH2 C6H5	LiC(CH3)3	DEE	+25	68%, (H5C2)2NCOCl	[1788]
M–(ring), MO– ...–N–CH3	LiC(CH3)3	DEE	+25	63%, (H3C)3SiCl	[1789]

[a] DEE = diethyl ether; PET = petroleum ether (pentanes, hexanes, heptanes). [b] El−X′ = electrophilic trapping reagent.
[c] In the presence of N,N,N',N'-tetramethylethylenediamine (TMEDA).

Table 129. Metalation of fluorobenzenes using reagents M–R'.

M–R	M–R'	Sv [a]	T[°C]	Product, El–X' [b]	Reference
(M, fluorobenzene, F)	LiC_4H_9	THF	−50	60%, CO_2	[1790]
	$LiC_4H_9/KOC(CH_3)_3$	THF	−75	98%, CO_2	[1791]
(M, difluorobenzene, F, F)	$LiC_4H_9/KOC(CH_3)_3$	THF	−75	83%, CO_2	[1791]
(M, difluorobenzene, F, F)	$LiC_4H_9/KOC(CH_3)_3$	THF	−75	65%, CO_2	[1791]
(M, difluorobenzene, F, F)	LiC_4H_9	THF	−65	88%, CO_2	[1792]
(M, trifluorobenzene, F, F, F)	$LiCH(CH_3)C_2H_5$	THF	−75	94%, CO_2	[1793]
(M, trifluorobenzene, F, F, F)	$LiCH(CH_3)C_2H_5$	THF	−75	87%, CO_2	[1793]
(M, trifluorobenzene, F, F, F)	$LiCH(CH_3)C_2H_5$	THF	−75	88%, I_2	[1793]
(M, tetrafluorobenzene, F, F, F, F)	LiC_4H_9	THF	−70	82%, CO_2	[1792]
(M, tetrafluorobenzene, F, F, F, F)	LiC_4H_9	DEE	−70	45%, $(H_3C)_3CNO$	[1794]
(M, tetrafluorobenzene, F, F, F, F)	LiC_4H_9	DEE	−65	85%, CO_2	[1792,1795]
(M, pentafluorobenzene, F, F, F, F, F)	$BrMgC_2H_5$	THF	+25	85%, CO_2	[1796]
	LiC_4H_9	DEE	−65	92%, CO_2	[1792,1796]

[a] DEE = diethyl ether; THF = tetrahydrofuran; in the case of mixtures (containing for example paraffinic cosolvents) only the most polar component is indicated. [b] El–X' = electrophilic trapping reagent.

Table 130. M−R′ Promoted metalation of trifluoromethyl substituted benzenes and naphthalenes.

M−R	M−R′	Sv [a]	T[°C]	Product, $El-X'$ [b]	Reference
	LiC₄H₉ LiCH₃/KOC(CH₃)₃	DEE THF	+35 −75	40%[c], CO_2 83%, CO_2	[1753,1797] 1791,1798
	Li−N⟨	THF	−75	80%[d], CO_2	[1798,1799]
	Li−N⟨	THF	−75	94%, CO_2	[1798,1800]
	Li−N⟨	THF	−75	93%[e], CO_2	[1798,1801]
	LiCH₃/KOC(CH₃)₃	THF	−75	92%, CO_2	[1791,1798]
	Li−N⟨	DEE	−75	94%, CO_2[f]	[1798]
	Li−N⟨	DEE	−25	94%, CO_2	[1798,1802]
	LiC₄H₉	DEE	+35	76%, CO_2	[1803]
	LiC₄H₉/KOC(CH₃)₃	THF	−25	42%, CO_2	[1804]
	LiC₄H₉/KOC(CH₃)₃	THF	−75	82%, CO_2	[1804]

[a] DEE = diethyl ether; THF = tetrahydrofuran; in the case of mixtures (containing for example paraffinic cosolvents) only the most polar component is indicated. [b] $El-X'$ = electrophilic trapping reagent. [c] As an 83:16 *o-/m-/p*-mixture of regioisomers. [d] A mixture of 2,3- and 3,4-bis(trifluoromethyl)benzoic acids is formed in a ratio of 2:1. [e] A yield of 85% is obtained when methyllithium in the presence of potassium *tert*-butoxide is used as the metalation reagent. [f] Contaminated with trace amounts (<3%) of a regioisomer.

Table 131. Metalation of chlorobenzenes using reagents M–R'.

M–R	M–R'	Sv [a]	T[°C]	Product, El–X' [b]	Reference
M–(2-chlorophenyl)	$LiCH(CH_3)C_2H_5$	THF	−105	88%, $H_5C_6SO_2CQ{=}CH_2$ [c]	[1805]
M–(2,3-dichlorophenyl)	$LiCH(CH_3)C_2H_5$	THF	−100	82%, CO_2	[1806]
M–(2,5-dichlorophenyl)	LiC_4H_9	THF	−70	90%, H_5C_6CN	[1806–1808]
M–(2,5-dichlorophenyl)	$Li{-}N$(piperidide)	THF	−75	82%, CO_2	[1806]
M–(2,3,5-trichlorophenyl)	$Li{-}N$(piperidide)	THF	−75	76%, CO_2	[1806]
M–(2,3,6-trichlorophenyl)	$LiCH(CH_3)C_2H_5$	THF	−75	77%, CO_2	[1806]
M–(2,4,6-trichlorophenyl)	LiC_4H_9	THF	−50	94%, CO_2	[1809]
M–(2,3,4,6-tetrachlorophenyl)	$Li{-}N$(piperidide)	THF	−75	53%, CO_2	[1806]
M–(2,3,5,6-tetrachlorophenyl)	$Li{-}N$(piperidide)	THF	−75	63%, CO_2	[1806]
M–(2,3,4,5-tetrachlorophenyl)	$Li{-}N$(piperidide)	THF	−75	95%, CO_2	[1806]
M–(pentachlorophenyl)	LiC_4H_9	THF	−65	91%, CO_2	[1806,1810]

[a] THF = tetrahydrofuran; in the case of mixtures (containing for example paraffinic cosolvents) only the most polar component is indicated. [b] El–X' = electrophilic trapping reagent. [c] Q = $Si(CH_3)_3$.

Table 132. M–R′ Promoted metalation of pyrroles at the 2-position.

M–R	M–R′	Sv [a]	T[°C]	Product, $El–X′$ [b]	Reference
M–pyrrole, N–CH$_3$	LiC$_4$H$_9$[c]	DEE	+40	70%, CO$_2$	[1811–1813]
M–pyrrole, N–SEM [d]	LiC(CH$_3$)$_3$	THF	−10	75%, H$_5$C$_6$CH$_2$Br	[1814]
M–pyrrole, N–COOC(CH$_3$)$_3$	Li–N (piperidide)	THF	−80	45%, H$_5$C$_6$COCl	[1815]
M–pyrrole, N–O=C–NC(CH$_3$)$_3$ M	LiC(CH$_3$)$_3$	THF	−75	64%, (H$_3$C)$_2$NCHO	[1816]
M–pyrrole, N–COOLi	LiC(CH$_3$)$_3$	THF	−70	55%, 4-H$_3$CC$_6$H$_4$SO$_2$F	[1817]
M–pyrrole, N–SO$_2$C$_6$H$_5$	LiC(CH$_3$)$_3$	THF	−80	69%, H$_3$COD	[1815]
M–pyrrole, N–C$_6$H$_5$	LiC$_4$H$_9$/KOC(CH$_3$)$_3$	THF	−75	75%, CO$_2$	[1818–1821]
M–pyrrole, N–aryl–OCH$_3$	LiC$_4$H$_9$[e]	THF	−75	68%, CO$_2$	[1822]
M–pyrrole, N–aryl–CH$_3$	LiC$_4$H$_9$[e]	THF	−75	32%[f], CO$_2$	[1823]
M–pyrrole, N–aryl–CF$_3$	LiC$_4$H$_9$[e]	THF	−75	6%[g], CO$_2$	[1823]

[a] DEE = diethyl ether; THF = tetrahydrofuran; in the case of mixtures (containing for example paraffinic cosolvents) only the most polar component is indicated. [b] $El–X′$ = electrophilic trapping reagent. [c] In the presence of $N,N,N′,N′$-tetramethylethylenediamine (TMEDA). [d] SEM = (2-trimethylsilylethyloxy)methyl. [e] In the presence of $N,N′,N′,N″,N″$-pentamethyldiethylenetriamine (PMDTA); in contrast, only the 2-position of the anisyl ring is metalated when the reaction is carried out in the presence of TMEDA. [f] In addition, 32% of metalation at the *para*-position. [g] Along with 22% of *meta*-metalation and 2% of α, *ortho*-dimetalation.

Table 133. M–R′ Promoted metalation of indoles at the 2-position.

M–R	M–R′	Sv [a]	T[°C]	Product, El–X′ [b]	Reference
M–indole, N–CH₃	LiC₄H₉	DEE	~+40	59%, $(H_5C_2OCO)_2$	[1824,1825]
M–indole, 5-OCH₃, N–CH₃	LiC(CH₃)₃	THF	0	39%, pyridyl–CHO	[1826]
R [c]–indole, N–CH₃	LiC₄H₉	THF	0	74%, D_2O	[1827]
M–indole, N–CH₂OCH₃	LiC(CH₃)₃	DEE	+25	80%, CO_2	[1828]
M–indole, N–CH₂N(CH₃)₂	LiC₄H₉	THF	0	85%, $(H_5C_6)_2CO$	[1829,1830]
M–indole, N–COOC(CH₃)₃	LiC(CH₃)₃	THF	−75	66%, $H_3COCO)_2$	[1815]
M–indole, N–C(=O)–NC(CH₃)₃, M	LiC(CH₃)₃	THF	−75	64%, $(H_3C)_2NCOC_6H_5$	[1816]
M–indole, N–COOM	LiC(CH₃)₃	THF	−70	55%, H_5C_6NCO	[1831]
H₃C–indole, M–, N–COOM	LiC(CH₃)₃	THF	−20	54%, $(H_3C)_2NCHO$	[1832]
M–indole, N–SO₂C₆H₅	LiC(CH₃)₃	THF	+25	63%, CO_2	[1826,1833,1834]

[a] DEE = diethyl ether; THF = tetrahydrofuran; in the case of mixtures (containing for example paraffinic cosolvents) only the most polar component is indicated. [b] El–X′ = electrophilic trapping reagent. [c] R = (H₃C)₂NCH₂CH₂.

Table 134. Metalation of pyrazoles at 4- and 5-positions using reagents M−R′.

M−R	M−R′	Sv [a]	T[°C]	Product, El−X′ [b]	Reference
(pyrazole, N−M)	LiC_4H_9	DEE	−30	9%, CO_2	[1835]
(4-Br pyrazole, N−M)	LiC_6H_5	DEE	+25	35%, CO_2	[1835]
(pyrazole, N−CH_3)	LiC_4H_9	DEE	+25	87%, $(H_5C_6)_2CO$	[1835,1836]
(pyrazole, N−C_3H_7)	LiC_4H_9	DEE	+25	81%, H_5C_6CHO	[1837]
(pyrazole, N−$CH_2C_6H_5$)	LiC_6H_5	DEE	+25	57%, CO_2	[1835,1837]
(pyrazole, N−C_6H_5)	LiC_4H_9	THF	−65	39%[c], CO_2	[1818,1836,1838]
(pyrazole, N−CH_2OM)	$LiN(^iC_3H_7)_2$	THF	−20	56%, H_5C_6NCO	[1839]
(3,N-dimethyl pyrazole)	LiC_4H_9	DEE	~+40	60%[d], H_5C_6CHO	[1837]
(H_3C, N−C_6H_5, CF_3 pyrazole)	LiC_4H_9	THF	0	11%[e], $(H_3C)_3SiCl$	[1840]
(F_3C, N−C_6H_5, CF_3 pyrazole)	LiC_4H_9	THF	0	95%, $(H_3C)_3SiCl$	[1840]

[a] DEE = diethyl ether; THF = tetrahydrofuran; in the case of mixtures (containing for example paraffinic cosolvents) only the most polar component is indicated. [b] El−X′ = electrophilic trapping reagent. [c] Along with 10% of an isomeric acid resulting from metalation at the *ortho*-position. [d] Concomitant metalation of the 5-position and at the *N*-methyl group in a 2:1 ratio. [e] Concomitant metalation of the 4-position and the phenyl *ortho*-position in a ratio of 1:3.

Table 135. Metalation of imidazoles using reagents M−R′.

M−R	M−R′	Sv [a]	T[°C]	Product, El−X′ [b]	Reference
(imidazole, N-CH$_3$)	LiC$_4$H$_9$	DEE	−25	48%, (pyridyl)−CHO	[1841−1846]
(imidazole, 2-CH$_3$, N-CH$_3$)	LiC$_4$H$_9$	DEE	−80	30%, H$_3$CCHO	[1843]
(imidazole, N-CH$_2$C$_6$H$_5$)	LiC$_4$H$_9$	DEE	−60	67%, CO$_2$	[1841,1842,1846]
(imidazole, N-C$_6$H$_5$)	LiC$_4$H$_9$	DEE	+25	76%, (H$_5$C$_6$)$_2$CO	[1841]
(imidazole, N-CH$_2$OCH$_3$)	LiC$_4$H$_9$	DEE	+25	56%, (cyclohexanone)	[1842]
(imidazole, N-SEM) [c]	LiC$_4$H$_9$	THF	−75	94%, H$_3$CI	[1847]
(imidazole, 2-CH$_3$, N-SEM) [c]	LiC$_4$H$_9$	THF	−40	89%, (H$_3$C)$_2$NCHO	[1847]
(imidazole, 2-CH$_2$CH(CH$_3$)$_2$, N-SEM) [c]	LiC$_4$H$_9$	THF	−75	85%, (H$_3$C)$_2$NCHO	[1848]
(imidazole, 2-C$_6$H$_5$, N-SEM) [c]	LiC$_4$H$_9$	THF	−75	86%, (cyclohexyl)CHO	[1848]
(imidazole, N-SO$_2$N(CH$_3$)$_2$)	LiC$_4$H$_9$	THF	−75	60%, (H$_5$C$_6$)$_2$CO	[1849,1850]

[a] DEE = diethyl ether; THF = tetrahydrofuran; in the case of mixtures (containing for example paraffinic cosolvents) only the most polar component is indicated. [b] El−X′ = electrophilic trapping reagent. [c] SEM = [2-(trimethylsilyl) ethoxy]methyl.

Table 136. Metalation of furans using reagents M−R′.

M−R	M−R′	$S\nu$ [a]	T[°C]	Product, El−X′ [b]	Reference
M—furan	LiC_4H_9	THF	−15	70%, $(H_5C_2O)_2CHCH_2Br$	[1851–1855]
M—furan—CH_3	LiC_4H_9	DEE	0	58%, [structure H_5C_2O cyclohexenone]	[1856,1857]
M—furan (CH_3)	$LiC(CH_3)_3$	THF	−75	83%, CO_2	[1858]
M—furan—$Si(CH_3)_3$	LiC_4H_9	DEE	~+40	62%, CO_2	[1851]
M—furan—CH_2NHBz [c]	LiC_4H_9	THF	−30	66%, H_3CI	[1859]
M—furan—$CH=N−OM$	LiC_4H_9 [d]	THF	−75	77%, $(H_9C_4)_3SnCl$	[1860]
M—furan—CH_2CHOM (CH_3)	LiC_4H_9	THF	−15	80%, $(H_2C)_2O$	[1855]
M—furan—CH_2O—cyclohexyl	LiC_4H_9	THF	−75	56%, $H_{31}C_{15}Br$	[1861]
M—furan—$CH(OC_2H_5)_2$	LiC_4H_9	DEE	+25	80%, $(H_3C)_3SiCl$	[1862]
M—furan—dioxolane	LiC_4H_9 [e]	[e]	[e]	67%, $(H_5C_6)_2CO$	[1863]
H_3CO OCH_3 M—furan	LiC_4H_9 [d]	DEE	0	78% [f], H_3CI	[1864]
M—furan—COOM	$LiN(^iC_3H_7)_2$	THF	−75	80%, $(H_3C)_2CO$	[1865]
Cl M—furan	$LiN(^iC_3H_7)_2$	THF	−75	41%, $(H_3C)_2C=CHCH_2Br$	[1866]
Br M—furan	$LiN(^iC_3H_7)_2$	THF	−75	66%, $(H_3C)_2C=CHCH_2Br$	[1866]

[a] DEE = diethyl ether; THF = tetrahydrofuran; in the case of mixtures (containing for example paraffinic cosolvents) only the most polar component is indicated. [b] El−X′ = electrophilic trapping reagent. [c] Bz = COC_6H_5 (benzoyl). [d] In the presence of $N,N,N′,N′$-tetramethylethylenediamine (TMEDA). [e] Not specified. [f] Along with about 10% of dimetalation.

Table 137. Metalation of thiophenes using reagents M–R'.

M–R	M–R'	$S\nu$ [a]	$T[°C]$	Product, $El–X'$ [b]	Reference
M–thiophene (2-)	LiC$_4$H$_9$	DEE	0	49%, FClO$_3$	[1867–1869]
	NaC$_2$H$_5$ [c]	DEE	+10	60%, CO$_2$	[738]
M–thiophene–CH$_3$	LiC$_2$H$_5$	DEE	+40	53%, FClO$_3$	[1868,1870]
M–thiophene (3-CH$_3$)	LiC$_4$H$_9$	DEE	+40	59%[d], (H$_3$C)$_2$NCHO	[1871]
M–thiophene (3-CH(CH$_3$)$_2$)	LiC$_4$H$_9$ [e]	DEE	+40	66%[f], (H$_3$C)$_2$SO$_2$	[1872]
M–thiophene (3-C$_6$H$_5$)	LiC$_4$H$_9$	DEE	0	59%[g], CO$_2$	[1873]
M–thiophene–Si(CH$_3$)$_3$	LiC$_4$H$_9$	DEE	+40	62%, CO$_2$	[1851]
M–thiophene–pyridyl	LiC(CH$_3$)$_3$	THF	−60	87%[h], (H$_5$C$_6$)$_2$CO	[1874]
M–thiophene–CH$_2$N(CH$_3$)$_2$	LiC$_4$H$_9$	THF	+45	50%, H$_5$C$_6$CHO	[1875]
M–thiophene–CH(OC$_2$H$_5$)$_2$	LiC$_4$H$_9$	DEE	−35	56%, H$_3$COCH$_2$Cl	[1876]
M–thiophene–COOM	LiN(^{i}C$_3$H$_7$)$_2$	THF	−75	83%, H$_5$C$_2$I	[1865]
M–thiophene–OCH$_3$	LiC$_6$H$_5$	DEE	+25	77%, H$_2$C=CH–CH$_2$Cl	[1877]
M–thiophene (3-F)	LiC$_4$H$_9$	DEE	+40	75%, CO$_2$	[1878]
M–thiophene–Cl	Na–thiophene [c]	DEE	+40	92%, CO$_2$	[738]
M–thiophene (3-Br)	LiC$_6$H$_5$	DEE	+25	72%, CO$_2$	[1879]

[a] DEE = diethyl ether; THF = tetrahydrofuran; in the case of mixtures (containing for example paraffinic cosolvents) only the most polar component is indicated. [b] $El–X'$ = electrophilic trapping reagent. [c] Generated *in situ*. [d] Concomitant metalation at the 2- and 5-position in the approximate ratio 1:5. [e] In the presence of *N,N,N',N'*-tetramethylethylenediamine (TMEDA). [f] Concomitant metalation at the 2- and the 5-position in the ratio 1:100. [g] Concomitant metalation at the 2- and the 5-position in the ratio roughly 1:1. [h] In DEE at 0 °C, LiC$_4$H$_9$ attacks the 2- and 5-position in a ratio of about 5:1 (total yield: 75% of trimethylsilylated derivatives).

Table 138. Metalation of isoxazoles using reagents M−R'.

M−R	M−R'	Sv [a]	T[°C]	Product, El−X' [b]	Reference
isoxazole, 4-CH$_2$OCH$_3$, 3-CH$_3$	LiC$_4$H$_9$	THF	−60	18%[c], H$_3$C−I	[1880]
isoxazole, 4-OCH$_3$, 3-CH$_3$	LiC$_4$H$_9$	THF	−75	10%[d], CO$_2$	[1881]
isoxazole, 4-OCH$_3$, 3-C$_6$H$_5$	LiC$_4$H$_9$	THF	−70	88%, I$_2$	[1882]
isoxazole, 4-OCH$_3$, 3-o-C$_6$H$_4$Cl	LiC$_4$H$_9$	THF	−70	62%, CO$_2$	[1882]
isoxazole, 4-C$_6$H$_5$, 3-OC$_2$H$_5$	LiC$_4$H$_9$	THF	−70	81%, CO$_2$	[1882]
isoxazole, 4-C$_6$H$_5$, 3-OCH(CH$_3$)$_2$	LiC$_4$H$_9$	THF	−70	77%, CO$_2$	[1882]
isoxazole, 4-C$_6$H$_5$, 3-OC(CH$_3$)$_3$	LiC$_4$H$_9$	THF	−70	75%, CO$_2$	[1882]
isoxazole, 4-C$_6$H$_5$, 3-SC$_2$H$_5$	LiC$_4$H$_9$	THF	−30	73%, CO$_2$	[1882]
isoxazole, 4-C$_6$H$_5$, 3-Cl	LiC$_4$H$_9$	THF	−70	80%, CO$_2$	[1882]

[a] THF = tetrahydrofuran; in the case of mixtures (containing for example paraffinic cosolvents) only the most polar component is indicated. [b] El−X' = electrophilic trapping reagent. [c] The main product (72%) is derived from the deprotonation of the methyl group at the 5-position. [d] Along with considerable amounts (47%) of an isomer derived from the deprotonation of the methyl group at the 5-position.

Table 139. Metalation of oxazoles using reagents M−R′.

M−R	M−R′	Sv [a]	T[°C]	Product, El−X′ [b]	Reference
M–(oxazole)	LiC₄H₉	THF	−75	50%, (H₃C)₂NCHO	[1883–1885]
M–(2-methyloxazole)	LiC₄H₉	DEE	−75	60%, (H₃C)₃SnCl	[1886]
M–(2-phenyloxazole)	Li–N⟨ [c]	THF	0	75%, H₅C₆CHO	[1887]
M–(5-phenyloxazole)	LiC₄H₉	THF	−60	96%, D₂O	[1888]
M–(COOM-, CH₃-oxazole)	LiC₄H₉	THF (?)	−75	92%, D₂O	[1889]
M–(benzoxazole)	LiC₄H₉	DEE	−75	85%, (H₃C)₃SnCl	[1890]

[a] DEE = diethyl ether; THF = tetrahydrofuran; in the case of mixtures (containing for example alkanic cosolvents) only the most polar component is indicated. [b] El−X′ = electrophilic trapping reagent. [c] Or butyllithium in THF at −75 °C.

Table 140. Metalation of isothiazoles using reagents M−R′.

M−R	M−R′	Sv [a]	T[°C]	Product, El−X′ [b]	Reference
M–(isothiazole)	LiC₄H₉	THF	−65	75%, (H₃C)₂NCHO	[1891]
M–(3-methylisothiazole)	LiC₄H₉	THF	−65	50%, CO₂	[1891,1892]
M–(4-methylisothiazole)	LiC₄H₉	THF	−65	40%, CO₂	[1891,1892]
M–(MOOC, CH₃-isothiazole)	LiC₄H₉	THF	−65	52%, Br₂	[1891]
M–(Cl-isothiazole)	LiC₄H₉	THF	−65	68%, CO₂	[1891]
M–(Br-isothiazole)	LiC₄H₉	THF	−65	73%, (H₃C)₂NCHO	[1891]
M–(I-isothiazole)	LiC₄H₉	THF	−65	33%, (H₃C)₂NCHO	[1891]

[a] DEE = diethyl ether; THF = tetrahydrofuran; in the case of mixtures (containing for example paraffinic cosolvents) only the most polar component is indicated. [b] El−X′ = electrophilic trapping reagent.

Table 141. Metalation of thiazoles using reagents M−R′.

M−R	M−R′	Sv [a]	T[°C]	Product, El−X′ [b]	Reference
M–(thiazol-2-yl)	LiC_4H_9	DEE	−60	87%, $(H_3C)_2CHCHO$	[1893,1894]
M–(4-CH_3-thiazol-2-yl)	LiC_4H_9	DEE	−60	26%, H_3CI	[1893]
M–(4-$C(CH_3)_3$-thiazol-2-yl)	$LiCH_3$	MEGME	−75	18%, crystalline	[1895]
M–(4,5-$(CH_3)_2$-thiazol-2-yl)	LiC_6H_5	DEE	+25	58%, H_2CO	[1896]
M–(4-Br-thiazol-2-yl)	LiC_4H_9	DEE	−75	95%, $(H_3C)_3SiCl$	[1894]
M–(benzothiazol-2-yl)	LiC_4H_9	DEE	−70	59%, $F_2C=CCl_2$ [c]	[1869,1897–1899]
M–(2-CH_3-thiazol-5-yl)	LiC_4H_9	DEE	−70	77% [d], H_5C_6CHO	[1900,1901]
M–(2-CH_3-4-C_6H_5-thiazol-5-yl)	LiC_4H_9	THF	−75	86%, H_5C_2I	[1902]
M–(2-CH_3-4-(o-C_6H_4-OCH_3)-thiazol-5-yl)	LiC_4H_9	THF	−75	86%, H_3CI	[1902]
M–(2-$Si(CH_3)_3$-4-$Si(CH_3)_3$-thiazol-5-yl)	$LiC(CH_3)_3$	THF	−65	40%, $(H_3C)_3SiCl$	[1894]
M–(2-Cl-thiazol-5-yl)	LiC_4H_9	DEE	−80	88%, H_3CCHO	[1903]
M–(2-Cl-4-Cl-thiazol-5-yl)	$LiN(^{i}C_3H_7)_2$	THF	−75	66%, H_5C_2OOCCl	[1904]
M–(2-Br-4-Br-thiazol-5-yl)	$LiN(^{i}C_3H_7)_2$	THF	−75	68%, CO_2	[1904]

[a] DEE = diethyl ether; THF = tetrahydrofuran; MEGME = monoethylene glycol dimethyl ether; in the case of mixtures only the most polar component is indicated. [b] El–X′ = electrophilic trapping reagent. [c] One of the chlorine atoms is replaced by the 2-benzothiazolyl entity. [d] Concomitant deprotonation of the methyl group occurring only to a small extent (~5%).

Table 142. M−R′ Promoted metalation of nitrogen-functionalized pyridines.

M−R	M−R′	Sv [a]	T[°C]	Product, El−X′ [b]	Reference
(pyridine structure) MNCOC(CH₃)₃	LiC₄H₉	THF	0	65%, H₅C₂COOCCl	[1905]
(pyridine structure) MNCOC(CH₃)₃	LiC₄H₉	THF	0	46%, CO₂	[1905,1906]
(pyridine structure) MNCOC(CH₃)₃	LiC₄H₉	THF	0	74%, H₃CI	[1905]
(pyridine structure) MNCOOC(CH₃)₃	LiC(CH₃)₃	THF	−20	28%[c], Cl(CH₂)₃I	[1907]
(pyridine structure) CH₃ / MN—OCH₃ / COOC(CH₃)₃	LiC₄H₉[d]	THF	−25	68%, I₂	[1908]
(pyridine structure) H₃C / MN—OCH₃ / COOC(CH₃)₃	LiC₄H₉	THF	−75	87%, I₂	[1909]
(pyridine structure) N=O	Li-N (structure)	THF	0	50%, H₅C₆CHO	[1910]
(pyridine structure) O=C-N(ⁱC₃H₇)₂	Li-N (structure)	MEGME	−75	63%[e], (cyclopentanone structure)=O	[1911,1912]
(pyridine structure) O=C-N-C₆H₅ / M	LiC₄H₉[d]	THF	−25	70%, H₅C₆CON(CH₃)₂	[1913]

[a] THF = tetrahydrofuran; MEGME = monoethylene glycol dimethyl ether ('glyme'); in the case of mixtures only the most polar component is indicated. [b] El−X′ = electrophilic trapping reagent. [c] The iodine and chlorine bearing tails reacting consecutively, C-alkylation is immediately followed by N-alkylation thus affording a tetrahydroisoquinoline derivative. [d] In the presence of $N,N,N′,N′$-tetramethylethylenediamine (TMEDA). [e] Product isolated after acid treatment as a lactone.

Table 143. M−R′ Promoted metalation of methoxy and carboxy pyridines.

M−R	M−R′	$S\nu$ [a]	$T[°C]$	Product, $El−X'$ [b]	Reference
OCH₃	LiN(^{i}C₃H₇)₂ [c]	THF	0	45%, (H₃C)₂CO	[1914]
OCH₃	LiC₄H₉ [d]	THF	−40	49%, H₃CCHO	[1915]
OCH₃	Li—⟨⟩	THF	−25	65%, I₂	[1916–1917]
OCH₃	LiC₄H₉ [e]	PET	0	84%[f], (H₃C)₃SiCl	[1918]
OCH₃—OCH₃ OCH₃	LiC₄H₉	THF	−40	45%[g], PCl₃	[1919]
OCH₂OCH₃	LiC(CH₃)₃	DEE	−75	90%, ClCH₂CH₂I	[1920]
OCSN(C₂H₅)₂	Li-N⟨⟩	THF	−75	91%, ⟨O⟩-CHO	[1921]
COOM	Li-N⟨⟩	THF	0	85%, CO₂	[1922]
COOM	Li-N⟨⟩	THF	−50	73%, CO₂	[1922]
COOM	Li-N⟨⟩	THF	−50	65%, CO₂	[1922]
Cl COOM	Li-N⟨⟩	THF	−75	73%, CO₂	[1922]
O=C Cl OM	Li-N⟨⟩	THF	−75	69%, CO₂	[1922]

[a] DEE = diethyl ether; THF = tetrahydrofuran; in the case of mixtures (containing for example paraffinic cosolvents) only the most polar component is indicated; PET = petroleum ether (pentanes, hexanes, heptanes). [b] $El−X'$ = electrophilic trapping reagent. [c] Generated *in situ* from methyllithium and small amounts of diisopropylamine. [d] In the presence of *N*,*N*,*N*′,*N*′-tetramethylethylenediamine (TMEDA). [e] Adduct of butyllithium and lithium 2-(dimethylamino)ethoxide ('Caubère's base') in large excess. [f] Metalation at the 3-position and nucleophilic addition of butyllithium as side reactions (5–10% each). [g] Tri(hetaryl)phosphine.

Table 144. Metalation of halogenated pyridines using reagents M−R′.

M−R	M−R′	Sv [a]	T[°C]	Product, El−X′ [b]	Reference
(2-fluoropyridine, M at 3)	LiN(^{i}C$_3$H$_7$)$_2$	THF	−75	87%, (H$_3$C)$_3$SiCl	[1923,1924]
(3-fluoropyridine, M at 4)	LiC$_4$H$_9$[c]	DEE	−40	65%[d], (H$_5$C$_2$)$_2$CO	[1925]
(3-fluoropyridine, M at 2)	LiC$_4$H$_9$[c]	THF	−40	65%, (H$_5$C$_2$)$_2$CO	[1926]
(2-chloropyridine, M at 3)	LiN(^{i}C$_3$H$_7$)$_2$	THF	−75	74%, (H$_3$C)$_3$SiCl	[1923,1927]
(2-chloropyridine, M at 3)	LiN(^{i}C$_3$H$_7$)$_2$	THF	−75	96%[e], (H$_3$C)$_3$SiCl	[1923,1928]
(3-chloropyridine, M at 2)	LiN(^{i}C$_3$H$_7$)$_2$	THF	−75	92%, (H$_3$C)$_3$SiCl	[1923,1926]
(2-bromopyridine, M at 3)	LiN(^{i}C$_3$H$_7$)$_2$	THF	−70	55%, (H$_3$C)$_3$SiCl	[1929]
(2-bromopyridine, M at 3)	LiN(^{i}C$_3$H$_7$)$_2$	THF	−75	61%, (H$_5$C$_6$S)$_2$	[1923,1928]
(3-bromopyridine, M at 2)	LiN(^{i}C$_3$H$_7$)$_2$	THF	−60	[f], (H$_5$C$_2$)$_2$CO	[1930]
(2,3-dichloropyridine, M at 4)	LiN(^{i}C$_3$H$_7$)$_2$	THF	−80	85%, H$_5$C$_6$CHO	[1931]
(2,6-dichloropyridine, M at 4)	LiN(^{i}C$_3$H$_7$)$_2$	THF	−80	76%[g], H$_3$CI	[1927,1932]
(2,3-dichloropyridine, M at 5)	LiN(^{i}C$_3$H$_7$)$_2$	THF	−65	29%, CO$_2$	[1933]

[a] DEE = diethyl ether; THF = tetrahydrofuran; in the case of mixtures (containing for example paraffinic cosolvents) only the most polar component is indicated. [b] El−X′ = electrophilic trapping reagent. [c] In the presence of N,N,N′,N′-tetramethylethylenediamine (TMEDA). [d] Along with 10% of an isomer resulting from metalation at the 2-position. [e] When LiC$_4$H$_9$/TMEDA is used as the reagent, metalation occurs mainly at the 2-position. [f] Main product; yield not specified; *in situ* trapping. [g] Concomitant attack at the 3- and 4-position in a ratio of 9:1.

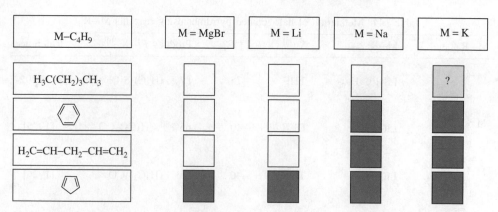

Figure 10. Metalation thresholds as a function of the basicity gradient between the reagent (butylmagnesium bromide, -lithium, -sodium or -potassium) and hydrogen/metal interconversion product (cyclopentadienyl-, 2,4-pentadienyl-, phenyl- and pentylmetal) and of the electropositivity of the metal employed. The shaded and white boxes mean that the reaction does and does not take place, respectively.

It would be quite erroneous to assume that butylpotassium, the most powerful reagent, gives the best results in terms of yields and purity of metalation product. In fact its reactions are neither clean nor quantitative. The lack of controlled reactivity has to be attributed to a large extent of the insolubility of butylpotassium and most other organopotassium and -sodium compounds in paraffinic media. Ethereal solvents cannot be used as they would undergo α- or β-elimination (see pp. 289–292). For example, butylpotassium is instantaneously decomposed in tetrahydrofuran containing solvents at temperatures even below $-100\,°C$.

Thus it was by virtue of necessity that systematic attempts were undertaken to pep up butyllithium by coordinating it with chelating auxiliaries, in particular N,N,N',N'-tetramethylethylenediamine (TMEDA). The metalation of toluene (at $+25\,°C$) and benzene (at $+50\,°C$) belong to the most remarkable achievements in this respect[1935–1936], although in other events TMEDA was suspected to cause nothing but 'placebo effects'[1937]. The activation provided by such a diamine remains modest. Moreover, aliphatic amines are not inert towards organometallic reagents (see Table 81) and a considerable fraction of the latter may get lost by reacting with the complexand.

To activate butyllithium more efficaciously, the auxiliary had to fulfil two requirements simultaneously: to dispose of an unrivalled electron-donor capacity and to be sterically or electronically protected against proton abstraction. On this rational ground, the superbasic mixture of butyllithium and potassium *tert*-butoxide was conceived[228,444]. Unlike butylpotassium, it is stable in tetrahydrofuran at $-75\,°C$ and momentarily even at $-50\,°C$. Although a large number of superbase-promoted metalation reactions can be advantageously performed in petroleum ether, ethereal solvents frequently offer the additional benefit of selectivity. No reliable information being available, one can only speculate about the structure and composition of the superbase. A butylpotassium/lithium *tert*-butoxide mixed aggregate is possibly the major, though not sole, component of the reagent in solution. The alcoholate would then be responsible for attenuation of the overall reactivity and stabilization towards solvent attack. At this point it should be mentioned that substoichiometric amounts (*e.g.* as little as 10%) of phenyllithium suffice to protect phenylsodium against degradation by diethyl ether[1262].

A variety of other mixed-metal reagents has also been developed and tested. Among these are combinations of methyllithium, *sec*-butyllithium, *tert*-butyllithium, pentylsodium and lithium diisopropylamide with potassium *tert*-butoxide and butyllithium with sodium *tert*-butoxide. They outperform pentylsodium or butylpotassium in the metalation of cyclopropylic, olefinic or allylic positions (see Table 145).

● *Resonance and Dipole Stabilization of Organometallics*

The local accumulation of electron excess at a metal-bearing carbon atom can be effectively attenuated by mesomeric delocalization of the fractional negative charge. Archetypical model cases are allylic or benzylic metal compounds and heteroanalogs thereof such as enolates and phenolates. This *p-orbital resonance* concentrates electron density at backbone centers which are next but one with respect to each other and alternate with nodal points.

$$H_2C=CH-CH=CH-\overset{\ominus}{\overset{..}{C}H} \quad \rightleftharpoons \quad H_2C=CH-\overset{\ominus}{\overset{..}{C}H}=CH-CH_2 \quad \rightleftharpoons \quad H_2\overset{\ominus}{\overset{..}{C}}=CH-CH=CH-CH_2$$

Any heteroatom stabilizes carbanions and organometallics first of all by its inductive electron withdrawal. Despite this apparent equality, second-row and heavier elements nevertheless outperform the corresponding first-row elements by orders of magnitudes. This has been attributed to an additional effect, the so-called *d-orbital resonance*[1938–1939]. It invokes the overlap of the carbanionic lone pair with an empty d orbital which should be relatively low-lying in the case of heavy elements such as silicon, tin, phosphorus, chlorine, bromine and iodine.

$$\underset{/}{\overset{\backslash}{-}}Q-\overset{\ominus}{\overset{..}{C}H_2} \quad \rightleftharpoons \quad \underset{/}{\overset{\backslash}{-}}Q=CH_2$$

According to this formalism, charge transfer from the carbon to a heteroatom imposes a partial double bond character on the connection between the two elements. However, one does not observe any significant hindrance of free rotation around such a bond, for example the $P-C^\alpha$ axis of a phosphorus ylid[1940]. Therefore, it deems appropriate to describe the phenomenon of carbanion-stabilization by heavy elements in a less sophisticated and more pictorial fashion by *polarization*[1941] (spatial deformation) of the carbanionic or organometallic orbital. The lengthening of nuclear distances to second-row or higher-period elements increases the gap between vicinal bonds and dilutes the electron density per volume unit. Electron–electron repulsions diminish heavily and, as a

Text continues on page 251

Table 145. Metalation of model hydrocarbons either with standard organosodium and -potassium reagents or with mixed metal superbases: reaction conditions and yields after trapping with the electrophiles specified in brackets.

Metalation reaction	Previous work [a]	Superbase results [a]
	NaC_5H_{11} PET, 100 h, 25 °C 3%, CO_2[1489]	$NaC_5H_{11}/KOC(CH_3)_3$[b] PET, 50 h, 25 °C 30%, CO_2[1489]
	NaC_5H_{11} PET, 22 days, 25 °C 5%, CO_2[1488]	$NaC_5H_{11}/KOC(CH_3)_3$[b] PET, 24 h, 25 °C 60%, $(H_3C)_3SiCl$[1489]
	NaC_5H_{11} PET, 7 days, 25 °C 20%[c], CO_2[1487]	[d]
	NaC_5H_{11} PET, 69 days, 25 °C 18%[e], CO_2[1488]	$NaC_5H_{11}/KOC(CH_3)_3$[b] PET, 1 h, 25 °C 68%, H_3CI[440]
	NaC_5H_{11} PET, 17 days, 25 °C 60%, CO_2[1488]	$LiC_4H_9/NaOC(CH_3)_3$[b] THF, 15 h, −50 °C 88%, $(H_3C)_3SiCl$[1638]
	NaC_5H_{11} PET, 5 h, 25 °C 0%[f], CO_2[1488]	$LiC_4H_9/NaOC(CH_3)_3$[b] THF, 15 h[g,i], −50 °C 92%, $(H_3C)_3SiCl$[1638]
	NaC_5H_{11} PET, 24 h, 25 °C 0%, CO_2[499]	$LiC_4H_9/KOC(CH_3)_3$ THF, 1 h, −55 °C 74%, CO_2[1560,1568]
	NaC_5H_{11} PET, 30 days, 25 °C 38%[h], CO_2[1937]	$Li(CH_3)C_2H_5/KOC(CH_3)_3$ PET, 24 h, 25 °C[b] 77%, $(H_2C)_2O$[440]
$H_{17}C_8$ → $H_{17}C_8$—M	NaC_5H_{11}[j] PET, 30 days, 25 °C 52%[k], CO_2[434,436]	$LiC_4H_9/KOC(CH_3)_3$ THF, 6 h, −50 °C[1564] 70%, $(H_3CO)_2BF/H_2O_2$

[a] THF = tetrahydrofuran; PET = petroleum ether (pentanes, hexanes, heptanes). [b] Similar results, although with lower yields, with $LiC_4H_9/KOC(CH_3)_3$. [c] Cyclopropane used as a cosolvent. [d] Not attempted. [e] Impure material. [f] Only fragmentation products were identified. [g] Virtually the same yield was obtained after 2 h. [h] With 'aged' (*e.g.* 4 weeks old) pentylsodium and sodium isopropylalcoholate about 80% yield. [i] With pentylsodium (PET, 18 h, 25 °C) and a fourfold excess of the alkene about 45% yield. [j] Mixtures of 2-nonyl-3-butenoic acid (main component), (*E*)-2-tridecenoic acid and (*Z*/*E*)-3-tridecenoic acid. [k] Mixture of regio- and stereoisomers.

consequence, the lone pair or metal-holding pair can expand well towards the heavy element before encountering resistance.

O-α-lithiated esters (see Table 89) and *N*-α-lithiated amides, amidines, carbamates or nitrosamines (see Tables 82–86) are neither p- nor d-orbital resonance stabilized. However, they obviously form quite readily. This may reflect merely an acceleration due to strong metal coordination by the carbonyl oxygen atom at the transition state[1942]. However, quantum mechanical calculations point to strong dipole stabilization of the metal-free carbanionic ground state species[1943]. As a prerequisite, the lone pair must lie in the amide plane (**273b**) rather than to be parallely aligned with the amide π-orbitals (**273a**).

273a   273b

Carbonyl groups can also provide thermodynamic stability to many other species by mere dipole interactions and metal coordination without exerting resonance delocalization. Noteworthy classes of compounds are 'pseudo-enolates'[1944,1945], the carbanionic and carbonyl π-orbitals which occupy perpendicular planes, imidoylmetals[1946,1947] and acylmetals[1948,1949] including congeners such as carbamoyllithiums[1950–1952] and thiocarbamoyllithiums[1953,1954].

• *Neighboring-Group Assistance to Hydrogen/Metal Permutations*

Neighboring-group controlled ('directed') *ortho*-selective metalation and subsequent electrophilic substitution of arenes occupies a prominent place in the toolbox of modern organic synthesis. The concept emerged from the systematic studies of G. Wittig[834], H. Gilman[1239] and C. R. Hauser[1955] and found numerous fervent disciples, notably H. W. Gschwend[1956] and V. Snieckus[1957]. Despite the maturity gained by the method since then, the fundamental question of how a substituent modulates CH acidities in its vicinity continues to attract unabated curiosity.

Any heteroatom or functional group facilitates hydrogen/metal exchange at nearby sites. The substituents have been classified according to their metalation-promoting power as strong, moderate or weak activators[1957]. Carbonyl, sulfonyl and phosphinoyl groups rank highest, α-metallooxyalkyls and halogens lowest in this hierarchy (see Table 146).

Most of the efficiency comparisons were performed with aromatic substrates carrying different substituents in 1,2- or 1,4-positions[1958]. This approach relies on two premises. The two substituents should not perturb each other. This will be generally the case unless a strong electron donor is matched with a strong electron acceptor. Furthermore, each substituent should display its electronic effect exclusively in the *ortho*-position and not in the *meta*-position. This second tacit assumption is incorrect in virtually every instance. Therefore *inter*molecular comparisons are more conclusive. To this end, two differently monosubstituted substrates are juxtaposed either in independent kinetic runs or in competition experiments[1959,1960].

Table 146. Neighboring group assistance to hydrogen/metal permutation processes: various substituents ordered, from top to bottom, by their decreasing efficiency.

Hetero-linked substituents [a]	Carbon-linked substituents [a]
$-SO_2NR_2$	$-CONR_2$
$-SO_2N(M)R$	$-CON(M)R$
$-PO(NR)_2$	$-CH_2N(M)R$
$-N(M)COR$	$-CH=NR$
$-N(M)COOR$	$-C(OR')=NR$ [b]
$-OCONR_2$	$-C\equiv N$
$-OR'$ [c]	$-CH_2OM$
$-SR'$ [c]	$-CH(OM)NR_2$
$-F$	$-CH(NR_2)_2$
$-Cl(-Br)$ [d]	$-CH(OR)_2$
$-I$	$-CF_3$
$-NR_2$	$-OM$

[a] R = alkyl, rarely also aryl. [b] For example, oxazolines. [c] R' = CH_3, C_6H_5, OCH_2OCH_3. [d] Chlorine and bromine have virtually equal activator strengths.

Whatever the relevance and accuracy of the technique employed, the numbers found are only valid for a given set of chemical reactions under given experimental conditions. Any change of reaction parameters such as solvent, concentration, temperature and, in particular, reagent may profoundly alter both absolute and relative rates.

In a simplified description one can differentiate between two types of substituent effects on metalation rates. Any electronegative group will acidify its environment by inductive electron withdrawal, and thus stabilize carbon centers carrying a partial negative charge at both the product and at the transition-state level. The substituent effect increases with the polarity of the organometallic bond, hence with the donor capacity of the solvent (THF > DEE > PET). Alternatively, the substituent may facilitate the permutational exchange occurring at the *ortho*-position by metal coordination. To become effective, the latter must be more intense at the transition state than at the preceding ground state. Small metals (Li, Mg) can optimally exploit the coordination potential due to their high local charge density, whereas big ones are fairly insensitive to this kind of interaction.

The interplay between inductive and coordinative effects can be demonstrated in a simple experiment[971]. When an equimolar mixture of anisole and benzotrifluoride is treated with butyllithium in the presence of potassium *tert*-butoxide, only the halogenated substrate undergoes metalation. In contrast, only the ether is attacked when *sec*-butyllithium is used as the reagent.

● *Competition between Reaction Sites and Reaction Modes*

The choice of the reagent may determine the outcome of not only an intermolecular but also an intramolecular competition. If the same substrate contains two different activating substituents, one can, in fortunate instances, metalate at will either the position adjacent to the more electronegative one or the better coordinating one by appropriately matching the reagents. Many examples of such *optional site selectivity* have been documented[1961], for instance with 2- and 4-fluoroanisole[1962], *N-tert*-butoxy-2- and -4-anisidine[1963], *N-tert*-butoxy-4-fluoroaniline[1964], 3-chlorobenzotrifluoride[1965] and 1,3-bis(trifluoromethyl)benzene[1798, 1966].

LiC$_4$H$_9$/PMDTA

(structure: fluoro-methoxybenzene, F ortho to OCH$_3$; metalation positions indicated by LiC$_4$H$_9$/PMDTA and LiC$_4$H$_9$)

LiC$_4$H$_9$

(structure: 4-fluoroanisole, F para to OCH$_3$; LiC$_4$H$_9$/PMDTA and LiC$_4$H$_9$)

LiC$_4$H$_9$/KOC(CH$_3$)$_3$

(structure: OCH$_3$ and NHCOOC(CH$_3$)$_3$ ortho; LiC(CH$_3$)$_3$)

(structure: OCH$_3$ para to NHCOO(CH$_3$)$_3$; LiC$_4$H$_9$/KOC(CH$_3$)$_3$ and LiC(CH$_3$)$_3$)

(structure: F para to NHCOOC(CH$_3$)$_3$; LiC$_4$H$_9$/KOC(CH$_3$)$_3$ and LiC(CH$_3$)$_3$)

LiC$_4$H$_9$

(structure: 3-chlorobenzotrifluoride, Cl and CF$_3$; LiCH(CH$_3$)C$_2$H$_5$)

LiC$_4$H$_9$/KOC(CH$_3$)$_3$

(structure: 1,3-bis(trifluoromethyl)benzene, F$_3$C and CF$_3$; LiCH(CH$_3$)C$_2$H$_5$/PMDTA or LITMP, LiC(CH$_3$)$_3$/THP (in part))

Benzylamines are generally converted into *N*-acyl derivatives before being submitted to metalation. This introduces a further reaction variable. The proper choice of the protective group may dictate the regioselectivity of the hydrogen/metal exchange process. *N,N*-Dimethylcarbamoyl-protected methoxy- or fluorobenzylamines tend to be deprotonated at the *ortho*-position nearest to the nitrogen-functional side chain, whereas metalation of *tert*-butyloxycarbonyl (BOC)-protected benzylamines occurs at an amidomethyl-remote, but oxygen- or halogen-adjacent site[1967]. The unsubstituted benzyl urea is still attacked at the *ortho*-position, but the unsubstituted *N*-benzyl *tert*-butylcarbamate at the α-position rather than at any aromatic center[1967]. In all such cases, two equivalents of the reagent are required, the acidic NH group being deprotonated beforehand (M=H → M=Li).

X = OC(CH₃)₃:
LiC₄H₉/KOC(CH₃)₃   OCH₃   X = N(CH₃)₂:
↖   ⤻ LiCH(CH₃)C₂H₅
CH₂NMCOX

X = N(CH₃)₂:
H₃CO   ⤻ LiCH(CH₃)C₂H₅
X = OC(CH₃)₃: ⤻
LiC₄H₉/KOC(CH₃)₃   CH₂NMCOX

X = N(CH₃)₂:
⤻ LiCH(CH₃)C₂H₅
CH₂NMCOR
⇑
X = OC(CH₃)₃:
LiCH(CH₃)C₂H₅

X = OC(CH₃)₃:
LiC₄H₉/KOC(CH₃)₃
↓   F
⇑   CH₂NMCOX
X = N(CH₃)₂:
LiC(CH₃)₃

X = N(CH₃)₂:
F   ⤻ LiCH(CH₃)C₂H₅
⤻
CH₂NMCOX
X = OC(CH₃)₃:
LiC₄H₉/KOC(CH₃)₃

It is by no means a trivial task to discriminate rigorously between aromatic and benzylic positions. Only butylpotassium[445] and the butyllithium/potassium *tert*-butoxide mixture[444] are capable of converting toluene cleanly into benzylpotassium. TMEDA-activated butylsodium[445] and butyllithium[446] also significantly (>10%) attack the ring. It needs trimethylsilylmethylpotassium to generate pure α-cumylpotassium (2-phenyliso-propylpotassium)[440], while all other reagents concomitantly promote hydrogen/metal exchange at the α-, *meta*- and *para*-positions of isopropylbenzene (cumene)[441–443].

N,N,2-Trimethylaniline reacts with butyllithium in the presence of TMEDA to afford 2-(dimethylamino)phenyllithium[450], whereas the *para*-isomer is metalated at the nitrogen-adjacent position under the same conditions[450]. However, the latter substrate undergoes exclusive hydrogen/metal exchange at the methyl group when treated with butyllithium/potassium *tert*-butoxide in tetrahydrofuran[451]. *N,N*-Dimethylbenzylamine reacts with butyllithium[1968], *tert*-butyllithium[1969] or butylsodium[1968] at the *ortho*-position, with phenylsodium[1968] or butyllithium/potassium *tert*-butoxide[1969] at the benzylic site. Depending on the base, *N*-benzoylaniline is deprotonated at either the *ortho*-position of the benzoyl part[1750] or at the benzylic α-position[1970].

LiC₄H₉/KOC(CH₃)₃ ⇒ CH₃

⤻ LiC₄H₉
LiC₄H₉/TMEDA ⤻
N(CH₃)₂
CH₂—N(CH₃)₂
⇑
LiC₄H₉/KOC(CH₃)₃

Treatment of 2-methylanisole (2-methoxytoluene) with pentylsodium leads to pure 2-methoxybenzylsodium, whereas TMEDA-activated butyllithium produces mixtures of *ortho-* and α-intermediates[448]. 4-Methylanisole is regioselectively deprotonated by butyl-lithium at the oxygen-adjacent position[449], whereas superbasic reagents attack the *ortho-* and α-positions randomly[444]. Also alkyl aryl sulfides may be preferentially metalated at an aromatic or benzylic position, again depending on the thioether structure and the reagent[1971]. A comparison between lithium benzylalcoholate and its carbon dioxide adduct, lithium benzylcarbonate, demonstrates once more how a protective group can completely remodel regioselectivities. The former substrate is exclusively metalated at the *ortho-*[1785], the latter at the α-position[1972]. The macrocycle 1,3-xylylene-15-crown-4 stages an impressive case of metal-dependent optional site selectivity. Unlike butyllithium, which inevitably abstracts a benzylic proton, ethereal phenylmagnesium bromide transfers, in exchange against hydrogen, its metal to the aromatic position flanked by the two ansa attachments[1973].

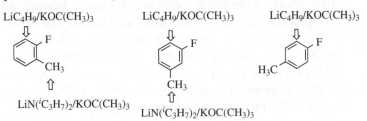

Finally 2- and 3-fluorotoluenes and, by extension, other 2- and 3-alkylated fluoro-benzenes undergo hydrogen/metal exchange at the (least congested) halogen-adjacent or the benzylic position depending on whether butyllithium/potassium *tert*-butoxide or lithium diisopropylamide/potassium *tert*-butoxide ('LIDA-KOR') are employed as the base[1974,1975]. However, due to electron−electron repulsion, only trace amounts of a 4-fluorobenzylpotassium can be generated in this way[1365].

<table>
<tr><td>LiC4H9/KOC(CH3)3</td><td>LiC4H9/KOC(CH3)3</td><td>LiC4H9/KOC(CH3)3</td></tr>
</table>

LiN(^{i}C$_3$H$_7$)$_2$/KOC(CH$_3$)$_3$ LiN(^{i}C$_3$H$_7$)$_2$/KOC(CH$_3$)$_3$

There are many other kinds of reaction sites that may compete with each other for bases. With 2-(*N*-BOC-ω-aminoalkyl)toluenes **274** as substrates, the α- and α'-positions rival with each other. If the side chain is short (n = 0 − 1), the benzylic methyl group, otherwise the benzylic methylene group, is deprotonated[1976]. 1,2,3,4-Tetrahydro-2-methylisoquinoline and derivatives thereof (**275a**) are metalated at the 4-position[1977], their adducts (**275b**) with boron trifluoride at the 1-position, however[1978]. Lithiation of the spiro-*N*-BOC-piperidine **276** occurs at a cyclopropanic rather the *N*-adjacent methylene group[1979]. A proton is abstracted from the 3-position of 1-methyl-1,2,3,6-tetrahydropyridine (**277a**)[1980], but from the 6-position of its amidine analog **277b**[1981]. The hydrogen/metal interconversion is oriented towards the side chain-adjacent position of 1-(dimethylamino)ethylferrocene (**278a**, R=X=CH₃), but mainly towards the unsubstituted bottom ring of the *O*-lithio hemiaminal **278b** (NR₂= 4-methylpiperazyl, X = OLi)[1982].

There are also numerous examples of competing typoselectivities some of which have already been mentioned above (see pp. 56–58). Organolithiums react with *N,N*-dimethylcarboxamides under nucleophilic addition, affording ketones after hydrolysis, but with *N,N*-diisopropylcarboxamides under *ortho*-metalation. The *N,N*-diethyl analogs undergo addition with butyllithium and *ortho*-lithiation with *sec*-butyllithium (see Table 123)[1983]. In the same way, the carbonyl groups in the methyl esters of 2,4,6-triisopropyl- and 2,4,6-tri-*tert*-butylbenzoic acid proved sterically inaccessible. Once more the organometallic reagent cannot perform a nucleophilic addition, as usual, but must content itself with proton abstraction from the methyl group[1512,1513,1540].

Polychloroarenes often exhibit ambivalent reactivity. 1,2,3,4-Tetrachlorobenzene mainly undergoes metalation with methyllithium and only marginally halogen/metal exchange, affording the intermediates **279** and **280** in a 13:1 ratio, whereas *tert*-butyllithium exclusively elicits the latter reaction mode[1984]. Phenyllithium and butyllithium give product mixtures in the ratios 8:1 and 1:3, respectively[1984].

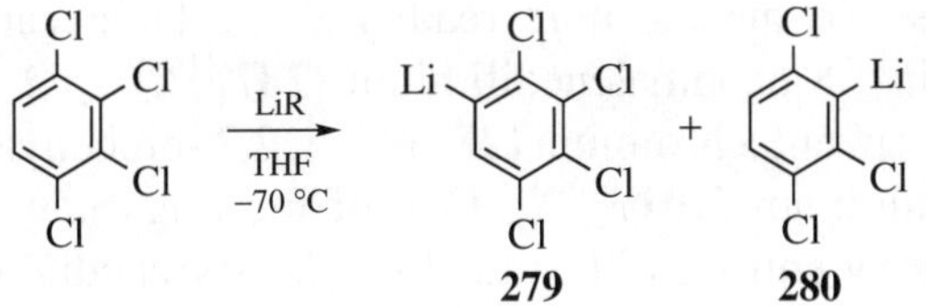

279 **280**

● *Dimetalation*

The outcome of a trapping experiment can be uncertain. As demonstrated above (see p. 120), the conversion of 1-isopropyloxy-2,2-diido-1-phenylethylene into 1-isopropyloxy-2-methyl-1-phenylpropene by consecutive treatment with methyllithium and methyl iodide does not involve a *gem*-dilithiated species but is brought about by two successive sequences of halogen/metal exchange and electrophilic substitution[1133].

Likewise, no trilithiated intermediate **281** is formed when an excess of *tert*-butyllithium and subsequently chlorotrimethylsilane are added to 1,3,5-trifluorobenzene in tetrahydro-furan at $-100\,^{\circ}$C. The two reagents being inert towards each other at such low temperatures, the triple substitution leading to 1,3,5-trifluoro-2,4,6-tris(trimethylsilyl)benzene (**282**) is accomplished stepwise, each monometalation being followed by instantaneous *in situ* trapping[1985]. At $-75\,^{\circ}$C, the same reaction mixture produces 2,4,6-tri-*tert*-butyl-1,3-phenylenedilithium (**283**) in a cascade which is unleashed by a second lithiation and in which lithium fluoride eliminations alternate with *tert*-butyllithium additions to the 'arynes' thus set free[1985].

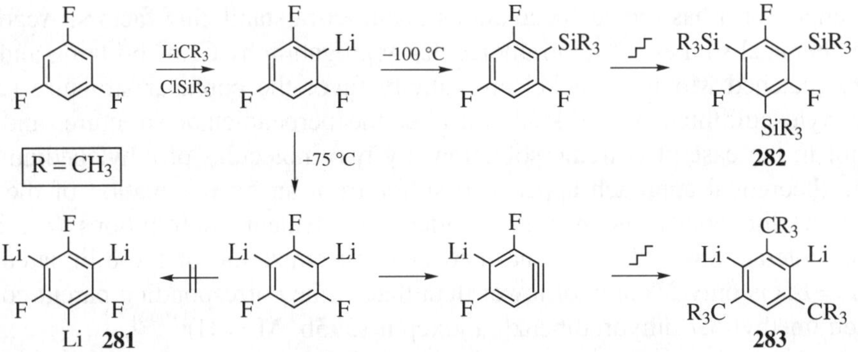

R = CH₃

282

Li 281 **283**

Unequivocal cases of dimetalation are nevertheless documented. 2,5-Dimethoxy-1,4-phenylenedilithium (**284**)[1986], 2,4,6-trimethoxy-1,3-phenylenedilithium (**285**)[1987], 1-methyl-2,5-pyrrolediyldilithium (**286**, Y = NCH₃, M = Li)[1988], 2,5-furanediyldilithium (**286**, Y = S, M = Li)[1988] and 2,5-thiophenediyldisodium (**286**, Y = S, M = Na)[1989] are typical examples.

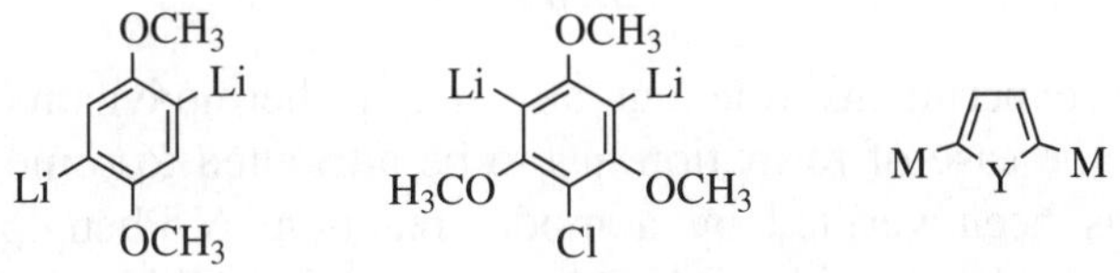

284 **285** **286**

A substrate tolerates two metals more readily if the latter can be accommodated in two different rings as in 1,8-naphthalenedilithium (**287**)[1990], 1,1′-ferrocenediyldilithium (**288**)[1991], bis(1-lithiobenzene)chromium (**289**)[1992], 2,2′-biphenylenedilithium (**290**)[1993] and 1,12-triphenylenedilithium (**291**)[1994]. One of the rings may even be replaced by a vinyl, styryl or ethenyloxy entity as illustrated by β,*o*-styrenediyldilithium (**292**)[1995], α-butyl-β,*o*-stilbenediyldilithium (**293**)[1996–1997] and 2-(1-potassioethenyloxy)phenylpotassium (**294**)[236].

The ease with which such *o,o′*- or *o,β*-dilithiated derivatives of styrenes, biphenyls and congeners form has fueled speculations about extra stabilizing factors. According to quantum chemical studies[1998], intramolecular aggregation by CLiC bridging and orbital-symmetry matched Möbius–Hückel aromaticity favor the coplanar structure (**295a**) of 2,2′-biphenylenedilithium by 17 kcal/mol over the perpendicular structure, and still by 9 kcal/mol in the case of extreme solvation (by two molecules of ethylenediamine). As often, the theoretical approach appears to suffer from an overestimation of the ionicity of organometallic bonds and hence a neglect of covalent contributions (*i.e.* spatially directional interactions). The torsional barrier to racemization of the dilithiated species **295b** (M = Li) is only 2 kcal/mol lower than that of the corresponding parent compound 5,5,7,7-tetramethyl-5,7-dihydrodibenz[*c,e*]oxepin (**295b**, M = H)[1999].

This and further experiments rule out any major thermodynamical bonus for 1,4-dilithiation. If real, the ease of formation has to be attributed to some kinetic advantage. This hypothesis has been verified by a model reaction. *N*-Phenylpyrrole is the only substrate known to undergo with TMEDA-activated butyllithium in diethyl ether or hexanes a second metalation faster than the preceding first metalation[2000]. However, when unconsumed *N*-phenylpyrrole is still present or fresh one is added, the α,o-dilithiated species **297** initially generated slowly reverts to the α-monometal compound **296**, so far only identified as a transient intermediate on the way to the dimetalated product. As revealed by this equilibration process, the dimetal species **297** is just kinetically favored but more basic, hence thermodynamically less stable than the monometal derivative **296**[2000].

In this way, the lithium atom reveals itself as a delicate, though potentially powerful hydrogen/metal exchange-promoting neighboring group. It does not exert the slightest acidifying effect by inductive electron withdrawal. Its only mode of action is to aggregate the metalating reagent and, by amplifying this coordinative interaction at the transition state, to lower the energy of activation for the exchange process.

● *Basicity Gradient-Driven Secondary Transformations*

A metal, at the moment of its attachment to an organic backbone, may not immediately end up at the most electronegative carbon atom. Its dislocation to another position may lower its basicity and thus improve its thermodynamic stability. Whenever this goal becomes practically attractive, there are two universally applicable mechanisms to achieve it, hydrogen/metal and halogen/metal permutation.

Permutational hydrogen/metal exchange is generally a slow process. Fortunate circumstances such as intramolecular reaction may, however, speed it up sufficiently. The internal proton transfer of ω-phenylalkyllithiums **298** is not only merely fast but its outcome depends also crucially on the tail length. 4-Phenylbutyllithium (n = 2) and 5-phenylpentyllithium (n = 3) afford resonance-stabilized benzylic species, whereas 3-phenylpropyllithium can only attack the *ortho*-position[2001].

Intramolecularity has been further established for the isomerization of 2-(2-furyl)ethyllithium (**299**) to 3-ethyl-2-furyllithium[2002] and seems probable for the conversion of 1-(allyldiphenylsilyl)hexyllithium (**300**), the adduct of butyllithium to the corresponding vinylsilane, into 1-(hexyldiphenylsilyl)allyllithium[2003].

Most secondary transmetalations occur intermolecularly. They are in general mediated by catalytic amounts of incidentally formed hydrocarbons which are the C—H acids common to each pair of final organometallic product and its precursor. In this sense such reactions can be conceived as counterparts of base-catalyzed isomerizations of alkenes[2004] which involve a common carbanion as the key intermediate (*e.g.* allylbenzene *via* phenylallylpotassium to predominantly *trans*-1-phenylpropene[2005]). Thus, when 2-methyl-1-phenyl-1-propenyllithium (**301**) reacts with 2-methyl-1-phenylpropene it converts the latter into the more stable 2-methyl-1-phenylpropene[2006]. Analogously, prenylpotassium (**302**) is reincarnated as (Z)-2-methyl-2-butenylpotassium (at $\geq -60\,^{\circ}$C)[399] and 5-methyl-2,4-hexadienyllithium (**303**) as 2-methyl-2,4-hexadienyllithium (at $+40\,^{\circ}$C)[2007].

The permanent regeneration of the hydrocarbon template is the prerequisite for and the common feature of all such isomerizations. To avoid it, work at low temperatures is recommended using an excess of the organometallic reagent.

Such spontaneous transformations are of course a nuisance. If the isomerized species were desired, one would access it by direct metalation of the appropriate hydrocarbon rather than take the roundabout route by way of a labile organometallic precursor. The same argument holds for the preparation of the intermediate **305** which is straightforward when *N*-cyclohexylpiperonylideneimine is metalated with butyllithium at −75 °C[2008]. The isomeric species **304** is readily made from the corresponding bromo derivative by halogen/metal exchange. To preserve its structural identity, one has to store it at low temperatures because at +25 °C it isomerizes rapidly to species **305**[2008]. A close analogy was simultaneously found in the heterocyclic series. 2-Chloro-3-thienyllithium (**306**), generated by halogen/metal permutation, can be trapped as such, but upon standing it slowly forms 5-chloro-2-thienyllithium (**307**) by transmetalation[2009].

However, a secondary transmetalation may sometimes offer a synthetically valuable option. When 2-(3,4-dimethoxyphenyl)ethyl-*N*,*N*-dimethylamine is treated with TMEDA-activated butyllithium, the sterically least-hindered site is deprotonated instead of the most acidic one[2010]. Thus, intermediate **308** is initially formed, which gradually metamorphoses to species **309** in which the organometallic bond is flanked by both heterosubstituents[2010].

Similar observations were made with (2,5-dimethoxyphenyl)diphenylphosphine oxide as the substrate. Bulky bases being mandatory for the metalation of phosphine oxides[2011], *tert*-butyllithium was employed. It was found to promote hydrogen/metal exchange preferentially at one of the unsubstituted phenyl rings (ratio of intermediates **310** and **311** after 2 min 75:25)[2012]. It was necessary to wait 4 h to find the metal mainly, though not exclusively, attached to the methoxy-bearing aryl ring (**310** : **311** = 10:90)[2012].

Once again, very similar examples exist in the heterocyclic series. 2-[2-(4,4-Dimethyl-Δ^2-oxazolinyl)]thiophene[2013], 2-(2-Δ^2-imidazolinyl)furan[2014] and 2-(2-Δ^2-imidazolinyl)thiophene[2014] are all metalated by butyllithium at the 3-position (species **312** and **314b,c**), but afford the 5-lithiated isomers **313** and **315b,c** under equilibration conditions (more polar solvent and higher temperature or amide bases). Only the 3-lithiation of 2-(2-Δ^2-imidazolinyl)-1-methylpyrrol (**314a**)[2014] is definitive.

314a: Y = NCH$_3$
314b: Y = O   **315b**: Y = O
314c: Y = S   **315c**: Y = S

Bromine/lithium and iodine/lithium interconversions proceed much more rapidly than transmetalations and should hence be more efficient if it is a matter of helping metals swap position. This may be experienced involuntarily. When 1,2,4-tribromobenzene **316** is treated with sodium amide in liquid ammonia[2015] or potassium *tert*-butoxide in *N,N*-dimethylformamide[2016], a product mix results which contains, along with starting material, 1,3,5-tribromobenzene, 1,4-dibromobenzene and 1,2,3,5- and 1,2,4,5-tetrabromo-benzene as the major components. J. F. Bunnett[2017] coined the euphemistic term 'halogen dance' for this randomization process, although more prosaic minds may rather feel tempted to call it 'halogen scramble' or 'halogen mess'.

Clean transformations can be achieved if a strong base is employed in stoichiometric amounts. For example, when 2-bromobenzotrifluoride is treated with lithium 2,2,6,6-tetramethylpiperidide in tetrahydrofuran at $-100\,°C$, deprotonation occurs at the position next to the heavy halogen. Intermediate **317** is stable enough to be trapped at $-100\,°C$ but isomerizes at $-75\,°C$ to the new species **318** in which the lithium is flanked by both substituents. The switching over of the metal to a more electronegative site provides a driving force strong enough to make the reaction unidirectional. The positional permutation of the lithium and bromine atom does not involve a transient 'aryne' as one might suspect at a first glance, but is mediated by trace amounts of incidentally formed 2,3-dibromobenzotrifluoride which acts as a self-regenerating intermediate for quick halogen/metal exchange[2018].

Often the organometallic intermediate isomerizes faster than it is formed. This is the case with 1,2,3-tribromobenzene and 3-bromo-4-iodobenzotrifluoride (2-bromo-1-iodo-4-trifluoromethylbenzene) which upon treatment with lithium diisopropylamide at $-75\,°C$ instantaneously afford the species **319** and **320**[2019].

Basicity-lowering halogen and lithium migrations have been well studied for heterocycles. The first synthetically useful examples were published by H. C. van der Plas *et al.* Under optimal conditions, 5-bromo-3-methylisothiazole[2020] and 5-bromo-3-methyl-1-phenylpyrazole[2021] were converted *via* intermediates **321–324** into the corresponding 4-bromo isomers in satisfactory yields (85 and 70%, respectively), although inevitably some starting material was also recovered and dehalogenated products were isolated in addition.

| Y = S | M = Na | **321** | **322** |
| Y = NC$_6$H$_5$ | M = K | **323** | **324** |

The extensive investigations[2022,2031] of bromothiophenes can be summarized in a few conclusions. Whereas 2-bromothiophene may be concomitantly deprotonated at the 3- and 5-positions[2023,2024,2029], one or the other of these sites are blocked in 2,3-dibromothiophene[2022,2027,2028], 2-bromo-3-(methylthio)thiophene[2026], 2,5-dibromothiophene[2025] and 2-bromo-5-methylthiophene[2023]. Proton abstraction occurs at the remaining most acidic position, thus anticipating the course of the subsequent shift of the halogen and the metal. With 2,3,5-tribromothiophene[2031] and 2,3-dibromo-5-methylthiophene[2029] there is no choice left and the site of deprotonation as well as the outcome of the secondary transformation is predictable. Whereas the isomerized organometallic species (**326, 328** or **330**) can in general be derivatized without difficulties[2025,2029], interception of the primary intermediates (**325, 327** or **329**) requires fortunate circumstances and carefully elaborated protocols[2026,2030].

Similar basicity-lowering halogen migrations have been accomplished with 2,3- and 2,5-dibromofuran and 2-bromo-5-methylfuran involving intermediates **331–334**[2030]. As in all previous cases, the chemical system aims at attaining the deepest thermodynamical sink it can reach.

The work disclosed by G. Quéguiner *et al.* on the isomerization of bromo-[2032] and iodopyridines[2033] is unsurpassed as far as the practical importance is concerned. Upon amide-promoted deprotonation at the 4-position, 2-chloro- and 2-fluoro-3-bromopyridines (**335**, R = H, H₃C) instantaneously undergo a halogen–metal exchange generating the 3-lithiated 4-halopyridines. If the 4-position is already occupied, the 2-position, as in the case of 3-fluoro-4-iodopyridine (**336**)[2033], or the 5-position, as in the case of 2-chloro-3-fluoro-4-iodopyridine (**337**)[2033], will be lithiated.

Relatively unexplored are base-driven halogen migrations in the quinoline series. The only known example exploits the superior acidity of the 4-position. 3-Fluoro-4-iodoquinoline (**338**), readily prepared from 3-fluoroquinoline by consecutive treatment with lithium diisopropylamide, is attacked by the same reagent at the 2-position to undergo instantaneous halogen–metal permutation and to produce 3-fluoro-2-iodo-4-quinolyllithium[2034,2035].

c) Working procedures

The order in which the working procedures are presented mirrors the arrangement of the Tables of reactive intermediates. Species resulting from the deprotonation of resonance-inactive saturated, then allylic and benzylic sp^3 centers are covered first. Next come olefinic and arylic metal derivatives, followed by, metalated 5- and 6-membered heterocycles. Within each subsection, pure hydrocarbons are treated as substrates first, followed by nitrogen-, oxygen- and halogen-substituted congeners.

The selection made neglects a bit structure types already covered elsewhere (in particular in *Organic Syntheses*) and emphasizes methodical novelties. Thus, the bulk of the protocols deals with allylic, benzylic and olefinic metal species without heterosubstituents (see pp. 267–276) and alkoxy-, aroxy- or fluorine-bearing arylmetals (see pp. 278–282).

Several examples highlight the phenomenon of optional site selectivity. *N,N*,4-Trimethylaniline can be either metalated at the benzylic or an aromatic position next to the nitrogen atom (see pp. 274 and 278). Both 2- and 4-fluoroanisole can undergo clean proton abstraction at either an oxygen or a halogen adjacent position (see pp. 280 and 281). Other substrates such as diphenyl ether (see pp. 279 and 280) and 1-phenylpyrrole (see pp. 282 and 283) illustrate the possibility of performing mono- or dimetalations selectively at will.

1-Tricyclo[2.2.1.0^{2,6}]heptylpotassium ('nortricyclylpotassium')[440]

Tricyclo[2.2.1.0^{2,6}]-heptane (1.9 g, 20 mmol) is added to a vigorously stirred suspension of pentylsodium (20 mmol) and potassium *tert*-butoxide (2.2 g, 20 mmol) in pentanes (40 mL). After 24 h, the mixture is treated with chlorotrimethylsilane (3.2 mL, 2.7 g, 25 mmol). 1-(Trimethylsilyl)tricyclo[2.2.1.0^{2,6}]heptane is isolated by distillation under reduced pressure [b.p. 52–55 °C/17 mmHg; yield: 60%]. Starting material (about 35%) is recovered.

Methoxymethylpotassium[1533]

Dibutylmercury (1.6 g, 5.0 mmol) is slowly added to a vigorously stirred suspension of potassium/sodium alloy (78:22 w/w; 1.0 mL, 0.86 g, 18 mmol) in pentanes (20 mL). After 30 min of stirring, the solvent is stripped off and the residue dissolved in precooled (−75 °C) dimethyl ether (5.0 mL, 4.0 g, 86 mmol). The mixture is kept for 2 h at 60 °C before being treated with benzaldehyde (1.0 mL, 1.1 g, 10 mmol). The excess of metal is cautiously destroyed by dropwise addition of ethanol (5 mL) followed by water (10 mL). Extraction with hexanes (3 × 10 mL) and distillation give 2-methoxy-1-phenylethanol [b.p. 123–124 °C/15 mmHg; yield: 75%]. A slightly modified protocol for the preparation of butylpotassium is given in Section 4.3.1 (see p. 158).

Dichloromethyllithium[1550]

$$CH_3Cl_2 \xrightarrow{\text{LiC}_4\text{H}_9} Li-CHCl_2 \xrightarrow{\text{HgCl}_2} Hg(CHCl_2)_2$$

Butyllithium (41 mmol) in hexanes (25 mL) is added dropwise over 45 min to dichloromethane (2.6 mL, 3.2 g, 40 mmol) in tetrahydrofuran (90 mL) at −75 °C. The mixture is treated with a solution of mercuric chloride (5.4 g, 20 mmol) in tetrahydrofuran (50 mL). After hydrolysis with a saturated aqueous solution of ammonium chloride (50 mL), bis(dichloromethyl)mercury is extracted with diethyl ether and precipitated with ethanol [m.p. 152–154 °C; yield 93%].

Trichloromethyllithium[1553]

$$CHCl_3 \xrightarrow{\text{LiC}_4\text{H}_9} Li-CCl_3 \xrightarrow{\text{HgCl}_2} Hg(CCl_3)_2$$

Butyllithium (10 mmol) in hexanes (15 mL) is added dropwise over 15 min to chloroform (1.6 mL, 2.4 g, 20 mmol) in tetrahydrofuran (32 mL) and diethyl ether (8 mL) at −110 °C. The mixture is treated with a solution of mercuric chloride (1.4 g, 5.0 mmol) in tetrahydrofuran (10 mL). After hydrolysis with a saturated aqueous solution of ammonium chloride (10 mL), bis(trichloromethyl)mercury is extracted with diethyl ether and, after evaporation of the solvents, crystallized from cyclohexane [m.p. 146.5–148.5 °C; yield: 96%].

Allylpotassium[1558]

Without inert gas protection, a 250 mL round-bottom flask is cooled to −75 °C and filled with pentanes (80 mL) a concentrated (about 11 M) solution of butyllithium (0.25 mol) in hexanes (23 mL) and potassium *tert*-butoxide (28 g, 0.25 mol). Propene (20 mL, 11 g, 0.25 mol) is added being transferred from a Schlenk tube kept at −75 °C by means of a precooled pipet. The flask is closed with a rubber stopper in the middle of which a hole is drilled to fix a thermometer. The mixture is vigorously stirred for 15 min at −25 °C (salt-ice/methanol bath), 15 min at 0 °C (ice-bath) and 30 min at 25 °C. At −25 °C, trimethylsilylchloride (42 mL, 38 g, 0.25 mol) is added dropwise. Without any further workup, the product is isolated by distillation [b.p. 166–168 °C; n_D^{20} 1.4437; yield: 69%].

endo-2-Butenylpotassium[2036]

(Z)-2-Butene (14 mL, 8.5 g, 0.15 mol) is condensed into a solution of butyllithium (0.10 mol) and potassium *tert*-butoxide (11 g, 0.10 mol) in neat tetrahydrofuran (75 mL) at −75 °C. After 15 min of stirring at −50 °C, the mixture is transferred into a −75 °C cold dry-ice/methanol bath and treated with fluoro-dimethoxyborane diethyl etherate (50 mL, 45 g, 0.27 mol). The orange-yellow mixture immediately decolorizes. After 30 min at −75 °C, propanol (8.0 mL, 6.4 g, 0.11 mol) is added dropwise. After 30 min at −75 °C, the mixture is poured into 2.0 M aqueous sodium hydroxide (0.2 L) and vigorously stirred 2 h at 25 °C before being extracted with hexane (3 × 25 mL). Distillation affords 4-methyl-5-hexen-3-ol having an *erythro/threo* composition of 96:4 [b.p. 66−68 °C/50 mmHg; yield: 62%].

exo-2-Butenylpotassium[2036]

The same procedure applied to (E)-2-butene (14 mL, 8.5 g, 0.14 mol) furnishes 4-methyl-5-hexen-3-ol with an *erythro/threo* ratio of 3:97 [b.p. 64−67 °C/50 mmHg; yield: 61%].

2-Phenylallylpotassium[1318,1377]

Butyllithium (40 mmol) in neat tetrahydrofuran (25 mL) is added dropwise, over 30 min, to a well stirred solution of α-methylstyrene (5.2 mL, 4.7 g, 40 mmol) and potassium *tert*-butoxide (4.5 g, 40 mmol) in tetrahydrofuran (120 mL) at −75 °C. After 1 h at −50 °C, the mixture is added portionwise to chlorotrimethylstannane (8.0 g, 40 mmol) in tetrahydrofuran (30 mL) at −75 °C. The turbid liquid is poured into water (0.2 L). The aqueous phase is extracted with hexanes (3 × 40 mL) and the combined organic layers are washed with brine (2 × 0.10 L) before being dried and evaporated. Distillation through a short (10 cm) Vigreux column gives colorless trimethyl(2-phenyl-2-propenyl)stannane [b.p. 68−69 °C/0.1 mmHg; n_D^{20} 1.5542; yield: 31%].

2-(Trimethylsilyl)allylpotassium[1318,1377]

Trimethyl-(1-methylethenyl)silane (2.3 g, 20 mmol) and potassium *tert*-butoxide (2.2 g, 20 mmol) are consecutively added to butyllithium (20 mmol) in neat tetrahydrofuran (50 mL) at −75 °C. After 5 h at −50 °C, the mixture is treated with chlorotrimethylstannane and worked up as described in the preceding protocol. Trimethyl[1-(trimethylstannylmethyl)ethenyl]silane is obtained as a colorless liquid [b.p. 33–34 °C/0.7 mmHg; n_D^{20} 1.4816; yield: 45%].

2-Pinen-10-ylpotassium[151,1229]

(1*S*)-α-Pinene (7.9 mL, 6.8 g, 50 mmol) and potassium *tert*-butoxide (5.6 g, 50 mmol) are added to a solution of butyllithium (50 mL) in hexanes (35 mL). Under stirring, the suspension is heated for 30 min to 50 °C. At −75 °C, the brown mixture is consecutively treated with precooled tetrahydrofuran (25 mL), fluorodimethoxyborane dietherate (11 mL, 10 g, 60 mmol), 35% aqueous hydrogen peroxide (5.2 mL, 5.8 g, 60 mmol) and 3.0 M aqueous sodium hydroxide (20 mL). After stirring for 1 h at 25 °C, virtually pure (1*R*)-(−)-myrtenol is obtained by extraction with diethyl ether (2 × 30 mL) and distillation [b.p. 105–109 °C/14 mmHg; n_D^{20} 1.4920, $[\alpha]_D^{20}$ − 46.4° (*c* = 1.0, methanol); yield: 49%].

endo-5-(2,3-Dimethyltricyclo[2.2.1.0^{2,6}]hept-3-yl)-2-methyl-2-pentenylpotassium ('α-santalenylpotassium')[1567]

Potassium *tert*-butoxide (2.5 g, 22 mmol) is added to a solution of butyllithium (24 mmol) and 5-(2,3-dimethyltricyclo[2.2.1.0^{2,6}]hept-3-yl)-2-methyl-2-pentene ('α-santalene'; 4.1 g, 20 mmol) in neat tetrahydrofuran (20 mL) at −75 °C. The mixture is stirred until homogeneous, then kept for 2 h at −50 °C. At −75 °C, it is consecutively treated with fluorodimethoxyborane diethyl etherate (7.5 mL, 6.6 g, 40 mmol) and 35% aqueous hydrogen peroxide (4.0 mL, 4.6 g, 47 mmol). After 1 h of stirring at 25 °C, the aqueous phase is saturated with sodium chloride before being extracted with diethyl ether (2 × 25 mL). The combined organic layers are absorbed on silica gel (25 mL). The dried powder is placed on top of a chromatography column filled with fresh silica gel (100 mL) and hexanes. Elution with a 15:85 (v/v) mixture of ethyl acetate and hexanes followed by evaporation gives a colorless oil having the characteristic smell of α-santalol [b.p. 92–93 °C/0.1 mmHg; n_D^{20} 1.5022; $[\alpha]_D^{20} + 7°$ (CHCl₃); yield: 46%]. No trace of the unnatural (*E*) isomer is detected by gas chromatography.

α-Santalene (77%) is readily prepared from 7-iodomethyl-1,7-dimethyltricyclo-[2.2.1.0²,⁶]heptane[1567] and 3-methyl-2-butenylpotassium ('prenylpotassium')[440] in tetrahydrofuran. Without isolation of the hydrocarbon, the reaction can be continued applying the procedure given above to afford α-santalol with 35% overall yield.

(E)-2,4-Pentadienylpotassium, suspended in hexanes[230]

1,4-Pentadiene (4.1 mL, 2.7 g, 40 mmol) and butyllithium (40 mmol) in hexanes (25 mL) are rapidly added to a well stirred slurry of potassium *tert*-butoxide (4.5 g, 40 mmol) in hexanes (25 mL) kept in an ice bath. After 20 min at 0 °C, the mixture is cooled to −75 °C before being treated with fluorodimethoxyboron dietherate (8.0 mL, 7.1 g, 43 mmol) in diethyl ether (12 mL). Aqueous sodium hydroxide (10 mL of a 5 M solution) and 35% aqueous hydrogen peroxide (5.0 mL, 5.7 g, 58 mmol) are added and the two-phase mixture is vigorously stirred during 2 h. The organic layer is decanted, briefly dried and concentrated. Distillation of the residue gives 2,4-pentadien-1-ol having a *cis/trans* ratio of 3:97 [b.p. 55–57 °C/12 mmHg; yield: 88%].

(Z)-2,4-Pentadienylpotassium, dissolved in tetrahydrofuran[230]

1,4-Pentadiene (40 mmol) is metalated as described above. The hexane is stripped off and the residue is dissolved in precooled (−75 °C) tetrahydrofuran (0.25 L). The mixture is treated with fluorodimethoxyborane etherate (8.0 mL, 7.1 g, 43 mmol) and 35% aqueous hydrogen peroxide (5.0 mL, 5.7 g, 58 mmol). 2,4-Pentadien-1-ol is isolated having a *cis/trans* ratio of 98:2 [b.p. 62–63 °C/18 mmHg; yield: 59%].

1,3-Cyclohexadienylpotassium[398]

At −75 °C, 1,4-cyclohexadiene (1.9 mL, 1.6 g, 20 mmol) is added dropwise to a solution of trimethylsilylmethylpotassium (20 mmol; see page 158) in tetrahydrofuran (40 mL). After 15 h at −50 °C, the mixture is poured on freshly crushed dry ice. The solid material is neutralized with ethereal hydrogen chloride (10 mL of a 2.3 M solution) before being treated with ethereal diazomethane until the yellow color persists. The solution is filtered and concentrated. Bulb-to-bulb ('Kugelrohr') distillation (bath temperature 130–140 °C/10 mmHg) affords 44% of methyl 2,5-cyclohexadienylcarboxylate besides small amounts (approx. 5%) of methyl benzoate.

3,7-Dimethyl-2,,4,6-octadienyllithium[1227]

3,7-Dimethyl-1,3,6-octatriene (ocimene; mixture of (Z) and (E) isomers: 1.4 g, 10 mmol) is added to a solution of lithium diisopropylamide (prepared from diisopropylamine and butyllithium) (10 mmol) in hexamethylphosphoric triamide (1.0 mL), tetrahydrofuran (7.0 mL) and hexanes (7.0 mL) at −75 °C. After 5 h at −75 °C, the reddish-black mixture decolorizes when fluorodimethoxyborane diethyl etherate (2.2 mL, 2.0 g, 12 mmol) and 35% aqueous hydrogen peroxide (1.0 mL, 1.1 g, 12 mmol) are consecutively added. After 15 h of stirring at 25 °C, the organic phase is diluted with diethyl ether (10 mL) and, protected by nitrogen against contact with air. It is washed with brine (3 × 10 mL), dried and evaporated. Distillation affords (2E,4E)-3,7-dimethyl-2,4,6-octatrien-1-ol ('*trans*-dehydrogeraniol') [b.p. 84–85 °C/2 mmHg; mp 47–49 °C (from pentanes); yield: 51%]. The yield increases (to 68%) if the metalation of ocimene is carried out with a 0.5 M solution of trimethylsilylmethylpotassium (see pp. 158 and 159) for 1 h at −75 °C.

(2E,4E)-3-Methyl-5-(2,6,6-trimethyl-1-cyclohexenyl)-2,4-pentadienylpotassium[233]

Butyllithium (16 mL, 25 mmol) in hexanes (15 mL) and (E)-1-(3-methyl-2,4-pentadienyl)-2,6,6-trimethylcyclohexene (5.1 g, 25 mmol; from 4-(2,6,6-trimethyl-1-cyclohexenyl)-2-methyl-2-pentenyl[2037–2038] and the methyltriphenylphosphonium bromide/sodium amide 'instant ylid'[2039–2040] in 87% yield) are added to a suspension of potassium *tert*-butoxide (2.8 g, 25 mmol) vigorously stirred at 0 °C. After 60 min, the solvent is evaporated under reduced pressure while the reaction mixture is kept in an ice bath. The dark red residue is dissolved in precooled (−75 °C) tetrahydrofuran (0.10 L) and consecutively treated with fluorodimethoxyborane diethyl etherate (8.0 mL, 7.1 g, 43 mmol), 35% aqueous hydrogen peroxide (5.0 mL, 5.6 g, 65 mmol) and 3.0 M aqueous sodium hydroxide

(15 mL, 45 mmol). After 30 min of stirring at +25 °C, the mixture is saturated with sodium chloride, the organic layer decanted and absorbed on silica gel (15 mL). The dry powder is poured on top of a column filled with more silica (85 mL) and eluted with a 2:3 (v/v) mixture of diethyl ether and hexanes to afford (2*E*,4*E*)-3-methyl-5-(2,6,6-trimethyl-1-cyclohexenyl)-2,4-pentadiene-1-ol[2041] [b.p. 132–134 °C/0.1 mmHg; n_D^{20} 1.4756; yield: 54%].

(2E,4E,6E,8E)-3,7-Dimethyl-9-(2,6,6-trimethyl-1-cyclohexenyl)-2,4,6,8-nonatetraenylpotassium[233]

At −75 °C, butyllithium (25 mmol) in hexanes (15 mL), hexamethylphosphoric triamide (10 mL) and (2*E*,4*E*,6*E*)-2-(3,7-dimethyl-2,4,6,8-nonatetraenyl)-1,3,3-trimethylcyclohexene (6.8 g, 25 mmol) were consecutively added to a solution of diisopropylamine (3.5 mL, 2.5 g, 25 mmol) in tetrahydrofuran (0.10 L). After 2 h at −75 °C, the mixture is treated with fluorodimethoxyborane diethyl etherate (7.5 mL, 6.6 g, 40 mmol) and, 5 min later, with a 35% aqueous solution of hydrogen peroxide (3.5 mL, 4.0 g, 41 mmol) and a 3.0 M aqueous solution of sodium hydroxide (30 mL, 0.10 mol). After stirring for 1 h at 25 °C, the aqueous phase is saturated with sodium chloride, the organic layer decanted and evaporated. The residue is dissolved in dichloromethane (50 mL) and pyridine (50 mL), acetyl chloride (3.6 mL, 3.9 g, 50 mmol) is added and the mixture kept 1 h at 25 °C before being concentrated and absorbed on silica gel (20 mL). The powder is dried under reduced pressure and poured on top of a column filled with more silica gel (0.2 L) which had been thoroughly air degassed and nitrogen loaded. Elution with a 1:9 (v/v) mixture of oxygen-free diethyl ether and hexanes followed by evaporation affords a light-yellow oil which becomes colorless upon crystallization from pentanes [m.p. 60–62 °C; yield: 41%].

(2*E*,4*E*,6*E*)-2-(3,7-dimethyl-2,4,6,8-nonatetraenyl)-1,3,3-trimethylcyclohexene is readily accessible from (2*E*)-2-methyl-4-(2,6,6-trimethylcyclohexenyl)-2-butenyl by addition of 3-methyl-2,4-pentadienyllithium and subsequent dehydration of the resulting (2*E*,6*E*)-3,7-dimethyl-1-(2,6,6-trimethylcyclohexenyl)-2,6,8-nonatrien-4-ol [m.p. −41 to −38 °C; n_D^{20} 1.5203; yield: 84%] using (methoxycarbonylsulfamoyl)trimethylammonium hydroxide[2042] (Burgess' reagent). The C_{20}-pentene (m.p. −8 to −6 °C; b.p. 138–139 °C/0.1 mmHg) is obtained in 81% yield (relative to the C_{20}-tetraenol precursor).

1-Naphthylmethylpotassium[2043]

A solution of butyllithium (0.10 mol) in hexanes (60 mL) is evaporated to dryness and the residue is taken up in precooled (−75 °C) tetrahydrofuran (0.10 L). 1-Methylnaphthalene (14 mL, 14 g, 0.10 mol) and potassium *tert*-butoxide (11 g, 0.10 mol) are consecutively added. The mixture is stirred 3 min at −50 °C until the alcoholate has dissolved and is kept 2 h at −50 °C. At −75 °C, oxetane (trimethylenoxide; 7.8 mL, 7.0 g, 0.12 mol) is introduced. The mixture is allowed to reach 25 °C and the solvent is evaporated. Brine (50 mL) is added and the product is extracted with hexanes (3 × 25 mL). Upon distillation 4-(1-naphthyl)-1-butanol is collected [b.p. 150–152 °C/3 mmHg; yield: 81%].

4-Ethylbenzylpotassium[1622]

A mixture of *p*-ethyltoluene (2.8 mL, 2.4 g, 20 mmol), butyllithium (22 mmol), potassium *tert*-butoxide (2.5 g, 22 mmol) and hexanes (15 mL) is vigorously stirred for 2 h at 25 °C. At −25 °C, isopropyl bromide (2.1 mL, 2.7 g, 22 mmol) is dropwise added over the course of 15 min. 1-Ethyl-4-isobutylbenzene is isolated by distillation [b.p. 87–89 °C/10 mmHg; yield: 87%].

1-(4-Isobutylphenyl)ethylpotassium[1622]

Under vigorous stirring, a mixture of 1-ethyl-4-isobutylbenzene (3.2 g, 20 mmol), butyllithium (25 mmol), potassium *tert*-butoxide (2.8 g, 25 mmol) and hexanes (20 mL) is heated for 3 h to 60 °C before being poured on crushed dry ice covered with anhydrous tetrahydrofuran (50 mL). After evaporation of the volatiles, the residue is dissolved in water (0.10 L). The aqueous phase is washed with diethyl ether (3 × 30 mL), acidified to a pH of 1 and extracted again with diethyl ether (3 × 30 mL). The organic layer is washed with a 1% aqueous solution of sodium hydrogen carbonate (1 × 10 mL, then 1 × 15 mL) and dried. The oily material left behind upon evaporation of the solvent is crystallized from hexanes to afford pure 2-(4-isobutylphenyl)propanoic acid [m.p. 70–72 °C; yield: 79%].

α-Cumylpotassium (1-methyl-1-phenylethylpotassium)[440]

Isopropylbenzene (cumene; 0.50 mL, 0.43 g, 3.6 mmol) is added to a precooled solution of trimethylsilylmethylpotassium (see pp. 158 and 159, 3.6 mmol) in tetrahydrofuran (5.0 mL). After 24 h at −50 °C, the reaction is quenched with methyl iodide (0.30 mL, 0.68 g, 4.8 mmol). *tert*-Butylbenzene (74%) is identified by gas chromatograph as the sole product, no trace of cymols (isopropyltoluenes) being present. When the metalation time is reduced to 2 h without increasing simultaneously the concentration, the yield of *tert*-butylbenzene drops to ≤10%.

4-(Dimethylamino)benzylpotassium[451]

At −75 °C, butyllithium (15 mmol) in hexanes (10 mL) and potassium *tert*-butoxide (1.7 g, 15 mmol) are added to a solution of *N,N,p*-trimethylaniline (*N,N*-dimethyl-*p*-toluidine; 2.2 mL, 2.0 g, 15 mmol) in tetrahydrofuran (15 mL). The mixture is vigorously stirred until it becomes homogeneous, then kept 48 h at −75 °C using a cryogenic unit. After the addition of dimethyl sulfate (1.4 mL, 1.9 g, 15 mmol), it is allowed to reach 25 °C. Gas chromatographic analysis by means of an internal reference substance ('standard') reveals the presence of 65% of *N,N*-dimethyl-*p*-ethylaniline besides some unconsumed starting material. The main product is isolated by distillation (b.p. 96–97 °C/10 mmHg).

(E)-3,3-Dimethyl-1-butenylsodium[1489]

3,3-Dimethyl-1-butene (3.0 mL, 2.0 g, 23 mmol) is added to the suspension of pentylsodium (20 mmol) and disodium pinacolate (20 mmol) in pentane (50 mL). After 50 h of vigorous stirring, the sealed Schlenk tube is opened and the mixture is treated with a 0.8 M solution of formaldehyde[2044] (10 mmol) in tetrahydrofuran (25 mL). After neutralization, 4,4-dimethyl-2-penten-1-ol having a (*Z/E*) ratio of 0.2:99.8 is isolated [b.p. 76–77 °C/20 mmHg; yield: 88%].

(E)-2-(3,3-Dimethylbicyclo[2.2.1]heptylidene)methylpotassium
[(E)-10-camphenylpotassium][440,1146]

At 25 °C, a mixture of camphene (8.7 g, 65 mmol), pentylsodium (see page 101; 65 mmol), potassium *tert*-butoxide (11 g, 0.10 mol) in pentanes (0.15 L) was stirred for 2 h at high speed (5000 rpm[2045]). While stirring continues, dimethyl sulfate (9.5 mL, 13 g, 0.10 mol) is added dropwise at −75 °C. The mixture is centrifuged and the supernatant liquid decanted. Upon distillation some unconsumed camphene (6%; b.p. 84–86 °C/12 mmHg) and *(E)*-3-ethylidene-2,2-dimethylbicyclo[2.2.2]heptane are collected [(*Z*)/(*E*) ≤1:99; b.p. 90–91 °C/12 mmHg; n_D^{20} 1.4816; yield 78%].

2-Bicyclo[2.2.2]oct-2-enylpotassium[1489]

Bicyclo[2.2.2]oct-2-ene (27 g, 25 mmol) is added to the suspension of pentylsodium (25 mmol) and potassium *tert*-butoxide (2.8 g, 25 mmol) in pentanes (25 mL). After 10 h of vigorous stirring, the mixture is consecutively treated with fluorodimethoxyborane etherate[1229] (5.0 mL, 4.5 g, 27 mmol) and 35% aqueous hydrogenperoxide (3.0 mL, 3.4 g, 35 mmol) to which a few pellets of sodium hydroxide have been added. 2-Bicyclo[2.2.2]octanone is isolated by bulb-to-bulb ('Kugelrohr') distillation [b.p. 130–135 °C/10 mmHg; m.p. 177–178 °C; yield 70%].

2-Bicyclo[2.2.1]hepta-2,5-dienylsodium[1638]

Butyllithium (50 mmol), from which the commercial hexanes solvent has been stripped off, is dissolved in precooled (−75 °C) tetrahydrofuran (0.10 L). Sublimed sodium *tert*-butoxide (5.3 g, 55 mmol) and bicyclo[2.2.1]hepta-2,5-diene ('norbornadiene'; 5.1 mL, 4.5 g, 50 mmol) are added. The reaction mixture is stirred until it becomes homogeneous. After 15 h at −50 °C, it is treated with chlorotrimethylsilane (7.6 mL, 6.5 g, 60 mmol). Upon distillation under reduced pressure, 2-(trimethylsilyl)bicyclo[2.2.1]hepta-2,5-diene is obtained [b.p. 55–57 °C/10 mmHg; yield: 92%].

2-Spiro[4.4]nona-1,3-dienylsodium[1489]

Spiro[4.4]nona-2,4-diene (3.0 g, 25 mmol) is added to the suspension of pentyl-sodium (25 mmol) in pentanes (25 mL). After 100 h of vigorous stirring, the reaction mixture is consecutively treated with fluorodimethoxyborane etherate[1229] (5.0 mL, 4.5 g, 27 mmol), 35% aqueous hydrogen peroxide (3.0 mL, 3.4 g, 35 mmol) and 5 M aqueous sodium hydroxide (5 mL). Spiro[4.4]non-4-en-3-one is isolated by distillation [b.p. 92–94 °C/10 mmHg; yield: 45%]. Trapping with chlorotrimethylsilane gives (2-spiro[4.4]nona-1,3-dienyl)trimethylsilane; yield: 66%.

2-Cyclohepta-1,3,5-trienylsodium[1638]

Butyllithium (30 mmol), from which the commercial hexanes solvent has been stripped off, is taken up in precooled (−90 °C) tetrahydrofuran (50 mL). At −50 °C, cycloheptatriene (3.1 mL, 2.8 g, 30 mmol) and sodium *tert*-butoxide (2.9 g, 30 mmol) are added and dissolved under stirring. After having been allowed to stand 15 h at −50 °C, the reaction mixture is treated with chlorotrimethylsilane (4.0 mL, 3.4 g, 32 mmol) and the 2-(trimethylsilyl)cyclohepta-1,3,5-triene formed is isolated without any further workup by distillation [b.p. 100–102 °C/10 mmHg; yield: 57%].

1-Ethoxyvinyllithium[1670,2046]

At −30 °C, ethyl vinyl ether (9.6 mL, 7.2 g, 0.10 mol) and *N,N,N′,N′*-tetramethyl-ethylenediamine (15 mL, 12 g, 0.10 mol) are added to *tert*-butyllithium (0.10 mol) in pentanes (0.10 L). The reaction is complete after 1 h at 25 °C. Addition of benzaldehyde (1.0 mL, 11 g, 0.10 mol) and neutralization with acetic acid (6.0 mL, 6.3 g, 0.11 mol), immediately followed by distillation, give 2-ethoxy-1-phenyl-2-propen-1-ol [b.p. 93–95 °C/1 mmHg; yield: 57%].

1-(2-Tetrahydropyranyloxy)-1-propenyl potassium(lithium)[237]

Under stirring, potassium *tert*-butoxide (2.8 g, 25 mmol) and *sec*-butyllithium (25 mmol) in hexanes (25 mL) are consecutively added to a solution of *cis*-1-propenyl-2-tetrahydropyranyl ether (3.6 g, 25 mmol) in tetrahydrofuran (50 mL). After 1 h at −75 °C, the reaction mixture is treated with methyl iodide (1.8 mL, 4.0 g, 28 mmol). (Z)-1-Methyl-1-propenyl 2-tetrahydropyranyl ether is isolated by distillation [b.p. 87–89 °C/15 mmHg; yield: 66%] (83% formed according to gas chromatographic analysis).

5-Chloro-3,4-dihydro-2H-pyran-6-yllithium[1687]

A 1.5 M solution of butyllithium (0.11 mol) in hexanes (70 mL) and 5-chloro-3,4-dihydro-2*H*-pyran (12 g, 0.10 mol) in tetrahydrofuran (30 mL) are mixed at −75 °C. After 5 min at 25 °C, the solution is placed in an ice-salt bath (−20 °C) and diethyl sulfate (14 mL, 17 g, 0.11 mol) is added over the course of 10 min under stirring. The mixture is washed with brine (3 × 25 mL) and concentrated. Distillation affords 5-chloro-3,4-dihydro-2*H*-pyran [b.p. 56–58 °C; n_D^{20} 1.4688; yield: 79%]. The latter product undergoes clean ring opening when a 2.0 M solution of it in tetrahydrofuran is treated for 15 min at 25 °C with finely dispersed sodium sand (3.0 molar equivalent). After careful hydrolysis with methanol and water, 4-heptyn-1-ol is isolated [b.p. 93–95 °C/14 mmHg; yield: 84%].

3,5-Di-tert-butylphenylpotassium[1712]

1,3-Di-*tert*-butylbenzene (11 mL, 9.5 g, 50 mmol) and potassium *tert*-butoxide are added to a 1.5 M solution of butyllithium (53 mmol) in hexanes (35 mL). After 48 h of stirring at 25 °C, the mixture is poured on dry ice covered with tetrahydrofuran (25 mL). After evaporation of the solvents, the residue is acidified with formic acid (5.0 mL, 6.1 g, 0.13 mol) and the 3,5-di-*tert*-butylbenzoic acid crystallized from 50% aqueous ethanol [m.p. 173–174 °C; yield 89%]. Only 67% are obtained if the metalation time is shortened to 16 h.

5-(1,1,3,3-Tetramethyl-1,3-dihydroisobenzofuryl)potassium[1712,2047]

Potassium *tert*-butoxide (2.5 g, 22 mmol) is added to the solution of 1,1,3,3-tetra-methyl-1,3-dihydroisobenzofuran (3.5 g, 20 mmol) and *tert*-butyllithium (22 mmol) in hexanes (45 mL). The mixture is sonicated for 2 h at 25 °C before being consecutively treated with lithium bromide (25 mmol) in tetrahydrofuran (50 mL) and dimethylformamide (3.0 mL, 2.9 g, 39 mmol). Elution from silica gel with a 1:5 (v/v) mixture of ethyl acetate and hexanes affords 5-(1,1,3,3-tetramethyl-1,3-dihydroisobenzofuran)carbaldehyde [m.p. 83–85 °C; yield: 52%].

5-Methyl-2-(dimethylamino)phenyllithium[450]

A 2.3 M solution (20 mL) of butyllithium (46 mmol) in hexanes was added to *N,N*,4-trimethylaniline (4.3 mL, 4.0 g, 30 mmol) and *N,N,N′,N′*-tetramethylethylenediamine (6.8 mL, 5.2 g, 45 mmol) in hexanes (0.25 L). The resulting suspension was stirred for 4 h at +25 °C before being treated with benzophenone (8.2 g, 45 mmol) in diethyl ether (25 mL). The 2-(α-hydroxydiphenylmethyl)-*N,N*,4-trimethylaniline is first extracted with 1.0 M hydrochloric acid (3 × 50 mL) and, after basification of the combined aqueous layers, with diethyl ether (3 × 50 mL). After washing (2 × 25 mL of brine), drying and evaporation of the solvent, the residue is recrystallized from a benzene/hexanes mixture [m.p. 169–173 °C; yield: 55%; 81% crude].

2-[N-(Dimethylcarbamoyl)-N-lithio]aminomethyl-5-fluorophenyllithium[2048]

The commercial solvent (pentanes) is stripped off from *sec*-butyllithium (40 mL) and the residue is dissolved in precooled tetrahydrofuran (20 mL). *N*-(Dimethylcarbamoyl)-4-fluorobenzylamine (3.9 g, 20 mmol) is added. After 4 h at −75 °C, the mixture is poured on an excess of freshly crushed dry ice. The semi-solid material left behind after the evaporation of the organic solvent is partitioned between water (50 mL) and diethyl ether (25 mL). The aqueous phase is washed with chloroform (2 × 25 mL) and acidified to pH 3 before being extracted with dichloromethane (3 × 25 mL). The combined organic layers are dried and evaporated. Recrystallization from toluene affords colorless 2-[*N*-(dimethylcarbamoyl)aminomethyl]-5-fluorobenzoic acid [m.p. 148–149 °C (decomp.); yield: 86%].

5-[N-1-Lithiooxy-2,2-dimethylpropylidene)]-1,3-benzodioxolan-4-yllithium[2049]

At $-75\,^\circ$C, butyllithium (0.24 mol) in hexanes (0.15 L) is added to *N*-pivaloylpiperonylamine (24 g, 0.10 mol) in tetrahydrofuran (0.25 L). After the mixture is allowed to stand 1 h at $0\,^\circ$C, a suspension has formed which is treated, at $-75\,^\circ$C, with fluorodimethoxyborane diethyl etherate (28 mL, 25 g, 0.15 mol). After evaporation of the solvents, the residue is dissolved in methanol (0.15 L) and 35% aqueous hydrogen peroxide (25 mL, 28 g, 0.29 mol) is added. A white pasty precipitate deposits in the course of 30 min. The methanol is evaporated and the remaining mixture is partitioned between saturated aqueous ammonium chloride (0.25 L) and dichloromethane (0.25 L). The aqueous layer is again extracted with dichloromethane (2 × 0.10 L) and the combined organic layers are washed with water (0.10 L) and evaporated. Recrystallization of the residue from ethanol (or 2-propanol) affords 4-hydroxy-5-pivaloylamidomethyl-1,3-benzodioxolane [m.p. 197–198 $^\circ$C; yield: 68%].

2-Phenoxyphenylpotassium[2050]

Potassium *tert*-butoxide (2.8 g, 25 mmol) is dissolved in a solution containing butyllithium (25 mmol) and diphenyl ether (4.3 g, 25 mmol) in tetrahydrofuran (30 mL) at $-75\,^\circ$C. The mixture is allowed to stand 5 h at $-75\,^\circ$C, before sulfur (0.96 g, 30 mmol) is added. At $+25\,^\circ$C, the mixture is treated with 5 M hydrogen chloride in diethyl ether and is then evaporated to dryness. The residue is extracted with hot hexanes (3 × 25 mL). Upon concentration and cooling of the combined organic phases, 2-phenoxybenzenethiol crystallizes as colorless needles [m.p. 68–69 $^\circ$C; yield: 66%].

2,2'-Oxybis(phenyllithium)[2050,2051]

A solution of diphenyl ether (17 g, 0.10 mol), *N,N,N',N'*-tetramethylethylenediamine (TMEDA; 30 mL, 23 g, 0.20 mol) and butyllithium (0.20 mol) in diethyl ether

(0.10 L) and hexanes (0.15 L) is allowed to stand for 3 h at +25 °C. A brownish red precipitate settles out. Under cooling with an ice bath, *P,P*-dichlorophenylphosphine (benzenephosphonous dichloride; 27 mL, 36 g, 0.20 mol) is added dropwise. The mixture is evaporated to dryness and the residue extracted with hot hexane. Upon concentration and cooling of the solution, 10-phenylphenoxaphosphine crystallizes as colorless platelets [m.p. 95–96 °C; yield 59%].

3-Fluoro-2-methoxyphenyllithium[1962]

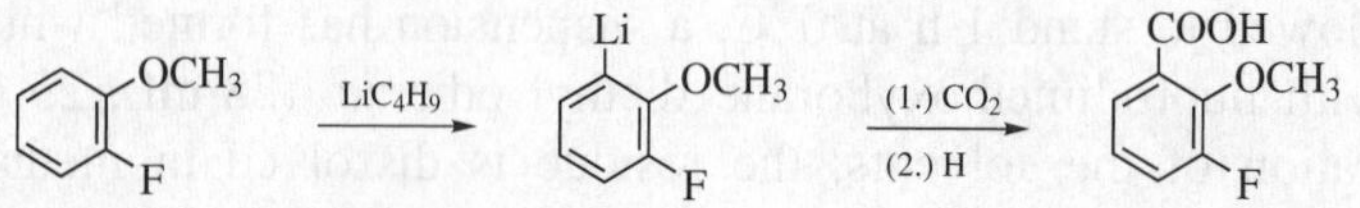

2-Fluoroanisole (5.6 mL, 6.3 g, 50 mmol) is added to a solution of butyllithium (50 mmol) in neat tetrahydrofuran kept at −75 °C. After 50 h, the mixture is poured on crushed dry ice. Evaporation of the solvent, neutralization of the residue with ethereal hydrogen chloride (2.5 M, 70 mL) and recrystallization from toluene affords pure 3-fluoro-2-methoxybenzoic acid [m.p. 154–156 °C; yield 50%].

5-Fluoro-2-methoxyphenyllithium[1962]

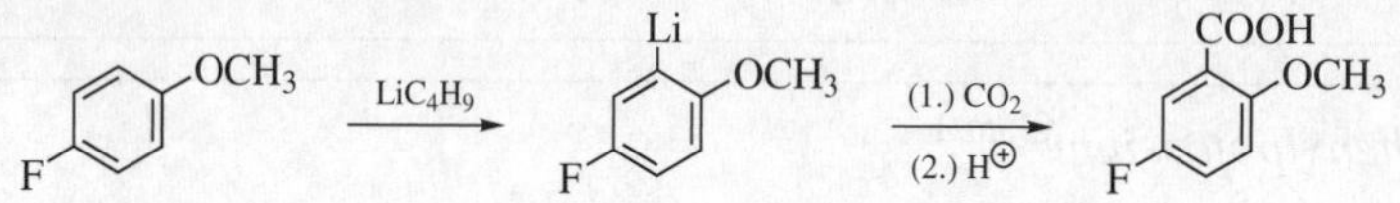

The same protocol applied to the *para* isomer of fluoroanisole (50 mmol) gives 5-fluoro-2-methoxybenzoic acid [m.p. 84–85 °C (from hexanes); yield 50%].

2-Fluoro-3-methoxyphenyllithium[1962]

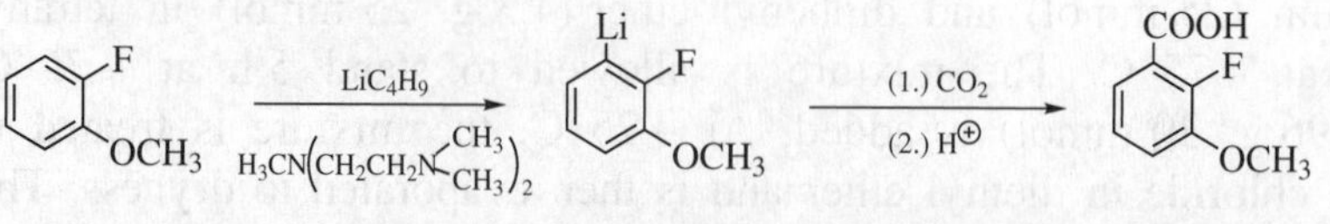

A solution containing 2-fluoroanisole (5.6 mL, 6.3 g, 50 mmol), butyllithium (50 mmol) and *N,N,N′,N″,N″*-pentamethylethylene-diamine (PMDTA; 10.2 mL, 8.7 g, 50 mmol) in a mixture of tetrahydrofuran (60 mL) and hexane (40 mL) is kept 2 h at −75 °C. The 2-fluoro-3-methoxybenzoic acid is again isolated by crystallization after neutralization with ethereal hydrogen chloride [m.p. 154–156 °C (from water); yield 87%].

2-Fluoro-5-methoxyphenyllithium[1962]

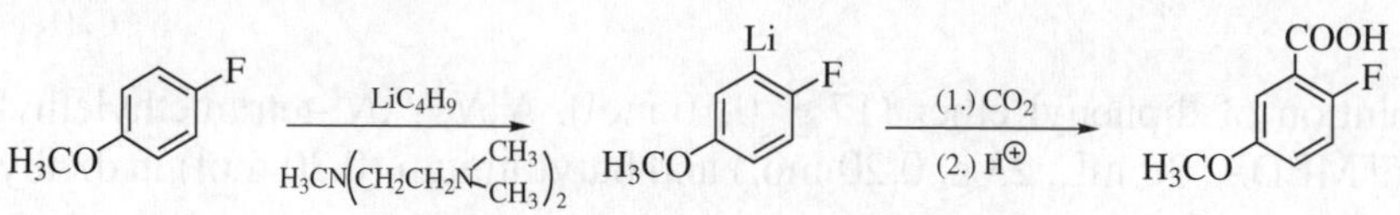

When the same protocol, as detailed in the preceding box, is applied to the *para*-isomer of fluoroanisole, 2-fluoro-5-methoxybenzoic acid is obtained [m.p. 144–145 °C (from toluene); yield: 85%].

5-[N-(Dimethylcarbamoyl)-N-lithio]aminomethyl-2-fluorophenyllithium[2048]

The commercial solvent (pentanes) is stripped off from butyllithium (40 mmol) and the residue is dissolved in precooled tetrahydrofuran (20 mL). *N,N,N′,N″,N″*-Pentamethyldiethylenetriamine (PMDTA; 8.4 mL, 6.9 g, 40 mmol) and *N*-dimethylcarbamoyl-4-fluorobenzylamine (3.9 g, 20 mmol) are added. After 4 h at −75 °C, the mixture is poured on an excess of freshly crushed dry ice. The semi-solid material left behind after the evaporation of the organic solvent is partitioned between water (50 mL) and diethyl ether (25 mL). The aqueous phase is washed with chloroform (2 × 25 mL) and acidified to pH 3 before being extracted with dichloromethane (3 × 25 mL). The combined organic layers are dried and evaporated. Recrystallization from water affords colorless 5-[*N*-(dimethylcarbamoyl)aminomethyl]-2-fluorobenzoic acid [m.p. 186–197 °C (decomp.); yield: 78%].

2-(Trifluoromethyl)phenylpotassium[1798]

At −75 °C, potassium *tert*-butoxide (2.8 g, 25 mmol) is added to a solution of methyllithium (25 mmol) and (trifluoromethyl)benzene (benzotrifluoride; 3.0 mL, 3.7 g, 25 mmol) in tetrahydrofuran (50 mL) and diethyl ether (20 mL; the solvent in which methyllithium is commercially supplied). Stirring is interrupted after 10 min when the mixture has become homogeneous. After 2 h at −75 °C it is poured on an excess of freshly crushed dry ice. The semi-solid material obtained after neutralization with an excess of ethereal hydrogen chloride and evaporation of the solvents is extracted with hot toluene (50 mL). Upon concentration and cooling of the filtered solution, 2-(trifluoromethyl)benzoic acid crystallizes [m.p. 111–113 °C; yield: 83%].

2,6-Bis(trifluoromethyl)phenylpotassium[1798]

When the protocol of the preceding preparation is applied to 1,3-bis(trifluoromethyl)benzene (hexafluoro-*m*-xylene; 3.9 mL, 5.4 g, 25 mmol), 2,6-bis(trifluoromethyl)benzoic acid is obtained [m.p. 133–135 °C (from hexanes); yield: 92%].

2,4-Bis(trifluoromethyl)phenyllithium[1798]

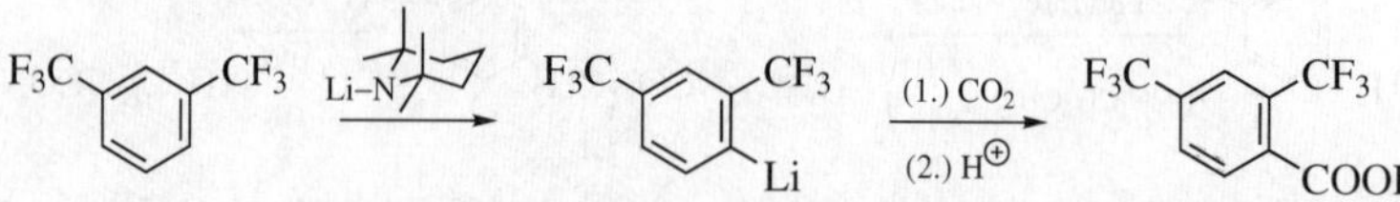

The commercial solvent (hexanes) is stripped off from butyllithium (25 mmol) and the residue dissolved in precooled tetrahydrofuran (25 mL). At −75 °C, 2,2,6,6-tetramethylpiperidine (4.2 mL, 3.5 g, 25 mmol) and 1,3-bis(trifluoromethyl)benzene (3.9 mL, 5.4 g, 25 mmol) are added consecutively. The mixture is kept at this temperature for 2 h before being poured on an excess of freshly crushed dry ice. Neutralization, extraction and crystallization [see the isolation of 2-(trifluoromethyl)benzoic acid above] afford 2,4-bis(trifluoromethyl)benzoic acid [m.p. 109–111 °C; yield: 94%].

1-Phenyl-2-pyrrylpotassium[1821]

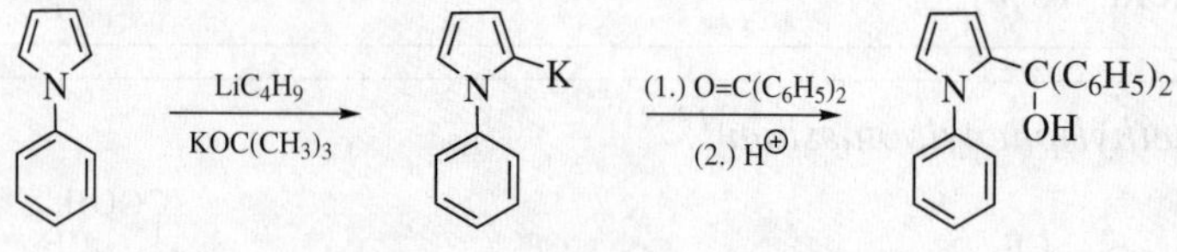

Potassium *tert*-butoxide (2.8 g, 25 mmol) and butyllithium (25 mmol in 15 mL hexanes) are consecutively added to a solution of 1-phenylpyrrole (3.6 g, 25 mmol) in tetrahydrofuran (30 mL) immersed in a methanol/dry ice bath. The mixture is vigorously stirred until homogeneous. After 90 min at −75 °C, it is treated with benzophenone (4.6 g, 25 mmol). After addition of brine (50 mL), the organic phase is decanted and the aqueous phase extracted with diethyl ether (2 × 20 mL). The combined organic layers are washed with brine (2 × 20 mL), dried and evaporated. Recrystallization from hexanes gives colorless 2-(diphenylhydroxymethyl)-1-phenylpyrrole [m.p. 108–109 °C (decomp.); yield: 74%].

1-(o-Lithiophenyl)-2-pyrryllithium[1821]

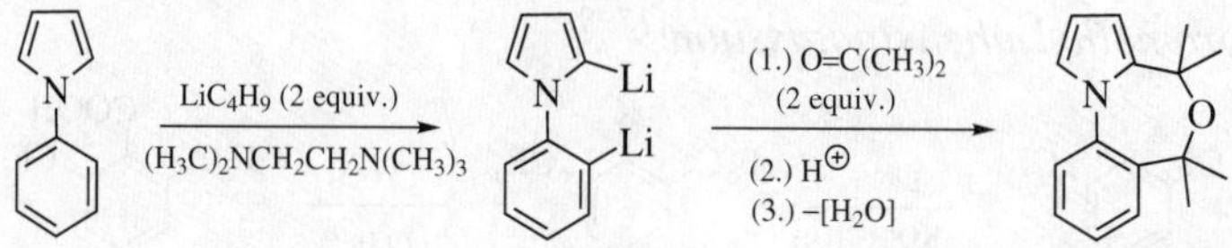

At 0 °C, butyllithium (55 mmol) in hexanes (40 mL) and *N,N,N',N'*-tetramethylethylenediamine (TMEDA; 8.3 mL, 6.4 g, 55 mmol) are added to 1-phenylpyrrole (3.6 g, 25 mmol) in diethyl ether (40 mL). After 30 min at 0 °C, the mixture is treated with acetone (4.0 mL, 3.2 g, 55 mmol) and, again 30 min later, washed with water (10 mL) and brine (20 mL). Upon evaporation of the organic solvents, crude 1-phenylpyrrole-2,2'-bis(1-methyl-1-ethanol) is left behind [m.p. 33–38 °C (from hexanes); yield: 90%]. The diol is dissolved in toluene (0.10 L). After the addition of silica gel (45 mL), the slurry was stirred for 3 h at 65 °C before being evaporated. Elution from a column filled with more silica gel (125 mL) with a 1:9 mixture of diethyl ether and hexanes gave analytically pure 4,4,6,6-tetramethyl-4*H*,6*H*-pyrrolo-[1,2-*a*][4.1]benzooxazepine [m.p. 48–49 °C; b.p. 123–125 °C/0.4 mmHg; yield: 58% (relative to 1-phenylpyrrole].

2-Furyllithium[2052]

Butyllithium (55 mmol) in hexanes (35 mL) is added to furan (4.4 mL, 4.1 g, 60 mmol) in tetrahydrofuran (60 mL). After 3 h at 0 °C, the mixture is treated with benzophenone (9.1 g, 50 mmol) before being poured, 15 min later, onto brine (50 mL). The organic phase was decanted and the aqueous one extracted with diethyl ether (2 × 25 mL). The combined organic layers were dried and evaporated. Upon recrystallization from diethyl ether/hexanes, α-(furyl)-α,α-diphenylmethanol was isolated quantitatively [m.p. 86–87 °C; yield: 99%].

1-Phenyl-5-trifluoromethyl-4-pyrazolyllithium[2053]

Diisopropylamine (2.8 mL, 2.0 g, 20 mmol), *N,N,N',N'',N''*-pentamethyldiethylenetriamine (PMDTA; 4.2 mL, 3.5 g, 20 mmol) and 1-phenyl-5-(trifluoromethyl)-pyrazole were consecutively added to a solution of butyllithium (20 mmol) in tetrahydrofuran (35 mL) and hexanes (15 mL) kept at −75 °C. After 5 min the reaction mixture was poured on freshly crushed dry ice. After evaporation of the solvents, the residue was portioned between hexanes (25 mL) and a 1.0 M aqueous solution (0.10 L) of sodium hydroxide. The alkaline layer was washed with diethyl ether (2 × 25 mL), acidified to pH 1, saturated with sodium chloride and extracted with diethyl ether (3 × 50 mL). 1-Phenyl-5-(trifluoromethyl)pyrazole-4-carboxylic acid was left behind upon evaporation of the dried organic layers [m.p. 132–133 °C (sublimed); yield: 62%].

4-Bromo-2-trifluoromethyl-3-quinolyllithium[2054]

At $-75\,°C$, diisopropylamine (4.3 mL, 3.1 g, 30 mmol) in tetrahydrofuran (15 mL) and 4-bromo-2-(trifluoromethyl)quinoline[2055] are consecutively added to butyllithium (25 mmol) in hexanes (15 mL). After 2 h at $-75\,°C$, the mixture is poured on an excess of freshly crushed dry ice. When the organic solvents are evaporated, the residue is dissolved in water (50 mL), washed with diethyl ether (3 × 40 mL), acidified with demi-concentrated hydrochloric acid to pH 1 and extracted with ethyl acetate (5 × 40 mL). The combined organic layers were washed with brine (2 × 40 mL), dried and concentrated. 4-Bromo-2-trifluoromethyl-3-quinolinecarboxylic acid crystallized from ethyl acetate in form of white needles [m.p. 207–208 °C; yield: 85%].

The acid can be easily converted into its methyl ester [m.p. 90–92 °C; yield: 95%] and subsequently be reduced to methyl 2-trifluoromethyl-3-quinolinecarboxylate [m.p. 51–52 °C; yield: 90%] by the treatment of its 0.5 M solution in dichloromethane with tributyltin hydride.

5   Handling Polar Organometallic Compounds

The practical aspects of synthetically oriented organometallic chemistry often turn out to be an unsurmountable hurdle for beginners. A first and major concern regards real or alleged hazards. Opinions diverge concerning the most appropriate equipment to ensure rigorous exclusion of oxygen and humidity during the reactions. Although convenient, is it advisable to use commercial organometallic reagents? No matter whether such reagents have been purchased or prepared by oneself, how long can they be stored without decomposition? Finally, how can one easily and reliably check the identity, purity and concentration of an organometallic intermediate in solution? All these issues are addressed in this last section.

5.1   Hazards

Organometallics are neither explosives nor witchcraft. The experimenter in the laboratory faces just one serious risk. Organometallic reagents like any other alkaline material may cause irremediable damage to the eyes. Therefore, it is essential to wear goggles with sideshields before entering any organometallic laboratory. If the eyes are inadvertently brought into contact with an alkaline substance, they should be immediately rinsed with 3% aqueous boric acid and a physician should be consulted urgently. If organolithium solutions are spilled over parts of the skin, diluted (*e.g.* 5%) acetic acid can be used for neutralization.

Some reagents such as *tert*-butyllithium are pyrophoric, especially if concentrated. Potassium/sodium alloy catches fire particularly rapidly. Self-ignitions are nevertheless extremely rare. If such an accident still happens, extinguishers operating with watery foam, halocarbons or carbon dioxide (!) must not be used. Containers filled with limestone powder and shovels should be kept everywhere in reach to cover spilled solutions. Even sand, if dry and uncontaminated with mud or butts, will do for fighting a small fire.

The situation changes completely when organolithium compounds are employed on an industrial scale. Many issues deserve careful consideration: the transportation in pressure vessels or tank cars, inert gas distribution systems and safety valves, non-corrosive reactors equipped with cooling devices down to $-100\,°C$, the exothermicity of the reaction, the elimination of by-products (butane !) and the disposal of waste, the recycling of solvents and metals, the protective wear for workers, the compliance with regulations and so forth. Reagent suppliers are willing to provide competent advice. The two main manufacturers of organolithiums can be contacted at the following addresses:

Chemetall Lithium Division
Trakehner Str. 3
D-604897 Frankfurt / Germany
(Fax: ++49 / 69 / 81 76 20 53)

FCM Corporation Lithium Division
Commercial Road
Bromborough L 62 2NL / UK
(Fax: ++44 / 151 / 482 73 61

Chemetall Foote Corp.
348 Holiday Inn Drive
Kings Mountain, NC 28086 / USA
(Fax: ++1 / 704 / 734 27 18)

FCM Corporation Lithium Division
449 North Cox Road
Gastonia, NC 28054 / USA
(Fax: ++1 / 704 / 868 53 30)

Chemetall Japan
1-7-10 Kanda Surugaida (Bldg. 6F)
Chiyoda-ku, Tokyo 101-0062 / Japan
(Fax: ++881 / 3 / 32 59 20 65)

Asia Lithium Corporation
5-24 Miyahara 3-Chome (Bldg. 11F)
Yodogawa-ku, Osaka / Japan
(Fax: ++81 / 66 / 399 23 45)

5.2   Equipment

Syringes and rubber septa may suffice to ensure 'anaerobic conditions'[2056] for mini-scale manipulations. Synthetically oriented work (on a 0.01–1.0 mol scale) can be more efficiently performed with special equipment and under an atmosphere of at least 99.99% pure nitrogen. Inert gas/vacuum line apparatus (see Fig. 11) and third-generation Schlenk glassware (Fig. 12) can be ordered from Pfeifer Gerätebau, Promenade 15, D-98711 Frauenwald, Germany.

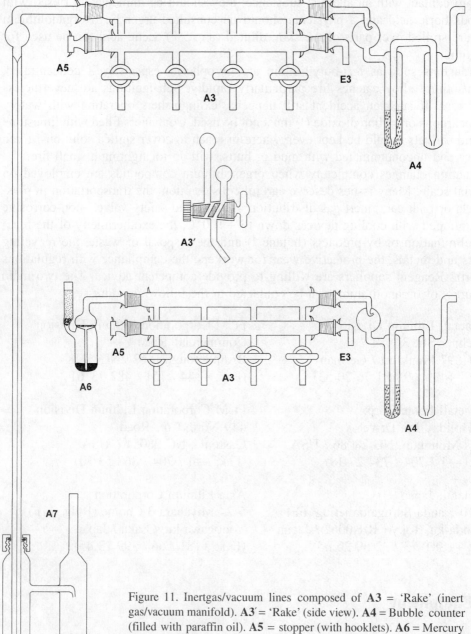

Figure 11. Inertgas/vacuum lines composed of **A3** = 'Rake' (inert gas/vacuum manifold). **A3′** = 'Rake' (side view). **A4** = Bubble counter (filled with paraffin oil). **A5** = stopper (with hooklets). **A6** = Mercury seal with mercury-tight frit and outlet tube filled with sulfur. **A7** = Mercury gauge with 80 cm long capillary. **E3** = Tubing adapter, right angle, male joint.

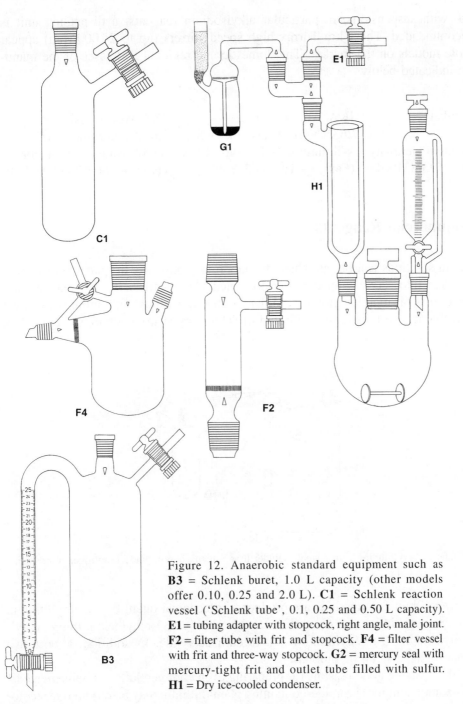

Figure 12. Anaerobic standard equipment such as
B3 = Schlenk buret, 1.0 L capacity (other models
offer 0.10, 0.25 and 2.0 L). **C1** = Schlenk reaction
vessel ('Schlenk tube', 0.1, 0.25 and 0.50 L capacity).
E1 = tubing adapter with stopcock, right angle, male joint.
F2 = filter tube with frit and stopcock. **F4** = filter vessel
with frit and three-way stopcock. **G2** = mercury seal with
mercury-tight frit and outlet tube filled with sulfur.
H1 = Dry ice-cooled condenser.

For work with suspensions, in particular alkylsodium reagents, a dispersing unit is strongly recommended. The Ultra-Turrax high speed stirrers (up to 24 000 rpm) appear to be the sole models on the market. The domestic and regional addresses of the manufacturer are indicated below.

Janke & Kunkel   IKA Works   IKA Works Asia
Postfach   2635 North Chase Pkwy.SE   Lot 2 Japan Indah 1/2
D-79219 Staufen/Germany Wilmington, NC 28405/USA   48000 Rawang-Selangor
(Fax: ++49/7633/831901) (Fax: ++1/910/4527693)   (Fax: ++60/603/60933940)

5.3   Commercial Reagents

Alkyllithium solutions can be purchased for laboratory purposes in quantities as small as 100 mL filled into screw-capped or serum-sealed glass bottles. If such reagents are regularly used, it is far more economic to order 5–10 L quantities. Siphoning is the most convenient technique for transfer of the liquid to Schlenk-type storage burets (Figure 13).

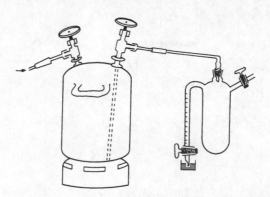

Figure 13. Transfer of butyllithium from a steel cylinder to a Schlenk burette under a nitrogen atmosphere.

Several organolithium compounds are produced in technical quantities: methyllithium, butyllithium, *sec*-butyllithium, *tert*-butyllithium, hexyllithium, 2-ethylhexyllithium, octyllithium, phenyllithium, 2-thienyllithium and lithium acetylide. In addition, almost two dozen Grignard reagents can be purchased.

However, there can be good reasons for preparing even commercially available reagents oneself. For example, lithium bromide-containing phenyllithium may be required in stereo-controlled Wittig reactions[2057]. Whereas the salt remains in solution when the reagent is prepared from bromobenzene in diethyl ether (see page 100), it precipitates from the technically used 3:7 ether/cyclohexane mixture. Upon request, lithium bromide-containing methyllithium can be obtained, but the reagent/salt stoichiometry, tacitly assumed to be 1:1, may vary between 0.7 and 1.5. The quality of salt-free alkyllithiums again leaves much to be desired. In extreme cases, 1 L of a commercial solution may contain up to 100 g of non-volatile hydrocarbons which progressively accumulate as the manufacturers recycle the solvents returned by their customers. These by-products may not cause much harm if one wishes to perform a metalation or addition reaction that ultimately leads to a product that is polar enough to be easily separated and purified. On the other hand, it is not advisable to use commercial reagents in mechanistic, in particular kinetic, studies without prior purification. The results may prove totally erratic.

5.4   Storability of Reagents

If tailor-made for synthetic purposes, organometallic intermediates for synthetic pur-poses are used as soon as prepared. Commercial reagents are generally purchased in large quantities and are consumed only over a longer period of time. In such cases, the researcher has to know how to conserve the material.

Two modes of decomposition tend to shorten the shelf life of organometallic species. They may react with the solvent or undergo spontaneous β-elimination of metal hydride (or other metal-bearing fragments).

5.4.1   Stability Towards Hydride Elimination

Hydride elimination is difficult to control. Fortunately, with most standard reagents this occurs to a significant extent only at elevated temperatures $(50-150\,°C)$[2058–2060]. There is, however, an important exception. Commercial solutions of *sec*-butyllithium should be stored in the cold (around $0\,°C$ or $-25\,°C$). This avoids a loss of about 1% of the organometallic titer per week due to formation of lithium hydride, as well as 1- and 2-butene. This process is considerably accelerated if more than trace amounts of lithium alkoxide are present[2061,2062]. Alkoxides are inevitably formed when air enters into contact with an organometallic solution.

5.4.2   Stability Towards Solvent Attack

Hexane and other alkanes are virtually inert towards standard organometallics. However, whenever an ethereal reaction medium is employed, one has to worry about reactions between the reagent and the solvent. For example, it is unreasonable to carry out metalation reactions in tetrahydrofuran with butyllithium at room temperature or with *tert*-butyllithium at $0\,°C$, if one knows how sensitive the reagent is towards the solvent under such conditions. To be on the safe side for metalation reactions one should choose a temperature range that ensures a reagent half-life of at least 10 h. The losses will be negligible if the reaction can be terminated after 60 min and are still tolerable if it extends overnight (some 15 h). The half-life should attain, or even exceed 100 h (*i.e.* 6 weeks) if the intention is to store the ethereal solution of a reagent over a longer period of time.

Numerous studies have been undertaken to quantify the chemical stability of methyllithium[704,2061], primary alkyllithium[684,2061,2063], secondary alkyllithium[684,1431], cyclopropyllithium[684], cyclohexyllithium[1431], tertiary alkyllithiums[1432], vinyllithium[1310] and phenyllithium[704,2061]. Butyllithium, *sec*-butyllithium and *tert*-butyllithium have received particular attention[2064]. However, the accuracy of the reported data should not be overestimated. So far, rate constants and reaction orders have been measured in only a single case[2065] and a profound kinetic analysis of solvent-mediated decomposition of organometallic reagents is still lacking. For practical purposes it is sufficient to compile half-lives $\tau_{1/2}$ of organometallic reagents as a function of the solvent and the temperature in a semi-quantitative way (see Tables 147 and 148). The relevant trends become thus even more clearly visible.

Table 147. Chemical stability of standard organolithium compounds in *d*iethyl *e*ther (DEE), *tetrah*ydro*f* uran (THF) and (*mono*)*e*thylene *g*lycol di*m*ethyl *e*ther (MEGME; 1,2-dimethoxyethane): temperature ensuring approximate reagent half-lives of 10 h.

Temperature [°C]	DEE	THF	MEGME
+100			
+75	Li−CH₃		
+50	Li−⬡	Li−CH₃	
+25	Li−C₄H₉	Li−⬡	Li−CH₃
0	Li−CH(CH₃)(C₂H₅)	Li−C₄H₉	Li−⬡
−25	Li−C(CH₃)₃	Li−CH(CH₃)(C₂H₅)	Li−C₄H₉
−50		Li−C(CH₃)₃	Li−CH(CH₃)(C₂H₅)
−75			Li−C(CH₃)₃
−100			

Table 148. Chemical stability of prominent alkylalkali compounds in tetrahydrofuran (THF): temperature ensuring approximate reagent half-lives of 10 h.

Temperature [°C]	Li–R	Na–R	K–R
+50	Li–CH$_3$		
+25	Li–CH$_2$Si(CH$_3$)$_3$		
0	LiCH$_2$C$_3$H$_7$		
–25		Na–CH$_3$	
–50			
–75		Na–CH$_2$C$_3$H$_7$	K–CH$_2$Si(CH$_3$)$_3$
–100			K–CH$_3$
–125			
–150			Na–CH$_2$C$_3$H$_7$

It makes sense to vary the temperature thresholds by 25 °C steps. According to a well-established rule-of-thumb this should increase or reduce reaction rates by a factor of 10. This means that the replacement of a hydrogen atom in methyllithium by a first alkyl group (as in butyllithium) enhances the aggressiveness of the reagent towards ethereal solvents by two powers of 10, and by a second and third alkyl group (as in *sec*- and *tert*-butyllithium) by an additional power of 10 each time (see Table 147). The loosely aggregated phenyllithium (see page 22) proves to be considerably more labile in ethereal solvents than expected in view of its basicity (which is much inferior than that of methyllithium, not to speak of any other alkyllithium). The replacement of lithium by a heavier alkali metal makes the organometallic species much more vulnerable. Butylsodium and butylpotassium can be conserved in tetrahydrofuran at −75 °C and −150 °C, respectively, only for a short while (see Table 148). In contrast, the superbasic mixed-metal complex composed of butyllithium and potassium *tert*-butoxide as trimethylsilylmethylpotassium survive in tetrahydrofuran at −75 °C for an almost indefinite period of time.

Polar organometallic compounds are less stable in tetrahydrofuran (or dimethyl ether) than in diethyl ether (or tetrahydrofuran) and again less stable in (mono)ethylene glycol dimethyl ether (or diethylene glycol dimethyl ether). Dilution of the neat ethereal solvent with pentanes or hexanes may retard the organometallic degradation by up to 5 times, the addition of an equivalent amount of TMEDA may accelerate it by up to 10 times. The solvent polarity effect (see Table 147) reflects differences in the coordinating, and hence activating, power of the various ethers and, not less importantly, differences in the modes of reaction.

When treated with strong bases (M–R), diethyl ether simply undergoes a direct, possibly *syn*-periplanar[509] β-elimination of metal ethoxide (transition state **339**). Unlike benzyl ethyl ether[2066–2068], diethyl ether is not acidic enough to be susceptible to α-deprotonation. Otherwise an α-metalated species **340** would be generated that could have become engaged in either a Wittig rearrangement[2069] producing 2-butanol or in an α',β-elimination[2066] affording again ethylene and metal ethoxide. The intermediacy of species such as **340** has been ruled out on the basis of deuterium-labeling experiments under certain conditions[2067], but has been demonstrated under others[2068].

$$H_3C-CH-O-CH_2 \quad \cdots \cdots \rightarrow \quad H_3C-CH-CH_2-CH_3$$
$$\underset{M}{|} \quad \underset{CH_2}{\diagdown} \qquad\qquad\qquad \underset{MO}{|}$$
$$\diagdown H$$

340

$$+ M-R$$

$$H_3C-CH_2-O-CH_2-CH_3 \quad \longrightarrow \quad \left[\begin{array}{c} H_2C\cdots CH \\ H_3C-CH_2-O \qquad H \\ M-R \end{array} \right]^{\ddagger} \longrightarrow \quad H_3C-CH_2 \ + \ H_2C=CH_2$$
$$\underset{MO}{|}$$

339

β-Elimination is once more the dominant reaction mode when alkyllithium reagents interact with the chelating solvent ethyleneglycol dimethyl ether[2070]. In addition, a substitution process can intervene that transfers a methyl group to the organometallic reagent (M–R)[444].

$$\overset{S_N2}{\nearrow} \quad H_3CO-CH_2-CH_2-OM + H_3C-R$$

$$H_3CO-CH_2-CH_2-O-CH_3 + M-R$$

$$\underset{E2}{\searrow} \quad H_3CO-M + H_2C=CH-OCH_3 + H-R$$

A similar alkylation of the organometallic reagent may occasionally occur if tetrahydrofuran acts as the solvent. This was found to be the case with allyl type organopotassium species[2071], with lithium dimethylphosphide[2072] and with organocuprates[2073], all of them being excellent nucleophiles. The more basic alkylmetal reagents, however, effect a hydrogen/metal exchange at the α-position. The resulting *gem*-counterpolarized intermediate **341** (M=K) can be trapped at −75 °C with electrophiles[1533]. Above −60 °C, species **341** (M = K, Li) performs a fragmentation and decomposes to ethylene and acetaldehyde enolate **342**[2074,2075]. The latter component may be intercepted by carbonyl compounds, its nucleophilic addition affording aldol-type products[2075–2077].

5.5 Analysis

To find out what happens with a given reagent in a given solvent under given conditions, reliable methods for the determination of organometallic concentrations must be available. In general, titration techniques are used that allow distinction between the 'reagent basicity', *i.e.* the alkalinity originating from the hydrolysis of the organometallic, and the 'residual basicity', *i.e.* the alkalinity that results from the presence of metal oxides, hydroxides and carbonates. The latter compounds form when reagent solutions experience air contact.

The same methods can be used to evaluate the concentration of an organometallic species generated as a key intermediate in a synthesis sequence. Frequently, however, the presence of the expected intermediate is determined in a more pragmatic way. An aliquot of the reaction mixture is withdrawn and is poured on dry ice or is added to an ethereal solution of benzaldehyde. The carboxylation and α-arylhydroxymethylation of organometallics being virtually quantitative transformations, they provide a very meaningful estimate of the concentration with which the crucial intermediate is formed.

If more detailed information is requested, NMR studies may provide the answer. Of course, 'invisible' solvents such as perdeuterated tetrahydrofuran, dioxane, dimethyl ether, diethyl ether or benzene have to be employed, and the sample-containing tubes have to be well stoppered or, better, flame sealed. The findings can be quantified if a known amount of an inert reference compounds (*e.g.* cyclohexane or neopentane) is added as an internal standard.

In some situations only qualitative information is sought. Often one simply wants to know whether the organometallic reagent has been essentially consumed. Color tests are most convenient to obtain such information.

5.5.1   Qualitative Tests

Inspired by earlier work dealing with the addition of Grignard reagents to diarylketones[2078–2080], H. Gilman *et al.* devised color tests for easy determination of the presence or absence of organometallic reagents. The most widely used Gilman–Schulze test (*Color Test I*)[2081] responds to practically all polar organometallics. A small sample (0.1–0.5 mL) of the dissolved reagent is transferred by pipet into a 1.0 M solution of 4,4′-bis(dimethylamino)benzophenone (Michler's ketone) in benzene (1–2 mL). Upon addition of water (1–2 mL) and a few drops of a 0.2% solution of iodine in glacial acetic acid, the most intense malachite green-blue color of a benzhydryl-type cation **343** appears.

The characteristic color is associated with the formation of a di- or triarylmethyl cation depending on whether an alkyl- or arylmetal compound was present. In the latter case, ionization of the carbinol precursor does not require acetic acid; it can be brought about in a buffered medium (*e.g.* with 20% aqueous pyrocatechol[2082]) and even in the absence of iodine.

A distinction between aliphatic and other organometallic reagents may be very valuable, for example when primary, secondary or tertiary butyllithium is used in a neighboring group-assisted metalation of an aromatic position. The Gilman–Swiss test (*Color Test II*)[2083] is specific for just alkyllithium reagents. The key step is a halogen–metal inter-conversion. A small sample of the presumed organometallic solution (0.5 mL) is intro-duced into a 15% solution (1 mL) of 4-bromo-*N,N*-dimethylaniline in benzene. Then a 15% solution (1 mL) of benzophenone in benzene and, a few minutes later, concentrated hydrochloric acid (1 mL) are added. A bright red color, characteristic of the triarylmethyl cation **344**, results only if an alkyllithium reagent was present.

To perform a Gilman test requires seconds rather than minutes. It is strongly recommended to follow the evolution of organometallic reactions by making use of this extremely simple and instructive means of verification.

5.5.2 Titration

Organometallic reagents inevitably contain some metal oxides, hydroxides, alkoxides (or aroxides) and even carbonates inadvertently produced by reaction with oxygen, moisture and atmospheric carbon dioxide. Simple titration with acid would therefore overestimate the true titer of the reagent by monitoring simultaneously the organometallic and the inorganic basicity. K. Ziegler et al.[1226] were the first to suggest a double titration method which allows determination of the organometallic titer as the difference between the 'total alkalinity' and the 'residual alkalinity', the latter being due to the inorganic material left over after the selective destruction of the organometallic reagent (for example, with butyl bromide and dibenzylmercury).

G. Wittig and G. Harborth[2084] presented an improved protocol with 1,2-dibromoethane as the killer reagent. A few months later H. Gilman and A. H. Haubein[2085] published another version using benzyl chloride to neutralize selectively the organometallic compound. This method became quite popular until H. Gilman and F. K. Cartledge[2086] suggested replacing benzyl chloride by allyl bromide, which gave more reliable results.

In the meanwhile the 1,2-dibromoethane-based double titration has become the standard method. Simple in execution, it can be applied to primary, secondary, tertiary, allylic and benzylic organolithium compounds. It can even be adapted to the titer evaluation of organomagnesiums, metal dialkylamides (such as lithium diisopropylamide) and 'radical-anions' (such as the naphthalene–lithium 1:1 adduct)[2087].

Wittig–Harborth Double Titration[2084]

$$OH^{\ominus} \xleftarrow{\ H_2O\ } M\text{--}R \xrightarrow[\text{(2.) } H_2O]{\text{(1.) } BrCH_2CH_2Br} Br^{\ominus} + Br\text{--}R \ (\text{or } H\text{--}R)$$

$$+$$

$$OH^{\ominus} \xleftarrow{\ H_2O\ } M\text{--}OR' \xrightarrow[\text{(2.) } H_2O]{\text{(1.) } BrCH_2CH_2Br} OH^{\ominus}$$

Total alkalinity: The organometallic solution (10 mL) is rapidly transferred into an Erlenmeyer flask filled with water (50 mL) and titrated with 1.0 M hydrochloric acid against phenolphthalein to neutral.

Residual alkalinity: The organometallic solution (10 mL) is placed in a Schlenk 'tube' and a slight excess of neat 1,2-dibromoethane (e.g. 2.0 mL, 4.4 g, 23 mmol) is added by means of a pipet. After 30 min at +25 °C, the mixture is poured into water (20 mL) and the vessel carefully rinsed with water (3 × 10 mL). The combined aqueous layers are titrated as specified above.

The residual alkalinity of commercial reagents usually falls in the range 5–15% of the organometallic titer. If it significantly exceeds the upper limit, one has to worry about the quality of the reagent and try to identify the origin of the problem. Thus, the result of a double titration can warn the researcher that something is out of control.

Despite this obvious advantage, much effort has been devoted to supplant the double by single titration techniques. The principle is always the same. One seeks as the titrant an acid that only reacts with the organometallic but not with the weaker mineral bases. All one needs then in addition is a method with which to visualize the neutralization endpoint.

Benzyl-type organometallics commonly exhibit a characteristic and intense orange-red or cherry red color. Hence they may be taken as self-indicators and be titrated directly with a normalized solution of benzoic acid in an ethereal or aromatic solvent[2088]. A decisive extension was realized by using small amounts of benzylic metal compounds as indicators while titrating colorless organometallics. For example, if one dissolves a known amount, say 5.0 mmol, of benzoic acid and a spatula tip of triphenylmethane (0.01–0.05 mol) in a mixture of anhydrous dimethyl sulfoxide (8.0 mL) and ethylene glycol dimethyl ether (2.0 mL) and adds an organolithium solution dropwise, the appearance of the cherry-red color of triphenylmethyllithium marks the complete neutralization of the acid[2089]. Lithium alkoxides, if present, do not interfere in the deprotonation of benzoic acid (or acetophenone) as blind tests have demonstrated[2089]. The next step was to combine the acid and the indicator properties in one and the same molecule. For example, when an organometallic solution of unknown titer is added dropwise to diphenylacetic acid, the latter is progressively converted into the salt until all the acid is consumed. One drop more and deprotonation at the α-position of the carboxylate will occur, thus generating the vividly yellow-colored enediolate **345**[2090]. The use of 1-pyreneacetic acid[2091] (color change to red), 2,5-dimethoxybenzyl alcohol[2092] (color change to deep red), 4-biphenylylmethanol[2093] (color change to orange), 1,3-diphenylacetone p-toluenesulfonyl-hydrazone[2094] (color change to orange), N-phenyl-N-2-naphthylamine[2095] (color change to yellow-orange) and 2-benzyl-N-(tert-butoxycarbonyl)aniline[2096] (color change to yellow-orange) is based on the same principles.

All the methods mentioned so far mark the endpoint of a titration by the appearance of color. There are a few others where decoloration signals neutralization. Addition of a spatula tip (e.g. 2 mg) of 1,10-phenanthroline to the solution of an organolithium species produces a characteristic rust-red charge-transfer complex. On titration with a standardized solution of 2-butanol in xylene, sharp endpoints can be observed[2097,2098]. Grignard reagents, giving rise to a bright violet color if diethyl ether is present, can be analyzed in the same way[2097]. Alkyllithiums and aryllithiums, and also lithium N,N-dialkylamides, deprotonate N-benzyl-p-biphenylmethylimine efficaciously, converting its into an intensively blue azaallyl species[2099]. Titration with a 1.0 M solution of 2-butanol until complete decoloration allows one to determine the organometallic concentration. A single method avoids the use of an acid and replaces it by a colored electrophile[2100]. The brick-red diphenyl ditelluride (**346**) reacts smoothly with butyllithium and other polar organometallics, including alkynyllithium and organomagnesium halides, to afford an organylphenyltelluride and a metal phenyltelluride, both colorless. Again the disappearance of color indicates the endpoint of titration.

346 (red)　　　　　(colorless)　　　(colorless)

Although attractive *per se*, the price of the key reagent (**346**) of approx. 6500 €/mol compromises the dissemination of the method. Other single-titration protocols suffer from other drawbacks. As already mentioned, the required inertness of mineral bases under the test conditions may not always be rigorously fulfilled. If solvents have to be added, or standardized solutions of organic acids (*e.g.* 2-butanol) have to be employed, the danger of bringing in moisture and oxygen should not be underestimated. Finally, some of these methods are impaired by their limited scope of applicability, failing frequently with organomagnesium compounds, metal amides, acetylides and even aryl- or methyllithium. Therefore, the good old double titration procedure may well remain the most popular tool for organometallic titer evaluation.

Acknowledgments

Some of the research work presented in this chapter was supported by the University of Lausanne, the Swiss National Science Foundation, Bern (grants 20-49′307-96 and 20-55′303-98) and the Federal Office for Education and Science, Bern (grant 97.0083 linked to the TMR-project FMRXCT970120 and grant C 98.0110 linked to the COST-project D12/004). The author is indebted to all collaborators mentioned in the literature references for their fine contributions and in particular to his senior assistant Dr. Frédéric Leroux for proof-reading and countless valuable hints. The manuscript has been processed and illustrated by Elsbeth Schlosser who has thus demonstrated once more her skills and self-sacrifice.

6　References

[1] Bugge, G.; (ed.), *Das Buch der großen Chemiker, Vol. 2*, Verlag Chemie, Weinheim, **1930** (reissue **1979**).
[2] Partington, J. R.; *A History of Chemistry, Vol. 4*, McMillan, London; **1964**.
[3] *Poggendorff: Biographisch-literarisches Handwörterbuch der exakten Naturwissenschaften, Vol. 7b*, Part 6, Akademie-Verlag, Berlin, **1980**.
[4] Wöhler, F.; *Ann. Phys. (Leipzig)* **1828**, *12*,253.
[5] Prior to publication, Wöhler sent on 22 February 1828 a letter to his friend and former teacher Berzelius informing him with unconcealed triumph that '*ich Harnstoff machen kann, ohne dazu Nieren oder überhaupt ein Thier, sei es Mensch oder Hund, nöthig zu haben*' (I can make urea without needing kidneys or an animal at all, be it man or dog).

[6]   Wöhler, F.; *Ann. Chem. Pharm. (Heidelberg)* **1840**, *35*, 111; see also: Mallet, J. E.; *Ann. Chem. Pharm. (Heidelberg)* **1851**, *79*, 223.

[7]   Bunsen, R.; *Ann. Chem. Pharm. (Heidelberg)* **1843**, *42*, 14.

[8]   Frankland, E.; *Ann. Chem. Pharm. (Heidelberg)* **1849**, *71*, 171, 213; *J. Chem. Soc.* **1849**, *11*, 263, 297.

[9]   Wanklyn, J. A.; *Ann. Chem. Pharm. (Leipzig)* **1858**, *108*, 67; *Proc. Roy. Soc.* **1858**, *9*, 341.

[10] Wittig, G.; Davis, P.; Koenig, G.; *Chem. Ber.* **1951**, *84*, 627; and later publications, *e.g.* Wittig, G.; Closs, G.; Mindermann, F.; *Liebigs Ann. Chem.* **1955**, *594*, 89, spec. 101.

[11] Seebach, D.; *Angew. Chem.* **1969**, *81*, 690; *Angew. Chem. Int. Ed. Engl.* **1969**, *8*, 639; Seebach, D.; *Angew. Chem.* **1979**, *91*, 259; *Angew. Chem. Int. Ed. Engl.* **1979**, *18*, 239.

[12] Schlosser, M.; *Struktur und Reaktivität polarer Organometalle*, Springer, Berlin, **1973**.

[13] Wittig, G.; *Experientia* **1958**, *14*, 389.

[14] Buncel, E.; *Carbanions: Mechanistic and Isotope Aspects*, Elsevier, Amsterdam, **1975**.

[15] Weast, R. C. (ed.); *Handbook of Chemistry and Physics*, CRC Press, Boca Raton (Florida), 67th edition, p. E-76–E-77.

[16] Coppens, P.; *Angew. Chem.* **1977**, *89*, 33; *Angew. Chem. Int. Ed. Engl.* **1977**, *16*, 32.

[17] Angermund, K.; Claus, K. H.; Goddard, R.; Krüger, C.; *Angew. Chem.* **1985**, *97*, 241; *Angew. Chem. Int. Ed. Engl.* **1985**, *24*, 237.

[18] Coppens, P.; Hall M. B. (eds.), *Electron Distributions and the Chemical Bond*, Plenum Press, New York, **1982**; Coppens, P.; *X-Ray Charge Densities and Chemical Bonding*, University Press, Oxford, **1997**.

[19] Toriumi, K.; Saito, Y.; *Adv. Inorg. Chem. Radiochem.* **1983**, *27*, 27.

[20] Coppens, P.; *Coord. Chem. Rev.* **1985**, *65*, 285.

[21] Hirshfeld, F. L.; *Cryst. Rev.* **1991**, *2*, 169.

[22] Weyrich, W. (ed.); Proceedings of the 10th Sagamore Conference on Charge, Spin and Momentum Densities, *Z. Naturforsch.* **1993**, *48a*, 11–462.

[23] Eades, R. A.; Gassman, P. G.; Dixon, D. A.; *J. Am. Chem. Soc.* **1981**, *103*, 1066.

[24] Catlow, R. A.; Stoneham, A. M.; *J. Phys. C* **1983**, *16*, 4321; *Chem. Abstr.* **1983**, *99*, 200'617u.

[25] Maslen, E. N.; Spackman, M. A.; *Austr. J. Phys.* **1985**, *38*, 273; *Chem. Abstr.* **1985**, *103*, 147'349z.

[26] Seiler, P.; Dunitz, J.; *Helv. Chim. Acta* **1986**, *69*, 1107.

[27] Seiler, P.; *Acta Cryst. B*, **1993**, *49*, 223.

[28] Escudero, F.; Yanez, M.; *Mol. Phys.* **1982**, *45*, 617.

[29] This argument was developed by Professor J. D. Dunitz, Zurich.

[30] Power, P. P.; *Acc. Chem. Res.* **1988**, *21*, 147.

[31] Al-Juaid, S. S.; Eaborn, C.; Hitchcock, P. B.; McGeary, C. A.; Smith, J. D.; *J. Chem. Soc., Chem. Commun.* **1989**, 273.

[32] Belluš, D; Klingert, B.; Lang, R. W.; Rihs, A.; *J. Organomet. Chem.* **1988**, *339*, 17.

[33] Belch, A. C.; Berkowitz, M.; McCammon, J. A.; *J. Am. Chem. Soc.* **1986**, *108*, 1755.

[34] Neupert-Laves, K.; Dobler, M.; *Helv. Chim. Acta* **1975**, *58*, 432.

[35] Pocker, Y.; Buchholz, R. F.; *J. Am. Chem. Soc.* **1981**, *93*, 2905.

[36] Schwesinger, R.; Piontek, K.; Littke, W.; Prinzbach, H.; *Tetrahedron Lett.* **1985**, *2*, 1201.

[37] Sarma, R.; Ramirez, F.; McKeever, B.; Chaw, Y. F.; Marecek, J. F.; Nierman, D.; MacCaffrey, T. M.; *J. Am. Chem. Soc.* **1977**, *99*, 5289.

[38] Owen, J. D.; *J. Chem. Soc., Dalton Trans.* **1980**, 1066.

[39] Veya, P.; Floriani, C.; Chiesi-Villa, A.; Guastini, C.; *Organometallics* **1991**, *10*, 1652.

[40] Pajerski, A. D.; Bergstresser, G. L.; Parvez, M.; Richey, H. G.; *J. Am. Chem. Soc.* **1988**, *110*, 4844.

[41] Riche, C.; Pascard-Billy, C.; *J. Chem. Soc., Chem. Commun.* **1977**, 183.

[42] Dobler, M.; Phizackerley, R. P.; *Helv. Chim. Acta* **1974**, *57*, 664.

[43] Norman, J. A. T.; Pez, G. P.; *J. Chem. Soc., Chem. Commun.* **1991**, 971.

[44] Mappes, G. W.; Friedman, S. A.; Fehlner, T. P.; *J. Phys. Chem.* **1970**, *74*, 3307; see also: Ashcroft, S. J.; Beech, G.; *Inorganic Thermodynamics*, Van Nostrand Reinhold, New York, **1973**; Ahlrichs, R.;

Theor. Chim. Acta **1974**, *35*, 59; Akkerman, O. S.; Schat, G.; Evers, E. A. I. M.; Bickelhaupt, F.; *Recl. Trav. Chim. Pays-Bas* **1983**, *102*, 109.

[45] Liebman, J. F; Greenberg, A.; Williams R. E. (eds.); *Advances in Boron and the Boranes*, Verlag Chemie VCH, Weinheim, **1988**.

[46] Schleyer, P. V. R.; Bühl, M.; Fleischer, U.; Koch, W.; *Inorg. Chem.* **1990**, *29*, 153.

[47] Olah, G. A.; Wade, K.; Williams R. E.; (eds.), *Electron Deficient Boron and Carbon Clusters*, Wiley, New York **1991**.

[48] Bell, R. P.; Emeleus, H. J.; *Quart. Rev., Chem. Soc.* **1948**, *2*, 141.

[49] Caroll, B. L.; Bartell, L. S.; *J. Chem. Phys.* **1965**, *42*, 1135, 3076; *Inorg. Chem.* **1968**, *7*, 219.

[50] Hoffmann, E. G.; *Liebigs Ann. Chem.* **1960**, *629*, 104.

[51] Laubengayer, A. W.; Gilliam, W. F.; *J. Am. Chem. Soc.* **1941**, *63*, 477.

[52] Henrickson, C. H.; Eyman, D. P.; *Inorg. Chem.* **1967**, *6*, 1461.

[53] Pitzer, K. S.; Gutowsky, H. S.; *J. Am. Chem. Soc.* **1946**, *68*, 2204.

[54] Lewis, P. H.; Rundle, R. E.; *J. Chem. Phys.* **1953**, *21*, 986.

[55] Coates, G. E.; Glockling, F.; Huck, N. D.; *J. Chem. Soc.* **1952**, 4496.

[56] Goubeau, J.; Walter, K.; *Z. Anorg. Allg. Chem.* **1963**, *322*, 58.

[57] Snow, A. I.; Rundle, R. E.; *Acta Cryst.* **1951**, *4*, 348.

[58] Kaufmann, E.; Raghavachari, K.; Reed, A. E.; Schleyer, P. V. R.; *Organometallics* **1988**, *7*, 1597.

[59] For early examples of organolithium structures, see: Hubberstey, P.; *Coord. Chem. Rev.* **1982**, *40*, 1–63, spec. 48–49.

[60] Brown, T. L.; Rogers, M. T.; *J. Am. Chem. Soc.* **1957**, *79*, 1859.

[61] West, R.; Glaze, W.; *J. Am. Chem. Soc.* **1961**, *83*, 3580.

[62] Chinn, J. W.; Lagow, R. J.; *Organometallics* **1984**, *3*, 75.

[63] Plavsic, D.; Srzic, D.; Klasinc, L.; *J. Phys. Chem.* **1986**, *90*, 2075.

[64] Andrews, L.; *J. Chem. Phys.* **1967**, *47*, 4834.

[65] Atwood, J. L.; Fjeldberg, T.; Lappert, M. F.; Luong-Thi, N. T.; Shakair, R.; Thorne, A. J.; *J. Chem. Soc., Chem. Commun.* **1984**, 1163.

[66] Haaland, A.; *Topics Curr. Chem.* **1975**, 53, 1.

[67] Beagley, B.; in *Molecular Structure by Diffraction Methods* (Sim, G. A.; Sutton, L. E.; eds.), Vol. *1*, 111, Specialist Periodical Report, Chemical Society, London **1973**; *Chem. Abstr.* **1974**, *83*, 69 628s.

[68] Dietrich, H.; *Acta Cryst.* **1963**, *16*, 681; *J. Organomet. Chem.* **1981**, *205*, 291.

[69] Whine, W. E.; Stucky, G. D.; Peterson, S. W.; *J. Am. Chem. Soc.* **1975**, *97*, 6401.

[70] Hughes, R. P.; Powell, J.; *J. Chem. Soc., Chem. Commun.* **1971**, 275.

[71] Armstrong, D. R.; Clegg, W.; Colquhoun, H. M.; Daniels, J. A.; Mulvey, R. E.; Stephenson, I. R.; Wade, K.; *J. Chem. Soc., Chem. Commun.* **1987**, 630.

[73] Brookhart, M.; Green, M. L. H.; *J. Organomet. Chem.* **1983**, *250*, 395; Novoa, J. J.; Whangbo, M. - h.; Stucky, G. D.; *J. Org. Chem.* **1991**, *56*, 3181.

[74] See in this respect: Opitz, A.; Koch, R.; Katritzky, A. R.; Fan, W.-q.; Anders, E.; *J. Org. Chem.* **1995**, *60*, 3743.

[75] Kottke, T.; Stalke, D.; *Angew. Chem.* **1993**, *105*, 619; *Angew. Chem. Int. Ed. Engl.* **1993**, *32*, 580.

[76] Teclé, B.; Maqsudur Rahman, A. F. M.; Oliver, J. P.; *J. Organomet. Chem.* **1986**, *317*, 267.

[77] Zerger, R.; Rhine, W.; Stucky, G. D.; *J. Am. Chem. Soc.* **1974**, *96*, 6048.

[78] Weiss, E.; Lambertsen, T.; Schubert, B.; Cockcroft, J. K.; Wiedemann, A.; *Chem. Ber.* **1990**, *123*, 79.

[79] See also: Weiss, E.; Lucken, E. A. C.; *J. Organomet. Chem.* **1964**, *2*, 197; Weiss, E.; Hencken, G.; *J. Organomet. Chem.* **1969**, *21*, 265.

[80] Weiss, E.; Lambertsen, T.; Schubert, B.; Cockcroft, J. K.; *J. Organomet. Chem.* **1988**, *358*, 1.

[81] See also: Weiss, E.; Sauermann, G.; *Chem. Ber.* **1979**, *103*, 265.

[82] Weiss, E.; Köster, H.; *Chem. Ber.* **1977**, *110*, 717.

[83] Weiss, E.; Corbelin, S.; Cockcroft, J. K.; Fitch, A. N.; *Chem. Ber.* **1990**, *123*, 1629.

[84] See also: Weiss, E.; Sauermann, G.; Thirase, G.; *Chem. Ber.* **1983**, *116*, 74.

[85] Markies, P. R.; Schat, G.; Akkerman, A. S.; Bickelhaupt, F.; *Organometallics* **1990**, 9, 2243.

[86] Schiemenz, B.; Power, P. P.; *Angew. Chem.* **1996**, *108*, 2288; *Angew. Chem. Int. Ed. Engl.* **1996**, *35*, 2150.

[87] Bauer, W.; Hampel, F. M.; *J. Chem. Soc., Chem. Commun.* **1992**, 903.

[88] Goldfuss, B.; Schleyer, P. V. R.; Hampel, F.; *J. Am. Chem. Soc.* **1997**, *119*, 1072.

[89] Geissler, M.; Kopf, J.; Schubert, B.; Weiss, E.; Neugebauer, W.; Schleyer, P. V. R.; *Angew. Chem. Int. Ed. Engl.* **1987**, *99*, 569; *Angew. Chem.* **1987**, *26*, 587; see also: Schubert, B.; Weiss, E.; *Chem. Ber.* **1983**, *116*, 3212.

[90] Weiss, E.; Plass, H.; *Chem. Ber.* **1968**, *101*, 2947.

[91] Üffing, C.; Köppe, R.; Schnöckel, H.; *Organometallics* **1998**, *17*, 3512.

[92] Almenningen, A.; Haaland, A.; Lusztyk, J.; *J. Organomet. Chem.* **1975**, *85*, 279.

[93] Nugent, K. W.; Beatti, J. K.; Hambley, T. W.; Snow, M. R.; *Aust. J. Chem.* **1984**, *37*, 1601.

[94] Haaland, A.; Lusztyk, J.; Brunvoll, J.; Starowieyski, K. B.; *J. Organomet. Chem.* **1975**, *85*, 279.

[95] Andersen, R. A.; Blohm, R.; Boncella, J. M.; Burns, C. J.; Volden, H. V.; *Acta Chem. Scand. A* **1987**, *41*, 24.

[96] Bünder, W.; Weiss, E.; *J. Organomet. Chem.* **1975**, *92*, 1.

[97] Nichols, M. A.; Williard, P. G.; *J. Am. Chem. Soc.* **1993**, *115*, 1568.

[98] Toney, J.; Stucky, G. D.; *J. Chem. Soc., Chem. Commun.* **1967**, 1168.

[99] Köster, H.; Thoeness, D.; Weiss, E.; *J. Organomet. Chem.* **1979**, *160*, 1.

[100] See also: Barnett, N. D. R.; Mulvey, R. E.; Clegg, W.; O'Neil, P. A.; *J. Am. Chem. Soc.* **1993**, *115*, 1573.

[101] Zerger, R. P.; Stucky, G. D.; *J. Am. Chem. Soc.* **1973**, 44.

[102] Stucky, G.; Rundle, R. E.; *J. Am. Chem. Soc.* **1964**, *86*, 4825.

[103] Weidenbruch, M.; Herrndorf, M.; Schäfer, A.; Pohl, S.; Saak, W.; *J. Organomet. Chem.* **1989**, *361*, 139.

[104] Hope, H.; Power, P. P.; *J. Am. Chem. Soc.* **1983**, *105*, 5320.

[105] Thoeness, D.; Weiss, E.; *Chem. Ber.* **1978**, 111, 3157.

[106] Schümann, U.; Kopf, J.; Weiss, E.; *Angew. Chem.* **1985**, *97*, 222, *Angew. Chem. Int. Ed. Engl.* **1985**, *24*, 215; see also: Schümann, U.; Weiss, E.; *Angew. Chem.* **1988**, *100*, 573; *Angew. Chem. Int. Ed. Engl.* **1988**, *27*, 584.

[107] Beno, M. A.; Hope, H.; Olmstead, M. M.; Power, P. P.; *Organometallics* **1985**, *4*, 2117.

[108] Bartlett, R. A.; Dias, H. V. R.; Power, P. P.; *J. Organomet. Chem.* **1988**, *341*, 1.

[109] Schümann, U.; Behrens, U.; Weiss, E.; *Angew. Chem.* **1989**, *101*, 481; *Angew. Chem. Int. Ed. Engl.* **1989**, *28*, 476.

[110] Schümann, U.; Weiss, E.; Dietrich, H.; Mahdi, W.; *J. Organomet. Chem.* **1987**, *322*, 299.

[111] Zarges, W.; Marsch, M.; Harms, K.; Boche, G.; *Chem. Ber.* **1989**, *122*, 2303.

[112] Schade, C.; Schleyer, P. V. R.; Dietrich, H.; Mahdi, W.; *J. Am. Chem. Soc.* **1986**, *108*, 2484; see also: Corbelin, S.; Lorenzen, N. P.; Kopf, J.; Weiss, E.; *J. Organomet. Chem.* **1970**, *96*, 1150.

[113] Patterman, S. P.; Karle, I. L.; Stucky, G. D.; *J. Am. Chem. Soc.* **1970**, *92*, 1150.

[114] Zarges, W.; Marsch, M.; Harms, K.; Koch, W.; Frenking, G.; Boche, G.; *Chem. Ber.* **1991**, *124*. 543.

[115] Diphenyl-2-pyridylmethylalkalimetal compounds: Pieper, U.; Stalke, D.; *Organometallics* **1993**, *12*, 1201.

[116] Hoffmann, D.; Bauer, W.; Schleyer, P. V. R.; Pieper, U.; Stalke, D.; *Organometallics* **1993**, *12*. 1193.

[117] Köster, H.; Weiss, E.; *J. Org. Chem.* **1988**, *341*, 1.

[118] Brooks, J. J.; Stucky, G. D.; *J. Am. Chem. Soc.* **1972**, *94*, 346.

[119] Olmsteadt, M. M.; Power, P. P.; *J. Am. Chem. Soc.* **1985**, *107*, 2174.

[120] Hein, F.; Schramm, H.; *Z. Phys. Chem. (Leipzig)* **1930**, *151*, 234.

[121] Bauer, W.; Seebach, D.; *Helv. Chim. Acta* **1984**, *67*, 1972.

[122] Bauer, W.; Winchester, W. R.; Schleyer, P. V. R.; *Organometallics* **1987**, *6*, 2371.

[123] Wittig, G.; Meyer, F. J.; Lange, G.; *Liebigs Ann. Chem.* **1951**, *571*, 167.

[124] West, P.; Waack, R.; *J. Am. Chem. Soc.* **1967**, *89*, 4395.

[125] Fraenkel, G.; Beckenbaugh, W. E.; Yang, P. P.; *J. Am. Chem. Soc.* **1976**, *98*, 6878.

[126] McKeever, L. D.; Waack, R.; Doran, M. A.; Baker, E. B.; *J. Am. Chem. Soc.* **1969**, *91*, 1057; see also: Waack, R.; *J. Am. Chem. Soc.* **1968**, *90*, 3244.

[127] Bywater, S.; Lachance, P.; Worsfold, D. J., *J. Phys. Chem.* **1975**, *79*, 2148.

[128] Fraenkel, G.; Fraenkel, A. M.; Geckle, J. J.; Schloss, F.; *J. Am. Chem. Soc.* **1979**, *101*, 4745.

[129] Fraenkel, G.; Henrichs, M.; Hewitt, J. M.; Su, B. M.; Geckle, M. J.; *J. Am. Chem. Soc.* **1980**, *102*, 3345.

[130] Seebach, D.; Hässig, R.; Gabriel, J.; *Helv. Chim. Acta* **1983**, *66*, 2148.

[131] Jackman, L. M.; Scarmoutzos, L. M.; *J. Org. Chem.* **1984**, *40*, 4627.

[132] Günther, H.; Moskau, D.; Bast, P.; Schmalz, D.; *Angew. Chem.* **1987**, *99*, 1242; *Angew. Chem. Int. Ed. Engl.* **1987**, *26*, 1212.

[133] Brown, T. L.; Rogers, M. T.; *J. Am. Chem. Soc.* **1957**, *79*, 1859.

[134] Brown, T. L.; Gerteis, R. L.; Bafus, D. A.; Ladd, J. A.; *J. Am. Chem. Soc.* **1964**, *86*, 2135.

[135] Lewis, H. L.; Brown, T. L.; *J. Am. Chem. Soc.* **1970**, *92*, 4664.

[136] West, P.; Waack, R.; Purmort, J. I.; *J. Am. Chem. Soc.* **1970**, *92*, 840.

[137] Baney, R. H.; Krager, R. J.; *Inorg. Chem.* **1964**, *3*, 1657.

[138] Hartwell, G. E.; Brown, T. L.; *Inorg. Chem.* **1966**, *5*, 1257.

[139] Panek, E. J.; Whitesides, G. M.; *J. Am. Chem. Soc.* **1972**, *94*, 8769.

[140] McKeever, L. D.; Waack, R.; Doran, M. A.; Baker, E. B.; *J. Am. Chem. Soc.* **1968**, *90*, 3244.

[141] Glaze, W. H.; Adams, G. M.; *J. Am. Chem. Soc.* **1966**, *88*, 4653.

[142] Bywater, S.; Worsfold, D. J.; *J. Organomet. Chem.* **1967**, *10*, 1.

[143] Weiner, M.; Vogel, G.; West, R.; *Inorg. Chem.* **1962**, *1*, 654.

[144] Waack, R.; Doran, M. A.; *J. Am. Chem. Soc.* **1969**, *91*, 2456.

[145] Schlosser, M.; Ladenberger, V.; *Chem. Ber.* **1967**, *100*, 3877.

[146] Wehman, E.; Jastrzebski, J. T. B. H.; Ernsting, J. M.; Grove, D. M.; Koten, G. V.; *J. Organomet. Chem.* **1988**, *133*, 353.

[147] Eppers, O.; Günther, H.; *Helv. Chim. Acta* **1992**, *75*, 2553.

[148] Fraenkel, G.; Pramanik, P.; *J. Chem. Soc., Chem. Commun.* **1983**, 1527.

[149] Morton, M.; Fetters, L. J.; *J. Polymer Sci.* **1964**, A2, 3311; *Chem. Abstr.* **1964**, *61*, 9585d.

[150] Stähle, M.; Schlosser, M.; *J. Organomet. Chem.* **1981**, *220*, 277.

[151] Desponds, O.; PhD thesis, Université de Lausanne, **1991**.

[152] Hartwell, G. E.; Brown, T. L.; *J. Am. Chem. Soc.* **1966**, *88*, 4625.

[153] Darensbourg, M. Y.; Kimura, B. Y.; Hartwell, G. E.; Brown, T. L.; *J. Am. Chem. Soc.* **1970**, *92*, 1236.

[154] McGarrity, J. F.; Ogle, C. A.; Brich, Z.; Loosli, H. R.; *J. Am. Chem. Soc.* **1985**, *107*, 1810.

[155] Weiner, M. A.; West, R.; *J. Am. Chem. Soc.* **1963**, *85*, 485.

[156] Applequist, D. E.; O'Brien, D. F.; *J. Am. Chem. Soc.* **1963**, *85*, 743.

[157] Brown, T. L.; *Acc. Chem. Res.* **1968**, *1*, 23.

[158] Hurd, D. T.; *J. Org. Chem.* **1948**, *13*, 711.

[159] Seitz, L. M.; Brown, T. L.; *J. Am. Chem. Soc.* **1966**, *88*, 4140.

[160] House, N. O.; Latham, R. A.; Whitesides, G. M.; *J. Org. Chem.* **1967**, *32*, 2481.

[161] Wittig, G.; Ludwig, R.; Polster, R.; *Chem. Ber.* **1955**, *88*, 294.

[162] Schlosser, M.; Stähle, M.; *Angew. Chem.* **1980**, *92*, 497; *Angew. Chem. Int. Ed. Engl.* **1980**, *19*, 487.

[163] Nordlander, J. E.; Roberts, J. D.; *J. Am. Chem. Soc.* **1959**, *81*, 1769.

[164] Nordlander, J. E.; Young, W. G.; Roberts, J. D.; *J. Am. Chem. Soc.* **1961**, *83*, 494.

[165] Whitesides, G. M.; Nordlander, J. E.; Roberts, J. D.; *J. Am. Chem. Soc.* **1962**, *84*, 2010.

[166] Brownstein, S.; Bywater, S.; Worsfold, D. J.; *J. Organomet. Chem.* **1980**, *199*, 1.

[167] Hogen-Esch, T. E.; Smid, J.; *J. Am. Chem. Soc.* **1966**, *88*, 307, 318.

[168] Reich, H. J.; Borst, J. P.; Dykstra, R. R.; Green, D. P.; *J. Am. Chem. Soc.* **1993**, *115*, 8728.

[169] Reich, H. J.; Borst, J. P.; *J. Am. Chem. Soc.* **1991**, *113*, 1835.

[170] Reich, H. J.; Kulicke, K. J.; *J. Am. Chem. Soc.* **1995**, *117*, 6621; **1996**, *118*, 273.

[171] Reichardt, C.; *Solvents and Solvent Effects in Organic Chemistry*, 2nd ed., Verlag Chemie VCH, Weinheim, **1988**, *Chem. Rev.* **1994**, *94*, 2319–2358; see also: Dimroth, K.; Reichardt, C.; Siepmann, T.; Bohlmann, F.; *Liebigs Ann. Chem.* **1963**, *661*, 1; Reichardt, C.; Harbusch-Bornert, E.; *Liebigs Ann. Chem.* **1983**, 721.

[172] Bhattacharyya, D. N.; Lee, C. L.; Smid, J.; Szwarc, M.; *J. Phys. Chem.* **1965**, *69*, 612.

[173] Slates, R. V.; Szwarc, M.; *J. Phys. Chem.* **1965**, *69*, 608.

[174] Smid, J.; unpublished work; quoted according to ref. [167b], foonote (24).

[175] Battacharyya, D. N.; Lee, C. L.; Smid, J.; Szwarc, M.; *J. Phys. Chem.* **1965**, *69*, 608.

[176] Strohmeier, W.; Seifert, F.; Landsfeld, H.; *Z. Elektrochem., Ber. Bunsenges.* **1962**, *66*, 312

[177] Strohmeier, W.; Landsfeld, H.; Gernert, F.; *Z. Elektrochem., Ber. Bunsenges.* **1962**, *66*, 823.

[178] Kraus, C. A.; Fuoss, R. M.; *J. Am. Chem. Soc.* **1933**, *55*, 2387.

[179] Swift, E.; *J. Am. Chem. Soc.* **1938**, *60*, 1403.

[180] Hill, D. C.; Burkus, J.; Luck, S. M.; Hauser, C. R.; *J. Am. Chem. Soc.* **1959**, *81*, 2787.

[181] Brown, J. M.; Elliot, R. J.; Richards, W. G.; *J. Chem. Soc., Perkin Trans. 2* **1982**, 485; Christl, M.; Leininger, H.; Brückner, D.; *J. Am. Chem. Soc.* **1983**, *105*, 4843; Köhler, F. H.; Hertkorn, N.; *Chem. Ber.* **1983**, *116*, 3274; Washburn, W. N.; *J. Org. Chem.* **1983**, *48*, 4287; Hertkorn, N.; Köhler, F. H.; Müller, G.; Reber, G.; *Angew. Chem.* **1986**, *98*, 462; *Angew. Chem. Int. Ed. Engl.* **1986**, *25*, 468; but see also: Grutzner, J. B.; Jorgensen, W. L.; *J. Am. Chem. Soc.* **1981**, *103*, 1372; Kaufmann, E.; Mayr, H.; Chandrasekhar, J.; Schleyer, P. V. R.; *J. Am. Chem. Soc.* **1981**, *103*, 1375.

[182] Hertkorn, N.; Köhler, F. H.; *J. Organomet. Chem.* **1988**, *355*, 19; *Z. Naturforsch. B: Chem. Sci.* **1990**, *45*, 848.

[183] Harder, S.; Prosenc, M. H.; *Angew. Chem.* **1994**, *106*, 1830; *Angew. Chem. Int. Ed. Engl.* **1994**, *33*, 1744; see also: Bauer, W.; Sivik, M.R.; Friedrich, D.; Schleyer, P. V. R.; Paquette, L. A.; *Organometallics* **1992**, *11*, 4178.

[184] Harder, S.; Prosenc, M. H.; Rief, U.; *Organometallics* **1996**, *15*, 118.

[185] Cotton, F. A.; *J. Organomet. Chem.* **1975**, *100*, 29.

[186] Rhee, W. Z. M.; Zuckerman, J. J.; *J. Am. Chem. Soc.* **1975**, *97*, 2291.

[187] Flood, T. C.; Rosenberg, E.; Sarhangi, A.; *J. Am. Chem. Soc.* **1977**, *99*, 4334.

[188] Bercaw, J. E.; Rosenberg, E.; Roberts, J. D.; *J. Am. Chem. Soc.* **1974**, *96*, 612.

[189] Gansow, O. A.; Burke, A. R.; Vernon, W. D.; *J. Am. Chem. Soc.* **1976**, *98*, 5817.

[190] Cotton, F. A.; Hanson, B. E.; Jamerson, J. D.; Stults, B. R.; *J. Am. Chem. Soc.* **1977**, *99*, 3293.

[191] Kreißl, F. R.; Eberl, K.; Uedelhoven, W.; *Chem. Ber.* **1977**, *110*, 3782.

[192] Faller, J. W.; Rosan, A. M.; *J. Am. Chem. Soc.* **1977**, *99*, 4858.

[193] Albright, T. A.; *Acc. Chem. Res.* **1982**, *15*, 149.

[194] Harland, L.; Stephenson, G. R.; Whittaker, M. J., *J. Organomet. Chem.* **1984**, *263*, C30.

[195] Benn, R.; *Org. Magn. Reson.* **1983**, *21*, 723.

[196] Benn, R.; Mynott, R.; Topalović, I.; Scott, F.; *Organometallics* **1989**, *8*, 2299; see also: Stanger, A.; *Organometallics* **1991**, *10*, 2979.

[197] Elschenbroich, C.; Bär, F.; Bilger, E.; Mahrwald, D.; Nowotny, M.; Metz, B.; *Organometallics* **1993**, *12*, 3373.

[198] Sanger, M. J.; Angelici, R. J.; *Organometallics* **1994**, *13*, 1821.

[199] Ceccon, A.; Gambaro, A.; Agostini, G.; Venzo, A.; *J. Organomet. Chem.* **1981**, *217*, 79.

[200] Rerek, M. E.; Basolo, F.; *Organometallics* **1984**, *3*, 647.

[201] Kündig, E. P.; Spichiger, C. P.; Bernardinelli, G.; *J. Organomet. Chem.* **1985**, *286*, 183.

[202] Kurosawa, H.; Ogoshi, S.; Kawasaki, Y.; Murai, S.; *J. Am. Chem. Soc.* **1990**, *112*, 2813.

[203] Granberg, K. L.; Bäckvall, J. E.; *J. Am. Chem. Soc.* **1992**, *114*, 6858.

[204] Starý, I.; Zajícek, J.; Kocovský, O.; *Tetrahedron* **1992**, *48*, 7229.

[205] Steiner, U.; Hansen, H. J.; Bachmann, K.; von Philipsborn, W.; *Helv. Chim. Acta* **1977**, *60*, 643.

[206] Schleyer, P. V. R.; Kos, A. J.; Kaufmann, E.; *J. Am. Chem. Soc.* **1983**, *105*, 7617.

[207] Neugebauer, W.; Kos, A. J.; Schleyer, P. V. R.; *J. Organomet. Chem.* **1982**, *228*, 107.

[208] Kranz, M.; Dietrich, H.; Mahdi, W.; Müller, G.; Hampel, F.; Clark, T.; Hacker, R.; Neugebauer, W.; Kos, A. J.; Schleyer, P. V. R.; *J. Am. Chem. Soc.* **1993**, *115*, 4698.

[209] Brown, K. J.; Berry, M. S.; Lingenfelter, D.; Murdoch, J. R.; *J. Am. Chem. Soc.* **1984**, *106*, 4717.

[210] Brown, K. J.; Murdoch, J. R.; *J. Am. Chem. Soc.* **1984**, *106*, 7843.

[211] Brown, K. J.; Berry, M. S.; Murdoch, J. R.; *J. Org. Chem.* **1985**, *50*, 4345.

[212] Desponds, O.; Schlosser, M.; *Tetrahedron* **1994**, *50*, 5881.

[213] Solladié-Cavallo, A.; Solladié, G.; Tsamo, E.; *J. Org. Chem.* **1979**, *44*, 4189.

[214] Jaouen, G.; Top, S. Vessières, A.; Sayer, B. G.; Frampton, C. S.; McGlinchey, M. J.; *Organometallics* **1992**, *11*, 4061; Jaouen, G.; Vessières, A.; Butler, I. S.; *Acc. Chem. Res.* **1993**, *2*, 361.

[215] See footnote (21) of ref. [230].

[216] The NMR spectra of cyclohexenylpotassium recorded of ethereal solutions at −150 °C do not differentiate between methylene hydrogens on the potassium-bearing and the opposite face (Stähle, M.; Schlosser, M.; unpublished, **1979**).

[217] Lias, S. G.; Bartmess, J. E.; Liebman, J. F.; Holmes, J. L.; Levin, R. D.; Mallard, W. G.; *J. Phys. Chem. Ref. Data* **1988**, *17*, Suppl. 1, upgraded by the National Institute of Standards and Technology, Gaithersburg, MD (database **1993** package 19B, version 3.0).

[218] Bywater, S.; Worsfold, D. J.; *J. Organomet. Chem.* **1971**, *33*, 273; Brownstein, S.; Worsfold, D. J.; *Can. J. Chem.* **1972**, *50*, 1246.

[219] Sandel, V. R.; McKinley, S. V.; Freedman, H. H.; *J. Am. Chem. Soc.* **1968**, *90*, 495; Fraenkel, G.; *J. Am. Chem. Soc.* **1973**, *95*, 3208; Bushweller, C. H.; Sturgess, J. S.; Cipullo, M.; Hoogasian, S.; Gabriel, M. W.; Bank, S.; *Tetrahedron Lett.* **1978**, *19*, 1359.

[220] Schmidt, H.; Schweig, A.; Manuel, G.; *J. Chem. Soc., Chem. Commun.* **1975**, 667.

[221] Schlosser, M.; Desponds, O.; Lehmann, R.; Moret, E.; Rauchschwalbe, G.; *Tetrahedron* **1993**, *49*, 10175.

[222] West, P.; Purmort, J. I.; McKinley, S. V.; *J. Am. Chem. Soc.* **1968**, *90*, 797.

[223] Zieger, H. E.; Roberts, J. D.; *J. Org. Chem.* **1969**, *34*, 1976.

[224] van Eikema Hommes, N. J. R.; Bühl, M.; Schleyer, P. V. R.; Wu, Y.-d.; *J. Organomet. Chem.* **1991**, *409*, 307.

[225] Schlosser, M.; Hartmann, J.; David, V.; *Helv. Chim. Acta* **1974**, *57*, 1567.

[226] Schlosser, M.; Hartmann, J.; *J. Am. Chem. Soc.* **1976**, *98*, 4674.

[227] Schlosser, M.; *Pure Appl. Chem.* **1988**, *60*, 1627.

[228] Schlosser, M.; *Modern Synth. Meth.* **1992**, *6*, 227.

[229] Kotthaus, M.; Schlosser, M.; *Tetrahedron Lett.* **1998**, *39*, 4031.

[230] Schlosser, M.; Rauchschwalbe, G.; *J. Am. Chem. Soc.* **1978**, *100*, 3258.

[231] Yasuda, H.; Yamauchi, M.; Ohnuma, Y.; Nakamura, A.; *Bull. Chem. Soc. Jpn.* **1981**, *54*, 1481; see also: Yasuda, H.; Ohnuma, Y.; Yamauchi, M.; Tani, H.; Nakamura, A.; *Bull. Chem. Soc. Jpn.* **1979**, *52*, 2036; Yasuda, H.; Yamauchi, M.; Nakamura, A.; Sei, T.; Kai, Y.; Yasuoka, N.; Kasai, N.; *Bull. Chem. Soc. Jpn.* **1980**, *53*, 1089; Yasuda, H.; Ohnuma, Y.; Nakamura, A.; Kai, Y.; Yasuoka, N.; Kasai, N.; *Bull. Chem. Soc. Jpn.* **1980**, *53*, 1101.

[232] Bosshardt, H.; Schlosser, M.; *Helv. Chim. Acta* **1980**, *63*, 2393.

[233] Rauchschwalbe, G.; Jenny, T.; Zellner, A.; Schlosser, M.; unpublished results (**1977, 1987, 1997**).

[234] Evans, D. A.; Andrews, G. C.; Buckwalter, B.; *J. Am. Chem. Soc.* **1974**, *96*, 5560.

[235] Still, W. C.; MacDonald, T. L.; *J. Am. Chem. Soc.* **1974**, *96*, 5561.

[236] Hartmann, J.; Muthukrishnan, R.; Schlosser, M.; *Helv. Chim. Acta* **1974**, *57*, 2261.

[237] Hartmann, J.; Stähle, M.; Schlosser, M.; *Synthesis* **1974**, 888.

[238] Hoppe, D.; *Angew. Chem.* **1984**, *96*, 930; *Angew. Chem. Int. Ed. Engl.* **1984**, *23*, 932.

[239] Hoffmann, R. W.; Kemper, B.; *Tetrahedron Lett.* **1981**, *22*, 5263.

[240] Moret, E.; Schlosser, M.; *Tetrahedron Lett.* **1984**, 25, 4491.

[241] Schlosser, M.; Margot, C.; Maccaroni, P.; Leroux, F.; *Tetrahedron* **1998**, *54*, 12853.

[242] Lehn, J. M.; *Top. Curr. Chem. (Fortschr. Chem. Forsch.)* **1970**, *15*, 311.

[243] Lambert, J. B.; *Top. Stereochem.* **1971**, *6*, 19.

[244] Lambert, J. B.; Takeuchi, Y.; *Acyclic and Cyclic Organonitrogen Stereodynamics* (2 volumes), VCH Publ., New York **1992**.

[245] Skaarup, S.; *Acta Chem. Scand.* **1972**, *26*, 4190.

[246] Nakanishi, H.; Yamamoto, O.; *Tetrahedron* **1974**, *30*, 2115.

[247] Bach, R. D.; Wolber, G. J.; *J. Org. Chem.* **1982**, *47*, 245.

[248] Christen, D.; Minkwitz, R.; Nass, R.; *J. Am. Chem. Soc.* **1987**, *109*, 7020.

[249] Mack, H. G.; Christen, D.; Oberhammer, H.; *J. Mol. Struct.* **1988**, *190*, 215.

[250] Montanari, F.; Moretti, I.; Torre, G.; *J. Chem. Soc., Chem. Commun.* **1968**, 1694; **1969**, 1086.

[251] Boyd, D. R.; *Tetrahedron Lett.* **1968**, *9*, 4561.

[252] Mannschreck, A.; Linß, J.; Seitz, W.; *Liebigs Ann. Chem.* **1969**, *727*, 224.

[253] Felix, D.; Eschenmoser, A.; *Angew. Chem.* **1968**, *80*, 197; *Angew. Chem. Int. Ed. Engl.* **1968**, 7, 224.

[254] Driessler, F.; Ahlrichs, R.; Staemmler, V.; Kutzelnigg, W.; *Theor. Chim. Acta* **1973**, *30*, 315.

[255] Ahlrichs, R.; Driessler, F., Lischka, H.; Staemmler, V.; Kutzelnigg, W; *J. Chem. Phys.* **1975**, *62*, 1235.

[256] Salzner, U.; Schleyer, P. V. R.; *Chem. Phys. Lett.* **1992**, *199*, 267.

[257] Schöllkopf, U.; in *Houben-Weyl: Methoden der organischen Chemie* (E. Müller, ed.), Thieme, Stuttgart **1970**, Vol. *13/1*, p. 9.

[258] Clark, T.; Schleyer, P. V. R.; Pople, J. A.; *J. Chem. Soc., Chem. Commun.* **1978**, 137.

[259] Schlosser, M.; Jan, G.; Byrne, E.; Sicher, J.; *Helv. Chim. Acta* **1973**, *56*, 1630.

[260] Schlosser, M.; Tran Dinh, A.; *Angew. Chem.* **1981**, *93*, 1114; *Angew. Chem. Int. Ed. Engl.* **1981**, *20*, 1039.

[261] Mori, S.; Kim, B. H.; Nakamura, M.; Nakamura, E.; *Chem. Lett.* **1997**, 1079.

[262] Cram, D. J.; *Fundamentals of Carbanion Chemistry*, Academic Press, New York, **1965**.

[263] Letsinger, R. L.; *J. Am. Chem. Soc.* **1950**, 72, 4842.

[264] Witanowski, M.; Roberts, J. D.; *J. Am. Chem. Soc.* **1966**, *88*, 737.

[265] Fraenkel, G.; Dix, D. T.; *J. Am. Chem. Soc.* **1966**, *88*, 979.

[266] Jensen, F. R.; Nakamaye, K. L.; *J. Am. Chem. Soc.* **1968**, *90*, 3248.

[267] Ford, W. T.; Buske, G.; *J. Am. Chem. Soc.* **1974**, *96*, 621.

[268] Stille, J. K.; Sannes, K. N.; *J. Am. Chem. Soc.* **1972**, *94*, 8489.

[269] Yoshikawa, K.; Bekki, K.; Karatsu, M.; Toyoda, K.; Kamia, T.; Morishima, I.; *J. Am. Chem. Soc.* **1976**, *98*, 3272.

[270] Bergbreiter, D. E.; Reichert, O. M.; *J. Organomet. Chem.* **1977**, *125*, 119.

[271] Maercker, A.; Schumacher, R.; Buchmeier, W.; Lutz, H. D.; *Chem. Ber.* **1991**, *124*, 2489.

[272] Glaze, W. H.; Selman, C. M.; *J. Organomet. Chem.* **1968**, *11*, P3; *J. Org. Chem.* **1968**, *33*, 1987.

[273] Eliel, E. L.; Hartmann, A. A.; Abatjoglou, A. G.; *J. Am. Chem. Soc.* **1974**, *96*, 1807.

[274] Cohen, T.; Matz, J. R.; *J. Am. Chem. Soc.* **1980**, *102*, 6900; Cohen, T.; Lin, M.-t.; *J. Am. Chem. Soc.* **1984**, *106*, 1130; Rychnovsky, S. D.; Mickus, D. E.; *Tetrahedron Lett.* **1989**, *30*, 3011.

[275] Krief, A.; Defrère, L.; *Tetrahedron Lett.* **1996**, *37*, 8011.

[276] Krief, A.; Defrère, L.; *Tetrahedron Lett.* **1996**, 37, 8015.

[277] McDougal, P. G.; Condon, B. D.; Laffosse, M. D.; Lauro, A. M.; VanDerweer, D.; *Tetrahedron Lett.* **1988**, *29*, 2547.

[278] Schlosser, M.; Christmann, K. F.; *Angew. Chem.* **1966**, *78*, 115; *Angew. Chem. Int. Ed. Engl.* **1966**, *5*, 126.

[279] Schlosser, M.; *Top. Stereochem.* **1970**, *5*, 1.

[280] Schlosser, M.; Christmann, K. F.; *Synthesis* **1969**, 38.

[281] Walborsky, H. M.; Impastato, F. J.; *J. Am. Chem. Soc.* **1959**, *81*, 5835.

[282] Walborsky, H. M.; Impastato, F. J.; Young, A. E.; *J. Am. Chem. Soc.* **1964**, *86*, 3283.

[283] Walborsky, H. M.; Periasamy, H. P.; *J. Am. Chem. Soc.* **1974**, *96*, 3711.

[284] Applequist, D. E.; Peterson, A. H.; *J. Am. Chem. Soc.* **1961**, *83*, 862.

[285] Welch, J. G.; Magid, R. M.; *J. Am. Chem. Soc.* **1967**, *89*, 5300.

[286] Tanaka, K.; Minami, K.; Funaki, I.; Suzuki, H.; *Tetrahedron Lett.* **1990**, *31*, 2727.

[287] Lehmkuhl, H.; Mehler, K.; *Liebigs Ann. Chem.* **1982**, 2244.

[288] Freeman, P. K.; Hutchinson, L. L.; *Tetrahedron Lett.* **1976**, *17*, 1849.

[289] Nesmeyanov, A. N.; Borisov, A. E.; Volkenau, N. A.; *Izvest. Akad. Nauk S. S. S. R., Otdel. Khim. Nauk* **1954**, 992; *Chem. Abstr.* **1955**, *49*, 6892d.

[290] Curtin, D. Y.; Koehl, W. J.; *J. Am. Chem. Soc.* **1962**, *84*, 1967.

[291] Seyferth, D.; Vaughan, L. G.; *J. Am. Chem. Soc.* **1964**, *86*, 883.

[292] Seyferth, D.; Vaughan, L. G.; Suzuki, R.; *J. Organomet. Chem.* **1964**, *1*, 437.

[293] Neumann, H.; Seebach, D.; *Chem. Ber.* **1978**, *111*, 2785.

[294] Martin, G. J.; Martin, M. L.; *Bull. Chem. Soc. Fr.* **1966**, 1636.

[295] Georgoulis, C.; Meyet, J.; Smadja, W.; *J. Organomet. Chem.* **1976**, *121*, 271.

[296] Kumar, A.; Singh, A.; Devaprobhakata, D.; *Tetrahedron Lett.* **1975**, *16*, 3343.

[297] Panek, E. J.; Neff, B. L.; Chu, H.; Panek, M. G.; *J. Am. Chem. Soc.* **1975**, *97*, 3996.

[298] Zweifel, G; Rajagopalan, S.; *J. Org. Chem.* **1985**, *107*, 700.

[299] Knorr, R; von Roman, T.; *Angew. Chem.* **1984**, *96*, 349; *Angew. Chem. Int. Ed. Engl.* **1984**, *23*, 366.

[300] Negishi, E.-i.; Takahashi, T.; *J. Am. Chem. Soc.* **1986**, *108*, 3402.

[301] Mahler, H.; Braun, M.; *Tetrahedron Lett.* **1987**, *28*, 5145; *Chem. Ber.* **1991**, *124*, 1379.

[302] Feit, B. A.; Melamed, U.; Schmidt, R. R.; Speer, H.; *Tetrahedron* **1981**, *37*, 2143.

[303] Ficini, J.; Depezay, J. C.; *Tetrahedron Lett.* **1968**, *9*, 937.

[304] Cahiez, G.; Bernard, D.; Normant, J. F.; *Synthesis* **1976**, 245.

[305] Shinokubo, H.; Miki, H.; Yokoo, T.; Oshima, K.; Utimoto, K.; *Tetrahedron* **1995**, *43*, 11681.

[306] Still, W. C.; Sreekumar, C.; *J. Am. Chem. Soc.* **1980**, *102*, 1201.

[307] Chan, P. C.-m.; Chong, J. M.; *J. Org. Chem.* **1988**, *53*, 5584.

[308] Chong, J. M.; Mar, E. K.; *Tetrahedron* **1989**, *45*, 7709.

[309] Chan, P. C.-m.; Chong, J. M.; *Tetrahedron Lett.* **1990**, *31*, 1985.

[310] Marshall, J. A.; Welmaker, G. S.; Gung, B. W.; *J. Am. Chem. Soc.* **1991**, *113*, 647.

[311] Hoppe, D.; Krämer, T.; Schwark, J. R.; Zschage, O.; *Pure Appl. Chem.* **1990**, *62*, 1999.

[312] Hoppe, D.; Ahrens, H.; Guarnieri, W.; Helmke, H.; Kolczewski, S.; *Pure Appl. Chem.* **1996**, *68*, 613.

[313] Hoppe, D.; Hense, T.; *Angew. Chem.* **1997**, *109*, 2376; *Angew. Chem. Int. Ed. Engl.* **1997**, *36*, 2282.

[314] Paetow, M.; Kotthaus, M.; Grehl, M.; Fröhlich, R.; Hoppe, D.; *Synlett* **1994**, 1034.

[315] Paulsen, H.; Graeve, C.; Hoppe, D.; *Synthesis* **1996**, 141; Paulsen, H.; Graeve, C.; Fröhlich, R.; Hoppe, D.; *Synthesis* **1996**, 145.

[316] Kerrick, S. T.; Beak, P.; *J. Am. Chem. Soc.* **1991**, *113*, 9708.

[317] Beak, P.; Basu, A.; Gallagher, D. J.; Park, Y. S.; Thayumanavan, S.; *Acc. Chem. Res.* **1996**, *29*, 552.

[318] Elworthy, T. R.; Meyers, A. I.; *Tetrahedron* **1994**, *50*, 6089.

[319] Gawley, R. E.; Zhang, Q.-h.; *Tetrahedron* **1994**, *50*, 6077.

[320] Gawley, R. E.; *J. Am. Chem. Soc.* **1987**, *109*, 1265.

[321] Rein, K.; Goicoechea-Pappas, M.; Anklekar, T. V.; Hart, G.; Smith, G. A.; Gawley, R. E.; *J. Am. Chem. Soc.* **1989**, *111*, 2211.

[322] Meyers, A. I.; Guiles, J.; Warmus, J. S.; Gonzales, M. A.; *Tetrahedron Lett.* **1991**, *32*, 5505.

[323] Hoppe, D.; Zschage, O.; *Angew. Chem.* **1989**, *101*, 67; *Angew. Chem. Int. Ed. Engl.* **1989**, *28*, 67.

[324] Hoppe, D.; Zschage, O.; *Tetrahedron Lett.* **1992**, *48*, 5657.

[325] Hoppe, D.; Brönneke, A.; *Synthesis* **1982**, 1045; *Tetrahedron Lett.* **1983**, *24*, 1687.

[326] Basu, A.; Beak, P.; *J. Am. Chem. Soc.* **1996**, *118*, 1575.

[327] Beak, P.; Du, H.; *J. Am. Chem. Soc.* **1993**, *115*, 2516.

[328] Lutz, G. P.; Du, H.; Gallagher, D. J.; Beak, P.; *J. Org. Chem.* **1996**, *61*, 4542.

[329] Gallagher, D. J.; Du, H.; Long, S. A.; Beak, P.; *J. Am. Chem. Soc.* **1996**, *118*, 11391.

[330] Schlosser, M.; Limat, D.; *J. Am. Chem. Soc.* **1995**, *117*, 12342.

[331] Hoppe, I.; Marsch, M.; Harms, K.; Boche, G.; Hoppe, D.; *Angew. Chem.* **1995**, *107*, 2328; *Angew. Chem. Int. Ed. Engl.* **1995**, *34*, 2158.

[332] Thayumanavan, S.; Lee, S.; Liu, C.; Beak, P.; *J. Am. Chem. Soc.* **1994**, *116*, 9755.

[333] Peoples, P. R.; Grutzner, J. B.; *J. Am. Chem. Soc.* **1980**, *102*, 4709.

[334] Hoffmann, R. W.; Rühl, T.; Chemla, F.; Zahneisen, T.; *Liebigs Ann. Chem.* **1992**, 719.

[335] Hoppe, D.; Hintze, F.; Tebben, P.; *Angew. Chem.* **1990**, *102*, 1457; *Angew. Chem. Int. Ed. Engl.* **1990**, *29*, 1422.

[336] Hintze, F.; *Doctoral Dissertation*, University of Kiel, **1993**, quoted according to ref. [313].

[337] Marsch, M.; Harms, K.; Zschage, O.; Hoppe, D.; Boche, G.; *Angew. Chem.* **1991**, *103*, 338; *Angew. Chem. Int. Ed. Engl.* **1991**, *30*, 321.

[338] Zschage, O.; Hoppe, D.; *Tetrahedron* **1992**, *48*, 8389.

[339] Carstens, A.; Hoppe, D.; *Tetrahedron* **1994**, *50*, 6097.

[340] Park, Y. S.; Boys, M. L.; Beak, P.; *J. Am. Chem. Soc.* **1996**, *118*, 3757; see also: Faibish, N. C.; Park, Y. S.; Lee, S.; Beak, P.; *J. Am. Chem. Soc.* **1997**, *119*, 11561; Weisenburger, G. A.; Faibish, N. C.; Pippel, D. J.; Beak, P.; *J. Am. Chem. Soc.* **1999**, *121*, 9522.

[341] Darwent, B. de B.; *Bond Dissociation Energies in Simple Molecules*, National Bureau of Standards, Washington, 1970.

[342] Wu, C. H.; *J. Chem. Phys.* **1976**, *65*, 2040.

[343] Zmbov, K. F.; Wu, C. H.; Ihle, H. R.; *J. Chem. Phys.* **1977**, *67*, 4603.

[344] Partridge, H.; Bauschlicher, C. W.; Pettersson, L. G. M.; McLean, A. D.; Liu, B.; Yoshimine, M.; Komornicki, A.; *J. Chem. Phys.* **1990**, *92*, 5377.

[345] On the basis of Pauling's formula and electronegativities (Pauling, L.; *The Nature of the Chemical Bond*, 3rd ed., Cornell University Press, Ithaca N.Y., **1960**) these bond ionicities can be derived: $C-H$ 4%, $C-Hg$ 10%, $C-Zn$ 18%, $C-Mg$ 30%, $C-Li$ 43%, $C-Na$ 47%, $C-K$ 51%; see also: Chan, T. H.; Fleming, I.; *Synthesis* **1979**, 761.

[346] DePuy, C. H.; Gronert, S.; Barlow, S. E.; Bierbaum, V. M.; Damrauer, R.; *J. Am. Chem. Soc.* **1989**, *111*, 1968.

[347] Ellison, G. B.; Engelking, P. C.; Lineberger, W. C.; *J. Am. Chem. Soc.* **1978**, *100*, 2556.

[348] Graul, S. T.; Squires, R. R.; *J. Am. Chem. Soc.* **1990**, *112*, 2517.

[349] Peerboom, R. A. L.; Rademaker, G. J.; Dekoning, L. J.; Nibbering, N. M. M.; *Rapid Commun. Mass Spectrom.* **1992**, *6*, 394; *Chem. Abstr.* **1992**, *117*, 111071c.

[350] Ervin, K. M.; Gronert, S.; Barlow, S. E.; Gilles, M. K.; Harrison, A. G.; Bierbaum, V. M.; DePuy, C. H.; Lin, W. C.; *J. Am. Chem. Soc.* **1990**, *112*, 5750.

[351] MacKay, G. J.; Hemsworth, R. S.; Bohme, D. K.; *Can. J. Chem.* **1976**, *54*, 1624.

[352] Lykke, K. R.; Murray, K. K.; Lineberger, W. C.; *Phys. Rev. A* **1991**, *43*, 6104; *Chem. Abstr.* **1991**, *115*, 37907v.

[353] Davico, G. E.; Bierbaum, V. M.; DePuy, C. H.; Ellison, G. B.; Squires, R. R.; *J. Am. Chem. Soc.* **1995**, *117*, 2590.

[354] Schulz, P. A.; Mead, R. D.; Jones, P. L.; Lineberger, W. C.; *J. Chem. Phys.* **1982**, *77*, 1153.

[355] Bartmess, J. E.; Scott, J. A.; McIver, R. T.; *J. Am. Chem. Soc.* **1979**, *101*, 6047.

[356] Wetzel, D. M.; Brauman, J. I.; *J. Am. Chem. Soc.* **1988**, *110*, 8333.

[357] Zimmerman, A. H.; Gygax, R.; Brauman, J. I.; *J. Am. Chem. Soc.* **1978**, *100*, 5595.

[358] Taft, R. W.; Bordwell, F. G.; *Acc. Chem. Res.* **1988**, *21*, 463.

[359] Hanstorp, D.; Gustafsson, M.; *J. Phys. B* **1992**, *25*, 1773; *Chem. Abstr.* **1992**, *116*, 262872v.

[360] Bartmess, J. E.; *Negative Ion Energetics Data, Chemistry WebBook: NIST Standard Reference Data Base Number 69* (Mallard W. G.; Linstrom, P. J.; eds), Nat. Inst. of Standards and Technol., Gaithersburg, Nov. **1998**.

[361] Moret, E.; Desponds, O.; Schlosser, M.; unpublished results (**1987**); see also: ref. [221] (pp. 10200–10201) and ref. [227] (p. 1632).

[362] Howden, M. E. H.; Maercker, A.; Burdon, J.; Roberts, J. D.; *J. Am. Chem. Soc.* **1966**, *88*, 1732.

[363] Maercker, A.; Roberts, J. D.; *J. Am. Chem. Soc.* **1966**, *88*, 1742.

[364] Roberts, J. D.; Mazur, R. H.; *J. Am. Chem. Soc.* **1951**, *73*, 2509.

[365] Patel, D. J.; Hamilton, C. L.; Roberts, D. J.; *J. Am. Chem. Soc.* **1965**, *87*, 5144.

[366] Lansbury, P. T.; Pattison, V. A.; Clement, W. A.; Sidler, J. D.; *J. Am. Chem. Soc.* **1964**, *86*, 2247.

[367] Hanack, M.; Schneider, H. J.; *Angew. Chem.* **1967**, *79*, 709; *Angew. Chem. Int. Ed. Engl.* **1967**, *6*, 666.

[368] Sarel, S.; Yovell, J.; Sarel-Imber, M.; *Angew. Chem.* **1968**, *80*, 592; *Angew. Chem. Int. Ed. Engl.* **1968**, *7*, 577.

[369] Bowry, V. W.; Lusztyk, J.; Ingold, K. U.; *J. Am. Chem. Soc.* **1991**, *113*, 5697.

[370] Bogdanović, B.; Liao, S.; Mynott, R.; Schlichte, K.; Westeppe, U.; *Chem. Ber.* **1984**, *117*, 1378; see also: Bogdanović, B.; *Angew. Chem.* **1985**, 97, 253, spec. 256; *Angew. Chem. Int. Ed. Engl.* **1985**, *24*, 262, spec. 265.

[371] X-Ray structure: Engelhardt, L. M.; Harvey, S.; Raston, C. L.; White, A. H.; *J. Organomet. Chem.* **1988**, *341*, 39.

[372] Kaiser, E. T.; Kevan, L.; (eds.), *Radical Ions*, Wiley, New York, **1968**.

[373] Szwarc, M.; *Prog. Phys. Org. Chem.* **1968**, *6*, 322.

[374] Holy, N. L.; *Chem. Rev.* **1974**, *74*, 243.

[375] Gerson, F.; Huber, W.; *Acc. Chem. Res.* **1987**, *20*, 85.

[376] Rautenstrauch, V.; *J. Am. Chem. Soc.* **1976**, *98*, 5035; **1977**, *99*, 6280.

[377] Hirota, N.; *J. Am. Chem. Soc.* **1968**, *90*, 3603; see also: Hirota, N.; Carraway, R.; Schook, W.; *J. Am. Chem. Soc.* **1968**, *90*, 3611; Takeshita, T.; Hirota, N.; *J. Chem. Phys.* **1973**, *58*, 3745.

[378] Freeman, P. K.; Hutchingson, L. L.; *J. Org. Chem.* **1980**, *45*, 1924.

[379] Cohen, T.; Sherbine, J. P.; Matz, J. R.; Hutchins, R. R.; McHenry, B. M.; Willey, P. R.; *J. Am. Chem. Soc.* **1984**, *106*, 3245.

[380] Katz, T. J.; *J. Am. Chem. Soc.* **1960**, *82*, 3784, 3785; Heinz, W.; Langensee, P.; Müllen, K.; *J. Chem. Soc., Chem. Commun.* **1986**, 947; Cox, R. H.; Harrison, L. W.; Austin, W. K.; *J. Phys. Chem.* **1973**, *77*, 200.

[381] Hsieh, H. L.; *J. Organomet. Chem.* **1967**, *7*, 1; Huynh Ba, G.; Jérôme, R.; Teyssié, P.; *J. Organomet. Chem.* **1980**, *190*, 107; Narita, T.; Hagiwara, T.; Hamana, H.; Yanagisawa, H.; Akazawa, Y.; *Makromol. Chem.* **1986**, *187*, 739; *Chem. Abstr.* **1986**, *105*, 6879b; Benken, R.; Finneiser, K.; von Puttkamer, H.; Günther, H.; Eliasson, B.; Edlund, U.; *Helv. Chim. Acta* **1986**, *69*, 955; Benken, T.; Günther, H.; *Helv. Chim. Acta* **1988**, *71*, 694.

[382] Selman, S.; Eastham, J. F.; *J. Org. Chem.* **1965**, *30*, 3804.

[383] Garst, J. F.; Pacifici, J. A.; *J. Am. Chem. Soc.* **1975**, *97*, 1802.

[384] Hückel, W.; Bretschneider, *Liebigs Ann. Chem.* **1939**, *540*, 157; see also: Hückel, W.; *Fortschr. Chem. Forsch.* **1966**, *6*, 197; *Chem. Abstr.* **1967**, *66*, 10299c.

[385] Wooster, C. B.; Godfrey, K. L.; *J. Am. Chem. Soc.* **1937**, *59*, 596.

[386] Birch, A. J.; Subba Rao, G.; *Adv. Org. Chem.* **1972**, *8*, 1.

[387] Birch, A. J.; *J. Chem. Educ.* **1975**, *52*, 458.

[388] Harvey, R. G.; *Synthesis* **1979**, 161.

[389] Rabideau, P. W.; *Tetrahedron* **1989**, 1579.

[390] Grovenstein, E.; Longfield, T. H.; Quest, D. E.; *J. Am. Chem. Soc.* **1977**, *99*, 2800.

[391] Hackspill, L.; *Ann. Chim. Phys. (Paris)* [8] **1913**, *28*, 653.

[392] Hackspill, L.; *Helv. Chim. Acta* **1928**, *11*, 1003, spec. 1026.

[393] de Postis, J.; *C.R. Séances Acad. Sci.* **1946**, *222*, 398; *Chem. Abstr.* **1946**, *40*, 3104[1].

[394] Collignon, N.; *J. Organomet. Chem.* **1975**, *96*, 139.

[395] Yang, S. C.; Nelson, D. D.; Stwalley, W. C.; *J. Chem. Phys.* **1983**, *78*, 4541.

[396] Bates, R. B.; Gosselink, D. W.; Kaczynski, J. A.; *Tetrahedron Lett.* **1967**, *8*, 199, 205; Bates, R. B.; Brenner, S.; Cole, C. M.; Davidson, E. W.; Forsythe, G. D.; McCombs, D. A.; Roth, A. S.; *J. Am. Chem. Soc.* **1973**, *95*, 926.

[397] Paul, R.; Tchelitcheff, S.; *C.R. Séances Acad. Sci.* **1954**, *239*, 1222; *Chem. Abstr.* **1955**, *49*, 13'915d.

[398] Lehmann, R.; Schlosser, M.; unpublished results (**1979**).

[399] For a working procedure, see pp. 270–271.

[400] Schlosser, M.; Bosshardt, H.; Walde, A.; Stähle, M.; *Angew. Chem.* **1980**, *92*, 302; *Angew. Chem. Int. Ed. Engl.* **1980**, *19*, 303.

[401] Gilman, H.; Bradley, C. W.; *J. Am. Chem. Soc.* **1938**, *60*, 2333.

[402] Approximate relative rates of reaction between hydrogen and liquid metals at 250 °C: Li > Cs > K > Na ~ 1.0:0.8:0.1:0.01 (Hill, S. E.; Pulham, R. J.; *J. Chem. Soc., Dalton Trans.* **1982**, 217).

[403] Schlenk, W.; Marcus, E.; *Ber. Dtsch. Chem. Ges.* **1914**, *47*, 1664.

[404] Conant, J. B.; Garvey, B. S.; *J. Am. Chem. Soc.* **1927**, *49*, 2599.

[405] Grovenstein, E.; Bhatti, A. M.; Quest, D. E.; Sengupta, D.; VanDerveer, D.; *J. Am. Chem. Soc.* **1983**, *105*, 6290.

[406] 1,2-Diphenylethane does undergo cleavage if dissolved in glycol dimethyl ether and brought in contact with a potassium mirror at low temperatures (Pearson, J. M.; Williams, D. M.; Levy, M.; *J. Am. Chem. Soc.* **1971**, *93*, 5478).

[407] Maercker, A.; Passlack, M.; *Chem. Ber.* **1983**, *116*, 710.

[408] Fraenkel, G.; Cooper, J. W.; *Tetrahedron Lett.* **1968**, *9*, 599.

[409] Jensen, F. R.; Bedard, R. L.; *J. Org. Chem.* **1959**, *24*, 874.

[410] Grovenstein, E.; Williamson, R. E.; *J. Am. Chem. Soc.* **1975**, *97*, 646.

[411] First example: Ziegler, K.; Crößmann, F.; *Ber. Dtsch. Chem. Ges.* **1929**, *62*, 1768.

[412] Grovenstein, E.; *J. Am. Chem. Soc.* **1957**, *79*, 4985.

[413] Zimmermann, H. E.; Smentowski, F. J.; *J. Am. Chem. Soc.* **1957**, *79*, 5455.

[414] Grovenstein, E.; Williams, L. P.; *J. Am. Chem. Soc.* **1961**, *83*, 412.

[415] Grovenstein, E.; Wentworth, G.; *J. Am. Chem. Soc.* **1963**, *85*, 3305.

[416] Grovenstein, E.; Rogers, L. C.; *J. Am. Chem. Soc.* **1964**, *86*, 854.

[417] Zimmerman, H. E.; Zweig, A.; *J. Am. Chem. Soc.* **1961**, *83*, 1196.

[418] Grovenstein, E.; Beres, A.; Cheng, Y.-m.; Pegolotti, J. A.; *J. Org. Chem.* **1972**, 37, 1281.

[419] Grovenstein, E.; *Angew. Chem.* **1978**, *90*, 317; *Angew. Chem. Int. Ed. Engl.* **1978**, *17*, 313.

[420] Grovenstein, E.; Ku, P.-c.; *J. Am. Chem. Soc.* **1982**, *104*, 6681; *J. Org. Chem.* **1982**, *47*, 2928.

[421] Grovenstein, E.; Singh, J.; Patil, B. B.; VanDerveer, D.; *Tetrahedron* **1994**, *50*, 5971.

[422] Blaise, C. C.; *C.R. Séances Acad. Sci.* **1901**, *132*, 38.

[423] Nützel, K.; in *Houben-Weyl: Methoden der organischen Chemie*, (Müller, E.; ed.) Vol. *13/2a*, Thieme, Stuttgart, **1973**, 47, spec. 352–366.

[424] Ziegler, K.; Ohlinger, H.; *Liebigs Ann. Chem.* **1932**, *495*, 84.

[425] Bockmühl, M.; Ehrhart, G.; *Liebigs Ann. Chem.* **1949**, *561*, 52.

[426] Raphael, R. A.; *Acetylenic Compounds in Organic Synthesis*, Butterworth, London, **1955**.

[427] Brandsma, L.; Verkruijsse, H. D.; *Synthesis of Acetylenes, Allenes and Cummulenes*, Elsevier, Amsterdam, **1981**.

[428] Gilman, H.; Jacoby, A. L.; Ludeman, H.; *J. Am. Chem. Soc.* **1938**, *60*, 2336; see also: Jones, R. G.; Gilman, H.; *Chem. Rev.* **1954**, *54*, 835.

[429] Swain, C. G.; *J. Am. Chem. Soc.* **1947**, *69*, 2306.

[430] O'Sullivan, W. I.; Swamer, F. W.; Humphlett, W. J.; Hauser, C. R.; *J. Org. Chem.* **1961**, *26*, 2306; see also: Hauser, C. R.; Puterbaugh, W. H.; *J. Am. Chem. Soc.* **1953**, *75*, 4756; Puterbaugh, W. H.; Hauser, C. R.; *J. Org. Chem.* **1959**, *24*, 416; Hauser, C. A.; Dunnavant, W. R.; *J. Org. Chem.* **1960**, *25*, 1296.

[431] Brown, C. A.; *J. Org. Chem.* **1974**, *39*, 1324, 3913.

[432] Stähle, M.; Schlosser, M.; unpublished results (**1978**).

[433] Fahaut, J.; Miginiac, P.; *Helv. Chim. Acta* **1978**, *61*, 2275.

[434] Broaddus, C. D.; *J. Org. Chem.* **1964**, *29*, 2689.

[435] Morton, A. A.; Lanpher, E. J.; *J. Org. Chem.* **1955**, *20*, 839.

[436] Broaddus, C. D.; Logan, T. J.; Flaut, T. J.; *J. Org. Chem.* **1963**, *28*, 1174.

[437] Hubert, A. J.; Reimlinger, H.; *Synthesis* **1969**, *97*, spec. 100.

[438] Bourguel, M.; *C.R. Séances Acad. Sci.* **1925**, *179*, 686; *Chem. Abstr.* **1925**, *19*, 966.

[439] Broaddus, C. D.; Muck, D. L.; *J. Am. Chem. Soc.* **1967**, *89*, 6533.

[440] Hartmann, J.; Schlosser, M.; *Helv. Chim. Acta* **1976**, *59*, 453.

[441] Schlosser, M.; Strunk, S.; *Tetrahedron Lett.* **1984**, *225*, 741.

[442] Bryce-Smith, D.; *J. Chem. Soc.* **1954**, 1079; **1963**, 5983.

[443] Benkeser, R. A.; Liston, T. V.; *J. Am. Chem. Soc.* **1960**, *82*, 3221; Benkeser, R. A.; Trevillyan, A. E.; Hooz, J.; *J. Am. Chem. Soc.* **1962**, *84*, 4971.

[444] Schlosser, M.; *J. Organomet. Chem.* **1967**, *8*, 9.

[445] Broaddus, C. D.; *J. Am. Chem. Soc.* **1966**, *88*, 4174.

[446] Chalk, A. J.; Hoogeboom, T. J.; *J. Organomet. Chem.* **1968**, *11*, 615.

[447] Broaddus, C. D.; *J. Org. Chem.* **1970**, *35*, 10.

[448] Harmon, T. E.; Shirley, D. A.; *J. Org. Chem.* **1974**, *39*, 3164.

[449] Letsinger, R. L.; Schnizer, A. W.; *J. Org. Chem.* **1951**, *16*, 869.

[450] Ludt, R. E.; Crowther, G. P.; Hauser, C. R.; *J. Org. Chem.* **1970**, *35*, 1288.

[451] Moret, E.; Schlosser, M.; unpublished results (**1978**).

[452] For working procedures, see pp. 274–278.

[453] Schlenk, W.; Hilleman, H.; Rodloff, I.; *Liebigs Ann. Chem.* **1931**, *487*, 135.

[454] Müller, E.; Gawlick, H.; Kreutzmann, W.; *Liebigs Ann. Chem.* **1934**, *515*, 109.

[455] Hudson, B. E.; Hauser, C. R.; *J. Am. Chem. Soc.* **1941**, *63*, 3156, 3163.

[456] Tomboulian, P.; *J. Org. Chem.* **1959**, *24*, 229.

[457] Gilman, H.; Gaj, B. J.; *J. Org. Chem.* **1963**, *28*, 1725.

[458] House, H. O.; Kramar, V.; *J. Org. Chem.* **1963**, *28*, 3362.

[459] House, H. O.; Gall, M.; Olmstead, H. D.; *J. Org. Chem.* **1971**, *36*, 2361.

[460] Vedejs, E.; *J. Am. Chem. Soc.* **1974**, *96*, 5944.

[461] Lee, R. A.; McAndrews, C.; Patel, K. M.; Reusch, W.; *Tetrahedron Lett.* **1973**, *14*, 965.

[462] Woodward, R. B.; Singh, T.; *J. Am. Chem. Soc.* **1950**, *72*, 494.

[463] Ringold, H. J.; Malhotra, S. K.; *Tetrahedron Lett.* **1962**, *3*, 669; see also: Woodward, R. B.; Patchett, A. A.; Barton, D. H. R.; Ives, D. A. J.; Kelly, R. B.; *J. Am. Chem. Soc.* **1954**, *76*, 2852; Ringold, H. J.; Rosenkrantz, G.; *J. Org. Chem.* **1957**, *22*, 602.

[464] Stork, G.; Hudrlik, P. F.; *J. Am. Chem. Soc.* **1968**, *90*, 4462, 4464.

[465] Kharasch, M. S.; Tawney, P. O.; *J. Am. Chem. Soc.* **1941**, *63*, 2308.

[466] Gilman, H.; Kirby, R. G.; *J. Am. Chem. Soc.* **1941**, *63*, 2046.

[467] Schultz, A. G.; Yee, Y. K.; *J. Org. Chem.* **1976**, *41*, 4044.

[468] Prévost, C.; *Bull. Soc., Chim. Fr.* **1931**, 1372; Miginiac-Groizeleau, L.; Miginiac, P.; Prévost, C.; *Bull. Soc. Chim. Fr.* **1965**, 3560; Agami, C.; Prévost, C.; Brun, M.; *Bull. Soc., Chim. Fr.* **1967**, 706.

[469] Young, W. G.; Prater, A. N.; *J. Am. Chem. Soc.* **1932**, *54*, 404; Roberts, J. D.; Young, W. G.; *J. Am. Chem. Soc.* **1945**, *67*, 148; Young, W. G.; Roberts, J. D.; *J. Am. Chem. Soc.* **1945**, *67*, 319; **1946**, *68*, 649.

[470] Benkeser, R. A.; *Synthesis* **1971**, 347.

[471] Inhoffen, H. H.; Burkhardt, H.; Quinkert, G.; *Chem. Ber.* **1959**, *92*, 1564.

[472] Felkin, H.; Frajerman, C.; Roussi, G.; *Bull. Soc., Chim. Fr.* **1970**, 3704.

[473] Rautenstrauch, V.; *J. Chem. Soc., Chem. Commun.* **1978**, 519.

[474] Hartmann, J.; Schlosser, M.; *Synthesis* **1975**, 328.

[475] Margot, C.; *Doctoral Dissertation*, Ecole Polytechnique Fédérale de Lausanne, **1985**; Koch, K.; *Diploma Work*, Université de Lausanne, **1988**.

[476] Yanagisawa, A.; Yasue, K.; Yamamoto, H.; *Synlett* **1996**, 842; see also: Yanagisawa, A.; Hibino, H.; Habaue, S.; Hisada, Y.; Yamamoto, H.; *J. Org. Chem.* **1992**, *57*, 6386.

[477] Ioffe, D. B.; Mostova, M. I.; *Russ. Chem. Rev.* **1973**, *42*, 56.

[478] Russell, G. A.; *J. Am. Chem. Soc.* **1959**, *81*, 2017.

[479] Eastham, J. F.; Cannon, D. Y.; *J. Org. Chem.* **1960**, *25*, 1504.

[480] Castagnetti, E.; Schlosser, M.; unpublished results (**1999**).

[481] Guggisberg, Y.; Faigl, F.; Schlosser, M.; *J. Organomet. Chem.* **1991**, 415, 1.

[482] Tiffeneau, M.; Delange, R.; *C.R. Acad. Sci.* **1903**, *137*, 573; Raaen, V. F.; Eastham, J. F.; *J. Am. Chem. Soc.* **1960**, *82*, 1349; Bernadon, C.; Deberly, A.; *J. Chem. Soc., Perkin Trans. 1* **1980**, 2631; Benkeser, R. A.; Snyder, D. C.; *J. Org. Chem.* **1982**, *47*, 1243.

[483] Gaudemar, M.; *Tetrahedron* **1976**, *32*, 1689.

[484] Bhupathy, M.; Cohen, T.; *Tetrahedron Lett.* **1985**, *26*, 2619; see also: Jefford, C. W.; Jaggi, D.; Boukouvalas, J.; *Tetrahedron Lett.* **1986**, *27*, 4011.

[485] Bandermann, F.; Sinn, H.; *Makromol. Chem.* **1966**, *96*, 150; *Chem. Abstr.* **1966**, *65*, 12287d.

[486] Szwarc, M.; *Nature (London)* **1956**, *178*, 1168; Szwarc, M.; Levy, M.; Milkovich, R.; *J. Am. Chem. Soc.* **1956**, *78*, 2656.

[487] Tobolsky, A. V.; Rogers, C. E.; *J. Polymer Sci.* **1959**, *38*, 205; **1959**, *40*, 73; *Chem. Abstr.* **1960**, *54*, 4021f, 12632f.

[488] Stearns, R. S.; Forman, L. E.; *J. Polymer Sci.* **1959**, *41*, 381; *Chem. Abstr.* **1960**, *54*, 17934d.

[489] *Br. Pat.* 848065 (to Phillips Petroleum, filed on 17 October **1956**), *Chem. Abstr.* **1961**, *55*, 15982d.

[490] Wittig, G.; *Naturwissenschaften* **1942**, *30*, 696; *Experientia* **1958**, *14*, 389.

[491] Wittig, G.; Köbrich, G.; *Endeavor* **1969**, *28* (*105*), 123.

[492] Han, C.-C.; Brauman, J. I.; *J. Am. Chem. Soc.* **1989**, *111*, 6491.

[493] Grutzner, J. B.; Lawlor, J. M.; Jackman, L. M.; *J. Am. Chem. Soc.* **1972**, *94*, 2306.

[494] O'Brien, D. H.; Russell, C. R.; Hart, A. J.; *J. Am. Chem. Soc.* **1979**, *101*, 633.

[495] Evans, A. G.; Hamid, M. A.; Rees, N. H.; *J. Chem. Soc. B* **1971**, 1110, 2164.

[496] Hamid, M. A.; *Can. J. Chem.* **1972**, *50*, 3761.

[497] Nazzal, A.; Proessdorf, H.; Mueller-Westerhoff, U.; *J. Am. Chem. Soc.* **1981**, *103*, 7678.

[498] Ahlberg, P.; Davidsson, Ö.; *J. Chem. Soc., Chem. Commun.* **1987**, 623.

[499] Davidsson, Ö.; Löwendahl, M.; Ahlberg, P.; *J. Chem. Soc., Chem. Commun.* **1992**, 1004; Ahlberg, P.; Davidsson, Ö.; Löwendahl, M.; Hilmersson, G.; Karlsson, A.; Håkansson, M.; *J. Am. Chem. Soc.* **1997**, *119*, 1745; Ahlberg, P.; Karlsson, A.; Davidsson, Ö.; Hilmersson, G.; Löwendahl, M.; *J. Am. Chem. Soc.* **1997**, *119*, 1751.

[500] Huisgen, R.; Sauer, J.; *Angew. Chem.* **1960**, *72*, 91, spec. 100.

[501] West, P.; Waack, R.; Purmort, J. I.; *J. Organomet.* **1969**, *19*, 267.

[502] Yamataka, H.; Matsuyama, T.; Hanafusa, T.; *J. Am. Chem. Soc.* **1989**, *111*, 4912.

[503] Yamataka, H.; Nishikawa, K.; Hanafusa, T.; *Chem. Lett.* **1990**, 1711.

[504] Yamataka, H.; Kawafuji, Y.; Nagareda, K.; Miyano, N.; Hanafusa, T.; *J. Org. Chem.* **1989**, *54*, 4706.

[505] Yamataka, H.; Miyano, N.; Hanafusa, T.; *J. Org. Chem.* **1991**, *56*, 2573.

[506] Yamataka, H.; Fujimura, N.; Kawafuji, Y.; Hanafusa, T.; *J. Am. Chem. Soc.* **1987**, *109*, 4305.

[507] Swain, C. G.; Kent, L.; *J. Am. Chem. Soc.* **1950**, *72*, 518.

[508] Ashby, E. C.; Laemmle, J.; Neumann, H. M.; *Acc. Chem. Res.* **1974**, *7*, 272.

[509] Letsinger, R. L.; Bobko, E.; *J. Am. Chem. Soc.* **1953**, *75*, 2649.

[510] Sicher, J.; *Angew. Chem.* **1972**, *84*, 177; *Angew. Chem., Int. Ed. Engl.* **1972**, *11*, 200.

[511] Schlosser, M.; in *Houben-Weyl: Methoden der organischen Chemie* (Müller, E.; ed.), Vol 5/1*b*, Thieme, Stuttgart **1972**, 9, spec. 40.

[512] Schlosser, M.; Jan, G.; Byrne, E.; Sicher, J.; *Helv. Chim. Acta* **1973**, *56*, 1630.

[513] Schlosser, M.; Tran Dinh, A.; *Helv. Chim. Acta* **1979**, *62*, 1194; Schlosser, M.; Tran Dinh, A.; *Angew. Chem.* **1981**, *93*, 1114; *Angew. Chem. Int. Ed. Engl.* **1981**, *20*, 1039.

[514] Schlosser, M.; Tarchini, C.; Tran Dinh, A.; Ruzziconi, R.; Bauer, P. J.; *Angew. Chem.* **1981**, *93*, 1116; *Angew. Chem. Int. Ed. Engl.* **1981**, *20*, 1041; Bauer, P. J.; Exner, O.; Ruzziconi, R.; Tran Dinh, A.; Tarchini, C.; Schlosser, M.; *Tetrahedron* **1994**, *50*, 1707.

[515] Matsuda, H.; Hamatani, T.; Matsubara, S.; Schlosser, M.; *Tetrahedron* **1988**, *44*, 2855.

[516] Nakamura, M.; Nakamura, E.; Koga, N.; Morokuma, K.; *J. Am. Chem. Soc.* **1993**, *115*, 11016.

[517] Harder, S.; van Lenthe, J. H.; van Eikema Hommes, N. J. R.; Schleyer, P. V. R.; *J. Am. Chem. Soc.* **1994**, *116*, 2508.

[518] Kaufmann, E.; Sieber, S.; Schleyer, P. V. R.; *J. Am. Chem. Soc.* **1989**, *111*, 121.

[519] Dixon, R. E.; Streitwieser, A.; Laidig, K. E.; Bader, R. F. W.; Harder, S.; *J. Phys. Chem.* **1993**, *97*, 3728.

[520] Nagase, S.; Uchibori, Y.; *Tetrahedron Lett.* **1982**, *23*, 2585.

[521] Kaufmann, E.; Sieber, S.; Schleyer, P. V. R.; *J. Am. Chem. Soc.* **1989**, *111*, 4005.

[522] Houk, K. N.; Rondan, N. G.; Schleyer, P. V. R.; Kaufmann, E.; Clark, T.; *J. Am. Chem. Soc.* **1985**, *107*, 2821.

[523] Wittig, G.; Schöllkopf, U.; *Tetrahedron* **1958**, *78*, 3, 91.

[524] Wittig, G.; *Angew. Chem.* **1958**, *70*, 65; *Quart. Rev.* **1966**, *20*, 191.

[525] Tochtermann, W.; *Angew. Chem.* **1966**, *78*, 355; *Angew. Chem. Int. Ed. Engl.* **1966**, *5*, 351.

[526] Reich, H. J.; Philips, N. H.; Reich, I. L.; *J. Am. Chem. Soc.* **1985**, *107*, 4101.

[527] Reich, H. J.; Green, D. P.; Philips, N. H.; *J. Am. Chem. Soc.* **1989**, *111*, 3444; *J. Am. Chem. Soc.* **1991**, *113*, 1414.

[528] Boche, G.; Schimeczek, M.; Cioslowski, J.; Piskorz, P.; *Eur. J. Org. Chem.* **1998**, 1851.

[529] Wakefield, B. J.; *The Chemistry of Organolithium Compounds*, Pergamon Press, New York, **1974**, pp. 51–52.

[530] Farnham, W. B.; Calabrese, J. C.; *J. Am. Chem. Soc.* **1986**, *108*, 2449.

[531] Bailey, W. F.; Patricia, J. J.; *J. Organomet. Chem.* **1988**, *352*, 1.

[532] Bergbreiter, D. E.; Reichert, O. M.; *J. Organomet. Chem.* **1977**, *125*, 119.

[533] Jedlicka, B.; Crabtree, R. H.; Siegbahn, E. M.; *Organometalllics* **1997**, *16*, 6021.

[534] Trogler, W. C.; *Organometallic Radical Processes*, Elsevier, Amsterdam, **1990**.

[535] Kornblum, N.; Michel, R. E.; Kerber, R. C.; *J. Am. Chem. Soc.* **1966**, *88*, 5660.

[536] Russell, G. A.; Danen, W. C.; *J. Am. Chem. Soc.* **1966**, *88*, 5563.

[537] Kim, J. K.; Bunnett, J. F.; *J. Am. Chem. Soc.* **1970**, *92*, 7463.

[539] Russell, G. A.; *Adv. Phys. Org. Chem.* **1987**, *23*, 271.

[540] Chen, K. S.; Bertini, F.; Kochi, J. K.; *J. Am. Chem. Soc.* **1973**, *95*, 1340.

[541] Bargon, J.; Fischer, H.; *Z. Naturforsch.* **1967**, *22A*, 1156.

[542] Ward, H. R.; Lawler, R. G.; *Acc. Chem. Res.* **1972**, *5*, 18.

[543] Lawler, R. G.; Ward, H. R.; *Acc. Chem. Res.* **1972**, *5*, 25.

[544] Kaptein, R.; *Adv. Free-Radical Chem.* **1975**, *5*, 319; see also: Kapstein, R.; *J. Chem. Soc., Chem. Commun.* **1971**, 732.

[545] Closs, G. L.; Czeropski, M. S.; *J. Am. Chem. Soc.* **1977**, *99*, 6127.

[546] Schaart, B. J.; Bodewitz, H. W. H. J.; Blomberg, C.; Bickelhaupt, F.; *J. Am. Chem. Soc.* **1976**, *98*, 3712.

[547] Garst, J. F.; Cox, R. H.; *J. Am. Chem. Soc.* **1970**, *92*, 6389.

[548] Agami, C.; Chauvin, M.; Levisalles, J.; *Bull. Soc., Chim. Fr.* **1970**, 2712.

[549] Molle, G.; Bauer, P.; Dubois, J. E.; *J. Org. Chem.* **1982**, *47*, 4120.

[550] Schmulbach, C. D.; Hinckley, C. C.; Wasmund, D.; *J. Am. Chem. Soc.* **1968**, *90*, 6600.

[551] Schlenk, W.; Appendrodt, J.; Michael, A.; Thal, A.; *Ber. Dtsch. Chem. Ges.* **1914**, *47*, 473; see also: Schlenk, W.; Bergmann, E.; *Liebigs Ann. Chem.* **1928**, *464*, 22.

[552] Wooster, C. B.; *Chem. Rev.* **1932**, *11*, 1.

[553] Ashby, E. C.; Lopp, I. G.; Buhler, J. D.; *J. Am. Chem. Soc.* **1975**, *97*, 1964.

[554] Kumar, A.; Singh, A.; Devaprabhakara, D.; *Tetrahedron Lett.* **1975**, *16*, 3343.

[555] Ingold, K. U.; *Acc. Chem. Res.* **1980**, *13*, 317.

[556] Beckwith, A. L.; *Chem. Soc. Rev.* **1993**, *22*, 143.

[557] Krusic, P. J.; Fagan, P. J.; San Filippo, J.; *J. Am. Chem. Soc.* **1977**, *99*, 250.

[558] Newcomb, M.; Williams, W. G.; *Tetrahedron Lett.* **1985**, *26*, 1179; Newcomb, M.; Williams, W. G.; Crumpacker, E. L.; *Tetrahedron Lett.* **1985**, *26*, 1183.

[559] Hill, E. A.; Chen, A. T.; Doughty, A.; *J. Am. Chem. Soc.* **1976**, *98*, 167.

[560] Arai, S.; Sato, S.; Shida, S.; *J. Chem. Phys.* **1960**, *33*, 1277.

[561] Griller, D.; Ingold, K. U.; *Acc. Chem. Res.* **1980**, *13*, 317.

[562] Ashby, E. C.; *Acc. Chem. Res.* **1988**, *21*, 414.

[563] Ashby, E. C.; Argyropoulos, J. N.; *J. Org. Chem.* **1985**, *50*, 3274.

[564] Garst, J. F.; Hines, J. B.; Bruhnke, J. D.; *Tetrahedron Lett.* **1986**, *27*, 1963.

[565] Ross, G. A.; Koppang, M. D.; Bartak, D. E.; Woolsey, N. F.; *J. Am. Chem. Soc.* **1985**, *27*, 1963.

[566] Ward, H. R.; *J. Am. Chem. Soc.* **1967**, *89*, 5517.

[567] Dessy, R. E.; Kandil, S. A.; *J. Org. Chem.* **1965**, *30*, 3857; *J. Am. Chem. Soc.* **1966**, *88*, 3027.
[568] Wooster, C. B.; Mitchell, N. W.; *J. Am. Chem. Soc.* **1930**, *52*, 1042.
[569] Bodewitz, H. W. H. J.; Blomberg, C.; Bickelhaupt, F.; *Tetrahedron* **1973**, *29*, 719.
[570] Garst, J. F.; *Acc. Chem. Res.* **1991**, *24*, 95.
[571] Walling, C.; *Acc. Chem. Res.* **1991**, *24*, 255; *Tetrahedron* **1981**, *37*, 1625.
[572] Garst, J. F.; Ungváry, F.; Baxter, J. T.; *J. Am. Chem. Soc.* **1997**, *119*, 253.
[573] Walborsky, H. M.; *Acc. Chem. Res.* **1990**, *23*, 286.
[574] Walborsky, H. M.; Zimmermann, C.; *J. Am. Chem. Soc.* **1992**, *114*, 4996.
[575] Hamdouchi, C.; Topolski, M.; Goedken, V.; Walborsky, H. M.; *J. Org. Chem.* **1993**, *58*, 3148.
[576] Garst, J. F.; *Acc. Chem. Res.* **1971**, *4*, 400; Garst, J. F.; Barton, F. E:; *J. Am. Chem. Soc.* **1974**, *96*, 523.
[577] Johnston, L. J.; Scaiano, J. C.; Ingold, K. U.; *J. Am. Chem. Soc.* **1984**, *106*, 4877.
[578] Johnston, L. J.; Ingold, K. U.; *J. Am. Chem. Soc.* **1986**, *108*, 2343.
[579] Deycard, S.; Hughes, L.; Lusztyk, J.; Ingold, K. U.; *J. Am. Chem. Soc.* **1987**, *109*, 4954.
[580] Boche, G.; Schneider, D. R.; Wintermayr, H.; *J. Am. Chem. Soc.* **1980**, *102*, 5697.
[581] Jacobus, J.; Pensak, D.; *J. Chem. Soc. D* **1969**, 400.
[582] Schlenk, W.; Ochs, R.; *Ber. Dtsch. Chem. Ges.* **1916**, *49*, 608.
[583] Panek, E. J.; Whitesides, G. M.; *J. Am. Chem. Soc.* **1972**, *94*, 8769.
[584] Panek, E. J.; Kaiser, L. R.; Whitesides, G. M.; *J. Am. Chem. Soc.* **1977**, *99*, 3708.
[585] Whitesides, G. M.; Newirth, T. L.; *J. Org. Chem.* **1975**, *40*, 3448.
[586] Okazaki, R.; Shibata, K.; Tokitoh, N.; *Tetrahedron Lett.* **1991**, *32*, 6601.
[587] Screttas, C. G.; *J. Chem. Soc., Chem. Commun.* **1971**, 406.
[588] Buck, P.; *Angew. Chem.* **1969**, *81*, 136; *Angew. Chem., Int. Ed. Engl.* **1969**, *8*, 120.
[589] Guthrie, R. D.; Weisman, G. R.; Burdon, L. G.; *J. Am. Chem. Soc.* **1974**, *96*, 6955.
[590] Schlenk, W.; Ochs, R.; *Ber. Dtsch. Chem. Ges.* **1916**, *49*, 608.
[591] Blomberg, C.; Mosher, H. S.; *J. Organomet. Chem.* **1968**, *13*, 519.
[592] Olah, G. A.; Wu, A.-h.; Farooq, O.; *Synthesis* **1991**, 1179.
[593] Holm, T.; Crossland, I.; *Acta Chem. Scand.* **1971**, *25*, 59.
[594] Ashby, E. C.; Laemmle, J.; Neumann, H. M.; *Acc. Chem. Res.* **1974**, *7*, 272; Zhang, Y.; Wenderoth, B.; Su, W.-y.; Ashby, E. C.; *J. Organomet. Chem.* **1985**, *292*, 29.
[595] Ashby, E. C.; *Pure Appl. Chem.* **1980**, *52*, 545.
[596] Holm, T.; *Acta Chem. Scand., Ser. B* **1983**, B37, 569.
[597] Kharasch, M. S.; Kleiger, S. C.; Martin, J. A.; Mayo, F. R.; *J. Am. Chem. Soc.* **1941**, *63*, 2305.
[598] Ashby, E. C.; Buhler, J. D.; Lopp, J. G.; Weisemann, T. L.; Bowers, J. S.; Laemmle, J. T.; *J. Am. Chem. Soc.* **1976**, *98*, 6561.
[599] Ashby, E. C.; Wiesemann, T. L.; *J. Am. Chem. Soc.* **1978**, *100*, 189.
[600] Walling, C.; *J. Am. Chem. Soc.* **1988**, *110*, 6846.
[601] Ziegler, K.; Schäfer, O.; *Liebigs. Ann. Chem.* **1930**, *479*, 150.
[602] Schlenk, W.; Bergmann, E.; *Liebigs Ann. Chem.* **1928**, *463*, 1, spec. 3–4; **1930**, *479*, 78.
[603] Newcomb, M.; Burchill, M. T.; *J. Am. Chem. Soc.* **1984**, *106*, 8276.
[604] Ashby, E. C.; Pham, T. N.; Park, B.; *Tetrahedron Lett.* **1985**, *26*, 4691.
[605] Bailey, W. F.; Patricia, J. J.; Nurmi, T. T.; Wang, W.; *Tetrahedron Lett.* **1986**, *27*, 1861.
[606] Bailey, W. F.; Patricia, J. J.; Nurmi, T. T.; *Tetrahedron Lett.* **1986**, *27*, 1865.
[607] Juaristi, E.; Jiménez-Vásquez, H. A.; *J. Org. Chem.* **1991**, *56*, 1623.
[608] Bailey, W. F.; Carson, M. W.; *Tetrahedron Lett.* **1999**, *40*, 5433.
[609] Applequist, D. E.; O'Brien, D. F.; *J. Am. Chem. Soc.* **1963**, *85*, 743.
[610] Ziegler, K.; Bähr, K.; *Ber. Dtsch. Chem. Ges.* **1928**, *61*, 253.
[611] Mulvaney, J. E.; Groen, S.; Carr, L. J.; Garlund, Z. G.; Garlund, S. L.; *J. Am. Chem. Soc.* **1969**, *91*, 388.
[612] For an analogy, see: Ziegler, K.; Schäfer, O.; *Liebigs Ann. Chem.* **1930**, *479*, 150.
[613] Panek, E. J.; *J. Am. Chem. Soc.* **1973**, *95*, 8460.

[614] Wittig, G.; Wittenberg, D.; *Liebigs Ann. Chem.* **1957**, *606*, 8; see also: Katz, T. J.; *J. Am. Chem. Soc.* **1960**, *82*, 3784.

[615] Lwow, M.; *Z. Chem.* **1871**, *[2] 7*, 257.

[616] Noller, C. R.; *J. Am. Chem. Soc.* **1929**, *51*, 594.

[617] Bott, K.; *Tetrahedron Lett.* **1994**, *35*, 555.

[618] Tanaka, J; Nojima, M.; Kusabayashi, S.; *J. Am. Chem. Soc.* **1987**, *109*, 3391.

[619] Wohl, A.; Mylo, B.; *Ber. Dtsch. Chem. Ges.* **1912**, *45*, 322.

[620] Dillon, R. T.; Lucas, H. J.; *J. Am. Chem. Soc.* **1928**, *50*, 1712.

[621] Wibaut, J. P.; Huls, R.; van der Voort, H. G. P.; *Recl. Trav. Chim. Pays-Bas* **1952**, *71*, 798, 1012.

[622] Dornfeld, C. A.; Coleman, G. H.; *Org. Synth., Coll. Vol.* **1955**, *3*, 701.

[623] Stetter, H.; Reske, R.; *Chem. Ber.* **1970**, *103*, 643.

[624] Eliel, E. L.; Nader, F.; *J. Am. Chem. Soc.* **1969**, 91, 536; **1970, 92**, 584.

[625] Bailey, W. F.; Croteau, A. A.; *Tetrahedron Lett.* **1981**, *22*, 545.

[626] Blomberg, C.; Vreughenhil, A. D.; Homsma, T.; *Recl. Trav. Chim. Bays-Bas* **1963**, *82*, 355; see also: Giusti, G.; *Bull. Soc. Chim. Fr.* **1972**, 4335; Westera, G.; Blomberg, C.; Bickelhaupt, F.; *J. Organomet. Chem.* **1974**, *82*, 291.

[627] Takano, S.; Ohkawa, T.; Ogasawara, K.; *Tetrahedron Lett.* **1988**, *29*, 1823.

[628] Ponaras, A. A.; *Tetrahedron Lett.* **1976**, *17*, 3105.

[629] Cheng, W.-L.; Yeh, S.-M.; Luh, T.-Y.; *J. Org. Chem.* **1993**, *58*, 5576.

[630] Axiotis, G. P.; *Tetrahedron Lett.* **1981**, *22*, 1509.

[631] Bruylants, P.; *Bull. Soc., Chim. Belg.* **1924**, *33*, 467; **1926**, *35*, 139.

[632] Ahlbrecht, H.; Dollinger, H.; *Synthesis* **1985**, 743.

[633] Courgois, G.; Harama, M.; Miginiac, L.; *J. Organomet. Chem.* **1980**, *198*, 1.

[634] Pollak, I. E.; Grillot, G. F.; *J. Org. Chem.* **1967**, *32*, 2892.

[635] Wasserman, H. H.; Dion, R. P.; *Tetrahedron Lett.* **1982**, *23*, 785.

[636] Seebach, D.; Betschart, C.; Schiess, M.; *Helv. Chim. Acta* **1984**, *67*, 1593.

[637] Morimoto, T.; Takahashi, T.; Sekiya, M.; *J. Chem. Soc., Chem. Commun.* **1984**, 794.

[638] Bestmann, H. J.; Wölfel, G.; *Angew. Chem.* **1984**, *96*, 52; *Angew Chem. Int. Ed. Engl.* **1984**, *23*, 53.

[639] Bockmühl, M.; Ehrhart, G.; *Liebigs Ann. Chem.* **1949**, *561*, 52.

[640] Knorr, L.; *Ber. Dtsch. Chem. Ges.* **1904**, *37*, 3507; Salomon, G.; *Helv. Chim. Acta* **1934**, *17*, 851; Gibbs, C. F.; Marvel, C. S.; *J. Am. Chem. Soc.* **1935**, *57*, 1137; Golumbic, C.; Friton, J. S.; Bergmann, M.; *J. Org. Chem.* **1946**, *11*, 518; Bartlett, P. D.; Ross, S. D.; Swain, C. G.; *J. Am. Chem. Soc.* **1947**, *69*, 2971; Capon, B.; McManus, S. P.; *Neighboring Group Participation*, Vol. *1*, Plenum, New York, **1976**, pp. 231–243.

[641] Schlosser, M.; *Angew. Chem.* **1964**, *76*, 124, 258; *Angew. Chem. Int. Ed. Engl.* **1964**, *3*, 287, 362.

[642] Dettmeier, U.; Grosskinsky, O. A.; Mack, K. -E.; Wirtz, R.; in *Winnacker-Küchler: Chemische Technologie*, Vol. *6* (Harnisch H.; Steiner R.; Winnacker K. eds), Hanser, München, **1982**, p. 133; see also: US-Pat. 2891977 (Ciba, 1955).

[643] Nützel, K.; in *Houben-Weyl: Methoden der organischen Chemie* (Müller E., ed.), Vol. *13/2a*, Thieme, Stuttgart, **1973**, 553, spec. 589; 636–642.

[644] Nützel, K.; in *Houben-Weyl: Methoden der organischen Chemie* (Müller E., Ed.), Vol. *13/2a*, Thieme, Stuttgart, **1973**, 47; spec. 63, 78.

[645] Callen, J. E.; Dornfield, C. A.; Coleman, G. H.; *Org. Synth., Coll. Vol.* **1955**, *3*, 26; Coburn, E. R.; *Org. Synth., Coll. Vol.* **1955**, *3*, 696; Paul, R.; Riobé, O.; Maumy, M.; *Org. Synth., Coll. Vol.* **1988**, *6*, 676.

[646] Schöllkopf, U.; in *Houben-Weyl: Methoden der organischen Chemie* (Müller E., ed.), Vol. *13/1*, Thieme, Stuttgart. **1970**, 87, spec. 134–148.

[647] Lusch, M. J.; Phillips, W. V.; Sieloff, R. F.; Nomura, G. S.; House, H. O.; *Org. Synth., Coll. Vol.* **1990**, *7*, 346; Wittig, G.; Hesse, A.; *Org. Synth., Coll. Vol.* **1988**, *6*, 901, spec. 903, note 5.

[648] Gilman, H.; Zoellner, E. A.; Selby, W. M.; *J. Am. Chem. Soc.* **1933**, *55*, 1252.

[649] Fischer, H.; *Org. Synth., Coll. Vol.* **1955**, *2*, 198; Moyer, W. W.; Marvel, C. S.; *Org. Synth., Coll. Vol.* **1955**, *2*, 26.

[650] Gilman, H.; Moore, F. W.; Baine, O.; *J. Am. Chem. Soc.* **1941**, *63*, 2479.

[651] Anker, R. M.; Cook, A. H.; *J. Am. Chem. Soc.* **1941**, 323, spec. 328.

[652] Karrer, P.; Benz, J.; *Helv. Chim. Acta* **1948**, *31*, 1048, spec. 1052.

[653] Eid, C. N.; Konopelski, J. P.; *Tetrahedron* **1991**, *47*, 975.

[654] Kim, Y.-J.; Bernstein, M. P.; Galiano Roth, A. S.; Romesberg, F. E.; Williard, P. G.; Fuller, D. J.; Harrison. A.T.; Collum, D. B.; *J. Org. Chem.* **1991**, *56*, 4435.

[655] Holm, T.; *J. Chem. Soc., Perkin Trans. 2* **1981**, 464.

[656] Luche, J.-L.; Damiano, J.-C.; *J. Am. Chem. Soc.* **1980**, *102*, 7926.

[657] Morton, A. A.; Richardson, G. M.; Hallowell, A. T.; *J. Am. Chem. Soc.* **1941**, *63*, 327.

[658] Coleman, G. H.; Craig, D.; *Org. Synth., Coll. Vol.* **1943**, *2*, 179.

[659] Munch-Petersen, J.; *Org. Synth., Coll. Vol.* **1973**, *5*, 762.

[660] Kumada, M.; Tamao, K.; Sumitami, K.; *Org. Synth., Coll. Vol.* **1988**, *6*, 407.

[661] Holmes, A. B.; Sporikou, C. N.; *Org. Synth., Coll. Vol.* **1993**, *8*, 606.

[662] Bryce-Smith, D.; Turner, E. E.; *J. Chem. Soc.* **1953**, 861

[663] Gilman, H.; Beel, J. A.; Brannen, C. G.; Bullock, M. W.; Dunn, G. E.; Miller, L. S.; *J. Am. Chem. Soc.* **1949**, *71*, 1499.

[664] Morton, A. A.; Davidson, J. B.; Newey, H. A.; *J. Am. Chem. Soc.* **1942**, *64*, 2240.

[665] Bachmann, G. B.; *Org. Synth., Coll. Vol.* **1943**, *2*, 323.

[666] Morton, A. A.; Marsh, F. D.; Coombs, R. D.; Lyons, A. L.; Penner, S. E.; Ramsden, H. E.; Baker, V. B.; Little, E. L.; Letsinger, R. L.; *J. Am. Chem. Soc.* **1950**, *72*, 3785.

[667] Morton, A. A.; Brown, M. L.; Holden, M. E. T.; Letsinger, R. L.; Magat, E. E.; *J. Am. Chem. Soc.* **1945**, *67*, 2224.

[668] Schwindeman, J. A.; Morrison, R. C.; Dover, B. T.; Engel, J. F.; Kamienski, C. W.; Hall, R. W.; Sutton, D. E.; *US Pat.* 5'332'533 (to FMC Corp., filed on 6 July 1993, issued on 26 July 1994); *Chem. Abstr.* **1994**, *121*, 231044.

[669] Morton, A. A.; Davidson, J. B.; Best, R. J.; *J. Am. Chem. Soc.* **1942**, *64*, 2239; Morton, A. A.; Davidson, J. B.; Hakan, B. L.; *J. Am. Chem. Soc.* **1942**, *64*, 2242.

[670] Fiandanese, V.; Marchese, G.; Martina, V.; Ronzini, L.; *Tetrahedron Lett.* **1984**, *25*, 4805.

[671] Meals, R. N.; *J. Org. Chem.* **1944**, *9*, 211; Damico, R.; *J. Org. Chem.* **1964**, *29*, 1971.

[672] Andersen, R. A.; Wilkinson, G.; *Inorg. Synth.* **1979**, *19*, 262; Shioiri, T.; Aoyama, T.; Mori, S.; *Org. Synth., Coll. Vol.* **1993**, *8*, 612.

[673] Tessier-Youngs, C.; Beachley, O. T.; *Inorg. Synth.* **1986**, *24*, 95.

[674] Rossi, F. M.; McCusker, P. A.; Hennion, G. F.; *J. Org. Chem.* **1967**, *32*, 1233; Blomberg, C.; Salinger, R. M.; Mosher, H. S.; *J. Org. Chem.* **1969**, *34*, 2385.

[675] Zook, H. D.; March, J.; Smith, D. F.; *J. Am. Chem. Soc.* **1959**, *81*, 1617; Chang, B.-H.; Tung, H.-S.; Brubaker, C. H.; *Inorg. Chim. Acta* **1981**, *51*, 143.

[676] Drake, N. L.; Cooke, G. B.; *Org. Synth., Coll. Vol.* **1943**, *2*, 406.

[677] Gilman, H.; Kirby, R. H.; *Org. Synth., Coll. Vol.* **1941**, *1*, 524.

[678] Glaze, W. H.; Lin, J.; Felton, E. G.; *J. Org. Chem.* **1965**, *30*, 1258.

[679] Puntambeker, S. V.; Zoellner, E. A.; *Org. Synth., Coll. Vol.* **1941**, *1*, 524.

[680] Kamienski, G. W.; Esmay, D. L.; *J. Org. Chem.* **1960**, *25*, 1807.

[681] Smith, W. N.; *J. Organomet. Chem.* **1974**, *82*, 1.

[682] Giancaspro, G.; Sleiter, G.; *J. Prakt. Chem.* **1979**, *321*, 876.

[683] Smith, L. I.; McKenzie, S.; *J. Org. Chem.* **1950**, *15*, 74; Roberts, J. D.; Chambers, V. C.; *J. Am. Chem. Soc.* **1951**, *73*, 3176.

[684] Seyferth, D.; Cohen, H. M.; *J. Organomet. Chem.* **1963**, *1*, 15.

[685] Gilman, H.; Cathin, W. E.; *Org. Synth., Coll. Vol.* **1932**, *1*, 182.

[686] Johnson, O. H.; Nebergall, W. H.; *J. Am. Chem. Soc.* **1949**, *71*, 1720.

[687] Rieke, D.; Bales, S. E.; *Org. Synth., Coll. Vol.* **1988**, *6*, 845.

[688] Bixler, R. L.; Niemann, C.; *J. Org. Chem.* **1958**, *23*, 742.

[689] Molle, G.; Bauer, P.; Dubois, J. E.; *J. Org. Chem.* **1983**, *48*, 2975.

[690] Normant, H.; Ficini, J.; *Bull. Soc., Chim. Fr.* **1956**, 1441.

[691] Ramsden, H. E.; Leebrick, J. R.; Rosenberg, S. D.; Miller, E. H.; Walburn, J. J.; Balint, A. E.; Cserr, R.; *J. Org. Chem.* **1957**, *22*, 1602.

[692] Foster, D. J.; *US Pat.* 298569 (23 May 1961, to Union Carbide Corp.); *Chem. Abstr.* **1961**, *55*, 22134a.

[693] Anderson, R. G.; Silverman, M. B.; Ritter, D. M.; *J. Org. Chem.* **1958**, *23*, 750.

[694] Babler, J. H.; Olsen, D. O.; Arnold, W. H.; *J. Org. Chem.* **1974**, *39*, 1656; Dasgupta, S. K.; Crump, D. R.; Gut, M.; *J. Org. Chem.* **1974**, *39*, 1658.

[695] Braude, E. A.; Timmons, C. J.; *J. Chem. Soc.* **1950**, 2207; Braude, E. A.; Coles, J. A.; *J. Chem. Soc.* **1950**, 2012, 2014.

[696] Threlkel, R. S.; Bercraw, J. E.; Seidler, P. F.; Stryker, J. M.; Bergman, R. D.; *Org. Synth., Coll. Vol.* **1993**, *8*, 505.

[697] Broaddus, T. J.; Logan, T. J.; Flautt, T. J.; *J. Org. Chem.* **1963**, *28*, 1174.

[698] Normant, H.; Maitte, P.; *Bull. Soc. Chim. Fr.* **1960**, 1424.

[699] Maitte, P.; *Bull. Soc. Chim. Fr.* **1959**, 499.

[700] Ramsden, H. E.; Balint, A. E.; Whitford, W. R.; Walburn, J. J.; Cserr, R.; *J. Org. Chem.* **1957**, *22*, 1202.

[701] Pickard, P. L.; Tolbert, T. L.; *Org. Synth., Coll. Vol.* **1973**, *5*, 520; Schwartz, A.; Madan, P.; Whitesell, J. K.; Lawrence, R. M.; *Org. Synth., Coll. Vol.* **1993**, *8*, 516.

[702] Wittig, G.; *Angew. Chem.* **1940**, *53*, 242; **1949**, *53*, 242.

[703] Gilman, H.; Morton, J. W.; *Org. React.* **1954**, *8*, 258, spec. 286.

[704] Gilman, H.; Gaj, B. J.; *J. Org. Chem.* **1957**, *22*, 1165.

[705] Schöllkopf, U.; in *Houben-Weyl: Methoden der organischen Chemie*, vol. *13/1* (Müller, E.; ed.), Thieme, Stuttgart, **1970**, p. 147.

[706] Gilman, H.; Zoellner, E. A.; Selby, W. M.; *J. Am. Chem. Soc.* **1932**, *54*, 1957.

[707] Gilman, H.; Pacevitz, H. A.; Baine, O.; *J. Am. Chem. Soc.* **1940**, *62*, 1514.

[708] Ziegler, K.; *Angew. Chem.* **1936**, *49*, 459.

[709] Boekelheide. V.; Morrison, G. C.; *J. Am. Chem. Soc.* **1958**, *80*, 3905.

[710] Gilman, H.; Gainer, G. C.; *J. Am. Chem. Soc.* **1947**, *69*, 877.

[711] Wittig, G.; Hellwinkel, D.; *Chem. Ber.* **1964**, *97*, 784.

[712] Gilman, H.; Dunn, G. E.; *J. Am. Chem. Soc.* **1951**, *73*, 5078.

[713] Truce, W. E.; Lyons, J. F.; *J. Am. Chem. Soc.* **1951**, *73*, 126.

[714] Smith, L. I.; *Org. Synth., Coll. Vol.* **1943**, *2*, 360; Bowen, D. M.; *Org. Synth., Coll. Vol.* **1955**, *3*, 553; Barnes, R. P.; *Org. Synth., Coll. Vol.* **1955**, *3*, 555.

[715] Bähr, G.; Gelius, R.; *Chem. Ber.* **1958**, *91*, 818.

[716] Fuson, R. C.; Hammann, W. C.; Jones, P. R.; *J. Am. Chem. Soc.* **1957**, *79*, 928.

[717] Gilman, H.; Nelson, R. D.; *J. Am. Chem. Soc.* **1948**, *70*, 3316.

[718] Pearson, D. E.; Cowan, D.; Beckler, J. D.; *J. Org. Chem.* **1959**, *24*, 504.

[719] Gilman, H.; John, N. B. S.; Schulze, F.; *Org. Synth., Coll. Vol.* **1955**, *2*, 425.

[720] Foster, D. J.; *Ger. Pat.* 1114193 (28 Sept. 1961, to Union Carbide Corp.); *Chem. Abstr.* **1962**, *57*, 2166c.

[721] Gilman, H.; Zoellner, E. A.; Dickey, J. B.; *J. Am. Chem. Soc.* **1929**, *51*, 1583.

[722] Gilman, H.; Brannen, C. G.; *J. Am. Chem. Soc.* **1950**, *72*, 4280.

[723] Bell, F., Waring, D. H.; *J. Chem. Soc.* **1949**, 267; Hemetsberger, H.; Neustern, F. -U.; *Tetrahedron* **1982**, *38*, 1175; Kikuchi, H.; Seki, S.; Yamamoto, G.; Mitsuhashi, T.; Nakamura, N.; Ōki, M.; *Bull. Chem. Soc. Jpn.* **1982**, *55*, 1514; Wellman, D. E.; Lassila, K. R.; West, R.; *J. Org. Chem.* **1984**, *49*, 965; Lee, S.; Arita, K.; Kajimoto, O.; Tamao, K.; *J. Phys. Chem. A* **1997**, *101*, 5228.

[724] Dornfeld, C. A.; Coleman, G. H., *Org. Synth., Coll. Vol.* **1955**, *3*, 701.

[725] Levason, W.; Smith, K. G.; McAuliffe, C. A.; McCullough, F. P.; Sedgwick, R. D.; Murray, S. G.; *J. Chem. Soc., Dalton Trans.* **1979**, 1718.

[726] Gilman, H.; Stuckwisch, C. G.; *J. Am. Chem. Soc.* **1950**, *72*, 4553.

[727] Gilman, H.; Melvin, H. W.; *J. Am. Chem. Soc.* **1950**, *72*, 995; Gilman, H.; Summers, L.; *J. Am. Chem. Soc.* **1950**, *72*, 2767.

[728] Taber, D. F.; Meagley, R. P.; Supplee, D.; *J. Chem. Educ.* **1996**, *73*, 259.

[729] Wright, J. B.; Gutsell, E. S.; *J. Am. Chem. Soc.* **1959**, *81*, 5193.

[730] Rehnberg, N.; Magnusson, G.; *Tetrahedron Lett.* **1988**, *29*, 3599.

[731] Brooklyn, R. J.; Finar, I. L.; *J. Chem. Soc. [C]* **1968**, 466.

[732] Grootaert, W. M.; Mijngheer, R.; De Clerq, P. J.; *Tetrahedron Lett.* **1982**, *23*, 3287.

[733] Gilman, H.; Wright, G. F.; *J. Am. Chem. Soc.* **1933**, *55*, 2893.

[734] Takeda, A.; Shinhama, K.; Tsuboi, T.; *Bull. Chem. Soc. Jpn.* **1977**, *50*, 1903.

[735] Krause, E.; Renwanz, G.; *Ber. Dtsch. Chem. Ges.* **1929**, *62*, 1710.

[736] Frisell, C.; Lawesson, S.-O.; *Org. Synth., Coll. Vol.* **1973**, *5*, 642.

[737] Cymerman-Craig, J.; Loder, J. W.; *Org. Synth., Coll. Vol.* **1963**, *4*, 667.

[738] Schick, J. W.; Hartough, H. D.; *J. Am. Chem. Soc.* **1948**, *70*, 286.

[739] Peters, A. T.; Walker, D.; *J. Chem. Soc.* **1957**, 1525.

[740] Hurd, C. D.; Kreuz, K. L.; *J. Am. Chem. Soc.* **1952**, *74*, 2965.

[741] Lawesson, S.-O.; *Arkiv Kemi* **1957**, *11*, 317.

[742] Lawesson, S.-O.; *Arkiv Kemi* **1957**, *11*, 337.

[743] Gattermann, L.; *Liebigs Ann. Chem.* **1912**, *393*, 230.

[744] Gronowitz, S.; Petterson, K.; *J. Heterocycl. Chem.* **1976**, *13*, 1099.

[745] Kurkjy, R.; Brown, E. V.; *J. Am. Chem. Soc.* **1952**, *74*, 6260.

[746] *Brit. Pat.* 779100 (**1954**, to Metal and Thermit Corp.); *Chem. Abstr.* **1958**, *52*, 2084a.

[747] Kochetkov, N. K.; Sokolov, S. D.; Vagurtova, N. M.; Nifantev, E. E.; *Dokl. Akad. Nauk S. S. S. R.* **1960**, *133*, 598; *Chem. Abstr.* **1960**, *54*, 24656f.

[748] Normant, H.; *Bull. Soc., Chim. Fr.* **1957**, 728.

[749] Overhoff, J.; Proost, W.; *Recl. Trav. Chim. Pays-Bas* **1938**, *57*, 179.

[750] Wibaut, J. P.; Huls, R.; *Recl. Trav. Chim. Pays-Bas* **1952**, *71*, 1021.

[751] Wibaut, J. P.; van der Voort, H. G. P.; *Recl. Trav. Chim. Pays-Bas* **1952**, *71*, 798.

[752] Wibaut, J. P.; Heeringa, L. G.; *Recl. Trav. Chim. Pays-Bas* **1955**, *74*, 1003.

[753] Chambers, R. D.; Hutchinson, J.; Musgrave, W. K. R.; *J. Chem. Soc.* **1965**, 5040.

[754] Dua, S. S.; Gilman, H.; *J. Organomet. Chem.* **1968**, *12*, 234.

[755] Berry, D. J.; Wakefield, B. J.; *J. Chem. Soc. [C]* **1969**, 2342.

[756] Linnemann, E.; *Ber. Dtsch. Chem. Ges.* **1877**, *10*, 1111.

[757] Shriner, R. L.; Neumann, F. W.; *Org. Synth., Coll. Vol.* **1955**, *3*, 73.

[758] Frank, R. L.; Smith, P. V.; *Org. Synth., Coll. Vol.* **1955**, *3*, 73.

[759] Wilkinson, R.; *J. Chem. Soc.* **1931**, 3057.

[760] Noller, C. R.; *Org. Synth., Coll. Vol.* **1943**, *2*, 184.

[761] Hennion, G. F.; Sheehan, J. J.; *J. Am. Chem. Soc.* **1949**, *71*, 1964.

[762] Suslick, K. S. (ed.); *Ultrasound: Its Chemical, Physical and Biological Effects*, VCH, New York, **1988**.

[763] Skell, P. S.; McGlinchey, M. J.; *Angew. Chem.* **1975**, *87*, 215; *Angew. Chem. Int. Ed. Engl.* **1975**, *14*, 195.

[764] Reichelt, W.; *Angew. Chem.* **1975**, *87*, 239; *Angew. Chem. Int. Ed. Engl.* **1975**, *14*, 218.

[765] Klabunde, K. J.; *Angew. Chem.* **1975**, *87*, 309; *Angew. Chem. Int. Ed. Engl.* **1975**, *14*, 287.

[766] Kündig, E. P.; Moskovits, M.; Ozin, G. A.; *Angew. Chem.* **1975**, *87*, 314; *Angew. Chem. Int. Ed. Engl.* **1975**, *14*, 292.

[767] Rieke, R. D.; *Acc. Chem. Res.* **1977**, *10*, 301.

[768] Klabunde, K. J.; *Chemistry of Free Atoms and Particles*, Academic Press, New York, **1980**.

[769] Oppolzer, W.; Schneider, P.; *Tetrahedron Lett.* **1984**, *25*, 3305.

[770] Cox, D. N.; Roulet, R. R.; *Organometallics* **1985**, *4*, 2001; **1986**, 5, 1886; *J. Organomet. Chem.* **1988**, *342*, 87.

[771] Fürstner, A.; *Angew. Chem.* **1993**, *105*, 171; *Angew. Chem. Int. Ed. Engl.* **1993**, *32*, 164.

[772] Rieke, R. D.; Hanson, M. V.; *Tetrahedron* **1997**, *53*, 1925.

[773] Bare, W. D.; Andrews, L.; *J. Am. Chem. Soc.* **1998**, *120*, 7293.

[774] Fürstner, A. (ed.); *Active Metals: Preparation, Characterization, Applications*, VCH, Weinheim, **1995**.

[775] Bönnemann, H.; Bogdanović, B.; Brinkmann, R.; Spliethoff, B.; *J. Organomet. Chem.* **1993**, *451*, 23.

[776] Van den Ancker, T. R.; Harvey, S.; Raston, C. L.; *J. Organomet. Chem.* **1995**, *502*, 35.

[777] Wilson, S. E.; *Tetrahedron Lett.* **1975**, *16*, 4651.

[778] Freeman, P. K.; Hutchinson, L. L.; *J. Org. Chem.* **1980**, *45*, 1924.

[779] Freeman, P. K.; Hutchinson, L. L.; *J. Org. Chem.* **1983**, *48*, 4705.

[780] Nützel, K.; in *Houben/Weyl: Methoden der organischen Chemie* (Müller, E.; ed.), Vol *13/2a*, G. Thieme, Stuttgart, **1973**, p. 47–527.

[781] Garst, J. F.; Ungváry, F.; Batlaw, R.; Lawrence, K. E.; *J. Am. Chem. Soc.* **1991**, *113*, 6697; Pearson, D. E.; Cowan, D.; Beckler, J. D.; *J. Org. Chem.* **1959**, *24*, 504.

[782] Ashby, E. C.; Yu, S. H.; Bearch, R. G.; *J. Am. Chem. Soc.* **1970**, *92*, 433.

[783] Kündig, E. P.; Perret, C.; *Helv. Chim. Acta* **1981**, *64*, 2606.

[784] Silver, M. S.; Shafer, P. R.; Nordlander, J. E.; Rüchardt, C.; Roberts, J. D.; *J. Am. Chem. Soc.* **1960**, *82*, 2646; Howden, M. E. H.; Maercker, A.; Burdon, J.; Roberts, J. D.; *J. Am. Chem. Soc.* **1966**, *88*, 1732.

[785] Nunomoto, S.; Yamashita, Y.; *J. Org. Chem.* **1979**, *44*, 4788.

[786] Wada, E.; Kanemasa, S.; Fujiwara, I.; Tsuge, O.; *Bull. Chem. Soc. Jpn.* **1985**, *58*, 1942.

[787] Fleming, F. F.; Jiang, T.; *J. Org. Chem.* **1997**, *62*, 7890.

[788] Michel, E.; Troyanowsky, C.; *Tetrahedron Lett.* **1973**, *52*, 5157.

[789] Dasgupta, S. K.; Gut, M.; *J. Org. Chem.* **1975**, *40*, 1475.

[790] Normant, H.; Crisan, C.; *Bull. Soc. Chim. Fr.* **1959**, 459.

[791] de Botton, M.; *Bull. Soc., Chim. Fr.* **1966**, 2212, 2466; **1969**, 3719.

[792] Schmeichel, M.; Redlich, H.; *Synthesis* **1996**, 1002.

[793] Bagli, J. F.; Bogri, T.; *J. Org. Chem.* **1972**, *37*, 2132.

[794] Alvarez, F. S.; Wren, D.; Prince, A.; Kluge, A. F.; Untch, K. G.; Fried, J. H.; *J. Am. Chem. Soc.* **1972**, *94*, 7823.

[795] Kluge, A. F.; Untch, K. G.; Fried, J. H.; *J. Am. Chem. Soc.* **1972**, *94*, 9256.

[796] Anderson, C. B.; Geis, M. P.; *Tetrahedron* **1975**, *31*, 1149.

[797] Eaton, P. E.; Mueller, R. H.; Carlson, G. R.; Cullison, D. A.; Cooper, G. F.; Chou, T.-C.; Krebs, E.-P.; *J. Am. Chem. Soc.* **1977**, *99*, 2755.

[798] Bal, S. A.; Marfat, A.; Helquist, P.; *J. Org. Chem.* **1982**, *47*, 5045.

[799] Gil, J. F.; Ramón, D. J.; Yus, M.; *Tetrahedron* **1993**, *49*, 4923.

[800] Ficini, J.; Dépezay, J.-C.; *Bull. Soc. Chim. Fr.* **1966**, 3878.

[801] Steinborn, D.; *J. Organomet. Chem.* **1979**, *182*, 313.

[802] Ficini, J.; Sarrade-Loucheur, G.; Normant, H.; *Bull. Soc. Chim. Fr.* **1962**, 1219.

[803] Ghosez, L.; *Angew. Chem.* **1972**, *84*, 901; *Angew. Chem. Int. Ed. Engl.* **1972**, *11*, 852.

[804] Wiaux-Zamar, C.; Dejonghe, J.-P.; Ghosez, L.; Normant, J. F.; Villieras, J.; *Angew. Chem.* **1976**, *88*, 417; *Angew. Chem. Int. Ed. Engl.* **1976**, *15*, 371.

[805] Madelmont, J. C.; Caubère, P.; *J. Organomet. Chem.* **1976**, *116*, 15.

[806] Nguyen, T.; Wakselman, C.; *J. Fluorine Chem.* **1975**, *6*, 311.

[807] von Werner, K.; Blank, H.; Gisser, A.; Manhart, E.; *J. Fluorine Chem.* **1980**, *16*, 193.

[808] Ziegler, K.; Colonius, H.; *Liebigs Ann. Chem.* **1930**, *479*, 136.

[809] Stiles, M.; Mayer, R. P.; *J. Am. Chem. Soc.* **1959**, *81*, 1501, footnote 38b; Wright, J. B.; Gutsell, E. S.; *J. Am. Chem. Soc.* **1959**, *81*, 5193, footnote 10.

[810] Borkowski, W. L.; *Am. Pat.* 3'293'313 (filed 1 May 1962, issued 20 Dec. 1966 to Foote Mineral Co.); *Chem. Abstr.* **1967**, *66*, 46 477b.

[811] Stapersma, J.; Klumpp, G. W.; *Tetrahedron* **1981**, *37*, 187.

[812] Stapersma, J.; Kuipers, P.; Klumpp, G. W.; *Recl. Trav. Chim. Pays-Bas* **1982**, *101*, 213.

[813] Marx, J. N.; *Tetrahedron Lett.* **1975**, *31*, 1251.

[814] Edwards, P. J.; Entwistle, D. A.; Genicot, C.; Kim, K. S.; Levy, S. V.; *Tetrahedron Lett.* **1994**, *40*, 7443.

[815] Yus, M.; *Chem. Soc. Rev.* **1996**, 155.

[816] Eisenbach, C. D.; Schnecko, H.; Kern, W.; *Eur. Polym. J.* **1975**, *11*, 699; *Chem. Abstr.* **1975**, *83*, 147766b.

[817] Dépezay, J. -C.; Le Merrer, Y.; *Tetrahedron Lett.* **1974**, 2751, 2755.

[818] Barluenga, J.; Flórez, J.; Yus, M.; *J. Chem. Soc., Perkin Trans. 1* **1983**, 3019; Barluenga, J.; Fernández, J. R.; Yus, M.; *J. Chem. Soc., Perkin Trans. 1* **1988**, 302; Barluenga, J.; Foubelo, F.; Fañanás, F. J.; *Tetrahedron* **1989**, *45*, 2183.

[819] Caine, D.; Frobese, A. S.; *Tetrahedron Lett.* **1978**, *19*, 883.

[820] Barluenga, J.; Montserrat, J. M.; Flórez, J.; *Tetrahedron Lett.* **1992**, *33*, 6183.

[821] Foubelo, F.; Yus, M.; *Tetrahedron Asymmetry* **1996**, *7*, 2911.

[822] Hoffmann, M.; Kessler, H.; *Tetrahedron Lett.* **1994**, *35*, 6067.

[823] Bockmühl, M.; Ehrhart, G.; *DRP* 622'875 (7 Dec. 1935) and 633'083 (2 July 1936) to Farbwerke Hoechst.

[824] Ruschig, H.; Fugmann, R.; Meixner, W.; *Angew. Chem.* **1958**, *70*, 71.

[825] Reimschuessel, H. K.; *J. Org. Chem.* **1960**, *25*, 2256.

[826] Ziegler K.; Nagel, K.; Patheiger, M.; *Z. Anorg. Allg. Chem.* **1955**, *282*, 345.

[827] Gilman, H.; Geel, J. A.; Brannen, C. G.; Bullock, M. W.; Dunn, G. E.; Miller, R.; *J. Am. Chem. Soc.* **1949**, *71*, 1499.

[828] West, R.; Rochow, E. G.; *J. Org. Chem.* **1953**, *18*, 1739.

[829] West, R.; Glaze, W. H.; *J. Org. Chem.* **1961**, *26*, 2096.

[830] Gilman, H.; Soddy, T. S.; *J. Org. Chem.* **1957**, *22*, 565; Waack, R.; Doran, M. A.; *J. Org. Chem.* **1967**, *32*, 3395.

[831] Maercker, A.; Theis, M.; in *Organometallic Syntheses, Vol. 3* (King, R. B.; Eisch, J. J.; eds.), Elsevier, Amsterdam, **1986**, 378.

[832] Schöllkopf, U.; Küppers, H.; Traencker, H. J.; Pitteroff, W.; *Liebigs Ann. Chem.* **1967**, *704*, 120.

[833] Morton, A. A.; Hechenbleikner, I.; *J. Am. Chem. Soc.* **1936**, *58*, 1697; Mixer, R. Y.; Young, W. G.; *J. Am. Chem. Soc.* **1956**, *78*, 3379.

[834] Wittig, G.; Pockels, U.; Dröge, H.; *Ber. Dtsch. Chem. Ges.* **1938**, *71*, 1903.

[835] Gilman, H.; Langham, W.; Jacoby, A. L.; *J. Am. Chem. Soc.* **1939**, *61*, 106.

[836] Aberhart, D. J.; Lin, L. J.; *J. Chem. Soc., Perkin Trans. 1* **1974**, 2320; Parry, R. J.; Mizusawa, A. E.; Chiu, I. C.; Naidu, M. V.; Ricciardone, M.; *J. Am. Chem. Soc.* **1985**, *107*, 2512, spec. 2519.

[837] Bailey, W. F.; Punzalan, E. R.; *J. Org. Chem.* **1990**, *55*, 5404.

[838] Wieringa, J. H.; Wynberg, H.; Strating, J.; *Synth. Commun.* **1971**, *1*, 7.

[839] Negishi, E.-i.; Swanson, D. R.; Rousset, C. J.; *J. Org. Chem.* **1990**, *55*, 5406.

[840] Bailey, W. F.; Nurmi, T. T.; Patricia, J. J.; Wang, W.; *J. Am. Chem. Soc.* **1987**, *109*, 2442.

[841] Bailey, W. F.; Ovaska, T. V.; Leipert, T. K.; *Tetrahedron Lett.* **1989**, *30*, 3901.

[842] Dondyl, B.; Doussot, P.; Portella, C.; *Synthesis* **1992**, 995.

[843] Gassman, P. G.; O'Reilly, N. J.; *J. Org. Chem.* **1987**, *52*, 2481.

[844] Pierce, O. R.; McBee, E. T.; Judd, G. F.; *J. Am. Chem. Soc.* **1954**, *76*, 474; Lagowski, J. J.; *Quart. Rev.* **1959**, *13*, 23, spec. 238.

[845] Johncock, P.; *J. Organomet. Chem.* **1969**, *19*, 257.

[846] Denson, D. D.; Smith, C. F.; Tamborski, C.; *J. Fluorine Chem.* **1973**, *3*, 247; Smith, C. F.; Soloski, E. J.; Tamborski, C.; *J. Fluorine Chem.* **1974**, *4*, 35; Nguyen, T.; *J. Fluorine Chem.* **1975**, *5*, 115.

[847] Dua, S. S.; Howells, R. D.; Gilman, H.; *J. Fluorine Chem.* **1974**, *4*, 409.

[848] Chambers, R. D.; Musgrave, W. K. R.; Savory, J.; *J. Chem. Soc.* **1962**, 1993.

[849] Johncock, P.; *J. Organomet. Chem.* **1966**, *6*, 433.

[850] Sadhu, K. M.; Matteson, D. S.; *Tetrahedron Lett.* **1986**, *27*, 795; Matteson, D. S.; *Tetrahedron* **1998**, *54*, 10555; *Chemtech* **1999**, *29*, 6; *J. Organomet. Chem.* **1999**, *581*, 51.

[851] Köbrich, G.; Fischer, R. H.; *Tetrahedron* **1968**, *24*, 4343.

[852] Cainelli, G.; Umani-Ronchi, A.; Bertini, F.; Grasselli, P.; Zubiani, G.; *Tetrahedron* **1971**, *27*, 6109; Tarhouni, R.; Kirschleger, B.; Rambaud, M.; Villieras, J.; *Tetrahedron Lett.* **1984**, *25*, 835.

[853] Villiéras, J.; *C.R. Seances Acad. Sci.* **1965**, *261*, 4137; *Chem. Abstr.* **1966**, *64*, 6673e.

[854] Villiéras, J.; *Bull. Soc. Chim. Fr.* **1967**, 1520; Bolm, C.; Pupowicz, D.; *Tetrahedron Lett.* **1997**, *38*, 7349.

[855] Concellón, J. M.; Llavona, L.; Bernard, P. L.; *Tetrahedron* **1995**, *51*, 5573.

[856] Tarhouni, R.; Kirschleger, B.; Villieras, J.; *J. Organomet. Chem.* **1984**, *272*, C1.

[857] Hoeg, D. F.; Lusk, D. I.; Crunbliss, A. L.; *J. Am. Chem. Soc.* **1965**, *87*, 4147.

[858] Köbrich, G.; *Angew. Chem.* **1967**, *79*, 15; *Angew. Chem. Int. Ed. Engl.* **1967**, *6*, 41.

[859] Köbrich, G.; *Angew. Chem.* **1972**, *84*, 557; *Angew. Chem. Int. Ed. Engl.* **1972**, *11*, 473.

[860] Seyferth, D.; Lambert, R. L.; *J. Organomet. Chem.* **1973**, *54*, 123.

[861] Seebach, D.; Hässig, R.; Gabriel, J.; *Helv. Chim. Acta* **1983**, *66*, 308.

[862] Fischer, R. H.; Köbrich, G.; *Chem. Ber.* **1968**, *101*, 3230.

[863] Rausis, T.; Schlosser, M.; unpublished results (**1999**).

[864] Schmidbaur, H.; Schier, A.; Schubert, U.; *Chem. Ber.* **1983**, *116*, 1938.

[865] Garst, J. F.; Ungváry, F.; Batlaw, R.; Lawrence, K. E.; *J. Am. Chem. Soc.* **1991**, *113*, 5392.

[866] Walborsky, H. M.; Zimmermann, C.; *J. Am. Chem. Soc.* **1992**, *114*, 4996.

[867] Applequist, D. E.; Peterson, A. H.; *J. Am. Chem. Soc.* **1961**, *83*, 862.

[868] Walborsky, H. M.; Impastato, F. J.; Young, A. E.; *J. Am. Chem. Soc.* **1964**, *86*, 3283.

[869] Walborsky, H. M.; Hamdouchi, C.; *J. Org. Chem.* **1993**, *58*, 1187.

[870] Schmidt, A.; Köbrich, G.; *Tetrahedron Lett.* **1974**, *15*, 2561.

[871] Lansbury, P. T.; Sidler, J. D.; *Tetrahedron Lett.* **1965**, *6*, 691.

[872] Lansbury, P. T.; Sidler, J. D.; *J. Chem. Soc., Chem. Commun.* **1965**, 373.

[873] Schöllkopf, U.; Wittig, G.; *Tetrahedron* **1958**, *3*, 91; Wittig, G.; Tochtermann, W.; *Liebigs Ann. Chem.* **1962**, *660*, 23.

[874] Corey, E. J.; Ulrich, P.; *Tetrahedron Lett.* **1975**, *16*, 3685.

[875] Dowd, P.; Kaufmann, C.; Abeles, R. H.; *J. Am. Chem. Soc.* **1984**, *106*, 2703.

[876] Gadwood, R. C.; Rubino, M. C.; Nagarajan, S. C.; Michel, S. T.; *J. Org. Chem.* **1985**, *50*, 3255.

[877] Moss, R. A.; Hui, H. K.; *Synth. Commun.* **1984**, *14*, 305.

[878] Ishihara, T.; Hayashi, K.; Ando, T.; Yamanaka, H.; *J. Org. Chem.* **1975**, *40*, 3264.

[879] Köbrich, G.; Goyert, W.; *Tetrahedron* **1968**, *24*, 4327.

[880] Seyferth, D.; Lambert, R. L.; Massol, M.; *J. Organomet. Chem.* **1975**, *88*, 255; see also: ref. [1360].

[881] Loozen, H. J. J.; Castenmiller, W. A.; Buter, E. J. M.; Buck, H. M.; *J. Org. Chem.* **1976**, *41*, 2965; Loozen, H. J.; Robben, W. M. M.; Buck, H. M.; *Recl. Trav. Chim. Pays-Bas* **1976**, *95*, 245.

[882] Oku, A.; Harada, T.; Homoto, Y.; Iwamoto, M.; *J. Chem. Soc., Chem. Commun.* **1988**, 1490.

[883] Braun, M.; Dammann, R.; Seebach, D.; *Chem. Ber.* **1975**, *108*, 2368.

[884] Kitatani, K.; Yamamoto, H.; Hiyama, T.; Nozaki, H.; *Bull. Chem. Soc. Jpn.* **1977**, *50*, 2158; Kitani, K.; Hiyama, T.; Nozaki, H.; *J. Am. Chem. Soc.* **1975**, *97*, 949; *Bull. Chem. Soc. Jpn.* **1977**, *50*, 3288.

[885] Banwell, M. G.; Lambert, J. N.; Gravatt, G. L.; *J. Chem. Soc., Perkin Trans. 1* **1993**, 2817.

[886] Weber, A.; Galli, R.; Sabbioni, G.; Stämpfli, U.; Walter, S.; Neuenschwander, M.; *Helv. Chim. Acta* **1989**, *72*, 41; Morizawa, Y.; Kanakura, A.; Yamamoto, H.; Hiyama, T.; Nozaki, H.; *Bull. Chem. Soc. Jpn.* **1984**, *57*, 1935.

[887] Taylor, K. G.; Hobbs, W. E.; Saquet, M.; *J. Org. Chem.* **1971**, *36*, 369.

[888] Baird, M. S.; Baxter, A. G. W.; *J. Chem. Soc., Perkin Trans. 1* **1979**, 2317.

[889] Neumann, H.; Seebach, D.; *Chem. Ber.* **1978**, *111*, 2785.

[890] Rottländer, M.; Boymond, L.; Cahiez, G.; Knochel, P.; *J. Org. Chem.* **1999**, *64*, 1080.

[891] Peterson, M. A.; Polt, R.; *Synth. Commun.* **1992**, *22*, 477.

[892] Yokoo, T.; Shinokubo, H.; Oshima, K.; Utimoto, K.; *Synlett* **1994**, 645.

[893] Cahiez, G.; Bernard, D.; Normant, J. F.; *Synthesis* **1976**, 245.

[894] Ashe, A. J.; Savla, P. M.; *J. Organomet. Chem.* **1993**, *461*, 1.

[895] Bertrand, M.; Leandri, G.; Meou, A.; *Tetrahedron Lett.* **1979**, *20*, 1841.

[896] Köbrich, G.; Stöber, I.; *Chem. Ber.* **1970**, *103*, 2744.

[897] Atwell, W. H.; Weyenberg, D. R.; Gilman, H.; *J. Org. Chem.* **1967**, *32*, 885.

[898] Eshelby, J. J.; Parsons, Crowley, P. J.; *J. Chem. Soc., Perkin Trans. 1* **1996**, 191.

[899] Brook, A. G.; Duff, J. M.; *Can. J. Chem.* **1973**, *51*, 2024.

[900] Baird, M. S.; Hussain, H. H.; Nethercott, W.; *J. Chem. Soc., Perkin Trans. 1* **1986**, 1845.

[901] Baird, M. S.; Grehan, B.; *J. Chem. Soc., Perkin Trans. 1* **1993**, 1547.

[902] Seebach, D.; Neumann, H.; *Chem. Ber.* **1974**, *107*, 847.

[903] Cope, A. C.; Burg, M.; Fenton, S. W.; *J. Am. Chem. Soc.* **1952**, *74*, 173.

[904] Gwynn, D. E.; Whitesides, G. M.; Roberts, J. D.; *J. Am. Chem. Soc.* **1965**, *87*, 2862.

[905] Askani, R.; Kirsten, R.; *Tetrahedron Lett.* **1979**, *20*, 1491.

[906] Barluenga, J.; Canteli, R.-M.; Flórez, J.; *J. Org. Chem.* **1994**, *59*, 1586.

[907] Barluenga, J.; Canteli, R.-M.; Flórez, J.; *J. Org. Chem.* **1994**, *59*, 602.

[908] Duhamel, L.; Poirier, J. M.; *Bull. Soc. Chim. Fr. II* **1982**, 297.

[909] Duhamel, L.; Poirier, J. M.; Tedga, N.; *J. Chem. Res.* **1983**, 222.

[910] Lau, K. S. Y.; Schlosser, M.; *J. Org. Chem.* **1978**, *43*, 1595.

[911] Schlosser, M.; Wei, H.-x.; *Tetrahedron* **1997**, *53*, 1735.

[912] Wei, H.-x.; Schlosser, M.; *Tetrahedron Lett.* **1996**, *37*, 2771.

[913] Kluge, A. F.; Untch, K. G.; Fried, J. H.; *J. Am. Chem. Soc.* **1972**, *94*, 7827, 9256.

[914] Meyers, A. I.; Spohn, R. F.; *J. Org. Chem.* **1985**, *50*, 4872.

[915] Ficini, J.; Dépezay, J.-C.; *J. Org. Chem.* **1989**, *54*, 2629.

[916] Ficini, J.; Dépezay, J.-C.; *Tetrahedron Lett.* **1969**, *10*, 4797.

[917] Smith, A. B.; Branca, S. J.; Pilla, N. N.; Guaciaro, M. A.; *J. Org. Chem.* **1982**, *47*, 1855.

[918] Smith, A. B.; Branca, S. J.; Guaciaro, M. A.; Wovkulich, P. M.; Korn, A.; *Org. Synth, Coll. Vol.* **1990**, *7*, 271.

[919] Janssen, C. G. M.; Simons, L. H. J. G.; Godefroi, E. F.; *Synthesis* **1982**, 389.

[920] Manning, M. J.; Raynolds, P. W.; Swenton, J. S.; *J. Am. Chem. Soc.* **1976**, *98*, 5008.

[921] Ramondenc, Y.; Plé, G.; *Tetrahedron* **1993**, *49*, 10855.

[922] Soullez, D.; Plé, G.; Duhamel, L.; *J. Chem. Soc., Perkin Trans. 1* **1997**, 1639.

[923] Duhamel, L.; Duhamel, P.; Gallic, Y.; *Tetrahedron Lett.* **1993**, *34*, 319.

[924] Duhamel, L.; Duhamel, P.; Lecouve, J. P.; *J. Chem. Res. (S)* **1986**, 34.

[925] Wang, Q.; Wei, H.-x.; Schlosser, M.; *Eur. J. Org. Chem.* **1999**, 3268.

[926] Schlosser, M.; Wei, H.-x.; unpublished results (**1997**).

[927] Ashimori, A.; Takaharu, M.; Overman, L. E.; Poon, D. J.; *J. Org. Chem.* **1993**, *58*, 6949.

[928] Corey, E. J.; Beames, D. J.; *J. Am. Chem. Soc.* **1972**, *94*, 7210; Corey, E. J.; Sachdev, H. S.; *J. Am. Chem. Soc.* **1973**, *95*, 8483.

[929] Skotnicki, J. S.; Schaub, R. E.; Bernardy, K. F.; Siuta, G. J.; Poletto, J. F.; Weiss, M. J.; *J. Med. Chem.* **1977**, *20*, 1551.

[930] Borer, B. C.; Taylor, R. J. K.; *Synlett* **1992**, 117.

[931] Kowalski, C. J.; Fields, K. W.; *J. Am. Chem. Soc.* **1982**, *104*, 1777.

[932] Kowalski, C. J.; Weber, A. E.; Fields, K. W.; *J. Org. Chem.* **1982**, *47*, 5088.

[933] Wei, H.-x.; *Doctoral Dissertation*, Université de Lausanne, **1998**, pp. 28–35, 92–96.

[934] Schlosser, M.; Hammer, E.; *Helv. Chim. Acta* **1974**, *57*, 276.

[935] Caine, D.; Frobese, A. S.; *Tetrahedron Lett.* **1978**, *19*, 5167; Caine, D.; Ukachukwu, V. C.; *Tetrahedron Lett.* **1983**, *24*, 3959.

[936] Sullivan, R.; Lacher, J. R.; Park, J. D.; *J. Org. Chem.* **1964**, *29*, 3664.

[937] Drakesmith, F. G.; Richardson, R. D.; Stewart, O. J.; Tarrant, P.; *J. Org. Chem.* **1968**, *33*, 286.

[938] Moreau, P.; Albadri, R.; Redwane, N.; Commeyras, A.; *J. Fluorine Chem.* **1980**, *15*, 103; Redwane, N.; Moreau, P.; Commeyras, A.; *J. Fluorine Chem.* **1982**, *20*, 699.

[939] Martin, S.; Sauvêtre, R.; Normant, J. F.; *Tetrahedron Lett.* **1982**, *23*, 4329.

[940] Normant, J. F.; *J. Organomet. Chem.* **1990**, *400*, 19.

[941] Denson, D. D.; Smith, C. F.; Tamborski, C.; *J. Fluorine Chem.* **1973**, *3*, 247.

[942] Normant, J. F.; Foulon, J. P.; Masure, D.; Sauvêtre, R.; Villiéras, J.; *Synthesis* **1975**, 122.

[943] Sauvêtre, R.; Masure, D.; Chuit, C.; Normant, J. F.; *C. R. Acad. Sci.* **1979**, *288*, 335; *Chem. Abstr.* **1979**, *91*, 91124m; Gillet, J. F.; Sauvêtre, R.; Normant, J. F.; *Synthesis* **1986**, 355.

[944] Tarrant, P.; Johncock, P.; Savory, J.; *J. Org. Chem.* **1963**, *28*, 839; see also: Knunyants, I. L.; Sterlin, R. N.; Yatsenko, R. D.; Pinkina, L. N.; *Izvest. Akad. Nauk S.S.S.R., Otdel. Khim. Nauk* **1958**, 1345; *Chem. Abstr.* **1959**, *53*, 6987g; Sterlin, R. N.; Knunyants, I. L.; Pinkina, L. N.; Yatsenko, R. D.; *Izvest. Akad. Nauk S.S.S.R., Otdel. Khim. Nauk* **1959**, 1492; *Chem. Abstr.* **1960**, *54*, 1270g; Kaez, H. D.; Stafford, S. L.; Stone, F. G. A.; *J. Am. Chem. Soc.* **1959**, *81*, 6336.

[945] Drakesmith, F. G.; Stewart, O. J.; Tarrant, P.; *J. Org. Chem.* **1968**, *33*, 280.

[946] Masure, D.; Sauvêtre, R.; Normant, J. F.; Villiéras, J.; *Synthesis* **1976**, 761.

[947] Masure, D.; Chuit, C.; Sauvêtre, R.; Normant, J. F.; *Synthesis* **1978**, 458.

[948] Köbrich, G.; Flory, K.; *Chem. Ber.* **1966**, *99*, 1773.

[949] Köbrich, G.; Wagner, E.; *Angew. Chem.* **1970**, *82*, 548; *Angew. Chem. Int. Ed. Engl.* **1970**, *9*, 524.

[950] Wittig, G.; Weinlich, J.; Wilson, R. W.; *Chem. Ber.* **1965**, *98*, 458.

[951] Gilman, H.; Langham, W.; Moore, F. W.; *J. Am. Chem. Soc.* **1940**, *62*, 2327; Gilman, H.; Jones, R. G.; *J. Am. Chem. Soc.* **1941**, *63*, 1443.

[952] Schlosser, M.; Ladenberger, V.; *J. Organomet. Chem.* **1967**, *8*, 193.

[953] Trepka, W. J.; Sonnenfeld, R. J.; *J. Organomet. Chem.* **1969**, *16*, 317.

[954] Wittig, G.; Braun, H.; Cristau, H.; *Liebigs Ann. Chem.* **1971**, *751*, 17.

[955] Schlosser, M.; Kadibelban, T.; unpublished results (**1964**).

[956] Mulvaney, J. E.; Carr, L. J.; *J. Org. Chem.* **1968**, *33*, 3286.

[957] Sita, L. R.; Bickerstaff, R.; *J. Am. Chem. Soc.* **1989**, *111*, 3769.

[958] Hellwinkel, D.; Aulmich, G.; Melan, M.; *Chem. Ber.* **1981**, *114*, 86.

[959] Bähr, G.; Gelius, R.; *Chem. Ber.* **1958**, *91*, 818; Sharp, P. R.; Astruc, D.; Schrock, R. R.; *J. Organomet. Chem.* **1979**, *183*, 477; Seebach, D.; Weller, T.; Protschuk, G.; Beck, A. K.; Hoekstra, M. S.; *Helv. Chim. Acta* **1981**, *64*, 716; Zigler, S. S.; Johnson, L. M.; West, R.; *J. Organomet. Chem.* **1988**, *341*, 187; Vedejs, E.; Lee, N.; *J. Am. Chem. Soc.* **1995**, *117*, 891.

[960] Betts, E. E.; Barclay, L. R. C.; *Can. J. Chem.* **1955**, *33*, 1768.

[961] Weiss, H.; Oehme, H.; *Z. Anorg. Allg. Chem.* **1989**, *572*, 186.

[962] Olmstead, M. M.; Power, P. P.; *J. Organomet. Chem.* **1991**, *408*, 1.

[963] Gilman, H.; Moore, F. W.; *J. Am. Chem. Soc.* **1940**, *62*, 1843.

[964] Gilman, H.; Cook, T. H.; *J. Am. Chem. Soc.* **1940**, *62*, 2813.

[965] Neugebauer, W.; Clark, T.; Schleyer, P. V. R.; *Chem. Ber.* **1983**, *116*, 3283.

[966] Eaborn, C.; Lasocki, Z.; Sperry, J. A.; *J. Organomet. Chem.* **1972**, *35*, 245.

[967] Takahashi, M.; Hatano, K.; Kimura, M.; Watanabe, T.; Oriyama, T.; Koga, G.; *Tetrahedron Lett.* **1994**, *35*, 579.

[968] Gilman, H.; Marrs, O. L.; *J. Org. Chem.* **1960**, *25*, 1194.

[969] Creary, X.; *J. Org. Chem.* **1987**, *52*, 5026; van Doorn, J. A.; Meijboom, N.; *Recl. Trav. Chim. Pays-Bas* **1992**, *111*, 170; Ishii, A.; Okazaki, R.; Inamoto, N.; *Bull. Chem. Soc. Jpn.* **1987**, *60*, 1037; Fraenkel, G.; Subramanian, S.; Chow, A.; *J. Am. Chem. Soc.* **1995**, *117*, 6300.

[970] Okazaki, R.; Unno, M.; Inamoto, N.; *Chem. Lett.* **1987**, 2293.

[971] Ginanneschi, A.; Schlosser, M.; unpublished results (**1999**).

[972] Gilman, H.; Banner, I.; *J. Am. Chem. Soc.* **1940**, *62*, 344.

[973] Giumanini, A. G.; Lercker, G.; *J. Org. Chem.* **1970**, *35*, 3756.

[974] Walton, D. R. M.; *J. Chem. Soc. [C]* **1966**, 1706.

[975] Cartoon, M. E. K.; Cheesman, G. W. H.; *J. Organomet. Chem.* **1981**, *212*, 1.

[976] Gilman, H.; Spatz, S. J.; *J. Am. Chem. Soc.* **1941**, *63*, 1553.

[977] Hellwinkel, D.; Wilfinger, H. J.; *Chem. Ber.* **1974**, *107*, 1428; see also: ref. [1020].

[978] Gilman, H.; Murray, A.; Stuckwisch, C. G.; Summers, L.; unpublished results; quoted according to Jones, R. G.; Gilman, H.; *Org. React.* **1951**, *6*, 339, refs. 38 and 59.

[979] Gilman, H.; Stuckwisch, C. G.; *J. Am. Chem. Soc.* **1941**, *63*, 2844; **1942**, *64*, 1007.

[980] Murray, A.; Foreman, W. W.; Langham, W.; *J. Am. Chem. Soc.* **1948**, *70*, 1037; Murray, A.; Langham, W.; *J. Am. Chem. Soc.* **1952**, *74*, 6289; see also: ref. [1009].

[981] Parham, W. E.; Egberg, D. C.; Sayed, Y. A.; Thraikill, R. W.; Kayser, G. E.; Neu, M.; Montgomery, W. C.; Jones, L. D.; *J. Org. Chem.* **1976**, *41*, 2628.

[982] Gilman, H.; Melstrom, D. S.; *J. Am. Chem. Soc.* **1948**, *70*, 4177.

[983] Köbrich, G.; Buck, P.; *Chem. Ber.* **1970**, *103*, 1412; Buck, P.; Köbrich, G.; *Chem. Ber.* **1970**, *103*, 1420.

[984] Kemmitt, T.; Levason, W.; *Organometallics* **1989**, *8*, 1303.

[985] Tunney, S. E.; Stille, J. K.; *J. Org. Chem.* **1987**, *52*, 748.

[986] Gilman, H.; Brown, G. E.; *J. Am. Chem. Soc.* **1945**, *67*, 824.

[987] Wittig, G.; Fuhrmann, G.; *Ber. Dtsch. Chem. Ges.* **1940**, *73*, 1197; Glaze, W. H.; Ranade, A. C.; *J. Org. Chem.* **1971**, *36*, 3331; Liebeskind, L. S.; Zhang, J.; *J. Org. Chem.* **1991**, *56*, 6379.

[988] Harder, S.; Boersma, J.; Brandsma, L.; van Mier, G. P. M.; Kanters, J. A.; *J. Organomet. Chem.* **1989**, *364*, 1.

[989] Ladd, J. A.; Parker, J.; *J. Chem. Soc., Dalton Trans.* **1972**, 930; Dodge, J. A.; Chamberlin, A. R.; *Tetrahedron Lett.* **1988**, *29*, 4827; see also: ref. [1028].

[990] Langham, W.; Brewster, R. Q.; Gilman, H.; *J. Am. Chem. Soc.* **1941**, *63*, 545.

[991] Klumpp, G. W.; Sinnige, M. J.; *Tetrahedron Lett.* **1986**, *27*, 2247.

[992] Gilman, H.; Willis, H. B.; Swislowsky, J.; *J. Am. Chem. Soc.* **1939**, *61*, 1371; Gilman, H.; Swislowsky, J.; Brown, G. E.; *J. Am. Chem. Soc.* **1940**, *62*, 348.

[993] Fuson, R. C.; Freedman, B.; *J. Org. Chem.* **1958**, *23*, 1161; Letsinger, R. L.; Nazy, J. R.; Hussey, A. S.; *J. Org. Chem.* **1958**, *23*, 1806.

[994] Ayer, W. A.; McCaskill, R. H.; *Can. J. Chem.* **1981**, *59*, 2161.

[995] Lovely, C. J.; Bhat, A. S.; Coughenour, H. D.; Gilbert, N. E.; Brueggemeier, R. W.; *J. Med. Chem.* **1997**, *40*, 3756.

[996] Foster, A. B.; Jarman, M.; Leung, O.-T.; McCague, R.; Leclerq, G.; Devleeschouwer, N.; *J. Med. Chem.* **1985**, *28*, 1491.

[997] Phadke, A. S.; Butler, C.; Robinson, B. C.; Morgan, A. R.; *Tetrahedron Lett.* **1993**, *34*, 6359.

[998] Meyers, A. I.; Roth, G. P.; Hoyer, D.; Barner, B. A.; Laucher, D.; *J. Am. Chem. Soc.* **1988**, *110*, 4611.

[999] Law, S.-J.; Morgan, J. M.; Masten, L. W.; Borne, R. F.; Arana, G. W.; Kula, N. S.; Baldessarini, R. J.; *J. Med. Chem.* **1982**, *25*, 213.

[1000] Lamza, L.; *J. Prakt. Chem.* **1964**, *25*, 294; Lee, K.; Turnbull, P.; Moore, H. W.; *J. Org. Chem.* **1995**, *60*, 461.

[1001] Getahun, Z.; Jurd, L.; Chu, P. S.; Lin, C. M.; Hamel, E.; *J. Med. Chem.* **1992**, *35*, 1058.

[1002] Hartman, G. D.; Halczenko, W.; Phillips, B. T.; *J. Org. Chem.* **1985**, *50*, 2423, 2427.

[1003] Gilman, H.; Arntzen, C. E.; Webb, F. J.; *J. Org. Chem.* **1945**, *10*, 374; Gilman, H.; Arntzen, C. E.; *J. Am. Chem. Soc.* **1947**, *69*, 1537; Davidsohn, W.; Laliberté, B. R.; Goddard, C. M.; Henry, M. C.; *J. Organomet. Chem.* **1972**, *36*, 283; Talley, J. J.; *Synthesis* **1983**, 845; Talley, J. J.; Evans, I. A.; *J. Org. Chem.* **1984**, *49*, 5267; Heinicke, J.; Kadyrov, R.; Kindermann, M. K.; Koesting, M.; Jones, P. G.; *Chem. Ber.* **1996**, *129*, 1547.

[1004] Neville, R. G.; *J. Org. Chem.* **1961**, *26*, 3031.

[1005] Selnick, H. G.; Bourgeois, M. L.; Butcher, J. W.; Radzilowski, E. M.; *Tetrahedron Lett.* **1993**, *34*, 2043.

[1006] Harrer, W.; Kurreck, H.; Reusch, J.; Gierke, W.; *Tetrahedron* **1975**, *31*, 625, spec. 627; Kurato, H.; Tanaka, T.; Sauchi, T.; Kawase, T.; Oda, M.; *Chem. Lett.* **1997**, 947.

[1007] Bailey, F. B.; Taylor, R.; *J. Chem. Soc. B* **1971**, 1446; Meek, D. W.; Dyer, G.; Workman, M. O.; *Inorg. Synth.* **1976**, *16*, 168.

[1008] Toyota, S.; Matsuda, Y.; Nagaoka, S.; Ōki, M.; Akashi, H.; *Bull. Chem. Soc. Jpn.* **1996**, *69*, 3115.

[1009] Gilman, H.; Gainer, G. G.; *J. Am. Chem. Soc.* **1947**, *69*, 1946.

[1010] Gilman, H.; Gorsich, R. D.; *J. Am. Chem. Soc.* **1955**, *77*, 3919.

[1011] Gilman, H.; Gorsich, R. D.; Gai, B. J.; *J. Org. Chem.* **1962**, *27*, 1023.

[1012] Nad, M. M.; Talalaeva, T. V.; Kazennikova, G. V.; Kocheshkov, K. A.; *Izvest. Akad. Nauk S.S.S.R., Otdel. Khim. Nauk* **1959**, 65; *Chem. Abstr.* **1959**, *53*, 14976f; Bridges, A. J.; Patt, W. C.; Stickney, T. M.; *J. Org. Chem.* **1990**, *55*, 773; Coe, P. L.; Waring, A. J.; Yarwood, T. D.; *J. Chem. Soc., Perkin Trans. 1* **1995**, 2729.

[1013] Talalaeva, T. V.; Kazennikova, G. V.; Kocheshkov, K. A.; *Zh. Obshch. Khim.* **1959**, *29*, 1593; *Chem. Abstr.* **1960**, *54*, 8677i; Ong, H. H.; Profitt, J. A.; Anderson, V. B.; Kruse, H.; Wilker, J. C.; Geyer, H. M.; *J. Med. Chem.* **1981**, *24*, 74.

[1014] Tamborski, C.; Moore, G. J.; *J. Organomet. Chem.* **1971**, *26*, 153.

[1015] Coe, P. L.; Stephens, R.; Tatlow, J. C.; *J. Chem. Soc.* **1962**, 3227; see also: Fenton, D. E.; *J. Organomet. Chem.* **1964**, *2*, 437.

[1016] Gilman, H.; Woods, L. A.; *J. Am. Chem. Soc.* **1944**, *66*, 1981.

[1017] Yu, Y.; Sun, J.; Chen, J.; *J. Organomet. Chem.* **1997**, *533*, 13.

[1018] Aeberli, P.; Houlihan, W. J.; *J. Organomet. Chem.* **1974**, *67*, 321.

[1019] Porwisiak, J.; Schlosser, M.; *Chem. Ber.* **1996**, *129*, 233.

[1020] Hellwinkel, D.; Lindner, W.; Wilfinger, H. J.; *Chem. Ber.* **1974**, *107*, 1428.

[1021] Gilman, H.; Gaj, B. J.; *J. Org. Chem.* **1957**, *22*, 447; Franzen, V.; Joschek, H.-I.; *Liebigs Ann. Chem.* **1967**, *703*, 90.

[1022] Rausch, M. D.; Tibbetts, F. E.; Gordon, H. B.; *J. Organomet. Chem.* **1966**, *5*, 493.

[1023] Moser, G. A.; Fischer, E. O.; Rausch, M. D.; *J. Organomet. Chem.* **1971**, *27*, 379.

[1024] Mickailov, B. M.; Bronovitskaya, V. P.; *Zh. Obshch. Khim.* **1952**, *22*, 157, 159; *Engl. Transl.* 195, 198; *Chem. Abstr.* **1952**, *46*, 11169c.

[1025] Eaborn, C.; Pande, K. C.; *J. Chem. Soc.* **1960**, 3200.

[1026] Nefedov, O. M.; Dyachenko, A. I.; *Dokl. Akad. Nauk. S.S.S.R.* **1971**, *198*, 593; *Chem. Abstr.* **1971**, *75*, 88225j; Gilman, H.; Chen, L. S.; Chen, G. J.; Tamborski, C.; *J. Organomet. Chem.* **1980**, *193*, 283.

[1027] Muke, B.; Kauffmann, T.; *Chem. Ber.* **1980**, *113*, 2739.

[1028] Roger, H. R.; Houk, J.; *J. Am. Chem. Soc.* **1982**, *104*, 522.

[1029] Chen, L. S.; Chen, G. J.; Tamborski, C.; *J. Organomet. Chem.* **1983**, *251*, 139; Chen, G. J.; Tamborski, C.; *J. Organomet. Chem.* **1983**, *251*, 149.

[1030] Nishiyama, H.; Isaka, K.; Itoh, K.; Ohno, K.; Nagase, H.; Matsumoto, K.; Yoshiwara, H.; *J. Org. Chem.* **1992**, *57*, 407.

[1031] Murphy, R. A.; Kung, H. F.; Kung, M.-p.; Billings, J.; *J. Med. Chem.* **1990**, *33*, 171.

[1032] Chen, L. S.; Chen, G. J.; Tamborski, C.; *J. Organomet. Chem.* **1981**, *215*, 281.

[1033] Smith, C. F.; Moore, G. J.; Tamborski, C.; *J. Organomet. Chem.* **1971**, *33*, C21.

[1034] Berry, D. J.; Wakefield, B. J.; *J. Organomet. Chem.* **1970**, *23*, 1.

[1035] Hedberg, F. L.; Rosenberg, H.; *Tetrahedron Lett.* **1969**, *10*, 4011.

[1036] Dong, T.-Y.; Lai, L.-l.; *J. Organomet. Chem.* **1996**, *509*, 131.

[1037] Chen, W.; Stephenson, E. K.; Cava, M. P.; Jackson, Y. A.; *Org. Synth.* **1992**, *70*, 151.

[1038] Muchowski, J. M.; Naef, R.; *Helv. Chim. Acta* **1984**, *67*, 1168.

[1039] Kozikowski, A. P.; Cheng, X.-m.; *J. Org. Chem.* **1984**, *49*, 3239.

[1040] Amat, M.; Halida, S.; Sathyanarayana, S.; Bosch, J.; *J. Org. Chem.* **1994**, *59*, 10.

[1041] Amat, M.; Halida, S.; Bosch, J.; *Tetrahedron Lett.* **1994**, *35*, 793.

[1042] Amat, M.; Halida, S.; Sathyanarayana, S.; Bosch, J.; *J. Org. Chem.* **1997**, *74*, 248.

[1043] Pavlik, J. W.; Kurzweil, E. M.; *J. Heterocycl. Chem.* **1992**, *29*, 1357.

[1044] Hüttel, R.; Schön, M. E.; *Liebigs Ann. Chem.* **1959**, *625*, 55.

[1045] Butler, D. E.; DeWald, H. A.; *J. Org. Chem.* **1971**, *36*, 2542.

[1046] Kirk, K. L.; *J. Heterocycl. Chem.* **1985**, *22*, 57.

[1047] Apen, P. G.; Rasmussen, P. G.; *J. Heterocycl. Chem.* **1992**, *29*, 1091.

[1048] Iddon, B.; Petersen, A. K.; Becher, J.; Christensen, N. J.; *J. Chem. Soc., Perkin Trans. 1* **1995**, 1475.

[1049] Iddon, B.; Khan, N.; *J. Chem. Soc., Perkin Trans. 1* **1982**, 735.

[1050] Iddon, B.; Khan, N.; *J. Chem. Soc., Perkin Trans. 1* **1987**, 1445.

[1051] Fukuyama, Y.; Miwa, T.; Tokoroyama, T.; *Synthesis* **1974**, 443; see also: Gronowitz, S.; Sörlin, G.; *Acta Chem. Scand.* **1961**, *15*, 1419.

[1052] Florentin, D.; Roques, B. P.; Fournie-Zaluski, M. C.; *Bull. Soc. Chim. Fr.* **1976**, 1999; see also: Zaluski, M. C.; Robba, M.; Bonhomme, M.; *Bull. Soc. Chim. Fr.* **1970**, 1838; Sornay, R.; Meunier, J. M.; Fournari, P.; *Bull. Soc. Chim. Fr.* **1971**, 990.

[1053] Bury, P.; Hareau, G.; Kocieński, P.; Dhanak, D.; *Tetrahedron* **1994**, *50*, 8793.

[1054] Nazarova, Z. N.; Tertov, B. A.; Gabaraeva, Y. A.; *Zhur. Org. Khim.* **1965**, *5*, 190; **1970**, *6*, 1346; *Chem. Abstr.* **1968**, *68*, 13021m; **1970**, *73*, 66646d.

[1055] Martin, G. J.; Mechin, B.; Leroux, Y.; Paulmier, C.; Meunier, J. C.; *J. Organomet. Chem.* **1974**, *67*, 327.

[1056] Campaigne, E.; Foye, W. O.; *J. Am. Chem. Soc.* **1948**, *70*, 3941.

[1057] Lawesson, S.-O.; *Arkiv Kemi* **1957**, *11*, 387; *Chem. Abstr.* **1962**, *56*, 10173c.

[1058] Moses, P.; Gronowitz, S.; *Arkiv Kemi* **1961**, *18*, 119.

[1059] Campaigne, E.; Rogers, R. B.; *J. Heterocycl. Chem.* **1973**, *10*, 963.

[1060] Haiduc, I.; Gilman, H.; *Rev. Roum. Chem.* **1971**, *16*, 305; *Chem. Abstr.* **1971**, *75*, 36230n.

[1061] Rausch, M. D.; Criswell, T. R.; Ignatowicz, A. K.; *J. Organomet. Chem.* **1968**, *13*, 419.

[1062] Yurev, Y. K.; Sadovaya, N. K.; Grekova, E. A.; *Zhur. Obshch. Khim.* **1964**, *34*, 847; *Chem. Abstr.* **1964**, *60*, 15817d.

[1063] Gronowitz, S.; Frejd, T.; *Acta Chim. Scand.* **1970**, *24*, 2656.

[1064] Furukawa, N.; Shibutani, T.; Fujihara, H.; *Tetrahedron Lett.* **1987**, *28*, 5845.

[1065] Trécourt, F.; Breton, G.; Bonnet, V.; Mongin, F.; Marsais, F.; Quéguiner, G.; *Tetrahedron Lett.* **1999**, *40*, 4339.

[1066] Gilman, H.; Spatz, S. M.; *J. Org. Chem.* **1951**, *16*, 1485; see also: ref. [976].

[1067] Wibaut, J. P.; de Jonge, A. P., van der Voort, H. G. P.; Otto, P. P. H. L.; *Recl. Trav. Chim. Pays-Bas* **1951**, *70*, 1054.

[1068] Malmberg, H.; Nilsson, M.; *Tetrahedron* **1986**, *42*, 3981.

[1069] French, H. E.; Sears, K.; *J. Am. Chem. Soc.* **1951**, *73*, 469.

[1070] Wibaut, J. P.; Heeringa, L. G.; *Recl. Trav. Chim. Pays-Bas* **1955**, *74*, 1003.

[1071] Grewe, R.; Mondon, A.; Nolte, E.; *Liebigs Ann. Chem.* **1949**, *564*, 161.

[1072] Brown, E. V.; Shambhu, M. B.; *Org. Prep. Proc.* **1970**, *2*, 285; *Chem. Abstr.* **1971**, *75*, 20130u.

[1073] Parks, J. E.; Wagner, B. E.; Holm, R. H.; *J. Organomet. Chem.* **1973**, *56*, 53; Utimoto, K.; Sakai, N.; Nozaki, H.; *J. Am. Chem. Soc.* **1974**, *96*, 5601; *Tetrahedron Lett.* **1976**, *32*, 769.

[1074] Cai, D.-w.; Hughes, D. L.; Verhoeven, T. R.; *Tetrahedron Lett.* **1996**, *37*, 2537.

[1075] Peterson, M. A.; Mitchell, J. R.; *J. Org. Chem.* **1997**, *62*, 8237.

[1076] Mallet, M.; Quéguiner, G.; *Tetrahedron* **1979**, *35*, 1625.

[1077] Parham, W. E.; Piccirilli, R. M.; *J. Org. Chem.* **1977**, *42*, 257.

[1078] Berry, D. J.; Wakefield, B. J., Cook, J. D.; *J. Chem. Soc. [C]* **1971**, 1227.

[1079] Banks, R. E.; Haszeldine, R. N.; Phillips, E.; Young, I. N.; *J. Chem. Soc. [C]* **1967**, 2091.

[1080] Cook, J. D.; Wakefield, B. J.; Clayton, C. J.; *J. Chem. Soc., Chem. Commun.* **1967**, 150; Dua, S. S.; Gilman, H.; *J. Organomet. Chem.* **1968**, *12*, 299; Cook, J. D.; Wakefield, B. J.; *J. Organomet. Chem.* **1968**, *13*, 15.

[1081] Berry, D. J.; Wakefield, B. J.; *J. Chem. Soc. [C]* **1969**, 2342.

[1082] Hertz, H. S.; Kabacinski, F. F.; Spoerri, P. E.; *J. Heterocycl. Chem.* **1969**, *6*, 239.

[1083] Hirschberg, A.; Peterkoski, A.; Spoerri, P. E.; *J. Heterocycl. Chem.* **1965**, *2*, 209.

[1084] Rho, T.; Abuh, Y. F.; *Synth. Commun.* **1994**, *24*, 253.

[1085] Ulbricht, T. L. V.; *Tetrahedron* **1959**, *6*, 225.

[1086] Goudgeon, N. M.; Naguib, F. N. M.; El Kouni, M. H.; Schinazi, R. F.; *J. Med. Chem.* **1993**, *36*, 4250.

[1087] Arukwe, J.; Undheim, K.; *Acta Chem. Scand. B* **1986**, *40*, 588; Undheim, K.; Benneche, T.; *Acta Chem. Scand.* **1993**, *47*, 102.

[1088] Alauddin, M. M.; Conti, P. S.; *Tetrahedron* **1994**, *50*, 1699.

[1089] Pierce, J. B.; Walborsky, H. M.; *J. Org. Chem.* **1968**, *33*, 1962.

[1090] Wittig, G.; Wahl, V.; unpublished results (**1961**); Wahl, V.; *Doctoral Dissertation*, University of Heidelberg, **1962**, pp. 15–17 and 55–57.

[1091] Foulger, N. J.; Wakefield, B. J.; *J. Organomet. Chem.* **1974**, *69*, 321.

[1092] McBee, E. T.; Sanford, R. A.; *J. Am. Chem. Soc.* **1950**, *72*, 5574.

[1093] Winkler, H. J. S.; Winkler, H.; *J. Am. Chem. Soc.* **1964**, *88*, 964, 969.

[1094] Narasimhan, N. S.; Ammanamanchi, R.; *J. Chem. Soc., Chem. Commun.* **1985**, 1368.

[1095] Boatman, R. J.; Whitlock, B. J.; Whitlock, H. W.; *J. Am. Chem. Soc.* **1977**, *99*, 4822.

[1096] Gallagher, D. J.; Beak, P.; *J. Am. Chem. Soc.* **1991**, *113*, 7984; see also: Beak, P.; Chen, C.-W.; *Tetrahedron Lett.* **1985**, *26*, 4979; Beak, P.; Musick, T. J.; Chen, C.-W.; *J. Am. Chem. Soc.* **1988**, *110*, 3538.

[1097] Pierce, O. R.; Meiners, A. F.; McBee, E. T.; *J. Am. Chem. Soc.* **1953**, *75*, 2516.

[1098] Dua, S. S.; Howells, R. D.; Gilman, H.; *J. Fluorine Chem.* **1974**, *4*, 409.

[1099] Moreau, P.; Dalverny, G.; Commeyras, A.; *J. Fluorine Chem.* **1975**, *5*, 265.

[1100] Avolio, S.; Malan, C.; Marek, I.; Knochel, P.; *Synlett* **1999**, 1820.

[1101] Boymond, L.; Rottländer, M.; Cahiez, G.; Knochel, P.; *Angew. Chem.* **1998**, *110*, 1801; *Angew. Chem. Int. Ed. Engl.* **1998**, *37*, 1701.

[1102] Gassman, P. G.; Fentiman, A. F.; *Tetrahedron Lett.* **1970**, *11*, 1021.

[1103] Parham, W. E.; Bradsher, C.; *Acc. Chem. Res.* **1982**, *15*, 300.

[1104] Parham, W. E.; Jones, L. D.; *J. Org. Chem.* **1976**, *41*, 2704.

[1105] Parham, W. E.; Jones, L. D.; Sayed, Y. A.; *J. Org. Chem.* **1976**, *41*, 1184; Parham, W. E.; Jones, L. D.; *J. Org. Chem.* **1976**, *41*, 1187; Parham, W. E.; Egberg, D. C.; Sayed, Y. A.; Thraikill, R. W.; Keyser, G. E.; Neu, M.; Montgomery, W. C.; Jones, L. D.; *J. Org. Chem.* **1976**, *41*, 2628.

[1106] Hergruter, C. A.; Brewer, P. D.; Tagat, J.; Helquist, P.; *Tetrahedron Lett.* **1977**, *18*, 4146.

[1107] Hardcastle, I. R.; Quayle, P.; Ward, E. L. M.; *Tetrahedron Lett.* **1994**, *35*, 1747; Hardcastle, I. R.; Quayle, P.; *Tetrahedron Lett.* **1994**, *35*, 1749.

[1108] Koźlecki, T.; Syper, L.; Wilk, K. A.; *Synthesis* **1997**, 681.

[1109] Stanetty, P.; Krumpak, B.; Rodler, I. K.; *J. Chem. Res. (S)* **1995**, 342.

[1110] Fuson, R. C.; Hammann, W. C.; Smith, W. E.; *J. Org. Chem.* **1954**, *19*, 674.

[1111] Wender, P. A.; Glass, T. E.; *Synlett* **1995**, 516.

[1112] Wender, P. A.; Wessjohann, L. A.; Peschke, B.; Rawlins, D. B.; *Tetrahedron Lett.* **1995**, *36*, 7181.

[1113] Comins, D. L.; Brown, J. D.; *Tetrahedron Lett.* **1981**, *22*, 4213; *J. Org. Chem.* **1984**, *49*, 1078; **1989**, *54*, 3730; Comins, D. L.; Killpack, M. O.; *J. Org. Chem.* **1987**, *52*, 104.

[1114] Comins, D. L.; *Synlett* **1992**, 615.

[1115] Saward, C. J.; Vollhardt, K. P. C.; *Tetrahedron Lett.* **1975**, *16*, 4539.

[1116] Wiberg, K. B.; Pratt, W. E.; Bailey, W. F.; *J. Am. Chem. Soc.* **1977**, *99*, 2297.

[1117] Bailey, W. F.; Gagnier, R. P.; *Tetrahedron Lett.* **1982**, *23*, 5123.

[1118] Sauers, R. R.; Schlosberg, S. B.; Pfeffer, P. E.; *J. Org. Chem.* **1968**, *33*, 2175.

[1119] Wong, T.; Cheung, S. S.; Wong, H. N. C.; *Angew. Chem.* **1988**, *100*, 716; *Angew. Chem. Int. Ed. Engl.* **1988**, *27*, 705.

[1120] Nishiyama, H.; Isaka, K.; Itoh, K.; Ohno, K.; Nagase, H.; Matsumoto, K.; Yoshiwara, H.; *J. Org. Chem.* **1992**, *57*, 407.

[1121] Boatman, R. J.; Whitlock, B. J.; Whitlock, H. W.; *J. Am. Chem. Soc.* **1977**, *99*, 4822.

[1122] Toth, J. E.; Fuchs, P. L.; *J. Org. Chem.* **1987**, *52*, 473.

[1123] Gilman, H.; Gaj, B. J.; *J. Org. Chem.* **1957**, 22, 447.

[1124] Snyder, H. R.; Weaver, C.; Marshall, C. D.; *J. Am. Chem. Soc.* **1949**, *71*, 289.

[1125] Hellwinkel, D.; Schenk, W.; *Angew. Chem.* **1969**, *81*, 1049; *Angew. Chem. Int. Ed. Engl.* **1969**, *8*, 987.

[1126] Tanaka, S.; Iso, T.; Doke, Y.; *J. Chem. Soc., Chem. Commun.* **1997**, 2063.

[1127] Märkl, G.; Mayr, A.; *Tetrahedron Lett.* **1974**, *15*, 1817.

[1128] Rot, N.; Bickelhaupt, F.; *Organometallics* **1997**, *16*, 5027.

[1129] Ferede, R.; Noble, M.; Cordes, A. W.; Allison, N. T.; Lay, J.; *J. Organomet. Chem.* **1988**, *339*, 1.

[1130] Duerr, B. F.; Chung, Y.-S.; Czarnik, A. W.; *J. Org. Chem.* **1988**, *53*, 2120.

[1131] Rahman, M. T.; *Monatsh. Chem.* **1983**, *114*, 249.

[1132] Barluenga, J.; Rodriguez, M. A.; Campos, P. J.; Asensio G.; *J. Am. Chem. Soc.* **1988**, *110*, 5567.

[1133] Maercker, A.; Bös, B.; *Main Group Metal Chem.* **1991**, *14*, 67; *Chem. Abstr.* **1993**, *119*, 226014v.

[1134] Murphy, R. A.; Kung, H. F.; Kung, M.-P.; Billings, J.; *J. Med. Chem.* **1990**, *33*, 171.

[1135] Voß, G.; Gerlach, H.; *Chem. Ber.* **1989**, *122*, 1199.

[1136] Iddon, B.; *Heterocycles* **1985**, *23*, 417.

[1137] Iddon, B.; Khan, N.; *Tetrahedron Lett.* **1986**, *27*, 1635; *J. Chem. Soc., Perkin Trans. 1* **1987**, 1445, 1453; Iddon, B.; Petersen, A. K.; Becher, J.; Christensen, N. J.; *J. Chem. Soc., Perkin Trans. 1* **1995**, 1475.

[1138] Groziak, P.; Wei, L.; *J. Org. Chem.* **1991**, *56*, 4296.

[1139] Lipshutz, B. H.; Hagen, W.; *Tetrahedron Lett.* **1992**, *33*, 5865.

[1140] Groziak, P.; Wei, L.; *J. Org. Chem.* **1992**, *5*, 3776.

[1141] Mahler, H.; Braun, M.; *Tetrahedron Lett.* **1987**, *28*, 5145; *Chem. Ber.* **1991**, *124*, 1379; Braun, M.; Opdenbusch, K.; *Liebigs Ann. Chem.* **1997**, 141.

[1142] Granjean, D.; Pale, P.; *Tetrahedron Lett.* **1993**, *34*, 1155.

[1143] Mühlebach, M.; Neuenschwander, M.; Engel, P.; *Helv. Chim. Acta* **1993**, *76*, 2089.

[1144] Taylor, K. G.; Chaney, J.; *J. Am. Chem. Soc.* **1972**, *94*, 8924.

[1145] Seyferth, D.; Lambert, R. L.; *J. Organomet. Chem.* **1975**, *88*, 287.

[1146] Garamszegi, L.; Schlosser, M.; *Chem. Ber.* **1997**, *130*, 77.

[1147] Wei, H.-x.; Schlosser, M.; *Chem. Eur. J.* **1998**, *4*, 1738.

[1148] Wang, X.-h.; Zhong, G.-f.; Garamszegi, L.; Schlosser, M.; unpublished results (**1993**).

[1149] Lau, K. S. Y.; Schlosser, M.; *J. Org. Chem.* **1978**, *43*, 1595.

[1150] Hoye, T. R.; Martin, S. J.; Peck, D. R.; *J. Org. Chem.* **1982**, *47*, 331; see also: Gilman, H.; Zoellner, E. A.; Selby, W. M.; *J. Am. Chem. Soc.* **1932**, *54*, 1957.

[1151] Dougherty, T. K.; Lau, K. S. Y.; Hedberg, F. L.; *J. Org. Chem.* **1983**, *48*, 5773.

[1152] Desponds, O.; Schlosser, M.; *J. Organomet. Chem.* **1996**, *507*, 257.

[1153] Boykin, D. W.; Patel, A. R.; Lutz, R. E.; Burger, A.; *J. Heterocycl. Chem.* **1967**, *4*, 459.

[1154] Miller, T. M.; in *Handbook of Chemistry and Physics*, 76th edition (D. R. Lide; ed.), CRC Press, Boca Raton, **1995**, pp. 10/180–10/191.

[1155] Azzena, U.; Dessanti, F.; Melloni, G.; Pisano, L.; *Tetrahedron Lett.* **1999**, *40*, 8291.

[1156] Almena, J.; Foubelo, F.; Yus, M.; *Tetrahedron Lett.* **1993**, *34*, 1649.

[1157] Almena, J.; Foubelo, F.; Yus, M.; *Tetrahedron* **1994**, *50*, 5775.

[1158] Almena, J.; Foubelo, F.; Yus, M.; *Tetrahedron* **1996**, *52*, 8545.

[1159] Guijarro, D.; Yus, M.; *Tetrahedron Lett.* **1994**, *50*, 3447.

[1160] Wittenberg, D.; Gilman, H.; *J. Org. Chem.* **1958**, *23*, 1063; Aguiar, A. M.; Giacin, J.; Mills, A.; *J. Org. Chem.* **1962**, *27*, 674; Aguiar, A. M.; Beisler, J.; Miller, A.; *J. Org. Chem.* **1962**, *27*, 1001.

[1161] Tamborski, C.; Ford, E.; Lehn, W. L.; Moore, G. J.; Soloski, E. J.; *J. Org. Chem.* **1962**, *27*, 619.

[1162] Maercker, A.; *Angew. Chem.* **1987**, *99*, 1002; *Angew. Chem. Int. Ed. Engl.* **1987**, *26*, 972.

[1163] Cohen, T.; Bhupathy, M.; *Acc. Chem. Res.* **1989**, *22*, 152.

[1164] Bartmann, E.; *Angew. Chem.* **1986**, *98*, 629; *Angew. Chem. Int. Ed. Engl.* **1986**, *25*, 653.

[1165] Cohen, T.; Jeong, I.-h.; Mudryk, B.; Bhupathy, M.; Awad, M. M. A.; *J. Org. Chem.* **1990**, *55*, 1528.

[1166] Bachki, A.; Foubelo, F.; Yus, M.; *Tetrahedron: Asymmetry* **1995**, *6*, 1907; **1996**, *7*, 2997.

[1167] Rothberg, I.; Schneider, L.; Kirsch, S.; O'Fee, R.; *J. Org. Chem.* **1982**, *47*, 2675.

[1168] Müller, E.; Bunge, W.; *Ber. Dtsch. Chem. Ges.* **1936**, *69*, 2164; Morton, A. A.; Lanpher, E. J.; *J. Org. Chem.* **1958**, *23*, 1636, spec. 1638; see also: Schorigin, P.; *Ber. Dtsch. Chem. Ges.* **1923**, *56*, 176; Lüttringhaus, A.; in *Houben-Weyl: Methoden der organischen Chemie* (Müller, E.; ed.), Thieme, Leipzig, **1924**, Vol. 2, 424; Eargle, D. H.; *J. Org. Chem.* **1963**, *28*, 1703.

[1169] Lazana, M. C.; Franco, M. L.; Herold, B. J.; Maercker, A.; *IX Encontro Anual da Sociedade Portuguesa de Quimica, Estrutura e Reactividade Molecular*, Coimbra (Portugal), **1986**, Commun. 8 CP 15; quoted according to ref. [1162].

[1170] Maercker, A.; Schaller, G.; Jansen, H.; unpublished results (**1975–1978**); quoted according to ref. [1162].

[1171] Evans, A. G.; Roberts, P. B.; Tabner, B. J.; *J. Chem. Soc. B* **1966**, 269.

[1172] Kadyrov, R.; Heinicke, J.; Kindermann, M. K.; Heller, D.; Fischer, C.; Selke, R.; Fischer, A. K.; Jones, P. G.; *Chem. Ber.* **1997**, *130*, 1663.

[1173] Maercker, A.; Jaroschek, H.-J.; *J. Organomet. Chem.* **1976**, *108*, 145.

[1174] Maercker, A.; *Organomet. Synth.* **1986**, *3*, 398; *J. Organomet. Chem.* **1969**, *18*, 249.

[1175] Bartmann, E.; *J. Organomet. Chem.* **1987**, *332*, 19.

[1176] Eisch, J. J.; Jacobs, A. M.; *J. Org. Chem.* **1963**, *28*, 2145.

[1177] Letsinger, R. L.; Traynham, J. G.; *J. Am. Chem. Soc.* **1948**, *70*, 3342.

[1178] Morton, A. A.; Magat, E. E.; Letsinger, R. L.; *J. Am. Chem. Soc.* **1947**, *69*, 950, spec. 959.

[1179] Miginiac, P.; Bouchoule, C.; *Bull. Soc. Chim. Fr.* **1968**, 4156; Miginiac, P.; *Bull. Soc. Chim. Fr.* **1970**, 1077.

[1180] Rautenstrauch, V.; *Helv. Chim. Acta* **1974**, *57*, 496.

[1181] Burns, T. P.; Rieke, R. D.; *J. Org. Chem.* **1987**, *52*, 3674.

[1182] Gilman, H.; Schwebke, G. L.; *J. Org. Chem.* **1962**, *27*, 4259.

[1183] Ziegler, K.; Schnell, B.; *Liebigs Ann. Chem.* **1924**, *437*, 227; see also: ref. [1218].

[1184] Ziegler, K.; Dislich, H.; *Chem. Ber.* **1957**, *90*, 1107.

[1185] Wittig, G.; Happe, W.; *Liebigs Ann. Chem.* **1947**, *557*, 205.

[1186] Schlenk, W.; Bergmann, E.; *Liebigs Ann. Chem.* **1928**, *463*, 1, spec. 199–200.

[1187] Cohen, T.; Sherbine, J. P.; Matz, J. R.; Hutchins, R. R.; McHenry, B. M.; Willey, P. R.; *J. Am. Chem. Soc.* **1984**, *35*, 3245.

[1188] Tanaka, K.; Minami, K.; Funaki, I.; Suzuki, H.; *Tetrahedron Lett.* **1990**, *31*, 2727.

[1189] Strohmann, C.; Abele, B. C.; *Angew. Chem.* **1996**, *108*, 2514; *Angew. Chem. Int. Ed. Engl.* **1996**, *35*, 2378.

[1190] Cohen, T.; Matz, J. R.; *J. Am. Chem. Soc.* **1980**, *102*, 6900.

[1191] Cohen, T.; Matz, J. R.; *Tetrahedron Lett.* **1981**, *22*, 2455.

[1192] Amano, S.; Fujiwara, K.; Murai, A.; *Synlett* **1997**, 1300.

[1193] Cohen, T.; Lin, M.-T.; *J. Am. Chem. Soc.* **1984**, *106*, 1130.

[1194] Rychnovsky, S. D.; Skalitzky, D. J.; *J. Org. Chem.* **1992**, *57*, 4336.

[1195] Rychnovsky, S. D.; Mickus, D. E.; *Tetrahedron Lett.* **1989**, *30*, 3011; Rychnovsky, S. D.; *J. Org. Chem.* **1989**, *54*, 4982; Rychnovsky, S. D.; Plzak, K.; Pickering, D.; *Tetrahedron Lett.* **1994**, *35*, 6799.

[1196] Cohen, T.; Daniewski, W. M.; Weisenfeld, R. B.; *Tetrahedron Lett.* **1978**, *19*, 465.

[1197] Maercker, A.; Jaroschek, H.-J.; *J. Organomet. Chem.* **1976**, *116*, 21.

[1198] Guo, B.-S.; Doubleday, W.; Cohen, T.; *J. Am. Chem. Soc.* **1987**, *109*, 4710.

[1199] McCullough, D. W.; Bhupathy, M.; Piccolino, E.; Cohen, T.; *Tetrahedron* **1991**, *47*, 9727.

[1200] Cohen, T.; Guo, B.-S.; *Tetrahedron Lett.* **1986**, *42*, 2803.

[1201] Deleris, G.; Kowalski, J.; Dunoguès, J.; Calas, R.; *Tetrahedron Lett.* **1977**, *33*, 4211; Pillot, J. -P.; Deleris, G.; Dunoguès, J.; Calas, R.; *J. Org. Chem.* **1979**, *44*, 3397.

[1202] Fraenkel, G.; Cabral, J.; Lanter, C.; Wang, J.-h.; *J. Org. Chem.* **1999**, *64*, 1302.

[1203] Cabral, J. A.; Cohen, T.; Doubleday, W. W.; Duchelle, E. F.; *J. Org. Chem.* **1999**, *64*, 1302; Fraenkel, G.; Guo, B. S.; Yü, S. H.; *J. Org. Chem.* **1992**, *57*, 3680.

[1204] Guijarro, D.; Mancheño, B.; Yus, M.; *Tetrahedron Lett.* **1992**, *33*, 5597; *Tetrahedron* **1992**, *48*, 4593.

[1205] Guijarro, D.; Guillena, G.; Mancheño, B.; Yus, M.; *Tetrahedron* **1994**, *50*, 3427; Guijarro, D.; Yus, M.; *Tetrahedron* **1994**, *50*, 3447.

[1206] Mudryk, B.; Cohen, T.; *J. Org. Chem.* **1991**, *56*, 5760.

[1207] Jedlinski, Z.; Kowalczuk, M.; Misiolek, A.; *J. Chem. Soc., Chem. Commun.* **1988**, 1261.

[1208] Mudryk, B.; Cohen, T.; *J. Am. Chem. Soc.* **1991**, *113*, 1866.

[1209] Ramón, D.; Yus, M.; *Tetrahedron* **1992**, *48*, 3585.

[1210] Bartmann, E.; *J. Organomet. Chem.* **1985**, *284*, 149.

[1211] Freudenberg, K.; Lautsch, W.; Piazolo, G.; *Ber. Dtsch. Chem. Ges.* **1941**, *74*, 1879, spec. 1886.

[1212] Hurd, C. D.; Oliver, G.; *J. Am. Chem. Soc.* **1959**, *81*, 2795.

[1213] Dewald, R. R.; Conlon, N. J.; Song, W. M.; *J. Org. Chem.* **1989**, *54*, 261; Zimmerman, H. E.; Wang, P. A.; *J. Am. Chem. Soc.* **1990**, *112*, 1280; Azzena, U.; Denurra, T.; Melloni, G.; Fenude, E.; Rassu, G.; *J. Org. Chem.* **1992**, *57*, 1444.

[1214] Araki, S.; Butsugan, Y.; *Chem. Lett.* **1988**, 457.

[1215] Maccaroni, P.; Leroux, F.; Schlosser, M.; unpublished results (**1998**); Maccaroni, P.; PhD thesis, Université de Lausanne **1998**.

[1216] Almena, J.; Foubelo, F.; Yus, M.; *Tetrahedron* **1995**, *51*, 3351; *Tetrahedron* **1995**, 51, 3365.

[1217] Gil, J. F.; Ramón, D. J.; Yus, M.; *Tetrahedron* **1993**, *49*, 9535; see also: Azzena, U.; Melloni, G.; Nigra, C.; *J. Org. Chem.* **1993**, *58*, 6707.

[1218] Ziegler, K.; Thielmann, F.; *Ber. Dtsch. Chem. Ges.* **1923**, *56*, 1740.

[1219] Gerdil, R.; Lucken, E. A. C.; *J. Chem. Soc.* **1963**, 2857, 5444; **1964**, 3916.

[1220] Screttas, C. G.; Micha-Screttas, M.; *J. Org. Chem.* **1978**, *43*, 1064; *J. Org. Chem.* **1979**, *44*, 713.

[1221] Screttas, C. G.; Smonou, I. C.; *J. Organomet. Chem.* **1988**, *342*, 143.

[1222] Rücker, C.; Prinzbach, H.; *Tetrahedron Lett.* **1983**, *24*, 4099.

[1223] Hoffmann, R. W.; Kemper, B.; *Tetrahedron Lett.* **1981**, *22*, 5263.

[1224] Alonso, D. A.; Alonso, E.; Nájera, C.; Ramón, D. J.; Yus, M.; *Tetrahedron* **1997**, *53*, 4835.

[1225] Beau, J.-M.; Sinaj, P.; *Tetrahedron Lett.* **1985**, *26*, 6185; *Tetrahedron Lett.* **1985**, *26*, 6189; *Tetrahedron Lett.* **1985**, *26*, 6193.

[1226] Ziegler, K.; Crössmann, F.; Kleiner, H.; Schäfer, O.; *Liebigs Ann. Chem.* **1929**, *473*, 1, spec. 19.

[1227] Rauchschwalbe, G.; Zellner, A.; unpublished results (**1977**; **1993**); Zellner, A.; *Doctoral Dissertation*, Université de Lausanne, **1997**, spec. pp. 53–54 and 122–123; see also: ref. [221], spec. p. 10197.

[1228] Eisch, J. J.; Kaska, W. C.; *J. Org. Chem.* **1962**, 27, 3745.

[1229] Rauchschwalbe, G.; Schlosser, M.; *Helv. Chim. Acta* **1975**, *58*, 1094.

[1230] For a revision of the ether stoichiometry, see: Schlosser, M.; Franzini, L.; *Synthesis* **1998**, 150, footnote (14).

[1231] Chamberlin, A. R.; Bloom, S. H.; *Org. React.* **1990**, *39*, 1.

[1232] Brown, P. A.; Jenkins, P. R.; *Tetrahedron Lett.* **1982**, *23*, 3733.

[1233] Chamberlin, A. R.; Liotta, E. L.; Bond, F. T.; *Org. Synth., Coll. Vol.* **1990**, *7*, 77.

[1234] Butikofer, P. A.; Eugster, C. H.; *Helv. Chim. Acta* **1983**, *66*, 1148.

[1235] Hojo, M.; Masuda, R.; Saeki, T.; Fujimori, K.; Tsutsumi, S.; *Synthesis* **1977**, 789.

[1236] Satoh, T.; Sato, T.; Oohara, T.; Yamakawa, K.; *J. Org. Chem.* **1989**, *54*, 3130; Satoh, T.; Mizu, Y.; Kawashima, T.; Yamakawa, K.; *Tetrahedron* **1995**, *51*, 703; Satoh, T.; Horiguchi, K.; *Tetrahedron Lett.* **1995**, *36*, 8235; Satoh, T.; Takano, K.; *Tetrahedron* **1996**, *52*, 2349; Satoh, T.; Takano, K.; Ota, H.; Someya, H.; Matsuda, K.; Yamakawa, K.; *Tetrahedron* **1998**, *54*, 5557.

[1237] Cardellicchio, C.; Fiandanese, V.; Naso, F.; *J. Org. Chem.* **1992**, *57*, 1718; Cardellicchio, C.; Iacuone, A.; Naso, F.; *Tetrahedron Lett.* **1996**, *37*, 6017; Capozzi, M. A. M.; Cardellicchio, C.; Fracchiolla, G.; Naso, F.; Tortorella, P.; *J. Am. Chem. Soc.* **1999**, *121*, 4708.

[1238] Hoffmann, R. W.; Nell, P. G.; *Angew. Chem.* **1999**, *111*, 354; *Angew. Chem. Int. Ed. Engl.* **1999**, *38*, 338.

[1239] Gilman, H.; Bebb, R. L.; *J. Am. Chem. Soc.* **1939**, *61*, 109.

[1240] Gilman, H.; Webb, F. J.; *J. Am. Chem. Soc.* **1949**, *71*, 4062.

[1241] Seebach, D.; Peleties, N.; *Angew. Chem.* **1969**, *81*, 465; *Angew. Chem. Int. Ed. Engl.* **1969**, *8*, 450; *Chem. Ber.* **1972**, *105*, 511; Gröbel, B. T.; Seebach, D.; *Chem. Ber.* **1976**, *110*, 852, 867.

[1242] Ogawa, S.; Masutomi, Y.; Furukawa, N.; Erata, T.; *Heteroat. Chem.* **1992**, *3*, 423; *Chem. Abstr.* **1993**, *118*, 169224s.

[1243] Maercker, A.; Bodenstedt, H.; Brandsma, L.; *Angew. Chem.* **1992**, *104*, 1387; *Angew. Chem. Int. Ed. Engl.* **1992**, *31*, 1339.

[1244] Kondo, Y.; Shilai, M.; Uchiyama, M.; Sakamoto, T.; *J. Chem. Soc., Perkin Trans. 1* **1996**, 1781.

[1245] Dumont, W.; Bayet, P.; Krief, A.; *Angew. Chem.* **1974**, *86*, 857; *Angew. Chem. Int. Ed. Engl.* **1974**, *13*, 804.

[1246] Clarembeau, M.; Krief, A.; *Tetrahedron Lett.* **1984**, *25*, 3629.

[1247] Clarembeau, M.; Krief, A.; *Tetrahedron Lett.* **1985**, *26*, 1093.

[1248] Krief, A.; Bousbaa, J.; *Tetrahedron Lett.* **1997**, *38*, 6289.

[1249] Seebach, D.; Meyer, N.; Beck, A. K.; *Liebigs Ann. Chem.* **1977**, 846.

[1250] Krief, A.; Dumont, W.; Clarembeau, M.; Bernard, G.; Badaoui, E.; *Tetrahedron Lett.* **1989**, *45*, 2005.

[1251] Krief, A.; Nazih, A.; *Tetrahedron Lett.* **1995**, *36*, 8115.

[1252] Hoffmann, R.; Brückner, R.; *Chem. Ber.* **1992**, *125*, 1957.

[1253] Krief, A.; Dumont, W.; *Tetrahedron Lett.* **1997**, *38*, 657.

[1254] Ryter, K.; Livinghouse, T.; *J. Org. Chem.* **1997**, *62*, 4842.

[1255] Clive, D. L. J.; *J. Chem. Soc., Chem. Commun.* **1973**, 695; Reich, H. J.; Reich, I. L.; Renga, J. M.; *J. Am. Chem. Soc.* **1973**, *95*, 5813; Sharpless, K. B.; Lauer, R. F.; *J. Org. Chem.* **1974**, *39*, 429.

[1256] Halazy, S.; Krief, A.; *J. Chem. Soc., Chem. Commun.* **1979**, 1136.

[1257] Krief, A.; Bousbaa, J.; *Tetrahedron Lett.* **1997**, *38*, 6291.

[1258] Krief, A.; Derouane, D.; Dumont, W.; *Synlett* **1992**, 907; see also: Krief, A.; Bousbaa, J.; *Synlett* **1996**, 1007.

[1259] Krief, A.; Badaoui, E.; Dumont, W.; *Tetrahedron Lett.* **1993**, *34*, 8517.

[1260] Krief, A.; Mercier, J.; unpublished results (**1999**); see also: Krief, A.; Remacle, B.; Mercier, J.; *Synlett* **2000**, 1443.

[1261] Seyferth, D.; Suzuki, R.; Murphy, C. J.; Sabet, C. R.; *J. Organomet. Chem.* **1964**, *2*, 431.

[1262] Wittig, G.; Ludwig, R.; Polster, R.; *Chem. Ber.* **1955**, *88*, 294.

[1263] Wittig, G.; Bickelhaupt, F.; *Chem. Ber.* **1958**, *91*, 865.

[1264] Wittig, G.; Benz, E.; *Chem. Ber.* **1958**, *91*, 873.

[1265] Wittig, G.; Bickelhaupt, F.; unpublished results (**1955**); quoted according to Schöllkopf, U.; in *Houben-Weyl: Methoden der organischen Chemie* (Müller, E.; ed.), **1970**, Vol. *13/1*, 93, spec. 127.

[1266] Carothers, W. H.; Coffman, D. D.; *J. Am. Chem. Soc.* **1930**, *52*, 1254.

[1267] Schlenk, W.; Holtz, J.; *Ber. Dtsch. Chem. Ges.* **1917**, *50*, 262.

[1268] Schorigin, P.; *Ber. Dtsch. Chem. Ges.* **1908**, *41*, 2711, 2717, 2723.

[1269] Carothers, W. H.; Coffman, D. D.; *J. Am. Chem. Soc.* **1929**, *51*, 588.

[1270] Gilman, H.; Kirby, R. H.; *J. Am. Chem. Soc.* **1936**, *58*, 2074; see also: Gilman, H.; Young, R. V.; *J. Org. Chem.* **1936**, *1*, 315, spec. 326.

[1271] Whitmore, F. C.; Zook, H. D.; *J. Am. Chem. Soc.* **1942**, *64*, 1783.

[1272] Braun, D.; Betz, W.; Kern, W.; *Makromol. Chem.* **1960**, *42*, 89.

[1273] Leroux, F.; Schlosser, M.; unpublished results (**1999**).

[1274] Benoit, P.; Collignon, N.; *Bull. Soc. Chim. Fr.* **1975**, 1302.

[1275] Hartmann, J.; Schlosser, M.; unpublished results (**1976**).

[1276] Stähle, M.; Hartmann, J.; Schlosser, M.; *Helv. Chim. Acta* **1977**, *60*, 1730.

[1277] Nesmeyanov, A. N.; Borisov, A. E.; Saveleva, I. S.; Golubeva, E. I.; *Izvest. Akad. Nauk S.S.S.R., Otdel. Khim. Nauk* **1958**, 1490; *Chem. Abstr.* **1959**, *53*, 7973a, 13045c.

[1278] Bartocha, B.; Douglas, C. M.; Gray, M. Y.; *Z. Naturf. B* **1959**, *14*, 809.

[1279] Lühder, K.; Lobitz, P.; Beyer, A.; Strege, S.; Anderson, H.; *J. Prakt. Chem.* **1988**, *330*, 650.

[1280] Waack, R.; Doran, M. A.; *J. Am. Chem. Soc.* **1963**, *85*, 1651; *J. Org. Chem.* **1967**, *32*, 3395.

[1281] Morton, A. A.; Letsinger, R. L.; *J. Am. Chem. Soc.* **1947**, *69*, 172.

[1282] Waack, R.; Doran, M. A.; Glatzke, A. L.; *J. Organomet. Chem.* **1972**, *46*, 1.

[1283] Ziegler, K.; *Ber. Dtsch. Chem. Ges.* **1931**, *64*, 445.

[1284] Wittig, G.; Bickelhaupt, F.; *Chem. Ber.* **1958**, *91*, 883.

[1285] Vecchiotti, L.; *Ber. Dtsch. Chem. Ges.* **1930**, *63*, 2275.

[1286] Wittig, G.; Herwig, W.; *Chem. Ber.* **1955**, *88*, 962, spec. 972.

[1287] Courtot, C.; Castani, M. G.; *C. R. Séances Acad. Sci.* **1936**, *203*, 197.

[1288] Gilman, H.; Brown, R. E.; *J. Am. Chem. Soc.* **1930**, *52*, 3314.

[1289] Ziegler, K.; Crössmann, F.; Kleiner, H.; Schäfer, O.; *Liebigs Ann. Chem.* **1929**, *473*, 1, spec. 19.

[1290] Gilman, H.; Young, R. V.; *J. Org. Chem.* **1936**, *1*, 315; Gilman, H.; Wu, T. C.; *J. Org. Chem.* **1953**, *18*, 753.

[1291] Seyferth, D.; Freyer, W.; *J. Org. Chem.* **1961**, *26*, 2604.

[1292] Cainelli, G.; Dal Bello, G.; Zubiani, G.; *Tetrahedron Lett.* **1965**, *6*, 3429.

[1293] Dickhaut, J.; Giese, B.; *Org. Synth.* **1992**, *70*, 164.

[1294] Seyferth, D.; Welch, D. E.; Raab, G.; *J. Am. Chem. Soc.* **1962**, *84*, 4266.

[1295] Smith, M. R.; Gilman, H.; *J. Organomet. Chem.* **1972**, *37*, 35.

[1296] Mauzé, B.; *J. Organomet. Chem.* **1979**, *170*, 265; Doucoure, A.; Mauzé, B.; Miginiac, L.; *J. Organomet. Chem.* **1982**, *236*, 139.

[1297] Seyferth, D.; Murphy, G. J.; Woodruff, R. A.; *J. Am. Chem. Soc.* **1974**, *96*, 5011.

[1298] Warner, C. M.; Noltes, J. G.; *J. Organomet. Chem.* **1970**, *24*, C4.

[1299] Talalaeva, T. V.; Kocheshkov, K. A.; *Izvest. Akad. Nauk S.S.S.R., Otdel. Khim. Nauk* **1953**, 126; *Chem. Abstr.* **1954**, *48*, 3285h.

[1300] Seitz, L. M.; Brown, T. L.; *J. Am. Chem. Soc.* **1966**, *88*, 2174.

[1301] Goldstein, M. J.; Wenzel, T. T.; *Helv. Chim. Acta* **1984**, *67*, 2029.

[1302] Gilman, H.; Moore, F. W.; Jones, R. G.; *J. Am. Chem. Soc.* **1941**, *63*, 2482; Dessy, R. E.; Kitching, W.; Psarras, T.; Salinger, R.; Chen, A.; Civers, T.; *J. Am. Chem. Soc.* **1966**, *88*, 460.

[1303] Huffman, J. C.; Nugent, W. A.; Kochi, J. K.; *Inorg. Chem.* **1980**, *19*, 2749.

[1304] Reck, C. E.; Winter, C. H.; *Organometallics* **1997**, *16*, 4493.

[1305] Chung, C.; Lagow, R. J.; *J. Chem. Soc., Chem. Commun.* **1972**, 1078.

[1306] Maercker, A.; Theis, M.; *Angew. Chem.* **1984**, *96*, 990; *Angew. Chem. Int. Ed. Engl.* **1984**, *23*, 995.

[1307] Seitz, D. E.; Zapata, A.; *Synthesis* **1981**, 557.

[1308] Seyferth, D.; Cohen, H. M.; *Inorg. Chem.* **1963**, *2*, 625.

[1309] Corey, E. J.; De, B.; *J. Am. Chem. Soc.* **1984**, *106*, 2735.

[1310] Seyferth, D.; Weiner, M. A.; *J. Am. Chem. Soc.* **1961**, *83*, 3583; Seyferth, D.; Vaughan, L. G.; *J. Am. Chem. Soc.* **1964**, *86*, 883; Maercker, A.; Graule, T.; Demuth, W.; *Angew. Chem.* **1987**, *99*, 1075; *Angew. Chem. Int. Ed. Engl.* **1987**, *26*, 1032.

[1311] Corey, E. J.; Eckrich, T. M.; *Tetrahedron Lett.* **1984**, *25*, 2415, 2419.

[1312] Corey, E. J.; Wollenberg, R. H.; *J. Am. Chem. Soc.* **1974**, *96*, 5581; see also: Seyferth, D.; Vick, *J. Organomet. Chem.* **1978**, *144*, 1; Keck, G. E.; Byers, J. H.; Tafesh, A. M.; *J. Org. Chem.* **1988**, *54*, 1127; Farina, V.; Hauck, S. I.; *J. Org. Chem.* **1991**, *56*, 4317.

[1313] Märkl, G.; Hofmeister, P.; *Tetrahedron Lett.* **1976**, *18*, 3419.

[1314] Ashe, A. J.; Lohr, L. L.; Al-Taweel, S. M.; *Organometallics* **1991**, *10*, 2424.

[1315] Wada, E.; Kanemasa, S.; Fujiwara, I.; Tsuge, O.; *Bull. Chem. Soc. Jpn.* **1985**, *58*, 1942.

[1316] Seyferth, D.; Weiner, M. A.; *J. Org. Chem.* **1959**, *24*, 1395; **1961**, *26*, 4797.

[1317] Desponds, O.; Schlosser, M.; *J. Organomet. Chem.* **1991**, *409*, 93.

[1318] Seyferth, D.; Jula, T. F.; *J. Organomet. Chem.* **1974**, *66*, 195.

[1319] Gilman, H.; Rosenberg, S. D.; *J. Org. Chem.* **1959**, *24*, 2063; Seyferth, D.; Suzuki, R.; Murphy, C. J.; Sabet, C. R.; *J. Organomet. Chem.* **1964**, *2*, 431.

[1320] Hoffmann, R. W.; Rühl, T.; Chemla, F.; Zahneisen, T.; *Liebigs Ann. Chem.* **1992**, 719.

[1321] Chong, J. M.; MacDonald, G. K.; Park, S. B.; Wilkinson, S. H.; *J. Org. Chem.* **1993**, *58*, 1266.

[1322] Peterson, D. J.; *J. Am. Chem. Soc.* **1971**, *93*, 4027.

[1323] Peterson, D. J.; Ward, J. F.; *J. Organomet. Chem.* **1974**, *66*, 209.

[1324] Pearson, W. H.; Lindbeck, A. C.; *J. Am. Chem. Soc.* **1991**, *113*, 8546.

[1325] Ahlbrecht, H.; Weber, P.; *Synthesis* **1992**, 1018.

[1326] Pearson, W. H.; Szura, D. P.; Harter, W. G.; *Tetrahedron Lett.* **1988**, *29*, 761; Pearson, W. H.; Szura, Postich, M. J.; *J. Am. Chem. Soc.* **1992**, *114*, 1329.

[1327] Pearson, W. H.; Jacobs, V. A.; *Tetrahedron Lett.* **1994**, *35*, 7001.

[1328] Goswami, R.; Corcoran, D. E.; *Tetrahedron Lett.* **1982**, *23*, 1463.

[1329] Tanaka, K.; Minami, K.; Funaki, I.; Suzuki, H.; *Tetrahedron Lett.* **1990**, *31*, 2727.

[1330] Imanieh, H.; MacLeod, D.; Quayle, P.; Davies, G. M.; *Tetrahedron Lett.* **1989**, *30*, 2693.

[1331] Corey, E. J.; Eckrich, T. M.; *Tetrahedron Lett.* **1983**, *24*, 3163.

[1332] Duchene, A.; Mouko-Mpegna, D.; Quintard, J. P.; *Bull. Soc. Chim. Fr.* **1985**, 787.

[1333] Still, W. C.; *J. Am. Chem. Soc.* **1978**, *100*, 1481.

[1334] Jones, T. K.; Denmark, S. E.; *J. Org. Chem.* **1985**, *50*, 4037.

[1335] Hutchinson, D. K.; Fuchs, P. L.; *J. Am. Chem. Soc.* **1987**, *109*, 4930.

[1336] Johnson, C. R.; Medich, J. R.; *J. Org. Chem.* **1988**, *53*, 4131.

[1337] Sawyer, J. S.; Macdonald, T. L.; McGarvey, G. J.; *J. Am. Chem. Soc.* **1984**, *106*, 3376.

[1338] Sato, T.; *Synthesis* **1990**, 259, spec. 260–262.

[1339] McGarvey, G. J.; Kimura, M.; *J. Org. Chem.* **1985**, *50*, 4655.

[1340] Burke, S. D.; Shearouse, S. A.; Burch, D. J.; Sutton, R. W.; *Tetrahedron Lett.* **1980**, *21*, 1285; Quintard, J. P.; Elissondo, B.; Hattich, T.; Pereyre, M.; *J. Organomet. Chem.* **1985**, *285*, 149.

[1341] Gadwood, R. C.; *Tetrahedron Lett.* **1984**, *25*, 5851.

[1342] Gadwood, R. C.; Nagarajan, S. C.; Michel, S. T.; *J. Org. Chem.* **1985**, *50*, 3255.

[1343] Sawyer, J. S.; Kucerovy, A.; Macdonald, T. L.; McGarvey, G. J.; *J. Am. Chem. Soc.* **1988**, *110*, 842.

[1344] Parrain, J. L.; Veaudet, I.; Cintrat, J.-C.; Duchêne, A.; Quintard, J. P.; *Bull. Soc. Chim. Fr.* **1994**, *131*, 304.

[1345] Colombo, L.; DiGiacomo, M.; Brusotti, G.; Delogu, G.; *Tetrahedron Lett.* **1994**, *35*, 2063.

[1346] Taylor, O. J.; Wardell, J. L.; *J. Chem. Res. [S]* **1989**, 98.

[1347] Prandi, J.; Audin, C.; Beau, J.-M.; *Tetrahedron Lett.* **1991**, *32*, 769.

[1348] Lesimple, P.; Beau, J.-M.; Sinaÿ, P.; *Carbohydr. Res.* **1987**, *171*, 289.

[1349] Hoffmann, M.; Kessler, H.; *Tetrahedron Lett.* **1994**, *35*, 6067; Hoffmann, M.; Burkhart, F.; Hessler, G.; Kessler, H.; *Helv. Chim. Acta* **1996**, *79*, 1519.

[1350] Hanessian, S.; Martin, M.; Desai, R. C.; *J. Chem. Soc., Chem. Commun.* **1986**, 926.

[1351] McGarvey, G. J.; Bajwa, J. S.; *J. Org. Chem.* **1984**, *49*, 4091.

[1352] Booth, C.; Imanieh, H.; Quayle, P.; Lu, S.-Y.; *Tetrahedron Lett.* **1992**, *33*, 413.

[1353] Casson, S.; Kocieński, P.; *Synthesis* **1993**, 1133, spec. 1140.

[1354] Ficini, J.; Falou, S.; Touzin, A. M.; d'Angelo, J.; *Tetrahedron Lett.* **1977**, *40*, 3589.

[1355] Wollenberg, R. H.; Albizati, K. F.; Peries, R.; *J. Am. Chem. Soc.* **1977**, *99*, 7365.

[1356] Corey, E. J.; Wollenberg, R. H.; *J. Org. Chem.* **1975**, *40*, 2265.

[1357] Zhao, Y.-k.; Quayle, P.; Kuo, E. A.; *Tetrahedron Lett.* **1994**, *35*, 3797.

[1358] Gill, M.; Bainton, H. P.; Rickards, R. W.; *Tetrahedron Lett.* **1981**, *22*, 1437.

[1359] Sato, T.; Kawase, A.; Hirose, T.; *Synlett* **1992**, 891.

[1360] Seyferth, D.; Lambert, R. L.; *J. Organomet. Chem.* **1973**, *55*, C53; see also: ref. [1145].

[1361] Piers, E.; Karunaratne, V.; *Tetrahedron* **1989**, *45*, 1089.

[1362] Matthews, D. P.; Waid, P. P.; Sabol, J. S.; McCarthy, J. R.; *Tetrahedron Lett.* **1994**, *35*, 5177.

[1363] Seyferth, D.; Welch, D. E.; Raab, G.; *J. Am. Chem. Soc.* **1962**, *84*, 4266.

[1364] Seyferth, D.; Wursthorn, K. R.; *J. Organomet. Chem.* **1979**, *183*, 455.

[1365] Faigl, F.; Marzi, E.; Schlosser, M.; *Chem. Eur. J.* **2000**, *6*, 771.

[1366] de Boor, H. J. R.; Akkerman, O. S.; Bickelhaupt, F.; *Organometallics* **1990**, *9*, 2898.

[1367] Sandosham, J.; Undheim, K.; *Tetrahedron* **1994**, *50*, 275.

[1368] Cottet, F.; Schlosser, M.; unpublished results (**1999**).

[1369] Yang, Y.; Wong, H. N. C.; *J. Chem. Soc., Chem. Commun.* **1992**, 1723; see also: Song, Z. Z.; Wong, H. N. C.; Yang, Y.; *Pure Appl. Chem.* **1996**, *68*, 723.

[1370] Lee, S. H.; Hanson, R. N.; Weitz, D. E.; *Tetrahedron Lett.* **1984**, *25*, 1751.

[1371] Strohmann, C.; *Angew. Chem.* **1996**, *108*, 600; *Angew. Chem. Int. Ed. Engl.* **1996**, *35*, 528.

[1372] Meyer, N.; Seebach, D.; *Chem. Ber.* **1980**, *113*, 1290.

[1373] Zidani, A.; Vaultier, M.; *Tetrahedron Lett.* **1986**, *27*, 857.

[1374] Newman-Evans, R. H.; Carpenter, B. K.; *Tetrahedron Lett.* **1985**, *26*, 1141.

[1375] Tamborski, C.; Ford, F. E.; Soloski, E. J.; *J. Org. Chem.* **1963**, *28*, 237.

[1376] Prakash, H.; Sisler, H. H.; *Inorg. Chem.* **1972**, *11*, 2258.

[1377] Desponds, O.; *Doctoral Dissertation*, University of Lausanne, **1991**, pp. 29–33, 80–92, 100–102 and 141–142.

[1378] Seyferth, D.; Vaughan, L. G.; *J. Organomet. Chem.* **1963**, *1*, 138.

[1379] Eisch, J. J.; *J. Org. Chem.* **1963**, *28*, 707.

[1380] Wittig, G.; Leo, M.; *Ber. Dtsch. Chem. Ges.* **1930**, *63*, 943.

[1381] Salzberg, P. L.; Marvel, C. S.; *J. Am. Chem. Soc.* **1928**, *50*, 1737.

[1382] Wittig, G.; Stahnecker, E.; *Liebigs Ann. Chem.* **1957**, *605*, 69; see also: Wittig, G.; Schlör, H. H.; *Suom. Kemistil. B* **1958**, *31*, 2.

[1383] Bunz, U.; Szeimies, G.; *Tetrahedron Lett.* **1990**, *31*, 651.

[1384] Oeffner, K.; *Doctoral Dissertation*, Universität Siegen (**1993**); quoted according to Maercker, A.; in *Houben-Weyl: Methoden der organischen Chemie*, Vol. *E19d* (Hanack, M.; ed.), Thieme, Stuttgart, **1993**, p. 510.

[1385] Pauling, L.; Kekule Lecture, London, **1958** (15 Sept.); *The Nature of the Chemical Bond*, 3rd ed., Cornell University Press, Ithaca, **1960**, pp. 136–142, spec. 138.

[1386] Hackspill, L.; Rohmer, R.; *C.R. Séances Acad. Sci.[2]* **1943**, *217*, 152.

[1387] Rautenstrauch, V.; *Angew. Chem.* **1975**, *87*, 254; *Angew. Chem. Int. Ed. Engl.* **1975**, *14*, 259.

[1388] van Eikema Hommes, N. J. R.; Bickelhaupt, F.; Klumpp, G. W.; *Angew. Chem.* **1988**, *100*, 1100; *Angew. Chem. Int. Ed. Engl.* **1988**, *27*, 1083.

[1389] Maercker, A.; Grebe, B.; *J. Organomet. Chem.* **1987**, *334*, C21.

[1390] Bogdanović, B.; Wermeckes, B.; *Angew. Chem.* **1981**, *93*, 691; *Angew. Chem. Int. Ed. Engl.* **1981**, *20*, 684; Bogdanović, B.; Cordi, A.; Stepowska, H.; Locatelli, P.; Wermeckes, B.; Wilczok, U.; Haertel, K.; Richtering, H.; *Chem. Ber.* **1984**, *117*, 42; Bogdanović, B.; *Angew. Chem.* **1985**, *97*, 253; *Angew. Chem. Int. Ed. Engl.* **1985**, *24*, 262.

[1391] Maercker, A.; Klein, K.-D.; *Angew. Chem.* **1989**, *101*, 63; *Angew. Chem. Int. Ed. Engl.* **1989**, *28*, 83.

[1392] Maercker, A.; Girreser, U.; *Angew. Chem.* **1990**, *102*, 718; *Angew. Chem. Int. Ed. Engl.* **1990**, *29*, 667.

[1393] Goldstein, M. J.; Wenzel, T. T.; Whittacker, G.; Yates, S.F.; *J. Am. Chem. Soc.* **1982**, *104*, 2669.

[1394] Goldstein, M. J.; Wenzel, T. T.; *J. Chem. Soc., Chem. Commun.* **1984**, *1654*, 1655.

[1395] Trinks, R.; Müllen, K.; *Chem. Ber.* **1987**, *120*, 1481.

[1396] Goldstein, M. J.; Tomoda, S.; Whittacker, G.; *J. Am. Chem. Soc.* **1974**, *96*, 3676.

[1397] Maercker, A.; Graule, T.; Girreser, U.; *Angew. Chem.* **1986**, *98*, 174; *Angew. Chem. Int. Ed. Engl.* **1986**, *25*, 167.

[1398] Maercker, A.; Graule, T.; Girreser, U.; *Organomet. Synth.* **1988**, *4*, 366.

[1399] Maercker, A.; Girreser, U.; *Tetrahedron* **1994**, *50*, 8019.

[1400] Levin, G.; Jagur-Grodzinski, J.; Szwarc, M.; *J. Am. Chem. Soc.* **1970**, *92*, 2268.

[1401] Schlenk, W.; Bergmann, E.; *Liebigs Ann. Chem.* **1928**, *463*, 1, spec. 71–83.

[1402] Braye, E. H.; Hübel, W.; Caplier, I.; *J. Am. Chem. Soc.* **1961**, *83*, 4406.

[1403] Laguerre, M.; Dunoguès, J.; Calas, R.; *Tetrahedron Lett.* **1978**, *19*, 57.

[1404] Dowd, P.; *J. Chem. Soc., Chem. Commun.* **1965**, 568; see also: Bernard, J.; Schnieders, C.; Müllen, K.; *J. Chem. Soc., Chem. Commun.* **1985**, 12; Rajca, A.; Tolbert, L. M.; *J. Am. Chem. Soc.* **1985**, *107*, 2969.

[1405] Smith, L. I.; Hoehn, H. H.; *J. Am. Chem. Soc.* **1941**, *63*, 1184; see also: Schlenk, W.; Bergmann, E.; *Liebigs Ann. Chem.* **1928**, *463*, 106.

[1406] O'Brien, D. H.; Breeden, D. L.; *J. Am. Chem. Soc.* **1981**, *103*, 3237.

[1407] Smith, J. G.; Oliver, E.; Boettger, T. J.; *Organometallics* **1983**, *2*, 1577.

[1408] Yokoyama, Y.; Takahashi, K.; *Chem. Lett.* **1987**, 589.

[1409] Fujita, K.; Ohnuma, Y.; Yasuda, H.; Tani, H.; *J. Organomet. Chem.* **1976**, *113*, 201.

[1410] Yang, M.; Yamamoto, K.; Otake, N.; Ando, M.; Takase, K.; *Tetrahedron Lett.* **1970**, *11*, 3843; see also: Yasuda, H.; Nakano, Y.; Natsukawa, K.; Tani, H.; *Macromolecules* **1978**, *11*, 586; *Chem. Abstr.* **1978**, *89*, 60087f.

[1411] Yasuda, H.; Kajihara, Y.; Mashima, K.; Nagasuna, K.; Lee, K.; Nakamura, A.; *Organometallics* **1982**, *1*, 388.

[1412] Baker, R.; Cookson, R. C.; Saunders, A. D.; *J. Chem. Soc., Perkin Trans. 1* **1976**, 1809.

[1413] Akutagawa, S.; Otsuka, S.; *J. Am. Chem. Soc.* **1976**, *98*, 7420.

[1414] Xiong, H.; Rieke, R. D.; *J. Org. Chem.* **1992**, *57*, 6560; Sell, M. S.; Xiong, H.; Rieke, R. D.; *Tetrahedron Lett.* **1993**, *34*, 6007, 6011.

[1415] Bogdanović, B.; Liao, S.-t.; Schlichte, K.; Westeppe, U.; *Organomet. Synth.* **1988**, *4*, 410.

[1416] Laguerre, M.; Dunoguès, J.; Calas, R.; *Tetrahedron Lett.* **1980**, *21*, 831.

[1417] Maercker, A.; Berkulin, W.; Schiess, P.; *Tetrahedron Lett.* **1984**, *25*, 1701.

[1418] Benken, R.; Finneiser, K.; von Puttkamer, H.; Günther, H.; Eliasson, B.; Edlund, U.; *Helv. Chim. Acta* **1986**, *69*, 955.

[1419] Bausch, J. W.; Gregory, P. S.; Olah, G. A.; Prakash, G. K. S.; Schleyer, P. V. R.; Segal, G. A.; *J. Am. Chem. Soc.* **1989**, *111*, 3633.

[1420] Braye, E. H.; Hübel, W.; Caplier, I.; *J. Am. Chem. Soc.* **1961**, *83*, 4406.

[1421] Sekiguchi, A.; Zigler, S. S.; Haller, K. J.; West, R.; *Recl. Trav. Chim. Pays-Bas* **1988**, *107*, 197.

[1422] Yasuda, H.; Kajihara, Y.; Mashima, K.; Nagasuna, K.; Nakamura, A.; *Chem. Lett.* **1981**, 671; Yasuda, H.; Kajihara, Y.; Nagasuna, K.; Mashima, K.; Nakamura, A.; *Chem. Lett.* **1981**, 719.

[1423] Bogdanović, B.; Liao, S.-t.; Schwickardi, M.; Sikorsky, P.; Spliethoff, B.; *Angew. Chem.* **1980**, *92*, 845; *Angew. Chem., Int. Ed. Engl.* **1980**, *19*, 818.

[1424] Crandall, J. K.; Clark, A. C.; *Tetrahedron Lett.* **1969**, *10*, 325; *J. Org. Chem.* **1972**, *26*, 4236; see also: Crandall, J. K.; Rojas Vargas, A. C.; *Org. Synth., Coll. Vol.* **1988**, *6*, 786.

[1425] Felkin, H.; Swierczewski, G.; Tambute, A.; *Tetrahedron Lett.* **1969**, *10*, 707.

[1426] Chérest, M.; Felkin, H.; Frajerman, C.; Lion, C.; Roussi, G.; Swierczewski, G.; *Tetrahedron Lett.* **1966**, *7*, 875.

[1427] Eisch, J. J.; Husk, G. R.; *J. Am. Chem. Soc.* **1965**, *87*, 4194.

[1428] Richey, H. G.; Szucs, S. S.; *Tetrahedron Lett.* **1971**, *12*, 3785; Richey, H. G.; Von Rein, F. W.; *Tetrahedron Lett.* **1971**, *12*, 3781.

[1429] Eisch, J. J.; Merkley, J. H.; *J. Organomet. Chem.* **1969**, *20*, P27; Richey, H. G.; Von Rein, F. W.; *J. Organomet. Chem.* **1969**, *20*, P32; Von Rein, F. W.; Richey, H. G.; *Tetrahedron Lett.* **1971**, *12*, 3777.

[1430] Ziegler, K.; Gellert, H.-G.; *Liebigs Ann. Chem.* **1950**, *567*, 195.

[1431] Maercker, A.; Theysohn, W.; *Liebigs Ann. Chem.* **1971**, *747*, 70.

[1432] Bartlett, P. D.; Friedman, S.; Stiles, M.; *J. Am. Chem. Soc.* **1953**, *75*, 1771; Bartlett, P. D.; Tauber, S. J.; Weber, W. P.; *J. Am. Chem. Soc.* **1969**, *91*, 6362; Bartlett, P. D.; Goebel, C. V.; Weber, W. P.; *J. Am. Chem. Soc.* **1969**, *91*, 7425.

[1433] Spialter, L.; Harris, C. W.; *J. Org. Chem.* **1966**, *31*, 4263.

[1434] Jung, M. E.; Blum, R. B.; *Tetrahedron Lett.* **1977**, *18*, 3791.

[1435] Kottke, T.; Lagow, R. J.; Hoffmann, D.; Thomas, R. D.; *Organometallics* **1997**, *16*, 789.

[1436] Richey, H. G.; Rees, T. C.; *Tetrahedron Lett.* **1966**, *7*, 4297.

[1437] Felkin, H.; Umpleby, J. D.; Hagaman, E. W.; Wenkert, E.; *Tetrahedron Lett.* **1972**, *13*, 2285.

[1438] Bailey, W. F.; Punzalan, E. R.; *J. Org. Chem.* **1990**, *55*, 5404; Bailey, W. F.; Khanolkar, A. D.; *J. Org. Chem.* **1990**, *55*, 6058.

[1439] Bailey, W. F.; Khanolkar, A. D.; *Organometallics* **1993**, *12*, 239.

[1440] Analogous retro-addition (ring-opening) reactions: Hill, E. A.; Richey, H. G.; Rees, T. C.; *J. Org. Chem.* **1963**, *28*, 2161; Hill, E. A.; Davidson, J. A.; *J. Am. Chem. Soc.* **1963**, *85*, 1866; **1964**, *86*,

4663; Lansbury, P. T.; Pattison, V. A.; Clement, W. A.; Sidler, J. D.; *J. Am. Chem. Soc.* **1964**, *86*, 2247.

[1441] Richey, H. G.; Kossa, W. C.; *Tetrahedron Lett.* **1969**, *10*, 2313.

[1442] Derocque, J.-L.; Sundermann, F.-B.; *J. Org. Chem.* **1974**, *39*, 1411; Dessy, R. E.; Kandil, S. A.; *J. Org. Chem.* **1965**, *30*, 3857; Kandil, S. A.; Dessy, R. E.; *J. Am. Chem. Soc.* **1966**, *88*, 3027; Richey, H. G.; Rothman, A. M.; *Tetrahedron Lett.* **1968**, *9*, 1457.

[1443] Magid, R. M.; Welch, J. G.; *J. Am. Chem. Soc.* **1966**, *88*, 5681; Welch, J. G.; Magid, R. M.; *J. Am. Chem. Soc.* **1967**, *89*, 5300; see also: Wawzonek, S.; Studnicka, B.; Bluhm, H. J.; Kallio, R. E.; *J. Am. Chem. Soc.* **1965**, *87*, 2069; Wawzonek, S.; Bluhm, H. J.; Studnicka, B.; Kallio, R. E.; McKenna, E. J.; *J. Org. Chem.* **1965**, *30*, 3028.

[1444] Mulvaney, J. E.; Gardlund, Z. G.; *J. Org. Chem.* **1965**, *30*, 917.

[1445] Wittig, G.; Otten, J.; *Tetrahedron Lett.* **1963**, *4*, 601.

[1446] Wittig, G.; *Angew. Chem.* **1962**, *74*, 479; *Angew. Chem. Int. Ed. Engl.* **1962**, *1*, 415.

[1447] Worsfold, D. J.; Bywater, S.; *Can. J. Chem.* **1960**, *38*, 1891; Bywater, S.; Worsfold, D. J.; *Can. J. Chem.* **1962**, *40*, 1564; Worsfold, D. J.; Bywater, S.; *Can. J. Chem.* **1964**, *42*, 2884; Hsieh, H. L.; *J. Polymer Sci.* **1965**, *3A*, 163, 173; Johnson, A. F.; Worsfold, D. J.; *J. Polymer Sci.* **1965**, *3A*, 449.

[1448] Köbrich, G.; Stöber, I.; *Chem. Ber.* **1970**, *103*, 2744; see also: Waack, R.; Doran, M. A.; Stevenson, P. E.; *Organomet. Chem.* **1965**, *3*, 481; Grovenstein, E.; Wentworth, G.; *J. Am. Chem. Soc.* **1967**, *89*, 1852.

[1449] Ziegler, K.; Schäfer, W.; *Liebigs Ann. Chem.* **1934**, *511*, 101.

[1450] Knox, G. R.; Pauson, P. L.; *J. Chem. Soc.* **1961**, 4610; see also: Little, W. F.; Koestler, R. C.; *J. Org. Chem.* **1961**, *26*, 3247; Keana, J. F. W.; Guzikowski, A. P.; Morat, C.; Volwerk, J. J.; *J. Org. Chem.* **1983**, *48*, 2661.

[1451] Djakovitch, L.; Herrmann, W. A.; *J. Organomet. Chem.* **1997**, *545*, 399.

[1452] Hafner, K.; Weldes, H.; *Liebigs Ann. Chem.* **1957**, *606*, 90; Hafner, K.; Bernhard, C.; Müller, R.; *Liebigs Ann. Chem.* **1961**, *650*, 35.

[1453] Cason, L. F.; Brooks, H. G.; *J. Am. Chem. Soc.* **1952** *74*, 4582; *J. Org. Chem.* **1954**, *19*, 1278.

[1454] Seyferth, D.; Weiner, M. A.; *J. Am. Chem. Soc.* **1962**, *84*, 361.

[1455] Peterson, D. J.; *J. Org. Chem.* **1966**, *31*, 950.

[1456] Carlson, R. M.; Helquist, P. M.; *Tetrahedron Lett.* **1969**, *10*, 173.

[1457] Franzen, V.; Joschek, H.-I.; *Liebigs Ann. Chem.* **1967**, *703*, 90.

[1458] Mulvaney, J. E.; Garlund, Z. G.; Garlund, S. L.; Newton, D. J.; *J. Am. Chem. Soc.* **1966**, *88*, 476; see also: Mulvaney, J. E.; Carr, L. J.; *J. Org. Chem.* **1968**, *33*, 3286.

[1459] Bauer, W.; Feigel, M.; Müller, G.; Schleyer, P. V. R.; *J. Am. Chem. Soc.* **1988**, *110*, 6033.

[1460] Gaudemar, M.; *C. R. Séances Acad. Sci. [C]* **1971**, *273*, 1669; *Chem. Abstr.* **1972**, *76*, 72623x.

[1461] Normant, J.-F.; Marek, I.; Lefrançois, J.-M.; *Pure Appl. Chem.* **1992**, *64*, 1857.

[1462] Marek, I.; Normant, J.-F.; *Chem. Rev.* **1996**, *96*, 3241.

[1463] Suzuki, K.; Imai, T.; Yamanoi, S.; Chino, M.; Matsumoto, T.; *Angew. Chem.* **1997**, *109*, 2578; *Angew. Chem. Int. Ed. Engl.* **1997**, *36*, 2469.

[1464] Nakamura, E.; Kubota, K.; Sakata, G.; *J. Am. Chem. Soc.* **1997**, *119*, 5457.

[1465] Mangano, G.; Schlosser, M.; unpublished results (**1999**).

[1466] Mochida, K.; Kojima, K.; Yoshida, Y.; *Bull. Chem. Soc. Jpn.* **1987**, *60*, 2255.

[1467] Way, K. R.; Stwalley, W. C.; *J. Chem. Phys.* **1973**, *59*, 5298; Hussein, K.; Effantin, C.; D'Incan, J.; Verges, J.; Barrow, R. F.; *Chem. Phys. Lett.* **1986**, *124*, 105; Zemke, W. T.; Stwalley, W. C.; *Chem. Phys. Lett.* **1988**, *143*, 84.

[1468] Matteson, G. W.; Whaley, T. P.; *Inorg. Synth.* **1957**, *5*, 10; see also: Bank, S.; Lois, T. A.; *J. Am. Chem. Soc.* **1968**, *90*, 4505.

[1469] Milas, N. A.; Lee, S. W.; Sakal, E.; Wohlers, H. C.; MacDonald, N. S.; Grossi, F. X.; Wright, N. A.; *J. Am. Chem. Soc.* **1948**, *70*, 1584, spec. 1589; Milas, N. A.; Sakal, E.; Plati, J. T.; Rivers, J. T.; Gladding, J. K.; Grossi, F. X.; Weiss, Z.; Campbell, M. A.; Wright, H. F.; *J. Am. Chem. Soc.* **1948**, *70*, 1597, spec. 1602; Oroshnik, W.; Mebane, A. D.; *J. Am. Chem. Soc.* **1949**,

71, 2062; Cheeseman, G. W. H.; Heilbron, I.; Jones, E. R. H.; Sondheimer, F.; Weedon, B. C. L.; *J. Chem. Soc.* **1949**, 1516, 2031; Inhoffen, H. H.; Bohlmann, F.; Bartram, K.; Rummert, G.; Pommer, H.; *Liebigs Ann. Chem.* **1950**, *570*, 54; Beumel, O. F.; Harris, R. F.; *J. Org. Chem.* **1963**, *28*, 2775; **1964**, *29*, 1872; **1965**, *30*, 814.

[1470] Hess, K.; Munderloh, M.; *Ber. Dtsch. Chem. Ges.* **1918**, *51*, 377; Rudledge, T. F.; *J. Org. Chem.* **1957**, *22*, 649; Tedeschi, R. J.; Moore, G. L.; *J. Org. Chem.* **1969**, *34*, 435.

[1471] Ruzicka, L.; Hoffmann, K.; *Helv. Chim. Acta* **1937**, *20*, 1280.

[1472] Moureu, C.; Desmots, H.; *Bull. Soc. Chim. Fr. [3]* **1902**, *27*, 360.

[1473] Henne, A. L.; Greenlee, K. W.; *J. Am. Chem. Soc.* **1943**, *65*, 2020.

[1474] Glaser, C.; *Liebigs Ann. Chem.* **1870**, *154*, 137, 161; Johnson, J. R.; Schwartz, A. M.; Jacobs, T. L.; *J. Am. Chem. Soc.* **1938**, *60*, 1882.

[1475] Yasuda, H.; Ohnuma, Y.; Yamauchi, M.; Tani, H.; Nakamura, A.; *Bull. Chem. Soc. Jpn.* **1979**, *52*, 2036; Yasuda, H.; Yamauchi, M.; Ohnuma, Y.; Nakamura, A.; *Bull. Chem. Soc. Jpn.* **1981**, *54*, 1481.

[1476] Claff, C. E.; Morton, A. A.; *J. Org. Chem.* **1955**, *20*, 440.

[1477] House, H. O.; Kramar, V.; *J. Org. Chem.* **1962**, *27*, 4146; see also: Hanriot, M.; Saint-Pierre, O.; *Bull. Soc. Chim. Fr. [3]* **1889**, *1*, 774; Werner, A.; Grob, A.; *Ber. Dtsch. Chem. Ges.* **1904**, *37*, 2898.

[1478] Gilman, H.; Gorsich, R. D.; *J. Org. Chem.* **1958**, *23*, 550.

[1479] Weißgerber, R.; *Ber. Dtsch. Chem. Ges.* **1908**, *41*, 2913.

[1480] Meier, R.; *Chem. Ber.* **1953**, *86*, 1483; Scherf, G. W.H.; Brown, R. K.; *Can. J. Chem.* **1960**, *38*, 697.

[1481] Weißgerber, R.; *Ber. Dtsch. Chem. Ges.* **1909**, *42*, 569.

[1482] Barber, W. A.; *Inorg. Synth.* **1960**, *6*, 11; see also: Saito, T.; *J. Chem. Soc., Chem. Commun.* **1977**, 1422.

[1483] Hafner, K.; Kaiser, H.; *Liebigs Ann. Chem.* **1958**, *618*, 140.

[1484] Ziegler, K.; Foitzheim-Kühlhorn, H.; Hafner, K.; *Chem. Ber.* **1956**, *89*, 434.

[1485] Gilman, H.; Breuer, F.; *J. Am. Chem. Soc.* **1934**, *56*, 1123.

[1486] Schönberg, A.; Petersen, E.; Kaltschmitt, *Ber. Dtsch. Chem. Ges.* **1933**, *66*, 233.

[1487] Lanpher, E. J.; Redmen, L. M.; Morton, A. A.; *J. Org. Chem.* **1958**, *23*, 1370.

[1488] Finnegan, R. A.; McNees, R. S.; *J. Org. Chem.* **1964**, *29*, 3234, 3241.

[1489] Schlosser, M.; Hartmann, J.; Stähle, M.; Kramař, J.; Walde, A.; Mordini, A.; *Chimia* **1986**, *40*, 306.

[1490] Gassman, P. G.; Mansfield, K. T.; *J. Org. Chem.* **1967**, *6*, 287.

[1491] Cohen, H. M.; *J. Organomet. Chem.* **1967**, *9*, 375.

[1492] See also: Wiberg, K. B.; Lampman, G. M.; Ciula, R. P.; Connor, D. S.; Schertler, P.; Lavanish, J.; *Tetrahedron* **1965**, *21*, 2749.

[1493] Closs, G. L.; Closs, L. E.; *J. Am. Chem. Soc.* **1963**, *85*, 2022.

[1494] Closs, G. L.; Larrabee, R. B.; *Tetrahedron Lett.* **1965**, *6*, 287.

[1495] Klumpp, G. W.; Kool, M.; Schakel, M.; Schmitz, R.F.; Boutkan, C.; *J. Am. Chem. Soc.* **1979**, *101*, 7065.

[1496] Lepley, A. R.; Giumanini, A. G.; *Chem. Ind. (London)* **1965**, 1035.

[1497] Ahlbrecht, H.; Dollinger, H.; *Tetrahedron Lett.* **1984**, *25*, 1353.

[1498] Smith, W. N.; *Adv. Chem. Ser.* **1974**, *130*, 23; *Am. Chem. Soc.*, Washington.

[1499] Peterson, D. J.; *J. Organomet. Chem.* **1970**, *21*, P63.

[1500] Köhler, F. H.; Hertkorn, N.; Blümel, J.; *Chem. Ber.* **1987**, *120*, 2081; see also: Harder, S.; Lutz, M.; *Organometallics* **1994**, *13*, 5173.

[1501] Schakel, M.; Aarnts, M. P.; Klumpp, G. W.; *Recl. Trav. Chim. Pays-Bas* **1990**, *109*, 305.

[1502] Luitjes, H.; Schakel, M.; Aarnts, M. P.; Schmitz, R. F.; de Kanter, F. J. J.; Klumpp, G. W.; *Tetrahedron* **1997**, *53*, 9977.

[1503] Meyers, A. I.; Jagdmann, G. E.; *J. Am. Chem. Soc.* **1982**, *104*, 877.

[1504] Dieter, R. K.; Alexander, C. W.; *Tetrahedron Lett.* **1992**, *33*, 5693.

[1505] Meyers, A. I.; Hellring, S.; Ten Hoeve, W.; *Tetrahedron Lett.* **1981**, *22*, 5115; Meyers, A. I.; Hellring, S.; *Tetrahedron Lett.* **1981**, *22*, 5119.

[1506] Meyers, A. I.; Edwards, P. D.; Rieker, W. F.; Bailey, T. R.; *J. Am. Chem. Soc.* **1984**, *106*, 3270.

[1507] Gawley, R. E.; Hart, G.; Goicoechea-Pappas, M.; Smith, A. L.; *J. Org. Chem.* **1986**, *51*, 3076; Gawley, R. E.; Hart, G.; Bartolotti, L. J.; *J. Org. Chem.* **1989**, *54*, 175; Rein, K. S.; Chen, Z.-H.; Perumal, P. T.; Echegoyen, L.; Gawley, R. E.; *Tetrahedron Lett.* **1991**, *23*, 1941.

[1508] Meyers, A. I.; Rieker, W. F.; Fuentes, L. M.; *J. Am. Chem. Soc.* **1983**, *105*, 2082.

[1509] Shawe, T. T.; Meyers, A. I.; *J. Org. Chem.* **1991**, *56*, 2751.

[1510] Gonzales, M. A.; Meyers, A. I.; *Tetrahedron Lett.* **1989**, *21*, 43, 47.

[1511] Monn, J. A.; Rice, K. C.; *Tetrahedron Lett.* **1989**, *21*, 911.

[1512] Beak, P.; McKinnie, B. G.; Reitz, D. B.; *Tetrahedron Lett.* **1977**, *18*, 1839; see also: Beak, P.; Brubaker, G. R.; Farney, R. F.; *J. Am. Chem. Soc.* **1976**, *98*, 3621.

[1513] Schlecker, R.; Seebach, D.; Lubosch, W.; *Helv. Chim. Acta* **1978**, *61*, 512.

[1514] Schlecker, R.; Seebach, D.; *Helv. Chim. Acta* **1977**, *60*, 1459.

[1515] Beak, P.; Zajdel, W. J.; *J. Am. Chem. Soc.* **1984**, *106*, 1010.

[1516] Wykypiel, W.; Lohmann, J.-J.; Seebach, D.; *Helv. Chim. Acta* **1981**, *64*, 1337.

[1517] Lubosch, W.; Seebach, D.; *Helv. Chim. Acta* **1980**, *63*, 102.

[1518] Hassel, T.; Seebach, D.; *Helv. Chim. Acta* **1978**, *61*, 2237.

[1519] Reitz, D. B.; Beak, P.; Tse, A.; *J. Org. Chem.* **1981**, *46*, 4316.

[1520] Park, Y. S.; Beak, P.; *Tetrahedron* **1996**, *52*, 12333.

[1521] Beak, P.; Wu, S.-d.; Yum, E. K.; Jun, Y. M.; *J. Org. Chem.* **1994**, *59*, 276.

[1522] Kerrick, S. T.; Beak, P.; *J. Am. Chem. Soc.* **1991**, *113*, 9708.

[1523] Beak, P.; Lee, W. K.; *J. Org. Chem.* **1990**, *55*, 2578; *J. Org. Chem.* **1993**, *58*, 1109.

[1524] Pfammatter, E.; Seebach, D.; *Liebigs Ann. Chem.* **1991**, 1323.

[1525] Beak, P.; Yum, E. K.; *J. Org. Chem.* **1993**, *58*, 823.

[1526] Katritzky, A.; Sengupta, S.; *J. Chem. Soc., Perkin Trans. 1* **1989**, 17.

[1527] Ahlbrecht, H.; Kornetzky, D.; *Synthesis* **1988**, 775; see also: Ahlbrecht, H.; Schmitt, C.; *Synthesis* **1994**, 719.

[1528] Ahlbrecht, H.; Schmitt, C.; Kornetzky, D.; *Synthesis* **1991**, 637.

[1529] Seebach, D.; Enders, D.; *Angew. Chem.* **1972**, *84*, 350, 1186; *Angew. Chem. Int. Ed. Engl.* **1972**, *11*, 301, 1101; *J. Med. Chem.* **1974**, *17*, 1225; *Angew. Chem.* **1975**, *87*, 1; *Angew. Chem. Int. Ed. Engl.* **1975**, *14*, 15.

[1530] Saavedra, J. E.; *J. Org. Chem.* **1983**, *48*, 2388.

[1531] Fraser, R. R.; Grindley, T. B.; Passannanti, S.; *Can. J. Chem.* **1975**, *53*, 2473; Fraser, R. R.; Passannanti, S.; *Synthesis* **1976**, 540.

[1532] Fraser, R. R.; Boussard, G.; Postescu, I. D.; Whiting, J. J.; Wigfield, Y. Y.; *Can. J. Chem.* **1973**, *51*, 1109.

[1533] Lehmann, R.; Schlosser, M.; *Tetrahedron Lett.* **1984**, *25*, 745.

[1534] Corey, E. J.; Eckrich, T. M.; *Tetrahedron Lett.* **1983**, *24*, 3165.

[1535] Eisch, J. J.; Galle, J. E.; *J. Organomet. Chem.* **1976**, *121*, C10.

[1536] Eisch, J. J.; Galle, J. E.; *J. Am. Chem. Soc.* **1976**, *98*, 4646.

[1537] Molander, G. A.; Mautner, K.; *J. Org. Chem.* **1989**, *54*, 4042.

[1538] Mani, N. S.; Townsend, C. A.; *J. Org. Chem.* **1997**, *62*, 636,

[1539] Murthi, K. K.; Salomon, R. G.; *Tetrahedron Lett.* **1994**, *35*, 517.

[1540] Beak, P.; McKinnie, B. G.; *J. Am. Chem. Soc.* **1977**, *99*, 5213.

[1541] Reitz, D. B.; Beak, P.; Farney, R. F.; Helmick, L. S.; *J. Am. Chem. Soc.* **1978**, *100*, 5428.

[1542] Beak, P.; Baillargeon, M.; Carter, L. G.; *J. Org. Chem.* **1978**, *43*, 4255.

[1543] Beak, P.; Carter, L. G.; *J. Org. Chem.* **1981**, *46*, 2363.

[1544] Hoppe, D.; Paetow, M.; Hintze, F.; *Angew. Chem.* **1993**, *105*, 430; *Angew. Chem. Int. Ed. Engl.* **1993**, *32*, 394.

[1545] Würthwein, E.-U.; Behrens, K.; Hoppe, D.; *Chem. Eur. J.* **1999**, *5*, 3459.

[1546] Schwerdtfeger, J.; Hoppe, D.; *Angew. Chem.* **1992**, *104*, 1547; *Angew. Chem. Int. Ed. Engl.* **1992**, *31*, 1505.

[1547] Guarnieri, W.; Grehl, M.; Hoppe, D.; *Angew. Chem.* **1994**, *106*, 1815; *Angew. Chem. Int. Ed. Engl.* **1994**, *33*, 1734.

[1548] Boie, C.; Hoppe, D.; *Synthesis* **1997**, 176.

[1549] Adcock, J. L.; Luo, H.-m.; *J. Org. Chem.* **1993**, *58*, 1999.

[1550] Köbrich, G.; Trapp, H.; Flory, K.; Drischel, W.; *Chem. Ber.* **1966**, *99*, 689; Köbrich, G.; Merkle, H. R.; *Chem. Ber.* **1966**, *99*, 1782.

[1551] Villiéras, J.; Perriot, P.; Normant, J. F.; *Bull. Soc. Chim. Fr.* **1977**, 765.

[1552] Villiéras, J.; *Bull. Soc. Chim. Fr.* **1967**, 1511, 1520; see also: Villiéras, J.; *J. Organomet. Chem. Rev.* **1971**, *A7*, 81.

[1553] Köbrich, G.; Flory, K.; Fischer, R. H.; *Chem. Ber.* **1966**, *99*, 1793.

[1554] Taguchi, H.; Yamamoto, H.; Nozaki, H.; *J. Am. Chem. Soc.* **1974**, *96*, 3010.

[1555] Reeve, W.; McKee, J. R.; Brown, R.; Lakshmanan, S.; McKee, G. A.; *Can. J. Chem.* **1980**, *58*, 485.

[1556] Köbrich, G.; Fischer, R. H.; *Chem. Ber.* **1968**, *101*, 3208, 3219.

[1557] Morton, A. A.; Brown, M. L.; Holden, M. E. T.; Letsinger, R. L.; Magat, E. E.; *J. Am. Chem. Soc.* **1945**, *67*, 2224.

[1558] Desponds, O.; Franzini, L.; Schlosser, M.; *Synthesis* **1997**, 150.

[1559] Schlosser, M.; Hartmann, J.; *Angew. Chem.* **1973**, *85*, 544; *Angew. Chem. Int. Ed. Engl.* **1973**, *12*, 508.

[1560] Giner, J. L.; Margot, C.; Djerassi, C.; *J. Org. Chem.* **1989**, *54*, 2117.

[1561] Klusener, P. A. A.; Tip, L.; Brandsma, L.; *Tetrahedron* **1991**, *47*, 2041.

[1562] Broaddus, C. D.; *J. Am. Chem. Soc.* **1972**, *94*, 4298.

[1563] Adrianome, M.; Delmond, B.; *Tetrahedron Lett.* **1985**, *26*, 6341; *J. Chem. Soc., Chem. Commun.* **1985**, 1203.

[1564] Franzini, L.; Moret, E.; Schlosser, M.; *Chem. Ber.* **1997**, *130*, 83.

[1565] Moret, E.; Fürrer, J.; Schlosser, M.; *Tetrahedron* **1988**, *44*, 3539.

[1566] Schlosser, M.; Dahan, R.; Cottens, S.; *Helv. Chim. Acta* **1984**, *67*, 284.

[1567] Schlosser, M.; Zhong, G.-f.; *Tetrahedron Lett.* **1993**, *34*, 5441.

[1568] Zhong, G.-f.; Schlosser, M.; *Synlett* **1994**, 173.

[1569] Thomas, E. W.; *Tetrahedron Lett.* **1983**, *24*, 1467; Sternberg, E.; Binger, P.; *Tetrahedron Lett.* **1985**, *26*, 301.

[1570] Wilson, S. R.; Phillips, L. R.; *Tetrahedron Lett.* **1975**, *16*, 3047; Wilson, S. R.; Phillips, L. R.; Natalie, K. J.; *J. Am. Chem. Soc.* **1979**, *101*, 3340.

[1571] Garner, C. M.; Thomas, A. A.; *J. Org. Chem.* **1995**, *60*, 7051.

[1572] Moret, E.; Schneider, P.; Margot, C.; Stähle, M.; Schlosser, M.; *Chimia* **1985**, *39*, 231.

[1573] Rauchschwalbe, G.; Schlosser, M.; *Helv. Chim. Acta* **1975**, *58*, 1094.

[1574] Leroux, F.; Galland, A.; Schlosser, M.; unpublished results (**1999**).

[1575] Kotthaus, M.; Schlosser, M.; *Tetrahedron Lett.* **1998**, *39*, 4031.

[1576] Schlosser, M.; Franzini, L.; unpublished results (**1993**); Brunner-Franzini, L.; *Doctoral Dissertation*, Université de Lausanne, **1996**, pp. 15–18, 74–78 and 81–82.

[1577] Linstrumelle, G.; Michelot, D.; *J. Chem. Soc., Chem. Commun.* **1975**, 561.

[1578] Michelot, D.; Linstrumelle, G.; *Tetrahedron Lett.* **1976**, *17*, 275.

[1579] Clinet, J.-C.; Linstrumelle, G.; *Nouv. J. Chim.* **1977**, *1*, 373; *Synthesis* **1981**, 875.

[1580] Michelot, D.; Clinet, J.-C.; Linstrumelle, G.; *Synth. Commun.* **1982**, *12*, 739.

[1581] Audin, P.; Doutheau, A.; Ruest, L.; Goré, J.; *Bull. Soc. Chim. Fr. II* **1981**, 313.

[1582] Hooz, J.; Cazada, J. G.; McMaster, D.; *Tetrahedron Lett.* **1985**, *26*, 271; *Org. Synth., Coll. Vol.* **1993**, *8*, 226.

[1583] Rajagopalan, S.; Zweifel, G.; *Synthesis* **1984**, 111.

[1584] Bauer, D.; Köbrich, G.; *Chem. Ber.* **1976**, *109*, 2185.

[1585] Köbrich, G.; Merkel, D.; *Liebigs Ann. Chem.* **1972**, *761*, 50, *Chem. Ber.* **1973**, 2040.

[1586] Ford, W. T.; Newcomb, M.; *J. Am. Chem. Soc.* **1974**, *96*, 309.

[1587] Ernst, R. D.; Freeman, J. W.; Swepston, P. N.; Wilson, D. R.; *J. Organomet. Chem.* **1991**, *402*, 17.

[1588] Ayalon-Chass, D.; Ehlinger, E.; Magnus, P.; *J. Chem. Soc., Chem. Commun.* **1977**, 722; Ehlinger, E.; Magnus, P.; *Tetrahedron Lett.* **1980**, *211*, 11.

[1589] Tsai, D. J. S.; Matteson, D. S.; *Tetrahedron Lett.* **1981**, *22*, 2751.

[1590] Koumaglo, K.; Chan, T. H.; *Tetrahedron Lett.* **1984**, *25*, 717.

[1591] Corriu, R. J. P.; Masse, J.; Samate, D.; *J. Organomet. Chem.* **1975**, *93*, 71; Corriu, R. J. P.; Lanneau, G. F.; Masse, J.; Samate, D.; *J. Organomet. Chem.* **1977**, *127*, 281; Corriu, R. J. P.; Guerin, C.; M'Boula, J.; *Tetrahedron Lett.* **1981**, *22*, 2985.

[1592] Li, L.-H.; Wang, D.; Chan, T. H.; *Tetrahedron Lett.* **1991**, *32*, 2879.

[1593] Horvath, R. F.; Chan, T. H.; *J. Org. Chem.* **1989**, *54*, 317.

[1594] Mordini, A.; Palio, G.; Ricci, A.; Taddei, M.; *Tetrahedron Lett.* **1988**, *29*, 4991.

[1595] Moret, E.; Franzini, L.; Schlosser, M.; *Chem. Ber.* **1997**, *130*, 335.

[1596] Martin, S. F.; DuPriest, M. T.; *Tetrahedron Lett.* **1977**, *18*, 3925.

[1597] Ahlbrecht, H.; Eichler, J.; *Synthesis* **1974**, 672.

[1598] Eisch, J. J.; Shah, J. H.; *J. Org. Chem.* **1991**, *56*, 2955.

[1599] Julia, M.; Schouteeten, A.; Baillarge, M.; *Tetrahedron Lett.* **1974**, *15*, 3433.

[1600] Katritzky, A. R.; Li, J.-Q.; Malhotra, N.; *Liebigs Ann. Chem.* **1992**, 843.

[1601] Quast, H.; Weise Vélez, C. A.; *Angew. Chem.* **1978**, *90*, 224; *Angew. Chem. Int. Ed. Engl.* **1978**, *17*, 213.

[1602] Quast, H.; Weise Vélez, C. A.; *Angew. Chem.* **1974**, *86*, 380; *Angew. Chem. Int. Ed. Engl.* **1974**, *13*, 342.

[1603] Resek, J. E.; Beak, P.; *Tetrahedron Lett.* **1993**, *34*, 3043.

[1604] Evans, D. A.; Andrews, G. C.; Buckwalter, B.; *J. Am. Chem. Soc.* **1974**, *96*, 5560.

[1605] Still, W. C.; Macdonald, T. L.; *J. Am. Chem. Soc.* **1974**, *96*, 5561; *J. Org. Chem.* **1976**, *41*, 3620.

[1606] Seyferth, D.; Mammarella, R. E.; Klein, H. A.; *J. Organomet. Chem.* **1980**, *194*, 1.

[1607] Rautenstrauch, V.; *Helv. Chim. Acta* **1972**, *55*, 594, 3064.

[1608] Oakes, F. T.; Saylor, R. W.; Sebastian, J. F.; *Synth. Commun.* **1982**, *12*, 607.

[1609] Funk, R. L.; Bolton, G. L.; *Tetrahedron Lett.* **1988**, 29, 1111; *J. Am. Chem. Soc.* **1988**, *110*, 1290.

[1610] Kozikowski, A. P.; Isobe, K.; *Tetrahedron Lett.* **1979**, *20*, 833.

[1611] Eisch, J. J.; Galle, J. E.; *J. Org. Chem.* **1990**, *55*, 4835.

[1612] Carlson, R. M.; *Tetrahedron Lett.* **1978**, *19*, 111; Carlson, R. M.; White, L. L.; *Synth. Commun.* **1983**, *13*, 237; Liu, T.-l.; Carlson, R. M.; *Synth. Commun.* **1993**, *23*, 1437.

[1613] Cardillo, G.; Contento, M.; Sandri, S.; *Tetrahedron Lett.* **1974**, *15*, 2215.

[1614] Wang, W.-B.; Roskamp, E. J.; *Tetrahedron Lett.* **1992**, *33*, 7631.

[1615] Hosomi, A.; Ando, M.; Sakurai, H.; *Chem. Lett.* **1984**, 1385.

[1616] Julia, M; Verpeaux, J.-N.; Zahneisen, T.; *Bull. Soc. Chim. Fr.* **1994**, *131*, 539.

[1617] Angeletti, E.; Baima, R.; Canepa, C.; Degani, I.; Tonachini, G.; Venturello, P.; *Tetrahedron* **1989**, *45*, 7827.

[1618] Hiyama, T.; Shinoda, M.; Tsukanaka, M.; Nozaki, H.; *Bull. Chem. Soc. Jpn.* **1980**, *53*, 1010.

[1619] Bryce-Smith, D.; Turner, E. E.; *J. Chem. Soc.* **1950**, 1975.

[1620] Bach, R. D.; Klix, R. C.; *Tetrahedron Lett.* **1986**, *27*, 1983.

[1621] Engelhardt, L. M.; Leung, W.-P.; Raston, C. L.; White, A. H.; *J. Chem. Soc., Dalton Trans.* **1985**, 337.

[1622] Faigl, F.; Schlosser, M.; *Tetrahedron Lett.* **1991**, *32*, 3369.

[1623] Hart, A. J.; O'Brien, D. H.; Russell, C. R.; *J. Organomet. Chem.* **1974**, *72*, C19.

[1624] Thurner, A.; Ágai, B.; Faigl, F.; *J. Chem. Res. (S)* **1998**, 158; see also: Ólier-Reuchet, C.; Aitken, D. H.; Bucourt, R.; Husson, H.-P.; *Tetrahedron Lett.* **1995**, *36*, 8221.

[1625] Schlosser, M.; Schneider, P.; *Helv. Chim. Acta* **1980**, *63*, 2404.

[1626] Maercker, A.; Berkulin, W.; Schiess, P.; *Angew. Chem.* **1983**, *95*, 248; *Angew. Chem. Int. Ed. Engl.* **1983**, *22*, 246; see also: ref. [1417].

[1627] Fuhrer, W.; Gschwend, H. W.; *J. Org. Chem.* **1979**, *44*, 1133.

[1628] Houlihan, W. J.; Parrino, V. A.; Uike, Y.; *J. Org. Chem.* **1981**, *46*, 4511.

[1629] Stabler, S. R.; quoted according to Clark, R. D.; Jahangir, A.; *Org. React.* **1995**, *47*, 1.

[1630] Clark, R. D.; Muchowski, J. M.; Fisher, L. E.; Flippin, L. A.; Repke, D. B.; Souchet, M.; *Synthesis* **1991**, 871.

[1631] Schlosser, M.; Maccaroni, P.; Marzi, E.; *Tetrahedron* **1998**, *54*, 2763.

[1632] Michaelides, M. R.; Schoenleber, R.; Thomas, S.; Yamamoto, D. M.; Britton, D. R.; MacKenzie, R.; Kebabian, J. W.; *J. Med. Chem.* **1991**, *34*, 2946.

[1633] Kun, K. A.; Cassidy, H. G.; *J. Org. Chem.* **1962**, *27*, 841.

[1634] Braun, M.; Ringer, E.; *Tetrahedron Lett.* **1983**, *24*, 1233.

[1635] Sano, H.; Ohtsuka, H.; Migita, T.; *J. Am. Chem. Soc.* **1988**, *110*, 2014.

[1636] Takagishi, S.; Schlosser, M.; *Synlett* **1991**, 119.

[1637] Brandsma, L.; Verkruijsse, H. D.; Schade, C.; Schleyer, P. V. R.; *J. Chem. Soc., Chem. Commun.* **1986**, 260.

[1638] Stähle, M.; Lehmann, R.; Kramař, J.; Schlosser, M.; *Chimia* **1985**, 229.

[1639] Finnigan, R. A.; McNees, R. S.; *Chem. Ind. (London)* **1996**, 1450.

[1640] Nativi, C.; Radivà, N.; Ricci, A.; Seconi, G.; Taddei, M.; *J. Org. Chem.* **1991**, *56*, 1450.

[1641] Kramař, J.; *Doctoral Dissertation*, Université de Lausanne, **1978**, pp. 20–30 and 47–72.

[1642] Schipperijn, A. J.; Smael, P.; *Recl. Trav. Chim. Pays-Bas* **1973**, *92*, 1121; Schipperijn, A. J.; *Recl. Trav. Chim. Pays-Bas* **1986**, *105*, 66.

[1643] Nesmeyanova, O. A.; Rudashevskaya, T. Y.; Lukina, M. Y.; *Izvest. Akad. Nauk S.S.S.R., Otdel. Khim. Nauk* **1965**, 1510; *Chem. Abstr.* **1965**, *63*, 16221g; Isaka, M.; Matsuzawa, S.; Yamago, S.; Ejiri, S.; Miyachi, Y.; Nakamura, E.; *J. Org. Chem.* **1989**, *54*, 4727.

[1644] Dumont, C.; Vidal, M.; *Bull. Soc. Chim. Fr.* **1973**, 2301; see also: Closs, G. L.; in *Adv. Alicycl. Chem., Vol. 1* (Hart, H.; Karabatsos, G. I.; eds.), Acad. Press, New York, **1966**, p. 102.

[1645] Kumagai, T.; Aga, M.; Okada, K.; Oda, M.; *Bull. Chem. Soc. Jpn.* **1991**, *64*, 1428.

[1646] Stork, G.; Shiner, C. S.; Cheng, C.-W.; Polt, R. L.; *J. Am. Chem. Soc.* **1986**, *108*, 304.

[1647] Katritzky, A. R.; Li, J.-q.; Malhotra, N.; *Liebigs Ann. Chem.* **1992**, 843.

[1648] Beak, P.; Lee, W. K.; *J. Org. Chem.* **1993**, *58*, 1109.

[1649] See also: Paquette, L. A.; Tae, J.-s.; *Tetrahedron Lett.* **1997**, *38*, 3151.

[1650] Schlosser, M.; Schneider, P.; *Angew. Chem.* **1979**, *91*, 515; *Angew. Chem. Int. Ed. Engl.* **1979**, *18*, 489.

[1651] Stout, D. M.; Takaya, T.; Meyers, A. I.; *J. Org. Chem.* **1975**, *40*, 563.

[1652] Meyers, A. I.; Edwards, P. D.; Bailey, T. R.; Jagdmann, G. E.; *J. Org. Chem.* **1985**, *50*, 1019.

[1653] Comins, D. L.; *Tetrahedron Lett.* **1983**, *24*, 2807.

[1654] Comins, D. L.; Weglarz, M. A.; *J. Org. Chem.* **1988**, *53*, 4437;**1991**, 2506.

[1655] Schmidt, R. R.; Talbiersky, J.; *Angew. Chem.* **1976**, *88*, 193; *Angew. Chem. Int. Ed. Engl.* **1976**, *15*, 171; *Angew. Chem.* **1977**, *89*, 891; *Angew. Chem. Int. Ed. Engl.* **1977**, *16*, 853; *Angew. Chem.* **1978**, *90*, 220; *Angew. Chem. Int. Ed. Engl.* **1978**, *17*, 204.

[1656] Schmidt, R. R.; Talbiersky, J.; *Synthesis* **1977**, 869.

[1657] Schmidt, R. R.; Berger, G.; *Chem. Ber.* **1976**, *109*, 2936.

[1658] Meghani, P.; Joule, J. A.; *J. Chem. Soc., Perkin Trans. 1* **1988**, 1.

[1659] Alvarez, M.; Salas, M.; Rigat, L.; Veciana, A., de; Joule, J. A.; *J. Chem. Soc., Perkin Trans. 1* **1992**, 351.

[1660] Urata, H.; Tanaka, M.; Fuchikami, T.; *Chem. Lett.* **1987**, 751.

[1661] Katritzky, A. R.; Fan, W.-Q.; Koziol, A. E.; Palenik, G. J.; *Tetrahedron* **1987**, *43*, 2343.

[1662] Barluenga, J.; Foubelo, F.; Gonzáles, R.; Fañanás, F. J.; Yus, M.; *J. Chem. Soc., Chem. Commun.* **1990**, 587; Barluenga, J.; Gonzáles, R.; Fañanás, F. J.; *Tetrahedron Lett.* **1992**, *33*, 831.

[1663] Barluenga, J.; Fañanás, F. J.; Foubelo, F.; Yus, M.; *J. Chem. Soc., Chem. Commun.* **1988**, 1135; *Tetrahedron Lett.* **1988**, *29*, 4859.

[1664] Hänssgen, D.; Odenhausen, E.; *Chem. Ber.* **1979**, *112*, 2389.

[1665] Burns, S. A.; Corriu, R. J. P.; Huynh, V.; Moreau, J. J. E.; *J. Organomet. Chem.* **1987**, *333*, 281.

[1666] Schulze, J.; Boese, R.; Schmid, G.; *Chem. Ber.* **1981**, *114*, 1297.

[1667] Baldwin, J. E.; Höfle, G. A.; Lever, O. W.; *J. Am. Chem. Soc.* **1974**, *96*, 7125.

[1668] Baldwin, J. E.; Lever, O. W.; Tzodikov, N. R.; *J. Org. Chem.* **1976**, *41*, 2312, 2874.

[1669] Levy, A. B.; Schwartz, S. J.; *Tetrahedron Lett.* **1976**, *17*, 2201; Levy, A. B.; Schwartz, S. J.; Wilson, N.; Christie, B.; *J. Organomet. Chem.* **1978**, *156*, 123.

[1670] Schöllkopf, U.; Hänßle, P.; *Liebigs Ann. Chem.* **1972**, *763*, 208; see also: Hoppe, I.; Schöllkopf, U.; *Liebigs Ann. Chem.* **1980**, 1474.

[1671] Dexheimer, E. M.; Spialter, L.; *J. Organomet. Chem.* **1976**, *107*, 229.

[1672] Soderquist, J. A.; Hassner, A.; *J. Am. Chem. Soc.* **1980**, *102*, 1577.

[1673] Venturello, P.; Deagostino, A.; unpublished results (**1999**); see also: Balma Tivola, P.; Deagostino, A.; Fenoglio, C.; Mella, M.; Prandi, C.; Venturello, P.; *Eur. J. Org. Chem.* **1999**, 2143.

[1674] O'Connor, B. R.; *J. Org. Chem.* **1968**, *33*, 1991.

[1675] Boeckman, R. K.; Bruza, K. J.; *Tetrahedron Lett.* **1977**, *18*, 4187; *Tetrahedron* **1981**, *37*, 3997; see also: Boeckman, R. K.; Charette, A. B.; Asberom, T.; Johnston, B. H.; *J. Am. Chem. Soc.* **1991**, *113*, 5337.

[1676] Oakes, F. T.; Yang, F. A.; Sebastian, J. F.; *J. Org. Chem.* **1982**, *47*, 3094.

[1677] Hedtmann, U.; Welzel, P.; *Tetrahedron Lett.* **1985**, *26*, 2773.

[1678] Paul, R.; Tchelitcheff, S.; *Bull. Soc. Chim. Fr.* **1952**, 808.

[1679] Contradictory results: Barber, C.; Jarowicki, K.; Kocieński, P.; *Synlett* **1991**, 197.

[1680] Weber, G. F.; Hall, S. S.; *J. Org. Chem.* **1979**, *44*, 364.

[1681] Saylor, R. W.; Sebastian, J. F.; *Synth. Commun.* **1982**, *12*, 579.

[1682] Fétizon, M.; Goulaouic, P.; Hanna, I.; Prangé, T.; *J. Org. Chem.* **1988**, *53*, 5672; Fétizon, M.; Hanna, I.; Rens, J.; *Tetrahedron Lett.* **1985**, *26*, 3453; Fétizon, M.; Goulaouic, P.; Hanna, I.; *Synthesis* **1987**, 503.

[1683] Cox, P.; Mahon, M. F.; Molloy, K. C.; Lister, S.; Gallagher, T.; *Tetrahedron Lett.* **1988**, *29*, 1993; **1989**, *30*, 2437.

[1684] Friesen, R. W.; Sturino, C. F.; *J. Org. Chem.* **1990**, *55*, 2572.

[1685] Schmidt, R. R.; Preuss, R.; Betz, R.; *Tetrahedron Lett.* **1987**, *28*, 6591.

[1686] Schlosser, M.; Schaub, B.; Spahić, B.; Sleiter, G.; *Helv. Chim. Acta* **1973**, *56*, 2166.

[1687] Riobé, O.; Lebouc, A.; Delaunay, J.; *C. R. Séances Acad. Sci. C* **1977**, *284*, 281; *Chem. Abstr.* **1977**, *86*, 189646e; Hassan, M. A.; *J. Chem. Soc. Pak.* **1983**, *5*, 103; *Chem. Abstr.* **1983**, *99*, 212107j.

[1688] Drakesmith, F. G.; Richardson, R. D.; Stewart, O. J.; Tarrant, P.; *J. Org. Chem.* **1968**, *33*, 286; see also: Drakesmith, F. G.; Stewart, O. J.; Tarrant, P.; *J. Org. Chem.* **1968**, *33*, 472.

[1689] Taguchi, T.; Morikawa, T.; Kitagawa, O.; Mishima, T.; Kobayashi, Y.; *Chem. Pharm. Bull.* **1985**, *33*, 5137.

[1690] Martinet, P.; Sauvêtre, R.; Normant, J.-F.; *J. Organomet. Chem.* **1989**, *367*, 1.

[1691] Burton, D. J.; Spawn, T. D.; Heinze, P. L.; Bailey, A. R.; Shin-ya, S.; *J. Fluorine Chem.* **1989**, *44*, 167.

[1692] Morken, P. A.; Lu, H.-y.; Nakamura, A.; Burton, D. J.; *Tetrahedron Lett.* **1991**, *32*, 4271.

[1693] Campbell, S. F.; Stephens, R.; Tatlow, J. C.; *J. Chem. Soc., Chem. Commun.* **1967**, 151.

[1694] Huebner, C. F.; Puckett, R. T.; Brzechffa, M.; Schwartz, S. L.; *Tetrahedron Lett.* **1970**, *11*, 359.

[1695] Kenndoff, J.; Polborn, K.; Szeimies, G.; *J. Am. Chem. Soc.* **1990**, *112*, 6117.

[1696] Köbrich, G.; Flory, K.; *Tetrahedron Lett.* **1964**, *5*, 1137; *Chem. Ber.* **1966**, *99*, 1773.

[1697] Kasatkin, A.; Whitby, R. J.; *Tetrahedron Lett.* **1997**, *38*, 4857.

[1698] Köbrich, G.; Drischel, W.; *Angew. Chem.* **1965**, *77*, 95; *Angew. Chem. Int. Ed. Engl.* **1965**, *4*, 74; *Tetrahedron* **1966**, *22*, 2621; see also: Nelson, D. J.; *J. Org. Chem.* **1984**, *49*, 2059.

[1699] Schlosser, M.; Ladenberger, V.; *Chem. Ber.* **1967**, *100*, 3893.

[1700] Köbrich, G.; Trapp, H.; Hornke, I.; *Tetrahedron Lett.* **1964**, *5*, 1131; Köbrich, G.; Trapp, H.; *Chem. Ber.* **1966**, *99*, 670, 680.

[1701] Nesmeyanov, A. N.; Perevalova, E. G.; Golovnya, R. V.; Nesmeyanova, O. A.; *Dokl. Akad. Nauk S.S.S.R.* **1954**, *97*, 459; *Chem. Abstr.* **1956**, *49*, 9633f; see also: Benkeser, R. A.; Goggin, D.; Schroll, G.; *J. Am. Chem. Soc.* **1954**, *76*, 4025.

[1702] Sanders, R.; Mueller-Westerhoff, U. T.; *J. Organomet. Chem.* **1996**, *512*, 219.

[1703] Arguarch, G.; Samuel, O.; Kagan, H. B.; *Eur. J. Org. Chem.* **2000**, 2885; Arguarch, G.; Samuel, O.; Riant, O.; Daran, J.-C.; Kagan, H. B.; *Eur. J. Org. Chem.* **2000**, 2893.

[1704] Slocum, D. W.; Rockett, B. W.; Hauser, C. R.; *J. Am. Chem. Soc.* **1965**, *87*, 1241.

[1705] Marquarding, D.; Klusacek, H.; Gokel, G.; Hoffmann, P.; Ugi, I.; *J. Am. Chem. Soc.* **1970**, *92*, 5389.

[1706] Hayashi, T.; Mise, T.; Fukushima, M.; Kagotani, M.; Nagashima, N.; Hamada, Y.; Matsumoto, A.; Kawakami, S.; Kanishi, M.; Yamamoto, K.; Kumada, M.; *Bull. Chem. Soc. Jpn.* **1980**, *53*, 1138; see also: Butler, I. R.; Cullent, W. R.; Rettig, S. J.; *Organometallics* **1986**, *5*, 1320.

[1707] Aratani, T.; Gonda, T.; Nozaki, H.; *Tetrahedron* **1970**, *26*, 5453.

[1708] Booth, D. J.; Rockett, B. W.; *Tetrahedron Lett.* **1967**, *8*, 1483.

[1709] Slocum, D. W.; Stonemark, F. E.; *J. Org. Chem.* **1973**, *38*, 1677.

[1710] Slocum, D. W.; Engelmann, T. R.; Jennings, C. A.; *Austr. J. Chem.* **1968**, *21*, 2319; Slocum, D. W.; Jennings, C. A.; Engelmann, T. R.; Rocket, B. W.; Hauser, C. R.; *J. Org. Chem.* **1971**, *36*, 377; Slocum, D. W.; Jones, W. E.; Ernst, C. R.; *J. Org. Chem.* **1972**, *37*, 4278.

[1711] Schlosser, M.; Choi, J. H.; Takagishi, S.; *Tetrahedron* **1990**, *46*, 5633.

[1712] Baston, E.; Wang, Q.; Schlosser, M.; *Tetrahedron Lett.* **2000**, *41*, 667.

[1713] Brandsma, L.; Hommes, H.; Verkruijsse, H. D.; de Jong, R. L. P.; *Recl. Trav. Chim. Pays-Bas* **1985**, *104*, 226.

[1714] Wittig, G.; Merkle, W.; *Ber. Dtsch. Chem. Ges.* **1942**, *75*, 1491.

[1715] Lepley, A. R.; Khan, W. A.; Giumanini, A. B.; Giumanini, A. G.; *J. Org. Chem.* **1966**, *31*, 2047.

[1716] Ludt, R. E.; Growther, G. P.; Hauser, C. R.; *J. Org. Chem.* **1970**, *35*, 1288.

[1717] Hay, J. V.; Harris, T. M.; *Org. Synth., Coll. Vol.* **1988**, *6*, 478.

[1718] Jung, I. N.; Jones, P. R.; *J. Am. Chem. Soc.* **1975**, *97*, 6102.

[1719] Harder, S.; Boersma, J.; Brandsma, L.; Kanters, J. A.; Bauer, W.; Schleyer, P. V. R.; *Organometallics* **1989**, *8*, 1696.

[1720] Loppnow, G. R.; Melamed, D.; Hamilton, A. D.; Spiro, T. G.; *J. Phys. Chem.* **1993**, *97*, 8957.

[1721] Schlosser, M.; Takagishi, S.; unpublished results (**1990**).

[1722] Gilman, H.; Gray, S.; *J. Org. Chem.* **1958**, *23*, 1476.

[1723] Cauquil, G.; Casadevall, A.; Casadevall, E.; *Bull. Soc. Chim. Fr.* **1960**, 1049.

[1724] Muchowski, J. M.; Venuti, M. C.; *J. Org. Chem.* **1980**, *45*, 4798.

[1725] Hillis, L. R.; Gould, S. J.; *J. Org. Chem.* **1985**, *50*, 718.

[1726] Stanetty, P.; Koller, H.; Mihovilovic, M.; *J. Org. Chem.* **1992** , *57*, 6833.

[1727] Ubeda, J. I.; Villacampa, M.; Avendaño, C.; *Synlett* **1997**, 2833.

[1728] Cho, I.-s.; Gong, L.-y.; Muchowski, J. M.; *J. Org. Chem.* **1991**, *56*, 7288.

[1729] Reed, J. N.; Rotchford, J.; Strickland, D.; *Tetrahedron Lett.* **1988**, *29*, 5725.

[1730] Takagishi, S.; Katsoulos, G.; Schlosser, M.; *Synlett* **1992**, 360.

[1731] Maggi, R.; Ginanneschi, A.; Schlosser, M.; unpublished results (**1996**).

[1732] Reuter, D. C.; Flippin, L. A.; McIntosh, J.; Caroon, J. M.; Hammaker, J.; *Tetrahedron Lett.* **1994**, *35*, 4899.

[1733] Macdonald, J. E.; Poindexter, G. S.; *Tetrahedron Lett.* **1987**, *28*, 1851.

[1734] Jones, F. N.; Vaulx, R. L.; Hauser, C. R.; *J. Org. Chem.* **1963**, *28*, 3461.

[1735] Klein, K. P.; Hauser, C. R.; *J. Org. Chem.* **1967**, *32*, 1479.

[1736] Narasimhan, N. S.; Bhide, B. H.; *Tetrahedron* **1971**, *27*, 6171.

[1737] Slocum, D. W.; Bock, G.; Jennings, C. A.; *Tetrahedron Lett.* **1970**, *11*, 3443; Slocum, D. W.; Jennings, C. A.; *J. Org. Chem.* **1976**, *41*, 3653.

[1738] Narasimhan, N. S.; Mali, R. S.; Kulkarni, B. K.; *Tetrahedron Lett.* **1981**, *22*, 2797; see also: Ager, D.; *Tetrahedron Lett.* **1981**, *22*, 2923; Hlasta, D. J.; Bell, M. R.; *Tetrahedron Lett.* **1985**, *26*, 2151.

[1739] Dai-Ho, G.; Mariano, P. S.; *J. Org. Chem.* **1987**, *52*, 704.

[1740] Beak, P.; Brown, R. A.; *J. Org. Chem.* **1977**, *42*, 1823.

[1741] Wang, X.; Snieckus, V.; *Synlett* **1990**, 313.

[1742] de Silva, S. O.; Reed, J. N.; Billedeau, R. J.; Wang, X.; Norris, D. J.; Snieckus, V.; *Tetrahedron* **1992**, *48*, 4863.

[1743] de Silva, S. O.; Reed, N.; Snieckus, V.; *Tetrahedron Lett.* **1978**, *19*, 5099.

[1744] See also: Beak, P.; Snieckus, V.; *Acc. Chem. Res.* **1982**, *15*, 306; Watanabe, M.; Sahara, M.; Furukawa, S.; Billedeau, R. J.; Snieckus, V.; *Tetrahedron Lett.* **1982**, *23*, 1647; Potts, K. T.; Bhattacharjee, D.; Walsh, E. B.; *J. Org. Chem.* **1986**, *51*, 2011.

[1745] Mills, R. J.; Snieckus, V.; *Tetrahedron Lett.* **1984**, *25*, 483.

[1746] Zani, C. L.; de Olivéira, A. B.; Snieckus, V.; *Tetrahedron Lett.* **1987**, *28*, 6561; see also: de Silva, S. O.; Ahmad, I.; Snieckus, V.; *Can. J. Chem.* **1979**, *57*, 1598.

[1747] Bindal, R. D.; Katzenellenbogen, J. A.; *J. Org. Chem.* **1987**, *52*, 3181.

[1748] Iwao, M.; Watanabe, M.; de Silva, S. O.; Snieckus, V.; *Tetrahedron Lett.* **1981**, *22*, 2349.

[1749] Puterbaugh, W. H.; Hauser, C. R.; *J. Org. Chem.* **1964**, *29*, 853.

[1750] Mao, C. L.; Barnish, I. T.; Hauser, C. R.; *J. Heterocycl. Chem.* **1969**, *6*, 475.

[1751] Bhide, B. H.; Parekh, H. J.; *Chem. Ind. (London)* **1974**, 733; Bhide, B. H.; Gupta, V. P.; *Indian J. Chem.* **1977**, *15B*, 30; *Chem. Abstr.* **1977**, *87*, 68077g.

[1752] Fitt, J. J.; Gschwend, H. W.; *J. Org. Chem.* **1976**, *41*, 4029.

[1753] Shirley, D. A.; Johnson, J. R.; Hendrix, J. P.; *J. Organomet. Chem.* **1968**, *11*, 209.

[1754] Vetter, W.; Schill, G.; Zürcher, C.; *Chem. Ber.* **1974**, *107*, 424.

[1755] Finnegan, R. A.; Altschuld, J. W.; *J. Organomet. Chem.* **1967**, *9*, 193.

[1756] Shirley, D. A.; Hendrix, J. P.; *J. Organomet. Chem.* **1968**, *11*, 217.

[1757] Wittig, G.; Fuhrmann, G.; *Ber. Dtsch. Chem. Ges.* **1940**, *73*, 1197.

[1758] Muthukrishnan, R.; Schlosser, M.; *Helv. Chim. Acta* **1976**, *59*, 13.

[1759] Oita, K.; Gilman, H.; *J. Org. Chem.* **1956**, *21*, 1009; *J. Am. Chem. Soc.* **1957**, *79*, 339; Gilman, H.; Trepka, W. J.; *J. Org. Chem.* **1961**, *26*, 5202; **1962**, *27*, 1418; Gilman, H.; Miles, D.; *J. Org. Chem.* **1958**, *23*, 1363.

[1760] Lüttringhaus, A.; von Sääf, G.; *Liebigs Ann. Chem.* **1939**, *542*, 241; **1947**, *557*, 25.

[1761] Wittig, G.; Pohmer, L.; *Chem. Ber.* **1956**, *89*, 1334.

[1762] Gilman, H.; Gorsich, R. D.; *J. Org. Chem.* **1957** , *22*, 687.

[1763] Gilman, H.; Gray, S.; *J. Org. Chem.* **1958**, *23*, 1476.

[1764] Gilman, H.; Young, R. V.; *J. Am. Chem. Soc.* **1934**, *56*, 1415; **1935**, *57*, 1121.

[1765] Shirley, D. A.; Cheng, C. F.; *J. Organomet. Chem.* **1969**, *20*, 251; Shirley, D. A.; Harmon, T. E.; Cheng, C. F.; *J. Organomet. Chem.* **1974**, *69*, 327.

[1766] Narasimhan, N. S.; Paradkar, M. V.; *Indian J. Chem.* **1969**, *7*, 536; Narasimhan, N. S.; Mali, R. S.; *Tetrahedron Lett.* **1973**, *14*, 843; *Tetrahedron* **1975**, *31*, 1005.

[1767] Houlihan, W. J.; Pieroni, A. J.; *J. Heterocycl. Chem.* **1973**, *10*, 405.

[1768] Wittig, G.; Uhlenbrock, W.; Weinhold, P.; *Chem. Ber.* **1962**, *95*, 1692.

[1769] Bergmann, E. D.; Pappo, R.; Ginsburg, D.; *J. Chem. Soc.* **1950**, 1369; see also: Barnes, R. A.; Reinhold, D. F.; *J. Am. Chem. Soc.* **1952**, *74*, 1327; Bruce, J. M.; *J. Chem. Soc.* **1959**, 2366, spec. 2374; Braun, D.; Seelig, E.; *Chem. Ber.* **1964**, *97*, 3098; Evans, D. A.; Mitch, C. H.; *Tetrahedron Lett.* **1982**, *23*, 285; Bunce, N. J.; Stephenson, K. L.; *Can. J. Chem.* **1989**, *67*, 220; Trauner, D.; Bats, J. W.; Werner, A.; Mulzer, J.; *J. Org. Chem.* **1998**, *63*, 5908.

[1770] Hoffmann, A. K.; Feldman, A. M.; Gelblum, E.; *J. Am. Chem. Soc.* **1964**, 646; Boltze, K.-H.; Dell, H.-E.; Jansen, H.; *Liebigs Ann. Chem.* **1967**, *709*, 63; Zweig, A.; Maurer, A. H.; Roberts, B. G.; *J. Org. Chem.* **1967**, *32*, 1322; Ishizu, K.; Mukai, K.; Shibayama, A.; Kondo, K.; *Bull. Chem. Soc. Jpn.* **1977**, *50*, 2269; Brachi, J.; Rieker, A.; *Synthesis* **1977**, 708; Reed, M. W.; Pollart, D. J.; Perri, S. T.; Foland, L. D.; Moore, H. W.; *J. Org. Chem.* **1988**, *53*, 2477; Vaillancourt, V.; Albizati, K. F.; *J. Org. Chem.* **1992**, *57*, 3627; Wada, M.; Mishima, H.; Watanabe, T.; Natsume, S.; Konishi, H.; Kirishima, K.; Hayase, S.; Erabi, T.; *Bull. Chem. Soc. Jpn.* **1995**, *68*, 243; Nagel, U.; Krink, T.; *Chem. Ber.* **1995**, *128*, 309.

[1771] Horner, L.; Simons, G.; *Phosphorus Sulfur* **1983**, *14*, 253; *Chem. Abstr.* **1983**, *98*, 179525y; Fujita, M.; Hiyama, T.; *J. Am. Chem. Soc.* **1984**, *106*, 4629; Kraus, G. A.; Hagen, M. D.; *J. Org. Chem.* **1985**, *50*, 3252; Turnbull, P.; Moore, H. W.; *J. Org. Chem.* **1995**, *60*, 3274.

[1772] Schill, G.; Logemann, E.; *Chem. Ber.* **1973**, *106*, 2910.

[1773] Maggi, R.; Gorgerat, S.; Schlosser, M.; unpublished results (**1999**).

[1774] van Koten, G.; Leusink, A. J.; Noltes, J. G.; *J. Organomet. Chem.* **1975**, *85*, 105; Baker, L.-J.; Bott, R. C.; Bowmaker, G. A.; Graham, A.; Healy, P. C.; Skelton, B. W.; Schwerdtfeger, P.; White, A. H.; *J. Chem. Soc., Dalton Trans.* **1995**, 1341; Nagel, U.; Nedden, H. G.; *Chem. Ber.* **1997**, *130*, 989.

[1775] Mathison, I. W.; Gueldner, R. C.; Carroll, D. M.; *J. Pharm. Sci.* **1968**, *57*, 1820; *Chem. Abstr.* **1969**, *70*, 3470j.

[1776] Schaefer, W.; Leute, R.; Schlude, H.-J.; *Tetrahedron Lett.* **1967**, *8*, 4303; Böhme, H.; Bomke, U.; *Arch. Pharm. (Weinheim)* **1970**, *303*, 779; *Chem. Abstr.* **1970**, *73*, 120232f.

[1777] Schill, G.; Murjahn, K.; *Chem. Ber.* **1971**, *104*, 3587.

[1778] Ellison, R. A.; Kotsonis, F. N.; *J. Org. Chem.* **1973**, *38*, 4192; see also: Young, S. D.; Coblens, K. E.; Ganem, B.; *Tetrahedron Lett.* **1981**, *22*, 4887; Wada, A.; Kanatomo, S.; Nagai, S.; *Chem. Pharm. Bull.* **1985**, *33*, 1016.

[1779] Townsend, C. A.; Bloom, L. M.; *Tetrahedron Lett.* **1981**, *22*, 3923.

[1780] Jeganathan, S.; Tsukamoto, M.; Schlosser, M.; *Synthesis* **1990**, 109.

[1781] Parham, W. E.; Anderson, E. L.; *J. Am. Chem. Soc* **1948**, *70*, 4187.

[1782] Stern, R.; English, J.; Cassidy, H. G.; *J. Am. Chem. Soc.* **1957**, *79*, 5797.

[1783] Townsend, C. A.; Davis, S. G.; Christensen, S. B.; Link, J. C.; Lewis, C. P.; *J. Am. Chem. Soc.* **1981**, *103*, 6885.

[1784] Matsumoto, T.; Kakigi, H.; Suzuki, K.; *Tetrahedron Lett.* **1991**, *32*, 4337.

[1785] Meyer, N.; Seebach, D.; *Chem. Ber.* **1980**, *113*, 1304.

[1786] Uemura, M.; Tokuyama, S.; *Chem. Lett.* **1975**, 1195.

[1787] House, H. O.; Bare, T. M.; Hanners, W. E.; *J. Org. Chem.* **1969**, *34*, 2209.

[1788] Schlosser, M.; Geneste, H.; *Chem. Eur. J.* **1998**, *4*, 1969.

[1789] Schlosser, M.; Simig, G.; Geneste, H.; *Tetrahedron* **1998**, *54*, 9023.

[1790] Gilman, H.; Soddy, T. S.; *J. Org. Chem.* **1957**, *22*, 1715.

[1791] Schlosser, M.; Katsoulos, G.; Takagishi, S.; *Synlett* **1990**, 747.

[1792] Tamborski, C.; Soloski, E. J.; *J. Org. Chem.* **1966**, *31*, 743, 746.

[1793] Büker, H. H.; Nibbering, N. M. M.; Espinosa, D.; Mongin, F.; Schlosser, M.; *Tetrahedron Lett.* **1997**, *38*, 8519.

[1794] Petersen, T. E.; Torssell, K.; Becher, J.; *Tetrahedron* **1973**, *29*, 3833, spec. 3842.

[1795] Tamborski, C.; Burton, W. H.; Breed, L. W.; *J. Org. Chem.* **1966**, *31*, 4229.

[1796] Harper, R. J.; Soloski, E. J.; Tamborski, C.; *J. Org. Chem.* **1964**, *29*, 2385.

[1797] Roberts, J. D.; Curtin, D. Y.; *J. Am. Chem. Soc.* **1946**, *68*, 1658.

[1798] Schlosser, M.; Mongin, F.; Porwisiak, J.; Dmowski, W.; Büker, H. H.; Nibbering, N. M. M.; *Chem. Eur. J.* **1998**, *4*, 1281.

[1799] Lukmanov, V. G.; Alekseeva, L. A.; Burmakov, A. I.; Yagupolskii, L. M.; *Zhur. Org. Khim.* **1973**, *9*, 1019; *Chem. Abstr.* **1973**, *79*, 42101v.

[1800] Aeberli, P.; Houlihan, W. J.; *J. Organomet. Chem.* **1974**, *67*, 321.

[1801] Bartle, K. D.; Hallas, G.; Hepworth, J. D.; *Org. Magn. Res.* **1973**, *5*, 479.

[1802] Carr, G. E.; Chambers, R. D.; Holmes, T. F.; Parker, D. G.; *J. Organomet. Chem.* **1987**, *325*, 13; Filler, R.; Gnandt, W. K.; Chen, W.; Lin, S.; *J. Fluorine Chem.* **1991**, *52*, 99.

[1803] Grocock, D. E.; Jones, T. K.; Hallas, G.; Hepworth, J. D.; *J. Chem. Soc. C* **1971**, 3305.

[1804] Schlosser, M.; Castagnetti, E.; unpublished results (**1999**).

[1805] Iwao, M.; *J. Org. Chem.* **1990**, *55*, 3622.

[1806] Marzi, E.; Schlosser, M.; unpublished results (**1998**).

[1807] Kress, T. H.; Leanna, M. R.; *Synthesis* **1988**, 803; Bennetau, B.; Rajarison, F.; Dunoguès, J.; Babin, P.; *Tetrahedron* **1993**, *49*, 10843.

[1808] Reich, H. J.; Wollowitz, S.; *J. Am. Chem. Soc.* **1982**, *104*, 7051.

[1809] Bochis, R. J.; Chabala, J. C.; Harris, E.; Peterson, L. H.; Barash, L.; Beattie, T.; Brown, J. E.; Graham, D. W.; Waksmunski, F. S.; Tischler, M.; Joshua, H.; Smith, J.; Colwell, L. F.; Wyvratt, M. J.; Fisher, M. H.; Tamas, T.; Nicolich, S.; Schleim, K. D.; Wilks, G.; Olson, G.; *J. Med. Chem.* **1991**, *34*, 2834; see also: Haiduc, I.; Gilman, H.; *J. Organomet. Chem.* **1968**, *12*, 394.

[1810] Tamborski, C.; Soloski, E. J.; Dills, C. E.; *Chem. Ind. (London)* **1965**, 2067.

[1811] Shirley, D. A.; Gross, B. H.; Roussel, P. A.; *J. Org. Chem.* **1955**, *20*, 225.

[1812] Gjøs, N.; Gronowitz, S.; *Acta. Chem. Scand.* **1971**, *25*, 2596.

[1813] Wrackmeyer, B.; Nöth, H.; *Chem. Ber.* **1976**, *109*, 1075.

[1814] Muchowski, J. M.; Solas, D. R.; *J. Org. Chem.* **1984**, *49*, 203; see also: Chadwick, D. J.; Hodgson, S. T.; *J. Chem. Soc., Perkin Trans. 1* **1982**, 1833.

[1815] Hasan, I.; Marinelli, E. R.; Lin, L.-C. C.; Fowler, F. W.; Levy, A. B.; *J. Org. Chem.* **1981**, *46*, 157.

[1816] Gharpure, M.; Stoller, A.; Bellamy, F.; Firnau, G.; Snieckus, V.; *Synthesis* **1991**, 1079.

[1817] Katritzky, A. R.; Akutagawa, K.; *Org. Prep. Proc. Int.* **1988**, *20*, 585; *Chem. Abstr.* **1989**, *111*, 173934w.

[1818] Marxer, A.; Siegrist, M.; *Helv. Chim. Acta* **1974**, *57*, 1988, spec. 1999.

[1819] Chadwick, D. J.; Cliffe, I. A.; *J. Chem. Soc., Perkin Trans. 1* **1979**, 2845.

[1820] Minato, A.; Tamao, K.; Hayashi, T.; Suzuki, K.; Kumada, M.; *Tetrahedron Lett.* **1981**, *22*, 5319.

[1821] Faigl, F.; Schlosser, M.; *Tetrahedron* **1993**, *49*, 10271.

[1822] Faigl, F.; Fogassy, K.; Thurner, A.; Töke, L.; *Tetrahedron* **1997**, *53*, 4883.

[1823] Faigl, F.; Fogassy, K.; Szücz, E.; Kovács, K.; Keserü, G. M.; Harmat, V.; Böcskei, Z.; Töke, L.; *Tetrahedron* **1999**, *55*, 7881.

[1824] Shirley, D. A.; Roussel, P. A.; *J. Am. Chem. Soc.* **1953**, *75*, 375.

[1825] Ziegler, F. E.; Spitzner, E. B.; *J. Am. Chem. Soc.* **1970**, *92*, 3492; **1973**, *95*, 7146.

[1826] Sundberg, R. J.; Parton, R. L.; *J. Org. Chem.* **1976**, *41*, 163.

[1827] Bosin, T. R.; Rogers, R. B.; *J. Labelled Compds.* **1973**, *9*, 395; *Chem. Abstr.* **1974**, *80*, 47761r.

[1828] Sundberg, R. J.; Russel, H. R.; *J. Org. Chem.* **1972**, *38*, 3324.

[1829] Hlasta, D. J.; Bell, M. R.; *Heterocycles* **1989**, *29*, 840.

[1830] Katritzky, A. R.; Lue, P.; Chen, Y. X.; *J. Org. Chem.* **1990**, *55*, 3688.

[1831] Katritzky, A. R.; Akutagawa, K.; *Tetrahedron Lett.* **1985**, *26*, 5935; Bergman, J.; Venemalm, L.; *Tetrahedron Lett.* **1987**, *28*, 3741; **1988**, *29*, 2993; *J. Org. Chem.* **1992**, *57*, 2495; see also: Buttery, C. D.; Jones, R. G.; Knight, D. W.; *J. Chem. Soc., Perkin Trans. 1* **1993**, 1425.

[1832] Katritzky, A. R.; Akutagawa, K.; Jones, R. A.; *Synth. Commun.* **1988**, *18*, 1151.

[1833] Sundberg, R. J.; Broome, R.; Walters, C. P.; Schnur, D.; *J. Heterocycl. Chem.* **1981**, *18*, 807.

[1834] Saulnier, M. G.; Gribble, G. W.; *J. Org. Chem.* **1983**, *48*, 2690; Gribble, G. W.; Saulnier, M. G.; Obaza-Nutaitis, J. A.; Ketcha, D. M.; *J. Org. Chem.* **1992**, *57*, 5891.

[1835] Hüttel, R.; Schön, M. E.; *Liebigs Ann. Chem.* **1959**, *625*, 55.

[1836] Alley, P. W.; Shirley, D. A.; *J. Am. Chem. Soc.* **1958**, *80*, 6271.

[1837] Butler, D. E.; Alexander, S. M.; *J. Org. Chem.* **1972**, *37*, 215.

[1838] Katritzky, A. R.; Jayaram, C.; Vassilatos, S. N.; *Tetrahedron* **1983**, *39*, 2023.

[1839] Katritzky, A. R.; Lue, P.; Akutagawa, K.; *Tetrahedron* **1989**, *45*, 4253.

[1840] Iwata, S.; Qian, C.-P.; Tanaka, K.; *Chem. Lett.* **1992**, 357.

[1841] Shirley, D. A.; Alley, P. W.; *J. Am. Chem. Soc.* **1957**, *79*, 4922.

[1842] Roe, A. M.; *J. Chem. Soc.* **1963**, 2195.

[1843] Noyce, D. S.; Stowe, G. T.; *J. Org. Chem.* **1973**, *38*, 3762.

[1844] Jutzi, P.; Sakriss, W.; *Chem. Ber.* **1973**, *106*, 2815.

[1845] Bedford, C. D.; Harris, R. N.; Howd, R. A.; Miller, A.; Nolen, H. W.; Kenley, R. A.; *J. Med. Chem.* **1984**, *27*, 1431.

[1846] Iversen, P. E.; Lund, H.; *Acta Chem. Scand.* **1966**, *20*, 2649.

[1847] Whitten, J. P.; Mathews, D. P.; McCarthy, J. R.; *J. Org. Chem.* **1986**, *51*, 1891.

[1848] Lipshutz, B. H.; Vaccaro, W.; Huff, B.; *Tetrahedron Lett.* **1986**, *27*, 4095.

[1849] Chadwick, D. J.; Ngochindo, R. I.; *J. Chem. Soc., Perkin Trans. 1* **1984**, 481.

[1850] Carpenter, A. J.; Chadwick, D. J.; *Tetrahedron* **1986**, *42*, 2351.

[1851] Benkeser, R. A.; Currie, R. B.; *J. Am. Chem. Soc.* **1948**, *70*, 1780.

[1852] Ramanathan, V.; Levine, R.; *J. Org. Chem.* **1962**, *27*, 1216; see also: Gschwend, H. W.; Rodriguez, H. R.; *Org. React.* **1979**, *26*, 1, spec. 97.

[1853] Niwa, E.; Aoki, H.; Tanaka, H.; Munakata, K.; *Chem. Ber.* **1966**, *99*, 712.

[1854] Gombos, J.; Haslinger, E.; Zak, H.; Schmidt, U.; *Monatsh. Chem.* **1975**, *106*, 219; Schmidt, U.; Gombos, J.; Haslinger, E.; Zak, H.; *Chem. Ber.* **1976**, *109*, 2628.

[1855] Arco, M. J.; Trammell, M. H.; White, J. D.; *J. Org. Chem.* **1976**, *41*, 2075.

[1856] Büchi, G.; Wuest, H.; *J. Org. Chem.* **1966**, *31*, 977; see also: Goldfarb, Y. L.; Danyushevskii, Y. L.; Vinogradova, M. A.; *Dokl. Akad. Nauk S.S.S.R.* **1963**, *151*, 332; *Chem. Abstr.* **1963**, *59*, 8681c; Kametani, T.; Nemoto, H.; Takeuchi, M; Hibino, S.; Fukumoto, K.; *Chem. Pharm. Bull.* **1976**, *24*, 1354; *Chem. Abstr.* **1976**, *85*, 94514a.

[1857] Cederlund, B.; Lantz, R.; Hörnfeldt, A. B.; Thorstad, O.; Undheim, K.; *Acta Chem. Scand.* **1977**, *B31*, 198.

[1858] Kilchherr, S.; Leroux, F.; Schlosser, M.; unpublished results (**1999**).

[1859] Ohno, K.; Machida, M.; *Tetrahedron Lett.* **1981**, *22*, 4487; see also: Carpenter, A. J.; Chadwick, D. J.; *J. Org. Chem.* **1985**, *50*, 4362.

[1860] Ager, D. J.; *Tetrahedron Lett.* **1983**, *24*, 5441.

[1861] Milberg, A. N.; Wiley, R. A.; *J. Med. Chem.* **1978**, *21*, 245.

[1862] Thames, S. F.; Odom, H. C.; *J. Heterocycl. Chem.* **1966**, *3*, 490.

[1863] Kulnevich, V. G.; Zelikman, Z. I.; Shkrebets, A. I.; Tertov, B. A.; Ketslakh, M. M.; *Khim. Geterosikl. Soedin.* **1973**, *5*, 595; *Chem. Abstr.* **1973**, *79*, 42424c.

[1864] Iten, P. X.; Hofmann, A. A.; Eugster, C. H.; *Helv. Chim. Acta* **1978**, *61*, 430, spec. 435.

[1865] Knight, D. W.; Nott, A. P.; *J. Chem. Soc., Perkin Trans. 1* **1981**, 1125; **1983**, 791.

[1866] Nguyên, D. L.; Schlosser, M.; *Helv. Chim. Acta* **1977**, *60*, 2085.

[1867] Adamson, D. W.; *J. Chem. Soc.* **1950**, 885.

[1868] Schultz, R. D.; Taft, D. D.; O'Brien, J. P.; Shea, J. L.; Mork, H. M.; *J. Org. Chem.* **1963**, *28*, 1420.

[1869] Okuhara, K.; *J. Org. Chem.* **1976**, *41*, 1487.

[1870] Goldfarb, Y. L.; Kirmalova, M. L.; Kalik, M. A.; *Zh. Obshch. Khim.* **1959**, *29*, 2034; *Chem. Abstr.* **1960**, *54*, 8775h; see also: Gronowitz, S.; Gestblom, B.; *Arkiv Kemi* **1962**, *18*, 513; *Chem. Abstr.* **1962**, *57*, 3000c.

[1871] Gronowitz, S.; Moses, P.; Hörnfeldt, A. B.; Håkansson, R.; *Arkiv Kemi* **1961**, *17*, 165; *Chem. Abstr.* **1962**, *57*, 8528h; see also: Sicé, J.; *J. Org. Chem.* **1954**, *19*, 70.

[1872] Gronowitz, S.; Cederlund, B.; Hörnfeldt, A. B.; *Chem. Scr.* **1974**, *5*, 217; *Chem. Abstr.* **1974**, *81*, 49013t.

[1873] Leardini, R.; Martelli, G.; Spagnolo, P.; Tiecco, M.; *J. Chem. Soc. C* **1970**, 1464.

[1874] Kauffmann, T.; Wienhofer, E.; Wöltermann, A.; *Angew. Chem.* **1971**, *83*, 795; *Angew. Chem. Int. Ed. Engl.* **1971**, *10*, 741; Kauffmann, T.; Mitschker, A.; *Tetrahedron Lett.* **1973**, *14*, 4039.

[1875] Hawkins, R. T.; Stroup, D. B.; *J. Org. Chem.* **1969**, *34*, 1173.

[1876] Goldfarb, Y. L.; Fabrichnyi, B. P.; Rogovik, V. I.; *Isvest. Akad. Nauk S.S.S.R., Otdel. Khim. Nauk* **1965**, 515; *Chem. Abstr.* **1965**, *63*, 562f.

[1877] Sicé, J.; *J. Am. Chem. Soc.* **1953**, *75*, 3697.

[1878] Gronowitz, S.; Rosen, U.; *Chem. Scr.* **1971**, *1*, 33; *Chem. Abstr.* **1971** , *75*, 20080e.

[1879] Gronowitz, S.; *Arkiv Kemi* **1954**, 7, 361; *Chem. Abstr.* **1955**, *49*, 13216b.

[1880] Garner, J.; Howarth, G. A.; Hoyle, W.; Roberts, S. M.; Suschitzky, H.; *J. Chem. Soc., Perkin Trans. 1* **1976**, 994.

[1881] Bowden, K.; Crank, G.; Ross, W. J.; *J. Chem. Soc. [C]* **1968**, 172.

[1882] Shirley, D. A.; Gilmer, J. C.; *J. Org. Chem.* **1962**, *27*, 4421.

[1883] Edwards, P. D.; Lewis, J. J.; Perkins, C. W.; Trainor, D. A.; Wildonger, R. A.; *Eur. Pat. Appl.* EP 291234 (filed 17 Nov. 1988, to ICI Americas); *Chem. Abstr.* **1989**, *110*, 232092z; Baker, R.; Snow, R. J.; Saunders, J.; Showell, G. A.; *Eur. Pat. Appl.* EP 307141 (filed 15 March 1989, to Merck, Sharp & Dohme); *Chem. Abstr.* **1989**, *111*, 153786w.

[1884] Hodges, J. C.; Patt, W. C.; Connolly, C. J.; *J. Org. Chem.* **1991**, *56*, 449.

[1885] Hilf, C.; Bosold, F.; Harms, K.; Marsch, M.; Boche, G.; *Chem. Ber.* **1997**, *130*, 1213.

[1886] Dondoni, A.; Fantin, G.; Fogagnolo, M.; Medici, A.; Pedrini, P.; *Synthesis* **1987**, 693.

[1887] Whitney, S. E.; Rickborn, B.; *J. Org. Chem.* **1991**, *56*, 3058.

[1888] Schröder, R.; Schöllkopf, U.; Blume, E.; Hoppe, I.; *Liebigs Ann. Chem.* **1975**, 533.

[1889] Meyers, A. I.; Lawson, J. P.; *Tetrahedron Lett.* **1981**, *22*, 3163.

[1890] Jutzi, P.; Gilge, U.; *J. Organomet. Chem.* **1983**, *246*, 159.

[1891] Caton, M. P. L.; Jones, D. H.; Slack, R.; Wooldridge, K. R. H.; *J. Chem. Soc.* **1964**, 446.

[1892] Horning, D. E.; Muchowski, J. M.; *Can. J. Chem.* **1974**, *52*, 2950.

[1893] Metzger, J.; Koether, B.; *Bull. Soc. Chim. Fr.* **1953**, 708; Beraud, J.; Metzger, J.; *Bull. Soc. Chim. Fr.* **1962**, 2072.

[1894] Medici, A.; Pedrini, P.; Dondoni, A.; *J. Chem. Soc., Chem. Commun.* **1981**, 655; Dondoni, A.; Fantin, G.; Fogagnolo, M.; Medici, A.; Pedrini, P.; *J. Org. Chem.* **1988**, *53*, 1748.

[1895] Boche, G.; Hilf, C.; Harms, K.; Marsch, M.; Lohrenz, J. C. W.; *Angew. Chem.* **1995**, *107*, 509; *Angew. Chem. Int. Ed. Engl.* **1995**, *34*, 487.

[1896] Erne, M.; Erlenmeyer, H.; *Helv. Chim. Acta* **1948**, *31*, 652.

[1897] Pinkerton, F. H.; Thames, S. F.; *J. Heterocycl. Chem.* **1971**, *8*, 257.

[1898] Jutzi, P.; Hoffmann, H. J.; *Chem. Ber.* **1973**, *106*, 594.

[1899] Jutzi, P.; Hoffmann, H. J.; Beier, K.; Wyes, K. H.; *J. Organomet. Chem.* **1974**, *82*, 209.

[1900] Crousier, J.; Metzger, J.; *Bull. Soc. Chim. Fr.* **1967**, 4134.

[1901] Meyers, A. I.; Knaus, G. N.; *J. Am. Chem. Soc.* **1973**, *95*, 3408; Hamana, H.; Sugasawa, T.; *Chem. Lett.* **1983**, 333.

[1902] Knaus, G.; Meyers, A. I.; *J. Org. Chem.* **1974**, *39*, 1192.

[1903] Noyce, D. S.; Fike, S. A.; *J. Org. Chem.* **1973**, *38*, 3316.

[1904] Athmani, S.; Bruce, A.; Iddon, B.; *J. Chem. Soc., Perkin Trans. 1* **1992**, 215.

[1905] Turner, J. A.; *J. Org. Chem.* **1983**, *48*, 3401.

[1906] Güngör, T.; Marsais, F.; Quéguiner, G.; *Synthesis* **1982**, 499.

[1907] Reed, J. N.; Rotchford, J.; Strickland, D.; *Tetrahedron Lett.* **1988**, *29*, 5725.

[1908] Godard, A.; Rocca, P.; Pomel, V.; Thomas-dit-Dumont, L.; Rovera, J. C.; Thaburet, J. F.; Marsais, F.; Quéguiner, G.; *J. Organomet. Chem.* **1996**, *517*, 25.

[1909] Pomel, V.; Rovera, J. C.; Godard, A.; Marsais, F.; Quéguiner, G. *J. Heterocycl. Chem.* **1996**, *33*, 1995.

[1910] Meyers, A. I.; Gabel, R. A.; *J. Org. Chem.* **1982**, *47*, 2633.

[1911] Iwao, M.; Kuraishi, T.; *Tetrahedron Lett.* **1983**, *24*, 2649.

[1912] Watabe, M.; Shinoda, E.; Shimizu, Y.; Furukawa, S.; Iwao, M.; Kuraishi, T.; *Tetrahedron* **1987**, *43*, 5281; Bolitt, V.; Mioskowski, C.; Reddy, S. P.; Falck, J. R.; *Synthesis* **1988**, 388.

[1913] Epsztajn, J.; Jóźwiak, A.; Czech, K.; Szcześniak, A.; *Monatsh. Chem.* **1990**, *121*, 909.

[1914] Trécourt, F.; Mallet, M.; Marsais, F.; Quéguiner, G.; *J. Org. Chem.* **1988**, *53*, 1367.

[1915] Marsais, F.; Le Nard, G.; Quéguiner, G.; *Synthesis* **1982**, 235.

[1916] Comins, D. L.; Morgan, L. A.; *Tetrahedron Lett.* **1991**, *32*, 5919.

[1917] Trécourt, F.; Mallet, M.; Mongin, O.; Gervais, B.; Quéguiner, G.; *Tetrahedron* **1993**, *49*, 8373.

[1918] Gros, P.; Fort, Y.; Quéguiner, G.; Caubère, P.; *Tetrahedron Lett.* **1995**, *36*, 4791; Gros, P.; Fort, Y.; Caubère, P.; *J. Chem. Soc., Perkin Trans. 1* **1998**, 1685.

[1919] Chan, A. S. C.; Chen, C.-C.; Cao, R.; Lee, M.-R.; Peng, S.-M.; Lee, G. H.; *Organometallics* **1997**, *16*, 3469.

[1920] Winkle, M. R.; Ronald, R. C.; *J. Org. Chem.* **1982**, *47*, 2101; Ronald, R. C.; Winkle, M. R.; *Tetrahedron* **1983**, *39*, 2031.

[1921] Beaulieu, F.; Snieckus, V.; *Synthesis* **1992**, 112.

[1922] Mongin, F.; Trécourt, F.; Quéguiner, G.; *Tetrahedron Lett.* **1999**, *40*, 5483.

[1923] Gribble, G. W.; Saulnier, M. G.; *Tetrahedron Lett.* **1980**, *21*, 4137; *Heterocycles* **1993**, *35*, 151.

[1924] Estel, L.; Marsais, F.; Quéguiner, G.; *J. Org. Chem.* **1988**, *53*, 2740.

[1925] Marsais, F.; Quéguiner, G.; *Tetrahedron* **1983**, *39*, 2009.

[1926] Marsais, F.; Trécourt, F.; Bréant, P.; Quéguiner, G.; *J. Heterocycl. Chem.* **1988**, *25*, 81.

[1927] Mallet, M.; *J. Organomet. Chem.* **1991**, *406*, 49.

[1928] Marsais, F.; Bréant, P.; Ginguène, A.; Quéguiner, G.; *J. Organomet. Chem.* **1981**, *216*, 139.

[1929] Marsais, F.; Laperdrix, B.; Güngör, T.; Mallet, M.; Quéguiner, G.; *J. Chem. Res. (S)* **1982**, 278.

[1930] Mallet, M.; Quéguiner, G.; *Tetrahedron* **1982**, *38*, 3035.

[1931] Radinov, R.; Chanev, C.; Haimova, M.; *J. Org. Chem.* **1991**, *56*, 4793; Radinov, R.; Haimova, M.; Simova, E.; *Synthesis* **1986**, 886.

[1932] Güngör, T.; Marsais, F.; Quéguiner, G.; *J. Organomet. Chem.* **1981**, *215*, 139.

[1933] Cale, A. D.; Gero, T. W.; Walker, K. R.; Lo, Y. S.; Welstead, W. J.; Jaques, L. W.; Johnson, A. F.; Leonard, C. A.; Nolan, J. C.; Johnson, D. N.; *J. Med. Chem.* **1989**, *32*, 2178; spec. 2198.

[1934] Finnegan, R. A.; *Tetrahedron Lett.* **1963**, *4*, 429.

[1935] Eberhardt, G. G.; *US Pat.* 3206519 (to Sun Oil Co., appl. 15 June 1962, issued 14 Sept. **1965**); *Chem. Abstr.* **1965**, *63*, 14757; Eberhardt, G. G.; Butte, W. A.; *J. Org. Chem.* **1964**, *29*, 2928; Eberhardt, G. G.; *Organomet. Chem. Rev.* **1966**, *1*, 491.

[1936] Langer, A. W.; *Trans. N.Y. Acad. Sci.* **1965**, *27*, 741; *Chem. Abstr.* **1965**, *63*, 18131f.

[1937] Morton, A. A.; Finnegan, R. A.; *J. Polymer Sci.* **1959**, *38*, 19.

[1938] Collum, D. B.; *Acc. Chem. Res.* **1992**, *25*, 448.

[1939] Kimball, G. E.; *J. Chem. Phys.* **1940**, *8*, 188; Doering, W. V. E.; Hoffmann, A. K.; *J. Am. Chem. Soc.* **1955**, *77*, 521.

[1940] Liu, Z.-p.; Schlosser, M.; *Tetrahedron Lett.* **1990**, *31*, 5753.

[1941] Rank, A.; Allen, L. C.; Mislow, K.; *Angew. Chem.* **1970**, *82*, 453, spec. 455; *Angew. Chem. Int. Ed. Engl.* **1970**, *9*, 400, spec. 402; Lischka, H.; *J. Am. Chem. Soc.* **1977**, *99*, 353; Francl, M. M.; Pellow, R. C.; Allen, L. C.; *J. Am. Chem. Soc.* **1988**, *110*, 3723.

[1942] Beak, P.; Zajdel, W. J.; Reitz, D. B.; *Chem. Rev.* **1984**, *84*, 471, spec. 515–516.

[1943] Rondan, N. G.; Houk, K. N.; Beak, P.; Zajdel, W. J.; Chandrasekhar, J.; Schleyer, P. V. R.; *J. Org. Chem.* **1981**, *46*, 4108; Bach, R. D.; Braden, M. L.; Wolber, G. J.; *J. Org. Chem.* **1983**, *48*, 1509.

[1944] Eastwood, F. W.; Gunawardana, D.; Wernert, G. T.; *Austr. J. Chem.* **1982**, *35*, 2289.

[1945] Shiner, C. S.; Berks, A. H.; Fisher, A. M.; *J. Am. Chem. Soc.* **1988**, *110*, 957.

[1946] Walborsky, H. M.; Niznik, G. E.; *J. Am. Chem. Soc.* **1969**, *91*, 7778; Niznik, G. E.; Morrison, W. H.; Walborsky, H. M.; *J. Org. Chem.* **1974**, *39*, 600; Yamamoto, Y.; Kondo, K.; Moritani, I.; *J. Org. Chem.* **1975**, *40*, 3644.

[1947] Ugi, I.; Fetzer, U.; *Chem. Ber.* **1961**, *94*, 2239.

[1948] Staudinger, H.; *Ber. Dtsch. Chem. Ges.* **1908**, *41*, 2217.

[1949] Lapkin, I. I.; Anvarova, G. Y.; Povarnitsyna, T. Y.; *Zh. Obshch. Khim.* **1966**, *36*, 1952; *Chem. Abstr.* **1967**, *66*, 76048e.

[1950] Bánhidai, B.; Schöllkopf, U.; *Angew. Chem.* **1973**, *85*, 861; *Angew. Chem. Int. Ed. Engl.* **1973**, *12*, 836; Schöllkopf, U.; Beckhaus, H.; *Angew. Chem.* **1976**, *88*, 296; *Angew. Chem. Int. Ed. Engl.* **1976**, *15*, 293.

[1951] Fraser, R. R.; Hubert, P. R.; *Can. J. Chem.* **1974**, *52*, 185.

[1952] Smith, K.; Swaminathan, K.; *J. Chem. Soc., Chem. Commun.* **1976**, 387.

[1953] Enders, D.; Seebach, D.; *Angew. Chem.* **1972**, *84*, 1186; *Angew. Chem. Int. Ed. Engl.* **1972**, *11*, 1101; *Angew. Chem.* **1973**, *85*, 1104; *Angew. Chem. Int. Ed. Engl.* **1973**, *12*, 1014; *J. Med. Chem.* **1974**, *17*, 1225.

[1954] Seebach, D.; Lubosch, W.; Enders, D.; *Chem. Ber.* **1976**, *109*, 1309.

[1955] Hauser, C. R.; Puterbaugh, W. H.; *J. Org. Chem.* **1964**, *29*, 853.

[1956] Gschwend, H. W.; Rodriguez, H. R.; *Org. React.* **1979**, *26*, 1.

[1957] Snieckus, V.; *Chem. Rev.* **1990**, *90*, 879.

[1958] Slocum, D. W.; Jennings, C. A.; *J. Org. Chem.* **1976**, *41*, 3653; Beak, P.; Brown, R. A.; *J. Org. Chem.* **1979**, *44*, 4463; **1982**, *47*, 34; Beak, P.; Tse, A.; Hawkins, J.; Chen, C.-w.; Mills, S.; *Tetrahedron* **1983**, *39*, 1983.

[1959] Ingold, C. K.; Shaw, F. R.; *J. Chem. Soc.* **1927**, 2918.

[1960] Schlosser, M.; Ladenberger, V.; *Chem. Ber.* **1967**, *100*, 3901.

[1961] Mongin, F.; Maggi, R.; Schlosser, M.; *Chimia* **1996**, *50*, 650.

[1962] Katsoulos, G.; Takagishi, S.; Schlosser, M.; *Synlett* **1991**, 731.

[1963] Maggi, R.; Schlosser, M.; *J. Org. Chem.* **1996**, *61*, 5430.

[1964] Takagishi, S.; Katsoulos, G.; Schlosser, M.; *Synlett* **1992**, 360.

[1965] Mongin, F.; Desponds, O.; Schlosser, M.; *Tetrahedron Lett.* **1996**, *37*, 2767.

[1966] Schlosser, M.; Katsoulos, G.; Takagishi, S.; *Synlett* **1990**, 747.

[1967] Katsoulos, G.; Schlosser, M.; *Tetrahedron Lett.* **1993**, *34*, 6263.

[1968] Puterbaugh, W. H.; Hauser, C. R.; *J. Am. Chem. Soc.* **1963**, *85*, 2467; *J. Org. Chem.* **1963**, *28*, 3461.

[1969] Ahlbrecht, H.; Harbach, J.; Hauck, T.; Kalinowski, H.-O.; *Chem. Ber.* **1992**, *125*, 1753.

[1970] Tischler, A. N.; Tischler, M. H.; *Tetrahedron Lett.* **1978**, *19*, 3.

[1971] Gilman, H.; Webb, F. J.; *J. Am. Chem. Soc.* **1940**, *62*, 987; Shirley, D. A.; Reeves, B. J.; *J. Organomet. Chem.* **1969**, *16*, 1; Cabiddu, S.; Fattuoni, C.; Floris, C.; Gelli, G.; Melis, S.; *J. Organomet. Chem.* **1992**, *441*, 197.

[1972] Katritzky, A. R.; Fan, W.-Q.; Akutagawa, K.; *Synthesis* **1987**, 415.

[1973] Markies, P. R.; Nomoto, T.; Akkerman, O. S.; Bickelhaupt, F.; Smeets, W. J. J.; Spek, A. L.; *Angew. Chem.* **1988**, *100*, 1143; *Angew. Chem. Int. Ed. Engl.* **1988**, *27*, 1084; see also: Markies, P. R.; Nomoto, T.; Schat, G.; Akkerman, O. S.; Bickelhaupt, F.; Smeets, W. J. J.; Spek, A. L.; *Organometallics* **1991**, *10*, 3826.

[1974] Takagishi, S.; Schlosser, M.; *Synlett* **1991**, 119.

[1975] Schlosser, M.; Geneste, H.; *Chem. Eur. J.* **1998**, *4*, 1969.

[1976] Clark, R. D.; Johangir; *Tetrahedron* **1993**, *49*, 1351.

[1977] Ito, Y.; Nakatsuka, M.; Saegusa, T.; *J. Am. Chem. Soc.* **1982**, *104*, 7609.

[1978] Kessar, S. V.; Singh, P.; Vohra, R.; Pal Kaur, N.; Nain Singh, K.; *J. Chem. Soc., Chem. Commun.* **1991**, 568.

[1979] Park, Y. S.; Beak, P.; *Tetrahedron* **1996**, *52*, 12333.

[1980] Moret, E.; Schneider, P.; Margot, C.; Stähle, M.; Schlosser, M.; *Chimia* **1985**, *39*, 231.

[1981] Meyers, A. I.; Edwards, P. D.; Rieker, W. F.; Bailey, T. R.; *J. Am. Chem. Soc.* **1984**, *106*, 3270.

[1982] Iftime, G.; Moreau-Bossuet, C.; Manoury, E.; Balavoine, G. G. A.; *J. Chem. Soc., Chem. Commun.* **1996**, 527.

[1983] Bindel, R. D.; Katzenellenbogen, J. A.; *J. Org. Chem.* **1987**, *52*, 3181.

[1984] Haiduc, I.; Gilman, H.; *Chem. Ind. (London)* **1968**, 1278.

[1985] Schlosser, M.; Guio, L.; Leroux, F.; *J. Am. Chem. Soc.* **2001**, *123*, 3822.

[1986] Slocum, D. W.; Hayes, G.; Kline, N.; *Tetrahedron Lett.* **1995**, *36*, 8175.

[1987] Cabiddu, S.; Contini, L.; Fattuani, C.; Floris, C.; Gelli, G.; *Tetrahedron* **1991**, *47*, 9279.

[1988] Chadwick, D. J.; Willbe, C.; *J. Chem. Soc., Perkin Trans. 1* **1977**, 887.

[1989] Morton, A. A.; Claff, C. E.; *J. Am. Chem. Soc.* **1954**, *76*, 4935.

[1990] Boche, G.; Decker, G.; Etzrodt, H.; Mahdi, W.; Kos, A. J.; Schleyer, P. V. R.; *J. Chem. Soc., Chem. Commun.* **1984**, 1483.

[1991] Nesmeyanov, A. N.; Perevalova, E. G.; Golovnya, R. V.; Nesmeyanova, O. A.; *Dokl. Akad. Nauk S.S.S.R.* **1954**, *97*, 459; *Chem. Abstr.* **1956**, *49*, 9633f; Benkeser, R. A.; Goggin, D.; Schroll, G.; *J. Am. Chem. Soc.* **1954**, *76*, 4025; Rausch, M. D.; Ciapenelli, D. J.; *J. Organomet. Chem.* **1967**, *10*, 127.

[1992] Elschenbroich, C.; *J. Organomet. Chem.* **1968**, *14*, 157.

[1993] Neugebauer, W.; Kos, A. J.; Schleyer, P. V. R.; *J. Organomet. Chem.* **1982**, *228*, 107.

[1994] Ashe, A. J.; Kampf, J. W.; Savla, P. M.; *Heteroat. Chem.* **1994**, *5*, 113; *Chem. Abstr.* **1994**, *121*, 255737c.

[1995] Maercker, A.; Bodenstedt, H.; Brandsma, L.; *Angew. Chem.* **1992**, *104*, 1387; *Angew. Chem. Int. Ed. Engl.* **1992**, *31*, 1339; Günther, H.; Eppers, O.; Hausmann, H.; Hüls, D.; Mons, H.-E.; Klein, D.; Maercker, A.; *Helv. Chim. Acta* **1995**, *78*, 1913.

[1996] Mulvaney, J. E.; Garlund, Z. G.; Garlund, S. L.; Newton, D. J.; *J. Am. Chem. Soc.* **1966**, *88*, 476.

[1997] Curtin, D. Y.; Quirk, R. P.; *Tetrahedron* **1968**, *24*, 5791.

[1998] Kos, A. J.; Schleyer, P. V. R.; *J. Am. Chem. Soc.* **1980**, *102*, 7928; Kranz, M.; Dietrich, H.; Mahdi, W.; Müller, G.; Hampel, F.; Clark, T.; Hacker, R.; Neugebauer, W.; Kos, A. J.; Schleyer, P. V. R.; *J. Am. Chem. Soc.* **1993**, *115*, 4698.

[1999] Desponds, O.; Schlosser, M.; *Tetrahedron* **1994**, *50*, 5881.

[2000] Faigl, F.; Schlosser, M.; *Tetrahedron* **1993**, *49*, 10271.

[2001] Maercker, A.; Passlack, M.; *Chem. Ber.* **1982**, *115*, 540.

[2002] Liotta, D.; Saindane, M.; Waykole, L.; Stephens, J.; Grossman, J.; *J. Am. Chem. Soc.* **1988**, *110*, 2667.

[2003] Corriu, R. J. P.; Lanneau, G. F.; Masse, J. P.; Samate, D.; *J. Organomet. Chem.* **1977**, *127*, 281.

[2004] Hubert, A. J.; Reimlinger, H.; *Synthesis* **1969**, 97.

[2005] Ela, S. W.; Cram, D. J.; *J. Am. Chem. Soc.* **1966**, *88*, 5791.

[2006] Knorr, R.; Lattke, E.; *Tetrahedron Lett.* **1977**, *18*, 4655; *Chem. Ber.* **1981**, *114*, 2116; see also: Knorr, R.; Hintermeyer-Hilpert, M.; Böhrer, P.; *Chem. Ber.* **1990**, *123*, 1137.

[2007] Bates, R. B.; Brenner, S.; Deines, W. H.; McCombs, D. A.; Potter, D. E.; *J. Am. Chem. Soc.* **1970**, *92*, 6345.

[2008] Ziegler, F. E.; Fowler, K. W.; *J. Org. Chem.* **1976**, *41*, 1564.

[2009] Gronowitz, S.; Frejd, T.; *Acta Chem. Scand., Ser. B* **1976**, *30*, 485; Gronowitz, S.; Holm, B.; *Acta Chem. Scand., Ser. B* **1976**, *30*, 505.

[2010] Liang, C. D.; *Tetrahedron Lett.* **1986**, *27*, 1971.

[2011] Schaub, B.; Jenny, T.; Schlosser, M.; *Tetrahedron Lett.* **1984**, *25*, 4097; Schaub, B.; Schlosser, M.; *Tetrahedron Lett.* **1985**, *26*, 1623.

[2012] Brown, J. M.; Woodward, S.; *J. Org. Chem.* **1991**, *56*, 6803.

[2013] Ribéreau, P.; Quéguiner, G.; *Tetrahedron* **1984**, *40*, 2107.

[2014] Chadwick, D. J.; Ennis, D. S.; *Tetrahedron* **1991**, *47*, 9901.

[2015] Wotiz, J. H.; Huba, F.; *J. Org. Chem.* **1959**, *24*, 595; see also: Vaitiekunas, A.; Nord, F. F.; *J. Am. Chem. Soc.* **1953**, *75*, 1764.

[2016] Bunnett, J. F.; Moyer, C. E.; *J. Am. Chem. Soc.* **1971**, *93*, 1183; Bunnett, J. F.; Feit, I. N.; *J. Am. Chem. Soc.* **1971**, *93*, 1205; Mach, M. H.; Bunnett, J. F.; *J. Org. Chem.* **1980**, *45*, 4660.

[2017] Bunnett, J. F.; *Acc. Chem. Res.* **1972**, *5*, 139.

[2018] Mongin, F.; Desponds, O.; Schlosser, M.; *Tetrahedron Lett.* **1996**, *37*, 2767.

[2019] Desponds, O.; Mongin, F.; Schlosser, M.; unpublished results (**1997**).

[2020] de Bie, D. A.; van der Plas, H. C.; *Tetrahedron Lett.* **1968**, *9*, 3905.

[2021] de Bie, D. A.; van der Plas, H. C.; Geurtsen, G.; Nijdam, K.; *Recl. Trav. Chim. Pays-Bas* **1973**, *92*, 245; see also: de Bie, D. A.; van der Plas, H. C.; Geurtsen, G.; *Recl. Trav. Chim. Pays-Bas* **1971**, *90*, 594.

[2022] Moses, P.; Gronowitz, S.; *Arkiv Kemi* **1961**, *18*, 119; *Chem. Abstr.* **1962**, *56*, 10173c; Gronowitz, S.; Holm, B.; *Acta Chem. Scand.* **1969**, *23*, 2207.

[2023] Reinecke, M. G.; Adickes, H. W.; *J. Am. Chem. Soc.* **1968**, *90*, 511; Reinecke, M. G.; Adickes, H. W.; Pyun, C.; *J. Org. Chem.* **1971**, *36*, 2690, 3820.

[2024] van der Plas, H. C., de Bie, D. A., Geurtsen, G.; Reinecke, M. G.; Adickes, H. W.; *Recl. Trav. Chim. Pays-Bas* **1974**, *93*, 33.

[2025] Kano, S.; Yuasa, Y.; Yokomatsu, T.; Shibuya, S.; *Heterocycles* **1983**, *20*, 2035.

[2026] Taylor, E. C.; Vogel, D. E.; *J. Org. Chem.* **1985**, *50*, 1002.

[2027] Sauter, F.; Fröhlich, H.; Kalt, W.; *Synthesis* **1989**, *10*, 771.

[2028] Fröhlich, H.; Kalt, W.; *J. Org. Chem.* **1990**, *55*, 2993.

[2029] Fröhlich, J.; *Progr. Heterocycl. Chem.* **1994**, *6*, 1.

[2030] Fröhlich, J.; Hametner, C.; Kalt, W.; *Monatsh. Chem.* **1996**, *127*, 325; Fröhlich, J.; Hametner, C.; *Monatsh. Chem.* **1996**, *127*, 435.

[2031] Hawkins, D. W.; Iddon, B.; Longthorne, Darren, S.; Rosyk, P. J.; *J. Chem. Soc., Perkin Trans. 1* **1994**, 2735.

[2032] Mallet, M.; Branger, G.; Marsais, F.; Quéguiner, G.; *J. Organomet. Chem.* **1990**, *382*, 3119.

[2033] Rocca, P.; Cochennec, C.; Marsais, F.; Thomas-dit-Dumont, L.; Mallet, M.; Godard, A.; Quéguiner, G.; *J. Org. Chem.* **1993**, *58*, 7832.

[2034] Arzel, E.; Rocca, P.; Marsais, F.; Godard, A.; Quéguiner, G.; *Tetrahedron Lett.* **1998**, *39*, 6465.

[2035] Arzel, E.; Rocca, P.; Marsais, F.; Godard, A.; Quéguiner, G.; *Tetrahedron* **1999**, *55*, 12149.

[2036] Fujita, K.; Schlosser, M.; *Helv. Chim. Acta* **1982**, *65*, 1258.

[2037] Heilbron, I. M.; Johnson, A. W.; Jones, E. R. H.; Spinks, A.; *J. Chem. Soc.* **1942**, 727.

[2038] Isler, O.; Huber, W.; Ronco, A.; Kofler, M.; *Helv. Chim. Acta* **1947**, *30*, 1911.

[2039] Schlosser, M.; Schaub, B.; *Chimia* **1982**, *39*, 396; El-Khoury, M.; Wang, Q.; Schlosser, M.; *Tetrahedron Lett.* **1996**, *37*, 9047.

[2040] Schlosser, M.; Christmann, K. F.; *Liebigs Ann. Chem.* **1967**, 708, 1; *Synthesis* **1969**, 38.

[2041] Wendler, N. L.; Slates, H. L.; Trenner, N. R.; Tishler, M.; *J. Am. Chem. Soc.* **1951**, *73*, 719; see also: Reddy, A. M.; Rao, V. J.; *J. Org. Chem.* **1992**, *57*, 6727.

[2042] Burgess, E. M.; Penton, H. R.; Taylor, E. A.; *J. Org. Chem.* **1973**, *38*, 26.

[2043] Schlosser, M.; Hartmann, J.; *Angew. Chem.* **1973**, *85*, 544; *Angew. Chem. Int. Ed. Engl.* **1973**, *12*, 508.

[2044] Schlosser, M.; Jenny, T.; Guggisberg, Y.; *Synlett* **1990**, 704.

[2045] Ultra-Turrax (drive T-50, shaft S50KG/HH-G40G, dispersion head G40G), Janke & Kunkel, D-79217 Staufen (see also page 288).

[2046] Schöllkopf, U.; in *Houben-Weyl: Methoden der organischen Chemie*, Vol. *13/1* (Müller, E.; ed.), Thieme, Stuttgart, **1970**, p. 116.

[2047] Shi, G.-Q., Cottens, S.; Shiba, S. A.; Schlosser, M.; *Tetrahedron* **1992**, *48*, 10569.

[2048] Katsoulos, G.; Schlosser, M.; unpublished results (**1993**); Katsoulos, G.; *Doctoral Dissertation*, Université de Lausanne, **1993**, pp. 81–86 and 115–116.

[2049] Simig, G.; Schlosser, M.; *Tetrahedron Lett.* **1988**, *29*, 4277; **1990**, *31*, 5213.

[2050] Wang, X.-h.; Schlosser, M.; unpublished results (**1992**).

[2051] Oita, K.; Gilman, H.; *J. Am. Chem. Soc.* **1957**, *79*, 339; Gilman, H.; Trepka, W. J.; *J. Org. Chem.* **1961**, *26*, 5202; Granoth, I.; Levy, J. B.; Symmes, C.; *J. Chem. Soc., Perkin Trans. 2* **1972**, 697.

[2052] Gschwend, H. W.; unpublished results; quoted according to Gschwend, H. W.; Rodriguez, H. R.; *Org. React.* **1979**, *26*, 1, spec. 97.

[2053] Volle, J.-N.; Schlosser, M.; unpublished results (**1999**); see also: Okada, E.; Masuda, R.; Hojo, M.; *Heterocycles* **1992**, *34*, 791.

[2054] Marull, M.; Schlosser, M.; unpublished results (**1999**).

[2055] Pinder, R. M.; Burger, A.; *J. Med. Chem.* **1968**, *11*, 267.

[2056] Shriver, D. F.; Drezdzon, M. A.; *The Manipulation of Air-Sensitive Compounds*, Wiley, New York, **1986**.

[2057] Schlosser, M.; Christmann, K. F.; *Angew. Chem.* **1965**, *77*, 682; *Angew. Chem. Int. Ed. Engl.* **1965**, *4*, 689; Schlosser, M.; Huynh Ba, T.; Schaub, B.; *Tetrahedron Lett.* **1985**, *26*, 311.

[2058] Carothers, W. H.; Coffman, D. D.; *J. Am. Chem. Soc.* **1929**, *51*, 588; **1930**, *52*, 1254.

[2059] Morton, A. A.; Newey, H. A.; *J. Am. Chem. Soc.* **1942**, *64*, 2247; Morton, A. A.; Lanpher, E. J.; *J. Org. Chem.* **1956**, *21*, 93.

[2060] Ziegler, K.; Gellert, H.-G.; *Liebigs Ann. Chem.* **1950**, *567*, 179, 185; Ziegler, K.; Nagel, K.; Patheiger, M.; *Z. Anorg. Allg. Chem.* **1955**, *285*, 345.

[2061] Gilman, H.; Haubein, A. H.; Hartzfeld, H.; *J. Org. Chem.* **1954**, *19*, 1034.

[2062] Glaze, W. H.; Lin, J.; Felton, E. G.; *J. Org. Chem.* **1966**, *31*, 2643.

[2063] Gilman, H.; Schwebke, G. L.; *J. Organomet. Chem.* **1965**, *4*, 483.

[2064] Stanetty, P.; Mihovilovic, M. D.; *J. Org. Chem.* **1997**, *62*, 1514.

[2065] Honeycutt, S. C.; *J. Organomet. Chem.* **1971**, *29*, 1.

[2066] Wittig, G.; Polster, R.; *Liebigs Ann. Chem.* **1956**, *599*, 13.

[2067] Letsinger, R. L.; Pollart, D. F.; *J. Am. Chem. Soc.* **1956**, *78*, 6179; Letsinger, R. L.; *Angew. Chem.* **1958**, *70*, 153.

[2068] Maercker, A.; Demuth, W.; *Liebigs Ann. Chem.* **1977**, 1909.

[2069] Wittig, G.; Löhmann, L.; *Liebigs Ann. Chem.* **1956**, *599*, 13.

[2070] Fitt, J. J.; Gschwend, H. W.; *J. Org. Chem.* **1984**, *49*, 209.

[2071] Moret, E.; Desponds, O.; Schlosser, M.; *J. Organomet. Chem.* **1991**, *409*, 83.

[2072] Goldberg, R. E.; Lewis, D. E.; Cohn, K.; *J. Organomet. Chem.* **1968**, *15*, 491.

[2073] Millon, J.; Linstrumelle, G.; *Tetrahedron Lett.* **1976**, *17*, 1095; Huynh, C.; Derguini-Boumechal, F.; Linstrumelle, G.; *Tetrahedron Lett.* **1979**, *20*, 1503.

[2074] Rembaum, A.; Siao, S. P.; Indictor, N.; *J. Polymer Sci.* **1962**, *56*, 517; *Chem. Abstr.* **1962**, *56*, 13'083a; Bates, R. B.; Kroposki, L. M.; Potter, D. E.; *J. Org. Chem.* **1972**, *37*, 560; Maercker, A.; Theysohn, W.; *Liebigs Ann. Chem.* **1971**, *747*, 70.

[2075] Tomboulian, P.; Anick, D.; Beare, S.; Dumke, K.; Hart, D.; Hites, R.; Metzger, A.; Nowak, R.; *J. Org. Chem.* **1973**, *38*; see also: Bates, R. B.; Kroposki, L. M.; Potter, D. E.; *J. Org. Chem.* **1972**, *37*, 560.

[2076] Jung, M. E.; Blum, R. B.; *Tetrahedron Lett.* **1977**, *18*, 3791.

[2077] Kamata, K.; Terashima, M.; *Heterocycles* **1980**, *14*, 205.

[2078] von Baeyer, A.; Villiger, V.; *Ber. Dtsch. Chem. Ges.* **1903**, *36*, 2775.

[2079] Ehrlich, P.; Sachs, F.; *Ber. Dtsch. Chem. Ges.* **1903**, *36*, 4296.

[2080] Sachs, F.; Sachs, L.; *Ber. Dtsch. Chem. Ges.* **1904**, *37*, 3088.

[2081] Gilman, H.; Schulze, F.; *J. Am. Chem. Soc.* **1925**, *47*, 2002.

[2082] Gaidis, J. M.; *J. Organomet. Chem.* **1967**, *8*, 385.

[2083] Gilman, H.; Swiss, J.; *J. Am. Chem. Soc.* **1940**, *62*, 1847.

[2084] Wittig, G.; Harborth, G.; *Ber. Dtsch. Chem. Ges.* **1944**, *77*, 315, spec. 322.

[2085] Gilman, H.; Haubein, A. H.; *J. Am. Chem. Soc.* **1944**, *66*, 1515.

[2086] Gilman, H.; Cartledge, F. K.; *J. Organomet. Chem.* **1964**, *2*, 447.

[2087] Ager, D. J.; *J. Organomet. Chem.* **1982**, *241*, 139.

[2088] Conant, J. B.; Wheland, G. W.; *J. Am. Chem. Soc.* **1932**, *54*, 1212.

[2089] Eppley, R. L.; Dixon, J. A.; *J. Organomet. Chem.* **1967**, *8*, 176.

[2090] Kofron, W. G.; Baclawski, L. M.; *J. Org. Chem.* **1976**, *41*, 1879.

[2091] Kiljunen, H.; Hase, T. A.; *J. Org. Chem.* **1991**, *56*, 6950.

[2092] Winkle, M. R.; Lansinger, J. M.; Ronald, R. C.; *J. Chem. Soc., Chem. Commun.* **1980**, 87.

[2093] Juaristi, E.; Martinez-Richa, A.; Garcia-Rivera, A.; Cruz-Sanchez, J. S.; *J. Org. Chem.* **1983**, *48*, 2603.

[2094] Lipton, M. F.; Sorensen, C. M.; Sadler, A. C.; Shapiro, R. H.; *J. Organomet. Chem.* **1980**, *186*, 155.

[2095] Bergbreiter, D. E.; Pendergras, E.; *J. Org. Chem.* **1981**, *46*, 219.

[2096] Suffert, J.; *J. Org. Chem.* **1989**, *54*, 509.

[2097] Watson, S. C.; Eastham, J. F.; *J. Organomet. Chem.* **1967**, *9*, 165.

[2098] For a modified version, see: Lipshutz, B. H.; in *Organometallics in Synthesis: A Manual* (Schlosser, M.; ed.), Wiley, Chichester, **1994**, p. 188.

[2099] Duhamel, L.; Plaquevent, J.-C.; *J. Org. Chem.* **1979**, *44*, 3404.

[2100] Aso, Y.; Yamashita, H.; Otsubo, T.; Ogura, F.; *J. Org. Chem.* **1989**, *54*, 5627.

II

Organotin Chemistry

JAMES A. MARSHALL

Department of Chemistry, University of Virginia Charlottesville, USA

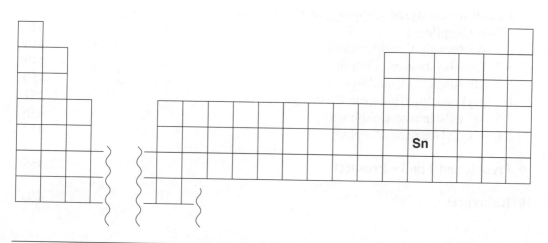

Organometallics in Synthesis: A Manual. Edited by Manfred Schlosser.
© 2002 John Wiley & Sons Ltd

Table of Contents

1   Introduction

This chapter covers the most useful synthetic applications of organotin compounds. Each section describes methods for preparing a given class of tin compounds, followed by a discussion of uses of that class as reagents in organic synthesis. Coverage focuses on ionic rather than free radical reactions.

Before we begin, a brief summary of the potential hazards associated with organotin compounds is in order. Tin, like most heavy metals, is toxic. However, the metal itself, and its inorganic salts, are relatively nontoxic compared to lead, for example. The toxicity of organotin compounds depends on the size and number of organic substituents. Tin compounds with larger substituents are less toxic than those with smaller substituents. Toxicity increases with an increasing number of substituents. Thus $RSnX_3 < R_2SnX_2 < R_3SnX$. Tetraorganotin compounds are degraded in vivo to triorgano derivatives leading to a delayed effect. Furthermore, the formed compounds are not readily assimilated, so their net toxicity is lower than that of the trialkyltin counterparts. Table 1 summarizes data from animal experiments for methyl, butyl, and phenyltin chlorides, the most commonly used sources of organotin compounds of interest to synthetic chemists[1]. Of these, the methyl derivatives clearly pose the greatest potential hazard.

This fact, combined with their relatively higher volatility, would indicate that extra caution should be exercised when working with methyl tin compounds. For this reason, and in part because of economic considerations, much of the organotin chemistry to be discussed employs a tributyltin derivative as the reagent. In some cases, however, reactions proceed more efficiently with the trimethyl tin analogue, possibly for steric reasons.

A pervasive problem in organotin chemistry is the removal of tin byproducts. These byproducts are typically tributyl or trimethylstannyl halides, stannoxanes, or ditin species. The tin halides can be removed by stirring the organic extracts with triethylamine or by adding the amine to the reaction mixture. Alternatively, washing the extracts with dilute ammonium hydroxide will solubilize tributyltin halides. Tributyltin chloride is soluble in hexane but insoluble in acetonitrile. Thus partitioning the reaction mixture between these two immiscible solvents can be used to remove Bu_3SnCl in the hexane layer. Aqueous KF reacts with tributyltin halides to form an insoluble tin fluoride. Trimethyltin chloride is water soluble and rather volatile, so its removal presents no special problems. However, care should be exercised because of its toxicity.

Table 1. Ranges of LD_{50} Values (mg/kg) for Selected Organotin Chlorides Based on Oral Administration to Rats.

R	R_3SnCl	R_2SnCl_2	$RSnCl_3$
Me	9–20	74–237	575–1350
Bu	122–349	112–219	2200–2300
Ph	118–135	—	—

The removal of stannoxanes is less well defined. Treatment of the reaction mixture with amines or aqueous KF is reasonably effective but several cycles may be needed. Hexabutylditin is removable by chromatography. On silica gel it is eluted with hexanes. Accordingly, separation from nonpolar products is problematic. In some cases it is possible to carry inseparable ditin byproducts into the next reaction without adverse effects.

2   Tetraalkyltin Compounds

2.1   Preparation

2.1.1   Nucleophilic Substitution

Tetraalkyltin compounds can be prepared by direct substitution of organotin halides with Grignard or organolithium reagents[1,2]. Alternatively a trialkyl or triaryltin hydride can be converted to a nucleophilic species with a strong base. Such anionic tin reagents readily displace halides to form the tetraalkyltin[3]:

$$R^1{}_nSnX_{4-n} \xrightarrow{R^2M} R^1{}_nSnR^2{}_{4-n} + MX$$

$$X = Cl, Br, I; M = MgX, Li; R^1, R^2 = \text{alkyl or aryl}$$

$$R^1{}_3SnH \xrightarrow{LiN(^iPr)_2} R^1{}_3SnLi \xrightarrow{R^2X} R^1{}_3SnR^2$$

Reactions of Me_3SnLi and Me_3SnNa with (*R*)-2-octyl tosylate, chloride, and bromide proceed with predominant inversion of configuration. Displacement of the bromide with the stannylsodium affords the product of lowest ee. Competing S_N2 and free radical processes are thought to be responsible. However, the corresponding displacements by Ph_3SnM occur with complete inversion of configuration:

$$\underset{C_6H_{13} \quad Me}{\overset{X}{\bigwedge}} \xrightarrow{MSnMe_3} \underset{C_6H_{13} \quad Me}{\overset{SnMe_3}{\bigwedge}}$$

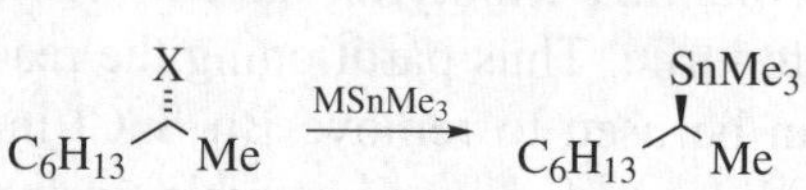

$$M = \text{Na or Li}$$

Displacement of a Tosylate with Me_3SnLi

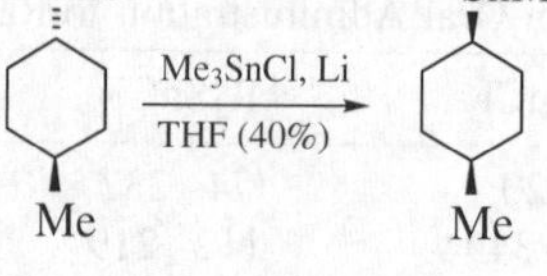

(a) Preparation of Trimethyltinlithium

Lithium metal (3.36 g, 0.48 mol) was cut into small pieces which were then protected and flattened with a hammer. The flattened Li pieces (now about 2 cm in diameter) were then cut into smaller (approximately 2 mm wide) pieces and placed in a 250-mL round-bottom flask containing THF and fitted with a condenser, drying tube, N_2 inlet, and pressure equalizing dropping funnel. A solution of $(CH_3)_3SnCl$ (9.58 g, 0.048 mol) in dry THF (~30 mL) was placed in the dropping funnel, the reaction vessel was cooled to 0 to about −5 °C and blanketed with N_2. The Li/THF was stirred vigorously and the $(CH_3)_3SnCl$ solution was added dropwise. A color change to dark olive green usually appears after about 15 min. Stirring was continued for about 2 h. The unreacted Li metal was removed by filtering the solution (under N_2 pressure) through a fitted bent side arm into an attached 250-mL three-neck round-bottom flask.

(b) *cis*-4-Methylcyclohexyltrimethylstannane

trans-4-Methylcyclohexyl tosylate (11.5 g, 0.043 mol) in dry THF (~30 mL) was added dropwise to the preformed $(CH_3)_3SnLi$ solution cooled to 0 °C under N_2. After 5 h, the reaction was quenched with 20% NH_4Cl solution (~20 mL). The ethereal layer was separated and the aqueous layer extracted with ether. The combined organic extracts were dried over $MgSO_4$ and ether was removed under reduced pressure. Distillation yielded an oil (bp 95–100 °C (20 mm), 40% yield).

The reaction of Me_3SnLi or Ph_3SnLi with cyclohexene oxide follows an S_N2 pathway[4]. 1-Bromoadamantane and 1-chlorocamphane afford substitution products with Me_3SnLi[5]. These reactions may take place by a free radical pathway:

R = Me or Ph

2.1.2 1,4 Addition

α,β-Unsaturated ketones yield 1,4 adducts with trialkylstannyllithium compounds[6]. With cyclohexenones the addition proceeds by axial attack. The intermediate enolates can be trapped with alkyl halides to give β-stannyl α-alkyl ketones. If the addition of Me_3SnLi to cyclohexenone is carried out in ether rather than THF the unstable 1,2 adduct is formed:

2.2   Reactions

2.2.1   Oxidation

Treatment of secondary tetraorgano tin compounds with CrO_3-pyridine affords ketones. Tertiary organostannanes yield mixtures of alcohols and elimination products:

Oxidation of a Tributylalkylstannane to a Ketone

4-Isopropyl-4-cycloheptenone

A mixture of the chloride (1.0 g, 6.0 mmol), granular Mg (0.43 g, 18 mmol), and ether (8.5 mL) was refluxed for 20 min, as a solution of tributylstannyl chloride (3.3 g, 10 mmol) in ether (3 mL) was added. After 20 min at reflux, the mixture was poured into saturated aq NH_4Cl and extracted with ether. The crude product in dichloromethane (25 mL) was added to a mixture of pyridine (19 mL, 0.24 mol) and chromium(VI) oxide (12 g, 0.12 mol) in dichloromethane (300 mL). After 12 h at 25 °C, the mixture was filtered through a pad of Celite 545, the filtrate was washed with 5% NaOH solution, 1 M HCl, and brine. Purification by silica gel colum chromatography gave the ketone (0.60 g, 65% yield; bp 80 °C bath temp, 3 mmHg).

2.2.2 Fragmentation

Various cyclic γ-stannyl alcohols undergo β-fragmentation to ω-unsaturated ketones or aldehydes upon treatment with PhIO[8]. A high degree of stereocontrol is observed in fragmentations of β-substituted γ-stannyl cyclohexanols. A concerted pathway can be postulated for these reactions:

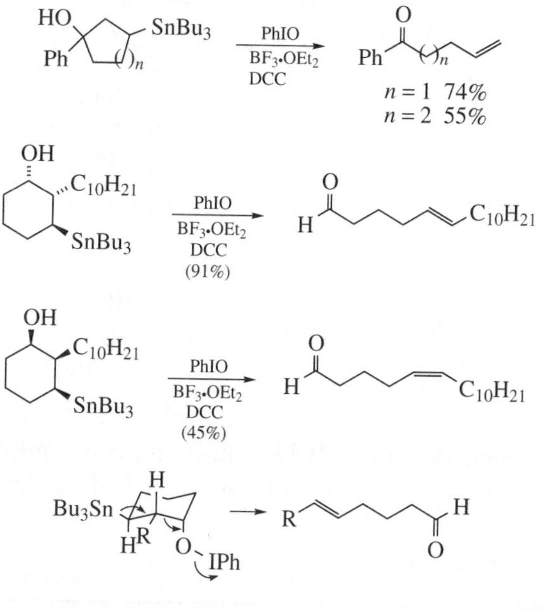

2.2.3 Intramolecular Additions

In certain circumstances the carbon–tin sigma bond can function as a latent carbanionic nucleophile. A particularly favorable case involves intramolecular addition to a stabilized carbocation leading to a five- or six-membered ring[9]. Allylic alcohols can also serve as precursors to the cationic center, but with tertiary allylic alcohols β-hydride transfer is observed:

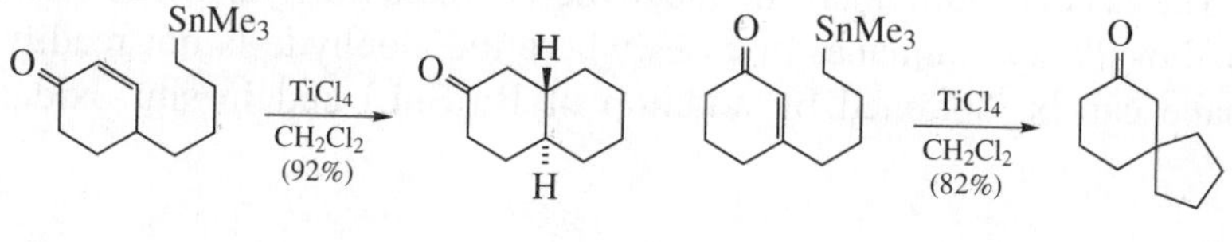

Cyclization of a secondary δ-trimethylstannyl aldehyde has been shown to occur with retention of configuration at the carbon–tin bond[10]. In contrast, analogous reactions leading to cyclopropanes proceed with inversion of stereochemistry[11,12]. The trialkyltin nucleofugal carbon can also serve as a terminator in cationic alkene cyclizations[13].

R = H   83%
R = Me   58%

BF₃-promoted Cyclization of a δ-Stannyl Aldehyde

A mixture of the aldehyde (0.1 g, 0.33 mmol) and triphenylphosphine (0.033 g) was treated with BF₃•OEt₂, and the product was purified by flash chromatography. Elution with light petroleum (bp 30–40 °C)/ether, 5 : 1 v/v, afforded the cyclopentanol (0.024 g, 52%) as an oil.

3   Acylstannanes

3.1   Preparation

Acylstannanes can be prepared by addition of Bu₃SnMgBr to two equivalents of an aldehyde[14]. The second equivalent of aldehyde serves as a hydride acceptor from the initially formed α-alkoxy stannane. In cases where the aldehyde is not readily available, the acylstannane can be obtained by addition of Bu₃SnLi and in situ oxidation of the lithium alkoxide with 1,1'-(azodicarbonyl)dipiperidine (ADD)[15]:

$$2RCHO \xrightarrow{BrMgSnBu_3} \underset{\substack{\| \\ O}}{RC}-SnBu_3 + RCH_2OMgX$$

$$R = Me, Et, {}^iPr, C_5H_{11}, {}^tBu$$

$$RCHO \xrightarrow{LiSnBu_3} \begin{bmatrix} OLi \\ | \\ RCSnBu3 \end{bmatrix} \xrightarrow{ADD} \underset{\substack{\| \\ O}}{RC}-SnBu_3$$

$$R = C_6H_{13}, (E)\text{-}MeCH=CH, (E)\text{-}BuCH=CH, (E)\text{-}c\text{-}C_6H_{11}CH=CH$$

Addition of Bu₃SnLi to Crotonaldehyde and in situ Oxidation with ADD

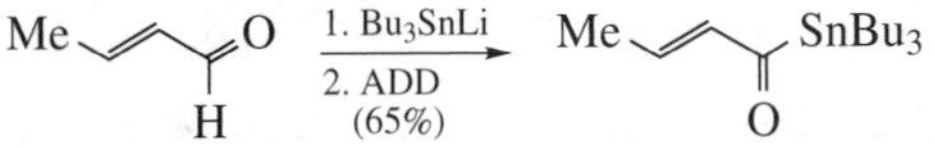

(E)-1-(Tri-n-butylstannyl)-2-buten-1-one

To a stirred, cooled (0 °C) solution of 35 mL (18 mmol) of 0.5 M LDA in THF was added 4.7 mL (18 mmol) of Bu₃SnH. After 15 min, the resulting solution was cooled to −78 °C and a solution of 1.1 g (16 mol) of crotonaldehyde in 15 mL of THF was introduced. The reaction solution was stirred for 10 min before 4.5 g (18 mmol) of 1,1'-(azodicarbonyl)dipiperidine (ADD) was added, and the reaction mixture was warmed to 0 °C. After stirring for 1 h at 0 °C, the dark orange reaction mixture was quenched with saturated aqueous NH₄Cl. The resulting mixture was extracted with ether, and the organic layer was washed with 3% HCl, saturated NaHCO₃, and brine and dried over MgSO₄. After removal of the solvent under reduced pressure and column chromatography, the acyl stannane (3.7 g, 65%) was obtained as a light yellow oil.

Direct acylation of Me₃SnLi or Bu₃SnLi can be effected with thiol esters (Table 2). Best results are obtained with aroyl esters[16].

Acylstannanes from Thiolesters

Benzoyltrimethylstannane

To a cooled solution ($-78\,°C$) of Me$_3$SnLi (0.60 mmol) was added dropwise 0.091 g of PhCOSMe (0.60 mmol, 73 µL) in 1 mL of THF over 3 min. The reaction mixture was stirred for 1 h at $-78\,°C$, diluted with 5 mL of ether, quenched with saturated NH$_4$Cl, and then slowly warmed to 25 °C. The aqueous layer was removed with a syringe, and the organic phase was washed three times with water and brine. The yellow solution obtained was dried over Na$_2$SO$_4$, and the solvent was removed under vacuum, giving 0.250 g of material, which was purified by elution under nitrogen on Florisil with hexane/ether 10 : 1 as the eluant, yielding 0.156 g (58%) of benzoyltrimethylstannane.

Table 2. Acylation of Lithiostannanes with Thiol Esters.

$$R^1 \overset{O}{\underset{}{\overset{\|}{C}}} X \xrightarrow{\text{LiSnR}_3} R^1 \overset{O}{\underset{}{\overset{\|}{C}}} SnR_3$$

R^1	R	X	Yield(%)
Ph	Bu	SMe	58
Ph	Me	SMe	58
Ph	Bu	SPh	62
m-MeC$_6$H$_4$	Bu	SPh	49
2-furyl	Bu	SPh	47
Pr	Bu	SMe	39

A second direct method involves Pd^0-catalyzed addition of Me$_3$SnSnMe$_3$ to acyl halides (Table 3). The method fails with Bu$_3$SnSnBu$_3$ and with branched acyl chlorides[17]. Propenoylstannanes can be prepared from allenyl ethers by a three-step sequence involving lithiation, stannylation and acid catalyzed hydrolysis[18].

Table 3. Pd-catalyzed Acylation of Hexamethyldistannane.

$$R \overset{O}{\underset{}{\overset{\|}{C}}} Cl + Me_3SnSnMe_3 \xrightarrow[\text{B(Pd(PPh}_3)_2\text{Cl}_2)}{\text{A(Pd(PPh}_3)_4)} R^1 \overset{O}{\underset{}{\overset{\|}{C}}} SnMe_3$$

R^1	Method	Yield(%)
Me	A	70
Et	B	70
^{t}Bu	A or B	0
Ph	A or B	80
2-furyl	B	80

Note: Bu$_3$SnSnBu$_3$ gave incomplete reaction even with excess RCOCl.

Acylstannanes by Hydroylsis of Enol Ethers

Propenoyltrimethylstannane

To a solution of 4.0 g (31 mmol) of 1-(1-ethoxyethoxy)-1,2-propadiene in 24 mL of anhydrous THF was added at −78 °C 21.6 mL of BuLi (1.6 M in hexane, 34.4 mmol), and the mixture was quenched with 6.10 g (30.5 mmol) of Me$_3$SnCl in 8 mL of THF. The bright yellow solution was stirred 20 min at −78 °C, let slowly warm to 25 °C, then poured onto aqueous NaHCO$_3$ and a 1 : 1 mixture of ether and pentane. The organic layer was separated, dried and the solvent was removed to afford a yellow oil (7.6 g, 84%) which, upon distillation under vacuum, afforded 5.1 g of stannane. A nitrogen-purged flask was charged with 0.5 mL of 2 N H$_2$SO$_4$ in 4 mL of THF and treated with 1.14 g (3.9 mmol) of 1-(1-ethoxyethoxy)-1-trimethylstannyl-1,2-propadiene. The mixture was stirred at 25 °C for 1 h, during which time it turned yellow-orange. It was then diluted with water and worked up under strictly inert atmosphere. Removal of the solvent under vacuum afforded 730 mg (91%) of propenoyltrimethylstannane.

3.2 Reactions

3.2.1 Oxidation/Reduction

Acylstannanes are rapidly oxidized to tin carboxylates upon exposure to air. The carbonyl group of acyl stannanes is highly electrophlic. Additions of hydride take place with ease. The most useful application of this reaction involves reduction with (R)- or (S)-BINAL-H to give nonracemic α-hydroxy stannanes of >90% ee:

Synthetic applications of these stannanes are described in later sections of this chapter.

3.2.2   Acylation

Acylstannanes undergo Pd^0-catalyzed coupling with acyl chlorides to yield α-diketones. Symmetrical α-diketones can be prepared similarly from acyl chlorides and Bu$_3$SnSnBu$_3$. In this case the acyl stannane is formed in situ [19]:

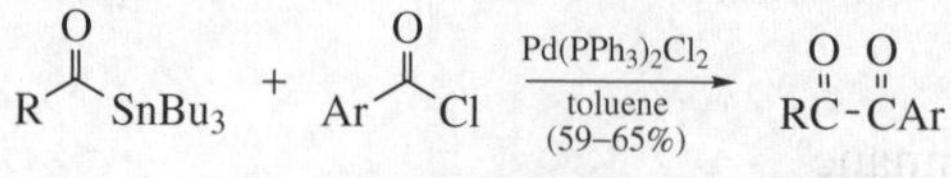

R = ^{i}Pr, Et; Ar = Ph, *p*-MeOC$_6$H$_4$, *p*-MeC$_6$H$_4$

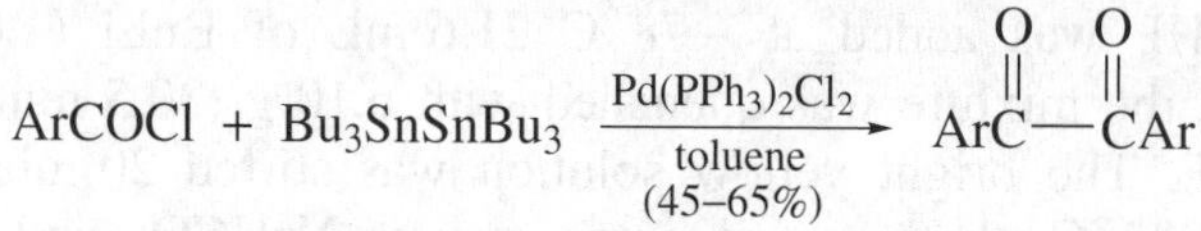

$$\text{ArCOCl} + \text{Bu}_3\text{SnSnBu}_3 \xrightarrow[\substack{\text{toluene} \\ (45\text{–}65\%)}]{\text{Pd(PPh}_3)_2\text{Cl}_2} \text{ArC—CAr}$$

Ar = Ph, *p*-ClC$_6$H$_4$, *p*-MeC$_6$H$_4$, *p*-MeOC$_6$H$_4$

4   α-Oxygenated Organostannanes

4.1   Preparation of Racemic α-Oxygenated Organostannanes

The first general synthesis of α-alkoxystannanes involved addition of Bu$_3$SnLi to an aldehyde followed by treatment of the resulting alcohol adduct with ethyl chloroethyl ether in the presence of *N,N*-dimethylaniline. The ethers are converted to stable organolithium derivatives upon treatment with BuLi. These, in turn, react with carbonyl compounds to give substitution products[20]:

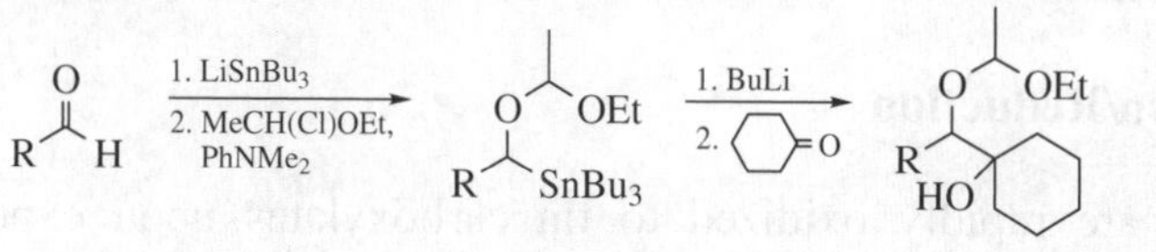

R = C$_6$H$_{13}$ (97%)   R = C$_6$H$_{13}$ (81%)
R = *c*-C$_6$H$_{13}$ (95%)   R = *c*-C$_6$H$_{11}$ (80%)

Conversion of Aldehydes into Ethoxyethyl Ethers of α-Hydroxy Stannanes

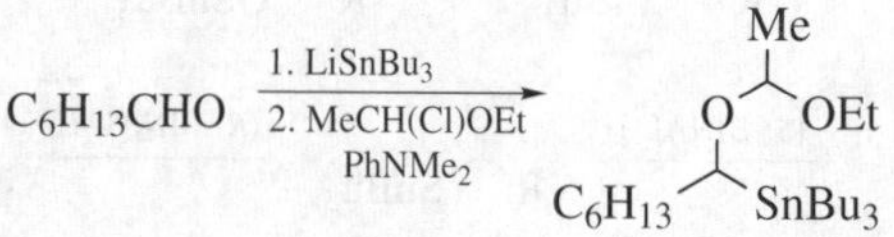

1-(1-Ethoxyethoxy)-1-tributylstannylheptane

Anhydrous THF (10 mL) and diisopropylamine (0.8 mL) were stirred under nitrogen at 0 °C while BuLi (2 mL of a 2.5 M hexane solution, 5 mmol) was added dropwise. The resulting solution was stirred for an additional 5 min and tributyltin hydride (1.45 g, 1.32 mL, 5.00 mmol) was added by syringe. After about 15 min at 0 °C the reaction was complete.

To a solution of Bu$_3$SnLi (10 mmol), prepared as described and chilled to -78 °C, was added a solution of the aldehyde (1.14 g, 1.34 mL, 10 mmol) in about 2 mL of anhydrous THF. After stirring for 5 min, the cold reaction mixture was quenched with dilute NH$_4$Cl and partitioned between petroleum ether and water. The organic phase was dried over sodium sulfate and the solvent was removed at reduced pressure below 30 °C. The product α-hydroxy organostannane was immediately converted to the *O*-ethoxyethyl derivative as described below.

The crude product in 25 mL of CH$_2$Cl$_2$ containing 2.5 mL of *N*,*N*-dimethylaniline was cooled to 0 °C under a drying tube. The solution was stirred and α-chloroethyl ethyl ether (1.7 mL, 15 mmol) was added. After 1 h, the reaction mixture was poured into 100 mL of petroleum ether and washed successively with ice-cold 0.5 N HCl and water. The organic phase was dried over sodium sulfate. The solvent was finally removed at reduced pressure to yield the ethoxyethyl-protected α-alkoxy organostannane (4.6 g, 97%). The material was essentially homogeneous by thin-layer chromatography (TLC) and was used without purification for preparation of the corresponding organolithium reagent. Analytically pure material was obtained by short-column chromatography on silica gel G with 0.5% ethyl acetate/petroleum ether.

4.2 Reactions of Racemic α-Oxygenated Organostannanes

4.2.1 Alkylation

Treatment of the α-alkoxy stannanes with BuLi and alkylation of the intermediate organolithium with reactive halides leads to the expected ethers. This methodology was applied to a synthesis of 9-hydroxydendrolasin and dendrolasin. In these transformations the α-alkoxy stannane functions as a carbinol anion equivalent:

9-hydroxydendrolasin   dendrolasin

Alkylation of α-Ethoxyethyl Ethers of α-Hydroxy Stannanes

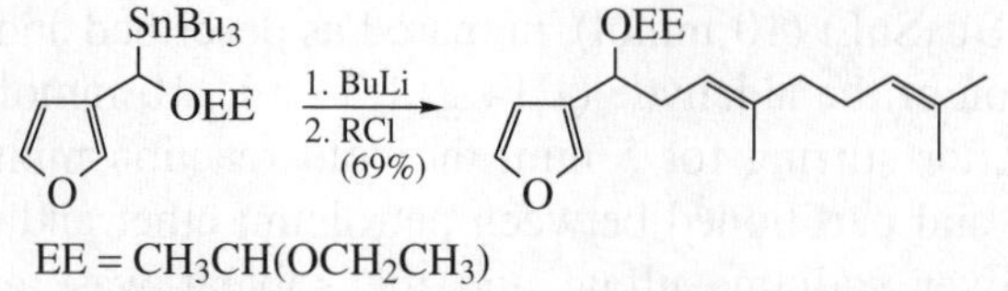

EE = CH₃CH(OCH₂CH₃)

9-(1-Ethoxyethoxy)dendrolasin

A mixture of anhydrous THF (93 mL) and 1,2-dimethoxyethane (3 mL) was chilled to −78 °C under nitrogen. Butyllithium (1 mL of a 2.5 M hexane solution, 2.5 mmol) was added. This mixture was stirred while the alkoxystannane (0.918 g, 2 mmol) in 2 mL of anhydrous THF was added dropwise. After about 1 min, geranyl chloride, freshly and rapidly filtered through a 1-in. Florisil column with cold pentane, (0.431 g, 2.5 mmol) was injected. The mixture was stirred for 1 h at −78 °C and then poured into 50 mL of petroleum ether. The solution was washed with water, dried, and solvent was removed to yield a light yellow oil (1.41 g). TLC (5% ethyl acetate/petroleum ether) showed a single major spot ($R_f = 0.44$) in addition to several minor spots. Short-column chromatography (60 g of silica gel G, 2.5% ethyl acetate/petroleum ether) allowed isolation of the major product as a colorless oil (0.425 g, 69%).

The intermediate α-alkoxy organolithium species were found to be configurationally stable[21]. Both additions to aldehydes or ketones and alkylations proceed with retention of stereochemistry. Similarly, the axial or equatorial cyclohexyllithium species from cis- or trans-1-methoxymethyl-4-tert-butylcyclohexyl tributyltin adds to benzaldehyde or propanal with retention[22]:

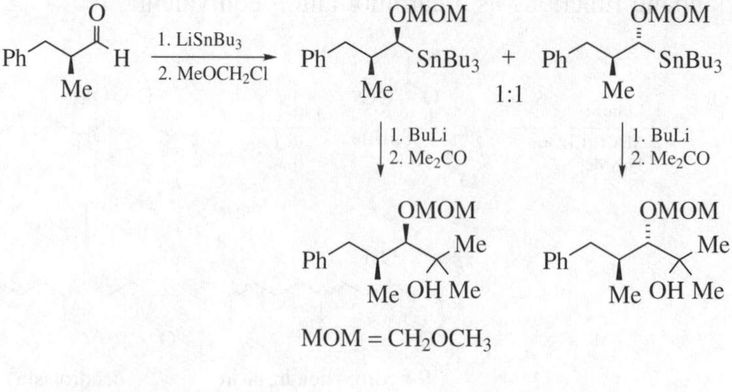

MOM = CH₂OCH₃

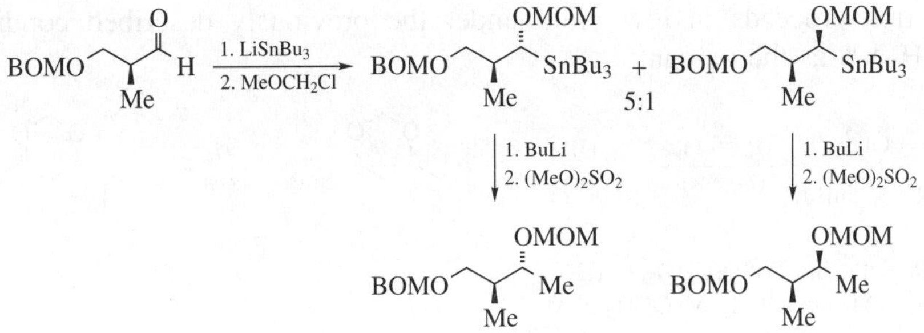

BOM = C₆H₅CH₂OCH₂– or PhCH₂OCH₂–

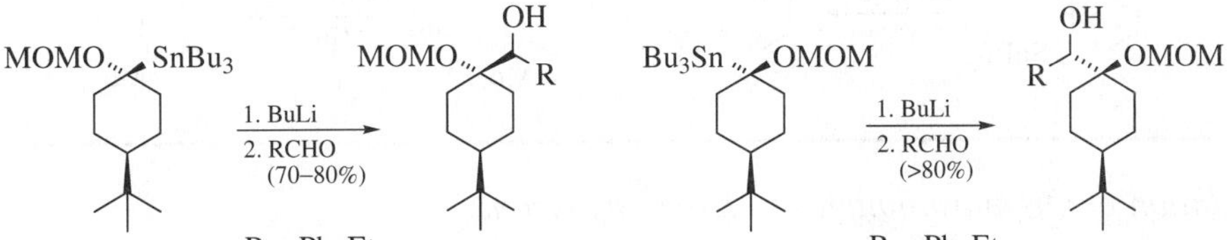

R = Ph, Et

R = Ph, Et

MOM = CH₃OCH₂–

4.2.2 Acylation

Acylation of α-alkoxy organolithium reagents can be effected with *N,N*-dimethyl carboxamides[23]. An intramolecular variant of this reaction leads to four- and five-membered rings. In these examples tin–lithium exchange is markedly faster than addition of BuLi to the amide carbonyl:

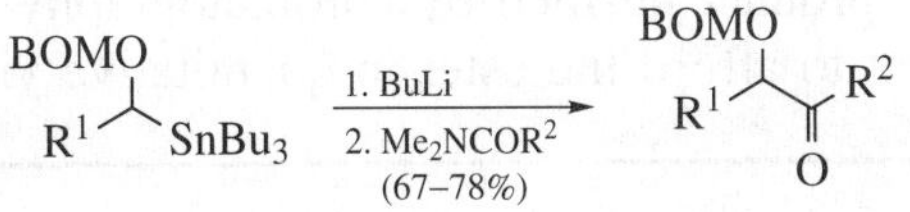

R¹ = Et, ᶦPr; R² = Ph, Et, H
BOM = PhCH₂OCH₂

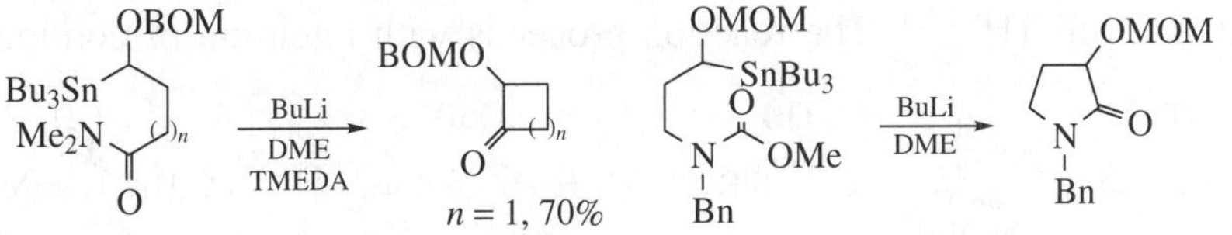

n = 1, 70%
n = 2, 75%

BOM = PhCH₂OCH₂
MOM = CH₂OCH₃

4.2.3 Oxidation

α-Alkoxyorganostannanes are oxidized to the corresponding esters by ozone[24]. The reaction also shows promise for secondary and tertiary tetraalkylstannanes. The latter reaction proceeds in low yield under the previously described conditions with $CrO_3 \bullet (C_5H_5N)_2$ as the oxidant:

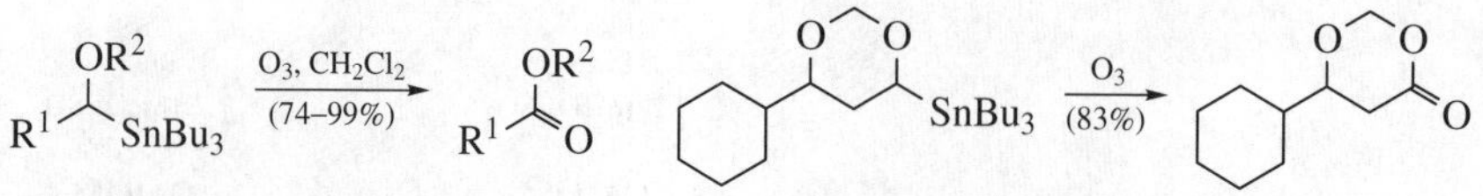

$R^1 = {}^i\text{Pr, Bu, TBSO(CH}_2)_5\text{, C}_{11}\text{H}_{23}$
$R^2 = \text{MenthylOCH}_2\text{, MeOCH}_2\text{, SiMe}_3$

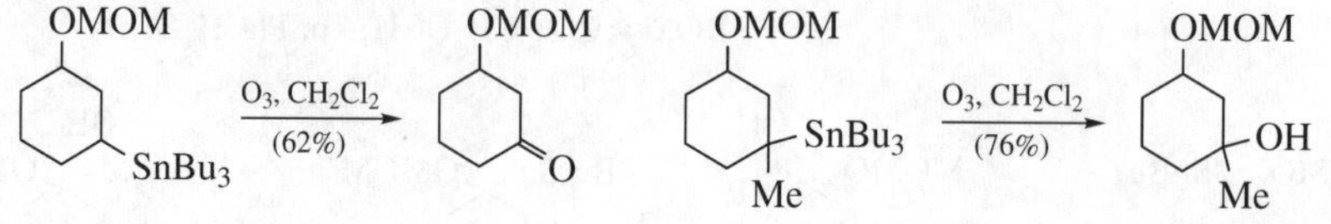

Oxidation of Organostannanes to Esters by Ozone

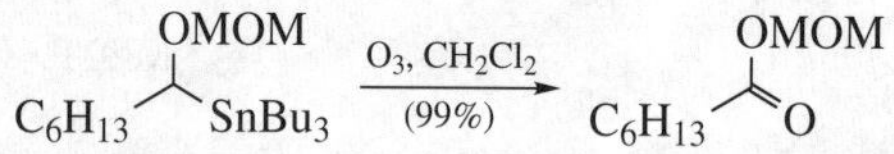

Methoxymethyl Heptanoate

The stannane (3.0 mmol) was dissolved in CH_2Cl_2 (50 mL) and cooled to $-78\,°C$. Ozone was bubbled through the solution until a blue color persisted. Ar gas was then bubbled through the solution for 5 min to dispel excess ozone, and the reaction mixture was allowed to warm to $25\,°C$. The solvent was removed under reduced pressure and the crude product purified by chromatography (silica gel) using ethyl acetate/hexane as eluent to afford the ester in quantitative yield.

4.2.4 Reverse Brook Rearrangement

α-Silyloxy organostannanes undergo a reverse Brook rearrangement upon treatment with excess BuLi in THF[25]. The reaction proceeds with retention of configuration:

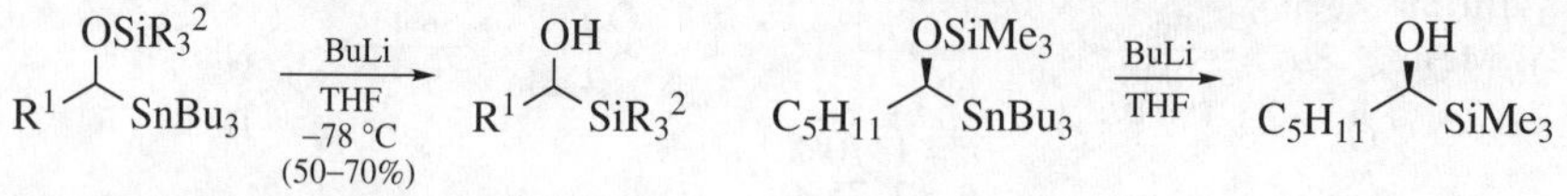

$R^1 = \text{Ph, }c\text{-C}_6\text{H}_{11}\text{, C}_5\text{H}_{11}\text{, C}_7\text{H}_{15}\text{, }{}^t\text{Bu}$
$R^2 = \text{Me, Et, }{}^i\text{Pr}$

Reverse Brook Rearrangement

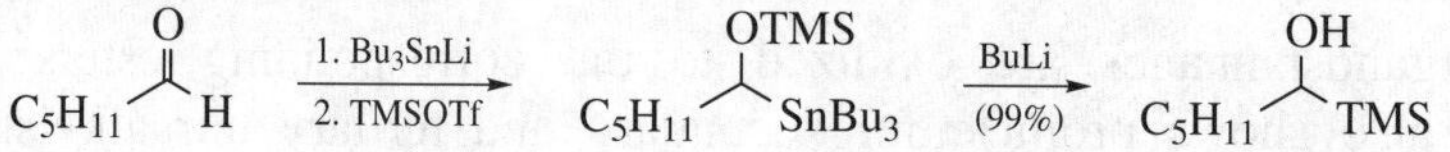

1-(Trimethylsilyl)-1-hexanol

A THF solution of the stannyllithium (0.95 equiv of 0.1 M) was cooled to −78 °C and hexanal (1.0 equiv) was added dropwise. The reaction mixture was stirred at −78 °C for 15 min and then quenched at −78 °C by the dropwise addition of trimethylsilyl (TMS) triflate (1.5 equiv). The mixture was allowed to gradually warm to 25 °C and then stirred for 1 h. The solution was then diluted with 100 mL of petroleum ether, washed with saturated aqueous NaCl solution, and dried over anhydrous sodium sulfate. The solvents were removed under reduced pressure to provide the crude stannyl silyl ether.

The crude stannyl silyl ether in 25 mL of dry THF was cooled to −78 °C (under Ar). A hexane solution of BuLi (3.0 equiv) was added dropwise by syringe. The reaction mixture was then allowed to stir at −78 °C for 15 min and quenched at −78 °C by the rapid addition of 5 mL of water. After warming to 25 °C, the reaction mixture was worked up by extraction with petroleum ether. The (α-hydroxyalkyl)trialkylsilane product (91% yield) was purified by flash chromatography on silica gel with hexane to first elute the tetraalkylstannane byproduct, followed by gradient elution of 1%, 2%, and then 3% ethyl acetate hexane.

4.2.5 Cupration

α-Methoxymethoxyorganocuprates can be prepared from the corresponding stannanes via the lithiated intermediates[26]. Best results are obtained with highly purified stannanes and CuCN to give the higher-order cyanocuprate. These cuprates add to cyclohexenone and cyclopentenone to afford diastereomeric mixtures of 1,4 adducts. Optimal yields are obtained when the reaction is performed in the presence of a five-fold excess of TMS-Cl. The intermediate silyl ethers can be hydrolyzed during work-up:

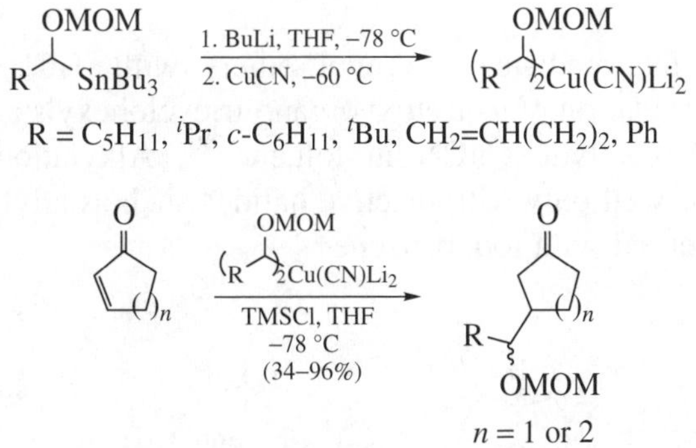

$R = C_5H_{11}, {}^iPr, c\text{-}C_6H_{11}, {}^tBu, CH_2=CH(CH_2)_2, Ph$

$n = 1$ or 2

Transmetalation and 1,4 Addition of an α-OMOM Stannane to Cyclohexenone

3-(1-Methoxymethoxy-2-methylpropyl)cyclohexanone

A solution of the stannane (1.0 mmol) in 5 mL of THF was cooled to −78 °C then 0.50 mL of 2.6 M BuLi in hexane (1.3 mmol) was added and the solution was stirred for 5 min at −78 °C. A 25-mL round bottom flask containing 0.045 g (0.5 mmol) of CuCN suspended in 2 mL of THF was cooled to −78 °C. The α-alkoxylithio species was transferred by cannula to the suspension of CuCN at −78 °C. The cuprate mixture was gradually allowed to warm to −60 °C (bath temperature) over a period of 0.5 h. A clear, homogeneous solution was obtained. A second 25-mL round bottom flask containing a solution of the enone (0.5 mmol) in 3 mL of THF was cooled to −78 °C. Trimethylsilyl chloride (0.32 mL, 2.5 mmol) was added to the enone solution. The enone/TMS-Cl mixture was then added to the cuprate solution (at −78 °C) by cannula. The resulting mixture was stirred for 1 h at −78 °C, and gradually warmed to 0 °C over 2.5 h. The reaction mixture was quenched by the addition of 1 mL of 1.0 N aqueous HCl, stirred for 10 min, and diluted with 100 mL of ether. The mixture was washed sequentially with a 1 : 1 mixture of aqueous NH_4Cl/1.0 N HCl, saturated aqueous NaCl, and $NaHCO_3$. The layers were separated and the organic phase was dried over anhydrous $MgSO_4$. After removal of the solvent under reduced pressure, the crude reaction product was purified by flash chromatography on silica gel with 15–20% ethyl acetate/petroleum ether as eluent, (96% yield).

Treatment of the intermediate silyl enol ethers with $TiCl_4$ leads to fused ring tetrahydrofurans[27]. Acylation of α-acetoxyorgano tricyclohexylstannanes by acyl halides can be effected with catalytic CuCN in toluene[28]. Alkylation of the intermediate cyanocuprate proceeds well only with reactive halides such as allyl or cinnamyl bromide. No coupling was observed with iodobenzene:

TMS = Si(CH₃)₃ or SiMe₃   R = Ph, Me, C₅H₁₁, ⁱPr, ᵗBu

4.3   Preparation of Nonracemic α-Oxygenated Organostannanes

4.3.1   Reduction of Acylstannanes

As noted earlier, reduction of acylstannanes with (R) or (S)-BINAL-H affords (S)-or (R)-α-hydroxy organostannanes of >90% ee. These alcohols are converted to the MOM = CH_3OCH_2 (MOM), BOM = $PhCH_2OCH_2$ (BOM), or TBS = Si(tert-Bu)Me$_2$ (TBS) ethers by reaction with the appropriate halides in the presence of tertiary amine bases. Attempted alkylation with stronger bases (NaH, BuLi) and less reactive halides (MeI, allyl bromide) affords Bu$_3$SnH and the starting aldehyde:

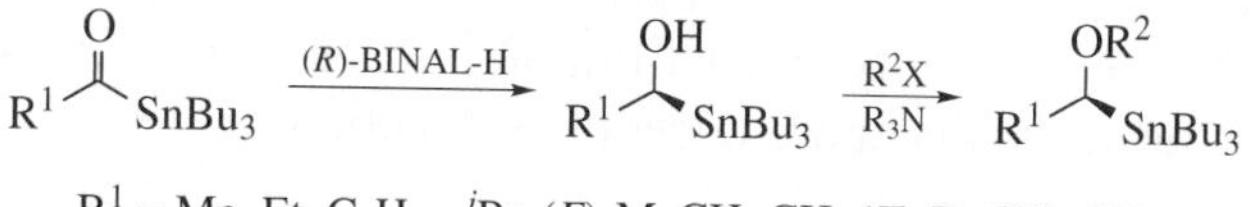

R^1 = Me, Et, C_5H_{11}, ^iPr, (E)-MeCH=CH, (E)-BuCH=CH
R^2X = MeOCH$_2$Cl, PhCH$_2$OCH$_2$Cl, ^{t}BuMe$_2$SiCl
R_3N = (^iPr)$_2$NEt (50–60% yield)

Reduction of an Acylstannane by BINAL-H and Formation of the MOM Ether

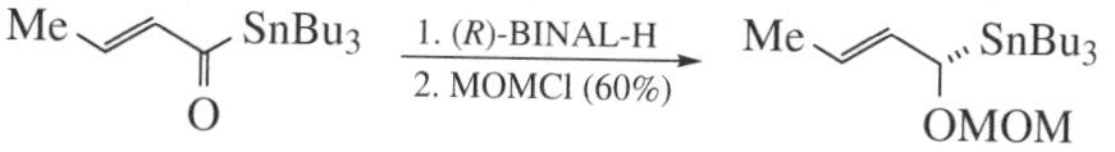

(E)-1-(Methoxymethoxy)-1-tributylstannyl-2-butene

A solution of 25 mL (25 mmol) of 1.0 M LiAlH$_4$ in THF was added to 50 mL of THF with stirring, and then 25 mL (25 mmol) of 1.0 M EtOH in THF was added over 30 min. The reaction mixture was stirred for 30 min. To this mixture was added a solution of 7.2 g (25 mmol) of (R)-1,1'-bi-2-naphthol in 50 mL of THF over 1 h. The milky white reaction mixture was heated to reflux for 50 min. It was then allowed to reach ambient temperature and cooled to −78 °C. To this suspension was added the acyl stannane in 17 mL of THF over 1 h. The reaction mixture was stirred for 24 h at −78 °C and then quenched with MeOH, followed by saturated aqueous NH$_4$Cl. The phases were separated, and the aqueous phase was treated with 3% HCl and extracted with ether. The organic layer was dried over MgSO$_4$ and concentrated under reduced pressure. This residue was triturated with 300 mL of hexanes and filtered, affording 6.9 g (96%) of recovered binaphthol.

The filtrate was concentrated under reduced pressure, affording the crude hydroxy stannane. This material was dissolved in 10 mL of CH$_2$Cl$_2$, and 4.4 mL (25 mmol) of (^iPr)$_2$NEt was added, followed by 1.0 mL (12 mmol) of MOMCl. After stirring overnight, the reaction mixture was quenched with saturated aqueous NH$_4$Cl. The phases were separated, and the aqueous phase was extracted with ether. The combined organic phases were dried over MgSO$_4$, concentrated under reduced pressure, and chromatographed slowly (elution with hexanes) through silica gel to yield

2.6 g (45% from starting aldehyde) of the α-(alkoxy) stannane ($[\alpha]_D - 56(c\ 1.4,$ $CH_2Cl_2)$). ^{1}H NMR analysis of the _O_-methylmandelate derivative of the hydroxy stannane indicated an ee of $>95\%$ for this material.

4.3.2 From Nonracemic Acetals

Several alternative routes to nonracemic α-alkoxyorganostannanes have been devised. In the first of these, a chiral acetal is prepared from (_S,S_)-2,4-pentanediol and the diethyl acetal of formyl tributyltin (Table 4)[29]. Addition of a Grignard reagent to this acetal in the presence of TiCl$_4$ affords the (_S_)-α-alkoxyorganostannane with modest to excellent enantioselectivity. Stereoselectivity arises from S$_N$2 attack of the Grignard reagent on the acetal carbon from the sterically more accessible direction.

Table 4. Synthesis of Nonracemic α-Oxygenated Stannanes from a Chiral Formylstannane Acetal.

R	Yield (%)	de
Me	71	30
Et	85	>95
Bu	54	92
CH$_2$=CHCH$_2$	30	>95
Ph	67	20

Nonracemic α-Alkoxystannanes from a Chiral Acetal

(_S,S_)-3-Hydroxy-1-methylbutyl (_S_)-1-(Tributylstannyl)propyl Ether

To a solution of acetal (5.0 g, 12.7 mmol) in benzene (80 mL) were added (_S,S_)-2,4-pentanediol (2.0 g, 19.2 mmol) and _p_-toluenesulfonic acid monohydrate (250 mg). After being stirred for 1 h at 25 °C, the solution was heated under reflux and the ethanol/benzene azeotrope (40 mL) was collected with a Dean–Stark apparatus. The resulting mixture was poured into aqueous pH 7 buffer and extracted with ether. The combined extracts were washed with brine, dried over sodium sulfate, filtered and concentrated. Purification by flash chromatography gave 3.7 g of acetal (74% yield) as a colorless oil ($[\alpha]_D + 14.7$ (c 1.11, $CHCl_3$)).

A 1.0 M solution of $TiCl_4$ in CH_2Cl_2 (0.2 mmol) was added to a solution of acetal (0.040 g, 0.1 mmol) in CH_2Cl_2 (5 mL) at $-78\,°C$ under nitrogen. After 5 min, EtMgBr (0.94 M in THF, 0.54 mL, 0.5 mmol) was slowly added to the mixture. After an additional 5 min, HCl (0.5 N) was added, and the mixture was warmed to $25\,°C$. The resulting mixture was extracted with ether. The combined extracts were washed with brine, dried over sodium sulfate, filtered, and concentrated. Purification by flash chromatography gave 45 mg of stannane (85% yield) as a colorless oil.

4.3.3 Resolution

A second route to nonracemic α-oxygenated organostannanes involves resolution of the menthyloxymethyl ethers of α-hydroxy organostannanes[30]. The 1 : 1 mixture of diastereomeric ethers can be separated by column chromatography. The resolved menthyl ethers are cleaved with Me_2BBr to yield the (R)- or (S)-α-hydroxy derivatives. These can be reprotected with an appropriate alkoxymethyl or silyl halide. Alternatively, an acetal interchange reaction can be effected with TMSOMe and TMSOTf to yield the MOM ethers directly:

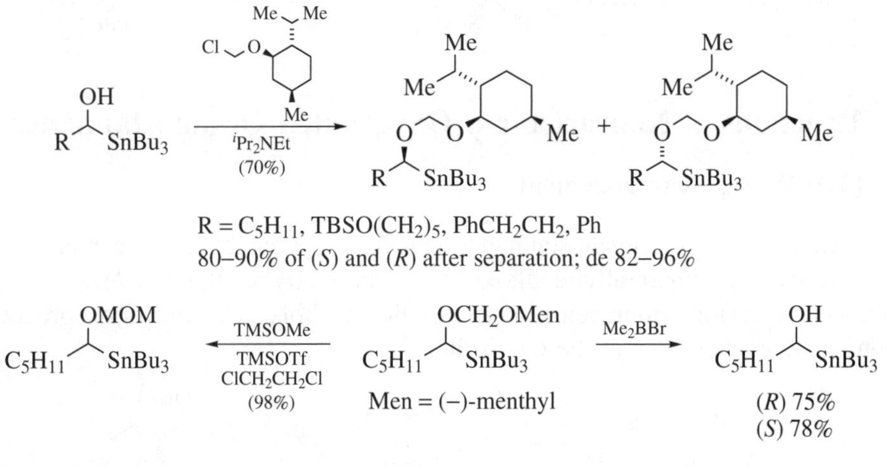

R = C_5H_{11}, $TBSO(CH_2)_5$, $PhCH_2CH_2$, Ph
80–90% of (S) and (R) after separation; de 82–96%

Men = (−)-menthyl

(R) 75%
(S) 78%

4.3.4 Asymmetric Stannylation

Chelation-directed deprotonation of primary carbamates in the presence of (−)-sparteine followed by addition of Me_3SnCl affords the (R) stannyl derivatives of >95% ds[31a]. If (+)-sparteine is employed as an additive (mismatched case), a 28 : 72 mixture of the (R) and (S) stannane diastereomers is obtained. In the absence of sparteine, the (R)

diastereomer predominates 98 : 2. Evidently, the proximal acetonide grouping significantly directs the deprotonation reaction:

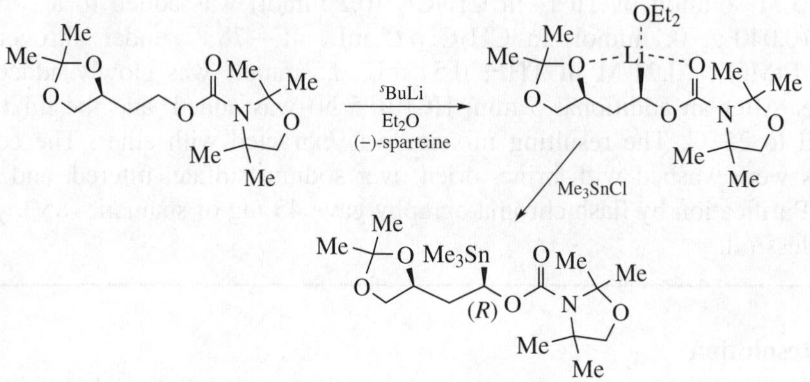

Deprotonation of (R)-1-phenylethyl carbamate followed by addition of Me₃SnCl leads
to the (R) stannane with net inversion of C−H stereochemistry[31b]. Lithiation of this
stannane and subsequent addition of Me₃SnCl affords the (S) enantiomer. The apparent
inversion of stereochemistry of the stannylation steps is only observed with benzylic
stannanes. The low inversion barrier of the benzyl system appears responsible for this
unusual case[31b]:

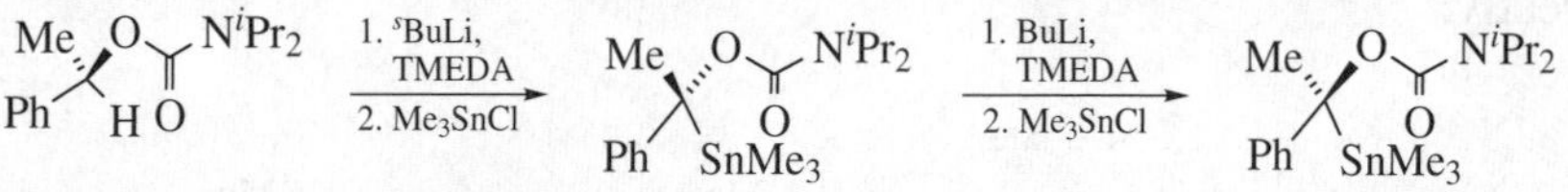

4.4   Reactions of Nonracemic α-Oxygenated Organostannanes

4.4.1   [2,3] Wittig Rearrangement

Nonracemic α-allyloxy organostannanes can be prepared by displacement of the α-
mesyloxy derivative with an allylic alkoxide[32]. These allylic ethers undergo [2,3] Wittig
rearrangement to afford nonracemic homoallylic alcohols. The reactions proceed with
inversion of stereochemistry at the C−Sn bond:

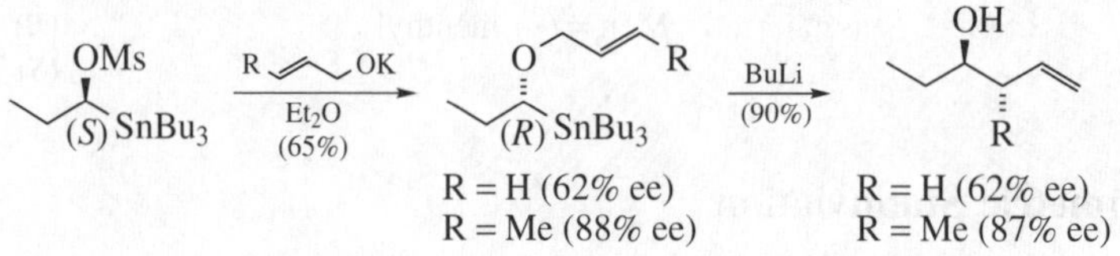

4.4.2   Mitsunobu Inversion

α-Imido organostannanes can be prepared by Mitsunobu displacement of α-hydroxy
precursors with imides[33]. The derived N-methyl Boc = tert-BuOCO (Boc) derivatives

are converted to *N*-methyl α-amino acid derivatives by lithiation and carboxylation. The α-amino organolithium intermediates are more prone to racemize than their α-alkoxy counterparts. As a result, lower temperatures are recommended for optimal retention of chirality:

$$DEAD = EtO_2CN=NCO_2Et$$

R = Et, 93% (92% ee)
R = ^{i}Pr, 75% (92% ee)

4.4.3 Cupration

Higher-order cyanocuprates derived from nonracemic α-alkoxy organostannanes add to enones, as previously described for the racemic counterparts[34]. However, the reaction occurs with significant racemization. Furthermore, the results are not reproducible. It is proposed that the racemization is initiated by trace amounts of oxygen.

The TMEDA-stablized cuprates, obtained from diastereomeric nonracemic stannyl dioxanes, are configurationally stable. Addition to ethyl propiolate affords the 1,4 adducts without any detectable racemization. In contrast, addition of the corresponding higher order cyanocuprates yields a 70 : 30 mixture of diastereomeric 1,4 adducts:

$$TMEDA = Me_2NCH_2CH_2NMe_2$$

The configurational stability of the foregoing TMEDA complexes of the dioxanyl-cuprates must reside, in part, with their cyclic nature. The analogous cuprate derived from a simple α-alkoxyorganostannane gives the 1,4 adduct of ethyl propiolate with appreciable racemization[35]:

18% ee

Coupling of the cyanocuprates derived from α-alkoxy or α-acyloxy organostannanes with allylic or acyl halides generally proceeds in low yield. However, the analogous thionocarbamates afford the coupled products in high yield with retention of configuration[36]. Furthermore the reaction is catalytic in CuCN (8 mol%):

$$\underset{R^1}{\overset{\overset{\displaystyle S}{\overset{\|}{OCNR_2}}}{\diagdown}}\underset{}{\diagup}SnBu_3 \quad \xrightarrow[\substack{(80-95\%)}]{\substack{CuCN \\ R^2X}} \quad \underset{R^1}{\overset{\overset{\displaystyle S}{\overset{\|}{OCNR_2}}}{\diagdown}}\underset{}{\diagup}R^2$$

$$82-84\% \; ee$$

$R^1 = C_9H_{19}, C_7H_{15}, PhCH_2CH_2, Ph$

$R = Me, (CH_2)_4$

$R^2X = \diagup\!\!\!\diagdown^{Br} \; C_7H_{15}C=CCH_2Br \; HC=CCH_2Br, \; BuCOCl, \; EtSCOCl$

Copper-Catalyzed Allylation of a Nonracemic α-Hydroxy Stannane Thiocarbamate

$$\underset{C_9H_{19}}{\overset{\overset{\displaystyle S}{\overset{\|}{O\diagup\!\!C\diagdown N\diagup}}}{}}\!SnBu_3 \quad \xrightarrow[\substack{Br\diagup\!\!\!\diagdown \\ (95\%)}]{CuCN} \quad C_9H_{19}\overset{\overset{\displaystyle S}{\overset{\|}{O\diagup\!\!C\diagdown N\diagup}}}{\diagdown}\!\diagup\!\!\!\diagdown$$

(*R*)-4-(Pyrrolidinethiocarbamoyloxy)tridec-1-ene

A THF solution (1.5 mL) of decanoyltributylstannane (0.373 g, 0.83 mmol) was added dropwise at −78 °C during 20 min to 12 mL of THF solution of (*R*)-BINAL-H prepared from LiAlH₄ (0.095 g, 2.5 mmol), absolute EtOH (0.115 mg, 2.5 mmol), and (*R*)-(+)-1,1′-bi-2-naphthol (0.717 mg, 2.5 mmol). After 2 h, the reaction was quenched with 6 mL of saturated NH₄Cl and the mixture was diluted with ether (60 mL). The organic layer was washed with water and brine, dried, and concentrated in vacuo. The residue was suspended in hexane (12 mL), and the precipitated binaphthol was removed by filtration. Concentration of the filtrate afforded the crude α-hydroxy stannane which was immediately converted to the corresponding pyrrolidine carbamate with thiocarbonyldiimidazole (0.222 g, 1.25 mmol) and DMAP = 4-Me₂NC₆H₄N (DMAP) (8 mg) in CH₂Cl₂ (2 mL), followed by pyrrolidine (2 mL). Extractive isolation and chromatographic purification (silica gel) with EtOAc/hexane (3 : 97) yielded 0.323 g (69% yield) of stannane as a colorless oil ($[\alpha]_D +$ 40.1 (*c* 1.67, CHCl₃)). Its stereochemical purity (95% ee) was determined by ¹H NMR analysis of the (*R*)-MTPA ester ((*R*)-(+)-MTPA, DCC/DMAP, CH₂Cl₂, 5 h) of the hydroxy stannane precursor. (MTPA = PhC(OMe)CF₃CO = *m*ethoxy*t*rifluoromethyl*p*henyl*a*cetic, DCC = dicyclohexylcarbodiimide)

Allyl bromide (0.014 g, 0.12 mmol) was added to the above stannane (0.072 g, 0.13 mmol) and copper cyanide (0.0009 g, 0.01 mmol) in 1.5 mL of anhydrous THF. The reaction mixture was heated at 45 °C under an argon atmosphere in a sealed tube for 14 h. Concentration in vacuo and chromatographic purification (silica gel) with 5% EtOAc/hexane gave 0.034 g (94% yield) of adduct as a colorless oil ($[\alpha]_D + 8.8$ (*c* 1.08, CHCl₃)). The absolute stereochemistry and %ee were determined after hydrolysis by comparisons with an authentic standard of (*R*)-1-tridecen-4-ol and the corresponding Mosher esters.

5 Vinylstannanes

5.1 Preparation

5.1.1 Hydrostannation[1]

Addition of Bu₃SnH to alkynes at elevated temperatures in the presence of azoisobutyronitrile (AIBN) leads to vinylstannanes. Unsymmetrical internal alkynes give rise to (*E*)/(*Z*) mixtures of regioisomers and therefore find little use in synthesis. Terminal alkynes afford mainly the terminal vinylstannanes, also as (*E*)/(*Z*) mixtures. It has been shown that the Bu₃Sn radicals involved in these additions can cause isomerization of the initially formed (*Z*) isomers:

Propargylic alcohols and ethers show a remarkable regioselectivity in free radical hydrostannations[37]. The Bu₃Sn grouping is introduced adjacent to the OR (alkoxy) function in a wide range of substrates:

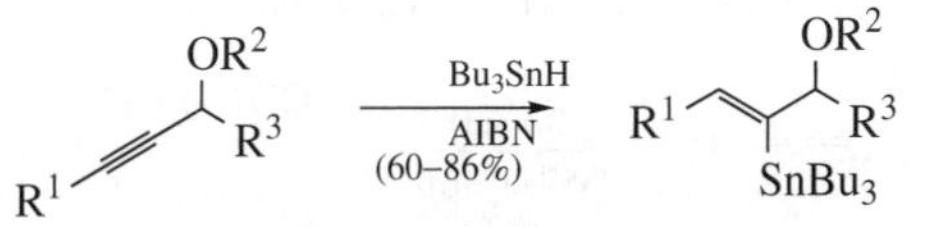

1:1 to 12:1 (*Z*):(*E*)

R^1 = CH₂SiMe₃, Me, Bu
R^2 = H, Me, ᵗBuMe₂Si, Ac
R^3 = H, Me, Bu, Ph, CH₃(CH₂)₅

Hydroxyl-directed Hydrostannation of Alkynes

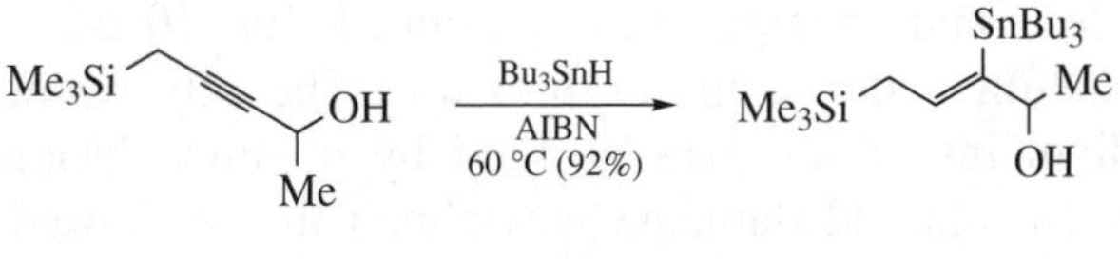

(*E*)-3-(Tributylstannyl)-5-(trimethylsilyl)-3-buten-2-ol

The reaction was performed in a sealed tube. Tributyltin hydride (0.7 mmol), the propargyl alcohol (1.44 g, 9.2 mmol), and AIBN (3 mg, 0.02 mmol) were heated at 60 °C for 1 h. NMR analysis of the crude product mixture showed quantitative conversion. The vinylstannane, 3.00 g (92% yield), was isolated by column chromatography on silica gel with hexane/ethyl acetate (11 : 1).

Under carefully controlled conditions, it is possible to prepare the (Z) isomers with complete regiocontrol and with high stereocontrol[1]. Hydrostannations catalyzed by Pd^0 and Mo^0 complexes are complete within minutes and afford the (E) isomers. Propargylic alcohols and ethers give rise to mixtures of regioisomers[38].

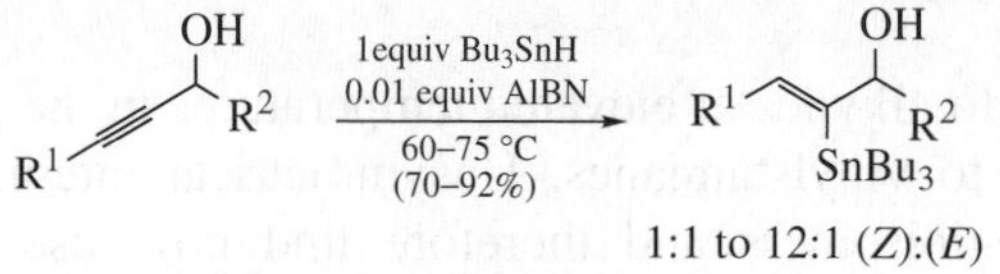

$$R^1 = Me_3SiCH_2, Bu$$
$$R^2 = H, Me, {}^iPr, C_3H_7, C_6H_{13}$$

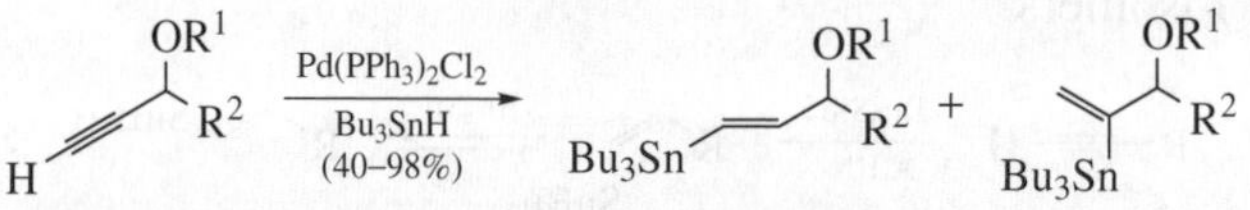

1:1 to 12:1 (Z):(E)

$$R^1 = H, Ph, TBS$$
$$R^2 = H, Me, C_5H_{11}$$

50:50 to 100:0

Palladium(0)-catalyzed Hydrostannation of Alkynes

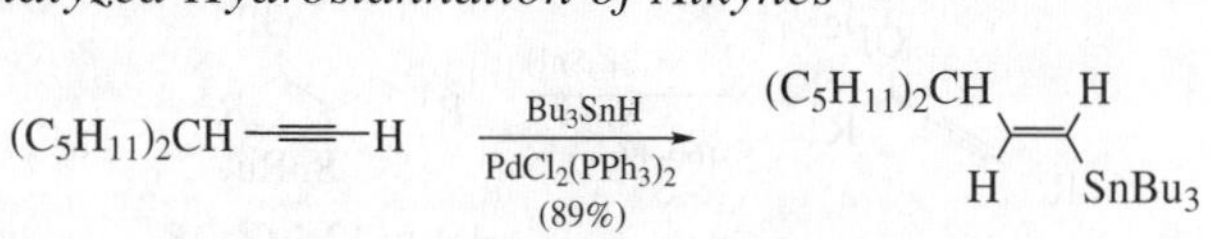

(E)-1-(Tributylstannyl)-3-pentyl-1-pentene

To a THF solution (3 mL) of 3-pentyl-1-octyne (0.180 g, 1.00 mmol) containing $PdCl_2(PPh_3)_2$ (0.014 g, 0.02 equiv) was added Bu_3SnH dropwise with a syringe over a period of 1–2 min. After about 1.1 equiv had been added, the originally light yellow solution abruptly turned orange-brown. An additional 0.1 equiv of Bu_3SnH was added, and stirring was continued for 10 min. THF was then removed by distillation under aspirator pressure. The oily residue, contaminated with black palladium impurities, was purified by column chromatography (silica gel, cyclohexane). Hexabutyldistannane was eluted first, followed immediately by (E)-1-(tributylstannyl)-3-pentyl-1-pentene (yield 0.423 g, 90%). The molybdenum-catalyzed reaction (0.4 equiv of catalyst) was conducted in a similar way (yield 0.422 g, 89%). In this case, the end of the reaction is signaled by a sudden color change of the solution, from dark purple to black.

Hydrostannation of internal alkynes proceeds poorly with the Pd^0 catalyst. However, satisfactory results are achieved with $Mo(CO)_2(MeCN)_2$. In contrast, 1-bromoalkynes afford (*E*)-terminal vinylstannanes with high regio and stereoselectivity. Interestingly, 1-chloro-1-octyne yields the (*E*)-terminal vinylstannane with retention of the Cl substituent:

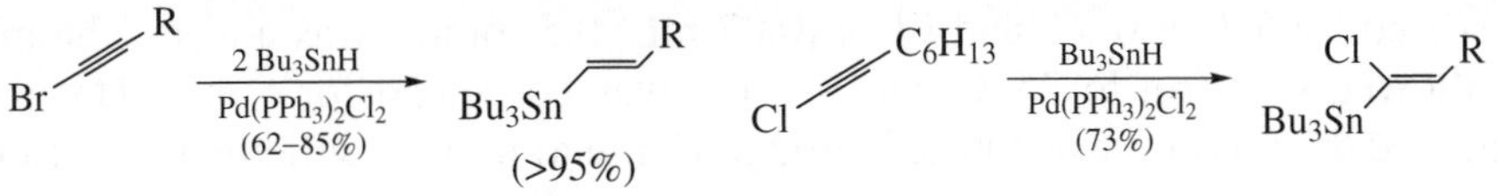

R = $SiMe_3$, CH_2OTHP, C_4H_9

Hydrostannation of alkynoic esters catalyzed by $Pd(PPh_3)_4$ leads to (*E*)-α-stannylated acrylic esters[39]. Alkynyl ketones are hydrostannylated under these conditions to afford the (*Z*)-α-stannylated products:

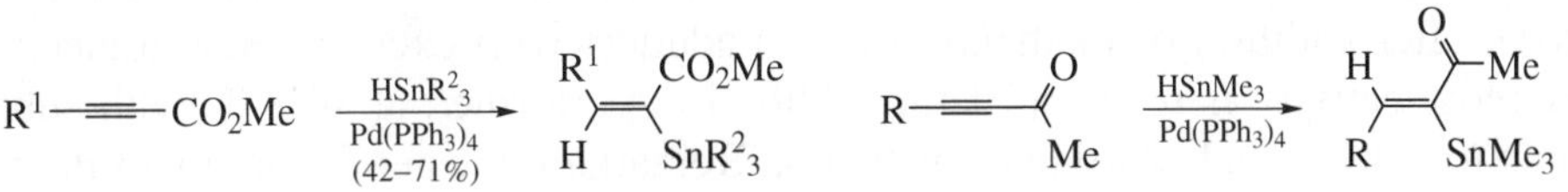

R^1 = H, Me, Ph, CO_2Me
R^2 = Me, Ph

R = H, 51%
R = Et, 58%

These methods can be applied to more complex systems as illustrated in the following examples[40]. Note the OH-directing influence on the regioselectivity of the first example:

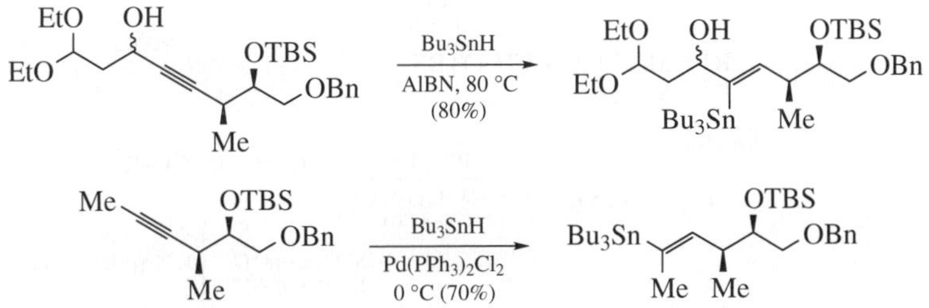

A Lewis acid-catalyzed hydrostannation of alkynes has been described[41]. In the presence of $ZrCl_4$ or $HfCl_4$, Bu_3SnH and Bu_2SnH_2 add to terminal alkynes to afford (*Z*)-vinylstannanes with high regioselectivity:

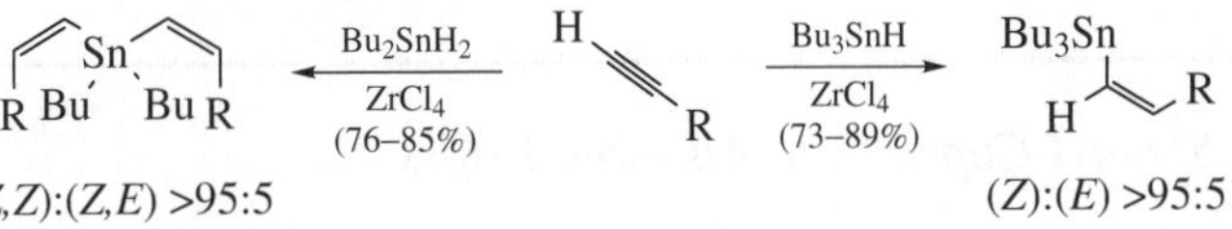

(*Z*,*Z*):(*Z*,*E*) >95:5 (*Z*):(*E*) >95:5

R = C_6H_{13}, $PhCH_2$, 1-cyclohexenyl R = C_6H_{13}, Ph, $TMSO(CH_2)_3$, $Bno(CH_2)_3$

Lewis Acid-catalyzed Hydrostannation of Alkynes

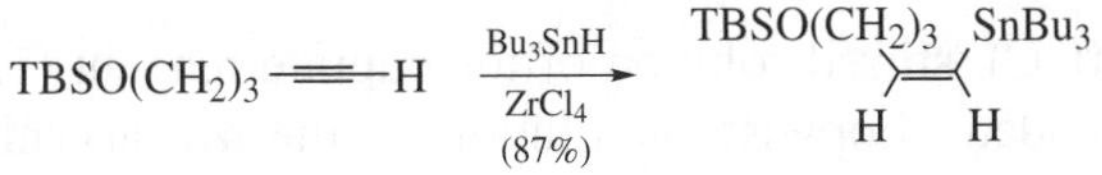

(Z)-1-(Tributylstannyl)-5-(tert-butyldimethylsilyl)oxy-1-pentene

To a suspension of ZrCl₄ (0.047 g, 0.2 mmol) in toluene (0.5 mL) was added the alkyne (0.24 mL, 1.0 mmol) at 0 °C under an Ar atmosphere. The mixture was stirred for 5 min, and then Bu₃SnH (0.42 mL, 1.5 mmol) was added. The mixture was stirred for 1 h at 0 °C, and Et₃N (0.07 mL, 0.5 mmol) was added. The mixture was allowed to warm to 25 °C, and stirring was continued for 5 min. Hexane was added, and the mixture was filtered through Celite to remove solid material. Removal of the solvents under reduced pressures gave an oily material in 87% yield.

5.1.2   Stannylmetalation

Alkynoic esters undergo facile 1,4 addition of PhS(Me₃Sn)CuLi at −100 °C in THF to afford, after addition of methanol, the (E) adducts with excellent stereoselectivity[42]. The stereoselectivity is reversed if the addition is performed at −78 °C with subsequent warming to −48 °C and addition of methanol. Remarkably, when 2.5 equiv of the cuprates Me₃SnCu(SMe₂)BrLi or Me₃SnCuC≡CC(OMe)Me₂Li are employed, the (E)-α,β-bis-stannanes are obtained[43]:

$R^1 = Me, Et, CH_2CH_2OTBS$
$R^2 = Me, Et$

$R^1 = Me, Et, {}^iPr, C_6H_{13}, c\text{-}C_3H_5, TBSOCH_2$
$R^2 = Me, Et$
$RCu = Me_3SnCu(SMe_2)BrLi \text{ or}$
$Me_3SnCuC≡CC(OMe)Me_2Li$

1,4 Addition of Stannyl Cuprates to Alkynoic Esters

(a) (E)-Ethyl 3-Trimethylstannyl-2-pentenoate

To a cold (−100 °C), stirred solution of the cuprate reagent (1.0 mmol) in 10 mL of dry THF was added, dropwise, a solution of the α,β-alkynic ester (0.5 mmol) in 0.5 mL of dry THF containing 0.85 mmol of dry MeOH. The reaction mixture was stirred at $-100\,^{\circ}$C for 15 min and at $-78\,^{\circ}$C for 3 h. MeOH (0.2 mL) and Et$_2$O (30 mL) were added and the mixture was allowed to warm to 25 $^{\circ}$C. The resulting yellow slurry was filtered through a short column of silica gel (10 g, elution with 30 mL of Et$_2$O). The oil obtained by concentration of the combined eluate was chromatographed on silica gel ($\sim$3 g). Elution with petroleum ether ($\sim$10 mL) gave Me$_3$SnSnMe$_3$. Further elution with Et$_2$O ($\sim$8 mL), followed by distillation of the material thus obtained, provided the product (79% yield, bp 110–125 $^{\circ}$C at 20 mmHg).

(b) (Z)-Ethyl 3-Trimethylstannyl-2-pentenoate

To a cold ($-78\,^{\circ}$C), stirred solution of the cuprate reagent (0.39 mmol) in 5 mL of dry THF was added a solution of the α,β-alkynic ester (0.30 mmol) in 0.5 mL of dry THF. The reaction mixture was stirred at $-78\,^{\circ}$C for 15 min and at $-48\,^{\circ}$C for 4 h. MeOH or EtOH (0.2 mL) and Et$_2$O (30 mL) were added and the mixture was allowed to warm to 25 $^{\circ}$C. The yellow slurry was treated with anhydrous MgSO$_4$ and then was filtered through a short column of Florisil (elution with 30 mL of Et$_2$O). Concentration of the combined eluate gave an oil, which was either distilled directly to give the product, or was chromatographed on silica gel (8–15 g, elution with petroleum ether and Et$_2$O), and then distilled (76% yield, bp 103–115 $^{\circ}$C at 20 mmHg).

The cuprate derived from Ph$_3$SnLi and CuBr•SMe$_2$ adds to acetylene to afford the (Z)-Ph$_3$Sn vinyl cuprate. Addition of a reactive halide gives the (Z)-vinylstannane in high yield[44]. Studies have shown that the stannylcupration of terminal alkynes does not occur below $-35\,^{\circ}$C. Furthermore, the additions are reversible but the equilibrium strongly favors the adducts[45]:

"Bu$_3$SnCu" = BuSnCu(CN)Li•2LiCl or (Bu$_3$Sn)$_2$Cu(CN)Li$_2$

It is possible to prepare R$_3$Sn cuprates directly from R$_3$SnH by reaction with the higher-order cyanocuprate Bu$_2$Cu(CN)Li$_2$[46]. The R$_3$Sn cuprate species thus prepared adds to terminal alkynes to afford vinylstannanes. Addition of $Bu_3SnMgMe$ to several terminal alkynes in the presence of 5 mol% CuCN yields (E)-terminal vinylstannane as major or exclusive products:

$$2\ R_3SnH \xrightarrow[\text{THF, } -78\,°C]{Bu_2Cu(CN)Li_2} R_3SnCu(Bu)(CN)Li_2 + Bu_4Sn + H_2$$

$$HO\text{—}\!\!\equiv\!\!\text{—}H \xrightarrow[\substack{\text{THF, } -78\,°C \\ (87\%)}]{Bu_3SnCu(Bu)(CN)Li_2} HO\diagup\diagdown SnBu_3$$

$$HO\diagup\!\!\equiv\!\!\text{—}H \xrightarrow[\substack{\text{THF, } -78\,°C \\ (90\%)}]{Bu_3SnCu(Bu)(CN)Li_2} HO\diagdown\diagup\diagdown SnBu_3 \quad + \quad HO\diagdown\diagup SnBu_3$$

$$50:50$$

$$R\text{—}\!\!\equiv\!\!\text{—}H \xrightarrow[\substack{\text{CuCN, THF} \\ (70-88\%)}]{Bu_3SnMgMe} \substack{R\diagdown\diagup \\ SnBu_3}$$

Stannylcupration of Alkynes

$$H\text{—}\!\!\equiv\!\!\text{—}\!\overset{Me}{\underset{Me}{C}}\!\text{—}OH \xrightarrow[(87\%)]{Bu_3SnCu(Bu)CNLi_2} \substack{HO\diagdown Me \\ \diagup\diagdown Me \\ Bu_3Sn}$$

3-Tributylstannyl-2-methyl-3-buten-2-ol

To a slurry of CuCN (0.067 g, 0.72 mmol) in THF (2 mL) at $-78\,°C$, BuLi (0.63 mL, 1.50 mmol) was added dropwise. The mixture was allowed to warm slightly to yield a colorless, homogeneous solution which was recooled to $-78\,°C$ and Bu_3SnH (0.40 mL, 1.5 mmol) was added by syringe. Stirring was continued and over about 10 min, the solution yellowed and H_2 gas was liberated. 2-Methyl-3-butyn-2-ol (0.66 mL, 0.68 mmol) was added neat by syringe and the reaction mixture was stirred for 5 min before being poured into 10 mL of 10% NH_4OH/90% saturated NH_4Cl. Extraction with ether was followed by combining the extracts and drying over Na_2SO_4. The solvent was removed in vacuo and the residue chromatographed on silica gel. Elution with hexanes/ethyl acetate (95 : 5) + 1% Et_3N gave the stannane (0.221 g, 87%) as a colorless oil.

The stannylmetal reagents $Bu_3SnAlEt_2$, and $(Bu_3Sn)_2Zn$ give mixtures of regioisomers with catalytic CuCN, or $Pd(PPh_3)$[47]. The intermediate vinylmagnesium intermediates react with various electrophiles to afford substitution products with retention of stereochemistry. In contrast, the Pd^0-catalyzed addition of $Bu_2SnSiMe_2Ph$ to terminal alkynes leads to the internal vinylstannane after desilylation with Bu_4NF[48]:

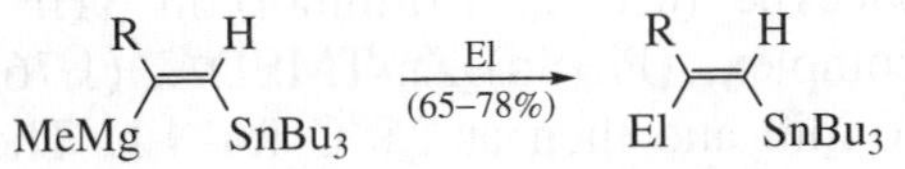

$$R = BnOCH_2CH_2 \ (100:0), \ Ph \ (>95:5), \ C_{10}H_{21} \ (70:30)$$

$$El = Me(MeI), \ Et(EtI), \ allyl \ (CH_2=CHCH_2Br), \ PhCHOH(PhCHO)$$

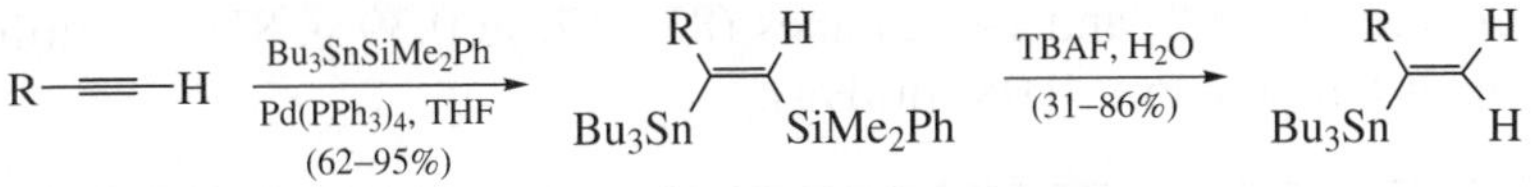

$$R = Ph, \ Bu, \ CH_2OBn, \ CH_2CH_2OTHP, \ CH_2CH_2OH, \ CH_2NMe_2$$

Copper(I)-Catalyzed Stannylmagnesiation of Alkynes

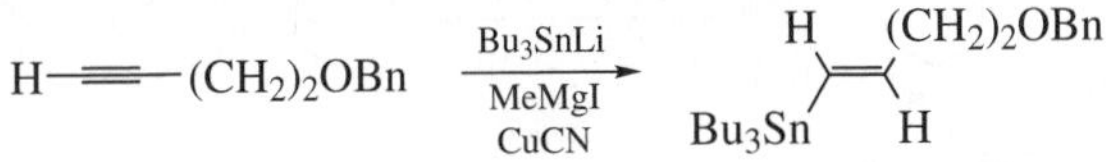

(E)-1-Tributylstannyl-4-benzyloxy-1-butene

An ethereal solution of MeMgI (1.0 M, 3.0 mL, 3.0 mmol) was added to a THF solution of Bu₃SnLi, prepared from SnCl₂ (0.58 g, 3.0 mmol) and BuLi (1.5 M, 6.0 mL, 9.0 mmol) at 0 °C under argon. After 15 min, CuCN (0.004 g, 5 mol%) and 4-benzyloxy-1-butyne (0.16 g, 1.0 mmol) in THF (5 mL) were added and the whole was stirred for 30 min at 0 °C. Aqueous work-up and alumina column chromatography gave the stannane (0.38 g, 88% yield) as a single product.

Copper(I)-catalyzed Stannylzincation of Alkynes

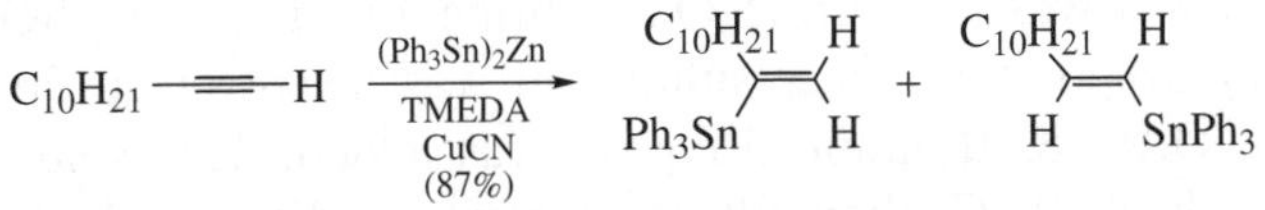

2-Triphenylstannyl-1-dodecene and (E)-1-Triphenylstannyl-1-dodecene

A hexane solution of Et₂Zn (1.5 M, 6.7 mL, 10 mmol) and TMEDA (1.28 g, 11 mmol) were combined in THF (10 mL) under an argon atmosphere. The reaction mixture was cooled to −78 °C, and a solution of Ph₃SnH (7.7 g, 22 mmol) in THF (10 mL) was added dropwise. The resulting mixture was warmed to 0 °C and kept there for 2 h. After being stirred for another 5 h at 25 °C, the mixture was cooled in ice, and pentane (20 mL) was added. The precipiated white solid was filtered, washed with 1 : 2 THF/pentane, and dried in vacuo to give the complex (7.2 g, 82% yield; mp 172–175 °C).

A solution of 1-dodecyne (0.16 g, 1.0 mmol) in THF (2 mL) was added to a suspension of the complex, $(Ph_3Sn)_2Zn \cdot TMEDA$, (1.76 g, 2.0 mmol) in THF (10 mL) at 0 °C for 30 min and then at 25 °C for 4 h. The reaction mixture was poured into ice water and extracted with ethyl acetate. The combined organic layers were washed with saturated aq NaCl solution and dried over Na_2SO_4. Removal of the solvent afforded a crude oil, which was chromatographed over silica gel (hexane) to give a mixture of the 2- and 1-stannanes (83 : 17; in 0.39 g. 87% combined yield; bp 200 °C bath temperature, 0.08 mmHg).

Vinylstannanes from Vinyl Triflates and Stannyl Cuprates[49]

1-(Tributylstannyl)cyclohexene

To 1.06 mL of $^iPr_2NH$ (7.54 mmol, 2.6 equiv) in 50 mL of THF at −60 to −80 °C was added 4.71 mL of 1.6 M BuLi (7.54 mmol, 2.6 equiv) and the resulting solution was stirred for 30 min. To this solution at −60 to −80 °C was added 1.72 mL of Bu_3SnH (6.38 mmol, 2.2 equiv) and the reaction mixture was stirred for an additional 20 min. At this point 0.286 g of solid CuCN (3.19 mmol, 1.1 equiv) was added and the solution was slowly warmed to −60 to −50 °C. The solid dissolved resulting in a translucent green-yellow solution to which 0.668 g of triflate (2.90 mmol, 1 equiv) was added dropwise, whereupon the color changed to a deep red-orange. The reaction mixture was further warmed to −25 to −20 °C (but not above −20 °C) and the progress was monitored by GC. After 2 h, the solution was poured into a mixture of 50 mL of pentane and 30 mL of 10% aqueous NH_4Cl. The pentane layer was washed with H_2O and brine, dried over $MgSO_4$, and concentrated on a rotary evaporator. The resultant oil was dissolved in 50 mL of EtOAc and 1.45 g of AgOAc (8.70 mmol, 3 equiv) was added. This solution was stirred open to air for 2 h at 25 °C, then filtered through Celite, washed with H_2O and brine, dried over $MgSO_4$, and concentrated on a rotary evaporator. The product was purified by gravity colum chromatography on silica gel (pretreated with 10–15% by weight of Et_3N) by elution with hexanes ($R_f = 0.68$). In this manner the vinyltin was obtained in 97% yield (1.04 g) as a colorless oil of analytical purity.

The complex $(Ph_3Sn)_2Zn \cdot TMEDA$ reacts with vinyl halides and triflates to yield vinyl-stannanes in the presence of catalytic $Pd(PPh_3)_4$. (*E*)-Vinyl halides are converted to (*E*)-vinylstannanes but the (*Z*) isomers afford mixtures of (*E*) and (*Z*) products. Cyclo-hexenyl triflates yield the corresponding stannanes through treatment with higher-order stannyl cyanocuprates:

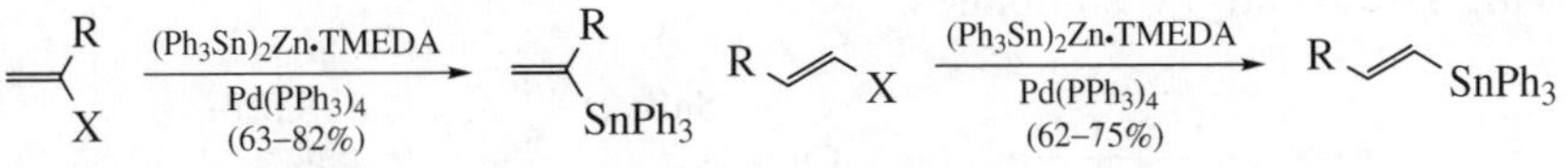

R = C$_{10}$H$_{21}$, C$_5$H$_{11}$, Me
X = Br, OTf

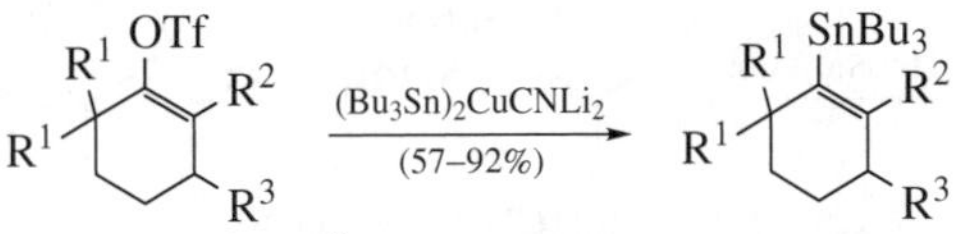

R^1 = H or Me, R^2 = H or Me, R^3 = H or Me

Alkynylboronates undergo regioselective stannylation with migration of an alkyl substituent from boron to carbon giving rise to (*E*)-vinylstannanes. These intermediates can be converted to vinylcuprates with loss of the boron substituent. Addition of various electrophiles affords the vinylstannanes[50]:

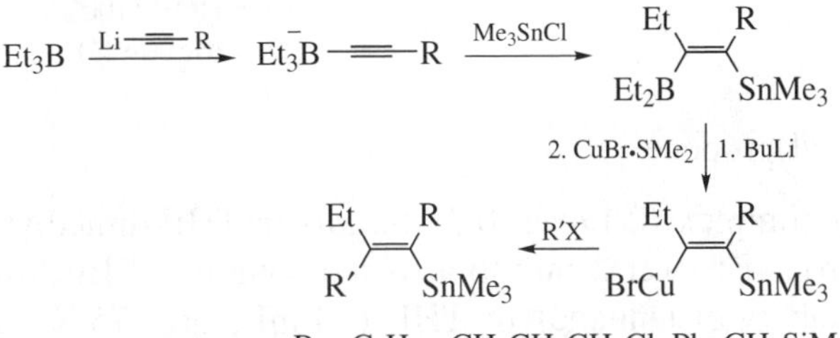

R = C$_6$H$_{13}$, CH$_2$CH$_2$CH$_2$Cl, Ph, CH$_2$SiMe$_3$
R'X = MeI, CH$_2$=CHCH$_2$Br

When the intermediate cuprate species is treated with MeOH or HOAc a mixture of regioisomeric stannanes is produced. Evidently retrostannylcupration and readdition are competitive with protonolysis, but not with electrophilic substitution:

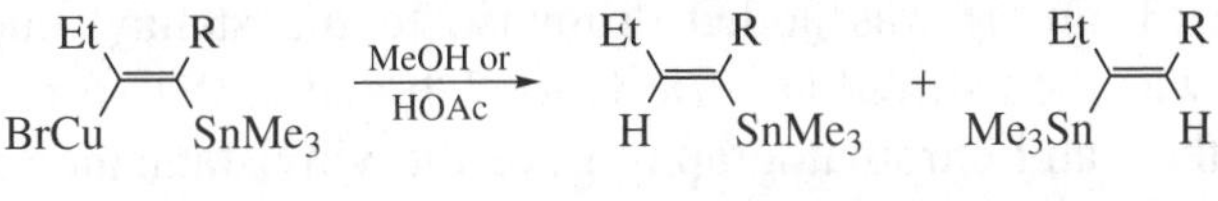

2-Stannylated 1,3-dienes undergo Diels–Alder cycloaddition with a variety of dienophiles to give cyclohexenylstannanes[51]. More reactive dienophiles such as maleic anhydride, dimethyl fumarate, and dimethyl acetylenedicarboxylate afford the adducts in 70–80% yield:

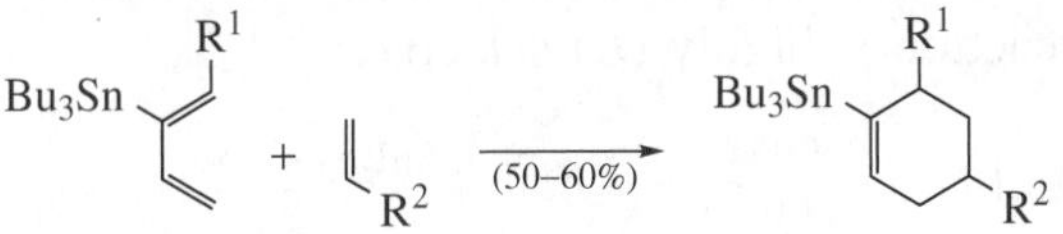

R^1 = H, Me, Ph
R^2 = COCH$_3$, OEt, CO$_2$Et, NO$_2$

Stannylcupration of allene followed by addition of reactive electrophiles affords 2-stannylalkenes. The reaction takes a different regiochemical course with substituted allenes, giving rise to allylic stannanes:

$$=\!\cdot\!= \xrightarrow{\text{(Bu}_3\text{Sn)}_2\text{CuLi}} \text{Bu}_3\text{SnCu}\diagdown\!\!\!\overset{\text{SnBu}_3}{=} \xrightarrow{\text{RX}} \text{R}\diagdown\!\!\!\overset{\text{SnBu}_3}{=}$$

RX = CH$_2$=CHCH$_2$Br, Me$_2$C=CHCH$_2$Br, PhCOCl

$$\diagdown\!\!\!\!=\!\cdot\!= \xrightarrow{\text{(Bu}_3\text{Sn)}_2\text{CuLi}} \overset{\text{CuSnBu}_3}{\diagdown\!\!\!\diagup}\text{SnBu}_3 \xrightarrow{\text{RX}} \overset{\text{R}}{\diagdown\!\!\!\diagup}\text{SnBu}_3$$

RX = MeI, CH$_2$=CHCH$_2$Br

Stannylcupration of Allenes

$$\overset{\text{H}}{\underset{\text{H}}{>}}\!=\!\cdot\!=\!\overset{\text{H}}{\underset{\text{H}}{<}} \xrightarrow{\text{(Bu}_3\text{Sn)}_2\text{CuLi}\cdot\text{SMe}_2} \text{Bu}_3\text{SnCu}\diagdown\!\!\!\overset{\text{SnBu}_3}{=} \xrightarrow{\text{RX}} \text{E}\diagdown\!\!\!\overset{\text{SnBu}_3}{=}$$

RX = HOH (100%)
RX = MeI (69%)

(a) 2-(Tributylstannyl)propene

The CuBr•SMe$_2$ complex (0.175 g, 0.85 mmol) in THF-dimethyl sulfide (2 mL, 1 : 1) was added dropwise over 3 min to a stirred solution of Bu$_3$SnLi (1.73 mmol, prepared from hexabutyldistannane) in THF (10 mL) at −75 °C, and the orange solution was stirred for 15 min. Excess allene (2 mmol) was added, and the mixture was stirred for 15 min, affording the stannylcuprate. Addition of water afforded the vinylstannane.

(b) 2-Tributylstannyl-1-butene

Methyl iodide (3 mmol) was added dropwise to the stannyl cupration mixture (2 mmol) which was then stirred at −78 °C for 1 h and at 0 °C for 1 h. An aqueous work-up using ether, and chromatography gave the vinylstannane (69%).

5.1.3 Miscellaneous Methods

Terminal vinylic stannanes can be prepared by Cr^{+2}-mediated addition of Bu$_3$SnCHBr$_2$ to aldehydes[52]. The reaction is highly (*E*) selective:

$$\text{RCHO} \xrightarrow[\substack{\text{CrCl}_2,\,\text{LiI} \\ \text{DMF, THF} \\ (53–63\%)}]{\text{Br}_2\text{CHSnBu}_3} \text{R}\diagup\!\!\!\diagdown\!\!\!\diagup\text{SnBu}_3$$

R = MeO$_2$C(CH$_2$)$_4$,
NC(CH$_2$)$_6$, MeCO(CH$_2$)$_{10}$

Vinylzirconium compounds, obtained through hydrozirconation of enynes, undergo transmetalation with Bu$_3$SnCl to afford terminal stannyl dienes[53]. Esters of α-hydroxy allylic stannanes are converted through Ireland–Claisen rearrangement to terminal (*E*)-vinylstannanes[54]:

$$R^1, R^2, R^3 = H: R^1, R^2 = H;$$
$$R^3 = OMe, R^1: R^3 = H;$$
$$R^2 = OMe, R^1: R^2 = (CH_2)_4; R^3 = H$$

R = H (60%), R = Me (62%)

γ-Hydroxy vinylstannanes can be prepared by titanium-catalyzed hydrometalation of propargylic alcohols followed by transmetalation with Bu_3SnCl[55]. These alcohols undergo Eschenmoser–Claisen rearrangement to allylic stannanes[56]:

$$R^1 = {}^iPr, c\text{-}C_6H_{11}$$
$$R^2 = TMS, DPS, SnBu_3, Ph, C_4H_9$$

Transmetalation of a Vinyltitanium Intermediates with Bu$_3$SnCl

(*E*)-1-Cyclohexyl-3-(tributylstannyl)-2-hepten-1-ol

A Schlenk tube was charged with 2.1 mmol of ethereal ^{i}BuMgCl and ether (2 mL). The solution was cooled in an ice-bath and dicyclopentadienyltitantium dichloride (10 mol%) was added in one portion causing a metallic green precipitate to form. The propargylic alcohol (1 mmol) was added dropwise and the resulting mixture was heated to reflux until TLC indictated the hydromagnesiation was complete (6 h). The solvent was removed under vacuum and the resulting solid dissolved in THF. Tributyltin chloride was added dropwise and the reaction mixture was heated to reflux for 2 h. Extraction with ether and flash chromatography gave the vinylstannanes in 79% yield.

Addition of Bu_3Sn radicals to alkenes is faster than addition to alkynes. Both additions are reversible. However, when an alkene and an alkyne are connected by an appropriate tether, cyclization takes place leading to a vinylstannane[57]. The reaction course is dictated by the greater reactivity of vinyl versus alkyl radicals. The vinylstannane products can be converted to carboxylic acid derivatives by lithiation and treatment with CO_2:

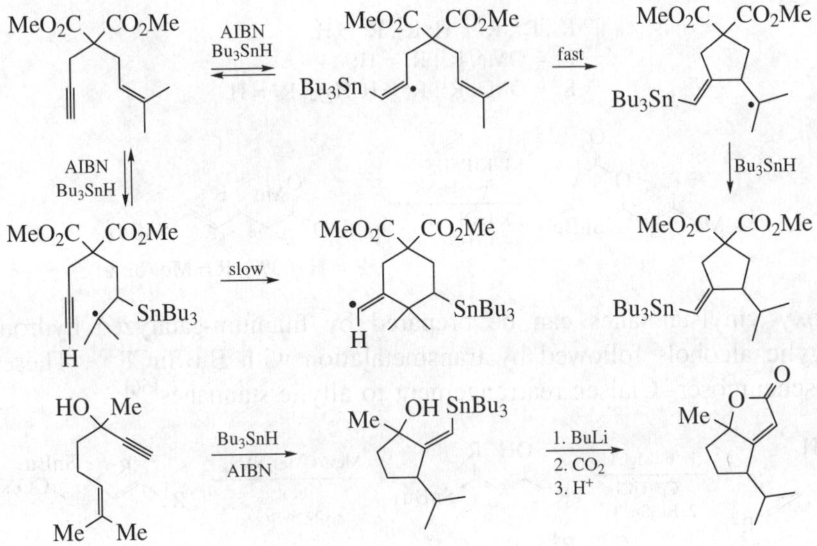

5.1.4   *α*-Oxygenated Vinylstannanes

Lithiation of enol carbamates followed by addition of Me_3SnCl has been used to prepare *α*-oxygenated vinylstannanes. Addition of cyanocuprates leads to cuprates with migration of the cuprate substituent and loss of the carbamate leaving group. The cuprates can be converted to vinylstannanes by treatment with Me_3SnCl[58]:

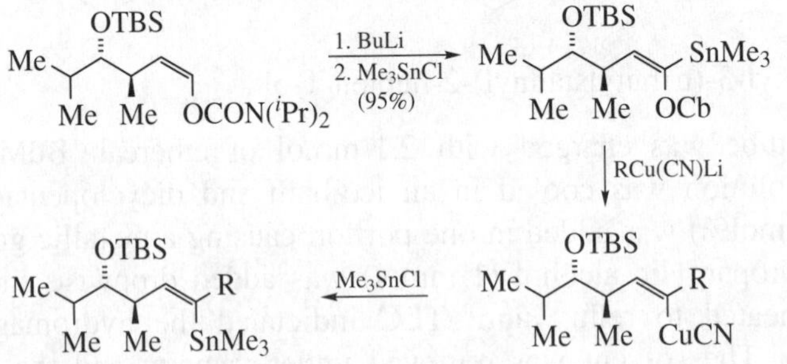

An alternative route to *α*-oxygenated vinylstannanes starts with norbornenones and proceeds by addition of Bu_3SnLi to the ketone and conversion of the resulting adduct to an ether. Thermolysis effects a retro Diels–Alder reaction to yield the vinylstannane[59]:

$$R^1 = H, Me, CH_2OBn$$
$$R^2 = Me, Bn, (CH_2)_2OMe$$

5.2 Reactions

5.2.1 Lithiation

Vinylstannanes are readily converted to vinyllithium intermediates upon treatment with butyllithium in tetrahydrofuran. The reaction proceeds with retention of configuration. An early application of this methodology was used to prepare vinylcuprates. Conjugate addition of the β-stannyl vinylcuprate reagent derived from (E)-1,2-bis- (tributylstannyl)ethene to cyclohexenones affords the β-stannyl vinyl adduct[60]. Oxidative destannylation leads to the β-alkynyl ketones. The direct preparation of such ketones through 1,4 addition of alkynylcuprates cannot be achieved owing to the unreactivity of these reagents:

1,4 Addition of a β-Stannylvinylcuprate and Subsequent Oxidative Elimination

1-Ethynylbicyclo[4.3.0]nonan-3-one

To a solution of 6.04 g (10.0 mmol) of *trans*-1,2-bis-(tributylstannyl)ethylene in 25 mL of dry THF at −78 °C was added 5.00 mL (10.6 mmol) of 2.13 M BuLi. Following slow warming to −40 °C during 30 min, the clear, light yellow solution was transferred through metal tubing (18 gauge) to another flask containing a suspension of 1.61 g (10.2 mmol) of 1-pentynyllithium in 6 mL of THF and

5.0 mL (27 mmol) of hexamethylphosphorous triamide over 30 min. After recooling to −78 °C the solution was stirred for 45 min and 0.727 g (5.34 mmol) of enone in 5.0 mL of THF was added over 12 min. After stirring for 30 min at this temperature, the reaction mixture was warmed to −40 °C during 15 min and then quenched by pouring into ice-cold saturated aqueous ammonium sulfate. The organic layer was separated and the aqueous layer was extracted with ether. The combined ethereal layers were washed with 2% sulfuric acid, filtered through a pad of filteraid, and dried (MgSO$_4$) to afford, after removal of solvent, 7.77 g of yellow-brown oil. Column chromatography on alumina with hexane as eluent gave 3.14 g (91%) of tetrabutyltin. Further elution with CHCl$_3$ yielded 4.1 g (93%) of the ketone.

A solution of 0.23 g (0.40 mmol) of the above ketone in 5 mL of dry acetonitrile was treated with 0.23 g (0.52 mmol) of lead tetraacetate. The reaction mixture became homogeneous after 3 min and a brown precipitate formed. After 3 h, TLC analysis (CHCl$_3$) showed no starting material. Dilution of the mixture with pentane and filtration through Celite and alumina afforded almost pure ethynyl ketone (0.051 g, 64%), homogeneous by TLC analysis.

A method for the synthesis of (Z,Z) skipped dienes entails conversion of a (Z)-vinylstannane to the cuprate followed by coupling with (Z)-3-tributylstannyl-2-propenyl acetate[61]. The process can be carried out reiteratively to form skipped trienes:

Reiterative Coupling of β-Stannylvinyl Grignard Reagents with γ-Stannyl Allylic Acetates to form Skipped Dienes

(Z,Z)-1-Tributylstannyl-1,4-decadiene

A solution of vinylstannane (10.71 g, 27.7 mmol) in 30 mL of THF was treated at −70 °C over 5 min with BuLi (27.7 mmol, hexane solution). The green solution was stirred for 1 h at −40 °C and then treated over 5 min at −70 °C with 30 mmol of ethereal magnesium bromide (from dibromoethane and magnesium turnings in

25 mL of ether). The slurry was stirred at −40 to −30 °C for 1 h and then treated successively at −70 °C with the acetate (30.5 mmol, 11.85 g in 30 mL of THF) over 15 min and Li_2CuCl_4 (0.15 mmol in 3.0 mL of THF). The bright yellow slurry was stirred at −40 to −30 °C for 6 h, then warmed to −15 °C and poured into 150 mL of aqueous NH_4Cl/NH_3 buffer (pH 8). The crude product (23.62 g) was isolated by extractive workup followed by filtration through a plug of basic alumina (2% Et_3N in hexane as eluant). Distillation (129–136 °C, 0.08 mmHg) gave 10.46 g (89%) of diene as a colorless liquid, which was homogeneous by TLC.

Various 1,3-dienes have been prepared by reaction of dienyllithium intermediates with electrophiles. α,β-Distannylated acrylic esters react with MeLi to give allenolates. Subsequent addition of electrophiles affords the α-substitution products:

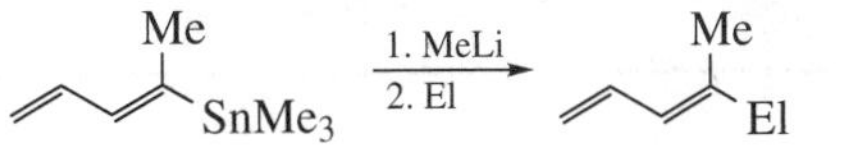

El = C_8H_{17} ($C_8H_{17}Br$),

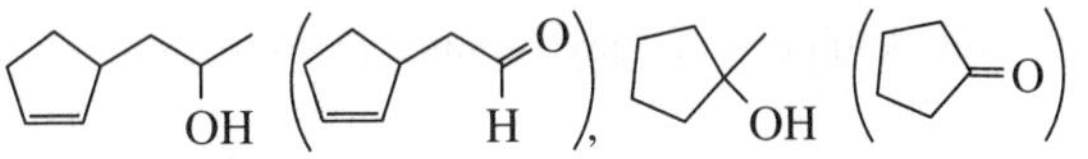

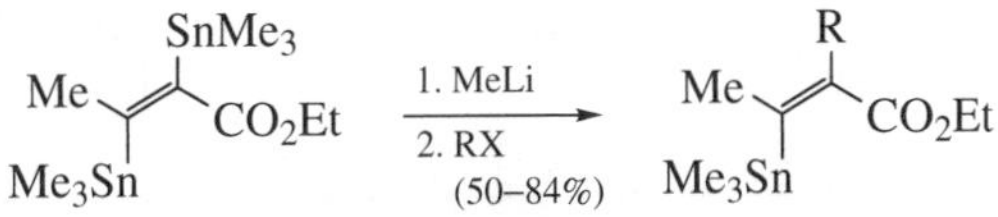

RX = MeI, CH_2=CHCH$_2$Br, BnBr, BuI

Vinylcuprates can be prepared from vinylstannanes by lithiation and subsequent treatment with Cu^I complexes. These cuprates afford 1,4 adducts with conjugated carbonyl compounds. This sequence was employed in the synthesis of a terpenoid pheromone of the square-necked grain beetle[62]:

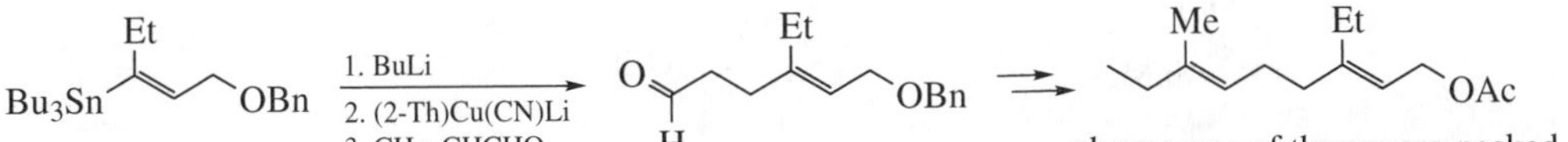

pheromone of the square-necked grain beatle

Conversion of a Vinylstannane to a Cyanocuprate and Subsequent 1,4 Addition

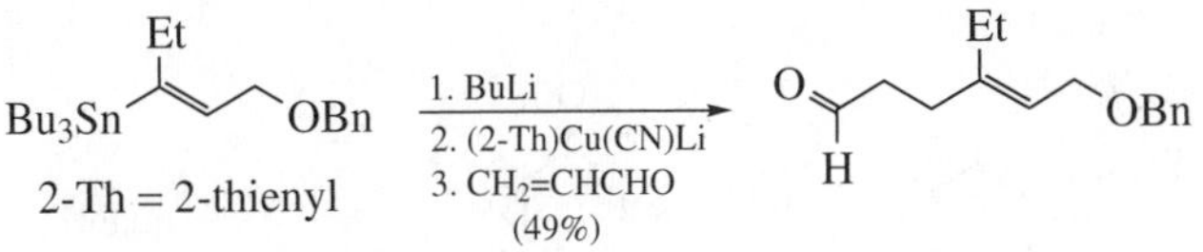

2-Th = 2-thienyl

(*E*)-6-Benzyloxy-4-ethyl-4-hexenal

The vinylstannane (4.85 g, 10.0 mmol) under an argon atmosphere, deoxygenated by two cycles of evacuation of the flask with oil pump vacuum and purging with argon, was dissolved in THF (50 mL) and cooled to −78 °C. BuLi (4.86 mL, 11 mmol, 2.5 M in hexane) was added dropwise, and after 2 h (2-thienyl)Cu(CN)Li (46.0 mL, 11.5 mmol, 0.25 M in THF) was added by syringe during 30 min, then HMPA (3.5 mL, 20 mmol) was added followed by the dropwise addition of acrolein (1.0 mL, 15 mmol) and TMS-Cl (1.9 mL, 15 mmol) in THF (10 mL). After 2 h the reaction contents were poured into a mixture of 1 N HCl (50 mL) and Et₂O (50 mL) and stirred for 0.5 h. The Et₂O layer was separated and the aqueous layer extracted with Et₂O. Removal of solvent followed by chromatography with ethyl acetate/hexane (1 : 9) as the eluant gave the adduct (1.2 g, 49%) as a slightly yellow oil.

Table 5. Conjugate Additions of Vinylcuprates Derived from Vinylstannanes.

Stannane	1,4 adduct	Yield (%)
		93
		76
		94
		quant

5.2.2 Cupration

Vinylcuprates can also be prepared directly from vinylstannanes through metal exchange with a higher-order cyanocuprate[63]. The cuprates thus formed undergo 1,4 additions to enones (Table 5).

This methodology was employed in a synthesis of the prostaglandin E (PGE) analogue misoprostol. The requisite vinylstannane can be prepared by a tin–copper exchange reaction with (E)-1,2-bis(tributylstannyl)ethene and the higher-order methyl 2-thienyl cyanocuprate and subsequent reaction of this reagent with a terminal epoxide[64]. Other terminal epoxides react similarly:

Direct Conversion of a Vinylstannane to a Mixed Cyanocuprate and Subsequent Addition to an Epoxide

(E)-1-Tributylstannyl-4-methyl-1-octen-4-ol

To 2.0 mL of a 0.5 M solution of dilithio-2-thienyl[2-(E)-tributylstannylethenyl]-cyanocuprate in ether was added MeLi (0.76 mL, 1.45 M in ether, 1.10 mmol). The cooling bath was removed and to the homogeneous dilithio methyl(2-thienyl)cyano-cuprate solution was added, by syringe, (E)-bis(tributylstannyl)ethylene (0.53 mL, 0.061 g, 1.0 mmol). The solution was allowed to warm to 25 °C over 30 min. An aliquot (0.01 mL) was withdrawn by syringe and added to 0.5 mL of a 1 : 1 mixture of hexane/saturated NH$_4$Cl/NH$_4$OH (9 : 1). After being vigorously shaken for 5 min the hexane layer was withdrawn, dried over K$_2$CO$_3$, and analyzed by gas chromatography for disappearance of the starting stannane (R_t = 9.78 min) and the formation of methyltributylstannane (R_t = 1.38 min) and tributylvinylstannane (R_t = 1.76 min). The dark red solution of cuprate was then cooled (−78 °C) and the epoxide (0.88 mmol) was added by syringe. The reaction mixture was stirred at −78 °C for 1 h, warmed to 0 °C for 1 h and then quenched by pouring into a vigorously stirred solution of saturated NH$_4$Cl/NH$_4$OH (9 : 1; 10 mL). After 30 min the dark blue aqueous mixture was extracted with EtOAc, the layers separated and the aqueous layer re-extracted with EtOAc. The organic extracts were combined, washed with saturated NaCl, dried (Na$_2$SO$_4$), concentrated, in vacuo, to an oil which was purified by medium pressure chromatography on silica gel (pretreated with triethylamine) using hexane/ethyl acetate (95 : 5) as the eluent to provide the β-hydroxy-(E)-vinylstannane (74% yield).

An intramolecular variant of the 1,4 addition has been achieved through use of CuCl for the transmetalation[65]:

$$R^1 = H, Me, Et; R^2 = H, Me, CO_2Me$$

5.2.3   Condensations

Both (E)- and (Z)-2-ethoxyvinyl tributylstannanes have been found to function as an acetaldehyde equivalent in Lewis acid-catalyzed additions to aldehydes[66]. Ketones fail to react. A likely pathway for this reaction involves attack on the aldehyde–Lewis acid complex by the enol ether, followed by elimination of the stannane moiety and rearrangement of the resulting β-hydroxy enol ether after addition of water:

$$R = Ph, \quad Ph \diagup \diagdown, \quad Me \diagup \diagdown$$

Aldehyde Homologation with EtOCH=CHSnBu₃

Cinnamaldehyde

To a cold ($-78\,°C$) CH_2Cl_2 solution (6 mL) of benzaldehyde (0.20 mL, 2 mmol) was added by syring 0.25 mL (2 mmol) of $BF_3•OEt_2$. A solution of (*E*)- or (*Z*)-2-ethoxyvinyl tributylstannane (0.72 g, 2 mmol) in 3 mL of CH_2Cl_2 was added dropwise and the mixture was stirred at $-78\,°C$ for 1 h. A 1 : 1 mixture of $MeOH/H_2O$ was added at $-78\,°C$ and the mixture was allowed to reach $25\,°C$ with stirring. The organic phase was washed with water and dried over $MgSO_4$. The product was obtained in 86% yield after chromatography on silica gel.

6   Allylic Stannanes[67]

6.1   Preparation

Allylic tin compounds can be prepared through metal–metal exchange of allylic Grignard reagents with triphenyltin chloride in tetrahydrofuran[68]. These triphenyltin derivatives are used to prepare allylic lithium compounds by transmetalation with PhLi in ether. The tetraphenyltin thus produced is insoluble and can be removed by filtration:

$$\text{Ph}_3\text{SnCl} + \underset{\substack{\text{R} = \text{H, X} = \text{Br}\\ \text{R} = \text{Me, X} = \text{Cl}}}{\overset{\text{R}}{\diagup\!\!\!\diagdown}\text{X}} \xrightarrow[\text{THF}]{\text{Mg}} \text{Ph}_3\text{Sn}\overset{\text{R}}{\diagup\!\!\!\diagdown} \xrightarrow{\text{PhLi}} \text{Ph}_4\text{Sn} + \text{Li}\overset{\text{R}}{\diagup\!\!\!\diagdown}$$

Transmetalation of Ph₃SnCl with Allylmagnesium Bromide

$$\text{Ph}_3\text{SnCl} + \diagup\!\!\!\diagdown\text{MgBr} \longrightarrow \text{Ph}_3\text{Sn}\diagup\!\!\!\diagdown$$

Allyl Triphenyltin

Magnesium turnings (24.3 g, 1.0 g atom) were placed in a 2-L three-necked flask equipped with a water condenser, mechanical stirrer, and addition funnel. THF (320 mL) was added, and the reaction was initiated by adding a small amount of allyl bromide. The rapidly stirred mixture then was heated to reflux, and a solution of 96 g (0.25 mol) of triphenyltin chloride and 47 g (0.39 mol) of allyl bromide in 240 mL of THF was added slowly over the course of 3 h. After the addition had been completed, the reaction mixture was heated at reflux overnight, cooled, and hydrolyzed with saturated NH_4Cl solution. The supernatant organic phase was separated and the aqueous phase was extracted with ether. The combined organic extracts were distilled to remove solvent. The residue was dissolved in ether and shaken with a solution of potassium fluoride. The ether layer was separated and the ether removed at reduced pressure. The solid residue was recrystallized from petroleum ether to give 67 g of allyltriphenyltin, mp 72.5–74 °C. On concentrating the filtrate, an additional 9 g of product, mp 70–72 °C, was obtained. The total yield was 78%.

Direct coupling of allylic halides with Bu_3SnCl can be effected with magnesium in the presence of catalytic lead bromide[69]:

$$R\diagup\diagdown X \xrightarrow[\substack{THF, Bu_3SnCl \\ (71–94\%)}]{Mg, PbBr_2} R\diagup\diagdown SnBu_3$$

$$R = Ph, Me, H; X = Br, Cl$$

An alternative coupling procedure employs zinc in aqueous THF containing NH_4Cl (Table 6)[70]. The use of Bu_2SnCl_2 leads to diallyltin species. When crotyl bromide is employed, a mixture of (E) and (Z) isomers is produced in nearly equal amounts:

$$Bu_2SnCl_2 + R\diagdown\diagup Br \xrightarrow[H_2O, THF]{Zn} \left(R\diagup\diagdown\right)_2 SnBu_2 + R\diagup\diagdown\underset{R}{Sn}\diagdown\diagup + \diagup\diagdown\underset{R}{Sn}\diagdown\diagup$$

$$\underset{\mathbf{1}}{} \qquad \underset{\mathbf{2}}{} \qquad \underset{\mathbf{3}}{}$$

$$R = H, 78\%; R = Me^a, 89\%^b$$

(a) 76% (E), 9% (Z), 15% methallyl (b) 6% **1**, 42% **2**, 52% **3**

Table 6. In situ Preparation of Allylic and Allenic Stannanes from Halides.

$$Bu_3SnCl + RBr \xrightarrow[H_2O, THF]{Zn} Bu_3SnR$$

RBr	Yield (%)
$CH_2=CHCH_2Br$	77
$CH_3CH=CHCH_2Br^a$	82[b]
$HC≡CCH_2Br$	80[c]

[a]76% (E), 9% (Z), 15% methallyl. RBr
[b]59% (E), 41% (Z). Bu_3SnR
[c]Allenic stannane.

Zinc-mediated Coupling of Allylic Bromides and Bu₃SnCl

$$\diagup\diagdown Br + Bu_3SnCl \xrightarrow[H_2O–NH_4Cl]{Zn} \diagup\diagdown\diagdown SnBu_3$$

Crotyl Tributyltin

In a two-necked flask equipped with a condenser and a dropping funnel, tributyltin chloride and a 30% excess of zinc powder were suspended in a H_2O-NH_4Cl saturated THF mixture. Under vigorous stirring, crotyl bromide (in a 1 : 1 stoichiometric ratio with respect to zinc) was added dropwise at a rate sufficient to maintain a gentle reflux. The addition took about 30 min. The heterogeneous mixture was stirred for a further 30 min then extracted with petroleum ether (35–60 °C). The organic layer was separated and solvent was removed leaving the pure crotyltin compound.

Transmetalation of allylic tin compounds with $SnCl_4$ proceeds by an S_E2' process (Table 7)[71]. The intially formed chlorostannanes undergo subsequent 1,3 isomerization to afford a mixture favoring the linear isomer as an $(E):(Z)$ mixture. The initial products are enriched in the (Z) isomer but, after equilibration a roughly 2 : 1 mixture favoring the (E) isomer is formed. This equilibration is thought to be an intramolecular process, but an intermolecular pathway has not been excluded.

6.1.2 Stannylation of Allylic Mesylates

Primary allylic alcohols are converted to allylic stannanes regioselectively and stereoselectively through in situ formation of the mesylate at -78 °C followed by addition of Bu_3SnLi[72]:

$R = H, Me, C_5H_{11}, CH_2ODPS$

$R^1 = H, Me, CH_2CH_2CH=CMe_2$
$R^2 = Me, CH_2CH_2CH=CMe_2, CH_2CH(ODPS)Ph$
$R^3 = H, Me$

Table 7. Transmetalation of (Crotyl)tributyltin with Tin Tetrachloride.

Equivalents of $SnCl_4$	4	5	6	7
0.5	0	0	16[a]	84
1.0	45[b]	55	0	0
3.0	88[c]	12	0	0

[a] 65 : 35 $(E):(Z)$ [b] 40 : 60 $(E):(Z)$ [c] 50 : 50 $(E):(Z)$.
Note: for **4** $(E):(Z)$ is 66 : 33 after 12 h in all runs.

Preparation of Allylic Stannanes from Allylic Alcohols via the Mesylates and Bu₃SnLi

(2-Pentyl-2-propenyl)tributylstannane

A solution of 1.78 M BuLi in cyclohexane (2.81 mL, 5.0 mmol) was added at −75 °C to a solution of alcohol (0.64 g, 5.0 mmol, 1.0 equiv) in THF (4 mL). After 20 min MsCl (0.39 mL, 0.57 g, 5.0 mmol) was added. After 35 min a solution of Bu₃SnLi was added dropwise by cannula—prepared from ⁱPr₂NH (0.77 mL, 0.56 g, 5.5 mmol, 1.1 equiv), 1.78 M BuLi in cyclohexane (2.95 mL, 5.25 mmol, 1.05 equiv), and Bu₃SnH (1.32 mL, 1.46 g, 5.0 mmol) in THF (4 mL). After 2 h at −78 °C the solution was allowed to warm to 25 °C and quenched after 12 h with H₂O (20 mL). Extraction with ⁱBuOMe, evaporation of the solvent, and filtration over a pad of flash silica gel (deactivated with Et₃N) with petroleum ether (100 mL) gave the stannane (1.70 g, 85%).

6.1.3   Free Radical Stannylation of Allylic Xanthates

Terminally substituted allylic stannanes are obtained through free radical stannylation of terminal allylic xanthates (Table 8)[73]. The reaction involves an initial thermal rearrangement to the dithiocarbonate and subsequent addition and homolytic elimination:

Table 8. Synthesis of Allylic Tributylstannanes from Allylic Methyl Xanthates.

R¹	R²	Yield of **8** (%)	Yield of **9** (%)
Me	H	79	78
H	Me	64	84
Pr	H	86	73
Pr	Me	70	59

6.2   Reactions

6.2.1   Lewis Acid-promoted Additions to Aldehydes

The most extensively studied reaction of allylic tin derivatives is their addition to aldehydes to afford homoallylic alcohols. First reported in 1967 as a thermal reaction between triethylallyltin and mainly aromatic aldehydes, numerous variations have since been developed[74]:

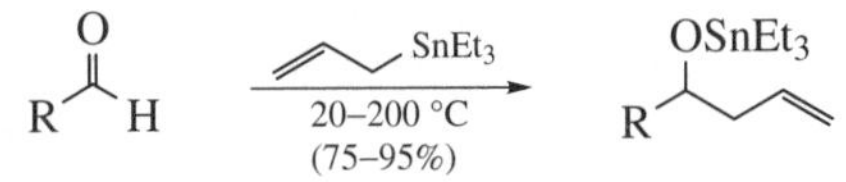

R = Ph, *p*-ClC$_6$H$_4$, *p*-O$_2$NC$_6$H$_4$, *p*-Cl$_3$CC$_6$H$_4$, PhCH=CH, C$_7$H$_{15}$

Some years later it was discovered that the allylation is considerably more facile in the presence of excess (2 mol) BF$_3$•OEt$_2$. For aldehydes with OH or NO$_2$ substituents, 3–4 equivalents of BF$_3$•OEt$_2$ are required[75]:

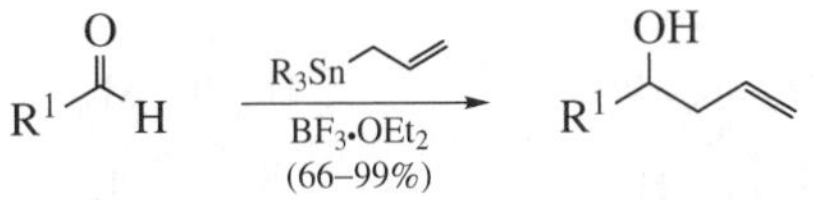

R^1 = Ph, *p*-MeC$_6$H$_4$, *p*-MeOC$_6$H$_4$, C$_6$H$_{13}$, 2-furyl, 3-pyridyl
R = Me or Bu

4-*tert*-Butylcyclohexanone affords a large predominance of the axial alcohol:

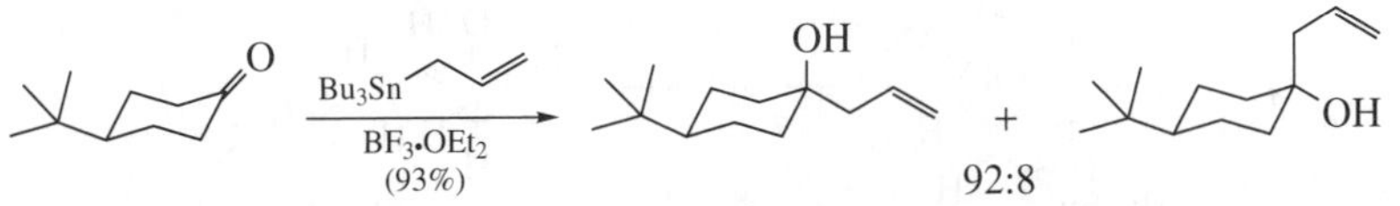

BF$_3$ -Promoted Addition of Bu$_3$ SnCH$_2$ CH=CH$_2$ to Aldehydes

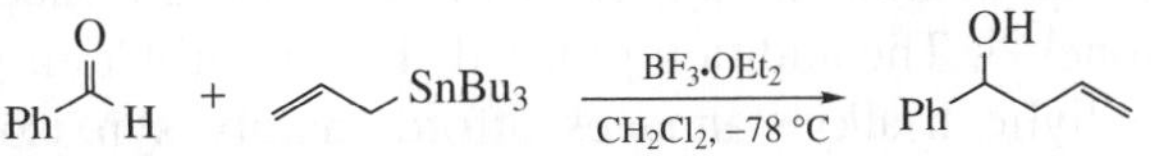

1-Phenyl-3-buten-1-ol

To a CH$_2$Cl$_2$ solution of benzaldehyde (1.0 mmol) was added BF$_3$•OEt$_2$ (2.0 mmol) and allyltributyltin (1.1 mmol) at −78 °C under N$_2$. The resulting mixture was allowed to warm slowly to 0 °C (1–2 h). Extractive work-up and purification by preparative TLC gave the homoallyl alcohol in 92% yield.

When these additions are carried out with cis- or trans-crotyltin derivatives it is found that syn adducts are obtained as major products from both stannanes[76]. An acyclic transition state was proposed for these additions. The assumption was made that the aldehyde–BF$_3$ complex should show little affinity for the Bu$_3$Sn grouping and the orientation of the two pi systems would be determined by steric interactions between the aldehyde substituent and the crotyl methyl group. Accordingly the double bond geometry would play no role in product stereochemistry. Several transition state orientations of the carbonyl and double bond can lead to the observed syn adducts. The outcome of most additions is consistent with the antiperiplanar arrangement, although for an intramolecular addition leading to a bicyclo[2.2.2.]octanol, the synclinal arrangement is preferred[77]:

R^{1}CHO + Me⎓⎓SnR$^2_3$ $\xrightarrow[\substack{-78\ °C,\ CH_2Cl_2 \\ (82–92\%)}]{BF_3 \cdot OEt_2}$

(E) or (Z)

R^1 = Ph, ^{i}Pr, ^{i}Bu, Et$_2$CH, Et, Me
R^2 = Bu, Me

>90:10 syn:anti

(E)   → syn ←   (Z)

antiperiplanar   → syn ←   synclinal

An early synthetic application of the crotyltin addition afforded a precursor of the Prelog–Djerassi lactone[78]. The addition proceeds by an anti-Cram pathway. It has been found that whereas allylic trialkylstannanes afford mainly syn isomers in addition to aldehydes, the triphenyl cinnamyl analogues give a predominance of the anti products[79]:

$\Longrightarrow$ Prelo–Djerassi lactone

RCHO + Ph⎓⎓SnPh$_3$ $\xrightarrow[\substack{CH_2Cl_2,\ -78\ °C \\ (69–84\%)}]{BF_3 \cdot OEt_2}$

R = Me, PhCH$_2$CH$_2$, Ph, c-C$_6$H$_{11}$

>99:1

Additions of allylstannanes to α-oxygenated aldehydes are sensitive to the Lewis acid, the solvent, and the nature of the α-substituent[80]. A major factor in determining aldehyde facial selectivity is the extent to which chelation of the α-oxygen and the carbonyl oxygen can take place. In nonpolar solvents with $MgBr_2$ or $TiCl_4$ as the Lewis acids, the syn adducts are highly favored as a result of chelation control (Table 9). Additions promoted by $BF_3 \bullet OEt_2$ proceed by Felkin-Ahn, or Conforth (dipolar) control and yield mainly anti adducts.

R = Bn or TBS

Addition of the crotylstannane analogue to cyclohexanecarboxaldehyde is also influenced by the choice of Lewis acid[81]. A strong preference for the syn adduct is seen with $BF_3 \bullet OEt_2$ or $TiCl_4$ in equimolar amounts. Preequilibration of the tin reagent with 2 equiv of $TiCl_4$ before addition of the aldehyde leads to the anti adduct in large predominance. Both $MgBr_2$ and $SnCl_4$ give mixtures of syn, and anti, and linear products (Table 10).

Similar trends are seen with the crotylstannane and α-oxygenated aldehydes[82]. The syn,syn and syn,anti products from **16** and **18** are the result of nonchelated transition states. These products predominate when $BF_3 \bullet OEt_2$ is employed as the Lewis acid promoter. The syn,syn product from **16** can also arise through a chelated transition state. This adduct along with the anti,syn isomer from **17** is formed with $MgBr_2$ or $TiCl_4$ as the Lewis acid promoter (Table 11).

Table 9. Addition of (Allyl)tributyltin to an α-Alkoxy Aldehyde Promoted by Lewis Acids (LA).

R	LA	Solvent	**10 : 11**
Bn	$BF_3 \bullet OEt_2$	CH_2Cl_2	39 : 61
Bn	$MgBr_2$	CH_2Cl_2	100 : 0
Bn	$MgBr_2$	THF	20 : 80
Bn	$TiCl_4$	CH_2Cl_2	100 : 0
TBS	$BF_3 \bullet OEt_2$	CH_2Cl_2	5 : 95
TBS	$MgBr_2$	CH_2Cl_2	21 : 79

Table 10. Additions of (Crotyl)tributyltin to Cyclohexanecarboxaldehyde Promoted by Lewis Acids (LA).

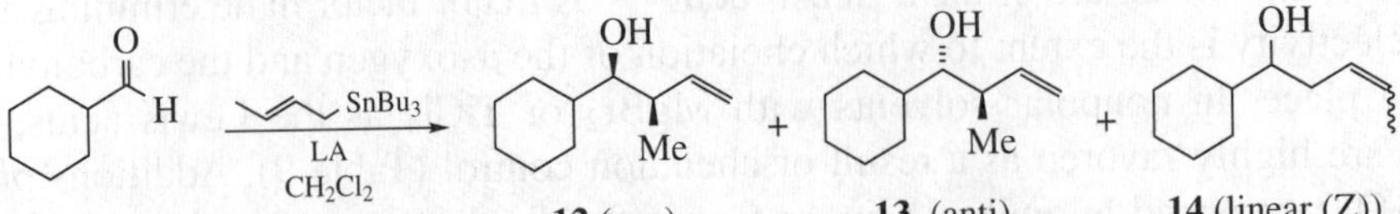

12 (syn)　　　**13** (anti)　　　**14** (linear (Z))
　　　　　　　　　　　　　　　　　　　　15 (linear (E))

LA	**12 : 13 : 14 : 15**
$BF_3 \cdot OEt_2$	96 : 4 : 0 : 0
$MgBr_2$	52 : 36 : 12 : 0
$SnCl_4$	23 : 26 : 36 : 15
$TiCl_4$	90 : 7 : 2 : 1
$TiCl_4^a$	4 : 91 : 0 : 5

^{a}2.1 equiv.

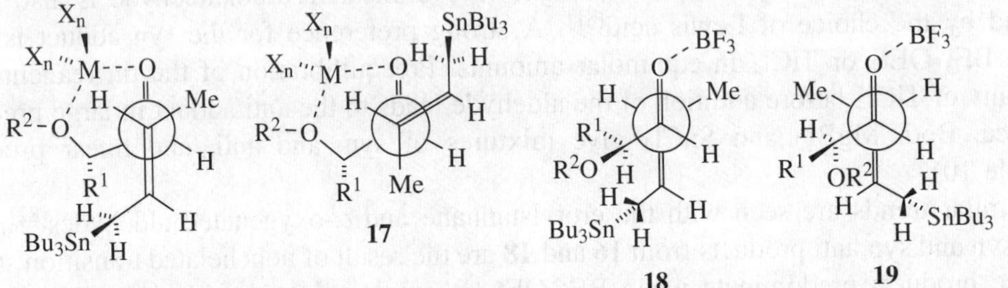

16　　M = Ti or Mg
　　　　X = Cl or Br
　　　　n = 1 or 2

17　　**18**　　**19**

Table 11. Addition of (Crotyl)tributyltin to α-Oxygenated Aldehydes Promoted by Lewis Acids (LA).

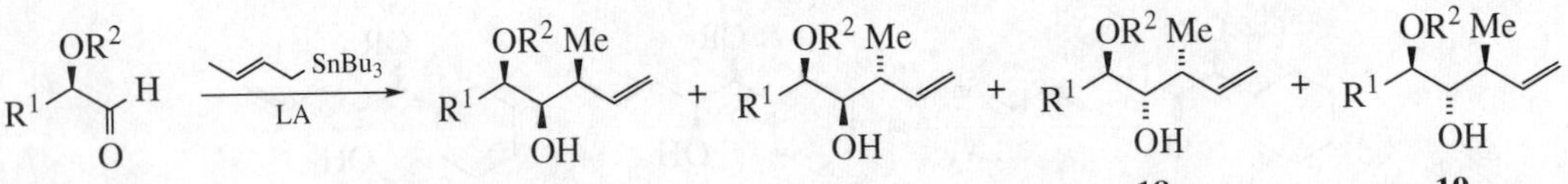

16　　　**17**　　　**18**　　　**19**

R^1	R^2	LA	**16 : 17 : 18 : 19**
c-C_6H_{11}	Bn	$BF_3 \cdot OEt_2$	66 : 1 : 26 : 7
c-C_6H_{11}	Bn	$TiCl_4$	63 : 37 : 0 : 0
c-C_6H_{11}	Bn	$MgBr_2$	92 : 8 : 0 : 0
Bu	Bn	$BF_3 \cdot OEt_2$	39 : 4 : 45 : 12
Bu	Bn	$MgBr_2$	93 : 7 : 0 : 0
Bu	CH_2OBn	$BF_3 \cdot OEt_2$	23 : 3 : 66 : 8
Bu	CH_2OBn	$MgBr_2$	91 : 9 : 0 : 0

Table 12. Addition of (Crotyl)tributyltin to β-Oxygenated Aldehydes Promoted by Lewis Acids (LA).

R	LA	20 : 21 : 22 : 23
DPS	BF$_3$•OEt$_2$	10 : 0 : 90 : 0
DPS	TiCl$_4$	33 : 0 : 67 : 0
Bn	MgBr$_2$	81 : 7 : 10 : 2
Bn	MgI$_2$	78 : 22 : 0 : 0

Chiral α-methyl-β-oxygenated aldehydes also show a preference for Cram diastereoselection with BF$_3$•OEt$_2$ and allylstannane nucleophiles (Table 12)[83]. Chelation is not important for β-silyloxy aldehydes so reactions promoted by TiCl$_4$ favor the Cram product **22**, as well. However the benzyloxy analogues afford mainly the syn,syn adduct **20** with MgX$_2$ Lewis acids via a chelated transition state.

A subtle change in stereochemistry is observed with β-methyl allylic stannanes and α-oxygenated aldehydes[84]. The typical syn adduct is formed in the BF$_3$•OEt$_2$-promoted reaction. However the chelation controlled addition (MgBr$_2$), favors the anti adduct. The crotylstannane reagent, lacking the β-methyl substituent, affords the expected syn adduct. Evidently, the β-methyl group causes the synclinal to be favored over the usual antiperiplanar arrangement in the transition state.

A recent examination of the proposal that cis- and trans-crotylstannanes both show nearly identical preferences for syn adducts has provided a somewhat modified conclusion (Table 13)[85]. It was found that in BF$_3$•OEt$_2$-promoted reactions, the trans-crotylstannane reacts faster than the cis-isomer and shows higher syn selectivity. It is proposed that these reactions proceed mainly by synclinal transition state arrangements which are stabilized by favorable orbital interactions between the lowest unoccupied molecular orbital (LUMO) of the carbonyl−BF$_3$ complex and the highest occupied molecular orbital (HOMO)

Table 13. Addition of cis- and trans-(Crotyl)tributyltin to Aldehydes.

R^1	Cis:trans	Syn:anti
c-C_6H_{11}	90 : 10	94 : 6
c-C_6H_{11}	74 : 26	86 : 14
c-C_6H_{11}	12 : 88	48 : 42
Ph	90 : 10	98 : 2
Ph	74 : 26	95 : 5
Ph	12 : 88	81 : 19

of the allylstannane. This is best achieved with the trans-stannane. The corresponding arrangement with the cis-stannane is less favorable because of steric interactions between the crotyl CH_3 substituent and the BF_3. Consequently these additions proceed more slowly and less selectively.

trans-Crotylstannanes also show higher rates and greater syn selectivities than their cis- counterparts in $MgBr_2$-promoted additions to aldehydes (Table 14). However, the differences are much smaller than those found for $BF_3 \cdot OEt_2$-promoted additions. A major consideration in these chelation-controlled reactions is steric in nature. Orientations in which the vinylic hydrogen of the stannane occupies the most crowded position with regard to the chelated carbonyl are highly preferred. The lowest energy transition states

Table 14. Addition of trans- and cis-(Crotyl)tributyltin to α-Benzyloxypropanal Promoted by MgBr₂.

Trans:cis	Syn,syn:Anti,syn
90 : 10	91 : 9
74 : 26	88 : 12
12 : 88	85 : 15

are usually those with an antiperiplanar orientation of the carbonyl and allylic stannane double bond.

BF₃-Promoted Addition of Bu₃SnCH₂CH=CHCH₃ to Aldehydes

syn-2-Methyl-1-cyclohexyl-3-buten-1-ol

   To a stirring solution of cyclohexanecarboxyaldehyde (0.030 g, 20 μL, 0.28 mmol) in CH₂Cl₂ (2.9 mL) at −78 °C was added BF₃•OEt₂ (44 mg, 38 μL, 0.31 mmol) dropwise. The reaction became slightly yellow and, after 15 min, the crotylstannane (0.197 g, 0.57 mmol) was added down the side of the flask to provide for ample cooling. The color of the mixture faded slightly to afford a clear and faintly yellow solution. The reaction was judged complete after 1 h and was quenched cold with saturated aqueous NaHCO₃ (3 mL) as the bath was removed. After warming to 25 °C the mixture was extracted with CH₂Cl₂ dried over Na₂SO₄ and filtered through a plug of Celite (0.5 cm) and silica gel (2 cm) and the solvent was distilled under reduced pressure to give the crude product as a clear colorless liquid, in 88% yield.

$MgBr_2$-Promoted Addition of $Bu_3SnCH{=}CHCH_3$ to an α-Alkoxy Aldehyde

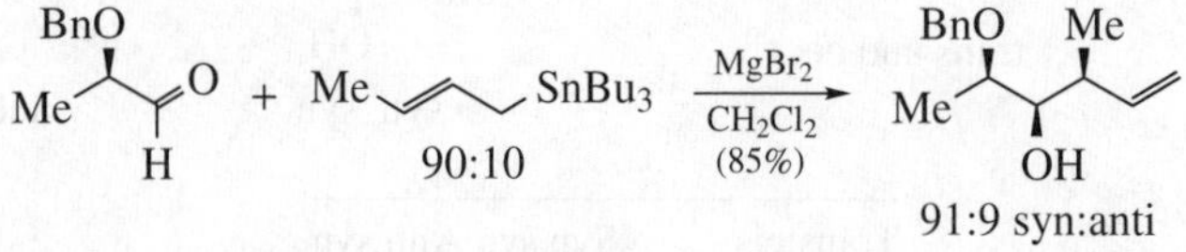

(2*R*,3*R*,4*S*)-2-(Benzyloxy)-4-methyl-5-hexen-3-ol

A solution of the aldehyde (0.450 g, 2.74 mmol) in 30 mL of CH_2Cl_2 was cooled to −23 °C and $MgBr_2$ etherate (1.14 g, 5.48 mmol) was added in one portion. The mixture became cloudy and then, over the next 10 min, the solid material partially dissolved and the solution became slightly yellow. After 15 min, tributylcrotylstannane (1.42 g, 4.11 mmol) was added dropwise by syringe down the side of the flask to allow for ample cooling. The mixture was stirred for an additional 2 h at −23 °C before the bath was allowed to expire. After 12 h, saturated aqueous $NaHCO_3$ (30 mL) was added, and stirring was continued for 25 min. The mixture was extracted with CH_2Cl_2, dried over Na_2SO_4 and filtered through a plug of Celite and silica gel. The solvent was removed and the crude product was purified by radial plate liquid chromatography. The adduct eluted with hexanes, 5% EtOAc/hexanes, 10% EtOAc/hexanes and 15% EtOAc/hexanes as a clear colorless liquid (0.51 g, 85%).

Various α-oxygenated aldehydes react with allylic stannanes in the presence of 5.0 M $LiClO_4{\bullet}OEt_2$[86]. Syn adducts are highly favored (> 95 : 5). The reaction proceeds through a Li chelated aldehyde. Diastereoselectivity is significantly diminished when α-silyloxy aldehydes are employed:

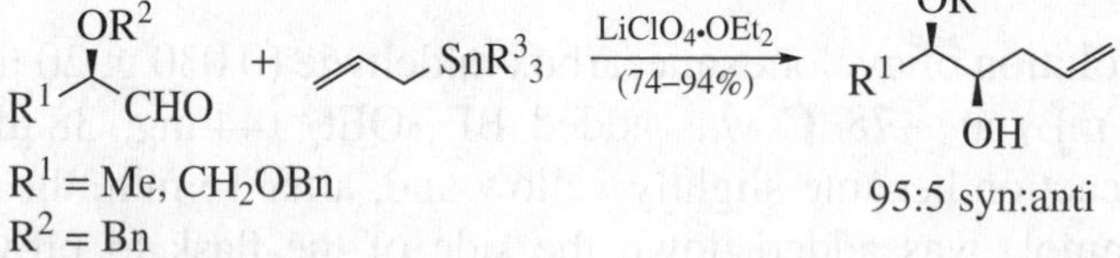

R^1 = Me, CH_2OBn
R^2 = Bn
R^1, R^2 = pyranose, furanose
R^3 = Me, Bu

The reaction of allylic stannanes with aldehydes can be promoted by $MeSiCl_3$ as the Lewis acid[87]. The initial products are silyl ethers which are hydrolyzed to the homoallylic alcohols. The reaction was shown not to involve tin–silicon exchange. A regioreversed version of the crotylstannane addition was found to be promoted by $CoCl_2$[88]. This reaction may involve a cobalt–tin exchange and subsequent S_E2' addition:

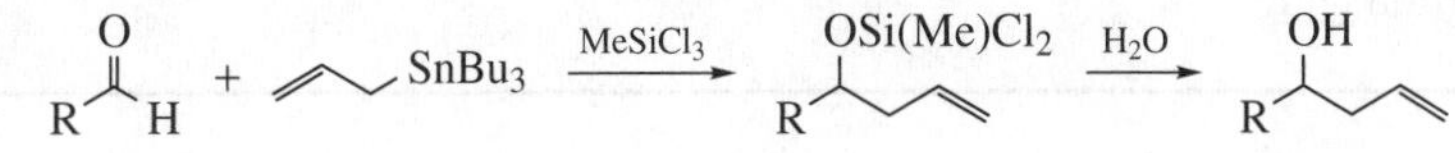

R = Ph, Pr

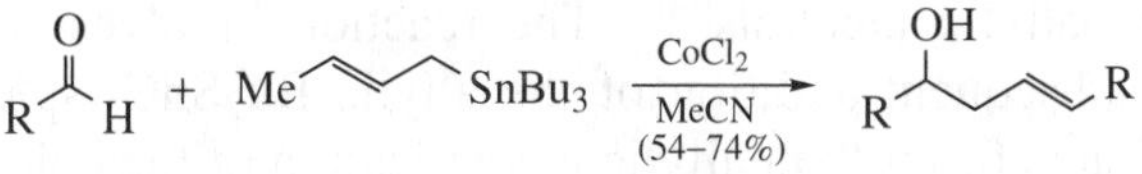

R = Ph, C_6H_{13}, c-C_6H_{11}, (E)-MeCH=CH

Additions of cinnamyl or crotyl tributyltin to aldehydes can be promoted by $SnCl_2$[89]. The stereochemistry of the addition is solvent dependent. In acetonitrile anti adducts are favored, whereas syn products predominate in CH_2Cl_2. It is suggested that transmetalation takes place in acetonitrile leading to a divalent allylic tin chloride which then reacts with the aldehyde by a cyclic six-membered transition state. The addition can also be performed with imines leading to homoallylic amines:

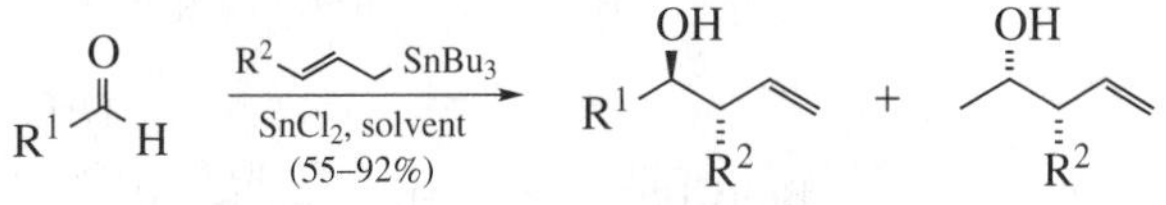

R^1 = Ph, Me, ^iPr, c-C_6H_{11},   100:0–52:48 anti:syn (MeCN)
    (E)-MeCH=CH   82:18 syn:anti (CH_2Cl_2)
R^2 = Ph, Me
solvent = MeCN, CH_2Cl_2

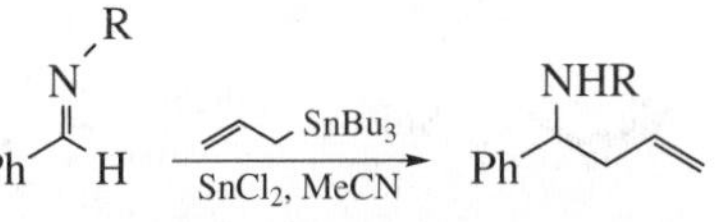

R = Bn 91%
R = Me 89%

Addition of Allyltributyltin to Benzaldehyde in the Presence of $SnCl_2$

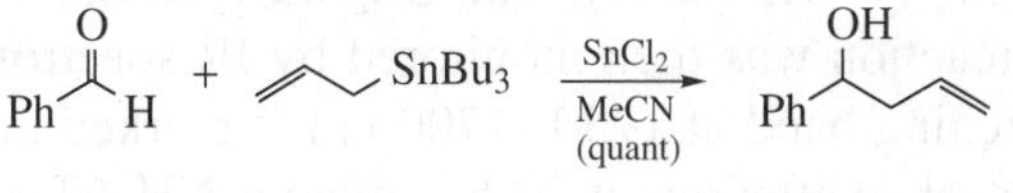

1-Phenyl-3-buten-1-ol

Allyltributyltin (1.0 mmol) was added to a stirred suspension of $SnCl_2$ (1.0 mmol) and benzaldehyde (1.0 mmol) in dry MeCN (1.0 mL). The color of the suspension changed to dark brown in 30 min. No striking difference in the yield was observed when the aldehyde was added to a mixture of stannane and $SnCl_2$. After 2 h, diethyl ether (100 mL) and aqueous NH_4F (15%, 40 mL) were added, the organic layer was washed with water, dried ($MgSO_4$) and the solvent was removed.

> The homoallyl alcohol was purified by flash chromatography (hexane/diethyl ether, 9 : 1) on silica gel.

Crotyltributyltin reacts with conjugated aldehydes in the presence of Bu_2SnCl_2 to afford unbranched (Z)-homoallylic alcohols[90]. The reaction involves transmetalation of the crotylstannane and subsequent addition of the allylic Bu_2SnCl species to the aldehyde. Evidently this addition is faster than allylic isomerization of the chlorostannane, a process that would lead to the branched products:

$$R^1 \overset{CHO}{\underset{R^2}{\diagup}} \quad \xrightarrow[\substack{Bu_2SnCl_2 \\ (82–90\%)}]{\substack{Bu_3Sn \diagdown\diagup Me \\ (E)/(Z)\ 3:1}} \quad R^1 \overset{OH}{\underset{R^2}{\diagup}} Me$$

$$R^1 = R^2 = H;\ R^1 = Me,\ R^2 = H;\ R^1 = H,\ R^2 = Me;\ R^1 = Pr,\ R^2 = H$$

$$Bu_3Sn \diagdown\diagup Me \xrightarrow{Bu_2SnCl_2} \overset{\diagup Me}{\underset{Bu_2SnCl}{\diagup}} \xrightarrow[fast]{RCHO} R \overset{OH}{\diagup} Me$$

$$\downarrow slow$$

$$Bu_2(Cl)Sn \diagdown\diagup Me \xrightarrow{RCHO} R \overset{OH}{\underset{Me}{\diagup}}$$

Addition of Crotyltributyltin to Enals in the Presence of $Bu_2 SnCl_2$

$$\overset{Me}{\underset{CHO}{\diagup}} \quad + \quad Me \diagdown\diagup SnBu_3 \quad \xrightarrow[(90\%)]{Bu_2SnCl_2} \quad \overset{Me}{\diagup} \underset{OH}{\diagup} Me$$

(Z)-2-Methylhepta-1,5-dien-3-ol

Equimolar amounts (25 mmol) of (E/Z)-$Bu_3SnCH_2CH=CHCH_3$ (typically 3 : 1) and methacrolein were mixed and the mixture added as quickly as possible with stirring to solid Bu_2SnCl_2 (1–1.5 equiv). The Bu_2SnCl_2 dissolved almost at once and the progress of the reaction was then monitored by IR spectroscopy. Disappearance of the carbonyl stretching band at 1750–1700 cm^{-1} marked the end of the reaction, which was usually reached after about 20 h. Aqueous NH_4Cl was then added and the carbinol and organotin compounds extracted with ether. Subsequently the carbinol was separated by distillation in yields of 80–90%. Only the cis-crotyl carbinol was obtained, together with small amounts of branched methylallyl carbinol.

The formation of the (Z) product can be understood in terms of a cyclic transition state in which steric interactions with the butyl substituents on tin require the allylic Me to assume an axial orientation. A similar preference is found for thermal reactions of

α-oxygenated allylic stannanes[91], and SnCl$_4$-promoted reactions of δ-oxygenated allylic stannanes[92]:

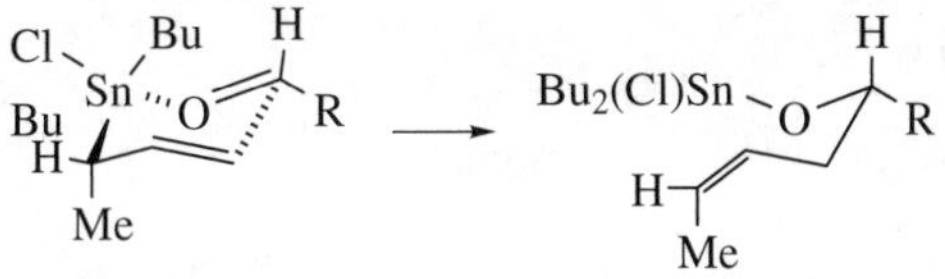

In a related process, BuSnCl$_3$ promotes the addition of crotyltributyltin to aldehydes[93]. The course of these additions depends upon the stereochemistry of the starting crotylstannane. The (Z) isomer affords the linear (Z) adducts nearly exclusively, whereas a 60 : 40 mixture of trans- and cis-crotylstannanes leads to a mixture of linear (Z) and branched syn and anti adducts (Table 15). Evidently the cis-crotylstannane undergoes transmetalation with BuSnCl$_3$ to yield the γ-BuSnCl$_2$ intermediate which reacts with the aldehyde to afford the linear (Z) adduct. As the trans-crotylstannane is more reactive than the cis isomer, it is possible that the trans component of the 60 : 40 mixture reacts with the aldehyde without prior transmetalation. In that case, the BuSnCl$_3$ could serve as the Lewis acid promoter. Alternatively, it has been suggested that the trans-crotylstannane undergoes transmetalation, without allylic inversion, leading to the α-BuSnCl$_2$ intermediate which then reacts with the aldehyde through the usual cyclic six-centered transition state. This possibility seems less likely in view of the known propensity of allylic stannanes for S$_E$2' reactions with electrophiles.

When the crotylstannane and BuSnCl$_3$ are premixed for a short time before addition of aldehyde, the reaction outcome is similar to that in which the BuSnCl$_3$ is added last. However, an extended preequilibration of the stannane with BuSnCl$_3$ leads to a large preponderance of the branched adducts, favoring the syn isomer. Presumably, these adducts are formed by way of a cyclic transition state with the syn isomer arising mainly

Table 15. Addition of trans- and cis-(Crotyl)tributyltin to Aldehydes Promoted by BuSnCl$_3$.

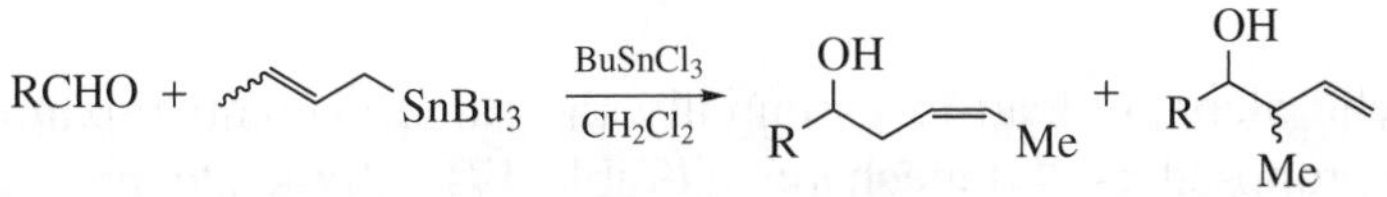

R	trans:cis	Method[a]	T (°C)	Time (h)	Yield (%)	Linear (%)	syn:anti (%)
Ph	0 : 100	A	−78	—	96	96	3 : 1
Ph	60 : 40	A	−78	—	96	44	34 : 22
C$_6$H$_{13}$	0 : 100	A	−78	—	94	95	5[b]
C$_6$H$_{13}$	60 : 40	A	−78	—	90	69	31[b]
Ph	0 : 100	B	−10	0.02	92	96	3 : 1
Ph	60 : 40	B	−10	0.02	95	54	23 : 23
Ph	0 : 100	B	0	8	93	6	80 : 14
Ph	60 : 40	B	0	8	68	—	69 : 31

[a]A = BuSnCl$_3$ added last; B = BuSnCl$_3$ and crotylstannane premixed for the indicated time before addition of aldehyde.
[b]Inseparable mixture of syn and anti products.

from the cis-crotyl dichlorobutyltin intermediate:

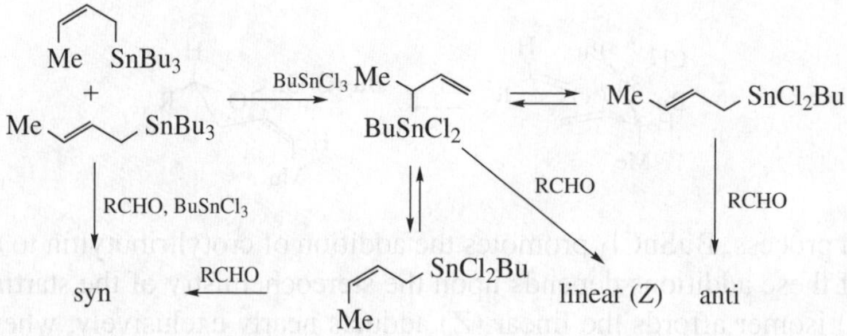

π-Allylpalladium intermediates are involved in a process for converting allylic alcohols to allylic chlorostannanes, which react in situ with aldehydes to yield homoallylic alcohols (Table 16)[94]. Stannylation is achieved by reaction of the electrophilic π-allyl palladium intermediates with excess $SnCl_2$ in various donor solvents. The actual allylating reagent is the trichlorostannane.

Table 16. In situ Preparation of (Allyl)trichlorotin from Allyl Alcohol and Addition to Aldehydes.

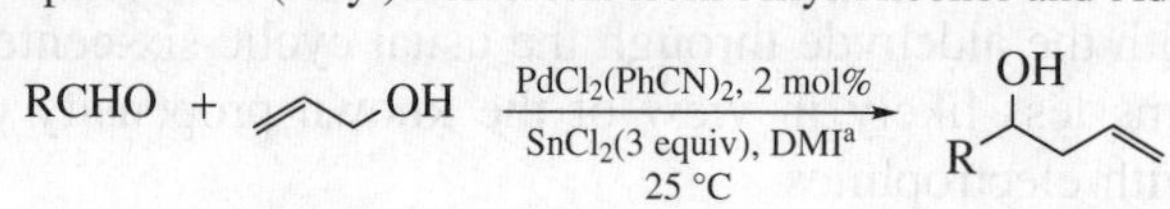

R	Time (h)	Yield (%)
C_6H_{13}	38	77
(E)-MeCH=CH	26	36
(E)-PhCH=CH	24	81
Ph	25	74

[a]DMI = 1,3-dimethyl-2-imidazolidinone.

Isomeric allylic alcohols lead to essentially the same syn:anti mixture of adducts in accord with the proposed π-allyl mechanism (Table 17). Allylic alcohols with substituents at the C-1 or C-3 positions afford mainly anti adducts. Larger substituents favor increased anti selectivity as expected for a cyclic six-membered transition state in which the (E)-allylic stannane affords the anti adduct and vice versa.

Palladium(0)-catalyzed Formation of an Allylic Stannane from an Allylic Alcohol and SnCl₂ and Subsequent Addition to an Aldehyde

2-Methyl-1-phenyl-3-buten-1-ol

To a mixture of SnCl$_2$ (0.57 g, 3 mmol), benzaldehyde (0.11 g, 1 mmol), and (*E*)-2-buten-1-ol (0.11 g, 1.5 mmol) in 1,3-dimethyl-2-imidazolidinone (3 mL) was added PdCl$_2$(PhCN)$_2$ (0.02 mmol) at ambient temperature under nitrogen. After being stirred for 25 h, the mixture was diluted with 120 mL of ether/dichloromethane (2 : 1) and washed successively with aqueous 10% HCl solution, aqueous NaHCO$_3$ solution, water, and brine. The extracts were dried over anhydrous MgSO$_4$. Then removal of solvent and purification by column chromatography on silica gel (hexane/ethyl acetate 7 : 1) afforded 0.12 g (0.74 mmol, 74%) of adduct.

Table 17. In situ Preparation of (Allylic)trichlorotin from Allylic Alcohols and Addition to Aldehydes.

R^1	R^2	R^3	Time (h)	Yield (%)	syn:anti
C$_6$H$_{13}$	Me	H	72	65	45 : 55
C$_6$H$_{13}$	H	Me	96	75	45 : 55
c-C$_6$H$_{11}$	Me	H	72	76	14 : 86
c-C$_6$H$_{11}$	H	Me	96	97	18 : 82
Ph	Pr	H	96	83	11 : 89
Ph	Ph	H	73	79	7 : 93
c-C$_6$H$_{11}$	Ph	H	70	59	0 : 100

[a]DMI = 1,3-dimethyl-2-imidazolidinone.

2,4-Pentadienylstannanes have the potential to react with aldehydes at the 3- or 5-position (Table 18). Reactions promoted by ZnCl$_2$ in ether give mainly the adducts derived from attack by C-3 on the aldehyde[95]. When CH$_2$Cl$_2$ is employed as the solvent the selectivity for C-3 versus C-5 decreases from 88 : 12 to 45 : 54 with benzaldehyde. Electron-donating solvents enhance the C-3 selectivity to 96 : 4 (MeCN) and > 99 : 1 (DMF). The stronger Lewis acids ZnBr$_2$ and ZnI$_2$ give decreased ratios of the two adducts (49 : 51 and 17 : 83, respectively). The stannyl dienes CH$_2$=CH−CH=CH−CH$_2$SnMe$_3$ and MeCH=CH−CH=CH−CH$_2$SnMe$_3$ afford C-3 adducts almost exclusively with benzaldehyde and cyclohexanecarboxaldehyde in the presence of ZnCl$_2$ in ether.

These additions are thought to involve a cyclic transition state with the weaker Lewis acid and an acyclic transition state with the stronger Lewis acids. The role of solvent is to diminish the coordinating ability of the Lewis acid thereby enabling the aldehyde carbonyl oxygen to more strongly associate with the tin atom. A similar argument has been advanced to explain the regiochemical preference at C-3 for the reaction of benzaldehyde

Table 18. Additions of a (2,4-Pentadienyl)trimethylstannane to Benzaldehyde Promoted by Lewis Acids.

mixture of (*E*) and (*Z*)

(24) (mixture of syn and anti)

Solvent	**24 : 25**	Lewis acid[a]	**24 : 25**
CH_2Cl_2	46 : 54	$ZnBr_2$	49 : 51
Et_2O	88 : 12	ZnI_2	17 : 83
MeCN	96 : 4		
DMF	> 99 : 1		

[a] In CH_2Cl_2.

with 2,4-pentadienyl tributyltin catalyzed by AgOTf•PPh$_3$ in 1 : 1 THF/H$_2$O[96].

6.2.2   Lewis Acid Catalyzed Additions to Aldehydes

Additions of allylic stannanes to aldehydes with promotors such as BF$_3$•OEt$_2$, MgBr$_2$, TiCl$_4$, and SnCl$_4$ typically require one or more equivalents of the Lewis acid. In such cases, complexation of the Lewis acid with the product renders it ineffective as a catalyst. Several approaches have been developed to circumvent this problem. One of these involves regeneration of the Lewis acid from the product complex through reaction with an appropriate electrophile. This approach has been successfully applied to reactions employing Bu$_2$SnCl$_2$ as the Lewis acid (Table 19) [97]. The first step of this sequence involves redistribution of the catalyst with the allylic stannane to afford the intermediate, allyldibutyltin chloride, which reacts with the aldehyde to yield the stannoxane adduct. This adduct reacts with an added acid chloride or TMS-Cl to produce the ester or silyl ether of the homoallylic alcohol and the regenerated catalyst.

The addition of cinnamyl tributyltin to isobutyraldehyde can be effected with 10 mol% of InCl$_3$ in the presence of one equivalent of Me$_3$SiCl[98]. In acetonitrile the *anti* adduct is favored, whereas the syn product predominates in CH$_2$Cl$_2$. It is proposed that transmetalation takes place in acetonitrile with addition occurring through a cyclic six-center transition state. In methylene chloride, transmetalation is slow and the major product

Table 19. Catalyzed Addition of (Allyl)butyldichlorotin to Aldehydes.

R^1	R^2	Yield (%)	Time (h)
Ph	Me_3Si	76	4
Ph	PhCO	89	2
Ph	MeCO	86	2
Ph	BnOCO	85	4
Ph	MeOCO	89	3
C_7H_{15}	MeOCO	83	3
c-C_6H_{11}	MeOCO	87	3
^tBu	MeOCO	48	58

is formed by way of an acyclic transition state with $InCl_3$ serving as a Lewis acid. In both cases the Me_3SiCl liberates the $InCl_3$ from the oxygen of the alcohol product. The combination of catalytic $InCl_3$ and stoichiometric Me_3SiCl can also be employed for additions of alkynylstannanes to aldehydes. In that case transmetalation preceeds addition.

88:12 in CH_3CN (82% yield)
12:88 in CH_2Cl_2 (85% yield)

$R^1 = $ Ph, p-MeC_6H_4, p-$NO_2C_6H_4$, (E)-MeCH=CH, ^tBu, Et
$R^2 = $ Ph, Pr, H

Various Pd and Pt phosphine complexes catalyze the addition of allylic stannanes to aldehydes and imines [99]. The Pt complex $PtCl_2(PPh_3)_2$ is the most effective of the three examined (Table 20). Allyl and methallyl stannanes are more reactive than cis-crotyl. The latter stannane affords nearly 1 : 1 mixtures of syn and anti adducts. Interestingly, imines are more reactive than the related aldehydes, whereas the opposite reactivity is seen for

Table 20. Palladium-catalyzed Additions of (Allyl)tributyltin to Benzaldehyde.

$$PhCHO \ + \ \diagup\!\!\diagdown\!\!\diagup SnBu_3 \ \xrightarrow[\substack{25\,°C \\ 4\,days}]{cat} $$

Catalyst	Yield (%)
$PtCl_2(PPh_3)_2$	90
$PdCl_2(PPh_3)_2$	64
$Pd(PPh_3)_4$	40

additions performed with $BF_3 \bullet OEt_2$ as the promotor.

$$RCHO \ + \ \underset{Me}{\diagup\!\!\diagdown\!\!\diagup} SnBu_3 \ \xrightarrow[THF,\ reflux]{PtCl_2(PPh_3)_2} $$

R = *p*-BrC$_6$H$_4$, 1 h, 48% yield   ca. 1:1 syn:anti
R = *p*-NO$_2$C$_6$H$_4$, 8 days, 65% yield

$$\underset{R^1}{\overset{N\diagdown R^2}{\diagup\!\!\diagup}}\!\!\!H \ + \ \diagup\!\!\diagdown\!\!\diagup SnBu_3 \ \xrightarrow{PtCl_2(PPh_3)_2} $$

R^1 = Ph, R^2 = *p*-NO$_2$C$_6$H$_4$, 91% yield
R^1 = Bn, R^2 = *p*-NO$_2$C$_6$H$_4$, 98% yield
R^1 = *p*-MeO$_2$CC$_6$H$_4$, R^2 = Ph, 98% yield

The reactions with $PdCl_2(PPh_3)_2$ proceed via a bis-π-allyl palladium complex formed in situ from the allylic stannane (Table 21). Addition of an aldehyde to this complex leads to a mixed π-allyl σ-allyl complex which undergoes a metallo ene reaction. The resulting π-allyl complex reacts with additional allylic stannane to regenerate the bis-π-allyl Pd

Table 21. Addition of (Allyl)tributyltin to Aldehydes Catalyzed by$PtCl_2(PPh_3)_2$.

$$R^1CHO \ + \ \underset{\ }{\overset{R^2}{\diagup\!\!\diagdown}} SnBu_3 \ \xrightarrow[THF,\ reflux]{PtCl_2(PPh_3)_2} $$

R^1	R^2	Time (h)	Yield (%)
o-BrC$_6$H$_4$	H	8	94
p-BrC$_6$H$_4$	Me	10	63
c-C$_6$H$_{11}$	H	10	99
c-C$_6$H$_{11}$	Me	14	67
C$_5$H$_{11}$	H	24	40
C$_5$H$_{11}$	Me	16	49

Table 22. Addition of Tetraallyltin to Aldehydes Catalyzed by Sc(OTf)$_3$.

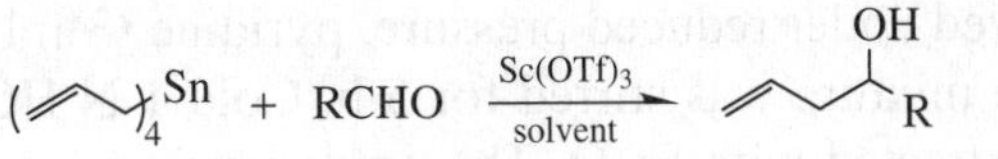

$$(\diagdown\!\!\diagdown\!\!\diagup\,Sn)_4 \;+\; RCHO \;\xrightarrow[\text{solvent}]{\text{Sc(OTf)}_3}\; \diagup\!\!\diagdown\!\!\diagup\overset{\text{OH}}{\diagdown}R$$

R	Solvent	Yield (%)
PhCH$_2$CH$_2$	THF/H$_2$O (9 : 1)	92
	EtOH/H$_2$O (9 : 1)	96
	MeCN/H$_2$O (9 : 1)	96
	EtOH	86
	MeCN	94
c-C$_6$H$_{11}$	THF/H$_2$O (9 : 1)	83
D-Arabinose	MeCN/H$_2$O (9 : 1)	93
o-HOC$_6$H$_4$	THF/H$_2$O (9 : 1)	99
2-Pyridyl	THF/H$_2$O (9 : 1)	99

complex and liberate the homoallylic alcohol adduct as its Bu$_3$Sn ether:

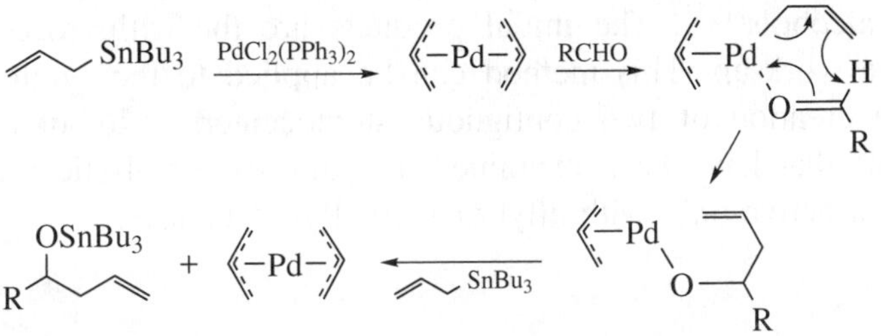

A second approach to catalyzed additions utilizes Lewis acids that do not form strong complexes with the homoallylic alcohol products. Lanthanide triflates fulfil this condition. Thus, as little as 5 mol% of Sc(OTf)$_3$ catalyzes the addition of tetraallyltin to various aldehydes in a variety of solvents[100]. All four allyl groups participate in the addition. The reaction can be carried out on unprotected carbohydrates (Table 22).

Addition of Tetraallyltin to D-Arabinose in an Aqueous Medium Catalyzed by Sc(OTf)$_3$

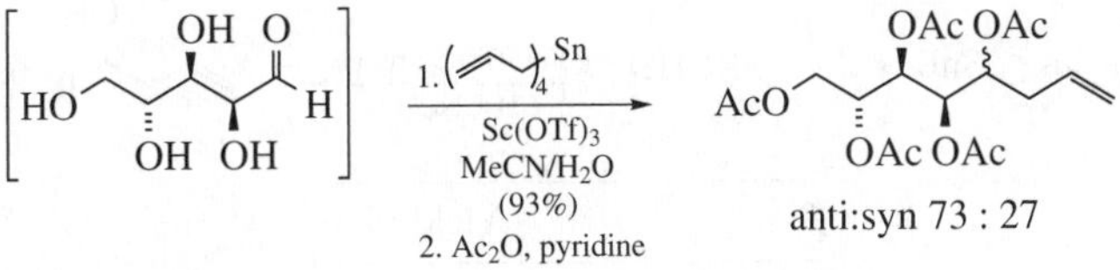

Addition of Tetraallyltin to D-Arabinose Catalyzed by Sc(OTf)$_3$

To a mixture of Sc(OTf)$_3$ (0.015 mmol, 5 mol%) and D-arabinose (0.30 mmol) in aqueous CH$_3$CN (1 : 4; 1.5 mL) was added tetraallyltin (0.15 mmol) in CH$_3$CN

(1.5 mL) at 25 °C. The mixture was stirred for 60 h at this temperature. After the solvents were removed under reduced pressure, pyridine (3 mL) and Ac$_2$O (1.5 mL) were added, and the mixture was stirred for 5 h. Cold 1 N HCl was added, and the organic layer was extracted with Et$_2$O. The crude product was chromatographed on silica gel to afford the pentaacetylated adduct (93%).

Additions of allyltributyltin to aldehydes are also catalyzed by Yb(OTf)$_3$ (Table 23). In this case best results are obtained under anhydrous conditions. Here too, 5 mol% of catalyst is effective[101].

6.2.3   Chiral Lewis Acid-catalyzed Additions to Aldehydes[102]

The addition of allylic stannanes to aldehydes in the presence of a 0.2–1.0 equivalents of a tartrate-derived acyloxyborane and (CF$_3$CO$_2$)$_2$O leads to enantioenriched syn-homoallylic alcohols[103]. The initial products are the trifluoroacetates which are hydrolyzed in the work-up. This method can be applied to the synthesis of branched adducts with the creation of two contiguous stereocenters. All but one of the other chiral Lewis acids that have been examined as catalysts for allylic stannane additions to aldehydes are effective only with allyl or methallyl stannanes:

R = Ph, (E)-MeCH=CH, BuC≡C, C$_3$H$_7$

Table 23. Addition of (Allyl)tributyltin to Aldehydes Catalyzed by Sc(OTf$_3$).

R	Yield (%)
Ph	85
p-NO$_2$C$_6$H$_4$	93
C$_5$H$_{11}$	93
c-C$_6$H$_{11}$	94
2-furyl	85

The exception is a catalyst prepared from (S)-BINOL and (^{i}PrO)$_2$TiCl$_2$ which effects additions of γ-methylated allylic stannanes to methyl glyoxylate leading to the homoallylic alcohol adducts as syn/anti mixtures of modest ee[104]. However, this catalyst is highly selective in additions of allyl tributyltin to various aldehydes[105]. An analogous catalyst, from 1,1′-bi-2-naphthol (BINOL) and Ti(O-^{i}Pr)$_4$, gives similar results[106]:

R = C$_5$H$_{11}$, C$_7$H$_{15}$, c-C$_6$H$_{11}$, (E)-PhCH=CH, Ph

The combination of BINOL (10 mol%) and Ti(O-^{i}Pr)$_4$ (5 mol%) catalyzes the addition of allyl and methallyl tributyltin to various aldehydes in the presence of 4A molecular sieves with high efficiency[107]. The reactions require several days to reach completion. Additions of allyl and methallyl tributyltin to aldehydes are also catalyzed by a complex of BINAP and AgOTf (5–15 mol%) in THF[108]:

R = C$_5$H$_{11}$, C$_7$H$_{15}$, c-C$_6$H$_{11}$, (E)-PhCH=CH, Ph
R^1 = H, Me

R = Ph, p-BrC$_6$H$_4$, 2-furyl, (E)-PhCH=CH, (E)-PrCH=CH

Chiral bis-oxazoline–metal salt complexes have also been employed as catalysts for enantioselective additions of allyltributyltin to aldehydes[109]. Of the metal salts examined—FeI$_2$, MgI$_2$, MgBr$_2$, InCl$_3$, Mg(ClO$_4$)$_2$, SnCl$_2$, Sn(OTf)$_2$, Cu(OTf)$_2$, ZnCl$_2$, ZnBr$_2$, and ZnI$_2$—only InCl$_3$, Sn(OTf)$_2$, ZnCl$_2$, ZnBr$_2$, and ZnI$_2$ gave measureable enantioselectivity. The latter three perform equally well (40% ee with octanal). A number of bis-oxazoline ligands were examined with various zinc salts but in no case did the ee of the adduct exceed 40% (Table 24). Somewhat better results were realized with cyclohexanecarboxaldehyde and a bis-oxazoline-Zn(OTf) catalyst (45% yield of adduct with 46% ee).

Table 24. Addition of (Allyl)tributyltin to Octanal Catalyzed by Chira bis-Oxazoline–Zinc Complexes.

R^1	R^2	R^3	X	Time (h)	Yield (%)	ee(%)
Ph	H	H	I	18	78	40
Ph	H	H	Br	18	54	40
Ph	H	H	Cl	18	10	40
Ph	Ph	H	I	20	45	38^a
Ph	H	Me	I	18	68	23
^{i}Pr	H	Me	Br	20	35	5
^{t}Bu	H	H	OTf	18	15	14

^{a}The enantiomeric auxiliary was employed.

6.2.4   SnCl$_4$-promoted Additions of 4-, 5-, and 6-Oxygenated Allylic Stannanes to Aldehydes[110]

4-Oxygenated allylic stannanes undergo 1,4 elimination upon treatment with Lewis acids such as BF$_3$•OEt$_2$[111]. However, with SnCl$_4$ a transmetalation takes place and the intermediate chlorostannane adds to aldehydes affording (Z)-homoallylic alcohols:

R = Pr, ^{i}Pr, c-C$_6$H$_{11}$, p-O$_2$NC$_6$H$_4$, p-MeOC$_6$H$_4$

The 3-stannylated intermediate is thought to be stabilized by coordination to the 4-oxygen substituent. Subsequent reaction with aldehydes proceeds through a chair transition state. With chiral α-substituted aldehydes the major products are formed via Felkin–Ahn transition states[112]:

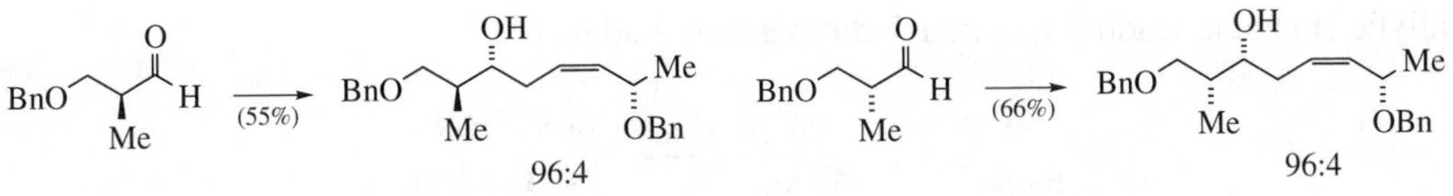

Allylic stannanes with a 5-oxygen substituent also afford linear (Z)-homoallylic alcohols in $SnCl_4$-promoted additions to aldehydes. Stereocontrol is imposed by a 4-methyl substituent via a cyclic chelated intermediate. Subsequent reaction with the aldehyde proceeds through a bicyclic transition state to afford the 1,5-anti adduct[113]:

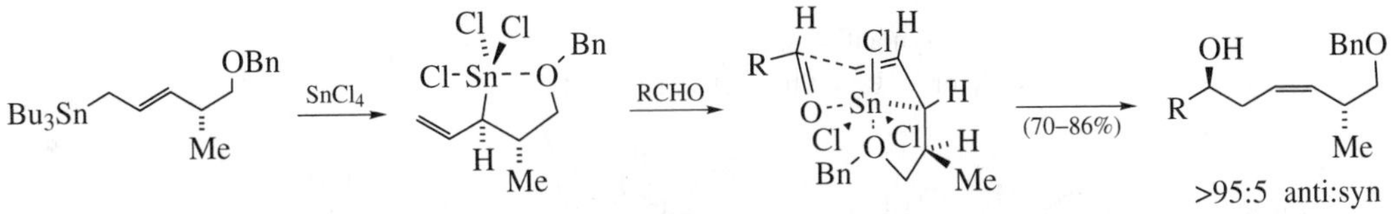

R = Et, ^{i}Pr, Ph, p-ClC$_6$H$_4$, p-MeOC$_6$H$_4$, (S)- or (R)-MeCH(OBn)

Addition of [(E)-4-Benzyloxy-2-pentenyl]tributylstannane to Benzaldehyde Following Transmetalation with SnCl$_4$

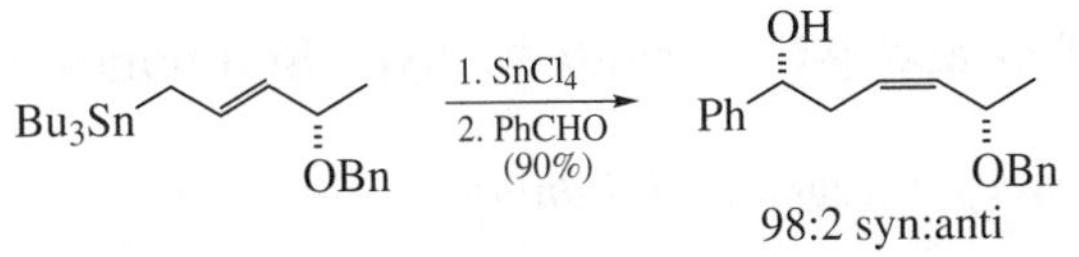

(Z)-syn-1-Phenyl-5-benzyloxy-3-hexen-1-ol

A cooled solution of $SnCl_4$ (0.028 g, 0.11 mmol) in anhydrous CH_2Cl_2 (0.3–0.5 mL) was added dropwise to a solution of the (4-benzyloxy-2-pentenyl)tributylstannane (0.050 g, 0.11 mmol) in anhydrous CH_2Cl_2 (3 mL) at −78 °C, and the mixture was stirred for 5 min. A cooled solution of the aldehyde (0.11 mmol) in CH_2Cl_2 was then added, and the mixture was stirred at −78 °C for 1 h. Saturated aqueous $NaHCO_3$ was added, the mixture was allowed to warm to 25 °C and was then extracted with CH_2Cl_2. The organic extracts were combined and washed twice with water and brine, and dried ($MgSO_4$). After concentration under reduced pressure, the residue was chromatographed on silica gel, using Et_2O/petroleum ether (1 : 3) as eluant, to give the product, a 98 : 2 mixture of syn and anti isomers in 90% yield.

Allylic stannanes with secondary 5-alkoxy groups afford syn-1,6-dioxygenated adducts with $SnBr_4$ and aldehydes[114]. In this case only the methoxy or hydroxy-substituted stannanes give high levels of diastereoselection. With α-substituted chiral aldehydes, Felkin–Ahn facial selectivity is observed. A similar trend is seen with a 6-hydroxy allylic stannane leading to syn-1,7-dioxygenated adducts[115]:

syn:anti = 85:15 to 96:4
R[1] = Me or H; R[2] = Ph, *p*-ClC$_6$H$_4$, Me, Et, ^{i}Pr

R = Me (36%), Et (61%), ^{i}Pr (63%), ^{t}Bu (38%), Ph (72%)

6.2.5 Additions of α- and γ-Oxygenated Allylic Stannanes to Aldehydes

6.2.5.1 Racemic α- and γ-Oxygenated Stannanes

α-Oxygenated allylic stannanes can be prepared by addition of Bu$_3$SnLi to conjugated enals followed by etherification of the alcohol adducts with reactive halides[116]. These stannanes add to aldehydes at elevated temperatures to afford anti adducts via a cyclic six-membered transition state. The reaction is most effective with aromatic aldehydes. Unbranched aliphatic aldehydes tend to self-condense under the reaction conditions. The exclusive formation of (Z)-enol ethers is thought to arise from a preferential axial disposition of the OCH$_2$OMe substituent in the transition state. The alternative equatorial arrangement places this substituent into closer proximity to the Bu substituents of the tin:

R = Ph, PhCH=CH, *p*-O$_2$NC$_6$H$_4$, *p*-ClC$_6$H$_4$, Et, ^{i}Pr, C$_6$H$_{13}$

γ-Oxygenated allylic stannanes can be prepared by lithiation of allylic ethers with sec-butyllithium and subsequent addition of Bu₃SnCl or Ph₃SnCl[117]. These stannanes add to α- or β-oxygenated aldehydes in the presence of MgBr₂ to afford syn adducts as major or exclusive products. The stereochemistry of the products is consistent with an acyclic chelated transition state:

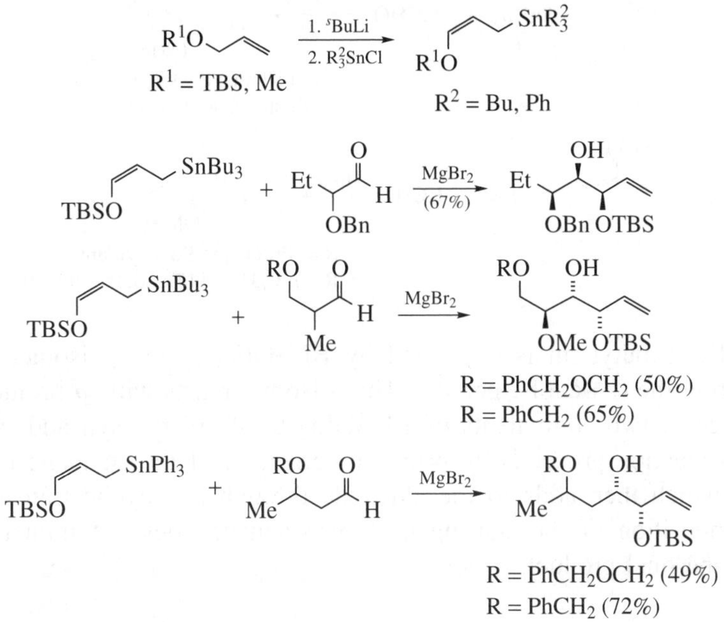

$R^1 = TBS, Me$

$R^2 = Bu, Ph$

R = PhCH₂OCH₂ (50%)
R = PhCH₂ (65%)

R = PhCH₂OCH₂ (49%)
R = PhCH₂ (72%)

Sequential Lithiation and Stannylation of an Allylic tert-Butyldimethylsilyl (TBS) Ether

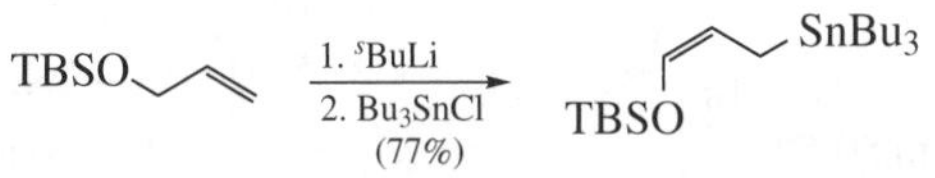

(Z)-[3-(tert-Butyldimethylsilyl)oxy]-2-propenyl Tributylstannane

To a solution of tert-butyldimethylsilyl allyl ether (3.21 g, 0.02 mmol) in THF (50 mL) at −78 °C was added sec-butyllithium (1.2 equiv) followed by 5 mL of HMPA. After 15 min, tributyltin chloride (5.45 mL, 0.022 mmol) was added. After an additional 15 min, the mixture was allowed to warm to 25 °C and then poured into 100 mL of hexane. The resulting solution was washed with saturated aqueous NH₄Cl solution, water, then dried over MgSO₄ and concentrated in vacuo. Kugelrohr distillation gave 7.12 g (77%) of stannane as a colorless liquid (bp 195–200 °C oven temp, 0.01 mmHg).

Hydrostannantion of methoxyallene leads to a separable 1 : 1 mixture of (*E*) and (*Z*)-γ-methoxy allylic stannanes. These stannanes afford syn adducts as major products in BF₃-promoted additions to aldehydes[118]. This result is consistent with an acyclic transition state analogous to that of the crotylstannane:

MeO⌒⌒SnBu₃ + PhCHO $\xrightarrow[\text{(86\%)}]{\text{BF}_3\cdot\text{OEt}_2}$ Ph⌒⌒⌒ OH, OMe

(*E*)-stannane 93:7 syn:anti
(*Z*)-stannane 91:9 syn:anti

MeO⌒⌒SnBu₃ + RCHO $\xrightarrow{\text{BF}_3\cdot\text{OEt}_2}$ R⌒⌒⌒ OH, OMe

R = ^{i}Pr, 60%; >96:4 syn:anti
R = *c*-C₆H₁₁, 44%; 85:15 syn:anti

α-Ethoxyallyl tributyl tin is converted by BF₃•OEt₂ to the γ isomer, as an 85 : 15 mixture favoring the (*Z*)-enol ether[119]. The α isomer reacts with *p*-bromobenzaldehyde in the presence of two equivalents of BF₃•OEt₂ to afford the syn adduct as the major product. This reaction proceeds by prior isomerization of the stannane to the γ-ethoxy intermediate which then adds to the aldehyde through an acyclic transition state. The (*E*) : (*Z*) composition of the starting α-ethoxystannane does not influence the stereochemistry of the final product:

OEt, SnBu₃ $\xrightarrow{\text{BF}_3\cdot\text{OEt}_2}$ Bu₃Sn⌒⌒ OEt + Bu₃Sn⌒⌒ OEt

85:15

Bu₃Sn⌒⌒Me, OEt + Br⌒CHO $\xrightarrow[\text{(80–95\%)}]{\text{BF}_3\cdot\text{OEt}_2}$ Br⌒⌒ OH, Me, OEt

(*Z*):(*E*) = 80:20 to 50:50   syn:anti = 93:7

Thermolysis of a 50 : 50 (*E*) : (*Z*) mixture of the foregoing stannane with *p*-bromobenzaldehyde affords the anti (*Z*) adduct as the major product in 50% yield. Presumably this adduct is formed by preferential reaction of the (*E*)-stannane through a cyclic transition state as is found for other analogous reactions:

Bu₃Sn⌒⌒Me, OEt + Br⌒CHO $\xrightarrow[\text{(50\%)}]{100\ °C}$ Br⌒⌒ OH, Me OEt

1:1 (*E*):(*Z*)   98:2 anti:syn
                                                    98:2 (*Z*):(*E*)

6.2.5.2 *Nonracemic α- and γ-Oxygenated Stannanes*

The first nonracemic α-alkoxy allylic stannane was obtained by resolution through the (−)-menthyloxymethyl ether[91b,120]. This ether was prepared by amine-catalyzed addition of chloromethyl (−)-menthyl ether to racemic α-hydroxy crotyl stannane. Subsequent separation of the diastereomeric ethers yielded the enantiomerically pure stannanes. Thermolysis of the (R) stannane with benzaldehyde affords the (S,S) adduct, whereas the (S) stannane gives the (R,R) adduct. These products are the expected ones from a cyclic transition state as seen for the analogous reactions of racemic α-alkoxy allylic stannanes. Thermolysis of the (R) stannane with cinnamaldehyde and cyclohexanecarboxaldehyde affords the (S,S) adducts in comparable yield:

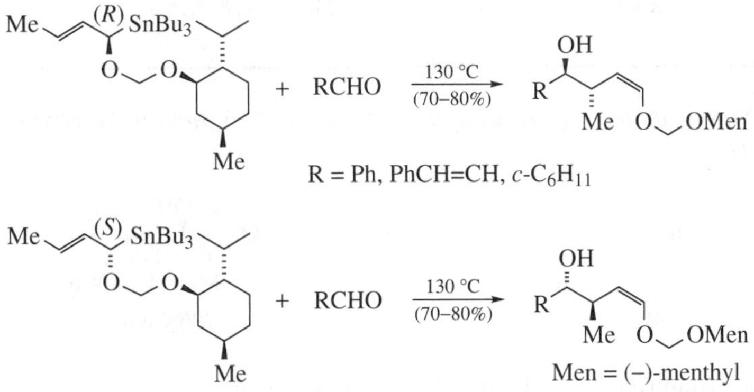

R = Ph, PhCH=CH, c-C$_6$H$_{11}$

Men = (−)-menthyl

Thermolysis of a Nonracemic α-Alkoxy Crotylstannane and Benzaldehyde

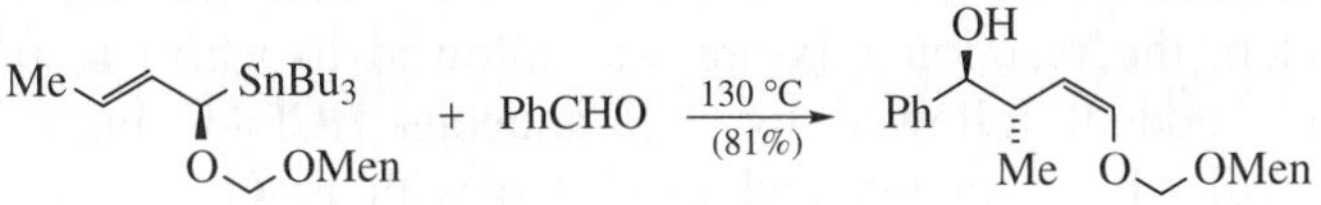

(1S,2S,Z)-4-[(−)-(Menthyloxy)methoxy]-2-methyl-1-phenyl-3-buten-1-ol

A mixture of the (R)-1-alkoxy-2-butenylstannane (0.102 g, 0.19 mmol) and benzaldehyde (98 μL, 1 mmol) was heated at 130 °C for 14.5 h under an atmosphere of argon. Short-column chromatography of the mixture with ether/petroleum ether (1 : 5) as eluant gave the adduct (0.053 g, 81%), as a colorless oil ($[\alpha]_D^{20}$ − 129.2 (c 1.03, CHCl$_3$)).

Nonracemic α- and γ-carbamoyloxy crotylstannanes can be prepared by enantioselective lithiation of crotyl N,N-diisopropyl carbamate with BuLi in the presence of (−)-sparteine[121]. Addition of tributyl or trimethyltin chloride leads to a mixture of chromatographically separable γ- and α-stannylated products of relatively high ee. Both stannanes are formed with inversion of configuration. Thermolysis of the (R)-α-OCb allylic stannane of 80% ee with benzaldehyde affords the (*S*,*S*) anti adduct of 79% ee in 79% yield:

Me⏦⏦OCb —BuLi→ (−)-sparteine Me⏦⏦—H Li · (−)-sparteine
Cb = CON(^{i}Pr)$_2$

│ R$_3$SnCl

Me⏦⏦SnR$_3$ +   Me⏦⏦
         │OCb   R$_3$Sn   OCb

R = Bu (58%, 90% ee)   R = Bu (22%, 88% ee)
R = Me (21%, 82% ee)   R = Me (48%, 82% ee)

Addition of Nonracemic γ-Carbamoyloxy Allylic Stannanes to Aldehydes in the Presence of TiCl$_4$

Me⏦⏦   Me
Bu$_3$Sn   OCb   +   Me⏦—O   —TiCl$_4$→   Me⏦⏦—OH
                          H   (96%)   Me Me OCb
(96% ee)   (96% ee)

(3*R*,4*S*,*Z*)-6-Carbamoyloxy-2,4-dimethyl-5-hexen-3-ol

To the solution of the enantioenriched stannane (1.0 mmol) and isobutyraldehyde (1.0 mmol) in CH$_2$Cl$_2$ (6 mL), cooled under Ar to below −78 °C, 1 M TiCl$_4$ in dry CH$_2$Cl$_2$ (1.2 mL) was added by a syringe over 5 min. Stirring was continued at −78 °C for 1 h, the reaction mixture was allowed to warm to 0 °C and Et$_2$O (10–20 mL) was added, followed by 2 N aqueous HCl (10 mL). The aqueous phase was immediately separated and extracted with Et$_2$O. The combined ether solutions were washed with saturated aqueous NaHCO$_3$ and dried (Na$_2$SO$_4$). After removal of the solvents, the residue was purified by flash chromatography (silica gel, Et$_2$O/pentane 1 : 3), to give the adduct in 96% yield.

An alternative route to the γ-OCb allylic stannanes entails reaction of the lithium complex with Ti(O − ^{i}Pr)$_4$ and subsequent treatment of the resulting α-OCb titanium intermediate with stannyl chlorides. This process affords the γ-OCb allylic stannanes of opposite configuration from those obtained in the direct stannylation:

Me⏦⏦—H Li   —Ti(O-^{i}Pr)$_4$→   Me⏦⏦Ti(O-^{i}Pr)$_3$ —R$_3$SnCl→   Me⏦⏦
       │O   OCb   R$_3$Sn   OCb
      O   R = Bu (80%, 95% ee)
    N(^{i}Pr)$_2$ R = Me (73%, 88% ee)

Previous findings indicated that these stannanes do not undergo BF_3-promoted additions to aldehydes[122]. However, upon treatment with $TiCl_4$ in the presence of aldehydes or ketones, transmetalation and subsequent addition leads to the anti products:

$$R^1 = {}^iPr, {}^tBu, EtO_2C(CH_2)_2, EtO_2C(CH_2)_3$$
$$R^2 = H, Me$$

The S_N2' reaction of cyanocuprates derived from Grignard reagents to the (R,R)-2,4-pentanediol acetal of (E)-3-(tributyltin)propenal provides an alternative route to enantioenriched γ-alkoxy allylic stannanes[123]. The addition requires three equivalents of $BF_3 \cdot OEt_2$. The assigned configuration of the products is based on comparison of their optical rotation with analogous MOM ethers of known configuration:

R = Bu [61%, 94:6 *(Z):(E)*, 78% de]
R = ^iPr [65%, 94:6 *(Z):(E)*, 68% de]

Possibly the most general method for obtaining α-oxygenated allylic stannanes of high ee is through asymmetric reduction of conjugated acylstannanes with (R)- or (S)-BINAL-H (Table 25). The acylstannanes are readily available from conjugated aldehydes by addition of Bu_3SnLi and *in situ* oxidation with 1,1'-(azodicarbonyl)dipiperidine (ADD). The (R)-BINAL-H reagent affords the (S)-α-hydroxy allylic stannanes. The hydroxy stannanes are protected as $MeOCH_2$, $BnOCH_2$ or $^tBuMe_2Si$ ethers by addition of the appropriate chloride in the presence of $^iPr_2NEt$.

The first Lewis acid-promoted addition of a nonracemic α-alkoxy allylic stannane to an aldehyde was carried out in connection with the synthesis of cembranolides[124]. This intramolecular S_E2' addition afforded a 14-membered alkynol in 88% yield as a single cis-(Z) enantiomer. Less than 5% of the trans isomer was produced:

OMOM = OCH_2OCH_3

Table 25. Synthesis of Enantioenriched α-Hydroxy Allylic Stannes.

R	ee(%)	Configuration	Hydride
Me	>95	(S)	(R)-BINAL-H
Bu	>95	(S)	(R)-BINAL-H
c-C6H11	90	(S)	(R)-BINAL-H

Intermolecular additions to achiral aldehydes are less selective. The major products are syn-(E) and syn-(Z) adducts, favoring the former (Table 26). Anti-(Z) and anti-(E) isomers are formed as minor products[125].

Although mixtures of diastereomers are obtained, each isomer reflects the enantiomeric purity of the starting allylic stannane, typically 95% or better. Thus the additions are stereospecific, though not totally stereoselective. These results can be understood by consideration of acyclic transition states in which electrophilic attack on the double bond occurs anti to the allylic tin substituent. Carbonyl facial selectivity is governed by steric

Table 26. Addition of an α-Oxygenated Allylic Stannane to Aldehydes.

OBOM = OCH2OCH2Ph

R	Yield(%)	**26 : 27 : 28 : 29**
C6H13	80	70 : 27 : 0 : 3
(E)-C4H9CH=CH	72	80 : 17 : 1 : 2
C4H9C≡C	88	51 : 25 : 7 : 17

interactions between the aldehyde and double-bond substituents:

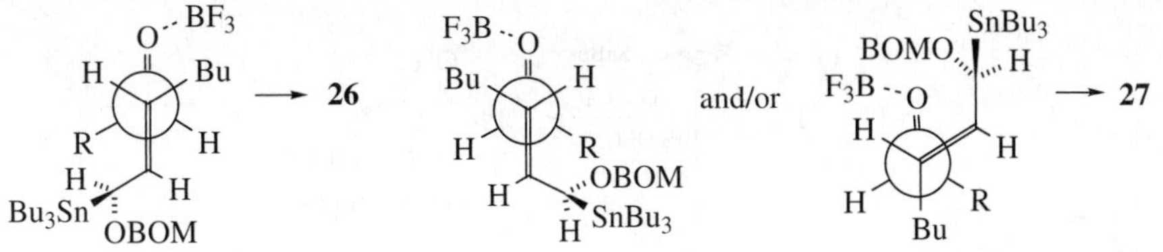

Additions to α-branched aldehydes are significantly more diastereoselective[126]. In studies directed toward the synthesis of macrolide antibiotics, it was found that chirality matched additions to nonracemic α-methyl substituted aldehydes exhibit significant substrate control leading to syn-(*Z*) or syn-(*E*) adducts with high efficiency. It is also possible to effectively discriminate between two diastereomeric allylic stannanes in such additions, resulting in a kinetic resolution:

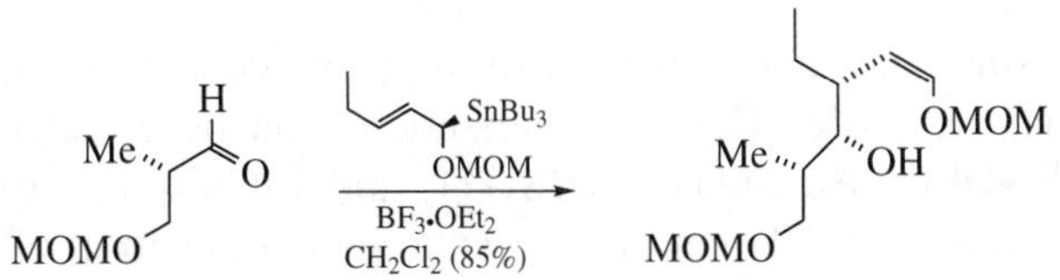

>95:5 syn:anti

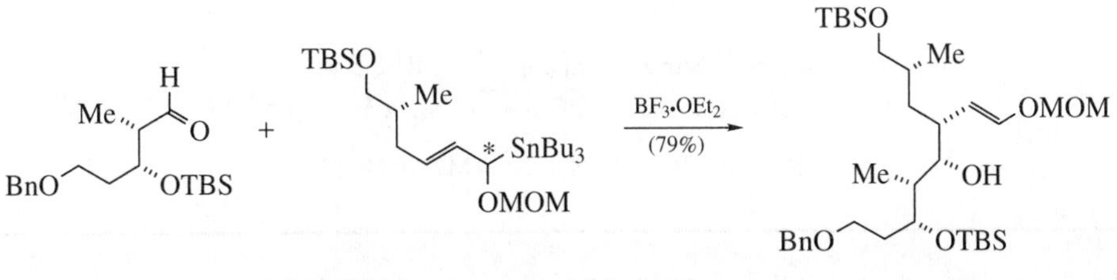

2.5 equivalents (racemic at C*) sole adduct

A β-methyl substituent significantly increases the diastereoselectivity of BF_3-promoted additions to aliphatic aldehydes[127]. The syn-(*E*) adducts are obtained nearly exclusively. However, such stannanes fail to react with aromatic aldehydes:

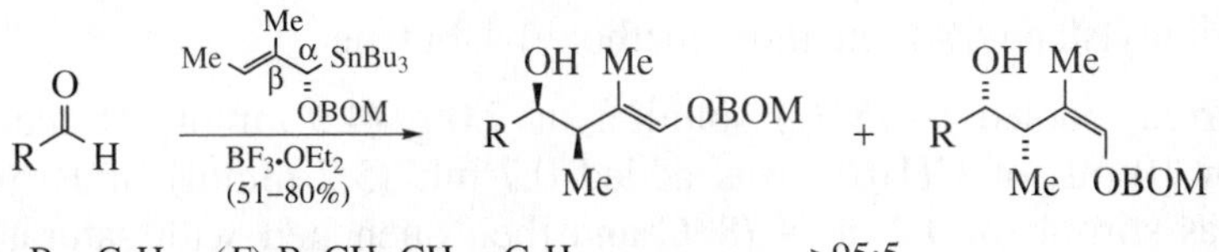

R = C_6H_{13}, (*E*)-BuCH=CH, *c*-C_6H_{11} >95:5

The β-unsubstituted analogues of the foregoing stannanes yield syn-(*Z*) adducts in BF_3-promoted additions to benzaldehyde. Mixtures of syn-(*E*) and syn-(*Z*) products are obtained with aliphatic aldehydes. The differing selectivities for aliphatic and aromatic aldehydes have been attributed to the stronger BF_3 complex of the former which increases the ionic character of the reaction and leads to an enhanced inside alkoxy effect. Aliphatic aldehydes form looser complexes with BF_3 and the product distribution of those reactions is controlled mainly by steric factors:

> outside alkoxy   inside alkoxy

α-Alkoxy allylic stannanes are converted to the γ-oxygenated isomers by Lewis acids in the absence of an aldehyde. These isomerizations can be effected by various Lewis acids including $BF_3 \cdot OEt_2$, Bu_3SnOTf, TBSOTf, and $LiClO_4$ in Et_2O. Stronger Lewis acids such as $TiCl_4$, $SnCl_4$, $AlCl_3$, and $ZnCl_2$ cause decomposition of the stannane. With the former Lewis acids the isomerizations appear to be irreversible leading to the (Z)-enol ethers with inversion of stereochemistry:

$$R^1 = Me, Bu, c\text{-}C_6H_{11}; \ R^2 = MOM, BOM$$

Rearrangement of an α-Alkoxy Allylic Stannane to the γ Isomer

(S,Z)-3-(Tributylstannyl)-1-(methoxymethyl)-1-butene

To a stirred, cooled ($-78\,°C$) solution of 2.0 g (4.9 mmol) of the α-(alkoxy) stannane in 10 mL of CH_2Cl_2 was added 0.7 mL (5.7 mmol) of $BF_3 \cdot Et_2O$. The solution was stirred for 1 h at $-78\,°C$ and then quenched with saturated aqueous $NaHCO_3$ and warmed to $25\,°C$. The phases were separated, and the aqueous phase was extracted with ether. The combined organic phases were washed with brine, dried over anhydrous $MgSO_4$, and concentrated under reduced pressure. Chromatography (elution with hexanes) of the crude product gave 1.6 g (80%) of the γ-(alkoxy) stannane ($[\alpha]_D + 135$ (c 2.0, CH_2Cl_2)).

The isomerization was shown to be an intermolecular reaction. Thus exposure of an equal mixture of an α-OMOM/α-Me$_3$Sn and an α-OCH$_2$OCH$_2$C$_6$H$_4$OMe/α-Bu$_3$Sn allylic stannane to BF$_3$•OEt$_2$ led to an equal mixture of the four possible γ-oxygenated stannanes[128]. This process is essentially irreversible. Treatment of the γ-oxygenated allylic stannanes with BF$_3$•OEt$_2$ gave none of the α-isomers. Furthermore exposure of a mixture of two different γ-oxygenated stannanes led to no exchange in the recovered γ isomers:

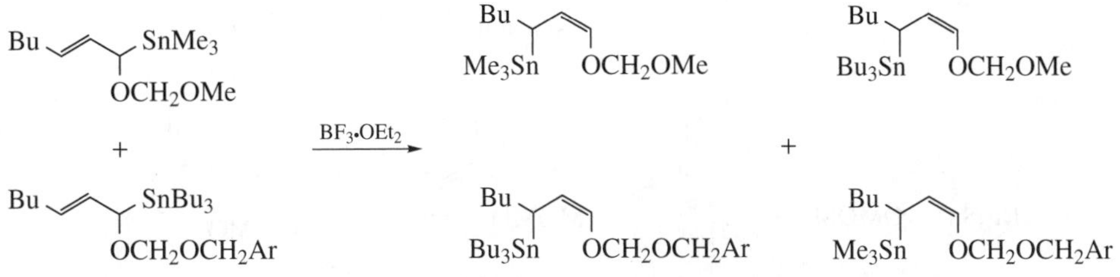

Ar = p-MeOC$_6$H$_4$

(Z)-γ-Silyloxy allylic stannanes can be prepared by Lewis acid-catalyzed isomerization of the α-silyloxy isomers[129]. The (E)-γ-silyloxy isomers are secured through 1,4 addition of the higher order Bu$_3$Sn cuprate to an enal and trapping of the enolate with TBS-Cl.

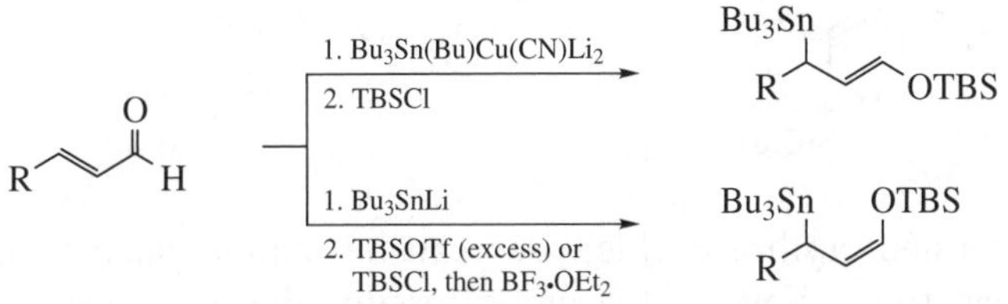

Nonracemic (E)-γ-silyloxy allylic stannanes are available by equilibration of the enantioenriched (Z) isomers with Yb(OTf)$_3$ followed by separation of the (E)/(Z) mixture by chromatography[130]. When BF$_3$•OEt$_2$ is employed as the catalyst the α-OTBS stannane affords only the γ-(Z) isomer (80% yield). Attempted isomerization of the γ-(Z) isomer with BF$_3$•OEt$_2$ leads to decomposition:

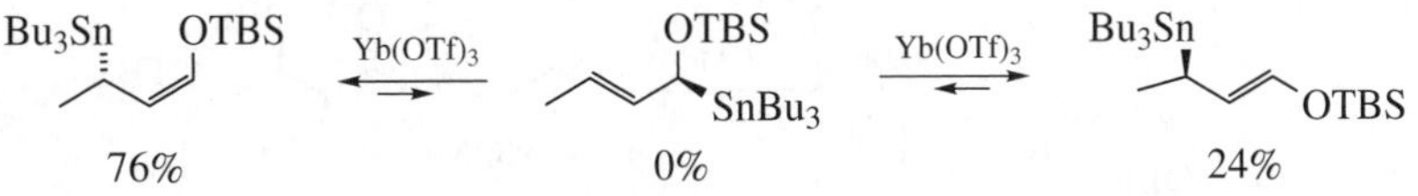

The S$_E$2' additions of γ-oxygenated allylic stannanes to aldehydes are promoted by BF$_3$•OEt$_2$ and, in some cases, MgBr$_2$•OEt$_2$. These additions proceed with high syn:anti diastereoselectivity. The ee of the addition product is equal to that of the starting stannane:

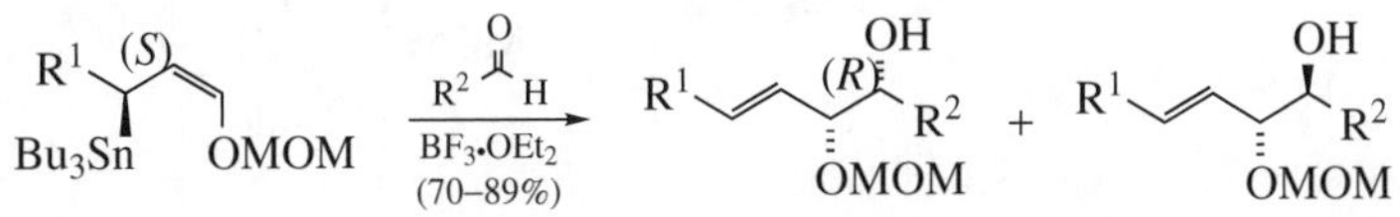

R^1 = Me, Bu

85:15 to 98:2

R^2 = Ph, C$_6$H$_{13}$, c-C$_6$H$_{11}$, (E)-BuCH=CH, BuC≡C

α-Oxygenated aldehydes afford mainly syn,anti adducts with BF₃•OEt₂ and syn,syn adducts with MgBr₂ as the promoter[131]. The former reactions involve Felkin–Ahn transition states, whereas the latter proceed via chelated transition states:

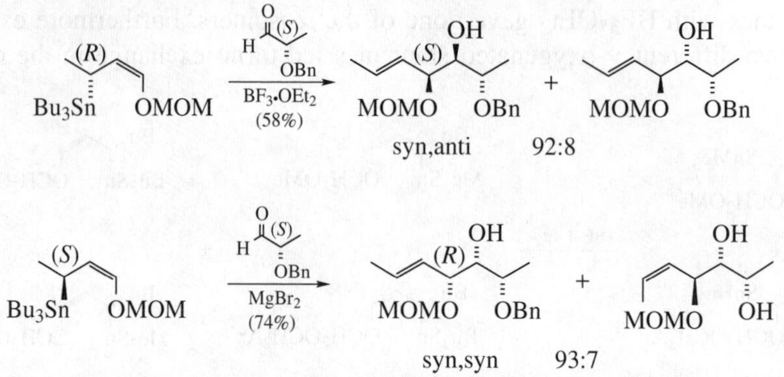

syn,anti   92:8

syn,syn   93:7

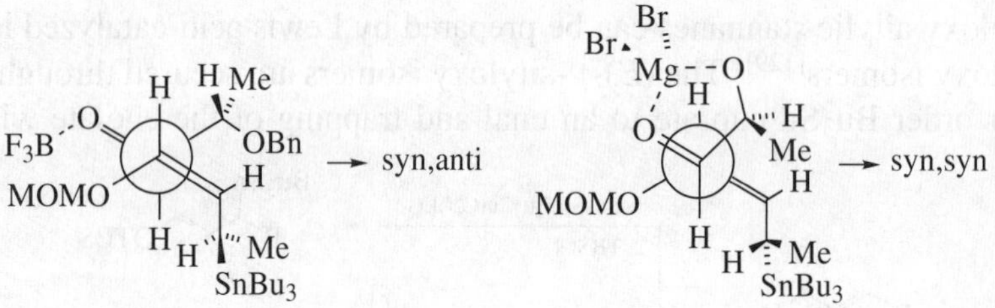

Protected threose and erythrose aldehydes can be homologated to hexose precursors with high stereoselectivity. Four of the eight possible diastereomers can be prepared in this manner[132]. In the MgBr₂-promoted additions, the racemic stannane can be employed in excess. The chelation-controlled transition state leads to a highly favored preference for the matched (R) stannane enantiomer with recovery of enantiomerically enriched (S) stannane:

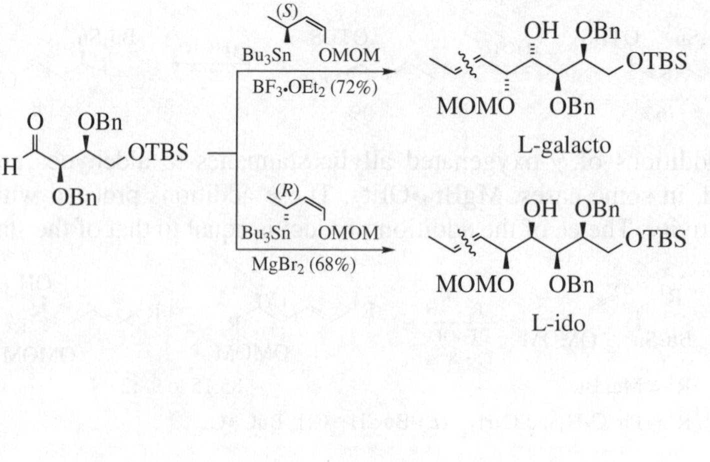

L-galacto

L-ido

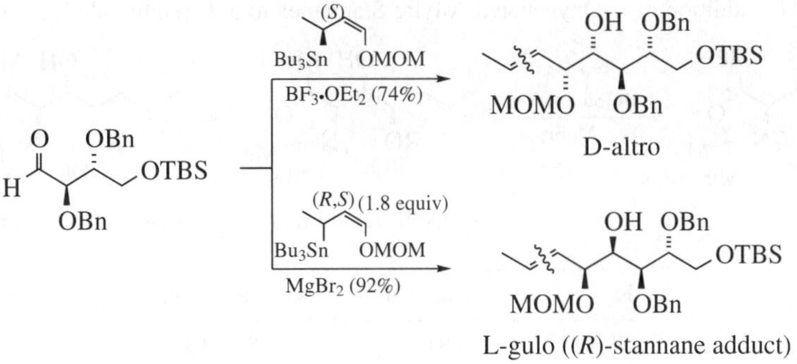

A similar enantioselectivity is observed in related α-oxygenated and α-N-acyl aldehydes. In the former case, a 2.6 : 1 mixture favoring the matched adduct is obtained when the racemic stannane is used in a 3 : 1 molar ratio[133]. The major adduct was converted to bengamide E, a novel marine natural product:

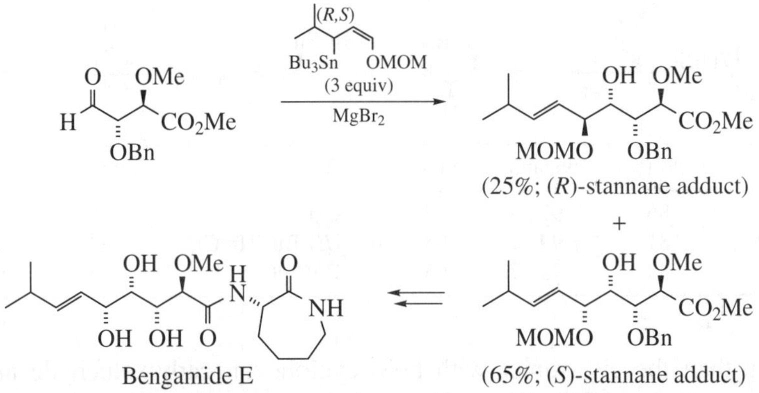

Certain α-N-acyl aldehydes show excellent kinetic discrimination for γ-oxygenated allylic stannanes. With the threonine-derived aldehyde, selectivity is significantly higher for the OTBS derivative compared to the OMOM analogue (Table 27)[134].

The γ-silyloxy allylic stannanes are more highly syn selective than their OMOM or OBOM counterparts (Table 28). The (E) isomers give syn adducts almost exclusively, even with the sterically less demanding alkynyl aldehydes. The additions are stereospecific, with the exception of the (E)-stannane and an alkynyl aldehyde. This reaction appears to proceed by both a syn and anti S_E2' pathway.

6.2.6 Relative Reactivities

Little is known about the relative reactivity of various allylic stannanes toward aldehydes with Lewis acids promoters. trans-Crotyl tributyltin reacts faster and is more

Table 27. Addition of α-Oxygenated Allylic Stannanes to a Threonine-derived Aldehyde.

syn,syn (*R*)-stannane   syn,anti (*S*)-stannane

R	Yield (%)	syn,syn:syn,anti
MOM[a]	84	87 : 13
TBS[a]	87	>99 : 1
TBS[b]	84	>99 : 1

[a]Racemic stannane (2 equivalents).
[b](*R*)-stannane (95% ee).

Table 28. syn:anti Selectivity for Additions of (*Z*)- and (*E*)-γ-Silyloxy Allylic Stannanes to Aldehydes.

R	Yield (%)	syn:anti	ee (%)	R	Yield (%)	syn:anti	ee (%)
C$_6$H$_{13}$	86	97 : 3	95	C$_6$H$_{13}$	79	>99 : 1	95
(*E*)-BuCH=CH	81	>99 : 1	95	(*E*)-BuCH=CH	83	>99 : 1	95
BuC≡C	84	93 : 7	95	BuC≡C	87	97 : 3	20

syn selective than the cis isomer with both cyclohexanecarboxaldehyde and benzaldehyde in the presence of BF$_3$•OEt$_2$. Although less dramatic, this trend is also seen in
MgBr$_2$-promoted additions to α-benzyloxypropanal. Competitive rate studies in which
one equivalent each of two different allylic stannanes is mixed with one equivalent of
cyclohexanecarboxyaldehyde and BF$_3$•OEt$_2$ at −78 °C in CH$_2$Cl$_2$ reveals the reactivity
order given in Table 29[135].

Evidently, oxygenation at the α or γ position leads to diminished reactivity. This trend
can also be expected for other aldehydes, although the ordering of the various oxygenated
species may differ.

6.2.7   InCl$_3$-promoted Additions to Aldehydes

In contrast to BF$_3$ and MgBr$_2$-promoted reactions, InCl$_3$-promoted additions of crotyl
tributyltin to aldehydes afford mainly anti adducts[136]. The yields and selectivities are
highest with nonconjugated aldehydes. Nonracemic α-oxygenated allylic stannanes afford

Table 29. Relative Reactivity of Allylic Stannanes Toward c-$C_6H_{11}CHO/BF_3\bullet OEt_2$ in CH_2Cl_2 at $-78\,°C$.

Stannane	Relative Reactivity
Me⌒⌒$SnBu_3$	100
⌒⌒$SnBu_3$	32
TBSO Me ⌒$SnBu_3$	14
TBSO⌒⌒$SnBu_3$ Me	11
Me⌒⌒$SnBu_3$ OTBS	6
MOMO Me ⌒$SnBu_3$	3
Me⌒⌒$SnBu_3$ OMOM	2

anti-1,2-diol derivatives. The reaction proceeds with allylic retention and stereochemical inversion, in contrast to the crotyl case which reacts with allylic inversion:

anti:syn 84:16 to 98:2

$R = C_6H_{13}$, ^iPr, c-C_6H_{11}, (E)-BuCH=CH $R = c$-C_6H_{11}, (E)-BuCH=CH

The foregoing results are consistent with a reaction pathway involving allylic transmetalation with allylic and configurational inversion and subsequent addition of a transient allylic indium species to the aldehyde through a cyclic six-center transition state. By this scenario, the crotyl system must undergo allylic isomerization of the initial indium species to the crotyl derivative prior to ensuing addition to the aldehyde. Application of this reaction to protected threose and erythrose aldehydes leads to precursors of four of the eight hexose diastereomers with high enantio and diastereoselectivity[137]:

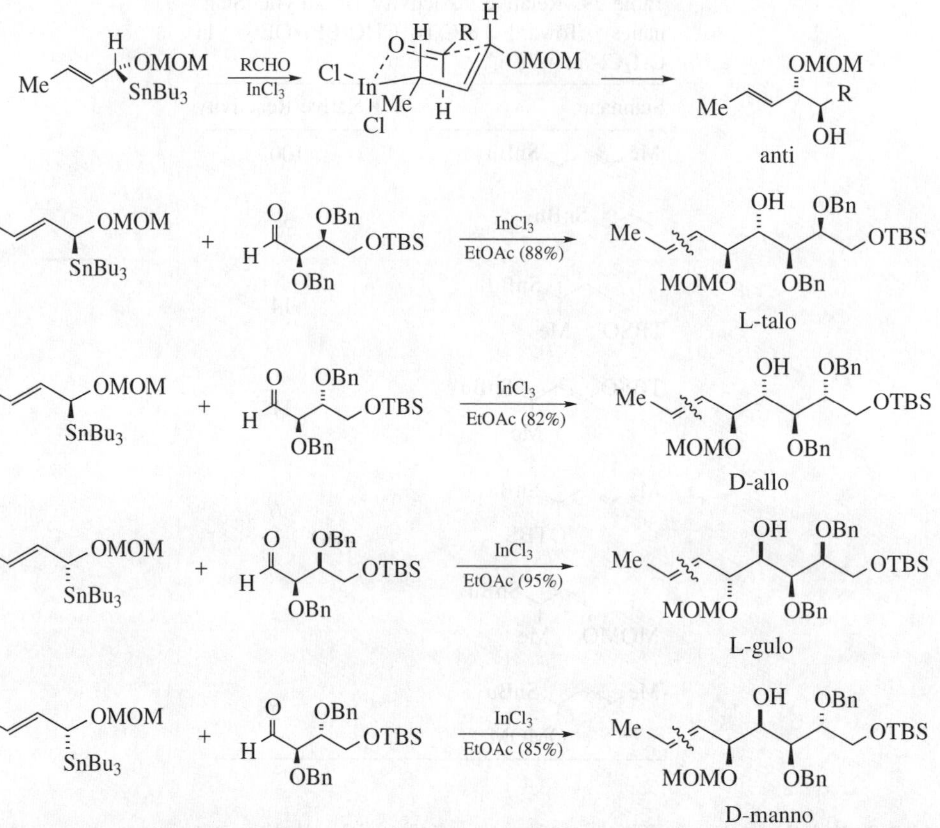

anti

L-talo

D-allo

L-gulo

D-manno

Addition to a tartrate-derived γ-oxygenated dialdehyde forms the basis of a bidirectional synthesis of the C_2 symmetric core unit of certain Annonaceous acetogenins[138]:

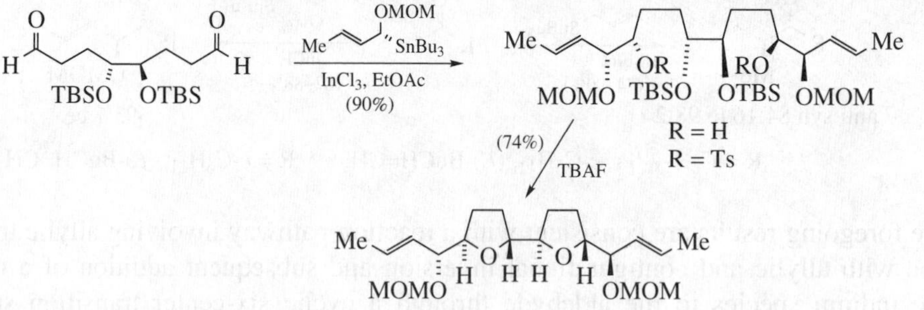

6.2.8 Oxidation

The oxidation of allylic stannanes with peroxycarboxylic acids leads to allylic alcohols with allylic transposition[139]. As the allylic stannanes can be prepared from allylic alcohols, the overall conversion represents a net 1,3-alcohol interchange. An epoxide intermediate (not isolated) has been suggested although a concerted mechanism cannot be ruled out:

$$R\diagdown\diagup OH \xrightarrow{[ref.73]} R\diagdown\diagup SnBu_3 \xrightarrow[(61-66\%)]{ArCO_3H} R\diagdown\diagup\diagdown OH$$

$$R = C_6H_{13}, C_8H_{17}, PhCH_2, PhCH(CH_3)$$

Oxidation of an Allylic Stannane to a Transposed Allylic Alcohol with MCPBA

$$Ph\diagdown\diagup\overset{Me}{\diagdown}\diagup SnBu_3 \xrightarrow[\substack{CH_2Cl_2 \\ (66\%)}]{m\text{-}ClC_6H_4CO_3H} Ph\diagdown\diagup\underset{OH}{\diagdown}\diagup\overset{Me}{\diag=}$$

3-Methyl-1-phenyl-3-buten-2-ol

To a solution of *m*-chloroperoxybenzoic acid (0.35 g, 2.03 mmol) in CH$_2$Cl$_2$ (10 mL) was added dropwise at 25 °C a solution of tributyl-2-methyl-4-phenyl-2-butenylstannane (0.65 g, 1.49 mmol; exothermic reaction). The resultant mixture was heated at reflux for 3 h. The solvent was removed under reduced pressure and the residue was treated with aqueous 1 N HCl (5 mL), and ether (10 mL) with stirring overnight at 25 °C. The ether layer was separated, washed with aqueous NaHCO$_3$, dried over MgSO$_4$, and evaporated. The crude product was purified by column chromatography (silica gel, eluting with hexane and subsequently with ether) and further purified by Kugelrohr distillation (yield 0.16 g, 66%; bp 125 °C, 2 mmHg).

Photooxygenation of nonracemic allylic stannanes proceeds with loss of the allylic hydrogen to afford nonracemic hydroperoxides[140]. The stereochemistry is consistent with a concerted ene reaction. Destannylated products are not observed:

$$C_5H_{11}\diagdown\diagup\overset{SnBu_3}{\diagdown}\diagup CO_2Me \xrightarrow[\substack{Rose\ Bengal \\ (45-58\%)}]{O_2,\ h\nu} C_5H_{11}\diagdown\diagup\overset{HOO\quad SnBu_3}{\diagdown}\diagup CO_2Me$$

$$C_5H_{11}\diagdown\diagup\overset{SnPh_3}{\diagdown}\diagup CO_2Me \xrightarrow[\substack{Rose\ Bengal \\ (60-74\%)}]{O_2,\ h\nu} C_5H_{11}\diagdown\diagup\overset{HOO\quad SnPh_3}{\diagdown}\diagup CO_2Me$$

Direct oxidation of *β*-carbamoyloxy allylic stannanes has been effected with SeO$_2$ or SeO$_2$ as catalyst in ^{t}BuOOH[141]. The reaction proceeds without allylic transposition and affords the (*E*)-vinylic carbamates predominantly. The transformation most likely involves a tin ene reaction followed by 2,3 rearrangement:

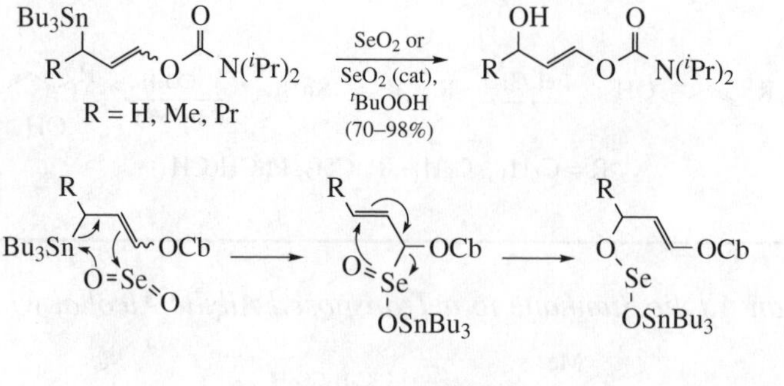

7   Allenic, Propargylic, and Alkynyl Stannanes

7.1   Preparation

7.1.1 S_N2' Displacement of Propargylic Mesylates

The first practical synthesis of allenic tin compounds involved the S_N2' displacement of sulfonates with organotin cuprates[142]. Primary propargylic mesylates favor propargylic stannanes. These may be formed by direct S_N2' displacement:

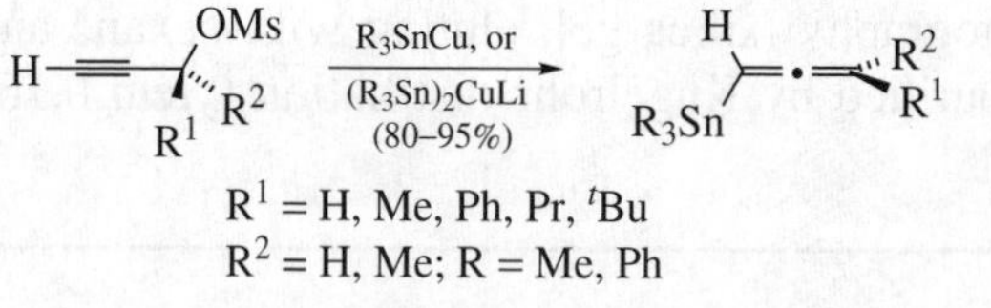

Secondary propargylic mesylates with internal alkynes afford equilibrating mixtures of allenic and propargylic stannanes. The product distribution reflects the relative stabilities of the two isomers which, in turn, is at least partly determined by steric effects. The reaction proceeds with inversion of stereochemistry:

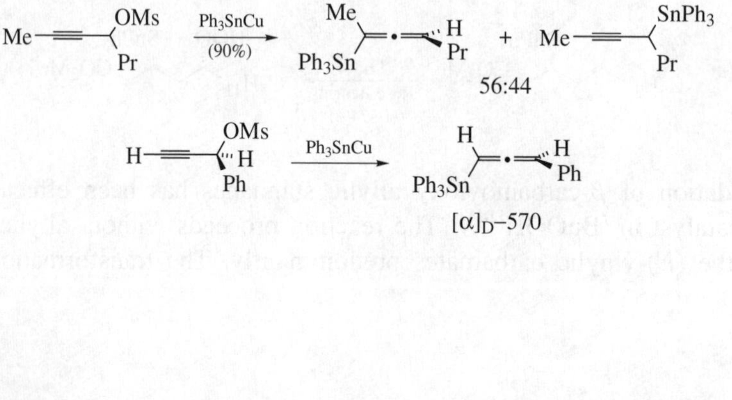

Preparation of an Enantioenriched Allenic Stannane from a Propargylic Alcohol[143]

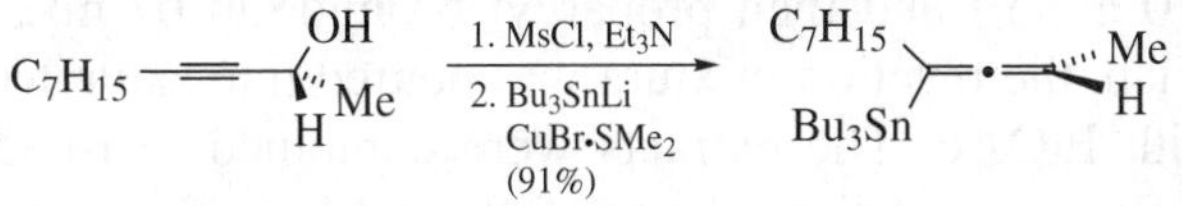

(*P*)-4-(Tributylstannyl)-2,3-undecadiene

To a mixture of 0.079 g (0.47 mmol) of (*R*)-3-undecyn-2-ol and 0.13 mL (0.94 mmol) of Et_3N in 3 mL of CH_2Cl_2 was added 0.06 mL (0.71 mmol) of methanesulfonyl chloride at $-78\,^{\circ}C$. The resulting mixture was stirred at $-78\,^{\circ}C$ for 1 h, then quenched with saturated $NaHCO_3$, and extracted with ether. The ether layer was washed with brine and dried over $MgSO_4$. Concentration under reduced pressure yielded the crude mesylate, which was dried in vacuo and directly used for the next reaction without further purification.

To a solution of 0.09 mL (0.65 mmol) of diisopropylamine in 3 mL of HMPA/THF (1 : 1) was added dropwise 0.24 mL (0.61 mmol) of 2.5 M BuLi at $0\,^{\circ}C$. After 30 min, 0.16 mL (0.60 mmol) of Bu_3SnH was added to the mixture. The resulting mixture was stirred at $0\,^{\circ}C$ for 15 min, then 0.150 g (0.47 mmol) of the mesylate in 3 mL of THF was added dropwise during 10 min. The reaction mixture was quenched with saturated $NaHCO_3$ and extracted with ether. The extracts were dried over $MgSO_4$ and concentrated. The residue was chromatographed on silica gel (hexane) to yield 188 mg (91%) of allenylstannane ($[\alpha]_D + 88.2$ ($CHCl_3$, c 0.90)).

7.1.2 In situ Coupling of Allenic Grignard Reagents with Alkyltin Halides

Direct coupling of propargylic bromides with Bu_3SnCl can be effected with Mg in the presence of catalytic $PbBr_2$. The reaction is compatible with an alkynyl ester function:

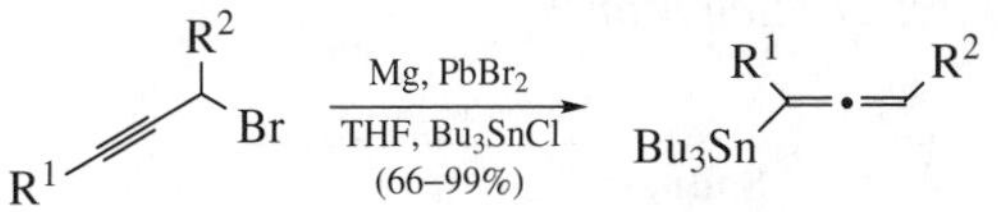

$R^1 = H, CO_2Me; R^2 = H, Me$

Magnesium-promoted Coupling of a Propargylic Bromide with Bu_3SnCl

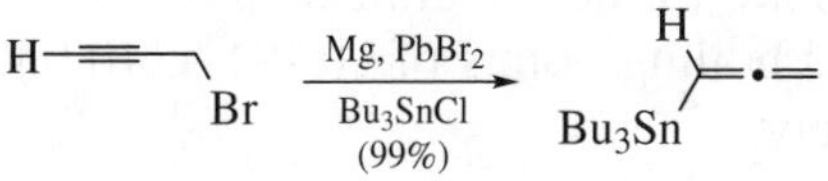

Tributyl-1,2-propadienylstannane

To a stirred suspension of Mg (0.29 g, 12 mmol) and $PbBr_2$ (5 mol%, 0.22 g, 0.6 mmol) in THF (20 mL) were added at ambient temperature tributylstannyl chloride (2.71 mL, 10 mmol) and then propargyl bromide (1.07 mL, 12 mmol). After being stirred for 1 h, the reaction mixture was poured into saturated aqueous NH_4Cl and extracted with EtOAc. The extracts were combined, washed with brine, and dried (Na_2SO_4). Removal of the solvents followed by column chromatography on silica gel using hexane as an eluent afforded the stannane (3.26 g, 99%) as a colorless liquid.

Transmetalation of Allenylmagnesium Bromide with Bu_2SnCl_2 [144]

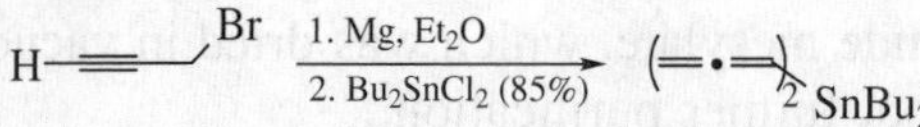

Dibutyl Diallenyltin

Allenylmagnesium bromide was obtained by dropwise addition (4 h) of an ether solution of propargyl bromide (165 g, 1.38 mol; 400 mL of diethyl ether) to Mg turnings (70 g, 2.86 mol) amalgamated with mercury chloride. The Grignard reagent solution was separated and a solution of dibutyltin dichloride (70 g, 0.23 mol) in 250 mL of diethyl ether was added to it during 2 h. After hydrolysis with ice the ethereal layer was separated and dried over Na_2SO_4. Most of the solvent was distilled and 100 mL of methanol was added. The solution was refluxed for 10 min in order to isomerize the product to the allenic form. Distillation of the residue under reduced pressure gave 61 g (85%) of diallenyldibutyltin (bp 88–90 °C at 0.1 mmHg).

Redistribution of Dibutyl Diallenyltin[•]

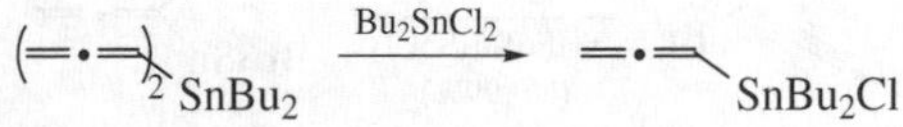

Dibutyl Allenyltin Chloride

Equimolar amounts (30 mmol) of dibutyl diallenyltin and dibutyltin dichloride were stirred together at 25 °C for 2 h. Several batches of this compound were prepared and used without further treatment. Two batches were distilled under reduced pressure and had boiling points of 90–91 °C (0.05 mmHg) and 83–83.5 °C (0.04 mmHg), respectively.

Alkynylstannanes can be prepared from terminal alkynes through lithiation with BuLi followed by addition of R_3SnCl (R = Me or Bu). A milder approach entails heating the terminal alkyne with Bu_3SnOMe[•] or Bu_3SnNMe_2[147]:

$$Bu_3SnX + H\text{---}\!\!\equiv\!\!\text{---}R \xrightarrow[\text{reflux}]{\text{toluene}} Bu_3Sn\text{---}\!\!\equiv\!\!\text{---}R + HX$$

$$X = OMe \text{ or } NMe_2$$

7.2 Reactions

7.2.1 Uncatalyzed Additions to Chloral

Propargyl and allenylstannanes react with chloral in CCl_4 to afford the S_N2' products[148]. The relative rates are dependent upon the stannane substituent with $Me_3Sn > Et_3Sn > Bu_3Sn > Ph_3Sn$. For the Ph_3Sn derivatives the following relative reactivities are observed: $Ph_3SnCH(Me)C\equiv CH$ (15) > $Ph_3SnCH_2C\equiv CH$ (6.5) > $Ph_3SnCH=C=CH_2$ (1.0). Under the foregoing reaction conditions the Me-substituted derivatives, $Ph_3SnCH_2C\equiv CMe$ and $Ph_3SnCH=C=CHMe$, give no adducts with chloral in CCl_4 after 8 days. An evaluation of the reaction parameters indicates that the additions proceed by a cyclic concerted transition state. Interestingly, high concentrations of chloral cause isomerization of the propargyl and allenic stannanes. The equilibrium favors the unbranched isomer with allenic preferred over propargylic when both are unbranched:

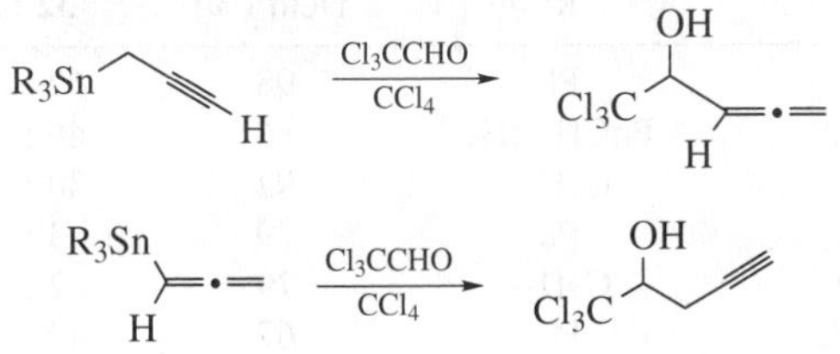

Addition of various mixtures of allenic and propargylic stannane isomers to chloral under equilibrating conditions leads to the same mixture of propargylic and allenic products. Thus equilibration is fast relative to addition. However, under nonequilibrating conditions (low concentration), it is possible to produce a single isomeric adduct derived from either the allenic or propargylic stannane:

$$Me_3SnCH_2C\equiv CH \qquad\qquad CH_2=C=CH(OH)CHCCl_3$$
$$\textbf{28} \qquad\qquad\qquad\qquad\qquad \textbf{30}$$

$$\Big\updownarrow \qquad \xrightarrow{Cl_3CCHO} \qquad +$$

$$CH_2=C=CHSnMe_3 \qquad\qquad Cl_3CH(OH)CH_2C\equiv CH$$
$$\textbf{29} \qquad\qquad\qquad\qquad\qquad \textbf{31}$$

$$\textbf{28:29} = 5:95 \text{ to } 90:10 \qquad\qquad \textbf{30:31} = 33:67$$

7.2.2 Additions of Propargyl and Allenyltin Halides to Aldehydes

Various procedures have been developed for the in situ formation of allenic and propargylic halostannanes from propargylic halides and stannous salts. Subsequent addition of aldehydes leads to the adducts[149]. The ratio of regioisomeric products depends on the structure of the starting halide and the aldehyde (Table 30). In most cases, the allenic and propargylic stannanes are interconverting and the ratio of the derived products will thus depend upon the relative transition state energies of the two addition reactions.

Table 30. Addition of Halostannanes Derived from Primary Propargylic Iodides to Aldehydes.

R^1	R^2	Yield (%)	**32 : 33**
H	Ph	98	52 : 48
H	$PhCH_2CH_2$	86	46 : 54
H	C_8H_{17}	92	30 : 70
Me	Ph	79	3 : 97
Me	C_8H_{17}	79	2 : 98
TMS	Ph	67	40 : 60
TMS	C_8H_{17}	76	0 : 100
CO_2Me	Ph	55	0 : 100

[a]DMI = 1,3-dimethyl-2-imidazolidinone.

Formation of a Halostannylallene and Addition to Benzaldehyde

1-Phenyl-3-pentyn-1-ol

A suspension of stannous chloride (0.209 g, 1.1 mmol), 1-bromo-2-butyne (0.133 g, 1.0 mmol) and sodium iodide (0.165 g, 1.1 mmol) in 2 mL of 1,3-

dimethyl-2-imidazolidinone (DMI) was stirred for 1 h at 25 °C. The mixture was then cooled to 0 °C and benzaldehyde (0.085 g, 0.8 mmol) in DMF (1 mL) was added dropwise. After the mixture was stirred at this temperature for 11 h, water was added and the mixture was extracted with ether. The organic layer was washed with water, dried over anhydrous MgSO₄ and the solvent was removed under reduced pressure. The residue was subjected to silica gel column chromatography yielding a 3 : 97 mixture of 1-phenyl-2-methylbuta-2,3-dien-1-ol and 1-phenyl-3-pentyn-1-ol in 79% yield.

A second in situ method involves the use of metallic tin and aluminum to convert propargyl bromide to the diallenic tin dibromide which reacts subsequently with added aldehydes and ketones to afford homopropargylic alcohols (Table 31)[150]. Similarly, TMS propargyl iodide is converted to the propargylic stannane with tin and aluminum (Table 32). Subsequent addition of aldehyde affords a mixture of propargyl and allenyl adducts whose composition is highly solvent dependent.

Table 31. Addition of bis-Allenyltin Dibromide to Aldehydes and Ketones.

R¹	R²	Yield (%)
Ph	H	94
C₆H₁₃	H	77
(E)-MeCH=CH	H	72
CH₂CH₂CH₂CH₂CH₂		90
Me	C₆H₁₃	91

Table 32. Addition of bis-TMS Allenyl/Propargyltin Diiodide to Aldehydes.

R	Solvent	Yield (%)	34 : 35
C₆H₁₃	MeCN/DMSO	89	95 : 5
C₆H₁₃	MeCN	66	30 : 70
C₆H₁₃	(MeOCH₂)₂	75	8 : 92
PhCHO	MeCN/DMSO	88	90 : 10
PhCHO	(MeOCH₂)₂	69	11 : 89
(E)-MeCH=CH	MeCN/DMSO	60	92 : 8
(E)-MeCH=CH	(MeOCH₂)₂	64	9 : 91

Formation of TMS Propargyltin Iodide and Addition to Heptanal

$$TMS-\!\!\!\equiv\!\!\!-\diagup^{I} \xrightarrow[\text{2. } C_6H_{13}CHO\ (88\%)]{\substack{\text{1. Sn, Al} \\ \text{MeCN, DMSO}}} TMS\!\!\!=\!\!\!\bullet\!\!\!=\!\!\!\diagdown_{HO}^{TMS}\!\!\!-C_6H_{13} \quad + \quad TMS-\!\!\!\equiv\!\!\!\diagdown_{HO}\!\!\!-C_6H_{13}$$

$$95:5$$

3-(Trimethylsilyl)-1,2-decadien-4-ol

To a stirred suspension of metallic tin (1.6 mmol) and aluminum, (1.6 mmol) in a mixed solvent (4 mL of MeCN, 0.3 mL of DMSO) was added the propargylic iodide (2.1 mmol) at 25 °C, and stirring was continued for 1 h. Heptanal (1.0 mmol) was added to the reaction mixture, and stirring was continued for an additional 1 h at 25 °C. Water was added and the product was extracted with ether. After removal of solvent, the crude allenylcarbinol (88% of a 95 : 5 mixture) was purified by column chromatography on silica gel.

Allenylmagnesium bromide is transmetalated by Bu_2SnCl_2 to yield the diallenyltin compound. This undergoes a redistribution with a second mole of Bu_2SnCl_2, resulting in chloro dibutyl allenyltin. Addition of this reagent to aldehydes can be effected in water to afford homopropargylic alcohols as the major adducts (Table 33):

$$\equiv\!\!\!\diagup^{Br} \xrightarrow[\text{2. } Bu_2SnCl_2\ (85\%)]{\text{1. Mg, Et}_2O} Bu_2Sn\!\!\diagdown\!(\!=\!\!\bullet\!\!=\!)_2 \xrightarrow{Bu_2SnCl_2} 2\ Bu_2(Cl)Sn\diagup\!\!=\!\!\bullet\!\!=$$

The addition of allenyl tributyltin to aldehydes in the presence of Bu_2SnCl_2 can also be carried out in water (Table 34). The products are mixtures of homopropargyl and allenyl isomers whose composition depends upon the reactivity of the aldehyde. Reactive aldehydes favor the allenic product and vice versa.

This outcome can be understood in terms of a rapid transmetalation followed by a slower isomerization. The more reactive aldehydes intercept the initially formed propargylic stannane before it appreciably isomerizes to the more stable allenic isomer. With

Table 33. Addition of (Allenyl)dibutyltin Chloride to Aldehydes in Water.

$$\underset{R}{\overset{O}{\|}}\!\!\diagdown_{H} + Bu_2(Cl)Sn\diagup\!\!=\!\!\bullet\!\!= \xrightarrow{H_2O} R\diagdown^{OH}\!\!\!\diagup\!\!\!\equiv + R\diagdown^{OH}\!\!\!\diagup\!\!=\!\!\bullet\!\!=$$

$$\textbf{36} \qquad\qquad \textbf{37}$$

R	Yield (%)	**36** (%)	**37** (%)
Et	100	90	10
^iPr	100	95	5
^tBu	98	95	5
(E)-MeCH=CH	98	90	10

Table 34. Transmetalation of (Allenyl)tributyltin with Bu_2SnCl_2 and in situ Addition to Aldehydes.

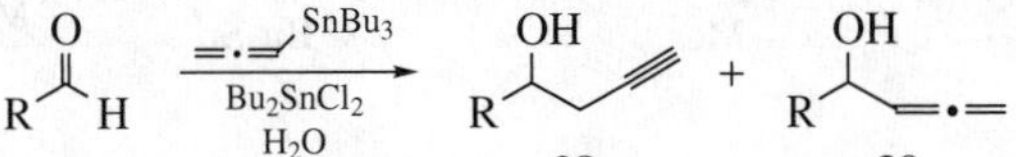

R	Yield (%)	**38 : 39**
Me	95	25 : 75
^iPr	97	24 : 76
^tBu	95	40 : 60
$CH_2=CH$	98	90 : 10
$CH_2=C(CH_3)$	90	65 : 35
(E)-MeCH=CH	95	95 : 5

the less reactive aldehydes, the major product is derived from the more stable allenic stannane:

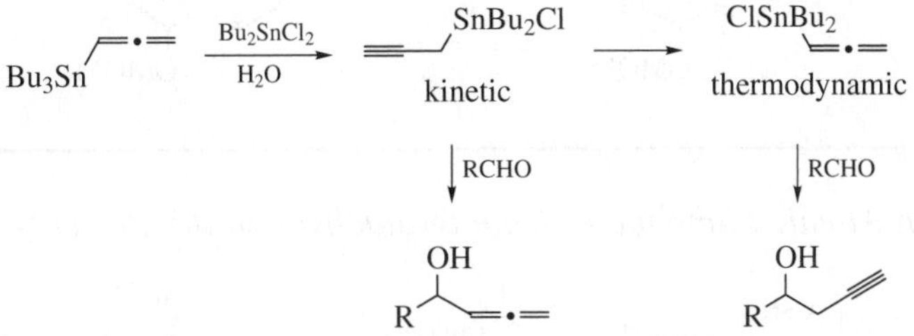

7.2.3 Additions of Preformed Allenyltin Derivatives to Aldehydes

Addition of tributyltin cuprates to propargylic tosylates or mesylates affords allenic tin reagents. These add to aldehydes in the presence of $BF_3 \cdot OEt_2$ or $MgBr_2$ to produce mixtures of syn and anti adducts (Table 35). Diastereoselectivity is highest with α-branched aldehydes:

Table 35. Addition of an (Allenyl)tributyltin Derivative to Aldehydes Promoted by $BF_3 \cdot OEt_2$ and $MgBr_2$.

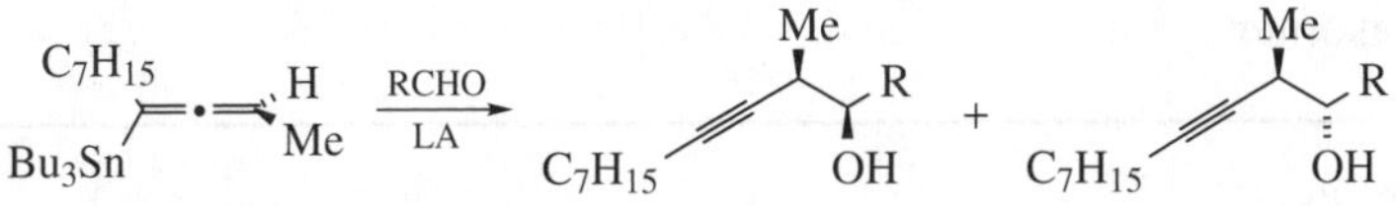

R	LA	Yield (%)	**40 : 41**
C_6H_{13}	$BF_3 \cdot OEt_2$	83	37 : 63
C_6H_{13}	$MgBr_2$	56	69 : 31
^iPrCHO	$BF_3 \cdot OEt_2$	80	99 : 1
^iPrCHO	$MgBr_2$	48	88 : 12
Bu	$BF_3 \cdot OEt_2$	92	99 : 1

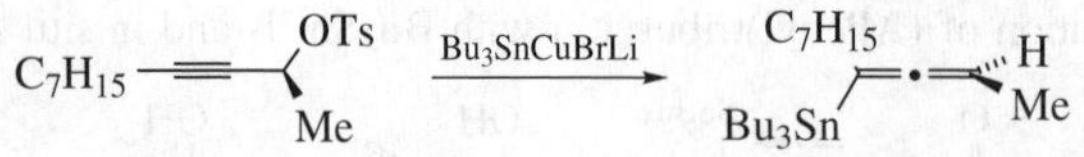

An intramolecular version of this reaction leads to cyclic homopropargylic alcohols of 12–14 members in high yield[151]:

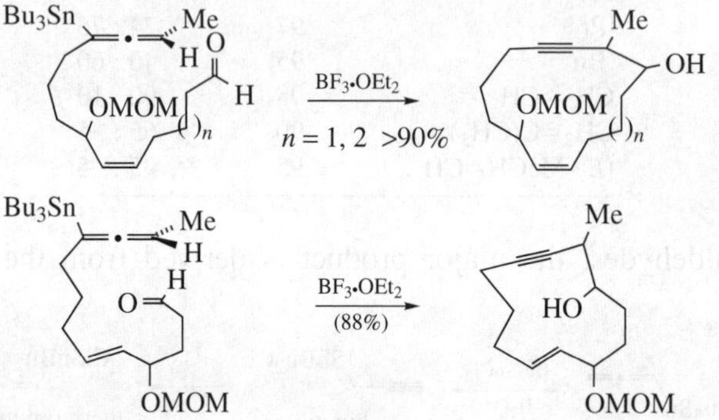

An intramolecular version of this reaction leads to cyclic homopropargylic alcohols of 12–14 members in high yield:

Addition of an Allenic Stannane to Isobutyraldehyde in the Presence of BF₃•OEt₂

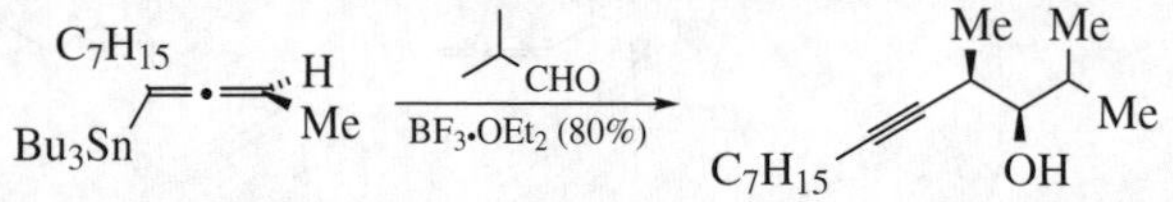

syn-2,4-Dimethyl-5-tridecyn-3-ol

To a solution of 0.20 mL (1.68 mmol) of $BF_3 \cdot OEt_2$ in 8 mL of CH_2Cl_2 was added dropwise a mixture of 0.180 g (0.40 mmol) of allenic stannane and 0.10 mL (1.1 mmol) of isobutyraldehyde in 3 mL of CH_2Cl_2 at $-78\,°C$. The mixture was stirred at $-78\,°C$ for 30 min, quenched with saturated $NaHCO_3$, and extracted with ether. The ether layer was dried over $MgSO_4$ and concentrated. The residue was chromatographed on silica gel (hexane/ether, 4 : 1) to yield 0.072 g (80%) of alcohol as a single isomer.

Enantioenriched propargylic mesylates are converted to allenic stannanes with inversion of configuration upon S_N2' displacement with a tributyltin cuprate[152]. Subsequent additions of these stannanes to enantioenriched 2-benzyloxypropanal show double stereodifferentiation with $BF_3 \cdot OEt_2$ or $MgBr_2$ as the promoter. In the latter case, the facial selectivity is the result of chelation control. A similar effect is seen with enantioenriched 3-benzyloxy-2-methylpropanal. The matched pairing affords the syn,anti adduct **48** nearly exclusively for both BF_3- and $MgBr_2$-promoted additions[153]:

These additions, like their allylic counterparts, proceed through acyclic transition states. The syn,anti adduct **48** could result from chelation control in the MgBr$_2$ reaction with the (*P*)-allenic stannanes. However, the syn,syn adduct **46**, from the (*M*)-allenic stannane, must be formed through a Felkin–Ahn transition state:

Addition of an Allenic Stannane to an Aldehyde Promoted by BF$_3$•OEt$_2$

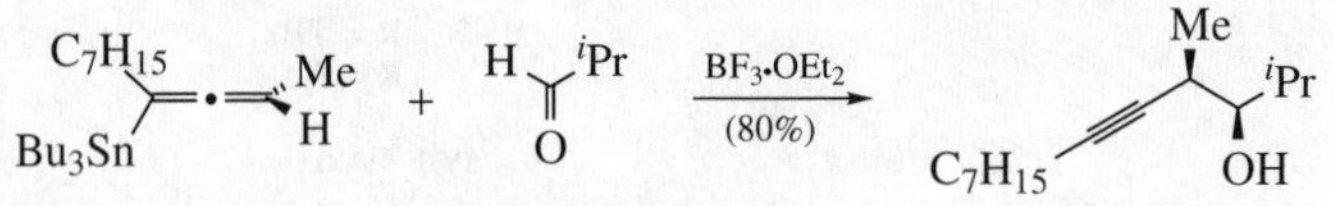

(*R,R*)-2,4-Dimethyl-5-tridecyn-3-ol

To a solution of 0.20 mL (1.35 mmol) of BF$_3$•OEt$_2$ in 3 mL of CH$_2$Cl$_2$ was added dropwise a mixture of 0.180 g (0.40 mmol) of stannane and 0.10 mL (1.12 mmol) of isobutyraldehyde in 3 mL of CH$_2$Cl$_2$ at $-78\,°$C. The mixture was stirred at $-78\,°$C for 30 min, then quenched with saturated NaHCO$_3$, and extracted with ether. The ether layer was dried over MgSO$_4$ and concentrated. The residue was chromatographed on silica gel (hexane/ether, 4 : 1) to yield 0.072 g (80%) of alcohol adduct as a single isomer ($[\alpha]_D + 3.2$ (CHCl$_3$, *c* 1.05)).

Transmetalation of allenic tributyltin reagents with SnCl$_4$ leads initially to the propargylic trichlorotin derivative[154]. If an aldehyde is present, subsequent addition affords the allenylcarbinol adduct. In the absence of an aldehyde, the propargylic tin intermediate converts to the allenic isomer which reacts with added aldehyde to yield the homo-propargylic alcohol product. In both cases the addition reaction proceeds through a cyclic transition state:

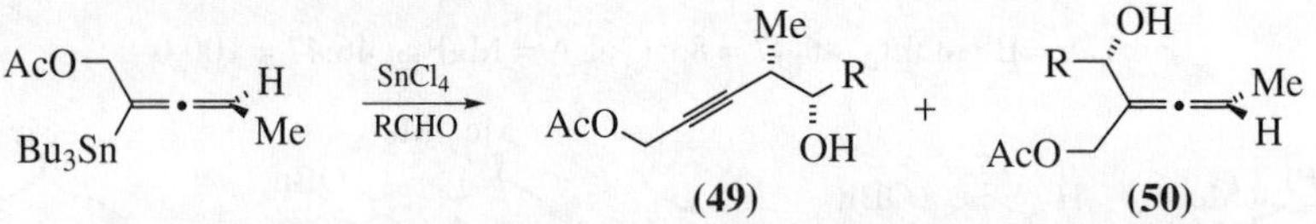

Add SnCl$_4$ at 0 °C, cool to -78 °C, then add aldehyde, **49:50** = 100:0
Add SnCl$_4$ at -78 °C, wait 10 min., then add aldehyde, **49:50** = 0:100

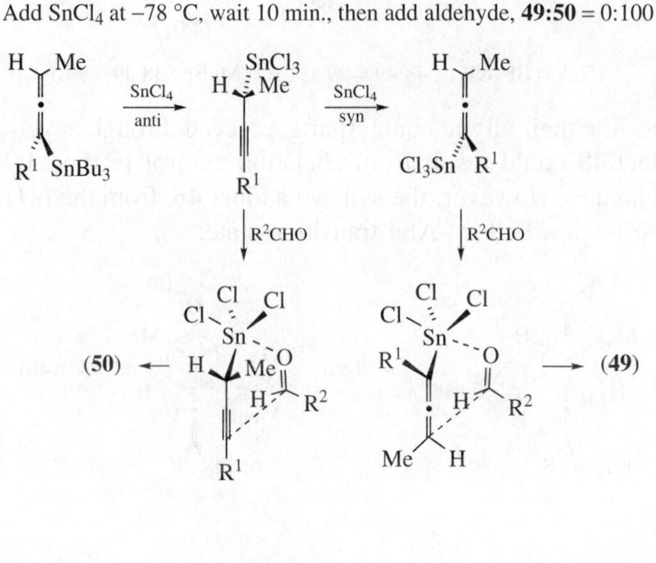

Lithiation of the methyl ether of TMS propargyl alcohol with ^{t}BuLi followed by quenching with Ti(O-^{i}Pr)$_4$ and subsequent addition of Bu$_3$SnCl affords the propargyl-stannane derivative in 89% yield[155]. Quenching the lithiated species with Et$_2$AlCl then addition of Bu$_3$SnCl gives the allenyltin product in 83% yield. Both stannanes react with cyclohexanecarboxaldehyde after conversion to the allenyltitanium derivatives, to afford the anti propargylic 1,2-diol derivative:

$$R = c\text{-}C_6H_{11}$$

7.2.4 Additions of Alkynyltin Derivatives to Aldehydes

Additions of alkynylstannanes to aldehydes can be carried out with InCl$_3$ as a catalyst (Table 36). Regeneration of the InCl$_3$ is effected by stoichiometric Me$_3$SiCl. Of the three solvents examined, MeCN, toluene, and CH$_2$Cl$_2$, the former proved most effective. Both aromatic and aliphatic aldehydes can be alkynylated by this procedure. The propargylic alcohol products were isolated as their TMS ethers by distillation.

Table 36. Transmetalation of (Alkynyl)tributyltin with InCl$_3$ and in situ Addition toAldehydes.

R^1	R^2	T (°C)	Yield (%)
Ph	Ph	25	92
Ph	Pr	−10	100
Et	Ph	25	88
^{t}Bu	Pr	25	100
p-NO$_2$C$_6$H$_4$	Ph	25	68
p-MeC$_6$H$_4$	Ph	25	76
(E)-MeCH=CH	Ph	25	79

Addition of a Bu₃Sn Alkyne to an Aldehyde Catalyzed by InCl₃

$$PhCHO \ + \ Bu_3Sn\!\!-\!\!\equiv\!\!-\!\!Ph \ \xrightarrow[\text{MeCN, TMS-Cl}]{\text{InCl}_3 \ (10 \text{ mol \%})} \ Ph\overset{\text{OTMS}}{\underset{Ph}{\diagdown}}$$

1,3-Diphenyl-2-propyn-1-ol TMS Ether

A mixture of InCl₃ (0.2 mmol) and (phenylethynyl)tributyltin (2.0 mmol) in aceto-nitrile (2 mL) was stirred for 5 min at 25 °C and benzaldehyde (2.0 mmol) and Me₃SiCl (2.0 mmol) were subsequently added at ambient temperature. After 30 min the reaction mixture was washed with aqueous NH₄F to remove Bu₃SnCl, and extracted with ether. The product was isolated in 92% yield by distillation (100 °C at 0.03 mmHg).

8   Palladium-catalyzed Coupling of Organostannanes (Stille Couplings)

A recent volume of *Organic Reactions* provides an in depth review of this subject[156]. The present coverage will therefore only focus on the main categories of these coupling reactions.

8.1   Mechanistic Considerations

Palladium-catalyzed couplings of organic halides and organotin compounds proceed by a three-step oxidative addition, transmetalation, reductive-elimination cycle, typical of other transition metal-mediated cross-coupling reactions. In step 1, Pd^0Ln_n inserts into the R−X bond to form a Pd^{II} species. In the ensuing transmetalation step, one of the tin substituents is transferred to Pd^{II} with exchange of X, forming R_3SnX and a diorgano Pd^{II} species, which undergoes reductive elimination with regeneration of Pd^0Ln_n:

$$1. \quad PdLn_n \ + \ R^1X \ \xrightarrow{\text{oxidative-addition}} \ R^1PdXLn_n$$

$$2. \quad R^1PdXLn_n \ \xrightarrow[\text{transmetalation}]{Bu_3SnR^2} \ Bu_3SnX \ + \ R^1PdR^2Ln_n$$

$$3. \quad R^1PdR^2Ln_n \ \xrightarrow{\text{reductive-elimination}} \ R^1-R^2 \ + \ PdLn_n$$

A Pd^{II} species can also serve as the catalyst. In that case, reduction to Pd^0 by a portion of the R_3SnR^1 reactant generates the active Pd^0 catalyst:

$$2R_3SnR^1 + PdX_2 \xrightarrow{\text{Ln}} R^1 \text{—} R^1 + 2R_3SnX + PdLn_n$$

8.2 Tetraalkylstannane Couplings

Benzyl and aryl halides couple with Me_4Sn and Bu_4Sn to afford the alkylated benzyl or phenyl derivatives[157]. The reaction is conducted in hexamethylphosphoric triamide (HMPA) or *N*-methylpyrrolidone (NMP) with $BnPd(PPh_3)_2Cl$ as the catalyst:

$$R_4Sn + BrCH_2Ar \xrightarrow[\text{HMPA}]{BnPd(PPh_3)_2Cl} RCH_2Ar + R_3SnBr$$

R = Me, Ar = Ph (82%); R = Bu, Ar = Ph (42%)

$$R_4Sn + BrAr \xrightarrow[\text{HMPA}]{BnPd(PPh_3)_2Cl} RAr + R_3SnBr$$

R = Me; Ar = Ph (89%), *p*-MeC_6H_4 (84%),

p-$MeOC_6H_4$ (85%), *p*-AcC_6H_4 (95%)

R = Bu; Ar = Ph (42%)

A considerable improvement in coupling efficiency is realized with alkyl-substituted 1-aza-5-stannabicyclo[3.3.3]undecane[158]. Both diphenylphosphinoferrocene•$PdCl_2$ and $Pd(PPh_3)_4$ serve as catalysts for couplings with aryl bromides:

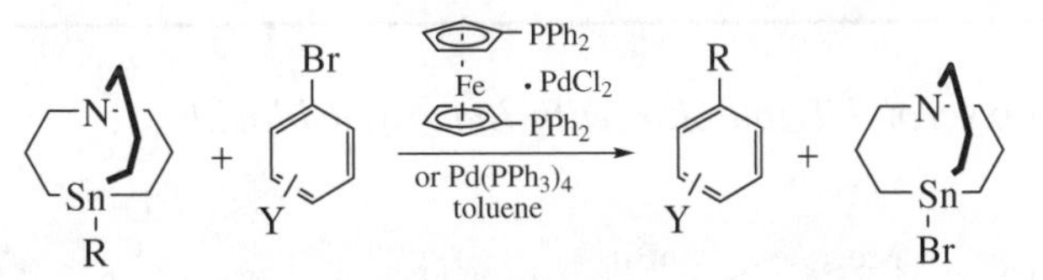

R = Me; Y = *p*-MeO (94%), *p*-Me_2N (56%), *m*-NO_2 (93%)
R = Bu; Y = *p*-MeO (64%), *m*-NO_2 (86%)
R = CH_2OMOM; Y = *p*-MeO (61%), *p*-Me_2N (63%), *m*-NO_2 (80%)

Both (*E*)- and (*Z*)-1-iodo-1-heptene also undergo efficient coupling. The reaction proceeds with retention of the alkene stereochemistry. The enhanced reactivity of these tin reagents can be attributed to a transannular interaction between the electron donor nitrogen and the tin which serves to increase the length of the Sn—R bond:

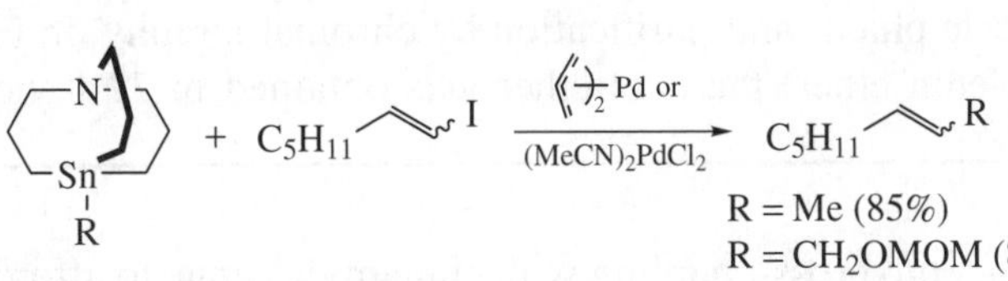

R = Me (85%)
R = CH_2OMOM (81%)

Cyanomethylation of aryl bromides can be effected through Pd-catalyzed coupling with cyanomethyl tributyltin[159]. The reaction is not successful with aryl bromides bearing strongly electron-withdrawing groups (COMe, CN, NO_2):

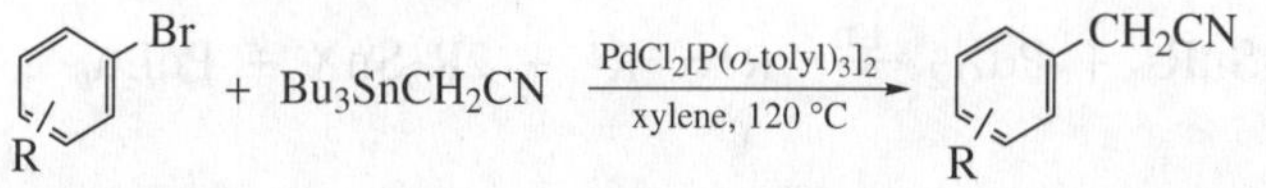

R = o-Me (74%), m-Me (74%), p-Me (78%), o-MeO (70%), p-MeO (77%),
o-Cl (67%), p-Cl (66%), p-CN (trace), p-NO$_2$ (trace)

8.3   Allylstannane Couplings

Cross-coupling of allylic stannanes with allylic halides proceeds in only moderate yield and affords mixtures of regioisomers[160]. Vinyl and arylstannanes, however, couple efficiently and regioselectively with allylic halides[161]. The reaction is compatible with numerous functional groups including, ester, aldehyde, nitrile and hydroxyl. The vinylic couplings proceed with retention of stereochemistry. Acylation of α-ethoxy crotyl tributyltin occurs with allylic inversion to afford the acylated enol ethers as mixtures of (E) and (Z) isomers[162]:

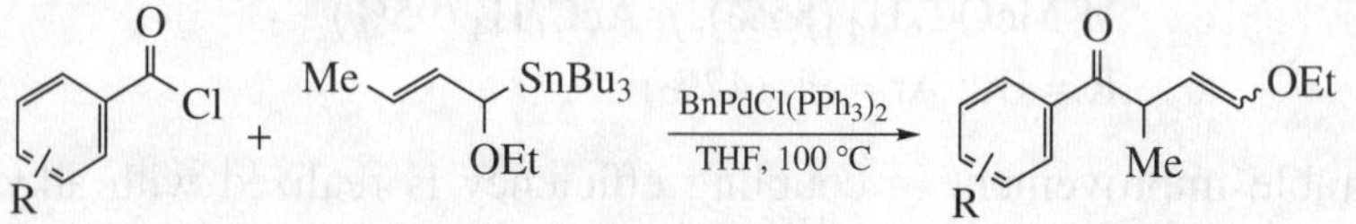

R = H (72%), m-MeO (67%), o-MeO (52%), p-Br (69%)

Coupling of α-Ethoxycrotyl Tributyltin with Benzoyl Chloride

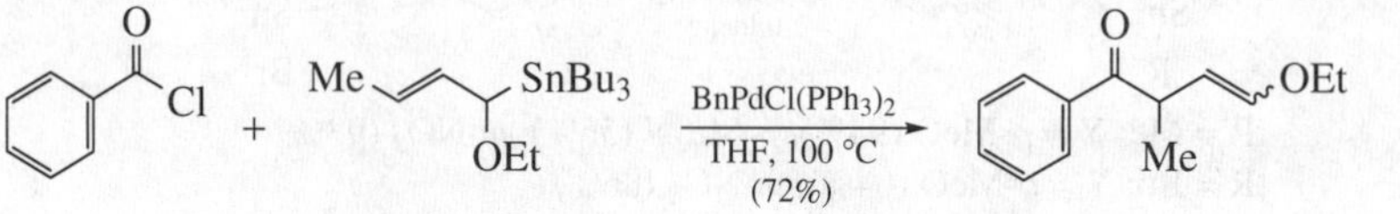

4-Ethoxy-2-methyl-1-phenyl-3-buten-1-one

Benzoyl chloride (0.70 g, 5.0 mmol), α-ethoxycrotyltributyltin (2.5 g, 6.4 mmol) and BnClPd(PPh$_3$)$_2$ (0.030 g, 0.04 mmol) in dry THF (5 mL) were heated in a sealed tube at 100 °C over a period of 16 h. After hexane−acetonitrile partition, concentration of the acetonitrile phase, and purification by chromatography on Florisil (eluent 15% ether in petroleum ether) the enol ether was obtained in 72% yield.

Allylic stannanes undergo cross-coupling with cinnamyl acetate to afford 1,5-dienes[163]. The reactions take place by a S_E2'/S_N2' pathway. The coupling is quite sensitive to steric effects:

$$R^1 = R^2 = H, R^3 = Me \ (69\%)$$
$$R^1 = Me, R^2 = R^3 = H \ (32\%)$$
$$R^1 = R^2 = Me, R^3 = H \ (4\%)$$

Coupling between allylic stannanes and allylic bromides is more efficient. These reactions also proceed by a S_E2'/S_N2' pathway:

$R = Me, R^1 = R^2 = H \ (43\%)$; $R = H, R^1 = R^2 = Me \ (48\%)$;
$R = R^1 = H, R^2 = Me \ (58\%)^a$

^{a}20% of the linear product is also formed

8.4   Aryl and Vinylstannane Couplings

Aryl and vinylstannanes undergo facile Pd^0-catalyzed coupling with allylic halides. The reaction is tolerant of a variety of functional groups including alcohols, esters and nitriles. Coupling proceeds with retention of vinyl stereochemistry[164]:

Palladium-catalyzed Coupling of a Vinyltin with an Allylic Halide

(Z,Z)-Methyl 3-Methoxy-2,5-heptadienoate

To a solution of 3.0 mol% of bis-(dibenzylideneacetone)palladium(0) in 5 mL of THF was added 6.0 mol% of triphenylphosphine, followed by the allylic halide and 1.0 equiv of organotin. The reaction solution was stoppered and heated at 50 °C until palladium metal precipitated (24–48 h). The reaction was cooled, partitioned between 100 mL of ether and 100 mL of one-third saturated potassium fluoride, and vigorously stirred for 30 min. The resulting precipitate of tributylstannyl fluoride was removed by gravity filtration, and the organic layer was separated, washed with brine, and dried (MgSO₄). The product was isolated in 86% yield by medium-pressure liquid chromatography on a silica gel column.

When the foregoing coupling reaction is conducted under three or more atmospheres of CO, carbonyl insertion takes place affording allylic ketones:

Palladium-catalyzed Carbonylative Coupling of a Vinyltin with an Allylic Halide

(E)-1-(3-Furyl)-8-methyl-3,7-nonadien-1-one

A solution of 3 mol% (based on the allylic halide) of bis-(dibenzylideneacetone) palladium(0) in 5 mL of THF and 6.0 mol% of triphenylphosphine was introduced into a pressure tube under nitrogen, followed by 2–5 mmol each of allyl halide and tin reagent. The tube was fitted with a pressure gauge and pressurized to 50 psig with carbon monoxide, and the pressure was released. This was repeated two more times, and the tube was repressurized to 45–55 psig. The tube was then placed in a constant temperature bath at 50 °C, and the contents were stirred for 24 h. The reaction mixture was cooled and the product was isolated in 69% yield by extraction with ether.

Vinyl halides also serve as useful coupling partners for vinylstannanes. A series of β-iodo cyclopentenones were converted to the β-vinyl enones in this manner[165]. The coupling reaction is compatible with alcohol and even carboxylic acid functionality:

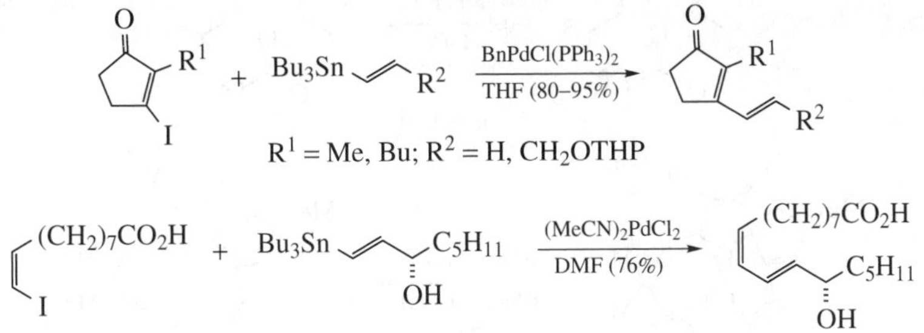

$R^1 = Me, Bu; R^2 = H, CH_2OTHP$

Vinyl triflates can also be used as coupling partners for vinylstannanes. These reactions are catalyzed by Pd^0–phosphine complexes and proceed best when 2–3 equivalents of LiCl are present[166]:

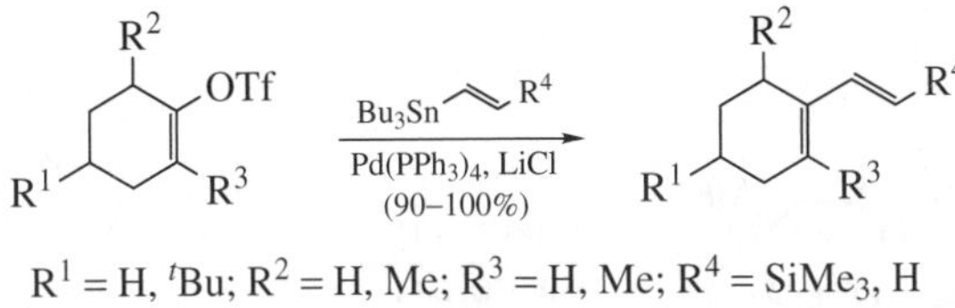

$R^1 = H, \,^tBu; R^2 = H, Me; R^3 = H, Me; R^4 = SiMe_3, H$

Palladium-catalyzed Coupling of a Vinylic Stannane and a Vinylic Triflate

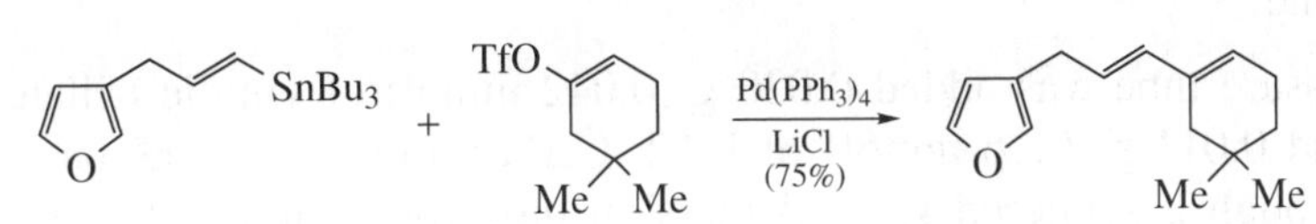

1-[(E)-3-(3-Furyl)-1-propenyl]-5,5-dimethylcyclohexene

A mixture of LiCl (0.20 g, 4.7 mmol), Pd(PPh$_3$)$_4$ (0.31 g, 1.8 mol%), triflate (0.38 g, 1.5 mmol), and vinyltin (0.61 g, 1.5 mmol) in THF (20 mL) was heated at reflux for 24 h, cooled to 25 °C, and diluted with 20 mL of pentane. The resulting mixture was washed with 5% NH$_4$OH solution. The extracts were washed with brine, filtered through a small plug of silica gel, and concentrated under reduced pressure to give a green oil. Column chromatography (silica gel, hexane) afforded 0.24 g of diene as a colorless oil (75% yield).

An intramolecular version of the coupling leads to macrocyclic lactones[167]. Intramolecular vinyl–vinyl coupling with CO insertion was employed as a key step in the total synthesis of jatraphone and epijatraphone[168]:

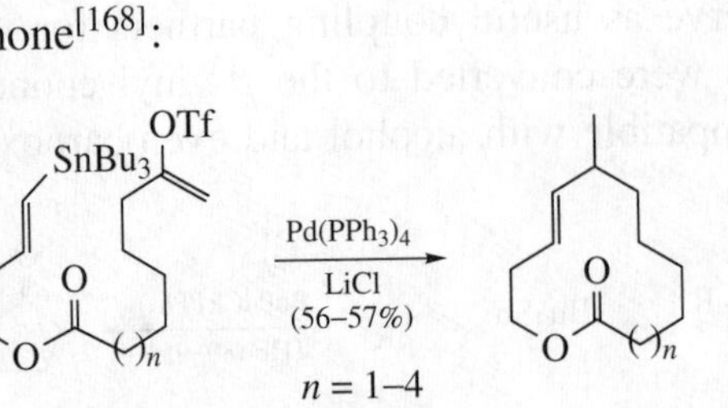

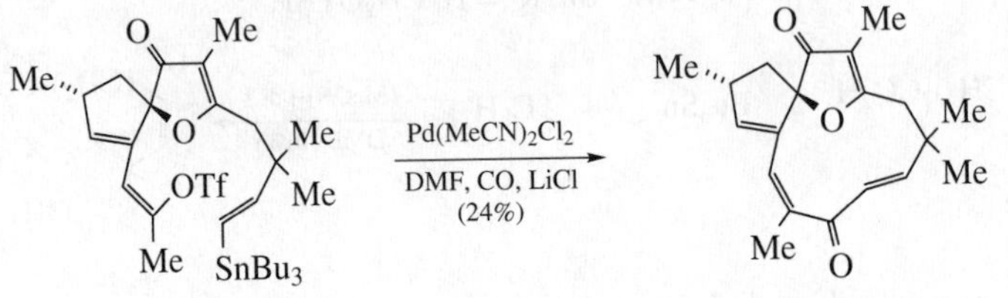

jatraphone

Intramolecular Pd-catalyzed Coupling of a Vinylic Stannane and a Vinylic Triflate with CO Insertion

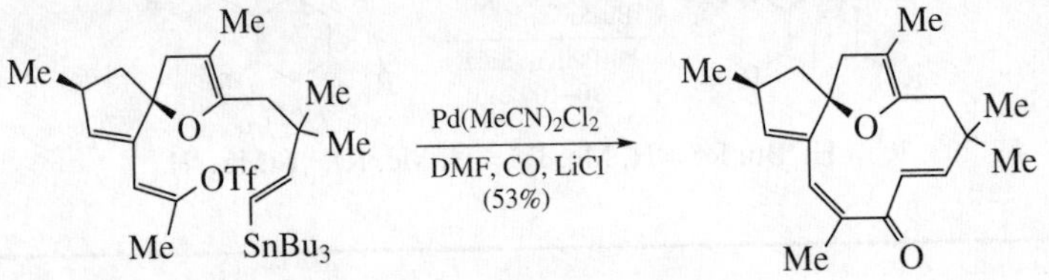

Epijatraphone

To a pressure tube was added 0.030 g (0.042 mmol) of vinylic triflate in 10 mL of DMF and 0.011 g (0.26 mmol) of LiCl. Carbon monoxide was passed into the solution through a scintered glass dispersion tube for 30 min. This was followed by the addition of 0.001 g (0.004 mmol) of bis(acetonitrile)palladium(II) chloride in 3 mL of DMF. The tube was then pressurized with CO to 50 psi and the mixture stirred at 25 °C for 13 h, after which time palladium black had precipitated. The tube was vented and the solution taken up in ether and washed with water. The organic layer was dried over Na_2SO_4. Removal of solvent under reduced pressure afforded an oily residue that was purified by column chromatography on silica gel with 25% ethyl acetate/hexanes to give 0.007 g (53%) of product as a white crystalline solid.

Various acid chlorides can be coupled to α-methoxyvinyl trimethylstannane in the presence of a Pd^0 catalyst leading to α-methoxyvinyl ketones[169]. α-Stannylated tetronic acid derivatives undergo Pd^0-catalyzed acylation to afford the α-acyl derivatives[170]. This structural unit is found in a number of ionophore natural products:

Me$_3$Sn~OMe $\xrightarrow[\text{C}_6\text{H}_6\ (44-86\%)]{\text{RCOCl}\ \text{BnPdCl(PPh}_3)_2}$

R = Me, C$_7$H$_{15}$, c-C$_6$H$_{11}$, ^{t}Bu, (CH$_2$)$_4$Cl, Ph, 2-furyl

R = Ph, (E)-PhCH=CH, Et, 2-furyl

3-Substituted furans are abundant in nature but there are few synthetic routes to these compounds. Furans can be acylated at the 3-position by cross-coupling of the 3-tributylstannyl derivative with various acid chlorides[171]. The procedure has been applied to perilla ketone, a fragrance component of mint oil:

R = C$_8$H$_{19}$ (86%), Ph (91%), 2-thienyl (95%), 2-furyl (95%)
Me$_2$CHCH$_2$CH$_2$ (perilla ketone, 74%)

Aryl–aryl coupling is best effected with Pd(AsPh$_3$)$_4$ as the catalyst in the presence of excess LiCl[172]. These reaction conditions also lead to efficient transfer of Bu from Bu$_4$Sn:

R = Bu (95%), Ph (82%), p-HOCH$_2$C$_6$H$_4$ (97%),
o-MeOC$_6$H$_4$ (88%), p-MeOC$_6$H$_4$ (92%)

Coupling of an Arylstannane and a Vinyl Triflate

1-(p-Methoxyphenyl)-4-tert-butylcyclohexene

The triflate (0.263 g, 0.92 mmol), triphenylarsine (0.023 g, 0.073 mmol), and Pd$_2$dba$_3$ (8.3 g, 0.018 mmol) were dissolved in 5 mL of anhydrous degassed N-methylpyrrolidone, and after the purple color was discharged (5 min) (p-methoxyphenyl)tributyltin (0.430 g, 1.08 mmol) in N-methylpyrrolidone (NMP) (2 mL) was added. After 16 h at 25 °C, the solution was treated with 1 M aqueous KF solution (1 mL) for 30 min, diluted with ethyl acetate, and filtered. The filtrate was extensively washed with water. Drying, followed by removal of solvent, gave a crude oil, that was purified by reversed-phase flash chromatography (C-18, 10% CH$_2$Cl$_2$ in acetonitrile) to yield 0.201 g (89%) of a white solid. Recrystallization from methanol gave white needles, mp 78–9 °C.

Coupling of an Arylstannane and an Aryl Triflate

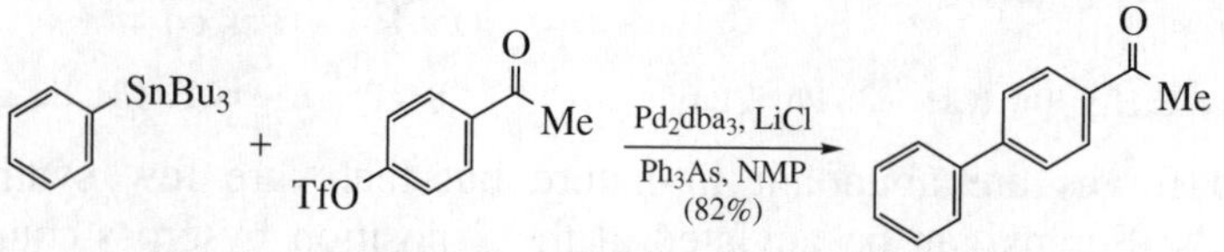

p-Phenylacetophenone

The triflate (0.308 g, 1.15 mmol), triphenylarsine (0.028 g, 0.092 mmol), Pd$_2$dba$_3$ (0.0105 g, 0.023 mmol), and lithium chloride (0.146 g, 3.44 mmol) were placed in a dry flask and stirred in anhydrous degassed NMP (5 mL) for 10 min. Phenyl-tributyltin (0.450 mL, 1.38 mmol) was then added neat by syringe, and the solution was stirred at 25 °C for 70 h. Addition of 1 M aqueous KF (2 mL), with stirring for 30 min, and dilution with ethyl acetate was followed by filtration. The filtrate was extensively washed with water, dried, and solvent was removed. Flash chromatography (silica, 10% ethyl acetate in hexane) gave the biphenyl as a white solid (0.185 g, 82%).

8.5   Allenylstannane Couplings

Only a few examples of allenylstannane cross-couplings have been recorded. Aryl iodides afford the arylated allenes in modest yield with Pd(PPh$_3$)$_4$ generated in situ from Pd$_2$dba$_3$ and PPh$_3$[173]. In some cases the addition of excess LiCl facilitates the reaction. Aryl triflates couple with tributylstannylallene in the presence of tris(2-furyl)phosphine, generated in situ from Pd$_2$dba, and 10% CuI in DMF[174]. Coupling is appreciably less efficient in the absence of CuI. Excess LiCl accelerates the reaction:

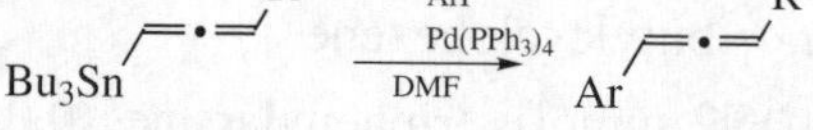

Ar = Ph, R = H (40%); Ar = 2-thienyl, R = H (42%);
Ar = Ph, R = Me (45%); Ar = R = Ph (45%)

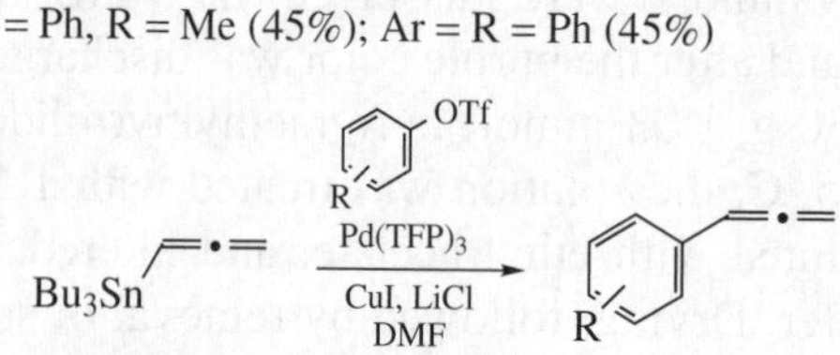

R = *p*-MeCO (60%), *o*-Ph (20%), *p*-Ph (31%),
o-CO$_2$Me (71%), *m*-MeO (70%)

TFP = (2-furyl)$_3$P

Coupling of an Allenylstannane and an Aryl Iodide

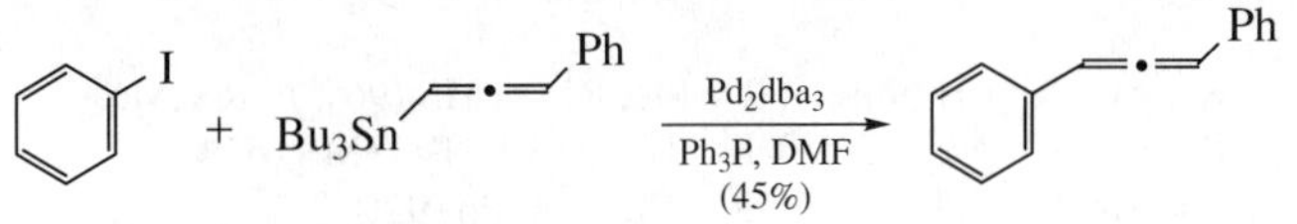

1,3-Diphenyl-1,2-propadiene

   To a mixture of 3-phenyl-1-(tributylstannyl)-1,2-propadiene (0.040 g, 0.1 mmol) and iodobenzene (0.020 g, 0.1 mmol) in 0.4 mL of degassed anhydrous DMF was added palladium dibenzylideneacetone chloroform complex (0.0031 g, 0.003 mmol) and triphenylphosphine (0.0062 g, 0.024 mmol). The resulting yellow solution was stirred at 25 °C. After 20 h, the reaction was quenched by addition of 0.5 mL of 10% NH_4OH solution. After stirring for 20 min, the mixture was extracted with hexane, washed with water, dried over sodium sulfate, filtered, and concentrated under vacuum. Column chromatography through silica gel with hexane as the elutant provided 0.007 g (45% yield) of 1,3-diphenylallene identical to an authentic sample. Approximately 20% of the starting stannyl allene was recovered in an earlier fraction.

8.6   Alkynylstannane Couplings

   The cross-coupling of alkynylstannanes with vinylic iodides can be achieved efficiently under mild reaction conditions to afford conjugated enynes in high yield. Various Pd^0 catalysts can be used including $Pd(PPh_3)_4$, $(MeCN)_2PdCl_2$, and $(Ph_3P)_2PdCl_2$. The latter two tend to be most effective. As expected, double-bond geometry is retained, although prolonged reaction times lead to mixtures of double-bond isomers. Trimethylstannanes are appreciably more reactive than their tributyl counterparts, but they are unstable and must be prepared immediately prior to use. However, the tributyl derivatives can be stored. Furthermore they are less toxic:

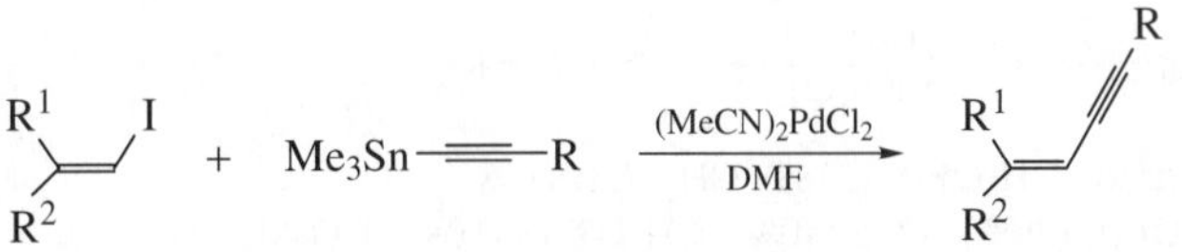

$R^1 = H$, $R^2 = R = Bu$ (88%); $R^1 = H$, $R^2 = Bu$, $R = (CH_2)_4OAc$ (97%); $R^1 = Bu$, $R^2 = H$, $R = (CH_2)_4OAc$ (91%); $R^1 = H$, $R^2 = (E)$-$CH=CH(CH_2)_2OTHP$, $R = H$ (68%)[a]

[a]The Bu_3Sn derivative was employed.

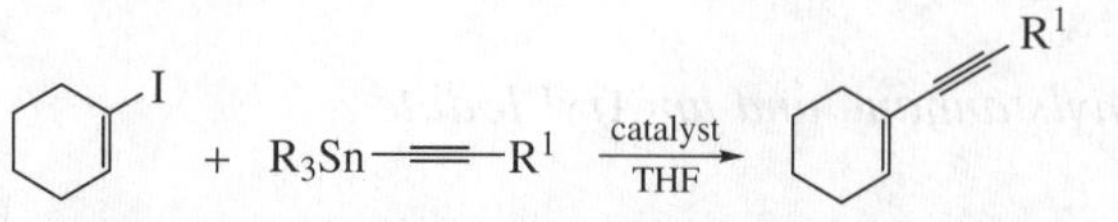

R = Me, R^1 = H (90%)^a; R = Me, R^1 = TMS (96%)^a; R = Me,
R^1 = CH$_2$OTMS (92%)^b; R = Bu, R^1 = CH$_2$OTMS (67%)^b;
R = Me, R^1 = Ph (90%)^a; R = Bu, R^1 = Ph (92%)^b

^{a}Catalyst = Pd(PPh$_3$)$_4$. ^{b}Catalyst = (Ph$_3$P)$_2$PdCl$_2$.

Coupling of an Alkynylstannane with a Vinylic Iodide

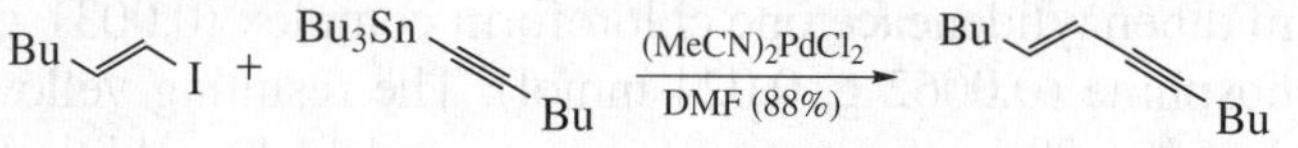

(E)-5-Dodecen-7-yne

To a dry and degassed solution of 0.013 g (0.050 mmol) of (MeCN)$_2$PdCl$_2$ in 10 mL of *N,N*-dimethylformamide (DMF) was added 0.211 g (1.00 mmol) of (*E*)-1-iodo-1-hexene and 0.316 g (1.29 mmol) of 1-hexynyltrimethylstannane at −50 °C. The homogeneous yellow reaction mixture turned black within 3 min. Analysis by GLC indicated complete consumption of the vinyl iodide, The black reaction mixture was added to 50 mL of water in a separatory funnel, and this aqueous mixture was extracted with ether. The combined ethereal extracts were washed with water and brine and dried over potassium carbonate. The dried extracts were filtered through a plug of alumina and concentrated under reduced pressure. Purification was accomplished by radial chromatography with pentane, followed by bulb-to-bulb distillation, bp 63–64 °C (0.3 mmHg) to yield 0.155 g (88%) of a colorless liquid.

Alkynylstannanes couple with acid chlorides to give alkynyl ketones in modest yield. The reactions are conducted in 1,2-dichloroethane with (Ph$_3$P)$_2$PdCl$_2$ as the catalyst. The variability in yields is due, in part, to the lability of the alkynones:

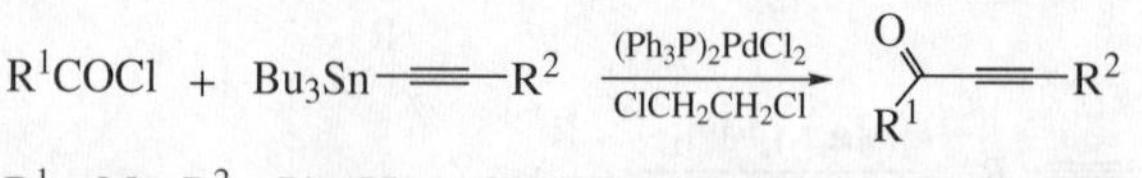

R^1 = Me; R^2 = Ph (55%), CH(OEt)$_2$ (31%), CH$_2$OTBS (48%)
R^1 = ^{i}Pr; R^2 = CH(OEt)$_2$ (70%), TMS (40%), CH$_2$OTBS (60%), CO$_2$Me (67%)
R^1 = Ph; R^2 = CH(OEt)$_2$ (68%), TMS (64%), CH$_2$OTBS (66%)

Arylation of ethoxyacetylene can be achieved through coupling of the aryl iodide with tributylstannyl ethoxyacetylene[175]. Attempted coupling of aryl iodides and alkynes, in the presence of CuCl and catalytic (PPh$_3$)$_2$PdCl$_2$ affords resins. With the same Pd catalyst and the stannyl alkyne coupling is successful. The reaction fails with *N*-acetamido or

Table 37. Palladium-catalyzed Coupling of Aryl Halides with Phenylacetylene.

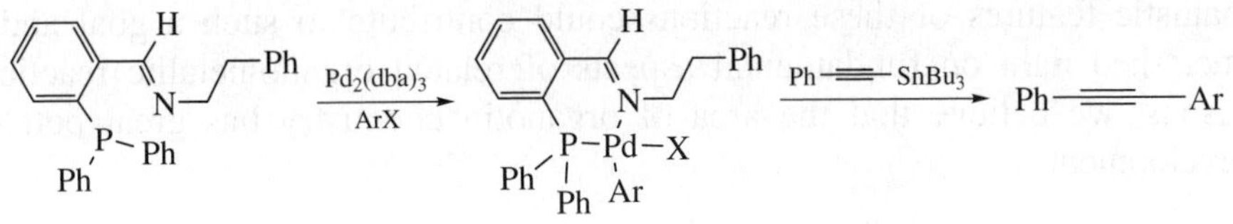

ArX	Solvent	Yield (%)
PhI	THF	93
p-EtO$_2$CC$_6$H$_4$I	THF	86
p-MeO$_2$C$_6$H$_4$I	THF	91
o-CF$_3$C$_6$H$_4$I	toluene	92
p-MeCOC$_6$H$_4$Br	toluene	90
PhOTf	toluene	93

N,N-dimethyl substituted aryl iodides. The alkynic products are converted to arylacetic esters through acid hydrolysis:

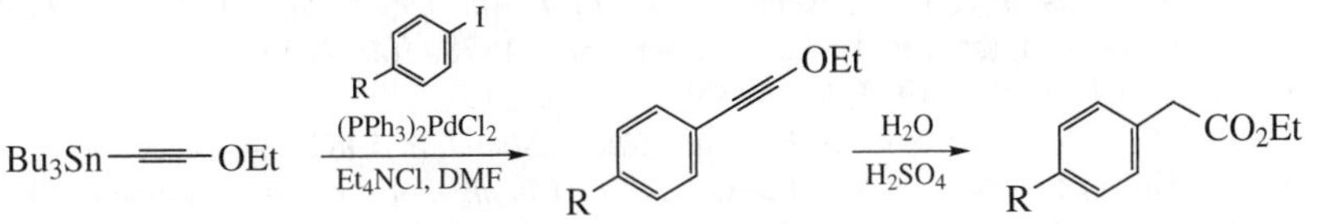

R = H (60%), NO$_2$ (52%), CO$_2$Et (59%), Me (45%), MeO (60%), AcNH (0%), Me$_2$N (0%)

Coupling of tributylstannyl phenylacetylene with various aryl halides proceeds in high yield when catalyzed by a phosphinobenzylidene amine catalyst (Table 37). The reaction is thought to proceed via a six-membered palladacycle intermediate. Coupling with vinyl and phenyltributyltin is also facile[176]:

9 Present and Future Prospects

Organotin compounds can be employed in a wide range of useful synthetic applications. Unlike many other organometallic compounds, they are isolable and stable. Moreover, they retain their stereochemical integrity. Typical reactions of allylic, allenic, propargylic, vinyl and aryltin reagents with a variety of electrophilic substrates and appropriate catalysts result in efficient and often highly stereoselective carbon–carbon bond formation. A number of these reactions can be conducted in aqueous solvents, and some without solvent. The production of toxic tin byproducts is a cause for concern, but as these generally contain tin–halogen bonds, they are effectively detoxified through treatment with amines to form insoluble materials easily removed by filtration. Of course, these materials ultimately require safe disposal which makes large-scale industrial applications of this chemistry problematic. Nonetheless, the methodology can serve as a useful starting point in the discovery and screening of new materials and potential drug candidates. Clearly, the discovery of catalytic processes could markedly change this scenario. Continued studies on mechanistic features of these reactions could contribute to such a goal and, at the same time, shed light on fundamental aspects of related organometallic reactions. For these reasons, we believe that the area of organotin chemistry has great potential for future development.

10   References

[1] Review: Pereyre, M.; Quintard, J. -P.; Rahm, A.; *Tin in Organic Synthesis*, Butterworths, London, **1987**, p 7.

[2] Other reviews of organotin compounds: (a) Harrison, P. G.; *Chemistry of Tin*, **1989**, Chapman and Hall, New York; (b) Marshall, J. A.; Jablonowski, J. A.; *The Chemistry of Organic Germanium, Tin, and Lead Compounds*, (S. Patai, Ed.), J Wiley, New York, 1995, Chapter 3.

[3] (a) San Filippo, J.; Silberman, J.; *J. Am. Chem. Soc.* **1981**, *103*, 5588. (b) Kitching, W.; Olszowy, H.; Waugh, J.; Doddell, D.; *J. Org. Chem.* **1978**, *43*, 898. (c) Koermer, G.S.; Hall, M. L.; Traylor, T. G.; *J. Am. Chem. Soc.* **1972**, *94*, 7205.

[4] Fish, R. H.; Broline, B. M.; *J. Organomet. Chem.* **1977**, *136*, C41.

[5] (a) Kuivila, H. G.; Considine, J. L.; Kennedy, J. D.; *J. Am. Chem. Soc.* **1972**, *94*, 7206. (b) San Filippo, J.; Silberman, J.; Fagan, P. J.; *J. Am. Chem. Soc.* **1978**, *100*, 4834.

[6] Still, W. C.; *J. Am. Chem. Soc.* **1977**, *99*, 4836.

[7] Itoh, A.; Saito, T.; Oshima, K.; Nozaki, H.; *Bull. Chem. Soc. Jpn.* **1981**, *54*, 1456.

[8] (a) Ochiai, M.; Ukita, T.; Nagao, Y.; Fujita, E.; *J. Chem. Soc. Chem. Commun.* **1984**, 1007. (b) Ochiai, M.; Ukita, T.; Nagao, Y.; Fujita, E.; *J. Chem. Soc. Chem. Commun.* **1985**, 637. (c) Nakatani, K.; Isoe, S.; *Tetrahedron Lett.*, **1985**, *26*, 2209.

[9] (a) Macdonald, T. L.; Mahalingam, S.; *J. Am. Chem. Soc.* **1980**, *102*, 2113. (b) Macdonald, T. L.; Mahalingam, S.; *Tetrahedron Lett.*, **1981**, *22*, 2077.

[10] Fleming, I.; Rowley, M.; *Tetrahedron*, **1986**, *42*, 3181.

[11] (a) Davis, D. D.; Johnson, H. T.; *J. Am. Chem. Soc.* **1974**, *96*, 7576. (b) Fleming, I.; Urch, C. J.; *J. Organomet. Chem.* **1985**, *285*, 173.

[12] Review of anionic organotin compounds: Sato, T.; *Synthesis* **1990**, 259.

[13] Macdonald, T. L., Mahalingam, S.; O'Dell, D. E.; *J. Am. Chem. Soc.* **1990**, *103*, 6767.

[14] (a) Quintard, J. P.; Elissondo, B.; Mouko, D.; Pegna, M.; *J. Organomet. Chem.* **1983**, *251*, 175. (b) Chan, P. C. -M.; Chong, J. M.; *J. Org. Chem.* **1988**, *53*, 5584.

[15] (a) Marshall, J. A.; Welmaker, G. S.; Gung, B. W.; *J. Am. Chem. Soc.* **1991**, *113*, 647. (b) Degl'Innocenti, A.; Stucchi, E.; Capperucci, A.; Mordini, A.; Reginato, G.; Ricci, A.; *Synlett* **1992**, 332.

[16] Capperucci, A.; Degl'Innocenti, A.; Faggi, C.; Reginato, G.; Ricci, A.; Dembech, P.; Seconi, G.; *J. Org. Chem.* **1989**, *54*, 2966.

[17] Mitchell, T. N.; Kwetkat, K.; *Synthesis* **1990**, 1001.

[18] Ricci, A.; Degl'Innocenti, A.; Capperucci, A.; Reginato, G.; Mordini, A.; *Tetrahedron Lett.* **1991**, *32*, 1899.

[19] (a) Verlhac, J. -B.; Chanson, E.; Jousseaume, B.; Quintard, J. -P.; *Tetrahedron Lett.* **1985**, *26*, 6075. (b) Kosugi, M.; Naka, H.; Haruda, S.; Sano, H.; Migita, T.; *Chem. Lett.* **1987**, 1371.

[20] Still, W. C.; *J. Am. Chem. Soc.* **1978**, *100*, 1481.

[21] Still, W. C.; Sreekumar, C.; *J. Am. Chem. Soc.* **1980**, *102*, 1201.

[22] Sawyer, J. S.; Kucerovy, A.; Macdonald, T. L.; McGarvey, G. J.; *J. Am. Chem. Soc.* **1988**, *110*, 842.

[23] (a) McGarvey, G. J.; Kimura, M.; *J. Org. Chem.* **1985**, *50*, 4657. (b) McGarvey, G. J.; Kimura, M.; *J. Org. Chem.* **1982**, *47*, 5422.

[24] Linderman, R. J.; Jaber, M.; Tetrahedron Lett. **1994**, *35*, 5993.

[25] Linderman, R. J.; Ghannam, A.; *J. Am. Chem. Soc.* **1990**, *12*, 2392.

[26] Linderman, R. J.; Godfrey, A.; Horne, K.; *Tetrahedron* **1989**, *45*, 495.

[27] Linderman, R. J.; Godfrey, A.; *J. Am. Chem. Soc.* **1988**, *110*, 6249.

[28] Linderman, R. J.; Siedlecki, J. M.; *J. Org. Chem.* **1996**, *61*, 6492.

[29] Tomooka, K.; Igarashi, T.; Nakai, T.; *Tetrahedron Lett.* **1994**, *35*, 1913.

[30] (a) Linderman, R. J.; Cusack, K. P.; Jaber, M. R.; *Tetrahedron Lett.* **1996**, *37*, 6649. (b) Jephcote, V. J.; Pratt, A. J.; Thomas, E. J.; *J. Chem. Soc., Chem. Commun.* **1984**, 800.

[31] (a) Helmke, H.; Hoppe, D.; *Synlett* **1995**, 978. (b) Hoppe, D.; Carstens, A.; Kramer, T.; Angew. Chem., Int. Ed. Engl. **1990**, *29*, 1424. (c) Hoppe, D.; Hintze, F.; Tebben, P.; *Angew. Chem., Int. Ed. Engl.* **1990**, *29*, 1422.

[32] Tomooka, K.; Igarashi, T.; Watanabe, M.; Nakai, T.; *Tetrahedron Lett.* **1992**, *33*, 5795.

[33] Chong, J. M.; Park, S. B.; *J. Org., Chem.* **1992**, *57*, 2220.

[34] Linderman, R. J.; Griedel, B. D.; *J. Org. Chem.* **1990**, *55*, 5428.

[35] Linderman, R. J.; Griedel, B. D.; *J. Org. Chem.* **1991**, *56*, 5491.

[36] Falck, J. R.; Bhatt, R. K.; Ye, J.; *J. Am. Chem. Soc.* **1995**, *117*, 5973.

[37] Nativi, C.; Taddei, M.; *J. Org. Chem.* **1988**, *53*, 820.

[38] Zhang, H. X.; Guibé, F.; Balavoine, G.; *J. Org. Chem.* **1990**, *55*, 1857.

[39] Cochran, J. C.; Bronk, B. S.; Terrence, K. M.; Phillips, H. K.; *Tetrahedron Lett.* **1990**, *31*, 6621.

[40] Benechie, M.; Skrydstrup, T.; Khuong-Huu, F.; *Tetrahedron Lett.* **1991**, *32*, 7535.

[41] Asao, N.; Liu, J. -X.; Sudoh, T.; Yamamoto, Y.; *J. Org. Chem.* **1996**, *61*, 4568.

[42] (a) Piers, E.; Morton, J. E.; *J. Org. Chem.* **1980**, *45*, 4263. (b) Piers, E.; Chong, J. -M.; Morton, H. E.; *Tetrahedron* **1989**, *45*, 363.

[43] Piers, E.; Chong, J. -M.; *J. Org. Chem.* **1982**, *47*, 1602.

[44] Westmijze, H.; Ruitenberg, K.; Meijer, J.; Vermeer, P.; *Tetrahedron Lett.* **1982**, *23*, 2797.

[45] Hutzinger, M. W.; Singer, R. D.; Oehlschlager, A. C.; *J. Am. Chem. Soc.* **1990**, *112*, 9397.

[46] (a) Lipshutz, B. H.; Ellsworth, E. L.; Dimoch, S. H.; Reuter, D. C.; *Tetrahedron Lett.* **1989**, *30*, 2065. (b) Lipshutz, B. H.; Sharma, S.; Reuter, D. C.; *Tetrahedron Lett.* **1990**, *31*, 7253.

[47] (a) Hibino, J.; Matsubara, S.; Morizawa, Y.; Oshima, K.; Nozaki, H.; *Tetrahedron Lett.* **1984**, *25*, 2151. (b) Beaudet, I.; Parrain, J. -L.; Quintard, J. -P.; *Tetrahedron Lett.* **1991**, *32*, 6333.

[48] Nonaka, T.; Okuda, Y.; Matsubara, S.; Oshima, K.; Utimoto, K.; Nozaki, H.; *J. Org. Chem.* **1986**, *51*, 4716.

[49] Gilbertson, S. R.; Challener, C. A.; Bos, M. E.; Wulff, W. D.; *Tetrahedron Lett.* **1988**, *29*, 4795.

[50] Wang, K. K.; Chu, K. -H.; Lin, Y.; Chen, J. -H.; *Tetrahedron* **1989**, *45*, 1105.

[51] Nativa, C.; Taddei, M.; *Tetrahedron* **1989**, *45*, 1131.

[52] Hodgson, D. M.; Boulton, L. T.; Maw, G. N.; *Tetrahedron Lett.* **1994**, *35*, 2231.

[53] Fryzuk, M. D.; Bates, G. S.; Stone, C.; *Tetrahedron Lett.* **1986**, *29*, 1537.

[54] Ritter, K.; *Tetrahedron Lett.* **1990**, *31*, 869.

[55] Lautens, M.; Huboux, A. H.; *Tetrahedron Lett.* **1990**, *31*, 3105.

[56] Lautens, M.; Huboux, A. H.; Chin, B.; Downer, J.; *Tetrahedron Lett.* **1990**, *31*, 5829.

[57] Stork, G.; Mook, R.; *J. Am. Chem. Soc.* **1987**, *109*, 2829.

[58] Kocienski, P.; Dixon, N. J.; *Synlett* **1989**, 52.

[59] McGarvey, G. J.; Bajwa, J. S.; *J. Org. Chem.* **1984**, *49*, 4091.

[60] Corey, E. J.; Wollenberg, R. H.; *J. Am. Chem. Soc.* **1974**, *96*, 5581.

[61] Corey, E. J.; Eckrich, T. M.; *Tetrahedron Lett.* **1984**, *25*, 2419.

[62] Dodd, D. S.; Pierce, H. D.; Oehlschlager, A. C.; *J. Org. Chem.* **1992**, *57*, 5250.

[63] Behling, J. R.; Babiak, K. A.; Ng, J. S.; Campbell, A. L.; *J. Am. Chem. Soc.* **1988**, *110*, 2641.

[64] (a) Behling, J. R.; Ng, J. S.; Babiak, K. A.; Campbell, A. L.; Elsworth, E.; Lipshutz, B. H.; *Tetrahedron Lett.* **1989**, *30*, 27. (b) Lipshutz, B. H.; Ellsworth, E. L.; Dimock, S. H.; Reuter, D. C.; *Tetrahedron Lett.* **1989**, *30*, 2065.

[65] Piers, E.; McEachern, E. J.; Burns, P. A.; *J. Org. Chem.* **1995**, *60*, 2322.

[66] Cabezas, J. A.; Oehlschlager, A. C.; *Tetrahedron Lett.* **1995**, *36*, 5127.

[67] Reviews: (a) Nishigaichi, Y.; Takuwa, A.; Naruta, Y.; Maruyama, K.; *Tetrahedron* **1993**, *49*, 7395. (b) Yamamoto, Y.; Asao, N.; *Chem. Rev.* **1993**, *93*, 2207. (c) Yamamoto, Y.; Shida, N.; *Advances in Detailed Reaction Mechanisms*, Vol. 3, JAI Press, Greenwich, CT *1994*, 1. (d) Marshall, J. A.; *Chemtracts-Organic Chemistry* **1992**, *5*, 75. (e) Marshall, J. A.; *Chem. Rev.* **1996**, *96*, 31.

[68] Seyferth, D.; Weiner, M. A.; *J. Org. Chem.* **1961**, *61*, 4797.

[69] Tanaka, H.; Abdul Hai, A. K. M.; Ogawa, H.; Torii, S.; *Synlett* **1993**, 835.

[70] Carofiglio, T.; Marton, D.; Tagliavini, G.; *Organometallics* **1992**, *11*, 2964.

[71] Naruta, Y.; Nishigaichi, Y.; Maruyama, K.; *Tetrahedron* **1989**, *45*, 1067.

[72] Weigand, S.; Brückner, R.; *Synthesis* **1996**, 475.

[73] Ueno, Y.; Sano, H.; Okawara, M.; *Tetrahedron Lett.* **1980**, *21*, 1767.

[74] (a) König, K.; Neumann, W. P.; *Tetrahedron Lett.* **1967**, 495. (b) Pratt, A. J.; Thomas, E. J.; *J. Chem. Soc., Chem. Commun.* **1982**, 1115.

[75] Naruta, Y.; Ushida, S.; Maruyama, K.; *Chem. Lett.* **1979**, 919.

[76] Yamamoto, Y.; Yatagai, H.; Naruta, Y.; Maruyama, K.; *J. Am. Chem. Soc.* **1980**, *102*, 7107.

[77] (a) Denmark, S. E.; Weber, E. J.; *J. Am. Chem. Soc.* **1984**, *106*, 7970. (b) Denmark, S. E.; Weber, E. J.; Wilson, T. M.; Willson, T. M.; *Tetrahedron* **1989**, *45*, 1053.

[78] Maruyama, K.; Ishihara, Y.; Yamamoto, Y.; *Tetrahedron Lett.* **1981**, *22*, 4235.

[79] Koreeda, M.; Tanaka, Y.; *Chem. Lett.* **1982**, 1299.

[80] Keck, G. E.; Boden, E. P.; *Tetrahedron Lett.* **1984**, *25*, 265.

[81] Keck, G. E.; Abbott, D. E.; Boden, E. P.; Enholm, E. J.; *Tetrahedron Lett.* **1984**, *25*, 3927.

[82] Keck, G. E.; Boden, E. P.; *Tetrahedron Lett.* **1984**, *25*, 1879.

[83] Keck, G. E.; Abbott, D. E.; *Tetrahedron Lett.* **1984**, *25*, 1883.

[84] (a) Mikami, K.; Kawamoto, K.; Loh, T. -P.; Nakai, T.; *J. Chem. Soc., Chem. Commun.* **1990**, 1161. (b) Fleming, I.; *Chemtracts-Organic Chemistry*, **1990**, 21.

[85] Keck, G. E.; Savin, K. A.; Cressman, E. N. K.; Abbott, D. E.; *J. Org. Chem.* **1994**, *59*, 7889.

[86] Henry, K. J.; Grieco, P. A.; Jagoe, C. T.; *Tetrahedron Lett.* **1992**, *33*, 1817.

[87] Marshall, R. L.; Young, D. J.; *Tetrahedron Lett.* **1992**, *33*, 1365.

[88] Iqbal, J.; Joseph, S. J.; *Tetrahedron Lett.* **1989**, *30*, 2421.

[89] Yasuda, M.; Sugawa, Y.; Yamamoto, A.; Shibata, I.; Baba, A.; *Tetrahedron Lett.* **1996**, *37*, 5951.

[90] Boaretto, A.; Marton, D.; Tagliavini, G.; Gambaro, A.; *Inorg. Chem. Acta* **1983**, *77*, L196.

[91] (a) Pratt, A. J.; Thomas, E. J.; *J. Chem. Soc., Perkin Trans. 1* **1989**, 1521. (b) Jephcote, V. J.; Pratt, A. J.; Thomas, E. J.; *J. Chem. Soc., Perkin Trans. 1* **1989**, 1529.

[92] (a) McNeill, A. H.; Thomas, E. J.; *Synthesis* **1994**, 322. (b) McNeill, A. H.; Thomas, A. J.; *Tetrahedron Lett.* **1993**, *34*, 1699.

[93] Miyake, H.; Yamamura, K.; *Chem. Lett.* **1992**, 1369.

[94] Takahara, J. P.; Masuyama, Y.; Karusu, Y.; *J. Am. Chem. Soc.* **1992**, *114*, 2577.

[95] Nishigaichi, Y.; Fujimoto, M.; Takuwa, A.; *Synlett* **1994**, 731.

[96] Yanagisawa, A.; Nakatsuka, Y.; Nakashima, H.; Yamamoto, H.; *Synlett* **1997**, 993.

[97] (a) Whitesell, J. K.; Apoduca, R.; *Tetrahedron Lett.* **1996**, *37*, 3955. (b) Yano, K.; Baba, A.; Matsuda, H.; *Bull. Chem. Soc. Jpn.* **1992**, *65*, 66.

[98] Yasuda, M.; Miyai, T.; Shibata, I.; Baba, A.; Nomura, R.; Matsuda, H.; *Tetrahedron Lett.* **1995**, *36*, 9497.

[99] Nakamura, H.; Iwama, J.; Yamamoto, Y.; *J. Am. Chem. Soc.* **1996**, *118*, 6641.

[100] Hachiza, I.; Kobayashi, S.; *J. Org. Chem.* **1993**, *58*, 6958.

[101] Aspinall, H. C.; Browning, A. F.; Grieves, N.; Ravenscroft, P.; *Tetrahedron Lett.* **1994**, *35*, 4639.

[102] Review: Marshall, J. A.; *Chemtracts-Organic Chemistry*, **1996**, *9*, 280.

[103] Marshall, J. A.; Tang, Y.; *Synlett* **1992**, 653.

[104] Aoki, S.; Mikami, K.; Terada, M.; Nakai, T.; *Tetrahedron* **1993**, *49*, 1783.

[105] Costa, A. L.; Piazzo, M. G.; Tagliavini, E.; Trombini, C.; Umani-Ronchi, A.; *J. Am. Chem. Soc.* **1993**, *115*, 7001.

[106] Bedeschi, P.; Casolari, S.; Costa, A. L.; Tagliavini, E.; Umani-Ronchi, A.; *Tetrahedron Lett.* **1995**, *36*, 7897.

[107] Keck, G. E.; Tarbet, K. H.; Geraci, L. S.; *J. Am. Chem. Soc.* **1993**, *115*, 8467. Keck, G. E.; Geraci, L. S.; *Tetrahedron Lett.* **1993**, *34*, 7827. Keck, G. E.; Krishnamurthy, D.; Grier, M. C.; *J. Org. Chem.* **1993**, *58*, 6543.

[108] Yanagisawa, A.; Nakashima, H.; Ishiba, A.; Yamamoto, H. *J. Am. Chem. Soc.* **1996**, *118*, 4723.

[109] Cozzi, P. G.; Orioli, P.; Tagliavini, E.; Umani-Ronchi, A.; *Tetrahedron Lett.* **1997**, *38*, 145.

[110] Review: Thomas, E. J.; *Chemtracts-Organic Chemistry*, **1994**, *7*, 207.

[111] McNeill, A. H.; Thomas, E. J.; *Tetrahedron Lett.* **1990**, *31*, 6239.

[112] McNeill, A. H.; Thomas, E. J.; *Tetrahedron Lett.* **1992**, *33*, 1369.

[113] Carey, J. S.; Thomas, E. J.; *Synlett* **1992**, 585.

[114] Carey, J. S.; Thomas, E. J.; *Tetrahedron Lett.* **1993**, *34*, 3935.

[115] Carey, J. S.; Thomas, E. J.; *J. Chem. Soc., Chem. Commun.* **1994**, 285.

[116] Pratt, A. J.; Thomas, E. J.; *J. Chem. Soc., Chem. Commun.* **1982**, 1115.

[117] Keck, G. E.; Abbott, D. E.; Wiley, M. R.; *Tetrahedron Lett.* **1987**, *28*, 139.

[118] Koreeda, M.; Tanaka, Y.; *Tetrahedron Lett.* **1987**, *28*, 143.

[119] (a) Quintard, J. -P.; Dumartin, G.; Elissondo, B.; Rahm, A.; Pereyre, M.; *Tetrahedron* **1989**, *45*, 1017. (b) Quintard, J. -P.; Elissando, B.; Pereyre, M.; *J. Org. Chem.* **1983**, *48*, 1559.

[120] Jephcote, V. J.; Pratt, A. J.; Thomas, E. J. *J. Chem. Soc., Chem. Commun.* **1984**, 800.

[121] Paulsen, H.; Graeve, C.; Hoppe, D.; *Synthesis* **1996**, 141.

[122] Zschage, O.; Schwark, J. -R.; Kramer, T.; Hoppe, D.; *Tetrahedron* **1992**, *48*, 8377.

[123] (a) Parrain, J. -L.; Cintrat, J. -C.; Quintard, J. -P.; *J. Organomet. Chem.* **1992**, *437*, C19. (b) Watrelot, S.; Parrain, J. -L.; Quintard, J. -P.; *J. Org. Chem.* **1994**, *59*, 7959.

[124] (a) Marshall, J. A.; Gung, W. Y.; *Tetrahedron Lett.* **1988**, *29*, 1657. (b) Marshall, J. A.; Crooks, S. L.; DeHoff, D. S.; *J. Org. Chem.* **1988**, *53*, 1616. (c) Marshall, J. A.; Gung, W. Y.; *Tetrahedron Lett.* **1988**, *29*, 3899.

[125] Marshall, J. A.; Gung, W. Y.; *Tetrahedron* **1989**, *45*, 1043.

[126] Marshall, J. A.; Yashunsky, D. V.; *J. Org. Chem.* **1991**, *56*, 5493.

[127] Gung, B. W.; Smith, D. T.; Wolf, M. A.; *Tetrahedron* **1992**, *48*, 5455.

[128] Marshall, J. A.; Gung W.Y.; *Tetrahedron Lett.* **1989**, *30*, 7349.

[129] (a) Marshall, J. A.; Welmaker, G. S.; *J. Org. Chem.* **1992**, *57*, 7158. (b) Marshall, J. A.; Beaudoin, S.; Lewinski, K.; *J. Org. Chem.* **1993**, *58*, 5876.

[130] Marshall, J. A.; Jablonowski, J. A.; Elliott, L. M.; *J. Org. Chem.* **1995**, *60*, 2662.

[131] Marshall, J. A.; Luke, G. P.; *J. Org. Chem.* **1991**, *56*, 483.

[132] Marshall, J. A.; Seletsky, B. M.; Luke, G. P.; *J. Org. Chem.* **1994**, *59*, 3413.

[133] Marshall, J. A.; Luke, G. P.; *J. Org. Chem.* **1993**, *58*, 6229.

[134] Marshall, J. A.; Seletsky, B. M.; Coan, P. S.; *J. Org. Chem.* **1994**, *59*, 5139.

[135] Marshall, J. A.; Jablonowski, J. A.; Welmaker, G. S.; *J. Org. Chem.* **1996**, *61*, 2904.

[136] Marshall, J. A.; Hinkle, K. W.; *J. Org. Chem.* **1995**, *60*, 1920.

[137] Marshall, J. A.; Hinkle, K. W.; *J. Org. Chem.* **1996**, *61*, 105.

[138] Marshall, J. A.; Hinkle, K. W.; *J. Org. Chem.* **1996**, *61*, 4247.

[139] Ueno, Y.; Sano, H.; Okawara, M.; *Synthesis* **1980**, 1011.

[140] Dussault, P. H.; Lee, R. J.; *J. Am. Chem. Soc.* **1994**, *116*, 4485.

[141] Madec, D.; Férézou, J. -P.; *Synlett* **1996**, 867.

[142] (a) Ruitenberg, K.; Westmijze, H.; Meijer, J.; Elsevier, C. J.; Vermeer, P.; *J. Organomet. Chem.* **1983**, *241*, 417. (b) Ruitenberg, K.; Westmijze, H.; Kleijn, H.; Vermeer, P.; *J. Organomet. Chem.* **1984**, *277*, 227.

[143] Marshall, J. A.; Wang, X. -J.; *J. Org. Chem.* **1990**, *55*, 6246.

[144] Boaretto, A.; Marton, D.; Tagliavini, G.; Gambaro, A.; *J. Organomet. Chem.* **1985**, *286*, 9.

[145] Boaretto, A.; Marton, D.; Tagliavini, G.; *J. Organomet. Chem.* **1985**, *297*, 149.

[146] Logue, M. W.; Teng, K.; *J. Org. Chem.* **1982**, *47*, 2549.

[147] (a) Stille, J. K.; Simpson, J. H.; *J. Am. Chem. Soc.* **1987**, *109*, 2138. (b) Stille, J. K.; *Angew. Chem., Int. Ed. Engl.* **1986**, *25*, 508.

[148] Lequan, M.; Guillerm, G.; *J. Organomet. Chem.* **1973**, *54*, 153.

[149] Mukaiyama, T.; Harada, T.; *Chem. Lett.* **1981**, 621.

[150] Nokami, J.; Tamaoka, T.; Koguchi, T.; Okawara, R.; *Chem. Lett.* **1984**, 1939.

[151] Marshall, J. A.; Wang, X. -J.; *J. Org. Chem.* **1991**, *56*, 6264.

[152] Marshall, J. A.; Wang, X. -J.; *J. Org. Chem.* **1991**, *56*, 3211.

[153] Marshall, J. A.; Wang, X. -J.; *J. Org. Chem.* **1992**, *57*, 1242.

[154] Marshall, J. A.; Perkins, J.; *J. Org. Chem.* **1994**, *59*, 3509.

[155] Anies, C.; Lallemand, J. -Y.; Pancrazi, A.; *Tetrahedron Lett.* **1996**, *37*, 5519.

[156] Farina, V.; Krishnamurthy, V.; Scott, W. J.; *Org. React.* **1997**, *50*, 1.

[157] Milstein, D.; Stille, J. K.; *J. Am. Chem. Soc.* **1979**, *101*, 4992.

[158] Vedejs, E.; Haight, A. R.; Moss, W. O.; *J. Am. Chem. Soc.* **1992**, *114*, 6556.

[159] Kosugi, M.; Ishiguro, M.; Negishi, Y.; Sano, H.; Migita, T.; *Chem. Lett.* **1984**, 1511.

[160] Godschalx, J.; Stille, J. K.; *Tetrahedron Lett.* **1980**, *21*, 2599.

[161] Sheffy, F. K.; Stille, J. K.; *J. Am. Chem. Soc.* **1983**, *105*, 7173.

[162] Verlhac, J. B.; Pereyre, M.; Quintard, J. -P.; *Tetrahedron* **1990**, *46*, 6399.

[163] Trost, B. M.; Keinan, E.; *Tetrahedron Lett.* **1980**, *21*, 2595.

[164] (a) Sheffy, F. K.; Godschalx, J. P.; Stille, J. K.; *J. Am. Chem. Soc.* **1984**, *106*, 4833. (b) Merrifield, J. H.; Godschalx, J.; Stille, J. K.; *Organometallics* **1984**, *3*, 1108. (c) Baillargeon, V. P.; Stille, J. K.; *J. Am. Chem. Soc.* **1986**, *108*, 452.

[165] Stille, J. K.; Sweet, M. P.; *Tetrahedron Lett.* **1989**, *30*, 3645.

[166] Scott, W. J.; Stille, J. K.; *J. Am. Chem. Soc.* **1986**, *108*, 3033.

[167] (a) Stille, J. K.; Tanaka, M.; *J. Am. Chem. Soc.* **1987**, *109*, 3785. (b) Labadie, J. W.; Stille, J. K.; *J. Am. Chem. Soc.* **1983**, *105*, 6129.

[168] Gyorkos, A. C.; Stille, J. K.; Hegedus, L. S.; *J. Am. Chem. Soc.* **1990**, *112*, 8465.

[169] Soderquist, J. A.; Leong, W. W. -H.; *Tetrahedron Lett.* **1983**, *24*, 2361.

[170] Ley, S. V.; Wadsworth, D. J.; *Tetrahedron Lett.* **1989**, *30*, 1001.

[171] Bailey, T. R.; *Synthesis* **1991**, 242.

[172] Farina, V.; Krishnam, B.; Marshall, D. R.; Roth, G. P.; *J. Org. Chem.* **1993**, *58*, 5434.

[173] Aidhen, I. S.; Braslau, R.; *Synth. Commun.* **1994**, *24*, 789.

[174] Badone, D.; Cardamone, R.; Guzzi, U.; *Tetrahedron Lett.* **1994**, *35*, 5477.

[175] Sakamoto, T.; Yasuhara, A.; Kondo, Y.; Yamanaka, H.; *Synlett* **1992**, 502.

[176] Shirakawa, E.; Yoshida, J.; Takaya, H.; *Tetrahedron Lett.* **1997**, *38*, 3759.

III

Organoboron Chemistry

KEITH SMITH

Department of Chemistry, University of Wales Swansea, Swansea, UK

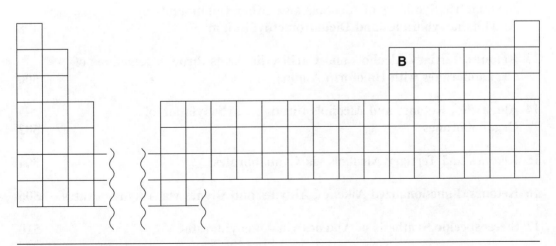

Organometallics in Synthesis: A Manual. Edited by Manfred Schlosser.
© 2002 John Wiley & Sons Ltd

Contents

1   Introduction

The discovery of the hydroboration reaction[1] in the late 1950s signaled a new era in synthetic organic chemistry by providing convenient access for the first time to organo-boranes, a class of reagents which was later to prove to be of unrivalled versatility. During the 1960s much effort, almost entirely on the part of the group of H. C. Brown, was put into exploration of the scope of the hydroboration reaction for the synthesis of organoboranes, despite the fact that such compounds had shown few signs of being synthetically useful. This latter situation changed dramatically in the 1970s as a number of specialist organoborane research groups joined Brown's group in exploring the reactions of organoboranes. They discovered that organoboranes, despite low reactivity in tradi-tional reactions of organometallic reagents such as Grignard reagents, were possessed of properties that gave rise to an enormous diversity and range of other synthetically useful reactions. Against this background, many nonspecialist synthetic groups began in the 1980s to utilize and indeed to extend the useful reactions of organoboranes, and an article predicted that the 1990s would begin to see the use of such reactions commercially[2]. This prediction has been vindicated. As the twenty-first century begins, organoboranes must now pass into the realm of standard reagents.

Despite the evident utility of organoboranes as reagents and the inexorable trend in their utilization, however, many practicing organic chemists have remained reluctant to take advantage of their potential. It was in an attempt to rectify this situation that the first edition of this contribution was made. There are emerging signs of more widespread utilization of organoboranes by practitioners of organic chemistry, but there are still many practitioners who are either unaware of the advantages offered by organoboranes or continue to lack confidence in their handling. Also, a number of new and useful developments have emerged since the first edition appeared. This second edition attempts to update the first edition, while retaining the emphasis on generally useful reactions and procedures.

The reasons for reluctance to embrace the new opportunities may be several:

1. fear of possible dangers associated with unfamiliar materials that might be toxic, pyrophoric or in other ways hazardous;
2. uncertainty about how to handle unfamiliar chemicals;
3. anecdotal accounts of difficulties experienced by others in getting reactions to work as reported;
4. lack of knowledge of reactions that would be useful for the synthetic challenges being undertaken;
5. insufficient confidence in ability to choose the best reagent for the job.

1.1   Safety

The sorts of boron compounds of interest for organic synthesis carry no extreme toxic hazards. Naturally, like any reactive materials they should be treated with care and ingestion or contact should be avoided, but most of the compounds are readily decomposed into boric acid, which is a relatively minor toxic hazard. The toxicity of any more stable compounds that might be produced would most likely be unknown, but normal safety precautions associated with handling organic chemicals should be sufficient in most cases.

1.2 Handling

Most boron−carbon bonds are quite stable to water, although some unsaturated boron compounds, especially allylboranes, provide exceptions. However, boron−hydrogen, boron−oxygen, boron−nitrogen and boron−halogen bonds are generally fairly easily hydrolyzed; as almost all precursors of organoboranes involve one or other of these types of bonds, it is usual to carry out reactions in dry solvents.

Very few organoboranes are pyrophoric. Those that are have significant volatility, such as trimethylborane, triethylborane and triallylborane, and even they can be handled safely enough under an inert atmosphere. Less volatile compounds, like the higher trialkylboranes, oxidize readily in air but without ignition. Increasing hindrance around the boron atom or attachment of electron-releasing substituents like hydroxy groups to the boron atom decrease the sensitivity to oxygen, and some compounds can even be handled in air without detriment. Nevertheless, it is usual for reactions to be carried out routinely under an atmosphere of nitrogen or, exceptionally, argon. Mixtures are usually opened to air only after an oxidizing mixture has been admitted, in order to avoid potential side reactions resulting from oxygen-induced radical chain reactions.

The apparatus used for organoborane reactions can be extremely simple—a dry round-bottom flask equipped with a magnetic follower and a septum is adequate for many purposes. The system is flushed with nitrogen using syringe needles, safety-vented by a paraffin-oil bubbler or by use of a rubber balloon connected via a needle, charged or sampled by means of syringes, and stirred magnetically. Although somewhat more sophisticated apparatus is required for reflux, filtration, or other types of manipulation, there is generally no requirement for apparatus that is more complex than that used for handling other organometallic reagents.

1.3 Reproducibility

Some early reports of hydroboration reactions emphasized the *in situ* generation of borane from sodium tetrahydroborate (borohydride) and, for example, boron trifluoride etherate. Such a method is adequate for simple hydroboration−oxidation of an unsubstituted alkene, but is not recommended for anything more complex because of possible side reactions involving the tetrahydroborate, the Lewis acid or any other component of the mixture. Unjustified use of the *in situ* procedure may have led to some disappointments, particularly in the early days. Similarly, when understanding was less than it is today, there may have been attempts to carry out reactions that had little real chance of success. Such cases can lead to anecdotal accounts of difficulties that carry weight beyond their significance. However, the numerous publications by nonspecialists that demonstrate successful application of organoborane reactions testify to the fact that reported reactions are reproducible. Some attention to detail, such as use of dry solvents and control of stoichiometry, may be necessary, but given such attention the reactions are just as reliable as any other organic reactions.

1.4   Selection of Reaction and Reagent

Whereas the range of reactions of most classes of organometallic reagents is very narrow and easily assimilated with the help of a simple mechanistic rationale, the situation for organoboranes is much more complex. First, there are many mechanistically diverse types of reaction, involving simple organic group transfers, intramolecular rearrangements, pericyclic processes, transition metal catalyzed cross-couplings, radical reactions, boron-stabilized carbanions, *etc*. Second, there can be up to four different organic groups attached to boron, some of which may be involved in the reaction whereas others are merely disposable blocking groups. Alternatively, halogen, alkoxy or other functionalities may replace some of the organic groups. The nonspecialist is therefore confronted with a bewildering array of possibilities.

For this reason, the following sections are highly selective and substantially free of mechanistic discussion. Only well-tried and tested reactions are included and the emphasis is on providing a simple procedure that can be used immediately by a nonspecialist. The literature contains more extensive works that include further details, including mechanistic discussion[3,4]. Recent works have dealt in detail with current topics in boron chemistry[5], hydroborations catalyzed by transition metal complexes[6], stereodirected synthesis with organoboranes[7], and reductions, which includes reductions with substituted borohydrides[8].

2   Simple Hydroboration–Oxidation

Simple hydroboration–oxidation can be carried out with any hydroborating agent[3,4]. The most convenient and cost-effective is borane–dimethyl sulfide, which is commercially available as a neat liquid or as a solution in dichloromethane. It offers possibilities for use of various solvents and for large-scale work.

Preparation of (–)-cis-Myrtanol by Hydroboration–Oxidation of (–)-β-Pinene[4]

A dry, 2 L, three-necked flask fitted with a mechanical stirrer, a septum-capped pressure-equalizing dropping funnel and a reflux condenser vented *via* a bubbler is flushed with nitrogen and then charged with $(-)$-β-pinene (238 mL, 1.5 mol) and hexane (500 mL). It is cooled in an ice bath (to dissipate the heat generated during reaction) and borane–dimethyl sulfide (52.5 mL, neat liquid, 0.55 mol) is added from the dropping funnel, with stirring, over 30 min. The cooling bath is removed and the mixture is stirred for 3 h at 25 °C to ensure complete reaction, by which time the flask contains a solution of tri-*cis*-myrtanylborane.

Ethanol (500 mL) is added cautiously (on account of initial evolution of hydrogen arising from excess borane) followed by aqueous sodium hydroxide (185 mL of 3 mol L^{-1} solution). The flask is again immersed in an ice-water bath and hydrogen peroxide (185 mL of 30% aqueous solution) is added at such a rate that the temperature does not exceed 40 °C. The cooling bath is removed and the reaction completed by heating at 50 °C for 1 h. The mixture is poured into ice-water (5 L), mixed with diethyl ether (2 L) and then separated. The organic layer is washed with water (2×1 L) and saturated sodium chloride (1 L), dried over potassium carbonate, filtered and evaporated to give a light yellow oil (230 g). Short-path distillation gives $(-)$-*cis*-myrtanol (196 g, 85%).

The above procedure is satisfactory for many alkenes. However, more hindered alkenes may require longer reaction times or more forcing conditions at either stage of the reaction. If necessary, more concentrated peroxide (50% or even 60% solution) may be used in the oxidation stage. Some alkenes react only as far as the dialkylborane (R_2BH) or monoalkylborane (RBH_2) stage and in such cases the ratio of borane to alkene is raised from 1 : 3 to 1 : 2 or 1 : 1, respectively, or an excess of borane is used. Caution should be exercised as a much larger quantity of hydrogen will be evolved in such cases.

If regioselectivity of hydroboration should be a problem (hydroboration of 1-hexene gives 94% attachment of B to C-1; styrene gives 80% attachment to B to C-1; internal alkenes give poor regioselectivity) the most generally useful reagent for giving high regioselectivity is the commercially available dialkylborane, 9-borabicyclo-[3.3.1]nonane (9-BBN-H)[9]. This must be used at 1 : 1 stoichiometry with the alkene. In the case of hydroboration of alkynes, the most regioselective reagent is dimesitylborane[10], which is also commercially available. Oxidation of a vinylborane, such as is produced in the reaction of an alkyne, requires a buffered peroxide solution in order to minimize hydrolysis of the organoborane intermediate. The product is an aldehyde or ketone.

Finally, if enantioselective hydroboration is required, the reagent of choice is diisopinocampheylborane, dilongifolylborane or monoisopinocampheylborane, depending on the nature of the alkene[11]. These organoboranes are prepared freshly before use (see Section 6).

3   Simple Trialkylboranes and Other Triorganylboranes

The most convenient and most versatile method for synthesis of symmetrical trialkyl-boranes involves the hydroboration of alkenes with borane. The experimental procedure is as described for the preparation of tri-*cis*-myrtanylborane in Section 2. It is important that the stoichiometry be close to the theoretical 3 : 1 (alkene/borane) in order to avoid contamination by dialkylborane species. Alternatively, an excess of alkene may be used provided that the excess is easily removed following reaction. For most purposes, the solution of trialkylborane obtained can be used directly for further reactions, but if neces-sary the solvent can be removed under reduced pressure and the product can be distilled under nitrogen or under reduced pressure.

Many triorganylboranes cannot be prepared by hydroboration, including aryl-, allyl- and most tertiary alkylboranes[12]. In such cases the most useful general method for synthesis of the organoborane involves the reaction of a reactive organometallic reagent with trifluoroborane etherate or some similar reagent[13].

Preparation of Triphenylborane from Phenylmagnesium Bromide and Trifluoroborane–Diethyl Etherate[13]

$$3PhMgBr + BF_3 \cdot OEt_2 \longrightarrow Ph_3B + 3MgBrF + Et_2C$$

A dry, 4 L, three-necked flask equipped with a mechanical stirrer, a 1 L pressure-equalizing dropping funnel capped with a septum, and a still-head assembled for distillation with a condenser and 2 L receiver is flushed with nitrogen, vented through a bubbler connected to the receiver adapter. The flask is charged (double-ended needle) with trifluoroborane–diethyl etherate (142 g, 1 mol) and xylene (1 L). A freshly prepared and estimated ethereal solution of phenylmagnesium bromide (3 mol in about 1 L of diethyl ether) is added dropwise, *via* the dropping funnel, with stirring, at 25–35 °C over a period of 3 h (slow addition so as to minimize the production of tetraphenylborate). A little ether distils during the addition and as soon as the addition is complete the temperature is raised to allow the rest of the ether to distil out (stopped when the xylene begins to distil at 138 °C). The distilla-tion set-up is replaced by a bent sinter tube attached to a 3 L two-necked receiver vented through a bubbler and the still-hot solution is forced through the sinter under slight nitrogen pressure. The residual salts are extracted with hot (120–130 °C) xylene (2 × 500 mL) and the combined xylene extracts are distilled, without a fractionating column, under reduced pressure. After removal of the solvent and a small forerun boiling below 155 °C at 0.1 mmHg, crude triphenylborane distils at 155–166 °C (0.1 mmHg). A single recrystallization from heptane under nitrogen gives triphenylborane (217 g, 90%; mp 148 °C, which is essentially pure).

4   Monoorganylboranes and Diorganylboranes

Hydroboration of some alkenes with borane proceeds rapidly only as far as the mono-alkylborane or dialkylborane stage and more slowly thereafter[3]. Thus, by careful control of the stoichiometry and reaction temperature it is possible, in such cases, to produce the appropriate mono- or dialkylborane cleanly. In many other cases this direct production of mono- or dialkylboranes by hydroboration is not possible.

Perhaps the most important dialkylborane of all is 9-borabicyclo[3.3.1]nonane (9-BBN-H; **1**)[14], which is commercially available as a crystalline solid. It is sufficiently stable in air to be transferred quickly without special precautions, though it should always be stored under an inert atmosphere. The preparation of this dialkylborane is slightly more complicated than other simple dialkylboranes and nonspecialists are therefore advised to use the commercial material:

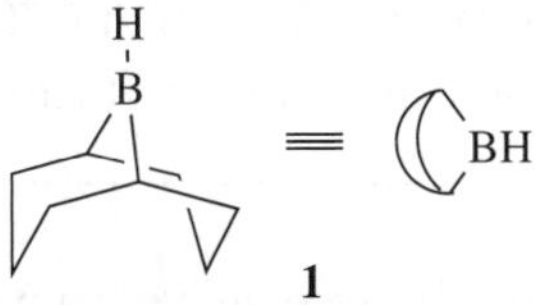

The types of alkenes which readily give dialkylboranes are trialkylethenes such as 2-methyl-2-butene and relatively hindered 1,2-dialkylethenes such as cyclohexene. 2-Methyl-2-butene gives the so-called disiamylborane, whereas cyclohexene gives dicyclohexylborane[15]. It is usual to produce them *in situ* immediately prior to their application in further reactions.

Preparation of Dicyclohexylborane[15]

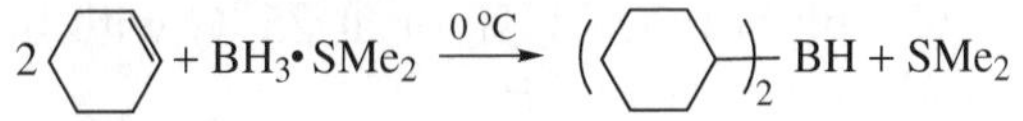

A 200 mL, round-bottomed, two-necked flask equipped with a pressure-equalizing dropping funnel fitted with a septum, a magnetic follower and a reflux condenser vented through a mercury bubbler, is flushed with nitrogen. Cyclohexene (16.4 g, 0.2 mol) and dry diethyl ether (75 mL) are added and the mixture is cooled to 0 °C. Borane–dimethyl sulfide (7.7 g, 0.1 mol) is introduced to the dropping funnel and added dropwise to the stirred solution over 30 min. The dropping funnel is washed through with diethyl ether (25 mL) and the mixture is stirred for 3 h at 0 °C. Dicyclohexylborane (as its dimer) precipitates as white crystals and can be used directly as a suspension for most purposes. If isolation is required, the ether and dimethyl sulfide can be removed by distillation in a slow stream of nitrogen. The product can also be sublimed in vacuum. Its melting point is 103–105 °C.

The most commonly utilized monoalkylborane is thexylborane, which is not very stable over prolonged periods and therefore has to be carefully and freshly prepared.

Preparation of Thexylborane[4]

$$Me_2C=CMe_2 + BH_3 \cdot THF \longrightarrow Me_2CHCMe_2BH_2$$

A dry, two-necked 200 mL flask fitted with a septum inlet, a magnetic follower and an outlet leading to a mercury bubbler is flushed with nitrogen. Borane–tetrahydrofuran (100 mL of 1 mol L^{-1} solution) is added using a double-ended needle and the flask is cooled in an ice-salt bath at -10 to $-15\,°C$. A 2 mol L^{-1} solution of 2,3-dimethyl-2-butene in tetrahydrofuran (50 mL) is added dropwise, with stirring, over 30 min, by syringe, while maintaining the temperature at or below $0\,°C$. The mixture is then stirred for 2 h at $0\,°C$ to complete the reaction. The clear solution so produced is used directly for further reactions.

2,3-Dimethyl-2-butene is relatively expensive, but the dimer of isobutene (2,4,4-trimethylpentane or Dib-2) is cheaper. The influence of the bulky *tert*-butyl group means that hydroboration of this trialkylethene can be controlled to provide the corresponding monoalkylborane, known as DIBborane:

Dib-2   DIBborane

The hydroboration step is carried out in a manner similar to that used in the synthesis of thexylborane, but at $0\,°C$ with borane–THF or at $25\,°C$ with borane–dimethyl sulfide. The product exists as a bridged dimer in THF[16].

The preparations of chiral mono- and dialkylboranes, also available by hydroboration, are described in Section 6. Some diorganylboranes cannot be prepared by hydroboration and in such cases organometallic reagents are generally used. An example is the preparation of dimesitylborane, although this compound may also be available commercially.

Preparation of Dimesitylborane[3,10]

$$2MesBr \xrightarrow{2Mg} 2MesMgBr \xrightarrow{BF_3 \cdot OEt_2} Mes_2BF \xrightarrow{1/4\ LiAlH_4} Mes_2BH$$

(Mes = mesityl = 2,4,6-trimethylphenyl)

A dry, 250 mL, two-necked flask equipped with a septum-capped pressure-equalizing dropping funnel and a reflux condenser leading to a mercury bubbler is charged with magnesium turnings (2.73 g, 114 mmol) and then flushed with nitrogen. A solution of mesityl bromide (22.3 g, 112 mmol) in tetrahydrofuran (THF; 56 mL) is transferred to the dropping funnel by double-ended needle and then the magnesium turnings are heated gently with an air gun. The mesityl bromide solution is added dropwise at such a rate as to give constant reflux. After completion of the addition the reaction mixture is heated at 80–90 °C for 3 h, cooled to ambient temperature, diluted with THF (30 mL) and transferred by double-ended needle to a graduated flask. The solution is made up to 100 mL with further THF and an aliquot is standardized by titration against 0.2 mol L^{-1} HCl. The yield is about 98%.

Another 250 mL, two-necked flask equipped with a septum-capped pressure-equalizing dropping funnel and a mechanical stirrer is flushed with nitrogen. The Grignard reagent prepared as above (94 mL of 1.1 mol L^{-1} solution, 104 mmol) is transferred to the funnel and the flask is charged with trifluoroborane–etherate (7.44 g, 52 mmol) that has been distilled from calcium hydride. The flask is immersed in an ice-water bath and the Grignard reagent is added, at such a rate as to maintain the temperature of the reaction mixture below about 30 °C, with rapid stirring. After completion of the addition the mixture is stirred for a further 1 h at 25 °C, the flask is disconnected, and the mixture is left overnight in a deep-freeze, stoppered with a septum and under nitrogen. The supernatant liquid is transferred by double-ended needle to another flask and the solvent is removed under reduced pressure (with protection from moisture by a drying tube). Light petroleum (30–40 °C, 30 mL) is added to precipitate magnesium fluoride and after the mixture has settled the supernatant liquid is transferred to another nitrogen-flushed flask. The residue is washed with further light petroleum (3 × 30 mL) and the combined supernatant liquids are concentrated to about 25 mL, then set aside in a deep-freeze for 18 h. The product crystallizes out and the mother liquor is removed and further concentrated to produce a second crop. Solvent is removed from the combined crystals under reduced pressure (oil pump), to give dimesitylfluoroborane (9.08 g, 70%; mp 70–72 °C).

A dry 500 mL flask equipped with a magnetic follower and a septum is connected by a needle to a mercury bubbler and flushed with nitrogen. Dimesitylfluoroborane (8.4 g, 50 mmol), prepared as described above, in 1,2-dimethoxyethane (50 mL), is charged by syringe or double-ended needle. This solution is stirred while a solution of lithium aluminum hydride in 1,2-dimethoxyethane (63 mL of 0.2 mol L^{-1} solution, 12.6 mol) is added dropwise through a double-ended needle. A white precipitate forms and stirring is maintained for 1 h. Dry benzene (100 mL) is added, the mixture is stirred for 30 min, and the mixture is then allowed to settle. The supernatant liquid is removed *via* a double-ended needle and the residue is washed with benzene (3 × 100 mL) in a similar manner. The combined supernatant liquids are allowed to settle overnight at ambient temperature, then filtered rapidly through a sintered glass funnel containing a 5 cm thick layer of Celite under a stream of nitrogen. The clear filtrate is concentrated under reduced pressure (protected by a drying tube) and the crude product (9 g) is crystallized from 1,2-dimethoxyethane (about 60 mL) to give colorless crystals of dimesitylborane (5.2 g, 69%; mp 164–166 °C).

Bis(2,4,6-triisopropylphenyl)borane (ditripylborane, $Trip_2BH$), which is even more hindered, is prepared from ditripylfluoroborane that has been made in a manner similar to that described above for dimesitylfluoroborane.

Bis(pentafluorophenyl)borane has been prepared in a somewhat different manner by reacting dimethylbis(pentafluorophenyl)stannane with trichloroborane to give chlorobis (pentafluorophenyl)borane, followed by reduction with chlorodimethylsilane[17]. It remains to be seen how much use will be made of this dimeric solid, which is stable for months in an inert atmosphere.

2,4,6-Triisopropylphenylborane ($TripBH_2$)[18] resembles the bulky monoalkylborane, thexylborane, in some of its properties. However, unlike thexylborane it is a solid that is quite stable to decomposition. It shows even greater selectivity than thexylborane in further hydroboration reactions[19].

Preparation of 2,4,6-Triisopropylphenylborane (TripBH$_2$)[18]

A dry, 250 mL, round-bottomed flask equipped with a magnetic follower and a septum is charged with dimethoxytripylborane (2.02 g, 7.32 mmol) and then flushed with nitrogen. Dry diethyl ether (60 mL) and dry pentane (15 mL) are added and the mixture is stirred at 0 °C.

Another dry, 100 mL, round-bottomed flask equipped with a magnetic follower and a septum is charged with lithium aluminum hydride (1.0 g) and then flushed with nitrogen. Dry diethyl ether (40 mL) is added and the suspension is stirred at 25 °C overnight. The solid residue is removed by filtration under nitrogen through a sintered funnel charged with a little Celite to give a clear solution of lithium aluminum hydride in diethyl ether. A solution of lithium aluminum hydride (0.87 mol L^{-1}, 8.5 mL), prepared as described before, is added *via* a syringe to dimethoxytripylborane (prepared by reaction of the tripyl Grignard reagent with trimethoxyborane) in solution at 0 °C, whereon a white precipitate immediately results. The ice-bath is removed and the mixture is stirred at 25 °C for 3 h, then the supernatant is removed through a double-ended needle under nitrogen. The solid residue is washed twice with a mixture of diethyl ether (60 mL) and pentane (15 mL). The supernatants are combined and the solvent is removed under reduced pressure to give $LiTripBH_3$ as a white solid (1.42 g, 87%; ν_{max} 2220 cm^{-1} (B–H)).

Another dry, 100 mL, round-bottomed flask equipped with a magnetic follower and a septum is charged with $LiTripBH_3$ (1.26 g, 5.63 mmol) and then flushed with nitrogen. Dry diethyl ether (20 mL) and dry pentane (5 mL) are added and the mixture is stirred at 25 °C. Trimethylsilyl chloride (0.55 ml, 5.8 mmol, distilled from CaH_2 under nitrogen) is added dropwise, and the mixture is stirred at room temperature for 3 h. The solvent is removed under reduced pressure to give tripylborane (1.14 g, 94%) as a white solid (ν_{max} 2503 cm^{-1} (B–H), 1580 cm^{-1} (B...H...B)).

5 Mixed Triorganylboranes

Early attempts to synthesize mixed trialkylboranes of the type $R^1R^2R^3B$ were dogged by failure, with the result that there was a view that such compounds were not stable and that they were prone to redistribution to give mixtures of the symmetrical trialkylboranes and other mixed trialkylboranes (e.g. $R^1_2R^2B$). However, it is now well established that this is not the case[3,12]. When redistribution products are obtained it is the result of failure to produce the intermediate mono- and dialkylboron compounds cleanly. The key to success is therefore to ensure that the intermediate compounds are produced cleanly, and then to control the conditions carefully during their subsequent conversion into mixed trialkylboranes.

It is often not possible to produce a particular mono- or dialkylborane directly by hydroboration, in which case it must be made in other ways (see the preparation of dimesitylborane in Section 4 and of other products in Section 8). Section 4 gives procedures for clean preparation of examples of mono- and dialkylboranes that are amenable to the direct approach.

If the subsequent conversion into a mixed trialkylborane involves hydroboration of a single alkene, as in the preparation of dicyclohexyl-1-octylborane by reaction of dicyclohexylborane with 1-octene or of thexyldicyclopentylborane by reaction of thexylborane with cyclopentene, the procedure is very simple. It consists simply of adding the alkene, in stoichiometric amount or slight excess, to the solution or suspension of the mono- or dialkylborane under nitrogen and allowing the mixture to stir at -10 to $+20\,°C$ for a period of time. Once reaction is complete, modest temperatures cease to be any problem.

Preparation of Dicyclohexyl-1-octylborane

Dicyclohexylborane (0.1 mol) is prepared as a suspension in diethyl ether as described in Section 4. The reaction flask is cooled in ice-water to dissipate heat of reaction and the mixture is stirred while 1-octene (11.2 g, 100 mmol) is added dropwise. After completion of the addition the reaction mixture is stirred for 3 h at 25 °C. (With more hindered alkenes it may be necessary to heat the mixture to 50 °C at this stage.) The suspended dicylcohexylborane dissolves during the reaction to produce a clear solution of dicyclohexyl-1-octylborane, which can be used directly in subsequent reactions.

For similar reactions using thexylborane it is necessary to keep the reaction mixture at about −10 °C during addition of the alkene[20]. Even under such conditions, however, it is not possible to control the stoichiometry to produce a thexylmonoalkylborane cleanly from the least hindered alkenes such as 1-octene. Fortunately, it is possible to achieve this in the case of most internal alkenes and a further hydroboration reaction can then be used to give a fully mixed trialkylborane.

Preparation of thexylcyclopentyl(6-acetoxyhexyl)borane[20]

A 200 mL flask equipped with a magnetic follower and a septum and vented by a needle leading to a paraffin-oil bubbler is flushed with nitrogen. Thexylborane (100 mmol) is prepared as described in Section 4 and the reaction flask is then immersed in a cooling bath set at −10 °C. The mixture is stirred at −10 °C during the dropwise addition, by syringe, of cyclopentene (6.8 g, 100 mmol), for a further 1 h thereafter, and then during the subsequent dropwise addition of 6-acetoxy-1-hexene (14.2 g, 100 mmol). The mixture is stirred for a further 1 h at −10 °C and then allowed to warm to 25 °C to give a solution containing the desired organoborane ready for direct utilization in subsequent reactions.

The preceding procedure is applicable only to cases in which the first alkene is a 1,2-disubstituted ethene or a trisubstituted ethene and the second alkene is a 1-alkene, a 1,1-disubstituted ethene or a 1,2-disubstituted ethene. If other fully mixed trialkylboranes are required a more complicated procedure is needed[21,22].

Preparation of Thexyldecyloctylborane[21,22]

A dry, 100 mL flask equipped with a magnetic follower and a septum and vented *via* a needle leading to a paraffin-oil bubbler is flushed with nitrogen. Thexylchloro-borane–dimethyl sulfide (20 mmol) in dichloromethane (20 ml) is prepared *in situ* (see Section 8), then cooled to 0 °C and stirred during addition of 1-octene (2.24 g, 20 mmol). The mixture is warmed to 25 °C and stirred for 2 h, then cooled to −10 °C (ice–salt bath). A solution of 1-decene (2.80 g, 20 mmol) in THF (20 mL) is added, followed by the dropwise addition of potassium triisopropoxyhydroborate (21 mL of a commercial 1 mol L^{-1} solution in THF, 21 mmol), with vigorous stirring. The mixture is stirred for 2 h at 0 °C to produce a solution of the desired product, which can be used directly.

Completely mixed organoboranes of the type $TripBR^1R^2$ (Trip = tripyl = 2,4,6-triisopropylphenyl) can be obtained from tripylborane via stepwise hydroborations in the same way as with thexylborane. However, it is easier to obtain such compounds where both R^1 and R^2 are primary groups, even when they are different, than in the case with thexylborane[19].

Some higher crotylboranes can be obtained cleanly by hydroboration of allenes with dialkylboranes (Section 7).

In some cases it is either necessary or more convenient to introduce the final organic group(s) via an organometallic reagent rather than by a hydroboration reaction. The organometallic reagents generally chosen are either Grignard reagents or organolithium reagents, although others are also possible[3,12]. The choice of the leaving group from boron can be critical. In the case of fairly hindered compounds it is reasonable to use halogenoboron compounds[3,23]:

$$Mes_2BF \xrightarrow{EtMgBr} Mes_2BEt + FMgBr$$

In the case of relatively unhindered compounds, however, the triorganylborane formed may react rapidly with further organometallic reagent, resulting in formation of some tetraorganylborate salt, which can be difficult to remove, and causing lowering of yield. In such cases it is advantageous to use a poorer leaving group, such as methoxide. The immediate product is then a methoxyborate that has to be broken down, for example by use of boron trifluoride. An example of this approach is given in Section 7, with the preparation of a dialkylalkynylborane.

6   Chiral Organoboranes

The simplest way of preparing useful chiral organoboranes is by hydroboration of a chiral alkene with borane–dimethyl sulfide. If the alkene is sufficiently unreactive that it can be cleanly converted into a dialkylborane (Section 4), this latter species can act as a chiral hydroborating agent for the asymmetric hydroboration of prochiral alkenes. The most widely used example of such a dialkylborane is diisopinocampheylborane (Ipc_2BH)[24,25], prepared from optically active α-pinene, which is readily available. High enantioselectivity is achieved in its reactions with (Z)-1,2-disubstituted ethenes such as cis-2-butene[3,4]. This is therefore a very useful method for preparation of chiral organoboranes containing the corresponding asymmetric units.

Preparation of (R)-2-Butyldiisopinocampheylborane[3,4]

A 250 mL, two-necked flask equipped with a septum inlet, a magnetic follower and a distillation head leading to a condenser with a cooled ($-78\,^\circ$C), bubbler-vented receiver is flushed with nitrogen. The flask is charged with THF (15 mL) and neat borane–dimethyl sulfide (5.05 mL, 50 mmol), cooled to $0\,^\circ$C and stirred during the dropwise addition of $(-)$-α-pinene (15.9 mL, 100 mmol; $\alpha_D^{23} - 48.7^\circ$, corresponding to 95% ee). After a further 3 h at $0\,^\circ$C, a mixture of the solvent and dimethyl sulfide (total 13 ml) is removed under reduced pressure (about 30 mmHg; protection from moisture by drying tube). (If α-pinene of 99% ee is used, the Ipc$_2$BH produced at this point is ready for direct further reaction with (Z)-2-butene, but with the lower-purity material described here an equilibration step is now required.) The distillation set-up is rapidly replaced with a stopper and further $(-)$-α-pinene (2.4 mL, 15 mmol) and THF (18 mL) are added to the reaction flask. The mixture is left to equilibrate for 3 days at $0\,^\circ$C to give a white suspension of diisopinocampheylborane of about 99% ee. This suspension can be used directly or the excess α-pinene can be removed (with THF) by syringe and fresh THF added.

The flask is cooled to about $-10\,^\circ$C by immersion in an ice–salt bath and *cis*-2-butene (3.1 g, 55 mmol) is added, in solution if necessary. After 4 h stirring at $0\,^\circ$C the temperature is allowed to rise to $25\,^\circ$C and the solution contains the desired asymmetric organoborane, diisopinocampheyl-(R)-2-butylborane (50 mmol), available for further reaction. Oxidation (see Section 2) gives (R)-2-butanol in very high enantiomeric purity.

Diisopinocampheylborane also readily hydroborates 1,1-disubstituted ethenes, but in such cases the chiral induction is rather low[26]. It does not successfully hydroborate more hindered alkenes such as *trans*-1,2-disubstituted ethenes or trisubstituted ethenes because displacement of α-pinene competes with hydroboration. As a result, several different hydroborating agents are present in the mixture, each one exhibiting different stereo-selectivity, which can lead overall to very low asymmetric induction. Dilongifolyborane is less prone to alkene displacement and less hindered than diisopinocampheylborane and consequently gives better asymmetric induction than the latter in reactions with the more hindered alkenes. Its preparation from optically active longifolene is straightforward, like that of dicyclohexylborane (Section 4).

An alternative reagent for hydroboration of relatively hindered alkenes is monoisopino-campheylborane (IpcBH$_2$)[27]. This reagent presents an additional advantage because it can be reacted with an equimolar amount of such a hindered alkene to give the corresponding alkylisopinocampheylborane which, like other dialkylboranes, tends to crystallize from the solution as its hydrogen-bridged dimer. By recrystallization, these compounds can be obtained with close to 100% ee and in substantial yield. Furthermore, α-pinene can then be displaced by reaction with acetaldehyde to give a chiral alkyldiethoxyborane[28]. These compounds are useful for conversion into many other compounds containing one chiral organic group attached to boron[29].

Preparation of Optically Pure Monoisopinocampheylborane

$$2Ipc_2BH \xrightarrow{\text{TMEDA}} (IpcBH)_2 \cdot TMEDA \xrightarrow{2BF_3 \cdot OEt_2} IpcBH_2$$

(−)-Diisopinocampheylborane (100 mmol) in diethyl ether (65 mL) is prepared as described above in a two-necked flask fitted with a magnetic follower, a nitrogen inlet and a reflux condenser leading to a mercury bubbler. The mixture is brought to reflux and tetramethylethylenediamine (TMEDA; 7.54 mL, 50 mmol) is added dropwise. The mixture is held at reflux for 30 min and then a small aliquot is withdrawn into a syringe and pushed back into the mixture (this process helps to induce crystallization). The mixture is allowed to cool to 25 °C and then kept at 0 °C overnight, after which time the supernatant liquid is removed by means of a double-ended needle. The crystalline TMEDA complex of monoisopinocampheylborane is washed with pentane (3 × 25 mL) and the solid is dried for 1 h at 15 mmHg (drying tube needed to protect from moisture) and 2 h at 1 mmHg to give optically pure $(IpcBH_2)_2 \cdot TMEDA$ (16.4 g, 79%; mp 140.5–141 °C; $[\alpha]_D^{23} + 69.03°$ (THF)). This solid can be stored for prolonged periods and free monoisopinocampheylborane can be liberated when required, as described in the following paragraph.

The complex (14.6 g, 35 mmol) is charged to a 250 mL flask equipped with a magnetic follower and a septum and the flask is flushed with nitrogen. THF (50 mL) is added and the mixture is stirred until the solid dissolves. Trifluoroborane-etherate (8.6 mL, 70 mmol) is added with constant stirring and the mixture is stirred for a further 1.25 h at 25 °C and then filtered under nitrogen through a sinter tube (transfer by double-ended needle). The solid residue is washed with ice-cold THF (3 × 9 mL) and the washings are added to the original supernatant liquid. This combined solution contains monoisopinocampheylborane (typically 80–84% yield) ready for further reaction. It is advisable to check the quantity by gas titration[4].

Preparation of Optically Pure Diethoxy(2-phenylcyclopentyl)borane[28]

A 25 mL, two-necked flask equipped with a magnetic follower, a septum inlet and a tube leading to a mercury bubbler is flushed with nitrogen and then charged with a solution of monoisopinocampheylborane (from (+)-α-pinene) in diethyl ether (52.6 mL of 0.95 mol L^{-1}, 50 mmol). The flask is cooled to $-35\,°C$, 1-phenylcyclopentene (7.2 g, 50 mmol) in diethyl ether (10 ml), precooled to $-35\,°C$, is added dropwise and with stirring, and the mixture is then kept at $-35\,°C$ for 36 h without stirring. A white solid separates and the supernatant liquid is removed by double-ended needle. The crystals are washed with cold diethyl ether (3 × 20 mL) at $-35\,°C$. Diethyl ether (50 ml) is added to the solid and the mixture is warmed to $20\,°C$ for 5–10 min, whereupon a homogeneous solution is obtained. It is then cooled to $0\,°C$ for 15 h and white crystalline needles separate. The supernatant solution is removed by double-ended needle and the crystalline solid is washed with ice-cold diethyl ether (3 × 10 mL). The solid is then dried at 15 mmHg for 1 h to yield isopinocampheyl-(1*S*,2*S*)-*trans*-2-phenylcyclopentyl)borane (10.3 g, 70%) of very high optical purity (>99% ee).

The solid dialkylborane (10.3 g, 35 mmol) is suspended in diethyl ether (30 ml), acetaldehyde (7.84 mL, 140 mmol) is added, and the mixture is stirred at room temperature for 6 h. Excess aldehyde and diethyl ether are pumped off and the residue is distilled under reduced pressure to give diethoxy-(1*S*,2*R*)-*trans*-2-phenylcyclopentyl)borane (8.0 g, 32.5 mmol, 65% based on original monoisopinocampheylborane and alkene; bp 80–82 °C/0.01 mmHg). Oxidation of an aliquot of this compound in the standard way (Section 2) gives (+)-*trans*-2-phenylcyclopentanol of >99% ee.

The optically active alkyldiethoxyboranes prepared as described above can be converted into other compounds by any of the other organoborane reactions described herein, ranging from reduction to the monoalkylborane and further hydroboration to direct cleavage by oxidation. They are therefore extremely important intermediates[29].

Although among chiral monoalkylboranes monoisopinocampheylborane has received most attention, on account of its historical precedence and the cheap availability of α-pinene, it is not necessarily the ideal choice in a number of cases. In particular, with relatively unhindered alkenes it leads to relatively low enantiomeric excesses during hydroboration.

(2-Alkylapopinanyl)boranes offer improved selectivity in some such cases. The alkenes required for the synthesis of these monoalkylboranes are readily prepared by metalation of α-pinene using Schlosser's base (butyllithium plus potassium *tert*-butoxide,) followed by reaction with an alkyl halide, or by further modifications of initial products obtained in this way[30]:

Slow addition of these higher homologues of α-pinene to a slight excess of borane–dimethyl sulfide leads to mixtures of organoboranes in which the monoisopino-campheylborane homolog is greatly predominant (generally over 90%). Furthermore, treatment with TMEDA allows precipitation of the pure bis(monoalkylborane) adducts[31].

Preparation of TMEDA-Bis(2-isopropylisopinocampheylborane)[31]

A dry, 500 mL, round-bottomed flask equipped with a magnetic follower and a pressure-equalizing dropping funnel capped by a septum is flushed with nitrogen and then charged with borane–dimethyl sulfide (10.0 mol L^{-1}, 42 mmol). Dry THF (32 ml) is added, then 2-isopropylapopinene (50.0 mmol) is added dropwise with stirring while the reaction temperature is maintained at 35 °C. The reaction is continued at 25 °C for 24 h. The THF and volatiles are removed under reduced pressure (12 mmHg, 50–55 °C, 1 h). The residue is redissolved in diethyl ether (32 mL) and the solution is refluxed for 0.5 h. TMEDA (21 mmol) is added dropwise to the refluxing mixture of boranes. Once addition is complete, the solution is maintained at reflux for an additional 0.5 h. The solution is cooled to 25 °C and then kept at 0 °C overnight. The supernatant liquid is removed *via* a double-ended needle, and the white solid is washed with cold pentane (50 mL) and then filtered. The solid is dried under vacuum to give TMEDA–bis(2-isopropylapoisopinocampheylborane) in 72–78% isolated yield, with chemical and optical purity each approaching >99%.

If the enantiomeric purity of the original alkylapopinene is high then the enantiomeric purity of the monoalkylborane is also high. However, the enantiomeric purity of the organoboranes can be raised to levels higher than those of the original alkenes by the simple expedient of controlling changes in the steric requirements of the organyl group in the 2-position of apopinene[32].

After liberation from their TMEDA adducts, the free homochiral monoalkylboranes are powerful hydroborating agents and achieve asymmetric hydroboration of relatively hindered prochiral alkenes, *trans*-disubstituted ethenes and trisubstituted ethenes, in higher optical purities than IpcBH$_2$[33]. These are then available for the same kinds of reactions as are found to be useful with alkylmonoisopinocampheylboranes:

An alternative way of modifying the selectivity is to use halogeno(isopinocampheyl)-boranes, which are effective hydroborating agents but have somewhat greater steric requirements than monoisopinocampheylborane itself. In some cases this leads to more favorable chiral induction during hydroboration reactions[34]. Such reagents can be prepared by a number of different methods, of which the most convenient may be the reduction of the corresponding dihalogeno(isopinocampheyl)borane with lithium aluminum hydride[35].

Preparation of Isopinocampheylchloroborane by Reduction of Isopinocampheyldichloroborane with Lithium Aluminium Hydride[35]

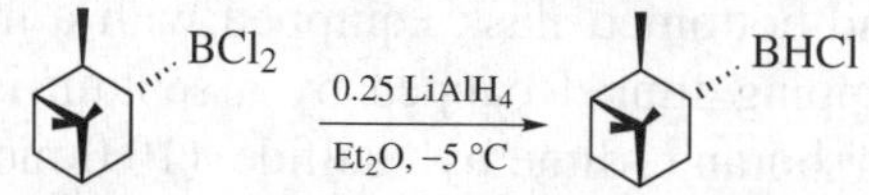

A 25 mL round-bottom flask equipped with a magnetic follower and a septum is flushed with argon and a solution of isopinocampheyldichloroborane (1.33 g, 6.12 mmol) in diethyl ether (5.0 mL) is added. The solution is cooled to $-5\,^{\circ}$C and a solution of lithium aluminum hydride (LAH) in ether (1.05 M, 1.46 mL, 1.53 mmol) is slowly added. After 15 min the solution is ready for use, and ^{11}B NMR shows that the solution is primarily (88%) IpcBHCl, with about 6% each of IpcBH$_2$ and IpcBCl$_2$.

Chiral dialkylhalogenoboranes are increasingly important reagents for enantioselective reductions of ketones (Section 23). The most widely used, known as DIP-Chloride^{TM} in its commercially available form, is diisopinocampheylchloroborane. If it is necessary to make this or a related compound on a laboratory scale, the most convenient procedure may be to reduce the appropriate trihalogenoborane with a trialkylsilane in the presence of the desired alkene, as illustrated below for the preparation of bromobis-(2-isobutylapoisopinocampheyl)borane[36]:

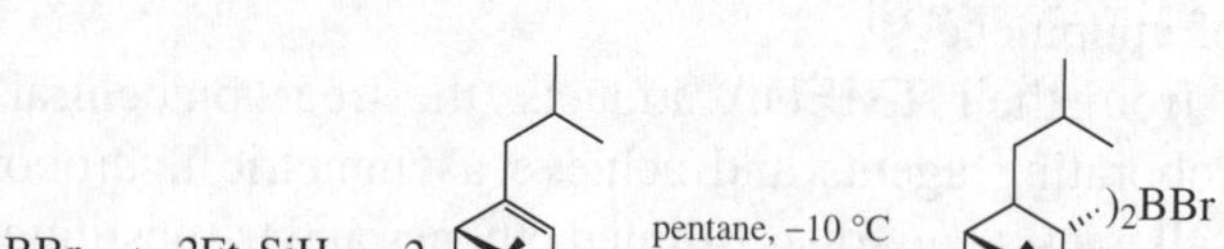

An alternative way of obtaining chiral organoboranes involves reaction of an alkyldialkoxyborane derived from a chiral diol, such as pinanediol, with dichloromethyllithium[37]. The initially formed α-chloroalkylboron compounds are frequently of very high enantiomeric purity and the chloride can be easily displaced in a stereodefined way to give more complex optically active species. This reaction has been developed into a very useful and general approach to asymmetric synthesis (Section 12).

Preparation of (1R)-1-Phenylpentyl-((+)-pinanediyldioxy)borane by Homologation of the Corresponding Butyl Derivative[37]

(a) LiCHCl$_2$, −100 °C
(b) ZnCl$_2$, 20 °C
(c) PhMgBr, −78 to 25 °C

Butyl((+)-pinanediyldioxy)borane

(+)-Pinanediol (341 g, 2 mol) in diethyl ether (500 mL) and light petroleum (bp 30–40 °C, 500 mL) is stirred with boric acid (62.5 g, 1 mol), and a solution of KOH (65 g of 85%, 1 mol) is added in portions, resulting in an exothermic reaction and formation of a voluminous white precipitate, which is collected, washed with ether, dried and recrystallized twice from acetone/water (90 : 10). This gives potassium bis(pinanediol)borate (about 45% yield) of about 100% ee even from pinanediol of 92% ee. To a solution of this solid (23 g) in ice-cold water (75 mL) is added a mixture of diethyl ether and light petroleum (1 : 1, 150 mL), followed by ice-cold hydrochloric acid (65 mL of 2 mol L^{-1}) in small portions, with stirring. After the cloudiness in the aqueous phase disappears, the layers are separated, and the aqueous phase is saturated with NaCl and extracted with ether (50 mL). The combined organic phases are treated with butyldihydroxyborane (butylboronic acid); (11.5 g, 113 mmol) and kept at 25 °C for 2 h, then dried over magnesium sulfate, evaporated and distilled under reduced pressure to give butyl((+)-pinanediyldioxy)borane (about 20 g, 85 mmol, 76%; bp 68–70 °C/0.1 mmHg).

(1R)-1-Phenylpentyl((+)-pinanediyldioxy)borane

A 250 mL flask equipped with a magnetic follower and a septum is flushed with argon, charged with dry THF (50 mL) and pure dichloromethane (3.25 mL, 60 mmol), cooled to −100 °C in a 95% ethanol/liquid nitrogen slush bath, and stirred as butyllithium in hexane (25.5 mL of 2.0 mol L^{-1}, 51 mmol) is added dropwise by syringe down the side of the flask (so that it is already cold on reaching the reaction solution) over a period of 15 min. (The solution should remain colorless or pale yellow although a white precipitate forms; darkening is a sign of overheating and decomposition.) The solution is stirred at −100 °C for a further 15 min. The butyl(pinanediyldioxy)borane (9.98 g, 42 mmol), prepared as above, in anhydrous ether (25 mL) is injected and, under a rapid stream of argon, the septum is briefly removed so that rigorously anhydrous powdered zinc chloride (3.8 g, 28 mmol) can be admitted. The septum is replaced and the flask is flushed thoroughly with argon. The mixture and cooling bath are allowed to warm slowly to 25 °C and stirring is maintained overnight. (It is sometimes possible to use the crude solution directly at this stage, but isolation of the intermediate product is described here.) The mixture is then concentrated on a rotary evaporator (bath temperature below 30 °C) and the thick residue is stirred with light petroleum (bp 30–40 °C, 100 mL) and then saturated aqueous ammonium chloride (25 mL). The phases are separated, the aqueous phase is washed with light petroleum (2 × 50 mL) and the combined organic extracts are filtered through a bed of anhydrous magnesium sulfate. Concentration yields a residue that is chromatographed on silica, eluted with 4% ethyl acetate in hexane. The product at this stage is usually sufficiently pure for further reaction, but it can be distilled if required to give pure (1*S*)-1-chloropentyl((+)-pinanediyldioxy)borane (10.3 g, 86%; bp 113–115 °C/0.2 mmHg).

A 100 mL flask equipped with a magnetic follower and a septum is flushed with argon and charged with dry THF (35 mL) and the chloropentyl(pinanediyldioxy) borane (2.85 g, 10 mmol), prepared as described above. The solution is cooled to −78 °C and stirred as a solution of phenylmagnesium bromide in THF or ether (10 mL of 1 mol L^{-1}, 10 mmol) is added by syringe. The mixture is allowed to warm to 25 °C and stirred for 20 h more. Hydrochloric acid (10 mL of 1.2 mol L^{-1}) is added and the product is extracted into ether (3 × 25 mL). The combined ether extracts are dried over magnesium sulfate, concentrated, and then distilled under reduced pressure to give (1*R*)-1-phenylpentyl((+)-pinanediyldioxy)borane (2.62 g, 80%; bp 125–128 °C/0.1 mmHg).

7   Preparation of Unsaturated Organoboranes

Vinylboranes can in some cases by synthesized by direct hydroboration of an alkyne. This is only possible if the desired product has the right geometry (*i.e.* if boron and hydrogen are *cis* to each other) and if the nature of the other groups is such that the regioselectivity is appropriate (*i.e.* terminal alkenyl groups can be obtained from 1-alkynes, whereas symmetrical internal alkynes offer no regioselectivity problems). Even then, attempts to produce alkenylboranes from unhindered hydroborating agents such as borane–dimethyl sulfide or even 9-BBN-H may be unsuccessful because of dihydroboration or polymerization. Therefore, in reality the method is quite restricted in scope[3]. Nevertheless, for those cases where it is appropriate it is usually the simplest method, and is illustrated by the preparation of dicyclohexyl-1-octenylborane.

Preparation of Dicyclohexyl-(E)-1-octenylborane

Dicyclohexylborane (200 mmol) in THF (100 mL) is prepared in a 500 mL flask as described in Section 4 (except for solvent). The flask is immersed in an ice/salt bath until the temperature is reduced to about $-10\,°C$ and a solution of 1-octyne (22.0 g, 200 mmol) in THF (20 mL) is added as rapidly as possible (to minimize any dihydroboration) while maintaining the temperature below $10\,°C$. The reaction is then allowed to warm to $25\,°C$ and stirred for 3 h to complete the hydroboration (the solid dicyclohexylborane dissolves during this process). The solution thus obtained contains the dicyclohexyloctenylborane, which may be used directly for further transformations.

For aryl-, alkynyl-, most allyl- and many alkenylboron compounds hydroboration cannot be used to put in the unsaturated group. In these cases the most versatile method is to use the corresponding organolithium or organomagnesium reagents. The procedure is similar whichever class of unsaturated organoborane is required, and is illustrated by the preparation of dicyclohexyl-1-octynylborane.

Preparation of Dicyclohexyl-1-octynylborane

Dicyclohexylborane (100 mmol) in THF (50 mL) is generated *in situ* (Section 4), under nitrogen, in a 500 mL flask equipped with a gas inlet with stopcock, a septum inlet and a magnetic follower. Methanol (3.2 g, 100 mmol) is added, dropwise and with stirring, and the hydrogen evolved is allowed to escape through a bubbler connected via a needle through the septum. The solution is stirred for a further 1–2 h at $25\,°C$, during which time hydrogen evolution ceases and the dicyclohexylborane solid dissolves to form dicyclohexylmethoxyborane.

Meanwhile, a separate 250 mL flask equipped with a magnetic follower and a septum is flushed with nitrogen and then charged with 1-octyne (11 g, 100 mmol) and THF (75 mL). The mixture is cooled to $-78\,°C$ and *n*-butyllithium (40 mL of 2.5 mol L^{-1} solution in hexane, 100 mmol) is added dropwise to give 1-lithiooctyne.

The flask containing the dicylcohexylmethoxyborane is cooled in a $-78\,°C$ bath and the 1-lithiooctyne solution is added to it by means of a double-ended needle. The 250 mL flask is rinsed with additional THF (5 mL) to complete the transfer of 1-lithiooctyne and the bulk solution is stirred for 30 min at $-78\,°C$. Trifluoroborane-etherate (18.8 g, 133 mmol) is added by syringe and the mixture is stirred at $-78\,°C$ for a further 15 min and then warmed to $25\,°C$. The reaction mixture can be used directly for many purposes, but if isolation is required the mixture is first concentrated under reduced pressure and then mixed with pentane. The precipitated material is removed under nitrogen by filtration through a glass sinter and the product can then be obtained from the filtrate by removal of the solvent and distillation under reduced pressure.

1,2-Dienes are hydroborated by dicyclohexylborane to give higher crotylboranes with predominantly (*E*) geometry about the double bond[38]. Such compounds are useful for the synthesis of *anti* homoallylic alcohols by reactions with aldehydes and ketones (Section 19):

$$\underset{H}{\overset{R}{>}}C{=}C{=}CH_2 \ + \ Chx_2BH \ \longrightarrow \ \overset{R}{\diagdown}\diagdown\diagup BChx_2$$

8   Preparation of Other Organoboron Compounds

For some purposes it is necessary to have a boron compound of the type R_2BX or RBX_2, where X is a methoxy or halogeno group. Such compounds can be readily obtained by addition of dry HX to the corresponding RBH_2 or R_2BH (as in the procedure for dicyclohexyl-1-octynylborane in Section 7) or, in the case of alkoxy compounds, from the corresponding halogeno compounds by reaction with dry alcohol. Compounds of the type $RB(OEt)_2$ can be obtained from IpcBRH by reaction with acetaldehyde (Section 6) and such compounds can be chain-extended by reaction with the anion of dichloromethane (see Sections 6 and 12)[3]. However, it is also possible to insert the organyl group directly by hydroboration using HBX_2 or H_2BX derivatives. Of these, the most conveniently available are catecholborane, dibromoborane–dimethyl sulfide[39–41] and monochloroborane–dimethyl sulfide[42,43]. The first two are available commercially, but are easily prepared from catechol and borane[3,4,44] or tribromoborane–dimethyl sulfide and borane–dimethyl sulfide[3] if desired.

Preparation of B-((E)-2-Cyclohexylethenyl)catecholborane[44]

A 100 mL, two-necked flask equipped with a reflux condenser vented to a mercury bubbler, a PTFE stopcock capped by a silicone rubber septum and a magnetic follower is flushed with nitrogen. Cyclohexylethyne (10.8 g, 100 mmol) and catecholborane (12.1 g, 100 mmol) are added by syringe and the mixture is stirred and heated to 70 °C for 1 h. After cooling the product is ready for further

> reaction, but if required it can be distilled (114 °C/2 mmHg) to provide pure *B*-((*E*)-2-cyclohexylethenyl)catecholborane (18.7 g, 82%).

Reactions of catecholborane, such as that described above, are intrinsically rather slow. Indeed, internal alkynes require longer reaction times (about 4 h at 70 °C) and those of alkenes require even higher temperatures (about 100 °C)[44,45]. Fortunately, the reactions can be performed at moderate temperatures if catalyzed by rhodium complexes[46] and this even allows the possibility of asymmetric hydroboration by use of homochiral catalysts[47]. The use of catalyzed hydroboration has been reviewed[6].

Preparation of Hexyldibromoborane[39–41]

$$Br_3B \cdot SMe_2 \xrightarrow{BH_3 \cdot SMe_2} Br_2BH \cdot SMe_2 \xrightarrow{\text{1-hexene}} Br_2B(CH_2)_5CH_3 \cdot SMe_2$$

$$\downarrow Br_3B$$

$$Br_2B(CH_2)_5CH_3$$

A 100 mL flask equipped with a septum and a magnetic follower is flushed with nitrogen and then charged with dimethyl sulfide (14.9 mL, 12.4 g, 200 mmol) and pentane (50 mL). The flask is cooled in an ice bath and fitted with a bubbler connected *via* a needle and the contents are stirred vigorously while tribromoborane (9.5 mL, 25.3 g, 100 mmol) is added dropwise by syringe (exothermic reaction). The mixture is brought to 25 °C and the volatile materials are removed at the pump (protection from moisture) to leave tribromoborane–dimethyl sulfide (31.5 g, 99%; mp 106–107 °C) as a white powder.

Dimethyl sulfide (10 mL) and borane–dimethyl sulfide (4.75 mL, 47.5 mmol) are added, with stirring, and the mixture is held at 40 °C for 12 h. Excess dimethyl sulfide is removed at the pump (protection from moisture) to leave dibromoborane–dimethyl sulfide (35.1 g, about 100%) as a clear, viscous liquid.

A two-necked flask equipped with a reflux condenser vented through a mercury bubbler, a PTFE stopcock capped by a silicone rubber septum and a magnetic follower is flushed with nitrogen and charged with dichloromethane (75 mL) and 1-hexene (12.5 mL, 100 mmol) by syringe. The mixture is stirred at 25 °C during the dropwise addition of dibromoborane–dimethyl sulfide (12.8 mL, 100 mmol), prepared as above, and then heated under reflux for 3 h. On cooling to 25 °C the product, hexyldibromoborane–dimethyl sulfide, is ready for many further manipulations.

If required, the product can be distilled (bp 99–100 °C/1 mmHg) to give the pure complex (29 g, 91%). If the product free of dimethyl sulfide is needed, the mixture after the hydroboration step is cooled to 0 °C and stirred during addition of dibromoborane (10.0 mL, 105 mmol). The mixture is stirred at 25 °C for 1 h and the solvent is removed to leave a mixture of liquid hexyldibromoborane and solid tribromoborane–dimethyl sulfide. This mixture is distilled directly under reduced pressure at a bath temperature less than $100\,^{\circ}\text{C}$ ($Br_3B{\cdot}SMe_2$ melts at $108\,^{\circ}\text{C}$) to give pure hexyldibromoborane (18.0 g, 71%; bp $56\text{--}58\,^{\circ}\text{C}/0.9$ mmHg).

Preparation of Chlorodicyclopenylborane[3,42,43]

$$BH_3{\cdot}SMe_2 \xrightarrow{\ CCl_4\ } BH_2Cl{\cdot}SMe_2 \longrightarrow \left(\langle\rangle\right)_2 BCl$$

A two-necked flask equipped with a septum, a reflux condenser leading to a mercury bubbler and a magnetic follower is flushed with nitrogen and charged with borane–dimethyl sulfide (10.1 mL, 100 mmol). Tetrachloromethane (9.7 mL, 15.4 g, 100 mmol) is added, with stirring, and the mixture is held under reflux for 20 h, by which time the product is predominantly $BH_2Cl{\cdot}SMe_2$. Estimation of an aliquot by gas titration shows the presence of two mole equivalents of hydrogen per mole of boron and hydrolysis and titration of the liberated HCl show one mole equivalent of labile chlorine per mole of boron. However, ^{11}B NMR shows the presence of equilibrium quantities (about 8% each) of $Cl_2BH{\cdot}SMe_2$ ($\delta - 2.0$ ppm) and $BH_3{\cdot}SMe_2$ ($\delta - 19.8$ ppm) in addition to $Cl_2BH{\cdot}SMe_2$ ($\delta - 6.7$ ppm).

A 200 mL flask equipped with a septum and a magnetic follower is flushed with nitrogen and charged with cyclopentene (14.7 g, 216 mmol) in pentane or diethyl ether (90 mL). The mixture is stirred at $0\,^{\circ}\text{C}$ while the monochloroborane–dimethyl sulfide, prepared as described above, is added slowly by syringe. The mixture is then allowed to warm to $25\,^{\circ}\text{C}$ and stirred for 2 h at this temperature. The solution thus obtained can be used directly for further transformations or the solvent may be removed under reduced pressure to leave chlorodicyclopentylborane which is about 93% pure (containing some $RBCl_2$ and R_3B). Distillation under reduced pressure gives the pure product (about 80% yield); bp $69\text{--}70\,^{\circ}\text{C}/1.2$ mmHg). Unhindered dialkylchloroboranes may retain the dimethyl sulfide during solvent removal but it is lost during distillation.

Thexylchloroborane–dimethyl sulfide is prepared in a manner similar to that used for chlorodicyclopentylborane, but using monochloroborane–dimethyl sulfide (100 mmol), dimethyl sulfide (1.5 mL, 20 mmol), dichloromethane (24 mL) and 2,3-dimethyl-2-butene (14.8 mL, 110 mmol). The alkene is added dropwise to the stirred mixture of the other components at $0\,^{\circ}\text{C}$, over a period of 1 h, and the resultant clear solution is stirred for 30 min at $0\,^{\circ}\text{C}$ and 3 h at $25\,^{\circ}\text{C}$ to complete the preparation. The solution obtained can be used directly or the solvent and excess dimethyl sulfide can be removed under reduced pressure (protection from moisture).

9 Replacement of Boron by a Functional Group

The simplest applications of organoboron compounds involve replacement of boron by a functional group such as OH, halogen or NH_2[3,4]:

$$\text{B—R} \longrightarrow \text{R—X}$$
$$(X = \text{OH, Cl, Br, I, } NH_2)$$

The most common application is oxidation with alkaline hydrogen peroxide, which converts alkylboron compounds into the corresponding alcohols (Section 2). All stereochemical features of the organoborane are retained in the alcohol and all three groups attached to boron can be utilized.

In some cases the sensitivity of other functionalities present in the organic product may dictate that oxidants other than alkaline hydrogen peroxide must be used to convert an organoborane into the corresponding alcohol. In such cases, buffered hydrogen peroxide, *meta*-chloroperbenzoic acid or trimethylamine *N*-oxide may be used with advantage[3].

Oxidation of vinylboranes gives aldehydes or ketones. However, if the oxidant used is (diacetoxyiodo)benzene the product is the corresponding enol acetate, with stereochemistry inverted with respect to the initial alkenylborane[48]:

$$R^1 \diagdown\!\!\!\diagup\!\!\!\diagdown \text{B(OR}^2)_2 \xrightarrow[\text{NaI, DMF, 20 °C}]{\text{PhI(OCOCH}_3)_2} R^1 \diagdown\!\!\!\diagup\text{OCOCH}_3$$

There are other types of reactions that bring together a group initially attached to boron (particularly an unsaturated group) and an oxygen-based moiety, but these have closer similarity to C–C bond-forming cross-coupling reactions and are therefore considered together with such reactions in Section 18.

Replacement of boron by bromine or iodine is best carried out by use of sodium methoxide in the presence of the free halogen. In the case of iodinolysis only two of the three alkyl groups on a trialkylborane are readily converted into iodoalkane[49], whereas in the case of brominolysis all three groups can be utilized[50]. In both cases the process occurs with complete inversion of stereochemistry at the displaced carbon atom[51]:

$$\left(\diagup\!\!\!\diagdown\right)_3 \text{B} + 2I_2 + 3\text{NaOMe} \longrightarrow 2 \underset{\text{I}}{\diagup\!\!\!\diagdown} + 2\text{NaI} + \text{Na}^+ \; \diagup\!\!\!\diagdown\text{B(OMe)}_3^-$$

Aryldihydroxyboranes can be converted into aryl bromides[52] and vinyldihydroxyboranes can be converted into vinyl bromides, again with inversion[53]:

$$\underset{R}{\overset{H}{\diagdown}}C\!=\!C\underset{H}{\overset{B(OH)_2}{\diagup}} \xrightarrow{\text{Br}_2,\ \text{NaOH}} \underset{R}{\overset{H}{\diagdown}}C\!=\!C\underset{Br}{\overset{H}{\diagup}}$$

The procedure for preparation of methyl 11-bromoundecenoate is representative.

Preparation of Methyl 11-Bromoundecenoate[50]

$$3H_2C=CH(CH_2)_8CO_2Me \xrightarrow[0\,°C]{BH_3 \cdot THF} B[(CH_2)_{10}CO_2Me]_3 \xrightarrow[0\,°C]{Br_2, NaOMe} 3Br(CH_2)_{10}CO_2Me$$

A two-necked flask equipped with a magnetic follower, a stopcock-protected septum and a pressure-equalizing dropping funnel connected to a bubbler is flushed with nitrogen. THF (75 mL) and methyl 10-undecenoate (29.7 g, 33.5 mL, 150 mmol) are added and the mixture is cooled to 0 °C. Borane–THF (50.0 mL of 1.0 mol L^{-1} solution in THF, 50 mmol) is added dropwise with stirring and the mixture is then stirred for 30 min at 0 °C and 30 min at 25 °C to complete the hydroboration. Methanol (1 mL) is added and the temperature is again lowered to 0 °C. Bromine (10.0 mL, 200 mmol) is added dropwise at such a rate as to maintain a temperature of 0 °C. A freshly prepared solution of sodium methoxide in methanol (60 mL of a 4.16 mol L^{-1} solution, 250 mmol) is then added dropwise and with stirring over 45 min in order that the temperature remains below 5 °C. The reaction mixture is allowed to warm to 20 °C and treated with pentane (50 mL), water (20 mL) and saturated aqueous potassium carbonate (20 mL). The pentane layer is separated and the aqueous layer is extracted with further pentane (3 × 50 mL). The combined pentane extracts are washed with water (2 × 50 mL) and then brine (50 mL), dried over K$_2$CO$_3$, filtered and evaporated under reduced pressure to give a colorless oil (41.2 g, 98%). Reduced pressure distillation at 126–128 °C/0.65 mmHg gives methyl 11-bromoundecanoate (35.4 g, 85%) contaminated by a small amount of methyl 10-bromoundecanoate.

There are several ways of converting trialkylboranes into primary amines, of which the use of hydroxylamine-*O*-sulfonic acid[54] and use of *in situ*-generated chloramine[55] are worthy of particular note, though a maximum of only two of the alkyl groups can be utilized. A recent application resulted in the production of enantiomerically pure isopinocampheylamine from methyldiisopinocampheylborane[56].

Preparation of Methyl 11-Aminoundecanoate[55]

$$B[(CH_2)_{10}CO_2Me]_3 \xrightarrow{aqueous\ NH_3,\ NaOCl} H_2N(CH_2)_{10}CO_2Me$$

Methyl 10-undecenoate (30 mmol) is hydroborated in THF solution as described in the preceding procedure. The solution is cooled to 0 °C and aqueous ammonium hydroxide (4.9 mL of 2.05 mol L^{-1} solution, 10 mmol) is added, followed by commercial bleach (15.4 mL of a 0.78 mol L^{-1} solution, 12 mmol), dropwise. A precipitate forms and the suspension is stirred for 15 min at 0 °C and then allowed to warm to 25 °C. The mixture is made acidic by addition of 10% hydrochloric acid and then extracted with diethyl ether (2 × 50 mL). The aqueous layer is made alkaline with aqueous NaOH (3 mol L^{-1}) and the product is extracted into ether (2 × 75 mL). The combined ether layers are washed with brine and then dried over KOH. Removal of the solvent gives methyl 11-aminoundecanoate (1.64 g, 76% based on transformation of just one alkyl group). The method can be applied for incorporation of ^{15}N and the authors also state that use of two mole equivalents of ammonium hydroxide permits utilization of two of the three alkyl groups[55].

Alkyldichloroboranes react with alkyl azides to give dialkylamines, with one of the alkyl groups emanating from the borane and one from the azide[57]. The reaction can be particularly useful for the synthesis of cyclic amines[58]:

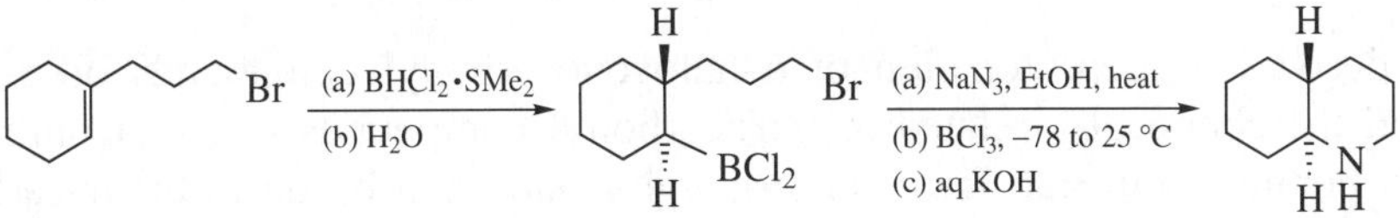

There are also possibilities for converting organoboranes into *N*-alkylsulfonamides[59] and other nitrogen derivatives[3]. Some of the latter reactions are mechanistically quite different and are more akin to cross-coupling reactions, involving coupling of a boron-bound group with an external amine or related compound, under the influence of a catalyst. They are therefore considered along with C–C bond-forming cross-coupling reactions in Section 18.

Another important simple cleavage reaction is protonolysis, which is included here although there is no replacement by a functional group. With a few exceptions (such as allylboranes), boron–carbon bonds are not readily cleaved by water. Rather, carboxylic acids are more commonly used to cleave C–B bonds protonolytically, and even then it is necessary to raise the temperature to about 160 °C for effective removal of all three groups[60]. The reaction is capable of maintaining all the stereochemical features of the organoborane.

Preparation of cis-Pinane by Protonolysis of Trimyrtanylborane[3,4,60]

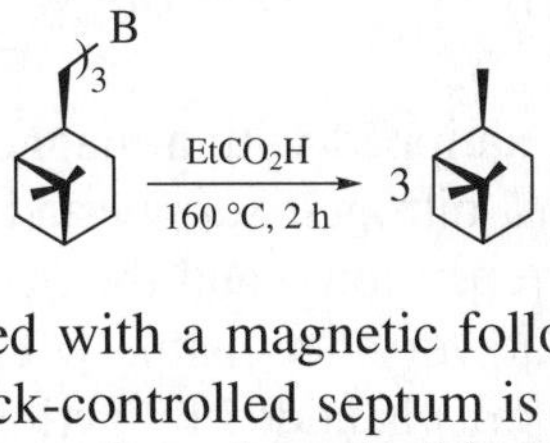

A two-necked flask equipped with a magnetic follower, a condenser leading to a mercury bubbler and a stopcock-controlled septum is flushed with nitrogen, Trimyrtanylborane (33 mmol) is prepared as described in Section 2 and then the solvent is removed under reduced pressure and replaced by diglyme (33 mL). Degassed propanoic acid (11 mL, about 50% excess) is added and the mixture is stirred and heated under reflux (about 160 °C) for 2 h, then cooled. Excess of aqueous NaOH (3 mol L^{-1}) is added and the diglyme phase is diluted with pentane (50 mL). The organic phase is separated, washed with ice-water (5 × 50 mL) to remove diglyme, dried over MgSO$_4$ and evaporated to give *cis*-pinane (12.4 g, 90%).

Vinylboranes are generally protonolyzed more readily than alkylboranes and still yield products with complete retention of configuration[3,4].

10   α-Alkylation of Carbonyl and Other Compounds by Organoboranes

Anions derived by deprotonation of α-halogenocarbonyl compounds or α-halogeno-nitriles, and the chemically related α-diazocarbonyl compounds or α-diazonitriles, react with organoboranes with transfer of an organyl group from boron to the α-carbon atom. Hydrolysis produces the corresponding α-alkyl- or α-aryl-carbonyl compounds or -nitriles[3].

$$\begin{array}{c} \text{(a) } \overset{-}{\text{C}}\text{HCOY} \\ | \\ \diagdown \\ \diagup \text{B-R} \xrightarrow[\text{(b) } H_2O]{\quad X \quad} RCH_2COY \end{array}$$

$$(X = Cl, N_2{}^+; \; Y = OEt, CH_3, \text{etc.})$$

Only one of the three alkyl groups of a trialkylborane is transferred and for optimum utilization of organic residues it is preferable to use organyldichloroboranes in the case of diazocarbonyl compounds[61] or B-alkyl-9-BBN derivatives in the case of α-halogeno-carbonyl compounds[62,63]. Presumably, organyldibromoboranes will behave in a similar way to organyldichloroboranes.

Preparation of Ethyl p-Chlorophenylacetate from Ethyl Diazoacetate and p-Chlorophenyldichloroborane[61]

$$\text{Cl}-\!\!\left\langle\!\!\bigcirc\!\!\right\rangle\!\!-\text{BCl}_2 \xrightarrow[\text{(b) } H_2O]{\text{(a) } N_2CHCO_2Et} \text{Cl}-\!\!\left\langle\!\!\bigcirc\!\!\right\rangle\!\!-\text{CH}_2\text{CO}_2\text{Et}$$

A 50 mL, two-necked flask equipped with a magnetic follower, a septum-capped stopcock and a line leading to a nitrogen supply and a vacuum pump is evacuated and filled with nitrogen (three repetitions) and then cooled to −25 °C by immersion in an acetone/tetrachloromethane/dry ice bath. *p*-Chlorophenyldichloroborane[64] (1.94 g, 10 mmol) in THF (10 mL) is added by syringe and then stirred while ethyl diazoacetate (1.25 g, 11 mmol) in THF (10 mL) is added at a rate which allows smooth liberation of nitrogen (about 1 mL every 4 min). With the mixture still stirring at −25 °C, water (5 mL) and methanol (5 mL) are added and the cooling bath is then removed. The mixture is poured into saturated aqueous ammonium carbonate (75 mL) and extracted with diethyl ether (3 × 50 mL). The combined ether extracts are dried over magnesium sulfate and the solvent is then removed under reduced pressure. Distillation under reduced pressure gives ethyl *p*-chlorophenylacetate (1.80 g, 91%; bp 106–107 °C/3.5 mmHg). Alkyldichloroboranes, as opposed to aryldichloroboranes, are reacted at lower temperatures, around −62 °C, and generally give lower yields.

Preparation of a Steroid Nitrile by α-Alkylation of α-Chloroacetonitrile[62,63]

Solid 9-BBN-H (0.13 g, 1.05 mmol) is placed in a 50 mL flask equipped with a magnetic follower and a stopcock fitted with a septum and connected to a nitrogen supply and a vacuum pump. The system is repeatedly evacuated and refilled with nitrogen and then the steroid **2** (0.35 g, 1.03 mmol) dissolved in THF (2 mL) is added by syringe, with stirring. The hydroboration step is allowed to proceed for 15 h at 20 °C, and the mixture is then cooled to 0 °C. A slurry of freshly prepared potassium 2,6-di-*tert*-butyl-4-methylphenoxide in THF (2.19 mL of 0.47 mol L^{-1} slurry, 1.03 mmol) is added via a syringe fitted with a wide-bore needle, followed by chloroacetonitrile (65 μL, 1.03 mmol). The mixture is stirred for 1 h at 0 °C, then ethanol (0.4 mL) is added and the whole is stirred for 15 min at 25 °C. Hexane (5 mL) is added and the solution is extracted with aqueous sodium hydroxide (3 × 25 mL of 1 mol L^{-1}) and water (2 × 20 mL). The organic phase is dried over magnesium sulfate and concentrated under reduced pressure. Chromatography on silica gel gives pure **3** (0.26 g, 65% yield); mp 179–181 °C.

Similar reactions can be carried out with the anion derived from allyl chloride. The anion is generated by reaction with a hindered lithium dialkylamide base at low temperature, and reacts with alkoxyboracyclanes, for example, to give ring-expanded products[65]:

As the immediate products are allylboranes, the utility of the process can be extended further by subsequent reaction with carbonyl compounds (Section 19). Alternatively, the initial products can be encouraged to undergo an allylic rearrangement by decomplexation of the amine (if this is a problem) and being allowed to heat up[65]:

In any of these reactions the product is finally fixed when the boron is removed by oxidation or hydrolysis.

These reactions all have in common the availability of a leaving group α to boron, which is displaced by a group migrating from boron, typically with inversion of configuration at the recipient carbon and retention at the migrating center. When the reagent possesses more than one leaving group it is possible to bring about two or even three such rearrangements. Such reactions have no parallel with other types of organometallic reagents, thereby offering unique synthetic opportunities. Some such reactions are discussed in the following sections.

11   Ketones and Tertiary Alcohols Through Reactions of Organoboranes with Acyl Carbanion Equivalents

Acyl carbanion equivalents, or anions with two α-leaving groups, can lead to transfer of two organyl groups from boron to carbon. Subsequent oxidation leads to the appropriate boron-free product, typically a secondary or tertiary alcohol. Of the various types of acyl carbanion equivalents available, anions of 1,1-bis(phenylthio)alkanes[66] or 2-alkyl-1,3-benzodithioles[67] are the most useful of the readily prepared ones for this application. The former anions are more hindered than the latter and are chosen for reactions with relatively unhindered organoboranes, whereas the latter are used for more hindered organoboranes[67].

Preparation of Cyclohexyldicyclopentylmethanol from Thexyldicyclopentylborane and 2-Cyclohexyl-1,3-benzodithiole[67]

A 100 mL, three-necked flask equipped with a magnetic follower, a septum-capped stopcock, an angled, rotatable side arm and a line to a nitrogen supply and vacuum pump is charged with 2-cyclohexyl-1,3-benzodithiole (0.472 g, 2 mmol) in the flask and mercury(II) chloride (1.63 g, 6 mmol) in the side arm and then repeatedly evacuated and refilled with nitrogen. THF (5 mL) is added to dissolve the dithiole and the solution is cooled to $-30\,^{\circ}$C. Butyllithium (1.37 mL of 1.6 mol L^{-1} solution in hexane, 2.2 mmol) is added by syringe and the mixture is stirred for 75 min at $-30\,^{\circ}$C.

Meanwhile, in a separate flask thexyldicyclopentylborane (2 mmol) in THF (5 mL) is prepared as described for other mixed organoboranes in Section 5. This solution is transferred to the flask containing the organolithium solution by syringe and further THF (2 mL) is used to transfer last traces. The mixture is allowed to warm to $25\,^{\circ}$C over 1 h and then stirred for 3 h. (Note that for ketone synthesis (as in the following preparation) the mixture is directly oxidized at this point.)

The mixture is cooled to $-78\,^{\circ}$C and stirred vigorously as the side arm is rotated to release the mercury(II) chloride into the flask. The mixture is allowed to warm to $25\,^{\circ}$C and stirred overnight. (Note that for ketone synthesis the following oxidation procedure is carried out directly on the mixture without introduction of mercury(II) chloride.)

The mixture is cooled to $0\,^{\circ}$C and then aqueous sodium hydroxide (10 mL of 5 mol L^{-1}) and 50% aqueous hydrogen peroxide (7 mL) are successively added, the latter dropwise and with care. The mixture is stirred at $25\,^{\circ}$C for 5 h and the product is then extracted into pentane (2×100 mL), washed with water (2×100 mL), dried over sodium sulfate, filtered and evaporated under reduced pressure. The syrupy residue is chromatographed on alumina (activity III), eluted successively with pentane (100 mL), dichloromethane/pentane (1 : 1, 150 mL) and dichloromethane (250 mL). The dichloromethane fraction is evaporated under reduced pressure and then pumped overnight to remove residual 2,3-dimethyl-2-butanol, leaving cyclohexyldicyclopentylmethanol (0.40 g, 80%).

If the addition of mercury(II) chloride is omitted from the above procedure only a single migration takes place. Oxidation then gives an aldehyde or ketone, depending upon the 2-substituent of the benzodithiole moiety. The optimum utilization of boron-bound alkyl groups is achieved by use of *B*-alkyl-9-BBN derivatives, but to avoid unnecessary duplication of procedures, the example below uses a thexyldialkylborane[68].

Preparation of Cyclohexyl Cyclopentyl Ketone by Reaction of
Thexyldicyclopentylborane with 2-Cyclohexyl-1,3-benzodithiole[68]

The procedure is as described in the preceding procedure, except that there is no need for the angled side arm charged with mercury(II) chloride, until the point when the mercury chloride would have been added. At that point oxidation as described later in the same procedure is carried out. Isolation of the product is also very similar, but silica rather than alumina is used in the chromatography. The yield of the ketone is about 76%.

The anion of dichloromethane has been used in a way similar to that described in this section for benzodithiole anions but on alkyldialkoxyboranes rather than trialkylboranes. Furthermore, if the dialkoxyborane part is generated from an optically active diol the rearrangement step can take place in a highly stereoselective manner to give chiral products with high enantiomeric purities. This process has been developed into one of the most powerful general approaches known for asymmetric synthesis and therefore justifies a separate section (Section 12).

12 Asymmetric Synthesis of Alcohols and Other Compounds by Alkyldialkoxyboranes and Dichloromethyllithium

Although dichloromethyllithium reacts readily with triphenylborane, resulting in migrations of two phenyl groups from boron to carbon in a manner similar to that described for other reagents in Section 10 but under milder conditions[69], its more important reactions are with alkyldialkoxyboranes, where only one alkyl group is available for migration. The immediate product is an α-chloroalkyldialkoxyborane, which reacts with nucleophiles to yield other α-substituted products[3].

$$\text{RB(OR}^1)_2 \xrightarrow[\text{low temp.}]{\text{LiCHCl}_2} \text{RCHClB(OR}^1)_2 \xrightarrow{\text{R}^2\text{MgCl}} \text{RR}^2\text{CHB(OR}^1)_2$$

Furthermore, if the alkoxy groups are chiral the reaction is useful for asymmetric synthesis. A simple example of the use of this procedure for synthesis of a chiral organoborane is given in Section 6. The reaction has been well reviewed by Matteson, whose group developed it[70]. The success of the reaction depends on a number of special features, as discussed in the following paragraphs.

First, for optimum stereoselectivity the two alkoxy groups are replaced by a single, chiral alkylenedioxy group. In early reports cyclic pinanediol derivatives were used, but in more recent reports use of cyclic products derived from C_2-symmetric chiral diols has been more common. In particular, much use has been made of products derived from diisopropylethanediol (DIPED)[71] or dicyclohexylethanediol (DICHED)[72]. Very high enantioselectivities can be achieved with all three types under appropriate conditions.

The second important feature is the use of zinc chloride as a coreactant. This complexes to a chlorine atom, thereby promoting displacement of the chloride group and allowing the reaction to be conducted at a lower temperature. Lower temperatures enhance the selectivity of the process. Furthermore, by complexing the liberated chloride ion the zinc chloride also inhibits epimerization of the intermediate α-chloroalkylboron compound, a reaction that is catalyzed by chloride ion. This leads to further enhancement of the selectivity.

Finally, the two epimeric intermediates may undergo different reactions, with the major epimer proceeding as described thus far, but with the conformation of the minor epimer leading to preferential migration of an alkoxy group rather than the alkyl group. In this way, even the small amount of the minor diastereoisomer at the intermediate stage fails to produce a similar level of the minor epimer following rearrangement. Consequently, after removal of boron following oxidation the alcohol is obtained with exceptionally high enantiomeric purity. Furthermore, prior to oxidation the intermediate organoborane is capable of undergoing the entire sequence of reactions a second time, thereby permitting the introduction of two contiguous asymmetric centers[73]. The process can also be repeated multiple times to generate several contiguous asymmetric centers:

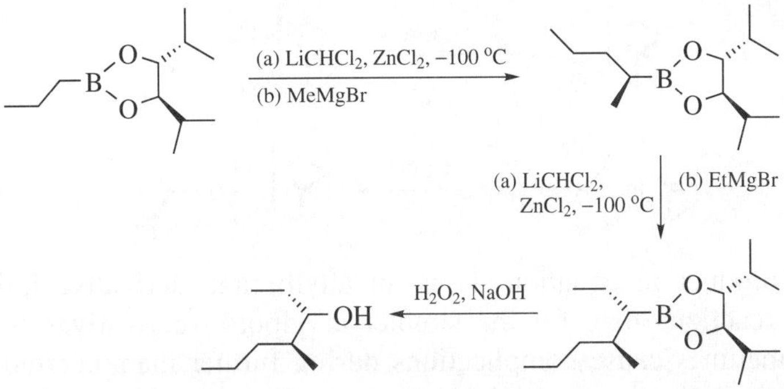

In the example illustrated above, the nucleophiles used in both stages to displace chloride from the intermediate α-chloroalkylboron compounds are reactive alkylmetal reagents. However, this need not be the case, and use of other types of nucleophilic reagents adds to the general utility of the reaction.

For example, use of an enolate as the nucleophile leads (after oxidation of the intermediate boron compound) to aldol-like products, though not always in high enantiomeric purity unless special precautions are taken[74]. Use of lithium benzyloxide as the nucleophile allows introduction of an oxygen-bound substituent that can eventually be cleaved by hydrogenation to give an alcohol, and repetition of the process allows the build-up of a polyol with several contiguous asymmetric centers, such as are found in simple sugars[70]. Introduction of an amino group can be achieved by use of lithium heamethyldisilazide or, provided the bromo analog rather than the chloro compound is used, sodium azide, under phase transfer conditions. The synthesis of serine illustrates how both OH and NH_2 groups can be incorporated into the same pathway[70]:

Deuterium has also been introduced stereoselectively *via* a deuteriated hydride[70]. The groups originally bound to boron in the above examples were also alkyl groups, but this is another unnecessary restriction. For example, the usual reactions with dichloro-methyllithium occur with the pinanediol derivatives of vinylboronic acid (equation 1)[70] and allylboronic acid (equation 2)[75]:

$$(1)$$

$$(2)$$

Because the product in equation (1) is an allylborane derivative it is subject to a wider range of reactions than for the simpler alkylboron derivatives (see Section 19), and this can sometimes cause complications during further manipulation. However, the homoallylic boron derivative formed in equation (2) behaves normally in most cases. Thus the Matteson reaction is an extremely powerful and general method.

13   Ketones, Tertiary Alcohols and Carboxylic Acids through Reactions of Organoboranes with Haloform Anions

Anions derived from dichloromethyl methyl ether (DCME), chloroform or related halo-forms react with trialkylboranes with spontaneous migration of all three alkyl groups from boron to carbon. Oxidation yields a tertiary alcohol in which all of the structural features of the trialkylborane have been riveted into the corresponding hydroxycarbon compound[76]:

With small modifications the procedure can be used to convert even very hindered organoboranes into the corresponding tertiary alcohols[77].

Preparation of Cyclohexylcyclopentylthexylmethanol

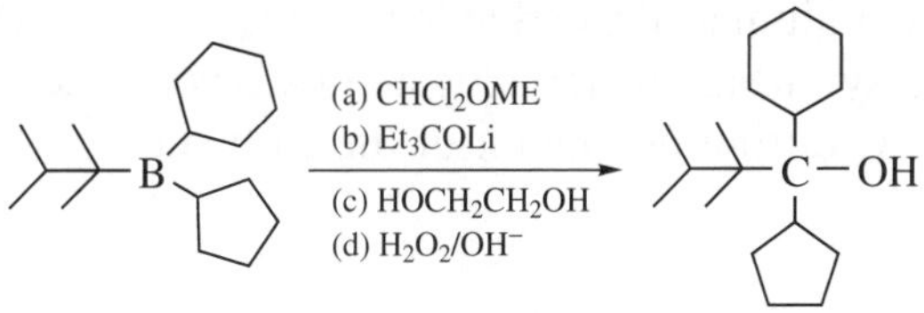

A 500 mL, three-necked flask equipped with a magnetic follower, septum-capped stopcock, reflux condenser and mercury bubbler is flushed with nitrogen. Thexyl-cyclohexylcyclopentylborane (100 mmol) in THF (100 mL) is prepared in the flask by a procedure analogous to that described for other thexyldialkylboranes in Section 5. The solution is cooled to 0 °C and stirred while purified DCME (25.3 g, 0.22 mol, excess) is added, followed by a solution of lithium triethylmethoxide (111 mL of a 1.8 mol L^{-1} solution in hexane, 0.20 mol, excess) dropwise over a period of 20–30 min. The mixture is brought to 25 °C and stirred for a further 30 min, during which time a heavy precipitate of lithium chloride forms. Ethylene glycol (12.4 g, 0.20 mol, excess) is added and the volatile materials are removed at the pump. Ethanol (95%, 50 mL) and THF (25 mL) are added and then solid NaOH (24 g, 0.6 mol). When most of the solid has dissolved, hydrogen peroxide (30%, 50 mL) is added, cautiously and with stirring, over 2 h, while the temperature of the mixture is kept below 50 °C. To complete the oxidation the mixture is held at 55–60 °C for 2 h, then cooled and saturated with NaCl. The organic layer is separated and the aqueous layer is extracted with diethyl ether (3 × 50 mL). The combined organic extracts are dried over magnesium sulfate and evaporated under reduced pressure. Rapid kugelrohr distillation (in two portions) at 0.5 mmHg gives cyclohexylcyclopentylthexylmethanol (17.4 g, 66%; bp 115–120 °C/0.5 mmHg).

If the same procedure is applied to dialkylmethoxyboranes the two alkyl groups are transferred from boron to carbon and oxidation gives the corresponding ketone[3,78]:

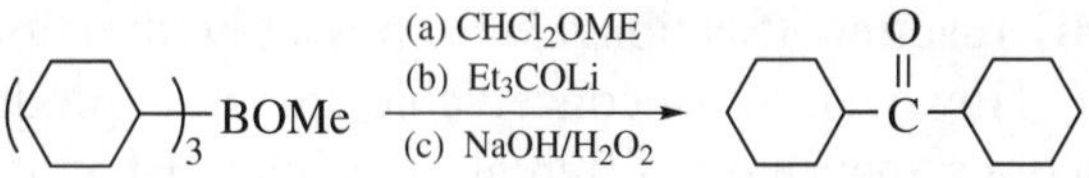

The reaction can be extended to the synthesis of unsaturated ketones from the corresponding organoboranes, and if the alkyl group on boron is chiral the chirality is carried through into the product[79]:

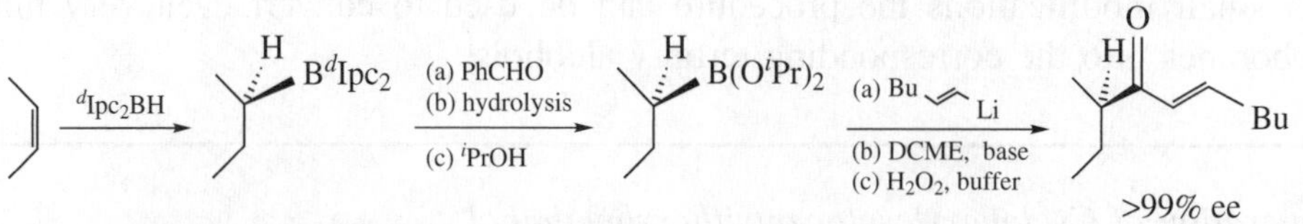

Use of trichloromethyllithium as the carbanion source and 2-alkyl-1,3,2-dithiaborolanes as substrates allows the synthesis of homologated carboxylic acids, but the temperature has to be kept at $-100\,°C$ during utilization of the carbanion[3,80]:

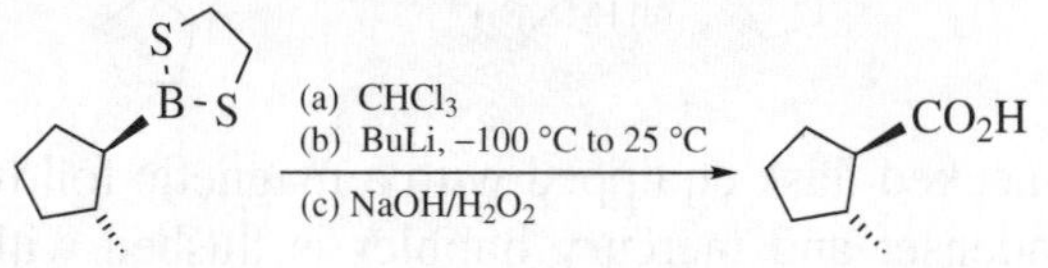

14   Aldehydes, Ketones and Alcohols Through Carbonylation of Organoboranes

The reactions of organoboranes with carbon monoxide (carbonylation reactions) are extremely versatile[3,81]. The reaction is rather slow and requires heating to about $100\,°C$ and/or the use of elevated pressures in order to achieve a reasonable rate. Oxidation of the mixture after such a process gives rise to ketones[3,4,82]:

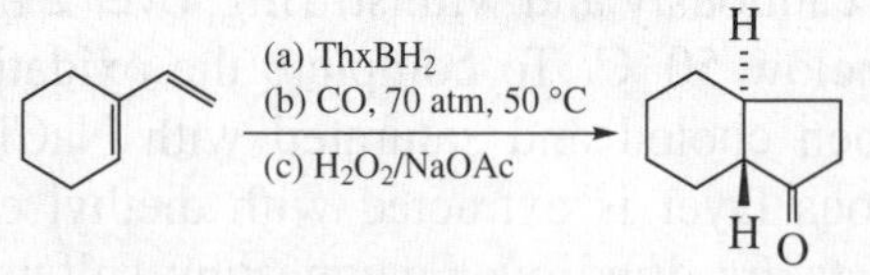

In practice, the use of the cyanoborate reaction (Section 15) achieves the same overall result under rather more convenient conditions, but the carbonylation reaction preceded the cyanoborate reaction and has been applied to a wider range of organoborane substrates[3,81]. The use of thexyldialkylboranes conserves valuable alkyl groups, and has the overall effect of replacement of a thexylboron unit by a carbonyl group.

If the temperature is raised to about $150\,°C$, especially if ethylene glycol is added, a third rearrangement may occur, giving rise to the riveted product after oxidation[81]. In practice, the DCME reaction (Section 13) is probably the first choice method for such a transformation. However, it is conceivable that the product from the DCME and carbonylation reactions may have different stereochemistry at the newly generated hydroxycarbon center, so that the two methods may not always be interchangeable. A third possible method involves the cyanoborate reaction (Section 15) and in this case a difference has been observed with one substrate[83]. Therefore, it is of interest to have a procedure, as illustrated by the preparation of a perhydrophenalenol[84].

Preparation of cis,cis,trans-Perhydro-9B-phenalenol from
trans,trans,trans-1,5,9-Cyclododecatriene[3,4,84]

A three-necked, 1 L flask equipped with a magnetic follower, a septum inlet, a mercury bubbler and a short Vigreux column connected to a distillation apparatus is flushed with nitrogen and charged with borane–triethylamine (57.6 g, 0.5 mol) and diglyme (300 mL). The temperature is raised to 140 °C and a solution of *trans,trans,trans*-1,5,9-cyclododecatriene (81 g, 0.5 mol) in diglyme (100 mL) is added over 2 h by means of a syringe pump. Most of the diglyme is removed at atmospheric pressure and the residue is then heated at 200 °C (internal temperature) for 6 h (this isomerizes other organoboranes). Spinning band distillation under reduced pressure gives the perhydroboraphenalene **4** in 98% purity (bp 115–117 °C/10 mmHg).

Compound **4** (17.6 g, 0.1 mol), THF (50 mL) and ethylene glycol (16.8 mL, 18.6 g, 0.3 mol) are charged to an autoclave under a flow of nitrogen. The autoclave is filled with carbon monoxide at 1000 psi pressure and the temperature is raised to 150 °C for 2 h. The autoclave is cooled and opened and pentane (100 mL) is added to help remove the mixture. The solution is washed with water, dried over magnesium sulfate and evaporated.

THF (100 mL) and ethanol (95%, 100 mL) are added and the mixture is stirred during addition of aqueous NaOH (37 mL of 6 mol L^{-1}, 0.22 mol, excess) and then 30% aqueous hydrogen peroxide (37 mL, excess), the latter added dropwise while the solution is maintained at a temperature below 40 °C. Once the vigorous reaction is over the temperature is raised to 50 °C for 3 h. The mixture is then cooled to 25 °C, an equal volume of pentane is added, and the organic phase is separated, washed with water (3 × 50 mL), dried over magnesium sulfate and evaporated to yield *cis,cis,trans*-perhydro-9*b*-phenalenol. Recrystallization from pentane gives the pure product (13.6 g, 70%; mp 78–78.5 °C).

The carbonylation reaction is at its most useful for the synthesis of aldehydes via a single migration. In order to achieve this a hydride reducing agent is needed for the process and this also has the benefit of enhancing the rate of carbon monoxide uptake. As a result, the reaction can be carried out at 0 °C and at atmospheric pressure. The most useful hydrides are lithium trimethoxyaluminium hydride (which results in a gelatinous precipitate during work-up and can present separation difficulties) and potassium triisopropoxyborohydride (which can lead to polymerization if precautions are not taken to stop the stirring during addition of the hydride). The procedure given below is appropriate for the latter hydride. *B*-Alkyl-9-BBN derivatives allow maximum utilization of alkyl residues[3,85].

Preparation of Cyclopentanecarboxaldehyde by Hydride-Induced Carbonylation of B-Cyclopentyl-9-BBN[3,85]

The apparatus used in this procedure is a Brown automatic gasimeter, which allows monitoring of the uptake of carbon monoxide. However, if this information is not specifically required the carbon monoxide can be introduced by direct feed from a cylinder or more conveniently by means of a rubber balloon connected to a syringe needle.

The Brown automatic gasimeter is set up for carbonylation as described elsewhere[4], and fitted with a 500 mL reaction flask equipped with a magnetic follower and a septum inlet. The entire system is flushed with nitrogen and then a solution of 9-BBN-H (40 mL of 0.5 mol L^{-1} in THF, 20 mmol) and cyclopentene (1.36 g, 20 mmol) are added by syringe. The mixture is stirred for 2 h at 25 °C, then cooled to 0 °C. The stirrer is stopped, a solution of potassium triisopropoxyborohydride (20 mL of 1.0 mol L^{-1} in THF, 20 mmol) is added and the system is then flushed with carbon monoxide (by injecting about 8 mL of formic acid into the generator flask containing hot sulfuric acid). Stirring is then recommenced and uptake of carbon monoxide can be monitored. After 15 min (uptake long complete) the system is flushed with nitrogen and the mixture is stirred vigorously during addition of a pH 7 buffer (40 mL) followed by 30% hydrogen peroxide (8 mL). After the initial vigorous reaction, the cooling bath is removed and stirring is maintained for 15 min. Potassium carbonate (about 50 g) is added to saturate the mixture and the organic layer is removed. The aqueous layer is extracted with diethyl ether (2 × 50 mL) and the combined organic layers are dried over potassium carbonate and then carefully evaporated to remove solvent. Distillation at 50 mmHg gives cyclopentanecarboxaldehyde (1.84 g, 94%; bp 80–81 °C/50 mmHg).

---

15   Ketones and Tertiary Alcohols *via* Cyanoborates

Addition of solid sodium or potassium cyanide to a solution of a trialkylborane in an ether solvent results in dissolution of the solid with formation of a cyanoborate salt[86]. Addition of an electrophile induces rearrangement and acylating agents lacking α-hydrogen atoms are particularly useful. Trifluoroacetic anhydride permits reaction under mild conditions and use of one mole equivalent leads to two migrations, giving a ketone after oxidation[87]. Optimal utilization of alkyl groups is achieved by use of thexyl-dialkylboranes and this also permits the synthesis of mixed or cyclic ketones from the appropriate organoboranes[87,88].

Preparation of 3-Acetoxyphenylpropyl Cyclopentyl Ketone via the Cyanoborate Reaction[87]

5

A 100 mL, three-necked flask is fitted with a magnetic follower, a rotatable angled side-tube, a septum-capped stopcock and a line leading to a nitrogen supply and vacuum pump. The side tube is charged with dry, powdered sodium cyanide (0.54 g, 11 mmol) and the apparatus is then repeatedly evacuated and refilled with nitrogen. A solution of the trialkylborane **5** (10 mmol) in THF (about 22 mL) is prepared as described for a related example in Section 5 by successive addition of THF, borane–THF, 2,3-dimethyl-2-butene, cyclopentene and 3-acetoxyphenyl-1-propene. The mixture is then stirred at 25 °C and the side arm is rotated to introduce the sodium cyanide. Stirring is maintained for 1 h, by which time most of the cyanide dissolves. The mixture is cooled in a bath at −78 °C, trifluoroacetic acid (2.53 g, 12 mmol) is added dropwise with vigorous stirring and the cooling bath is then removed. The mixture is stirred for 1 h at 25 °C and then cooled to 10 °C. *m*-Chloroperbenzoic acid (3.9 g, 22.5 mmol, enough to cleave two B–C bonds) in dichloromethane (15 mL) is added. The mixture is stirred for 30 min at 25 °C and the products are then extracted into pentane (150 mL). The extract is successively washed with aqueous solutions of sodium carbonate (25 mL of 1 mol L^{-1}), sodium thiosulfate (25 mL of 1 mol L^{-1}) and hydrochloric acid (25 mL of 0.01 mol L^{-1}), dried over magnesium sulfate and evaporated. The crude product is transferred to a column packed with dry silica (100 g) with the aid of a little pentane and the column is eluted with pentane and then dichloromethane. The dichloromethane is removed to provide the essentially pure product, which can be distilled to give pure 3-acetoxyphenylpropyl cyclopentyl ketone (2.19 g, 80%; bp 105 °C/5 mmHg).

By use of excess trifluoroacetic anhydride and a period of heating the reaction can be encouraged to proceed further, resulting in a third migration and the riveting of the original organoborane[89] in a manner analogous to the methods using carbon monoxide or DCME. The DCME method occurs under milder conditions, but there is a possibility of different diastereoisomeric products in some cases.

Preparation of Tricyclohexylmethanol from Tricyclohexylborane by the Cyanoborate Reaction[89]

$$\text{(a) KCN}$$
$$\text{(b) } (CF_3CO)_2O, \text{ excess, 40 °C}$$
$$\text{(c) } H_2O_2/OH^-$$

The apparatus described in the preceding procedure is set up, charged with potassium cyanide (0.72 g, 11 mmol) in the side-arm and flushed with nitrogen. Tricyclohexylborane (10 mmol) is prepared as described for other trialkylboranes in Sections 2 and 5, from cyclohexene (3.04 mL, 30 mmol) and borane–THF (6.7 mL of 1.5 mol L^{-1} solution, 10 mmol), allowing 3 h at 50 °C for completion of the hydroboration step. The THF is removed under reduced pressure and diglyme (10 mL) is added. The side arm is rotated to introduce the cyanide and stirring is maintained for 1 h, by which time most of the solid dissolves. The solution is cooled to 0 °C and trifluoroacetic anhydride (TFAA; 12.6 g, 60 mmol) is added. The temperature is then raised to 40 °C for 6 h. After cooling, excess TFAA is removed under reduced pressure and the flask is immersed in an ice-water bath. Aqueous NaOH (12 mL of 3 mol L^{-1}) is added, followed by hydrogen peroxide (8 mL of 50%), slowly and with care. Once the initial vigorous reaction has subsided, the cooling bath is removed and the oxidation is completed by stirring at 25 °C for 3 h and at 50 °C for 15 min.

The mixture is extracted with pentane (150 mL) and the extract is washed with aqueous NaOH (2 × 25 mL of 2 mol L^{-1}) and water (2 × 25 mL), dried over magnesium sulfate and evaporated. The crude product is transferred with a little pentane to a column filled with dry, neutralized (with triethylamine) silica gel and eluted with pentane (50 mL) and then dichloromethane (300 mL). The dichloromethane fractions are evaporated to yield tricyclohexylmethanol (2.39 g, 86%), which is pure by GC. Recrystallization from pentane gives very pure product (2.21 g, 79%; mp 93–94 °C).

16   Ketones, Functionalized Alkenes, Alkynes and Diynes *via* Alkynylborates

Addition of alkynyllithiums to trialkylboranes gives rise to lithium trialkylalkynyl-borates, which are susceptible to attack by electrophiles on the triple bond. Such attack leads to rearrangement involving migration of an alkyl group from boron to the adjacent carbon atom. The exact nature of the product obtained depends upon the nature of the electrophile and the method of work-up. If the electrophile is a protonic acid or an alkylating agent the intermediate is a vinylborane which is generally a mixture of *E* and *Z* isomers. However, both isomers give rise to the same ketone on oxidation, making this a useful synthesis of ketones[3,90].

Preparation of 8-Allyl-7-Tetradecanone by Allylation of Lithium Trihexyloctynylborate[3,90]

$$Hex_3B \xrightarrow[\substack{(b)\ H_2C=CHCH_2Br \\ (c)\ H_2O_2/OH^-}]{(a)\ LiC\equiv CHex} \underset{\underset{CH_2CH=CH_2}{|}}{HexCOCHHex}$$

A 100 mL, three-necked flask equipped with a magnetic follower, a septum-capped stopcock, a septum-capped pressure-equalizing dropping funnel and a line leading to a nitrogen supply/vacuum system is flushed with nitrogen. Trihexylborane (5 mmol) is prepared in the dropping funnel from borane–THF (3.4 mL of 1.47 mol L^{-1}, 5 mmol) and 1-hexene (1.26 g, 15 mmol), with swirling to mix the reagents thoroughly, and the mixture is left for 1 h. Meanwhile, the reaction flask is immersed in an ice-water bath and charged with 1-octyne (0.55 g, 5 mmol), light petroleum (bp 40–60 °C, 5 mL) and butyllithium (3.2 mL of 1.56 mol L^{-1} in hexane, 5 mmol), with stirring. The cooling bath is removed and the mixture is stirred for 30 min at 25 °C. The cooling bath is replaced and the mixture is stirred during addition of the trihexylborane solution from the dropping funnel. The funnel is rinsed out with diglyme (5 mL) and the mixture is stirred until all of the precipitated octynyllithium dissolves (only a few minutes). The volatiles are then removed under reduced pressure, leaving a solution of lithium trihexyloctynylborate in diglyme.

Stirring is maintained for 15 min at 25 °C and then allyl bromide (0.61 g, 5 mmol) is added by syringe. The reaction mixture is stirred at 40 °C for 2 h, cooled in ice, and then oxidized by successive addition of aqueous NaOH (2 mL of 5 mol L^{-1}) and 50% hydrogen peroxide (1.5 mL of 50%), the latter dropwise (from the dropping funnel) and with vigorous stirring and after removal of the septa from the apparatus. After the initial exothermic reaction subsides the cooling bath is removed and the mixture is stirred for 3 h at 25 °C. The product is extracted into diethyl ether (2 × 20 mL) and the extract is washed with water, dried over magnesium sulfate and evaporated to give a syrupy residue which is transferred with a little pentane to a column packed with dry silica gel (100 g). Elution is with pentane (100 mL) and then dichloromethane (2 × 150 mL) and removal of the dichloromethane provides almost pure 8-allyl-7-tetradecanone (1.11 g, 88%), which can be further purified by distillation if required (bp 96–98 °C/0.8 mmHg).

When the electrophile added to the alkynylborate is an α-bromocarbonyl compound, iodoacetonitrile or even propargyl bromide, the vinylborane intermediate is formed highly stereoselectively. Thus, by hydrolysis of the intermediate it is possible to obtain (Z)-alkenyl ketones, carboxylates, nitriles and alkynes[3,91]. Of course, oxidation of the intermediate to give a ketone (see above) is also possible.

Preparation of (Z)-4-Hexyl-4-undecen-2-one from Trihexyloctynylborate and Bromoacetone[3,91]

$$\text{Li}^+\text{HexB}\bar{\text{C}}\equiv\text{CHex} \xrightarrow[\text{(--LiBr)}]{\text{BrCH}_2\text{COCH}_3} \underset{\text{Hex}_2\text{B}}{\overset{\text{Hex}}{\diagdown}}\text{C}=\text{C}\underset{\text{Hex}}{\overset{\text{CH}_2\text{COCH}_3}{\diagup}} \xrightarrow{{}^i\text{PrCO}_2\text{H}} \underset{\text{H}}{\overset{\text{Hex}}{\diagdown}}\text{C}=\text{C}\underset{\text{Hex}}{\overset{\text{CH}_2\text{COCH}_3}{\diagup}}$$

Lithium trihexyloctynylborate (5 mmol) is prepared in diglyme (5 mL) exactly as described in the preceding procedure. The mixture is cooled in a $-78\,^\circ$C bath and stirred while bromoacetone (0.75 g, 5.5 mmol) is added by syringe. The mixture is allowed to warm to 25 $^\circ$C and then heated at 55 $^\circ$C for 6 h to produce the intermediate vinylborane, followed by cooling to 25 $^\circ$C.

Degassed 2-methylpropanoic acid (1 mL) is added and the mixture is stirred for 3 h at 25 $^\circ$C. The mixture is neutralized by addition of aqueous NaOH (5 mol L^{-1}) and then a further 1.5 mL of the same NaOH solution is added. Hydrogen peroxide (3 mL of 50%, large excess) is added cautiously and the mixture is stirred overnight (this oxidizes residual B–C bonds). The product is extracted into diethyl ether and the extract is washed with water, dried over magnesium sulfate and evaporated. The product is transferred with the aid of a little pentane to a column of dry silica gel and successively eluted with light petroleum (bp 40–60 $^\circ$C) and then pentane/dichloromethane (1 : 1). Evaporation of the latter fractions yields (Z)-4-hexyl-4-undecen-2-one (0.95 g, 75%), which can be further purified by distillation if required (bp 100–102 $^\circ$C/1.5 mmHg).

$\alpha\beta$-Unsaturated ketones are obtained stereospecifically in 50–70% yields when the electrophile is a 2-alkyl-1,3-dioxolanium salt[92]:

$$\text{Li}^+\text{R}^1{}_3\bar{\text{B}}\text{C}\equiv\text{CR}^2 + \text{R}^3\!-\!\!\overset{\text{O}}{\underset{\text{O}}{\diagup\hspace{-0.3em}\diagdown}}\!\!\overset{+}{} \quad \text{FSO}_3^- \longrightarrow \underset{\text{H}}{\overset{\text{R}^1}{\diagdown}}\!\!=\!\!\underset{\text{COR}^3}{\overset{\text{R}^2}{\diagup}}$$

When the electrophile added to a trialkylalkynylborate is iodine, the intermediate 2-iodoalkenylborane eliminates dialkyliodoborane to produce an alkyne. This can be a useful synthesis of unsymmetrical alkynes[93]:

$$\text{Ph}_3\text{B} \xrightarrow[\text{(b) I}_2]{\text{(a) CLi}\equiv\text{CMe}_3} \text{PhC}\equiv\text{CCMe}_3$$

Use of unsymmetrical triorganylboranes in order to minimize the wastage of potentially valuable organic residues is not very successful in most cases, on account of the similarity in the migratory aptitudes of different groups in this reaction. However, alkynyl groups show significantly higher relative migratory aptitudes than secondary alkyl groups and so di-*sec*-alkyldialkynylborates give rise to conjugated diynes[94]. In order to make the reaction useful for the synthesis of unsymmetrical diynes, all that is required is a means of synthesis of the appropriate dialkyldialkynylborate. This can be achieved by addition of an alkynyllithium to an isolated dialkylalkynylborane[95], prepared as described in Section 7, or *in situ* by use of dicyclohexyl(methylthio)borane as the starting material[96].

Preparation of 5,7-Tetradecadiyne from a Dialkynylborate[96]

$$\left(\left\langle \bigcirc \right\rangle \right)_2 BH \xrightarrow[\substack{(c)\ LiC\equiv CBu \\ (d)\ I_2}]{\substack{(a)\ MeSH \\ (b)\ LiC\equiv CHex}} Hex\equiv CC\equiv CBu$$

A 100 mL, two-necked flask equipped with a magnetic follower, a septum and a septum-capped pressure equalizing dropping funnel is flushed with nitrogen. Dicyclohexylborane (5 mmol) is prepared in THF (10 mL) as described in Section 4 and methanethiol (0.75 mL of 8.2 mol L^{-1} in THF, 6 mmol) is then added, with stirring and use of a bubbler connected *via* a needle to vent liberated hydrogen. The mixture is stirred for 2 h to complete the formation of the methylthioborane (the dicyclohexylborane dissolves) and the flask is briefly pumped (via a needle) to remove excess methanethiol. If necessary, some additional THF is added to bring the volume back to about 10 mL.

Meanwhile, in each of two separate, nitrogen-flushed 50 mL flasks equipped with a magnetic follower and a septum, the two alkynyllithium reagents are prepared. The flask is cooled in an ice-water bath and charged with the appropriate alkyne (5 mmol), pentane (5 mL) and butyllithium solution (3.9 mL of 1.29 mol L^{-1} in hexane, 5 mmol). The mixture is stirred for 30 min at 25 °C and THF (4 mL) is then added. The solution is cooled to −78 °C before use in the next step.

The main flask, containing dicyclohexyl(methylthio)borane, is cooled in a −78 °C bath and the cooled solution of 1-lithiooctyne is added by syringe, residual traces being transferred with the aid of additional THF (2 mL). The reaction mixture is allowed to warm to 25 °C over 15 min and then recooled to −78 °C for addition, in the same way, of the cooled 1-lithiohexyne solution. The mixture is again warmed to 25 °C and stirred for 1 h, then most of the hydrocarbon solvent is removed under reduced pressure. If necessary, a little THF is added back to bring the final volume to about 10 mL. The solution is cooled to −78 °C and a solution of iodine (2.54 g, 10 mmol) in THF (20 mL) is added dropwise from the dropping funnel over 15 min. The mixture is stirred at −78 °C for a further 15 min then allowed to warm to 25 °C over 1 h. Saturated aqueous sodium thiosulfate solution (5 mL) and aqueous NaOH (5 mL of 5 mol L^{-1}) are added and the mixture is stirred for 1 h. The products are extracted into pentane (3 × 40 mL) and the extract is washed with water (3 × 50 mL), dried over magnesium sulfate and evaporated. The product is purified on a column of silica gel by elution with pentane and then 5% dichloromethane in pentane. The fractions containing the pure product are combined and evaporated to give 5,7-tetradecadiyne (0.584 g, 61%), which can be distilled if required (bp 66 °C/0.02 mmHg).

17   Stereospecific Synthesis of Alkenes *via* Alkenylborates

Most of the reactions of alkynylborates (Section 16) can probably be achieved with alkenylborates, although the products will obviously be at a lower oxidation level. Some of the reactions have been demonstrated[3], but in general the reactions have been less widely studied. One reaction that has found extensive use is the reaction with iodine, which produces alkenes in a highly stereoselective way[3].

The alkenylborate used in such procedures can be obtained by addition of an alkenyllithium to a trialkylborane[97], but this results in the wastage of two boron-bound alkyl groups. This problem can be overcome by utilization of alkyldimethoxyboranes as substrates[98]. The stereochemistry around the double bond is inverted during the process:

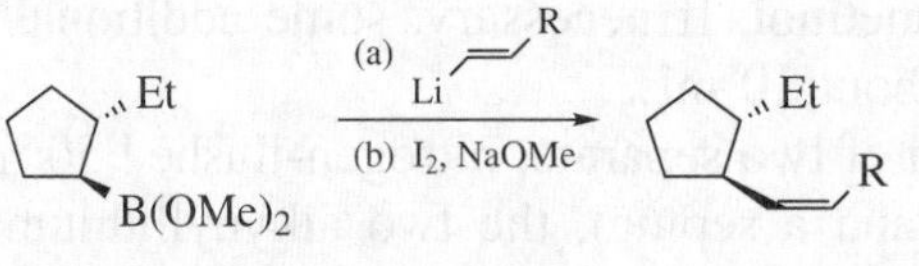

Alternatively, alkylalkenylboron compounds obtained *via* hydroboration reactions can be converted into borate salts by addition of methoxide and then reacted with iodine. For optimal utilization of organic residues an alkyldibromoborane can be reduced to an alkylmonobromoborane, which is then used to hydroborate an alkyne[99]:

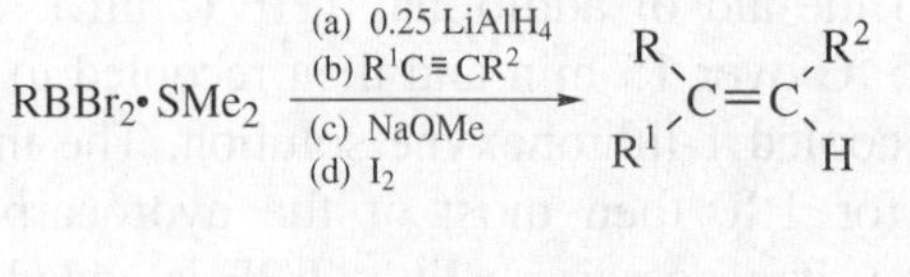

The simplest procedure of all involves the direct hydroboration of an alkyne with an easily formed dialkylborane, followed by reaction with a base to generate the borate and iodine to induce the reaction[100]. This reaction is used to illustrate the procedure because the other cases differ primarily only in how the organoborane is generated.

Preparation of (Z)-1-Cyclohexyl-1-hexene[3,100]

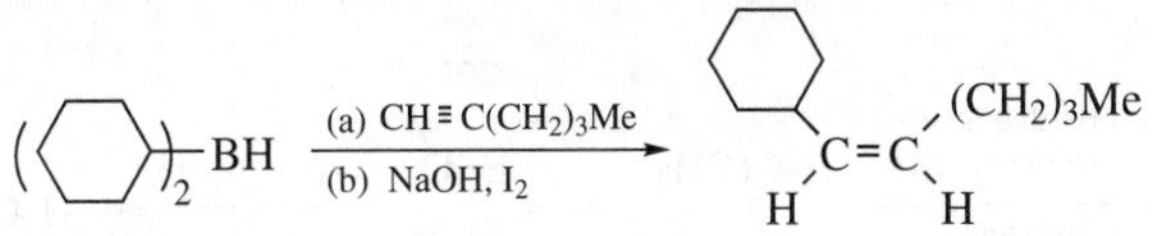

A 100 mL, three-necked flask equipped with a magnetic follower, a thermometer, a septum and a septum-capped pressure-equalizing dropping funnel is flushed with nitrogen. Dicyclohexylborane (25 mmol) is prepared as described in Section 4 from cyclohexene (4.1 g, 50 mmol) in THF (20 mL) and borane–THF (13.9 mL of 1.8 mol L^{-1}, 25 mmol) at 0–5 °C. The alkenyldialkylborane is generated, as described in Section 7 for a related example, by addition of 1-hexyne (2.05 g, 25 mmol) at −10 °C and then stirring at 25 °C until the precipitate dissolves, followed by 1 h more. The mixture is then cooled to −10 °C and aqueous NaOH (15 mL of 6 mol L^{-1}) is added, followed by the dropwise addition, with stirring, of a solution of iodine (6.35 g, 25 mmol) in THF (20 mL) over a period of 15 min. The mixture is allowed to warm to 25 °C and excess iodine is decomposed by addition of a small amount of aqueous sodium thiosulfate solution. The product is extracted into pentane (2 × 25 mL) and the extract is washed with water (2 × 10 mL), dried over magnesium sulfate and evaporated. Fractional distillation under reduced pressure gives (Z)-1-cyclohexyl-1-hexene (3.11 g, 75%; bp 44–45 °C/1 mmHg).

Conjugated enynes can be obtained from alkenylalkynylborates[101], whereas conjugated dienes can be obtained by treatment of dialkenylalkoxyboranes with methoxide and iodine[102]:

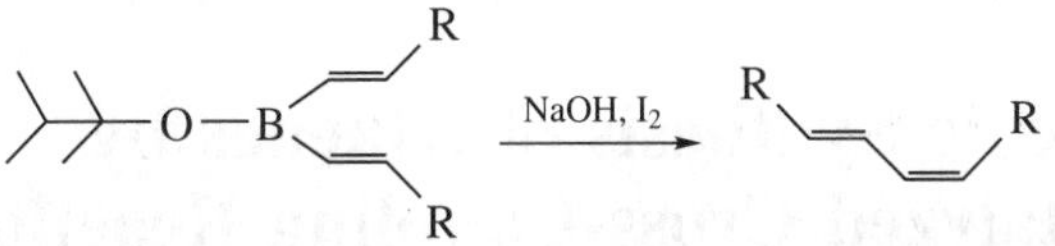

1-Bromoalkenyl(organyl)boranes, obtainable in stereospecific form by hydroboration of 1-bromoalkynes with appropriate hydroborating agents, undergo stereospecific rearrangement on addition of a nucleophile, without the need for added iodine[3]. However, in these cases the intermediate formed does not have the possibility to eliminate an iodoborane unit, so that the product is a substituted vinylborane. Oxidation then gives a ketone or protonolysis gives an alkene, again stereospecifically. The principal variables in designing a reaction are the nature of the organyl group that is to migrate, and the means by which it is to be incorporated. In the example shown in the following preparation, the migrating group is an allyl group introduced by use of allylmagnesium chloride. The final product in such a case, following oxidative work-up, is a β,γ-unsaturated ketone[103].

Synthesis of 1-Nonen-4-one by Reaction of a 1-Bromoalkenylborane with Allylmagnesium Chloride[103]

$$CH_3(CH_2)_3C \equiv CBr \xrightarrow[\text{(c) allylMgCl}]{\substack{\text{(a) } Br_2BH \cdot SMe_2 \\ \text{(b) } HO(CH_2)_3OH}} \quad \xrightarrow{H_2O_2, \text{NaOAc}} \quad$$

1-Bromo-1-hexyne is hydroborated in a manner similar to that described for the preparation of hexyldibromoborane in Section 8, and is converted into 1-bromo-1-hexenyl-1,3,2-dioxaborolane[104] by reaction with 1,3-propanediol. The cyclic boronate (2.47 g, 10 mmol) is dissolved in dry THF (10 mL) in an argon-filled, 50 mL, round-bottom flask equipped with a magnetic follower and a septum. The solution is cooled to $-78\,^\circ$C and stirred vigorously during addition of allylmagnesium chloride (5.5 mL of a 2 M solution in THF, 11 mmol). After 5 min the cooling bath is removed and the mixture is allowed to warm to $25\,^\circ$C. Dry HMPA (4 mL) is added, and the temperature is raised to $50\,^\circ$C and maintained for 6 h or until boron NMR shows that the reaction is complete (disappearance of the borate signal at $\delta = 0$). (Note that the reaction is faster and gives a higher yield in the presence of HMPA, but does occur in its absence.) The mixture is then cooled and diluted with pentane (20 mL) and the solution is then washed with water (2×25 mL), dried (MgSO$_4$) and concentrated under reduced pressure. The crude product is oxidized under acetate-buffered conditions by addition of aqueous NaOAc and H$_2$O$_2$ (Section 2 discusses oxidation), the product is extracted into pentane, and the extract is washed with water (2×25 mL), dried (MgSO$_4$) and concentrated again. The product is then distilled immediately to avoid isomerization, yielding 1-nonen-4-one (1.12 g, 77%; bp 65–66 $^\circ$C/15 mmHg).

18 Stereospecific Synthesis of Alkenes by Metal-Catalyzed Cross-Coupling Reactions

The cross-coupling between an organic halide and an organoboron compound catalyzed by palladium compounds has become a widely used and versatile method for the generation of carbon–carbon bonds[105]. At its simplest, the reaction can be represented as:

$$R^1\text{-}X + R^2\text{-}BY_2 \longrightarrow R^1\text{-}R^2$$

The reaction is at its most efficient when the reacting partners are an alkenyl bromide and an alkenylboron compound, such as an alkenyldihydroxyborane (alkenylboronic acid)[106]. The reaction will tolerate many functional groups and the stereochemistry is retained in both portions of the diene.

Preparation of Bombykol by Palladium-Induced Coupling of a Bromoalkene with an Alkenyldihydroxyborane[106]

$$HO(CH_2)_9C\equiv CH \xrightarrow[\text{(b) } H_2O]{\text{(a) } 2 \text{ catecholborane}} HO(CH_2)_9-C \overset{H}{\underset{H}{\diagdown}} C-B(OH)_2 \xrightarrow[\text{Pd(PPh}_3)_4]{\underset{H}{\overset{Br}{\diagup}} C-H \atop Me(CH_2)_2-C} HO(CH_2)_9-C \overset{H}{\underset{H}{\diagdown}} C-C \overset{H}{\underset{Me(CH_2)_2}{\diagdown}} C-H$$

A 100 mL, two-necked flask equipped with a magnetic follower, a septum inlet and a reflux condenser connected to a bubbler is flushed with nitrogen and charged with 10-undecyn-1-ol (1.51 g, 9 mmol) in THF (3 mL). Catecholborane (2.0 mL, 18 mmol) is added dropwise, with stirring, and the mixture is stirred at 25 °C until hydrogen evolution ceases and then brought to reflux for 5 h. The mixture is cooled and water (60 mL) is added. The whole is stirred for 2 h and then cooled to 0 °C. The solid is collected by filtration, washed with water (3 × 20 mL) and dried to give (*E*)-11-hydroxy-1-undecenyldihydroxyborane (1.6 g, 83%), which is used without further purification in the next stage.

A 50 mL, two-necked flask equipped with a magnetic follower, septum inlet and reflux condenser leading to a bubbler is flushed with nitrogen and charged successively with a solution of tetrakis(triphenylphosphine)palladium (0.29 g, 0.25 mmol) in benzene (20 mL), (*Z*)-1-bromo-1-pentene (0.75 g, 5 mmol), 11-hydroxy-1-undecenyldihydroxyborane (1.18 g, 5.5 mmol) and a solution of sodium ethoxide in ethanol (5 mL of 2 mol L^{-1}). The solution is heated at reflux for 2.5 h, with stirring, and then cooled to 25 °C and oxidized by successive addition of aqueous NaOH (0.5 mL of 3 mol L^{-1}) and hydrogen peroxide (0.5 mL of 30%), to destroy any residual organoborane. The whole is stirred for 1 h and the products are then extracted into diethyl ether (30 mL) and the extract is washed with saturated NaCl (2 × 15 mL), dried over magnesium sulfate and evaporated to yield crude bombykol (0.96 g, 82%), which can be further purified by kugelrohr distillation (bp 125 °C/0.1 mmHg).

The reaction, known as the Suzuki–Miyaura coupling, has been reviewed[107]. It is an extremely versatile reaction and tolerates many variations. The method has been used, for example, to prepare stereochemically defined vinylsilanes[108]. In certain cases nickel salts rather than palladium compounds have been used as catalysts[109–111], The coupling partners can include aryl groups on the boron moiety[109–111], or as the halide partner[112]. A mechanistic study has been carried out for the palladium-catalyzed biaryl coupling case[111].

The organic halide can be replaced in some cases by other compounds, such as allylic acetates or epoxides[109,110]. The reaction can be applied to alkynylboron derivatives and

with some modifications to *B*-alkyl-9-BBN derivatives[105]. It is even possible to couple
the latter compounds to iodoalkanes in the presence of potassium phosphate as well as a
palladium catalyst, but the yields are only moderate in such cases[113].

In a somewhat related reaction, arylboronic acids can be coupled with a variety of
heteroatom-linked groups under the influence of copper(II) acetate as catalyst[114–116].
The reaction, which also resembles the Ullmann diaryl ether synthesis, can indeed be
applied to the synthesis of diaryl ethers[114], but can also be applied to the synthesis
of *N*-aryl heteroaromatic compounds[115] or to *N*- or *O*-arylation of a range of other
compounds[116].

---

*Synthesis of 1-(4-Tolyl)-1H-benzimidazole by Copper(II) Acetate-Catalyzed
Coupling of p-Tolylboronic Acid with Benzimidazole*

A 25 mL round-bottom flask equipped with a magnetic follower is charged
under air with *para*-tolylboronic acid (0.090 g, 0.67 mmol), benzimidazole (0.039 g,
0.33 mmol), anhydrous copper(II) acetate (0.091 g, 0.50 mmol), activated 4A molecular sieves (0.250 g), pyridine (1.0 mL of a 0.67 M solution in dichloromethane,
0.67 mmol), and dichloromethane (4 mL). A stopper is loosely fitted and the mixture
is stirred for 2 days at 25 °C. When TLC indicates that the reaction is complete, the
mixture is filtered through Celite, the Celite is washed with a little methanol and
the filtrate is evaporated under reduced pressure. The crude product is purified by
column chromatography (silica, eluted by hexane/ethyl acetate/methanol 84 : 15 : 1)
to give 1-*para*-tolyl-1*H*-benzimidazole (0.046 g, 67%).

---

19   Homoallylic Alcohols *via* Allylboranes

Allylic boron compounds generally react rapidly with aldehydes, or somewhat less
rapidly with ketones, to give homoallylic alcohols[3]. The reaction takes place with transposition of the allylic group, but many allylic organoboranes themselves undergo allylic
rearrangement to interconvert the isomers so care must be taken to ensure that the desired
allylic organoborane is indeed the one undergoing reaction:

$$X_2BCHR^1CH{=}CHR^2 \xrightarrow[\text{(b) H}_2\text{O}]{\text{(a) R}^3COR^4} R^1CH{=}CHCHR^2\overset{\overset{\displaystyle OH}{|}}{C}R^3R^4$$

Organoboranes possessing more than one allylic group should be avoided because the rates of reaction will differ as each one reacts in turn. This may lead to complications. The nature of the other boron-bound groups (X in the above equation) has a strong influence on both the rate of allylic rearrangement of the organoborane and the rate of reaction of the organoborane with carbonyl compounds. The more powerfully electron-donating groups (dialkylamino > alkoxy > alkyl) slow down the reactions substantially. Thus, dialkylboryl derivatives must be prepared and reacted at low temperatures if it is necessary to avoid allylic rearrangement of the allylic organoborane prior to reaction with the carbonyl compound, whereas dialkoxyboryl derivatives are relatively stable up to ambient temperatures. The latter compounds are also less reactive towards carbonyl compounds, however, and rearrangement may precede reactions with ketones. Reactions with aldehydes are usually free of such complications.

There is the possibility of substantial control over the various stereochemical features in the product homoallylic alcohols, namely over the geometry about the double bond, the *syn* or *anti* relationship between the two newly created sp^3 centers, and the absolute configuration of the product, when chiral. For example, in reactions of dialkoxy(1-methylallyl)boranes with aldehydes the $(Z)/(E)$ ratio for the double bond in the product can be varied from $3:1$ to $1:2$ simply by changing the alkoxy groups, bulky ones favoring the (Z) product[117].

The *syn/anti* selectivity is primarily determined by the geometry of the double bond in the initial allylborane (which can also be altered by reversible allylic rearrangement). (Z)-Allylboranes give *syn* alcohols[118], whereas (E)-allylboranes give *anti* alcohols[3].

Chiral induction is achieved by having optically active groups on boron, such as two isopinocampheyl groups[119] or an optically active cyclic alkylenedioxy group[120]. By using a chiral auxiliary with the appropriate configuration, it is possible to obtain the appropriate enantiomer of the product which, when coupled with the right choice of geometry of the double bond in the allylic organoborane, allows synthesis of the desired enantiomer of either diastereoisomer (*syn* or *anti*), rendering this a very versatile method[3,119].

Symmetrical dialdehydes react with *B*-allyldiisopinocampheylborane to give homochiral C_2-symmetric diols[121], whereas *anti*-1-alkene-2,3-diols are obtained from ((*E*)-γ-(1,3,2-dioxaborinanyl)allyl)diisopinocampheylborane[122]:

(de > 95%, ee > 90%)

Allyldiisopinocampheylboranes incorporating a diphenylamino group can be used to generate β-diphenylamino alcohols[123].

In terms of the details of how the reaction should be carried out, the only differences required involve the preparation and handling of the initial organoborane. Simple dialkylallylboranes can be prepared and purified by procedures similar to that described in Section 7 for a diakylalkynylborane, but if isomerization of the allylic organoborane is to be avoided it may be necessary to use the organoborane directly as generated *in situ* and to maintain a low temperature[119]. Appropriate higher crotylboranes may also sometimes be obtained by hydroboration of allenes (Section 7). From the point of obtaining the required allylboron compound onwards, the reaction procedure given below for a simple case[124] is satisfactory.

Preparation of 5,5-Dimethyl-1-hexen-4-ol by Reaction of
B-Allyl-9-borabicyclo[3.3.1]nonane with Pivaladehyde[124]

A 50 mL flask equipped with a magnetic follower and a septum connected via a needle to a mercury bubbler is flushed with nitrogen through a needle and then charged with *B*-allyl-9-BBN (3.775 g, 23.3 mmol) and purified (alkene-free) pentane (25 mL). The mixture is cooled to 0 °C (lower temperatures throughout are recommended for geometrically or positionally labile organoboranes) and stirred during dropwise addition, from a syringe, of freshly distilled pivaldehyde (2.60 mL, 2.00 g, 23.3 mmol). The mixture is allowed to warm up and stirred for 1 h at 25 °C then neat ethanolamine (1.40 mL, 1.42 g, 23.3 mmol) is added in order to free the product and precipitate the 9-BBN byproduct. The slurry thus obtained is stirred for 30 min and the contents of the flask are then poured into a centrifuge tube, the residue being washed through with a little pentane (10 mL). The mixture is centrifuged and the clear supernatant liquid is removed with a syringe. The precipitate is washed repeatedly with pentane (3 × 15 mL), each time with thorough mixing prior to centrifugation, and the combined organic solutions are concentrated by blowing with a stream of nitrogen (reduced pressure is avoided at this stage because of the volatility of the product). The residual oil is distilled under reduced pressure to give 5,5-dimethyl-1-hexen-4-ol (2.53 g, 85%; bp 55.5–56 °C/19 mmHg).

The process can be extended to allenylboranes, homoallenylboranes and propargyl-boranes. As with allylboranes, allylic rearrangement accompanies these processes, which result in the formation of homopropargyl alcohols from allenylboranes[125], alkyl(1,3-butadien-2-yl)methanols from homoallenylboranes[126, 127], and homoallenyl alcohols from propargylboranes[128, 129]:

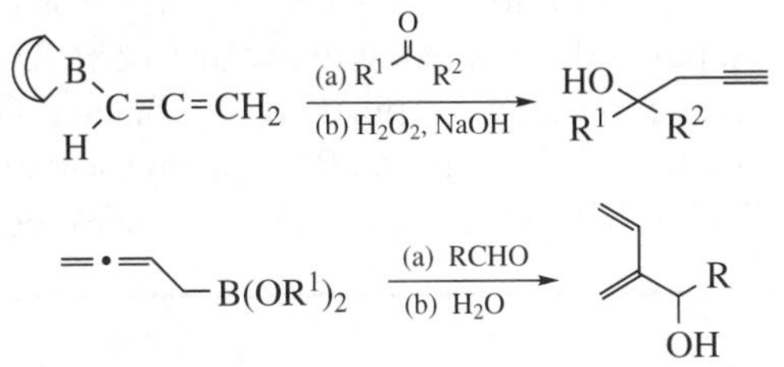

Given that there is an appropriate chiral moiety in the residual groups attached to boron, the reactions are enantioselective. Cyclic dialkoxyboranes derived from hindered dialkyl tartrates provide suitable chiral auxiliaries in the homoallenylborane reactions, and high enantiomeric excesses can be obtained, particularly in reactions with homochiral aldehydes when the matched enantiomer of the organoborane is used[127]. Diisopinocam-pheylboron derivatives have been recommended for the enantioselective synthesis of chiral α-allenic alcohols[127, 128]. The required propargylic organoboranes can be obtained by several methods, of which one is illustrated in the following preparation.

Synthesis of (R)-2-Propyl-4-methylhexa-1,2-dien-4-ol by Propargylboration[128]

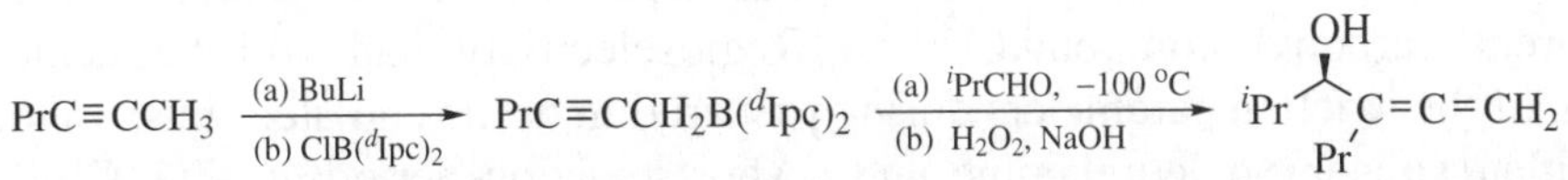

A 50 mL round-bottom flask is fitted with a magnetic follower and a septum and is thoroughly flushed with argon through a needle and a paraffin oil bubbler. The flask is charged, *via* syringes, with butyllithium (4.0 mL of 2.5 M solution in hexane, 10.0 mmol) and THF (4 mL) and cooled to −10°C. 2-hexyne (1.18 mL, 10.5 mmol) is added and the mixture is stirred for 10 min at −10°C and then allowed to warm to 25°C and stirred for a further 30 min. It is then cooled in a −78°C bath and a solution of ^{d}(Ipc)$_2$BCl (the isomer from (+)-α-pinene, available as the (−) enantiomer of DIP-Chloride™; 3.53 g, 11.0 mmol) in diethyl ether (20 mL), that has been previously chilled to 0°C, is added dropwise *via* a double ended needle. (Note: it is important that the temperature of the solution is close to −78°C before the DIP-Cl is added or the THF can be cleaved by the reagent under catalysis by LiCl liberated during the reaction.) The mixture is stirred for 10 min, the cooling bath is then removed, the mixture is allowed to warm to 25°C, and stirring is continued for a further 30 min. The solvents are removed under reduced pressure and diethyl ether (20 mL) is added (the reaction occurs better in pure ether than in the mixture with THF and hexane). The mixture is then cooled to $-100\,^{\circ}C$ (ethanol/liquid N_2) and stirred as a solution of 2-methylpropanal (0.91 mL, 10.0 mmol) in ether (5 mL) that has previously been cooled to $-78\,^{\circ}C$ is added slowly *via* a double ended needle. After 2 h the mixture is allowed to warm to $25\,^{\circ}C$ and oxidation is effected by addition of H_2O_2 (5 mL of 30%, 44 mmol) and NaOH (5.0 mL of 3 M, 15.0 mmol), followed by vigorous stirring for 4 h. The organic layer is then separated, washed with water (2×25 mL), dried ($MgSO_4$) and evaporated. The crude product is purified by flash column chromatography (silica, hexane/ethyl acetate 98 : 2) to give (*R*)-2-propyl-4-methylhexa-1,2-dien-4-ol (1.23 g, 80%; $\alpha_D - 19.1\,^{\circ}$ ($CHCl_3$), corresponding to 96% ee).

20   Aldol Reactions of Vinyloxyboranes

The aldol reaction has long been an important weapon in the armory of the synthetic organic chemist, but the traditional reaction involving alkali metal enolates suffers from a number of severe disadvantages that limit its general utility. In particular, crossed aldol reactions are often problematic, and ability to control regioselectivity, diastereoselectivity, and enantioselectivity is limited. The use of boron enolates (vinyloxyboranes) allows many of the disadvantages to be overcome, and this modification of the aldol reaction has in consequence become an extremely important synthetic method[3,130].

Vinyloxyboranes can be formed regiospecifically by a number of routes[131], but the recent rise in importance of these species has depended on their direct generation from their parent carbonyl compounds[3,130]. Regioselectivity can still be achieved by variation of the reaction parameters during preparation of the enolate. Thus, use of dibutyl-boryl triflate, diisopropylethylamine and a short reaction period at $-78\,^{\circ}C$ allows total conversion of 2-pentanone into its kinetic boron enolate, whereas use of 9-BBN triflate, 2,6-lutidine and a long reaction period at $-78\,^{\circ}C$ gives exclusively the thermodynamic enolate[3,130,132]:

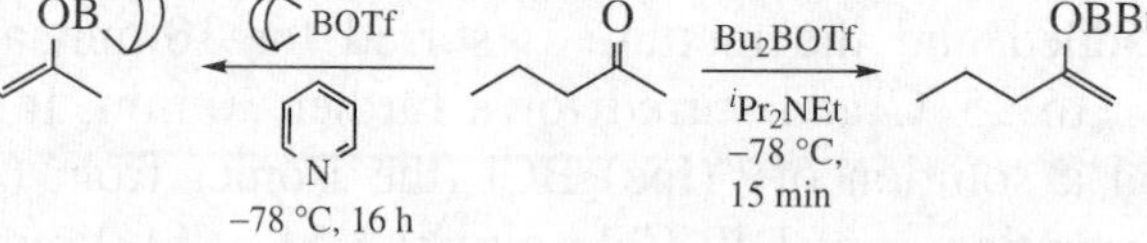

Subsequent aldol reactions with aldehydes take place without loss of regiochemical integrity, but reactions with ketones are slower and preservation of regiochemical integrity can then be more of a problem.

An advantage of boron enolates over most other kinds of enolates is the high diastereoselectivity displayed in their reactions with aldehydes. There is a very good correlation between the geometry of the enolate and the diastereoisomer formed in its aldol reactions. (*Z*)-Enolates give almost 100% stereoselectivity for formation of *syn*-aldols, whereas

(*E*)-enolates strongly favor the *anti*-aldol products, especially if the groups on boron are relatively bulky[3,133,134]:

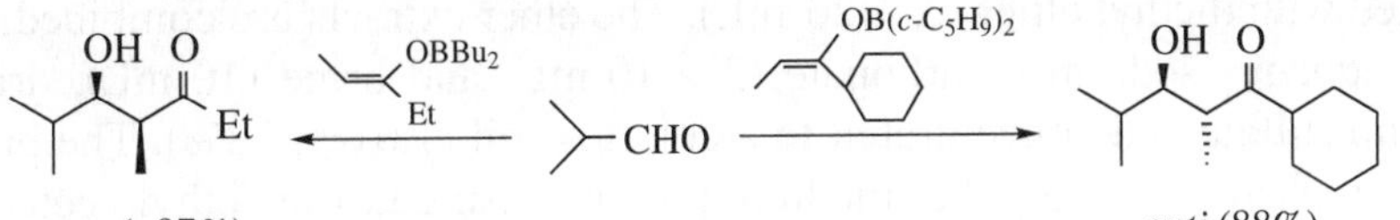

In order to take advantage of the stereoselectivity of the reaction it is necessary to be able to generate boron enolates with the appropriate geometry. Use of a hindered dialkylboryl triflate (*e.g.* dicyclopentyl) and diisopropylethylamine at 0 °C yields predominantly the (*E*) isomer, whereas use of less hindered reagents (e.g. dibutyl) and low temperature (−78 °C) gives almost exclusively the (*Z*) isomer[133].

Preparation of syn-1-Hydroxy-2-methyl-1-phenyl-3-pentanone via a Boron Enolate (Vinyloxyborane)[132,133]

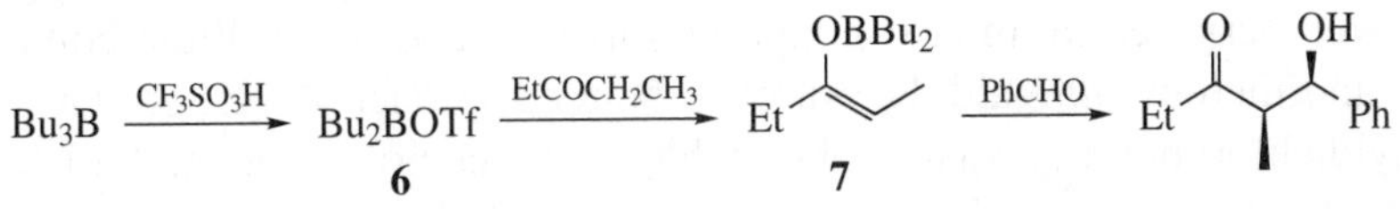

Dibutylboryl triflate **6** is prepared as follows. A 100 mL flask equipped with a magnetic follower and a septum connected *via* a needle to a bubbler is flushed with argon. Tributylborane (15.16 g, 83.3 mmol) is charged by syringe, followed by a small amount of trifluoromethanesulfonic acid (1.0 g). The mixture is stirred and warmed to 50 °C until evolution of butane begins (there is an induction period), and is then cooled back to 25 °C. The remaining trifluoromethanesulfonic acid (11.51 g, total of 83.3 mmol) is added dropwise at such a rate as to maintain a temperature between 25 and 50 °C. The mixture is then stirred for a further 3 h at 25 °C. Short-path distillation under reduced pressure in an atmosphere of argon gives pure dibutylboryl triflate (**6**; 19.15 g, 84%; bp 60 °C/2 mmHg).

A 100 mL flask equipped as described above is flushed with argon, charged with dry diisopropylethylamine (0.85 g, 6.6 mmol), dibutylboryl triflate (**6**; 1.81 g, 6.6 mmol) and dry diethyl ether (15 mL) and then cooled to −78 °C. 3-Pentanone (0.52 g, 6.0 mmol) is added, dropwise and with stirring, and the mixture is stirred for a further 30 min at −78 °C, during which the boron enolate **7** is formed along with a white precipitate of diisopropylethylammonium triflate.

Benzaldehyde (0.64 g, 6.0 mmol) is added dropwise and the mixture is stirred for a further 30 min at −78 °C and 1 h at 0 °C. The reaction is quenched by addition to a pH 7 phosphate buffer solution (50 mL) and the product is extracted into diethyl ether (2 × 30 mL). The combined ether extracts are washed with brine (2 × 10 mL) and concentrated under reduced pressure. The oil thus obtained is dissolved in methanol (20 mL), the solution is cooled to 0 °C, and hydrogen peroxide solution (6.5 mL of 30%) is added. The mixture is stirred at 25 °C for 2 h and water (50 mL) is then added. Most of the methanol is removed under reduced pressure (note that the mixture contains peroxide and should not be evaporated to dryness), and the residue is extracted with diethyl ether (2×20 mL). The ether extracts are combined, washed with 5% aqueous sodium bicarbonate (2×10 mL) and brine (10 mL), dried over magnesium sulfate and concentrated to a colorless oil (1.01 g, 88%). The product is chromatographed on silica gel at medium pressure using hexane/ethyl acetate (8 : 1) to give *syn*-1-hydroxy-2-methyl-1-phenyl-3-pentanone (0.89 g, 77%).

A systematic study of the factors governing the stereoselective formation of (*E*)- or (*Z*)-boron enolates from ethyl ketones ($R^1COCH_2CH_3$), using various R^2_2BX reagents (X = Cl, Br, I, OSO_2CH_3, and OSO_2CF_3), in the presence of various *tert*-amines, has led to several generalizations[135]. (*Z*)-Boron enolates, leading to *syn*-aldols on reaction with aldehydes, are favored by lower steric requirements of R^2, a better leaving group (X = I or OSO_2CF_3), a more hindered *tert*-amine, a polar solvent, and relatively concentrated medium. Conversely, more hindered R^2, relatively poor leaving groups, *tert*-amines of lower steric requirements, nonpolar solvents, and more dilute conditions favor (*E*)-enolates and thereby *anti*-aldols. An even better approach to (*Z*)-enolates makes use of cyclohexyldichloroborane, easily obtainable from cyclohexene and dichloroborane, or other alkyldichloroboranes. This approach is much less sensitive to other reaction parameters and consistently gives good *syn* selectivity in subsequent aldol reactions[136].

Introduction of chirality features into the vinyloxyborane unit allows the reaction to be extended to the synthesis of nonracemic products[3,137]. Furthermore, by appropriate choice of a homochiral aldehyde in reaction with a chiral vinyloxyborane it is possible to make use of double asymmetric induction to maximize the enantioselectivity of the reaction[3,138].

One of the simplest approaches to the incorporation of chirality is to use diisopinocampheylboryl enolates, obtained by reaction of homochiral DIP-chloride™ with an appropriate ketone[139–142]. Unfortunately, the degree of asymmetric induction in the aldol reaction is not always large in such cases, though double asymmetric induction with matched reagents helps[142]. Asymmetric induction is quite good also when a diisopinocampheylboryl enolate of a relatively unhindered alkyl methyl ketone is reacted with a relatively hindered aldehyde[143]:

(89% ee)

Dialkoxyvinyloxyboranes, available through a number of procedures[144,145], also take part in aldol reactions. However, the reactions are slower and *syn* products are obtained predominantly irrespective of the geometry of the enolate[146].

Boron enolates derived from esters[147] or from *N,N*-dialkylamides[148] can be obtained by use of dicyclohexyliodoborane, and either diastereoisomer may be favored, depending on the particular reaction parameters. A chiral amide enolate reaction that gives high diastereoselectivity and high enantioselectivity is the following[149]:

$$
\begin{array}{c}
\text{(a) Bu}_2\text{BOTf} \\
\text{(b) }^i\text{Pr}_2\text{NEt} \\
\xrightarrow{\hspace{1.5cm}} \\
\text{(c) PhCHO} \\
\text{(d) H}_2\text{O}_2
\end{array}
$$

(84% yield, >97% de)

The product is easily hydrolyzed to yield the corresponding β-hydroxy acid in high enantiomeric purity.

21 Alcohols and Alkenes *via* Boron-Stabilized Carbanions

Although a dialkylboryl group can be expected to provide about as much stabilization as a carbonyl group to a carbanion center, the generation of such boron-stabilized anions is much more problematical[3]. Nevertheless, methods have been developed that make the anions available for synthetic reactions. Probably the most useful method involves the direct deprotonation of a hindered organoborane such as an alkyldimesitylborane with a moderately hindered base such as mesityllithium[150]. The anions thus generated can then be reacted with various electrophiles to give the corresponding products[3].

For example, dimesitylboryl-stabilized carbanions react readily with primary-alkyl bromides and iodides to give the corresponding alkylated organoboranes, which can be oxidized to yield alcohols[151]. Although reactions with secondary halides are less efficient, it is possible to make more highly branched derivatives by successive deprotonations followed by alkylations with dimethyl sulfate[152].

Preparation of 2-Octanol from Ethyldimesitylborane and 1-Iodohexane[151]

$$
\text{Mes}_2\text{BCH}_2\text{Me} \xrightarrow{\text{MesLi}} \underset{\mathbf{8}}{\text{Mes}_2\text{B}\overset{-}{\text{C}}\text{HMe}} \xrightarrow{\text{HexI}} \underset{\text{HexI}}{\overset{\text{Hex}}{\text{Mes}_2\text{BCHMe}}} \xrightarrow{[\text{O}]} \overset{\text{OH}}{\text{HexCHMe}}
$$

A 50 mL flask equipped with a magnetic follower is charged with dry mesityl bromide (MesBr, Mes = 2,4,6-trimethylphenyl; 1.095 g, 5.5 mmol), then fitted with a septum connected *via* a needle to a bubbler and flushed with nitrogen *via* a second needle. Dry THF (10 mL) is added and the mixture is cooled to −78 °C for addition of *tert*-butyllithium (7.33 mL of 1.5 mol L^{-1} solution in hexane, 11 mmol), dropwise and with stirring. The mixture, which develops a cloudy yellow color, is stirred for 15 min at −78 °C and then for 15 min at 25 °C. The bubbler is removed so that the solution of mesityllithium so obtained can be withdrawn into a wide-needle syringe when required.

Meanwhile, a 100 mL flask equipped as above is charged with ethyldimesityl-borane (see Sections 5 and 7; 1.39 g, 5 mmol) and flushed with nitrogen. Dry THF (5 mL) is added and the mixture is stirred as the solution of mesityllithium is added by syringe. The mixture is stirred for 1 h at 25 °C (longer for more hindered cases) to generate the anion **8**. The contents of the flask are cooled to 0 °C for addition, by syringe, of 1-iodohexane (1.17 g, 5.5 mmol). The cooling bath is then removed and the pink mixture is stirred for 30 min at 25 °C. The septum is removed and the organoborane is oxidized by successive addition of methanol (5 mL), aqueous NaOH (2.5 mL of 5 mol L^{-1}) and hydrogen peroxide (5 mL of 50%), followed by stirring overnight at 25 °C and then bringing to reflux for 3 h (longer for more hindered cases). The mixture is cooled and saturated with potassium carbonate. The organic layer is separated and the aqueous layer is extracted with diethyl ether (2 × 30 mL). The combined organic extracts are washed with 10% aqueous citric acid (30 mL) and water (20 mL), dried over magnesium sulfate and evaporated below 40 °C. The crude product is chromatographed on silica gel by gradient elution with dichloromethane through chloroform and diethyl ether. 2-Octanol (0.442 g, 68%) is obtained from the fractions containing chloroform or chloroform with a small amount of ether following removal of the solvent under reduced pressure at less than 40 °C.

A particularly useful type of reaction of anions such as **8** is that with carbonyl compounds, known as the boron-analogous Wittig reaction[153]. In the reaction with aromatic aldehydes it is possible to vary the work-up procedure to obtain either (*E*)- or (*Z*)-alkenes[3,154].

Preparation of (E)- or (Z)-1-Phenyl-1-nonene by the Boron-analogous Wittig Reaction[3,154]

The anion **9** is prepared from dimesityloctylborane (3 mmol) in a manner anal-ogous to that described for **8** in the preceding procedure. The solution is cooled to −78 °C. Meanwhile, a 10 mL Wheaton bottle equipped with a septum is flushed with argon, charged with freshly distilled benzaldehyde (0.223 g, 2.1 mmol) in THF (3 mL) and cooled to −78 °C. The benzaldehyde solution is added to the stirred solu-tion of anion **9** via a double-ended needle and the mixture is stirred for 2 h at −78 °C to give intermediate **10**. This solution is used for preparation of either alkene, as detailed below.

(*E*)-1-Phenyl-1-nonene

A solution of chlorotrimethylsilane (0.337 g, 3.1 mmol) in THF (3 mL) in an argon-flushed Wheaton bottle is cooled to −78 °C and transferred *via* a double-ended needle to the stirred solution of intermediate **10**. The mixture is stirred at −78 °C for 1 h and then gradually allowed to warm to 20 °C and stirred for a further 16 h. The volatile components are removed under reduced pressure and dry light petroleum (bp 30–40 °C, 30 mL) is added. The mixture is stirred and then allowed to settle and the petroleum solution is decanted from the precipitated solid, concentrated and chromatographed on alumina, eluted with pentane. An oil consisting of mesitylene and the trimethylsilyl ether of **10** is obtained. This oil is cooled to −78 °C and a solution of aqueous HF (1 mL of 40%) in HPLC grade acetonitrile (20 mL) is added. The cooling bath is removed and the solid reaction mixture is allowed to warm to 20 °C and then stirred for a further 30 min. The mixture is poured into pentane (30 mL) and the pentane phase is separated. The aqueous phase is extracted a second time with pentane (30 mL) and the combined extracts are washed with water (3 × 20 mL), dried over magnesium sulfate and evaporated to give a crude product (1.68 g). Chromatography on silica, eluted with pentane, gives a fraction containing the product and mesitylene, which on pumping overnight to remove mesitylene leaves pure (*E*)-1-phenyl-1-nonene (0.36 g, 84%). The product can be distilled under reduced pressure if required (bp 80–82 °C/0.1 mmHg).

(*Z*)-1-Phenyl-1-nonene

The solution of intermediate **10** is cooled to −110 °C and a precooled (−78 °C) solution of trifluoroacetic anhydride (0.55 g, 2.6 mmol) in THF (3 mL) is added *via* a cooled double-ended needle. The mixture is stirred for 1 h at −110 °C and 4 h at −78 °C then left to warm to 25 °C overnight. Volatile materials are removed under reduced pressure and then light petroleum (bp 30–40 °C, 30 mL) is added. The mixture is stirred and then allowed to settle and the petroleum layer is decanted from the precipitated solid. The extract is evaporated and the residue is chromatographed on silica, eluted with pentane. The fractions containing mesitylene and the product are pumped overnight to remove mesitylene and leave pure (Z)-1-phenyl-1-nonene (0.33 g, 77%). The product can be distilled if required (bp 79–83 °C/0.1 mmHg).

The reaction with aliphatic aldehydes is less reliable under the procedures described above. Instead, alkenes are more reliably obtained by using an acid in admixture with the aldehyde during addition to the anion **9**[155]. The stereochemistry can be controlled by the choice of acid, strong acids such as trifluoromethanesulfonic acid or HCl giving predominantly the (*E*)-alkene, and weak acids such as acetic acid generally giving predominantly the (*Z*)-alkene.

22 Reductions with Trialkylhydroborates

There are two features of trialkylhydroborates (trialkylborohydrides) that justify their use instead of simple reagents such as sodium borohydride in certain circumstances: they are much more reactive and can therefore accomplish reactions that are very slow or low yielding with the simple reagents; and they are much more hindered and may therefore give rise to more selective reactions, particularly more stereoselective reactions, when steric factors are important[3].

Probably the most useful reaction that takes account of the high reactivity of trialkylhydroborates is the conversion of alkyl halides and tosylates into the corresponding alkanes[156]. Several trialkylhydroborates are commercially available and can be used directly. Otherwise, unhindered trialkylhydroborates are relatively easily synthesized by stirring THF solutions of the corresponding trialkylboranes with solid LiH, NaH or KH, whereas hindered trialkylboranes are easily converted into their lithium hydroborates by reaction with *tert*-butyllithium[3]. Lithium aminoborohydrides such as $LiPr_2NBH_3$ appear to be generally useful as agents for the transfer of hydride to unhindered or hindered organoboranes alike[157].

Preparation of Cyclooctane by Reduction of Cyclooctyl Tosylate with Lithium Triethylhydroborate[3,156]

A 300 mL, two-necked flask equipped with a magnetic follower, a reflux condenser connected to a bubbler and a septum-capped side arm is flushed with nitrogen, charged with dry THF (20 mL) and cyclooctyl toluenesulfonate (7.05 g, 25 mmol) by syringe, and cooled to 0 °C. Lithium triethylhydroborate (33.3 mL of 1.5 mol L^{-1} in THF, 50 mmol; note that as commercial solutions are less concentrated, longer reaction times may be necessary with them unless concentrations are adjusted) is added to the stirred solution, the ice-bath is removed, the mixture is stirred for 2 h at 25 °C, and excess hydride is then destroyed by cautious addition of water (hydrogen is evolved). Oxidation of the triethylborane byproduct is achieved by successive addition of aqueous NaOH (20 mL of 3 mol L^{-1}) and hydrogen peroxide (20 mL of 30%, added dropwise; Section 2), followed by stirring at 25 °C for 1 h. The mixture is allowed to separate, the aqueous layer is extracted with pentane (2 × 20 mL) and the combined organic extracts are washed with water (4 × 15 mL), dried over magnesium sulfate and concentrated by distillation at atmospheric pressure (because of the volatility of the product). The residue is transferred to a small distillation apparatus and distilled at atmospheric pressure to give cyclooctane (2.27 g, 81%) as a colorless oil (bp 142–146 °C), contaminated by about 3% of cyclooctene.

Other reactions for which trialkylhydroborates may have advantages include demethylation of methyl aryl ethers[158] and chemoselective reduction of *N*-Boc-protected lactams[159].

The ability of trialkylhydroborates to effect diastereoselective reductions is well illustrated by the reduction of substituted cyclohexanones[160]. Use of lithium tri-*sec*-butylhydroborate (L-Selectride™) at −78 °C allows the production of 90% of the less stable *cis* isomer on reduction of 4-methylcyclohexanone and with lithium trisiamylhydroborate the proportion is 99%. With 4-*tert*-butylcyclohexanone the proportion is even higher[160]. Recently, results comparable with those achieved using lithium trisiamylhydroborate at −78 °C have been achieved at 0 °C with the even more hindered reagent, lithium ethylbis(2,4,6-triisopropylphenyl)hydroborate[161]. In this case there is the additional advantage that the triorganylborane byproduct is air stable and can easily be recovered and reused.

Preparation of cis-4-tert-Butylcyclohexanol by Reduction of
4-tert-Butylcyclohexanone with Lithium Trisiamylhydroborate[3,160]

A 250 mL, two-necked flask equipped with a magnetic follower, a reflux condenser connected to a bubbler and a septum-capped stopcock is flushed with nitrogen, charged with lithium trisiamylhydroborate solution (70 mL of 0.4 mol L^{-1} in THF, 28 mmol) and immersed in a −78 °C cooling bath. A solution of 4-*tert*-butylcyclohexanone (3.7 g, 24 mmol) in THF (25 mL) is cooled to 0 °C and then added by syringe to the trisiamylhydroborate solution as rapidly as is consistent with keeping the reaction solution cold (for maximum selectivity). The mixture is stirred vigorously for 2 h at −78 °C and then allowed to warm to 25 °C over 1 h. Water (4 mL) and ethanol (10 mL) are added and the trisiamylborane byproduct is oxidized by successive addition of aqueous NaOH (10 mL of 6 mol L^{-1}) and hydrogen peroxide (15 mL of 30%), followed by warming at 40 °C for 30 min once the initial vigorous reaction subsides (Section 2). The mixture is cooled and the aqueous phase is saturated with potassium carbonate. The organic phase is separated, the aqueous phase is further extracted with diethyl ether/tetrahydrofuran (2 × 20 mL, 1 : 1 mixture), and the combined extracts are dried over magnesium sulfate. The volatile solvents and 3-methyl-2-butanol are removed under reduced pressure to leave *cis*-4-*tert*-butylcyclohexanol (3.65 g, 98%) as a white solid (mp 80 °C), which is at least 99.5% *cis*-isomer by GC.

In reductions of acyclic ketones by trialkylhydroborates the nature of substituents on the ketones, as well as their size, can be important in determining the diastereoselectivity[162].

In general, homochiral trialkylhydroborates have not proved to be as useful for enantio-selective reductions of ketones as certain types of tricoordinate organoboranes (Section 23). However, some success has been achieved recently with some substituted derivatives of *B*-isopinocampheyl-9-BBN hydride[163].

23   Stereoselective Reductions with Organoboranes

Trialkylboranes generally display little reactivity towards aldehydes or ketones. However, organoboranes possessing an isopinocampheyl group are more reactive, effecting reduction of some compounds with concomitant displacement of α-pinene. Furthermore, as such organoboranes can be obtained in nonracemic form, the reagents can be used to induce asymmetry during reduction. *B*-Isopinocampheyl-9-BBN (Alpine-borane™) is one such reagent. It reduces aldehydes readily at 65 °C and if its deuterio derivative is used, or if a 1-deuterioaldehyde is used, the corresponding deuteriated primary alcohol is obtained in very high optical purity[3,164].

Preparation of (S-Benzyl-1-d Alcohol by Reduction of 1-Deuteriobenzaldehyde with Alpine-Borane™

A 1 L, two-necked flask equipped with a magnetic follower, a reflux condenser connected to a bubbler and a septum-capped stopcock is flushed with nitrogen and then charged with *B*-isopinocampheyl-9-BBN (52.9 g, 205 mmol, from (+)-α-pinene of 93% ee) in THF (400 mL) by a double-ended needle. Benzaldehyde-1-*d* (19.0 mL, 185 mmol) is added and the mixture is stirred for 10 min at 25°C and then heated at reflux for 1 h. The mixture is cooled to 20 °C and acetaldehyde (5 mL) is added to destroy residual trialkylborane. THF is removed under reduced pressure and the mixture is then pumped at oil-pump vacuum using a bath temperature of 40 °C to remove α-pinene. Nitrogen is readmitted, diethyl ether (150 mL) is added, the solution is cooled to 0 °C and 2-aminoethanol (12.5 g, 205 mmol) is added in order to precipitate the 9-BBN derivative. The precipitate is filtered off and washed with diethyl ether (2 × 20 mL). The combined organic extracts are washed

with water (25 mL), dried over magnesium sulfate and evaporated to give a product
that is further purified by distillation under reduced pressure. (*S*)-Benzyl-1-*d* alcohol
(16.5 g, 82%) of 88% ee is collected at 110 °C/30 mmHg.

Alpine-borane[TM] reacts sluggishly with simple ketones, but reductions can still be
effected with good enantioselectivity if the reactions are carried out without solvent
and/or under pressure[165]. Alternatively, alkynyl ketones, which are much less sterically
demanding, are reduced more readily to the corresponding alkynylmethanols, and the
latter are easily converted into saturated alcohols. 'NB-Enantrane', the organoborane
derived by reaction of 9-BBN-H with nopol benzyl ether, reduces alkynyl ketones with
even greater enantioselectivity[166].

For enantioselective reduction of simple ketones, including aromatic ketones
such as acetophenone (98% ee for the corresponding alcohol), hindered aliphatic
ketones such as pinacolone (95% ee), and hindered alicyclic ketones such as
2,2-dimethylcyclopentanone (98% ee), the best reagent appears to be diisopinocampheylchloroborane (DIP-Chloride[TM]; commercially available, but see Section 8 for
preparation of similar reagents)[167]. However, some hindered ketones such as pinacolone
react much more sluggishly than relatively unhindered analogs such as acetophenone and
the reaction requires 12 days at 25 °C without solvent for completion[167]. Several diacyl-
aromatics cleanly produce C_2-symmetric diols under the standard conditions[168].

*Preparation of (S)-1-Phenylethanol by Reduction of Acetophenone with
Diisopinocampheylchloroborane*[167]

A 250 mL flask equipped with a magnetic follower and a septum is flushed
with nitrogen and charged with a solution of diisopinocampheylchloroborane (9.0 g,
28 mmol; from (+)-α-pinene) in THF (20 mL). The solution is cooled to −25 °C
and acetophenone (3.05 mL, 26 mmol) is added by syringe, whereupon the mixture
turns yellow. The mixture is stirred for 7 h at −25 °C (^{11}B NMR of a methanolysis
aliquot shows the reaction to be complete) and the volatiles are then removed at
aspirator pressure. α-Pinene is removed under reduced pressure (0.1 mmHg, 8 h) and
the residue is dissolved in diethyl ether (100 mL). Diethanolamine (6.0 g, 57 mmol)
is added and after 2 h the solid is removed by filtration. The solid is washed with
pentane (2 × 30 mL) and the combined organic solutions are concentrated. The
residue is distilled to give (*S*)-1-phenylethanol (2.3 g, 72%; bp 118 °C/22 mmHg,
with 97% ee.

DIP-Chloride™ is not a very satisfactory reagent for stereoselective reduction of simple unhindered aliphatic ketones, but reductions of α- or β-hydroxyketones are much more successful, giving high enantioselectivity with just one equivalent of the reagent via internal reduction of the intermediate diisopinocampheylboryl ketoalkoxides[169].

The reduction of ketones with DIP-Chloride™ is tolerant of several other functional groups, such as ester and halogenoalkyl groups, which increases the value of the process. It can be used, for example, in the synthesis of either enantiomer of the potential anti-psychotic agent, α-(4-fluorophenyl)-4-(2-pyrimidinyl)-1-piperazinebutanol, via enantio-selective reduction of the intermediate 4-chloro-4′-fluorobutyrophenone[170]. The reduction occurs without affecting either the fluoroaryl or chloroalkyl groups:

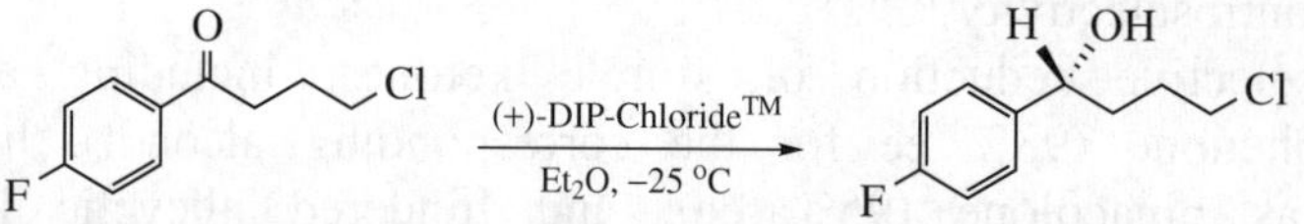

α-Fluoroalkyl ketones can also be reduced enantioselectively and without problem[171]. Indeed, perfluoroalkyl groups appear always to act as the groups that control the enantio-selectivity. If the reduction is applied to trifluoromethyl bromomethyl ketone, followed by base-catalyzed cyclization of the bromohydrin obtained, the result is a useful synthesis of α-(trifluoromethyl)oxirane[172]:

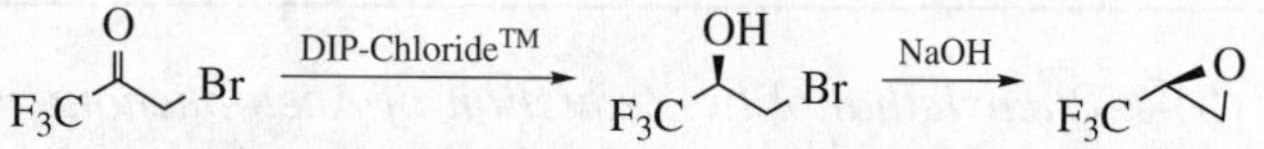

For the racemic form of the specific ketone shown in the above equation, use of 0.65 equivalents of (−)-DIP-Chloride™ allows recovery of 61% of the (R)-ketone having over 99% ee[173].

Ketones that already possess an asymmetric center react with the two different enan-tiomers of DIP-Chloride™ to give different selectivity. With matched reacting pairs the selectivity can be very impressive, as is the case in the reduction of ethyl 1-methyl-2-oxocarboxylate[174]:

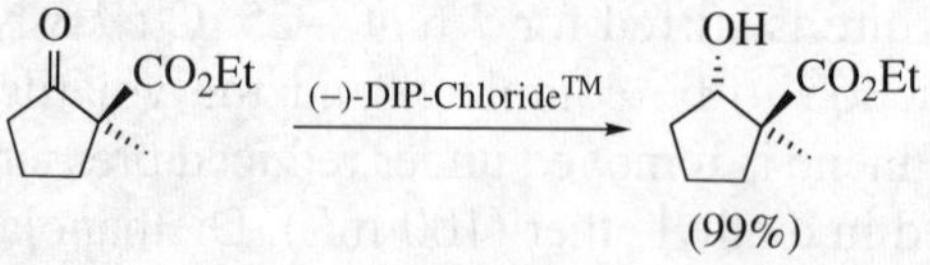

(99%)

The substantial differences in rates between the matched and unmatched pairs allow the reactions to be used for kinetic resolution of racemic α-tertiary ketones by the simple expedient of using a deficiency of the reducing agent.

Ketones that contain an asymmetric dialkoxyboryl group can be reduced with consid-erable asymmetric induction by borane–dimethyl sulfide, which offers opportunities for enantioselective synthesis of various diols and their derivatives[175]:

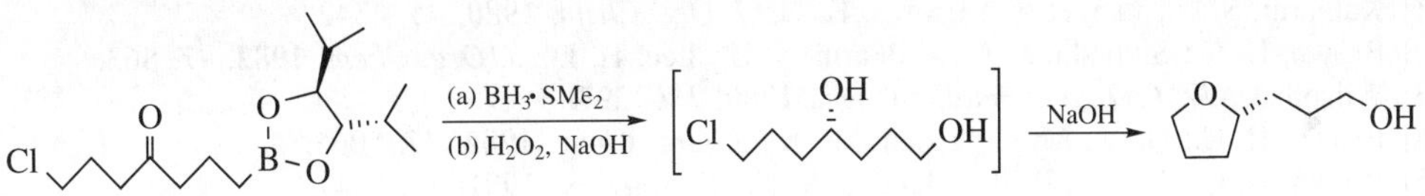

24 Conclusions

This chapter provides well-tried procedures for some of the more important synthetic methods based on organoboron compounds. In particular, it provides a good starting point for anyone interested in making use of the enormous synthetic potential of organo-boranes, especially those experiencing reluctance on account of lack of familiarity with the handling of such reagents. Naturally, in order to try to keep the material to a reasonable quantity, the choice of reactions for inclusion has had to be highly selective. Also, procedures may need to differ, even for the same reaction type, depending on factors such as the degree of steric hindrance in the substrate. Such variations can only be alluded to in a work of this length. For additional procedures and more extensive discussions of reactions of organoboranes, specialized monographs are available[3,4].

25 References

[1] Brown, H. C.; *Hydroboration*, Benjamin, New York, **1962**; reprinted with Nobel lecture, Benjamin/ Cummings, Reading, Massachusetts, **1980**.

[2] Smith, K.; *Chem. Ind. (London)* **1987**, 603.

[3] Pelter, A.; Smith, K.; Brown, H. C.; *Borane Reagents*, Academic Press, London, **1988**.

[4] Brown, H. C.; Kramer, G. W.; Levy, A. B.; Midland, M. M.; *Organic Syntheses via Boranes*, Wiley, New York, **1975**.

[5] Kabalka, G. W.; *Current Topics in the Chemistry of Boron*, Royal Society of Chemistry, Cambridge, **1994**.

[6] Beletskaya, I.; Pelter, A.; *Tetrahedron* **1997**, *53*, 4957.

[7] Matteson, D. S.; *Stereodirected Synthesis with Organoboranes*, Springer-Verlag, Berlin, **1995**.

[8] Seyden-Penne, J.; *Reductions by the Alumino- and Borohydrides in Organic Synthesis*, Wiley-VCH, New York, **1997**.

[9] Brown, H. C.; Liotta, R.; Scouten, C. G.; *J. Am. Chem. Soc.* **1976**, *98*, 5297.

[10] Pelter, A.; Singaram, S.; Brown, H. C.; *Tetrahedron Lett.* **1983**, *24*, 1433.

[11] Brown, H. C.; Jadhav, P. K.; Mandal, A. K.; *Tetrahedron* **1981**, *37*, 3547.

[12] Smith, K.; *Chem. Soc. Rev.* **1974**, *3*, 443.

[13] Köster, R.; Binger, P.; Fenzl, W.; *Inorg. Synth.* **1974**, *15*, 134.

[14] Brown, H. C.; Lane, C. F.; *Heterocycles* **1977**, *7*, 453.

[15] Brown, H. C.; Desai, M. C.; Jadhav, P. K.; *J. Org. Chem.* **1982**, *47*, 5065.

[16] Schwier, J. R.; Brown, H. C.; *J. Org. Chem.* **1993**, *58*, 1546.

[17] Parks, D. J.; von H. Spence, R. E.; Piers, W. E.; *Angew. Chem., Int. Ed. Engl.* **1995**, *34*, 809.

[18] Pelter, A.; Smith, K.; Buss, D.; Jin, Z.; *Heteroat. Chem.* **1992**, *3*, 275.

[19] Smith, K.; Pelter, A.; Jin, Z.; *J. Chem. Soc., Perkin Trans. 1* **1993**, 395.

[20] Negishi, E.; Brown, H. C.; *Synthesis* **1974**, 77.

[21] Kulkarni, S. U.; Lee, H. D.; Brown, H. C.; *J. Org. Chem.* **1980**, *45*, 4542.

[22] Brown, H. C.; Sikorski, J. A.; Kulkarni, S. U.; Lee, H. D.; *J. Org. Chem.* **1982**, *47*, 863.

[23] Wilson, J. W.; *J. Organometallic Chem.* **1980**, *186*, 297.

[24] Brown, H. C.; Desai, M. C.; Jadhav, P. K.; *J. Org. Chem.* **1982**, *47*, 5065.

[25] Brown, H. C.; Jadhav, P. K.; Mandal, A. K.; *Tetrahedron* **1981**, *37*, 3547.

[26] Zweifel, G.; Ayyangar, N. R.; Munekata, T.; Brown, H. C.; *J. Am. Chem. Soc.* **1964**, *86*, 1076.

[27] Brown, H. C.; Mandal, A. K.; Yoon, N. M.; Singaram, B.; Schwier, J. R.; Jadhav, P. K.; *J. Org. Chem.*, **1982**, *47*, 5069.

[28] Brown, H. C.; Vara Prasad, J. V. N.; Gupta, A. K.; Bakshi, R. K.; *J. Org. Chem.* **1987**, *52*, 310.

[29] (a) Brown, H. C.; Singaram, B.; *Pure Appl. Chem.* **1987**, *59*, 879; (b) Brown, H. C.; Singaram, B.; *Acc. Chem. Res.* **1988**, *21*, 287.

[30] (a) Rauchschwalbe, G.; Schlosser, M.; *Helv. Chim. Acta* **1975**, *58*, 1094; (b) Brown, H. C.; Dhokte, U. P.; *J. Org. Chem.* **1994**, *59*, 2025.

[31] Brown, H. C.; Dhokte, U. P.; *J. Org. Chem.* **1994**, *59*, 2365.

[32] Brown, H. C.; Dhokte, U. P.; *J. Org. Chem.* **1994**, *59*, 5479.

[33] Dhokte, U. P.; Brown, H. C.; *Tetrahedron Lett.* **1994**, *35*, 4715.

[34] (a) Dhokte, U. P.; Brown, H. C.; *Tetrahedron Lett.* **1996**, *37*, 9021; (b) Brown, H. C.; Mahindroo, V. K.; Dhokte, U. P.; *J. Org. Chem.*, **1996**, *61*, 1906.

[35] Dhokte, U. P.; Kulkarni, S. U.; Brown, H. C.; *J. Org. Chem.* **1996**, *61*, 5140.

[36] Dhokte, U. P.; Soundararajan, R.; Ramachandran, P. V.; Brown, H. C.; *Tetrahedron Lett.* **1996**, *37*, 8345.

[37] (a) Matteson, D. S.; Ray, R.; Rocks, R. R.; Tsai, D. J.; *Organometallics* **1983**, *2*, 1536; (b) Matteson, D. S.; Sadhu, K. M.; Peterson, M. L.; *J. Am. Chem. Soc.* **1986**, *108*, 810; (c) review: Matteson, *Acc. Chem. Res.* **1988**, *21*, 294.

[38] Narla, G.; Brown, H. C.; *Tetrahedron Lett.* **1997**, *38*, 219.

[39] Brown, H. C.; Ravindran, W.; *Inorg. Chem.* **1977**, *16*, 2938.

[40] (a) Brown, H. C.; Ravindran, N.; Kulkarni, S. U.; *J. Org. Chem.* **1980**, *45*, 384; (b) Brown, H. C.; Racherla, U. S.; *J. Org. Chem.* **1986**, *51*, 895.

[41] Brown, H. C.; Campbell, J. B.; *J. Org. Chem.* **1980**, *45*, 389.

[42] Paget, W. E.; Smith, K.; *J. Chem. Soc., Chem. Commun.* **1980**, 1169.

[43] Brown, H. C.; Ravindran, N.; *J. Org. Chem.* **1977**, *42*, 2533.

[44] (a) Brown, H. C.; Gupta, S. K.; *J. Am. Chem. Soc.* **1972**, *94*, 4370; (b) Brown, H. C.; Chandrasekharan, J.; *J. Org. Chem.* **1983**, *48*, 5080.

[45] (a) Brown, H. C.; Gupta, S. K.; *J. Am. Chem. Soc.* **1971**, *93*, 1816; (b) Brown, H. C.; Gupta, S. K.; *J. Am. Chem. Soc.* **1975**, *97*, 5249.

[46] Manning, D.; Nöth, H.; *Angew. Chem., Int. Ed. Engl.* **1985**, *24*, 878.

[47] Sato, M.; Miyaura, N.; Suzuki, A.; *Tetrahedron Lett.* **1990**, *31*, 231.

[48] Murata, M.; Satoh, K.; Watanabe, S.; Masuda, Y.; *J. Chem. Soc., Perkin Trans. 1* **1998**, 1465.

[49] De Lue, N. R.; Brown, H. C.; *Synthesis* **1976**, 114.

[50] (a) Brown, H. C.; Lane, C. F.; *J. Am. Chem. Soc.* **1970**, *92*, 6660.

[51] (a) Brown, H. C.; DeLue, N. R.; Kabalka, G. W.; Hedgecock, H. C.; *J. Am. Chem. Soc.* **1976**, *98*, 1290; (b) Kabalka, G. W.; Gooch, E. E.; *J. Org. Chem.* **1981**, *46*, 2582.

[52] Kuivila, H. G.; Hendrickson, A. R.; *J. Am. Chem. Soc.* **1952**, *74*, 5068.

[53] Brown, H. C.; Hamaoka, T.; Ravindran, N.; *J. Am. Chem. Soc.* **1973**, *95*, 6456.

[54] (a) Brown, H. C.; Heydkamp, W. R.; Breuer, E.; Murphy, W. S.; *J. Am. Chem. Soc.* **1964**, *86*, 3565; (b) Rathke, M. W.; Inoue, N.; Varma, K. R.; Brown, H. C.; *J. Am. Chem. Soc.* **1966**, *88*, 2870.

[55] (a) Kabalka, G. W.; Sastry, K. A. R.; McCollum, G. W.; Yoshioka, H.; *J. Org. Chem.* **1981**, *46*, 4296; (b) Kabalka, G. W.; Sastry, K. A. R.; McCollum, G. W.; Lane, C. F.; *J. Chem. Soc. Chem. Commun.* **1982**, 62.

[56] Ramachandran, P. V.; Rangaishenvi, M. V.; Singaram, B.; Goralski, C. T.; Brown, H. C.; *J. Org. Chem.* **1996**, *61*, 341.

[57] Brown, H. C.; Salunkhe, A. M.; Singaram, B.; *J. Org. Chem.* **1991**, *56*, 1170.

[58] Brown, H. C.; Salunkhe, A. M.; *Tetrahedron Lett.* **1993**, *34*, 1265.

[59] Jigajinni, V. B.; Pelter, A.; Smith, K.; *Tetrahedron Lett.* **1978**, 181.

[60] (a) Brown, H. C.; Murray, K.; *J. Am. Chem. Soc.* **1959**, *81*, 4108; (b) Zweifel, G.; Brown, H. C.; *J. Am. Chem. Soc.* **1964**, *86*, 393.

[61] Hooz, J.; Bridson, J. N.; Calzada, J. G.; Brown, H. C.; Midland, M. M.; Levy, A. B.; *J. Org. Chem.* **1973**, *38*, 2574.

[62] Brown, H. C.; Nambu, H.; Rogić, M. M.; *J. Am. Chem. Soc.* **1969**, *91*, 6854.

[63] Midland, M. M.; Kwon, Y. C.; *J. Org. Chem.* **1981**, *46*, 229.

[64] Hooz, J.; Calzada, J. G.; *Org. Prep. Proc. Int.* **1972**, *4*, 219.

[65] (a) Brown, H. C.; Jayaraman, S.; *Tetrahedron Lett.* **1993**, *34*, 3997; (b) Brown, H. C.; Jayaraman, S.; *J. Org. Chem.* **1993**, *58*, 6791.

[66] Hughes, R. J.; Ncube, S.; Pelter, A.; Smith, K.; Negishi, E.; Yoshida, T.; *J. Chem. Soc., Perkin Trans. 1* **1977**, 1172.

[67] Ncube, S.; Pelter, A.; Smith, K.; *Tetrahedron Lett.* **1979**, 1895.

[68] Ncube, S.; Pelter, A.; Smith, K.; *Tetrahedron Lett.* **1979**, 1893.

[69] Köbrich, G.; Merkle, H. R.; *Chem. Ber.* **1967**, *100*, 3371.

[70] See reference 7, Chapter 5.

[71] Matteson, D. S.; Kandil, A. A.; *Tetrahedron Lett.* **1986**, *27*, 3831.

[72] Hoffmann, R. W.; Ditrich, K.; Köster, G.; Stürmer, R.; *Chem. Ber.* **1989**, *122*, 1783.

[73] Tripathy, P. B.; Matteson, D. S.; *Synthesis* **1990**, 200.

[74] (a) Matteson, D. S.; Michnick, T. J.; *Organometallics* **1990**, *9*, 3171; (b) Matteson, D. S.; Man, H. W.; *J. Org. Chem.* **1994**, *59*, 5734.

[75] Matteson, D. S.; Campbell, J. D.; *Heteroat. Chem.* **1990**, *1*, 109.

[76] Brown, H. C.; Carlson, B. A.; *J. Org. Chem.* **1973**, *38*, 2422.

[77] Brown, H. C.; Katz, J.-J.; Carlson, B. A.; *J. Org. Chem.* **1973**, *38*, 3968.

[78] (a) Carlson, B. A.; Brown, H. C.; *J. Am. Chem. Soc.* **1973**, *95*, 6876; (b) Carlson, B. A.; Brown, H. C.; *Synthesis* **1973**, 776; (c) Carlson, B. A.; J.-Katz, J.; Brown, H. C.; *J. Organomet. Chem.* **1974**, *67*, C39; (d) Brown, H. C.; Katz, J.-J.; Carlson, B. A.; *J. Org. Chem.* **1975**, *40*, 813.

[79] Brown, H. C.; Mahindroo, V. K.; *Tetrahedron: Asymmetry* **1993**, *4*, 59.

[80] Brown, H. C.; Imai, T.; *J. Org. Chem.* **1984**, *49*, 892.

[81] Brown, H. C.; *Acc. Chem. Res.* **1969**, *2*, 65.

[82] (a) Brown, H. C.; Negishi, E.; *J. Am. Chem. Soc.* **1967**, *89*, 5477; (b) Brown, H. C.; Negishi, E.; *J. Chem. Soc., Chem. Commun.* **1968**, 594.

[83] Pelter, A.; Maddocks, P. J.; Smith, K.; *J. Chem. Soc., Chem. Commun.* **1978**, 805.

[84] (a) Brown, H. C.; Negishi, E.; *J. Am. Chem. Soc.* **1967**, *89*, 5478; (b) Brown, H. C.; Dickason, W. C.; *J. Am. Chem. Soc.* **1969**, *91*, 1226.

[85] Brown, H. C.; Hubbard, J. L.; Smith, K.; *Synthesis* **1979**, 701.

[86] Pelter, A.; Hutchings, M. G.; Smith, K.; *J. Chem. Soc., Chem. Commun.* **1970**, 1529.

[87] Pelter, A.; Smith, K.; Hutchings, M. G.; Rowe, K.; *J. Chem. Soc., Perkin Trans. 1* **1975**, 129.

[88] Pelter, A.; Hutchings, M. G.; Smith, K.; *J. Chem. Soc., Chem. Commun.* **1971**, 1048.

[89] Pelter, A.; Hutchings, M. G.; Rowe, K.; Smith, K.; *J. Chem. Soc., Perkin Trans. 1* **1975**, 138.

[90] Pelter, A.; Bentley, T. W.; Harrison, C. R.; Subrahmanyam, C.; Laub, R. J.; *J. Chem. Soc., Perkin Trans. 1* **1976**, 2419.

[91] Pelter, A.; Gould, K. J.; Harrison, C. R.; *J. Chem. Soc., Perkin Trans. 1* **1976**, 2428.

[92] Pelter, A.; Colclough, M. E.; *Tetrahedron* **1995**, *51*, 811.

[93] (a) Suzuki, A.; Miyaura, N.; Abiko, S.; Itoh, M.; Brown, H. C.; Sinclair, J. A.; Midland, M. M.; *J. Am. Chem. Soc.* **1973**, *95*, 3080; (b) *J. Org. Chem.* **1986**, *51*, 4507.

[94] Pelter, A.; Smith, K.; Tabata, M.; *J. Chem. Soc., Chem. Commun.* **1975**, 857.

[95] Sinclair, J. A.; Brown, H. C.; *J. Org. Chem.* **1976**, *41*, 1078.

[96] Pelter, A.; Hughes, R. J.; Smith, K.; Tabata, M.; *Tetrahedron Lett.* **1976**, 4385.

[97] (a) Utimoto, K.; Uchida, K.; Yamaya, M.; Nozaki, H.; *Tetrahedron* **1977**, *33*, 1945; (b) Miyaura, N.; Tagami, H.; Itoh, M.; Suzuki, A.; *Chem. Lett.* **1974**, 1411; (c) LaLima, N. J.; Levy, A. B.; *J. Org. Chem.* **1978**, *43*, 1279.

[98] (a) Evans, D. A.; Thomas, R. C.; Walker, J. A.; *Tetrahedron Lett.* **1976**, 1427; (b) Evans, D. A.; Crawford, T. C.; Thomas, R. C.; Walker, J. A.; *J. Org. Chem.* **1976**, *41*, 3947.

[99] (a) Brown, H. C.; Basaviah, D.; *J. Org. Chem.* **1982**, *47*, 1792, 3806, 5407; (b) Kulkarni, S. U.; Basaviah, D.; Zaidlewicz, M.; Brown, H. C.; *Organometallics* **1982**, *1*, 212.

[100] Zweifel, G.; Arzoumanian, H.; Whitney, C. C.; *J. Am. Chem. Soc.* **1967**, *89*, 3652.

[101] Negishi, E.; Lew, G.; Yoshida, T.; *J. Chem. Soc., Chem. Commun.* **1973**, 874.

[102] Zweifel, G.; Polston, N. L.; Whitney, C. C.; *J. Am. Chem. Soc.* **1968**, *90*, 6243.

[103] Brown, H. C.; Soundararajan, R.; *Tetrahedron Lett.* **1994**, *35*, 6963.

[104] Brown, H. C.; Imai, T.; *Organometallics* **1984**, *3*, 1392.

[105] Article giving most of the important background references: Miyaura, N.; Ishiyama, T.; Sasaki, H.; Ishikawa, M.; Satoh, M.; Suzuki, A.; *J. Am. Chem. Soc.* **1989**, *111*, 314.

[106] (a) Miyaura, N.; Suginome, H.; Suzuki, A.; *Tetrahedron* **1983**, *39*, 3271; (b) Yatagai, H.; Yamamoto, Y.; Maruyama, K.; Sonoda, A.; Murahashi, S.-I.; *J. Chem. Soc., Chem. Commun.* **1977**, 852; (b) Uenishi, J.; Beau, J. M.; Armstrong, R. W.; Kishi, Y.; *J. Am. Chem. Soc.* **1987**, *109*, 4756.

[107] Miyaura, N.; Suzuki, A.; *Chem. Rev.* **1995**, *95*, 2457.

[108] Soderquist, J. A.; Leon, G.; *Tetrahedron Lett.* **1998**, *39*, 3989.

[109] Kobayashi, Y.; Takahisa, E.; Usmani, S. B.; *Tetrahedron Lett.* **1998**, *39*, 597.

[110] Usmani, S. B.; Takahisa, E.; Kobayashi, Y.; *Tetrahedron Lett.* **1998**, *39*, 601.

[111] Moreno-Manas, M.; Perez, M.; Pleixats, R.; *J. Org. Chem.* **1996**, *61*, 2346.

[112] Firooznia, F.; Gude, C.; Chan, K.; Satoh, Y.; *Tetrahedron Lett.* **1998**, *39*, 3985.

[113] Ishiyama, T.; Abe, S.; Miyaura, N.; Suzuki, A.; *Chem. Lett.* **1992**, 691.

[114] Evans, D. A.; Katz, J. L.; West, T. R.; *Tetrahedron Lett.* **1998**, *39*, 2937.

[115] Lam, P. Y. S.; Clark, C. G.; Saubern, S.; Adams, J.; Winters, M. P.; Chan, D. M. T.; Combs, A.; *Tetrahedron Lett.* **1998**, *39*, 2941.

[116] Chan, D. M. T.; Monaco, K. L.; Wang, R.-P.; Winters, M. P.; *Tetrahedron Lett.* **1998**, *39*, 2933.

[117] Hoffmann, R. W.; Weidmann, U.; *J. Organomet. Chem.* **1980**, *195*, 137.

[118] Hoffmann, R. W.; Zeiss, H.-J.; *Angew. Chem., Int. Ed. Engl.* **1979**, *18*, 306.

[119] Brown, H. C.; Bhat, K. S.; *J. Am. Chem. Soc.* **1986**, *108*, 5919.

[120] Tsai, D. J. S.; Matteson, D. S.; *Tetrhaedron Lett.* **1981**, *22*, 2751.

[121] Ramachandran, P. V.; Chen, G.-M.; Brown, H. C.; *Tetrahedron Lett.* **1997**, *38*, 2417.

[122] Brown, H. C.; Narla, G.; *J. Org. Chem.* **1995**, *60*, 4686.

[123] Barrett, A. G. M.; Seefeld, M. A.; *Tetrahedron* **1993**, *49*, 7857.

[124] (a) Kramer, G. W.; Brown, H. C.; *J. Org. Chem.* **1977**, *42*, 2292; (b) Kramer, G. W.; Brown, H. C.; *J. Organomet. Chem.* **1977**, *132*, 9.

[125] (a) Brown, H. C.; Khire, U. R.; Racherla, U. S.; *Tetrahedron Lett.* **1993**, *34*, 15; (b) Brown, H. C.; Khire, U. R.; Narla, G.; Racherla, U. S.; *J. Org. Chem.* **1995**, *60*, 544.

[126] Soundararajan, R.; Li, G.; Brown, H. C.; *Tetrahedron Lett.* **1995**, *36*, 2441.

[127] Soundararajan, R.; Li, G.; Brown, H. C.; *J. Org. Chem.* **1996**, *61*, 100.

[128] Kulkarni, S. V.; Brown, H. C.; *Tetrahedron Lett.* **1996**, *37*, 4125.

[129] Brown, H. C.; Khire, U. R.; Narla, G.; *J. Org. Chem.* **1995**, *60*, 8130.

[130] (a) Review: Cowden, C. J.; Paterson, I.; *Org. React.* **1997**, *51*, 1; (b) Mukaiyama, T.; *Org. React.* **1982**, *28*, 203.

[131] Köster, R.; Fenzl, W.; *Angew. Chem., Int. Ed. Engl.* **1968**, *7*, 735.

[132] Inoue, T.; Mukaiyama, T.; *Bull. Chem. Soc. Japan* **1980**, *53*, 174.

[133] Evans, D. A.; Nelson, J. V.; Vogel, E.; Taber, T. R.; *J. Am. Chem. Soc.* **1981**, *103*, 3099.

[134] Van Horn, D. E.; Masamune, S.; *Tetrahedron Lett.* **1979**, 2229.

[135] (a) Brown, H. C.; Ganesan, K.; Dhar, R. K.; *J. Org. Chem.* **1992**, *57*, 3767; (b) Brown, H. C.; Ganesan, K.; Dhar, R. K.; *J. Org. Chem.* **1993**, *58*, 147; (c) Ganesan, K.; Brown, H. C.; *J. Org. Chem.* **1993**, *58*, 7162.

[136] Ramachandran, P. V.; Xu, W.; Brown, H. C.; *Tetrahedron Lett.* **1997**, *38*, 769.

[137] Masamune, S.; Choy, W.; Kerdesky, F. A. J.; Imperiali, B.; *J. Am. Chem. Soc.* **1981**, *103*, 1566.

[138] Masamune, S.; Choy, W.; Petersen, J. S.; Sita, L. R.; *Angew. Chem., Int. Ed. Engl.* **1985**, *24*, 1.

[139] Paterson, I.; Lister, M. A.; McClure, C. K.; *Tetrahedron Lett.* **1986**, *26*, 4787.

[140] (a) Paterson, I.; Goodman, J. M.; Lister, M. A.; Schumann, R. C.; McClure, C. K.; Norcross, R. D.; *Tetrahedron* **1990**, *46*, 4663; (b) Paterson, I.; *Pure Appl. Chem.* **1992**, *64*, 1821.

[141] Paterson, I.; Goodman, J. M.; *Tetrahedron Lett.* **1989**, *30*, 997.

[142] (a) Paterson, I.; Smith, J. D.; *Tetrahedron Lett.* **1993**, *34*, 5351; (b) Paterson, I.; Cumming, J. G.; Smith, J. D.; Ward, R. A.; *Tetrahedron Lett.* **1994**, *35*, 441.

[143] Ramachandran, P. V.; Xu, W.; Brown, H. C.; *Tetrahedron Lett.* **1996**, *37*, 4911.

[144] Hoffmann, R. W.; Ditrich, K.; *Tetrahedron Lett.* **1984**, *25*, 1781.

[145] Gennari, C.; Colombo, L.; Poli, G.; *Tetrahedron Lett.* **1984**, *25*, 2279.

[146] Gennari, C.; Cardani, S.; Colombo, L.; Scolastico, C.; *Tetrahedron Lett.* **1984**, *25*, 2283.

[147] Ganesan, K.; Brown, H. C.; *J. Org. Chem.* **1994**, *59*, 2336.

[148] Ganesan, K.; Brown, H. C.; *J. Org. Chem.* **1994**, *59*, 7346.

[149] Davies, S. G.; Doisneau, G. J.-M.; Prodger, J. C.; Sanganee, H. J.; *Tetrahedron Lett.* **1994**, *35*, 2373.

[150] Pelter, A.; Singaram, B.; Williams, L.; Wilson, J. W.; *Tetrahedron Lett.* **1983**, *24*, 623.

[151] Pelter, A.; Williams, L.; Wilson, J. W.; *Tetrahedron Lett.* **1983**, *24*, 627.

[152] Wilson, J. W.; *J. Organomet. Chem.* **1980**, *186*, 297.

[153] Pelter, A.; Singaram, B.; Wilson, J. W.; *Tetrahedron Lett.* **1983**, *24*, 635.

[154] Pelter, A.; Buss, D.; Colclough, E.; *J. Chem. Soc., Chem. Commun.* **1987**, 297.

[155] Pelter, A.; Smith, K.; Elgendy, S.; Rowlands, M.; *Tetrahedron Lett.* **1989**, *30*, 5647, 5643.

[156] (a) Krishnamurthy, S.; Brown, H. C.; *J. Org. Chem.* **1976**, *41*, 3064; (b) Krishnamurthy, S.; Brown, H. C.; *J. Org. Chem.* **1983**, *48*, 3085.

[157] Harrison, J.; Alvarez, S. G.; Godjoian, G.; Singaram, B.; *J. Org. Chem.* **1994**, *59*, 7193.

[158] Majetich, G.; Zhang, Y.; Wheless, K.; *Tetrahedron Lett.* **1994**, *35*, 8727.

[159] Pedregal, C.; Ezquerra, J.; Escribano, A.; Carreno, M. C.; Ruano, L. G.; *Tetrahedron Lett.* **1994**, *35*, 2053.

[160] Krishnamurthy, S.; Brown, H. C.; *J. Am. Chem. Soc.* **1976**, *98*, 3383.

[161] Smith, K.; Pelter, A.; Norbury, A.; *Tetrahedron Lett.* **1991**, *32*, 6243.

[162] Evans, D. A.; Dart, M. J.; Duffy, J. L.; *Tetrahedron Lett.* **1994**, *35*, 8541.

[163] Weissman, S. A.; Ramachandran, P. V.; *Tetrahedron Lett.* **1996**, *37*, 3791.

[164] Midland, M. M.; Greer, S.; Tramontano, A.; Zderic, S. A.; *J. Am. Chem. Soc.* **1979**, *101*, 2352.

[165] (a) Brown, H. C.; Pai, G. G.; *J. Org. Chem.* **1985**, *50*, 1384; (b) Brown, H. C.; Pai, G. G.; *J. Org. Chem.* **1983**, *48*, 1784.

[166] Midland, M. M.; Kazubski, A.; *J. Org. Chem.* **1982**, *47*, 2814.

[167] (a) Chandrasekharan, J.; Ramachandran, P. V.; Brown, H. C.; *J. Org. Chem.* **1985**, *50*, 5446; (b) Chandrasekharan, J.; Ramachandran, P. V.; Brown, H. C.; *J. Org. Chem.* **1986**, *51*, 3394.

[168] Ramachandran, P. V.; Chen, G.-M.; Lu, Z.-H.; Brown, H. C.; *Tetrahedron Lett.* **1996**, *37*, 3795.

[169] Ramachandran, P. V.; Lu, Z.-H.; Brown, H. C.; *Tetrahedron Lett.* **1997**, *38*, 761.

[170] Ramachandran, P. V.; Gong, B.; Brown, H. C.; *Tetrahedron Asymmetry*, **1993**, *4*, 2399.

[171] Ramachandran, P. V.; Gong, B.; Teodorovic, A. V.; Brown, H. C.; *Tetrahedron Asymmetry*, **1994**, *5*, 1061, 1075.

[172] Ramachandran, P. V.; Gong, B.; Brown, H. C.; *J. Org. Chem.* **1995**, *60*, 41.

[173] Ramachandran, P. V.; Chen, G.-M.; Brown, H. C.; *J. Org. Chem.* **1996**, *61*, 88.

[174] Ramachandran, P. V.; Chen, G.-M.; Brown, H. C.; *J. Org. Chem.* **1996**, *61*, 95.

[175] (a) Molander, G. A.; Bobbitt, K. L.; *J. Am. Chem. Soc.* **1993**, *115*, 7517; (b) Conole, G.; Mears, R. J.; De Silva, H.; Whiting, A.; *J. Chem. Soc., Perkin Trans. 1* **1995**, 1825.

IV

Organoaluminum Chemistry

HISASHI YAMAMOTO

Graduate School of Engineering, Nagoya University Nagoya, Japan

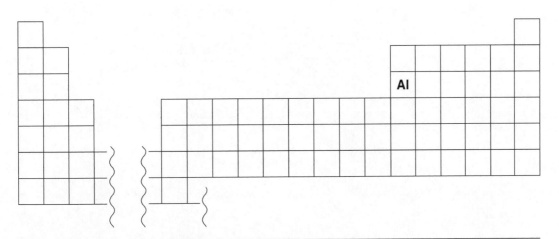

Organometallics in Synthesis: A Manual. Edited by Manfred Schlosser.
© 2002 John Wiley & Sons Ltd

Contents

1   Introduction

The list of the most abundant elements by weight in the Earth's layers is headed by oxygen (48.9%), followed by silicon (26.3%), aluminum (7.7%), iron (4.7%) and calcium (3.4%). The three most abundant elements are often present as a group within the Earth's crust in the form of widely-distributed aluminum silicate minerals. Thus, aluminum is one of the most plentiful metal elements in our planet. The object of this chapter, organoaluminum compounds, are known to be the commonest organometallics. These compounds are used as highly-active catalysts and co-catalysts for the polymerization of a wide range of organic monomers. The reagent of organoaluminum compounds in organic synthesis is, however, relatively limited and used only as hydride reagents and some Lewis acid catalysts. This Chapter essentially deals with all aspects of organoaluminum reagents in organic synthesis to encourage wider and more important usage of the reagent in future.

Aluminum is known as an element which is toxic to humans, and which can cause anemia, dementia and death. Aluminum can replace magnesium (Mg^{2+}), since it binds more strongly to the same ligand. These properties are probably due to the high reactivity of the aluminum compounds towards oxygen- and nitrogen-containing functional groups.

2   General Properties of Organoaluminum Compounds

Although their role in the alkene-oriented world of petrochemicals is well-established[1], organoaluminum reagents are relative newcomers as tools in selective organic syntheses[2]. Organoaluminum compounds are the cheapest of the active metals, and thus are gradually replacing other organometallic derivatives as more economical reducing and alkylating agents.

The characteristic properties of aluminum reagents derive mainly from the high Lewis acidity of the organoaluminum monomers, which is directly related to the tendency of the aluminum atom to complete electron octets. Bonds from aluminum to electronegative atoms such as oxygen or the halogens are extremely strong—the energy of the Al−O bond is estimated to be 138 kcal mol^{-1}. Because of this strong bond, nearly all organoaluminum compounds are particularly reactive with oxygen and often ignite spontaneously in air. These properties, commonly identified as 'oxygenophilicity', are of great value in the designing of selective synthetic reactions.

The strong Lewis acidity of organoaluminum compounds appears to account for their tendency to form 1 : 1 complexes, even with neutral bases such as ethers. In contrast to lithium and magnesium derivatives, ether and related solvents may even retard the reactivity of organoaluminum compounds. The relatively belated appreciation of organoaluminum chemistry may be traced to the longstanding confinement to ethereal solvents. The coordinated group may be activated or deactivated depending upon the type of reaction. Furthermore, on coordination of an organic molecule an auxiliary bond can become coupled to the reagent and promote the desired reaction.

The major difference between organoaluminum compounds and more common Lewis acids, such as aluminum chloride and bromide, is attributable to the structural flexibility of organoaluminum reagents. Thus, the structure of an aluminum reagent is easily modified by changing one or two of its ligands.

The following questions should be posed when seeking the most effective use of organoaluminum compounds.

1. What kind of stereochemical or electrochemical reactivity difference in substrates can we expect by coordination with the aluminum reagent?
2. Among the three ligands of organoaluminum compounds, which one has the highest reactivity?
3. What are the oxygenophilicities? What kind of reaction can we design with this unusual reactivity?
4. An organoaluminum reagent is a typical Lewis acid. However, once coordinated with a substrate, the aluminum reagent behaves as a typical ate complex and one of the ligands behaves as a nucleophile. This is an acid–base complexed-type reagent. What are the characteristic features of these reagents?
5. By changing from a trivalent organoaluminum reagent to tetravalent ate complex, the reactivity of the reagent is changed significantly. What is the chemistry of the ate complex of the aluminum reagent in the presence of donor ligands?

The chemistry of organoaluminum reagent is discussed here with the intention of answering some of these questions.

3   Organoaluminum Reactivity After Coordination with a Substrate

Most trivalent organoaluminum-promoted reactions were initiated by coordination of the substrate to the aluminum atom. For example, when cyclohexanone **1** was treated with trimethylaluminum, the carbonyl oxygen was first coordinated with aluminum atom[3]. If the reaction was performed in ether or tetrahydrofuran, axial alcohol **2** was produced through a four-centered transition state. A similar complex was generated in hydrocarbon solvents with a 1 : 1 ratio of the carbonyl compound and aluminum reagent. However, if the ratio was changed to 1 : 2, the reaction proceeded through the six-membered transition state as depicted in Scheme 1 to yield equatorial alcohol as the major product[4,5]. This example provides a typical illustration of the solvent effect of an aluminum reagent-promoted reaction. In ethereal solution, the oxygen atom of the solvent molecules strongly coordinates with the aluminum atom and the self-association of aluminum reagent becomes much weaker. The resulting dimethylaluminum alkoxide forms a stable dimer and the rate of methylation decreases significantly[6].

Pronounced solvent effects in the course of organoaluminum-induced reactions are frequently observed. Thus unimolecular decomposition of the 1 : 1 complex of citronellal-trimethylaluminum at −78 °C to room temperature yielded the acyclic compound **3** in

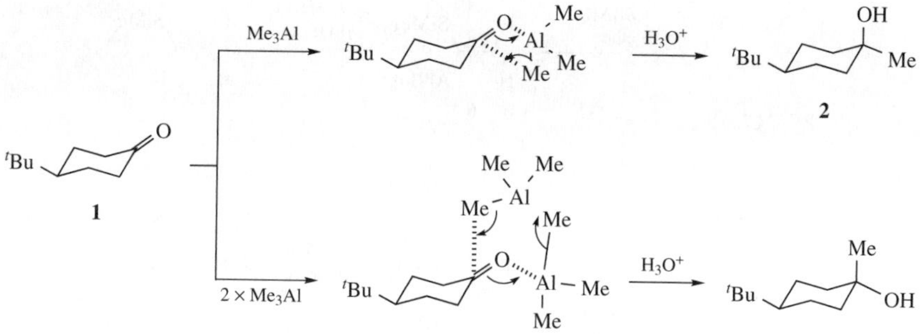

Scheme 1.

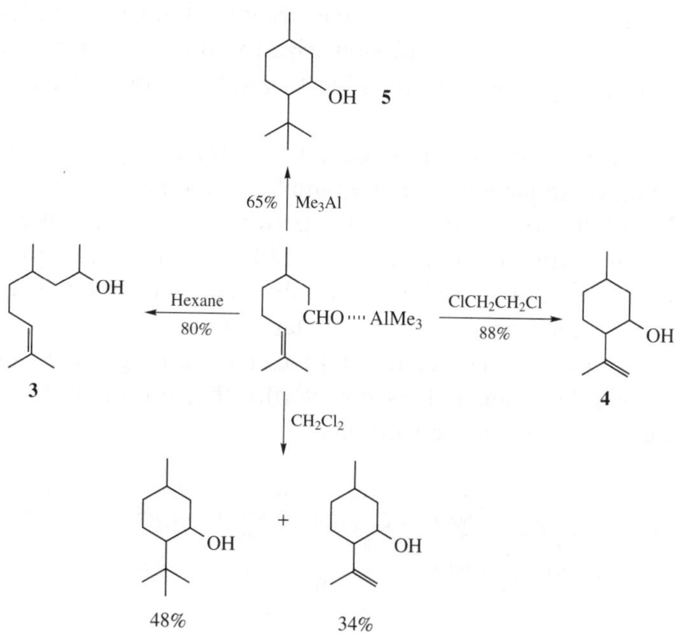

Scheme 2.

hexane, whereas isopregol (**4**) was produced exclusively in 1,2-dichloroethane. Furthermore, the cyclization–methylation product **5** was formed with high selectivity using excess trimethylaluminum in dichloromethane at low temperature (Scheme 2)[7].

Hydroalumination of acetylene using diisobutylaluminum hydride (DIBAH) was known to occur through a cis addition mechanism similar to that of hydroboration[8]. If silyl acetylene was used as substrate and if the reaction was performed in hexane, the initially formed cis addition product **6** was isomerized to trans isomer spontaneously (Scheme 3)[9]. The alkenic π electron may coordinate with the vacant orbital of aluminum atom and also with the silicon atom through pπ–dπ conjugation, and these coordinations may accelerate the rotation with C—C bond[10]. However, if the reaction occurs in hexane in the presence

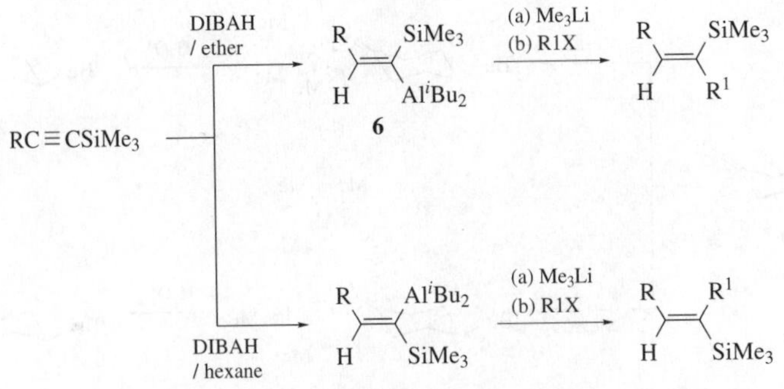

Scheme 3.

of triethylamine or in an ethereal solvent, the vacant orbital of the aluminum metal is occupied by these donor molecules and isomerization does not easily occur. Therefore, by choosing the solvent system, it is possible to produce either the cis or trans alkene stereoselectively.

A significant solvent effect was also observed during the study of the Beckmann rearrangement alkylation sequence[11]. For example, when the reaction of cyclopentanone oxime tosylate (**7**) with tripropylaluminum was performed in hexane at a low temperature, the resulting imine was directly reduced with DIBAH and *N*-propylcyclopentylamine **8** was produced (Scheme 4). However, if the same reaction was performed in dichloromethane, the major rearrangement–alkylation product, turned out to be piperidine. Furthermore, if the reaction was conducted at 40 °C, the rearrangement–alkylation product coninine (**9**) was the sole product. It is noteworthy that no amine **8** or piperidine was produced under the above reaction conditions[12].

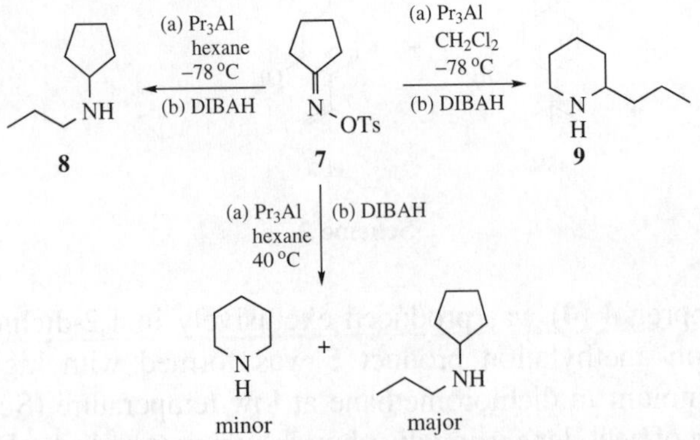

Scheme 4.

With hydrocarbon solvents, coordination of the solvent to the aluminum atom may be minimal. Thus, in these solvent systems self-association of the aluminum atoms may be more significant. As an example, the association of trimethylaluminum in toluene at various temperatures is shown in Fig. 1[13]. At −55 °C, the ratio between bridged

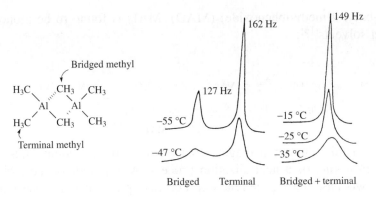

Figure 1. NMR spectrum of trimethylaluminum in toluene.

methyl and terminal methyl was found to be 1 : 2, so that most of the trimethylaluminum would have the dimeric structure[14]. By increasing the temperature, one of the trimethyl-aluminum methyl groups was rapidly exchanged, randamizing the NMR signals of the two types of methyl groups. At low temperatures, the vacant organoaluminum atom orbital may therefore be occupied by the methyl group of other molecules and be unavailable for reaction. At higher temperature, however, the existence of monomeric trimethyla-luminum can be expected. The product distribution reflects these general features of aluminum reagents.

The reducing agents R_2AlH or R_3Al, which are capable of facile elimination of alkenes, upon epoxide linkage leads to alcohols if the reaction proceeds in a regio- and stereo-selective manner. Eisch *et al.* examined the reduction of alkyl-, aryl- and silyl-substituted epoxides by such neutral hydride sources as $^{i}Bu_2AlH$ and $^{i}Bu_3Al$[15]. Although these reagent formed etherate complexes when employed in ethereal media, they are freely soluble in hydrocarbons without complexation. In fact, by the reaction of either $^{i}Bu_2AlH$ or $^{i}Bu_3Al$ with terminal epoxides in the presence or the absence of Lewis bases such as ethers or amines, a highly regioselective cleavage into either the primary alcohol **10** or the secondary alcohol **11** was observed:

$$C_8H_{17}\diagdown\diagup OH \xleftarrow[\substack{(b)\ H_2O \\ 93\%}]{\substack{(a)\ ^{i}Bu_3Al \\ (in\ THF) \\ 60\ °C,\ 6\ h}} C_8H_{17}\diagdown\triangleleft_O \xrightarrow[\substack{(b)\ H_2O \\ 95\%}]{\substack{(a)\ ^{i}Bu_2AlH \\ (in\ C_7H_{16}) \\ 25\ °C,\ 6\ h}} C_8H_{17}\diagup\diagdown_{OH}$$

10   **11**

The bulky ligand effectively solved this association–dissociation problem. With an exceedingly bulky ligand, organoaluminum reagents no longer associate with each other but remain as monomers in solution. In other words, the aluminum reagent no longer shares the alkyl or halogen group with an adjacent molecule because of the steric bulkiness of the ligand.

2,6-Di-*tert*-butyl-4-methylphenol is a typical bulky ligand for an aluminum reagent. This readily available phenol reacts with trimethylaluminum to generate methylaluminum bis(2,6-di-*tert*-butyl-4-methylphenoxide) (MAD). MAD is found to be monomeric even in hydrocarbon solvents[16]:

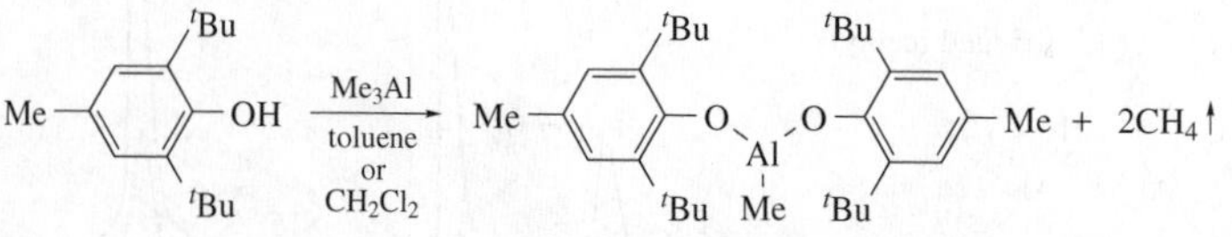

MAD

Highly Lewis acidic aluminum reagents react with carbonyl compounds to form the 1 : 1 complex, in a kind of a neutralization process. A lone pair of the carbonyl group behaves as a Lewis base during the reaction. For example, benzophenone reacts with an equivalent amount of trimethylaluminum at low temperatures to generate the orange 1 : 1 complex **12** almost instantaneously[3,4,5]. Although similar complexes can be generated between other organometallics and carbonyl compounds, they are usually quite unstable and are further transformed to alcohol by successive hydride reduction or alkylation processes. In a highly Lewis acidic aluminum reagent, the formed complex is sufficiently stable and has a long enough lifetime for a variety of reactions:

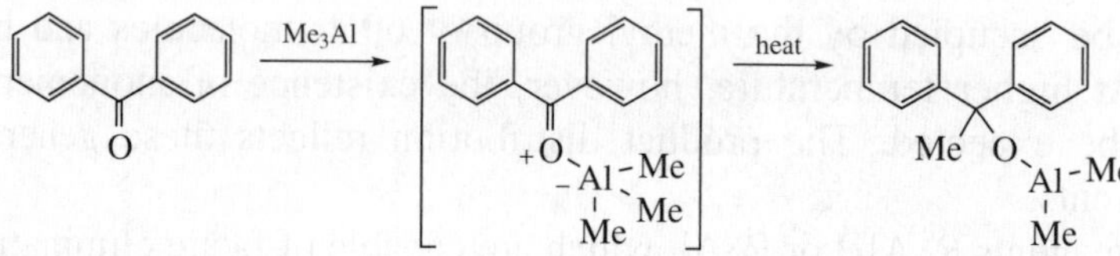

12

The bulky organoaluminum reagent MAD was mixed with the carbonyl compound 4-*tert*-butylcyclohexanone and gave the stable 1 : 1 complex **13**. This complex was treated with methyllithium at low temperature to yield an equatorial alcohol **14** whose stereochemistry was the opposite of the product from the reaction of a simple treatment of cyclohexanone and methyllithium. The observed selectivity was found to be nearly perfect (Scheme 5)[16].

The interaction of MAD with ketones yields the Lewis acid–base complexes which have been characterized by spectroscopic analysis. All the complexes show a decrease in the carbonyl stretching frequency in the IR spectrum and downfield shift in the CMR (carbon NMR) for the carbonyl α carbon, consistent with coordination of the ketone to aluminum. Recently, the X-ray crystallographic determination of the MAD–benzophenone complex was reported by Barron, which showed the benzophenone is coordinated such that aluminum atom is in the π-nodal plane of the ketone carbonyl and oriented syn with respect to the aluminum methyl[17]. Although benzophenone should be different from cyclohexanone, the result is highly interesting and could give us the possibility of an electron transfer rather than nucleophilic attack mechanisim for the MAD–benzophenone complex reaction with alkyllithium.

MAD can also be successfully utilized as a highly efficient nonchelating Lewis acid for achieving high stereoselectivity in 1,*n* asymmetric induction in both cyclic and acyclic-systems. Thus, addition of α-benzyloxycyclohexanone to MAD on subsequent treatment

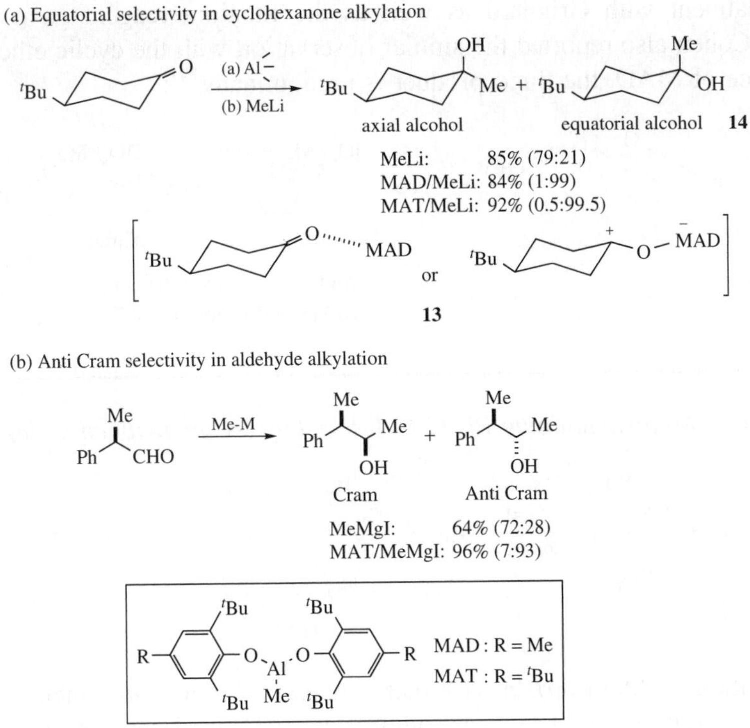
(a) Equatorial selectivity in cyclohexanone alkylation
(a) Al
(b) MeLi
axial alcohol
equatorial alcohol 14
MeLi: 85% (79:21)
MAD/MeLi: 84% (1:99)
MAT/MeLi: 92% (0.5:99.5)
or
13
(b) Anti Cram selectivity in aldehyde alkylation
Me-M
Cram
Anti Cram
MeMgI: 64% (72:28)
MAT/MeMgI: 96% (7:93)
MAD : R = Me
MAT : R = ^tBu

Scheme 5.

with MeMgBr in ether afforded a mixture of chelation and nonchelation products, **15** and **16**, in a ratio of 5 : 95 (81% yield)[18]:

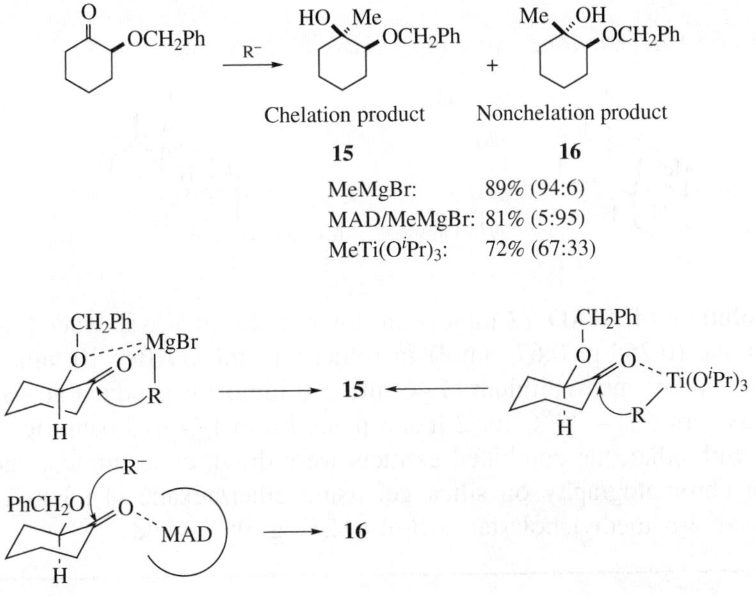
Chelation product
Nonchelation product
15
16
MeMgBr: 89% (94:6)
MAD/MeMgBr: 81% (5:95)
MeTi(O^iPr)3: 72% (67:33)

Simple treatment with Grignard as well as the titanium reagent gave the opposite selectivities. Cohen also reported the similar observation with the cyclic ethers **17**. Again in the presence of MAD, the threo product is predominant[19]:

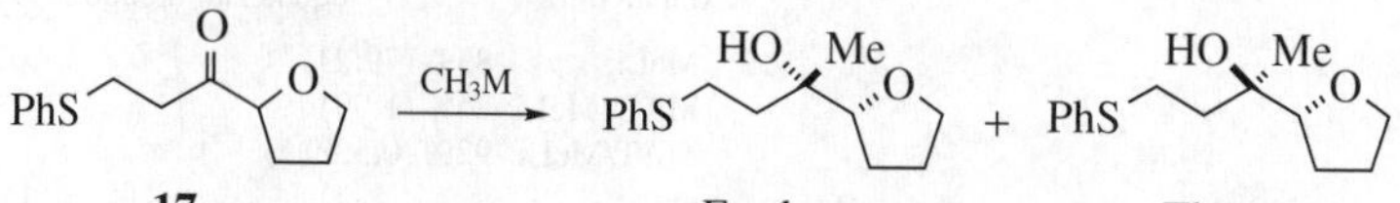

Erythro   Threo

MeLi: 78% (6.9:1)
MAD/MeLi: 80% (1:10.2)

Preparation of Methyleluminum Bis (2,6-di-tert-butyl-4-methylphenoxide) (MAD)

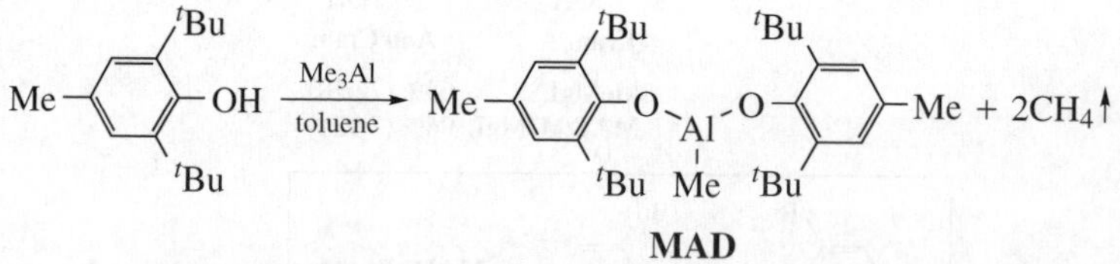

MAD

To a solution of 2,6-di-*tert*-butyl-4-methylphenol (2 equiv) in toluene was added at room temperature a 2 M hexane solution of trimethylaluminum (1 equiv). The methane gas evolved immediately. The resulting mixture was stirred at 25 °C for 1 h and used as a solution of MAD in toluene without further purification.

Alkylation of 3-Cholestanone using the MAD/CH$_3$Li System

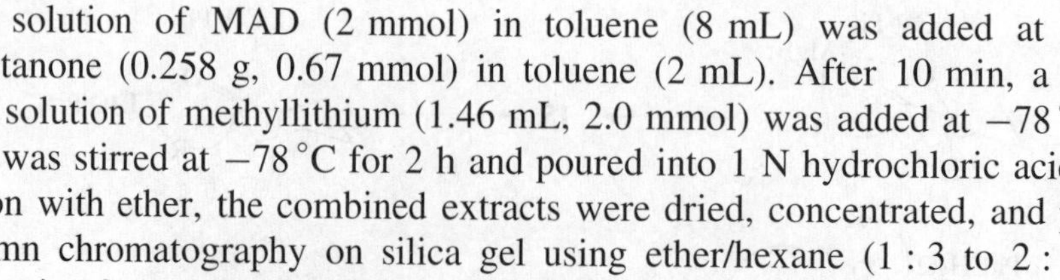

To a solution of MAD (2 mmol) in toluene (8 mL) was added at −78 °C 3-cholestanone (0.258 g, 0.67 mmol) in toluene (2 mL). After 10 min, a 1.37 M ethereal solution of methyllithium (1.46 mL, 2.0 mmol) was added at −78 °C. The mixture was stirred at −78 °C for 2 h and poured into 1 N hydrochloric acid. After extraction with ether, the combined extracts were dried, concentrated, and purified by column chromatography on silica gel using ether/hexane (1 : 3 to 2 : 1) as a eluant to give 3-*α*-methylcholestan-3 *β*-ol (0.257 g, 96% yield).

Amphiphilic Alkylation of α-Alkoxy Cyclic Ketone

threo:erythro = 92:8

To a solution of MAD (1.5 mmol) in CH_2Cl_2 (5 mL) was added at $-78\,°C$ 2-benzyloxycyclohexanone (0.5 mmol) followed by an ethereal solution of CH_3Li (1.5 mmol). The solution was maintained at $-78\,°C$ for 2 h. The reaction mixture was poured into 1 N HCl, and extracted with CH_2Cl_2. The combined extracts were, after drying over Na_2SO_4 and concentration, purified by column chromatography on silica gel (ether/hexane as eluent) to furnish *threo*-2-benzyloxy-1-methyl-1-cyclohexanol (75% yield).

4   Relative Mobility of Organoaluminum Ligands

An organoaluminum compound is sometimes the reagent of choice for the introduction of the alkynyl group to the substrate. In fact, the alkynylaluminum reagent can be readily prepared from the corresponding lithium or magnesium compound.

When the reactivities of these alkynylaluminum reagents were compared with those of the corresponding organocopper reagents, the reactivity trend for the ligands was found to be completely different. When attached to copper, the alkynic ligand is stabilized by interaction with the copper d orbital and cannot be transferred easily[20]. Such stabilization does not occur with the aluminum reagent and the less basic alkynic ligand is transferred preferentially[21,22]:

With α,β-unsaturated ketones the possibility of 1,4 addition of R_3Al also exists, and such conjugate addition can become the principal mode of reaction. It is interesting that addition of Me_2AlI or Ph_2AlI to enones, occurs principally in a 1,4 manner[23]:

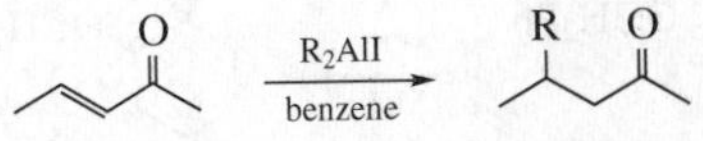

R=Ph: 66% (1,4/1,2 = 100:0)
R=Me: 39% (1,4/1,2 = 61:39)

Even more striking was the observation reported by Kunz for the similar 1,4 addition processes. Thus, diethylaluminum chloride reacts smoothly with α,β-unsaturated *N*-acyloxazolidinones providing chiral β-branched carboxylic acid derivatives. An unexpected contrast between the mode of reaction of dimethylaluminum chloride and that of the highly homologs is observed: whereas diethylaluminum chloride and its higher homologs reacts with the acceptors at low temperature via a polar pathway, dimethylaluminum chloride requires activation by UV light or radical initiation[24]:

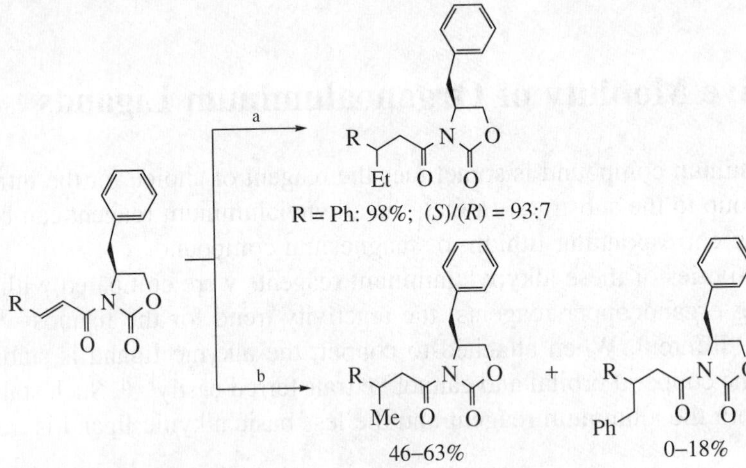

R = Ph: 98%; (S)/(R) = 93:7

46–63%   0–18%

Path a: Et_2AlCl (4 equiv, 1 M) in hexane/toluene at −78 °C under argon.
Path b: Me_2AlCl (4 equiv) in hexane/toluene, 5 h irradiation. Total reaction time 30–48 h.

The heterosubstituted ligand of the organoaluminum compound is transferred much more easily than the alkyl ligands. For example, treatment of diethylaluminum amide with methanol produced ammonia by cleaving the Al−N bond. No formation of ethane was detected in this experiment[25]. Similarly, cleavage of the Al−S bond or Al−P bond occurred preferentially by treatment of dimethylaluminum sulfide or phosphite with methanol, respectively. In general, R_2AlX-type reagents revealed a similar reactivity profile. This reactivity of the aluminum reagents is in sharp contrast to that of the organocopper reagents where the PhS− and PhO− groups are frequently utilized as nontransfer groups[19]. The reaction between diethylaluminum amide and methanol is shown below. The reaction may be understood as a ligand exchange process in which the exchange process from **18** to **19** required a much smaller activation energy than the evolution of ethane gas:

$$(CH_3CH_2)_2AlNH_2 + CH_3OH \longrightarrow (CH_3CH_2)_2AlOCH_3 + NH_3$$

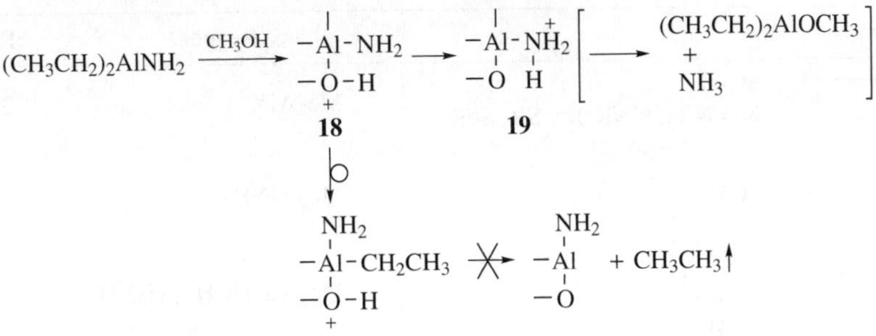

$$(CH_3CH_2)_2AlNH_2 \xrightarrow{CH_3OH} \underset{\underset{+}{\overset{|}{-O-H}}}{\overset{|}{-Al-NH_2}} \longrightarrow \underset{\underset{}{\overset{|}{-O\ H}}}{\overset{|}{-Al-NH_2^+}} \left[\longrightarrow \begin{array}{c} (CH_3CH_2)_2AlOCH_3 \\ + \\ NH_3 \end{array} \right]$$

18   **19**

$$\begin{array}{c} NH_2 \\ | \\ -Al-CH_2CH_3 \\ | \\ -O-H \\ + \end{array} \quad \cancel{\longleftarrow} \quad \begin{array}{c} NH_2 \\ | \\ -Al \\ | \\ -O \end{array} + CH_3CH_3\uparrow$$

This characteristic may be utilized over a broad range of organic syntheses. Thus Table 1 shows several examples of the reactions of AlR_2X, all of which depend upon this feature.

The dialkylaluminum reagents derived from RSH, RSeH, and RTeH have proved to be valuable for preparing heteroatom-substituted carbonyl compounds by conjugate additions to α,β-unsaturated carbonyl compounds, which are otherwise difficult to prepare[42]. The similar Michael addition reactions were oberved for azide and cyanide reagents[43]:

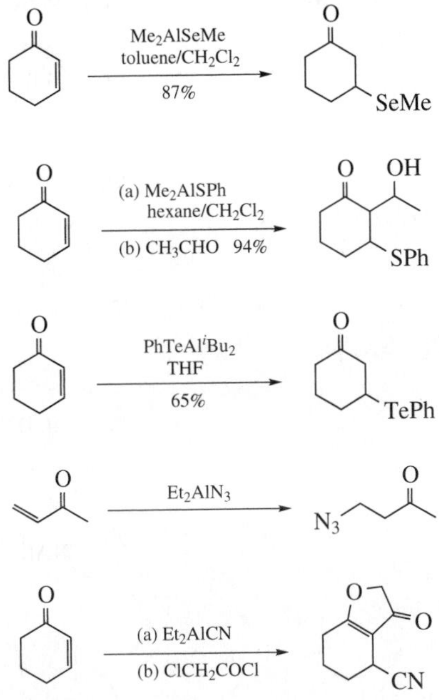

Reaction of Methyl Heptanoate with $Me_2AlSeMe$

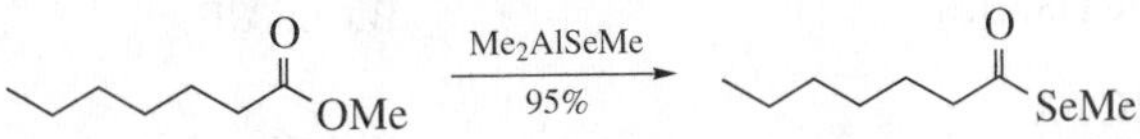

To a 50-mL side-arm flask containing 4.1 g (0.052 mol) of selenium powder was added 25.2 mL (0.05 mol) of a 17.0% solution of trimethylaluminum in

Table 1. Nucleophilic substitution reactions of aluminum ligands.

Transformation	Reagent	References
$-COOR \longrightarrow -COX$, $X = NR_2, NNR^1R^2, SR, SeR$	Me_2AlX	26, 27, 28 29
$-COOR \longrightarrow -CN$	Me_2AlNH_2	30
$-COOR \longrightarrow$ (imidazoline ring)	$Me_2AlNHCH_2CH_2NH_2$	31
$\overset{H}{\underset{COOR}{>}} \longrightarrow$ (dithiolane)	$(Me_2AlSCH_2)_2CH_2$	32
epoxide $\longrightarrow \overset{OH}{\underset{X}{}}$, $X = NR_2, SeR$	Et_2AlNR_2 / Me_2AlSeR	33 / 29
epoxide $\longrightarrow$ $PhS\overset{}{\diagup}\diagdown OH$	Et_2AlSPh	34
$\overset{}{\diagup}OPO(OR)_2 \longrightarrow \overset{}{\diagup}X$, $X = OPh, NHPh, SR$	Me_2AlX	35
$\underset{OSO_2R}{N} \longrightarrow \overset{}{N}\overset{}{X}$, $X = SR, SeR$	R'_2AlX	36
$RCN \longrightarrow \overset{NH}{\underset{R-C-NR'R''}{\parallel}}$	$Me_3Al(Cl)NR'R''$	37
$H\overset{O}{\diagup} \longrightarrow \diagdown OH$	(cyclic) NAlEt$_2$	38
$\overset{H}{\diagdown}O \longrightarrow \overset{}{\diagdown}OAlR_2$	(cyclic) NAlEt$_2$	39
$EtCHO \longrightarrow EtCOSnBu_3$	$Bu_3SnAlEt_2$	40
$RCHO \longrightarrow RCOX$, $X = SR, SeR, TeR$	R'_2AlX	41

toluene. The reaction mixture was heated at reflux for 2 h, then cooled to room temperature, allowing the unreacted selenium to settle from solution. A 1.1 mL (2.2 mmol) aliquot of dimethylaluminum methyl selenolate was transferred by syringe to a solution of methyl heptanoate (0.288 g, 2.0 mmol) in 5 mL of argon-degassed dichloromethane at 0 °C. After 30 min at this temperature, the yellow solution was warmed to room temperature over 30 min and then treated with moist sodium sulfate. The resulting mixture was extracted with ether and the isolated crude product purified by bulb-to-bulb distillation (95 °C over temperature, 30 mm) to yield 0.393 g (95%) of the desired selenol ester as a yellow oil.

Reaction of Benzaldehyde with $^{i}Bu_2AlTeBu$

In a flame-dried flask equipped with an Ar inlet and a rubber septum was placed BuTeTeBu (0.185 g, 0.5 mmol). After $^{i}Bu_2AlH$ (1 N in hexane, 1 mL) was added under Ar, the solution was stirred at 25 °C for 1 h and then 2 mL of THF was added. The solution was cooled to −23 °C and 2 mmol of benzaldehyde and 0.5 mL of Et_2AlCl (1 N in hexane) were injected. The mixture was gradually warmed to 25 °C, stirred for 1 h, and poured into saturated NH_4Cl solution. The products were extracted with Et_2O (30 mL × 3), dried over $MgSO_4$, and concentrated *in vacuo*. Medium-pressure liquid chromatography of the residue gave purebutyl tellurobenzoate in 71% yield (0.205 g) in a hexane/Et_2O (100 : 1) fraction.

---

5 Oxygenophilicity of Organoaluminum Reagents

Organoaluminum reagents are highly reactive compounds and those with alkyl groups of four or less carbon atoms are usually pyrophoric; that is, they ignite spontaneously in air at ambient temperature. All aluminum alkyl compounds react violently with water. Increasing the molecular weight by increasing the number of carbons in the alkyl groups, or by substituting halogens for these groups, generally reduces pyrophoric reactivity.

Whatever the detail, it seems probable by analogy with other organometallic oxygenations that the initial step in oxidation must be formation of a peroxide **20** which either undergoes reaction with additional organoaluminum to form dialkylaluminum alkoxide **21** or else undergoes rearrangement to methylaluminum dialkoxide[44]:

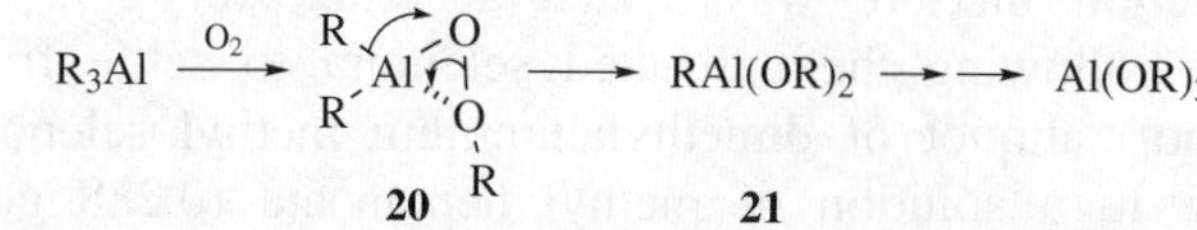

$$R_3Al \xrightarrow{O_2} \underset{\underset{R}{|}}{\overset{R}{\underset{20}{\text{Al}}}} \cdots \longrightarrow RAl(OR)_2 \longrightarrow \longrightarrow Al(OR)_3$$
$$21$$

Recently, Lewinski *et al.* reported the isolation of the (alkylperoxo)aluminum compound **22** derived from the oxidation of di-*tert*-butylaluminum O,O'-chelate complex. Thus, reaction of the $^tBu_3Al$–ether complex with methyl salicylate generates the chelate compound, which upon exposure with dry oxygen at $-15\,°C$ leads to the formation of aluminum peroxide **22** in essentially quantitative yield[45]:

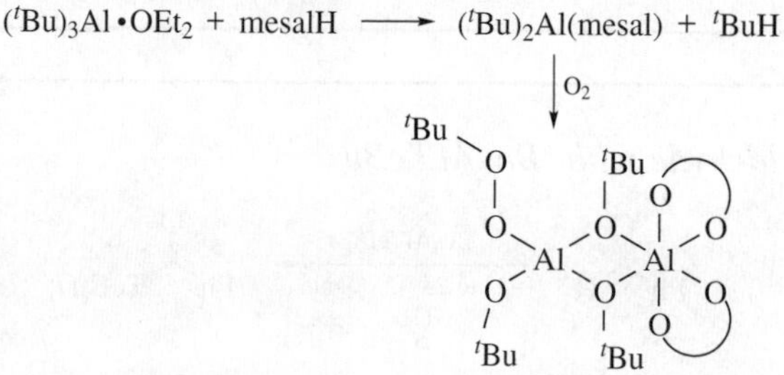

$$(^tBu)_3Al \cdot OEt_2 + \text{mesalH} \longrightarrow (^tBu)_2Al(\text{mesal}) + {}^tBuH$$

22

On an industrial scale, the oxidation of aluminum alkyls with dry air to aluminum alkoxides can be made to proceed smoothly in high yields. In the laboratory such air oxidation can readily be carried out on unsymmetrical aluminum alkyls resulting from the hydroalumination of alkenes with aluminum hydrides[46]:

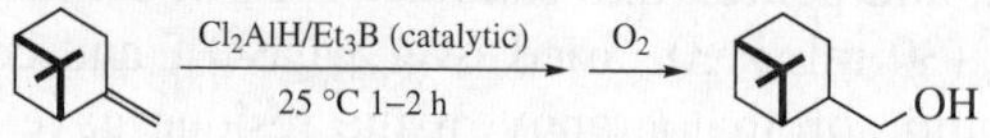

$$\xrightarrow[\text{25 °C 1–2 h}]{Cl_2AlH/Et_3B \text{ (catalytic)}} \xrightarrow{O_2}$$

A unique utilization of the oxygenophilicity of the aluminum reagent was reported by Lautens *et al.* Thus, nickel-catalyzed hydroalumination of oxabicyclic alkene **23** proceeds smoothly at low temperatures. The hydroaluminated intermediate was converted to the ring-opened product by simply heating in the presence of diisobutylaluminum chloride. Even more interesting is the asymmetric hydroalumination–β-elimination sequence which achieved up to 97% ee[47]:

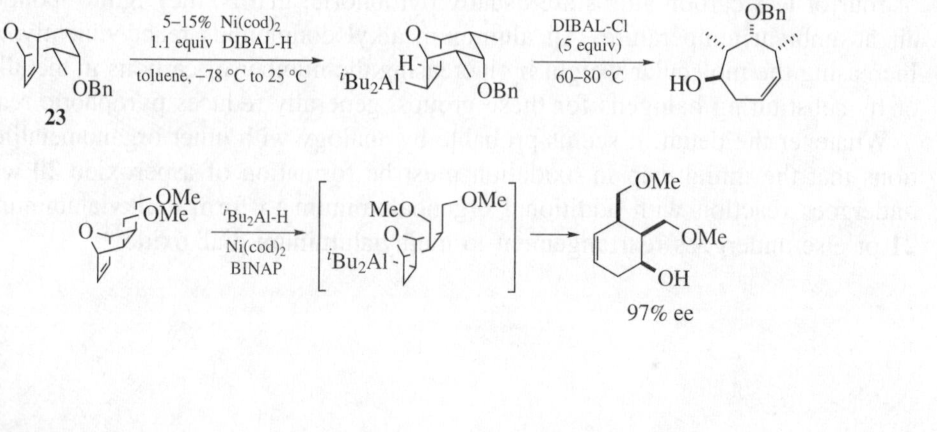

97% ee

The extraordinarily high reactivity of the organoaluminum reagent must be carefully controlled when using this reagent for selective reactions—the reagent's high oxygenophilicity should be conserved.

Combination with $Cp_2TiCl_2-Me_3Al$ (Tebbe's complex) resulted in an excellent reagent for the transformation of an ester to a vinyl ether. This reaction clearly utilized the oxygenophilicity of the aluminum reagent. In this particular case, the intermediate oxyanion was eliminated before its conversion to the ketone. Thus, a sufficiently oxygenophilic metal is necessary for the success of this transformation[48]:

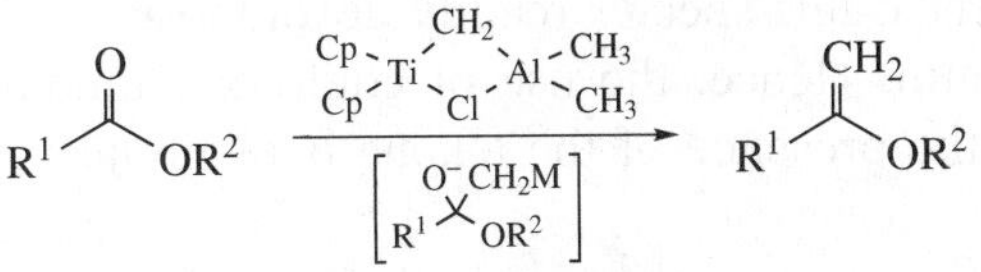

Oxidation reactions catalyzed by aluminum alkoxides are also useful in organic synthesis. The following examples reveal mild oxidation reactions using organoaluminum reagents[49]. When the aluminum reagent was treated with *tert*-butylhydroperoxide, the intermediate **24** is generated. Strong coordination of oxygen to the aluminum atom results in the highly polarized structure **24**, in which the oxygen is electrophilic. The first reaction shown below is similar to the titanium-catalyzed oxidation by Sharpless and co-workers, and the transition state should be the three-membered ring structure **25**[50]. If there is an appropriate alkene in the system, oxidation takes place smoothly with this oxygen. In the case of simple secondary alcohols the oxidation produces ketone instead, through the possible transition state of **26**.

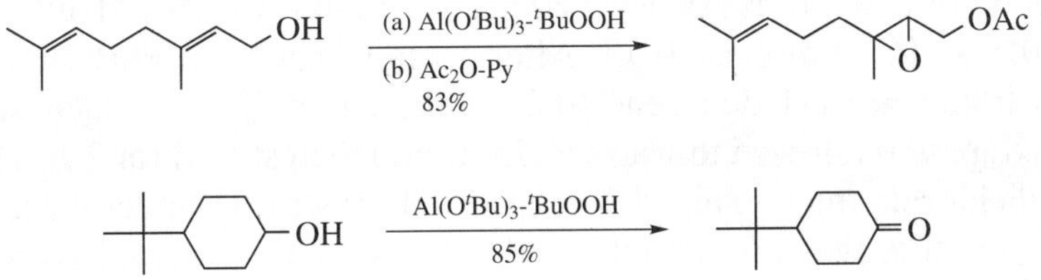

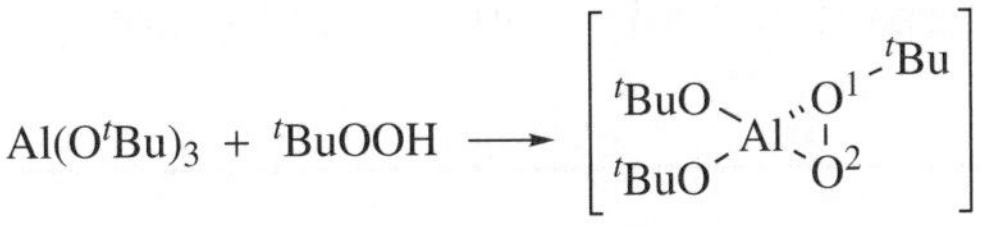

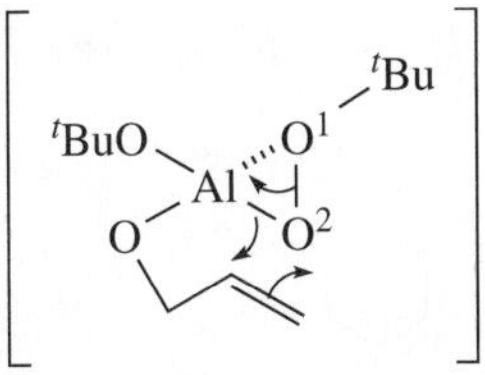

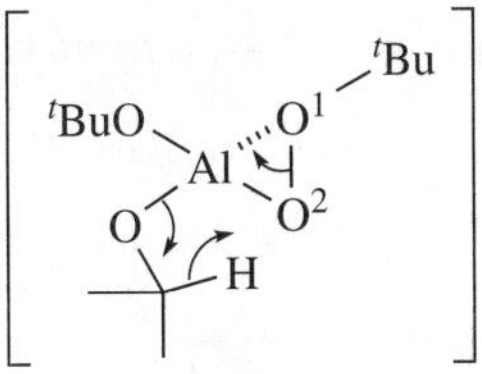

Pronounced solvent effects as well as temperature effects can alter the course of organoaluminum-induced reactions. Halogenated solvents can coordinate with aluminum metal and the aluminum reagent behaves as an ate complex in these solvent system. Such halogenophilicity of the aluminum reagent can be utilized in many synthetic processes including the following cyclopropanation reaction[51]: Thus, treatment of alkenes with various organoaluminum compounds and alkylidene iodide under mild conditions produced the corresponding cyclopropanes highly efficiently. In this reaction the intermediate dialkyl(iodomethyl)aluminum species **27** is responsible for the cyclopropanation of alkenes; this species readily decomposes in the absence of alkene or by excess trialkylaluminum. Hence, the use of equimolar amounts of trialkylaluminum and methyleneiodide in the presence of the alkene is essential for reproducible results:

$$RCHI_2 \ + \ R'_3Al \ \longrightarrow \ \left[\begin{array}{c} I \\ | \\ CH\text{-}AlR'_2 \\ | \\ R \end{array} \right] \ \xrightarrow[84-99\%]{C_{10}H_{21}\diagup\!\!\diagup} \ C_{10}H_{21}\triangle R$$

27

R = H and Me
R′ = Me, Et and *i*Bu

Hydroalumiation Using the $Cl_2AlH-PhB(OH)_2$ System

$$CH_{10}H_{21}CH=CH_2 \ \xrightarrow[PhB(OH)_2]{Cl_2AlH} \ \xrightarrow[91\%]{O_2} \ C_{11}H_{23}CH_2OH$$

To a solution of anhydrous $AlCl_3$ (0.200 g, 1.5 mmol) in ether (1 mL) was added $LiAlH_4$ (0.019 g, 0.5 mmol) at 0 °C. After 15 min, catalytic $PhB(OH)_2$ (0.006 g, 0.05 mmol) followed by 1-dodecene (0.222 mL, 1 mmol) was added at 0 °C. The resulting mixture was allowed to warm to 25 °C and then stirred for 2 h. The dodecylaluminum dichloride thus obtained was oxidized by stirring under oxygen at 25 °C for 3 h. The mixture was poured into iced 1 N HCl and extracted with ether. The combined extracts were dried, concentrated, and purified by column chromatography on silica gel (AcOEt/hexane = 1 : 3) to give 1-dodecanol (0.167 g, 91% yield) as a colorless oil.

Cyclopropanation using Organoaluminum Reagent-1-hydroxymethyl-4-(1-methylcyclopropyl)-1-cyclohexene

A dry 1-L, three-necked, round-bottomed flask is equipped with a gas inlet, a 50-mL pressure-equalizing dropping funnel, a rubber septum, and a Teflon-coated magnetic stirring bar. The flask is flushed with argon, after which 10.7 g (0.07 mol) of (S)-perillyl alcohol followed by 350 mL of dichloromethane is injected through the septum into the flask. The solution is stirred and 37.3 mL (0.147 mol) of triisobutylaluminum is added from the dropping funnel over a period of 20 min at 25 °C. After stirring the mixture at 25 °C for 20 min, 7.3 mol (0.091 mol) of diiodomethane is added dropwise with a syringe over a 10-min period. The mixture is stirred at 25 °C 4 h, and poured into 400 mL of ice-cold 8% aqueous sodium hydroxide. The organic layer is separated, and the aqueous layer is extracted twice with 100 mL portions of dichloromethane. The combined extracts are dried over anhydrous sodium sulfate, and concentrated with a rotary evaporator at ca. 20 mm Hg. The residual oil is distilled under reduced pressure to give 11.1 g (96%) of 1-hydroxymethyl-4-(1-methylcyclopropyl)-1-cyclohexene as a colorless liquid (bp 132–134 °C at 24 mmHg).

6 Organoaluminum Reagents as an Acid–Base Complex System

Epoxide is isomerized to afford an allylic alcohol under basic conditions. With a strong base, one of the protons adjacent to the epoxide is removed, resulting in a high yield of an allylic alcohol. If several protons can be removed, the product becomes a mixture.

For effective rearrangement, the reagent requires two reaction sites, a Lewis acid site to coordinate oxygen of the epoxide and a Lewis base site to remove the proton from the system. Organoaluminum amide was found to be the reagent of choice for this purpose. With the aluminum atom strongly coordinated to the oxygen (it forms an ate complex) a lone pair of nitrogen electrons becomes essential for removal of the proton, and a well defined transition state results. With such a rigid transition structure, a highly regioselective deprotonation proceeds[38]:

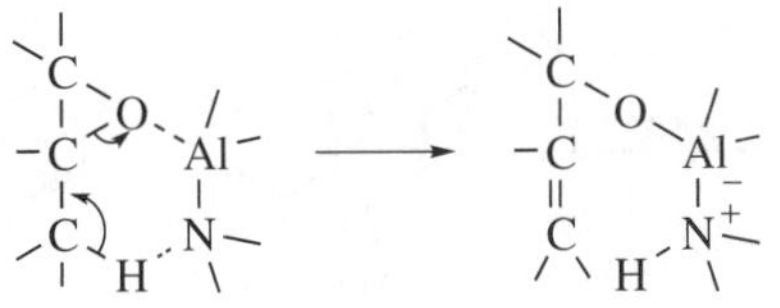

A pair of stereoisomeric epoxides was chosen for the reaction. As shown below, epoxide **28** has open space on the upper side of the molecule and epoxide **29** has the open space on the lower side of the structure. With conventional reagents, no discrimination was possible between the two structures. If the aluminum reagent attacks from the less-hindered side of the molecule, the adjacent nitrogen would approach the proton closest to the aluminum metal. For epoxide **28**, this should be a proton from methyl group and for epoxide **29** the methylene proton would be removed. Consequently, the two isomers gave rise to two different products[38]:

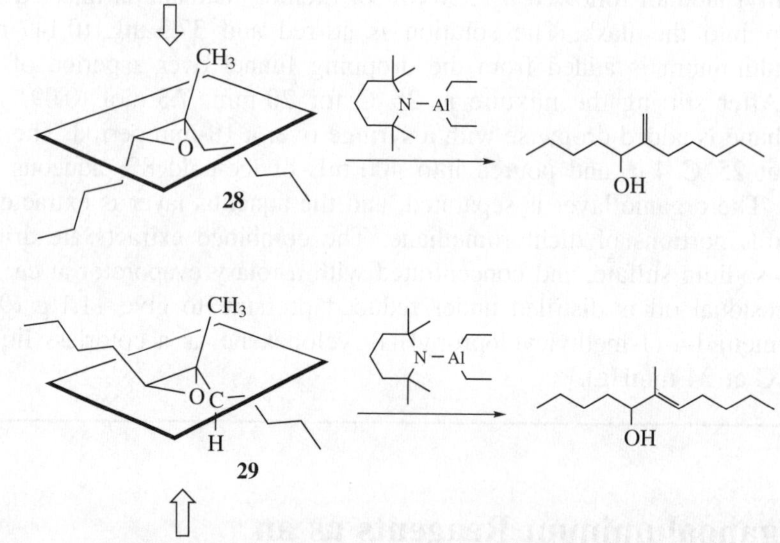

A similar acid–base-type reaction is shown below. Thus, treatment of the (E)-N-methyl-N-phenylhydrozone **30** of 5-methyl-3-heptanone with DATMP (diethylaluminum 2,2,6,6-tetramethylpyperidide) affords 3-*sec*-butyl-2-ethyl-1-methylinodol as a sole isolable product and its (Z) isomer **31** produced 1,3-dimethyl-2-(2-methylbutyl)indol with high regioselectivity under similar reaction conditions[52]:

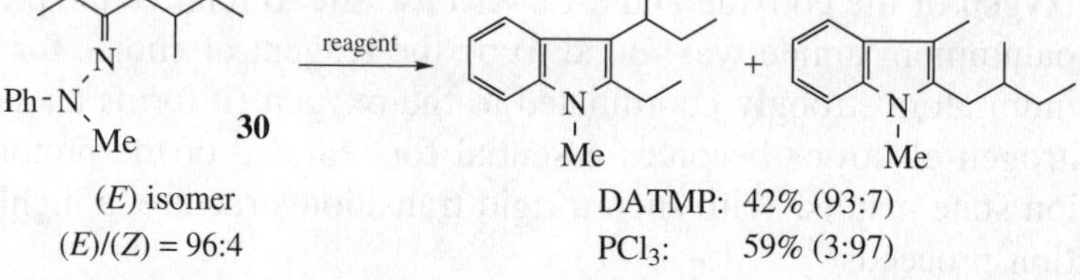

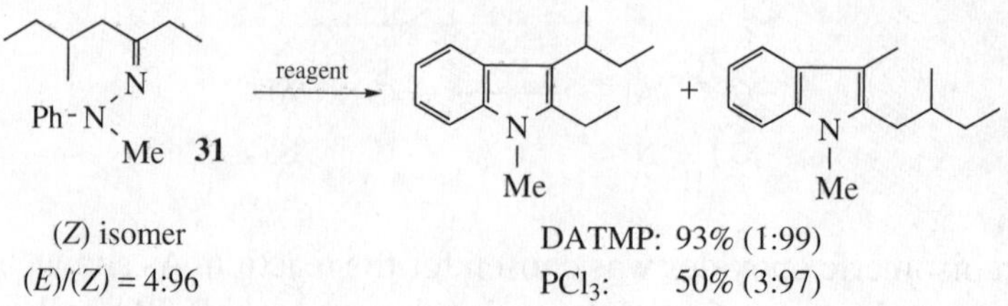

The Beckmann rearrangement is the skeletal rearrangement of ketoximes in the presence of Brøensted or Lewis acids, to give amides or lactams. The reaction has found broad application in the manufacture of synthetic polyamides. It is a preferred way to incorporate the nitrogen atom efficiently in both acyclic and alicyclic systems, thereby providing a powerful method for a variety of alkaloid syntheses. The reaction proceedes through a carbenium–nitrilium ion **32** after being captured at the oxime oxygen by Lewis acid. The intermediate cation adds water to generate an amide or lactam[53]:

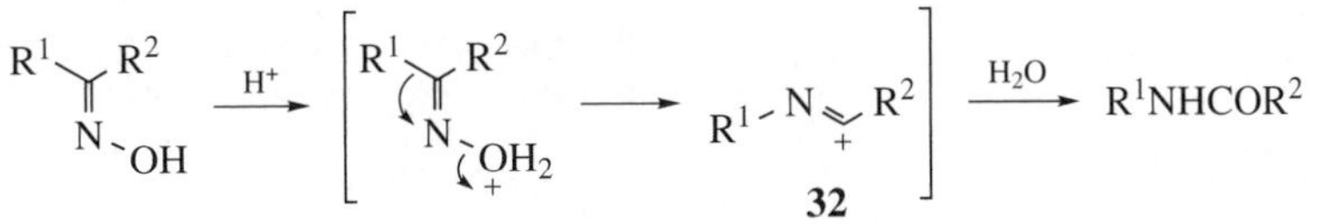

If the reaction is performed in the absence of water, the resulting cation is not efficiently trapped and subsequently decomposes in various ways. Use of the organoaluminum reagent as an acid–base complex system was found to be highly useful in this case. Thus, after the rearrangement process, the resulting ate complex was able to supply the alkyl group to the cationic intermediate. The resulting imine **33** can be reduced to the amine by DIBAH[11,12]:

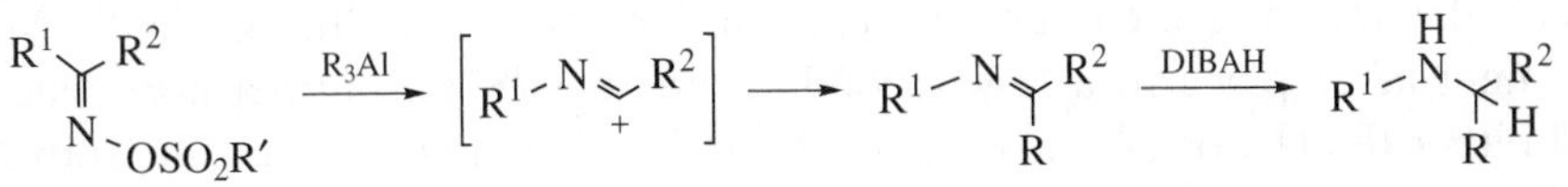

An example of the reaction for the synthesis of pumiliotoxine C is shown below. The starting ketone was selectively prepared from the corresponding bicyclic cyclopentenone derivative **34**, and was hydrogenated in the presence of palladium black to give a cis-fused ring system. The saturated ketone was exposed with hydroxyamine and then with *p*-toluenesulfonyl chloride to give the tosylate. After recrystallization followed by the Beckmann rearrangement–alkylation process, pumiliotoxin C was obtained stereoselectively[11,12]:

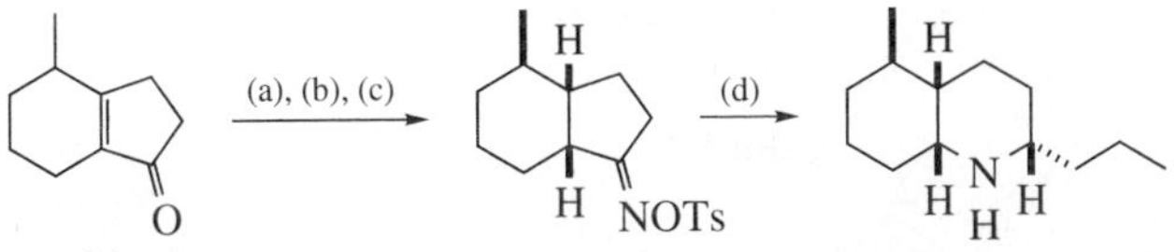

Pumiliotoxin C

(a) H_2/Pd–CH_3CH_2COOH
(b) H_2NOH HCl–NaOAc
(c) TsCl–Et_3N
(d) Pr_3Al–DIBAH

The Claisen rearrangement of an allyl vinyl ether generally requires high temperatures. However, in the presence of an organoaluminum catalyst, the reaction can be done at a much lower temperature. After the reaction, the corresponding aldehyde can be alkylated by an organoaluminum reagent. Diethylaluminum mercaptide or diethylaluminum chloride–triphenylphosphine may be used for the preparation of unsaturated aldehyde. It should be noted that other, stronger Lewis acids such as BF_3 and $TiCl_4$ could not be used for this reaction. An acid–base complex system plays an important role in this particular reaction[54]:

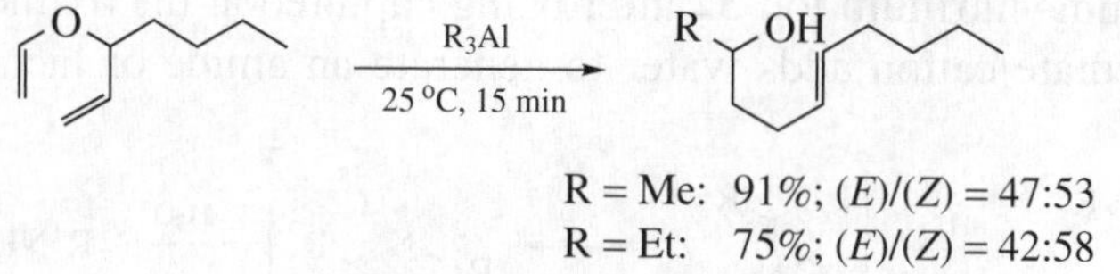
R_3Al
25 °C, 15 min
R = Me: 91%; (E)/(Z) = 47:53
R = Et: 75%; (E)/(Z) = 42:58

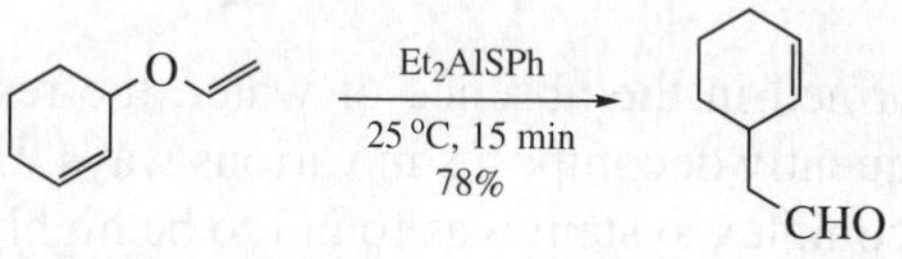
Et_2AlSPh
25 °C, 15 min
78%
CHO

Bulky aluminum reagents can also used in the Claisen rearrangement. For example, the bulky methylaluminum bis(2,6-diphenylphenoxide) (MAPH) or methylaluminum bis(4-bromo-2,6-*tert*-butylphenoxide) (MABR) are highly effective reagents which able to catalyze the rearrangement under very mild reaction conditions. With MABR the rearrangement takes place in a few seconds even at −78 °C. Furthermore, each catalyst produced either the (*E*) or (*Z*) alkenic aldehyde by changing the reagent from MAPH to MABR. The observed selectivity may come from precise molecular recognition of the substrate conformer by the bulky aluminum reagent[55]:

(*Z*) selectivity

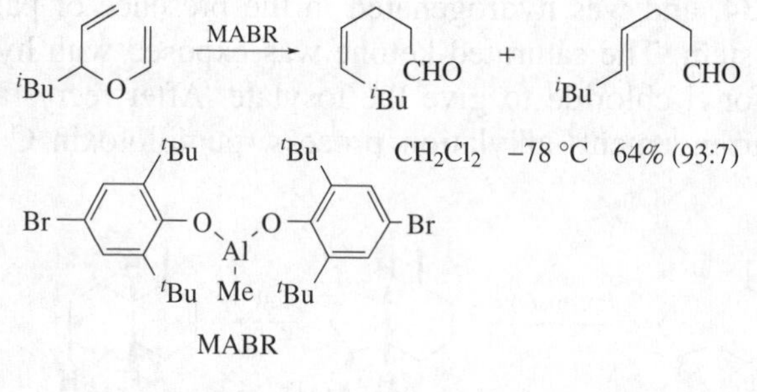
MABR
CHO
CHO
CH_2Cl_2 −78 °C 64% (93:7)
MABR

(*E*) selectivity

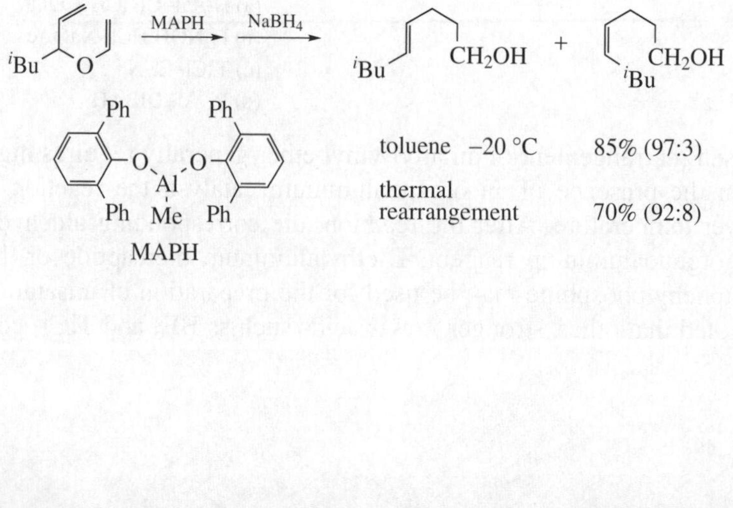
MAPH
$NaBH_4$
CH_2OH
CH_2OH
toluene −20 °C 85% (97:3)
thermal rearrangement 70% (92:8)
MAPH

Possible mechanism

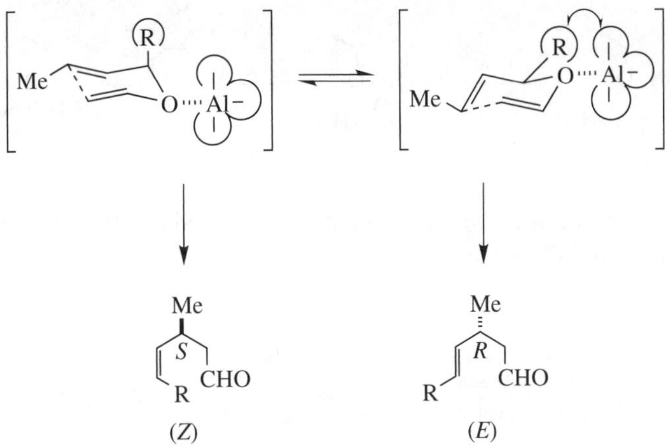

Reductive cleavage of acetals with organoaluminum hydride reagents affords stereoselectively syn-reduced products. The observed high diastereoselectivity was ascribed to the stereospecific coordination of the organoaluminum reagent of one of the acetal oxygens followed by hydride attack syn to the cleaved carbon–oxygen bond. Thus, preferential complexation of a Lewis acid with an acetal oxygen and subsequent anti substitution to the cleaved C–O bond take place in the reaction of an acetal using the Lewis acid–nucleophilic system (invertive reaction)[56]. This retentive subsitution of an acetal is exceptional and may be explained by the tight ion pair between the aluminum ate complex and the oxocarbenium ion:

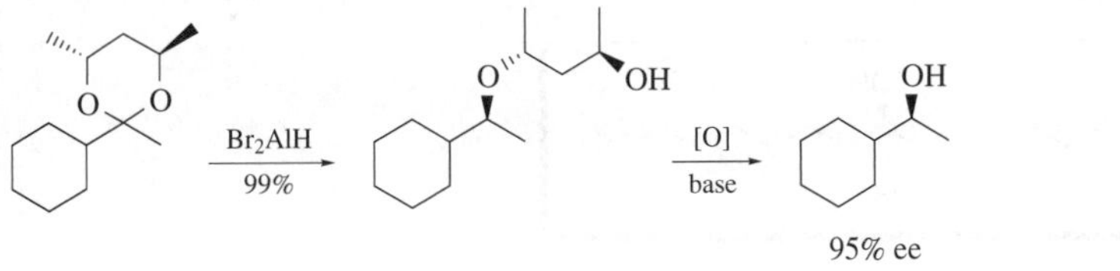

Possible mechanism

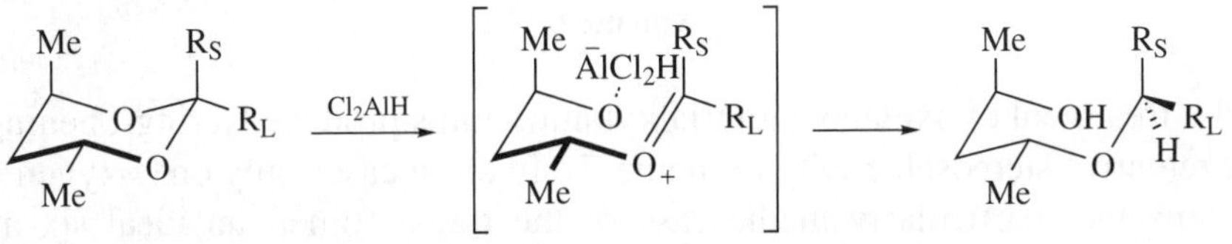

A novel methodology of diastereoselective acetal cleavage reactions was achieved using trialkylaluminum and dialkylaluminum aryloxide methods which are complementary to each other; when an appropriate reagent system and reaction conditions are chosen, better than 95% selectivity can be achieved in most cases[57]:

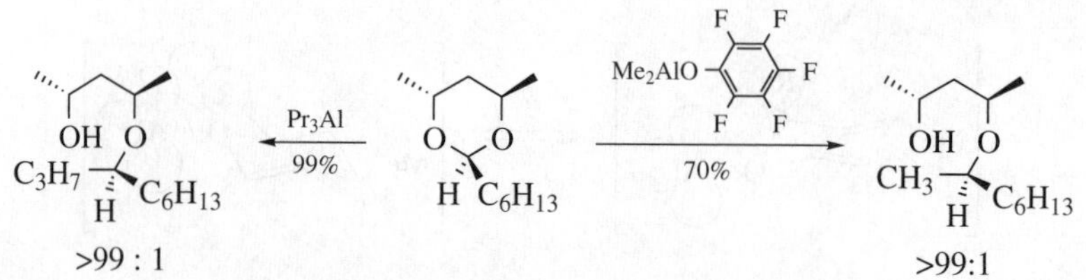

>99 : 1   99%   70%   >99:1

These methods were applied to stereoselective synthesis of the natural product (S)-(+)-8-hydroxypamitic acid (**35**; Scheme 6)

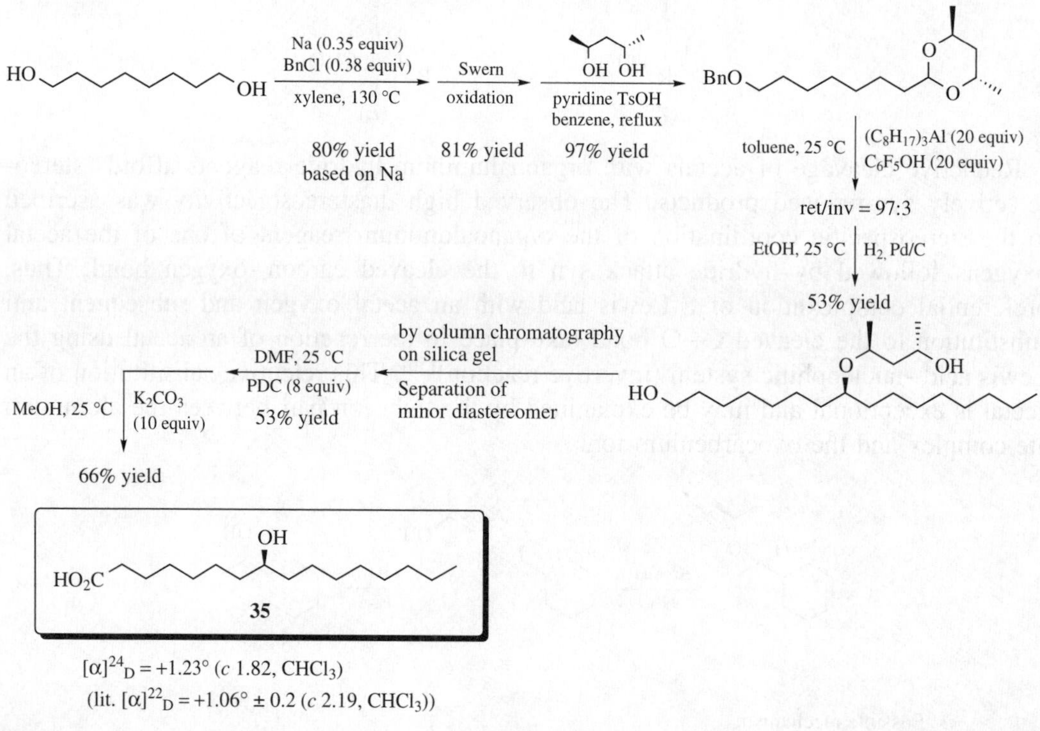

$[\alpha]^{24}_D = +1.23°$ (c 1.82, CHCl$_3$)
(lit. $[\alpha]^{22}_D = +1.06° \pm 0.2$ (c 2.19, CHCl$_3$))

Scheme 6.

Similarly, treatment of oxetane with trialkylaluminums produced a ring-opening product with strict retentive stereoselectivity (Scheme 7). In these cases only one oxygen is present in the system and, particularly in the case of the trans isomer, an ideal six-membered transition state can be expected. A slight decrease of selectivity and reactivity for the cis isomer may caused by the axial methyl or phenyl group for the transition structure.

Another example of the cleavage of acetals by elimination was observed in the reaction between cyclic acetals and triisobutylaluminum or an aluminum amide. Thus, triisobutyl-aluminum acts as a Lewis acid and as a base for the abstraction of a proton to afford the enol ether. The method was utilized to the total synthesis of laudolure (**36**; Scheme 8)[58].

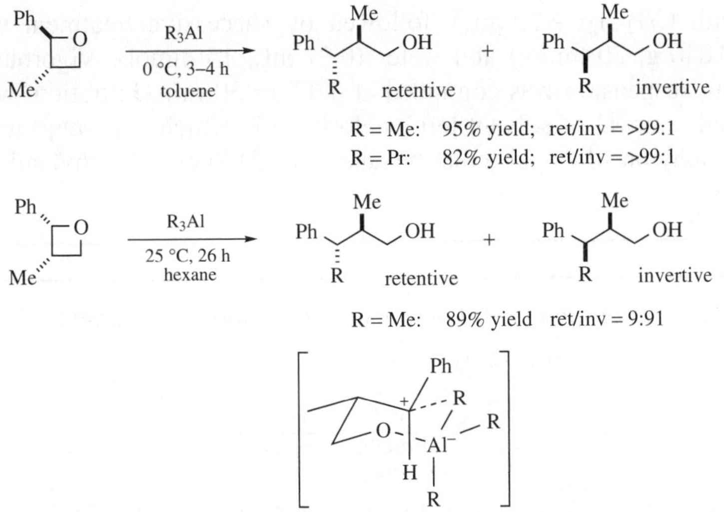

Scheme 7.

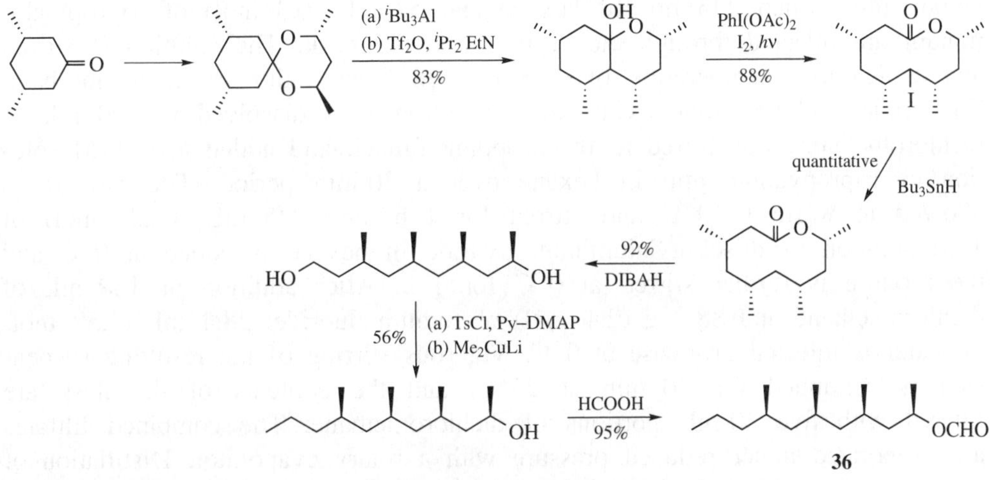

Scheme 8.

Beckmann Rearrangement–Reduction Sequence

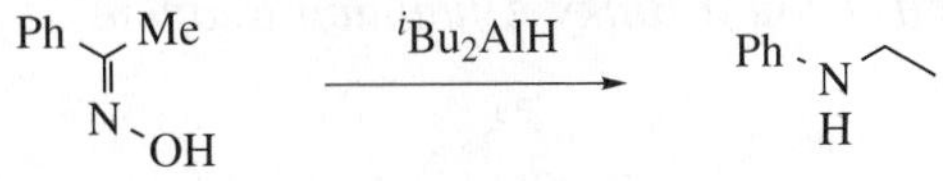

DIBAH (5 mL of a 1 M hexane solution, 5 mmol) was added dropwise to a solution of acetophenone oxime (0.135 g, 1 mmol) in dry CH$_2$Cl$_2$ (10 mL) at 0 °C, and the mixture was stirred at 0 °C for 2 h. Then the reaction was quenched by dilution with CH_2Cl_2 (~ 20 mL), followed by successive treatment with sodium fluoride (0.840 g, 20 mmol) and water (0.27 mL, 15 mmol). Vigorous stirring of the resulting suspension was continued at $0\,°C$ for 30 min. Filtration, washing with CH_2Cl_2, and removal of solvent left a colorless oil, which was subjected to column chromatography on silica gel (ether/hexane $= 1 : 3$) to give *N*-ethylaniline (0.111 g, 92% yield).

Beckmann Rearrangement–Alkylation using an Aluminum Reagent

A dry, 2-L, three-necked, round-bottomed flask is equipped with a variable-speed mechanical stirrer, 300-mL pressure-equalizing dropping funnel bearing a gas inlet at its top, and a rubber septum. The apparatus is flushed with argon, after which 243 mL of hexane and 57 mL (0.3 mol) of tripropylaluminum are injected through the septum into the flask. The solution is stirred and cooled to a temperature of -73 to $-78\,°C$ in a dry ice–methanol bath. The crude cyclohexanone oxime methanesulfonate is dissolved in 100 mL of dichloromethane, transferred to the dropping funnel, and added to a 1 M solution of tripropylaluminum in hexane over a 30-min period. The mixture is allowed to warm to $0\,°C$ and stirred for 1 h, and 225 mL (0.225 mol) of 1 M solution of diisobutylaluminum hydride in hexane is added at $0\,°C$ and the mixture is further stirred at $0\,°C$ for 1 h. After addition of 100 mL of dichloromethane and 88.2 g (2.1 mol) of sodium fluoride, 28.4 mL (1.58 mol) of water is injected dropwise at $0\,°C$. Vigorous stirring of the resulting suspension is continued for 30 min at $25\,°C$, and the contents of the flask are filtered with five 30-mL portions of dichloromethane. The combined filtrates are evaporated under reduced pressure with a rotary evaporator. Distillation of the residual liquid under reduced pressure affords 11.3–12.2 g (53–58%) of 2-propyl-1-azacycloheptane as a colorless liquid (bp 79–81$\,°C$, 18 mmHg).

Epoxide Rearrangement using a Bulky Aluminum Reagent

A dry, 1-L, three-necked, round-bottomed flask is equipped with a gas inlet, rubber septum, pressure-equalizing dropping funnel, and a magnetic stirring bar. The flask is charged with 3.42 g (12 mmol) of 4-bromo-2, 6-di-*tert*-butylphenol and flushed with argon, after which 600 mL of freshly distilled dichloromethane is added. The mixture is stirred, degassed under vacuum, flushed with argon, and 3 mL (6 mmol) of a 2 M hexane solution of trimethylaluminum is injected through the septum to the flask at 25 °C. The resulting solution is stirred at this temperature for 1 h to give methylaluminum bis(4-bromo-2,6-di-*tert*-butylphenoxide) almost quantitatively. The reaction vessel is cooled to a temperature of −20 °C in a dry ice/ *o*-xylenebath. Then 11.8 g (60 mmol) of *trans*-stilbene oxide is dissolved in 25 mL of dry dichloromethane, transferred to the dropping funnel, and added over 15–20 min at −20 °C. The mixture is stirred at −20 °C for 4 h. After addition of 1.01 g (24 mmol) of sodium fluoride, 324 μL (18 mmol) of water is injected dropwise at −20 °C. The entire mixture is vigorously stirred at −20 °C for 5 min and at 0 °C for 30 min. The contents of the flask are filtered with the aid of three 50-mL portions of dichloromethane. The combined filtrates are concentrated to about 100 mL under reduced pressure with a rotary evaporator. Silica gel is added and the remainder of the dichloromethane is removed using a rotary evaporator. The residue is layered on a column of silica gel and eluted (ether/dichloromethane/hexane, 1 : 2 : 20 to 1 : 1 : 10 as eluent) to give 10.3 g (yield, 87%) of diphenylacetaldehyde as a colorless oil.

Alkylative Cleavage of Acetals with Trialkylaluminum

To a solution of acetal (0.5 mmol) in toluene (10 mL) was added dropwise tripropylalminum (1.5 mL of 2.0 M in hexane, 3.0 mmol) at 0 °C. The reaction mixture was stirred at a 60 °C. After complete conversion (5–6 h), the resulting mixture was poured into 2 N aqueous sodium hydroxide (20 mL) and extracted with hexane (20 mL × 3). The combined organic layers were dried over magnesium sulfate and concentrated *in vacuo*. The residue was purified by column chromatography on silica gel (hexane/EtOAc as eluent) to give alkylation product (99% yield) as a colorless oil.

Alkylative Cleavage of Acetals with the Combined use of Trialkylaluminum and Pentafluorophenol

To a solution of trimethylaluminum (1.5 mL of a 2.0 M solution in hexane, 3.0 mmol) in toluene (10 mL) was added pentafluorophenol (1.5 mL of a 2.0 M solution in toluene, 3.0 mmol) at 25 °C under argon. The reaction mixture was stirred for 1 h at that temperature, during which period the evolution of alkane gas ceased. To this was introduced acetal (0.5 mmol) in toluene (1 mL) at 25 °C. After being stirred for 3 h, the solution was poured into 2 N aqueous sodium hydroxide (20 mL), and the product was extracted with hexane (20 mL × 3). The combined organic layers were dried over magnesium sulfate. The solvent was evaporated, and the residue was purified by column chromatography on silica gel (hexane/EtOAc as eluent) to give alkylation product (70% yield) as a colorless oil.

Alkylative Cleavage of Oxetanes with Trimethylaluminum

To a solution of trimethylaluminum (2.0 M in hexane, 1.5 mL, 3.0 mmol) in solvent (15 mL) was added dropwise oxetane (0.5 mmol) at 0 °C. The reaction solution was stirred at 0 °C. After 3 h, the resulting solution was poured into cooled 2 M aqueous sodium hydroxide (20 mL), and extracted with hexane (20 mL × 3). The combined organic layers were dried over magnesium sulfate, concentrated *in vacuo*. The residue was purified by column chromatography on silica gel (hexane/EtOAc as eluent) to give the corresponding alcohol (95% yield) as a colorless oil.

7   Ate Complex and its Reactivity

As described previously, the initial reaction of a trivalent organoaluminum compound with organic molecules is always the coordination of the substrate with an aluminum atom.

In contrast, a preformed aluminum ate complex behaves quite differently: the complex has already satisfied the octet rule and behaves as a nucleophile rather than a Lewis acid. For example, reduction of α,β-unsaturated epoxide gave a completely different product when the aluminum reagent was changed from a neutral trivalent species to an ate complex[59]. With DIBAH, the 1,4-addition-type reduction product was found to be the major product through a possible six-membered transition state, whereas with lithium aluminum hydride, a direct reduction product was formed preferentially. This is a typical example of the difference in reactivity between a trivalent aluminum reagent and an ate complex:

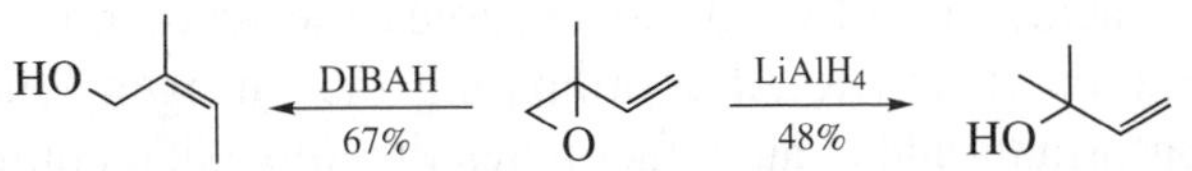

Alkenylaluminum reagent was used in a key synthetic process in construction of the prostaglandin structure. The preparation of alkenylaluminum **37** is straightforward, using simple hydroalumination of a terminal alkyne. Preparation of the ate complex **38** from this alkenylaluminum reagent is not difficult and a 1,4 rather than a 1,2 addition product can be anticipated by the reaction with cyclopentenone derivative[60]:

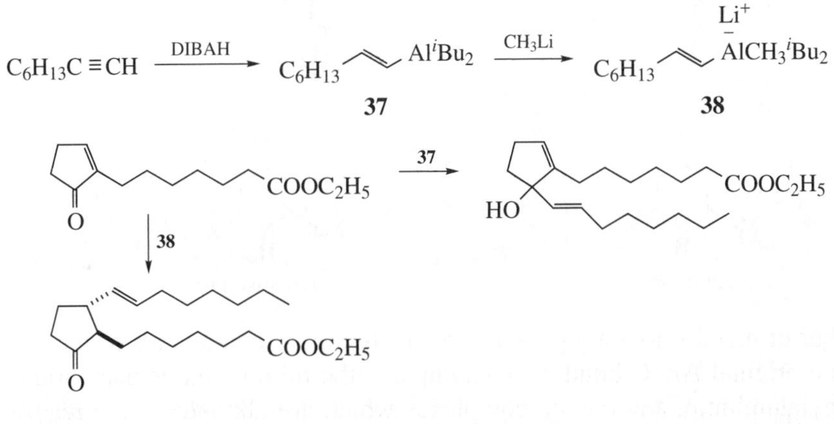

Shown below is the synthesis of solenopsin using organoaluminum methodology, in which the anti oxime **39** was converted into the corresponding mesylate in quantitative yield. Treatment of the mesylate in methylene chloride with excess trimethylaluminum in toluene resulted in the formation of imine. The final step of the synthesis should be simple reduction; unfortunately, however, the usual reduction with DIBAH gave the cis isomer almost exclusively. Moderate selectivity for the desired trans isomer was observed with lithium aluminum hydride. Excellent stereoselectivity was finally attained in formation of the trans form using lithium aluminum hydride in ethereal solvent in the presence of an equimolar amount of trialkylaluminum at low temperature. Thus, solenopsin was obtained almost exclusively using lithium aluminum hydride–trimethylaluminum in THF at a low temperature[61]:

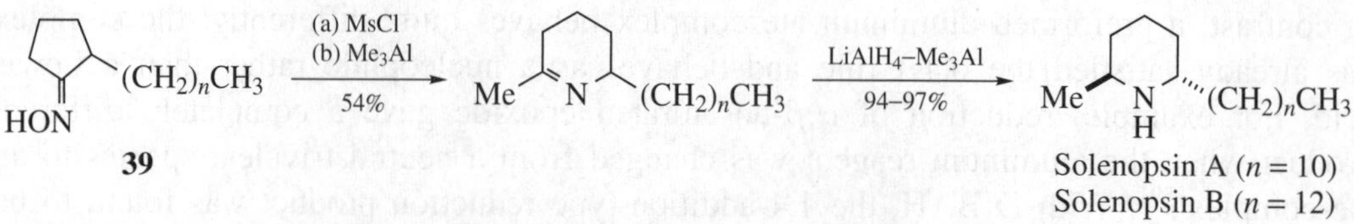

Solenopsin A (n = 10)
Solenopsin B (n = 12)

The observed selectivity can be predicted. In the hydride reduction of the imine, the underside approach of the hydride ion toward the imino π bond is preferred by stabilization of the σ* orbital through electron delocalization from the σ C−H bond into the σ* orbital, producing the cis isomer[62]. However, in the presence of a trialkylaluminum species as the Lewis acid, the alkyl group occupies the axial position because of the steric interaction between R and R₃Al coordinating the nitrogen lone-pair electrons by A-strain[63]. Such conformational change facilitates the upperside approach of the hydride ion toward the imino π bond and furnishes the desired trans isomer:

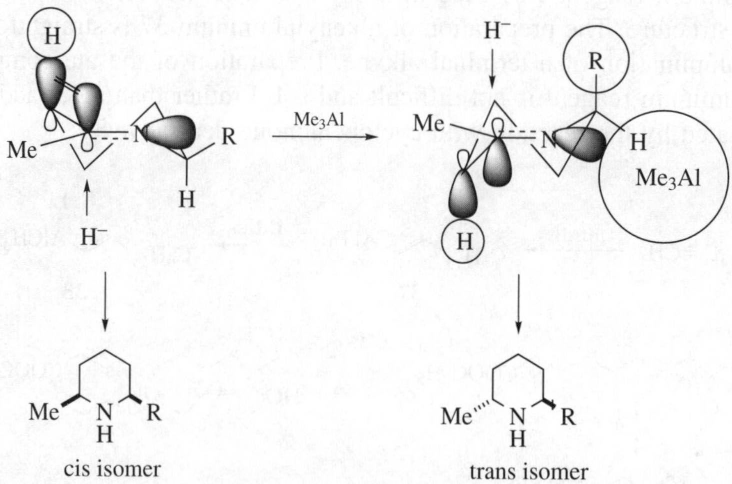

cis isomer   trans isomer

A number of metal salts interact with organoaluminum compounds to modify the reactivity of the original Al−C bond. One example is the lithium alkyls that form aluminates with alkenylaluminum, anionic ate complexes which are alkylated more readily than the neutral aluminum derivatives:

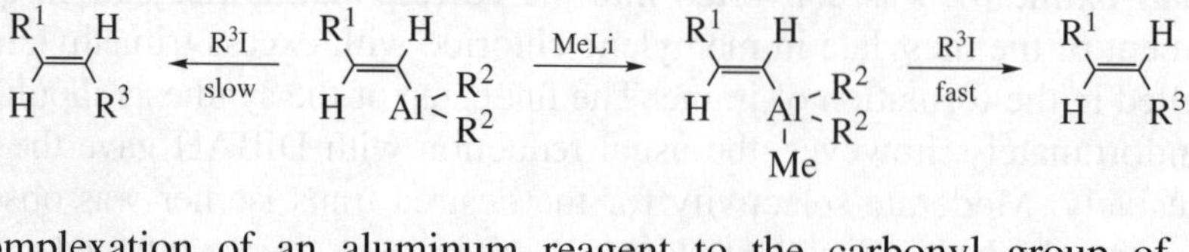

Prior complexation of an aluminum reagent to the carbonyl group of an unsaturated ester to generate aluminum ate complex would decrease the electron density of its alkenic moiety and thus function to increase the electron-withdrawing properties of this substituent. This methodology for activation of α,β-unsaturated carbonyl compounds is well established in Diels–Alder reactions. Thus, unusual exo selectivity was observed in some cases[64]:

$R^1 = Ph, R^2 = H$: 81% (73:27)
$R^1 = Ph, R^2 = Me$: 81% (96:4)
$R^1 = R^2 = Me$: 87% (87:13)

ATPH

Using the bulky reagent aluminum tris(2,6-diphenylphenoxide) (ATPH), ionic reactions such as 1,4 addition of nucleophiles could be effectively achieved[64]:

RLi = MeLi (83%), PhLi(86%), PhC≡CLi (99%)

Radical acceptor activation by an aluminum reagent has also been achieved[65]: Thus, addition of butyl radical to methyl crotonate (**40**) was accelerated in the presence of an aluminum reagent:

LA:	
TiCl$_4$	<2%
BF$_3$	5%
Bu$_2$BOTf	7%
Et$_2$AlCl	18%
EtAlCl$_2$	33%

Similarly, intramolecular addition of alkenyl radical to the β position of [(dimethyl-phenyl)methyl] cyclohexyl α,β-unsaturated esters showed good to excellent diastereo-selctivity in the presence of aluminum reagents. It is well known that acrylate is fixed in the (S)-trans conformation in the presence of a Lewis acid. Thus, it is not surprising to see that a suitable Lewis acid capable of surviving under the radical conditions would allow reaction to occur predominantly from the (*S*)-trans conformer[66]:

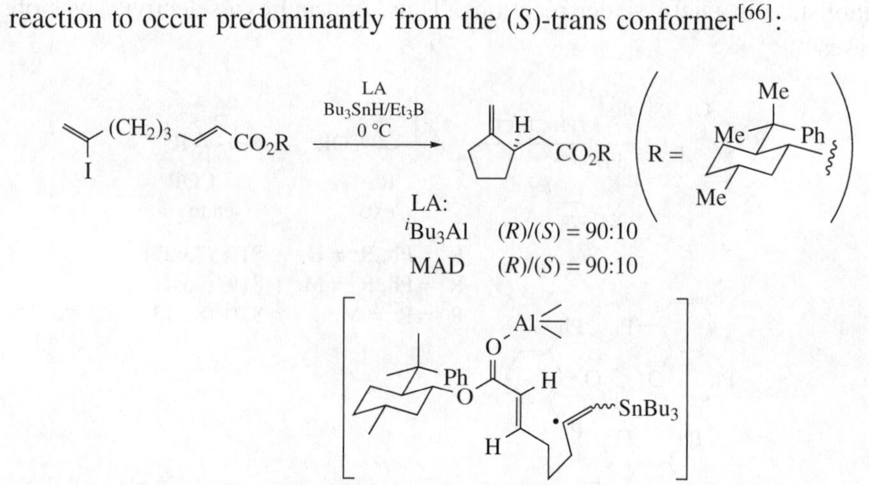

LA:
*i*Bu₃Al   (*R*)/(*S*) = 90:10
MAD   (*R*)/(*S*) = 90:10

Nucleophilic reactions are promoted by Lewis acids through coordination and activation of substrates as shown above. However, a side reaction such as degradative attack of nucleiphiles to Lewis acid is usually inevitable. Inoue and Aida proposed a Lewis acid-assisted high-speed living anionic polymerization, in which the degradative nucleophilic Lewis acid neutralization is sterically suppressed by the combination of bulky Lewis acids with aluminum porphyrins as nucleophiles. Thus, the nucleophile and Lewis acid are both so large that they are unable to react directly with each other, whereas the monomer is able to coordinate to the Lewis acid and is activated for nucleophilic attack. A solution of methylaluminum tetraphenylprophyrin (TPP)AlMe and methyl methacrylate was irradiated with visible light. After a 2.5 h irradiation, methyl methacrylate polymerization occurred with only 6% conversion. However, when three equivalents of bulky aluminum reagent was added to this reaction mixture, under diffuse light, a strikingly vigorous reaction took place with heat evolution to attain 100% conversion within three seconds[67]:

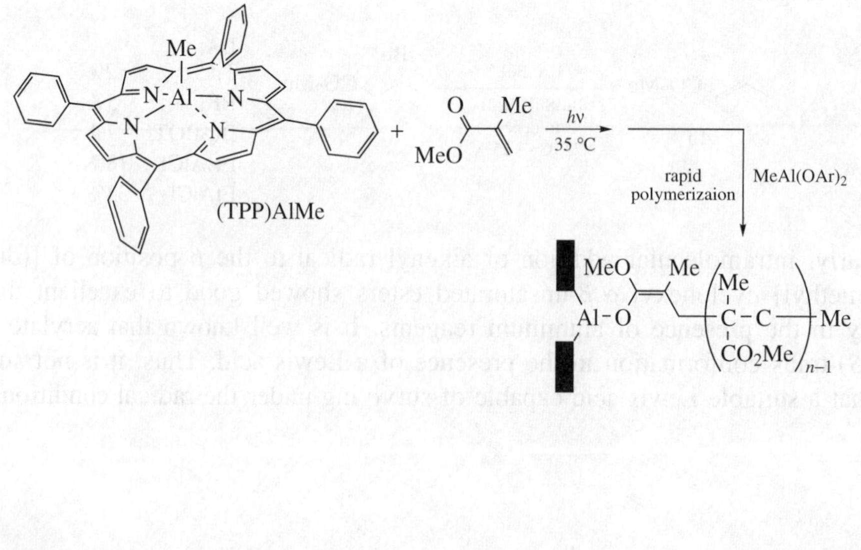

Diels–Alder Reaction using Dimenthyl Fumarate

A solution of (−)-dimenthyl fumarate (9.42 g, 24 mmol) in toluene (144 mL) was treated with diethylaluminum chloride (24 mL of a 1 M hexane solution) at −78 °C under nitrogen. The mixture was stirred for 15 min and to the resulting orange solution was added cyclopentadiene (5.85 mL, 72 mmol) dropwise at the same temperature. Acidic work-up was conducted and the product was isolated by column chromatography on silica gel (11.0 g, 100% yield, 99% de).

Michael Addition of Lithium Acetylide in the Presence of ATPH

A 1-M hexane solution (1.5 mL, 1.5 mmol) of Me_3Al was added dropwise to a stirred solution of 2,6-diphenylphenol (1.1 g, 4.5 mmol) in CH_2Cl_2 (10 mL) at room temperature and the resulting colorless solution was stirred there for 30 min to furnish aluminum tris(2,6-diphenylphenoxide) (ATPH) in hexane/CH_2Cl_2. This solution was cooled to −78 °C and 2-cyclohexenone (103 μL, 1 mmol) was added at this temperature, giving an enone–ATPH complex. Then, lithium phenylacetylide (2 mmol), which was prepared by treatment of phenylacetylene (231 μL, 2.1 mmol) in THF (6 mL) with a 1.6 M hexane solution (1.25 mL, 2 mmol) of BuLi at 0 °C for 20 min, was transferred by a cannula to the enone–ATPH complex in hexane–CH_2Cl_2 at −78 °C. The whole mixture was stirred at −78 °C for 1 h. The reaction mixture was worked up with diluted HCl and extracted with ether. The combined organic extracts were dried over Na_2SO_4, concentrated under reduced pressure, and purified by column chromatography on silica gel (ether/hexane = 1 : 5 to 1 : 1) to furnish 3-(phenylethynyl)cyclohexanone (0.196 g, 99% yield).

Michael Addition of Alkylbarium Reagent in the Presence of ATPH

$$Ph \diagup CHO \xrightarrow[\substack{(b)\ BuBaI \\ 97\%}]{(a)\ ATPH} \underset{\substack{Bu}}{Ph \diagup CHO}$$

$$(1,4/1,2 = 97:3)$$

ATPH is conveniently prepared by treatment of 2,6-diphenylphenol (3 equiv) in CH_2Cl_2 with a 1 M hexane solution of Me_3Al at 25 °C for 30 min. Initial complexation of *trans*-cinnamaldehyde with ATPH (1.1 equiv) in CH_2Cl_2 and subsequent addition of BuBaI (1.1 equiv) in ether at −78 °C gave rise to a conjugate adduct, 3-phenylheptanal, predominantly in 97% yield along with 3% of the 1,2 adduct.

1,6 Addition of ^tBuLi to Acetophenone in the Presence of ATPH

$$\xrightarrow[\substack{(b)\ ^tBuLi}]{(a)\ ATPH}$$

A 2-M hexane solution (0.75 mL, 1.5 mmol) of Me_3Al was added dropwise to a stirred solution of 2,6-diphenylphenol (1.1 g, 4.5 mmol) in toluene (10 mL) at 25 °C and the resulting colorless solution was stirred there for 30 min to furnish aluminum tris(2,6-diphenylphenoxide)(ATPH) in toluene. This solution was cooled to −78 °C and acetophenone (117 µL, 1 mmol) was added at this temperature, giving a ketone–ATPH complex. Then, the mixture was diluted with THF (10 mL) and a 1.57 M pentane solution (1.27 mL, 2 mmol) of ^tBuLi was added dropwise to the ketone–ATPH complex in toluene/THF at −78 °C. The whole mixture was stirred at −78 °C for 3 h. The reaction mixture was worked up by addition of concentrated HCl (1 mL) followed by vigorous stirring of the mixture, and extracted with CH_2Cl_2. The combined organic extracts were dried over Na_2SO_4, concentrated under reduced pressure, and purified by column chromatography on silica gel (CH_2Cl_2/hexane = 1 : 1) to furnish the 1,6 adduct (0.166 g, 93% yield).

Alkylation of Unsymmetrical Ketones by the Combined use of ATPH and LDA

$$\xrightarrow[LDA]{ATPH} \xrightarrow{MeOTf}$$

To a solution of ATPH (1.1 equiv) in toluene (5.0 mL) was added 2-methylcyclohexanone (61 µL, 0.50 mmol) at −78 °C under argon atmosphere, followed by a transfer of a THF (5.0 mL) solution of LDA, prepared from diisopropylamine (77 μL, 0.55 mmol) and a 1.60 M hexane solution of BuLi (0.34 mL, 0.55 mmol) at −78 °C for 30 min, by a steel cannula. After stirring at −78 °C for 1 h, to the mixture was added MeOTf (566 μL, 5.0 mmol) dropwise, and the mixture was maintained at this temperature for 7.5 h, poured into 1 N HCl, and was extracted with ether. The organic layer was dried over Na_2SO_4, and concentrated by rotary evaporator. The residue was purified by column chromatography on silica gel (CH_2Cl_2/pentane = 1 : 1 as eluent) to give 0.033 g of a mixture of 2,2-dimethylcyclohexanone and 2,6-dimethylcyclohexanone (53% yield) as a colorless oil. The ratio of 2,2-dimethyl-cyclohexanone and 2,6-dimethylcyclohexanone was determined by GC analysis.

Butyl Radical Addition under the Influence of Et_2AlCl

(syn/anti = 87:13)

tert-Butyl-2-methylene-3-hydroxypentanonate (0.020 g, 0.107 mmol) was dissolved in toluene (0.8 mL) and the solution was cooled to −78 °C. Diethylaluminum chloride (0.11 mL, 1 M solution in hexane, 0.11 mmol) was added, followed in order by BuI (0.040 mL, 0.321 mmol), Bu_3SnH (0.090 mL, 0.321 mmol), and finally Et_3B (0.030 mL, 1 M solution in hexane, 0.030 mmol). If the progress of the reaction was not seen at the initial stage, a small volume of air (ca 0.050 mL) was bubbled into the solution by syringe. After 6 h at −78 °C, complete consumption of Bu_3SnH was observed by TLC analysis. The reaction was terminated at −78 °C by the addition of 1 N HCl and ether. After warming up to 25 °C with stirring, the organic layer was separated, washed twice with 1 N NaOH (to convert Bu_3SnX to Bu_3SnOH), dried over Na_2SO_4, and concentrated to an oil. ^1H NMR showed the ratio of syn and anti isomers to be 87 : 13. The crude oil was purified on silica gel to give the title compound (0.0196 g, 76% yield) again as an 87 : 13 mixture of the syn and anti isomers.

8 General Hints Concerning the Use of Organoaluminum Compounds: Handling and Cautions

Alkylaluminum compounds are typically clear, colorless, mobile liquids with low melting points (except trimethylaluminum, mp 15 °C) and low vapor pressures at ambient temperatures. Aluminum alkyls are miscible in all proportions and compatible with saturated and aromatic hydrocarbons. Organoaluminum compounds are also soluble in ethers and tertiary amines, accompanied with exothermic complex formation.

Aluminum alkyls react very vigorously with air, water or other hydroxylic reagents. The violence of this reaction depends on molecular weight of the aluminum alkyl. Diluted solutions of aluminum alkyls in hydrocarbon solvents are nonpyrophoric and could be used in the same manner as more common organometallic reagents, such as butyllithium. Aromatic solvents such as benzene and toluene, and saturated aliphatic solvents such as hexane can be used. Generally, unsaturated aliphatic and aromatic compounds should not be used. Some chlorinated or oxygenated solvents may be used but others, such as carbon tetrachloride, may react violently with the formation of toxic gases. The nonpyrophoric concentration varies with the aluminum alkyl compound, the temperature, and the solvent: for C_4 compounds and below concentrations up to $10-20$ wt % alkyls are generally nonpyrophoric; for C_5 and above, the concentration can be increased to over 20%. Thus, organoaluminum reagents in hydrocarbon solvents under these concentrations can be used as easily as butyllithium in hexane. However, the use of neat aluminum alkyls should be undertaken with extreme caution.

Operations involving organoaluminum compounds thus should be carried out under an inert gas such as nitrogen or argon. This is quite important not only because of the pyrophoric character of the reagent, but also to eliminate losses due to hydrolysis and oxidation of the organoaluminum species.

Organoaluminum compounds are available as 1 M hydrocarbon solutions. However, if aluminum compounds are available only in lecture bottles, the following procedures can be used to extract the compound and to prepared a solution.

Procedure for Removing Organoaluminum Compounds from a Lecture Bottle

1. With the lecture bottle clamped in an upright position (Fig. 2), remove the valve unit ① so that only the bottle remains. Quickly attach the adapter ② to the opening of the bottle and apply a slow stream of nitrogen or argon.
2. Use a Luer-lock hypodermic syringe with a needle, the plunger ④ of which should be well greased for safe removal of an ignitable organoaluminum compound. The Luer-lock part should also be slightly greased and well sealed with tape before commencing the procedure.
3. Open the ball valve ③ with a wrench, put the needle into the open neck of the adapter as above, and withdraw the desired amount of organoaluminum compound. Then carefully and quickly remove the syringe from the bottle.
4. Close the valve ③, followed by stopping the flow of nitrogen or argon.

Another type of a lecture bottle shown in Fig. 3 is also available, and similar handling for this has been described in the literature[68].

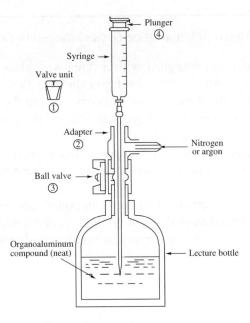

Figure 2. Removal of organoaluminum compound from a lecture bottle.

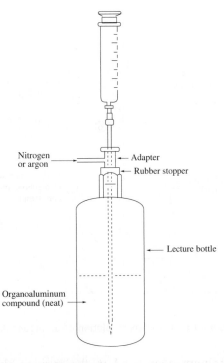

Figure 3. Removal of organoaluminum compound from a lecture bottle.

Procedure for Preparing a Solution of an Organoaluminum Compounds

1. A round-bottomed flask, equipped with a three-way glass and Teflon stopcock ⑤, is flame dried or dried with a heat gun under a flow of nitrogen or argon (Fig. 4). The joint of the flask and the stopcock ⑥, and the joint of the glass and the teflon cock ⑦ should be well greased. The joint ⑥ should be well sealed both with parafilm and scotch tape.
2. While a gentle stream of nitrogen or argon is maintained into the flask from the vent ⑧, an appropriate amount of hydrocarbon as solvent is added from the vent ⑧ using a syringe.
3. The requisite amount of organolaluminum compound, obtained using the previous procedure, is transferred from the vent ⑨. It is recommended that the needle end of the syringe is immersed in the hydrocarbon during draining for safety reason.

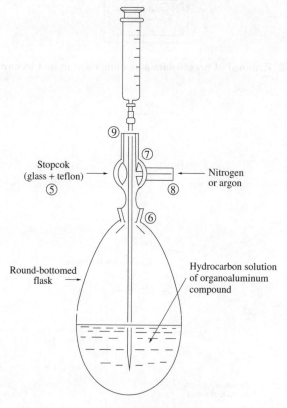

Figure 4. Preparation of a hydrocarbon solution of an organoaluminum compound.

4. The remaining aluminum compound in the syringe should be diluted before discarding it safely by pouring into a flask containing plenty of hexane in which

the needle end is immersed. Rinse the syringe several times by withdrawing hexane, and wash with diluted HCl, H_2O and then with acetone. The resulting hexane solution of the aluminum residue should be disposed of by pouring slowly into a flask filled with ice.

5. The system should be kept out of line when not in use by closing the three-way stopcock, followed by covering the vents ⑧ and ⑨ with rubber caps to ensure temporary protection from air and moisture.

The concentration of a hydrocarbon solution of an aluminum compound can be calculated from its own density (Table 2). Titration is also possible, but is not generally used to determine the concentration of this kind of species.

Table 2. Density of organoaluminum compounds. The density decreases at about 0.001–0.0015 per degree centigrade.

Compound	Density $(g \bullet mL^{-1}, {}^\circ C)^a$
Trimethylaluminum	0.743, 30
Triethylaluminum	0.835, 25
Triisobutylaluminum	0.781, 25
Diisobutylaluminum hydride	0.798, 25
Diethylaluminum chloride	0.961, 25
Ethylaluminum dichloride	1.207, 50

[a]Cited from a handbook by Texas Alkyls, Stauffer Chemical Company, 1976.

Although it is possible to store the solution at 25 °C, storage is strongly recommended at below 0 °C, because aluminum compounds are sometimes volatile enough to corrode the greased parts of the storage apparatus.

9 Specialized Topics

9.1 Ziegler–Natta Catalysis

Although it is not the purpose of this manuscript to review the complete details of organoaluminum chemistry, it is still appropriate to mention some of the recent advances in the Ziegler–Natta catalyst. As originally described[1], the Ziegler–Natta catalyst is a combination of a transition metal compound and an organometallic compound of Group 1, 2 or 3. It has been referred to as a coordination catalyst, reflecting the view that the catalysis occurs via coordination of the alkene to the active metal center, followed by insertion into a metal–alkyl bond. A variety[69] of reagent systems have been developed

and some of them are proving quite useful in industry. A combined use of trialkylaluminums and cyclopentadienyltitanium catalysts is unique among them[69]: The catalyst is
not particularly active, but is soluble in organic solvents and thus is popular for academic
research studies. The activity of this catalyst is relatively low; however, activity an order
of magnitude higher is obtained when small amounts of water are added to the trialkylaluminum. Peak activities are found with between 0.2 and 0.5 moles of water per mole
of aluminum alkyl[70]. Although the detailed structure of this aluminum reagent is not
known, the system seems to have a vast potential not only for industrial use, but also for
more special use in organic sysnthesis:

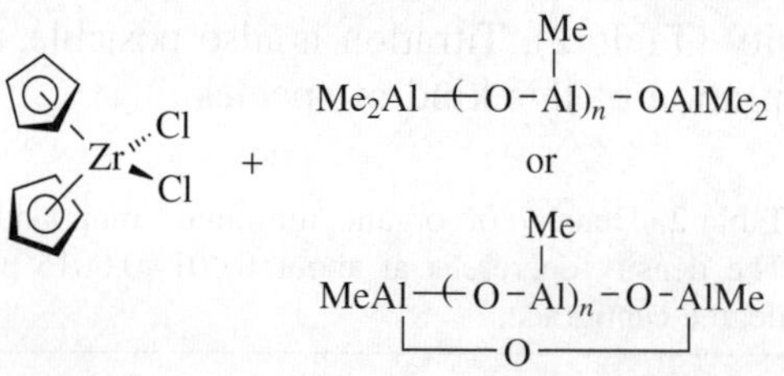

A clear explanation for the catalytic activity of MAO has been sought through iden-
tifying special structural features in MAO that could cause such cocatalysis. As a huge
excess of MAO is required per equivalent of transition metal component, such an MAO
structural unit might constitute a small percentage of the various components in an
MAO sample. An illustrative example of this approach in trying to understand the catalytic
activity of an MAO is a recent investigation that concludes that the active units are
the terminal Me_2Al groups of the aluminoxanes chain, although what the electronic
basis would be for the activity of such a unit remains unclear. The recent findings by
Barron and co-workers on the structure of oligomers of tetra-*tert*-butyldialuminoxane
and *tert*-butylaluminoxanes are probably best viewed in this light. The interaction of
a Zr catalyst with these aluminoxanes has been investigated and the Lewis acid–base
complexes isolated and characterized. This type of cage species could be responsible for
zirconocene polymerization of ethylene[71]:

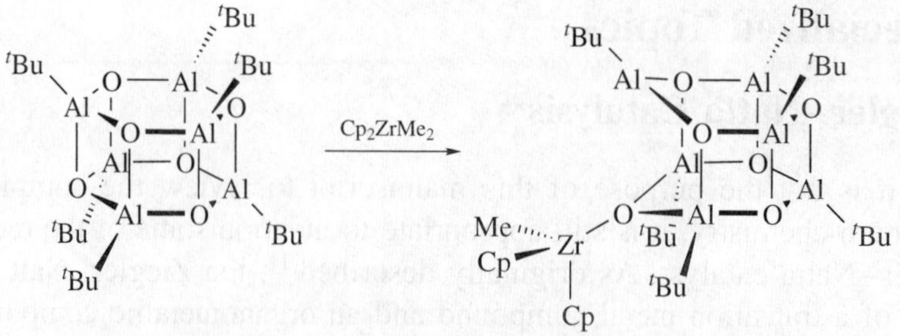

At present the true structure and the cocatalytic activity of methylaluminoxanes remain
unsolved problems, and represent significant challenges for future research.

9.2 Complexation Chromatography

The ready availability of different types of hindered polyphenols enables the molecular designing of various polymeric organoaluminum reagents. For example, this chemistry allows the realization of a complexation chromatography, i.e. separation of heteroatom-containing solutes by complexation with stationary, insolubilized organoaluminum reagents[72]. Accordingly, treatment of sterically hindered triphenol **41** (2 mmol) in CH_2Cl_2 with Me_3Al (3 mmol) at room temperature for 1 h gave rise to the polymeric monomethylaluminum reagent **42**. After evaporation of solvent, the residual solid was ground to a powder and mixed with silanized silica gel (1.7 g) in an argon box. This was packed in a short-path glass column (10 mm × 150 mm) as a stationary phase and washed once with dry, degassed hexane to remove unreacted free triphenol. Then a solution of methyl 3-phenylpropyl ether and ethyl 3-phenylpropyl ether (0.5 mmol each) in degassed hexane was charged on this short-path column and eluted with hexane/ether to realize a complete separation. This technique allows the surprisingly clean separation of structurally similar ether substrates. Ethyl 3-phenylpropyl ether and isopropyl 3-phenylpropyl ether, or the THF and THP ethers of 4-(*tert*-butyldiphenylsiloxy)-1-butanol can be separated equally well with this short-path column chromatography.

41 → **42**

Me_3Al (1.5 equiv) CH_2Cl_2

10 References

[1] (a) Ziegler, K.; in *Organometallic Chemistry* (Zeiss, H.; Ed.), Reinhold, New York, **1960**, 194; (b) Boor, J.; *Ziegler-Natta Catalysts and Polymerizations*, Academic Press, New York, **1979**.

[2] (a) Mole, T.; Jeffery, E. A.; *Organoaluminum Compounds*, Elsevier, Amsterdam, **1972**; (b) Bruno, G.; *The Use of Aluminum Alkyls in Organic Synthesis*, Ethyl Corporation, Baton Rounge, (1970, 1973, 1980); (c) Negishi, E.; *J. Organometal. Chem. Libr.*, **1976**, *1*, 93; (d) Yamamoto, H.; Nozaki, H.; *Angew. Chem., Int. Ed. Engl.* **1978**, *17*, 169; (e) Negishi, E.; *Organometallics in Organic Synthesis* Vol. 1, **1980**, 286, Wiley, New York; (f) Eisch, J. J.; *Comprehensive Organometallic Chemistry* (Wilkinson, G.; Stone, G. G. A.; Abel, E. W.; Ed.), Vol. 1, **1982**, 555, Pergamon Press, Oxford; (g) Zietz, J. R.; Robinson, G. C.; Lindsay, K. L.; in *Comprehensive Organometallic Chemistry*

(Wilkinson, G.; Stone, G. G. A.; Abel, E. W.; Ed.), Vol. 7, **1982**, 365, Pergamon Press, Oxford;
(h) Zeifel, G.; Miller, J. A.; *Org. React.* **1984**, *32*, 375; (i) Maruoka, K.; Yamamoto, H.; *Angew. Chem.,
Int. Ed. Engl.* **1985**, *24*, 668; (j) Maruoka, K.; Yamamoto, H.; *Tetrahedron* **1988**, *44*, 5001.

[3] (a) Ashby, E. C.; Laemmle, J.; Parris, G. E.; *J. Organomet. Chem.* **1969**, *19*, 24; (b) Mole, T.; Surtees,
J. R.; *Aust. J. Chem.* **1964**, *17*, 961.

[4] (a) Asby, E. C.; Yu, S. H.; *J. Chem. Soc. D* **1971**, 351; (b) Ashby, E. C.; Yu, S. H.; Roling, P. V.; *J.
Org. Chem.* **1972**, *37*, 1918; (c) Laemmle, J.; Ashby, E. C.; Roling, P. V.; *J. Org. Chem.* **1974**, *38*,
2526.

[5] Ashby, E. C.; Laemmle, J. T.; *Chem. Rev.* **1975**, *75*, 521.

[6] Mole, T.; *Aust. J. Chem.* **1966**, *19*, 381.

[7] Sakane, S.; Maruoka, K.; Yamamoto, H.; *J. Chem. Soc. Jpn.* **1985**, 324.

[8] (a) Wilke, G.; Muller, H.; *Chem. Ber.* **1956**, *89*, 444; (b) Wilke, G.; Muller, H.; *Liebigs Ann. Chem.*
1958, *618*, 267; (c) Wilke, G.; Muller, H.; *Liebigs Ann. Chem.* **1960**, *629*, 222; (d) Wilke, G.;
Schneider, W.; *Bull. Soc. Chim. Fr.* **1963**, 1462.

[9] (a) Eisch, J. J.; Damasevitz, G. A.; *J. Org. Chem.* **1976**, *41*, 2214; (b) Uchida, K.; Utimoto, K.;
Nozaki, H.; *J. Org. Chem.* **1976**, *41*, 2215.

[10] Eisch, J. J.; Rhee, S.; *J. Am. Chem. Soc.* **1975**, *97*, 4673.

[11] Hattori, K.; Matsumura, Y.; Miyazaki, T.; Maruoka, K.; Yamamoto, H.; *J. Am. Chem. Soc.* **1981**, *103*,
7368.

[12] Maruoka, K.; Miyazaki, T.; Ando, M.; Matsumura, Y.; Sakane, S.; Hattori, K.; Yamamoto, H.; *J. Am.
Chem. Soc.* **1983**, *105*, 2831.

[13] Ham, N. S.; Mole, T.; *Prog. Nucl. Magn. Reson. Spectrosc.* **1961**, *4*, 91.

[14] (a) Ramey, K. C.; Brien, J. F. O.; Hasegawa, I.; Borchert, A. E.; *J. Phys. Chem.* **1965**, *69*, 3418;
(b) Williams, K. C.; Brown, T. L.; *J. Am. Chem. Soc.* **1966**, *88*, 5460; (c) Jeffery, E. A.; Mole, T.;
Aust., J. Chem. **1969**, *22*, 1129.

[15] Eisch, J. J.; Liu, Z-R.; Singh, M.; *J. Org. Chem.* **1992**, *57*, 1618.

[16] Maruoka, K.; Itoh, T.; Sakurai, M.; Nonoshita, K.; Yamamoto, H.; *J. Am. Chem. Soc.* **1988**, *110*, 3588.

[17] Power, M. B.; Bott, S. G.; Atwood, J. L.; Barron, A. R.; *J. Am. Chem. Soc.* **1990**, *112*, 3446.

[18] Maruoka, K.; Oishi, M.; Shinohara, K.; Yamamoto, H.; *Tetrahedron Lett.* **1994**, *50*, 8983.

[19] Ahn, Y.; Cohen, T.; *J. Org. Chem.* **1994**, *59*, 3142.

[20] (a) Corey, E. J.; Beames, D. J.; *J. Am. Chem. Soc.* **1972**, *94*, 7210; (b) Collins, P. W.; Dajani, E. Z.;
Bruhn, M. S.; Brown, C. H.; Palmer, J. R.; Pappo, R.; *Tetrahedron Lett.* **1975**, *13*, 4217; (c) Posner,
G. H.; Whitten, C. E.; Sterling, J.; *J. Am. Chem. Soc.* **1973**, *95*, 7788; (d) Posner, G. H.; Brunelle, D. J.;
Sinoway, L.; *Synthesis* **1974**, 662.

[21] (a) Fried, J.; Lin, C.; Ford, S. H.; *Tetrahedron Lett.* **1969**, *7*, 1379; (b) Fried, J.; Lin, C.; Mehra, M.;
Kao, W.; Dalven, P.; *Ann. N. Y. Acad. Sci.* **1971**, *180*, 36.

[22] Hanse, R. T.; Carr, D. B.; Schwartz, J.; *J. Am. Chem. Soc.* **1978**, *100*, 2244.

[23] Ashby, E. C.; Noding, S. A.; *J. Org. Chem.* **1979**, *44*, 4792.

[24] Rück, K.; Kunz, H.; Synthesis, **1993**, 1018, *Angew. Chem., Int. Ed. Engl.* **1991**, *30*, 694.

[25] Gosling, K.; Smith, J. D.; Wharmby, D. H. W.; *J. Chem. Soc. A* **1969**, 1738.

[26] (a) Basha, A.; Lipton, M.; Winreb, S. M.; *Tetrahedron Lett.* **1977**, *15*, 4171; (b) Corey, E. J.; Beames,
D. J.; *J. Am. Chem. Soc.* **1973**, *95*, 5829.

[27] Benderly, A.; Stavchansky, S.; *Tetrahedron Lett.* **1988**, *29*, 739.

[28] Hatch, R. P.; Weinreb, S. M.; *J. Org. Chem.* **1977**, *42*, 3960.

[29] Kozikowski, A. P.; Ames, A.; *J. Org. Chem.* **1978**, *43*, 2735.

[30] Wood, J. L.; Khatri, N. A.; Weinreb, S. M.; *Tetrahedron Lett.* **1979**, *17*, 4907.

[31] Neef, G.; Eder, U.; Sauer, G.; *J. Org. Chem.* **1981**, *46*, 2824.

[32] Corey, E. J.; Kozikowski, A. P.; *Tetrahedron Lett.* **1975**, *13*, 925.

[33] Overman, L. E.; Flippin, L. A.; *Tetrahedron Lett.* **1981**, *22*, 195.

[34] Yasuda, A.; Takahashi, M.; Takaya, H.; *Tetrahedron Lett.* **1981**, *22*, 2413.

[35] Kitagawa, Y.; Hashimoto, S.; Iemura, S.; Yamamoto, H.; Nozaki, H.; *J. Am. Chem. Soc.* **1976**, *98*, 5030.

[36] Maruoka, K.; Miyazaki, T.; Ando, M.; Matsumura, Y.; Sakane, S.; Hattori, K.; Yamamoto, H.; *J. Am. Chem. Soc.* **1983**, *105*, 2831.

[37] Garigipati, R. S.; *Tetrahedron Lett.* **1990**, *31*, 1969.

[38] Yasuda, A.; Tanaka, S.; Oshima, K.; Yamamoto, H.; Nozaki, H.; *J. Am. Chem. Soc.* **1974**, *96*, 6513; Tanaka, S.; Yasuda, A.; Yamamoto, H.; Nozaki, H.; *J. Am. Chem. Soc.* **1975**, *97*, 3252.

[39] Nozaki, H.; Oshima, K.; Takai, K.; Ozawa, S.; *Chem. Lett.* **1979**, 379.

[40] Kosugi, M.; Naka, H.; Sano, H.; Migita, T.; *Bull Chem. Soc. Jpn.* **1987**, *67*, 3462.

[41] Inoue, T.; Takeda, T.; Kambe, N.; Ogawa, A.; Ryu, I.; Sonoda, N.; *J. Org. Chem.* **1994**, *59*, 5824.

[42] (a) Kozikowski, A. P.; Ames, A.; *J. Org. Chem.* **1978**, *43*, 2735; (b) Itoh, A.; Ozawa, S.; Oshima, K.; Nozaki, H.; *Tetrahedron Lett.* **1980**, *21*, 361; (c) Sasaki, K.; Aso, Y.; Otubo, T.; Ogura, F.; *Chem. Lett.* **1989**, 607.

[43] (a) Chung, B. Y.; Park, Y. S.; Cho, I. S.; Hyun, B. C.; *Bull. Korean Chem. Soc.* **1988**, *9*, 269. (b) Bernasconi, S.; Jommi, G.; Montanari, S.; Sisti, M.; *Synthesis* **1987**, 1126.

[44] (a) Davies, A. G.; Hall, C. D.; *J. Chem. Soc.* **1963**, 1192; (b) Hock, H.; Kropf, H.; Ernst, F.; *Angew. Chem.* **1959**, *71*, 541.

[45] Lewinski, J.; Zachara, J.; Grabska, E.; *J. Am. Chem. Soc.* **1996**, *118*, 6794.

[46] Maruoka, K.; Sno, H.; Shinoda, K.; Nakai, S.; Yamamoto, H.; *J. Am. Chem. Soc.* **1986**, *108*, 6036.

[47] Lautens, M.; Chiu, P.; Ma, S.; Rovis, T.; *J. Am. Chem. Soc.* **1995**, *117*, 532.

[48] Pine, S. H.; Zahler, R.; Evans, D. A.; Grubbs, R. H.; *J. Am. Chem. Soc.* **1980**, *102*, 3270.

[49] Takai, K.; Oshima, K.; Nozaki, H.; *Tetrahedron Lett.* **1980**, *21*, 1657.

[50] (a) Katsuki, T.; Sharpless, K. B.; *J. Am. Chem. Soc.* **1980**, *102*, 5974; (b) Rossiter, B. E.; Katsuki, T.; Sharpless, K. B.; *J. Am. Chem. Soc.* **1981**, *103*, 464; (c) Matin, V. S.; Woodard, S. S.; Katsuki, T.; Yamada, Y.; Ikeda, M.; Sharpless, K. B.; *J. Am. Chem. Soc.* **1981**, *103*, 6237.

[51] (a) Maruoka, K.; Fukutani, Y.; Yamamoto, H.; *J. Org. Chem.* **1985**, *50*, 4412. (b) Maruoka, K.; Sakane, S.; Yamamoto, H.; *Org. Synth.* **1989**, *67*, 176.

[52] Maruoka, K.; Oishi, M.; Yamamoto, H.; *J. Org. Chem.* **1993**, *58*, 7638.

[53] McCarty, C. G.; in *Chemistry of the Carbon-Nitrogen Double Bond* (Patai, S.; Ed.), Wiley-Interscience, New York, **1970**, 408.

[54] Takai, K.; Mori, I.; Oshima, K.; Nozaki, H.; *Tetrahedron Lett.* **1981**, *22*, 3985.

[55] Nonoshita, K.; Banno, H.; Maruoka, K.; Yamamoto, H.; *J. Am. Chem. Soc.* **1990**, *112*, 316.

[56] Ishihara, K.; Mori, A.; Yamamoto, H.; *Tetrahedron* **1990**, *46*, 4595.

[57] Ishihara, K.; Hanaki, N.; Yamamoto, H.; *J. Am. Chem. Soc.* **1993**, *115*, 10695.

[58] Kaino, M.; Naruse, Y.; Ishihara, K.; Yamamoto, H.; *J. Org. Chem.* **1990**, *55*, 5814.

[59] Lenox, R. S.; Katzenellenbogen, J. A.; *J. Am. Chem. Soc.* **1973**, *95*, 957.

[60] Bernady, K. F.; Floyd, M. B.; Poletto, J. F.; Weiss, M. J.; *J. Org. Chem.* **1979**, *44*, 1438.

[61] Matsumura, Y.; Maruoka, K.; Yamamoto, H.; *Tetrahedron Lett.* **1982**, *23*, 1929.

[62] Cieplak, A. S.; *J. Am. Chem. Soc.* **1981**, *103*, 4540.

[63] Narula, A. S.; *Tetrahedron Lett.* **1981**, *22*, 2017.

[64] Saito, S.; Yamamoto, H.; *Chem. Commun.* **1997**, 1585.

[65] Urabe, H.; Ymashita, K.; Suzuki, K.; Kobayashi, K.; Sato, F.; *J. Org. Chem.* **1995**, *60*, 3576.

[66] Nishida, M.; Ueyama, E.; Hayashi, H.; Ohtake, Y.; Yamaura, Y.; Yanaginuma, E.; Yonemitsu, O.; Nishida, A.; Kawahara, N.; *J. Am. Chem. Soc.* **1994**, *116*, 6455.

[67] Kuroki, M.; Watanabe, T.; Aida, T.; Inoue, S.; *J. Am. Chem. Soc.* **1991**, *113*, 5903.

[68] Nagata, W.; Yoshioka, M.; *Org. Synth. Coll.* **1988**, *6*, 14.

[69] Young, J. R.; Stille, J. R.; *J. Am. Chem. Soc.* **1992**, *114*, 4936; and references cited therein.

[70] Sinn, H.; Kaminsky, W.; Vollmer, H.-J.; Woldt, R.; *Angew. Chem., Int. Ed. Engl.* **1980**, *19*, 390.

[71] Harlan, C. J.; Bott, S. G.; Barron, A. R.; *J. Am. Chem. Soc.* **1995**, *117*, 6465.

[72] Maruoka, K.; Nagahara, S.; Yamamoto, H.; *J. Am. Chem. Soc.* **1990**, *112*, 6115.

V

Organozinc Chemistry

EIICHI NAKAMURA

Department of Chemistry, University of Tokyo, Japan

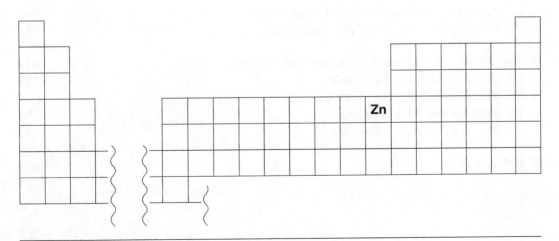

Organometallics in Synthesis: A Manual. Edited by Manfred Schlosser.
© 2002 John Wiley & Sons Ltd

Contents

1 Structures and Reactivities of Organozinc Reagents

1.1 History and Current Status of Organozinc Reagents in Synthesis

Of the numerous organometallic compounds bearing a metal–carbon σ bond, organolithium and magnesium compounds represent the most popular polar organometallic reagents serving as particularly useful reagents for organic synthesis (see Chapter 1). The reagents possess a countercationic metal of low electronegativity, and show high basicity and nucleophilicity, deprotonating an acidic substrate or transferring an alkyl group to an electrophilic carbon center. They are frequently used also as a source of alkyl groups in the preparation of organometallics through transmetalation reactions. With these powerful rivals around, organozinc(II) reagents have long received rather scant attention of synthetic chemists. However, since the mid 1980s there has been a renaissance of organozinc chemistry.

Organozinc compounds were the first group of compounds possessing a metal–carbon σ bond. Its preparation dates back to the year 1849[1], when Frankland discovered that heating a mixture of metallic zinc with either methyl iodide or ethyl iodide in a sealed tube affords self-inflammable dialkylzinc. The reaction, which involves a Schlenk-type equilibrium and which can now be formulated as given below, is still commonly used. An ate complex Et$_3$ZnNa was also prepared as early as in 1858[2]. There have been several comprehensive reviews on organozinc compounds[3,4]:

$$2RI + 2Zn \longrightarrow 2RZnI \rightleftarrows R_2Zn + ZnI_2$$

Most of the basic reactivities of organozinc compounds toward acid halides and carbonyl compounds were reported before the end of 1870s, leading to the discovery of the synthetically versatile Reformatsky reagent in 1887[5]. By the end of the nineteenth century, the status of organozinc reagents in classical organic synthesis was firmly established. However, the discovery of Grignard reagents at the turn of the century put an end to this first flourishing period of organozinc chemistry. Grignard reagents, being easier to prepare and much more reactive than the zinc reagents, quickly acquired widespread popularity. Functionalized organozinc compounds were known as early as in the 1940s[6], but, then, without the assistance of modern transition metal chemistry, they did not receive much attention. The Reformatsky reagent has lost its importance after the discovery of a more convenient preparation of reactive lithium enolates[7]. The major problem of organozinc chemistry was the low nucleophilic reactivities of the reagent. Interestingly enough, the revival of organozinc reagents during the 1980s depends on this very drawback, which has become an advantage over the Grignard and lithium reagents.

One of the recent major advances in organozinc chemistry is the development of 'functionalized organozinc reagent' bearing functional groups, in particular, electrophilic ones such as alkoxycarbonyl and carbonyl. The report of Negishi and co-workers on the use of organozinc reagents in nickel- and palladium-catalyzed substitution reactions[8] was the first step toward this goal. Organozinc reagents were subsequently found to be superior to Grignard reagents in Pd-catalyzed substitution (see Chapter 10) reactions at an sp^2 carbon center[9,10]. In these reactions, the zinc reagent substitutes the X group of an RMX catalytic intermediate (M = Pd etc.), which then gives the product R^1–R^2 upon reductive elimination. The poorly reactive alkylzinc reagent is just reactive enough to undergo this transmetalation process without touching the R^1 and R^2 group:

$$R^2X + cat.\ Pd(0) \longrightarrow R^2\text{–}Pd\text{–}X \xrightarrow{R^1\text{–}ZnX} R^2\text{–}Pd\text{–}R^1 \longrightarrow R^1\text{–}R^2 + Pd(0)$$

R^1 = alkyl, vinyl, aryl; R^2 = vinyl, aryl; X = halogen, CF$_3$SO$_3$

Pioneering works on functionalized organozinc reagents by Boersma[11b] and Thiele *et al.*[12] showed the compatibility of nonelectrophilic groups (*e.g.* halogen) in organozinc compounds. The discovery of nucleophilically reactive β-metalloesters (metal homoenolates) by Kuwajima and Nakamura[13] marked a breakthrough in the chemistry of functionalized organometallics by showing that an electrophilic group such as an ester group is compatible with the organozinc reagent. Many previous attempts to use alkali or alkali earth metal homoenolates had failed because of the intrinsic instability of such functionalized nucleophiles which undergo rapid intramolecular reactions. However, stable β-metalloesters, for instance β-stannylesters, were simply unreactive nucleophiles. It was first reported in 1983[14] that a titanium homoenolate is a stable (isolable) yet sufficiently reactive nucleophile to be used for organic synthesis[15]. A zinc homoenolate, reported the following year[16], turned out to be synthetically more versatile. ZnCl$_2$ activation of the C–C bond of 1-alkoxy-1-siloxycyclopropane generated the zinc homoenolate of alkyl propionate (see below). The compound is remarkably stable and can even be distilled, yet is reactive under catalyzed conditions undergoing rapid Me$_3$SiCl-accelerated conjugate addition and acylation in the presence of a catalytic amount of copper(I) halide and Me$_3$SiCl. This report also demonstrated, for the first time, the remarkable accelerating effects of Me$_3$SiCl[17] and the usefulness of organozinc reagent in organocopper chemistry (see Chapter 6). BF$_3$ was also found to accelerate the copper-catalyzed conjugate addition of R$_2$Zn with a different stereochemical outcome[18]. Palladium and nickel catalysts catalyze the substitution reaction of the homoenolate with vinyl and aryl halides and tosylates, and acid chlorides[19,20]. The otherwise unreactive organozinc reagent was found to undergo addition to carbonyl compounds in the presence of Lewis acids such as Me$_3$SiCl[21] and BF$_3$[22]. The basic synthetic transformations of the zinc homoenolate illustrates the synthetic versatility of functionalized organozinc compounds (Scheme 1)[23].

The synthetic inconvenience of the cyclopropane route to zinc homoenolate was subsequently circumvented by the use of Frankland's classical synthesis of organozinc reagents. Tamaru, Yoshida and co-workers reported in 1985 that not only the homoenolate but also its higher homologs can be prepared by the reaction of the corresponding functionalized alkyl iodide and zinc metal[24]. In 1987, their work culminated in the first preparation of a nonstabilized carbanionic species containing a carbonyl group[25]. Knochel *et al.* starting their studies on functionalized zinc species in 1988 also utilized the same approach[26]

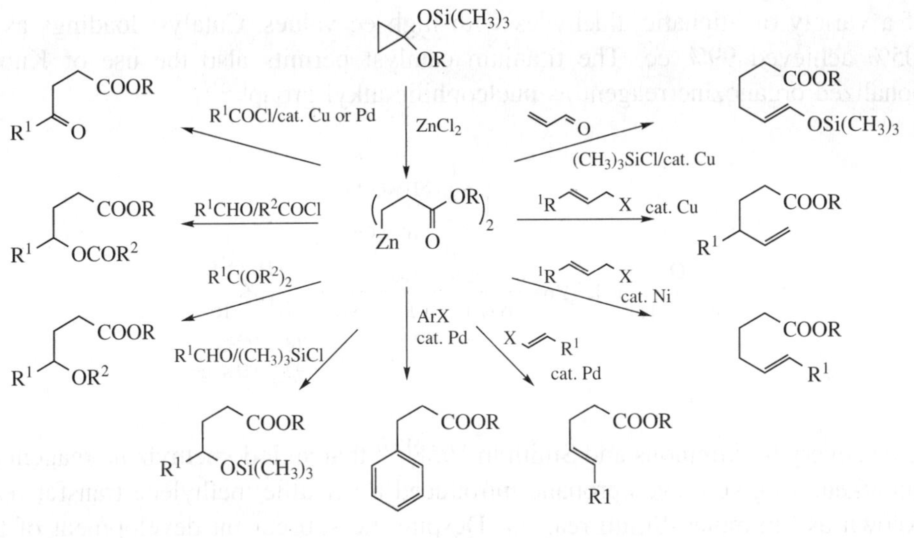

Scheme 1.

and later invented many useful preparative routes, which are discussed in the next section. As a result of extensive investigations in recent years, the chemistry of functionalized organozinc reagents is now firmly established[27]:

$$ \text{FG-RI} \xrightarrow{\text{Zn}} \text{FG-R-ZnI} \xrightarrow[\text{cat.}]{R^{1+}} \text{FG-R-R}^1 $$

Another important breakthrough in recent organozinc chemistry was Oguni's discovery of catalytic asymmetric carbonyl addition[28] in 1984[29]. A dialkylzinc reagent, which does not react with the carbonyl compound by itself, was found to add to an aldehyde in the presence of a small amount of a chiral amino alcohol via a six-centered transition state[30]. Thus, when a suitable amino alcohol such as the one shown below[31] is present in a mixture of R_2Zn and an aldehyde, the zinc reagent reacts only through complex formation with the chiral ligand to give an optically active secondary alcohol in nearly 100% ee. During the course of this research, it was also noted that even a catalyst of low enantiomeric purity was found to give very high ee values[32], convincingly demonstrating the principle of enantio-amplification[33]:

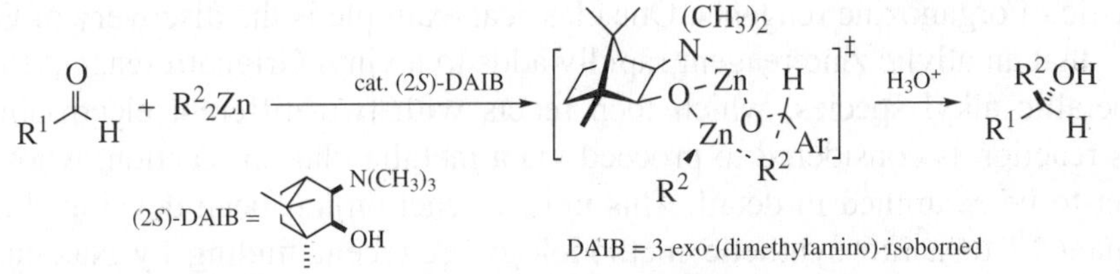

DAIB = 3-exo-(dimethylamino)-isoborned

The original amino alcohol system permitted only aromatic aldehydes to serve as electrophiles in the reaction. Development of a chiral titanium catalyst[34] allowed the use of a variety of aliphatic aldehydes with high ee values. Catalyst loadings as small as 0.05% achieved 99% ee. The titanium catalyst permits also the use of Knochel's functionalized organozinc reagent as nucleophilic alkyl group[35]:

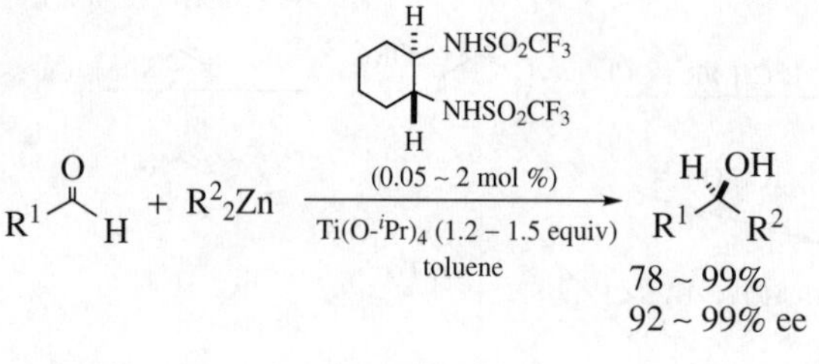

The discovery by Simmons and Smith in 1958[36] that an iodomethylzinc reagent reacts with an alkene to give a cyclopropane introduced a versatile methylene transfer reagent, now known as Simmons–Smith reagent. Despite the subsequent development of transition metal catalyzed cyclopropanation reactions using α-diazo carbonyl compounds, the Simmons–Smith reaction still retains its importance as a reliable route to cyclopropanes. It's merit resides in the ability to transfer alkylidene groups (CRR') as opposed to the necessary use of acylmethylene group (CH$_2$COX) in the transition metal catalyzed routes. In a modification of the Simmons–Smith reaction Furukawa et al.[37] utilized the reaction of Et$_2$Zn with CH$_2$I$_2$ and significantly broadened the scope of the alkylidenation methodology. It also provided the first beneficial use of a metal–halogen exchange reaction in organozinc chemistry. The catalytic asymmetric cyclopropanation reaction of Kobayashi et al. using a chiral titanium catalyst[38] spurred the research in asymmetric reaction, and reliable synthetic protocols are being established in this field of research:

$$Zn + CH_2I_2 \xrightarrow{Et_2O} ICH_2ZnI \quad \text{Simmons–Smith reagent}$$

$$Et_2Zn + CH_2I_2 \xrightarrow[\substack{\text{aprotic} \\ \text{solvents}}]{\text{various}} ICH_2ZnEt \quad \text{Furukawa reagent}$$

High affinity towards unactivated alkenes relative to carbonyl compounds is a useful characteristic of organozinc reagents. One classical example is the discovery of Gaudemar in 1971[39] that an allylic zinc reagent rapidly adds to a vinyl Grignard reagent to generate a 1,1-dimetallic alkyl species, which then reacts with two different electrophiles. This analogous reaction is considered to proceed via a metalla-claisen reaction, whose mechanism is yet to be examined in detail. This unique reaction has been developed by Marek and Normant[40] to a full synthetic methodology. A recent finding by Nakamura et al. is that the reaction of zincated hydrazones exhibits formal similarity to the Gaudemar reaction, and provides a new synthesis of ketone derivatives by one-pot four-component coupling reactions[41]:

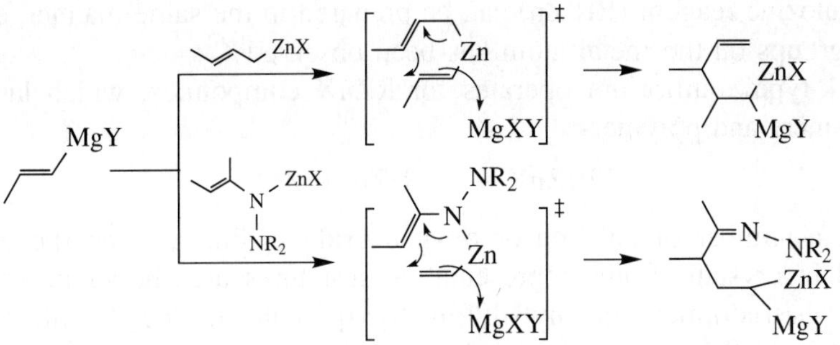

In a similar vein, Kubota and Nakamura have demonstrated the addition of a zinc enolate or an azaenolate to simple alkenes including ethylene and styrene[42]. Despite its apparent counterthermodynamic nature, this alkenic analog of the aldol reaction has been shown to be reasonably general[43,44]. Not unexpectedly, the corresponding lithium enolates do not undergo such a reaction. This alkenic aldol reaction has enabled multi-component coupling synthesis of carbonyl compounds, opening up a new avenue to the derivatization of carbonyl compounds:

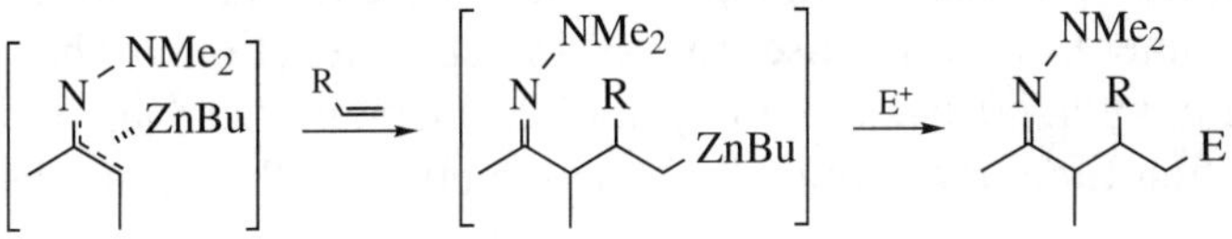

In the background of recent advances of organozinc chemistry is the improvement of preparative methods of organozinc compounds, as detailed in the next section. The rather low reactivity of metallic zinc has posed various problems of reproducibility. Both trial-and-error and analytical approaches have led to the development of the reactive zinc slurry known as Rieke zinc[45] and other chemical and physical "cleaning" methods. Organozinc reagents can now be prepared in a manner as reliable as in Grignard chemistry. Transmetalation and other methods, which are unavailable for lithium and Grignard reagents, also offer synthetically useful possibilities for organozinc compounds.

1.2   Structures of Organozinc Compounds

Organozinc(II) compounds can be classified, according to the number of alkyl groups on the metal, into three classes: $RZnX$, R_2Zn and $R_3Zn^- M^+$, where R is an sp^3, sp^2 or sp carbon group, enolate, or aza-enolate, X is a halogen or chalcogen atom, and M is an alkali or alkali earth metal. These compounds can be correlated by the sequence of reactions of zinc halide with an alkyllithium reagent:

$$ZnX_2 \xrightarrow{RLi} RZnX \xrightarrow{RLi} R_2Zn \xrightarrow{RLi} R_3Zn^{\ominus} Li^{\oplus}$$

Mixed diorganozinc reagent (RR'Zn) can be prepared in the same manner, and exchange of the alkyl groups on the metal atom has been observed[46].

A Schlen k-type equilibrium operates for RZnX compounds, which latter may also exist as oligomers and polymers:

$$2RZnX \longrightarrow R_2Zn + ZnX_2$$

For instance, a number of halogen or oxygen bridged dimeric zinc species have been characterized in crystals. Some representative structures are shown in Scheme 2. The structure of bis[tris(dimethylphenylsilyl)methyl-(μ^2-chloro)-zinc][47] illustrates such a structure. EtZnI exists in crystals as a polymer involving a tetracoordinated zinc atom[48]. In the absence of coordinating ligands, dialkylzinc compounds possess a linear structure as shown for bis(2,4-di-*tert*-butylpentadienyl)-zinc[49]. The C–Zn–C bond is bent in the presence of a bidentate ligand, as seen the structure of dimethyl-(*N,N,N',N'*-tetramethylethylenediamine) zinc[50]. The crystal structures of ate complexes are also reported in the literature. Trialkylzincate species have a planar tricoordinate structure, and the countercation stays nearby to neutralize the charge. A typical structure is shown for sodium tri(neopentyl)zincate[51].

Structural studies on the Simmons–Smith reagent have seen much progress owing to recent interests in its use for asymmetric reactions. The crystal structure of (*R*)-2,3-dimethoxy-4,7,7-trimethylbicyclo(2.2.1)heptane-*O,O'*-bis(iodomethyl)zinc[52] revealed, for the first time, the iodomethylzinc structure of the Simmons–Smith reagent.

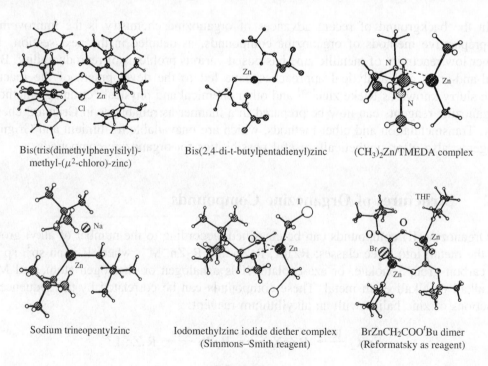

Bis(tris(dimethylphenylsilyl)-
methyl-(μ^2-chloro)-zinc)

Bis(2,4-di-t-butylpentadienyl)zinc

$(CH_3)_2$Zn/TMEDA complex

Sodium trineopentylzinc

Iodomethylzinc iodide diether complex
(Simmons–Smith reagent)

BrZnCH$_2$COO'Bu dimer
(Reformatsky as reagent)

Scheme 2.

It has a typical structure of R_2Zn species complexed with a bidentate ligand. The crystallographic analysis of $[CF_3CCl_2ZnCl\bullet Et_2O]_2$[53] revealed a chloride-bridged dimeric structure for this deactivated Simmons–Smith reagent. This observation suggests the possibility that the reactive Simmons–Smith reagent ICH_2ZnI may well exist as an iodide-bridged dimer.

The structure of the Reformatsky reagent is generally depicted as an α-metalated ester ($BrZnCH_2COOR$), yet its exact structure has remained elusive until recently. It is now known to exist as a dimer in solution, except in strongly polar DMSO wherein it exists as a monomer. The structure of the dimer in crystals[54], shown in Scheme 2 (on page 2) for $R = {}^tBu$, has solved this long-standing puzzle. The compound possesses a carbon–zinc bond which is conjugated to the carbonyl group, while the zinc atom is simultaneously bound to the carbonyl oxygen atom of another enolate structure in the dimer complex. Such an ambivalent structure is not a norm, however. For instance, the glycine enolate takes the form of an archetypal O-bound zinc enolate structure, in which the carbonyl oxygen and the α nitrogen atom form a stable bidentate chelate with the zinc atom[55].

2   Preparation of Organozinc Compounds

2.1   Outline of Preparative Methods

Low chemical reactivities characterize the chemistry of organozinc compounds as reagents for organic synthesis. This property has long been regarded as a disadvantage as compared to the highly reactive organolithium and -magnesium reagents. However, this changed during the 1980s, when a new emphasis on selectivities and functional tolerance changed the image of organozinc reagents dramatically. These stable and mildly reactive reagents now enjoy the status of key reagents for the synthesis of polyfunctionalized compounds, transition metal-mediated reactions, and enantio- and diastereoselective C–C bond formation.

The preparative methods for organozinc compounds may be classified into three major categories. The classical and still most popular method involves direct preparation of a carbon–zinc bond by oxidative addition of an alkyl halide to metallic zinc. Metal–metal exchange reaction constitutes the second most useful category, including the reaction of organolithium or magnesium compounds with zinc halide (ZnX_2) or organozinc halide (RZnX). The third category involves metal–halogen exchange reaction using a low-molecular-weight dialkylzinc compound and an organic halide.

As stated in the previous section, organozinc compounds can be classified by the numbers of organic groups covalently attached to the center metal, and each class of compounds has its own best set of preparations. For instance, RZnX-type reagents are best prepared by oxidative zincation reactions under mild reductive conditions. However, this method cannot be applied to the synthesis of R_2Zn reagents. Thus R_2Zn compounds bearing functional groups on the R moiety are best prepared by transmetalation between zinc and boron. With the progress of recent organozinc chemistry, the diversity of zinc reagents has became much greater than that available in organolithium and magnesium chemistry. For simplicity of classification, this section describes the various preparations according to the nature of the organic groups on the metal: (1) alkylzinc reagents; (2) alkenyl, aryl, and alkynylzinc reagents; (3) allylic, benzylic, and propargylic reagents. Each sub-section is further divided according to the preparative methods.

2.2 Alkylzinc Reagents

2.2.1 Oxidative Zincation

The reaction of metallic zinc and an alkyl halide is the direct and synthetically most useful method for zinc–carbon bond formation. The method is applicable to the preparation of a number of functionalized organozinc reagents. Alkyl iodides are the best substrate, whereas alkyl bromides and chlorides may be employed for the reaction with highly active metallic zinc.

For a long time, reproducibility had frequently posed problems in oxidative zincation reactions. The most common initial step for activation of zinc metal involves washing commercial zinc power with aqueous HCl (activated zinc power). Further activation is achieved by the addition of a metallic additive (Cu and Ag salts) or an organic activating agent (1,2-dibromoethane or Me₃SiCl or both). Alkali metal reduction of zinc halide, as studied extensively by Rieke, also provides a reliable route to reactive zinc metal.

2.2.1.1 Activated Zinc Powder[56]

Numerous procedures have been developed to obtain metallic zinc suitable for the oxidative zincation of alkyl halides. Chemical activation method includes washing with aqueous HCl, addition of iodine or 1,2-dibromoethane, as well as making alloys by addition of an inorganic copper(I), copper(II)[57], or silver[58] salt to zinc dust, granules or powder. Physical activation may be achieved by ultrasound, metal vaporization, and electrochemistry. The main purpose of activation is believed to be the removal of oxidized material on the metal surface to expose an active surface. Requiring no special apparatus, chemical activation provides convenient methods for laboratory-scale experiments. Ultrasound irradiation in an ultrasound bath is also easy to carry out and can be combined with chemical activation methods.

The solid–liquid biphasic system used in oxidative zincation reactions often poses a problem of reproducibility. Utimoto, Takai et al.[59] found that a small amount of lead contaminant greatly affects the reducing ability of the bulk zinc metal. Thus, zinc powder obtained by pyrometallurgy contains 0.04–0.07 mol % of lead with respect to zinc. In contrast, electrolytic zinc produced by hydrometallurgy is >99.998% pure. The lead impurity accelerates the methylenation of carbonyl compounds, whereas it almost stops the formation of the Simmons–Smith reagent and oxidative zincation of alkyl iodides. (Suppliers may differ in the preparative method used for metallic zinc.) Me₃SiCl, often used as an activator, may in fact simply suppress the adverse effects of lead impurity. On the basis of these findings, the following procedure may be recommended as a standard preparation of activated zinc. Zinc powder activated by aqueous HCl washing is kept under argon, and is further activated *in situ* with Me_3SiCl or a combination of $Me_3SiCl/1,2$-dibromoethane.

Activation of Commercial Zinc Powder by HCl[60]

$$Zn \text{ dust} \xrightarrow{\text{dilute HCl}} \text{activated Zn dust}$$

About 250 g of 20-mesh zinc was covered with 300 mL of 5% HCl (aq) and stirred vigorously, for three times. The metal was washed by decantation three times with 300 mL of distilled water, twice with 200 mL of acetone, and twice with 200 mL of ether. The metal was dried rapidly in a vacuum desiccator. Five pounds of zinc was processed, combined to ensure uniformity, and kept in a stoppered bottle in a desiccator.

Activation of Commercial Zinc Powder by Me_3SiCl[61]

$$Zn \text{ dust} \xrightarrow{Me_3SiCl} \text{activated Zn dust}$$

Me_3SiCl (0.30 mL, 0.26 g, 2.4 mmol) was added via a syringe to a suspension of zinc powder (2.10 g, 32 mmol) in anhydrous ether (or tetrahydrofuran, 50.0 mL THF). The mixture was stirred for 15 min at 25 °C.

Combined use of Me_3SiCl and 1,2-Dibromoethane[62]

$$Zn \text{ dust} \xrightarrow[\text{(b) } Me_3SiCl]{\text{(a) } BrCH_2CH_2Br} \text{activated Zn dust}$$

A suspension of 1.70 g (26 mmol) of zinc (99.99% purity) in 2.0 mL of THF containing 0.17 g (1.0 mmol) of 1,2-dibromoethane was heated to 65 °C for 1 min and cooled to 25 °C. Me_3SiCl (0.10 mL, 0.087 g, 0.80 mmol) was added, and then the resulting suspension was stirred at 25 °C for 15 min.

The preparation of alkylzinc reagents with zinc powder is described next. Excellent yields are obtained with primary and secondary alkyl iodides, but not with alkyl bromides or chlorides.

Preparation of Secondary (and Primary) Alkylzinc Iodide[26]

$$FG\text{-}RI + Zn \xrightarrow{THF} FG\text{-}RZnI$$
90% yield

$$FG\text{-}R = C_4H_9, CH(CH_3)_2, c\text{-}C_6H_{13}, NC(CH_2)_3,$$
$$C_2H_5OCO(CH_2)_3, \ t\text{-}C_4H_9OCOCH(CH_3)(CH_2)_3$$

A suspension of 1.70 g (26 mmol) of zinc (99.99% purity) in 2.0 mL of THF containing 0.19 g (1.0 mmol) of 1,2-dibromoethane was heated to 65 °C for 1 min, cooled to 25 °C, and 0.10 mL (0.8 mmol) of Me_3SiCl was added. After 15 min at 25 °C, a solution of the iodide (25 mmol) in 10 mL of THF was slowly added (at 25 °C in the case of a secondary iodide; at 30 °C in the case of a primary iodide). After the addition, the reaction mixture was stirred for 12 h at 25–30 °C (or 35–40 °C in the case of a primary iodide). Usually less than 0.1 g of zinc remains, indicating a yield of 90%.

2.2.1.2   Rieke Zinc

In 1973, Rieke reported that metallic zinc generated by reduction of zinc chloride with alkali metal shows much higher reactivities than commercial zinc powder[45,63]. Interestingly, some organozinc reagents prepared by the use of Rieke zinc sometimes exhibit reactivities that differ from those of the reagents prepared from commercial metallic zinc. Treatment of zinc halide with alkali metal in a hydrocarbon or ethereal solvent produces a reactive slurry of metallic zinc, which is useful for the preparation of alkylzinc reagents and Reformatsky reagents. In addition, the use of Rieke zinc allows the use of alkyl bromides and aryl bromides in oxidative zincation reactions. The method can be applied to alkyl groups bearing ketone, ester, cyano, and chloro groups:

$$ZnCl_2 + 2\,K \longrightarrow Zn^* + 2\,KCl$$

$$ZnCl_2 + 2\,LiNp \xrightarrow{THF\ or\ DME} Zn^* + 2\,LiCl +$$

$$Zn(CN)_2 + 2\,LiNp \longrightarrow Zn^* + 2\,LiCN +$$

LiNp = lithium naphthalenide

The first generation of Rieke zinc employed extremely reactive metallic sodium and potassium as the reducing agent and hence is experimentally inconvenient, especially, on a large scale. In the 1991 improvement, the use of a small amount of naphthalene allowed the use of nonpyrophoric lithium metal, and the resulting zinc slurry was found to be even more reactive. In situations where the naphthalene or lithium halide hampers the subsequent reaction, the slurry is washed several times with dry THF[64]. Two methods have been proposed: in the one-pot procedure, a mixture of ZnCl₂, lithium metal and naphthalene is mixed together and stirred vigorously; and in the two-pot procedure, a THF solution of ZnCl₂ is added dropwise to a solution of lithium naphthalenide. The latter procedure allows the preparation of either reactive zinc metal in a micropowder form or in a sponge form by regulating the rate of the addition. When one needs to use a solvent other than THF, the sponge form is used. The following procedure illustrates the method, using a stoichiometric amount of naphthalene.

Preparation of Rieke zinc (Zn*) from ZnCl₂ by Lithium Naphthalenide Reduction[65]

$$\underset{\text{(1 equiv)}}{\text{ZnCl}_2} \xrightarrow[\text{THF or DME}]{\overset{\text{Li (2 equiv)}}{\text{naphtalene (2 equiv)}}} \text{Zn}^*$$

Two 50-mL flasks, A and B, were equipped with rubber septa, condensers topped with argon inlets, and Teflon-coated magnetic stir bars. Flask A was charged with freshly cut lithium (0.21 g, 30.6 mmol) and naphthalene (4.0 g, 31.15 mmol). Flask B was charged with anhydrous ZnCl₂ (2.09 g, 15.4 mmol). Both of these operations were performed in an argon atmosphere dry box. The flasks were transferred to the manifold system and the argon inlet fitted. Freshly distilled THF (15 mL) was added to both flask A and B via a syringe, and the mixtures were stirred at 25 °C. The solution in flask A changed from colorless to dark green almost immediately. The lithium was consumed in about 2 h, and the ZnCl₂ solution was transferred dropwise to the flask A by a cannula over 15 min. (A slow addition, about 3 s per drop with no excess of zinc halide, results in an extremely fine black slurry of active zinc. This slurry takes several hours to settle and can easily be transferred by a cannula. With faster addition, about 1 s or less per drop, the active zinc formed is sponge shaped. The solvent can be changed very easily by a cannula if the subsequent reaction needs a different solvent. The active zinc, prepared by both methods described above, has similar reactivity.) The active zinc was typically used at this point, but it can be washed several times with fresh solvent if naphthalene presents a problem with product isolation or if the solvent needs to be changed.

Ultrasound irradiation greatly reduces the reaction time and avoids the use of naphthalene for the one-pot preparation of the Rieke zinc. The method is more convenient than the original procedure, which needs careful control of stirring because of the solid/liquid nature of the reaction.

Preparation of Ethyl 3-Phenyl-3-hydroxypropionate from Benzaldehyde, Ethyl Bromoacetate and Zn[66]*

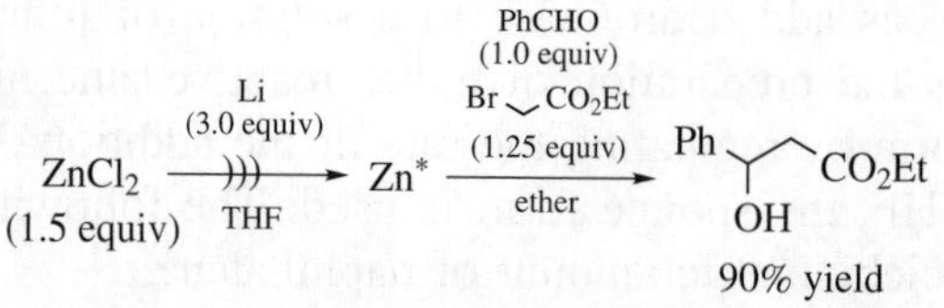

90% yield

In a clean and dry single-neck round-bottom flask equipped with an N_2 inlet and a condenser were placed $ZnCl_2$ (1.65 g, 12 mmol), THF (10 mL) and lithium dispersion (0.17 g, 24 mmol, 30% in mineral oil). The mixture was sonicated. The reduction was very vigorous, so sonication was stopped intermittently to maintain control. After 1 h, sonication was stopped, and a fine gray, dispersion of zinc was obtained. The THF was removed *in vacuo* and replaced with diethyl ether (10 mL) and then the flask cooled to 0 °C. Ethyl bromoacetate (1.67 g, 10 mmol) and benzaldehyde (0.85 g, 8.0 mmol) were weighed separately. One-tenth of the bromo ester was added to the zinc/ether dispersion. The remaining bromo ester and benzaldehyde were mixed and added dropwise to the dispersion. After addition was complete, the mixture was warmed to 25 °C and stirred vigorously for 1 h, then poured into cold water, and stirred for 15 min. Extraction with ether, drying with $MgSO_4$, and vacuum evaporation left a pale yellow liquid (1.39 g, 90% yield based on benzaldehyde).

The following example illustrates the unique reactivities of the organometallics prepared using Rieke zinc. For instance, the zinc reagent prepared in this manner undergoes 1,4 addition in the presence of Me_3SiCl[67]. The reason for this curious reactivity is yet to be clarified.

Conjugate Addition of Zinc Homoenolate Prepared Using Rieke Zinc[64]

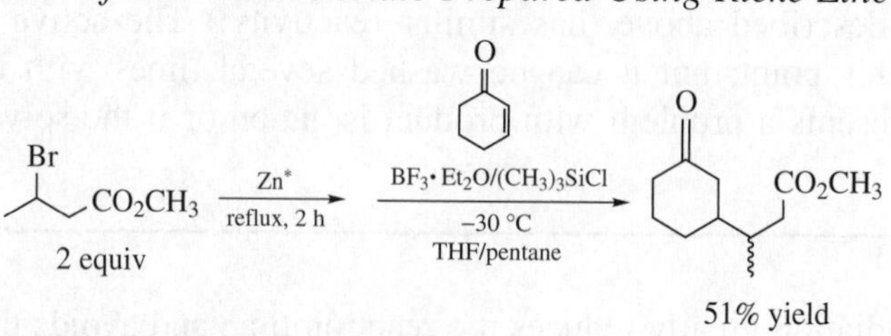

51% yield

To a flask charged with finely cut lithium (0.172 g, 24.8 mmol), naphthalene (0.320 g, 2.5 mmol), and THF (10 mL) under argon was transferred by cannula a THF solution of zinc chloride (1.61 g, 11.8 mmol) dropwise such that addition was complete in about 1.5 h with moderate stirring. The stirring was stopped when the lithium was totally consumed, and the active zinc was allowed to settle. The supernatant was then removed. Fresh THF (25 mL) was added, the mixture briefly stirred and then allowed to settle, and the supernatant subsequently removed. The active zinc was then washed one more time. THF (10 mL) was added to the flask, and the active zinc was ready for use. To the stirring mixture of active zinc was added methyl 3-bromobutyrate (1.37 g, 7.59 mmol), and the mixture was refluxed for 2 h. The reaction mixture was allowed to cool to 25 °C, and the zinc settled in 4 h. The alkylzinc bromide reagent was added dropwise to a solution of 2-cyclohexen-1-one (0.40 g, 4.14 mmol), BF₃•Et₂O (1.56 g, 11 mmol), and Me₃SiCl (15 mmol) in pentane (90 mL) at −30 °C, such that the addition was complete in 20 min. The heterogeneous reaction mixture was stirred for 3.5 h at −30 °C. The reaction mixture was then quenched by addition to saturated NH₄Cl (30 mL). Methyl 3-(3'-oxocyclohexyl)butanoate (0.416 g, 2.10 mmol, 51%) was isolated from the crude reaction mixture by flash chromatography (hexane/ethyl acetate).

Conversion of an alkyl chloride to the corresponding organozinc chloride is difficult but not impossible. For instance, Rieke zinc converts both alkyl iodides and bromides to the zinc reagents. When an alkyl chloride is used, excess KI is added to convert the chloride *in situ* to the corresponding iodide[65]. Further improvement came with the use of zinc cyanide/lithium naphthalenide[67], which converts an alkyl chloride directly into an alkylzinc reagent. The conditions are compatible with a bulky tertiary amide and cyano groups, but not with ester and ketone groups:

Organozinc Chloride Prepared from Primary Alkyl Chloride and Rieke Zn* Starting with Zn(CN)₂

Caution: cyanides are highly toxic, and care must be taken in the handling of the reagent and the waste.

Lithium (0.18 g, 26.02 mmol), zinc cyanide (1.49 g, 12.69 mmol), and naphthalene (0.33 g, 2.60 mmol) in freshly distilled THF (20 mL) were stirred for 5 h at 25 °C under argon. To the resulting black slurry of zinc at 25 °C was added *N,N*-diisopropyl-3-chlorobutamide (4.15 mmol), and the mixture stirred for 12 h. The excess zinc was then allowed to settle for about 4 h. The resulting clear supernatant solution was transferred carefully via cannula to a solution of copper(I) cyanide (2.60 mmol) and lithium bromide (2.60 mmol) in THF (10 mL) at −45 °C. A gummy white precipitate immediately formed in the reaction mixture. The resulting mixture was allowed to warm to 25 °C over 3 h and the reaction mixture was quenched by addition to a saturated ammonium chloride solution (40 mL). (**Caution:** additional acidification may be necessary by addition of 3 N HCl to dissolve precipitated metal salts and adequate measures should be taken to prevent hydrogen cyanide inhalation.) 5-Oxo-5-phenyl-(*N,N*-diisopropyl)pentamide (0.49 g, 1.78 mmol, 61% yield) was isolated from the crude residue by flash chromatography (ethyl acetate/hexanes).

A unique feature of the Rieke zinc is its applicability to the preparation of tertiary alkylzinc halide from alkylbromide. The rate of alkylzinc halide formation decreases in an order of tertiary, secondary and primary, which is the same as that observed with the Me$_3$SiCl/1,2-bromoethane activated zinc power[27]. The *tert*-alkylzinc reagent undergoes copper-catalyzed conjugate addition and acylation:

Preparation of Rieke Zinc from ZnCl$_2$ using a Catalytic Amount of Naphthalene. Generation and ^{13}C NMR Measurement of tert-butyllzinc Bromide

To a flask charged with finely cut (about $0.75 \times 1 \times 5$ mm^3) Li (0.11 g, 16 mmol), naphthalene (0.20 g, 1.6 mmol), and THF (10 mL) under argon was transferred via cannula a solution of zinc chloride (1.09 g, 8.0 mmol) in THF (15 mL) dropwise so addition was complete in 1.5 h. The mixture was then vigorously stirred for an additional 0.5 h until the lithium metal was consumed. The stirring was then stopped and zinc settled in 10 min. The supernatant was removed via cannula. The Rieke zinc was then washed with two consecutive portions of dry THF (15 mL). A final portion of THF was added and the Rieke zinc was ready for use.

To a slurry of Rieke zinc (1.03 g, 15.8 mmol) in THF (10 mL) under argon was added from a disposable syringe, 1.97 g (14.4 mmol) of *tert*-butyl bromide. The exothermic reaction was completed in 1 h. The zinc was allowed to settle, and 0.50 mL of the supernatant was cannulated to an NMR tube capped with septum. A sealed capillary tube containing CDCl$_3$ inside the NMR tube was used to gain NMR lock (^{13}C NMR δ 32.3, 21.7; THF: δ 25.3).

Preparative entries to reactive zinc metal other than the Rieke one are also known. For instance, metallic zinc deposited on titanium oxide is reported to convert secondary alkyl and benzyl bromide into the corresponding zinc bromide reagents.

Preparation of Activated Zinc Deposited on Titanium Oxide[68]

$$Na \xrightarrow{\text{TiO}_2} Na/TiO_2 \xrightarrow{\text{ZnCl}_2} Zn^*/TiO_2$$

A three-necked 100 mL flask equipped with an argon inlet, a glass stopper and a septum cap was charged with TiO$_2$ (18 g, 380 mmol) and heated for 2 h at 150 °C under vacuum (0.1 mmHg). The glass stopper was replaced by a mechanical stirrer, the reaction flask was flushed with argon, and sodium (1.50 g, 65 mmol) was added immediately. Alternatively, the Na could be added at 25 °C to the dry TiO$_2$ (TiO$_2$ can be stored for several weeks after drying, as described above, without noticeable changes). The reaction mixture was vigorously stirred at 150 °C for 15 min and cooled to 0 °C leading to a gray homogeneous powder. A solution of dry ZnCl$_2$ (4.57 g, 35.5 mmol) in THF (20 mL) was added with stirring. After 15 min, the activated Zn on TiO$_2$ was ready to use. Storage of the Zn powder leads to a loss of activity.

Potassium–graphite laminate (C$_8$K) can be regarded as an aggregate of polymeric alkali naphthalenide, and hence is expected to be a good reducing agent. Indeed, it acts as an excellent substitute for potassium and has been found to be useful for the preparation of Reformatsky and allylic zinc reagents.

Reduction of ZnCl$_2$ with Potassium Graphite[69]

$$2\,C_8K + ZnCl_2 \longrightarrow C_{16}Zn + 2\,KCl$$

Graphite powder (Roth, 3.1 g) was poured in a 100 mL two-necked flask and heated under argon at 150 °C. Freshly cleaned potassium (1.2 g, 30.7 mmol) was added in pieces; when the potassium melts, the mixture was vigorously stirred with a Teflon-covered magnetic stirring bar, and heating is stopped. Potassium–graphite (C_8K) so prepared was a bronze-colored powder (pyrophoric). C_8K was covered with THF (30 mL) and solid anhydrous $ZnCl_2$ (2.05 g, 15 mmol) was added. A very exothermic reaction takes place, causing the solvent to reflux vigorously. When the temperature subsided, the reaction mixture was stirred at 70 °C (external oil bath) for an additional 30 min.

The presence of a silver salt further activates the potassium graphite. Thus, Zn/Ag–graphite converts an α-bromoacetate into the corresponding Reformatsky reagent at -78 °C, as opposed to 25 °C and 0 °C required by Rieke zinc and Zn/graphite, respectively. In addition, conversion of α-substituted α-bromoacetate and α-choloroacetate into the corresponding Reformatsky reagent also takes place at 0 °C.

Reduction of $ZnCl_2$ with Zn/Ag–graphite[70]

$$ZnX_2 + \left(\begin{array}{c} AgOCOCH_3 \\ 8 \text{ mol}\% \end{array}\right) \xrightarrow{C_8K} Zn^*/Ag\text{–graphite}$$

Potassium (1.2 g, 30.6 mmol) was added in pieces to graphite (3.0 g, 250 mmol), which was previously degassed for 15 min under argon at 150–160 °C. When the potassium melts, the mixture was vigorously stirred with a magnetic stirring bar, yielding the bronze-colored pyrophoric C_8K within 10–15 min. After cooling, the C_8K was suspended in anhydrous THF (30 mL), and a mixture of freshly fused $ZnCl_2$ (2.0 g, 14.7 mmol) and AgOAc (0.2 g, 1.2 mmol) was added in portions causing the solvent to boil. The black mixture was refluxed for 30 min to ensure complete reduction and was used immediately.

2.2.2　Transmetalation

Transmetalation from organolithium, magnesium, aluminum, boron, mercury, and cadmium reagents to zinc halide generates a variety of organozinc reagents. If functional group tolerance is an issue, the use of organoboranes is the synthetic choice. This subsection first describes transmetalation from lithium, magnesium reagent, and then from alkyl boranes, especially those bearing functional groups.

2.2.2.1　Preparation of Alkylzinc Reagent from RLi or RMgX

It is a great merit of this approach that one can make RZnX, R_2Zn, and $R_3Zn^-M^+$ species simply by using stoichiometric amounts of the alkyl anion donor. In contrast to the variety of RZnX reagents available by oxidative zincation methods, the availability of R_2Zn species is rather limited. Except for distillable R_2Zn compounds (R $\leq$ butyl), R_2Zn reagents are prepared by transmetalation methods (this section) or by halogen–metal exchange reactions (next subsection). There are certain reactions, wherein the metal halide byproducts (LiX or MgXX′) interfere with the subsequent organic transformations and hence must be removed (such as an enantioselective carbonyl addition which may loose the selectivity)[71]. When starting with an alkyllithium reagent, one can remove lithium halide byproduct by precipitating it by dilution with a hydrocarbon solvent. Magnesium halide byproducts can be removed by complexation with dioxane.

Preparation of Salt-free Dialkylzinc from Grignard Reagent in the Presence of Dioxane[72]

$$2\,RCH_2MgX \; + \; ZnCl_2 \; \rightleftharpoons \; (RCH_2)_2Zn \; + \; 2\,MgXCl$$

A 1.0 M $ZnCl_2$ solution in Et_2O (Aldrich; 20 mL, 20 mmol) was diluted with 10 mL of Et_2O and 0.040 mol alkylmagnesium halide (in ether, 2.2 M, freshly prepared as usual from the alkyl halide and magnesium tunings and titrated shortly before use) was added. The resulting suspension was stirred at 25 °C for 2 h, then treated with 12 mL of 1,4-dioxane (freshly distilled from sodium metal) and stirred for an additional 45 min. Subsequent filtration under an inert atmosphere (using a Schlenk filter) yielded a clear solution of the zinc reagent.

The following example illustrates physical removal of the magnesium salt by precipitation and centrifuge. The group selectivity demonstrated in this example is also notable; namely, transfer of the methyl group was not observed (<1%). Competition between butyl and *tert*-butyl results in selective transfer of the butyl group.

Preparation of Butyl(methyl)zinc and its Reaction with Benzaldehyde[73]

$$CH_3MgCl + C_4H_9MgCl + ZnCl_2 \xrightarrow[\text{(b) 1,4-dioxane filteration}]{\text{(a) centrifuging ether}} CH_3(C_4H_9)Zn$$

82.4% ee
85% yield

A thick glass centrifuge tube was charged with 3 mL of a 1.0 M $ZnCl_2$ solution in ether. A 3.0 M MeMgCl solution in THF (1.0 mL) was added slowly followed by the slow addition of 1.5 mL of 2.0 M BuMgCl in ether. After stirring for 15 min the solution was centrifuged at 4000 rpm for 5 min. The supernatant was then transferred with double-ended needle through a filter into a 50 mL round-bottom flask. The remaining salts in the centrifuge tube were washed with 10 mL of distilled hexanes and the washings, which were also centrifuged, transferred to the round bottom flask. 1,4-Dioxane (12.8 μL) was added to the clear solution and stirred for 15 min.

The resulting suspension was then filtered and transferred to another 50 mL flask containing 0.0425 g of Chirald®. The salt-free solution was then cooled to 0 °C and reacted with 0.30 mL of benzaldehyde (3.0 mmol) for 8 h. The reaction mixture was finally quenched with saturated NH_4Cl, dried and volatiles removed *in vacuo*. After chromatography on silica gel pure (R)-(+)-1-phenyl-1-pentanol (0.42 g, 2.6 mmol 85% yield) was obtained with 82.4% ee.

2.2.2.2   Preparation of Alkylzinc Reagent from Alkylborane Compounds

Despite the use of pyrophoric dimethylzinc or diethylzinc, transmetalation between an organoborane reagent and an alkylzinc reagent is a synthetically viable reaction, because the highly volatile alkylborane byproduct can be removed easily. The low reactivities of dialkylzinc, combined with the availability of various organoborane compounds through various methods including hydroboration, makes this synthetic approach quite attractive on a laboratory scale. The following example illustrates a sequence involving hydroboration, boron/zinc transmetalation, and conjugate addition. The organozinc reagents thus prepared have been applied to various reactions including coupling reaction and enantioselective carbonyl additions[74]. Functional group compatibility of the process with ester, imide, cyano, nitro, borinate, tert-butyl dimethyl siloxy and alkyl halide groups has been demonstrated[75]. Starting with dienol silyl ether, the hydroboration/transmetalation sequence affords a diorganozinc compound possessing an enol silyl ether functionality[76].

Chemoselective Hydroboration, Transmetalation and Conjugate Addition[75]

Ethyl 3-butenylacrylate (1.85 g, 12.0 mmol) was cooled to 0 °C, and Et_2BH (12.0 mmol, 1 equiv; prepared from $BH_3 \cdot Me_2S$ (3.80 g, 50 mmol), Et_3B (9.80 g, 100 mmol), and Et_2O (14.8 g)) was slowly added via syringe. After 3 h at 25 °C, the solvents were removed under vacuum (0.1 mmHg, 0 °C, 0.5 h) affording the expected diethyl(alkyl)borane (2.30 g, 86% yield; 95% purity by ^{1}H and ^{13}C NMR analysis). The organoborane (1.01 g, 4.5 mmol) was transferred to a 50 mL Schlenk flask and cooled to 0 °C, and Et$_2$Zn (9.0 mmol, 0.92 mL, 2 equiv) was added. After 0.5 h at 0 °C, the excess Et$_2$Zn and formed Et$_3$B were pumped off (0.1 mmHg, 0 °C, 3 h). The resulting oil was diluted with THF (3 mL) and cooled to −80 °C, and a THF solution (4.5 mL) of CuCN•2LiCl (prepared from CuCN (0.40 g, 4.5 mmol) and LiCl (0.83 g, 9.0 mmol) was added. The reaction mixture was warmed to 0 °C and immediately cooled back to −80 °C. Diethylbenzylidene malonate (1.00 g, 4.05 mmol, 0.9 equiv) was added, and the reaction mixture was allowed to warm to 25 °C. After 2 h, the conversion was complete, as shown by GC analysis of the reaction aliquot. The reaction mixture was quenched with a saturated aqueous NH$_4$Cl solution (50 mL). A crude product which was purified by chromatography and desired compound (1.36 g, 82%; overall yield 71% based on the starting alkene) was obtained.

Transmetalation between diethylzinc and a borane reagent bearing a secondary alkyl group is a very inefficient reaction (excess diethylzinc and a reaction time of a few days). This problem can be resolved by the use of diisopropylzinc. This study also showed that the overall transformation takes place with retention of stereochemistry or the metal–carbon bond. Experimental details are given in Chapter 6[77]:

44% yield
96% trans

2.2.3 Halogen–Zinc Exchange

The halogen–metal exchange method is the second general route to dialkylzinc compounds. In spite of extensive use of this method for the synthesis of organolithium compounds, it has been underutilized until recently, since the preparation of Furukawa reagent by the reaction of Et$_2$Zn and CH$_2$I$_2$ has been preferred. It is only lately that the utility of the method was rediscovered as a route to functionalized organozinc reagents bearing ester, cyano, borinate, and alkyl chloride functionalities. The absence of metal halide byproducts allows the use of the *in situ* generated organozinc reagent for catalytic enantioselective carbonyl addition reactions:

$$FG-RI + Et_2Zn \xrightarrow[\substack{45-55°C \\ (3-5\ equiv)\ 1-12\ h}]{neat} FG-RZnEt + EtI \xrightarrow[\substack{40-50°C,\ 2\ h \\ (-Et_2Zn, -EtI)}]{0.1\ mmHg} (FG-R)_2Zn$$

Preparation of Dialkyzinc by Iodine–Zinc Exchange and its Cu-Mediated Michael Addition[78]

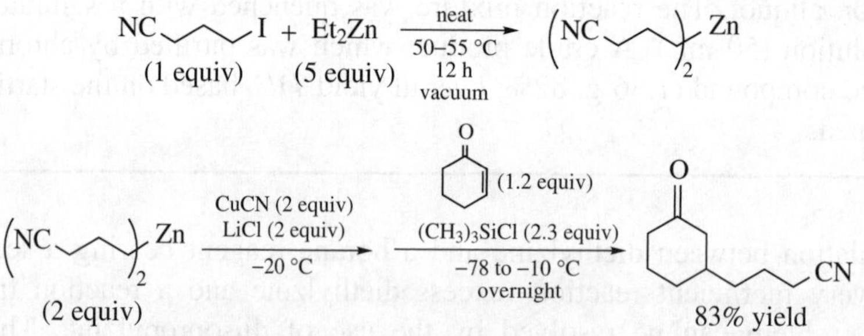

A Schlenk tube equipped with a septum cap and an argon outlet was charged with 4-iodobutyronitrile (1.20 g, 6 mmol) and diethylzinc (3.0 mL, 30 mmol). The reaction mixture was warmed to 50–55 °C and was stirred for 12 h at this temperature. GLC analysis of a hydrolyzed reaction aliquot indicates the completion of the reaction. The ethyl iodide formed, and excess diethylzinc was removed in vacuum (0.1 mmHg, 50 °C, 2 h). The resulting oil of bis(3-cyanopropyl)zinc was dissolved in dry THF (3 mL) and added to a THF solution (6 mL) of CuCN (0.270 g, 3.0 mmol) and LiCl (0.025 g, 5.9 mmol, dried for 2 h at 150 °C under 0.1 mmHg) at −20 °C. The resulting light-green solution was cooled to −78 °C, and Me_3SiCl (0.8 g, 7 mmol) and 2-cyclohexenone (0.335 g, 35 mmol) were added successively. The reaction mixture was slowly allowed to warm to −10 °C overnight and was worked up in the usual way to afford, after purification by flash chromatography (30% AcOEt in hexane), the ketone (0.483 g, 2.9 mmol, 83% yield).

The recent findings that CuI[79], MnBr_2/CuCl[80] and Ni(acac)_2[281] catalyze this halogen–metal exchange reaction greatly increased the synthetic utility of the transformation by allowing the use of a nearly stoichiometric amount of Et_2Zn, and the conversion of alkyl bromides and chlorides into the corresponding dialkylzinc compounds. One example of each is given below.

Preparation of Di-(5-acetoxypentyl)zinc in the Presence of a Catalytic Amount of CuI and its Enantioselective Addition to (Z)-2-Bromo-2-hexenal

$$AcO\diagdown\diagup\diagdown\diagup Br + Et_2Zn \xrightarrow[\substack{50\,°C \\ 8\,h}]{\substack{CuI\,(cat.) \\ neat}} \left(AcO\diagdown\diagup\diagdown\diagup\right)_2 Zn$$

(1.25 equiv)

$$\left(AcO\diagdown\diagup\diagdown\diagup\right)_2 Zn + \underset{\substack{Br \\ (1\,equiv)}}{\diagup\diagdown\diagup\diagdown\!\!\overset{O}{\underset{}{\diagdown}}H} \xrightarrow[\substack{toluene,\,-20\,°C,\,10\,h}]{\substack{chiral\,ligand^*\,(0.075\,equiv) \\ Ti(O^iPr)_4\,(2\,equiv)}} \xrightarrow{H^+} \underset{\substack{Br \\ 95\%\,yield,\,94\%\,ee}}{\overset{OH}{\diagup\diagdown\diagup\diagdown\diagup\diagdown\!OAc}}$$

(2 equiv)

A Schlenk flask equipped with an argon inlet and a septum cap was charged with CuI (0.002 g, about 0.01 mmol), 5-iodopentyl acetate (4.1 g, 16 mmol) and Et_2Zn (2.0 mL, 20 mmol). The reaction mixture was warmed to 50 °C and stirred for 8 h at this temperature. The Schlenk flask was connected to vacuum source (0.1 mmHg) and the excess Et_2Zn and formed EtI were collected in a trap cooled with liquid N_2. This operation required about 2 h at 50 °C. The resulting dialkylzinc was diluted in toluene (8 mL) and was ready to use.

A 100 mL three-neck flask equipped with an argon inlet, a thermometer, and a septum cap was charged with dry toluene (3 mL), $Ti(O^iPr)_4$ (2.4 mL, 8.0 mmol) and (1R,2R)-1,2-bis(trifluorosulfamido)cyclohexane (0.120 g, 0.3 mmol). The reaction mixture was heated to 40–45 °C for 0.5 h and cooled to −60 °C. Di(5-acetoxypentyl) zinc (8 mmol prepared as described above) in toluene and then (Z)-2-bromo-2-hexenal (0.710 mg, 4.0 mmol) were slowly added and stirred for 10 h at −20 °C. After the usual work-up, the crude resulting oil was purified by flash chromatography affording the pure allylic alcohol (1.15 g, 91% yield, 94% ee) determined by the ^{1}H NMR analysis of its derivative with (S)-(−)-O-acetylmandelic acid.

Preparation of Dialkylzinc from Alkylbromide in the Presence of MnCl$_2$ and CuCl and its Pd-catalyzed Coupling Reaction with 4-Chloroiodobenzene

$$C_2H_5O_2C\diagdown\diagup\diagdown Br + Et_2Zn \xrightarrow[\substack{DMPU \\ 25\,°C,\,4\,h}]{\substack{MnBr_2\,(0.05\,equiv) \\ CuCl\,(0.03\,equiv)}} \left(C_2H_5O_2C\diagdown\diagup\diagdown\right)_2 Zn$$

(1 equiv)　(0.9 equiv)

$$\left(C_2H_5O_2C\diagdown\diagup\diagdown\right)_2 Zn + I\!-\!\!\bigcirc\!\!-Cl \xrightarrow[\substack{THF/DMPU \\ -30\,°C\,to\,25\,°C\,(0.5\,h) \\ to\,65\,°C\,overnight}]{\substack{Cl_2Pd(dppf) \\ (0.04\,equiv)}} C_2H_5O_2C\diagdown\diagup\diagdown\!\!\bigcirc\!\!-Cl$$

(<1.1 equiv)　(1 equiv)　73% yield

A three-necked flask equipped with a thermometer, a gas inlet, and a magnetic stirring bar was charged under argon with MnBr$_2$ (0.064 g, 0.30 mmol) in DMPU

(5 mL). CuCl (0.017 g, 0.17 mmol), ethyl-4-bromobutyrate (1.17 g, 6.0 mmol) and Et$_2$Zn (0.54 mL. 5.4 mmol) were successively added. The reaction mixture turned dark red and was stirred for 4 h at 25 °C. GC analysis of a reaction aliquot showed a complete conversion to the zinc reagent. After cooling to −30 °C, a solution of PdCl$_2$(dppf) (0.185 g, 0.2 mmol) and 4-chloroiodobenzene (1.19 g, 5.0 mmol) in THF (5 mL) was slowly added. The reaction mixture was warmed up to 25 °C for 0.5 h and was then stirred at 65 °C overnight and quenched with an aqueous 2 M HCl solution (20 mL). After work-up and evaporation of the solvent, the crude residue obtained was purified by flash chromatography, yielding 0.82 g (73% yield) of the coupling product.

Preparation of Dialkylzinc from Bromo- or Chloroalkane by Halogen–Zinc Exchange in the Presence of Ni(acac)$_2$

$$RCH_2X \xrightarrow[\text{neat, 50–60 °C}]{\substack{(C_2H_5)_2Zn \\ Ni(acac)_2 \ (cat.)}} (RCH_2)_2Zn + C_2H_5X\uparrow$$
$$X = Br, Cl$$

A 50 mL two-necked flask equipped with an argon inlet, a magnetic stirring bar, and a septum cap was charged with Ni(acac)$_2$ (0.051 g, 0.20 mmol, 5 mol %) followed by 1-bromoalkane (4.0 mmol). The mixture was cooled to 0 °C, and Et$_2$Zn (0.82 mL, 8.0 mmol, 2 equiv) was added dropwise. After completion of the addition, the cooling bath was removed and the reaction mixture was stirred in a preheated oil bath at 55 °C for 2 h. **Caution:** large amounts of gas are produced by the reaction, especially for large-scale preparations! The excess of Et$_2$Zn was distilled off *in vacuo*. The black residue was dissolved in THF (2 mL) and cooled to −60 °C, and a solution of CuCN•2LiCl (CuCN 0.322 mg, 3.6 mmol; LiCl 0.305 g, 7.2 mmol) in THF was added. The electrophile (2.80 mmol, 0.7 equiv) was added, and the reaction mixture was warmed to the temperature described. It was diluted with ether and was quenched by addition of a saturated aqueous NH$_4$Cl solution. The crude product was purified by flash chromatography to afford the pure product.

2.2.4   C−C Bond Activation

ZnCl$_2$ cleaves the electron-rich C−C bond in α-siloxy cyclopropane to generate a zinc homoenolate. The reaction must have generated an RZnCl reagent which, however, affords the R$_2$Zn reagent through Schlenck equilibrium.

Cyclopropane Cleavage Route to Zinc Homoenolate

In a L three-necked flask, two necks of which were covered with rubber septa and the other connected to a nitrogen-vacuum manifold, was placed 17.2 g of zinc chloride. The flask was evacuated to approximately 2 mmHg and heated with a burner with swirling until practically all of the salt melted. The flask was cooled and filled with nitrogen. The dried salt weighed 16.4–17 g (about 0.12 mol). An efficient magnetic stirring bar and a Dimroth condenser in place of a rubber septum were set in position, and the flask was again flushed with nitrogen. Ether (300 mL) was introduced via the septum, and stirring was initiated and maintained throughout the reaction. The mixture was refluxed gently for 1 h to aid dissolution of the solid state. The flask was cooled, and 1-trimethylsilyloxy-1-ethoxycyclopropane was introduced with the aid of a hypodermic syringe over 5 min. The cloudy mixture was stirred at 25 °C for 1 h; the more dense lower layer should have mostly disappeared at this point. The mixture was refluxed for 30 min to complete homoenolate formation.

2.2.5 Hydrozincation

Nickel-catalyzed hydrozincation of terminal alkene provides a simple and convenient route to dialkylzinc reagents. Thus, warming a mixture of Et_2Zn and the alkene in the presence of a catalytic amount of Ni(II) generates a dialkylzinc compound[82]. The reaction is considered to involve a nickel hydride species generated by a transmetalation–β-elimination sequence, which undergoes hydronickelation and transmetalation from nickel back to zinc. The reaction is limited to monosubstituted alkenes and hence to the preparation of primary alkylzinc species. Because of the reversible nature of the hydronickelation reaction, the reaction of a simple alkene stops at 40–60% conversion. The reaction can be driven to completion with allylic alcohols, homoallylic alcohols and allylic amines, wherein the products stabilize by chelate formation. The dialkylzinc compounds prepared in this manner undergo conjugate addition and titanium-mediated enantioselective carbonyl addition[83].

Preparation of Dialkylzinc Reagent by Hydronickelation–Transmetalation and its Use for Cu-mediated Coupling Reactions

To a stirred suspension of 4-hydroxyhex-5-enyl pivalate (0.50 g, 5.0 mmol), 1,5-cyclooctadiene (0.05 mL, 10 mol %) and $Ni(acac)_2$ (0.060 g, 0.25 mmol, 5 mol %) was added at $-78\,°C$ neat Et_2Zn (1.1 mL, 11 mmol). The reaction mixture was slowly warmed to $40\,°C$ leading to a black suspension. The reaction was completed by stirring at $40\,°C$ for 4 h. Excess Et_2Zn was pumped off at high vacuum (0.1 mmHg at $25\,°C$). The residual solid was dissolved in THF (5 mL) and treated with Me_3SiCl (1.2 mL, 10 mmol) at $-78\,°C$. The mixture was allowed to warm to $25\,°C$ overnight. Then a THF solution (10 mL) of CuCN (0.89 g, 10 mmol) and LiCl (0.84 g, 20 mmol) was added at $-60\,°C$. After stirring at $0\,°C$ for 30 min, the reaction mixture was cooled again to $-78\,°C$ and ethyl (2-bromoethyl)acrylate (1.3 g, 6.5 mmol)was added. The allylation reaction was completed after stirring at $0\,°C$ for 30 min. The reaction mixture was then poured into ice-cooled saturated aqueous NH_4Cl (100 mL) and was extracted with ether (3 × 100 mL). The crude product was purified by flash chromatography affording the product (0.99 g, 3.35 mmol, 67% yield).

2.3 Aryl-, Alkenyl-, and Alkynylzinc Reagents

Whereas the oxidative zincation route provides a versatile preparative entry to sp^3 alkylzinc compounds, it is less efficient for the halides possessing an sp^2 carbon–halogen bond which is intrinsically more stable than the sp^3 carbon–halogen bond. Activated commercial zinc converts only the iodides into the corresponding zinc compounds in good yield. There are recent reports on useful alternative routes involving rearrangement of vinylzincate[84] and transmetalation from aryl iodides[85].

2.3.1 Oxidative Zincation

Whereas the alkenyl and arylzinc halides (RZnX) can be prepared by the oxidative zincation of the corresponding iodides and zinc metal activated with $Me_3SiCl/$ 1,2-dibromoethane, these halides are much less reactive than alkyl halides. Here, Rieke zinc reveals it's advantages over commercial zinc powder. The preparation of diorganozinc species (R_2Zn) is even more difficult for the alkenylzinc and arylzinc classes, and a special method such as ultrasound irradiation of a mixture of lithium, metallic lithium and $ZnCl_2$ is necessary:

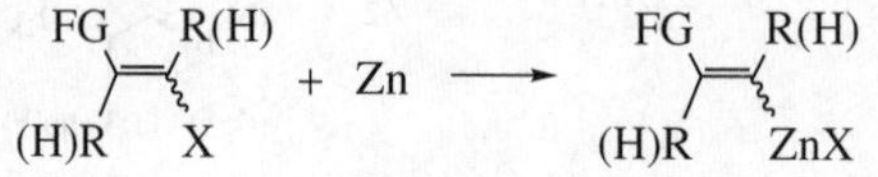

X = Cl, Br, I
Zn = activated or reactive zinc
solvent: THF, DMF
FG: H, C_6H_{13}, C_6H_5, RCOO, RCO, $2\text{-}(C_2H_5OCO)C_6H_4$, $p\text{-}C_6H_4(CH_3)SO_2$, $C_6H_5SO_2NH$

Preparation of Rieke Zn and Generation of Organozinc Halides, and their Copper-mediated Coupling Reaction with Acid Chloride[65]

$$ZnCl_2 \xrightarrow[\text{THF or DME}]{\substack{\text{Li (2 equiv)} \\ \text{Naphthalene (2 equiv)}}} Zn^*$$

(1 equiv)

$$C_2H_5O_2C-\langle\rangle-I \xrightarrow[\text{THF}]{\substack{Zn^* \\ (2\ \text{equiv})}} C_2H_5O_2C-\langle\rangle-ZnI$$

(1 equiv)   25 °C, 3 h, and
then stand for 3 h
(for excess zinc to settle)

(b) C$_4$H$_9$COCl (1 equiv) | (a) CuCN· 2LiBr

$$C_2H_5O_2C-\langle\rangle-COC_4H_9$$

(83% yield)

Two 50 mL flasks, A and B, were equipped with rubber septa, condensers topped with argon inlets, and Teflon-coated magnetic stir bars. Flask A was charged with freshly cut lithium (0.213 g, 30.6 mmol) and naphthalene (3.987 g, 31.2 mmol). Flask B was charged with anhydrous ZnCl$_2$ (2.09 g, 15.4 mmol). Both of these operations were performed in the argon-atmosphere dry box. The flasks were transferred to the manifold system and the argon inlet fitted. Freshly distilled THF (15 mL) was added to both flasks by syringe, and the mixtures were stirred at 25 °C. The solution in flask A changed from colorless to dark green almost immediately. The lithium was consumed in about 2 h, and the ZnCl$_2$ solution was transferred dropwise to flask A by a cannula over 15 min. (A slow addition, about 3 s per drop with no excess of zinc halide, results in an extremely fine black slurry of active zinc. This slurry takes several hours to settle and can easily be transferred by a cannula. With faster addition, about 1 s or less per drop, the active zinc formed is sponge shaped. The solvent can be changed very easily by a cannula if further reaction needs a different solvent. The active zinc, prepared by both methods described above, has similar reactivity.) The active zinc was typically used at this point, but it can be washed several times with fresh solvent if naphthalene presents a problem with product isolation or if the solvent needs to be changed.

Ethyl 4-iodobenzoate (1.934 g, 7.0 mmol) was added neat, via a syringe, to the active zinc (15.4 mmol) at 25 °C. The reaction mixture was stirred for 3 h at 25 °C. The solution was left to stand for about 3 h to allow the excess zinc to settle from the dark brown organozinc iodide solution. The top solution was then transferred carefully using a cannula to another two-necked flask under argon atmosphere and cooled to −20 °C. (In the cross-coupling reaction with acid chlorides, excess zinc must be removed form the organozinc solution as it will react with the acid chlorides, resulting in homocoupling of the acid chloride.) A solution prepared by mixing

CuCN (0.651 g, 7.27 mmol) and anhydrous LiBr (1.273 g, 14.7 mmol) in THF (10 mL) was added at −20 °C. The reaction mixture was gradually warmed to 0 °C and stirred at 0 °C for about 15 min. The solution was then cooled to −25 °C, and valeryl chloride (0.851 g, 7.02 mmol) was added neat via a syringe. The mixture was then worked up by pouring into a saturated NH_4Cl solution (20 mL) and extracting with diethyl ether (3 × 20 mL). The combined organic layers were dried over anhydrous $CaCl_2$. The resultant crude product was chromatographed on silica gel using gradient elution to give ethyl 4-(1-oxopentyl)benzoate (1.360 g, 5.31 mmol) as a white crystalline solid in 83% isolated yield.

Preparation of 2-(2-tert-Butoxycarbonyl-2-propenyl)phenyl Cyclohexylketone

Zn: Zinc dust activated by dibromoethane was used.

A 50 mL three-necked flask equipped with a dropping funnel, a thermometer and argon inlet was charged with zinc dust (Aldrich 325 mesh; 0.820 g, 12.5 mmol) in 2 mL of dry DMF. After zinc activation with dibromoethane[26] (about 0.200 g) the zinc suspension was heated to 45–50 °C (internal temperature) with an oil bath and a solution of 2-iodophenyl cyclohexyl ketone (1.50 g, 4.77 mmol) in 4 mL of DMF containing 0.050 g of undecene (internal standard) was slowly added over 40 min. After 4 h at 45–50 °C, GLC analysis of a hydrolyzed aliquot showed that less than 5% of the iodide remained, while an iodonolysis of a reaction aliquot gave 64% reformation of the iodide. The reaction mixture was allowed to settle, and the supernatant solution of the organozinc iodide was cannulated into a solution of dry LiCl (dried for 1 h at 120 °C at 0.1 mmHg; 0.424 g, 10.0 mmol) and CuCN (0.45 g, 5.0 mmol) in 6 mL of THF at −70 °C. A solution of *tert*-butyl α-(bromomethyl)acrylate (0.35 g, 1.73 mmol) was added and the reaction was allowed to warm to 0 °C and worked up as usual after 1 h at this temperature. Flash-chromatographical purification of the residue provided 0.467 g (72%) of the coupling product as an analytically pure compound.

There is a recent report that Zn*/Ag–graphite (described in section 3.2.2.1) effectively converts several aryl and heteroaromatic iodides into the corresponding arylzinc iodides:

F_5C_6-ZnI

[structure: N-methylpyridine with ZnI]

[structure: N-methylimidazole with NC and NC groups, ZnI]

[structure: benzothiazole with ZnI]

C_6H_5CO—[thiophene]—ZnI

The merit of this preparation resides in it's simple operation and short reaction time (>80% yield, THF, 25 °C, several hours), while the demerit is the use of potassium metal[86].

Zn/Ag–graphite-mediated Oxidative Zincation*

C_6H_5CO—[thiophene]—I (1.3 equiv) $\xrightarrow[\substack{\text{THF, 25 °C} \\ \text{15 min}}]{\text{Zn*/Ag–graphite}}$ $\xrightarrow[\substack{-10 °C, 0.5 \text{ h}}]{\substack{\text{CuCN·2LiCl} \\ (1.3 \text{ equiv})}}$ $\xrightarrow[\substack{-60 \text{ to } -10 °C, \\ 14 \text{ h}}]{\substack{C_6H_5COCl \\ (1.0 \text{ equiv})}}$ C_6H_5CO—[thiophene]—COC_6H_5

76% yield

A 100 mL three-necked flask equipped with an argon inlet, a glass stopper and a septum cap was charged with graphite (1.65 g, 137 mmol) and heated to 160 °C. Potassium (0.67 g, 17.1 mmol) was added in small pieces under a steady stream of argon with vigorous stirring, resulting in the formation of bronze-colored C_8K within 15 min as a fine powder. A second three-necked flask was charged with zinc chloride (1.17 g, 8.6 mmol) which was dried at 140 °C for 2 h under high vacuum. After cooling to 25 °C, THF (10 mL) was added affording a solution to which silver acetate (0.140 g, 0.85 mmol) was added. The resulting heterogeneous slurry was transferred, at 25 °C, with a syringe to the previously prepared C_8K while vigorously stirring (alternatively, a mixture of $ZnCl_2$ and AgOAc can be added as a solid to a suspension of C_8K in THF). A slightly exothermic reaction occurs. After stirring for 1 h 5-benzoyl-2-iodothiophene (0.900 g, 2.9 mmol) was added as a solid. GC analysis of an iodolyzed and hydrolyzed reaction aliquot indicated that a complete insertion had occurred after 15 min reaction time. The excess zinc/silver–graphite was allowed to settle for 1–2 h and the supernatant solution of the zinc reagent was transferred to a THF (2 mL) solution of CuCN (0.26 g, 3.0 mmol) and LiCl (0.24 g, 5.7 mmol) at −60 °C. Benzoyl chloride (0.30 g, 2.14 mmol, 0.75 equiv for the starting iodothiophene) was added and the reaction mixture was warmed to −10 °C and stirred for about 14 h at −10 °C and worked up in the usual way. The crude light brown residue was purified by flash chromatography affording 0.470 g (1.64 mmol, 76% yield) of the analytically pure 2,5-dibenzoylthiophene.

The preparation of alkenylzinc reagents by oxidative zincation is equally difficult. For instance, the oxidative zincation of 1-iodooctene with zinc powder needs a reaction time of 14 h at 70 °C in DMF, with loss of alkene stereochemistry[87]. An exception is the reaction of haloalkene connected to an electron-withdrawing group, which facilitates one-electron transfer from zinc metal to the alkene. Even vinyl chlorides, which are generally unreactive substrates, can be converted to the corresponding vinylzinc chloride. The vinylzinc reagents as shown in the following preparation serve as vinylogous acyl anion synthons[88].

Preparation of Functionalized Alkenylzinc Halide

85% yield

Other examples:

45 °C, 4 h,
80% yield

(E)/(Z) 1:9
25 °C, 2–3 h,
90% yield

(E)/(Z) 1:1
−15 to 10 °C, 2 h,
50% yield

(E)/(Z) 9:1
25 °C, 4 h,
50% yield

A dry, three-necked, 50 mL flask equipped with an argon inlet, magnetic stirring bar and a low temperature thermometer was charged with zinc dust (1.95 g, 30 mmol, Aldrich, 325 mesh) and flushed with argon. 1,2-Dibromoethane (0.200 g, 1.1 mmol) in THF (3 mL) was added. The zinc suspension was heated with a heat gun to emulation, allowed to cool and was heated again. This process was repeated three times. Then trimethylchlorosilane (0.15 mL, 1.2 mmol) was added, and after 10 min of stirring, a solution of 3-iodo-2-cyclohexen-1-one (2.22 g, 10 mmol) in THF (3 mL) was added dropwise over 15–20 min. During the addition, the temperature rose to 55 °C. The reaction mixture was stirred for 1 h at 25 °C and the progress of the reaction was monitored by GC analysis of hydrolyzed reaction aliquots. After completion of the reaction, THF (8 mL) was added and the zinc was allowed to settle for 1–2 h at 25 °C. GC analysis of a reaction aliquot indicates complete conversion of the alkenyl iodide to the organozinc as well as the formation of less than 8% of dimer. The yield of 3-oxo-1-cyclohexen-1-ylzinc iodide was estimated to be 85%.

2.3.2   Transmetalation

Reagents of the type Ar_2Zn are difficult to prepare other than by the reaction of aryllithium and ZnX_2 or $ZnX_2/TMEDA$, which is easy to handle. RZnX and R_2Zn reagents can be prepared simply by adjusting the stoichiometry of the reagent (i.e. 1 and 2 equiv of RLi to ZnX_2, respectively). This route was employed in the palladium-catalyzed coupling between an arylzinc reagent and an sp^2 halide where the presence of metal halide byproduct (e.g. LiBr) does not affect the reaction course[8].

Preparation of the Complex $ZnCl_2 \cdot TMEDA$[89]

$$ZnCl_2 + Me_2NCH_2CH_2NMe_2 \longrightarrow ZnCl_2 \cdot Me_2NCH_2CH_2NMe_2$$

The complex was prepared by mixing 19 mL of a saturated $ZnCl_2$/THF solution and 5 mL of TMEDA and allowing to stand for several hours at 25 °C. The crude crystals separated were collected and recrystallized from THF.

Enantioselective addition of an alkenylzinc reagent illustrates the transmetalation method. The LiBr byproduct does not affect the enantioselectivity of the reaction induced by a stoichiometric amount of the lithium salt of the optically active amino alcohol.

Preparation of Alkenylzinc Bromide from Alkenyl Lithium Reagent and Asymmetric Alkenylation of Aldehyde[90]

Lithium, containing 2% Na (0.490 g, 70 mmol) was obtained from the commercial 15% suspension in hexane (5.6 mL) by evaporation of the solvents under vacuum at 25 °C. To the lithium powder under argon was added dry ether (20 mL) and the suspension was cooled to −35 °C. With stirring a solution of pure (>98%) (Z)-1-bromo-1-propene (1.7 g, 14 mmol) in ether (2 mL) was added dropwise. The mixture was stirred at −35 °C for 2 h and filtered under argon. The 0.6 M (Z)-1-propenyllithium solution (17 mL, 10 mmol) was cooled to −35 °C and treated dropwise with a 0.6 M solution of commercial zinc bromide in ether (25 mL, 15 mmol). The reaction mixture was stirred for an additional 1 h at 0 °C and then a solution of lithium (1S,2R)-N-methylephedrate, prepared by the addition of butyllithium 1.6 M in hexane (6.4 mL, 10 mmol) to a solution of (+)-N-methylephedrine 81.8 g, 10 mmol) in toluene (60 mL) at 0 °C, was cannulated. The clear colorless solution was stirred for 1 h at 0 °C and cyclohexanecarboxaldehyde (1.20 mL, 10 mmol) was added neat. After stirring for 1 h at 0 °C the reaction was quenched by the addition of saturated ammonium chloride solution, the organic phase was separated and the aqueous phase was extracted with ether. The organic extracts were washed with a second portion of the ammonium chloride solution, dried (Na$_2$SO$_4$) and concentrated. Flash chromatography and bulb-to-bulb distillation (130 °C at 0.01 mmHg) gave 1.20 g (78% yield) of (R)-(Z)-a-(1-pro-penyl)cyclohexanemethanol with 93% ee.

Catalytic enantioselective alkenylation of aldehydes necessitates the use of a salt-free alkenylzinc reagents. Oppolzer *et al.* used boron/zinc exchange between equimolar amounts of alkenyldicyclohexylborane and dimethyl or diethylzinc, and the resulting alkenylzinc reagent undergoes a catalytic aminoalcohol-catalyzed carbonyl addition. The utility of the transmetalation route is illustrated by (−)-muscone synthesis[91]. At temperatures below 0 °C, an alkenylalkylzinc compound exists as it is and not in equilibrium with a mixture of a dialkenylzinc and a dialkylzinc compound:

Preparation of Alkenylalkylzinc and Catalytic Enantioselective Addition to Aldehyde[92]

Cyclohexene (205 µL, 2.0 mmol) was added under Ar at 0 °C to a stirred solution of borane–(dimethylsulfide) complex (100 µL, 1.1 mmol) in hexane (1 mL). After 3 h at 0 °C, 1-octyne (150 µL, 1.0 mmol) was added, and the mixture was stirred at 25 °C for 1 h. Then, the solution was cooled to −78 °C. Addition (over 10 min) of a 1.0 M solution of (CH$_3$CH$_2$)$_2$Zn in hexane (1.05 mL, 1.05 mmol) followed by addition of DAIB (0.002 g, 0.01 mmol), immersion of the mixture into an ice bath (0 °C), addition (over 20 min) of propionaldehyde (72 µL, 1.0 mmol) in hexane (4 mL), stirring the mixture at 0 °C for 1 h, addition of saturated aqueous NH$_4$Cl solution, extraction (ether), washing, drying, and evaporation of the extracts, and chromatography of the residue (silica gel, hexane/ether) yielded allyl alcohol (0.155 g, 91% yield, 84% ee).

Hydrozirconation of a terminal alkyne followed by transmetalation from the resulting alkenylzirconium to dimethylzinc provides a useful route to alkenylzinc reagents.

Preparation of Alkylalkenylzinc Reagent by Hydrozirconation of a Terminal Alkyne[93]

94% yield

A solution of 1-hexyne (0.053 g, 0.64 mmol) in 2 mL of CH_2Cl_2 was kept under an atmosphere of N_2 and treated at 25 °C with zirconocene hydrochloride (0.160 g, 0.64 mmol). The mixture was stirred at 25 °C until a homogeneous solution formed and cooled to −65 °C. Dimethylzinc (0.33 mL, 2.0 M solution in toluene) was added over 5 min. The reaction mixture was immersed in an ice bath and a solution of *trans*-cinnamaldehyde (0.070 g, 0.53 mmol) in 2 mL of CH_2Cl_2 was added over 10 min. The mixture was extracted with ether twice. Evaporation of the solvents and chromatography of the residue on silica gel yielded allyl alcohol (0.108 g, 94% yield).

2.3.3 Halogen–Zinc Exchange

An arylzincate species is a good nucleophile toward carbonyl compounds, as opposed to arylzinc halides, which undergo carbonyl addition only in the presence of a Lewis acid. A conventional route to arylzincates involves addition of two equivalents of alkyllithium to an arylzinc halide. A convenient and chemoselective metalation route was recently reported, in which treatment of an aryl iodide with lithium trimethylzincate generates an aryldimethylzincate, which adds to a carbonyl compound with exclusive transfer of the aryl group. In the example below, the conventional halogen–lithium exchange method affords a mixture of the 2- and 3-lithioindols, whereas the present method exclusively affords the 3-zincated indol. Upon reaction of aryl iodide and bromide, Me_4ZnLi_2 generates the corresponding arylzinc reagents[94].

Generation of Lithium Indol-3-yl(dimethyl)zincate and its Reaction with Benzaldehyde[94]

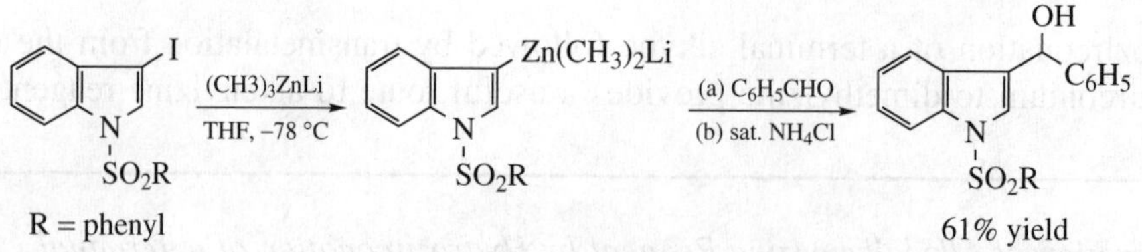

R = phenyl

61% yield

CH$_3$Li (1.03 M solution in Et$_2$O; 3.21 mL, 3.3 mmol) was added at 0 °C to a solution of ZnCl$_2$–TMEDA (0.253 g, 1.0 mmol) in THF (5 mL) and the mixture was stirred at 0 °C for 30 min. It was then cooled to −78 °C and treated with 3-iodo-1-phenylsulfonylindole (0.383 g, 1.0 mmol) in THF (2 mL). The resulting mixture was then gradually warmed to 25 °C over 12 h after which it was treated with aqueous NH$_4$Cl (2 mL), concentrated by removal of THF under reduced pressure, diluted with water (20 mL) and extracted with CHCl$_3$ (3 × 30 mL). The combined organic extracts were dried over MgSO$_4$ and evaporated. The crude material was purified by silica gel column chromatography to give a viscous oil (0.221 g, 61% yield).

An interesting alkylative preparation of alkenylzinc reagents involves the reaction of 1,1-dibromoalkene and a lithium trialkylzincate at −85 °C. A bromine–zinc exchange reaction gives first 1-bromoalkenylzincate, which then at 0 °C undergoes alkyl group migration and elimination of LiBr[95].

Preparation of Alkenylzincate by the Reaction of 1,1-Dihaloalkenes with Trialkylzincate

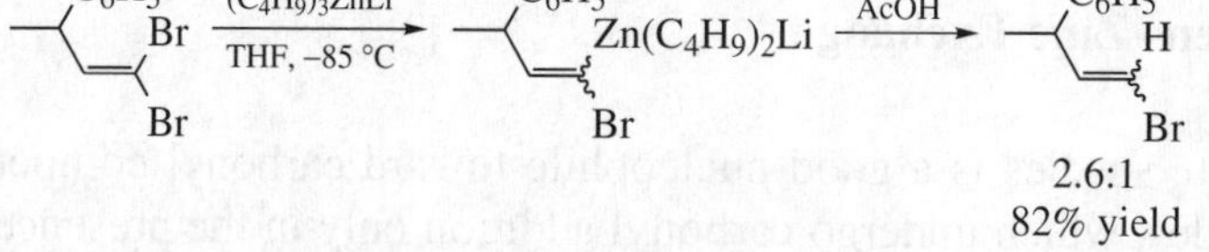

2.6:1
82% yield

To a solution of ZnCl$_2$ (0.0827 g, 0.61 mmol) in THF (2 mL) was added BuLi (1.62 M in hexane, 1.12 mL, 1.8 mmol) at 0 °C. After being stirred for 15 min, the mixture was cooled to −85 °C. To the cooled mixture was slowly added a THF (1.2 mL) solution of 1,1-dibromo-3-phenyl-1-butene (0.142 g, 0.49 mmol), and the mixture was stirred for 3 h at the same temperature. The reaction was quenched with 10% AcOH in THF. The mixture was poured into 1 M HCl(aq) and was extracted twice with ether. The combined organic layers were washed with aqueous NaHCO$_3$

and dried. Analysis of the mixture by GC using an internal standard method showed the formation of (E)- and (Z)-1-bromo-3-phenyl-1-butene (2.6 : 1) in 82% yield.

Preparation of Alkylalkenylzinc by the Reaction of 1,1-Dihaloalkenes with Trialkylzincate at 0 °C

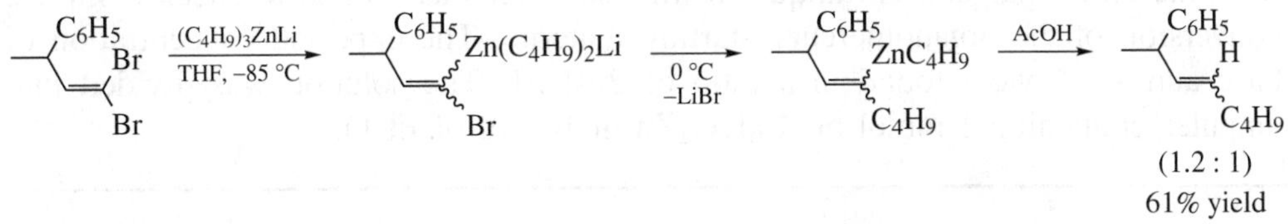

(1.2 : 1)
61% yield

To a solution of $ZnCl_2$ (0.243 g, 1.78 mmol) in THF (6 mL) at 0 °C was added BuLi (1.62 M in hexane, 3.16 mL, 5.12 mmol). After being stirred for 15 min, the mixture was cooled to −85 °C. To the cooled mixture, a THF (2 mL) solution of 1,1-dibromo-3-phenyl-1-butene (0.339 g, 1.17 mmol) was slowly added. After being stirred for 2 h at the same temperature, the mixture was allowed to warm to 0 °C over a 2.5 h period. The reaction was quenched with 10% AcOH in THF. The mixture was poured into aqueous $NaHCO_3$ and dried. Analysis of the mixture by GC using an internal standard method showed the formation of (E)- and (Z)-2-phenyl-3-octene (1 : 1.9) in 61% yield.

2.3.4 Other Methods

Carbozincation (Section 3.5) and hydrozincation of an alkyne takes place in a cis fashion, providing stereodefined routes to alkenylzinc compounds. In the presence of Cp_2TiCl_2, a mixture of alkali metal hydride and ZnI_2 reacts with diene and internal alkyne, presumably via ZnHX, to give allylzinc and alkenylzinc reagents, respectively. When the substituents on the alkyne are two alkyl groups, the regioselectivity of the addition is poor. Using combinations of aryl/alkyl and alkyl/silyl groups, regioselectivities of about 9 : 1 favoring attachment of the zirconium atom next to the aryl and silyl groups resulted[96]:

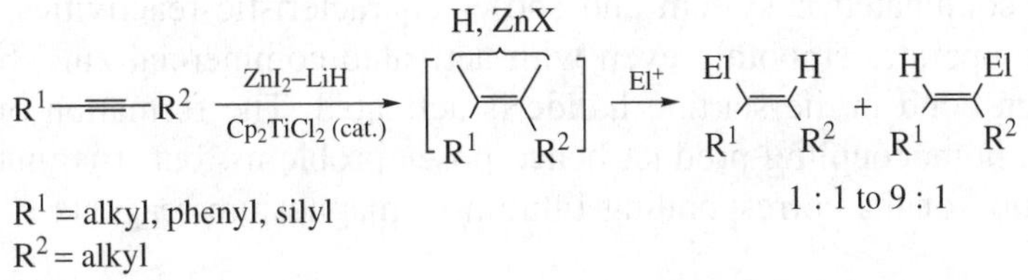

R^1 = alkyl, phenyl, silyl
R^2 = alkyl

1 : 1 to 9 : 1

Transmetalation between organomercury compounds and zinc metal generates salt-free R_2Zn reagents. The merit of this method resides in the generally poor availability of organomercury compounds, and their handling.

Preparation of Diphenylzinc by Oxidative–Reductive Transmetalation[97]

$$(C_6H_5)_2Hg + Zn \text{ (large excess)} \longrightarrow (C_6H_5)_2Zn + Zn\text{–Hg amalgam}$$

Diphenylmercury (7.1 g, 20 mmol) was stirred with zinc (13 g, 200 mmol) for 2 weeks in diethyl ether (200 mL). The clear, light-brown solution was decanted from the zinc amalgam. An aliquot of this solution was titrated to check complete conversion of the organomercury starting material. The expected concentration of base and Zn^{2+} were found in a ratio of 2.01 : 1. The solution was divided into ampules containing 1 mmol of $(C_6H_5)_2Zn$ in 10 mL of Et_2O.

With alkynylzinc compounds, the alkynyl proton is reasonably acidic and, in a polar solvent (DMF, DMSO, HMPA), is susceptible to deprotonation with diethylzinc[98], which is not conducive for the preparation of alkyl, aryl and alkenylzinc compounds. Deprotonation of phenylacetylene with diphenylzinc generates bis(phenylethynyl)zinc in about 80% yield as colorless crystals[99].

Preparation of Bis(phenylethynyl)zinc

$$C_6H_5\!\!-\!\!\equiv \xrightarrow{(C_6H_5)_2Zn} \left(C_6H_5\!\!-\!\!\equiv\right)_{\!2}Zn$$

A solution of pure diphenylzinc (2.2 g, 10.2 mmol) in 50 mL of ether is added to a solution of freshly distilled phenylacetylene (2.17 mL, 21.2 mmol) in 50 mL of ether. Colorless microcrystalline bis(phenylethynyl)zinc precipitated after a few hours and was collected by filtration (about 80% yield).

2.4   Allylic, Benzylic, Propargylic, and Allenic Zinc Reagents

The carbon–zinc bond in allylic, benzylic, and propargyllic zinc reagents is involved as a part of a conjugated π system and shows characteristic reactivities. The oxidative zincation route operates smoothly even with activated commercial zinc, because the sp^3 carbon–halogen bond in the starting halide is activated. The formation of allyl–allyl or benzyl–benzyl homocoupling product hence poses problems, but it is much less so than in the preparation of the corresponding lithium or magnesium reagents.

2.4.1   Oxidative Zincation

Allyllic zinc bromide can be prepared in nearly quantitative yield from the corresponding allylic bromide in THF with activated commercial zinc powder (1.1–1.2 equiv).

The unreacted zinc powder can be removed by filtration, or centrifuge and decantation. The allylic zinc reagent thus prepared in THF is colorless to pale green and stable for a few weeks at 0 °C in a tightly closed Schlenk flask. Dimerization and precipitation of metallic zinc result when the solution is kept at 25 °C. A solution in a solvent other than THF can be prepared by removal of THF at reduced pressure[100]. The concentration of reagent can be determined by acid–base titration as in Grignard reagent chemistry. Diallylzinc can be made by the transmetalation method.

Preparation of Allylzinc Bromide

$$\text{Br} \xrightarrow[\text{THF}]{\text{Zn}} \text{ZnBr}$$

A suspension of zinc dust (0.784 g, 12 mmol) in 3 mL of THF was added to 1,2-dibromoethane (30 µL, 0.35 mmol, 3 mol %). The reaction mixture was heated with a heat gun to reflux for a few minutes and cooled to 25 °C. Chlorotrimethylsilane (15 µL, 0.12 mmol. 1 mol %) was added and the reaction mixture was stirred at 25 °C for about 20 min. The zinc dust became spongy during the stirring at 25 °C. Allylbromide (0.87 mL, 10 mmol) in dry THF (7 mL) was added dropwise at 0 °C over 1 h. After the addition, the ice-bath was removed and the reaction mixture was stirred at 25 °C for 1 h. Filtration through Celite gave clear light-green (sometimes almost colorless) solution of allylzinc bromide in about 90–95% yield.

The Zn*/TiO$_2$ method described in the section on alkylzinc is also an effective oxidative zincation route to benzylzinc halides. The homocoupling byproduct forms in only 1% yield, which is 1/20 of the amount formed by the use of commercial zinc powder.

Preparation of Secondary Benzylic Zinc Bromide[68]

(83% yield)

A THF solution (10 mL) of 1-bromo-1-phenylpentane (2.37 g, 10.5 mmol) was slowly added (one drop every 5 s) at 0 °C to the suspension of Zn*/TiO$_2$ (prepared as described above). Formation of the benzylic zinc reagent was complete after 1 h, as indicated by GC analysis of hydrolyzed aliquots. The solution of zinc reagent was filtered under argon over Celite providing a clear solution of the benzylic reagent. This was added at −60 °C to a THF solution (10 mL) of CuCN (0.93 g, 10 mmol) and LiCl (0.90 g, 21 mmol). The reaction mixture was warmed to −20 °C, stirred for 5 min, and cooled back down to −78 °C. A THF solution (2 mL) of iodo-2-cyclohexen-1-one (1.63 g, 7.4 mmol) was added and the reaction mixture was warmed to −30 °C and stirred for 8 h at this temperature. The reaction mixture was worked up, as above, leading to a crude oil which was purified by flash chromatography giving the coupling product as a clear oil (1.48 g, 6.1 mmol, 83% yield).

2.4.2   Transmetalation

The preparation of diallylzinc reagents is best achieved by transmetalation. Preparations starting with an allylic Grignard reagent and with a borane reagent are known. It has been reported that the yield of the parent diallylzinc reagent is about 30% by the former method, whereas it is nearly quantitative by the latter[101]:

$$\left(\diagup\!\!\diagdown\right)_3 B \;+\; Me_2Zn \;\longrightarrow\; \left(\diagup\!\!\diagdown\right)_2 Zn \;+\; Me_3B \uparrow$$

2.4.3   Other Methods

A useful new technology involves the palladium-catalyzed synthesis of allylic zinc reagents starting from allyl ether or allyl carbonate. The reagents are prepared *in situ* and will add to a carbonyl compound[102]. Propargylic reagents are also available by the same method[103]:

$$\diagup\!\!\diagdown OR \;+\; Et_2Zn \;+\; R'CHO \xrightarrow{\;Pd\,(cat.)\;} \diagup\!\!\diagdown\!\!\overset{OH}{\underset{R'}{\diagup}}$$

R = Ph, Piv

The reaction of the Simmons–Smith reagent with a carbon nucleophile provides a useful method for one-carbon homologation of the nucleophile. For instance, the reaction of a lithium enolate Generates a ketone homoenolate, and a vinylcopper and alkynylcopper generates allylic[104] and propargylic metallic reagents, respectively[105]:

$$\underset{}{\overset{OSi(CH_3)_3}{\bigcirc}} \xrightarrow[\substack{(b)\ (ICH_2)_2ZnI\\ -40\ to\ 25\ °C}]{(a)\ CH_3Li} \underset{}{\overset{O\quad ZnI}{\bigcirc}}$$

3   Addition to Carbon–Oxygen Double Bonds

A zinc reagent prepared by reduction of an α-haloester with metallic zinc undergoes addition to carbonyl compounds. Reported more than 100 years ago by Reformatsky[5], this reaction has played the central role in the textbook description of organozinc reactions.

Whereas the high reactivity of Reformatsky, allyllic and propargyllic zinc reagents toward carbonyl groups has been known for a long time, it is only very recently that alkylzinc reagents have been found to undergo carbonyl addition in the presence of a Lewis acid, especially when complexed with a chiral organic ligand. This method of enantioselective carbonyl addition now represents an important synthetic methodology for C—C bond formation, surpassing the conventional lithium- and magnesium-based methods with its ability to incorporate sensitive functionalities with the nucleophilic organometallic reagent.

3.1 Reformatsky Reaction

Conventional solvents for the preparation of the Reformatsky[7] reagent are benzene and ether, in which a mixture of an α-haloester and a carbonyl compound is reduced *in situ* with metallic zinc to prepare the desired β-hydroxyester, posing problems of side reactions. However, in an improved procedure using dimethoxymethane, one can prepare the reagent in high yield prior to the carbonyl addition, realizing a much higher overall yield and with reproducible results. Ultrasound irradiation has proved useful in the preparation of the Reformatsky reagent[106].

With all these variations, the scope of the Reformatsky reaction may now be fully explored, subsequently giving way to more convenient methods that use lithium enolates prepared by direct deprotonation of the corresponding ester (i.e. not from an α-haloester) with a strong lithium base such as lithium diisopropylamide. However, there are several important cases reported for the beneficial use of the Reformatsky reaction under modified conditions. Organoaluminum additives act as Lewis acids to enhance the reactivity of the Reformatsky reagent. The next preparation illustrates this use in a macrocyclic intramolecular cyclization performed in the presence of Et$_2$AlCl, in which the presence of an aldehyde moiety precludes the application of lithium-based chemistry.

Macrocyclization route to 3-Hydroxypentadecanolide[107]

A solution of diethylaluminum chloride (1.5 mmol, 1.5 mL of a 1 M solution) in hexane was added to a slurry of zinc–silver couple (3.27 g, 50 mmol) in anhydrous THF (50 mL) under argon at 20 °C. After 10 min, the suspension was warmed to 35 °C and a solution of bromoacetate (0.335 g, 1.0 mmol) in THF (10 mL) was added slowly over a period of 4 h. Stirring was continued for an additional 20 min, and the reaction was terminated by the addition of pyridine (0.9 mL). Then the suspension was diluted with ether, washed with iced 2 M HCl, and brine. The organic layer and ether extracts of the aqueous washings were dried and concentrated to afford lactone (0.175 g, 48% yield) as a colorless oil after purification by column chromatography on silica gel.

The aluminum-activation method is especially useful in the Reformatsky reaction of a fluorinated ester[58]:

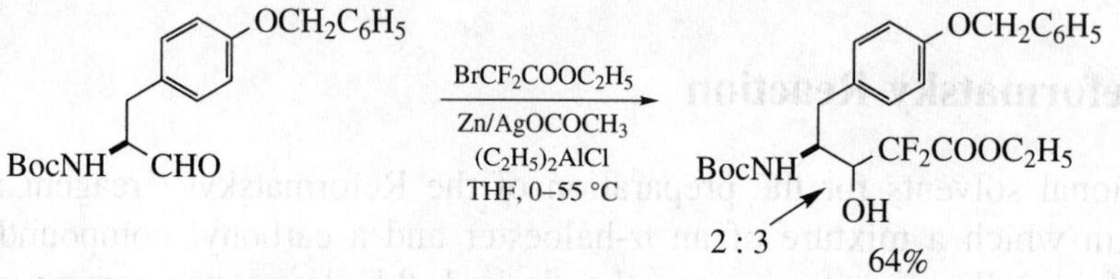

The utility of the Reformatsky reagent is not limited to carbonyl addition reactions. It has been used to prepare alkyl α-silylacetates, which then serve as precursors to a modified Reformatsky reagent useful for synthetic transformations such as silylation[108] and Wittig-like carbon homologation[109]:

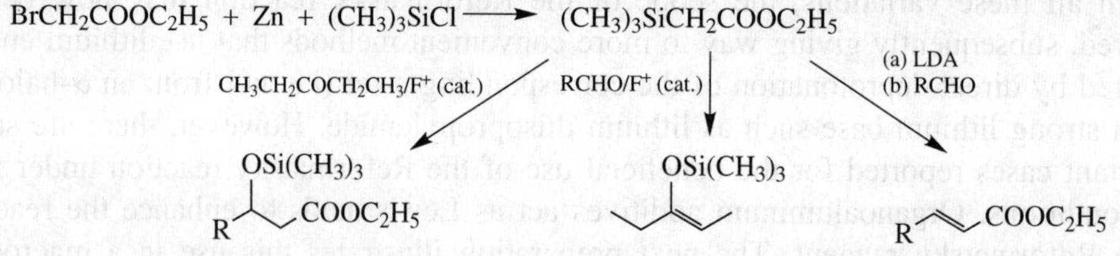

Synthesis of Ethyl Trimethylsilylacetate[110]

In a 3 L, three-necked flask fitted with a 1 L pressure-equalizing dropping funnel, mechanical stirrer, and efficient condenser connected to a nitrogen source were placed 97.5 g (1.5 mol) of zinc powder and 14.9 g (0.15 mol) of cuprous chloride. After the reaction vessel was flushed with nitrogen, a static nitrogen atmosphere was maintained for the remainder of the reaction. A mixture of 150 mL of ether and 550 mL of benzene was added to the flask, and the resulting mixture was refluxed with stirring for 30 min with the aid of an electric heating mantle. Heating was discontinued and a solution of 109 g (128 mL, 1.0 mol) of chlorotrimethylsilane and 184 g (123 mL, 1.1 mol) of ethyl bromoacetate in a mixture of 90 mL of ether and 350 mL of benzene was promptly added through the dropping funnel at such a rate as to maintain the reaction at gentle reflux. When the addition was complete, after about 1 h, the mixture was heated at reflux for 1 h and then cooled in an ice-bath. While the mixture was stirred, 300 mL of aqueous 5% HCl was added through the dropping funnel over a 10 min period. The mixture was extracted with ether.

The organic phases were combined and washed with saturated aqueous NaCl, saturated aqueous NaHCO$_3$, and finally with saturated aqueous NaCl. The organic layer was dried over MgSO$_4$, the mixture was filtered, and the filtrate was concentrated on a rotary evaporator to a volume of about 400 mL. The residual yellow liquid was distilled in a 30 cm vacuum-jacketed Vigreux column at atmospheric pressure until the boiling point was 90 °C. The remaining liquid was distilled at reduced pressure to give, after a small forerun, 118 g (74%) of ethyl trimethylsilylacetate (bp 93–94 °C, 104 mmHg).

3.2   Addition Reactions with and without Lewis Acid Assistance

Allylic, propargylic, and allenyl zinc reagents can be viewed as carbon analogs of the Reformatsky reagent and are very reactive in carbonyl addition reactions[111]. It may appear strange that an alkylzinc reagent bearing a potentially more reactive alkyl anionic moiety does not add normally to a carbonyl compound, and the reason for this low reactivity is unclear at present. However, alkylzinc compounds undergo carbonyl addition in the presence of a Lewis acid additive including that bearing a chiral ligand.

Reduction of an allylic halide with lithium or magnesium metal is not a good synthetic reaction because of a competing self-coupling reaction of the allylic moiety. This side reaction is not serious in the case of zinc, and the reduction makes a good synthetic route to allylic zinc reagents. The ease of preparation and the high reactivity of the allylic zinc reagents makes the organozinc route complementary to the lithium and magnesium routes. The reaction has conventionally been carried out in ethereal solvents, but a recent report recommend the use of DMF. When carried out in a polar solvent (DMA or DMPU) in the presence of a catalytic amount of lithium iodide, conversion of allylic mesylate and phosphate to an allylic zinc reagent is also possible[112].

Allylation of Aldehyde with Allylic Zinc Reagent in DMF[113]

1 : 1

90%

Into a stirred solution of aldehyde (0.50 g, 5.1 mmol) and allylicbromide (0.90 g, 7.4 mmol) in 5 mL of DMF, some pieces of small zinc plate (0.50 g, 7.7 mmol) were added at 25 °C under ambient atmosphere. An exothermic reaction started within 10 min and had ceased in 30 min. Then, the reaction mixture was poured into a saturated aqueous solution of NH$_4$Cl (100 mL), extracted with ether (30 mL × 3), and the combined organic phase was dried over MgSO$_4$. After it was worked up as usual, the product was purified by Kugelrohr distillation under reduced pressure and obtained in 90% yield.

It is remarkable that the allylation of carbonyl compounds can be achieved by the *in situ* reaction of an allylic bromide with metallic zinc in an aqueous medium[114]. In a recent modification, aqueous NH$_4$Cl and an organic phase (THF or reversed phase C$_{18}$ silica gel) may be employed to form a two-phase reaction medium, wherein the reaction is carried out at 25 °C and under open air. In light of the water sensitivity of allylic zinc compounds, the reaction may appear to proceed via a radical intermediate. However, the ordinary carbonyl addition product was formed:

Allylzincation of Aldehyde in Aqueous Medium[115]

   A mixture was prepared with 1.0 mmol of the aldehyde or ketone, 1 mL of satu-rated aqueous ammonium chloride solution, 0.100 or 0.200 g of reversed phase silica gel (or other organicpolymer), 1.2–2.0 mmol of the allyl halide, and 1.2–2.0 mmol of zinc dust. The mixture was stirred for 0.5–16 h at 25 °C open to the air. The reaction mixture was filtered through a fritted-glass funnel with the help of a water aspirator, washing first with 5 mL of water to remove the inorganic materials. The product was washed off the polymer with diethyl ether, and the ether layer was dried (MgSO$_4$) and concentrated under vacuum to yield the product.

A synthetic problem frequently observed is S$_E$2/S$_E$2′ regioselectivity in the carbonyl addition of allylmetal and propargylic reagents. That is, the site of the C–C bond forma-tion relative to the position of the metal atom in starting material is often difficult to control because of a facile 1,3 metal shift in the starting material and the consequent diversity of mechanisms for the following carbonyl addition. When a bulky acylsilane (the best being acyltriisopropylsilylane) is used as a carbonyl acceptor, the steric bulk of the silyl group dictates the regioselectivity in such a way that the less substituted end of the allylic group become attached to the carbonyl carbon. An application to prostaglandin synthesis illustrates the functional group tolerance of the zinc methodology:

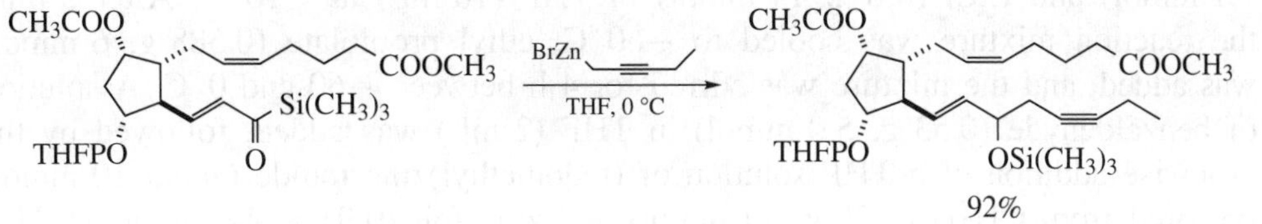

Propargylation of Acylsilane[116]

A suspension of zinc dust (0.131 g, 2.0 mmol) in dry THF (1 mL) containing 1.2-dibromoethane (0.020 mL, 0.23 mmol) was heated to reflux for a few minutes and cooled to 0 °C. To this mixture was added a solution of the propargylic bromide (1 mmol) in dry THF (1 mL) and stirred for 1 h at this temperature. After addition of the acylsilane (0.5 mmol) at 0 °C or 20 °C, the resulting mixture was stirred for 10–45 min and poured into saturated aqueous NH_4Cl (5 mL). Extractive work-up with ether followed by column chromatography on silica gel gave the homopropargylic α-hydroxysilane. Conceptually related control can also be made by placing a silyl group on the propargylic metal.

An interesting variation of the allylic zinc chemistry starts from methylene homologation reaction of a vinyl zinc/copper reagent with the Simmons–Smith reaction. The resulting allylic zinc/copper complex underwent S_E2' regioselective addition to a carbonyl compound (and also an imine) to give an α-methylene lactone in good yield.

One-carbon Homologation of Vinylcopper with Simmons–Smith Reagent[104]

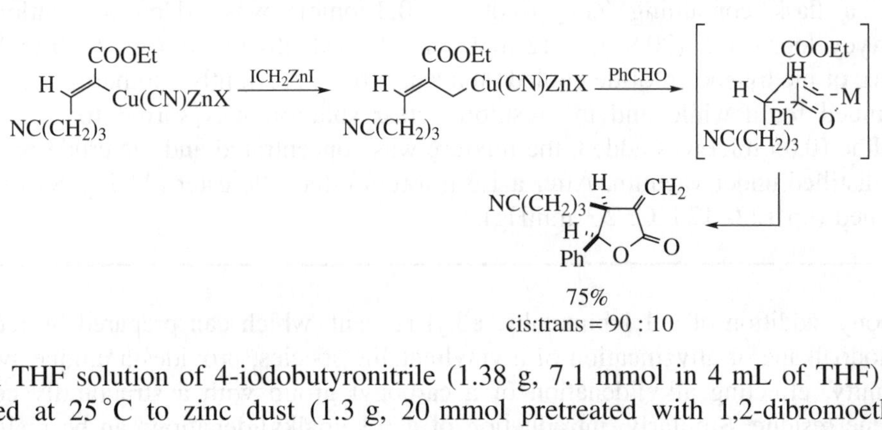

75%
cis:trans = 90 : 10

A THF solution of 4-iodobutyronitrile (1.38 g, 7.1 mmol in 4 mL of THF) was added at 25 °C to zinc dust (1.3 g, 20 mmol pretreated with 1,2-dibromoethane (0.3 g, 1.5 mmol) and Me_3SiCl (0.1 mL)). An exothermic reaction occurred, and the reaction mixture was stirred for 1.5 h at 40 °C. The excess of zinc was allowed to settle, and the supernatant solution was added to a solution of CuCN (0.635 g,

7.1 mmol) and LiCl (0.6 g, 14 mmol) in THF (10 mL) at $-10\,°C$. After 5 min, the reaction mixture was cooled to $-60\,°C$, ethyl propiolate (0.588 g, 6 mmol) was added, and the mixture was stirred for 4 h between -60 and $0\,°C$. A solution of benzaldehyde (0.53 g, 5.0 mmol) in THF (2 mL) was added, followed by the dropwise addition of a THF solution of (iodomethyl)zinc iodide (about 10 mmol) prepared from CH_2I_2 (3.75 g, 14 mmol) and zinc foil (0.91 g, 14 mmol) in THF (8 mL) at $25–26\,°C$ (3 h). The reaction mixture was allowed to warm to $0\,°C$ and was stirred for 0.5 h at this temperature. After the usual work-up, the residual oil was purified by flash chromatography, affording products as a cis/trans mixture (90 : 10, 0.9 g, 75% yield).

Although alkylzinc compounds do not react with carbonyl compounds, they do undergo carbonyl addition in the presence of a Lewis acid. Me_3SiX-mediated addition of zinc homoenolate represents an early example. The zinc homoenolate is inert to an aldehyde by itself, but smoothly reacts with it in the presence of Me_3SiCl in methylene chloride. Alternatively, if one starts with a siloxycyclopropane and a catalytic amount of ZnI_2 to generate the homoenolate and Me_3SiI *in situ*, then the γ-siloxyester adduct is obtained in high yield. Stoichiometric reaction of the zinc homoenolate with $TiCl(O^iPr)_3$ generates a titanium homoenolate, which adds to an aldehyde by itself[117].

ZnI$_2$-catalyzed Addition of Zinc Homoenolate[21]

To a flask containing ZnI_2 (0.094 g, 0.3 mmol) was added a solution of ethoxycyclopropane (20.9 g, 0.12 mol) and benzaldehyde (10.6 g, 0.10 mol) in 60 mL of methylene chloride cooled on a water-bath. A slightly exothermic reaction continued for a while and the resulting clear solution was stirred for 4 h. Dry pyridine (0.05 mL) was added, the mixture was concentrated and the crude product was distilled under vacuum. After a 1.5 g forerun, the title ester (24.7 g, 88%) was obtained (bp $122–125\,°C$, 2.3 mmHg).

Carbonyl addition of a 1,1-bimetallic alkyl reagent, which can prepared by reduction of 1,1-iodoalkane or allyzincation of a vinylmetallic species, provides a unique synthetic opportunity, effecting alkylidenation of a carbonyl group with a structurally complex alkylidene residue. Similarly, introduction of a γ-ketoalkylidenation can be realized by the use of the 1,1-bimetallic species generated by the coupling of zincated hydrazone and a vinyl Grignard reagent.

Alkylidenation with a 1,1-Bimetallic Reagent[118]

To a solution of (Z)-γ-iodo allylic ether (0.500 g, 1.77 mmol) in Et$_2$O (30 mL) was added at $-78\,^\circ$C, 2 equiv of ^{t}BuLi (1.6 M solution in hexane, 2.2 mmol, 3.5 mL). This solution was warmed to $-65\,^\circ$C for 10 min to complete the lithium–iodine exchange and then at $-65\,^\circ$C, 1.5 equiv of crotyl magnesium bromide was added (1 M solution in Et$_2$O, 2.6 mL, 2.6 mmol) followed by the slow addition of 1.5 equiv of ZnBr$_2$ at $-50\,^\circ$C (1.0 M solution in Et$_2$O, 2.6 mL, 2.6 mmol). The reaction mixture was stirred at $-20\,^\circ$C for 5 h and the quantitative formation of the adduct was checked by GC. To this solution of the bismetallic reagent was added in one portion 0.190 g of freshly distilled benzaldehyde (1.77 mmol) in 5 mL of Et$_2$O at $-50\,^\circ$C. The reaction mixture was stirred at $-50\,^\circ$C for 1 h and the temperature was warmed to 20 $^\circ$C. The mixture was hydrolyzed with aqueous 1 M HCl (10 mL) and, after usual treatment, the residue was chromatographed on SiO$_2$ (cyclohexane/ethyl acetate 98 : 2; yield 0.350 g, 68%).

Synthesis of γ,δ-Unsaturated Ketone Hydrazone[41]

76% yield

To a solution of the cyclohexanone *N,N*-dimethylhydrazone (65 µL, 0.40 mmol) in Et$_2$O (0.50 mL) was added ^{t}BuLi (1.52 M in pentane, 0.26 mL, 0.40 mmol) at $-78\,^\circ$C, and the mixture was warmed to 0 $^\circ$C. After 4 h, ZnBr$_2$ (0.33 M in Et$_2$O, 1.20 mL, 0.40 mmol) was added to the solution of the lithiated hydrazone. After 1 h, vinylmagnesium bromide (1.09 M in THF, 0.37 mL, 0.40 mmol) was added to the suspension of zincated hydrazone. The mixture slowly turned to a pale yellow suspension. After 20 min, benzaldehyde (81 µL, 0.80 mmol) was added.

The reaction mixture was stirred for 1 h at 0 °C, and then a 1/15 N phosphate buffer solution was added to the reaction mixture. The mixture was separated into an aqueous layer and an organic layer. The aqueous layer was extracted three times with Et_2O. The combined organic extracts were successively washed with 30% potassium sodium tartrate and saturated sodium chloride solution, dried over sodium sulfate, and evaporated *in vacuo*. The residual oil was purified on silica gel (8 g, first deactivated with 2 mL *N,N*-dimethylaniline) to obtain the title compound (0.0775 g, 76%, as a yellow oil) as an 82 : 18 mixture of (*E*) and (*Z*) isomers.

3.3   Enantioselective Additions

Although the enantioselective alkylation of carbonyl compounds has been studied extensively for a long time by the use of various organometallic reagents, a truly useful and highly selective version of this reaction was realized only recently for the catalytic addition of an alkylzinc reagent to an aldehyde[28,29]. In the fully optimized prototypal procedure[33], only 2 mol % of an optically active amino alcohol effects highly enantioselective carbonyl addition of an alkylzinc compound to an aromatic or an α,β-unsaturated aldehyde. Most noteworthy is the fact that a chiral ligand of rather low ee (e.g. 15% ee) effects highly enantioselective addition (e.g. 95% ee). This (positive) nonlinear relationship between the purity of the catalyst turned out to provide the first practical case of chiral amplification phenomenon[32], allowing the use of an only partially resolved catalyst to achieve near 100% selectivity[119].

Enantioselective Alkylation of Aldehyde Catalyzed by an Aminoalcohol[120]

>95% ee, 80% yield

A dry Schlenk tube containing a Teflon-coated stirring bar and argon atmosphere was charged with optically active (−)-DAIB (0.0111 g, 0.056 mmol) and toluene (6 mL). A 4.19 M toluene solution of Et_2Zn (0.81 mL, 3.4 mmol) was added at 15 °C. The mixture was stirred for 15 min and then cooled to −78 °C. To this was added furfural (0.30 g, 3.1 mmol) in one portion. The reaction mixture was stirred at 0 °C for 6 h. Then saturated aqueous NH_4Cl solution (3 mL) was added. The mixture was extracted three times with ether. The combined organic layers were washed with 1 M HCl solution, water, and brine, dried over anhydrous Na_2SO_4, and concentrated under reduced pressure. The crude mixture was distilled to obtain (*S*)-2-(furyl)-1-hexanol (>95% ee; 0.28 g, 80% yield).

The use of a mixed reagent system consisting of an alkylzinc reagent, Ti(O*i*Pr)$_4$, and a chiral sulfonamide ligand[121] removed some of the important limitations of the above procedure, allowing highly enantioselective addition of functionalized alkylzinc reagents to aliphatic as well as aromatic aldehydes.

Chiral Titanium Catalyst for Symmetric Alkylation of an Aldehyde

To a cooled (0 °C) solution of (1*R*,2*R*)-1,2-diaminocyclohexane (4.6 g, 40 mmol) in CH$_2$Cl$_2$ (120 mL) was added diisopropylethylamine (31 mL, 0.18 mol), and the mixture was stirred for 10 min. Then, *para*-nitrobenzenesulfonylchloride (17.7 g, 80 mmol) was added at −40 °C. The mixture was allowed to warm to 25 °C. After being stirred for 30 min, the mixture was poured into 1 M HCl (300 mL), and the product was extracted with Et$_2$O. The organic phase was washed with saturated NaCl, dried over anhydrous Na$_2$SO$_4$, and concentrated. The residue was recrystallized from CH$_2$Cl$_2$/hexane to afford the desired compound (14.3 g, 74% yield) as colorless needles.

In a flame-dried round-bottom flask was placed the bis-sulfonylamide catalyst (0.189 g, 0.5 mmol) under an argon atmosphere. To this was added degassed toluene (10 mL) and Ti(O*i*Pr)$_4$ (8.53 g, 30 mmol), and the mixture was stirred at 40 °C for 20 min. After being cooled to −78 °C, Et$_2$Zn (1.0 M hexane solution, 30 mL, 30 mmol) was added to the solution; the solution soon turned orange. To the resulting solution was added benzaldehyde (2.12 g, 25 mmol) in toluene (2 mL) and the mixture was warmed to −20 °C, and was stirred at that temperature for 2 h. The reaction was quenched by adding 2 M HCl, and the product was extracted with Et$_2$O. The organic phase was washed with saturated NaCl, dried over anhydrous Na$_2$SO$_4$, and concentrated. The residue was chromatographed on a silica gel column chromatography to obtain crude 1-phenylpropanol. Distillation gave the pure alcohol (3.27 g, yield 98%; bp 103 °C/15 mmHg).

Perhaps of little importance when viewed from the current demands of synthetic efficiency, an extremely interesting example of asymmetric autocatalysis was discovered recently in the asymmetric carbonyl addition of diisopropylzinc to an amino aldehyde[122].

Asymmetric Autocatalysis[123]

57% ee   89% **ee**

Pyrimidyl alcohol (0.0317 g (0.208 mmol; 57% ee) containing the (*S*) isomer (0.249 g) and (*R*) isomer (0.0068 g)) in toluene (49 mL) and diisopropylzinc (1.2 mL of 1 M toluene solution, 1.2 mmol) was stirred for 20 min at 0 °C, and then a toluene solution (1.8 mL) of pyrimidine-5-carboxaldehyde (0.1127 g, 1.04 mmol) was added at 0 °C. This reaction mixture was stirred for 96 h at 0 °C, and was then quenched by the addition of 1.0 M HCl (5 mL) and saturated aqueous NaHCO₃ (15 mL) at 0 °C. The mixture was filtered using celite and the filtrate was extracted with ethyl acetate. The extract was dried over anhydrous sodium sulfate and allowed to evaporate. Purification of the crude product using thin-layer chromatography yielded the pyrimidyl alcohol (0.1273 g) as a mixture of the newly formed alcohol and the catalyst alcohol (0.0317 g). Analysis of the product by high-performance liquid chromatography using a chiral column showed it has 81% ee; it therefore must consist of the (*S*) isomer (0.1154 g) and the (*R*) isomer (0.0119 g). The amount of newly formed alcohol is $0.1273 - 0.0317 = 0.0956$ g (0.628 mmol, 60% yield), which consists of the (*S*) isomer ($0.1154 - 0.0249 = 0.0905$ g) and the (*R*) isomer ($0.0119 - 0.0068 = 0.0051$ g). The newly formed alcohol thus has an enantiomeric purity of 89% ee.

---

4   Addition to Carbon–Nitrogen Multiple Bonds

Addition of organozinc reagents to C–N multiple bonds is much less studied than carbonyl additions, because of low reactivity of such bonds and competitive deprotonation of acidic α protons. Whereas organozinc reagents including alkyl, alkenyl and arylzinc reagents do not add to simple imines, they do add to more electrophilic compounds such as iminium salts[124] and *N*-acylimines[125]. Addition to simple imines can be achieved in the presence of a Lewis acid such as $MgBr_2$[126]:

Allylic zinc reagents are more reactive than alkylzinc reagents, but the reaction may suffer complication due to competing S_E2 and S_E2' pathways giving rise to branched and linear products, respectively[127]. Diastereoselectivity issues arise in the latter product. Allylic zinc reagents bearing an allyic functional group, such as sulfonyl[128], phenoxy[129], ester and silyl groups, are known.

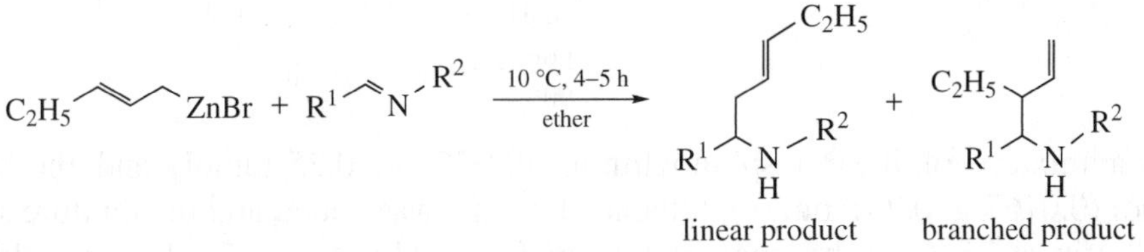

linear product   branched product

The Reformatsky reagent reacts with an imine to provide a synthetic route to β-amino-esters and β-lactams[130]. The stereoselectivity depends both on the substituents and the solvent:

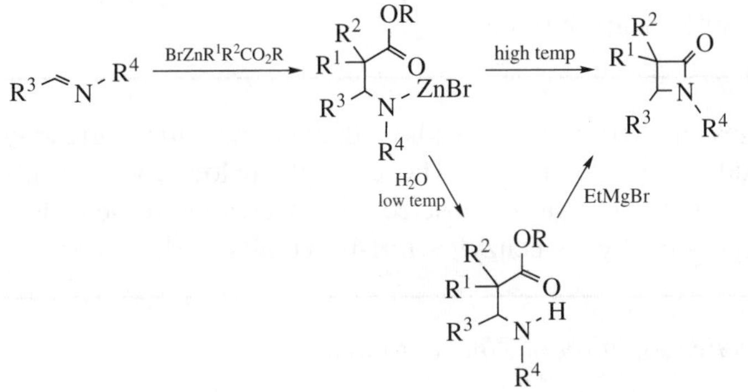

4.1 Addition to Imines

In spite of recent advances in enantioselective carbonyl additions, enantiocontrolled nucleophilic addition to a C=N double bond is still a difficult task. The amino alcohol-mediated protocol successfully employed for carbonyl addition does not effect conversion of the imine substrates, even under forcing conditions. The examples with acyl[125] and phosphoryl imine[131] illustrate the current status of the development.

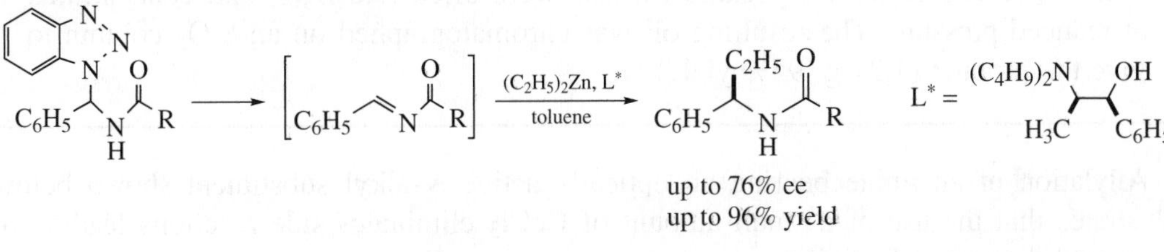

up to 76% ee
up to 96% yield

Enantioselective Alkylation of Phosphoryl Imines[132]

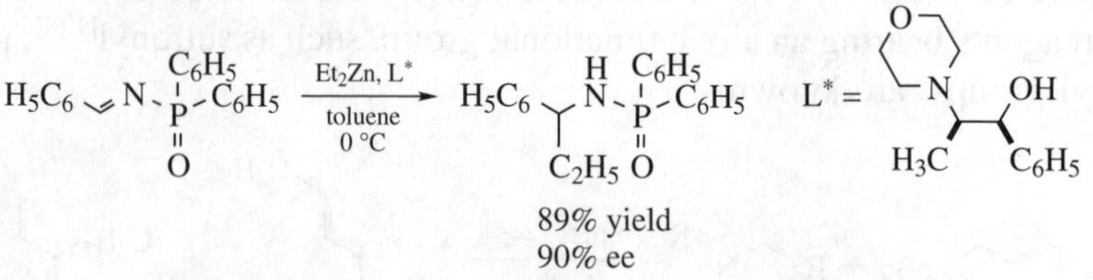

89% yield
90% ee

To a mixture of the phosphinoylimine (0.0767 g, 0.25 mmol) and the amino alcohol (0.0767 g, 0.251 mmol), toluene (1.5 mL) was added and the mixture stirred for 10 min at 25 °C. A hexane solution of Et_2Zn (1.0 M, 0.75 mL) was added at 0 °C, and the mixture was stirred for 22 h. The reaction was quenched with saturated aqueous NH_4Cl (5 mL). The mixture was extracted with dichloromethane and dried (Na_2SO_4) and the solvent was evaporated. Purification of the residue by silica gel TLC (acetone/hexane 1 : 2) afforded optically active (*S*)-phosphoramide (0.0745 g, 0.222 mmol, 89% yield; 90% ee).

Trialkylzincates are more reactive alkyl donors than the corresponding dialkylzinc reagents and add to simple imines. The example below for 2-pyridylcarboxaldehyde illustrates the reactivity. The diastereoselectivity decreases in the order of vinyl > ethyl, isopropyl > butyl > methyl > benzyl > *tert*-butyl, allyl[133].

Diastereoselective Addition of Zincate to Imines

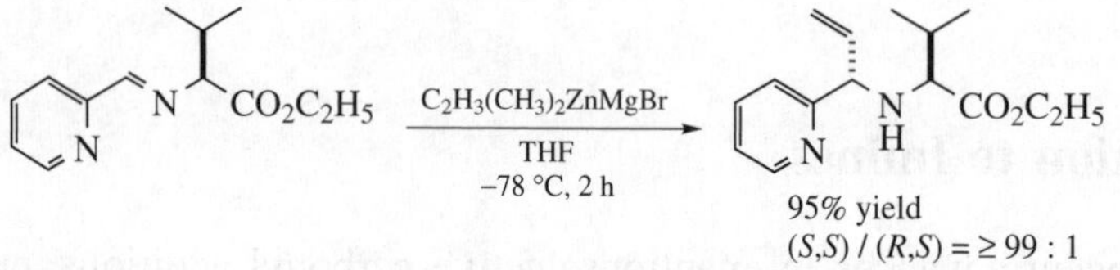

95% yield
(*S*,*S*) / (*R*,*S*) = ≥ 99 : 1

A solution of the imine (1.10 g, 5 mmol) in anhydrous THF (5 mL) was added over 30 min to a magnetically stirred solution at −78 °C of the triorganozincate (5 mmol). The reaction mixture was stirred for 2 h, then quenched with 10% aqueous $NaHCO_3$ (10 mL). The organic layer was separated and the aqueous phase extracted with Et_2O. The combined organic extracts were dried (Na_2SO_4) and concentrated at reduced pressure. The resulting oil was chromatographed on an SiO_2 column to give the product (1.24 g, 95% yield).

Allylation of an imine bearing an optically active *N*-alkyl substituent shown below illustrates that the use of a small amount of $CeCl_3$ eliminates side reactions leading to decreased diastereoselectivity.

Diastereoselective Allylation of Imines[134]

R = C₆H₅, 3-pyridyl, CH(CH₃)₂

$R = C_6H_5$, 3-pyridyl, $CH(CH_3)_2$

~ 100% yield
~ 100% selectivity

To a suspension of Zn powder (0.13 g, 2.0 mmol) in THF (3 mL), stirred magnetically and cooled with an ice bath, was added $CeCl_3 \cdot 7H_2O$ (0.037 g, 0.1 mmol) and then a solution of the imine (1.0 mmol) and allyl bromide (0.182 g, 1.6 mmol) in THF (3 mL) was added. The reactions in the presence of anhydrous salts were carried out as follows. The salts (about 0.1 mmol) were dried and weighed in the same apparatus used for the subsequent reaction: anhydrous $CeCl_3$ was obtained by heating $CeCl_3 \cdot 7H_2O$ at 150 °C for 2 h. To the salt was added THF (3 mL), Zn, and the organic reagents as above. The progress of the reactions was followed by GC-MS analysis and after usual quenching and work-up the products were isolated generally in quantitative yield with satisfactory purity for subsequent use.

Allylic zinc reagents bearing a chiral bisoxazoline undergo highly enantioselective addition to cyclic imines[135] and the oxime of an α-ketoester[136].

Enantioselective Allylation of Cyclic Imines

88% yield
97.5% ee

The lithio bis-oxazoline was prepared in THF by the treatment of the corresponding protio compound (1.45 g, 6.1 mmol) in THF (7.0 mL) with BuLi (1.61 M, hexane) at 0 °C. A 1.06 M THF solution of methallylzinc bromide (5.35 mL, 5.7 mmol) was then added. After 30 min, 6,7-dimethoxy-3,4-dihydro-isoquinoline (0.96 g, 5.0 mmol) was added at −70 °C. After 6 h, the reaction mixture was quenched with 0.5 mL of MeOH, and then diluted with Et₂O. The organic layer was washed with 0.2 M NaOH and dried (Na₂SO₄) and filtered. Concentration of the filtrate afforded a mixture (2.92 g) of the desired product (95.5% yield by NMR) and a 2 : 1 complex of bis-oxazoline and zinc (BOX₂−Zn). A 2.62 g portion of the crude product was purified by silica gel chromatography to obtain the desired amine

(1.09 g, 4.4 mmol, 88% yield) as well as the protio bis-oxazoline (30% recovery) and BOX$_2$–Zn (40% recovery), which can be quantitatively converted to the protio compound without racemization.

Addition of *N,N*-disubstituted glycine enolate to an imine provides an efficient one-pot synthesis of amino-β-lactams. High levels of trans selectivity can be achieved by judicious choice of solvent and steric bulk of the substituents.

β-Lactam Synthesis from Glycine Enolates[137]

$$(C_2H_5)_2NCH_2CO_2C_2H_5 \xrightarrow[\text{Et}_2\text{O, 25 °C}]{\text{LDA}} \xrightarrow{\text{ZnCl}_2}$$

To a stirred solution containing 1.01 g (10 mmol) of ^{i}Pr$_2$NH in 30 mL of solvent (benzene, toluene, or THF) was added 10 mmol of BuLi (6.67 mL of a 1.6 M solution in hexane). The resulting solution was stirred for 10 min, and then 10 mmol of a *N,N*-disubstituted glycine ester was added at 25 °C. The reaction mixture was stirred for an additional 15 min, and then 10 mmol of ZnCl$_2$ (10.0 mL of a 1.0 M solution in Et$_2$O) was added. After the mixture was stirred for 30 min, 10 mmol of an appropriate imine was added. Then, the reaction mixture was refluxed until no further formation of 2-azetidinone could be detected by ^{1}H NMR. The reaction mixture was allowed to cool to 25 °C and then quenched with 20 mL of a saturated aqueous NH$_4$Cl solution. The precipitated salts were filtered through sintered glass. The aqueous layer was separated and extracted with two 30 mL portions of Et$_2$O. The combined organic extracts were dried over Na$_2$SO$_4$ and concentrated *in vacuo* to afford the crude 2-azetidinone products. The products were purified by recrystallization, flash chromatography, or preparative HPLC techniques.

4.2   Addition to Nitriles

Nitriles are even less reactive than imines. Alkylzinc reagents are unreactive. Allyllic zinc reagents and Reformatsky reagents do react with nitriles but used not be so synthetically useful. Modifications of the conditions (use of THF and slow addition of bromoester) have led to successful addition of the Reformatsky reagent to a nitrile. Ultrasound irradiation may be beneficial.

Addition of Reformatsky Reagent to Nitriles[138]

A suspension of 16.0 g (5 equiv) of activated zinc dust in 150 mL of THF was heated to reflux under N_2. Several 0.1 mL portions of methyl bromoacetate were added with vigorous stirring to initiate the reaction. When the green color appeared, 14.55 g (49.0 mmol) of cyano mesylate in 50 mL of THF were added. Then 18.9 mL (4 equiv) of methyl bromoacetate were added dropwise over 45 min to the refluxing mixture. The mixture was refluxed 10 min longer, cooled to 25 °C, diluted with 430 mL of THF, and quenched with 70 mL of 50% aqueous K_2CO_3. Rapid stirring for 45 min gave two distinct layers. The THF layer was decanted, and the residue was rinsed with THF. The combined THF layers were dried over $MgSO_4$ and concentrated. This crude intermediate was then stirred with 14 g of powdered K_2CO_3 in 150 mL of DMF for 14 h. The reaction mixture was diluted with 150 mL of Et_2O, filtered through Celite, concentrated, and adsorbed on Florisil. The Et_2O eluate was concentrated and purlfied by MPLC, giving 10.66 g (38.8 mmol, 79%) of the product as colorless crystals from hexanes/EtOAc. (mp 117–118 °C).

4.3 Addition to Iminium Salts

Iminium salts and pyridinium salts are good acceptors of alkylzinc reagents. Regioselectivity issues (1,2 versus 1,4 addition) arise in the latter case, and the selectivity is moderate[139], supposedly conforming to the HSAB (hard and soft acids and bases) principle[140]. Copper catalysis accelerates the reaction and results in selective 1,4 addition[141].

Copper-catalyzed 1,4 Addition to Acylpyridium Salts[142]

To a solution of benzylbromide (13 mmol) in THF (20 mL) is added at 0 °C activated Zn (15.6 mmol). After stirring at 0–5 °C under N_2 atmosphere for 3 h, this solution is added to dry THF (20 mL) containing CuCN (0.9 g, 10 mmol) and LiCl (0.9 g, 10 mmol) at −78 °C. After warming up to −20 °C for 5 min, the mixture is cooled again to −7 °C. This solution is added to a preformed solution of pyridinium chloride (from ethyl chloroformate (10 mmol), functionalized pyridine (10 mmol), THF (40 mL), at −25 °C for 30 min). The mixture is warmed up slowly to 25 °C and is allowed to stand overnight (12 h). After quenching with 5% NH_4OH and evaporation of THF, the intermediate is then subjected to oxidation by S_8 (1.3 equiv) in decalin (15 mL) under reflux. The product is worked up by evaporation of decalin and purified by column chromatography to afford the desired pyridine.

5   Carbometalation of Carbon–Carbon Multiple Bonds

Organozinc compounds are unique among main group organometallics for their high affinity to alkenes, leading to the generation of stable organometallic products[143] which can be further exploited as synthetically useful nucleophilic reagents. The reaction takes place with a variety of organozinc compounds, including alkyl, allylic, propargylic, allenyl, and silyl compounds as well as, most uniquely, enolates and azaenolates. Intramolecular applications of the reactions leads to the formation of carbocyclic compounds. Transition metal catalysts (e.g. Cu or Ni) have been successfully employed to improve the synthetic efficiency of the reactions. Addition of an alkyl or allylic zinc compound to an alkene or an alkyne generally takes place in a syn fashion, suggesting that a cyclic transition state may be involved. However, like various other organometallic reactions, details of the reaction mechanisms are often unclear:

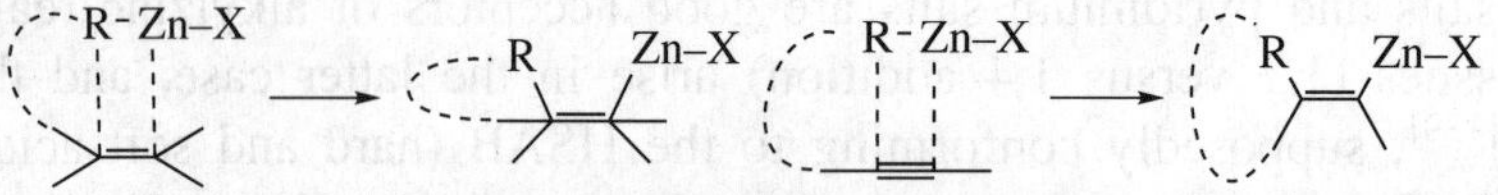

The carbometalation reaction of isolated alkenes and alkynes has mechanistic features that cannot be understood by simple anionic addition. Rather, the reaction must be considered as the result of cooperative action of the Lewis acidic zinc(II) metal and the nucleophilic alkyl anion moiety. For instance, the regioselectivity of the addition of allylic zinc reagent to styrene is considerably influenced by the nature of the alkyl ligand of the metal, as shown below[144]. If the reaction were simply a nucleophilic addition of a free anion, the reaction would have afforded the branched product due to the anion-stabilizing effect of the phenyl group and would remain unaffected by the nature of the R group on the zinc atom. The exact role of the zinc metal–alkene interaction in the carbometalation reaction is yet to be understood.

R = $^tC_4H_9$, linear/branched = 1 : 1
R = Cl, linear/branched = 1 : 43~90

5.1 Carbometalation of Alkenes

Intermolecular carbometalation of an unactivated alkene with an alkylzinc reagent is not a good synthetic reaction. However, the intramolecular version takes place smoothly to afford five-membered carbocycles in good yield[145]. The starting alkylzinc reagent may be synthesized *in situ* from a suitable alkenyl halide by reduction with metallic zinc or by palladium-catalyzed trasmetalation with Et_2Zn. The alkylzinc reagent formed by the cyclization can be further trapped by electrophiles, such as an electron-deficient alkene[83].

Palladium-catalyzed Intramolecular Cyclization and Trapping[83]

A three-necked flask equipped with a magnetic stirring bar, a thermometer, and a gas inlet was charged with $PdCl_2(dppf)$ (0.07 g, 2 mol %) in THF (5 mL) and cooled to −78 °C. The iodide (1.33 g, 5.0 mmol) and Et_2Zn (1.23 g, 1.0 mL, 10 mmol) were added. After the mixture was warmed to 25 °C and stirred for 4 h, the solvent and excess Et_2Zn were removed (0.1 mm Hg, 25 °C, 1 h). After addition of THF (5 mL) and cooling of the mixture to −40 °C, CuCN•2LiCl (5.0 mmol) in THF (5 mL) was added, and the reaction mixture was warmed to 0 °C (5 min) and cooled to −78 °C. Nitrostyrene (1.12 g, 7.5 mmol) in THF (3 mL) was added, and the reaction mixture was slowly warmed to 0 °C and stirred for 2 h. After the usual work-up, the residual oil was purified by flash column chromatography to yield 4 g as a clear oil (1.16 g, 81% yield).

As in the carbonyl addition reaction, allylic zinc reagents are more reactive than alkyl zinc reagents, and undergo addition with a high level of S_E2' regioselectivity forming a new C–C bond at the carbon γ to the starting C–Zn bond:

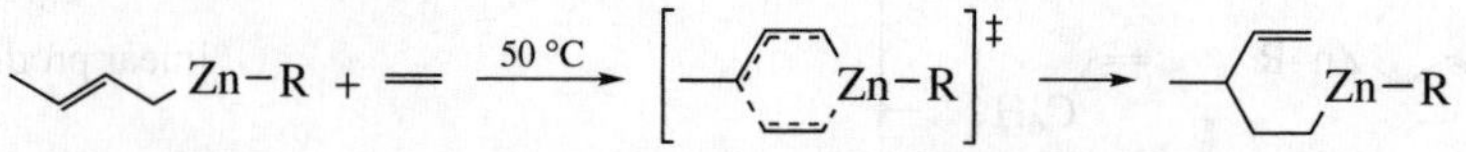

In order to promote carbozincation of an alkenic bond, it is useful to include a substituent capable of coordinating to the zinc metal. A notable example is the nickel-catalyzed carbozincation of acrolein acetal, which affords a protected zinc homoenolate (3-zincio acetal). Note that the resulting regioselectivity of the addition reaction is opposite to the normal conjugate addition to α,β-unsaturated aldehyde.

Allylzincation of Allylic Ethers[146]

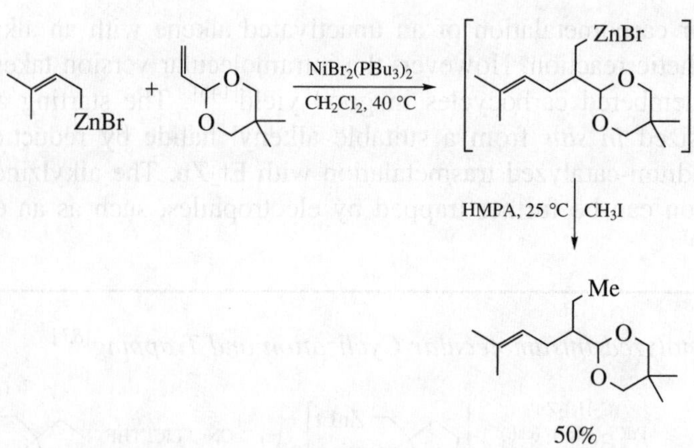

50%

To a solution of 3-methyl-2-butenylzinc bromide (0.88 mmol) in CH_2Cl_2 (2 mL) was added catalytic $NiBr_2(PBu_3)_2$ (0.0156 g, 0.025 mmol) at 25 °C. After 10 min, 5,5-dimethyl-2-vinyl-1,3-dioxane (0.0356 g, 0.25 mmol) was added at 25 °C, and the resulting mixture was stirred at 40 °C for 40 min. The protected zinc homoenolate, thus obtained was methylated by adding a mixture of CH_3I (0.156 mL, 2.5 mmol) and HMPA (1.04 mL, 6.0 mmol) at 25 °C for 16 h. The mixture was poured into a saturated NH_4Cl aqueous solution and extracted with ether. The combined extracts were dried and concentrated, and the product was purified by column chromatography on silica gel to give 5,5-dimethyl-2-(1'-ethyl-4'-methyl-3'-pentenyl)-1,3-dioxane (0.028 g, 50% yield) as a colorless oil.

Allenyl compounds also act as powerful electrophilic acceptors toward organozinc compounds. Regioselectivities of carbozincation with an organozinc reagent alone, and with a mixed organozinc(II)/copper(I) complex differ from each other, but reasons remain unclear[147]:

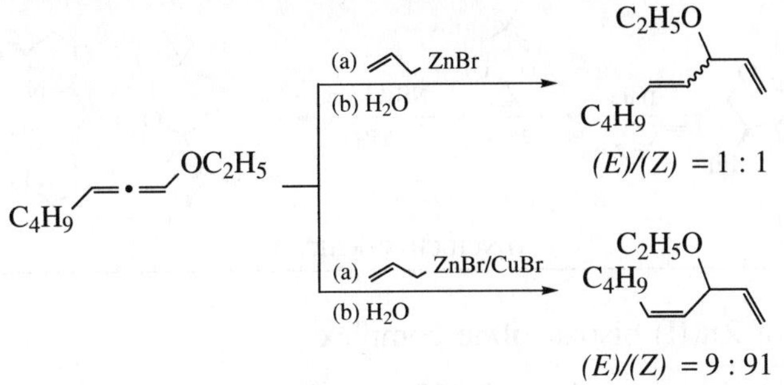

Cyclopropenone acetal is a stable and readily available compound[148] and has been studied to obtain basic information on the stereochemisty of allylmetalation reactions. Choice of the reaction conditions allows chelate and nonchelate pathways to operate, realizing high levels of remote asymmetric induction as well as diastereoselectivity as to the forming C–C bond. Thus, the reaction of cinnamylzinc bromide with an optically active acetal in methylene chloride takes place with good overall selectivity to give one isomer, which accounts for 92% of the product isomers. However, the reaction in a mixture of HMPA and THF gives another isomer, with 91% selectivity[149]. The isomer obtained in methylene chloride is due to a chelation effect that places the metal on the equatorially disposed cyclopropene carbon, whereas the one obtained in a polar solvent mixture is due to a nonchelation reaction[150]:

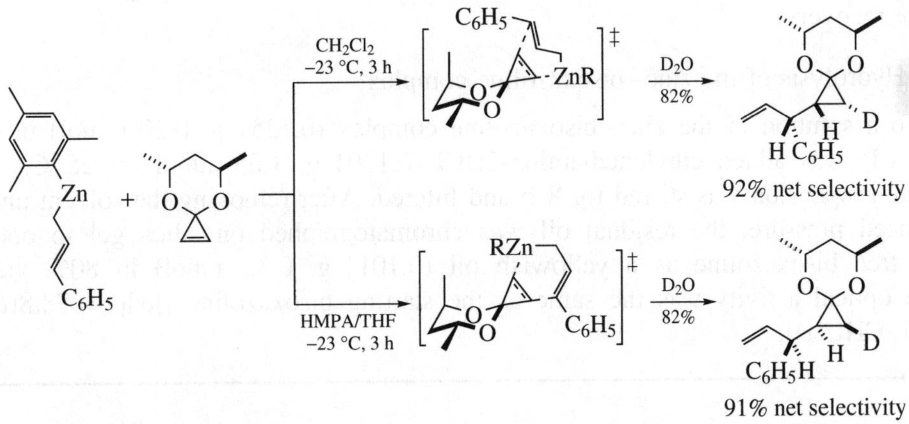

The stereochemical environment created by a rigid and optically active bis-oxazoline provides a rare example of ligand-induced enantioselective carbometalation of an alkenic bond[151]. The resulting cyclopropanone acetal serves as a useful precursor to optically active compounds, for instance after cyclopropane ring opening.

Asymmetric Allylzincation of Cyclopropenone Acetal

(a) Preparation of Zn(II) bisoxazoline complex

To a solution of bisoxazoline (0.422 g, 1.38 mmol) and 2,2'-dipyridyl (about 0.001 g) as an indicator in dry THF (1.8 mL) was added a 1.6 M hexane solution of BuLi, at −78 °C to 0 °C. The colorless clear solution changed to a red-brown suspension immediately after completion of the addition of BuLi. The suspension was stirred at 0 °C for 30 min and at 25 °C for 30 min The brownish suspension was cooled to 0 °C and then a 1.0 M THF solution of allylzinc bromide (1.25 mL, 1.25 mmol) was added. Then the reaction mixture turned a clear red solution. After stirring at 0 °C for 30 min, cyclopropenone acetal (165 mL, 1.1 mmol) was added to the solution. After stirring for 12 h at 25 °C, 50 mL of saturated NH$_4$Cl was added. The resulting white precipitate was filtered and the solvent was evaporated under reduced pressure. The residual oil was chromatographed on silica gel to obtain the allylation product as a colorless oil (0.169 g, 85% yield) and a 2 : 1 white crystalline complex of the bisoxazoline and Zn(II) (0.416 g, 0.62 mmol, 80% recovery).

(b) Hydrolysis of the zinc–bisoxazoline complex

To a solution of the zinc–bisoxazoline complex (0.1351 g, 0.20 mmol) in dry CH$_2$Cl$_2$ was added ethylenediamine•2HCl (0.1391 g, 1.05 mmol) at 25 °C. The white suspension was stirred for 8 h and filtered. After removing the solvent under reduced pressure, the residual oil was chromatographed on silica gel to obtain the free bisoxazoline as a yellowish oil (0.1011 g, 0.32 mmol) in 80% yield. The optical activity was the same as the starting bisoxazoline $\{[\alpha]_D = 78.8(c = 1.01, \text{CHCl}_3)\}$.

5.2   Carbometalation of Alkynes

An organocopper reagent prepared by mixing an organozinc reagent with a CuX reagent (X = halogen, CN) is a good reagent for the carbometalation of alkynes. This type of reagent often shows chemical reactivities different from those of the organocopper reagents that bear the same R group but prepared from an organolithium reagent. The mixed Cu/Zn reagents are generally less reactive, thermally more stable, and more selective than the Cu/Li reagents and their reactions may be carried out at 25 °C. When prepared by a suitable method, the R group may bear a variety of functionality. The addition reaction of a functionalized alkyne makes a useful synthetic entry to trans alkenes bearing functional groups.

Although these reagents may be depicted in a simplified formula as shown below, their true structures are unknown. Recent theoretical studies[152,153] on Li-based organocopper clusters (Me_2CuLi complexes) showed that the countercation metal (e.g. Li^+) and Zn metals play much more active roles than previously considered:

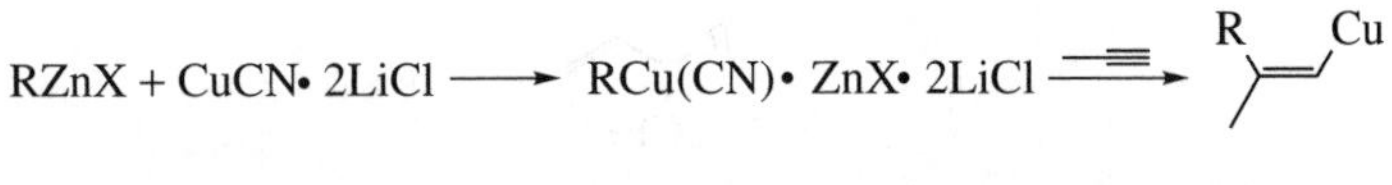

$$RZnX + CuCN \cdot 2LiCl \longrightarrow RCu(CN) \cdot ZnX \cdot 2LiCl \xrightarrow{\quad} \underset{CF_3}{\overset{R \quad Cu}{\diagup\!=\!\diagup}}$$

Generally speaking, perfluoroalkylmetallic species are not synthetically useful reagents. For instance, perfluoroalkyllithium and magnesium reagents are unstable due to elimination of metal halide. The high stability of the zinc–carbon bond was successfully exploited to remedy this problem. Thus, a perfluoroalkyl reagent was generated *in situ* by reduction of the corresponding iodide or bromide with zinc power and added to a terminal alkyne in 32–66% yield. The profile of the reaction was very different from the usual chemistry of alkyl zinc compounds. Thus, ultrasound irradiation was mandatory and, unlike ordinary copper-mediated carbozincation reactions, the reaction took place nonstereoselectively.

Trifluoromethylzincation of Alkynes[154]

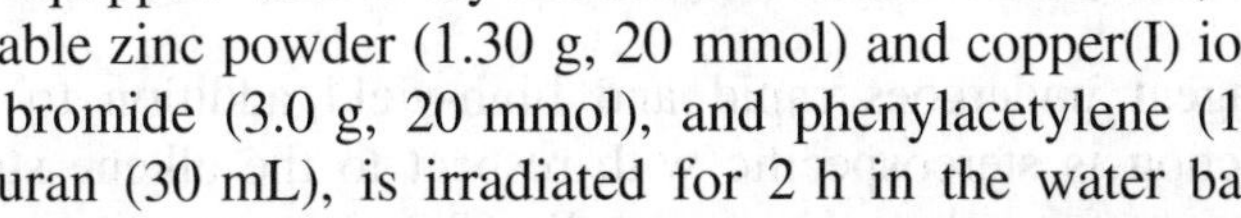

$$CF_3I + H\!\!=\!\!\!=\!\!R \xrightarrow[\substack{THF \\ ultrasound \\ bath, 2 h}]{Zn\ dust} [\,CF_3ZnI\,] \longrightarrow \underset{CF_3 \quad R}{\overset{H}{\diagup\!=\!\diagup}}$$

R = C_6H_5, C_4H_9, $HOCH_2$
65%, 71%, 61%, respectively
(E)/(Z) = ca. 7 : 3

A flask, equipped with a dry-ice/acetone reflux condenser, containing commercially available zinc powder (1.30 g, 20 mmol) and copper(I) iodide (0.72 g), trifluoromethyl bromide (3.0 g, 20 mmol), and phenylacetylene (1.02 g, 10 mmol) in tetrahydrofuran (30 mL), is irradiated for 2 h in the water bath of an ultrasound laboratory cleaner. Then, the solution was poured into a 2% HCl solution and an oily material was extracted with ether. After the ethereal solution was dried over $MgSO_4$, the solvent was removed. Distillation gave 3,3,3-trifluoro-1-phenylpropene [55% yield, 1.11 g; bp 71–73 °C (25 mmHg)].

In a similar manner, the perfluoroalkylzinc compounds undergoes stereospecific substitution reaction with (*E*)-2-bromostyrene derivatives in the presence of Pd(PPh$_3$)$_4$, and S$_N$2'-selective allylation reaction with allylic halides in the presence of Pd(OAc)$_2$. Addition across a double bond of isoprene takes place in the presence of Cp$_2$TiCl$_2$:

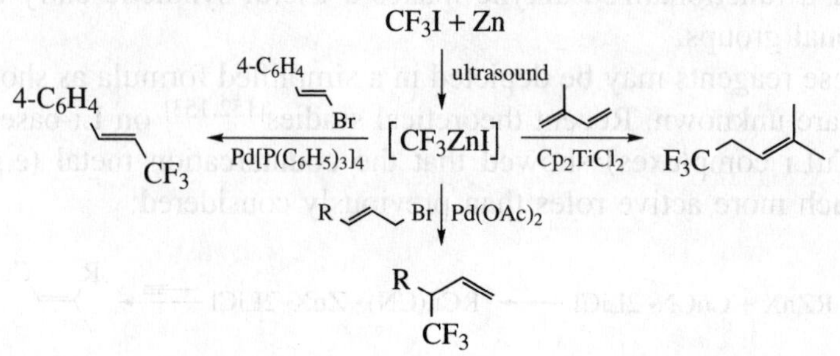

Silicon-substituted alkynes serve as excellent acceptors of allylic zinc reagents. The reaction can also be catalyzed by a Zr catalyst, taking place in a syn-addition manner[155]:

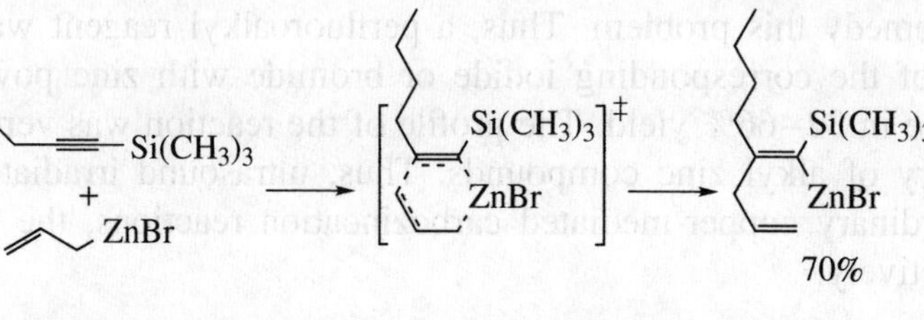

When a terminal alkyne reacts with an allylic zinc reagent, deprotonation of the acidic alkynic proton takes place, and the allylic zinc reagent onto this alkynyl metal species. An intermolecular variation to make five-membered ring compounds[157]:

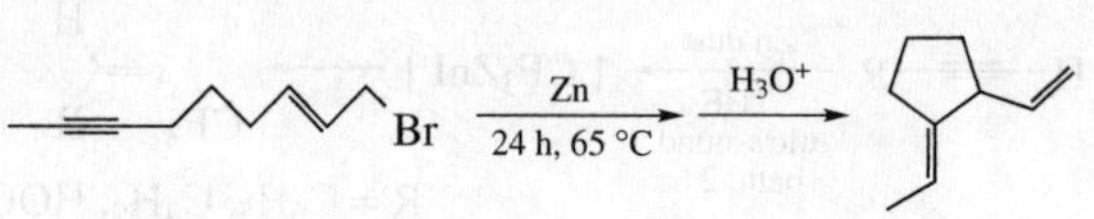

5.3   Coupling of Allylic Zinc Reagents with Vinyl Metallic Species

An allylic zinc reagent undergoes rapid and high-yield addition to a vinyl Grignard reagent. The reaction is stereospecific with respect to the alkene stereochemistry. The reaction has been suggested to be a metalla-Claisen rearrangement rather than carbometalation reaction, but there are so far no firm mechanistic support studies. The product of the addition reaction is a unique 1,1-dimetallic species, which reacts with a carbonyl compound to give an alkene. Treatment of the dimetallic species with two different electrophiles results in the introduction of two different substituents in one operation:

In the reaction using an allylic alcohol as the substrate, there occurs asymmetric induction due to this chiral center, generating three chiral centers in one operation. Assuming a chair six-centered transition state, the sense of the stereoselectivity requires isomerization of the crotyl double bond from trans to cis in the transition state.

Diastereoselective Allylzincation of Vinyl Grignard Reagents[156]

75%
>95% ds

To a solution of (Z)-γ-iodo allylic ether (0.500 g, 1.8 mmol) in Et$_2$O (30 mL) was added at $-78\,°$C, 2 equiv of ^{t}BuLi (1.6 M solution in hexane, 2.2 mmol, 3.5 mL). This solution was warmed to $-65\,°$C for 10 min to complete the lithium–iodine exchange and then at $-65\,°$C, 1.5 equiv of crotyl magnesium bromide was added (1 M solution in Et$_2$O, 2.6 mL, 2.6 mmol) followed by the slow addition of 1.5 equiv of ZnBr$_2$ at $-50\,°$C (1 M solution in Et$_2$O, 2.6 mL, 2.6 mmol). The reaction mixture was stirred at $-20\,°$C for 5 h and the quantitative formation of the adduct was checked by GC. The hydrolysis was done with an aqueous solution of hydrochloric acid (1 M solution, 20 mL). The aqueous phase was extracted twice with ether and the combined phases were washed with HCl (1 M). The organic layer was treated overnight with an aqueous solution of Na$_2$S, washed with Na$_2$CO$_3$, dried over MgSO$_4$ and then concentrated *in vacuo*. The residue was chromatographed on SiO$_2$ to give 0.280 g (75%) of the product.

5.4 Addition of Zinc Enolates to Alkenes and Vinyl Metallic Species

An enolate anion adds smoothly to a carbonyl compound but not to an alkene. It was recently revealed that this latter alkenic analog of the aldol reaction, namely addition of an enolate to an alkene such as ethylene and stryrene, is possible if a Zn cation is used as a countercation of an enolate. The use of BuZn cation rather than BrZn is often crucial for the success of intermolecular reactions. Thus, a zincated hydrazone of a ketone bearing a BuZn cation undergoes smooth addition to ethylene at 25 °C and the resulting 'bis-homoenolate' can be trapped with an external electrophile. As the alkenic aldol reaction is an energetically disfavored reaction, the role of the zinc atom must be very significant. The regioselectivity of the reaction with substituted alkenes suggests that the reaction is not an ordinary conjugate addition reaction.

Addition of Zincated Azaenolate to Ethylene[42]

$R^1 = C_6H_5CH_2CH_2$; $R^1 = C_6H_5CH_2$;

R = H (branched product; R ≠ H)

'BuLi in pentane (1.48 M, 2.0 mL, 3.0 mmol) at −70 °C was added to a solution of *N,N*-dimethylhydrazone of 1,5-diphenylpentan-3-one (0.86 mL, 3.0 mmol) in Et₂O (3.0 mL), and the mixture was warmed to 0 °C. After 4 h, ZnBr₂ in Et₂O (0.33 M, 9.0 mL, 3.0 mmol) and, after another 1 h, butyllithium in hexane (1.67 M, 1.8 mL, 3.0 mmol) were added, and the mixture was warmed to 25 °C. After 1 h, the prepared zincated hydrazone was transferred to an autoclave, and pressurized with ethylene (8 atm) for 4 days with stirring. The resulting adduct was quenched with aqueous 30% potassium sodium tartrate, and the mixture was extracted with ethyl acetate. After the organic extract had been washed with aqueous sodium bicarbonate and aqueous sodium chloride, dried over sodium sulfate, and evaporated *in vacuo*, 1.5 g of an oily crude product was obtained. Purification on silica gel afforded a small amount of the starting hydrazone and the protonated adduct (0.833 g, 90%).

Instead of treatment of the adduct with water, it was allylated *in situ* in 81% overall yield by addition of CuBr•Me₂S (1.5 equiv) and then allyl bromide (3 equiv) followed by 3 h stirring and aqueous work-up as above. The hydrazone product was smoothly hydrolyzed in 85% isolated yield by stirring a mixture of the hydrazone (1 mmol), CuCl₂ (1 mmol), 3 mL of 0.05 M phosphate buffer, 5 mL of water and 15 mL of THF at 25–35 °C for 9 h.

This alkenic analog of the aldol reaction has proved to be useful in the asymmetric synthesis of carbonyl compounds. When an optically active hydrazone is employed, the reaction shows considerable levels of diastereoselectivity. The selectivity is moderate for unhindered alkenes, but it is quite high for a bulky cyclopropenone acetal. The resulting cyclopropanone structure can be converted to various ester derivatives by cyclopropane ring opening.

SAMP–Hydrozone-induced Asymmetric Alkenic Aldol Reaction[43]

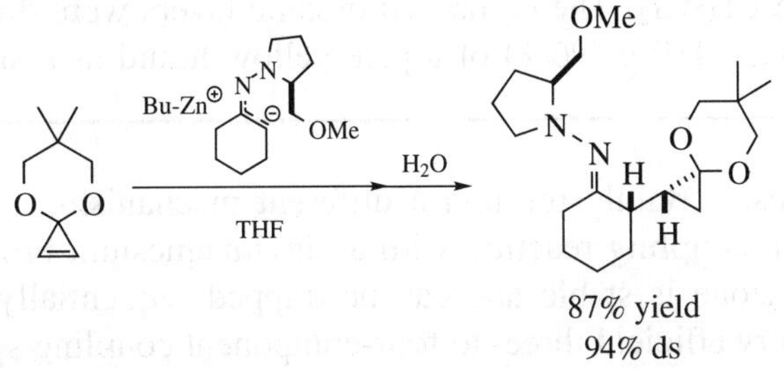

87% yield
94% ds

To a solution in SAMP–hydrazone of cyclohexanone (86 mL, 0.40 mmol) in THF (0.40 mL) was added ^{t}BuLi (1.70 M in pentane, 0.26 mL, 0.44 mmol). After 4 h, ZnCl$_2$ (1.03 M in THF, 0.40 mL, 0.41 mmol) was added to the lithium azaenolate. After 30 min at 0 °C BuLi (1.60 M in hexane, 0.26 mL, 0.40 mmol) was added at −70 °C. The mixture was warmed to 0 °C and after 30 min, cyclopropenone acetal (28 mL, 0.20 mmol) was added. The reaction mixture was warmed to 25 °C. After 2 h, a 1/15 M phosphate buffer solution was added, and the mixture was separated into an aqueous layer and an organic layer. The aqueous layer was extracted with ethyl acetate three times. The combined organic extracts were washed successively with a saturated sodium bicarbonate solution and with saturated sodium chloride solution, dried over sodium sulfate, and evaporated *in vacuo*. The residual oil was purified on silica gel to obtain the desired adduct (0.0613 g, 87%) as a 94 : 2 : 4 mixture of diastereomers.

An intramolecular alkenic aldol reaction also takes place readily. The enolate of glycin bearing a ZnBr counter cation undergoes ready addition to the internal alkenic bond to give pyrrolidine derivatives. The less-reactive enolate of a ketone hydrazone also undergoes similar cyclization when the countercation is a ZnBu group.

Pyrrolidine Synthesis by a Intramolecular Alkenic Aldol Reaction[44]

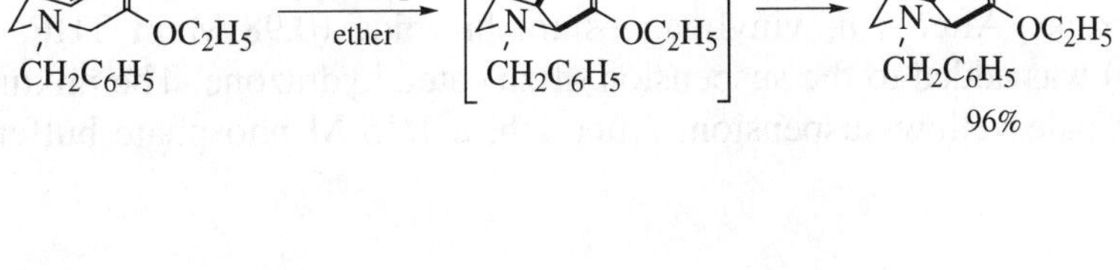

96%

Ethyl-*N*-benzylglycinate (0.500 g, 2.0 mmol) was dissolved in dry ether (5 mL), under argon, and cooled to −78 °C. LDA (2.2 mL, 2.2 mmol) was added and the reaction stirred for 5 min at −78 °C. A 1.0-M solution of ZnBr$_2$ in ether ZnBr$_2$ (6.0 mL, 6.0 mmol) was added at −90 °C, and the temperature was raised to 25 °C. After 20 min stirring, the reaction was quenched with saturated aqueous NH$_4$Cl, diluted with ether, and washed with aqueous NH$_4$Cl. The aqueous layers were then extracted with CH$_2$Cl$_2$. The combined organic layers were dried (MgSO$_4$) and concentrated to give 0.480 g (96%) of a pale yellow liquid as a single product.

Similarly, but almost certainly through a different mechanism, a zincated hydrazone also undergoes a rapid coupling reaction with a vinylmagnesium bromide. The resulting *γ,γ*-dimetalated hydrazone is stable and can be trapped sequentially with two different electrophiles. Thus, very efficient three- to four-component coupling synthesis of carbonyl compounds have been realized:

Coupling of Zincated Hydrazone with Vinyl Grignard Reagents[41]

To a solution of 1,5-diphenyl-3-pentanone *N,N*-dimethylhydrazone (2.86 mL, 10.0 mmol) in Et$_2$O (10.0 mL) was added *t*BuLi (1.48 M in pentane, 6.76 mL, 10.0 mmol) at −78 °C, and the mixture was warmed to 0 °C. After 4 h, ZnBr$_2$ (0.33 M in Et$_2$O, 30.0 mL, 10.0 mmol) was added to the solution of the lithiated hydrazone. After 1 h, vinylmagnesium bromide (0.98 M in THF, 10.2 mL, 10.0 mmol) was added to the suspension of zincated hydrazone. The mixture slowly turned to a pale yellow suspension. After 1 h, a 1/15 M phosphate buffer solution was added to the reaction mixture. The mixture was separated into an aqueous layer and an organic layer. The aqueous layer was extracted three times with Et$_2$O. The combined organic extracts were successively washed with 30% potassium sodium tartrate and saturated NaCl, dried over Na$_2$SO$_4$, and evaporated *in vacuo*. The residual oil (3.59 g) was purified on silica gel to obtain the desired compound (2.86 g, 93%, as a yellow oil) together with recovery of the starting hydrazone (7%).

The coupling of the vinylmagnesium bromide and the zincated hydrazone was achieved as above. After 1 h, dimethyldisulfide (0.99 mL, 11.0 mmol) and allylbromide (4.33 mL, 50.0 mmol) were added at 0 °C for 1 h and the reaction mixture was warmed to 25 °C. After 40 h. a 1/15 M phosphate buffer solution was added. The mixture was separated into an aqueous layer and an organic layer. The aqueous layer was extracted three times with Et$_2$O. The combined organic extracts were successively washed with 30% potassium sodium tartrate and saturated sodium chloride solution, dried over sodium sulfate, and evaporated *in vacuo*. The residual oil (3.55 g) was purified on silica gel to obtain the desired compound (2.64 g, 67%, as a yellow oil) as an isomeric mixture of isomers owing to the stereoisomerism of the hydrazone moiety and chiral centers. The same reaction starting with 114 mL (0.40 mmol) of the hydrazone afforded the desired compound in 84% yield (0.134 g).

5.5   Conjugate Addition to Electron-deficient Alkenes

The conjugate addition of dialkylzinc reagents R$_2$Zn to electron-deficient alkenes is not a good synthetic reaction, whereas triorganozinc species R$_3$Zn$^\ominus$ may be used as a donor of a non-stabilized carbanion in a conjugate addition. The true synthetic utility of organozinc compounds was discovered by the facility of zinc homoenolate species to undergo conjugate addition in the presence of a catalytic amount of Cu(I) halide, Me$_3$SiCl, and a strongly coordinating solvent (HMPA, DMPU)[158] or BF$_3$•Et$_2$O[159]. Stoichiometric use of CuCN allowed the use of RZnI reagents for conjugate addition. Use of a dummy methyl group in the formation of RZnMe$_2$$^\ominus$ (R = higher alkyl group) allowed the use of a catalytic amount of a Cu(I) catalyst and more economical use of the precious R group[160]. Nickel catalysis also promotes conjugate addition of the organozinc reagent[161]. The utility of the RZnX/Cu(I)/Lewis acid combination is now known to be extensive, as discussed in the organocopper chapter. Triorganozincate species are better alkyl donors and may be of some synthetic use:

$$R_3Zn^- \text{ or } [R_2Zn\bullet Cu(I) + MX_n] \; + \; \text{\includegraphics{}} \longrightarrow \text{\includegraphics{}}$$

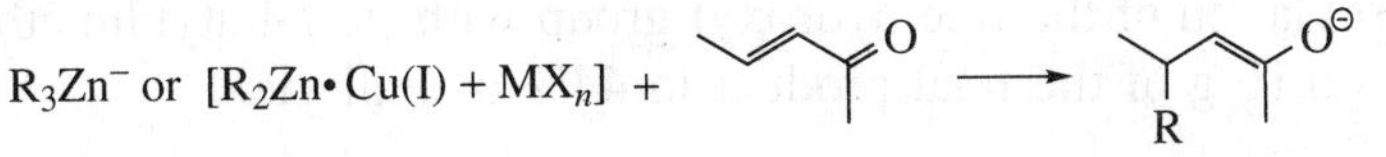

The conversion of a steroidal CD ring intermediate into a BCD ring precursor shown below illustrates the BF$_3$-mediated conjugate addition of dialkylzinc reagent in the presence of a catalytic amount of CuBr. The reaction of a mixture of zinc homoenolate/ Me₃SiCl/HMPA with the enone takes place in quantitative yield but with no stereoselectivity as to the C$_8$ chirality created by the reaction. Interestingly enough, when Me₃SiCl is removed and BF₃•Et₂O is employed as an accelerator, the reaction became almost totally β-selective. Subtle influence of added Lewis acid on the stereoselectivity of cuprate conjugate addition has been observed in various examples.

Stereoselective Conjugate Addition using the R₂Zn/Cu(I)(cat)/BF₃Et₂O system[162]

44% overall yield

To an acetonirile solution of NaI (19.1 g, 127 mmol) were added an acetonitrile solution (20 mL) of the hydindanone (6.7 g, 32 mmol), triethylamine (22.7 mL, 163 mmol) and chlorotrimethylsilane (16.0 mL, 126 mmol). The mixture was stirred at 25 °C for 40 min and extracted four times with hexane. The extract was washed with aqueous sodium bicarbonate and then with brine and dried. The crude oily enol silyl ether was dissolved in acetonitrile (50 mL), to which palladium acetate (7.0 g, 31 mmol) was added, and the mixture was stirred at room temperature for 5 h. Most of the solvent was removed *in vacuo*, and the residue was extracted with ethyl acetate/ether. The enone thus obtained was dissolved in 100 mL of THF containing CuBr•Me₂S (0.813 g, 3.95 mmol), to which a solution of zinc homoenolate (58 mmol) followed by HMPA (10 mL, 57.5 mmol) and BF₃•Et₂O (13 mL, 106 mmol) was added. After stirring for 3 h at 25 °C, the reaction mixture was poured into a column of alumina and kept there for 5 h to remove most of the silyl group. Elution with ether followed by aqueous HCl treatment afforded the hydroxy keto ester. Silylation of the free hydroxyl group with (*tert*-butyldimethyl)silyl chloride afforded 6.16 g of the final product in 44% overall yield.

Expeditious one-step synthesis of highly functionalized cyclopentenones has been achieved by using the copper-assisted carbozincation reaction of a zinc homoenolate as key step. An intermediary ester enolate species is acylated *in situ* to form the cyclopen-tenone skeleton. Very mild conditions allowed the presence of functional groups such as epoxide and acetals.

2-Carbalkoxy and 2-Carboxamidocyclopent-2-en-1-ones from Alkynic Esters or Amides[163]

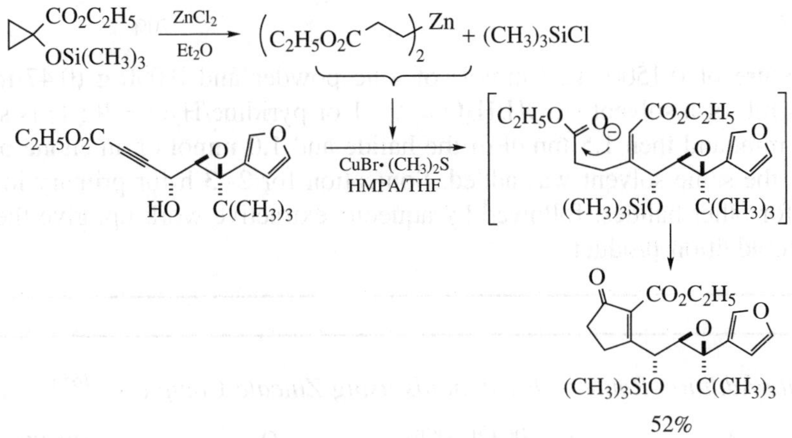

52%

To a stirring solution of 4.8 mL (24 mmol) of (1-ethoxycyclopropyl)oxy trimethyl-silane in 18 mL of diethyl ether was added 18 mL (18 mmol) of zinc chloride (1.0 M in ethyl ether) via syringe in one portion at 25 °C. This mixture was then sonicated for 40 min followed by stirring at 25 °C for an additional 10 min. At this point the heterogeneous mixture was cooled to 0 °C, and to this was added succes-sively 0.308 g (1.5 mmol) of copper(I) bromide dimethyl sulfide complex, 5 mmol of alkynic ester or amide in 18 mL of THF, and 4.2 mL (24 mmol) of HMPA. After addition, the mixture was allowed to stir for 5 min at 0 °C, and then the ice bath was removed and the stirring was continued for 4 h. The reaction was quenched with saturated ammonium chloride solution, and the organic layer was washed with half-saturated ammonium hydroxide solution until no blue color appeared in the wash. The resulting organic layer was washed with water and saturated sodium chloride solution and dried over magnesium sulfate. Concentration followed by chromatography of the crude oil gave the corresponding trimethylsilyl-protected hydroxy cyclopentenone.

The uncatalyzed conjugate reaction of organozinc compounds generally is not the method of choice if the same operation can be carried out with organocopper chemistry, which is usually general vastly superior. However, in some cases the zinc chemistry may be of considerable use. As mentioned previously, an alkylzinc reagent prepared by oxidative zincation with the Rieke zinc undergoes Me_3SiCl-mediated conjugate addition.

Zinc-mediated conjugate addition of an alkyl halide can be achieved in an aqueous medium under ultrasound irradiation.

Zinc-mediated Coupling of an Alkyl Halide and Enone or Enal[164]

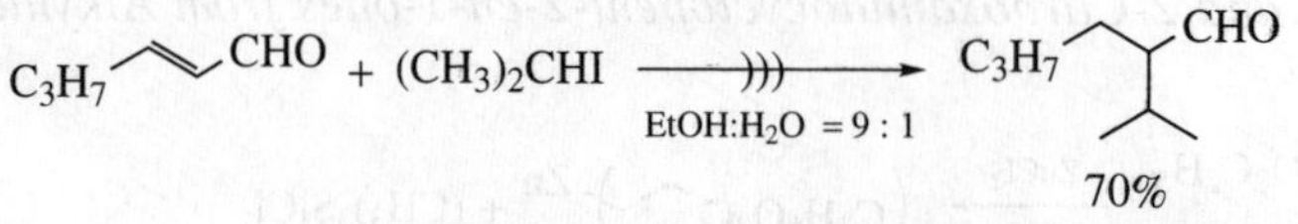

$$C_3H_7 \diagup\!\!\diagup CHO + (CH_3)_2CHI \xrightarrow[\text{EtOH:H}_2\text{O} = 9\,:\,1]{)))} C_3H_7 \diagup\!\!\diagup^{CHO}$$

70%

A mixture of 0.150 g (2.3 mmol) of zinc powder and 0.090 g (0.47 mmol) of CuI in 4 mL of a solvent (EtOH/H_2O = 9 : 1 or pyridine/H_2O = 9 : 1) is sonicated for 1–3 min, and then 1.5 mmol of the halide and 1.0 mmol of an enone or enal in 1 mL of the same solvent was added. Sonication for 2–3 h for primary iodide and 40 min for other halides, followed by aqueous extractive work-up, give the desired conjugate addition product.

Conjugate Addition Route to Prostanoids using Zincate Complexes[165]

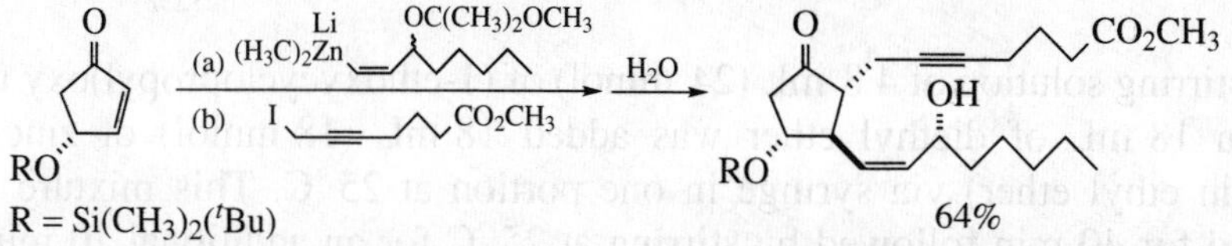

R = Si(CH₃)₂(ᵗBu)

64%

The vinyl iodide corresponding to the zincate structure above (1.16 mmol) was metalated (−78 °C, 1 h) in dry hexane with ᵗBuLi (1.20 mmol). Dimethylzinc was prepared separately from ZnCl₂•TMEDA (1.17 mmol) and MeLi (2.43 mmol) in dry THF at −20 °C. The (±) zincate was prepared by addition of the vinyllithium to a solution of ZnMe₂ in THF (in one portion at −78 °C, then at −50 °C for 1 h). The chiral enone (*R*)-4-hydroxycyclopent-Z-eno-1-one (0.39 mmol) in THF was added to the zincate at −78 °C (dropwise for 30 min, then at −40 °C for 30 min) and HMPA (3.89 mmol) was added at −30 °C. After cooling the mixture to −78 °C, the propargyllic iodide (1.99 mmol) in THF was added dropwise for 30 min and the reaction mixture was stirred for another 1 h (−78 °C to −50 °C). Usual work-up and hydrolysis of the alkylated product (50% aqueous AcOH/THF, 30 min at 25 °C) gave the product in 64% yield.

Organozinc compounds undergo nickel-catalyzed conjugated addition to certain enones[166], which has been made enantioselective by the ligand system using an optically active amino alcohol and an achiral diamine. The earliest example illustrates the sythetic utility and high ee of such a process.

Enantioselective Nickel-catalyzed Conjugate Addition[167]

$$R_2Zn + \quad \underset{Ar}{\overset{C_6H_5}{\diagdown}} =O \quad \xrightarrow{\text{Ni(acac)}_2/\text{ligand}/2,2\text{-Dipyridyl}} \quad \underset{Ar}{\overset{C_6H_5}{\underset{R}{\diagup}}} =O$$

ligand (14 mol %) = $\underset{HO}{\overset{C_6H_5}{H}} \underset{N(C_4H_9)_2}{\overset{CH_3}{H}}$ R = C$_2$H$_5$, C$_4$H$_9$; Ar = C$_6$H$_5$, 4-CH$_3$OC$_6$H$_4$

A mixture of Ni(acac)$_2$ (0.0360 g, 0.14 mmol) and (1S, 2R)-N,N-dibutylnor-ephedrine (DBNE; 0.0896 g, 0.34 mmol) in CH$_3$CN (2 mL) was heated with stirring at 80 °C for 1 h, then the solvent was removed in vacuo. 2,2′-Dipyridyl (0.0219 g, 0.14 mmol) and CH$_3$CN (2 mL) was added to the residue and the mixture was stirred at 80 °C for 1 h. The resulting green solution was cooled to 25 °C. An acetonitrile solution (4 mL) of chalcone (benzylideneacetophenone; 0.4165 g, 2.0 mmol) was added and the mixture was stirred at 25 °C for 20 min, then cooled to −30 °C. A toluene solution of diethylzinc (1.0 M, 2.4 mL, 2.4 mmol) was added and the reaction mixture was stirred for 12 h at −30 °C. The reaction was quenched by adding 1 M HCl and the mixture was extracted with dichloromethane. The extract was dried over anhydrous sodium sulfate and the solvent was evaporated under reduced pressure. The residue was purified by silica gel TLC (developing solvent, chloroform/hexane = 1 : 1) and the conjugated adduct with 90% ee was obtained in 47% yield. The ee was determined by HPLC using a chiral column.

6   Simmons–Smith Reaction

The Simmons–Smith reagent, which delivers an alkylidene unit to an ene, making a cyclopropane ring, has been the most representative class of zinc reagent[168]. In its original preparation reported in 1958[36], the reagent was prepared in the presence of an alkenic substrate by oxidative addition of a geminal diiodide and zinc–copper couple in refluxing ether. In the following years a number of modifications have been reported among which Furukawa modification reported in 1966 is the most notable[37]. The Furukawa reagent involves a metal exchange reaction between Et$_2$Zn and an iodide and can be performed in a variety of solvents including aromatic solvents and haloalkanes:

$$Zn + CH_2I_2 \xrightarrow[\text{Et}_2O]{} ICH_2ZnI \quad \text{Simmons-Smith reagent}$$

$$Et_2Zn + CH_2I_2 \xrightarrow[\substack{\text{various} \\ \text{aprotic} \\ \text{solvents}}]{} ICH_2ZnEt \quad \text{Furukawa reagent}$$

For a long time after the discovery of the reaction, the structure of the reagent and hence the mechanism of the reaction remained unclear. This was largely due to the instability of the reagent and the presence of Schlenk equilibrium of various species. Two decomposition pathways of the Furukawa reagent are shown below[169]. The cyclopropanation reaction must always compete with these side reactions, making mechanistic studies difficult and requiring synthetic studies to be carried out by the use of excess reagent:

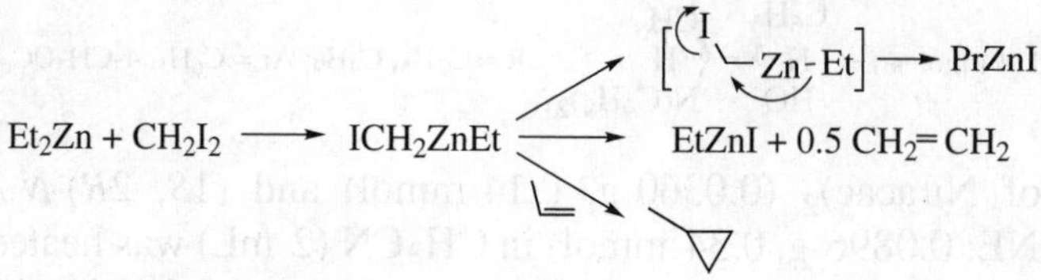

$$\text{Et}_2\text{Zn} + \text{CH}_2\text{I}_2 \longrightarrow \text{ICH}_2\text{ZnEt} \longrightarrow \text{EtZnI} + 0.5\ \text{CH}_2{=}\text{CH}_2$$

Despite these difficulties, considerable progress has been made in the structural elucidation of the reaction. The first key step was the structure determination of ((R)-2,3-dimethoxy-4,7,7-trimethylbicyclo[2.2.1]heptane-O,O′)-bis(iodomethyl)zinc in crystals[52], which revealed the iodomethylzinc structure of the Simmons–Smith reagent. More recent experimental[169,170] and theoretical studies[171] have resulted in the proposed mechanistic profile shown below, wherein the crucial activating effects of the Lewis acid are through aggregate formation (such as the binuclear cluster shown or possibly higher clusters of zinc alkoxides). Coordination of a suitable chiral ligand on the Lewis acid leads to the formation of enantiomerically enriched cyclopropanes:

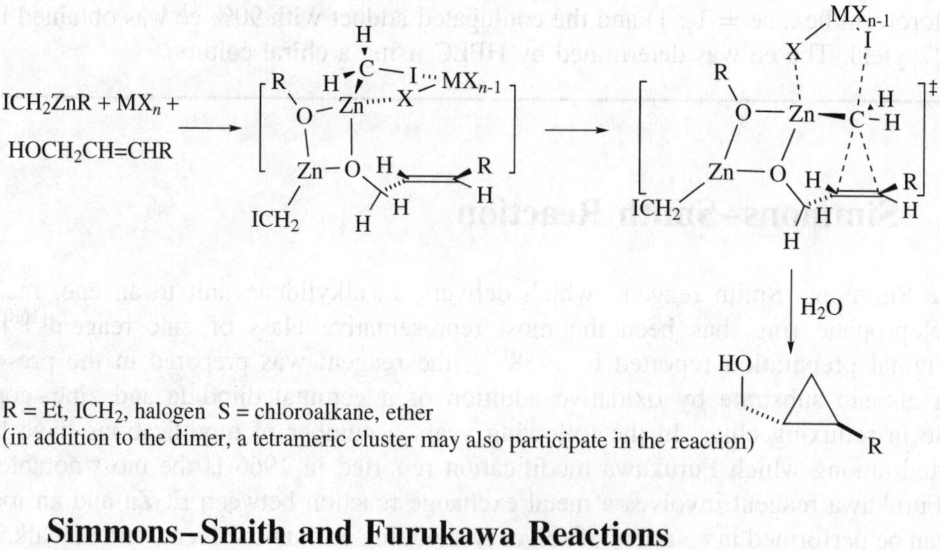

R = Et, ICH$_2$, halogen   S = chloroalkane, ether
(in addition to the dimer, a tetrameric cluster may also participate in the reaction)

6.1   Simmons–Smith and Furukawa Reactions

The Simmons–Smith reagent is typically prepared by heating a mixture of the alkenic substrate and Zn/Cu couple in ether. Under the optimum conditions for reactive alkenes, one can achieve good yields with near stoichiometric amounts of Zn/Cu, CH$_2$I$_2$ and the alkene, whereas excess reagent may be necessary for less reactive substrates to compensate the loss of the Simmons–Smith reagent.

Cyclopropanation of Cyclohexene using the Standard Simmons–Smith Reagent[172]

$$CH_2I_2 + Zn/Cu + \underset{Et_2O, \text{ reflux } 15 \text{ h}}{\bigcirc \longrightarrow} \bigcirc\hspace{-3pt}\triangleright$$

A mixture of methylene iodide (190 g, 0.71 mol), zinc–copper couple (46.8 g, 0.72 g atom), cyclohexene (53.3 g, 0.65 mol), anhydrous ether (250 mL), and a crystal of iodine was heated under reflux for 15 h. The decanted ether solution was washed with saturated ammonium chloride solution, sodium bicarbonate solution, and water. Distillation afforded 36 g (58% yield) of norcarane (bp 116–117 °C).

Although the Furukawa modification requires the use of pyrophoric Et_2Zn, it enjoys several assets over the Simmons–Smith reagent. First, it allows the use of aprotic solvents other than ethereal ones. This has proved to be especially important for asymmetric cyclopropanation, where the ether solvent reduces the selectivity as well as the reaction rate. The reaction conditions are milder (as exemplified by the following preparation) than the conventional Zn(Cu) procedure, which results in total destruction of the vinyl ether by cationic polymerization reaction. Although appearing hazadous in view of the pyrophoric nature of Et_2Zn, the presence of molecular oxygen accelerates the reaction[173].

Furukawa Cyclopropanation of Enol Ethers[174]

$$\overset{}{\diagup}\hspace{-4pt}O\hspace{-4pt}\diagdown\hspace{-2pt}\diagup \xrightarrow[\text{hexane, reflux}]{1/2\ Et_2Zn + CH_2I_2} \triangleright\hspace{-4pt}O\hspace{-4pt}\diagdown\hspace{-2pt}\diagup$$

A 500 mL round-bottomed flask with four necks was equipped with a large magnetic stirring bar, a stopper, a dropping funnel, a thermometer, and a reflux condenser connected to a nitrogen source through a three-way stopcock. Isobutyl vinyl ether (20 g, 0.20 mol) and 1.0 M Et_2Zn in hexane (150 mL, 0.15 mol) are introduced with a syringe via an open neck while nitrogen is introduced gently from the three-way stopcock. The flask is stopped and heated at 30 °C, while CH_2I_2 (70.6 g, 0.25 mol) was added dropwise from the funnel over 30 min to 1 h, while gentle refluxing is maintained. The reaction mixture became turbid. After 30 min heating the reaction was complete (GLC analysis), and the content of the flask is poured into saturated NH_4Cl. The aqueous layer was extracted several times, and the organic extracts were washed with 10% $NaHCO_3$ and then with water. After drying over $MgSO_4$ and distillation (bp 112 °C), the cyclopropane (19 g) was obtained in 84% yield. Ethyl iodide (bp 72 °C) can be removed by fractional distillation.

Careful studies of the practical and mechanistic aspects of the Simmons–Smith reaction have led recently to findings that the $ClCH_2I/Et_2Zn$ reagent is more reactive than the conventional CH₂I₂/Et₂Zn reagent and that the reaction takes place much faster in 1,2-dichoroethane than in a conventional ethereal solvent. It was also found that, in the cyclopropanation of an allylic alcohol, the reagent should be prepared prior to the addition of the alcohol rather than premixing Et₂Zn and the alcohol.

Preparation of (R,S)-trans-2-Phenyl-1-cyclopropylmethanol[175]

$$C_6H_5 \diagdown\diagup OH \xrightarrow[\text{1,2-dichloroethane}]{ClCH_2I + Et_2Zn} C_6H_5 \diagdown\triangleright\diagup OH$$

In a 25 mL, two-neck flask, a solution of Et₂Zn (410 µL, 4.0 mmol, 2.0 equiv) in dichloroethane (7 mL) was cooled to 0 °C, and ClCH₂I (585 µL, 8.0 mmol, 4.0 equiv) was added via syringe. The solution was stirred for 5 min at 0 °C, and solution of *trans*-cinnamyl alcohol (0.268 g, 2.0 mmol) in 1,2-dichloroethane (3 mL) was added slowly via syringe. The reaction mixture was stirred for 20 min at 0 °C, and then quenched with saturated NH₄Cl (20 mL). The reaction was allowed to warm to 25 °C, stirred vigorously for 10 min, and extracted with ether. The extracts were washed with H₂O and brine, combined, dried (K₂CO₃), filtered through a pad of silica gel, and concentrated at aspirator pressure. The crude product was purified by silica gel chromatography and bulb-to-bulb distillation to afford 0.275 g (93%) of the product as a clear colorless oil [bp 145–150 °C (2 torr)].

The Simmons–Smith cyclopropanation of alkenes bearing a nearby free alcohol takes place much faster than those without such a functionality, and the stereochemistry of the product suggests that the reagent is anchored by the oxygen functionality. As shown in Table 1, the rate of the Simmons–Smith reaction is much more influenced by the proximity of an oxygen functionality rather than by its electronic influence (e.g. enol ether)[176].

Table 1. Methylenation, with CH₂I₂/Zn–Cu, of cyclic allylic alcohols, one homoallyl alcohol and one ether: relative rate and configuration (position of the methylene bridge) relative to OH..

Compound	OH	OH	OCH₃	OH	OCH₃
k(rel)	1.00	0.46	0.50	0.091	0.059
cis =	100%	—	100%	100%	45%

A novel generation of a substituted zinc carbenoid by deoxygenation of a carbonyl compound with zinc metal and disilaethane has been reported, in which aromatic aldehydes, α,β-unsaturated aldehydes and ketones serve as precursors of the carbenoid, and lead to construction of novel cyclopropane structures.

Deoxygenative Generation of Zinc Carbenoids[177]

A solution of distilled 4,4-dimethyl-2-cyclohexen-1-one (247 µL, 1.9 mmol) in dry ether (1.8 mL) was added slowly using a syringe pump over 36 h to a vigorously stirred mixture of flame-dried zinc amalgam (1.3 g, 19 mmol, 10 equiv), dry ether (2 mL), styrene (0.43 mL, 3.8 mmol, 2 equiv) and 1,2-bis(chlorodimethylsilyl) ethane (2.20 mL of a 1.28 M solution in ether, 2.8 mmol, 1.5 equiv) under nitrogen at reflux. The cooled mixture was filtered through Celite and the separated zinc washed with ether (50 mL). The ethereal solution was washed with saturated aqueous sodium bicarbonate solution and brine, dried over $MgSO_4$ and then concentrated *in vacuo*. The crude residue was chromatographed to afford the spirooctene (0.221 g, 55% yield) as a colorless oil.

6.2   Stereoselective Cyclopropanation Reactions

Remarkable progress has been made in the preparation of enantiomerically enriched cyclopropanes. Earlier examples include the cyclopropanation of alkenes bearing a chiral auxiliary, such as chiral acetal, enol ether, boronic acid, and sugar moieties[178]. These must act as a group that anchors the zinc reagent.

Diastereoselective Cyclopropanation of Tartaric Ester Acetal[179]

$R = CH_3, C_3H_7, C_6H_5$
$R' = C_2H_5, {}^iC_3H_7$

88–94% ee

To a solution of the acetal, $R = Ph, R' = Et$ (6.4 g, 20 mmol), in dry hexane (220 mL) is added Et_2Zn (100 mmol, 32.3 mL of 3.1 M hexane solution) at $-20\,°C$. Methylene iodide (16.2 mL, 0.20 mmol) was added dropwise to the resulting stirred solution and the mixture was vigorously stirred at $-20\,°C$ for 6 h and $0\,°C$ for 6 h. The reaction mixture was poured into cold aqueous ammonium chloride and the product was repeatedly extracted with ether. The ether layer was washed with sodium thiosulfate and water. The combined ether layers were dried over sodium sulfate and concentrated *in vacuo*. Chromatography on silica gel afforded the pure cyclopropane, $R = Ph, R' = Et$, as a colorless oil (6.08 g, 91% yield; $[\alpha]^{25}D - 94.2°$ (c 1.03, EtOH)).

Several research groups have reported enantioselective Simmons–Smith cyclopropanation reactions using a chiral ligand[180]. Practically useful levels of enantioselectivity has been reported for allylic alcohol substrates. These reactions necessitate the use of more than one equivalent of the ligand and excess Furukawa reagent (partly because of decomposition), and hence are not yet in their ultimate ideal form. However, they have been applied to the synthesis of cyclopropane natural products[181].

Asymmetric Cyclopropanation of (E)-5-phenyl-2-penten-1-ol[182]

Ph⌇⌇⌇OH $\xrightarrow{\text{Et}_2\text{Zn, CH}_2\text{I}_2}$ Ph⌇△⌇OH

NHSO₂CF₃ / NHSO₂CF₃ (12 mol %)

To a solution of (1R,2R)-1,2-bis(p-nitrobenzenesulfonylamino)cyclohexane (0.354 g, 0.73 mmol, 12 mol %) and (E)-5-phenyl-2-penten-1-ol (0.988 g, 6.1 mmol) in 200 mL of anhydrous CH_2Cl_2 were added successively a hexane solution of Et_2Zn (0.98 M, 12.4 mL, 12.2 mmol) and CH_2I_2 (4.89 g, 18.3 mmol) in 20 mL of CH_2Cl_2 at −23 °C. The reaction mixture was stirred at that temperature for 5 h, then 40 mL of 2 M NaOH solution was added, and the product was extracted with Et_2O. The organic phase was washed with saturated aqueous NaCl, dried over anhydrous Na_2SO_4, and condensed under reduced pressure. The residue was chromatographed on silica gel to afford (2R,3R)-5-phenyl-2,3-methano-1-pentanol as a colorless oil (1.06 g, 100%; $[a]_D^{20}$ − 20.3° (c 1.14, $CHCl_3$)). The sulfonamide was recovered quantitatively from the combined aqueous solution after acidification with HCl solution. The enantiomeric excess of 5-phenyl-2,3-methano-1-pentanol was determined to be 82% ee by HPLC analysis by the use of a chiral stationary phase.

The reagent reported by Charette *et al.* has become the most popular so far. To increase its stability, the Furukawa reagent is complexed with DME, but is still used in excess because a smaller amount resulted in lower enantioselectivity (84% ee). Care must be taken as to the details of the conditions such as the rate of addition of the reagent, because they affect the yield and the selectivity.

Stoichiometric Simmons–Smith Cyclopropanation of Cinnamyl Alcohol[183]

Me₂NOC   CONMe₂

O   O
 \ /
  B
  |
  Bu
  (a)

C_6H_5⌇OH $\xrightarrow[\text{Ch}_2\text{Cl}_2, -10\,°\text{C, 2 h}]{\text{(b) (ICH}_2)_2\text{Zn· DME}}$ C_6H_5⌇△⌇OH

100%, 93% ee

(a) Preparation of dioxaborolane

To a solution of 30.6 g (0.15 mol) of (+)-N,N,N',-tetramethyltartaric acid diamide in 100 mL of anhydrous toluene was added 18.3 g (0.18 mol) of 1-butaneboronic acid. The mixture was heated under reflux to remove the H_2O produced in the reaction (Dean–Stark, 15 h). The reaction mixture was cooled to 25 °C and concentrated under reduced pressure. The residue was dissolved in a minimum of CH_2Cl_2, filtered to remove excess 1-butaneboronic acid, and concentrated under reduced pressure to produce 37.8 g (93%) of the desired dioxaborolane.

(b) Preparation of the $Zn(CH_2I)_2 \cdot DME$ complex solution in CH_2Cl_2

To a solution of 3.8 mL (37 mmol) of Et_2Zn in 37 mL of CH_2Cl_2 and 3.9 mL (37 mmol) of freshly distilled DME at −15 °C was added 6.0 mL (75 mmol) of CH_2I_2 at a rate to keep the internal temperature below −10 °C (about 20 min). This clear colorless solution was used directly in the cyclopropanation reaction.

(c) General procedure for the enantinoselective cyclopropanation of cinnamyl alcohol: (2S,3S)-*trans*-(3-Phenylcyclopropyl)methanol

To a mixture of 2.20 g (8.2 mmol, 1.1 equiv) of dioxaborolane, 1.0 g (7.5 mmol, 1.0 equiv) of cinnamyl alcohol, and 0.300 g of 4 A molecular sieve in 37 mL of CH_2Cl_2 at −15 °C was added the previously prepared solution of $Zn(CH_2I)_2$ DME complex at a rate to keep the internal temperature below −10 °C (about 20 min). The resulting mixture was stirred at −10 °C for 2 h, after which time TLC analysis showed complete consumption of the starting material. A saturated aqueous NH_4Cl was added, and the layers were separated. The aqueous layer was washed three times with ether. The combined organic layers were then stirred vigorously for 12 h with aqueous KOH (5 M), and the layers were separated. The organic layer was then successively washed with 10% aqueous HCl, saturated aqueous $NaHCO_3$, H_2O, saturated aqueous NaCl, and concentrated under reduced pressure. The residue was chromatographed on silica gel to afford 1.10 g (100%) of the desired cyclopropylmethanol that was identical in all respects to known material ($[\alpha]_D$+66° (*c* 1.9, EtOH)).

7   Substitution at Carbon by Organozinc Reagents

Organozinc compounds are rather unreactive nucleophiles toward organic halides and need nickel or palladium catalysts to achieve such cross-coupling reactions. Although the basic reactivities of the organozinc reagents are similar to those of the Grignard reagents, the former are much milder so that various functional groups may be compatible with the reaction. In certain cases, transmetalation from magnesium to zinc improves the yield of the overall transformation. The proposed mechanism of the $Pd(PPh_3)_4$-catalyzed reaction is shown below. As the characteristics of the substitution reaction are more influenced by the nature of the organic halide rather than the organozinc reagent, this section is classified according to the structure of the organic halides.

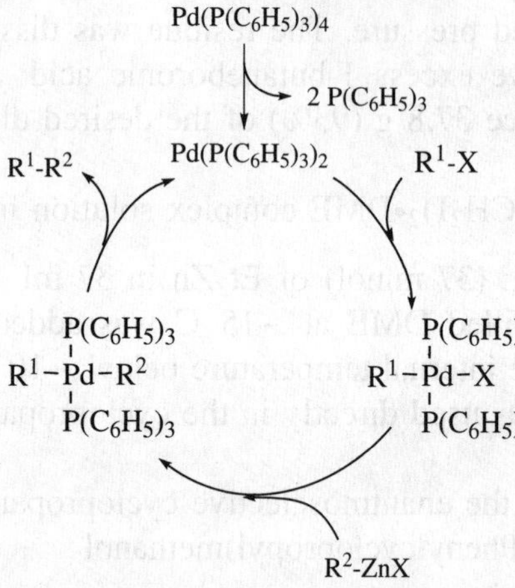

$Pd(P(C_6H_5)_3)_4$

2 $P(C_6H_5)_3$

R^1-R^2 $Pd(P(C_6H_5)_3)_2$ R^1-X

$\begin{array}{c} P(C_6H_5)_3 \\ | \\ R^1-Pd-R^2 \\ | \\ P(C_6H_5)_3 \end{array}$ $\begin{array}{c} P(C_6H_5)_3 \\ | \\ R^1-Pd-X \\ | \\ P(C_6H_5)_3 \end{array}$

R^2-ZnX

7.1   Reaction with Allylic Halides

Substitution reactions of alkyl halides are scarce, because of the low reactivity of organozinc reagents under noncatalyzed conditions and of the poor efficiency of transition metal catalysis. The latter is due to the slow rate of oxidative addition of alkyl halides to transition metal complexes, and the instability of the alkyl transition metal intermediates (*e.g.* β-hydride elimination). Allylic halides are also unreactive under ordinary conditions. However, a unique case reported that they do react with the alkylzinc reagents at 25 °C in the presence of a strong donor ligand such as DMF, HMPA, and TMEDA. The reaction is highly S_N2' selective and may generate a chiral center, whose stereochemistry can be controlled by the chirality in a nearby center with high selectivity.

Diastereoselective and S_N2'-Selective Alkylation of Allylic halides[184]

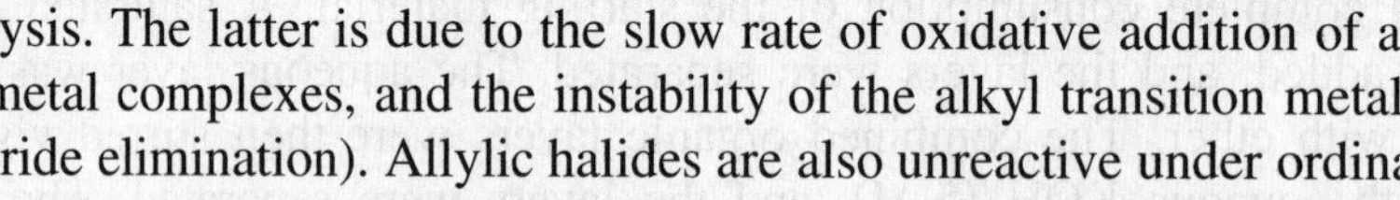

$(C_4H_9)_2Zn \cdot 2LiCl$ + C$_6$H$_5$—CH$_3$ allyl Cl $\xrightarrow[\substack{\text{hexane/THF} \\ -70\,°C \text{ to } 25\,°C, 4\text{ h}}]{\text{HMPA (3 equiv)}}$ C$_6$H$_5$—CH$_3$ allyl C$_4$H$_9$

87%
97% S_N2', 89% ds

To a solution of $ZnCl_2$ (1.0 M THF, 1.5 mmol, 1.5 mL) was added BuLi (1.58 M hexane, 3.0 mmol) at −70 °C. After addition of HMPA (3 mmol), an allylic chloride (1.0 mmol) was added at −70 °C. The reaction mixture was warmed to 25 °C, and stirred for 4 h. After addition of ca. 2 mL of hexane saturated with water, the reaction mixture was passed through a pad of silica gel. Capillary GC analysis of the filtrate indicated the S_N2'/S_N2 ratio. Purification of the crude product by silica gel column chromatography afforded the allylation product.

A Cu(I) catalyst can catalyze the coupling of an organozinc reagent with an allylating agent (halide and phosphates) with a high level of S_N2' selectivity[185]. When a nickel or palladium catalyst is employed, the alkylation reaction takes place with the less hindered side of the allylic system[186].

7.2 Reaction with Alkenyl Halides

Coupling of an alkenyl halide and an organozinc compound can be achieved in the presence of a nickel or palladium catalyst. Nickel is more reactive but palladium is more selective, higher yielding, and shows wider functional group tolerance. The palladium-catalyzed reactions are applicable to alkyl, allyl, alkenyl, and arylzinc reagents, and to chloro-, bromo-, and iodoalkenes, whose reactivities increase in this order. The use of a nickel catalyst may be preferred for the less-reactive alkenyl chlorides. Iodoalkenes tend to give byproducts (*e.g.* homocoupling products) under nickel catalysis, which can be avoided under palladium catalysis. Alkenyl triflates, which can be prepared readily from carbonyl compounds, also couple smoothly with organozinc reagents under palladium catalysis. In all cases, the stereochemistry of the alkenylating agents is retained in the product.

In the reaction of an alkylzinc reagent bearing a hydrogen at the β position, an intermediary alkyl transition metal may undergo metal hydride elimination to give various side products. For instance, the reaction of a secondary butylzinc halide with a haloalkene produces a normal butylalkene and reduced alkene. The latter product may form in the reaction of primary alkylzinc halide. Homoallylic and homopropargylic zinc halides are less prone to β elimination and afford the 1,5-diene and 1,5-enyne in high yields, even in the presence of a simple phosphine such as $Pd(PPh_3)_4$[187]. Alkenyl-, alkynyl-, and arylzinc reagents react with alkenyl halides and triflates under nickel or palladium catalysis to give 1,3-dienes, 1,3-enynes and styrenes with retention of the stereochemistry of the zinc reagents.

Pd-catalyzed Cross-coupling of Arylzinc Halides with Alkenyliodide[188]

80%

A solution of (4-methoxyphenyl)zinc chloride was prepared by adding (4-methoxyphenyl)lithium (prepared by reacting 4-bromoanisole (1.68 g, 9.0 mmol, 3 equiv) with lithium metal (0.249 g, 36 mmol, 12 equiv) in 6 mL of anhydrous ether at 25 °C for 2 h) to a solution of anhydrous zinc chloride (1.23 g, 9.0 mmol, 3 equiv) in anhydrous THF (12 mL). The resultant solution was refluxed for 30 min and then cooled to 25 °C. This solution was reacted, in a separate flask, with a mixture of $Pd(P(C_6H_5)_3)_4$ (0.173 g, 0.05 equiv) and (E)-1,2-diphenyl-1-iodo-1-butene (1.00 g, 3.0 mmol) in anhydrous THF (7 mL). The resultant mixture was refluxed for 1 h and cooled to 25 °C, quenched with 3 M hydrochloric acid, and extracted with hexane. The combined organic layers were washed with saturated sodium bicarbonate solution and dried over anhydrous sodium sulfate. After removal of the solvent, the residue was chromatographed on silica gel to give 0.754 g (80%) of white crystalline product.

7.3   Reactions of Aryl Halides

Aryl halides also serve as good substrates for the reaction with organozinc reagents under nickel or palladium catalysis. The reactivity increases in the order of chloride, bromide, and iodide. The compound $PdCl_2(dppf)$ is a suitable catalyst for the reaction of alkylzinc halides, especially secondary alkylzinc reagents, which tend to undergo β-elimination as discussed above.

Pd-catalyzed Cross-coupling of sec-Butylzinc Chloride with Bromobenzene[9,10]

In a 50 mL two-necked flask equipped with a stirring bar, a serum cap, and a three-way stopcock, the catalyst (0.0293 g, 0.04 mmol) and anhydrous $ZnCl_2$ (1.23 g, 9.0 mmol) were placed. The reaction vessel was filled with argon after evacuation and then charged, at −78 °C, with anhydrous THF (10 mmol) and ^sBuMgCl (8.0 mmol) in diethyl ether with a syringe through the serum cap. To the mixture was then added bromobenzene (0.628 g, 4.0 mmol) at −78 °C. The reaction mixture was stirred at 25 °C overnight and hydrolyzed with 10% aqueous HCl. The organic layer and diethyl ether extracts from the aqueous layer were combined, washed with saturated sodium hydrogen carbonate solution, and then with water, and dried over Na_2SO_4. After evaporation of the solvent, bulb-to-bulb distillation afforded 0.51 g (95%) of sec-butylbenzene.

Functionalized alkylzinc reagents prepared by ZnCl₂-mediated ring opening of siloxycyclopropanes[23] or by oxidative zincation of alkyl iodides with Zn(Cu)[189] react smoothly with aryl bromides and iodides in the presence of PdCl₂[P(2-CH₃-C₆H₄)₃]₂. The reaction is especially useful for functionalization of heteroaromatic compounds.

Pd-catalyzed Cross-coupling of Zinc Homoenolate with o-Tolylbromide[23]

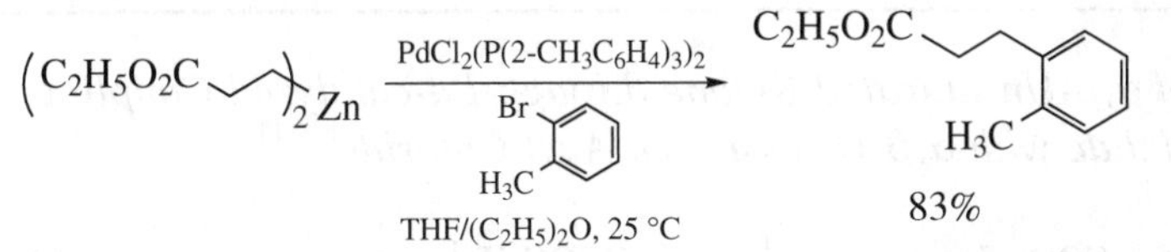

A crude approximately 0.3 M ethereal solution of the zinc homoenolate[23a] was concentrated *in vacuo* (1 mmHg) at 30–40 °C. The yield was assumed to be about 80%. The residue was dissolved in THF to make a 0.5 M solution. A halide and PdCl₂(P(2-CH₃-C₆H₄)₃)₂ (5 mol %) was weighed into another flask under nitrogen, and to it was added 2.5 equiv of the homoenolate at 0 °C. The mixture was stirred overnight at 25 °C. About 5 equiv each of powdered KF and water were added, the suspension was stirred for 3 h, and the organic phase was separated, dried over Na₂SO₄, and concentrated under a reduced pressure. The residue was purified by chromatography on silica gel to afford the desired product (83%).

Pd-catalyzed Cross-coupling of a Functionalized Alkylzinc Halide with a Heteroaryl Halide[190]

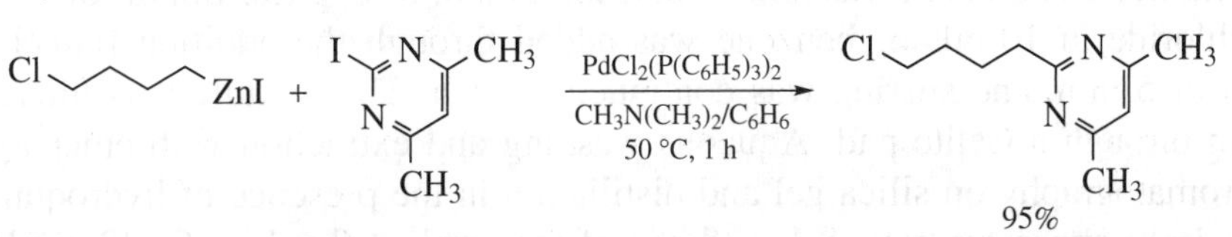

A mixture of zinc–copper couple (0.75 g, 11.5 mmol) and 4-chloro-1-iodobutane (7.5 mmol) in dry benzene (13 mL) and dry *N,N*-dimethylacetamide (1 mL) was stirred at 70 °C for 3–4 h under a nitrogen atmosphere. Then, PdCl₂(P(C₆H₅)₃)₂ (0.13 g, 0.19 mmol) and a solution of 2-iodo-4,6-dimethylpyrimidine (5.0 mmol) in dry benzene (2 mL) were added at 25 °C. The mixture was stirred at 50 °C for 1 h, then washed with water, and dried over MgSO₄. The solvent was removed and the residue was purified by column chromatography on silica gel. After concentration, bulb-to-bulb distillation gave the pure product in 95% yield.

7.4 Reactions of Acyl Chlorides

Whereas the ketone synthesis by the reaction of an organozinc reagent with acid halide was a classical synthetic operation, its efficiency has been dramatically improved by palladium catalysis. Both organozinc halides[24,191] and diorganozinc reagents[192] serve as nucleophiles. Alkenylzinc and alkynylzinc reagents give α,β-unsaturated and alkynic ketones, respectively. β- and γ-Zinco esters give the respective ketoesters in high yields.

Synthesis of α,β-Unsaturated Ketone through Pd-catalyzed Coupling of an Alkylzinc Halide with α,β-Unsaturated Acyl Chloride[193]

A 300 mL, four-necked round-bottomed flask containing a magnetic stirring bar was fitted with a serum cap, a thermometer, a 100 mL serum-capped pressure-equalizing addition funnel, and a reflux condenser equipped at the top with a nitrogen inlet. The dry apparatus was flushed with nitrogen and 5.6 g (85.5 mmol) of zinc–copper couple and 20 mL of benzene were introduced. A mixture of 13.8 g (57 mmol) of ethyl 4-iodobutyrate, 9 mL of N,N-dimethylacetamide, and 70 mL of benzene was transferred into the addition funnel by cannulation techniques and added to the stirred Zn(Cu) suspension over 3 min at 25 °C. The mixture was vigorously stirred for 1 h at 25 °C and then heated at gentle reflux with an oil bath for 4.5 h. After the mixture was cooled to 60 °C, a solution of benzene was added over 1 min through the addition funnel and stirring was continued for 5 min at the same temperature. The oil bath was removed, a solution of 5.23 g (50 mmol) of methacryloyl chloride in 10 mL of benzene was added through the addition funnel over a period of 5 min, and stirring was continued for 1 h. The mixture was filtered with suction through a Celite pad. Aqueous washing and extraction with ether followed by chromatography on silica gel and distillation in the presence of hydroquinone in a Kugelrohr apparatus gave 8.1 g (88%) of the product (bp 185 °C, 20 mmHg).

---

8 References

[1] Frankland, E.; *Liebigs Ann. Chem.* **1849**, *71*, 171.

[2] Wanklyn, J. A.; *Liebigs Ann. Chem.* **1858**, *107*, 125.

[3] Nützel, K.; in *Houben-Weyl: Methoden der Organis Chemie*, Vol. 1 3/2a, (ed.: Müller, E.) Thieme, Stuttgart, **1973**, 553.

[4] (a) Boersma, J.; in *Comprehensive Organometallic Chemistry*, Vol. 2, (eds: G. Wilkinson, F. G. A. Stone, E. W. Abel), Pergamon, New York, **1982**, Chap. 16; (b) O'Brien, P.; in *Comprehensive Organometallic Chemistry* Vol 3, (J. R. Wardell, Ed.), Elsevier, Oxford, **1995**, 175. (c) Erdik, E.; *Organozinc Reagents in Organic Synthesis*, CRC Press, Boca Raton, **1996**.

[5] Reformatsky, S. N., *Ber. Dtsch. Chem. Ges.* **1887**, *20*, 1210.

[6] Hunsdiecker, H.; Erlbach, H.; Vogt, E.; German Patent 722467, **1942**; *Chem. Abstr.* **1943**, *37*, p5080.

[7] Rathke, M. W.; *Org. React.* **1975**, *22*, 423.

[8] (a) Negishi, E.; King, A. O.; Okukado, N.; *J. Org. Chem.* **1977**, *42*, 1821; (b) Negishi, E.; Bagheri, V.; Chatterjee, S.; Luo, F. T.; *Tetrahedron Lett.* **1983** *24*, 5181.

[9] Hayashi, T.; Konishi, M.; Kobori, Y.; Kumada, K.; Higuchi, T.; Hirotsu, K.; *J. Am. Chem. Soc.* **1984**, *106*, 158.

[10] Minato, A.; Tamao, K.; Hayashi, T.; Suzuki, K.; Kumada, M.; *Tetrahedron Lett.* **1980**, *21*, 845.

[11] (a) Hofstee, H. K.; van der Meulen, J. D.; van der Kerk, G. J. M.; *J. Organomet. Chem.* **1978**, *153*, 245; (b) Dekker, J.; Münningghoff, J. K.; Boersma, J.; Spek, A. L.; *Organometallics* **1987**, *6*, 1236.

[12] (a) Thiele, K. H.; Heinrich, M.; Brüser, W.; Schröder, S, Z.; *Anorg. Allg. Chem.* **1977**, *221*, 432; (b) Thiele, K. H.; Langguth, E.; Muller, D. E. Z.; *Anorg. Allg. Chem.* **1980**, *462*, 152.

[13] (a) Kuwajima, I.; Nakamura, E.; *Topics Curr. Chem.*, **1990**, *155*, 1; (b) Kuwajima, I.; Nakamura, E.; in *Comprehensive Organic Synthesis*, Vol 2, (eds: B. M. Trost, I. Fleming), Pergamon Press, Oxford, **1991**, Chapter 1.14.

[14] (a) Nakamura, E.; Kuwajima, I.; *J. Am. Chem. Soc.* **1983**, *105*, 651; (b) Nakamura, E.; Kuwajima, I.; *J. Am. Chem. Soc.* **1977**, *99*, 7360; (c) Ryu, I.; Ando, M.; Ogawa, A.; Murai, S.; Sonoda, N.; *J. Am. Chem. Soc.* **1983**, *105*, 7192.

[15] Nakamura, E.; Kuwajima, I.; *J. Am. Chem. Soc.* **1985**, *107*, 2138.

[16] Nakamura, E.; Kuwajima, I.; *J. Am. Chem. Soc.* **1984**, *106*, 3368.

[17] Nakamura, E.; in *Organocopper Reagents*, (ed.: R. J. K. Taylor), Oxford University Press, Oxford, **1994**, Chapter 6, 129.

[18] Horiguchi, Y.; Nakamura, E.; Kuwajima, I.; *J. Org. Chem.* **1986**, *51*, 4323.

[19] Nakamura, E.; Kuwajima, I.; *Tetrahedron Lett.* **1986**, *27*, 83.

[20] (a) Sekiya, K.; Nakamura, E,; *Tetrahedron Lett.* **1988**, *29*, 5155; (b) Nakamura, E.; Sekiya, K.; Kuwajima, I.; *Tetrahedron Lett.* **1987**, *28*, 337.

[21] Oshino, H.; Nakamura, E.; Kuwajima, I.; *J. Org. Chem.* **1985**, *50*, 2802.

[22] Yeh, M. C. P.; Knochel, P.; Santa, L. E.; *Tetrahedron Lett.* **1988**, *29*, 3887.

[23] (a) Nakamura, E.; Aoki, S.; Sekiya, K.; Oshino, H., Kuwajima, I.; *J. Am. Chem. Soc.* **1987**, *109*, 8056; (b) Nakamura, E.; Kuwajima I.; *Org. Synth. Coll.* **1993**, *8*, 277.

[24] (a) Tamaru, Y.; Ochiai, H.; Sanda, F.; Yoshida, Z.; *Tetrahedron Lett.* **1985**, *26*, 5559. (b) Tamaru, Y.; Ochiai, H.; Nakamura, T.; Tsubaki, K.; Yoshida, Z.; *Tetrahedron Lett.* **1985**, *26*, 5559.

[25] Tamaru, Y.; Ochiai, H.; Nakamura, T.; Yoshida, Z.; *Angew. Chem. Int. Ed. Engl.* **1987**, *26*, 1157.

[26] Knochel, P.; Yeh, M. C. P.; Berk, S. C.; Talbert, J.; *J. Org. Chem.* **1988**, *53*, 2390.

[27] Knochel, P.; Singer, R. D.; *Chem. Rev.* **1993**, *93*, 2117.

[28] (a) Noyori, R.; Kitamura, M.; *Angew. Chem., Int. Ed. Engl.* **1991**, *30*, 49; (b) Soai, K.; Niwa, S.; *Chem. Rev.* **1992**, *92*, 833.

[29] Oguni, N.; Omi, T.; *Tetrahedron Lett.* **1984**, *25*, 2823.

[30] (a) Yamakawa, M.; Noyori, R.; *J. Am. Chem. Soc.* **1995**, *177*, 6327; (b) Nakamura, M.; Nakamura, E.; Koga, N.; Morokuma, K.; *J. Am. Chem. Soc.* **1993**, *115*, 11016.

[31] Kitamura, M.; Suga, S.; Kawai, K.; Noyori, R.; *J. Am. Chem. Soc.* **1986**, *108*, 6071.

[32] Oguni, N.; Matsuda, Y.; Kaneko, T.; *J. Am. Chem. Soc.* **1988**, *110*, 7877.

[33] Noyori, R.; Suga, S.; Kawai, K.; Okada, S.; Kitamura, M.; *Pure Appl. Chem.* **1988**, *60*, 1597.

[34] Yoshioka, M.; Kawakita, T.; Ohno, M.; *Tetrahedron Lett.* **1989**, *30*, 1657.

[35] Rozema, M. J.; Achyatha Rao, S.; Knochel, P. *J. Org. Chem.*; **1992**, *57*, 1956.

[36] (a) Simmons, H. E.; Smith, R. D.; *J. Am. Chem. Soc.* **1958**, *80*, 5323; (b) Simmons, H. E.; Smith, R. D.; *J. Am. Chem. Soc.* **1959**, *81*, 4256.

[37] (a) Furukawa, J.; Kawabata, N.; Nishimura, J.; *Tetrahedron* **1968**, *24*, 53; (b) Furukawa, J.; Kawabata, N.; Nishimura, J.; *Tetrahedron Lett.* **1968**, 3495.

[38] Kobayashi, S.; Takahashi, H.; Yoshioka, M.; Ohno, M.; *Tetrahedron Lett.* **1992**, *33*, 2575.

[39] Gaudemar, M.; *C.R. Séances Acad. Sci. Série C*, **1971**, *273*, 1669; *Chem. Abstr. 76*: 72623.

[40] Marek, I.; Normant, J. F.; *Chem. Rev.* **1996**, *96*, 3241.

[41] Nakamura, E.; Kubota, K.; Sakata, G.; *J. Am. Chem. Soc.* **1997**, *119*, 5457.

[42] Kubota, K.; Nakamura, E.; *Angew. Chem., Int. Ed. Engl.* **1997**, *36*, 2491.

[43] (a) Nakamura, E.; Kubota, K.; *J. Org. Chem.* **1997**, *62*, 792; (b) Nakamura, E.; Kubota, K.; *Tetrahedron Lett.* **1997**, *38*, 7099.

[44] (a) Karoyan, P.; Chassaing, G.; *Tetrahedron Lett.* **1997**, *38*, 85; (b) Lorthiosis, E.; Marek, I.; Normant, J. F.; *Tetrahedron Lett.* **1997**, *38*, 89.

[45] Rieke, R. D.; Hanson, M. V.; *Tetrahedron*, **1997**, *53*, 1925.

[46] Nehl, H.; Scheidt, W. R.; *J. Organomet. Chem.* **1985**, *289*, 1.

[47] Al-Juaid, S. S.; Eaborn, C.; Habtemariam, A.; Hitchcock, P. B.; Smith, J. D.; Tavakkoli, K.; Webb, A. D.; *J. Organomet. Chem.* **1993**, *462*, 45.

[48] Moseley, P. T.; Shearer, H. M. M.; *J. Chem. Soc. Dalton Trans.* **1973**, 64.

[49] Ernst, R. D.; Freeman, J. W.; Swepston, P. N., Wilson, D. R.; *J. Organomet. Chem.* **1991**, *402*, 17.

[50] O'Brien, P.; Hursthouse, M. B.; Motevalli, M.; Walsh, J. R.; Jones, A. C.; *J. Organomet. Chem.* **1993**, *449*, 1.

[51] Purdy, A. P.; George, C. F.; *Organometallics.* **1992**, *11*, 1955.

[52] Denmark, S. E.; Edwards, J. P.; Wilson, S. R.; *J. Am. Chem. Soc.* **1992**, *114*, 2592.

[53] Behm, J.; Lotz, S. D.; Herrmann, W. A.; *Z. Anorg. Allg. Chem.* **1993**, *619*, 849.

[54] Dekker, J.; Budzelaar, P. H. M.; Boersma, J.; van der Kerk, G. J. M.; Spek, A. L.; *Organometallics* **1984**, *3*, 1403.

[55] van der Steen, F. H.; Boersma, J.; Spek, A. L.; van Koten, G.; *Organometallics* **1991**, *10*, 2467.

[56] Erdik, E.; *Tetrahedron*, **1987**, *43*, 2203.

[57] (a) LeGoff, E.; *J. Org. Chem.* **1964**, *29*, 2048; (b) Benefice-Malouet, S.; Blancou, H.; Commeyras, A.; *J. Fluorine Chem.* **1985**, *30*, 171.

[58] (a) Curran, T. T.; *J. Org. Chem.* **1993**, *58*, 6360; (b) Jiang, B.; Xu, Y.; *J. Org. Chem.* **1991**, *56*, 7336.

[59] (a) Takai, K.; Kakiuchi, T.; Kataoka, Y.; Utimoto, K.; *J. Org. Chem.* **1994**, *59*, 2668; (b) Takai, K.; Kakiuchi, T.; Utimoto, K.; *J. Org. Chem.* **1994**, *59*, 2671.

[60] Newman, M. S.; Evans, F. J.; *J. Am. Chem. Soc.* **1995**, *117*, 946.

[61] Picotin, G.; Miginiac, P.; *J. Org. Chem.* **1987**, *52*, 4796.

[62] (a) Yeh, M. C. P.; Chen, H. G.; Knochel, P.; *Org. Synth.* **1992**, *70*, 195; (b) Achyutha Rao, S.; Knochel, P.; *J. Am. Chem. Soc.* **1991**, *113*, 5735.

[63] (a) Rieke, R. D.; Hudnall, P. M.; Uhm, S. J.; *J. Chem. Soc., Chem. Commun.* **1973**, 269; (b) Rieke, R. D.; Uhm, S. J.; *Synthesis* **1975**, 452; (c) Arnold, R. T.; Kulenovic, S. T.; *Synth. Commun.* **1977**, *7*, 223.

[64] Hanson, M. V.; Rieke, R. D.; *J. Am. Chem. Soc.* **1995**, *117*, 10775.

[65] (a) Zhu, L.; Wehmeyer, R. N.; Rieke, R. D.; *J. Org. Chem.* **1991**, *56*, 1445; (b) Rieke, R. D.; Li, P. T.-J.; Burns, T. M.; Uhm, S. J.; *J. Org. Chem.* **1981**, *46*, 4323.

[66] Boudjouk, P.; Thompsom, D. P.; Ohrbom, W. H.; Han, B.-H.; *Organometallics* **1986**, *5*, 1257.

[67] (a) Hanson, M.; Rieke, R. D.; *Synth. Commun.* **1995**, *25*, 101; (b) Rieke, R. D.; Hanson, M. V.; Brown, J. D.; Niu, Q. J.; *J. Org. Chem.* **1996**, *61*, 2726; (c) Hanson, M. V.; Brown, J. D.; Rieke, R. D.; Niu, Q. J.; *Tetrahedron Lett.* **1994**, *35*, 7205.

[68] Stadtmüller, H.; Greve, B.; Lennick, K.; Chair, A.; Knochel, P.; *Synthesis* **1995**, 69.

[69] Boldrini, G. P.; Savoia, D.; Tagliavini, E.; Trombini, C.; Umani-Ronchi, A.; *J. Org. Chem.* **1983**, *48*, 4108.

[70] (a) Fürstner, A.; *Synthesis* **1989**, 517; (b) Csuk, R.; Fürstner, A.; Weidmann, H.; *J. Chem. Soc., Chem. Commun.* **1986**, 775.

[71] Corey, E. J.; Hannon, F. J.; *Tetrahedron Lett.* **1987**, *28*, 5233.

[72] von dem Bussche-Hünnefeld, J. L.; Seebach, D.; *Tetrahedron*, **1992**, *48*, 5719.

[73] Laloë, E.; Srebnik, M.; *Tetrahedron Lett.* **1994**, *35*, 5587.

[74] Langer, F.; Waas, J.; Knochel, P.; *Tetrahedron Lett.* **1993**, *34*, 5261.

[75] Langer, F.; Schwink, L.; Devasagayaraj, A.; Chavant, P.-Y.; Knochel, P.; *J. Org. Chem.* **1996**, *61*, 8229.

[76] Devasagayaraj, A.; Schwink, L.; Knochel, P.; *J. Org. Chem.* **1995**, *60*, 3311.

[77] Micouin, L.; Oestreich, M.; Knochel, P.; *Angew. Chem., Int. Ed. Engl.* **1997**, *36*, 245.

[78] Rozema, M. J.; Sidduri, A.; Knochel, P.; *J. Org. Chem.* **1992**, *57*, 1956.

[79] Rozema, M. J.; Eisenberg, C.; Lütjens, H.; Ostwald, R.; Belyk, K.; Knochel, P.; *Tetrahedron Lett.* **1993**, *34*, 3115.

[80] Klement, I.; Knochel, P.; Chau, K.; Cahiez, G.; *Tetrahedron Lett.* **1994**, *35*, 1177.

[81] Vettel, S.; Vaupel, A.; Knochel, P.; *J. Org. Chem.* **1996**, *61*, 7473.

[82] Vettel, S.; Vaupel, A.; Knochel, P.; *Tetrahedron Lett.* **1995**, *36*, 1023.

[83] (a) Stadtmüller, H.; Lentz, R.; Tucker, C. E.; Stüdemann, T.; Dörner, W.; Knochel, P.; *J. Am. Chem. Soc.* **1993**, *115*, 7027; (b) Stadtmüller, H.; Tucker, C. E.; Vaupel, A.; Knochel, P.; *Tetrahedron Lett.* **1993**, *34*, 7911; (c) Vaupel, A.; Knochel, P.; *Tetrahedron Lett.* **1994**, *35*, 8349.

[84] Harada, T.; Katsuhira, T.; Hara, D.; Kotani, Y.; Maejima, K.; Kaji, R.; Oku, A.; *J. Org. Chem.* **1993**, *58*, 4897.

[85] Kondo, Y.; Takazawa, N.; Yamazaki, C.; Sakamoto, T.; *J. Org. Chem.* **1994**, *59*, 4717.

[86] Fürstner, A.; Singer, R.; Knochel, P.; *Tetrahedron Lett.* **1994**, *35*, 1047.

[87] Majid, T. N.; Knochel, P.; *Tetrahedron Lett.* **1990**, *31*, 4413.

[88] (a) Rao, C. J.; Knochel, P.; *J. Org. Chem.* **1991**, *56*, 4593; (b) Knochel, P.; Rao, C. J.; *Tetrahedron* **1993**, *49*, 29.

[89] Isobe, M.; Kondo, S.; Nagasawa, N.; Goto, T.; *Chem. Lett.* **1977**, 679.

[90] Oppolzer, W.; Radinov, R. N.; *Tetrahedron Lett.* **1991**, *32*, 5777.

[91] (a) Oppolzer, W.; Radinov, R. N.; Brabander, J. D.; *Tetrahedron Lett.* **1995**, *36*, 2607; (b) Oppolzer, W.; Radinov, R. N.; *J. Am. Chem. Soc.* **1993**, *115*, 1593.

[92] Oppolzer, A.; Radinov, R. N.; *Helv. Chim. Acta* **1992**, *75*, 170.

[93] Wipf, P.; Xu, W.; *Tetrahedron Lett.* **1994**, *35*, 5197.

[94] (a) Konda, Y.; Takazawa, N.; Yoshida, A; Sakamoto, T. *J. Chem. Soc. Perkin Trans. 1*, **1995**, 1207. (b) Kondo, Y.; Takazawa, N.; Yamazaki, C.; Sakamoto, T.; *J. Org. Chem.* **1994**, *59*, 4717.

[95] Harada, T.; Katsuhira, T.; Hara, D.; Kotani, Y.; Maejima, K.; Kaji, R.; Oku, A.; *J. Org. Chem.* **1993**, *58*, 4897.

[96] Gao, Y.; Harada, K.; Hata, T.; Urabe, H.; Sato, F.; *J. Org. Chem.* **1995**, *60*, 290.

[97] Markies, P. R.; Schat, G.; Akkerman, O. S.; Bickelhaupt, F.; *Organometallics*, **1991**, *10*, 3538.

[98] Okhlobystin, O. Y.; Zakhartin, L. I.; *Organometal. Chem.* **1965**, *3*, 259.

[99] Nast, R.; Künzel, O.; Müller, R.; *Chem. Ber.* **1962**, *95*, 2155.

[100] Yanagisawa, A.; Habaue, S.; Yamamoto, H.; *J. Am. Chem. Soc.* **1989**, *111*, 366.

[101] (a) Thiele, K.-H.; Zdunneck, P.; *J. Organomet. Chem.* **1965**, *4*, 10; (b) Thiele, K.-H., Engelhardt, G.; Köhler, J.; Arnstedt, M.; *J. Organomet. Chem.* **1967**, *9*, 385.

[102] Yasui, K.; Goto, Y.; Yajima, T.; Taniseki, Y.; Fugami, K.; Tanaka, A.; Tamaru, Y.; *Tetrahedron Lett.* **1993**, *34*, 7619.

[103] (a) Tamaru, Y.; Goto, S.; Tanaka, A.; Shimizu, M.; Kimura, M.; *Angew. Chem., Int. Ed. Engl.* **1996**, *35*, 878; (b) Tamaru, Y.; Tanaka, A.; Yasui, K.; Goto, S.; Tanaka, S.; *Angew. Chem., Int. Ed. Engl.* **1995**, *34*, 787.

[104] Knochel, P.; Rao, S. A.; *J. Am. Chem. Soc.* **1990**, *112*, 6146.

[105] Rozema, M. J.; Knochel, P.; *Tetrahedron Lett.* **1991**, *32*, 1855.

[106] Han, B.-H.; Boudjouk, P.; *J. Org. Chem.* **1982**, *47*, 5030.

[107] Maruoka, K.; Hashimoto, S.; Kitagawa, Y.; Yamamoto, H.; Nozaki, H.; *Bull. Chem. Soc. Jpn.* **1980**, *53*, 3301.

[108] Nakamura, E.; Murofushi, T.; Shimizu, M.; Kuwajima, I.; *J. Am. Chem. Soc.* **1976**, *98*, 2346.

[109] Taguchi, H.; Shimoji, K.; Yamamoto, H.; Nozaki, H.; *Bull. Chem. Soc. Jpn.* **1974**, *47*, 2529.

[110] Kuwajima, I.; Nakamura, E.; Hashimoto, K.; *Org. Synth., Coll.* **1990**, *7*, 512.

[111] Yamamoto, Y.; Asao, N.; *Chem. Rev.* **1993**, *93*, 2207.

[112] (a) Jubert, C.; Knochel, P.; *J. Org. Chem.* **1992**, *57*, 5425; (b) Masuyama, Y.; Kinugawa, N.; Kurusu, Y.; *J. Org. Chem.* **1987**, *52*, 3702.

[113] Shono, T.; Ishifune, M.; Kashimura, S.; *Chem. Lett.* **1990**, 449.

[114] Petrier, C.; Luche, J. L.; *J. Org. Chem.* **1985**, *50*, 910.

[115] Wilson, S. R.; Guazzaroni, M. E.; *J. Org. Chem.* **1989**, *54*, 3087; (b) Killinger, T. A.; Boughton, N. A.; Runge, T. A.; Wolinsky, J.; *J. Organomet. Chem.* **1977**, *124*, 131.

[116] Yanagisawa, A.; Habaue, S.; Yamamoto, H.; *Tetrahedron* **1992**, *48*, 1969.

[117] DeCamp, A. E.; Kawaguchi, A. T.; Volante, R. P.; Shinkai, I.; *Tetrahedron Lett.* **1991**, *32*, 1867.

[118] Marek, I.; Lefrançois, J.-M.; Normant, J. F.; *Bull. Soc. Chim. Fr.* **1994**, *131*, 910.

[119] Kitamura, M.; Okada, S.; Suga, S.; Noyori, R.; *J. Am. Chem. Soc.* **1989**, *111*, 4028.

[120] Noyori, R.; Suga, S.; Kawai, K.; Okada, S.; Kitamura, M.; Oguni, N.; Hayashi, M.; Kaneko, T.; Matsuda, Y.; *J. Organomet. Chem.* **1990**, *382*, 19.

[121] (a) Takahashi, H.; Kawakita, T.; Ohno, M.; Yoshioka, M.; Kobayashi, S.; *Tetrahedron* **1992**, *48*, 5691; (b) Takahashi, H.; Yoshioka, M.; Shibasaki, M.; Ohno, M.; Imai, N.; Kobayashi, S.; *Tetrahedron* **1995**, *51*, 12013; (c) Whitney, T. A.; *J. Org. Chem.* **1980**, *45*, 4214.

[122] (a) Shibata, T.; Morioka, H.; Hayase, T.; Choji, K.; Soai, K.; *J. Am. Chem. Soc.* **1996**, *118*, 471; (b) Shibata, T.; Choji, K.; Morioka, H.; Hayase, T.; Soai, K.; *Chem. Commun.* **1996**, 751; (c) Shibata, T.; Choji, K.; Hayase, T.; Aizu, Y.; Soai, K.; *Chem. Commun.* **1996**, 1235.

[123] Soai, K.; Shibata, T.; Morioka, H.; Choji, K.; *Nature (London)* **1995**, *378*, 767.

[124] Shono, T.; Hamaguchi, H.; Sasaki, M.; Fujita, S.; Nagami, K.; *J. Org. Chem.* **1983**, *48*, 1621.

[125] Katritzky, A. R.; Harris, P. A.; *Tetrahedron: Asymmetry* **1992**, *3*, 437.

[126] Thomas, J.; *Bull. Soc. Chim. Fr.* **1973**, 1300.

[127] Miginiac, L.; Mauze, B.; *Bull. Soc. Chim. Fr.* **1968**, 3832.

[128] Auvray, P.; Knochel, P.; Normant, J. F.; *Tetrahedron* **1988**, *44*, 4495.

[129] van der Louw, J.; van der Baan, J. L.; Stichter, H.; Out, G. J. J.; Bickelhaupt, F.; Klumpp, G. W.; *Tetrahedron Lett.* **1988**, *29*, 3579.

[130] (a) Dardoize, F.; Moreau, J.-L.; Gaudemar, M.; *Bull. Soc. Chim. Fr.* **1972**, 3841; (b) Luche, J.-L.; Kagan, H. B.; *Bull. Soc. Chim. Fr.* **1969**, 3500.

[131] Suzuki, T.; Narisada, N.; Shibata, T.; Soai, K.; *Tetrahedron: Asymmetry* **1996**, *7*, 2519.

[132] Soai, K.; Hatanaka, T.; Miyazawa, T.; *J. Chem. Soc., Chem. Commun.* **1992**, 1097.

[133] (a) Alvaro, G.; Pacioni, P.; Savoia, D.; *Chem. Eur. J.* **1977**, *3*, 726; (b) Alvaro, G.; Savoia, D.; *Tetrahedron: Asymmetry* **1996**, *7*, 2083.

[134] (a) Bocoum, A.; Savoia, D.; Umani-Ronchi, A.; *J. Chem. Soc., Chem. Commun.* **1993**, 1542; (b) Basile, T.; Bocoum, A.; Savoia, D.; Umani-Ronchi, A.; *J. Org. Chem.* **1994**, *59*, 7766.

[135] Nakamura, M.; Hirai, A.; Nakamura, E.; *J. Am. Chem. Soc.* **1996**, *118*, 8489.

[136] Hanessian, S.; Yang, R.-Y.; *Tetrahedron Lett.* **1996**, *37*, 8997.

[137] van der Steen, F. H.; Klejin, H.; Jastrzebski, J. T. B. H.; van Koten, G.; *J. Org. Chem.* **1991**, *56*, 5147.

[138] Hannick, S.; Kishi, Y.; *J. Org. Chem.* **1983**, *48*, 3833.

[139] Comins, D. L.; O'Connor, S.; *Tetrahedron Lett.* **1987**, *28*, 1843.

[140] Yamaguchi, R.; Nakazono, Y.; Matsuki, T.; Hata, E.; Kawanishi, M.; *Bull. Chem. Soc. Jpn.* **1987**, *60*, 215.

[141] Shiao, M.-J.; Liu, K.-H.; Lin, L.-G.; *Synlett* **1992**, 655.

[142] Shing, T.-L.; Chia, W.-L.; Shiao, M.-J.; Chau, T.-Y.; *Synthesis* **1991**, 849.

[143] Knochel, P., in *Comprehensive Organic Synthesis*, Vol. *4*, (B. M. Trost, Ed.) Pergamon, New York, **1991**, p. 865.

[144] (a) Lehmkuhl, H.; Nehl, H.; *J. Organomet. Chem.* **1981**, *221*, 131; (b) Lehmkuhl, H.; McLane, R.; *Liebigs Ann. Chem.* **1980**, 736.

[145] Meyer, C.; Marek, I.; Courtemanche, G.; Normant, J. F.; *Synlett* **1993**, 266.

[146] Yanagisawa, A.; Habaue, S.; Yamamoto, H.; *J. Am. Chem. Soc.* **1989**, *111*, 366.

[147] (a) Oppolzer, N.; Schröder, F.; *Tetrahedron Lett.* **1994**, *35*, 7939; (b) Normant, J. F.; Quirion, J.-C.; Masuda, Y.; Alexakis, A.; *Tetrahedron Lett.* **1990**, *31*, 2879; (c) Marek, I.; Alexakis, A.; Mangeney, P.; Normant, J. F.; *Bull. Soc. Chim. Fr.* **1992**, *129*, 171.

[148] (a) Isaka, M.; Ejiri, S.; Nakamura, E.; *Tetrahedron* **1992**, *48*, 2045; (b) Nakamura, M.; Wang, X. Q.; Isaka, M.; Yamago, S.; Nakamura, E.; *Org. Synth.* submitted; (c) Nakamura, E.; *J. Synth. Org. Chem., Jpn.* **1994**, *52*, 935; (d) Nakamura, E.; *Pure Appl. Chem.* **1996**, *68*, 123.

[149] Kubota, K.; Nakamura, M.; Isaka, M.; Nakamura, E.; *J. Am. Chem Soc.* **1993**, *115*, 5867.

[150] Nakamura, E.; Mori, S.; Nakamura, M.; Nakamura, E.; *J. Am. Chem. Soc.* **1998**, *120*, 13334.

[151] Nakamura, M.; Arai, M.; Nakamura, E.; *J. Am. Chem. Soc.* **1995**, *117*, 1179.

[152] Nakamura, E.; Mori, S.; Nakamura, M.; Morokuma, K.; *J. Am. Chem. Soc*, **1997**, *119*, 4887.

[153] Nakamura, E.; Mori, S.; Morokuma, K.; *J. Am. Chem. Soc*, **1997**, *119*, 4900.

[154] Kitazume, T.; Ishikawa, N.; *J. Am. Chem. Soc.* **1985**, *107*, 5186; (b) Kitazume, T.; Ishikawa, N.; *Chem. Lett.* **1982**, 1453.

[155] (a) Negishi, E.; Miller, J. A.; *J. Am. Chem. Soc.* **1983**, *105*, 6761; (b) Molander, G. A.; *J. Org. Chem.* **1983**, *48*, 5409.

[156] (a) Marek, I.; Lefrancois, J.-M.; Normant, J. F.; *Bull. Soc. Chim. Fr.* **1994**, *131*, 910; (b) Normant, J. F.; Marek, I.; LeFran, Lois, J.-M.; *Synlett* **1992**, 633.

[157] Courtois, G.; Masson, A.; Miginiac, L.; *C. R. Séances Acad. Sci. Serie C* **1978**, *286*, 265; *Chem. Abstr.* *89*: 24482.

[158] Nakamura, E.; Kuwajima, I.; *Org. Synth. Coll. Vol.* **1987**, *66*, 43.

[159] Nakamura, E.; *Synlett*, **1991**, 539.

[160] Lipshutz, B. H.; Wood, M. R.; Tirado, R.; *J. Am. Chem. Soc.* **1995**, *117*, 6126.

[161] Petrier, C.; de Souza Barbosa, J. C.; Dupuy, C.; Luche, J.-L.; *J. Org. Chem.* **1985**, *50*, 5761.

[162] Horiguchi, Y.; Nakamura, E.; Kuwajima, I.; *J. Am Chem. Soc.* **1989**, *111*, 6257.

[163] Crimmins, M. T.; Nantermet, P. G.; Trotter, B. W.; Vallin, I. M.; Watson, P. S.; McKerlie, L. A.; Reinhold, T. L.; Cheung, A. W.-H.; Steton, K. A.; Dedopoulou, D.; Gray, J. L.; *J. Org. Chem.* **1993**, *58*, 1038.

[164] Petrier, C.; Dupuy, C.; Luche, J. L.; *Tetrahedron Lett.* **1986**, *27*, 3149.

[165] Takahashi, T.; Nakazawa, M.; Kanoh, M.; Yamamoto, K.; *Tetrahedron Lett.* **1990**, *31*, 7349.

[166] Greene, A. E.; Lansard, J. P.; Luche, J. L.; Petrier, C.; *J. Org. Chem.* **1984**, *49*, 931.

[167] Soai, K.; Hayasaka, T.; Ugajin, S.; *J. Chem. Soc., Chem. Commun.* **1989**, 516.

[168] Simmons, H. E.; Cairns, T. L.; Vladuchick, S. A.; *Org. React.* **1973**, *20*, 1.

[169] Charette, A. B.; Marcoux, J.-F.; *J. Am. Chem. Soc.* **1996**, *118*, 4539.

[170] Denmark, S. E.; O'Connor, S. P.; *J. Org. Chem.*, **1997**, *62*, 3390.

[171] Nakamura, E.; Hirai, A.; Nakamura, M.; *J. Am. Chem. Soc.* **1998**, *120*, 5844.

[172] Smith, R. D.; Simmons, H. E.; *Org. Synth. Coll.* **1973**, *5*, 855.

[173] Miyano, S.; Hashimoto, H.; *J. Chem. Soc., Chem. Commun.* **1971**, 1418.

[174] Nishimura, J.; Okada, Y., in *Experimental Organometallic Chemistry (in Japanese)*, (eds: Sato, F. Yamamoto, K, Imamoto T.), Kodansha Scientific, Tokyo, **1992**, 59–61; *Chem. Abstr. 117*: 90494.

[175] Denmark, S. E.; Edwards, J. P.; *J. Org. Chem.* **1991**, *56*, 6974.

[176] (a) Winstein, S.; Sonnenberg, J.; *J. Am. Chem. Soc.* **1961**, *83*, 3235; (b) Chan, J. H.-H.; Rickborn, B.; *J. Am. Chem. Soc.* **1968**, *90*, 6406; (c) Staroscik, J. A.; Rickborn, B.; *J. Org. Chem.* **1972**, *37*, 738.

[177] (a) Motherwell, W. B.; Roberts, L. R.; *J. Chem. Soc., Chem. Commun.* **1992**, 1582; (b) Afonso, C. A. M.; Motherwell, W. B.; O'Shea, D. M.; Roberts, L. R.; *Tetrahedron Lett.* **1992**, *33*, 3899.

[178] Mash, E. A.; Nelson, K. A.; *J. Am. Chem. Soc.* **1985**, *107*, 8256.

[179] (a) Arai, I.; Mori, A.; Yamamoto, H.; *J. Am. Chem. Soc.* **1985**, *107*, 8254; (b) Mori, A.; Arai, I.; Yamamoto, H.; *Tetrahedron*, **1986**, *42*, 6447.

[180] (a) Imai, N.; Takahashi, H.; Kobayashi, S.; *Chem. Lett.* **1994**, 177; (b) Takahashi, H.; Yoshioka, M.; Kobayashi, S.; *J. Synth. Chem. Soc. Jpn.* **1997**, *55*, 714; *Chem. Abstr. 127*: 175984; (c) Denmark, S. E.;

Christenson, B. L.; Coe, D. M.; O'Connor, S. P.; *Tetrahedron Lett.* **1995**, *36*, 2215; (d) Kitajima, H.; Ito, K.; Aoki, Y.; Katsuki, T.; *Bull. Chem. Soc. Jpn.* **1997**, *70*, 207.

[181]   (a) Barrett, A. G. M.; Kasdorf, K.; *J. Chem. Soc., Chem. Commun.* **1996**, 325; (b) Barrett, A. G. M.; Kasdorf, K.; *J. Am. Chem. Soc.* **1996**, *118*, 11030.

[182]   Takahashi, H.; Yoshioka, M.; Shibasaki, M.; Ohno, M.; Imai, N.; Kobayashi, S.; *Tetrahedron* **1995**, *51*, 12013.

[183]   Charette, A. B.; Prescott, S.; Brochu, C.; *J. Org. Chem.* **1995**, *60*, 1081.

[184]   Arai, M.; Kawasuji, T.; Nakamura, E.; *Chem. Lett.* **1993**, 357.

[185]   Sekiya, K.; Nakamura, E.; *Tetrahedron Lett.* **1988**, *29*, 5155.

[186]   Negishi, E.; Chatterjee, S.; Matsushita, H.; *Tetrahedron Lett.* **1981**, *22*, 3737.

[187]   Negishi, E.; Valente, L. F.; Kobayashi, M.; *J. Am. Chem. Soc.* **1980**, *102*, 3298.

[188]   Miller, R. B.; Al-Hassan, M. I.; *J. Org. Chem.* **1985**, *50*, 2121.

[189]   Tamaru, Y.; Ochiai, H.; Nakamura, T.; Yoshida, Z.; *Tetrahedron Lett.* **1986**, *27*, 955.

[190]   Sakamoto, T.; Nishimura, S.; Kondo, Y.; Yamanaka, H.; *Synthesis* **1988**, 485.

[191]   Negishi, E.; Bagheri, V.; Chatterjee, S.; Luo, F.-t.; Miller, J. A.; Stoll, A. T.; *Tetrahedron Lett.* **1983**, *24*, 5181.

[192]   Grey, R. A.; *J. Org. Chem.* **1984**, *49*, 2288.

[193]   Tamaru, Y.; Ochiai, H.; Nakamura, T.; Yoshida, Z.; *Org. Synth. Coll.* **1993**, *8*, 274.

VI
Organocopper Chemistry

BRUCE H. LIPSHUTZ

Department of Chemistry, University of California, Santa Barbara, USA

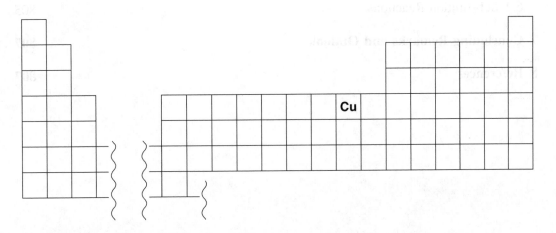

Organometallics in Synthesis: A Manual. Edited by Manfred Schlosser.
© 2002 John Wiley & Sons Ltd

Contents

1 Organocopper Chemistry in Focus

In the first edition of this Manual just over 100 procedures were provided in the chapter on organocopper chemistry. Since 1990 several new and useful procedures have been developed, many of which are now included in this chapter and represent most of the major classes of organocopper reagents. Much of the accent of late has been in the area of transmetalations, a phenomenon which has blossomed to include both stoichiometric as well as catalytic amounts of Cu^I. Such processes provide routes to copper reagents using other organometallic species as precursors, arrived at in ways characteristic of each metal. A separate section has now been created in response to this rapidly evolving chemistry.

Notwithstanding these advances, there is no denying the observation that usage of organocopper reagents remains greatest at the grass-roots level. Thus, all of the fundamental types of couplings involving traditional cuprates derived from either organolithium or Grignard reagents have been strictly maintained, and in most cases expanded in scope.

With new advances come more options for preparing and applying copper chemistry to synthetic problems. However, as pointed out in the earlier version of this Manual, one already is confronted with a list of considerations that need be addressed prior to selecting a copper reagent for use. When environmental concerns are added to this list the inertial barrier can seem high indeed.

Taken for granted in organocopper chemistry is the requirement for careful exclusion of air and moisture. Other experimental decisions that must be made prior to carrying out a reaction include:

- Which copper(I) precursor should be chosen and what level of purity is required?
- Although ethereal solvent is the norm, which is best for a particular reaction type?
- With so many variations of copper reagents from which to choose, is there one that is likely to be best suited for the goal in mind?
- Under what circumstances should additives be relied upon to assist with a selected coupling reaction?
- When and how can the number of equivalents of a potentially valuable organolithium, or other organometallic, be minimized without sacrificing efficiency?
- Is there a process catalytic in copper that saves on reagents and, in particular, waste disposal costs?

Although these and other questions arise with each circumstance surrounding the selection of a copper reagent, there is no shortage of usage of copper chemistry, judging from the continuous flow of reviews on this subject[1]. The explanations behind the intense usage of these reagents are not cryptic. Copper reagents, although utilized for several unrelated types of couplings, tend to react under very mild conditions and generally afford synthetically useful yields[2]. The reagents themselves, formally at least, are soft nucleophiles, but in the majority of cases are actually electrophilically driven.

That is, without the presence of gegenions capable of Lewis acidic character (e.g., Li$^+$, MgX$^+$), most couplings do not take place[3]. Such fundamental properties impart highly valued chemoselectivity patterns not witnessed with other organometallic species, especially those composed of far more ionic metals (e.g., RLi, RMgX, R$_3$Al, RZnX, etc.). Thus, few would consider introducing, in any direct way, an alkyl, vinylic, or aromatic moiety in a Michael sense by means other than copper chemistry (although alternatives do exist). The remarkable penchant for copper reagents to add to conjugated carbonyl groups in a 1,4 manner[2c], together with their relative inertness towards carbonyl 1,2 additions[4], combine to enhance their worth as selective agents. In substitution reactions, their low basicity encourages displacement over competing elimination[2d]. While these two processes (i.e., 1,4 addition and substitution) make up the lion's share of organocopper reagent usage, other characteristic reactions such as carbocupration and metallocupration of an acetylene are extremely valuable. Although 1,2 additions are rare, nonconjugated ketones and especially aldehydes can also be appropriate reaction partners[4].

In light of the comments above, this chapter strives to provide not only a starting point for researchers with little experience in organocopper chemistry, but also a central resource for those working in the field at any level. It still seems that an argument could be made that manipulation of copper reagents may well be less technically demanding than the successful preparation and use of a Grignard reagent, thus involving skills normally taken for granted beyond the early stages of one's career in organic chemistry. However, the successful handling of such powerful tools of synthesis does not imply an understanding of the mechanistic details surrounding their use, nor an appreciation of their composition or structure. Although this field of chemistry began in the mid-1950s[5], it is only since the mid-1980s that mechanistic insight has begun to accrue, brought forth by both experimental and theoretical contributions, concerning what are still often considered black-box phenomena. Thus, although the emphasis here is mainly on how to prepare and utilize copper reagents to form carbon–carbon and carbon–heteroatom bonds, there is much yet to be learned about the nature of organocopper complexes from the physical organic perspective.

2   Organocopper Complexes Derived from Organolithium Reagents

2.1   Lower-order Gilman Cuprates, R$_2$CuLi

Historically, Gilman's original recipe for combining two equivalents of an organolithium with either CuI or CuSCN was envisaged to yield cuprate R$_2$CuLi (**1**), proceeding via an intermediate (presumed polymeric) organocopper RCu[5]. However, it is now known that for X = I, Br, and Cl that this reaction does hold, whereas thiocyanate ligand loss does not occur but leads to a completely different reagent (R$_2$Cu(SCN)Li$_2$;

see below)[6]. Species **1** may consist of virtually any sp^3-, or sp^2-, or sp-based ligands originating from the corresponding organolithium (RLi) precursors:

$$2RLi + CuX \longrightarrow [(RCu)_n + RLi + LiX] \longrightarrow \boxed{R_2CuLi} + LiX$$

1

Generation of a homocuprate **1** (i.e., where both R groups are the same) is a very straightforward operation, as long as good quality CuI, CuBr, or CuBr•SMe$_2$ is used. There are a number of methods for purifying these salts[7], and of course all cuprous halides are readily available commercially. When using CuI, care must be exercised to protect it from light and excessive moisture. When pure, it is a nicely flowing white powder. Tinges of yellow/orange suggest traces of I$_2$ are present, although it is the accompanying product of disproportionation, Cu^{II}, that is the more serious impurity. Hence, although Soxhlet extraction will remove halogen, only recrystallization frees the (CuI)$_n$ from other materials. One procedure that works well is as follows.

Purification of (CuI)$_n$[8]

A dark brown solution of CuI (Fisher, 13.167 g, 69.1 mmol), KI (Fisher, 135 g, 813 mmol), and 100 mL water were treated with charcoal and filtered through a Celite pad. The yellow filtrate was then diluted with 300 mL water causing a fine gray/white precipitate to form. The resulting mixture was cooled in an ice/water bath and additional water was added portionwise to ensure complete precipitation. The precipitate was then suspended and filtered on a medium sintered glass frit under a cone of purging dry N$_2$. The filtrant was triturated under purging dry N$_2$ successively with four 100 mL portions of water, four 80 mL portions of acetone, and four 80 mL portions of distilled ether. The remaining off-white solid was allowed to dry with suction on the frit under a purge of dry N$_2$ and transferred to a 50 mL round-bottom flask, wrapped in foil, and dried under vacuum (≤ 0.20 mmHg) overnight. The evacuated flask was then heated to 90 °C for 4 h. Yield: 9.33 g (49.0 mmol, 70%) of grayish white powdery CuI.

As for CuBr, its use in the uncomplexed state is rare, and can be problematic. As the dimethyl sulfide complex, however, it is an excellent cuprate precursor[9]. Unfortunately, though, aged material tends to lose percentages of volatile Me$_2$S, and can easily lead to improper stoichiometries. That is, with losses of Me$_2$S, excess CuBr will be present and the two equivalents of RLi added cannot fully form R$_2$CuLi[10]. Hence, relatively fresh bottles of CuBr•SMe$_2$ should be employed, and once opened, they should be well-wrapped with parafilm and stored under a blanket of Ar. Alternatively, the CuBr•SMe$_2$ complex can be prepared fresh following the standard procedure given below.

Purification of the CuBr•SMe₂ Complex[9]

To 40.0 g (279 mmol) of pulverized CuBr (Fisher) was added 50 mL (42.4 g, 682 mmol) of Me₂S (Eastman, bp 36–38 °C). The resulting mixture, which warmed during dissolution, was stirred vigorously and then filtered through a glass-wool plug. The residual solid was stirred with an additional 30 mL (25 g, 409 mmol) of Me₂S to dissolve the bulk of the remaining solid and this mixture was filtered. The combined red solutions were diluted with 200 mL of hexane. The white crystals that separated were filtered with suction and washed with hexane until the washings were colorless. The residual solid was dried under N₂ to leave 51.6 g (90%) of the complex as white prisms that dissolved in an Et₂O•SMe₂ mixture to give a colorless solution. For recrystallization, a solution of 1.02 g of the complex in 5 mL of Me₂S was slowly diluted with 20 mL of hexane to give 0.96 g of the pure complex as colorless prisms, mp 124–129 °C (dec.). The complex is essentially insoluble in hexane, Et₂O, acetone, CHCl₃, CCl₄, MeOH, EtOH, and H₂O. Although the complex does dissolve in DMF and in DMSO, the observation that heat is evolved and that the resulting solutions are green in color suggests that the complex has dissociated and that some oxidation (or disproportionation) to give Cu^{II} species has occurred.

Copper(I) triflate (Fluka, CF₃SO₃Cu•0.5C₆H₆) is also a good source from which to generate lower-order cuprates, the subsequent chemistry from which works well[11]. However, it is a far more expensive and somewhat more sensitive precursor than the halides, and as such has not found widespread acceptance.

The other major component required in forming R₂CuLi is the organolithium. One of the surest pitfalls in all of cuprate chemistry, irrespective of reagent type, is a lack of knowledge concerning the quality and quantity of RLi being used to form the copper reagent. When relying on commercially available RLi, it is *never* a good practice to assume that a molarity as given on a bottle is correct. Usually they are sufficiently off the mark which, irrespective of direction (i.e., a high or low titre), is detrimental. Thus, with a high titre, insufficient RLi will be introduced, leading to mixtures of copper species, whereas with a low titre, excess RLi (i.e., above the 2.0 equivalents needed) will inadvertently be present, which can have disastrous consequences.

Several titration procedures can be followed to ensure that correct amounts of RLi are being combined with CuX[12]. One which is frequently called on relies on complexation between lithium and 1,10-phenanthroline, which is brown/red in color. Neutralization with ^{s}BuOH leads to a lemon yellow endpoint, from which the molarity of an RLi can be determined, as outlined for the titration of BuLi. At least three consistent measurements should be made to ensure accuracy. (For titration of ^{t}BuLi, benzene was used in place of ether at 25 °C.)

Titration of Organolithium Reagents: Normal Addition[13]

An over-dried 50 mL Erlenmeyer flask and stirring bar were cooled under an Ar stream. A few crystals of 1,10-phenanthroline (Fisher) were placed in the flask and dissolved in 15–20 mL of diethyl ether. The flask was then capped with a rubber septum, maintained under an Ar atmosphere, and stirred at 0 °C. The organolithium was added dropwise to this colorless solution until a colored endpoint was observed (e.g., for butyllithium a reddish-brown color was noted). Exactly 1 mL of organolithium was then added. To this solution, ^{s}BuOH was added dropwise until the colored endpoint was obtained (for butyllithium red/brown → pale yellow). This method was then repeated, again adding 1 mL organolithium (first drop again indicating endpoint) followed by titration with ^{s}BuOH. The molarity was determined from the equations

$$\text{Molarity (RLi)} = \frac{\text{mmol (RLi)}}{\text{mL (RLi) used}}$$

$$\text{mmol (RLi)} = \text{mmol (}^s\text{BuOH)} = \frac{\text{density (}^s\text{BuOH)}}{\text{MW (}^s\text{BuOH)}} \times \text{mL (}^s\text{BuOH)}$$

$$= \frac{806 \text{ mg mL}^{-1}}{74.12 \text{ mg mmol}^{-1}} \times \text{mL (}^s\text{BuOH)}$$

Still more accurate is the reverse of the above procedure, i.e., addition of the RLi to the ^{s}BuOH. The control realized by adding a solution of RLi to the alcohol, as opposed to introducing neat ^{s}BuOH to a solution of RLi, accounts for the improvement.

Titration of Organolithium Reagents: Inverse Addition

Methyllithium and butyllithium were purchased as solutions, 1.4 M in Et$_2$O and 2.5 M in hexane respectively, from the Aldrich Chemical Co. and were titrated as follows. An oven-dried 25 mL Erlenmeyer flask equipped with a magnetic stirrer bar was cooled under Ar and charged with 10 mL of freshly distilled Et$_2$O and one crystal of 1,10-phenanthroline. At 25 °C, 50 μL of 2-pentanol (distilled from CaH$_2$) were added via 100 μL syringe to provide a clear/colorless solution. The solution of organolithium reagent in question was then added dropwise via a 0.5 mL syringe to the vigorously stirred alcoholic Et$_2$O solution until an endpoint was realized. This was evidenced by a persistent tan/light-maroon colored solution. Beware that if too much base has been added a brown/maroon solution will result, thus indicating the proper endpoint has been surpassed. This protocol was repeated in the same Erlenmeyer flask until three trials were within 95% agreement of organometallic solution added. The molarity of organolithium reagent may be calculated following the formula given in the alternative procedure above.

With high-quality CuI or CuBr•SMe$_2$ in hand and a firm, accurate grip on the molarity of an RLi of interest, forming the corresponding Gilman cuprate **1** is relatively trivial. Before preparing an organocopper reagent of any type, however, there are a few pointers that must be considered:

1. All reactions involving organocopper complexes should be run under either an argon or nitrogen atmosphere with a slightly positive pressure applied being vented to a bubbler. Argon, although more expensive than nitrogen, is strongly endorsed for two major reasons. First, it is heavier than air, and hence blankets any reaction that may have residual amounts of other lighter gases present. Second, its use permits (although not recommended) removal of stoppers for purposes of introducing solids to the reaction vessel.

2. As essentially every organocopper reaction is sensitive to some degree to moisture, the educt to be added to the reagent must be dry. It is therefore good technique to routinely dry a substrate azeotropically with toluene at room temperature under a high vacuum prior to use.

3. Following similar lines of thought, and as is normally done prior to other organometallic chemistry, the glassware should either be stored in an oven at temperatures above 120 °C, or flame-dried just before use.

To prepare a lower-order cuprate R$_2$CuLi, then, the copper salt is slurried in an ethereal solvent and cooled to −78 °C, where two equivalents of the RLi are added. Warming to dissolution completes the process. Depending upon the type of reaction to be performed and the specific R$_2$CuLi being utilized, the cuprate may be recooled to −78 °C prior to addition of the substrate. The cornerstone of modern cuprate chemistry, Gilman's Me$_2$CuLi, is prepared in this manner, the subsequent chemoselective 1,4 addition of which, e.g., to a bromo enone, is described below.

Preparation of Me$_2$CuLi; Chemoselective 1,4 Addition to a Primary Bromo Enone[14]

To a cold (0 °C) solution of Me$_2$CuLi, prepared from 0.365 g (1.78 mmol) of CuBr•SMe$_2$ in 8 mL of Et$_2$O and 5 mL of Me$_2$S at 0 °C to which had been added 3.56 mmol of MeLi (halide free), was introduced a solution of 0.295 g (1.07 mmol) of the bromo enone in 5 mL of Et$_2$O. The resulting mixture, from which an orange precipitate separated, was stirred at 0–3 °C for 1.5 h and then siphoned into a cold aqueous solution (pH 8) of NH$_3$ and NH$_4$Cl. The ethereal extract of this mixture was dried and concentrated and the residual crude product (0.35 g of yellow liquid) was chromatographed on silica gel with an Et$_2$O/hexane eluent (1 : 39 v/v). The bromo ketone was collected as 0.28 g (92%) of colorless liquid (n^{25}D = 1.4687). From a comparable reaction in Et$_2$O at 0–5 °C for 2 h, the yield of the bromo ketone was 83%.

For lithium reagents that require prior generation by metal–halogen (usually Li–I or Li–Br) or metal–metal (e.g., Li–Sn) exchange, warming the 2RLi + CuX mixture to effect dissolution may present complications due to ligand isomerization and/or cuprate decomposition. To avoid this potential problem, the Cu^I salt can be solubilized by admixture with any of several possible additives, such as (MeO)$_3$P[15], LiX[16], Me$_2$S, etc. Of course, this trick can be applied to uses of commercial RLi, and is one way to avoid β-hydride elimination[17] in alkyl cuprates (e.g., BuLi). This trivial modification is exemplified by the formation and ultimate addition of a (Z)-2-ethoxyethenyl group to cyclohexenone.

Conjugate Addition of a Divinylic Cuprate to Cyclohexenone[18]

(82%)

A solution of cis-2-ethoxyvinyllithium was prepared from 2.18 g (6.04 mmol) of cis-1-ethoxy-2-tri-butylstannylethylene and butyllithium (1.1 equiv) in 15 mL of THF at −78 °C over 1 h. A solution of 0.577 g (3.03 mmol) of purified cuprous iodide and 0.89 mL (12.1 mmol) of Me$_2$S in 5 mL of THF was then added over 5 min. After stirring for 1 h at −78 °C, 0.264 g (2.75 mmol) of cyclohexenone in 5 mL of THF was added over 10 min. After stirring for 1 h, the mixture was warmed to −40 °C over 30 min, quenched with aqueous 20% NH$_4$Cl solution, and extracted with ether. The product was purified by column chromatography (silica gel, CHCl$_3$) to afford 0.379 g (82%) of the desired product.

Conversion of the lithium enolate of acetone dimethylhydrazone to the (presumed) corresponding azacuprate occurs upon addition of predissolved CuI in diisopropyl sulfide at −78 °C. Thus, 1,4 rather than 1,2 addition of the equivalent of acetone enolate, for instance to methyl vinyl ketone, occurs in excellent yield.

Conjugate Addition of a Cuprate Derived from a Lithiated N,N-Dimethylhydrazone to Methyl Vinyl Ketone[19]

(85%)

A precooled (ca. −30 °C) clear solution of 0.96 g (5 mmol) of cuprous iodide in 2.88 mL (20 mmol) of diisopropyl sulfide and 10 mL of THF was added dropwise with stirring at −78 °C to a suspension of 6-lithio-2-methylcyclohexanone dimethyl-hydrazone (10 mmol, generated from 1.54 g (10 mmol) of 2-methylcyclohexanone dimethylhydrazone and lithium diisopropylamide (10 mmol) in 40 mL of THF. The lithium compound dissolved during warming of the orange reaction mixture from −78 °C to −20 °C over 30 min and from −20 °C to 0 °C over 10 min, resulting in a clear golden yellow solution. It was cooled again to −78 °C, and 0.41 mL (5 mmol) of methyl vinyl ketone was added dropwise. After 2 h, the reaction was slowly warmed to room temperature over a period of 12 h. The black/brown reaction mixture was poured into a solution of saturated NH$_4$Cl containing NH$_4$OH (pH 8) and repeatedly extracted with CH$_2$Cl$_2$. The organic phase was shaken several times with NH$_4$Cl/NH$_4$OH solution until the aqueous phase was no longer blue. The combined aqueous phase was again extracted with CH$_2$Cl$_2$ and the combined organic phases were then dried over sodium sulfate. After removal of the solvent by rotary evaporation in vacuo, the crude product (1.19 g, spectroscopic yield 100%) was purified by distillation to give 0.42 g (85%) of a light yellow oil, bp 100 °C (0.05 mmHg).

Conjugate addition reactions by Gilman cuprates can also occur on extended alkenic networks[20], initially shown by Yamamoto using an ester of sorbic acid[21]. Although the organocopper species 'RCu•BF$_3$' (see below) adds mainly in a 1,4 sense to give **3**, Bu$_2$CuLi affords the 1,6 adduct **4**. Other substrate types, such as enynoates, have also been examined and the regioselectivity noted. However, these latter 1,6 additions give rise to an allenyl enoate, which is subject to further trapping by an electrophile. Here again, a regiochemical issue arises, although in practice the α-site is strongly favored, thereby generating allenic products. Quenching with aldehydes (PhCHO, CH$_2$=CHCHO, HCHO) or ketones (e.g., acetone) works well[22]. The combination of enynoate and cuprate produces initially an orange/red coloration, starting from the clear, colorless cuprate. As the reaction progresses, the color dissipates as a yellow suspension forms indicative of (MeCu)$_n$. Strongly basic or acidic conditions should be avoided during work-up, as double-bond isomerization and/or retro aldol may occur.

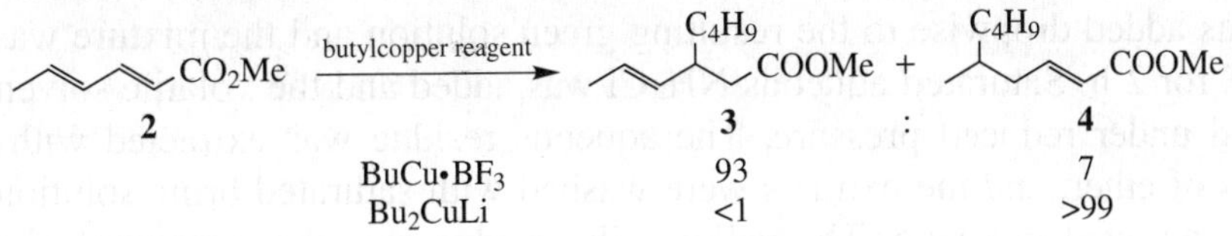

	2	**3**	:	**4**
BuCu•BF$_3$		93		7
Bu$_2$CuLi		<1		>99

1,6 Addition of R$_2$CuLi to an Enynoate–Electrophile Trapping (A Three-component Coupling)[23]

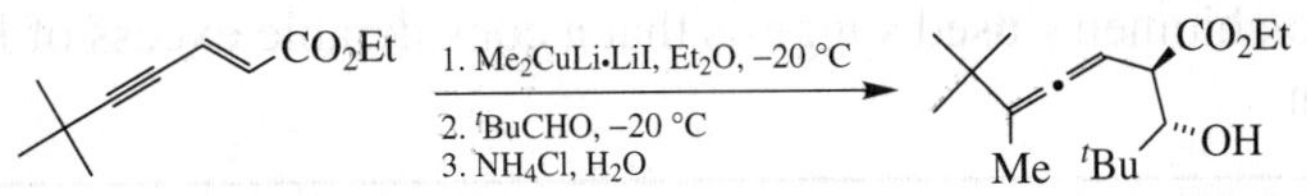

To a suspension of 2.86 g (15.0 mmol) of copper(I) iodide in 40 mL of diethyl ether was added dropwise at 0 °C 20.0 mL (30.0 mmol) of methyllithium (1.5 M solution in diethyl ether). The mixture was stirred for 15 min at 0 °C, then cooled to −20 °C, and a solution of 1.80 g (10.0 mmol) of ethyl 6,6-dimethyl-2-hepten-4-ynoate in 20 mL of diethyl ether was added dropwise. Stirring at −20 °C was continued for 1 h prior to the addition of 1.72 g (2.2 mL, 20.0 mmol) of pivaldehyde. After stirring for another 1 h at −20 °C the mixture was poured into 50 mL of a saturated NH$_4$Cl solution, and the salts and excess of water were removed by filtration through Celite. The crude product was purified by column chromatography (SiO$_2$, diethyl ether/hexane, 1 : 2) to yield 2.30 g (82%) as a colorless liquid. The diastereomeric composition was determined by GC (SE 30) to be 87 : 13.

Displacement reactions with R$_2$CuLi are also extremely valuable, yet equally as straightforward to conduct. The most cooperative electrophiles are primary centers bearing iodide, bromide, or tosylate. Epoxides of varying substitution patterns likewise couple quite well, especially less-hindered examples (e.g., monosubstituted oxiranes). Tosylates (in Et$_2$O) and iodides (in THF) are rated roughly comparable in terms of leaving group ability toward R$_2$CuLi[24], and relatively unreactive cuprates such as Ph$_2$CuLi can nonetheless be used to prepare functionalized aromatic systems, as illustrated below[25]. Differences in leaving group ability can also be used to advantage, as in the case of displacement in a chloro bromide by a dicyclopropyl cuprate[26].

Double Displacement of a Ditosylate with Ph$_2$CuLi[25]

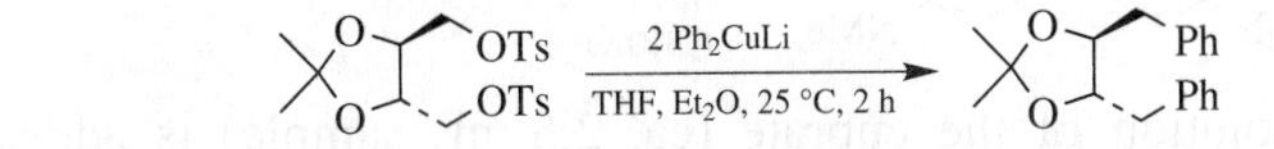

To a solution of (3.0 g, 15.75 mmol) of cuprous iodide in 10 mL of dry ether, stirred at 0 °C under dry Ar was added dropwise 20 mL of 2.1 M phenyllithium (42 mmol) as a solution in 75% benzene to 25% hexane. A solution of (1.93 g,

4.10 mmol) of 2,3-*o*-isopropylidene-L-threitol ditosylate in 12 mL ether and 3 mL THF was added dropwise to the resulting green solution and the mixture was stirred at 25 °C for 2 h. Saturated aqueous NH$_4$Cl was added and the volatile solvents were removed under reduced pressure. The aqueous residue was extracted with several portions of ether, and the extracts were washed with saturated brine solution, dried, and concentrated in vacuo. The yellow oily residue was chromatographed on 20 g of silica gel, eluting first with hexane to remove biphenyl, then with hexane/ethyl acetate (3 : 1) to elute the product. Distillation at 140 °C (0.1 mmHg) yielded 0.650 g (47%) of the colorless product.

Note: the stoichiometry used suggests that a considerable excess of PhLi is present in this reaction.

Selective Displacement of a Bromo Chloride by a Cyclopropylcuprate[26]

A solution of 1.1 M cyclopropyllithium in ether (660 mL) was added over 45 min at −35 °C to a slurry of 73 g (0.38 mol) of cuprous iodide in 660 mL of THF. After a Gilman test[27] (see below) was negative, 1-bromo-4-chlorobutane (54 g, 0.32 mol) was rapidly added to the mixture which was held at −35 °C for 1.5 h. Aqueous saturated NH$_4$SO$_4$ was then added and the mixture was filtered. The product was extracted with 2 L of ether/pentane (1 : 1). The organic layer was washed several times with water, then with brine. After drying over calcium sulfate, the extract was distilled through a 45 cm Vigreux column to remove solvents. The pot residue was then short-path distilled to yield 37.4 g (90%) of the product, bp 58–59 °C (17 mmHg).

The Gilman test[27] referred to in the procedure above is a standard means of establishing the presence of a free (usually unwanted), highly basic organolithium or Grignard species in a reaction solution.

Gilman Test for Free RLi (or RMgX)[27]

To an ethereal solution of the cuprate (ca. 2.5 mL sample) is added an equal volume of a 1% solution of Michler's ketone in toluene and the mixture is allowed to warm to room temperature. H$_2$O (1.5 mL) is then added and the mixture stirred vigorously for 5 min. A 0.2% solution of I$_2$ in glacial acetic acid is then added dropwise. If a true blue solution results, this is indicative of a positive test (i.e., there is free RLi/RMgX in the cuprate solution). A solution which develops a yellow, tan, or greenish coloration implies a negative test.

Ring opening of an epoxide occurs under conditions reflecting both the nature of the substrate and the relative reactivity of the cuprate. In general, although THF is an acceptable solvent, Et_2O is preferred as the Lewis acidity of lithium cations associated with R_2CuLi is maximized in this medium[28]. Mixed solvent systems, when necessary, are certainly acceptable. The compatibility of functional groups with soft R_2CuLi is oftentimes used to advantage.

Opening of a Cyclohexene Epoxide with Me_2CuLi[29]

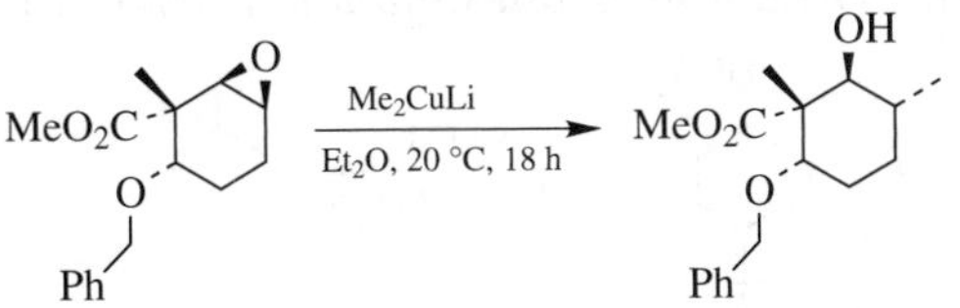

$$\text{(85\%)}$$

To a solution of lithium dimethylcuprate (from 6.4 mL of 0.75 M methyllithium (4.8 mmol) and 0.490 g (2.58 mmol) of cuprous iodide) in ether under N_2 at 0 °C was added methyl *c*-6-benzyloxy-*t*-2,3-epoxy-1-methylcyclohexane-*r*-1-carboxylate in ether and the mixture was stirred at 20 °C for 18 h. Addition of saturated aqueous NH_4Cl and extraction with ether gave the product as a colorless oil (0.125 g, 85%).

Another attractive feature of lower-order cuprates R_2CuLi is their ability to induce coupling at an sp^2 carbon center bearing an appropriate leaving group. Although procedures exist which augment the action of R_2CuLi alone in this regard (see Section 2.1.3), there are many circumstances where additives are not required, such as for vinyl bromides and vinyl triflates.

Substitution of a Vinylic Bromide Using Ph_2CuLi[30]

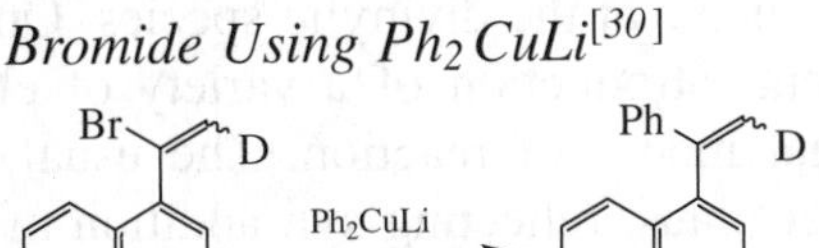

$$\text{(60\%)}$$

(4:1 (*E*):(*Z*))   (4:1 (*Z*):(*E*))

Lithium diphenylcuprate was prepared at 0 °C by slowly adding 25 mL of 1.86 M (46.5 mmol) phenyllithium solution to a suspension of 5.03 g (24.4 mmol) of cuprous bromide–dimethyl sulfide complex in 20 mL of dry ether. A yellow precipitate formed initially which changed to a homogeneous green solution after complete addition. After 40 min at 0 °C, a solution of 1.36 g (5.81 mmol) of 1-(1-bromo-2-deuterioethenyl)naphthalene $((E):(Z) = 4:1)$ in 3 mL of dry ether was then added. After 4.5 h at 0 °C, the reaction mixture was poured into aqueous saturated NH_4Cl solution (pH 9 by addition of NH_4OH), and this was stirred for 1.5 h. The ether layer was separated, washed twice with brine, and then dried. Removal of solvent afforded a light yellow oil which was purified by short-path distillation, collecting the fraction with bp 124–134 °C (1 mmHg). The yield was 0.80 g (60%) of the product which was crystallized from methanol, mp 57.5–58.5 °C; the (Z) isomer predominated 4:1.

Substitution of a Vinylic Trifluoromethanesulfonate (Triflate) with Me$_2$CuLi[31]

(75%)

A solution of 2.0 M methyllithium in hexane (5.5 mL, 10.8 mmol) was added to a stirred slurry of cuprous iodide (1.43 g, 7.5 mmol) in 15 mL of THF at 0 °C. A solution of 1-trifluoromethanesulfonyloxy-4-*tert*-butylcyclohexene (0.625 g, 2.2 mmol) in 5 mL of THF was added and the reaction mixture was stirred at −15 °C for 12 h. It was then diluted with hexane, filtered through a pad of Florisil, and concentrated on a rotary evaporator in vacuo. Chromatography of the residue on silica gel provided the product (0.250 g, 75%).

Note: The reported stoichiometry is 1.4 : 1 MeLi:CuI, which suggests that "Me$_2$CuLi" was not fully formed.

Carbocupration of alkynes with R_2CuLi (as well as with magnesio cuprates; see below) is an excellent inroad to vinylic cuprates[2j,32]. A 2 : 1 stoichiometry of the alkyne to R_2CuLi is required so as to arrive at the divinylic species. Once the intermediate reagent has been formed, subsequent introduction of a variety of electrophiles is possible and couplings follow the normal modes of reaction. The usual regiochemistry of addition places the R group at the terminus, reflecting syn addition of the elements 'R−Cu'.

Carbocupration Followed by Acid Quench of a 1-Alkyne Using Bu$_2$CuLi[2j]

(91%)

Butyllithium (50 mmol) was added to a suspension of copper iodide (5.3 g, 28 mmol) in ether (50 mL) at −40 °C. The mixture was stirred at −35 °C to −25 °C for 30 min and to the resulting solution was added at −55 °C 2-propynal diethyl acetal (6.4 g, 50 mmol) in ether (30 mL). After stirring for 30 min at −40 °C, the vinylcuprate (ready for further use with $E^+ \neq H^+$, if desired), was hydrolyzed with a saturated aqueous NH_4Cl solution (80 mL) admixed with a concentrated NH_4OH solution (20 mL). After filtration on Celite and separation of the layers, the organic phase was dried with potassium carbonate, the solvents were evaporated in vacuo, and the crude product was distilled (8.4 g, 91%; bp 96–97 °C (15 mmHg)).

Some of the most useful carbocuprations involve acetylene gas (ca. 2 equiv) and give rise to bis-(Z)-alkenylcuprates, which can then go on to transfer both vinyl groups in many reactions. A generalized procedure is as follows:

Carbocupration of Acetylene Gas: Preparation of a Lithium Bis-Z-alkenylcuprate[2j]

$$2 \ HC \equiv CH \xrightarrow[\text{−50 °C to −20 °C, 15 min}]{R_2CuLi, \ Et_2O} \quad \underset{R}{\overset{R}{\diagdown}} CuLi \xrightarrow{El^+} R \diagup El$$

The organolithium reagent (50 mmol) prepared in ether (from the corresponding bromide and lithium) was added to a suspension of copper(I) iodide (5.35 g, 28 mmol) or copper(I) bromide–dimethyl sulfide (1 : 1 complex; 5.75 g, 28 mmol) in ether (100 mL) at −35 °C. The mixture was stirred for 20 min at −35 °C to effect dissolution. Acetylene, cleared from acetone (through a −78 °C trap) was measured in a water gasometer (50 mmol; 1.2 L) and bubbled into the reaction mixture after being dried over a column packed with calcium chloride. The temperature was allowed to rise from −50 °C (at the start) to −25 °C. Stirring of the pale green solution was maintained for 20 min at −25 °C, at which point the cuprate is ready for use.

Although acetylene carbocupration with R_2CuLi leads to (Z)-alkenylcuprates (see above), it is also possible in the presence of excess acetylene to effect a double carbocupration to (Z,Z)-dienylcuprates[33]. The first equivalent of acetylene reacts at −50 °C; conversion to the dienylcuprate requiring warming to 0 °C but no higher as decomposition ensues. A range of electrophiles (e.g., enones, aldehydes, CO_2, X_2, activated halides) can then be introduced affording the expected dienes in fair to good yields. Preparation of the navel orangeworm pheromone (hexadeca-(11Z,13Z)-dienal) is illustrative of the method.

Double Carbocupration of Acetylene: Synthesis of a Pheromone[34]

(33%)

A stock solution of ethyllithium was prepared by addition of bromoethane (2.61 mL, 25 mmol) to a suspension of finely cut lithium wire (1.00 g, 143 mmol) in hexane (30 mL) at −20 °C. After being stirred at this temperature for 1 h, the mixture was allowed to warm to room temperature during an additional 2 h. Titration gave the molarity as 0.85 M (73%). A portion of this ethyllithium solution (8 mL, 6.8 mmol) was added dropwise to a stirred suspension of CuBr•SMe$_2$ (0.689 g, 3.36 mmol) in ether (25 mL) at −40 °C. A homogeneous blue/black solution of lithium diethylcuprate formed. After being stirred at −35 °C for 30 min, the solution was cooled to −50 °C and treated with acetylene (165 mL, 7.4 mmol). The mixture was allowed to warm to −25 °C for 30 min and then to −10 °C. More acetylene (300 mL, 13.4 mmol) was added during ca. 10 min, while the temperature was maintained at −10 °C. Once addition was complete, the solution was cooled to −40 °C and treated with the iodo acetal (1.0 g, 2.79 mmol) and HMPA (Me$_2$N)$_3$P = O (0.5 mL, 2.8 mmol). After being stirred at −40 to 0 °C for 3 h, the reaction was quenched with water (30 mL), diluted with brine (50 mL), and the product was extracted with ether (2 × 30 mL). The combined extracts were washed with brine (50 mL) and dried over MgSO$_4$. After chromatography on silica (CH$_2$Cl$_2$), the product acetal was treated with oxalic acid (2 g) in water (20 mL). THF was added until the mixture became homogeneous and the solution was stirred for 2 h at 60 °C. The product was extracted with light petroleum (3 × 100 mL), and the extract was dried over MgSO$_4$ and concentrated under reduced pressure. Rapid distillation from a Kugelrohr oven (125 °C, 0.3 mmHg) gave hexadeca-(11Z,13Z)-dienal (0.22 g, 33%). The ^{1}H NMR data were consistent with published values[35].

Addition of R$_2$CuLi across simple alkenes, as opposed to alkynes, is an unknown reaction. However, with cyclopropenone ketal **2**, homo- and mixed Gilman cuprates react at −70 °C in 1 min to afford a product of cis addition[36]. Both activated and unactivated electrophiles (in the presence of HMPA) can be employed to trap the intermediate vinylic cuprate, which occurs with retention of stereochemistry. The ketal can be stored (in an ampoule) at −20 °C, although it is handled at ambient temperatures. A slight excess of

R_2CuLi is usually used; greater quantities of cuprate do not improve yields. Given the rapidity of the carbocupration step, the ketal should be added as a solution in Et_2O or THF, rather than as a neat liquid. If vinylic cuprates are used, the vinylcyclopropane products can be rearranged to cyclopentenone derivatives. Likewise, trapping the intermediate of such an addition with a vinylic electrophile to afford a divinylcyclopropane can ultimately lead to seven-membered ring formation. Should these latter two types of transformations be of interest, it should be noted that vinylcyclopropanone ketals are sensitive to acid, opening via a cyclopropylcarbinyl cation to the corresponding β,γ-unsaturated ester.

Carbocupration of a Cyclopropenone Acetal with R_2CuLi[37]

To a THF (13 mL) solution of 1-bromocyclooctene (0.63 mL, 4.4 mmol) was added at −70 °C ^{t}BuLi (5.8 mL of a 1.53 M solution in pentane, 8.8 mmol) over 30 s. After stirring for 3 min, the vinyllithium solution was added via a cannula to a suspension of CuBr•SMe$_2$ (0.451 g, 2.2 mmol) in Et_2O (3 mL). The mixture was stirred at −40 °C for 20 min, then cooled to −70 °C. The cyclopropenone ketal **5** (0.28 mL, 2 mmol)[38] in Et_2O (1.5 mL) was added to the cuprate solution over 1 min and stirred for 5 min. An Et_2O (2 mL) solution of iodomethane (0.62 mL, 10 mmol) and HMPA (0.38 mL, 2.2 mmol) was added and the mixture was warmed slowly to 0 °C over 3 h, and then stirred for 1 h at 0 °C. The solution was poured into saturated NH$_4$Cl (15 mL) and the water layer was extracted with Et_2O. The combined organic layer was filtered through a short column of silica gel and concentrated in vacuo. The residual oil was chromatographed on silica gel (2% ethyl acetate in hexane) to obtain the 2,3-disubstituted cyclopropanone ketal as a colorless oil (0.423 g, 79%).

An unusual indolyl Gilman cuprate has been used, together with a nonracemic *N*-acylated pyridine carboxaldehyde derivative, to prepare 1,4-dihydropyridines with high diastereoselectivities[39,40]. The chirality derives from animals of type **6**, which when N-activated add Gilman cuprates to the 4-position preferentially, as in **7**. Attack at C-6 can compete depending upon the reagent and solvents used, with THF apparently essential for obtaining the C-4 adduct with good stereocontrol. Methyl chloroformate and acetyl chloride are the most effective activating groups, rather than *N*-alkyl moieties (e.g., *N*-benzyl). Cuprate additions take place at low temperatures. Curiously, acylation of the pyridine ring is faster than reaction of the acylating reagent with the cuprate, and hence the cuprate/pyridine aminal are admixed to which is then added the acid halide or chloroformate.

6   **7 (R = Me, Et, ...)**

Addition of a Gilman Cuprate to a Nonracemic Pyridine Aminal in the Presence of a Chloroformate[39]

(50%), de = 95%

A solution of BuLi in hexane (1.52 mmol) was added to a solution of indole (0.178 g, 1.52 mmol) in THF at −80 °C. The mixture was stirred for 30 min and the temperature allowed to warm to −40 °C and then cooled to −60 °C. CuI (0.145 g, 0.76 mmol) was added and the reaction mixture stirred for 1 h at −40 °C and then cooled to −80 °C. During this time, the reaction mixture became a deep blue solution. The pyridine aminal (0.100 g, 0.3 mmol) in THF (10 mL) and methyl chloroformate (0.12 mL, 1.52 mmol) were added, and the reaction mixture was stirred for 12 h and then allowed to warm to room temperature. The reaction was quenched by addition of an aqueous solution of NH_4OH/NH_4Cl (1 : 1). The mixture was diluted with Et_2O and washed with aqueous NH_4Cl solution. The organic layer was dried over anhydrous Na_2CO_3 and concentrated in vacuo to afford a yellow oil which was purified by column chromatography (SiO_2, Cyclohexane/dichloromethane/-ether = 70 : 10 : 20) to give 0.076 g (50%, de = 95%) of pure product.

2.1.1   Reactions in the Presence of $BF_3 \cdot Et_2O$

Exposure of the Gilman dimeric cuprate $(Me_2CuLi)_2$ in THF to distilled $BF_3 \cdot Et_2O$ at −78 °C has been shown to give rise to $Me_3Cu_2Li_2$, together with $MeLi \cdot BF_3$, the former species likely to be responsible (along with $BF_3 \cdot Et_2O$) for the chemistry of this cuprate–Lewis acid pair[41]:

$$[R_2CuLi]_2 + 2BF_3 \rightleftharpoons R_3Cu_2Li + RLi \cdot BF_3 + BF_3$$

Regardless of the events that occur rapidly upon mixing R_2CuLi with $BF_3 \cdot Et_2O$, of necessity carried out at low temperatures of $-78\,°C$ up to about $-50\,°C$, this combination is responsible for remarkable accelerations in numerous situations where R_2CuLi alone is either too unreactive or incompatible with increasing temperatures needed for a reaction to ensue[42]. Little or no visible changes in solutions of R_2CuLi occur upon addition of $BF_3 \cdot Et_2O$ at $-78\,°C$, and there is no induction period prior to introduction of a substrate. Thus, as is true for other additives, e.g., Me_3SiCl (see below), the cuprates are prepared exactly as described earlier, for these additives figure in only after the initial preparation. At least one equivalent of $BF_3 \cdot Et_2O$ is required for the full benefits to be realized, in part due to the probable formation of boron ate complexes, rather than lithium salts, as the initial products[43]. In each of the examples that follow, it is noteworthy that $BF_3 \cdot Et_2O$ is essential for the desired chemistry to take place. The reactions include the conjugate addition to a highly hindered propellane skeleton[44], displacement of an aziridine to afford a secondary amine (see also Section 2.2)[45], and a highly diastereoselective bond formation involving a chiral, nonracemic acetal[46].

$BF_3 \cdot Et_2 O$-assisted Conjugate Addition of $Me_2 CuLi$ to a Hindered α,β-unsaturated Ketone[44]

(71%)

To a cold ($-30\,°C$) slurry of 0.870 g (4.6 mmol) of purified cuprous iodide and 8.0 mL of ether was added under Ar 6.0 mL (9.6 mmol) of methyllithium (1.2 M). The clear solution of lithium dimethylcuprate was stirred for 5 min and cooled to $-78\,°C$, and 0.19 mL (1.54 mmol) of freshly distilled $BF_3 \cdot Et_2O$ was added. After 5 min stirring of the mixture, a solution of 0.310 g (1.6 mmol) of isomeric enones in 2 mL of ether was added dropwise; an immediate precipitation of methylcopper was observed. The mixture was stirred at $-78\,°C$ for 15 min, an additional 0.08 mL (0.75 mmol) of $BF_3 \cdot Et_2O$ was added, and the mixture was stirred at $-78\,°C$ for 1 h. After the mixture had slowly warmed to room temperature, the organic material was extracted into ether and washed with saturated NH_4Cl, water, and brine, and then dried. Evaporation of the solvent in vacuo afforded 0.277 g of crude product which was shown by IR to consist of approximately a 60 : 40 mixture of saturated ketone and enone, respectively. Without prior isolation of the ketone, the crude product mixture was recycled in the same manner as described above. Kugelrohr distillation (bp 90–100 °C) (0.4 mmHg), afforded 0.233 g (71%) of product as a 2 : 1 mixture of anti and syn isomers, respectively; VPC (195 °C) cleanly separated the mixture. The first component off the column was the anti isomer.

BF$_3$•Et$_2$O-promoted Coupling of an Aryl Cuprate with a Protected Aziridine[45]

A 50 mL round-bottom flask charged with CuI (1.5 mmol) and THF (6 mL) was cooled under Ar to −40 °C and treated with phenyllithium (3 mmol in 7 : 3 cyclohexane/ether). The resulting black mixture was stirred for 15 min, then cooled to −78 °C. To it was rapidly added the 4,4-dimethoxybenzhydryl (DMB)-protected aziridine (0.5 mmol) in THF (0.5 mL) followed by BF$_3$•Et$_2$O (1.5 mmol). After warming the mixture to room temperature, 14% NH$_4$OH (15 mL) was added together with ether (10 mL) and solid NH$_4$Cl (1 g). The resulting dark blue aqueous layer was extracted three times with 1 : 1 hexane/ether. The combined extracts were dried (K$_2$CO$_3$), filtered and concentrated to afford the *N*-DMB derivative of β-phenethyl-amine in 95% yield after flash chromatography (4 : 1 hexane/ethyl acetate). This sample was deprotected by stirring in 88% formic acid (5 mL) at 80–85 °C for 90 min. After removing the solvent in vacuo at 15 mmHg and then at 0.5 mmHg, the amine was partitioned between 5% aqueous HCl and ether to furnish pure β-phenethylamine (0.044 g, 80%).

BF$_3$•Et$_2$O-assisted Cuprate Alkylation of a Chiral, Nonracemic Acetal[46]

To a slurry of CuI (2.86 g, 15 mmol) in Et$_2$O (70 mL) was slowly added, at −40 °C, an ethereal solution of hexyllithium•LiBr (30 mL of a 1 M solution, 30 mmol; prepared from hexylbromide and Li metal). The blue solution of the cuprate was ready after complete dissolution of CuI (30–60 min). After cooling to −78 °C, the chiral acetal (0.58 g, 5 mmol) dissolved in 10 mL of Et$_2$O, was slowly added. If the addition is too rapid it will result in the formation of a yellow/orange precipitate which redissolves upon warming. Under vigorous stirring, a solution of BF$_3$•Et$_2$O (1.9 mL, 15 mmol) in Et$_2$O (10 mL) was added. The reaction was exothermic and was allowed to warm to −55 °C (internal temperature) for 15 min, whereupon no starting material remained (GC analysis). The orange/red mixture was hydrolyzed by addition of 30 mL of aqueous NH$_4$Cl and 20 mL of aqueous ammonia. The salts were filtered off and the aqueous layer extracted twice

$(2 \times 50$ mL $Et_2O)$. The combined organic phases were dried over Na_2SO_4 and the solvents were removed in vacuo. The residue was chromatographed on silica gel ($80 : 20$ cyclohexane/EtOAc) to afford 0.897 g of a pale yellow oil (89%). The diastereomer at C-2 may be prepared in the same manner from Me_2CuLi and 2-hexyl-(R,R)-4,5-dimethyl dioxolane.

2.1.2 Reactions in the Presence of Me₃SiCl

Although Gilman reagents are modified under the influence of $BF_3 \bullet Et_2O$[41], just how TMS-Cl alters the reaction pathway of lower order cuprates remains a hot topic of debate[47]. Arguments favoring both trapping of intermediate Cu^{III} adducts[48] and trace quantities of carbonyl–TMS-Cl Lewis acid–Lewis base activation[49] have been advanced, although the latter proposal seems unlikely in light of recent high-level calculations[50]. Whatever the role, this simple synthetic maneuver usually pays off handsomely, and should be considered whenever hindered unsaturated carbonyl-containing molecules are involved. An illustrative procedure is cited below, the process (apart from the introduction of Me_3SiCl to the cuprate) being otherwise essentially identical to the standard mode of reagent use. Usually, 2–10 equivalents of TMS-Cl are used given its inexpensive nature, and the silyl enol ether, rather than ketone, is the initially expected product. A subsequent hydrolysis under very mild conditions, or fluoride treatment, returns the carbonyl compound. Interestingly, TMS-CN is regarded to be as effective as TMS-Cl as an activating additive for cuprate 1,4 additions[51]. An investigation in some detail (Michael additions, 1,2 additions, ^7Li NMR) supports this notion, although whether these two silanes act along similar lines is open for discussion[52].

Me₃SiCl-accelerated 1,4 Addition of Bu₂CuLi to Acrolein[53]

$$\text{CHO} \xrightarrow[\text{HMPA, Me}_3\text{SiCl, } -70\,^\circ\text{C, 2.5 h}]{\text{Bu}_2\text{CuLi, THF}} \text{Bu} \diagup\!\!\diagdown\!\!\diagup\text{OTMS}$$

(80%)

(98:2(*E*):(*Z*))

To a stirred suspension of $CuBr \bullet SMe_2$ (0.0617 g, 0.30 mmol) in 0.8 mL of THF at $-70\,°C$ was added dropwise butyllithium in hexane (0.60 mmol). The mixture was stirred at $-40\,°C$ for 30 min, and then cooled to $-70\,°C$ after which HMPA (174 μL, 1.0 mmol) was added. After several minutes, a mixture of acrolein (33.4 μL, 0.50 mmol), chlorotrimethylsilane (120 μL, 1.0 mmol) and decane (internal standard) in 0.3 mL of THF was added dropwise. After 2.5 h at $-70\,°C$, 80 μL of triethylamine and pH 7.4 phosphate buffer were added. GLC analysis of the resulting mixture indicated 80% yield and an (*E*) : (*Z*) ratio of 98 : 2.

Recent work also suggests that TMS-Br, rather than the silyl chloride, may provide even a greater boost in reagent reactivity[54,55]. This additive has been used more frequently, however, with lower-order cyanocuprates RCu(CN)Li than with Gilman cuprates R_2CuLi, as the latter reagents are only likely to have good compatibility characteristics with this highly reactive additive at low temperatures for short periods of time. An example of RCu(CN)Li + TMS-Br use can be found in Section 2.2.5.

2.1.3   Reactions in the Presence of Other Additives

Cuprate couplings effected in the presence of highly polar solvents, such as DMF or HMPA[14], or modified by inclusion of other metals can lead to changes in chemoselectivity or product formation that would not otherwise occur. For example, 1,4-diene formation by attachment of a vinylic cuprate (in particular when derived via carbocupration) to an sp^2 center is best done in the presence of a catalytic amount of Pd(Ph$_3$P)$_4$ and $ZnBr_2$ (1 equiv)[56]. Thus, following preparation of a (Z)-vinylic cuprate (see above), a typical procedure is exemplified by the preparation of (7E,9Z)-dodecadien-1-yl acetate.

Coupling of a Vinylic Iodide with a Vinylic Cuprate: 1,3-Diene Synthesis[56]

To a solution of (Z)-dibutenyl cuprate (prepared from 6 mmol of EtLi in 30 mL ether and acetylene via carbocupration; see above) were added at $-40\,^\circ$C in 20 mL THF, then a solution of 0.700 g $ZnBr_2$ in 10 mL THF. After stirring for 30 min at $-20\,^\circ$C, a mixture of 3 mmol 1-iodo-1-(E)-octen-8-yl acetate (94% (E) purity) and 0.15 mmol Pd(PPh$_3$)$_4$ in 10 mL THF were added. The reaction mixture was slowly warmed to $+10\,^\circ$C and after 30 min at this temperature, 40 mL of saturated NH_4Cl solution were added. The organic phase was concentrated in vacuo and 50 mL of pentane were added to precipitate the inorganic salts which were filtered off. The organic solution was washed twice with 20 mL of saturated NH_4Cl solution, dried over $MgSO_4$ and the solvent removed in vacuo. The crude residue was purified by preparative TLC to afford a 78% yield of the pheromone (97% (E/Z) purity).

On occasion, silicon derivatives that depart from the commonly employed trimethylsilyl-X (see above) are used, such as Et$_3$SiCl. The bulkier additive usually leads to a more stable silyl enol ether product. Another alternative reagent, bis(trimethylsilyl)acetamide (BSA), has also been examined and found to be comparable to Et$_3$SiCl in its activation of enones[57]. In the case of doubly activated Michael acceptors, as with the enone below, BSA was particularly noteworthy in that competitive net reduction was not observed.

Use of Bis(trimethylsilyl)acetamide (BSA) in Cuprate 1,4 Additions[57]

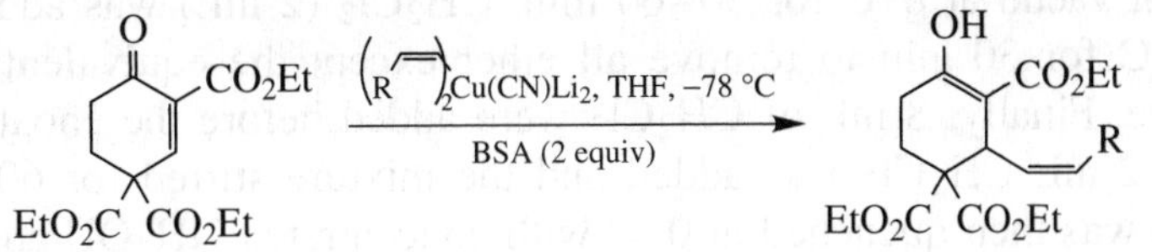

(R = SnMe₃, 92%)

The cuprate (3.36 mmol), was cooled to $-78\,°C$ and treated rapidly with BSA (1.11 mL, 4.48 mmol), followed by a solution of the keto ester (0.700 g, 2.24 mmol) in dry THF (3 mL). After 15 min, TLC (hexane/CH_2Cl_2/EtOAc = 10 : 2 : 1) indicated complete reaction. After a further 15 min the solution was poured into a mixture of ice-cold Et_2O and 1 : 4 NH_4OH/NH_4Cl. The mixture was stirred at $25\,°C$ for 15 min. The resulting deep blue mixture was extracted with ether and purified by flash column chromatography using (hexane/CH_2Cl_2/EtOAc = 10 : 2 : 1 + 1% Et_3N) to afford the product as a wax (1.03 g, 92%).

2.1.4   Reactions in Nonethereal Media

Although not routine, cuprate reactions can be run in solvents other than Et_2O or THF[58]. The impetus may lie, for example, in changes expected where stereochemical issues are present, or where increased interactions between substrate and cuprate are desirable. Solvents such as hexane, benzene, toluene, or (as in the case below) CH_2Cl_2, may be used to replace Et_2O in which the cuprates are initially formed. NMR analyses confirm, at least in the case of CH_2Cl_2, that all of the Et_2O is not removed, the residual solvent serving as ligands on Li^+[59]. As long as the cuprate is used immediately after solvent exchange (in vacuo), decomposition does not appear to be a problem. Occasionally, manipulations of this type can have multiple benefits, such as increasing both the rate and stereoselectivity of 1,4 additions. Another example is given in Section 2.3.

Conjugate Addition of Me₂CuLi to an α,β-Unsaturated Ester in CH₂Cl₂[59]

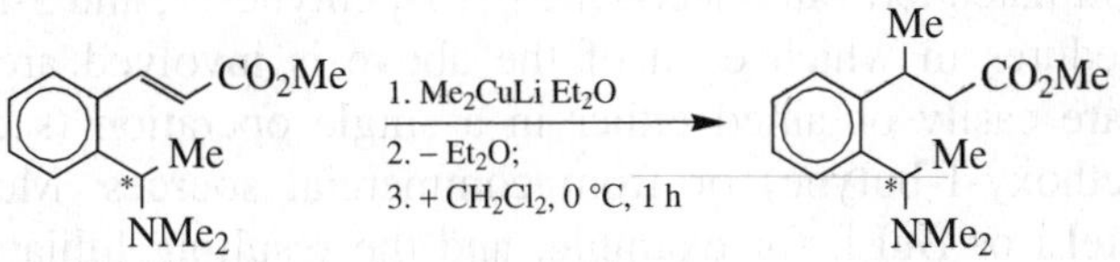

results in: Et_2O (40%), de <35

(89%), de 0   (+TMS-I)

THF (<1%) de –

CH_2Cl_2 (92%), de 71

Methyllithium (1.1 mL, 1.76 mmol, 1.6 M) was added to a slurry of CuI (0.167 g, 0.88 mmol) in 3 mL diethyl ether at 0 °C. After stirring for 10 min the ether was evaporated in vacuo at 0 °C for 30–60 min. CH_2Cl_2 (2 mL) was added and evaporated at 0 °C for 30 min to remove all ether except the equivalent coordinating to the cuprate. Finally, 8 mL of CH_2Cl_2 were added before the enoate (0.4 mmol) dissolved in 2 mL CH_2Cl_2 was added and the mixture stirred for 60 min at 0 °C. The reaction was then quenched at 0 °C with concentrated NH_4OH/aqueous NH_4Cl and extracted with CH_2Cl_2 (3 × 10 mL). The organic layer was washed with brine, dried (Na_2SO_4), and the solvent evaporated in vacuo (yield 92%, de 71%). The same procedure was used for lithium diphenylcuprate (phenyllithium, 0.8 mL, 2 M).

2.2   Lower-order Mixed Cuprates, R_tR_rCuLi

In a reaction of a Gilman homocuprate, transfer of only one R group usually takes place leading to a byproduct organocopper $(RCu)_n$ which is lost on work-up (as RH and copper salts). Thus, based on RLi invested, for every equivalent of cuprate (which requires 2RLi to form), the maximum realizable yield with one equivalent of substrate is 50%. Although tolerable for most commercially obtained RLi, those cuprates whose precursors must be synthetically prepared and then lithiated are too costly to sacrifice. To conserve valued RLi, mixed cuprates have been developed which derive from two different organolithiums: one is the organometallic of interest, R_tLi, the 'transferable' group; the other, R_rLi, consists of a 'residual' ligand R_r which is less prone to release by copper. When R_tLi and R_rLi combine with CuX (X = I, Br), R_tR_rCuLi is formed. Subsequent reaction with a suitable substrate leads to transfer of the R_t group in preference to R_r, with loss of the byproduct $(R_rCu)_n$ being of no chemical consequence.

2.2.1 R_r = alkynic

Many different residual ligands (R_r) have been developed over the years[60]. Perhaps the most widely used are alkynic derivatives, first introduced in 1972[61]. Examples include lithiated tert-butylacetylene[62], 1-pentyne[61], and 3-methyl-3-methoxy-1-butyne[63]. Procedures in which each of the above is involved are given below. All of these alkynes are easily obtained either in a single operation (see procedure below for 3-methyl-3-methoxy-1-butyne) or from commercial sources. Metalation is readily performed with MeLi or BuLi, for example, and the resulting lithiated alkyne is ready for use. When added to CuI, the product of metathesis, a cuprous acetylide, tends to have reasonable solubility in Et_2O and is readily dissolved in THF as solutions bearing a red-to-orange color. Owing to their relative volatility, reformation of the neutral alkynes on work-up does not complicate this methodology.

Michael Addition of a Mixed Alkynic Cuprate to Isophorone[62]

$$\text{(enone)} \xrightarrow[\text{Et}_2\text{O, 5–7 °C, 20 min}]{\text{[tert-butylacetylide]}-\text{Cu(Me)Li}} \text{(product ketone)}$$

(76%)

A solution of lithiated tert-butylacetylene, prepared from 0.269 g (3.28 mmol) of tert-butylacetylene and 3.06 mmol of MeLi in 2.4 mL of Et_2O, was added with stirring to a cold (10–13 °C) slurry of 0.573 g (3.02 mmol) of purified CuI in 2.0 mL of Et_2O. To the resulting cold (5–7 °C) red/orange solution of the acetylide was added 1.7 mL of an Et_2O solution containing 2.74 mmol of MeLi. This addition resulted in a progressive color change from red/orange to yellow to green. To the resulting cold (5–7 °C) solution of the cuprate was added 2 mL of an Et_2O solution containing 2.08 mmol of isophorone. The color of the reaction mixture changed progressively from green to yellow (1–2 min) to red/orange (20 min), after which the mixture was partitioned between Et_2O and an aqueous solution (pH 8) of NH_4Cl and NH_3. The resulting orange Et_2O solution was washed with 3×25 mL of aqueous 28% NH_3 to complete hydrolysis and the remaining colorless Et_2O solution was washed with H_2O, dried, and concentrated in vacuo. After the residual yellow liquid (0.288 g) had been mixed with a known weight of $\text{C}_{14}\text{H}_{30}$ (as an internal standard), analysis (GLPC, silicone fluid QF_1 on Chromosorb P, apparatus calibrated with known mixtures) indicated the presence of the tetramethyl ketone (76% yield), and the unchanged enone (12% recovery). None of the reduced ketone was detected. Collected samples of the ketones were identified with authentic samples by comparison of GLPC retention times and IR and NMR spectra.

Conjugate Addition of a Lower-order Alkynic Butyl Cuprate[61]

$$\text{(cyclohexenone)} \xrightarrow[\text{Et}_2\text{O, HMPA, −78 °C, 15 min}]{\text{C}_3\text{H}_7-\text{Cu(Bu)Li}} \text{(3-butylcyclohexanone)}$$

(97%)

A slurry of 0.64 g of dry propylethynylcopper (4.90 mmol) in 10 mL of anhydrous ether was treated with 1.80 mL of dry hexamethylphosphorus triamide (9.80 mmol), and the mixture was stirred at room temperature under Ar until a clear solution was obtained (5–10 min). To the cooled (−78 °C) solution was then added 3.10 mL of a 1.49 M solution of butyllithium (4.62 mmol) in hexane, and the resulting yellow solution was stirred for 15 min at −78 °C.

The solution of mixed cuprate so formed was then treated with 2.50 mL of a 1.80 M solution of 2-cyclohexenone (4.50 mmol) in anhydrous ether, stirred for 15 min at −78 °C, quenched by pouring into ice-cold aqueous ammonium sulfate solution, and extracted with ether. The ethereal layers were extracted with ice-cold 2% (v/v) sulfuric acid, then filtered through Celite, and washed with aqueous sodium bicarbonate (5%). The dried (Na_2SO_4) extracts afforded almost pure 3-butylcyclohexanone (0.675 g, 97%), homogeneous by TLC analysis (R_f 0.40; ether/benzene, 1 : 10), and >99% pure by GLPC analysis (GLPC retention time, SE-30, 10 ft, 10% column, 170 °C, 5.2 min). The IR and NMR spectra were also satisfactory.

Preparation of 3-Methyl-3-methoxy-1-butyne[63]

(84%)

A slurry of sodium hydride (7.2 g, 150 mmol; 50% in mineral oil) in 150 mL of DMF was cooled to 0 °C, and 8.4 g (100 mmol) of 2-methyl-3-butyn-2-ol dissolved in 100 mL of DMF was added dropwise over 30 min. The reaction mixture was stirred for an additional 30 min, and dimethyl sulfate (19 g, 14.3 mL, 150 mmol) was slowly added over a 20 min period. After stirring for an additional 5 min at 0 °C, the flask was allowed to warm to 25 °C and stirring was continued for 45 min. Excess sodium hydride was then destroyed by the dropwise addition of glacial acetic acid to the cooled (0 °C) reaction mixture. Direct distillation through a 30 cm Vigreux column afforded 8.2 g (84%) of pure material (bp 77–80 °C literature bp 80 °C).

Displacement of a Benzylic Halide with a Mixed Lower-order Alkynic Cuprate[63]

(92%)

A 1 M solution of 3-methoxy-3-methyl-1-butyne in THF was treated at 0 °C with 1 equiv of butyllithium. The clear, colorless solution was stirred for 5–10 min and transferred into a slurry of cuprous iodide in THF (1 mmol mL^{-1}), precooled to 0 °C. The resulting red/orange solution was then stirred at this temperature for 30 min and subsequently transferred either by syringe or cannula to a −78 °C solution of the lithio reagent (0.5–1 M). Under the above conditions, a virtually instantaneous reaction occurred, yielding a pale yellow to colorless solution of the mixed cuprate. The use of more concentrated conditions led to the appearance of a white precipitate during cuprate formation (presumably lithium iodide) which readily dissolved at around −30 °C to give homogeneous solutions. Addition of the substrate (0.985 equiv) as a solution in THF at −78 °C was followed by warming to −20 °C and stirring for several hours. Standard extractive work-up employing pH 8 aqueous ammonium hydroxide in saturated NH$_4$Cl gave the product (92%) which was virtually pure (by NMR). The formation of these mixed Gilman reagents can also be accomplished satisfactorily by the addition of an R$_t$Li reagent to a solution of the cuprous acetylide (i.e., normal addition).

Other nonalkynic groups have also found favor as residual ligands (R$_r$) in lower-order cuprate reactions. The strong association of copper with sulfur, as well as with phosphorus, has led to the development of the thiophenoxide (PhS−)[64], 2-thienyl (2-Th−)[65], and dicyclohexylphosphido [(C$_6$H$_{11}$)$_2$P−][16,66] ligands in this capacity. Each in lithiated form is readily prepared from commercial materials and has proven to serve in a manner similar to that of the alkynic unit (see above).

2.2.2   R$_r$ = thiophenyl

Acylation of a Mixed Phenylthio Cuprate[64a]

(87%)

A stirred suspension of 4.19 g (22.0 mmol) of cuprous iodide in 45 mL of THF was treated at 25 °C with 18.3 mL of 1.20 M (22.0 mmol) lithium thiophenoxide in 1 : 1 THF/hexane. A clear, yellow solution formed within 5 min but became a cloudy suspension upon cooling to −78 °C. Dropwise addition of 10.6 mL of 2.06 M (21.8 mmol) tert-butyllithium in pentane to the cold (−78 °C) suspension gave a fine, nearly white precipitate. Into this cold (−78 °C) suspension was injected after 5 min 15.0 mL of a precooled (−78 °C) solution containing 2.81 g (20.0 mmol) of benzoyl chloride in THF. Addition of the substrate regenerated the cloudy, yellow suspension, and the reaction was stirred for 20 min before quenching was effected by injection of 5.0 mL (125 mmol) of absolute methanol. The reaction mixture was allowed to warm to room temperature and poured into 200 mL of saturated, aqueous NH$_4$Cl, and the yellow precipitate thus formed was removed by suction filtration. The aqueous phase was extracted with 3 × 100 mL of ether and the combined ether phases were washed twice with 50 mL of 1 N NaOH and dried with $MgSO_4$. Solvent was removed in vacuo to afford 3.21 g (99%) of a slightly yellow oil with spectral properties essentially identical with those of pure pivalophenone. Short-path distillation gave 2.82 g (87% yield based on benozyl chloride) of colorless pivalophenone (bp 105–106 °C at 15 mmHg (literature bp = 103–104 °C at 13 mmHg); n_D^{20} 1.5092 (literature n_D^{20} 1.5090).

2.2.3 R_r = 2-thienyl

In the case of mixed thienylcuprates $[R_t(2-Th)CuLi]$, the thiophene obtained from various vendors is usually not of satisfactory quality to use as received. At least one distillation (bp 84 °C) is strongly recommended, and clean thiophene should appear as a water-white liquid. Metalation in ethereal media should afford solutions pale yellow (when done in THF at *ca.* −25 °C) to yellow (at 0 °C to *ca.* 25 °C).

Michael Addition of a Mixed 2-Thienyl Cuprate to an α,β-*Unsaturated Ester Assisted by* Me_3SiCl[65]

Butyllithium (2.5 mmol) was added to a solution of thiophene (3 mmol) in ether (5 mL) at 0 °C and the solution was stirred at 25 °C for at least 40 min. Then another 2.5 mL ether was added, the mixture was cooled in an ice bath, and finally powdered copper iodide (2.5 mmol) was added. 2-Thienylcopper formed immediately as a yellowish suspension. The mixture was stirred for about 5 min and then methyllithium (4.85–4.95 mmol) was added. Then the mixture was stirred until the Gilman test (see page 676) for free alkyllithium was negative (about 5 min). The color of the cuprate solution is either yellow or light green. The reaction mixture was then cooled to about −50 °C and methyl cinnamate (2 mmol) in ether (2.5 mL) was added. The addition resulted in a shiny yellow color. Within 1 min from the substrate addition trimethylchlorosilane (5 mmol) was added. The temperature was allowed to rise to 0 °C and the reaction was followed by GLC. After work-up the crude product, dissolved in pentane, was chromatographed through silica gel to separate trimethylsilylthiophene from the conjugate adduct. The silica gel was then eluted with ether. After filtration, drying with sodium sulfate and evaporation the yield was 0.268 g (75%) of methyl 3-phenylbutanoate.

2.2.4 R_r = dialkylphosphido

The use of dicyclohexylphosphine, en route to the mixed phosphido lower-order cuprates $R_t(C_6H_{11})_2PCuLi$, requires some care in terms of handling (i.e., minimized exposure to air). The cuprates once formed, however, are particularly stable reagents, although they participate readily in substitution and Michael additions.

Conjugate Addition of a Lower-order Phosphido Cuprate to Cyclohexenone[66]

(80%)

A 10.7 g (54.0 mmol) quantity of dicyclohexylphosphine (K & K Labs or Organometallics) dissolved in 30 mL of dry, oxygen-free ether in a septum-sealed 100 mL pear-shaped flask was cooled to 0 °C (ice bath), and 37.5 mL (54.0 mmol) of 1.44 M (0.21 M residual base) butyllithium (Aldrich, in hexane) was added. The resulting suspension was stirred at 0 °C for 1 h and then transferred by cannula into a suspension of 10.9 g (53.0 mmol) of CuBr•SMe₂ (Aldrich) in 60 mL of ether in a septum-sealed 500 mL round-bottom flask, which was also at 0 °C. (The flask containing the phosphide was rinsed with 10 mL of ether). The homogeneous brown solution was stirred for 15 min at 0 °C and then cooled to −50 °C for 15 min. A 36.8 mL (53.0 mmol) quantity of 1.44 M butyllithium was added, and the homogeneous brown solution was stirred for 15 min at −50 °C. It was then cooled to −75 °C for 15 min, and 4.80 g (49.9 mmol) of 2-cyclohexen-1-one (Aldrich, distilled and refrigerated) in 20 mL of ether in a septum-sealed 50 mL pear-shaped flask cooled to −75 °C was added by cannula. (A further 5 mL of ether was used to rinse the 2-cyclohexen-1-one flask.) After 45 min at −75 °C, a 2.0 mL aliquot (out of a total of 200 mL) was withdrawn by syringe and added to 1 mL of 3 M aqueous NH₄Cl in a septum-sealed 2 dram vial, which also contained 0.0297 g of tetradecane (internal standard). Calibrated GLC analysis indicated a 91% yield of product; no starting material remained. After a total of 1 h, 200 mL of 3 M aqueous NH₄Cl (deoxygenated with N₂) was added to the reaction mixture, which was allowed to warm to ambient temperature. The final pH of the aqueous layer was 8. The mixture was filtered through Celite 545, and the filter deposit was washed with 200 mL of ether. The organic layer was separated and back-extracted with 250 mL of 0.2 M aqueous sodium thiosulfate, 200 mL of NH₄Cl solution, 250 mL of 0.4 M aqueous sodium thiosulfate, and finally 100 mL of NH₄Cl. The aqueous layers were sequentially extracted with 50 mL of ether, which were added to the original organic layer. Drying over anhydrous sodium sulfate (Baker, granular) and evaporation under reduced pressure (<30 °C, 30 torr) gave 9.31 g of a yellow oil, which was purified by flash chromatography on 180 g of Florisil (Fisher) slurry-packed in a 3.5 × 40 cm column and eluted with 3.5 L of 5% ether/hexane followed by 0.5 L of 10% ether/hexane. All fractions collected were of 50 mL (the column volume was 350 mL), and 5.61 g (98% pure by GLC) of product was recovered from fractions 11–76. An additional 0.53 g (98% pure) was obtained by stripping the column with 20% ether/hexane (~750 mL). The total yield of pure product was 6.14 g (80%).

2.2.5　R_r = cyano

All of the above dummy ligands R_r have one feature in common: they are introduced in lithiated form (R_rLi) either to react with CuX to initially generate $R_rCu + LiX$ and thence, with R_tLi, R_tR_rCuLi, or with R_tCu, to form R_tR_rCuLi directly. Another alternative does exist, however, in which the R_r is already bound to copper as the Cu^I salt. That is, CuCN is unique in that treatment with R_tLi affords directly a lower-order cyanocuprate $R_tCu(CN)Li$, rather than the products of metathesis, $R_tCu + LiCN$ (Scheme 1)[67]. The strength of the copper–cyanide bond is presumably responsible for this behavior, but what is gained in simplicity of preparation and stability is paid for in reactivity. Hence, although lower-order cyanocuprates are easily formed by this simple 1 : 1 correspondence (Scheme 1), the cyano ligand serving as the nontransferable R_r, they are best used in reactions with activated electrophiles, such as allylic epoxides.

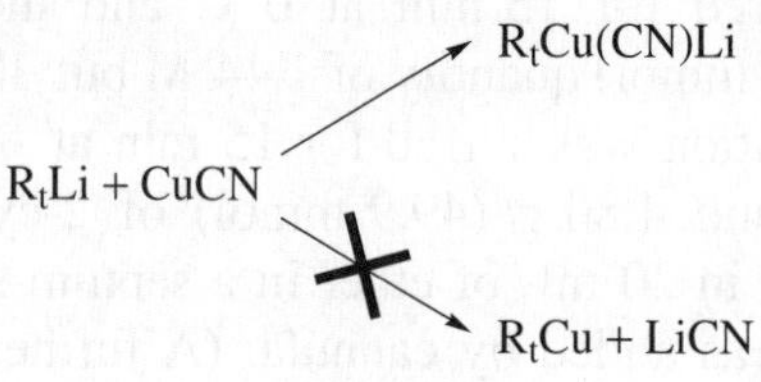

Scheme 1.

Ring Opening of an Allylic Epoxide with a Lower-order Cyanocuprate[67a]

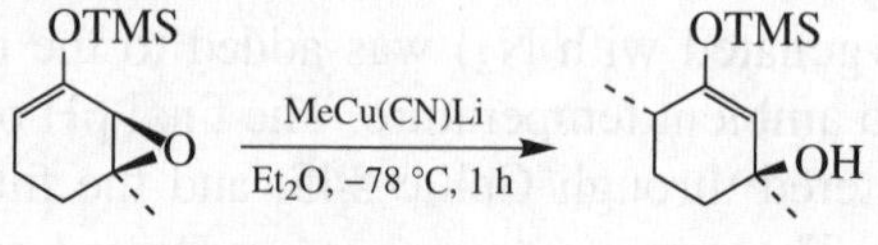

(93%)

(Caution: All cuprate reactions using CuCN should never be quenched with acidic aqueous solutions so as to prevent generation of HCN.) CuCN (0.720 g, 8 mmol) was placed into a flame-dried round-bottom flask, which was then filled with *ca.* 40 mL of dry ether and cooled to $-40\,^\circ$C under N_2. Then, 4.67 mL of methyllithium in ether (1.71 M, 8 mmol) was added and the yellowish suspension stirred for *ca.* 30 min at $-40\,^\circ$C, until no CuCN was visible at the bottom of the flask. After cooling to $-78\,^\circ$C, a solution of 0.400 g (2.0 mmol) of the epoxy enol ether in 5 mL of dry ether was added dropwise, with an intensification of the yellow color. The mixture was allowed to warm to $25\,^\circ$C over 5 h and then quenched with 30 mL of a saturated NH_4Cl solution. After filtration through a Celite pad and washing of the ether layer with brine solution, the organic phase was dried over sodium sulfate and concentrated in vacuo to yield 0.390 g (93%) of adduct. The crude reaction product was of high purity, as determined by 360 MHz ^{1}H NMR and ^{13}C NMR. Only one product was detectable by the aforementioned spectroscopic techniques.

Cyanocuprate ring-opening reactions of aziridines, whether activated (in an allylic sense), or by an electron-withdrawing group on nitrogen, present interesting opportunities for establishing amino functionality in selected targets. For example, the sulfonylated aziridine **8** has been coupled along S_N2 lines with diaryl cuprate **9a** to give the desired trans arrangement in **10** crucial to the synthesis of (+)-7-deoxypancratistatin[68]. Likewise, the ortho-metallated aromatic in the form of derived cuprate **9b** reacts with the same educt to afford **10b**, which was ultimately converted to (+)-pancratistatin[69].

(**a**) R = R′ = H   (**10a**, 32%)
(**b**) R = OTBS, R′ = CONMe$_2$ (**10b**, 75%)

Alternative anti-S_N2', rather than S_N2, attack by lower-order cyanocuprates has been noted with acyclic allylic aziridines (see also the reactions of magnesio cuprates with the corresponding γ-mesyloxy derivatives, Section 3.2.2). Both the (*E*)-cis and (*Z*)-trans isomers lead to the same (*E*)-product **11**, whereas the (*E*)-trans and (*Z*)-cis isomers afford the corresponding diastereomer **12**[70]. Diastereoselectivity of the order of >98 : 2 is to be expected. Reactions are rapid at $-78\,^\circ$C, requiring only minutes prior to quenching. Ligands in RCu(CN)Li that effectively transfer, in addition to methyl, include Et, Bu, ^{i}Pr, (^{i}PrO)Me$_2$SiCH$_2$ and p-FC$_6$H$_4$CH$_2$. The presence or absence of lithium salts is of no consequence. Other reagents, such as R$_2$Cu(CN)Li$_2$ or R$_2$CuLi•LiI, gave unwanted reduced esters **11** and **12**, R = H. Only aziridines participate in these couplings, as other rings of larger size (**13** and **14**) were recovered unchanged from exposure to these reaction conditions. Nonallylic aziridines of general structures **15** and **16** are capable alkylating agents as well, the former toward $R_2Cu(CN)Li_2$ (2 equiv)[71], the latter reacting with R_2CuLi[72].

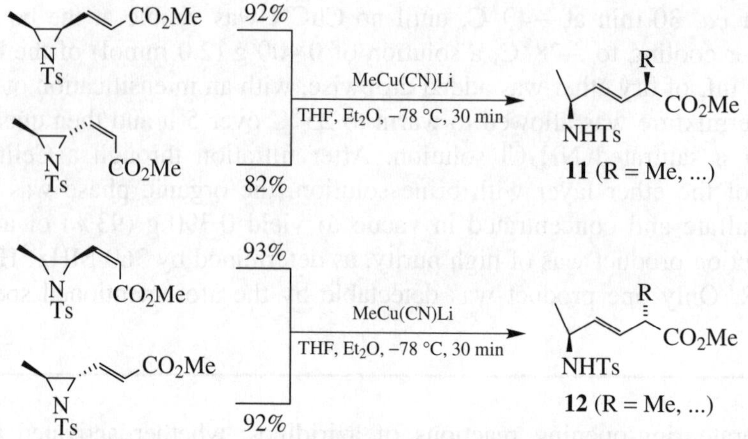

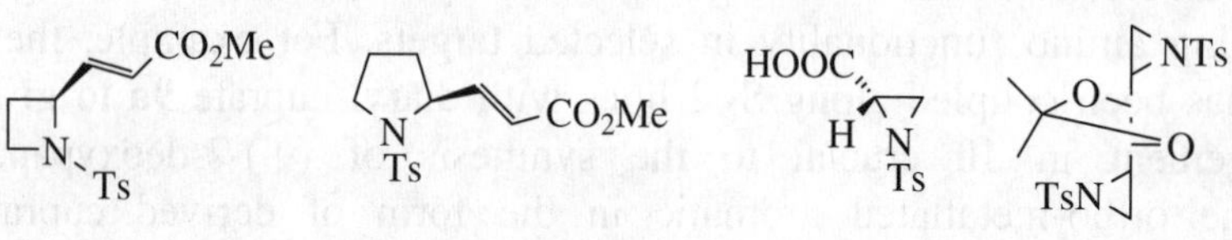

92%

MeCu(CN)Li

THF, Et$_2$O, -78 °C, 30 min

11 (R = Me, ...)

82%

93%

MeCu(CN)Li

THF, Et$_2$O, -78 °C, 30 min

12 (R = Me, ...)

92%

13   **14**   **15**   **16**

Reaction of a Nonracemic N-Tosyl-γ,δ-epimino Enoate with a Lower-order Cyanocuprate[70a]

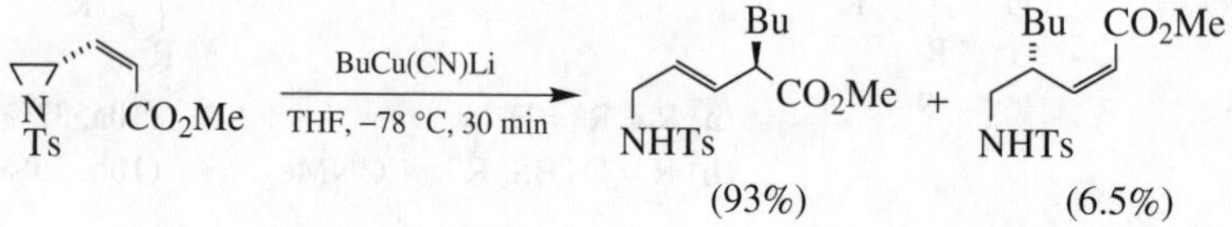

BuCu(CN)Li

THF, -78 °C, 30 min

(93%)   **(6.5%)**

To a stirred slurry of CuCN (0.1791 g, 2 mmol, 4 equiv) in dry THF (2 mL) under argon at -78 °C was added via syringe butyllithium (1.63 mol dm^{-3} solution in hexane; 1.23 mL, 2 mmol, 4 equiv), and the mixture was allowed to warm to 0 °C. After being stirred at this temperature for 10 min a solution of the α, β-enoate (0.1407 g, 0.5 mmol, 1 equiv) in dry THF (2 mL) was added dropwise at -78 °C with stirring. Stirring was continued for 30 min, followed by quenching with aqueous saturated NH$_4$Cl–28% NH$_4$OH (1 : 1, 1 mL). Usual work-up led to a mixture of products as a colorless oil, which was separated by flash chromatography over silica gel eluting with hexane/EtOAc (2 : 1), yielding, in order of elution, a minor product (0.011 g, 6.5%) and a major product (0.158 g, 93%; colorless oil; $[a]_D^{20} - 30.7$ (c 1.40 in CHCl$_3$); $\Delta\varepsilon - 0.313$ (216.3 nm in isooctane).

For hindered enones where 1,4 additions of cuprates RCu(CN)Li are especially difficult, TMS-Cl may not provide the panacea so frequently observed with this additive. However, it has been found that TMS-Br can have a remarkable impact on the extent of conversion and hence, overall yield of ketone formed[54,55]. For example, in the case of the bicyclic system **17**, the reaction progresses initially even at −78°, whereas in the presence of TMS-Cl only 34% of the product was realized. Thus, the yield was more than doubled by simply switching to TMS-Br.

Conjugate Addition of RCu(CN)Li to a Hindered Enone in the Presence of TMS-Br[55]

17   (83%)

To a cold (−97 °C), rapidly stirred solution of freshly distilled 4-iodo-2-trimethylgermyl-1-butene (0.893 g, 2.99 mmol) in dry THF (30 mL) was quickly added a solution of tert-butyllithium in pentane (3.8 mL, 5.68 mmol). The resultant bright yellow, cloudy solution was stirred at −97 °C for 10 min and then warmed to −78 °C. Copper(I) cyanide (0.2946 g, 3.29 mmol) was added in one portion and the suspension became pale yellow after ~5 min. Warming to −30 °C for 3 min provided a homogeneous solution, either tan or pale yellow in color. The solution of the vinylgermane cyanocuprate was recooled immediately to −78 °C to avoid decomposition, at which point the solution became heterogeneous again. A mixture of freshly distilled enone (0.363.4 g, 1.89 mmol) and TMS-Br (1.5 mL, 11.3 mmol) in THF (4 mL) was added dropwise. As the addition proceeded, the reaction mixture turned bright yellow, then bright orange and eventually brownish and homogeneous. Stirring was continued for 8 h at −78 °C and at −48 °C for 2 h. The solution was poured into water (20 mL) and the resultant mixture was stirred for 15 min. Ether (30 mL) and aqueous NH_4Cl/NH_4OH (pH 8–9) (20 mL) were added, and the mixture was stirred vigorously (open to the atmosphere) until the aqueous layer was blue (overnight). The phases were separated and the aqueous layer was extracted with ether (3 × 20 mL). The combined organic extracts were washed with brine (20 mL), dried over anhydrous magnesium sulfate and the solvent was removed in vacuo. The resulting crude product was purified by flash chromatography (35 g of silica gel, 95 : 5 petroleum ether/ether) and the oil thus obtained was distilled (air-bath temperature 132–139 °C at 0.1 torr) to afford 0.5717 g (83%) of the keto germane, a mixture of epimers, as a colorless oil. The ratio of epimers varied somewhat from experiment to experiment but was found to be, in this case, 4 : 1.

The presence of an equivalent of CuCN can have a major impact on softer organo-lithiums, such as that obtained upon deprotonation of nonracemic amide **18**[73], prepared from the 'Merck ligand' (1*S*,2*R*)-cis-aminoindanol[74]. Attempts at direct amination of lithi-ated **18** using lithium tert-butyl-*N*-tosyloxycarbamate (LiBTOC) as a synthon for 'NHBOC' (NH−C(O)O−t−Bu) were unsuccessful, as were the corresponding zinc halide, diorgan-iozinc, and zincate combinations. Introduction of CuCN (1.1 equiv), however, led to the rapid consumption of both educt and LiBTOC at −78 °C to give **19** in good yield and excel-lent de. Hydrolysis to the corresponding amino acid was accomplished with 6 N HCl, and the chiral auxiliary recovered (80%) upon adjustment of pH and extraction with CH_2Cl_2.

α-Amination of a Chiral, Nonracemic Amide Derived from Nonracemic 1-Amino-2-indanol[73]

1. BuLi, THF, −78 °C
2. CuCN, −78 °C to 0 °C
3. TsON(Li)BOC, −78 °C

18   **19** (77%) ≥ 99% de

To a solution of the amide (1.0 g, 3.11 mmol) in THF (10 mL) was slowly added 2.5 M BuLi (1.33 mL, 3.33 mmol) such that $T_i \leq -70\,°C$ ($T_{bath} = -78\,°C$). The resulting solution (light yellow) was aged at −78 °C for 1 h and gradually trans-ferred via cannula into a slurry of CuCN (0.2785 g, 3.11 mmol) in THF (10 mL) such that $T_i \leq -60\,°C$ ($T_{bath} = -78\,°C$). The dry ice bath was removed and the resulting slurry was allowed to warm until it became homogenous (*ca.* − 5 °C), after which the yellow solution was cooled back to −78 °C. In another flask, 2.5 M BuLi (1.33 mL, 3.33 mmol) was added to a solution of tert-butyl-*N*-tosyloxycarbamate (0.8439 g, 3.11 mmol) in THF (10 mL) such that $T_i \leq -70\,°C$ ($T_{bath} = -78\,°C$). The resulting solution (light yellow) was aged at −78 °C for 1 h and then slowly transferred via cannula into the previously prepared amide cuprate solution such that $T_i \leq 60\,°C$ ($T_{bath} = -78\,°C$). The reaction mixture immediately turned dark purple. Upon completion of the transfer, the reaction mixture was stirred at −78 °C for 1 h, and then MeOH (0.63 mL) was added to the reaction mixture such that $T_i \leq -70\,°C$ ($T_{bath} = -78\,°C$) followed by the addition of saturated aqueous NH$_4$Cl solution (40 mL). The mixture was allowed to warm to 25 °C followed by the addition of EtOAc (40 mL). The organic layer was separated and the aqueous layer extracted with EtOAc (40 mL × 2). The organic layers were combined, washed with brine (80 mL), dried over MgSO$_4$ and then concentrated via rotary evaporation. The resul-tant crude product ($R_f = 0.68$, 33% EtOAc in hexane) was further purified by silica gel chromatography (silica gel, 80 : 1 versus-crude product, isocratic, 15% EtOAc in hexane) to give the purified material.

2.3 Higher-order Cyanocuprates R$_2$Cu(CN)Li$_2$

Retention of the cyano group on copper when CuCN is exposed to R$_t$Li (Scheme 1) presumably reflects an element of π basicity between the filled d orbitals on Cu^I and π acidity of the vacant π^* orbitals in the nitrile. These interactions, together with the polarization of the C≡N ligand, may be responsible for the lower-order species (RCu(CN)Li) being receptive toward a second equivalent of RLi, thereby forming 'R$_2$Cu(CN)Li$_2$':

$$RLi + CuCN \longrightarrow RCu(CN)Li \xrightarrow{\text{RLi}} \boxed{R_2Cu(CN)Li_2}$$

These presumed Cu^I dianions are far more reactive than the corresponding lower-order cyanocuprates, and compare quite favorably with Gilman reagents as well in this regard[2a,e−g]. And yet, while more robust toward primary halides and epoxides, for instance, they are at the same time more stable than lower-order species R$_2$CuLi[75], another fringe benefit of the cyano ligand. Although the existence of the higher-order concept as it applies to CuCN has been challenged from both the experimental[76] and theoretical[77] perspectives, the outcome of this ongoing debate has no bearing on the use of cyanocuprate technology. What does seem clear from the arguments presented on both sides[78] is that there is something unique about the cyano ligand within the cuprate cluster, relative to halide ion. Whether the cyanide group is on copper initially or at some later stage of a reaction does not enhance nor diminish the value of these reagents. Thus, the term 'higher-order cyanocuprate' is used for consistency herein, because the expression 'cyano-Gilman cuprate', meant to be synonymous and indicative of a 'lower-order' species[79], could be confused with the RLi/CuCN ratio of 1 : 1 (which is a lower-order, or Gilman-like cyanocuprate, RCu(CN)Li).

The precursor to both RCu(CN)Li and R$_2$Cu(CN)Li$_2$ is CuCN, which offers several advantages over CuI or CuBr•SMe$_2$. CuCN is far less costly, and requires no special handling (such as recrystallization or protection from light), although it is good practice to store it in an Abderhalden at 56 °C (refluxing acetone) over KOH. It comes in several different forms (according to the Merck Index) which is responsible for the different colors to ethereal solutions of the cuprates derived therefrom. They range from an apple-juice light brown (at *ca.* 0.3 M) with CuCN in the tan form (Mallinckrodt) to yellowish green (green form, Fluka), to almost water-white solutions (white powder, Aldrich). Fortunately, there is no distinction between them in terms of the subsequent chemistry of the derived R$_2$Cu(CN)Li$_2$[80].

With both CuCN and the RLi of interest in hand, admixture in THF or Et$_2$O at −78 °C followed by slight warming effects dissolution to the higher-order cuprate. Care should be exercised at this point so as not to slosh or overvigorously stir the mixture, thereby placing the CuCN too high up on the walls of the flask, which if not avoided could engender slight reagent decomposition (observed as a black ring around the flask). Once homogeneity has been reached in THF (or cloudy solutions on occasion in Et$_2$O), recooling to −78 °C sets the stage for substrate addition. Unsaturated enones tend to react at these low temperatures fairly quickly[81], as do unhindered epoxides[82] and primary bromides[83]. Et$_2$O is the preferred solvent for most conjugate additions and oxirane couplings, mainly because of the enhanced Lewis acidity of the Li$^+$ ion in solution. However, halide displacements are not push–pull events, and therefore THF is the solvent of choice[80]. **As with the case above involving a lower-order cyanocuprate, work-up procedures should always avoid the use of highly acidic solutions to prevent generation of noxious HCN.**

1,4 Addition of a Higher-order Cyanocuprate to Mesityl Oxide[81]

(83%)

Ph$_2$Cu(CN)Li$_2$ was prepared by the addition of PhLi (0.65 mL, 1.44 mmol) to CuCN (0.066 g, 0.74 mmol) in 0.95 mL of Et$_2$O at −78 °C. Warming this mixture to 0 °C produces a yellowish but not quite homogeneous solution. The temperature was returned to −78 °C at which point mesityl oxide (57 µL, 0.5 mmol) was added neat via syringe. Stirring was continued at −78 °C. After 45 min the solution became viscous and further stirring was difficult. Quenching after 1 h and work-up in the usual manner were followed by column chromatography on SiO$_2$ with 3 : 1 pentane/Et$_2$O, yielding 0.0726 g (83%) of product (R_f (3 : 1 pentane/Et$_2$O) = 0.43).

Coupling of a Divinyl Higher-order Cyanocuprate with a 1,1-Disubstituted Oxirane[82]

(94%)

To (vinyl)$_2$Cu(CN)Li$_2$ at 0 °C, formed via addition of vinyllithium (0.96 mL, 2.0 mmol) to CuCN (0.089 g, 1.0 mmol) in 1.5 mL of THF, was added 92 µL (0.77 mmol) of the epoxide. The solution was stirred at 0 °C for 5 h, then quenched, and worked up in the usual fashion. Chromatographic purification on SiO$_2$ with 10% Et$_2$O/pentane yielded 0.0923 g (94%) as a clear oil (R_f (1 : 1 Et$_2$O/pentane) = 0.73).

Displacement of a Primary Bromide by a Higher-order Cyanocuprate[83]

Br−(CH$_2$)$_4$−CN $\xrightarrow[\text{THF, −50 °C, 2.5 h}]{\text{Bu}_2\text{Cu(CN)Li}_2}$ CH$_3$−(CH$_2$)$_7$−CN

(92%)

Copper cyanide (0.0896 g, 1.0 mmol) was placed in a 25 mL, two-necked round-bottom flask, evacuated with a vacuum pump, and purged with Ar, and the procedure was repeated three times. THF (1.0 mL) was injected by syringe and the resulting slurry cooled to $-75\,^\circ$C where butyllithium (0.8 mL, 2.0 mmol) was added dropwise. Subsequent warming to $0\,^\circ$C produced a tan solution, which was immediately recooled to $-50\,^\circ$C, followed by dropwise addition, via syringe, of 5-bromovaleronitrile (0.089 mL, 0.77 mmol). The reaction was stirred at this same temperature for 2.5 h followed by quenching with 5 mL of 9 : 1 saturated NH_4Cl/concentrated NH_4OH solution. Extraction with Et_2O (3 × 5 mL), drying over Na_2SO_4, and evaporation of the solvent in vacuo resulted in a light yellow oil, which was chromatographed on silica gel with 15% Et_2O/pentane to yield 0.099 g (92%) of a clear liquid (R_f (1 : 4 Et_2O/pentane) = 0.5; NMR and IR data were identical when compared with that of pelargononitrile (Aldrich)).

Leaving groups other than epoxides and halides have been used together with cyanocuprates[84]. One interesting example involves alkynyl(phenyl)iodonium tosylates, which couple with bis-alkynic cyanocuprates to afford diynes, as in Scheme 2[85]. Attempts to use an insoluble alkynylcopper reagent, rather than a cuprate, gave significant amounts of homocoupling products. This is yet another situation where a normally nontransferable ligand, (i.e., an alkynic group) can be transferred from copper when part of a homo-rather than mixed cuprate (see Section 2.4.5)[86]. These phenyliodonium tosylates are also responsive to Gilman cuprates R_2CuLi to give disubstituted alkynes. When this leaving group is part of an alkenyl system, it reacts with cuprates to give stereodefined alkenes. When applied to double displacements *en route* to endiynes, it was necessary to avoid THF as reaction solvent and switch to CH_2Cl_2[87]. Most unusual is the formation of the cuprate in CH_2Cl_2, rather than in THF followed by solvent exchange (see Section 2.1.4).

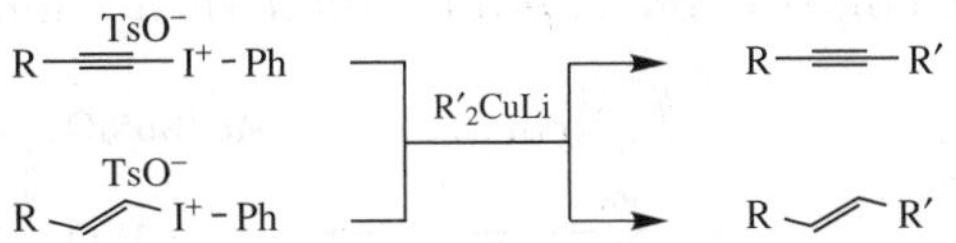

Scheme 2.

Alkylation of a Dialkynic Cyanocuprate with an Alkynic Phenyliodonium Tosylate[85]

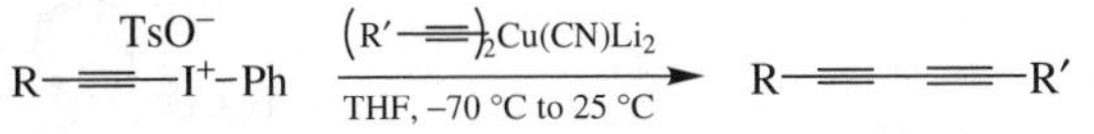

To a solution of phenylacetylene (0.31 g, 3.0 mmol) in THF (20 mL) was added dropwise BuLi (1.5 mol dm^{-3} in hexane solution, 3.0 mmol) at $-70\,^\circ$C under a nitrogen atmosphere. CuCN (0.13 g, 1.5 mmol) was then added and the mixture was stirred at $-40\,°C$ for 2 h; it was then cooled to $-70\,°C$. Solid hex-1-ynyl(phenyl)iodonium tosylate (0.46 g, 1.0 mmol) was added to the cooled mixture which was then allowed to warm to room temperature after which it was poured into saturated aqueous ammonium chloride and the resulting precipitate filtered off. The filtrate was extracted with ether and the extract was washed, dried, and concentrated in vacuo. The products were analyzed by GC (column, OV-17) and separated by column chromatography on silica gel with hexane/dichloromethane eluent.

Higher-order cyanocuprates composed of N-protected α-aminoalkyl groups have been developed as routes to several nitrogen-containing structured arrays[88]. Both acyclic (**20**) as well as cyclic (**21**) reagents are available, and undergo a variety of couplings (e.g., acylation[89], 1,4 additions[90], etc.). Included in these schemes is the coupling with a vinyl iodide which, as in the example below[91], was ultimately used to make (±)-norruspoline. Importantly, lithiation, which is accomplished with ^{s}BuLi, need be carried out with clear solutions of this organolithium (from FMC Lithium Corporation). Those that contain particulate material should be filtered and freshly titrated, with best results ultimately being obtained with titres between 1.3 and 1.5 M. Lithiations should be given no longer than the 1 h prescribed, as carbamate deprotonation may occur and lead to decomposition of the lithiated educt in THF.

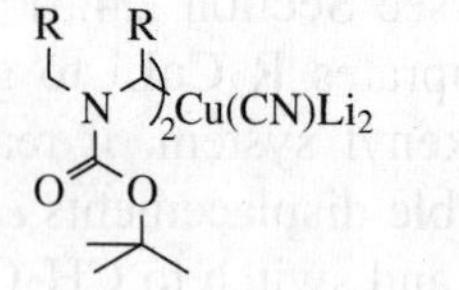

20 **21**

Coupling of an α-Aminoalkyl Cyanocuprate with a Vinyl Iodide[91]

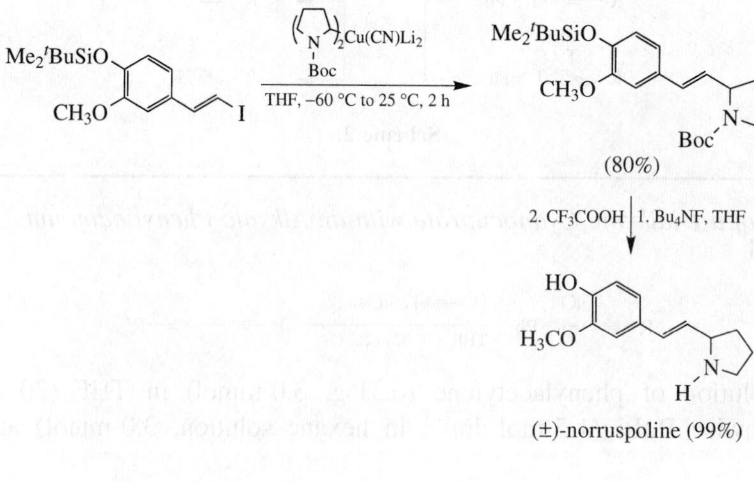

(80%)

2. CF₃COOH | 1. Bu₄NF, THF

(±)-norruspoline (99%)

(−)-Sparteine (1.2 mmol) was added to a stirred solution of *N*-tert-butoxycarbonyl (*N*-Boc)-protected pyrrolidine (1 mmol) in THF (3 mL), under nitrogen at −78 °C, followed by dropwise addition of ^{s}BuLi (1.2 mmol). The reaction mixture was stirred for 1.0 h followed by addition of CuCN•2LiCl (CuCN (0.5 mmol) solubilized with LiCl (1 mmol)) in THF (3 mL) via cannula at −78 °C. The solution turned dark yellow and the reaction mixture was slowly warmed to −60 °C over 50 min. The *E*-vinyl iodide derivative of vanillin (1.5 mmol) in THF (2 mL) was added via cannula. The reaction mixture was warmed to 25 °C after 10 min and then stirred at 25 °C for 2 h. The reaction was quenched with saturated aqueous NH$_4$Cl (15 mL), the combined phases filtered through Celite, and the aqueous phase extracted with ether (3 × 6 mL). The combined ether extracts were washed with saturated aqueous NH$_4$Cl, dried over anhydrous MgSO$_4$, and concentrated in vacuo. Purification by flash chromatography (silica gel) gave the vinylic substitution product in 80% yield. The product was treated with Bu$_4$NF (2.5 mmol) to remove the silyl protecting group and then with triflic acid to remove the Boc protecting group. Purification of the crude material using medium pressure liquid chromatography (MPLC) (silica gel, 15% ether/petroleum ether) gave (±)-norruspoline (99% yield).

All of the previous reactions of 'R$_2$Cu(CN)Li$_2$' have in common an R group which is carbon based. Although generation of C−C bonds via cuprates is of paramount importance, various copper reagents bearing other types of ligands (e.g., Si, Sn, or H, see below) are also of considerable value. To be included in this group is the chemistry of bis-amidocuprates, such as Li$_2$Cu(CN)[NBn(TMS)]$_2$, prepared in an analogous fashion using CuCN and two equivalents of the lithiated amine[92]. When added to a dienoate, the presumed cuprate delivers nitrogen in a 1,4 fashion. Although similar results are realized with the corresponding Gilman amino cuprate (i.e., from CuI rather than CuCN), subsequent trapping of an intermediate enolate by an aldehyde procedes more smoothly and cleanly with the higher-order cuprate. The sequence is tantamount to a three-component coupling which produces an intermediate readily parlayed into a β-lactam such as **22**, amenable to conversion to the 1-β-methylcarbapenems. When applied to sultam educt **23**, adduct **24** was formed as a single diastereomer, the TMS moiety being lost on work-up[93].

22

Three-component Coupling with an Amido Cyanocuprate[93]

1. Li$_2$[Cu(CN)(N(Bn)TMS]$_2$
   THF, −100 °C, 20 min

2. CH$_3$CHO, −100 °C to −70 °C, 1 h
3. TBS-Cl, Imidazole, CH$_2$Cl$_2$, 25 °C

23   **24** (71%)

To a solution of *N*-(trimethylsilyl)benzylamine (6.08 mL, 31.0 mmol) in THF (40 mL) at −78 °C was slowly added BuLi (18.6 mL, 1.61 M in hexane, 80.0 mmol). After stirring for 25 min, CuCN (1.34 g, 15.0 mmol) was added and the mixture was allowed to warm to −45 °C. After stirring for 5 min, the solution was cooled to −78 °C. To the mixture was added a cold (−100 °C) solution of the sultam (1.86 g, 5.0 mmol) in THF (15 mL) over a period of 20 min. After 40 min, acetaldehyde (10 mL, 5 M in THF, 50 mmol) was added. After the mixture had been allowed to gradually warm to −70 °C over a period of 1 h, it was poured into a mixture of aqueous saturated NH$_4$Cl and 28% aqueous NH$_3$ (1 : 1, 50 mL) with vigorous stirring, and then diluted with Et$_2$O. The organic phase was separated, washed with a mixture of aqueous saturated NH$_4$Cl and 28% aqueous NH$_3$ (1 : 1, 50 mL) twice, and brine, dried (K$_2$CO$_3$) and concentrated in vacuo to leave the crude product, which was dissolved in CH$_2$Cl$_2$ (100 mL). To this solution was added imidazole (3.4 g, 49.9 mmol) and then tert-butyldimethylsilyl chloride (5.7 g, 37.8 mmol) at 0 °C. The mixture was stirred overnight at 25 °C, and then quenched with H$_2$O. The organic phase was separated, washed with brine, dried (K$_2$CO$_3$), and concentrated in vacuo. The residue was purified by chromatography on silica gel using a mixture of hexane and EtOAc (5 : 1) as eluent to give the product silyl ether (2.26 g, 71% yield) as a viscous oil.

2.3.1   Reactions in the Presence of BF$_3$•Et$_2$O

When higher-order cyanocuprates are exposed to an equivalent or more of BF$_3$•Et$_2$O in THF, two significant events occur immediately, in spite of no visible change in appearance of the solution: (a) an equilibrium is established with the lower order cyanocuprate, and

$$R_2Cu(CN)Li_2 + BF_3 \rightleftharpoons R_2Cu(CN\text{-}BF_3)Li_2 \rightleftharpoons RCu(CN)Li + RLi\text{•}BF_3$$

(b) the BF$_3$ situates itself to a significant degree on the nitrile ligand in R$_2$Cu(CN)Li$_2$[94]. As control experiments demonstrated that the reactive species is surely the higher order reagent, it now seems likely that the boost in cuprate reactivity may, at least in part, be ascribed to the rapid inclusion of this potent Lewis acid into the cuprate cluster. Thus, as the enone is added, it sees a species bearing a far stronger carbonyl-activating moiety than would otherwise be the case (i.e., BF_3 rather than, or in addition to, Li^+). Although the ^1H and ^{11}B NMR data unequivocally attest to the equilibrium as shown, whether the BF_3-complexed cuprate species is the actual reagent responsible for the chemistry is unknown. What is not a matter for conjecture, however, is the frequent extraordinary difference this combination of reagents can make in otherwise challenging substrates[42]. The argument is particularly convincing, for example, with isophorone, where only with $BF_3 \cdot Et_2O$ present could a phenyl group be delivered to the β site in high yield[43].

$BF_3 \cdot Et_2O$-assisted 1,4 Addition of a Higher-order Cyanocuprate to a Hindered Enone[95]

$$\text{Ph}_2\text{Cu(CN)Li}_2, \text{THF}, \text{BF}_3\cdot\text{Et}_2\text{O}$$
$$-78\,°C \text{ to } -50\,°C, 1\,h, -15\,°C, 0.75\,h$$

(>95%)

Ph$_2$Cu(CN)Li$_2$ was prepared as a yellow solution in THF (0.6 mL)/Et$_2$O (0.6 mL) using PhLi (2.0 mmol, 0.90 mL) and CuCN (1 mmol, 0.0896 g). BF$_3$•Et$_2$O (1.0 mmol, 0.13 mL) was added to the cuprate at $-78\,°C$ with no visible change seen. Isophorone (0.50 mmol, 0.074 mL) was added, neat, at $-78\,°C$ followed by stirring at $-50\,°C$ for 1 h, and then at $-15\,°C$ for 0.75 h. Quenching followed by quantitative VPC analysis indicated that phenyl transfer had occurred to the extent of >95%. Filtration through SiO$_2$ afforded pure material.

Usually this additive, if successful in turning an otherwise sluggish coupling into a facile process, is compatible with R$_2$Cu(CN)Li$_2$ up to about $-50\,°C$. Temperatures much above this limit start to seriously erode the percentage of educt consumption, presumably due to side reactions between 'RLi' and BF$_3$ (see page 704), as well as BF$_3$-mediated opening of THF by the RLi•BF$_3$ present[96]. Fortunately, S$_N$2 displacements of epoxides with the R$_2$Cu(CN)Li$_2$/BF$_3$ mixture also tend to occur at low temperatures and at greater rates relative to reactions in the absence of this additive[97]. Such is also true for openings of oxetanes.

$BF_3 \cdot Et_2O$-mediated Opening of Oxetane by a Higher-order Cyanocuprate[97]

1. Ph$_2$Cu(CN)Li$_2$, Et$_2$O, BF$_3$-Et$_2$O
   $-78\,°C$ to $-50\,°C$, 1 h
2. Ac$_2$O, pyr

OAc

(87%)

An ethereal solution of PhLi•LiBr (49.2 mL of a 1.3 M solution in Et$_2$O, 64 mmol) was rapidly added to a suspension of CuCN (3 g, 33 mmol) in Et$_2$O (100 mL) at −30 °C. Stirring was continued for 10 min at −15 °C until a grey solution was obtained. This solution was cooled to −40 °C and trimethylene oxide (1.74 g, 30 mmol) in Et$_2$O was added. The yellowish solution was again cooled to −78 °C, whereupon BF$_3$•Et$_2$O (4.2 mL, 32 mmol) in Et$_2$O (20 mL) was slowly added. The reaction was complete after 1 h at −50 °C. The yellow turbid solution was then hydrolyzed by addition of 60 mL of aqueous NH$_4$Cl and 40 mL of aqueous ammonia. The salts are filtered off and the aqueous layer extracted twice (2 × 100 mLEt$_2$O). The combined organic phases were concentrated in vacuo. The residue was directly acetylated by dissolving in pyridine (50 mL) and addition, at 0 °C, of Ac$_2$O (8.55 mL, 90 mmol). After stirring overnight at room temperature, MeOH (5 mL) was added to destroy excess Ac$_2$O. After 1 h, Et$_2$O (250 mL) was added and this solution was washed once with aqueous NaHCO$_3$ solution (100 mL), then four times with aqueous NH$_4$Cl (4 × 100 mL) to remove most of the pyridine, then, once with 1 N HCl (100 mL). The organic phase was dried over MgSO$_4$ and concentrated in vacuo. The residue was distilled through a 15 cm Vigreux column affording the desired product (4.65 g, 87% yield), bp 74 °C (0.05 mmHg).

2.3.2   Reactions in the Presence of Me$_3$SiCl

Although the extent of impact of Me$_3$SiCl on lower-order homocuprates R$_2$CuLi has yet to be fully delineated[47−50], admixture of this silyl halide with R$_2$Cu(CN)Li$_2$ even at −100 °C has a dramatic effect on these species[98]. Remarkably, it is the cyano ligand that is sequestered, giving rise to Me$_3$SiCN along with the lower-order cuprate:

$$R_2Cu(CN)Li_2 + 2Me_3SiCl \xrightarrow{\text{THF, } < -78\,°C} Me_3SiCN + R_2CuLi + Me_3SiCl + LiCl$$

The spectroscopically established presence of TMS-CN raises questions as to its role in this chemistry; nevertheless, the use of TMS-Cl (in excess) appears to offer benefits similar to those noted for reactions of lower-order cuprates with α,β-unsaturated carbonyl systems (vide supra)[47−50]. Improvements in rates and diastereoselectivities of 1,2 additions observed due to the in situ formation of TMS-CN, together with residual TMS-Cl, have also been noted[98].

Conjugate Addition of an α-Alkoxy Higher-order Cyanocuprate[99,100]

(96%)

A solution of the stannane (1.0 mmol) in 5 mL THF was cooled to $-78\,^\circ$C (CO_2/acetone). A 0.50 mL sample of a 2.6 M solution of butyllithium in hexane (1.3 mmol) was then added and the solution was stirred for 5 min at $-78\,^\circ$C. A second 25 mL round-bottom flask containing 0.045 g (0.5 mmol) copper(I) cyanide suspended in 2 mL THF was then cooled to $-78\,^\circ$C (CO_2/acetone). The α-alkoxylithio species was transferred via cannula to the suspension of copper cyanide at $-78\,^\circ$C. The cuprate mixture was gradually allowed to warm to $-60\,^\circ$C (bath temperature) over a period of 30 min. A clear, homogeneous solution was obtained. A third 25 mL round-bottom flask containing a solution of the enone (0.5 mmol) in 3 mL THF was cooled to $-78\,^\circ$C (CO_2/acetone). Trimethylsilyl chloride (0.32 mL, 2.5 mmol) was then added to the enone solution. The enone/TMS-Cl mixture was then added to the cuprate solution (at $-78\,^\circ$C) via cannula. The resulting mixture was stirred for 1 h at $-78\,^\circ$C, and then gradually warmed to $0\,^\circ$C (ice bath) over an additional 2.5 h time period. The reaction mixture was quenched by the addition of 1 mL of 1.0 N aqueous HCl, stirred for 10 min, and then diluted with 100 mL ether. The mixture was then washed sequentially with a 1 : 1 mixture of aqueous NH_4Cl/1.0 N HCl (1×40 mL), saturated aqueous NaCl (1×40 mL), and saturated Na_2CO_3 (1×40 mL). The layers were separated and the organic phase was dried over anhydrous $MgSO_4$. After removal of the solvent under reduced pressure, the crude reaction product was purified by flash chromatography on silica gel using 15–20% ethyl acetate/petroleum ether as eluent; (yield 96%).

The procedure cited above[99], which effectively enables introduction of a protected α-hydroxyalkyl appendage in a 1,4 sense to an enone, is best carried out with fresh, purified α-alkoxystannane (by column chromatography). Clear solutions of the higher-order cuprate should be used; cloudy mixtures imply impure stannane and lead to inferior results. The use of the MOM derivative in this case is not essential, as the MEM analog works equally as well. However, the benzyloxy methyl (BOM) derivative is not an acceptable choice.

2.4 Higher-order Mixed Cyanocuprates $R_tR_rCu(CN)Li_2$

The impetus behind the development of more highly mixed higher order cuprates was exactly the same as that which led to the development of the lower-order analogs R_tR_rCuLi; that is, to conserve potentially valuable organolithium reagents, two equivalents of which normally go to form $R_2Cu(CN)Li_2$, whereas most often only one is transferred to the educt. Extensive trials have led to several observations of a general nature. It is now appreciated that for Michael additions[81], alkyl and vinylic ligands transfer selectively over acetylenic, thienyl, dimsyl, N-imidazoyl, trimethylsilylmethyl, and even a simple methyl moiety[101]. Several procedures for these residual ligand types (R_r) are illustrated below. The simplest, of course, is that which allows for reagent formation using materials out of a bottle, and in this sense only the latter two (e.g., $R_t(Me)Cu(CN)Li_2$) meet this criterion.

2.4.1 $R_r = 2$ – thienyl

With the advent of the 'cuprate in a bottle', (2-Th)Cu(CN)Li[101b], sold by Aldrich (catalog No. 32417-5), mixed higher-order cuprates can be easily prepared by simply adding an R_tLi of one's choosing to this precursor to form $R_t(2\text{-}Th)Cu(CN)Li_2$. Alternatively, 2-lithiothiophene can be generated and then added to CuCN to arrive at this reagent in situ, followed by higher-order cuprate formation[102].

Michael Addition of a Mixed Thienyl Higher-order Cyanocuprate to an Enoate[102]

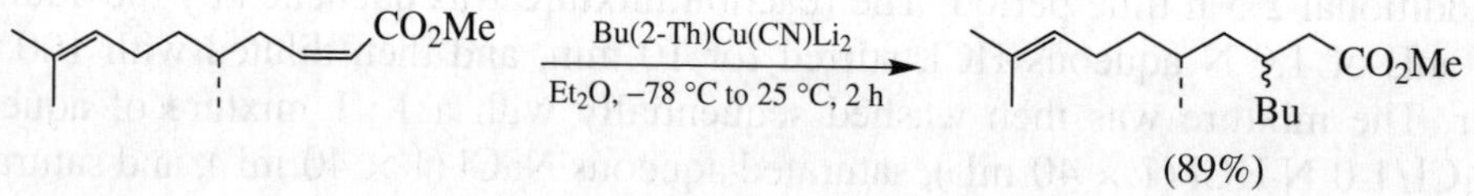

The mixed cuprate was formed using CuCN (0.054 g, 0.61 mmol), 2-lithiothiophene (0.61 mmol), and BuLi (0.24 mL, 2.53 M, 0.61 mmol) in 1.4 mL of Et_2O. The unsaturated ester (121 μL, 0.55 mmol) was added to the cold (−78 °C) cuprate, the solution from which was slowly warmed to 25 °C and stirred there for 2 h. Quenching with 10% NH_4OH/saturated aqueous NH_4Cl and extractive work-up with Et_2O followed by solvent removal in vacuo and chromatography on SiO_2 with 5% Et_2O/pentane afforded 0.131 g (89%) product (R_f(10%Et_2O/petroleum ether) = 0.60).

The diminished reactivity of higher-order thienyl cuprates toward hindered (e.g., β,β-disubstituted) enones can usually be overcome by simply adding $BF_3\cdot Et_2O$ to the performed, cold cuprate. Thus, in the case of isophorone, an otherwise sluggish coupling of a vinyl cuprate occurs at −78 °C in less than 1 h.

BF$_3$•Et$_2$O-assisted 1,4 Addition of a Mixed 2-Thienyl Higher-order Cyanocuprate $R_t(2\text{-}Th)Cu(CN)Li_2$[43]

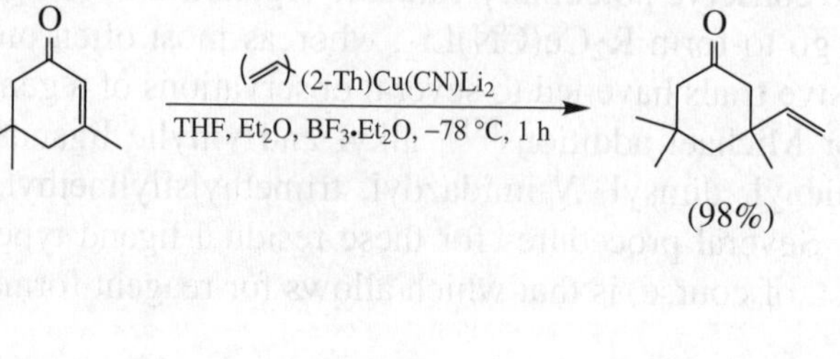

Thiophene (0.082 mL, 1.02 mmol) was added to THF (0.6 mL) in a 10 mL two-neck pear flask at $-78\,^\circ$C, followed by BuLi (0.315 mL, 1.0 mmol). Stirring was continued at the same temperature for 15 min, then at $0\,^\circ$C for 30 min. The solution was transferred, via cannula, into a slurry of CuCN (89.6 mg, 1.0 mmol) and Et$_2$O (0.6 mL), with a wash of 0.6 mL Et$_2$O. Warming to $0\,^\circ$C gave a light tan solution which was recooled to $-78\,^\circ$C where vinyllithium (0.60 mL, 1.0 mmol) was introduced. After addition of neat BF$_3\bullet$Et$_2$O (0.13 mL, 1.0 mmol), isophorone (0.104 mL, 0.70 mmol) was added, neat, followed by stirring at $-78\,^\circ$C for 1 h and quenching. VPC analysis indicated that vinyl transfer had occurred to the extent of 98%. Chromatography on SiO$_2$ with Et$_2$O/pentane (15 : 85) gave pure material (TLC R_f(15:85 pentane/Et$_2$O) = 0.32).

Insofar as substitution reactions are concerned, the selectivity of transfer using R$_t$Li = MeLi is not satisfactory[82]. Hence, it is necessary to resort to other choices, such as the thienyl-containing system R$_t$(2 − Th)Cu(CN)Li$_2$. Unhindered, monosubstituted epoxides couple very nicely with these mixed cuprates, even when less robust vinylic groups are undergoing transfer, as in the case of opening a chiral, nonracemic glycidol ether.

Opening of a Chiral, Nonracemic Epoxide with a Higher-order Cyanocuprate Containing the 2-Thienyl Ligand[102]

Thiophene (88 µL, 1.1 mmol) was added to THF (1 mL), at $-78\,^\circ$C followed by butyllithium (0.39 mL, 1.1 mmol). The cooling bath was removed and the temperature raised to $0\,^\circ$C over 5 min and stirred for an additional 30 min. The faint yellow anion was then transferred, via cannula, into a two-neck flask containing cuprous cyanide (0.0896 g, 1 mmol) and THF (1 mL), which was previously purged with Ar and cooled to $-78\,^\circ$C. Warming to $0\,^\circ$C produced a light tan solution which was cooled to $-78\,^\circ$C and vinyllithium (0.5 mL, 1 mmol) was injected, with immediate warming to $0\,^\circ$C (no visible change). It was then cooled to $-78\,^\circ$C and to it was added, via cannula, a precooled solution of (2S)-benzyl-2-epoxypropyl ether (0.149 g, 0.91 mmol) in THF (1 mL). After 2.5 h at $0\,^\circ$C, the reaction was quenched with 5 mL of a 90% saturated NH$_4$Cl/concNH$_4$OH solution, extracted with ether (2 $\times$ 10 mL) and dried over sodium sulfate. Concentration in vacuo, followed by chromatography on silica gel (230–400 mesh) with ether/petroleum ether (2 : 3) afforded 0.161 g (92%) of a clear liquid (bp $90\,^\circ$C at 0.1 mmHg; R_f (1 : 1 ether/petroleum ether) = 0.33; $[\alpha]_D = -2.2^\circ$ ($c = 3$, CHCl$_3$)).

2.4.2 R_r = dimsyl

Generation and Use of "DMSO Cuprates" $Li_2[CH_3SOCH_2Cu(CN)R]$[60e]

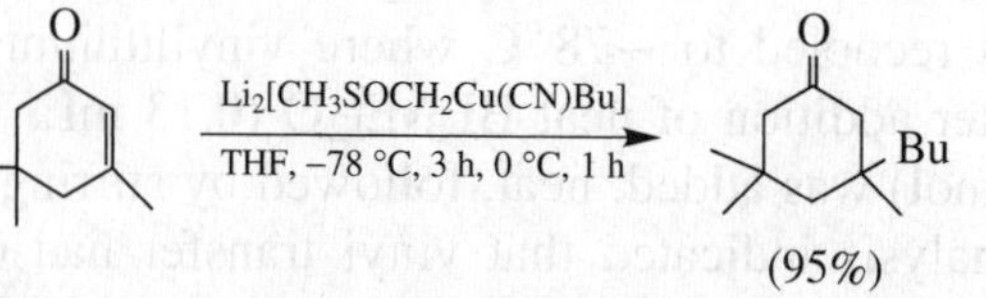

(95%)

The lithio anion of DMSO was generated as a 0.2 M solution in THF by treatment of DMSO with butyllithium (1 equiv) at 0 °C for 15 min. This was then transferred to a slurry of cuprous cyanide (1 equiv) in THF at −78 °C via cannula. The mixture was warmed to 0 °C resulting in a light green slurry which was recooled to −78 °C and butyllithium (1 equiv) was added and allowed to warm to 0 °C to ensure cuprate formation. It was then cooled to −78 °C and a solution of 3,5,5-trimethylcyclohexen-1-one (0.45 equiv) in THF was added via syringe. After 3 h at −78 °C and an additional 1 h at 0 °C, the reaction was quenched with a saturated NH_4Cl solution containing 10% NH_4OH. After stirring for 15 min, it was suction filtered through Celite; the filter cake was washed with ether and the aqueous phase extracted with more ether. Analysis of the combined organic phases by VPC showed the product had formed in 95% yield.

2.4.3 R_r = methyl

Selective Ligand Transfer From a Mixed Dialkyl Higher-order Cyanocuprate to 2-Cyclopentenone[81]

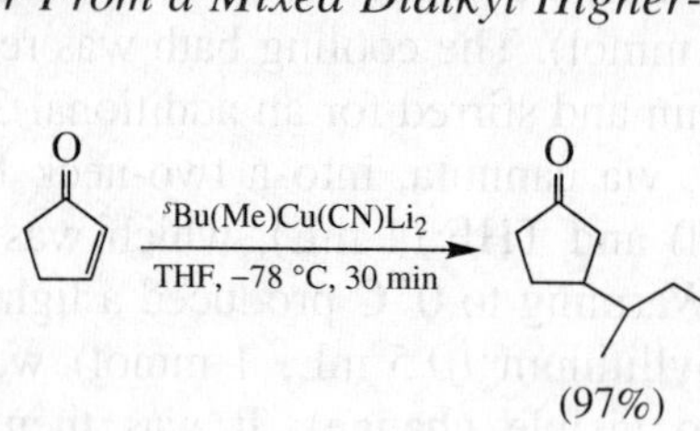

(97%)

To CuCN (0.0672 g, 0.75 mmol) in 1.25 mL of cold (−78 °C) THF was added MeLi (0.47 mL, 0.75 mmol) and the mixture was warmed to 0 °C. Recooling to −78 °C and addition of ^{s}BuLi (0.61 mL, 0.75 mmol) was followed by injection, via syringe, of cyclopentenone (42 μL, 0.5 mmol) and stirring for 0.5 h. Quenching and extractive (Et_2O) work-up followed by chromatography on silica gel (40% Et_2O/pentane) afforded 0.067 g (97%) of the product as a light oil (R_f (1 : 1Et_2O/pentane) = 0.69).

2.4.4 $R_r = N - imidazoyl$

An attractive alternative to the 2-thienyl ligand is the nitrogen-to-copper bound imidazole moiety[103]. The driving force behind its development lies in the invariable occurrence of homocoupled material formed upon work-up of mixed cuprate reactions, which for $R_t(2\text{-Th})Cu(CN)Li_2$ implies the formation of 2, 2'-dithiophene, **25**. Although this byproduct is usually not chemically problematic, it is nonpolar and frequently requires chromatographic separation. No such issues exist for the corresponding cuprate derived from *N*-lithioimidazole plus CuCN, to which is added an R_tLi.

$$R_t(2\text{-Th})Cu(CN)Li_2 \xrightarrow{\;E^+\;} R_t - E + \text{[25]}$$

25

⇑

byproduct

The combination of CuCN with N-lithiated imidazole leads to a pale green solution of the lower-order cyanocuprate **26**. Although when cooled to $-78\,°C$ there is some precipitation, introduction of an R_tLi affords a homogenous yellow solution upon slight warming. The stability of **27** as a 0.2 M solution in THF has been examined, and it appears that storage at about $4\,°C$ (in a refrigerator) for at least 1 week does not change the titre. A potentially even more practical method involves mixing powdered *N*-lithioimidazole with CuCN in THF at $25\,°C$ to obtain the lower-order cuprate. Thus, the lithiation step at the time of cuprate generation is avoided, and no difference in yields was noted.

$$\text{[imidazole-Li]} + CuCN \xrightarrow[25\,°C]{THF} \underset{\textbf{26}}{N\text{-Cu(CN)Li}} \xrightarrow{R_tLi} \underset{\textbf{27}}{R_t\text{-Cu(CN)Li}_2}$$

1,4 Addition of a Mixed Imidazoyl Cyanocuprate to an Enone[103]

$$\xrightarrow[\text{THF, BF}_3\bullet\text{OEt}_2,\ -78\,°C,\ 15\ \text{min}]{\text{Ph(Imid)Cu(CN)Li}_2}$$

(83%)

Imidazole (0.100 g, 1.469 mmol), was dissolved in 5 mL of THF, cooled to $-78\,°C$ and BuLi (2.79 M, 0.526 mL, 1.469 mmol), was added dropwise via syringe. CuCN (0.132 g, 1.469 mmol), and LiCl (0.125 g, 2.938 mmol) were placed in 5 mL of THF and stirred until dissolved. The solubilized CuCN•2LiCl was then added via cannula to the previously formed 1-lithioimidazole at $-78\,°C$. The resulting suspension was allowed to warm to $0\,°C$ and stirred until a light green solution formed.

The solution was cooled to $-78\,^{\circ}$C (precipitation occurred), and PhLi (1.88 M, 0.781 mL, 1.469 mmol), was added. The solution was carefully warmed until a clear, faint yellow solution formed. The solution was then cooled to $-78\,^{\circ}$C, BF$_3$•Et$_2$O (0.181 mL, 1.469 mmol) was added via syringe, then 4-isopropylcyclohexenone (0.166 mL, 1.130 mmol) was added via syringe, stirred for 15 min at $-78\,^{\circ}$C, and then poured into 20 mL of water saturated with NH$_4$Cl and 10% NH$_4$OH. Extraction with 3×30 mL of ethyl acetate, drying with Na$_2$SO$_4$, and then solvent removal in vacuo gave a residue which was purified by flash chromatography (silica gel, 8 : 2 hexane/ethyl acetate), to give 0.203 g (83%), of a clear oil (TLC (ethyl acetate/hexane, 10 : 90) $R_f = 0.60$).

2.4.5   R$_r$ = trimethylsilylmethyl(TMSM)

As alluded to above and first noted by Whitesides[17], mixed cuprates bearing dummy ligands R$_r$, whether of a lower- or higher-order blend, tend to react more sluggishly than their homocuprate analogs (i.e., R$_2$CuLi or R$_2$Cu(CN)Li$_2$). However, it has been found that the ligand Me$_3$SiCH$_2$ (TMSM) is not only nontransferable in its mixed cuprates R$_t$(TMSM)CuLi and R$_t$(TMSM)Cu(CN)Li$_2$, but it imparts an excellent reactivity profile relative to the use of homocuprates[104]. Such is the case in both THF and Et$_2$O solvent. The main byproduct of the selective transfer of R$_t$ is the proton-quenched TMSM group, i.e., tetramethylsilane. Importantly, TMSM mixed cuprates display impressive stability data even at $25\,^{\circ}$C, presumably due to a β-silicon effect on metal centers. The precursor organolithium is known[105] and can be combined with CuCN (or CuI) in the usual 1 : 1 stoichiometry, followed by introduction of R$_t$Li.

Conjugate Addition of Bu(TMSM)CuLi•LiI to Cyclohexenone[104]

Commercial trimethylsilylmethyllithium (1.0 M in THF, Aldrich) was used to prepare CuCH$_2$Si(CH$_3$)$_3$, according to a literature procedure. Thus, 0.1905 g of CuI, suspended in 8 mL of dry (Na/benzophenone) THF at $-78\,^{\circ}$C, was treated with 1.02 mL of 0.98 M LiCH$_2$Si(CH$_3$)$_3$ (1.00 mmol). After 0.1 h at $-78\,^{\circ}$C, the reaction mixture was transferred to an ice bath and stirred for 0.1 h at $0\,^{\circ}$C. The reaction mixture was cooled to $-78\,^{\circ}$C, and 0.41 mL of 2.44 M BuLi (1.00 mmol, Aldrich) was added. After stirring at $-78\,^{\circ}$C for 0.1 h, the cuprate was annealed at $0\,^{\circ}$C for 0.1 h. The brown solution was cooled to $-78\,^{\circ}$C and treated with 0.100 mL of 2-cyclohexenone (0.100 g, 1.00 mmol, Aldrich 95%) and 0.110 g of undecane (internal standard) in 0.1 mL of the same solvent. After 6 min, the reaction mixture was quenched with 3 mL of nitrogen-sparged, saturated aqueous sodium bicarbonate.

The organic layer was diluted with an equal volume of ether and dried over anhydrous sodium sulfate before GLC analysis.

Although TMSM mixed cuprates $R_t(TMSM)CuLi$ and $R_t(TMSM)Cu(CN)Li_2$ transfer R_t selectively, it had been shown years earlier that TMSM homocuprates $(TMSM)_2Cu(CN)Li_2$ are excellent reagents for converting epoxysilanes **28** to allyl silanes via adducts **29**[106]. Cuprates derived from either $TMSCH_2MgCl$ and 10 mol% CuI, or notably $TMSCH_2Li/CuI$ (2 : 1, i.e., the Gilman cuprate $(TMSM)_2CuLi$) were unacceptable alternatives. Interestingly, the latter reagent afforded only 10% reaction after 12 h at 0° C, whereas the higher-order species gave product in excellent yield in 6 h under otherwise identical conditions. No mixed reagents were tested in this study. More hindered epoxides (i.e., **28**, R = ^iPr, ^tBu, Ph) took slightly longer at higher temperatures (7 h at 25 °C).

Representative Coupling of $(TMSM)_2 Cu(CN)Li_2$ with an Epoxide[106]

R = Me, 90%; Pr, 86%

Bu, 91%; ^iPr, 81%

^tBu, 90%; Ph, 89%

To CuCN (1.79 g, 20 mmol) in THF (50 mL) at −78 °C was added trimethylsilylmethyllithium in hexanes (44.6 mL, 0.87 M, 40 mmol) dropwise. The cold bath was removed and dissolution occurred after *ca.* 15 min at which time the mixture was recooled to −78 °C and the epoxide (20 mmol) was added dropwise via syringe. After slowly warming to 0 °C, the mixture was allowed to stir for an additional 6 h. The reaction was quenched with a 9 : 1 $NH_4Cl(sat)/NH_4OH$ (conc) solution (50 mL). After stirring for 15 min, extraction with ether (50 mL) gave an organic layer which was dried (Na_2SO_4), concentrated in vacuo and distilled to give the desired erythro disilylated alcohol.

2.4.6 $R_r = R_t$ or R_t' (cuprate oxidations)

One of the pitfalls associated with cuprate chemistry is the sensitivity of these reagents to molecular oxygen, or traces of other oxidizing agents. Usually, for a mixed cuprate such as RR′CuLi, its unwanted oxidation leads to as many as three byproducts often produced in statistical ratios, i.e., about $1 : 2 : 1$:

$$RR'CuLi \xrightarrow{[O]} R-R + R-R' + R'-R'$$

Control of these ratios can be achieved, however, for mixed aryl cyanocuprates ArAr′Cu(CN)Li$_2$ when the higher-order reagent is formed under kinetic conditions[107]. That is, initial formation of the lower-order cyanocuprate ArCu(CN)Li (or Ar′Cu(CN)Li) is followed by cooling to $-125\,^\circ$C prior to introduction of the other Ar′Li (or ArLi). The cuprate must be formed in 2-methyltetrahydrofuran (2-MeTHF) which allows for the temperature to be maintained at this level without freezing the solvent, otherwise not possible in THF. Bubbling dry gaseous O$_2$ through such a species at $-125\,^\circ$C results in its conversion to the unsymmetrical biaryl Ar–Ar′ with selectivities usually greater than 93%. Of paramount importance is that the temperature of ArCu(CN)Li not be raised significantly above $-125\,^\circ$C during addition of Ar′Li, the second aryllithium also having been made in 2-MeTHF. Other critical experimental parameters include:

1. The quality of ′BuLi being used to generate ArLi and Ar′Li; if not relatively fresh, products containing tert-butylated aromatic rings, and those reflecting coupling of Ar′ and Ar to the CN ligand may be found[108].

2. Replacement of CuCN by either CuI (in the form of the solubilized LiI, LiBr, or Bu$_3$P salts), or CuBr (itself or as CuBr•SMe$_2$) do not afford results that are in any way related to those from cyanocuprate oxidations[109].

These couplings apply likewise to dinaphthyl cuprates[108], and provide similar selectivities from mixed phenyl–naphthyl or heteroaryl–naphthyl[110] cyanocuprates.

General Procedure for the Intermolecular Synthesis of a Mixed Biaryl[107]

> 93% selectivity

All joints were greased and wrapped with parafilm. A dry three-necked 50 mL or 100 mL flask equipped with a rubber septum (because of pressure changes during cooling, the septum should be secured with copper wire), a hose adapter for argon/vacuum, and a thermometer adapter with septum were used (if problems with foaming during the oxidation with oxygen occur, a 100 mL flask should be used). A gas dispersion tube (Aldrich, porosity 40–60 μm) was put through the septum of the thermometer adapter and could be moved for the oxidation by pushing it into the solution. The open end of the gas dispersion tube was sealed with a small septum. In the flask were placed a stirring bar (2.5 cm long) and 0.090 g (1 mmol) of dry CuCN. The flask was then evacuated to about 0.01 torr for at least half an hour and heated twice with a heat gun. The flask was then filled with argon. Evacuation and filling with argon were each repeated twice. With a syringe, 18 mL of dry MeTHF was transferred to the flask, and the suspension was cooled to about $-78\,°C$. To the resulting suspension was added exactly 1 mmol (*ca.* 1 mL of solution) of aryllithium (ArLi). The cooling bath was removed, and the suspension was allowed to warm until all of the CuCN had dissolved. The speed of stirring was adjusted so that the solution was well stirred (this is important in order to get good mixing at these very low temperatures). The resulting water-white to pale yellow solution (a bright yellow color is an indication of excess CuCN, i.e., hydrolyzed ArLi) was then cooled to $-125\,°C$ using a pentane/liquid nitrogen cooling bath in a Dewar flask. The whole flask was smothered with pentane (up to the beginning of the necks), and the Dewar was kept covered with a layer of frozen pentane by periodically adding liquid nitrogen during the reaction. A thermocouple was placed in the cooling bath to monitor the temperature. To this solution of the lower order cuprate was added exactly 1 mmol (*ca.* 1 mL of solution) of Ar′Li via syringe to the well-stirred solution. Addition should be dropwise and slow, at a speed of *ca.* 1 drop per second; 5–7 min are required for this addition. The solution was stirred for an additional 8–10 min. Next, 0.5 mL of TMEDA (3.3 mmol) was added dropwise over *ca.* 3 min, and the solution was stirred for an additional 7 min. The argon was shut off, and the flask was connected to a bubbler via a syringe needle. A strong flow of dry oxygen was passed into the small septum of the gas dispersion tube via a syringe needle, and the gas dispersion tube was pushed into solution. The color of the solution changes within minutes from light yellow to black. After 1 h, the flask was evacuated under vacuum and purged with argon, and the reaction was then quenched with 2 mL of a 1 : 1 mixture of methanol/concentrated aqueous $NaHSO_3$ solution. The reaction mixture was then allowed to warm to $25\,°C$ and acidified with concentrated hydrochloric acid (0.5 mL). A sample of this solution was taken, washed with brine, and analyzed by GC. The reaction solution (and the GC sample) were poured into a separatory funnel and extracted 3–4 times from water (50 mL) with CH_2Cl_2. The combined organic phases were dried over magnesium sulfate and filtered, and the solvent was evaporated in vacuo. The residue was flash chromatographed with CH_2Cl_2/hexane (1 : 9) on silica gel. The pure fractions were combined, and the solvents were evaporated in vacuo to afford the pure biaryl.

When the aryl ligands on copper are the same (Ar′ = Ar), the kinetic conditions of the process are no longer crucial and, hence, reagent formation and oxidation can be done at much higher temperatures. This has found use in the preparation of biphenyl **30**[111] and the dibromobisilole **31**[112].

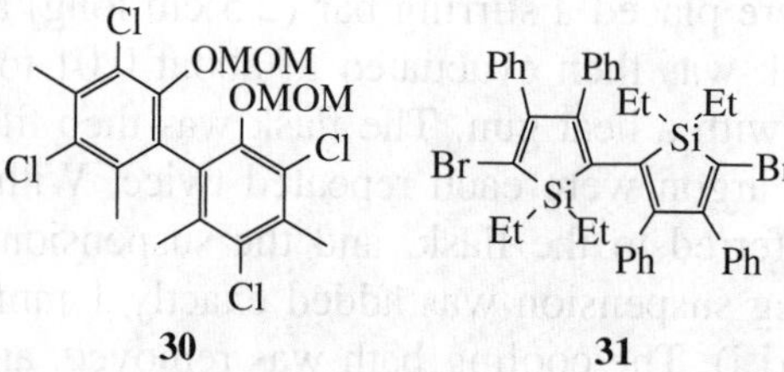

30

31

Further applications of cyanocuprate oxidations have appeared, including a key biaryl coupling between nonracemic partners **32** to produce **33** en route to calphostin A[113]. The issue of atropediastereoselection now comes into play, in this case an 8 : 1 ratio being formed favoring the desired (S) isomer.

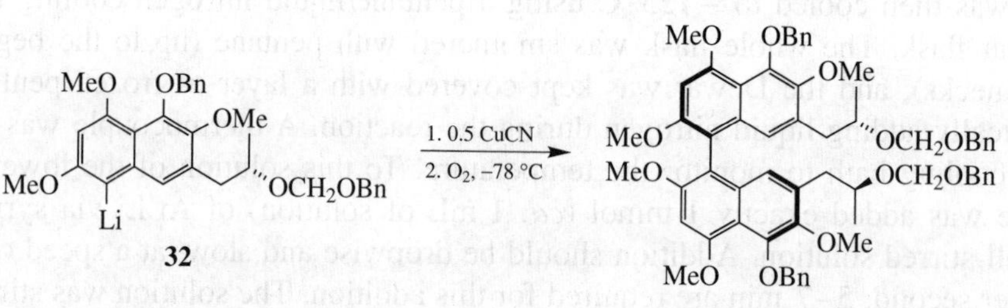

32

1. 0.5 CuCN
2. O$_2$, −78 °C

33 (70%)

Enhancement of an atropisomer distribution can potentially be realized using an intramolecular protocol. Proper selection of a nonracemic tether not only allows for nonkinetic conditions (i.e., higher temperatures) but imparts high enantioselectivity to the biaryl upon removal of the tether. Nonracemic (S)-BINOL has been prepared in this fashion (e.g., **34**)[114]. Related intramolecular oxidations of heteroaromatics, albeit devoid of axial chirality, have also been successfully carried out[110].

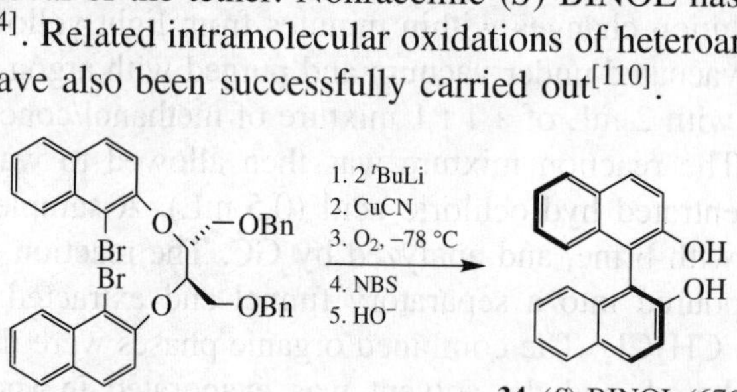

1. 2 ^{t}BuLi
2. CuCN
3. O$_2$, −78 °C
4. NBS
5. HO$^-$

34, (S)-BINOL (67%)

When the ligands on copper are not both carbon-based, such as with amidocuprates R(R$_2'$N)Cu(CN)Li$_2$, oxidation leads to product amines, thereby effecting a net electrophilic amination[115]. Copper is required for this process, not surprisingly, as use of either RLi or RMgX in place of RCu(CN)Li did not lead to products **35**. Molecular oxygen as oxidant is the reagent of choice, as yields with RuO$_2$, CuCl$_2$, VoCl$_3$, and (TMSO)$_2$ were inferior. Alkyl, vinyl, and heteroaryl groups can be coupled to nitrogen. Aliphatic substituents on the secondary amine are the best educts. Substituted hydrazines follow from the corresponding use of lithiated materials, although best yields were obtained upon oxidation of the zinc halocuprate $R(R'_2NNR'')Cu(CN)Zn$ using O_2 together with 20 mol% *o*-dinitrobenzene[116].

$$RCu(CN)Li + R'_2NLi \longrightarrow \text{Amidocuprate} \xrightarrow{[O]} R{-}N\begin{smallmatrix} R' \\ R' \end{smallmatrix}$$

35

Typical Procedure for Amine Formation via Oxidation of an Amidocuprate[115]

(62%)

Butyllithium (2.00 mmol) was added to 10 mL of THF at −40 °C. CuCN (2 mmol) was introduced quickly under a nitrogen flow, and the reaction mixture was stirred with dissolution of the salt, which was complete generally after about 20 min. To a brown clear solution of BuCuCNLi (2.00 mmol), cooled to −40 °C, was added a THF solution of *N*-lithioanilide (2.00 mmol). The mixture was allowed to react at this temperature for 15 min. Into the reaction mixture cooled to −78 °C was introduced a vigorous stream of oxygen for an additional 20 min. A dark precipitate was formed during this time. The reaction mixture was allowed to rise to 25 °C, filtered through a Celite pad, and concentrated in vacuo. The crude material was purified by flash chromatography on silica gel with pentane/ether (8 : 2) to give 0.185 g (62%) of the product as a pale yellow liquid.

2.5 Organocopper (RCu) Reagents

The preparation of organocopper complexes generally scribed as 'RCu', with or without an additive, traditionally follows along lines noted earlier for cuprate formation but with one important distinction: the stoichiometry of RLi (or RMgX, see below) to CuX (X = I, Br) is 1 : 1, rather than 2 : 1:

$$RLi + CuX \longrightarrow RCu + LiX \text{ or } R(X)CuLi$$

$$(X = Br, I)$$

Although Lewis basic additives, such as sulfides (e.g., Me_2S, see below) and phosphines (e.g., Bu_3P) are typically present presumably to stabilize polymeric 'RCu' formed (as $RCu \leftarrow L$)[2c,d], there are a number of situations where a species derived solely from metathesis is useful in its own right. When using 'RCu', it is important to appreciate that isolated, ***dry reagent may be explosive***, especially where R = alkyl. Hence, 'RCu' is best prepared and utilized in situ[117].

Asymmetric Conjugate Addition of MeCu to a Chiral, Nonracemic Vinyl Sulfoximine[118]

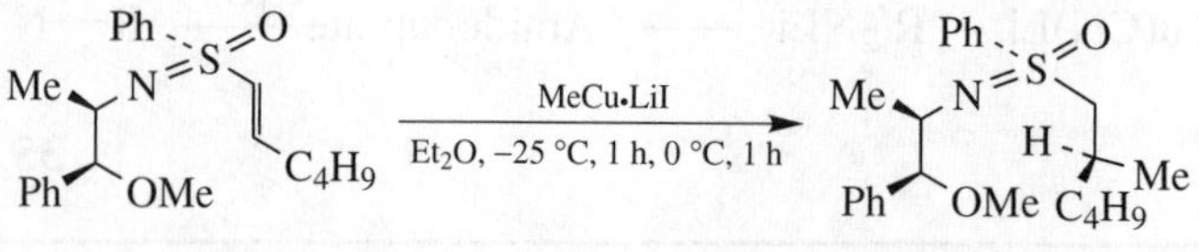

(72%)

To a stirred suspension of cuprous iodide (0.486 g, 2.56 mmol) in ether (12.8 mL) at −25 °C, was added methyllithium (2.56 mmol). After 30 min, (SR,1S,2R)-N-(1-methoxy-1-phenyl-2-propyl)-S-(1-hexenyl)-S-phenylsulfoximine (0.190 mg, 0.511 mmol) in ether (2 mL) was added, and the mixture was stirred at −25 °C for 1 h. It was then allowed to warm to 0 °C over a period of 1 h and, after an additional 1 h at 0 °C, the reaction was quenched with aqueous NH$_4$Cl (20 mL). The layers were separated and the ether layer dried and concentrated in vacuo. Analysis of the crude reaction mixture by HPLC indicated two compounds in a ratio of 96.5 to 3.5. Purification of the crude material by preparative thin layer chromatography (ethyl acetate/hexane 2 : 3) gave the product as a colorless oil.

More commonly, as alluded to above, an equimolar amount of a trialkylphosphine is used in conjunction with CuI, which together afford a THF-soluble Cu^I salt. Once the RLi (1 equiv) has been added, the 'RCu•PR$_3'$' is ready for use, although the price for using this additive must be figured in terms of its pyrophoric nature (for R = butyl), effect on work-up, toxicity, and potential chromatographic separation.

Michael Addition of a Phosphine-stabilized Organocopper Complex (RCu•PR$_3'$)[119]

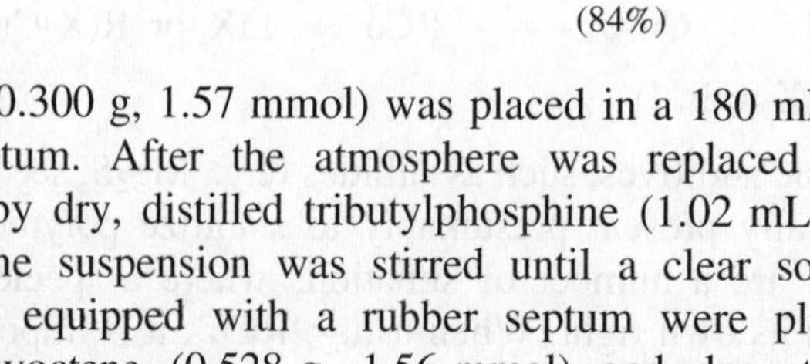

(84%)

Cuprous iodide (0.300 g, 1.57 mmol) was placed in a 180 mL ampule equipped with a rubber septum. After the atmosphere was replaced by Ar, dry THF (20 mL) followed by dry, distilled tributylphosphine (1.02 mL, 4.10 mmol) were added at 25 °C. The suspension was stirred until a clear solution resulted. In a 30 mL test tube equipped with a rubber septum were placed (E)-1-iodo-3-tetrahydropyranyloxyoctene (0.528 g, 1.56 mmol) and dry ether (6 mL). After cooling to −95 °C, tert-butyllithium (1.68 mL, 3.12 mmol) in pentane was added to this solution, with stirring, over 1 min. The mixture was stirred at $-78\,^{\circ}C$ for 2 h. The resulting white suspension was added at $-78\,^{\circ}C$, with stirring, to the above prepared ethereal solution of the cuprous iodide–phosphine complex through a stainless steel cannula under a slight Ar pressure. After the mixture was stirred at $-78\,^{\circ}C$ for 10 min, to this solution was then added slowly, along the cooled wall of the reaction vessel, a solution of cyclopentenone (0.103 g, 1.25 mmol) in cold $(-78\,^{\circ}C)$ THF (10 mL) through a stainless steel cannula under a slight Ar pressure over 50 min. The mixture was stirred at $-78\,^{\circ}C$ for 1 h. A saturated aqueous solution of NH_4Cl (15 mL) was added at $-78\,^{\circ}C$ and the mixture shaken vigorously. The organic layer was separated and the aqueous layer extracted with ether (30 mL). The combined extracts were dried over $MgSO_4$, evaporated, and chromatographed on triethylamine-treated silica gel (30 g) using 2000 : 100 : 1 hexane/ethyl acetate/triethylamine mixture as eluent to give the adduct (0.310 g, 84%, mixture of diastereomers) as a colorless oil.

Perhaps the most reactive form of $RCu{\cdot}PR_3'$ is arrived at via lithium naphthalide reduction of $CuI{\cdot}PR_3'$, which generates highly active $Cu{\cdot}PR_3'^{[120a]}$. Although both Ph_3P and Bu_3P have been employed, the trialkylphosphine complexes ultimately give reagents of higher reactivity. Preformed $Cu{\cdot}PR_3'$, or this species generated in situ, gives similar results en route to $RCu{\cdot}PR_3'$. Typical procedures are given below, in these cases for the coupling with an acid chloride to generate a ketone, and the 1,4 addition to an enone. It is especially noteworthy that highly functionalized organocopper reagents can be made in this fashion, wherein carboalkoxy, cyano, halo, and even epoxy groups can be tolerated. More recently, lithium naphthalide reduction of $(2\text{-Th})Cu(CN)Li$ at $-78\,^{\circ}C$ has led to a process which does not involve phosphines[120b]. **These reactions should be carried out under Ar, as lithium naphthalide gradually reacts with N_2.**

Acylation of an Organocopper Species Prepared via Lithium Naphthalide Reduction of CuI/PPh₃: Ketone Formation[120a]

Br⌒⌒CO₂Et → 1. Li⁰/Naphthalene, CuI·PPh₃ → Ph⌒⌒⌒CO₂Et
2. PhCOCl, THF, $-35\,^{\circ}C$, 1.5 h, 25 °C, 30 min
(81%)

Lithium (0.708 g, 10.2 mmol) and naphthalene (1.588 g, 12.39 mmol) in freshly distilled THF (10 mL) were stirred under Ar until the Li was consumed (*ca.* 2 h). CuI (1.751 g, 9.194 mmol) and PPh_3 (2.919 g, 11.13 mmol) in THF (15 mL) were stirred for 30 min giving a thick white slurry which was transferred via cannula to the dark green solution of lithium naphthalide at 0 °C. (Later experiments showed that slightly better results were obtained if the lithium naphthalide solution was added to the CuI/PPh_3 mixture.) The resultant reddish-black solution of active copper was stirred for 20 min at 0 °C. Ethyl 4-bromobutyrate (0.3663 g, 1.888 mmol) and the GC internal standard decane (0.1566 g, 1.101 mmol) were added neat via syringe to the active copper solution at −35 °C. The solution was allowed to stir for 10–15 min at −35 °C followed by addition of benzoyl chloride (0.7120 g, 5.065 mmol) neat to the organocopper solution at −35 °C. The reaction was allowed to stir for 90 min at −35 °C followed by warming to 25 °C for 30 min. (GC analysis showed the reaction to be essentially complete after stirring at −35 °C.) The reaction was then worked up by pouring into saturated aqueous NH₄Cl, extracting with Et₂O, and drying over anhydrous sodium sulfate. (For compounds not sensitive to base, the ether layer was also washed with 5% aqueous NaOH solution.) Silica gel chromatography (hexanes, followed by mixtures of hexanes/ethyl acetate) and further purification by preparative thin-layer chromatography (2 mm plate) provided 4-carboethoxy-1-phenyl-1-butanone in 81% isolated yield (93% GC yield after quantitation using the isolated product for the preparation of GC standards).

1,4 Addition of an Organocopper Complex Prepared from Lithium Naphthalide Reduction of CuI•PPh₃[120a]

$$C_8H_{17}Br \quad \xrightarrow[\substack{\text{2. Cyclohexenone} \\ \text{THF, } -78\,°C \text{ to } 25\,°C, 4\,h}]{\substack{\text{1. Li}^0/\text{Naphthalene, CuI•PPh}_3}} \quad \text{(93\%)}$$

Lithium (71.2 mg, 10.3 mmol) and naphthalene (1.592 g, 12.42 mmol) in freshly distilled THF (10 mL) were stirred under Ar until the Li was consumed (*ca.* 2 h). A solution of CuI•PBu₃ (3.666 g, 9.333 mmol) and PBu₃ (2.89 g, 14.3 mmol) in THF (5 mL) was added via cannula to the dark green lithium naphthalide solution at 0 °C and the resultant reddish-black active copper solution was stirred for 20 min. 1-Bromooctane (0.9032 g, 4.677 mmol) and the GC internal standard decane (0.1725 g, 1.212 mmol) in THF (5 mL) were added rapidly via cannula to the active copper solution at −78 °C. The organocopper formation was typically complete within 20 min at −78 °C. 2-Cyclohexen-1-one (0.1875 g, 1.950 mmol) in THF (10 mL) was added slowly dropwise over 20 min to the organocopper species at −78 °C. The reaction was allowed to react at −78 °C, −50 °C, −30 °C, and 25 °C for 1 h each. The reaction was then worked up by pouring into saturated aqueous NH₄Cl, extracting with Et₂O, and drying over anhydrous sodium sulfate. Silica gel chromatography (hexanes followed by mixtures of hexanes/ethyl acetate) and further purification by preparative thin-layer chromatography provided 3-octylcyclohexanone (93% GC yield after quantitation using the isolated product for the preparation of GC standards).

Although ethereal solvents are standard fare for reactions of 'RCu', another solvent of increasing popularity is dimethyl sulfide (SMe_2). Historically, sulfides have served mainly as additives, especially valued for their ability to solubilize CuI and CuBr, thereby obviating the usual call for warming slurries of $2RLi + CuX$ to effect dissolution[121]. In some situations, greater percentages of Me_2S were involved, but always no more than as co-solvent[9]. When used to the exclusion of Et_2O and THF, for instance, it appears that not only are these reagents of greater thermal stability, they are also more reactive toward enones, epoxides, acid halides, etc. than in traditional ethereal media[122]. This may be due to changes in solubility properties, or possibly differences in reagent constitution. From the experimental point of view, its low boiling point (38 °C) simplifies work-up.

Conjugate Addition of 'PhCu•LiI in Me_2S[122]

$CuI•SMe_2$ → (ca. 90%)

1. PhLi, Et_2O/C_6H_{14}, −50 °C, 30 min
2. $\bigcirc$=O, Me_2S −75 °C to 0 °C, 2.5 h

A 500 mL recovery flask was charged with 20.00 g of CuI (105.0 mmol, Alfa ultrapure) which was dissolved in 40 mL of deoxygenated (Ar sparge) Me_2S (Aldrich, gold label) at 25 °C under N_2. Upon cooling the solution to −50 °C, a white solid precipitated; therefore, an additional 120 mL of Me_2S was added to the cold suspension in order to redissolve the CuI. A 55.5 mL volume of 1.86 M PhLi (103 mmol, 0.17 M residual base, Aldrich) solution (ether/cyclohexane) was added via syringe over *ca.* 1 h. The dark greenish-yellow solution was stirred at −50 °C for 30 min. It was then cooled to −75 °C (*ca.* 15 min) and 9.63 g of 2-cyclohexenone (100.2 mmol, Aldrich, freshly opened bottle) dissolved in 15 mL of Me_2S was added to the rapidly stirred solution over *ca.* 3 min via cannula from a 50 mL strawberry-shaped flask cooled in a dry ice/2-propanol bath. After 2 h at −75 °C and 30 min at 0 °C, the reaction was complete. Work-up consisted of the addition of 100 mL of 3 M aqueous NH_4Cl, separation of phases, and extraction of the organic phase with 4×100 mL of 3 M aqueous NH_4Cl. The combined aqueous phases were back-extracted with 100 mL of ether. The combined organic layers were dried over anhydrous Na_2SO_4 and the solvent was removed by rotary evaporation in vacuo. (A dry ice trap was inserted between the rotary evaporator and the aspirator to which it was connected.) The residue was treated with 100 mL of hexane and filtered; the filter cake was washed with a total of 75 mL of fresh hexane. The hexane was removed in vacuo and the residue, which still contained some solid, was dissolved in 100 mL of ether, which was extracted with 2×100 mL of 0.5 M sodium thiosulfate. The combined thiosulfate layers were back-extracted with 100 mL of ether and the combined ether layers were dried over anhydrous Na_2SO_4. Rotary evaporation left 16.8 g of crude 3-phenylcyclohexanone (94% pure by GLC). Flash chromatography on a 30 mm × 60 mm column of basic alumina (50 g, Woelm act. I) eluted with hexane afforded 13.7 g of product in the first four 50 mL fractions. Further elution with 200 mL of ether yielded 2.3 g of product. The purity was not improved by this chromatography; therefore, 15.9 g of the chromatographed material was distilled at 0.01 mmHg. Three fractions were collected: 0.8 g (65–88 °C, 49% pure by GLC), 3.0 g (88–92 °C, 91% pure), and 8.3 g (92–94 °C, 99% pure). The main impurity was biphenyl from the commercial PhLi solution. (Little biphenyl was observed in the small-scale reactions, which employed solid PhLi, free of biphenyl.)

2.5.1   With $BF_3 \cdot Et_2O$

Switching the exposure of organocopper complexes from Lewis bases to Lewis acids, most notably $BF_3 \cdot Et_2O$, brings about some major changes not only in the chemistry observed but in the composition of the reagents themselves. Introduction of ≥ 1 equivalent of $BF_3 \cdot Et_2O$ to the products of metathesis between RLi and CuI (see page 717) leads to a reagent mixture highly prone toward both displacements of allylic leaving groups[123], as well as Michael additions to enones, enoates and unsaturated acids[21,124]. The standard protocol involves little more than cooling the RCu•LiI formed initially and introducing the Lewis acid followed by the substrate. With the former class of substrates, complete allylic rearrangement is the norm.

Allylic Alkylation of 'RCu•BF$_3$'[123]

In a 200 mL flask, equipped with a magnetic stirrer and maintained under N_2, were placed 1.9 g (10 mmol) of CuI and 20 mL of dry THF. Butyllithium in hexane (1.3 M, 10 mmol) was added at −30 °C, and the resulting mixture was stirred at this temperature for 5 min. The mixture was then cooled to −70 °C, and $BF_3 \cdot OEt_2$ (47%, 1.3 mL, 10 mmol) was added. After the mixture was stirred for a few minutes, cinnamyl chloride (1.53 g, 10 mmol) was added, and the mixture was allowed to warm slowly to room temperature with stirring. The product was filtered through the column of alumina using petroleum ether. The alkene thus obtained in essentially pure form was distilled under reduced pressure (1.64 g, 94%, bp 65–66 °C at 5 mmHg).

Conjugate Addition of 'RCu•BF₃' to an α,β-Unsaturated Acid[121,124]

$$\text{COOH} \xrightarrow[\text{Et}_2\text{O},\ -70\,°\text{C to } 25\,°\text{C}]{\text{BuCu•BF}_3} \underset{\text{Bu}}{\text{COOH}}$$

(74%)

In a 200 mL flask, equipped with a magnetic stirrer and maintained under N₂, were placed 60 mL of dry ether and 5.7 g (30 mmol) of purified CuI. Butyllithium in hexane (1.3 M, 30 mmol) was slowly added at −30 °C to −40 °C, and the resulting dark brown suspension was stirred for 5 min. The mixture was then cooled to −70 °C, and BF₃•OEt₂ (47%, 3.9 mL, 30 mmol) was slowly added. The color changed from dark brown to black and BuCu•BF₃ seemed to be present as a precipitate. After the mixture was stirred for a few minutes, an ether solution of crotonic acid (0.86 g, 10 mmol) was added at −70 °C. The color immediately changed to deep black. The mixture was allowed to warm slowly to 25 °C with stirring. Addition of water, separation, and distillation yielded the desired carboxylic acid (1.16 g, 74%, bp 75–76 °C at 1 mmHg).

Although the RLi + CuLi + BF₃ combination was originally described as RBF₃⁻Cu⁺[123] and now commonly as 'RCu•BF₃', more recent evidence has shown that the LiI present (see page 717) plays a critical role[125]. In fact, these reagents clearly involve iodocuprates, perhaps of general form RCu(I)Li or R(I₂)Cu₂Li, which arise as a result of the metathesis and/or by the action of BF₃ on the initially formed R(I)CuLi dimer:

$$2\text{RCu•LiI} \rightleftharpoons [\text{R(I)CuLi}]_2 \underset{\text{THF, }-80\,°\text{C}}{\overset{2\text{BF}_3}{\rightleftharpoons}} 2\text{RLi•BF}_3 + 2\text{CuI}$$

Irrespective of these subtle changes, the reagent system 'RCu•BF₃' provides, in some instances, the only alternative for successful 1,4 additions (e.g., with unsaturated acids; see above).

2.5.2 With Me₃SiCl

As an alternative to BF₃•Et₂O as co-reagent with 'RCu' in Michael additions, Me₃SiCl has also been found to function admirably in this regard. Usually, to drive a reaction to completion forming the product silyl enol ether, other additives such as HMPA[49] and/or basic amines (e.g., Et₃N)[48d] are deemed necessary. The combination of reagents 'RCu•Me₃SiCl', where R = an allylic ligand, is very successful in effecting 1,4 delivery of allylic ligands to α,β-unsaturated ketones[126]. The allylic copper species derives from lithiation of an allylic stannane, followed by treatment with precooled (to −78 °C)

LiI-solubilized CuI in THF, to which is then added Me$_3$SiCl:

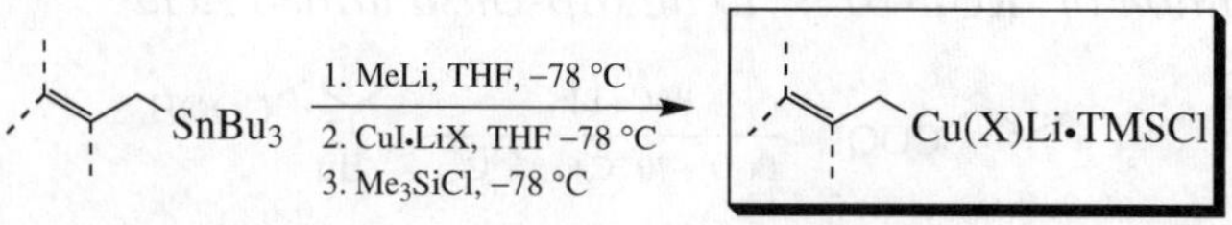

Allylic systems examined include allyl, methallyl, crotyl, and prenyl. Crotylcopper reacts virtually exclusively at its α site, as is true for prenylcopper. However, mixtures of (E) and (Z) isomers are to be expected from the former $((E) : (Z) \approx 3 : 1)$. This method gives good to excellent yields of conjugate adducts, but only moderate results for highly hindered cases.

1,4 Addition of Allylcopper•Me$_3$SiCl to an α,β-Unsaturated Ketone[126]

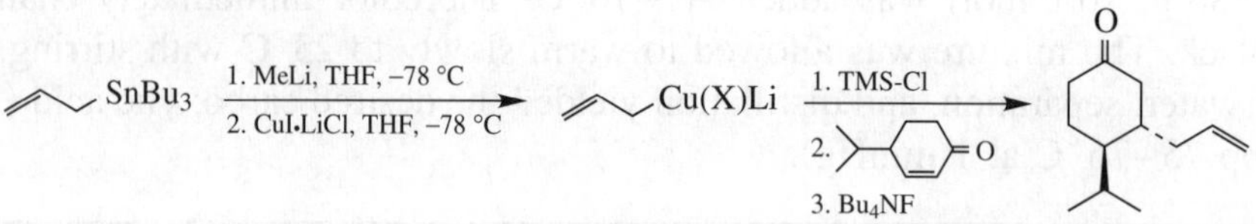

(87%)

CuI (0.400 g, 2.10 mmol) and dry LiCl (0.89 g, 2.10 mmol) were placed in a 10 mL round bottom flask equipped with a stir bar and sealed with a septum. The flask was evacuated and purged with Ar; the process was repeated three times. THF (1.5 mL) was injected, and the mixture was stirred for 5 min to yield a yellow, homogeneous solution which was then cooled to $-78\,^{\circ}$C. Concurrently, a solution of allyllithium (2.0 mmol) was prepared from allyltributylstannane (0.62 mL, 2.0 mmol) and MeLi (1.25 mL, 2.00 mmol) in THF (1.0 mL) at $-78\,^{\circ}$C (15 min). This solution was then transferred via a dry ice-cooled cannula to the CuI/LiCl solution ($-78\,^{\circ}$C) to yield a tan solution. TMS-Cl (0.17 mL, 2.1 mmol) was added followed immediately by the neat addition of 4-isopropyl-2-cyclohexenone (0.11 mL, 0.75 mmol). The reaction was allowed to proceed for 30 min before being quenched with 5 mL of a saturated NH$_4$Cl solution. Extraction with 4×20 mL of ether was followed by combining the organic layers and drying over Na$_2$SO$_4$. The solvent was then removed in vacuo, and the resulting oil was treated with THF (5 mL) and Bu$_4$NF (2.0 mL, 2.0 mmol) for 15 min. The solvent was again removed in vacuo, and the residue was subjected to flash chromatography (9 : 1 petroleum ether/EtOAc) to yield 0.118 g (87%) of 3-(1-propen-3-yl)cyclohexanone as a colorless oil: (R_f (9 : 1 petroleum ether/EtOAC) = 0.28).

2.5.3   With TMS-I

Although the precise role(s) of TMS-Cl in reactions of 'RCu' as well as of cuprates[48–50,98] is of current interest[127], a theoretical study suggests that it is not functioning simply as a Lewis acid toward a Michael acceptor[50]. However, the corresponding iodide (i.e., TMS-I) could be viewed as serving in this capacity[128]. Organocopper reagents admixed with an equivalent of TMS-I are highly effective in delivering various 'R' groups to α,β-unsaturated ketones[55]. The 'RCu' species can be solubilized with Bu_3P, but in most cases where R $\neq$ Me, additives such as phosphines, HMPA, DMAP, or TMEDA are not necessary. Most reactions occur at $-78\,^{\circ}C$ with unhindered enones, whereas less-reactive partners such as enoates (as with the example of **36** below) require somewhat elevated temperatures. Upon completion, it is important for optimal yields to quench the reaction with dry pyridine (from CaH_2), which presumably consumes excess TMS-I and neutralizes any HI generated from introduction of water. The TMS-I used should be stored cold ($-25\,^{\circ}C$) under argon and over copper chips in a flask or bottle protected from light. Use of this reagent combination in THF, as well as in Et_2O, is certainly permissable, and has been studied in some detail[127].

Conjugate Addition of BuCu•TMS-I to an Enoate[129]

BuCu•TMS-I
Et_2O, $-60\,^{\circ}C$, 20 h

36 (R = 1-naphthyl, phenyl, methyl)

(93%) de = 98%, (S) configuration,
R = 1-naphthyl

(91%) de = 70%, (S) configuration,
R = phenyl

(79%) de = 37%, (R) configuration,
R = methyl

Butyllithium (3.1 mL, 5.0 mmol, 1.6 M) was added dropwise to a slurry of copper(I) iodide (1.05 g, 5.5 mmol) in diethyl ether (10 mL) at $-78\,^{\circ}C$. The slurry was stirred for 40 min at $-60\,^{\circ}C$. Colorless iodotrimethylsilane (0.71 mL, 5.0 mmol) was added dropwise and the slurry stirred for 5 min. The temperature was lowered to $-78\,^{\circ}C$ and a solution of the bornyl ester (0.5–1.5 mmol) in diethyl ether (10–15 mL) was added dropwise (2 mL min^{-1}). The reaction was quenched with 10 mL dry pyridine/dry ether (1 : 9) and the suspension stirred (1–2 h) at low temperature. Diethyl ether saturated with ammonia/ammonium chloride buffer was added and the mixture stirred an additional 4 h. The temperature was raised to $20\,^{\circ}C$ and aqueous ammonium chloride/ammonia (5 mL) was added. The excess heterogeneous material was filtered over Celite, washed with several portions of ether and the organic layer separated. The aqueous phase was extracted with ether (3×50 mL) and the combined ether extracts washed with dilute copper(II) sulfate, brine, dried over anhydrous sodium sulfate, and the solvent evaporated in vacuo. Flash chromatography (5–10% diethyl ether/pentane) gave pure product.

2.6   Reactions of Silyl and Stannyl Cuprates ($R_3MCu \cdot L_n$, M = Si, Sn)

2.6.1   Si−Cu Reagents

Organosilicon intermediates occupy a rather prominent position in synthetic organic chemistry, as they are extremely versatile for constructing carbon−carbon as well as carbon−heteroatom bonds[130]. Other than the simplest of these materials, which may be purchased from commercial sources, organosilanes such as those bearing vinylic and allylic appendages must be prepared prior to use. One versatile approach utilizes organocuprate chemistry, based on $CuCN$[131]. Starting with $PhMe_2SiCl$, the red lithiosilane ($PhMe_2SiLi$) is formed using Li^0 (lithium shot)[132]. This organometallic is stable for a few days at $0\,^\circ C$, and requires titration (usually 0.9–1.2 M) to ensure accuracy in the subsequent cuprate-forming step.

Preparation of $Me_2PhSiLi$[131 a,b]

$$Me_2PhSiCl \xrightarrow[\text{THF, 0 °C, 18 h}]{Li^0} \boxed{Me_2PhSiLi}$$

Chlorodimethyl(phenyl)silane (5 mL, Aldrich or home-made, in both cases contaminated with about 10% of the bromide, coming from its preparation using phenylmagnesium bromide and dichlorodimethylsilane) was vigorously stirred with lithium shot (1 g) in dry THF (30 mL) under N_2 or Ar at $0\,^\circ C$ for 18 h to give a deep red solution of the silyllithium reagent. In general, the red color was persistent only after 0.5 to 1 h. The molarity of the solution was measured by adding a 0.8 mL aliquot to water (10 mL) and titrating the resulting mixture against HCl (0.1 M) using phenolphthalein as indicator. The solution was generally found to be 0.9–1.2 M.

Alternatively, $(Me_2PhSi)_2$ can be cleaved with Na-containing lithium wire to arrive at halide-free $PhMe_2SiLi$, although the presence of LiX salts has no effect on the outcome of the cuprate reactions. Once in hand, addition to CuCN (0.5 equiv) affords reddish solutions of the higher order disilylcuprate:

$$Me_2PhSiCl \xrightarrow[\text{THF, 0 °C, 18 h}]{Li^0} PhMe_2SiLi \xrightarrow[\text{THF, 0 °C, 20 min}]{0.5\ CuCN} \boxed{(Me_2PhSi)_2Cu(CN)Li_2}$$

Much of the value of these particular silyl cuprates (Fleming reagents) lies in the facility with which the $C-SiPhMe_2$ bond can be converted to a $C-OH$ bond with retention of stereochemistry[133]. Thus, this silyl residue is a masked hydroxyl group.

These reagents smoothly add Michael-wise to α,β-unsaturated ketones[131b], readily displace allylic acetates (the regio- and stereochemistry of which is substrate and conditions dependent)[131c], and effect silylcupration of 1-alkynes to afford (*E*)-vinylsilanes.

Silylcupration of a 1-Alkyne Using $(PhMe_2Si)_2Cu(CN)Li_2$ [131 a]

$$CH_3(CH_2)_{10}C\equiv CH \xrightarrow[\substack{THF,\ Et_2O,\ -78\ °C,\ 2\ h \\ then\ 0\ °C,\ 30\ min}]{(Me_2PhSi)_2Cu(CN)Li_2} CH_3(CH_2)_{10}\diagup\!\!\!\diagdown SiMe_2Ph$$

(95%)

The silyllithium reagent (see above) (73.3 mL of a 1.14 M solution in THF, 83.6 mmol) was added by syringe to a slurry of copper(I) cyanide (3.74 g, 41.9 mmol) in THF (50 mL) at 0 °C over 2 min. The mixture was stirred at 0 °C for 20 min and cooled to −78 °C. The color of the solution changed little, and vigorous stirring was essential for all the cyanide to react. Tridecyne (5.7 g, 31.7 mmol) was added in ether (15 mL) and the mixture stirred for 2 h, allowed to warm to 0 °C, and stirred for a further 30 min, by which time the solution had turned black. The reaction was quenched with aqueous NH_4Cl solution at 0 °C, stirred for 10 min at this temperature, brought to room temperature, and filtered over Celite. The layers were separated and the aqueous layer extracted with ether. The organic layers were dried (Na_2SO_4) and evaporated in vacuo. Chromatography (SiO_2, hexane) gave the vinylsilane (10.5 g, 95%) as an oil (R_f(hexane) = 0.51).

Silyl ligands containing other substitution patterns, e.g. tert-butyldimethyl- and thexyldimethylsilyl-, can be delivered in a manner analogous to the chemistry of the Me_2PhSi group. The preparation of mixed cuprates $^tBuMe_2Si(R)Cu(CN)Li_2$ and (thexyl)$Me_2Si(R)Cu(CN)Li_2$, R = Me or Bu, however, relies on the transmetalation of the corresponding silyltrimethylstannanes, $R'Me_2Si-SnMe_3$ (R' = thexyl or ^tBu)[134]. These water-white, stable liquids are easily prepared from the corresponding silyl chlorides and Me_3SnLi. By simply mixing $R'Me_2Si-SnMe_3$ with, for example $Bu_2Cu(CN)Li_2$ at 25 °C, the desired higher-order cuprate is formed directly without prior lithiation of the silane. Once formed, the higher-order silyl cuprates can be anticipated to react quite readily with the usual assortment of coupling agents.

$$R'Me_2Si\text{-}SnMe_3 + Me_2Cu(CN)Li_2 \xrightarrow[25\ °C,\ 1\ h]{THF} \boxed{R'Me_2Si(Me)Cu(CN)Li_2} + Me_4Sn$$

Preparation of Thexyldimethylsilyltrimethylstannane[135]

$$Me_3SnH \xrightarrow[THF]{LDA} Me_3SnLi \xrightarrow[0\ °C,\ 3\ h]{\overset{\displaystyle \vert\!\!\!-\!SiMe_2Cl}{}} \vert\!\!\!-\!Si\text{-}SnMe_3$$

(>60%)

This reaction should be performed in an efficient hood due to the extreme toxicity and volatility of Me₃SnH! To 40 mL of THF was added diisopropylamine (3 mL, 29.7 mmol) via syringe and the solution was cooled to −78 °C, after which was added BuLi in hexanes (12 mL, 26.4 mmol). Warming this solution by removal of the ice bath for 10 min and then recooling to −78 °C was followed by addition of trimethyltin hydride (7.8 mL, 25.9 mmol)[136]. The reaction was warmed to 0 °C for 1 h. At this temperature thexyldimethylsilyl chloride (7.4 mL, 37.6 mmol) was added and this solution was then allowed to stir at 0 °C for 3 h. After this time, the heterogeneous mix was filtered using a Schlenck funnel and distilled (72 °C, 0.2 mmHg) yielding a clear water-white liquid (>60%; R_f(hexanes) = 0.9).

Ring Opening of Isoprene Oxide with (thexyl)Me₂Si(Bu)Cu(CN)Li₂ [134,135]

(74%)
((E):(Z)9:1)

To a two-necked flask equipped with a two-way valve connected to a source of Ar and a vacuum pump was added CuCN (0.0456 g, 0.51 mmol). It was gently flame-dried under vacuum followed by Ar addition; this evacuation and Ar reentry process was repeated three times. THF (1 mL) was then added and the slurry was cooled to −78 °C, at which point BuLi in hexanes (0.46 mL, 1.06 mmol) was added. The reaction was then warmed via removal of the bath and stirred until a pale yellow homogeneous solution resulted (approximately 3–5 min). While warming it is important that all the CuCN be dissolved by occasionally tipping the flask to immerse any remaining salt. The flask was then cooled to −78 °C and thexyldimethylsilyltrimethylstannane (0.18 mL, 0.73 mmol) was introduced neat via syringe. The reaction was warmed to 22 °C for 1.5 h to complete the transmetalation, following which the yellow homogeneous cuprate was then recooled to −78 °C and the epoxide (0.04 mL, 0.404 mmol) was added also neat via syringe. The solution was then stirred for an additional 30 min, after which it was quenched at −78 °C via addition of 10% NH₄OH/saturated aqueous NH₄Cl. Work-up involved warming the reaction to 22 °C, addition of Et₂O, extraction and drying of the organic layer (brine/Na₂SO₄) followed by rotary evaporation in vacuo and chromatography (silica gel, 2 : 40 : 58 TEA/Et₂O/hexanes) to yield the (E)- and (Z)-vinylsilanes as a clear oil (0.068 g, 74%) in a 9 : 1 ratio (GC; R_f (1 : 1 Et₂O/hexanes) = 0.42).

2.6.2 Sn−Cu Reagents

Recent advances in organotin chemistry have provided considerable incentive for development of new technologies which position a trialkylstannyl moiety into organic substrates[137,138]. Several reagents exist from which to choose, depending upon the particular transformation of interest. The following equations show some of these, together with the derivation of each reagent.

$$\text{Me}_3\text{Sn}-\text{SnMe}_3 \xrightarrow[-20\,°\text{C, 15 min}]{\text{MeLi, THF}} \text{Me}_3\text{SnLi} \xrightarrow[\text{THF, }-48\,°\text{C, 10 min}]{\text{CuBr·SMe}_2} \boxed{\text{Me}_3\text{SnCu·SMe}_2}$$

37

$$\text{Me}_3\text{Sn}-\text{SnMe}_3 \xrightarrow[0\,°\text{C, 15 min}]{\text{MeLi, THF}} \text{Me}_3\text{SnLi} \xrightarrow[-20\,°\text{C, 15 min}]{\text{PhSCu, THF}} \boxed{\text{Me}_3\text{Sn(PhS)CuLi}}$$

38

$$\text{Me}_3\text{Sn}-\text{SnMe}_3 \xrightarrow[\substack{\text{2. 2MeLi, }-20\,°\text{C,}\\ 50\text{ min}}]{\substack{1.\ \overset{\displaystyle \langle\!\langle\ \rangle\!\rangle}{\text{S}}\ \text{THF}\\-20\,°\text{C}}} \left[\overset{\displaystyle \langle\!\langle\ \rangle\!\rangle}{\text{S}}\!\!-\text{Li} + \text{Me}_3\text{SnLi}\right] \xrightarrow{\text{CuCN}} \boxed{\text{Me}_3\text{Sn(2-Th)Cu(CN)Li}_2}$$

39

$$2\text{Bu}_3\text{SnH} + \text{Bu}_2\text{Cu(CN)Li}_2 \xrightarrow[10\text{ min}]{\text{THF, }-78\,°\text{C}} \boxed{\text{Bu}_3\text{Sn(Bu)Cu(CN)Li}_2}$$

40

$$\text{Me}_3\text{Si}-\text{SiMe}_3 \xrightarrow[\text{2. Me}_3\text{SnCl}]{1.\ \text{MeLi, THF, HMPA}} \text{Me}_3\text{Si}-\text{SnMe}_3 \xrightarrow{\text{Bu}_2\text{Cu(CN)Li}_2} \boxed{\text{Me}_3\text{Sn(Bu)Cu(CN)Li}_2}$$

41

Reagent **37**, prepared from hexamethylditin (see procedure below), is especially effective at stannylcupration of terminal unactivated alkynes[139]. When used (2 equiv) in tandem with a proton source (MeOH, 60 equiv) good to excellent yields of the product 2-trimethylstannyl-1-alkenes are obtained, accompanied by small amounts of the isomeric (*E*)-1-trimethylstannyl-1-alkene. These byproducts tend to be unstable on chromatographic separation, and are not easily isolated in small-scale work.

Preparation of Me₃SnCu·SMe₂ from Me₃Sn−SnMe₃[140]

$$\text{Me}_3\text{Sn}-\text{SnMe}_3 \xrightarrow[\text{2. CuBr·SMe}_2\text{, THF, }-48\,°\text{C, 10 min}]{1.\ \text{MeLi, THF, }-20\,°\text{C, 15 min}} \text{Me}_3\text{SnCu·SMe}_2$$

37

(a) Me₃SnLi

To a cold (−20 °C), stirred solution of hexamethylditin in dry THF (∼10 mL mmol⁻¹ of Me₃Sn−SnMe₃) was added a solution of methyllithium (1.0 equiv) in ether. The mixture was stirred at −20 °C for 15 min to afford a pale yellow solution of trimethylstannyllithium.

(b) Me₃SnCu•SMe₂

To a cold (−48 °C), stirred solution of trimethylstannyllithium (0.65 mmol) in 5 mL of dry THF was added, in one portion, solid cuprous bromide–dimethyl sulfide complex (113 g, 0.65 mmol). The mixture was stirred at −48 °C for 10 min to give a dark red solution of the cuprate reagent.

Stannylcupration of a 1-Alkyne with Me₃SnCu•SMe₂ [139]

To a cold (−78 °C), stirred solution of Me₃SnCu•SMe₂ (0.4 mmol) in 3 mL of dry THF (Ar atmosphere) was added sequentially a dry THF solution (0.5 mL) of the 1-alkyne (0.039 g, 0.2 mmol) and dry MeOH (0.5 mL, 12 mmol). The dark red reaction mixture was stirred at −78 °C for 10 min and at −63 °C for 12 h. Saturated aqueous NH₄Cl (pH 8; 5 mL) and ether (20 mL) were added and the mixture was allowed to warm to room temperature with vigorous stirring. Stirring was continued until the aqueous phase became deep blue. The layers were separated and the aqueous phase was extracted with ether (2 × 5 mL). The combined organic extracts were washed with saturated aqueous NH₄Cl (pH 8; 2 × 5 mL) and dried (MgSO₄). Removal of the solvent afforded a crude oil that, on the basis of GC analysis, consisted of a 95 : 5 mixture of two products. Subjection of the crude oil to flash chromatography on silica gel (2 × 15 cm column, elution with 1 : 20 ether/petroleum ether), followed by distillation (air-bath temperature 115–120 °C, 0.7 mmHg) of the oil thus obtained gave 0.059 g (84%) of the vinylstannane as a colorless oil.

Mixed Gilman cuprate **38**, prepared as illustrated, undergoes 1,4 additions quite easily, on occasion with assistance from in situ HOAc or MeOH. Although an alternative to reagent **37**, it offers particular advantages in reactions with alkynic esters, where careful control of conditions can lead to excellent (E) versus (Z) stereoselectivities in the resulting enoates. Thus, when conducted in THF containing MeOH at $-78\,^{\circ}$C for 3 h, the product of (E) stereochemistry results.

Warming to $-48\,^{\circ}$C for 4 h, followed by addition of MeOH, reverses the isomeric ratio (Scheme 3).

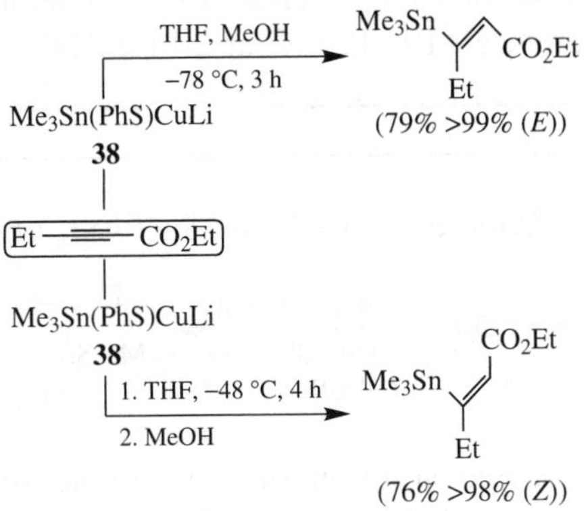

Scheme 3.

Preparation of $Me_3Sn(SPh)CuLi$[140]

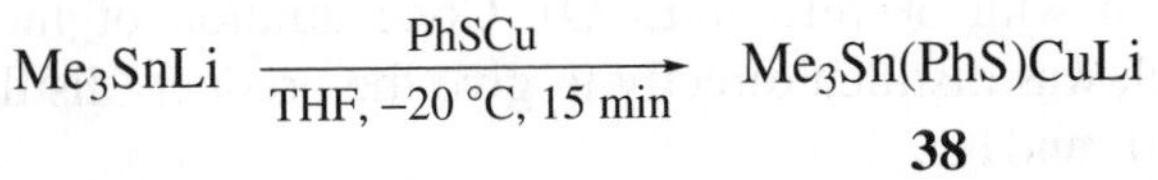

$$Me_3SnLi \xrightarrow[\text{THF}, -20\,^{\circ}\text{C}, 15\text{ min}]{\text{PhSCu}} Me_3Sn(PhS)CuLi$$
$$\textbf{38}$$

To a cold ($-20\,^{\circ}$C), stirred solution of trimethylstannyllithium (0.75 mol) in 10 mL of dry THF was added in one portion solid phenylthiocopper(I) (0.132 g, 0.75 mmol). The slurry was stirred at $-20\,^{\circ}$C for 15 min to afford a dark red solution of the cuprate reagent.

Stannylcupration of an Alkynic Ester with a Mixed Lower-order Stannylcuprate at $-78\,^{\circ}$C in the Presence of MeOH[141]

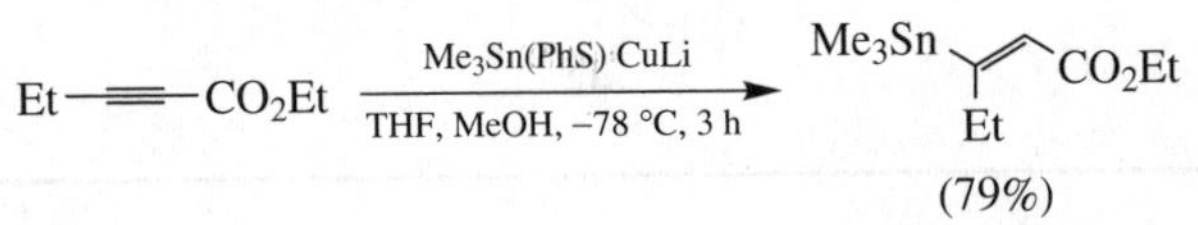

To a cold ($-100\,^{\circ}$C), stirred solution of the cuprate reagent $Me_3Sn(PhS)CuLi$ (1.0 mmol) in 10 mL of dry THF was added, dropwise, a solution of the α,β-alkynic ester (0.5 mmol) in 0.5 mL of dry THF containing 0.85 mmol of dry MeOH. The reaction mixture was stirred at $-100\,^{\circ}$C for 15 min and at $-78\,^{\circ}$C for 3 h. MeOH

(0.2 mL) and Et_2O (30 mL) were added and the mixture was allowed to warm to room temperature. The resulting yellow slurry was filtered through a short column of silica gel (10 g, elution with 30 mL of Et_2O). The oil obtained by concentration of the combined eluate was chromatographed on silica gel (*ca.* 3 g). Elution with petroleum ether (*ca.* 10 mL) gave $Me_3Sn-SnMe_3$. Further elution with Et_2O (*ca.* 8 mL), followed by distillation of the material thus obtained, provided the product (79%; distillation temperature 110–125 °C at 20 mmHg).

---

*Stannylcupration of an Alkynic Ester with $Me_3Sn(PhS)CuLi$ (**8**) at -48 °C*[141]

$$Et\!\!=\!\!=\!\!CO_2Et \quad \xrightarrow[\substack{THF, -48\,°C, 4\,h \\ 2.\ MeOH}]{1.\ Me_3Sn(PhS)CuLi} \quad \overset{Et}{\underset{Me_3Sn}{>\!\!=\!\!<}}CO_2Et$$

(76%)

To a cold (-78 °C), stirred solution of the cuprate reagent $Me_3Sn(PhS)CuLi$ (0.39 mmol) in 5 mL of dry THF was added a solution of the α,β-alkynic ester (0.3 mmol) in 0.5 mL of dry THF. The reaction mixture was stirred at -78 °C for 15 min and at -48 °C for 4 h. MeOH or EtOH (0.2 mL) and Et_2O (30 mL) were added and the mixture was allowed to warm to room temperature. The yellow slurry was treated with anhydrous $MgSO_4$ and then was filtered through a short column of Florisil (elution with 30 mL of Et_2O). Concentration of the combined eluate gave an oil, which was distilled directly to give the product (distillation temperature 120–125 °C at 20 mmHg).

Cyanocuprates **39**, **40**, and **41** are of more recent vintage, and it is clear that whether a thienyl or alkyl ligand is part of their make-up, the trialkylstannyl moiety is always transferred (virtually exclusively) over the remaining ligand. All three routes-involve one pot protocols, not an insignificant feature, especially for large-scale processes. Cuprates **39** are also useful for generating vinylstannanes of general structure **42** (compare the use of mixed cuprate **38** in Scheme 3), as well as several other product types.

$$\overset{Me_3Sn}{\underset{OH}{>\!\!=\!\!<}}CO_2Et$$

42

Preparation of $Me_3Sn(2\text{-}Th)Cu(CN)Li_2$[142]

$$(Me_3Sn)_2 \quad \xrightarrow[\substack{2.\ 2\ MeLi, -20\,°C, 50\,min \\ 3.\ CuCN, -78\,°C, 10\,min \\ 4.\ -48\,°C, 10\,min}]{1.\ thiophene, THF, -20\,°C} \quad Me_3Sn(2\text{-}Th)Cu(CN)Li_2$$

39

To a cold (−20 °C), stirred solution of (Me₃Sn)₂ (0.164 g, 0.5 mmol) in 10 mL of dry THF were added, successively, thiophene (0.042 g, 0.5 mmol) and a solution of MeLi (1.0 mmol, low halide or LiBr complex) in Et₂O. After the pale yellow solution had been stirred at −20 °C for 50 min, it was cooled to −78 °C and CuCN (0.045 g, 0.5 mmol) was added. The resulting suspension was stirred for 5 min at −78 °C and for 10 min at −48 °C to provide a bright yellow solution of the cuprate reagent. The solution was cooled to −78 °C and used immediately.

Michael Addition of the Mixed Higher-order Stannylcuprate
Me₃Sn(2-Th)Cu(CN)Li₂ [142]

(90%)

To a cold (−78 °C), stirred solution of Me₃Sn(2-Th)Cu(CN)Li₂ (0.5 mmol; see above) in 10 mL of dry THF (Ar atmosphere) was added 0.037 g (0.33 mmol) of the enone. After the solution had been stirred at −78 °C for 5 min and at −20 °C for 4 h, it was treated with saturated aqueous NH₄Cl/NH₄OH (pH 8) (10 mL) and Et₂O (10 mL). The vigorously stirred mixture was exposed to air and allowed to warm to 25 °C. The phases were separated, and the aqueous phase was extracted with Et₂O (3 × 10 mL). The combined organic extracts were dried (MgSO₄) and concentrated. Flash chromatography (1 : 4 Et₂O/petroleum ether) of the residual oil, followed by distillation (90 °C, 2.0 mmHg) of the material thus obtained gave 0.083 g (90%) of the product.

An especially simple route to Bu₃Sn-incorporated higher-order cuprates involves a presumed transmetalation between Bu₃SnH and Bu₂Cu(CN)Li₂, which occurs upon mixing of these components at −78 °C[143]. The resulting reagent behaves as a mixed cuprate **40** and transfers the Bu₃Sn moiety selectively to several types of educts. A 2 : 1 stoichiometry of tin hydride to Bu₂Cu(CN)Li₂ is essential, and the commercially obtained Bu₃SnH should be dry and from a relatively new bottle (or if not, then freshly distilled).

Stannylcupration of a 1-Alkyne Using Bu₃Sn(Bu)Cu(CN)Li₂ Generated via
Transmetalation of Bu₃SnH [143]

(87%)

CuCN (0.672 g, 0.75 mmol) was added to a dry 10 mL round-bottom flask equipped with a stir bar and rubber septum. The flask was evacuated with a vacuum pump, and purged with Ar. This process was repeated three times. THF (2 mL) was injected and the slurry cooled to −78 °C where BuLi (0.63 mL, 1.50 mmol) was added dropwise. The mixture was allowed to warm slightly to yield a colorless, homogeneous solution which was recooled to −78 °C where Bu₃SnH (0.40 mL, 1.5 mmol) was added via syringe. Stirring was continued and over *ca.* 10 min, the solution yellowed and H_2 gas was liberated. 2-Methyl-3-butyn-2-ol (0.066 mL, 0.68 mmol) was added neat via syringe and the reaction mixture stirred for 5 min before being quenched into a 10 mL bath of 10% NH_4OH/90% saturated NH_4Cl. Extraction with 3 × 20 mL ether was followed by combining the extracts and drying over Na_2SO_4. The solvent was removed in vacuo and the residue chromatographed on silica gel. Elution with petroleum ether/ethyl acetate $(95 : 5) + 1\%Et_3N$ gave 2-methyl-3(tributylstannyl)-3-buten-2-ol (0.221 g, 87%) as a colorless oil (TLC R_f (10 : 90 EtOAc/petroleum ether) = 0.38).

Although the above transmetalation concept has been applied to the formation of the trimethyltin analog (i.e., $Me_3Sn(Me)Cu(CN)Li_2$), another procedure which does not involve trimethyltin hydride[136] has been developed[144]. The in situ generation of $Me_3Sn-SiMe_3$, which uses the Me_3Si group as a 'bulky proton', likewise transmetalates with $R_2Cu(CN)Li_2$ to afford a trimethyltin-containing mixed higher-order cuprate **41**[145]. Solutions of $Me_3Sn-SiMe_3$ in THF/HMPA are stable at room temperature for months, as long as they are protected from air and moisture. All operations are conducted in a single flask, and the overall process leading to **41** appears to proceed virtually in quantitative yield.

Stannylcupration of a 1-Alkyne Using $Me_3Sn(Bu)Cu(CN)Li_2$ Formed in situ from Transmetalation of $Me_3Sn-SiMe_3$[144]

Me₃Si−SiMe₃ →
1. MeLi, THF, HMPA, −78 °C, to −30 °C, 1 h
2. Me₃SnCl, −78 °C to −50 °C, 1.5 h
3. Bu₂Cu(CN)Li₂, −78 °C, to −50 °C, 1 h
4. HOCH₂CH₂C≡CH, MeOH, −78 °C to −25 °C
→ HO⌁SnMe₃ (74%)

Ethereal methyllithium (6.6 mL, 1.5 M) was added to a solution of hexamethyl-disilane (2.0 mL, 10.0 mmol, Aldrich) in 24 mL of THF/HMPA (3 : 1 v/v) at −78 °C under Ar. The resulting deep red solution was stirred for 1 h while allowing it to warm to −30 °C. The reaction was cooled to −78 °C after which Me₃SnCl (1.99 g, 10.0 mmol, available commercially from Aldrich) in 2 mL of THF was added. The reaction mixture was further stirred for 1.5 h while warming to −50 °C. In a separate vessel, Bu₂Cu(CN)Li₂ (10.0 mmol, prepared from 8.7 mL of 2.3 M BuLi and 0.89 g of CuCN) in 10 mL of THF was prepared at −45 °C. After stirring for 30 min, this solution was transferred via cannula to a solution of $Me_3Sn-SiMe_3$ at −78 °C. The resulting lemon-yellow solution was warmed to −50 °C and stirred for 1 h to ensure complete transmetalation. 3-Butyn-1-ol (0.63 g, 9.0 mmol) was then added neat via syringe followed by 5 mL of MeOH. The reaction immediately turned red in color. After 30 min the bath was removed and the solution warmed to room temperature. Usual work-up followed by chromatography on silica gel (hexanes/ethyl acetate, 8 : 1) gave 1.56 g (74%) of 4-hydroxy-2-trimethylstannyl-1-butene and 0.17 g (8%) of 4-hydroxy-1-trimethylstannyl-1-butene. GC analysis revealed a purity of >95% for both isomers, 4-hydroxy-2-trimethylstannyl-1-butene.

2.7 Other Cuprate Aggregates ($R_5Cu_3Li_2$)

The 2 : 1 ratio of organolithium(s) to CuX (X = halogen, CN) resulting in either R_2CuLi or $R_2Cu(CN)Li_2$ is by no means the only stoichiometry leading to discrete reagents. Higher ratios, 3RLi:CuI, give higher-order cuprates R_3CuLi_2[146,147], whereas addition of 1.5 equivalents RLi to CuI in THF leads to R_3Cu_2Li[146a,b] and 1.66 equivalents of RLi plus CuI (in Et_2O) gives $R_5Cu_3Li_2$[146a,148]. The higher-order species Me_3CuLi_2 is in equilibrium with Me_2CuLi and free MeLi in Me_2O (although apparently not in THF or Et_2O)[149], which tends to limit its use[146a,b]. R_3Cu_2Li is relatively unreactive[146c], although in the presence of $BF_3 \cdot Et_2O$[41] and Me_3SiCl[150] it works well in 1,4 additions to enones. The species $R_5Cu_3Li_2$, with R = Me, is particularly effective (versus Me_2CuLi, for instance) for conjugate methylation of α,β-unsaturated aldehydes[148]. The quenching process is experimentally important, with best yields obtained by either rapidly adding acetic acid to the −75 °C reaction mixture, or Me_3SiCl in the presence of amines Et_3N and HMPA. A typical case involves the addition of $Me_5Cu_3Li_2$ to the β,β-disubstituted enal cyclohexylideneacetaldehyde.

Conjugate Methylation of an α,β-Unsaturated Aldehyde with $Me_5Cu_3Li_2$ [148]

Purified cuprous iodide (3 mmol) was placed in a dry 50 mL three-necked flask containing a magnetic stirring bar. Two necks of the flask were closed by rubber septa and the other by a vacuum take-off equipped with a stopcock. The flask was alternately evacuated and filled with N_2 (three cycles), and dry ether (10 mL) was then injected. The slurry was stirred at *ca.* 0 °C (ice bath), and commercial ethereal MeLi (*ca.* 1.8 M, 5 mmol) was injected over 2–3 min. A dark yellow precipitate of methylcopper was deposited and then dissolved. Five minutes after the end of the addition the colorless (or faintly yellow) solution was cooled to $-75\,°C$ and the enal (1 mmol) in ether (1 mL plus 2×1 mL rinse) was added over 5 min. The reaction mixture was stirred at this temperature for 2 h. The temperature was then allowed to rise to $-40\,°C$ over 1.5 h, and the mixture was recooled to $-75\,°C$ and quenched with acetic acid. Work-up and Kugelrohr distillation (125–$130\,°C$, 10 mmHg) gave the product (90%) of better than 97% purity (VPC (vapor phase chromatography) = GC (gas chromatography) = GLC (gas-liquid chromatography), diethyleneglycol succinate, $120\,°C$). The material contained 1% (VPC) of the 1,2 addition product, 1-cyclohexylidene-2-propanol.

2.8   Hydrido Cuprates

The first complex metal hydride of copper, $LiCuH_2$, was reported back in 1974[151]. This species, along with several related hydrido cuprates Li_2CuH_3, Li_3CuH_4, and Li_4CuH_5, are all prepared by LAH reduction of preformed methyl cuprates resulting from various MeLi to CuI ratios[152]:

$$5MeLi + CuI \longrightarrow Li_4CuMe_5 + LiI$$

$$2Li_4CuMe_5 + 5LiAlH_4 \longrightarrow Li_4CuH_5 + 4LiAlH_2Me_2$$

These species have been individually examined in terms of their potential to reduce haloalkanes, ketones, and enones[153]. Among the many observations made was one concerning the remarkable reactivity of Li_4CuH_5 towards alkyl halides, a reagent which was shown to be more powerful than LAH in this capacity. Li_4CuH_5 is soluble in THF and is utilized at room temperature.

Preparation of Li$_4$CuH$_5$[152]

$$5MeLi + CuI \longrightarrow [Li_4CuMe_5 + LiI]$$

$$2Li_4CuMe_5 + 5LiAlH_4 \longrightarrow Li_4CuH_5 + 4LiAlH_2Me$$

An Et_2O solution of CH_3Li (10.0 mmol) was added dropwise to a well stirred CuI (2.0 mmol) slurry in Et_2O at $-78\,°C$. A clear solution resulted in a few minutes. To this solution was added $LiAlH_4$ (5.0 mmol) in ether and the reaction mixture was stirred at room temperature for 1 h during which time a white crystalline solid formed. The insoluble solid was filtered, washed with ether, and dried under vacuum. The product was analyzed and the X-ray powder-diffraction pattern was recorded. Calculated analysis for $Li_4CuH_5LiCuH = 4.00 : 1.00 : 5.00$; found $= 4.10 : 1.00 : 5.09$. The white solid was stable at room temperature for over a week.

Another reagent for carrying out either halide/sulfonate reductions[154] or conjugate reductions of enones[155] is that derived from addition of BuLi to CuH, presumably forming LiCuH(Bu). Pure, anhydrous copper(I) hydride decomposes to hydrogen and metallic copper above −20 °C; it is indefinitely stable at −78 °C. Suspensions of copper(I) hydride in ether are relatively air insensitive; the dry solid is pyrophoric. Tributylphosphine and copper(I) hydride form a 1 : 1 complex, the high solubility of which has prevented its isolation. The procedure for preforming CuH from DIBAL reduction of CuBr, along with some relevant information on its properties, is given below[156]. Use of LAH in place of DIBAL leads to inferior results.

Preparation of CuH[156]

$$CuBr \xrightarrow[\text{2. Et}_2\text{O (300 equiv), } -78\,°\text{C}]{\text{1. DIBAL, } -50\,°\text{C, pyridine (100 equiv)}} CuH$$
$$(>90\%)$$

Copper(I) hydride was prepared by treating 1 equiv of copper(I) bromide dissolved in 100 equiv of pyridine with 1.1 equiv of diisobutylaluminum hydride (20% in heptane) at −50 °C. Vigorous mixing produced a homogeneous, dark brown solution, from which copper(I) hydride could be precipitated by dilution with *ca.* 300 equiv of ether. Centrifugation, separation of the supernatant liquid, and repeated washing of the precipitate with ether, all at −78 °C, permitted isolation of copper(I) hydride as a brown solid in greater than 90% yield. The ratio of hydride to copper in this material was 0.96 ± 0.04; it contains less than 0.5% aluminum or bromine but retains *ca.* 25% pyridine, based on copper.

With CuH in hand, best conditions for using the ate complex (i.e., from the introduction of an RLi) involve Et$_2$O at −40 °C for conjugate reductions, whereas halides and sulfonates usually require ambient temperature over about 2 h.

Reduction of a Mesylate Using Hydrido Cuprate LiCuH(Bu)[154]

$$\underset{\text{CH}_3(\text{CH}_2)_5\text{CH(CH}_2)_{10}\text{CO}_2\text{Et}}{\overset{\text{OSO}_2\text{CH}_3}{|}} \xrightarrow[\text{Et}_2\text{O, } -40\,°\text{C to 25}\,°\text{C, 2 h}]{\text{LiCuH(Bu)}} CH_3(CH_2)_{16}CO_2Et$$
$$(85\%)$$

CuH (6.0 mmol; see above) was prepared under Ar at −50 °C in a 50 mL round-bottom flask, equipped with a magnetic stirrer and sealed with a rubber septum. After the CuH was washed with four 20 mL portions of cold (−50 °C) ether, 15 mL of cold (−40 °C) ether was added with stirring and then 6.0 mmol of cold BuLi in hexane was syringed into the flask over 1 min. The resulting dark brown solution (reagent partially insoluble) was stirred for 10 min. After addition of 0.610 g (1.5 mmol) of ethyl 12-mesyloxystearate in 1 mL of ether, the cooling bath was removed and stirring was continued for 2 h at room temperature. The reaction mixture was poured into aqueous saturated NH₄Cl solution, the ethereal layer was decanted and the aqueous layer was washed with 50 mL of ether. The combined ether extracts were dried, filtered, and the solvents removed in vacuo to provide 400 mg (85%) of ethyl stearate following silicic acid chromatography.

More recently, two newer methods for conjugate reduction have appeared which offer elements of convenience, efficiency, and chemoselectivity. In one, the known copper hydride hexamer [(Ph₃P)CuH]₆[157], routinely used in benzene or toluene solution at 25 °C, was found to be an extremely effective conjugate reductant toward α,β-unsaturated ketones and esters[158]. The reagent is a stable material, fully compatible with Me₃SiCl and insensitive to the presence of water. In fact, water is added to benzene or toluene solutions of this reagent in cases where short-lived intermediates are formed to suppress byproduct formation. Isolated, unactivated alkenes, however, are completely inert. The corresponding deuteride, [(Ph₃P)CuD]₆, is also available, with both reacting to deliver all six hydrides or deuterides per cluster to the substrate. The procedure for preparing [(Ph₃P)CuH]₆ as a red crystalline solid is given below[159]. Alternatively, this reagent may be purchased from Aldrich.

Preparation of [(Ph₃P)CuH]₆[159]

$$\text{NaO-}^t\text{Bu} + \text{CuCl} + \text{PPh}_3 \xrightarrow[\substack{\text{PhCH}_3/\text{PhH} \\ 25\,°\text{C}}]{1\text{ atm H}_2} \boxed{[(\text{Ph}_3\text{P})\text{CuH}]_6} + {}^t\text{BuOH} + \text{NaCl}$$

(a) NaO-ᵗBu

Toluene (150 mL) was added via cannula to a dry, N₂-flushed 500 mL Schlenk flask equipped with a Claisen head, a condenser topped with a gas inlet, and a 50 mL pressure-equalizing addition funnel. Under positive pressure of N₂, sodium (3.52 g, 0.153 mol, cut in thin slices) was added. The toluene was heated with an oil bath to 70–80 °C. tert-Butanol (45.4 mL, 35.4 g, 0.478 mol) was delivered dropwise. The vigorouslystirred mixture was heated until all of the sodium had reacted (12–24 h). Under a positive N₂ flush, the condenser was replaced with a rubber septum, and for ease of subsequent cannula transfer of the resultant suspension, large clumps of the alkoxide which may be present were broken up with a glass rod.

(b) [(Ph₃P)CuH]₆

Triphenylphosphine (100.3 g, 0.383 mol) and copper(I) chloride (15.14 g, 0.153 mol) were added to a dry, septum-capped 2 L Schlenk flask and placed under N₂. Benzene (approximately 800 mL) was added via cannula, and the resultant suspension was stirred. The NaO-^{t}Bu/toluene suspension was transferred via a wide-bore cannula to the reaction flask, washing if necessary with additional toluene or benzene, and the yellow nearly homogeneous mixture was placed under positive pressure (1 atm) of H_2 and stirred vigorously for 15–24 h. During this period the residual solids dissolved, the solution turned red, typically within 1 h, then dark red, and some gray or brown material precipitated. The reaction mixture was transferred under N_2 pressure through a wide-bore Teflon cannula to a large Schlenk filter containing several layers of sand and Celite. The reaction flask was rinsed with several portions of benzene, which were then passed through the filter. The very dark red solution was concentrated under vacuum to approximately one-third of its volume and acetonitrile (300 mL) was layered onto the benzene, promoting crystallization of the product. The yellow/brown supernatant was removed via cannula, and the product was washed several times with acetonitrile and dried under high vacuum to give 25.0–32.5 g (50–65%) of bright red to dark red crystals.

The degree of crystallinity and purity of the product varies somewhat with the degree of care exerted in the crystallization procedure. The major impurity present in the product was observed in the ^{1}H NMR as two broad resonances at approximately 7.6 and 6.8 ppm, and has not been identified. Small amounts of this byproduct have no perceptible effect on subsequent reduction chemistry. Crystallization can also be induced by addition of hexane or pentane with no effect on product purity or yield.

Conjugate Reduction of the Wieland–Miescher Ketone with [(Ph$_3$P)CuH]$_6$[158]

(85%)
(17:1 cis/trans)

[(Ph$_3$P)CuH]$_6$ (1.61 g, 0.82 mmol), weighted out under an inert atmosphere, and the keto enone (0.400 g, 2.26 mmol) were added to a 100 mL, two-necked flask under positive N_2 pressure. Deoxygenated benzene (60 mL) containing 100 μL of H_2O (deoxygenated by N_2 purge for 10 min) was added via cannula, and the resultant red solution was allowed to stir at room temperature until starting material had been consumed as judged by TLC analysis (8 h). The cloudy red/brown reaction mixture was opened to air, and stirring was continued for 1 h, during which time copper-containing decomposition products precipitated. Filtration through Celite and removal of the solvent in vacuo gave crude product which was purified by flash chromatography (yield was 82% of a 17 : 1 mix of cis/trans isomers).

Another method for conjugate reduction of enones/enals entails the proposed in situ formation of a halohydrido cuprate H(X)CuLi via transmetalation of Cl(I)CuLi (i.e., CuI•LiCl) with Bu₃SnH[160].

$$CuI + LiCl \longrightarrow Cl(I)CuLi \xrightarrow[\text{THF}]{Bu_3SnH} \boxed{H(X)CuLi} + Bu_3SnX$$

$$(X = Cl \text{ or } I)$$

Whatever the reagent's composition, it is compatible with Me₃SiCl at low temperatures (<0 °C) and can be applied to substrates containing unprotected keto groups, esters, allylic acetates, and sulfides. The procedure is best performed using excess LiCl.

Conjugate Reduction of an α,β-Unsaturated Aldehyde Using an in situ Generated Hydrido Cuprate H(X)CuLi[160]

To a −60 °C solution of CuI (0.1904 g, 1.00 mmol) and LiCl (0.1008 g, 2.38 mmol) in THF (4.5 mL) was added 8-acetoxy-2,6-dimethyl-2,6-octadienal (0.080 g, 0.381 mmol) followed by Me₃SiCl (0.27 mL, 2.11 mmol). After 10 min, Bu₃SnH (0.30 mL, 1.12 mmol) was added dropwise producing a cloudy yellow slurry. The reaction mixture was then allowed to warm to 0 °C gradually over 2 h during which darkening to a reddish-brown color was observed. Quenching was carried out with 10% aqueous KF solution (3 mL) leading to an orange precipitate. The organic layer was filtered through Celite and evaporated in vacuo, and the residue rapidly stirred with additional amounts of 10% KF for *ca.* 30 min before diluting with ether. The organic layer was then washed with saturated aqueous NaCl solution and dried (Na₂SO₄). The solvent was then removed in vacuo and the material chromatographed on silica gel. Elution with EtOAc/hexanes (10 : 90) gave 0.081 g (quantitative yield) of the product as a colorless oil (TLC R_f (15% EtOAc/hexanes) = 0.22).

---

3   Organocopper Complexes Derived from Grignard Reagents

Reactions of Grignard reagents under the influence of Cu^I salts enjoy a rich history of service in organic chemistry. From the early observations of Kharasch using catalytic quantities of CuCl with RMgX[161], to the various stoichiometric recipes developed over time, few would argue today with the value of such time-honored chemistry. Undoubtedly much of the popularity stems from a combination of factors, not the least of which include the general availability of Grignards, and the facile nature of magnesiocopper-based addition to, for example, α,β-unsaturated ketones and terminal alkynes. As with lithio copper reagents (see above), many of the same reactivity patterns are observed with Grignard-derived species, both in the catalytic as well as stoichiometric modes of use. In addition, most of the reaction variables and parameters to be addressed in cuprate couplings using $Cu^I/RMgX$ are identical to those associated with Cu^I/RLi reagents. Thus, all reactions should be run under an inert atmosphere of N_2 or Ar, preferably the latter. Ethereal solvents are again the norm, and Grignards obtained from commercial sources should be titrated. Although early work focused on the use of CuCl as the Cu^I salt[161], more recent developments have shifted significantly toward usage of purified CuI, CuBr as its Me_2S complex, or the Cu^{II} chloride species Li_2CuCl_4 (Kochi's catalyst)[162].

3.1 Reactions Using Catalytic Cu^I Salts

Conjugate additions of a functionalized Grignard reagent, as with simpler analogs, is a popular means of setting the stage for subsequent manipulations (such as annulation)[2a]. When noncommercially available Grignards are to be employed, the entire process can fortunately still be run in one pot, in spite of the fact that information concerning the precise amount of Grignard present in the medium is unknown. Given the truly catalytic nature of the process, it is likely that the reactive species is R_2CuMgX, formed under the conditions. A typical procedure for preparing the Grignard reagent of, in this case, a protected chloro aldehyde, and its eventual Michael addition to cyclohexenone is given below. In general, with only small percentages of copper salts being necessary, work-up procedures are somewhat simplified.

Copper Bromide–Dimethylsulfide Catalyzed 1,4 Addition of a Grignard to Cyclohexenone[163]

(73%)

Magnesium turnings (0.60 g, 25 mmol) were ground for a few minutes with a mortar and pestle and were immediately placed into an N_2-filled flask. A solution of 2-(3-chloropropyl)-1,3-dioxolane (1.2 mL, 8.3 mmol), 1,2-dibromoethane (0.05 mL), and THF (1.6 mL) were added at 25 °C, and the mixture was stirred in a 70 °C bath at which temperature Grignard formation began. The reaction flask was then placed in a 25 °C bath and was stirred for 30 min, diluted with additional THF (5 mL), stirred for 1.25 h and then cooled to −78 °C. A solution of cuprous bromide–SMe$_2$ complex (0.41 g, 2.0 mmol) and Me$_2$S (4 mL) was then added dropwise and the mixture was stirred at $-78\,°$C for 1 h. A solution of cyclohexenone (0.66 mL, 6.8 mmol) and ether (7 mL) was then introduced dropwise over a 7-min period, and the mixture stirred at $-78\,°$C for 2.5 h and then warmed in an ice-water bath. After being stirred at $0\,°$C for 5 min, the mixture was quenched by the addition of a saturated aqueous solution (5 mL) of NH$_4$Cl (adjusted to pH 8 with aqueous NH$_4$OH) and stirred at $25\,°$C for 1.5 h. The dark-blue aqueous layer was removed, the ether layer washed with two additional 10 mL portions of water and a saturated aqueous solution (15 mL) of NaCl, and dried over MgSO$_4$. Concentration by rotary evaporation in vacuo gave 1.28 g of the crude product which was purified by flash chromatography (silica gel, 1 : 1 hexanes/ethyl acetate) to give 1.05 g (73%) of the product as a colorless oil. An analytical sample was obtained by bulb-to-bulb distillation (oven temperature $80\,°$C, 0.2 mmHg).

Due to the catalytic nature of the copper salt present, the resulting enolate must necessarily be mainly associated with a magnesio halide counterion (MgX$^+$), suggesting that in situ trapping with an electrophile may occur. Usually, more reactive alkylating agents work best, one example of which employs tert-butyl bromoacetate as the quenching agent following a copper-catalyzed 1,4 addition.

1,4 Addition/in situ Trapping of an Enone Using Catalytic CuI/Vinyl Grignard[164]

(79%)

To a slightly brownish slurry of vinylmagnesium bromide (1.5 M THF solution, 119 mmol) in Me$_2$S (16 mL) and THF (100 mL) was added cuprous iodide (0.658 g, 3.2 mol % relative to substrate) at $-78\,°$C and then 2-methyl-2-cyclopentenone (10.4 g, 108 mmol) in 20 mL of THF over 40 min at $-50\,°$C to $-60\,°$C to give a dark brown thick solution. After being stirred at the same temperature for 50 min, the reaction mixture was cooled down to $-78\,°$C, 47 mL (270 mmol) of HMPA was added slowly, and then 44 mL (270 mmol) of tert-butyl bromoacetate was added. The reaction mixture was warmed very slowly to $0\,°$C over 6 h, and stirring was continued for 18 h at room temperature, resulting in a brown solution with a white precipitate. The reaction mixture was quenched with an aqueous solution of NH$_4$Cl and extracted with ether three times, and the combined extracts were washed with water twice and then brine, dried over MgSO$_4$, and evaporated in vacuo. Fractional distillation of the residue afforded 22.2 g (114 mmol) of *tert*-butyl bromoacetate and 20.3 g (79%) of the product with 96% stereoselectivity (bp 80–82 °C at 0.4 mmHg).

The relative rapidity and selectivity of the addition (1,4 versus 1,2) by catalytic Cu^I/RMgX is further exemplified in the case of an α,β-unsaturated aldehyde. With substrates of this type, it is essential to use both Me_3SiCl (2 equiv) and HMPA (2–3 equiv), otherwise the process is slowed considerably at −70 °C, and ratios of 1,4 to 1,2 adduct drop from above 200 : 1 to 4 : 1. The expected product in all cases is the TMS enol ether, most notably of the predominantly *E* configuration.

Copper(I)-catalyzed Conjugate Addition of RMgX to an α,β-Unsaturated Aldehyde/Me_3SiCl Mix[165]

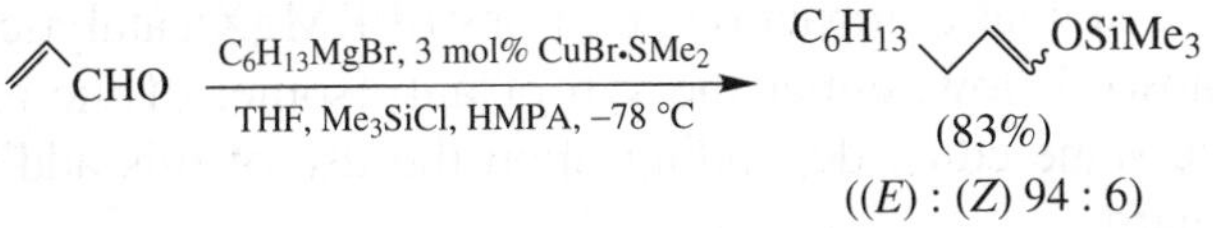

(83%)

((*E*) : (*Z*) 94 : 6)

To a cooled (−78 °C) THF solution (60 mL) of hexylmagnesium bromide (prepared from 35 mmol of 1-bromohexane and 37.5 mmol of magnesium in 85–90% yield), hexamethylphosphoric triamide (10.5 mL, 60 mmol; **CAUTION– potent carcinogen**), and cuprous bromide–SMe_2 complex (0.257 g, 1.25 mmol) was added dropwise a mixture of acrolein (1.67 mL, 25 mmol) and chlorotrimethylsilane (6.4 mL, 50 mmol) in 20 mL of THF over 30 min. After 3 h, triethylamine (7 mL) and hexane (100 mL) were added. The organic layer was washed with water to remove hexamethylphosphoric triamide and dried over $MgSO_4$. The product (3.86 g, 83%; 94% (*E*) by GLC analysis) was obtained by distillation (74 °C, 1 mmHg).

Copper-catalyzed Grignard ring openings of epoxides are commonly used, mild reactions. Transdiaxial opening of cyclohexene oxide, even with the usually less robust phenyl Grignard/catalytic Cu^I mix, takes place cleanly at −30 °C.

Opening of an Epoxide with Catalytic CuI/RMgX[166]

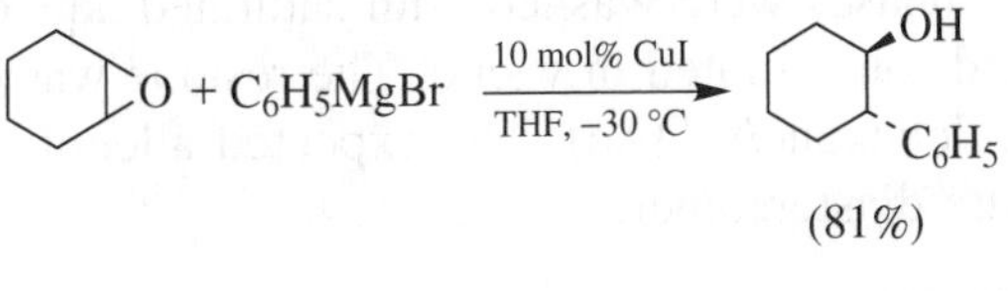

(81%)

To 10.9 g (0.45 mol) of magnesium in 100 mL of THF was added 73.0 g (0.47 mol) of bromobenzene in 100 mL of THF over 1 h. The resulting mixture was stirred for 30 min and then 8.85 g (46.5 mmol) of cuprous iodide was added and the mixture cooled to $-30\,°C$. A solution of 29.45 g (0.30 mol) of cyclohexene oxide in 50 mL of THF was then added dropwise. After the addition was complete, the mixture was stirred for 3 h and then quenched by being poured into 100 mL of cold saturated aqueous NH_4Cl solution. The solution was extracted with ether and the organic layers were combined, dried, and concentrated in vacuo to afford a liquid that was distilled at $80\,°C$ (0.23 mmHg) to afford 43.1 g (81%) of a yellow solid which was recrystallized from pentane, mp $56.5-57.0\,°C$.

Although the effect of added TMS-Cl is frequently most pronounced in conjugate addition schemes, it has a dramatic impact on additions of RMgX/catalytic Cu^I to propargylic oxiranes[167]. As shown below, either the syn or anti isomer of the allenic products can be realized from the same educt depending upon the use of this additive and the halide present in the Grignard.

Copper-catalyzed Addition of a Grignard (RMgBr) to a Propargylic Epoxide: anti Allenol Formation[167]

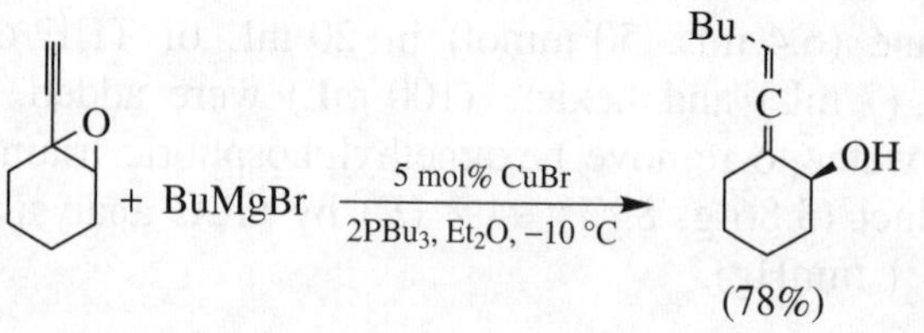

To an ethereal solution (30 mL Et_2O) of ethynyl cyclohexene oxide (0.400 g, 3.28 mmol), were successively added: (1) at $0\,°C$, a solution of $CuBr\cdot2PBu_3$ (1.65 mL of a 0.1 M solution in Et_2O, 0.165 mmol, 0.05 equiv); and (2) at $-50\,°C$ and slowly, a solution of BuMgBr (6.55 mL of a 1 N solution in Et_2O, 6.55 mmol). The stirred solution was allowed to warm to $-10\,°C$ over 30 min, then hydrolyzed with 20 mL of a saturated aqueous solution of NH_4Cl admixed with 5 mL of aqueous NH_4OH. The aqueous phase was extracted twice with 30 mL Et_2O, then the combined organic phases were washed with saturated aqueous NH_4Cl (20 mL), dried over $MgSO_4$ and concentrated in vacuo. The residue was chromatographed on silica gel (70 : 30 cyclohexane/EtOAc). The expected allenol (0.460 g, 78% yield) was obtained as a pure diastereomer.

CuBr-catalyzed Addition of a Grignard (RMgCl) to a Propargylic Epoxide in the Presence of Me₃SiCl: syn Allenol Formation[167]

To a suspension of CuBr (0.024 mg, 0.167 mmol) in Et$_2$O (20 mL) and pentane (20 mL), at 0 °C, were added ethynyl cyclohexene oxide (0.400 g, 3.28 mmol), then Me$_3$SiCl (0.42 mL, 3.28 mmol). The mixture was cooled to −50 °C and a solution of BuMgCl (6.55 mL of a 1 N solution in Et$_2$O, 6.55 mmol) was slowly added. The mixture warmed to 0 °C for 30 min, then hydrolyzed and worked up as in the example above. After column chromatography 0.545 g (92% yield) of the syn allenol were collected as an inseparable mixture of two diastereomers (88 : 12 by GC, after acetylation; IR and ^{1}H NMR are identical to the anti diastereomer).

β-Lactones are highly susceptible to attack by RMgX modified by Cu^I, with carbon–carbon bond formation occurring at the β-carbon to afford alkylated carboxylic acids[168]. The procedure is attractively straightforward and can be applied to α,β- and α,α-substituted β-propiolactones.

Catalytic CuCl/RMgX Opening of a β-Lactone: a Carboxylic Acid Synthesis[169]

Butylmagnesium bromide (1 M in ether, 2.4 mL, 2.4 mol) was slowly added to a suspension of cuprous chloride (0.004 g, 0.04 mmol) in 6 mL of THF at 0 °C under Ar. β-Propiolactone (0.144 g, 2 mmol) in 2 mL of THF was next added dropwise. The mixture was stirred at 0 °C for 15 min and quenched by adding 3 N HCl solution. From the organic layer, heptanoic acid was extracted with 3 N NaOH solution. The alkaline solution was acidified, extracted with ether and concentrated to give pure heptanoic acid in 90% yield, bp 65 °C (1.0 mmHg).

Similar chemistry when performed on chiral, nonracemic (protected) α-amino-β-lactones provides a quick, clean entry to α-amino acids[170]. Although excess RMgX is present (>5 equiv), virtually no racemization is seen under the reaction conditions.

*Ring Opening of an α-Amino-β-propiolactone via Catalytic CuBr•SMe₂/RMgX:
Synthesis of a Chiral, Nonracemic Amino Acid*[170]

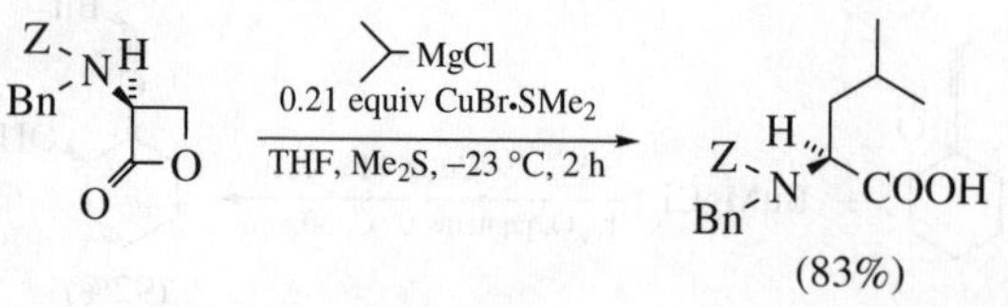

(83%)

Isopropylmagnesium chloride in Et₂O (3.0 mmol, 1.0 mL) was added dropwise over 5 min to the β-lactone (0.180 g, 0.578 mmol) and CuBr•SMe₂ (0.025 g, 0.122 mmol) in THF (6 mL)/SMe₂ (0.3 mL) at −23 °C. The mixture was stirred 2 h at −23 °C and quenched by addition to cold degassed 0.5 N HCl (20 mL). Extraction and washing of the ethereal phases followed by reversed phase MPLC (55% MeCN/H₂O, 3.3 mL min⁻¹) yielded 0.170 g (83%) of product as an oil ([α]$^{25}_D$ = −44.7° (c 2.5, CHCl₃); optical purity analysis by GC showed no detectable (R) isomer, i.e., ≥99.4% ee).

Displacement reactions of primary halides by RMgX are usually assisted by catalytic amounts of lithium tetrachlorocuprate, Li₂CuCl₄[2d,162]. Admixture of a Grignard with this salt (1–5 mol %) leads rapidly to the Cu^I oxidation state which then participates in a manner similar to other Cu^I salts, in this case effecting a substitution event. Shown below are two representative procedures; the first leads to formation of substituted butadienes[171], whereas the second demonstrates the generation of a trisubstituted aromatic product[172].

*Cross-coupling of a Vinylic Grignard with an Alkyl Halide Assisted by Catalytic
Li₂CuCl₄*[171]

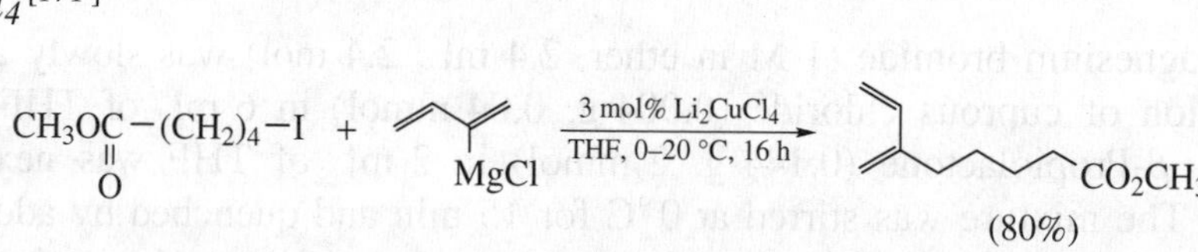

(80%)

To a mixture of halide (0.1 mol), Li₂CuCl₄ (3 mol % relative to halide), and THF (50 mL) in a 300 mL four-necked flask was added the vinyl Grignard (0.1 mol) in THF (100 mL) dropwise with stirring at 0 °C under a N₂ atmosphere. An exothermic reaction occurred during the addition and the color of the contents gradually changed from reddish brown to black. After the completion of the addition, stirring was continued at 20 °C for 16 h. The organic layer was separated after hydrolyzing the reaction mixture with 6 N HCl, and the aqueous layer was extracted with two portions of diethyl ether (100 mL). The combined organic extracts were washed first with 5% aqueous $NaHCO_3$ and then with water, dried (Na_2SO_4), and distilled at 71–76°C (2 mmHg). The reaction product (80%) was identified by comparing its IR, MS and NMR spectra with the reported data. The product gave reasonable elemental analyses.

Li_2CuCl_4-catalyzed Coupling of an Aryl Grignard with an Alkyl Halide[172]

Under dry N_2, 5-chloro-1,3-dimethoxybenzene (40 g, 0.23 mol), magnesium (6 g, 0.25 mol) and a small amount of 1,2-dibromoethane in THF (80 mL) were heated under reflux for 6 h. The solution was cooled in ice and a mixture of 1-iodopentane (42.6 mL, 0.325 mol) and Li_2CuCl_4 (30 mL of a 0.2 M solution in THF, 6 mmol) was added dropwise over a period of 30 min. The resulting black mixture was stirred at 0°C for 90 min and at 20°C for an additional 16 h. The almost solid reaction mixture was acidified with 6 N HCl (160 mL) and extracted with ether (2 × 200 mL). The organic extract was washed with 15% aqueous NH_4OH (60 mL) and water (60 mL), dried with $MgSO_4$, and evaporated in vacuo. According to the ^1H NMR spectrum of the residual product, olivetol dimethyl ether was formed in 74% yield. Distillation afforded the pure product (31.9 g, 66%) as a colorless liquid, bp 152–156°C (12 mmHg).

In the former study, use of CuI in place of Li_2CuCl_4 afforded inferior results. Other substrates tested included 3-bromopropyl chloride, which reacted selectively at the center bearing bromine. Aryl halides also couple under roughly comparable conditions. Perhaps most noteworthy is the finding that hydroxyl, ester, ether, and cyano functions remained unaffected throughout the reaction.

Both simple and more functionalized Grignard reagents, together with a Cu^I source, can afford products derived from couplings with allylic centers. For example, the mesitoate of a cyclohexenol, prepared specifically for stereochemical studies, clearly indicates that a CuCN-catalyzed S_N2' displacement takes place with inversion of stereochemistry[173]. The example below not only addresses the manner in which couplings of this sort can be performed, but also highlights the use of CuCN in a catalytic role. Although this salt is not used nearly as frequently as CuI, CuBr•SMe$_2$, or Li_2CuCl_4 for this type of cuprate reaction, it can afford completely different outcomes from the other Cu^I salts owing to the lack of metathesis between RMgX and CuCN, as discussed previously for RLi/CuCN mixtures. Hence, with CuCN in the presence of excess RMgX, the reactive species may be either RCu(CN)MgX, or possibly the higher-order magnesio cuprate '$R_2Cu(CN)(MgX)_2$'[174].

Substitution of an Allylic Carboxylate with a Grignard Reagent Catalyzed by CuCN[173]

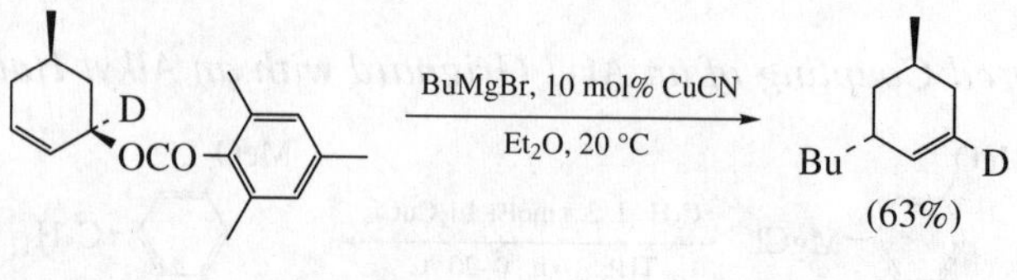

A flask equipped with a magnetic stirrer and septum was charged with 0.054 g (0.6 mmol) of cuprous cyanide. After flushing with dry N_2, 2 mL of anhydrous ether was added and the suspension was chilled to −10 °C. An ether solution of butylmagnesium bromide (6 mmol, prepared from 0.987 g of 1-bromobutane and 0.146 g of magnesium in 8 mL of ether) was added through a cannula, and after stirring the mixture for 10 min, a solution of 0.778 g (3 mmol) of α-duterio-cis-5-methyl-2-cyclohexenyl mesitoate in 2 mL of ether was added. The cooling bath was removed and the mixture stirred at room temperature for 6.5 h after which it was quenched with 2 mL of aqueous NH_4Cl solution. The resulting mixture was filtered, the precipitate washed with ether, and the ether solution dried over $MgSO_4$. Removal of solvent by fractionation followed by column chromatography (silica gel, pentane/ether) and vacuum distillation gave 0.289 g (63% yield) of a clear mobile oil, bp 58–60 °C (7.4 mmHg).

The Grignard of 3-chloromethylfuran, formed quantitatively and under the influence of Kochi's catalyst, smoothly displaces chloride ion from prenyl chloride in an S_N2 fashion almost instantaneously at 0 °C.

Coupling of an Allylic Halide with RMgX Catalyzed with Li_2CuCl_4[175]

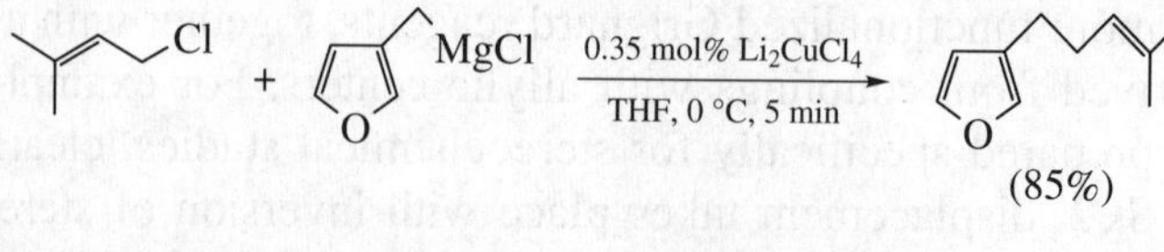

To 0.104 g (4.29 mmol) of magnesium turnings covered with 3 mL of THF under Ar was added 0.5 g (4.29 mmol) of 3-chloromethylfuran in 2 mL of THF in one portion. The mixture was allowed to stir for 30 min at 25 °C, then warmed in a preheated 50 °C oil bath for 30 min to provide a golden-yellow solution. The solution was chilled in an ice-water bath and 0.448 g (4.29 mmol) of freshly distilled

1-chloro-3-methyl-2-butene in 2 mL of THF was added in one portion followed immediately by the addition of 0.15 mL of a 0.1 M solution of Li_2CuCl_4 in THF. The resulting black suspension was stirred for 5 min at $0\,^{\circ}C$, poured into petroleum ether (50 mL), washed with 5% aqueous Na_2CO_3 solution (50 mL), water (50 mL), and dried over Na_2SO_4. Concentration in vacuo provided a pale yellow liquid which was purified by bulb-to-bulb distillation to give 0.547 g (85%) of perillene as a colorless liquid, bp $80\,^{\circ}C$ (20 mmHg).

Attack by a magnesio cuprate at the primary position in prenyl and related systems appears to be a general phenomenon[176]. Thus, in the case of a (Z)-trisubstituted allylic acetate and a Grignard reagent derived from a primary bromide, copper bromide-catalyzed coupling gave the expected product with only a small quantity of the competing S_N2' product (19 : 1). By switching to Li_2CuCl_4 as catalyst, only the product of straight substitution at the carbon bearing the leaving group is afforded.

Coupling of a (Z)-Allylic Acetate with a Grignard in the Presence of Li_2CuCl_4 [176]

A mixture of magnesium (0.255 g, 10.5 mmol) and 1,2-dibromoethane (26 µL, 0.3 mmol) in THF (10 mL) was heated at reflux. To the activated magnesium was added a solution of 4-bromo-3-methylbutyl benzyl ether (2.57 g, 10 mmol) in THF (2 mL) at $20\,^{\circ}C$ and the mixture was refluxed for 15 min. The Grignard reagent was added dropwise at $0\,^{\circ}C$ under Ar to a mixture of the acetate (0.98 g, 5 mmol) in THF (8 mL) and a 0.1 M solution of Li_2CuCl_4 in THF (2.0 mL, 0.2 mmol). After stirring for 1 h at $0\,^{\circ}C$, the mixture was partitioned between ether (50 mL) and saturated aqueous NH_4Cl (50 mL). The ether layer was washed with saturated aqueous NH_4Cl (30 mL), dried with $MgSO_4$, concentrated, and distilled under reduced pressure to give the product (1.38 g (79%); bp $134\,^{\circ}C$ (0.2 mmHg)). The product was further purified by silica gel column chromatography using 5 : 1 hexane/isopropyl ether as eluent.

3.2   Reactions of Stoichiometric Species

Admixture of a Grignard reagent with an equivalent of a copper halide leads to a metathesis reaction affording an organocopper species in the presence of a magnesium halide salt:

$$RMgX + CuX \xrightarrow[<0\,°C]{\text{ethereal solvent}} RCu \bullet MgX_2$$

This representation is precisely analogous to the conversion of an organolithium to $RCu \bullet LiX$, more accurately scribed as a halo cuprate $R(X)CuLi$[126]. Spectroscopic studies on the corresponding reagent $RCu \bullet MgX_2$, however, suggest that the composition of this mixture of $RMgX/CuX$ is not as simple as drawn (i.e., '$RCu \bullet MgX_2$')[177]. Studies of this sort are often hampered in part due to limited reagent solubility in common ethereal media.

When two equivalents of $RMgX$ relative to CuX are used, the magnesium halide analog of the Gilman reagent is formed. In this case, as with the 1 : 1 ratio, an equivalent of MgX_2 is generated relative to the lower order cuprate R_2CuMgX:

$$2RMgX + CuX \xrightarrow[<0\,°C]{\text{THF and/or Et}_2O} R_2CuMgX + MgX_2$$

The chemistry of these two species, just as with the reagents themselves, is quite different. The lower reactivity and basicity of the neutral organocopper $RCu \bullet MgX_2$ is used to advantage especially in carbocuprations and displacements of highly reactive electrophiles (e.g., allylic leaving groups). However, magnesio cuprates serve well in situations where a more robust, ate complex is needed, as in substitutions with less-reactive centers and in conjugate additions.

3.2.1 $RCu \bullet MgX_2$

One of the most valued reactions of $RCu \bullet MgX_2$ is their facile addition across terminal alkynes of the elements 'R$^-$' and 'Cu$^+$', namely carbocupration[178]. The regiochemistry is predictably such that the copper atom resides at the least-hindered alkyne carbon, thereby forming a new organocopper•MgX$_2$ complex of the vinylic type:

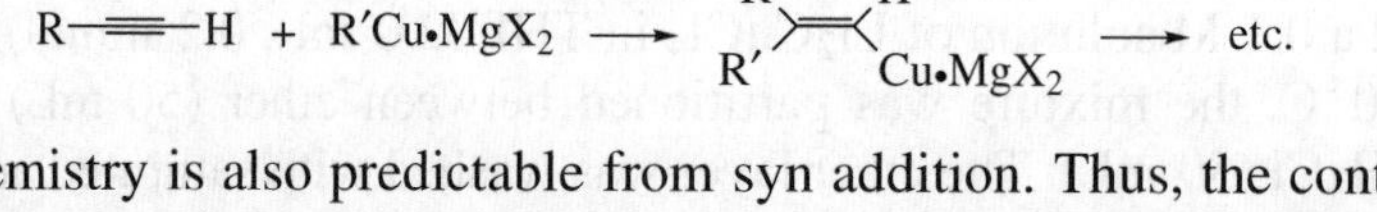

The stereochemistry is also predictable from syn addition. Thus, the control offered by this chemistry for stereodefined alkene preparation is especially noteworthy, and has been utilized extensively in the area of pheromone total synthesis when even trace amounts of isomeric impurities can be detrimental to potency.

Carbocupration of a 1-Alkyne Using RCu•MgX$_2$ in Et$_2$O[179]

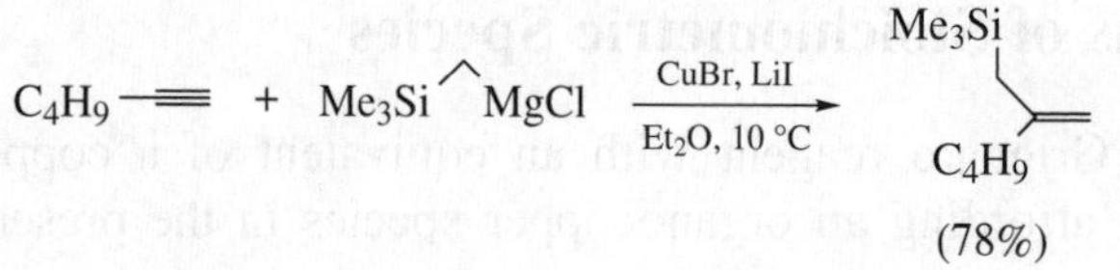

To a suspension of cuprous bromide (2.2 g, 15 mmol) and lithium iodide (1 N solution in ether, 20 mL, 20 mmol) in ether (50 mL) was added, at 0 °C, a solution of trimethylsilylmethylmagnesium chloride (0.9 M in ether, 17 mL, 15 mmol). The mixture first gave a yellow precipitate and then a homogeneous pale green solution which was stirred at −5 °C for 1 h. After addition of 1-hexyne (1.0 g, 12.2 mmol), the mixture was allowed to warm to 10 °C and stirred at this temperature for 18 h (brown solution) and then hydrolyzed with 100 mL of buffered ammonia solution. The mixture was filtered, decanted, and then the organic layer was washed with brine (10 mL) and dried over $MgSO_4$. The solvent was evaporated under vacuum and the residue distilled through a 10 cm Vigreux column to afford 1.8 g (78%) of pure product, bp 70 °C (10 mmHg).

Conjugated alkenes, starting with an enyne, can be realized in short order via this chemistry. Starting with a system of defined double-bond geometry and with knowledge of the expected cis delivery of R−Cu, a product of (*Z,E*) disposition can be obtained in good yields.

Carbocupration of a Functionalized, Terminal Enyne[180]

$$\text{SEt} + \text{EtCu} \cdot \text{MgBr}_2 \xrightarrow[-20\,°C,\,2\,h]{Et_2O} \text{Et} \diagup\!\!\diagdown\!\!\diagup \text{SEt}$$

(82%)

To a cooled (−40 °C) suspension of purified CuBr (7.9 g, 55 mmol) or CuBr•SMe_2 complex (11.3 g, 55 mmol) in 100 mL Et_2O, was added EtMgBr (25 mL of a 2 N solution in Et_2O). After stirring for 30 min at −30 °C to −35 °C, a yellow/orange precipitate of ethylcopper was formed. To this suspension was added (*Z*)-1-ethylthio-1-buten-3-yne (5.6 g, 50 mmol) dissolved in 20 mL Et_2O. The mixture was slowly warmed up to −20 °C, whereupon it dissolved, and was stirred for 2 h at this temperature. The dark red solution was hydrolyzed with aqueous NH_4Cl (50 mL), admixed with 5 N HCl (30 mL), the salts were filtered off, and the aqueous phase extracted once with Et_2O (50 mL). The combined organic phases were washed twice with aqueous NH_4Cl and then dried over $MgSO_4$ and the solvents were removed in vacuo. The residue was distilled through a 15 cm Vigreux column to afford 5.8 g (82% yield) of isomerically pure product, bp 87 °C (15 mmHg).

Although in the example above the intermediate vinyl copper species was hydrolyzed to the 1,1-disubstituted alkene, the electrophile used need not be limited to H^+. Further elaborations are indeed possible, as the following two cases demonstrate, using ethylene oxide in the former[178c], and allyl bromide in the latter[181]. Note that prior to introducing the oxirane, the vinylcopper was necessarily treated with an equivalent of pentynyllithium to form an ate species of sufficient reactivity to open even this simple coupling partner.

CuBr/RMgX-mediated Carbocupration of an Alkyne: Trapping with an Epoxide[178c]

$$C_6H_{13} \underset{120\ h}{\overset{\underset{CuBr\cdot SMe_2}{\overset{MeMgBr}{}}}{\xrightarrow{Et_2O,\ -23\ °C}}} \left[\underset{C_6H_{13}}{\overset{Me}{\diagdown}} Cu\cdot MgBr_2 \right] \overset{1.\ C_3H_7 \!\equiv\! Li}{\underset{2.\ \triangle}{\xrightarrow{Et_2O,\ HMPT}}} \underset{C_6H_{13}}{\overset{Me}{\diagdown}} OH$$

(75%)

To a mixture of cuprous bromide–SMe₂ (0.82 g, 4.0 mmol), ether (5 mL), and Me₂S (4 mL) at −45 °C under N₂ was added a 2.90 M solution (1.39 mL, 4.0 mmol) of methylmagnesium bromide in ether over 2 min. After 2 h, 1-octyne (0.52 mL, 3.5 mmol) was added over 1 min to the yellow/orange suspension. The mixture was stirred at −23 °C for 120 h, and then the resulting dark green solution was cooled to −78 °C. A solution of 1-lithio-1-pentyne (prepared from 4.0 mmol of butyllithium and 4.0 mmol of 1-pentyne), ether (5 mL), and hexamethylphosphoric triamide (1.4 mL, 8.0 mmol; **Caution: potent carcinogen**) was transferred to the green solution. After 1 h, ethylene oxide (0.21 mL, 4.0 mmol), which had been condensed at −45 °C, was added with a dry-ice-cooled syringe over 30 s. The resulting mixture was stirred at −78 °C for 2 h, allowed to stand at −25 °C for 24 h, quenched at 0 °C by addition of an aqueous solution (5 mL) of NH₄Cl (adjusted to pH 8 with ammonia), and then partitioned between ether and water. The crude product (90% pure by GC) was purified by column chromatography on silica gel (CH₂Cl₂) to give a colorless oil (0.44 g, 75%).

Carbocupration of Ethoxyacetylene with Br₂CuLi/RMgX Followed by Alkylation with an Allylic Halide[181]

$$EtO \!\equiv\! \underset{THF,\ -20\ °C}{\overset{\underset{Br_2CuLi}{\overset{C_6H_5MgBr}{}}}{\xrightarrow{\hspace{1cm}}}} \left[\underset{EtO}{\overset{C_6H_5}{\diagdown}} Cu\cdot MgBr_2 \right] \overset{\diagup\!\!\diagdown Br}{\underset{-20\ °C}{\xrightarrow{\hspace{1cm}}}} \underset{EtO}{\overset{C_6H_5}{\diagdown}}$$

(96%)

To a stirred solution of phenylcopper (prepared in situ by stirring phenyl magnesium bromide (0.01 mol) with 0.01 mol of the THF-soluble complex lithium dibromocuprate at −50 °C for 1 h) in THF (35 mL) was added 0.01 mol of ethoxy-acetylene at −50 °C. The mixture was then stirred for 1 h at −20 °C. Subsequently, allyl bromide (0.01 mol) was added and the mixture stirred for 3 h after which it was poured into an aqueous NH₄Cl solution (200 mL) containing NaCN (2 g) and extracted with pentane (3 × 50 mL). The combined extracts were washed with water (6 × 100 mL) to remove THF and dried over MgSO₄. The solvent was removed in vacuo and the residue purified by column chromatography eluting with pentane, to afford the product (96%) of 95% purity by GC.

Two other valuable reactions of organocopper complexes derived from an initial carbocupration step are iodination[182] and carboxylation[183]. The vinylic iodides resulting from I_2 quenching lead to strictly defined stereochemistries (over 99.9% pure), in good yields. These halides are useful precursors to the corresponding organolithiums[184], or are electrophiles for various stereospecific substitution reactions[185].

Preparation of Vinylcopper Reagents Prior to Iodination/Carboxylation[182]

To a suspension of CuBr (50 mmol) in ether (50 mL) is added dropwise, at −35 °C, an ethereal solution of RMgBr (50 mmol). After 30 min, a yellow or brownish suspension (according to the nature of the R group) of RCu is obtained. A solution of 1-alkyne (50 mmol) in ether (30 mL) is then added dropwise and the reaction mixture is allowed to warm slowly to −15 °C. The temperature must be carefully kept for 1.5 h between −15 °C and −12 °C (very important!). The vinylcopper reagent is thus quantitatively obtained as a dark green solution.

To effect iodination, the vinylic copper is simply treated with one equivalent of molecular iodine in the cold and then warmed. For large-scale (over 50 mmol) reactions, the iodine should be added portion-wise. For purification of products by distillation, a small amount of copper powder added to the crude material is recommended.

Vinylic Iodides via Iodination of Vinylic Copper Reagents[182]

(64–76%)

Finely crushed solid iodine (50 mmol) is added at once at −50 °C to a solution of vinylcopper reagent (50 mmol) prepared as above. The reaction mixture is then allowed to warm to 0 °C. Stirring is continued for 30 min to 1 h, until formation of a precipitate of CuI and discoloration of the supernatant is noted. Hydrolysis is performed at −10 °C with a mixture of saturated aqueous NH_4Cl and $NaHSO_3$ (80 mL/10 mL). Next, the precipitate is filtered off and washed twice with ether

(2 × 50 mL). After decantation, the aqueous layer is extracted twice with pentane or cyclohexane. The combined organic layers are then washed with a diluted solution of NaHSO₃ (if necessary to eliminate free iodine) and then dried over MgSO₄. The solvents are removed by distillation in vacuo and the product is isolated by distillation.

Exposure of a vinylic copper•MgX₂ species to dry CO_2 results in the nearly quantitative, stereospecific (>99.9%) conversion to α,β-unsaturated carboxylic acid salts, which ultimately give the acids upon quenching with aqueous HCl[183]. This particular reaction is unusual in that it requires a catalytic amount of triethylphosphite [$(EtO)_3P$], but otherwise is very much akin to the follow-up procedures offered above once carbocupration has occurred.

α,β-Unsaturated Carboxylic Acids via Carboxylation of Vinylic Copper Reagents[183]

$$R \diagup \overset{}{\underset{R'}{=}} Cu\cdot MgX_2 \quad \xrightarrow[\substack{2.\ CO_2,\ -40\ ^{\circ}C,\ 2\ h,\ then\ warm\ to\ -25\ ^{\circ}C \\ 3.\ 3N\ HCl,\ -30\ ^{\circ}C}]{1.\ P(OEt)_3\ (10\ mol\%),\ HMPT,\ -40\ ^{\circ}C} \quad R \diagup \overset{}{\underset{R'}{=}} COOH$$

(60–96%)

To a solution of vinylcopper reagent (50 mmol) prepared as above is added, at −40 °C, HMPT (40 mL) and a catalytic amount of P(OEt)₃ (5 mmol). A slow stream of dried CO_2 is then bubbled into the reaction mixture for 2 h. During the carbonation, the temperature is slowly allowed to warm to room temperature. The hydrolysis is performed at −30 °C by adding a 3 N HCl solution (80 mL). After decantation and extraction twice with cyclohexane (2 × 50 mL), the combined organic layers are washed twice with 2 N HCl (2 × 30 mL), water (80 mL), and then dried over MgSO₄. The solvents are removed under vacuum and the product is isolated by distillation.

Displacements of other allylic leaving groups by organocopper•MgX₂ reagents not derived from an initial carbocupration can also be effected. The stereochemistry of the process is such that the reagent attacks the nucleofuge in an S_N2', anti fashion.

S_N2' *Opening of a Vinylic Lactone with* $RCu•MgX_2$ [186]

$$\overset{O}{\underset{}{\bigcirc}}\!\!=\!\!O \quad \xrightarrow[THF,\ -20\ ^{\circ}C]{MeMgBr,\ CuBr\cdot SMe_2} \quad Me \diagdown\!\!\bigcirc\!\!\diagup COOH$$

(97%)

To a solution of cuprous bromide–SMe$_2$ complex (71.0 g, 0.35 mol) in Me$_2$S (300 mL) and THF (700 mL) at −20 °C was added methylmagnesium bromide (125 mL, 2.85 M in THF, 0.35 mol). After stirring at −20 °C for 1 h, a solution of 2H-cyclopenta[b]furan-2-one (21.5 g, 0.18 mol) in THF (200 mL) was added dropwise via an addition funnel. The mixture was stirred at −20 °C for 5 h, poured into 1 N NaOH and stirred for 2 h. The organic layer was separated and the aqueous layer was acidified to pH $ca.$ 2 with 1 N HCl. After extraction with ether, the organic phase was washed with water and brine, dried over MgSO$_4$, and concentrated in vacuo to provide a yellow oil (23.65 g, 97.6%), which was characterized as the methyl ester (prepared by standard diazomethane treatment).

RCu•MgX$_2$-based couplings with propargylic electrophiles occur readily under mild conditions to afford allenic products. For the educt below, the stereochemical outcome reflects approach of the copper species from the face of the alkyne anti to the departing group[187a], although the product was originally believed to be derived from cuprate syn delivery[187c]. This observation, therefore, is not an exception to the rule (i.e., expected anti addition)[188]. Other factors, such as temperature, solvent, and leaving group, can also affect the regiochemistry of these couplings[189].

Displacement of a Steroidal Propargylic Sulfinate with RCu•MgX$_2$ [187a]

A solution of the methylmagnesium halide (0.03 mol) in THF (30 mL) was added cautiously to a stirred suspension of cuprous bromide (0.03 mol) in THF (50 mL) at −50 °C and stirred at −30 °C for 30 min. 17α-Ethynyl-17β-methanesulfinyloxy-3-methoxy-1,3,5(10)-estratriene (5.58 g, 15 mmol) in THF (10 mL) was then added at −50 °C over 10 min. The reaction mixture was raised within 10 min to 20 °C. After 45 min, it was poured into a saturated solution of NH$_4$Cl in water (200 mL) containing NaCN (2 g). It was then extracted with hexane (3 × 50 mL), the combined extracts were washed with water and then dried over MgSO$_4$. Evaporation of solvent in vacuo afforded the product (4.55 g, 98%) which was recrystallized from ethanol (mp 71.0–71.5 °C; $[\alpha]_D^{23} = −16.05°$ (CH$_2$Cl$_2$)).

Although an allylic copper reagent prepared from an allylic lithium, CuI, and TMS-Cl is very effective for allylic ligand transfer to enones in a Michael sense (see p. 724)[127], replacing an allylic lithium by an allylic Grignard is also an option[190]. The same 1 : 1 : 1 stoichiometry applies, and CuBr•SMe$_2$ can be used in place of CuI. The halide in RMgX (R = an allylic moiety) is of no consequence in this chemistry, although ample precedent exists where this variable can play a pivotal role (see above). Allylic groups successfully transferred include allyl, crotyl, methallyl, and prenyl; yields are routinely good and the couplings are rapid at −78 °C. The initially formed TMS enol ether is routinely cleaved to the ketone with TBAF in THF as part of the work-up.

Conjugate Addition of Methallylcopper to an Enone[190]

(71%)

CuBr•SMe$_2$ (0.411 g, 2.00 mmol) and dry LiCl (0.0846 g, 2.0 mmol) were placed in a two-neck 25 mL round-bottom flask equipped with a stir bar and sealed with two septa. The flask was evacuated with a vacuum pump and then purged with argon while drying with a heat gun, this process being repeated three times. THF (3.00 mL) was injected and the mixture stirred for 3 min to yield a yellow, homogeneous solution which was then cooled to −78 °C. Concurrently, the Grignard solution (6.43 mL, 1.80 mmol) was transferred via syringe dropwise to the copper complex. TMS-Cl (253 µL, 2.00 mmol) was added followed immediately by the neat addition of the enone (0.90 mmol). The reaction was allowed to proceed for ≥5 min before being quenched at −78 °C with a saturated aqueous NH$_4$OH/NH$_4$Cl solution (1 : 9). Extraction with 4 × 25 mL of ether was followed by combining the organic layers and drying over anhydrous NaSO$_4$. The solvent was removed in vacuo and the resulting liquid was dissolved in THF (3.00 mL) and treated with TBAF (2.00 mL, 2.00 mmol) for 10 min. The solvent was again removed in vacuo, and the oil was subjected to flash chromatography (pentane/ether, 8 : 1) to afford the product (0.123 g, 71%) as a colorless oil (TLC R_f = 0.21 (pentane/ether, 8 : 1)).

3.2.2   RCu(CN)MgX

The allylic mesylate present in γ-oxygenated (E)-α, β-unsaturated esters, **43**, has also been found to be susceptible to (anti) S$_N$2′ displacement with lower-order cyanocuprates derived from Grignard reagents, RCu(CN)MgX, so long as BF$_3$ is present in equimolar amounts[191]. Competing 1,4 addition to the enoate or S_N2 displacement is not significant. THF, THF/Et$_2$O (5 : 1), or THF/hexanes (5 : 1) mixtures are the media of choice. Use of Et$_2$O alone is not likely to afford high yields due to generation of what is effectively the product of 1,4 reduction (i.e., **44**, R′ = H). Use of the acetate in place of mesylate as leaving group is also unproductive, starting material being recovered. Tosylates were not studied but are assumed to be acceptable substitutes for mesylates. The corresponding lithio lower-order cyanocuprate mix, RCu(CN)Li•BF$_3$, can be used with equal success. The products **44** are usually realized with de values of the order of 99 : 1. Using nonracemic enoates **43**, highly valued dipeptide isosteres are readily obtained, thus accounting for the use of this chemistry by several major pharmaceutical companies.

S_N2' Displacement of an Allylic Mesylate: Synthesis of a Dipeptide Isostere[191]

44, R = Me, R′ = ^{i}Pr, (85%)
other R, R′ groups:
R = Me, R′ = Me (93%)
R = Ph, R′ = ^{i}Bu, (96%)

To a stirred slurry of CuCN (1.89 g, 21 mmol) in 20 mL of dry THF was added by syringe 15 mL (21 mmol) of 1.4 M ^{i}PrMgCl in THF, and the mixture was allowed to warm to 0 °C and to stir at this temperature for 15 min. BF$_3$•Et$_2$O (2.58 mL, 21 mmol) was added to the above mixture at −78 °C and the mixture was stirred for 5 min. A solution of the α,β-enoate **52** (1.77 g, 5.25 mmol) in dry THF (8 mL) was added dropwise to the above reagent at −78 °C with stirring, and the stirring was continued for 30 min followed by quenching with 8 mL of a 1 : 1 saturated NH$_4$Cl/28% NH$_4$OH solution. The mixture was extracted with Et$_2$O and the extract was washed with water and dried over MgSO$_4$. Concentration under reduced pressure gave a colorless oil, which was purified by flash chromatography over silica gel with hexane/EtOAc (3 : 1) followed by recrystallization from hexane to yield the product (1.28 g, 85% yield) as colorless crystals of better than 99% optical purity as determined by capillary gas chromatography and ^{1}H NMR (mp 50 °C; $[\alpha]^{20}_D$ + 21.5° (c 0.90, CHCl$_3$)).

The RCu•MgX$_2$ formulation has found use in a remarkable five-fold addition of PhCu•MgBr$_2$ to buckminsterfullerene (C$_{60}$)[192]. Unlike the corresponding organolithium or Grignard reagents alone, which undergo monoaddition, multiple addition occurs in THF/toluene solution between −70 °C and 25 °C. No intermediate products were detected by HPLC.

Procedure for the Addition of 'PhCu•MgBr$_2$' to C$_{60}$: Synthesis of C$_{60}$Ph$_5$H[192]

C$_{60}$Ph$_5$H

A solution of C$_{60}$ (0.330 g, 0.458 mmol) in 150 mL of toluene was cooled to −78 °C and cannulated over 15 min to a magnetically stirred solution of organocopper reagent prepared from PhMgBr (0.94 M in THF, 7.81 mL, 7.33 mmol) and CuBr•Me$_2$S (1.51 g, 7.35 mmol) at −78 °C. The resulting dark green suspension was gradually warmed to −20 °C over 3 h. At this point, the purple color of the supernatant disappeared completely (quenching an aliquot with NH$_4$Cl solution resulted in complete recovery of C$_{60}$, suggesting that the solid materials are C$_{60}$−copper complexes). The mixture was then allowed to warm to 25 °C with continuous stirring, the progress of the reaction monitored by HPLC (nakalai tesque, Buckyprep, 10 × 250 mm, toluene/hexane = 7.3, no intermediate was observed during the reaction). After 2 h stirring at room temperature, the color of the solid became bright wine red, with the supernatant red colored. The reaction was complete after 2 days and then quenched with aqueous NH$_4$Cl solution. The mixture was separated into two phases, and the aqueous phase extracted with toluene three times. The combined organic phase was washed with brine, dried by passing through a column of Na$_2$SO$_4$, and evaporated in vacuo. After washing the residual powder with hexane (×4) to remove biphenyl (0.776 g, 100%), analytically pure C$_{60}$Ph$_5$H was obtained in 94% yield (0.496 g). The ^{1}H and ^{13}C NMR spectra are in good agreement with the values reported in the literature. Analytical data was obtained for the 0.5 toluene complex.

3.2.3   R$_2$CuMgX

Magnesio cuprates R$_2$CuMgX are the reagents of choice for openings of β-propiolactones to afford carbon homologated carboxylic acids. These displacements are characterized by short reaction times, which is fortunate owing to the relative thermal instability of the cuprates. Effective stirring and cooling (internal monitoring) are essential for both reagent formation and the subsequent introduction of educt, as these are exothermic processes. A nice feature of this chemistry is the color changes associated with completion of the reaction. Thus with R = Me the initial slightly yellow coloration becomes a deeper yellow, also: R = Bu, gray → yellow; R = ^{t}Bu, grey → greenish white; R = Ph, lemon yellow → yellow; R = vinyl, dark green → purple; R = allyl, reddish yellow → brown. Although the example below gives details for unsubstituted β-propiolactone[193], this chemistry applies to other related unsaturated substrates[194], including lactones **45–50**. In all cases, the products are those resulting from S_N2' additions.

| **45** | **46** | **47** | **48** | **49** | **50** |

Carboxylic Acid Formation From Ring Opening of β-Propiolactone with
$R_2 CuMgX$[193]

A flask equipped with a magnetic stirring bar and a septum was charged with CuI (0.420 g, 2.20 mmol). After flushing with dry Ar, 10 mL of anhydrous THF and 1 mL of Me$_2$S were added and the solution was chilled to −30 °C. Butylmagnesium bromide (1.00 M in THF, 4.4 mmol) was slowly added to this solution and the mixture was stirred for 30 min at this temperature. Then a solution of β-propiolactone 0.144 g (2.00 mmol) in 2 mL of THF was added dropwise to the flask. The mixture was stirred at −30 °C for 1 h and then allowed to warm to 0 °C for 1 h. After the reaction was quenched by addition of 2 mL of 3 M HCl, heptanoic acid was extracted with three 5 mL portions of 3 M NaOH from the organic layer. The alkaline solution was acidified with 3 mL of 6 M HCl and then extracted with ether. The ethereal extracts were washed with brine and dried (MgSO$_4$). Concentration gave pure heptanoic acid. An analytical sample was obtained by bulb-to-bulb distillation.

Ketones are easily realized from acid chlorides using magnesio diorganocuprates. As with lithio cuprates, mixed magnesio reagents generally represented as R_tR_rCuMgX can also be anticipated to form and show a selectivity of transfer profile related to their lithio analogs. In the case described below, initial preparation of (MeCu)$_n$, from CuI and MeLi, can be followed by the introduction of a Grignard reagent (1 equiv) in the usual ethereal media[195]. After appropriate manipulation of temperature to ensure reagent generation (i.e., warming to get the MeCu into solution), the substrate is introduced. Unlike R_2CuLi or R_tR_rCuLi, Grignard-based reagents such as $R(Me)CuMgX$ usually exist as suspensions at $-78\,°C$. They do, however, qualitatively have somewhat better thermal stability.

Ketone Formation From Reaction of an Acid Halide with a Mixed Magnesio Cuprate R_tR_rCuMgX[195]

$$\text{Me}-\!\!\!\bigcirc\!\!\!-\text{MgBr} \quad \xrightarrow[\text{THF, Et}_2\text{O, }-78\,°\text{C to }25\,°\text{C}]{\text{PhCOCl, MeCu}}$$

(79%)

In a 1 L, flame-dried, three-necked round-bottom flask equipped with an overhead stirrer and low-temperature thermometer, a bright yellow suspension of methyl-copper was prepared by the reaction of 30 mL of a 1.73 M (51.9 mmol) ether solution of methyllithium (0.11 M in residual base) with a $-78\,°C$ suspension of CuI (9.6 g, 50.8 mmol) in 100 mL of THF. The bright yellow color characteristic of methylcopper formed when this reaction mixture was warmed to $25\,°C$. It was then cooled down to $-70\,°C$ and 26 mL of a 1.96 M (51.0 mmol) ether solution of 4-methylphenylmagnesium bromide was added with a syringe. The resulting suspension was allowed to warm to $25\,°C$ and after cooling the deep purple solution to $-78\,°C$, a solution of benzoyl chloride (13.0 mL, 112 mmol) in THF (30 mL) was added dropwise by syringe. The reaction mixture was then warmed to $25\,°C$ and allowed to stir for 30 min. It was quenched with 8 mL of absolute methanol and then added to 600 mL of saturated aqueous NH_4Cl solution. Stirring for 2 h dissolved the copper salts, the ethereal phase separated and the aqueous portion was washed with two 100 mL portions of ether. The combined organic fractions were washed once with 100 mL of 0.1 N aqueous sodium thiosulfate, 3×100 mL of 1.0 N NaOH, and 1×200 mL saturated NaCl, and then dried over potassium carbonate. The product 4-methylbenzophenone was isolated by distillation (7.8 g; 79% yield; bp $120-130\,°C$ at 0.6 mmHg).

Replacement of halogen by various alkyl groups at the 2, 3 and 4 positions in the pyridine, quinoline, and 1,10-phenanthroline series can be accomplished using excess R_2CuMgX[196]. Although $CuBr/2RMgX$ is recommended for methyl- and ethylations, insertion of a tert-butyl moiety is most efficiently carried out with ^tBuMgCl in the presence of CuCN.

Double Displacement on a Heteroaromatic System Using R₂CuMgX[196]

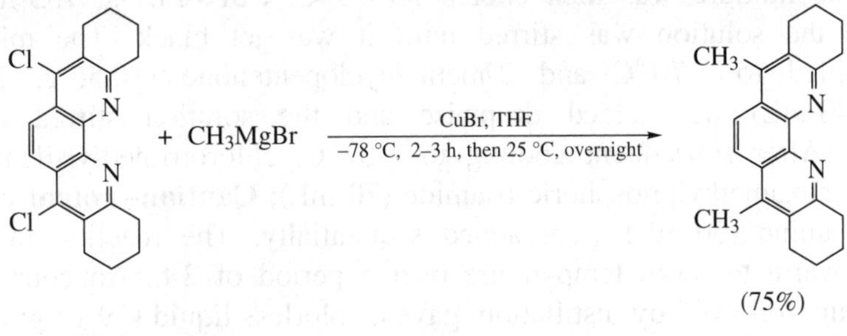

(75%)

A mixture of 8.0 g (56 mmol) of anhydrous CuBr, 250 mL of anhydrous THF, and 40 mL of ethereal methylmagnesium bromide (2.9 M) was stirred under N₂ at −78 °C for 20 min. Dichlorophenanthroline (2.5 g, 7.0 mmol) was added, and the reaction mixture was stirred for 2–3 h at −78 °C and then overnight at 25 °C. The reaction mixture was quenched by dropwise addition of saturated aqueous NH₄OH and then extracted with CHCl₃ (3 × 75 mL). The combined extracts were stirred for 20 min with 50 mL of ethylenediamine, and then 250 mL of water were added cautiously. The aqueous layer was extracted with CHCl₃ (3 × 75 mL), and the combined CHCl₃ solutions were dried (MgSO₄) and evaporated in vacuo. Column chromatography of the residue (4 : 1 EtOAc/CH₃OH) gave 75% of spectroscopically pure product as a white solid (mp 220 °C dec).

Michael additions based on stoichiometric magnesio cuprates are also very popular, useful processes, in particular where cyclopentenones are concerned. The initial adduct of this 1,4 event can be parlayed into other derivatives, such as regiospecifically generated silyl enol ethers, or α-alkylated products[197]. Both types of 1,4 addition/electrophilic trapping procedures are described below, in these cases used to provide intermediates for steroid total synthesis.

Conjugate Addition of R₂CuMgX: in situ Trapping with Me₃SiCl to a Cyclopentenone[197]

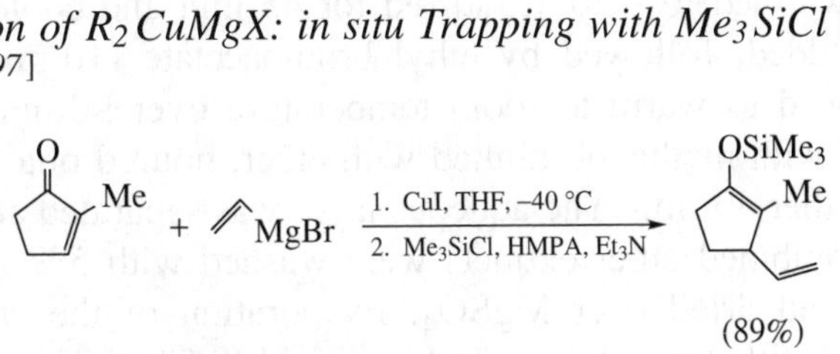

(89%)

To magnesium (6.07 g, 250 mmol) and one crystal of iodine in THF (100 mL) was added vinyl bromide (70.5 mL, 1 mol) in THF (60 mL) at a rate so as to maintain the reaction temperature at 45 °C. After all the magnesium had disappeared, the solution was heated at 45 °C under a stream of N_2 to remove excess vinyl bromide. The mixture was then cooled to −5 °C, CuI (25.7 g, 135 mmol) was added, and the solution was stirred until it was jet black. The mixture was quickly chilled to −70 °C and 2-methylcyclopentenone (10.56 g, 110 mmol) in THF (40 mL) was added dropwise and the solution stirred at −40 °C for 45 min. After subsequent cooling to −60 °C, chlorotrimethylsilane (34 mL, 365 mmol), hexamethylphosphoric triamide (70 mL); **Caution−potent carcinogen** and triethylamine (50 mL) were added sequentially. The reaction mixture was allowed to warm to room temperature over a period of 2 h. Aqueous petroleum ether work-up, followed by distillation, gave a colorless liquid (19.19 g; 89% yield; bp 64−66 °C at 3.1 mmHg).

1,4 Addition of $R_2 CuMgX$ to a Cyclopentenone Followed by in situ Alkylation[197]

(81%)
[1:3 cis:trans]

To magnesium (2.66 g, 109 mmol) and one crystal of I_2 in THF (100 mL) was added vinyl bromide (29.5 mL, 418 mmol) in THF (60 mL) at such a rate as to maintain the reaction temperature at 45 °C. After all the magnesium had disappeared, the solution was heated at 45 °C under a stream of N_2 to remove excess vinyl bromide. The mixture was then cooled to −5 °C, copper(I) iodide (10.44 g, 54.8 mmol) was added, and the solution stirred until it was jet black. The mixture was quickly chilled to −70 °C and 2-methylcyclopentenone (4.79 g, 49.9 mmol) in THF (45 mL) was added dropwise. After the addition was complete (30 min) the solution was warmed to −30 °C, stirred for 45 min and cooled to −70 °C, and HMPA (50 mL) added, followed by ethyl bromoacetate (10 mL, 91 mmol). The solution was allowed to warm to room temperature over 90 min, then stirred for 30 min, quenched with methanol, diluted with ether, poured onto saturated NH_4Cl, and stirred for another 30 min. The aqueous layer was separated and extracted twice with ether. The combined ether extracts were washed with 5% aqueous $Na_2S_2O_3$, water, and brine and dried over $MgSO_4$. Evaporation of the ether left a yellow liquid which was distilled to give a colorless liquid (8.52 g, 81%), a single peak by GC and a single spot by TLC (R_f (1 : 4 ether/petroleum ether) = 0.27; bp 85−87 °C (0.5 mmHg)).

With a β-halocyclopentenone, the conjugate addition of R_2CuMgX leads to reformation of the enone, a net substitution reaction. When carried out by adding the Grignard to a slurry of the chloro enone/CuI in cold THF, an excellent yield, at least with the functionalized substrate and Grignard shown below, was obtained.

Conjugate Addition–Elimination on a β-Chlorocyclopentenone Mediated by R_2CuMgX[198]

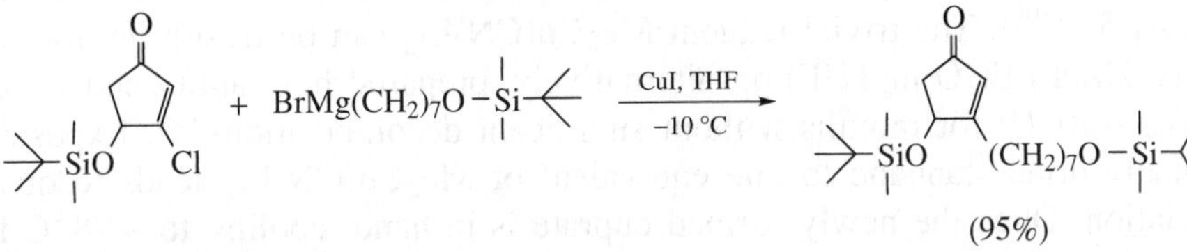

(95%)

The required Grignard reagent was prepared by adding a solution of 7-bromo-1-(*tert*-butyldimethyl-silyloxy)heptane (6.19 g, 20 mmol) in THF (15 mL), over 1.5 h to magnesium (0.491 g, 20.2 mmol) in refluxing THF (15 mL). The consumption of magnesium was complete after heating at reflux for a further 3 h. The concentration of reagent was measured by standard titration of an aliquot (1 mL) after hydrolysis. A suspension of CuI (0.078 g, 0.4 mmol) in THF (2 mL) containing 3-chloro-4-(*tert*-butyldimethylsilyloxy)cyclopent-2-en-1-one (0.100 g, 0.4 mmol) was stirred vigorously at $-10\,^\circ$C under Ar. Dropwise addition of the above Grignard reagent (0.41 M in THF, 1.85 mL, 0.76 mmol) produced a green solution which was stirred at $-10\,^\circ$C for 10 min. The reaction was rapidly quenched with saturated aqueous NH_4Cl solution (5 mL), and after addition of ether (5 mL), the mixture was stirred at room temperature for 1 h before dilution with water (10 mL) and extraction with ether (5 $\times$ 10 mL). The combined extracts were washed with brine (2 $\times$ 5 mL), dried over $MgSO_4$ and evaporated. Preparative TLC (silica gel, 50 : 1 CH_2Cl_2/methanol, v/v) gave the product as a colorless oil (0.167 g, 95%; bp (Kugelrohr) 135 $^\circ$C (0.2 mmHg)).

---

4 Copper Reagents by Transmetalation

4.1 Stoichiometric Ligand Exchanges

In a departure from prior art as a route to mixed cyanocuprates, selected higher-order reagents can be formed directly by ligand exchange processes between non-copper-containing organometallics and trivial higher-order cuprates, generalized as follows:

$$R_t\text{–}M + R'_2Cu(CN)Li_2 \longrightarrow R'\text{–}M + R_tR'Cu(CN)Li_2$$

Specifically, this chemistry has now been applied to vinylic and allylic stannanes, in situ generated vinylic zirconocenes and alanes, (Z)-vinyl tellurides, and zinc reagents. Note that the precursor transferrable ligand R_t in these schemes does not rely on highly basic organolithiums.

4.1.1   Vinyl and Allyl Stannanes

Vinylic stannanes, prepared in a variety of ways (such as hydrostannylation)[137], are converted to cuprates **51** upon treatment with $Me_2Cu(CN)Li_2$ in THF at room temperature for 1–1.5 h[199]. The trivial reagent $Me_2Cu(CN)Li_2$ can be freshly preformed (from CuCN plus $2MeLi/Et_2O$ in THF) or, alternatively, prepared in quantity and stored in the freezer (*ca.* −20 °C) for months without significant decomposition[200]. Exposure of the azeotropically dried stannane to one equivalent of $Me_2Cu(CN)Li_2$ leads to quantitative transmetalation. Once the newly formed cuprate is in hand, cooling to −78 °C followed by introduction of an α,β-unsaturated ketone effects the desired 1,4 addition. Some key points concerning this overall one-pot process include: (1) for both vinylic and allylic (see below) stannanes, the extent of transmetalation is dependent upon the purity of the stannanes; (2) stannanes should *not* be stored under rubber septa, as exposure to these (rather than glass) stoppers dramatically reduces the extent of ligand exchange; (3) with vinyl stannanes, there is no obvious color change during the transmetalation; (4) the process is quite amenable to scale-up, having been run successfully (for the example below) on a mole scale in pilot plant equipment[199]; (5) for substitution reactions, cuprates $Me(vinylic)Cu(CN)Li_2$ are not useful because, unlike their reactions with enones, they transfer the methyl ligand in preference to the vinyl group[82]. Hence, the transmetalation should be conducted with $Me(2\text{-Th})Cu(CN)Li_2$ resulting in ligand swapping to give $(vinylic)(2\text{-Th})Cu(CN)Li_2$, which then selectively transfers the vinylic moiety[102]. Below, a procedure for the preparation of the antisecretory agent misoprostol is given, the lower side chain being appended to the cyclopentenone using this transmetalation approach.

51

Transmetalation of a Vinylstannane with $Me_2Cu(CN)Li_2$ Followed by Michael Addition: Preparation of Misoprostol[199]

Copper cyanide (1.21 g, 13.5 mmol, flame-dried under Ar) in THF (15 mL) was treated with methyllithium (20.6 mL, 1.44 M in Et$_2$O, 29.7 mmol) at 0 °C. The cooling bath was removed, and vinylstannane (7.65 g, 15.2 mmol) in THF (15 mL) was added. After 1.5 h at ambient temperature the mixture was cooled to −64 °C and the enone (3.2 g, 9.63 mmol) in THF (15 mL) was added rapidly via cannula. The temperature rose to −35 °C. After 3 min, the mixture was quenched into a 9 : 1 saturated NH$_4$Cl/NH$_4$OH solution. Ether extraction followed by solvent removal in vacuo provided 11 g of residue which was solvolyzed (3 : 1 : 1 acetic acid/THF/water, 100 mL) and chromatographed (silica gel, EtOAc/hexane eluent) to provide 3.15 g of product (8.2 mmol, 91%) which was identical in all respects with an authentic standard of misoprostol[201].

With allylic stannanes, the transmetalation is far more facile[202], not an unexpected observation judging from the order of release of ligands from tin[203]. Under standard conditions of THF (0 °C, 30 min), higher-order diallylic cuprates are formed to which can then be added various electrophiles, including vinylic triflates[204a], alkyl halides[202], and epoxides[202]:

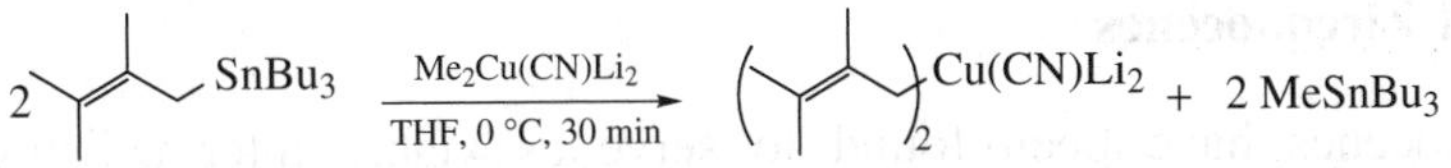

Under a CO atmosphere 1,4 additions (i.e., allylic acylations) are also facile[204b] Starting with virtually colorless solutions of Me$_2$Cu(CN)Li$_2$ and an allylic tin, these in situ generated cuprates take on a bright yellow coloration. For best results from these transmetalations, again, it is crucial that the initial stannanes be kept free from exposure to rubber septa. Also, it should be appreciated that vinylic iodides and bromides appear to be unacceptable reaction partners[204a]. The former lead to (frequently large) percentages of the product of reduction, whereas the latter do not retain their stereochemical integrity (i.e., double-bond geometry). Insofar as alkyl halides are concerned, as allylic cuprates are among the most reactive cuprates known, they can be used to displace even primary chlorides at −78 °C in minutes[202]. Thus, whereas bromides are perfectly acceptable, less stable, usually light-sensitive iodides are unnecessary and are likely to afford lower yields of coupling product.

In situ Diallylcuprate Formation from an Allylstannane: Displacement of a Primary Chloride[202]

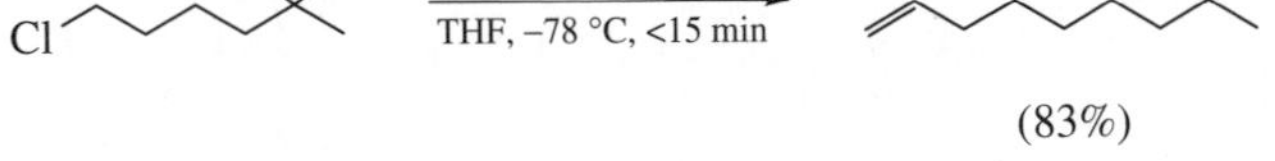

(83%)

CuCN (0.112 g, 1.25 mmol) was gently flame-dried under vacuum, followed by flushing with Ar. This process was repeated two times. THF (1.5 mL) was added via syringe and the resulting slurry was cooled to $-78\,^{\circ}$C. MeLi (1.18 M, 2 equiv, 2.11 mL, 2.5 mmol) was added dropwise, and the mixture was warmed to $0\,^{\circ}$C and stirred until homogeneous. After recooling to $-78\,^{\circ}$C, allyltributylstannane (2 equiv, 0.66 g, 0.62 mL, 2.5 mmol) was added dropwise and the solution was warmed to $0\,^{\circ}$C, stirred for 30 min (during which time it became yellow), and recooled to $-78\,^{\circ}$C. A cooled solution ($0\,^{\circ}$C) of the 5-chloropent-2-one acetal (0.8 equiv, 0.17 g, 0.15 mL, 1.0 mmol) in THF (1 mL) was transferred via cannula to the cuprate solution. The reaction was stirred 15 min, followed by quenching with 3 mL of aqueous saturated NH$_4$Cl. Water (4 mL) was added and the aqueous mixture was poured into a separatory funnel and extracted with ether (2×20 mL). The combined organic layers were washed with water (15 mL), brine (15 mL), dried over Na$_2$SO$_4$ and the solvent was removed in vacuo giving a clear liquid. Chromatography (silica gel, 13 : 1 petroleum/ether) afforded 0.113 g (0.66 mmol, 83%) of a clear liquid product (R_f (10% ether in petroleum ether) = 0.35).

4.1.2   Vinyl Zirconocenes

Vinylzirconocenes have been found to serve as viable intermediates en route to mixed vinylic cuprates[205]. Hydrozirconation of a 1-alkyne produces vinyl organometallic **52**, which is converted to the methylated species **53** at low temperature. Addition of Me$_2$Cu(CN)Li$_2$ presumably induces transmetalation to the desired mixed reagent **51**, which transfers the vinyl ligand in a Michael fashion to enones (Scheme 4). The initial hydrozirconation step, normally run in aromatic hydrocarbon (e.g., benzene) as solvent[206], is best performed here in THF thereby speeding up the process dramatically (due to greater solubility of the reagent in THF). Introduction of both the MeLi and

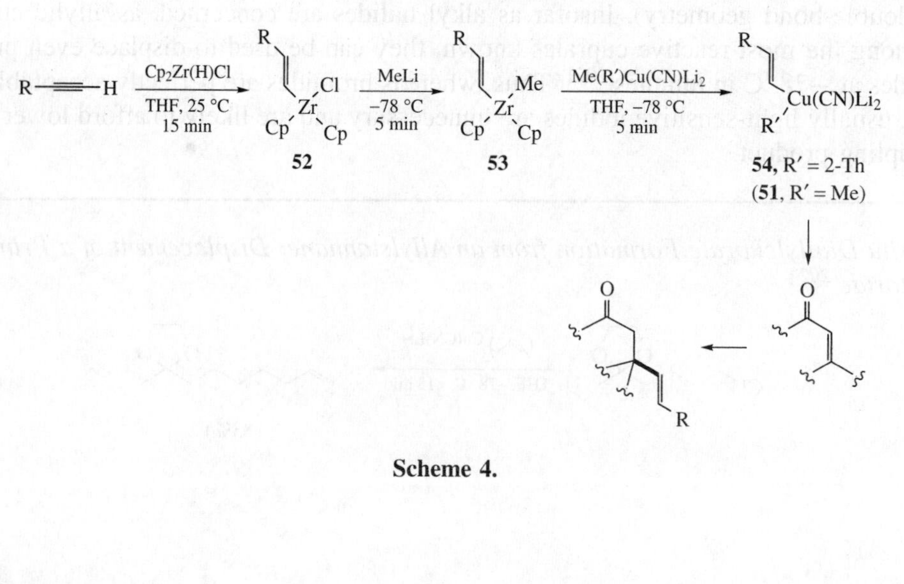

Scheme 4.

$Me_2Cu(CN)Li_2$ to **52** should be done with precooled (to $-78\,^\circ C$) reagents. Their addition to yellow solutions of **52** does not lead to any noticeable differences in color during ligand exchange. The extent of transmetalation can be sensitive to the quality of the MeLi; bottles containing noticeable amounts of particulates may prove problematic. Should incomplete transmetalation be observed, use of about 2 equiv of dimethoxyethane (DME) usually negates the deleterious effects of dissolved extraneous lithium salts. Use of Et_2O as cosolvent, introduced along with the educt, also enhances the overall process. Final concentrations of about 0.1 M are recommended. The synthesis of misoprostol (in bis-silylated form) using this protocol is illustrative.

Higher-order Mixed Cyanocuprate Formation/1,4 Addition via Transmetalation of a Vinyl Zirconocene with $Me_2\,Cu(CN)Li_2$[205a]

A 10 mL round bottom flask equipped with a stir bar was charged with zirconocene chloride hydride (0.129 g, 0.50 mmol) and sealed with a septum. The flask was evacuated with a vacuum pump and purged with Ar, the process being repeated three times. THF (1.50 mL) was injected and the mixture stirred to generate a white slurry which was treated with trimethyl-[(1-methyl-1-(2-propynyl)-pentyl)-oxy]-silane (0.124 mL, 0.50 mmol). The mixture was stirred for 15 min to yield a nearly colorless solution which was cooled to $-78\,^\circ C$ and treated via syringe with ethereal MeLi (0.35 mL, 0.50 mmol) to generate a bright yellow solution. Concurrently, CuCN (0.045 g, 0.50 mmol) was placed in a 5 mL round bottom flask equipped with a stir bar, and sealed under septum. The flask was evacuated and purged with Ar as above and ether (0.50 mL) was added via syringe. The resulting slurry was cooled to $-78\,^\circ C$ and treated with MeLi in ether (0.70 mL, 1.0 mmol). The mixture was warmed to yield a suspension of $Me_2Cu(CN)Li_2$ which was recooled to $-78\,^\circ C$ and added via cannula to the zirconium solution. The mixture was stirred for 15 min at $-78\,^\circ C$ to yield a bright yellow solution which was treated via cannula with methyl 7-(5-oxo-3-[(triethylsilyl)-oxy]-1-cyclopenten-1-yl)heptanoate (0.088 mL, 0.25 mmol) in ether (0.50 mL). After 10 min the mixture was quenched with 20 mL of 10% NH_4OH in saturated aqueous NH_4Cl. The product was extracted with 3×30 mL of ether and dried over Na_2SO_4. The solution was then filtered through a pad of celite and the solvent removed in vacuo. The resulting residue was submitted to flash chromatography on silica gel (9 : 1 petroleum ether EtOAc) to yield the protected form of misoprostol (0.132 g, 92%) as a colorless oil which displayed spectral characteristics identical with authentic material[200].

Using the same vinylzirconocene intermediate **52**, two alternative procedures may be utilized which arrive at the same type of cuprate (e.g., **51**)[205a]. Thus, in the first scenario, treatment of **52** with 2MeLi followed by (2-thienyl)Cu(CN)Li (available from Aldrich, catalog number 32,417-5) presumably affords **54**, R′ = 2-Th, which then goes on to deliver the vinylic moiety in a 1,4 sense. The simplest procedure, however, is one which calls for addition of 3MeLi to **52**, to which is then added LiCl-solubilized CuCN in cold THF. Subsequent introduction of the unsaturated ketone completes this straightforward sequence.

Hydrozirconation/Transmetalation Using 2MeLi/(2-Th)Cu(CN)Li on a Vinyl Zirconocene: Michael Addition of a Mixed Cuprate to Isophorone[205a]

(71%)

A 10-mL round-bottom flask equipped with a stir bar was charged with zirconocene chloride hydride (0.258 g, 1.0 mmol) and sealed with a septum. The flask was evacuated with a vacuum pump and purged with Ar, the process being repeated three times. THF (3.0 mL) was injected and the mixture stirred to generate a white slurry which was treated via syringe with phenyl acetylene (0.110 mL, 1.0 mmol). The mixture was stirred for 15 min to yield a bright red solution which was cooled to −78 °C and treated via syringe with ethereal MeLi (1.40 ml, 2.0 mmol). Concurrently, CuCN (0.0895 g, 1.0 mmol) was placed in a 5 mL round-bottom flask equipped with a stir bar, and sealed under septum. The flask was evacuated and purged with Ar as above and THF (1.0 mL) added via syringe. The resulting slurry was cooled to −78 °C and treated via cannula with a solution of 2-thienyllithium prepared from the metalation of thiophene (0.080 mL, 1.0 mmol) with BuLi (0.43 mL, 1.0 mmol) in THF (1.50 mL) at −30 °C (25 min). The mixture was warmed to yield a suspension of (2-Th)Cu(CN)Li which was recooled to −78 °C and added via cannula to the vinyl zirconate solution. The mixture was stirred for 30 min at −78 °C to yield a bright red solution which was treated with BF$_3$•Et$_2$O (0.12 mL, 1.0 mmol) followed by the addition of isophorone (0.075 mL, 0.5 mmol). After 1 h the mixture was quenched with 10 mL of 10% NH$_4$OH in saturated aqueous NH$_4$Cl. The product was extracted with 3 × 50 mL of ether and dried over Na$_2$SO$_4$. The solution was then filtered through a pad of celite and the solvent removed in vacuo. The resulting residue was submitted to flash chromatography on silica gel (9 : 1 petroleum ether/ethyl acetate), to give a 71% yield (0.086 g) of 3-(1-phenylethen-2-yl)-3,5,5-trimethylcyclohexanone as a thick yellow oil. The above procedure can alternatively be carried out with commercially available (2-Th)Cu(CN)Li (2.94 mL, 1.0 mmol) which when cooled to $-78\,^\circ$C can be added directly to a vinyl zirconate solution which has been treated with one equivalent of MeLi (1.01 mL, 1.0 mmol) at $-78\,^\circ$C (R_f (9 : 1 petroleum ether/ethyl acetate) = 0.37).

Mixed Cyanocuprate Generation from Treatment of a Vinyl Zirconocene with 3MeLi and then CuCN•2LiCl: Preparation of Protected 15-Methyl-PGE$_1$ Methyl Ester[205 a]

(84%)

A 10 mL round-bottom flask equipped with a stir bar was charged with zirconocene chloride hydride (0.258 g, 1.0 mmol) and sealed with a septum. The flask was evacuated with a vacuum pump and purged with Ar, the process being repeated three times. THF (0.5 mL) was injected and the mixture stirred to generate a white slurry which was treated with trimethyl [(1-methyl-1-(1-ethynyl)-hexyl)-oxy]silane (0.148 g, 0.70 mmol) as a solution in THF (0.5 mL). The mixture was stirred at room temperature for 30 min to yield a yellow/orange solution which was cooled to $-78\,^\circ$C and treated via cannula with a diethyl ether (3.0 mL) solution of MeLi (1.91 mL, 2.10 mmol, in THF/cumene), which had been precooled to $-78\,^\circ$C, to generate a bright yellow solution. The mixture was stirred for 10 min. Concurrently, CuCN (0.063 g, 0.7 mmol) and LiCl (0.059 g, 1.4 mmol) were placed in a 5 mL round bottom flask equipped with a stir bar, and sealed under septum. The flask was evacuated and purged with Ar as above and THF (1.0 mL) added via syringe. The mixture was stirred for 5 min at room temperature to generate a colorless homogeneous solution which was cooled to $-78\,^\circ$C and added via cannula to the zirconate solution. The mixture was stirred for 5 min at $-78\,^\circ$C to yield a bright yellow solution which was treated with a precooled ($-78\,^\circ$C) solution of methyl-7-(5-oxo-3-[(triethylsilyl)-oxy]-1-cyclopenten-1-yl)heptanoate (0.0883 g, 0.25 mmol) in Et$_2$O (0.50 mL). After 20 min the mixture was quenched with 10 mL of 10% NH$_4$OH in saturated aqueous NH$_4$Cl. The product was extracted with 3 × 35 mL of ether and dried over Na$_2$SO$_4$. The solution was then filtered through a pad of celite and the solvent removed in vacuo. The resulting residue was submitted to flash chromatography on silica gel (9 : 1 petroleum ether/EtOAc) to give an 84%

yield (0.120 g) of the protected form of 15-methyl-PGE$_1$ as a yellow oil (R_f (9 : 1 petroleum ether/EtOAc) = 0.41).

The transmetalation approach to lithio vinyl cuprate formation, due to the rapid and especially mild conditions of the ligand exchange between cyanocuprates and vinyl zirconocenes (e.g., **55**) tolerates electrophilic groups within the cuprate cluster. Thus, terminal alkynes bearing an ω-functional group such as nitrile, ester, and chloride undergo conversion to the mixed vinylic methyl cyanocuprates **56**, which then behave in the usual way toward enones[207]. Noteworthy is the finding that an equivalent of MeLi, used previously to replace chloride on zirconium in **52**, is not required. Simple (e.g., ethyl) esters are unacceptable functional groups, presumably due to carbonyl complexation with the vinyl zirconocene which negates its Lewis acidity. Triisopropylsilyl (TIPS) esters, however, eliminate this problem. With more sterically hindered enones, an equivalent of BF$_3$•OEt$_2$ is needed. Attempts to introduce ketones (i.e., FG = R−CO−) were not made owing to the facility with which Schwartz' reagent reduces this functional group.

FG ══ →[Cp$_2$Zr(H)Cl][THF, rt] FG⌒ZrCp$_2$Cl →[Me$_2$Cu(CN)Li$_2$][THF, −78 °C] FG⌒Cu(CN)Li$_2$ (Me) →[enone] (structure)

(*E*)-**55**   **56** [FG = CN, Cl, CO$_2$TIPS]

Hydrozirconation–Transmetalation: Preparation and Conjugate Addition of a Functionalized Vinylic Lithio Cyanocuprate[207]

(reaction scheme)
1. Cp$_2$Zr(H)Cl, THF, 25 °C
2. Me$_2$Cu(CN)Li$_2$, THF, −78 °C
3. cyclohexenone, −78 °C, 3 h

(75%)

A 25 mL flask was charged with 0.300 g (1.06 mmol) of zirconocene chloride hydride, 3 mL of THF, and a stir bar. 5-Cyanohex-1-yne (0.086 g, 1.04 mmol) was added with a 100 µL syringe and the suspension was stirred for 10 min. The yellow solution was cooled to −78 °C and 2.08 mmol of Me$_2$Cu(CN)Li$_2$ (prepared by adding 1.50 mL of 1.45 M MeLi in ether to a stirred suspension of 0.046 g CuCN in 3 mL of THF at −78 °C, followed by gentle warming to ambient temperature and then recooling to −78 °C) were added by cannula. The solution was stirred for 15 min and 0.050 g (0.052 mmol) of cyclohex-2-en-1-one added. After 3 h, the cold solution was poured into a beaker containing 20 mL of ether and 20 mL of 10% (by weight) aqueous ammonium hydroxide/saturated ammonium chloride. The mixture was transferred to a separatory funnel and the organic layer was removed. The

> aqueous layers were washed with 10 mL of saturated sodium bicarbonate, 10 mL of brine and dried over anhydrous magnesium sulfate. The organic layer was concentrated in vacuo and the resulting oil purified by silica gel chromatography (10% EtOAc/hexanes) to give 0.077 g (75%) of 3-(5-cyanopent-1-enyl)cyclohexane as a clear oil (R_f (10% EtOAc/hexanes) = 0.30).

Alkylations of vinylzirconocenes cannot be accomplished using the same ligand exchange phenomena based on $Me_2Cu(CN)Li_2$, as the mixed methyl vinyl cyanocuprate formed will preferentially transfer the methyl rather than vinyl ligand[81]. To circumvent this unusual inversion of ligand release from copper relative to conjugate addition schemes, the same method used for vinylic stannanes (see page 764) has been applied using the mixed thienyl methyl cuprate $Me(2\text{-}Th)Cu(CN)Li_2$ in place of the dimethyl analog. Presumably, ligand exchange now leads to the mixed vinylcuprate known to retain the 2-thienyl moiety on copper[102], allowing for the desired alkylation[208]. Epoxides, activated halides, and unactivated triflates are good reaction partners, although the latter electrophiles require use of MeMgCl in place of MeLi in the sequence. Unactivated bromides or iodides do not couple.

Alkylation of a Vinylzirconocene Mediated by $Me(2\text{-}Th)Cu(CN)Li_2$ [208]

1. Cp$_2$Zr(H)Cl, THF, 25 °C
2. 2MeLi, −78 °C
3. (2-Th)Cu(CN)Li, −78 °C
4. PhCH$_2$Br, −78 °C to 25 °C

(95%)

A 10 mL two-necked round-bottom flask equipped with a stir bar was charged with zirconocene chloride hydride (0.126 g, 0.48 mmol). The flask was evacuated and purged with argon, the process being repeated three times. THF (3 mL) was injected followed by the addition of 1-dimethyl-*tert*-butylsilyloxy-3-butyne (0.095 mg, 0.51 mmol). The mixture was stirred for 15 min to yield a clear yellow solution which was cooled to −78 °C and treated with ethereal MeLi (0.95 mL, 1.04 mmol). Concurrently, thiophene (46 mL, 0.048 g, 0.57 mmol) was placed in a round-bottom flask with a stir bar. The flask was evacuated and purged with argon as above and THF (1 mL) was introduced via syringe. The solution was cooled to −20 °C to which was added BuLi (0.25 mL, 0.50 mmol). The solution was then stirred for 1 h at −20 °C and transferred via cannula to a suspension of CuCN (0.045 g, 0.51 mmol) in THF (1 mL) which was precooled to −78 °C. The bath was subsequently removed and the suspension warmed to 25 °C to obtain a clear orange solution which was cooled again to −78 °C and transferred via cannula to the solution containing the vinyl zirconocene. After being stirred for 30 min at −78 °C, the resulting solution was treated with benzyl bromide (0.040 g, 0.23 mmol) and then kept for 1 h at −78 °C followed by warming to 25 °C and stirring at this temperature for an additional 3 h. Quenching was carried out using 10% concentrated NH_4OH in saturated aqueous NH_4Cl, and was followed by extraction with ether. The extracts were washed with water, dried, and concentrated in vacuo. Column chromatography with silica gel (10% EtOAc in hexane) gave 1-dimethyl-*tert*-butylsilyloxy-6-phenyl-3-pentene (0.059 g, 95%).

4.1.3   Vinyl Alanes

An initial carboalumination of a terminal alkyne gives rise to a vinylalane, which transfers its ligand from aluminum to copper upon treatment with a dialkynic higher-order cyanocuprate[209]. Prior to the transmetalation step, the toluene/dichloroethane are removed in vacuo and exchanged for Et_2O.

Carboalumination/Transmetalation of a 1-Alkyne to a Mixed Higher-order Cuprate Followed by 1,4 Addition[209]

(95%)

A suspension of 0.053 g (0.18 mmol) of zirconocene dichloride in 2 mL of dry 1,2-dichloroethane was treated at 0 °C with 0.78 mL (1.56 mmol) of a 2.0 M solution of trimethylaluminum in toluene, followed by addition of a solution of 0.046 g (0.56 mmol) of 1-hexyne in 0.3 mL of 1,2-dichloroethane. The reaction mixture was stirred for 3 h at 25 °C, the solvent was removed in vacuo, and 3 mL of dry Et_2O was added. The ethereal solution of the vinylalane was added to the solution of 0.050 g (0.56 mmol) of flame-dried CuCN in 2 mL of THF, which had previously been treated at −23 °C with 2.24 mL (1.12 mmol) of a 0.5 M solution of 1-hexynyllithium in THF/hexane (5 : 1). After the reaction mixture was stirred for 5 min at −23 °C, a solution of 0.048 g (0.50 mmol) of cyclohexenone in 1 mL of Et_2O was added dropwise, and stirring at −23 °C was continued for another 30 min. The mixture was quenched into a 9 : 1 saturated NH_4Cl/NH_4OH solution and extracted three times with Et_2O, and the combined organic layers were dried ($MgSO_4$), filtered through silica gel, and chromatographed (silica gel, 1 : 5 EtOAc/hexane) to yield 0.092 g (95%) of product ketone.

4.1.4   Vinyl Tellurides

Transmetalations of vinyl stannanes, alanes, and zirconocenes, by virtue of the metal chemistry leading to these species, all afford (*E*)-vinyl cuprates. A route to (*Z*)-vinyl cuprates is available via the corresponding vinyl tellurides, taking advantage of the stereoselectivity of hydrotelluration across acetylenes[210]. Reduction of a ditelluride (RTeTeR) in the presence of an alkyne occurs with NaBH$_4$, easily monitored at ambient temperatures by the disappearance of the red ditelluride color. Heating under an inert atmosphere leads directly to the vinyl telluride, usually as stable, isolable materials:

$$\text{RTeTeR} \xrightarrow[\text{EtOH, N}_2,\ 25\,°\text{C}]{\text{NaBH}_4} [\ 2\,\text{RTe}^- \ ] \xrightarrow{R'\equiv} \underset{\textbf{57}}{R'\diagup\text{TeR}}$$

Treatment of vinyl tellurides **57** (R = Bu, but not Ph) with Me$_2$Cu(CN)Li$_2$ in THF at 25 °C leads to a mixed methyl vinyl cyanocuprate which selectively delivers the (*Z*)-vinyl ligand to α,β-unsaturated ketones, as expected[210]. Although the same result can be obtained initially using the cuprate Bu(Imid)Cu(CN)Li$_2$ (Imid = *N*-imidazoyl; see above)[103], more convenient is the reagent Bu(2-Th)Cu(CN)Li$_2$ given the commercial availability of (2-Th)Cu(CN)Li[101b]. The product of transmetalation in this case is the mixed (*Z*)-vinyl (2-thienyl) cuprate, which is also selective for vinyl transfer to enones. Work-up follows the usual protocols (see above), and volatile tellurium byproducts (e.g., Bu$_2$Te) are easily evaporated at reduced pressures. **All reactions, due to the volatility of these simple tellurides, should be conducted in a well-ventilated hood.**

General Procedure for the Transmetalation/1,4 Addition of a (Z)-Vinyl Telluride with Bu(2-Th)Cu(CN)Li$_2$ [210]

$$R\diagup\text{TeBu} \xrightarrow[\text{2. α,β-unsaturated ketone, }-78\,°\text{C to }25\,°\text{C}]{\text{1. Bu(2-Th)Cu(CN)Li}_2,\ \text{THF, }25\,°\text{C, 1 h}}$$

Butyllithium (1.35 mL, 1.84 M in hexane, 2.5 mmol) was added to a solution of thiophene (0.21 g, 2.5 mmol) in THF (2 mL) previously cooled to −78 °C under nitrogen. The temperature was raised to −10 °C, and the solution was stirred for 30 min. The yellow solution was transferred via cannula to another flask containing a suspension of CuCN (0.18 g, 2.0 mmol) in THF (3 mL) previously cooled to −78 °C. Heating the mixture to room temperature produced a homogeneous solution which was then cooled to −78 °C and treated dropwise with butyllithium (1.1 mL of a 1.84 M solution in hexane, 2.0 mmol). The stirring was maintained for 15 min at −78 °C, and then the mixture was heated to room temperature. A solution of the vinylic telluride (2.1 mmol) in THF (2 mL) was added. After being stirred for 1 h at room temperature, the solution containing the vinylic cuprate was ready to be reacted with the electrophile.

To a solution of the vinylic telluride/$R_tR_rCu(CN)Li_2$ prepared as above and cooled to $-78\,°C$ was added the appropriate enone (2.0 mmol) via syringe. The cooling bath was removed, and the mixture was stirred for 20 min at room temperature, then treated with a mixture of saturated aqueous solutions of NH_4Cl and NH_4OH (4 : 1), and extracted with ethyl acetate (30 mL). The organic phase was separated, washed with brine (2 × 50 mL), and dried over anhydrous $MgSO_4$. The solvent was evaporated in vacuo and the residue was chromatographed on a silica gel column eluting first with hexane to remove the diorganotelluride and then with a mixture of hexane/ethyl acetate (5 : 1) to afford the pure product.

As with mixed higher-order cyanocuprates in general, these reagents are also reactive toward epoxides, and hence transfer the (Z)-vinyl ligand in this type of coupling[211]. Monosubstituted oxiranes react regioselectively above $-10\,°C$, whereas allylic epoxides are opened at $-78\,°C$ to give regioisomeric products. There same cuprates can be used to make (Z)-endiynes via couplings with 1-bromoalkynes[212].

Coupling of (Z)-Vinylic, 2-Thienylcyanocuprate with an Epoxide[211]

To a solution of the vinylic telluride/(2-Th)Cu(Bu)(CN)Li_2 system prepared as described above, cooled to $-78\,°C$, was added the appropriate epoxide (2.0 mmol) via syringe. In the case of the allylic epoxides, the mixture was stirred for 1 h at $-78\,°C$ and then worked up. In the other cases the cooling bath was removed and the mixture was stirred for 2 h at $25\,°C$. In both cases the work-up was performed by adding a mixture of saturated solutions of NH_4Cl and NH_4OH (4 : 1) and extracting with ethyl acetate (30 mL). The organic phase was separated, washed with brine (2 × 50 mL), and dried with anhydrous $MgSO_4$. The solvent was evaporated in vacuo, and the residue was chromatographed on silica gel, eluting first with petroleum ether to remove the tellurium containing byproduct and then with petroleum ether/ethyl acetate (4 : 1). In some cases, $BF_3\bullet Et_2O$ (0.25 mL, 2.0 mmol) was added at $-78\,°C$ prior to the addition of the epoxide, and the mixture was stirred at $-78\,°C$ for 2 h before performing the work-up described above.

4.1.5   Organozinc Reagents

Several processes involving stoichiometric amounts of Cu^I en route to zinc cuprates have appeared[213]. One case in point concerns the generation and use of a zinc homoenolate precursor, derived from a cyclopropanone ketal[214a]. Intermediates of this type are extremely versatile, participating in many different carbon–carbon bond-forming events. When exposed to catalytic quantities of a Cu^I salt, they can be effectively utilized as Michael donors, as well as regioselective nucleophiles in allylations. The organozinc precursor is initially formed in situ using dried ZnCl$_2$ of high quality which can be tolerated in slight excess. HMPA is also an essential ingredient in both reaction types, as is Me$_3$SiCl. DMPU can be substituted for the former additive in conjugate additions, although lower yields of 1,4 adduct are to be expected[214].

1,4-Addition of a Zinc Homoenolate Catalyzed by CuBr•SMe$_2$[214a]

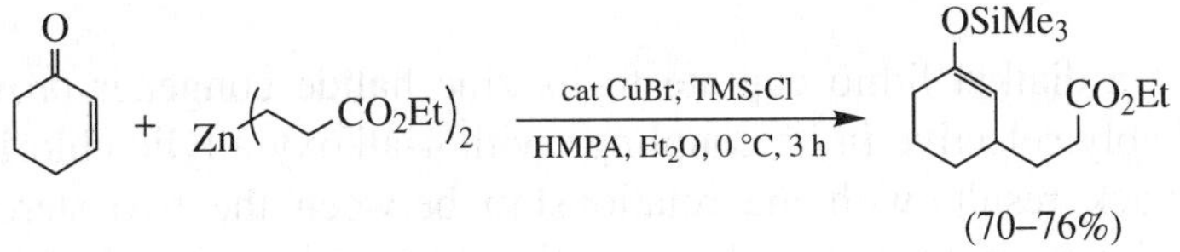

(70–76%)

To a solution of ZnCl$_2$ (16.4–17.0 g), freshly fused under vacuum, in 500 mL of ether was added 1-(trimethylsiloxy)-1-ethoxycyclopropanone (41.80 g, 240 mmol) over 5 min. The cloudy mixture was stirred at room temperature for 1 h and then refluxed for 30 min. The clear colorless solution of the zinc homoenolate and chlorotrimethylsilane was cooled with an ice bath, and to this was added CuBr•SMe$_2$ (0.4 g, 2 mmol). Cyclohexenone (9.62 g, 100 mmol) and then HMPA (34.8 mL, 200 mmol) were added over 5 min. A slightly exothermic reaction occurred initially and the bath was removed after 20 min. After 3 h at 25 °C, 40 g of silica gel and 300 mL of dry hexane were added while the mixture was stirred vigorously. The supernatant was decanted, and the residue was extracted twice with a mixture of ether and hexane. HMPA was collected as a low-boiling fraction (about 50–80 °C), and after about 1 g of forerun, the desired product (18.9–20.5 g, 70–76%) was obtained as a fraction boiling at 122–125 °C (2.40 mmHg). This product contains 0.2% of the double-bond regioisomer.

CuBr•SMe$_2$-catalyzed Allylation of a Zinc Homoenolate[214a]

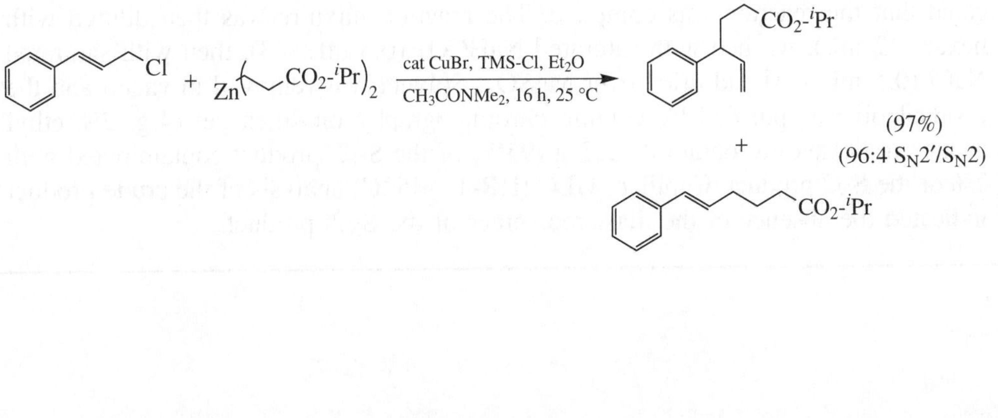

(97%)

(96:4 S$_N$2'/S$_N$2)

To a mixture of cinnamyl chloride (144 µL, 1.0 mmol), CuBr•SMe$_2$ (0.010 g, 0.05 mmol), and HMPA (2–5 equiv) or dimethylacetamide (4.5 mL), was added a 0.37 M ethereal solution of homoenolate (1.7 mmol, 4.5 mL; containing 2 equiv of Me$_3$SiCl), and the solution was stirred for 16 h at room temperature. The reaction mixture was diluted with ether and washed five times with water. After washing with saturated NaCl, drying, and concentration, the crude product was purified by chromatography (2% ethyl acetate in hexane) to obtain 0.148 g (97%) of the allylated product as a 96 : 4 mixture of the S$_N$2′ and the S$_N$2 isomers (GLC retention times on OV-1, 23 m 194 °C were 5.4 and 8.1 min, respectively). The isomers were separated by medium-pressure chromatography for analysis (bp 80 °C at 0.3 mmHg).

Conversion of a dialkyl lithio cuprate to its zinc halide congener provides a reagent which is remarkably selective in its couplings with 4-alkoxy allylic chlorides[215a]. Products of S$_N$2′ attack result, with the relationship between the two stereogenic centers predominantly, if not exclusively, of the anti configuration. Given the availability of chiral, nonracemic allylic and propargylic alcohols and the fact that cuprate attack is not limited by alkene disubstitution α to the alkoxy moiety, these factors argue well for future applications of this chemistry to the construction of quaternary centers of defined absolute stereochemistry.

S$_N$2′ Anti Allylation of R$_2$CuZnX[215a]

To a suspension of CuBr•SMe$_2$ (0.1028 g, 0.5 mmol) in THF was added 1.57 M BuLi in hexane (0.57 mL, 1.00 mmol) at −70 °C. The resulting solution was warmed to between −50 °C and −40 °C, stirred for 40 min, then cooled to −70 °C. A 1 M THF solution of fused ZnCl$_2$ (0.5 mL, 0.5 mmol) was added, turning the solution dark brown. After stirring for 10 min, 4-benzyloxy-5-methyl-2-hexenyl chloride (117 µL, 0.50 mmol) was added dropwise. After stirring for 15 h, TLC analysis indicated that the reaction was complete. The reaction mixture was then diluted with hexane (2 mL), washed with saturated NaHCO$_3$ (0.5 mL × 3), then with saturated NaCl (0.5 mL × 3) and dried over MgSO$_4$. Solvent was removed in vacuo and the residual oil was purified by column chromatography on silica gel (4 g, 2% ethyl acetate in hexane) to obtain 0.1235 g (95%) of the S$_N$2′ product contaminated with 2% of the S$_N$2 product. Capillary GLC (HR-1, 145 °C) analysis of the crude product indicated the absence of the diastereoisomer of the S$_N$2′ product.

Related lower-order zinc cuprates can be derived from CuCN and a preformed organozinc halide, rather than via transmetalations between lithio cuprates and ZnX_2 (see above). They enjoy a considerable level of functional group tolerance within these organometallic species due to a less reactive carbon–zinc bond associated with the cuprate precursors RZnX. Hence, cuprates of general stoichiometric formula RCu(CN)ZnX (X = halide) may contain within R functionality which includes esters, nitriles, halides, ketones, phosphonates, thioethers, sulfoxides, sulfones, and amines[213]. The initial phase of reagent preparation calls for generation of the organozinc iodide, using cut zinc foil (Alfa, 0.25 mm thick, 30 cm wide, 99.9% purity) or zinc dust (Aldrich), −325 mesh. A detailed procedure is given below which includes the subsequent formation of the cyanocuprate. For benzylic systems, three slightly different protocols have been developed (procedures 1, 2, and 3 below)[216]. Once formed, where a yield of 90% of RZnX has been assumed, the reagent is inversely added to solubilized CuCN•2LiCl to form the lower-order cyanocuprate (see Procedure 4 below).

Formation of $RCH_2Cu(CN)ZnI$ via RCH_2ZnI, Prepared from $RCH_2I + Zn^0$ [217]

$$RCH_2\text{-}I \xrightarrow[\text{2. CuCN•2LiCl, }-40\,°C\text{ to }0\,°C]{\text{1. }Zn^0,\text{ THF, TMS-Cl, 4 h, }BrCH_2CH_2Br,\text{ 35–45 }°C} \boxed{RCH_2Cu(CN)ZnI}$$

A dry 100 mL, three-necked, round bottom flask was equipped with a magnetic stirring bar, a 50 mL pressure equalizing addition funnel bearing a rubber septum, a three-way stopcock and a thermometer. The air in the flask was replaced by dry Ar and charged with 4.71 g (72 mmol) of cut zinc (*ca.* 1.5 × 1.5 mm). The flask was again flushed three times with Ar. 1,2-Dibromoethane (0.2 mL, 2.3 mmol) and 3 mL of THF were successively injected into the flask which was then heated gently with a heat gun until ebullition of the solvent was observed. The zinc suspension was stirred a few minutes and heated again. The process was repeated three times after which 0.15 mL (1.2 mmol) of chlorotrimethylsilane was injected through the addition funnel. The cut zinc foil turned to a gray color and was ready to use after 10 min of stirring. The reaction mixture was heated to 30 °C with an oil bath and 60 mmol of the iodide dissolved in 30 mL of THF was added dropwise over 40 min. After addition, the reaction mixture was stirred for 4 h at 43 °C to give a dark brown/yellow solution of the zinc reagent. A second dry 100 mL three-necked round bottom flask was equipped with a magnetic stirring bar, a three-way stopcock and two glass stoppers. The flask was charged with 4.59 g (108 mmol) of lithium chloride. The flask was heated with an oil bath at 130 °C (oil bath temperature under vacuum (0.1 mmHg)) for 2 h in order to dry the lithium chloride. The reaction flask was then cooled to 25 °C and flushed with Ar. The two glass stoppers were replaced by a low temperature thermometer and a rubber septum and 4.84 g (54 mmol) of copper cyanide was added. The flask was flushed three times with Ar and 40 mL of freshly distilled THF was added which led, after 15 min, to a clear yellow/green solution of the complex CuCN•2LiCl. This solution was cooled to about −40 °C and the two 100 mL, three-necked flasks were connected via a Teflon cannula or a stainless steel needle. The solution of the zinc reagent was then transferred to the THF solution of copper cyanide and lithium chloride. The resulting dark green solution was warmed to 0 °C within 5 min and was ready to use for the next step after 5 min of stirring at this temperature.

Preparation of Benzylic Zinc Halides: From Zinc Foil (Procedure 1)[216]

Cut zinc foil (approximately 5 mm^2 pieces, 2.30 g, 36 mmol) in 3 mL of dry THF was added to a dry, three-necked flask equipped with an Ar inlet, a thermometer, and an addition funnel. 1,2-Dibromoethane (0.150 g) was then added, and the mixture was heated with a heat gun until evolution of soap-like bubbles of ethylene and darkening of the zinc surface indicated activation. The mixture was cooled to 0–5 °C (ice bath), and a solution of the benzylic bromide (30 mmol) in 15 mL of THF was added dropwise (1 drop every 5–10 s). The reaction mixture was stirred at 5 °C until GLC analysis showed that the starting material was completely consumed (1–4 h).

Preparation of Benzylic Zinc Halides: From Zinc Dust (Procedures 2 and 3)[216]

Zinc dust (0.67 g, 10.5 mmol) in 1 mL of dry THF was added to a three-necked flask equipped with an Ar inlet, a thermometer, and an addition funnel. The mixture was cooled in an ice bath, and a solution of the benzylic bromide (7 mmol) in 7 mL of dry THF was added dropwise (1 drop every 5–10 s). The reaction mixture was stirred at 5 °C until GLC analysis showed that the starting material was completely consumed (1–4 h).

Zinc dust (1.34 g, 21 mmol) in 1.5 mL of dry THF was added to a three-necked flask equipped with an Ar inlet, a thermometer, and an addition funnel. A solution of the benzylic chloride (7 mmol) in 5.5 mL of dry THF and 1.5 mL of DMSO was added dropwise at room temperature. The reaction mixture was stirred at room temperature until GLC analysis showed complete conversion of the starting material to the zinc organometallic (22–24 h).

Preparation of Benzylic Copper Derivatives from Benzylic Zinc Halides by Transmetalation (Procedure 4)[216]

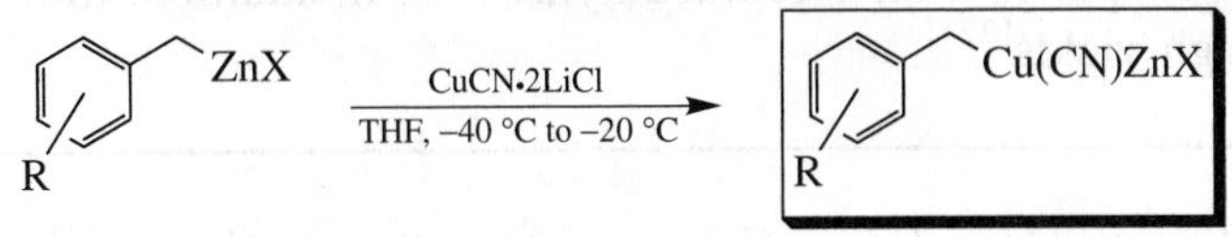

In a three-necked flask equipped with an Ar inlet, a thermometer, and a rubber septum, heat-dried LiCl (1.14 g, 27 mmol) was combined with CuCN (1.21 g, 13.5 mmol) in 10.5 mL of dry THF. The mixture was cooled to $-40\,^\circ$C and the benzylic organozinc reagent (14.7 mmol, prepared as described above) was added to the copper solution by cannula. The mixture was allowed to warm to $-20\,^\circ$C for 5 min, then cooled to $-78\,^\circ$C (dry ice/acetone bath). The copper reagent was then ready to react with various organic electrophiles, as described below.

As only small percentages of Wurtz-like coupling occurs with, for example, benzylic halide conversions to benzylic zinc halides, this type of organometallic can be prepared and used in many subsequent couplings. In addition to the example shown below for a Michael reaction[216], these cuprates are applicable to substitutions with allylic-, trialkylstannyl-, and acid halides, as well as 1,2 additions to aldehydes, and addition–eliminations to β-halo enones[216].

1,4-Addition of a Benzylic Lower-order Cyanocuprate to an Ynoate[216a]

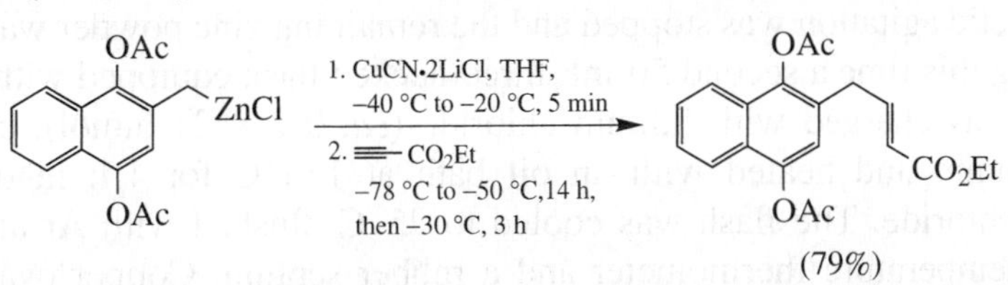

1,4-Diacetoxynaphthylzinc chloride was prepared from 1,4-diacetoxy-2-chloro-methylnaphthalene (13 mmol) according to Procedure 3 (above) with heating at $45\,^\circ$C for 3 h. GLC analysis showed complete conversion with less than 3% Wurtz coupling. After the transmetalation to the corresponding copper reagent by the addition of CuCN•2LiCl (as indicated above), ethyl propiolate (1.02 mL, 10 mmol) was added at $-78\,^\circ$C. The reaction was allowed to stir at $-50\,^\circ$C for 14 h and then at $-30\,^\circ$C for 3 h. Work-up and purification of the residue by flash chromatography (hexane/ethyl acetate 3 : 1, then 2 : 1) afforded 0.51 g (14% yield) of the double-bond isomerized product and 2.55 g (79% yield) of the desired product.

Unactivated primary iodides are also easily converted to the corresponding function-alized zinc iodides. These can be used to form ketones via treatment with an acid chloride[218] or α,β-unsaturated ketones in the presence of Me$_3$SiCl[219], or to synthesize alkynes by way of couplings with 1-bromoalkynes[220]. In addition, they add smoothly to nitro alkenes in high yields[221].

Coupling of a Functionalized Zinc Cuprate with a Bromoalkyne[220]

The cuprate can be prepared using the general procedure above from cut zinc foil, or alternatively using zinc dust, as follows. A 50 mL three-necked flask equipped with a thermometer, a 25 mL pressure-equalizing dropping funnel sealed with a septum, a vacuum/argon inlet and a magnetic stirring bar was charged with zinc dust (Aldrich, −325 mesh; 2.1 g, 18 mmol) and evacuated by means of a pump and refilled with Ar three times, 1,2-Dibromoethane (0.150 g, 0.8 mmol) in 2 mL of THF was added, and the suspension was heated gently to ebullition with a heat gun. The zinc suspension was stirred *ca.* 1 min and heated again. This process was repeated three times and chlorotrimethylsilane (0.1 mL, 0.8 mmol) was then added. After a few minutes, 4-iodobutyronitrile (1.66 g, 8.5 mmol) in 3.5 mL of dry THF was slowly added at 35–40 °C. After 3 h of stirring at 40 °C, GLC analysis of a hydrolyzed aliquot indicated completion of the reaction. The yield was estimated to be *ca.* 85%.

The magnetic agitation was stopped and the remaining zinc powder was allowed to settle. During this time a second 50 mL three-necked flask equipped with a magnetic stirring bar was charged with lithium chloride (*ca.* 0.9 g, 21 mmol), connected to a vacuum pump and heated with an oil bath at 130 °C for 1 h in order to dry the lithium chloride. The flask was cooled to 25 °C, flushed with Ar and equipped with a low-temperature thermometer and a rubber septum. Copper cyanide (0.9 g, 10 mmol) was added, followed by 10 mL of THF. A yellow/green solution formed after a few minutes of stirring. The reaction mixture was cooled to about −40 °C and the above prepared THF solution of 3-cyanopropylzinc iodide was added via syringe. The reaction mixture was allowed to warm to 0 °C and was ready for use in the next step.

The solution of NC(CH$_2$)$_3$Cu(CN)ZnI was cooled to −78 °C and 1-(2-bromoethynyl)-cyclohexene (0.915 g, 5 mmol) was added. The reaction mixture was stirred at −70 °C to −65 °C for 20 h and then warmed slowly to −10 °C. No starting material was left, as indicated by GLC analysis. The reaction mixture was poured into an Erlenmeyer flask containing ether (200 mL) and a saturated solution of aqueous NH₄Cl (50 mL). The insoluble copper salts were removed by vacuum filtration and the two layers were separated. The organic layer was washed with a saturated solution of aqueous NH₄Cl (2 × 50 mL), and the aqueous layer was extracted with ether (2 × 50 mL). The combined organic layers were then washed with a saturated aqueous solution of NaCl and dried over MgSO₄. After filtration, the solvent was removed and the resulting crude oil was purified by flash chromatography (6 : 1 hexane/EtOAc), affording 0.670 g of analytically pure 6-(1-cyclohexenyl)-5-hexyne-1-nitrile, as well as a small second fraction of impure product (0.030 g, 48% purity) in an overall yield of 79%.

Acylation of a Functionalized Zinc Cuprate[218]

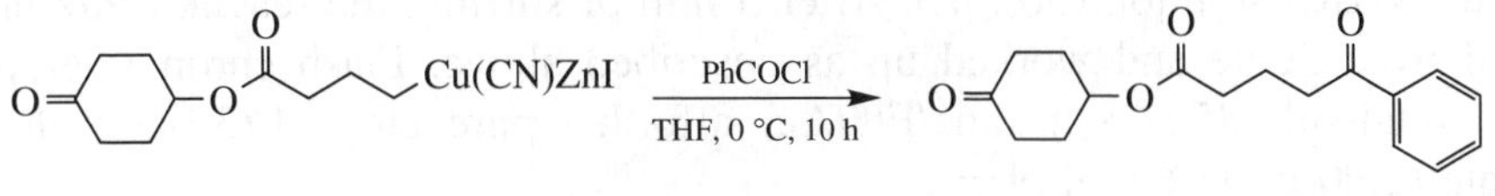

(80%)

To a suspension of zinc dust (activated with 1,2-dibromoethane and chlorotrimethylsilane as described above (1.63 g, 25 mmol)) in 1.5 mL of THF was slowly added 4-oxocyclohexyl 4-iodobutanoate (3.10 g, 10 mmol) in 4 mL of THF at 30 °C. After 4 h at this temperature, GLC analysis of a hydrolyzed aliquot indicated a conversion of 87%. The zinc reagent was cooled to 25 °C and added via syringe to a solution of copper cyanide (0.67 g, 7.5 mmol) and predried lithium chloride (0.67 g, 15 mmol) in 8 mL of THF at −30 °C. The resulting milky solution was warmed to 0 °C for 5 min and cooled back to −20 °C. Benzoyl chloride (0.914 g, 6.5 mmol) was added and the reaction mixture was stirred at −3 °C for 10 h. The reaction mixture was poured into a mixture of ether (200 mL) and saturated aqueous NH₄Cl solution (50 mL). After filtration and work-up as described above, the resulting crude oil was purified by flash chromatography (hexane/EtOAc) affording white crystals (mp 58–59 °C) of pure 4-oxocyclohexyl-5-oxo-5-phenylpentanoate (1.51 g, 80% yield).

Conjugate Addition of a Functionalized Zinc Cuprate to Cyclohexenone[219]

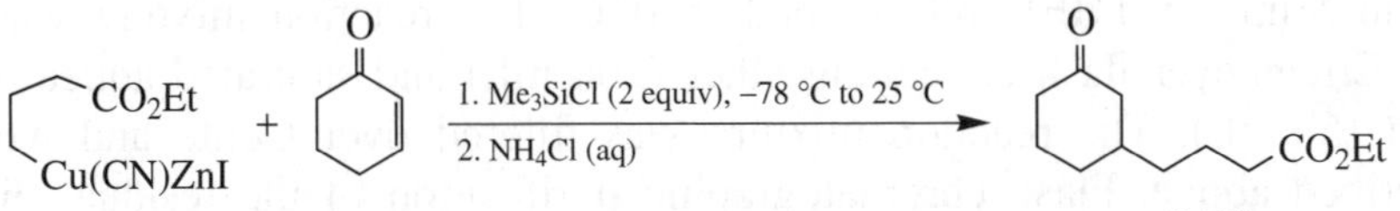

(94%)

To a suspension of zinc dust (activated with 1,2-dibromoethane and chlorotrimethylsilane as described above (*ca.* 3 g, 45 mmol)) in 2 mL of THF was slowly added, at 45 °C, a solution of ethyl 4-iodobutyrate (4.90 g, 20 mmol) in 8 mL of THF. After 4 h at this temperature, GLC analysis of a hydrolyzed aliquot indicated over 95% conversion. The zinc reagent was cooled to 25 °C and added via syringe to a solution of copper cyanide (1.44 g, 16 mmol) and predried lithium chloride (1.36 g, 32 mmol) in 16 mL of dry THF cooled at −30 °C. The resulting grey/greenish solution was warmed to 0 °C for a few minutes and cooled back to −78 °C. Chlorotrimethylsilane (4.1 mL, 32 mmol) was added, followed by a solution of cyclohexenone (1.39 g, 14.5 mmol) in 5 mL of dry ether. The reaction mixture was stirred for 3 h at −78 °C, then allowed to warm to 25 °C overnight. GLC analysis of a hydrolyzed reaction aliquot showed no cyclohexenone left. The reaction mixture was poured into an Erlenmeyer flask containing ether (50 mL) and saturated aqueous NH₄Cl solution (50 mL). After 5 min of stirring, the reaction mixture was filtered over Celite and worked up as described above. Flash chromatography of the residual oil (85 : 15 hexane/EtOAc) afforded pure ethyl 4-(3-oxocyclohexyl) butyrate (2.90 g, 94% yield)[222].

Michael Addition of a Functionalized Zinc Cuprate to a Nitroalkene[221]

To a suspension of zinc dust (activated with 1,2-dibromoethane and chlorotrimethylsilane as described above (*ca.* 2 g, 30 mmol)) in 2 mL of THF was slowly added, at 40 °C, a solution of 4-chloro-1-iodobutane (2.75 g, 12.5 mmol) in 5 mL of THF. After 3 h at 40 °C, GLC analysis of a hydrolyzed aliquot indicated over 95% conversion. The zinc reagent was cooled to 25 °C and added, via syringe, to a solution of copper cyanide (0.9 g, 10 mmol) and predried lithium chloride (0.85 g, 20 mmol) in 10 mL of dry THF cooled at −30 °C. The resulting greenish solution was warmed to 0 °C for a few minutes and then cooled to −78 °C. 1-Nitropentene (0.86 g, 7.5 mmol) was added dropwise at −78 °C. The cooling bath was removed after addition and the reaction mixture was allowed to warm slowly to 0 °C. After 4 h at this temperature, no nitropentene was left, as indicated by GLC analysis. The reaction mixture was cooled to −78 °C, quenched with acetic acid (2 mL in 5 mL of THF) and warmed to 0 °C. The reaction mixture was poured into an Erlenmeyer flask containing ether (200 mL) and saturated aqueous NH₄Cl solution (50 mL). The reaction mixture was filtered over Celite and worked up as described above. Flash chromatographic purification of the residual oil (20 : 1 hexane/ether) afforded pure 1-chloro-5-(nitromethyl) octane (1.40 g, 90% yield).

Most methods for preparing organozinc reagents RZnX involve conversion of lithio precursors RLi by way of ZnX_2, or direct insertion of Zn^0 into R−X. The former places obvious restrictions on functional-group tolerance, whereas the latter negates any option for maintenance of a stereocenter bearing metal, as insertion proceeds via a radical mechanism. As a means of retaining stereochemical integrity at secondary centers in cyclic systems which are to undergo C−C bond formation, a transmetalation scheme has been developed using an initial hydroboration followed by ligand exchange with $^iPr_2Zn$[223]. The resulting cyclic secondary zinc species subsequently transmetalates to copper in the usual way (i.e., with CuCN•2LiCl) and couples with various electrophiles, E^+. The ligand exchange from organozinc **58** to cuprate **59** proceeds with clean retention of stereochemistry so long as the temperature is carefully controlled.

Hydroboration/Tandem Transmetalation of a Cyclic Alkene: Alkynylation of a Zinc Cuprate[223]

A 50 mL, two-necked flask equipped with a rubber septum and an argon inlet was charged with 1-phenylcyclopentene (0.649 g, 4.50 mmol). After the alkene had been degassed, diethylborane (1.32 g, 13.5 mmol, 3 equiv; prepared from triethylborane and borane/methyl sulfide complex) was added dropwise by syringe at 25 °C. The resulting mixture was stirred for 4 days at 40 °C. Excess diethylborane was removed by distillation (25 °C, 2 h). $^iPr_2Zn$ (1.36 g, 9.00 mmol, 2 equiv) was carefully added by syringe at 25 °C. The reaction mixture was immediately cooled to −10 °C and kept at this temperature for 10 h with exclusion of light. Meanwhile, a mixture of CuCN (0.806 g, 9.00 mmol, 2 equiv) and LiCl (0.763 g, 18.0 mmol, 4 equiv) was dried in vacuo (130 °C, 3 h). The copper and lithium salts were dissolved in dry THF (9 mL) and the solution cooled to −78 °C. Then excess $^iPr_2Zn$ was removed by distillation (25 °C, 1 h); the flask was covered with aluminum foil to exclude light. The flask was equipped with an internal thermometer and the organozinc compound was diluted with dry THF (2 mL) and cooled to −78 °C. The previously prepared CuCN•2LiCl solution was added. The reaction mixture was stirred at −78 °C for 15 min, and 2-bromo-1-phenylacetylene (1.63 g, 9.00 mmol, 2 equiv) in dry THF (5 mL) was added dropwise within 20 min while the temperature of the reaction mixture was kept below −70 °C. The reaction mixture was warmed to 25 °C and worked up as usual. The crude material was purified by flash chromatography (hexanes), affording the product (0.459 g, 1.82 mmol, 41% yield) as a pale yellow liquid.

4.2   Processes Catalytic in Copper(I)

Ligand exchange between various organometallics and simple cuprates represents one mechanism by which cuprates of interest can be formed without recourse to traditional protocols. Hence, organometallics other than RLi and RMgX were presented in the previous section that are highly prone toward transmetalation, in particular, with cyanocuprates. However, these are stoichiometric events—a ligand-for-ligand swap driven by several factors. Efforts to decrease dependence on copper have been significant[224], one justification for which could be viewed as the reduced environmental impact, and another the associated costs of chemical waste disposal. Some of the same metals involved in the stoichiometric processes (see Section 4.1) can now be used with catalytic amounts of copper, and other metals have now been added to this important category.

4.2.1   Zirconocenes

Conjugate additions of (functionalized) alkyl zirconocenes can be realized using 10 mol % CuBr•SMe₂ in THF at 40 °C[225]. Lower temperatures tend to decrease yields. The presence of TMS-Cl (3 equiv) had a marginal impact, and the presence of other additives such as HMPA or excess Me₂S either completely inhibited the addition, or significantly dropped the overall yields. The presence of lithium (e.g., LiCl, LiClO₄) or magnesium (e.g., MgBr₂) salts also had a severely deleterious effect, implying that in situ procedures for generating Schwartz' reagent[226] which produce these byproduct salts should not be employed. Other copper salts (e.g., CuI•2LiCl, CuCN•2LiCl, CuI•P(OEt)₃, and CuSPh) gave results inferior to the CuBr•SMe₂ complex.

Copper-catalyzed Conjugate Addition of an Alkylzirconocene[225]

A solution of 0.150 g (0.59 mmol) of 4-triisopropylsilyloxycarbonyl-1-butene in 5 mL of THF was treated at 22 °C with 0.166 g (0.64 mmol) of Cp₂Zr(H)Cl. The mixture was stirred at 40 °C until a homogeneous yellow solution formed. After cooling to 22 °C, 0.065 g (0.59 mmol) of 3-methyl-2-cyclohexen-1-one and 0.012 g (0.06 mmol) of CuBr•SMe$_2$ were added. The dark green solution was stirred at 40 °C for 20 min, quenched with 25 mL of wet Et$_2$O and washed twice with 10 mL of saturated aqueous NaHCO$_3$ solution. The organic layer was dried over anhydrous Na$_2$SO$_4$, filtered through a pad of silica gel, and concentrated in vacuo. The residue was purified by chromatography on silica gel (EtOAc/hexane, 1 : 4) to give 0.170 g (78%) of 3-methyl-3-(4-triisopropylsilyloxycarbonyl)butylcyclohexan-1-one as a colorless oil.

Alkenylzirconocenes, prepared by hydrozirconation of 1-alkynes, are also subject to transmetalation–conjugate addition, in this case using catalytic amounts of the trivial higher-order cyanocuprate Me$_2$Cu(CN)Li$_2$[227]. To effect this process at −78 °C, the presence of Me$_3$ZnLi (from either 3MeLi + ZnCl$_2$•TMEDA, or MeLi + Me$_2$Zn) is needed to regenerate the cyanocuprate initially tied up as the gegenion associated with enolate **60**. Competitive methyl transfer from either the intermediate mixed cyanocuprate **51** or trimethylzincate is minor. Thus, the Me$_2$Zn, via Me$_3$ZnLi, is effectively a methyllithium "shuttle", allowing for its presence without complications due to acid–base chemistry or 1,2 addition. Only 5 mol % of the cuprate is needed for complete reaction. The entire process, following hydrozirconation, occurs at −78 °C which is fortunate because, at higher temperatures, 1,4 addition of Me$_3$ZnLi can become significant[228]. Slow addition of the enone is necessary for best results. Other soft sources of MeLi (e.g., Me$_3$NiLi, Me$_3$MnLi, Me$_3$Ti(O-^{i}Pr)$_2$Li) gave unacceptable yields. MeLi from FMC Lithium Corporation, which comes in THF/cumene, was somewhat preferable chemically to MeLi in Et$_2$O, and is less expensive. Use of commercial Me$_2$Zn in heptane (Texas Alkyls) is recommended, as the ZnCl$_2$•TMEDA[229] route introduces extra LiCl as well as Et$_2$O as co-solvent, both of which do not help with subsequent transmetalations[225]. The intermediate zinc enolates **61** may also be used further in either alkylations or aldol additions, and hence have led to an efficient three-component coupling route to prostaglandins[230].

Cyanocuprate-catalyzed 1,4 Addition of a Vinylzirconocene[230]

1. Cp₂Zr(H)Cl
2. cat Me₂Cu(CN)Li₂, MeLi, Me₃ZnLi
3. cyclopentenone

8:1 mix   (87%) 8:1 mix

To a 10 mL round bottom flame-dried flask equipped with a stir bar and septum and having been cooled to 25 °C under a stream of argon was added $Cp_2Zr(H)Cl$ (0.2985 g, 1.16 mmol) followed by 3.0 mL of THF, and an 8 : 1 mix of α- and β-anomers of 1-(3-butynyl)-4,6-dimethyl-D-glucal (225 μL, 1 mmol). The two-phase reaction mixture was shielded from light by wrapping in aluminum foil and allowed to stir for 30 min, after which the reaction was judged to be complete as evidenced by a homogeneous solution. This mixture was then cooled to −78 °C, and at this temperature MeLi was added in ether (0.49 mL, 1 mmol), dropwise over 1 min. At the same time, to another 10 mL round-bottom flask, equipped with a stir bar and septum and having been cooled under argon, was added CuCN (0.0052 g, 0.06 mmol), followed by 0.5 mL of THF. The mixture was cooled to −78 °C, Me_2Zn in heptane (0.61 mL, 0.52 mmol) was added dropwise to the stirred slurry, followed quickly by dropwise addition of MeLi in ether (0.30 mL, 0.63 mmol). After 5 min, this three-phase slurry was placed in an ice bath for 10 min during which time both remaining liquid phases became clear and homogeneous. This solution was then recooled to −78 °C and the solution containing the methyl vinyl zirconocene, still at −78 °C, was transferred via cannula into the flask containing the zincate/catalytic cuprate mixture. After 5 min of stirring 2-cyclopenten-1-one (43 μL, 0.5 mmol) in 1 mL THF at −78 °C was added portionwise over 35 min. After 1 h of additional reaction time the reaction was quenched at −78 °C with 1 mL of a 9 : 1 saturated aqueous NH_4Cl/concentrated NH_4OH solution ("cuprate quench"). The reaction mixture was then transferred to a separatory funnel containing an additional 14 mL of the 9 : 1 quench and 25 mL of ether, with the help of an additional 25 mL of ether. The combined organic layers were washed three times with 20 mL of brine and then dried over anhydrous $MgSO_4$. After concentration in vacuo, flash chromatography on silica gel using 2 : 1 (petroleum ether/ethyl acetate) afforded 0.1354 g (87%) of the product, an 8 : 1 mix of anomers, as a clear oil TLC (petroleum ether/ethyl acetate, 4 : 1) $R_f = 0.12$.

The corresponding alkylations of alkyl zirconocenes have also been pursued, and a method now exists for this type of C−C bond construction[231]. These organometallics are rendered active toward allylic electrophiles in the presence of 10 mol % CuCN (with or without 2LiCl) or 10 mol % CuBr•SMe₂. Although chlorides, bromides, and phosphates work well, acetates are not reactive, nor are benzylic systems. The S_N2' mode of attack is highly preferred. Primary unactivated halides (e.g., a bromide) can even be tolerated within the alkylzirconocene. Reaction times tend to be lengthy, and increasing the concentration did not make a noticeable difference in rate.

Copper(I)-catalyzed Hydrozirconation–Alkylation of a 1-Alkene[231]

93:7

(98%)

A dry 10 mL round bottom flask is charged with 0.100 g (0.39 mmol) $Cp_2Zr(H)Cl$. To this is added 3 mL of dry THF and 0.058 g (0.39 mmol) of 5-bromopentene. The mixture is stirred at room temperature for 1 h to give a clear yellow solution. At this point, 0.070 g (0.26 mmol) of diethyl cinnamylphosphate are added followed by 0.005 g (0.06 mmol) of CuCN. The suspension is stirred overnight and then poured into 10 mL of pentane/15 mL of 10% aqueous NaOH. The organic layer is removed and washed with 10 mL of 1N HCl and then 10 mL of brine. Further drying over anhydrous magnesium sulfate, filtration through a silica gel pad, and removal of solvent in vacuo yields 0.068 g (98%) of a 93 : 7 mix of S_N2' and S_N2 products.

4.2.2 Alanes

A catalytic amount of THF-soluble CuCN•2LiCl[218] has been found effective for delivering vinyl alanes to conjugated enones[232]. Thus, following a standard Negishi carboalumination[206], inverse addition of this in situ generated species to a THF/Et$_2$O solution of the substrate in the presence of a copper(I) salt at 0 °C leads to good yields of 1,4 adducts. The process is not sensitive to divalent sulfur, can accommodate unprotected alcohols (in their–$OAlMe_2$ form), and is applicable to other types of Michael acceptors (e.g., an Evans chiral auxiliary)[233].

Copper-catalyzed Addition of a Vinyl Alane to an Enone[232]

(82%)

A solution of zirconocene dichloride (0.0840 g, 0.287 mmol) in 2.25 mL of CH$_2$Cl$_2$ was treated with 1.75 mL (3.0 mmol) of a 2.0 M solution of trimethylaluminum in hexane at 0 °C followed by the careful dropwise addition of 3-butyn-1-ol (0.0806 g, 1.15 mmol) dissolved in 1.0 mL of CH_2Cl_2. The resulting solution was allowed to warm to room temperature with stirring and then taken up into a gas-tight 5 mL motor-driven syringe with a 0.50 mL CH_2Cl_2 wash. A 1.0 M solution of CuCN•2LiCl was prepared by dissolving a flame-dried admixture of CuCN (0.0850 g, 0.95 mmol) and LiCl (0.0817 g, 1.93 mmol) in 0.95 mL of THF at 25 °C with stirring. The contents of the syringe were added slowly, dropwise, to 4.0 mL of dry Et_2O with stirring at 0 °C. When the solution became visibly yellow, 0.10 mL of the 1.0 M CuCN•2LiCl solution was added dropwise followed by the dropwise addition of a solution of 2-methylcyclopentenone (0.0959 g, 0.998 mmol) in 1.0 mL of Et_2O at 0 °C. The vinylalane was added over 1 h and the resulting suspension stirred at 0 °C for an additional 1.5 h. The reaction mixture was then poured into a solution of 20 mL of 1.0 M tartaric acid and 20 mL of Et_2O at 0 °C. The biphasic suspension was allowed to warm to room temperature, with vigorous stirring continued until the solids had dissolved. The layers were then separated and the aqueous phase was extracted three times with Et_2O. The combined organic layers were washed with brine, dried over anhydrous Na_2SO_4, and chromatographed (silica gel, CH_2Cl_2/EtOAc 17 : 3) to yield 0.1488 g (82%) of the desired keto alcohol as a 2.8 : 1 mixture of diastereomers by capillary GC ($R_f = 0.16(9 : 1$ CH_2Cl_2/EtOAc)).

4.2.3   Organomanganese Reagents

By conversion of an RLi or RMgX to the corresponding organomanganese(II) reagents RMnCl, 1,4 additions can be effected in the presence of 1–3 mol % CuCl[234]:

$$MnCl_2 + \begin{matrix} RMgX \\ or \\ RLi \end{matrix} \longrightarrow RMnCl \xrightarrow[\text{cat CuCl}]{\text{enone}} \text{1,4 adduct}$$

In some cases, yields from reactions using this combination are better than those from either the corresponding copper-catalyzed Grignard or cuprate Michael-type couplings. Notably, hindered enones (e.g., pulegone 88%, isophorone 95%, mesityl oxide 94%; addition of a Bu group) are not problematic[234], and the reactions are conducted at a very convenient 0 °C over a 2–4 h period.

Due to the hygroscopic nature of $MnCl_2$, it must be handled quickly in air. It should be dried in vacuo before use for 2 h at 180 °C under vacuum (10^{-2} mmHg). A number of commercial vendors supply anhydrous $MnCl_2$ (purity $\geq$ 99%) at rather inexpensive prices. Material supplied by Chemetals, Inc., as low-nickel-grade flakes, works especially well for this chemistry.

Preformation of RMnCl can be accomplished using either an organolithium or Grignard reagent. They must be prepared and utilized under an inert atmosphere (N_2 or Ar).

Preparation of RMnCl from RLi and MnCl$_2$ [235]

$$RLi + MnCl_2 \xrightarrow[\text{2. 30 min to 3 h, 25 °C}]{\text{1. mix at }-35\text{ °C to 0 °C}} RMnCl + LiCl$$

$$\left\{ \text{in Et}_2\text{O} \right\} \left\{ \text{in THF} \right\} \qquad \left\{ \begin{array}{c} \text{brownish solution} \\ \text{or suspension} \end{array} \right\}$$

A solution of 50 mmol of an organolithium compound in ether or a hydrocarbon is added, under stirring, to 52 mmol of anhydrous manganese chloride in 80 mL of THF under N$_2$, between -35 °C and 0 °C. After 30 min of stirring at 25 °C, the organomanganese chloride reagent is quantitatively obtained as a brownish solution.

Importantly, when using reactive organolithiums (e.g., BuLi), the reaction mixture should be kept closer to -35 °C, whereas less basic RLi (e.g., MeLi, ArLi, or vinyl-lithiums) can be used at 0 °C. Although all RMnCl compounds have stability at ambient temperatures, those derived from *sec*-alkyl–M or tert-alkyl–M (M = Li or MgX) may show decomposition at 25 °C should MnCl$_2$ of poor quality be used. Starting with RMgX, RMnCl is formed as follows.

Preparation of RMnCl from RMgX and MnCl$_2$ [235]

Organomanganese chloride reagents are prepared by adding, between -10 °C to 0 °C under stirring, a solution of 50 mmol of an organomagnesium halide in ether or better in THF (RMgX, X = Cl, Br, I) to a suspension of 52 mmol of anhydrous manganese chloride in 80 mL of THF. The reaction is quantitative after stirring for 30 min to 3 h at room temperature. A brownish solution or suspension is thus obtained.

Recently, formation of organomanganese bromides has been achieved using Rieke manganese[236]. The direct oxidative addition of highly reactive manganese to alkyl bromides can be accomplished using the Rieke method of activation. Thus, reduction of anhydrous manganese halides with two equivalents of lithium and a catalytic amount of naphthalene as an electron carrier affords an extremely fine, divided black powder partially soluble in THF at ambient temperatures. Primary, secondary, and tertiary alkyl bromides all readily participate under mild conditions (between 0 °C and 25 °C) in high yields over about 2 h. Couplings with acid chlorides using organomanganese reagents formed via this procedure do not require copper catalysis:

$$MnCl_2 \xrightarrow[\text{THF, 25 °C 3 h}]{\text{Li}^0 (2 \text{ equiv}), \text{ naphthalene (0.31 equiv)}} \text{Rieke Mn}^*$$

Once the organomanganese derivative has been prepared, finely crushed CuCl (or CuCl$_2$) is introduced prior to the substrate. In addition to efficient Michael-type processes[234], the catalytic Cu^I/RMnCl system is highly effective for acylation reactions using acid chlorides[237]. Although it is usually possible to acylate RMnCl in good yields without resorting to copper catalysis, the presence of these salts clearly accelerates the process. Moreover, with secondary or tertiary alkyl-manganese chlorides, or MeMnCl, the yields are usually dramatically improved under the influence of Cu^I.

1,4 Addition of RMnCl Catalyzed by CuCl (or CuCl$_2$): General Procedure[234]

(86–98%)

To 52 mmol of organomanganese reagent (RMnCl) in 80–100 mL of THF (prepared as above) is added, at 0 °C, 1–3 mol % of pulverized copper chloride (CuCl or CuCl$_2$). After 3 min, 50 mmol of enone dissolved in 40 mL of anhydrous THF is added dropwise over 10 min. The reaction mixture is then allowed to warm to 25 °C and, after 1.5–4 h, depending upon the nature of the enone (for example, 1.5 h for cyclohexenone and 4 h for Bu(Pr)C=CHCOBu), hydrolyzed with 60 mL of a 1 N HCl solution. The aqueous layer is decanted and extracted with ether or cyclohexane (3 × 100 mL) and the combined organic layers are washed with a diluted aqueous NH$_4$OH/NH$_4$Cl solution to eliminate copper salts, and then with 50 mL of an aqueous Na$_2$CO$_3$ solution. After drying over MgSO$_4$, the solvents are removed in vacuo, and the ketone is isolated by distillation (86–98% yield).

Copper-catalyzed Acylation of Organomanganese Reagents with Carboxylic Acid Chlorides: General Procedure[237]

(69–95%)

To 52 mmol of organomanganese reagent in 80–100 mL of THF (see above for preparation) is added, at −10 °C, 1–3 mol % of pulverized copper chloride (CuCl or CuCl$_2$). After 3 min, 50 mmol of carboxylic acid chloride dissolved in 40 mL of anhydrous THF is added to the pot dropwise over 10 min. On occasion, the yield can be improved by adding the carboxylic acid chloride at $-30\,°C$ instead of at $-10\,°C$. The reaction mixture is then allowed to warm to $25\,°C$ and, after 20–30 min (no more!), hydrolyzed with 60 mL of a 1 N HCl solution. The aqueous layer is decanted and extracted with ether or cyclohexane (3 × 100 mL) and the combined organic layers are washed with a diluted aqueous NH$_4$OH/NH$_4$Cl solution to eliminate copper salts, and then with 50 mL of an aqueous Na$_2$CO$_3$ solution. After drying over MgSO$_4$, the solvents are removed in vacuo and the ketone is isolated by distillation (yield usually 69–95%).

Copper-catalyzed alkylations of organomanganese reagents have also been added to the repertoire associated with the chemistry of this group 7 metal[238]. Alkyl bromides, functionalized (e.g., ketone and ester moieties) or otherwise, are excellent substrates, although iodides and sulfonates work equally well. THF is the main solvent, but the presence of NMP is essential for good yields. Only 3 mol % CuCl$_2$, together with LiCl (2 equiv), is needed, and the couplings proceed at $25\,°C$. Best results are obtained with finely pulverized, anhydrous LiCl (99% from Aldrich), which is hygroscopic and should be handled quickly. Admixture of the LiCl and MnCl$_2$ in THF leads to a green/brown solution in time, the rate of dissolution being very dependent upon the purity and grain size of these two salts. Using analytically pure materials, total dissolution can occur in less than 10 min, as opposed to about 4 h with unpulverized samples.

Copper-catalyzed Alkylation of an Organomanganese Reagent[239]

$$C_7H_{15}CH_2Br \xrightarrow[\text{3 mol\% CuCl}_2\text{•2LiCl, }-10\,°C,\,2\,h]{^tBuMnCl,\ THF,\ NMP} C_7H_{15}CH_2 \!-\!\!\!<$$

A three necked, 500 mL round-bottomed flask is fitted with a mechanical stirrer, a pressure-equalizing dropping funnel (100 mL) and a Claisen head containing a low-temperature thermometer and a nitrogen inlet. To the flask is added anhydrous tetrahydrofuran (THF, 140 mL), 10.57 g (84 mmol) of anhydrous manganese chloride, and 7.12 g (168 mmol) of lithium chloride. The mixture is stirred at room temperature until the salts are fully dissolved (12–24 h), and then a solution of tert-butylmagnesium chloride (77 mmol) in THF is added dropwise at $-10\,°C$ over a 15 min period. Stirring is continued for 25 min at $0\,°C$ at which point 70 mL of NPM (*N*-methylpyrrolidinone), 2.1 mL of a 0.1 N THF solution of CuCl$_4$Li$_2$ (0.21 mmol), and 13.51 g (70 mmol) of 1-octylbromide in 20 mL of anhydrous THF are successively added dropwise. The reaction mixture is stirred for 1.5 h at ambient temperature and then hydrolyzed by adding 200 mL of 0.5 M aqueous hydrochloric acid over 10 min. After addition of 150 mL of petroleum ether (35–60 °C) and decantation, the aqueous layer is extracted three times with 150 mL portions of petroleum ether (35–60 °C). The combined organic layers are washed with 100 mL of water, then dried over anhydrous magnesium sulfate. After filtration, the solvents are removed in vacuo to give the product in crude form in 90% yield. Distillation under reduced pressure affords 10.7 g of pure 2,2-dimethyldecane (90%, bp 82–84 °C at 8–10 torr).

4.2.4   Alkyltitanium Reagents

A very effective approach to directing alkylations of an organolithium with allylic systems in an S_N2' manner is to convert the RLi to its trialkoxytitanium derivative RTi(O-$^iPr)_3$ or tetraalkoxy ate complex RTi(O-$^iPr)_4$Li[240]. The neutral species is essentially inert toward an allylic chloride at −70 °C in THF, but upon addition of 6 mol % CuBr•SMe$_2$, products of alkylation are formed in good yields and with regioselectivities strongly favoring S_N2' attack. Even better chemical yields are realized with the corresponding ate complex, using CuI•2LiCl as the Cu^I catalyst. The best leaving groups are chlorides and phosphates, both giving comparable results. Comparison reactions with Bu$_2$CuLi and BuLi afforded inferior regioselectivities. Both metals are essential for obtaining high levels of regiocontrol.

Copper-catalyzed S_N2' Alkylation of an Alkyltitanium Reagent[240]

To a solution of TiCl(O-$^iPr)_3$ (1.64 M in hexanes, 0.146 mL) in THF (0.6 mL) was added BuLi (1.64 M in hexanes, 0.4 mmol, 0.24 mL) at −70 °C. A THF solution of CuI•2LiCl (1 M, 0.012 mmol) was added at this temperature to the orange solution, which then turned brown. Cinnamyl chloride (0.25 mmol, 0.035 mL) was added, and the reaction mixture was stirred for 6 h at −70 °C. After addition of hexanes saturated with water, the reaction mixture was passed through a pad of silica gel. GC analysis (HR-1, 0.25 mm I.D. × 30 m, 100 °C) of the filtrate indicated an S_N2'/S_N2 ratio of 99.5 : 0.5. Purification of the crude product by silica gel column chromatography afforded 0.031 g (88%) of the allylation products. The product was identified by comparison (TLC, GC, and ^{1}H NMR) with an authentic sample.

4.2.5 Organozinc Reagents

A very convenient procedure for the copper-catalyzed conjugate addition of a dialkylzinc to an enone has been reported, the key to success being the use of trace amounts of *N*-benzylbenzenesulfonamide, **62**, along with a copper(I) salt[241]. Reactions tend to be rapid in toluene at 0 °C, and can be scaled up. A variety of copper salts (e.g., CuCN, CuOTf, CuBr, etc.) can be used interchangeably. Polar solvents such as DMF, THF, and CH₃CN should not replace toluene, although Et₂O is satisfactory. Thus far, R_2Zn, R = Me, Et, Bu, and Ph have been shown to work well with α,β-unsaturated ketones. However, unsaturated esters are not alkylated. The intermediate zinc enolates can be further alkylated to afford mixtures of products of three-component couplings.

Copper-catalyzed Conjugate Addition of a Dialkylzinc to an Enone[241]

A dry 500 mL Schlenk tube containing a Teflon-coated stirring bar was charged, under an argon stream, with CuCN (0.099 g, 1.1 mmol), the sulfonamide **62** (0.272 g, 1.1 mmol) and toluene (60 mL), and was cooled to 0 °C with an ice bath. Zn(C₂H₅)₂ (26.0 g, 0.21 mol) was added, and the mixture was stirred for 10 min. To this was added the enone (20.2 g, 0.21 mol). The resulting pale red suspension was stirred at 0 °C for 2 h, and then poured into saturated aqueous NH₄Cl solution (200 mL). The organic layer was removed, and the aqueous layer was extracted twice with ether (100 mL). The combined organic layers were washed with water (100 mL) and brine (100 mL), dried over anhydrous Na₂SO₄, and concentrated under reduced pressure to give a *ca.* 25% toluene solution of the product. The mixture was distilled at 191–193 °C and atmospheric pressure to give the product in 90% isolated yield (24.0 g). GC analysis of the reaction mixture showed the yield to be >99%; 1-ethyl-2-cyclohexen-1-ol was not detected by capillary GC.

Functionalized organozinc halides can be added to enones in a 1,4 manner, following conversion to the corresponding mixed alkyl zinc and transmetalation to copper at low temperatures with Me₂Cu(CN)Li₂[242]. In the presence of 2–3 equiv of TMS-Cl, the enolate is trapped so as to recycle copper. Functionality such as chlorides, nitriles, ketones, and simple esters are all usable, as is an acylsilane[243]. More hindered substrates require greater amounts of the silyl chloride (10 equiv). Work-up with Bu₄NF leads to the keto products.

Conjugate Addition of a Functionalized Organozinc Halide Catalyzed by a Cyanocuprate[242,244]

I-(CH₂)₄CO₂Et

1. Zn⁰
2. MeLi, −78 °C
3. 10% Me₂Cu(CN)Li₂
—————————————→
4. 3 TMS-Cl
5. cyclohexenone
6. TBAF

(structure of product: cyclohexanone bearing a -(CH₂)₅CO₂Et side chain)

(90%)

To a 50 mL round bottom flask equipped with a stir bar and septum and having been cooled under a stream of argon was added zinc powder (3.0 g, 46 mmol), followed by 16 mL of THF and 1,2-dibromoethane (155 μL, 1.80 mmol). This stirred slurry was heated to *ca.* 65 °C for 1 min using a heat gun. After this time, the flask was placed in a tap water bath (*ca.* 23 °C). TMS-Cl (155 μL, 1.22 mmol) was added, and the slurry was stirred for 15 min. Neat ethyl 5-iodovalerate (6.7 mL, 40.0 mmol) was added dropwise over 10 min. The flask was placed in a 35 °C oil bath for overnight stirring. The reaction was initially exothermic, and THF was seen slowly refluxing around the neck of the round bottom flask. Dry ice was used to cool the neck of the flask so that the septum was not extracted by the THF. Twelve hours later, when the reaction was judged complete by TLC, the oil bath was removed. As usual, the presence of a white dust, that never quite settles out, was noted. The reaction was allowed to cool to 25 °C, at which time 20 mL of THF were added to the stirred slurry. Stirring was then halted, and the solids were allowed to settle, such that the THF solution of the alkylzinc iodide could be transferred via cannula to a clean, dry 250 mL round-bottom flask equipped with a stir bar and septum. Two additional washes of THF (29 mL each) were carried out to ensure complete transfer and proper dilution of alkylzinc iodide. This solution was cooled to −78 °C, and MeLi in ether (23 mL, 36 mmol, 1.56 M) was added dropwise over 15 min. During the same time that the alkylmethylzinc was being formed, to a 25 mL round bottom flask, equipped with a stir bar and septum and having been cooled under argon, was added CuCN (0.1789 g, 2.0 mmol), followed by 10 mL of THF. The mixture was cooled to −78 °C, and MeLi in ether (2.6 mL, 4.1 mmol, 1.56 M) was added slowly. Once the addition of MeLi was complete, the slurry was gently warmed until the reaction mixture became homogeneous, and then it was recooled to −78 °C. With both solutions at −78 °C, the higher-order dimethyl cyanocuprate was transferred via cannula to the alkylmethylzinc reagent. After 5 min at −78 °C, neat TMS-Cl (7.77 mL, 61.2 mmol) was added dropwise over 10 min, followed by 10 more minutes of stirring. Finally, neat 2-cyclohexenone (2.02 mL, 20.9 mmol) was added dropwise over 15 min via syringe. After 4.5 h, the reaction was complete as determined by TLC and was quenched by pouring the reaction mixture into 200 mL of pH 7 buffer and 200 mL of Et₂O in a 1 L separatory funnel. An additional 25 mL of ether were used to complete the transfer. The aqueous layer was further extracted with an additional 50 mL of ether. The combined organic layers were shaken for 5 min with 50 mL of 1 M TBAF in THF (Aldrich), followed by three washes with 100 mL of brine. The organics were then dried over anhydrous $MgSO_4$. After concentration in vacuo, flash chromatography on silica gel using 10 : 1 petroleum ether/ethyl acetate afforded 4.09 g (90%) of the product as a clear, pale-yellow oil (TLC (petroleum ether/ethyl ether 10 : 1) $R_f = 0.11$).

Rather than using a mixed dialkylzinc (from FG—ZnCl + MeLi), conversion of the zinc halide precursor to a mixed zincate with two equivalents of MeLi now leads to a species susceptible to ligand exchange with catalytic MeCu(CN)Li at $-78\,°C$ in THF. This mixture likewise effects conjugate delivery of the functionalized ligand to an enone—but without recourse to excess TMS-Cl[245]. Introduction of the enone leads to a bright yellow coloration, assumed to be a cuprate-π-complex, which gradually fades as the reaction progresses. Reactions that were not successful failed to produce this initial color. As the zinc enolate is not converted to a silyl ether, it is available for secondary couplings with aldehydes (i.e., a three-component coupling; Scheme 5).

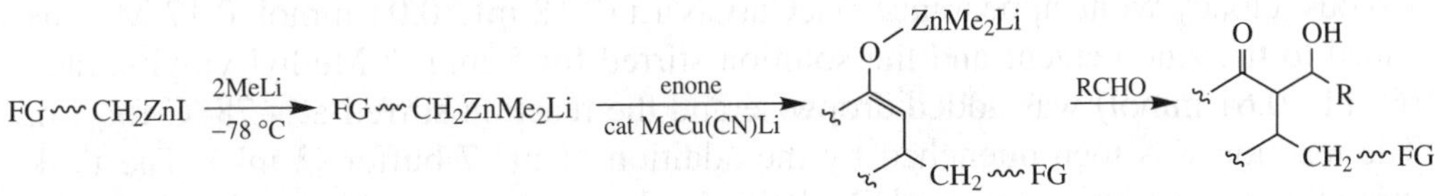

Scheme 5.

This same combination of reagents (i.e., mixed zincate + catalytic MeCu(CN)Li) is especially effective at delivering the functionalized ligand from zinc via copper to an allylic epoxide in an S_N2' sense[246]. Chlorides and nitriles as the functional group (FG) are well-behaved, but a triisopropylsilyl (TIPS) ester is again essential for this functional group, due to (presumed) competing ester carbonyl–zinc Lewis acid–Lewis base interactions which negate transmetalation[207]. Note that a highly valuable allylic alcohol is produced, an ideal precursor to further manipulations (e.g., Sharpless epoxidations, enal formation, etc.):

Alkylation of a Functionalized Mixed Zincate with an Allylic Epoxide Using Catalytic Amounts of MeCu(CN)Li[246]

TIPSO—C(=O)—(CH₂)₅ZnI
$\xrightarrow[\text{3. isoprene epoxide}]{\begin{array}{l}\text{1. 2 MeLi, THF, }-78\,°C\\\text{2. cat MeCu(CN)Li}\end{array}}$
TIPSO—C(=O)—(chain)—OH   (87%)

−78 °C, 4 h

1,2-Dibromoethane (7.5 µL, 0.085 mmol) was added to a stirred suspension of zinc dust (0.158 g, 2.42 mmol) in THF (1.0 mL). The solution was heated to reflux for 1 min, then allowed to cool to 25 °C. Trimethylsilylchloride (7.5 µL, 0.06 mmol) was added and the reaction was stirred for 30 min. Triisopropylsilyl 6-iodohexanoate (0.40 mL, 1.21 mmol) was added dropwise and the reaction was stirred overnight (12 h) at 40 °C shielded from light. The solution was transferred away from the residual zinc dust via cannula. Additional THF (0.5 mL) was added to the zinc dust, the solution was vigorously stirred, the zinc was allowed to settle and the THF was added via cannula to the reaction mixture. This process was repeated one additional time. The resulting THF solution was cooled to −78 °C and methyllithium (1.42 mL, 2.18 mmol, 1.53 M) was added. Within 10 min, the mixture took on a viscous, cloudy white appearance. MeCu(CN)Li (0.18 mL, 0.03 mmol, 0.17 M) was added to the zinc reagent and the solution stirred for 5 min. 2-Methyl vinyloxirane (60 µL, 0.61 mmol) was added dropwise and the reaction stirred at −78 °C for 4 h. The reaction was then quenched by the addition of pH 7 buffer (3 mL). The flask was allowed to warm, saturated Rochelles' salt (2 mL) was added to break up the emulsion and the solution extracted with ether (3 × 4 mL). The combined extracts were dried over anhydrous sodium sulfate, filtered and concentrated in vacuo. The product, (*E*)-triisopropylsilyl-10-hydroxy-9-methyldec-8-enoate, was isolated by column chromatography (9% EtOAc, 2% Et₃N in hexanes) to the extent of 0.1901 g, 0.53 mmol, 87% (R_f = 0.12 (9% EtOAc in hexanes)).

4.2.6   Alkynyl Silanes

Transmetalation between a trimethylsilylated alkyne and catalytic amounts of CuCl has recently led to a simple method for generating alkynylcopper reagents capable of undergoing acylations with acid chlorides[247]. The use of CuCl, rather than CuI, CuBr, or CuCN, is highly recommended for best yields. The medium plays a crucial role in this ligand exchange process, the solvent of choice being 1,3-dimethylimidazolidinone (DMI), although DMF is a not-too-distant alternative, yield-wise. The exchange is sensitive to steric demands of the group attached to the other end of the alkyne, suggesting an initial π-complexation with CuCl. Acid-sensitive residues within the alkyne are tolerated, such as acetate and tert-butyldimethylsilyloxy moieties. Yields in the 80–90% range are the norm.

Transmetalation-Acylation of an Alkynylsilane with CuCl[247]

$$Ph\text{---}\!\!\equiv\!\!\text{---}SiMe_3 + R'COCl \xrightarrow[\text{DMI, 80 °C, 5 h}]{\text{cat CuCl}} Ph\text{---}\!\!\equiv\!\!\text{---}C\overset{O}{\underset{R'}{\diagdown}}$$

To a mixture of 1-phenyl-2-(trimethylsilyl)ethyne (0.174 g, 1.0 mmol) and CuCl (0.020 g, 0.2 mmol) in DMI (0.5 mL) was added benzoyl chloride (0.155 g, 1.1 mmol) at room temperature. After stirring for 5 h at 80 °C, the reaction mixture was diluted with chloroform. Filtration through a short plug of Florisil followed by column chromatographic purification on silica gel (ethyl acetate/hexane = 1 : 20) gave 1,3-diphenylpropyn-1-one (0.177 g) in 85% yield. No precipitation of copper(I) acetylides was observed during the reaction.

---

5   Miscellaneous Copper Reagents

Although the vast majority of organocopper complexes in use today are of the forms RCu, RR'CuM, M = Li, MgX, and $R_2Cu(CN)Li_2$, several alternative reagents have also been developed. Changes in the gegenion(s) associated with a cuprate complex can exert considerable influence over several reaction parameters, such as (1) the regio- and stereo-chemistry of additions, (2) reagent reactivity, (3) stability of the reagent, (4) compatibility with functionality within the ligands on copper, and (5) tendency to form lower-order rather than higher-order reagents. In general, it is now well-established that replacement of lithium by another metal in a lower- or higher-order cuprate can be expected to decrease reagent reactivity[3,248]. Although magnesio cuprates are therefore usually less robust than their lithio analogs, organozinc-based reagents are less prone still to react. Moving to nonassociating counterions, such as R_4N^+, depresses reactivity still further (see below). These guidelines lend further support to the notion that many cuprate couplings are in fact electrophilically driven, with the hard Li^+ intimately involved in placing R_2CuLi or $R_2Cu(CN)Li_2$ at the top of the reactivity scale. Insight as to how to go about fine tuning the behavior of copper reagents by the judicious choice of the counterion(s), for instance, is now starting to accrue, to which the discussion and representative procedures that follow attest.

5.1 $[RCu(CN)_2]Li[N(C_4H_9)_4]$

Independent of the exact structure of higher-order cyanocuprates[76], it is likely that '$R_2Cu(CN)Li_2$' or, if written as '$R_2CuLi\cdot LiCN$' or '$R_2Cu^-Li_2CN^+$', contain two mono-valent metal ions associated with the complex. The two gegenions need not be identical, the formula thus being generalized as $R_2Cu(CN)MM'$, or $R_2CuM \cdot M'CN/R_2CuM' \cdot MCN$. As the ligands on copper also need not be identical, the stoichiometry generalizes further to $RR'Cu(CN)MM'$ or the various equivalent Gilman combinations. Moreover, the nature of the gegenions can be chosen to reflect a cationic species which has no obligation to be of the alkali metal series. Hence, cuprates composed of less commonly employed components have started to appear which, not surprisingly, have modified properties. One case where multiple variations have been found advantageous for selected synthetic transformations concerns the higher-order dicyano species derived from $CuCN$, Bu_4NCN, and an organolithium[249]:

$$CuCN + Bu_4CN \xrightarrow[25\,°C]{MeOH} Cu(CN)_2NBu_4 \xrightarrow[-40\,°C \text{ to } -25\,°C,\ 2\ h]{RLi,\ THF} \boxed{RCu(CN)_2LiNBu_4}$$

The precursor, $Cu(CN)_2NBu_4$, a colorless solid, is prepared in MeOH at ambient temperatures under an inert atmosphere.

Preparation of $Bu_4NCu(CN)_2$ [249]

To a slurry of 1.79 g (20 mmol) of cuprous cyanide in 100 mL of anhydrous methanol at 25 °C under N_2 was added a solution of 5.36 g (20 mmol) of tetrabutylammonium cyanide in 50 mL of methanol. After stirring for 10 min at 25 °C, the resulting homogeneous solution was rotary evaporated to dryness, and the resulting paste was dried by azeotropic evaporation with toluene (3 × 70 mL) at ambient temperature to furnish a white crystalline solid. The complex was dried under high vacuum at 25 °C for 24 h.

Treatment of a slurry of $Cu(CN)_2NBu_4$ in THF with one equivalent of an R_tLi ($R_t =$ alkyl, phenyl, vinyl, alkenyl) leads to nearly homogeneous solutions of $R_tCu(CN)_2LiNBu_4$, which participate in the selective delivery of the transferable ligand (R_t) to enones, as well as in displacements of primary halides[249]. Due to both the second nitrile group (presumably on copper) and replacement of a lithium with tetrabutylammonium as one of the two counterions, reactivity is diminished considerably. This is manifested in both reaction types, where conjugate additions to even unhindered α,β-unsaturated ketones are performed at $-50\,°C$ over 1 h, and primary iodides take 4 h at $-25\,°C$. By contrast, the corresponding cuprates $R_2Cu(CN)Li_2$ would react at $-78\,°C$ with either functional group within minutes. With lower reactivity, however, comes added selectivity, and indeed the incentive for development of these higher-order dicyanocuprates arose due to dissatisfaction with yields obtained using other copper reagents for the substitution reaction detailed below. A second procedure highlighting use of these cuprates as Michael donors is also given.

Displacement of a Primary Halide Using RCu(CN)$_2$LiNBu$_4$[249]

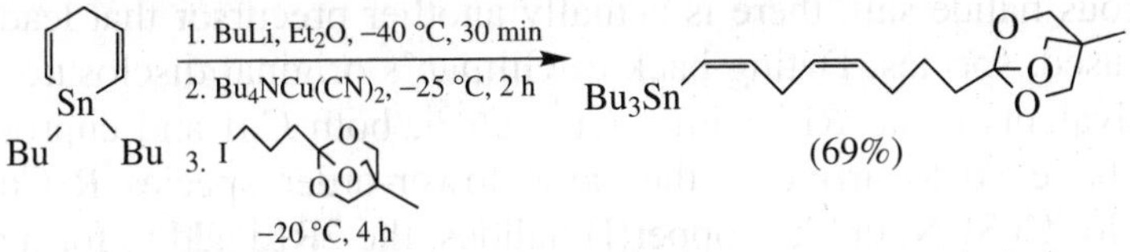

To a solution of 0.299 g (1.00 mmol) of the stannane in 2 mL of anhydrous ether at −78 °C was added 0.404 mL of 2.50 M (1.01 mmol) BuLi. The yellow solution was stirred for 1 h at −78 °C, then for 30 min at −40 °C. The vinyllithium solution was added to a suspension of 0.354 g (0.99 mmol) of tetrabutylammonium dicyanocopper(I) in 8 mL of anhydrous THF. The orange suspension was stirred for 15 min at −78 °C, then warmed to −25 °C and stirred an additional 2 h. To the above cuprate mixture at −25 °C was added a solution of 0.328 g (1.10 mmol) of the iodoorthoester in 5 mL of anhydrous THF. The mixture was stirred for 4 h at −25 °C, then quenched by the addition of 2 mL of 2 M aqueous ammonium chloride solution (pH 9.0). The crude mixture was extracted with 3 × 50 mL of 1 : 3 pentane/ether, washed with brine, and dried over anhydrous potassium carbonate. Evaporation of solvent in vacuo followed by chromatography on triethylamine-deactivated silica gel in 1 : 5 ether/hexane gave 0.363 g (69%) of the product (R_f (1 : 5ether/hexane) = 0.62).

Typical 1,4 Addition of R$_t$Cu(CN)$_2$LiNBu$_4$ to an Enone[249]

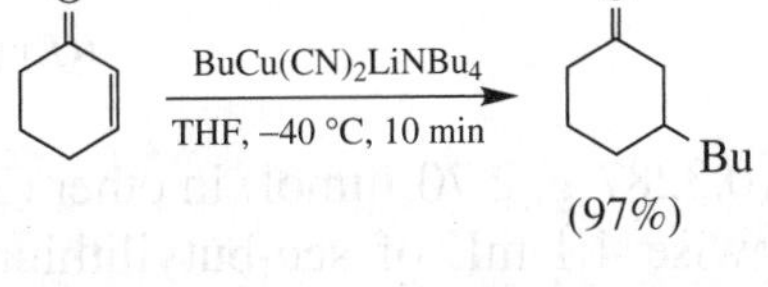

To a suspension of 0.179 g (0.5 mmol) of Bu$_4$NCu(CN)$_2$ in 4 mL of anhydrous THF at −78 °C under N$_2$ was added 0.202 mL of 2.5 M (0.505 mmol) BuLi. After stirring for 1 h at −40 °C, 0.053 g (0.55 mmol) of 2-cyclohexenone (neat) was added and the reaction mixture stirred for an additional 10 min at −40 °C. The reaction was quenched by the addition of 2 mL of saturated aqueous NH$_4$Cl, and the crude product extracted with ether. The organic phase was washed with brine and dried over anhydrous MgSO$_4$. Evaporation of solvent followed by chromatography on silica gel in 1 : 2 ether/petroleum ether afforded 0.0745 g (97%) of the product.

5.2   R$_2$Cu(SCN)Li$_2$

Although the term 'organocuprate' has come to imply a copper reagent arising from CuCN or a cuprous halide salt, there is actually another precursor that leads to a unique, albeit seldomly used, species. Dating back to Gilman's original disclosure on the admixture of two equivalents of an RLi with a Cu^I salt[5], both CuI and cuprous thiocyanate (CuSCN) were believed to arrive at the same lower-order species R$_2$CuLi. It is now appreciated that for CuSCN, unlike copper(I) halides, the 2RLi add to form a higher-order cuprate R$_2$Cu(SCN)Li$_2^{[6]}$, i.e., the thiocyano analog of R$_2$Cu(CN)Li$_2$:

$$2RLi \ + \ CuSCN \ \xrightarrow[\text{solvent}]{\text{ethereal}} \ \boxed{R_2Cu(SCN)Li_2}$$

63

However, they have been reported to afford results in certain contexts which are quite different from those obtained with other cuprates. Thus, for the displacement reaction of a configurationally defined vinylogous thioester (below), cuprates **63** were a welcomed alternative to R$_2$CuLi insofar as the stereoselectivities of these couplings (i.e., (E)/(Z) ratios) are concerned[250]. Cuprous thiocyanate, the precursor to **63**, is a commercially available, inexpensive, air and light-stable off-white solid. Given these attractive features, it is somewhat surprising that it has not been utilized to a greater extent.

Displacement on a Vinylogous Thioester Using R$_2$Cu(SCN)Li$_2$ [250a]

(86%)

((E):(Z) 11:89)

To a mixture of CuSCN (0.3287 g, 2.70 mmol) in ether (20 mL) cooled to $-10\,°C$ to $-20\,°C$ was added dropwise 4.1 mL of sec-butyllithium (1.33 M, 5.45 mmol). After 30 min, the light tan solution was chilled to $-78\,°C$, and the educt (0.2108 g, 1.35 mmol) was added via syringe. After 1.5 h, the temperature had risen to $-58\,°C$, and TLC indicated the reaction was complete. The reaction was quenched with saturated aqueous NH$_4$Cl and standard workup gave 0.2357 g of a yellow oil. Purification by TLC using a preparative plate (1000 μm SiO$_2$, 1% ethyl acetate/petroleum ether, v/v) afforded the pure (Z) isomer (0.1492 g, 66%), $R_f = 0.28$, and a minor fraction (which was a mixture of the (E) and (Z) isomers (0.0421 g), for an overall yield of 85%). VPC analysis of the crude mixture indicated an (E)/(Z) ratio of 11 : 89.

5.3   R₂Cu(CN)LiMgX

Another variation among higher-order cyanocuprates involves the mixing of gege-nions to reflect, for example, the combination of an organolithium and Grignard reagent, together with CuCN, to presumably form R₂Cu(CN)LiMgX[251]:

$$R_rLi + R_tMgX + CuCN \xrightarrow[\text{solvent}]{\text{ethereal}} \boxed{R_tR_rCu(CN)LiMgX}$$

64

The driving force behind this variation lies in the availability and ease of preparation of most Grignard reagents. When an equivalent is added to preformed (2-thienyl)Cu(CN)Li, either freshly made up or purchased in a bottle, a reagent results which is capable of effecting substitution and 1,4 additions. The solubility at low temperatures, however, of lithio magnesiohalide cuprates is not nearly that of dilithio reagents, with those derived from RMgCl (rather than RMgBr) seemingly of greater tendency to dissolve in THF/Et₂O at −78 °C. The actual outcome of the reaction, however, appears not to be dependent upon the halide present in the Grignard precursor (and, hence, in **64**). Although the reactivity of cuprates **64** is lowered due to the switch from 2Li⁺ to Li⁺MgX⁺, even congested enones are acceptable educts. On occasion, however, BF₃•Et₂O (1 equiv) must be present in solution (which together, should not be warmed above about −50 °C). With unhindered systems, the Lewis acid is not needed. THF is the best medium solubility-wise, although the reagents frequently appear as greyish slurries. In Et₂O as the major or sole solvent, a far less soluble, brown, sticky material is commonly noted at the bottom of the flask.

Reaction of a Monosubstituted Epoxide with $R_t(2\text{-}Th)Cu(CN)LiMgX$[251]

(81%)

2-Thienyllithium (0.6 mmol) was prepared from thiophene (50 μL, 0.62 mmol) and BuLi (0.25 mL, 2.40 M) in THF (0.40 mL) at −30 °C for 30 min and then added to a precooled (−78 °C) slurry of CuCN (0.055 g, 0.61 mmol) in THF (1.5 mL) followed by BuMgCl (0.24 mL, 0.60 mmol). Upon warming to 25 °C, the mixture dissolved to a green/amber solution with traces of a fluffy white precipitate. Recooled to −78 °C, the mixture thickened to a tan slurry and to it was added the neat epoxide (56 μL, 0.5 mmol). The reaction was warmed to 0 °C and stirring continued for 4 h followed by quenching and work-up in the usual way. Chromatography on SiO₂ with 20% Et₂O/Skelly Solve afforded 0.0635 g (81%) of a clear oil ($R_f = 0.25$, identical spectroscopically with an authentic sample).

Conjugate Addition of R_t(2-Th)Cu(CN)LiMgX to an α,β-Unsaturated Ketone[251]

(85%)

CuCN (0.102 g, 1.14 mmol) was placed in an oven-dried two-neck round bottom flask equipped with a magnetic stir bar. The salt was gently flame dried (30 s) under vacuum and then purged with Ar. Dry THF (1.0 mL) was added and the stirred slurry was cooled to −78 °C. 2-Thienyllithium was prepared in a second two-neck round bottom flask from thiophene (91 μL, 1.14 mmol) in dry THF (1.0 mL) at −30 °C to which was added BuLi (0.47 mL, 2.44 M in hexanes, 1.14 mmol) and the clear colorless solution was stirred at 0 °C for 30 min. The preformed 2-thienyllithium was added to the CuCN slurry at −78 °C and warmed to 0 °C over 30 min until the tan/brown slurry became clear. The mixture was cooled to −78 °C and the Grignard reagent derived 2-(2-bromoethyl)-1,3-dioxane (80 μL, 1.42 M in THF, 1.14 mmol) was added dropwise and the resulting mixture was warmed to 0 °C for 2 min and cooled back to −78 °C. Cyclohexenone (freshly distilled, 100 μL, 1.03 mmol) was added and the reaction was stirred at −78 °C for 2.25 h and quenched by the addition of 5 mL of a 90% NH$_4$Cl (saturated)/10% NH$_4$OH (concentrated) solution. After stirring at 25 °C for 30 min, the solution was worked up in the usual way. Column chromatography on SiO$_2$ with 1 : 1 Et$_2$O/Skelly Solve afforded 0.186 g (85%) of product as a clear liquid.

6   Choosing an Organocopper Reagent

With well over 100 representative procedures furnished herein for making carbon–carbon, carbon–nitrogen, carbon–hydrogen, carbon–tin, and carbon–silicon bonds via a copper reagent, where does one start the process of deciding upon a specific reagent for a particular transformation? As most of the variations among the reagent pool from which to choose are for purposes of C–C bond construction, attention is directed toward this major subgroup rather than those composed of heteroatom or hydrido ligands.

The two main types of couplings that rely heavily on organocopper reagents are conjugate addition and substitution reactions. It is also true that carbocupration is an extremely valuable, regio- and stereoselective process, but the number of different copper species which effect this reaction is relatively limited. Examples provided herein cite the use of lower-order lithio (R$_2$CuLi) and magnesio (R$_2$CuMgX) cuprates, as well as

Grignard-based organocopper species RCu•MgX$_2$, which together account for essentially all of this chemistry. Relatively few cases involving, for example, mixed Gilman reagents have been used for carbocupration purposes, and additives such as Lewis acids do not figure in these additions to alkynes. Furthermore, higher-order cyanocuprates bearing all-carbon ligands are usually excluded from contention in that they are apparently too basic to be utilized in this context unless handled under special circumstances (e.g., intramolecular processes)[252], for they induce an acid–base reaction by abstraction of a terminal alkyne proton.

In anticipating conditions for a particular coupling, the question of competing cuprate decomposition may arise. It was pointed out early on that[2d], in particular, ligands containing β-hydrogens appear to be especially good candidates for syn elimination of the elements of CuH[17]. A more recent study comparing a homocuprate (Bu$_2$CuLi) with several mixed reagents (RR'CuLi) at $-50\,°$C, $0\,°$C, and $25\,°$C suggests that Gilman reagents are actually far more thermally stable than previously realized[75].

6.1   Conjugate Addition Reactions

Focusing first on the need to carry out a 1,4 addition, there are several initial, mainly visual (i.e., nonexperimental) observations that can quickly narrow the field of logical choices:

1. Other than the presence of an α,β-unsaturated carbonyl unit, what other (electrophilic or Lewis basic) functionality is located in the molecule? This question arises because, although catalytic copper-driven Grignard additions preferentially deliver in a 1,4 manner, the excess RMgX in the pot may not be tolerated elsewhere in the molecule. However, the cornerstone of stoichiometric cuprate use is their reluctance to add, for instance, in a 1,2 fashion to isolated ketones, especially in the presence of functionalities more compatible with soft copper complexes. Further insight along these lines can be gleaned from an extensive survey on the uses of organocopper reagents in natural products-related total syntheses[2a]. What seems to come from this inspection, perhaps not unexpectedly, is that far more researchers opt for stoichiometric lithio or magnesio cuprates than for the catalytic Cu^I/RMgX protocol. The majority of uses of the catalytic system are with simpler substrates, where there is little room for competing modes of addition. Thus, the build-up of intermediates early in a synthetic scheme might be one of the goals. Finer control within more advanced intermediates tends to be left to the domain of stoichiometric reagents, mostly lithium based. Until recently, these were the choices. However, since the advent of transmetalation processes which effect a net ligand exchange between a metal that does not do 1,4 addition and copper, a new era of opportunity has begun. Add the virtues of carrying along within the cuprate electrophilic centers and the value of cuprate chemistry reaches new heights. And most of the advances can still be augmented by the benefits associated with additives (see below).

2. How costly is the ligand to be introduced into the substrate? When a carbon fragment of interest (R_t) is easily obtained from, say, a commercial supplier of organolithium (R_tLi) or Grignard reagents (R_tMgX), or is a single step removed from readily available materials (e.g., R_tMgBr, from R_tBr, EtLi from EtBr, or (*E*)- and (*Z*)-propenyllithium from the respective halides[253], etc.), then it is wise to prepare the corresponding homocuprates: R_2CuLi, R_2CuMgX, or $R_2Cu(CN)Li_2$. Half of the R_tM (M = Li, MgX) will be lost upon work-up but the benefits associated with greater reactivity (compared to that of a mixed reagent R_tR_rCuM) and the elimination of any selectivity of transfer issue far outweigh the cost of one equivalent of R_tM. However, should the ligand R_tM require a number of steps to prepare or involve expensive reagents (e.g., a costly protecting group) such that sacrificing one full equivalent from a cuprate coupling is less desirable or even intolerable, then it is advantageous to select a mixed cuprate (R_tR_rCuM). As, by definition, these reagents are usually[104] less reactive than homocuprates[254], and as the nature of the gegenion M also affects reactivity, a lithio cuprate (rather than a Grignard-derived reagent) usually offers the best chance for success. As for R_rLi, there are many from which to choose (including, as examples, lithiated tert-butylacetylene, 1-pentyne, thiophene, tert-butoxide, thiophenoxide, trimethylsilylacetylene, and 3-methoxy-3-methyl-1-butyne). At the outset, however, there is no reason to select one which is either costly or requires special handling, preparation or manipulation. Based on these stringent requirements, thiophene and thiophenol should be seriously considered, as both easily metalate with BuLi in THF and can then be utilized directly for cuprate generation. Lithiated 3-methoxy-3-methyl-1-butyne is also an excellent choice for R_rLi, its precursor 2-methyl-3-butyn-2-ol being very inexpensive and O-methylated in high yield. The resulting 3-methoxy-3-methyl-1-butyne is a clear, colorless, room-temperature-stable liquid which in lithiated form is extremely soluble in THF. Thus, if sulfur compounds are (for whatever reason) to be avoided, this alkyne is a good alternative as long as THF, at least in part, is in the medium. Of course, it is possible that the newly introduced Me_3SiCH_2 (TMSM; see Section 2.4.5[104]) ligand may provide all of the benefits of reactivity and selectivity of ligand transfer such that use of a homocuprate is not necessary. As of this writing, however, the scope and generality of this ligand in these contexts have not been established, although it has already been applied in a related way to organozinc chemistry[255].

3. Will additives such as $BF_3\cdot Et_2O$ or Me_3SiCl help? Perhaps the best approach to answering this question is to assess the environment surrounding the chromophore. That is, although these additives do indeed have the potential to turn a completely unsuccessful 1,4 addition reaction into a remarkably efficient and rapid process (e.g., an α,β-unsaturated amide)[48c], in most cases they may not be necessary and should only be considered after the results of an initial small-scale reaction are known. Their impact is most noticeable in sterically congested situations, or where the substrate is normally not a satisfactory Michael acceptor toward cuprates due to its highly cathodic (i.e., ≤ -2.4 V in DMF) reduction potential (e.g., as with unsaturated nitriles)[256]. For these situations, as well as even more general uses, it pays to consider $R_tCu\cdot BF_3$, as the

mix of CuI plus R$_t$Li (1 equiv) followed by introduction of BF$_3$•Et$_2$O in THF has been shown to undergo 1,4 additions to α,β-unsaturated carboxylic acids. There are also small but possibly significant prices to be paid when employing these additives—with BF$_3$•Et$_2$O, Lewis acid-sensitive functionalities (e.g., ketals) may prohibit their use, and the selectivity of transfer in mixed cuprates is usually affected in a detrimental way (i.e., the dummy ligand is more competitively transferred). With Me$_3$SiCl, the initial product is the silyl enol ether, which requires a separate hydrolysis step (i.e., treatment with mild acid or fluoride ion) to arrive at the carbonyl stage. Thus, the best approach is the simplest approach—once a reagent has been decided upon, try the coupling without recourse to any other materials in the pot (including HMPA, phosphines, phosphites, BF$_3$•Et$_2$O, or Me$_3$SiX). Only when it is clear that special circumstances exist (such as with highly hindered substrates, thermally unstable or less reactive cuprates, problems due to limited solubility with the substrate and/or cuprate, etc.) should one resort to these modifications.

6.2 Substitution Reactions

Displacements by organocopper reagents, as with conjugate additions, cover a wide range of available reagents. Coming up with generalities for selecting one, however, can be far more challenging in that there is a greater diversity of electrophiles associated with this aspect of organocopper chemistry. In other words, although Michael additions define (in the vast majority of cases) the substrate as containing an α,β-unsaturated carbonyl unit, suitable partners for substitution reactions include: (a) primary, secondary, and vinylic halides, as well as numerous sulfonate derivatives (e.g., mesylate, tosylate, triflate); (b) epoxides of the (most often) mono- and disubstituted type; (c) allylic substrates such as halides, epoxides, and acetates; (d) propargylic leaving groups; (e) miscellaneous participants, including lactones, sulfates, heteroaromatics, heteroatoms, and other organometallic species, to name a few.

Of greatest interest, based solely on relative numbers, are reactions of primary leaving groups and monosubstituted epoxides. These two classes need be distinguished with respect to reagent choice. Additives that do not play a role in displacements, such as with halides, can significantly alter the chemistry of epoxides via Lewis acid coordination, for instance. Thus, although the former are essentially push reactions (i.e., nucleophilically driven), displacements involving oxiranes (with Li$^+$, MgX$^+$, or either in the presence of BF$_3$) involve a push–pull type of process.

For carrying out a displacement on a primary halide in THF or sulfonate in Et$_2$O, it is important to match the reactivity of the leaving group with the reactivity of the reagent. For simplicity, the following scale (Scheme 6) can be used as a guide, where each halide correlates with a particular lithio cuprate. Thus, if a primary iodide is chosen, a Gilman cuprate should lead to an excellent yield of the desired product at low temperatures (usually *ca.* $-78\,^\circ$C). If, however, an iodide is of limited stability, is difficult to prepare, or is light sensitive, then a bromide would suffice so long as either (a) the lower-order

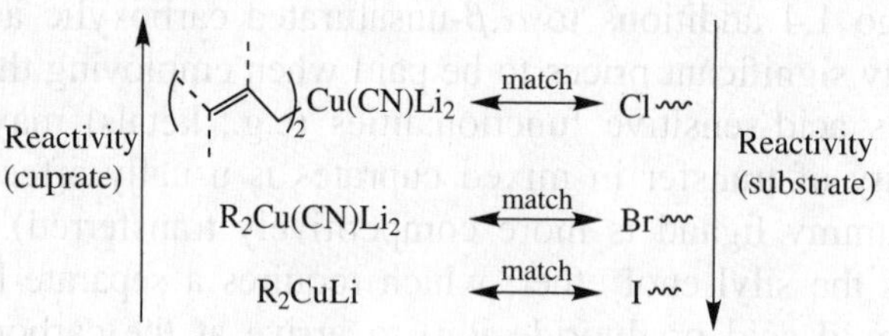

Reactivity patterns for cuprate substitution reactions at sp³ centers

Scheme 6.

cuprate is used at higher temperatures for longer times, or (b) a more reactive higher-order cyanocuprate is employed, which should do the same coupling at temperatures ≤ −50 °C. At the very bottom of the reactivity scale lie the primary chlorides, which are generally unacceptable educts toward lower-order cuprates, and slow-reacting with most higher-order cyanocuprates. By switching to allylic cyanocuprates, however, they become quite susceptible and can be displaced in minutes at −78 °C.

Although stoichiometric lithio cuprate-based substitutions are textbook reactions, strong competition comes in the form of copper-catalyzed Grignard processes. Many of the same arguments in favor of this system apply as were offered for conjugate additions (see above). Moreover, perhaps the best source of catalytic Cu^I, rarely used for 1,4 additions, is Kochi's catalyst, Li_2CuCl_4. Hence, if the starting material is not especially complex and the Grignard reagent is readily available, the catalytic Li_2CuCl_4/RMgX combination is hard to beat in terms of simplicity, and yields can be every bit as good (or better) than those realized from the corresponding lithium cuprate.

Ring openings of epoxides can also be accomplished in good yields via the catalytic Cu^I/RMgX or lithio cuprate route, and to a lesser extent, using R_2CuMgX. In general, as Et_2O is the solvent of choice over THF (in which the gegenion(s) present is (are) of greater Lewis acidity), lithio cuprates tend to have better solubility characteristics. With monosubstituted oxiranes, assuming no likely interference from other functionalities within the substrate, essentially any of the above reagents is a valid first choice. Distinctions will merely lay in the temperatures of the couplings which are tied not only to reagent reactivity but also reagent solubility (i.e., lithio > magnesio cuprates in both cases). It is also well worth noting, however, that potential Schlenk equilibria associated with Grignard-based reagents may lead to halohydrin byproducts (Scheme 7). Disubstituted epoxides (1,1 or 1,2) are more likely to be opened efficiently by lithio cuprates. When conservation of ligands (R_t) is called for, and hence reagent reactivity is decreased, use of ≥1 equiv BF_3•Et_2O may prove highly beneficial. For greatest reagent reactivity without recourse to additives, higher-order cyanocuprates $R_2Cu(CN)Li_2$ in Et_2O appear to offer the best opportunities, and on occasion can be successful with trisubstituted epoxides. For activated, allylic cases, the reagent of choice is clearly the lower-order cyanocuprate RCu(CN)Li. Of course, this particular substitution reaction can now be done catalytically in cuprate using a mixed zincate reagent (see above).

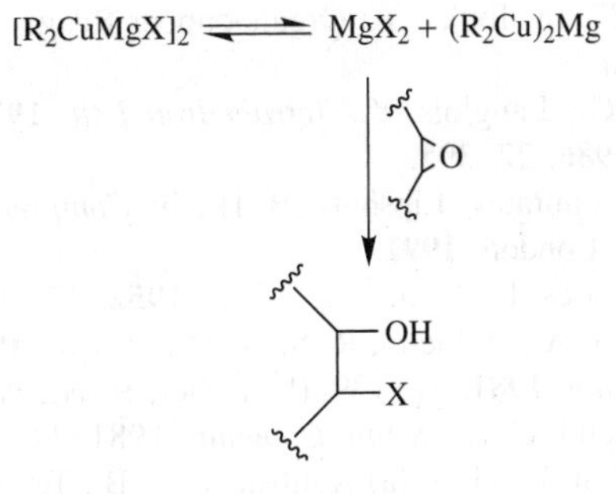

$$[R_2CuMgX]_2 \rightleftharpoons MgX_2 + (R_2Cu)_2Mg$$

Scheme 7.

7 Concluding Remarks and Outlook

Judging alone from the advances made in organocopper chemistry since the appearance of the first edition, it would be hard not to conclude that this field is still vibrant. The continuing stream of new discoveries and uses of copper reagents has increased significantly the options now available for applying this chemistry to synthetic problems. But there is more to be accomplished. Much, if not most, of the synthetic chemistry of copper is still stoichiometric in transition metal and, to an increasing degree, this is not in line with the times. Although not an expense-related issue, environmental concerns and waste-disposal costs will continue to dictate the extent of usage, especially at the industrial level. The extraordinary growth in organopalladium chemistry, catalytic by necessity, is strong testimony to the notion that, in time, catalysis is likely to envelope all transition metal-mediated transformations. Surely, this is a worthy goal for synthetic chemists. Thus, challenges for organocopper chemistry into the next millennium will be a mix of ongoing problems (e.g., asymmetric induction in 1,4 additions), and new concerns (e.g., converting all of the key reactions of cuprates to processes catalytic in copper). This chapter, therefore, not only pays extensive tribute to prior art, but offers a sampling of the most recent developments of increasing acceptance and importance, and even a glimpse of the future. What is clear is that more valuable methodology involving copper is on the horizon.

8 References

[1] Collman, J. P.; Hegedus, L. S.; Norton, J. R.; Finke, R. G.; *Principles and Applications of Organotransition Metal Chemistry*, University Science Books, Mill Valley, CA, **1987**, p. 682.

[2] (a) Lipshutz, B. H.; Sengupta, S.; *Org. React.* **1992**, *41*, 135. (b) Posner, G. H.; *An Introduction to Synthesis Using Organocopper Reagents*, Wiley, New York, **1980**. (c) Posner, G. H.; *Org. React.* **1972**, *19*, 1. (d) Posner, G. H.; *Org. React.* **1975**, *22*, 253. (e) Lipshutz, B. H.; *Synlett* **1990**, 119. (f) Lipshutz, B. H.; *Synthesis* **1987**, 325. (g) Lipshutz, B. H.; Wilhelm, R. S.; Kozlowski, J. A.; *Tetrahedron* **1984**, *40* 5005. (h) Taylor, R. J. K.; *Synthesis* **1985**, 364. (i) Erdik, E.; *Tetrahedron* **1984**, *40*, 641. (j) Normant, J. R.; Alexakis, A.; *Synthesis* **1981**, 841. (k) Kauffmann, T.; *Angew. Chem.* **1974**, *86*, 321. (l) Taylor, R. J. K.; (Ed) *Organocopper Reagents*, University Press, Oxford, 1994. (m)

Wipf, P.; *Synthesis* **1993**, 537. (n) Ibuka, T.; Organocopper Reagents in Organic Synthesis, Camellia and Rose Press, Osaka, 2000.

[3] (a) Ouannes, C.; Dressaire, G.; Langlois, Y.; *Tetrahedron Lett.* **1977**, 815. (b) Hallnemo, G.; Ullenius, C.; *Tetrahedron Lett.* **1986**, *27*, 395.

[4] Review on 1,2 additions of cuprates: Lipshutz, B. H.; in *Comprehensive Organic Synthesis*, (B.M. Trost, Ed.), Pergamon Press, London, **1991**.

[5] Gilman, H.; Jones, R. G.; Woods, L. A.; *J. Org. Chem.* **1952**, *17*, 1630.

[6] Lipshutz, B. H.; Kozlowski, J. A.; Wilhelm, R. S.; *J. Org. Chem.* **1983**, *48*, 546.

[7] (a) Wuts, P. G. M.; *Syn. Comm.* **1981**, *11*, 139. (b) Keller, R. N.; Wycoff, H. D.; *Inorg. Synth.* **1946**, *2*, 1 (c) Theis, A. B.; Townsend, C. A.; *Synth. Commun.* **1981**, *11*, 157.

[8] Based on a modification of that found in: (a) Kauffman, G. B.; Tetev, L. A.; *Inorg. Synth.* **1963**, *7*, 9. (b) Linstrumelle, G.; Krieger, J. K.; Whitesides, G. M.; *Org. Synth.* **1976**, *55*, 103.

[9] House, H. O.; Chu, C. -Y.; Wilkins, J. M.; *J. Org. Chem.* **1975**, *40*, 1460.

[10] Lipshutz, B. H.; Whitney, S.; Kozlowski, J. A.; Breneman, C. M.; *Tetrahedron Lett.* **1986**, *27*, 4273.

[11] Bertz, S. H.; Gibson, C. P.; Dabbagh, G.; *Tetrahedron Lett.* **1987**, *28*, 4251.

[12] Suffert, J.; *J. Org. Chem.* **1989**, *54*, 509 (and references therein).

[13] Watson, S. C.; Eastham, J. F.; *J. Organomet. Chem.* **1967**, *9*, 165.

[14] House, H. O.; Lee, T. V.; *J. Org. Chem.* **1978**, *43*, 4369.

[15] House, H. O.; Fischer, W. F.; *J. Org. Chem.* **1968**, *33*, 949.

[16] (a) Whitesides, G. M.; Fischer, W. F.; SanFilippo, J.; Bashe, R. W.; House, H. O.; *J. Am. Chem. Soc.* **1969**, *91*, 4871. (b) Westmijze, H.; Kleijn, H.; Vermeer, P.; *Tetrahedron Lett.* **1977**, 2023.

[17] Whitesides, G. M.; Stedronsky, E. R.; Casey, C. P.; SanFilippo, J.; *J. Am. Chem. Soc.* **1970**, *92*, 1426.

[18] Wollenberg, R. H.; Albizati, K. F.; Peries, R.; *J. Am. Chem. Soc.* **1977**, *99*, 7365.

[19] Corey, E. J.; Enders, D.; *Chem. Ber.* **1978**, *111*, 1362.

[20] Review: Krause, N.; Gerold, A.; *Angew. Chem. Int. Ed. Engl.* **1997**, *109*, 186.

[21] Yamamoto, Y.; Yamamoto, S.; Yatagai, H.; Ishihara, Y.; Maruyama, K.; *J. Org. Chem.* **1982**, *47*, 119.

[22] Koop, U.; Handke, G.; Krause, N.; *Liebigs Ann.* **1996**, 1487.

[23] (a) Laux, M.; Krause, N.; Koop, U.; *Synlett* **1996**, 87. (b) Arndt, S.; Handke, G.; Krause, N.; *Chem. Ber.* **1993**, *126*, 251.

[24] Johnson, C. R.; Dutra, G. A.; *J. Am. Chem. Soc.* **1973**, *95*, 7777.

[25] Hill, R. K.; Bradberg, T. F.; *Experientia* **1982**, *38*, 70.

[26] Willy, W. E.; McKean, D. R.; Garcia, B. A.; *Bull. Chem. Soc. Jpn.* **1976**, *49*, 1989.

[27] Gilman, H.; Schulze, F.; *J. Am. Chem. Soc.* **1925**, *47*, 2002.

[28] Johnson, C. R.; Herr, R. W.; Wieland, D. M.; *J. Org. Chem.* **1973**, *38*, 4263.

[29] Sirat, H. M.; Thomas, E. J.; Wallis, J. D.; *J. Chem. Soc., Perkin Trans. 1* **1982**, 2885.

[30] Nelb, R. G.; Stille, J. K.; *J. Am. Chem. Soc.* **1976**, *98*, 2834.

[31] McMurry, J. E.; Scott, W. J.; *Tetrahedron Lett.* **1980**, *21*, 4313.

[32] Normant, J. F.; *Pure Appl. Chem.* **1978**, *50*, 709.

[33] Casy, G.; Furber, M.; Lane, S.; Taylor, R. J. K.; Burford, S. C.; *Philos. Trans. R. Soc. London, Ser. A* **1988**, *326*, 565.

[34] Furber, M.; Taylor, R. J. K.; Burford, S. C.; *J. Chem. Soc., Perkin Trans. 1* **1986**, 1809.

[35] (a) Bishop, C. E.; Morrow, G. W.; *J. Org. Chem.* **1983**, *48*, 657. (b) Sonnet, P. E.; Heath, R. R.; *J. Chem. Ecol.* **1980**, *6*, 221.

[36] Nakamura, E.; Isaka, M.; Matsuzawa, S.; *J. Am. Chem. Soc.* **1988**, *110*, 1297.

[37] (a) Isaka, M.; Nakamura, E.; *J. Am. Chem. Soc.* **1990**, *112*, 7428. (b) Nakamura, E.; unpublished data (see ref. 36).

[38] (a) Baucom, K. B.; Butler, G. B.; *J. Org. Chem.* **1972**, *37*, 1730. (b) Breslow, R.; Pecorara, J.; Sugimoto, T.; *Org. Synth. Coll.* Vol. 6, **1988**, 361. (c) Boger, D. L.; Brotherton, C. E.; George, G. I.; *Org. Synth.* **1987**, *65*, 32.

[39] Mangeney, P.; Gosmini, R.; Raussou, S.; Commerçon, M.; Alexakis, A.; *J. Org. Chem.* **1994**, *59*, 1877.

[40] Raussou, S.; Urbain, N.; Mangeney, P.; Alexakis, A.; *Tetrahedron Lett.* **1996**, *37*, 1599.

[41] Lipshutz, B. H.; Ellsworth, E. L.; Siahaan, T. J.; *J. Am. Chem. Soc.* **1989**, *111*, 1351.

[42] Yamamoto, Y.; *Angew. Chem. Int. Ed. Engl.* **1986**, *25*, 947.

[43] Lipshutz, B. H.; Parker, D. A.; Kozlowski, J. A.; Nguyen, S. L.; *Tetrahedron Lett.* **1984**, *25*, 5959.

[44] Smith, A. B.; Jerris, P. J.; *J. Org. Chem.* **1982**, *47*, 1845.

[45] Eis, M. J.; Ganem, B.; *Tetrahedron Lett.* **1985**, *26*, 1153.

[46] Ghribi, A.; Alexakis, A.; Normant, J. F.; *Tetrahedron Lett.* **1984**, *25*, 3083.

[47] (a) Bertz, S. H.; Miao, G.; Rossiter, B. E.; Snyder, J. P.; *J. Am. Chem. Soc.* **1995**, *117*, 11023. (b) Lipshutz, B. H.; Dimock, S. H.; James, B.; *J. Am. Chem. Soc.* **1993**, *115*, 9283.

[48] (a) Corey, E. J.; Boaz, N. W.; *Tetrahedron Lett.* **1985**, *26*, 6019. (b) Alexakis, A.; Sedrani, R.; Mangeney, P.; *Tetrahedron Lett.* **1990**, *31*, 345. (c) Alexakis, A.; Berlan, J.; Besace, Y.; *Tetrahedron Lett.* **1986**, *27*, 1047. (d) Johnson, C. R.; Marren, T. J.; *Tetrahedron Lett.* **1987**, *28*, 27.

[49] (a) Hiroguchi, Y.; Komatsu, M.; Kuwajima, I.; *Tetrahedron Lett.* **1989**, *30*, 7087. (b) Nakamura, E.; Matsuzawa, S.; Horiguchi, Y.; Kuwajima, I.; *Tetrahedron Lett.* **1986**, *27*, 4025, 4029.

[50] Lipshutz, B. H.; Aue, D. H.; James, B.; *Tetrahedron Lett.* **1996**, *37*, 8471.

[51] Lipshutz, B. H.; James, B.; *Tetrahedron Lett.* **1993**, *34*, 6689.

[52] Bertz, S. H.; Smith, R. A. J.; *Tetrahedron* **1990**, *46*, 4091.

[53] Matsuzawa, S.; Hiroguchi, Y.; Nakamura, E.; Kuwajima, I.; *Tetrahedron* **1989**, *45*, 349.

[54] (a) Bergdahl, M.; Lindstedt, E. -L.; Nilsson, M.; Olsson, T.; *Tetrahedron* **1989**, *45*, 535. (b) Bergdahl, M.; Lindstedt, E. -L.; Nilsson, M.; Olsson, T.; *Tetrahedron* **1988**, *44*, 2055.

[55] Piers, E.; Renaud, J.; *J. Org. Chem.* **1993**, *58*, 11.

[56] (a) Gardette, M.; Jabri, N.; Alexakis, A.; Normant, J. F.; *Tetrahedron* **1984**, *40*, 2741. (b) Jabri, N.; Alexakis, A.; Normant, J. F.; *Bull. Soc. Chim. Fr. II* **1983**, 332.

[57] Pereira, O. Z.; Chan, T.; *J. Org. Chem.* **1996**, *61*, 5406.

[58] Hallnemo, G.; Ullenius, C.; *Tetrahedron* **1983**, *39*, 1621.

[59] Ullenius, C.; unpublished data quoted according to ref. 58.

[60] (a) Ledlie, D. B.; Miller, G.; *J. Org. Chem.* **1979**, *44*, 1006 (3-(dimethylamino)-1-propyne). (b) Tsuda, T.; Yazawa, T.; Watanabe, K.; Fujii, T.; Saegusa, T.; *J. Org. Chem.* **1981**, *46*, 192 (mesitylcopper). (c) Mandeville, W. H.; Whitesides, G. M.; *J. Org. Chem.* **1974**, *39*, 400 (alkyl, aryl groups). (d) Johnson, C. R.; Dhanoa, D. S.; *J. Chem. Soc., Chem. Commun.* **1982**, 358 (sulfonyl) anion). (e) Johnson, C. R.; Dhanoa, D. S.; *J. Org. Chem.* **1987**, *52*, 1885 (DMSO anion).

[61] Corey, E. J.; Beames, D. J.; *J. Am. Chem. Soc.* **1972**, *94*, 7210.

[62] House, H. O.; Umen, M. J.; *J. Org. Chem.* **1973**, *38*, 3893.

[63] Corey, E. J.; Floyd, D. M.; Lipshutz, B. H.; *J. Org. Chem.* **1978**, *43*, 3418.

[64] (a) Posner, G. H.; Whitten, C. E.; Sterling, J. J.; *J. Am. Chem. Soc.* **1973**, *95*, 7788. (b) Posner, G. H.; Brunelle, D. J.; Sinoway, L.; *Synthesis* **1974**, 662. (c) Posner, G. H.; Whitten, C. E.; *Org. Synth.* **1976**, *55*, 122.

[65] (a) More information about commercially available methyl thienyl cuprate is given in the FMC Lithium Division Brochure, *Organometallics in Organic Synthesis*, **1990** p 33. (b) Lindstedt, E. -L.; Nilsson, M.; Olsson, T.; *J. Organomet. Chem.* **1987**, *334*, 255. (c) Lindstedt, E. -L.; Nilsson, M.; *Acta Chem. Scand. B* **1986**, *40*, 466.

[66] (a) Bertz, S. H.; Dabbagh, G.; *J. Org. Chem.* **1984**, *49*, 1119. (b) Bertz, S. H.; Dabbagh, G.; Villacorta, G. M.; *J. Am. Chem. Soc.* **1982**, *104*, 5824.

[67] (a) Marino, J. P.; Jaen, J. C.; *J. Am. Chem. Soc.* **1982**, *104*, 3165. (b) Acker, R. D.; *Tetrahedron Lett.* **1977**, 3407. (c) Acker, R. D.; *Tetrahedron Lett.* **1978**, 2399. (d) Hamon, L.; Levisalles, J.; *Tetrahedron* **1989**, *45*, 489.

[68] Tian, X.; Maurya, R.; Konigsberger, K.; Hudlicky, T.; *Synlett* **1995**, 1125.

[69] Tian, X.; Hudlicky, T.; Konigsberger, K.; *J. Am. Chem. Soc.* **1995**, *117*, 3643.

[70] (a) Fujii, N.; Nakai, K.; Tamamura, H.; Otaka, A.; Mimura, N.; Miwa, Y.; Taga, T.; Yamamoto, Y.; Ibuka, T.; *J. Chem. Soc., Perkin Trans. 1* **1995**, 1359. (b) Ibuka, T.; Mimura, N.; Ohno, H.; Nakai, K.; Akaji, M.; Habashita, H.; Tamamura, H.; Miwa, Y.; Taga, T.; Fujii, N.; *J. Org. Chem.* **1997**, *62*, 2982.

(c) Wipf P.; Fritch, P. C.; *J. Org. Chem.* **1994**, *59*, 4875. For related reactions of propargylic aziridines, see Ohno, H.; Toda, A.; Fujii, N.; Takemoto, Y.; Tanaka, T.; Ibuka, T. *Tetrahedron*, **2000**, *56*, 2811.

[71] Church, N. J.; Young, D. W.; *Tetrahedron Lett.* **1995**, *36*, 151.

[72] (a) Bergmeier, S. C.; Seth, P. P.; *J. Org. Chem.* **1997**, *62*, 2671. (b) Dureault, A.; Tranchepain, I.; Depezay, J.; *J. Org. Chem.* **1989**, *54*, 5324.

[73] (a) Zheng, N.; Armstrong, J. D.; McWilliams, J. C.; Volante, R. P.; *Tetrahedron Lett.* **1997**, *38*, 2817. (b) Ryan, J. H.; Stang, P. J.; *Tetrahedron Lett.* **1997**, *38*, 5061.

[74] (a) Askin, D.; Eng, K. K.; Rossen, K.; Purick, R. M.; Wells, K. M.; Volante, R. P.; Reider, P. J.; *Tetrahedron Lett.* **1994**, *35*, 673. (b) Davies, I. W.; Senanayake, C. H.; Castonguay, L.; Larsen, R. D.; Verhoeven, T. R.; Reider, P. J.; *Tetrahedron Lett.* **1995**, *36*, 7619. (c) Ghosh, A. K.; Onishi, M.; *J. Am. Chem. Soc.* **1996**, *118*, 2527.

[75] Bertz, S. H.; Dabbagh, G.; *J. Chem. Soc., Chem. Commun.* **1982**, 1030.

[76] (a) Bertz, S. H.; *J. Am Chem. Soc.* **1990**, *112*, 4031. (b) Bertz, S. H.; *J. Am. Chem. Soc.* **1991**, *113*, 5470. (c) Stemmler, T.; Penner-Hahn, J. E.; Knochel, P.; *J. Am. Chem. Soc.* **1993**, *115*, 348. (d) Barnhart, T. M.; Huang, H.; Penner-Hahn, J. E.; *J. Org. Chem.* **1995**, *60*, 4310.

[77] Stemmler, T. L.; Barnhart, T.; Penner-Hahn, J. E.; Tucker, C. E.; Knochel, P.; Bohme, M.; Frenking, G.; *J. Am. Chem. Soc.*, **1995**, *117*, 12489.

[78] (a) Cabezas, J. A.; Oehlschlager, A. C.; *J. Am Chem. Soc.* **1997**, *119*, 3878. (b) Lipshutz, B. H.; Sharma, S.; Ellsworth, E. L.; *J. Am. Chem. Soc.* **1990**, *112*, 4032. (c) Lipshutz, B. H.; James, B.; *J. Org. Chem.* **1994**, *59*, 7585.

[79] Bertz, S. H.; Miao, G.; Eriksson, M.; *J. Chem. Soc., Chem. Commun.* **1996**, 815.

[80] Lipshutz, B. H.; Wilhelm, R. S.; Floyd, D. M.; *J. Am. Chem. Soc.* **1981**, *103*, 7672.

[81] Lipshutz, B. H.; Wilhelm, R. S.; Kozlowski, J. A.; *J. Org. Chem.* **1984**, *49*, 3938.

[82] Lipshutz, B. H.; Wilhelm, R. S.; Kozlowski, J. A.; *J. Org. Chem.* **1984**, *49*, 3928.

[83] Lipshutz, B. H.; Parker, D.; Kozlowski, J. A.; Miller, R. D.; *J. Org. Chem.* **1983**, *48*, 3334.

[84] Bonini, B. F.; Capperuci, A.; Comes-Franchini, M.; Degl'Innocenti, A.; Mazzanti, G.; Ricci, A.; Zani, P.; *Synlett* **1993**, 937.

[85] Kitamura, T.; Tanaka, T.; Taniguchi, H.; Stang, P. J.; *J. Chem. Soc., Perkin Trans. 1* **1991**, 2892.

[86] Kang, J.; Cho, W.; Lee, W. K.; *J. Org. Chem.* **1984**, *49*, 1838.

[87] (a) Stang, P. J.; Blume, T.; Zhdankin, V. V.; *Synthesis* **1993**, 35. (b) Kitamura, T.; Mihara, I.; Taniguchi, H.; Stang, P. J.; *J. Chem Soc., Perkin Trans. 1* **1990**, 614.

[88] Dieter, R. K.; Dieter, J. W.; Alexander, C. W.; Bhinderwala, N. S *J. Org. Chem.* **1996**, *61*, 2930.

[89] Dieter, R.K.; Sharma, R. R.; Ryan, W.; *Tetrahedron Lett.* **1994**, *38*, 783.

[90] (a) Dieter, R. K.; Alexander, C. W.; *Synlett* **1993**, 407. (b) Dieter, R. K.; Alexander, C. W.; *Tetrahedron Lett.* **1992**, *33*, 5693. Dieter, R. K.; Alexander, C. W.; Nice, L. E. *Tetrahedron*, **2000**, *56*, 2767.

[91] Dieter, R. K.; Sharma, R. R.; *Tetrahedron Lett.* **1997**, *38*, 5937.

[92] Yamamoto, Y.; Asao, N.; Uyehara, T.; *J. Am Chem. Soc.* **1992**, *114*, 5427.

[93] Asao, N.; Uyehara, T.; Tsukada, N.; Yamamoto, Y.; *Bull. Chem. Soc. Jpn.* **1995**, *68*, 2103.

[94] Lipshutz, B. H.; Ellsworth, E. L.; Siahaan, T. J.; *J. Am. Chem. Soc.* **1988**, *110*, 4834.

[95] Kozlowski, J. A.; Thesis, Ph.D.; University of California, Santa Barbara, 1985 (see ref. 43).

[96] Eis, M. J.; Wrobel, J. E.; Ganem, B.; *J. Am. Chem. Soc.* **1984**, *106*, 3693.

[97] Alexakis, A.; Jachiet, D.; Normant, J. F.; *Tetrahedron* **1986**, *42*, 5607.

[98] Lipshutz, B. H.; Ellsworth, E. L.; Siahaan, T. J.; Shirazi, A.; *Tetrahedron Lett.* **1988**, *29*, 6677.

[99] Linderman, R. J.; Godfrey, A.; Horne, K.; *Tetrahedron* **1989**, *45*, 495.

[100] Linderman, R. J.; McKenzie, J. R.; *J. Organomet. Chem.* **1989**, *361*, 31.

[101] (a) Selective transfer of a vinylic ligand over a methyl group in a Gilman cuprate: Posner, G. H.; Sterling, J. J.; Whitten, C. E.; Lentz, C. M.; Brunelle, D. J.; *J. Am. Chem. Soc.* **1975**, *97*, 107. (b) Leyendecker, F.; Drouin, J.; Debesse, J. J.; Conia, J. M.; *Tetrahedron Lett.* **1977**, 1591. (c) Lipshutz, B. H.; Koerner, M.; Parker, D. A.; *Tetrahedron Lett.* **1987**, *28*, 945.

[102] Lipshutz, B. H.; Kozlowski, J. A.; Parker, D. A.; McCarthy, K. E.; *J. Organomet. Chem.* **1985**, *285*, 437.

[103] Lipshutz, B. H.; Fatheree, P.; Hagen, W.; Stevens, K. L.; *Tetrahedron Lett.* **1992**, *33*, 1041.

[104] Bertz, S. H.; Eriksson, M.; Miao, G.; Snyder, J. P.; *J. Am. Chem. Soc.* **1996**, *118*, 10906.

[105] Jarvis, J. A. J.; Pearce, R.; Lappert, M. F.; *J. Chem. Soc., Dalton Trans.* **1977**, 999.

[106] Soderquist, J. A.; Santiago, B.; *Tetrahedron Lett.* **1989**, *30*, 5693.

[107] Lipshutz, B. H.; Siegmann, K.; Garcia, E.; *J. Am. Chem. Soc.* **1991**, *113*, 8161.

[108] Lipshutz, B. H.; Siegmann, K.; Garcia, E.; Kayser, F.; *J. Am. Chem. Soc.* **1993**, *115*, 9276.

[109] Lipshutz, B. H.; Kayser, F.; Siegmann, K.; *Tetrahedron Lett.* **1993**, *34*, 6693.

[110] Lipshutz, B. H.; Kayser, F.; Maullin, N.; *Tetrahedron Lett.* **1994**, *35*, 815.

[111] Corey, E. J.; Letavic, M. A.; Noe, M. C.; Sarshar, S.; *Tetrahedron Lett.* **1994**, *35*, 7553.

[112] Tamao, K.; Yamaguchi, S.; Shiro, M.; *J. Am. Chem. Soc.* **1994**, *116*, 11715.

[113] Coleman, R. S.; Grant, E. B.; *J. Am Chem. Soc.* **1995**, *117*, 10889.

[114] Lipshutz, B. H.; Kayser, F.; Liu, Z.; *Angew. Chem. Int. Ed. Engl.* **1994**, *33*, 1842.

[115] (a) Alberti, A.; Cane, F.; Dembech, P.; Lazzari, D.; Ricci, A.; Seconi, G.; *J. Org. Chem.* **1996**, *61*, 1677. (b) Casarini, A.; Dembech, P.; Lazzari, D.; Marini, E.; Reginato, G.; Ricci, A.; Seconi, G.; *J. Org. Chem.* **1993**, *58*, 5620.

[116] Cane, F.; Brancaleoni, D.; Dembech, P.; Ricci, A.; Seconi, G.; *Synthesis* **1997**, 545.

[117] van Koten, G.; University of Utrecht, personal communication.

[118] Pyne, S. G.; *J. Org. Chem.* **1986**, *51*, 81.

[119] Suzuki, M.; Suzuki, T.; Kawagishi, T.; Morita, Y.; Noyori, R.; *Isr. J. Chem.* **1984**, *24*, 118.

[120] (a) Rieke, R. D.; Wehmeyer, R. M.; Wu, T. -C.; Ebert, G. W.; *Tetrahedron* **1989**, *45*, 443. (b) Rieke, R. D.; Wu, T. -C.; Stinn, D. E.; Wehmeyer, R. M.; *Synth. Commun.* **1989**, *19*, 1833.

[121] Representative early uses: (a) Corey, E. J.; Carney, R. L.; *J. Am. Chem. Soc.* **1971**, *93*, 7318. (b) Clark, R. D.; Heathcock, C. H.; *Tetrahedron Lett.* **1974**, 1713. (c) Van Mourik, G. L.; Pabon, H. J. J.; *Tetrahedron Lett.* **1978**, 2705. (d) Mansson, J. -E.; *Acta Chem. Scand. B* **1978**, *32*, 543.

[122] Bertz, S. H.; Dabbagh, G.; *Tetrahedron* **1989**, *45*, 425.

[123] Maruyama, K.; Yamamoto, Y.; *J. Am. Chem. Soc.* **1977**, *99*, 8068.

[124] Yamamoto, Y.; Maruyama, K.; *J. Am. Chem. Soc.* **1978**, *100*, 3241.

[125] Lipshutz, B. H.; Ellsworth, E. L.; Dimock, S. H.; *J. Am. Chem. Soc.* **1990**, *112*, 5869.

[126] Lipshutz, B. H.; Ellsworth, E. L.; Dimock, S. H.; Smith, R. A. J.; *J. Am. Chem. Soc.* **1990**, *112*, 4404.

[127] Eriksson, M.; Johansson, A.; Nilsson, M.; Olsson, T.; *J. Am. Chem. Soc.* **1996**, *118*, 10904.

[128] Jung, M. E.; Lyster, M. A.; *J. Am. Chem. Soc.* **1977**, *99*, 968.

[129] Bergdahl, M.; Nilsson, M.; Olsson, T.; Stern, K.; *Tetrahedron* **1991**, *47*, 9691.

[130] (a) Colvin, E. W.; *Silicon in Organic Synthesis*, Butterworths, London, 1981. (b) Weber, W. P.; *Silicon Reagents for Organic Synthesis*, Springer-Verlag, Berlin, 1983. (c) Colvin, E. W.; *Silicon Reagents in Organic Synthesis*, Academic Press, Orlando, FL, 1988.

[131] (a) Fleming, I.; Newton, T. W.; Roessler, F.; *J. Chem. Soc., Perkin Trans. 1* **1981**, 2527. (b) Ager, D. J.; Fleming, I.; Patel, S. K.; *J. Chem. Soc., Perkin Trans. 1* **1981**, 2520. (c) Fleming, I.; Thomas, A. P.; *J. Chem. Soc., Chem. Commun.* **1985**, 411. (d) Fleming, I.; Rowley, M.; Cuadrado, P.; Gonzalez-Nogal, A. M.; Pulido, F. J.; *Tetrahedron* **1989**, *45*, 413. (e) Barbero, A.; Cuadrado, P.; Fleming, I.; Gonzalez, A. M.; Pulido, F. J.; *J. Chem. Soc., Chem. Commun.* **1992**, 351.

[132] George, M. V.; Peterson, D. J.; Gilman, H.; *J. Am. Chem. Soc.* **1960**, *82*, 403.

[133] (a) Fleming, I.; Henning, R.; Plaut, H.; *J. Chem. Soc., Chem. Commun.* **1984**, 351. (b) Fleming, I.; Sanderson, P. E. J.; *Tetrahedron Lett.* **1987**, *28*, 4229. (c) Tamao, K.; Kawachi, A.; Ito, Y.; *J. Am. Chem. Soc.* **1992**, *114*, 3989.

[134] Lipshutz, B. H.; Reuter, D. C.; Ellsworth, E. L.; *J. Org. Chem.* **1989**, *54*, 4975.

[135] Reuter, D. C.; Ph D Thesis, University of California, Santa Barbara, 1990.

[136] Lipshutz, B. H.; Reuter, D. C.; *Tetrahedron Lett.* **1989**, *30*, 4617.

[137] Pereyre, M.; Quintard, J. -P.; Rahm, A.; in *Tin in Organic Synthesis*, Butterworths, London, **1987**.

[138] Yamamoto, Y.; (Ed), *Organotin Compounds in Organic Synthesis*, Tetrahedron Symposia-in-Print No. 36, **1989**.

[139] (a) Piers, E.; Chong, J. M.; *Can. J. Chem.* **1988**, *66*, 1425. Other procedures involving stannylcuprations of alkynes: (b) Marino, J. P.; Emonds, M. V. M.; Stengel, P. J.; Oliveira, A. R. M.; Simonelli, F.; Ferrerra, J. T. B.; *Tetrahedron Lett.* **1992**, *33*, 49. (c) Barbero, A.; Cuadrado, P.; Fleming, I.; Gonzalez, A. M.; Pulido, F.; Rubio, R.; *J. Chem. Soc., Perkin Trans. 1* **1993**, 1657.

[140] Piers, E.; Morton, H. E.; Chong, J. M.; *Can. J. Chem.* **1987**, *65*, 78.

[141] (a) Piers, E.; Chong, J. M.; Morton, H. E.; *Tetrahedron* **1989**, *45*, 363. The corresponding tributyltin analogs of both reagents **37** and **38**, i.e., $Bu_3SnCu\bullet SMe_2$ and $(Bu_3Sn)(PhS)CuLi$, have also been prepared and utilized in a similar capacity: (b) Piers, E.; Chong, J. M.; Gustafson, K.; Andersen, R. J.; *Can. J. Chem.* **1984**, *62*, 1. (c) Piers, E.; Wong, T.; Ellis, K. A.; *Can. J. Chem.* **1992**, *70*, 2058.

[142] Piers, E.; Tillyer, R. D.; *J. Org. Chem.* **1988**, *53*, 5366.

[143] (a) Lipshutz, B. H.; Ellsworth, E. L.; Dimock, S. H.; Reuter, D. C.; *Tetrahedron Lett.* **1989**, *30*, 2065. (b) Gilbertson, S. R.; Challener, C. A.; Bos, M. E.; Wulff, W. D.; *Tetrahedron Lett.* **1988**, *29*, 4795.

[144] Lipshutz, B. H.; Sharma, S.; Reuter, D. C.; *Tetrahedron Lett.* **1990**, *31*, 7253.

[145] Oehlschlager, A. C.; Hutzinger, M. W.; Aksela, R.; Sharma, S.; Singh, S. M.; *Tetrahedron Lett.* **1990**, *31*, 165.

[146] (a) Ashby, E. C.; Watkins, J. J.; *J. Am. Chem. Soc.* **1977**, *99*, 5312. (b) Ashby, E. C.; Watkins, J. J.; *J. Chem. Soc., Chem. Commun.* **1976**, 784. (c) Ashby, E. C.; Lin, J. J.; *J. Org. Chem.* **1977**, *42*, 2805. (d) Ashby, E. C.; Lin, J. J.; Watkins, J. J.; *J. Org. Chem.* **1977**, *42*, 1009.

[147] Bertz, S. H.; Dabbagh, G.; *J. Am. Chem. Soc.* **1988**, *110*, 3668.

[148] Clive, D. L. J.; Farina, V.; Beaulieu, P. L.; *J. Org. Chem.* **1982**, *47*, 2572.

[149] Lipshutz, B. H.; Kozlowski, J. A.; Breneman, C. M.; *J. Am. Chem. Soc.* **1985**, *107*, 3197.

[150] Lipshutz, B. H.; Ellsworth, E. L.; unpublished results.

[151] Ashby, E. C.; Korenowski, T. F.; Schwartz, R. D.; *J. Chem. Soc., Chem. Commun.* **1974**, 157.

[152] Ashby, E. C.; Goel, A. B.; *Inorg. Chem.* **1977**, *16*, 3043.

[153] Ashby, E. C.; Lin, J. -J.; Goel, A. B.; *J. Org. Chem.* **1978**, *43*, 183.

[154] Masamune, S.; Bates, G. S.; Georghiou, P. E.; *J. Am. Chem. Soc.* **1974**, *96*, 3686.

[155] Boeckman, R. K.; Michalak, R.; *J. Am. Chem. Soc.* **1974**, *96*, 1623.

[156] Whitesides, G. M.; San Filippo, J.; Stredronsky, E. R.; Casey, C. P.; *J. Am. Chem. Soc.* **1969**, *91*, 6542.

[157] (a) Churchill, M. R.; Bezman, S. A.; Osborn, J. A.; Wormald, J.; *Inorg. Chem.* **1972**, *11*, 1818. (b) Bezman, S. A.; Churchill, M. R.; Osborn, J. A.; Wormald, J.; *J. Am. Chem. Soc.* **1971**, *93*, 2063. (c) Goeden, G. V.; Caulton, K. G.; *J. Am. Chem. Soc.* **1981**, *103*, 7354. (d) Lemmen, T. H.; Folting, K.; Huffman, J. C.; Caulton, K. G.; *J. Am. Chem. Soc.* **1985**, *107*, 7774.

[158] Mahoney, W. S.; Brestensky, D. M.; Stryker, J. M.; *J. Am. Chem. Soc.* **1988**, *110*, 291.

[159] (a) Brestensky, D. M.; Huseland, D. E.; McGettigan, C.; Stryker, J. M.; *Tetrahedron Lett.* **1988**, *29*, 3749. (b) Daeuble, J. F.; McGettigan, C.; Stryker, J. M.; *Tetrahedron Lett.* **1990**, *31*, 2397. (c) Koenig, T. M.; Daeuble, J. F.; Brestensky, D. M.; Stryker, J. M.; *Tetrahedron Lett.* **1990**, *31*, 3237. (d) Chen, J-X.; Daeuble, J. F.; Stryker, J. M. *Tetrahedron*, **2000**, *56*, 2789.

[160] Lipshutz, B. H.; Ung, C. S.; Sengupta, S.; *Synlett* **1989**, 64. Lipshutz, B. H.; Chrisman, W.; Noson, K.; Papa, P.; Sclafari, J. A.; Vivian, R. W.; Keith, J. M. *Tetrahedron*, **2000**, *56*, 2779.

[161] Kharasch, M. S.; Tawney, P. O.; *J. Am. Chem. Soc.* **1941**, *63*, 2308.

[162] (a) Tamura, M.; Kochi, J. K.; *J. Organomet. Chem.* **1972**, *42*, 205. (b) Tamwa, M.; Kochi, J. K.; *Synthesis* **1971**, 303.

[163] Bal, S. A.; Marfat, A.; Helquist, P.; *J. Org. Chem.* **1982**, *47*, 5045.

[164] Ito, Y.; Nakatsuka, M.; Saegusa, T.; *J. Am. Chem. Soc.* **1982**, *104*, 7609.

[165] Horiguchi, Y.; Matsuzawa, S.; Makamura, E.; Kuwajima, I.; *Tetrahedron Lett.* **1986**, *27*, 4025.

[166] Whitesell, J. K.; Lawrence, R. M.; Chen, H. H.; *J. Org. Chem.* **1986**, *51*, 4779.

[167] Alexakis, A.; Marek, I.; Mangeney, P.; Normant, J. F.; *Tetrahedron Lett.* **1989**, *30*, 2387.

[168] (a) Fujisawa, T.; Sato, T.; Kawara, T.; Naruse, K.; *Chem. Lett.* **1980**, 1123. (b) Fujisawa, T.; Sato, T.; Kawara, T.; Noda, A.; Obinata, T.; *Tetrahedron Lett.* **1980**, *21*, 2553.

[169] Sato, T.; Kawara, T.; Kawashima, M.; Fujisawa, T.; *Chem. Lett.* **1980**, 571.

[170] Arnold, L. D.; Drover, J. C. G.; Vederas, J. C.; *J. Am. Chem. Soc.* **1987**, *109*, 4649.

[171] Nunomoto, S.; Kawakami, Y.; Yamashita, Y.; *J. Org. Chem.* **1983**, *48*, 1912.

[172] Novak, J.; Salemink, C. A.; *Synthesis* **1981**, 597.

[173] (a) Kotsuki, H.; Kadota, I.; Ochi, M.; *J. Org. Chem.* **1990**, *55*, 4417; (b) Kotsuki, H.; Kadota, I.; Ochi, M.; *Tetrahedron Lett.* **1989**, *30*, 1281. (c) Kotsuki, H.; Kadota, I.; Ochi, M.; *Tetrahedron Lett.* **1989**, *30*, 3999.

[174] (a) Tseng, C. C.; Paisley, S. D.; Goering, H. L.; *J. Org. Chem.* **1986**, *51*, 2884; (b) Tseng, C. C.; Yen, S. -J.; Goering, H. L.; *J. Org. Chem.* **1986**, *51*, 2892. (c) Underiner, T. L.; Paisley, S. D.; Schmitter, J.; Lesheski, L.; Goering, H. L.; *J. Org. Chem.* **1989**, *54*, 2369.

[175] Tanis, S. P.; *Tetrahedron Lett.* **1982**, *23*, 3115.

[176] Suzuki, S.; Shiono, M.; Fujita, Y.; *Synthesis* **1983**, 804.

[177] (a) Four, P.; LeTri, Ph.; Riviere, H.; *J. Organomet. Chem.* **1977**, *133*, 385. (b) Ashby, E. C.; Goel, A. B.; Smith, R. S.; *J. Orgnomet. Chem.* **1981**, *212*, C47. (c) Ashby, E. C.; Goel, A. B.; *J. Org. Chem.* **1983**, *48*, 2125. (d) Ashby, E. C.; Smith, R. S.; Goel, A. B.; *J. Org. Chem.* **1981**, *46*, 5133. (e) Goel, A. B.; Ashby, E. C.; *Inorg. Chem. Acta* **1981**, *54*, L199. (f) Westmijze, H.; George, A. V. E.; Vermeer, P.; *Recl. Trav. Chim. Pays-Bas* **1983**, *102*, 322.

[178] (a) Gardette, M.; Alexakis, A.; Normant, J. F.; *Tetrahedron* **1985**, *41*, 5887. (b) Marfat, A.; McGuirk, P. R.; Helquist, P.; *J. Org. Chem.* **1979**, *44*, 3888. (c) Iyer, R. S.; Helquist, P.; *Org. Synth.* **1985**, *64*, 1.

[179] Foulon, J. P.; Commercon, M. B.; Normant, J. F.; *Tetrahedron* **1986**, *42*, 1389.

[180] (a) Alexakis, A.; Normant, J. F.; Villieras, J.; *J. Organomet. Chem.* **1975**, *96*, 471. (b) Normant, J. F.; Alexakis, A.; in: *Modern Synthetic Methods*, Vol. 3, (ed., Scheffold, R.;), Wiley, NY, **1983** p. 139.

[181] Wijkens, P.; Vermeer, P.; *J. Organomet. Chem.* **1986**, *301*, 247.

[182] Normant, J. F.; Cahiez, G.; Chuit, C.; Villieras, J.; *J. Organomet. Chem.* **1974**, *77*, 269.

[183] (a) Normant, J. F.; Cahiez, G.; Chuit, C.; Villieras, J.; *J. Organomet. Chem.* **1973**, *54*, C53. (b) *J. Organomet. Chem.* **1974**, *77*, 281.

[184] Cahiez, G.; Bernard, D.; Normant, J. F.; *Synthesis* **1976**, 245.

[185] Normant, J. F.; Commercon, A.; Cahiez, G.; Villieras, J.; *C.R. Acad. Sci. Paris, Ser. C* **1974**, *278*, 967.

[186] Curran, D. P.; Chen, M. -H.; Leszczweski, D.; Elliot, R. L.; Rakiewicz, D. M.; *J. Org. Chem.* **1986**, *51*, 1612.

[187] (a) Vermeer, P.; Westmijze, H.; Kleijn, H.; van Dyck, L. A.; *Recl. Trav. Chim. Pays-Bas* **1978**, *97*, 56. (b) Westmijze, H.; Vermeer, P.; *Tetrahedron Lett.* **1979**, 4101. (c) Elsevier, C. J.; Meijer, J.; Westmijze, H.; Vermeer, P.; van Dijck, L. A.; *J. Chem. Soc., Chem. Commun.* **1982**, 84.

[188] (a) Stork, G.; Kreft, A.; *J. Am. Chem. Soc.* **1977**, *99*, 3850. (b) Magid, R. M.; Fruchey, O. S.; *J. Am. Chem. Soc.* **1977**, *99*, 8368. (c) Olsson, L. -I.; Claesson, A.; *Acta Chem. Scand. B* **1979**, *33*, 679. (d) Marek, I.; Mangeney, P.; Alexakis, A.; Normant, J. F.; *Tetrahedron Lett.* **1985**, *27*, 5499. (e) Tadema, G.; Everhardus, R. H.; Westmijze, H.; Vermeer, P.; *Tetrahedron Lett.* **1978**, 3935.

[189] Macdonald, T. L.; Reagan, D. R.; Brinkmeyer, R. S.; *J. Org. Chem.* **1980**, *45*, 4740.

[190] Lipshutz, B. H.; Hackmann, C.; *J. Org. Chem.* **1994**, *59*, 7437.

[191] Ibuka, T.; Habashita, H.; Otaka, A.; Fujii, N.; *J. Org. Chem.* **1991**, *56*, 4370.

[192] Sawamura, M.; Iikura, H.; Nakamura, E.; *J. Am. Chem. Soc.* **1996**, *118*, 12850.

[193] Fujisawa, T.; Mori, T.; Kawara, T.; Sato, T.; *Chem. Lett.* **1982**, 569.

[194] (a) Fujisawa, T.; Sato, T.; Kawara, T.; Kawashima, M.; Shimizu, H.; Ito, Y.; *Tetrahedron Lett.* **1980**, *21*, 2181. (b) Fujisawa, T.; Sato, T.; Kawara, T.; Tago, H.; *Bull. Chem. Soc. Jpn.* **1983**, *56*, 345. (c) Kawashima, M.; Sato, T.; Fujisawa, T.; *Tetrahedron* **1989**, *45*, 403. (d) Tamaru, Y., Tanigawa, H., Yamamoto, T., Yoshida, Z.; *Angew. Chem., Int. Ed. Engl.* **1989**, *28*, 351.

[195] Bergbrieter, D. E.; Killough, J. M.; *J. Org. Chem.* **1976**, *41*, 2750.

[196] Bell, T. W.; Hu, L. -Y.; Patel, S. V.; *J. Org. Chem.* **1987**, *52*, 3847.

[197] Funk, R. L.; Vollhardt, K. P. C.; *J. Am. Chem. Soc.* **1980**, *102*, 5253.

[198] Gill, M.; Rickards, R. W.; *J. Chem. Soc., Perkin Trans. I* **1981**, 599.

[199] Behling, J. R.; Babiak, K. A.; Ng, J. S.; Campbell, A. L.; Moretti, R.; Koerner, M.; Lipshutz, B. H.; *J. Am. Chem. Soc.* **1988**, *110*, 2641.

[200] Lipshutz, B. H.; Koerner, M.; unpublished results.

[201] Collins, P. W.; *J. Med. Chem.* **1986**, *29*, 437 (and references therein).

[202] Lipshutz, B. H.; Crow, R.; Dimock, S. H.; Ellsworth, E. L.; Smith, R. A. J.; Behling, J. R.; *J. Am. Chem. Soc.* **1990**, *112*, 4063.

[203] (a) Milstein, D.; Stille, J. K.; *J. Org. Chem.* **1979**, *44*, 1613. (b) Labadie, J. W.; Tueting, D.; Stille, J. K.; *J. Org. Chem.* **1983**, *48*, 4634.

[204] (a) Lipshutz, B. H.; Elworthy, T. R.; *J. Org. Chem.* **1990**, *55*, 1695. (b) Lipshutz, B. H.; Elworthy, T. R.; *Tetrahedron Lett.* **1990**, *31*, 477.

[205] (a) Lipshutz, B. H.; Ellsworth, E. L.; *J. Am. Chem. Soc.* **1990**, *112*, 7440. (b) Babiak, K. A.; Behling, J. R.; Dygos, J. H.; McLaughlin, K. T.; Ng, J. S.; Kalish, V. J.; Kramer, S. W.; Shone, R. L.; *J. Am. Chem. Soc.* **1990**, *112*, 7441. (c) Dygos, J. H.; Adamek, J. P.; Babiak, K. A.; Behling, J. R.; Medich, J. R.; Ng, J. S.; Wieczorek, J. J.; *J. Org. Chem.* **1991**, *56*, 2549.

[206] (a) Schwartz, J.; Labinger, J. A.; *Angew. Chem., Int. Ed. Engl.* **1976**, *15*, 333. (b) Negishi, E.; Takahashi, T.; *Synthesis* **1988**, 1. (c) Negishi, E.; Takahashi, T.; *Aldrichimica Acta* **1985**, *18*, 31. (d) Negishi, E.; *Pure Appl. Chem.* **1981**, *53*, 2333. (e) Yoshifuji, M.; Loots, M.; Schwartz, J.; *Tetrahedron Lett.* **1977**, 1303. (f) Buchwald, S. L.; Lamaire, S. J.; Nielsen, R. B.; Watson, B. T.; King, S. M.; *Tetrahedron Lett.* **1987**, *28*, 3895.

[207] Lipshutz, B. H.; Keil, R.; *J. Am. Chem. Soc.* **1992**, *114*, 7919.

[208] Lipshutz, B. H.; Kato, K.; *Tetrahedron Lett.* **1991**, *32*, 5647.

[209] Ireland, R. E.; Wipf, P.; *J. Org. Chem.* **1990**, *55*, 1425.

[210] Tucci, F. C.; Chieffi, A.; Comasseto, J. V.; *J. Org. Chem.* **1996**, *61*, 4975.

[211] Marino, J. P.; Tucci, F.; Comasseto, J. V.; *Synlett* **1993**, 761.

[212] (a) de Araujo, M. A.; Comasseto, J. V.; *Synlett* **1995**, 1145. Displacements of the phenyltelluride moiety by cuprates: (b) Ogawa, A.; Tsuboi, Y.; Obayashi, R.; Yokoyama, K.; Ryu, I.; Sonoda, N.; *J. Org. Chem.* **1994**, *59*, 1600.

[213] (a) Knochel, P.; Singer, R.; *Chem. Rev.* **1993**, *93*, 2117. (b) Knochel, P.; *Synlett* **1995**, 393.

[214] (a) Nakamura, E.; Aoki, S.; Sekiya, K.; Oshino, H.; Kuwajima, I.; *J. Am. Chem. Soc.* **1987**, *109*, 8056. (b) Nakamura, E.; Kuwajima, I.; *Org. Syn.* **1988**, *66*, 43. (c) Tamaru, Y.; Tanigawa, H.; Yamamoto, T.; Yoshida, Z.; *Angew. Chem., Int. Ed. Engl.* **1989**, *28*, 351.

[215] (a) Nakamura, E.; Sekiya, K.; Arai, M.; Aoki, S.; *J. Am. Chem. Soc.* **1989**, *111*, 3091. (b) Sekiya, K.; Nakamura, E.; *Tetrahedron Lett.* **1988**, *29*, 5155.

[216] (a) Berk, S. C.; Yeh, M. C. P.; Jeong, N.; Knochel, P.; *Organometallics* **1990**, *9*, 3053. (b) Berk, S. C.; Knochel, P.; Yeh, M. C. P.; *J. Org. Chem.* **1988**, *53*, 5789.

[217] Yeh, M. C. P.; Chen, H. G.; Knochel, P.; *Org. Synth.* **1992**, *70*, 195.

[218] Knochel, P.; Yeh, M. C. P.; Berk, S. C.; Talbert, J.; *J. Org. Chem.* **1988**, *53*, 2390.

[219] Yeh, M. C. P.; Knochel, P.; Butler, W. M.; Berk, S. C.; *Tetrahedron Lett.* **1988**, *29*, 6693.

[220] Yeh, M. C. P.; Knochel, P.; *Tetrahedron Lett.* **1989**, *30*, 4799.

[221] Retherford, C.; Yeh, M. C. P.; Schipor, I.; Chen, H. G.; Knochel, P.; *J. Org. Chem.* **1989**, *54*, 5200.

[222] Alternative route to this product using organocopper chemistry: Wehmeyer, R. M.; Rieke, R. D.; *J. Org. Chem.* **1987**, *52*, 5056.

[223] Micouin, L.; Oestreich, M.; Knochel, P.; *Angew., Chem., Int. Ed. Engl.* **1997**, *36*, 245.

[224] Lipshutz, B. H.; *Acc. Chem. Res.* **1997**, *30*, 277.

[225] Wipf, P.; Xu, W. J.; Smitrovich, J. H.; Lehmann, R.; Venanzi, L. M.; *Tetrahedron* **1994**, *50*, 1935.

[226] (a) Lipshutz, B. H.; Keil, R.; Ellsworth, E. L.; *Tetrahedron Lett.* **1990**, *31*, 7257. (b) Swanson, D. R.; Nguyen, T.; Noda, Y.; Negishi, E. *J. Org. Chem.* **1991**, *56*, 2590.

[227] Lipshutz, B. H.; Wood, M. R.; *J. Am. Chem. Soc.* **1993**, *115*, 12625.

[228] (a) Tückmantel, W.; Oshima, K.; Nozaki, H.; *Chem. Ber.* **1986**, *119*, 104. (b) Kjonaas, R. A.; Vawter, E. J.; *J. Org. Chem.* **1986**, *5*, 3993.

[229] Isobe, M.; Kondo, S.; Nagasawa, N.; Goto, T.; *Chem. Lett.* **1977**, 679.

[230] Lipshutz, B. H.; Wood, M. R.; *J. Am. Chem. Soc.* **1994**, *116*, 11689.

[231] Venanzi, L. M.; Lehmann, R.; Keil, R.; Lipshutz, B. H.; *Tetrahedron Lett.* **1992**, *33*, 5857.

[232] Lipshutz, B. H.; Dimock, S. H.; *J. Org. Chem.* **1991**, *56*, 5761.

[233] Evans, D. A.; Chapman, K. T.; Bisaha, J.; *J. Am. Chem. Soc.* **1984**, *106*, 4261.

[234] Cahiez, G.; Alami, M.; *Tetrahedron Lett.* **1989**, *30*, 3541.

[235] (a) Friour, G.; Cahiez, G.; Normant, J. F.; *Synthesis* **1984**, 37. (b) Cahiez, G.; *L'Actualite Chimique* **1984**, *9*, 24. (c) Cahiez, G.; Alami, M.; *Tetrahedron Lett.* **1986**, *27*, 569.

[236] Kim, S. H.; Hanson, M. V.; Rieke, R. D.; *Tetrahedron Lett.* **1996**, *37*, 2197; Kim, S. H.; Rieke, R. D. *J. Org. Chem.* **2000**, *65*, 2322.

[237] Cahiez, G.; Laboue, B.; *Tetrahedron Lett.* **1989**, *30*, 7369.

[238] (a) Cahiez, G.; Marquais, S.; *Synlett* **1993**, 45. (b) Cahiez, G.; Marquais, S.; *Pure Appl. Chem.* **1996**, *68*, 53. (c) Klement, I.; Knochel, P.; Chau, K.; Cahiez, G.; *Tetrahedron Lett.* **1994**, *35*, 1177. (d) Riguet, E.; Klement, I.; Reddy, C. K.; Cahiez, G.; Knochel, P.; *Tetrahedron Lett.* **1996**, *37*, 5865.

[239] Cahiez, G.; *An. Quim.* **1995**, *91*, 561.

[240] Arai, M.; Lipshutz, B. H.; Nakamura, E.; *Tetrahedron* **1992**, *48*, 5709.

[241] Kitamura, M.; Miki, T.; Nakano, K.; Noyori, R.; *Tetrahedron Lett.* **1996**, *37*, 5141.

[242] Lipshutz, B. H.; Wood, M. R.; Tirado, R.; *J. Am. Chem. Soc.* **1995**, *117*, 6126.

[243] Lipshutz, B. H.; Lindsley, C.; Susfalk, R.; Gross, T.; *Tetrahedron Lett.* **1994**, *35*, 8999 (and references therein).

[244] Lipshutz, B. H.; Wood, M. R.; Tirado, R.; *Org. Synth.* **1998**, *76*, 252.

[245] Lipshutz, B. H.; Gross, T.; Buzard, D. J.; Tirado, R.; *J. Chin. Chem. Soc.* **1997**, *44*, 1.

[246] Lipshutz, B. H.; Woo, K.; Gross, T.; Buzard, D. J.; Tirado, R.; *Synlett* **1997**, 477.

[247] Ito, H.; Arimoto, K.; Sensui, H.; Hosomi, A.; *Tetrahedron Lett.* **1997**, *38*, 3977.

[248] (a) Lipshutz, B. H.; Ellsworth, E. L.; Behling, J. R.; Campbell, A. L.; *Tetrahedron Lett.* **1988**, *29*, 893. (b) Bertz, S. H.; Dabbagh, G.; *Organometallics* **1988**, *7*, 227.

[249] Corey, E. J.; Kyler, K.; Raju, N.; *Tetrahedron Lett.* **1984**, *25*, 5115.

[250] (a) Dieter, R. K.; Silks, L. A.; *J. Org. Chem.* **1986**, *51*, 4687. (b) Dieter, R. K.; Silks, L. A.; Fishpaugh, J. R.; Kastner, M. E.; *J. Am. Chem. Soc.* **1985**, *107*, 4679. (c) Dieter, R. K.; *Tetrahedron* **1986**, *42*, 3029.

[251] Lipshutz, B. H.; Parker, D. A.; Nguyen, S. L.; McCarthy, K. E.; Barton, J. C.; Whitney, S. E.; Kotsuki, H.; *Tetrahedron* **1986**, *42*, 2873.

[252] Achyutha Rao, S.; Knochel, P.; *J. Am. Chem. Soc.* **1991**, *113*, 5735.

[253] Whitesides, G. M.; Casey, C. P.; Krieger, J. K.; *J. Am. Chem. Soc.* **1971**, *93*, 1379.

[254] Mandeville, W. H.; Whitesides, G. M.; *J. Org. Chem.* **1974**, *39*, 400.

[255] Berger, S.; Langer, F.; Lutz, C.; Knochel, P.; Mobley, T. A.; Reddy, C. K.; *Angew. Chem., Int. Ed. Engl.* **1997**, *36*, 1496.

[256] House, H. O.; *Acc. Chem. Res.* **1976**, *9*, 59. Lipshutz, B. H.; Wilhelm, R. S.; Nugent, S. T.; Little, R. D.; Baizer, M. M., *J. Org. Chem.* **1983**, *48*, 3303.

VII

Organotitanium Chemistry

MANFRED T. REETZ

Max-Planck-Institut für Kohlenforschung, Mülheim/Ruhr, Germany

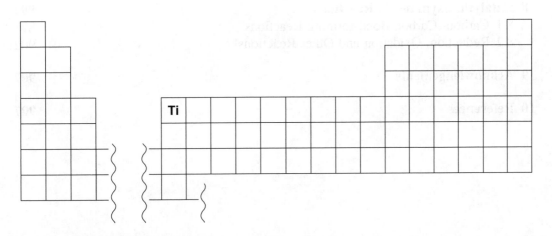

Organometallics in Synthesis: A Manual. Edited by Manfred Schlosser.
© 2002 John Wiley & Sons Ltd

Contents

1 Introduction

Titanium is the seventh most abundant metal on earth and has been used in a multitude of reactions in organic, inorganic, and polymer chemistry. The emphasis in this chapter is on the use of titanium(IV) in C–C bond formation, especially on the titaration of classical carbanions or CH-acidic compounds which generates selective reagents for Grignard, aldol and Michael additions, and substitution reactions. Another important aspect to be treated is the $TiCl_4$-mediated carbonyl addition and alkylation reactions of allylsilanes, enolsilanes, and enolizable C–H-acidic compounds, processes that may or may not involve organotitanium intermediates, depending on the conditions and substrates. A variety of titanium-promoted alkene-forming reactions are also included, as are enantioselective C–C bond-forming reactions induced by stoichiometric amounts of chiral titanium compounds. The use of low-valent titanium reagents in organic synthesis is also described, as are Ti-induced radical processes. Further attention is paid to recent developments of titanium-based enantioselective catalysts in Grignard-type and aldol additions, Diels–Alder and $[2 + 2]$-cycloadditions, ene reactions, and reductions. The Sharpless epoxidation has been presented in detail elsewhere, and only the essential aspects are reiterated here. Space limitation does not allow for the treatment of the Ziegler–Natta polymerization[1], TiX_4-mediated Friedel–Crafts reactions[2], $Ti(OR)_4$-induced transesterifications[3], and most of the inorganic chemistry associated with titanium-based sandwich compounds[4].

A number of titanium compounds such as CH_3TiX_3 have been known for decades, but were not tested in organic synthesis[4,5]. Such physical organic aspects as bond energies, structure, spectroscopy, quantum mechanical treatments[6] and kinetics have been delineated elsewhere[7]. Only a few central points are reiterated here.

The Ti–C bond is not particularly weak in a thermodynamic sense, as was assumed for a long time. However, kinetically low-energy decomposition pathways such as β-hydride elimination may limit the use of certain organotitanium compounds[7]. The Ti–O bond is fairly strong ($\sim$115 kcal mol^{-1}), which means that any reaction in which a Ti–C bond converts to a Ti–O entity is expected to have a strong driving force. The Ti–C bond is typically 2.1 Å long, which is similar to Li–C and Mg–C bonds, but shorter than Zr–C (2.2 Å). Importantly, the Ti–O bond is short ($\sim$1.7 Å) relative to Li–O ($\sim$2.0 Å), Mg–O ($\sim$2.1 Å) and Zr–O ($\sim$2.2 Å) bonds. This means that transition states in such reactions as Grignard or aldol additions are expected to be more compact in the case of titanium. The four ligands in Ti(IV) compounds are tetrahedrally arranged around the metal.

Reagents of the type $RTiCl_3$ and $RTi(NR_2)_3$ are monomeric in solution. Compounds such as $CH_3Ti(OR)_3$ are also monomeric or show little aggregation provided the alkoxy ligand is as bulky as isopropoxy (or bulkier). Analogs having smaller ligands (*e.g.*, ethoxy) are dimeric. The electronic nature of titanium changes greatly in the series $TiCl_4$, Cl_3TiO^iPr, $Cl_2Ti(O^iPr)_2$, $ClTi(O^iPr)_3$ and $Ti(O^iPr)_4$ in that Lewis acidity decreases as electron-donating alkoxy ligands are introduced. The same effect operates in reagents such as CH_3TiCl_3 versus $CH_3Ti(O^iPr)_3$. Amino ligands also decrease Lewis acidity dramatically, as do pentahaptocyclopentadienyl (Cp) groups[7]. For example, CH_3TiCl_3 readily forms octahedral bisetherates, whereas $CH_3Ti(NEt_2)_3$ or $CpTiCl_3$ do not even form stable monoetherates.

2   Synthesis, Stability and Chemoselective Grignard and Aldol Additions of Organotitanium Reagents

2.1   Titanation of Classical Carbanions

2.1.1   Titanating Agents

Organolithium and -magnesium compounds constitute the largest class of organometallic reagents used in organic synthesis[8]. They are usually referred to as carbanions and include not only simple alkyl, aryl, and vinyl reagents, but also the vast number of resonance-stabilized and heteroatom-substituted species used by organic chemists. The reagents are generally basic and reactive, which means that the degree of chemo- and stereoselectivity in reactions with carbonyl compounds may be limited. For example, in the 1940s it was noted that phenylmagnesium bromide hardly discriminates between aldehyde and ketone functionalities[9]. A similar lack of chemoselectivity has been observed for alkyllithium reagents and for lithium ester enolates, deprotonated nitriles, sulfones, and similar reactive species[10,11]. Many of the classical reagents also lead to undesired enolization and to indiscriminate attack at polyfunctional molecules[7,10–15].

In 1979–80, it was discovered that certain organotitanium reagents prepared by transmetalation of RMgX, RLi or R_2Zn undergo chemo- and stereoselective C–C bond-forming reactions with carbonyl compounds and certain alkyl halides[16–20]. These observations led to the working hypothesis that titanation of classical carbanions increases chemo- and stereoselectivity (Scheme 1)[21]. It became clear that steric and electronic properties of the reagents can be adjusted in a predictable way by proper choice of the ligands X at titanium[7,10,13].

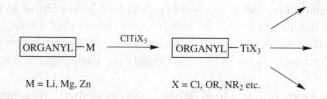

Scheme 1. Titanation as a means to control carbanion selectivity.

If the ligand X is Cl, the organyltitanium trichlorides $RTiCl_3$ are fairly Lewis acidic, forming octahedral bisetherates with diethyl ether or tetrahydrofuran (THF).

This property can be modulated by introducing alkoxy or amino ligands which increase the electron density at the metal. This electronic difference is important in chelation-controlled or non-chelation-controlled additions to alkoxy carbonyl compounds (Section 3) and in certain S_N1 substitution reactions of alkyl halides (Section 7). By varying the size of the alkoxy or amino ligands, steric properties are also easily tuned. Finally, chiral modification is possible (Sections 3.3 and 8). For the purpose of controlling chemoselectivity, the nature of the ligand is not crucial in most cases, whereas stereoselectivity depends critically on the type of ligand. An important parameter not apparent in Scheme 1 is the possibility of increasing the number of ligands in the reagents. For example, if $Ti(OR)_4$ is used as the titanating agent for RLi, ate complexes $RTi(OR)_4Li$ evolve which are bulkier than the neutral $RTi(OR)_3$ counterparts.

Common titanating agents are $TiCl_4$, $ClTi(O^iPr)_3$[10,11], $ClTi[N(CH_3)_2]_3$[22] and $ClTi[N(C_2H_5)_2]_3$[23]:

$$TiCl_4 + 3\ Ti(O^iPr)_4 \longrightarrow 4\ ClTi(O^iPr)_3$$

$$TiCl_4 + 3\ LiNR_2 \longrightarrow ClTi(NR_2)_3$$

$$R = CH_3,\ C_2H_5$$

General Remarks Concerning Experimental Procedures

All manipulations should be carried out in dry flasks under an atmosphere of an inert gas (nitrogen or argon). Generally, $TiCl_4$ taken from fresh bottles can be used without prior purification. However, if moisture has entered the bottle, the $TiCl_4$ should be distilled (bp 135–136 °C) and kept under nitrogen or argon. For manipulation, syringe techniques are best employed. The various sections contain information concerning the stability of organotitanium reagents and hints on what to avoid. Concerning work-up, simple quenching with water usually poses no problems. If problems with TiO_2-containing emulsions arise, saturated solutions of NH_4F or KF should be used, extraction times being kept short[7].

Synthesis of Chlorotriisopropoxytitanium, $ClTi(O^iPr)_3$[10,11]

A 1-L three-necked flask equipped with a dropping funnel, magnetic stirrer and nitrogen inlet is charged with tetraisopropoxytitanium (213 g, 0.75 mol). $TiCl_4$ (47.5 g, 0.25 mol) is then slowly added at about 0 °C. After reaching 25 °C the mixture is distilled (61–65 °C, 0.1 torr) to afford 247 g (95%) of product. The syrupy liquid is >98% pure and slowly solidifies at room temperature. Gentle warming with a heat gun results in liquid formation. This may be necessary during distillation to prevent clogging. Manipulation with a syringe and serum cap presents no problems. Alternatively, the product can be mixed with the appropriate solvent (*e.g.*, pentane, toluene, diethyl ether, THF, CH_2Cl_2) to provide stock solutions. The reagent is hygroscopic, but can be stored in pure form or in solution under nitrogen for months. The actual synthesis can also be performed in solvents. The reagent is commercially available.

Synthesis of Chlorotriphenoxytitanium, $ClTi(OC_6H_5)_3$ [11]

To a solution of chlorotriisopropoxytitanium (71.9 g, 276 mmol) in 500 mL of toluene are added phenol (77.9 g, 828 mmol). The deep red solution is concentrated by distillation through a 10-cm Vigreux column, which removes most of the isopropanol azeotropically. The residue is distilled in a Kugelrohr (250 °C, 0.001 torr) to give 95.8 g (96%) of product which crystallizes on cooling. A 0.26 M stock solution in THF can be kept under exclusion of air.

This method of exchanging OR groups is fairly general, including the *in situ* preparation of derivatives with chiral OR groups.

Synthesis of Tetrakis(diethylamino)titanium, $Ti(NEt_2)_4$ [23b]

To a stirred solution of ethylmagnesium bromide (1.5 mmol) in diethyl ether (500 mL; prepared from ethyl bromide (168.9 g, 1.55 mol) and Mg (37.7 g, 1.55 mol)), Et_2NH (109.7 g, 1.5 mol) dissolved in diethyl ether (200 mL) is added dropwise at 0 °C. When no more ethane is liberated, the mixture is stirred for 2 h at 20 °C and filtered through glass-wool. The solution is added dropwise at 0–5 °C to a vigorously stirred suspension of $TiCl_4 \cdot 2THF$ (83.0 g, 0.25 mol) in benzene (250 mL). After the addition, the stirring is continued for 1 h at 0 °C, 2 h at 25 °C, and 2 h at reflux. The precipitated magnesium salts are filtered and washed with diethyl ether (100 mL). To the combined organic solutions dioxane (150 mL) is added slowly to precipitate the magnesium halides as sparingly soluble dioxane addition complexes. The solvents are removed from the filtrate by distillation *in vacuo*. The residue is distilled at low pressure through a 20-cm Vigreux column, affording 33–42 g (40–50%) of product (bp 112 °C, 0.3 mbar; ^{13}C NMR (benzene-d_6): $\delta = 45.45$ ppm (CH_2), 15.67 ppm (CH_3)).

Synthesis of Chlorotris(diethylamino)titanium, $ClTi(NEt_2)_3$ [23a]

A mixture of diethylamine (73.1 g, 1.0 mol) and lithium (7.7 g, 1.1 mol) in dry diethyl ether (400 mL) is heated at reflux temperature under an atmosphere of nitrogen. A solution of styrene (52.1 g, 0.5 mol) in diethyl ether (150 mL) is slowly added over a period of 2 h. The mixture is refluxed for an additional 30 min, and then cooled to 0 °C. At 0 °C, a solution of $TiCl_4$ (63.3 g, 0.33 mol) in toluene (150 mL) is added, the mixture refluxed for 1 h and the solvent stripped off. Vacuum distillation affords the desired reagent, (yield 72–76 g, 73–77%; bp 94–96 °C, 0.02 torr).

Lithium diethylamide can also be conveniently prepared by treatment of diethylamine with butyllithium. Treatment with $TiCl_4$ as above then affords $ClTi(NEt_2)_3$ in similar yield. An analogous procedure using dimethylamine affords chlorotris(dimethylamino)titanium $ClTi[N(CH_3)_2]_3$ (bp 82–83 °C, 0.01 torr[22]).

Synthesis of Dichlorodiisopropoxytitanium, $Cl_2Ti(O^iPr)_2$ in Solution[10]

The solution of tetraisopropoxytitanium (71 g, 0.25 mol) in 350 mL of diethyl ether is treated with $TiCl_4$ (47.5 g, 0.25 mol) at 0 °C. After stirring for 1 h, diethyl ether is added to a total volume of 0.5 L. This stock solution is 1 M and can be kept in the refrigerator under nitrogen. The same reaction occurs in CH_2Cl_2. The compound can also be obtained as a solid (see procedure for catalytic asymmetric glyoxylate-ene reaction in Section 3.8.1).

One of the dramatic results of titanation is the pronounced increase in chemo- and stereoselectivity of Grignard and aldol additions. For example, most reagents $RTiX_3$ are aldehyde selective in the presence of ketone functions[10–14], this being due to steric factors[24].

The terms 'metal tuning' and 'ligand tuning' have become popular in describing the methodology summarized in Scheme 1. Indeed, it has stimulated the search for other metals in controlling carbanion selectivity. Analogs involving zirconium[25–28], cerium[29], chromium[30], hafnium[31], manganese[32], iron[33], and ytterbium[34] are prominent examples. More recently, organozinc reagents have attracted attention, especially those containing functional groups not tolerated by Li or Mg compounds[35]. The influence of ligands needs to be studied before final comparisons can be made. The important virtues of titanium reagents include low cost, high degrees of chemo- and stereoselectivity, and non-toxic waste on work-up (ultimately TiO_2).

2.1.2 Chlorotitanium Reagents

A few general rules should be kept in mind when attempting to titanate carbanions[7]. Approaches that do *not* work well include titanation of primary alkyllithium reagents with $TiCl_4$, because this leads to the undesired reduction of titanium[13,36]. However, by choosing dialkylzinc reagents R_2Zn as the precursor, $TiCl_4$-mediated addition reactions with carbonyl compounds are possible[37]. These may not necessarily involve titanium trichlorides of the type $RTiCl_3$, but rather activation of the carbonyl compound by $TiCl_4$ complexation followed by reaction of the zinc reagents.

The parent compound CH_3TiCl_3 is easily accessible in better than 95% yield either by the titanation of methyllithium in diethyl ether[12] or by Zn–Ti exchange using $(CH_3)_2Zn$

in CH_2Cl_2[4,16,38]. In the former case the reagent exists in equilibrium with the mono- and bisetherate[12,39]. Although ether-free CH_3TiCl_3 can be distilled prior to reactions with electrophiles, *in situ* reaction modes are preferred:

$$CH_3Li(etherate) \xrightarrow[\text{diethyl ether}]{TiCl_4} CH_3TiCl_3(etherate)$$

$$(CH_3)_2Zn \text{ or } (CH_3)_3Al \begin{cases} \xrightarrow[CH_2Cl_2]{TiCl_4} CH_3TiCl_3 \\ \\ \xrightarrow[CH_2Cl_2]{TiCl_4} (CH_3)_2TiCl_2 \end{cases}$$

As dimethylzinc is pyrophoric[40], the first reaction is the method of choice. However, in certain chelation-controlled carbonyl additions[41] and substitution reactions[16,17], ether-free solutions are required, which make the use of dimethylzinc or trimethylaluminum mandatory. Solutions of $(CH_3)_2Zn$ in CH_2Cl_2 are much less hazardous and can be handled like butyllithium[38,42]. $(CH_3)_2TiCl_2$ is readily accessible by adjusting the amount of $TiCl_4$ relative to the organometallic precursor.

Synthesis and Carbonyl Addition Reactions of Trichloromethyltitanium Etherate, CH_3TiCl_3–etherate[12]

$TiCl_4$ (1.9 g, 10 mmol) is added by a syringe to about 50 mL of cooled ($-78\,°C$) diethyl ether, resulting in partial precipitation of the yellow $TiCl_4$–etherate. The equivalent amount of methyllithium or methylmagnesium chloride in diethyl ether is slowly added, causing the color change to black–purple. This mixture of CH_3TiCl_3–etherate is allowed to warm to about $-30\,°C$. Then an aldehyde (9 mmol) is added and the mixture is stirred for 2 h. In the case of ketones, the mixture is stirred for 2–5 h, during which time the temperature is allowed to reach $0\,°C$. In the case of sensitive polyfunctional ketones, it may be better to add the reagent to a cooled solution of substrate. In all cases the cold reaction mixture is poured onto water. In rare cases a TiO_2-containing emulsion may form, in which case a saturated solution of NH_4F should be used. Following the usual work-up (diethyl ether extraction, washing with water, drying over $MgSO_4$), the products are isolated by standard techniques, generally chromatography or Kugelrohr distillation.

Synthesis of Ether-free Trichloromethyltitanium, CH_3TiCl_3[16]

In a dry 100 mL flask equipped with a nitrogen inlet clean $TiCl_4$ (8.35 g, 4.8 mL, 44 mmol) is mixed with 80 mL of dry CH_2Cl_2. After cooling to $-30\,°C$, 22 mmol of $(CH_3)_2Zn$ (*e.g.*, 5.5 mL of a 4 M CH_2Cl_2 solution[38]) are slowly added via a syringe with stirring. CH_3TiCl_3 is formed quantitatively together with precipitated $ZnCl_2$.

This solution is used for *in situ* reactions, although CH_3TiCl_3 can be distilled[4]. Neat dimethylzinc is pyrophoric, although solutions in dichloromethane can be handled safely[38].

Synthesis of Ether-free Dichlorodimethyltitanium, (CH₃)₂TiCl₂ [38]

The procedure as above is used except that 44 mmol of $(CH_3)_2Zn$ are employed. As $(CH_3)_3TiCl_2$ is less stable than CH_3TiCl_3, the solution should be used in reactions as soon as possible. The reaction can also be carried out in pentane[4c].

Currently it is not known how general the problem of undesired reduction in reactions of organolithium reagents with $TiCl_4$ actually is[36]. Cases in which *no* difficulties are encountered include the titanation of cyclopropyllithium **1** → **2**[43] and of resonance-stabilized carbanions of the type **3** with formation of reagents **4**[44]. Certain enolates can also be titanated by $TiCl_4$[45]. The reagents are stable in diethyl ether solutions at -78 to $-10\,°C$ and undergo smooth aldehyde additions in the presence of ketone functionality.

Occasionally, trichlorides $RTiCl_3$ are not completely chemoselective, specifically when the precursor is extremely reactive, such as the lithium reagent **3**. The trichloride **4** is formed quantitatively, but aldehyde selectivity is only 73%[44,46]. Problems of this kind are solved by using alkoxy ligands as delineated below.

2.1.3 Alkoxytitanium Reagents

The majority of titanation reactions of carbanions has been performed with $ClTi(O^iPr)_3$. This generates compounds of the type $RTi(O^iPr)_3$, usually in excellent yield[7]. Typical examples are reagents **5**[10,11,47], **6**[10], **7**[10,11], **8**[43], **9**[10], **10**[10], **11**[13,48,49,50], **12**[51], **13**[52,53,54], **14**[55,56], **15**[56], **16**[10,11,57], **17**[57], **18**[58], **19**[43], **20**[59], **21**[10], **22**[60], **23**[10], **24a**[43], **24b**[61], **25**[62], **26**[63], **27**[7,43], **28**[7,43], **29**[7,43], **30**[44], **31**[64], **32**[65], **33**[66], **34**[67], and **35**[68]. The precise structures have not always been elucidated. However, in the case of several ambident nucleophiles, ^{13}C NMR spectroscopy has been helpful in determining the position of titanium. For example, α-lithio nitriles and hydrazones are titanated at nitrogen (see **23**, **24** and **31**), whereas α-lithio sulfones are titanated at carbon (*e.g.*, **21**)[46].

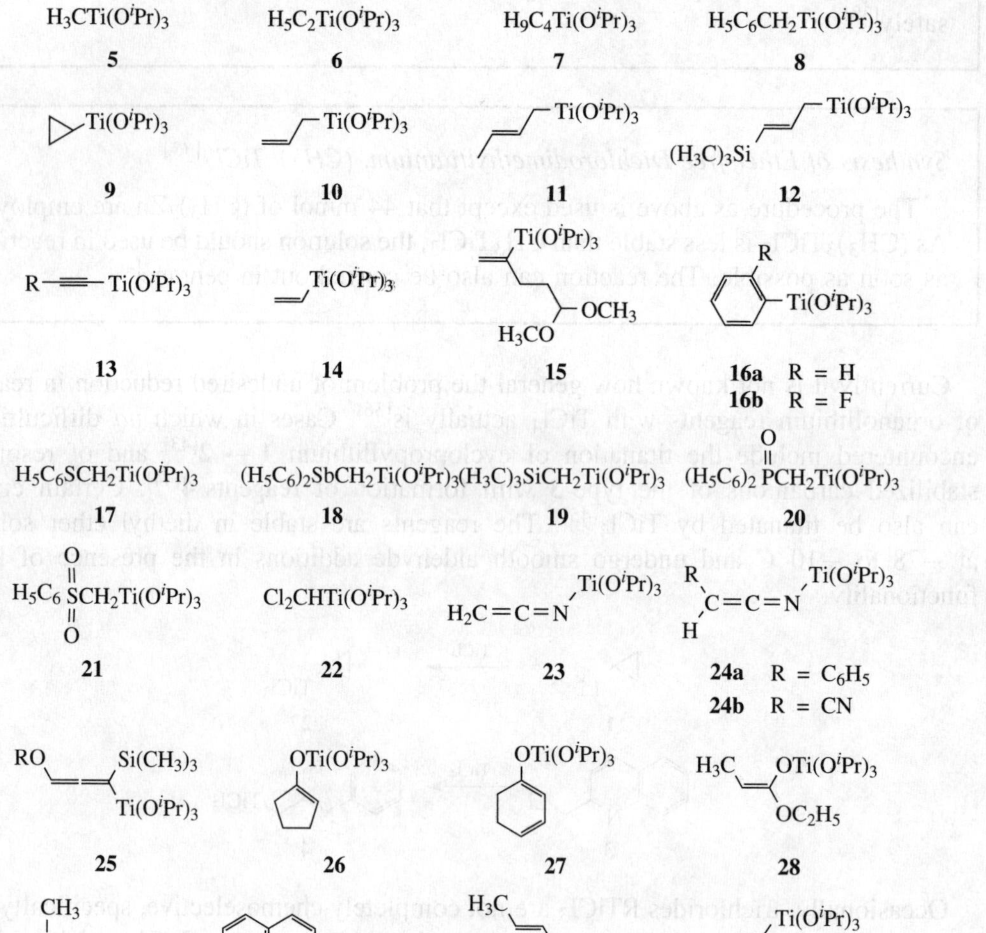

$$H_3CTi(O^iPr)_3 \qquad H_5C_2Ti(O^iPr)_3 \qquad H_9C_4Ti(O^iPr)_3 \qquad H_5C_6CH_2Ti(O^iPr)_3$$

5 **6** **7** **8**

▷–Ti(O^iPr)_3	∕=∖–Ti(O^iPr)_3	∕=∖–Ti(O^iPr)_3	(H_3C)_3Si∕=∖–Ti(O^iPr)_3
9	**10**	**11**	**12**

$$R-\!\!\equiv\!\!-Ti(O^iPr)_3 \qquad \text{=∕–Ti(O^iPr)_3}$$

13 **14**

(structure **15** with Ti(O^iPr)_3, OCH_3, H_3CO)

15

(aryl structures)

16a R = H
16b R = F

$$H_5C_6SCH_2Ti(O^iPr)_3 \qquad (H_5C_6)_2SbCH_2Ti(O^iPr)_3 \qquad (H_3C)_3SiCH_2Ti(O^iPr)_3 \qquad (H_5C_6)_2\overset{O}{\overset{\|}{P}}CH_2Ti(O^iPr)_3$$

17 **18** **19** **20**

$$H_5C_6\overset{O}{\underset{O}{\overset{\|}{\underset{\|}{S}}}}CH_2Ti(O^iPr)_3 \qquad Cl_2CHTi(O^iPr)_3 \qquad H_2C\!=\!C\!=\!N\!-\!Ti(O^iPr)_3 \qquad \underset{H}{\overset{R}{\diagdown}}C\!=\!C\!=\!N\!-\!Ti(O^iPr)_3$$

21 **22** **23** **24a** R = C_6H_5
24b R = CN

(structures)

25 **26** **27** **28**

(structures)

29 **30** **31** **32**

(structures)

33 **34** **35**

Reagents **24**, **25**, and **33–35** are chiral, although **24** and **25** are probably fluxional, thereby excluding the possibility of enantiomerically pure preparation. Further reagents are described in Section 3. The compound CH_3Ti(O^iPr)_3 (**5**) is so stable that it can be distilled[10,49], but not the alkyl homologs such as **6** and **7**. For synthetic purposes an *in* *situ* reaction mode is sufficient. This means that lithium salts are present in the reaction mixture. Pronounced differences in the selectivity of carbonyl addition reactions in the presence or absence of lithium or magnesium salts have not been observed to date, although this needs to be studied.

All of the reagents add smoothly to most aldehydes. The rate of addition to ketones is much lower, which means high degrees of aldehyde selectivity. Typical examples concern the reaction of the ketoaldehyde **36** with reagents **5** and **21** leading solely to products **37** and **38**, respectively. In contrast, the lithium precursors do not discriminate between the two carbonyl sites and lead to complex mixtures[10]:

Synthesis and Carbonyl Addition Reactions of Triisopropoxymethyltitanium,
$CH_3Ti(O^iPr)_3$ **(5)**[10,11,47]

Method A

A 2-L three-necked flask equipped with a dropping funnel and magnetic stirrer is charged with 250 mL of diethyl ether and chlorotriisopropoxytitanium (130.3 g, 0.50 mol) (see above) and cooled to −40 °C. The equivalent amount of methyl-lithium (e.g., 312.5 mL of a 1.6 M diethyl ether solution) is slowly added and the solution is allowed to warm to 25 °C within 1.5 h. The solvent is removed *in vacuo* and the yellow product distilled directly from the precipitated lithium chloride at 48–53 °C and 0.01 torr (113 g, (94%); ^{1}H NMR (CCl$_4$) δ (ppm) = 0.5 (s, 3H), 1.3 (d, 18H), 4.5 (m, 3H)). The compound is air sensitive, but can be kept under nitrogen in a refrigerator for weeks or months. On standing, slow crystallization begins, which can be reversed by gentle warming with a heat gun. Stock solutions can be prepared by mixing with the desired solvent.

Method B

The above procedure is applied on a smaller scale (10–100 mmol, as needed for immediate use), but the diethyl ether solution of $CH_3Ti(O^iPr)_3$ containing precipitated lithium chloride is used for carbonyl additions without any further treatment or purification (see below).

Addition of CH$_3$Ti(O^{i}Pr)$_3$ to 6-Oxo-6-phenylhexanal, 36[10]

The solution of 6-oxo-6-phenylhexanal (1.9 g, 10 mmol) in 40 mL of THF is treated with distilled CH$_3$Ti(O^{i}Pr)$_3$ (2.4 g, 10 mmol) at −78 °C. The cooling bath is removed and stirring is continued for 6 h. The reaction mixture is poured on-to dilute HCl, diethyl ether is added and the aqueous phase is extracted with diethyl ether. The combined organic phases are washed with water and dried over MgSO$_4$. The solvent is removed and the residue distilled in a Kugelrohr (220 °C, 0.02 torr) to yield 1.65 g (80%) of the adduct **37**. An *in situ* reaction mode leads to the same result.

Addition of CH$_3$Ti(O^{i}Pr)$_3$ to 3-Nitrobenzaldehyde[15]

A dry, 500-mL three-necked flask equipped with a pressure-equalizing 100 mL dropping funnel, argon inlet and magnetic stirrer is evacuated and flushed with argon (three cycles). The flask is charged with tetraisopropoxytitanium (16.0 mL, 57.7 mmol) via a plastic syringe and hypodermic needle and TiCl$_4$ (2.1 mL, 19.2 mmol) is added over 5 min, with gentle cooling of the flask in an ice-water bath, to give a viscous oil. After the addition of 70 mL of THF, the clear solution is stirred at 25 °C for 30 min. The dropping funnel is charged with methyllithium, (62 mL, 77 mmol, 1.24 M in hexane) which is added to the cooled (ice bath) THF solution over a period of 25–30 min. During the addition the resulting suspension changes from orange to bright yellow. After stirring at ice-bath temperature for 1 h, a solution of 3-nitrobenzaldehyde (10.6 g, 70 mmol) in 60 mL of THF is added from the dropping funnel within 20–25 min at the same temperature. The mixture is stirred at 0–5 °C for 1 h and then 60 mL of 2 M hydrochloric acid are added. The organic phase is separated in a separating funnel and the aqueous phase is extracted with three 150 mL portions of diethyl ether. The combined organic phases are washed with 100 mL of saturated NaHCO$_3$ solution and 100 mL of saturated NaCl solution and then dried over anhydrous MgSO$_4$. After filtration the solution is concentrated on a rotary evaporator and dried at 0.1 mm for 1 h. The residue, 11.0–11.1 g (94–95%) of an orange–brown viscous oil, sometimes solidifies on standing (mp 55–60 °C); the purity of the crude product is at least 95% (estimated by ^{1}H NMR). The product (3′-nitro-1-phenylethanol) can be purified by short-path distillation at 120–125 °C (0.15 mm Hg) to give 9.9–10.4 g (85–89%) of a yellow oil, which solidifies on standing at room temperature or at −30 °C in a freezer (mp 60.5–62.0 °C).

When performing reactions with polyfunctional compounds, it is generally important to add the organotitanium reagent to the organic substrate in order to obtain the maximum degree of chemoselectivity. The other reaction mode, *i.e.*, addition of the substrate to the reagent, means a local excess of the latter, which may lead to reactions at two or more sites[10,69].

A number of other points should be remembered when performing carbonyl addition reactions with RTi(O^{i}Pr)$_3$ reagents[7]. The reactivity varies widely, depending on the nature of the R group. Reagents derived from resonance-stabilized carbanions, such as **10–12**, **21**, **23–31**, and **33–35** are most reactive, rapidly adding to aldehydes at low temperatures (−78 °C to −20 °C, 1 h); addition to ketones is considerably slower, but generally fast enough for yields to be good under reasonable conditions (−50 °C to −10 °C, 3–4 h). The less reactive compound, CH$_3$Ti(O^{i}Pr)$_3$, adds smoothly to aldehydes at low temperatures, but ketones require 25 °C and long reaction times (6–72 h). Working in the absence of solvents speeds up the addition to bulky ketones dramatically[70]. Saturated analogs such as reagents **6–9** and **17–19** are even less reactive and generally fail to add to ketones or bulky aldehydes. When reacting the somewhat sluggish alkynyl compounds **13** with aldehydes, an excess of reagent is required[52–54].

Certain vinyltitanium reagents have been reported to decompose via reductive dimerization at −60 °C in THF, rendering them useless for synthetic purposes[25]. However, the rate of reductive elimination depends on the solvent, such undesired decomposition being negligible in diethyl ether[56]. Thus, besides the parent reagent **14**[55,56], a fairly wide variety of vinyltitanium reagents undergo Grignard-type addition to aldehydes in diethyl ether at −78 °C (4–8 h)[56]. The reactivity of the phenyltitanium reagent **16a**[11] is similar to that of the methyl analog **5**, but ortho-substituted derivatives such as **16b** require room temperature for effective aldehyde additions[11,57].

General Procedure for the Formation and Aldehyde Addition of Vinyltitanium Reagents[56]

A solution (1.5 M) of *tert*-butyllithium (5 mmol) in hexane is cooled to −78 °C under N$_2$ and the vinyl bromide (2.5 mmol) in 3 mL of freshly distilled diethyl ether is added dropwise over 5 min. An immediate precipitate of LiBr forms and the mixture is stirred at −78 °C for 20 min. A 1 M solution of ClTi(O^{i}Pr)$_3$ (2.5 mL) in hexane is then added and the solution, which turns brown immediately, is stirred for 20 min. The aldehyde (1 mmol) is added and the reaction mixture stirred at −78 °C until the disappearance of the aldehyde is apparent by TLC (8 h maximum). After quenching with saturated NH$_4$Cl/Et$_2$O (33 mL, 1 : 10), the ether layer is decanted. Two fresh portions of diethyl ether are similarly employed to complete extraction of the product(s). The combined ethereal solutions are dried, concentrated and the residue purified by chromatography (SiO$_2$), to afford the expected allylic alcohol.

In most cases an equivalent amount or a slight excess (10–20%) of titanating agent suffices[10,11]. However, it has been observed that in the titanation of the extremely reactive quinoline reagent **3** a double or even triple excess of ClTi(O^{i}Pr)$_3$ is necessary to ensure maximum chemoselectivity[44]. The reason for this is not clear, but may involve ate complexes **39**. Collapse of this intermediate with formation of the reagent **30** may be promoted by additional $ClTi(O^iPr)_3$ acting as a Lewis acid[44]. These observations suggest that whenever similar problems arise in other situations, an excess of titanating agent should be tried. In fact, cases in which stereoselectivity is influenced favorably by an excess of $ClTi(O^iPr)_3$ have been reported[71] (Section 3.3).

Secondary and tertiary alkyllithium and -magnesium reagents cannot be used in the titanation/carbonyl addition sequence owing to undesired β-hydride elimination, reduction and/or rearrangement[7]. This applies to titanating agents of the type $ClTi(O^iPr)_3$, $Cl_2Ti(O^iPr)_2$ and $TiCl_4$. However, sometimes these transition metal properties can be exploited in a useful manner, as in the titanium mediated hydromagnesiation of alkenes and alkynes[72-74] (see also Section 5.3).

Although zinc reagents of the type R_2Zn and $RZnX$ generally do not add smoothly to carbonyl compounds, amino alcohols[75] or Lewis acids such as $TiCl_4$[37] lead to rapid Grignard-type additions. A synthetically important variation concerns generation of zinc reagents $RZnBr$ followed by addition of $ClTi(O^iPr)_3$[76] or $Cl_2Ti(O^iPr)_2$[77]. This generates the corresponding titanium reagents which add smoothly to aldehydes. As the R group in the zinc reagent may contain additional functionality (as in Zn–homoenolates), this approach is very powerful[77]. Another potentially important development concerns the addition of Et_2Zn to aldehydes mediated by $Ti(O^iPr)_4$[35,37]; chiral versions occur enantioselectively (Section 8.1).

The reactivity of reagents of the type $CH_3Ti(OR)_3$ varies according to the nature of the alkoxy group[7,14,24]. For example, the reaction of $CH_3Ti(O^iPr)_3$ with heptanal at $-30\,^\circ C$ is about 40–50 times faster than that of the triethoxy analog $CH_3Ti(OC_2H_5)_3$[24]. This is because the latter is largely dimeric via Ti–O bridging and therefore less reactive, whereas the bulkier $CH_3Ti(O^iPr)_3$ is only partially aggregated[7,24]. Such phenomena should be useful in designing catalytic processes. The triphenoxy reagent $CH_3Ti(OPh)_3$ is considerably less reactive than $CH_3Ti(O^iPr)_3$ due to steric reasons[49]. Further variation of alkoxy groups needs to be examined, especially with respect to stereoselectivity (Section 3). Acyloxy ligands have yet to be tested systematically, although preliminary result using CH_3CO_2 ligands appear promising[43a].

Dialkyl- and dialkynyltitanium dialkoxides, such as **40**[10], **41**[78], and **42**[79], are easily prepared via $Cl_2Ti(O^iPr)_2$, which is accessible by mixing equivalent amounts of $TiCl_4$ and $Ti(O^iPr)_4$[10]. The X-ray structural analysis of **41** shows the presence of a C–Ti bond (not O–Ti)[78]:

$$TiCl_4 + Ti(O^iPr)_4 \longrightarrow Cl_2Ti(O^iPr)_2$$

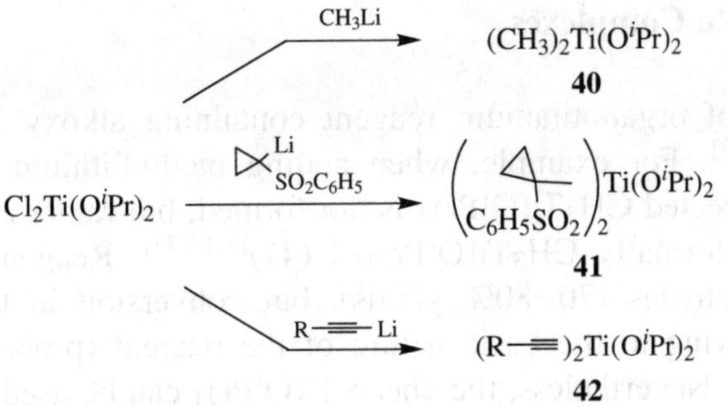

$$CH_3Li \longrightarrow (CH_3)_2Ti(O^iPr)_2$$
40

$$Cl_2Ti(O^iPr)_2 \xrightarrow{\quad \text{(Li, SO}_2\text{C}_6\text{H}_5\text{)} \quad} \left(C_6H_5SO_2 \right)_2 Ti(O^iPr)$$
41

$$\xrightarrow{R\!\!-\!\!\!\equiv\!\!-Li} (R\!\!-\!\!\!\equiv)_2Ti(O^iPr)_2$$
42

Reactivity toward carbonyl compounds increases dramatically in the series $CH_3Ti(O^iPr)_3 < (CH_3)_2Ti(O^iPr)_2 < Ti(CH_3)_4$[10,26]. Thus, reagent **40** reacts smoothly with sterically hindered ketones such as **43** (22 °C, 4 h; 98% yield of adduct **44**), whereas $CH_3Ti(O^iPr)_3$ requires 2 days for 90–94% conversion[10]. A more recent example of the virtue of reagent **40** concerns its reaction with the trifunctional compound **45**, which proceeds with complete chemo- and stereoselectivity[80]:

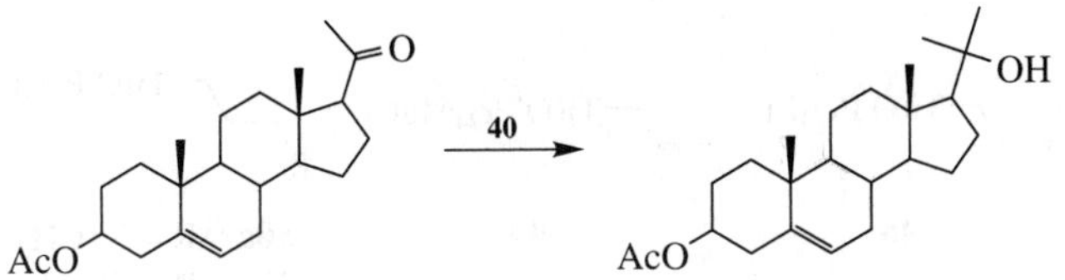

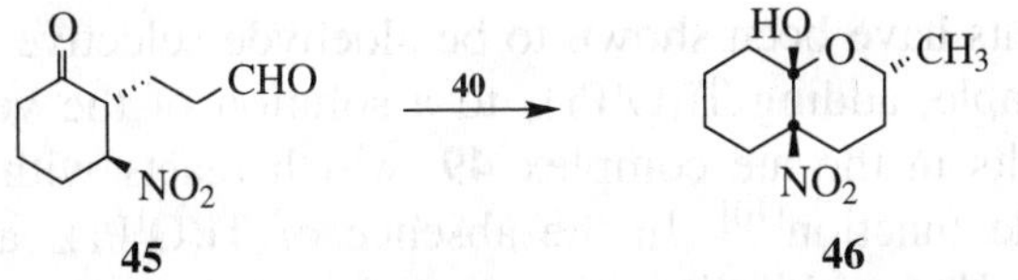

Similarly, dialkynyltitanium reagents **42** provide much higher yields in aldehyde additions than the alkynyltrialkoxytitanium counterparts[79].

A powerful method for the generation of functionalized carbon nucleophiles concerns the reaction of the corresponding iodides with zinc (activated by 4% 1,2-dibromoethane and 3% chlorotrimethylsilane) to produce zinc reagents, which can be transmetalated with $CuCN/2LiCl$[81] or $Cl_2Ti(O^iPr)_2$[82]. In the latter case the reagents readily add to aldehydes in high yield[82]. A disadvantage of this method is the fact that an excess of reagent (1.5 equiv) is required, in contrast to an alternative process involving three metals (1.3 equiv of $RCu(Cn)ZnI$ and 2 equiv of BF_3/OEt_2)[81].

Although compounds of the type $(CH_3)_3TiO^iPr$[4] have not been utilized in synthetic organic chemistry, the extremely reactive and non-basic tetramethyltitanium $(CH_3)_4Ti$ adds to sterically hindered enolizable ketones which normally fail to undergo Grignard reactions with CH_3Li, CH_3MgX or $(CH_3)_2Ti(O^iPr)_2$[10]. Organocerium reagents are ideal reagents for easily enolizable ketones[29], but highly sterically hindered substrates have not been tested. Tetramethylzirconium surpasses even $(CH_3)_4Ti$ as a supermethylating agent in such extreme situations[10].

2.1.4   Titanium Ate Complexes

A different type of organotitanium reagent containing alkoxy ligands concerns ate complexes[7,10,13,49,63]. For example, when adding methyllithium to $Ti(O^iPr)_4$ at low temperatures, the expected $CH_3Ti(O^iPr)_3$ is not formed, but rather an ate complex of yet undefined structure, formally $CH_3Ti(O^iPr)_4Li$ (**47**)[7,13,43]. Reagent **47** adds Grignard-like to aromatic aldehydes (70–80% yields), but conversion in the case of aliphatic aldehydes is poor owing to the basic nature of the reagent (probably competing aldol condensation)[7,13,43]. Nevertheless, the cheap $Ti(O^iPr)_4$ can be used as a titanating agent if the carbon nucleophile is particularly reactive, as in the case of resonance-stabilized precursors such as allylic lithium or magnesium reagents, lithium enolates and other heteroallylic species. Typical examples are **48**[10,51,83], **49**[10,51], **50a**[10,49,51], **50b**[84], and **51**[63] (see also Section 3). The structure of titanium ate complexes of enolates has been studied by NMR spectroscopy[85]. In some cases the ate complex may transform into the neutral reagent $RTi(O^iPr)_3$, especially if an excess of $Ti(O^iPr)_4$ is used[44,71].

$H_3CTi(O^iPr)_4Li$ $\diagdown\!\!\!\diagup Ti(O^iPr)_4Li$ $\diagdown\!\!\!\diagup Ti(O^iPr)_4MgCl$

47   **48**   **49**

$\diagdown\!\!\!=\!\!\!\diagup^{Ti(O^iPr)_4Li}_{R}$ $\bigcirc\!\!\!-OTi(O^iPr)_4Li$

50a $R = Si(CH_3)_3$ **51**
50b $R = SC_2H_5$

Many of the reagents have been shown to be aldehyde selective and/or stereoselective (Section 3). For example, adding $Ti(O^iPr)_4$ to a solution of the very reactive allylmagnesium chloride results in the ate complex **49**, which reacts with the ketoaldehyde **36** solely at the aldehyde function[10]. In the absence of $Ti(O^iPr)_4$ a complex mixture of products is obtained. Hence this titanation is a fast and cheap way to tame the very reactive allylmagnesium chloride.

---

*General Procedure for Formation and Carbonyl Addition of the Allyltitanium Ate Complex $CH_2 = CHCH_2Ti(O^iPr)_4MgCl$, **49**[10]*

A solution of 18 mmol of allylmagnesium chloride in about 40 mL of THF is treated with tetraisopropoxytitanium (5.68 g, 20 mmol) at $-78\,°C$. The formation of the orange ate complex **49** is complete after 30 min. Addition of aldehydes ($-30\,°C$, 1 h) or ketones ($-10\,°C$, 1–2 h) results in 85–95% conversion. The reaction mixtures are poured onto dilute HCl, diethyl ether is added, and the aqueous phase is extracted three times with diethyl ether. The combined organic phases are washed with water and dried over $MgSO_4$. In the case of di- or polyfunctional molecules, the solution containing **49** should be added to the substrate (reversed reaction mode).

2.1.5   Aminotitanium Reagents

Most alkyl-, aryl- and vinyllithium (or -magnesium) reagents can be titanated with $ClTi(NR_2)_3$ to form distillable compounds of the type $R^1Ti(NR_2)_3$[86]. Surprisingly, even the *tert*-butyltitanium analogs are thermally stable. Steric factors appear to prevent undesired β-hydride elimination. Unfortunately, many of the trisamino reagents do not undergo Grignard-type additions, because the amino groups are transferred onto the carbonyl moiety[10,13,87]. Nevertheless, these ligands can be used if the carbon nucleophile is very reactive. This requirement is fulfilled in the case of resonance-stabilized carbanions, such as **52–54**[10,51,63,64]. All of them are accessible by titanation of the Li or Mg precursor with $ClTi(NEt_2)_3$ and transfer the carbon nucleophile in preference to undesired amino addition. As chemoselectivity is achieved using $TiCl_4$, $Cl_2Ti(O^iPr)_2$ or $ClTi(O^iPr)_3$, the more expensive titanating agents $ClTi(NR_2)_3$ should be employed only if stereoselectivity is to be controlled (Section 3).

General Procedure for the Titanation of Carbanions and in situ Carbonyl Addition Reactions[10]

Standard techniques employing butyllithium or lithium diisopropylamide (LDA) are used to lithiate CH-acidic compounds. For example, the solution of acetonitrile (0.82 g, 20 mmol) in 50 mL of THF is treated with 11 mL of a 1.8 M butyllithium solution at $-78\,°C$. After 1 h, chlorotriisopropoxytitanium (0.52 g, 20 mmol) or chlorotris(diethylamino)titanium (0.66 g, 22 mmol) is added and the mixture is stirred for about 45 min at the same temperature, producing reagent **23** in above 90% yield. An aldehyde (19 mmol) is added and the mixture is stirred for 2 h and worked up in the usual way. Methyl phenyl sulfone is lithiated by slow addition of butyllithium at $0\,°C$. Stirring is continued for 30 min and the solution cooled to $-78\,°C$, followed by treatment with $ClTi(O^iPr)_3$; the temperature is then allowed to reach $-40\,°C$, which leads to reagent **21**. After cooling to $-78\,°C$, an aldehyde or a ketone is added and the stirred solution is allowed to thaw to room temperature over a period of 3 h. The usual workup affords the carbonyl addition product in 80–95% yield. Ethyl acetate is lithiated in THF using LDA at $-78\,°C$; titanation by $ClTi(O^iPr)_3$ occurs (0.5 h) at the same temperature to produce the triisopropoxytitanium enolate of ethyl ester. Aldol addition to aldehydes generally sets in at $-78\,°C$ to $-30\,°C$ within 2 h; ketones require 3–4 h (at $-78\,°C$ to $0\,°C$).

A 10% excess of organometallic reagent is used in all cases. Sometimes a two- or threefold excess of titanating agent, $Ti(O^iPr)_4$ or $ClTi(O^iPr)_3$, is necessary to maximize chemo- or stereoselectivity[44,71].

2.1.6   Cyclopentadienyltitanium Reagents

As mentioned previously, the Cp group is a strong electron-donating ligand in titanium chemistry. Indeed, the properties of mixed reagents of the type $CpTi(CH_3)Cl_2$,
$CpTi(CH_3)(OR)_2$, $CpTi(CH_3)(NR_2)_2$, $Cp_2Ti(CH_3)Cl$, $Cp_2Ti(CH_3)OR$ or $Cp_2Ti(CH_3)NR_2$
are dominated to a large degree by the Cp ligands. The concept of titanating reactive
carbanions in order to increase chemo- and stereoselectivity in carbonyl addition reactions
or in substitution processes has not been tested systematically using such Cp-containing
transmetalating agents as $CpTiCl_3$ (**55**), $Cp_2TiCl(OR)_2$, $Cp_2TiCl(NR_2)_2$, Cp_2TiCl_2 (**58a**),
$Cp_2TiCl(OR)$, or $Cp_2TiCl(NR_2)$, although a few select cases are known. This also pertains
to $CpTi^{III}$ species, which means that a great deal of research remains to carried out in this
area. Most of the applications of Cp-bearing titanium reagents, besides Ziegler–Natta
polymerization, involve alkenation, stereoselective reduction of alkenes, ketones, and
imines, hydrosilylation of the ketones, esters, lactones and imines, hydrometalation of
alkenes, dienes, and alkynes as well as dehalogenation, deoxygenation, and epoxidation
reactions. A recent review summarizes this area comprehensively[88], and only the major
aspects are delineated here. The most important developments concern catalytic processes.
Reagents **55**[4], **56**[89], **57**[90], **58**[4,88] and **59**[91] are prominent titanating agents or catalysts, whereas compounds **60**[4], **61**[92], **62**[89], **63**[88] and **64**[93] represent typical reagents
prepared by transmetalation.

The presence of Cp groups in methyltitanium reagents sharply reduces their reactivity in Grignard-type reactions. Thus, **60** reacts slowly with aldehydes, but not with ketones[7]. The introduction of two Cp ligands reduces carbonylophilicity even more, **63** not reacting readily in Grignard-type processes. Replacing the methyl group in these compounds by such C-nucleophiles as allyl or enolate residues increases reactivity toward aldehydes dramatically. Nevertheless, as chemoselectivity is most easily achieved by chloro, alkoxy, or amino ligands, there is no need to revert to Cp ligands, unless stereoselectivity is the goal, as in the aldol addition of **62**[89], oxidation of **64**[93], and the Grignard addition of reagent **65** having the stereogenic center at titanium[94]. These and other enantioselective reactions are discussed in Section 3.

2.1.7   Further Types of Chemoselectivity

Organotitanium reagents are also chemoselective in reactions with diketones, dialdehydes and carbonyl compounds containing additional functional groups[10,12,14,24]. In the case of aldehyde/aldehyde and ketone/ketone discrimination, competition experiments point to surprisingly high degrees of molecular recognition[10,13,24,44]. Small differences in steric environment can tip the balance in favor of the less-shielded reaction site. The mechanism of all of the Ti-based reactions probably involves reversible complexation of the carbonyl group by the metal followed by irreversible C–C bond formation[24].

The simplest case concerns the titanation of CH_3Li or CH_3MgX with $TiCl_4$, leading to the etherate of CH_3TiCl_3 which is a very selective reagent in cases in which the classical precursors fail. A case in point is the Grignard-type addition to *p*-nitroacetophenone which fails completely using CH_3Li or CH_3MgCl due to undesired electron-transfer processes[12].

Other than steric factors, chelation effects in the case of α- and β-alkoxy and amino carbonyl compounds also influence the sense and degree of molecular recognition[13,24,41]. In fact, detailed kinetic experiments[24] show that steric effects are completely overridden in the case of α-alkoxy and amino ketones which react much faster than analogous carbonyl compounds devoid of alkoxy or amino functionality.

In stereochemically relevant cases the chelation-controlled diastereomer is formed preferentially[24] (see also Section 3), corroborating the hypothesis that chelation enhances the rate of addition. α-Alkoxy aldehydes do *not* react faster, and in fact lead to non-chelation-controlled adducts[24] (Section 3.1.3). Similar observations have been made in related cases involving titanium and other metal reagents[61b,95,96]. Other activating factors such as field effects have been invoked in the aldol addition of lithium enolates[97].

Many other types of functionality are tolerated in carbonyl addition reactions of Ti reagents, including ester, cyano, bromo, and iodo groups[7,10,11]. An unusual case of chemo- and regioselectivity concerns the reaction of the novel Ti reagent **67** with the anhydride **66** leading to compound **68**, which is a precursor of the alkaloid ellipticine[98]. The Li precursor of **67** fails in this reaction:

66 + 67 → 68

Saturated reagents of the type $RTiCl_3$ and $RTi(O^iPr)_3$ generally do not add to aldimines. Exceptions are more reactive reagents such as enolates[99] or allylic compounds[100], or substrates with reactive C−N double bonds such as pyrimidones[52].

2.2   Alternative Syntheses of Organotitanium Reagents

In addition to the titanation of traditional carbanions, several other routes to organotitanium reagents are available. Some involve metal–metal exchange reactions of silicon, tin or lead compounds, specifically with the reactive $TiCl_4$ to form trichlorotitanium reagents, such as **69**[101], **70**[102], **71**[103], or **72**[104a]. Adding $Ti(O^iPr)_4$ or $Ti(NR_2)_4$ to such reagents has not been tried, which would make ligand tuning possible. All of the trichlorotitanium reagents add smoothly to aldehydes. $Et_4Pb/TiCl_4$ ethylates chiral aldehydes stereoselectively[104b−e].

In the case of enolsilanes, rapid Si−Ti exchange using $TiCl_4$ is possible only with (Z)-configured compounds[101]. The trichlorotitanium enolate derived from cyclopentanone decomposes as it is formed. An exception is the enolsilane from cyclohexanone which undergoes fairly fast Si−Ti exchange with formation of **69**[101]. Reaction with aldehydes is related to the Mukaiyama aldol addition[105] (Section 2.3), in which enolsilanes are treated with aldehydes in the presence of $TiCl_4$, a process which does *not* involve titanium enolates[105d].

Titanium homoenolates such as **74** are accessible by $TiCl_4$-mediated ring cleavage of the silicon compounds **73**[106]. The reagent is dimeric and adds smoothly to aldehydes to form adducts **75**, but reactions with ketones are sluggish. Treatment of the titanium homoenolate **74** with $Ti(O^iPr)_4$ results in the dichloroisopropoxy analog, which is monomeric and therefore more reactive, adding to ketones and sterically hindered aldehydes[107]. In a synthetically important extension, siloxycyclopropanes containing further functionalities have been shown to undergo Ti-mediated homoaldol additions[108a,b]. β-Stannylated ketones react similarly[108c].

General Procedure for Homoaldol Additions Based on Alkoxytitanium Homoenolates Derived from Siloxycyclopropanes[107]

Formation of alkyl-3-(trichlorotitanio)propionates **74**

To a water-cooled solution of $TiCl_4$ (110 μL, 1.0 mmol) in 2.0 mL of hexane is added a 1-alkoxy-1-siloxycyclopropane, (*e.g.*, **73** (R = ^iPr), 1.0 mmol) at 21 °C during 20 s. The initially formed milky white mixture turns brown in 10 s, and finally deep purple microcrystals precipitate. Heat evolution continues for several minutes. The mixture is allowed to stand for 30 min; NMR analysis of the supernatant reveals the quantitative formation of chlorotrimethylsilane (with 1,1,2,2-tetrachloroethane as an internal standard). The supernatant is removed with a syringe and the crystals are washed three times with hexane. The homoenolate **74** (R = ^iPr) weighs 0.223 g (83%). This procedure can be scaled up to the 10 g level without modification. Recrystallization from CH_2Cl_2 hexane gives an analytical sample as thin, deep purple needles. The titanium alkyl melts at 90–95 °C with a color change to reddish brown, and sublimes with some decomposition at 90–110 °C (0.005 torr).

Reaction of the alkoxide-modified homoenolate

The purified homoenolate **74** (1.5 mmol) is dissolved in 3 mL of CH_2Cl_2 at 0 °C and $Ti(O^iPr)_4$ (0.75 mmol) is added. After 5 min, 1 mmol of a carbonyl compound is added. The mixture is stirred for 1 h (up to 18 h in the case of bulky ketones). The mixture is poured into a stirred mixture of diethyl ether and water. After 10 min, the ethereal layer is separated and the aqueous layer is extracted three times with diethyl ether. The combined extract is washed with water, aqueous $NaHCO_3$, and saturated NaCl. After drying and concentration, the product is purified to obtain the desired 4-hydroxy ester or lactone. For unreactive substrates, 2 equivalent of **74** and 1 equivalent of $Ti(O^iPr)_4$ are used. For unreactive ketones, $Ti(O^tBu)_4$ should be employed.

Along a different line, isonitriles were reported to react with $TiCl_4$ to form the α,α-adducts **76**[109], which add chemoselectively to aldehydes[110]. Following aqueous work-up, the products are *N*-methyl-α-hydroxycarboxamides **77**:

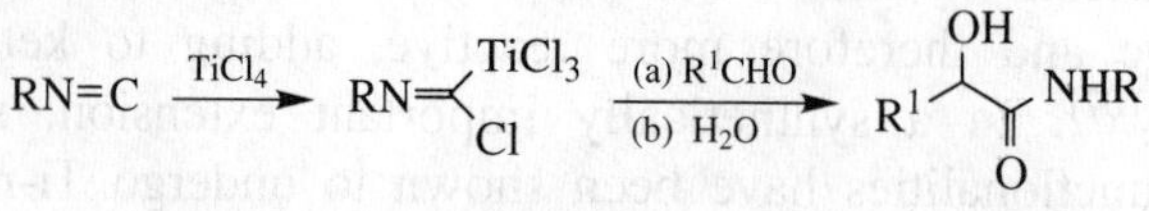

$$RN{=}C \xrightarrow{TiCl_4} RN{=}\!\!\begin{array}{c} TiCl_3 \\ \diagdown \\ Cl \end{array} \xrightarrow[\text{(b) }H_2O]{\text{(a) }R^1CHO} R^1\!\!\begin{array}{c} OH \\ | \\ \diagup \diagdown \\ \| \\ O \end{array}\!\!NHR$$

76   **77**

In a completely different approach, CH-acidic compounds are treated with $TiCl_4$ and a tertiary amine, thereby inducing titanation. An early example pertains to the Knoevenagel condensation of malonates with aldehydes in the presence of $TiCl_4$ and triethylamine (Sections 3.2.2 and 5.1)[111], a process which probably involves intermediate chlorotitanium enolates. A useful way to generate trichlorotitanium enolates involves the reaction of carbonyl compounds with $TiCl_4$ in the presence of amines[112], such as **78** → **79** (Evans method)[113]:

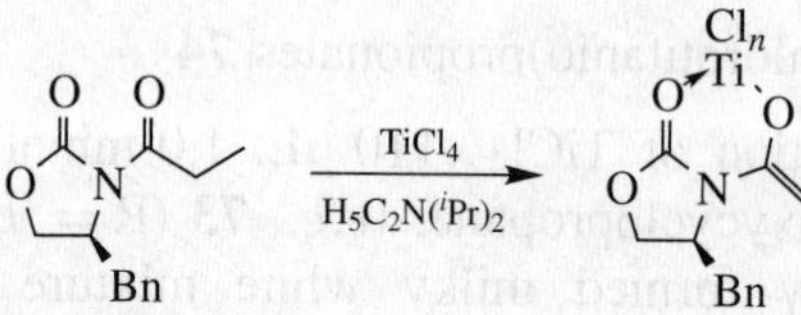

78   **79**

$$Bn = CH_2C_6H_5$$

The titanium enolates undergo S_N1 substitution reactions (Section 6) as well as aldol (Section 3) and Michael additions (Section 4), chiral versions being possible. Claisen condensations utilizing $Cl_2Ti(OTf)_2$ or $TiCl_4$ are also based on the titanation of carbonyl compounds[114].

Interestingly, even $Ti(OR)_4$ induces enolization of aldehydes and ketones, albeit at higher temperatures (20–140 °C). A one-pot procedure for aldol condensations has been devised[115]. For example, heating a mixture of acetone and benzaldehyde in the presence of $Ti(O^iPr)_4$ affords 75% of the aldol condensation product. In order to avoid undesired Meerwein–Ponndorf side-reactions which occur in some cases, $Ti(O^tBu)_4$ should be used.

In what appears to be an important recent development, ketones can be induced to undergo regio- and stereoselective aldol additions to aldehydes simply by reacting the components in the presence of *catalytic* amounts of $TiCl_4$ (Mahrwald method)[116] (Section 3.2.2). It is likely that Ti enolates are involved. A completely different strategy involves the reaction of ketene with $Ti(OR)_4$, leading to intermediate Ti enolates which undergo smooth aldol additions[117]. Chiral versions have not been reported.

In the area of Cp–titanium chemistry several ways to access new Ti reagents have been described which do not involve the titanation of organolithium compounds. One of them concerns the generation of π-allylic Ti^{III} reagents of the type **80** by hydrotitanation of butadiene or other 1,3-dienes[118]. They react with carbonyl compounds or CO_2, excellent to mediocre levels of stereoselectivity being observed in relevant cases (Section 3)[119,120].

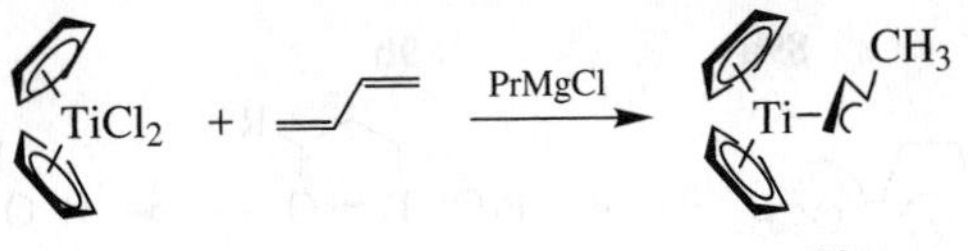

The Tebbe reagent **81**, prepared by treating Cp_2TiCl_2 **58a** with $Al(CH_3)_3$[121], represents a special case in that it is a precursor for the reactive nonisolable titanocene methylidene **84**, which can be trapped readily with a variety of unsaturated compounds to form new titanium compounds or catalysts as shown in Scheme 2[88]. Alkylidene analogs $Cp_2Ti = C(R)R^1$ are most easily generated by the reaction of dithiocacetals with $Cp_2Ti[P(OR)_3]_2$[122] (see also Section 5).

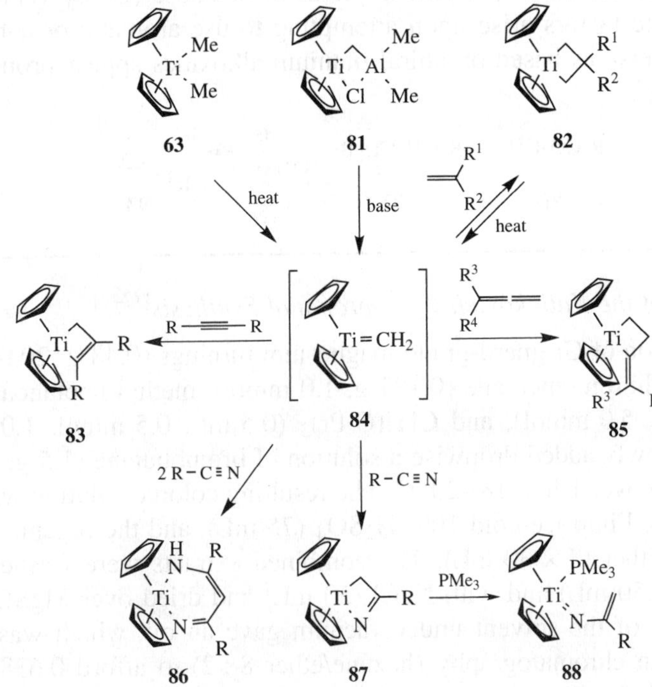

Scheme 2. Reactions of titanocene methylidene **84**.

A novel approach to a special class of organotitanium reagents is based on the reduction of $Ti(O^iPr)_4$ by CH_3CH_2MgBr with formation of ethylene, leading to the titanacyclopropane **89** in good yields (Kulinkovich method)[123]. The reagent, which can be considered to be either a Ti^{IV} species (**89a**) or a Ti^{II}–alkenes complex (**89b**), interacts with esters to form cyclopropanols **90**. In a formal sense ethylene dianions ($^-CH_2CH_2^-$) are involved:

$$2CH_3CH_2MgBr \xrightarrow[\substack{-2^iPrOMgBr \\ -C_2H_6}]{Ti(O^iPr)_4} \triangleright Ti(O^iPr)_2 \longleftrightarrow \| \cdots Ti(O^iPr)_2$$

89a   **89b**

$$RCO_2CH_3 \xrightarrow{89} (^iPrO)_2Ti \overset{O}{\underset{R}{\diagdown}} OCH_3 \longrightarrow (^iPrO)_2Ti \leftarrow O \overset{R}{\diagup} \longrightarrow R \overset{OCH_3}{\underset{\triangle}{OTi(O^iPr)_2}} \xrightarrow{H_2O/H^+} R \overset{OH}{\triangle}$$

90

This interesting method has been extended to include a variety of derivatives and substrates[124]. A significant improvement involves the use of $ClTi(O^iPr)_3$ instead of $Ti(O^iPr)_4$, particularly if homologs of CH_3CH_2MgBr are employed. For example, a one-pot procedure using the reaction components **91** and **92** in the presence of magnesium turnings and $ClTi(O^iPr)_3$ not only results in excellent yields of the corresponding cyclo-propanols **93**, but also in complete diastereoselectivity in favor of the products having the R^1/R^2 groups in the cis disposition in yields of 60–88% (Corey version)[124c]. Limitations due to steric factors arise upon attempting to use aromatic or α-branched esters. Enantioselective versions based on chiral titanium alkoxides appear promising[124c].

$$R^1CO_2CH_3 + R^2CH_2CH_2Br \xrightarrow[ClTi(O^iPr)_3]{Mg} R^2 \overset{R^1 \quad OH}{\triangle}$$

91   **92**   **93**

Corey Version of the Kulinkovich Cyclopropanol Synthesis[124c]

To a suspension of Grignard-grade magnesium turnings (0.48 g, 20.0 mmol) activated with 1,2-dibromomethane (0.187 g, 1.0 mmol), methyl heptanoate (**91**, $R^1 = C_6H_{13}$; 0.721 g, 5.0 mmol), and $ClTi(O^iPr)_3$ (0.5 mL, 0.5 mmol, 1.0 M) in THF (30 mL) was slowly added dropwise a solution of bromobutane (1.5 g, 11.0 mmol,) in THF (10 mL) over 1 h at 18–20 °C. The resulting colored solution was stirred for 2–3 h and poured into ice-cold 10% H_2SO_4 (75 mL), and the organic product was extracted with ether (3 × 30 mL). The combined extracts were washed with saturated $NaHCO_3$ (50 mL) and with NaCl (50 mL) and dried over $MgSO_4$. Filtration and evaporation of the solvent under vacuum gave an oil, which was purified by silica gel column chromatography (hexane/ether 8 : 2) to afford 0.638 g (75%) of *cis*-1-hexyl-2-ethylcyclopropanol (**93**, $R^1 = C_6H_{13}$, $R^2 = C_2H_5$) as a colorless oil.

In a similar reaction cyclopropylamines are accessible from the corresponding amides, making it the most versatile synthetic method for this class of compounds[125]. In the case of imides, cyclopropane formation is prevented due to the lower basicity of nitrogen, which means that the initial titanacycles can be trapped by protons or O_2[126]. This novel approach allows for the synthesis of functionalized pyrrolizidine, indolizidine and mitomycen alkaloids. The synthetic utility of titanacyclopropanes or of Ti^{II} complexes of alkenes goes far beyond these reactions. In particular, the Sato system comprising Ti(O^{i}Pr)$_4$/2^{i}PrMgBr (*i.e.*, the propene complex of Ti(O^{i}Pr)$_2$) has been used successfully in many different types of reactions[127]. For example, Ti(O^{i}Pr)$_4$/2^{i}PrMgBr reacts with unsaturated substrates such as alkenes, alkynes, or allenes, forming the corresponding titanium π-complexes, which in turn can be made to react with a variety of electrophiles[128]. For example, intramolecular Kulinkovich-type reactions are possible (**94** → **95**)[129]:

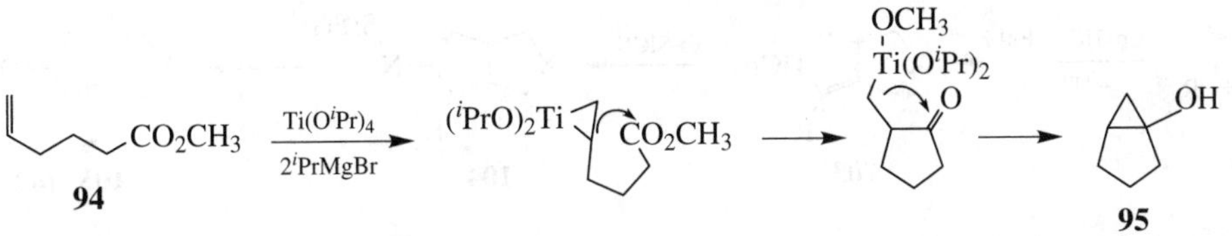

The reaction of Ti(O^{i}Pr)$_4$/2^{i}PrMgBr with alkenic substrates **96** having neighboring leaving groups leads to prochiral allylic titanium reagents which react stereoselectively with aldehydes or aldimines (Section 3.2.1)[127]. Here again the process is initiated by intermediate formation of titanacyclopropanes **97**:

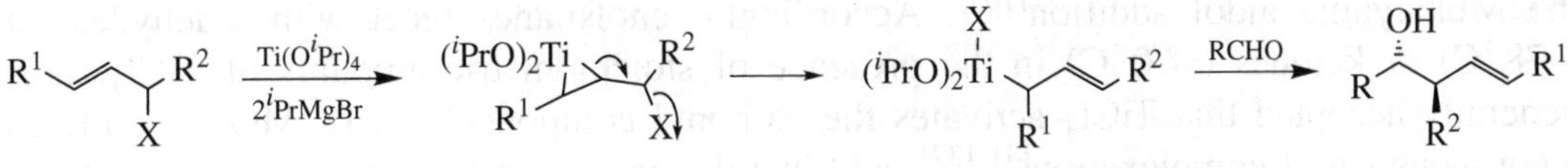

Along related lines, 1,6- and 1,7-dienes, enynes and diynes react with Ti(O^{i}Pr)$_4$/2^{i}PrMgBr to form titanabicycles which can be intercepted by CO, H$^+$ or aldehydes. In the case of substrates containing alkynic functionality, the reactions are initiated by the formation of titanacyclopropenes (or Ti–alkyne complexes). These and other variations demonstrate the utility and diversity of this chemistry[130].

Related are certain zirconocene- and titanocene-mediated reductive cyclizations of unsaturated substrates **100** such as enynes[131], diynes[132], and dienes[133]. Sometimes titanium is the preferred metal[134]. Precursors of the highly reactive titanocene 'Cp$_2$Ti' (which cannot be isolated) are such complexes as Cp$_2$TiCl$_2$/RMgX, Cp$_2$Ti(CO)$_2$, or Cp$_2$Ti(PMe$_3$)$_2$. Mechanistically, 'Cp$_2$Ti' first adds to the alkene or alkyne to form a three-membered titanocycle which then reacts to generate the products **101**. These can then be protonated, carbonylated, iminylated, halogenated, or converted into a variety of main group heterocycles and substituted benzene derivatives[134].

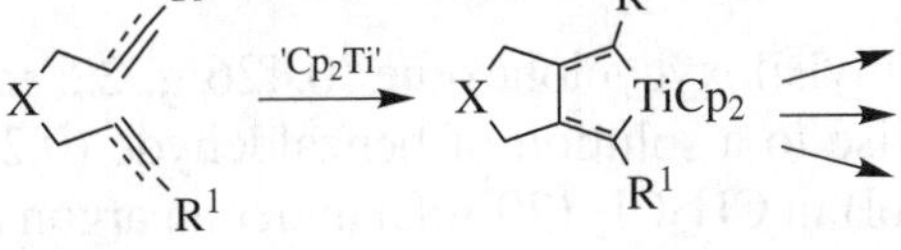

A disadvantage of such reactions has to do with the necessity of using stoichiometric amounts of Cp$_2$Ti precursors. However, catalytic versions are possible in certain cases, as in the synthesis of bicyclic cyclopentenones **105** and allylic amines. The trick is to use catalytic amounts of Cp_2TiCl_2/BuLi and stoichiometric amounts of Et_3SiCN which reacts with the intermediate titanacycles **103**, expelling 'Cp_2Ti' needed for a catalytic cycle[134]. A number of mechanistically related stoichiometric and catalytic processes have been reported, often involving three-membered titanocycles in the first step as in the reductive cyclization of enones[135] and in hetero Pauson–Khan reactions[136].

102   103   104   105 (42–82%)

2.3   TiCl₄-mediated Additions of Enol- and Allylsilanes to Carbonyl Compounds

One of the synthetically reliable methods for performing crossed aldol reactions is the Mukaiyama aldol addition[105]. Accordingly, enolsilanes react with aldehydes (at $-78\,°C$) or ketones (at $0\,°C$) in the presence of stoichiometric amounts of $TiCl_4$. It is generally accepted that $TiCl_4$ activates the carbonyl component via LUMO-lowering as a consequence of complexation[41, 137], and that the enolsilane then undergoes C–C bond formation. In mechanistic studies it was demonstrated that the silyl groups do not become bonded to the aldehyde oxygen atoms, *i.e.*, open transition states are involved[105b,d]. However, in rare cases prior Si–Ti exchange may occur (Section 2.2). In some cases it was possible to characterize $TiCl_4(ArCHO)_2$ complexes[138] and the trichlorotitanium aldolates by X-ray crystallography[139]. Various aspects of stereoselectivity are discussed in Section 3.2.2. Perhaps one of the disadvantages is the fact that stoichiometric amounts of $TiCl_4$ are needed. Catalytic amounts of $Cp_2Ti(OTf)_2$[140] and other Lewis acids have been shown to the effective[105b,e,141,142] (Section 7). *O*-Silyl ketene ketals undergo $TiCl_4$-promoted additions to imines with formation of β-lactams[105b,c].

Typical Procedure for the Mukaiyama Aldol Addition[105a]

A solution of 1-trimethylsilyloxycyclohexene (0.426 g, 2.5 mmol) in 10 mL of dry CH_2Cl_2 is added dropwise to a solution of benzaldehyde (0.292 g, 2.75 mmol) and $TiCl_4$ (0.55 g, 2.75 mmol) in CH_2Cl_2 (20 mL) under an argon atmosphere at $-78\,°C$. The mixture is stirred for 1 h and hydrolyzed at that temperature. Following extraction with diethyl ether, washing with water and drying over Na_2SO_4, the solution is concentrated and the residue is purified by column chromatography on SiO_2 using

CH$_2$Cl$_2$ to afford 115 mg (23%) of *erythro*-2-(hydroxyphenylmethyl)cyclohexanone (mp 103.5–105.5 °C from 2-propanol) and 0.346 g (69%) of the threo isomer (mp 75 °C from *n*-hexane/diethyl ether).

Closely related are allylsilane additions (Sakurai–Hosomi reaction)[143]. They too have been exploited widely, especially in diastereofacial additions to chiral aldehydes (Section 3.1.2). The use of chiral allylic silanes in TiCl$_4$-mediated aldehyde additions has been reviewed[144]. Although in the original procedure ambient temperature was chosen, addition to aldehydes is in fact rapid at −78 °C, whereas ketones require higher temperatures (−10 to 0 °C). This means that ketoaldehydes react completely aldehyde selectively[69]. According to an NMR study of the mechanism of the reaction, Si–Ti metathesis plays no role, and the initial products are titanium alkoxides[145]. A similar conclusion has been reached in a rapid injection NMR study of the TiCl$_4$-mediated chelation-controlled addition of allylsilanes to chiral α-alkoxyaldehydes[105d]. Various enantioselective catalytic versions has been devised (Section 8.1). Stereoselective intramolecular allylsilane additions to aldehydes[146, 147] and imines[148] constitute a useful way to synthesize carbocycles and heterocycles, respectively.

General Procedure for TiCl$_4$-mediated Allylsilane Additions to Aldehydes and Ketones[143,145]

To a solution of a carbonyl compound (2 mmol) and CH$_2$Cl$_2$ (3 mL) is added TiCl$_4$ (1 mmol) dropwise with a syringe at room temperature under nitrogen. An allylsilane (2 mmol) is then added rapidly and the mixture is stirred for 1 min (aldehyde additions) or 3 min (ketone additions). The usual aqueous workup affords 50–91% of the homoallylic alcohols. The reaction can also be carried out at low temperatures[145].

2.4 Reversal of Chemoselectivity via Selective *in situ* Protection of Carbonyl Functions

Although titanium reagents allow for high levels of chemoselectivity in Grignard and aldol additions to ketoaldehydes, dialdehydes, and diketones as delineated above, selective C–C bond formation at the sterically more hindered functionality may well be required in certain synthetic sequences. Classically, this problem can be solved to some extent by a sequence of reactions involving protection, reaction, and deprotection. Unfortunately, protection is not always completely site selective, even if modern reagents are chosen, as in the protective ketalization of the less-hindered ketone functionality in steroidal diketones using silicon reagents of the type $R_3SiOCH_2CH_2OSiR_3/R_3SiOTf$[149]. Titanium chemistry provides a partial solution to such problems. It was discovered that the amides $Ti(NR_2)_4$ ($R = CH_3$, C_2H_5) add rapidly to aldehydes at $-78\,^\circ$C to form adducts **106**, whereas ketones react only with the less bulky reagent $Ti[N(CH_3)_2]_4$ at $-40\,^\circ$C with formation of analogs **107**[150]:

$$Ti(NR_2)_4 \quad \begin{array}{c} R = CH_3 \\ R = C_2H_5 \end{array}$$

$$\xrightarrow[-78\,^\circ C]{R^1CHO} \quad R^1 \underset{NR_2}{\overset{OTi(NR_2)_3}{\diagup}}$$

106a R = CH₃
106b R = C₂H₅

$$\xrightarrow[-40\,^\circ C]{R^1 \overset{O}{\diagup} R^2} \quad \underset{R^1}{R_2N}\underset{R^2}{\overset{OTi(NR_2)_3}{\diagup}}$$

107 R = CH₃

This difference in reactivity can be exploited to protect aldehydes selectively in the presence of ketones. Following *in situ* protection, the addition of carbanions such as reactive alkyllithium reagents or enolates forces C—C bond formation to occur at the lessreactive ketone site. Aqueous workup regenerates the aldehyde function, for instance **36 → 108**. Ketone selectivity in this one-pot sequence is better than 99%, the isolated yield of the aldol adduct **108** being 88%[150]:

$$\underset{Ph}{\overset{O}{\diagdown}}\cdots\overset{O}{\diagdown}\overset{H}{} \quad \xrightarrow[\text{(b)}]{\text{(a) Ti(NEt}_2)_4} \quad \underset{Ph}{HO}\overset{COOC_2H_5}{\diagup}\cdots\overset{H}{\underset{O}{\diagdown}}$$

36   **108**

Differentiation between two ketone sites is also possible, as is the addition to ester functions in the presence of aldehyde moieties[150]. On the basis of this methodology selective alkenation (Section 5.1) of ketoaldehydes at the ketone function has been accomplished[151].

This methodology is powerful, but limitations have become apparent. Only very reactive carbanions which add rapidly at temperatures below $-30\,^\circ$C can be used. This is because decomposition of the amino adducts **106** and **107** with formation of enamines occurs at temperatures above $-25\,^\circ$C[150,152]. A related method based on the *in situ* protection by $TiCl_4/PPh_3$ is also simple to perform, but its synthetic scope remains to be explored[153].

3   Stereoselective Grignard and Aldol Additions

3.1   Diastereofacial Selectivity

As the two π-faces of an aldehyde or ketone having a stereogenic center at the α-position are diastereotopic, the addition of Grignard-like reagents or enolates may lead to unequal amounts of diastereomeric products. In the case of carbonyl compounds devoid of additional heteroatoms the problem of Cram/*anti*-Cram selectivity pertains, for which various theoretical models have been proposed[154,155]. If one of the substituents involves a heteroatom (alkoxy, amino groups), the problem of chelation against nonchelation is relevant[41]. A great deal of progress has been made in these areas, titanium chemistry often playing a pivotal role[7,41]. However, a few problems persist.

3.1.1 Cram Selectivity

Although Cram selectivity is generally acceptable in addition reactions of bulky Grignard reagents (*e.g.*, *tert*-butyl, isopropyl), mediocre diastereoselectivities are often observed in the case of smaller analogs such as methyl, alkyl, allyl, and phenyl reagents[154]. In a number of reactions the power of metal and ligand tuning has become apparent, as in the case of allyl additions to 2-phenylpropanal **109**[21,49]:

	110	:	111
CH_2=$CHCH_2MgCl$	65	:	35
CH_2=$CHCH_2Ti[N(CH_3)_2]_3$	88	:	12
CH_2=$CHCH_2Ti[N(Et_2)]_3$	94	:	6

The results show that increasing the size of the ligands at titanium increases diastereofacial selectivity. It should be mentioned that in a definitive study the reaction of methyl- and butyllithium (and Grignard analogs) with 2-phenylpropanal **109** was carefully reexamined[43b]. In contrast to earlier reports in the literature, Cram selectivities of 89–94% were observed at −78 °C. Hence 2-phenylpropanal is not a good probe to test Cram-type stereoselectivity with methyl- or butylmetal reagents.

An impressive application of titanium chemistry concerns the regio- and stereoselective reaction of **109** with the homoenolate **112**[156]. Of the eight possible diastereomers of γ-attack, only two are formed in a ratio of **113/114** = 93 : 7. This means high Cram preference and complete simple diastereoselectivity (see Section 3.2), in addition to exclusive formation of the cis form of the enol moiety. The lithium precursor affords a complex mixture of products.

Side-chain extending reactions of steroidal aldehydes and ketones are often fairly selective using classical organometallics[157], but titanation may increase asymmetric induction markedly[49]. A clear limitation concerns alkyltitanium reagents $RCH_2Ti(O^iPr)_3$ which are too sluggish to add to ketones[7]. Alkynyltin reagents add stereoselectively to steroidal aldehydes in the presence of $TiCl_4$[158]. Several interesting cases of Cram-selective additions of alkynyltitanium reagents in prostaglandin intermediates have been reported[54], although simple chiral aldehydes react nonselectively[53]. Sometimes vinyl-lithium reagents are more selective than the titanium analogs[55b].

3.1.2   Chelation-controlled Grignard and Aldol Reactions

In contrast to Cram selectivity in which the carbonyl compounds may react in a variety of conformations, alkoxy carbonyl compounds are capable of chelation. In such cases the number of degrees of freedom are greatly reduced, making asymmetric induction much easier. A systematization is shown in Scheme 3[41]. The problem is to find the right metal for each particular situation.

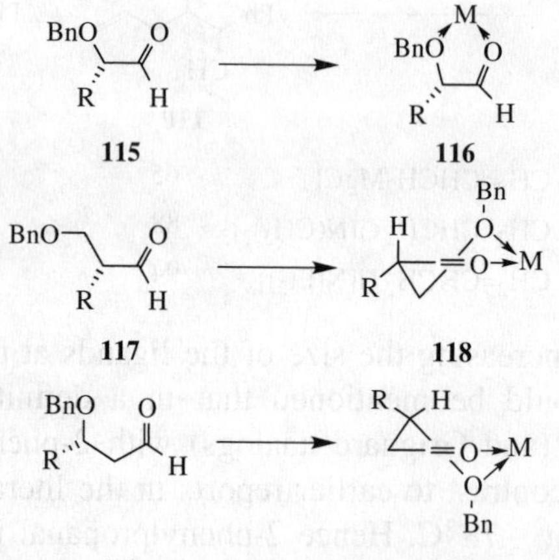

Scheme 3. Chelation modes of chiral alkoxy aldehydes.

Efforts up to 1980 provided solutions to only two general problems. According to Cram's classical work on α-alkoxy ketones, simple Grignard reagents RMgX react with high levels of chelation control[159]. Secondly, α-chiral β-alkoxyaldehydes **117** react with cuprates R_2CuLi to form the chelation-controlled Grignard-type adducts[160]. However, these methods do not generally extend to such aldehydes as **115** or **119**. Furthermore, aldol additions of lithium enolates usually afford mixtures of diastereomers in reactions with aldehydes **115**, **117** and **119**[41]. Finally, the classical reagents cannot be used to induce nonchelation control. In many cases metal and ligand tuning based on titanium chemistry solve such problems[41].

As CH_3TiCl_3 forms bisetherates (Section 2.1.2), it was speculated that aldehydes such as **115** should form intermediate Cram-like chelates **121** (R = Bn)[161]. Indeed, the ratio of chelation- to nonchelation-controlled adducts turned out to be **122**/**123** = 93 : 7[161].

Later the intermediacy of octahedral complexes of the type **121** was demonstrated by ^{13}C NMR spectroscopy[162]. They are the only recorded cases of direct spectroscopic evidence of Cram-type chelates. Chelation-controlled reactions of RTiCl$_3$ need to be performed in an ether-free medium (*e.g.*, CH$_2$Cl$_2$) because ether competes for the Lewis acidic titanium, thereby breaking up the chelate. A number of chelates involving TiCl$_4$ or CH$_3$TiCl$_3$ have been studied by quantum mechanical methods[41c].

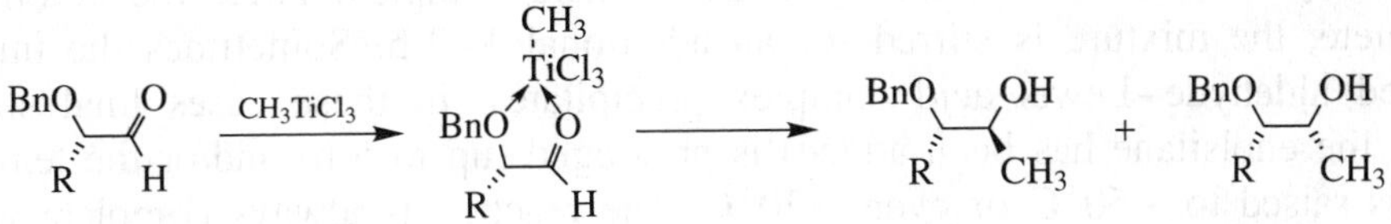

Although other highly Lewis-acidic titanium reagents also show excellent levels of chelation control[163], the range of reagents of the type RTiCl$_3$ is limited (Section 2.1.2), especially because ether-free solvents such as CH$_2$Cl$_2$ are mandatory[41]. Therefore, a different methodology had to be developed. Accordingly, the alkoxyaldehydes **115** are activated and 'tied up' by Lewis acids such as TiCl$_4$ to form chelates of the type **124**, which are then reacted with allyl- and enolsilanes[161], dialkylzinc reagents[161], tetraalkyl-lead compounds[104b,e] or (CH$_3$)$_3$SiCN[164] (Scheme 4). More recent examples[165] include the chelation controlled addition of methylenecyclopropane to aldehyde **115**[166]. In all of these reactions conversion is greater than 90% with chelation control being 90–100%.

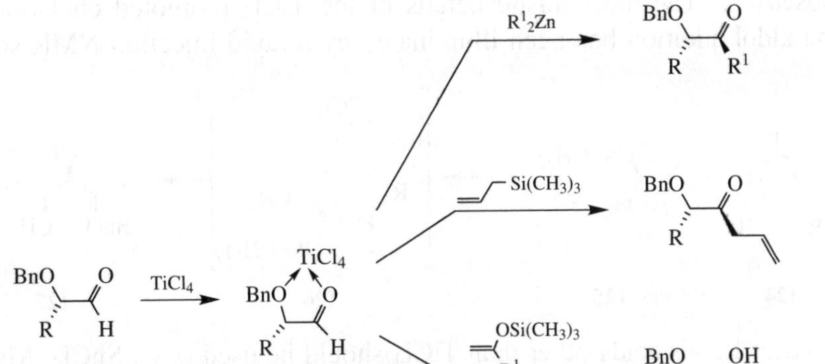

Scheme 4. Chelation-controlled reactions of **124**.

General Procedure for the TiCl₄-mediated Chelation-controlled Aldol Addition to Alkoxyaldehydes 115, 117, or 119[161,167a]

5.0 mmol of an alkoxyaldehyde in 50 mL of dry CH_2Cl_2 are treated with 5 mmol of $TiCl_4$ at $-78\,^{\circ}C$. After 5–10 min, 5 mmol of an enolsilane are added using a syringe. In the case of the more reactive ketene ketals, cooled ($-78\,^{\circ}C$) CH_2Cl_2 solutions are added to the aldehyde–Lewis acid complex. After the addition is complete, the mixture is stirred for an additional 1–2 h. Sometimes the initially formed aldehyde–Lewis acid complex precipitates. In these cases final stirring (after the enolsilane has been added) is prolonged (up to 5 h) and/or the temperature is raised to $-50\,^{\circ}C$ or even $-20\,^{\circ}C$. The reaction is always complete within 30 min after the solution has become homogeneous (very often within 1–2 min!). The mixture is then poured onto water, extracted twice with diethyl ether and the combined organic phases are neutralized with $NaHCO_3$ solution. In the case of acid-sensitive products, saturated $NaHCO_3$ solution is poured onto the reaction mixture. After drying over $MgSO_4$ the solvent is stripped off and the crude product worked up in the usual way. Conversion is at least 85% in all cases. The same applies to reactions on a 10–20 mmol scale.

If the carbon nucleophile is prochiral as in **125**, the problem of simple diastereo-selectivity (Section 3.2) is also relevant, i.e. four diastereomeric products are possible. In some cases complete control with formation of a single product **127** is possible, simple diastereoselectivity being syn[167]. Open transition states of the type **126** have been proposed[167]. The mechanistic details of the $TiCl_4$-promoted chelation-controlled Mukaiyama aldol addition has been illuminated by a rapid injection NMR study[105d]:

| 124 | 125 | 126 | 127 |

In some cases Lewis acids other than $TiCl_4$ should be used (e.g., $SnCl_4$, MgX_2)[41,167]. Perhaps a disadvantage is the fact that stoichiometric amounts of Lewis acid have to be employed. The observation that catalytic amounts of $LiClO_4$ induce chelation-controlled additions in some cases is a promising development[105d,e]. Nevertheless, the enormous generality observed in the case of $TiCl_4$-induced chelation-controlled additions remains to be shown for the catalytic processes. The concept of using Lewis acids capable of chelation in combination with carbon nucleophiles has evolved as a general principle in stereoselective C—C bond formation[41,168]. Synthetic, mechanistic and theoretical aspects have been reviewed[41]. The principle has been extended to α-chiral

β-alkoxyaldehydes (**117**)[167,169], β-chiral β-alkoxyaldehydes (**119**)[170] and α-chiral α,β-dialkoxyaldehydes[171]. This means that in Scheme 4 the metal M does not participate in the form of an organometallic reagent such as RMgX or RTiCl$_3$, but rather as a normal Lewis acid. Nuclear magnetic resonance[7,161,172] and X-ray structural data[161b] have been presented, for instance in the case of TiCl$_4$ and SnCl$_4$ chelates. Similarly, α- and β-alkoxy acid nitriles react to form tertiary cyanohydrins stereoselectively[164].

A typical example is the reaction of β-benzyloxyaldehydes **119** with allylsilane in the presence of TiCl$_4$ in CH$_2$Cl$_2$ at $-78\,^{\circ}$C. A chelate **128** is involved which is attacked at the less-hindered π-face, 1,3-asymmetric induction being greater than 95%[170]. Intramolecular versions lead to the reversal of diastereoselectivity (**129** versus **132**)[173a]. In these cases the aldehydes **130** become chelated in the form of intermediates **131**, which spontaneously react to form the products **132** with diastereoselectivities of 90%. Chelation-controlled intramolecular addition of allylsilanes in the presence of TiCl$_4$ affords carbocycles in other cases[173b].

119

Bn = CH$_2$C$_6$H$_5$

128

129

130

131

132

The types of prochiral C-nucleophiles capable of participating in chelation controlled reactions vary widely, (*e.g.*, **133**[174], **134**[167], **135**[175], **136a**[176a], **136b**[176b], **137**[177], and **138**[167c]). Silylallenes can also be used[178]. Further examples continue to be reported, including functionalized chiral crotylsilanes[179]. In most cases chelation control is better than 95% and simple diastereoselectivity is syn.

133

134

135

136

137

138

a: R^1 = CH$_3$, R^2 = S^{t}Bu

b: R^2 = Si(CH$_3$)$_3$, R^2 = OC$_2$H$_5$

Transition states such as **126** (or synclinal analogs) are believed to be involved[105b,167]. In a synthetically substantial extension, chelation-controlled Mukaiyama aldol additions of chiral β-formyl esters have been shown to proceed with greater than 90% diastereofacial selectivity[176c]. The products are γ-lactones having additional keto functionality, as in the synthesis of the pheromone (+)-eldanolide[180]. TiCl$_4$-mediated reactions of chiral β-hydroxy ketones are also stereoselective[181].

N,N-Dibenzylaminoaldehydes (**139**) undergo chelation-controlled allylsilane and (CH$_3$)$_2$Zn additions in the presence of TiCl$_4$, although diastereoselectivity is not consistently acceptable (the **140/141** ratio ranges between 65 : 35 and 95 : 5)[182]. This is due to steric inhibition of chelation. However, if the nitrogen has only one protective group, as in **142**, chelation-controlled allyl- and enolsilane additions mediated by TiCl$_4$ or SnCl$_4$ are highly effective[183], as are TiCl$_4$-promoted ene reactions[184]. The chelation-controlled addition of alkyl nucleophiles to **142** is possible by using cuprates or manganese reagents[185].

Titanium enolates from thioesters prepared by the Evans method (TiCl$_4$/NR$_3$) add to chiral α-alkoxyaldimines to produce the chelation-controlled adducts[186]. This recent development is of considerable interest in the stereoselective synthesis of β-lactams. Chelation controlled additions of allylstannanes and Grignard reagents to α,β-epoxyaldehydes mediated by TiCl$_4$ or MgBr$_2$ have also been reported (diastereoselectivity, ds > 95%)[187]. Related are TiCl$_4$-controlled hydride reductions of ketones[188,189] and (^{i}PrO)$_2$TiBH$_4$-induced reductions of α,β-epoxy ketones[189].

3.1.3   Non-chelation-controlled Grignard and Aldol Additions

Non-chelation control is more difficult to achieve because there are no general ways to reduce the number of rotamers of noncomplexed alkoxycarbonyl compounds[41]. Reagents incapable of chelation must be used. Electronic and/or steric factors, notably those defined by the Felkin−Anh model[155], must then be relied on. As Lewis acidity of organotitanium reagents decreases drastically on going from RTiCl$_3$ to RTi(OR^1)$_3$ (Section 2), the latter might be expected to be candidates in non-chelation-controlled reactions. Indeed, the parent compound CH$_3$Ti(O^{i}Pr)$_3$ adds to a variety of α-alkoxyaldehydes (*e.g.*, **115**) in exactly this manner[161]. Application to carbohydrate aldehydes is possible, but homologs such as butyltriisopropoxytitanium (**7**) are too sluggish to react with α-alkoxyaldehydes[43a]. However, they add to α,β-dialkoxyaldehydes with non-chelation control[55].

$$\begin{array}{ccccc} \textbf{115} & & \textbf{122} & 8:92 & \textbf{123} \\ R = CH_3 & & & & \end{array}$$

---

*Non-chelation-controlled Addition of $CH_3\,Ti(O^i Pr)_3$ (**5**) to α-Alkoxyaldehydes* **115**[161]

A solution of an α-alkoxyaldehyde **115** (5.0 mmol) in 25 mL of THF, diethyl ether, CH_2Cl_2 or hexane is treated with distilled $CH_3Ti(O^iPr)_3$ (1.32 g, 5.5 mmol) at $-30\,°C$ for 3 days. The mixture is poured onto 50 mL of saturated NH_4F solution, extracted several times with diethyl ether, washed with NaCl solution and dried over $MgSO_4$. The solvents are removed and the residue is Kugelrohr distilled, delivering 80–90% of product. Non-chelation control is highest if hexane is the solvent.

Titanium enolates derived from acetic acid esters do not add diastereoselectively to α-alkoxyaldehydes[190]. The method also fails in the case of crotyltitanium reagents[191]. In contrast, prochiral titanium enolates such as **145** having alkoxy or amino ligands show good degrees of non-chelation control[190], as do certain homoenolate equivalents based on titanium chemistry[156]. Simple diastereoselectivity in the aldol additions is syn (Section 3.2 discusses this type of diastereoselectivity):

$$\begin{array}{cccccc} \textbf{115} & & \textbf{145} & & \textbf{146} & 13:87 & \textbf{147} \\ R = CH_3 & & & & & \end{array}$$

As α-alkoxy ketones have a pronounced tendency to undergo efficient chelation control even with simple Grignard reagents, reversing the diastereoselectivity is expected to be difficult. Indeed, even $CH_3Ti(O^iPr)_3$ reacts with 99% chelation control[70]. However, the combination of metal, ligand and protective group tuning[192] can tip the balance completely. By using bulky silyl protective groups, addition of methyl- and allyltitanium reagents or titanium enolates of low Lewis acidity results in better than 99% non-chelation control, as in the reaction of the enolate **149** with α-siloxy ketones of the type **148**[70].

$$R = Si(CH_3)_2{}^t Bu$$

The use of bulky silyl protective groups in combination with classical reagents RMgX results in low degrees of selectivity. Similarly, the trimethylsilyl protective group is generally inefficient. Thus, steric factors are more important than possible electronic effects originating from silicon[193]. Clearly, it is the combination of metal tuning, ligand tuning and protective group tuning that ensures success! A limitation of this method concerns the less reactive alkyltitanium reagents that do not add to ketones.

N,N-Dibenzylaminoaldehydes (**139**) react with a variety of classical reagents RMgX and RLi with surprisingly high degrees of non-chelation control (ds > 90%)[182]. An exception is the very reactive allylmagnesium chloride (3 : 1 diastereomer mixtures). In this case prior addition of ClTi(NEt$_2$)$_3$ to the Grignard reagent results in greater than 90% non-chelation control[182]. The possible reasons for the general behavior of these aldehydes have been delineated elsewhere[194]. In complete contrast, there are no general ways to obtain consistently high levels of non-chelation control in reactions of α-alkoxyaldehydes or -ketones, in spite of progress achieved with a number of titanium reagents. Perhaps the combination of metal and protective group tuning will provide a solution to this problem. Reagent control[195], albeit more expensive, has been successful in the case of boron[196] and titanium[197–199] reagents and in the ene reaction catalyzed by chiral titanium complexes[200] (Sections 3.3 and 8.1).

A number of chelation and non-chelation-controlled titanium mediated reactions of chiral lactols[201], 2-alkoxy-1-(1,3-dithian-2-yl)-1-propanones[202a], α-ketoamides[202b], arylsulfinylacetophenones[203], and α-alkoxyimines[186d] have been reported. In another interesting development, titanium phenolates were shown to arylate glyceraldehyde acetonide and α-aminoaldehydes selectively with non-chelation control[204]. The titanium phenolate behaves as an enolate. Similarly, titanated pyrroles react stereoselectively with glyceraldehyde acetonide and with arabinofuranose and glucopyranose derivatives to form pyrrole *C*-glycoconjugates[204c]. The method constitutes an elegant route to heteroarylated sugars in a highly stereoselective manner (ds > 98%). In certain cases the cerium analogs afford diastereomers having the opposite relative configuration (chelation control), whereas the bromomagnesium salt of pyrrole produces mixtures[204].

3.2   Reactions Involving Simple Diastereoselectivity

3.2.1   Prochiral Allylic Titanium Reagents

The reaction of an aldehyde with a prochiral reagent such as a (*Z*)- or (*E*)-configured crotyl metal reagent affords syn and anti adducts, respectively[205], each in racemic form (only one enantiomer shown in Scheme 5). Among the most efficient reagents are crotylboron compounds, the (*E*)-configured form resulting in >95% anti selectivity and the (*Z*)-analog showing opposite simple diastereoselectivity (>95% syn selectivity)[205]. Compact chair-like transition states are the source of stereoselectivity (Scheme 5). Nevertheless, other metal systems have been tested, some of which are more easily accessible and still fairly stereoselective[205]. One of the drawbacks of boron reagents has to do with the fact that addition to ketones fails chemically, in which case titanium reagents should be considered.

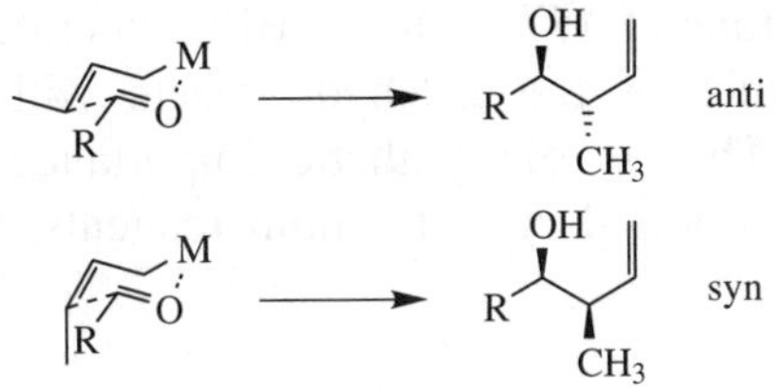

Scheme 5. Transition states of crotylmetal additions.

The titanation of crotylmagnesium chloride (which is a mixture of stereo- and regio-isomers) with ClTi(O^{i}Pr)$_3$, Ti(O^{i}Pr)$_4$ or ClTi(NEt$_2$)$_3$ provides stereoconvergently the (*E*)-configured reagents **11**, **152**, and **153**, respectively[13,49]:

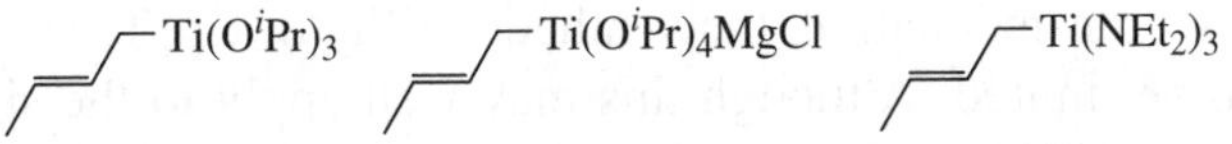

The addition of these reagents to aldehydes occurs with moderate to good degrees of anti selectivity[13,14,49]. Real benefits of the crotyltitanium reagents become apparent in reactions with prochiral ketones **154**, diastereoselectivity in favor of **155** generally being above 90%[13,14,49]. In the case of acetophenone, the ate complex **152** is best suited, but purely aliphatic ketones require the trisamide **153** for best results. Crotyltriphenoxytitanium also adds stereoselectively to ketones, but is less easily accessible[206]. In all cases chair-like transition states have been proposed.

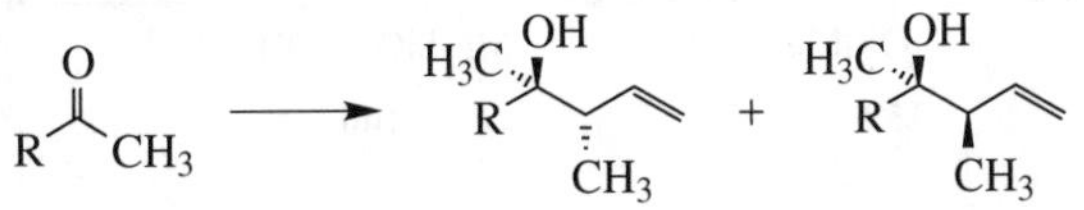

General Procedure for the Formation and Ketone Addition Reactions of Crotyltris(diethylamino)titanium, 153[10]

A solution of ClTi(NEt$_2$)$_3$ (3.29 g, 11 mmol; see above) in 50 mL of THF is treated with 10 mmol of crotylmagnesium chloride as a mixture of regio- and

stereoisomers in THF at 0 °C. The solution is allowed to reach 25 °C and is then
cooled to −78 °C. A ketone (9 mmol) is added, the mixture is stirred for 2–3 h
and then poured on dilute HCl. Extraction with diethyl ether followed by the usual
workup affords the desired alcohols in 75–85% yield.

Bis(cyclopentadienyl)titanium(IV) reagents **157**[207] and the related titanium(III)
compounds **80**[208] add to aldehydes with excellent anti selectivity. Reversal of
diastereoselectivity in reactions of **157** occurs if BF$_3$–etherate is used as an additive[209].
In this case the BF$_3$ adduct of the aldehyde reacts with the titanium reagent via
an open transition state. This is of synthetic importance because it has not been
possible to generate (Z)-configured crotyltitanium reagents, which are required for syn
selectivity.

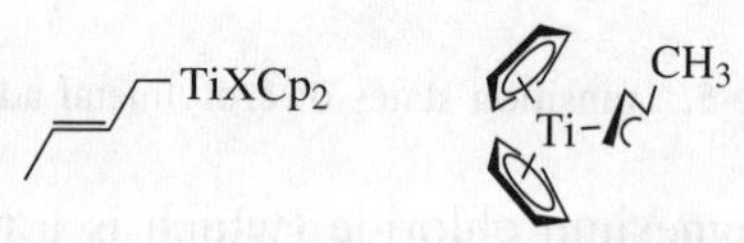

157 X = Cl, Br, I   **80**

In view of the ready availability of reagents **11**, **152**, and **153**, and of other crotylmetal
reagents prepared by transmetalation, the synthetic utility of Cp$_2$Ti reagents of the type **80**
and **157** appears to be limited. Although this may well apply to the simplest case (crotyl
Ti reagents), it does not necessarily pertain to more complicated allylic systems, because
these are often easily accessible by regioselective hydrotitanation of dienes[118−120] (see
Section 2.2). A particularly interesting case involves siloxy-substituted dienes of the type
158 which allow for the so-called allyltitanation–Mukaiyama aldol sequence with stereoselective creation of five contiguous stereogenic centers, as shown in Scheme 6[120].
The first step is regioselective hydrotitanation with formation of the Ti^{III} species **159**. It

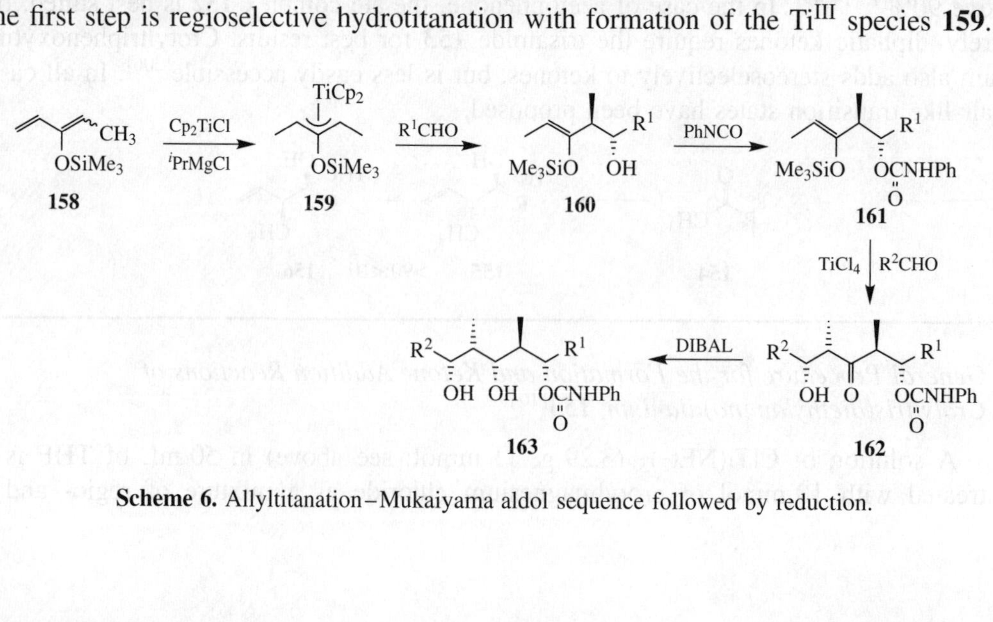

Scheme 6. Allyltitanation–Mukaiyama aldol sequence followed by reduction.

reacts with aldehydes R^1CHO in the expected anti selective manner (anti/syn > 98/2), (Z)-selectivity of enolsilane formation (compare **160**) being complete. After protection of the hydroxy function, a Mukaiyama aldol addition with complete syn selectivity is performed (Section 3.2.2 describes this type of reaction) followed by stereoselective reduction of the ketone function. The method is likely to be of utility in the synthesis of polyproprionate-derived natural products, especially if the stereoselectivity of the aldol process can be reversed on an optional basis and if an enantioselective variant can be developed.

A large number of heteroatom-substituted allyltitanium reagents have been generated and reacted with aldehydes to form the corresponding anti adducts, generally with more than 95% diastereoselectivity[7]. In most cases a very simple experimental protocol is used: generation of the allylic lithium precursor by deprotonation of the neutral compound followed by titanation with $Ti(O^iPr)_4$ and *in situ* aldehyde addition. Reductive cleavage of thioethers followed by titanation is an alternative procedure[210]. Titanium ate complexes (Section 2.1.4) have been postulated, although structural ambiguities remain. An early example is the silyl-substituted titanium ate complex **50a**, which adds regio- and stereoselectively to aldehydes with exclusive formation of the adducts **164**[49,51]. This one-pot procedure is simpler than the multistep process based on boron chemistry[211]. Titanation of the Li precursor using $ClTi(O^iPr)_3$[49], $ClTi(NEt_2)_3$[49] or Cp_2TiCl[212] also results in species which react anti selectively, but the cheap $Ti(O^iPr)_4$ is clearly the titanating agent of choice. Reactions with ketones are also stereoselective. The products are synthetically useful because stereospecific conversion into dienes via the Peterson elimination under basic or acidic conditions is possible[49].

50a   **164**

Using the above techniques, reagents **50b**[84], **165**[84], **166**[213], **167**[214], **168**[215], **169**[216], **170**[216] and **171**[7,43] were prepared and reacted with aldehydes. Again, the simplicity and high anti selectivity of this procedure make it more efficient than alternative methods based on other metals. For example, reagent **167** has been used in a short synthesis of the diterpene (±)-aplysin-20[214]. Compounds of the type **166/167** are excellent reagents for anti-selective homoaldol additions[215]. It is important to point out that some of the reagents are chiral, but were used in racemic form, such as **166**, **169**, and **170**. Therefore, enantiomerically pure forms should provide diastereomerically *and* enantiomerically pure adducts. This interesting aspect has been considered[216] in the case of **170** (Section 3.3). If syn-selective homo-aldol additions are required, a different approach is neccessary. Genuine titanium ester or amide homoenolates react with aldehydes with greater than 90% syn selectivity[217].

$$\underset{H_5C_2S}{\diagdown}\diagup Ti(O^iPr)_4Li$$

50b

$$\underset{H_5C_2S}{\diagup}\diagup Ti(O^iPr)_4Li$$

165

$$\underset{(H_3C)_3Si}{\diagup}\diagup \underset{S^tBu}{\overset{Ti(O^iPr)_4Li}{\diagup}}$$

166

$$\underset{(H_3C)_3Si}{\overset{RO}{\diagup}}\diagup Ti(O^iPr)_4Li$$

167

$$\underset{(H_5C_6)_2P}{\diagup}\diagup Ti(O^iPr)_4Li$$

168

$$[(H_3C)_2N]_3Ti\underset{(H_3C)_3Si}{\overset{O}{\diagdown}}\diagup \overset{O}{\diagdown}N(^iPr)_2$$

169

$$(^iPrO)_3Ti\overset{O}{\diagdown}\diagup\overset{O}{\diagdown}N(^iPr)_2$$

170

$$\underset{H_5C_6SO_2}{\diagup}\diagup Ti(O^iPr)_3$$

171

Many of the above reagents are useful in the synthesis of pheromones, terpenes, and 4-butanolides. Variously substituted prochiral allenyltitanium reagents **172**[218a,b] and **174**[219] react with aldehydes anti selectively with formation of β-alkynyl alcohols **173** and **175**, respectively. Triisopropoxy analogs of **172** (R^1 = CH$_3$, R^2 = (CH$_3$)$_3$Si) react stereoselectively with imines[218c].

$$\underset{Ti(O^iPr)_4Li}{\overset{R^1 \diagup R^2}{\diagdown}} \xrightarrow{RCHO} R\underset{\overset{|}{R^1}}{\overset{OH}{\diagup}}\diagup R^2$$

172   **173**

$$\underset{H_3C}{\overset{H_3CO}{\diagdown}}=\underset{Ti(O^iPr)_3}{\overset{H}{\diagup}} \xrightarrow{\overset{}{\diagdown}CHO} \underset{CH_3}{\overset{OH}{\diagup}}\underset{OCH_3}{\overset{H}{\diagup}}$$

174   **175**

3.2.2   Titanium-based Aldol Additions

Aldol additions using prochiral titanium enolates (and heteroatom analogs) are useful reagents, particularly when they are complementary to existing methodologies based on other metals[220]. The titanation of lithium enolates with ClTi(O^{i}Pr)$_3$ or ClTi(NR$_2$)$_3$ affords titanium enolates which add syn selectively to aldehydes, irrespective of the geometry of the enolate[63]. As diastereoselectivity is not consistently above 90%, the method is limited in scope. However, it is useful in the case of cyclic ketones[63], because it is complementary to anti-selective boron enolates[220]. A prime example is the enolate **176** derived from cyclohexanone, a case which also illustrates the power of ligand tuning[63]. Certain tin enolates also react syn selectively[221a]. The triisopropoxytitanium enolate derived from isobutyric acid methyl ester reacts stereoselectively with chirally modified imines to produce (4R)-β-lactams[221b]. The use of the lithium enolate results in the formation of the (4S)-β-lactams. Chirally modified triisopropoxytitanium enolates react highly selectively (Section 3.3). In fact, this is their primary area of application. Internally chelated (^{i}PrO)$_3$Ti enolates add either syn or anti selectively to aldehydes, depending upon the structure of the reagent[222].

176   177   178

X = O^{i}Pr
X = N(CH$_3$)$_2$ 86:14
X = N(Et)$_2$ 92:8
               97:3

The Mukaiyama crossed aldol addition, i.e. the reaction of an aldehyde and an enol-silane in the presence of TiCl$_4$, does not involve titanium enolates (Section 2.3) and generally affords mixtures of syn and anti adducts, although progress has been made[105].

The proper choice of substrates and conditions may result in high levels of simple diastereoselectivity irrespective of the geometry of the double bond[105b]. If R^1 is small (methyl), good to excellent anti selectivity results, provided R^2 is a bulky group as in sterically demanding ketones[223]. Some *O*-silyl ketene ketals (R^1 = CH$_3$, R^2 = OEt) also react anti selectively, as do enolsilanes from thioesters[224]. However, the latter add to cobalt-complexed propynal completely syn selectively[225]. In the case of bulky R^1 groups (silyl, *tert*-butyl), syn selectivity results[224]. Open transition states in which the enolsilane attacks the TiCl$_4$-activated aldehyde were proposed to explain the stereochemical results. The enolsilanes from 2-pyridyl thioesters undergo TiCl$_4$-mediated aldol additions with exclusive formation of syn adducts[226]. Sometimes other Lewis acids are more effective[223,227]. One of the important applications of the Mukaiyama reaction is the chelation-controlled addition of enolsilanes to TiCl$_4$-chelated forms of chiral α- and β-alkoxyaldehydes in which simple diastereoselectivity is of the syn type[41] (Section 3.1.2).

As mentioned in Section 2.2, some but not all enolsilanes derived from ketones react smoothly with TiCl$_4$ to form the corresponding titanium trichlorides which add to alde-hydes with moderate simple diastereoselectivity (65–89% syn selectivity)[101]. Cases of chiral enolsilanes undergoing the Si−Ti exchange reaction are known, leading to Cl$_3$Ti enolates which react diastereo- and enantioselectively[105b] (Section 3.3).

Significant progress was made with the discovery that ketones (*e.g.*, 179) and certain carboxylic acid derivatives can be titanated by treatment with TiCl$_4$ in the presence of tertiary amines (Evans methodology)[112,113,228]. Formally, this generates the trichloroti-tanium enolate and the amine hydrochloride, which may actually interact to form the tetrachlorotitanium ate complex with the ammonium salt as the counterion, such as 180. It is also not clear whether monomers or aggregates are involved. In any event, the reagents add to aldehydes with high degrees of syn selectivity (*e.g.*, 181 with 92% ds)[228]. This methodology has been used successfully in the aldol addition of thioesters to aldehydes and aldimines[186] and in a variety of different applications[229].

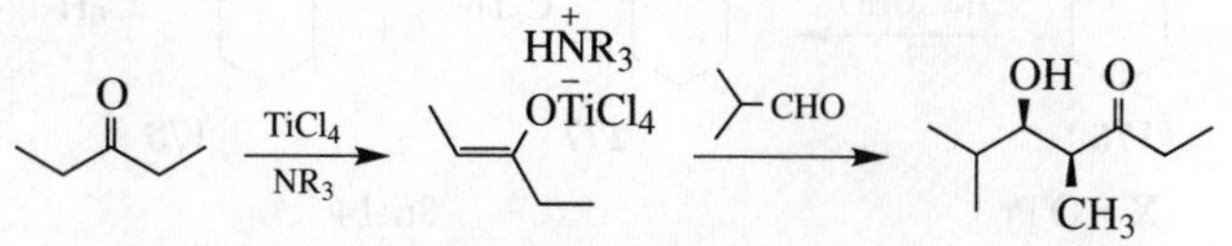

179   **180**   **181**

Typical Procedure for the Syn-selective TiCl4/NR3-mediated Aldol Addition of Carbonyl Compounds to Aldehydes[228]

TiCl4 (1.1 equiv) is added dropwise to a 0.2 M solution of 1.0 equiv of the ketone or carboxylic acid derivative in CH_2Cl_2 at $-78\,°C$ under N_2, giving a yellow slurry. After 2 min, 1.2 equiv of either Et_3N or $EtN(^iPr)_2$ are added dropwise, and the resulting deep red solution is stirred at $-78\,°C$ under N_2 for 1.5 h. After the dropwise addition of isobutylaldehyde (1.2 equiv), stirring is continued at $-78\,°C$ for 1.5 h. The reaction is terminated by addition of 1 : 1 v/v of saturated aqueous NH_4Cl, the mixture is warmed to $25\,°C$ and the product is isolated by a conventional extraction. The syn selectivity is 90–99%.

The sense and degree of stereoselectivity is comparable to those of the well-known boron-mediated processes[220], but the yields are often higher. Zimmerman–Traxler transition states accommodate the results[228]. Chirally modified enolates react diastereo- and enantioselectively[228] (Section 3.3). Hence this methodology has turned out to be a standard procedure.

Surprisingly, the TiCl4-mediated aldol addition can be carried out in the absence of amines[230]. A wide variety of cyclic and acyclic ketones react stereoselectively with benzaldehyde (93–100% syn), whereas syn selectivity in the case of aliphatic aldehydes is lower. These aldol additions need to be monitored, because prolonged reaction times lead to dehydration products. In the case of **179**, catalytic amounts of TiF4 were shown to be effective, although with lower simple diastereoselectivity (77% syn)[230]. In a significant extension, unsymmetrical ketones such as **182** were reacted with aldehydes in a regioselective manner in favor of the more substituted position in the ketone[231]. As only catalytic amounts of TiCl4 are needed and syn selectivity is high, this simple protocol constitutes a useful method:

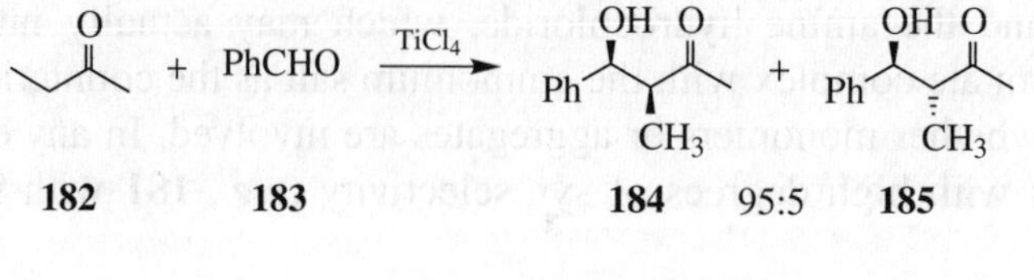

182   **183**   **184**   95:5   **185**

Another development which deserves attention is the use of titanium ate complexes of the type $Ti(O^iPr)_5Li$ to promote aldol reactions with high anti selectivity, especially in the case of sterically hindered aliphatic aldehydes[232]. If the titanium ate complex contains a hydride source as in $BuTi(O^iPr)_4Li$ (prepared from $Ti(O^iPr)_4$ and BuLi), a one-pot stereoselective aldol-Tishenko reaction is possible[233]. Following anti-selective aldol addition, the ketone function is reduced under chelation control with exclusive formation of anti-configurated 1,3-diols **186**:

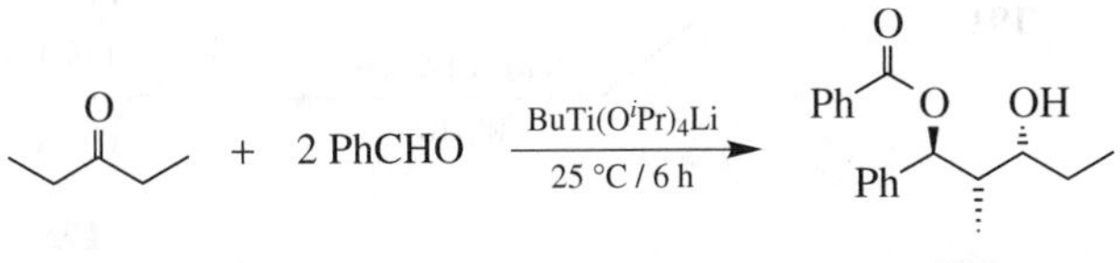

Preformed titanium enolates derived from lactams show moderate to good levels of simple diastereoselectivity[7,43]. Ligand tuning needs to be looked at. No method for the diastereoselective addition of aldehyde–enolates to aldehydes is known. Aldehyde titanium enolates likewise provide syn/anti mixtures, but titanated hydrazones **188** derived from aldimines add to aldehydes syn selectively (**189**/**190** = 10 : ≥90)[64]. The reagents **188** are (*E*)-configured, which means that syn selectivity is difficult to explain (boat transition state?)[64]. Titanated chiral bislactims react enantio- and diastereoselectively (Section 3.3).

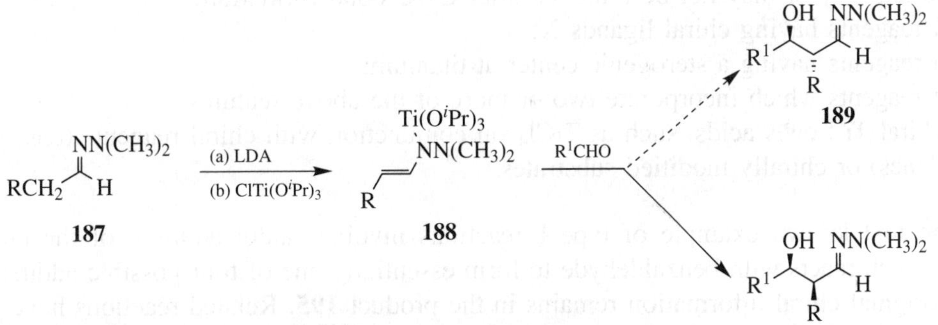

$TiCl_4/NEt_3$-mediated Knoevenagel condensations using active methylene carbonyl compounds probably proceed via chlorotitanium enolates[111]. In the case of the phosphonate **191**, the thermodynamically more stable (*E*)-configured condensation products **192** are formed preferentially[234]. In contrast, the sodium salt of **191** reacts with $ClTi(O^iPr)_3$ to form an isolable mixture of (*E*)/(*Z*) enolates[234] which condense stereoconvergently with aldehydes to form the (*Z*)-configured alkenes **193** selectively with concomitant transesterification[235]. Triethylamine as the base is also efficient[235]. Mechanistically, a kinetically controlled syn-selective aldol addition is involved followed by stereospecific *O*-titanate elimination[234]. Malonates and other CH-acidic compounds also undergo this type of Knoevenagel condensation[235,111]. The method is mild and works well in cases in which traditional Knoevenagel conditions fail[236].

$$(H_5C_2O)_2\overset{\overset{\displaystyle O}{\|}}{P}CH_2CO_2C_2H_5$$

191

3.3   Enantioselective Aldol and Grignard Additions Using Stoichiometric Amounts of Chiral Reagents

Chiral modification of titanium-mediated reactions is possible in one of several ways[7], using:

1. reagents $RTiX_3$ in which chirality occurs in the organyl moiety R – the chiral information may or may not be removed after C−C bond formation;
2. Ti reagents having chiral ligands X;
3. Ti reagents having a sterogenic center at titanium;
4. Ti reagents which incorporate two or more of the above features;
5. achiral Ti Lewis acids, such as $TiCl_4$, in conjunction with chiral reagents (*e.g.*, enolsilanes) or chirally modified substrates.

The first known example of type 1 reactions involves aldol addition of the enolate **194**, which reacts with benzaldehyde to form essentially one of four possible adducts[26]. The original chiral information remains in the product **195**. Related reactions have been reported for titanium enolates derived from β-lactams[237a] and titanated terpenes[237b].

This methodology has been extended to the enolate **197**, which adds to aldehydes to form adducts **198** preferentially (ds $\geq$ 98%)[71], which means essentially complete syn selectivity and complete diastereofacial selectivity. As the original chiral information can be cleaved off oxidatively, enolate **197** is a 'chiral propionate'. It competes well with the boron analog[238]. As a result of a systematic study it became clear that an excess of titanating agent $ClTi(O^iPr)_3$ enhances stereoselectivity dramatically. It is possible that lessselective ate complexes are formed in the initial titanation step which rapidly lose chloride ions in the presence of additional $ClTi(O^iPr)_3$. Based on earlier work on the titanation of lithium enolates using $Ti(O^iPr)_4$ with formation of titanium ate complexes[63], the lithium enolate derived from **196** was also reacted with $Ti(O^iPr)_4$[71]. The use of an excess of this titanating agent resulted in an enolate which showed the same selectivity as **197**. The influence of excess $ClTi(O^iPr)_3$ or $Ti(O^iPr)_4$[44,239] needs to be studied in other systems as well.

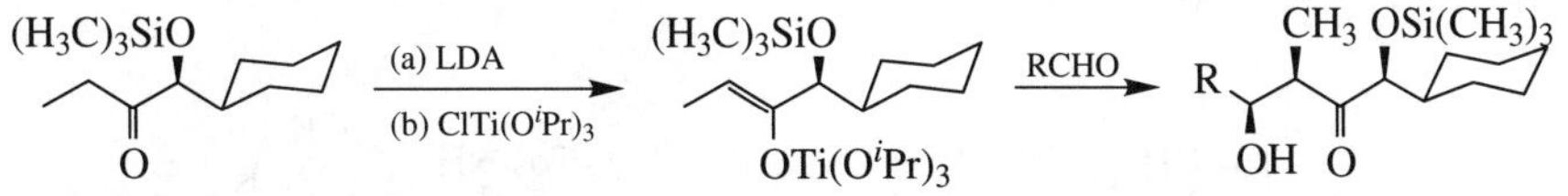

196   **197**   **198**

Along related lines, the enolate **200** was prepared and reacted with benzaldehyde[239]. Using a threefold excess of $ClTi(O^iPr)_3$ in diethyl ether, a 92 : 5 : 3 : 0 ratio of diastereomers was observed, adduct **201** being the major product. The result is synthetically significant because the analogous boron enolate derived from **199** shows opposite diastereofacial selectivity[220]. This is because the titanium reagent undergoes internal chelation (cf. **200**), in contrast to the boron analog:

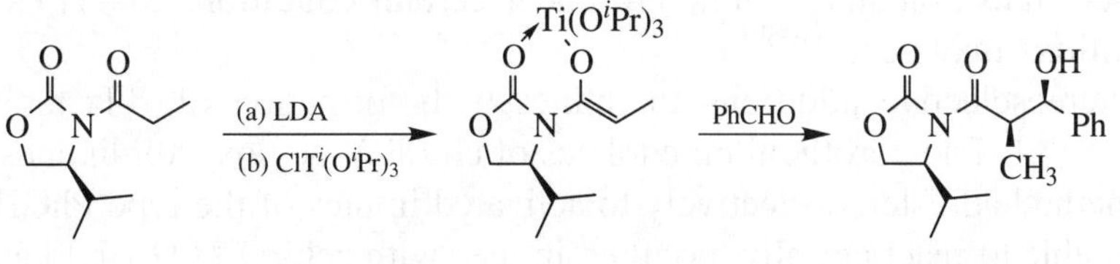

199   **200**   **201**

The previously mentioned $TiCl_4/EtN(^iPr)_2$-induced titanation and aldol reaction of carbonyl compounds (Section 3.2.2) is particularly effective in the case of chiral compounds[228,240]. Typical enolate precursors are **202–204**:

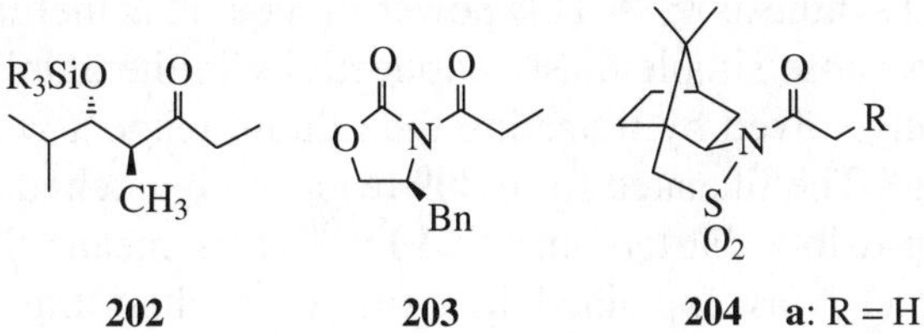

202   **203**   **204**   **a**: R = H

              Bn = $CH_2C_6H_5$ **b**: R = CH_3

The yields and stereoselectivities of these aldol reactions are impressive. For example, the Cl_3Ti–enolate of **203** affords syn-configured aldol adducts in enantiomerically pure form[228,241]. The absolute configuration ('Evans-syn' versus 'non-Evans-syn') can be reversed simply by varying the nature of the amine base and the Lewis acid stoichiometry[242]. In order to obtain the corresponding anti adducts, a different Ti-based method was developed[243]. Titanation of the ester **205** with $TiCl_4/^iPr_2NEt$ generates Ti–enolates which fail to react with aldehydes, presumably due to intramolecular coordination of titanium by the sulfonamide (which enables enolization of the ester in the first place). However, if the aldehyde component is precomplexed with additional $TiCl_4$, the Ti–enolates undergoes smooth aldol addition, anti selectivity ranging between 85% and 99%[243]. A transition state involving two titanium moieties was postulated for this synthetically important reaction.

Enantioselective versions of the Mukaiyama aldol additions based on chirally modified *O,O*-silyl ketene ketals have also been described[244]. In some cases the results are excellent, although a really general procedure encompassing high yield as well as above 95% stereoselectivity remains to emerge. In most of these reactions, the *O*-silyl reagents react via open transition states, although under certain conditions Si–Ti exchange prior to aldol additions may occur[105b].

Various stereoselective additions to imines with formation of *β*-lactams have been reported[105b,c,245]. Trichlorotitanium enolates of chiral *N*-acyloxazolidinones prepared by the Evans method add stereoselectively to activated imines of the type $PhCH = NTs$[246]. It is also possible to react chirally modified imines with achiral *O,O*-silyl ketene ketals in the presence of $TiCl_4$ or TiF_4, the nature of the Lewis acid determining the stereochemical outcome[247]. Alternatively, achiral Ti–enolates of thioesters or *α*-alkoxy acetates add stereoselectively to chiral imines[248].

Lithiated bis-lactim ethers **208** are known to be alkylated stereoselectively trans to the isopropyl group, hydrolysis then affording amino acids in above 98% enantiomeric purity (Schöllkopf amino acid synthesis)[249]. This powerful method is inefficient if aldehydes are used as electrophiles because simple diastereoselectivity in the aldol-type process is low. This problem was nicely solved by titanating the lithium reagent with $ClTi[(CH_3)_2]_3$ (or the *N,N*-diethyl analog). The titanated form **209** reacts with aldehydes to form essentially only one of the four possible diastereomers **210**[250]. This means that simple diastereoselectivity is anti, which was explained in terms of a chair transition state[250]. The adducts **210** can be hydrolyzed, cleaving off the recyclable chiral auxiliary (valine). Reagents of the type **209** have also been added to chiral aldehydes with complete reagent control[251]. The reagent also undergoes stereoselective Michael additions to nitroalkenes and *α,β*-unsaturated carbonyl compounds (Section 4).

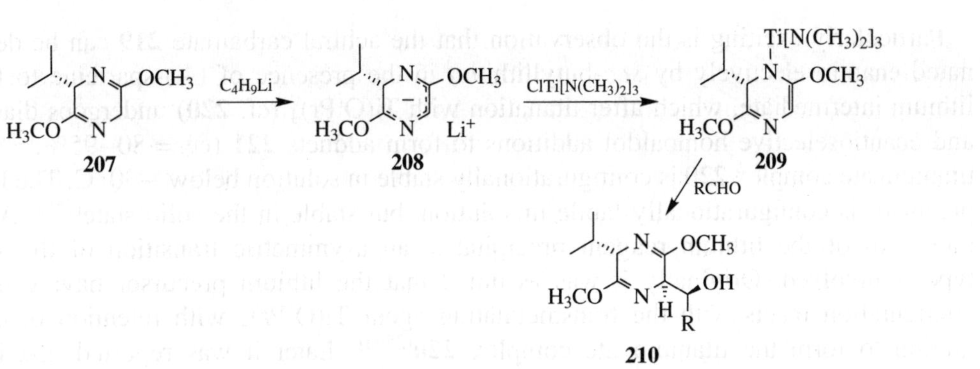

One of the most efficient methods for enantioselective homoaldol addition concerns the reaction of the titanium reagent **211**, which adds to aldehydes to produce essentially a single adduct **212**[252]. Cleavage of the chiral auxiliary affords furans **213**, which are readily oxidized to the enantiomerically pure lactones **214**:

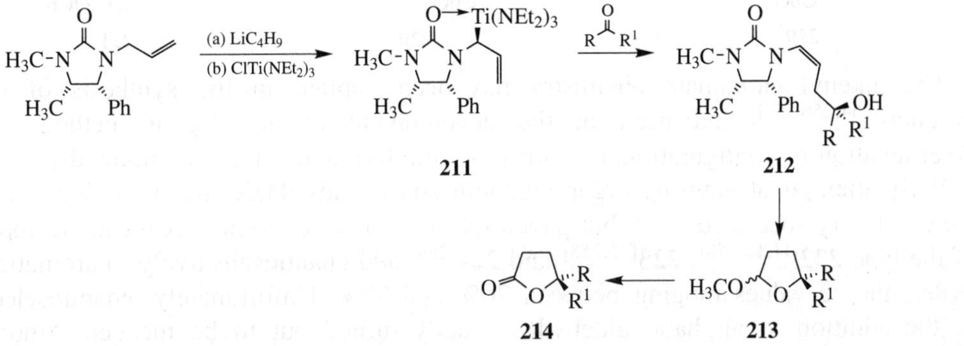

Prochiral homoenolates have also been prepared in enantiopure form[253] (Section 3.2 describes racemic versions). For example, lithiation of the enantiomerically pure alkenyl carbamate **215** results in the (unselective) lithium reagent **216** which can be titanated by $Ti(O^iPr)_4$ with retention of configuration to form the ate complex **217**. The latter reacts with 2-methylpropanal to form essentially a single diastereomer **218**, having an enantiomeric excess (ee) of 86%[253]. Interestingly, titanation with $ClTi(O^iPr)_3$ leads to racemization, whereas $ClTi(NEt_2)_3$ results in inversion of configuration. The reasons for the dramatic differences are not fully understood[254]. The stannylated form of **216** undergo stereoselective $TiCl_4$-mediated aldehyde additions which involve prior Sn–Ti exchange[103,255].

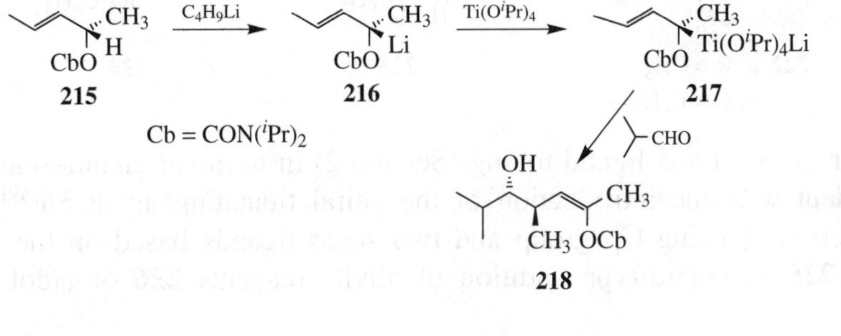

Particularly exciting is the observation that the achiral carbamate **219** can be deprotonated enantioselectively by *sec*-butyllithium in the presence of (−)-sparteine to form a lithium intermediate, which after titanation with Ti(O^{i}Pr)$_4$ (cf. **220**) undergoes diastereo- and enantioselective homoaldol additions to form adducts **221** (ee = 80–95%)[256]. The titanium ate complex **220** is configurationally stable in solution below −30 °C. The lithium precursor is configurationally labile in solution, but stable in the solid state[256]. As only one form of the lithium reagent precipitates, an asymmetric transition of the second type is involved. Originally it was assumed that the lithium precursor having the (*R*) configuration reacts with the transmetalating agent Ti(O^{i}Pr)$_4$ with retention of configuration to form the titanium ate complex **220**[256a]. Later it was reported that in fact the (*S*)-configured lithium precursor is transmetalated by Ti(O^{i}Pr)$_4$ with inversion of configuration to produce reagent **220**[256b]:

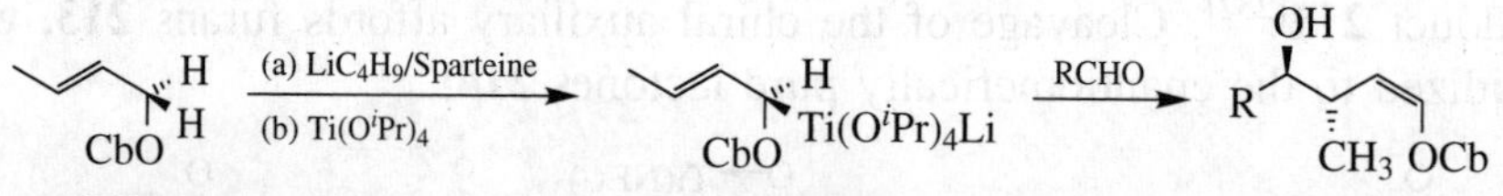

The alkenyl carbamate chemistry has been applied in the synthesis of natural products[254,256,257] and used in the development of an elegant method for the determination of configurational stability of chiral organometallic compounds[219].

Early attempts at devising organotitanium compounds RTiX$_3$ having chiral ligands X were not very successful[7,14], but paved the way to more efficient reagents. Compounds of the type **222**[11,13,258], **223**[11,258] and **224**[259] add enantioselectively to aromatic aldehydes, the ee values ranging between 70% and 95%. Unfortunately, enantioselectivity in the addition to aliphatic aldehydes usually turned out to be meager. Among the most efficient reagents are compounds of the type **223** based on α,α,α′,α′-tetraaryl-1,3-dioxolane-4,5-dimethanols (TADDOLs), which are prepared from tartrates. This ligand system is also useful in catalytic reactions (Section 8). The 'magic' effect of the four aryl groups has been discussed in detail[11,258,260]. An improved procedure based on the use of TADDOL, Ti(O^{i}Pr)$_4$, and RMgX has been described, the best results being obtained in reactions of alkyl aryl ketones[261].

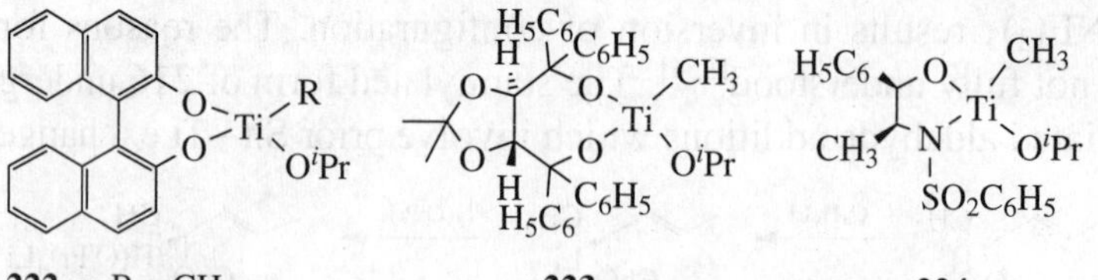

222 a: R = CH$_3$
    b: R = C$_6$H$_5$

223

224

The power of metal and ligand tuning (Section 2) in terms of enantioselectivity really became evident with the introduction of the chiral titanating agent **56**[89]. It contains a bulky, electron-donating Cp group and two sugar ligands based on the cheap diacetoneglucose **225**. Grignard-type addition of allylic reagents **226** or aldol additions of analogous enolates **62** are highly enantioselective (ee > 90%)[89]. Structural and mechanistic features have been delineated elsewhere, including correlations with ^{47}Ti/^{49}Ti NMR data[262].

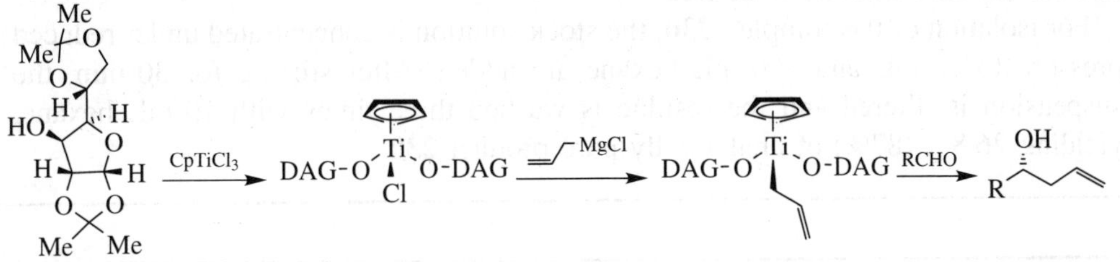

Reagent **228** adds to aldehydes to produce β-hydroxyamino acid derivatives **229**, simple diastereoselectivity (syn) being 96–98% and the ee values ranging between 87% and 98%[89,262]:

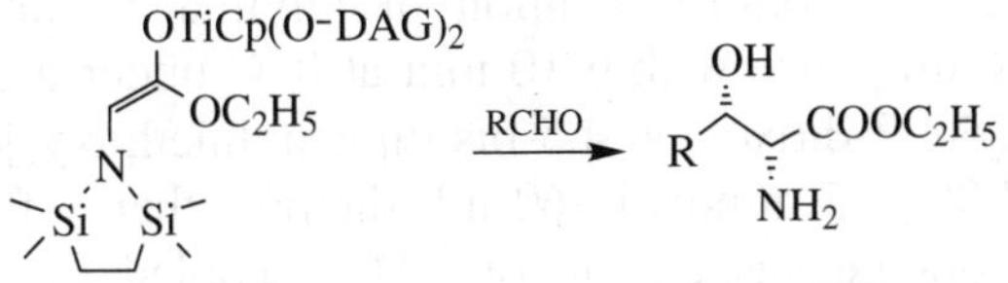

A stereochemically complementary allyl transfer reagent is the TADDOLate **231** which readily adds to aldehydes (ee > 95%)[89c]. In the case of analogous crotyltitanium reagents (and other substituted allylic and enolate compounds), enantioselectivity is also essentially complete. The reagents are so powerful that complete reagent control is observed in reactions with chiral aldehydes. TADDOLs have been used to prepare chiral cyanotitanium reagents which add enantioselectively to aldehydes (ee = 68–96%)[89d].

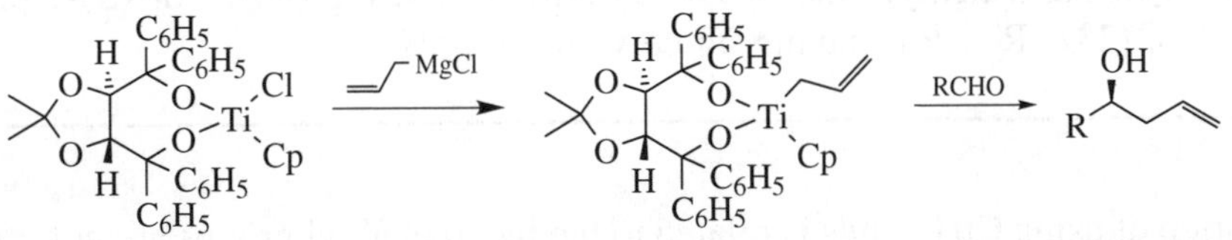

Preparation of the Chiral Titanating Agent [(4R,5R)-2,2-Dimethyl-1,3-dioxolan-4,5-bis(diphenylmethoxy)]cyclopentadienyl Chlorotitanate, 230[89,262]

A solution/suspension of freshly sublimed cyclopentadienyltrichlorotitanium (11 g, 50 mmol) in 400 mL diethyl ether (distilled over Na–benzophenone), is treated with (4*R*, 5*R*)-2,2-dimethyl-4,5-[bis(diphenylhydroxymethyl)]-1,3-dioxolane (23.3 g, 50 mmol) under argon in the absence of moisture. After 2 min at 25 °C, a solution of NEt$_3$ (12.65 g, 110 mmol) in 125 mL of diethyl ether is added dropwise to the stirred mixture within 1 h (efficient stirring is essential). After further stirring for 12 h, Et₃N·HCl (13.8 g) is filtered off under argon and washed three times with 50 mL of diethyl ether. This stock solution (610 mL) is assumed to be 82 mmol and can be used directly if desired.

For isolation of the complex **230**, the stock solution is concentrated under reduced pressure to 75 mL and 300 mL hexane are added. After stirring for 30 min, the suspension is filtered and the residue is washed three times with 10 mL hexane, yielding 26.8 g (87%) of analytically pure product **230**.

Typical Procedure for the Enantioselective Allylation of Aldehydes using [(4R,5R)-2,2-Dimethyl-1,3-dioxolan-4,5-bis(diphenylmethoxy)]allylcyclopentadienyl Chlorotitanate, 231[89]

A 5.3-mL volume of a 0.8 M solution of allylmagnesium chloride in THF (4.25 mmol) is added dropwise within 10 min at 0 °C under argon to a solution of [(4R, 5R)-2,2-dimethyl-1,3-dioxolane-4,5-bis(diphenylmethoxy)]cyclopentadienyl chlorotitanate **230** (3.06 g, 5 mmol) in 60 mL diethyl ether. After stirring for 1.5 h at 0 °C the slightly orange suspension of **231** is cooled to −74 °C and treated within 2 min with benzaldehyde (403 mg, 3.8 mmol, dissolved in 5 mL of diethyl ether). The mixture is stirred for 3 h at −74 °C. After hydrolysis with 20 mL of aqueous 45% NH₄F solution and stirring for 12 h at 25 °C, the reaction mixture is filtered over Celite and extracted twice with diethyl ether (50 mL). The combined extracts are washed with brine, dried with MgSO₄ and concentrated. The solid residue is stirred with 50 mL of pentane. Subsequent filtration furnishes 1.54 g of crude alcohol and 1.68 g of white crystalline material. Chromatography (200 g silica gel, CH₂Cl₂/hexane/diethyl ether 4 : 4 : 1) affords 0.521 g (93%) of (S)-1-phenyl-3-buten-1-ol (**232**; R = Ph) having an ee value of 95%.

A limitation of using Cp ligands is related to the fact that alkyl groups are not transferred chemically to aldehydes very well[7], owing to the electron-releasing and steric effects of such ligands. In these cases zirconium and hafnium analogs are better suited because they are more Lewis acidic[262]. Titanium reagents having two Cp groups are even less reactive. However, allyl transfer reactions are possible. The first case of such a chiral titanium reagent having two different Cp groups and the stereogenic center at the metal has been described[94]. As the two Cp groups which were used are too 'similar', enantioselectivity in aldehyde additions turned out to be poor. Nevertheless, reagents with a stereogenic center at titanium (or zirconium) constitute an intriguing goal for the future (see also reagent **224**[259]). The use of Brintzinger-type bridged titanocenes in the crotylation of aldehydes results in varying degrees of diastereo- and enantioselectivity[263]. These and other types of chiral titanocenes as well as other metallocenes in stoichiometric and catalytic processes have been reviewed[264].

4   Michael Additions

Two outstanding Michael-type reactions are $TiCl_4$-mediated additions of allyl-silanes[143,265] and enolsilanes[266,267], the Sakurai–Hosomi and Mukaiyama–Michael reactions, respectively:

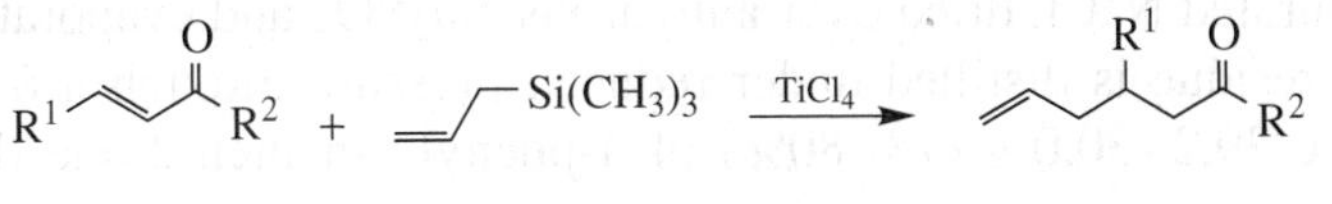

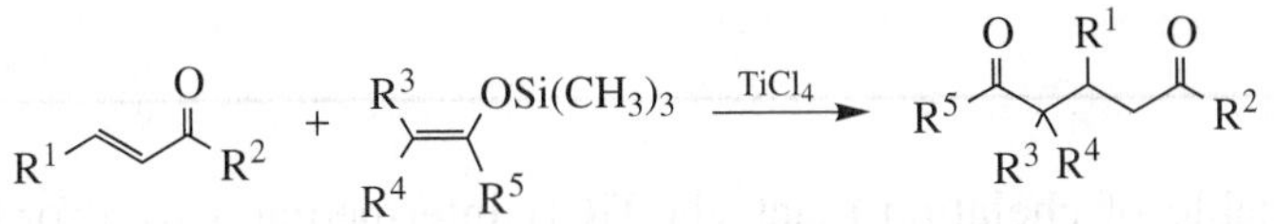

Neither of the reactions involve Si–Ti exchange prior to C–C bond formation[145d,266,267]. Rather, as illustrated below for the allylsilane addition, complexation of the carbonyl group by $TiCl_4$ initiates the process with subsequent formation of cationic species **233** (or Si-bridged siliranium ions) which undergo rapid chloride-induced desilylation:

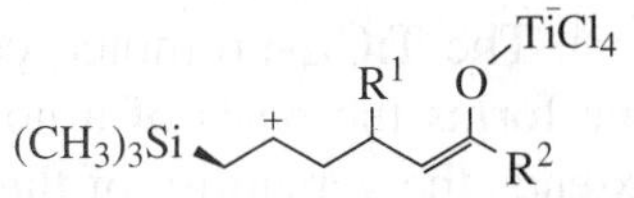

233

Such conjugate additions of allylsilanes to enones are quite general, whereas α,β-unsaturated esters react poorly[268]. If the enone is chiral, diastereofacial selectivity is possible, provided one of the π-faces is sterically shielded as in the case of the enone **234**, affording the adduct **235** exclusively in 85% yield[265]:

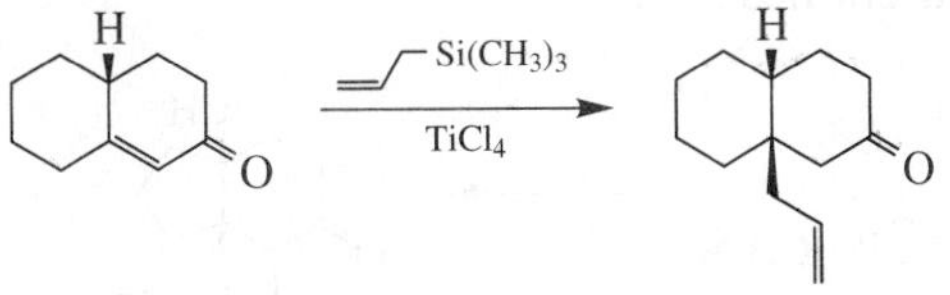

234   **235**

Typical Procedure for the TiCl$_4$-mediated Conjugate Addition of Allylsilanes[265]

A 2-L three-necked round-bottomed flask is fitted with the dropping funnel, mechanical stirrer, and reflux condenser attached to a nitrogen inlet. In the flask is placed benzalacetone (29.2 g, 0.20 mol) and 300 mL of CH_2Cl_2. The flask is immersed in a dry-ice/methanol bath ($-40\,°C$) and TiCl$_4$ (22 mL, 0.20 mol) is slowly added by syringe to the stirred mixture. After 5 min, a solution of allyltri-methylsilane (30.2 g, 0.26 mol) in 300 mL of CH$_2$Cl$_2$ is added dropwise with stirring over a 30-min period. The resulting red–violet reaction mixture is stirred for 30 min at $-40\,°C$, hydrolyzed by addition of 400 mL of water and, after the addition of 500 mL of diethyl ether with stirring, allowed to warm to $25\,°C$. The nearly colorless organic layer is separated and the aqueous layer is extracted with three 500-mL portions of diethyl ether. The organic layer and ether extracts are combined and washed successively with 500 mL of saturated sodium hydrogencarbonate and 500 mL of saturated NaCl, dried over anhydrous Na$_2$SO$_4$ and evaporated at reduced pressure. The residue is distilled under reduced pressure through a 6 inch Vigreux column to give 29.2–30.0 g (78–80%) of 4-phenyl-6-hepten-2-one (bp 69–71 °C, 0.2 mmHg).

Substrates capable of chelation react via TiCl$_4$ intermediates to afford good yields of chelation-controlled adducts[269]. In an interesting development, crotylsilanes were shown to undergo TiCl$_4$-promoted stereoselective conjugate additions to chiral α,β-unsaturated keto esters and to 2-(S)-tolylsulfinyl-2-cyclopentenone[270]. Stereoselective TiCl$_4$-induced additions of allyltrimethylsilane to chiral α,β-unsaturated N-acyloxazolidinones and N-enoyl-sultams constitute another example of conjugate C$-$C bond formation[271].

Conjugate additions to α,β-unsaturated acid nitriles are also possible[272], as are selective 1,6-additions to dienones[273]. The TiCl$_4$-promoted conjugate addition of allylsilane to N-acyl-2,3-dihydro-4-pyridone forms the basis of a novel synthesis of ($-$)-N-methyl-coniine, a process which also extends the versatility of the Sakurai–Hosomi reaction[274]. Various intramolecular allylsilane additions with formation of polycyclic and/or spiro compounds have been reported[275], as in the cyclization of enone 236[275c]. Sometimes undesired protodesilylation occurs (proton source?), in which case Lewis acids such as EtAlCl$_2$ (proton sponge!) or fluoride ions are the promoters of choice[275]. Little is known concerning high levels of simple diastereoselectivity in intermolecular conjugate additions of crotylsilanes to achiral enones[265].

| | TiCl$_4$ | 80 : 20 |
| | F$^-$ | 14 : 86 |

Related are TiCl$_4$-promoted allenylsilane conjugate additions[276]. In this case, however, the initial species is a silyl-substituted vinyl cation which undergoes rapid silyl group migration followed by ring closure and formation of cyclopentenes in good yields (e.g.,

239[276]). Although a variety of enones react similarly, the method cannot be extended to the unsubstituted parent allenyl reagent. Allenylstannanes substituted in the 3-position do not induce cyclization, the alkynyl adducts **240** being the products[277]:

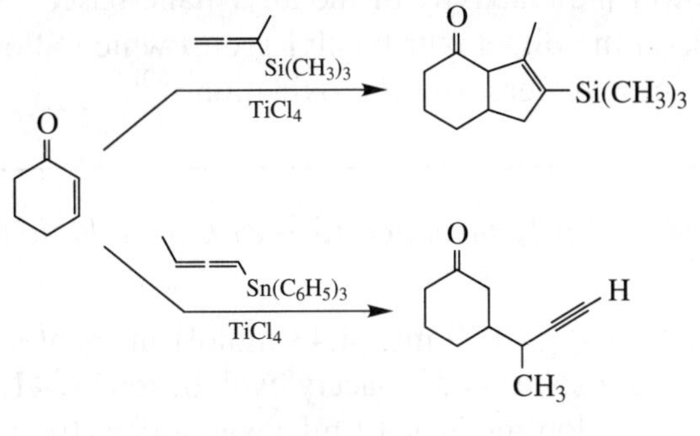

In a synthetically significant development concerning allylsilane additions, it was discovered that the use of sterically hindered silyl groups such as triisopropylsilyl or *tert*-butyldiphenylsilyl in place of the usual $(CH_3)_3Si$–entity leads to different products, namely cyclopentane derivatives in yields of 50–86% (*e.g.*, **243**[278]):

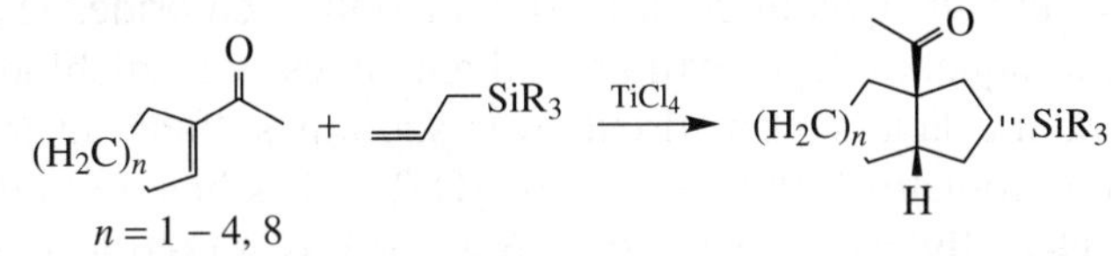

Diastereoselectivity in favor of the products having the silyl groups anti to the angular acetyl moiety is very high. Due to the steric hindrance at silicon, desilylation of the initial intermediate **245** is largely suppressed, allowing for reversible formation of two diastereomeric siliranium cations **244** (syn) and **246** (anti). Attack of the titanium enolate involves a 5-exo-tet cyclization (Baldwin rules: exocyclic sp^3 bond formation) and occurs preferentially with **246** due to stereoelectronic reasons:

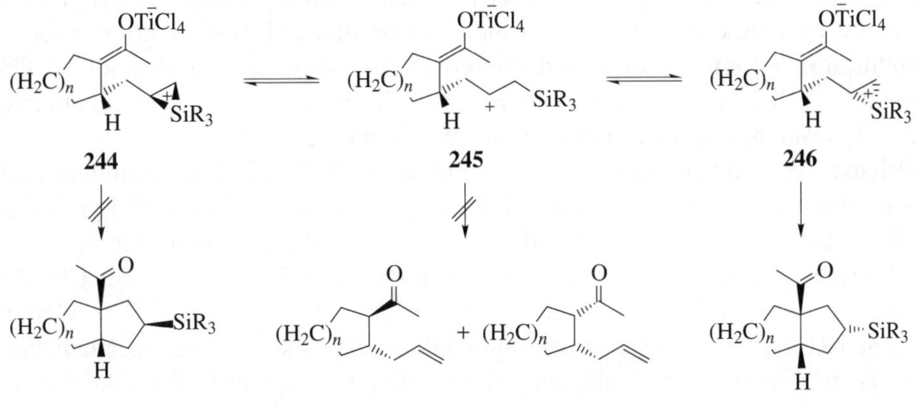

This convenient [3 + 2] cycloaddition has been applied to different substrates[278,279]. It also provides the possibility of replacing the silyl group by an alcohol function via Tamao–Fleming oxidation, provided silicon bears activating groups such as phenyl[278]. However, such groups lower the reactivity of the allylsilane-based [3 + 2] cycloaddition. An excellent compromise is the dimethyltritylsilyl group which allows for rapid [3 + 2] cycloaddition as well as smooth stereospecific oxidation[280].

Optimized Procedure of the TiCl$_4$-mediated [3 + 2] Cycloaddition of Allylsilanes to Enones[278]

A solution of TiCl$_4$ (0.840 g, 0.49 mL, 4.43 mmol) in dichloromethane (5 mL) was cooled to −20 °C and a solution of 1-acetylcyclohexene (**241**, $n = 2$) (0.500 g, 0.52 mL, 4.03 mmol) in dichloromethane (2 mL) was added (formation of the Lewis acid–enone complex as a yellow suspension). After cooling to −78 °C, a solution of allyltriisopropylsilane (1.20 g, 1.46 mL, 6.04 mmol) in dichloromethane (6 mL) was added. The reaction temperature was raised to −20 °C over a period of 4 h (the color changed to red–violet), the mixture was stirred for an additional 16 h at this temperature, and poured into an aqueous solution of ammonium chloride. The organic layer was separated, the aqueous layer extracted three times with dichloromethane, and the combined organic layers were dried over sodium sulfate. Evaporation of the solvent and flash chromatography (pentane/Et$_2$O 20 : 1, silica gel) afforded first the less-polar 1-acetyl-2-allylcyclohexane (0.016 g, 2%) as a colorless oil and then 1-acetyl-8-triisopropylsilylbicyclo[4.3.0]nonane (**243**, $n = 2$, R = ^{i}Pr; 1.12 g, 86%) as an oil.

Upon attempting to react α,β-unsaturated esters, a third reaction mode was discovered, namely the stereoselective formation of cyclobutane derivatives[280]. In this case the corresponding titanium ester enolates react preferentially at the more substituted position of the siliranium ions in a 4-exo-tet cyclization. In the case of alkynyl esters the products are cyclobutene derivatives, which can be induced to undergo another [2 + 2] cycloaddition in a domino process with stereoselective formation of bicyclo[2.2.0]hexane compounds[280]. It is interesting to note that allylsilanes add to α-keto esters anomalously in a [2 + 2] fashion, provided bulky groups at silicon are used[281].

Enolsilanes derived from ketones or esters undergo TiCl$_4$-mediated conjugate additions to α,β-unsaturated ketones and esters (Mukaiyama–Michael addition), the yields ranging between 55 and 95%[266]. If the enolsilane and/or the carbonyl component is sensitive to TiCl$_4$, mixtures of TiCl$_4$ and Ti(O^{i}Pr)$_4$ (which form the milder Lewis acids Cl$_3$TiO^{i}Pr or Cl$_2$Ti(O^{i}Pr)$_2$) or other promoters such as trityltriflate should be used[282a]. Compounds of the type (RO)$_2$Ti = O can also be employed as catalysts[282b]. The mechanism of this reaction is more complicated and thus more difficult to define than in the Mukaiyama aldol addition. This is due to several reasons:

- The Lewis acid can complex in a syn or anti manner[283].
- The complex can take on several different conformations[283].
- In some cases electron transfer prevails, as in the reaction of *O,O*-silyl ketene ketals with certain enones[284].

Typical Procedure for the TiCl₄-mediated Conjugate Addition of Enolsilanes[266]

A 500-mL three-necked flask is fitted with a mechanical stirrer, rubber septum and a two-way stopcock which is equipped with a balloon of argon gas. To the flask is added 100 mL of dry CH_2Cl_2, and the flask is cooled in a dry-ice-acetone bath. $TiCl_4$ (7.7 mL) is added by syringe through the septum. The septum is removed and replaced by a 100-mL pressure-equalizing dropping funnel containing a solution of 11.2 g of isopropylideneacetophenone in 30 mL of CH_2Cl_2. This solution is added over 3 min and the mixture is stirred for 4 min. A solution of 13.5 g of the enolsilane of acetophenone in 40 mL of CH_2Cl_2 is added dropwise with vigorous stirring over 4 min and the mixture is stirred for 7 min. The reaction mixture is poured into a solution of 22 g of Na_2CO_3 in 160 mL of water with vigorous magnetic stirring. The resulting white precipitate is removed by filtration through a Celite pad and the precipitate is washed with CH_2Cl_2. The organic layer of the filtrate is separated and the aqueous layer is extracted with two 40 mL portions of CH_2Cl_2. The combined organic extracts are washed with 60 mL of brine and dried over sodium sulfate. The CH_2Cl_2 solution is concentrated in a rotary evaporator and the residue is passed through a short column of silica gel (Baker 200 mesh, 400 mL) using 1.5 L of a 9 : 1 v/v mixture of hexane and ethyl acetate. The eluent is condensed and distilled: the first fraction (bp 81–85 °C, 0.6 mmHg, 2.04 g) is a mixture of isopropylideneacetophenone and acetophenone; the second fraction (bp 85–172 °C, 0.6 mmHg, 0.42 g) is a mixture of the above substances and the desired product; the third fraction (bp 172–178 °C, 0.6 mmHg) gives 14.0–15.2 g (72–78%) of 3,3-dimethyl-1,5-diphenylpentane-1,5-dione.

Although the problem of simple diastereoselectivity was not addressed in the early publications[266], mixtures of diastereomers are probably formed in most of the relevant cases. Later work using similar substrates and $TiCl_4$ or $SnCl_4$ showed that syn/anti mixtures are indeed formed, but that in some cases the reaction can be manipulated in a stereoselective sense[285]. Enolsilanes **250** derived from ketones result in moderate to high anti selectivity (cf. **251**), regardless of the geometry of the enolsilane. The (Z)-enolsilanes from propiophenone and related aromatic ketones show excellent anti selectivity if $SnCl_4$ is used as the Lewis acid (ds = 90–98%). (Z)- and (E)-configured *O*-silyl ketene ketals react highly syn selectively with *tert*-butyl enones to form adducts **252** preferentially (ds = 96–99%), but stereorandomly with most other enones[285].

In contrast, O,S-silyl ketene ketals **250** ($R = SR'$) show significantly higher degrees of diastereoselectivity[286]. In some cases the geometry of the O,S-silyl ketene ketal governs the stereochemical outcome, the (E)-configured reagent affording a 95 : 5 anti/syn product ratio, and the (Z) analog leading to a 9 : 91 ratio. However, the reaction is sensitive to the nature of the R groups in the enone as well as to the bulkiness of the silyl group.

Chiral enones have been shown to react with remarkable degrees of simple diastereoselectivity *and* diastereofacial selectivity[287]. Finally, chiral O,O-silyl ketene ketals undergo stereoselective TiCl$_4$-mediated conjugate additions (de = 72–75%)[288]. Acyclic transition states have been proposed for most of these reactions, but it is unlikely that a simple well-defined transition state pertains to all cases. More work is needed, especially in catalytic enantioselective versions (Section 8.1). In a novel approach Cl$_3$TiO^{i}Pr was used to promote the anti-selective conjugate addition of alkenylsulfides[289]. Additions to nitroalkenes are promoted by TiCl$_4$, the products being 1,4-diketones following Nef-type work-up[290].

Bona fide organotitanium reagents such as CH$_3$Ti(O^{i}Pr)$_3$[7,14] but also triisopropoxytitanium enolates derived from esters[13] generally add to enones in a 1,2 manner (Section 2). A notable exception is tetrabenzyltitanium[99]. Reagents RTi(O^{i}Pr)$_3$ and RTi(O^{i}Pr)$_4$Li undergo Cu-catalyzed 1,4-additions[291,292]. Ni(acac)$_2$ catalyzes the conjugate addition of CH$_3$Ti(O^{i}Pr)$_4$MgCl to sterically hindered enones, although (CH$_3$)$_3$Al is sometimes a better reagent in these otherwise difficult reactions[292]. In the case of unsaturated esters (*e.g.*, methyl methacrylate) the triisopropoxytitanium enolate derived from isobutyric acid ester induces group transfer polymerization (GTP) involving repetitive Michael additions[168b], as does the analogous titanium ate complex[168c].

The well-known propensity of many metal ate complexes (*e.g.*, cuprates, zincates, manganates) to show a preference for the 1,4-addition mode also shows up in other titanium enolate reactions. The first cases reported refer to the Michael addition of the titanium ate complexes prepared from thioamides by reaction of the lithium enolate with Ti(O^{i}Pr)$_4$[293]. Later a more systematic study showed that the Ti/Li ate complexes derived from propionic acid *tert*-butyl ester add to (E)-configured α,β-unsaturated esters and ketones to afford anti adducts with stereoselectivities up to 95%, whereas the reaction of the parent Li enolate is 90–95% syn selective[294]. This dramatic difference has been explained on the basis of an eight-membered transition state for the Li enolate and of an inverse electron demand Diels–Alder type of cycloaddition in the case of the Ti ate complex[294]. The Ti ate complexes also react stereoselectively with chiral enones[294].

Allyltitanium reagents add to enones cleanly in a 1,2 manner[43a]. However, by increasing the Michael acceptor propensity and by making the 1,2-addition mode less favorable (esters versus ketones), 1,4-additions may prevail[7]. The question of regioselective addition of carbon nucleophiles to 1-acylpyridinium salts has been addressed fairly often, especially as lithium enolates react unselectively to produce 1 : 1 mixtures of 1,2 and 1,4 adducts[295]. Simply treating the enolates with Ti(O^{i}Pr)$_4$ results in high degrees of 1,4 preference[295].

Of the large number of other types of Michael acceptors known in the literature, few have been reacted with organotitanium reagents. One of the early examples involves chiral α,β-unsaturated sulfoxides, which undergo chelation-controlled Michael additions of titanium reagents RTi(O^{i}Pr)$_3$ (R = CH$_3$, C$_2$H$_5$)[296]. The Schöllkopf bis-lactim methodology[249] (Section 3.3) is highly successful in Michael additions to prochiral nitroalkenes and α,β-unsaturated esters[297]. For example, the titanium reagent **209** adds to nitroalkenes such as **253** to produce essentially one of four possible diastereomers **254**. The more reactive lithium precursor is considerably less selective[297a].

Several other Ti-promoted Michael-type reactions are known. Reagents Ti(NR$_2$)$_4$ add to α,β-unsaturated ketones and esters in a conjugate manner; the intermediate trisaminotitanium enolates can be reacted with aldehydes in a tandem manner[298]. Finally, imines of α-amino esters undergo cycloaddition reactions with acrylates in the presence of triethylamine and ClTi(O^{i}Pr)$_3$ or Cl$_2$Ti(O^{i}Pr)$_2$[299]. Titanium enolates are involved which undergo Michael-type addition/cyclization.

5 Alkene-Forming Reactions

5.1 Wittig-type and Knoevenagel Alkylidenations

The Wittig alkylidenation is one of the most widely used synthetic reactions[300]. However, in the case of enolizable and/or highly functionalized ketones, the yields are often poor. Epimerization at stereogenic centers may also occur due to the basic nature of the ylides. In many of these cases titanium chemistry provides a solution to such problems[301].

The addition of TiCl$_4$ (0.7 parts) in dichloromethane to a mixture of CH$_2$Br$_2$ (1 part) and zinc dust (3 parts) in THF at room temperature leads within 15 min to a reagent which smoothly methylenates a variety of ketones, such as **255**[301a] and **257**[302]. The yield of methylenecycloheptane (**256**) is 83%, compared to the 10% yield obtained by the reaction of the Wittig reagent Ph$_3$P = CH$_2$. The sensitive ketone **257** yields no olefinic product using classical methodology, whereas the titanium-based procedure affords 60% of the desired alkene **258**:

255   256

257   258

The use of CH_2I_2 results in a more reactive and in some cases more selective reagent[301b]. A related reagent based on $CH_2I_2/Zn/Ti(O^iPr)_4$ reacts with ketoaldehydes solely at the aldehyde function[151].

On attempting to apply the titanium-mediated reaction to the synthesis of gibberellins, difficulties were encountered[303]. However, letting the reagent age in THF at 5 °C for 3 days prior to reaction resulted in excellent yields[303]. In fact, it was claimed that this version is the method of choice in other cases also, including the conversion of (+)-isomenthone into (+)-2-methylene-*cis*-*p*-menthane, which proceeds without any epimerization[304]. Several other impressive applications of the $CH_2X_2/Zn/TiCl_4$ reagents have been described[305].

Typical Procedure for the Methylenation of Ketones using $CH_2Br_2/TiCl_4/Zn$[303,304]

   Into a 1-L round-bottomed flask fitted with a magnetic stirrer and a pressure-equalizing dropping funnel connected to a nitrogen line are placed activated zinc powder, (28.75 g, 0.44 mol) dry THF (250 mL), and dibromomethane (10.1 mL, 0.144 mol). The mixture is stirred and cooled in a dry-ice/acetone bath at −40 °C. To the stirred mixture is added $TiCl_4$ dropwise (11.5 mL, 0.103 mol) over 15 min. The cooling bath is removed and the mixture is stirred at 5 °C (cold room) for 3 days under a nitrogen atmosphere. A reasonable rate of stirring must be maintained as the mixture thickens, but too fast a rate causes splashing up to the neck of the flask. The dark grey slurry is cooled with an ice-water bath and 50 mL of dry CH_2Cl_2 are added. To the stirred mixture are added 15.4 g (0.1 mol) of (+)-isomenthone in 50 mL of dry CH_2Cl_2 over 10 min. The cooling bath is removed and the mixture is stirred at room temperature (20 °C) for 1.5 h. The mixture is diluted with 300 mL of pentane and a slurry of 150 g of $NaHCO_3$ in 80 mL of water is added cautiously over 1 h. It is necessary to add the slurry dropwise at the beginning, allowing the effervescence to subside after each drop. After the inital vigorous effervescence, larger portions can be added. During this part of the addition, the stirrer becomes ineffective and gentle shaking by hand is continued until effervescence ceases. The residue is washed three times with 50-mL portions of pentane. The combined organic solutions are dried over a mixture of 100 g of Na_2SO_4 and 20 g of $NaHCO_3$, filtered through a sintered glass funnel and the solid desiccant is thoroughly washed with pentane. The solvent is removed at atmospheric pressure by flash distillation through a column (40 cm × 2.5 cm internal diameter) packed with glass helices. The liquid residue is distilled to give (+)-3-methylene-*cis*-*p*-menthane as a clear, colorless liquid (bp 105–107 °C, 90 mmHg; 13.6 g, 89% yield).

The reagent must be kept cold at all times because at room temperature the active reagent slowly decomposes and the mixture darkens considerably. Once prepared, the reagent can be stored at −20 °C (freezer) in a well-sealed flask without a significant loss of activity. A sample stored in this way for 1 year showed only a slight (5–10%) loss of activity. The molar activity of the active reagent is equivalent to the $TiCl_4$ molarity (determined by reaction with excess ketone followed by GLC analysis); however, an increase in the proportion of $TiCl_4$ makes no difference to the molar activity.

In an important extension of these reactions, alkylidenation of carboxylic acid esters **259** by means of $RCHBr_2/Zn/TiCl_4/TMEDA$ (tetramethylethylene diamine) to produce predominantly (Z)-alkenyl ethers **260** has been reported[306]:

$$RCO_2R^1 \xrightarrow[\text{Zn / TiCl}_4\text{ / TMEDA}]{R^2CHBr_2} \quad \underset{R^1O}{\overset{R}{\diagup}}\!\!=\!\!\underset{R^2}{}$$

$$\textbf{259} \qquad\qquad\qquad \textbf{260}$$

The method has unprecedented generality and can be extended to the synthesis of (Z)-configurated enolsilanes from trimethylsilyl esters[307], enamines from amides and alkenyl sulfides from thioesters[308]. Classical Wittig reactions fail in all of these transformations. A limitation of the titanium-based method is the poor yields in the case of alkylidenation reactions of aldehydes; in these cases low-valent chromium reagents are better suited[309].

Typical Procedure for the Alkylidenation of Carboxylic Acid Esters using
$RCHBr_2/Zn/TiCl_4/TMEDA$[306]

A solution of $TiCl_4$ (1.0 M, 4.0 mmol) in CH_2Cl_2 is added at 0 °C to THF (10 mL) under an argon atmosphere. To the yellow solution at 25 °C is added tetramethylethylenediamine (TMEDA; 1.2 mL, 8.0 mmol) and the mixture is stirred at 25 °C for 10 min. Zinc dust (0.59 g, 9.0 mmol) is added to the mixture. The color of the suspension turns from brownish yellow to dark greenish blue in a slightly exothermic process. After stirring at 25 °C for 30 min, a solution of methyl pentanoate (**259**, R = C_4H_9, R^1 = CH_3 0.12 g, 1.0 mmol) and 1,1-dibromohexane (0.54 g, 2.2 mmol) in THF (2 mL) is added to the mixture. The color of the resulting mixture gradually turns dark brown while stirring at 25 °C for 2 h. Saturated K_2CO_3 solution (1.3 mL) is added at 0 °C to the mixture. After it has been stirred at 0 °C for another 15 min, the mixture is diluted with diethyl ether (20 mL) and then passed rapidly through a short column of basic alumina (activity III) using diethyl ether–triethylamine (200 : 1, 100 mL). The resulting clear solution is concentrated and the residue is purified by column chromatography on basic alumina (activity III) with pentane to give the desired 5-methoxy-5-undecene (**260**, R = C_4H_9, R^1 = CH_3, R^2 = C_5H_{11} 0.18 g, 96% yield; (Z)/(E) = 91 : 9).

The titanation of Wittig–Horner compounds prior to the reaction with aldehydes does not result in the expected alkenes. Rather, stereoselective Knoevenagel condensations are observed[234,235] (Section 3.2.2). This type of reaction had been observed previously upon mixing active methylene compounds such as malonates with $TiCl_4/R_3N$[111]. The $ClTi(O^iPr)_3/NEt_3$ version is milder and even works in cases in which classical Knoevenagel conditions fail[236]:

$$H_2C \overset{A^1}{\underset{A}{\diagup\diagdown}} + RCHO \xrightarrow[\text{or} \atop ClTi(O^iPr)_3 \, / \, NR_3]{TiCl_4 \, / \, NR_3} R-CH=C \overset{A^1}{\underset{A}{\diagup\diagdown}}$$

A, A^1 = COR, COOR, CN, PO(OR)$_2$, NO$_2$

Typical Procedure for the TiCl$_4$-mediated Knoevenagel Condensation[111a]

To a stirred solution of 200 mL of dry THF is added TiCl$_4$ (11 mL, 100 mmol) in 25 mL of dry CCl$_4$ at 0 °C, which leads to a flaky yellow precipitate. An aldehyde (50 mmol) and a CH-acidic methylene compound such as malonic acid ester, acetoacetic ester or nitroacetic acid ester (50 mmol) are added in pure form or as a THF solution. Dry pyridine (16 mL, 200 mmol) or N-methylmorpholine (22 mL) in 30 mL of THF is slowly added to the stirred mixture within 1–2 h. The mixture is stirred at 0 °C (or 23 °C) for 16–24 h and diluted with 50 mL of diethyl ether. The usual work-up and purification (distillation or recrystallization) afford the Knoevenagel products in yields of 40–95%. Ketones react similarly at 23 °C (1 day)[111b].

Typical Procedure for the ClTi(O^{i}Pr)$_3$-mediated Knoevenagel Condensation[235]

Procedure A

A suspension of 0.25 g (10.5 mmol) of freshly washed (THF) NaH containing no NaOH or Na$_2$O in 30 mL of THF is treated with 2.24 g (10.0 mmol) of ethyl(diethoxyphosphoryl) acetate (**191**) at 25 °C and the mixture stirred until an almost clear solution forms. It is refluxed for 1 h, cooled to −78 °C and treated with chlorotriisopropoxytitanium (2.73 g, 10.5 mmol). The cooling bath is removed and the mixture stirred for 1.5 h at 25 °C. After the addition of an aldehyde (9.5 mmol), the mixture is stirred at room temperature for 4 h (overnight in the case of 4-methoxybenzaldehyde) and poured onto dilute hydrochloric acid. The aqueous phase is extracted twice with diethyl ether and the combined organic phase is washed with water and dried over MgSO₄. After stripping off the solvent, the crude product is distilled using a Kugelrohr. (*Z*)-Configured vinyl phosphonates **193** are obtained in 57–62% yield. Transesterification occurs under the reaction conditions.

Procedure B

A mixture of chlorotriisopropoxytitanium (5.2 g, 20 mmol), ethyl(diethoxy-phosphoryl)acetate (**191**; 2.24 g 10.0 mmol) and 10.0 mmol of an aldehyde in 30–40 mL of THF is treated with triethylamine (2.02 g, 20 mmol) at 0 °C. The amine hydrochloride precipitates and the mixture is stirred at 0 °C for 2–3 h; followed by work-up according to Procedure A. Procedure B can be applied to the Knoevenagel condensation of malonic acid esters with aldehydes.

The Tebbe reagent **81** (Section 2.2) is an isolable and well characterized compound which in the presence of bases methylenates carbonyl compounds, including enolizable and sensitive aldehydes, ketones, esters (*e.g.*, **261**), lactones and amides[121]. The reactive species is actually the carbene complex Cp₂Ti = CH₂ (**84**), which can also be generated by [2 + 2] cycloreversion of titanacyclobutanes **82** or, more conveniently, by thermolysis of Cp₂Ti(CH₃)₂ (**63**) at 60–80 °C[88,310]. The latter method constitutes a nearly neutral process, in contrast to the Tebbe reagent which contains a Lewis acidic aluminum component. Consequently, **63** can be used in the methylenation of acid-sensitive substrates, for example in the preparation of certain extremely labile enol ether products in which the Tebbe reagent fails[311]. Both methods have been used successfully in tandem methylenation/Claisen rearrangement processes[312], as in **263** → **264** → **265**[313]:

Typical Procedure for the Methylenation of Carboxylic Acid Esters using the Tebbe Reagent, 81[121 d]

To a 250 mL round-bottom flask equipped with a magnetic stirring bar are added bis(cyclopentadienyl)dichlorotitanium (**58a**; 5.0 g, 20.0 mmol). The flask is fitted with a rubber septum through which a large-gauge needle is passed to flush the system with dry nitrogen. After the vessel has been thoroughly purged, the nitrogen line flowing to the needle is opened to a mineral oil bubbler and 20 mL of trimethylaluminum solution (2.0 M in toluene, 40 mmol) are added by a nitrogen-purged syringe (**caution:** pyrophoric material). Methane gas evolved by the reaction is allowed to vent as the resulting red solution is stirred at 25 °C for 3 days.

The Tebbe reagent **81** thus formed is used *in situ* by cooling the mixture in an ice-water bath, then adding phenyl benzoate (**261**; 4.0 g, 20 mmol) of dissolved in 20 mL of dry THF by syringe or cannula to the cooled stirring solution over 5–10 min. After the addition, the reaction mixture is allowed to warm to 25 °C and is stirred for about 30 min. The septum is removed and 50 mL of anhydrous diethyl ether are added. To the stirred reaction mixture are gradually added 50 drops of 1 M sodium hydroxide over 10–20 min. Stirring is continued until gas evolution essentially ceases; then to the resulting orange slurry are added a few grams of anhydrous Na₂SO₄ to remove excess water. The mixture is filtered through a Celite pad on a large coarse frit using suction and liberal amounts of diethyl ether to transfer the product and rinse the filter pad. Concentration of the filtrate with a rotary evaporator to 5–8 mL provides crude product, which is purified by column chromatography on basic alumina (150 g) eluting with 10% diethyl ether in pentane. Fractions which contain the product are combined and evaporated to give 2.69–2.79 g (68–70%) of 1-phenoxy-1-phenylethene (**262**) as a pale yellow oil. The acid lability of enol ether products requires rigorous treatment of all glassware used for the reaction in order to avoid migration of the double bond in susceptible cases. Satifactory results are obtained by treating the glassware sequentially with ethanolic 0.5 M solutions of hydrogen chloride and potassium hydroxide for about 1 h, thoroughly rinsing with distilled water after each treatment and finally oven drying. This protocol is also effective for removing stubborn deposits on the glassware after the reaction. Methylenation must be carried out at −78 °C in the case of extremely sensitive substrates or products.

Recently, an unusual and synthetically important entry to titanium–alkylidene chemistry based on the reaction of dithioacetals **266** with the titanocene source Cp₂T[P(OEt)₃]₂ (**267**) has been described[314]. Subsequent reaction with aldehydes, ketones or esters affords the corresponding alkenes as a mixture of (*E*)/(*Z*) isomers **269**:

$$
\underset{\textbf{266}}{\overset{RS}{\underset{R^1}{\diagdown}}\hspace{-0.3em}\overset{SR}{\underset{R^2}{\diagup}}} + \underset{\textbf{267}}{Cp_2Ti[P(OEt)_3]_2} \longrightarrow \underset{\textbf{268}}{\overset{TiCp_2}{\underset{R^1}{\diagdown}\hspace{-0.3em}\underset{R^2}{\diagup}}} \xrightarrow{\overset{O}{\underset{R^3}{\diagdown}\hspace{-0.3em}\overset{\|}{\underset{R^4}{\diagup}}}} \underset{\textbf{269}}{\overset{R^1}{\underset{R^2}{\diagdown}}\hspace{-0.3em}\overset{R^3}{\underset{R^4}{\diagup}}}
$$

In summary, several highly efficient Ti-based alkylidenation reagents[315] are available which are complementary to Wittig technology. It remains to be seen which reagents will be used most often in the future[316]. The chemistry of titanacyclobutanes is fascinating, particularly in their role in alkene metathesis as applied to ring-opening polymerization[88,121b,c]. They also react with acid chlorides to form titanium enolates[317]. Carbene complexes of the type $Cp_2Ti = C = CH_2$ can also be generated from the appropriate precursors and undergo a variety of interesting reactions[318].

5.2   Deoxygenative Coupling of Carbonyl Compounds

In 1973–74, three different groups independently discovered that low-valent titanium induces the deoxygenative coupling of ketones and aldehydes to form alkenes. The methods differ in the source of low-valent titanium: $TiCl_3/3THF/Mg$[319], $TiCl_3/LiAlH_4$[320], and $Zn/TiCl_4$[321]. Subsequently, the McMurry reagent ($TiCl_3/LiAlH_4$) has become the most used. Titanium as a fine dispersion on graphite also turned out to be very effective and was successfully applied in cases in which other reagents had failed[322].

$$
\underset{R}{\overset{O}{\underset{}{\diagup}}}\hspace{-0.3em}\overset{\|}{\underset{R^1}{}} \xrightarrow{\text{'Ti'}} \underset{R\;R^1R\;R^1}{\overset{L_nTiO\quad OTiL_n}{\diagdown\hspace{-0.3em}\diagup}} \xrightarrow{\text{'Ti'}} \underset{R^1\quad R}{\overset{R\quad R^1}{\diagdown=\diagup}}
$$

The substrates can be saturated or unsaturated, and the reaction works in an intermolecular sense to yield acylic alkenes and intramolecularly on dicarbonyl compounds to yield cycloalkenes. Aromatic and α,β-unsaturated carbonyl compounds even react with commercial titanium powder activated by chlorosilanes[323]. Mechanistically, the reaction proceeds via a low-valent titanium-induced pinacol coupling followed by deoxygenation. Recent studies show that the actual nature of the low-valent titanium is highly dependent on the mode of preparation – thus, a polymeric HTiCl turned out to be the active species formed from $TiCl_2/LiAlH_4$, whereas bimetallic complexes (specifically $[TiMgCl_2(THF)_x]$ and $[Ti(MgCl)_2(THF)_x]$) are obtained on reaction of $TiCl_3$ with Mg[324]. Zero-valent titanium, which had previously been assumed to be the actual coupling reagent, has never been identified beyond doubt, although it may pertain in specific cases. Several extensive reviews on McMurry couplings and related processes covering the preparative and mechanistic aspects have appeared[325].

Unfortunately, in the early version reproducibility depended on the age, history and source of trichlorotitanium ($TiCl_3$). Although improvements based on a number of other low-valent titanium sources were reported[325], some confusion as to the best procedure persisted. This uncertainty ended with a definitive study of an optimized procedure based on TiCl$_3$/Zn−Cu in dimethoxyethane (DME)[326]. Accordingly, intermolecular couplings and intramolecular processes leading to small-ring cycloalkenes are best performed by rapid addition of the carbonyl component to a reagent prepared by using three equivalents of TiCl$_3$(DME)$_{1.5}$ per equivalent of carbonyl compound. More difficult intramolecular couplings require four or more equivalents of titanium reagent per carbonyl moiety and very slow addition to achieve high dilution[326]. The value of this new version is illustrated by the smooth coupling of diisopropyl ketone **270** to form tetraisopropylethylene **271** in 87% yield[326], compared to 12% by the original TiCl$_3$/LiAlH$_4$ method and 37% by the TiCl$_4$/Zn−Cu procedure in the absence of DME:

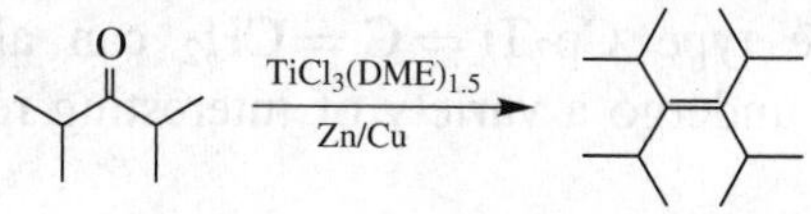

270   **271**

Optimized McMurry Coupling of Ketones[326]

Preparation of TiCl$_3$(DME)$_{1.5}$

TiCl$_3$ (25.0 g, 0.162 mol) is suspended in 350 mL of dry dimethoxyethane (DME) and the mixture is refluxed for 2 days under argon. After the mixture has cooled to 25 °C, filtration under argon, washing with pentane and drying under vacuum give the fluffy, blue crystalline TiCl$_3$(DME)$_{1.5}$ (32.0 g, 80%) that is used in the coupling reaction. The solvate is air sensitive but can be stored indefinitely under argon at 25 °C.

Preparation of zinc−copper couple

The zinc−copper couple is prepared by adding zinc dust (9.8 g, 150 mmol) to 40 mL of nitrogen-purged water, purging the slurry with nitrogen for 15 min and then adding CuSO$_4$ (0.75 g, 4.7 mmol). The black slurry is filtered under nitrogen, washed with deoxygenated (nitrogen-purged) water, acetone and diethyl ether, and then dried under vacuum. The couple can be stored for months in a Schlenk tube under nitrogen.

Typical procedure for ketone coupling

TiCl$_3$(DME)$_{1.5}$ (5.2 g, 17.9 mmol) and Zn−Cu (4.9 g, 69 mmol) are transferred under argon to a flask containing 100 mL of DME, and the resulting mixture is refluxed for 2 h to yield a black suspension. Cyclohexanone (0.44 g, 4.5 mmol) in 10 mL of DME is added and the mixture is refluxed for 8 h. After being cooled to room temperature, the reaction mixture is diluted with pentane (100 mL), filtered through a pad of Florisil, and concentrated in a rotary evaporator to yield cyclo-hexylidenecyclohexane (0.36 g, 97%) as white crystals (mp 52.5–53.5 °C. If a 3 : 1 ratio of TiCl$_3$(DME)$_{1.5}$ to carbonyl compound instead of 4 : 1 is used, the yield decreases from 97% to 94%; if a 2 : 1 ratio is used, the yield is 75%.

It has been demonstrated that low-valent titanium efficiently promotes intramolecular cross-coupling reactions of oxoesters or oxoamides **272** to yield aromatic heterocycles **273** such as furans, pyrroles, benzofurans, and indoles. These reactions are best effected in a one-pot fashion by mixing the substrate, TiCl$_3$, and Zn (the instant method)[327] and can even be achieved with catalytic amounts of TiCl$_3$ when carried out in the presence of chlorosilanes[323]. The power of this approach is illustrated by the successful synthesis of various alkaloids[327,328].

$$\textbf{272} \xrightarrow{\text{'Ti'}} \textbf{273}$$

272   **273**

(X = O, NR)

5.3   Hydrometalation and Carbometalation of Alkynes

Hydrometallation of alkynes using metals such as boron, aluminum or zirconium are part of standard synthetic organic methodology. In the hydromagnesiation of alkynes based on isobutyl- or isopropylmagnesium bromide as the hydride source, Cp$_2$TiCl$_2$ has been found to be an efficient catalyst[73,88,329]. The Ti^{III} species Cp$_2$TiH is the actual intermediate which hydrotitanates the alkyne in the catalytic cycle. Disubstituted alkynes **274** react at room temperature to produce (*E*)-alkenyl Grignard reagents **275** in excellent yields. After hydrolysis, (*Z*)-alkenes **276** of high purity are obtained. The reaction occurs with low regioselectivity for unsymmetrical dialkylacetylenes (which is of no consequence in the hydrolysis)[74,330]:

$$R\!-\!\!\!\equiv\!\!\!-R^1 \xrightarrow[\text{Cp}_2\text{TiCl}_2 \text{ (cat.)}]{(\text{H}_3\text{C})_2\text{CHCH}_2\text{MgBr}} \quad \textbf{275} \longrightarrow \textbf{276}$$

274   **275**   **276**

Regioselectivity is high in the case of alkylarylalkynes **277**, which means that quenching with electrophiles other than protons becomes meaningful[74,330]:

$$\text{Ph}\!-\!\!\!\equiv\!\!\!-R \xrightarrow[\text{Cp}_2\text{TiCl}_2 \text{ (cat.)}]{^i\text{C}_4\text{H}_9\text{MgBr}} \quad \textbf{278} \xrightarrow{\text{El}-\text{X}} \textbf{279}$$

277   **278**   **279**

The hydromagnesiation of 1-trimethylsilyl-1-alkylenes proceeds with complete regioselectivity if the reaction is carried out at 25 °C for 6 h. The intermediate 1-(silyl)vinylmagnesium reagents react with a variety of electrophiles, such as protons, RI/CuI, I$_2$, or RCHO. This constitutes a powerful method for the synthesis of prochirally pure vinylsilanes[330,331]. Hydrozincation is also possible[332].

Titanium has also played a role in the carbometalation of alkynes[333], although zirconium seems to be more versatile, for instance in Cp_2ZrCl_2-catalyzed additions of R_3Al[27,334]. However, titanium-based reactions appear to be superior in reactions of homopropargylic alcohols in which the hydroxy group dictates regioselectivity[335].

6   Titanium-Mediated Radical Processes

Some of the reactions mediated by low-valent titanium involve radical intermediates, as in the alkene-forming McMurry reaction (Section 5.2). If alkene formation is not the goal, the reaction can be stopped at the diol stage by using Ti^{III} reagents[325] or other low-valent metals[336]. A practical stereoselective pinacol coupling in an aqueous medium (H_2O/THF) has been devised in which stoichiometric amounts of Cp_2TiCl are reacted with aromatic aldehydes 280 to form high yields of diols 281/282, the d, l/meso ratio typically being 95 : 5[337]. An excess of NaCl is necessary for the reaction to proceed, presumably due to the necessity of displacing H_2O from the coordination sphere of titanium. Aldimines undergo reductive coupling with formation of d, l vicinal diamines in the presence of $TiCl_4$/Mg–Hg or $TiCl_4$/Mg/$BrCH_2CH_2Br$, processes that also occur with intermediate formation of azaketyl radicals[338].

With the aim of developing a pinacol coupling procedure which is catalytic in titanium and at the same time chemo- and diastereoselective, a mixture of stoichiometric amounts of Zn, $(CH_3)_3SiCl$ and $MgBr_2$ in the presence of catalytic amounts of a titanocene such as *rac*-ethylenebis(η^5-tetrahydroindenyltitanium)dichloride (59a) was tested[339]. This resulted in excellent yields of diols 281, diastereoselectivity being better than 96%. The method allows for considerable functional group tolerance. The catalytically active species for diastereoselective dimerization of the ketyl radicals bound to the Ti catalyst has been postulated to be 283[339], which suggests that an enantioselective version using the optically active form of the metallocene should be possible:

283

Another area with bright prospect for the future concerns the Ti-mediated one-electron reduction of epoxides as a means to generate highly functionalized radical intermediates[340]. Whereas the explosive development of chemo- and stereoselective radical reactions has provided organic chemists with valuable new tools, a limiting factor is the relatively low number of synthetically useful radical precursors[341]. The discovery that the reaction of an epoxide **284** with Cp_2TiCl leads to a one-electron reduction with concomitant ring-opening and formation of radicals **285/286** is therefore synthetically significant[340]. Regioselectivity of ring-opening is generally opposite to traditional S_N2 processes and is determined mainly by the stability of the intermediate radicals, *i.e.*, **285** is usually preferred:

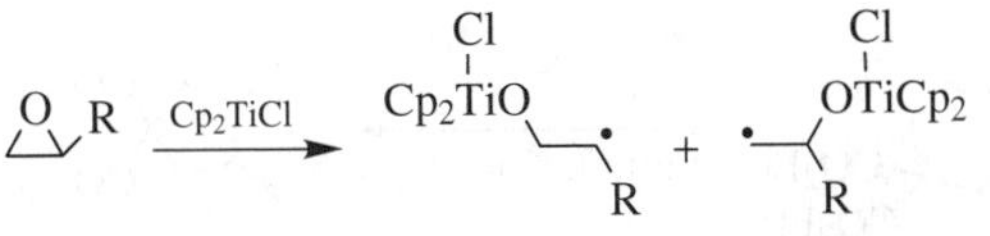

The radicals can be reduced by hydrogen-atom donors such as cyclohexa-1,4-diene with regioselective formation of alcohols, a reaction that has been applied in carbohydrate and nucleotide chemistry[340]. The method is regiochemically complementary to ring-opening reduction based on super hydride ($LiBEt_3H$). The radicals can also be trapped intra- or intermolecularly by alkenes. The reaction of epoxide **287** is a typical example, which also shows that the eventual radical **289** can be trapped by excess Cp_2TiCl to form a stable Ti^{IV}) species **290**, capable of undergoing such reactions as iodination. An expeditious approach to the synthesis of achiral and chiral butadienyl alcohols is also based on Cp_2TiCl-mediated radical ring-opening of epoxides[342].

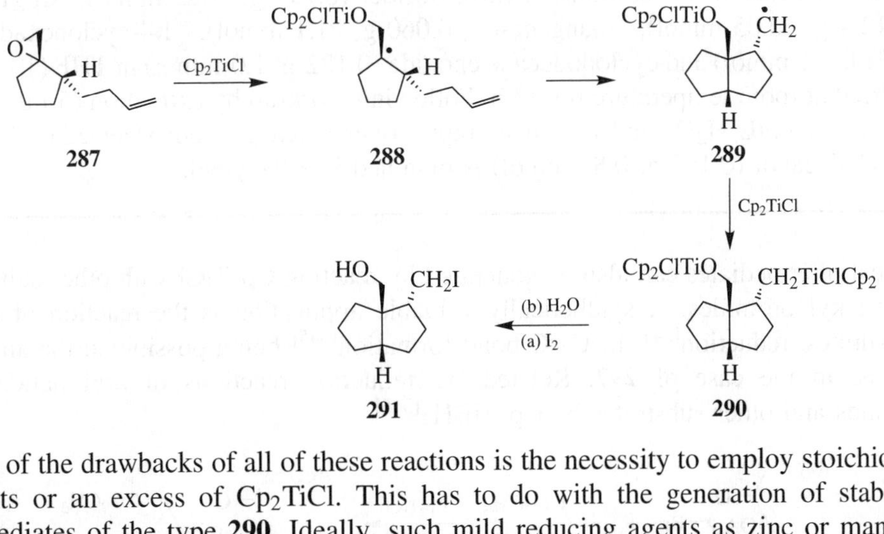

One of the drawbacks of all of these reactions is the necessity to employ stoichiometric amounts or an excess of Cp_2TiCl. This has to do with the generation of stable Ti^{IV} intermediates of the type **290**. Ideally, such mild reducing agents as zinc or manganese powder could allow for a catalytic process via the reduction of Ti^{IV} back to Ti^{III}. However, this is not possible, unless Ti^{IV} can be released from the intermediates in the form of

Cp$_2$TiCl$_2$. Although Me$_3$SiCl might fulfill this trick, it cannot be utilized because it reacts with the epoxides themselves. This difficult problem was recently solved by using a mild proton source which cleaves the strong Ti−O bond, but which does not interfere with the redox system. The combination of stoichiometric amounts of manganese (or zinc) powder as the reducing agent and 2,4,6-trimethylpyridine hydrochloride as the Ti−O cleaving agent and catalytic amounts of Cp$_2$TiCl$_2$ fulfills all of the requirements[343]. Typical examples are the reactions of epoxides **292** and **294**. This powerful method is expected to gain further significance if an enantioselective catalytic version can be devised.

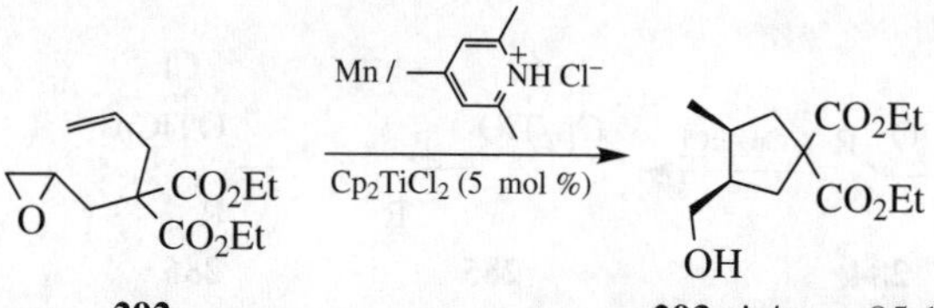

292   **293** cis/trans 85:15

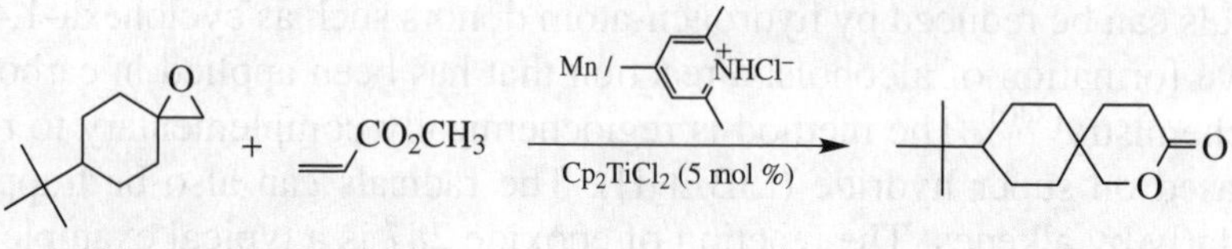

294   **295**   **296**

Typical Procedure for the Cp$_2$TiCl$_2$-catalyzed Reduction of Epoxides[343]

The suspension of collidine hydrochloride (0.197 g, 1.25 mmol), Cp$_2$TiCl$_2$ (0.012.4 g, 0.05 mmol), manganese (0.060 g, 1.1 mmol), 1,4-cyclohexadiene (425 μl, 4.5 mmol) and cyclododecane epoxide (0.182 g, 1.0 mmol) in THF (10 mL) is stirred at room temperature for 32 h. Following work-up by extraction (5 mL 2 M HCl, 2 × 20 mL H$_2$O) and chromatography over silica gel (ether/pet ether 1 : 3), cyclododecanol (0.153 g, 0.83 mmol) is obtained in 83% yield.

Radical intermediates can also be generated by reacting Cp$_2$TiCl with other substrates, such as alkyl bromides. A synthetically valuable application is the reaction of certain carbohydrates, reduction[344] or C−C bond formation[345] being possible at the anomeric center, as in the case of **297**. Related are reduction reactions of aryl halides, azo compounds and other substrates by Cp$_2$TiBH$_4$[346].

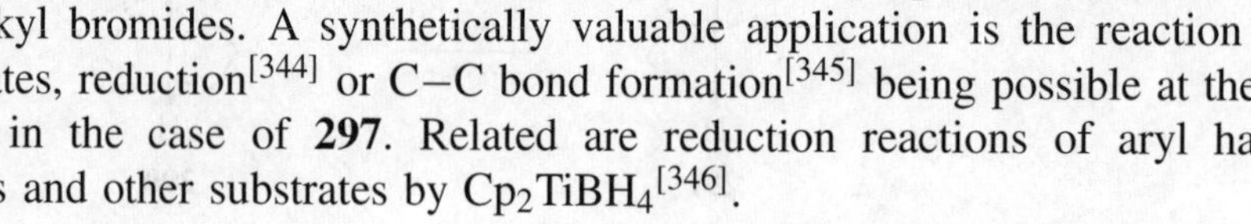

297   **295**   **298**

Finally, the titanocene-catalyzed stereoselective cyclocarbonylation of *o*-allyl aryl ketones to γ-butyrolactones, a kind of hetero-Pauson–Khand reaction also appears to involve radical intermediates[347].

7   Substitution Reactions

Substitution reactions involving carbon nucleophiles are synthetically important processes. In carbanion chemistry deprotonated carbonyl compounds (and their nitrogen analogs), sulfones, sulfoxides, thioesters, nitriles, etc. undergo smooth reactions with S_N2-active alkyl halides[8]. A serious synthetic gap becomes apparent when attempting to perform these classical reactions with tertiary alkyl halides and other base-sensitive alkylating agents not amenable to S_N2 substitution. Similarly, cuprates R_2CuLi and higher order analogs do not undergo substitution reactions with tertiary alkyl halides. Many of these problems can be solved using titanium reagents of high Lewis acidity which induce S_N1 reactions. Some of these reagents also allow for a combination of two C–C bond-forming reactions in a one-pot sequence, namely addition to carbonyl compounds followed by S_N1-type substitution of the oxygen function. Conversely, titanium reagents, regardless of the type of ligands, are generally not nucleophilic enough to undergo S_N2 reactions with primary alkyl halides. For example, compounds of the type CH_3TiX_3 (X = Cl, OR) do not react with RCH_2X (X = Cl, Br, I), nor do titanium enolates or α-titanated sulfones[7,13,43].

The Lewis acidic character of trichloromethyltitanium CH_3TiCl_3 (Section 2) is so pronounced that the reagent will ionize S_N1-active alkyl halides to form carbocations such as **300**, which are spontaneously captured by the non-basic carbon nucleophile[16,17]. Essentially all tertiary **299** and aryl-activated secondary alkyl halides **302** are methylated at −78 °C to 0 °C to form high yields of products **301** and **303**, respectively.

$$R_3C-Cl + H_3CTiCl_3 \longrightarrow [R_3C]^+ [H_3CTiCl_4]^- \longrightarrow R_3C-CH_3$$

299		**300**	**301**	**302**

303

Ether-free solvents such as CH_2Cl_2 are mandatory, as diethyl ether or THF will complex with CH_3TiCl_3, making then incapable of inducing S_N1 ionization. This means that precursors such as $(CH_3)_2Zn$ or $(CH_3)_3Al$ must be used in conjunction with $TiCl_4$, either stoichiometrically with formation of CH_3TiCl_3 or $(CH_3)_2TiCl_2$, or catalytically[16,17].

Typical Procedure for the Methylation of Tertiary Alkyl Chlorides[16,19]

A solution of 6 mmol $(CH_3)_2Zn$ (*e.g.*, 1.5 mL of a 4 M CH_2Cl_2 solution)[38] in 30 mL of dry CH_2Cl_2 is treated with $TiCl_4$ (0.180 g) and subsequently with *trans*-9-chlorodecalin (1.72 g, 10 mmol) at −30 °C. After 15 min the mixture is poured on ice-water. Following extraction with diethyl ether, washing with $NaHCO_3$ solution and drying over MgSO$_4$, the solvent is removed and the residue is distilled (90 °C, 12 torr) to provide 1.31 g (82%) of 9-methyldecalin as a 1 : 1 cis/trans mixture.

The fact that in some cases cationic rearrangements occur is in line with the proposed S$_N$1 mechanism[17]. Functional groups such as primary and secondary alkyl halides and esters are tolerated. (CH$_3$)$_3$Al in the presence of catalytic amounts of TiCl$_4$ is also effective[19]. Tertiary ethers and alcohols are methylated by excess (CH$_3$)$_2$TiCl$_2$[348].

Homologs (e.g., BuTiCl$_3$) do not induce efficient coupling because reduction of the carbocations via β-hydride abstraction prevails[19]. In this case dialkylzinc reagents in the presence of ZnCl$_2$ is the method of choice. Allylsilanes react smoothly with S$_N$1-active alkylating agents in the presence of TiCl$_4$[349]. Chromium-complexed benzylic alcohols undergo stereoselective substitution reactions with (CH$_3$)$_3$Al/TiCl$_4$[350]. The reaction of β-chlorosulfides such as **305** or **308** with (CH$_3$)$_2$Zn in the presence of catalytic amounts of TiCl$_4$ occurs via the intermediacy of epi-sulfonium ions which are captured stereospecifically by (CH$_3$)$_2$TiCl$_2$ in yields of 60–75%. Thus, substitution of chloride by methyl proceeds with retention of configuration[351]. Chiral alkenes show pronounced degrees of diastereofacial selectivity. The overall process, **304** → **306** and **307** → **309**, constitutes a simple means to carbosulfenylate alkenes regio- and stereoelectively[351]:

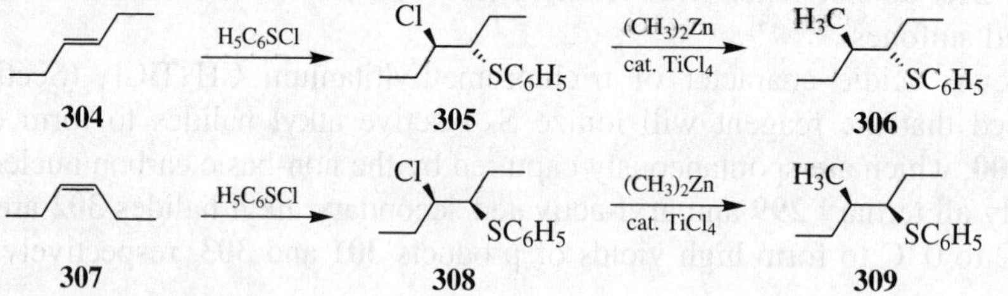

Direct geminal dimethylation of carbonyl compounds **310** → **312** using an excess of (CH$_3$)$_2$TiCl$_2$ works well for simple ketones[38,348] and aromatic aldehydes[42], a process which involves Grignard-type addition with formation of intermediates **311** followed by S$_N$1 ionization and further C−C bond formation. The yields generally range between 60% and 90%, but conversion may be lower in the case of highly hindered ketones.

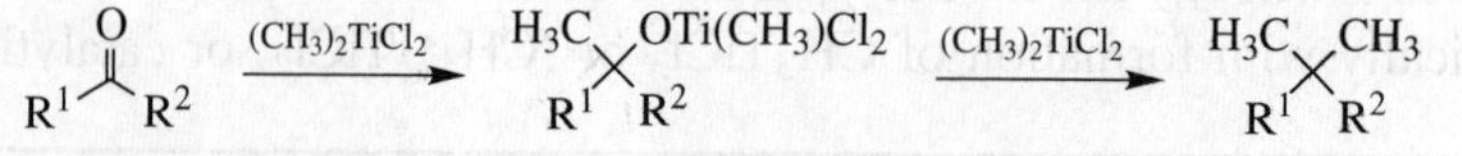

The limitation of this interesting method has to do with the fact that only a few types of additional functionalities are tolerated and that rearrangements may occur[352]. Nevertheless, in the synthesis of certain molecules with quaternary atoms[353], it is the method of choice[348] (e.g., **314** or **316**). (CH$_3$)$_2$Zn may be replaced by (CH$_3$)$_3$Al[38]. The conventional alternative is a three-step sequence based on Wittig methylenation,

Simmons–Smith cyclopropanation, and ring-opening hydrogenolysis.

313   314   315   316

General Procedure for the Geminal Dimethylation of Ketones using $(CH_3)_2 TiCl_2$ [38]

To a stirred solution of 40 mmol of $(CH_3)_2TiCl_2$ in dichloromethane (see above) is added a ketone (20 mmol) at $-30\,°C$. The mixture is slowly allowed to come to $25\,°C$ during a period of about 2 h and is then poured onto ice-water. The aqueous phase is extracted with ether and the combined organic phases are washed with H_2O and $NaHCO_3$. After drying over $MgSO_4$, the solvent is removed and the product distilled (*e.g.*, using a Kugelrohr) or crystallized. In the case of aryl or α,β-unsaturated ketones, addition is best performed at $-40\,°C$ and the mixture allowed to come to $0\,°C$ prior to the usual work-up.

Direct geminal dimethylation of ketones in which one of the newly introduced groups is an alkyl moiety and the other a methyl group is possible in some cases by a one-pot procedure in which an alkyllithium reagent (in ether-free solvent) is added to a ketone and the resulting tertiary alkoxide is treated with an excess of dichlorodimethyltitanium in CH_2Cl_2, as in the preparation of compounds **319** which are precursors of synthetic tetrahydrocannabinoids **320**[354]. The one-pot geminal dialkylation proceeds typically with yields of 70–90%.

317   318   319   320

Acetals, ketals and geminal dichlorides also undergo methylation reactions[7,355]. Certain acetals react with Grignard reagents in the presence of $TiCl_4$ to form the corresponding substitution products[356]. An interesting version pertains to the reaction of chiral acetals of the type **321** with $CH_3MgCl/TiCl_4$ to afford a 96 : 4 mixture of substitution products **322** and **323**, respectively[357]. S_N2-like processes have been postulated, but tight ion pairs in which 1,3-allylic strain is operating cannot be excluded[358].

$$\textbf{321} \xrightarrow[\text{TiCl}_4]{\text{H}_3\text{CMgCl}} \textbf{322} \;+\; \textbf{323}$$

Not only Grignard and alkyllithium reagents, but also allylsilanes[359], cyanotrimethylsilane[360], and silylated alkynes[361] function as efficient *C*-nucleophiles. Sometimes mixtures of TiCl$_4$ and Ti(O^{i}Pr)$_4$ are better suited. Many of these reactions may not involve the titanium intermediates RTiCl$_3$. However, CH$_3$TiCl$_3$, generated from (CH$_3$)$_3$Zn/TiCl$_4$, is in fact capable of undergoing such stereoselective substitution reactions[362]. The area of stereoselective substitution reactions of chiral nonracemic acetals using titanium reagents and other organometallics has been reviewed[363,364].

The classical alkylation of lithium enolates with S$_N$2-active alkyl halides is one of the important C–C bond-forming reactions[8]. However, tertiary and many base-labile secondary alkyl halides fail to react. This long-standing problem can be solved by reacting the enolsilanes derived from carbonyl compounds with tertiary and other S$_N$1-active alkyl halides in the presence of Lewis acids[364]. In the case of ketones, TiCl$_4$ is the Lewis acid of choice. The mechanism involves S$_N$1 ionization followed by addition of the carbocation to the enolsilane. The alternative mechanism, i.e. Si–Ti exchange followed by ionization of R$_3$CCl and alkylation of the tetrachlorotitanium enolate, may occur in rare systems in which alkylation of the enolsilane is slow relative to Si–Ti exchange[101b].

$$\text{R}_3\text{C-Cl} \xrightarrow{\text{TiCl}_4} \left(\text{R}_3\overset{+}{\text{C}} \;\; \overset{-}{\text{TiCl}_5}\right) \longrightarrow \longrightarrow$$

The yields of isolated products range between 60% and 95%. Regioselective *tert*-alkylation[365] is possible as the corresponding enolsilanes[366] are readily accessible, for example **324 → 325** and **326 → 327**:

$$\textbf{324} \xrightarrow[\text{TiCl}_4]{(\text{H}_3\text{C})_3\text{CCl}} \textbf{325} \qquad \textbf{326} \xrightarrow[\text{TiCl}_4]{(\text{H}_3\text{C})_3\text{CCl}} \textbf{327}$$

Undesired polyalkylation has never been observed. The reaction is best performed by treating the mixture of an enolsilane and a tertiary alkyl halide in CH$_2$Cl$_2$ with the equivalent amount of TiCl$_4$ at $-40\,^\circ$C to $-50\,^\circ$C. Alternative reaction modes, such as adding the enolsilane to a mixture of TiCl$_4$ and tertiary alkyl halide, result in poor yields (30–54%)[367]. In the case of enolsilanes which are sensitive to TiCl$_4$ (*e.g.*, those derived from aldehydes or esters), milder Lewis acids such as ZnCl$_2$ or BiCl$_3$ need to be employed[365].

Typical Procedure for the α-tert-Alkylation of Ketones via their Enolsilanes[365]

A dry 250-mL three-necked round-bottomed flask is fitted with an argon inlet, a gas bubbler, rubber septum, and magnetic stirrer. The apparatus is flushed with dry nitrogen or argon and charged with dry CH_2Cl_2 (120 mL), 1-trimethylsilyloxycyclopentene (15.6 g, 0.10 mol), and of 2-chloro-2-methylbutane (11.7 g, 0.11 mol). The mixture is cooled to $-50\,°C$ and a cold $(-50\,°C)$ solution of $TiCl_4$ (11 mL, 0.10 mol) of in 20 mL of CH_2Cl_2 is added within 2 min through the rubber septum with the aid of a syringe. During this operation rapid stirring and cooling is maintained. Direct sunlight should be avoided. The reddish brown mixture is stirred at the given temperature for an additional 2.5 h and then rapidly poured onto 1 L of ice-water. After the addition of 400-mL of CH_2Cl_2, the mixture is vigorously shaken in a separating funnel; the organic phase is separated and washed twice with 400 mL portions of water. The aqueous phase of the latter two washings is extracted with 200 mL of CH_2Cl_2, the organic phases are combined and dried over anhydrous Na_2SO_4. The mixture is concentrated using a rotary evaporator and the residue is distilled at $80\,°C$ (12 mmHg) to yield 9.2–9.5 g (60–62%) of 2-*tert*-pentylcyclopentanone as a colorless oil.

These and other Lewis acid-induced S_N1 alkylations are synthetically significant because they are complementary to the classical alkylations of lithium[8] or manganese[368] enolates using S_N2-active alkyl halides[364]. Any alkylating agent which has a higher S_N1 solvolysis activity than isopropyl halides is likely to be suitable. Neighboring group participation is an important stereochemical factor in relevant cases[369].

Acetals, ketals and chloro ethers can also be used in $TiCl_4$-mediated alkylations of enolsilanes[370]. Intramolecular versions result in carbocycles[371]. Again, chlorotitanium enolates are probably not involved. However, the previously mentioned enolization method based on $TiCl_4/NR_3$ (Sections 2.2, 3.2.2, and 3.3) can be used to induce α-alkoxy- and aminoalkylation of carbonyl compounds[113,372]. In the case of chiral chlorotitanium enolates, stereoselectivity often approaches 100%, as in the conversion of the amide **209** into the α-alkylated product **328**[113]. Less Lewis-acidic enolates undergo smooth aminoalkylation[66].

203 (Bn = benzyl)   328

Chirality can also be incorporated in the electrophile, as in the case of the chiral formyl cation equivalent **329** derived from ephedrine, which reacts stereoselectively with enolsilanes. Diastereoselectivity in the case of the substitution product **331** is 88%[373].

The reaction can also be induced by other Lewis acids such as SnCl$_4$, BF$_3$/OEt$_2$ and (CH$_3$)$_3$SiOTf[373]. Related approaches have been used in other enol- and allylsilane reactions[374].

329   **330**   **331**

Reports of epoxides as alkylating agents are rare. Undesired ring opening followed by a hydride shift to form aldehydes occurs if the medium is too Lewis acidic, as in the case of TiCl$_4$[43a]. Less Lewis-acidic reagents such as CH$_3$Ti(O^{i}Pr)$_3$ are not reactive enough to undergo smooth C−C bond formation with most epoxides[7,43a]. However, more reactive species such as allyltitanium reagents react with styrene oxide **332**[7,43a]. Regioselectivity in favor of ring-opened product **333** contrasts well with the nonselective behavior of the Grignard reagent. Conversion amounts to at least 80% in all cases. Other epoxides react less selectively. However, on going from isopropoxy to phenoxy ligands, a dramatic improvement results, ring opening of epoxides occurring consistently at the more substituted carbon atom in yields of 80−85%[375]. This is of practical importance because cuprates generally react at the less-substituted position.

332	**333**	**334**
H$_2$C=CHCH$_2$MgCl	70	30
H$_2$C=CHCH$_2$Ti(O^{i}Pr)$_3$	> 99	< 1
H$_2$C=CHCH$_2$Ti(O^{i}Pr)$_4$MgCl	> 99	< 1

Thus, the majority of Ti-mediated substitution reactions have S$_N$1-character, including Ti(OR)$_4$-induced stereoselective Pictet–Spengler cyclizations[376]. The intermediacy of cation-like species is also likely in iodocarbocyclizations[377]. Accordingly, alkenes that contain functionality having pronounced CH acidity, as in the case of the malonate **335**, undergo smooth cyclization in the presence of Ti(O^{t}Bu)$_4$/I$_2$. Titanium enolates are involved which nucleophilically attack intermediate iodonium ions formed at the alkenic moiety. Mechanistically, this is related to the classical iodolactonization in which iodonium ions are trapped intramolecularly by carboxy functions. The Ti-mediated iodocarbocyclications work well for a variety of different substrates, including alkynylmalonates with formation of iodomethylene cyclopentanes[377]. Catalytic enantioselective versions have also been reported[378,379].

CO₂CH₃ → **335** → Ti(O'Bu)₄ / I₂ → **336**

Along different lines, organotitanium reagents have been used in Pd- and Ni-catalyzed C–C bond formation (certain Ti enolates undergo Pd-catalyzed allylation using allylacetate[7,43]), as in the regioselective phenylation of propargyl acetates using PhTi(O^{i}Pr)$_3$[380]. Furthermore, titanium reagents RTi(O^{i}Pr)$_3$ and RTi(O^{i}Pr)$_4$Li undergo selective Cu(I)-catalyzed S$_N$2' allylation with allylic chlorides and phosphates[292]. Not enough data is available for a final evaluation, especially with respect to traditional alternatives.

A few reports of unusual substitution reactions of vinyl and aryl hydrogen atoms are known, such as **337** → **338** (65%)[381] and **339** → **340** (32%)[382]. The latter substitution reaction has been optimized and is general for homoallylic alcohols. In some cases trimethylaluminum (CH$_3$)$_3$Al can be replaced by CH$_3$Li. Homoallylic alcohols having a terminal alkenic double bond are converted stereoselectively into (E)-3-alken-1-ols, whereas 3-alken-1-ols with internal double bonds afford the corresponding branched 4-methyl-products (55–60% yields). In both cases an intriguing 'configurational inversion' of the double bond occurs[383].

Acylation reactions of organotitanium reagents have not been studied extensively. CH$_3$Ti(O^{i}Pr)$_3$ reacts with acid chlorides to form the isopropyl esters, not the ketones[7]. In contrast, imidazolides react with allylic titanium reagents to form the β,γ-unsaturated ketones in good yields[384], a process that fails using allylmagnesium chloride. A few cases of Ti-promoted Claisen condensations have been reported[114] which are also substitution processes (Section 2.2).

8 Catalytic Asymmetric Reactions

8.1 Carbon–Carbon Bond-forming Reactions

Prior to 1990 the role of titanium in catalytic enantioselective C–C bond formation[7,385] was limited. However, the number of Ti-based catalytic processes has increased considerably since then. The most important examples include ene reactions, Diels–Alder cycloadditions, Mukaiyama aldol additions, and Grignard-type processes. In most cases the Ti catalyst activates the carbonyl substrate, as in the $TiCl_4$-mediated Diels–Alder reaction of chiral alkenes or dienophiles[386]. In contrast to such reactions as hydrovinylation, hydrocyanation, hydroformylation and allylic substitution catalyzed by transition metals of the type nickel, rhodium, and palladium, processes involving Lewis acid catalysis often require fairly large amounts of catalyst (10–30 mol %). This inherent disadvantage, which translates into relatively low turnover numbers (TON), also applies to many Ti-catalyzed C–C bond-forming reactions. This means that the costs and efforts reside almost completely in the preparation of the chiral ligands. Indeed, the major challenge concerns the attainment of maximum enantioselectivity and high TON.

One of the most prominent Ti catalysts in asymmetric C–C bond formation is (*R*)- or (*S*)-BINOL–$TiCl_2$ **342**, first prepared in 1986 by the reaction of the dilithium salt **341** with $TiCl_4$[387]. Alternatively, BINOL **343** itself interacts with (^{i}PrO)$_2$TiCl$_2$ to form **342**[388]. Although ^{1}H and ^{13}C NMR spectra indicate high conversion and purity, in the absence of definitive structural proof formula **342** should be viewed with caution. Indeed, the X-ray structural analysis of a related chiral Ti complex derived from tetrachloro tetramethyl-(bis)phenol and $TiCl_4$ reveals a dimeric structure in which each phenolic oxygen from one bis(phenol) is bonded to a different titanium atom in a criss-cross manner[389]. The corresponding Ti complex prepared by treating the same (bis)phenol with Ti(O^{i}Pr)$_4$ is also dimeric, but has a different structure[389]. As **342** generally shows nonlinear effects in catalytic reactions, the catalytically active species must be a dimer or an aggregate[390].

Initially, enantioselectivity resulting from the use of (*R*)- or (*S*)-**342** in C–C bond-forming reactions turned out to be mediocre, as in certain Diels–Alder reactions[387]. However, a breakthrough was achieved upon applying the catalyst to the so-called glyoxylate–ene reaction[388,390]. Accordingly, 2–10 mol % of the catalyst **342** is first prepared from **343** in the presence of molecular sieves and then used to effect the ene reaction of methylglyoxylate **344** and alkenes of the type **345**. The ee values are generally very high (92–98%). Sometimes the bromo analog BINOL–TiBr$_2$ is more efficient, and as little as 1 mol % of catalyst can be used. Desymmetrization of divinyl substrates occurs with greater than 99% enantio- and diastereoselectivity.

$$\underset{344}{\overset{\overset{\displaystyle O}{\parallel}}{H}\overset{}{\diagdown}CO_2CH_3} \quad + \quad \underset{345}{\overset{}{\underset{R}{\diagup}}\hspace{-0.3em}=} \quad \xrightarrow{(R)\text{-}342} \quad \underset{346}{\overset{H_2C}{\underset{R}{\diagup}}\hspace{-0.3em}\overset{OH}{\diagdown}CO_2CH_3}$$

The role of the molecular sieves is not to trap any moisture (as has been claimed in the Sharpless epoxidation[391]), but to accelerate ligand exchange in the formation of the catalyst[388,390]. Indeed, if the alternative preparation of the catalyst is used based on the reaction of the dilithium salt **341** with TiCl₄[387], molecular sieves are not necessary[388]. The glyoxylate–ene reaction is fairly general for a variety of 1,1-disubstituted alkenes, including methylenecyclohexane, methylenecyclopentane and similar substrates[388]. A fascinating variation concerns a two-directional tandem glyoxylate–ene reaction of 2-trimethylsilylpropene[392]. In a mechanistic study, remarkable positive nonlinear effects were discovered which are also synthetically interesting[393]. For example, using 1 mol % of a catalyst derived from enantiomerically impure BINOL (ee = 33%), an ene reaction product having an ee value of 91% was obtained! This clearly indicates that the catalyst cannot be monomeric. If substrates such as isoprene are used, which can undergo either the ene or a hetero Diels–Alder reaction, ligand tuning enhances ene selectivity[394]. For example, 6-Br-BINOL–TiCl₂ increases ene selectivity and enantioselectivity. Variation of the bulkiness of the alkyl group in the glyoxylate can also be used to influence periselectivity in favor of the ene process, an instrument that was applied in a short enantioselective synthesis of ipsdienol[394].

General Procedure for the Asymmetric Glyoxylate–Ene Reaction Catalyzed by Dichloro-1,1'-bi-naphthalene-2,2'-dioxytitanium, 342[388]

Isolation of dichlorodiisopropoxytitanium

To a solution of tetraisopropoxytitanium (2.98 mL, 10 mmol) in hexane (10 mL) is added TiCl₄ (1.10 mL, 10 mmol) slowly at 25 °C. On addition of TiCl₄, heat evolves. After stirring for 10 min, the solution is allowed to stand for 6 h at 25 °C and the precipitate is then collected. The precipitate is washed with hexane (2 × 5 mL) and recrystallized from hexane (3 mL). The crystalline material is dried under reduced pressure and then dissolved in toluene to give a 0.3 M toluene solution.

Catalytic ene reaction

To a suspension of activated powdered molecular sieves 4 Å(0.500 g) in CH₂Cl₂ (5 mL) is added a 0.3 M toluene solution of the above dichlorodiisopropoxytita-nium (0.33 mL, 0.10 mmol) and (R)-(+)- or (S)-(−)-binaphthol (**343**; 0.0286 g, 0.10 mmol) at 25 °C under an argon atmosphere. After stirring for 1 h at 25 °C, the mixture is cooled to −70 °C. An excess of the isobutylene **345** (R = CH₃) is bubbled through the mixture (about 2 equiv), and freshly distilled methyl glyoxylate (**344**; 0.088 g, 1.0 mmol) is added. The mixture is then warmed to $-30\,°C$ and stirred for 8 h. The solution is poured onto saturated $NaHCO_3$ solution (10 mL). Molecular sieves 4 Å are filtered through a pad of Celite, and the filtrate is extracted with ethyl acetate. The combined organic layers are washed with brine. The extract is then dried over $MgSO_4$ and evaporated under reduced pressure. Separation by silica gel chromaography (hexane/ethyl acetate 20 : 1) gives a 72% yield of methyl 2-hydroxy-4-methyl-4-pentenoate (**346**, $R = CH_3$) having an ee value of 95%.

Inspite of the versatility of the $BINOL–TiX_2$-catalyzed ene reaction, it cannot be extended to normal saturated or aromatic aldehydes. Exceptions are aldehydes electronically related to glyoxylates, such as 3-methoxycarbonylpropynal. On the basis of these and other observations a transition-state model for the reaction has been proposed which predicts the absolute and relative stereochemistry of the reactions[395]. One of the key features is an intermediate complex with trigonal bipyramidal geometry in which the apical substituents are the most electronegative ligands and the coordinated aldehyde, the latter also undergoing formyl $CH\cdots O$ hydrogen bonding to the stereoelectronically most favorable oxygen lone pair of the BINOL ligand.

The $BINOL–TiX_2$ (X = Cl, Br) catalysts or derivatives thereof have also been applied in other reactions[396], although the generality which typifies the glyoxylate–ene reaction is usually not reached. Enantioselectivities of 92–96% have been reported for a few Diels–Alder and hetero Diels–Alder reactions[390,396]. These need to be compared to powerful alternative catalysts based on ligands other than BINOL and metals other than titanium[397]. $BINOL–TiCl_2$ (20 mol %) also catalyzes the addition of allylstannanes to aldehydes (ee = 80–97%)[398]. As stereoselective allylsilane additions are of great synthetic interest (Section 3), a catalytic enantioselective version would be desirable. However, allylsilanes are not very carbonylophilic, requiring fairly strong activation of the aldehyde by Lewis acids. Unfortunately, this invariably means that the Lewis acid remains attached to the oxygen atom of the addition product (rather than the silyl group), thereby preventing efficient catalysis. Indeed, the catalysts $BINOL–TiX_2$ (X = Cl, Br) fail in these reactions. A solution to this problem was discovered upon using a catalyst prepared from BINOL and TiF_4[399]. Fluorine not only has a higher electronegativity which should increase Lewis acidity, the unusually strong Ti−F bond having a mean bond dissociation energy (D_o) of 140 kcal mol^{-1} also leads to the expectation that the Me_3SiO-ether function will remain intact under the reaction conditions (D_o of Si−F is only 135 kcal mol^{-1}).Indeed, upon reacting 20 mol % of BINOL and 10 mol % of TiF_4 in CH_3CN a red–brown solid is formed which catalyzes the addition of allylsilanes to aldehydes with formation of adducts **347** having ee values of 60–94%. Although the structural and mechanistic details remain to be elucidated (nonlinear effects are observed), the results show that fluorine is an interesting ligand in titanium chemistry, as has been shown in other reactions[400].

$$RCHO + \underset{}{\overset{}{=\!\!\!-\!\!\!/\!\!\!-SiMe_3}} \xrightarrow{\text{(S)-BINOL/TiF}_4} R \overset{OSiMe_3}{\underset{\textbf{347}}{\wedge\!\!\wedge}}$$

The BINOL-TiX$_2$ system provides the opportunity to tune the ligands X in ways other than halogen variation, a strategy that has been used in the asymmetric reaction of allylstannanes[401] and enolsilanes[402] with aldehydes. Accordingly, BINOL is simply mixed with Ti(O^{i}Pr)$_4$, resulting in BINOL–Ti(O^{i}Pr)$_2$ which serves as the catalyst; ee values are generally 91–98%. The reaction can be run at 25 °C and, based on its simplicity, appears to be preferred over the BINOL–TiCl$_2$ version[403]. Although the BINOL Ti(O^{i}Pr)$_4$ ratio is usually 1 : 1, a ratio of 2 : 1 is sometimes beneficial. Structural variation in the ligand, as in the case of cyclo-BINOLS, may improve the ee in difficult cases (*e.g.*, cyclohexane carboxaldehyde)[404]. A different improvement makes use of B(OCH)$_3$ as an additive which speeds up the reaction[405]; in the case of allenylstannanes, propargylation of aldehydes is possible (ee = 92–97%)[405]. A different additive which also accelerates the reaction is ^{i}Pr$_3$SSi(CH$_3$)$_3$[406]. More work is necessary to define the scope and mechanism of these reactions.

In a definitive synthetic study concerning the use of 2 : 1 ratios of BINOL: Ti(OEt)$_4$ in the absence of additives or of molecular sieves it was discovered that for really superb results (ee > 96%) it is necessary to premix BINOL and Ti(OEt)$_4$ for about 2 h (not 1 h as previously recommended)[407]. This not only applies to allyl- and methallylstannanes, but also to functionalized stannanes which extend the scope of the methodology considerably. Excellent enantioselectivities are observed, which means that this variation constitutes the method of choice:

$$RCHO + \underset{}{\overset{}{=\!\!\!-\!\!\!/\!\!\!-SnBu_3}} \xrightarrow[\text{10 mol \% each}]{\text{(R)-BINOL/Ti(OEt)}_4} R \overset{OH}{\underset{\textbf{232}}{\wedge\!\!\wedge}}$$

General Procedure for the Optimized BINOL(TiOEt)$_4$-catalyzed Addition of Allylstannanes to Aldehydes[407]

A mixture of (R)-(+)-BINOL (**343**; 0.050 g, 0.17 mmol, 0.2 equiv) and Ti(OEt)$_4$ (0.018 mL, 0.020 g, 0.087 mmol, 0.1 equiv) in CH$_2$Cl$_2$ (0.5 mL) was stirred at 25 °C for 2 h. Hexanal (0.11 mL, 0.087 g, 0.87 mmol, 1.0 equiv) was added. After the mixture had been stirred for 10 min, it was cooled to −78 °C and allyltributyl-stannane (0.3180 g, 0.9605 mmol, 1.1 equiv) was added. The reaction proceeded for 12 days at −40 °C in the refrigerator. Addition of a saturated aqueous solution of NaHCO$_3$ (5 mL), 1 h of stirring, filtration through a frit filled with powdered NH$_4$Cl, extraction with CH$_2$Cl$_2$ (3 × 10 mL), and flash chromatography (2 cm, petroleum ether ^{t}BuOMe 10 : 1) yielded the pure homoallylic alcohol (S)-**232** (0.0898 g, 72%).

Although BINOL–Ti(O^{i}Pr)$_2$ is not a very efficient catalyst for the glyoxylate–ene reaction, complete hydrolysis leads to a moisture-tolerant catalyst, which is active[408]. Alternatively, addition of BINOL to the Ti complex speeds up the reaction considerably. This has been exploited in the development of a novel catalyst system in which racemic BINOL–Ti(O^{i}Pr)$_2$ as the Ti component and optically active BINOL as an activator are used to obtain ee values of 80–90%[409]. Such enantiomer-selective activation of racemic titanium complexes is based on a hydrogen bonded 1 : 1 complex of the two components which makes a kind of positive nonlinear effect possible. This type of asymmetric amplification is related to chiral poisoning[410], as in the BINOL–Ti(O^{i}Pr)$_2$-catalyzed allylstannane addition to aldehydes[411]. In the latter case racemic BINOL–Ti(O^{i}Pr)$_2$ is used in the presence of a chiral poison prepared from diisopropyl D-tartrate and Ti(O^{i}Pr)$_4$, resulting in ee values of up to 92%[411].

BINOL–Ti(O^{i}Pr)$_2$ catalyzes the addition of allenylstannanes to aldehydes with formation of homopropargyl alcohols (ee > 90%)[412]. BINOL–Ti(O^{i}Pr)$_2$ is also an excellent catalyst in certain Mukaiyama aldol additions, provided ether is chosen as the solvent[402]. Finally, H$_8$-BINOL–Ti(O^{i}Pr)$_2$ is an efficient catalyst in the addition of Et$_3$Al to aromatic aldehydes (ee > 90%)[413].

Another class of chiral diols used successfully in Ti-catalyzed C–C bond-forming reactions are TADDOLs (Section 3.3). The ketal part may vary as in the Seebach system **348**[258,260,414] or Narasaka system **350**[415]. Mechanistic and structural aspects have been discussed in detail[260,416]. The ligands are readily attached to titanium in the form of TADDOL–TiX$_2$ (**349** or **351**), for example by the reaction with Cl$_2$Ti(O^{i}Pr)$_2$ in the presence of molecular sieves.

348   **349** a: X = Cl   **350**   **351** a: X = Cl
                                   b: X = Br   b: X = O^{i}Pr
                                   c: X = OTs   c: X = OTs
                                   d: X = O^{i}Pr

Catalysts **349** and **351** have been used for a number of C–C bond-forming reactions including Diels–Alder reactions of acyloxazolidinone derivatives prepared from α,β-unsaturated carboxylic acids (*e.g.*, **352**)[416] and hetero Diels–Alder reactions[417] as well as Grignard-type reactions of R$_2$Zn[414,418] which occur with ee values of 60–98%. In a series of intriguing [2 + 2] cycloaddition reactions between dienophiles of the type **352** and alkenyl sulfides (*e.g.*, **353**) catalyzed by the Ti catalyst **351a** (10 mol %), high levels of enantioselectivity were observed (ee = 95–98%)[415]. These and other Lewis acid-catalyzed reactions of vinyl sulfides have been reviewed[419]. Related reactions of substituted 1,4-benzoquinones with substituted styrenes result in products having four contiguous stereocenters (ee = 86–92%)[420].

352 **a**: R = CH₃   **353**   **354**
      b: R = Ph
      c: R = CO₂CH₃

Typical Procedure for the Enantioselective Diels–Alder Reaction Mediated by the Narasaka Catalyst, 351a[415 b,c]

Preparation of the catalyst, **351a**

Under an argon atmosphere, to a toluene solution (5 mL) of dichlorodiiso-propoxytitanium (0.140 m, 0.59 mmol) is added a toluene solution (5 mL) of the chiral diol **350** (0.354 g, 0.67 mmol) at 25 °C, and the mixture is stirred for 1 h.

Diels–Alder reaction

To a toluene suspension (3 mL) of molecular sieves 4 Å(0.150 g) is added a toluene solution of the catalyst **351a** (about 0.07 mmol) and the mixture is cooled to 0 °C. A toluene solution (4 mL) of the fumaric acid derivative **352c** (0.140 g, 0.7 mmol) is added to the mixture and then hexane (5 mL) and isoprene (1 mL) are added. The mixture is stirred overnight at 0 °C, then pH 7 phosphate buffer is added, the organic materials are extracted with ethyl acetate and the combined extracts are dried over anhydrous MgSO₄. After evaporation of the solvent, the crude product is purfied by thin-layer chromatography (ethyl acetate/hexane 1 : 2) to give the pure Diels–Alder product in 92% yield (ee = 94%).

It has been pointed out that the direction of the stereochemical outcome of a series of different Ti-(R,R)-TADDOLate- and Ti-(S)-BINOLate-catalyzed reactions is the same, suggesting common mechanistic features[416]. Although the respective mechanisms require further studies, a useful mnemonic device has been proposed that allows for the prediction of rectus/sinister (*Re/Si*) topicity in each type of reaction (Scheme 7)[416]. Of course, with (S,S)-TADDOLS and (R)-BINOL the topicities are reversed.

The actual mechanism of these [4 + 2] and [2 + 2] cycloadditions has been studied to some extent[416,421]. A central hypothesis is the activation of substrates of the type **352** via Ti chelation at the dicarbonyl function, 10 possible diastereomeric octahedral complexes being possible. Insight into this complex state of affairs was gained by an X-ray analysis of the complex formed by reacting **352b** with **349a**[422]. The adduct **355** has the chloride ligands located trans to each other, and the alkene is in an s-cis geometry relative to the carbonyl functionality. It is likely that this is the reactive intermediate in

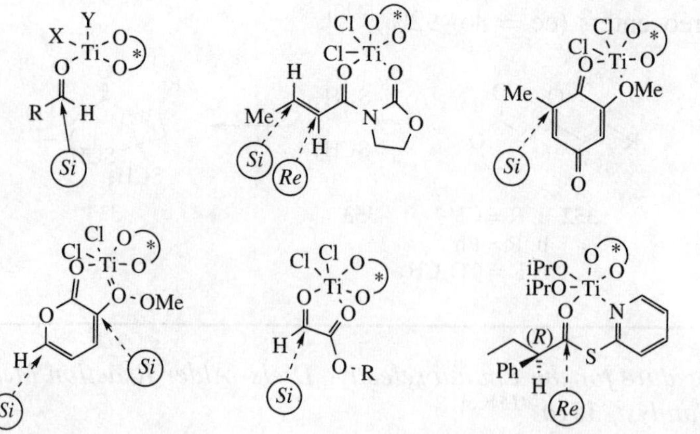

Scheme 7. Mnemonics for the stereochemical course of reactions mediated by Ti-(R,R)-TADDOLate or Ti-(S)-BINOLate.

the catalytic cycle, attack of the diene or of the alkenyl sulfide occurring preferentially from the less-hindered diastereotopic face. However, the situation may be more complex because derivatives of TADDOLates differing only in the nature of the ketal show different degrees of stereoselectivity[423]. Although not catalytic, the Diels–Alder reaction of 1-acetoxybutadiene and maleic acid imides mediated by TADDOL/Ti(O^{i}Pr)$_4$ also involves preorganization in the Ti–ligand sphere[424]. Dienophile–TiCl$_4$ complexation has been studied by extended X-ray absorption fine structure (EXAFS)[425].

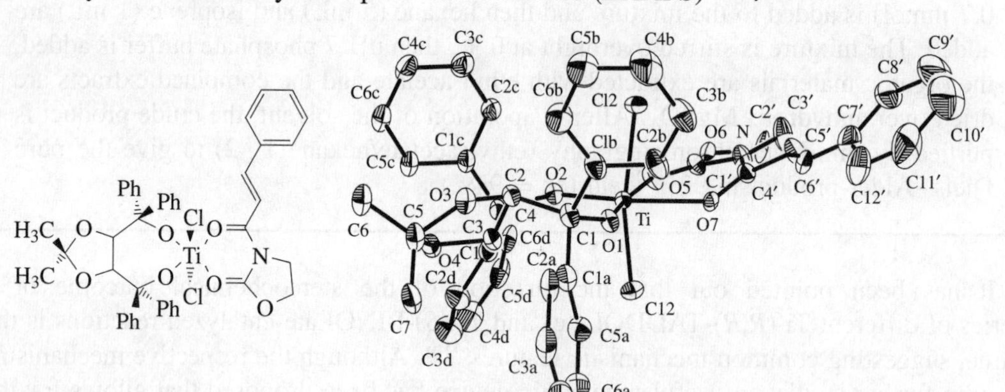

355

Although achiral[386,426] and chiral Ti-based Lewis acids have been used for a long time to mediate (or catalyze) Diels–Alder reactions, their use in analogous 1,3-dipolar cycloaddition reactions has only recently been described[427]. The challenge is to find a Lewis acid which does not bind exclusively to the 1,3-dipole, thereby killing the reaction. In the reaction of nitrones with substrates of the type **352a**, Ti catalysts of the type TADDOL–TiX$_2$ accelerate the process and also lead to pronounced degrees of stereoselectivity[428]. Of the four possible catalysts **349a–d**, the one with the tosylato ligand **349c** turned out to be the most effective. In the presence of 50 mol % of this catalyst nitrones having aryl substituents react with **352a** to afford the corresponding endo adducts exclusively, the ee values being 91–93%. In other cases stereoselectivity is lower.

A bewildering number of catalysts for the Grignard-type of addition of dialkylzinc reagents to carbonyl compounds has been described[75]. They are either chiral amino alcohols or metal coordination compounds, such as Ti complexes. The latter are based on the action of titanium as a Lewis acid, usually requiring fairly large amounts of catalysts (10–20 mol %). A notable exception is the ligand **356**, which in the presence of $Ti(O^iPr)_4$ catalyzes the enantioselective addition of Et_2Zn or Bu_2Zn to benzaldehyde and other aldehydes[429]. With as little as 0.0005 equivalents of chiral ligand, ee values of 98% and nearly quantitative conversions are possible. Importantly, the method tolerates the presence of functional groups in the organozinc reagents although larger amounts of catalyst (5–8 mol %) are usually required[430].

$$PhCHO \ + \ R_2Zn \ \xrightarrow[Ti(O^iPr)_4]{356} \ \underset{Ph}{HO \diagdown R}$$

Alkyltitanium intermediates formed by ligand exchange were originally believed to be intermediates[429]. However, based on more recent studies the ligand forms a Ti complex which binds R_2Zn and the aldehyde[431]. A wide variety of related vicinal diamo derivatives have been prepared as potential ligands, including some which form helical Ti complexes[431]. In some cases ligand tuning of this kind leads to even higher ee values, but sometimes at the expense of using more catalyst. Following the establishment of this catalyst system[429], other chiral ligands in combination with $Ti(O^iPr)_4$ were described including TADDOL[414] derivatives of, for example, amino alcohols[432], BINOL[433], and phosphoramides[434a], ee values of 90–98% being common. The first cases of additions to ketones appear promising[432c].

Typical Procedure for the Enantioselective Addition of $ZnEt_2$ to Aldehydes Catalyzed by Chiral Sulfonamides 356 in the Presence of $Ti(O^iPr)_4$[429]

In a flame-dried round-bottomed flask is placed (1R,2R)-1,2-N,N'-bis(trifluoromethylsulfonylamino)cyclohexane (**356**; 0.189 g, 0.5 mmol) under an argon atmosphere. To this are added degassed toluene (10 mL) and $Ti(O^iPr)_4$ (8.53 g, 30 mmol) and the mixture is stirred at 40 °C for 20 min. After being cooled to −78 °C, Et_2Zn (1.0 M hexane solution, 30 mL, 30 mmol) is added to the solution. The solution rapidly turns orange. To the resulting solution is added benzaldehyde (2.12 g, 25 mmol) in toluene (2 mL) and the mixture is warmed to −20 °C and stirred at that temperature for 2 h. The reaction is quenched by adding 2 M HCl and the product is extracted with diethyl ether. The organic phase is washed with saturated NaCl, dried over anhydrous Na_2SO_4 and concentrated. The residue is chromatographed on a silica gel column (ethyl acetate containing 2% hexane as eluent) to obtain crude 1-phenylpropanol. Distillation gives pure (S)-1-phenylpropanol (3.27 g, 98%; bp ~103 °C, 15 torr; [α] −48.6 °C (c 5.13, CHCl₃); ee = 99%).

Many different types of chiral Lewis acids are known to catalyze the Mukaiyama aldol addition[385b,434b,c], some of them being Ti complexes. One of the most efficient is based on the Carreira-catalyst 358 readily prepared from the ligand 357[435]:

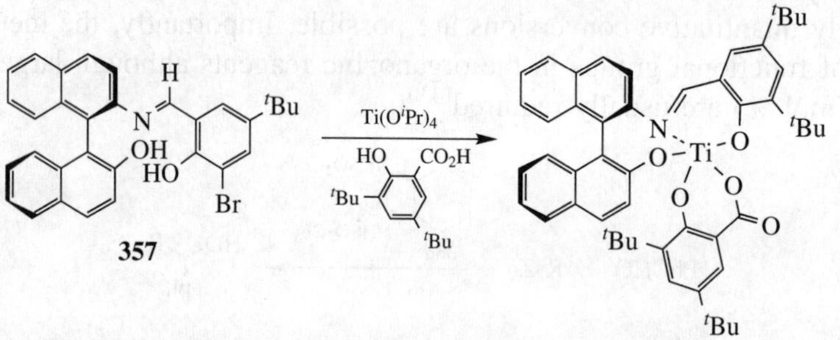

The success of this catalyst has to do with its ease of preparation and with the high levels of conversion and enantioselectivity even when using as little as 0.5 mol %. Thus, the Mukaiyama aldol addition of 359 to aliphatic, aromatic and α,β-unsaturated aldehydes proceeds with ee values of 94–97%, the yields of the products 360 following desilylation being 72–98 %[435]. Although prochiral (e.g., methyl-substituted) analogs of 359 do not behave analogously, certain dienolate additions to aldehydes are very stereoselective (ee = 80–94%)[436].

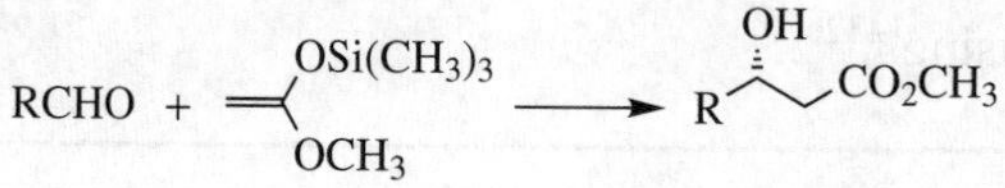

An in situ procedure which circumvents the azeotropic removal of isopropanal in the catalyst preparation makes use of (CH₃)₃SiCl and Et₃N[437]. It has been applied in a short synthesis of (R)-(−)-epinephrine. A useful variation is based on the application of 2-methoxypropene in acetone aldol additions catalyzed by 358[438]. Enantioselectivity is slightly lower (ee = 66–98%), but still outstanding in many cases (e.g., α,β-ynals).

In situ Procedure for Enantioselective Aldol Addition Catalyzed by the Carreira Catalyst, 358[437]

To a 5.5 mmol solution of the chiral Schiff base ligand (357; 0.044 equiv) in toluene at 23 °C was added Ti(OiPr)₄ (0.020 equiv). The orange solution was stirred for 1 h at 23 °C prior to adding via cannula a 10 mmol toluene solution of 3,5-di-*tert*-butylsalicylic acid (0.040 equiv, dried by dissolving in toluene and concentrating *in vacuo* twice). Stirring was continued for an additional hour at 23 °C and then the solution was cooled to −20 °C. To the cooled solution was added trimethylsilyl chloride (0.20 equiv) followed by triethylamine (1.0 equiv) and the aldehyde (1.0 equiv). After stirring the resulting solution for 15 min, the silyl ketene acetal $CH_2 = C(OSiMe_3)OCH_3$ (**359**) or $CH_2 = C(OSiMe_3)OCH_2Ph$ (2.0 equiv) was added dropwise neat. The reaction was allowed to gradually warm to 23 °C over 6 h before it was quenched by pouring onto water. The mixture was extracted with Et_2O; the combined organic extracts were washed with saturated aqueous NaCl solution and dried over anhydrous Na_2SO_4. The organic extracts were concentrated in vacuo and the residue was dissolved in 10% CF_3CO_2H/THF. Once the reaction was complete, the solution was then partitioned between Et_2O and water. The organic layer was washed with 1.0 M NaOH solution followed by brine. The separated organic layer was dried over anhydrous Na_2SO_4 and concentrated in vacuo. Purification by chromatography on silica gel gave the β-hydroxy ester adduct of the type **360**.

A number of other enantioselective Ti-catalyzed C−C bond-forming reactions have been reported, including cyclopropanation reactions[439], TADDOL−TiX$_2$-catalyzed iodocarbocyclizations[440] (compare Section 7) and Michael additions of *O,O*-silyl ketene ketals[441]. Chiral titanocenes as catalysts in enantioselective Pauson−Khand reactions as in the formation of **362** (85% yield; ee = 96%)[442] deserve special attention, as do similar chiral titanocenes in Diels−Alder reactions[443].

361   362

General Procedure for Enantioselective Catalytic Pauson−Khand Formation of Bicyclic Cyclopentenones[442]

In an argon filled glovebox, a dry sealable Schlenk flask is charged with (*S,S*)-(EBTHI)TiMe$_2$ (**59c**; 0.008 g, 0.025 mmol), toluene (2 mL), and a substrate (0.50 mmol) such as **361**. The Schlenk is removed from the glovebox, attached to a Schlenk line, evacuated, and backfilled with 14 psig CO.

Caution: Appropriate precautions should be taken when performing reactions under elevated CO pressure.

> The reaction mixture is heated to 90 °C for 12–16 h. After cooling the reaction mixture to 25 °C, the CO is cautiously released in the hood. The crude reaction mixture is filtered through a plug of silica gel with the aid of diethyl ether and purified by flash chromatography to provide product **362** in 85% yield (ee = 96%).

Several Ti-catalyzed addition reactions of Me₃SiCN to aldehydes have been reported[444], including the use of Ti(O^{i}Pr)₄ in the presence of chiral sufoximines resulting in ee values of 74–91%[445]. Ring-opening reactions of meso epoxides with (CH₃)₃SiCN or (CH₃)₃SiN₃ are catalyzed by chiral Ti complexes[446].

8.2   Reduction, Oxidation and Other Reactions

Enantioselective hydrogenation reactions of alkenes, ketones and imines are usually carried out with chiral rhodium, iridium or ruthenium catalysts[447]. Nevertheless, several Ti-catalyzed processes have been reported which are noteworthy because they are complementary to the traditional methods. One of them concerns the asymmetric reduction of imines catalyzed by the chiral titanium(III) hydride **364** which is prepared by *in situ* reduction of the titanocene **363** using sequentially an excess of butyllithium and PhSiH₃[448]:

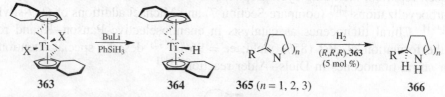

X + X = 1,1'-binaphth-2,2'-diolate

Although fairly high pressures were originally used, it was later discovered that the reactions can be carried out either at lower pressures (80 psig) and higher temperatures (65 °C) or at medium pressure (500 psig) and lower temperatures (21–45 °C) with little or no change in ee. The best results are obtained in the case of cyclic imines such as **365**, the ee values ranging between 95% and 99%. The process is so efficient that it is possible to kinetically resolve racemic disubstituted 1-pyrrolines (ee > 95%)[449]. Somewhat lower enantioselectivities are observed for the reduction of acyclic imines, probably because they exist as mixtures of (E) and (Z) isomers. The catalyst is also capable of mediating the hydrogenation of trisubstituted olefins enantioselectivity (ee = 90–96 %), albeit at higher pressures (2000 psig) and longer reaction times[450]. Likewise, the hydrogention of enamines of the type CH₂ = C(R)NR′₂ occurs with ee values of 91–95%[451]. The hydrosilylation of imines using (S,S)-(EBTHI)TiF₂ (**59b**) as the precatalyst is also possible[452]. Alternative methods are less effective in many of the examples reported. Thus, the efforts needed to prepare the titanocene catalyst **364** in optically active form are compensated by the positive catalyst performance. In other work ketones have been reduced by catecholborane, a process that is catalyzed by chiral Ti alkoxides (ee up to 97%)[453].

A fundamentally important development in organic synthesis is the Sharpless enantios-elective epoxidation of allylic alcohols using *tert*-butyl hydroperoxide, tetraisopropoxyti-tanium, and tartaric acid esters[391,454] (Scheme 8). In the original version, stoichiometric amounts of tartaric acid esters had to be used. However, by working in the presence of molecular sieves which bind water, only catalytic amounts are necessary.

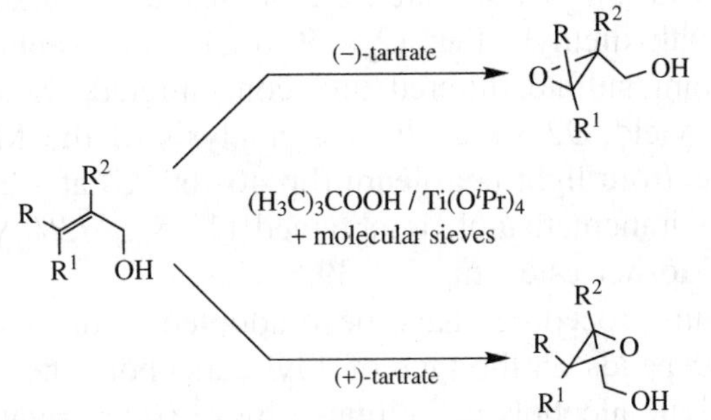

Scheme 8.

Typical Procedure for the Catalytic Sharpless Epoxidation of Allylic Alcohols[391]

Caution: Owing to possible explosions, *tert*-butyl hydroperoxide should not be used in pure form. Also, strong acids and transition metal salts known to be good autoxidation catalysts (Mn, Fe, Ru, Co, etc.) should never be added to high-strength *tert*-butyl hydroperoxide.

An oven-dried 1-L three-necked round-bottomed flask equipped with a magnetic stirbar, pressure equalizing addition funnel, thermometer, nitrogen inlet and bubbler is charged with 3.0 g of powdered, activated molecular sieves 4 Å and 350 mL of dry CH_2Cl_2. The flask is cooled to $-20\,°C$. L-(+)-Diethyl tartrate (1.24 g, 6.0 mmol) and $Ti(O^iPr)_4$ (1.49 mL, 1.42 g, 5.0 mmol, via a syringe) are added sequentially with stirring. The reaction mixture is stirred at $-20\,°C$ as *tert*-butyl hydroperoxide (39 mL, 200 mmol, 5.17 M in isooctane) is added through the addition funnel at a moderate rate (over about 5 min). The resulting mixture is stirred at $-20\,°C$ for 30 min. (E)-2-Octenol (12.82 g, 100 mmol, freshly distilled), dissolved in 50 mL of CH_2Cl_2, is then added dropwise through the same addition funnel over a period of 20 min, being careful to maintain the reaction temperature between $-20\,°C$ and $-15\,°C$. The mixture is stirred for an additional 3.5 h at $-20\,°C$ to $-15\,°C$.

The work-up procedure[391] is as follows. A freshly prepared solution of iron(II)sulfate heptahydrate (33 g, 0.12 mol) and tartaric acid (10 g, 0.06 mol), or 11 g (0.06 mol) of citric acid monohydrate instead of tartaric acid, in a total volume of 100 mL of deionized water is cooled to about $0\,°C$ by an ice-water bath. The epoxidation reaction mixture is allowed to warm to about $0\,°C$ and is then slowly poured into a beaker containing the precooled stirring iron(II)sulfate solution (external cooling is not essential during or after this addition). The two-phase mixture is stirred for 5–10 min and then transferred into a separating funnel. The phases are separated and the aqueous phase is extracted with two 30 mL portions of diethyl ether. The combined organic layers are treated with 10 mL of a precooled (0 °C) solution of 30% w/v NaOH in saturated brine. The two-phase mixture is stirred vigorously for 1 h at 0 °C. Following transfer to a separating funnel and dilution with 50 mL of water, the phases are separated and the aqueous layer is extracted with diethyl ether (2 × 50 mL). The combined organic layers are dried over sodium sulfate, filtered and concentrated, yielding a white solid (12.6 g, 88% crude yield; 92.3% ee by GC analysis of the Mosher ester). After two recrystallizations from light petroleum (bp 40–60 °C) at −20 °C, a white solid, (2S)-*trans*-3-pentyloxiranemethanol, is obtained (10.5 g, 73% yield; >98% ee by GC analysis of the Mosher ester; mp 38–39.5 °C).

Alternative work-up procedures have been adopted in other cases[391]. The best work-up procedure depends on the type of allylic alcohol. The epoxidation of low-molecular-weight allylic alcohols is facilitated by *in situ* derivatization[391].

Hundreds of successful examples have been reported by Sharpless and by other groups. As the epoxides can be used for further ring-opening reactions using cuprates, amines, etc., they constitute useful chiral nonracemic synthetic vehicles. Several extensive review have appeared[455]. Table 1. contains a number of typical examples which illustrate substrate scope. A definitive mechanistic study has appeared[456].

There are some limitations of the method, in that certain types of allylic alcohols cannot be successfully epoxidated, as shown in Table 2.[455]. First, some substrates react very

Table 1. Typical allylic alcohols successfully epoxidated by the Sharpless procedure (Scheme 8).

$$\underset{R}{\overset{R^1}{\diagdown}}\diagup\overset{}{\underset{R^2}{}}\text{--OH}$$

Substrate	ee (%)	Yield (%)
Unsubstituted $R = R^1 = R^2 = H$	95	15
trans-Substituted ($R^1 = R^2 = H$):		
$R = CH_3$	>95	45
$R = C_{10}H_{21}$	>95	79
$R = (CH_2)_3CH = CH_2)$	>95	80
$R = Me_3Si$	>95	60
$R = tert\text{-Bu}$	>95	65
$R = Ar$	95	0–95
$R = CH_2OCH_2Ph$	98	85

Table 1. (*continued*).

Substrate	ee (%)	Yield (%)
R =	>95	78–95
R = $H_5C_6CH_2O$	>95	70
R = $H_5C_6CH_2O$	>99	76
R = $H_5C_6CH_2O$	>99	70
cis-Substituted (R = R^2 = H):		
$R^1 = C_{10}H_{21}$	90	82
$R^1 = CH_2Ph$	91	83
$R^1 = CH_2OCH_2Ph$	92	84
1,1-Substituted (R = R^1 = H):		
R^2 = cyclohexyl	>95	81
$R^2 = C_{14}H_{29}$	>95	51
R^2 = *tert*-Bu	85	60
trans-1,1,2-Substituted (R^1 = H):		
R = $R^2 = C_6H_5$	<95	87
R = C_2H_5, $R^2 = CH_3$	<95	79
R = , $R^2 = CH_3$	<95	70
R = , $R^2 = CH_3$	<95	92
cis = 1, 1, 2 – Substituted (R = H):		
$R^1 = H_5C_6CH_2$, $R^2 = CH_3$	91	90
1,2,2-Substituted (R^2 = H):		
R = $(CH_3)_2C = CHCH_2CH_2$, $R^1 = CH_3$	<95	77
R = CH_3, $R^1 = (CH_3)_2C = CHCH_2CH_2$	94	79
1,1,2,2,-Substituted:		
R = C_6H_5, $R^1 = C_6H_5CH_2$, $R^2 = CH_3$	94	90

Table 2. Poor substrates for the Sharpless epoxidation.

Substrate	Result
	Slow epoxidation, 65% ee
	Slow epoxidation, 25% ee
	Slow epoxidation, 60% ee
	Slow epoxidation, no epoxy alcohol isolated
	Slow epoxidation using (−)-tatrate, 23% ee (2R-diastereomer)
	No reaction using (−)-tartrate
	Slow epoxidation, 67% ee using (+)-tartrate; 0% ee using (−)-tartrate
	No reaction using (+)-DET
	95% ee, 58% yield. Difficult to reproduce due to lactone diol formation
	Product epoxy alcohol unstable to reaction conditions; either no product or only very low yields obtained under standard conditions using stoichiometric Ti tartrate

slowly with poor to moderate enantioselectivity, such as certain (Z)-configurated allylic alcohols and some sterically hindered compounds of other substitution patterns. Second, some substrates lead to epoxy alcohols stereoselectively, but the products are unstable to the reaction conditions. Such problems can often be solved by *in situ* derivatization prior to final work-up[455]. Third, stereoselectivity is sensitive to stereogenic centers already present in the substrate. In fact, kinetic resolution is possible in many instances[455]. In terms of reagent control, this means that the mismatched cases may pose problems. In summary, the titanium–tartrate asymmetric epoxidation constitutes a significant synthetic procedure. In spite of several limitations, it is suprisingly versatile and has been applied in a multitude of natural product syntheses and other preparations. A complementary method is based on the use of chiral allylic alcohols which are first oxidized diastereoselectively to β-hydroperoxy alcohols, these serving as oxygen donors in the $Ti(O^iPr)_4$-catalyzed epoxidation with formation of epoxy diols[457].

The Sharpless catalyst system has been used in modified form in the asymmetric oxidation of thioethers to sulfoxides. If a stoichiometric amount of water is present, many thioethers are transformed into sulfoxides with excellent levels of enantioselectivity (ee > 90%)[458]. Ligands other than tartrates have also been used[459]. The best results in all of these efforts usually pertain to aryl methyl sulfides **367**:

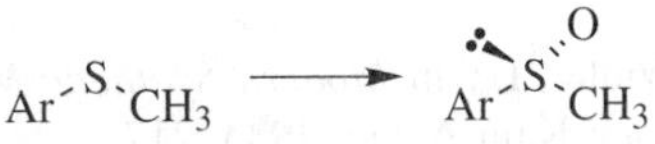

Several Ti-catalyzed reactions are known which do not involve reduction, oxidation, or C–C bond formation. For example, TADDOL–TiX$_2$ catalyzes the enantioselective Michael addition of *O*-benzyl hydroxylamines[460]. Various chiral titanium complexes have been used in the stereoselective ring opening of meso epoxides[446,461]. Finally, a highly enantioselective ring-opening reaction of prochiral meso anhydrides is possible by reaction with stoichiometric amounts of TADDOL–$Ti(O^iPr)_2$[462], although substoichiometric amounts of the titanium reagent also appear to be effective.

9 Acknowledgments

I wholeheartedly thank my co-workers whose name are listed in the references. Help in collecting literature on titanium chemistry was provided by H. Nehl. Special thanks are due to A. Rathofer, M. Lickfeld, and E. Enk for preparing the drawings, typing the manuscript and checking the references.

10 References

[1] (a) Sinn, H.; Kaminsky, W.; *Adv. Organomet. Chem.* **1980**, *18* 99; (b) Pino, P.; Mülhaupt, R.; *Angew. Chem., Int. Ed. Engl.* **1980**, *19* 857.

[2] (a) Rieche, A.; Gross, H.; Höft, E.; *Chem. Ber.* **1960**, *93* 88; (b) Cullinane, N. M.; Leyshon, D. M.; *J. Chem. Soc.* **1954**, 2942; (c) Izumi; Mukaiyama, T.; *Chem. Lett.* **1996**, 739; (d) Mikami, K.; Kotera, O.; Motoyama, Y.; Sakaguchi, H.; Maruta, M.; *Synlett* **1996**, 171.

[3] (a) Seebach, D.; Hungerbühler, E.; Naef, R.; Schnurrenberger, P.; Weidmann, B.; Züger, M.; *Synthesis* **1982**, 138; (b) Rehwinkel, H.; Steglich, W.; *Synthesis* **1982**, 826; (c) Imwinkelried, R.; Schiess, M.; Seebach, D.; *Org. Synth.* **1987**, *65* 230.

[4] (a) *Gmelin Handbuch, Titan-Organische Verbindungen*, Part 1 **1977**, Part 2 **1980**, Part 3 **1984**, Part 4 **1984**, Springer, Berlin; (b) Bottrill, M.; Gavens, P. D.; Kelland, J. W.; McMeeking, J.; in *Comprehensive Organometallic Chemistry*, (Wilkinson, G.; Stone, F. G. A.; Abel, E. W.; Eds), Pergamon Press, Oxford, **1982**, Chap. 22; (c) Thiele, K. H.; *Pure Appl. Chem.* **1972**, *30* 575; (d) Mintz, E. A.; in *Encyclopedia of Inorganic Chemistry*, (King, R. B.; Ed.) Vol. 8 Wiley, New York, **1994**, 4306.

[5] Segnitz, A.; in *Houben-Weyl-Müller, Methoden der Organischen Chemie*, Vol. 13/7, Thieme, Stuttgart, **1975**, 261.

[6] McGrady, G. S.; Downs, A. J.; Bednall, N. C.; McKean, D. C.; Thiel, W.; Jonas, V.; Frenking, G.; Scherer, W.; *J. Phys. Chem. A* **1997**, *101* 1951.

[7] Reetz, M. T.; *Organotitanium Reagents in Organic Synthesis*, Springer, Berlin, **1986**.

[8] (a) Stowell, J. C.; *Carbanions in Organic Synthesis*, Wiley, New York, **1979**; (b) Boche, G.; *Angew. Chem., Int. Ed. Engl.* **1989**, *28* 277; (c) Brandsma, L.; Verkruijsse, H.; *Preparative Polar Organometallic Chemistry 1*, Springer, Berlin, **1987**; (d) Brandsma, L.; *Preparative Polar Organometallic Chemistry 2*, Springer, Berlin, **1990**.

[9] Kharasch, M. S.; Cooper, J. H.; *J. Org. Chem.* **1945**, *10* 46.

[10] Reetz, M. T.; Westermann, J.; Steinbach, R.; Wenderoth, B.; Peter, R.; Ostarek, R.; Maus, S.; *Chem. Ber.* **1985**, *118* 1421.

[11] Seebach, D.; Weidmann, B.; Widler, L.; in *Modern Synthetic Methods*, (Scheffold, R.; Ed.) Vol. 3, Salle + Sauerländer, Frankfurt am Kain Aarau, **1983**, 217.

[12] Reetz, M. T.; Kyung, S. H.; Hüllmann, M.; *Tetrahedron* **1986**, *42* 2931.

[13] Reetz, M. T.; *Top. Curr. Chem.* **1982**, *106* 1.

[14] Weidmann, B.; Seebach, D.; *Angew. Chem., Int. Ed. Engl.* **1983**, *22* 31.

[15] Imwinkelried, R.; Seebach, D.; *Org. Synth.* **1988**, *67* 180.

[16] Reetz, M. T.; Westermann, J.; Steinbach, R.; *Angew. Chem., Int. Ed. Engl.* **1980**, *19* 900.

[17] Reetz, M. T.; Westermann, J.; Steinbach, R.; *Angew. Chem., Int. Ed. Engl.* **1980**, *19* 901.

[18] Reetz, M. T.; Steinbach, R.; Westermann, J.; Peter, R.; *Angew. Chem., Int. Ed. Engl.* **1980**, *19* 1011.

[19] Reetz, M. T.; Wenderoth, B.; Peter, R.; Steinbach, R.; Westermann, J.; *J. Chem. Soc., Chem. Commun.* **1980**, 1202.

[20] (a) Weidmann, B.; Seebach, D.; *Helv. Chim. Acta* **1980**, *63* 2451; (b) Weidmann, B.; Widler, L.; Olivero, A. G.; Maycock, C. D.; Seebach, D.; *Helv. Chim. Acta* **1981**, *64* 357.

[21] Reetz, M. T.; Steinbach, R.; Wenderoth, B.; Westermann, J.; *Chem. Ind. (London)* **1981**, 541.

[22] Benzing, E.; Kornicker, W.; *Chem. Ber.* **1961**, *94* 2263.

[23] (a) Reetz, M. T.; Urz, R.; Schuster, T.; *Synthesis* **1983**, 540; (b) Steinborn, D.; Wagner, I.; Taube, R.; *Synthesis* **1989**, 304.

[24] (a) Reetz, M. T.; Maus, S.; *Tetrahedron* **1987**, *43* 101; (b) Reetz, M. T.; Hugel, H.; Dresely, K.; *Tetrahedron* **1987**, *43* 109.

[25] Weidmann, B.; Maycock, C. D.; Seebach, D.; *Helv. Chim. Acta* **1981**, *64* 1552.

[26] Reetz, M. T.; Steinbach, R.; Westermann, J.; Urz, R.; Wenderoth, B.; Peter, R.; *Angew. Chem., Int. Ed. Engl.* **1982**, *21* 135; *Angew. Chem. Suppl.* **1982**, 257.

[27] Review of Zr reagents: Negishi, E.; Takahashi, T.; *Synthesis* **1988**, 1.

[28] Kauffmann, T.; Pahde, C.; Wingbermühle, D.; *Tetrahedron Lett.* **1985**, *26* 4059.

[29] (a) Imamoto, T.; Takiyama, N.; Nakamura, K.; Hatajima, T.; Kamiya, Y.; *J. Am. Chem. Soc.* **1989**, *111* 4392; (b) Kauffmann, T.; Pahde, C.; Tannert, A.; Wingbermühle, D.; *Tetrahedron Lett.* **1985**, *26* 4063; (c) Imamoto, T.; in *Comprehensive Organic Synthesis*, (Trost, B. M.; Fleming, I.; Eds) Vol. I, Pergamon Press, Oxford, **1991**, 231.

[30] (a) Okude, Y.; Hirano, S.; Hiyama, T.; Nozaki, H.; *J. Am. Chem. Soc.* **1977**, *99* 3179; (b) Kauffmann, T.; Abeln, R.; Wingbermühle, D.; *Angew. Chem., Int. Ed. Engl.* **1984**, *23* 729.

[31] For an informative comparison of Ti, Zr, Hf, Cr, Sm, Ce, Mn, Nb, Mo and other organometallics see: (a) Kauffmann, T.; in *Organometallics in Organic Synthesis*, (Werner, H.; Erker, G.; Eds) Vol. 2, Springer, Berlin, **1989**; (b) Kauffmann, T.; Kieper, H.; Pieper, H.; *Chem. Ber.* **1992**, *125* 899.

[32] (a) Cahiez, G.; Laboue, B.; *Tetrahedron Lett.* **1989**, *30* 3545; (b) Normant, J. F.; Cahiez, G.; in *Modern Synthetic Methods*, (Scheffold, R.; Ed.) Vol. 3, Salle + Sauerlander, Frankfurt am Kain, Aarau, **1983**, 173.

[33] (a) Kauffmann, T.; Laarmann, B.; Menges, D.; Voss, K. U.; Wingbermühle, D.; *Tetrahedron Lett.* **1990**, *31* 507; (b) Kauffmann, T.; Laarmann, B.; Menges, D.; Neiteler, G.; *Chem. Ber.* **1992**, *125* 163.

[34] (a) Molander, G. A.; Burkhardt, E. R.; Weinig, P.; *J. Org. Chem.* **1990**, *55* 4990; (b) Utimoto, K.; Nakamura, A.; Matsubara, S.; *J. Am. Chem. Soc.* **1990**, *112* 8189; (c) review of lanthanide reagents: Kagan, H.; Namy, J. L.; *Tetrahedron* **1986**, *42* 6573.

[35] Knochel, P.; Singer, R. D.; *Chem. Rev. (Washington, D.C.)* **1993**, *93* 2117.

[36] Raubenheimer, H. G.; Seebach, D.; *Chimia* **1986**, *40* 12.

[37] Reetz, M. T.; Steinbach, R.; Wenderoth, B.; *Synth. Commun.* **1981**, *11* 261.

[38] Reetz, M. T.; Westermann, J.; Kyung, S. H.; *Chem. Ber.* **1985**, *118* 1050.

[39] Reetz, M. T.; Westermann, J.; *Synth. Commun.* **1981**, *11* 647.

[40] Nützel, K.; in *Houben-Weyl-Müller, Methoden der Organischen Chemie*, Vol. 13/2, Thieme, Stuttgart, **1973**, 573.

[41] Reviews of chelation and non-chelation controlled additions to chiral alkoxy aldehydes and ketones: (a) Reetz, M. T.; *Angew. Chem., Int. Ed. Engl.* **1984**, *23* 556; (b) Reetz, M. T.; *Acc. Chem. Res.* **1993**, *26* 462; theoretical aspects of chelation: (c) Jonas, V.; Frenking, G.; Reetz, M. T.; *Organometallics* **1993**, *12* 2110.

[42] Reetz, M. T.; Kyung, S. H.; *Chem. Ber.* **1987**, *120* 123.

[43] (a) Reetz, M. T.; Peter, R.; Wünsch, T.; Steinbach, R.; Westermann, J.; Schuster, T.; unpublished results; (b) Reetz, M. T.; Stanchev, S.; Haning, H.; *Tetrahedron* **1992**, *48* 6813.

[44] Reetz, M. T.; Wünsch, T.; *J. Chem. Soc., Chem. Commun.* **1990**, 1562.

[45] Yu, K. L.; Handa, S.; Tsang, R.; Fraser-Reid, B.; *Tetrahedron* **1991**, *47* 189.

[46] Reetz, M. T.; *S. Afr. J. Chem.* **1989**, *42* 49.

[47] (a) Herman, D. F.; Nelson, W. K.; *J. Am. Chem. Soc.* **1953**, *75* 3882; (b) Rausch, M. D.; Gordon, H. B.; *J. Organomet. Chem.* **1974**, *74* 85.

[48] Widler, L.; Seebach, D.; *Helv. Chim. Acta* **1982**, *65* 1085.

[49] Reetz, M. T.; Steinbach, R.; Westermann, J.; Peter, R.; Wenderoth, B.; *Chem. Ber.* **1985**, *118* 1441.

[50] Dolle, R. E.; Nicolaou, K. C.; *J. Am. Chem. Soc.* **1985**, *107* 1691.

[51] Reetz, M. T.; Wenderoth, B.; *Tetrahedron Lett.* **1982**, *23* 5259.

[52] (a) Rise, F.; Undheim, K.; *J. Organomet. Chem.* **1985**, *291* 139; (b) Undheim, K.; Benneche, T.; *Heterocycles* **1990**, *30* 1155; (c) Rise, F.; Undheim, K.; *J. Chem. Soc., Perkin Trans. I* **1985**, 1997; (d) Gundersen, L. L.; Rise, F.; Undheim, K.; *Tetrahedron* **1992**, *48* 5647.

[53] Krause, N.; Seebach, D.; *Chem. Ber.* **1987**, *120* 1845.

[54] (a) Mahrwald, R.; Schick, H.; Vasileva, L. L.; Pivnitsky, K. K.; Weber, G.; Schwarz, S.; *J. Prakt. Chem.* **1990**, *332* 169; (b) Wu M.-J.; Yan, D.-S.; Tsai, H.-W.; Chen, S.-H.; *Tetrahedron Lett.* **1994**, *35* 5003.

[55] (a) Mead, K.; Macdonald, T. L.; *J. Org. Chem.* **1985**, *50* 422; (b) Schick, H.; Spanig, J.; Mahrwald, R.; Bohle, M.; Reiher, T.; Pivnitsky, K. K.; *Tetrahedron* **1992**, *48* 5579.

[56] Boeckman, R. K.; O'Connor, K. J.; *Tetrahedron Lett.* **1989**, *30* 3271.

[57] Weidmann, B.; Widler, L.; Olivero, A. G.; Maycock, C. D.; Seebach, D.; *Helv. Chim. Acta* **1981**, *64* 357.

[58] Kauffmann, T.; Antfang, E.; Ennen, B.; Klas, N.; *Tetrahedron Lett.* **1982**, *23* 2301.

[59] Kauffmann, T.; Schwartze, P.; *Chem. Ber.* **1986**, *119* 2150.

[60] Kauffmann, T.; Fobker, R.; Wensing, M.; *Angew. Chem., Int. Ed. Engl.* **1988**, *27* 943.

[61] (a) Kasatkin, A. N.; Biktimirov, R. K.; Tolstikov, G. A.; Khalilov, L. M.; *Zh. Org. Khim.* **1990**, *26* 1191; (b) Kauffmann, T.; Kieper, H.; *Chem. Ber.* **1992**, *125* 907.

[62] Anies, C.; Lallemand, J.-Y.; Pancrazi, A.; *Tetrahedron Lett.* **1996**, *37* 5519.

[63] Reetz, M. T.; Peter, R.; *Tetrahedron Lett.* **1981**, *22* 4691.

[64] Reetz, M. T.; Steinbach, R.; Keßeler, K.; *Angew. Chem., Int. Ed. Engl.* **1982**, *21* 864; *Angew. Chem. Suppl.* **1982**, 1899.

[65] Haarmann, H.; Eberbach, W.; *Tetrahedron Lett.* **1991**, *32* 903.

[66] Seki, M.; Miyake, T.; Izukawa, T.; Ohmizu, H.; *Synthesis* **1997**, 47.

[67] Kanemasa, S.; Mori, T.; Tatsukawa, A.; *Tetrahedron Lett.* **1993**, *34* 8293.

[68] (a) Hainz, R.; Gais, H.-J.; Raabe, G.; *Tetrahedron: Asymmetry* **1996**, *7* 2505; (b) Reggelin, M.; Weinberger, H.; Gerlach, M.; Welcker, R.; *J. Am. Chem. Soc.* **1996**, *118* 4765.

[69] (a) Aono, T.; Hesse, M.; *Helv. Chim. Acta* **1984**, *67* 1448; (b) Kostova, K.; Hesse, M.; *Helv. Chim. Acta* **1984**, *67* 1713.

[70] Reetz, M. T.; Hüllmann, M.; *J. Chem. Soc., Chem. Commun.* **1986**, 1600.

[71] (a) Siegel, C.; Thornton, E. R.; *J. Am. Chem. Soc.* **1989**, *111* 5722; (b) reactions of chiral α-benzoyl titanium enolates: Choudhury, A.; Thornton, E. R.; *Tetrahedron* **1992**, *48* 5701.

[72] (a) Cooper, G. D.; Finkbeiner, H. L.; *J. Org. Chem.* **1962**, *27* 1493; (b) Fell, B.; Asinger, F.; Sulzbach, R. A.; *Chem. Ber.* **1970**, *103* 3830.

[73] Review of titanium-catalyzed hydromagnesiation of alkenes and alkynes: Sato, F.; *J. Organomet. Chem.* **1985**, *285* 53.

[74] Sato, F.; Ishikawa, H.; Sato, M.; *Tetrahedron Lett.* **1981**, *22* 85.

[75] (a) Review of enantioselective additions of dialkylzinc reagents to aldehydes catalyzed by chiral β-amino alcohols: Noyori, R.; Kitamura, M.; *Angew. Chem., Int. Ed. Engl.* **1991**, *30* 49; (b) addition to ketones: Dosa, P. I.; Fu, G. C.; *J. Am. Chem. Soc.* **1998**, *120* 445.

[76] Knochel, P.; Normant, J. F.; *Tetrahedron Lett.* **1986**, *27* 4431.

[77] McWilliams, J. C.; Armstrong, J. D.; Zheng, N.; Bhupathy, M.; Volante, R. P.; Reider, P. J.; *J. Am. Chem. Soc.* **1996**, *118* 11970.

[78] Gais, H. J.; Vollhardt, J.; Lindner, H. J.; Paulus, H.; *Angew. Chem., Int. Ed. Engl.* **1988**, *27* 1540.

[79] Mukaiyama, T.; Suzuki, K.; Yamada, T.; Tabusa, F.; *Tetrahedron* **1990**, *46* 265.

[80] Hollenstein, S.; Hesse, M.; *Helv. Chim. Acta* **1996**, *79* 827.

[81] Yeh, M. C. P.; Knochel, P.; Santa, L. E.; *Tetrahedron Lett.* **1988**, *29* 3887.

[82] Yeh, M. C. P.; Knochel, P.; *Tetrahedron Lett.* **1988**, *29* 2395.

[83] Blaney, W. M.; Cuñat, A. C.; Ley, S. V.; Montgomery, F. J.; Simmonds, M. S. J.; *Tetrahedron Lett.* **1994**, *35* 4861.

[84] (a) Ikeda, Y.; Furuta, K.; Meguriya, N.; Ikeda, N.; Yamamoto, H.; *J. Am. Chem. Soc.* **1982**, *104* 7663; (b) Furuta, K.; Ikeda, Y.; Meguriya, N.; Ikeda, N.; Yamamoto, H.; *Bull. Chem. Soc. Jpn.* **1984**, *57* 2781.

[85] Bernardi, A.; Cavicchioli, M.; Marchionni, C.; Potenza, D.; Scolastico, C.; *J. Org. Chem.* **1994**, *59* 3690.

[86] (a) Bürger, H.; Neese, H. J.; *Chimia* **1970**, *24* 209; (b) Bürger, H.; Neese, H. J.; *J. Organomet. Chem.* **1972**, *36* 101.

[87] Schiess, M.; Seebach, D.; *Helv. Chim. Acta* **1982**, *65* 2598.

[88] Petasis, N. A.; Hu, Y.-H.; *Curr. Org. Chem.* **1997**, *1* 249.

[89] (a) Riediker, M.; Duthaler, R. O.; *Angew. Chem., Int. Ed. Engl.* **1989**, *28* 494; (b) Oertle, K.; Beyeler, H.; Duthaler, R. O.; Lottenbach, W.; Riediker, M.; Steiner, E.; *Helv. Chim. Acta* **1990**, *73* 353; (c) Duthaler, R. O.; Herold, P.; Wyler-Helfer, S.; Riediker, M.; *Helv. Chim. Acta* **1990**, *73* 659; (d) Minamikawa, H.; Hayakawa, S.; Yamada, T.; Iwasawa, N.; Narasaka, K.; *Bull. Chem. Soc. Jpn.* **1988**, *61* 4379; (e) Hafner, A.; Duthaler, R. O.; Marti, R.; Rihs, G.; Rothe-Streit, P.; Schwarzenbach, F.; *J. Am. Chem. Soc.* **1992**, *114* 2321.

[90] Okuda, J.; *Chem. Ber.* **1990**, *123* 1649.

[91] Yang, Q.; Jensen, M. D.; *Synlett* **1996**, 563.

[92] Cozzi, P. G.; Carofiglio, T.; Floriani, C.; *Organometallics* **1993**, *12* 2845.

[93] (a) Adam, W.; Müller, M.; Prechtl, F.; *J. Org. Chem.* **1994**, *59* 2358; (b) Schmittel, M.; Werner, H.; Gevert, O.; Söllner, R.; *Chem. Ber./Recl.* **1997**, *130* 195.

[94] Reetz, M. T.; Kyung, S. H.; Westermann, J.; *Organometallics* **1984**, *3* 1716.

[95] (a) Chen, X.; Hortelano, E. R.; Eliel, E.; *J. Am. Chem. Soc.* **1990**, *112* 6130; (b) Chen, X.; Hortelano, E. R.; Eliel, E. L.; Frye, S. V.; *J. Am. Chem. Soc.* **1992**, *114* 1778.

[96] Kauffmann, T.; Möller, T.; Rennefeld, H.; Welke, S.; Wieschollek, R.; *Angew. Chem., Int. Ed. Engl.* **1985**, *24* 348.

[97] Das, G.; Thornton, E. R.; *J. Am. Chem. Soc.* **1990**, *112* 5360.

[98] Miki, Y.; Tada, Y.; Yanase, N.; Hachiken, H.; Matsushita, K.; *Tetrahedron Lett.* **1996**, *43* 7753.

[99] Roulet, D.; Capéros, J.; Jacot-Guillarmod, A.; *Helv. Chim. Acta* **1984**, *67* 1475.

[100] (a) Yamamoto, Y.; Komatsu, T.; Maruyama, K.; *J. Chem. Soc., Chem. Commun.* **1985**, 814; (b) Fujisawa, T.; Hayakawa, R.; Shimizu, M.; *Tetrahedron Lett.* **1992**, *33* 7903.

[101] (a) Nakamura, E.; Kuwajima, I.; *Tetrahedron Lett.* **1983**, *24* 3343; (b) Heimbach, H.; thesis, Universität Bonn **1980**.

[102] Keck, G. E.; Abbott, D. E.; Boden, E. P.; Enholm, E. J.; *Tetrahedron Lett.* **1984**, *25* 3927.

[103] Krämer, T.; Schwark, J. R.; Hoppe, D.; *Tetrahedron Lett.* **1989**, *30* 7037.

[104] (a) Baker, W. R.; *J. Org. Chem.* **1985**, *50* 3942; (b) Yamamoto, Y.; Yamada, J.; *J. Am. Chem. Soc.* **1987**, *109* 4395; (c) Yamada, J.; Abe, H.; Yamamoto, Y.; *J. Am. Chem. Soc.* **1990**, *112* 6118; (d) Furuta, T.; Yamamoto, Y.; *J. Org. Chem.* **1992**, *57* 2981; (e) Yamamoto, Y.; Yamada, J.; Asano, T.; *Tetrahedron* **1992**, *48* 5587.

[105] Reviews of TiCl₄-mediated aldol additions of enolsilanes: (a) Mukaiyama, T.; *Org. React.* **1982**, *28* 203; (b) Gennari, C.; in *Comprehensive Organic Synthesis*, (Trost, B. M.; Ed.) Vol. 2, Pergamon Press, Oxford, **1990**, 629; (c) review of enolate additions to imines: Hart, D. J.; Ha, D. C.; *Chem. Rev. (Washington, D.C.)* **1989**, *89* 1447; mechanism of chelation controlled Mukaiyama aldol addition: (d) Reetz, M. T.; Raguse, B.; Marth, C. F.; Hügel, H. M.; Bach, T.; Fox, D. N. A.; *Tetrahedron* **1992**, *48* 5731; stereoselective Mukaiyama-type aldol additions under catalytic conditions: (e) Reetz, M. T.; Fox, D. N. A.; *Tetrahedron Lett.* **1993**, *34* 1119, and references cited therein.

[106] Nakamura, E.; Kuwajima, I.; *J. Am. Chem. Soc.* **1983**, *105* 651.

[107] Nakamura, E.; Oshino, H.; Kuwajima, I.; *J. Am. Chem. Soc.* **1986**, *108* 3745.

[108] (a) Reissig, H. U.; Holzinger, H.; Glomsda, G.; *Tetrahedron* **1989**, *45* 3139; (b) Kano, S.; Yokomatsu, T.; Shibuya, S.; *Tetrahedron Lett.* **1991**, *32* 233; (c) Sato, T.; Watanabe, M.; Murayama, E.; *Tetrahedron Lett.* **1986**, *27* 1621.

[109] Crociani, B.; Nicolini, M.; Richards, R. L.; *J. Organomet. Chem.* **1975**, *101* C1.

[110] Seebach, D.; Adam, G.; Gees, T.; Schiess, M.; Weigand, W.; *Chem. Ber.* **1988**, *121* 507.

[111] (a) Lehnert, W.; *Tetrahedron* **1972**, *28* 663; (b) Lehnert, W.; *Tetrahedron* **1973**, *29* 635; (c) Lehnert, W.; *Tetrahedron* **1974**, *30* 301; (d) Mukaiyama, T.; *Pure Appl. Chem.* **1982**, *54* 2455.

[112] (a) Harrison, C. R.; *Tetrahedron Lett.* **1987**, *28* 4135; (b) Brocchini, S. J.; Eberle, M.; Lawton, R. G.; *J. Am. Chem. Soc.* **1988**, *110* 5211.

[113] (a) Evans, D. A.; Urpi, F.; Somers, T. C.; Clark, J. S.; Bilodeau, M. T.; *J. Am. Chem. Soc.* **1990**, *112* 8215; (b) Evans, D. A.; Bilodeau, M. T.; Somers, T. C.; Clardy, J.; Cherry, D.; Kato, Y.; *J. Org. Chem.* **1991**, *56* 5750.

[114] (a) Tanabe, Y.; Mukaiyama, T.; *Chem. Lett.* **1986**, 1813; (b) Silveira, C. C.; Perin, G.; Braga, A. L.; *Synth. Commun.* **1995**, *25* 117.

[115] Mahrwald, R.; Schick, H.; *Synthesis* **1990**, 592.

[116] Mahrwald, R.; *Chem. Ber.* **1995**, *128* 919.

[117] Vuitel, L.; Jacot-Guillarmod, A.; *Synthesis* **1972**, 608.

[118] (a) Martin, H. A.; Jellinek, F.; *J. Organomet. Chem.* **1968**, *12* 149; (b) Lehmkuhl, H.; Fustero, S.; *Liebigs Ann. Chem.* **1980**, 1353.

[119] (a) Urabe, H.; Yoshikawa, K.; Sato, F.; *Tetrahedron Lett.* **1995**, *36* 5595; (b) Szymoniak, J.; Pagneux, S.; Felix, D.; Moïse, C.; *Synlett* **1996**, 46.

[120] (a) Szymoniak, J.; Thery, N.; Moïse, C.; *Bull. Soc. Chim. Fr.* **1997**, *134* 85; (b) Szymoniak, J.; Lefranc, H.; Moïse, C.; *J. Org. Chem.* **1996**, *61* 3926.

[121] (a) Tebbe, F. N.; Parshall, G. W.; Reddy, G. S.; *J. Am. Chem. Soc.* **1978**, *100* 3611; (b) Brown-Wensley, K. A.; Buchwald, S. L.; Cannizzo, L.; Clawson, L.; Ho, S.; Meinhardt, D.; Stille, J. R.; Strauss, D.; Grubbs, R. H.; *Pure Appl. Chem.* **1983**, *55* 1733; (c) Meinhart, J. D.; Anslyn, E. V.; Grubbs, R. H.; *Organometallics* **1989**, *8* 583; (d) Pine, S. H.; Kim, G.; Lee, V.; *Org. Synth.* **1990**, *69* 72.

[122] (a) Horikawa, Y.; Watanabe, M.; Fujiwara, T.; Takeda, T.; *J. Am. Chem. Soc.* **1997**, *119* 1127; (b) Horikawa, Y.; Nomura, T.; Watanabe, M.; Fujiwara, T.; Takeda, T.; *J. Org. Chem.* **1997**, *62* 3678.

[123] (a) Kulinkovich, O. G.; Sviridov, S. V.; Vasilevski, D. A.; *Synthesis* **1991**, 234; (b) Kulinkovich, O. G.; Sorokin, V. L.; Kel'in, A. V.; *Russ. J. Org. Chem.* **1993**, *29* 55.

[124] (a) Lee, J.; Kang, C. H.; Kim, H.; Cha, J. K.; *J. Am. Chem. Soc.* **1996**, *118* 291; (b) Cho, S. Y.; Lee, J.; Lammi, R. K.; Cha, J. K.; *J. Org. Chem.* **1997**, *62* 8235; (c) Corey, E. J.; Rao, S. A.; Noe, M. C.; *J. Am. Chem. Soc.* **1994**, *116* 9345.

[125] (a) Chaplinski, V.; de Meijere, A.; *Angew. Chem., Int. Ed. Engl.* **1996**, *35* 413; (b) Lee, J.; Cha, J. K.; *J. Org. Chem.* **1997**, *62* 1584.

[126] Lee, J.; Ha, J. D.; Cha, J. K.; *J. Am. Chem. Soc.* **1997**, *119* 8127.

[127] (a) Kasatkin, A.; Nakagawa, T.; Okamoto, S.; Sato, F.; *J. Am. Chem. Soc.* **1995**, *117* 3881; (b) Hikichi, S.; Gao, Y.; Sato, F.; *Tetrahedron Lett.* **1997**, *38* 2867.

[128] (a) Okamoto, S.; Kasatkin, A.; Zubaidha, P. K.; Sato, F.; *J. Am. Chem. Soc.* **1996**, *118* 2208; (b) Yoshida, Y.; Okamoto, S.; Sato, F.; *J. Org. Chem.* **1996**, *61* 7826.

[129] Takahashi, Y.; Nishioka, N.; Endoh, F.; Ikeda, H.; Miyashi, T.; *Tetrahedron Lett.* **1996**, *37* 1841.

[130] (a) Urabe, H.; Hata, T.; Sato, F.; *Tetrahedron Lett.* **1995**, *36* 4261; (b) Urabe, H.; Sato, F.; *J. Org. Chem.* **1996**, *61* 6756; (c) Gao, Y.; Shirai, M.; Sato, F.; *Tetrahedron Lett.* **1996**, *37* 7787.

[131] Negishi, E.; in *Comprehensive Organic Synthesis*, (Trost, B. M.; Fleming, I.; Eds) Vol. 5, Pergamon Press, Oxford, **1991**, 1163.

[132] Nugent, W. A.; Calabrese, J. C.; *J. Am. Chem. Soc.* **1984**, *106* 6422.

[133] Nugent, W. A.; Taber, D. F.; *J. Am. Chem. Soc.* **1989**, *111* 6435.

[134] Hicks, F. A.; Berk, S. C.; Buchwald, S. L.; *J. Org. Chem.* **1996**, *61* 2713.

[135] (a) Hewlett, D. F.; Whitby, R. J.; *J. Chem. Soc., Chem. Commun.* **1990**, 1684; (b) Kablaoui, N. M.; Buchwald, S. L.; *J. Am. Chem. Soc.* **1996**, *118* 3182; (c) Crowe, W. E.; Vu, A. T.; *J. Am. Chem. Soc.* **1996**, *118* 1557; (d) Schobert, R.; Maaref, F.; Dürr, S.; *Synlett* **1995**, 83.

[136] Kablaoui, N. M.; Hicks, F. A.; Buchwald, S. L.; *J. Am. Chem. Soc.* **1997**, *119* 4424.

[137] (a) Reetz, M. T.; Hüllmann, M.; Massa, W.; Berger, S.; Rademacher, P.; *J. Am. Chem. Soc.* **1986**, *108* 2405; (b) Shambayati, S.; Crowe, W. E.; Schreiber, S. L.; *Angew. Chem., Int. Ed. Engl.* **1990**, *29* 256; (c) Birney, D. M.; Houk, K. N.; *J. Am. Chem. Soc.* **1990**, *112* 4127; (d) Branchadell, V.; Oliva, A.; *J. Am. Chem. Soc.* **1992**, *114* 4357; (e) Corcoran, R. C.; Ma, J.; *J. Am. Chem. Soc.* **1992**, *114* 4536.

[138] Cozzi, P. G.; Solari, E.; Floriani, C.; Chiesi-Villa, A.; Rizzoli, C.; *Chem. Ber.* **1996**, *129* 1361.

[139] Cozzi, P. G.; Floriani, C.; Chiesi-Villa, A.; Rizzoli, C.; *Organometallics* **1994**, *13* 2131.

[140] Hollis, T. K.; Bosnich, B.; *J. Am. Chem. Soc.* **1995**, *117* 4570.

[141] Kobayashi, S.; Matsui, S.; Mukaiyama, T.; *Chem. Lett.* **1988**, 1491.

[142] Kobayashi, S.; Hachiya, I.; *Tetrahedron Lett.* **1992**, *33* 1625.

[143] Hosomi, A.; *Acc. Chem. Res.* **1988**, *21* 200.

[144] Masse, C. E.; Panek, J. S.; *Chem. Rev. (Washington, D.C.)* **1995**, *95* 1293.

[145] Mechanistic study of allylsilane additions: Denmark, S. E.; Almstead, N. G.; *Tetrahedron* **1992**, *48* 5565.

[146] (a) Wilson, S. R.; Price, M. F.; *Tetrahedron Lett.* **1983**, *24* 569; (b) Denmark, S. E.; Weber, E. J.; *J. Am. Chem. Soc.* **1984**, *106* 7970.

[147] Ihara, M.; Suzuki, S.; Tokunaga, Y.; Fukumoto, K.; *J. Chem. Soc., Perkin Trans. I* **1995**, 2811.

[148] Kercher, T.; Livinghouse, T.; *J. Am. Chem. Soc.* **1996**, *118* 4200.

[149] Hwu, J. R.; Wetzel, J. M.; *J. Org. Chem.* **1985**, *50* 3946.

[150] Reetz, M. T.; Wenderoth, B.; Peter, R.; *J. Chem. Soc., Chem. Commun.* **1983**, 407.

[151] Okazoe, T.; Hibino, J.; Takai, K.; Nozaki, H.; *Tetrahedron Lett.* **1985**, *26* 5581.

[152] Weingarten, H.; White, W. A.; *J. Org. Chem.* **1966**, *31* 4041.

[153] Kauffmann, T.; Abel, T.; Schreer, M.; *Angew. Chem., Int. Ed. Engl.* **1988**, *27* 944.

[154] (a) Cram, D. J.; Abd Elhafez, F. A.; *J. Am. Chem. Soc.* **1952**, *74* 5828; (b) Morrison, J. D.; Mosher, H. S.; *Asymmetric Organic Reactions*, Prentice-Hall, Englewood Cliffs, NJ, **1971**.

[155] Anh, N. T.; *Top. Curr. Chem.* **1980**, *88* 145.

[156] Review of homo-aldol additions: Hoppe, D.; *Angew. Chem., Int. Ed. Engl.* **1984**, *23* 932.

[157] Piatak, D. M.; Wicha, J.; *Chem. Rev. (Washington, D.C.)* **1978**, *78* 199.

[158] Yamamoto, Y.; Nishii, S.; Maruyama, K.; *J. Chem. Soc., Chem. Commun.* **1986**, 102.

[159] (a) Eliel, E. L.; in *Asymmetric Synthesis*, (Morrison, J. D.;Ed.) Vol. 2, Academic Press, New York, **1983**; (b) Still, W. C.; Schneider, J. A.; *Tetrahedron Lett.* **1980**, *21* 1035.

[160] Still, W. C.; McDonald, J. H.; *Tetrahedron Lett.* **1980**, *21* 1031.

[161] (a) Reetz, M. T.; Keßeler, K.; Schmidtberger, S.; Wenderoth, B.; Steinbach, R.; *Angew. Chem., Int. Ed. Engl.* **1983**, *22* 989; *Angew. Chem.* Suppl. **1983**, 1511; (b) Reetz, M. T.; Harms, K.; Reif, W.; *Tetrahedron Lett.* **1988**, *29* 5881.

[162] (a) CH_3TiCl_3 chelate of a chiral α-alkoxy ketone: Reetz, M. T.; Hüllmann, M.; Seitz, T.; *Angew. Chem., Int. Ed. Engl.* **1987**, *26* 477; (b) CH_3TiCl_3 chelate of aldehyde 115 (R = CH_3): Reetz, M. T.; Raguse, B.; Seitz, T.; *Tetrahedron* **1993**, *49* 8561.

[163] Fujii, H.; Taniguchi, M.; Oshima, K.; Utimoto, K.; *Tetrahedron Lett.* **1992**, *33*, 4579.

[164] Reetz, M. T.; Keßeler, K.; Jung, A.; *Angew. Chem., Int. Ed. Engl.* **1985**, *24* 989.

[165] (a) Hanessian, S.; Tehim, A.; Chen, P.; *J. Org. Chem.* **1993**, *58* 7768; (b) Franck-Neumann, M.; Bissinger, P.; Geoffrey, P.; *Tetrahedron Lett.* **1997**, *38* 4473; (c) Fujisawa, T.; Kooriyama, Y.; Shimizu, M.; *Tetrahedron Lett.* **1996**, *37* 3881; (d) Oikawa, H.; Oikawa, M.; Ueno, T.; Ichihara, A.; *Tetrahedron Lett.* **1994**, *35* 4809.

[166] Miura, K.; Takasumi, M.; Hondo, T.; Saito, H.; Hosomi, A.; *Tetrahedron Lett.* **1997**, *38* 4587.

[167] (a) Reetz, M. T.; Keßeler, K.; Jung, A.; *Tetrahedron* **1984**, *40* 4327; (b) Reetz, M. T.; Keßeler, K.; Jung, A.; *Tetrahedron Lett.* **1984**, *25* 729; (c) Keck, G. E.; Abbott, D. E.; *Tetrahedron Lett.* **1984**, *25* 1883.

[168] (a) Reetz, M. T.; in *Selectivities in Lewis Acid Promoted Reactions*, (Schinzer, D.; Ed.), NATO ASI Series C, Vol. 289, Kluwer, Dordrecht, **1989**, 107; (b) Reetz, M. T.; *Pure Appl. Chem.* **1985**, *57* 1781; (c) Yong, T.-M.; Holmes, A. B.; Taylor, P. L.; Robinson, J. N.; Segal, J. A.; *Chem. Commun. (Cambridge)* **1996**, 863.

[169] (a) Kiyooka, S.; Heathcock, C. H.; *Tetrahedron Lett.* **1983**, *24* 4765; (b) concerning cautionary remarks on this paper, see ref. [167b].

[170] Reetz, M. T.; Jung, A.; *J. Am. Chem. Soc.* **1983**, *105* 4833.

[171] Reetz, M. T.; Keßeler, K.; *J. Org. Chem.* **1985**, *50* 5434.

[172] Keck, G. E.; Castellino, S.; *J. Am. Chem. Soc.* **1986**, *108* 3847.

[173] (a) Reetz, M. T.; Jung, A.; Bolm, C.; *Tetrahedron* **1988**, *44* 3889; (b) Molander, G. A.; Andrews, S. W.; *Tetrahedron* **1988**, *44* 3869.

[174] Danishefsky, S. J.; Pearson, W. H.; Harvey, D. F.; Maring, C. J.; Springer, J. P.; *J. Am. Chem. Soc.* **1985**, *107* 1256.

[175] Uenishi, J.; Tomozane, H.; Yamato, M.; *Tetrahedron Lett.* **1985**, *26* 3467.

[176] (a) Gennari, C.; Bernardi, A.; Poli, G.; Scolastico, C.; *Tetrahedron Lett.* **1985**, *26* 2373; (b) Shirai, F.; Nakai, T.; *Chem. Lett.* **1989**, 445; (c) Angert, H.; Kunz, T.; Reissig, H.-U.; *Tetrahedron* **1992**, *48* 5681.

[177] Bernardi, A.; Cardani, S.; Gennari, C.; Poli, G.; Scolastico, C.; *Tetrahedron Lett.* **1985**, *26* 6509.

[178] Danheiser, R. L.; Carini, D. J.; Fink, D. M.; Basak, A.; *Tetrahedron* **1983**, *39* 935.

[179] Jain, N. F.; Takenaka, N.; Panek, J. S.; *J. Am. Chem. Soc.* **1996**, *118* 12 475.

[180] Angert, H.; Czerwonka, R.; Reißig H.-U.;   *Liebigs Ann.* **1996**, 259.

[181] Bartoli, G.; Bosco, M.; Sambri, L.; Marcantoni, E.; *Tetrahedron Lett.* **1997**, *38* 3785.

[182] Reetz, M. T.; Drewes, M. W.; Schmitz, A.; *Angew. Chem., Int. Ed. Engl.* **1987**, *26* 1141.

[183] (a) Vara Prasad, J. V. N.; Rich, D. H.; *Tetrahedron Lett.* **1990**, *31* 1803; (b) Takemoto, Y.; Matsumoto, T.; Ito, Y.; Terashima, S.; *Tetrahedron Lett.* **1990**, *31* 217; (c) D'Aniello, F.; Falorni, M.; Mann, A.; Taddei, M.; *Tetrahedron: Asymmetry* **1996**, *7* 1217.

[184] Mikami, K.; Kaneko, M.; Loh, T. P.; Terada, M.; Nakai, T.; *Tetrahedron Lett.* **1990**, *31* 3909.

[185] Reetz, M. T.; Rölfing, K.; Griebenow N.; *Tetrahedron Lett.* **1994**, *35* 1969.

[186] (a) Annunziata, R.; Cinquini, M.; Cozzi, F.; Cozzi, P. G.; *Tetrahedron Lett.* **1992**, *33* 1113; (b) Annunziata, R.; Cinquini, M.; Cozzi, F.; Cozzi, P. G.; Consolandi, E.; *Tetrahedron* **1991**, *47* 7897; (c) Cinquini, M.; Cozzi, F.; Cozzi, P. G.; Consolandi, E.; *Tetrahedron* **1991**, *47* 8767; (d) Annunziata, R.; Cinquini, M.; Cozzi, F.; Cozzi, P. G.; *J. Org. Chem.* **1992**, *57* 4155.

[187] Wang, S.; Howe, G. P.; Mahal, R. S.; Procter, G.; *Tetrahedron Lett.* **1992**, *33* 3351.

[188] (a) Maier, G.; Seipp, U.; Kalinowski, H.-O.; Henrich, M.; *Chem. Ber.* **1994**, *127* 1427; (b) Bartoli, G.; Bosco, M.; Sambri, L.; Marcantoni, E.; *Tetrahedron Lett.* **1996**, *37* 7421; (c) Sarko, C. R.; Collibee, S. E.; Knorr, A. L.; DiMare, M.; *J. Org. Chem.* **1996**, *61* 868.

[189] Ravikumar, K. S.; Chandrasekaran, S.; *Tetrahedron* **1996**, *52* 9137.

[190] Reetz, M. T.; Keßeler, K.; *J. Chem. Soc., Chem. Commun.* **1984**, 1079.

[191] Martin, S. F.; Li, W.; *J. Org. Chem.* **1989**, *54* 6129.

[192] For a definition of '*protective group tuning*', see: Reetz, M. T.; Binder, J.; *Tetrahedron Lett.* **1989**, *30* 5425.

[193] (a) Overman, L. E.; McCready, R. J.; *Tetrahedron Lett.* **1982**, *23* 2355; (b) Nakata, T.; Tanaka, T.; Oishi, T.; *Tetrahedron Lett.* **1983**, *24* 2653.

[194] (a) Reetz, M. T.; *Pure Appl. Chem.* **1988**, *60* 1607; (b) review of reactions of N,N-dibenzylaminoaldehydes, -aldimines and α,β-unsaturated esters: Reetz, M. T.; *Angew. Chem., Int. Ed. Engl.* **1991**, *30* 1531.

[195] Masamune, S.; Choy, W.; Petersen, J. S.; Sita, L. R.; *Angew. Chem., Int. Ed. Engl.* **1985**, *24* 1.

[196] Reetz, M. T.; Rivadeneira, E.; Niemeyer, C.; *Tetrahedron Lett.* **1990**, *31* 3863.

[197] Grauert, M.; Schöllkopf, U.; *Liebigs Ann. Chem.* **1985**, 1817.

[198] Hoppe, D.; Krämer, T.; Schwark, J. R.; Zschage, O.; *Pure Appl. Chem.* **1990**, *62* 1999.

[199] Duthaler, R. O.; Hafner, A.; Riediker, M.; *Pure Appl. Chem.* **1990**, *62* 631.

[200] Mikami, K.; Matsukawa, S.; Sawa, E.; Harada, A.; Koga, N.; *Tetrahedron Lett.* **1997**, *38* 1951.

[201] (a) Bloch, R.; Gilbert, L.; *Tetrahedron Lett.* **1987**, *28* 423; (b) Tomooka, K.; Matsuzawa, K.; Suzuki, K.; Tsuchihashi, G.; *Tetrahedron Lett.* **1987**, *28* 6339; (c) Reetz, M. T.; Schmitz, A.; Holdgrün, X.; *Tetrahedron Lett.* **1989**, *30* 5421.

[202] (a) Sato, T.; Kato, R.; Gokyu, K.; Fujisawa, T.; *Tetrahedron Lett.* **1988**, *29* 3955; (b) Fujisawa, T.; Ukaji, Y.; Funabora, M.; Yamashita, M.; Sato, T.; *Bull. Chem. Soc. Jpn.* **1990**, *63* 1894.

[203] Fujisawa, T.; Fujimura, A.; Ukaji, Y.; *Chem. Lett.* **1988**, 1541.

[204] (a) Casiraghi, G.; Cornia, M.; Casnati, G.; Gasparri Fava, G.; Ferrari Belicchi, M.; Zetta, L.; *J. Chem. Soc., Chem. Commun.* **1987**, 794; (b) Bigi, F.; Casnati, G.; Sartori, G.; Araldi, G.; Bocelli, G.; *Tetrahedron Lett.* **1989**, *30* 1121; (c) Casiraghi, G.; Cornia, M.; Rassu, G.; Del Sante, C.; Spanu, P.; *Tetrahedron* **1992**, *48* 5619.

[205] (a) Hoffmann, R. W.; *Angew. Chem., Int. Ed. Engl.* **1982**, *21* 555; (b) Schlosser, M.; *Pure Appl. Chem.* **1988**, *60* 1627.

[206] Seebach, D.; Widler, L.; *Helv. Chim. Acta* **1982**, *65* 1972.

[207] Sato, F.; Iida, K.; Iijima, S.; Moriya, H.; Sato, M.; *J. Chem. Soc., Chem. Commun.* **1981**, 1140.

[208] Kobayashi, Y.; Umeyama, K.; Sato, F.; *J. Chem. Soc., Chem. Commun.* **1984**, 621.

[209] Reetz, M. T.; Sauerwald, M.; *J. Org. Chem.* **1984**, *49* 2292.

[210] Cohen, T.; Guo, B. S.; *Tetrahedron* **1986**, *42* 2803.

[211] Tsai, D. J. S.; Matteson, D. S.; *Tetrahedron Lett.* **1981**, *22* 2751.

[212] Sato, F.; Suzuki, Y.; Sato, M.; *Tetrahedron Lett.* **1982**, *23* 4589.

[213] (a) Ukai, J.; Ikeda, Y.; Ikeda, N.; Yamamoto, H.; *Tetrahedron Lett.* **1984**, *25* 5173; (b) Ikeda, Y.; Ukai, J.; Ikeda, N.; Yamamoto, H.; *Tetrahedron Lett.* **1984**, *25* 5177.

[214] (a) Murai, A.; Abiko, A.; Shimada, N.; Masamune, T.; *Tetrahedron Lett.* **1984**, *25* 4951; (b) Murai, A.; Abiko, A.; Masamune, T.; *Tetrahedron Lett.* **1984**, *25* 4955.

[215] (a) van Hülsen, E.; Hoppe, D.; *Tetrahedron Lett.* **1985**, *26* 411; (b) Hanko, R.; Hoppe, D.; *Angew. Chem., Int. Ed. Engl.* **1982**, *21* 372; *Angew. Chem. Suppl.* **1982**, 961; (c) Hoppe, D.; Lichtenberg, F.; *Angew. Chem., Int. Ed. Engl.* **1982**, *21* 372.

[216] Krämer, T.; Hoppe, D.; *Tetrahedron Lett.* **1987**, *28* 5149.

[217] (a) Ochiai, H.; Nishihara, T.; Tamaru, Y.; Yoshida, Z.; *J. Org. Chem.* **1988**, *53* 1343; (b) Sakami, S.; Houkawa, T.; Asaoka, M.; Takei, H.; *J. Chem. Soc., Perkin Trans. I* **1995**, 285.

[218] (a) Furuta, K.; Ishiguro, M.; Haruta, R.; Ikeda, N.; Yamamoto, H.; *Bull. Chem. Soc. Jpn.* **1984**, *57* 2768; (b) Hiraoka, H.; Furuta, K.; Ikeda, N.; Yamamoto, H.; *Bull. Chem. Soc. Jpn.* **1984**, *57* 2777; (c) Yamamoto, Y.; Ito, W.; Maruyama, K.; *J. Chem. Soc., Chem. Commun.* **1984**, 1004.

[219] Hoffmann, R. W.; Lanz, J.; Metternich, R.; Tarara, G.; Hoppe, D.; *Angew. Chem., Int. Ed. Engl.* **1987**, *26* 1145.

[220] (a) Evans, D. A.; Nelson, J. V.; Taber, T. R.; *Top. Stereochem.* **1982**, *13* 1; (b) Heathcock, C. H.; in *Asymmetric Synthesis*, (Morrison, J. D.; Ed.), Vol. 3, Academic Press, Orlando, **1984**, 111.

[221] (a) Harada, T.; Mukaiyama, T.; *Chem. Lett.* **1982**, 467; (b) Fujisawa, T.; Ukaji, Y.; Noro, T.; Date, K.; Shimizu, M.; *Tetrahedron* **1992**, *48* 5629.

[222] (a) Kazmaier, U.; *Angew. Chem., Int. Ed. Engl.* **1994**, *33* 998; (b) Kazmaier, U.; Grandel, R.; *Synlett* **1995**, 945.

[223] (a) Banno, K.; Mukaiyama, T.; *Chem. Lett.* **1976**, 279; (b) Heathcock, C. H.; Davidsen, S. K.; Hug, K. T.; Flippin, L. A.; *J. Org. Chem.* **1986**, *51* 3027.

[224] (a) Gennari, C.; Molinari, F.; Cozzi, P.; Oliva, A.; *Tetrahedron Lett.* **1989**, *30* 5163; (b) Chan, T. H.; Aida, T.; Lau, P. W. K.; Gorys, V.; Harpp, D. N.; *Tetrahedron Lett.* **1979**, *20* 4029; (c) Matsuda, I.; Izumi, Y.; *Tetrahedron Lett.* **1981**, *22* 1805.

[225] Mukai, C.; Kataoka, O.; Hanaoka, M.; *J. Org. Chem.* **1993**, *58* 2946.

[226] Suh, K.-H.; Choo, D.-J.; *Tetrahedron Lett.* **1995**, *36* 6109.

[227] Gennari, C.; Beretta, M. G.; Bernardi, A.; Moro, G.; Scolastico, C.; Todeschini, R.; *Tetrahedron* **1986**, *42* 893.

[228] (a) Evans, D. A.; Rieger, D. L.; Bilodeau, M. T.; Urpi, F.; *J. Am. Chem. Soc.* **1991**, *113* 1047; (b) Xiang, Y.; Olivier, E.; Quimet, N.; *Tetrahedron Lett.* **1992**, *33* 457.

[229] Luke, G. P.; Morris, J.; *J. Org. Chem.* **1995**, *60* 3013.

[230] Mahrwald, R.; *Chem. Ber.* **1995**, *128* 919.

[231] Mahrwald, R.; Gündogan, B.; *J. Am. Chem. Soc.* **1998**, *120* 413.

[232] Mahrwald, R.; *Tetrahedron* **1995**, *51* 9015.

[233] Mahrwald, R.; Costisella, B.; *Synthesis* **1996**, 1087.

[234] Reetz, M. T.; von Itzstein, M.; *J. Organomet. Chem.* **1987**, *334* 85.

[235] Reetz, M. T.; Peter, R.; von Itzstein, M.; *Chem. Ber.* **1987**, *120* 121.

[236] Reetz, M. T.; Röhrig, D.; *Angew. Chem., Int. Ed. Engl.* **1989**, *28* 1706.

[237] (a) d'Angelo, J.; Pecquet-Dumas, F.; *Tetrahedron Lett.* **1983**, *24* 1403; (b) Hoppe, D.; in *Enzymes as Catalysts in Organic Synthesis*, (Schneider, M. P.; Ed.), Reidel, Dordrecht, **1986**, 177.

[238] Masamune, S.; Hirama, M.; Mori, S.; Ali, S. A.; Garvey, D. S.; *J. Am. Chem. Soc.* **1981**, *103* 1568.

[239] (a) Nerz-Stormes, M.; Thornton, E. R.; *Tetrahedron Lett.* **1986**, *27* 897; (b) camphor-derived Ti-enolate: Bonner, M. P.; Thornton, E. R.; *J. Am. Chem. Soc.* **1991**, *113* 1299.

[240] Yan, T.-H.; Lee, H.-C.; Tan, C.-W.; *Tetrahedron Lett.* **1993**, *34* 3559.

[241] Luke, G. P.; Morris, J.; *J. Org.Chem.* **1995**, *60* 3013.

[242] Crimmins, M. T.; King, B. W.; Tabet, E. A.; *J. Am. Chem. Soc.* **1997**, *119* 7883.

[243] Ghosh, A. K.; Onishi, M.; *J. Am. Chem. Soc.* **1996**, *118* 2527.

[244] (a) Helmchen, G.; Leikauf, U.; Taufer-Knöpfel, I.; *Angew. Chem., Int. Ed. Engl.* **1985**, *24* 874 (erratum 4057); (b) Oppolzer, W.; *Tetrahedron* **1987**, *43* 1969; (c) review of chiral 'acetate synthons':

Braun, M.; *Angew. Chem., Int. Ed. Engl.* **1987**, *26* 24; (d) Oppolzer, W.; Starkemann, C.; *Tetrahedron Lett.* **1992**, *33* 2439; (e) Gennari, C.; Bernardi, A.; Colombo, L.; Scolastico, C.; *J. Am. Chem. Soc.* **1985**, *107* 5812; (f) Gennari, C.; Colombo, L.; Bertolini, G.; Schimperna, G.; *J. Org. Chem.* **1987**, *52* 2754; (g) Cardani, S.; De Toma, C.; Gennari, C.; Scolastico, C.; *Tetrahedron* **1992**, *48* 5557.

[245] Gennari, C.; Schimperna, G.; Venturini, I.; *Tetrahedron* **1988**, *44* 4221.

[246] Abrahams, I.; Motevalli, M.; Robinson, A. J.; Wyatt, P. B.; *Tetrahedron* **1994**, *50* 12755.

[247] Shimizu, M.; Kume, K.; Fujisawa, T.; *Tetrahedron Lett.* **1995**, *36* 5227.

[248] (a) Annunziata, R.; Benagli, M.; Cinquini, M.; Cozzi, F.; Raimondi, L.; *Tetrahedron Lett.* **1993**, *34* 6921; (b) Annuziata, R.; Benaglia, M.; Chiovato, A.; Cinquini, M.; Cozzi, F.; *Tetrahedron* **1995**, *51* 10025; (c) Shimizu, M.; Ishida, T.; Fujisawa, T.; *Chem. Lett.* **1994**, 1403.

[249] Schöllkopf, U.; *Pure Appl. Chem.* **1983**, *55* 1799.

[250] (a) Schöllkopf, U.; Nozulak, J.; Grauert, M.; *Synthesis* **1985**, 55; (b) Bold, G.; Allmendinger, T.; Herold, P.; Moesch, L.; Schär, J. P.; Duthaler, R. O.; *Helv. Chim. Acta* **1991**, *75* 865.

[251] (a) Schöllkopf, U.; Beulshausen, T.; *Liebigs Ann. Chem.* **1989**, 223; (b) Beulshausen, T.; Groth, U.; Schöllkopf, U.; *Liebigs Ann. Chem.* **1991**, 1207.

[252] Roder, H.; Helmchen, G.; Peters, E. M.; Peters, K.; von Schnering, H. G.; *Angew. Chem., Int. Ed. Engl.* **1984**, *23* 898.

[253] Krämer, T.; Hoppe, D.; *Tetrahedron Lett.* **1987**, *28* 5149.

[254] Hoppe, D.; Zschage, O.; in *Organic Synthesis via Organometallics*, (Dötz, K. H.; Hoffmann, R. W.; Eds), Vieweg, Braunschweig, **1991**, 267.

[255] Paulsen, H.; Graeve, C.; Hoppe, D.; *Synthesis* **1996**, 141.

[256] (a) Hoppe, D.; Zschage, O.; *Angew. Chem., Int. Ed. Engl.* **1989**, *28* 65; (b) Zschage, O.; Hoppe, D.; *Tetrahedron* **1992**, *48* 5657; (c) Paulsen, H.; Hoppe, D.; *Tetrahedron* **1992**, *48* 5667.

[257] Hoppe, D.; Krämer, T.; Freire Erdbrügger, C.; Egert, E.; *Tetrahedron Lett.* **1989**, *30* 1233, and references cited therein.

[258] (a) Seebach, D.; Beck, A. K.; Schiess, M.; Widler, L.; Wonnacott, A.; *Pure Appl. Chem.* **1983**, *55* 1807; (b) Seebach, D.; Adam, G.; Gees, T.; Schiess, M.; Weigand, W.; *Chem. Ber.* **1988**, *121* 507; (c) Wang, J. T.; Fan, X.; Feng, X.; Qian, Y. M.; *Synthesis* **1989**, 291.

[259] Reetz, M. T.; Kükenhöhner, T.; Weinig, P.; *Tetrahedron Lett.* **1986**, *27* 5711.

[260] (a) Braun, M.; *Angew. Chem., Int. Ed. Engl.* **1996**, *35* 519; (b) Beck, A. K.; Dobler, M.; Plattner, A.; *Helv. Chim. Acta* **1997**, *80* 2073.

[261] Weber, B.; Seebach, D.; *Tetrahedron* **1994**, *50* 6117.

[262] Duthaler, R. O.; Hafner, A.; Riediker, M.; *Chem. Rev. (Washington, D.C.)* **1992**, *92* 807.

[263] Kuntz, B. A.; Ramachandran, R.; Taylor, N. J.; Guan, J.; Collins, S.; *J. Organomet. Chem.* **1995**, *497* 133.

[264] Halterman, R. L.; *Chem. Rev. (Washington, D.C.)* **1992**, *92* 965.

[265] (a) Hosomi, A.; Sakurai, H.; *J. Am. Chem. Soc.* **1977**, *99* 1673; (b) Sakurai, H.; Hosomi, A.; Hayashi, J.; *Org. Synth.* **1984**, *62* 86; (c) Tokoroyama, T.; Pan, L.-R.; *Tetrahedron Lett.* **1989**, *30* 197.

[266] Narasaka, K.; Soai, K.; Aikawa, Y.; Mukaiyama, T.; *Bull. Chem. Soc. Jpn.* **1976**, *49* 779.

[267] Narasaka, K.; *Org. Synth.* **1987**, *65* 12.

[268] Majetich, G.; Casares, A.; Chapman, D.; Behnke, M.; *J. Org. Chem.* **1986**, *51* 1745.

[269] Heathcock, C. H.; Kiyooka, S.; Blumenkopf, T. A.; *J. Org. Chem.* **1984**, *49* 4214; correction: *J. Org. Chem.* **1986**, *51* 3252.

[270] Pan, L.-R.; Tokoroyama, T.; *Tetrahedron Lett.* **1992**, *33* 1469.

[271] Wu, M. J.; Wu, C.-C.; Lee, P.-C.; *Tetrahedron Lett.* **1992**, *33* 2547.

[272] El-Abed, D.; Jellal, A.; Santelli, M.; *Tetrahedron Lett.* **1984**, *25* 1463.

[273] Nickisch, K.; Laurent, H.; *Tetrahedron Lett.* **1988**, *29* 1533.

[274] Sato, M.; Aoyagi, S.; Yago, S.; Kibayashi, C.; *Tetrahedron Lett.* **1996**, *37* 9063.

[275] (a) Review of intramolecular conjugate additions of allylsilanes: Schinzer, D.; *Synthesis* **1988**, 263; (b) Majetich, G.; Behnke, M.; Hull, K.; *J. Org. Chem.* **1985**, *50* 3615; (c) Schinzer, D.;

Allagiannis, C.; Wichmann, S.; *Tetrahedron* **1988**, *44* 3851; (d) Majetich, G.; Defauw, J.; *Tetrahedron* **1988**, *44* 3833.

[276] (a) Danheiser, R. L.; Carini, D. J.; Fink, D. M.; Basak, A.; *Tetrahedron* **1983**, *39* 935; (b) Danheiser, R. L.; Fink, D. M.; Tsai, Y. M.; *Org. Synth.* **1987**, *66* 8.

[277] Haruta, J.; Nishi, K.; Matsuda, S.; Tamura, Y.; Kita, Y.; *J. Chem. Soc., Chem. Commun.* **1989**, 1065.

[278] (a) Knölker, H.-J.; Baum, G.; Graf, R.; *Angew. Chem., Int. Ed. Engl.* **1994**, *33* 1612; (b) Knölker, H.-J.; Foitzik, N.; Goesmann, H.; Graf, R.; Jones, P. G.; Wanzl, G.; *Chem.-Eur. J.* **1997**, *3* 538.

[279] (a) Knölker, H.-J.; Foitzik, N.; Graf, R.; Pannek, J.-B.; *Tetrahedron* **1993**, *49* 9955; (b) Majetich, G.; Sing, J.-S.; Ringold, C.; Nemeth, G. A.; Newton, M. G.; *J. Org. Chem.* **1991**, *56* 3973; (c) Lee, Y.-G.; Ishimaru, K.; Iwasaki, H.; Ohkata, K.; Akiba, K.; *J. Org. Chem.* **1991**, *56* 2058; (d) Snider, B. B.; Zhang, Q.; *J. Org. Chem.* **1991**, *56* 4908; (e) Ipaktschi, J.; Heydari, A.; *Angew. Chem., Int. Ed. Engl.* **1992**, *31* 313; (f) Danheiser, R. L.; Takahashi, T.; Bertók, B.; Dixon, B. R.; *Tetrahedron Lett.* **1993**, *34* 3845; (g) Wu, M.- J.; Yeh, J.-Y.; *Tetrahedron* **1994**, *50* 1073.

[280] Brengel, G. P.; Meyers, A. I.; *J. Org. Chem.* **1996**, *61* 3230.

[281] Akiyama, T.; Kirino, M.; *Chem. Lett.* **1995**, 723.

[282] (a) Narasaka, K.; Soai, K.; Aikawa, Y.; Mukaiyama, T.; *Bull. Chem. Soc. Jpn.* **1976**, *49* 779; (b) Mukaiyama, T.; Hara, R.; *Chem. Lett.* **1989**, 1171.

[283] Bernardi, A.; Karamfilova, K.; Sanguinette, S.; Scolastico, C.; *Tetrahedron* **1997**, *53* 13 009.

[284] Sato, T.; Wakahara, Y.; Otera, J.; Nozaki, H.; Fukuzumi, S.; *J. Am. Chem. Soc.* **1991**, *113* 4028.

[285] Heathcock, C. H.; Norman, M. H.; Uehling, D. E.; *J. Am. Chem. Soc.* **1985**, *107* 2797.

[286] Fujita, Y.; Otera, J.; Fukuzumi, S.; *Tetrahedron* **1996**, *52* 9419.

[287] Heathcock, C. H.; Uehling, D. E.; *J. Org. Chem.* **1986**, *51* 279.

[288] Gennari, C.; Colombo, L.; Bertolini, G.; Schimperna, G.; *J. Org. Chem.* **1987**, *52* 2754.

[289] Takayanagi, M.; Tanino, K.; Kuwajima, I.; *Synlett* **1995**, 173.

[290] Miyashita, M.; Yanami, T.; Kumazawa, T.; Yoshikoshi, A.; *J. Am. Chem. Soc.* **1984**, *106* 2149.

[291] Arai, M.; Lipshutz, B. H.; Nakamura, E.; *Tetrahedron* **1992**, *48* 5709.

[292] Kabbara, J.; Flemming, S.; Nickisch, K.; Neh, H.; Westermann, J.; *Liebigs Ann.* **1995**, 401.

[293] Goasdoue, C.; Goasdoue, N.; Gaudemar, M.; *Tetrahedron Lett.* **1984**, *25* 537.

[294] (a) Bernardi, A.; Dotti, P.; Poli, G.; Scolastico, C.; *Tetrahedron* **1992**, *48* 5597; (b) Bernadi, A.; Marchionni, C.; Novo, B.; Karamfilova, K.; Potenza, D.; Scolastico, C.; Roversi, P.; *Tetrahedron* **1996**, *52* 3497; (c) Bernardi, A.; Capelli, A. M.; Cassinari, A.; Comotti, A.; Gennari, C.; Scolastico. C.; *J. Org. Chem.* **1997**, *57* 7029.

[295] Comins, D. L.; Brown, J. D.; *Tetrahedron Lett.* **1984**, *25* 3297.

[296] Posner, G. H.; Frye, L. L.; Hulce, M.; *Tetrahedron* **1984**, *40* 1401.

[297] (a) Schöllkopf, U.; Kühnle, W.; Egert, E.; Dyrbusch, M.; *Angew. Chem., Int. Ed. Engl.* **1987**, *26* 480; (b) Busch, K.; Groth, U. M.; Kühnle, W.; Schöllkopf, U.; *Tetrahedron* **1992**, *48* 5607; (c) Schöllkopf, U.; Pettig, D.; Busse, U.; *Synthesis* **1986**, 737.

[298] Hosomi, A.; Yanagi, T.; Hojo, M.; *Tetrahedron Lett.* **1991**, *32* 2371.

[299] Barr, D. A.; Grigg, R.; Sridharan, V.; *Tetrahedron Lett.* **1989**, *30* 4727.

[300] (a) Maercker, A.; *Org. React.* **1965**, *14* 270; (b) Schlosser, M.; Schaub, B.; de Oliveira-Neto, J.; Jeganathan, S.; *Chimia* **1986**, *40* 244.

[301] (a) Takai, K.; Hotta, Y.; Oshima, K.; Nozaki, H.; *Bull. Chem. Soc. Jpn.* **1980**, *53* 1698; (b) Hibino, J.; Okazoe, T.; Takai, K.; Nozaki, H.; *Tetrahedron Lett.* **1985**, *26* 5579.

[302] Minicione, E.; Pearson, A. J.; Bovicelli, P.; Chandler, M.; Heywood, G. C.; *Tetrahedron Lett.* **1981**, *22* 2929.

[303] Lombardo, L.; *Tetrahedron Lett.* **1982**, *23* 4293.

[304] (a) Lombardo, L.; *Org. Synth.* **1987**, *65* 81; (b) Fieser, L.F.; Fieser, M.; *Reagents for organic synthesis*, Vol. 1, Wiley, New York, **1967**, 1276.

[305] (a) Ogawa, Y.; Shibasaki, M.; *Tetrahedron Lett.* **1984**, *25* 1067; (b) Snowden, R. L.; Sonnay, P.; Ohloff, G.; *Helv. Chim. Acta* **1981**, *64* 25; (c) Kramer, A.; Pfander, H.; *Helv. Chim. Acta* **1982**, *65* 293.

[306] Okazoe, T.; Takai, K.; Oshima, K.; Utimoto, K.; *J. Org. Chem.* **1987**, *52* 4410.

[307] Takai, K.; Kataoka, Y.; Okazoe, T.; Utimoto, K.; *Tetrahedron Lett.* **1988**, *29* 1065.

[308] Takai, K.; Fujimura, O.; Kataoka, Y.; Utimoto, K.; *Tetrahedron Lett.* **1989**, *30* 211.

[309] Okazoe, T.; Takai, K.; Utimoto, K.; *J. Am. Chem. Soc.* **1987**, *109* 951.

[310] (a) Petasis, N. A.; Bzowej, E. I.; *J. Am. Chem. Soc.* **1990**, *112* 6392; (b) Petasis, N. A.; Hu, Y.-H.; Fu, D.-K.; *Tetrahedron Lett.* **1995**, *36* 6001; (c) Hughes, D. L.; Payack, J. F.; Cai, D.; Verhoeven, T. R.; Reider, P. J.; *Organometallics* **1996**, *15* 663.

[311] DeShong, P.; Rybczynski, P. J.; *J. Org. Chem.* **1991**, *56* 3207.

[312] (a) Ireland, R. E.; Varney, M. D.; *J. Org. Chem.* **1983**, *48* 1829; (b) Wilcox, C. S.; Long, G. W.; Suh, H.; *Tetrahedron Lett.* **1984**, *25* 395; (c) Kinney, W. A.; Coghlan, M. J.; Paquette, L. A.; *J. Am. Chem. Soc.* **1984**, *106* 6868.

[313] Petasis, N. A.; Patane, M. A.; *Tetrahedron Lett.* **1990**, *31* 6799.

[314] Horikawa, Y.; Watanabe, M.; Fujiwara, T.; Takeda, T.; *J. Am. Chem. Soc.* **1997**, *119* 1127.

[315] (a) Eisch, J. J.; Piotrowski, A.; *Tetrahedron Lett.* **1983**, *24* 2043; (b) Seetz, W. F. L.; Schat, G.; Akkerman, O. S.; Bickelhaupt, F.; *Angew. Chem., Int. Ed. Engl.* **1983**, *22* 248; *Angew. Chem. Suppl.* **1983**, 234.

[316] Reviews: (a) Breit, B.; *Angew. Chem., Int. Ed. Engl.* **1998**, *37* 535; (b) Pine, S. H.; *Org. React.* **1993**, *43* 1.

[317] Stille, J. R.; Grubbs, R. H.; *J. Am. Chem. Soc.* **1983**, *105* 1664.

[318] Beckhaus, R.; Oster, J.; Sang, J.; Strauß, I.; Wagner, M.; Synlett **1997**, 241.

[319] Tyrlik, S.; Wolochowicz, I.; *Bull. Soc. Chim. Fr.* **1973**, 2147.

[320] McMurry, J. E.; Fleming, M. P.; *J. Am. Chem. Soc.* **1974**, *96* 4708.

[321] Mukaiyama, T.; Sato, T.; Hanna, J.; *Chem. Lett.* **1973**, 1041.

[322] (a) Fürstner, A.; Weidmann, H.; *Synthesis* **1987**, 1071; (b) Fürstner, A.; *Angew. Chem., Int. Ed. Engl.* **1993**, *32* 164; (c) Clive, D. L. J.; Zhang, C.; Murthy, K. S. K.; Hayward, W. D.; Daigneault, S.; *J. Org. Chem.* **1991**, *56* 6447.

[323] Fürstner, A.; Hupperts, A.; *J. Am. Chem. Soc.* **1995**, *117* 4468.

[324] (a) Aleandri, L. E.; Becke, S.; Bogdanović, B.; Jones, D.; Rozière, J.; *J. Organomet. Chem.* **1994**, *472* 97; (b) Aleandri, L. E.; Bogdanović, B.; Gaidies, A.; Jones, D. J.; Liao, S.; Michalowisz, A.; Rozière, J.; Schott, A.; *J. Organomet. Chem.* **1993**, *459* 87; (c) Bogdanović, B.; Bolte, A.; *J. Organomet. Chem.* **1995**, *502* 109; (d) Eisch, J. J.; Shi, X.; Alila, J. R.; Thiele, S.; *Chem. Ber. Recl.* **1997**, *130* 1175.

[325] (a) Fürstner, A.; Bogdanović, B.; *Angew. Chem., Int. Ed. Engl.* **1996**, *35* 2442; (b) Lectka, T.; in *Active Metals. Preparation, Characterization, Applications,* (Fürstner, A.; Ed.), VCH, Weinheim, **1996**, 85; (c) McMurry, J. E.; *Chem. Rev. (Washington, D.C.)* **1989**, *89* 1513; (d) Lenoir, D.; *Synthesis* **1989**, 883.

[326] McMurry, J. E.; Lectka, T.; Rico, J. G.; *J. Org. Chem.* **1989**, *54* 3748.

[327] Fürstner, A.; Hupperts, A.; Ptock, A.; Janssen, E.; *J. Org. Chem.* **1994**, *59* 5215.

[328] Fürstner, A.; Ernst, A.; *Tetrahedron* **1995**, *51* 773; (b) Fürstner, A.; Weintritt, H.; Hupperts, A.; *J. Org. Chem.* **1995**, *60* 6637; (c) Fürstner, A.; Ernst, A.; Krause, H.; Ptock, A.; *Tetrahedron* **1996**, *52* 7329; (d) Fürstner, A.; Jumbam, D. N.; Seidel, G.; *Chem. Ber.* **1994**, *127* 1125.

[329] New applications of hydromagnesiation: Ito, T.; Yamakawa, I.; Okamoto, S.; Kobayashi, Y.; Sato, F.; *Tetrahedron Lett.* **1991**, *32* 371.

[330] (a) Hirao, T.; Yamada, N.; Ohshiro, Y.; Agawa, T.; *Chem. Lett.* **1982**, 1997; (b) Sato, F.; Tanaka, Y.; Sato, M.; *J. Chem. Soc., Chem. Commun.* **1983**, 165; (c) Yamamoto, K.; Kimura, T.; Tomo, Y.; *Tetrahedron Lett.* **1984**, *25* 2155; (d) Sato, F.; Kobayashi, Y.; *Org. Synth.* **1990**, *69* 106.

[331] Ito, T.; Yamakawa, I.; Okamoto, S.; Kobayashi, Y.; Sato, F.; *Tetrahedron Lett.* **1991**, *32* 371.

[332] Gao, Y.; Harada, K.; Hata, T.; Urabe, H.; Sato, F.; *J. Org. Chem.* **1995**, *60* 290.

[333] Negishi, E. J.; *Acc. Chem. Res.* **1987**, *20* 65.

[334] Rand, C. L.; Van Horn, D. E.; Moore, M. W.; Negishi, E.; *J. Org. Chem.* **1981**, *46* 4093.

[335] (a) Tweedy, H. E.; Coleman, R. A.; Thompson, D. W.; *J. Organomet. Chem.* **1977**, *129* 69; (b) Zitzelberger, T. J.; Schiavelli, M. D.; Thompson, D. W.; *J. Org. Chem.* **1983**, *48* 4781.

[336] (a) Konradi, A. W.; Kemp, S. J.; Pedersen, S. F.; *J. Am. Chem. Soc.* **1994**, *116* 1316; (b) Reetz, M. T.; Griebenow, N.; *Liebigs Ann.* **1996**, 335.

[337] Barden, M. C.; Schwartz, J.; *J. Am. Chem. Soc.* **1996**, *118* 5484.

[338] (a) Mangeney, P.; Tejero, T.; Alexakis, A.; Grosjean, F.; Normant, J.; *Synthesis* **1988**, 255; (b) Periasamy, M.; Reddy, M. R.; Kanth, J. V. B.; *Tetrahedron Lett.* **1996**, *37* 4767.

[339] (a) Gansäuer, A.; *Chem. Commun. (Cambridge)* **1997**, 457; (b) Gansäuer, A.; *Synlett* **1997**, 363.

[340] (a) Nugent, W. A.; RajanBabu, T. V.; *J. Am. Chem. Soc.* **1988**, *110* 8561; (b) RajanBabu, T. V.; Nugent, W. A.; *J. Am. Chem. Soc.* **1994**, *116* 986.

[341] Curran, D. P.; Porter, N. A.; Giese, B.; *Stereochemistry of Radical Reactions*, VCH, Weinheim, **1996**.

[342] Yadav, J. S.; Srinivas, D.; Shekharam, T.; *Tetrahedron Lett.* **1994**, *35* 3625.

[343] Gansäuer, A.; Pierobon, M.; Bluhm, H.; *Angew. Chem., Int. Ed. Engl.* **1998**, *37* 101.

[344] Cavallaro, C. L.; Schwartz, J.; *J. Org. Chem.* **1996**, *61* 3863.

[345] Spencer, R. P.; Schwartz, J.; *J. Org. Chem.* **1997**, *62* 4204.

[346] (a) Barden, M. C.; Schwartz, J.; *J. Org. Chem.* **1995**, *60* 5963; (b) Spencer, R. P.; Schwartz, J.; *Tetrahedron Lett.* **1996**, *37* 4357; (c) Liu, Y.; Schwartz J.; *J. Org. Chem.* **1994**, *59* 940; (d) reduction based on TiCl₄/Sm: Huang, Y.; Zhang, Y.; Wang, Y.; *Synth. Commun.* **1996**, *26* 2911.

[347] Kablaoui, N. M.; Hicks, F. A.; Buchwald, S. L.; *J. Am. Chem. Soc.* **1997**, *119* 4424.

[348] Reetz, M. T.; Steinbach, R.; Westermann, J.; *J. Chem. Soc., Chem. Commun.* **1981**, 237.

[349] (a) Sasaki, T.; Usuki, A.; Ohno, M.; *J. Org. Chem.* **1980**, *45* 3559; (b) Fleming, I.; Paterson, I.; *Synthesis* **1979**, 446.

[350] Uemura, M.; Isobe, K.; Hayashi, Y.; *Tetrahedron Lett.* **1985**, *26* 767.

[351] Reetz, M. T.; Seitz, T.; *Angew. Chem., Int. Ed. Engl.* **1987**, *26* 1028.

[352] Posner, G. H.; Kogan, T. P.; *J. Chem. Soc., Chem. Commun.* **1983**, 1481.

[353] Reviews of syntheses of compounds having quarternary C-atoms: (a) Martin, S. F.; *Tetrahedron* **1980**, *36* 419; (b) Rüchardt, C.; Beckhaus, H. D.; *Angew. Chem., Int. Ed. Engl.* **1980**, *19* 429.

[354] Reetz, M. T.; Westermann, J.; *J. Org. Chem.* **1983**, *48* 254.

[355] Reetz, M. T.; Steinbach, R.; Wenderoth, B.; *Synth. Commun.* **1981**, *11* 261.

[356] Ishikawa, H.; Mukaiyama, T.; Ikeda, S.; *Bull. Chem. Soc. Jpn.* **1981**, *54* 776.

[357] Lindell, S. D.; Elliott, J. D.; Johnson, W. S.; *Tetrahedron Lett.* **1984**, *25* 3947.

[358] Hoffmann, R. W.; *Chem. Rev. (Washington, D.C.)* **1989**, *89* 1841 (specifically p. 1855).

[359] Bartlett, P. A.; Johnson, W. S.; Elliott, J. D.; *J. Am. Chem. Soc.* **1983**, *105* 2088.

[360] Johnson, W. S.; Crackett, P. H.; Elliott, J. D.; Jagodzinski, J.; Lindell, S. D.; Natarajan, S.; *Tetrahedron Lett.* **1984**, *25* 3951.

[361] Johnson, W. S.; Elliott, R.; Elliott, J. D.; *J. Am. Chem. Soc.* **1983**, *105* 2904.

[362] Mori, A.; Maruoka, K.; Yamamoto, H.; *Tetrahedron Lett.* **1984**, *25* 4421.

[363] Reviews of Lewis-acid induced substitution reactions of acetals: (a) Alexakis, A.; Mangeney, P.; *Tetrahedron: Asymmetry* **1990**, *1* 477; (b) Mukaiyama, T.; Murakami, M.; *Synthesis* **1987**, 1043.

[364] Review of Lewis acid mediated α-alkylation of carbonyl compounds: Reetz, M. T.; *Angew. Chem., Int. Ed. Engl.* **1982**, *21* 96.

[365] (a) Reetz, M. T.; Maier, W. F.; Heimbach, H.; Giannis, A.; Anastassiou, G.; *Chem. Ber.* **1980**, *113* 3734; (b) Reetz, M. T.; Maier, W. F.; Chatziiosifidis, J.; Giannis, A.; Heimbach, H.; Löwe, U.; *Chem. Ber.* **1980**, *113* 3741.

[366] Review of enolsilane chemistry: Brownbridge, P.; *Synthesis* **1983**, 1 and 85.

[367] Chan, T. H.; Paterson, I.; Pinsonnault, J.; *Tetrahedron Lett.* **1977**, *18* 4183.

[368] (a) Reetz, M. T.; Haning, H.; *Tetrahedron Lett.* **1993**, *34* 7395; (b) Cahiez, G.; Figadère, B.; Cléry, P.; *Tetrahedron Lett.* **1994**, *35* 3065.

[369] Reetz, M. T.; Sauerwald, M.; Walz, P.; *Tetrahedron Lett.* **1981**, *22* 1101.

[370] (a) Review: Mukaiyama, T.; *Angew. Chem., Int. Ed. Engl.* **1977**, *16* 817; (b) Fleming, I.; Lee, T. V.; *Tetrahedron Lett.* **1981**, *22* 705 and references cited therein.

[371] Cockerill, G. S.; Kocienski, P.; Treadgold, R.; *J. Chem. Soc., Perkin Trans. 1* **1985**, 2101.

[372] (a) Saksena, A. K.; Girijavallabhan, V. M.; Wang, H.; Liu, Y.-T.; Pike, R. E.; Ganguly, A. K.; *Tetrahedron Lett.* **1996**, *37* 5657; (b) Barnett, C. J.; Wilson, T. M.; Evans, D. A.; Somers, T. C.; *Tetrahedron Lett.* **1997**, *38* 735.

[373] (a) Bernardi, A.; Cardani, S.; Carugo, O.; Colombo, L.; Scolastico, C.; Villa, R.; *Tetrahedron Lett.* **1990**, *31* 2779; (b) Palazzi, C.; Poli, G.; Scolastico, C.; Villa, R.; *Tetrahedron Lett.* **1990**, *31* 4223; (c) Pasquarello, A.; Poli, G.; Potenza, D.; Scolastico, C.; *Tetrahedron: Asymmetry.* **1990**, *1* 429.

[374] (a) Kondé-Friebold, K.; Hoppe, D.; *Synlett* **1990**, 99; (b) Basile, T.; Longobardo, L.; Tagliavini, E.; Trombini, C.; Umani-Ronchi, A.; *J. Chem. Soc., Chem. Commun.* **1990**, 759; (c) van Oeveren, A.; Feringa, B. L.; *J. Org. Chem.* **1996**, *61* 2920; (d) Hiemstra, H.; Speckamp, W. N.; in *Comprehensive Organic Synthesis*, (Trost, B. M.; Fleming, J.; Eds), Vol 2, Pergamon, Oxford, **1991**, 1047; (e) Allin, S. M.; Northfield, C. J.; Page, M. I.; Slawin, A. M. Z.; *Tetrahedron Lett.* **1997**, *38* 3627.

[375] Tanaka, T.; Inoue, T.; Kamei, K.; Murakami, K.; Iwata, C.; *J. Chem. Soc., Chem. Commun.* **1990**, 906.

[376] Schmidt, G.; Waldmann, H.; Henke, H.; Burkard, M.; *Chem.-Eur. J.* **1996**, *2* 1566.

[377] (a) Kitagawa, O.; Inoue, T.; Hirano, K.; Taguchi, T.; *J. Org. Chem.* **1993**, *58* 3106; (b) Inoue, T.; Kitagawa, O.; Oda, Y.; Taguchi, T.; *J. Org. Chem.* **1996**, *61* 8256.

[378] Inoue, T.; Kitagawa, O.; Kurumizawa, S.; Ochiai, O.; Taguchi, T.; *Tetrahedron Lett.* **1995**, *36* 1479.

[379] Inoue, T.; Kitagawa, O.; Ochiai, O.; Shiro, M.; Taguchi, T.; *Tetrahedron Lett.* **1995**, *36* 9333.

[380] Kasatkin, A. N.; Kulak, A. N.; Tolstikov, G. A.; *Bull. Acad. Sci. USSR, Div. Chem. Sci. (Engl. Transl.)* **1986**, *35* 871.

[381] Barber, J. J.; Willis, G.; Whitesides, G. M.; *J. Org. Chem.* **1979**, *44* 3603.

[382] (a) Schlosser, M.; Fujita, K.; *Angew. Chem., Int. Ed. Engl.* **1982**, *21* 309; *Angew. Chem., Suppl.* **1982**, 646.

[383] Moret, E.; Schlosser, M.; *Tetrahedron Lett.* **1985**, *26* 4423.

[384] Reetz, M. T.; Wenderoth, B.; Urz, R.; *Chem. Ber.* **1985**, *118* 348.

[385] (a) Hayashi, T.; Kumada, M.; in *Asymmetric Synthesis*, (Morrison, J. D.; Ed.), Vol. 5, Academic Press, Orlando, 1985, 147; (b) Narasaka, K.; *Synthesis* **1991**, 1.

[386] (a) Walborsky, H. M.; Barash, L.; Davis, T. C.; *Tetrahedron* **1963**, *19* 2333; (b) Springer, J. B.; Corcoran, R. C.; *J. Org. Chem.* **1996**, *61* 1443.

[387] Reetz, M. T.; Kyung, S. H.; Bolm, C.; Zierke, T.; *Chem. Ind.* **1986**, 824.

[388] Mikami, K.; Terada, M.; Nakai, T.; *J. Am. Chem. Soc.* **1990**, *112* 3949.

[389] Corey, E. J.; Letavic, M. A.; Noe, M. C.; Sarshar, S.; *Tetrahedron Lett.* *35* **1994**, 7553.

[390] (a) Mikami, K.; Shimizu, M.; *Chem. Rev. (Washington, D.C.)* **1992**, *92* 1021; (b) Mikami, K.; *Pure Appl. Chem.* **1996**, *68* 639.

[391] Gao, Y.; Hanson, R. M.; Klunder, J. M.; Ko, S. Y.; Masamune, H.; Sharpless, K. B.; *J. Am. Chem. Soc.* **1987**, *109* 5765.

[392] Mikami, K.; Matsukawa, S.; Nagashima, M.; Funabashi, H.; Morishima, H.; *Tetrahedron Lett.* **1997**, *38* 579.

[393] Mikami, K.; Terada, M.; *Tetrahedron* **1992**, *48* 5671.

[394] Terada, M.; Mikami, K.; *J. Chem. Soc., Chem. Commun.* **1995**, 2391.

[395] (a) Corey, E. J.; Barnes-Seeman, D.; Lee, T. W.; Goodman, S. N.; *Tetrahedron Lett.* **1997**, *38* 6513; (b) Corey, E. J.; Barnes-Seeman, D.; Lee, T. W.; *Tetrahedron Lett.* **1997**, *38* 1699.

[396] (a) Zimmer, R.; Suhrbier, J.; *J. Prakt. Chem.* **1997**, *339* 758; (b) Motoyama, Y.; Terada, M.; Mikami, K.; Synlett **1995**, 967.

[397] (a) Deloux, L.; Srebnik, M.; *Chem. Rev. (Washington, D.C.)* **1993**, *93* 763; (b) Kagan, H. B.; Riant, O.; *Chem. Rev. (Washington, D.C.)* **1992**, *92* 1007; (c) Evans, D. A.; Kozlowski, M. C.; Tedrow, J. S.; *Tetrahedron Lett.* **1996**, *37* 7481; (d) Sagasser, I.; Helmchen, G.; *Tetrahedron Lett.* **1998**, *39* 261.

[398] Costa, A. L.; Piazza, M. G.; Tagliavini, E.; Trombini, C.; Umani-Ronchi, A.; *J. Am. Chem. Soc.* **1993**, *115* 7001.

[399] Gauthier, D. R., Jr.; Carreira, E. M.; *Angew. Chem., Int. Ed. Engl.* **1996**, *35* 2363.

[400] Duthaler, R. O.; Hafner, A.; *Angew. Chem., Int. Ed. Engl.* **1997**, *36* 43.

[401] (a) Keck, G. E.; Tarbet, K. H.; Geraci, S.; *J. Am. Chem. Soc.* **1993**, *115* 8467; (b) Moris, F.; Gotor, V.; *J. Org. Chem.* **1993**, *58* 653.

[402] Keck, G. E.; Krishnamurthy, D.; *J. Am. Chem. Soc.* **1995**, *117* 2363.

[403] Mikami, K.; Takasaki, T.; Matsukawa, S.; Maruta, M.; *Synlett* **1995**, 1057.

[404] Lipshutz, B. H.; James, B.; Vance, S.; Carrico, I.; *Tetrahedron Lett.* **1997**, *38* 753.

[405] Yu, C.-M.; Choi, H.-S.; Yoon, S.-K.; Jung, W.-H.; *Synlett* **1997**, 889.

[406] Yu, C.-M.; Choi, H.-S.; Jung, W.-H.; Lee, S.-S.; *Tetrahedron Lett.* **1996**, *37* 7095.

[407] Weigand, S.; Brückner, R.; *Chem.-Eur. J.* **1996**, *2* 1077.

[408] Kitamoto, D.; Imma, H.; Nakai, T.; *Tetrahedron Lett.* **1995**, *36* 1861.

[409] Mikami, K.; Matsukawa, S.; *Nature (London)* **1997**, *385* 613.

[410] Faller, J. W.; Mazzieri, M. R.; Nguyen, J. T.; Parr, J.; Tokunaga, M.; *Pure Appl. Chem.* **1994**, *66* 1463.

[411] (a) Faller, J. W.; Sams, D. W. I.; Liu, X.; *J. Am. Chem. Soc.* **1996**, *118* 1217; (b) Faller, J. W.; Liu, X.; *Tetrahedron Lett.* **1996**, *37* 3449.

[412] Yu, C.-M.; Yoon, S.-K.; Choi, H.-S.; Baek, K.; *Chem. Commun. (Cambridge)* **1997**, 763.

[413] Chan, A. S. C.; Zhang, F.-Y.; Yip, C.-W.; *J. Am. Chem. Soc.* **1997**, *119* 4080.

[414] (a) Schmidt, B.; Seebach, D.; *Angew. Chem., Int. Ed. Engl.* **1991**, *30* 99; (b) Schmidt, B.; Seebach, D.; *Angew. Chem., Int. Ed. Engl.* **1991**, *30* 1321; (c) von dem Bussche-Hünnefeld, J. L.; Seebach, D.; *Tetrahedron* **1992**, *48* 5719.

[415] (a) Iwasawa, N.; Hayashi, Y.; Sakurai, H.; Narasaka, K.; *Chem. Lett.* **1989**, 1581; (b) Narasaka, K.; Iwasawa, N.; Inoue, M.; Yamada, T.; Nakashima, M.; Sugimori, J.; *J. Am. Chem. Soc.* **1989**, *111* 5340; (c) Narasaka, K.; Yamamoto, I.; *Tetrahedron* **1992**, *48* 5743.

[416] (a) Seebach, D.; Dahinden, R.; Marti, R. E.; Beck, A. K.; Plattner, D. A.; Kühnle, F. N. M.; *J. Org. Chem.* **1995**, *60* 1788; (b) Beck, A. K.; Dobler, M.; Plattner, D. A.; *Helv. Chim. Acta* **1997**, *80* 2073.

[417] Wada, E.; Yasuoka, H.; Kanemasa, S.; *Chem. Lett.* **1994**, 1637.

[418] (a) Seebach, D.; Beck, A. K.; Schmidt, B.; Wang, Y. M.; *Tetrahedron* **1994**, *50* 4363; (b) Oguni, N.; Satoh, N.; Fujii, H.; *Synlett* **1995**, 1043.

[419] Narasaka, K.; Hayashi, Y.; *Adv. Cycloadd it.* **1997**, *4* 87.

[420] Engler, T. A.; Letavic, M. A.; Reddy, J. P.; *J. Am. Chem. Soc.* **1991**, *113* 5068.

[421] (a) Haase, C.; Sarko, C. R.; DiMare, M.; *J. Org. Chem.* **1995**, *60* 1777; (b) Corey, E. J.; Matsumura, Y.; *Tetrahedron Lett.* **1991**, *32* 6289.

[422] Gothelf, K. V.; Hazell, R. G.; Jørgensen, K. A.; *J. Am. Chem. Soc.* **1995**, *117* 4435.

[423] Altava, B.; Burguete, M. I.; Fraile, J. M.; Garcia, J. I.; Luis, S. V.; Mayoral, J. A.; Royo, A. J.; Vicent, M. J.; *Tetrahedron: Asymmetry* **1997**, *8* 2561.

[424] Bienaymé, H.; *Angew. Chem., Int. Ed. Engl.* **1997**, *36* 2670.

[425] Assfeld, X.; Garcia, J.; Garcia, J. I.; Mayoral, J. A.; Proietti, M. G.; Ruiz-López, M. F.; Sánchez, M. C.; *J. Org. Chem.* **1996**, *61* 1636.

[426] (a) Oppolzer, W.; *Angew. Chem., Int. Ed. Engl.* **1984**, *23* 876; (b) Poll, T.; Metter, J. O.; Helmchen, G.; *Angew. Chem., Int. Ed. Engl.* **1985**, *24* 112; (c) Hollis, T. K.; Robinson, N. P.; Bosnich, B.; *J. Am. Chem. Soc.* **1992**, *114* 5464; (d) Motoyama, Y.; Tanaka, M.; Mikami, K.; *Inorg. Chim. Acta* **1997**, *256* 161.

[427] (a) Kanemasa, S.; Nishiuchi, M.; Kamimura, A.; Hori, K.; *J. Am. Chem. Soc.* **1994**, *116* 2324; (b) Murahashi, S.-I.; Imada, Y.; Kohno, M.; Kawakami, T.; *Synlett* **1993**, 395; (c) Gilbertson, S. R.; Dawson, D. P.; Lopez, O. D.; Marshall, K. L.; *J. Am. Chem. Soc.* **1995**, *117* 4431.

[428] Gothelf, K. V.; Thomsen, I.; Jørgensen, K. A.; *J. Am. Chem. Soc.* **1996**, *118* 59.

[429] (a) Yoshioka, M.; Kawakita, T.; Ohno, M.; *Tetrahedron Lett.* **1989**, *30* 1657; (b) Takahashi, H.; Kawakita, T.; Ohno, M.; Yoshioka, M.; Kobayashi, S.; *Tetrahedron* **1992**, *48* 5691; (c) Rozema, M. J.; Eisenberg, C.; Lütjens, H.; Ostwald, R.; Belyk, K.; Knochel, P.; *Tetrahedron Lett.* **1993**, *34* 3115.

[430] (a) Lütjens, H.; Nowotny, S.; Knochel, P.; *Tetrahedron: Asymmetry* **1995**, *6* 2675; (b) Lutz, C.; Knochel, P.; *J. Org. Chem.* **1997**, *62* 7895.

[431] (a) Guo, C.; Qiu, J.; Zhang, X.; Verdugo, D.; Larter, M. L.; Christie, R.; Kenney, P.; Walsh, P. J.; *Tetrahedron* **1997**, *53* 4145; (b) Qui, J.; Guo, C.; Zhang, X.; *J. Org. Chem.* **1997**, *62* 2665.

[432] (a) Ito, K.; Kimura, Y.; Okamura, H.; Katsuki, T.; *Synlett* **1992**, 573; (b) Soai, K.; Inoue, Y.; Takahashi, T.; Shibata, T.; *Tetrahedron* **1996**, *52* 13 355; (c) Ramon, D. J.; Yus, M.; *Tetrahedron Lett.* **1998**, *39* 1239.

[433] (a) Zhang, F.-Y.; Yip, C.-W.; Cao, R.; Chan, A. S. C.; *Tetrahedron: Asymmetry* **1997**, *8* 585; (b) Mori, M.; Nakai, T.; *Tetrahedron Lett.* **1997**, *38* 6233.

[434] (a) Soai, K.; Ohno, Y.; Inoue, Y.; Tsuruoka, T.; Hirose, Y.; *Recl. Trav. Chim. Pays-Bas* **1995**, *114* 145; (b) Hollis, T. K.; Bosnich, B.; *J. Am. Chem. Soc.* **1995**, *117* 4570; (c) Denmark, S. E.; Chen, C.-T.; *Tetrahedron Lett.* **1994**, *35* 4327.

[435] Carreira, E. M.; Singer, R. A.; Lee, W.; *J. Am. Chem. Soc.* **1994**, *116* 8837.

[436] Singer, R. A.; Carreia, E. M.; *J. Am. Chem. Soc.* **1995**, *117* 12 360.

[437] Singer, R. A.; Carreira, E. M.; *Tetrahedron Lett.* **1997**, *38* 927.

[438] Carreira, E. M.; Lee, W.; Singer, R. A.; *J. Am. Chem. Soc.* **1995**, *117* 3649.

[439] Charette, A. B.; Brochu, C.; *J. Am. Chem. Soc.* **1995**, *117* 11 367.

[440] Inoue, T.; Kitagawa, O.; Saito, A.; Taguchi, T.; *J. Org. Chem.* **1997**, *62* 7384.

[441] (a) Kobayashi, S.; Suda, S.; Yamada, M.; Mukaiyama, T.; *Chem. Lett.* **1994**, 97; (b) Bernardi, A.; Karamfilova, K.; Boschin, G.; Scolastico, C.; *Tetrahedron Lett.* **1995**, *36* 1363; (c) Bernadi, A.; Karamfilova, K.; Sanguinetti, S.; Scolastico, C.; *Tetrahedron* **1997**, *53* 13009.

[442] Hicks, F. A.; Buchwald, S. L.; *J. Am. Chem. Soc.* **1996**, *118* 11 688.

[443] Odenkirk, W.; Bosnich, B.; *J. Chem. Soc., Chem. Commun.* **1995**, 1181.

[444] (a) Hayashi, M.; Matsuda, T.; Oguni, N.; *J. Chem. Soc., Chem. Commun.* **1990**, 1364; (b) Hayashi, M.; Inoue, T.; Miyamoto, Y.; Oguni, N.; *Tetrahedron* **1994**, *50* 4385; (c) Pan, W.; Feng, X.; Gong, L.; Hu, W.; Li, Z.; Mi, A.; Jiang, Y.; *Synlett* **1996**, 337; (d) Yaozhong, J.; Xiangge, Z.; Wenhao, H.; Zhi, L.; Aiqiao, M.; *Tetrahedron: Asymmetry* **1995**, *6* 2915; (e) Belokon, Y.; Moscalenko, M.; Ikonnikov, N.; Yashkina, L.; Antonov, D.; Vorontsov, E.; Rozenberg, V.; *Tetrahedron: Asymmetry* **1997**, *8* 3245.

[445] Bolm, C.; Müller, P.; *Tetrahedron Lett.* **1995**, *36* 1625.

[446] (a) Cole, B. M.; Shimizu, K. D.; Krueger, C. A.; Harrity, J. P. A.; Snapper, M. L.; Hoveyda, A. H.; *Angew. Chem., Int. Ed. Engl.* **1996**, *35* 1668; (b) Eppley, A. W.; Totah, N. T.; *Tetrahedron* **1997**, *53* 16 545.

[447] Noroyi, R.; *Asymmetric Catalysis in Organic Synthesis*, Wiley, New York, **1994**.

[448] (a) Willoughby, C. A.; Buchwald, S. L.; *J. Org. Chem.* **1993**, *58* 7627; (b) Willoughby, C. A.; Buchwald, S. L.; *J. Am. Chem. Soc.* **1994**, *116* 8952.

[449] Viso, A.; Lee, N. E.; Buchwald, S. L.; *J. Am. Chem. Soc.* **1994**, *116* 9373.

[450] Broene, R. D.; Buchwald, S. L.; *J. Am. Chem. Soc.* **1993**, *115* 12 569.

[451] Lee, N. E.; Buchwald, S. L.; *J. Am. Chem. Soc.* **1994**, *116* 5985.

[452] Verdaguer, X.; Lange, U. E. W.; Reding, M. T.; Buchwald, S. L.; *J. Am. Chem. Soc.* **1996**, *118* 6784.

[453] (a) Gansäuer, A.; Pierobon, M.; Bluhm, H.; *Angew. Chem., Int. Ed. Engl.* **1995**, *34* 2005; (b) Almqvist, F.; Torstensson, L.; Gudmundsson, A.; Frejd, T.; *Angew. Chem., Int. Ed. Engl.* **1997**, *37* 376.

[454] Katsuki, T.; Sharpless, K. B.; *J. Am. Chem. Soc.* **1980**, *102* 5974.

[455] Reviews of the Sharpless epoxidation: (a) Rossiter, B. E.; in *Asymmetric Synthesis*, (Morrison J. D.; Ed.), Vol. 5, Academic Press, Orlando, **1985**, 193; (b) Finn, M. G.; Sharpless, K. B.; in *Asymmetric Synthesis*, (Morrison, J. D.; Ed.), Vol. 5, Academic Press, Orlando, **1985**, 247; (c) Pfenninger, A.; *Synthesis* **1986**, 89.

[456] (a) Woodard, S. S.; Finn, M. G.; Sharpless, K. B.; *J. Am. Chem. Soc.* **1991**, *113* 106; (b) Finn, M. G.; Sharpless, K. B.; *J. Am. Chem. Soc.* **1991**, *113* 113.

[457] Adam, W.; Peters, K.; Renz, M.; *J. Org. Chem.* **1997**, *62* 3183.

[458] (a) Brunel, J.-M.; Diter, P.; Duetsch, M.; Kagan, H. B.; *J. Org. Chem.* **1995**, *60* 8086; (b) Brunel, J. M.; Kagan, H. B.; *Synlett* **1996**, 404.

[459] (a) Furia, F. D.; Licine, G.; Modena, G.; Motterle, R.; *J. Org. Chem.* **1996**, *61* 5175; (b) Superchi, S.; Rosini, C.; *Tetrahedron: Asymmetry* **1997**, *8* 349; (c) Reetz, M. T.; Lohmer, G.; Schwickardi, R.; *Angew. Chem., Int. Ed. Engl.* **1997**, *36* 1526.

[460] Falborg, L.; Jørgensen, K. A.; *J. Chem. Soc., Perkin Trans. I* **1996**, 2823.

[461] Eppley, A. W.; Totah, N. I.; *Tetrahedron* **1997**, *53* 16545; and references cited therein.

[462] Seebach, D.; Jaeschke, G.; Wang, Y. M.; *Angew. Chem., Int. Ed. Engl.* **1995**, *34* 2395.

VIII

Organozirconium Chemistry

EI-ICHI NEGISHI

Department of Chemistry, Purdue University, West Lafayette, USA

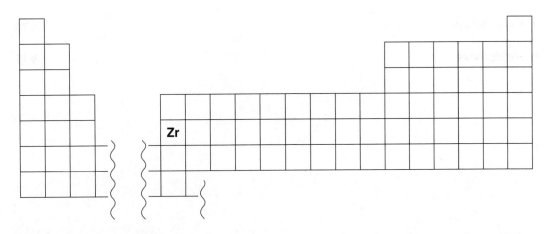

Organometallics in Synthesis: A Manual. Edited by Manfred Schlosser.
© 2002 John Wiley & Sons Ltd

Contents

1   Background

This Chapter deals with the chemistry of organozirconium compounds for organic synthesis. Both the preparation of organozirconium compounds and their reactions of synthetic utility are systematically discussed. Although the Zr chemistry is nearly 200 years old, one of the first, if not the first, organozirconium compounds of enormous synthetic significance, i.e., Cp_2ZrCl_2, was prepared only less than half a century ago in 1954, and the first systematic use of Zr in organic synthesis, with the exception of the use of Zr salts as Lewis acid catalysts in the Friedel–Crafts reaction and the Ziegler–Natta polymerization, probably was made with $HZrCp_2Cl$ in hydrozirconation in the early 1970s. In fact, essentially all of the topics discussed in this Chapter have been discovered and developed over the last quarter of a century.

Roughly 75–80% or possibly even more of the currently known organozirconium compounds are zirconocene and related derivatives. If one considers only those that have displayed some significant synthetic utilities, more than 90% of them would be zirconocene and related derivatives. This Chapter clearly reflects this fact in that nearly all compounds discussed herein are indeed zirconocene and related derivatives, although this may well change in the future.

In Section 8.2, several representative groups of reactions for the preparation of organozirconium compounds are discussed. This is followed by systematic discussions of the formation of carbon–heteroatom bonds including those bonds to hydrogen and metals (8.3), carbon–carbon bond formation by polar reactions (8.4), migratory insertion reactions including carbonylation (8.5), and more novel stoichiometric and catalytic carbozirconation and related addition reactions (8.6).

Due mainly to space limitations, no attempts have been made to provide a comprehensive coverage of organozirconium chemistry, and the coverage is indeed selective. For example, the Zr-catalyzed olefin polymerization is not discussed. The readers may supplement this text with a fairly large number of reviews and monographs listed as the references [1]–[50].

A combination of the IUPAC and some conventional nomenclatures is used. Some organozirconium compounds, such as $Cp_2ZrC_2H_4$, can be viewed as either Zr(II) π-complexes or Zr(IV) σ-complexes. It is judged that they are best represented as resonance hybrids of π- and σ-complexes, and that these representations may be used interchangeably. Thus, for example, $Cp_2ZrC_2H_4$ may be viewed as either an ethylene-Zr(II) complex or a zirconacyclopropane containing the $Cp_2Zr(IV)$ moiety.

Zirconium, which occurs to the extent of 0.022% in the lithosphere[1], is roughly as abundant as carbon. Although more expensive than Fe, Cu, Mn and Ti, it is one of the less expensive transition metals ($11 per mole of 98% pure $ZrCl_4$, Strem Chemicals). Despite these statistics, the use of Zr, especially as organozirconium compounds, in organic synthesis was virtually unknown until the mid-1970s. The subsequent discovery[2] and development of hydrozirconation as a synthetic tool[3] in the mid- to late-1970s clearly indicated some attractive synthetic potentials of Zr: carbon–carbon bond-forming reactions including carbonylation[4]; Ni- or Pd-catalyzed

cross-coupling[5]; Ni-catalyzed conjugate addition[6]; acylation by Zr-to-Al transmetal-
lation[7]; and Zr-catalyzed carboalumination of alkynes[8]. Explorations of $Cp_2Zr(II)$
chemistry ($Cp = \eta^5$-C_5H_5) in the 1980s and 1990s, together with further developments of the $Cp_2Zr(IV)$ chemistry mentioned above have now firmly established
Zr as one of the more useful and versatile transition metals in organic synthesis.
Although not as widely used as Cu and Pd, it rivals Ti, Cr, Co, Ni and Rh in
this respect. More importantly, its reactions complement those of the other transition
and main group metals. This chapter primarily deals with the preparation and reactions of organozirconium compounds with emphasis on their use in organic synthesis.
The great majority of organozirconium compounds used for organic synthesis today
are zirconocene derivatives, which therefore receive due attention, although other
synthetically useful derivatives will also be discussed. Despite the relatively short
history of Zr in organic synthesis, it is no longer practical to cover all the important
topics in detail here, and readers are referred to a few monographs[1,9] and other
reviews[10–48] for those topics that are not adequately. The most notable of these are
the Ziegler–Natta polymerization reactions[49].

2   Preparation of Organozirconium Compounds

The dozen or so basic patterns for the generation of organometals are generally applicable to the synthesis of organozirconium compounds[50]. In particular, transmetalation, hydrozirconation, and complexation provide the three most widely used routes
to organozirconiums. Table 1 illustrates these methods, together with oxidative addition. In principle, all of these methods can be applied to the synthesis of various types
of organozirconiums including zirconocene derivatives. In reality, the great majority of
synthetically useful organozirconium reactions known today involve Cp_2Zr derivatives.
Consequently, nearly all of the examples discussed in this chapter, including those in
Table 1, also deal with Cp_2Zr derivatives. Zirconocene derivatives may exist as Cp_2Zr^{IV},
Cp_2Zr^{III}, and Cp_2Zr^{II} complexes. At present, few reactions of Cp_2Zr^{III} derivatives are
known, and those that are synthetically interesting are fewer still. However, both Cp_2Zr^{IV}
and Cp_2Zr^{II} derivatives have been shown to be of considerable synthetic interest. In fact,
in many cases, the distinction between Cp_2Zr^{II} and Cp_2Zr^{IV} compounds is a nebulous
issue. Moreover, many of the "Cp_2Zr^{II} derivatives" may be best viewed as resonance
hybrids between Cp_2Zr^{II} and Cp_2Zr^{IV} derivatives, as exemplified by the products of
oxidative complexation or complexation in Table 1.

2.1   Transmetalation

Transmetalation represents the most widely applicable method for the synthesis of
organozirconium compounds. Synthesis of Cp_2Zr derivatives typically begins with the

Table 1. Preparation methods for Cp_2Zr compounds.

Reaction type	Transformation
Transmetallation	$Cp_2Zr^{IV} \longrightarrow Cp_2Zr^{IV}$
	$Cp_2ZrX_2 \xrightarrow{MR} Cp_2Zr(X)R \xrightarrow{MR} Cp_2ZrR_2$
Hydrozirconation	$Cp_2Zr^{IV} \longrightarrow Cp_2Zr^{IV}$

Oxidative addition	$Cp_2Zr^{II} \longrightarrow Cp_2Zr^{IV}$

Oxidative complexation (or complexation)	$Cp_2Zr^{II} \longrightarrow Cp_2Zr^{IV}$
	$Cp_2Zr^{II} \longrightarrow Cp_2Zr^{II}$

Complexation with donors	$Cp_2Zr^{II} \longrightarrow Cp_2Zr^{II}$
	$Cp_2\overset{II}{Zr}L_n \xrightarrow{mL} Cp_2\overset{II}{Zr}L_{n+m}$

reaction of $ZrCl_4$ with CpLi, CpNa, or CpMgBr (or Cl) to produce Cp_2ZrCl_2[51]. This compound is available from commercial sources (Boulder Chemical, Aldrich Chemical, Strem Chemical, etc.). It is not practical to synthesize $CpZrCl_3$ by the method mentioned above. However, an indirect route involving photochemical chlorination of Cp_2ZrCl_2[52] does provide a convenient route to $CpZrCl_3$.

Preparation of η^5-Cyclopentadienylzirconium Trichloride[52]

$$\text{Cp}_2\text{ZrCl}_2 \xrightarrow[\text{CCl}_4,\ 20-23\ °C]{\text{Cl}_2,\ h\nu} \text{CpZrCl}_3 + \text{C}_5\text{H}_5\text{Cl}_5$$
$$\qquad\qquad\qquad\qquad\qquad (96\%) \qquad (95\%)$$

In a 500-mL two-necked Schlenk tube, equipped with a thermometer, a gas inlet tube, and a magnetic stirrer, a suspension of 30 g (0.106 mol) of Cp_2ZrCl_2 in CCl_4 (300 mL) is saturated with chlorine gas. The reaction is initiated by short irradiation (1–2 min) with a 200-W Osram sunlight lamp. Chlorine is introduced at such a rate to maintain the exothermic reaction. Occasional external cooling may be necessary to keep the temperature of the reaction mixture within the optimal range of 20–23 °C. Chlorination is complete after about 2 h. A stream of argon is passed through the resulting white suspension to remove excess chlorine. The precipitate of pure CpZrCl_3 is separated and washed successively with chloroform (50 mL), CCl_4 (100 mL), and pentane (100 mL). After the precipitate is dried *in vacuo*, CpZrCl_3 is obtained as a white powder (yield 26 g 96%). Evaporation of the CCl_4 filtrate yields 24 g (95%) of pentachlorocyclopentane.

Zirconocene dichloride is relatively air and moisture stable, but protection from these and from light is advisable. The heavier analogues Cp_2ZrBr_2 and Cp_2ZrI_2, may be similarly prepared. They can also be prepared, perhaps more conveniently and economically, by direct halogen exchange or dialkylation–halogenolysis of Cp_2ZrCl_2. The general course of alkylation of Cp_2ZrCl_2 is shown in Scheme 1. However, these reactions can be complicated by a variety of side reactions, the nature of which depends strongly on the metal countercations among other factors.

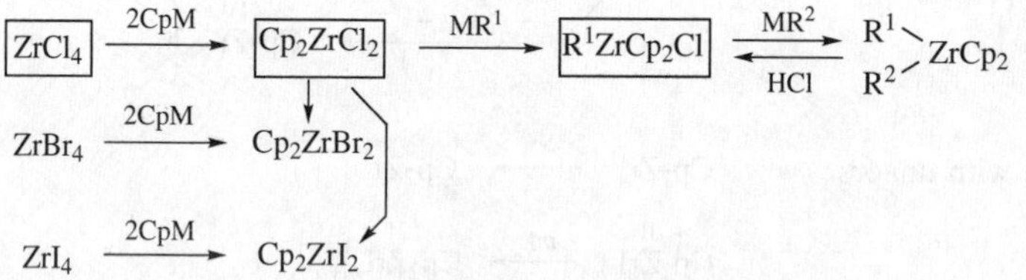

Scheme 1.

Both organolithiums and Grignard reagents tend to dialkylate Cp_2ZrCl_2 regardless of the reactant ratio, making the synthesis of $\text{R}^1\text{ZrCp}_2\text{Cl}$ by transmetalation a generally cumbersome and/or difficult task. For example, the most frequently used synthesis of MeZrCp_2Cl is the following two-step procedure.

Preparation of Chloro(methyl)zirconocene[2 a,53]

$$Cp_2ZrCl_2 \xrightarrow[\text{CH}_2\text{Cl}_2]{\substack{\text{aniline (1 equiv)} \\ \text{H}_2\text{O (0.7 equiv)}}} (Cp_2Zr)_2O \atop \underset{Cl}{|} \xrightarrow{Me_3Al} Cp_2Zr(Me)Cl$$

(a) Aniline (10 mL, 110 mmol) and water (1.3 mL, 72 mmol) were added to a solution of Cp_2ZrCl_2 (30 g, 103 mmol) in methylene dichloride (250 mL), and the mixture was shaken thoroughly. A white precipitate of aniline hydrochloride formed immediately. After chilling overnight in the refrigerator, the suspension was filtered cold. At times large crystals of the product are present at this stage, which must be redissolved by addition of more CH_2Cl_2. The filtrate was evaporated to a small volume, and petroleum ether (100 mL, bp 30–40°) was added to precipitate the product. Filtration and washing with petroleum ether gave white crystals of the oxo-bridged compound (26.5 g, 97% yield).

(b) To a slurry of $(Cp_2ZrCl)_2O$ in CH_2Cl_2 (80 mL) was added $Al_2(CH_3)_6$ (5.5 mL). The oxide rapidly dissolved to give a pale yellow solution which was stirred for a further 30 min before diethyl ether (30 mL) was added. Solvent was pumped off to lower the bulk and light petroleum ether (100 mL, bp 30–40°) was added with stirring to precipitate $Cp_2Zr(CH_3)Cl$ as pale yellow crystals (8 g, 75% yield) which were collected by filtration in an inert atmosphere, washed several times with light petroleum ether and dried by pumping under vacuum at 40°C (NMR (C_6D_6) δ 5.73 (s, 10 H) and 0.32 (s, 3 H) ppm). The compound is very soluble in aromatic solvents and is highly sensitive to moisture forming the bridged oxide again with elimination of methane.

Some bulky Grignard reagents do cleanly monoalkylate Cp_2ZrCl_2. Typically, its reaction with ^tBuMgCl gives $^tBuZrCp_2Cl$ as a transient species which then isomerizes to give $^iBuZrCp_2Cl$[54] (Scheme 2). The reaction of Cp_2ZrCl_2 with Me_3Al only partially monomethylates, with no indication of dimethylation[55]. Depending on the ligands, a carbon group may be transferred either from Zr to Al[56], or from Al to Zr[57] (Scheme 2).

2.2   Hydrozirconation

Hydrozirconation converts alkenes and alkynes into alkyl- and alkenylzirconium derivatives, respectively[2,3,10]. The most widely used reagent is $HZrCp_2Cl$, and its preparation as a pure substance has been accomplished by the treatment of Cp_2ZrCl_2 with various aluminum hydrides, such as $LiAlH_4$[2a], $LiAlH(O^tBu)_3$[2a], and

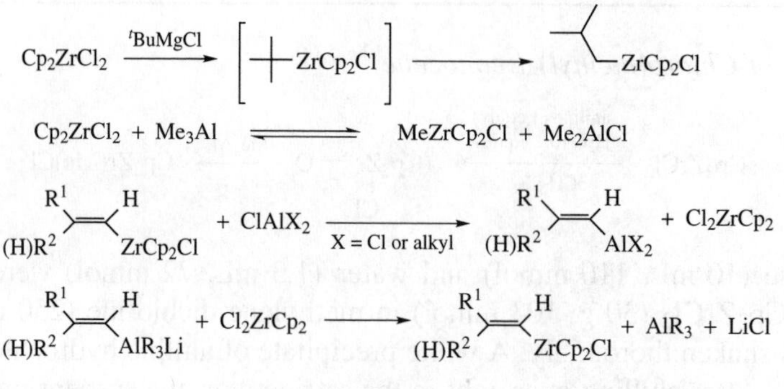

Scheme 2.

$NaAlH_2(OCH_2CH_2OCH_3)_2$[56], followed by purification. The low solubility of $HZrCp_2Cl$ in common organic solvents permits a simple, if rather tedious, method for its purification by filtration under an inert atmosphere. However, it is more convenient to generate $HZrCp_2Cl$ and its equivalents *in situ* by the treatment of Cp_2ZrCl_2 with $LiAlH_4$[54,58], $NaAlH_2(OCH_2CH_2OCH_3)_2$[54], $LiBEt_3H$[54,59], and ^tBuMgCl[54,60]. These reagents should be clearly distinguished from pure $HZrCp_2Cl$, as they may contain various impurities. They may be structurally different from $HZrCp_2Cl$, even though they act as its synthetic equivalents. For example, competitive formation of Cp_2ZrH_2, which can cause various unwanted side-reactions, is a problem in cases where $LiAlH_4$ is used. However, this problem can be largely avoided by treating Cp_2ZrH_2 with CH_2Cl_2 to produce $HZrCp_2Cl$[58].

Preparation of Chlorohydridozirconocene[58]

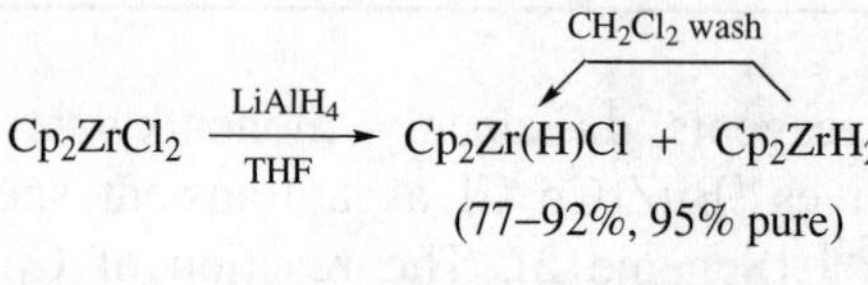

Zirconocene dichloride (100 g, 0.342 mol) was dissolved in dry THF (650 mL, heating required) in a 1-L Schlenk flask under argon. To this solution (at or slightly above 25 °C was added dropwise, over a 45-min period, a filtered solution of $LiAlH_4$ in diethyl ether (prepared from 3.6 g, 94.9 mmol of 95% $LiAlH_4$ and dry ether (100 mL) followed by filtration using a cannula fitted with a piece of glass fiber filter—a Schlenk-filtered or commercial clear solution would work as well). The resulting suspension was allowed to stir at 25 °C for 90 min. It was then Schlenk filtered under argon using a 'D' frit. The white solid was washed on the frit with THF (4 × 75 mL), CH_2Cl_2 (2 × 100 mL) with stirring or agitation of a stirbar immersed in the slurry and ether (4 × 50 mL). The resulting white solid was dried *in vacuo* to give a white powder (65.51–81 g, 77–92% yield). A small sample was assayed in a 5-mm NMR tube in C_6D_6 by treatment with a known amount of excess acetone, and the relative areas of the signal for the mono and diisopropoxides were determined by ^{1}H NMR (Cp$_2$Zr(H)Cl 94–96%, Cp$_2$ZrH$_2$ 4–6%).

A wide variety of solvents may be used for hydrozirconation. For the reaction of 1-hexene with HZrCp$_2$Cl at 0 °C, the rates of reaction in the decreasing order are: oxetane (37), morphorine (40), THF (113), and benzene (2210), the numbers in parentheses being $t_{1/2}$ in minutes[44].

Hydrozirconation with HZrCp$_2$Cl must involve a concerted syn addition of a H–Zr bond to C=C and C≡C bond *via* a transient 18-electron π-complex. The activation energy has been computationally estimated to be 0–15 kcal mol^{-1}[61].

Hydrozirconation of 1-Octyne with Chloro(hydrido)zirconocene to Generate (E)-1-Octenylzirconocene Chloride[10]

To a suspension of Cp$_2$Zr(H)Cl (1.52 g, 4.0 mmol) in 15 mL of THF was added 1-octyne (0.44 g, 4.0 mmol) at 25 °C. After stirring the mixture for 1 h, examination by ^{1}H NMR indicated the formation of (*E*)-1-octenylzirconocene chloride in >95% yield as judged by the integration of the Cp signal at 6.16 ppm relative to benzene added as an internal standard. This solution can be used in various further reactions. For example, protonolysis with 3 N HCl gives 1-octene in >90% yield which was identified by GLC coinjection of an authentic sample and NMR examination of the protonolysis product. In general, NMR examination of the reaction mixture and quantitative analysis of the protonolysis product provide two convenient means of quantitatively analyzing the hydrozirconation product.

The scope of hydrozirconation has been extensively delineated. As shown in Scheme 3, zirconium migrates along alkyl chains to occupy the terminal positions under the hydrozirconation conditions. In the reactions of alkynes, however, zirconium may only equilibrate between the two sp^2 carbon atoms without migrating further. Halogens, ethers, and metal-substituted alcohols are tolerated. Carbonyl groups and nitriles are competitively reduced.

Hydrogen-transfer hydrozirconation with ^{t}BuZrCp$_2$Cl generated *in situ* by the reaction of ^{t}BuMgCl with Cp$_2$ZrCl$_2$[54,60] is a particularly convenient alternative to hydrozirconation with HZrCp$_2$Cl. Alkynes generally react very satisfactorily, as exemplified by the reaction of HC≡CCH$_2$OTBS shown in Scheme 4[62].

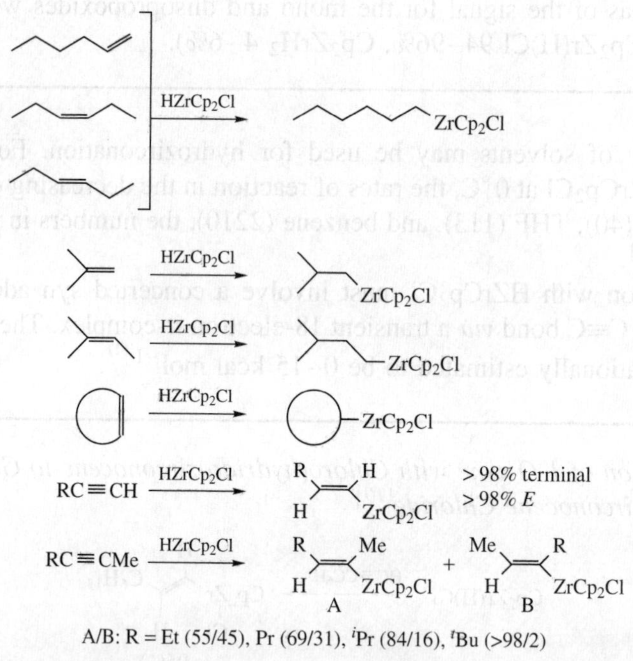

Scheme 3.

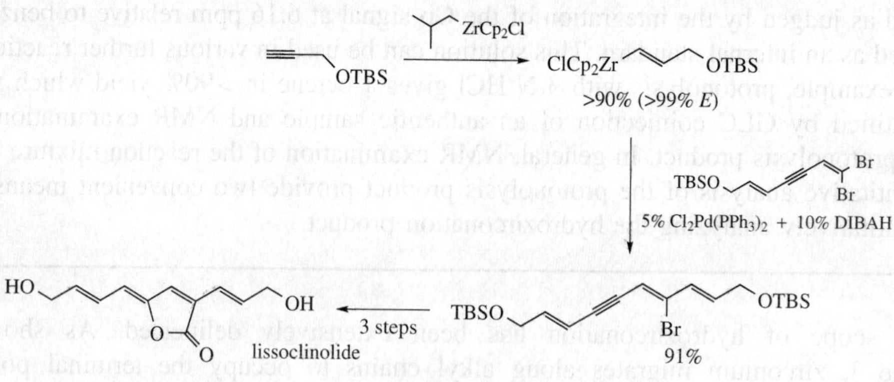

> 98% terminal
> 98% E

A/B: R = Et (55/45), Pr (69/31), ^{i}Pr (84/16), ^{t}Bu (>98/2)

ClCp$_2$Zr OTBS >90% (>99% E)

5% Cl$_2$Pd(PPh$_3$)$_2$ + 10% DIBAH

TBSO Br Br

HO OH 3 steps lissoclinolide TBSO Br OTBS 91%

Scheme 4.

Hydrogen-transfer hydrozirconation of alkenes with ^{i}BuZrCp$_2$Cl[54] not only is sluggish but also tends to be erratic. However, it has recently been shown that this reaction can be significantly accelerated by a wide variety of Lewis-acidic catalysts, such as AlCl$_3$, ZnCl$_2$, AgBF$_4$, Me$_3$SiI, and Cl$_2$Pd(PPh$_3$)$_2$[63] (Scheme 5). These results not only provide convenient procedures for hydrogen-transfer hydrozirconation, but also provide an explanation for the erratic behavior mentioned above.

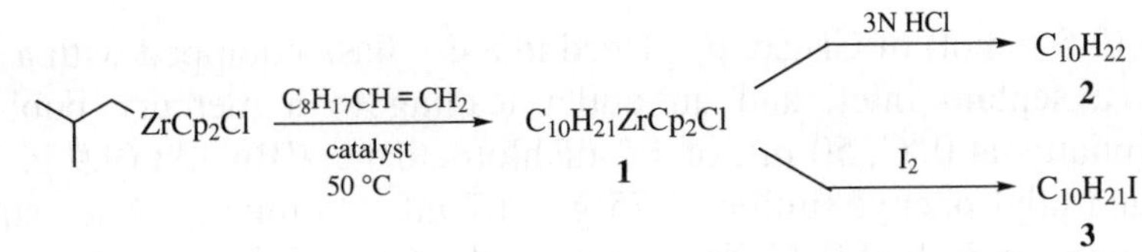

Scheme 5.

Catalyst (mol %)		1 (%)	2 (%)	3 (%)	Cp$_2$ZrCl$_2$ (%)
AlCl$_3$	(2)	87	91	84	5
ZnCl$_2$	(10)	79	81	71	8
AgBF$_4$	(20)	74	92	85	16
Me$_3$SiI	(5)	82	85	82	6
Cl$_2$Pd(PPh$_3$)$_2$	(5)	76	74	72	5

Hydrogen-transfer Hydrozirconation of 9-Decenyl Benzoate with ^{i}BuZrCp$_2$Cl in the Presence of Cl$_2$Pd(PPh$_3$)$_2$[63]

$$PhCOO(CH_2)_8CH=CH_2 \xrightarrow[\substack{5\% \ Cl_2Pd(PPh_3)_2 \\ PhH - Et_2O}]{^iBuZrCp_2Cl} PhCOO(CH_2)_{10}ZrCp_2Cl$$

To ^{i}BuZrCp$_2$Cl generated *in situ* from 2.2 mmol each of ^{t}BuMgCl and Cp$_2$ZrCl$_2$ were added Cl$_2$Pd(PPh$_3$)$_2$ (70 mg, 0.1 mmol) and 9-decenyl benzoate (520 mg, 2 mmol). The reaction mixture was heated at 50 °C for 3 h. Analysis of the mixture by ^{1}H NMR spectroscopy indicated the formation of the desired compound in 76% yield along with 5% of Cp$_2$ZrCl$_2$. After protonolysis with 3 N HCl and extractive work-up, column chromatography (5% EtOAc in hexane) provided 384 mg (74%) of decyl benzoate along with a minor amount (5%) of decanol.

Monosubstituted alkenes and certain alkynes can undergo hydroalumination with ^{i}Bu$_3$Al in the presence of a catalytic amount of Cp$_2$ZrCl$_2$, providing a convenient alternative to hydrozirconation.

Zirconium-catalyzed Hydroalumination–Deuterolysis of Allyl Phenyl Sulfide using Triisobutylalane and Zirconocene Dichloride[64]

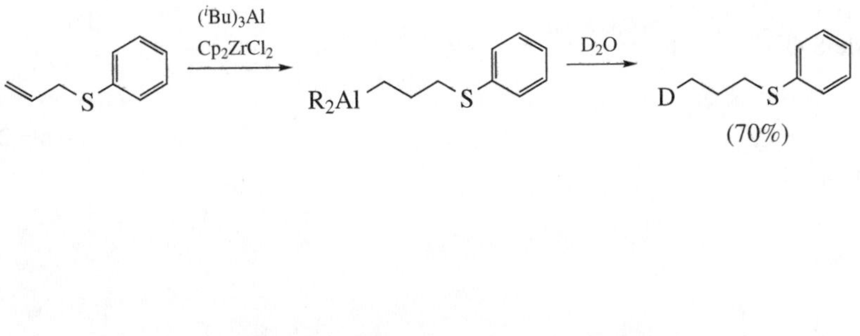

To 0.73 g (2.5 mmol) of Cl_2ZrCp_2 placed in a dry flask equipped with a magnetic stirring bar, a septum inlet, and an outlet leading to a mercury bubbler were added sequentially, at 0 °C, 50 mL of 1,2-dichloroethane, $(^iBu)_3Al$ (9.9 g, 12.6 mL, 50 mmol), and allyl phenyl sulfide (3.75 g, 3.67 mL, 25 mmol). After stirring the reaction mixture for 36 h at 0 °C, deuterium oxide (6.0 g, 5.44 mL, 300 mmol) was added. The product thus formed was extracted with 100 mL of pentane, washed, and dried. Distillation provided 2.68 g (70%) of a 97 : 3 mixture of 3- and 2-deuterio propyl phenyl sulfide: (bp 65–66 °C (1.7 mm Hg), $n_D^{22} = 1.4825$).

Although this reaction appears to involve hydrozirconation at the critical moment, it is discrete from the usual hydrozirconation. With alkynes, it undergoes regioisomerization restricted within a three carbon unit (Scheme 6).

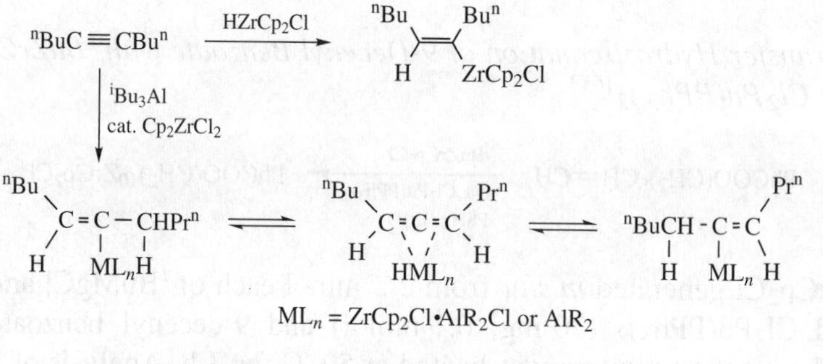

Scheme 6.

2.3   Oxidative Addition

Oxidative addition is a potentially important process for the preparation of organozirconium derivatives. One prototypical example was accidentally discovered in an unsuccessful attempt to bicyclize diallyl ether with Bu_2ZrCp_2. The reaction instead produced an oxidative addition product which must have been formed via oxidative complication–elimination[65]. This reaction has been developed into a convenient and useful method for generating allyl- and allenylzirconocene derivatives[66,67] (Scheme 7).

Preparation of (^{t}Butyldimethylsilyloxy)-3-phenylallylzirconocene[67e]

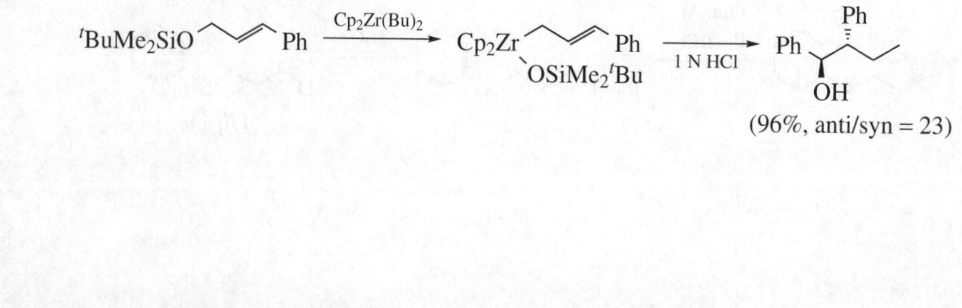

(96%, anti/syn = 23)

To a solution of Cp$_2$ZrCl$_2$ (248 mg, 0.85 mmol) in THF (3 mL) was added a solution of BuLi (1.45 M in hexane, 1.17 mL, 1.7 mmol) at −78 °C, and the mixture was stirred for 1 h. A solution of tert-butyldimethyl-3-phenylallyloxysilane (161 mg, 0.65 mmol) in THF (2 mL) was added to the above reaction mixture at −78 °C, and the resultant mixture stirred for 3 h at 25 °C. Examination by NMR indicated the formation of the desired product in essentially quantitative yield. This reaction mixture may be used directly for further reactions as exemplified by its reaction with benzaldehyde, described below.

Benzaldehyde (104 mg, 0.98 mmol) in THF (2 mL) was added at 0 °C to the above reaction mixture. After stirring for 2 h, 1 N HCl was added. The reaction mixture was extracted with ether, washed with brine, and dried over MgSO$_4$. Evaporation of the solvent under reduced pressure and purification of the residue by silica gel column chromatography afforded *anti*-1,2-diphenyl-3-buten-1-ol (96%, anti/syn = 23).

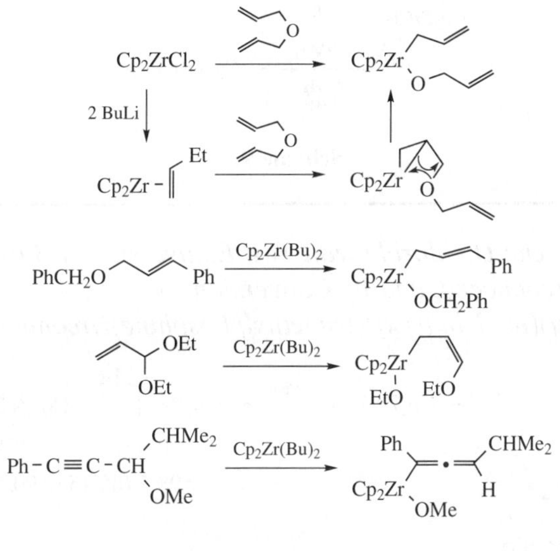

Scheme 7.

The scope of the oxidative addition of 'Cp$_2$Zr' derivatives has been significantly expanded by the development of oxidative addition of alkenyl chlorides to give alkenylzirconocene chlorides corresponding to formal Markovnikov hydrozirconation of alkynes[68].

2.4 Complexation (or Oxidative Complexation)

Reduction of Cp$_2$ZrCl$_2$ in the presence of π or non-bonding electron donors can provide Cp$_2$Zr^{II} π or n complexes (Scheme 8). Although 14-electron Cp$_2$Zr^{II} may be generated as a transient species, its generation as a free monomeric species does not appear to have been confirmed. Nor does it appear to be necessary. Electropositive metals, such as NaHg, Mg and HgCl$_2$, were mainly used in early studies[69–71] (Scheme 9).

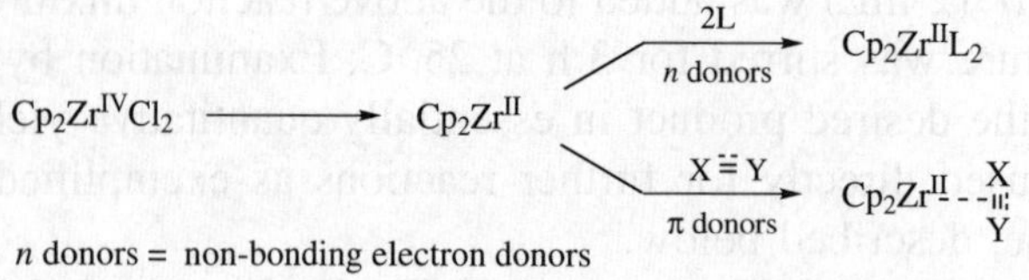

$$Cp_2Zr^{IV}Cl_2 \longrightarrow Cp_2Zr^{II} \xrightarrow[\substack{2L \\ n\ donors}]{} Cp_2Zr^{II}L_2$$

$$\xrightarrow[\substack{X \equiv Y \\ \pi\ donors}]{} Cp_2Zr^{II}\text{-}\text{-}\overset{X}{\underset{Y}{\|}}$$

n donors = non-bonding electron donors

Scheme 8.

More recently, β-hydrogen abstraction of diorganylzirconocenes has been developed to provide convenient routes to Cp$_2$Zr π complexes. In the Negishi–Takahashi protocol[48], β-hydrogen abstraction of dialkylzirconocenes yields Cp$_2$Zr–alkene complexes, which can be used to prepare other Cp$_2$Zr π complexes containing alkenes.

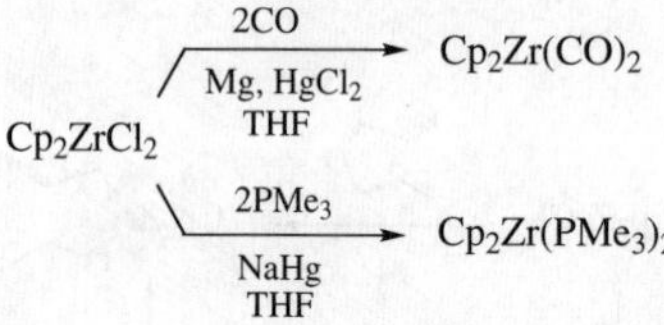

$$Cp_2ZrCl_2 \xrightarrow[\substack{Mg, HgCl_2 \\ THF}]{2CO} Cp_2Zr(CO)_2$$

$$\xrightarrow[\substack{NaHg \\ THF}]{2PMe_3} Cp_2Zr(PMe_3)_2$$

Scheme 9.

Reaction of Zirconocene Dichloride with Two Equivalents of Butyllithium to produce Di(butyl)zirconocene, and its Conversion to Bis(cyclopentadienyl)(η^2-1-butenyl)(trimethylphosphine)zirconium[72–74]

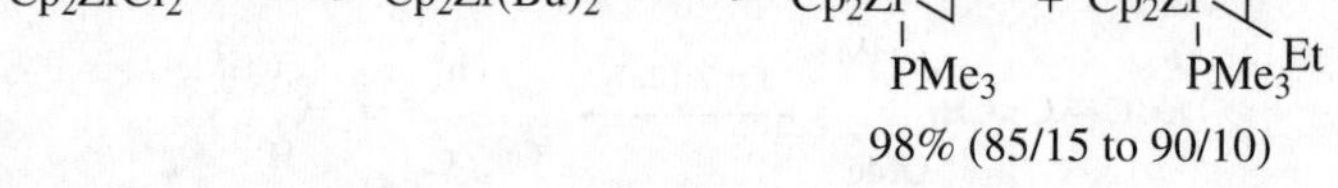

$$Cp_2ZrCl_2 \xrightarrow{2BuLi} Cp_2Zr(Bu)_2 \xrightarrow{PMe_3} Cp_2Zr\underset{PMe_3}{\overset{Et}{\diagup}} + Cp_2Zr\underset{PMe_3}{\overset{}{\diagdown}}Et$$

98% (85/15 to 90/10)

(a) Di(butyl)zirconocene

To 2.92 g (10 mmol) of Cl$_2$ZrCp$_2$ in 40 mL of THF was added 12.3 mL (1.63 M, 20 mmol) of BuLi at −78 °C, and the mixture was stirred for 1 h. An aliquot was quickly transferred to an NMR tube cooled at −78 °C. Its ^{1}H NMR spectrum taken after 30 min at −78 °C indicated the Cp singlet at δ 6.61 had completely shifted to δ 6.18. The yield of the product by ^{1}H NMR was essentially 100%. The mixture was warmed to 20 °C, and its decomposition was monitored by ^{1}H NMR. The first-order rate constant for this reaction was (4.6 ± 0.4) × 10^{-2} min^{-1} at 20 °C. The product was not identified. Another aliquot of the BuLi–Cl$_2$ZrCp$_2$ reaction mixture was treated with 2 equiv of I$_2$ in THF (−78 to 25 °C). Examination by ^{1}H NMR indicated the formation of I_2ZrCp_2 (δ 6.63) in 80% yield, and GLC analysis of the mixture, after hydrolytic work-up, indicated the formation of 2 equiv of BuI (ca. 100%), whose signal was discrete from that of octane present as a byproduct.

(b) Bis(cyclopentadienyl)(η^2-1-butenyl)(trimethylphosphine)zirconium

To a 2.0-mmol aliquot of $(Bu)_2ZrCp_2$ in THF was added PMe_3 (0.4 mL, 300 mg, 4 mmol). This mixture was warmed to 25 °C and stirred for 1 h. The ^1H NMR spectrum of the mixture indicated a signal at δ 5.29, appearing as an apparent triplet in 98% yield relative to benzene as an internal standard. Evaporation of the volatiles followed by dissolving the residue in C_6D_6 gave a sample for examination by NMR spectroscopy which indicated that the product consisted of a mixture of two stereoisomers in a ratio of 85/15 to 90/10.

Alkyne complexes of zirconocenes can be prepared similarly (see following procedure)[72,73,75–77] (Scheme 10).

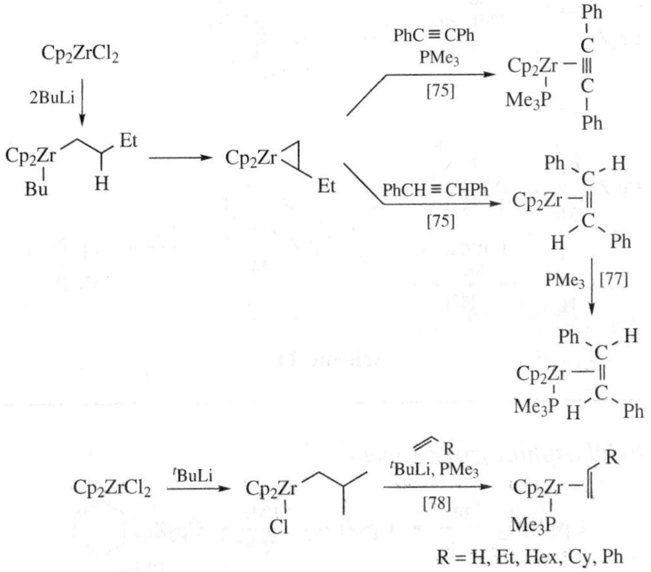

Scheme 10.

1,2-Diphenyl-3,3-bis(cyclopentadienyl)-3-(trimethylphosphine)-3-zirconacyclopropene[75]

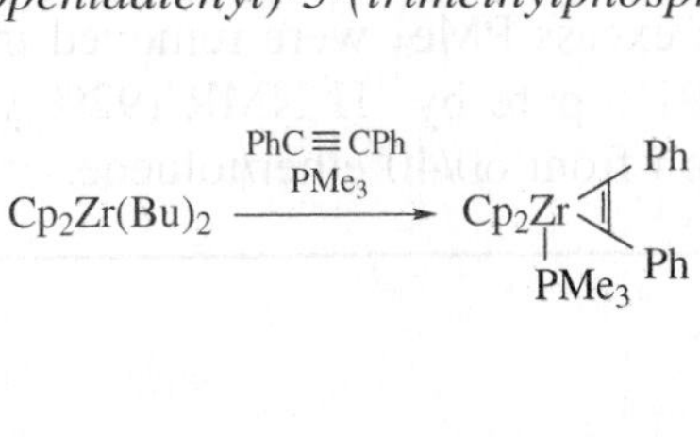

To $(Bu)_2ZrCp_2$ in THF generated at $-78\,°C$ by treating Cl_2ZrCp_2 (0.584 g, 2 mmol) with BuLi (2.0 M, 2 mL, 4 mmol) were added at $-78\,°C$ 0.356 g (2 mmol) of diphenylacetylene and trimethylphosphine (0.182 g, 2.4 mmol) at $-78\,°C$. After stirring the reaction mixture at $25\,°C$ for 3 h, the title compound was formed in 91% yield (^{1}H NMR). The mixture was evaporated until a yellow solid starting precipitating. A small amount of THF was added to dissolve the yellow solid. To this solution was added 15 mL of ether, which induced precipitation of a white solid. After filtering off the white solid, the resultant orange solution was kept at $-10\,°C$ overnight to give 0.524 g (55%) of yellow crystals of the title compound (mp 201–203 °C).

In the Erker–Buchwald protocol[24], the eventual π ligand is initially introduced either by transmetalation with organolithiums and Grignard reagents or by hydrozirconation[78–81] (Scheme 11). Examples are given in the following three procedures.

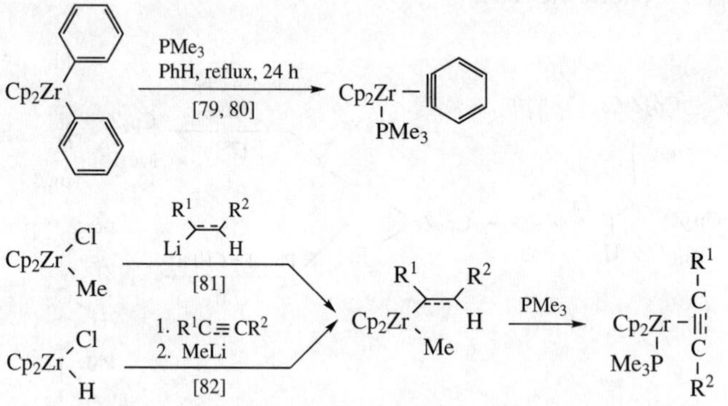

Scheme 11.

Benzyne(trimethylphosphine)zirconocene[79]

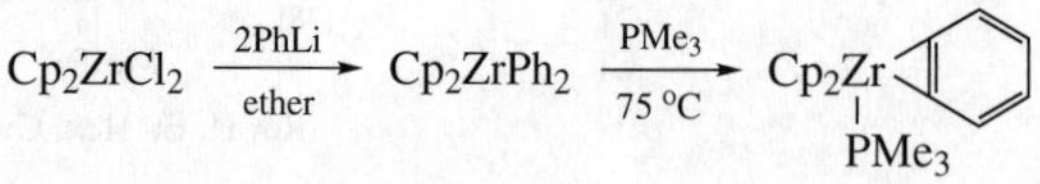

To a solution of diphenylzirconocene (2.08 g, 5.57 mmol), prepared by treatment of Cp_2ZrCl_2 with 2 equiv of PhLi in ether, was added under argon trimethylphosphine (4.41 g, 6.0 mL, 58 mmol), and the resulting solution was heated to $75\,°C$ for 24 h. Benzene and the excess PMe_3 were removed *in vacuo* to yield 2.08 g of a tan powder which was 92% pure by ^{1}H NMR (92% yield). A pure sample was obtained by recrystallization from 60/40 ether/toluene.

(s-trans-η^4-1,3-Butadiene)zirconocene by Photolysis of Diphenylzirconocene in the presence of 1,3-Butadiene[83,84]

$$Cp_2ZrPh_2 \xrightarrow{h\nu} \text{[structure]} ZrCp_2$$

A suspension of 7.7 g (20 mmol) of Cp_2ZrPh_2, prepared by the reaction of Cp_2ZrCl_2 with PhLi, in a solution of 1.7 g (31 mmol) of 1,3-butadiene in 250 mL of ether was irradiated for 15 h at $-20\,°C$ (HPK 125, Pyrex filter). The resultant red mixture was filtered and concentrated *in vacuo* at $0\,°C$. The residue was treated at $0\,°C$ with 120 mL of precooled pentane to give, after filtration, 3.0 g (50%) of the title compound as an orange powder. Low temperature recrystallization from toluene provided crystals for identification (mp 157–158 $°C$).

Hydrozirconation of Alkynes with Chlorohydridozirconocene and subsequent Generation of 1-Hexynezirconocene[81]

$$Cp_2Zr\underset{Cl}{\overset{H}{<}} \xrightarrow{HC\equiv CBu} Cp_2Zr\underset{Cl\ H}{\overset{H}{\diagup}}Bu \xrightarrow[PMe_3]{MeMgBr} Cp_2Zr\underset{PMe_3}{\diagdown}Bu$$

To a slurry of $Cp_2Zr(H)Cl$ (2.58 g, 10 mmol) in 30 mL of CH_2Cl_2 at $0\,°C$ under an argon atmosphere was added 1-hexyne (0.82 g, 10 mmol), and the mixture was stirred at $0\,°C$ for 1 h to give a clear solution of (E)-1-hexenylzirconocene chloride. This solution may be used for various further transformations. To the mixture obtained above were added MeMgBr in ether (1.9 M, 5.38 mL, 10 mmol) at $0\,°C$ and then an excess of PMe_3. After work-up and recrystallization, 1-hexyne(trimethylphosphine)zirconocene was obtained and identified by X-ray analysis.

Although most of the Cp_2Zr^{II} π complexes that have been fully identified by X-ray crystallography are 18-electron phosphine-stabilized derivatives, others containing ethers and amines, such as $Cp_2Zr(Me_3SiC\equiv CSiMe_3)(THF)$[46] and Cp_2Zr $(Me_3SiC\equiv CSiMe_3)\cdot Py$[81], have also been identified by X-ray crystallography and used as $ZrCp_2$ equivalents[46]. These procedures are, in principle, applicable to the synthesis of Cp_2Zr π complexes containing carbonyl compounds, imines, and other heteroatom π compounds[46,95].

*1-Trimethylsilyl-2,2-bis(cyclopentadienyl)-3-phenyl-2-zirconaaziridine–
Tetrahydrofuran Complex*[85]

To a solution of benzylamine (1.97 g, 18.4 mmol) in ether (20 mL) at 0 °C was
added butyllithium (11.2 mL of a 1.64 M solution in hexane, 184 mmol), and the
resulting suspension was allowed to stir for 5 min. Chlorotrimethylsilane (2.00 g,
18.4 mol) was added, and the reaction mixture was allowed to stir for 5 min. Butyllithium (11.2 mL of a 1.64 M solution in hexane, 18.4 mol) was added, and the
reaction mixture was allowed to stir for an additional 5 min. This solution was transferred by cannula to a solution of chloro(methyl)zirconocene (5.00 g, 18.4 mmol)
in THF (75 mL) at −78 °C, and the resulting solution was allowed to stir for 0.5 h,
warmed to 0 °C, and allowed to stir for 3 h. The reaction mixture was then warmed
to 25 °C and allowed to stir for 6 h. The solvent was removed to yield a yellow
solid. This was diluted with benzene (50 mL) and cannula filtered to remove LiCl.
The benzene was removed to yield a yellow solid which was washed with hexane
(3 × 10 mL) and dried *in vacuo* to yield 4.55 g of the title compound (53%).

3   Formation of Carbon–Heteroatom Bonds

As discussed in Section 2, monoorganylzirconocene derivatives represented by
RZrCp$_2$Cl, where R is alkyl, allyl, allenyl (or propargyl), and alkenyl, are now widely
available. Those containing a benzyl or aryl group are currently less readily available,
and their chemistry as well as that of RC≡CZrCp$_2$Cl remains largely unexplored. The
majority of known monoorganylzirconocene chlorides tend to be thermally stable at 25 °C.
However, they are sensitive to air and moisture, and they must be handled under an inert
atmosphere of argon or nitrogen. These compounds are formally 16-electron species with
one valence shell orbital empty. It is therefore reasonable to assume that, despite some
steric hindrance, the Zr atom can readily interact with single-electron or electron-pair
donors. In fact, the great majority of reactions appear to be initiated in this manner.

3.1   Bonds to Nonmetallic Heteroatoms

Cleavage of σ Zr–C bonds occurs readily upon treatment with H$_2$O or proton acids,
although the Zr–Cp bond usually survives mild protonolysis conditions. The use of D$_2$O or
DCl–D$_2$O permits replacement of Zr with D. Protonolysis of Zr–C$_{sp^2}$ bonds proceeds with

retention of configuration[86], whereas the stereochemistry of the corresponding reaction of Zr–Cp$_{sp^3}$ bonds does not appear to have been well established. In contrast with protonolysis, catalytic hydrogenolysis of Zr–C bonds is sluggish and not very clean[87]. Synthetically, it is useful to know that one or more D atoms can be introduced stereo- and regio-selectively through judicious use of deuterated alkynes, Cp$_2$Zr(D)Cl, and/or D$_2$O (Scheme 12).

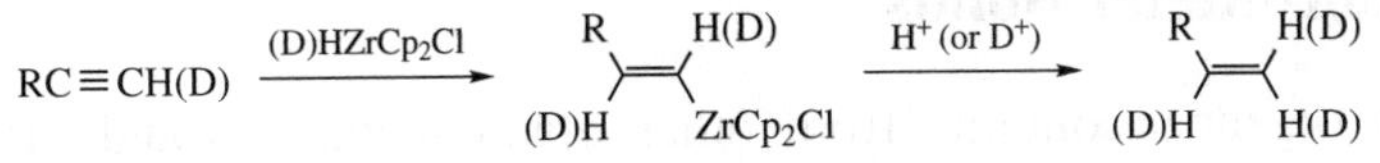

Scheme 12.

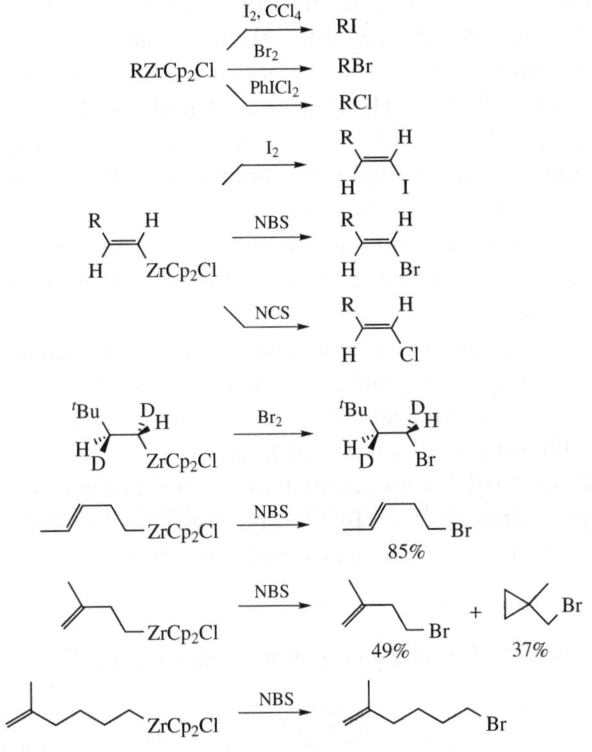

Scheme 13.

Iodinolysis of RZrCp$_2$Cl with I$_2$ in THF occurs readily and widely at or below 25 °C. However, it generally preserves the Cp–Zr bond[3,88]. Brominolysis with Br$_2$ is also facile, but it is often advantageous to use NBS in place of Br$_2$. Chlorinolysis is more sluggish, but favorable results may be obtained with NCS[89] or PhICl$_2$[3]. Iodinolysis and brominolysis of both Zr–C$_{sp^2}$ and Zr–C$_{sp^3}$ bonds have proceeded with retention of configuration. Brominolysis of some homoallyl derivatives suffers from skeletal rearrangement (Scheme 13)[89].

Alkylzirconocene derivatives can be converted to alcohols with H_2O_2–NaOH, ^tBuOOH, m-chloroperbenzoic acid, or O_2[90]. However, conversion of alkenylzirconocenes to aldehydes and ketones cannot be readily achieved[3]. Although oxidation reactions in some cases may proceed stereospecifically, those involving radicals do not.

3.2   Carbon–Metal Bonds

Synthesis of organozirconiums from other organometals through transmetalation is discussed in Section 2.1. In these reactions, zirconium should be as electronegative as or more electronegative than the other metal countercation. However, a carbon group may be transferred from Zr to more electronegative metals. If their electronegativities are comparable, as in the Zr versus Al case, transmetalation may proceed in either direction depending on the other ligands. In view of the electronegativity index of 1.2–1.4 for Zr[50], transfer of a carbon group from Zr to a variety of metals including B (electronegativity index 2.0), Si (1.8), Sn (1.7–1.8), Hg (1.4–1.9), Cu (1.75–1.9), Ni (1.75–1.8), and Pd (1.35–2.2) is expected to be favorable. In some cases, the products of transmetalation should be represented as 'ate' complexes containing both metals. Another important point to be noted is that, in cases where the organometallic reagents are generated and consumed in further steps under catalytic conditions, the required transmetalation needs not be thermodynamically favorable so long as the overall transformation is favorable. The following three scenarios are conceivable for the reactions of organozirconiums by transmetalation: (1) stoichiometric conversion into other organometals, (2) reactions of organozirconiums catalyzed by other metal compounds, and (3) reactions of other organometals catalyzed by zirconocene derivatives. In this section, only the first of these reactions discussed; the others are discussed later.

The treatment of $RZrCp_2Cl$ with metal halides and alkoxides containing Al[7,56,64], B (see following procedure)[91,92], Hg[93], and Sn[94,95] has led to the formation of organometals containing these metals as discrete and Zr-free species (Scheme 14).

Preparation of (Z)-(B-4-Methyl-2-pentenyl)catecholborane[92]

(68%)

To B-bromocatecholborane (0.994 g, 5 mmol) in methylene chloride at 0 °C was added 5 mmol of (4-methyl-2-pentenyl)zirconocene chloride in CH_2Cl_2. The same experiment was also conducted using 5 mmol of B-chlorocatecholborane (0.769 g) in 5 mL of methylene chloride (^{11}B NMR (CH_2Cl_2) δ +31.5 ppm (Br), +31.6 ppm (Cl); after methanolysis +31.7 ppm (Br) 79% conversion, +31.7 ppm (Cl); 76% conversion of starting material). In a similar reaction, methanolysis was not done but the

---

solids were separated from the supernate and were washed with 4 × 10 mL of pentane to remove any remaining boron species from the zirconium. These washings were combined with the supernate, and the solvents were removed under reduced pressure. An oil resulted which was distilled using short path distillation to afford 1.01 g (68%) of (*Z*)-(*B*-4-methyl-2-pentenyl)catecholborane (bp 85–90 °C (0.1 mmHg)).

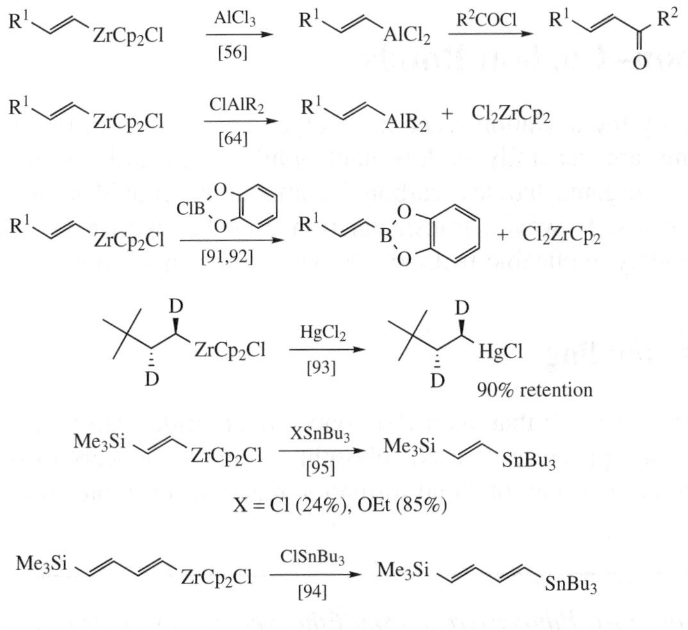

Scheme 14.

Once these compounds are formed, their subsequent reactions have little or nothing to do with Zr. Conversion of organylzirconocene chlorides into organyldichloroalanes with $AlCl_3$ promotes acylation with acyl chlorides to give ketones[7,56].

Octylaluminum Dichloride by Transmetalation from Zirconium to Aluminum and its Acetylation with Acetyl Chloride to produce 2-Decanone[7,56]

$$\text{OctZrCp}_2\text{Cl} \xrightarrow[\text{CH}_2\text{Cl}_2]{\text{AlCl}_3} \text{OctAlCl}_2 \xrightarrow[-30\,^\circ\text{C}]{\text{MeCOCl, CH}_2\text{Cl}_2} \text{OctCOMe}$$

Octylzirconocene chloride was generated from a mixture of linear octenes (1.12 g, 10 mmol) and $Cp_2Zr(H)Cl$ (2.58 g, 10 mmol) in CH_2Cl_2. This mixture was added to a suspension of 2.00 g (15 mmol) of $AlCl_3$, from which HCl had been removed by evacuation, in 10 mL of CH_2Cl_2 cooled to 0 °C. The resultant mixture was treated

with acetyl chloride (1.17 g, 15 mmol) at −30 °C for 1 h. After the usual aqueous
work-up, analysis by GLC indicated the formation of 2-decanone in 98% yield. The
product was identified by coinjection with an authentic sample.

This transmetalation is particularly important for transferring alkenyl groups, because
alkenylzirconocene chlorides do not undergo acylation with acyl chlorides.

4   Carbon–Carbon Bonds

As suggested by the acylation reaction of organozirconiums in the preceding section,
organozirconiums are generally of low nucleophilicity towards common organic electrophiles, such as organic halides, carbonyl compounds, epoxides, and α,β-unsaturated
carbonyl compounds. However, transmetalation, together with a few other techniques,
can provide a widely applicable method for overcoming this limitation.

4.1   Cross-coupling

The discovery in 1977[5] that alkenylzirconocene chlorides can react with aryl iodides
and bromides in the presence of a Ni–phosphine catalyst appears to represent the first
example of polar carbon–carbon bond formation reactions of organozirconiums catalyzed
by other metals.

*Preparation of (E)-β-Ethoxystyrene from Ethoxyethyne by Hydrozirconation and
Ni-catalyzed Cross-coupling*[5]

To 3.87 g (15 mmol) of Cl(H)ZrCp₂ suspended in 30 mL of dry benzene was
added 1.05 g (15 mmol) of ethoxyacetylene. After stirring the reaction mixture for
2 h at 25 °C, benzene and other volatile compounds were removed *in vacuo*, and
THF (10 mL) was added to the residue. In a separate flask Ni(PPh₃)₄ was prepared
from anhydrous Ni(acac)₂ (1 mmol), diisobutylaluminum hydride (1 mmol), triphenylphosphine (4 mmol), and 15 mL of THF. To the mixture containing the
β-ethoxyethenylzirconium derivative prepared above were added sequentially the
supernatant solution of Ni(PPh₃)₄ and 2.04 g (10 mmol) of iodobenzene. After stirring the reaction mixture for 12 h at 25 °C, GLC examination of a quenched aliquot
indicated the formation of (E)-β-ethoxystyrene in 99% yield. After extraction with

diethyl ether, washing with water, and drying, distillation provided 1.12 g (76%) of essentially pure (*E*)-*β*-ethoxystryene (bp 70–72 °C (1.6 mmHg).

Catalysis by Pd–phosphine complexes has been shown to be generally more satisfactory than that by Ni complexes, especially in alkenyl–alkenyl, alkynyl–alkenyl, and alkenyl–allyl coupling reactions (Schemes 15 and 16)[5,96–99]. In cases where alkenyl-metals are generated by carbometallation of alkynes with organoalanes and Cp_2ZrCl_2, alkenylalanes rather than alkenylzirconiums are the addition products. These alkenyl-lalanes may be used directly or used after conversion to alkenyl iodides in the Pd- or Ni-catalyzed cross-coupling reaction (Scheme 17). The use of Zn salts, such as $ZnBr_2$ and $ZnCl_2$, in conjunction with Pd or Ni catalysts can, in some cases, further promote the cross-coupling reaction of alkenylzirconiums and alkenylalanes[97]. However, in some of the most demanding cases, the use of preformed organozincs, typically generated by the Zr (or Al) $\rightarrow$ I $\rightarrow$ Li (or Mg) $\rightarrow$ Zn route, appears to provide highly satisfactory results[98]. Although Pd catalysts generally offer more satisfactory results for alkenyl–allyl coupling[99], Ni catalysts have been reported to be more satisfactory in the synthesis of coenzyme Q_n[100]. These features are demonstrated in the following three procedures.

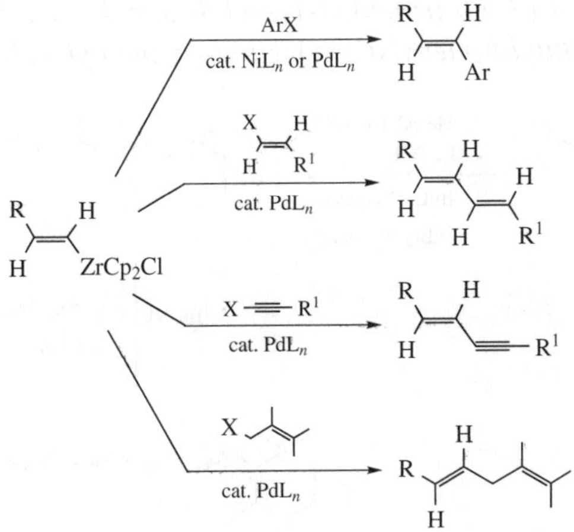

Scheme 15.

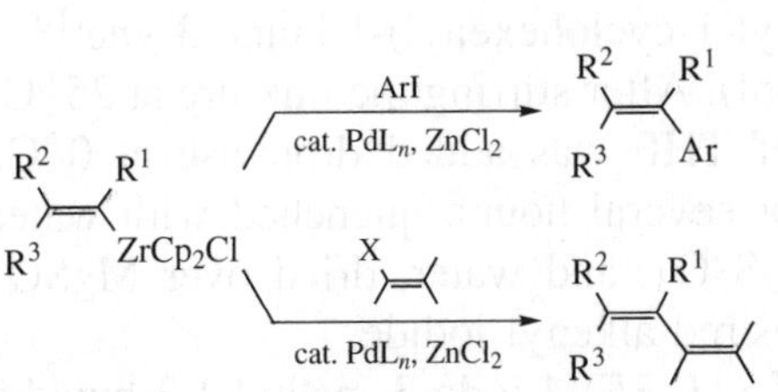

Scheme 16.

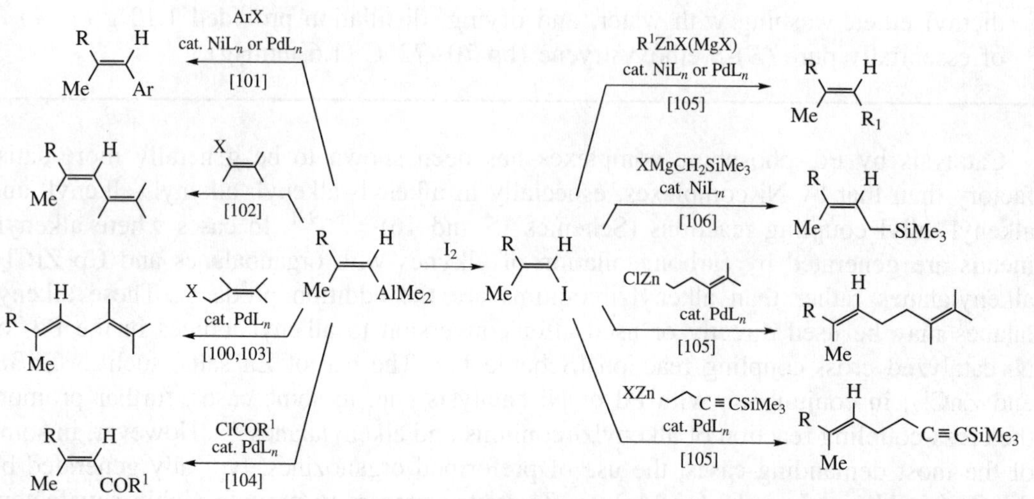

Scheme 17.

Preparation of Vitamin A by the Palladium-catalyzed Cross-coupling of (1E,3E)-Bis[2-methyl-4-(2,6,6-trimethyl-1-cyclohexen-1-yl)-1,3-butadiene-1-yl]zinc with (1E,3E)-5-(tert-Butyldiphenylsiloxy)-1-iodo-3-methyl-1,3-pentadiene[98]

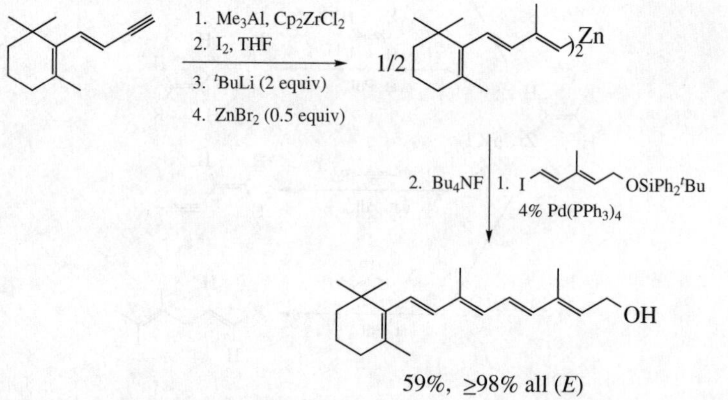

59%, ≥98% all (*E*)

(a) To Cp_2ZrCl_2 (2.92 g, 10 mmol) in 50 mL of 1,2-dichloroethane were added at 0 °C 1-(2,6,6-trimethyl-1-cyclohexenyl)-1-buten-3-yne[107] (1.74 g, 10 mmol) and Me_3Al (1.44 g, 20 mmol). After stirring the mixture at 25 °C for 3 h, iodine (3.04 g, 12 mmol) in 15 mL of THF was added dropwise at 0 °C. The reaction mixture was stirred at 25 °C for several hours, quenched with water/pentane, washed with aqueous $NaHCO_3$, $Na_2S_2O_3$, and water, dried over $MgSO_4$, and distilled to give 2.59 g (82%) of the desired alkenyl iodide.

(b) To a solution of (1*E*,3*E*)-4-iodo-3-methyl-1,3-butadien-1-yl-2,6,6-trimethyl-1-cyclohexene (0.174 g, 0.55 mmol) in 1.5 mL of THF was added ^{t}BuLi (0.64 mL,

1.1 mmol) at −78 °C. The mixture was stirred for 0.75 h at −78 °C, and then $ZnBr_2$ (1.0 M, 0.28 mL, 0.28 mmol) in THF was added. The resultant mixture was warmed to 25 °C, stirred for an additional 15 min and transferred to a mixture of (1*E*,3*E*)-(*tert*-butyldipehnylsiloxy)-1-iodo-3-methyl-1,3-pentadiene (0.231 g, 0.50 mmol), Pd(PPh₃)₄ (0.023 g, 0.02 mmol) and 0.5 mL of THF. After stirring for 0.75 h, the reaction mixture was quenched with water, extracted with pentane, dried over $MgSO_4$, concentrated, and examined by ^1H and ^{13}C NMR spectroscopy using DMF as an internal standard. The title compound of at least 98% stereoisomeric purity is formed in 87% NMR yield along with (1*E*,3*E*,5*E*,7*E*)-3,6-dimethyl-1,8-bis(2,6,6-trimethyl-1-cyclohexen-1-yl)-1,3,5,7-octatetraene, (2*E*,4*E*,6*E*,10*E*)-1,10-bis(*tert*-butyldipehnylsiloxy)-2,4,6,10-decatetraene, and (1*E*,3*E*)-3-methyl-1,3-butadienyl-2,6,6-trimethyl-1-cyclohexene in 4%, 1% and 18% yields, respectively. The starting iodide was consumed completely. Column chromatography (silica gel, hexane/ethyl acetate 20 : 1) provided 0.182 g (70%) of the *tert*-butyldiphenylsilyl ether of vitamin A.

(c) Treatment of the silyl-protected vitamin A obtained above with (Bu)₄NF in of THF provided, after work-up and evaporation, vitamin A which was at least 98% pure. Its NMR spectra were in good agreement with those of an authentic sample.

Generation of (E)-2-Methyl-1,3-butadienyldimethylalane by Zirconium-catalyzed Methylalumination and its Conversion to α-Farnesene by Palladium-catalyzed Cross-coupling[99]

(a) (*E*)-(2-Methyl-1,3-butadienyl)dimethylalane

An oven-dried, 1 L, two-necked, round-bottomed flask equipped with a magnetic stirring bar, a rubber septum, and an outlet connected to a mercury bubbler was charged with 7.01 g (24 mmol) of dichlorobis(η^5-cyclopentadienyl)zirconium and flushed with nitrogen. To this were added sequentially at 0 °C 100 mL of 1,2-dichloroethane, 12.48 g (120 mmol) of a 50% solution of 1-buten-3-yne in xylene, and 120 mL (240 mmol) of a 2 M solution of trimethylalane in toluene. The reaction mixture was stirred for 12 h at 25 °C.

(b) (3*E*,6*E*)-3,7,11-Trimethyl-1,3,6,10-dodecatetraene (α-farnesene)

To the solution of (*E*)-(2-methyl-1,3-butadienyl)dimethylalane prepared above were added 17.25 g (100 mmol) of geranyl chloride and 1.15 g (1 mmol) of tetrakis(triphenylphosphine)palladium dissolved in 100 mL of dry tetrahydrofuran, while the reaction temperature was controlled below 26–30 °C with a water bath. After the reaction mixture was stirred for 6 h at 25 °C, 250 mL of 3 N hydrochloric acid is slowly added at 0 °C. The organic layer was separated and the aqueous layer was extracted twice with pentane. The combined organic layer was washed with water, saturated aqueous sodium bicarbonate, and water. After the organic extract was dried over anhydrous magnesium sulfate, the solvent was removed thoroughly using a rotary evaporator (15–20 mm), and the crude product was passed through a short (15–20 cm) silica gel column (70–200 mesh) using hexane as an eluent. After the hexane was evaporated using a rotary evaporator, the residue was distilled using a 12-cm Vigreux column to give 16.70 g (83% based on geranyl chloride) of α-farnesene as a colorless liquid (bp 63–65° (0.05 mmHg)).

Coenzyme Q₄ via Zirconium-catalyzed Methylalumination and Nickel-catalyzed Alkenyl-benzyl Coupling[100]

(86%)

(a) To a 10 mL round-bottom Schlenk flask (equipped with a medium ground glass frit) was added zirconocene dichloride (0.073 g, 0.25 mmol) under an argon atmosphere. A solution of trimethylaluminum (0.75 mL, 2.0 M in hexane, 1.5 mmol) was added at 0 °C and stirred under reduced pressure until the hexane was removed. 1,2-Dichloroethane was added (1.0 mL), and the solution was allowed to stir and warm to 25 °C over 30 min. To this solution was added (5Z,9Z,13Z)-6,10,14-trimethyl-5,9,13-pentadecatrien-1-yne (0.224 g, 1.0 mmol), and the mixture was stirred at 0 °C for 30 min, after which carboalumination was complete, as determined by GC. The dichloroethane was pumped off *in vacuo*, and freshly distilled hexane (2 mL) was added, which was then also removed *in vacuo*. Additional hexane (5 mL) was then added to the flask to precipitate the zirconium salts. The hexane layer was removed by carefully decanting and filtering through the frit, with great care taken to avoid contamination by the zirconium salts. The left-over salts were not washed. The orange hexane solution was concentrated under reduced pressure and dissolved in THF (2.0 mL).

(b) To a 5 mL round-bottom flask were added bis(triphenylphosphine)nickel(II) chloride (Aldrich, 0.022 g, 0.033 mmol) and triphenylphosphine (0.018 g, 0.067 mmol) under an argon atmosphere at 25 °C. THF (1.0 mL) was added, followed by butyllithium (0.136 mL, 0.49 M in hexane, 0.067 mmol). The deep red solution was allowed to stir for 30 min, at which time 2,3,4,5-tetramethyoxy-6-methylbenzyl chloride (0.174 mg, 0.67 mmol) was added, and the subsequent dark blue solution was stirred for an additional 5 min. The solution containing the nickel catalyst was then transferred by cannula to the vinylalane at 25 °C, and the cross-coupling reaction was followed by GC analysis. When the reaction was complete (<15 min), the solution was diluted with diethyl ether (10 mL) and quenched at 0 °C by carefully adding 1.0 M HCl dropwise (3 mL). The mixture was allowed to stir for an additional 5 min and then extracted with diethyl ether. The combined organic layers were dried (Na$_2$SO$_4$/MgSO$_4$) and concentrated *in vacuo*. Silica gel column chromatography of the residue (petroleum ether) afforded the title compound (0.281 g, 87%) as a viscous, clear oil.

4.2 Conjugate Addition

Conjugate addition reactions of organozirconiums with α,β-unsaturated carbonyl compounds can be catalyzed by Cu (Scheme 18)[108–110], Zn[111] and Ni (Scheme 19)[6,112,113]. For the Ni-catalyzed conjugate addition reaction, a one-electron transfer mechanism was proposed.

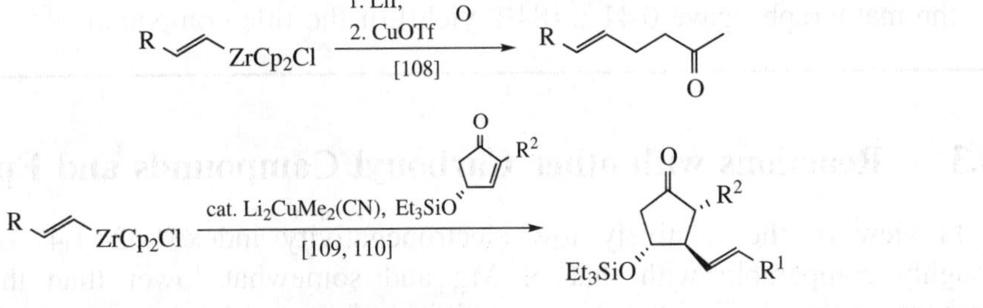

Scheme 18.

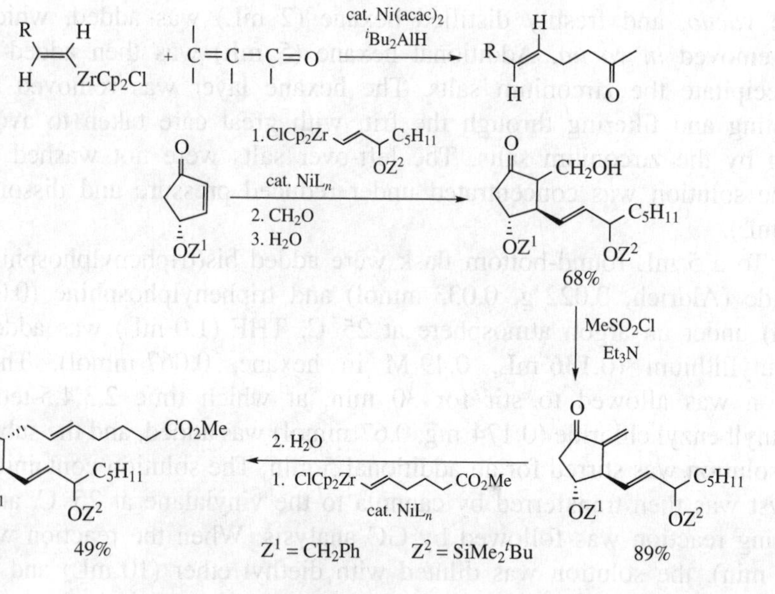

Scheme 19.

Preparation of
3-(trans-3-tert-Butyldimethylsiloxy-1-octen-1-yl)-4-cumyloxycyclopentanone[113]

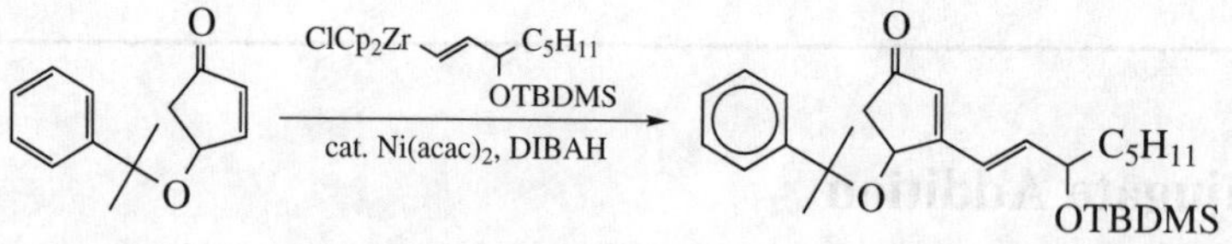

trans-3-*tert*-Butyldimethylsiloxy-1-octen-1-ylzirconocene chloride (1.217 g, 1.71 mmol) and 4-cumyloxycyclopentenone (0.245 g, 1.13 mmol) were dissolved in 15 mL of THF. This solution was added dropwise over 0.5 h to a second flask containing Ni(acac)$_2$ (0.066 g, 0.26 mmol) and DIBAH (0.27 mL of 0.94 M solution in THF, 0.25 mmol) dissolved in 4 mL of THF at 0 °C. The resulting solution was stirred for 4 h at 0 °C. Standard work-up, followed by preparative liquid chromatography gave 0.41 g (84% yield) of the title compound.

4.3   Reactions with other Carbonyl Compounds and Epoxides

In view of the relatively low electronegativity index (1.2–1.4) of Zr, which is roughly comparable with that of Mg and somewhat lower than those of Al and perhaps even of Zn, the low reactivity of organylzirconocene chlorides towards carbonyl compounds is puzzling. It is likely that the presence of two sterically demanding Cp groups is at least partially responsible for their low reactivity. In accord with this interpretation, alkyl(tributoxy)zirconiums react readily not only with aldehydes but also with ketones[114]. This reaction displays high addition/enolization ratios even with readily enolizable ketones.

Preparation of Methyltributoxyzirconium and its Reaction with β-Tetralone[114]

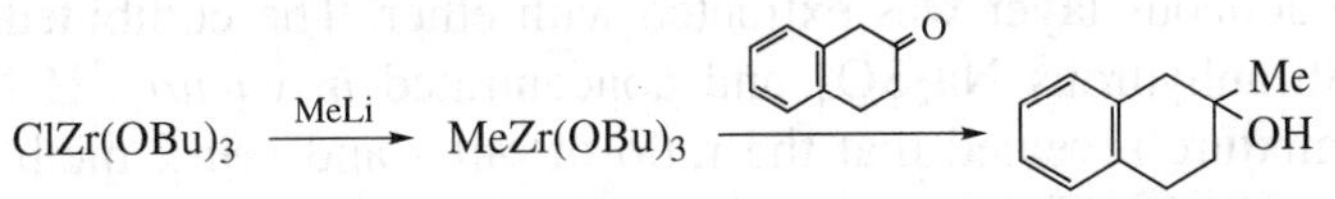

$$ClZr(OBu)_3 \xrightarrow{MeLi} MeZr(OBu)_3$$

(a) To a solution of 288 g (0.75 mol) of Zr(OBu)₄ in 600 mL of ether stirred at 0 °C under argon was slowly added 58.5 g (0.25 mol) of ZrCl₄. After stirring overnight, a brown solution was obtained. Concentration to a volume of 500 mL of partial evaporation furnished a 2 M stock solution of ClZr(OBu)₃.

(b) To a mixture of 6 mL of ether and 1.55 mL (3.1 mmol) of the 2 M solution of ClZr(OBu)₃ stirred at −10 °C under argon was added 2 mL of 1.5 M MeLi (3.0 mmol) in ether. After 1 h at 0 °C, a solution of 0.146 g (1.0 mmol) of β-tetralone in 2 mL of CH₂Cl₂ was added, and stirring was continued at 25 °C. The reaction mixture was worked up by pouring into 20% aqueous KF/ether. The ether phase was dried over MgSO₄ and evaporated. Chromatography provided 2-methyl-2-tetralol. The ratio of the extent of addition to that of enolization was 90 : 10.

Allylzirconocene derivatives are exceptionally reactive, and they can readily react with aldehydes, permitting an anti-selective synthesis of homoallyl alcohols (Scheme 20)[115].

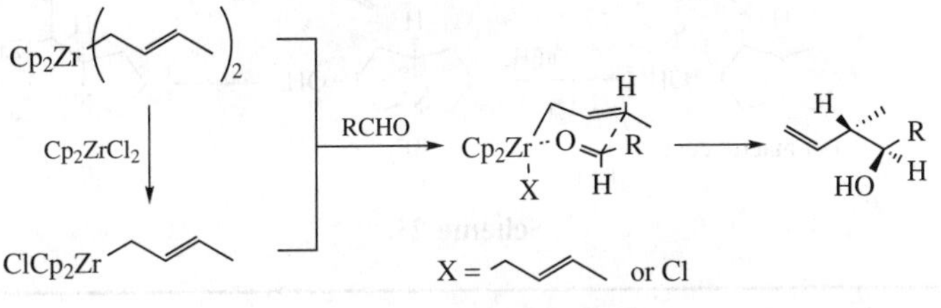

Scheme 20.

Reaction of Benzaldehyde with Crotylzirconocene Chloride Prepared in situ from Crotylmagnesium Chloride and Zirconocene Dichloride (anti-selective Synthesis of Homoallyl Alcohols)[115]

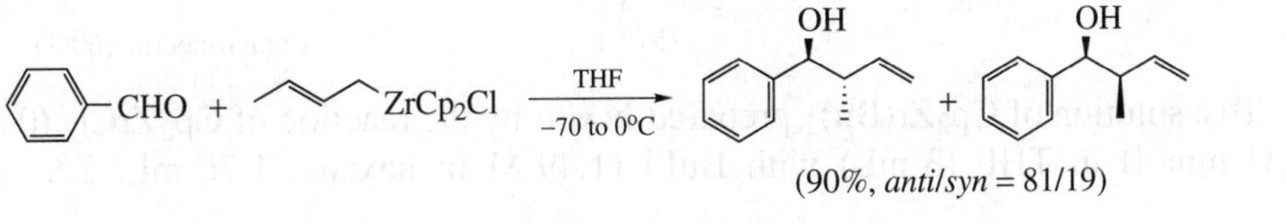

(90%, *anti/syn* = 81/19)

In a 50 mL flask equipped with a magnetic stirrer and maintained under N_2 was placed Cp_2ZrCl_2 (0.293 g, 1 mmol), and then dry THF (3 mL) was added to dissolve it. The flask was put in a dry ice/acetone bath, and the 0.25 M crotylmagnesium chloride in ether (1 mmol, 4 mL) was added. The mixture was stirred, and then benzaldehyde (1 mmol, 0.102 mL) was added. The mixture was stirred for 10 min, warmed to 0 °C, and quenched with aqueous NH_4Cl. The organic layer was separated, and the aqueous layer was extracted with ether. The combined organic layer was dried with anhydrous Na_2SO_4 and concentrated *in vacuo*. ^1H NMR analysis of the crude mixture revealed that the ratio of *anti*- and *syn*-2-methyl-1-phenyl-3-buten-1-ol was 81 : 19. The crude product was purified by a short silica-gel column chromatography using hexane/ether (20 : 1) as an eluent, giving 2-methyl-1-phenyl-3-buten-1-ol in 90% yield (0.145 g) (anti/syn = 81 : 19).

The synthesis of (−)-macronecine shown in Scheme 21 combines a novel oxidative addition procedure for generating allylzirconocene derivatives and an intramolecular addition to aldehydes to provide the crucial bicyclic precursor to the target compound.

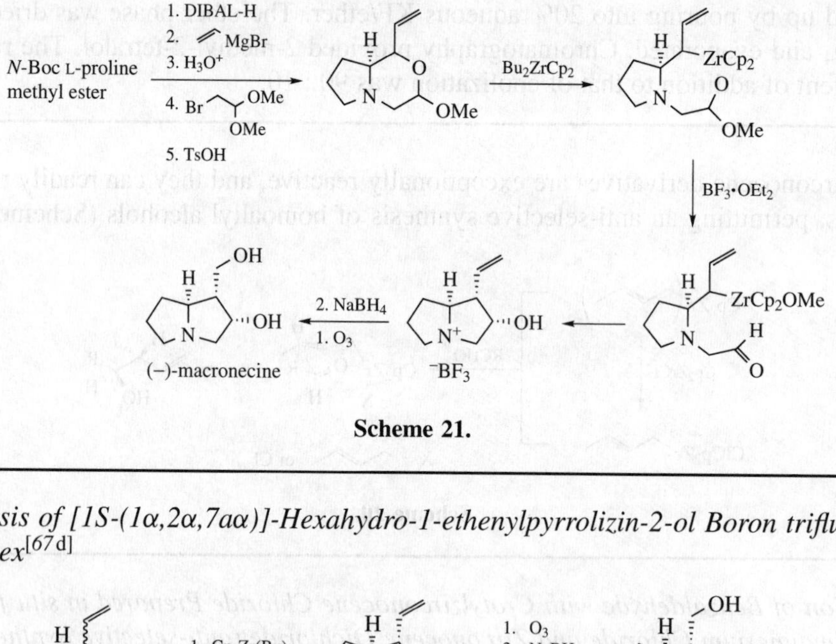

Scheme 21.

Synthesis of [1S-(1α,2α,7aα)]-Hexahydro-1-ethenylpyrrolizin-2-ol Boron trifluoride Complex[67d]

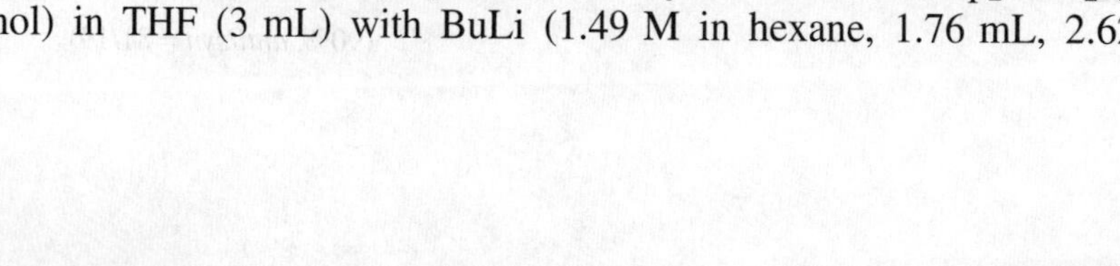

To a solution of $Cp_2Zr(Bu)_2$ prepared *in situ* by the reaction of Cp_2ZrCl_2 (0.383 g, 131 mmol) in THF (3 mL) with BuLi (1.49 M in hexane, 1.76 mL, 2.62 mmol)

at $-78\,^{\circ}$C for 1 h, was added a solution of the 1-aza-4-oxabicyclo[4.3.0]nanane derivative (a mixture of isomers; 0.200 g, 1.09 mmol) in THF (3 mL) at $-78\,^{\circ}$C, and the mixture was warmed to $25\,^{\circ}$C. After stirring for 4 h, a solution of BF$_3\bullet$OEt$_2$ (0.25 mL, 2 mmol) in THF was added to the reaction mixture at $0\,^{\circ}$C. Stirring was continued for 1 h at this temperature and then for an additional 1 h at $25\,^{\circ}$C. After addition of aqueous NH$_4$Cl, the mixture was extracted with CH$_2$Cl$_2$. The combined organic layers were washed with brine, dried over MgSO$_4$, and concentrated under reduced pressure. The residue was purified by silica gel column chromatography (hexane/ethyl acetate/10% NH$_3$–MeOH $= 5:5:1$) to give the title compound (0.138 g, 0.624 mmol, 57%) as a pale yellow oil ($[\alpha]_D^{24}-57.1^{\circ}$ (c 1.47, CHCl$_3$)). Ozonolysis, reduction of the resultant aldehyde, and neutralization with 10% NaOH converted this compound into $(-)$-macronecine in 60% yield (mp $124-127\,^{\circ}$C; $[\alpha]_D^{20} - 49.4^{\circ}$ (c 0.96, EtOH)).

1,3-Diene–zirconocene complexes may be viewed as bis(allyl)zirconocene derivatives. Indeed, they readily react not only with aldehydes and ketones but even with esters and nitriles. Furthermore, they can react with two different electrophiles in two discrete steps to permit the formation of two carbon–carbon bonds in a regio- and stereoselective manner (Scheme 22)[18,116].

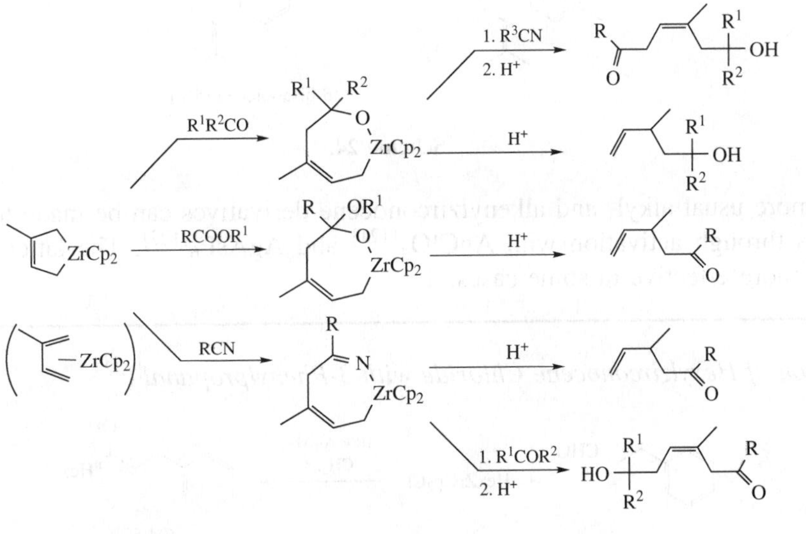

Scheme 22.

α-Metalloorganylzirconocene derivatives represent another class of organylzirconocene derivatives that display high reactivities towards carbonyl compounds[117] (Scheme 23). Even so, they appear to be less reactive than the corresponding organotitanium derivatives[118]. Related bimetallic complexes containing Zr and Zn[119] are also reactive towards ketones, and this reaction has been applied to the synthesis of β-bisabolene[120] (Scheme 24). It is not clear if these Zr–Al and Zr–Zn reagents react as bimetallic complexes, because they can, in principle, be converted *in situ* to metal–carbene complexes.

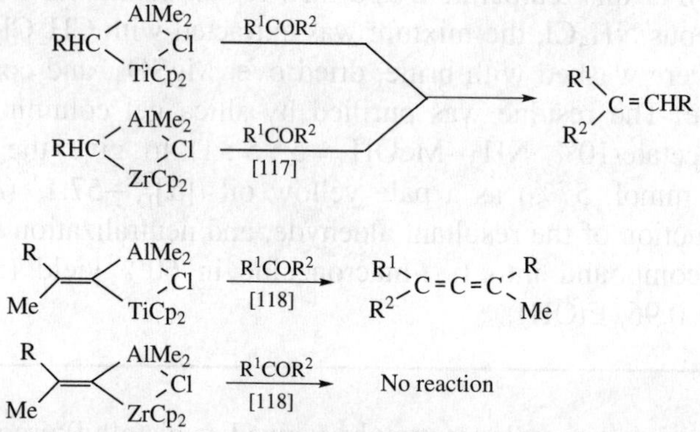

Scheme 23.

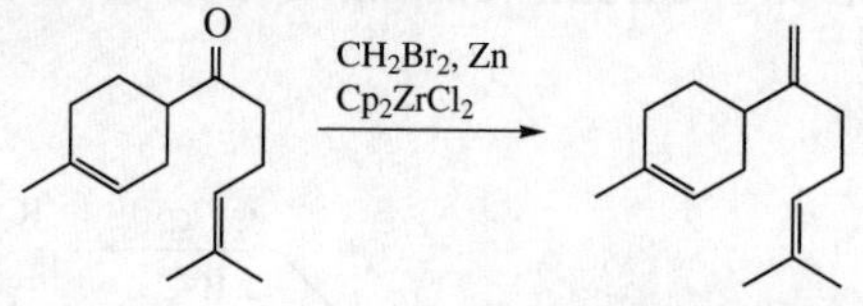

β-bisabolene (75%)

Scheme 24.

Even more usual alkyl- and alkenylzirconocene derivatives can be made to react with aldehydes through activation with AgClO$_4$[121] and AgAsF$_6$[122]. The latter is not only safer but more effective in some cases.

Reaction of Hexylzirconocene Chloride with 3-Phenylpropanal[122]

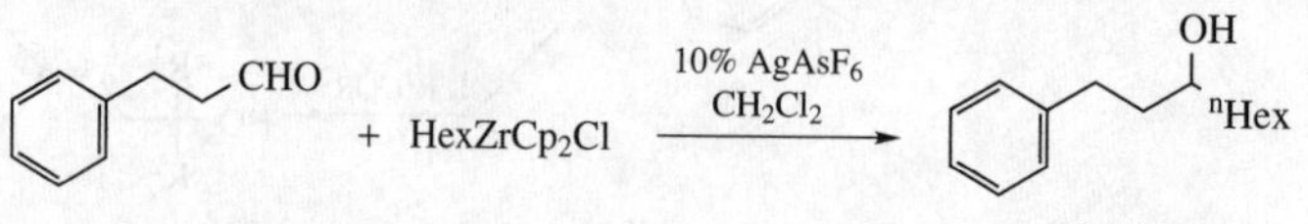

(94.5%)

A mixture of Cp$_2$Zr(H)Cl (0.270 g, 1.05 mmol) and 1-hexene (0.0931 g, 1.11 mmol) in CH$_2$Cl$_2$ (3 mL) was stirred for 10 min. To the resulting solution was added 3-phenylpropanol (0.0831 g, 0.619 mmol) in CH$_2$Cl$_2$ (3 mL) followed by AgAsF$_6$ (0.0201 g, 67.7 μmol, 10 mol %). The reaction mixture gradually turned dark brown. After 10 min, the mixture was poured into aqueous NaHCO$_3$. Extractive work-up (EtOAc) followed by purification with PTLC (heane/EtOAc = 4 : 1) gave 1-phenyl-3-nonanol as a colorless oil (0.129 g, 94.5%).

The same strategy is also effective in the reaction with epoxides.

Preparation of 1-Phenyl-3-octen-2-ol by the Reaction of (E)-1-Hexenylzirconocene Chloride with Styrene Oxide in the presence of Silver Perchlorate[123]

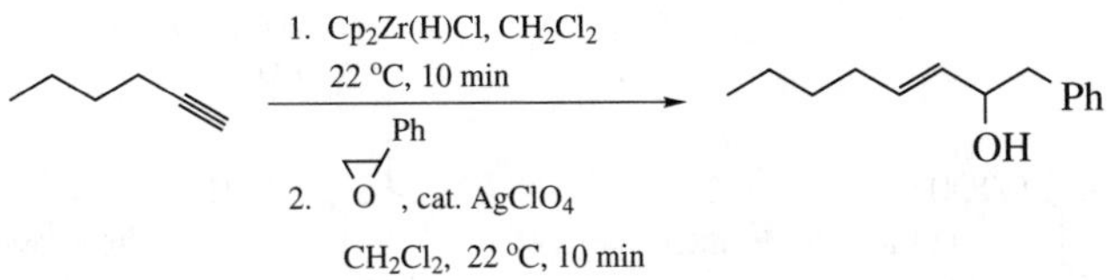

A solution of 0.200 g (1.22 mmol) of 1-hexyne in 4 mL of CH_2Cl_2 was treated at 22 °C with 0.313 g (1.22 mmol) of chlorohydridozirconocene. The mixture was stirred at 22 °C for 10 min. To the resulting yellow solution was added a solution of 0.122 g (1.02 mmol) of styrene oxide in 4 mL of CH_2Cl_2 followed by 0.011 g (0.05 mmol) of $AgClO_4$. After stirring for 10 min at 22 °C, the mixture was quenched with a saturated aqueous $NaHCO_3$ solution (5 mL), and extracted with 10 mL of EtOAc (3×). The combined organic layers were dried (Na_2SO_4), filtered through a pad of SiO_2 (EtOAc/hexanes = 1 : 4) to give 0.181 g (98%) of 1-phenyl-3-octen-2-ol as a colorless oil.

This reaction has been shown to involve Zr-catalyzed isomerization of epoxides to aldehydes (Scheme 25).

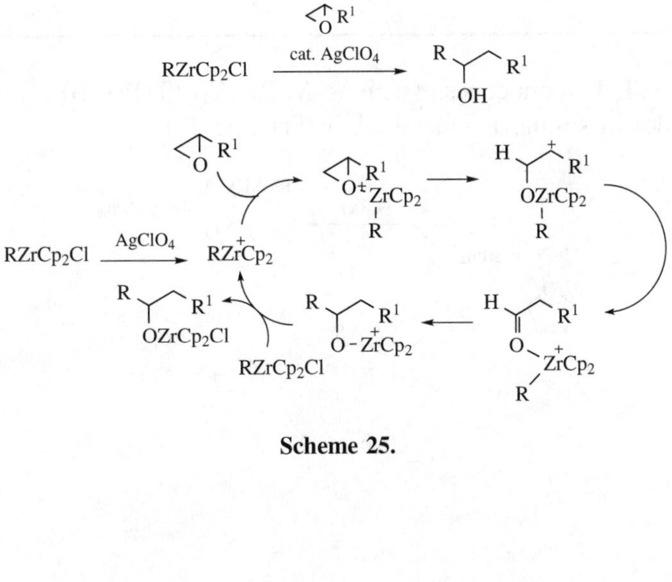

Scheme 25.

Generation of carbocationic species by the reaction of epoxides with Cp_2ZrCl_2 and $AgClO_4$ has been applied to the development of a highly regio- and stereoselective Diels–Alder reaction shown in Scheme 26.

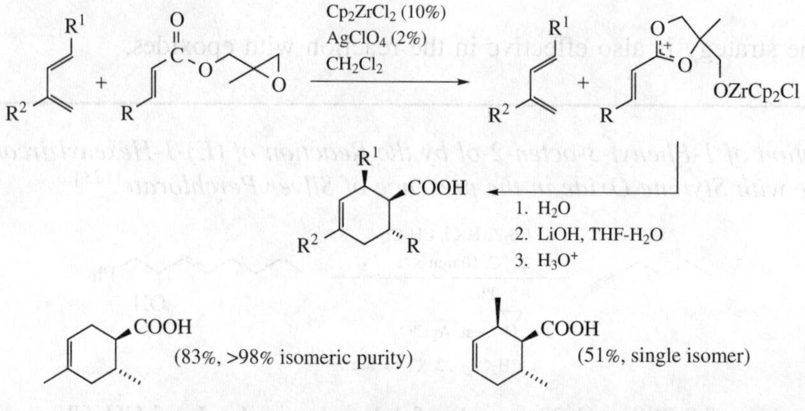

1. H_2O
2. LiOH, THF-H_2O
3. H_3O^+

(83%, >98% isomeric purity)

(51%, single isomer)

Scheme 26.

Preparation of trans-4,6-Dimethyl-3-cyclohexene-1-carboxylic Acid[124]

A solution of 0.030 g (0.19 mmol) of (E)-MeCH=CHCOOCH$_2$(Me)C$_2$H$_2$O in 1 mL of CH_2Cl_2 was treated at 0 °C with 0.130 g (1.9 mmol) of isoprene, 0.006 g (0.02 mmol) of Cp_2ZrCl_2 and finally 0.001 g (0.004 mmol) of $AgClO_4$. The reaction mixture was stirred for 7 h at 0 °C, treated with saturated aqueous NaHCO$_3$, and extracted with EtOAc (3×). The organic layers were dried (Na$_2$SO$_4$), filtered, evaporated *in vacuo*, and purified by chromatography (EtOAc/hexanes = 1 : 2). The product thus obtained was treated with 2 mL each of 1 M LiOH and THF at 25 °C overnight, concentrated, washed with Et$_2$O, acidified with 3 N HCl, extracted with Et$_2$O, dried, concentrated, and chromatographed on SiO$_2$ (EtOAc/hexanes = 1 : 3) to give 0.024 g (83%) of the title compound (mp 80–82.5 °C.

Cationic monoalkylzirconocenes, such as $MeZr^+Cp_2(THF)^-BPh_4$, react even with ketones and nitriles in similar manners[38,125] (Scheme 27).

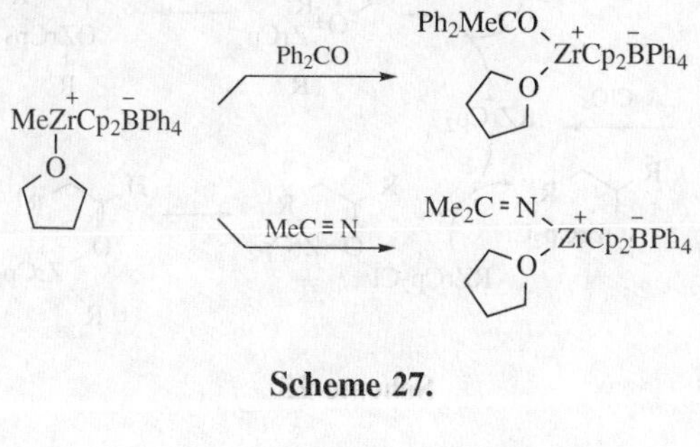

Scheme 27.

Transmetalation from Zr to Zn, originally used for the Ni- or Pd-catalyzed cross-coupling[97] provides a practical method for generating organometals of sufficient reactivity towards aldehydes. In most cases, this transformation itself may well be thermodynamically unfavorable. However, an equilibrium quantity of an organozinc or Zn-containing species may be generated *in situ* and consumed in subsequent reactions. Transmetalation reactions of alkenyl- and alkylzirconocene derivatives with Me$_2$Zn[44], Et$_2$Zn[44], ZnBr$_2$[111a], and LiZnMe$_3$[111c] have been reported (Scheme 28).

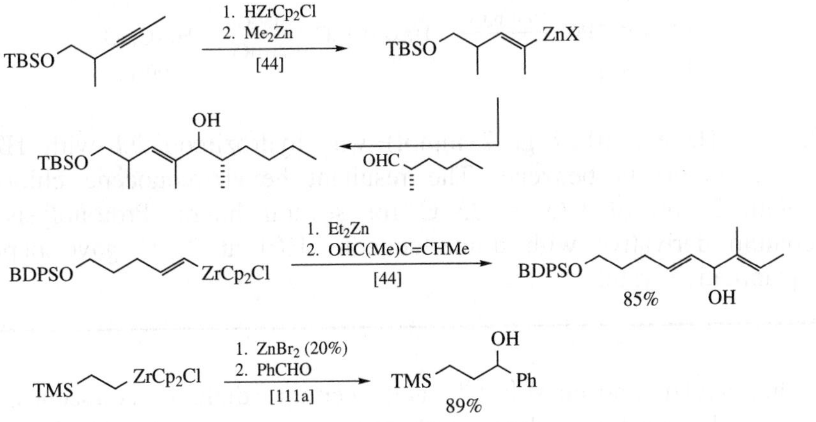

Scheme 28.

5 Carbonylation and Other Migratory Insertion Reactions

5.1 Carbonylation and Related Reactions

In contrast with common carbon electrophiles, which generally do not react readily with organylzirconocene derivatives, some carbon nucleophiles, such as CO and isonitriles, are quite reactive towards organylzirconocene derivatives, indicating that 16-electron RZrCp$_2$Cl can readily serve as electrophilic reagents. Thus, the reaction of RZrCp$_2$Cl with 1-3 atm of CO readily produces the corresponding RCOZrCp$_2$Cl that can be converted to

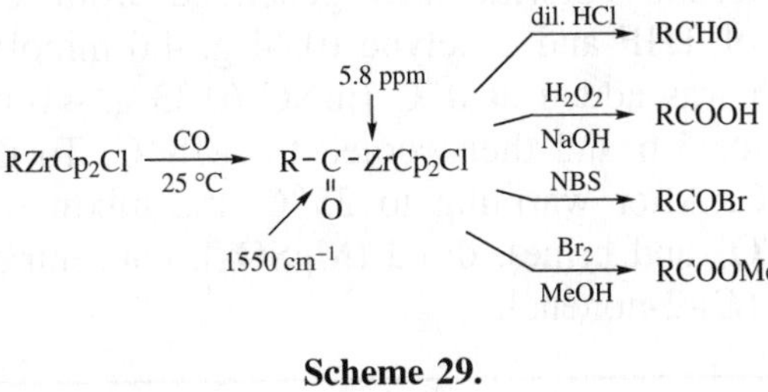

Scheme 29.

aldehydes and carboxylic acid derivatives (Scheme 29)[86,126]. The formation of aldehydes clearly indicates that the acyl group acts as an acyl anion equivalent rather than an acyl cation.

Conversion of Hexenes into Heptanal by Hydrozirconation–Carbonylation[126]

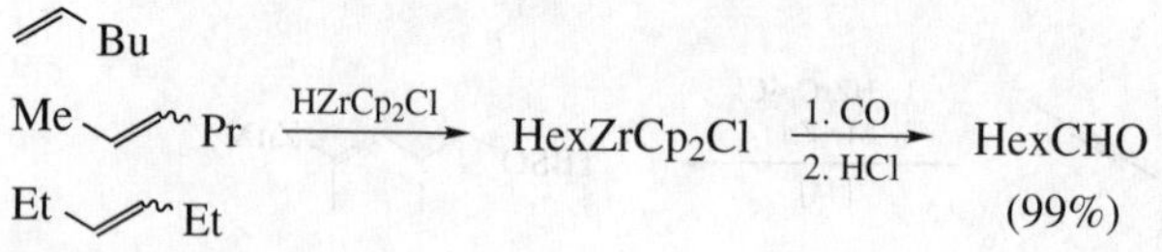

1, 2, or 3-Hexene (0.17 g, 2 mmol) was hydrozirconated with HZrCp$_2$Cl (0.515 g, 2 mmol) in benzene. The resultant hexylzirconocene chloride was treated with 2 atm of CO at 25 °C for several hours. Protonolysis of the acylzirconium derivative with dilute aqueous HCl at 25 °C gave heptanal in nearly quantitative yield.

Cyclic diorganylzirconocenes[16,127] and certain dialkylzirconocenes, such as Me$_2$ZrCp$_2$[128], also readily undergo carbonylation to give ketones. In some cases, the organozirconium products before protonolysis have been shown to be zirconium enolates[16,127].

Isonitriles are useful alternatives to CO for some one-carbon homologation reactions where CO is not very satisfactory. The preparation of α,β-unsaturated aldehydes is one such example.

Preparation of (E)-2-Nonenal from 1-Octyne by Hydrozirconation and One-carbon Homologation with Butyl Isocyanide[129]

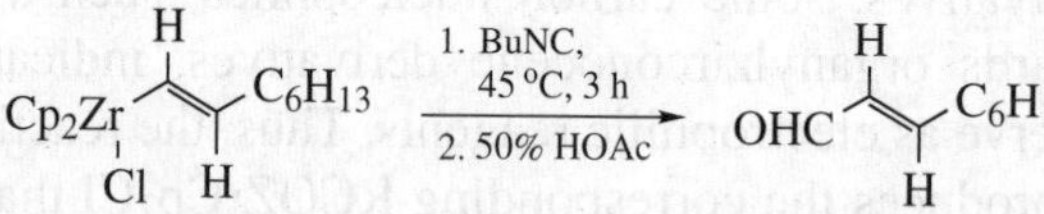

(E)-1-Octenylzirconocene chloride was generated from Cp$_2$Zr(H)Cl (1.52 g, 4.0 mmol) in 15 mL of THF and 1-octyne (0.44 g, 4.0 mmol) at 25 °C in >95% yield. To this solution was added at 0 °C BuNC (0.33 g, 4.0 mmol). The mixture was heated at 45 °C for 3 h and then cooled to −78 °C. To this was added 50% HOAc kept at −20 °C. After warming to 25 °C, the mixture was extracted with ether, washed (NaHCO$_3$ and brine), dried (MgSO$_4$), concentrated, and distilled to give 0.42 g (75%) of (E)-2-nonenal.

Another interesting reaction is the cyanation of alkenes shown in Scheme 30[130]. The aldehyde synthesis[129] shown in Scheme 30 has been applied to the synthesis of curacin A[44] (Scheme 31).

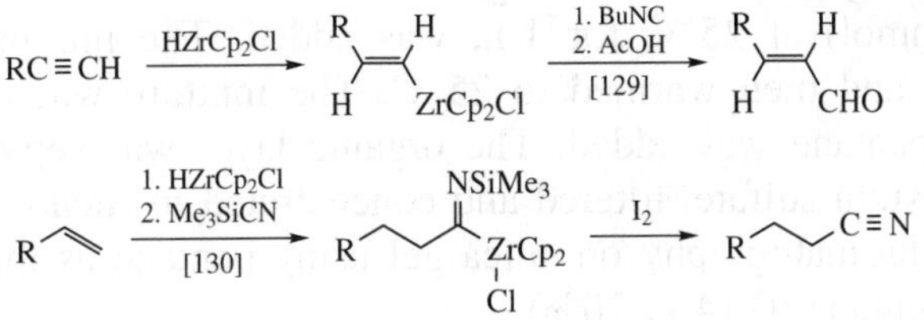

Scheme 30.

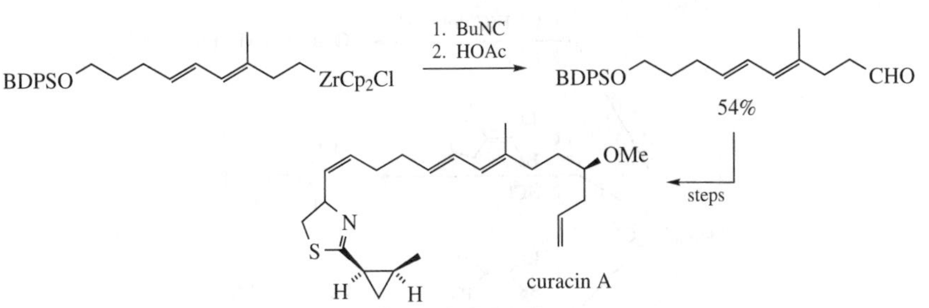

Scheme 31.

5.2 Migratory Insertion Reactions

Although the potential scope of the migratory insertion reactions of organozirconium compounds appears to be very broad, its current scope is largely limited to carbonylation and related reactions with isonitriles. Nevertheless, there are some scattered reports on other migratory insertion reactions, as shown in Schemes 32[131] and 33.

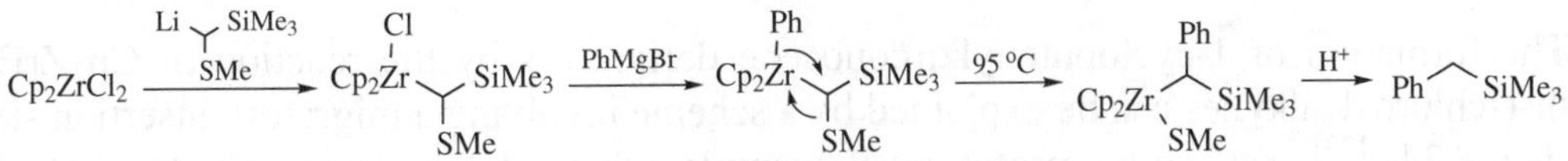

Scheme 32.

Preparation of Nonylidenecyclopentane[132]

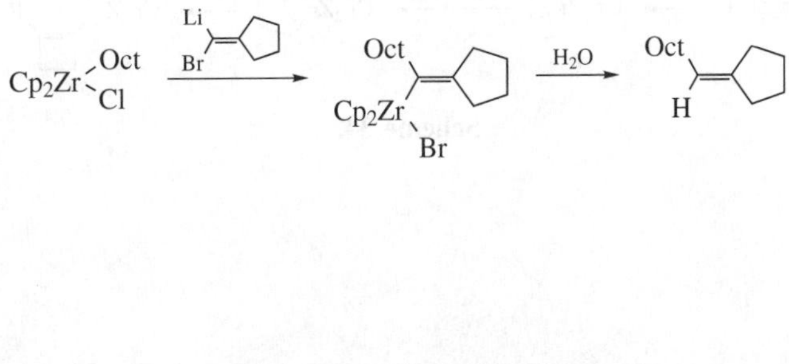

To a mixture of dibromomethylcyclopentane (0.24 g, 1.0 mmol) and 6 mL of Et$_2$O was added 0.4 mL (2.5 M, 1.0 mmol) of BuLi at $-110\,°C$ to $100\,°C$. After stirring at $-100\,°C$ for 0.5 h octylzirconocene chloride, prepared by the reaction of a suspension of Cp$_2$Zr(H)Cl (0.258 g, 1.0 mmol) in THF (3 mL) with 1-octene (0.156 mL, 1.0 mmol) at $25\,°C$ for 1 h, was added. The mixture was stirred at $-78\,°C$ for 0.5 h and then warmed to $25\,°C$. The mixture was quenched at $0\,°C$ with water, and pentane was added. The organic layer was separated, dried over anhydrous magnesium sulfate, filtered and concentrated *in vacuo*. The resulting oil was purified by chromatography on silica gel using pentane as the solvent to give nonylidenecyclopentane (0.14 g, 70%).

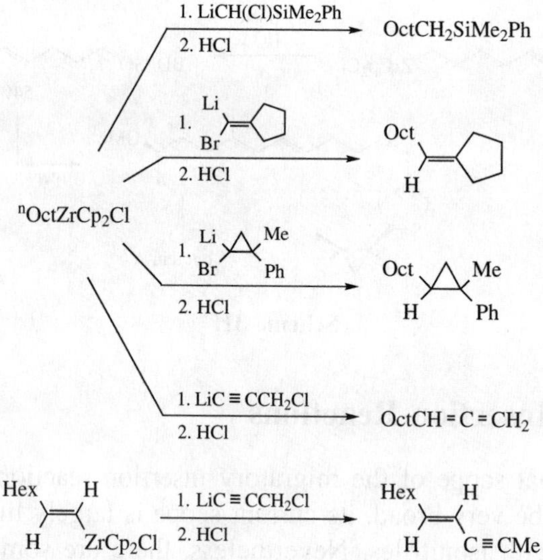

Scheme 33.

The formation of 1-cyclobutenylzirconocene derivatives by the reaction of Cp$_2$ZrEt$_2$ with 1-chloro-1-alkynes can be explained by a scheme involving a migratory insertion step (Scheme 34)[133]. Yet another prototypical example, which clearly deserves to be explored further, is the migratory insertion reaction of alkynylzirconium derivatives exemplified by the following procedure.

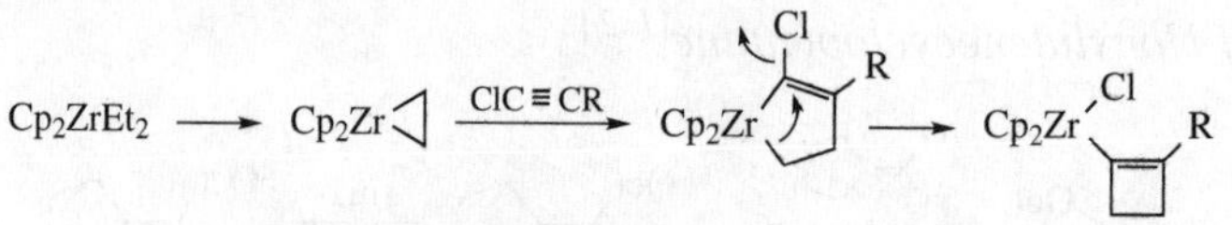

Scheme 34.

Reaction of Zirconocene Dichloride with Three Equivalents of 1-Octynyllithium Followed by Protonolysis to Produce (Z)-7-Hexadecen-9-yne[134]

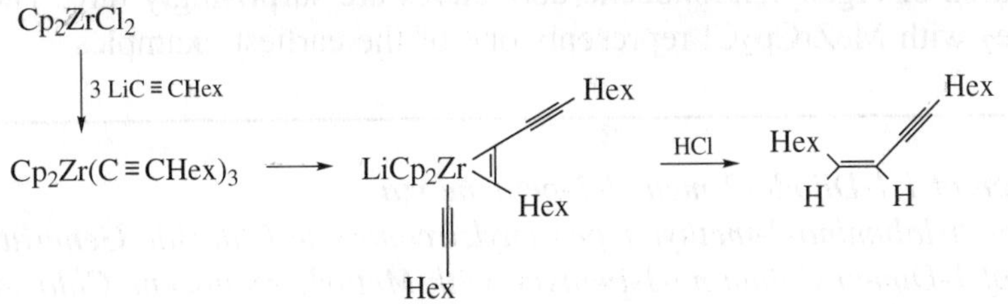

1-Octynyllithium, prepared from HexC≡CH (0.33 g, 0.44 mL, 3 mmol) and BuLi (2.3 M in hexane, 1.30 mL, 3 mmol) was added dropwise at −78 °C to a solution of Cp_2ZrCl_2 (0.292 g, 1 mmol) in THF (3 mL). The reaction mixture was warmed to 25 °C, stirred for an additional 24–48 h, quenched with 3 N HCl, extracted with hexane, washed with $NaHCO_3$, brine and dried over $MgSO_4$. GLC examination using a hydrocarbon internal standard indicated a 60–70% yield of (Z)-7-hexadien-9-yne, which was isolated in 55% yield (0.122 g), as an over 98% isomerically pure compound by concentration and silica gel chromatography (hexane).

6 Stoichiometric and Catalytic Carbometalation Reactions

6.1 Stoichiometric Carbozirconation

Organylzirconocene derivatives such as $RZrCp_2Cl$ are mostly 16-electron d^0 complexes with one empty valence shell orbital. As such they should, in principle, be capable of adding to alkenes and alkynes to undergo carbometalation. The key HOMO–LUMO interactions are those shown in Scheme 35, and it is essentially the same as those for hydrometalation except that an sp^n-hybridized carbon orbital participates in place of the 1s hydrogen orbital.

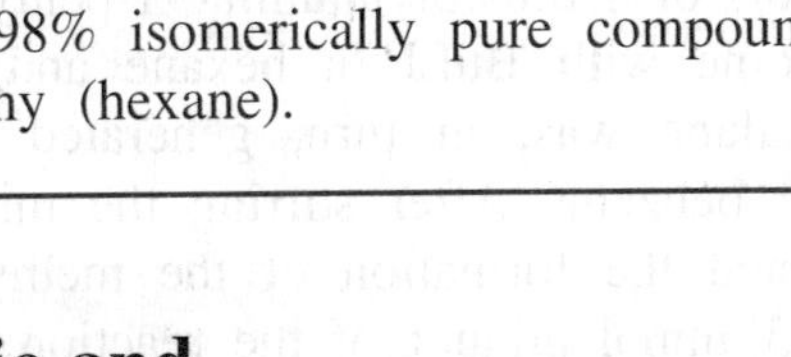

Scheme 35.

Indeed, carbozirconation has been implicated in a number of Zr-catalyzed carbometala-tion reactions, most notably Ziegler–Natta alkene polymerization[49,135] and Zr-catalyzed carboalumination[8,136], discussed later. However, clear-cut examples of the stoichiometric carbozirconation of organylzirconocene derivatives are surprisingly rare. The reaction of $PrC{\equiv}CAlMe_2$ with $MeZrCp_2Cl$ represents one of the earliest examples.

Preparation of 1,1-Diiodo-2-methyl-1-pentene via
(E)-1-Dimethylalumino-2-methyl-1-pentenylzirconocene Chloride Generated by the
Reaction of 1-Dimethylalumino-1-pentyne with Methylzirconocene Chloride[118]

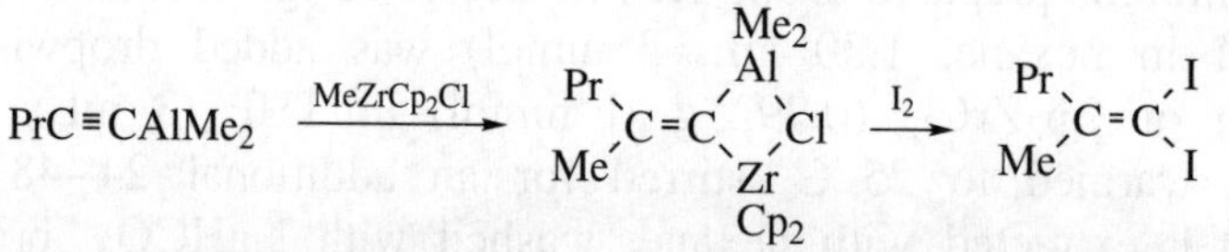

To 1.46 g (5.36 mol) of $MeZrCp_2Cl$ in 3 mL of methylene chloride was added 0.66 g (5.36 mmol) of 1-dimethylalumino-1-pentyne prepared by succes-sive treatment of 1-pentyne with BuLi in hexane and Me_2AlCl (1 equiv) in benzene. Chlorodimethylalane was, in turn, generated by mixing Me_3Al and $AlCl_3$ in a 2 : 1 ratio in benzene. After stirring the mixture for 3 h at 25 °C, analysis by NMR indicated the formation of the methylzirconation product in quantitative yield. To a 5 mmol aliquot of the reaction mixture obtained above was added 2.79 g (11 mmol) of iodine in 5 mL of THF at −78 °C. The reaction mixture was warmed to 25 °C and successively treated with water and pentane. The organic layer was treated with 3 N HCl and washed with aqueous $NaHCO_3$, $Na_2S_2O_3$, and water. After drying, distillation provided the title compound in 60% yield (92% by GLC).

Another significant example is the methylzirconation with a cationic reagent, $MeZr^+Cp_2(THF)^-BPh_4$ (Scheme 36)[125]. In both of these examples, cationic methylzirconocene species appear to be required for facile carbozirconation. Similar modes of activation of $RZrCp_2Cl$ towards carbonyl compounds have since been devised, as discussed in Section 4.3.

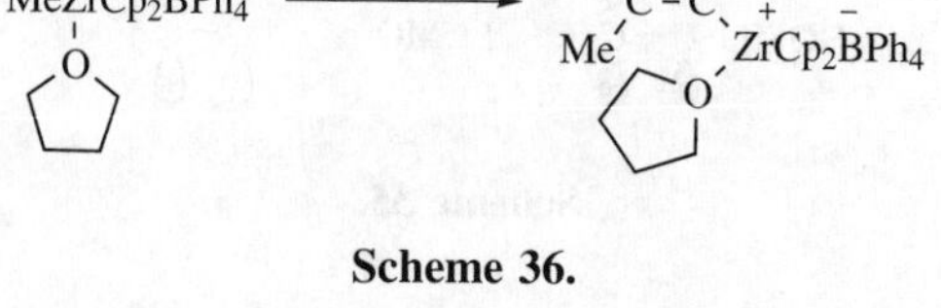

Scheme 36.

6.2   Zirconium–catalyzed Carbometalation

Apart from the Ziegler–Natta-type alkene polymerization catalyzed by zirconocene derivatives[49], most notably the Kaminsky protocol[136], the methylalumination of alkynes by Me$_3$Al and a catalytic amount of Cp$_2$ZrCl$_2$[8,137] (Scheme 37) appears to represent the first example of the organic synthetic reactions catalyzed by a zirconocene derivative. The reaction is generally high yielding, essentially 100% *syn* stereoselective, and about 95% regioselective for methylalumination of terminal alkynes[8,137,138].

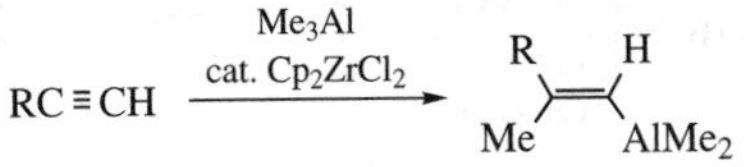

$$RC \equiv CH \xrightarrow[\text{cat. Cp}_2\text{ZrCl}_2]{\text{Me}_3\text{Al}} \underset{\underset{Me}{} \underset{AlMe_2}{}}{R \quad H}$$

Scheme 37.

Zirconium-catalyzed Methylalumination of Alkynes. Preparation of
(E)-2-Phenyl-1-propenyldimethylalane and (E)-1-Iodo-2-phenylpropene[138]

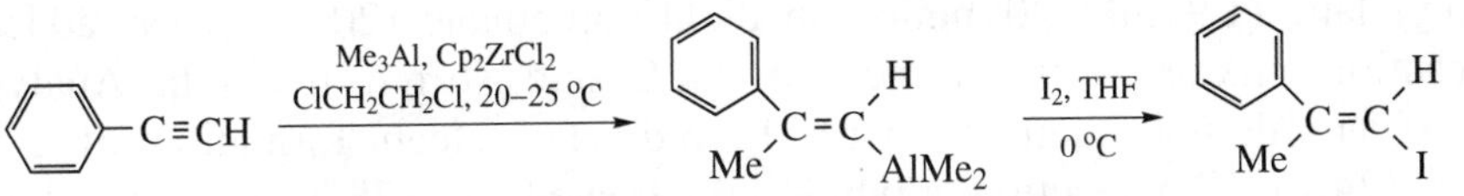

A dry 100-mL round-bottomed flask equipped with a magnetic stirring bar, a septum inlet, and an outlet connected to a mercury bubbler was charged with 2.92 g (10 mmol) of dichlorobis(η^5-cyclopentadienyl)zirconium, and the system was flushed with nitrogen. To this were added 25 mL of 1,2-dichloroethane and 1.44 g (12 mL, 20 mmol) of trimethylaluminum using syringes. The dichlorobis(η^5-cyclopentadienyl)zirconium dissolved within 10–15 min to produce a lemon-yellow solution. To this solution was added 1.02 g (1.11 mL, 10 mmol) of phenylethyne at 20–25 °C. After 24 h, examination by GLC (SE-30) of the reaction mixture after quenching with diluse. HCl/hexane indicated the formation of a 96 : 4 mixture of 2-phenylpropene and (E)-1-phenylpropene in 90–100% combined yield.

To the reaction mixture obtained above were added at 0 °C 3.04 g (12 mmol) of iodine dissolved in 15 mL of THF. After the iodine color had faded, the reaction mixture was quenched with a water/ether mixture. The organic layer was separated and washed with aqueous Na$_2$S$_2$O$_3$, dried over anhydrous magnesium sulfate, filtered, and distilled to give 1.79 g (73% yield) of (E)-1-iodo-2-phenylpropene (bp 72.5° (0.55 mmHg)). The stereoisomeric purity is ≥98%, and the regioisomeric purity is ≥95%.

Although some related reactions using Me$_3$Al–Cp$_2$TiCl$_2$[139] and R$_2$Zn–Cp$_2$ZrI$_2$[140] were also discovered, they are of limited synthetic utility. Various heterofunctionalities, such as halogens, alcohols, and amines, can be accommodated[141], but in some cases, they lead to side reactions including synthetically useful cyclobutenation[142–146] and *anti*-carbometalation[147] as the following procedure show.

1-Iodo-2-methylcyclobutene by Zirconium-catalyzed Methylalumination of 1-Alumino-substituted 4-Bromo-1-butyne[143]

$$HC \equiv CCH_2CH_2Br \quad \xrightarrow[\substack{3.\ Me_3Al \\ Cp_2ZrCl_2}]{\substack{1.\ BuLi \\ 2.\ Me_2AlCl}} \quad Me_2Al \diagup \square \diagdown Me \quad \xrightarrow{I_2} \quad I \diagup \square \diagdown Me$$

(59%)

To a solution of 4-bromo-1-butyne (1.33 g, 10.0 mmol) in pentane (20 mL) at −78 °C was added 2.71 N butyllithium in hexane (3.7 mL, 10.0 mmol) dropwise with stirring. After 30 min, chlorodimethylalane prepared by mixing trimethylalane (0.67 mL, 7 mmol) and $AlCl_3$ (0.47 g, 3.5 mmol) was added to the mixture. After 30 min, a solution of Cp_2ZrCl_2 (2.92 g, 10.0 mmol) and trimethylalane (1.9 mL, 20 mmol) in dichloromethane (20 mL) was added, and the reaction mixture was warmed to 25 °C and stirred for 3 h. Analysis by GLC of an aliquot quenched with I_2 indicated clean formation of a single product (74%). The reaction mixture was treated at −78 °C with a solution of I_2 (3.81 g, 15.0 mmol) in THF (10 mL), warmed to 0 °C, and quenched in a mixture of ice and 3 N HCl. The aqueous phase was extracted with diethyl ether, and the combined organic phases were washed with saturated aqueous $NaHCO_3$, aqueous $Na_2S_2O_3$, and saturated aqueous NaCl, dried ($MgSO_4$), and concentrated. The title iodide (1.14 g, 59%) distilled as a clear liquid (>95% pure by GLC; bp 80 °C (45 mm, Kugelrohr)).

Zirconium-catalyzed Anti Methylalumination of 3-Butyn-1-ol[147]

$$HC \equiv CCH_2CH_2OH \quad \xrightarrow[\substack{2.\ reflux,\ 3\ days}]{\substack{1.\ Me_3Al\ (3\ equiv) \\ Cp_2ZrCl_2\ (0.25\ equiv)}} \quad \substack{H \quad Me \\ MeAl \diagdown \diagup \\ O} \quad \xrightarrow[\substack{THF}]{I_2} \quad \substack{H \quad Me \\ I \diagdown \diagup \\ HO}$$

(60%, >98% Z)

A suspension of Cp_2ZrCl_2 (7.3 g, 25 mmol) in 1,2-dichloroethane (200 mL) was successively treated with trimethylalane (21.9 g, 28.8 mL, 300 mmol) and 3-butyn-1-ol (7.0 g, 100 mmol) in 1,2-dichloroethane (200 mL) at 25 °C. (**Caution**: Addition of the reagents must be done carefully and slowly. The reaction is exothermic.) The reaction mixture was refluxed for about 3 days to complete the desired stereoisomerization. The progress of the reaction may be monitored by NMR spectroscopy. Successive treatment of the resultant mixture with I_2 (38.0 g, 150 mmol) in THF (40 mL) at $-30\,^\circ$C, aqueous $Na_2S_2O_3$, and aqueous K_2CO_3 followed by the usual work-up and chromatography on silica gel (5 : 1 hexanes/EtOAc) afforded 12.6 g (60%) of (3Z)-4-iodo-3-methyl-3-buten-1-ol ($Z/E > 98/2$).

Once alkenylalanes are generated, they undergo essentially all of the known reactions of organoalanes and organoalanates, the latter of which can be readily generated by treating the former with alkyllithiums[148,149]. It suffices to point out here that various carbon–heteroatom bond formation reactions (Scheme 38)[57,150,151], conventional polar reactions for carbon–carbon bond formation (Scheme 39)[152,153], and the Ni- or Pd-catalyzed cross-coupling reactions (Scheme 17)[101–106] proceed satisfactorily in many cases. The procedure below illustrates carbon–carbon bond formation.

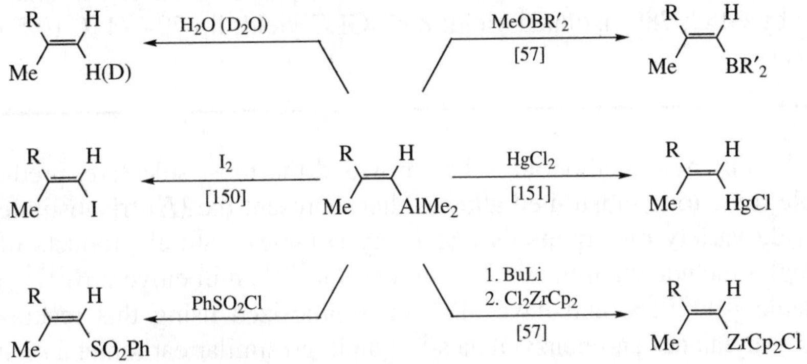

Scheme 38.

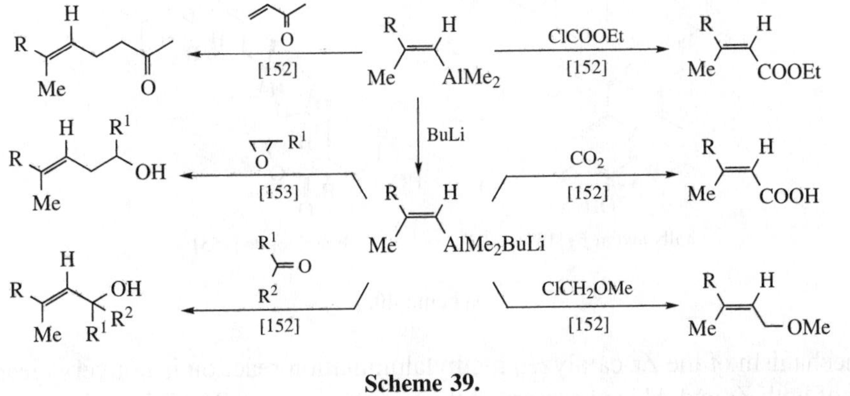

Scheme 39.

Preparation of Ethyl Geranate from 6-Methyl-5-hepten-1-yne by Zirconium-catalyzed Methylalumination and Treatment with Ethyl Chloroformate[152]

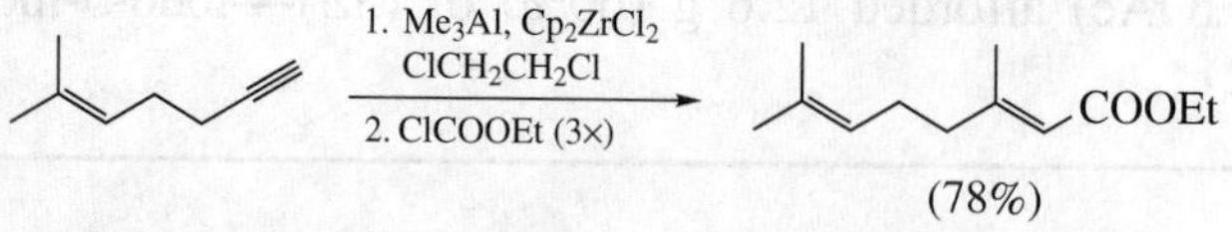

(78%)

To a solution of Cl_2ZrCp_2 (2.92 g, 10 mmol) and Me_3Al (1.22 g, 1.63 mL, 17 mmol) in 1,2-dichloroethane (25 mL), was added 6-methyl-5-hepten-1-yne (1.08 g, 10 mmol) at 25 °C, and the mixture was stirred for 2 h. After removal of the volatile components under reduced pressure (0.5 mmHg) at 50 °C, the carbo-metalated compound was extracted with hexane (5 × 6 mL), and transferred to another flask. To this extract was added ethyl chloroformate (3.26 g, 30 mmol), and the mixture was stirred for 1 h at 25 °C. After treatment with 3 N hydrochloric acid, ether, and aqueous sodium carbonate, distillation provided ethyl geranate (1.53 g, 98% pure by GLC; 78% isolated yield; 85% GLC yield; bp 72–74 °C (0.5 mmHg); n_D^{23} 1.4677).

Collectively, this new methodology has provided the most selective, predictable, and hence reliable route to trisubstituted alkenes that represent the (*E*)-trisubstituted isoprene units of a wide variety of terpenoids and many complex natural products of terpenoid origin. Examples include vitamin A[98], coenzyme Q*n*[100], milbemycin β_3[154], and brassi-nolide (Scheme 40)[155], which have all been synthesized using this reaction in some critical steps. Allylalanes and benzylalanes[156] undergo similar carbometalation reactions.

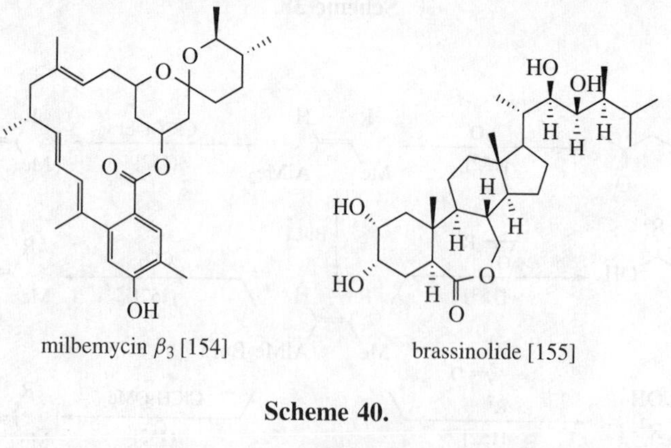

milbemycin β_3 [154]   brassinolide [155]

Scheme 40.

The mechanism of the Zr-catalyzed methylalumination reaction is not very clear, but the presence of both Zr and Al is necessary at the crucial moment. There have been indications

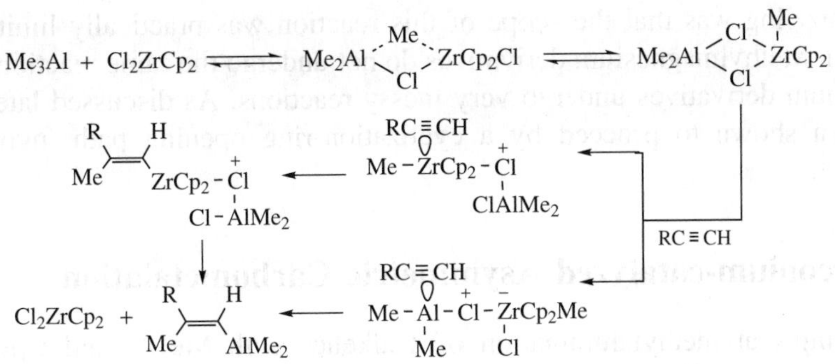

Scheme 41.

supporting a mechanism involving carbozirconation, followed by transmetalation and/or direct carboalumination (Scheme 41)[55,137].

The use of trialkylalanes containing ethyl and higher alkyl groups leads to competitive hydrometallation[8,137]. In fact, $^{i}Bu_3Al$–Cp_2ZrCl_2 can undergo hydroalumination, exclusively rather than carboalumination with either alkynes[157] or alkenes[64]. Zirconium-catalyzed carboalumination of terminal alkenes with Me_3Al–Cp_2ZrCl_2 may give isoalkyl-metals as the initial product. However, the product should then undergo competitive hydroalumination. In accord with this expectation, the reaction of 1-octene with Me_3Al and a catalytic amount of Cp_2ZrCl_2 gives 2-(n-hexyl)-1-decene in 59% yield along with an 18% yield of 2-methyl-1-octene without producing the desired 2-methyloctane in detectable amount[158]. In view of these results, it was somewhat puzzling that EtMgBr undergoes ethylmagnesiation in the presence of a catalytic amount of Cp_2ZrCl_2 (Procedure 36)[159,160].

Conversion of 1-Decene to 2-Ethyldecylmagnesium Bromide by Zirconium-catalyzed Ethylmagnesiation, Followed by Iodinolysis to Produce 3-Iodomethylundecane[160]

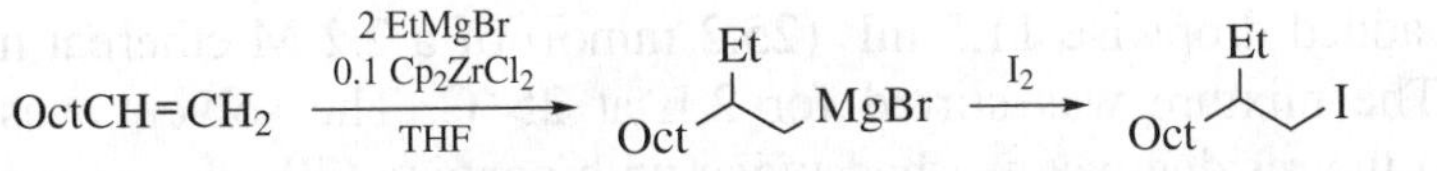

To Cp_2ZrCl_2 (0.292 g, 1 mmol) suspended in 20 mL of THF were added 1-decene (1.40 g, 10 mmol) and EtMgBr in THF (2.0 M, 15 L, 30 mmol) at 0 °C, and the reaction mixture was stirred for 1 day. Examination by NMR indicated the formation of 2-ethyldecylmagnesium bromide in 80% yield. Treatment of a 1 mmol aliquot (1/10 of the entire mixture) with I_2 (1.02 g, 4 mmol) in THF at 25 °C followed by quenching with 3 N HCl, extraction with pentane, washing with aqueous $Na_2S_2O_3$ and NaCl, drying ($MgSO_4$), and concentration provided 0.22 g (74%) of 3-iodomethylundecane which was ≥98% isomerically pure.

Equally puzzling was that the scope of this reaction was practically limited to ethyl-magnesiation. Methylmagnesium derivatives do not undergo the same reaction, and higher alkylmagnesium derivatives undergo very messy reactions. As discussed later, this reaction has been shown to proceed by a cyclization-ring opening path involving β-CH activation[160].

6.3   Zirconium-catalyzed Asymmetric Carbometalation

Early attempts at methylalumination of 1-alkenes with Me₃Al and Cp₂ZrCl₂ were largely unsuccessful. Evidently, the desired methylmetalation takes place, but the isoalkylzirconocene product in the presence of an alane readily hydroaluminates alkenes, as exemplified by the reaction shown in Scheme 42[64]. In view of these results, it is rather surprising that the reaction of 1-alkenes with one equivalent of Me₃Al, and a catalytic amount of dichlorobis(1-neomenthylindenyl)zirconium, **4**, in (CH₂Cl)₂, has produced in 70–90% yields the desired methylaluminated products readily oxidizable with O₂ to give the corresponding alcohols (Scheme 43)[158].

Preparation of Bis(1-neomenthylindenyl)zirconium Dichloride[161]

(a) (1-Neomenthylindenyl)lithium

To a solution of 6.4 g (25.2 mmol) of (+)-3-neomenthylindene in 100 mL of ether was added dropwise 11.5 mL (25.2 mmol) of a 2.2 M ethereal methyllithium solution. The mixture was stirred for 2 h at 25 °C. The solvent was removed *in vacuo*, and the residue was washed twice with pentane (30 mL each) to yield 5.4 g (82%) of the title compound.

(b) Bis(1-neomenthylindenyl)zirconium dichloride

To a suspension of 2.9 g (7.7 mmol) of ZrCl₄(THF)₂ in 50 mL of toluene was added at −78 °C a cold solution of 4.0 g (15.4 mmol) of (1-neonmenthylindenyl)lithium in 200 mL of tetrahydrofuran. The mixture was allowed to warm to ambient temperature during 6 h with stirring and then kept at room temperature for 12 h. The solvents were removed *in vacuo*. The yellow solid thus obtained was washed twice with pentane (50 mL each) and treated with 100 mL of methylene chloride, and the solution was filtered from the precipitated lithium chloride. The clear filtrate was concentrated *in vacuo* and allowed to crystallize at $-28\,°C$. The crystals obtained were suited for X-ray crystal structure analysis. The mother liquor was further concentrated and cooled to $-28\,°C$ for crystallization. Repeated recrystallization of the material from methylene chloride furnished the title compound in a diastereomerically pure state (3.15 g (61%); mp 146 °C; $[\alpha]_D = -77°$ ($c = 0.23$, toluene)).

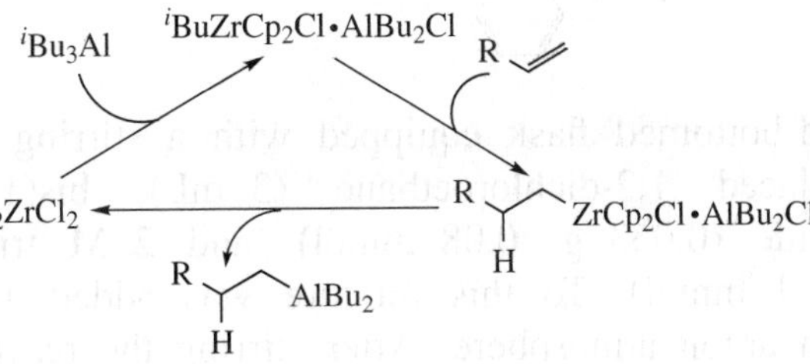

Scheme 42.

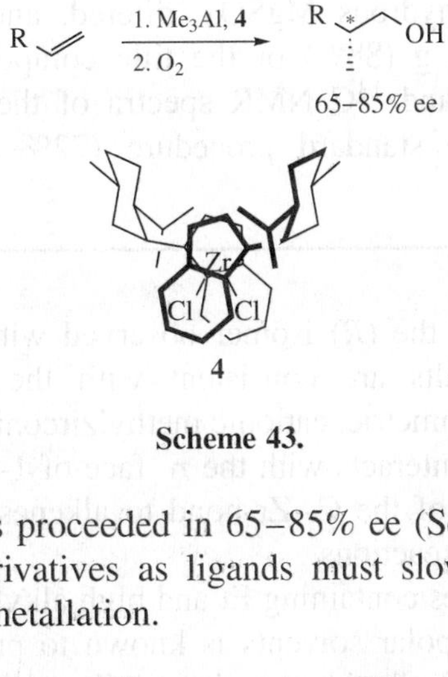

65–85% ee

Scheme 43.

Moreover, the reaction has proceeded in 65–85% ee (Scheme 43)[158]. Evidently, the use of certain bulky Cp derivatives as ligands must slow down β-dehydrometalation relative to the desired carbometalation.

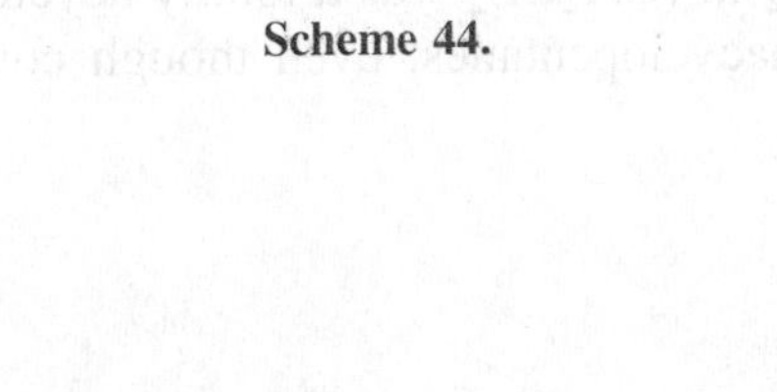

Scheme 44.

Conversion of 1-Octene to (2R)-2-Methyl-1-octanol by Zirconium-catalyzed Enantioselective Methylalumination[158]

$$C_6H_{13}CH=CH_2 \xrightarrow[\text{2. O}_2]{\text{1. Me}_3\text{Al, Cl}_2\text{ZrCp}_2^*} C_6H_{13}\overset{\overset{\text{Me}}{\underset{}{\big|}}}{\underset{}{}}\text{H}\ \text{OH}$$

(88%, 72% ee)

$Cp_2^* =$

In a dry, round-bottomed flask equipped with a stirring bar and a mercury bubbler were placed 1,2-dichloroethane (3 mL), bis(1-neomenthylindenyl) zirconium dichloride (0.053 g, 0.08 mmol), and 2 M trimethylaluminum in toluene (0.5 mL, 1 mmol). To this mixture was added 1-octene (0.156 mL, 1 mmol) under an argon atmosphere. After stirring the reaction mixture at 0 °C for 1 day, the mixture was treated with oxygen which was bubbled through for 30 min. The resultant mixture was further stirred under an oxygen atmosphere for 6 h, treated with 15% aqueous NaOH, extracted with Et$_2$O, washed with brine, dried over anhydrous MgSO$_4$, filtered, and concentrated. Kugelrohr distillation provided 0.125 g (88%) of the title compound. Enantiomeric excess was determined from ^{1}H and ^{13}C NMR spectra of the esters formed from (+)- and (−)-MTPA using the standard procedure (72% ee, $[\alpha]_D^{20}$ +7.4° (c 15.9, CH$_2$Cl$_2$)).

The selective formation of the (R) isomer observed with a wide range of 1-alkenes and other experimental results are consistent with the mechanism represented by Scheme 44, in which a C$_2$-symmetric, cationic methylzirconium derivative containing two 1-neomenthylindenyl groups interacts with the *re* face of 1-alkenes[158]. This mechanism clearly favors direct addition of the C−Zr bond to alkenes promoted by Al over direct C−Al bond addition in these reactions.

The reaction of trialkylalanes containing Et and high alkyl groups with 1-alkenes in the presence of Cp$_2$ZrCl$_2$ in nonpolar solvents is known to produce aluminacyclopentanes rather than the straightforward alkylalumination products[162] (Scheme 45). In principle, this cyclic carboalumination can proceed enantioselectively. However, the ee figures have been limited to 30–35%[163]. As suspected and desired, a complete mechanistic change was observed, when nonpolar solvents, such as hexanes, were replaced by polar solvents, for example CH$_2$Cl$_2$, (CH$_2$Cl)$_2$, and CH$_3$CHCl$_2$. Thus, the reaction of 1-decene with Et$_3$Al in the presence of Cp$_2$ZrCl$_2$ in (CH$_2$Cl)$_2$ was a totally acyclic process with no indication of the formation of aluminacyclopentanes, even though competitive dehydrometalation

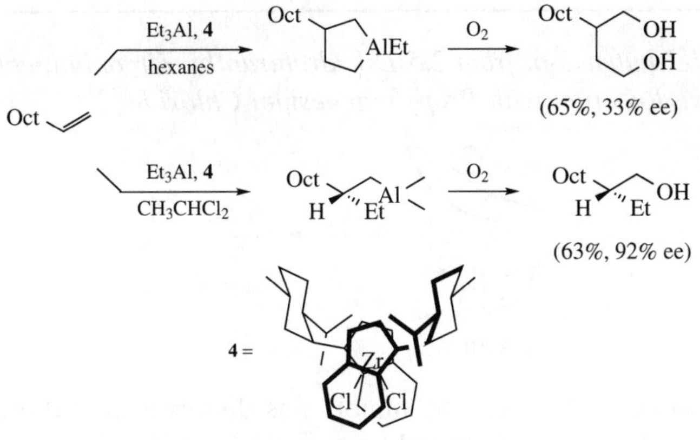

Scheme 45.

Table 2. Zirconium-catalyzed enantioselective alkylalumination of monosubstituted alkenes[a].

Alkene	Methylalumination		Ethylalumination	
	Yield (%)	ee (%)	Yield (%)	ee (%)
$HexCH = CH_2$	88	72	–	–
$BuCH = CH_2$	–	–	74	93
$^iBuCH = CH_2$	92	74	77	90
$^cHexCH = CH_2$	80	65	low	–
$^tBuCH = CH_2$	low	–	–	–
$PhCH_2CH = CH_2$	77	70	69	93
$HO(CH_2)_4CH = CH_2$	79	75	88	90
$Et_2N(CH_2)_3CH = CH_2$	68	71	56	95
$Me_2Si(CH_2CH = CH_2)_2^b$	81	74	66	96

[a]The reaction were run with 8 mol % of **4** in Scheme 43 in $(CH_2Cl)_2$, CH_3CHCl_2, or CH_2Cl_2, and the yield figures are for the oxidized alcohols which were uniformly (R) isomers.
[b]The structure of the oxidation product was:

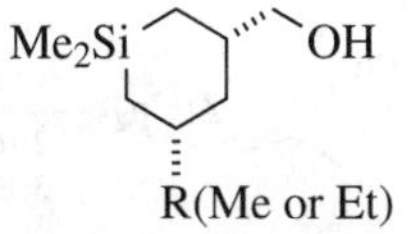

R(Me or Et)

and hydrometalation occurred to minor extents. Here again, the use of bulky and chiral 1-neomenthylindene in place of cyclopentadiene not only suppressed β-dehydrometalation but also led to the formation of the desired alkylmetallation products in 60–90% yields in 90–95% ee[163] (Scheme 45). Some representative results with Me_3Al and Et_3Al are summarized in Table 2.

Although the ee figures observed in the Zr-catalyzed cyclic carbometalation are generally low, certain allylic ethers[164] and allylic amines[165] have been enantioselectively ethyl- and propylmagnesiated in the presence of zirconocene complexes containing chiral Cp derivatives, such as [EBTHI]ZrCl_2.

(s)-2-Isopropyl-3-buten-1-ol from 2,5-Dihydrofuran by Zirconium-catalyzed Enantioselective Reaction with Propylmagnesium Chloride[164]

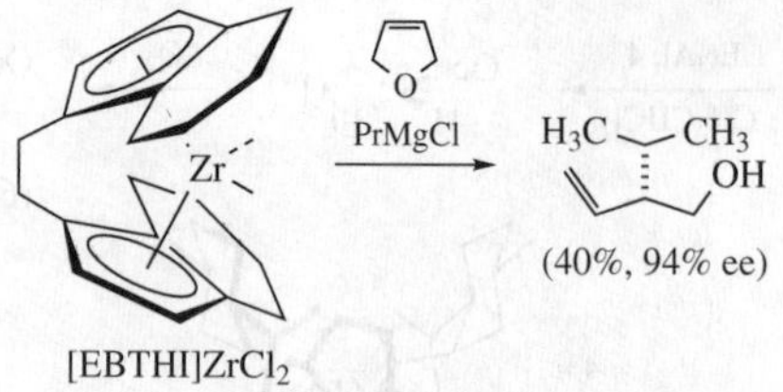

[EBTHI]ZrCl₂

2,5-Dihydrofuran (0.0600 g, 0.85 mmol) was dissolved in 1.0 mL of anhydrous THF in a flame-dried 10-mL round-bottom flask. After the addition of freshly prepared PrMgCl (3.29 mL, 4.28 mmol), the reaction mixture was allowed to stir for 5 min. At this time, 0.0365 g (0.08 mmol) of [EBTHI]ZrCl₂ was added. The reaction flask was then equipped with a reflux condenser and the mixture was allowed to stir at 70 °C for 12 h. After the solution was cooled to 0 °C, excess Grignard reagent was quenched through the dropwise addition of 2.0 mL of a 2.0 M solution of HCl. The mixture was diluted with 25 mL of distilled water and washed three times with 35-mL portions of diethyl ether. Combined organic layers were dried over anhydrous MgSO₄. Filtration of the drying agent and removal of solvent *in vacuo* afforded a pale yellow oil. Silica gel chromatography (2 : 1 pentane/ether) afforded 0.0390 g of (*S*)-2-isopropyl-3-buten-1-ol as a colorless oil (40% yield; 94% ee; 95% regioselective).

Asymmetric Conversion of N-Allylaniline to N-Phenyl-N-(2-methylthiomethyl)amine by Zirconium-catalyzed Enantioselective Ethylmagnesation[165]

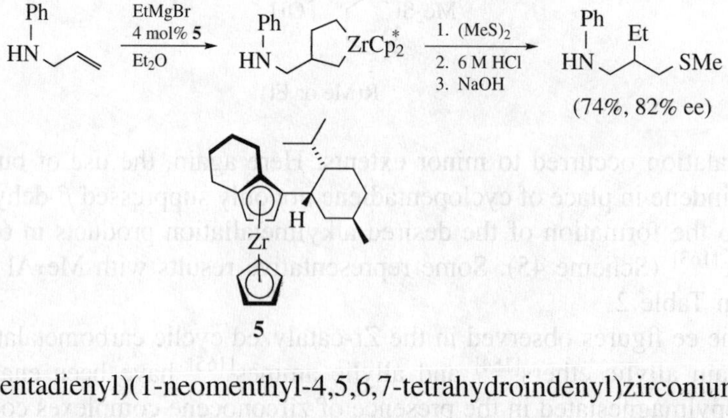

(74%, 82% ee)

5

(a) (Cyclopentadienyl)(1-neomenthyl-4,5,6,7-tetrahydroindenyl)zirconium dichloride

To   1-[(1*R*,2*S*,5*R*)-2-isopropyl-5-methylcyclohexyl]indene   (neomenthylindene) (3.45 g, 13.6 mmol) in THF (40 mL) in the dark at −78 °C was added butyllithium

(5.4 mL of a 2.5 M solution in hexanes, 13.6 mmol), and the orange solution was stirred at 25 °C for 2 h. With the exclusion of light, CpZrCl₃•2THF (6.63 g, 16.3 mmol) was slurried in cold (0 °C) THF (60 mL). To this was added the neomenthylindenyllithium solution *via* cannula to give a bright yellow solution which was stirred at 25 °C overnight. The solvent was removed *in vacuo*, and the residue was redissolved in dichloromethane (100 mL). Platinum oxide (0.090 g, 0.4 mmol) was added, and the mixture was stirred overnight under an atmosphere of hydrogen gas (1 bar). The colorless solution was filtered through Celite, and the solvent was removed to give the title complex as a pale solid. Recrystallization from hot toluene afforded colorless crystals (4.05 g (61%); mp 237–238 °C.

(b) *N*-Phenyl-*N*-(2-methylthiomethylbutyl)amine

To (cyclopentadienyl)(neomenthyltetrahydroindenyl)zirconium dichloride (0.020 g, 4 mol %) under argon was added *N*-allylaniline (0.133 g, 1.0 mmol) in diethyl ether (1 mL) followed by ethylmagnesium chloride (2.5 mL of a 2 M solution in diethyl ether, 5 mmol). The reaction mixture was stirred for 24 h at 25 °C before quenching with dimethyl disulfide (1 mL). After 3 h at 25 °C 6 M HCl (5 mL) was added, and the aqueous phase was washed with diethyl ether (3 × 20 mL). The aqueous layer was basified with aqueous NaOH, and the product was extracted into diethyl ether (3 × 20 mL). Drying (MgSO₄), solvent removal, and chromatography (silica, 5% diethyl ether in light petroleum as elutant) gave *N*-phenyl-*N*-(2-methylthiomethylbutyl)amine as a yellow oil (0.0154 mg, 74%). HPLC analysis on a Chiracel OD-H column (Diacel) (5 × 250 mm) showed an 82% ee.

Although these reactions show considerable promise, with some ee figures exceeding 80–90%, further efforts to eliminate the scope-limiting requirement for allylic heteroatoms are desirable.

6.4 Stoichiometric Bicyclization through Carbozirconation

Until recently, the chemistry of Zr^{II} compounds had lagged far behind that of Zr^{IV} complexes. Thus, for example, a comprehensive review of organozirconium chemistry published in 1974[1] allocated less than a page to the chemistry of Cp₂Zr^{II} derivatives. Despite some pioneering studies during the 1970s and early 1980s, the synthetic usefulness and versatility of the Cp₂Zr^{II} chemistry was not adequately recognized until the 1985–1987 period.

Monomeric Cp₂Zr^{II} would be a 14-electron species. In its singlet form, it would possess one filled nonbonding orbital and two empty valence shell orbitals. Consequently, in addition to forming complexes with π donors, such as alkenes, alkynes, carbonyl compounds, and nitriles, it should be capable of undergoing subsequent ring expansion with a second π donor to produce five-membered zirconacycles (Scheme 46). The latter process may be viewed as the carbozirconation of zirconacyclopropanes or zirconacyclopropenes (Scheme 47).

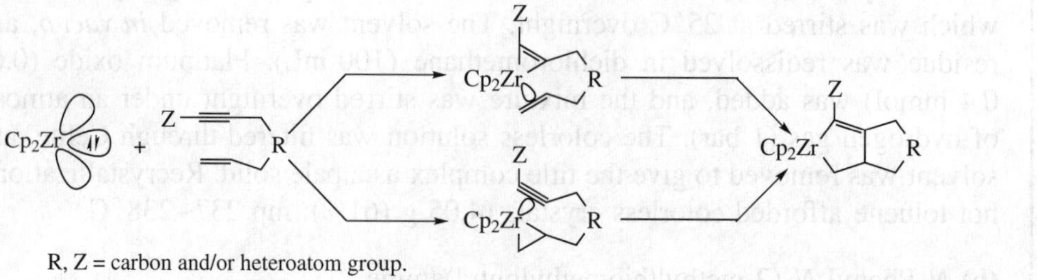

R, Z = carbon and/or heteroatom group.

Scheme 46.

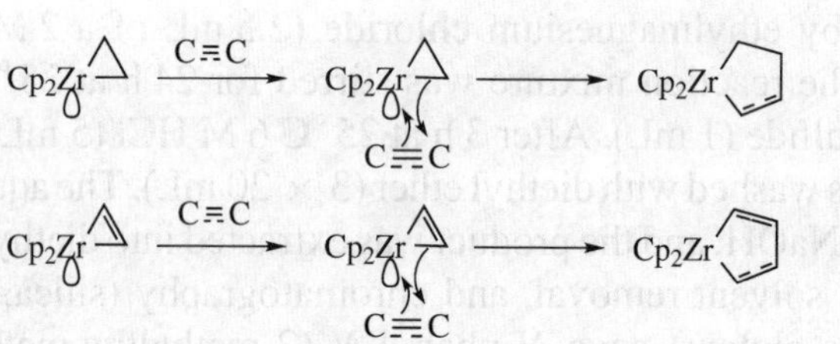

Scheme 47.

As suggested above, the reaction of various enynes[72,166-171] with 'Cp₂Zr' equivalents provided the desired zirconabicycles in high yields. Furthermore, carbonylation of the zirconabicycles produced the corresponding ketones.

---

*Generation of
3,3-Bis(cyclopentadienyl)-2-(trimethylsilyl)-3-zirconabicyclo[3.3.0]oct-1(2)-ene by
Zirconocene-promoted Enyne Bicyclization and Synthesis of
2-(Trimethylsilyl)bicyclo[3.3.0]oct-1(2)en-3-one by Carbonylation[72,166]*

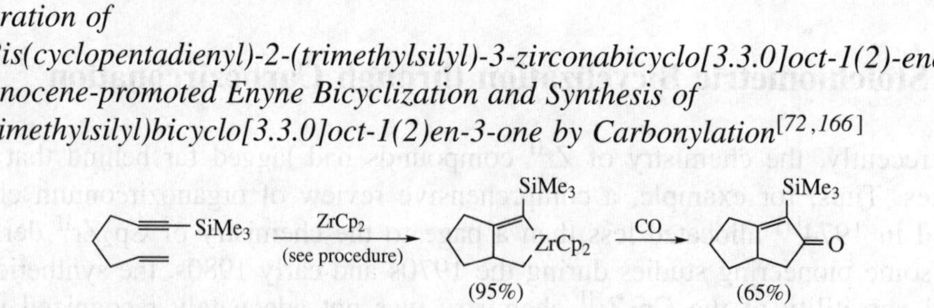

(a) Bicyclization procedure using Cl₂ZrCp₂ and BuLi

To Cl₂ZrCp₂ (0.614 g, 2.1 mmol) in THF (7 mL) was added dropwise at $-78\,^{\circ}$C BuLi (1.6 M, 2.63 mL, 4.2 mmol). After stirring for 1 h at $-78\,^{\circ}$C, 7-(trimethylsilyl)-1-hepten-6-yne (0.332 g, 2.0 mmol) in THF (3 mL) was added. The mixture was warmed to $25\,^{\circ}$C over 1–2 h and stirred for 3–6 h. ^{1}H NMR analysis of the signals due to the Cp (δ 6.17) and Me$_3$Si (δ 0.04) groups using benzene as an internal standard indicated the formation of the desired zirconabicycle in 95% yield. The resulting yellow-brown supernatant liquid was siphoned into a separate flask, and the volatile compounds were removed *in vacuo*. The residue was extracted with hexane and filtered through Celite under a nitrogen atmosphere. The filtrate was concentrated to dryness to provide about 1.1 g (95%) of 3,3-bis(cyclopentadienyl)-2-(trimethylsilyl)-3-zirconabicyclo[3.3.0]oct-1(2)-ene (90–95% pure by ^{13}C NMR).

(b) Bicyclization procedure using Cl$_2$ZrCp$_2$, Mg, and HgCl$_2$

Into a 100-mL flask equipped with a magnetic stirring bar, a septum inlet, and a mercury bubbler were introduced 0.81 g (3.0 mmol) of HgCl$_2$ and 0.73 g (30.0 mmol) of Mg turnings. The metals were dried at 70 °C *in vacuo* ($\leq$1 mmHg) for 2 h. After cooling to 0 °C, 30 mL of THF was added, followed by 0.50 g (3.0 mmol) of 7-(trimethylsilyl)-1-hepten-6-yne. Upon addition of 0.88 g (3.6 mmol) of Cp$_2$ZrCl$_2$, the reaction mixture was warmed to 25 °C and stirred for 12 h.

(c) Representative procedure for carbonylation of zirconabicycles 2-(Trimethylsily)bicyclo[3.3.0]oct-1-(2)-en-3-one

A THF solution of the zirconabicycle obtained above (10-mmol scale) was cooled to 0 °C and evacuated (10–20 mmHg). To this was introduced CO. This process was repeated three times. The initial pressure of 1.1 atm was attained by having a 76-mmHg bubbler. The reaction mixture was stirred for 2 h at 0 °C, quenched with 3 M HCl and pentane, extracted, washed with aqueous NaHCO$_3$ and NaCl, dried over MgSO$_4$, and distilled to give 1.26 g (65%) of the desired bicyclic ketone.

Iodinolysis of the zirconabicycles provided the corresponding diiodides (Scheme 48)[72, 166].

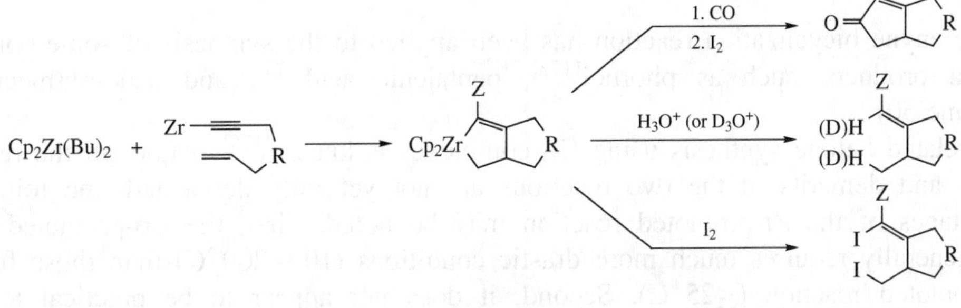

Scheme 48.

The Cp$_2$Zr-promoted enyne bicyclization reaction can proceed in a highly diastereoselective manner, as indicated by the results shown in Scheme 49 and the next procedure. The earlier examples involved enynes containing internal alkynes; however, enynes containing terminal alkynes may now be accommodated[171].

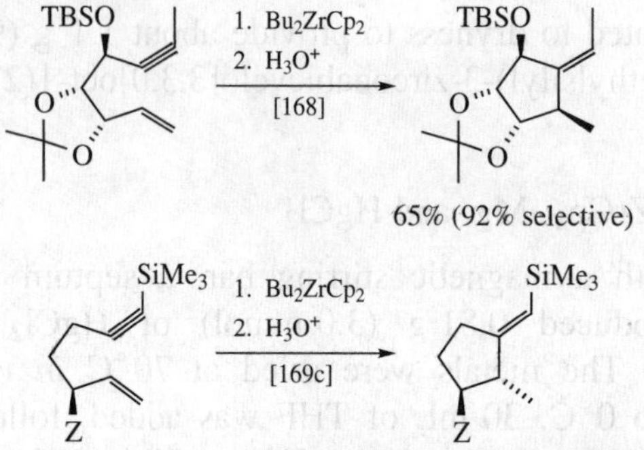

65% (92% selective)

Z	Yield (%)	Selectivity (%)
OTBS	78	98
OH	81	98
CONEt$_2$	93	>99
CH$_2$NEt$_2$	88	98

Scheme 49.

Zr-promoted Bicyclization of Heteroatom-substituted Enynes: Preparation of (1α,2β,3E)-N,N-Diethyl-2-methyl-3-[(trimethylsilyl)methylene]cyclopentane-1-carboxamide[169c]

A solution of BuLi (2.12 mmol) was added to Cp$_2$ZrCl$_2$ (0.307 g, 1.05 mmol) and DMAP (107 g, 1.08 mmol) in THF (10 mL) maintained at −78 °C. After 10 min, the mixture was warmed to −10 °C. The resulting yellow solution was recooled to −78 °C. A solution of N,N-diethyl-2-ethenyl-6-(trimethylsilyl)-5-hexynamide (0.132 g, 0.5 mmol) in THF (0.15 mL) was added dropwise, and the mixture was gradually warmed to 25 °C. After 10 h, the mixture was quenched with H$_2$O (1 mL), filtered through silica gel, extracted with Et$_2$O, washed with H$_2$O, aqueous H$_2$SO$_4$, H$_2$O, and aqueous NaHCO$_3$, brine, dried, and concentrated *in vacuo*. Chromatography (15% EtOAc/hexane) afforded 0.123 g (93%) of the title compound as a colorless oil.

The enyne bicyclization reaction has been applied to the synthesis of some complex natural products, such as phorbol[172], pentalenic acid[173], and iridomyrmecin[174] (Scheme 50).

A related ketone synthesis using Co complexes is known[175]. Although the relative merits and demerits of the two reactions are not yet fully delineated, the following advantages of the Zr-promoted reaction may be noted. First, the Co-promoted reaction generally requires much more drastic conditions (100–200 °C) than those for the Zr-promoted reaction (≤25 °C). Second, it does not appear to be practical to stop

phorbol[172]   pentalenic acid[173]   iridomyrmecin[174]

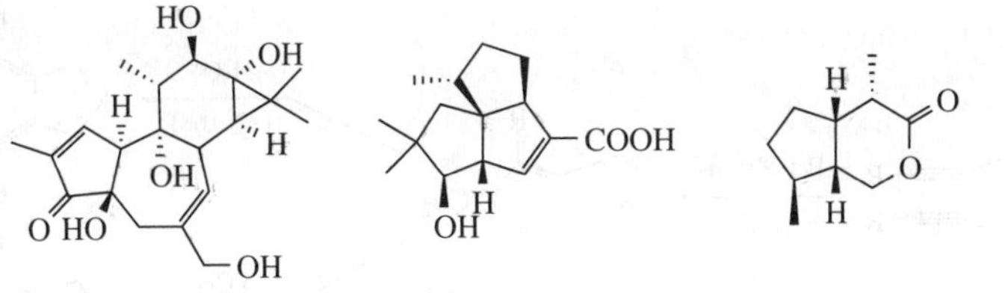

Scheme 50.

the Co-promoted reaction at the cyclic organometallic stage and convert the products into various types of organic compounds other than cyclopentenones. In contrast, zirconacyclopentenes are obtainable as discrete products, often in nearly quantitative yields, and they can be used for various purposes, as exemplified by the synthesis of phorbol[172] and of 7-*epi*-β-bulnesene[176].

Synthesis of (Z)-(1R*,5S*)-[2-(1′-Iodoethylidene)-5-methyl]cyclopentylethanal[176]

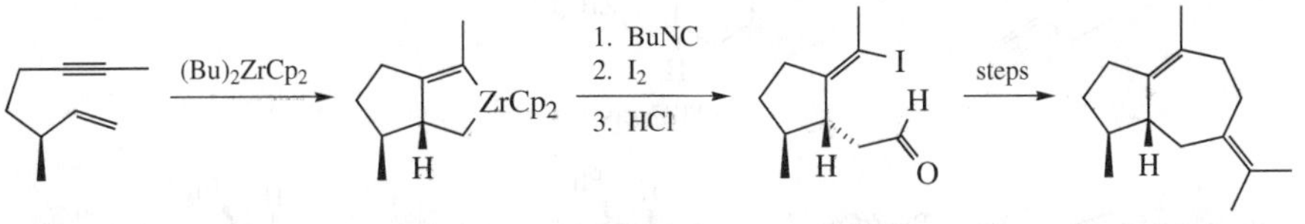

(68%)   7-*epi*-β-bulnesene

To Cp$_2$ZrCl$_2$ (1.45 g, 4.96 mmol) in THF (12 mL) were added successively BuLi (4 mL, 2.5 M in hexanes, 10 mmol; −78 °C, 1 h), 3 methyloct-1-en-6-yne (0.400 g, 3.28 mmol) in THF (1 mL; added at −78 °C, then warmed slowly from −78 °C to 25 °C within 1–2 h, followed by 5 h at 50 °C). The reaction mixture was quenched successively with BuNC (0.350 g, 97%, 4.09 mmol) in THF (1 mL −70 °C to −60 °C, 1 h), I$_2$ (1.56 g, 6.14 mmol) in THF (5 mL; added at −70 °C to −60 °C, then warmed to 25 °C), and HCl (4 mL, 3 N; 6 h at 25 °C). Extraction with ether, washing (Na$_2$S$_2$O$_3$/NaHCO$_3$/brine), drying over MgSO$_4$, evaporation of solvent, and chromatography on silica gel (10 : 1 hexanes/ethyl acetate) afforded 0.62 g (68%) of a 90 : 10 mixture of the title compound and its (1S*,5S*) epimer.

Third, the current scope of the Co-based methodology appears to be limited to cyclization by ene–yne coupling. In contrast, the Zr-based methodology is applicable to bicyclization of not only enynes but also diynes[72,74,177] (Scheme 51) and dienes[65,178–180] (Scheme 52). These reactions are exemplified by the following two procedures.

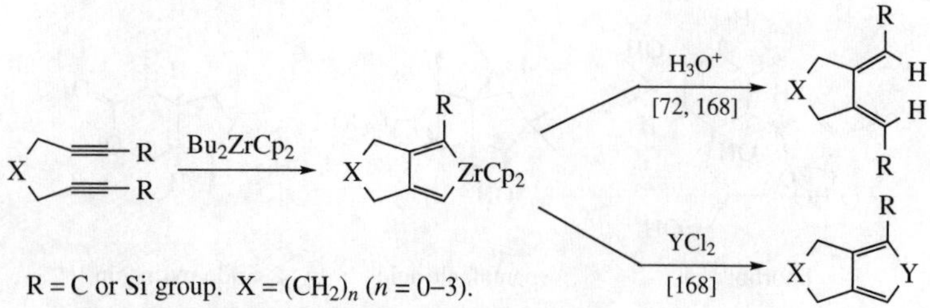

R = C or Si group. X = (CH$_2$)$_n$ (n = 0–3).
Y = Me$_2$Sn, PhP, PhAs, S, etc.

Scheme 51.

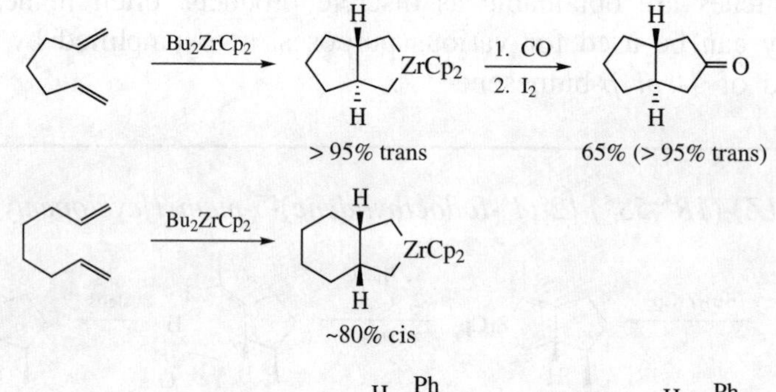

Scheme 52.

Conversion of 3,8-Undecadiyne to (E,E)-1,2-Bis(propylidene)cyclopentane by Zirconium-promoted Bicyclization[74]

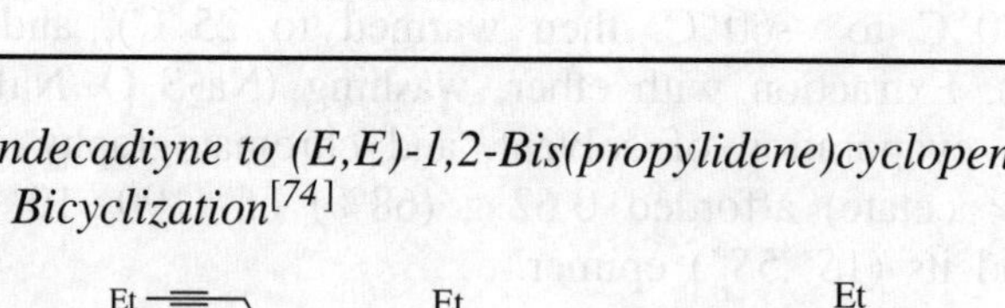

To Cp$_2$ZrCl$_2$ (2.92 g, 10 mmol) in 20 mL of THF was added BuLi in hexane (2.5 M, 8 mL, 20 mmol) at −78 °C. After 1 h, 3,8-undecadiyne (1.48 g, 10 mol) was added, and the mixture was slowly warmed to 23 °C and stirred for several hours. After quenching with 3 N HCl, extraction with pentane, washing with aqueous NaHCO$_3$ and brine, drying over MgSO$_4$, and chromatography, provided 1.20 g (80%) of the title compound.

Preparation of trans-Bicyclo[3.3.0]octan-3-one[65]

To a solution of Cp$_2$ZrCl$_2$ (0.876 g, 3 mmol) in THF (6 mL) at −78 °C was added dropwise 2.60 mL 92.3 M in hexane, 6 mmol) of BuLi. After stirring for 1 h at −78 °C, 1,6-heptadiene (0.41 mL, 3 mmol) was added dropwise. The mixture was warmed to 0 °C, stirred overnight, and then treated with CO (1.1 atm) at 0 °C. When CO uptake was completed, the carbonylated product was treated with I$_2$ (0.837 g, 3.3 mmol) in THF (3 mL) at 0 °C. The mixture was warmed to 25 °C, quenched with 3 N HCl, extracted with hexane/ether, washed with aqueous Na$_2$S$_2$O$_3$ and brine, and dried over MgSO$_4$. Concentration followed by purification by flash chromatography (85 : 15 hexane/ether) afforded 0.24 g (65%) of the title compound which was at least 95% trans.

The trans ring fusion observed in the diene bicyclization (Scheme 52) was unexpected, but synthetically useful. Using this and enyne bicyclization reactions, it is now feasible to readily and selectively synthesize all four possible diastereomers of 2-phenyl-bicyclo[3.3.0]octan-3-ones.

Preparation of (1S*,2S*,5R*)- and (1S,*2R*,5R*)-2-Phenylbicyclo[3.3.0]octan-3-ones by Carbonylation of the Corresponding Zirconabicyles[181]

(a) (1S*,2S*,5S*)-2-Phenyl-3-bis(cyclopentadienyl)zirconabicyclo[3.3.0]octane

To a mixture of Cl$_2$ZrCp$_2$ (0.292 g, 1 mmol) in THF (3 mL) was added at −78 °C BuLi (2.57 M in hexane, 0.78 mL, 2 mmol). After 1 h, (E)-1-phenyl-1,6-heptadiene (0.172 g, 1 mmol) was added, and the reaction mixture was warmed to 23 °C and stirred for 1.5 h. Analysis of the ^{1}H NMR Cp signals indicated the formation of an 81% yield of the title compound which was characterized by NMR and was shown to be stereochemically at least 95% pure.

(b) (1S*,2R*,5S*)-2-Phenyl-3-bis(cyclopentadienyl)zirconabicyclo[3.3.0]octane

This compound was generated in 93% yield from Cl$_2$ZrCp$_2$ (0.292 g, 1 mmol), BuLi (1.60 M in hexane, 1.2 mL, 2 mmol), and (Z)-1-phenyl-1,6-heptadiene

(0.172 g, 1 mmol). Examination by NMR spectroscopy indicated that the compound was contaminated with the C-2 epimer to the extent of about 10%. After 48 h at 23 °C, however, the initial 10 : 1 ratio changed to 1 : 9.

(c) Carbonylation of zirconabicycles: carbonylation of the $(1S^*,2S^*,5S^*)$isomer

The reaction mixture containing the $(1S^*,2S^*,5S^*)$ isomer (1 mmol) was cooled to 0 °C and treated with CO (1.1 atm). When CO uptake was complete (after 6 h), the reaction mixture was treated with iodine (0.3 g, 1.2 mmol) at 0 °C. The reaction mixture was kept at 23 °C for a few hours, quenched with 3 N HCl, extracted with Et_2O, washed with aqueous $Na_2S_2O_3$ and NaCl, and dried over $MgSO_4$. Concentration, followed by column chromatography (silica gel, hexane/Et_2O = 80 : 20) afforded 0.141 g (70%) of $(1S^*,2S^*,5R^*)$-2-phenylbicyclo[3.3.0]octane-3-one (diastereomeric purity ≥97%). Similarly, carbonylation of 1 mmol of a 9 : 1 mixture of the $(1S^*,2R^*,5S^*)$ and $(1S^*,2S^*,5S^*)$ isomers followed by the usual work-up and column chromatography (silica gel, pentane/ether = 90 : 10) provided 0.128 g (64%) of $(1S^*,2R^*,5R^*)$-2-phenylbicyclo[3.3.0]octan-3-one of 90–95% stereoisomeric purity.

The observed stereoselection is mostly thermodynamically controlled, although kinetically favored products may be obtained in cases where thermodynamic equilibration is sufficiently slow, as in the cyclization of 1,7-octadiene[65,178,179]. The thermodynamic preference for the formation of *trans*-β,β'-disubstituted cyclopentanones has been amply demonstrated in the cases of monocyclic zirconacyclopentanes, as discussed later. It should be noted, however, that *trans* fusion does not always lead to the thermodynamically favored diastereomers. Thus, in the synthesis of the tricyclic ketone shown in Scheme 53[182,183], both ring fusions are cis, although the two end rings are *anti* to each other with respect to the zirconacyclopentane ring. In the synthesis of dendrobine below[184], the desired all-cis stereochemistry is favored. So,

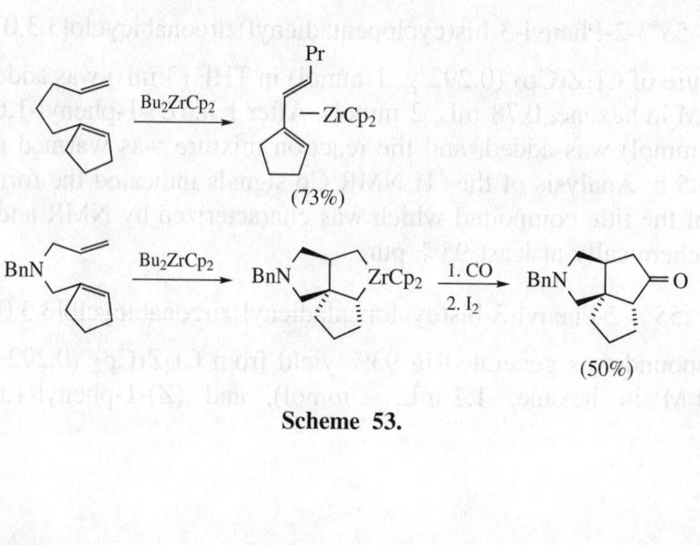

Scheme 53.

it is not *trans* fusion but the overall energetics that governs the stereochemistry, and *trans* fusion happens to be thermodynamically favored in some bicyclic cases.

(1S,4R,6S,8S,11S)-3-Benzyl-6-isopropenyl-11-methyl-3-azatricyclo[6.2.1.0⁴,¹¹]undecan-9-one[184] *(A key intermediate for the synthesis of dendrobine*

(47%)   Kende intermediate   (−)-dendrobine

To a stirred suspension of Cp_2ZrCl_2 (2.13 g, 7.28 mmol) in THF (30 mL) at −78 °C was added dropwise BuLi (1.80 M solution in hexane, 7.8 mL, 14 mmol), and the solution was stirred at −78 °C for 1 h. To this solution was added the starting diallylamine (1.58 g, 5.60 mmol) in THF (14 mL), and the solution was allowed to warm to 25 °C. After the solution was stirred for 1.5 h, the argon atmosphere was exchanged for carbon monoxide, and the solution was stirred for 15 h. After the solution was cooled to 0 °C, 10% HCl (10 mL) was added, and the solution was stirred at room temperature for 6 h. The resultant mixture was basified with K_2CO_3, and the aqueous layer was extracted with AcOEt. The organic layer was washed with brine, dried over Na_2SO_4, and concentrated. The residue was purified by column chromatography (1 : 10 AcOEt/hexane) to afford 0.8145 g (47%) of the title compound as colorless crystals (mp 51.5–52.0 °C (recrystallized from hexane, at −78 °C); $[\alpha]_D^{20} = +8.3°$ (*c* 0.980, $CHCl_3$); 90% ee).

It should be noted that the Zr-promoted bicyclization–carbonylation of 1-phenyl-6-hepten-1-yne gives 2-phenylbicyclo[3.3.0]oct-1(2)-en-3-one in 55% yield. Its conjugate reduction with CuBr (2 equiv) and $LiAlH(OMe)_3$ (3.6 equiv) in THF at −20 °C afforded (1S*,2S*,5S*)-2-phenylbicyclo[3.3.0]octan-3-one as the major product, which was isolated by column chromatography (silica gel, pentane/$Et_2O = 90 : 10$) in 36% yield, whereas catalytic hydrogenation catalyzed by $Pd/BaSO_4$ in EtOAc gave the (1S*,2R*,5S*) isomer of ≥95% stereoisomeric purity. Thus, all four possible diastereomers can be prepared by the Zr-promoted bicylization reaction[181].

Although less well developed, some heteroatom analogs of dienes, enynes, and diynes have also been bicyclized with Cp_2Zr reagents, as exemplified by the results shown in Scheme 54[185].

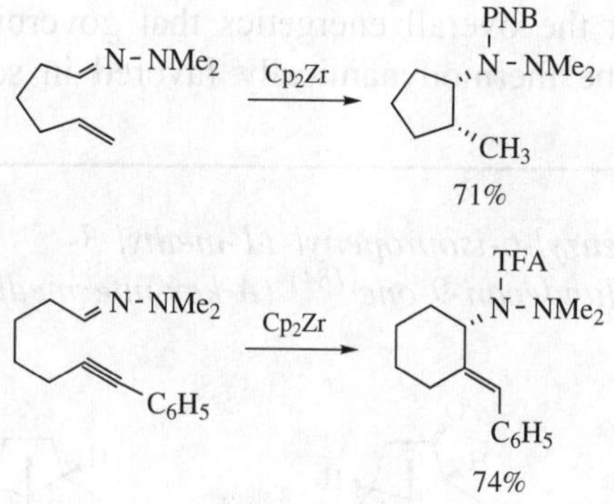

71%

74%

Scheme 54.

6.5   Cyclization and other Reactions of Zirconacyclopropanes and Zirconacyclopropenes

The initial successes discussed in the preceding section triggered a number of intensive studies in this area, which have not only made the chemistry of Cp_2Zr^{II} compounds synthetically useful and versatile but have also unraveled surprisingly intricate relationships between the chemistry of Cp_2Zr^{IV} compounds and that of Cp_2Zr^{II} derivatives.

In Zr-promoted $\pi–\pi$ coupling reactions, introduction of the first π compound (*e.g.* an alkene or an alkyne) is most conveniently achieved by β-H abstraction of diorganylzirconocenes using either the Negishi–Takahashi protocol[48] or the Erker–Buchwald protocol[24] discussed in Section 2.4. The β-H abstraction reaction of dialkylzirconocenes has been extensively delineated, and the following features have been clarified. Specifically, the reaction is a unimolecular, nondissociative process[186] in which β-agostic interaction between a β-H and the Zr empty orbital must play a crucial role (Scheme 55). This explains the seemingly irrational reactivity order $^sBu \geq$ $^tBu \geq$ Et > Bu > ^iBu. The relative reactivity order may be predicted according to the generalization β-CH_3 > β-CH_2 > β-CH[187]. Although Cp_2Zr π complexes are generally unstable, they can be stabilized by complexation with phosphines such as PMe_3 and then fully characterized[73,75–81]. Some heteroatom-containing zirconacyclopropanes and zirconacyclopropenes can also be prepared similarly[84,188,189].

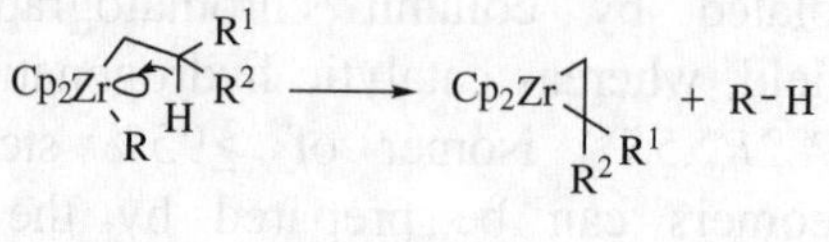

Scheme 55.

The β-H abstraction reaction is also applicable to the generation of three-membered zirconacycles containing heteroatoms such as N and S[24,84]. Those three-membered zirconacycles that contain O have been prepared by various methods, such as insertion and α-H abstraction of acylzirconium derivatives[190-193] (Scheme 56).

Once the Cp_2Zr^{II} π complex or three-membered zirconacycle has been generated, its treatment with the second π compound can lead to the formation of the corresponding five-membered zirconacycle (Scheme 57). Considering just the Zr-mediated coupling of alkenes, alkynes, carbonyl compounds, and imines or nitriles, there are about 10 binary combinations, most of which have been investigated since the late 1980s. However, those that have been extensively investigated deal with alkenes and alkynes. In each of these reactions, the 'pair' selectivity and regioselectivity can be significant issues, as detailed below for the alkenyl–alkenyl, alkenyl–alkynyl, and alkynyl–alkynyl coupling reactions. At least five different types of 'pair' selectivity patterns have been observed, as summarized in Table 3[34].

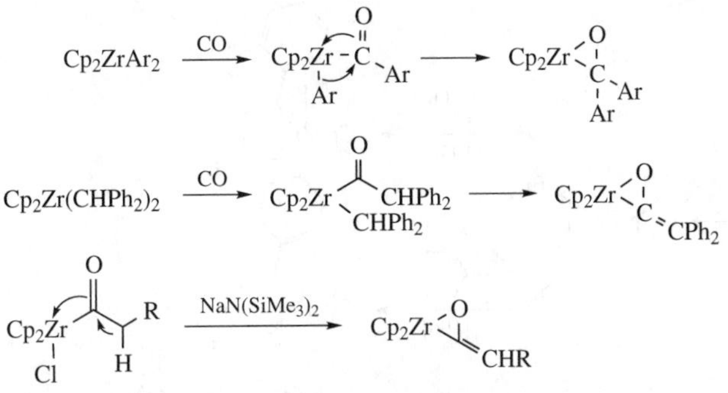

Scheme 56.

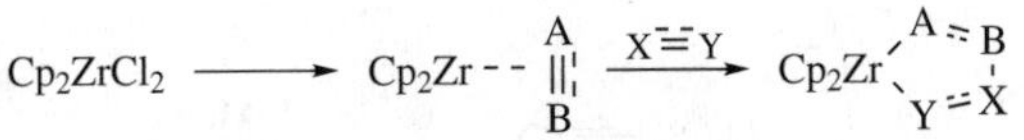

Scheme 57.

Some representative examples of the five different types of reactions involving alkenes and alkynes are shown in Scheme 58. The following common characteristics should

Table 3. Distinctive features of various types of ring expansion reactions of alkene–zirconocene complexes.

Type	Displacement of initial alkene	Pair selectivity	Equilibrating process or not
I	no	very high (~100%, kinetic)	no
II	no	high (thermodynamic)	yes
III	yes	homo-coupling	a
IV	yes	high	b
V	b	low	yes

[a] Not pertinent.
[b] Depends on the substrates.

also be noted. First, alkyl groups prefer to be β to Zr, presumably to minimize steric hindrance, whereas aryl, alkenyl, and alkynyl groups favor the α position, presumably due to favorable benzylic, allylic, and propargylic interaction with Zr. The α/β ratio for aryl is typically 90 : 10, whereas that for alkenyl is $\geq 95\%$. Second, two substituents of the two π compounds prefer to be trans to each other in the zirconacycles, presumably for steric reasons.

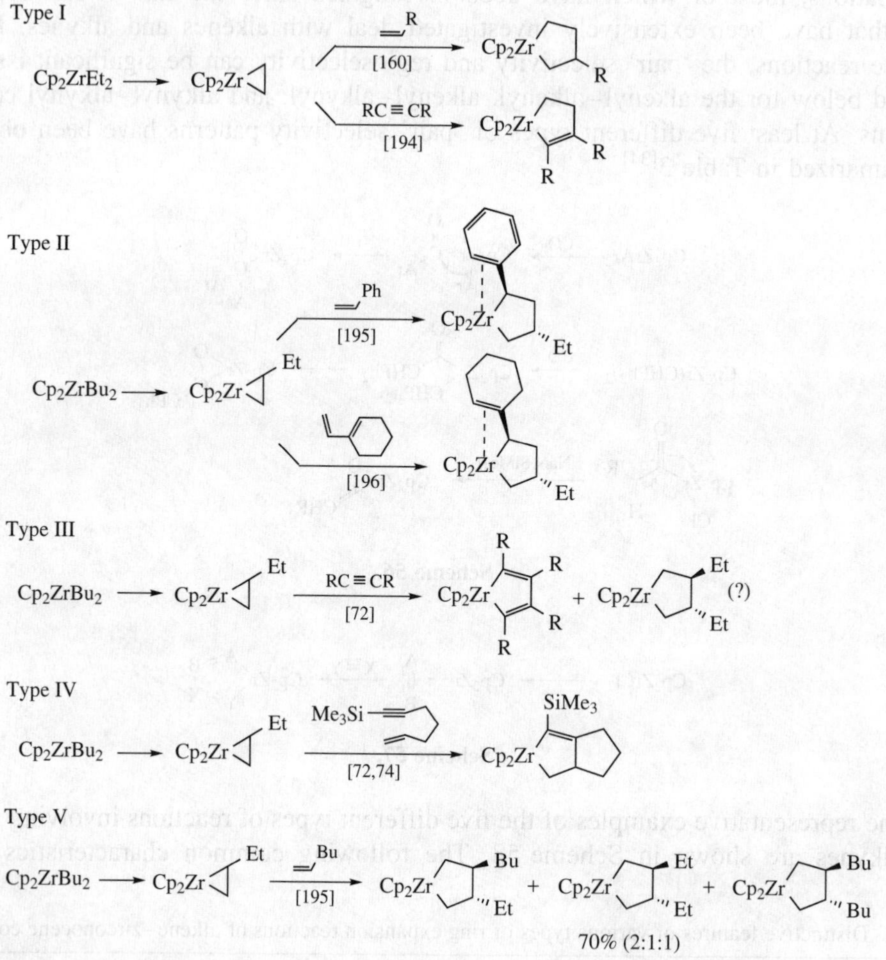

Scheme 58.

Although there has been little doubt about the five-membered cyclic structures for the organozirconium products in Scheme 58, it is reassuring that they have been confirmed by X-ray crystallography[197–199].

As in the cases of bicyclization, five-membered zirconacycles can be subjected to protonolysis, iodinolysis (following procedure)[160], carbonylation, and other reactions, to produce organic products.

Preparation of 1,1-Bis(cyclopentadienyl)-3-octyl-1-zirconacyclopentane and its Iodinolysis to 1,4-Diiodo-2-(n-octyl)butane[160]

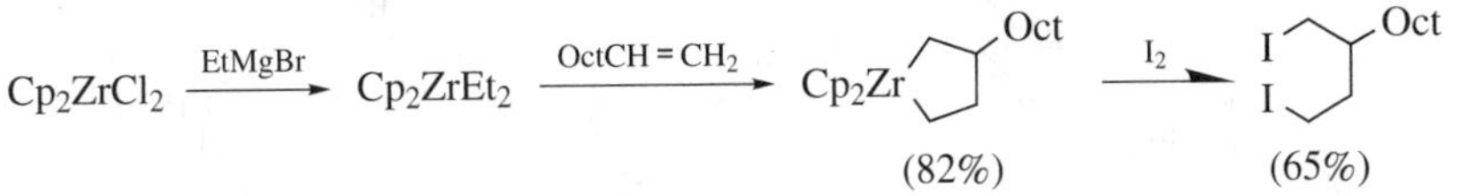

$$Cp_2ZrCl_2 \xrightarrow{EtMgBr} Cp_2ZrEt_2 \xrightarrow{OctCH=CH_2} Cp_2Zr\overset{Oct}{\diagdown} \xrightarrow{I_2} \overset{Oct}{\diagdown}$$

(82%)   (65%)

To Cp_2ZrCl_2 (2.92 g, 10 mmol) in 20 mL of THF was added at $-78\,^\circ$C EtMgBr in THF (2.0 M, 10 mL, 20 mmol). After 1 h, 1-decene (1.40 g, 10 mmol) was added, and the mixture was stirred at $0\,^\circ$C for several hours. Analysis of the reaction mixture by NMR indicated the formation of the expected zirconacyclopentane derivative in 82% yield. An aliquot (1 mmol) of the reaction mixture obtained above was treated with I_2 (0.76 g, 3 mmol) in THF. After the usual work-up, 1,4-diiodo-2-(octyl)butane was obtained in 65% yield based on 1-decene or Cp_2ZrCl_2.

Although generation of three-membered zirconacycles by β-H abstraction and their carbometallative ring expansion have been most widely used for the preparation of five-membered zirconacycles, the reaction of Cp_2ZrCl_2 with Na−naphthalene[200] or Mg and $HgCl_2$ in the presence of π compounds[201] were two of the earliest methods for the same purpose. A more recent development, namely generating $Cp_2Zr(Me_3SiC\equiv CSiMe_3)\cdot$THF by the treatment of activated Mg in THF in the presence of $Me_3SiC\equiv CSiMe_3$[46] and reacting it with the second π compound, is yet another method for the preparation of five-membered zirconacycles. Both Bu_2ZrCp_2 and $Cp_2Zr(Me_3SiC\equiv CSiMe_3)\cdot$THF have been extensively employed for the synthesis of polymeric arenes and heteroareanes[202–204].

Five-membered zirconacycles are often capable of reacting further with π compounds to produce a variety of products. One of the earliest examples is shown in Scheme 59[72]. In general, the C_β–$C_{\beta'}$ bonds of zirconacyclopentanes and zirconacyclopentenes are labile. This may present some difficulties as in the attempted bicyclization of 1-(4-pentenyl)cyclopentene which underwent regioisomerization of the terminal double bond to give a conjugated diene−Cp_2Zr complex (Scheme 53). This double-bond migration reaction presumably proceeds as summarized in Scheme 60[182,183]. Viewed from a different perspective, however, this regioisomerization reaction provides a novel route to Cp_2Zr−diene complexes. Even more useful is exploitation of this property to achieve highly selective cross-coupling of two π compounds (Scheme 61)[194,205,206].

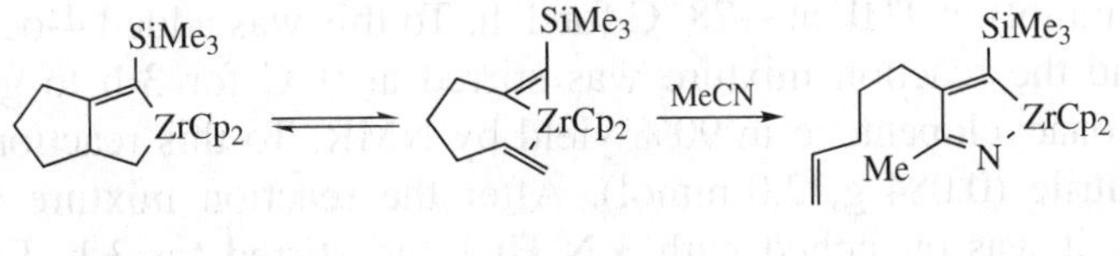

Scheme 59.

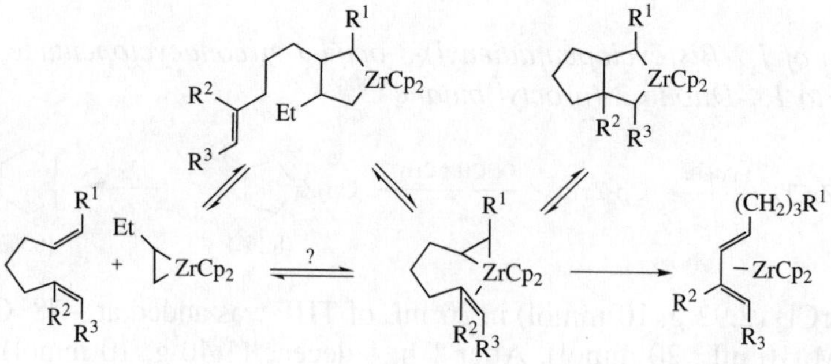

Scheme 60.

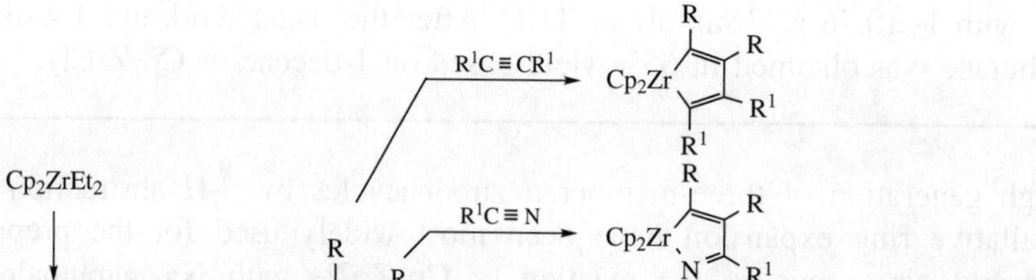

Scheme 61.

Preparation of (E)-3-(n-Propyl)-3-hepten-2-one from 4-Octyne and Acetonitrile by Zirconium-promoted Cyclization[194]

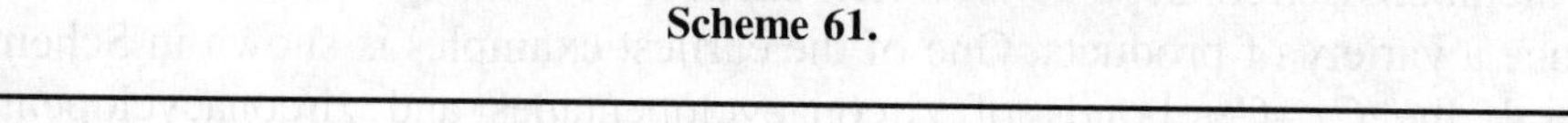

Diethylzirconocene was generated by treating Cp$_2$ZrCl$_2$ (0.584 g, 2.0 mmol) with EtMgBr (4.0 mmol) in THF at $-78\,^\circ$C for 1 h. To this was added 4-octyne (0.220 g, 20 mmol), and the reaction mixture was stirred at $0\,^\circ$C for 3 h to give the corresponding zirconacyclopentene in 90% yield by NMR. To this reaction mixture was added acetonitrile (0.084 g, 2.0 mmol). After the reaction mixture was stirred at $25\,^\circ$C for 3 h, it was quenched with 3 N HCl and stirred for 3 h. Extraction with ether, washing with aqueous NaHCO₃ and brine, drying with MgSO₄, concentration, and flash chromatography provided 0.170 g (60%) of the title compound, which was >98% (*E*).

This process also provides a plausible explanation for a facile skeletal rearrangement (Scheme 62)[207]. Metathesis of the σ bond in zirconacyclopentanes also leads to ring opening (Scheme 63)[160]. This stoichiometric σ-bond metathesis and β-H abstraction sequence coupled with yet another stoichiometric Type I cyclization (Scheme 58) provided a three-step catalytic cycle for the Dzhemilev ethylmagnesiation (Scheme 64)[160]. A similar mechanism may also be proposed for the Cp₂Zr-catalyzed ethylation of allyl ethers (Scheme 65)[208].

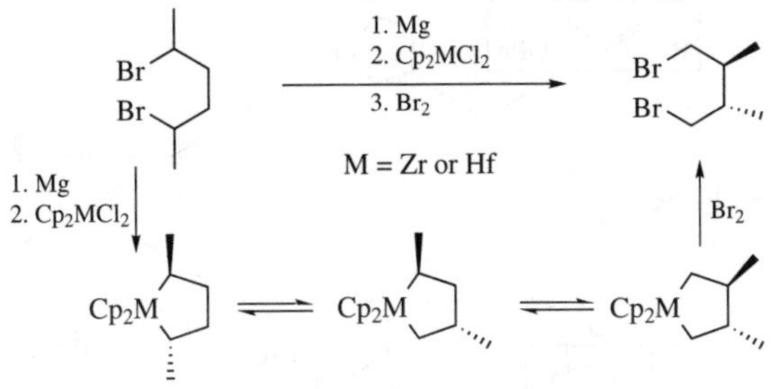

Scheme 62.

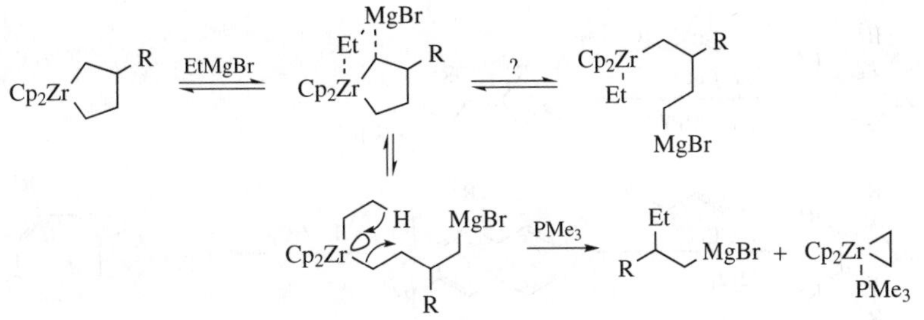

Scheme 63.

Five-membered zirconacylces can serve as nucleophiles under the catalytic influence of Cu and Ni compounds, such as CuCl. Under these conditions, they react readily with allyl chlorides[209–213] (Scheme 66), acyl chlorides[214,215] (Scheme 67) and alkynes[216,217] (Scheme 68).

Zirconacyclopropanes themselves can also undergo some reactions other than ring expansion reactions with π compounds. The alkene regioisomerization shown in Schemes 53 and 60 has been shown to proceed by a 1,3-H shift rather than a 1,2-H shift[183].

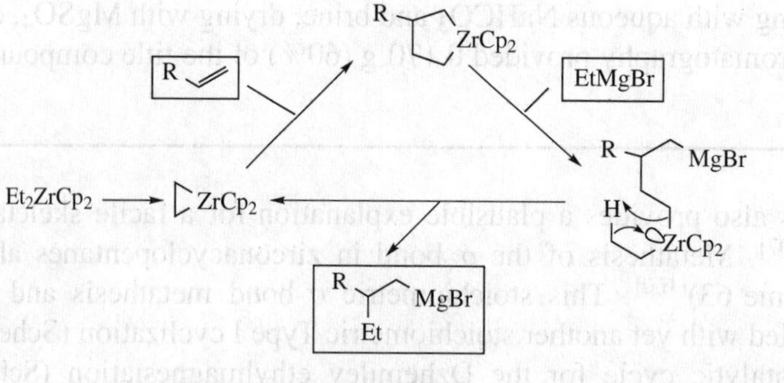

$Et_2ZrCp_2 \longrightarrow \triangleright ZrCp_2$

Scheme 64.

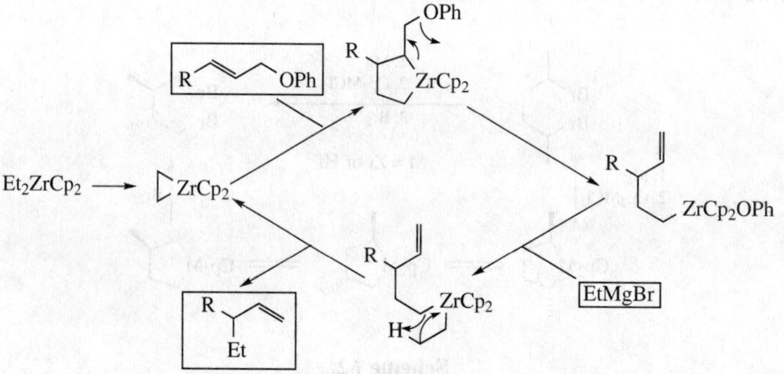

$Et_2ZrCp_2 \longrightarrow \triangleright ZrCp_2$

Scheme 65.

Cp_2Zr ··· $\xrightarrow[\text{[209, 210]}]{\text{cat. CuCl}}$ ··· $ZrCp_2Cl$

Cp_2Zr ··· $\xrightarrow[\text{[212]}]{\substack{2 \quad \diagdown Cl \\ \text{cat. CuCl}}}$ ··· $\xrightarrow{Bu_2ZrCp_2}$ Cp_2Zr ···

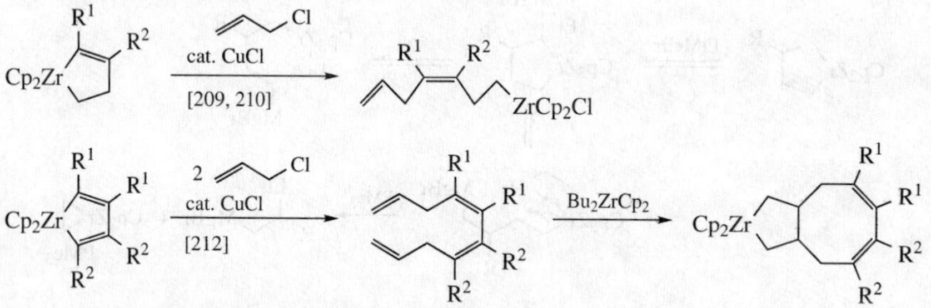

Scheme 66.

Stereoisomerization of stilbene catalyzed by Cp_2Zr^{II} derivatives has been shown to involve an intriguing polar mechanism (Scheme 69)[181], indicating that, in some cases, polar processes may take over usually dominant concerted processes. Formation of a novel bis(zirconocene) complex by the reaction of ethylene–Cp_2Zr and Me_2ZrCp_2 appears to involve a σ-bond metathesis between the two reagents (Scheme 70)[218]. Even the hydrosilation of alkenes is catalyzed by Bu_2ZrCp_2 (Scheme 71)[219].

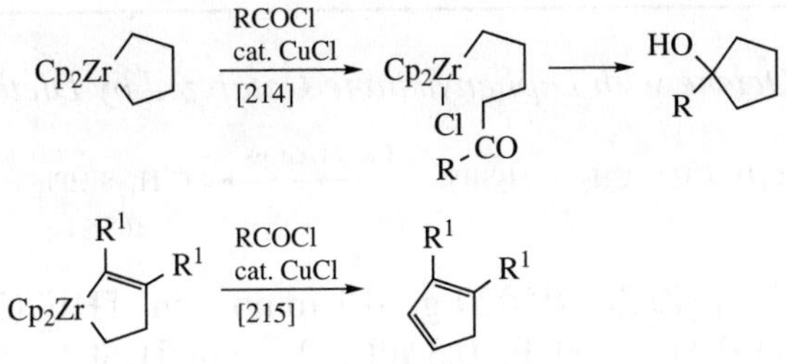

Scheme 67.

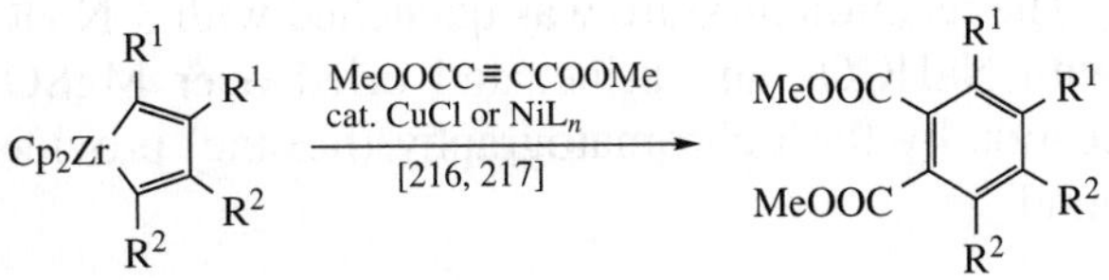

Scheme 68.

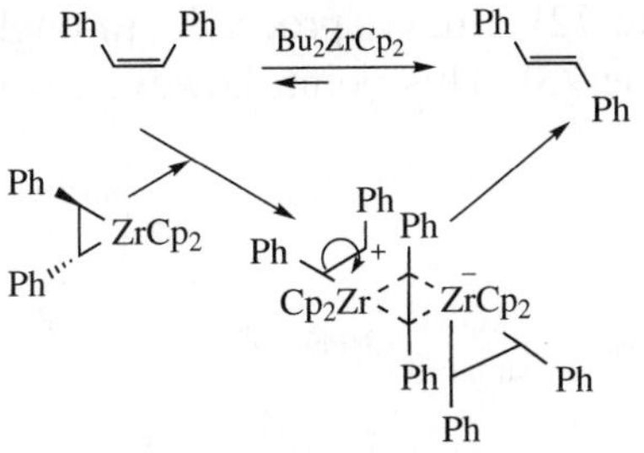

Scheme 69.

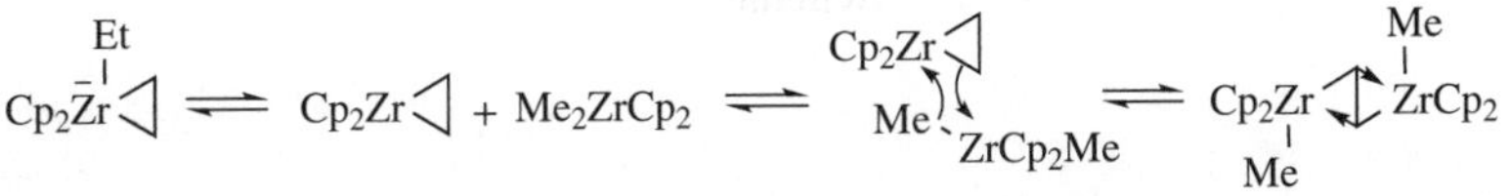

Scheme 70.

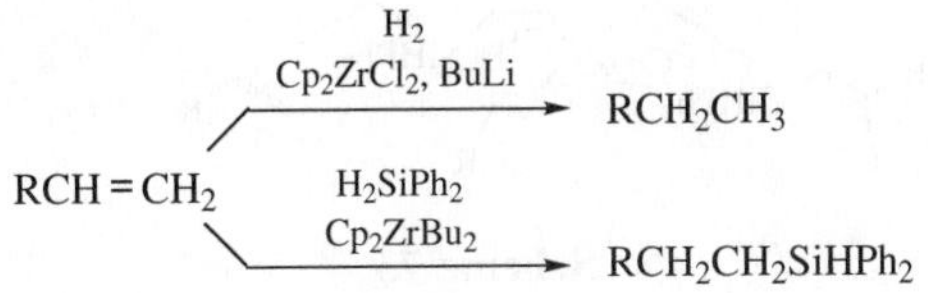

Scheme 71.

Hydrosilation of 1-Octene with Diphenylsilane Catalyzed by Diethylzirconocene[219]

$$C_6H_{13}CH = CH_2 + H_2SiPh_2 \xrightarrow{\text{Cat. Et}_2ZrCp_2} C_8H_{17}SiHPh_2$$

$$(65\%)$$

To a solution of Cp_2ZrCl_2 (0.029 g, 0.1 mmol) in THF (5 mL) was added dropwise EtMgBr (1.0 M in THF, 0.3 mL, 0.3 mmol) at $-78\,°C$. After 1 h at $-78\,°C$, diphenylsilane (0.207 mL, 1.1 mmol) and 1-octene (0.156 mL, 1 mmol) were sequentially added to the reaction mixture which was then warmed to $25\,°C$ and stirred for 1 h. The reaction mixture was quenched with 3 N HCl, extracted with hexane, washed with $NaHCO_3$ and brine, and dried over $MgSO_4$. Concentration followed by purification by flash chromatography (hexane) provided 0.191 g (65%) of the title compound.

A tentative oxidative addition–insertion–reductive elimination mechanism for this hydrosilation (Scheme 72) may proceed through σ-bond metathesis of zirconacyclopropanes (Scheme 73). This point, however, is under current investigation.

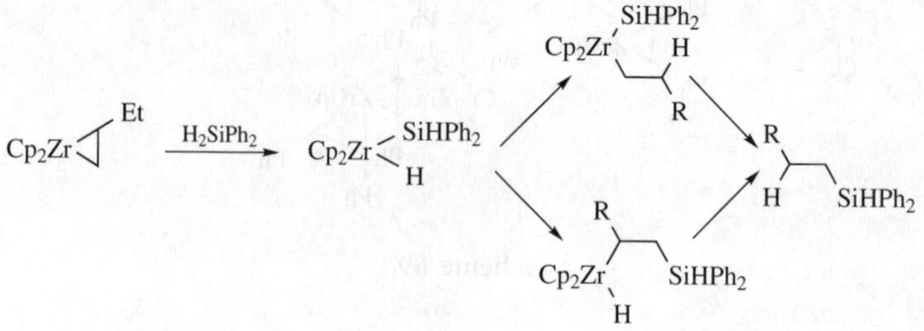

Scheme 72.

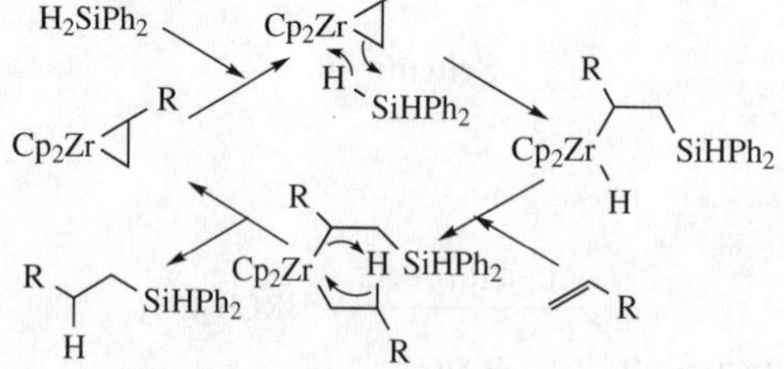

Scheme 73.

Five-membered zirconacycles and their derivatives have been shown to serve as catalysts for carbon–carbon and carbon–hetero atom bond-forming reactions. In this connection, clarification of the mechanism of the Zr-catalyzed ethylmagnesiation as a cyclic process involving C–H activation (Scheme 64) was a significant turning point, which necessitated reinvestigation of all Zr-catalyzed carbometalation reactions, as the overall equation showing the addition of the alkyl–metal bond to alkenes and alkynes should no longer be taken as an indication of straightforward four-centered addition reactions. Indeed, reinvestigation of the Zr-catalyzed carboalumination of alkynes with ethyl- and higher alkylalanes[118,127] has revealed a number of unexpected and highly intriguing results, as represented by the following results with ethylalanes.

First, the reaction of alkynes with Et$_2$AlCl and Cp$_2$ZrCl$_2$ shows no sign of cyclic processes. In polar solvents (such as (CH$_2$Cl)$_2$), which are essential to observing favorable results, this reaction is of synthetic utility, producing syn-addition products in high yields. This reaction can also accommodate higher alkyl groups[157]. Second, the corresponding reaction of Et$_3$Al and Cp$_2$ZrCl$_2$ is mostly or nearly exclusively cyclic[220] and proceeds better in nonpolar solvents (Scheme 74).

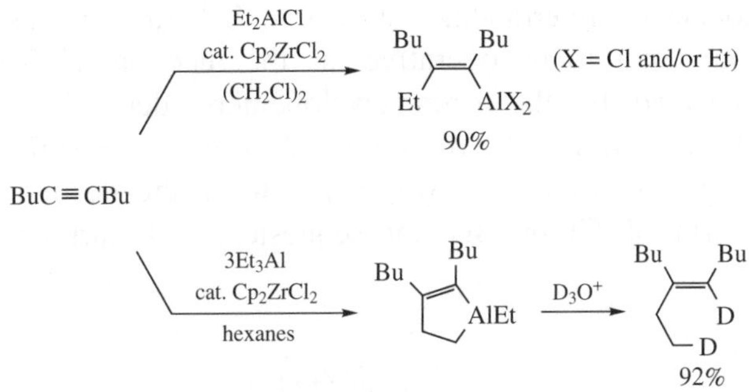

Scheme 74.

Zirconium-catalyzed Carboalumination of 5-Decyne with Triethylalane[157]

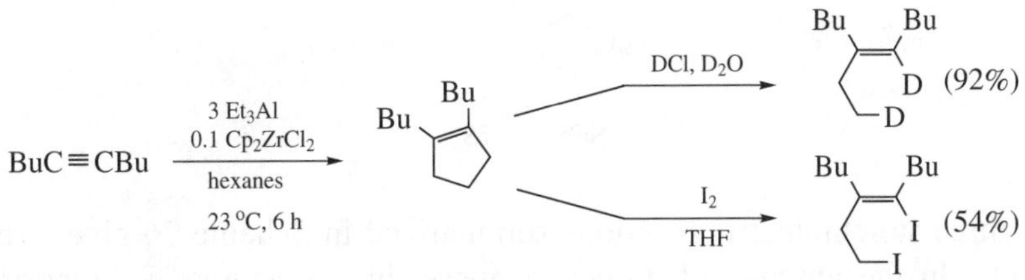

(a) Reaction of 5-decyne with Et$_3$Al in the presence of a catalytic amount of zirconocene dichloride in hexanes

A mixture of Cp$_2$ZrCl$_2$ (0.059 g, 0.2 mmol) and 5-decyne (0.36 mL, 2 mmol) in hexanes (4 mL), cooled to 0 °C was treated with Et$_3$Al (0.82 mL, 6 mmol) and warmed to 23 °C. After 6 h, analysis of a protonolyzed aliquot by GLC indicated that the reaction was complete.

(b) Deuterolysis

In another run, the reaction mixture was quenched at $0\,^{\circ}$C with DCl (20 wt % in D$_2$O, 5 mL) and THF (5 mL) and stirred for 2–3 h at 23 $^{\circ}$C. (Z)-5-Deuterio-6-(2-deuterioethyl)-5-decene was obtained in 92% yield, as determined by GLC ($\geq$98% D at the Me of the ethyl group and $\geq$98% D at C-5).

(c) Iodinolysis

In the third run (1 mmol scale), the reaction mixture was quenched with a solution of I$_2$ (3.1 g, 12.2 mmol) in THF (13 mL) at $0\,^{\circ}$C and stirred at 23 $^{\circ}$C for 8–10 h. After work-up, purification by column chromatography afforded 0.21 g (54%, 69% by NMR) of (E)-5-iodo-6-(2-iodoethyl)-5-decene.

Initially, a diethylation mechanism proceeding through Et$_2$ZrCp$_2$, ethylene–ZrCp$_2$, and the zirconacyclopentene intermediate shown in Scheme 75 was considered. This was subsequently found not to be operative, as the reaction of 5-decyne and Et$_3$Al (3 equiv) is not catalyzed by the zirconacyclopentene derived from 5-decyne and Et$_2$ZrCp$_2$, generated by treating Cl$_2$ZrCp$_2$ with 2 equiv of EtMgBr. However, addition of ClAlEt$_2$ (1 equiv relative to Cl$_2$ZrCp$_2$) induces the desired catalytic process. Clearly, an active form of Cl or Br, not sequestered as part of the Mg salts, is necessary.

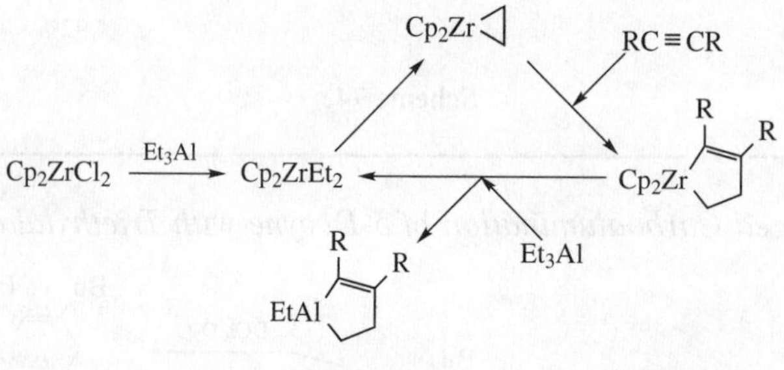

Scheme 75.

Third, the three stoichiometric reactions summarized in Scheme 76 give three discrete major products. In the absence of 5-decyne, these three reactions are reported to give discrete bimetallic complexes[221–225] (Scheme 77). It is thought that there is a 1 : 1 correspondence between the three equations in Scheme 76 and those in Scheme 77, and that the second one in each scheme represents the catalytic process.

Fourth, treatment of EtZrCp$_2$Cl, generated by hydrozirconation of ethylene with HZrCp$_2$Cl, with 1 equiv of Et$_3$Al provides a clean and high-yielding route to the five-membered bimetallic reagent (Scheme 78), which is thought to be the carbozirconating agent in the catalytic reaction. Indeed, its reaction with 5-decyne smoothly gives the same

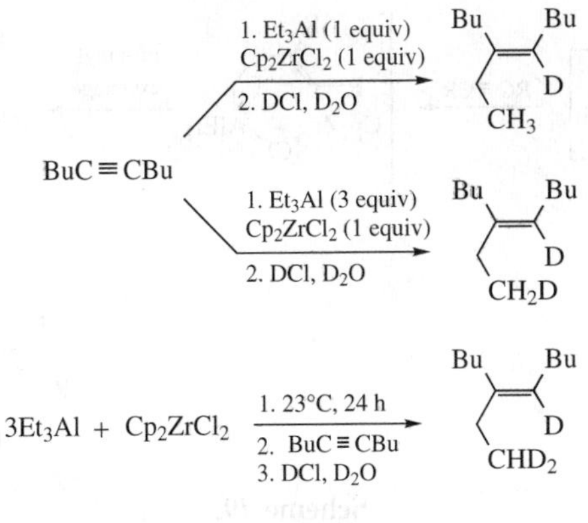

BuC≡CBu

1. Et₃Al (1 equiv)
Cp₂ZrCl₂ (1 equiv)
2. DCl, D₂O
→ (product with Bu, Bu, D, CH₃)

1. Et₃Al (3 equiv)
Cp₂ZrCl₂ (1 equiv)
2. DCl, D₂O
→ (product with Bu, Bu, D, CH₂D)

$3Et_3Al + Cp_2ZrCl_2$

1. 23°C, 24 h
2. BuC≡CBu
3. DCl, D₂O
→ (product with Bu, Bu, D, CHD₂)

Scheme 76.

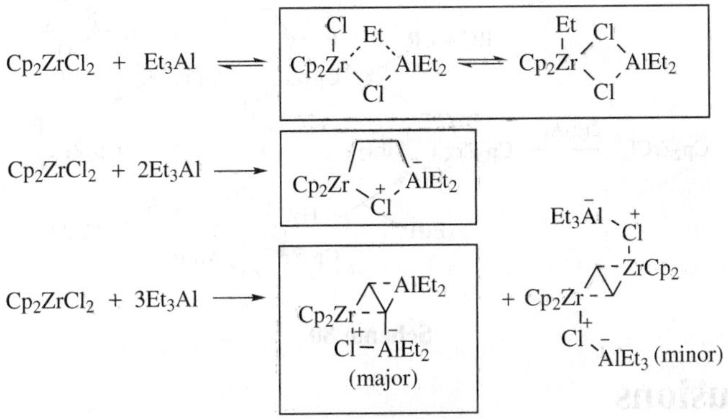

$Cp_2ZrCl_2 + Et_3Al \rightleftharpoons$ (bridged complexes)

$Cp_2ZrCl_2 + 2Et_3Al \longrightarrow$ (complex)

$Cp_2ZrCl_2 + 3Et_3Al \longrightarrow$ (major) + (minor)

Scheme 77.

$Cp_2Zr(H)Cl \xrightarrow[C_6D_6]{H_2C=CH_2}$ (complex) $\xrightarrow{Et_3Al}$ ... $\xrightarrow[-C_2H_4]{}$ ...

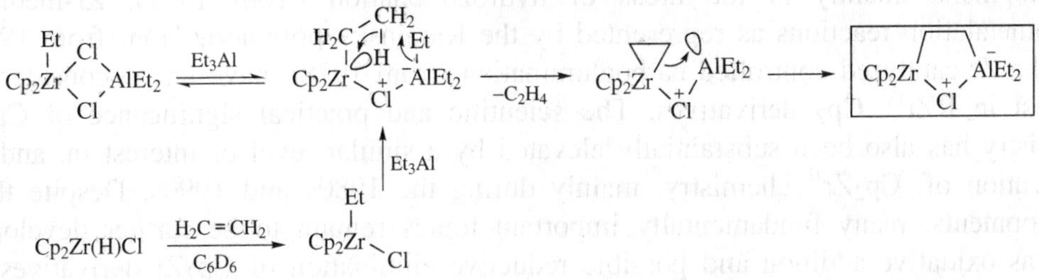

Scheme 78.

organoalane product and EtZrCp₂Cl under the stoichiometric conditions (Scheme 79), and it also induces the same catalytic reaction.

Together, these results provide strong support for the novel bimetallic and monoalkylative mechanism shown in Scheme 80. In the critical step producing the five-membered bimetallic complex, three components consisting of: (1) EtZrCp₂ for crucial agostic interaction; (2) Et₃Al but not Et₂AlCl to provide an Et base; and (3) one Cl atom not sequestered by Mg and hence available for linking Zr and Al are thought to be necessary.

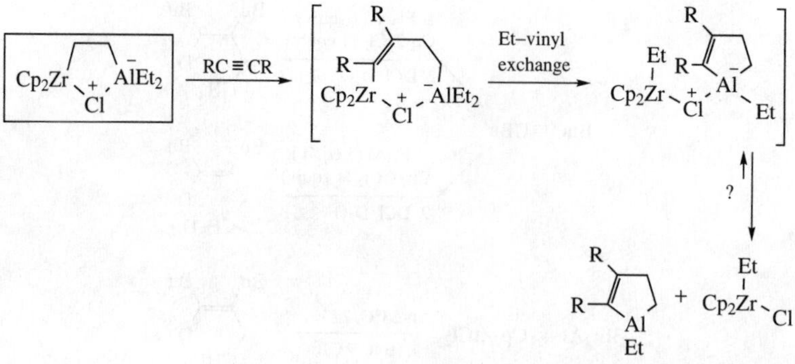

Scheme 79.

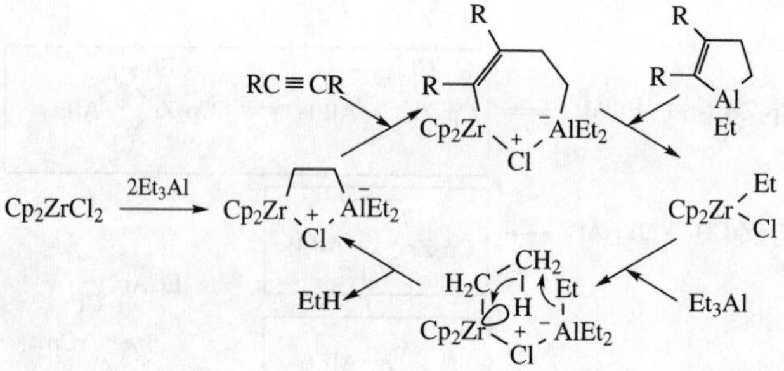

Scheme 80.

7   Conclusions

The chemistry of organozirconium compounds is now almost half a century old. Developments mainly in the areas of hydrozirconation (from 1971), Zr-mediated carbometalation reactions as represented by the Kaminsky polymerization (from 1974), and the Zr-catalyzed controlled carboalumination (from 1978), have led to considerable interest in $RZr^{IV}Cp_2$ derivatives. The scientific and practical significance of Cp_2Zr chemistry has also been substantially elevated by a similar level of interest in, and the application of, Cp_2Zr^{II} chemistry, mainly during the 1980s and 1990s. Despite these developments, many fundamentally important topics remain to be further developed, such as oxidative addition and possible reductive elimination of Cp_2Zr derivatives, σ-bond metathesis, and migratory insertion. The catalytic and/or asymmetric versions of the reactions discussed in this chapter will undoubtedly be developed further. It is reasonable to predict that zirconium will be widely used in the future by practising synthetic chemists.

Acknowledgments

Our research in this area has been supported primarily by the National Sciences Foundation, the National Institutes of Health, and Purdue University. The author is deeply indebted to a number of collaborators mentioned in the literature references for their contributions.

8 References

[1] Wailes, P. C.; Coutts, R. S. P.; Weigold, H.; *Organometallic Chemistry of Titanium, Zirconium, and Hafnium*, Academic Press, New York, **1974**, 302 pp.

[2] (a) Wailes, P. C.; Weigold, H.; *J. Organomet. Chem.*, **1970**, *24*, 405; (b) Wailes, P. C.; Weigold, H.; Bell, A. P.; *J. Organomet. Chem.*, **1971**, *27*, 373.

[3] Hart, D. W.; Schwartz, J.; *J. Am. Chem. Soc.*, **1974**, *96*, 8115.

[4] Bertelo, C. A.; Schwartz, J.; *J. Am. Chem. Soc.*, **1975**, *97*, 228.

[5] Negishi, E.; Van Horn, D. E.; *J. Am. Chem. Soc.*, **1977**, *99*, 3168.

[6] Loots, M. J.; Schwartz, J.; *J. Am. Chem. Soc.*, **1977**, *99*, 8045.

[7] Carr, D. B.; Schwartz, J.; *J. Am. Chem. Soc.*, **1977**, *99*, 638.

[8] Van Horn, D. E.; Negishi, E.; *J. Am. Chem. Soc.*, **1978**, *100*, 2252.

[9] Cardin, D. J.; Lappert, M. F.; Raston, C. L.; *Chemistry of Organozirconium and Hafnium Compounds*, John Wiley & Sons, New York, **1986**, 451 pp.

[10] Schwartz, J.; Labinger, J. A.; *Angew. Chem., Int. Ed. Engl.*, **1976**, *15,* 333.

[11] Cardin, D. J.; Lappert, M. F.; Raston, C. L.; Riley, R. I.; in *Comprehensive Organometallic Chemistry*, Vol. 1, (G. Wilkinson, F. G. A. Stone, E. W. Abel, Eds), Pergamon Press, Oxford, **1982**, 549.

[12] Cardin, D. J.; Lappert, M. F.; Raston, C. L.; Riley, R. I.; in *Comprehensive Organometallic Chemistry*, Vol. 1, (G. Wilkinson, F. G. A. Stone, E. W. Abel, Eds), Pergamon Press, Oxford, **1982**, 559.

[13] Negishi, E.; *Pure Appl. Chem.*, **1981**, *53*, 2333.

[14] Pez, G. P.; Armor, J. N.; *Adv. Organomet. Chem.*, **1981**, *19*, 1.

[15] Negishi, E.; *Acc. Chem. Res.*, **1982,** *15*, 340.

[16] Erker, G.; *Acc. Chem. Res.*, **1984**, *17*, 103.

[17] Negishi, E.; Takahashi, T.; *Aldrichim. Acta*, **1985**, *18*, 31.

[18] Yasuda, H.; Tatsumi, K.; Nakamura, A.; *Acc. Chem. Res.*, **1985**, *18*, 120.

[19] Erker, G.; Krüger, C.; Müller, G.; *Adv. Organomet. Chem.*, **1985**, *24*, 1.

[20] Yasuda, H.; Nakamura, A.; *Rev. Chem. Intermediates*, **1986**, *6*, 365.

[21] Dzhemilev, U. M.; Vostrikova, O. S.; Tolstikov, G. A.; *J. Organomet. Chem.*, **1986**, *304,* 17.

[22] Negishi, E.; *Acc. Chem. Res.*, **1987**, *20*, 65.

[23] Negishi, E.; Takahashi, T.; *Synthesis*, **1988**, 1.

[24] Buchwald, S. L.; Nielsen, R. B.; *Chem. Rev.*, **1988**, *88*, 1047.

[25] Buchwald, S. L.; Fisher, R. A.; *Chem. Scripta*, **1989**, *29*, 417.

[26] Nugent, W. A.; RajanBabu, T. V.; Taber, D. F.; *Chem. Scripta*, **1989**, *29,* 439.

[27] Negishi, E.; *Chem. Scripta*, **1989**, *29*, 457.

[28] Dzhemilev, U. M.; Vostrikova, O. S.; Tolstikov, G. A.; *Russ. Chem. Rev.*, **1990**, *59*, 1157.

[29] Negishi, E.; in *Comprehensive Organic Synthesis*, Vol. 5, (L. A. Paquette, Ed), Pergamon Press, Oxford, **1991**, 1163.

[30] Jordan, R. F.; *Adv. Organomet. Chem.*, **1991**, *32*, 325.

[31] Labinger, J. A.; in *Comprehensive Organic Synthesis*, Vol. 8, (B. M. Trost, I. Fleming, Eds), Pergamon Press, Oxford, **1991**, 667.

[32] Erker, G.; *Pure Appl. Chem.*, **1992**, *64*, 393.

[33] Broene, R. D.; Buchwald, S. L.; *Science*, **1993**, *261*, 1696.

[34] Negishi, E.; Takahashi, T.; *Acc. Chem. Res.*, **1994**, *27*, 124.

[35] Farina, V.; in *Comprehensive Organometallic Chemistry II*, Vol. 12, (E. W. Abel, F. G. A. Stone, G. Wilkinson, Eds), Pergamon Press, Oxford, **1995**, p. 161.

[36] Buchwald, S. L.; Broene, R. D.; in *Comprehensive Organometallic Chemistry II*, Vol. 12, (E. W. Abel, F. G. A. Stone, G. Wilkinson, Eds), Pergamon Press, Oxford, **1995**, 771.

[37] Jubb, J.; Fong, J.; Richeson, D.; Gambarotta, S.; in *Comprehensive Organometallic Chemistry II*, Vol. 4, (E. W. Abel, F. G. A. Stone, G. Wilkinson, Eds), Pergamon Press, Oxford, **1995**, 543.

[38] Guram, A. S.; Jordan, R. F.; in *Comprehensive Organometallic Chemistry II*, Vol. 4, (E. W. Abel, F. G. A. Stone, G. Wilkinson, Eds), Pergamon Press, Oxford, **1995**, 589.

[39] Binger, P.; Podubrin, S.; in *Comprehensive Organometallic Chemistry II*, Vol. 4, (E. W. Abel, F. G. A. Stone, G. Wilkinson, Eds), Pergamon Press, Oxford, **1995**, 439.

[40] Broene, R. D.; in *Comprehensive Organometallic Chemistry II*, Vol. 12, (E. W. Abel, F. G. A. Stone, G. Wilkinson, Eds), Pergamon Press, Oxford, **1995**, 313.

[41] Hanzawa, Y.; Ito, H.; Taguchi, T.; *Synlett*, **1995**, 299.

[42] Dzhemilev, U. M.; *Tetrahedron*, **1995**, *51*, 4333.

[43] Dzhemilev, U. M.; Sultanov, R. M.; Gaimaldinov, R. G.; *J. Organomet. Chem.*, **1995**, *491*, 1.

[44] Wipf, P.; Jahn, H.; *Tetrahedron*, **1996**, *52*, 12853.

[45] Negishi, E.; Kondakov, D. Y.; *Chem. Soc. Rev.*, **1996**, *26*, 417.

[46] Ohff, A.; Pulst, S.; Lefeber, C.; Peulecke, N.; Arndt, P.; Burkalov, V. V.; Rosenthal, U.; *Synlett*, **1996**, 111.

[47] Hoveyda, A. H.; Morken, J. P.; *Angew. Chem., Int. Ed. Engl.*, **1996**, *35*, 1262.

[48] Negishi, E.; Takahashi, T.; *Bull. Chem. Soc. Jpn.*, **1998**, *71*, 755.

[49] Boor, J.; *Ziegler-Natta Catalysis and Polymerization*, Academic Press, New York, **1978**, 670 pp.

[50] Negishi, E.; *Organometallics in Organic Synthesis*, Wiley-Interscience, **1980**, Chap. 2.

[51] Wilkinson, G.; Birmingham, J. M.; *J. Am. Chem. Soc.*, **1954**, *76*, 4281.

[52] Erker, G.; Berg, K.; Treschanke, L.; Engel, K.; *Inorg. Chem.*, **1982**, *21*, 1277.

[53] Wailes, P. C.; Weigold, H.; Bell, A. P.; *J. Organomet. Chem.*, **1971**, *33*, 181.

[54] Negishi, E.; Miller, J. A.; Yoshida, T.; *Tetrahedron Lett.*, **1984**, *25*, 3407.

[55] Negishi, E.; Yoshida, T.; *J. Am. Chem. Soc.*, **1981**, *103*, 4985.

[56] Carr, D. B.; Schwartz, J.; *J. Am. Chem. Soc.*, **1979**, *101*, 3521.

[57] Negishi, E.; Boardman, L. D.; *Tetrahedron Lett.*, **1982**, *23*, 3327.

[58] Buchwald, S. L.; La Maire, S. J.; Nielsen, R. B.; Watson, B. T.; King, S. M.; *Tetrahedron Lett.*, **1987**, *28*, 3895.

[59] Lipshutz, B. H.; Keil, R.; Ellsworth, E. L.; *Tetrahedron Lett.*, **1990**, *31*, 7257.

[60] Swanson, D. R.; Nguyen, T.; Noda, Y.; Negishi, E.; *J. Org. Chem.*, **1991**, *56*, 2590.

[61] Endo, J.; Koga, N.; Morokuma, K.; *Organometallics*, **1993**, *12*, 2777.

[62] Xu, C.; Negishi, E.; *Tetrahedron Lett.*, **1999**, *40*, 431.

[63] Makabe, H.; Negishi, E.; *Eur. J. Org. Chem.*, **1999**, 969.

[64] Negishi, E.; Yoshida, T.; *Tetrahedron Lett.*, **1980**, *21*, 1501.

[65] Rousset, C. J.; Swanson, D. R.; Lamaty, F.; Negishi, E.; *Tetrahedron Lett.*, **1989**, *30*, 5105.

[66] Ito, H.; Taguchi, T.; Hanzawa, Y.; *Tetrahedron Lett.*, **1992**, *33*, 7873.

[67] (a) Ito, H.; Nakamura, T.; Taguchi, T.; Hanzawa, Y.; *Tetrahedron Lett.*, **1992**, *33*, 3769; (b) Ito, H.; Taguchi, T.; Hanzawa, Y.; *Tetrahedron Lett.*, **1992**, *33*, 1295; (c) Ito, H.; Taguchi, T.; Hanzawa, Y.; *J. Org. Chem.*, **1993**, *58*, 774; (d) Ito, H.; Ikeuchi, Y.; Taguchi, T.; Hanzawa, Y.; Shiro, M.; *J. Am. Chem. Soc.*, **1994**, *116*, 5469; (e) Ito, H.; Nakamura, T.; Taguchi, T.; Hanzawa, Y.; *Tetrahedron*, **1995**, *51*, 4507.

[68] Takahashi, T.; Kotora, M.; Fischer, R.; Nishihara, Y.; Nakajima, K.; *J. Am. Chem. Soc.*, **1995**, *117*, 11039.

[69] Watt, G. W.; Drummond, F. O.; *J. Am. Chem. Soc.*, **1966**, *88*, 5926.

[70] Sikora, D. J.; Moriarty, K. J.; Rausch, M. D.; *Inorg. Synth.*, **1990**, *28*, 249.

[71] Fryzuk, M. D.; Hadad, T. S.; Berg, D. J.; *Coord. Chem. Rev.*, **1990**, *99*, 137.

[72] Negishi, E.; Holmes, S. J.; Tour, J. M.; Miller, J. A.; Cederbaum, F. E.; Swanson, D. R.; Takahashi, T.; *J. Am. Chem. Soc.*, **1989**, *111*, 3336.

[73] Binger, P.; Müller, P.; Benn, R.; Rufinska, A.; Gabor, B.; Krüger, C.; Betz, P.; *Chem. Ber.*, **1989**, *122*, 1035.

[74] Negishi, E.; Cederbaum, F. E.; Takahashi, T.; *Tetrahedron Lett.*, **1986**, *27*, 2829.

[75] Takahashi, T.; Swanson, D. R.; Negishi, E.; *Chem. Lett.*, **1987**, 623.

[76] Takahashi, T.; Murakami, M.; Kunishige,M.; Saburi, M.; Uchida, Y.; Kozawa, K.; Uchida, T.; Swanson, D. R.; Negishi, E.; *Chem. Lett.*, **1989**, 761.

[77] Swanson, D. R.; Negishi, E.; *Organometallics*, **1991**, *10*, 825.

[78] Erker, G.; Kropp, K.; *J. Am. Chem. Soc.*, **1979**, *101*, 3659.

[79] Buchwald, S. L.; Watson, B. T.; Huffman, J. C.; *J. Am. Chem. Soc.*, **1986**, *108*, 7411.

[80] Buchwald, S. L.; Lum, R. T.; Dewan, J. C.; *J. Am. Chem. Soc.*, **1986**, *108*, 7441.

[81] Buchwald, S. L.; Watson, B. T.; Huffman, J. C.; *J. Am. Chem. Soc.*, **1987**, *109*, 2544.

[82] Erker, G.; Wicher, J.; Engel, K.; Rosenfeldt, F.; Dietrich, W.; Kruger, C.; *J. Am. Chem. Soc.*, **1980**, *102*, 6346.

[83] Erker, G.; Wicher, J.; Engel, K.; Krüger, C.; *Chem. Ber.*, **1982**, *115*, 3300.

[84] Buchwald, S. L.; Watson, B. T.; Wannamaker, M. W.; Dewar, J. C.; *J. Am. Chem. Soc.*, **1989**, *111*, 4486.

[85] Rosenthal, U.; Ohff, A.; Baumann, W.; Tillack, A.; Görls, H.; Burlakov, V. V.; Shur, V. B.; *Z. Anorg. Allg. Chem.*, **1995**, *621*, 77.

[86] Labinger, J. A.; Hart, D. W.; Seibert, W. E.; Schwartz, J.; *J. Am. Chem. Soc.*, **1975**, *97*, 3851.

[87] Wailes, P. C.; Weigold, H.; Bell, A. P.; *J. Organomet. Chem.*, **1972**, *43*, C32.

[88] Hart, D. W.; Blackburn, T. F.; Schwartz, J.; *J. Am. Chem. Soc.*, **1975**, *97*, 679.

[89] Bertelo, C. A.; Schwartz, J.; *J. Am. Chem. Soc.*, **1976**, *98*, 262.

[90] Blackburn, T. F.; Labinger, J. A.; Schwartz, J.; *Tetrahedron Lett.*, **1975**, 3041.

[91] Cole, T. E.; Quintanilla, R.; Rodewald, S.; *Organometallics*, **1991**, *10*, 3777.

[92] Quintanilla, R.; Cole, T. E.; *Tetrahedron*, **1995**, *51*, 4297.

[93] Budnik, R. A.; Kochi, J. K.; *J. Organomet. Chem.*, **1976**, *116*, C3.

[94] Fryzuk, M. D.; Bates, G. S.; Stone, C.; *Tetrahedron Lett.*, **1986**, *27*, 1537.

[95] Kim, S.; Kim, K. H.; *Tetrahedron Lett.*, **1995**, *36*, 3725.

[96] Okukado, N.; Van Horn, D. E.; Klima, W. L.; Negishi, E.; *Tetrahedron Lett.*, **1978**, 1027.

[97] Negishi, E.; Okukado, N.; King, A. O.; Van Horn, D. E.; Spiegel, B. I.; *J. Am. Chem. Soc.*, **1978**, *100*, 2254.

[98] Negishi, E.; Owczarczyk, Z.; *Tetrahedron Lett.*, **1991**, *32*, 6683.

[99] Matsushita, H.; Negishi, E.; *J. Am. Chem. Soc.*, **1981**, *103*, 2882.

[100] Lipshutz, B. H.; Bulow, G.; Lowe, R. F.; Stevens, K. L.; *J. Am. Chem. Soc.*, **1996**, *118*, 5512.

[101] Negishi, E.; Baba, S.; *J. Chem. Soc., Chem. Comm.*, **1976**, 596.

[102] Baba, S.; Negishi, E.; *J. Am. Chem. Soc.*, **1976**, *98*, 6729.

[103] Negishi, E.; Matsushita, H.; Okukado, N.; *Tetrahedron Lett.*, **1981**, *32*, 2715.

[104] Negishi, E.; Bagheri, V.; Chatterjee, S.; Luo, F. T.; Miller, J. A.; Stoll, A. T.; *Tetrahedron Lett.*, **1983**, *24*, 5181.

[105] Negishi, E.; Valente, L. F.; Kobayashi, M.; *J. Am. Chem. Soc.*, **1980**, *102*, 3298.

[106] Negishi, E.; Luo, F. T.; Rand, C. L.; *Tetrahedron Lett.*, **1982**, *23*, 27.

[107] Negishi, E.; King, A. O.; Klima, W. L.; Patterson, W.; Silveira, A.; *J. Org. Chem.*, **1980**, *45*, 2526.

[108] Yoshifuji, M.; Loots, M. J.; Schwartz, J.; *Tetrahedron Lett.*, **1977**, 1303.

[109] Lipshutz, B. H.; Ellsworth, E. L.; *J. Am. Chem. Soc.*, **1990**, *112*, 7440.

[110] Lipshutz, B. H.; Keil, R.; *J. Am. Chem. Soc.*, **1992**, *114*, 7919.

[111] (a) Zheng, B.; Srebnik, M.; *J. Org. Chem.*, **1995**, *60*, 3278; (b) Pereira, S.; Srebnik, M.; *Organometallics*, **1995**, *14*, 3127; (c) Lipshutz, B. H.; Wood, M. R.; *J. Am. Chem. Soc.*, **1994**, *116*, 11689.

[112] Loots, M. J.; Schwartz, J.; *Tetrahedron Lett.*, **1978**, 4381.

[113] Schwartz, J.; Loots, M. J.; Kosugi, M.; *J. Am. Chem. Soc.*, **1980**, *102*, 1333.

[114] Weidmann, B.; Maycock, C. D.; Seebach, D.; *Helv. Chim. Acta*, **1981**, *64*, 1552.

[115] Yamamoto, Y.; Muruyama, K.; *Tetrahedron Lett.*, **1981**, *22*, 2895.

[116] Yasuda, H.; Kajihara, Y.; Mashima, K.; Nagasuna, K.; Nakamura, A.; *Chem. Lett.*, **1981**, 671.

[117] (a) Hartner, F. W.; Schwartz, J.; *J. Am. Chem. Soc.*, **1981**, *103*, 4979; (b) Hartner, F. W.; Schwartz, J.; Clift, S. M.; *J. Am. Chem. Soc.*, **1983**, *105*, 640; (c) Clift, S. M.; Schwartz, J.; *J. Am. Chem. Soc.*, **1984**, *106*, 8300.

[118] Yoshida, T.; Negishi, E.; *J. Am. Chem. Soc.*, **1981**, *103*, 1276.

[119] (a) Tour, J. M.; Redworth, P. V.; Wu, R.; *Tetrahedron Lett.*, **1989**, *30*, 3927; (b) Tucker, C. E.; Knochel, P.; *J. Am. Chem. Soc.*, **1991**, *113*, 9888.

[120] Argenti, L.; Bellina, F.; Carpita, A.; Dell'Amico, N.; Rossi, R.; *Synth. Commun.*, **1994**, *24*, 3167.

[121] Maeta, H.; Hashimoto, T.; Hasegawa, T.; Suzuki, K.;*Tetrahedron Lett.*, **1992**, *33*, 5965.

[122] Suzuki, K.; Hasegawa, T.; Imai, T.; Maeta, H.; Ohba, S.; *Tetrahedron*, **1995**, *51*, 4483.

[123] Wipf, P.; Xu, W.; *J. Org. Chem.*, **1993**, *58*, 825.

[124] Wipf, P.; Xu, W.; *Tetrahedron*, **1995**, *51*, 4551.

[125] Jordan, R. F.; Bajgur, C. S.; Willett, R.; Scott, B.; *J. Am. Chem. Soc.*, **1986**, *108*, 7410.

[126] Bertelo, C. A.; Schwartz, J.; *J. Am. Chem. Soc.*, **1975**, *97*, 228.

[127] Manriquez, J. M.; McAlister, D. R.; Sanner, R. D.; Bercaw, J. E.; *J. Am. Chem. Soc.*, **1978**, *100*, 2716.

[128] Erker, G.; Rosenfeldt, F.; *J. Organomet. Chem.*, **1980**, *188*, C1.

[129] Negishi, E.; Swanson, D. R.; Miller, S. R.; *Tetrahedron Lett.*, **1988**, *29*, 1631.

[130] Buchwald, S. L.; La Maire, S. J.; *Tetrahedron Lett.*, **1987**, *28*, 295.

[131] Mintz, E. A.; Ward, A. S.; Tice, D. S.; *Organometallics*, **1985**, *4*, 1308.

[132] Negishi, E.; Akiyoshi, K.; O'Connor, B.; Takagi, K.; Wu, G.; *J. Am. Chem. Soc.*, **1989**, *111*, 3089.

[133] Kasai, K.; Liu, Y.; Hara, R.; Takahashi, T.; *Chem. Commun.*, **1998**, 1989.

[134] Takagi, K.; Rousset, C. J.; Negishi, E.; *J. Am. Chem. Soc.*, **1991**, *113*, 1440.

[135] Kaminsky, W.; Vollmer, H.-J.; Heins, E.; Sinn, H.; *Makromol. Chem.*, **1974**, *175*, 443.

[136] Kaminsky, W.; Miri, M.; Sinn, H.; Woldt, R.; *Makromol. Chem. Rapid Commun.*, **1983**, *4*, 417.

[137] Negishi, E.; Van Horn, D. E.; Yoshida, T.; *J. Am. Chem. Soc.*, **1985**, *107*, 6639.

[138] Negishi, E.; Van Horn, D. E.; *Organomet. Synth.*, **1986**, *3*, 467.

[139] Van Horn, D. E.; Valente, L. F.; Idacavage, M. J.; Negishi, E.; *J. Organomet. Chem.*, **1978**, *156*, C20.

[140] Negishi, E.; Van Horn, D. E.; Yoshida, T.; Rand, C. L.; *Organometallics*, **1983**, *2*, 563.

[141] Rand, C. L.; Van Horn, D. E.; Moore, M. W.; Negishi, E.; *J. Org. Chem.*, **1981**, *46*, 4093.

[142] Negishi, E.; Boardman, L. D.; Tour, J. M.; Sawada, H.; Rand, C. L.; *J. Am. Chem. Soc.*, **1983**, *105*, 5344.

[143] Boardman, L. D.; Bagheri, V.; Sawada, H.; Negishi, E.; *J. Am. Chem. Soc.*, **1984**, *106*, 6105.

[144] Negishi, E.; Boardman, L. D.; Sawada, H.; Bagheri, V.; Stoll, A. T.; Tour, J. M.; Rand, C. L.; *J. Am. Chem. Soc.*, **1988**, *110*, 5383.

[145] Negishi, E.; Liu, F.; Choueiry, D.; Mohamud, M. M.; Silveira, A.; Reeves, M.; *J. Org. Chem.*, **1996**, *61*, 8325.

[146] Liu, F.; Negishi, E.; *Tetrahedron Lett.*, **1997**, *38*, 1149.

[147] Ma, S.; Negishi, E.; *J. Org. Chem.*, **1997**, *62*, 784.

[148] Mole, T.; Jeffery, E. A.; *Organoaluminum Compounds*, Elsevier, Amsterdam, **1972**, 465 pp.

[149] Zweifel, G.; Miller, J. A.; *Org. React.*, **1984**, *32*, 1.

[150] Negishi, E.; Van Horn, D. E.; King, A. O.; Okukado, N.; *Synthesis*, **1979**, 501.

[151] Negishi, E.; Jadhav, K. P.; Daotien, N.; *Tetrahedron Lett.*, **1982**, *23*, 2085.

[152] Okukado, N.; Negishi, E.; *Tetrahedron Lett.*, **1978**, 2357.

[153] Kobayashi, M.; Valente, L. F.; Negishi, E.; Patterson, W.; Silveira, A.; *Synthesis*, **1980**, 1034.

[154] Williams, D. R.; Barner, B. A.; Nishitani, K.; Phillips, J. G.; *J. Am. Chem. Soc.*, **1982**, *104*, 4708.

[155] Fung, S.; Siddall, J. B.; *J. Am. Chem. Soc.*, **1980**, *102*, 6580.

[156] Miller, J. A.; Negishi, E.; *Tetrahedron Lett.*, **1984**, *25*, 5863.

[157] Negishi, E.; Kondakov, D. Y.; Choueiry, D.; Kasai, K.; Takahashi, T.; *J. Am. Chem. Soc.*, **1996**, *118*, 9577.

[158] Kondakov, D. Y.; Negishi, E.; *J. Am. Chem. Soc.*, **1995**, *117*, 10771.

[159] Dzhemilev, U. M.; Vostrikova, O. S.; Sultanov, R. M.; *Izv. Akad. Nauk SSSR, Ser. Khim.*, **1983**, 218.

[160] Takahashi, T.; Seki, T.; Nitto, Y.; Saburi, M.; Rousset, C. J.; Negishi, E.; *J. Am. Chem. Soc.*, **1991**, *113*, 6266.

[161] Erker, G.; Aulbach, M.; Knickmeier, M.; Wingbermuhle, D.; Kruger, C.; Nolte, M.; Werner, S.; *J. Am. Chem. Soc.*, **1993**, *115*, 4590.

[162] Dzhemilev, V. M.; Ibragimov, A. G.; Zoltarev, A. P.; Muslukhov, R. R.; Tolstikov, G. A.; *Izv. Akad. Nauk SSSR, Ser. Khim.*, (Engl. Transl.), **1989**, 194.

[163] Kondakov, D. Y.; Negishi, E.; *J. Am. Chem. Soc.*, **1996**, *118*, 1577.

[164] (a) Morken, J. P.; Didiuk, M. T.; Hoveyda, A. H.; *J. Am. Chem. Soc.*, **1993**, *115*, 6997; (b) Didiuk, M. T.; Johannes, C. W.; Morken, J. P.; Hoveyda, A. H.; *J. Am. Chem. Soc.*, **1995**, *117*, 7097.

[165] Bell, L.; Whitby, R. J.; Jones, R. V. H.; Standen, M. C. H.; *Tetrahedron Lett.*, **1996**, *37*, 7139.

[166] Negishi, E.; Holmes, S. J.; Tour, J. M.; Miller, J. A.; *J. Am. Chem. Soc.*, **1985**, *107*, 2568.

[167] Negishi, E.; Swanson, D. R.; Cederbaum, F. E.; Takahashi, T.; *Tetrahedron Lett.*, **1987**, *28*, 917.

[168] (a) RajanBabu, T. V.; Nugent, W. A.; Taber, D. F.; Fagan, P. J.; *J. Am. Chem. Soc.*, **1988**, *110*, 7128; (b) Fagan, P. J.; Nugent, W. A.; *J. Am. Chem. Soc.*, **1988**, *110*, 2310.

[169] (a) Lund, E. C.; Livinghouse, T.; *J. Org. Chem.*, **1989**, *54*, 4487; (b) Van Wagenen, B. C.; Livinghouse, T.; *Tetrahedron Lett.*, **1989**, *30*, 3495; (c) Pagenkopf, B. L.; Lund, E. C.; Livinghouse, T.; *Tetrahedron*, **1995**, *51*, 4421.

[170] Miura, K.; Funatsu, M.; Saito, H.; Ito, H.; Hosomi, A.; *Tetrahedron Lett.*, **1996**, *37*, 9059.

[171] Barluenga, J.; Saz, R.; Fañanás, F. J.; *Chem. Eur. J.*, **1997**, *3*, 1324.

[172] Wender, P. A.; McDonald, F. E.; *J. Am. Chem. Soc.*, **1990**, *112*, 4956.

[173] Agnel, G.; Negishi, E.; *J. Am. Chem. Soc.*, **1991**, *113*, 7424.

[174] Agnel, G.; Owczarczyk, Z.; Negishi, E.; *Tetrahedron Lett.*, **1992**, *33*, 1543.

[175] (a) Khand, I. U.; Knox, G. R.; Pauson, P. L.; Watts, W. E.; Foreman, M. I.; *J. Chem. Soc., Perkin Trans. 1*, **1973**, 977; (b) Schore, N. E.; in *Comprehensive Organic Synthesis*, Vol. 5, (L. A. Paquette, Ed), Pergamon Press, Oxford, **1991**, 1037.

[176] Negishi, E.; Ma, S.; Sugihara, T.; Noda, Y.; *J. Org. Chem.*, **1997**, *62*, 1922.

[177] Nugent, W. A.; Thorn, D. L.; Harlow, R. L.; *J. Am. Chem. Soc.*, **1987**, *109*, 2788.

[178] Nugent, W. A.; Taber, D. F.; *J. Am. Chem. Soc.*, **1989**, *111*, 6435.

[179] Akita, M.; Yasuda, H.; Yamamoto, H.; Nakamura, A.; *Polyhedron*, **1991**, *10*, 1.

[180] Taber, D. F.; Louey, J. P.; *Tetrahedron*, **1995**, *51*, 4495.

[181] Negishi, E.; Choueiry, D.; Nguyen, T. B.; Swanson, D. R.; Suzuki, N.; Takahashi, T.; *J. Am. Chem. Soc.*, **1994**, *116*, 9751.

[182] Maye, J. P.; Negishi, E.; *Tetrahedron Lett.*, **1993**, *34*, 3359.

[183] Negishi, E.; Maye, J. P.; Choueiry, D.; *Tetrahedron*, **1995**, *51*, 4447.

[184] (a) Mori, M.; Uesaka, N.; Shibasaki, M.; *J. Org. Chem.*, **1992**, *57*, 3519; (b) Mori, M.; Saitoh, F.; Uesake, N.; Okamura, K.; Date, T.; *J. Org. Chem.*, **1994**, *59*, 4993; (c) Uesaka, N.; Saitoh, F.; Mori, M.; Shibasaki, M.; Okamura, K.; Date, T.; *J. Org. Chem.*, **1994**, *59*, 5633; (d) Mori, M.; Uesaka, N.; Saitoh, F.; Shibasaki, M.; *J. Org. Chem.*, **1994**, *59*, 5643; (e) Honda, T.; Satoh, S.; Mori, M.; *Organometallics*, **1995**, *14*, 1548; (f) Saitoh, F.; Mori, M.; Okamura, K.; Date, T.; *Tetrahedron*, **1995**, *51*, 4439.

[185] Jensen, M.; Livinghouse, T.; *J. Am. Chem. Soc.*, **1989**, *111*, 4495.

[186] Negishi, E.; Swanson, D. R.; Takahashi, T.; *J. Chem. Soc., Chem. Comm.*, **1990**, *18*, 1254.

[187] Negishi, E.; Nguyen, T.; Maye, J. P.; Choueiry, D.; Suzuki, N.; Takahashi, T.; *Chem. Lett.*, **1992**, 2367.

[188] Skibbe, V.; Erker, G.; *J. Organomet. Chem.*, **1983**, *241*, 15.

[189] Erker, G.; Dorf, U.; Czisch, P.; Petersen, J. L.; *Organometallics*, **1986**, *5*, 668.

[190] Straus, D. A.; Grubbs, R. H.; *J. Am. Chem. Soc.*, **1982**, *104*, 5499.

[191] Moore, E. J.; Straus, D. A.; Armantrout, J.; Sautarsiero, B. D.; Grubbs, R. H.; Bercaw, J. E.; *J. Am. Chem. Soc.*, **1983**, *105*, 2068.

[192] Ho, S. C. H.; Straus, D. A.; Armantrout, J.; Schaefer, W. P.; Grubbs, R. H.; *J. Am. Chem. Soc.*, **1984**, *106*, 2210.

[193] Waymouth, R. M.; Santarsiero, B. D.; Coots, R. J.; Bronikowski, M. J.; Grubbs, R. H.; *J. Am. Chem. Soc.*, **1986**, *108*, 1427.

[194] Takahashi, T.; Kageyama, M.; Denisov, V.; Hara, R.; Negishi, E.; *Tetrahedron Lett.*, **1993**, *34*, 687.

[195] Swanson, D. R.; Rousset, C. J.; Negishi, E.; Takahashi, T.; Seki, T.; Saburi, M.; Uchida, Y.; *J. Org. Chem.*, **1989**, *54*, 3521.

[196] Negishi, E.; Miller, S. R.; *J. Org. Chem.*, **1989**, *54*, 6014.

[197] Knight, K. S.; Wang, D.; Waymouth, R. M.; Ziller, J.; *J. Am. Chem. Soc.*, **1994**, *116*, 1845.

[198] Takahashi, T.; Fischer, R.; Xi, Z.; Nakajima, K.; *Chem. Lett.*, **1996**, 357.

[199] Fischer, R.; Walther, D.; Gebhardt, P.; Görls, H.; *Organometallics*, **2000** (in press).

[200] Watt, G. W.; Drummond, F. O.; *J. Am. Chem. Soc.*, **1970**, *92*, 826.

[201] Thanedar, S.; Farona, M. F.; *J. Organomet. Chem.*, **1982**, *235*, 65.

[202] Mao, S. S. H.; Tilley, T. D.; *J. Am. Chem. Soc.*, **1995**, *117*, 5365.

[203] Mao, S. S. H.; Tilley, T. D.; *Macromolecules*, **1997**, *30*, 5566.

[204] Jiang, B.; Tilley, T. D.; *J. Am. Chem. Soc.*, **1999**, *121*, 9744.

[205] Copéret, C.; Negishi, E.; Xi, Z.; Takahashi, T.; *Tetrahedron Lett.*, **1994**, *35*, 695.

[206] Takahashi, T.; Kondakov, D. Y.; Xi, Z.; Suzuki, N.; *J. Am. Chem. Soc.*, **1995**, *117*, 5871.

[207] Takahashi, T.; Fujimori, T.; Seki, T.; Saburi, M.;. Uchida, Y; Rousset, C. J.; Negishi, E.; *J. Chem. Soc., Chem. Comm.*, **1990**, *18*, 182.

[208] Suzuki, N.; Kondakov, D. Y.; Takahashi, T.; *J. Am. Chem. Soc.*, **1993**, *115*, 8485.

[209] Venanzi, L. M.; Lehman, R.; Keil, R.; Lipshutz, B. H.; *Tetrahedron Lett.*, **1992**, *33*, 5857.

[210] Kasai, K.; Kotora, M.; Suzuki, N.; Takahashi, T.; *J. Chem. Soc., Chem. Commun.*, **1995**, 109.

[211] Takahashi, T.; Kotora, M.; Kasai, K.; Suzuki, N.; *Tetrahedron Lett.*, **1994**, *35*, 5685.

[212] Takahashi, T.; Kotora, M.; Kasai, K.; Suzuki, N.; *Organometallics*, **1994**, *13*, 4183.

[213] Kotora, M.; Xi, Z.; Takahashi, T.; *J. Synth. Org. Chem. Jpn. (Engl.)*, **1997**, *55*, 958.

[214] Takahashi, T.; Kotora, M.; Xi, Z.; *J. Chem. Soc., Chem. Commun.*, **1995**, 1503.

[215] Takahashi, T.; Xi, Z.; Kotora, M.; Xi, C.; Nakajima, K.; *Tetrahedron Lett.*, **1996**, *37*, 7521.

[216] Takahashi, T.; Kotora, M.; Xi, Z.; *J. Chem. Soc., Chem. Commun.*, **1995**, 361.

[217] Takahashi, T.; Tsai, F.-Y.; Li, Y.; Nakajima, K.; Kotora, M.; *J. Am. Chem. Soc.*, **1999**, *121*, 11093.

[218] Takahashi, T.; Kasai, K.; Suzuki, N.; Nakajima, K.; Negishi, E.; *Organometallics*, **1994**, *13*, 3413.

[219] Takahashi, T.; Hasegawa, M.; Suzuki, N.; Saburi, M.; Rousset, C. J.; Fanwick, P. E.; Negishi, E.; *J. Am. Chem. Soc.*, **1991**, *113*, 8564.

[220] Dzhemilev, U. M.; Ibragimov, A. G.; Zoltarev, A. P.; Muslukov, R. R.; Tolstikov, G. A.; *Izv. Akad. Nauk SSSR, Ser. Khim.*, **1991**, 2570.

[221] Sinn, H.; Oppermann, G.; *Angew. Chem., Int. Ed. Engl.*, **1966**, *5*, 962.

[222] Sinn, H.; Kolk, E.; *J. Organomet. Chem.*, **1966**, *6*, 373.

[223] Kaminsky, W.; Sinn, H.; *Liebigs Ann. Chem.*, **1975**, 424.

[224] Kaminsky, W.; Sinn, H.; *Liebigs Ann. Chem.*, **1975**, 438.

[225] Kaminsky, W.; Kopf, J.; Sinn, H.; Vollmer, H.; *Angew. Chem., Int. Ed. Engl.*, **1976**, *15*, 629.

IX

Organoiron and Organochromium Chemistry

MARTIN F. SEMMELHACK

Department of Chemistry, Princeton University, Princeton, USA

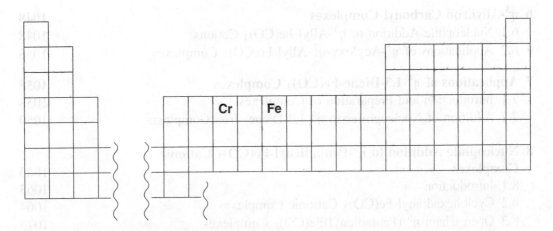

Organometallics in Synthesis: A Manual. Edited by Manfred Schlosser.
© 2002 John Wiley & Sons Ltd

Contents

1   Introduction

The emphasis in this chapter is the application in organic synthesis of organoiron and organochromium complexes. The coverage includes a diverse group of reagents and mechanisms, and each section contains a specific introduction. The space is limited, and the coverage is aimed at illustrating useful methods with generality and significant applications. It does not attempt to be comprehensive. Those methods not included are iron-promoted formation of oxyallyl cations[1,2], iron−carbene species in cyclopropane formation[3], iron−carbene cycloaddition chemistry[4], the peripheral effects on reactivity of 1,3-diene-Fe(CO)$_3$[5], CpFe(CO)$_2$[6], and arene-Cr(CO)$_3$ units[7]. A variety of pathways from the chromium−carbene complexes[8] are also not included.

As with conventional organic synthesis methods, the application of organometallic reagents in synthesis requires a solid footing in mechanism and an awareness of alternate pathways. The discussions begin with descriptions of each type of complex including preparation and handling. Then the mechanism and associated aspects of the scope and limitations are be covered, and lead to specific applications in synthesis. Finally, in selected cases representative examples of detailed procedures are reproduced.

It is possible to divide all organometallic reactions into two classes. The first class might be called 'polar' mechanisms and has strong parallels in organic chemistry. Polar reactions can be analyzed in terms of electrophile−nucleophile interactions, and the metal serves primarily as a complex functional group, interacting with one of the substrates (ligand) and modifying it's reactivity toward polar reagents. A widely used example is allylic alkylation, in which an allyl−palladium intermediate acts as a special allyl electrophile, with reactivity and selectivity influenced by the Pd[9]. A second mechanism class is 'nonpolar', epitomized by the Heck reaction, in which oxidative-addition, β-insertion, and β-elimination mechanisms occur in sequence[9]. Nonpolar processes take place within the coordination sphere of the metal, and are not in any direct way driven by polar features in the substrates. There is some analogy with concerted mechanisms in traditional organic chemistry. Such mechanistic distinctions have significance in recognizing parallels between methods and mechanisms.

The order of topics is based on ligand type, starting with several variations of η^1 ligands and proceeding in order to η^6-arene ligands. This ordering is chosen only for lack of a better organizational principle and does not correlate with mechanistic principles nor types of synthesis applications.

2   Reactions of Tetracarbonylferrate(−II), Collman's Reaction

2.1   Introduction and Preparation of the Reagent

Collman's reaction is illustrated in Scheme 1[10,11]. It is a three-step batch process, beginning with the reaction of tetracarbonylferrate(−II) **1** with an alkylating agent to

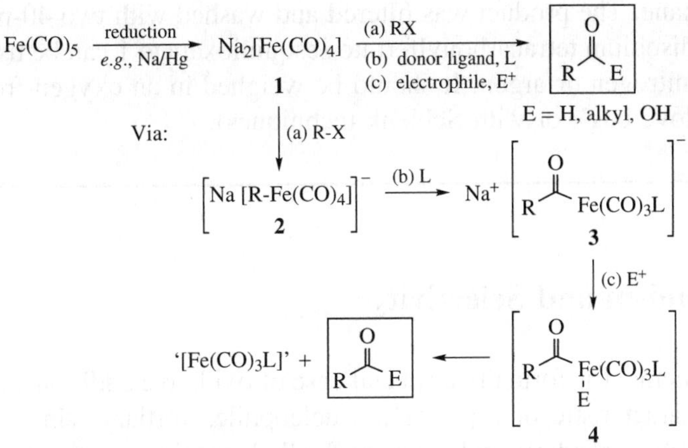

$$Fe(CO)_5 \xrightarrow[e.g., Na/Hg]{reduction} \underset{\mathbf{1}}{Na_2[Fe(CO)_4]} \xrightarrow[\substack{(b)\ \text{donor ligand, L} \\ (c)\ \text{electrophile, } E^+}]{(a)\ RX} \boxed{\underset{R \quad E}{\overset{O}{\diagup\!\!\diagdown}}}$$

E = H, alkyl, OH

Via: $\Big\downarrow$ (a) R-X

$$\underset{\mathbf{2}}{\Big[Na\,[R\text{-}Fe(CO)_4] \Big]^-} \xrightarrow{(b)\ L} Na^+ \underset{\mathbf{3}}{\left[\underset{R \quad Fe(CO)_3L}{\overset{O}{\diagup\!\!\diagdown}} \right]^-}$$

$\Big\downarrow$ (c) E^+

$$\text{'}[Fe(CO)_3L]\text{'} + \boxed{\underset{R \quad E}{\overset{O}{\diagup\!\!\diagdown}}} \longleftarrow \underset{\mathbf{4}}{\left[\underset{R \quad \underset{E}{Fe(CO)_3L}}{\overset{O}{\diagup\!\!\diagdown}} \right]}$$

Scheme 1. Collman's reaction.

give a transient η^1-alkyltetracarbonylferrate(0) intermediate **2**, which undergoes migratory insertion to produce an η^1-acyltricarbonylferrate(0) **3** with an additional ligand (L). Addition of another electrophile (E^+) leads through a transient alkylacyl$(CO)_3$(L)iron(II) intermediate **4** to an unsymmetrical carbonyl derivative and a $(CO)_3$(L)iron(0) byproduct. The overall result is the coupling of one alkyl group and one electrophile (alkyl, proton, alkene, oxygen) to one unit of carbon monoxide. Important mechanistic variations on the reaction include alternate procedures for the formation of the acyltricarbonylferrate(0) **3** and alternate substrates, such as alkenes, as the electrophile for the final step.

Tetracarbonylferrate($-$II) **1** is prepared by reduction of iron pentacarbonyl with, for example, sodium amalgam. It is readily isolated as the dioxane solvate and is available commercially. It should be handled to minimize exposure to air, and is exceedingly sensitive to oxygen when in solution. It is a moderate base and an exceptionally powerful nucleophile.

*Preparation of Disodium Tetracarbonylferrate **1***

$$Fe(CO)_5 \xrightarrow[e.g., Na/Hg]{reduction} \underset{\mathbf{1}}{Na_2[Fe(CO)_4]}$$

A mixture of dioxane (60 mL), sodium (1.06 g, 46.1 mg.-at), and benzophenone (0.91 g, 5.0 mmol) was prepared under argon. The solution was stirred vigorously and heated under reflux until the deep blue color of the benzophenone ketyl appeared. Iron pentacarbonyl (4.53 g, 2.98 mL, 23.1 mmol) was titrated into the blue solution to a white or slightly yellow end point over 2.5 h. The suspension was heated at reflux for another 45 min, then cooled to 25 °C. Precipitation of the disodium tetracarbonylferrate sesquidioxanate as a white powder is completed by adding

60 mL of hexane. The product was filtered and washed with two 40-mL portions of hexane. The disodium tetracarbonylferrate sesquidioxanate **1** can be used directly, or stored under nitrogen or argon. It should be weighed in an oxygen-free atmosphere ('drybox', 'glove bag', or with Schlenk techniques).

2.2   Mechanism and Selectivity

Step (a) in Scheme 1 is formally a special case of oxidative addition and follows a strict S_N2 pathway characteristic of a powerful nucleophile. Tertiary, vinyl, and aryl halides are inert but even secondary and neopentyl alkyl species react smoothly. Secondary alkyl halides lead to some elimination, but secondary alkyl tosylates give high yields of alkylation. Acyl halides also react, leading directly to the acyltetracarbonylferrate(0) **3** (see below). Allylic halides are reactive, but fail in the subsequent steps apparently due to elimination to the 1,3-diene-Fe(CO)$_3$ species[10]. For primary alkyl electrophiles, all of the typical S_N2 leaving groups are useful, including Cl, Br, I, and OTs.

Step (b) has been the subject of classic organometallic mechanistic investigations[12]. Added ligands promote the migration process, but even in the absence of an added ligand the acylferrate **3** is responsible for reaction with the electrophile; there is no evidence for direct reaction of the alkylferrate **2** with an added electrophile, except in the case where the electrophile is a proton. In that case, the simple reduction product R−H is produced.

Step (c) is again a special case of oxidative addition (via an S_N2 process in the case of alkyl halides), leading to the putative Fe^{II} species **4**, and then the final step of reductive elimination and formation of the unsymmetrical ketone (carbon electrophiles) or aldehyde (protons) proceeds rapidly. The byproduct is an iron(0) carbonyl derivative, which could be recycled but is generally discarded. There is no obvious possibility for a catalytic process based on these steps. The intermediate **3** is essentially a 'carbonyl anion', representing an Umpolung reagent[13], with no masking nor unmasking required.

The most serious limitation on the overall process is the relatively weak nucleophilic reactivity of the acylate−iron intermediate **3** toward the second oxidative addition (step (c)). The process is limited to highly reactive alkylating agents, such as methyl iodide and primary alkyl sulfonate esters. Allylic and benzylic halides should be reactive enough, but no example is reported.

Acyl halides are also sufficiently reactive, but a different pathway is observed. Remembering that the acylate−iron structure **3** is one contributing resonance structure, and the oxygen-localized anion structure **5** is also a good representation, it is easy to understand the formation of acylated alkoxy carbene complexes **6**, in close analogy with the conventional reactivity of enolate anions (O- versus C-alkylation). This process has been reported only briefly in a review article[8] and referenced to a thesis:

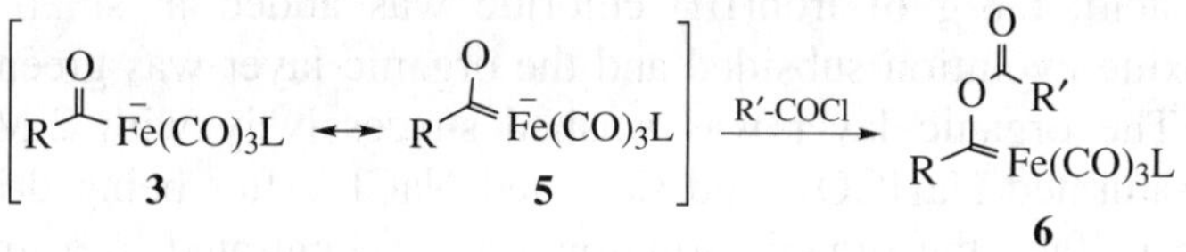

$$\left[R \underset{3}{\overset{O}{\underset{\text{Fe(CO)}_3\text{L}}{\parallel}}} \longleftrightarrow R \underset{5}{\overset{O}{\underset{\text{Fe(CO)}_3\text{L}}{\parallel}}} \right] \xrightarrow{R'\text{-COCl}} \underset{6}{\overset{O}{\underset{R}{\overset{O}{\underset{\text{Fe(CO)}_3\text{L}}{}}}}}R'$$

The limitation that only S_N2 substrates are successful in the first step can be circumvented by starting with an acyl halide. The acylate–iron intermediate **3** is formed directly, and can be taken through the second stages[10,14]. Another procedure for the preparation of the acylate iron intermediate **3** reverses the polarity of the reaction, using carbon nucleophiles (organo-Li and organo-MgX reagents) in reaction with an iron electrophile (iron pentacarbonyl)[15]. For this process, the limitations are determined by the functional group compatibility of the organo-Li (-MgBr) reagent, which are serious but well understood. Alternate processing is also possible from the acylate iron intermediate. Oxidizing agents can lead to the carboxylic acid (O_2/H_2O), ester (I_2, MeOH), or amide (I_2, RNH_2)[16]. The various options are summarized in Scheme 2.

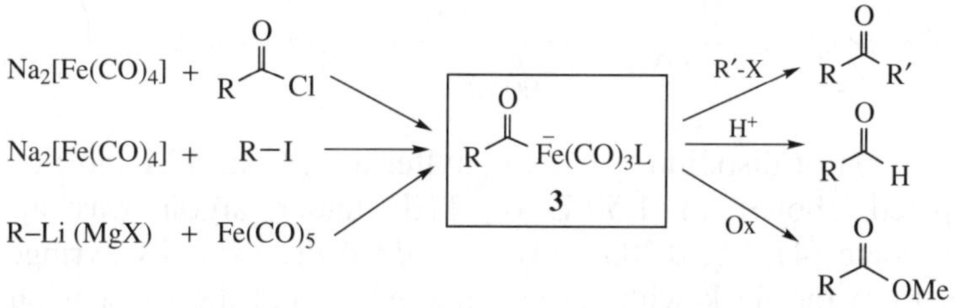

Scheme 2. Summary of conversions with Collman's reagent.

Procedure for Unsymmetrical Ketone Synthesis—Preparation of Methyl 7-Oxooctanoate[17]

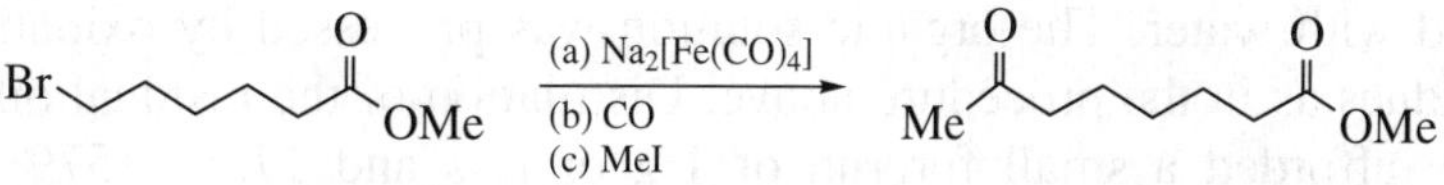

To a suspension of 7.2 g (0.021 mol) of disodium tetacarbonylferrate (as prepared above) in 60 mL of *N*-methylpyrrolidinone under argon was added dropwise methyl 6-bromohexanoate (4.18 g, 0.0200 mol). The resulting solution was stirred for 30 min at 25 °C and cooled in an ice bath before methyl iodide (6.4 g, 2.8 mL, 0.045 mol) was added over 20 min. The ice bath was removed and stirring was continued for 20–40 h. The dark-red mixture was poured into 3 L of saturated aqueous sodium chloride and extracted with three 40-mL portions of ether and one 40-mL portion of hexane. The combined organic solutions were washed with water and saturated NaCl. The organic solution was mixed with 40 mL of 2 M hydrochloric acid; 6.8 g of iron(III) chloride was added in small portions until carbon monoxide evolution subsided and the organic layer was green (triiron dodecacarbonyl). The organic layer was washed successively with 2 M hydrochloric acid, water, saturated NaHCO₃, and saturated NaCl. After being dried over anhydrous sodium sulfate, the organic solution was concentrated to a green oil with a rotary evaporator. Iron-containing byproducts such as triiron dodecacarbonyl and iron pentacarbonyl were removed by rapid chromatography on silica gel. The green triiron dodecacarbonyl was first eluted with hexane before the product was eluted with 1 : 1 ether–hexane. The solvent was evaporated, and the remaining liquid was distilled at reduced pressure, giving 2.40 g (70%) of methyl 7-oxooctanoate (bp 112–127 °C, 10 torr).

Procedure for Aldehyde Synthesis—Preparation of 7-Oxoheptanoate[15,18]

$$\text{Br}\diagdown\diagup\diagdown\diagup\overset{\displaystyle O}{\diagup}\text{OMe} \xrightarrow[\substack{\text{(b) CO}\\\text{(c) HOAC}}]{\text{(a) Na}_2[\text{Fe(CO)}_4]} \text{H}\overset{\displaystyle O}{\diagdown}\diagup\diagdown\diagup\overset{\displaystyle O}{\diagup}\text{OMe}$$

To a suspension of disodium tetracarbonylferrate (72 g, 0.21 mol) of sesquidioxanate (prepared above) in 1.50 L of THF under argon was added methyl 6-bromohexanoate (41.8 g, 0.200 mol) was added dropwise by syringe. The argon was flushed from the flask with carbon monoxide, and the suspension was stirred under 10 psi of carbon monoxide for at least 14 h, during which time the solid dissolves. Then a rapid flow of argon was swept through the flask and 50 mL of glacial acetic acid was added dropwise to the orange solution. Stirring was continued for 20 min, after which the deep red solution was concentrated to a volume of ca. 400 mL with a rotary evaporator and then poured into 2 L of water. The mixture was extracted with diethyl ether, and the combined organic solution was washed with water. The organic solution was processed by oxidative removal of iron residues as in the procedure above. Distillation of the residual oil after chromatography afforded a small forerun of 1 g or less and 17.9 g (57%) of methyl 7-oxoheptanoate (bp 65–80 °C, 0.1 torr).

2.3   Variation: Insertion of an Alkene with the Acylate Anion

Potentially useful in synthesis is the coupling of the acyl ligand with an alkene by β-insertion (Scheme 3). This requires that the acylate iron intermediate **7** be coordinatively

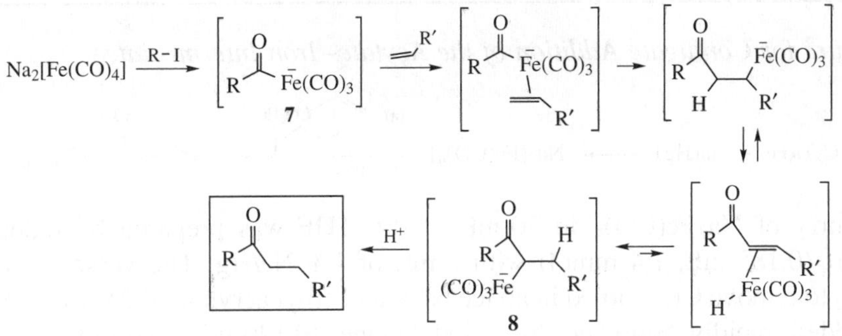

Scheme 3. Coupling of the acylate intermediate with an alkene.

unsaturated (*i.e.*, not further coordinated as in **3**), which is the situation immediately after migratory insertion, before an additional ligand has combined. The insertion process leads to an iron species which can undergo β-elimination of hydride and then re-addition to give the α-iron species **8** (an iron enolate)[19]. This species is stable toward further migratory insertion with CO and is decomposed by addition of acid in the isolation procedure. Although the method is severely limited, a few significant applications have been made, such as a key step in the synthesis of aphidicolin[20].

Electron-deficient alkenes are generally successful in inter- and intramolecular applications (Scheme 4)[10,21]. The reaction of BuI with $Fe(CO)_4(-II)$ gives the usual coordinatively unsaturated intermediate **7**, which then adds to ethyl methacrylate and to cyclopentenone to give, after protonation, the products from conjugate addition. The reaction proceeds well with electron-deficient alkenes, but is less efficient with substituted simple alkenes.

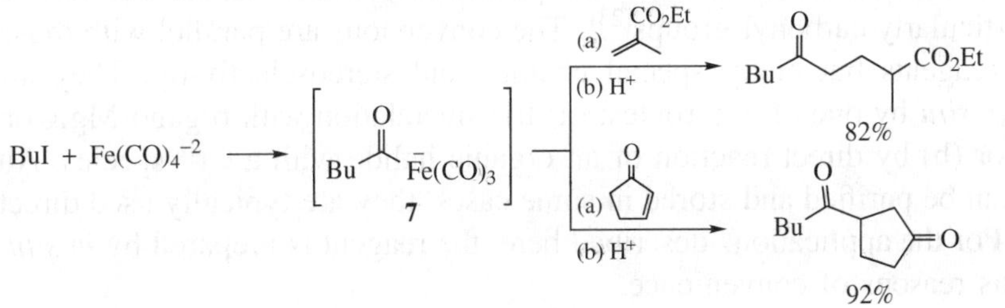

Scheme 4. Examples of alkene insertion.

The intramolecular example in the following equation shows that five-membered rings are formed efficiently; both electron deficient and simple alkenes participate[22]:

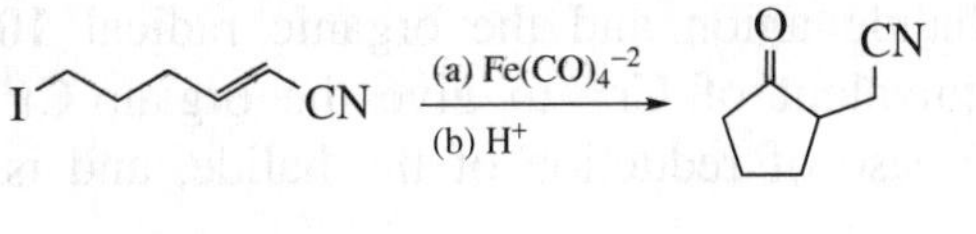

Procedure for Conjugate Addition of the Acylate–Iron Intermediate

$$(CO)_5Fe + Na(Hg) \longrightarrow Na_2[Fe(CO)_4] \xrightarrow[\text{(b) HOAc}]{\text{(a)}} $$

A slurry of $Na_2Fe(CO)_4$ in 20 mL of dry THF was prepared by reduction of $Fe(CO)_5$ (0.185 mL, 1.4 mmol) with 2 mL of 1% Na/Hg. The vessel was purged with N_2 to remove CO, cooled in an ice bath, and ethyl acrylate (0.25 mL, 2.5 mmol) was added rapidly followed by 1-iodobutane (0.120 mL, 1.0 mmol). The N_2 atmosphere was maintained throughout. The mixture was stirred at 23 °C for 4 h, and then treated with 0.12 mL of HOAc, stirred for 5 min more, and poured into water. The water layer was twice extracted with ether, and the extracts were combined, dried over anhydrous $MgSO_4$, filtered, and the filtrate was passed through a short plug of Al_2O_3 (neutral) to remove iron residues. The solution was concentrated by rotary evaporation and the residue was purified by short path distillation at aspirator vacuum. The colorless distillate was 156 mg (80%) of $C_4H_9COCH_2$ CH_2CO_2Et.

3   Addition of Allyl-, Alkenyl-, and Alkynylchromium(III) Reagents to Carbonyl Groups

3.1   Introduction

Organochromium(III) complexes are particularly useful in the addition to C=X pi bonds, particularly carbonyl groups[23]. The conversions are parallel with those of typical Grignard reagents but enjoy special chemo- and stereoselectivitiy. They are typically prepared *in situ* by one of two routes: (a) transmetalation with organo-MgX or organo-Li reagents, or (b) by direct reaction of an organic halide with a Cr^{II} species. Although the reagents can be purified and stored in some cases, they are typically used directly without isolation. For the applications described here, the reagent is prepared by *in situ* reduction, for obvious reasons of convenience.

Transmetalation with $CrCl_3$ produces mono-, di-, and triorgano-Cr^{III} species and was the method used to make the first organochromium complexes in the 1920s[24]. The intermediates typically are hexacoordinate with three neutral donor ligands such as THF and three anionic ligands (R or X). The direct reaction of Cr^{II} with R−X proceeds by a two-step electron transfer pathway (Scheme 5). The first step is one-electron reduction of the organic halide to produce a transient radical anion **9** which undergoes homolytic cleavage to give the halide anion and the organic radical **10**. The radical is further reduced by another equivalent of Cr^{II} to give the organo-Cr^{III} intermediate **11**. This process depends of the ease of reduction of the halide, and is particularly useful with

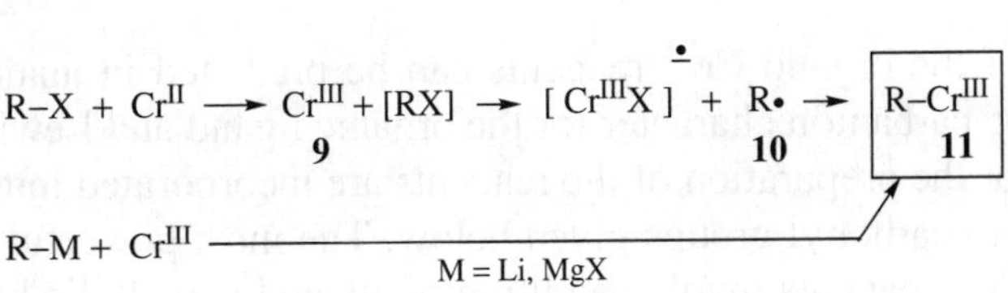

$$R-X + Cr^{II} \longrightarrow Cr^{III} + [RX] \longrightarrow [Cr^{III}X] + R\cdot \longrightarrow \boxed{R-Cr^{III}}$$

$$\underset{\mathbf{9}}{} \qquad\qquad \underset{\mathbf{10}}{} \qquad \underset{\mathbf{11}}{}$$

$$R-M + Cr^{III} \underset{M = Li, \, MgX}{\underline{\hspace{5cm}}}$$

Scheme 5. Mechanisms of formation of alkyl-Cr intermediates.

allylic, propargylic, and alkenyl halides. As the reagents are generated *in situ*, issues of handling are not critical, beyond the usual inert atmosphere techniques used with, for example, Grignard reagents.

Anhydrous $CrCl_2$ is available commercially, but it can also be generated by *in situ* reduction of $CrCl_3$ with lithium aluminum hydride[25] or sodium amalgam[26]. However, a critical feature in the reagent was uncovered during studies of the reaction of alkenyl iodides with ketones, promoted by Cr^{II}[27]. Trace amounts of nickel must be present in the $CrCl_2$ in order to have reproducible and efficient addition of the *alkenyl species* to a carbonyl substrate. This amount of Ni is typically present in commercial $CrCl_2$, and highly purified $CrCl_2$ is not recommended. After the recognition of the Ni effect, most procedures call for deliberate addition of 1 mol % Ni^{II} to the mixture, most commonly as $NiCl_2$ or nickel acetylacetonate, $Ni(AcAc)_2$. The nickel, presumably as Ni^0 or Ni^I[28], undergoes oxidative addition with the alkenyl halide to give alkenyl-Ni-X (or alkenyl-NiX_2), and the alkenyl group is then transferred to Cr^{III} producing the key intermediate (summarized for Ni^0-Ni^{II} in Scheme 6). The function of the Cr^{II} is to reduce Ni^{II} to Ni^0 (or Ni^I[28]), so that the Ni can function as a catalyst. This is expressed in the modified sequence, Scheme 6. Palladium(II), deliberately added to the medium, can also serve the role of the Ni, except that it is less effective with vinyl triflates[29]. It is assumed that the transmetalation from Pd to Cr is less effective in the case of the triflate counter ion. A significant consideration is the use of stoichiometric amounts (at least) of the chromium reagent and concomitant problems in disposal of the residues after reaction. Two recent improvements are promising: the use of electrochemical reduction to recycle the Cr[30] and the use of stoichiometric manganese salts with catalytic Cr[31].

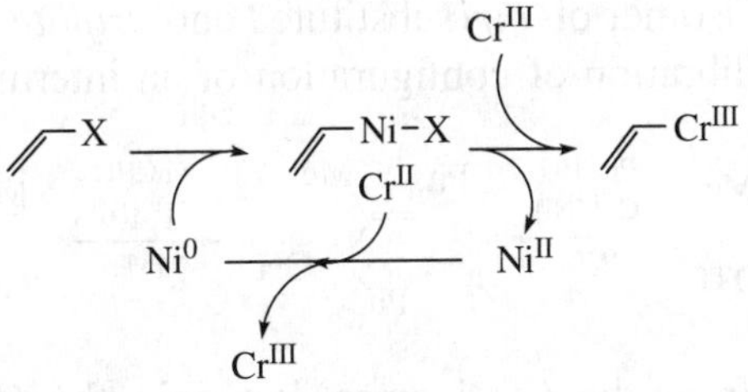

Scheme 6. Role of the Ni catalyst.

The reactivity of the organo-Cr^{III} reagents can be predicted in analogy with Grignard reagents, imagining carbanion character for the organic ligand and Lewis acid behavior for the Cr^{III}. Details for the preparation of the reagents are incorporated into the overall procedures for addition to carbonyl groups given below. The most powerful synthesis methods based on organo-Cr^{III} reagents involve addition of alkenyl and allylic halides to aldehydes and ketones. The most important advantage is the almost complete chemoselectivity for these functional groups.

3.2   Reactions with Alkenyl Halides and Triflates

The reaction is typically carried out by mixing the alkenyl halide or triflate with two or more mole equivalents of the Cr^{II}/Ni^{II} reagent (e.g., CrCl$_2$) in a polar aprotic solvent and adding the carbonyl component. The reactions proceed smoothly at 25 °C within minutes. The alkenyl species can be either the iodide, bromide, or triflate. Aldehydes give generally high yields and fast reactions, whereas ketones are not particularly good substrates. This leads to a useful chemoselectivity, in which substrates such as **12** bearing an aldehyde group and a ketone (or nitrile, ester, etc.) react completely selectively at the aldehyde[32]:

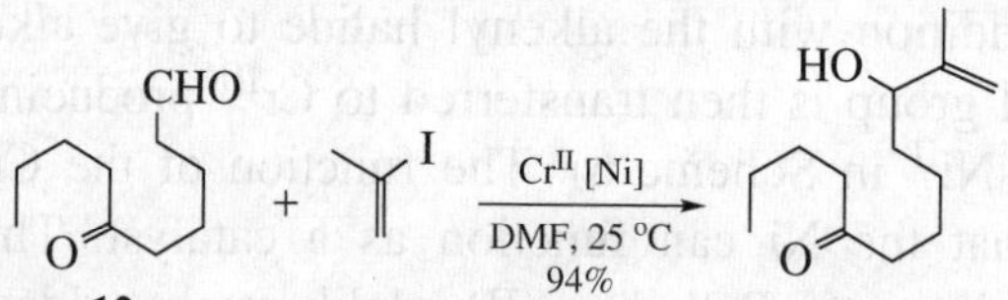

It is this chemoselectivity that strongly distinguishes the Cr^{III} methodology from conventional Grignard chemistry. Reactions with α,β-unsaturated aldehydes gives exclusively the 1,2-mode of addition[32]. Recently, improved conditions have been reported[33], which demonstrate that the addition of 4-tert-butylpyridine has a beneficial effect, including minimizing homocoupling of the vinyl halide. It assists in solubilizing the metal reagents.

The geometry of the double bond is maintained with 2-substituted-1-halo-alkenes (and triflates) where isomerization could give a different product[34]. However, as exemplified in eq 8, each geometrical isomer of a tri substituted enol triflate leads to the same product, presumably through equilibration of configuration of an intermediate[35]:

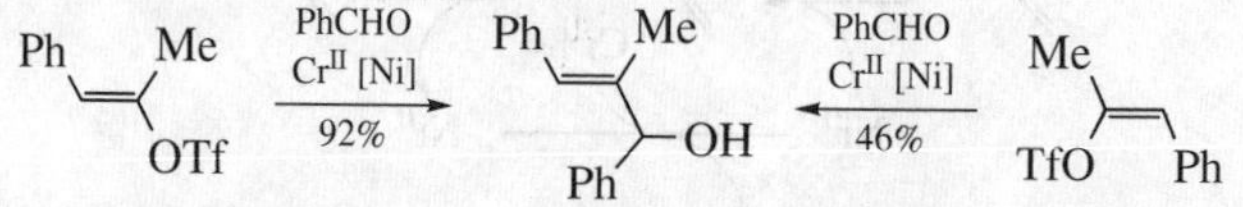

The yield is modest from the (Z) isomer, but only the (E) product was detected. Retention of geometry is a characteristic of the oxidative-addition mechanism involving catalytic Ni⁰; that step has been studied and found to proceed with complete retention[36]. An alternative mechanism involving a free vinyl radical from one-electron reduction of the vinyl species by Cr^{II} would be expected to include partial or complete isomerization due to the high rate of inversion of vinyl radicals. No comprehensive mechanistic picture has been presented to account for the variation in stereoselectivity. The process is not very sensitive to the substitution pattern of the vinyl derivative, and high yields are obtained with all substitution levels.

The usefulness of the Cr^{II}-promoted coupling of alkenyl halides was demonstrated particularly well in the synthesis work toward palytoxin, and also provided a test of the stereoselectivity with respect to the new stereogenic center created at the aldehyde carbonyl carbon[37]. Using the trisubstituted alkenyl iodide **13**, addition to the aldehyde with an α-stereogenic center (**14**) led to a 2 : 1 mixture of diastereomers, in 72% yield. It was also shown that the corresponding (Z) isomer of the starting alkenyl iodide gave the same mixture of diastereomers, but in lower yield, 15%:

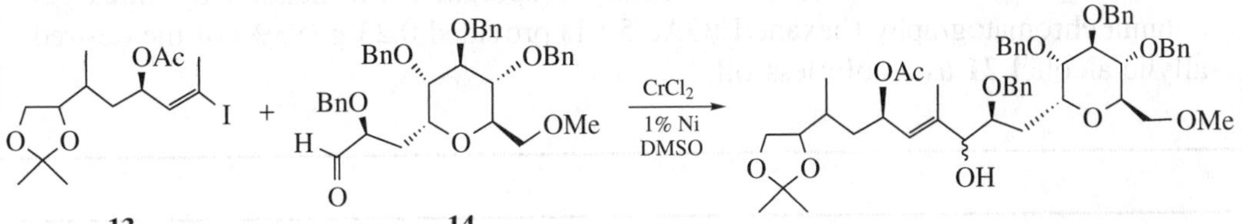

72% yield; 2:1 ratio of diastereomers

In a related example aimed at C-glycoside synthesis with a 1,2-disubstituted alkenyl iodide (**15**), addition to aldehyde **16** proceeded according to the Felkin–Ahn analysis and produces the diastereomer **17** in >94 : 6 ratio and complete retention of the (Z) geometry of the double bond[37]:

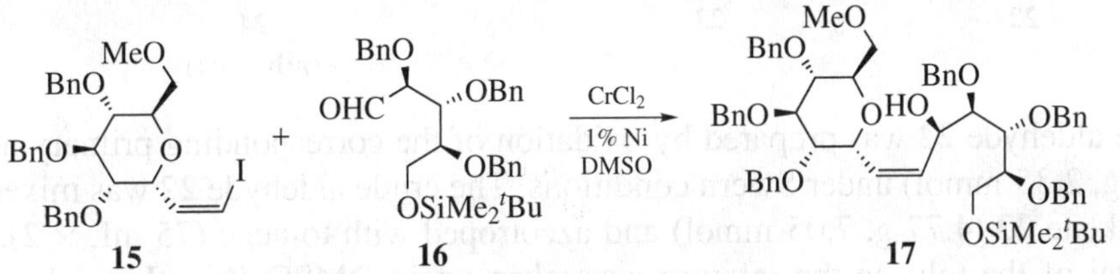

Intramolecular versions of the process are also highly successful. The structural analog **18** of brefeldin was prepared in 70% yield from the β-iodoenoate **19**; the epimer at C-4 was also detected but **18** was favored by a factor of >10 : 1[38]:

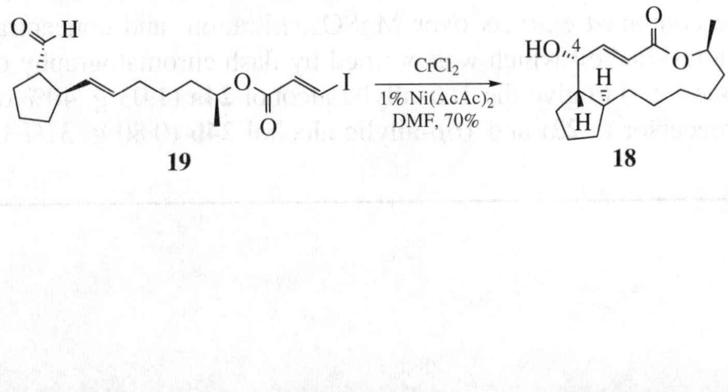

Procedure for Addition of Alkenyl Triflate to an Aldehyde[39]

PhCHO + C$_{10}$H$_{21}$ ⟶OTf $\xrightarrow[\text{DMF, 25 °C}]{\text{Cr}^{\text{II}}, \text{Ni}^{\text{II}}}$ C$_{10}$H$_{21}$ ⟶ Ph / OH

20   **21**

A mixture of anhydrous CrCl$_2$ (0.49 g, 4.0 mmol) and a catalytic amount of anhydrous NiCl$_2$ (0.0026 g, 0.020 mmol) in DMF (10 mL) was stirred at 25 °C for 10 min under an argon atmosphere. To this reagent at 25 °C was added sequentially a solution of benzaldehyde (0.11 g, 1.0 mmol) in DMF (5 mL) and a solution of alkenyl triflate **20** (0.63 g, 2.0 mmol) in DMF (5.0 mL). After being stirred at 25 °C for 1 h, the reaction mixture was diluted with ether (20 mL), poured into water (20 mL) and extracted with ether repeatedly. The combined organic extract was dried (Na$_2$SO$_4$) and concentrated by rotary evaporation. Purification by silica gel column chromatography (hexane/EtOAc 5 : 1) provided 0.23 g (83%) of the desired allylic alcohol **21** as a colorless oil.

Procedure for Addition of an Alkenyl Iodide to an Aldehyde[40]

22   **23**   **24**
(a:b = 5:4)

The aldehyde **22** was prepared by oxidation of the corresponding primary alcohol (1.28 g, 2.35 mmol) under Swern conditions. The crude aldehyde **22** was mixed with iodoalkene **23** (4.77 g, 7.05 mmol) and azeotroped with toluene (75 mL × 2). After removal of the toluene the mixture was taken up in DMSO (60 mL) under argon. By solid addition using Schlenk apparatus or in a 'dry box', solid anhydrous CrCl$_2$ containing 0.1% NiCl$_2$ (1.75 g, 14.2 mmol) was added portionwise. The dark green mixture was stirred under argon for 20 h at 23 °C, and then quenched by addition of saturated aqueous NH$_4$Cl solution. Extraction with EtOAc (3×), followed by drying of the combined extracts over MgSO$_4$, filtration, and concentration of the filtrate gave a residual oil which was purified by flash chromatography on silica gel (EtOAc/hexane 1 : 5) to give the 16α-allylic alcohol **24a** (1.03 g, 40% overall from the alcohol precursor to **22**) and 16β-allylic alcohol **24b** (0.80 g, 31% overall).

3.3 Addition of Allylchromium(II) Reagents to Aldehydes and Ketones

The coupling of allylic halides and sulfonate esters with aldehydes and ketones promoted by Cr^{II} is a very general and efficient process. There are important issues of regio-selectivity and stereoselectivity. The following equation summarizes the process, in which a dipolar aprotic solvent (DMF, DMSO) accelerates the reaction and is required for the reaction of allylic chlorides and tosylates.

$$R^1 \diagdown Br + H \overset{O}{\underset{R^3}{\diagdown}} R^2 \xrightarrow[\text{or DMF}]{\substack{CrCl_2 \\ DMSO}} \diagdown \overset{OH}{\underset{R^1 \ R^3}{\diagdown}} R^2$$

THF is often used with allylic bromides. With aldehydes, slightly more than two mole equivalents (one equivalent) of the Cr^{II} reagent is used, whereas in the slower reactions of ketones, an excess of Cr^{II} and the allylic halide are required for complete conversions. The reaction proceeds by allylic rearrangement and is invariably carried out with primary allylic halides (tosylates), leading to products with the alkene unit unsubstituted in the terminal position; there is no issue of alkene geometry in the product. In addition, the geometry of the alkene unit in the allylic halide has no determining effect on the relative configuration of the new stereogenic centers in the product; (*E*) and (*Z*) isomers give the same product mixture, as discussed below. Two new stereogenic centers are generated, often with excellent and predictable stereocontrol.

The addition to carbonyl groups is sensitive to steric and electronic effects. Complete regioselectivity is observed with substrates bearing both a ketone and aldehyde group[41]. In an intermolecular competition between 4-heptanone and 2-heptanone, the methyl ketone reacts fairly selectively (product ratio: 5 : 1)[41]. Ester groups are in general not reactive toward the allyl-Cr^{III} reagents:

$$\overset{O}{\diagdown}\diagdown\diagdown\underset{O}{\diagdown} Z + \diagdown Br \xrightarrow[\text{THF}]{CrCl_2} \diagdown\overset{OH}{\diagdown}\diagdown\diagdown\underset{O}{\diagdown} Z$$

Z = Bu: 66%
Z = OMe: 75%

Addition to simple ketones can be efficient, and offers one of the more effective means to deliver an allylic organometallic to cyclohexanones. Addition is preferred in the equatorial position by a factor of 7 : 1[41]:

$$^tBu\diagdown\diagdown\overset{O}{\diagdown} + \diagdown Br \xrightarrow[\text{THF}]{CrCl_2} {^tBu}\diagdown\diagdown\overset{OH}{\diagdown}\diagdown + \text{axial}$$

88:12
85% together

With α,β-unsaturated aldehydes, the addition is exclusively by the 1,2-mode, leading to 3-hydroxy-1,5-dienes[41]:

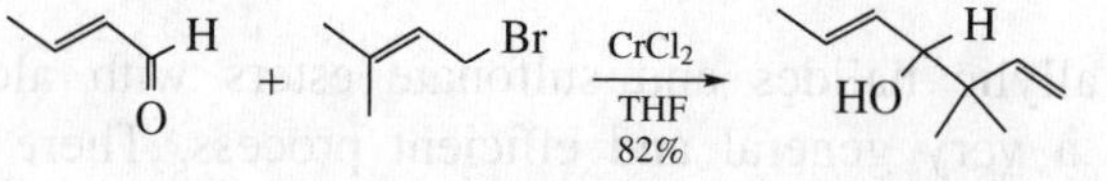

The addition of 3-substituted allylic halides, most commonly crotyl bromide, creates two new stereogenic centers. Numerous studies have been reported with various aldehydes, and a strong generalization can be made: both (E)- and (Z)-crotyl bromide leads to the anti addition product, and the selectivity is often > 10 : 1. An exception is pivaldehyde (2,2-dimethylpropionaldehyde) which leads to a small preference for the syn product[42]. The better solvent is THF, compared to DMF. This is one of the many examples of allylmetal reagents which add to carbonyl groups, and the stereochemical results depend strongly on the mechanism[43]:

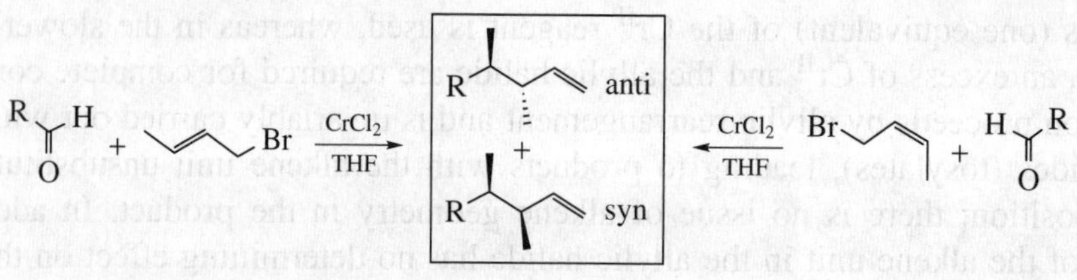

The crotylchromium reagents together with crotyl-Ti[IV] and crotyl-Zr[III] are referred to as belonging to mechanistic 'type 3'[44], and the anti selectivity is rationalized in terms of rapid equilibration of the double bond configuration in the crotyl-Cr reagent followed by a synclinal conformation in the complex leading to the transition state:

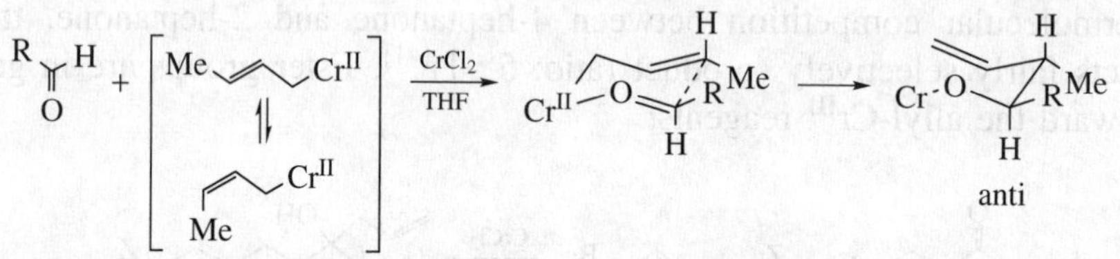

With an α-substituent on the aldehyde, an additional question of stereoselectivity arises, the classic question of facial selectivity. Assuming the usual preference for 1,2-anti selectivity, two diastereomers are likely, from addition according to Cram's rule or anti-Cram. The Felkin–Ahn analysis supports the prediction of the Cram's rule product being preferred, but the preference in simple cases is not strong[45]. Perhaps the least complicated case is the following, where crotyl bromide adds to 2-methylbutanal[42b]:

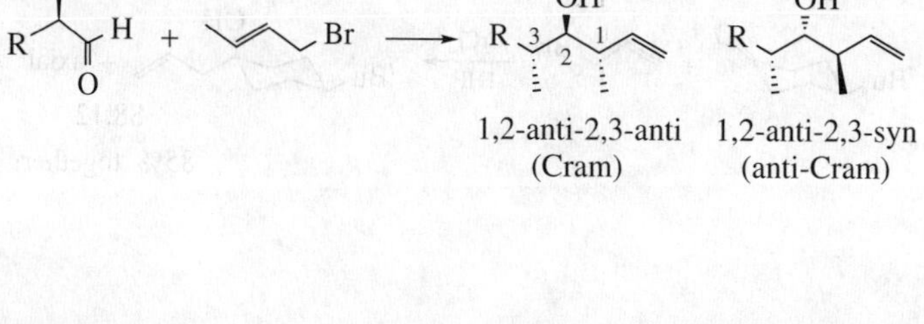

1,2-anti-2,3-anti 1,2-anti-2,3-syn
(Cram)   (anti-Cram)

The first two products add up to 93% of the total, with a 93 : 7 selectivity for 1,2-anti addition. The 2,3-selectivity is then 69 : 31, favoring the Cram (2,3-anti) product:

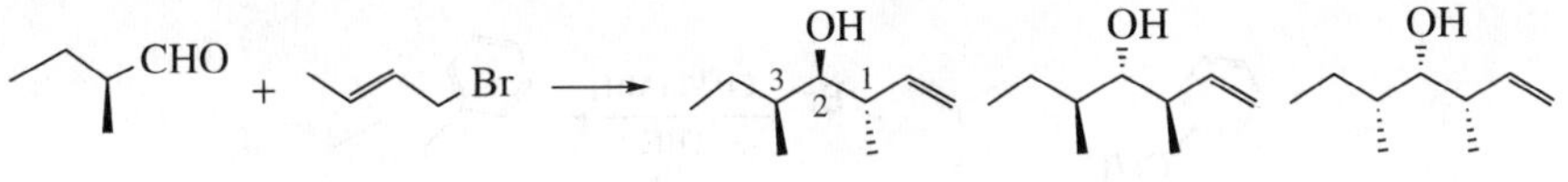

31%   62%   7%

With additional structural features to influence the selectivity, much better diastereoisomeric ratios are observed. For example, with an α-methyl group and β-oxygen (in **25**), the selectivity for the Cram product **26** is very high, with a yield of 86%[46]:

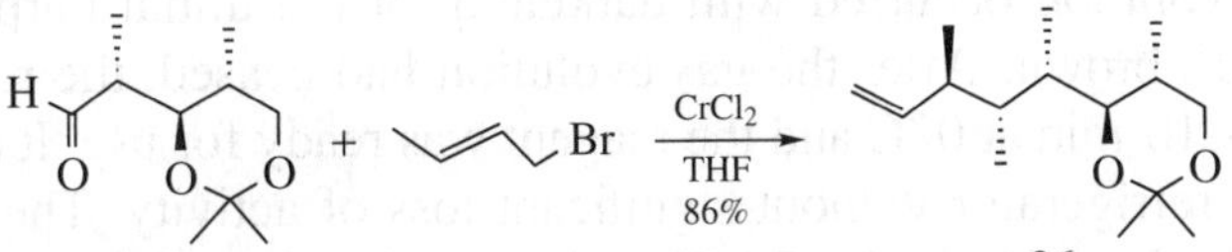

25

26
95:5 (other isomers)

It was demonstrated that this selectivity is not an example of chelation control, but is determined simply by the relative sizes of the α-substituents[47].

Simple α-alkoxy aldehydes, such as **27**, show the usual high 1,2-anti selectivity, and high selectivity in the 2,3 relationship, with some dependence on the protecting group; TBDMS, in the example below. Only one product was observed, attributed to anti addition in the 1,2 relationship and Cram (Felkin–Ahn) selectivity in the 2,3 relationship[48].

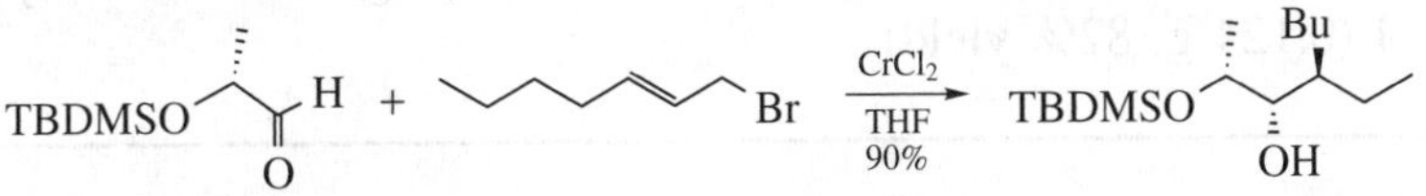

27

>99% this isomer

However, the α-alkoxy aldehyde **28** gave the usual high anti selectivity in the 1,2 relationship, but essentially equal amounts of the diastereoisomers with respect to 2,3[49]:

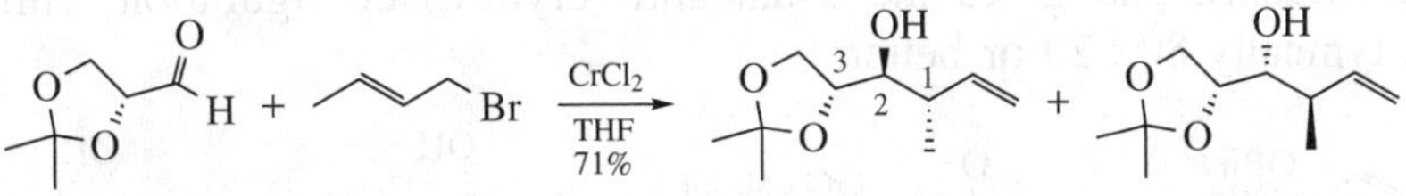

28

53%   :   45%

A certain lack of predictability is clear after considering that the closely related substrate **29** gives 100% Cram selectivity in setting the 2,3 relationship, and 96 : 4 anti/syn preference in the 1,2 relationship[50]:

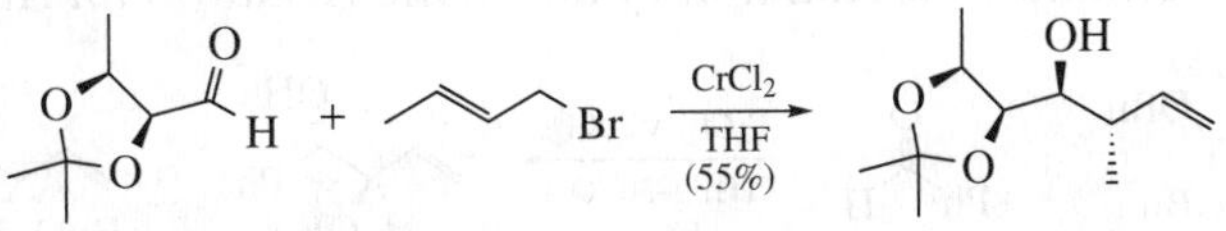

29

94% this isomer

Procedure for Intermolecular Addition of an Allylic Halide to an Aldehyde[51]

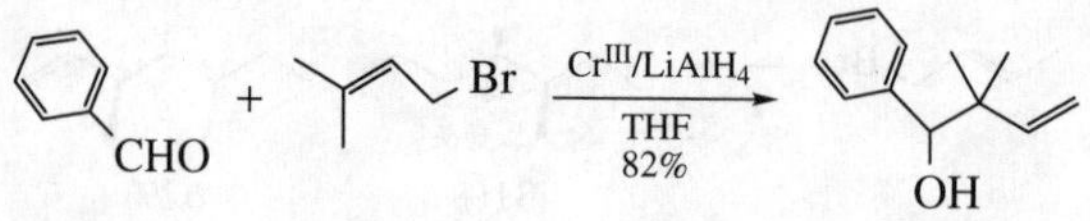

30

(a) Preparation of Cr^{II} from $LiAlH_4$ and $CrCl_3$. $LiAlH_4$ (0.044 g, 1.2 mmol) was added to $CrCl_3$ (0.37 g, 2.3 mmol) suspended in THF (5 mL) at 0 °C under argon. Immediate gas evolution occurred with darkening of the initial purple color which finally turned dark brown. After the gas evolution had ceased, the reaction mixture was stirred for 5–10 min at 0 °C and the reagent was ready for use. It could be stored for weeks in the refrigerator without significant loss of activity. The THF could be replaced by DMF by adding the DMF and removing the THF under vacuum at 25 °C. Equivalent results are obtained using commercial $CrCl_2$ (anhydrous).

(b) To the chromium reagent (prepared above, in THF) at 25 °C were added successively rapidly benzaldehyde (neat, 0.092 g, 0.86 mmol) and a solution of 1-bromo-3-methyl-2-butene (0.173 g, 1.2 mmol) in THF (5 mL). After being stirred for 2 h, the mixture was diluted with water (5 mL) and the resulting mixture was extracted thoroughly with ether. The ether solution was washed with brine, dried over Na_2SO_4, filtered, and concentrated. The residue was distilled (short path) with a bath temperature of 105–110 °C and 112 torr to give 2,2-dimethyl-1-phenyl-3-buten-1-ol (0.124 g, 82% yield).

A particularly useful variation is the special case where the allyl-Cr reagent is generated by cleavage of one C–O bond in an acrolein or methyl vinyl ketone dialkyl acetal, such as **31**[52]. The reaction proceeds with exclusive addition at the more substituted position (alkoxy substituent), and gives the usual anti (erythro) configuration with reasonable selectivity, typically 80 : 20 or better:

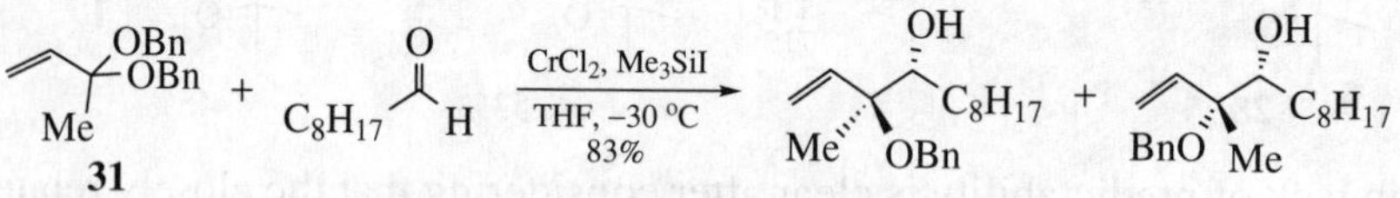

Procedure for Intermolecular Addition of an Allylic Acetal to an Aldehyde[52]

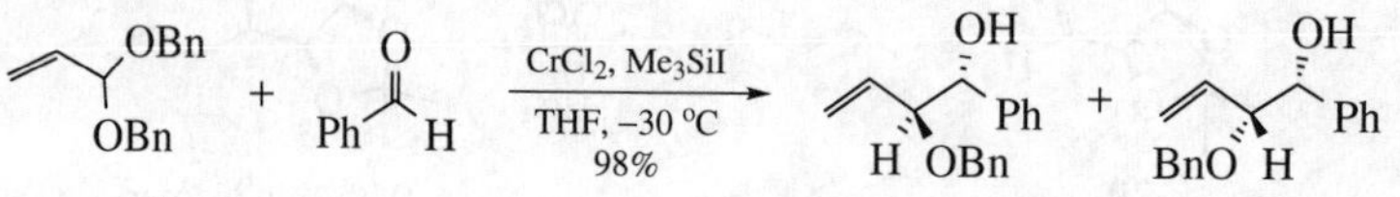

88:12

A suspension of CrCl$_2$ (0.74 g, 6.0 mmol) in THF (14 mL) was cooled to $-30\,°$C. To the suspension at $-30\,°$C was added successively a solution of acrolein dibenzyl acetal (0.51 g, 2.0 mmol) in THF (3 mL), a hexane solution of Me$_3$SiI (1.0 M, 2.0 mL, 2.0 mmol), and a solution of benzaldehyde (0.11 g, 1.0 mmol) in THF (3 mL). The mixture turned gradually from gray to brown-red. After being stirred at $-30\,°$C for 3 h, the suspension was poured into aqueous HCl (1 M, 15 mL) and extracted with ether. After being dried (MgSO$_4$), filtered, and concentrated, the combined organic solution yielded a residue which was purified by column chromatography on silica gel (hexane/EtOAc 10 : 1). The product, 2-benzyloxy-1-phenyl-3-buten-1-ol, was obtained in 98% yield (0.25 g) and an anti/syn diastereoisomer ratio of 88 : 12.

Intramolecular versions of the process have also been very successful, perhaps best exemplified by the formation of asperdiol **32**[53]. The yield is good, again displaying the high chemo- and stereoselectivity of this process. A mixture results from only moderate facial selectivity in the addition to the carbonyl with respect to the (remote) peripheral asymmetric centers:

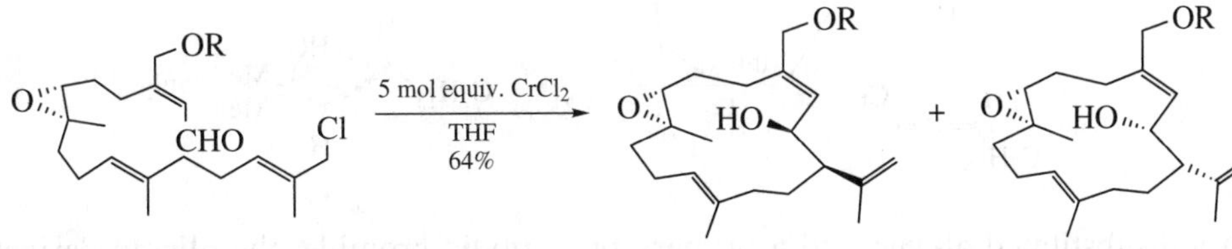

A recent report promises to expand the applications of allyl-Cr reactivity by demonstrating direct generation of the reactive intermediate from 1,3-dienes using catalytic Co−H reagents (*e.g.*, B$_{12}$-type with water as the source of hydride)[54]. The diene is simply mixed with an aldehyde, excess CrCl$_2$, and 10 mol % of B$_{12}$ in DMF, and then water was added to complete the process. The example here displays the high regio- and stereoselectivity obtained with 2-alkoxy-1,3-butadiene. The mechanism is not yet clear, but addition of H−Co^{III} to the 1,3-diene is proposed as the key activation step. Then allyl radical transfer to Cr^{II} produces the allyl-Cr^{III} intermediate. A related method involving three-component coupling of alkyl iodides, 1,3-dienes, and an aldehyde with Cr^{II} was also reported[55]:

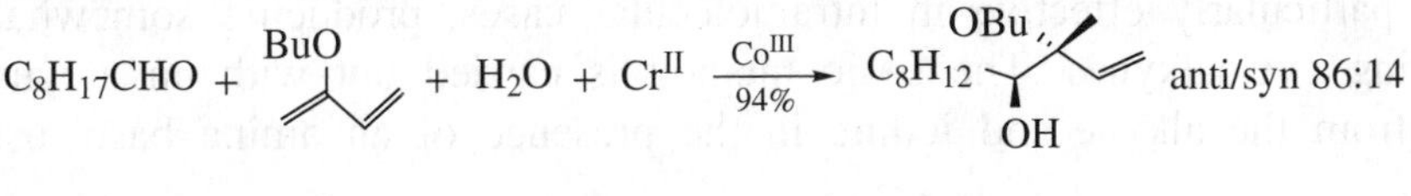

3.4   Extension to Propargyl Derivatives

Propargyl derivatives involve the usual complications of allene–Cr versus propargyl-Cr reactivity. Two isomeric chromium intermediates are formed, possibly at equilibrium, and the product mixture reflects the relative reactivity of the isomers[56]:

The ratio depends on the propargylic halide substitution pattern and on the carbonyl electrophile. With a terminal alkyne and a secondary bromide, the propargylic derivative is the favored product from aldehydes, interpreted as exclusive formation of the allenic-Cr intermediate and allylic transposition during addition to the aldehyde (path a). With the same alkynyl bromide and a ketone, the allenyl product is strongly favored (path b)[55]:

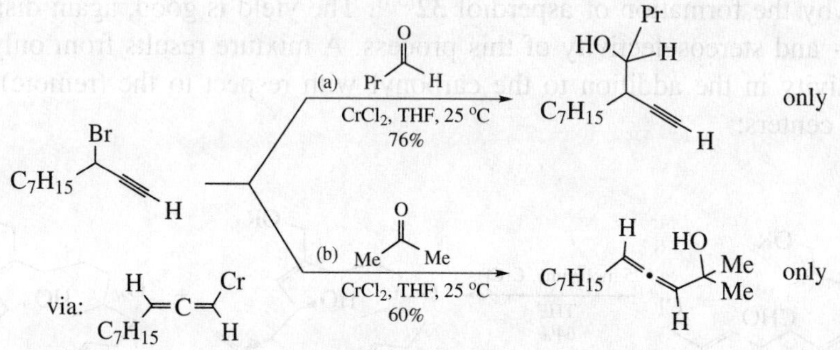

With a disubstituted alkyne and a primary propargylic bromide, the allenyl derivative is the only product, consistent with a propargylic-Cr intermediate[55]:

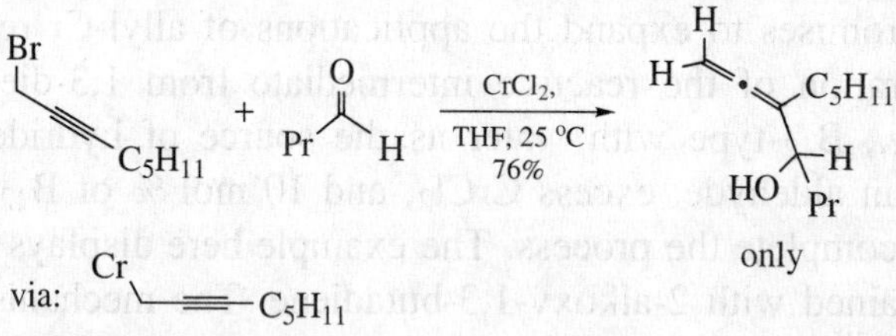

3.5   Extension to Alkynyl Derivatives

The generation of alkynylchromium reagents and their addition to carbonyl compounds has been particularly effective in intramolecular cases, producing somewhat strained medium-ring cycloalkynes. The basic method is carried out with an alkynyl iodide, prepared from the alkyne and iodine in the presence of an amine base, using DMF under mild conditions. The chemoselectivity is again excellent[57]:

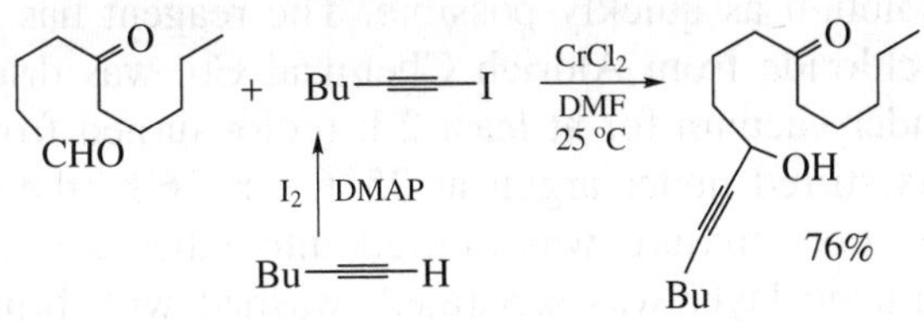

Selective 1,2 addition has also been demonstrated:

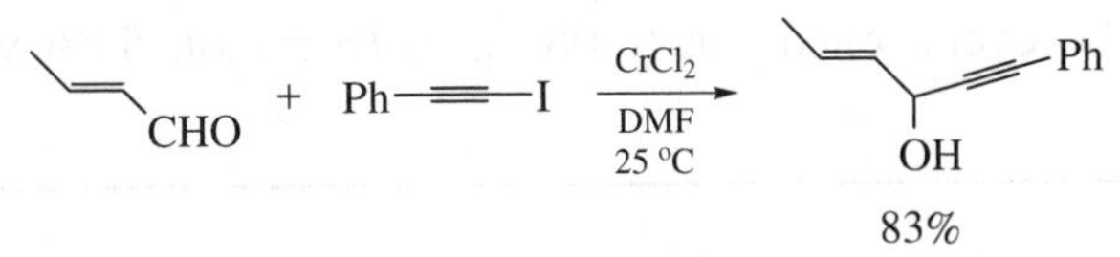

The corresponding alkynyl bromides also react well under essentially the same conditions, but the iodides are typically more accessible.

The development of interest in the synthesis of cyclic 1,5-diyne-3-enes has been driven since 1987 by the appearance of a family of natural products referred to as the enediyne toxins[58]. The Cr[II]-promoted coupling procedures have provided some of the more useful methods for the critical ring closure, especially in difficult cases[59]. The example in the following procedure illustrates the general idea; the alternative of inducing ring closure by treatment of the alkyne (before iodination) with strong base (*e.g.*, LDA) failed to give the desired ring closure.

Procedure for Cyclic Alkynes by Chromium(II)-promoted Ring Closure[59a]

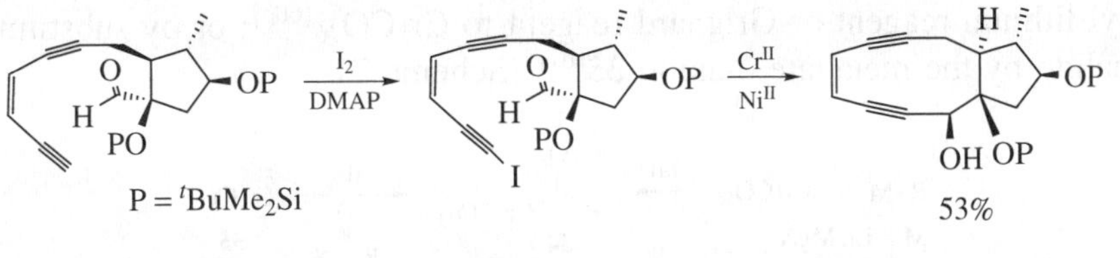

P = ^{t}BuMe$_2$Si

(a) A mixture of the alkyne (2.75 g, 5.97 mmol), 4-*N*,*N*-dimethylaminopyridine (6.88 g, 56.3 mmol), iodine (6.08 g, 24.0 mmol), and benzene (70 mL) was heated under argon at 49 °C for 1 h. The reaction mixture was cooled to 25 °C, transferred to a separatory funnel, and treated with saturated sodium bisulfite. The organic layer was separated and the aqueous layer was extracted with ether. The combined organic extracts were dried over magnesium sulfate, filtered, and the filtrate was concentrated by rotary evaporation. The residue from the organic layers was flash chromatographed on silica gel (0 → 5% ethyl acetate in hexane) to give the alkynyl iodide (3.05 g, 5.20 mmol, 87% yield) as a colorless oil.

(b) To a solution of the alkynyl iodide (0.201 g, 0.343 mmol) in THF (6 mL) under argon at 25 °C was added nickel(II) chloride (0.020 g, 0.15 mmol) all at once.

Anhydrous chromium(II) chloride (0.400 g, 3.42 mmol) was then quickly weighed and added to the solution as quickly possible. The reagent has to be dried before use; chromium(II) chloride from Aldrich Chemical Co. was dried under argon by heating (200 °C) under vacuum for at least 2 h (color turned from gray to brown). The suspension was stirred under argon at 25 °C for 16 h (the suspension should be brown). The reaction mixture was poured into ether and water in a separatory funnel. The organic layer was separated, washed with brine, and dried over magnesium sulfate. After filtration and concentration of the filtrate by rotary evaporation, the residue was flash chromatographed on silica gel (8% ethyl acetate in hexane) to give the cyclic enediyne (0.083 g, 0.18 mmol, 53% yield) as a colorless oil.

4 Applications of Fischer-type Carbene Complexes of Chromium

4.1 Introduction and Preparation of Complexes

Fischer-type carbene complexes[60] are those that are strongly stabilized by the presence of resonance electron donor substituents at the carbene carbon, as exemplified by the $Cr(CO)_5$ derivative **33**. The Cr derivatives were the first carbene complexes to be prepared and immediately elicited interest on the part of organic chemists to evaluate the reactivity of the carbene ligand. The primary method of preparation involves the alkylation (or acylation) of an acylate–metal anionic intermediate **34**. The acylate intermediates, in turn, can be prepared by two general routes: by addition of a carbon nucleophile such as an alkyl lithium reagent or Grignard reagent to $Cr(CO)_6$[60c]; or by substitution on an organic halide by the metallate dianion **35**[61] (Scheme 7).

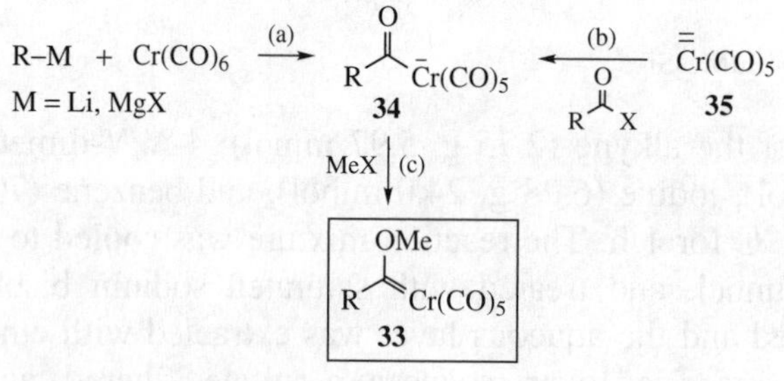

Scheme 7. Preparation of carbene–chromium complexes.

The primary route is path (a), utilizing organolithium reagents and alkylating with a reactive methylating agent such as Me_3OBF_4 or MeOTf. Alternatively, the very reactive and difficult $Cr(CO)_5(-II)$ can be brought into reaction with an acyl halide, leading directly to the acylate intermediate[61]. This latter method is particularly useful when functional group compatibility or other complications prohibit the use of the organolithium reagent required for path (a). The yields can be as high as 80–90%, and the products can be purified by conventional means, such as column chromatography. The carbene complexes are somewhat air sensitive, and handling in solution is best done under an inert gas atmosphere. The complexes are relatively stable thermally, and can be stored for long periods at ambient conditions.

Procedure for Preparation of a Fischer-type Carbene Complex of Chromium[62]

36   **37**

To a solution of BuLi (6.9 mL of 1.35 M solution in hexane, 9.32 mmol) was added into a solution of 2-bromo-1,4-dimethoxybenzene (2.00 g, 9.30 mmol) in ether (40 mL) via syringe over 15 min at −78 °C under argon. The solution turned from colorless to pale brown. After 1 h at −78 °C, the solution was transferred into a suspension of $Cr(CO)_6$ (2.1 g, 9.5 mmol) in ether (10 mL) at −78 °C via cannula. The mixture was allowed to warm with stirring to 23 °C over 20 min. The color changed from lemon yellow to deep brown. After another 2 h at 23 °C, the volatile material was removed at oil pump vacuum. The residue was dissolved in degassed water (60 mL) and trimethyloxonium tetrafluoroborate (17.4 g, 15 mmol) was added as a solid portionwise quickly. With vigorous stirring, a red precipitate formed quickly. The aqueous solution was neutralized with a saturated aqueous solution of $NaHCO_3$ (60 mL) and ether was added to dissolve the red solid. The aqueous layer was washed with ether, and the combined ether solution was dried over $MgSO_4$, filtered, and concentrated by rotary evaporation to leave a residual red oil. Flash chromatography (10% EtOAc in hexane) followed by crystallization from pentane at −78 °C gave the carbene complex **36** (1.92 g, 5.1 mmol, 55% yield) as red needles (mp 42–43 °C (dec)). The mother liquors contained a mixture (0.320 g, about 4 : 1) of complex **36** (minor) and the tetracarbonyl analog **37** (major).

4.2   General Reactivity and Mechanism

The pathways leading from the Fischer carbene complexes include polar mechanisms, where the carbon–metal double bond mimics a carbonyl group, behaving as a strong resonance and inductive electron withdrawing group and containing a very electrophilic carbon. There are also nonpolar mechanisms, driven by orbital interactions within the coordination sphere of the metal, especially cycloaddition reactions of pi bonds with the carbon–metal double bond.

Enolate/aldol reactivity:

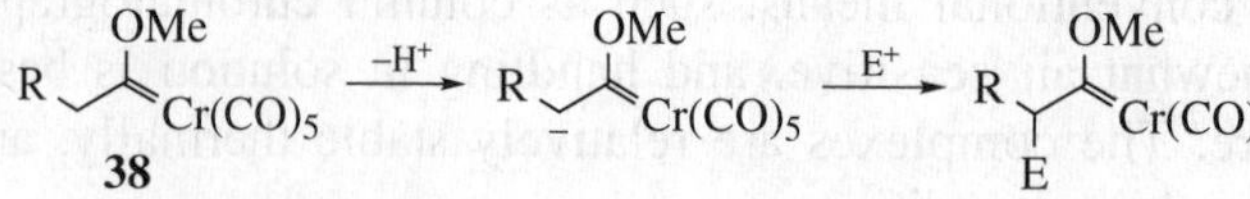

Acyl-transfer reactivity:

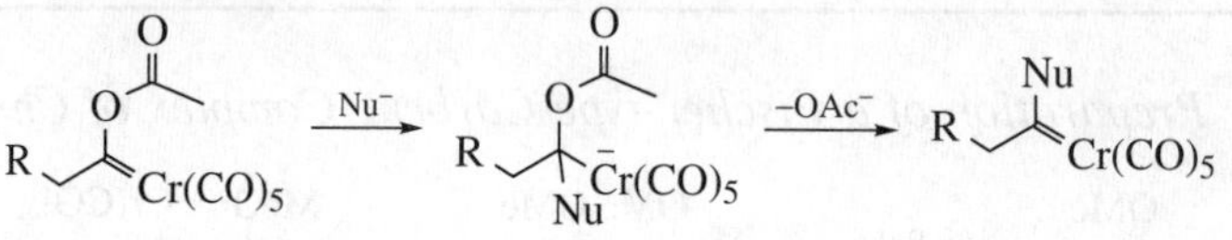

The effects are generally stronger compared to a simple carbonyl group. For example, the α-protons (in **38**) have a pKa of 12–15 in protic media or higher values (>20) in aprotic solvents such as acetonitrile[63], which makes the $Cr(CO)_5$ unit among the most powerful functional groups in acidifying the adjacent C–H bonds. Enolate- and aldol-type reactivity has been developed as potential synthesis methodology (Scheme 8), but is not surveyed here[60b,64,65,66]. An important aspect of aldol reactivity is the possibility of converting simple alkyl carbene complexes into vinyl carbene complexes, important reactants for the benzannulation reaction[67]. In addition to the base-promoted procedure, Lewis acids such as $SnCl_4$ are very effective and give the vinyl carbene complex directly[65]:

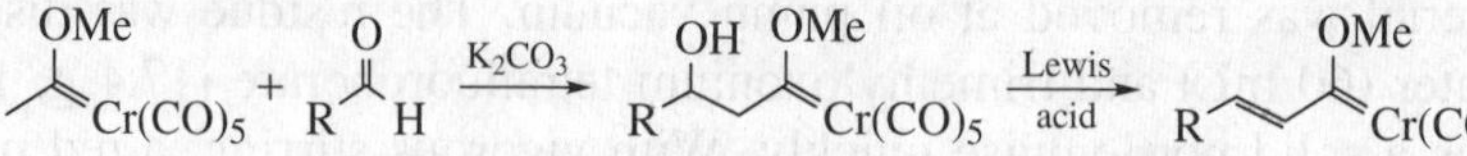

The carbene carbon also reacts as a pseudo-acyl transfer agent, especially in the case of the pseudo-anhydride **39** where the nucleophile is thought to give the tetrahedral intermediate **40**, followed by loss of acetate and formation of the new complex **41**. This process is useful when the nucleophile is a hetero atom with a stabilizing electron pair (O, N, S), but carbon nucleophiles lead to highly unstable products. It has been used to introduce reactive side chains for *intra*molecular coupling with the carbene ligand, as discussed below.

In direct analogy with conventional electron withdrawing groups, the $C=Cr(CO)_5$ group activates alkenes and alkynes as dienophiles toward the Diels–Alder reaction (Scheme 8).

This is particularly useful with alkynes, which are usually sluggish substrates in the Diels–Alder reaction, and leads to vinyl carbene complex **42**, again useful in benzannulation[68]:

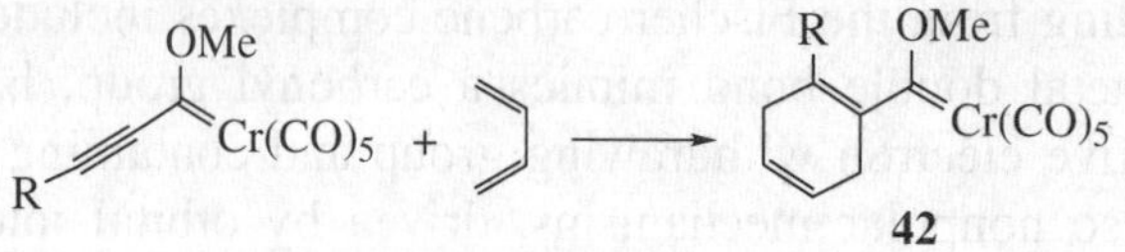

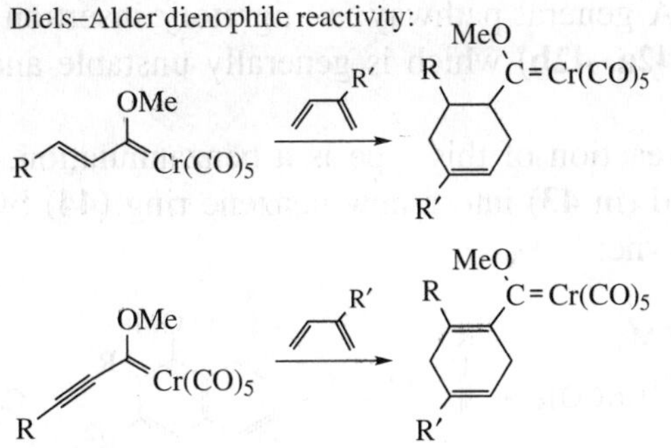

Diels–Alder dienophile reactivity:

Scheme 8. The carbene-chromium group activating the Diels–Alder reaction.

The oxidative sensitivity of the carbon–metal double bond makes possible very efficient replacement of the metal unit by an oxygen. Gentle reagents such as DMSO are sufficient, and lead to the corresponding ester (along with in this case, the dimethylsulfide–Cr(CO)$_5$ product which is a useful intermediate for recycling the chromium)[68]:

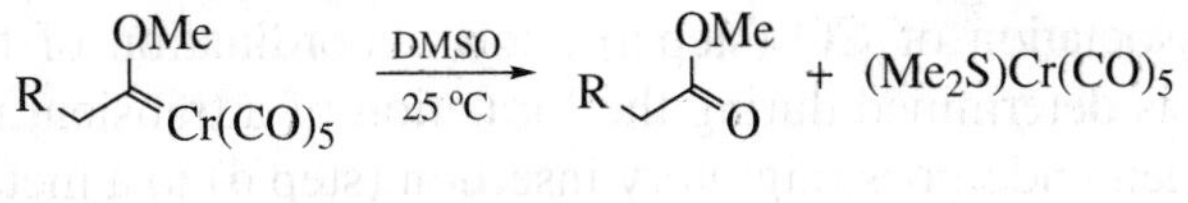

4.3   Benzannulation

4.3.1   Introduction

The primary nonpolar mechanism, briefly summarized in Scheme 9, is formally a [2 + 2] cycloaddition of an X=Y (or X≡Y) unit across the C=M bond to give reactive metallacyclobutane **41a** (or metallacyclobutene **41b**). Numerous pathways are open to this intermediate, depending on the structure of X–Y, the metal (Cr, Fe, or others), and

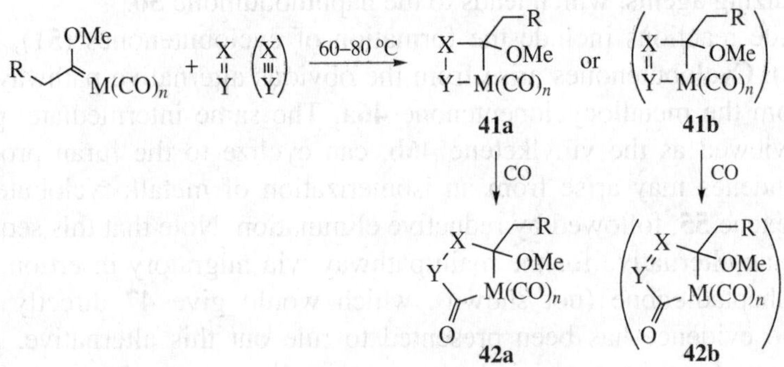

Scheme 9. General [2 + 2] cycloaddition/CO insertion pathway.

the peripheral ligands. A general pathway is migratory insertion to a CO ligand, giving a five-membered ring (**42a**, **42b**) which is generally unstable and leads to characteristic chemistry.

The most important reaction of this type is a benzannulation, which converts an aryl (or vinyl) carbene ligand (in **43**) into a new benzene ring (**44**) by incorporation of a CO ligand and an added alkyne:

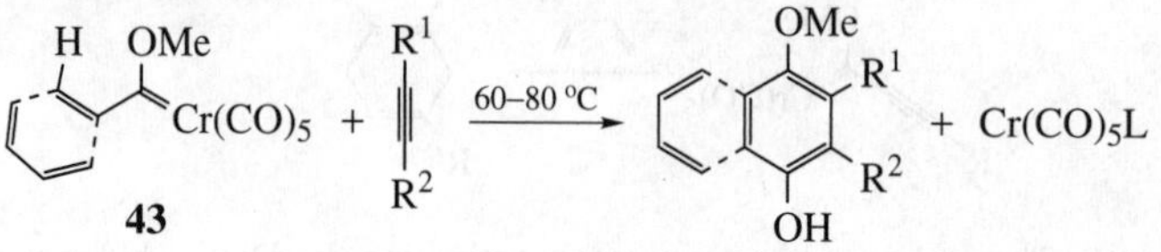

4.3.2   Mechanism

The mechanism of this process has been the subject of much study, and good support for the transient intermediates is now available, although direct observation has proved difficult[60b,69]. The central features of the mechanism are presented in Scheme 10. An initial rate-determining dissociation of CO (step a) allows coordination of the alkyne (step b). The regioselectivity is determined during the formation of a transient metallocyclobutene (**45**, step c), which then undergoes migratory insertion (step d) to a metallocyclopentenone complex (**46a**). This species can also be expressed as the resonance structure **46b**, a phenyl vinyl ketene complex. Even in the absence of the metal, phenyl vinyl ketenes are postulated to rearrange thermally to give naphthols via the keto tautomer (in **47**). In this case, the metal may play a role in the rearrangement to the metal-complexed benzannulated product, **47**, but no specific suggestion has been made. Then keto–enol tautomerization gives the first isolable intermediate, the naphthol complex **48**. The arene ligand in **48** is labile and easily displaced by weak donor ligands such as acetone solvent to give the free naphthohydro-quinone monomethyl ether **49**, which is somewhat air sensitive. If the reaction is run in the presence of acetic anhydride, the corresponding naphthol acetate is conveniently isolated at this stage[70]. A simple expedient to give a clean product and remove chromium residues as water-soluble Cr^{III} salts is to oxidize the crude product directly with iodine, 5% HNO_3, or other mild oxidizing agents, which leads to the naphthoquinone **50**.

Important side reactions include the formation of cyclobutenones (**51**), indenes (**53**), and furans (**52**). Cyclobutenones arise from the obvious alternative pathway of reductive elimination from the metallocyclopentenone **46a**. The same intermediate, perhaps more appropriately viewed as the vinylketene **46b**, can cyclize to the furan product **52**. The formation of indenes may arise from an isomerization of metallocyclobutene **45** to the metallocyclohexene **55**, followed by reductive elimination. Note that this sequence (**45** → **55**) also offers an alternative for the main pathway, via migratory insertion of **55** to give a metallocycloheptadienone (not shown), which would give **47** directly by reductive elimination. No evidence has been presented to rule out this alternative. Although the formation of cyclobutenones and indenes is interesting as synthesis methodology, no general process has been worked out.

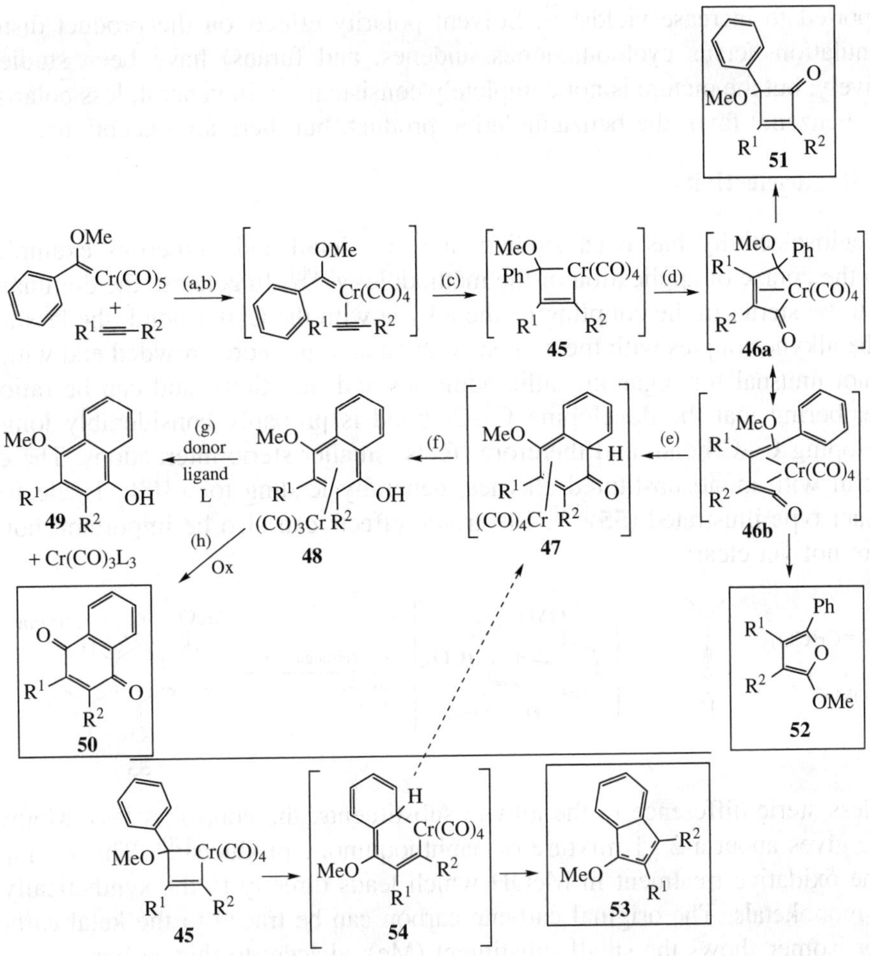

Scheme 10. Mechanism of the benzannulation reaction.

4.3.3 Scope and limitations

This remarkable process (benzannulation) is a powerful way to construct aromatic rings. Critical questions include the regioselectivity of alkyne incorporation with unsymmetrical alkynes, and functional group compatibility on the alkyne and on the carbene complex. General questions such as intramolecular applications also arise. An important question, so far unanswered, is: how can the Cr be recycled? This question transforms itself into the question: what other pathways can be imagined for preparation of the starting carbene–Cr complex, that would be more compatible with the other steps? The ultimate Cr product from the process can be a Cr^0 polycarbonyl complex (by simple ligand displacement of the arene ligand in the product) which is an appropriate starting material, but no gentle method is available for converting the Cr^0 byproduct to the key starting carbene–Cr reactant. There is no catalytic process. The influence of irradiation with a xenon lamp has been reported to increase yields[71]. Solvent polarity effects on the product distribution (benzannulation versus cyclobutenones, indenes, and furans) have been studied quite exhaustively, but the picture is not completely consistent[72]. In general, less polar solvents (hexane, benzene) favor the benzannulation product, but there are exceptions.

4.3.4   Regioselectivity

The regioselectivity has been studied in some detail and numerous examples have arisen in the course of application of the methodology[60b]. In general, the dominant effect appears to be steric. In the coupling of the alkyne with the C=Cr bond, the less hindered end of the alkyne couples with the carbene carbon and the more crowded end with the Cr. This is not unusual for organometallic additions and insertions, and can be rationalized by remembering that the developing C–Cr bond is probably considerably longer than the developing C–C bond, and therefore 'feels' smaller steric interactions. The effect is very useful with monosubstituted alkynes, generally leading to >100 : 1 selectivity for the product type illustrated (**55**)[73]. Electronic effects can also be important, but simple trends are not yet clear:

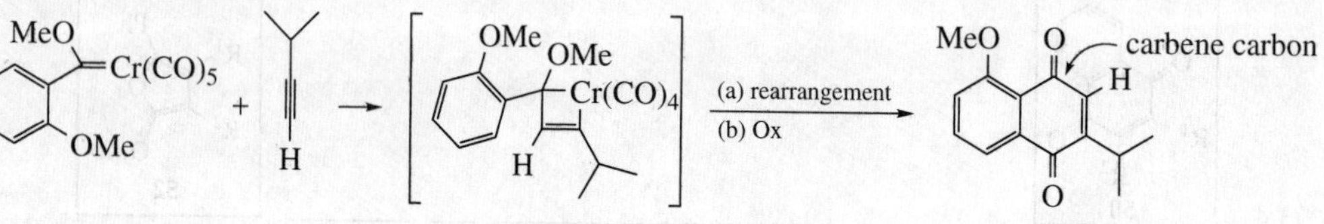

55

With less steric difference in the alkyne substituents, the control is less. Methylethyl-acetylene gives about a 2 : 1 mixture of naphthoquinone products[70]. This example also shows the oxidative treatment in MeOH which leads directly to the synthetically useful quinone monoketals. The original carbene carbon can be traced to the ketal carbon, and the major isomer shows the small substituent (Me) adjacent to this carbon:

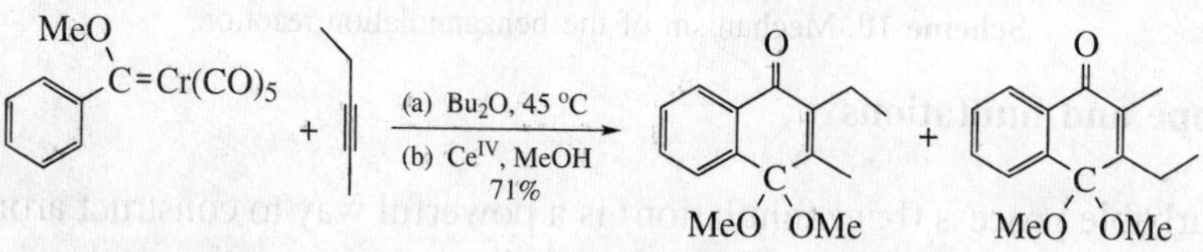

2.2 : 1

Procedure for Benzannulation with Arylcarbene Complexes—Formation of Naphthoquinones[74]

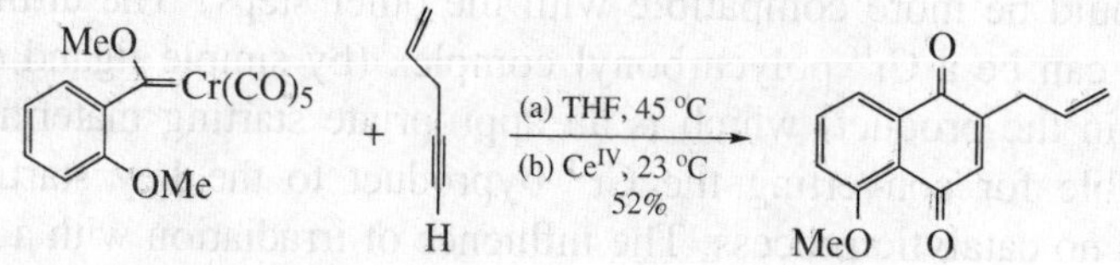

To a solution of the carbene complex (4.00 g, 11.7 mmol) in 100 mL of THF under argon at 25 °C was added pent-4-en-1-yne (1.5 mL, $\rho = 0.777$, 17.6 mmol) all at once. The mixture was heated at 45 °C for 36 h, cooled, and concentrated by rotary evaporation. The residue was dissolved in a mixture of acetonitrile (50 mL) and water (10 mL). To this mixture at 0 °C was added a solution of ceric ammonium nitrate (32.0 g, 58.4 mmol) in 50 mL of water. The mixture was warmed to 25 °C, stirred for 0.5 h, and concentrated by rotary evaporation to remove most of the acetonitrile. The residue was washed with ether five times, dried, and concentrated to leave a brown residue. Flash chromatography (silica gel, hexane/EtOAc) gave yellow crystals of the allyl naphthoquinone (1.40 g, 52% yield). Recrystallization from hexane/ether gave fine yellow crystals (mp 96.5–98 °C).

4.3.5   Functional Group Compatibility

Simple functional groups such as ester units and other carbonyl functionality are compatible with the benzannulations[75]. In general, the reaction is viewed as involving nonpolar driving forces, and is expected to be compatible with polar functionality as long as direct coordination effects are not severe. The example below also shows the convenient technique of displacing the arene ligand from the initially formed chromium complex by simple stirring in the presence of air and ambient light:

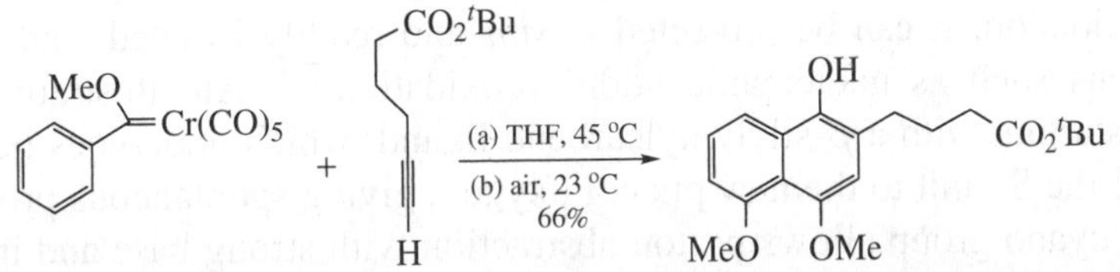

However, there are strong effects of substituents directly attached, or near to, the triple bond. For example, propargylic oxygen substituents can deflect the mechanism from benzannulation to related products such as furans and indenes[76]. Diphenylacetylene shows a very strong solvent effect on the distribution of products, whereas simple dialkylacetylenes give primarily the benzannulation product in all solvents:

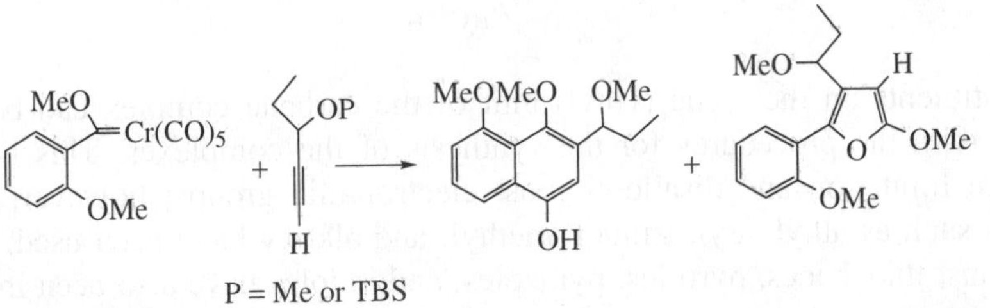

P = Me or TBS

There is an electronic effect on the regioselectivity, which can be complicated by the effect on product selectivity. For example, the alkenyl carbene complex **56** reacts with hex-3-yne-2-one to give one regioisomer of the benzannulation product along with a nearly equal amount of the indene-type product **57**:

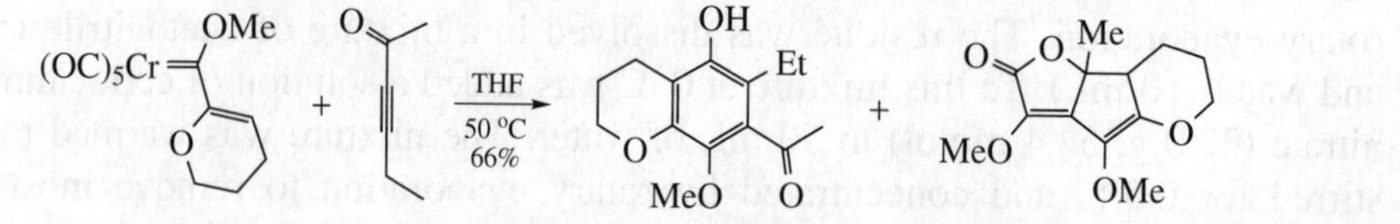

56   1:1   **57**

This process is an example of general benzannulations with alkenyl–carbene complexes, leading to benzohydroquinone complexes of chromium (**58**)[60]:

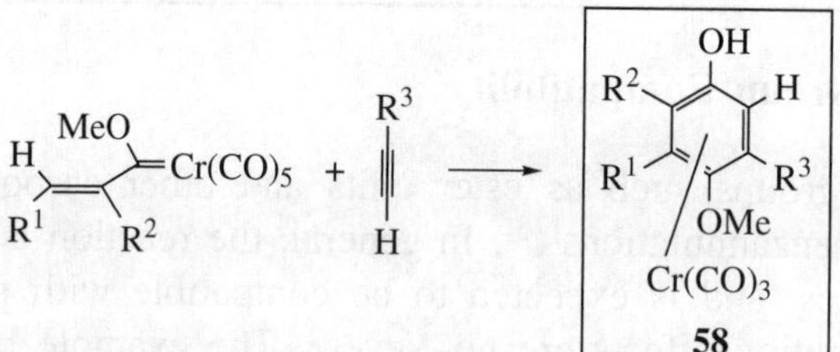

58

These simple arene complexes (*e.g.*, **58**) are much more stable toward ligand displacement compared to the naphthalene analogs, although the phenol ligand is particularly sensitive to oxidation. It can be protected *in situ* and readily isolated, and then used for further reactions such as nucleophile addition/oxidation[77]. An illustrative example is given below, starting with a *β*-silylvinylcarbene ligand, which undergoes benzannulation and transfer of the Si unit to the new phenol oxygen, giving spontaneous protection. Then the side-chain cyano group allows proton abstraction with strong base and intramolecular cyclization, giving a spirocyclohexenone in this case[77]:

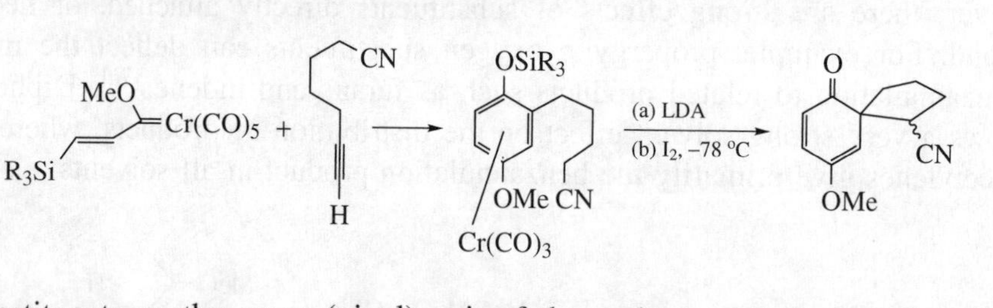

The substituents on the arene (vinyl) unit of the carbene complex can be any group compatible with the procedures for the synthesis of the complexes. This is typically a fairly severe limitation and disallows most electrophilic groups; however, many other substituents such as alkyl, aryl, trifluoromethyl, and alkoxy have been used. Heteroaryls such as furans, thiophenes, pyrroles, pyrazoles, and indoles have also been included. The aldol approach to vinyl carbene complexes should be more compatible.

The benzannulation process almost exclusively involves carbene complexes with an oxygen (methoxy) at the carbene carbon. A potentially useful diversion of the general mechanism occurs with the amino analogs. For example, the morpholino complex **59** reacts in DMF at elevated temperatures to give the indene **60** in 96% yield[78]. The usual regioselectivity obtains, and the amino indene derivatives can be converted directly to an indanone:

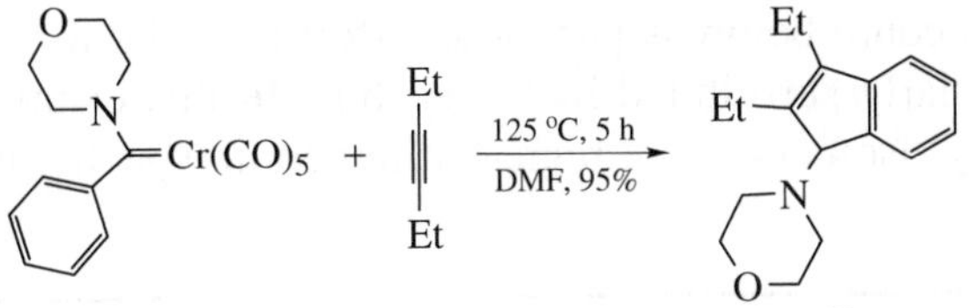

59   **60**

Procedure for Indanone Formation from Aminocarbene Complexes[78]

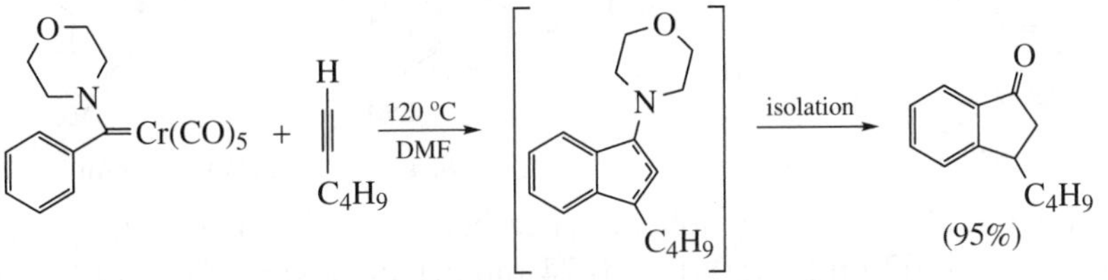

The phenyl morpholino carbene complex was prepared in 76% yield from the corresponding methoxyphenyl carbene complex with excess morpholine at 25 °C[79]. A DMF solution of the morpholino carbene complex (1.0 mmol) and 1-hexyne (2.0 mmol) under argon was heated at 120–125 °C for 5 h, and then for an additional 15 h at 90–95 °C. The mixture was cooled and diluted with ether. The ether layer was washed with brine, dried over Na_2SO_4, filtered, and concentrated by rotary evaporation. The residue was purified by flash chromatography on silica gel, giving 3-butylindanone in 95% yield as a colorless oil.

Even with alkoxy-substituted carbene ligands, the formation of indenes is often a significant side process, and becomes the major process under certain conditions. For example, the Mo and W analogs give indene products under conditions where the exact Cr derivatives give exclusively benzannulation[60b]. There is sometimes a poorly understood solvent effect, suggesting a general rule that benzannulation is favored by nonpolar solvents (hexane) and indene formation can become dominant in polar, coordinating solvents such as THF[80].

With vinylcarbene complexes (*e.g.*, **64**) where the vinyl group is β,β-disubstituted, the final proton shift to give an aromatic product is not available. In these cases, the intermediate cyclohexadienone can be isolated, as exemplified by the following spirocycle synthesis[81]:

This is potentially very useful methodology, and has already been employed in alkaloid synthesis[82]. The procedure below is part of an effort to probe for stereoselectivity in the formation of the angularly substituted hydrodecalins. In this example, the selectivity for *cis*- and *trans*-dimethyl isomers is not high, but in certain specific cases the selectivity is >90 : 10[83].

Procedure for Reaction of β,β-Disubstituted Carbene Ligands to give Cyclohexadienones[84]

A solution of 2,3-dimethylcyclohexenyl carbene complex **65** (0.128 g, 0.37 mmol) and pent-1-yne (0.070 mL, 0.051 g, 0.74 mmol) in hexane (7.4 mL) was deoxygenated by the freeze-pump-thaw method (three cycles) and then stirred at 60 °C under argon for 12 h. The solution was stirred in air at 25 °C for 1 h and then filtered through Celite. After removal of solvent on a rotary evaporator, the cyclohexadienone product (**66a/b**) was purified on silica gel with a 1 : 1 : 10 mixture of Et_2O/CH_2Cl_2/hexane as eluant to give 0.076 g (83% yield) of a 56 : 44 mixture of **66a** and **66b** (thin-layer chromatography (TLC) with Et_2O/CH_2Cl_2/hexane gave: **66b**, R_f 0.53; **66a**, R_f 0.48).

The intrinsic regioselectivity of the alkyne benzannulation reaction can be overridden by making the process intramolecular. This is particularly convenient when the reactant alkyne can be tethered via heteroatom exchange at the carbene carbon. For example, after addition of phenyl lithium to produce the metallate **67**, acylation gives the pseudo anhydride **68**[74]. Then reaction with an hydroxyalkyne (*e.g.*, **69**) proceeds by rapid oxygen exchange at below room temperature. Gentle heating of the tethered alkyne–carbene complex **70a** leads to benzannulation, where the regioselectivity is constrained by the tether dimensions. Alternatively, the alkyne can be tethered by means of silicon derivative **70b**[85] (Scheme 11).

The intramolecular benzannulations shown in Scheme 11 provide intermediates that are very useful in the synthesis of natural naphthoquinone antibiotics, such as frenolicin and nanaomycin. A synthesis of deoxyfrenolicin included the following procedure[74].

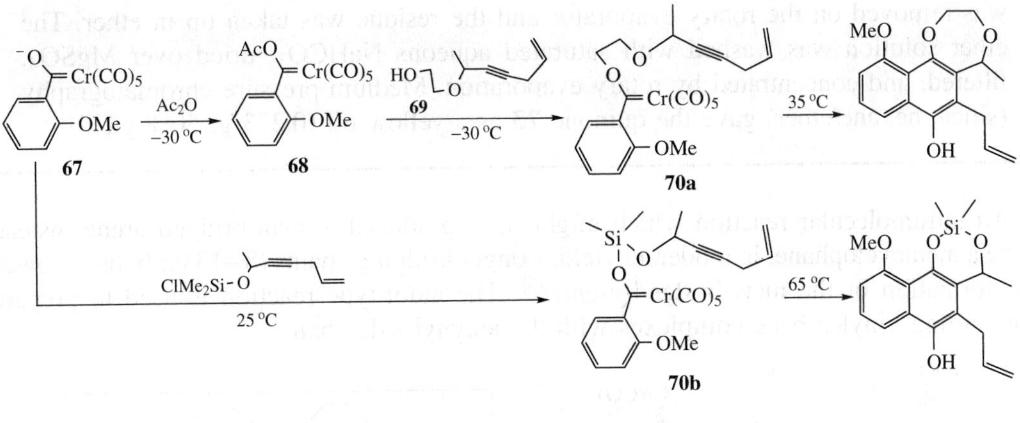

Scheme 11. Intramolecular benzannulation.

Procedure for Intramolecular Benzannulation in the Synthesis of Deoxyfrenolicin

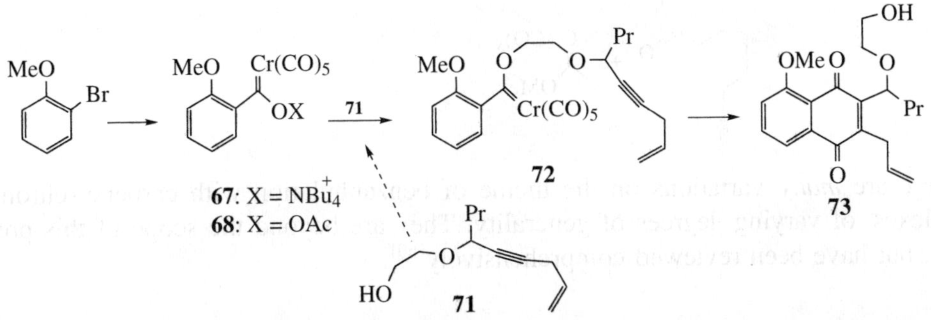

67: X = NBu₄⁺

The salt **67** (1.00 g, 2.49 mmol) was dissolved in CH₂Cl₂ (50 mL) under argon. The flask was covered with Al foil, cooled to −20 °C, and a solution of acetyl chloride (0.216 g, 2.74 mmol) in CH₂Cl₂ (25 mL) was added over 5 min to give a deep red soln of **68**. After addition, the mixture was warmed to −10 °C and stirred for 45 min. A solution of the alkynol **71** (0.563 g, 2.49 mmol) in CH₂Cl₂ (2.5 mL) was added and the mixture was stirred at 23 °C until the alkynol was consumed (TLC 2–3 h). The mixture was concentrated at aspirator pressure and the residue was washed with several small portions of pentane to dissolve carbene complex **72**, all the while maintaining an argon atmosphere. The combined pentane solutions were concentrated to leave the carbene complex as a red oil (0.806 g, 66% yield). This crude complex was dissolved in ether (50 mL) and heated at reflux for 64 h. The solvent was removed at aspirator vacuum, and the residue was dissolved in 20 mL of degassed acetonitrile/water (10 : 1). Dichlorodicyanoquinone (DDQ) (0.335 g, 1.5 mmol) was added as a solid in small portions under a slow flow of argon. The reaction is self-indicating, as excess DDQ is evidenced by the appearance of an intense brick-red color in the mixture. After addition was complete, the acetonitrile

was removed on the rotary evaporator and the residue was taken up in ether. The ether solution was washed with saturated aqueous NaHCO₃, dried over MgSO₄, filtered, and concentrated by rotary evaporation. Medium pressure chromatography (silica, hexane/ether) gave the quinone **73** as a yellow oil (0.273 g, 50% yield).

An intramolecular reaction which might have produced a meta-bridged arene instead gave a metacyclophane, in moderate yield. Longer bridging chains (8–13 carbons) allowed the formation of the meta-bridged arene[86]. The aldol-type reaction is used to prepare the starting vinylcarbene complexes with the alkynyl side chain:

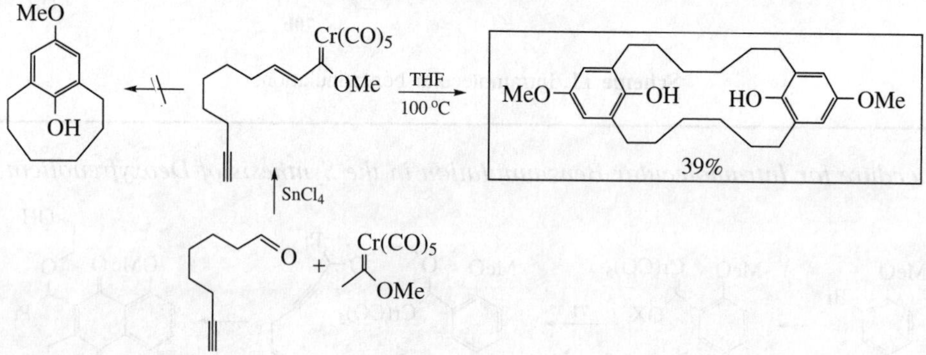

There are *many* variations on the theme of benzannulation with carbene–chromium complexes, of varying degrees of generality. They are beyond the scope of this presentation, but have been reviewed comprehensively[60].

4.4   Controlled Generation of Ketene–Chromium Complexes[87]

4.4.1   Introduction

An alternate pathway is followed with imines in the place of alkynes under visible light irradiation, giving β-lactams in an efficient and general procedure. Although the β-lactam product can be rationalized by initial cycloaddition of the imine in analogy with the general mechanism in Scheme 12, the actual mechanism is quite different and the irradiation is required. Irradiation with UV light is commonly used to promote ligand dissociation of carbene–chromium complexes, leading to ligand exchange. However, visible light selectively excites the MLCT (metal-to-ligand charge transfer) transition, corresponding to removal of an electron from the d-orbital-centered HOMO (highest occupied molecular orbital), and is referred to as a formal one-electron oxidation of the metal[87]. The process initiated by irradiation is not simple CO dissociation, but rather the formation of a ketene ligand (**74**, Scheme 12). Then coordination and a more-or-less conventional [2 + 2] cycloaddition of the ketene proceeds, leading to the β-lactam product directly. Compared to alternative methods of ketene generation, this process has

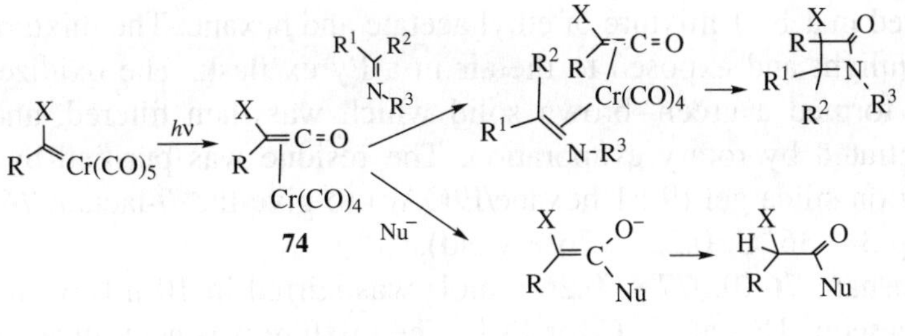

Scheme 12. Ketene generation and trapping reactions.

important advantages in efficiency and stereocontrol. The product β-lactams are useful structures and also undergo facile and predictable conversions into other equally useful intermediates. A special feature is the easy introduction of a chiral auxiliary and then the induction of asymmetry into the two new stereogenic centers in the product. Given the carbene–Cr complex as a new and controlled source of ketenes, it is not surprising to see trapping by other species such as simple nucleophiles. In essence, the discovery of this new method for the generation of ketenes re-opened the fertile area of ketene as a synthesis intermediate. The ketene complexes are generally not isolated, but are generated *in situ* and trapped by cycloaddition or nucleophile addition.

4.4.2 β-Lactam Formation from Imines

The [2 + 2] cycloaddition of imines with the ketene ligand generated from carbene–chromium complexes under irradiation is very general[87]. Even quinoline and isoquinoline give adducts. Most interesting are the adducts from thiazolines such as **75** which can lead to penicillin-like structures[88]. In the example given in the following procedure, one diastereoisomer was observed. A homochiral thiazolidine unit can serve as a chiral auxiliary unit, and is cleaved to give highly functionalized quaternary centers, as in **77**, which are also homochiral.

Procedure for Formation of a Homochiral β-Lactam and Cleavage to a Homochiral Quaternary Center[89]

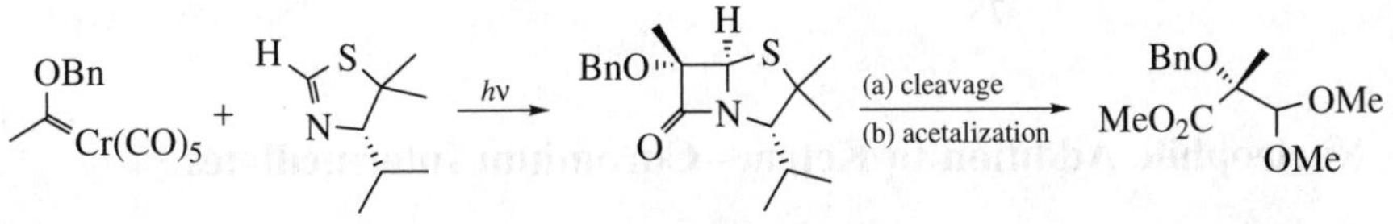

(a) A solution of homochiral thiazoline **75** (0.14 g, 1.10 mmol) in ether was added to the methylmethoxycarbene $Cr(CO)_5$ complex (0.33 g, 1.00 mmol) under argon at 23 °C. The yellow solution was irradiated with a 450-W Hanovia 7825 medium-pressure mercury lamp. After 3.5 day of irradiation (monitoring by TLC), the mixture was concentrated by rotary evaporation to give a green residue which was dissolved in a 1 : 1 mixture of ethyl acetate and hexane. The mixture was placed in bright sunlight and exposed to the air in a Pyrex flask. The oxidized chromium byproducts formed a green–brown solid which was then filtered, and the filtrate was concentrated by rotary evaporation. The residue was purified by radial chromatography on silica gel (9 : 1 hexane/EtOAc) to give the β-lactam **76** as colorless crystals (mp 34–36 °C; 0.22 g, 76% yield).

(b) The penam **76** (0.077 g, 0.26 mmol) was stirred in 10 mL of methanol saturated with gaseous HCl at 23 °C for 18 h. The mixture was neutralized with aqueous NaHCO₃ and extracted with ethyl acetate. The solvent was removed by rotary evaporation, and the residue was dissolved in 10 mL of 10% aqueous acetone and treated with 3 mol equivalents of I₂. The mixture was heated at reflux for 3 h, after which time the excess I₂ was decomposed with aqueous Na₂S₂O₃, and the mixture was extracted with ethyl acetate. Evaporation of the solution gave an aldehyde. It was dissolved in 10 mL of anhydrous methanol with 1 mL of trimethylorthoformate, and 3 mol equivalents of I₂ in a sealed tube and heated at 80 °C for 3 h. The mixture was cooled, poured into aqueous Na₂S₂O₃, and extracted with ethyl acetate. The ethyl acetate solution was concentrated and the residue was purified by preparative TLC (9 : 1 hexane/EtOAc) to give the dimethyl acetal **77** as a colorless oil ($[\alpha]_D + 1.22°$ (c 3.6, CH₂Cl₂)).

Using aminocarbene complexes (*e.g.*, **78**), asymmetry was introduced into the substituents on the nitrogen and then favored amino-substituted β-lactams with very high de. The next example also shows the use of a carbene ligand with a hydrogen substituent at the carbene carbon. Few such complexes were known before a method was developed for their synthesis, via the Cr(CO)₅ dianion and, for example, Vilsmeier's salts[90]:

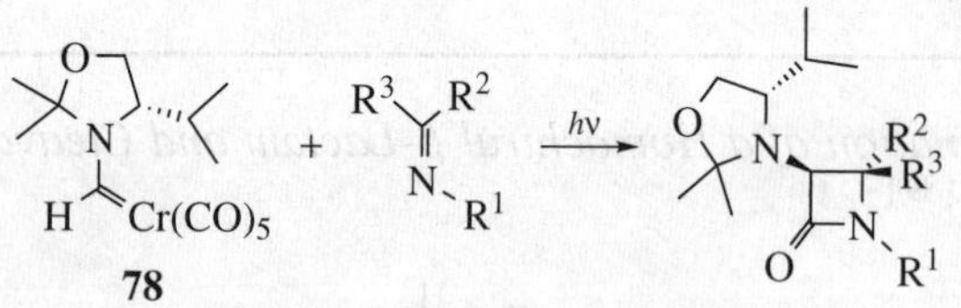

4.4.3   Nucleophile Addition to Ketene–Chromium Intermediates

With this convenient source of ketenes bearing a simple heteroatom substituent or a chiral auxiliary (the substituent on the heteroatom at the carbene carbon), other general reactions of ketenes have been explored[91]. With aminocarbene–Cr(CO)₅ complexes, irradiation and methanol addition leads to α-substituted α-amino acids. Many examples have been reported, with moderate yields in the preparation of the starting carbene complex, and very high yields in the ketene generation/trapping:

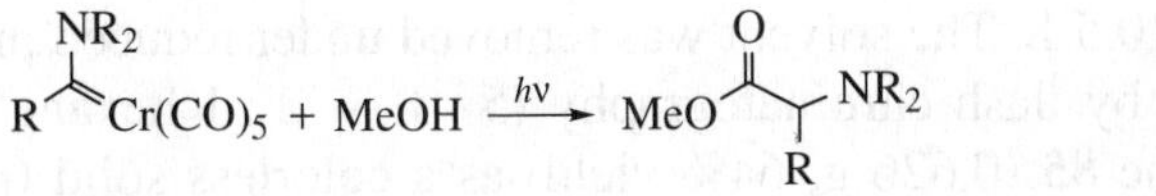

Using cleverly designed asymmetric amines, a collection of α-amino acids with various side chains was prepared in high enantiomeric excess, including asymmetric 2-^{2}H$_1$-glycine[92]. For example, the usual carbene anion **79** was acylated and then substitution by the homochiral amino alcohol **80** gave the asymmetric aminocarbene complex **81**[93]. Irradiation and intramolecular trapping (cyclization) leads to a lactone **82**, and then hydrogenolysis removes the directing group to give amino acid **83** (Scheme 13).

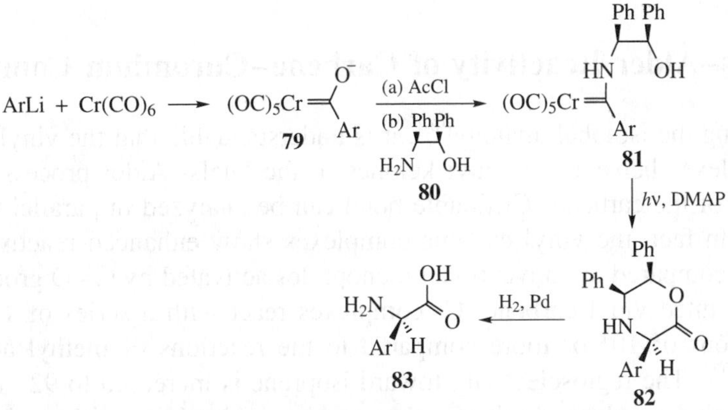

Scheme 13. Asymmetric α-amino acid synthesis.

This concept has been expanded to produce asymmetric templates for 4′-substituted nucleoside analogs[94], and in general asymmetric butenolide synthesis[95] through a [2 + 2] cycloaddition of the ketene ligand with an asymmetric alkene, as presented in the following Procedure.

Procedure for Trapping of the Ketene Ligand with Asymmetric Alkenes

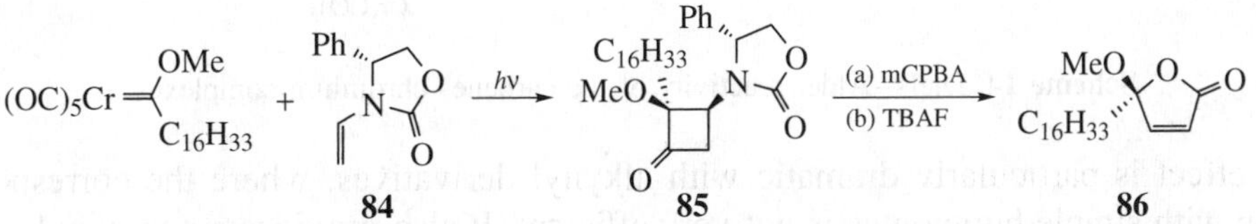

(a) The starting carbene complex was prepared by addition of C$_{16}$H$_{33}$Li to Cr(CO)$_6$, as usual, followed by alkylation with trimethyl oxonium fluoroborate. The carbene chromium complex (1.3 g, 2.83 mmol) and vinyl carbamate **84** (0.382 mg, 2.0 mmol) in degassed CH$_2$Cl$_2$ (50 mL) were placed in an Ace Glass pressure tube. The tube was charged with CO (60–90 psi, three cycles) and irradiated at 25 °C for 10.5 h. The solvent was removed under reduced pressure and the residue was purified by flash chromatography (5 : 1 → 2 : 1 hexane/EtOAc, SiO_2) to give cyclobutanone **85** (0.626 g, 64% yield) as a colorless solid (mp 74–75 °C).

(b) Compound **85** (0.300 g, 0.62 mmol), m-chloroperoxybenzoic acid (m-CPBA) (0.160 g, 0.74 mmol), and Li_2CO_3 (0.014 g, 0.2 mmol) in CH_2Cl_2 (50 mL) were stirred at 23 °C for 11 h. The mixture was washed sequentially with 10% $Na_2S_2O_3$ and Na_2CO_3 solutions, dried over $MgSO_4$, filtered, and concentrated by rotary evaporation to give essentially pure lactone **86** (0.309 g, 99% yield) as a white solid (mp 92–93 °C).

4.5   Diels–Alder Reactivity of Carbene–Chromium Complexes

Remembering the isolobal analogy[96], it is understandable that the vinyl (and alkynyl) carbene complexes behave like vinyl ketones in the Diels–Alder process (Scheme 14). The reactivity of the carbene–Cr double bond can be analyzed in parallel with the C=O double bond. In fact, the vinyl carbene complexes show enhanced reactivity and selectivity features compared to conventional dienophiles activated by C=O groups. The usual methoxy-substituted vinyl carbene–Cr complexes react with a series of 1,3-dienes with rate accelerations of 10^4 or more compared to the reactions of methyl acrylate with a similar series[97]. The regioselectivity toward isoprene is increased to 92 : 8 compared to 70 : 20 for methyl acrylate. Similarly, the endo/exo selectivity with cyclopentadiene is 94 : 6 versus 78 : 22 for methyl acrylate. The effect is comparable to that observed by Lewis acid catalysis, with $AlCl_3$.

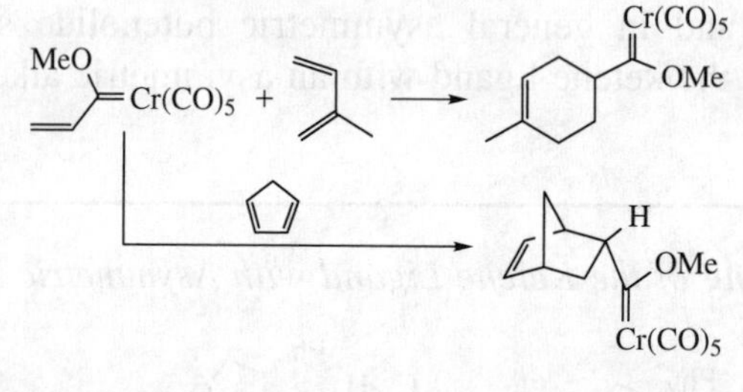

Scheme 14. Diels–Alder reactivity of the carbene–chromium complexes.

The effect is particularly dramatic with alkynyl derivatives, where the corresponding reaction with simple butynontes is not very efficient. It also provides a new vinyl carbene species of increased complexity, ready to enter into benzannulation[98]:

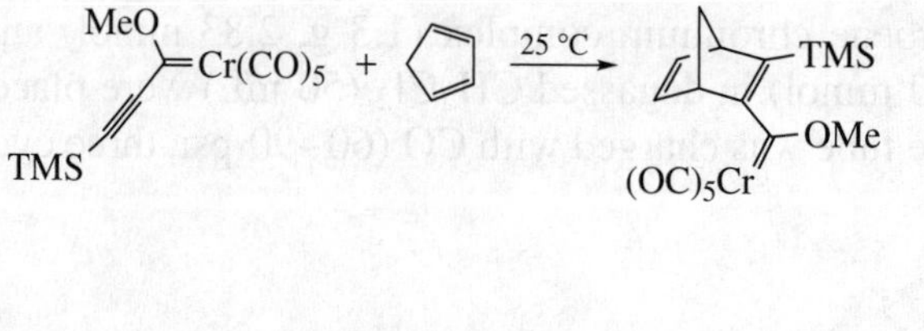

The general version in Scheme 15 illustrates the versatility of these two steps. The initially formed cyclohexenylcarbene–chromium complex with a β-H (A=H) leads to the usual hydroquinone monomethyl ether chromium complex, whereas the complex with a β-Me (A=Me) provides a cyclohexa-2,4-diene-1-one, free of the chromium. There are numerous variations on this theme, with other 1,3-dienes, other substituents on the vinyl group, and even other metals (*e.g.*, tungsten), but a complete review is beyond the scope of this chapter[99].

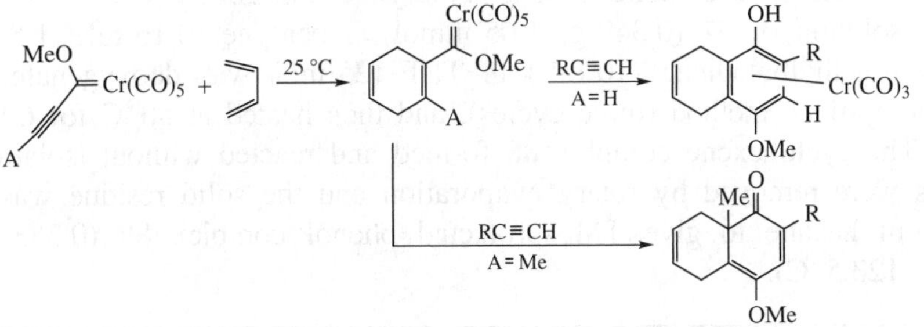

Scheme 15. The two-stage Diels–Alder/benzannulation process.

The tandem Diels–Alder reaction and benzannulation is a direct process for increasing complexity in organic substrates, as illustrated in the process below. The process can be run in one pot, with excess 1,3-diene and limited alkyne. It is particularly favored by the rate-enhancing effect of the $Cr(CO)_5$ unit on the Diels–Alder reaction involving alkynes, as in **87**. The intermediate vinylcarbene complex **88** reacts selectively with the alkyne to give the benzannulated product **89**. Gentle oxidative decomplexation (air/*hν* or I_2) can give the free arene.

Procedure for Formation of Alkynylcarbene Complexes, Diels–Alder Reaction, and Benzannulation of the 1,4-Cyclohexadiene Product[89]

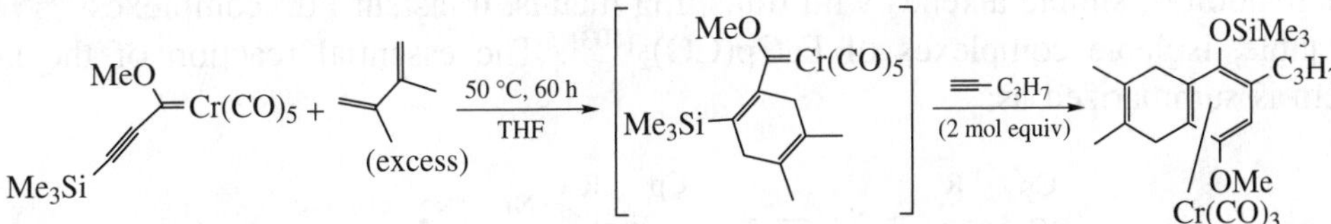

(a) A solution of TMS−C≡CLi was prepared by addition of BuLi (10.0 mL, 1.54 M in hexane, 15.4 mol) to a solution of TMS−C≡CH (1.47 g, 2.07 mL, 15.0 mmol) in 30 mL of ether under argon at −78 °C. After 30 min at this temperature, the solution was allowed to warm to 0 °C over 20 min, and then transferred via cannula to a slurry of $Cr(CO)_6$ (3.7 g, 17 mmol) in ether (150 mL) at 0 °C. The mixture was warmed to 25 °C and 25 mL of THF was added, resulting in a clear red solution. After 1.5 h, the solution was cooled to 0 °C and methyl fluorosulfonate (2.57 g, 22.5 mmol, 1.80 mL) was added dropwise. After 30 min, the mixture was quenched by addition of saturated aqueous. NaHCO$_3$ and stirred for 20 min. The organic layer was washed with brine, dried over anhydrous MgSO$_4$, filtered, and the filtrate was concentrated by rotary evaporation. The residue was purified by chromatography on silica gel (hexane, TLC R_f = 0.39). The fractions were concentrated by rotary evaporation and then at oil pump vacuum at 0 °C. The alkynyl complex **87** was obtained as a red solid (3.83 g, 77% yield; mp 26–27 °C).

(b) A solution of **87** (0.349 g, 1.05 mmol), 1-pentyne (0.16 mL, 1.58 mmol), and 2,3-dimethylbutadiene (10 mL) in THF (25 mL) was deoxygenated by the freeze-pump-thaw method (three cycles), and then heated at 50 °C for 60 h under argon. (The cyclohexene complex **88** formed and reacted without isolation.) The volatiles were removed by rotary evaporation and the solid residue was crystallized from hexane to give TMS-protected phenol complex **89** (0.383 g, 80%; mp 121–123.5 °C).

5   Applications of η^2-Alkene–Fe(Cp)(CO)$_2$ Complexes

5.1   Introduction

The coordination of alkenes and other polyenes to transition metals strongly increases the polarizibility of the pi bond system[100,101]. Depending on the oxidation state of the metal and the donor/acceptor properties of the peripheral ligands, the complex can behave as a more-or-less reactive electrophile or nucleophile. As simple alkenes, polyenes, and arenes react most commonly as nucleophiles in traditional organic chemistry, the *electrophilic* reactivity of coordinated alkenes and polyenes offers unique tactics for carbon–carbon and carbon–heteroatom coupling. There are two well-developed systems for activation of simple alkenes with transition metals, transient Pd^{II} complexes[102], and the stable, isolable complexes of FeCp(CO)$_2$[103]. The essential reaction of the latter system is summarized as:

$$\underset{OC}{\overset{Cp}{\underset{|}{OC-Fe}}}\!\!\!{\overset{+}{\diagup\!\!\!R}} \;+\; Nu^- \;\longrightarrow\; \underset{OC}{\overset{Cp}{\underset{|}{OC-Fe}}}\!\!\!{\overset{R}{\diagup}}\!-Nu \xrightarrow{etc.} \begin{array}{l}\text{organic}\\\text{products}\end{array}$$

Important questions for synthesis applications include: (a) source of the starting cationic complex, (b) range of nucleophiles which will add efficiently, (c) regioselectivity of the addition (which end of the double bond?), (d) stereoselectivity of the addition (configuration of the new stereogenic center), (e) processing routes for the product σ-alkyl-Fe complex to give simple organic products, and (f) opportunities for a catalytic process or easy recycling of the metal reagent.

The question of the range of nucleophiles is particularly interesting, as a number of potential side reactions is available. The Nu⁻ species can behave as a donor ligand, displacing the alkene from the iron before addition, or it can behave as a base and abstract an activated proton from an allylic position on the alkene ligand. The nucleophile might also function as an electron-transfer agent, reducing the starting cationic complex and causing rapid alkene dissociation from the iron. Regioselectivity is also complicated by the fact that the iron is bound more-or-less symmetrically with respect to the ends of the alkene; it provides an activating influence but not necessarily an orienting influence. The substituents on the alkene may play a dominant role even if de-activating (*e.g.*, Me or OMe). This contrasts strongly with conventional methods of activating alkenes toward nucleophile addition, by attaching an electron-withdrawing functional group via a sigma bond. The electron-withdrawing group then exerts both an activating and an orienting effect (*β*-addition). The stereochemical answer is simple: anti addition to the alkene is strongly preferred, consistent with the general path for nucleophile addition to coordinated polyenes[101].

5.2 Preparation and Properties of Cp(CO)₂Fe(Alkene)⁺ Complexes

There are three general routes for the preparation of the starting alkene complex. First, simple alkene exchange occurs at a useful rate even at 25 °C, and a more stable alkene complex is favored at equilibrium. In general, more substituted Cp(CO)₂Fe(alkene)⁺ complexes are more stable. It is convenient to prepare the complex of isobutene **90** and store it as a source of the Cp(CO)₂Fe unit. For example, reaction of Cp(CO)₂Fe(isobutene)⁺ with 4-vinylcyclohexene gives complex **91** with high selectivity under gentle thermal conditions[104]:

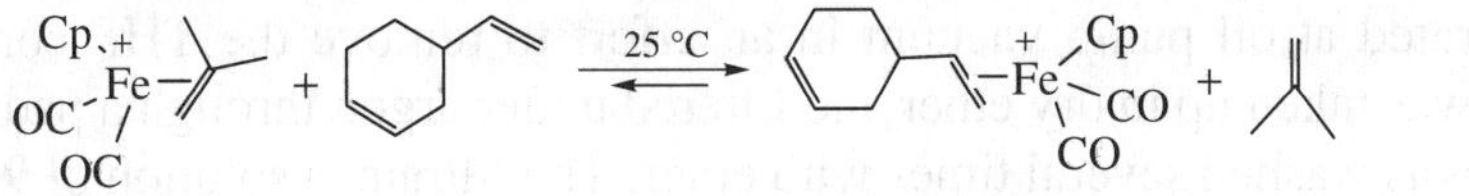

Second, the powerful anion nucleophile Cp(CO)₂Fe⁰ reacts rapidly in S_N2 fashion with epoxides to produce *β*-hydroxyalkyliron(0) species[104]. Gentle treatment with acid leads to rapid elimination of water and formation of the alkene–FeCp(CO)₂(II) cation complex **93**:

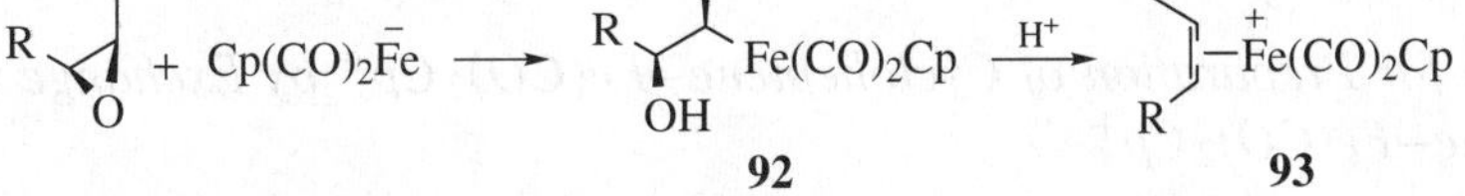

The reaction is highly stereospecific, requiring alkylation with inversion and a dominant anti transition-state conformation for elimination. Then starting with a cis alkene produces by this method the cis alkene product, assuming a cis epoxidation mechanism. In a related process, simple alkyl halides and sulfonate esters undergo S_N2 reaction with Cp(CO)$_2$Fe^0 to give alkyl-Fe^{II} species (*i.e.*, **94**). The common organometallic mechanism of β-hydride elimination would give the desired alkene–Fe complex, but is inhibited[105] by the stable, coordinatively saturated arrangement in **94**. Instead, β-hydride elimination must be induced intermolecularly by the addition of a strong hydride abstracting agent, such as trityl cation. The product cationic complexes (*e.g.*, **95**) are air-stable solids, and can be stored for long periods at ambient temperature:

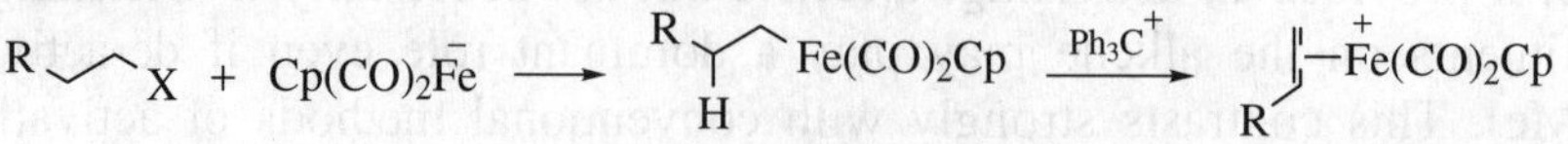

Third, as shown in the procedure below, an alternative is loss of a β-electronegative substituent, usually an alkoxy group induced by acid treatment.

Procedure for Preparation of Ethoxyethylene–Fe(CO)$_2$Cp$^+$ by S_N2 Reaction and β-Ethoxy Elimination[106]

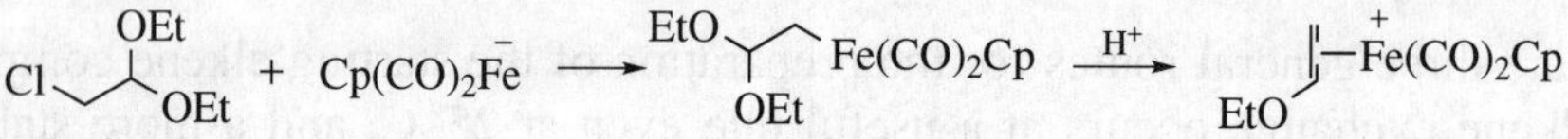

To a solution of the Cp(CO)$_2$Fe anion (0.01 mol, prepared from [Cp(CO)$_2$Fe]$_2$)[107] in 10 mL of THF under argon at 25 °C was added chloroacetaldehyde diethyl acetal (3.11 g, 0.020 mol) slowly by syringe. The mixture was heated at 50 °C for 2 h, then cooled to 25 °C and concentrated on a rotary evaporator. The residue was further concentrated at oil pump vacuum in an effort to remove the THF completely. The residue was taken up in dry ether and filtered under argon through a pad of Celite; the residue was washed several times with ether. The filtrate, a solution of **96**, was cooled to −78 °C and HBF$_4$•OEt$_2$ (3.82 g, 0.023 mol) was added dropwise over 0.5 h. After the solution was allowed to warm to 25 °C, the yellow precipitate was filtered, washed with ether, and dried under a stream of argon followed by oil pump vacuum, releasing with argon. The product, **97**, was a bright yellow salt (5.0 g, 70% yield).

Procedure for Preparation of Cycloheptene–Fe(CO)$_2$Cp$^+$ by Exchange with Isobutylene–Fe(CO)$_2$Cp$^+$

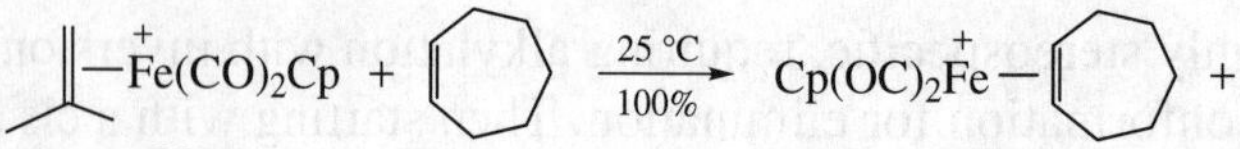

According to a general procedure[104], a solution of **98**•BF$_4$ (1 mmol) and cyclo-heptene (2.3 mol) in 1,2-dichloroethane under argon was heated at 60 °C. After 10 min, dry ether was added to induce precipitation of the cycloheptene salt **99**, which was isolated by filtration in quantitative yield. The product is a yellow, air-stable solid which decomposes rather than melting. The vinyl H appear in the ^{1}H NMR spectrum at δ 7.4 ppm.

5.3   Variation in the Nucleophile

The range of nucleophiles which will add successfully includes heteroatom species such as alkoxide, thiolate, and amide (R$_2$N$^-$) anions, as well as carbanions ranging in reactivity from nitro-stabilized anions to simple ketone enolates. However, anions derived from less acidic carbon acids (sulfur-stabilized carbanions, simple organo-Li and organo-MgBr species) lead to significant amounts of electron transfer reduction, and loss of the alkene ligand. The products from successful addition are new alkyl-Fe(Cp)(CO)$_2$ complexes (*e.g.*, **100**), and are relatively air stable, chromatographable substances. This is, in fact, the primary problem with the methodology. The intermediates are too stable to allow for a fast catalytic process. Various techniques are known for the further conversion of the intermediate **100**, such as oxidation (*e.g.*, to give **101**) and proteolysis[103]. No carbon–carbon coupling procedures from the intermediate have been developed, however. The overall stoichiometric feature of the method is clearly one of the more serious limitations, and has restricted the possible applications. The example below also demonstrates the clean anti addition pathway to the cyclopentene complex **102**:

| 102 | 100 | 101 |

5.4   Regioselectivity and Stereoselectivity

The reaction proceeds well with variously substituted alkenes, including those with alkoxy substituents, and these derivatives provide the examples with greatest potential for synthesis. A typical example is shown in Scheme 16, where the enolate anion from cyclohexanone adds to the complex **97** from ethyl vinyl ether[108]. This complex is viewed as an equivalent of the vinyl cation. The regioselectivity is perfect for addition at the more substituted end (typical of alkene–Fe^{II} complexes substituted with electron donor groups, and rationalized in analogy with Markovnikoff's rule for electrophilic addition to alkenes). The product **103** is a β-alkoxy ketone with an alkyl-Fe side chain. The compound can be

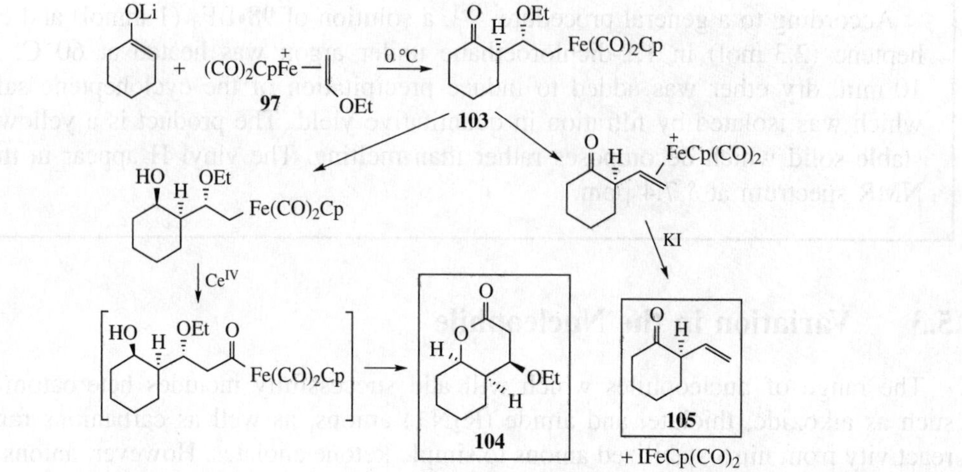

Scheme 16. Enolate addition to vinyl ether–FeCp(CO)$_2{}^+$ complexes.

manipulated by conventional methodology without disturbing the Fe group, until oxidation gives CO insertion and the product **104** at the carboxylic acid oxidation level, a lactone in this case. Alternatively, gentle acid treatment promotes anti elimination of alkoxy to give a new alkene–Fe^{II} complex, and then the iron can be detached by gentle ligand displacement with, for example, iodide anion. The resulting β,γ-unsaturated ketone **105** is difficult to produce by other methods because of the propensity for the double bond to isomerize into conjugation with the ketone.

The regioselectivity is particularly dramatic with methyl α-methoxyacrylate[109]. Whereas conventional nucleophilic addition would take the usual conjugate addition pathway, the Fe^{II} complex **106** leads to addition exclusively at the α-position. The product can be manipulated into an α-methylenelactone structure:

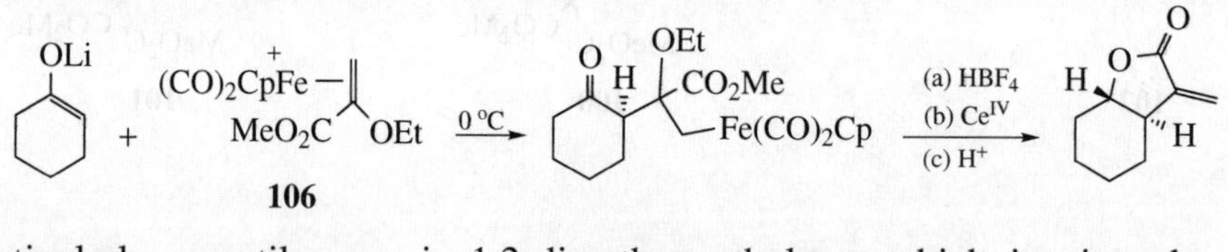

A particularly versatile case is 1,2-dimethoxyethylene, which is viewed as a 1,2-dication equivalent[100]. The cis version, prepared by exchange complexation with isobutylene–Cp(CO)$_2$Fe$^+$, undergoes nucleophile addition as usual, with the anti orientation, producing the alkyl-Fe^{II} species **107** (Scheme 17). Acid treatment induces anti elimination, giving the new alkene complex **108t**. The trans geometry is thermally unstable at elevated temperatures and can be equilibrated, a process which favors the cis geometry (**108c**). A second nucleophile addition, protonation, and ligand displacement gives the 1,2-disubstituted ethylene of either geometry.

The high stereoselectivity in the addition and elimination steps is very reliable, and adds significance to efforts to produce asymmetric alkene–FeCp(CO)$_2{}^+$ complexes[110].

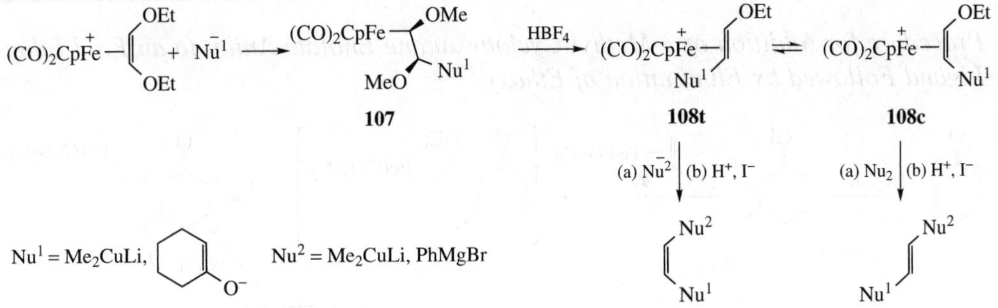

Scheme 17. Nucleophile addition to 1,2-dialkoxyethylene-FeCp(CO)$_2^+$ complexes.

A useful protocol involves diastereoselection in which a chiral diol is introduced as a chiral auxiliary. This process, shown in Scheme 18, produces the homochiral enantiomeric complexes **109** and **110**.

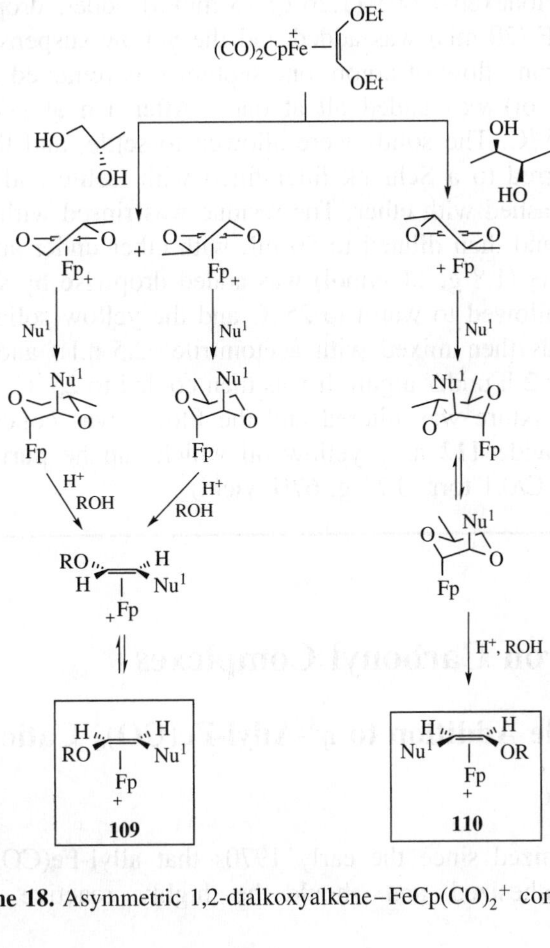

Scheme 18. Asymmetric 1,2-dialkoxyalkene–FeCp(CO)$_2^+$ complexes.

Procedure for Addition of 3-Methyl Cyclohexanone Enolate Anion to an Enol Ether Ligand Followed by Elimination of Ethoxy[108]

To a suspension of cuprous iodide (2.47 g, 13 mmol) in 15 mL of ether at 0 °C under argon was added dropwise over 5 min MeLi in ether (17.2 mL, 26 mmol) followed by 2-cyclohexen-1-one (1.26 g, 13 mmol) added dropwise over 2 min. After 15 min, THF (20 mL) was added and the yellow suspension was cooled to −78 °C. With a strong flow of argon, one septum was removed and solid complex **97** (4.37 g, 13 mmol) was added all at once. After 1 h at −78 °C, the mixture was warmed to 25 °C. The solids were allowed to settle, and the red supernatant liquid was transferred to a Schlenk filter filled with Celite and a pad of alumina which had been washed with ether. The residue was rinsed with ether. The filtrate was concentrated and then diluted to 50 mL with ether under argon and cooled to −78 °C. HBF₄•OEt₂ (1.8 g, 11 mmol) was added dropwise by syringe over 0.5 h. The mixture was allowed to warm to 25 °C and the yellow solid **111** was isolated by filtration. It was then mixed with acetonitrile (2.5 mL) and the mixture was heated at reflux for 2 h under argon. It was then cooled to 25 °C, and ether (50 mL) was added. The mixture was filtered and the filtrate was concentrated by rotary evaporation to provide **112** as a yellow oil which can be purified by short-path distillation (bp 30 °C/0.1 torr; 0.72 g, 67% yield).

6   η³-Allyliron Carbonyl Complexes

6.1   Nucleophile Addition to η³-Allyl-Fe(CO)₄ Cations

6.1.1   Introduction

It has been recognized since the early 1970s that allyl-Fe(CO)₄ cation complexes can be readily synthesized and should be highly reactive electrophiles[111,112].

Nucleophile coupling would give simple alkene species, loosely coordinated with the iron (Scheme 19). Obvious issues include: (a) scope of structural types for the allyl-Fe(CO)$_4$ cations; (b) scope of nucleophiles which add successfully; and (c) regio- and stereoselectivity in the nucleophile addition[113]. The Fe system has been overshadowed by the more general *catalytic* process for allylic alkylation involving, for instance, allylic acetates and Pd^0[114]. Nevertheless, in a limited way, the iron allyl complexes offer unique conversions. Two methods of preparation and the overall sequence are represented in Scheme 19.

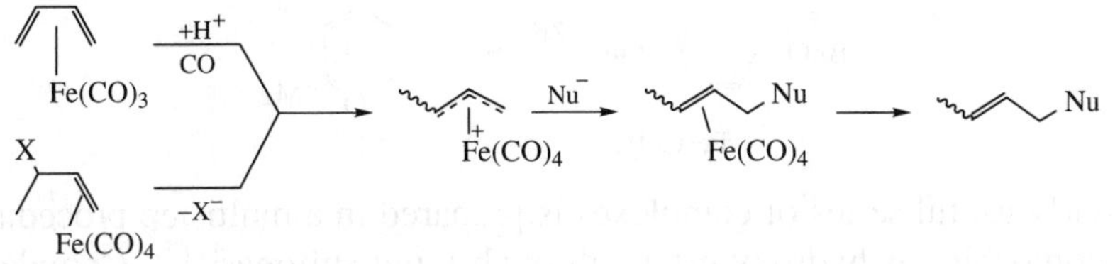

Scheme 19. Preparation and general reactions of the allyl-Fe(CO)$_4$ cations.

6.1.2 Preparation of Allyl-Fe(CO)$_4$ Cation Complexes

The easily synthesized η^4-(1,3-diene)Fe(CO)$_3$ complexes react at room temperature in the presence of CO at low pressure with mineral acids such as HBF$_4$ to give the syn-1-substituted-η^3-allyl-Fe(CO)$_4$ cation:

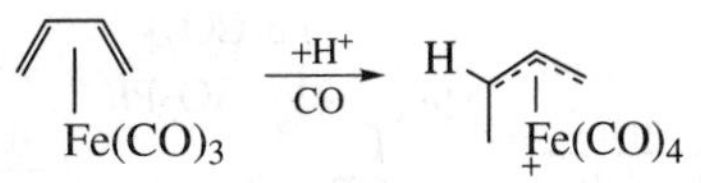

Allylic alcohols are convenient starting materials, using acid (*e.g.*, HBF$_4$) to promote loss of the hydroxyl group in the presence of the Fe(CO)$_4$ unit. The configuration of the allyl complex is determined by the starting allylic alcohol; no syn–anti isomerization is observed under these mild conditions[114]:

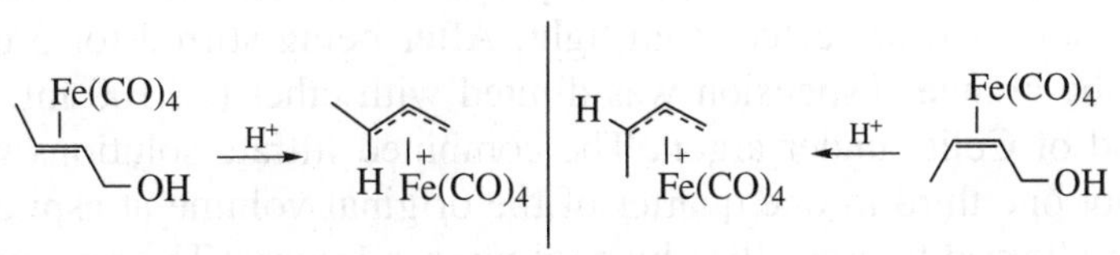

An important subtype is the vinyl-silane/allylic acetate. Preparation of the alkene complex by reaction with Fe$_2$(CO)$_9$ is followed by Lewis acid treatment (BF$_3$) to induce ionization. Again, the configuration of the alkene determines the configuration of the allyl ligand (syn or anti Si)[115]:

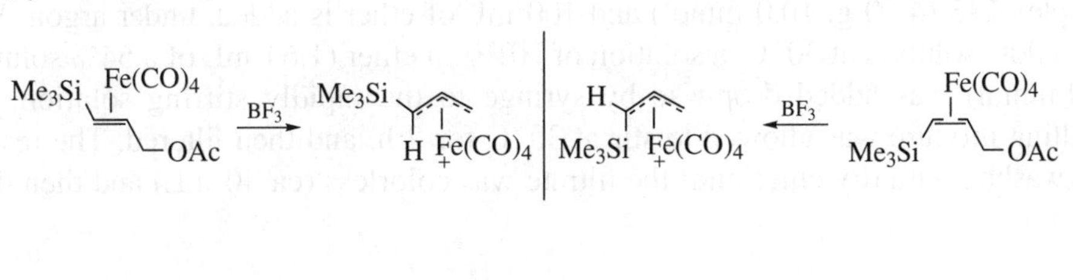

Even with α,β-unsaturated carbonyl compounds bearing a γ-alkoxy group, the configuration of the alkene unit is preserved[116]:

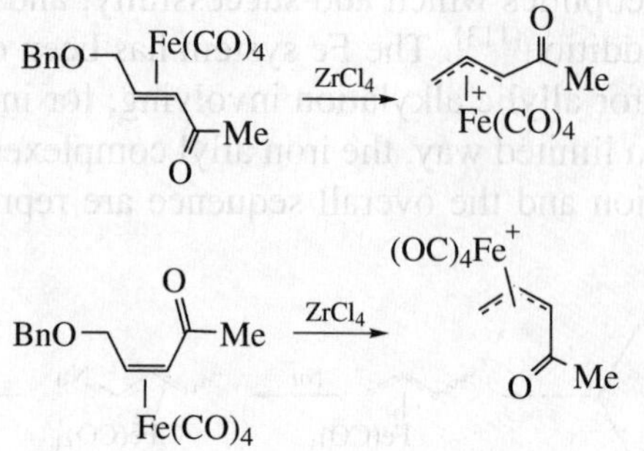

A particularly useful series of complexes is prepared in a multistep procedure from the convenient homochiral α-hydroxy esters, through vinyl sulfones[117]. Complexation with Fe$_2$(CO)$_9$ provides an alkene complex which then can be converted to the *homochiral* allyl-Fe(CO)$_4$ complex by acid treatment which promotes clean anti elimination.

Procedure for Preparation of Allyl-Fe(CO)$_4$ Cation Complexes: (+)-(1R,3R)-Tetracarbonyl[1-3h)-1-(phenylsulphonyl)but-1-ene-1-yl]iron(II) Tetrafluoroborate[108]

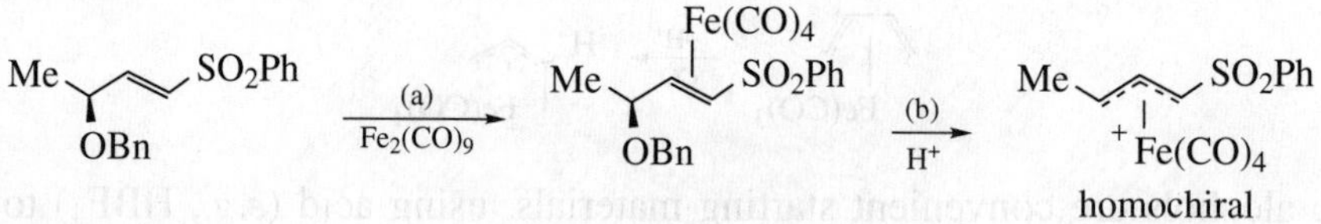

(a) A suspension of the homochiral vinyl sulfone (3.03 g, 10.0 mmol) and non-acarbonyldiiron (4.73 g, 13.0 mmol) was prepared in 150 mL of hexane under CO pressure (balloon) and protected from light. After being stirred for 3 days at 25 °C in the dark, the orange suspension was diluted with ether (ca 100 mL) and filtered through a pad of Celite under argon. The combined filtrate solutions were concentrated to about one third to one quarter of the original volume at aspirator pressure, and was then allowed to crystallize by cooling in a freezer. The pale yellow precipitate was washed with cold (−25 °C) hexane under argon and dried at oil pump vacuum at room temperature. The product 113 is a pale yellow crystalline solid (3.24 g, 69% yield). It is moderately air sensitive.

(b) To a 100-mL Schlenk flask is added a solution of vinyl sulfone Fe(CO)$_4$ complex 113 (4.70 g, 10.0 mmol) and 100 mL of ether is added, under argon. With the yellow solution at 30 °C, a solution of HBF$_4$ in ether (1.64 mL of a 54% solution, 12.0 mmol) was added dropwise by syringe to the rapidly stirring solution. The resulting mixture was allowed to stir at 30 °C for 2 h, and then filtered. The residue was washed with dry ether until the filtrate was colorless (ca 40 mL) and then dried at 25 °C under oil pump vacuum. The resulting pale yellow solid (**114**, 4.32 g, 96% yield) can be used without further purification. If desired, it can be further purified by addition of dry ether to a nitromethane solution. The final complex **114** is stored at −25 °C in a freezer under argon, but can be handled for short periods in air at room temperature.

6.1.3 Nucleophile Addition to the η^3-Allyl-Fe(CO)$_4$ Cations

The cationic allyl-Fe(CO)$_4$ complexes are highly electrophilic and react with a wide range of nucleophiles. In the early studies, stabilized carbanions such as malonate anion as well as simple phosphines and amines, including pyridine, were successful[111]. Regioselectivity rules were suggested, emphasizing a trend toward addition at the less substituted terminal position. Addition of triethylphosphite to **115** occurs rapidly at −20 °C to give adduct **116**:

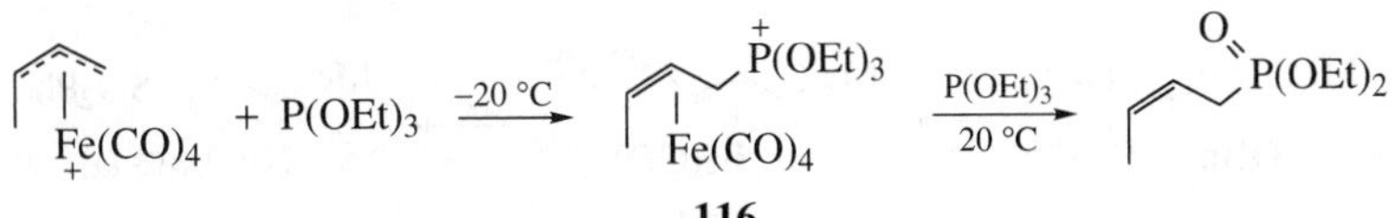

Simple electron-rich aromatics react spontaneously by electrophilic aromatic substitution[118], illustrating a significantly higher electrophilic reactivity for the allyl-Fe(CO)$_4$ cations compared to allyl-Pd^{II} complexes. The regioselectivity trend toward coupling at the less hindered end (with retention of double bond configuration) is clear, but a mixture is obtained:

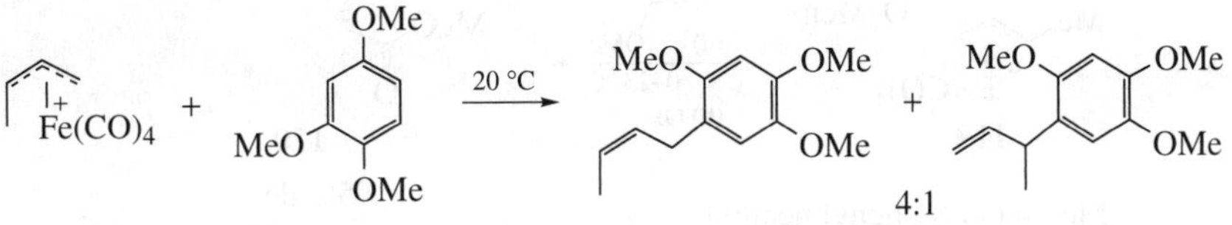

At the other end of the reactivity scale, simple alkylmetal species (LiCuX[119], ZnX[120], etc.) add in moderate yield, and appear more prone to give regioisomeric coupling products when a substituted allyl ligand is used. In addition, CO insertion can precede coupling, suggesting initial attack at the CO ligand or Fe, followed by syn migration to the allyl ligand:

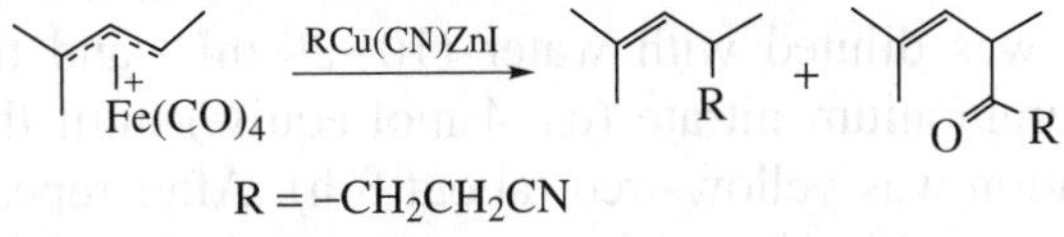

A most useful variation on this process is the reaction of enol acetates (e.g., with complex **117**)[121], enol ethers[122], and allyl-SnR$_3$ compounds[115], with allyl-Fe(CO)$_4$ complexes bearing an electron withdrawing group (keto, carboxylate, cyano, etc.) on one terminus. The iron complex can be generated *in situ*, by reaction of a Lewis acid with an $Fe(CO)_4$ complex of allylic ether:

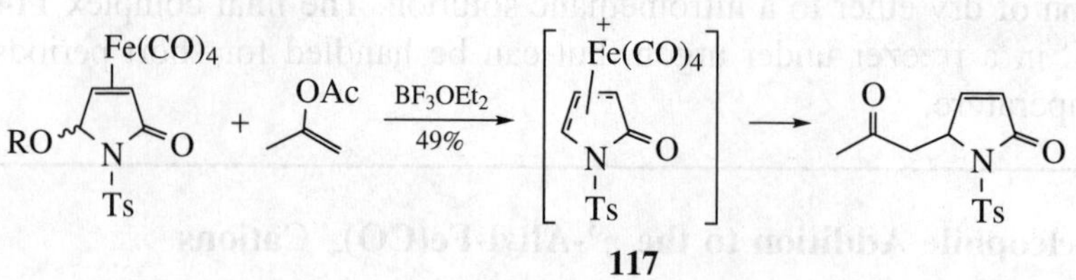

117

An important variation takes advantage of the relative ease of synthesis of homochiral γ-hydroxy-α,β-unsaturated carbonyl derivatives and their conversion into the cationic complexes, as discussed above. The sulfone analog has been particularly well worked out, with a series of nucleophiles[117]. The regioselectivity with these electronically asymmetric allyl ligands (*e.g.*, in **114**) is very high, favoring coupling away from the electron withdrawing substituent, to give **118**:

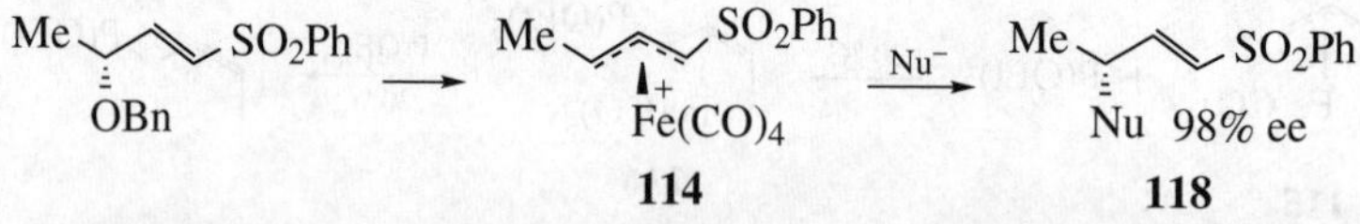

Nu = malonate anion, enol ether, secondary amine, aryl ether, etc.

Procedure for Addition of Enol Ethers to η³-Allyl-Fe(CO)₄ Complexes

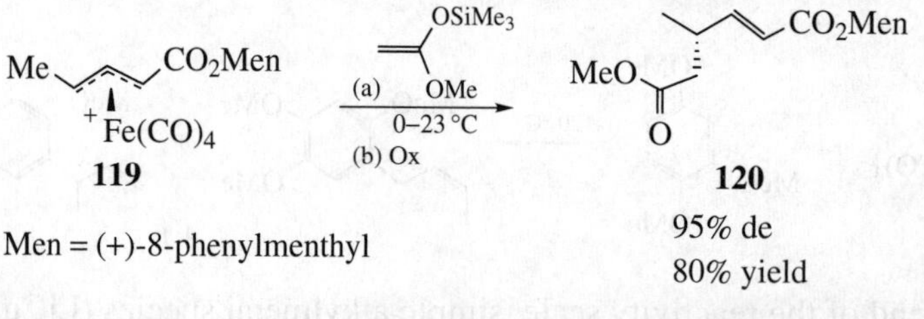

Men = (+)-8-phenylmenthyl

95% de
80% yield

To complex **119** (>99% one diastereoisomer, 1.66 g, 2.65 mmol) in a Schlenk flask under argon at 23 °C was added dichloromethane (10 mL) and the suspension was cooled to 0 °C. To the suspension was added a solution of the TMS ketene acetal from methyl acetate (0.88 g, 6.0 mmol) in dichloromethane (6 mL), and the mixture was allowed to warm to 23 °C. After 1–2 h, the solution was homogeneous yellow. It was diluted with water (10–20 mL) and treated at 0 °C with excess solid ceric ammonium nitrate (ca. 4 mol equiv.) until the evolution of CO ceased and the solution was yellow–red (about 8 h). After repeated extraction with dichloromethane, the combined organic layer was washed with aqueous ammonium fluoride and finally with pH 7 buffer. The organic solution was dried over $MgSO_4$, filtered, and concentrated by rotary evaporation. The residue was purified by flash column chromatography (SiO$_2$, ether/petroleum ether) to give **120** as a colorless oil (0.82 g, 80% yield; $[\alpha]_D^{28} - 9.1°$ (c 2.23, CHCl$_3$))[123].

6.2 Applications of η^1-Acyloxy-η^3-Allyl-Fe(CO)$_3$ Complexes

6.2.1 Introduction and Pathways

The oxidative addition of vinyl epoxides with Fe(CO)$_5$ or Fe$_2$(CO)$_9$ leads under mild conditions to acyloxyallyl complexes **121**, presumably via the zwitterion **122** or equivalent (Scheme 20). Such complexes can be induced oxidatively or thermally to undergo reductive-elimination and produce dihydropyrones or β-lactones. The oxygen can be exchanged for nitrogen as in **123**, which then lead to β-lactams[124]. The easy availability of substituted vinyl epoxides lends generality to this process, and the significance of the product types in medicinal chemistry increases the usefulness.

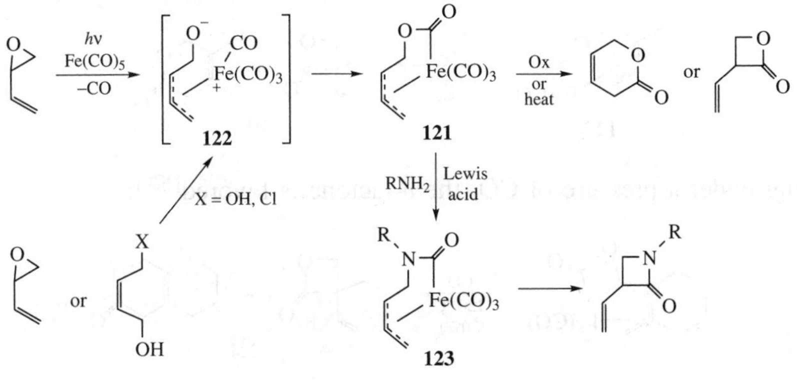

Scheme 20. Lactone and lactam synthesis.

The same acyloxy-allyl complexes can be obtained from vinyl cyclic sulfites (**124**)[125] or from 1,4-disubstituted cis-2-butenes (*e.g.*, **125**)[126]. The use of *cis*-butene-1,4-diols looks very attractive, but a thorough study[126] showed that although the yield of the complex is excellent for the parent case (1,4-dihydroxy-cis-2-butene), it is moderate to poor for a variety of substituted versions.

6.2.2 Preparation of Complexes

The initial oxidative addition is facilitated by creation of a coordinatively unsaturated Fe species, presumably Fe(CO)$_4$L, where L is a donor solvent such as THF. This process is promoted by irradiation if Fe(CO)$_5$ is used, and by simple thermal conditions or ultrasound treatment starting from Fe$_2$(CO)$_9$. The oxidative addition proceeds by inversion[127], and may produce the zwitterion species **122** or equivalent structure, before CO insertion and formation of the acyloxy allyl complex **121** (Scheme 20). The compounds are relatively air stable and can be purified by conventional chromatography. The regioselectivity of the epoxide ring opening is obviously controlled by the formation of the acyloxy-allyl ligand but, in general, two diastereoisomers (**126a** and **126b**) can form. They can be separated by chromatography, and often give different results in the ring-forming, 'demetalation' step:

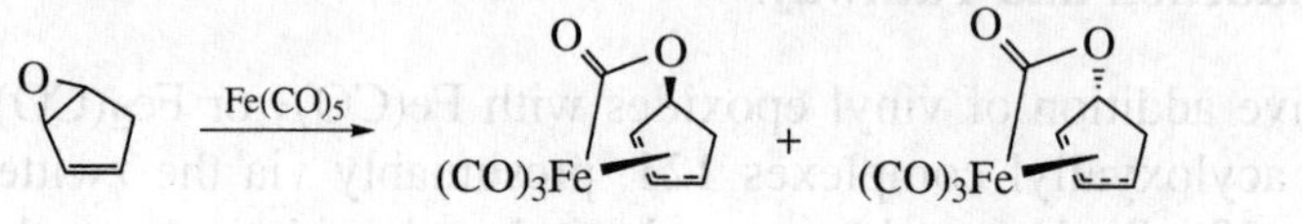

126a   **126b**

6.2.3   Formation of Lactones

Oxidation of an acyloxy allyl lactone Fe(CO)₃ complex (*e.g.*, **127**) favors formation of β-lactones, although a minor amount of the δ-lactone cannot be avoided:

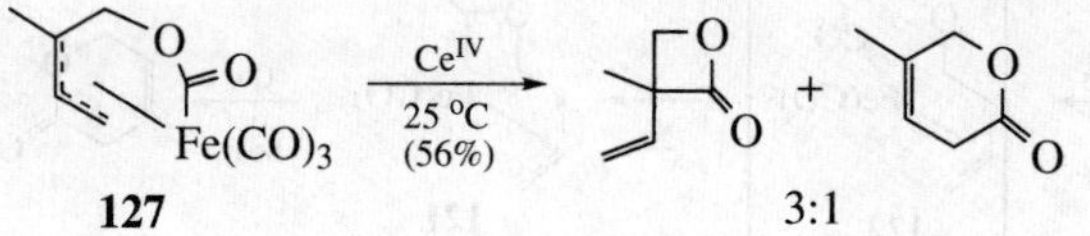

127   3:1

By heating under a pressure of CO, the δ-lactone is favored[128]:

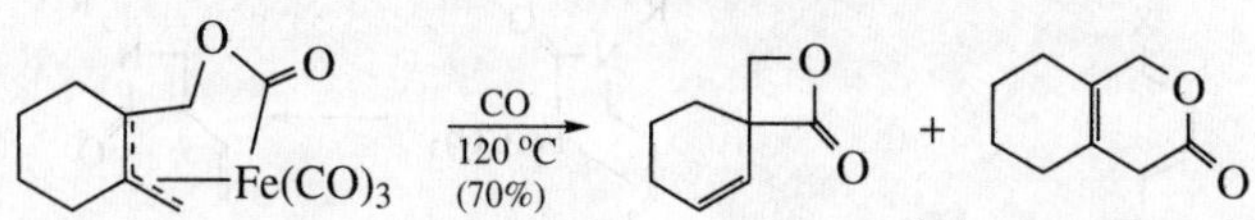

1:4

The ready availability of vinyl epoxides by selective epoxidation of 1,3-dienes opens useful synthesis routes to lactone natural products, such as malyngolide **128**[129]. Selective epoxidation of the allylic alcohol followed by reaction with diiron nonacarbonyl at room temperature produces **129**. Then heating under a pressure of CO leads to the δ-lactone **130** in 74% yield. Hydrogenation finishes off the synthesis of **128**:

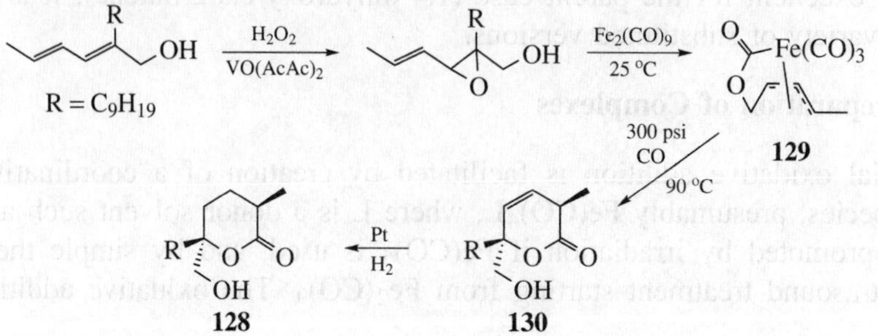

$R = C_9H_{19}$

128   **130**

The development of asymmetric epoxidation opens pathways to asymmetric vinyl epoxides and then homochiral acyloxy allyl iron complexes can be generated. Generation of the δ-lactone is a prelude to spiroketal formation, in natural products such as avermectin B$_{1a}$ and routiennocin[130]. To take advantage of the Sharpless asymmetric epoxidation (AE) technology, the starting material is allylic alcohol **131**. Conventional conversions from **132** produce the vinyl epoxide **133**. Complexation (**134**) followed by CO pressure (250 atm) produced a δ-lactone as a mixture of alkene isomers. Hydrogenation gave the desired homochiral lactone, **135**, which was elaborated to routiennocin:

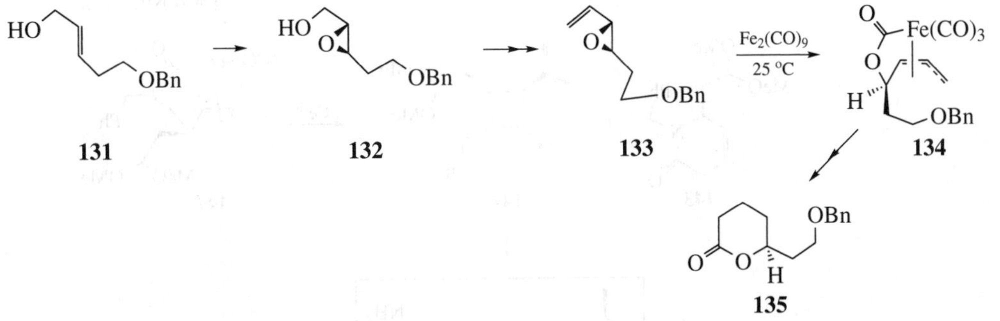

6.2.4 Formation of β-Lactams

The particular importance of β-lactam structures in antibiotic chemistry lends significance to the effort to divert the process from β-lactones to β-lactams. A possible approach, in analogy with that immediately above, is the reaction of vinyl aziridines with iron carbonyls. This process does produce β-lactams, but the relative difficulty in producing the starting vinyl aziridines has limited the applications[131]. A solution to the problem is based on the general reaction with nucleophiles of allyl metal complexes bearing α-leaving groups[132]. With allyl lactone iron complexes (*e.g.*, **136**), a mild Lewis acid is effective in promoting exchange of oxygen for nitrogen (to give **137**)[133]. The mechanism obviously involves several steps, and one sequence is discussed by Harrington[128]. Then oxidation with Ce^{III} induces β-lactam formation in yields of 70–90%:

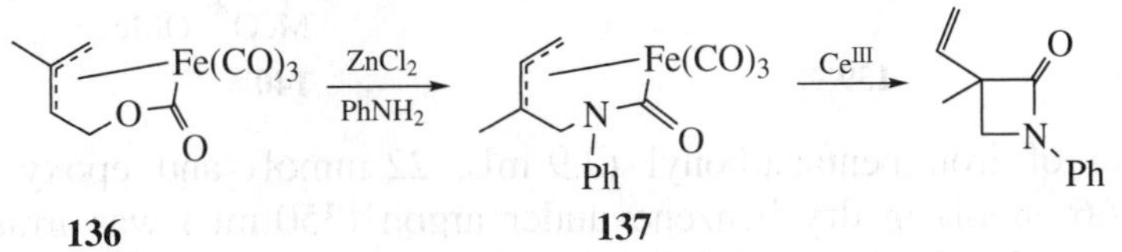

A centerpiece of this methodology development is the total synthesis of thienamycin[134]. The acetal ketone **138** is converted to the allylic epoxide **139** and then allowed to react with diiron nonacarbonyl to give the lactone complex **140** (Scheme 21). Then addition of benzyl amine produces the lactam complex **139** which can be oxidized to the β-lactam **142** (major) and the δ-lactam **143** (minor). Conventional methodology leads to thienamycin.

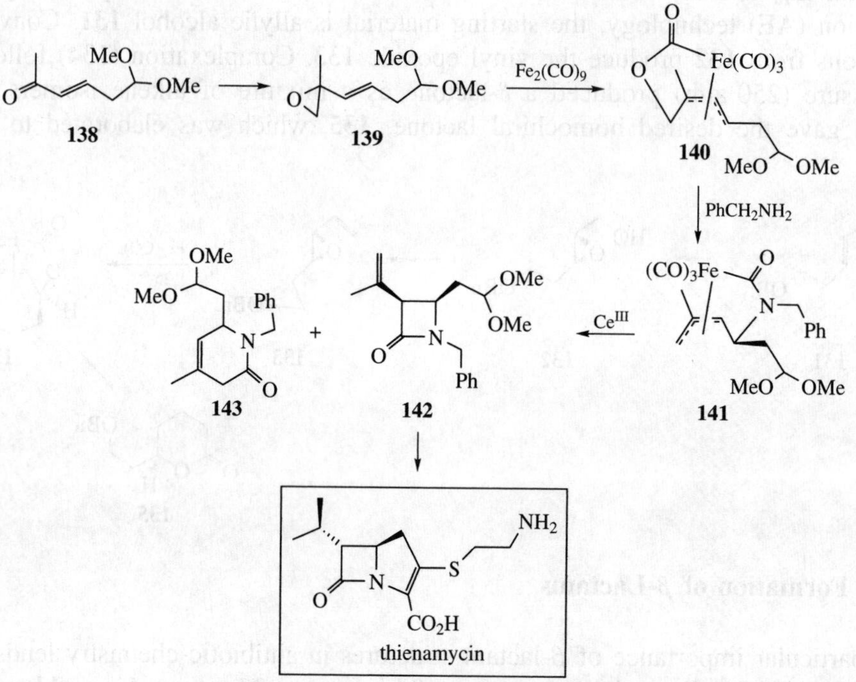

Scheme 21. Synthesis of thienamicin.

---

*Procedure for Preparation of the Allyl Lactone–Fe(CO)₃ Complex **140***

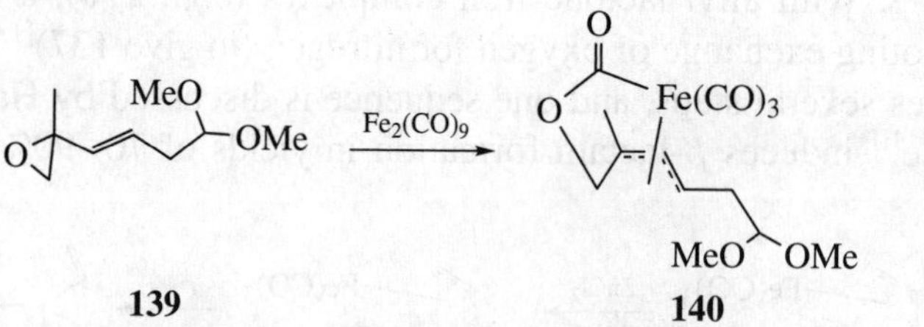

A solution of iron pentacarbonyl (2.9 mL, 22 mmol) and epoxy acetal **139** (0.673 g, 3.66 mmol) in dry benzene under argon (350 mL) was irradiated with a high pressure Hg lamp while following the progress by TLC (ether/hexanes). After 11.5 min, the mixture was filtered, frozen, and freeze dried. The residue was filtered through a Celite pad and chromatographed (SiO₂, ether/hexanes) to give complex **140**. The yield was 1.13 g (90%) as a yellow solid with mp 51–52 °C.

Procedure for Conversion of a Lactone Complex to a Lactam Complex

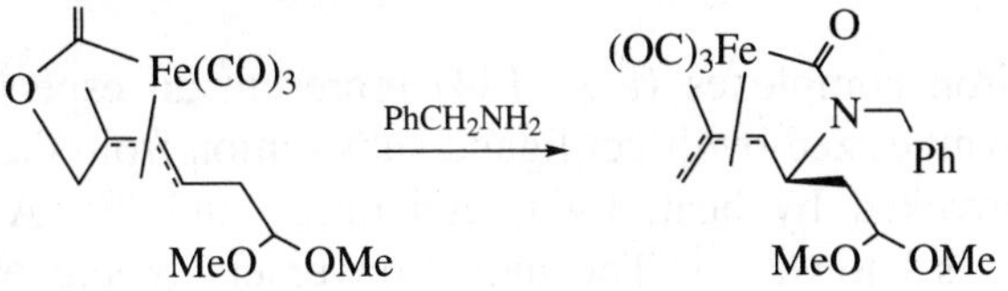

140 → **141**

To a solution of complex **140** (0.99 g, 2.9 mmol) in THF (50 mL) containing $ZnCl_2$ (25 mL of a 7.25 M solution in ether) under argon at 23 °C was added a solution of benzyl amine (3.2 g, 30 mmol) in ether (50 mL). After 6 h, the mixture was quenched with dilute aqueous citric acid and extracted with ether. After being dried over Na_2SO_4, the ether solution was filtered and concentrated by rotary evaporation. The residue was purified by column chromatography (SiO_2, ether/hexanes) to give complex **141** (0.72 g, 57%) as a yellow oil.

Procedure for Oxidative Decomplexation of ***141*** *to form β-Lactam* ***142***

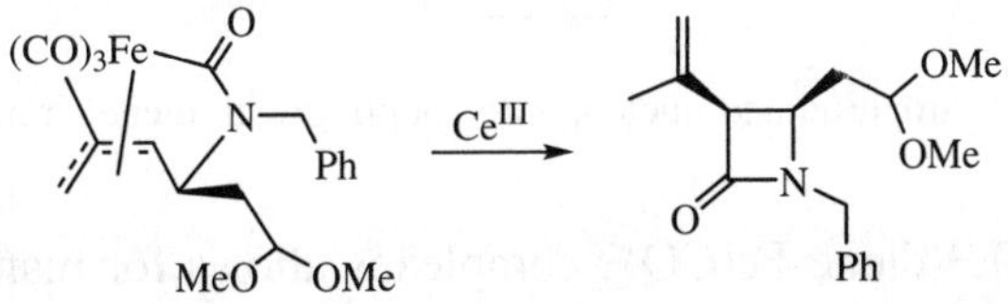

141 → **142**

A solution of complex **141** (0.68 g, 1.59 mmol) in MeOH (35 mL) was oxidized at −30 °C by adding a solution of ceric ammonium nitrate (8.66 g, 15.8 mmol) in MeOH (70 mL). After allowing the mixture to warm to 23 °C over 1 h, the solvent was removed under reduced pressure and the residue was partitioned between ether (100 mL) and water (20 mL). The organic layer was washed with brine, dried, and concentrated by rotary evaporation to leave a residue which was purified by column chromatography (SiO_2, ether/hexanes) to give the β-lactam **142** (0.24 g, 64%) as a colorless oil.

It should be noted that the allyltricarbonyliron lactone and lactam complexes are robust and can serve as activating and directing groups for synthesis transformations on side chains, such as Mukaiyama aldol reactions of adjacent carbonyl groups[135].

7   Applications of η^4-1,3-Diene-Fe(CO)$_3$ Complexes

7.1   Introduction and Preparation of the Complexes

The η^4-1,3-diene iron complexes (*e.g.*, **144**) represent an especially stable bonding mode and are easily synthesized by direct ligand substitution from the iron polycarbonyls, such as Fe(CO)$_5$, provoked by heat, light, and ultrasound[136]. A long list of cyclic and acyclic examples is known[137]. The most convenient precursor is Fe(CO)$_5$ which usually undergoes complexation by direct reaction at 100 °C or so. Lower temperatures are sufficient with concurrent irradiation of the Fe(CO)$_5$, and the more reactive sources of Fe(CO)$_3$, such as Fe$_2$(CO)$_9$ and Fe$_3$(CO)$_{12}$, also react at lower temperature. A new recipe which gives very high yields involves the addition of a 1-azabuta-1,3-diene as catalyst[138]. The pi orbital system is highly polarizible, and can react with electrophiles via electrophilic addition (**145**) and loss of a proton (as in classical electrophilic aromatic substitution) to give simple substitution (**147**), and with nucleophiles by addition to the pi system giving an η^3-allyl anionic intermediate **148** or an η^3-homoallyl anionic intermediate (**149**; Scheme 22).

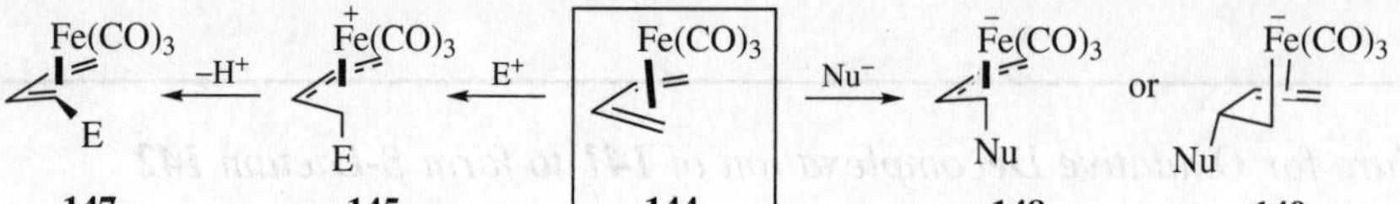

Scheme 22. Electrophilic and nucleophilic coupling with diene–iron complexes.

The stability of the 1,3-diene-Fe(CO)$_3$ complexes allows for many synthesis reactions to be carried out on the side chains of substituted versions, and the steric effect of the Fe(CO)$_3$ units can influence the introduction of new stereogenic centers at nearby positions[139].

Procedure for Photochemical Preparation of η^4-(1,3-Cyclohexadiene)Fe(CO)$_3$

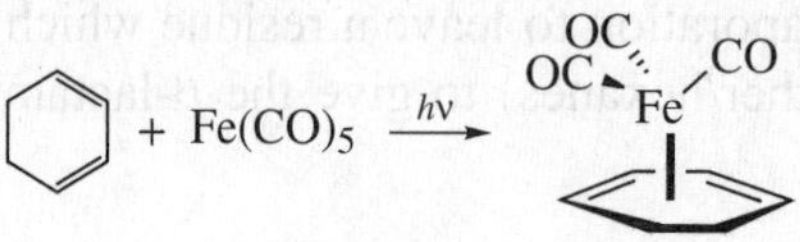

In a 100-mL flask was placed 1,3-cyclohexadiene (29.4 g, 0.15 mol) and Fe(CO)$_5$ (12.3 g, 0.15 mol) under argon at 23 °C. The mixture was irradiated by placing a Hanovia medium pressure Hg lamp (in a water-cooled jacket) close to the flask. After 8 h of irradiation, the mixture was directly distilled (8-cm Vigreux column)

and the fraction of bp 75–80 °C/5 torr was collected. The desired complex, η^4-(1, 3-cyclohexadiene)Fe(CO)₃, was obtained as an orange liquid (58% yield, 17.5 g). It solidified in the cold (mp 8–9 °C)[140].

Procedure for Thermal Preparation of η^4-(2-Methoxy-1,3-cyclohexadiene)Fe(CO)₃ **150** *by Rearrangement of a 1,4-Cyclohexadiene*

150

A mixture of 1-methoxy-1,4-cyclohexadiene (3.9 g, 35 mmol), iron pentacarbonyl (6.5 mL, 49 mmol), and dibutyl ether (35 mL) under argon was heated at reflux for 18 h. The mixture was then cooled to 23 °C, filtered, and the filtrate was concentrated to remove the solvent and excess reactants (heating on a rotary evaporator at aspirator vacuum). The residue was distilled through a short Vigreux column to give a center cut of yellow liquid iron diene complex **150** (bp 78–80 °C/0.2 torr; 5.5 g, 62% yield).

7.2 Addition of Nucleophiles to η^4-1,3-Diene–Iron Complexes

7.2.1 Mechanism and Scope of the Nucleophile

Quite reactive anions are required to convert the diene–Fe(CO)₃ complexes to anionic intermediates, consistent with the fact that the iron is in the zerovalent state. The only known examples involve carbanions derived from carbon acids with pK_a between about 22 and about 35[141]. Simple (unstabilized) carbon nucleophiles such as alkyl Grignard reagents and alkyllithium reagents show competitive addition to the CO ligand and are not useful. Useful carbanions are those stabilized by one cyano, one carboalkoxy, or two sulfur units (*e.g.*, 2-lithio-1,3-dithiane). The prototype reaction is shown with η^4-(1,3-cyclohexadiene)Fe(CO)₃ and 2-lithio-1,3-dithiane ion (Scheme 23). Three intermediates (**151, 152, 155**) are accessible, depending on the conditions. The kinetic product is the anionic homoallyl complex **151**, formed by a fast reaction at −78 °C via anti addition at an internal carbon. It is not stable, even with very reactive nucleophiles such as diphenylmethyllithium, and reverts to the reactants at higher temperatures, such as −30 °C[142]. Then a second (slower but more favorable) addition occurs at a terminal position to produce the η^3-allyl anionic complex **152**. The anionic intermediates are highly reactive toward oxygen and are not generally isolated nor characterized. They can be trapped

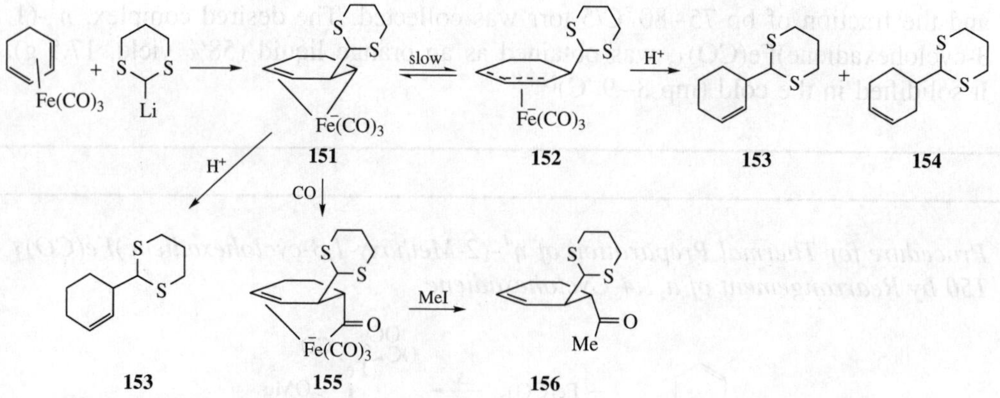

Scheme 23. Addition of nucleophiles and further processing of anionic intermediates.

by protonation, giving labile alkene–Fe complexes which spontaneously lose the alkene ligands. The predominant products are shown (**153** and **154**), but alkene positional isomers are also observed in minor amounts. The complex **151** can be seen as an analog of the first intermediate in Collman's reaction (above), and is observed to undergo parallel reactions. For example, with 1.1 atm (balloon) of CO, the acylate anionic complex **155** is formed by syn addition (analogous to the key intermediate in Collman's reaction). Then reaction with MeI (or other reactive alkylating agents; H$^+$ to give a CHO group; oxidizing agents to give –CO$_2$H) leads to the methyl ketone **156**, maintaining the trans relationship of the new substituents[143]. This latter process is potentially very useful. The equilibration between the homoallyl and allyl complexes appears to be general for all types of 1,3-diene ligands[144].

Procedure for Formation of 1,2-Disubstituted Cyclohexenes

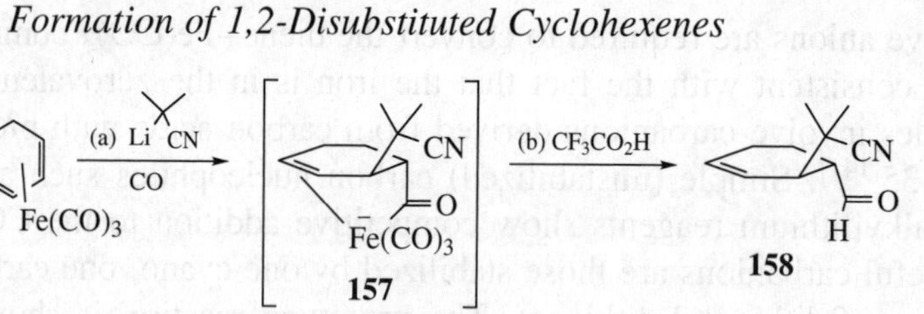

(a) To a solution of lithium diisopropylamide (2.75 mmol) in THF (8 mL) under argon at −78 °C was added neat dropwise via syringe 2-methylpropionitrile (2.95 mmol). The mixture was allowed to stir at −78 °C for 20 min, and HMPA was added (2 mL). Carbon monoxide was introduced into the solution via a syringe needle and was pressurized to about 2 psi above 1 atm. The flask was held at this positive pressure simply by tightening a wire around the septa. After a few minutes, the CO was released through another needle, and CO was bubbled through the solution for 30 s or so. The neat liquid complex, cyclohexa-1,3-diene-Fe(CO)$_3$ (0.500 g,

0.368 mmol, 2.3 mmol), was added via syringe over a period of 20 s. Then the gas exit needle was removed and the closed system was pressurized with CO to about 8–10 psi above 1 atm. The mixture was stirred at −78 °C for 1 h, and at 25 °C for 1 h, after which time it was again cooled to −78 °C, the CO needle was removed, and the system was depressurized by insertion of a needle into the septum. The needle was quickly removed when the gas efflux was over.

(b) To this solution containing **157**, trifluoroacetic acid (1.5 mL, 2.95 mmol) was added neat over 30 s, and the mixture was allowed to stir for 10 min at −78 °C. The mixture was poured into 20% aqueous acetone and ceric ammonium nitrate was added a solid in small portions until gas evolution was no longer evident. The mixture was poured into saturated aqueous sodium acetate solution, and the aqueous layer was washed three times with ether. The combined ether solution was washed with brine, dried over magnesium sulfate, filtered, and concentrated by rotary evaporation. The residue was purified by short path distillation to give **158** as a colorless oil (0.37 g, 92% yield).

7.2.2 Carbonylation of the Anionic Intermediate

There is another pathway open to simple 1,3-diene ligands[145]. For example, with (isoprene)Fe(CO)$_3$, the intermediate from internal addition and CO insertion is **159**, which is unstable toward cyclization via alkene insertion (Scheme 24). In this mechanistic rationalization, the first cyclized product is the β-Fe derivative **160**, which is coordinatively unsaturated and prone to β-hydrogen elimination, to give **161**. Then re-addition with the opposite regioselectivity gives **162**, which is effectively an iron-enolate anion. The structure is written arbitrarily as **162**, but the actual structure may involve coordination with the oxygen. Final treatment with mild acid cleaves the iron and gives the cis-3,4-disubstituted cyclopentanone **163**.

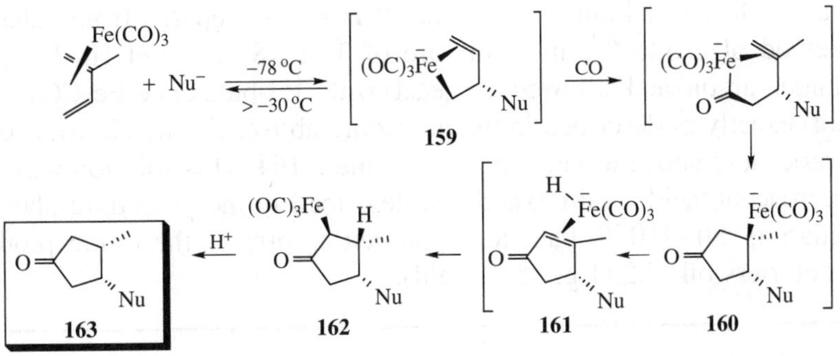

Scheme 24. Formation of cyclopentanones.

This pathway has been tested on a few substituted 1,3-dienes and is found to proceed efficiently when one alkene unit is monosubstituted. With 1,2-disubstituted alkenes, the addition/carbonylation proceeds without cyclization, presumably due to the inhibition of insertion into the heavily substituted alkene unit. If both internal positions are substituted, the nucleophile addition occurs only at the terminal positions, giving an η^3-allyl intermediate directly, with no tendency for carbonylation:

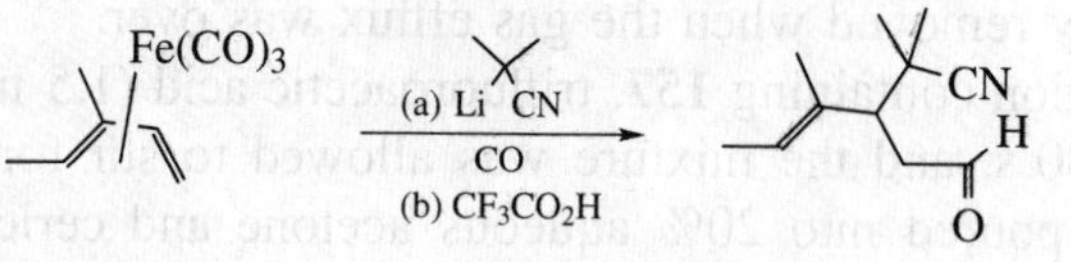

If both terminal positions are substituted, the alkene insertion/cyclization occurs to give a mixture of isomers:

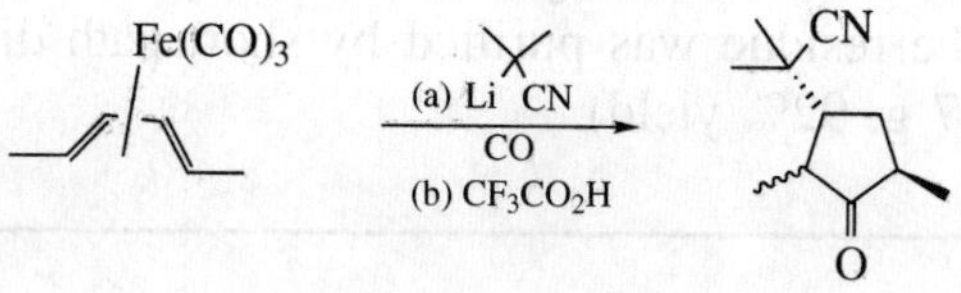

Procedure for Formation of a 3-Substituted Cyclopentanone

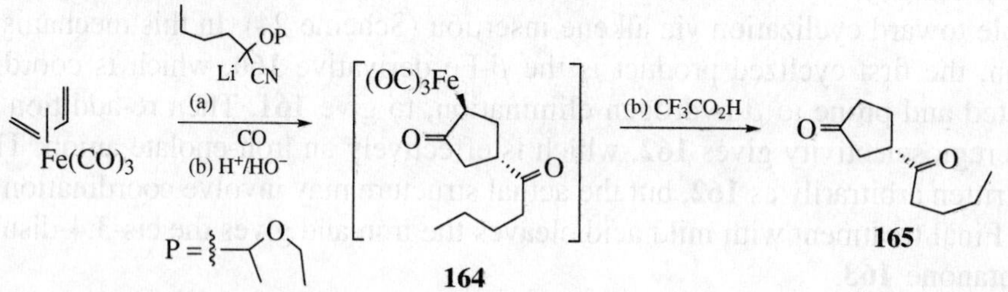

The cyanohydrin acetal anion (2.75 mmol) **164** was prepared from valeraldehyde by the method of Stork[146], in a mixture of THF (8 mL) and HMPA (2 mL) at $-78\,^\circ$C under argon and allowed to react with 1,3-butadiene-Fe(CO)$_3$ (0.444 g, 2.29 mmol) exactly as described in the procedure above. CO was introduced exactly as in the procedure above to generate intermediate **164**. This solution was quenched with trifluoroacetic acid, again exactly as described in the procedure above. Short-path distillation (80–110 $^\circ$C bath temperature/0.8 torr) of the crude product gave **165** as a colorless oil (0.294 g, 85% yield).

The general regioselectivity is dominated by steric effects, with nucleophile addition at the internal position away from the larger terminal substituents. The methoxy substituent is potentially useful, as the products are enol ethers which can be converted into ketones:

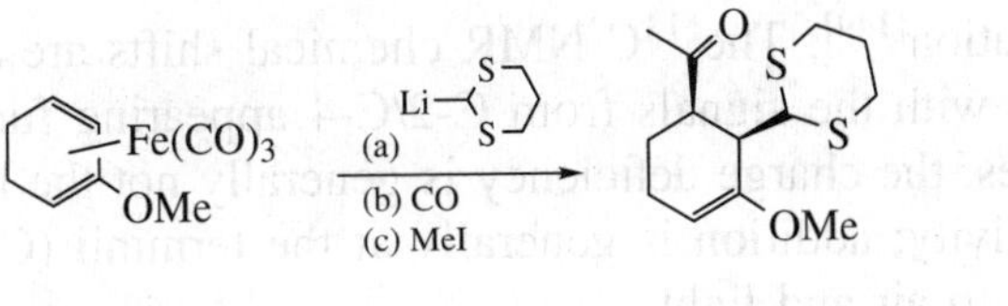

These reactions are stoichiometric, and it is difficult to see how a catalytic cycle could be constructed. At the same time, the iron reagents are inexpensive and relatively easy to dispose of from batch-wise processes. No significant applications in more complex synthesis have been reported.

8 Nucleophile Addition to η^5-Pentadienyl-Fe(CO)$_3$ Cationic Complexes

8.1 Introduction

Iron forms very stable cationic complexes with a pentadienyl ligand and it was established early in the study of the simplest derivatives that the complexes are highly electrophilic[147]. They are air-stable crystalline salts, often soluble in typical organic solvents. The parent pentadienyl-Fe(CO)$_3$ complex **166** reacts directly with electron-rich aromatic rings via electrophilic aromatic substitution at a terminal position to give the product 1,3-diene-Fe(CO)$_3$ species:

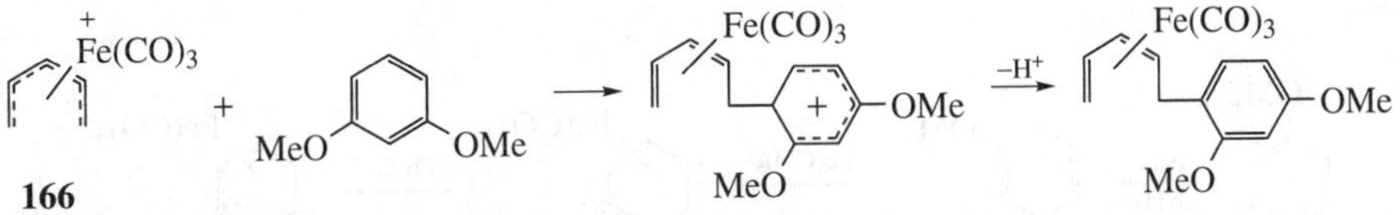

The complexes show a planar five-carbon unit bound to the iron, with one CO staggered at the open end of the dienyl ligand[148]. In the cyclohexadienyl case, the CH$_2$ group at C-6 is tilted away from the iron by 39°. The Me substituents in the open-chain case are tilted toward the metal by a small amount (9°). Substituents at C-1 and C-5 in the open chain case can exist in the exo (**167**) or endo (**168**) arrangement:

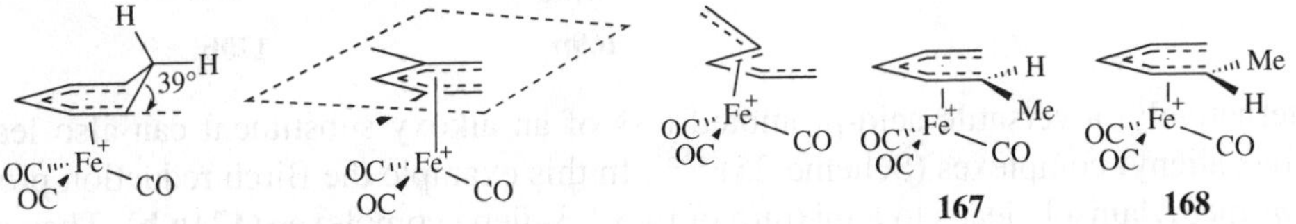

In general, endo is less stable than exo, but the barrier between them can be high, and both isomers can generally be obtained in pure form. Calculations (extended Huckle theory) suggest that the greatest charge deficiency is at C-2/C-4, with more electron density at C-3, and the least positive charge at C-1/C-5, consistent with the typical picture for a pentadienyl cation[149]. The ^{13}C NMR chemical shifts are also consistent with this charge distribution, with the signals from C-2/C-4 appearing furthest downfield[150]. As seen in the examples, the charge deficiency is generally not the rationale for nucleophile addition regioselectivity; addition is generally at the terminii (C-1/C-5). The complexes are generally stable to air and light.

Substituted versions of the open chain pentadienyl ligands have been prepared, evaluated as electrophiles, and applied in synthesis[151]. Their application has been more recent compared to the cyclohexadienyl-Fe(CO)$_3$ cationic complexes, and will be discussed in the second part.

8.2　Cyclohexadienyl-Fe(CO)$_3$ Cationic Complexes

8.2.1　Preparation of Complexes

The reactivity suggested in the above discussion is obviously very useful, but little development of the general process occurred until it was recognized that the starting complexes were readily available from Birch reduction products, which are themselves available in wide variety. Birch reduction products are 1,4-cyclohexadiene derivatives, and they react with Fe(CO)$_5$ with heating via hydrogen migration and formation of the 1,3-cyclohexadiene complex (*e.g.*, **169a,b**). Then hydride abstraction, with trityl cation, for example, produces the cyclohexadienyl-Fe(CO)$_3$ cationic complexes **170a** and **170b**[152]:

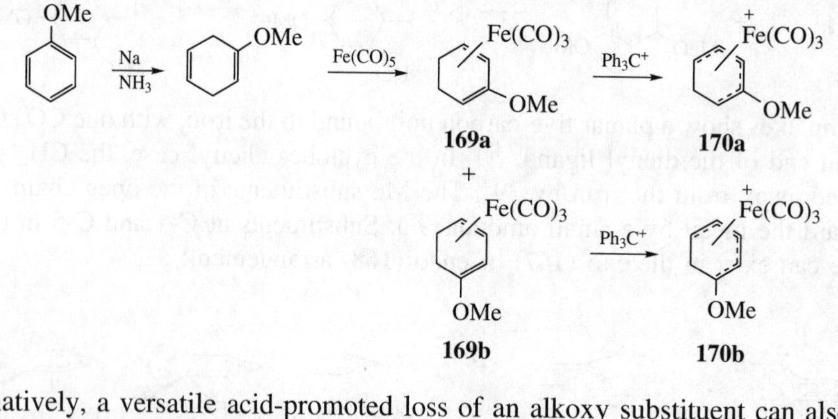

Alternatively, a versatile acid-promoted loss of an alkoxy substituent can also lead to cyclohexadienyl complexes (Scheme 25)[153]. In this example the Birch reduction product from *m*-methylanisole leads to a mixture of two 1,3-diene complexes (**171a,b**). Then addition of strong acid, such as sulfuric acid, initiates a protonation–deprotonation sequence which eventually leads to **172** with the MeO substituent at an sp^3 carbon; Loss of–OMe anion gives the methyl substituted cationic cyclohexadienyl-Fe(CO)$_3$ complex.

In a related process, a 1,3-cyclohexadiene-Fe(CO)$_3$ complex with an external hydroxymethyl substituent (*e.g.*, **173**) can undergo acid-promoted loss of −OH and hydrogen

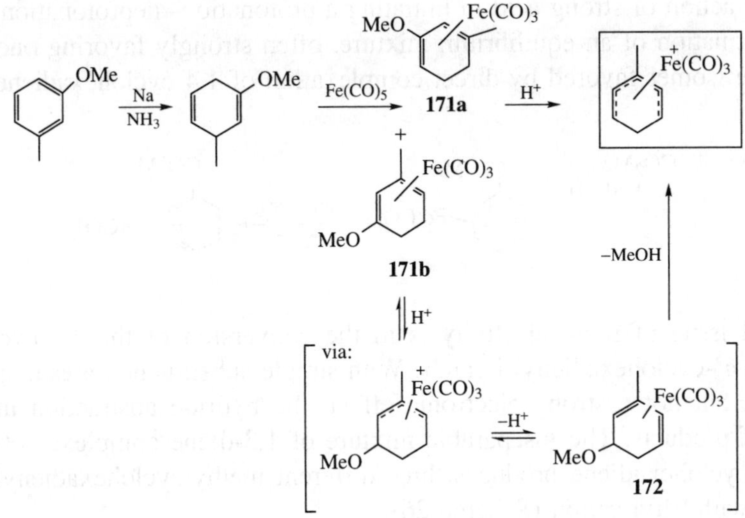

Scheme 25. Acid-promoted loss of MeO to prepare cyclohexadienyl complexes.

migration to give 1-methylcyclohexadienyl-Fe(CO)$_3$ complex **174**[154]:

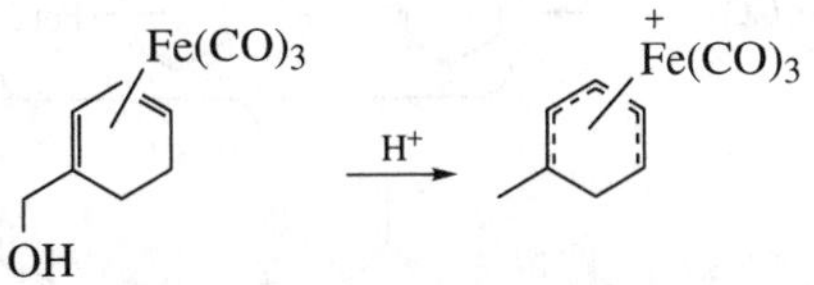

Regioselectivity is an important question at two stages of the complexation process. First, the conversion of Birch reduction products (substituted 1,3-cyclohexadienes) by direct reaction with Fe(CO)$_5$ generally leads to at least two positional isomers in significant amounts (as in Scheme 25). For example, the Birch reduction product from anisole reacts with Fe(CO)$_5$ to produce both the 1-methoxy and the 2-methoxy-1,3-cyclohexadiene ligands (**169a** : **169b** = 1 : 1). Similarly, the product from methyl benzoate leads to a 3 : 1 mixture from which the isomers can be separated by careful chromatography, as they are thermally and (moderately) air stable[154]:

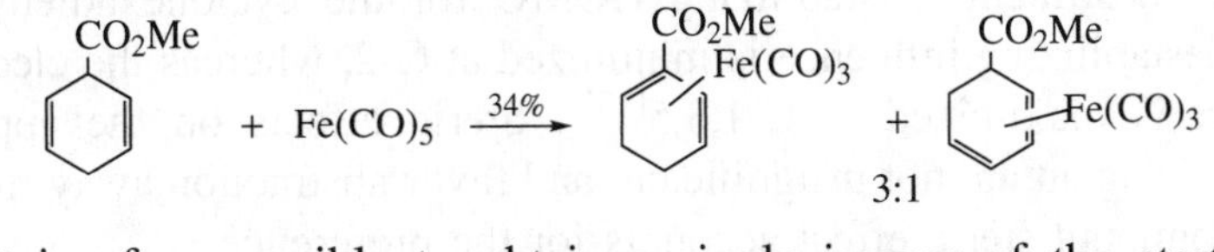

However, it is often possible to obtain a single isomer of the starting substituted 1,3-cyclohexadiene, and in that case complexation takes place without significant isomerization:

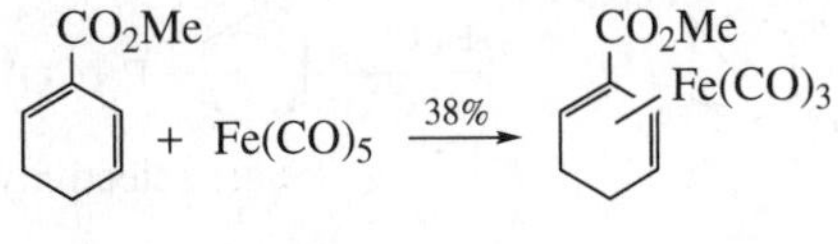

Finally, the action of strong acid in initiating a protonation–deprotonation equilibrium can induce formation of an equilibrium mixture, often strongly favoring one isomer and not always the isomer favored by direct complexation of 1,4-cyclohexadienes:

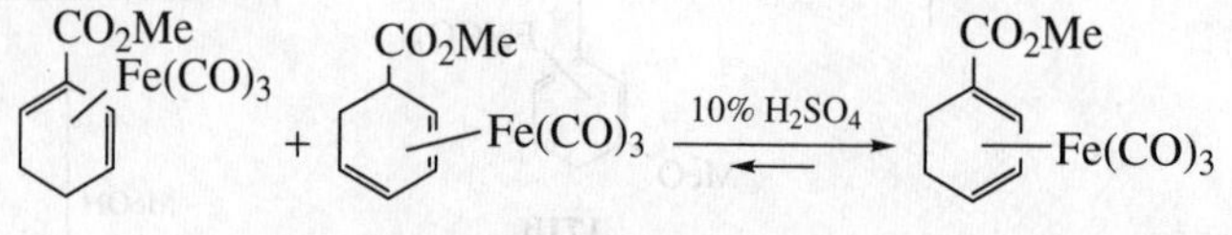

The second issue of regioselectivity is in the conversion of the 1,3-cyclohexadiene ligands to the η^5-cyclohexadienyl ligands. With simple substituents, presumably exerting a steric effect but not a strong electronic effect, the hydride abstraction method leads to mixtures of products. The inseparable mixture of 1,3-diene complexes obtained from 1-methyl-1,4-cyclohexadiene produces three different methylcyclohexadienyl complexes when treated with trityl cation (Scheme 26).

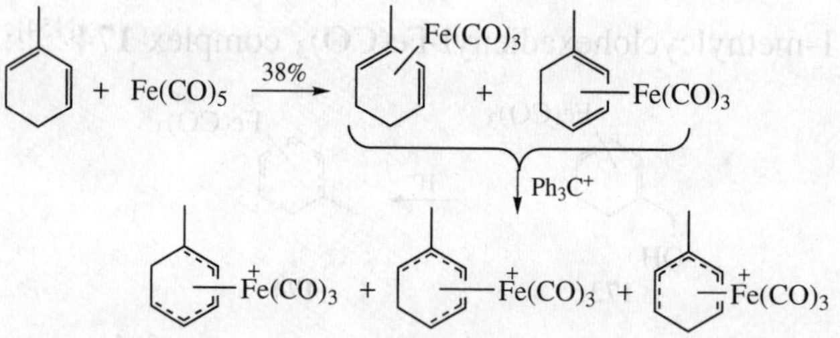

Scheme 26. Regio-isomeric complexes by hydride abstraction.

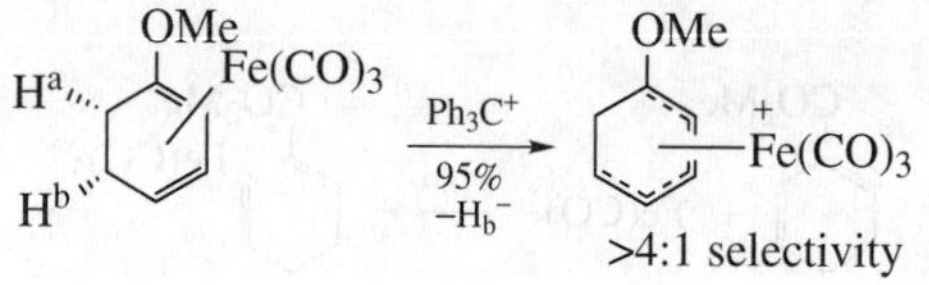

With strong electronic perturbation of one substituent, the regioselectivity of hydride abstraction can be high and very useful. In general, electron donors such as OMe will prefer to sit at the 2 position, whereas electron-accepting groups (CO$_2$Me, SiR$_3$) are favored in the 1,3,5 positions. An analysis has been made which emphasized the influence of the substituent in stabilizing HOMO for the cyclohexadienyl ligand: the electron donor destabilizing influence is minimized at C-2, whereas the electron acceptor stabilizing effect is maximized at C-1,3,5[155]. Steric effects on the approach of the hydride abstracting agent are not insignificant, and favor abstraction away from a terminal substituent. A dominant steric effect accounts for the preference whereas a dominant electron effect affords the following:

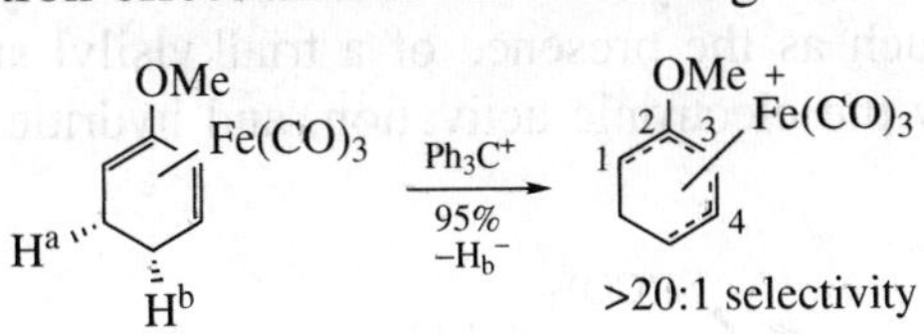

>20:1 selectivity

The electronic effect of a substituent is shown below, where the carbomethoxy group produces the complex with the substituent in the 3-position:

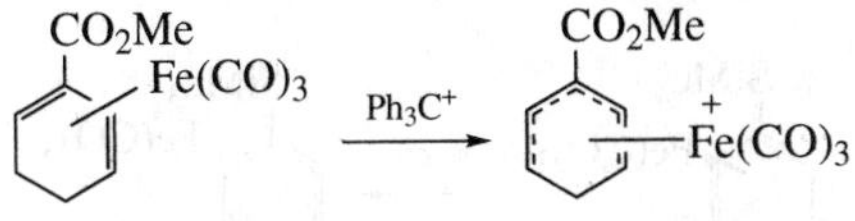

>4:1 selectivity

Again, the steric effect is used to rationalize the formation of the 1-substituted analog:

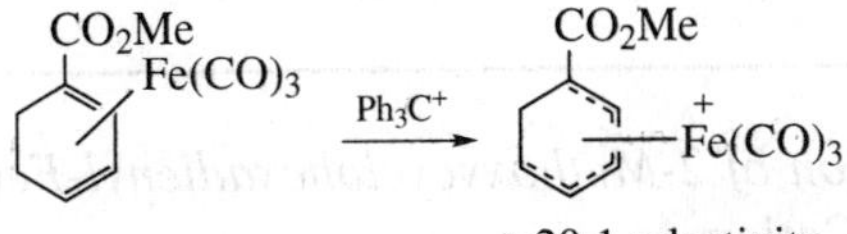

>20:1 selectivity

In these processes, the trialkylsilyl substituent behaves as an electron-withdrawing group, giving the 3-substituted product:

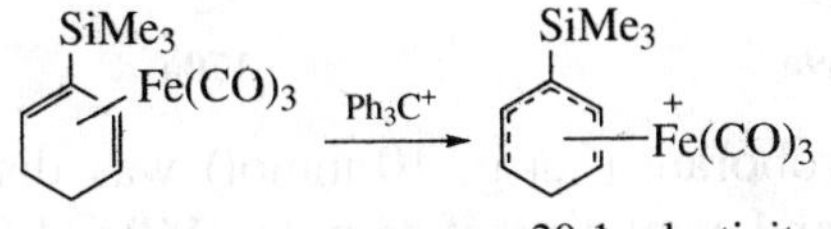

>20:1 selectivity

The alternate technique of acid-promoted loss of $-OH$ or $-OR$ is highly regiospecific. For example, *o*-methyl anisole (**175a**) proceeds through the usual Birch reduction/iron complexation/acid treatment to give the 2-methyl derivative **176a** exclusively. Then *m*-methylanisole (**175b**) follows the same sequence to give exclusively the 3-methyl derivative **176b**. By the same mechanism, the *p*-methyl isomer (**175c**) also gives complex **176a**:

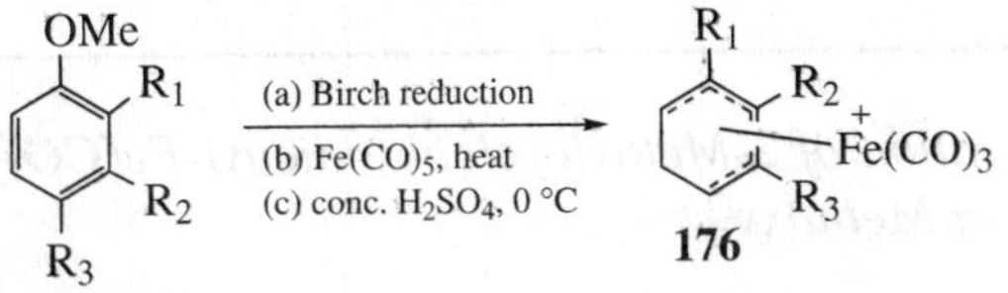

series (**a**) $R_1 = Me$; R_2, $R_3 = H$
series (**b**) R_1, $R_3 = H$; $R_2 = Me$
series (**c**) R_1, $R_2 = H$; $R_3 = Me$

It is worth noting that the hydride abstraction mechanism is sensitive to structural effects in the reactant. Generally, if the 1,3-cyclohexadiene ligand bears an exo substituent at

C-5 (the Me group in **177**), the hydride abstraction with trityl cation fails. However, with strong activation, such as the presence of a trialkylsilyl substituent **178**, the steric repulsion is overcome by the electronic activation, and hydride abstraction is successful to yield **179**:

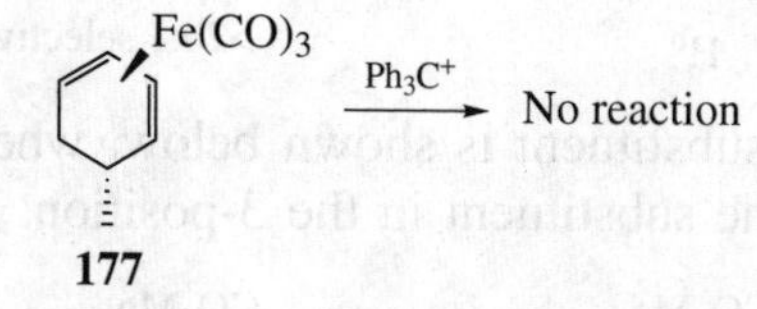

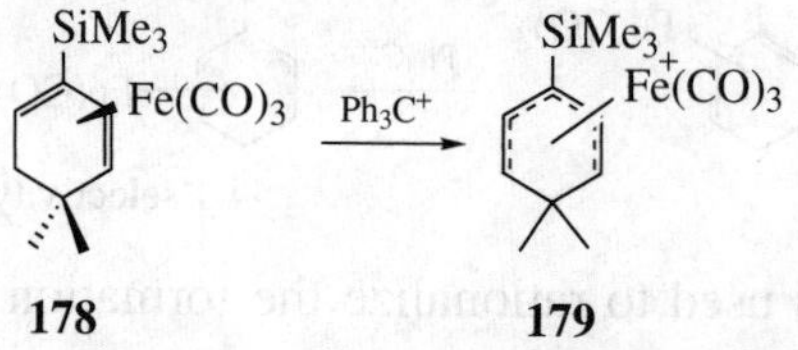

177

178   **179**

Procedure for Preparation of 2-Methoxycyclohexadienyl-Fe(CO)₃⁺ by Hydride Abstraction with Trityl Cation

169a   **170a**

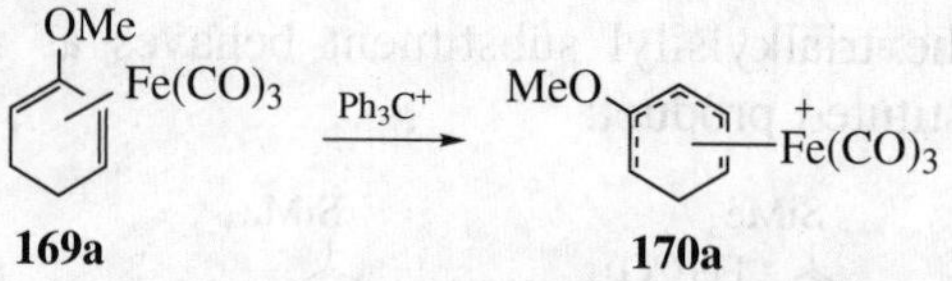

Triphenylmethyl fluoroborate (3.4 g, 10 mmol) was dissolved in 20 mL of dichloromethane at 23 °C and a solution of complex **169a** (1.8 g, 7.2 mmol) in 20 mL of dichloromethane was added. The mixture was allowed to stand for 30 min, and then it was poured into 120 mL of ether with stirring. The resulting precipitate was collected by filtration and washed with ether. The yellow solid residue (**170a**, 2.2 g, 91% yield) was pure enough for further applications, but can be purified by recrystallization from water[156].

Procedure for Preparation of 2-Methylcyclohexadienyl-Fe(CO)₃⁺ by Acid-promoted Loss of Methoxy[152]

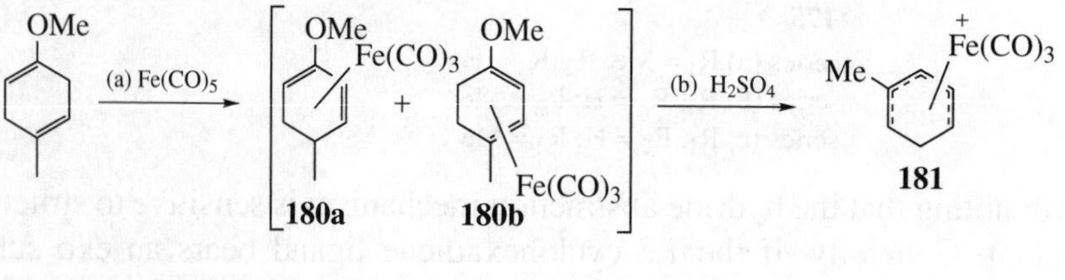

180a   **180b**   **181**

(a) A solution of the Birch reduction product from *p*-methylanisole (1-methoxy-5-methylcyclohexa-1,4-diene, 10 mmol) and Fe(CO)$_5$ (11 mmol) in 50 mL of dibutyl ether under argon was heated at reflux for 10 h. The solvent was removed under vacuum, and then the product was distilled (bp 78–80 °C/3.5 torr) to give a mixture of **180a** and **180b** as a yellow oil (3.3 mmol, 33% yield).

(b) The mixture of 1,3-diene complexes **180a/b** above was mixed with concentrated sulfuric acid (1.0 mL) and allowed to stand at 23 °C for 15 min. Ether was added (50 mL) giving a precipitate as an unfilterable gum. The ether was decanted and the residue was triturated with portions of ether. The residue was then dissolved in water (15 mL) and the solution was extracted with ether. Ammonium hexafluorophosphate (10% aqueous solution) was added to precipitate the desired salt as the hexafluorophosphate derivative. It was filtered and dried, 72% yield.

8.2.2 Nucleophile Addition to η^5-Cyclohexadienyl-Fe(CO)$_3$ Cationic Complexes

8.2.2.1 Range of Nucleophiles

The cyclohexadienyl-Fe(CO)$_3$ cationic complexes are highly versatile electrophiles, reacting with a wide range of nucleophiles including (Scheme 27):

- *heteroatom*—alkoxide, amines, phosphines, phosphites, sulfites (path a);
- *neutral carbon*—simple ketones (enol), enamines, silyl enol ethers, allyl silanes, silyl ketene acetals, electron-rich arenes (path b);
- *formally carbanions*—organo-Cu, -Zn, and -Cd, lithium and tin enolates of simple ketones, as well as highly stabilized enolates (malonate, β-ketoester), alkynylborates (path c).

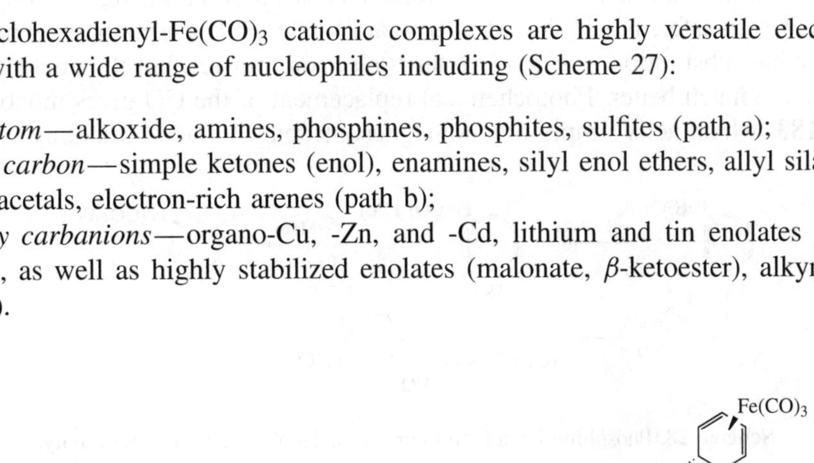

Scheme 27. Examples of nucleophile addition.

An interesting selectivity is shown by the following reaction of the electron-rich arene with the 3-methoxycyclo-hexadienyl complex. Reaction might have occurred via the nitrogen nucleophile, but the arene ring carbon is the reactive site, giving the adduct in 96% yield[157]:

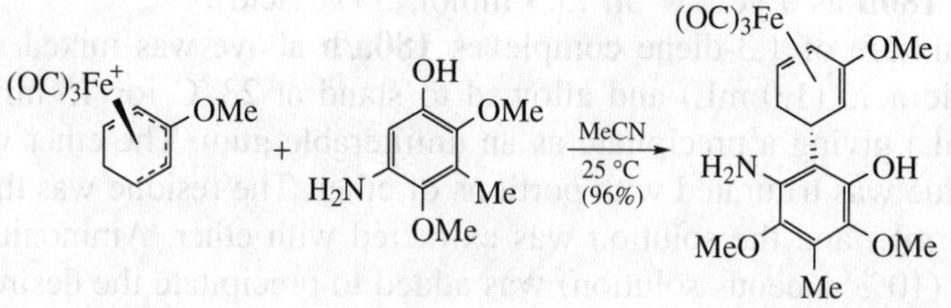

Successful addition of certain nucleophiles requires specific conditions of solvent or modified ligands. Significant side reactions can include addition to a CO ligand and electron transfer reduction. For example, the addition of simple organolithium and organomagnesium reagents in ether solvents gives low yields of nucleophile addition, and clear evidence of electron transfer reduction, such as product **182** (Scheme 28)[158]. Using other organometallic derivatives such as R$_2$Zn and R$_2$Cd, the yields can be better, but a more general solution is to replace one CO ligand with triphenylphosphine[159,160]. It is reasonable that the addition of a more strongly electron-donating ligand (Ph$_3$P) should retard electron transfer, but it should also retard nucleophile addition. It is not obvious why phosphine substitution *selectively* retards electron transfer, but the nucleophile addition results are much better. Photochemical replacement of the CO gives monophosphine complex **183**, which adds simple organomagnesium reagents very smoothly.

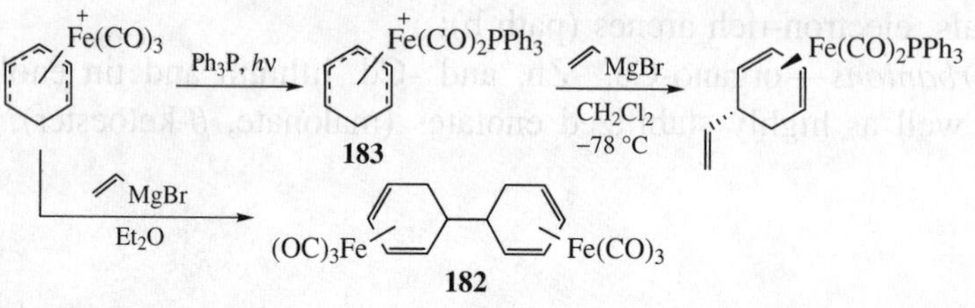

Scheme 28. Phosphine ligands to improve nucleophilic addition selectivity.

8.2.2.2 Regioselectivity in Nucleophile Addition

With terminal alkyl substituents, another potential side reaction is deprotonation, to give an exocyclic double bond in a neutral complex (**184**, Scheme 29). The relative rate of deprotonation compared to nucleophile addition is a strong function of the nature of the nucleophile[161]. A dramatic case is the comparison of a lithium enolate with a tin enolate in addition to complex **184**. This result (Scheme 29) is rationalized in terms of the high rate of proton abstraction by the hard lithium enolate.

Even with the softer malonate anion, a similar cyclohexadienyl complex **185** leads to a mixture of regioisomers as well as deprotonation. However, a strong ligand effect is

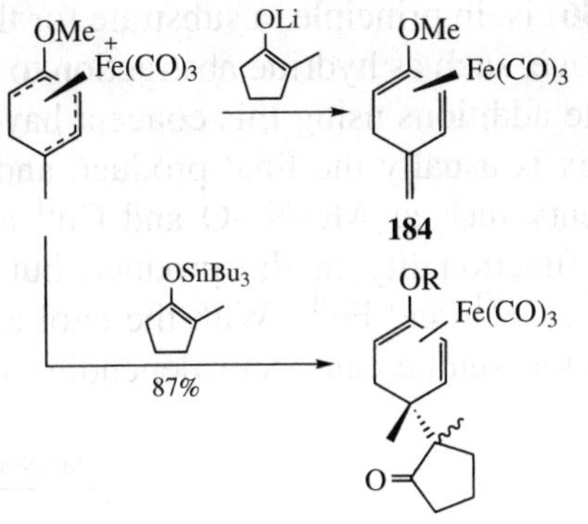

Scheme 29. Strong base reactivity of a ketone enolate.

observed when one CO is replaced with Ph₃P (**186**), resulting in very selective addition to the more substituted terminal position and no significant deprotonation (Scheme 30)[162].

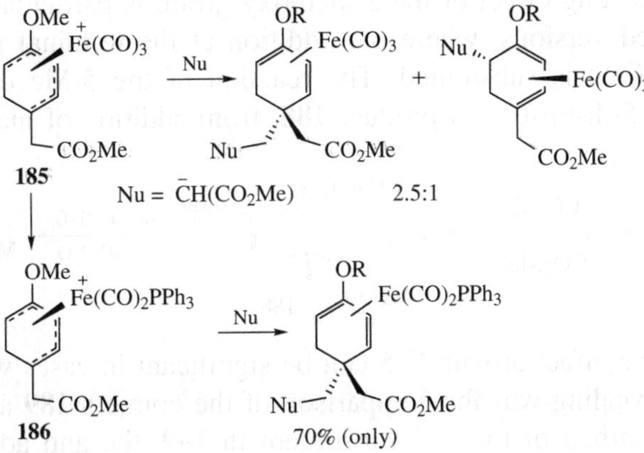

Scheme 30. Phosphine effect on deprotonation versus addition.

Certain aspects of selectivity are very reliable: the addition is always anti, from the side opposite the metal, and the addition is always at a terminal position. The issues are more interesting with substituted cyclohexadienyls in which the terminal positions are non-equivalent. Again, both steric and electronic effects are important in determining regioselectivity.

The directing influence of a methoxy substituent has been particularly well defined because of the ready availability of the starting complexes and the potential usefulness of the methoxy functional group, usually as an enol ether, in the product. The 2-methoxy complex **169a** shows very high selectivity for nucleophile addition at the terminal position away from the methoxy group, at C-5. Strategically, this is equivalent to nucleophile addition para to a methoxy group in anisole or to nucleophile substitution at the 4-position of a 2-cyclohexenone, and therefore complements traditional methodology[163]. The product

1,3-diene complex (*e.g.*, **186**) is, in principle, a substrate for the methodology discussed in the section immediately above, such as hydride abstraction to regenerate a new cyclohexadienyl ligand, but no double additions using this concept have been reported. Instead, the 1,3-cyclohexadiene complex is usually the final product, and is removed from the metal by oxidation. Gentle reagents such as Me$_3$N–O and Cu^{II} are sufficient and tend to be compatible with common functionality in the product, but numerous other agents are also effective, such as Ce^{IV}, Cr^{VI}, and Fe^{III}. With the enol ether function in the product, simultaneous hydrolysis to the ketone can occur depending on the oxidation conditions:

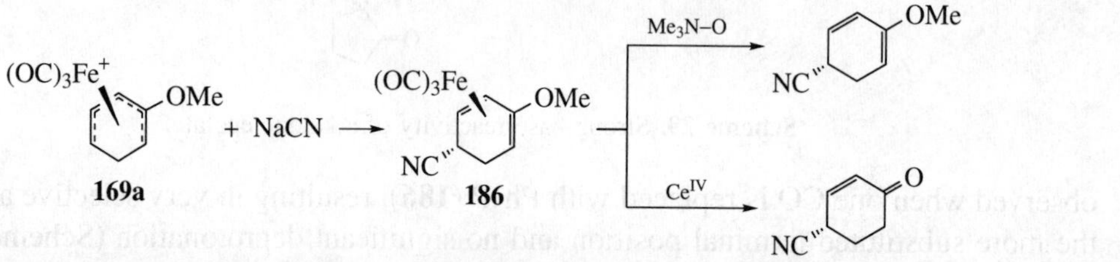

The electronic directing effect of the 2-methoxy group is particularly clear in the case of 5-alkyl substituted versions, where the addition at the terminal position away from the –OMe group (C-5) is substituted. The reaction of the 5-Me derivative **187** leads exclusively to the 5,5-disubstituted product **188**, from addition of malonate anion[164]:

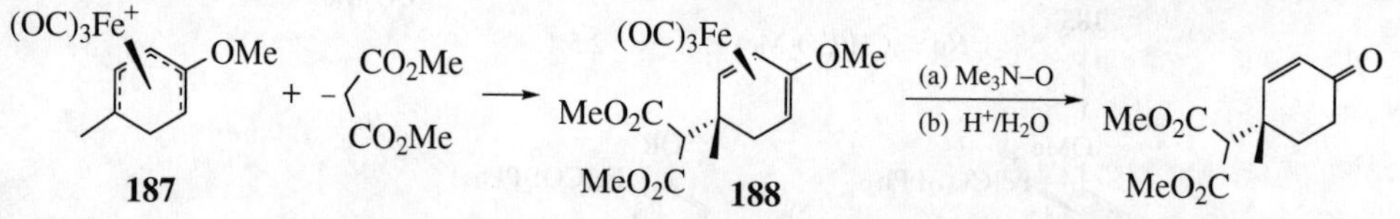

However, the steric effect around C-5 can be significant in cases with sufficient steric bulk. Particularly revealing was the comparison of the epimers **189** and **190**. With a syn Me group at the α-carbon of the C-5 substituent in **189**, the anti addition gives a 1 : 1 mixture of C-1 and C-5 attack:

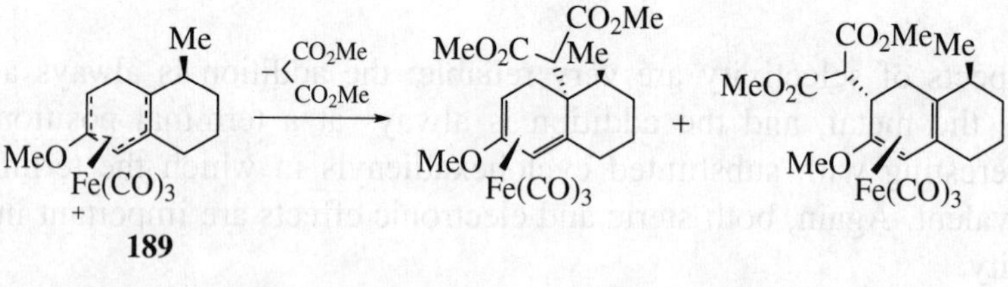

The epimer with the anti Me group (**190**) shows only addition at C-1:

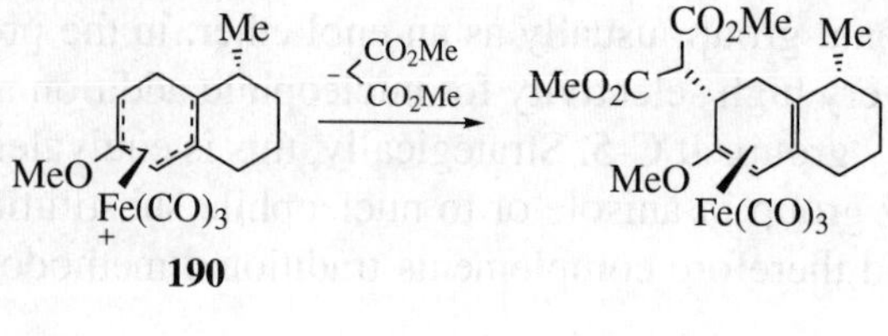

It is possible to tune the selectivity through steric effects at the alkoxy group. Replacing the MeO group with ^{i}PrO, for example, can inhibit addition at the terminus near the alkoxy group. In this case, the selectivity is still strong for the terminal position away from the MeO group by a factor of >3 : 1, but can be increased significantly to 9 : 1 by the simple expedient of adjusting the alkoxy group. The yields are 75–85%[165]:

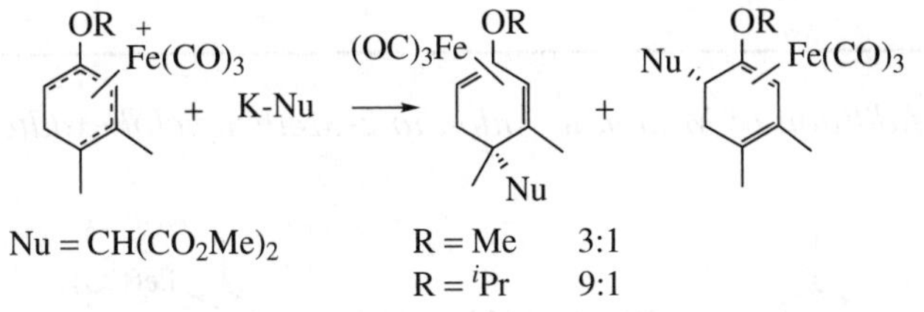

| Nu = CH(CO_2Me)_2 | R = Me | 3:1 |
| | R = ^{i}Pr | 9:1 |

The 3-methoxy substituted complex **191** gives smooth addition at one of the equivalent terminal positions[166]. However, related cases with larger (tertiary) exo substituents at C-4 are completely inhibited from undergoing substitution with malonate-type nucleophiles; with long reaction times or more vigorous conditions, the ligand is detached:

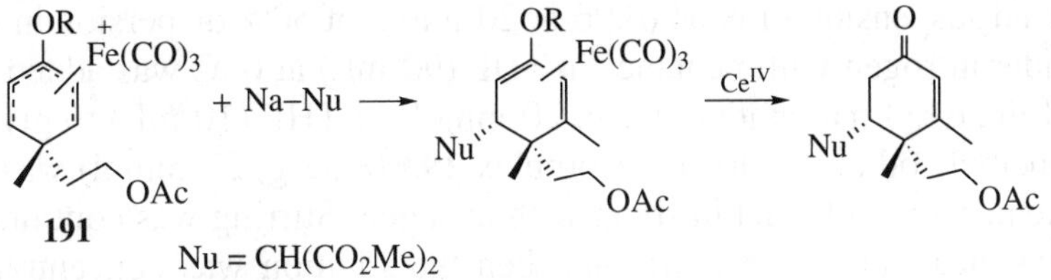

Nu = CH(CO_2Me)_2

With the 1-carboxy-substituted complex **192**, addition of a nitrogen nucleophile is completely selective for the remote terminal position, and with the usual 100% anti addition selectivity[167]. In this case, the starting complex **192** was a single enantiomer, and this asymmetry was transmitted to the final product, (−)-gabaculine:

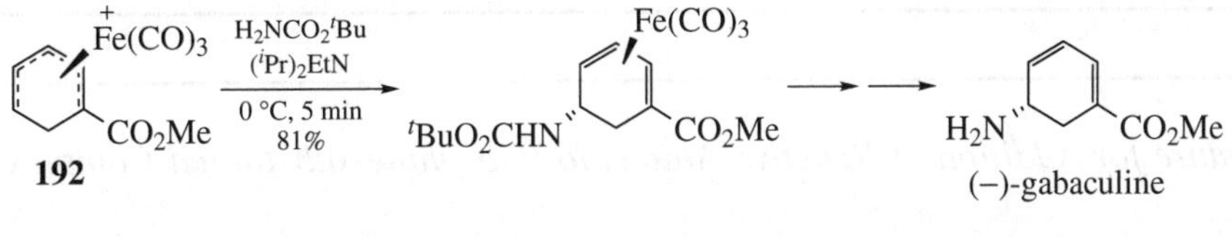

The reactivity of the cyclohexadienyl cationic complexes toward easily generated, stabilized nucleophiles means that intramolecular applications are simple to devise. Intramolecular reaction can add new restrictions to the regioselectivity. The selectivity for addition of the β-ketoester anion to the substituted terminal position is 100% in the following example, enforced by the opportunity to form a simple six-membered spirocycle rather than a more highly strained [5.3.1]bicyclic system (not shown):

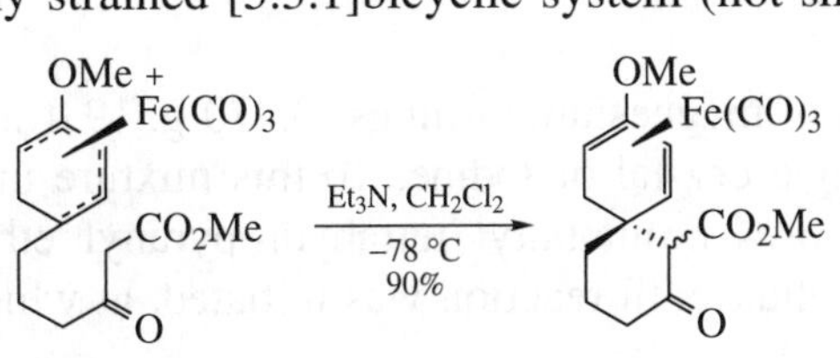

However, the success of this procedure depends on a delicate balance of reactivity. A side reaction is the removal of a proton from the side chain, to give an exocyclic double bond and a neutral complex. Malonate-type nucleophiles (in contrast to those from β-ketoesters) are not quite acidic enough to allow nucleophile generation compared to proton abstraction.

Procedure for Addition of Malonate Anion to 2-Methylcyclohexadienyl-Fe(CO)$_3$, **193**[168]

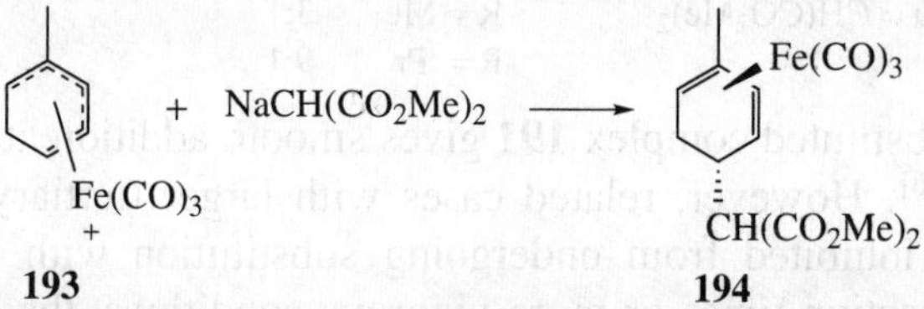

To a stirred suspension of NaH (0.96 g, 20 mmol of 50% dispersion in mineral oil; washed under nitrogen with pentane) in THF (90 mL) at 0 °C was added dropwise a solution of dimethyl malonate (2.64 g, 20 mmol) in THF (10 mL) to give a suspension of dimethyl sodium malonate. Complex **193** (8.32 g, 22 mmol) was added as a solid in one portion with backflushing with nitrogen. Stirring was continued until the mixture was clear (about 15 min), and then the solution was concentrated, diluted with ether (250 mL) and the resulting mixture was washed with water (3 × 100 mL), dried (MgSO$_4$), filtered, and the filtrate was concentrated to leave **194** (7.20 g, 99%) sufficiently pure for further applications. An analytical sample was obtained by crystallization from 5% ether in pentane.

Procedure for Addition of Reactive Anions to Phosphine-substituted Complexes[160]

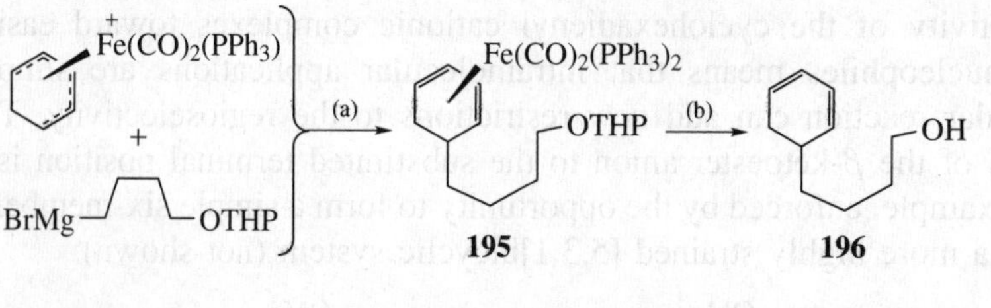

(a) To a suspension of magnesium turnings (0.470 g, 19.4 mmol) in 15 mL of dry THF was added a single crystal of iodine. To this mixture under argon was added 0.60 mL (3.2 mmol) of 4-bromobutyl tetrahydropyranyl ether. The mixture was stirred and heated at reflux until reaction was initiated, at which time the remaining bromide (3.0 mL, 16.1 mmol) was added dropwise via syringe over 5 min while the reaction temperature was maintained at 45 °C. After an additional 1 hr at 40 °C, the solution of the Grignard reagent was added dropwise via syringe over 10 min to a solution of dicarbonyl(cyclohexadienyl)triphenylphosphineiron complex (8.91 g, 14.9 mmol; PF_6 salt) in 100 mL of dry dichloromethane cooled at −78 °C. The reaction turned a deep red brown color. After being stirred for 2 h at −78 °C, the mixture was warmed to 0 °C and allowed to stir for an additional 5 min. The reaction was quenched by pouring into 100 mL of water. The organic layer was separated, and the aqueous layer was extracted with 50 mL of dichloromethane and of ether (2 × 50 mL). The combined organic layers were dried over anhydrous magnesium sulfate and evaporated under reduced pressure. The residue was purified on silica gel (25% ether/hexane → 50% ether/hexane) to give adduct **195** (8.19 g, 90% yield) as an amber viscous oil. This material was sufficiently pure to be demetalated and deprotected in the next step.

(b) The complex **195** (8.16 g, 13.4 mmol) was dissolved over 30 min in 180 mL of absolute EtOH. Solid $CuCl_2$ (16.8 g, 125 mmol) was added over a 5-min period, resulting in a mildly exothermic reaction (care must be taken to avoid excessive foaming). The reaction temperature was maintained at 25 °C for 5 h by means of a water bath. The mixture was then poured into 250 mL of water and extracted with 1 : 1 ether/hexane (4 × 75 mL). The combined extracts were dried over anhydrous $MgSO_4$, filtered, and the filtrate was concentrated to provide the crude product which was chromatographed on silica gel. Elution with 40% ether/hexane gave 1.29 g (63%) of the pure hydroxycyclohexadiene **196** as a colorless oil.

8.3 Open chain η^5-(Pentadienyl)Fe(CO)$_3$ Complexes[151, 169]

8.3.1 Preparation of Complexes

The abstraction of hydride from a 1-alkyl-1,3-diene-Fe(CO)$_3$ complex is not a general process. It requires the cisoid configuration of the alkyl substituent (Scheme 31, path a), an arrangement for which there are limited synthesis methods. Most general is acid-promoted loss of an oxygen substituent on a carbon adjacent to one of the termini of the 1,3-diene ligand (Scheme 31, path b)[170]. The mechanism appears to involve initial generation of the transoid cationic complex followed by rapid isomerization to the more stable cisoid version. A third method involves formation of a triene complexed as a 1,3-diene (path c)[171]. Again, the starting triene complex is most easily prepared as the transoid version, and isomerization to the cisoid dienyl complex is required (and usually fast). In general, the cisoid and transoid forms appear to be in equilibrium at room temperature, and in some sterically biased cases, such as the 1,1,5-trimethylpentadienyl-Fe(CO)$_3$ complex, the transoid form is favored at equilibrium[172].

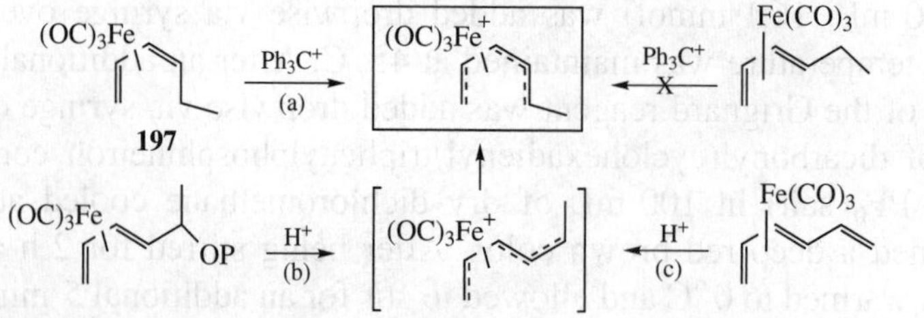

Scheme 31. Methods and selectivity in acyclic pentadienyl ligand formation.

Procedure for Preparation of Pentadienyl-Fe(CO)$_3$ Complexes by Acid-promoted Elimination[173]

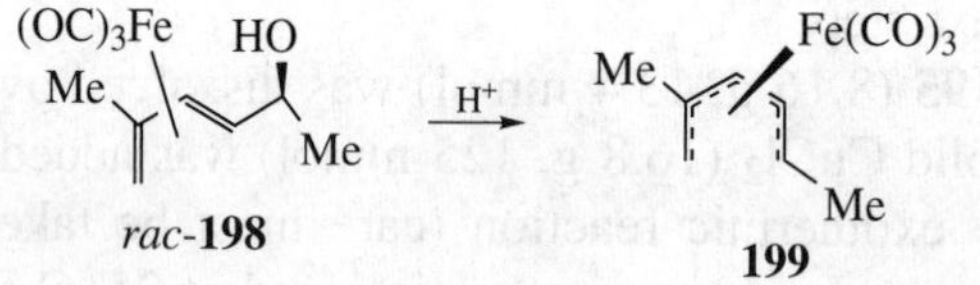

A cold solution of HPF$_6$ (3.5 mL, 60% in water) was added to acetic anhydride (3.5 mL) and allowed to stir for 30 min with cooling in an ice bath. To this solution was added dropwise a solution of tricarbonyl(dienyl)iron complex **198** (7 mmol) in acetic anhydride (3 mL) and ether (8 mL). The mixture was stirred for 10 min and then was added dropwise to a large excess of ether (ca 400 mL) at 25 °C. The precipitate was collected by suction filtration, dissolved in nitromethane (20 mL) and reprecipitated by addition of ether (400 mL). The precipitate was collected by suction filtration and dried in vacuo. The result was **199** as a yellow solid, essentially analytically pure, in 70% yield. (mp 153–159 °C).

8.3.2   Reactions with Nucleophiles

Addition of nucleophiles to the pentadienyl system has been observed from the cisoid complex at either termini or at C-2/C-4 (Scheme 32, path a), and from either of the isomeric transoid complexes (paths b and c) at either termini. Addition is not observed at the central carbon C-3. Considering this range of possibilities, the addition process nevertheless can give useful regioselectivities.

The high electrophilicity of the pentadienyl-Fe(CO)$_3$ complexes is reflected in their high reactivity even with simple electron-rich carbon compounds as well as conventional carbanion/organometallic derivatives. In the assessment as to whether the cisoid or transoid form will undergo the nucleophile addition, there are few generalizations. Less reactive nucleophiles such as electron-rich aromatics, allyl silane, and the alkynyl anion derived from TMS–alkyne and fluoride lead to transoid products (**200, 201**) from

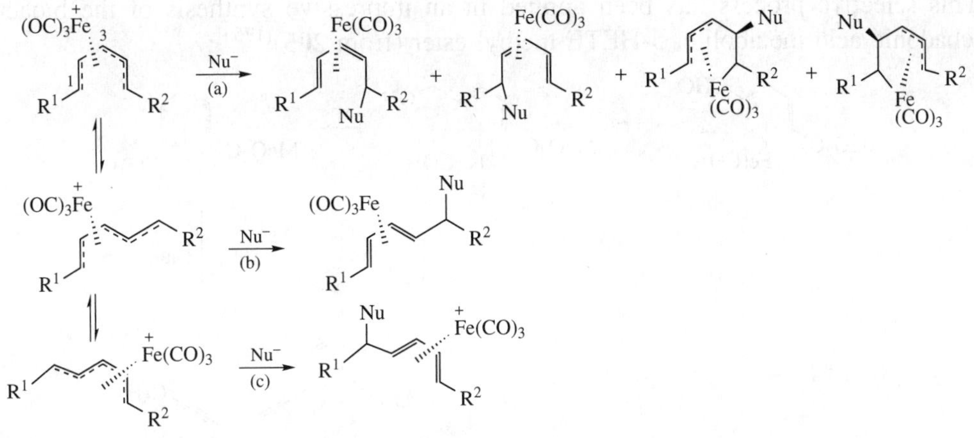

Scheme 32. Regioselectivity possibilities for the acyclic pentadienyl complexes.

addition at the less-substituted terminus, suggesting that the transoid dienyl complex is the reactive species:

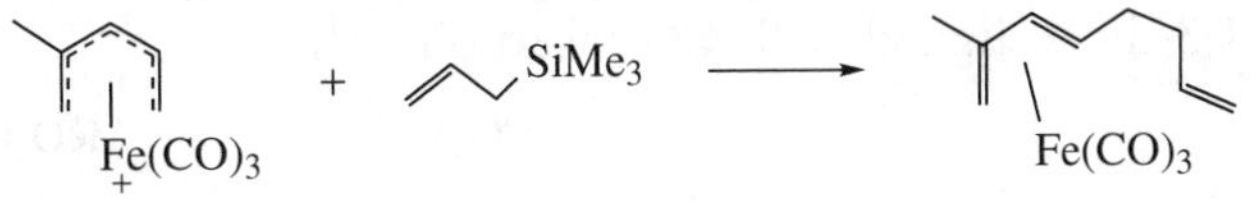

200

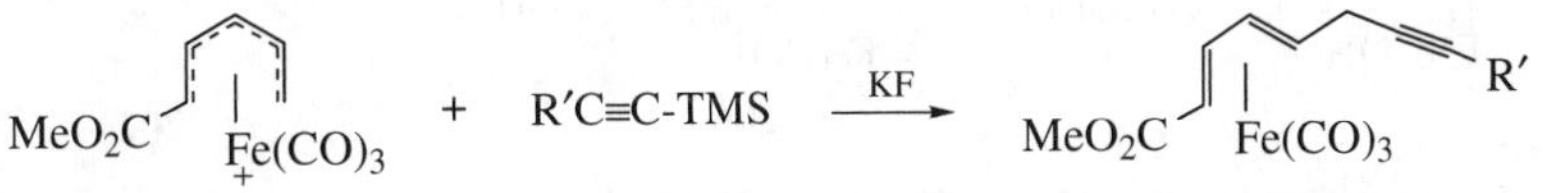

201

The same selectivity is seen when the cationic intermediate (transoid?) is generated by *in situ* treatment of an oxygen derivative (**202**) with a Lewis acid in the present of a nucleophilic species, to give **203**[174]:

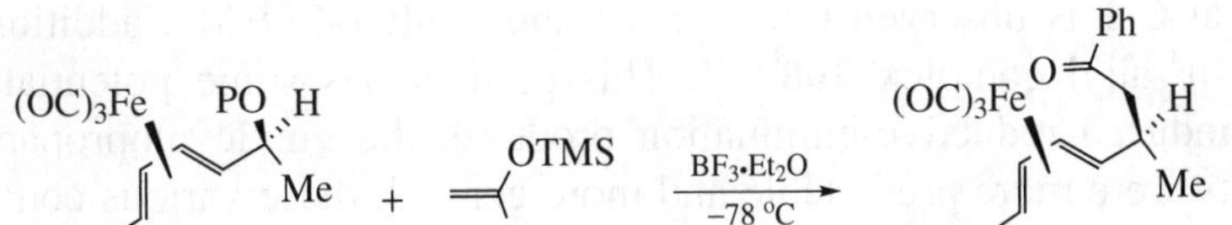

202 *rac*-**203**

Conversely, organo-Cd, -Cu/Li, and -Zn/Cu reagents produce the cisoid products. Attesting to the high electrophilicity of the dienyl cations, even an alkynyl (higher order) cuprate adds at −65 °C to give **204**[175]:

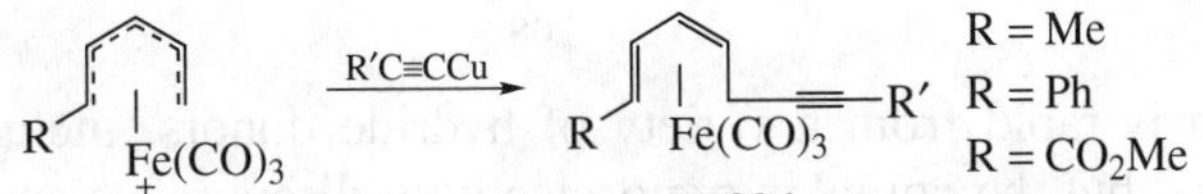

204

This selective process has been applied in an impressive synthesis of the bioactive arachadonic acid metabolite, 5-HETE methyl ester (from **205**)[175a]:

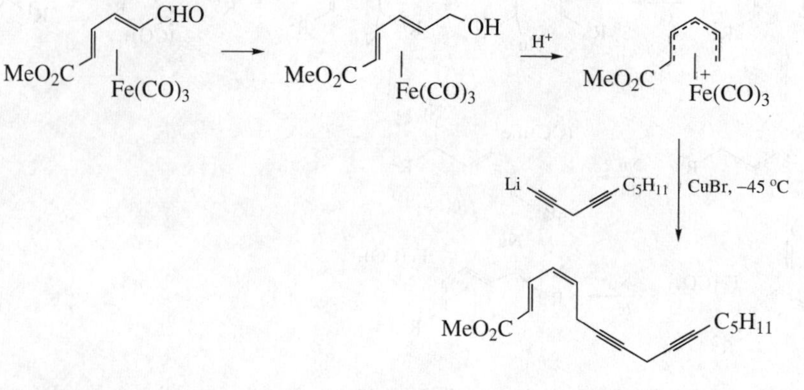

205

The zinc reagents are particularly useful as they can be prepared directly from the corresponding halide and can carry polar functionality. In the following example the initial product (**206**) was converted to the bicyclic system **207**, taking advantage of intramolecular addition to a 1,3-diene ligand with CO insertion[176]:

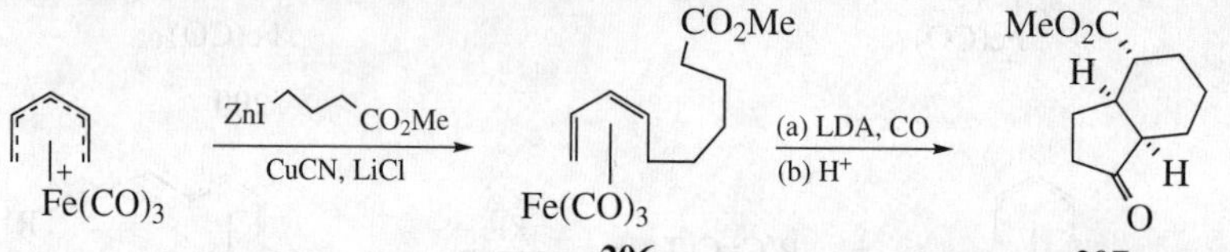

206   **207**

Reagents of the type R—Li (R = Ph, Me, Bu) give significant amounts of addition at both C-5 and C-4 with 1-Me pentadienyl-Fe(CO)$_3$ and the processes are not synthetically useful[177]. Carbon nucleophiles such as malonate anion give addition to 1-substituted pentadienyl ligands with regioselectivity that depends on the substituent. With a 1-Me substituent, a mixture of C-1 and C-5 adducts is obtained, whereas with 1-Ph, selective addition at C-1 is observed (50% yield) and, with 1-CO$_2$Me, addition is at C-2 to give η^1-alkyl-η^3-allyl complex **208**[178]. This product has some potential in synthesis, as oxidation-induced reductive-elimination produces the vinylcyclopropane **209**. If the regioselectivity were more predictable and more general, these various conversions could justify development as new methodology:

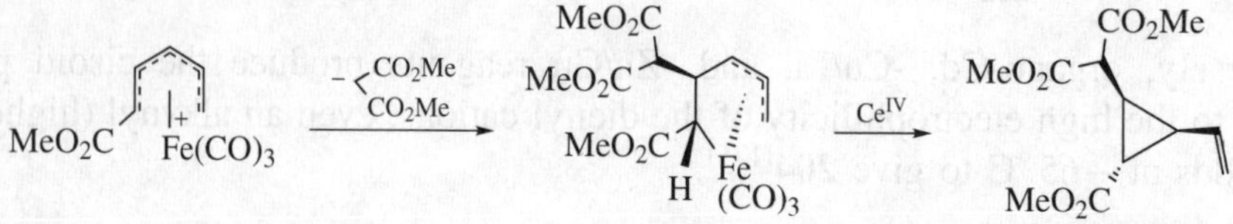

208   **209**

Hydride addition is rapid from a variety of hydride donors, including mild reagents such as Na(CN)BH$_3$, but the complex regioselectivity discourages applications. Addition can occur at C-1, to give both cisoid and transoid products, and at C-2, depending on the hydride reagent and the substituents on the pentadienyl system[169].

Heteroatom nucleophiles add under mild conditions, including O, N, S, and P derivatives[179]. With amine nucleophiles, addition at the less hindered terminal position is strongly favored, but both cisoid and transoid products can be obtained, depending on the amine. More basic alkyl amines tend to give the cisoid products **210**, whereas anilines tend to add to give transoid products **211**. Less basic anilines give more transoid products. Even a tertiary alkyl amine (*e.g.*, Et$_3$N) will add, giving an ammonium ion:

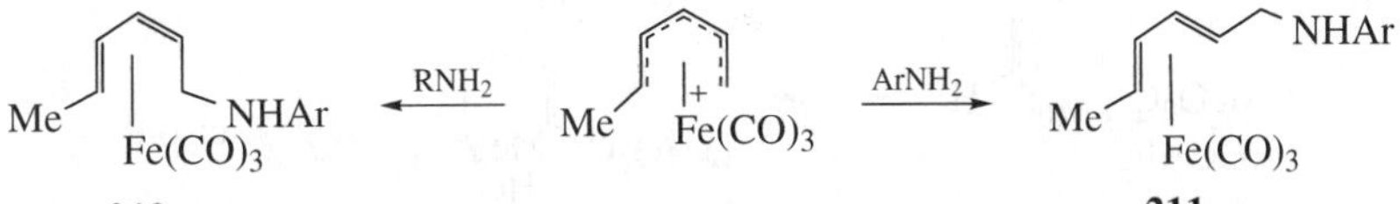

210

R = Et, Bn, etc.

211

Ar = *m*-NO$_2$, *p*-NO$_2$ or *p*-BrPh

Water and alcohols add at room temperature, and show useful regioselectivity[151, 180], Representative examples are shown below, where a competition between C-1 substituents and C-5−Me suggests the magnitude of the steric effect of the C-1 group; an apparent electronic effect (Ph strongly favors addition at the other terminus), is also suggested:

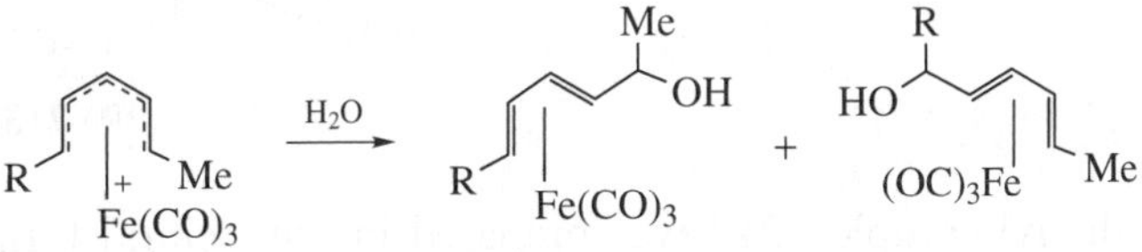

R = Et	61	39
R = ^{i}Pr	90	10
R = Ph	100	0

A C-2−Me group strongly favors addition away, at C-5, whereas a C-1−Me group favors addition at the adjacent (C-2) carbon:

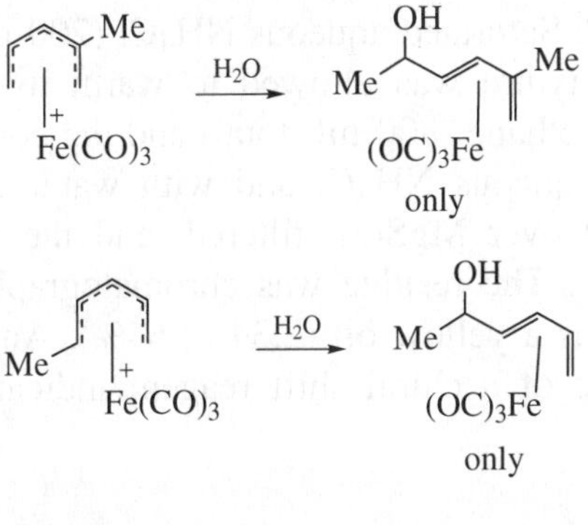

With an electron-withdrawing group (EWG) on the pentadienyl ligand at C-1, the reactivity is increased and perhaps that explains why the cisoid product is obtained (fast coupling with the more stable geometrical isomer). The reactions are stereospecific and occur at the terminus away from the EWG group, as illustrated below for a carbomethoxy group. The exo and endo isomers give different diastereoisomers as products, from addition at the side opposite the metal[181]:

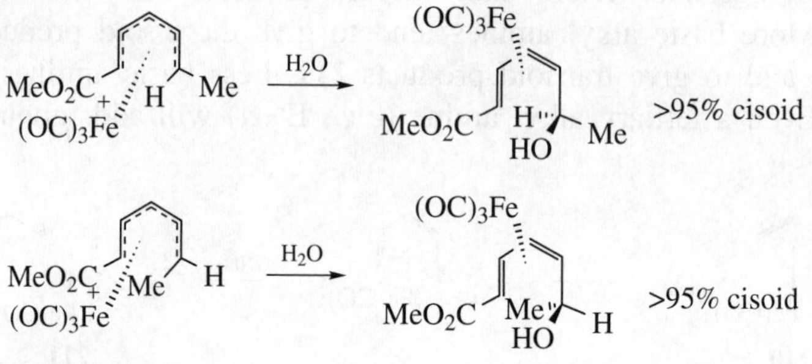

>95% cisoid

>95% cisoid

Procedure for Alkynyl Cuprate Addition to Homochiral
1-Carboxypentadienyl)Fe(CO)$_3$ $^{+[181]}$

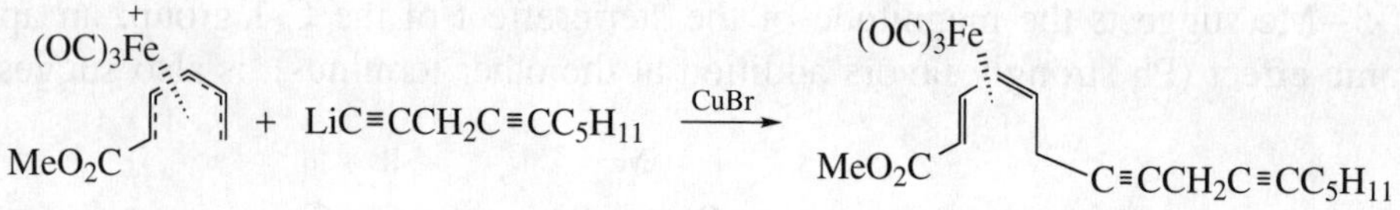

The starting dienyl complex **212** was prepared in homochiral form by resolution of a derivative[181]. To a solution of freshly distilled diyne (9.20 g, 68.6 mmol) in ether/THF (4 : 1, 300 mL) under argon at −78 °C was added dropwise butyllithium (1.6 M in hexane, 43 mL, 68.6 mmol) and the mixture was stirred for 15 min. The mixture was warmed to −45 °C and CuBr–Me$_2$S (4.70 g, 22.9 mmol) was added as a solid. The mixture was stirred for 4 h at −45 °C and the solid cation (2R)-**212** (3.75 g, 9.15 mmol) was added in one portion. The mixture was stirred for 4 h at −45 °C. Saturated aqueous NH$_4$Cl (200 mL) was added to quench the reaction, and the mixture was allowed to warm to 25 °C. The mixture was extracted with dichloromethane (800 mL total) and the combined organic layer was washed with saturated aqueous NH$_4$Cl and with water until it was neutral. The organic layer was dried over MgSO$_4$, filtered, and the filtrate was concentrated on the rotary evaporator. The residue was chromatographed (hexane/ether 20 : 1, SiO$_2$) to give (2R)-**213** as a yellow oil (2.34 g, 64%). Analysis by ^{1}H NMR spectroscopy in the presence of a chiral shift reagent indicated that the product was >92% ee.

9 Applications of η^6-Arene-Cr(CO)$_3$ Complexes

9.1 Preparation of (Arene)-Cr(CO)$_3$ Complexes

The general preparation of arene complexes is shown below. The L unit in the Cr(CO)$_3$L$_3$ can be CO (most common)[182–186], MeCN[187,188], alkylpyridine[189], ammonia[190,191], naphthalene[192], pyrrole[193], or other donor ligands. The rate (reaction temperature) is related to the nature of L; the most reactive readily available source of Cr(CO)$_3$ is (η^6-naphthalene)Cr(CO)$_3$, which undergoes favorable arene exchange under mild conditions with many substituted arenes. A technical problem in the use of Cr(CO)$_6$ is the crystallization of sublimed Cr(CO)$_6$ in the condenser. The original remedy is still used, a fairly complicated heated condenser apparatus known as the Strohmeier apparatus[182], but a simple air condenser is also effective[183]. The most general and convenient procedure also relies on continuous washing down of the Cr(CO)$_6$ with a mixture of THF and dibutyl ether at reflux[183,184]. A variety of polar and nonpolar aprotic solvents has been used; and, for some purposes such as complexation of α-amino acids with aromatic side chains, water/THF mixtures are effective[184].

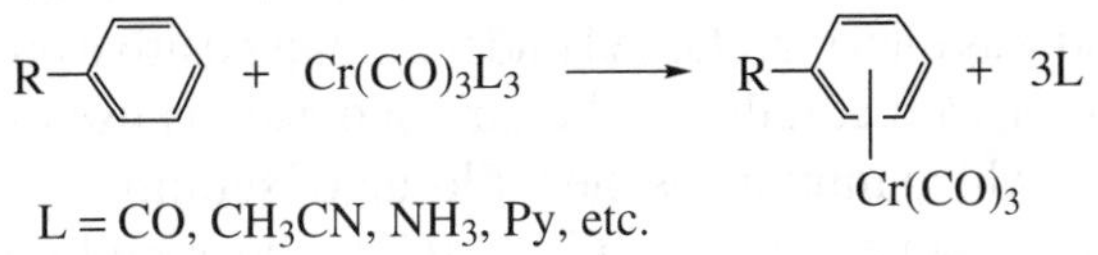

$$R-\!\!\bigcirc + \text{Cr(CO)}_3\text{L}_3 \longrightarrow R-\!\!\bigcirc\!\!\underset{\text{Cr(CO)}_3}{\big|} + 3L$$

$$L = \text{CO, CH}_3\text{CN, NH}_3\text{, Py, etc.}$$

The primary competing reaction is irreversible oligomerization of the coordinatively unsaturated Cr(CO)$_n$ species. Addition of small amounts of weak donor ligands, such as α-picoline[194,195] or THF[184] appear to lower the temperature necessary for the complexation step from 100 °C or more to 60 °C, presumably by assisting in the dissociation of one or more CO ligands and retarding the rate of oligomerization of the reactive source of Cr(CO)$_3$. After complexation, the yellow to red arene–Cr(CO)$_3$ complex can be purified by crystallization from nonpolar organic solvents or chromatographed on silica gel. The complexes are somewhat sensitive to air while in solution, but the solutions can be handled in air briefly and the crystalline solids can be stored without special precautions. Examples are known with a wide variety of arenes, including sensitive species such as (η^6-indole)Cr(CO)$_3$[196]. Most common synthesis operations, such as acid and base hydrolysis, hydride reduction, and carbanion addition to ketones, can be carried out on side-chain functional groups without disturbing the arene–Cr bond. However, even very mild oxidation conditions will detach the arene by oxidizing the metal. An important aspect of the complexation procedure is diastereoselectivity. Complexes of *o*- or *m*-disubstituted arenes (and more highly substituted) can have molecular asymmetry; a stereogenic center in a side chain leads to diastereoisomers (*e.g.*, **214a** and **214b**). Under favorable conditions, especially with low-temperature conditions for complexation, significant diastereoselectivity is observed in the complexation step[197].

214a   98 : 2   **214b**

Procedure for Synthesis of m-Fluoroanisolechromium Tricarbonyl

In a 250-mL one-neck RB flask equipped with a stir bar and an air condenser (straight 1-m by 3-cm glass tube), chromium hexacarbonyl (5.00 g, 2 mmol) was suspended in a mixture of *m*-fluoroanisole (10.0 mL, 11.0 g, 87.5 mmol), 80 mL of dibutyl ether (freshly distilled from LiAlH$_4$), and 10 mL of THF (freshly distilled from Na/benzophenone). The system was placed under argon by evacuating and filling with argon three times and was then maintained under a small positive pressure of argon. The mixture was heated for 24 h at reflux. (The air condenser is used so that the refluxing vapors remain warm, which minimizes the build-up of sublimed Cr(CO)$_6$ at the cooler parts of the condenser. Nevertheless, solid Cr(CO)$_6$ does build up in the condenser during the process. It can be returned to the reaction vessel simply by increasing the heat and allowing stronger reflux to wash it down; or the system can be cooled, opened, and the solid dislodged with a long glass rod. It is important that an argon atmosphere be maintained during the complexation process.) After the reaction period, the mixture was cooled to 23 °C and filtered quickly through a small path of Celite, exposed to the air. The filtrate was concentrated under vacuum and the residue was chromatographed on silica gel with ethyl acetate/hexanes (1 : 4) as eluent. The yellow band was collected yielding of the crystalline yellow η^6-(*m*-fluoroanisole)Cr(CO)$_3$ (4.47 g, 17.0 mmol, 75% yield); mp 62–64 °C.

9.2   Addition of Nucleophiles

9.2.1   Introduction

This section covers reactions in which coordination of a Cr(CO)$_3$ unit to the pi system of an arene ring activates the ring toward addition of nucleophiles, to give η^5-cyclohexadienyl-Cr(CO)$_3$ complexes (**215** in Scheme 33). If an electronegative atom is present in the ipso position, elimination of that atom (X in **215**) leads to nucleophilic aromatic substitution (**216**, path a). Mild oxidation of the arene complex gives the free

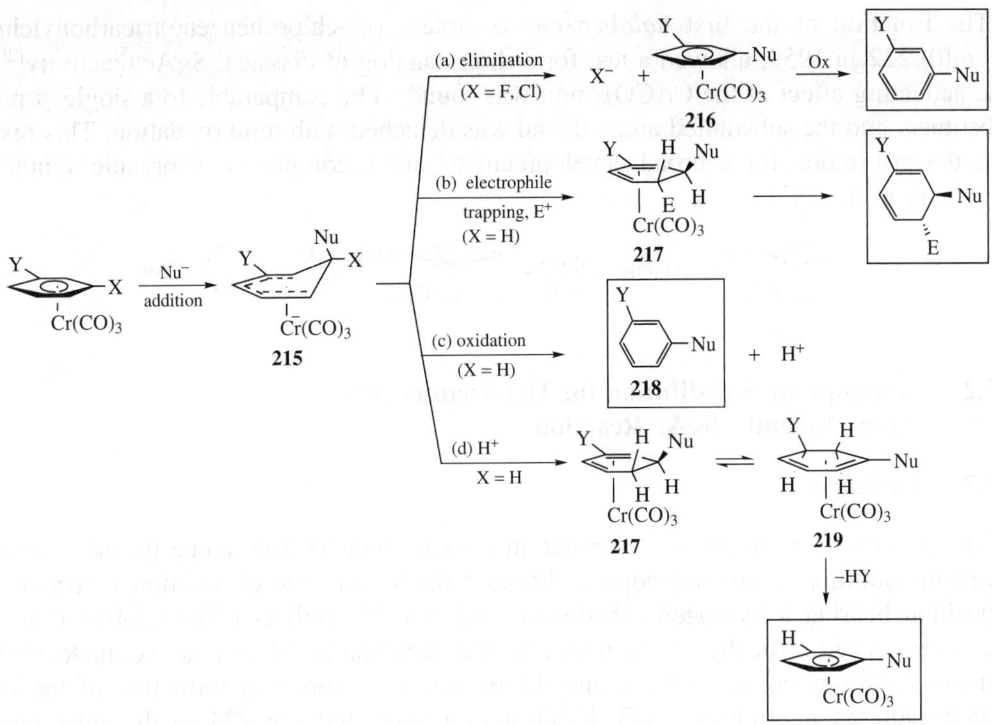

Scheme 33. Pathways following nucleophile addition.

arene. Reaction of the intermediate **215** with an electrophile (E$^+$) can give disubstituted 1,3-cyclohexadiene derivatives (**217**, path b). If a hydrogen occupies the ipso position (**215**, X=H), oxidation of the intermediate gives formal nucleophilic substitution for hydrogen (**218**, path c). An electronegative substituent, including (most usefully) methoxy in **215**, can allow H-rearrangement from the cyclohexadiene intermediate (**217**, path d) and then **219** can undergo elimination of HY, resulting in overall indirect (cine, tele) substitution under acid conditions (path d). General reviews have appeared[198–200] as well as others with an emphasis on η^6-arene-Cr(CO)$_3$ complexes[201–205].

The first arene–transition metal complexes were prepared in the 1950s[206,207] (**220**), and it was immediately recognized that the added polarizibility or electron deficiency would promote addition of nucleophiles to the arene ligand. A number of cyclohexadienyl complexes (**221**) were characterized following nucleophilic addition to the arene ligand, but the question of inducing the ipso hydrogen to depart was not answered until much later:

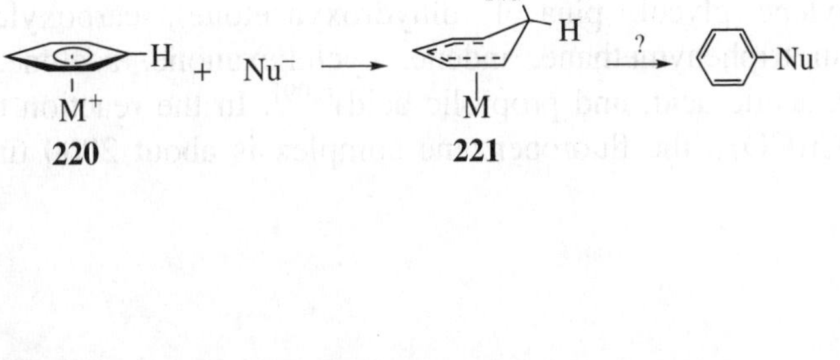

The isolation of the first *halo*benzene complex, (η^6-chlorobenzene)tricarbonylchromium(0) **222** in 1959, allowed a test for a direct analog of classical S_NAr reactivity[208]. The activating effect of the Cr(CO)$_3$ unit was found to be comparable to a single *p*-nitro substituent and the substituted arene ligand was detached with mild oxidation. This result was the inspiration for a broad development of arene complexes in organic synthesis beginning in the 1970s:

9.2.2   Nucleophilic Substitution for Heteroatoms on Arene Ligands, S$_N$Ar Reaction

9.2.2.1   Introduction

The smooth replacement of a heteroatom (usually halide) from arene ligands requires reversible addition of the nucleophile, because the kinetic site of addition is usually at a position bearing a hydrogen substituent. (Scheme 34, path k_1.) The relative rates of each step depend critically on the nature of the nucleophile. More reactive nucleophiles disfavor equilibration ($k_1 \gg k_{-1}$), and the process can stop with formation of the first cyclohexadienyl intermediate **223**. Equilibration leads through **224** to the substitution product.

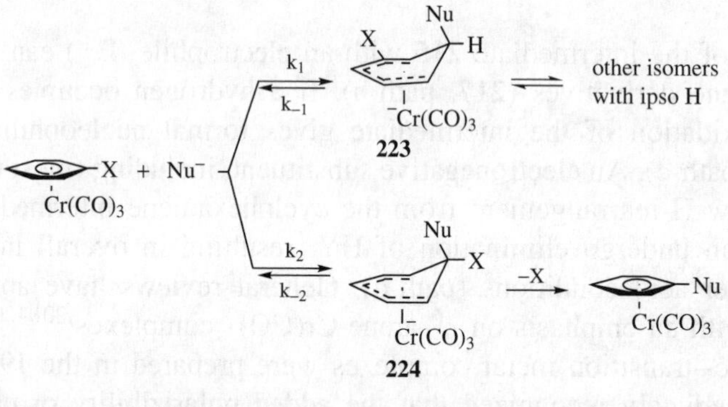

Scheme 34. Reversible addition of nucleophiles followed by irreversible loss of halide.

9.2.2.2   Intermolecular Direct Nucleophilic Substitution with Heteroatom Nucleophiles

A patent issued in 1965 claims substitution for fluoride on fluorobenzene–Cr(CO)$_3$ in DMSO by a long list of nucleophiles including: alkoxides (from simple alcohols, cholesterol, ethylene glycol, pinacol, dihydroxyacetone), carboxylates, amines, and carbanions (from triphenylmethane, indene, cyclohexanone, acetone, cyclopentadiene, phenylacetylene, acetic acid, and propiolic acid)[209]. In the reaction of methoxide with (halobenzene)–Cr(CO)$_3$, the fluorobenzene complex is about 2000 times more reactive than the chlorobenzene complex[210]. The difference is taken as evidence for a rate-limiting attack on the arene ligand followed by fast loss of halide; the concentration of the cyclohexadienyl anion complex (*e.g.*, **223/224** in Scheme 34) does not build up. In the reaction of (fluorobenzene)–Cr(CO)$_3$ with amine nucleophiles, the coordinated aniline product appears rapidly at 25 °C, and a mechanistic study suggests that the loss of halide is now rate limiting[211].

Hydroxide, alkoxide, and phenoxide nucleophiles react with (chlorobenzene)Cr(CO)$_3$ at 25–50 °C in polar aprotic media to give high yields of the phenol or aryl ether chromium complexes (**225**); oxidation with excess iodine at 25 °C for a few hours releases the free arene, CO, chromium(III), and iodide anion[212–214].

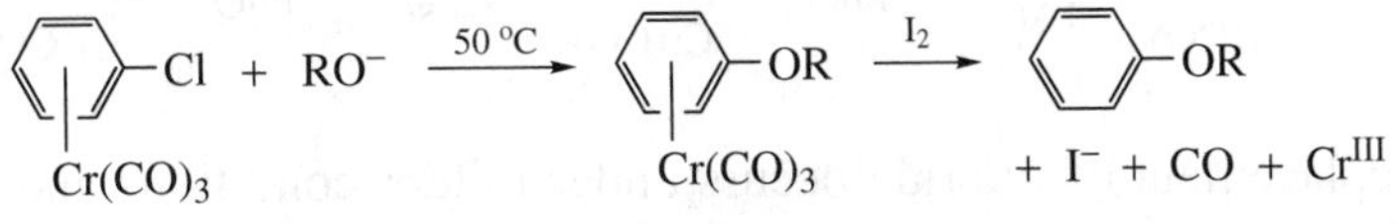

225

An alternative general procedure for detaching the arene ligand from the Cr(CO)$_3$ unit is to expose a solution of the complex to air in the presence of normal room light or sunlight. The solution soon becomes cloudy with oxidized chromium derivatives, and the organic solution can be processed to isolate the free arene[215].

The reaction proceeds almost exclusively by direct substitution (ipso), as shown by reactions of isomeric chlorotoluene complexes (**226** and **227**, Scheme 35)[212]. The relative rates of substitution of the isomeric ligands are similar, being of the order 1.0 : 1.4 : 2.4 for ortho/meta/para[212].

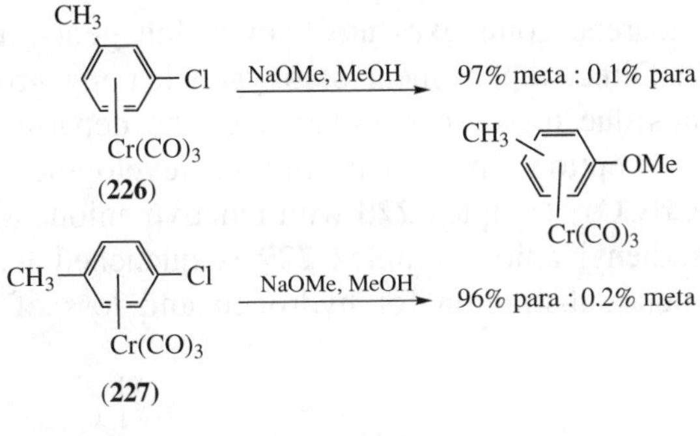

Scheme 35. Selectivity in halide replacement by methoxide.

Whereas polar protic solvents such as MeOH strongly retard reaction[216], phase-transfer catalysis using benzene[217] or addition of crown ethers to potassium alkoxides in benzene[216] allows substitution at 25 °C. Even with strong electron donors such as alkyl, methoxy, or dialkylamino in the ortho, meta, or para positions, substitution for chloride by potassium methoxide proceeds smoothly using the crown ether activation in benzene[216]:

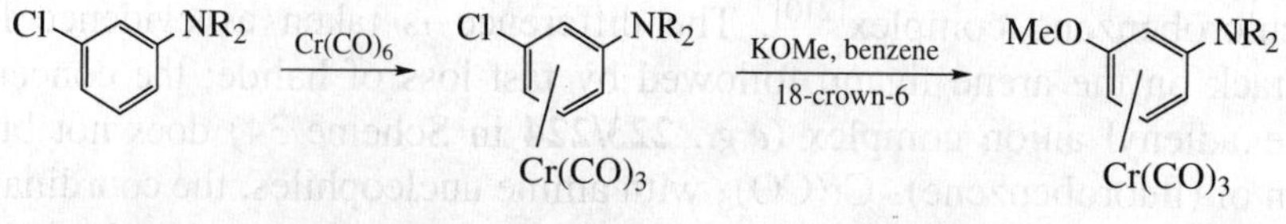

Cl—⟨ ⟩—NR$_2$ →[Cr(CO)$_6$]→ Cl—⟨ ⟩—NR$_2$ (Cr(CO)$_3$) →[KOMe, benzene / 18-crown-6]→ MeO—⟨ ⟩—NR$_2$ (Cr(CO)$_3$)

There is a suggestion that DMSO may be a particularly beneficial solvent, allowing sequential replacement of chloride from the *p*-dichlorobenzene ligand by oxygen nucleophiles[217]:

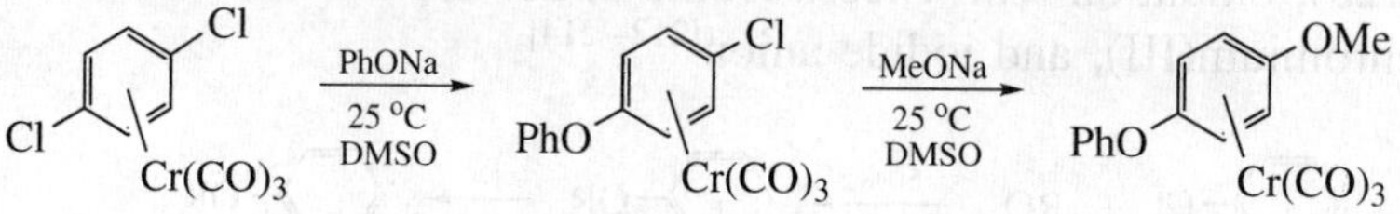

Cl / Cl–⟨ ⟩(Cr(CO)$_3$) →[PhONa / 25 °C / DMSO]→ Cl / PhO–⟨ ⟩(Cr(CO)$_3$) →[MeONa / 25 °C / DMSO]→ OMe / PhO–⟨ ⟩(Cr(CO)$_3$)

In general, replacement of fluoride occurs under milder conditions and in higher yield than chloride; however, compared to chloride, a fluoride substituent slows the complexation of arenes and often leads to lower yields in the formation of the complexes. An excellent procedure for formation of the parent fluorobenzene–Cr(CO)$_3$ complex has been reported[218]. Fluorobenzene was the substrate in substitution to produce aryl thioethers[219]:

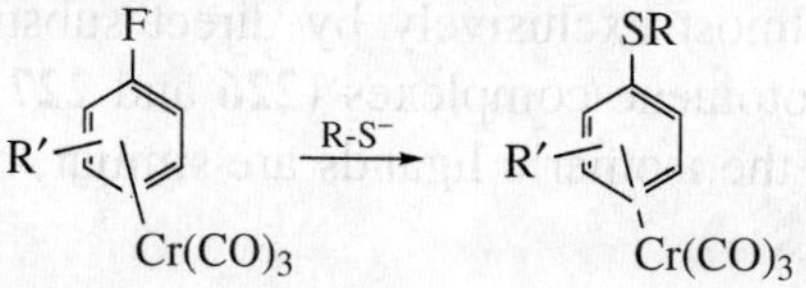

F / R'–⟨ ⟩(Cr(CO)$_3$) →[R-S⁻]→ SR / R'–⟨ ⟩(Cr(CO)$_3$)

The bromo- and iodoarene complexes are known, but generally are not effective in the S$_N$Ar reaction[208]. Direct replacement of oxygen leaving groups, such as tosylate, is not effective, perhaps due to steric retardation of the departure of the endo leaving group. One intriguing exception which bears further development is the reaction of the diphenyl ether mono-Cr(CO)$_3$ complex **228** with reactive anions (Scheme 36)[220]. If the intermediate cyclohexadienyl anion complex **229** is quenched with an oxidizing agent at low temperature, meta substitution for hydrogen and loss of the Cr(CO)$_3$ unit are

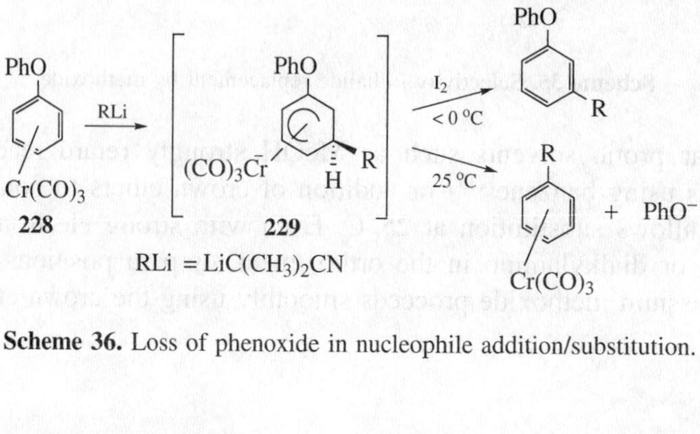

PhO—⟨ ⟩(Cr(CO)$_3$) **228** →[RLi]→ [PhO / (CO)$_3$Cr⟨ ⟩R H **229**] →[I$_2$ / < 0 °C]→ PhO—⟨ ⟩—R →[25 °C]→ R—⟨ ⟩(Cr(CO)$_3$) + PhO⁻

RLi = LiC(CH$_3$)$_2$CN

Scheme 36. Loss of phenoxide in nucleophile addition/substitution.

observed (a discussion of this pathway is given below). However, if the same intermediate is allowed to warm to room temperature, anion equilibration can occur even with the very reactive 2-methyl-1,3-dithianyl carbanions and direct (ipso) substitution for phenoxide is observed.

Thiolate anions[221,222] and oxime alkoxides[221] react under phase transfer conditions to give aryl thio ethers and *O*-aryl oximes, respectively; the *o*-dichlorobenzene complex **230** can be converted selectively to the monosubstitution product **231**:

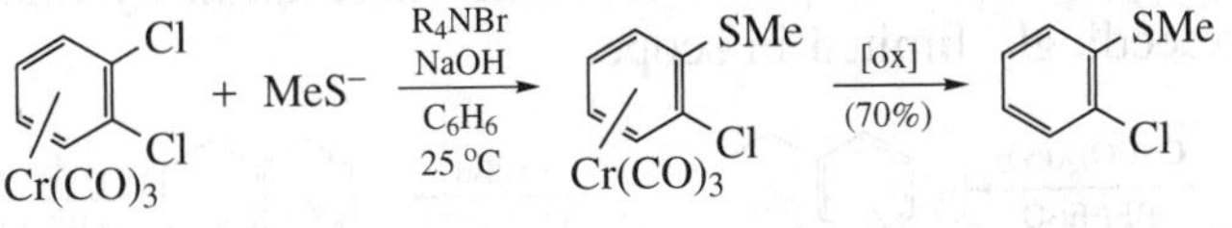

It has been established that simple primary and secondary amine nucleophiles react smoothly in the absence of added base, to produce aniline derivatives, but only a few examples have been reported[211]. To add an $-NH_2$ group, an effective procedure is reaction of the trifluoroacetamide anion with a (chlorobenzene)Cr(CO)$_3$ complex (*e.g.*, **232**) followed by gentle base treatment[223].

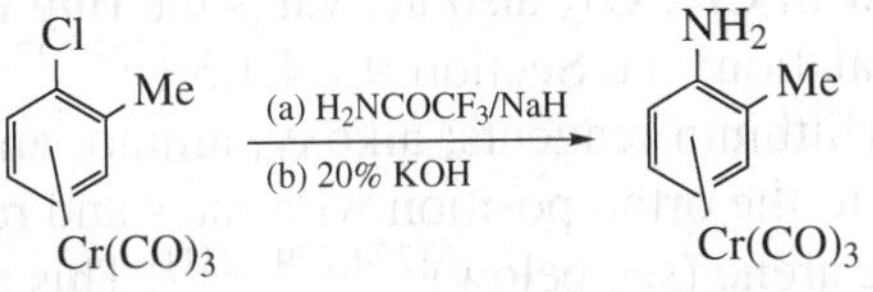

Complexation of chloroindole **234** with Cr(CO)$_6$ in diglyme/cyclohexane at 125 °C for 53 h using the Strohmeier apparatus gave the chromium complex (**235**) in 85% yield, based on 40% recovery of starting material. The unreacted tetrahydroisoquinoline can be separated from the complex by simple acid extraction, as the Cr(CO)$_3$ coordination renders the complexed amine relatively nonbasic. Indeed, the complexed amine is moderately acidic and must be protected (in situ) as the benzyl ether before methoxide treatment. The overall yield for protection, substitution, and oxidative decomplexation to **236** is 80%[216]:

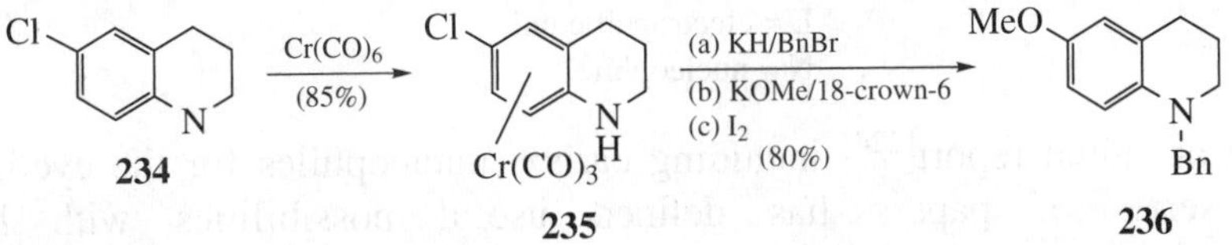

9.2.2.3 Cyclizations via Heteroatom Substitution for Halide

Intramolecular substitution for chloride or fluoride is particularly effective. Oxygen heterocycles with fused benzo rings are obtained from Cr(CO)$_3$ complexes of fluorobenzene with an *o*-(hydroxyalkyl) side chain[224,225]. For example, complexation of

3-(2-fluorophenyl)-1-propanol **237** with Cr(CO)$_3$(pyridine)$_3$ at 25 °C in ether (promoted by BF$_3$•Et$_2$O) gave complex **238**. Reaction in DMSO with excess potassium *tert*-butoxide for 3 h at 25 °C gave the chroman complex **239**[221]. The yield in the cyclization step was 75%, and iodine decomplexation to give benzopyran **240** was quantitative. Efforts to produce the dihydrobenzo*furan* under the same conditions failed; only intermolecular substitution products were obtained[226]. A related study demonstrated the same conversion (**237** → **240**) but using a Rh^{III} species as a *catalyst*[225]. This is a rare example of catalysis of nucleophilic aromatic substitution by transition metals and is unfortunately exceedingly limited in scope:

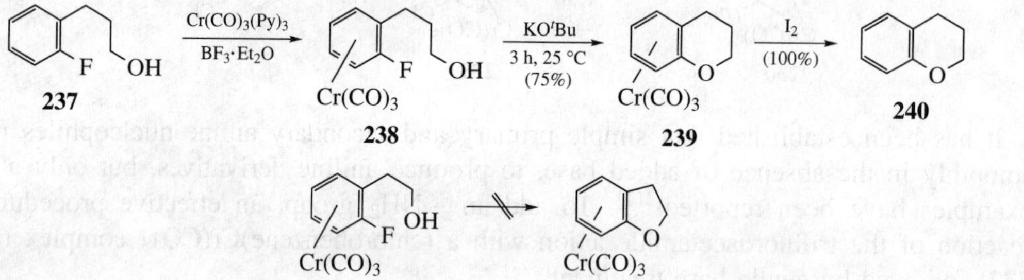

Coordination of arenes with Cr(CO)$_3$ also activates the ring hydrogens toward abstraction with strong base (metalation; see Section 9.2.4.1.3)[224,227,228]. Simple arene ligands can be metalated with alkyllithium reagents; alkoxy, amino, and halo substituents on the arene direct the metalation to the ortho position with rates and regioselectivity higher than with the corresponding free arene (see below)[224,229–231]. This allows a strategy for annulating aromatic rings via ortho lithiation (**241**), trapping with a bifunctional electrophile (to give **242**), and finally nucleophilic substitution for the electronegative substituent (usually F), to give the cycle **243**:

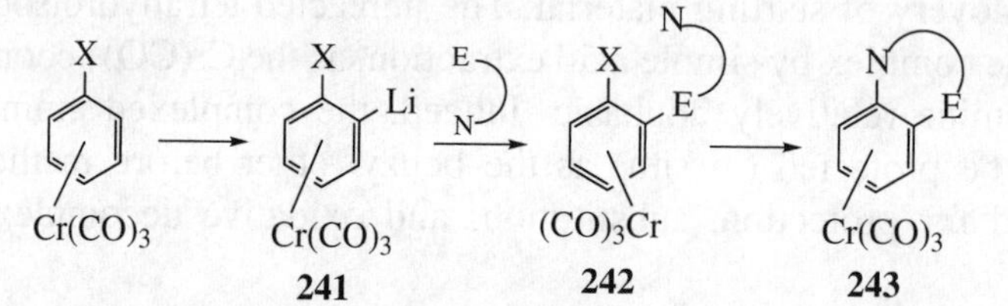

E = electrophilic unit
N = nucleophile

Following an initial report[224] including carbon nucleophiles for the cyclization (see below), a series of papers has defined useful possibilities with heteroatom nucleophiles[218,229–231]. Although chlorobenzene–Cr(CO)$_3$ undergoes lithiation to give a moderately stable species which can be trapped with electrophiles to produce the ortho-substituted chloroarene–Cr(CO)$_3$ complexes[232], the fluoro analog appears to be more efficient, in the same way. The ortho-lithio haloarene complexes are highly basic, and the electrophilic trapping species is restricted to those with low kinetic acidity and high electrophilicity. Good results are obtained with isocyanates, ketenes, and acyl derivatives with

α-protons of low acidity[229–231]. Examples include phenyl isocyanate, which reacts twice to give the six-membered heterocycle **244**, whereas the dimethyl-*N*-carboxy anhydride **245** proceeds directly to the five-membered ring **246** after spontaneous decarboxylation. It is possible that the *gem*-dimethyl group is important in favoring five-membered ring formation, considering the failure in forming furan rings by direct cyclization with a simple side-chain hydroxy group:

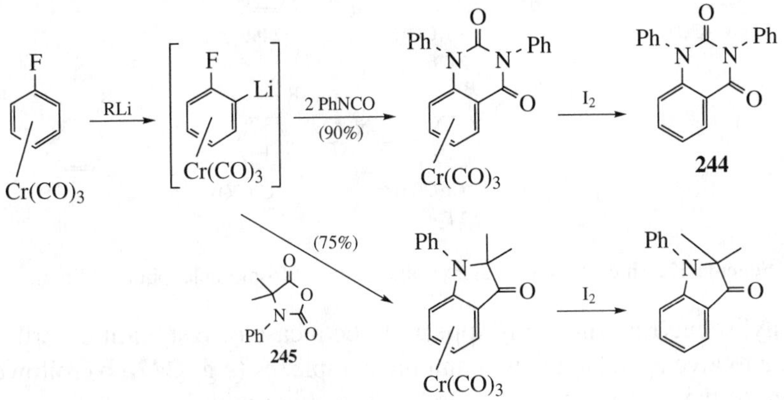

9.2.3 Nucleophilic Substitution with Carbon Nucleophiles

9.2.3.1 Addition/Elimination, S_NAr

Although carbon nucleophiles were suggested to be efficient in substitution for fluoride in the early patent[209], the first examples in the primary literature appeared in 1974[214,215]. It is now clear that there are three reactivity classes of carbon nucleophiles: (a) stabilized carbanions (from carbon acids with pK_a < about 18) which give unfavorable and easily reversible addition to the arene ligand; (b) more reactive carbanions (pK_a > 20) which give complete conversion to cyclohexadienyl addition products (**247a–c**) prior to slow equilibration via reversible anion addition; and (c) more reactive carbanions (pK_a > 20) which give irreversible addition to the arene ligand (Scheme 37). Reversibility also depends on the nature of the arene and the metal–ligand unit, as discussed below.

Sodio diethylmalonate is an example of type (**a**). Reaction with fluorobenzene–Cr(CO)$_3$ proceeds to completion after 20 h at 50 °C in HMPA to give the diethyl phenylmalonate complex in over 95% yield[213]. There is no evidence for an intermediate (*e.g.*, the cyclohexadienyl anion complex **247**). A satisfactory picture assumes that the anion adds reversibly and unfavorably ($k_1 < k_{-1}$, as in Scheme 34), slowly finding its way to the ipso position; then irreversible loss of fluoride gives the substitution product[213,214].

Lithio isobutyronitrile is an example of anion type (**b**). The initial addition to chlorobenzene–Cr(CO)$_3$ is complete within minutes at −78 °C, but the substitution product does not appear until the mixture is warmed to 25 °C[214]. Quenching with iodine after short reaction time leads to a mixture of phenylisobutyronitrile and *o*- and

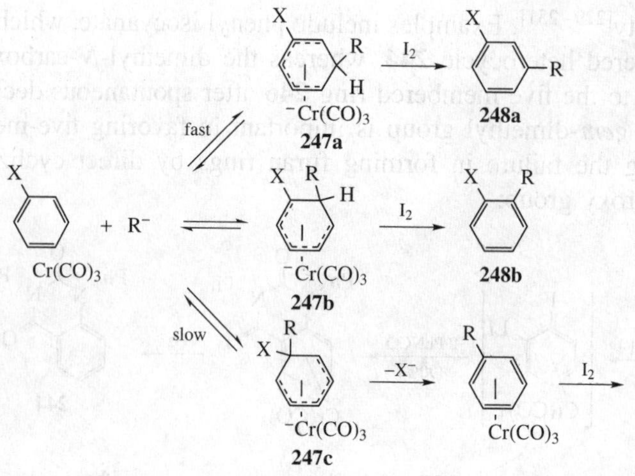

Scheme 37. Three cyclohexadienyl intermediates from nucleophile addition.

m-chlorophenylisobutyronitrile. This appears to be a case of fast addition ortho and meta to the chloride to give cyclohexadienyl anionic complexes (*e.g.*, **247a,b**) followed by slow rearrangement to the ipso intermediate **247c**. Quenching with iodine before equilibration is complete leads to oxidation of the intermediates **247a,b** and formal substitution for hydrogen, as in **248a,b**.

Carbon nucleophiles of type (**c**) add to the arene ligand and do not easily rearrange; examples include the very reactive anions such as 2-lithio-2-methyl-1,3-dithiane and the less sterically encumbered anions such as lithioacetonitrile and *tert*-butyl lithioacetate. In these cases, the anion adds to an unsubstituted position (mainly ortho or meta to Cl, as in **247a,b** and does not rearrange. Then iodine quenching, even after a long period at 25 °C, gives almost exclusively the products from formal substitution for hydrogen, as in **248a,b** in Scheme 36. Reversibility in the addition of the carbanion also depends on the nature of the substituents on the arene ligand and the solvent[233].

Hydrocarbon anions such as cyclopentadienyl and analogs (fluorenyl, indenyl, pentadienyl) substitute for fluoride, leading, for example, to phenylcyclopentadiene as a Cr(CO)$_3$ complex on the arene[234].

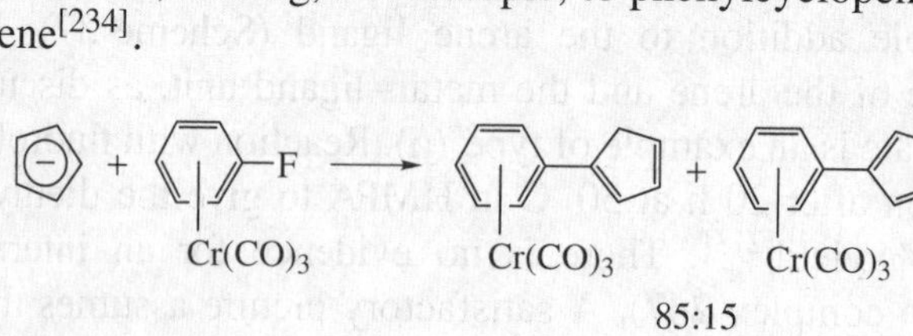

85:15

As before, fluoride is a better leaving group than chloride. The example shown below gives methyl-2-phenyl-2-(*tert*-butylthio) propionate in 94% yield after 15 h at 25 °C[235]. There is an important solvent effect on the rate of equilibration of the carbanions[233], and the conditions chosen for these early experiments are not necessarily the best; there may be room to expand the scope of useful anions by careful choice of media (less polar solvents should favor equilibration):

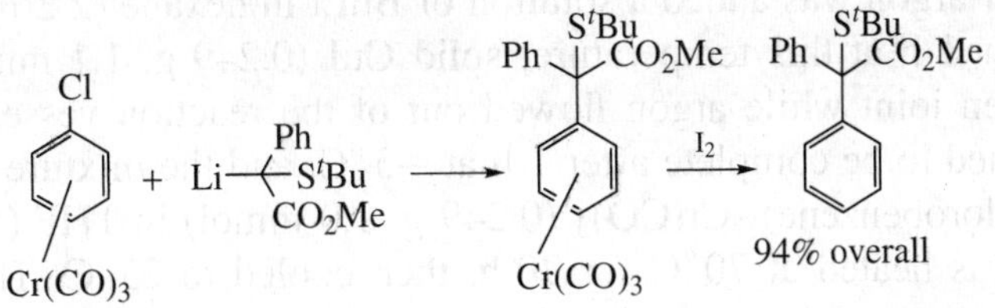

94% overall

An efficient synthesis of homochiral α-phenylalanine **249** utilizes an asymmetric enolate anion (**250**) and substitution for fluoride in the fluorobenzene–Cr(CO)₃ complex[236]:

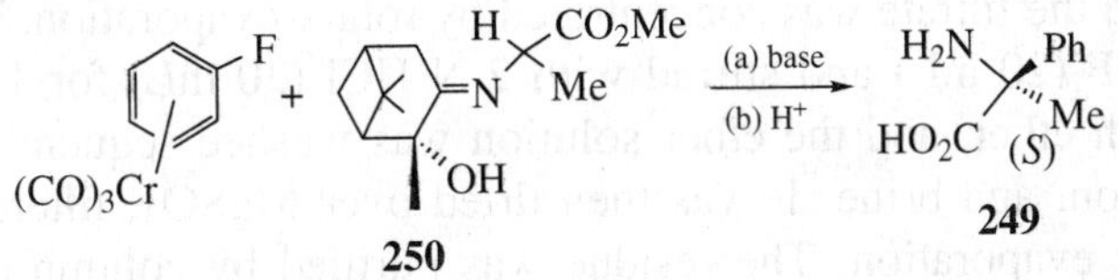

A recent example suggests that the substitution for chloride with primary ketone enolate anions can be accomplished via simple modifications[237]. The ketone is converted to *N,N*-dimethylhydrazone **251** and the lithio anion is mixed with CuI. The presumed copper enolate anion analog, **252**, gives smooth substitution for Cl at −5 °C in THF to give **253**.

Procedure for Substitution for Chloride using a Copper Lithium Azaenolate

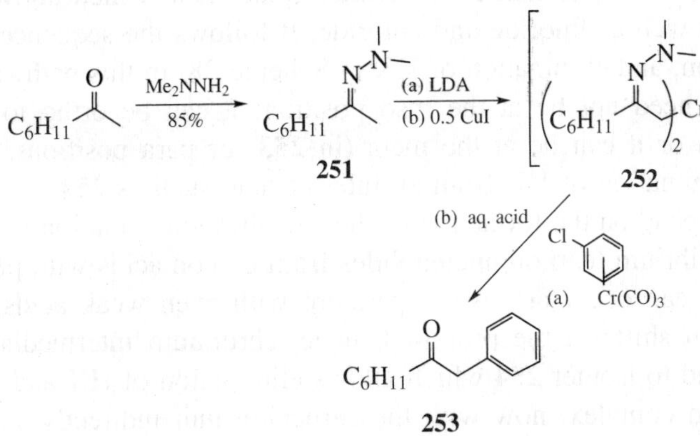

(a) Preparation of the hydrazone **251**. A mixture of 2-octanone (40 mmol), *N,N*-dimethylhydrazine (3.8 mL, 50 mmol) , trifluoroacetic acid (0.05 mL), and benzene (50 mL) was heated at reflux with a Dean–Stark trap employed to removed water. After 5 h, the mixture was cooled to 23 °C and diluted with ether and water. The organic layer was washed with brine, dried over MgSO₄, filtered, and concentrated by rotary evaporation. The residue was distilled to give the *N,N*-dimethylhydrazone of 2-octanone **251** in (84% yield; bp 84 °C/15 torr).

(b) To a solution of the hydrazone (2.1 mmol) in THF (2.0 mL) in a dried reaction flask under argon was added a solution of BuLi in hexane (2.2 mmol, 1.38 mL) at −5 °C. After 1 h at this temperature, solid CuI (0.249 g, 1.1 mmol) was added through an open joint while argon flowed out of the reaction vessel. Formation of **252** was assumed to be complete after 1 h at −5 °C, and the mixture was added to a solution of (chlorobenzene)–Cr(CO)$_3$ (0.249 g, 1.0 mmol) in THF (5 mL) at 23 °C. The mixture was heated at 70 °C for 20 h, then cooled to 23 °C. The mixture was quenched with water and extracted with ether. The ether solution was washed with brine, dried over MgSO$_4$, filtered, and the solution of complex **252** was exposed to air and sunlight for 3 h to detach the Cr from the substituted arene. The mixture was filtered, and the filtrate was concentrated by rotary evaporation. The residue was taken up in THF (20 mL) and stirred with 2 N HCl (30 mL) for 1 h. The mixture was diluted with ether, and the ether solution was washed sequentially with water, NaHCO$_3$ solution, and brine. It was then dried over MgSO$_4$, filtered, and concentrated by rotary evaporation. The residue was purified by column chromatography on SiO$_2$ (hexane/EtOAc 8 : 1) to give 1-phenyl 2-octanone **253** as a colorless oil in 72% yield.

9.2.3.2 Cine/Tele Substitution

There is an alternative mechanism for halide replacement which allows alkoxy to be a leaving group, as well as fluoride and chloride. It follows the sequence of nucleophile addition, protonation, and elimination of HX in Scheme 38. In this pathway, the addition of the nucleophile need not be at the ipso position; it can be ortho to halide leading to cine substitution or it can be at the meta (in **253**) or para positions, leading to tele substitution via elimination of HX from an intermediate such as **254**[238,239].

The processes depend on the formation of the cyclohexadienyl anion intermediates **253** in a favorable equilibrium (carbon nucleophiles from carbon acids with pK_a > 22 or so), protonation (which can occur at low temperature with even weak acids such as acetic acid), and hydrogen shifts in the proposed diene–chromium intermediates (*e.g.*, **255**). Hydrogen shifts lead to isomer **254** which allows elimination of HX and regeneration of an arene–chromium complex, now with the carbanion unit indirectly substituted for X (Scheme 38).

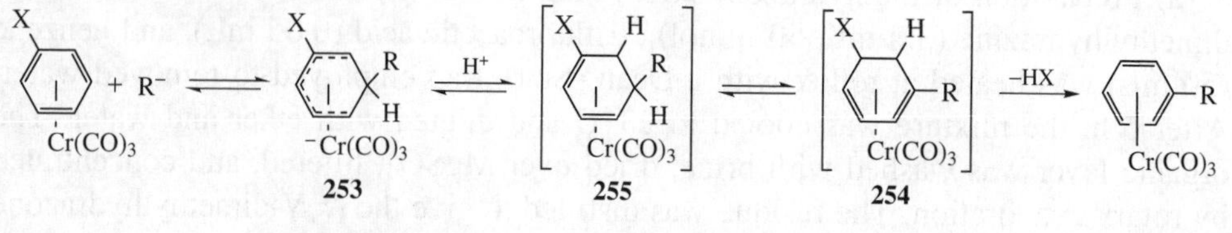

Scheme 38. Indirect substitution for halide and alkoxy.

A particularly useful example of indirect substitution utilizes alkoxy leaving groups; for example, the intermediate **256** from the reaction of the diphenyl ether–Cr(CO)₃ complex (**257**) and the 2-methyl-1,3-dithiane anion can be induced to eliminate phenol after protonation at low temperature. The result is tele substitution (scheme 39)[238].

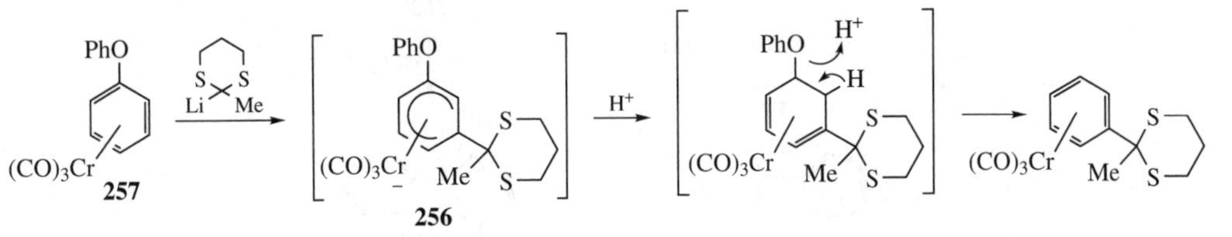

Scheme 39. Phenoxy as a leaving group in indirect substitution.

An impressive example is the reaction of (2,6-dimethyl-1-fluorobenzene)Cr(CO)₃ **258** with 2-lithio-2-methyl-1,3-dithiane at −78 °C followed by treatment with trifluoroacetic acid at −78 °C[239]. Loss of HF leads to the 1,2,4 (tele) substitution product **259** in 62% yield:

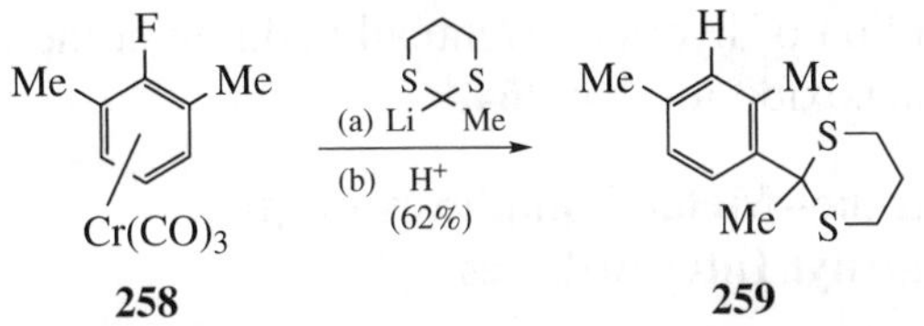

The directing effects of F (strong meta) versus Cl (weak ortho/meta, discussed below) allow control over the site of tele substitution; the chloro analog **260** leads to **261**, with the 1,3,5 substitution pattern[240]:

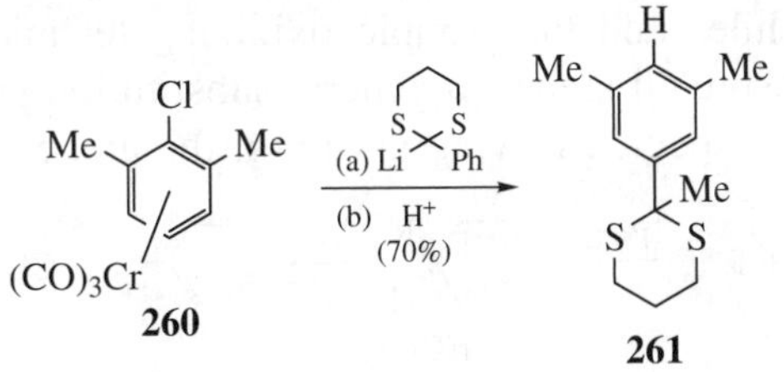

There is evidence for reversible addition of hydride[241] under mild conditions and substitution can occur indirectly (see immediately below). Addition of LiEt₃BH to (arene)Cr(CO)₃ complexes bearing a heteroatom leads, after further treatment with acid, to the product from cleavage of the heteroatom bond, to overall hydride substitution for the heteroatom via cine or tele pathways. Examples include cleavage of carbon–oxygen[242], carbon–halogen[243], and carbon–nitrogen[244] bonds. The regioselectivity and reversibility are shown clearly in the addition to the Cr(CO)₃ complex of dibenzofuran (Scheme 40)[241]. Reaction with LiEt₃BH (67 °C, 0.5 h, 55% yield) gave a yellow amorphous powder which was assigned structure **262**. Reaction with electrophiles such as

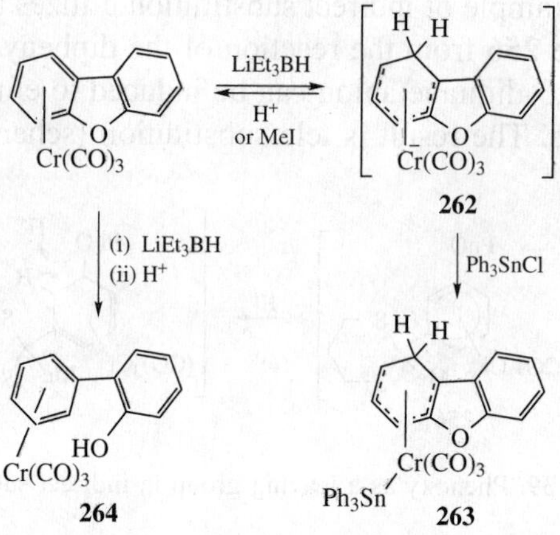

Scheme 40. Addition of hydride and substitution for oxygen.

H$^+$ and MeI leads to regeneration of the dibenzofuran complex, but triphenyltin chloride reacts at the metal, to give **263**. Longer reaction time with LiEt$_3$BH proceeds with apparent reversible addition of hydride, eventually adding at the position bearing oxygen and allowing loss of phenoxide to give **264**.

9.2.4   Addition to Arene–Metal Complexes to give η^5-Cyclohexadienyl Intermediates

9.2.4.1 *Addition/Oxidation: Formal Nucleophilic Substitution for Hydrogen*

9.2.4.1.1 General features   Once it is recognized that cyclohexadienyl anionic complexes of chromium[245] (and other metal–ligand systems) can be generated by addition of sufficiently reactive nucleophiles and that simple oxidizing techniques convert the anionic intermediates to free substituted arenes, a general substitution process becomes available that does not depend on a specific leaving group on the arene:

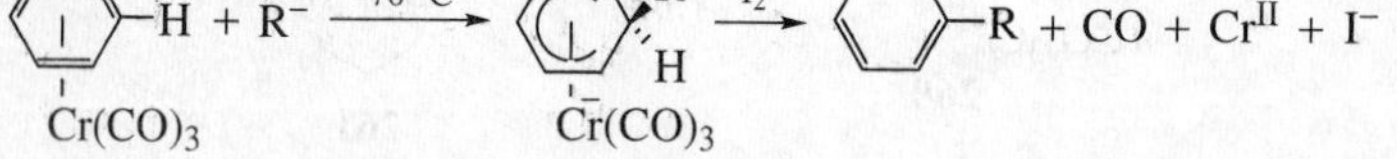

Although there is only one example of a heteroatom nucleophile[246], the process is general for carbanions derived from carbon acids with pK_a > 22 or so. Exceptions among the very reactive anions include organolithium reagents (deprotonation of the arene ligand and addition to the CO ligand are favored), and Grignard reagents (addition to the CO ligand is favored). With anions such as ester enolate anions, reaction occurs within minutes at $-78\,°$C and the intermediate cyclohexadienyl complex can be observed spectroscopically. The intermediates are exceedingly air sensitive and are generally quenched directly, without purification. In one case, from the addition of 2-lithio-1,3-dithiane, the adduct has been crystallized and fully characterized by X-ray diffraction analysis[245]. Oxidation with excess iodine (at least 2.5 mol-equiv of I_2) also proceeds at $-78\,^{\circ}C$ and is complete within hours below room temperature. The products are the substituted arene, HI, Cr^{III}, and CO. For acid-sensitive products such as trialkylsilyl-substituted arenes, an excess of amine (conveniently, diisopropylamine) can buffer the mixture.

9.2.4.1.2 Regioselectivity in addition/oxidation, substitution for hydrogen   Regioselectivity has been the subject of numerous studies with a variety of mono- and polysubstituted arene ligands. The addition of synthetically interesting carbanions to the typical arene ligand, especially those bearing electron-donating substituents, can be reversible at rates comparable to the rate of addition; it is not always obvious whether to assume kinetic or thermodynamic control[247,248-251]. However, correlations have been suggested that allow prediction or rationalization of regioselectivity with a modest degree of confidence. With some significant exceptions as discussed below, the difference between the kinetic and thermodynamic selectivity has not been determined or is small.

With arenes bearing a single resonance donor substituent (Do = NR_2, OMe, F), the addition is strongly preferred at the meta position, with small amounts of ortho substitution $(0-10\%)$[252,253]:

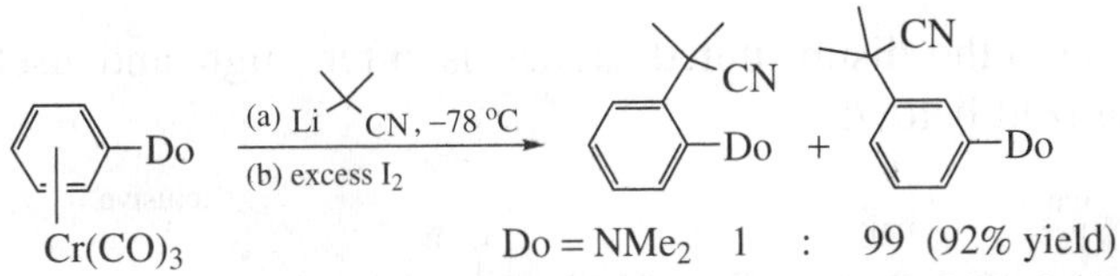

Do = NMe_2 1 : 99 (92% yield)
Do = F   2 : 98 (84% yield)

The meta acylation of the anisole ligand in **265**, using a carbonyl anion equivalent **266** as the nucleophile, illustrates the unique regioselectivity available with the $Cr(CO)_3$ activation, producing **267**[252]:

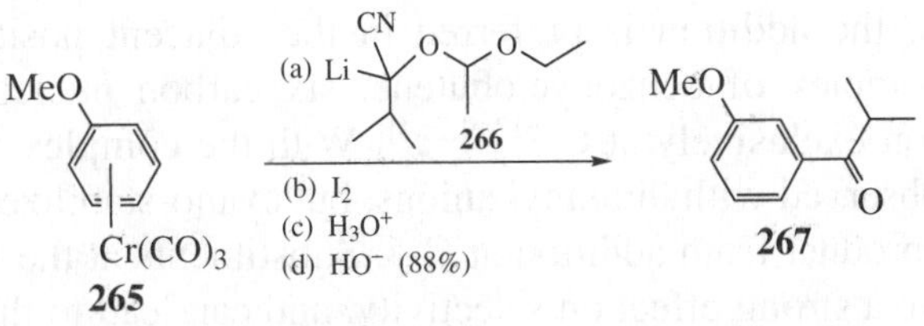

However, the selectivity is more complicated with a methyl or chloro substituent. Again, meta substitution is always significant, but ortho substitution can account for 50–70% of the mixture in some cases[251-253]. More reactive anions (1,3-dithianyl) and less substituted carbanions (*e.g.*, *tert*-butyl lithioacetate) tend to favor ortho substitution. The added activating effect of the Cl substituent allows addition of the pinacolone enolate anion, whereas no addition to the anisole nor toluene ligand is observed with the same anion:

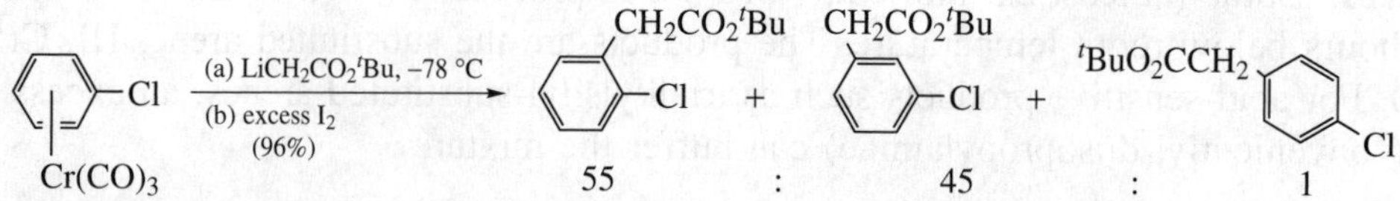

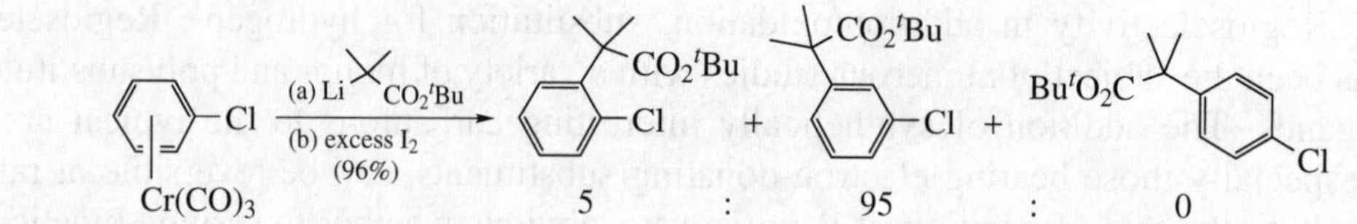

In general, electron-withdrawing substituents such as acyl and cyano are also activated toward nucleophile addition to the substituent (C=O or CN) and this process competes with addition to the ring itself. The CF$_3$ and SiMe$_3$ groups are also para directing:

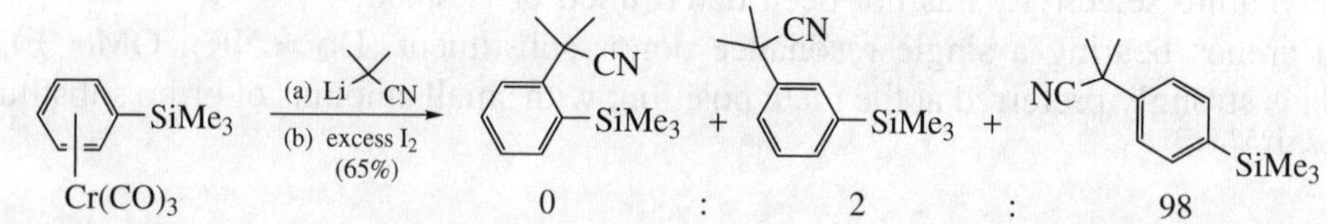

Regioselectivity in ortho-disubstituted arenes is often high and useful. A series of examples is summarized below:

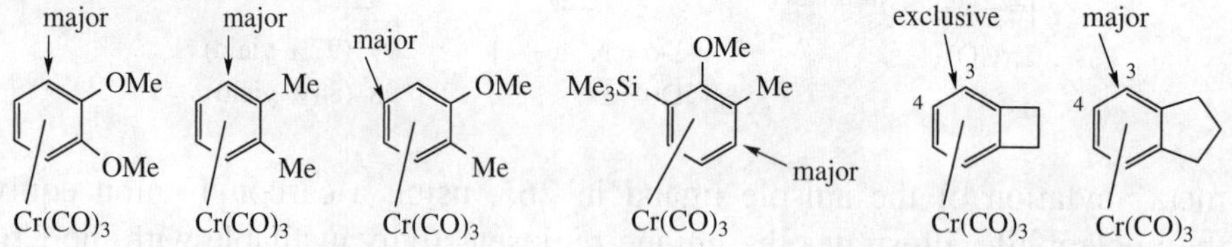

The resonance donor substituent (OMe) appears to dominate the directing influences, favoring addition at the less hindered position meta to itself[252]. However, with two identical substituents, the addition is preferred in the adjacent position[250,253,254]. For example, with the complex of benzocyclobutene, six carbon nucleophiles were tested and each gave addition exclusively at C-3[249,254]. With the complex of indane, selective addition at C-3 was observed with dithianyl anions, but cyano-stabilized anions gave up to 20% of the isomeric product from addition at C-4. Substituents at the benzylic carbons in the indane ligand have a strong effect on selectivity, and can lead to the 1,2,4-substitution product[250].

Addition to *o*-alkyl aniline complexes has been the subject of a detailed study, and presents a clear case of different products from kinetic versus thermodynamic control[249,250]. Addition of 2-lithio-2-cyanopropane to *N*-methyl(tetrahydroquinoline)-Cr(CO)$_3$ (**268**; Scheme 41) with variation in time, temperature, and solvent (THF or THF with 4 mol equiv of HMPA added) gives product distributions ranging from 2 : 1 for C-5/C-3 (**269a/b**) at −78 °C for 1 min to >96% addition at C-5 when the adducts (**270a,b**) are allowed to equilibrate (8.6 h at −78 °C). The data show that equilibration is occurring within minutes at −78 °C, and the product distribution does not change upon warming to 20 °C. The effect of added HMPA was also observed in halide substitution (addition/elimination)[213] and clearly established during addition to simple arenes[255]. With the less sterically demanding anion LiCH$_2$CN, addition is incomplete after 22 h at −78 °C, and favored at C-3 (82 : 18 for C-3/C-5).

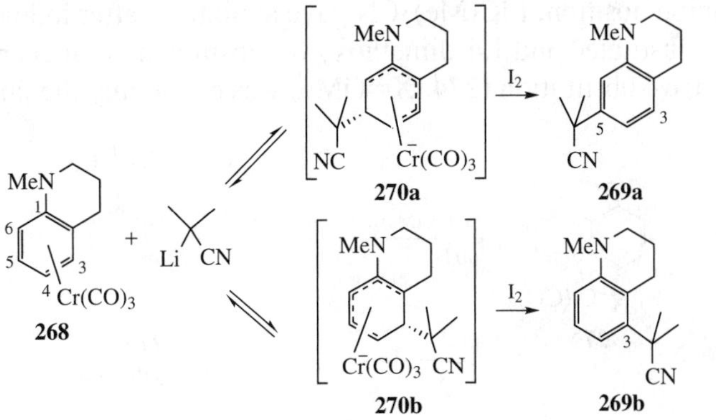

Scheme 41. Kinetic and thermodynamic control in regioselectivity.

The same study shows that with the complex of *N,N*-dimethyl-*o*-toluidine (**271**), the selectivity depends on time, temperature, and the nature of the anion[249]. Again, equilibration occurs with the tertiary nitrile-stabilized anion, favoring the C-4 substitution product, whereas lithioacetonitrile favors addition at C-3. The complex of *N*-methyl indoline (**272**) gives similar behavior, favoring addition at C-4 (indole numbering), but with significant addition at C-7, depending on the anion and opportunity for equilibration[249,256]. With the 2-methyl-1,3-dithianyl nucleophile, a high selectivity for C-4 is observed.

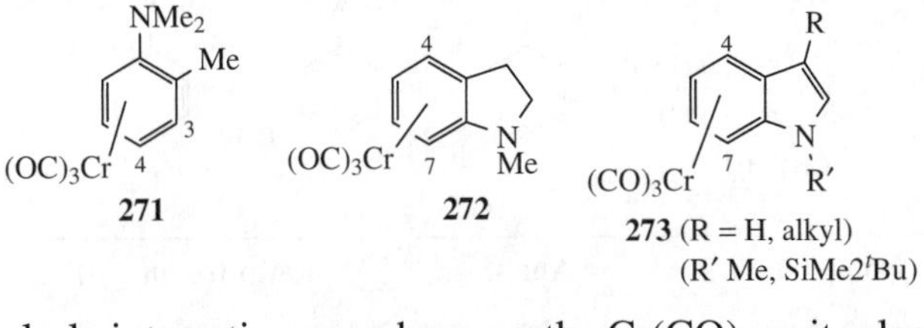

Indole is a particularly interesting case, because the Cr(CO)$_3$ unit selectively activates the six-membered ring (**273**)[256,257], whereas in free indole the five-membered ring dominates the (electrophilic addition) reactivity. The selectivity in addition to the Cr(CO)$_3$ complexes of indole derivatives shows a preference for addition at C-4 (indole numbering) and C-7, with steric effects (due to substituents at C-3 and N-1) as well as anion type influencing the selectivity[256]. A hydrogen substituent at C-3 and a tertiary carbanion leads to selective C-4 substitution. The same substrate adds 2-lithio-1,3-dithiane predominantly at C-7. With a trimethylsilylmethyl substituent at C-3, the addition is preferred at C-7.

Another example in which the regioselectivity of addition is different under kinetic rather than thermodynamic control is the naphthalene series. In the addition of

LiC(Me)$_2$CN to naphthalene-Cr(CO)$_3$ (**274**, X=H), a mixture of products is observed from addition at C-α and C-β in the ratio 42 : 58 under conditions where equilibration is minimized (0.3 h, $-65\,^\circ$C, THF/HMPA). With the same reactants, but in THF and at $0\,^\circ$C, the product is almost exclusively the α-substituted naphthalene (**275**, X=H)[247,253]. Using the standard procedures, 1,4-dimethoxynaphthalene is complexed at the less-substituted ring with high selectivity to give (**274**, H=OMe)[233]. Again, under conditions of minimum equilibration of anion position, LiC(Me)$_2$CN gave a mixture (after iodine oxidation) of the 1,4-dimethoxy-β-substituted and 1,4-dimethoxy-α-substituted products in the ratio 78 : 22. After equilibration, α-substitution (**274**, X=OMe) was essentially the only product found:

274   **275**
>98% selectivity

The steric effect on regioselectivity shows clearly in the series given in Table 1[258]. Comparing similar anion type, except for size, entries 1, 7, 8, and 9 show that ortho substitution is very significant with a primary carbanion, but (other entries) essentially absent with a tertiary cyano-stabilized anion. It is striking that as the size of the alkyl substituent on the arene increases, not only is ortho substitution disfavored, but meta is as well—compare entries 3–5, and compare 8 and 9. With the very large CH(^{t}Bu)$_2$ group (entry 6), only para substitution is observed. Regioselectivity is also dependent

Table 1. Steric effects on regioselectivity.

Entry	Complex	Anion	Ratio ($o : m : p$)	Combined yield
1	X=Me	LiCH$_2$CN	35 : 63 : 2	88
2	X=Me	LiC(OR)MeCN[a]	0 : 96 : 4	75
3	X=Et	LiC(OR)MeCN[a]	0 : 94 : 6	89
4	X=^{i}Pr	LiC(OR)MeCN[a]	0 : 80 : 20	88
5	X=^{t}Bu	LiC(OR)MeCN[a]	0 : 35 : 65	86
6	X=CH(^{t}Bu)$_2$	LiC(Me)$_2$CN	0 : 0 : 100	63
7	X=Me	LiCH$_2$SPh	52 : 46 : 2	96
8	X=^{i}Pr	LiCH$_2$SPh	47 : 46 : 7	86
9	X=^{t}Bu	LiCH$_2$SPh	35 : 32 : 33	88

[a]R=C(Me)OEt

on the electronic nature of the nucleophile. Most remarkably, addition of the primary sulfur-stabilized anion shows nearly equal amounts of ortho and meta substitution with the toluene ligand but, as the size of the arene substituent increases, the para substitution product increases at the expense of the meta product (entries 7–9).

The dependence of selectivity on anion type is understood as a change in the balance of charge versus orbital control[259]. Analysis of charge control requires a knowledge of charge distribution in the arene ligand. Analysis of orbital control can be approximated by emphasizing interaction of the frontier molecular orbitals, HOMO for the nucleophile and LUMO for the arene complex. It was recognized early that the ortho/para selectivity observed with weakly polar substituents such as Me and Cl correlated with the LUMO for the free arene, and the assumption was made that the LUMO for the arene ligand has a distribution of coefficients similar to the free arene[251,252]. This has been supported by computation at the level of extended Huckel theory[260]. However, the frontier orbital picture based on the free arene does not account for nearly exclusive meta selectivity in addition to anisole–Cr(CO)$_3$; LUMO for anisole shows essentially the same pattern as for toluene. With a strong resonance electron donor the traditional electronic picture (deactivation of the ortho and para positions) is sufficient to account for the observed meta selectivity. In this case the balance of charge control and orbital control is pushed toward charge control by strong polarization. The same argument applies to the aniline and fluorobenzene complexes. However, another rationale for polarization effects in the arene ligand is the based on the influence of substituent on the conformation of the Cr(CO)$_3$ unit[261–263].

Procedure for Selective Meta Addition/Oxidation, Substitution for Hydrogen

To a solution of LDA (3.3 mmol) in THF (30 mL) under argon at −78 °C was added 5-hexenenitrile (0.314 g, 3.30 mmol). After 1 h, HMPA (10.4 mL, 60 mmol) was added over 2 min, followed by a solution of **276** (1.19 g, 3.0 mmol) in THF (4 mL). After 30 min, the reaction mixture was warmed to 0 °C for 0.5 h, and then cooled again to −78 °C. A solution of iodine (3.0 g, 24 g-at.) in THF (10 mL) was added rapidly. The resulting black mixture was allowed to warm to 23 °C and was stirred at this temperature for 6 h. The mixture was diluted with ether and washed sequentially with saturated aqueous Na$_2$SO$_3$ and brine. It was dried over K$_2$CO$_3$, filtered, and concentrated by rotary evaporation. The residue was taken up in dioxane (50 mL) and 12 M aqueous HCl (5 mL) was added. The resulting mixture was heated at reflux for 3 h, and then cooled to 23 °C. After concentration by rotary evaporation, the residue was partitioned between ether and water, and the ether layer dried over K$_2$CO$_3$, filtered, and the filtrate was concentrated by rotary evaporation. The residual dark oil was purified by medium pressure LC to yield **277** (0.511 g, 60% yield) as a colorless oil.

9.2.4.1.3 Intramolecular addition/oxidation   Intramolecular addition/oxidation with reactive carbanions is generally successful; most of the examples involve cyano-stabilized carbanions. Formation of a six-membered fused ring is efficient, but five-membered fused ring formation is not[264]. The only addition/oxidation product is a cyclic dimer, the [3.3]-metacyclophane (Scheme 42).

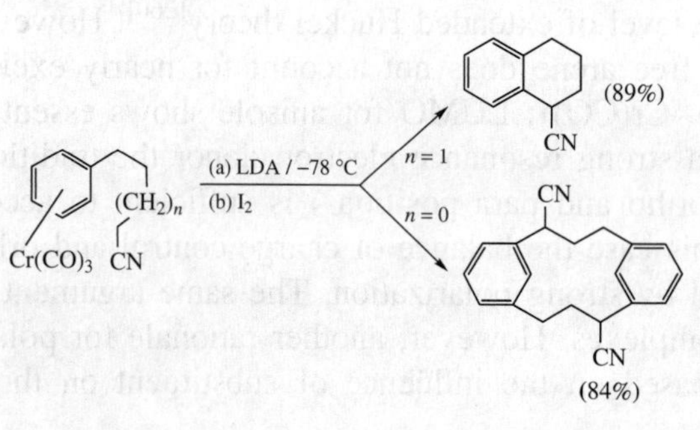

Scheme 42. Metacyclophane formation.

Reversibility again is apparent with the higher homolog **278**. At low temperature (0.5 h/−78 °C), a mixture of cyclohexadienyl anionic intermediates is formed with the spiro ring isomer **279** preferred by a factor of almost 3 : 1 (Scheme 43). An alternative quenching procedure, using trifluoroacetic acid (see below) retains the spiro ring and produces the spiro[5.5]cyclohexadiene product **280**. If the initial adduct is allowed to warm to 0 °C or above, essentially complete rearrangement to the fused ring adduct **281** occurs, and oxidative quenching gives the seven-membered ring **282** in good yield[263].

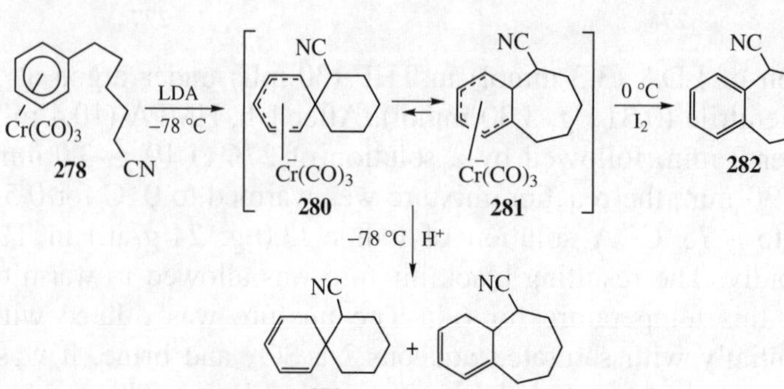

Scheme 43. Spiro versus fused ring formation.

9.2.4.2 Addition/Protonation: Synthesis of Substituted 1,3-Cyclohexadienes

The intermediate η^5-cyclohexadienyl anionic species (*e.g.*, **283**) from nucleophilic addition to the arene–Cr(CO)₃ complexes are obviously highly electron rich, and should be susceptible to reactions with electrophiles. Protonation is efficient at low temperature, and is suggested to produce a labile η^4-1,3-cyclohexadiene complex (**284**) which can undergo H-migration to give the more stable 1,3-cyclohexadiene isomer **285**; in simple systems, the major product is the 1-substituted 1,3-cyclohexadiene (**286**, Scheme 44)[245]. The products can be aromatized by reaction with 2,3-dichloro-5,6-dicyanobenzoquinone.

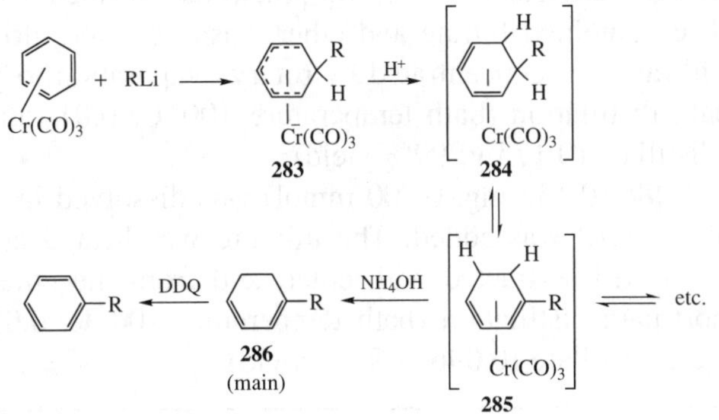

Scheme 44. Addition/protonation.

The reaction appears to be general, although only a limited number of examples have been reported. A particularly useful process begins with meta addition to anisole–Cr(CO)₃, followed by protonation and hydrolysis of the enol ether unit[264]. The result is the 5-substituted cyclohex-2-en-1-one **287**. The intermediate dienol ether **288** can be isolated in high yield, before the aqueous hydrolysis. Other alkene positional isomers of the product enone can be obtained selectively depending on the conditions of the acid treatment and the acid hydrolysis. The addition/protonation process does not change the oxidation state of the chromium, and a procedure has been defined for recovery of the Cr⁰ in order to allow direct recycling[265]. The addition of CO or other donor ligands during the protonation process does not influence the product distribution; CO insertion to produce formyl derivatives is not observed.

Procedure for Meta Substitution of Anisole and 3-Substituted Cyclohexenone Synthesis

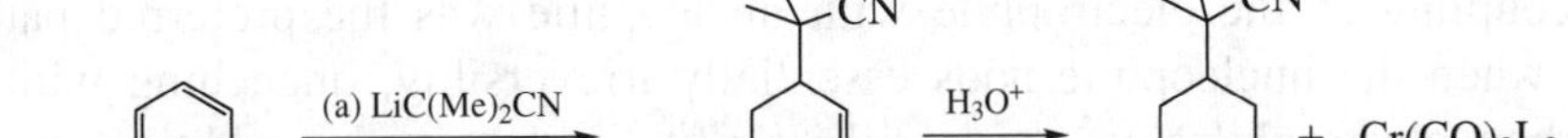

(a) To a sample of THF (10 mL) under argon at 0 °C was added diisopropyl amine (0.14 mL, 1.0 mmol) followed by a solution of BuLi in hexane (0.50 mL of 2.0 M solution). The mixture was stirred at 0 °C for 15 min and then cooled to −78 °C and isobutyronitrile (0.090 mL, 1 mmol) was added. The mixture was warmed to 0 °C for 15 min, and then cooled again to −78 °C. A solution of anisole–Cr(CO)$_3$ (0.245 g, 1.00 mmol) in THF (2 mL) was added dropwise over 1 min, and the solution was allowed to warm slowly to 0 °C over 1 h. The mixture was again cooled to −78 °C and trifluoroacetic acid (0.185 mL, 2.5 mmol) was added. The solution turned deep red. It was allowed to warm to 0 °C and was then partitioned between aqueous ammonium hydroxide and ether. A yellow precipitate formed which was filtered and discarded. The combined filtrate and ether washings were dried over MgSO$_4$, filtered, and the filtrate was concentrated by rotary evaporation to give a yellow oil residue. Short path distillation (bath temperature 100 °C, 0.01 torr) gave **286** as a colorless liquid distillate (0.175 g, 99% yield).

(b) A sample of **286** (0.132 mg, 0.800 mmol) was dissolved in THF (5 mL) and aqueous HCl (5%, 5 mL) was added. The mixture was heated at reflux for 14 h. The mixture was cooled, extracted with ether, and from the ether was isolated a colorless oil. Short-path distillation (bath temperature 100 °C, 0.01 torr) gave **287** as a colorless liquid distillate (0.096 g, 77% yield).

The intramolecular version of addition/protonation with (*m*-cyanoalkyl)anisole ligands produces spirocyclic enones such as **289**[263].

(a) LiNR$_2$, 1 h, −78 °C
(b) CF$_3$CO$_2$H
(c) NH$_4$OH
(d) H$_3$O$^+$
(96%)

289
(mixture of epimers)

The addition/protonation process has been coupled with meta addition of a carbonyl anion equivalent and the controlled exo addition of the incoming nucleophile to generate acorenone and acorenone B stereospecifically from *o*-methylanisole-Cr(CO)$_3$[266].

9.2.4.3 Addition/Acylation (allylation)

The efficient trapping of the cyclohexadienyl anionic intermediates with protons raises the possibility of quenching with carbon electrophiles. The process is not as general as the proton quench; early experiments suggested that reversal of the nucleophile addition and coupling of the electrophile with nucleophile was the preferred pathway[244]. However, when the nucleophile adds essentially irreversibly, quenching with a limited set of carbon electrophiles is successful[267–269]. For example, addition of 2-lithio-1, 3-dithiane to (benzene)Cr(CO)$_3$ followed by addition of methyl iodide and then oxidation or addition of a donor ligand (CO, Ph$_3$P) produces cyclohexa-1,3-diene **290** substituted in a trans relationship by both acetyl (Me + CO) and the nucleophile (Scheme 45). The insertion of CO occurs with a variety of electrophiles. The insertion is efficient without added CO, but somewhat higher yields are obtained under a modest pressure (<4 bar) of CO during electrophile coupling. The attachment of the electrophile is from the endo direction, consistent with initial addition to the metal (**291**) followed by CO insertion (**292**) and migration up to the arene unit; the products then have the trans arrangement of the new substituents.

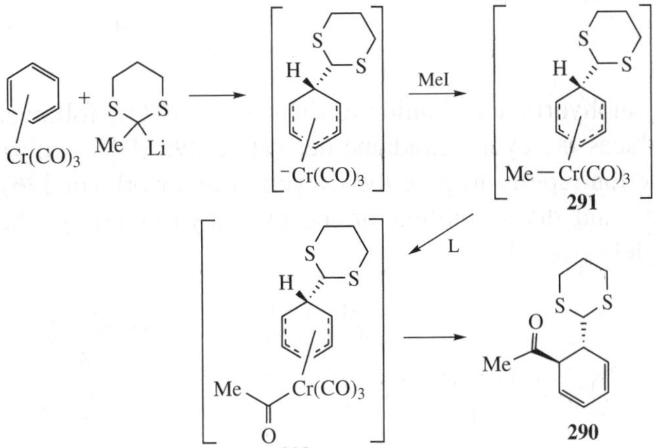

Scheme 45. Nucleophile addition/electrophile trapping with CO insertion.

More reactive anions such as the 2-lithio-1,3-dithiane derivatives, phenyllithium, and *tert*-butyllithium do not require a special solvent and proceed in high yield in THF. Although HMPA is known to suppress the migratory insertion to CO in anionic complexes[270], it does not deter the CO insertion in these cases. In the special case of a phenyl group substituted with an oxazoline, anion addition followed by trapping with allyl bromide leads to a 1,2-disubstitution product without CO insertion[271]. The only electrophile that adds without CO insertion is the proton, as discussed above. Good alkylating agents (primary iodides and triflates, allyl bromide, benzyl bromide) react below 0 °C, but ethyl bromide requires heating at 50 °C. The reaction is selective for a primary alkyl iodide in the presence of an ester or a ketone unit.

Although the addition/oxidation and the addition/protonation procedures are successful with ester enolates as well as more reactive carbon nucleophiles, the addition/acylation procedure requires more reactive anions and the addition of a polar aprotic solvent (HMPA has been used) to disfavor reversal of anion addition. Even under these conditions, cyano-stabilized anions and ester enolates fail (simple alkylation of the carbanion) but cyanohydrin acetal anions are successful. The addition of the cyanohydrin acetal anion **293** to 1,4-dimethoxynaphthalene-Cr(CO)$_3$ occurs by kinetic control at C-β in THF/HMPA and leads to the α,β-diacetyl derivative **294** after methyl iodide addition, and hydrolysis of the cyanohydrin acetal[266–268]. Surprisingly, acyl chlorides are unreactive. A simple

S$_N$2 process is proposed, based on the reactivity pattern of the electrophiles and the lack of rearrangement during alkylation with cyclopropyl carbinyl iodide (no long-lived radical intermediates)[267].

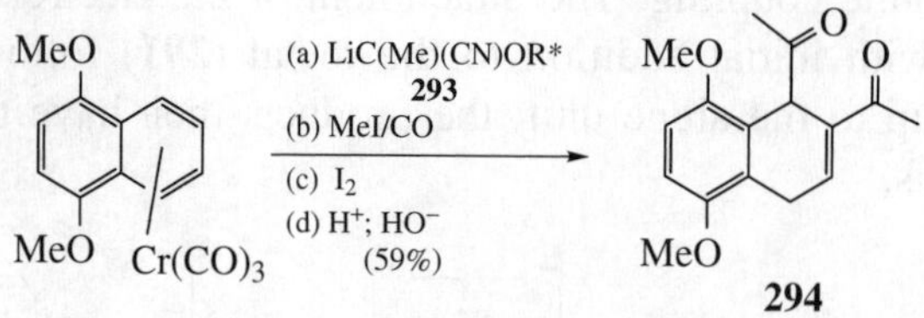

R* = CH(Me)(OEt)

Addition of a cyanohydrin acetal anion to (benzene)Cr(CO)$_3$ followed by reaction with allyl bromide produces the cyclohexadiene derivative 295 (94% yield) which undergoes a Diels–Alder reaction rapidly to give the tricyclic framework (in 296). After quenching with methyl iodide and disassembling of the cyanohydrin group, the diketone 296 is obtained in 50% yield overall[267]:

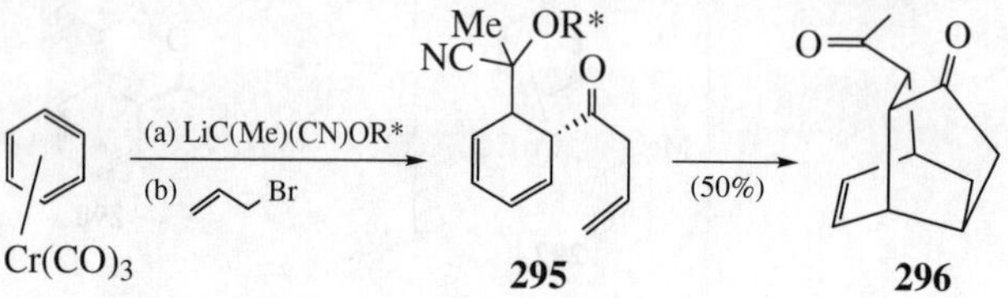

R* = CH(Me)(OEt)

9.2.5   Addition to Styrene-type Ligands Activated by Cr(CO)$_3$

Nucleophile addition to styrene derivatives (e.g., 297) coordinated with Cr(CO)$_3$ is another example of addition/electrophile trapping[272]. Addition of reactive anions is selective at the β-position of the styrene ligand, leading to the stabilized benzylic anion 298. The intermediate reacts with protons and a variety of carbon electrophiles to give substituted alkylbenzene ligands (in 299). In the dihydronaphthalene series, the selective exo addition of the nucleophile and exo addition of the electrophile (steric approach) results in exclusive formation of the cis-α,β-disubstituted tetralin 299[272]:

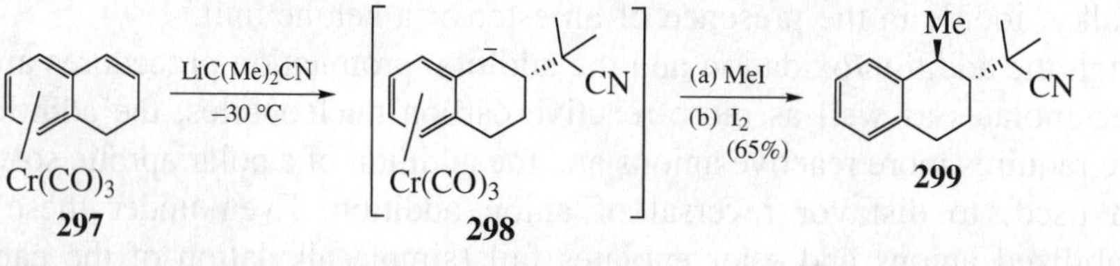

The addition/protonation procedure maintains the arene–chromium bond, and allows further application of the activating effect of the metal. In an approach to the synthesis of anthraquinone antibiotics, the dihydronaphthalene complex 300 was allowed to react with a cyanohydrin acetal anion and then quenched with acid[273]. The resulting tetralin complex **301** could be metalated effectively (see below) and carried on to key intermediate **302** in anthraquinone construction:

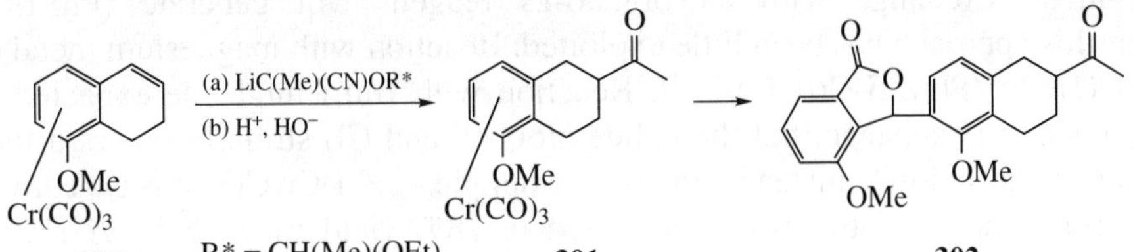

| **300** | R* = CH(Me)(OEt) | **301** | **302** |

9.3 Lithiation of the Arene Ligand and Trapping with Electrophiles

9.3.1 Introduction

There are three types of electrophilic sites in (benzene)Cr(CO)$_3$: the aromatic ring (Scheme 46, path a), the CO ligand (path b), and the ring protons (path c). Most common nucleophiles/bases will add to the ring (path a); a few organolithium reagents are known to add to the CO ligand (path b)[274,275]. Selective proton abstraction is highly demanding of the properties of the base, requiring high kinetic basicity and low nucleophilic reactivity. Other means of forming the aryllithium ligands parallel the methods of formation of traditional aryllithium reagents, halogen–metal exchange and transmetalation.

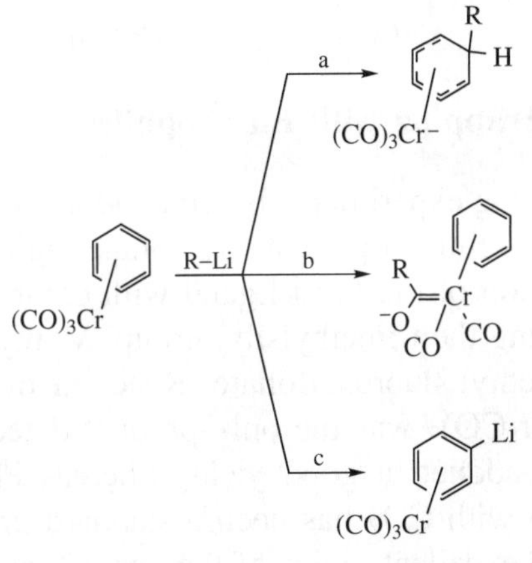

Scheme 46. Possible reactions with nucleophile/base.

In a quantitative study of the acidifying effect of coordination to Cr(CO)$_3$, equilibration measurements suggested that the pK_a of benzene–Cr(CO)$_3$ is about 34 (compared to 43 for benzene in cyclohexylamine) and that the acid-strengthening effect in a series of substituted derivatives is 6–7 pK_a units[276]. The same work shows that lithium diisopropylamide is sufficient to bring about complete and rapid proton removal.

The complexes (PhX)Cr(CO)$_3$ are known[277] for X = F, Cl, Br, and I, but only the fluoro and chloro examples are easily prepared. In principle, reaction with Li or halogen–metal exchange with an organo-Li reagent will generate (PhLi)Cr(CO)$_3$. However, this approach has been little exploited. Reaction with magnesium metal converts (PhI)Cr(CO)$_3$ to (PhMgI)Cr(CO)$_3$[278]. Reaction with BuLi might be expected to bring about halogen–Li exchange, but the halide atom (F and Cl) strongly favors ortho proton abstraction (see below) instead. In one example, (PhI)Cr(CO)$_3$ was converted to (PhLi)CrCO)$_3$ and the latter was methylated in 78% yield overall[279]. A recent paper takes advantage of KH-promoted desilylation of a trialkylsilylarene ligand to generate (and trap *in situ*) aryl anion coordinated with Cr(CO)$_3$[280].

One of the earliest techniques for preparation of (PhLi)Cr(CO)$_3$ is the reaction of diphenylmercury bis-Cr(CO)$_3$ with BuLi[281]. Quenching with CO$_2$ gave the benzoic acid complex in 50% yield. The method is limited primarily by low accessibility of the starting mercury compounds:

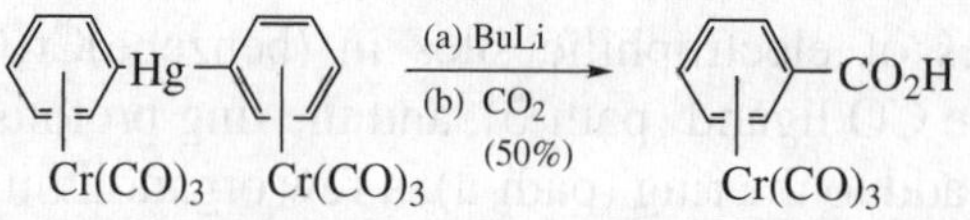

Related is the conversion of (PhLi)Cr(CO)$_3$ to other metal derivatives, such as a copper species, by metathesis with metal halides[282,283] (applications are discussed below):

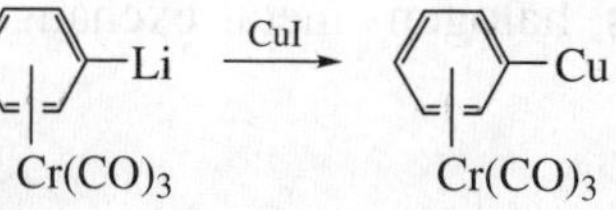

9.3.2 Generation and Trapping with Electrophiles

The only halogen-quenching experiments involve iodine, and the yield of (PhI)Cr(CO)$_3$ from (PhLi)Cr(CO)$_3$ is 26%, limited perhaps by competing oxidative decomplexation of the arene ligand[279]. Reaction of the PhLi ligand with chlorotrimethylsilane is an excellent technique for introducing the trimethylsilyl group. Methylation is also efficient, with either methyl iodide or methyl fluorosulfonate. Reaction of (PhLi)Cr(CO)$_3$ with BuBr and BuOTs failed; (PhH)Cr(CO)$_3$ was the only product detected in each case[282,283]. A chlorophosphine gives the adduct in good yield, whereas PhPCl$_2$ gives the bis adduct in 36% yield[281]. Reaction with CO$_2$ has been a standard probe for the presence of the aryl-Li ligand, and leads to excellent yields of the benzoic acid product. Quenching with acetyl chloride is complicated by apparent enolization, to give (PhH)Cr(CO)$_3$, and double addition. Reaction with ketones and aldehydes can be efficient, but is also complicated by enolization of the carbonyl derivative. Benzaldehyde and benzophenone give the corresponding adducts in moderate yield. However, acetone gives the adduct in only 29% yield, accompanied by 60% recovered (PhH)Cr(CO)$_3$[282,283]. To establish that enolization is the problem, the same reaction with perdeuteroacetone was shown to produce (PhD)Cr(CO)$_3$ (92% d$_1$). Interestingly, the formation of adduct was now much higher

(68%) and the recovered (PhD)Cr(CO)$_3$ was obtained in only 8% yield[283]. This appears to be the result of an isotope effect on the competing steps of addition to the carbonyl and enolization. With simple aldehydes, the yields are good to excellent, with minor amounts of enolization.

A special example of generation of nucleophilic aryl ligands and addition to carbonyls takes advantage of nucleophilic catalysis with aryl silane ligands[284]. An example is given in the following equation, and a series of substituted arene ligands and carbonyl electrophiles has been reported[285]. As the corresponding arylsilanes are likely to be prepared by metalation and silylation of an (arene)Cr(CO)$_3$ complex, it is not obvious what advantage the nucleophilic catalysis approach holds. Generally, direct quenching with the electrophile of choice is successful and saves the additional steps of silylation and silyl activation:

$$X = o\text{-Me},\ m\text{-Me},\ p\text{-Me},\ o\text{-Cl},\ m\text{-Cl},\ p\text{-Cl},\ H$$
$$R^1 = H,\ Me,\ Ph;\quad R^2 = Ph,\ Pr,\ Me$$

9.3.3   Regioselectivity in the Deprotonation of (Arene)Cr(CO)$_3$ Complexes

9.3.3.1 Alkyl Substituents

In free toluene, the kinetic acidity of the methyl protons is higher than the ring protons by a factor of 150[285]. Under equilibration conditions, the benzylic position of (PhMe)Cr(CO)$_3$ is strongly favored. However, kinetic deprotonation of the toluene complex with BuLi at low temperature followed by quenching with CO$_2$, produces a mixture of ortho (9%), meta (35%), and para (35%) toluic acids, along with phenyl acetic acid (20%). The apparent retardation in kinetic acidity for the benzylic position is attributed to a steric effect of the Cr(CO)$_3$ group, blocking the Me substituent with some selectivity[286]:

$$o: 9\%$$
$$m: 35\%$$
$$p: 35\%$$

In a systematic study of lithiation of alkylarene ligands[286] steric inhibition of ortho substitution is clear. The unhindered positions tend to be comparably reactive. With mesitylene, only benzylic (*i.e.*, α) lithiation/alkylation is observed. In general, the regio-selectivity of lithiation with alkylbenzene ligands is poor, and little development has occurred.

9.3.3.2   Halo Substituents

Treatment of uncomplexed haloaromatics using strong base produces selective ortho metalation[287], and opens a pathway to o-benzyne intermediates. The overall conversion to an o-lithiohaloarene and quenching with an electrophile is generally successful only with X=F[288]. For chloride and the heavier halides, halogen–metal exchange competes with proton abstraction, and benzyne formation for the o-lithiohaloarene is too fast to allow efficient electrophile trapping[289]. The effect of coordination of the haloarene to the Cr(CO)$_3$ group is to favor o-lithiation relative to benzyne formation, probably via acceleration of the deprotonation step.

With fluorobenzene, the initial proton abstraction is complete within 0.5 h at $-78\,°C$ to give a clear yellow solution; the mixture turns red with obvious decomposition when heated above $-20\,°C$[283,286]. A new feature arises which will be highly significant in development of this subject—the product from addition to benzaldehyde is a 60 : 40 mixture of diastereoisomers, based on the molecular asymmetry of a disubstituted arene–Cr(CO)$_3$ complex and the new stereogenic center at the benzylic position:

mixture

The special effect of metal coordination is clearer with (PhCl)Cr(CO)$_3$. Whereas o-lithiochlorobenzene (uncomplexed) has a half-life time of minutes at $-78\,°C$[290], the concentration of (o-lithiochlorobenzene)Cr(CO)$_3$ shows a unimolecular decay with a $t_{1/2}$ of 136 min at $0\,°C$[291]. Although the kinetics are consistent with formation of a coordinated benzyne, trapping experiments failed to show more than traces of an appropriate adduct[291]. If the solution is kept at low temperature prior to quenching, high yields of adducts with electrophiles are obtained. The reaction with CO_2 (98% yield) shows that (o-lithiochlorobenzene)Cr(CO)$_3$ was produced efficiently. Again, reaction with acetone suffers from significant protonation of the lithioarene, presumably due to enolization of the acetone. In these examples, there is no evidence for deprotonation at a site other than ortho.

E = PhCHO, Me$_3$SiCl, MeI, CO_2, etc.

From a systematic study of the addition of carbonyl electrophiles (aryl isocyanates, diphenyl ketene, N-carboxyanhydrides of amino acids, and an azlactone), it was concluded that successful electrophiles for quenching (o-lithiofluorobenzene)Cr(CO)$_3$ should have low kinetic acidity and high electrophilicity[292]. Another series of electrophiles failed due either to competing proton transfer to the aryl lithium (ketene thioacetals, succinimide, succinic anhydride, *N*-methylpyridinium iodide, ethylene oxide, aziridine *N*-toluenesulfonamide) or too low electrophilic reactivity (diphenylacetylene, PhN=S=O, PhCN). There has been no corresponding development of the chromium complexes of bromobenzene or iodobenzene.

9.3.3.3 *Lithiation of Anisole and Aniline Ligands*

Alkoxy substituents on arenes direct proton abstraction to the ortho position with high selectivity, apparently due to a combination of inductive effects and specific coordination of the base (Li counter ion) with the alkoxy substituent[287]. The effect is at least as powerful in the $Cr(CO)_3$ complexes of alkoxyarenes. (Anisole)$Cr(CO)_3$ shows a pK_a of 33, compared to 39 for anisole itself[276]. Proton abstraction is fast with either BuLi or $LiNR_2$ bases. The results of deprotonation of (anisole)$Cr(CO)_3$ and trapping with simple electrophiles are represented below. Substitution was observed only at the ortho positions, and 2,6-disubstitution was significant in several of the cases. With acetone, enolization is not significant (<10% recovered anisole complex):

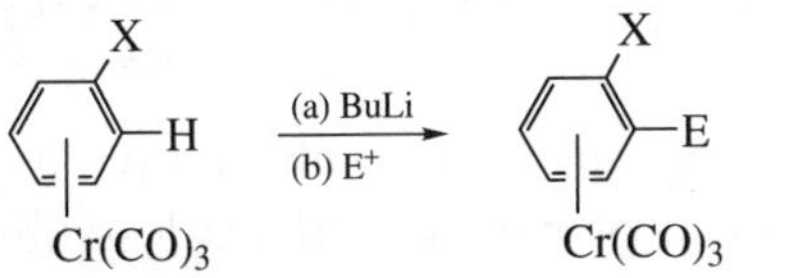

E = PhCHO, Me₃SiCl, MeI, CO₂, Me₂C=O, etc.

Transmetalation of the *o*-lithioanisole ligand to Cu^I or Pd^{II} allows coupling with allylic halides and vinyl halides (Scheme 47)[293].

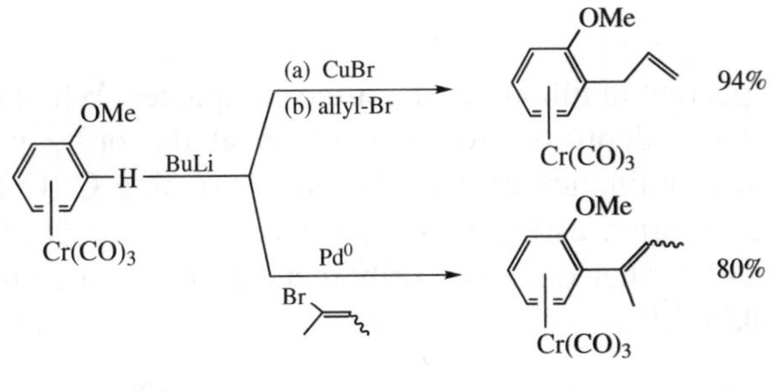

Scheme 47. Transmetalation and coupling.

The directing effect of the alkoxy group can be reversed by use of a sufficiently large protecting group. For example, the complex from phenol tri(isopropyl)silyl ether **303** was lithiated with ^{t}BuLi and then the aryl-Li intermediates were methylated to give predominately the meta substitution product[294]:

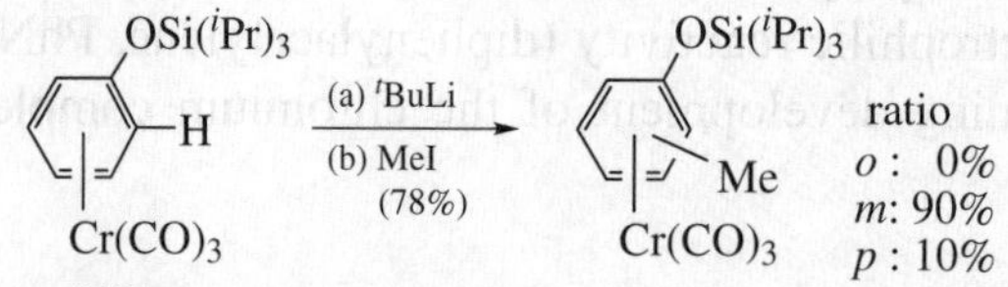

303

The following additional examples show selectivities for an unhindered position. The products from deprotonation and then quenching with electrophiles are shown. Only one isomer was obtained in each case. The electrophiles are MeI, ClSiMe$_3$, and ClCO$_2$Me. In the absence of the large group and in the absence of Cr(CO)$_3$, lithiation is directed by coordination to the heteroatom groups[293].

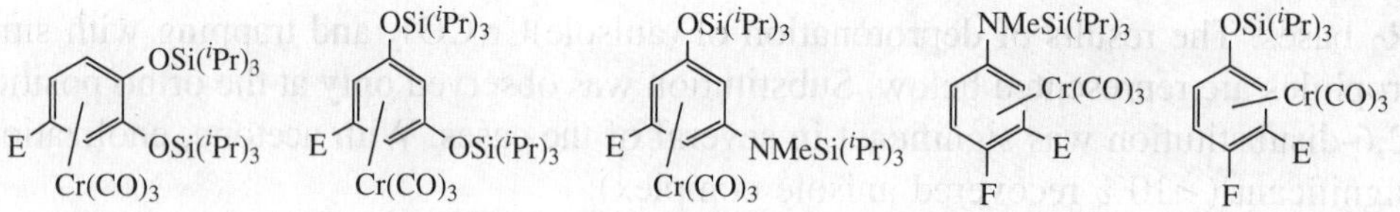

E = Me, 77%
E = SiMe$_3$, 87%
E = CO$_2$Me, 83%

E = Me, 77%
E = SiMe$_3$, 67%
E = CO$_2$Me, 80%

E = Me, 85%
E = SiMe$_3$, 85%

E = Me, 90%
E = SiMe$_3$, 83%
E = CO$_2$Me, 83%

E = Me, 65%
E = CO$_2$Me, 74%

Although the steric effect disfavoring ortho substitution is easy to appreciate, the resulting selectivity for meta instead of para is perhaps less obvious. These[293,294] and related cases[295] have been explained by consideration of the (assumed) favored eclipsed conformation of the Cr(CO)$_3$ tripod and resulting electron deficiency at the ring C—H eclipsing a CO ligand:

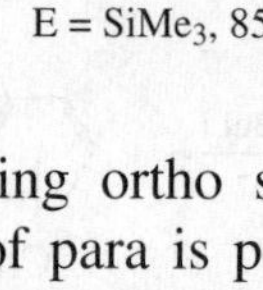

This effect is also apparent in lithiation of aniline complexes. While the free ligand *N,N*-dimethylaniline undergoes deprotonation selectively at the ortho position, presumably largely due to specific coordination effects, the corresponding Cr(CO)$_3$ complex gives a mixture with meta substitution as the major product (*o* : *m* : *p* 31:50:19)[296]. Installing a ^{t}BuMe$_2$Si group (**304**) further disfavors orthometalation and the process becomes very selective for meta (**305**)[295]:

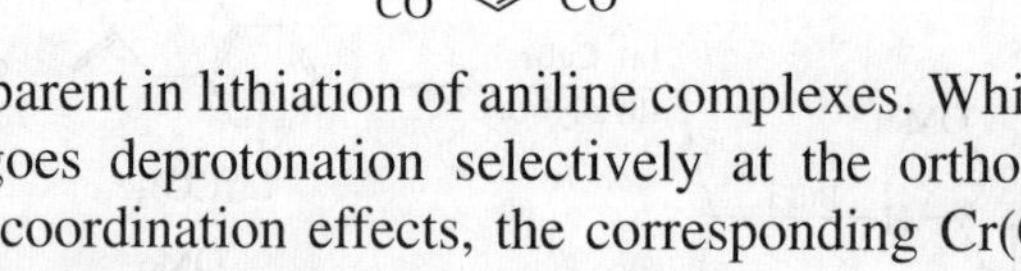

304

(a) BuLi
(b) PhCHO
(c) I$_2$, acid
(d) Ac$_2$O
(69% overall)

305

o:m:p = 0:98:2

Procedure for Lithiation, Transmetalation, and Alkylation

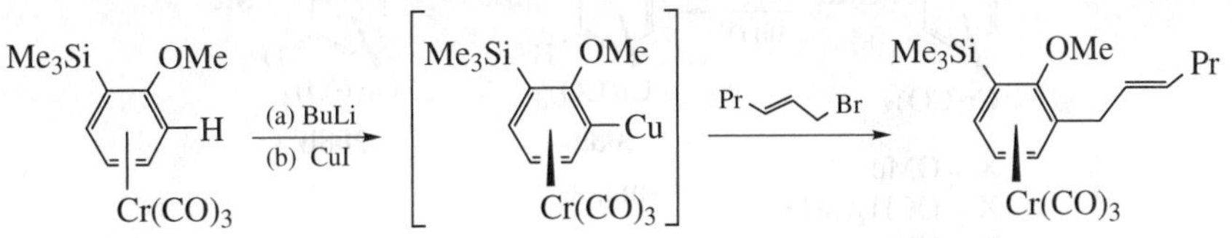

306 → **307**

To a mixture of yellow crystalline **306** (6.33 g, 20.0 mmol) in a flask equipped with a magnetic stir bar, vacuum adapter, serum stopper, and a solid addition funnel charged with cuprous iodide (4.19 g, 22 mmol) under argon was added THF (100 mL). The resulting yellow solution was cooled to $-30\,°C$ and BuLi (9.24 mL, 22 mmol, 2.38 M in hexane) was added dropwise over 2 min. After 3 h, the dark reaction mixture was warmed to $0\,°C$ for 5 min and then cooled again to $-30\,°C$. The cuprous iodide in the solid addition funnel was then added all at once. The reaction mixture was allowed to warm to $0\,°C$, stirred for 1 h, and then cooled again to $-30\,°C$. Then 1-bromo-(2*E*)-hexene (3.59 g, 22 mmol) was added all at once. The mixture was allowed to warm to $23\,°C$ and stirring was continued for 3 h. The mixture was partitioned between aqueous KI and ether, and the combined ether solution was dried over K_2CO_3, filtered, and concentrated on the rotary evaporator. The clear orange oil residue was purified by column chromatography (Florisil, ether) to give **307** as a colorless oil (7.69 g, 96% yield).

9.3.3.4 Ortholithiation Directed by Benzylic Nitrogen and Oxygen Substituents

Direct lithiation of benzylic alcohols and amines leads to efficient and selective ortholithiation[287]. Lithiation of (benzyl alcohol)Cr(CO)3 with BuLi leads to a directed nucleophilic addition to the ring, rather than ortholithiation[297]:

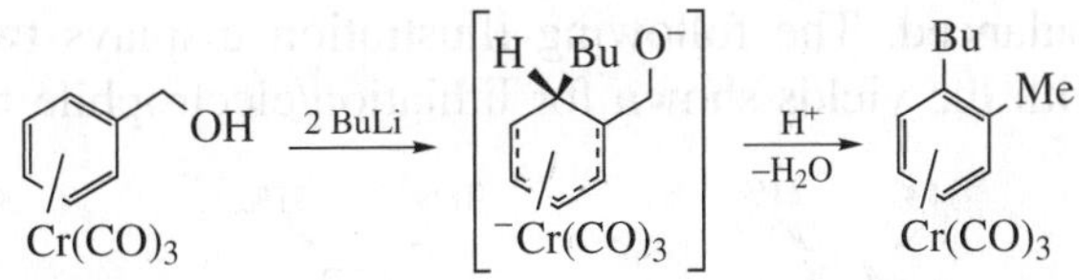

In contrast, the methyl or methoxymethyl *ethers* of 1-phenylethanol form Cr(CO)3 complexes which undergo ortholithiation with *s*BuLi; the ortho lithio derivative can be quenched with MeOD to give a monodeutero product (**308a,b**) in 54–60% yield[298]. Similarly, the Cr(CO)3 complex of *N,N*-dimethyl-1-phenylethylamine is deprotonated smoothly by *t*BuLi and the ortholithio derivative can be coupled with various electrophiles[298]. The most significant feature of these reactions is the high diastereoselectivity. Protons H^a and H^b in the starting complexes are diastereotopic. The products are formed with very high diastereoselectivity (94–100%) and in moderate yield:

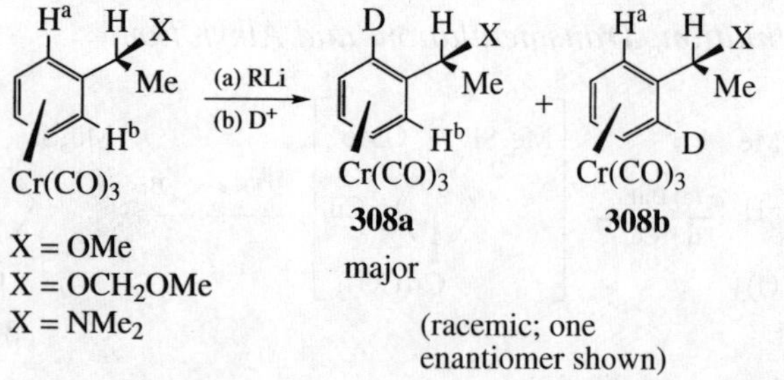
Hª H X
Me
(a) RLi
(b) D+
Hb
Cr(CO)3
X = OMe
X = OCH2OMe
X = NMe2
D H X
Me
Hb
Cr(CO)3
308a
major
+
Hª H X
Me
D
Cr(CO)3
308b
(racemic; one
enantiomer shown)

Diastereoselective deprotonation has also been observed with homochiral complexes (*e.g.*, **309**) prepared by acetalization of benzaldehyde[299]. In this case, the product **310** is homochiral. The selectivity was found to vary widely (diastereoisomer ratio from 34 : 66 to 93 : 7) depending on the alkyllithium base used and the reaction temperature:

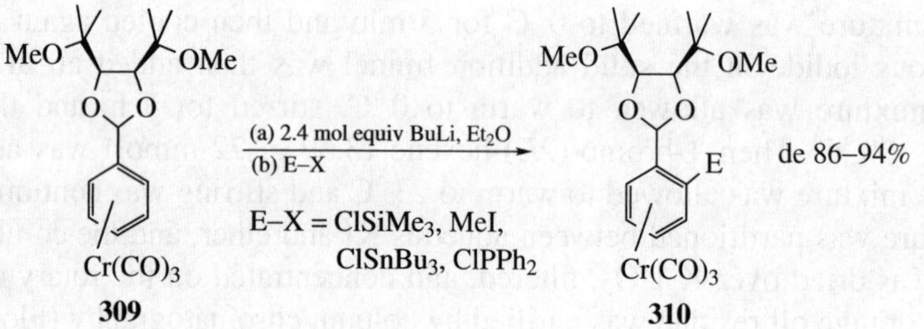
MeO
OMe
O O
Cr(CO)3
309
(a) 2.4 mol equiv BuLi, Et2O
(b) E–X
E–X = ClSiMe3, MeI,
ClSnBu3, ClPPh2
MeO
OMe
O O
E
Cr(CO)3
310
de 86–94%

9.3.3.5 *Correlation of Heteroatom-directing Effects*

A useful study demonstrated that the substituent effects on regioselectivity in deprotonation of free arenes is distinctly different from that for Cr(CO)$_3$-complexed arenes[300,301]. A series of para-disubstituted arenes were complexed and subjected to deprotonation conditions. From the ratio of the two possible products in each case, a reactivity series was generated. For example, the complex of *p*-methoxy-*N*,*N*-dimethylaniline (**311**) provides a comparison of the directing effects of OMe versus CH$_2$NMe$_2$. In this case, the effects are almost exactly balanced. The following illustration displays two cases, complexed and uncomplexed, with the yields shown for lithiation/electrophile trapping.

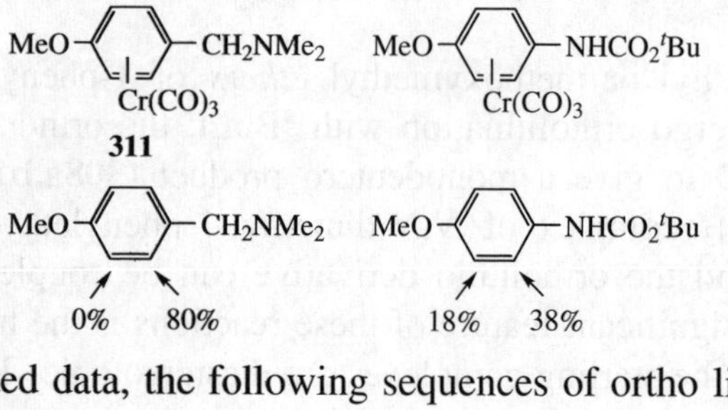
34%
31%
18%
51%
MeO
CH2NMe2
MeO
NHCO2tBu
Cr(CO)3
311
Cr(CO)3
MeO
CH2NMe2
MeO
NHCO2tBu
0%
80%
18%
38%

From these and related data, the following sequences of ortho lithiation reactivity can be defined:

- free arenes: $-CONR_2 > -SO_2NR_2 > -NHCOR > -CH_2NR_2 > -OMe > -NMe_2 = F$
- $Cr(CO)_3$ complexes: $-F > -CONHR > -NHCOR > -CH_2NR_2 = -OMe \gg -CH_2OMe > -NMe_2 = -SR$

The most dramatic difference is the case of fluoride. The implications are borne out by the example below. Whereas the secondary amide group is one of the most powerful ortho directors with free arenes, fluoride dominates in complex **312**. In fact, the arene ligand is so electron deficient that alkyllithium reagents add to the ring rather than remove a proton. But dialkyl amide bases rapidly remove the proton ortho to F and electrophilic trapping to give **313** proceeds as usual:

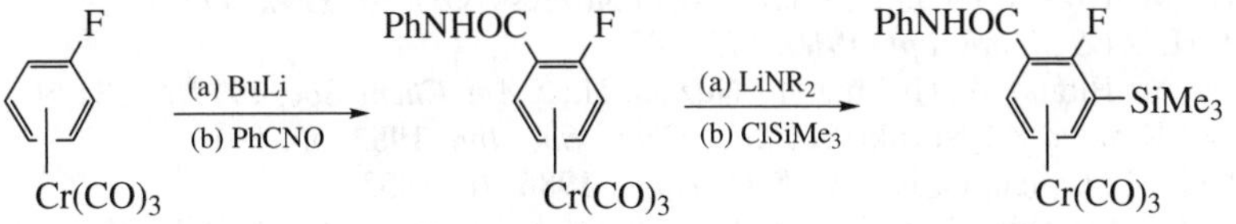

The collective results suggest that conformational effects are significant (best conformation of the $Cr(CO)_3$ tripod), and that inductive effects are relatively more important than specific coordination of the base (compared to the uncomplexed arene). The high inductive effect of F is critical in deprotonation of the fluorobenzene complexes.

10   References

[1] Noyori, R.; Hayakawa, Y.; *Org. React.* **1983**, *29*, 163.
[2] Noyori, R.; Hayakawa, Y.; *Tetrahedron* **1985**, *41*, 5879.
[3] Brookhart, M.; Studabaker, W. B.; *Chem. Rev.* **1987**, *87*, 411.
[4] Semmelhack, M. F.; Park, J.; Schnatter, W.; Tamura, R.; Steigerwald, M.; *Chem. Scr.* **1987**, *27*, 509.
[5] Semmelhack, M. F.; *Pure Appl. Chem.* **1981**, *53*, 2379.
[6] Paterson, I.; in *Comprehensive Organic Synthesis*, Vol. 2, (B. M. Trost, I. Fleming, Eds), Pergamon Press, Oxford, **1992**, 315.
[7] Davies, S. G.; McCarthy, T. D.; in *Comprehensive Organometallic Chemistry II*, Vol. *12*, (E. W. Abel, G. Wilkinson, F. G. A. Stone, Eds), Pergamon Press, New York, **1996**, 1039.
[8] Wulff, W. D.; in *Comprehensive Organometallic Chemistry II*, Vol. *12*, (E. W. Abel, F. G. A. Stone, G. Wilkenson, Eds), Pergamon Press, New York, **1996**, 1065.
[9] Examples are given in Chapter ● of this work.
[10] Collman, J. P.; *Acc. Chem. Res.* **1975**, *8*, 342.
[11] Bates, R. W.; in *Comprehensive Organometallic Chemistry II*, Vol. *12*, (E. W. Abel, G. Wilkinson, F. G. A. Stone, Eds), Pergamon Press, New York, **1996**, 351.
[12] (a) Collman, J. P.; Finke, R. G.; Cawse, H.; Brauman, J. N.; *J. Am. Chem. Soc.* **1977**, *99*, 2515; (b) Collman, J. P.; Finke, R. G.; Cawse, H.; Brauman, J. N.; *J. Am. Chem. Soc.* **1978**, *100*, 4766.
[13] Hassner, A.; Rai, K. M. L.; in *Comprehensive Organic Synthesis*, Vol. *1*, (B. M. Trost, I. Fleming, Eds), Pergamon Press, Oxford, **1992**, 542.
[14] Watanabe, Y.; Mitsubo, T.; Tanaka, M.; Yamamoto, K.; Okajima, T.; Takegami, Y.; *Bull. Chem. Soc. Jpn.* **1971**, *44*, 2569.

[15]   (a) Sawa, Y.; Ryang, M.; Tsutsumi, S.; *Tetrahedron Lett.* **1969**, *10*, 5188; (b) Yamashita, M.; Suemitsu, R.; *Tetrahedron Lett.* **1978**, *19*, 761.

[16]   Collman, J. P.; Winter, S. R.; Clark, D. R.; *J. Am. Chem. Soc.* **1972**, *94*, 1788.

[17]   Adapted from: Finke, R. G.; Sorrell, T. N.; in *Organic Syntheses*, Collected Vol VI, (W. A. Noland, Ed), John Wiley and Sons, NY, **1988**, 807.

[18]   Cooke, M. P.; *J. Am. Chem. Soc.* **1970**, *92*, 6080.

[19]   Cooke, M. P.; Parlman, R. M.; *J. Am. Chem. Soc.* **1977**, *99*, 5222.

[20]   McMurry, J. E.; Andrus, A.; Ksander, G. M.; Musser, J. W.; Johnson, M. A.; *Tetrahedron* **1981**, *Suppl. 1*, *37*, 19.

[21]   Masada, H.; Mizuno, M.; Suza, Y.; Wantanabe, Y.; Takegami, Y.; *Bull. Chem. Soc. Jpn.* **1970**, *43*, 3824.

[22]   Cooke, M. P.; Parlman, R. M.; *J. Am. Chem. Soc.* **1977**, *99*, 5222.

[23]   For a recent comprehensive review, see: Saccomano, N. A.; in *Comprehensive Organic Synthesis*, Vol. *1*, (B. M. Trost, I. Fleming, Eds), Pergamon Press, Oxford, **1992**, 173.

[24]   Zeiss, H. H.; *ACS Monograph* **1960**, *147*, 380.

[25]   (a) Okude, Y.; Hirano, ß.; Hiyama, T.; Nozaki, H.; *J. Am. Chem. Soc.* **1977**, *99*, 3179; (b) Hiyama, T.; Okude, Y.; Kimura, K.; Nozaki, H.; *Bull. Chem. Soc. Jpn.* **1982**, *55*, 561.

[26]   Wuts, P. G. M.; Callen, G. R.; *Synth. Commun.* **1986**, *16*, 1833.

[27]   (a) Takai, K.; Tagashira, M.; Kuroda, T.; Oshima, K.; Utimoto, K.; Nozaki, H.; *J. Am. Chem. Soc.* **1986**, *108*, 6048; (b) Jin, H.; Uenishi, J.; Christ, W. J.; Kishi, Y.; *J. Am. Chem. Soc.* **1986**, *108*, 5644.

[28]   For evidence implicating Ni[I] in related transformations, see: Tsou, T. T.; Kochi, J. K.; *J. Am. Chem. Soc.* **1979**, *101*, 7547.

[29]   Saccomano, N. A.; in *Comprehensive Organic Synthesis*, Vol. *1*, (B. M. Trost, I. Fleming, Eds), Pergamon Press, Oxford, **1992**, 193.

[30]   Grigg, R.; Putnikovic, B.; Urch, C. J.; *Tetrahedron Lett.* **1997**, *38*, 6307.

[31]   Furstner, A.; Shi, N.; *J. Am. Chem. Soc.* **1996**, *118*, 2533.

[32]   Hiyama, T.; Okude, Y.; Kimura, K.; Nozaki, H.; *Bull. Chem. Soc. Jpn.* **1982**, *55*, 561.

[33]   Stamos, D. P.; Sheng, X. C.; Chen, S. S.; Kishi, Y.; *Tetrahedron Lett.* **1997**, *38*, 6355.

[34]   Takai, K.; Kimura, K.; Kuroda, T.; Hiyama, T.; Nozaki, H.; *Tetrahedron Lett.* **1983**, *24*, 5281.

[35]   Takai, K.; Tagashira, M.; Kuroda, T.; Oshima, K.; Utimoto, K.; Nozaki, H.; *J. Am. Chem. Soc.* **1986**, *108*, 6048.

[36]   Semmelhack, M. F.; Helquist, P.; Gorzynski, J. D.; *J. Am. Chem. Soc.* **1972**, *94*, 9234.

[37]   Geokian, P. G.; Wu, T.-C.; Kang, H.-Y.; Kishi, Y.; *J. Org. Chem.* **1987**, *52*, 4823.

[38]   Schreiber, S. L.; Meyers, H. V.; *J. Am. Chem. Soc.* **1988**, *110*, 5198.

[39]   Takai, K.; Tagashira, M.; Kuroda, T.; Oshima, K.; Utimoto, K.; Nozaki, H.; *J. Am. Chem. Soc.* **1986**, *108*, 6048.

[40]   Based on a published procedure: Jin, H.; Uenishi, J.-I.; Christ, W. J.; Finan, J. M.; Fujioka, H.; Kishi, Y.; *J. Am. Chem. Soc.* **1986**, *108*, 644.

[41]   Okude, Y.; Hirano, S.; Hiyama, T.; Nozaki, H.; *J. Am. Chem. Soc.* **1977**, *99*, 3179.

[42]   (a) Buse, C. T.; Heathcock, C. H.; *Tetrahedron Lett.* **1978**, *19*, 1685; (b) Hiyama, T.; Kimura, K.; Nozaki, H.; *Tetrahedron Lett.* **1981**, *22*, 1037.

[43]   Roush, W.; in *Comprehensive Organic Synthesis*, Vol. 2, (B. M. Trost, I. Fleming, Eds), Pergamon Press, Oxford, **1992**, 19.

[44]   Denmark, S. E.; Weber, E.; *J. Am. Chem. Soc.* **1984**, *106*, 7970.

[45]   For a discussion, see: Hoffmann, R. W.; *Angew. Chem., Int. Ed. Engl.* **1987**, *26*, 489.

[46]   Nagaoka, H.; Kishi, Y.; *Tetrahedron Lett.* **1982**, *23*, 3873.

[47]   Lewis, M. D.; Kishi, Y.; *Tetrahedron Lett.* **1982**, *23*, 2343.

[48]   Mulzer, J.; Schultze, T.; Strecker, A.; Denzer, W.; *J. Org. Chem.* **1988**, *53*, 4098.

[49]   Jurczak, J.; Pikul, S.; Bauer, T.; *Tetrahedron* **1986**, *42*, 447.

[50]   Fronza, G.; Fuganti, P.; Graselli, G.; Petrocchi-Fantoni, G.; Zirotti, C.; *Chem. Lett.* **1984**, 335.

[51]   Hiyama, T.; Okude, Y.; Kimura, K.; Nozaki, H.; *Bull. Chem. Soc. Jpn.* **1982**, *55*, 561.

[52] Takai, K.; Nitta, K.; Utimoto, K.; *Tetrahedron Lett.* **1988**, *29*, 5263.

[53] Still, W. C.; Mobilio, D.; *J. Org. Chem.* **1983**, *48*, 4785.

[54] Takai, K.; Toratsu, C.; *J. Org. Chem.* **1998**, *63*, 6450.

[55] Takai, K.; Matsukawa, N.; Takahashi, A.; Fujii, T.; *Angew. Chem., Int. Ed. Engl.* **1998**, *37*, 152.

[56] Place, P.; Verniere, C.; Gore, J.; *Tetrahedron* **1981**, *37*, 1359.

[57] Takai, K.; Kuroda, T.; Nakatsukasa, S.; Oshima, K.; Nozaki, H.; *Tetrahedron Lett.* **1985**, *26*, 4371.

[58] (a) Lhermitte, H.; Grierson, D. S.; *Contemp. Org. Synth.* **1996**, *3*, 41; (c) Lhermitte, H.; Grierson, D. S.; *Contemp. Org. Synth.* **1996**, *3*, 93.

[59] (a) Semmelhack, M.; Gu, Y.; Ho, D. M.; *Tetrahedron Lett.* **1997**, *38*, 5583; (b) for a recent example, discussion, and leading references, see: Hartwig, C. W.; Py, S.; Fallis, A. G.; *J. Org. Chem.* **1997**, *62*, 7902.

[60] Reviews: (a) Wulff, W. D.; in *Comprehensive Organic Synthesis*, Vol. *5*, (B. M. Trost, I. Fleming, Eds), Pergamon Press, Oxford, **1992**, 469; (b) Wulff, W. D.; in *Comprehensive Organometallic Chemistry II*, Vol. *12*, (E. W. Abel, F. G. A. Stone, G. Wilkinson, Eds), Pergamon Press, New York, **1996**, 1103–1110; comprehensive discussions: (c) Dotz, K. H.; Fischer, H.; Hofman, P.; Kreissel, F. R.; Schubert, U.; Weiss, K.; *Transition Metal Carbene Complexes*, VCH, Deerfield Beach, FL, **1983**; (d) Harvey, D. F.; Sigano, D. M.; *Chem. Rev.* **1996**, *96*, 271.

[61] (a) Semmelhack, M. F.; Lee, G. R.; *Organometallics* **1987**, *6*, 1424; (b) Imwinkelried, R.; Hegedus, L. S.; *Organometallics* **1988**, *7*, 702.

[62] Nacheol Jeong, **1987**. PhD thesis *Carbene-Chromium Complexes in the Synthesis of Nogalarol*, Princeton University, 228.

[63] For a recent determination and leading references, see: Bernasconi, C. F.; Leyes, A. E.; *J. Am. Chem. Soc.* **1998**, *120*, 8632.

[64] Fruhauf, H.-W.; *Chem. Rev.* **1997**, *97*, 523.

[65] Anderson, B. A.; Wulff, W. D.; Rahm, A.; *J. Am. Chem. Soc.* **1993**, *115*, 4602.

[66] Wang, H.; Hsung, R. P.; Wulff, W. D.; *Tetrahderon Lett.* **1998**, *39*, 1849.

[67] For an improved procedure and leading references, see: Wang, H.; Hsung, R. P.; Wulff, W. D.; *Tetrahedron Lett.* **1998**, *38*, 1849.

[68] Wulff, W. D.; Bauta, W. E.; Kaesler, R. W.; Lankford, P. J.; Miller, R. A.; Murray, C. K.; Yang, D. C.; *J. Am. Chem. Soc.* **1990**, *112*, 3642.

[69] Waters, M. L.; Brfandvold, T. A.; Isaacs, L.; Wulff, W. D.; Rheingold, A. L.; *Organometallics* **1998**, *17*, 4298.

[70] Yamashita, A.; Timko, J. M.; Watt, W.; *Tetrahedron Lett.* **1988**, *29*, 2513.

[71] Choi, Y. H.; Rhee, K. S.; Kim, K. S.; Shin, G. C.; Shin, S. C.; *Tetrahedron Lett.* **1995**, *36*, 1871.

[72] For examples and discussion, see: Su, J.; Wulff, W. D.; Ball, R. G.; *J. Am. Chem. Soc.* **1998**, *63*, 8440 and references therein.

[73] Wulff, W. D.; Tang, P. C.; McCallum, J. S.; *J. Am. Chem. Soc.* **1981**, *103*, 7677.

[74] Semmelhack, M. F.; Bozell, J. J.; Keller, L.; Sato, T.; Spiess, E. J.; Wulff, W.; Zask, A.; *Tetrahedron* **1985**, *41*, 5803.

[75] Wulff, W. D.; Tang, P. D.; Chan, K. S.; McCallum, J. S.; Yang, D.; Gilbertson, S. R.; *Tetrahedron* **1985**, *41*, 5813.

[76] Semmelhack, M. F.; Jeong, N.; Lee, G. R.; *Tetrahedron Lett.* **1990**, *31*, 609.

[77] Chamberlin, S.; Wulff, W. D.; *J. Am. Chem. Soc.* **1992**, *114*, 10667.

[78] Yamashita, A.; *Tetrahedron Lett.* **1986**, *27*, 5915.

[79] Klabunde, U.; Fischer, E. O.; *J. Am. Chem Soc.* **1967**, *89*, 7141.

[80] Chan, K. S.; Peterson, G. A.; Brandvoldt, K. L.; Faron, K. L.; Challener, C. A.; Hyldahl, C.; Wulff, W. D.; *J. Organomet. Chem.* **1987**, *334*, 9.

[81] Wulff, W. D.; Gilbertson, S. R.; *J. Am. Chem. Soc.* **1985**, *107*, 503.

[82] Bauta, W. E.; Wulff, W. D.; Parkovic, S. F.; Saluzec, E. J.; *J. Org. Chem.* **1989**, *54*, 3249.

[83] Tang, P. C.; Wulff, W. D.; *J. Am. Chem. Soc.* **1984**, *106*, 1132.

[84] Hsung, R. P.; Wulff, W. D.; Challener, C. A.; *Synthesis* **1996**, 773.

[85] Gross, M. F.; Finn, M. G.; *J. Am. Chem. Soc.* **1994**, *116*, 10921.

[86] Wang, H.; Wulff, W. D.; *J. Am. Chem. Soc.* **1998**, *120*, 10573.

[87] Hegedus, L. S.; in *Comprehensive Organometallic Chemistry*, Vol. *II*, (E. W. Abel, F. G. A. Stone, G. Wilkinson, Eds), Pergamon Press, Oxford, **1995**, Chapter 5.4.

[88] Hegedus, L. S.; McGuire, M. A.; Schultze, L. M.; Yijun, A.; C. Anderson, O. P.; *J. Am. Chem. Soc.* **1984**, *106*, 2680.

[89] Thompson, D. K.; Suzuki, N.; Hegedus, L. S.; Satoh, Y.; *J. Org. Chem.* **1992**, *57*, 1461.

[90] Borel, C.; Hegedus, L. H.; Krebs, J.; Satoh, Y.; *J. Am. Chem. Soc.* **1987**, *109*, 1101.

[91] Hegedus, L. S.; *Acct. Chem. Res.* **1995**, *28*, 299.

[92] Hegedus, L. S.; Lastra, S. E.; Narukawa, Y.; Snustad, D. C.; *J. Am. Chem. Soc.* **1992**, *114*, 2991 and references therein.

[93] Hegedus, L.; deWeck, G.; D'Andrea, S.; *J. Am. Chem. Soc.* **1988**, *110*, 2122.

[94] Reed, A.D.; Hegedus, L. S.; *Organometallics* **1997**, *16*, 2313.

[95] Miller, M.; Hegedus, L. S.; *J. Org. Chem.* **1993**, *58*, 6779.

[96] Hoffmann, R.; *Angew. Chem., Int. Ed. Engl.* **1982**, *21*, 711.

[97] Wulff, W. D.; Yang, D. C.; *J. Am. Chem. Soc.* **1983**, *105*, 6726.

[98] Wulff, W. D.; Yang, D. C.; *J. Am. Chem. Soc.* **1984**, *106*, 7565.

[99] For a comprehensive review, see: Fruhauf, H.-W.; *Chem. Rev.* **1997**, *97*, 523 and 571.

[100] Rosenblum, M.; *Acc. Chem. Res.* **1974**, *7*, 122.

[101] A very influential paper laid out the issues of reactivity and selectivity in the addition of nucleophiles to metal-coordinated polyenes: Davies, S. G.; Green, M. L. H.; Mingos, D. M. P.; *Tetrahedron* **1978**, *34*, 3047.

[102] Review: Hegedus, L. S.; in *Comprehensive Organic Synthesis*, Vol. *4*, (B. M. Trost, I. Fleming, Eds), Pergamon Press, Oxford, **1992**, 551.

[103] Rosenblum, M.; Chang, T. C.; Foxman, B. M.; Samuels, S. B.; Stockman, C.; *Organic Synthesis Today and Tomorrow*, Proceedings of the 3rd IUPAC Symposium on Organic Synthesis, **1980**, Pergamon Press, Oxford, **1981**, 47.

[104] For the general procedure, see: Cutler, A.; Ehntholt, D.; Giering, W. P.; Lennon, P.; Raghu, S.; Rosan, A.; Rosenblum, M.; Tancrede, J.; Wells, D.; *J. Am. Chem. Soc.* **1976**, *98*, 3495.

[105] For a discussion, see: Collman, J. P.; Hegedus, L.; Norton, J. R.; Finke, R. G.; *Principles and Applications of Organotransition Metal Chemistry*, University Science Books, Mill Valley, California, **1987**, 386.

[106] For more details on this example and related discussion, see: Freeman, J., (Ed.) *Organic Syntheses*, Collected Vol. VIII, John Wiley & Sons, **1993**, 479.

[107] Eisch, J. J.; King, R. B.; (Eds), *Organometallic Syntheses*, Vol. *1*, Academic Press, New York, **1965**, 114.

[108] Zhen, W.; Chu, K.-H.; Rosenblum, M.; *J. Org. Chem.* **1997**, *62*, 3344.

[109] Chang, T. C. T.; Rosenblum, M.; *J. Org. Chem.* **1981**, *46*, 4626.

[110] Zhen, W.; Chu, K. H.; Rosenblum, M.; *J. Org. Chem.* **1997**, *62*, 3344.

[111] Whitesides, T. H.; *J. Am. Chem. Soc.* **1973**, *95*, 5792.

[112] Nicholas, K. M.; *J. Organomet. Chem.* **1981**, *212*, 107.

[113] For examples and leading references, see: (a) Enders, D.; Jandeleit, B.; Von Berg, S.; *Synlett* **1997**, 1; (b) Enders, D.; Jandeleit, B.; von Berg, S.; in *Organic Synthesais via Organometallics* OSM5, (G. Helmchem, Ed), Vieweg, Braunschweig, Germany, **1997**.

[114] Palladium-catalyzed allylic alkylation is reviewed in Chapter ● of this work.

[115] Green, J. R.; *Synthesis*, **1992**, 973.

[116] Enders, D.; Fey, P.; Schmnitz, T.; Lohray, B. B.; Jandeleit, B.; *J. Organomet. Chem.* **1996**, *514*, 227.

[117] Enders, D.; Jandeleit, B.; Raabe, G.; *Angew. Chem., Int. Ed. Engl.* **1994**, *33*, **1949**.

[118] Nicholas, K. M.; *Tetrahedron Lett.* **1987**, *28*, 845.

[119] Yeh, M.-C. P.; Tau, S.-I.; *J. Chem. Soc., Chem. Comm.* **1992**, 13.

[120] Pearson, A. J.; *Tetrahedron Lett.* **1975**, *16*, 3617.

[121] Hiemstra, H.; Speckamp, W. N.; *J. Chem. Soc., Chem. Comm.* **1995**, 617.

[122] Green, J. R.; *Tetrahedron Lett.* **1993**, *33*, 4497.

[123] Enders, D.; Frank, U.; Fey, P.; Jandeleit, B.; Loray, B.; *J. Organomet. Chem.* **1996**, *519*, 147.

[124] (a) Horton, A. M.; Hollinshead, D. M.; Ley, S. V.; *Tetrahedron* **1984**, *40*, 1737; (b) Ley, S. V.; Cox, L. R.; Meek, G.; *Chem. Rev.* **1996**, *96*, 423.

[125] Caruso, M.; Knight, J. G.; Ley, S. V.; *Synlett* **1990**, 224.

[126] Bates, R. W.; Dies-Martin, D.; Kerr, W. J.; Knight, J. G.; Ley, S. V.; Sakellaridis, A.; *Tetrahedron* **1990**, *46*, 4063.

[127] Moriarty, R. M.; DeBoer, B. G.; Churchill, M. R.; Yeh, J. J. S.; Chen, K. N.; *J. Am. Chem. Soc.* **1975**, *97*, 5602.

[128] For a survey, see: Harrington, P. J.; *Transition Metals in Total Synthesis*, John Wiley & Sons, NY, **1990**, 124.

[129] Horton, A. M.; Ley, S. V.; *J. Organomet. Chem.* **1985**, *285* C17.

[130] Kotecha, N. R.; Ley, S. V.; Mantegani, S.; *Synlett* **1992**, 395.

[131] Aumann, R.; Frohlich, K.; Ring, H.; *Angew. Chem., Int. Ed. Engl.* **1974**, *13*, 275.

[132] Becker, Y.; Eisenstadt, A.; Shvo, Y.; *Tetrahedron* **1976**, *32*, 2123.

[133] Annis, G. D.; Helllethwaite, E. M.; Hodgson, S. T.; Hollinshead, D. M.; Ley, S. V.; *J. Chem. Soc., Perkin Trans.* **1983**, *1*, 2851.

[134] Hodgson, S. T.; Hollinshead, D. M.; Ley, S. V.; *Tetrahedron* **1985**, 41, 5871.

[135] Ley, S. V.; Cox, L. R.; *J. Chem. Soc., Chem. Commun.* **1998**, 1339 and previous papers referenced therein.

[136] Ley, S. V.; Low, C. M. R.; White, A. D.; *J. Organomet. Chem.* **1986**, *302*, C13.

[137] For a compilation and leading references, see: Abel, E. W.; Stone, F. G. A.; Wilkinson, G. (Eds); *Comprehensive Organometallic Chemistry*, Vol. 7, Pergamon Press, New York, **1996**.

[138] Knolker, H.-J.; Baum, E.; Gonser, P.; Rohde, G.; Rottele, H.; *Organometallics* **1998**, *17*, 3916.

[139] Cox, L. R.; Ley, S. V.; *Chem. Soc. Rev.* **1998**, *27*, 301.

[140] King, R. B.; *Organometallic Syntheses* Vol. *1*, Academic Press, **1965**, 131.

[141] Semmelhack, M. F.; Herndon, J. W.; *Organometallics* **1983**, *2*, 363.

[142] Semmelhack, M. F.; Le, H. T. M.; *J. Am. Chem. Soc.* **1985**, *107*, 1455.

[143] Semmelhack, M. F.; Herndon, J. W.; Springer, J. P.; *J. Am Chem. Soc.* **1983**, *105*, 2497.

[144] Semmelhack, M. F.; Le, H. T. M.; *J. Am. Chem. Soc.* **1984**, *106*, 2715.

[145] Semmelhack, M. F.; Herndon, J. W.; Liu, J. K.; *Organometallics* **1983**, *2*, 1885.

[146] Stork, G.; Maldonado, L.; *J. Am. Chem. Soc.* **1971**, *93*, 5286.

[147] For background and a comprehensive review, see: Pearson, A. J.; in *Comprehensive Organic Synthesis*, Vol. *4*, (B. Trost, I. Fleming, Eds), Pergamon Press, **1992**, 663.

[148] (a) for the X-ray diffraction structure of a cyclohexadienyl complex, see: Eisenstein, O.; Butler, W. M.; Pearson, A. J.; *Organometallics* **1984**, *3*, 1150; (b) for an open chain complex, see: Reingold, A. L.; Haggarty, B. S.; Ma, H.; Ernst, R. D.; *Acta Cryst.* **1996**, *C52*, 1110.

[149] Dobosh, P. A.; Gresham, D. G.; Kowalski, D. J.; Lillya, C. P.; Magyar, E. S.; *Inorg. Chem.* **1978**, *17*, 775.

[150] For examples, see: Tao, C.; Donaldson, W. A.; *J. Org. Chem.* **1991**, *58*, 2134.

[151] Donaldson, W. A.; *Aldrichimica Acta* **1997**, *30*, 17.

[152] (a) Birch, A. J.; Cross, P. E.; Lewis, J.; White, D. A.; Wild, S. B.; *J. Chem. Soc. A* **1968**, 332; (b) Pearson, A. J.; Perrior, T. R.; *J. Chem. Soc., Perkin Trans. I* **1983**, 625.

[153] Pearson, A. J.; Ong, C. W.; *J. Org. Chem.* **1982**, *47*, 3780.

[154] Birch, A. J.; Williamson, D. H.; *J. Chem. Soc., Perkin Trans. I* **1973**, 1892

[155] Eisenstein, O.; Butler, W. M.; Pearson, A. J.; *Organometallics* **1984**, *3*, 1150.

[156] Birch, A. J.; Cross, P.; Lewis, J.; White, D.; Wild, S. B.; *J. Chem. Soc. A* **1968**, 332.

[157] Knolker, H.-J.; Frohner, W.; *Tetrahedron Lett.* **1997**, *38*, 4051.

[158] For examples and a useful solvent effect: Bandara, B. M. R.; Birch, A. J.; Khor, T. C.; *Tetrahedron Lett.* **1980**, *21*, 3625.

[159] Pearson, A. J.; Yoon, J.; *Tetrahedron Lett.* **1985**, *26*, 2399.

[160] Grieco, P. A.; Larsen, S. D.; *J. Org. Chem.* **1986**, *51*, 3553.

[161] Pearson, A. J.; in *Comprehensive Organic Synthesis*, Vol. *4*, (B. Trost, I. Fleming, Eds) Pergamon Press, Oxford, **1992**, 684.

[162] Millot, N.; Guillou, C.; Thal, C.; *Tetrahedron Lett.* **1998**, *39*, 2325.

[163] Birch, A. J.; Bandara, B. M. R.; Chamberlin, K.; Chauncy, B.; Dahler, Day, A. I.; Jenkins, I. D.; Kelly, L. F.; Khor, T.-C.; Kretschmer, G.; Liepa, A. J.; Narula, A. S.; Raverty, W. D.; Rizzardo, E.; Sell, C.; Stephenson, G. R.; Thompson, D. J.; Williamson, D. H.; *Tetrahedron Suppl. 9*, **1981**, *37*, 289.

[164] Pearson, A. J.; *J. Chem. Soc., Perkin Trans. 1* **1977**, 2069.

[165] Pearson, A. J.; Perrior, T.; Griffen, D. A.; *J. Chem. Soc., Perkin Trans. I*, **1983**, 625.

[166] Pearson, A. J.; Ong, C. W.; *J. Org. Chem.* **1982**, *47*, 3780.

[167] Bandara, B. M. R.; Birch, A. J.; Kelly, L. F.; *J. Org. Chem.* **1984**, *49*, 2496.

[168] Pearson, A. J.; Ray, T.; *Tetrahedron* **1985**, *41*, 5765.

[169] Gree, R.; Lellouche, J. P.; *Advances in Metal-Organic Chemistry*, Vol. *4*, (L. Liebeskind, Ed), JAI Press, NY, **1995**, 129.

[170] Mahler, J. E.; Pettit, R.; *J. Am. Chem. Soc.* **1963**, *85*, 3955.

[171] Johnson, B. F. G.; Lewis, J.; Parker, D. B.; Postle, S. R.; *J. Chem. Soc., Dalton Trans.* **1977**, *1977*, 794.

[172] Sorenson, T. S.; Jablonsky, C. R.; *J. Organomet. Chem.* **1970**, *25*, C62.

[173] Donaldson, W. A.; Jin, M.-J.; Bell, P. T.; *Organometallics* **1993**, *12*, 1174.

[174] Uemura, M.; Minami, T.; Yamashita, Y.; Hiyoshi, K.; Hayashi, Y.; *Tetrahedron Lett.* **1987**, *28*, 641.

[175] (a) Tao, C.; Donaldson, W. A.; *J. Org. Chem.* **1993**, *58*, 2134; (b) Donaldson, W. A.; Ramaswamy, M.; *Tetrahedron Lett.* **1989**, *30*, 1339.

[176] Yeh, M. C. P.; Sun, M. L.; Lin, S. K.; *Tetrahedron Lett.* **1991**, *32*, 13.

[177] Gree, R.; Lellouche, J. P.; *Advances in Metal-Organic Chemistry*, Vol. *4*, (L. Liebeskind Ed.), JAI Press, NY, **1995**, 149.

[178] (a) Donaldson, W. A.; Ramaswamy, M.; *Tetrahedron Lett.* **1989**, *30*, 1339; (b) Donaldson, W. A.; Ramaswamy, M.; *Tetrahedron Lett.* **1988**, *29*, 1343.

[179] For a summary, see: Donaldson, W. A.; *Aldrichimica Acta* **1997**, *30*, 19.

[180] For a recent example and leading references, see: Laabassi, M.; Toubet, L.; Gree, R.; *Bull. Soc. Chim. Fr.* **1992**, *129*, 47.

[181] Tao, C.; Donaldson, W. A.; *J. Org. Chem.* **1993**, *58*, 2134.

[182] Strohmeier, W.; *Chem. Ber.* **1961**, *94*, 2490.

[183] Semmelhack, M. F.; Keller, L.; Thebtaranonth, Y.; *J. Am. Chem. Soc.* **1977**, *99*, 959.

[184] Mahaffy, C. A. L.; Pauson, P. L., *Inorg. Synth.* **1979**, *19*, 154.

[185] Top, S.; Jaouen, G.; *J. Organomet. Chem.* **1979**, *182*, 381.

[186] Sergheraert, C.; Brunet, J.-C.; Tartar, A.; *J. Chem. Soc., Chem. Commun.* **1982**, 417.

[187] Ofele, K.; Dotzauer, E.; *J. Organomet. Chem.* **1971**, *30*, 211.

[188] Knox, G. R.; Leppard, D. G.; Pauson, P. L.; Watts, W. E.; *J. Organomet. Chem.* **1972**, *34*, 347.

[189] Ofele, K.; *Chem. Ber.* **1966**, *99*, 1732.

[190] Moser, G. A.; Rausch, M. D.; *Synth. React. Inorg. Met.-Org. Chem.* **1974**, *4*, 37.

[191] Verbrel, J.; Mercier, R.; Belleney, J.; *J. Organomet. Chem.* **1982**, *235*, 197.

[192] Kundig, E. P.; Perret, C.; Spichiger, S.; Bernardinelli, G.; *J. Organomet. Chem.* **1985**, *286*, 183.

[193] Semmelhack, M. F.; Goti, A.; *J. Organomet. Chem.* **1994**, *470*, C4.

[194] Pruett, R. L.; *Prep. Inorg. React.* **1965**, *2*, 187.

[195] Rausch, M. D.; *J. Org. Chem.* **1974**, *39*, 1787.

[196] Fischer, E. O.; Goodwin, H. A.; Kreiter, C. G.; Simmons, H. D.; Sonogashira, K.; Wild, S. B.; *J. Organomet. Chem.* **1968**, *14*, 359.

[197] Uemura, M.; Minami, T.; Hirotsu, K.; Hayashi, Y.; *J. Org. Chem.* **1989**, *54*, 469.

[198] Collman, J. P.; Hegedus, L.; Norton, J. R.; Finke, R. G.; *Principles and Applications of Organotransition Metal Chemistry*, University Science Books, Mill Valley, California, 1987, 424–431 and 921–940.

[199] (a) Watts, W. E.; in *Comprehensive Organometallic Chemistry*, Vol. *8*, (E. W. Abel, F. G. A. Stone, G. Wilkenson, Eds), Pergamon Press, Oxford, **1982**, 1013–1072; (b) Semmelhack, M. F.; in *Comprehensive Organometallic Chemistry II*, Vol. *12*, (E. W. Abel, F. G. A. Stone, G. Wilkenson, Eds), Pergamon Press, Oxford, **1995**, 979.

[200] Kane-Maguire, L. A. P.; Honig, E. D.; Sweigart, D. A.; *Chem. Rev.* **1984**, *84*, 525.

[201] Semmelhack, M. F.; *J. Organomet. Chem. Libr.* **1976**, *11*, 361.

[202] (a) Semmelhack, M. F.; *Ann, N. Y. Acad. Sci.* **1977**, *295*, 36; (b) Semmelhack, M. F.; *Pure Appl. Chem.* **1981**, *53*, 2379.

[203] Semmelhack, M. F.; Clark, G. R.; Garcia, J. L.; Harrison, J. J.; Thebtaranonth, Y.; Wulff, W.; Yamashita, A.; *Tetrahedron* **1981**, *37*, 3957.

[204] Uemura, M.; *Adv. Met.-Org. Chem.* **1991**, *2*, 195.

[205] Semmelhack, M. F.; in *Comprehensive Organic Synthesis*, Vol. *4*, (B. M. Trost,; I. Fleming, Eds), Pergamon Press, Oxford, **1992**, 517.

[206] Zeiss, H.; Wheatley, P. J.; Winkler, H. J. S.; *Benzenoid-Metal Complexes*, The Ronald Press Company, New York, **1966**.

[207] Silverthorn, W. E.; *Adv. Organomet. Chem.* **1975**, *13*, 47.

[208] Nichols, B.; Whiting, M. C.; *J. Chem. Soc.* **1959**, 551.

[209] Whiting, M. C.; (Ethyl Corp.), US Pat. **1966**, 3,225,071 (*Chem. Abstr.* **1966**,64, 6694h].

[210] Knipe, A. C., McGuinness, J.; Watts, W. E., *J. Chem. Soc., Chem. Commun.* **1979**, 842.

[211] Bunnett, J. F.; Hermann, H., *J. Org. Chem.* **1971**, *36*, 4081.

[212] Rosca, S. I.; Rosca, S.; *Revista de Chemie (Bucharest)*, **1964**, *25*, 461.

[213] Semmelhack, M. F.; Hall, H. T.; *J. Am. Chem. Soc.* **1974**, *96*, 7091.

[214] Semmelhack, M. F.; Hall, H. T.; *J. Am. Chem. Soc.* **1974**, *96*, 7092.

[215] Top, S.; Jaouen, G.; *J. Organomet. Chem.* **1979**, *82*, 381.

[216] Oishi, T.; Fukui, M.; Endo, Y.; *Heterocycles* **1977**, *7*, 947.

[217] Alemagna, A.; Baldoli, C.; Del Buttero, P.; Licandro, E.; Maiorana, S.; *J. Chem. Soc., Chem. Commun.* **1985**, 417.

[218] Ghavshou, M.; Widdowson, D. A.; *J. Chem. Soc. Perkin Trans. I* **1983**, 3065.

[219] Dickens, M. J.; Gilday, J. P.; Mowlem, T. J.; Widdowson, D. A.; *Tetrahedron* **1991**, *47*, 8621.

[220] Boutonnet, J.; Rose-Munch, F.; Rose, E.; *Tetrahedron Lett.* **1985**, 3899.

[221] Alemagna, A.; Del Buttero, P.; Gorini, C.; Landini, D.; Licandro, E.; Maiorana, S.; *J. Org. Chem.* **1983**, *48*, 605.

[222] Alemagna, A.; Cremonesi, P.; Del Buttero, P.; Licandro, E.; Maiorana, S.; *J. Org. Chem.* **1983**, *48*, 3114.

[223] Baldoli, C.; Buttero, P. D.; Maiorama, S.; *Tetrahedron Lett.* **1992**, *33*, 4049.

[224] Semmelhack, M. F.; Bisaha, J.; Czarny, M.; *J. Am. Chem. Soc.* **1979**, *101*, 768.

[225] Houghton, R. P.; Voyle, M.; Price, R.; *J. Organomet. Chem.* **1983**, *259*, 183.

[226] Houghton, R. P.; Voyle, M.; *J. Chem. Soc., Chem. Commun.* **1980**, 884.

[227] Card, R. J.; Trayhanovsky, W. S.; *J. Org. Chem.* **1980**, *45*, 2560.

[228] Semmelhack, M. F.; in *Comprehensive Organometallic Chemistry*, Vol. *12*, (E. W. Abel, F. G. A. Stone, G. Wilkenson, Eds), Pergamon Press, New York, **1996**, 1017.

[229] Gilday, J. P.; Widdowson, D. A.; *Tetrahedron Lett.* **1986**, 5525.

[230] Beswick, P. J.; Widdowson, D. A.; *Synthesis* **1985**, 492.

[231] Gilday, J. P.; Widdowson, D. A.; *J. Chem. Soc., Chem. Commun.* **1986**, 1235.

[232] Semmelhack, M. F.; Ullenius, C.; *J. Organomet. Chem.* **1982**, *235*, C10.

[233] Kundig, E. P.; Desobry, V.; Simmons, D. P.; Wenger, E.; *J. Am. Chem. Soc.* **1989**, *111*, 1804.

[234] Ceccon, A.; Gambaro, A.; Gotthardi, F.; Manoli, F.; Venzo, A.; *J. Organomet. Chem.* **1989**, *363*, 91.

[235] Clark, G.; Ph.D. Thesis, Cornell University, **1977**, 78.

[236] Chaari, M.; Jenki, A.; Lavergne, J.-P.; Viallefont, P.; *Tetrahedron* **1991**, *47*, 4619.

[237] Mino, T.; Matsuda, T.; Maruhashi, K.; Yamashita, M.; *Organometallics* **1997**, *16*, 3241.

[238] Rose-Munch, F.; Rose, E.; Semra, A.; Mignon, L.; Garcia-Oricain, A.; Knobler, C.; *J. Organomet. Chem.* **1989**, *363*, 297.

[239] Rose-Munch, F.; Rose, E.; Semra, A.; *J. Chem. Soc., Chem. Commun.* **1987**, 942.

[240] Rose-Munch, F.; Rose, E.; Semra, A.; Jeannin, Y.; Robert, F.; *J. Organomet. Chem.* **1988**, *353*, 53.

[241] Djukic, J. P.; Rose-Munch, F.; Rose, E.; *J. Am. Chem. Soc.* **1993**, *115*, 6434.

[242] Rose-Munch, F.; Djukic, J. P.; Rose, E.; *Tetrahedron Lett.* **1990**, *31*, 2589

[243] Djukic, J. P.; Geyermans, P.; Rose-Munch, F.; Rose, E.; *Tetrahedron Lett.* **1991**, *32*, 6703.

[244] Djukic, J. P.; Rose-Munch, F.; Rose, E.; *J. Chem. Soc. Chem. Commun.* **1991**, 1634.

[245] Semmelhack, M. F.; Hall, H. K.; Farina, R.; Yoshifuji, M.; Clark, G.; Bargar, T.; Hirotsu, K.; Clardy, J.; *J. Am. Chem. Soc.* **1979**, *101*, 3535.

[246] Keller, L.; Times-Marshall, K.; Behar, S.; Richards, K.; *Tetrahedron Lett.* **1989**, *30*, 3373.

[247] Kundig, E. P.; Cunningham, A. F.; Paglia, P.; Simmons, D. P.; *Helv. Chim. Acta* **1990**, *73*, 386.

[248] Ohlsson, B.; Ullenius, C.; *J. Organomet. Chem.* **1984**, *267*, C34.

[249] Ohlsson, B.; Ullenius, C.; *J. Organomet. Chem.* **1988**, *350*, 35.

[250] Ohlsson, B.; Ullenius, C.; Jagner, S.; Grivet, C.; Wegner, E.; Kundig, E. P.; *J. Organomet. Chem.* **1989**, *365*, 243 and references therein.

[251] Semmelhack, M. F.; Clark, G.; *J. Am. Chem. Soc.* **1977**, *99*, 1675.

[252] Clark, G.; Ph.D. Thesis, Cornell University, **1977**, 67.

[253] Semmelhack, M. F.; Clark, G. R.; Farina, R.; Saeman, M.; *J. Am. Chem. Soc.* **1979**, *101*, 217.

[254] Kundig, E. P.; *Pure Appl. Chem.* **1985**, *57*, 1855.

[255] Kundig, E. P.; Desobry, V.; Simmons, D. P.; *J. Am. Chem. Soc.* **1983**, *105*, 6962.

[256] Semmelhack, M. F.; Wulff, W.; Garcia, J. L.; *J. Organomet. Chem.* **1982**, *240*, C5.

[257] Kozikowski, A. P.; Isobe, K.; *J. Chem. Soc., Chem. Commun.* **1978**, 1076.

[258] Semmelhack, M. F.; Garcia, J. L.; Cortes, D.; Farina, R.; Hong, R.; Carpenter, B. K.; *Organometallics* **1983**, *2*, 467.

[259] Klopmann, G.; *Chemical Reactivity and Reaction Paths*, John Wiley & Sons, New York, **1974**.

[260] Albright, T. A.; Carpenter, B. K.; *Inorg. Chem.* **1980**, *19*, 3092.

[261] Jackson, W. R.; Rae, I. D.; Wong, M. G.; Semmelhack, M. F.; Garcia, J. N.; *J. Chem. Soc., Chem. Commun.* **1982**, 1359.

[262] Jackson, W. R.; Rae, I. D.; Wong, M. G.; *Aust. J. Chem.* **1986**, *39*, 303.

[263] Solladie-Cavallo, A.; Wipff, G.; *Tetrahedron Lett.* **1980**, 3047.

[264] Semmelhack, M. F.; Harrison, J. J.; Thebtaranonth, Y.; *J. Org. Chem.* **1979**, *44*, 3275.

[265] Boutonnet, J. C.; Levisalles, J.; Normant, J. M.; Rose, E.; *J. Organomet. Chem.* **1983**, *255*, C21–C23.

[266] Semmelhack, M. F.; Yamashita, A.; *J. Am. Chem. Soc.* **1980**, *102*, 5924.

[267] Kundig, E. P.; Simmons, D. P.; *J. Chem. Soc., Chem. Commun.* **1983**, 1320.

[268] Kundig, E. P.; Do Thi, N. P.; Paglia, P.; Simmons, D. P.; Spichiger, S.; Wenger, E.; in *Organometallics in Organic Synthesis*, (A. de Meijere, H. Tom Dieck, Eds), Springer-Verlag, Berlin, **1987**, 265–276.

[269] Cambie, R. C.; Clark, G. R.; Gallagher, S. R.; Rutledge, P. S.; Stone, M. J.; Woodgate, P. D.; *J. Organomet. Chem.* **1988**, *342*, 315.

[270] Collman, J. P.; *Acc. Chem. Res.* **1975**, *8*, 342.

[271] Kundig, E. P.; Bernardinelli, G.; Liu, R.; Ripa, A.; *J. Am. Chem. Soc.* **1991**, *113*, 9676.

[272] Semmelhack, M. F.; Seufert, W.; Keller, L.; *J. Am. Chem. Soc.* **1980**, *102*, 6584.

[273] Uemura, M.; Minami, T.; Hayashi, Y.; *J. Chem. Soc., Chem. Commun.* **1984**, 1193.

[274] Beck, H. J.; Fisher, E. O.; Kreiter, G. C.; *J. Organomet. Chem.* **1971**, *26*, C41.

[275] Fisher, E. O.; Stuckler, P.; Beck, H. J.; Kreisel, F. R.; *Chem. Ber.* **1976**, *109*, 3089.

[276] Fraser, R. R.; Mansour, T. S.; *J. Organomet. Chem.* **1986**, *310*, C60.

[277] Silverthorn, W. E.; *Adv. Organomet. Chem.* **1975**, *13*, 47.

[278] Moser, G. A.; Rausch, M. D.; *Synth. React. Inorg. Met.-Org. Chem.* **1974**, *4*, 37.

[279] Card, R. J.; Trayhanovsky, W. S.; *J. Org. Chem.* **1980**, *45*, 2555.

[280] Mandal, S. K.; Sarkar, A.; *J. Org. Chem.* **1998**, *63*, **1901**.

[281] Rausch, M. D.; Gloth, R. E.; *J. Organomet. Chem.* **1978**, *153*, 59.

[282] Bisaha, J. J.; PhD Dissertation, Princeton University, **1980**.

[283] Semmelhack, M. F.; Bisaha, J. J.; Czarny, M.; *J. Am. Chem. Soc.* **1979**, *101*, 768.

[284] Effenberger, F.; Schollkopf, K.; *Angew. Chem. Int. Ed. Engl.* **1981**, *20*, 266.

[285] Streitweiser, A.; Boerth, D.; *J. Am. Chem. Soc.* **1978**, *100*, 755.

[286] Card, R. J.; Trayhanovsky, W. S.; *J. Org. Chem.* **1980**, *45*, 2560.

[287] Gschwend, H. W.; Rodriguez, H. R.; *Org. Reactions*, **1979**, *26*, 1.

[288] Gilman, H.; Soddy, T. S.; *J. Org. Chem.* **1957**, 22, 1715.

[289] Wakefield, B. J.; *The Chemistry of Organolithium Compounds*, Pergamon Press, NY, **1974**, 38.

[290] Nefedov, O. M.; Dyanchenko, A. I.; *Dokl. Akad. Nauk SSSR* **1971**, *198*, 593.

[291] Semmelhack, M. F.; Ullenius, C.; *J. Organomet. Chem.* **1982**, *235*, C10.

[292] Gravshou, M.; Widdowson, D. A.; *J. Chem. Soc., Perkin I* **1983**, 3065.

[293] Mathews, N.; Widdowson, D. A.; *Synth. Lett.* **1990**, 467.

[294] Masters, N. F.; Widdowson, D. A.; *J. Chem. Soc., Chem. Commun.* **1983**, 955.

[295] For a general discussion, see: Semmelhack, M. F.; Garcia, J. L.; Cortes, D. A.; Farina, R.; Hong, R.; Carpenter, B. K.; *Organometallics* **1983**, *2*, 467.

[296] Fukui, M.; Ikeda, T.; Oishi, T.; *Tetrahedron Lett.* **1982**, *23*, 1605.

[297] Blagg, J.; Davies, S. G.; Goodfellow, C. L.; Sutton, K. H.; *J. Chem. Soc., Chem. Commun.* **1986**, 1283.

[298] Heppert, J. A.; Aube, J.; Thomas-Miller, M. E.; Milligan, M. L.; Takusagawa, F.; *Organometallics* **1990**, *9*, 727.

[299] Kondo, Y.; Green, J. R.; Ho, J.; *J. Org. Chem.* **1991**, *56*, 7199.

[300] Dickens, P. J.; Gilday, M.; Negri, J. T.; Widdowson, D. A.; *Pure Appl. Chem.* **1990**, *62*, 575.

[301] Gilday, C. C.; Negri, J. T.; Widdowson, D. A.; *Tetrahedron* **1989** *45*, 4605.

X
Organopalladium Chemistry

LOUIS S. HEGEDUS

Colorado State University, Fort Collins, Colorado, USA

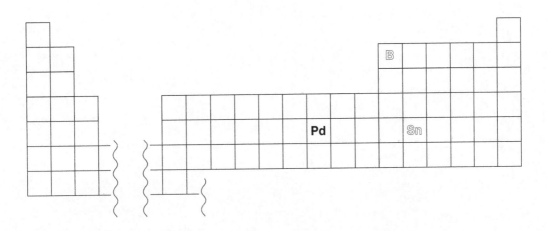

Organometallics in Synthesis: A Manual. Edited by Manfred Schlosser.
© 2002 John Wiley & Sons Ltd

Contents

1   Introduction

Although the majority of the fundamental processes which are catalyzed by palladium were well-established during the mid-1980s, their acceptance and use by synthetic organic chemists was minimal until the mid-1990s. However, as examples of palladium catalysis in the synthesis of highly functionalized complex molecules accumulated, its acceptance grew, until today, when it is one of the most commonly used metals in organic synthesis[1].

There are several reasons for this growing popularity. Palladium complexes have a very rich organic chemistry and are among the most readily available, easily prepared and easily handled of transition metal complexes. Their real synthetic utility lies in the very wide range of organic transformations promoted by palladium catalysts, and in the specificity and functional group tolerance of most of these processes. They permit unconventional transformations and give the synthetic chemist a wide choice of starting materials. When utilized with skill and imagination, exceptionally efficient total syntheses can be achieved. These points are the recurring themes of this chapter.

2   General Features

Palladium exists in two stable oxidation states, the $+2$ state and the zero-valent state, and it is the facile redox interchange between these two oxidation states which is responsible for the rich reaction chemistry that palladium complexes display. Each oxidation state has its own unique chemistry.

Palladium(II) complexes are electrophilic, and tend to react with electron-rich organic compounds, particularly alkenes and arenes. The most common starting material for most palladium complexes is palladium(II) chloride $[PdCl_2]_n$ a commercially available rust red–brown chloro-bridged polymer, insoluble in most organic solvents (Scheme 1). The polymeric structure is easily broken by donor ligands, resulting in monomeric $PdCl_2L_2$ complexes stable to air and soluble in most common organic solvents. Among the most useful are the nitrile complexes, $PdCl_2(RCN)_2$, prepared by stirring a suspension of $[PdCl_2]_n$ in the nitrile as solvent. These tan–golden solids are quite soluble in organic solvents, and stable to storage for years. The benzonitrile complex is most commonly used, as it is very soluble but, when used in quantity, elimination of the odorous water-insoluble relatively nonvolatile benzonitrile is inconvenient. Although slightly less soluble, the acetonitrile complex is more convenient to work with because acetonitrile is odorless,

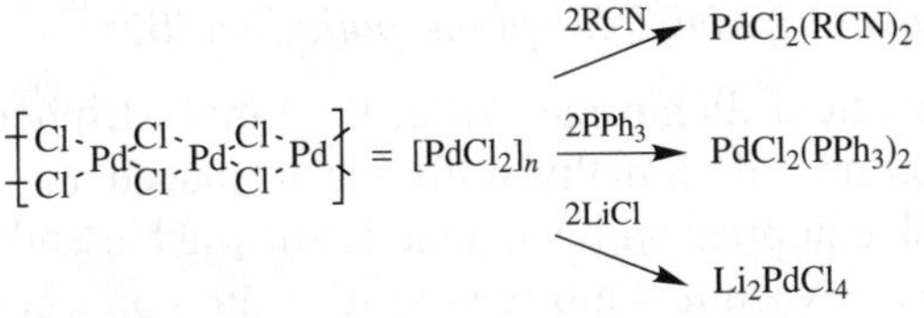

$$\left\{ \begin{array}{c} Cl \\ Cl \end{array} \!\! Pd \!\! \begin{array}{c} Cl \\ Cl \end{array} \!\! Pd \!\! \begin{array}{c} Cl \\ Cl \end{array} \!\! Pd \right\} = [PdCl_2]_n \quad \begin{array}{c} \xrightarrow{2RCN} \; PdCl_2(RCN)_2 \\ \xrightarrow{2PPh_3} \; PdCl_2(PPh_3)_2 \\ \xrightarrow{2LiCl} \\ \searrow \; Li_2PdCl_4 \end{array}$$

Scheme 1. Sources of palladium(II) catalysts.

water soluble, and volatile. Both nitriles are sufficiently labile to easily vacate coordination sites during reaction, making them excellent choices for catalysis.

Preparation of Dichlorobis(acetonitrile)palladium(II)

Anhydrous palladium(II) chloride was suspended in dry acetonitrile, and the resulting dark brown slurry was stirred for 12 h at 25 °C, producing an orange-gold slurry. Filtration followed by drying on the filter gave virtually a quantitative yield of the desired compound.

Treatment of $[PdCl_2]_n$ with triphenylphosphine produces the yellow crystalline $(Ph_3P)_2$ $PdCl_2$ complex, again, quite stable and easily stored and handled. In contrast to the nitrile ligand, the phosphines are much less labile and, as a consequence, $(Ph_3P)_2PdCl_2$ is infrequently used in systems requiring palladium(II) catalysis, although it is frequently the catalyst precursor of choice for palladium(0)-catalyzed processes.

Even chloride ion is able to break the $[PdCl_2]_n$ oligomer, treatment of which with two equivalents of LiCl in methanol produces Li_2PdCl_4, a red–brown hygroscopic solid relatively soluble in organic solvents. The final commonly used palladium(II) complex is palladium(II) acetate, $Pd(OAc)_2$, again commercially available and soluble in common organic solvents. It, too, is most commonly used as a catalyst precursor.

Palladium(0) complexes are strong nucleophiles and strong bases, and are usually used to catalyze reactions involving organic halides, acetates and triflates (see below). By far the most commonly used palladium(0) complex is $Pd(PPh_3)_4$, tetrakis(triphenylphosphine) palladium(0). This yellow, slightly air-sensitive (as a solid) crystalline solid is commercially available, but is expensive, and with four triphenylphosphines contributing 1048 to its molecular weight, one gram of $Pd(PPh_3)_4$ contains precious little expensive palladium, and lots of inexpensive phosphine. This, combined with the tendency of this complex to abruptly loose catalytic activity on standing, but with no accompanying change in appearance, make it advisable to prepare ones own material frequently and in small batches. Fortunately, it is easy to prepare by reducing almost any palladium(II) complex in the presence of excess phosphine (Scheme 2). In fact, in many instances palladium(0)–phosphine complexes are generated by reduction *in situ* and used without isolation, saving time as well as permitting the use of phosphine ligands other than triphenylphosphine.

Preparation of Tetrakis(triphenylphosphine)palladium(0)[2]

A mixture of palladium dichloride (18 g, 0.10 mol), triphenylphosphine (131 g, 0.50 mol), and 1200 mL of dimethylsulfoxide is placed in a single-necked, 2-L, round-bottomed flask equipped with a magnetic stirring bar and a dual-outlet adapter. A rubber septum and a vacuum–nitrogen system are connected to the outlets. The system is then placed under nitrogen with provision made for pressure relief through a mercury bubbler. The yellow mixture is heated by means of an oil bath with stirring until complete solution occurs (140 °C). The bath is then taken away, and the solution is rapidly stirred for 15 min. Hydrazine hydrate (20 g, 0.40 mol) is then rapidly added over about 1 min from a hypodermic syringe. A vigorous reaction takes place with evolution of nitrogen. The dark solution is then immediately cooled with a water bath; crystallization begins to occur at 125 °C. At this point the mixture is allowed to cool without external cooling. After the mixture has reached room temperature it is filtered under nitrogen on a coarse, sintered-glass funnel. The solid is washed successively with two 50-mL portions of ethanol and two 50-mL portions of ether. The product is dried by passing a slow stream of nitrogen through the funnel overnight. The resulting slightly air sensitive yellow crystalline product weighs 106 g (92% yield). Although it can be handled in air for short periods of time, it must be stored under an inert atmosphere. A change in color from yellow to green indicates oxidative decomposition. This procedure can be carried out on a much smaller scale with no dimunition of yield.

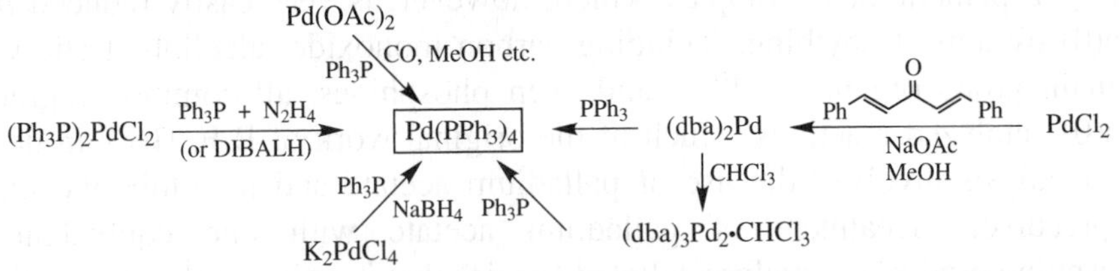

Scheme 2. Sources of palladium(0) catalysts.

Another exceptionally useful palladium(0) complex is $Pd(dba)_2$, the bis-dibenzylidine acetone complex. This is prepared by simply heating at reflux palladium(II) chloride and dba together in methanol, which acts as the reducing agent (getting oxidized to formaldehyde). A cherry-red precipitate, which is $Pd(dba)_2$, forms, is easily separated by filtration, and is quite air-stable, notwithstanding the fact that it is a palladium(0) complex. Purification (rarely necessary) by recrystallization from chloroform produces crystals of $Pd_2(dba)_3 \cdot CHCl_3$ which has the same activity as $Pd(dba)_2$.

Preparation of Bis(dibenzylideneacetone)palladium(0) and
Tris(dibenzylideneacetone)dipalladium(0)(chloroform)[3]

Palladium chloride (1.05 g, 5.92 mmol), was added to hot methanol (about 50 °C, 150 mL) containing dibenzylidene acetone (4.60 g, 19.6 mmol) and sodium acetate (3.90 g, 47.5 mmol). The mixture was stirred for 4 h at 40 °C to give a reddish-purple precipitate and allowed to cool to complete the precipitation. The precipitate was removed by filtration, washed successively with water and acetone and dried *in vacuo*. This material is bis(dibenzylideneacetone)palladium(0). The precipitate

(3.39 g) was dissolved in hot chloroform (120 mL) and filtered to give a deep violet solution. To the solution, diethyl ether (170 mL) was added slowly. Deep purple needles precipitated. These were removed by filtration, washed with diethyl ether, and dried *in vacuo*. The $Pd_2(dba)_3(CHCl_3)$ was obtained in 80% yield (mp 122–124 °C).

These are excellent palladium(0) catalyst precursors because they are very easily handled, can be stored without precaution for years, yet when dissolved and treated with a wide variety of phosphines, produce yellow solutions of the catalytically active PdL_n species in situ. It should be noted, however, that dba is a better ligand for palladium(0) than is triphenylphosphine. As a result, the solutions produced by addition of two equivalents of triphenylphosphine to $Pd(dba)_2$ contain less of the catalytically active species 'PdL_2' than do solutions of an equal concentration of PdL_4![4]. In spite of this, $Pd(dba)_2$ is a very convenient and generally useful catalyst.

Perhaps the most widely used catalyst precursor for palladium(0)-catalyzed processes is $Pd(OAc)_2$, a palladium(II) complex which, however, is very easily reduced *in situ* to palladium(0) by almost anything, including carbon monoxide, alcohols, tertiary amines, alkenes, main group organometallics, and even phosphines, all common ingredients in palladium(0)-catalyzed reactions. Much of the original work of R.F. Heck in developing palladium catalysis involved the use of palladium acetate and tri-*o*-tolylphosphine as a catalyst precursor. Treatment of palladium acetate with one equivalent of tri-*o*-tolylphosphine produces a cyclometallated species that is quite stable and easily isolated and handled, yet is a highly active and long-lived catalyst for a number of important reactions[5,6]:

$$Pd(OAc)_2 + \left(\bigcirc\right)_3 P \xrightarrow[50\,°C,\ 3\ min]{PhCH_3} 1/2 \left[\bigcirc\begin{matrix} P(otol)_2 \\ PdOAc \end{matrix}\right]_2$$

Preparation of trans-Di(μ-acetato)-bis[o-(di-o-tolylphosphino)benzyl] dipalladium(II)[5]

$Pd(OAc)_2$ (4.50 g, 20.0 mmol) was dissolved in toluene (500 mL). The red–brown solution was treated with tris(*o*-tolyl)phosphine (8.00 g, 26.3 mmol). The solution, which rapidly bleached to light orange, was heated to 50 °C for 3 min and then cooled to 25 °C. The solution was reduced under vacuum to a quarter of its volume. After addition of hexane (500 mL), the yellow precipitate was filtered off and dried under vacuum (yield 8.80 g, 93%). Crystallization from toluene/hexane or dichloromethane, hexane and filtration of the solution over Celite affords analytically pure, crystalline material.

Finally, soluble palladium clusters, made by reducing palladium(II) salts in the presence of tetraalkylammonium salts or polyvinylpyrrolidone[7], as well as palladium colloids in the presence of bases[8], catalyze a variety of useful processes. Thus, the synthetic chemist has a large number of palladium catalysts to choose from, and the best choice is not always obvious. As is the case with traditional organic synthetic methodology, the starting point is to find as close an analogy to the desired transformations as possible, and start with that catalyst and those conditions. However, as it is with traditional organic chemistry, the lack of a close analogy is common, and not catastrophic; although an impediment to progress, it is also an opportunity to discover new chemistry.

3 Palladium(II) Complexes in Organic Synthesis

3.1 Nucleophilic Attack on Palladium(II) Alkene Complexes

3.1.1 General Features

Alkenes rapidly and reversibly complex to soluble palladium(II) complexes, particularly $(RCN)_2PdCl_2$ and Li_2PdCl_4. For simple alkenes, the order of complexation is: $CH_2=CH_2 > RCH=CH_2 > cis\text{-}RCH=CHR > trans\text{-}RCH=CHR, \gg R_2C=CH_2, R_2C=CHR.$ ($R_2C=CR_2$ does not complex at all.) As palladium(II) is electrophilic, electron-rich alkenes such as enol ethers and enamides complex strongly, whereas electron-deficient alkenes complex poorly or not at all. Although the degree of complexation is related to the degree of activation of the alkene towards further reaction, stable alkene complexes are not only unnecessary, but actually a hindrance to catalysis, and alkenes that complex only very weakly can often be brought into reaction, particularly in intramolecular cases.

Once complexed to palladium(II), the alkene becomes generally reactive towards nucleophilic attack, this reversal of normal reactivity (electrophilic attack) being a consequence of coordination. Under appropriate conditions, nucleophiles ranging from chloride through phenyllithium—a range of about 10^{36} in basicity—can attack the metal-bound alkene. Attack usually occurs at the more substituted position of the alkene (that which would best stabilize the positive charge) and from the face opposite the metal. Nucleophiles that are poor ligands for palladium(II), such as halide, carboxylates, alcohols, and water, generally react without complication. Amines are substantially better ligands for palladium, and competitive attack at the metal with displacement of the alkene is sometimes a problem which can, however, often be managed. For this same reason, phosphines are rarely if ever used as spectator ligands in palladium(II)-catalyzed reactions of alkenes. Low-valent sulfur species, such as thiols and thioethers, are catalyst poisons and cannot be used as nucleophiles. Carbanions, particularly nonstabilized ones, are strong reducing agents, and reduction of palladium(II) to palladium(0) with concommitment oxidative coupling of the carbanion must be suppressed by the addition of appropriate ligands when carbanions are used as nucleophiles. Finally, remote functional groups in the substrate that strongly complex palladium(II) (e.g. phosphines, aliphatic amines, thiols) may irreversibly bind palladium(II), preventing the desired complexation and activation of the alkenic portion of the molecule.

Nucleophilic attack on the alkene produces a new carbon–nucleophile bond and a new palladium–carbon σ-bond. Regardless of how it is produced, this (usually) unstable σ-alkylpalladium(II) intermediate has a very rich reaction chemistry in its own right. Exposure of this class of complexes to subatmospheric pressures of carbon monoxide results in insertion of CO into the metal–carbon σ-bond, to give a σ-acyl complex. This insertion proceeds readily at temperatures above −20 °C, and effectively competes with β-elimination (see below). The resulting σ-acyl complexes are also unstable, and it is customary to generate them in the presence of a trap such as methanol, to result in cleavage to the free organic ester, and palladium(0). Exposure of solutions of the σ-alkylpalladium(II) complex to hydrogen, again at temperatures above −20 °C results in clean hydrogenolysis of the metal–carbon σ-bond, resulting in overall nucleophilic addition to the alkene. Again, palladium(0) is produced. Main group organometallics including Sn, Zr, Zn, Cu, B, Al, MgX, Si, but not Li readily transfer their R-group to σ-alkylpalladium(II) complexes at temperatures below −20 °C, producing dialkylpalladium(II) species which undergo reductive elimination upon warming, coupling the two R groups and again producing palladium(0). If the initial σ-alkylpalladium(II) complex is warmed above −20 °C, β-hydrogen elimination rapidly ensues, regenerating the alkene (e.g., overall nucleophilic substitution) and again giving palladium(0). The stereochemistry of β-elimination requires that the metal and the β-hydrogen be syn coplanar, and β-elimination can often be suppressed by preventing this. However, under normal circumstances, β-elimination is very facile, and is the primary competing side-reaction in processes attempting to utilize σ-alkylpalladium(II) species which have β-hydrogens. Indeed, alkene insertion, a very important processes for σ-alkylpalladium(II) complexes lacking β-hydrogens (e.g., aryl or vinyl complexes) can rarely be achieved with systems having β-hydrogens, because β-elimination is so facile. When β-hydrogens are lacking, simple alkenes, electron-rich alkenes such as enol ethers and enamides, and electron-deficient alkenes, all readily insert into palladium(II)–carbon σ bonds, resulting in alkylation of the alkene. The regiochemistry of the insertion is dominated by steric not electronic effects, and alkylation occurs predominately at the less-substituted alkene carbon.

Note that all of these reactions of σ-alkylpalladium(II) complexes produce palladium(0) in the final step, while palladium(II) is required to activate the alkene for the first step (nucleophilic attack). Thus, for catalysis, palladium(0) must be reoxidized to palladium(II) in the presence of substrate, nucleophile and product. A wide array of oxidants, including $O_2/CuCl_2$, $K_2S_2O_8$, and benzoquinone are available, and many successful catalytic systems have been developed. However carrying the palladium(0) to palladium(II) redox chemistry necessary for catalysis is often the most difficult aspect of developing new palladium(II)-based systems. All of the points discussed above are summarized in Scheme 3.

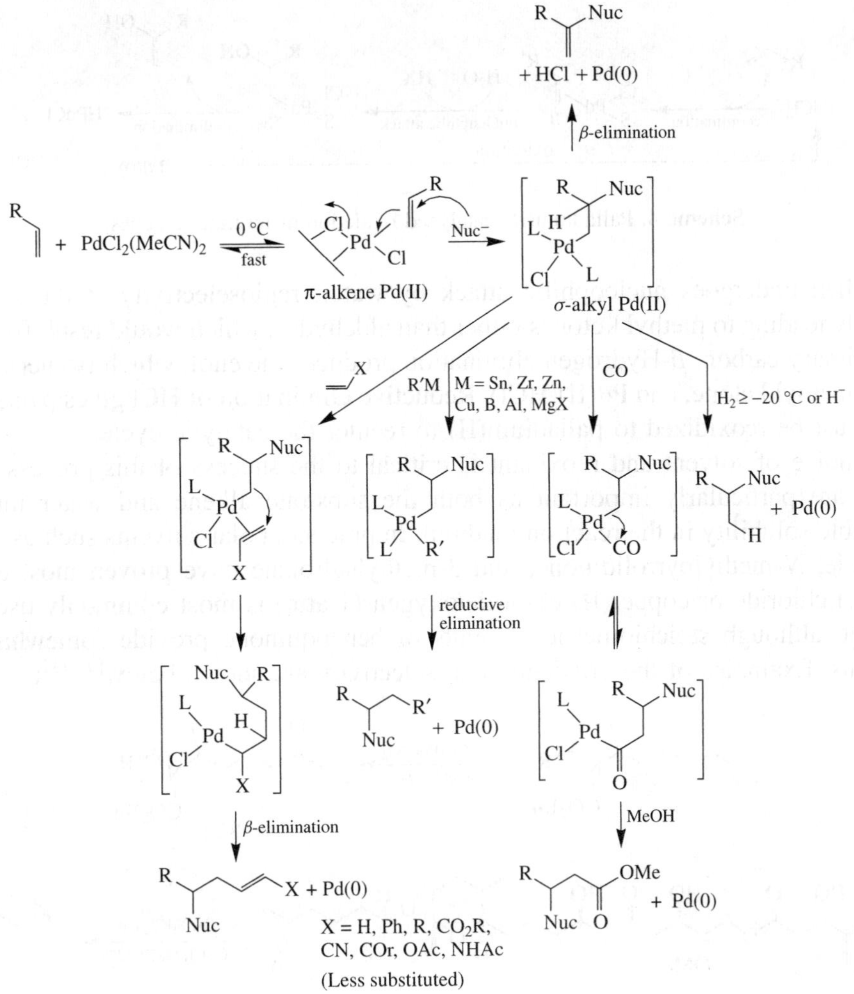

Scheme 3. Reactions of palladium(II)–alkene complexes.

3.1.2 Oxygen Nucleophiles

One of the earliest palladium(II)-catalyzed reactions of alkenes studied was the Wacker process, the formal oxidation of ethylene to acetaldehyde which involves, as the key step, palladium(II)-catalyzed nucleophilic addition of water to ethylene. Although this particular process is of little general use in organic synthesis, the closely related palladium(II)-catalyzed oxidation of terminal alkenes to methyl ketones is quite useful, primarily because this oxidation is quite selective for terminal alkenes and tolerant of a wide range of functional groups[9]. The reaction probably proceeds as in Scheme 4, and involves many of the processes discussed above.

The first step involves coordination of the alkene and, as terminal alkenes complex most strongly, the reaction is selective for terminal over internal alkenes. The complexed

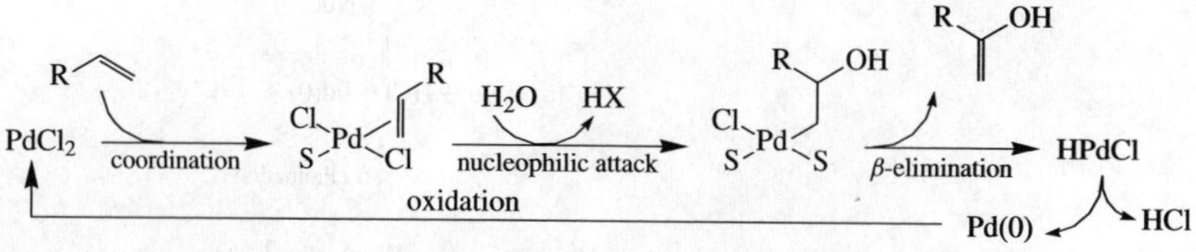

Scheme 4. Palladium(II)-catalyzed oxidation of terminal alkenes.

alkene then undergoes nucleophilic attack by water, regioselectivity at the 2-position, ultimately leading to methyl ketones rather than aldehydes, which would result from attack at the primary carbon. β-Hydrogen elimination produces the enol, which isomerizes to the product methyl ketone, and Pd(II)(HCl). Reductive elimination of HCl gives palladium(0), which must be reoxidized to palladium(II) to reenter the catalytic cycle.

The choice of solvent and reoxidant is critical to the success of this process. Solvent systems are particularly important as both the substrate alkene and water must have appreciable solubility in the reaction medium. In practice, polar solvents such as dimethyl formamide, *N*-methylpyrrolidinone, and 3-methylsulfolane have proven most effective. Copper(I) chloride or copper(II) chloride/oxygen (1 atm) is most commonly used as the reoxidant, although stoichiometric amounts of benzoquinone provide somewhat milder conditions. Examples of the efficiency and selectivity are shown below[9–11]:

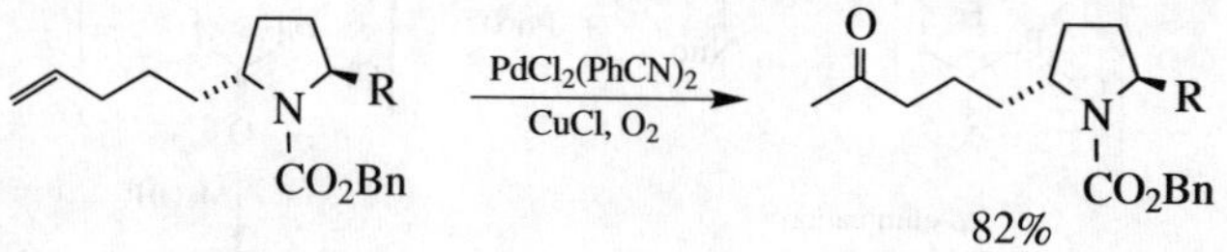

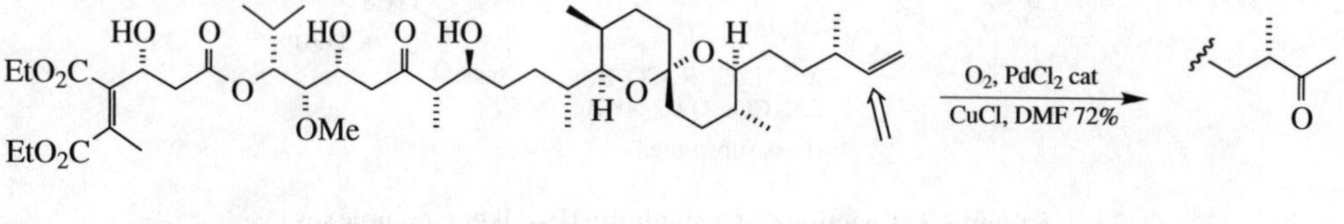

Oxidation of 1-(5-Hexenyl)-3,7-dimethylxanthine with Palladium(II) Chloride and Copper(II) Chloride[9]

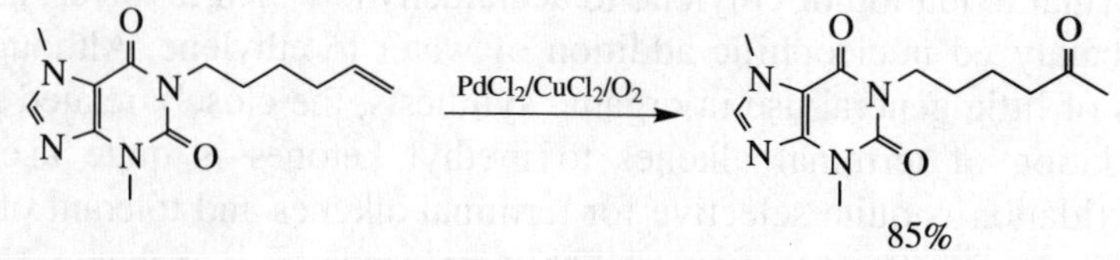

A solution of palladium(II) chloride (0.034 g) and copper(II) chloride (0.330 g) in dimethylformamide (100 mL) and water (100 mL) is prepared. To this solution, the xanthine (5 g), dissolved in dimethylformamide (50 mL) and water (50 mL) is added slowly during 4 h at 60–80 °C with passage of oxygen. After the addition, the mixture is stirred for 5 h at the same temperature. The solvent is evaporated under vacuum and the residue is extracted with chloroform. After the usual work-up, the crude material is recrystallized from benzene/hexane to give pentoxifylline as crystals (4.5 g (85% yield); mp 103 °C).

The regioselectivity of this process can be altered if there are adjacent functional groups that can coordinate to palladium, directing the nucleophilic attack to form stabilized five- or six-membered palladiacyclic intermediates. In the trans steroid, normal Wacker oxidation to form the methyl ketone was observed[12]. In contrast, the cis steroid gave a 1 : 1 mixture of ketone and aldehyde, the latter resulting from nucleophilic attack at the terminal position, placing the palladium within reach of the carbonyl oxygen to form a six-membered intermediate.

Oxidation of Steroidal Ketones[12]

BQ = benzoquinone   39%   38%

A solution of palladium acetate (0.022.4 g, 0.1 mmol, 0.1 equiv) and of benzo-quinone (0.54, 0.5 mmol, 0.5 equiv) in 10 mL of acetonitrile, 3.1 mL of water and 0.65 mL of perchloric acid (70%) were successively added. The mixture was stirred 0.5 h at 20 °C under argon. Then a solution of the steroid derivative (1 mmol) in 12 mL of acetonitrile was added and stirred at 20 °C. The mixture was poured in diethyl ether and washed with a solution of soda (30%). The aqueous layer was extracted with ether. The combined organic layers were dried over magnesium sulfate, filtered and concentrated under vacuum. The crude product was purified by chromatography on silica gel.

The directing effect was even more profound in the following case, for which the coordination of both oxygens of the ketal was invoked[13]:

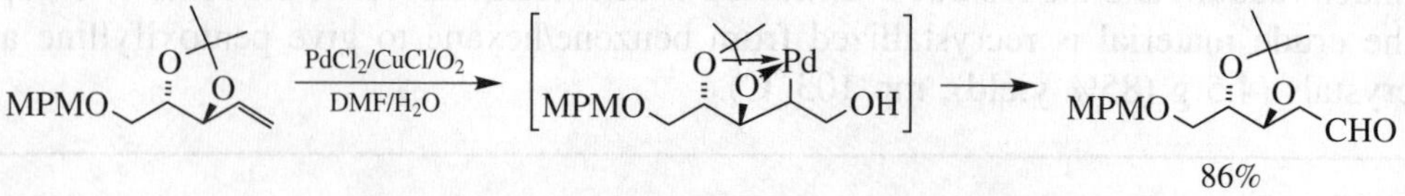

86%

In this case exclusively the aldehyde formed. With the corresponding diol as substrate, normal Wacker oxidation to form the methyl ketone was observed, implying that the OH groups coordinated less strongly than the acetonide. In contrast to most Wacker oxidations, even internal alkenes underwent efficient conversion to methyl ketones when they bore allylic ketal or even alcohol groups[13].

Oxidation of Allylic Diols[13]

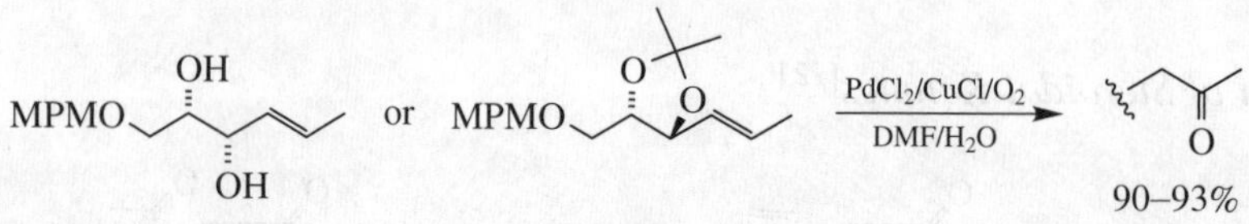

90–93%

(4S,5S)-4,5-Dihydroxy-6-(4-methoxybenzyloxy)-2-hexanone

To a stirred solution of PdCl$_2$ (0.007 g, 0.04 mmol) and CuCl (0.043 g, 0.40 mmol) in DMF and H$_2$O (7 : 1, 3 mL) under oxygen atmosphere (1 atm) was added (E)-(4S,5S)-4,5-dihydroxy-6-(4-methoxybenzyloxy)-2-hexene (0.100 g, 0.40 mmol). The resulting dark-brown solution was stirred at room temperature for 12 h and then extracted with ether. The organic layer was dried over anhydrous MgSO$_4$ and evaporated *in vacuo*. The crude product was separated by SiO$_2$ column chromatography (EtOAc, $R_f = 0.54$) to afford the product ketone (0.0998 g, 93%).

Internal alkenes lacking adjacent directing functional groups undergo reaction extremely slowly under normal reaction conditions, hence the specificity. Unfortunately, alternative conditions under which internal alkenes will react have not been found. The exception to this is conjugated enones, which can be efficiently converted to β-dicarbonyl compounds under very different conditions and almost certainly by a quite different mechanism. An example of this process, which requires the use of *tert*-butylhydroperoxide as a reoxidant, N-methylpyrrolidinone or isopropanol as solvent, and Na$_2$PdCl$_4$ as catalyst at 50–80 °C, is the following[9]:

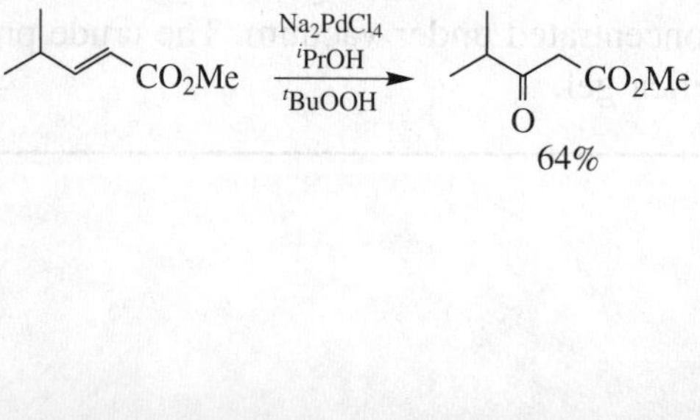

64%

Other oxygen nucleophiles, particularly alcohols and carboxylates[14], also efficiently attack alkenes in the presence of palladium(II) catalysts. Intermolecular versions have been little used, but intramolecular versions, particularly with highly functionalized complex substrates, have been extensively applied. Much of the early work predated the development of mild reoxidants for palladium (*e.g.*, benzoquinone) and many literature examples were carried out using stoichiometric quantities of palladium(II), because the substrates and/or products were unstable to the relatively harsh copper/oxygen redox conditions[15]:

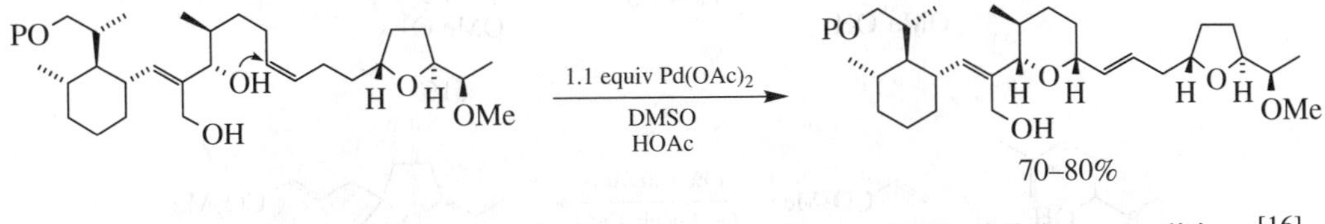

70–80%

However, when the substrate and product can tolerate the oxidizing conditions[16] or when milder oxidizing agents can be used[17], efficient catalysis can be achieved:

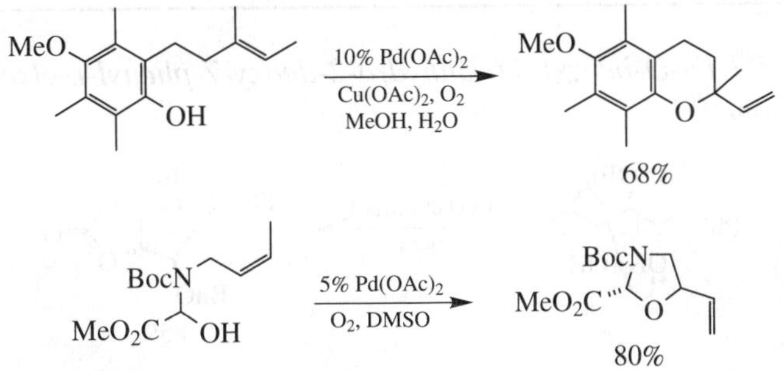

68%

80%

Reoxidation is necessary in these palladium(II)-catalyzed processes because 'HPdX' is produced in the β-elimination step (Scheme 4) and rapidly loses (reductive elimination) HX to produce palladium(0). If some group other than hydride, such as Cl⁻ or OH⁻, can be β-eliminated, subsequent reductive elimination to form palladium(0) (and Cl₂ or HOCl) does not occur, and the resulting PdCl₂ or Pd(OH)(Cl) can reenter the catalytic cycle directly, obviating the need for oxidants[18]:

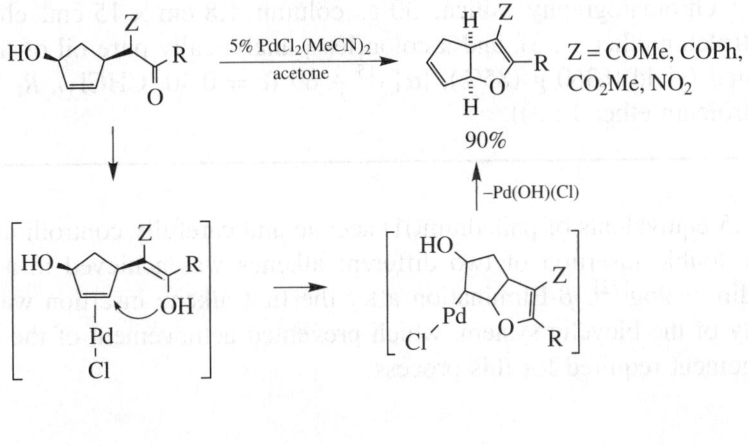

90%

The σ-alkylpalladium(II) intermediate resulting from nucleophilic attack on the alkene can be diverted from undergoing β-elimination by insertion of carbon monoxide[19,20] or alkenes[21], resulting in the very (if β-hydrides are absent) efficient formation of complex systems in a single process:

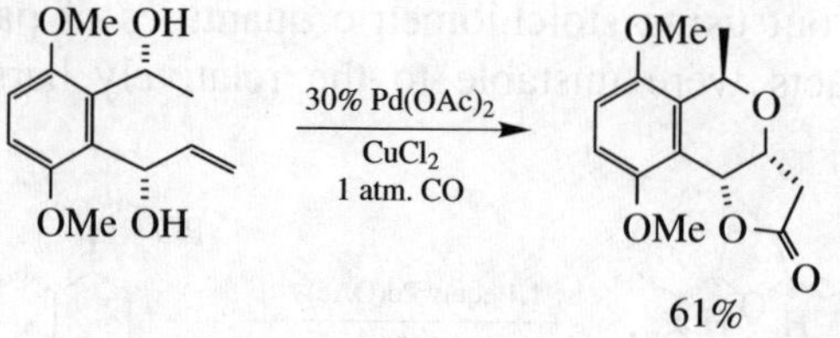

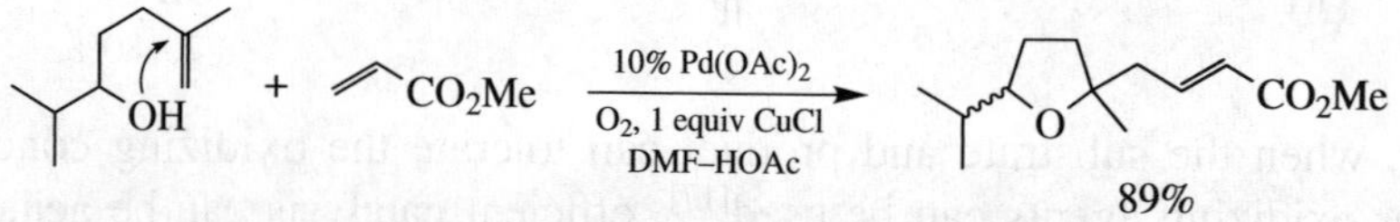

61%

89%

Preparation of 5,7-Di-O-benzyl-3,6-anhydro-2-deoxy-7-phenyl-L-glycero-D-ido-1,4-heptonolactone[20]

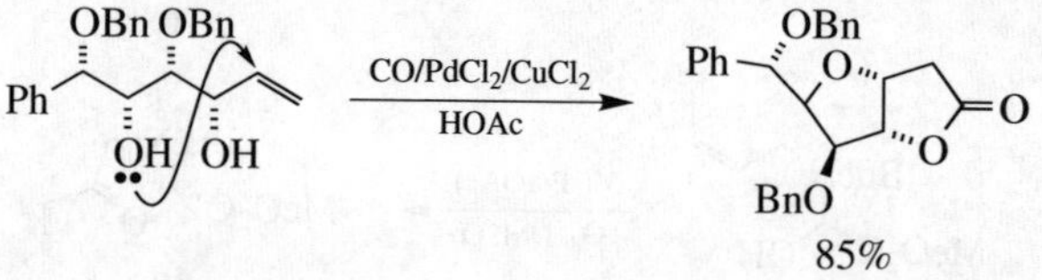

85%

(+)-5,7-Di-O-benzyl-7-epi-goniofufurone

A 50 mL flask purged with CO and connected to a balloon filled with CO was charged with PdCl$_2$ (0.017 g, 0.098 mmol), anhydrous CuCl$_2$ (0.397 g, 2.95 mmol), anhydrous NaOAc (0.242 g, 2.95 mmol), substrate tetraol (0.398 g, 0.984 mmol) and acetic acid (20 mL) and was stirred under CO at 25 °C for 8 h, then filtered through a short tube filled with cellulose. The crude product was purified by chromatography (silica, 30 g, column 1.8 cm × 15 cm, elution with EtOAc/petroleum ether 1 : 3), and a colourless, analytically pure oil of the product was obtained (yield 0.360 g (85%), $[\alpha]_D^{15} + 69$ ($c = 0.40$, CHCl$_3$), R_f 0.18 (with EtOAc/petroleum ether 1 : 3)).

By using 1.5 equivalents of palladium(II) acetate and carefully controlling the reaction conditions, a double insertion of two different alkenes was achieved in a synthesis of a prostaglandin analog[22]. β-Elimination after the first alkene insertion was suppressed by the rigidity of the bicyclic system, which prevented achievement of the syn coplanar Pd−H arrangement required for this process:

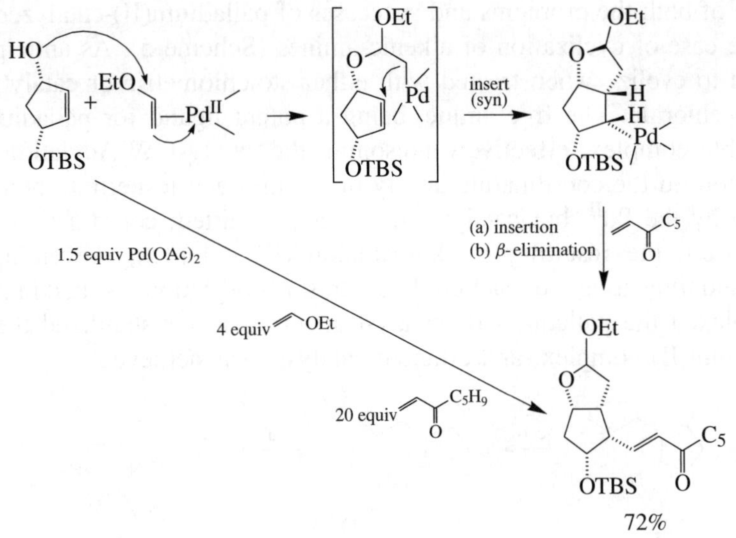

Carboxylic acids are also excellent nucleophiles for alkenes. *o*-Alkenic benzoic acids were cyclized to isocoumarins in excellent yield under both stoichiometric and catalytic conditions[23,24]:

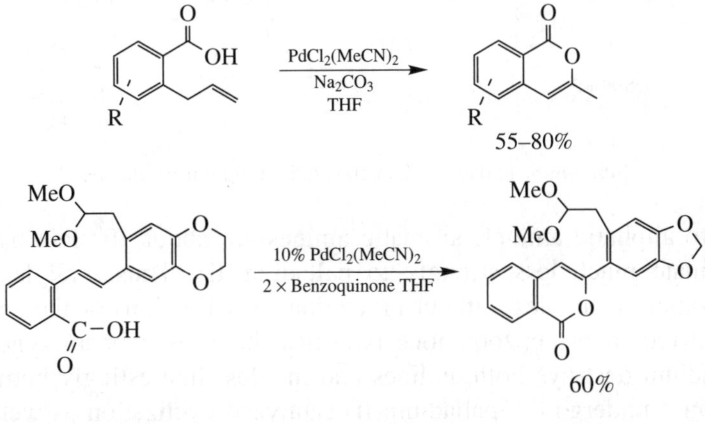

3.1.3 Nitrogen Nucleophiles

Although the palladium(II)-catalyzed reaction of oxygen nucleophiles with alkenes is a general and efficient process, nitrogen nucleophiles pose a number of serious problems, the major one being that aliphatic amines are better ligands for palladium(II) than are alkenes, and both the nucleophile and the product are catalyst poisons. Hence palladium(II)-catalyzed amination of alkenes by aliphatic amines has not been achieved, although a cumbersome process requiring one equivalent of palladium(II), three equivalents of a secondary amine, and a reductive removal of palladium has been developed[25]. The problem does indeed lie with the basicity of the amine nucleophile and, by reducing the basicity, catalysis can be achieved.

Illustrative of both the problems and successes of palladium(II)-catalyzed amination of alkenes is the case of cyclization of alkenic amines (Scheme 5). As anticipated, the free amines failed to cyclize when treated with either stoichiometric or catalytic amounts of palladium(II) chloride. The free amine, being a potent ligand for palladium(II), instead formed a stable complex, effectively poisoning the catalyst. *N*-Acylation of the amine sufficiently reduced the coordinating ability of the nitrogen to permit competitive alkene complexation by the Pd^{II}, but catalysis was again thwarted; coordination of the *N*-acyl oxygen stabilized the resulting σ-alkylpalladium(II) complex, preventing β-hydrogen elimination, and truncating the reaction. In contrast *N*-tosylation resulted in a substrate that neither complexed the palladium(II) catalyst at nitrogen nor stabilized the intermediate σ-alkylpalladium(II) complex, and efficient catalysis was achieved[26].

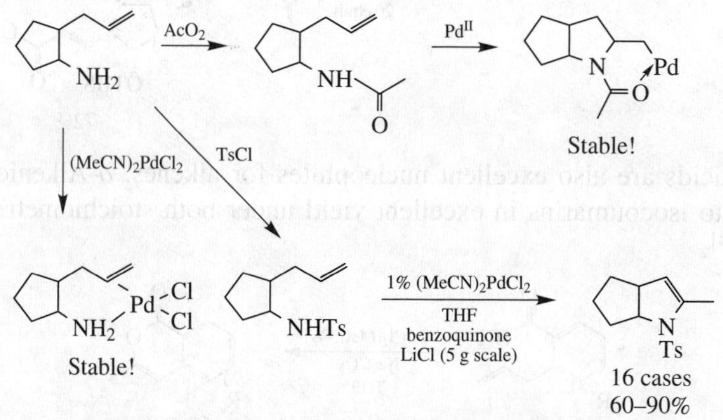

Scheme 5. Palladium(II)-catalyzed amination of alkenes.

In contrast to aliphatic amines, aromatic amines are about 10^6 less basic and consequently coordinate much less strongly to palladium(II). Thus *o*-allylanilines undergo catalytic cyclization to indoles without protection (deactivation) of the amine[27,28]. The use of the mild reoxidant benzoquinone is critical here, as copper/oxygen-based reoxidation of palladium destroys both anilines and indoles. Interestingly both *N*-acetyl, and *N*-sulfonylanilines undergo the palladium(II)-catalyzed cyclization as well. This process was efficiently utilized in the synthesis of indolactam V analogs[29].

Preparation of 4-Bromo-1-tosylindole[28]

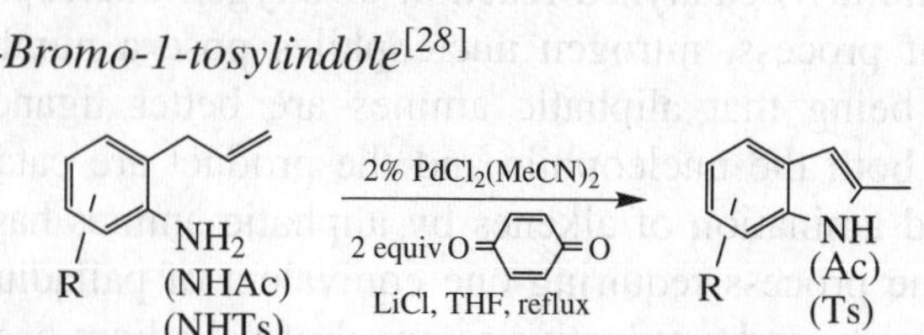

Method A

A 200-mL, one-necked recovery flask equipped with reflux condenser and vacuum adapter with argon balloon was charged with 3-bromo-2-ethenylaniline *p*-toluenesulfonamide (6.00 g, 17.03 mmol), *p*-benzoquinone (1.84 g, 17.03 mmol), LiCl (7.22 g, 0.17 mmol), PdCl$_2$(CH$_3$CN)$_2$ (0.44 g, 1.70 mmol, 10 mol %), and 85 mL of THF. The orange suspension was refluxed (bath 75 °C) for 18 h. The solvent was removed *in vacuo*. The residual brown, slightly gummy solid was transferred to a Soxhlet thimble and extracted with 250 mL of hexanes for 4 h. The resulting pot solution was treated with 0.5 g of charcoal and hot-gravity filtered. After cooling to 25 °C, the hexanes were removed *in vacuo*. The residual beige, slightly gummy solid was transferred to the top of a silica gel (10 g) column and dissolved-eluted with hexanes. Concentration of the eluent *in vacuo* afforded 4.70 g (78.8%) of the product as a colorless solid (mp 119–121 °C). This material may be recrystallized from hexanes to afford 4.58 g (76.7%) of shiny colorless crystals (mp 119–121 °C).

Method B

A 12-oz pressure bottle was charged with 3-bromo-2-ethenylaniline *p*-toluenesulfonamide (6.00 g, 17.03 mmol), *p*-benzoquinone (1.84 g, 17.03 mmol), LiCl (7.13 g, 0.170 mmol), PdCl$_2$(CH$_3$CN)$_2$ (0.22 g, 0.85 mmol, 5 mol %), and 50 mL of THF. The orange suspension was flushed with argon and then heated at 125 °C for 75 min. An identical work-up afforded 4.63 g (77.7%) of the product as a coloress solid (mp 118–121 °C). This material may be recrystallized from hexanes to afford 4.55 g (76.3%) of shiny colorless crystals (mp 120–122 °C).

Procedure of Palladium-assisted Intramolecular Amination[29]

PdCl$_2$(CH$_3$CN)$_2$ (0.0079 g, 30 µmol), benzoquinone (0.0338 g, 0.313 mmol), and LiCl (0.1303 g, 3.07 mmol) was dissolved in anhydrous THF (4.2 mL). The THF solution of the 14-*O*-acetate (0.1123 g, 0.304 mmol) was added to the solution. After being stirred for 2.5 h at 25 °C, the reaction mixture was concentrated *in vacuo* to dryness. The residue was purified by column chromatography on Wakogel C-100 using toluene and increasing amounts of acetone, followed by HPLC on μ-Bondasphere using 70% MeOH and on YMC A-023 using 85% hexane, 10% CHCl$_3$, and 5% ^{i}PrOH to give the product (0.0744 g, 0.203 mmol) in 67% yield ([α]$_D$ − 126.3° (c = 0.69, MeOH, 26 °C)).

By carefully controlling the conditions to promote alkene insertion and suppress β-hydride elimination (short reaction times, polar solvents, low temperatures) the intermediate σ-alkylpalladium(II) complex could be intramolecularly trapped by an adjacent alkene, resulting in bicyclization to the pyrroloindoloquinone system[30]:

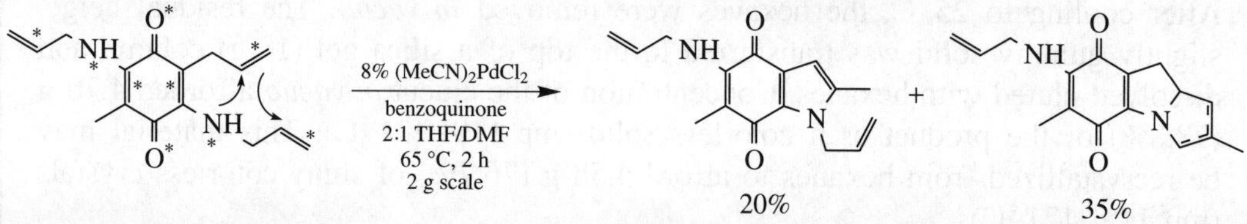

Note that although this substrate has nine potential functional groups (denoted by asterisks) which can coordinate to palladium(II) only coordination to one of these results in a situation where productive reaction can ensue. Provided the other sites do no irreversibly bind the catalyst, the catalytic process proceeds smoothly.

A very mild and general catalyst system for the cyclization of both aliphatic and aromatic alkenic tosamides has recently been developed[31]. It consists of palladium acetate in DMSO and relies upon oxygen as a reoxidant. In most cases the regioselectivity paralleled that of the previous systems, but *o*-allyl-*N*-tosyl anilines cyclized to give dihydroquinolines (attack at the less-substituted alkene terminus) rather than indoles.

Optically active vinylogous carbamates were efficiently prepared by palladium(II) catalyzed amination of methyl acrylate[32]:

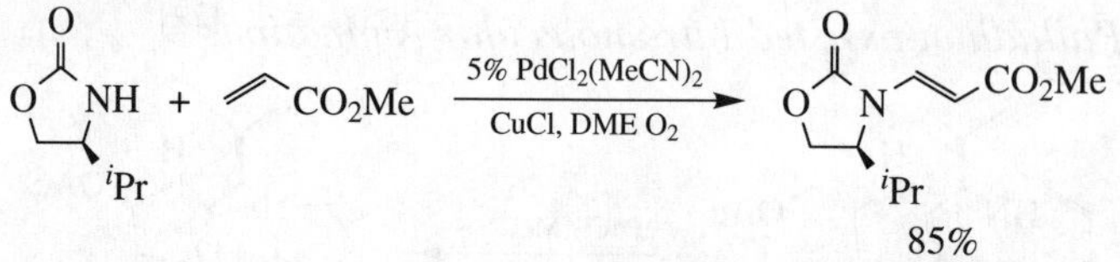

Although palladium(II) also catalyzes the amination of alkynes, the intermediate σ-alkenylpalladium(II) complex can insert alkynes leading to an homologated σ-alkenyl-palladium(II) complex, and ultimately alkyne oligomerization. However, if the reaction is carried out in the presence of carbon monoxide, efficient aminocarbonylation of alkynes can be achieved[33,34]:

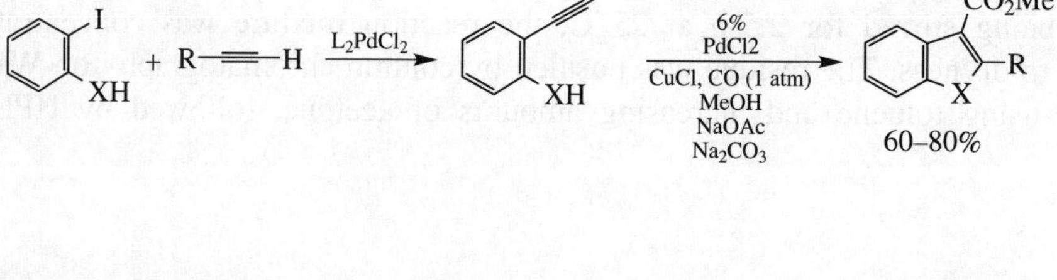

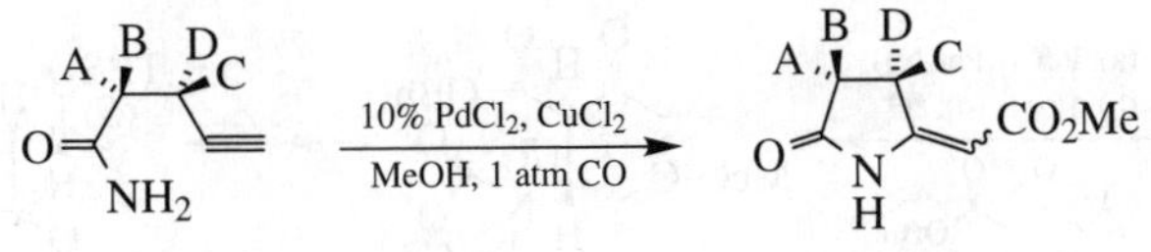

A, B, C, D = H, Me   50–90%

3.1.4 Carbon Nucleophiles

The use of carbon nucleophiles to attack alkenes coordinated to palladium(II) presents yet another set of problems. Palladium(II) salts are reasonably good oxidizing agents, and exposure of carbanions to them results in rapid reduction of palladium(II) to the metal, with concomitant oxidative coupling of the carbanion. This can be suppressed by protecting the palladium(II) from direct attack by the carbanion by the addition of triethyl amine to the alkene–palladium(II) complex at low temperature. Under these conditions, stabilized carbanions can be directed to attack the alkene rather than the metal, and alkylation of the alkene can be achieved. Again an unstable σ-alkylpalladium(II) complex is produced; this can β-eliminate to give the alkylated alkene, can be reduced to give the alkane, or can be carbonylated to give the ester (Scheme 6)[35].

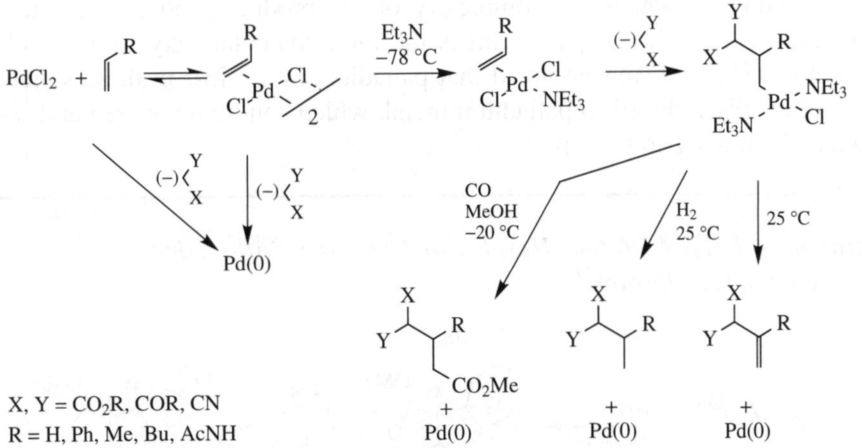

X, Y = CO$_2$R, COR, CN
R = H, Ph, Me, Bu, AcNH

Scheme 6. Palladium(II)-assisted alkylation of alkenes.

However, this process requires a stoichiometric amount of palladium, because reoxidation of palladium(0) in the presence of carbanions has not yet been achieved. Thus, the product must be economically justified before the chemistry is to be utilized. Just such a case is seen in the synthesis of a relay to (+)-thienamycin, wherein all of the functional groups as well as the absolute stereochemistry of the key stereogenic center are set in place in the initial palladium(II)-assisted carboacylation step[36], with good yield and with virtually complete stereoselectivity:

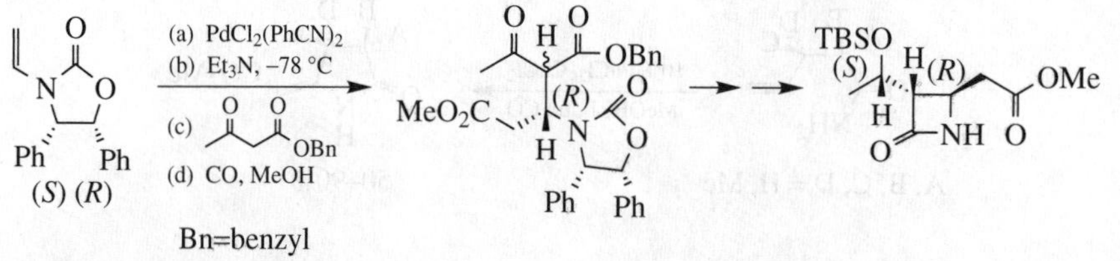

Bn=benzyl

This reaction takes advantage of the fact that CO insertion proceeds faster than β-hydrogen elimination at −20 °C, generating a σ-acyl complex which was cleaved by methanol to produce the ester. Cleavage of this same σ-acyl group with tin reagents, via transmetalation/reductive elimination (Scheme 3) permits the introduction of a wide variety of keto groups at the former alkene terminus[37]. Its use in the synthesis of (+)-Negamycin is shown below[38]:

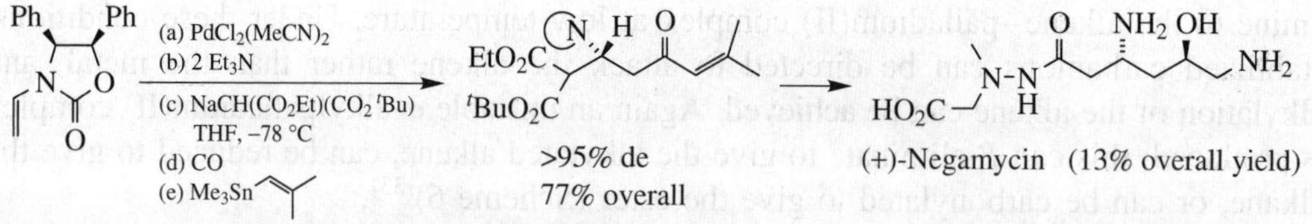

Another example in which the complexity of the product probably justifies the use of stoichiometric quantities of palladium is the intramolecular alkylation of alkenes by TMS enol ethers[39]. It should be noted that palladium is not lost in these stoichiometric reactions, but merely reduced to palladium metal, which can be recovered and reoxidized to palladium(II) in a separate step.

Preparation of 1,3bα,4,5,6,6aα-Hexahydro-3-methoxy-3aβ-methyl-5-oxopentaleno[1,2-c]furan[39]

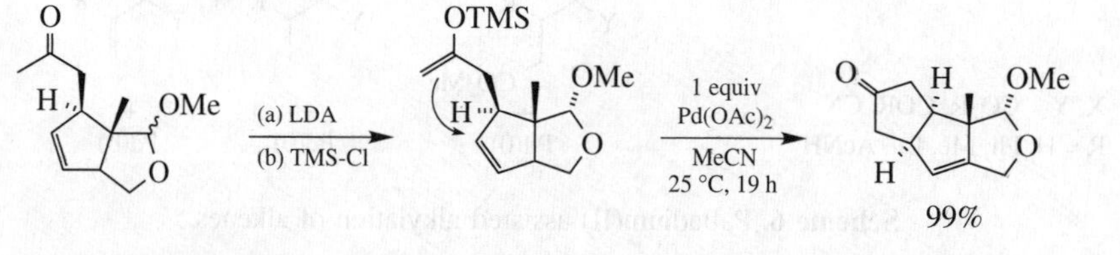

To a stirred solution of LDA, prepared from diisopropylamine (0.14 mL, 0.74 mmol) and butyllithium (10% solution in hexane, 0.48 mL, 0.74 mmol) in THF (5 mL), at −78 °C was added dropwise a solution of the starting keto-alkene (0.104 g, 0.495 mmol) in THF (3 mL), and the mixture was then stirred for 1 h. After addition of chlorotrimethylsilane (freshly distilled from CaH$_2$, 0.18 mL, 0.743 mmol) at −78 °C, the mixture was allowed to warm to 25 °C over a period of 1 h. A solution of this silyl enol ether dissolved in CH$_2$Cl$_2$/MeCN (6 mL, 1 : 1) was added dropwise to a stirred solution of Pd(OAc)$_2$ (0.167 g, 0.743 mmol) dissolved in MeCN (4 mL). The mixture was stirred at 25 °C for 19 h, then filtered through Celite. The filtrate was concentrated to give an oil, which was subjected to column chromatography with 4 : 1 hexane/EtOAc to give rise to the desired tricyclic product (0.0840 g, 99%) as an approximate 4 : 1 mixture of C-3 epimers. Each epimer was separated by flash column chromatography.

3.2 Palladium(II)-catalyzed Rearrangements

3.2.1 Rearrangements of Allylic Systems[40]

Palladium(II) complexes are capable of catalyzing rearrangements of all alkenic systems to which they complex, and, indeed, a competing reaction in the palladium(II)-catalyzed reactions of terminal alkenes discussed above is the rearrangement of these to internal alkenes, which are substantially less reactive[40]. However this ability to catalyze rearrangements has also been utilized in productive ways, particularly with allylic systems. The general transformation is shown below, together with a probable mechanism which involves coordination of the alkene to palladium(II), thereby activating it to nucleophilic attack: nucleophilic attack by Y giving a zwitterionic intermediate, then collapse of the intermediate to the rearranged product with ejection of palladium(II).

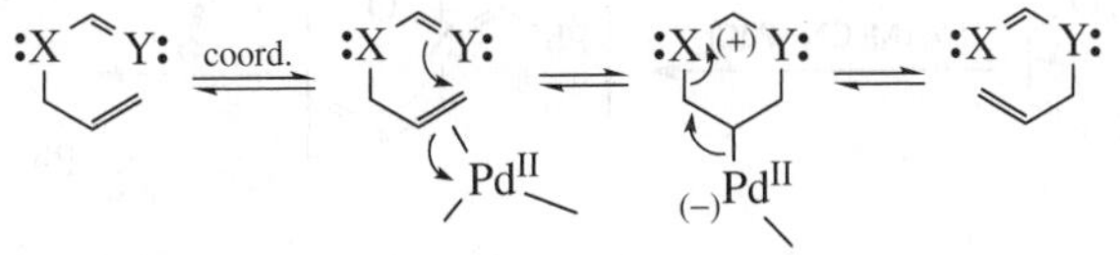

Note there is no net redox here and, in a sense, palladium(II) is behaving like a very selective Lewis acid. (Palladium(0) complexes catalyze similar rearrangements but both the mechanism and the selectivity are quite different from the same reactions catalyzed by palladium(II)!)

This rearrangement proceeds to give the thermodynamically most stable product. Thus tertiary acetates rearrange to primary acetates and rearrangements which result in conjugation of the allylic double bond are always favored. Even small steric differences between allylic isomers can strongly favor the less hindered product[41]:

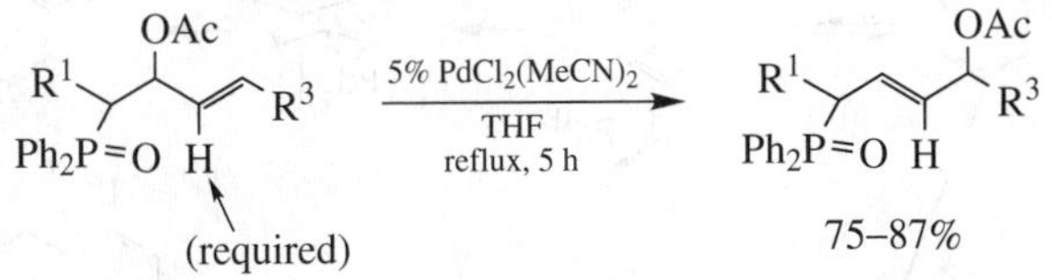

When optically active allyl acetates were rearranged, complete transfer of chirality was observed[42,43]: Note that the absolute stereochemistry observed at the newly formed chiral center must result from coordination of palladium to the face of the alkene opposite the acetate, and attack of the alkene from the face opposite the metal, as expected for the mechanism described above.

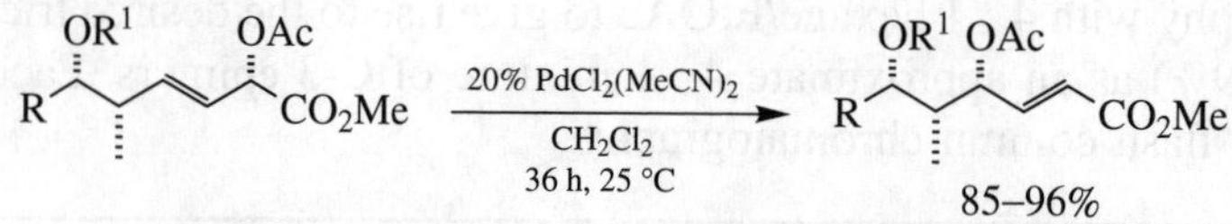

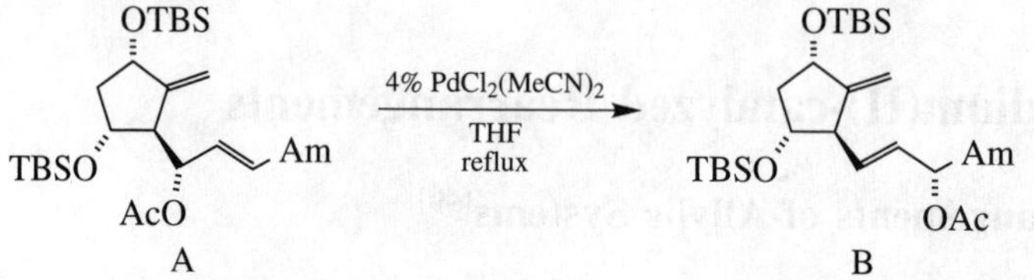

27:73 A/B (calculated thermodynamic ratio, 20:80)

This rearrangement is not restricted to allyl acetates, but is general for allylic systems containing heteroatoms, such as allyl formimidates[44], S-allylthioimidates[45], and allylic thionobenzoates[46], but these have found fewer applications in synthesis. A more synthetically useful variant involves the palladium(II)-catalyzed rearrangement of oxime *o*-allyl ethers to nitrones. When carried out in the presence of a 1,3-dipolarophile, combined rearrangement/1,3-dipolar cycloaddition can be achieved[47].

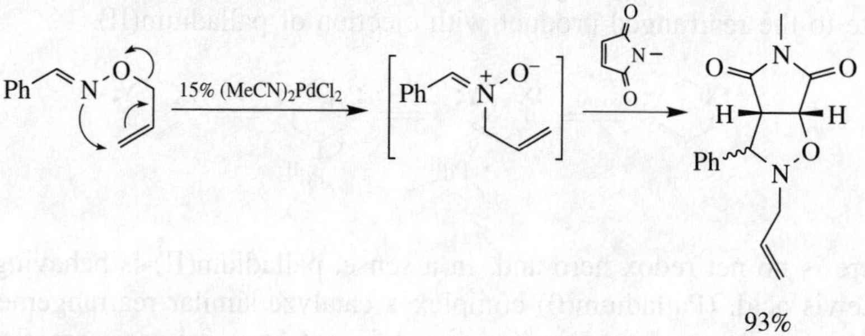

93%

Substantially more interesting is the palladium(II)-catalyzed Cope rearrangement[48], which enjoys rate enhancements of about 10^{10} over uncatalyzed processes, is more stereoselective than the thermal process, and proceeds with virtually complete transfer of chirality in a chain topographic sense:

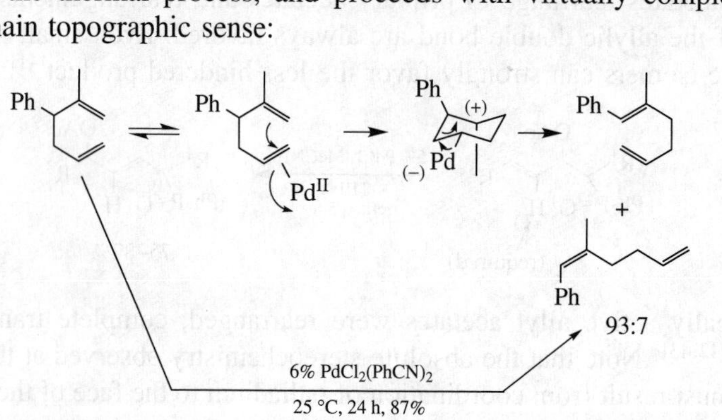

93:7

6% PdCl$_2$(PhCN)$_2$
25 °C, 24 h, 87%

Although useful, this process also suffers limitations consonant with the proposed mechanism. The diene must have a substituent either at C-2 or at C-5, but not both. That is, it must have a monosubstituted alkene to complex to palladium(II), as well as a substituent on the other alkene to stabilize positive charge in the intermediate. Within these confines, however, the reaction is quite useful. Oxy–Cope rearrangements have been catalyzed under similar conditions[49]:

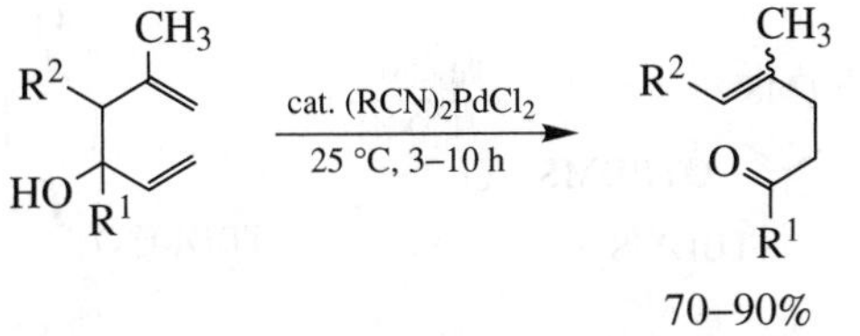

70–90%

3.2.2 Palladium(II)-catalyzed Cycloisomerization of Enynes

Treatment of suitably disposed enynes with a variety of palladium(II) catalysts produces cyclic dienes in excellent yield and under mild conditions (Scheme 7)[50]. Two possible mechanisms are shown, with the 'HPdX' route probably operative when using palladium(0) plus acetic acid as catalyst[51], whereas either is possible when using palladium acetate[52]. A wide array of functionality can be tolerated and the process has been used

Scheme 7.

to synthesize a number of relatively elaborate cyclic dienes[50,53,54]:

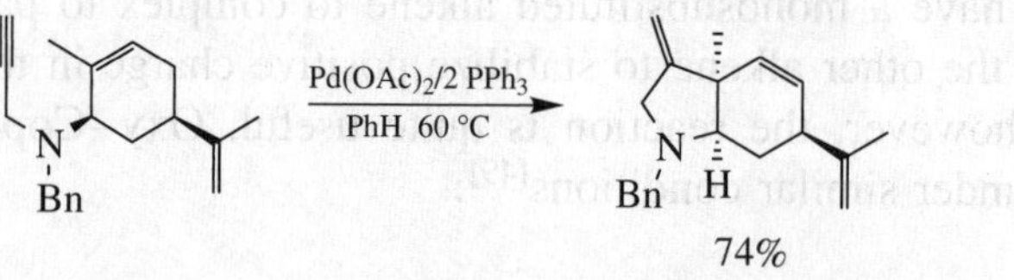

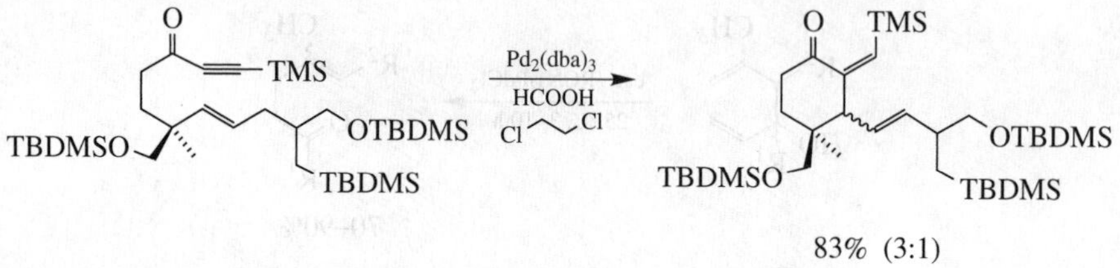

74%

83% (3:1)

Palladium-catalyzed Enyne Cyclization to a 1,4-Diene[53]

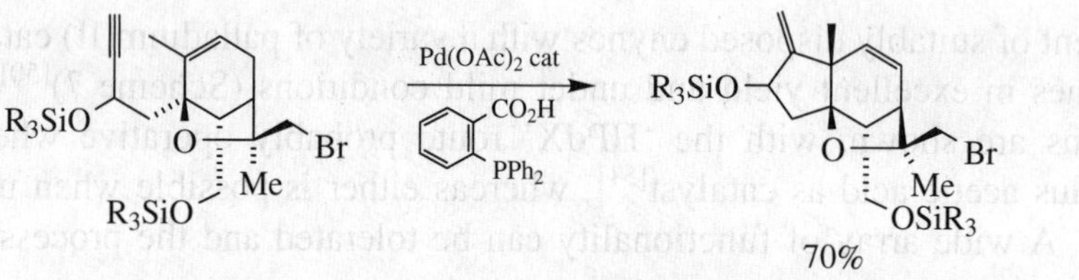

70%

Sequential addition of first 1,3-bis(dibenzophospalo)propane (0.5418 g, 1.32 mmol, 10 mol %) and then 2-diphenylphosphinobenzoic acid (0.4077 g, 1.32 mmol, 10 mol %) to a solution of the enyne (7.40 g, 13.3 mmol) in 175 mL of 1,2-dichloroethane at 25 °C was followed by addition of palladium acetate (0.297 g, 1.32 mmol, 10 mol %). The solution was stirred 20 min at 25 °C and then 21 h at reflux. After cooling, the solvent was removed *in vacuo* and the residue flash chromatographed (1 : 4 CH₂Cl₂/hexane) to give 5.17 g (70% yield) of the cyclized product ($R_f = 0.45$ (1 : 1, CH₂Cl₂/hexane); $[\alpha]_D = -71.5°$ (c 2.28, dichloromethane)).

The intermediate σ-alkylpalladium(II) species generated in the insertion step (path a, Scheme 7) can be intercepted before β-elimination by a number of reagents. Silanes reduced these complexes efficiently, producing alkenes from enynes[50]:

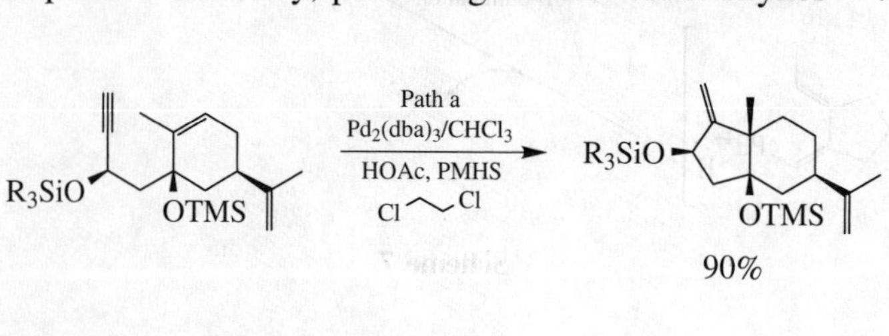

90%

and dienes from diynes[55]:

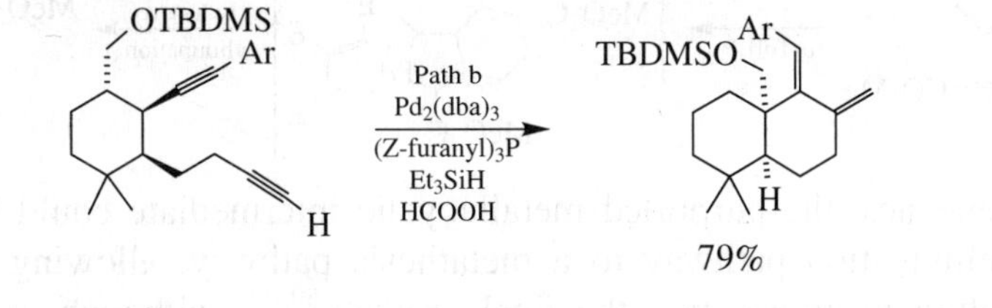

79%

Intramolecular trapping of this σ-alkylpalladium(II) complex by alkenes was also efficient, allowing polycyclizations[56], intramolecular cycloadditions[51]:

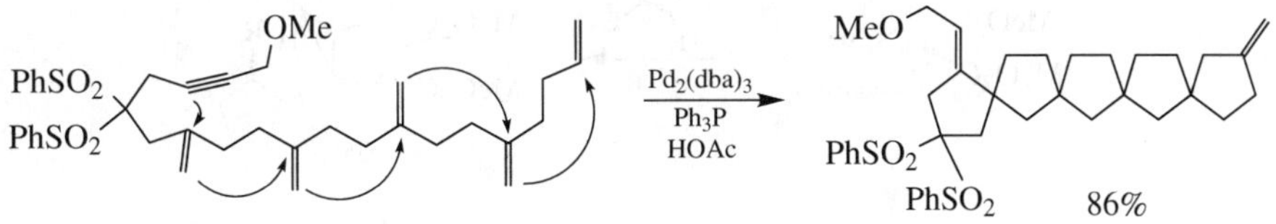

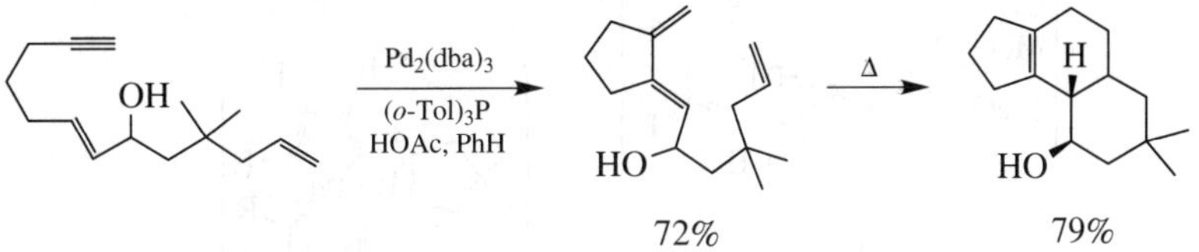

86%

72%   79%

and intramolecular electrocyclic reactions[57]:

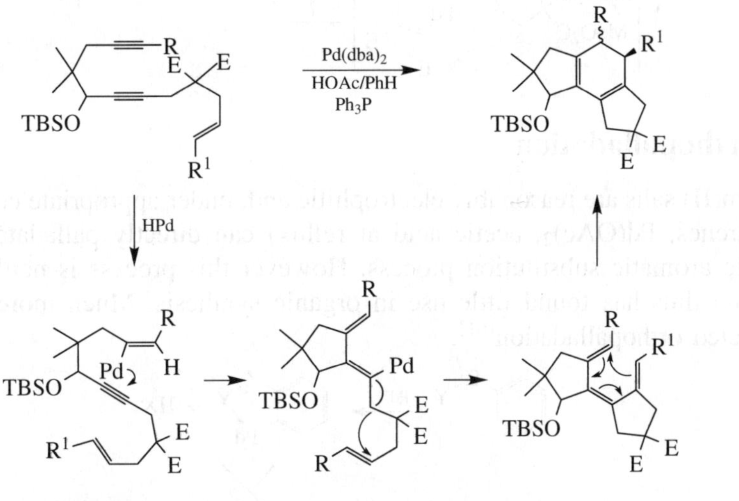

The complex tetrakis[(heptafluorobutoxy)carbonyl]-palladacyclopentadiene, PdTCPCHFB, catalyzed a somewhat different cycloisomerization of enynes, resulting in an overall 2 + 2 cycloaddition[58]:

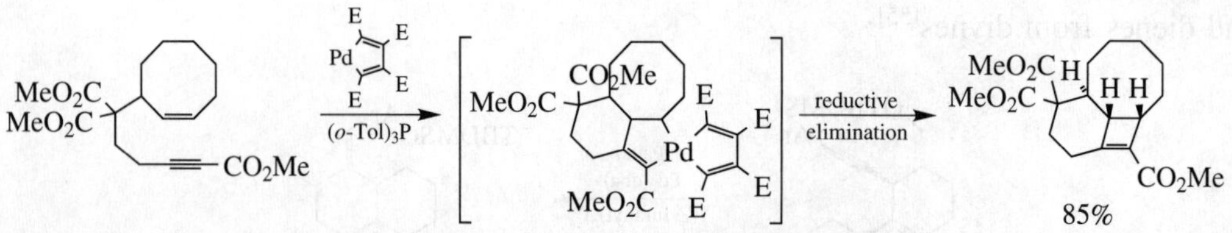

By using dieneynes, the purported metallacyclic intermediate could be diverted from the reductive elimination pathway to a metathesis pathway, allowing incorporation of additional unsaturated groups into the final product[59,60], although to date (1997) this process has found little additional application in synthesis:

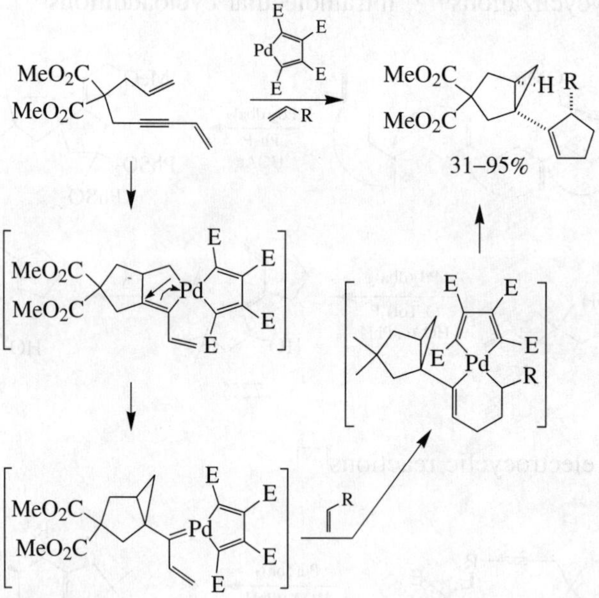

3.3   Orthopalladation

Palladium(II) salts are reasonably electrophilic and, under appropriate conditions, (electron rich arenes, Pd(OAc)$_2$, acetic acid at reflux) can directly palladate arenes via an electrophilic aromatic substitution process. However this process is neither general nor efficient, and thus has found little use in organic synthesis. Much more general is the ligand-directed orthopalladation[61]:

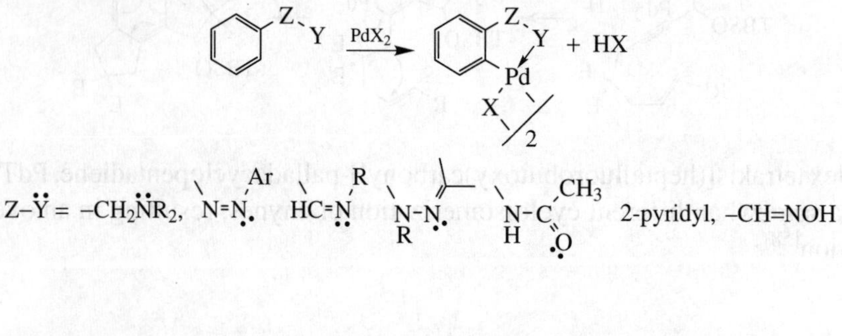

$$Z-\ddot{Y} = -CH_2\ddot{N}R_2, \quad N=\ddot{N}, \quad HC=\ddot{N}, \quad \underset{R}{N-\ddot{N}}, \quad \underset{H}{N-C}, \quad \text{2-pyridyl, } -CH=NOH$$

This process is very general, the only requirement being a lone pair of electrons in a benzylic position to precoordinate to the metal, and conditions conducive to electrophilic aromatic substitution (usually acetic acid solvent). The site of palladation is always ortho to the directing ligand and, when the two ortho positions are inequivalent, palladation results at the sterically less-congested position, regardless of the electronic bias. (This is in direct contrast to ortholithiation, which is electronically controlled and will usually occur at the more hindered position on rings having several electron-donating groups). The resulting chelate-stabilized, σ-arylpalladium(II) complexes are almost invariably quite stable, are easily isolated, purified and stored, and are quite unreactive.

This stability is clearly due to the chelate stabilization by Y, and is the major reason orthopalladation has not yet been developed for catalytic ortho functionalization of arenes. However, the reactivity intrinsic to σ-alkylpalladium(II) species (Scheme 3) can be activated by heating the *o*-palladated species in the presence of excess triethyl amine to disrupt the chelation by displacing Y, in the presence of an alkene. Under these conditions, facile insertion occurs[62]:

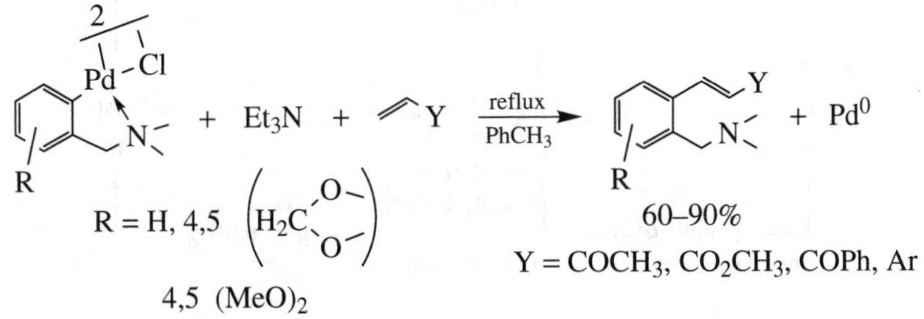

Related complexes of acetanilide undergo facile carbonylation[63]:

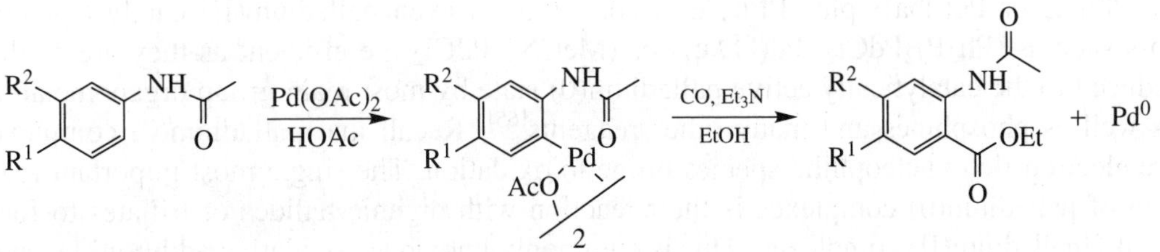

The unresolved problems for catalysis are the incompatible conditions for the orthopalladation step (HOAc) and the subsequent step (Et$_3$N), the stability of the orthopalladated product and, as always, the reoxidation of the palladium(0) produced in the final step. Until these are solved, orthopalladation will remain a stepwise, stoichiometric process of synthetic interest only when its regioselectivity or specificity make it uniquely appropriate for a desired transformation.

4   **Palladium(0) Complexes in Organic Synthesis**

4.1   **Oxidative Addition–Transmetalation**[64]

4.1.1   **General Features**

The palladium(0)-catalyzed coupling of aryl and alkenyl halides and triflates with main group organometallics via oxidative addition/transmetalation/reductive elimination sequences (Scheme 8) has been very broadly developed and has an overwhelming amount of literature associated with it[64]. However, relatively few fundamental principles are involved, the understanding of which will bring order out of this apparent chaos, and make the area both approachable and usable.

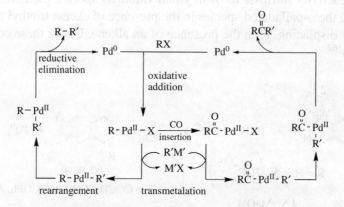

Scheme 8. Palladium(0)-catalyzed oxidative addition/transmetalation.

A very wide range of palladium catalyst precursors can be used in this system, and the choice is best made by analogy, although almost all will work reasonably well. Most often $Pd(PPh_3)_4$ or $Pd(dba)_2$ plus PPh_3 is used, although even palladium(II) catalyst precursors such as $(Ph_3P)_2PdCl_2$, $Pd(OAc)_2$ or $(MeCN)_2PdCl_2$ are efficient as they are readily reduced to the catalytically active palladium(0) state by most main group organometallics as well as phosphines and many other reagents[65]. Recall that palladium(0) complexes are electron-rich nucleophilic species prone to oxidation. The single most important reaction of palladium(0) complexes is their reaction with organic halides or triflates to form σ-alkylpalladium(II) complexes. This is commonly known as 'oxidative addition' because the metal is formally oxidized from palladium(0) to palladium(II) and the oxidizing agent, RX, adds to the metal.

The oxidative addition process has a number of general features. The order of reactivity is $I \gg OTf > Br \gg Cl$[66], such that chlorides are rarely useful in this reaction although some catalysts do have limited reactivity[5]. With aryl halides, added phosphines are required for the reaction of aryl bromides, but suppress the reaction of aryl iodides, permitting easy discrimination. With most substrates, the oxidative addition step proceeds readily at 25 °C. When higher temperatures are used, this is usually because some step other than the oxidative addition is sluggish at this temperature. From the standpoint of the substrate, the metal is getting oxidized, but the substrate is getting reduced. Thus electron-deficient halides are, in general, more reactive than electron-rich ones. Because β-hydrogen elimination is rapid above −20 °C, the halide or triflate substrate usually cannot have β-hydrogens, restricting this process to aryl and alkenyl substrates. With alkenyl substrates oxidative addition occurs with retention of alkene geometry, (as do all of the ensuing steps). The σ-aryl or σ-alkenylpalladium(II) complex formed by oxidative addition participates in the rich chemistry of this class of complexes presented in Scheme 3, including alkene and CO insertion (see Section 4.2), hydrogenolysis and transmetalation from main group organometallics.

A very wide range of main group organometallics 'transmetalate' to palladium(II); that is, they transfer their R group to palladium in exchange for the halide or triflate, generating a dialkylpalladium(II) complex and the main group metal halide or triflate. Transmetalations from Li, Mg, Zn, Zr, B, Al, Sn, Si, Ge, Hg, Tl, Cu, Ni and perhaps others have been reported but some (*e.g.*, Sn, B, Zn) are much more useful than others.

Transmetalation is favored from more electropositive to more electronegative metals, but this is of little use in assessing reactivity as electronegativities are only crudely known, are sensitive to spectator ligands, and if the next step in the process is irreversible, only a small equilibrium constant for transmetalation is required for the process to proceed. Transmetalation is almost invariably the rate-limiting step, and when oxidative addition/transmetalation sequences fail, it is this step that warrants attention. (Because transmetalation is rate-limiting, it is often possible to insert CO into the oxidative addition product prior to transmetalation, resulting in a carbonylative coupling (see below).) Organosilanes only transmetalate in the presence of added fluoride, whereas organoboranes require added alkoxide to transmetalate. These additives both generate the more reactive anionic -ate complexes (R_4SiF^-, $R_3BOR'^-$) for transmetalation and produce very stable neutral species (R_3SiF, R_2BOR') after transfer. Transmetalation occurs with retention of any stereochemistry in the organic group.

Once transmetalation has occurred, the remaining steps—rearrangement of the *trans*-dialkylpalladium(II) complex to the cis complex, and reductive elimination, producing the coupled products and regenerating the palladium(0) catalyst—are rapid. Because of this, β-hydrogens can be present in the R group transferred from main group organometallic, as reductive elimination is faster than β-hydride elimination from the dialkylpalladium(II) intermediate.

4.1.2   Transmetalation from Li, K, Mg, Zn (Negishi Coupling), and Cu

One of the earliest examples of coupling by oxidative addition/transmetalation was the palladium-catalyzed coupling of aryl or alkenyl halides with Grignard reagents[67]. A wide range of aromatic, heteroaromatic[68] and alkenyl halides[69] couple efficiently with most Grignard reagents, even those having β-hydrogens. Organolithium reagents are much less efficient at transmetalating to palladium, but also participate reasonably well in this coupling process[70]:

$$R = alkyl, aryl, heteroaryl, alkenyl$$

Configurationally unstable secondary alkyl Grignard reagents can be coupled to a range of aryl and alkenyl halides with high asymmetric induction when carried out in the presence of chiral (aminoalkylferrocenyl) phosphine ligands[71]. This is a kinetic resolution of the rapidly equilibrating racemic Grignard reagent, and asymmetric induction relies on equilibration being faster than coupling of the minor diastereoisomeric metal complex:

R = Ph, Et, hexyl
R′ = aryl, alkenyl

Quite a different sort of asymmetric induction was demonstrated in the palladium-catalyzed Grignard coupling with prochiral bis-triflates to generate optically active biaryls[72,73]:

R′ = Me, Ph
R = Ph, m-MePh, R″C≡C

Much more versatile are processes involving transmetalation from zinc reagents. These can be easily prepared from organolithium (and other main group organometallic) species by simple addition to these of one equivalent of anhydrous zinc chloride in THF. Organozinc reagents have at least three advantages: (1) they are among the most efficient reagents for transmetalation to palladium, (2) they tolerate a wide range of functional groups, including carbonyls, in the substrate, (3) they can be prepared directly from functionalized halides (*e.g.*, α-bromoesters) thus allowing functionality in both coupling partners[74–76]:

84%

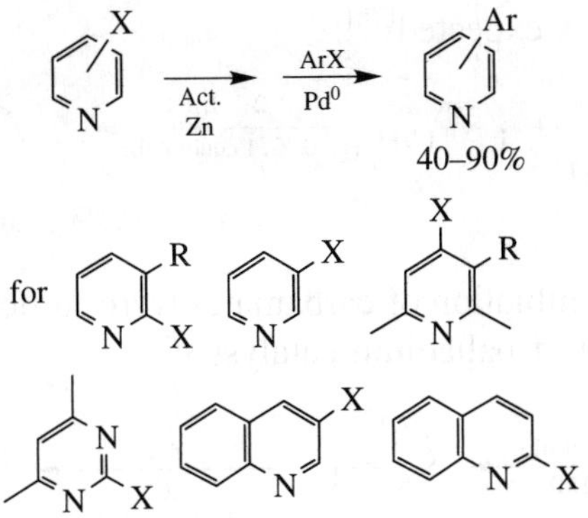

40–90%

for

This process has found use in combinatorial chemistry applications in the palladium(0)-catalyzed coupling of arylzinc halides with polymer-bound bromobenzoates[77]:

Palladium-catalyzed Oxidative Addition/Transmetalation from Zinc[74]

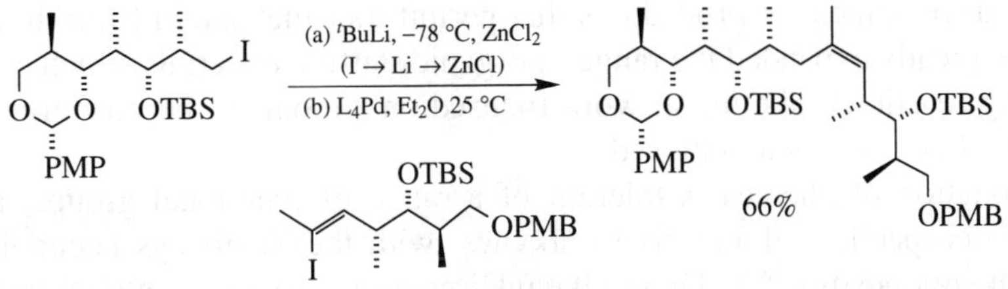

(a) ^{t}BuLi, –78 °C, ZnCl$_2$
(I → Li → ZnCl)

(b) L$_4$Pd, Et$_2$O 25 °C

66%

ZnCl$_2$ (1.32 g, 9.69 mmol) was dried at 160 °C under vacuum overnight and then treated with a solution of the starting primary iodide (5.25 g, 9.59 mmol) in dry Et$_2$O (50 mL) via a cannula (2 × 25 mL rinse). The mixture was stirred at room temperature until most of the ZnCl$_2$ dissolved and cooled to −78 °C. ^{t}BuLi (1.7 M in pentane, 17.0 mL) was added over 30 min, and the resultant solution was stirred 15 min further, warmed to 25 °C, and stirred for 1 h. The solution was added by cannula to a mixture of the substrate vinyl iodide (3.21 g, 6.19 mmol; 6 : 1 (Z)/(E) and Pd(PPh$_3$)$_4$ (0.3640 g, 0.315 mmol). The mixture was covered with aluminum foil, stirred overnight, and then diluted with EtOAc (100 mL), washed with brine (2 × 100 mL), dried over MgSO$_4$, filtered and concentrated *in vacuo*. Flash chromatography (5% ethyl acetate/hexane) gave the product (3.32 g, 66% yield) as a white semisolid ($[\alpha]_D^{23}$ − 28.6° (c 1.53, CHCl$_3$)).

Organocuprates transmetalate only slowly to palladium at temperatures below their decomposition points, greatly limiting their use. However, addition of one equivalent of zinc bromide promotes the coupling at about −10 °C, almost certainly via the organozinc intermediate. With alkenyl cuprates and alkenyl halides, retention of stereochemistry by both partners is observed, as expected[78]:

$$\text{(}\underbrace{}_{2}\text{CuLi)} + \text{I}\overset{}{\frown}\text{C}_5\text{H}_{11} \xrightarrow[\text{0 °C, 1 equiv ZnBr}]{\text{2% L}_4\text{Pd}} \overset{}{\frown\frown\frown}\text{C}_5\text{H}_{11}$$

86%
>99.5% *cis,cis*

Cuprates derived from α-lithiation of carbamates were sufficiently stable to couple with aryl halides in the presence of palladium catalysts[79]:

$$\underset{\substack{OO^t\text{Bu}}}{\overset{()_n}{\underset{N}{\bigvee}}} \xrightarrow[\text{TMEDA}]{^s\text{BuLi}} \underset{\substack{OO^t\text{Bu}}}{\overset{()_n}{\underset{N}{\bigvee}}}\text{-Li} \xrightarrow[\substack{\text{CuCN}\\ [(p\text{MeOPh})_3\text{P}]_4\text{Pd}}]{\text{ArI}} \underset{\substack{OO^t\text{Bu}}}{\overset{()_n}{\underset{N}{\bigvee}}}\text{-Ar}$$

$n = 1, 2$ 50–70%

4.1.3   Transmetalation from Zr, Al and B (Suzuki Coupling)

Alkyl- and alkenylzirconium, -aluminum and -boron complexes are readily available by the hydrometalation (hydrozirconation, hydroalumination, hydroboration) of alkenes and alkynes. The development of procedures that permit transmetalation from these metals to palladium has greatly expanded the range of organic starting materials which can enter into these coupling reactions. The initial work focused on zirconium, because transmetalation from boron had not yet been achieved.

Hydrozirconation of alkynes is tolerant of a range of functional groups, and is both regio- and stereospecific, adding cis to alkynes, with the Zr always occupying the less substituted alkenyl position[80]. These alkenylzirconium species couple efficiently to aryl and alkenyl halides in the presence of palladium(0) catalysts[81,82]:

$$\text{RC≡CH} \xrightarrow{\text{Cp}_2\text{Zr(H)Cl}} \overset{R}{\underset{[Zr]}{\frown}} \xrightarrow[\text{THF}]{\text{Pd(0) (from(PhCN)}_2\text{PdCl}_2)} $$

70–90%

Although internal alkynes are also hydrozirconated, the resulting internal alkenylzirconium complexes transmetalate only poorly to palladium(0). However, as was the case with alkyl cuprates, addition of zinc chloride—to produce the alkenyl zinc species which transmetalates more efficiently—solved the problem[83]. As is usually the case, the stereochemistry of both alkenic partners was maintained. Alkenyl alanes, from addition of diisobutylaluminum hydride to alkynes, underwent similar coupling reactions with similar yields and specificities[84].

Organoboranes are among the most common main group organometallics in organic synthesis, readily available by hydroboration of unsaturated species, or by boronation of organolithium reagents. Organoboron compounds are electrophilic at boron, but the organic groups are only weakly nucleophilic and lack sufficient reactivity to transmetalate to palladium. The realization that addition of an anionic base to neutral organoboranes dramatically increased the nucleophilicity of the organic groups led to successful transmetalation from boron to palladium by the simple expediency of carrying out the process in the presence of added base, and to the development of the Suzuki reaction[85]. To a large extent this chemistry has supplanted that of zirconium and aluminum.

The Suzuki reaction occurs under mild conditions and tolerates a wide range of functionality. The starting organoboranes are often more easily generated and handled than other main group organometallics, and the (water soluble) boron byproducts are both more easily separated and less toxic than those of other metals, particularly tin. Aryl- and 1-alkenylboranes couple with alkyl, alkenyl, aryl, allyl and alkynyl substrates (usually halides or triflates) and even alkyl boranes transmetalate and couple efficiently.

Alkyl boranes, from hydroboration of alkenes, couple efficiently to a range of aryl and alkenyl halides and triflates. A catalyst consisting of $PdCl_2$(dppf) and aqueous sodium hydroxide in THF is often used[86], but this metal complex in DMF with K_2CO_3 is also efficient as is $Pd(PPh_3)_4$ and aqueous NaOH in less polar solvents such as benzene and dioxane:

66–85%

The coupling of alkenyl halides or triflates with alkenyl boranes to make dienes is also efficient under a variety of conditions. With highly functionalized substrates, the use of aqueous TlOH in place of alkali metal hydroxides sometimes results in an increase in both rate and yield[87]. These conditions were used for perhaps the most spectacular of Suzuki couplings (Scheme 9)[88].

Suzuki Coupling of Alkenyl Halides with Alkenylboronic Acids[87]

82%

To 184 mg (0.74 mmol) of the alkenylboronic acid was added 222 mg (0.60 mmol) of the vinyl iodide, 5 mL of THF, and 43 mg (2.4 mmol) of water. The colorless solution was cooled to 0 °C, and 49 mg (0.07 mmol) of Pd[PPh$_3$]$_4$ in 1.5 mL of dry THF was added via cannula. After 5 min, 377 mg (1.51 mmol) of TlOEt was added to the orange solution over a 10 min period. After 2 h, the light brown suspension was partitioned between EtOAc and water, and the aqueous phase was extracted twice with EtOAc. The combined organic phases were washed twice with water and once with brine, dried over MgSO$_4$, filtered, and concentrated under vacuum to give a dark residue. The residue was filtered through silica (10 : 90 EtOAc:hexanes) to give an orange oil, which was then purified by HPLC (4 : 96 EtOAc:hexanes) to give 219 mg (82%) of the coupling product as a yellow oil: $R_f = 0.39$ (20 : 80 EtOAc:hexanes).

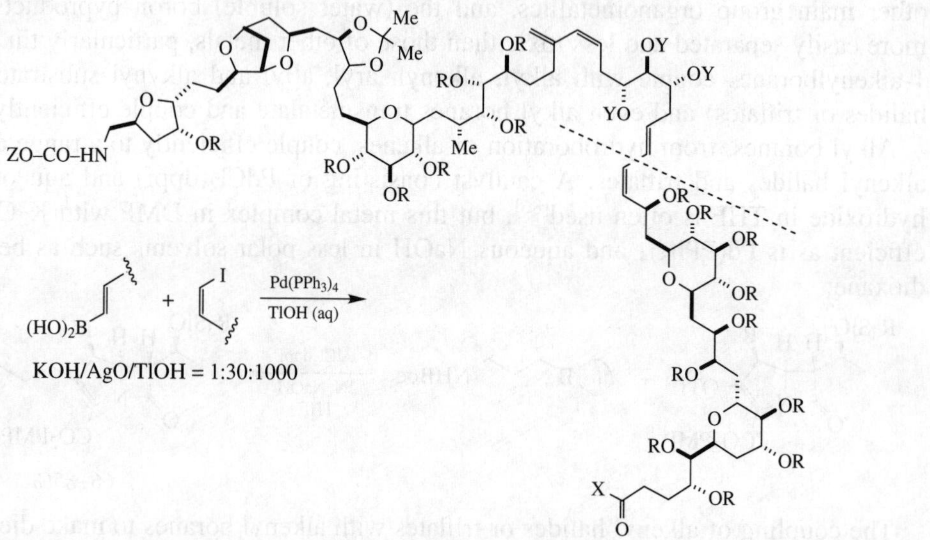

Scheme 9.

Arylboronic acids were among the first boron compounds to undergo Suzuki coupling, and this methodology is often the method of choice for biaryl coupling. The usual catalyst systems consist of Pd(PPh$_3$)$_4$ or PdCl$_2$(PPh$_3$)$_2$ and aqueous Na$_2$CO$_3$ in dimethoxyethane, although a variety of other bases are also effective including the very mild fluorides CsF and Bu$_4$NF[89]. Ligandless catalysts such as Pd(OAc)$_2$, Pd$_2$dba$_3$, and η^3-allyl PdCl are about 10 times more active than ligated catalysts but sometimes undergo deactivation before the reaction is complete.

Suzuki coupling to form biaryls is efficient with a wide range of highly substituted aryl triflates and aryl boronic acids, and is effective even on a very large scale[90]:

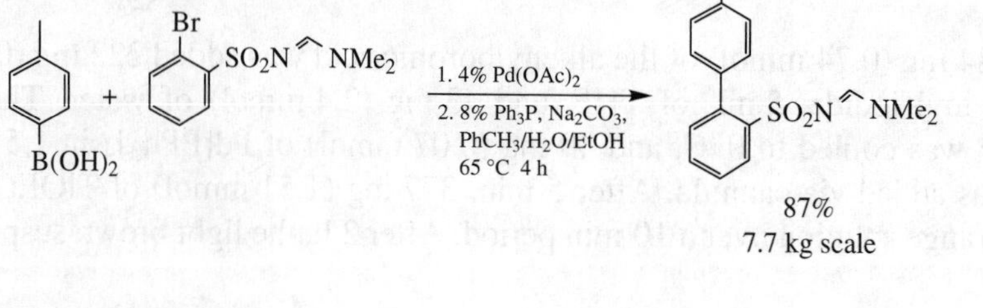

87%
7.7 kg scale

It has found applications in the synthesis of AB_2 building blocks for liquid crystal dendrimers (equation below[91]), oligophenyls with up to 15 aryl groups[92], and hexaoligophenyl-substituted arenes[93]. It proceeds efficiently with polymer-supported aryl halides, and provides the ability to generate combinatorial libraries effectively[94–97].

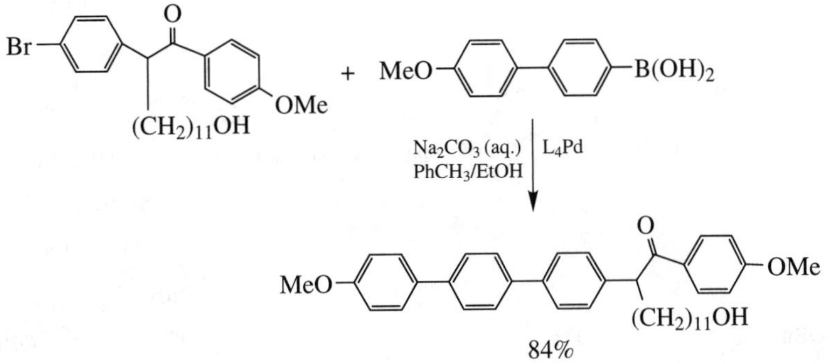

84%

The utility of transmetalation from boron has been further expanded by combining it with ligand-directed ortholithiation of aromatics[98]:

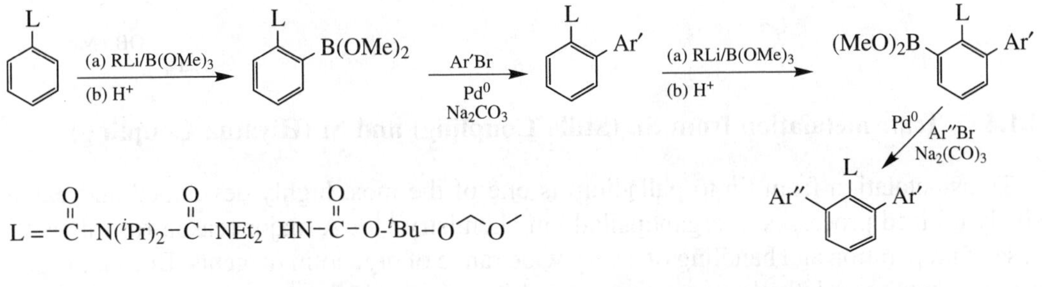

This combined process has found extensive application in the synthesis of complex aromatic compounds[99–102]:

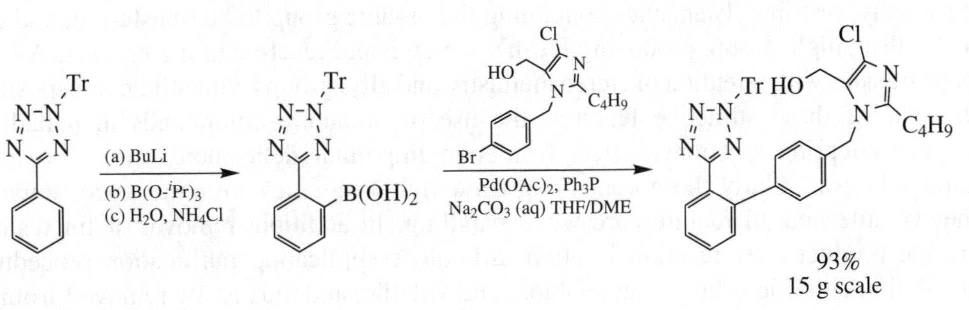

93%
15 g scale

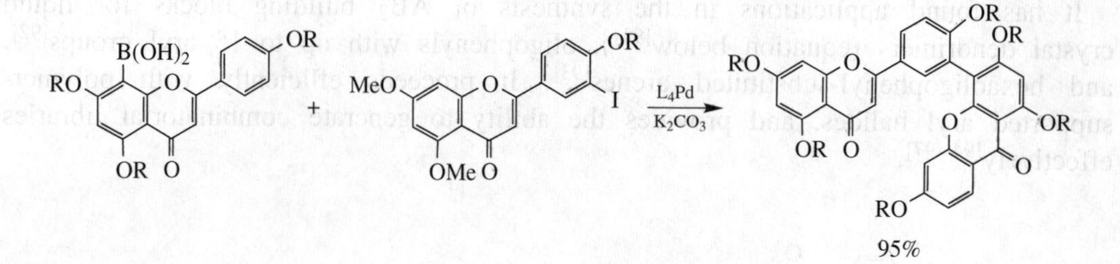

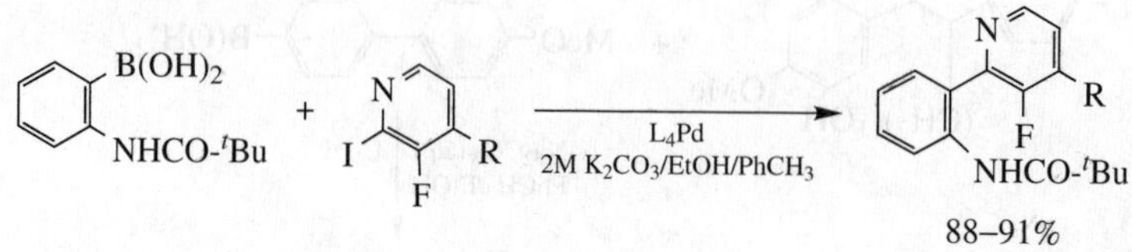

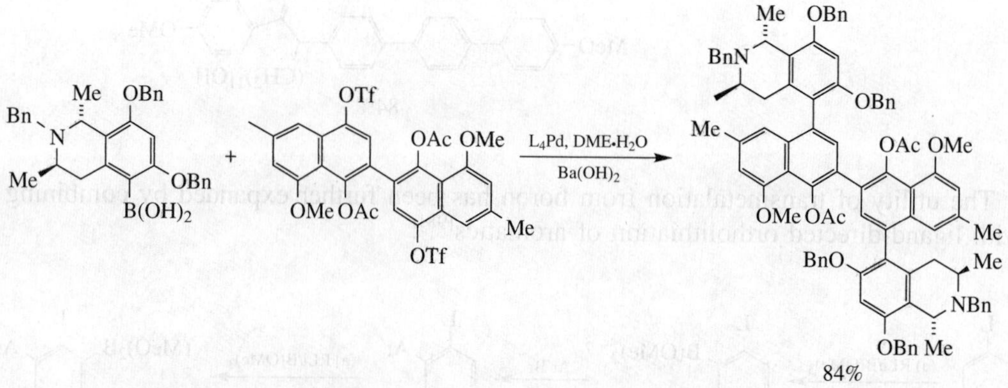

95%

88–91%

84%

4.1.4　Transmetalation from Sn (Stille Coupling) and Si (Hiyama Coupling)

Transmetalation from tin to palladium is one of the most highly developed and extensively utilized processes in organopalladium chemistry[103]. A major reason for this is the ease of preparation and handling of a very wide range of organotin reagents. Because organotin reagents are relatively unreactive towards most organic functional groups both the tin reagent and the substrate can contain extensive, unprotected functionality (as will be seen in the examples below). The rate of transfer of organic groups from tin to palladium is $RC \equiv C > RCH = CH > Ar > RCH = CHCH_2 \approx ArCH_2 \gg C_nH_{2n+1}$, making it feasible to use trimethyl or tributylstannanes containing the organic group to be transferred, and only transfer that single group, producing R_3SnX which is not reactive in the process. Alkenyl groups transfer with retention of stereochemistry and allyl groups with allylic transposition.

In spite of these attractive features, the use of organotin compounds in palladium-catalyzed coupling reactions suffers from two important deficiencies. Most organotin compounds, particularly those containing trimethyltin moieties are quite toxic, some are rather volatile and all require care when handling. In addition, removal of tin residues from the product after reaction is often difficult, complicating purification procedures. Trimethyltin chloride is both water soluble and volatile, and thus easily removed from the product by extraction or distillation. However, tributyltin chloride is relatively nonvolatile, is soluble in most organic solvents, and has a tendency to streak on silica gel chromatography, making its removal difficult. This feature has impeded the use of tin couplings in the manufacture of pharmaceuticals, where even traces of tin contaminants in the product are intolerable. Tributyltin chloride undergoes reaction with aqueous ammonia to form somewhat water-soluble complexes, and with aqueous KF to form insoluble oligomeric fluorides both of which can aid in its removal from reaction product mixtures. Another approach is the use of the $(C_6F_{13}CH_2CH_2)3Sn$ moiety in place of the tributyltin group. These reagents produce a tin residue that is easy to remove by extraction[104].

In most cases, it is the transmetalation step that is rate-limiting, and often accounts for the severe conditions used in early versions of this process. However, recent advances have resulted in the development of considerably milder conditions. For example, replacing triphenylphosphine, the ligand of choice in most of the earlier studies, with ligands of lower donicity, such as Ph_3As or $(2\text{-furyl})_3P$ led to up to a 10^3 increase in rate of transmetalation, allowing the use of substantially milder conditions[105].

The use of CuI as a co-catalyst can also dramatically improve palladium-catalyzed organotin cross-coupling reactions[106]. When triphenylphosphine complexes are used as catalysts, the CuI appears to facilitate the formation of coordinatively unsaturated palladium intermediates by scavenging triphenylphosphine, whereas in systems utilizing Ph_3As ligands and highly polar solvents, transmetalation from tin to copper may be occurring[107].

Virtually the entire array of substrates that can oxidatively add to palladium(0) complexes can be made to undergo coupling with organostannanes. Although acid chlorides are known to be alkylated by stannanes in the absence of palladium catalysts[108], the palladium-catalyzed process is efficient and has few limitations[109]. The normal catalyst system used is $(PhCH_2)Pd(PPh_3)_2Cl$ in chloroform and, again, use of Ph_3As or $(2\text{-furyl})_3P$ as ligands allows milder reaction conditions[105]. Aldehydes can be produced by using Bu_3SnH as the stannane[110].

Alkenyl bromides and iodides are generally reactive towards alkenylstannanes[111] and alkynylstannanes[112], to produce dienes and enynes, respectively, with retention of alkene geometry. The catalyst system of choice is the ligandless variety, $(MeCN)_2PdCl_2$ in THF or DMF, and the reaction generally proceeds under mild conditions. Again, extensive functionality is tolerated, making the process effective in the late stages of total synthesis[113,114]:

65%

Preparation of [4R-[4α(2E,4E),4aβ,7α(2S,3E,5E),8β,8aβ]-5-[Hexahydro-7-
[7-hydroxy-5-methyl-2-[[1,1-dimethylethyl)dimethylsilyl]oxy]-3,5-heptadienyl]-4a,
8-dimethyl-2,5-dioxo-2H,5H-pyrano[3,4-e]-1,3-oxazin-4-yl]-4-methyl-2,4-
pentadienal*[113]

To a solution of alkenyl iodide (0.120 g, 0.20 mmol) and alkenyl stannane
(0.070 g, 0.25 mmol) in 1.0 mL DMF at ambient temperature was added
(MeCN)$_2$PdCl$_2$ (0.0080 g, 0.03 mmol). The dark reaction mixture was stirred
at 25 °C for 2 h. The solvent was removed *in vacuo*. Purification by flash
chromatography (EtOAc) gave 0.098 g (90%) of coupled product as a solid
(mp 88–90 °C (from EtOAc); $[\alpha]_D^{25}$ + 175.1° (*c* 1.00, CHCl$_3$)).

The discovery that alkenyl (and aryl, see below) triflates participated in this palladium-
catalyzed coupling dramatically increases its utility in synthesis[115], as isomerically
pure alkenyl triflates are available directly from ketones[116]. When triphenyl phosphine
palladium catalysts were used in THF solvent, excess lithium chloride was required for the
coupling to proceed, and even under these conditions aryl stannanes were unreactive[115].
However, the use of As$_3$P in very polar solvents led to an optimized reaction that did
not require added lithium chloride[117]. This process is quite efficient even with highly
functionalized substrates[118,119]:

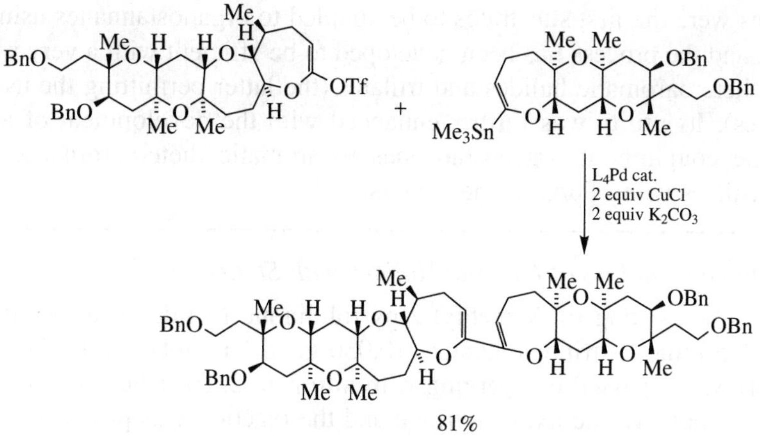

81%

Optimized Procedure for the Palladium-catalyzed Coupling between Alkenic Triflates and Arylstannanes

1-(4-Methoxyphenyl)-4-tert-butylcyclohex-1-ene[117]

4-*tert*-butyl-cyclohex-1-enyl triflate (0.2628 g, 0.918 mmol), triphenylarsine (0.023 g, 0.0734 mmol), and $Pd_2(dba)_3$ (0.0083 g, 0.0184 mmol Pd) were dissolved in anhydrous degassed *N*-methyl pyrrolidinone (NMP; 5 mL) and, after the purple color was discharged (5 min), (4-methoxyphenyl)tributyltin (0.430 g, 1.083 mmol) in NMP (2 mL) was added. After 16 h at 25 °C, the solution was treated with 1 M aqueous KF solution (1 mL) for 30 min, diluted with ethyl acetate, and filtered, and the filtrate was extensively washed with water. Drying, followed by evaporation, gave a crude oil, that was purified by reverse phase flash chromatography (C-18, 10% dichloromethane in acetonitrile; yield 0.201 g (89%) of a white solid). Recrystallization from methanol gave white needles (mp 78–9 °C).

Although rarely used, alkynyl halides also couple efficiently to organostannanes under palladium catalysis[120]:

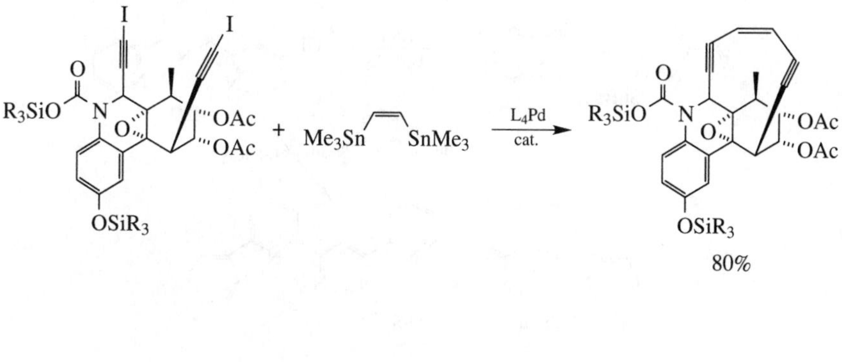

80%

Aryl halides were the first substrates to be coupled to organostannanes using palladium catalysis[121], and the process has been developed to be efficient with a very wide range of aromatic and heteroaromatic halides and triflates (the latter permitting the use of phenols as aryl sources). Its utility was further enhanced with the development of an optimized system for the coupling of organostannanes to aromatic, heteroaromatic, and alkenyl halides and triflates using Pd/C as the catalyst[122].

Pd/C-catalyzed Coupling of Organic Halides with Stannanes[122]

A solution consisting of *N*-methyl-2-pyrrolidinone (5 mL), 4-iodoacetophenone (0.121 g, 0.5 mmol), triphenylarsine (0.030 g, 0.1 mmol), and CuI (0.010 g, 0.05 mmol) was degassed by sparging with nitrogen. 2-Tri-*n*-butylstannylthiophene (0.310 g, 0.7 mol) was added by syringe and the reaction was placed in an oil bath set at 80 °C. Under positive nitrogen pressure, Pd/C (10%, 0.003 g, 0.003 mmol) was added and the mixture was allowed to stir at 80 °C for 24 h. The reaction was cooled, treated with a saturated KF solution and allowed to stir for 30 min. The mixture was passed through a pad of Celite and rinsed with ether. The filtrate was washed with water, then dried (MgSO$_4$) and concentrated to give a crude yellow solid. Chromatography (hexanes/ethyl acetate (10%), silica gel) furnished the desired product as white flakes (0.074 g, 60%; mp 207–208 °C).

Palladium-catalyzed tin couplings with aryl substrates have found broad application in the synthesis of highly functionalized compounds[123]:

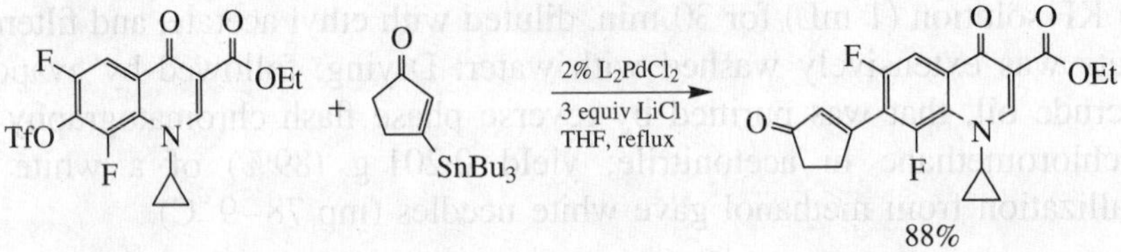

It has proven particularly useful for the coupling of heteroaromatic systems and has been used to synthesize extended polyheteroaromatic systems[124,125] and oligomeric systems[126,127]:

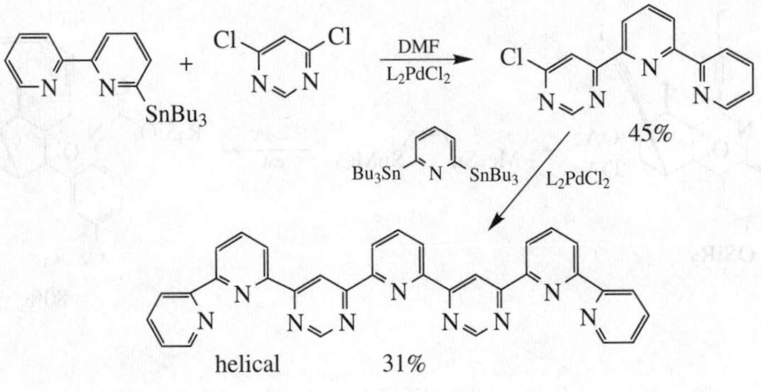

helical   31%

Although palladium-catalyzed amination of aryl halides by aminostannanes had been reported in 1983[128], little development took place until very recently, when two groups simultaneously readdressed the problem. Although the initial systems studied involved aminostannanes[129,130], subsequent studies showed that the use of chelating diphosphine ligands in the presence of sodium *tert*-butoxide resulted in the efficient amination of aryl halides by amines themselves, obviating the need for formation of aminostannane[130–133]:

$$\text{ArBr} \ + \ \text{RR'NH} \quad \xrightarrow[\substack{\text{or (DPPF)PdCl}_2 \\ {}^t\text{BuONa} \\ \text{PhCH}_3 \ 80°}]{\text{Pd}_2\text{dba}_3/\text{BINAP}} \quad \text{ArNRR'}$$

ArI   60–95%

Ar = *p*-NCPh, *p*-MePh, *p*-PhPh, *p*-PhCOPh, *p*-Et$_2$NCOPh, *o*-MeOPh, *m*-MeOPh, *o*-Me$_2$NPh, 3,5-Me$_2$Ph, *m*-^{t}BuO$_2$CPh, 2,5-Me$_2$Ph, *o*-Me, *p*-MeOPh

Amine = C$_6$H$_{13}$NH$_2$, BnNH$_2$, ⬡(NH$_2$) , (piperazine) , PhNHMe, H$_2$N—∕—N(morpholine)O

*p*ClPhNH$_2$, PhNH$_2$, ^{t}BuNH$_2$, ^{s}BuNH$_2$, (piperazine)—R

This methodology has been used in a combinatorial sense by coupling a range of amines to RINK-resin supported *p*-bromobenzamides and cinnamides[134]. Bromopyridines and bromoquinolines were also efficiently aminated by this system[135]. There is a recent indication that alkoxides will also attack aryl halides to give aryl ethers under similar conditions[136].

General Procedure for Arylation of Primary Amines[131]

A Schlenk tube was charged with aryl halide (1.0 mmol), amine (1.1 mmol), sodium *tert*-butoxide (1.4 mmol), tris(dibenzylideneacetone)dipalladium(0) (0.0025 mmol, 0.5 mol % Pd), BINAP (0.0075 mmol), and toluene (2 mL) under argon. The tube was immersed in an 80 °C oil bath with stirring until the starting material had been completely consumed as judged by GC analysis. The solution was then allowed to cool to 25 °C, taken up in ether (15 mL), filtered, and concentrated. The crude product was then purified further by flash chromatography on silica gel.

In contrast to stannanes, transmetalation from silicon to palladium has only recently been developed, and to date has found little application in complex organic synthesis. Again, the problem is to provide activation of the silicon starting material and stabilization of the silicon product of the transmetalation step. This is achieved by providing a source of fluoride[137], most easily in the form of the soluble tris(diethylamino)sulfonium difluorotrimethyl silicate (TASF)[138]:

$$RX + R'SiMe_3 \xrightarrow[\substack{HMPA \text{ or } THF/P(OEt)_3 \\ 50\,°C \text{ 1 equiv TASF}}]{2\% \langle\!\langle(PdCl)_{\!/2}} RR' \quad 60\text{–}90\%$$

$$RX = ArI, \;\; {}^{R}\!\!\diagdown\!\!\diagup\!I \;\; {}^{R}\!\!\diagdown\!\!\diagup\!Br \;\; {}^{R}\!\!\diagdown\!\!\diagup\!\!\diagdown\!{}^{Br}, \; ArOTf, \;\; \diagup\!\!\diagdown\!OTf$$

$$R' = \;\diagup\!\!\diagdown\!\!\diagup \;\; \diagup\!\!\diagdown\!\!\diagup\!\!\diagdown \;\; Ph\!\diagup\!\!\diagdown\!\!\diagup \;\; \diagdown\!\!\diagup\!\!\diagdown\!\!\diagup \quad R\!-\!\!\equiv\!\!\bullet\!-\!\{$$

$$\langle\!\langle(PdCl)_{\!/2} = [\eta^3\text{~allylPdCl}]_2$$

Under more forcing conditions (KF, DMF, 120 °C) even aryl chlorides are coupled to aryl and vinyl dichloroethylsilanes[139a]. Surprisingly, NaOH is a better promoter for Hiyama coupling than is KF[139b].

4.1.4 Transmetalation from Hg and Tl

Arylmercury and -thallium compounds are available by direct metalation of aromatic compounds, and participate in oxidative addition/transmetalation procedures as above. The toxicity of these metals and difficulties in removing traces of them from products has somewhat limited their use. However, they have been utilized to a greater extent in insertion chemistry, which is discussed below.

4.2 Palladium(0)-catalyzed Insertion Processes

4.2.1 Carbonylation

Carbon monoxide inserts very readily into palladium–carbon σ-bonds ($< -20\,°C$, <1 atm CO) regardless of how they were formed—nucleophilic attack on alkenes (Scheme 3), orthopalladation (Section 3.3), transmetalation (Scheme 8), or oxidative addition (this section)—and the resulting σ-acyl complexes are readily cleaved by a variety of nucleophiles, making palladium-catalyzed carbonylation of organic substrates the most versatile of reactions.

By far the most commonly used palladium-catalyzed carbonylation process is that involving oxidative addition (Scheme 10). As oxidative addition is the most difficult step, the scope of the overall process is limited by the *general* scope of oxidative addition discussed above; for example, β-hydrogens are not tolerated, and aryl, alkenyl, benzyl, and allyl halides (I > Br ≫ Cl) and triflates are reactive, etc. The reaction is widely tolerant of solvents and catalysts, although the presence of a ligand to stabilize palladium(0) is usually necessary to prevent the precipitation of metallic palladium, and the nucleophile (usually methanol, to produce esters, but others such as amines work as well) is normally used in excess. Again, structural complexity and functionality are tolerated in the substrate, and this is probably the method of choice for the conversion of organic halides and triflates to carboxylic acid derivatives. With the development of procedure to involve aryl[140] and vinyl triflates in palladium(0)-catalyzed oxidative additions, the scope of carbonylation

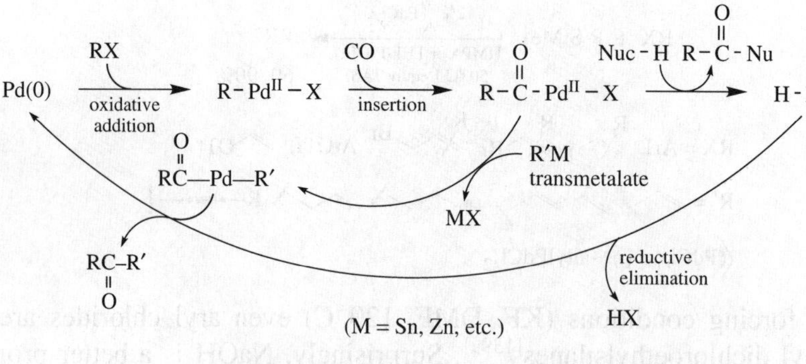

Scheme 10. Palladium(0)-catalyzed carbonylation of halides.

was greatly broadened, because phenols, and ketones, respectively, can be carbonylated via their triflates[141,142]:

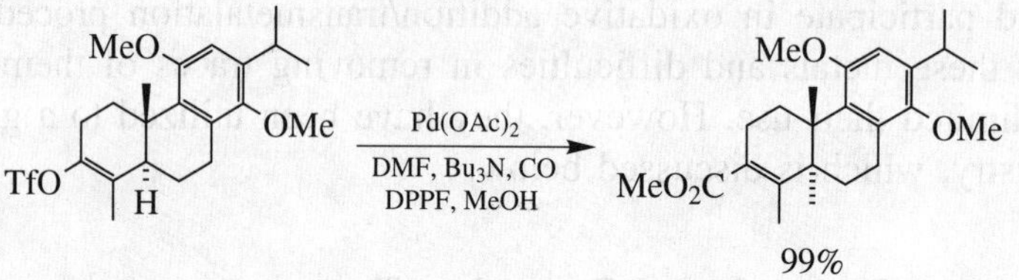

99%

Carbonylation is even efficient on a very large scale[143]:

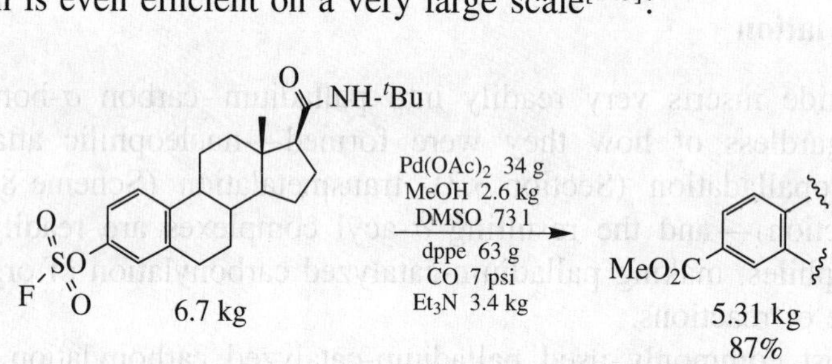

5.31 kg
87%

Palladium(0)-catalyzed carbonylation of vinyl triflates[142]

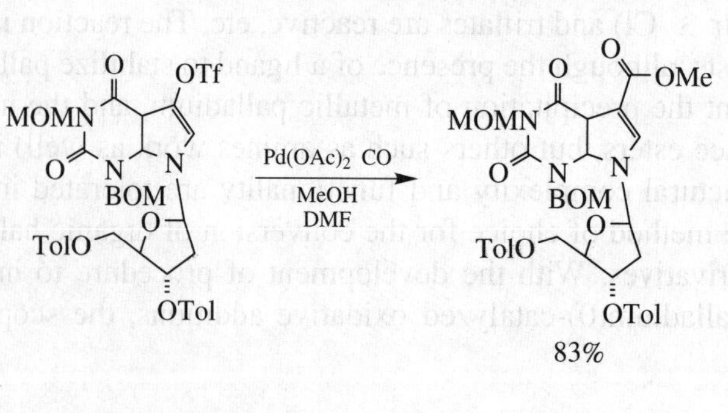

83%

1-[(Benzyloxy)methyl]-5-(methoxycarboxyl)-3-(methoxymethyl)-7-
(*p*-nitrophenethyl)pyrrolo[2,3-d]pyrimidine-2,4-dione

Triethylamine (1.0 mL) was added to a solution of the vinyl triflate,
p-nitrophenethyl in place of protected deoxyribose (1.00 g, 1.63 mmol) in DMF
(7.6 mL) and MeOH (3.4 mL) followed by Pd(OAc)$_2$ (0.0267 g, 0.12 mmol) and
PPh$_3$ (0.0762 g, 0.29 mmol). A stream of CO was passed into the solution for 5 min.
The resulting mixture was stirred at 70–72 °C for 4 h under a CO balloon and was
then evaporated to dryness under reduced pressure. The residue was purified by
flash chromatography using ethyl acetate/hexanes (7 : 3) as eluant to provide 0.715 g
(84%) of the carbonylated product as a light yellow foam. An analytical sample was
obtained by recrystallization from ethyl acetate/hexanes as a light yellow crystalline
solid (mp 140–142 °C).

Oxidative addition/carbonylation is used most frequently in an intramolecular manner,
forming lactones or lactams. It is quite efficient with aryl halides[144,145] as well as with
alkenyl halides[144,146]:

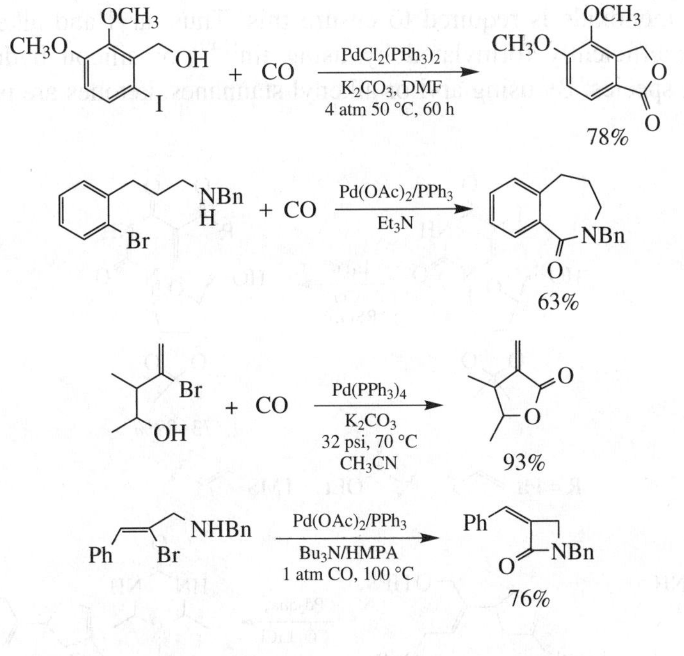

Again, this is a very general process and should be applicable to virtually any system
having appropriately situated nucleophiles to trap a σ-acylpalladium complex, and compatible with the requirements of oxidative addition enunciated above. That it is usable on a
large scale is seen in the equation below[147]:

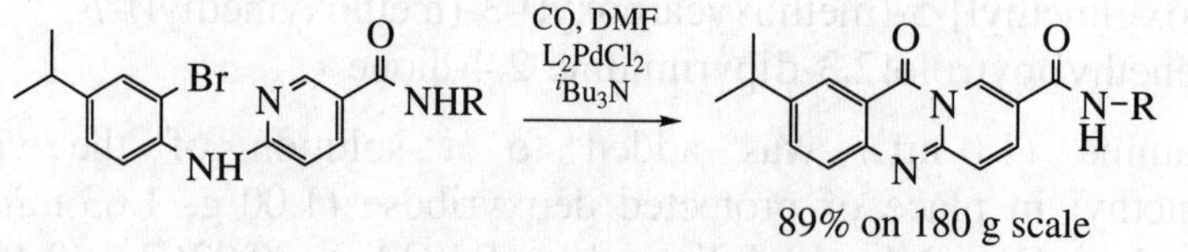

89% on 180 g scale

Under appropriate conditions, even enolate oxygens can trap the acylpalladium complex, giving isocoumarins[148]:

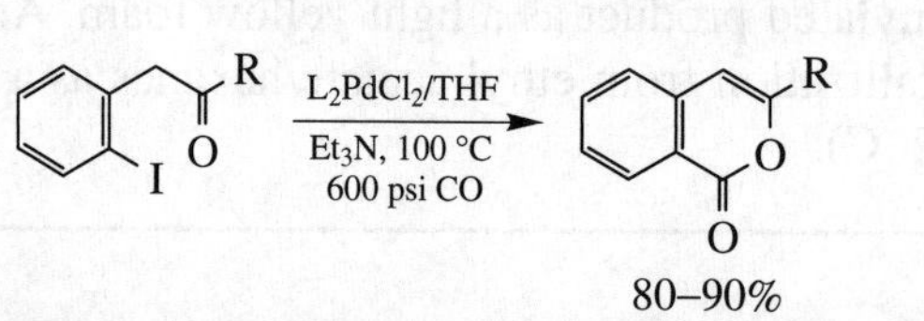

80–90%

By combining oxidative addition–CO insertion with transmetalation a very useful range of carbonylative coupling processes have been developed. The only complication is that CO insertion must preceed transmetalation and sometimes a moderate (50–90 psi) pressure of carbon monoxide is required to ensure this. Thus, aryl and alkenyl halides and triflates can be efficiently formylated by using tin[149] or silicon hydrides[150] as the transmetalating species. By using aryl or alkenyl stannanes, ketones are produced[151,152]:

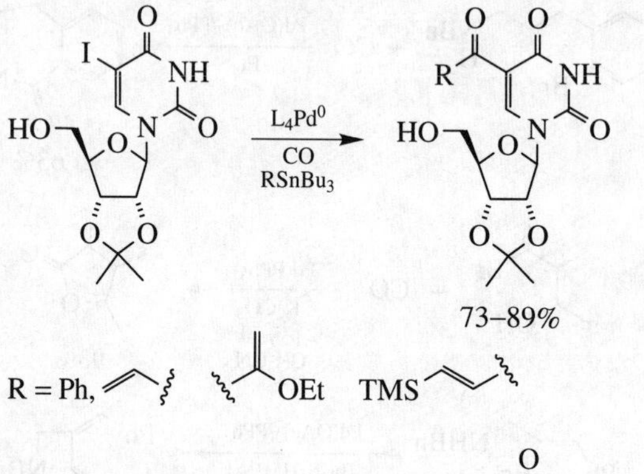

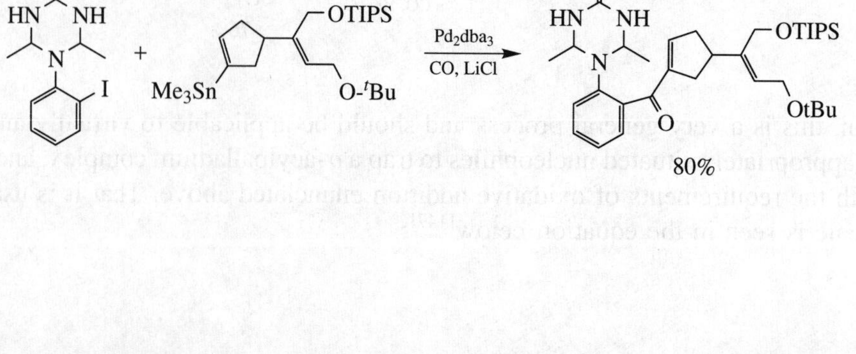

73–89%

R = Ph,

80%

General Procedure for Palladium-catalyzed Formylation of Aryl/Enol Triflates[150]

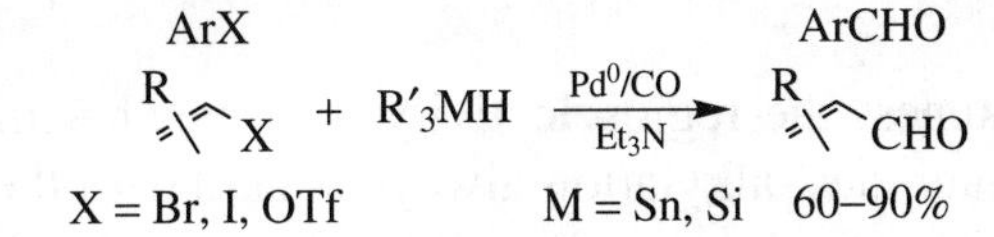

$$ArX \qquad\qquad ArCHO$$

$$\underset{X = Br,\ I,\ OTf}{\overset{R}{\underset{}{\diagup}}\diagdown X} \ + \ R'_3MH \ \xrightarrow[\ Et_3N\ ]{Pd^0/CO} \ \underset{M = Sn,\ Si \quad 60\text{--}90\%}{\overset{R}{\underset{}{\diagup}}\diagdown CHO}$$

A mixture of the triflate (1 mmol), Pd(OAc)$_2$ (0.0045 g, 0.02 mmol), and dppp (0.008 g, 0.02 mmol) in DMF (5 mL) was heated to 70 °C and CO was bubbled into the solution over 20 min. At this time the solution color sharply changed from light brown to deep brown. Then Et$_3$N (0.35 mL, 2.5 mmol) was introduced dropwise via microsyringe followed by addition of trioctylsilane (0.9 mL, 2.0 mmol). Stirring was continued at the same temperature until completion of the reaction. After dilution with H$_2$O, the mixture was extracted with Et$_2$O. The combined extracts were washed with H$_2$O, saturated NaHCO$_3$, and saturated NaCl, dried (Na$_2$SO$_4$), and evaporated. The crude product was purified by preparative TLC or column chromatography to give the desired aldehyde.

4.2.2 Palladium(0)-catalyzed Alkene Insertion Processes

One of the most useful of palladium(0)-catalyzed reactions is the oxidative addition/alkene insertion reaction—the Heck reaction[153] (Scheme 11). In this case, it is the insertion step which is the most difficult, and most limitations are imposed by structural features on the alkene. The limitations on the oxidative addition step are the same, and all of the substrates discussed above are viable candidates for alkene insertion chemistry. Again β-hydrogens are not tolerated in the substrate. A wide range of alkenes undergo this reaction, including unfunctionalized alkenes, electron-deficient alkenes such as acrylates and conjugated enones, and electron-rich alkenes such as enol ethers and enamides. With unfunctionalized or electron poor alkenes, insertion occurs to place the R group at the less-substituted position of the alkene, indicating that the regioselectivity of insertion is controlled by steric, not electronic effects, with these substrates. It is a cis insertion, with

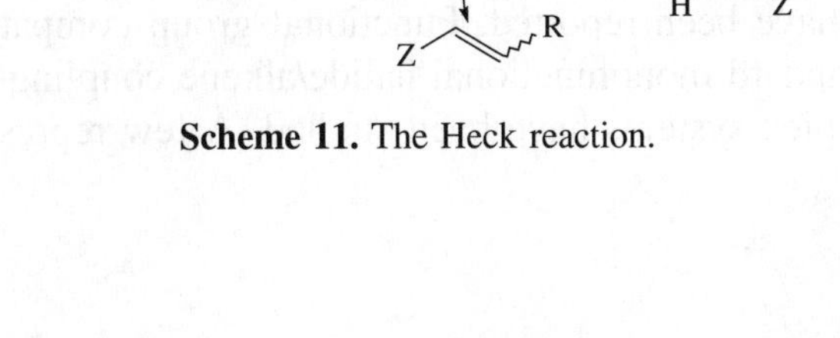

Scheme 11. The Heck reaction.

the alkyl group being delivered to the less-substituted carbon from the face of the alkene occupied by the metal. As might be expected, substitution on the alkene, particularly in the β-position, inhibits the reaction and, for intermolecular processes, β-disubstituted alkenes are unreactive.

With electron-rich alkenes, the regioselectivity is somewhat more complex[154]. With cyclic enol ethers and enamides, alkylation always occurs α to the heteroatom. However, with acyclic systems, mixtures of α- and β-alkylation are observed, the ratio of which is subject to electronic control. Electron-rich aryl halides lead to α-arylation, whereas electron-poor aryl halides lead to β-arylation of the alkene. Alkenyl triflates favor α-alkylations as does the addition of thallium(I) acetate or silver salts[155], and the use of bidentate phosphines. As shown in examples below, even aromatic 'double bonds' may insert into σ-alkylpalladium complexes, particularly in intramolecular versions of the process.

The classic conditions for the Heck reaction involve heating a mixture of the alkene, the halide, a catalytic amount of $Pd(OAc)_2$, and several equivalents of triethylamine in acetonitrile until the reaction is complete. (Triethylamine rapidly reduces $Pd(OAc)_2$ to the catalytically active palladium(0) state[156].) When aryl iodides are used, addition of phosphine is not required, and in fact inhibits the reaction. (Palladium on carbon can be used to catalyze reactions of aryl iodides[157].) With bromides, added phosphine is required for the oxidative addition step. Hindered phosphines such as (o-tolyl)₃P are used, because triphenylphosphine reacts with the halide substrate to form quaternary phosphonium salts.

However, with the dramatic increase in use of the Heck reaction, a number of improvements in catalyst systems have been made, resulting in a wider range of reactivity, higher turnovers and milder conditions. By using palladium acetate/sodium acetate/tri(o-tolyl)phosphine in DMF, catalyst loadings as low as 0.0005 mol % with turnover numbers of 134,000 have been achieved[158]. High reactivity under very mild conditions (20–50 °C) has been achieved by using quarternary ammonium chlorides and alkali metal carbonates[159] or acetates[160] as additives in DMF (Jeffery conditions). The quaternary salt is thought to promote the reaction by keeping the chloride concentration (required for catalyst activity) high[161], although quaternary ammonium sulfates also promote mild conditions[162] and allow the reaction to be run under anhydrous, aqueous/organic, and strictly aqueous conditions. Development of water-soluble phosphine ligands[163] has also permitted the use of aqueous solvent systems for the Heck reaction. Although most Heck reactions are run using catalysts generated *in situ* from palladium acetate and a reducing agent, some preformed catalysts are longer-lived and more reactive than those generated *in situ*[5,164]. Palladium on carbon will catalyze Heck reactions of aryl iodides[165], and a variety of polymer-bound palladium complexes also provide heterogeneous Heck conditions[165,166].

The Heck reaction is one of the most broadly used palladium-catalyzed processes, and myriad applications have been reported. Functional group compatibility is excellent. In addition to all the standard monofunctional halide/alkene couplings which abound in the literature, many complex systems have been studied. A few representative examples are shown in the equations below[167−169]:

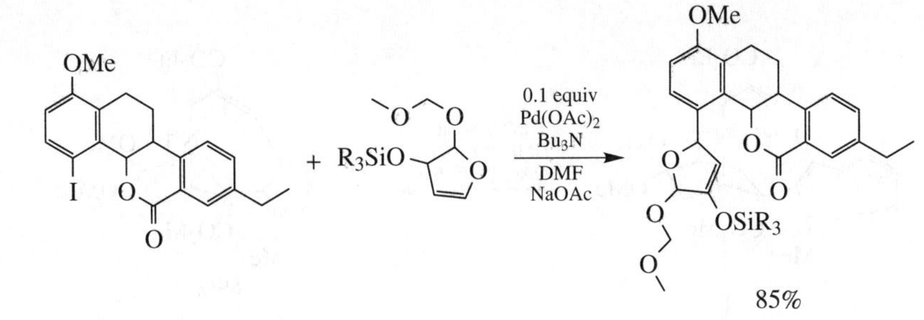

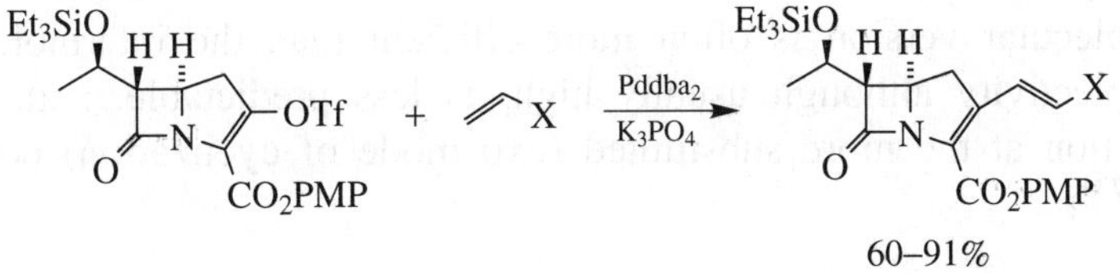

85%

60−91%

X = Ph, CO₂Me, CH₂OH, CH(OEt)₂, CH₂OAc, CH₂NHCO₂PMP

X = Ph, CO_2Me, CH_2OH, CH(OEt)_2, CH_2OAc, CH_2NHCO_2PMP

55−66%

85−99%

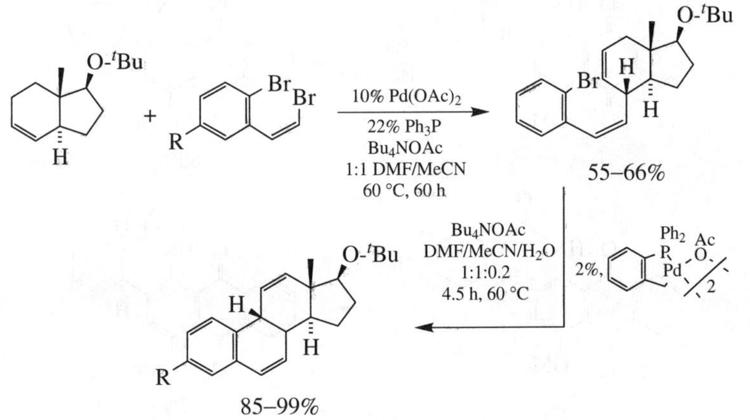

The process is efficient with solid-phase supported substrates, and the Heck reaction is making inroads in combinatorial chemistry[170]:

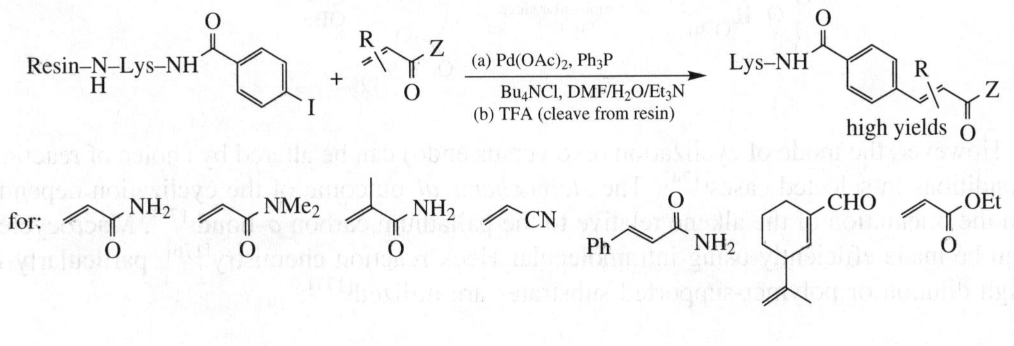

high yields

for:

Intramolecular versions[171] are particularly useful for the synthesis of complex ring systems[172a]:

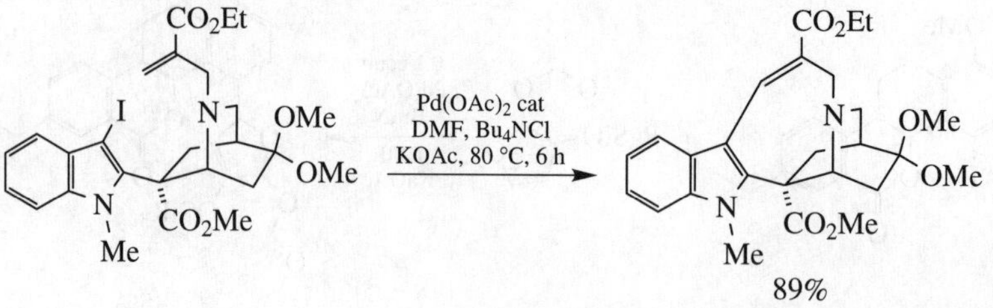

89%

This intramolecular version is often more efficient than the intermolecular version, but the regioselectivity although usually high, is less predictable, and, depending on ring size, insertion at the more substituted (exo mode of cyclization) position is often observed[172b,173]:

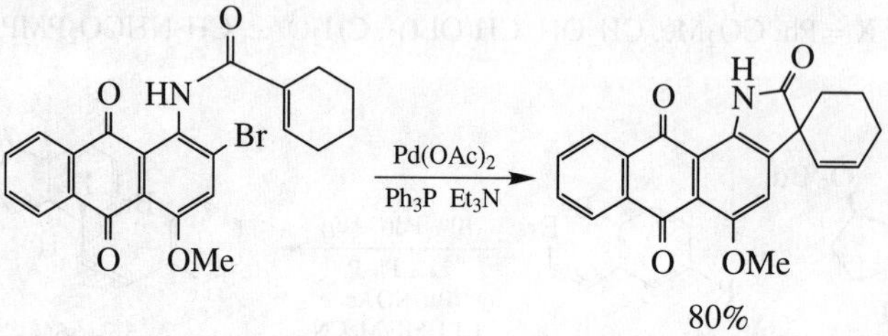

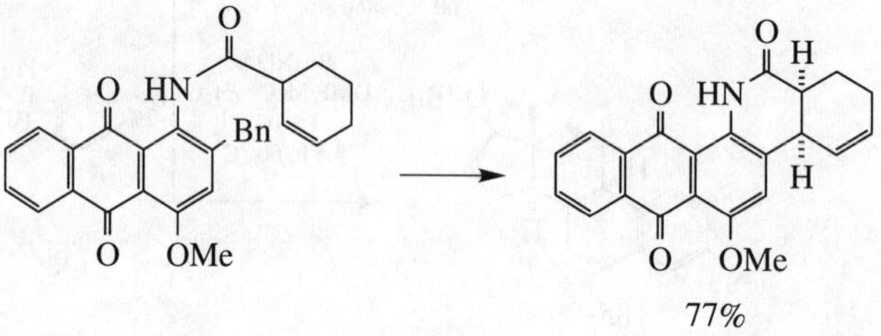

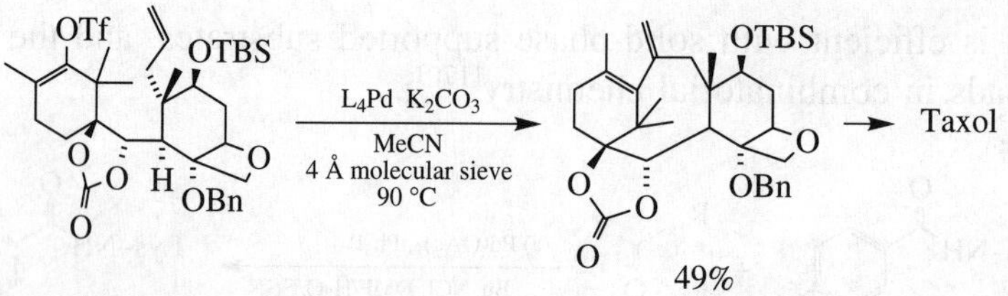

80%

77%

49%

However, the mode of cyclization (exo versus endo) can be altered by choice of reaction conditions in selected cases[174]. The *stereochemical* outcome of the cyclization depends on the orientation of the alkene relative to the palladium carbon σ-bond[175]. Macrocycles can be made efficiently using intramolecular Heck reaction chemistry[176], particularly if high dilution or polymer-supported substrates are utilized[177]:

Preparation of 1α,2β-Bis(tert-butyldimethylsiloxy)]-4-cyclohexylamino)-10,11-dimethoxy-1,2,7,8-tetrahydro-5-oxo-indolo[7,1-ab][3]-benzazepine[174]

A mixture of the starting iodoarene (0.200 g, 0.255 mmol), Pd(OAc)$_2$ (0.0034 g, 0.015 mmol); Bu$_4$NCl (0.154 g, 0.520 mmol), and KOAc (0.140 g, 1.43 mmol) in DMF (2.6 mL) was heated at 80 °C for 21.5 h. The reaction mixture was poured into ethyl acetate (25 mL), and water (25 mL) was added. The layers separated and the aqueous layer was extracted with ethyl acetate (3 × 25 mL). The combined organic layers were washed with brine (50 mL) and dried over anhydrous magnesium sulfate. Chromatography (silica gel, hexanes/ethyl acetate, 6 : 1) provided 0.098 g (58%) of the endo cyclized products as a yellow solid (mp 149–150 °C (acetonitrile)).

Preparation of 2,3β-[Bis(tert-butyldimethylsiloxy)]-5-(cyclohexylamino)-11,12-dimethoxy-1-dehydro-3,4,8,9-tetrahydro-6-oxo-1H-indolo[7a,1-a]isoquinoline[174]

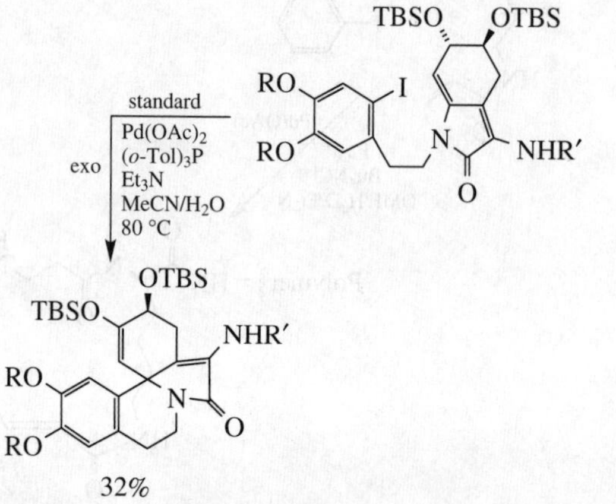

32%

A mixture of the starting iodoarene (0.118 g, 0.151 mmol), Pd(OAc)$_2$ (0.00075 g, 0.00334 mmol), PPh$_3$ (0.0025 g, 0.00953 mmol), and Et$_3$N (0.030.6 g, 0.302 mmol) in CH$_3$CN (1.15 mL) were heated at 82 °C for 20 h. Ethyl ether (20 mL) was added. The solution was washed with brine (25 mL) and dried over anhydrous magnesium sulfate. Chromatography (silica gel, hexanes/ethyl acetate, 6 : 1) provided 0.017 g (32% based on recovered starting material) of the exo product as a white solid (mp 170.5–171.5 °C (acetonitrile)).

By using chiral ligands, a high degree of asymmetric induction has been observed in selected cases[178–181]:

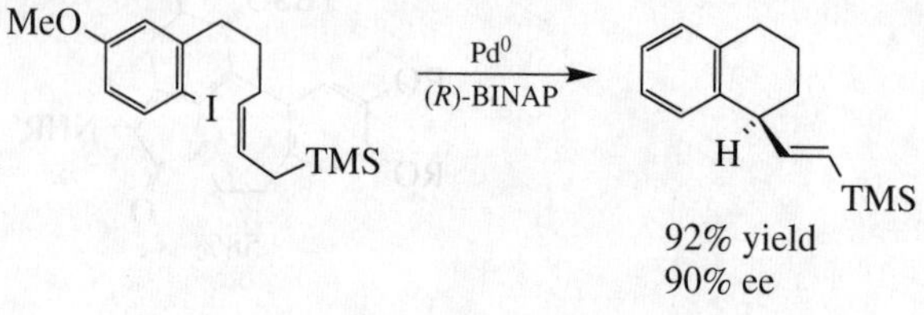

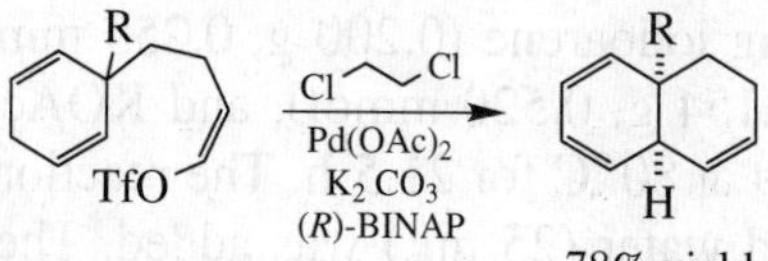

92% yield
90% ee

78% yield
95% ee

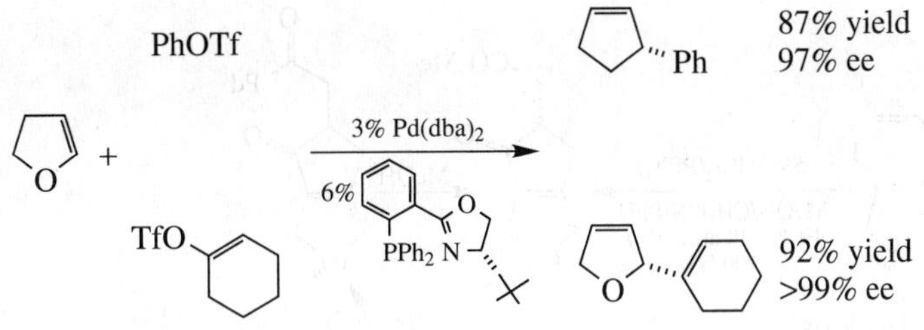

Heck oxidative addition–insertion sequences are not limited to simple alkenes. Even aromatic π-systems can insert[182,183]:

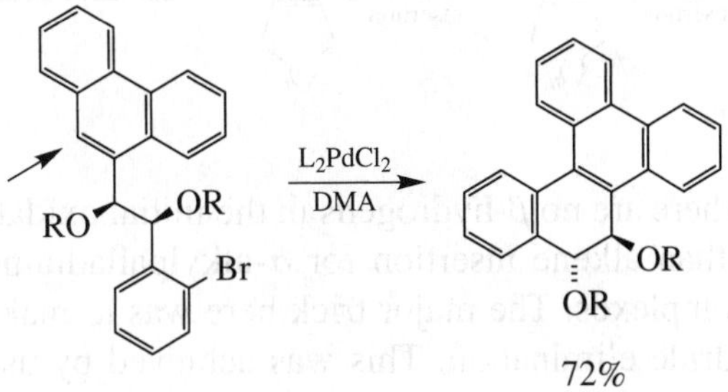

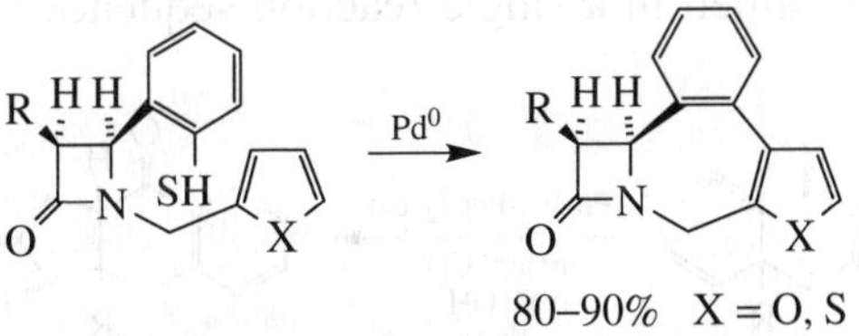

4.2.3 Palladium(0)-catalyzed Multiple (Cascade) Insertion Process

As presented above, σ-aryl- and alkenylpalladium(II) complexes have a very rich reaction chemistry, undergoing insertions, transmetalations, reductive eliminations, and β-hydrogen eliminations. In principle, these processes can be combined in any order and, as many of them generate another σ-bonded organopalladium complex, a sequence of these processes, forming several bonds, could be envisaged[184]. The only requirement is that the next step is more facile than reductive elimination or β-hydrogen elimination, which truncates the process. Thus, a sequence consisting of oxidative addition/CO insertion/alkene insertion/CO insertion and cleavage formed three carbon–carbon bonds per catalyst cycle[185]:

This works because there are no β-hydrogens in the initial oxidative adduct, and because CO insertion is faster than alkene insertion for σ-alkylpalladium complexes, but slower for σ-acyl palladium complexes. The major trick here was to make the final CO insertion competitive with β-hydride elimination. This was achieved by using high (40 atm) pressure of CO. This 'down-hill' sequence of events is critical but, when achieved, a large number of bonds can be formed in a single reaction sequence[186]:

Asymmetry can be induced in this process[187]:

Alkynes insert more readily than alkenes, and a variety of carbocyclization processes[188] have been developed[189–191]:

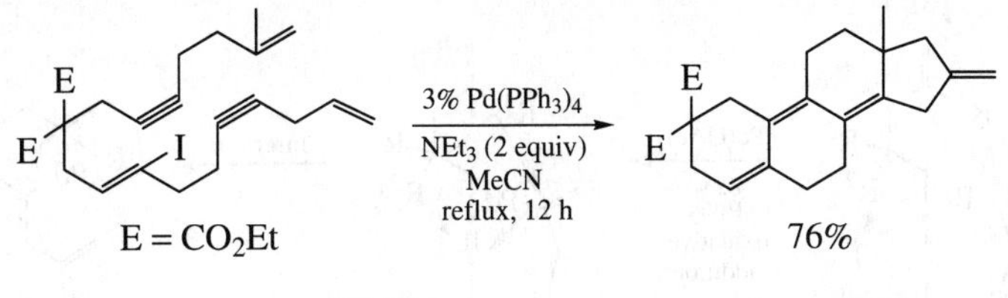

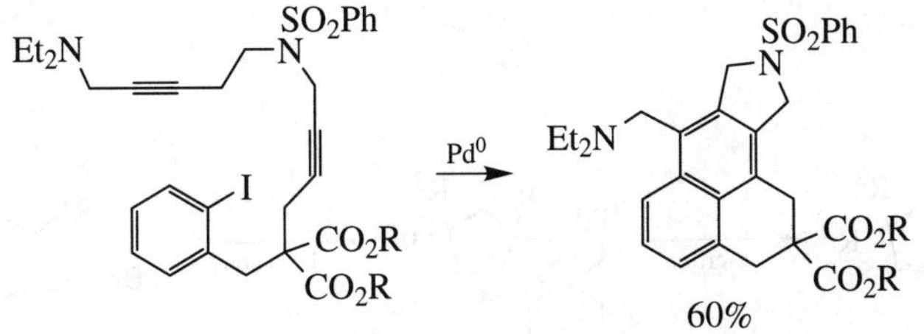

3% Pd(PPh$_3$)$_4$
NEt$_3$ (2 equiv)
MeCN
reflux, 12 h

E = CO$_2$Et

76%

Pd^0

60%

Again, any process in which σ-alkylpalladium(II) complexes participate can be used to either carry or truncate these 'cascade' processes. Thus, the reaction below[192] is truncated by a transmetalation step, whereas the one that follows[193] is completed by a nucleophilic attack. There are endless variations on these themes, and it is likely that all of them will be tried.

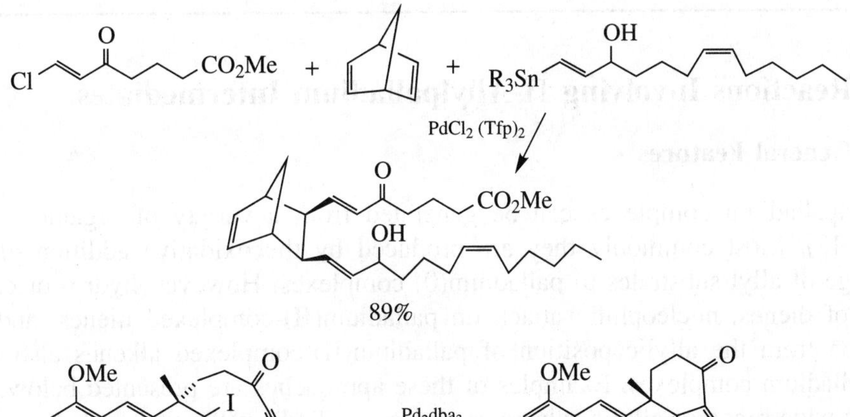

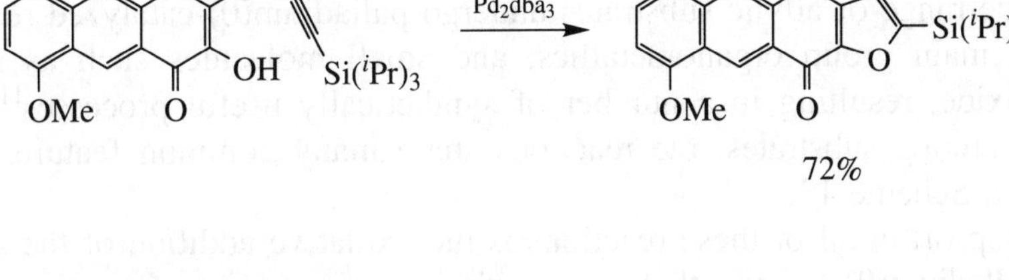

PdCl$_2$ (Tfp)$_2$

89%

Pd$_2$dba$_3$

72%

Preparation of Trimethyl-10-oxatetracyclo[6.4.1.0.0]tridec-2-ene-4,4,6-tri-carboxylate[190]

A mixture of bromodieneyne (0.5 g, 1.16 mmol), Pd(OAc)$_2$ (0.015 g, 0.058 mmol), PPh$_3$ (0.065 g, 0.232 mmol), and Ag$_2$CO$_3$ (0.638 g, 2.318 mmol) in 10 mL of MeCN was heated in a sealed tube at 130 °C for 3 h. Extractive work-up and flash column chromatography (silica gel; 1 : 12 ether–petroleum ether) provided 0.288 g (71%).

4.3   Reactions Involving Π-Allylpalladium Intermediates

4.3.1   General Features

π-Allylpalladium complexes can be generated from a variety of organic substrates (Scheme 12). Most commonly they are produced by the oxidative addition of a very wide range of allyl substrates to palladium(0) complexes. However, hydro- or carbopalladation of dienes, nucleophilic attack on palladium(II)-complexed dienes, and proton abstraction from the allylic position of palladium(II)-complexed alkenes also produce π-allylpalladium complexes. Examples of these approaches are presented below.

A very wide range of allylic substrates undergo palladium(0)-catalyzed reactions with nucleophiles, main group organometallics, and small molecules such as alkenes and carbon monoxide, resulting in a number of synthetically useful processes[194]. Despite the disparity among substrates, the reactions share many common features, which are summarized in Scheme 13.

The first step (a) in all of these reactions is the oxidative addition of the allylic C–X system to palladium(0), a step that goes with clean inversion of configuration at the allylic carbon and produces a σ-allylpalladium(II) complex. These are rarely detected (although often invoked) and rapidly collapse to the more stable π-allylpalladium(II)

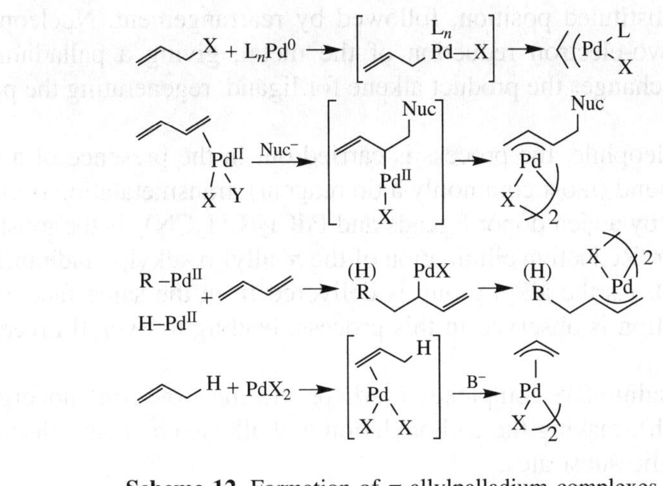

Scheme 12. Formation of π-allylpalladium complexes.

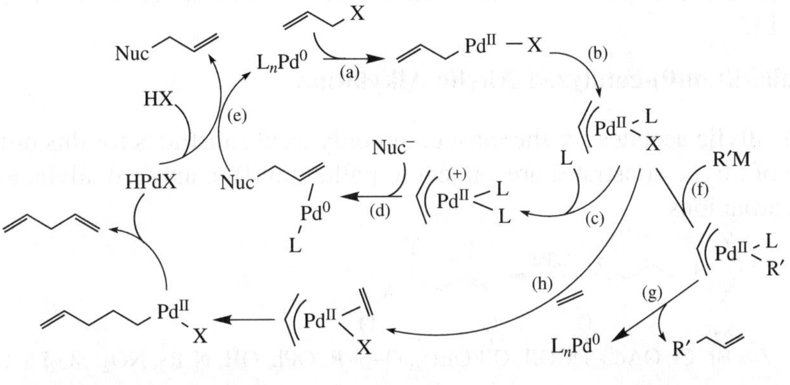

Scheme 13. Palladium(0)-catalyzed reactions of allylic substrates.

complex, step (b). These can be isolated, and are usually yellow, crystalline, air-stable solids. However, for catalysis, isolation is undesired and reactions are usually carried out under conditions which favor consumption of the π-allyl complex at a rate commensurate with its formation.

In the absence of added ligands, π-allylpalladium(II) complexes are relatively inert to nucleophilic attack. However, in the presence of phosphines, normally already there from the palladium(0) starting complex (most commonly Pd(PPh₃)₄), they undergo reaction with a wide variety of nucleophiles, most commonly amines or stabilized carbanions. The nucleophile attacks from the face *opposite* the metal, with clean inversion, resulting in net retention (two inversions) for the overall process (d). With unsymmetrical allyl complexes, attack usually occurs at the less-substituted position, but the regioselectivity is strongly dependent on the structural features of the substrate and the conditions of the reaction. Many nucleophiles, particularly amines but some stabilized carbanions as well, react *reversibly*; in these cases the kinetic product results from attack at the more substituted position, followed by rearrangement. Nucleophilic attack results in a formal two-electron reduction of the metal, giving a palladium(0) alkene complex (d) which exchanges the product alkene for ligand, regenerating the palladium(0) catalyst (e).

If, instead of a nucleophile, the process is carried out in the presence of a main group organometallic compound (most commonly a tin reagent), transmetalation occurs (f). This process is *suppressed* by added donor ligands and $PdCl_2(CH_3CN)_2$ is the most commonly used catalyst precursor. Reductive elimination of the π-allyl-σ-alkylpalladium(II) complex results in coupling (g). As the alkyl group is delivered from the same face as the metal, retention of configuration is observed in this process, leading to overall inversion for the total process.

Finally, π-allylpalladium(II) complexes (perhaps via their σ-form) undergo both CO and alkene insertion (h), making the carbonylation and alkene chemistry discussed above also accessible to allylic substrates.

π-Allylpalladium complexes are also produced by the addition of nucleophiles to palladium(II)–diene complexes. Once formed, these also undergo all of the reactions in Scheme 13.

4.3.2   Palladium(0)-catalyzed Allylic Alkylation

Although allylic acetates are the most commonly used substrates for this process, a very wide range of allylic substrates are subject to palladium(0)-catalyzed allylic alkylation by stabilized carbanions:

$Z = Br, Cl, OAc, -OCOR, OP(OEt)_2, O-S-R, OPh, OH, N^+R_3, NO_2, SO_2Ph, CN,$

$X, Y = CO_2R, COR, SO_2Ph, CN, NO_2$

The reactions are usually carried out by mixing the substrate with 1–2% of $Pd(PPh_3)_4$ catalyst in THF containing an additional 5–10 equivalents (based on the catalyst, not the substrate) of PPh_3, generating the anion in THF in a separate vessel, adding it to the substrate–catalyst mixture and heating to reflux for several hours. The reaction usually proceeds in excellent yield. There is an enormous number of examples of this process in the literature, with a very wide range of highly functionalized substrates and carbanions, so the reliability of this methodology is very well established. A few representative current examples are shown in the following equations[195–198]:

90%
complete chirality
transfer

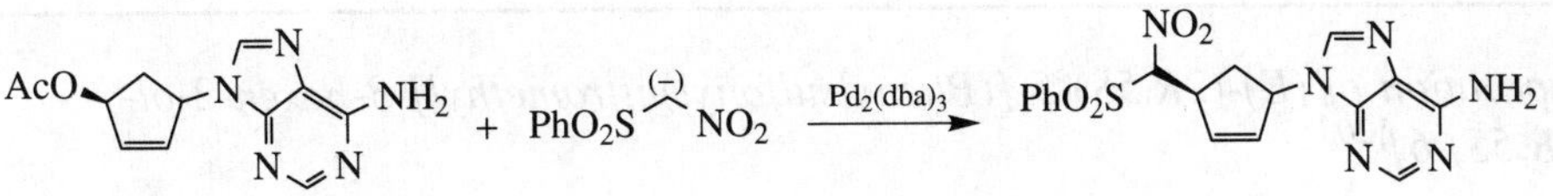

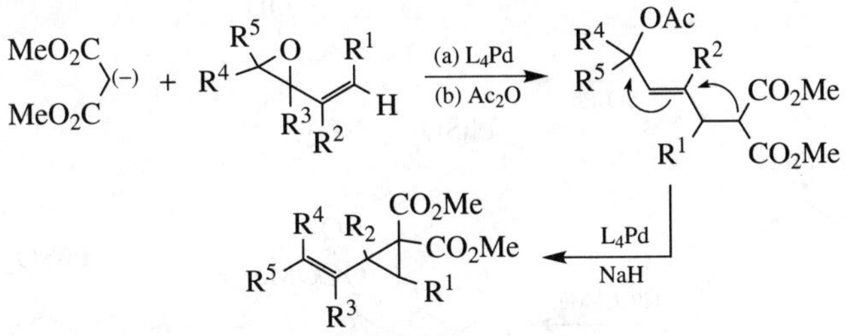

good yields

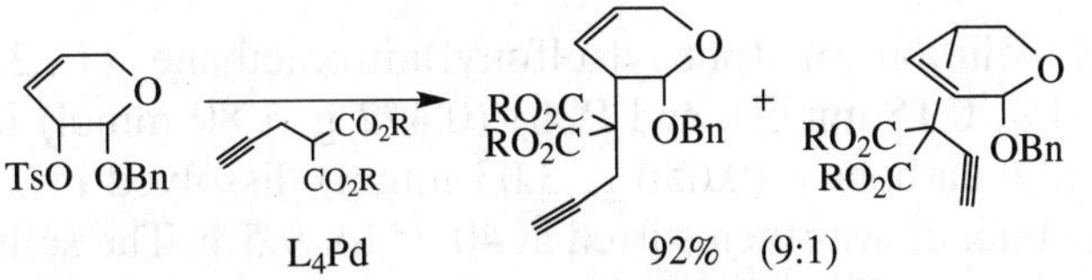

92% (9:1)

Intramolecular versions of this process are particularly useful for forming carbocyclic rings[199], and have been used to make from three-membered to 11-membered rings and macrocyclic rings successfully[200–202]:

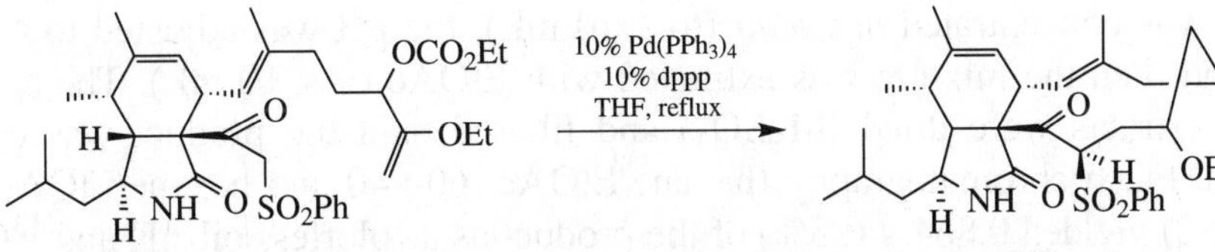

49%

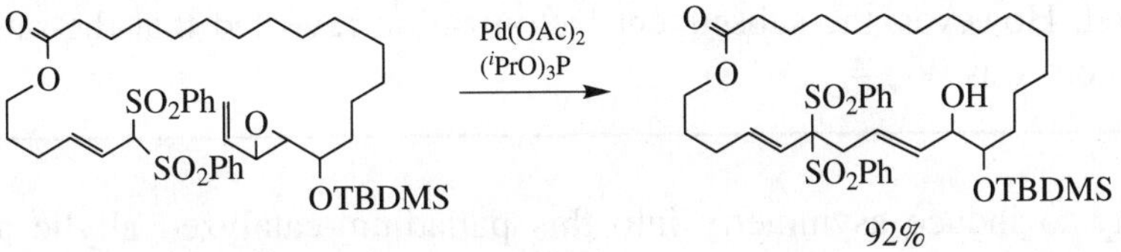

92%

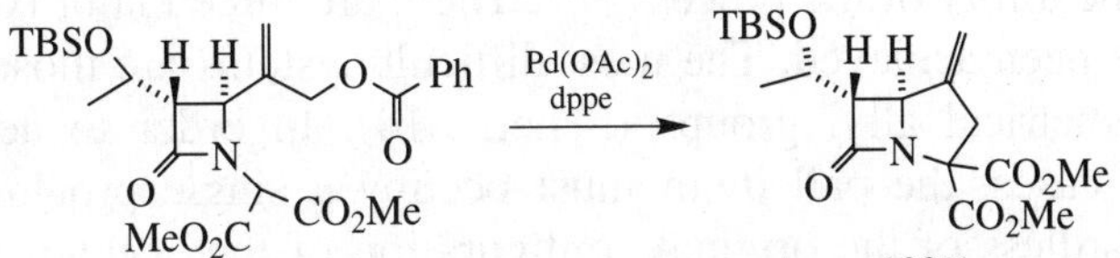

100%

Cl>OCO_2R>OAc>>>OH

The order of reactivity of leaving groups towards palladium(0) is Cl > OCO_2R > OAc ⋙ OH, allowing a high degree of discrimination in systems having multiple allylic sites of reactivity[203]. As systems such as these are easily generated from dienes (see below) many useful examples of this discrimination are extant.

Preparation of (E)-(2R,5S)-5-[(Phenylsulfonyl)nitromethyl]-3-hexen-2-ol-[(2R,5S)-6][203]

To a stirred solution of (phenylsulfonyl)nitromethane (1.22 g, 6.06 mmol), Pd(OAc)$_2$ (0.034 g, 0.15 mmol), and PPh$_3$ (0.472 g, 1.80 mmol) in THF (20 mL), was added the allyl carbonate (0.656 g, 3.03 mmol) dissolved in THF (5 mL). The yellow solution obtained was then stirred at 40 °C for 9.5 h. The solution was cooled, and the product was collected on silica. Flash chromatography (hexane/EtOAc, 80 : 20 and 70 : 30) afforded 1.67 g of a crude oil containing the desired product plus nucleophile. This oil was dissolved in CH$_3$OH (35 mL), 15% aqueous K$_2$CO$_3$ (25 mL) was added, and the solution was stirred at 40 °C for 2 h. The reaction mixture was concentrated *in vacuo* (to ≈10 mL), the pH was adjusted to ≈2 (2 M HCl), and then the mixture was extracted with EtOAc (4 × 10 mL). The combined organic extracts were dried (MgSO$_4$) and filtered, and the product was collected on silica. Flash chromatography (hexane/EtOAc, 60 : 40 and hexane/EtOAc/EtOH, 49 : 49 : 2) yielded 0.864 g (95%) of the product as a colorless oil. ^{1}H and ^{13}C NMR spectra were in accordance with those reported earlier, and no (2R,5R) isomer could be detected. However, the subsequent lactonization revealed that the ratio between these isomers was 96 : 4.

The ability to induce asymmetry into this palladium-catalyzed allylic alkylation has greatly expanded the utility of the process[204]. There are three major types of asymmetric induction that have been achieved. The most difficult systems are those involving achiral or racemic unsymmetrical allyl groups (Scheme 14). In order to achieve asymmetric induction in these cases, the palladium must occupy a single prochiral face of the π-allyl complex, regardless of the original configuration of the acetate. In acyclic systems this can be achieved by enantioface exchange of the metal via a σ−π−σ equilibration. Provided this kind of exchange can occur, and that one of the two diastereoisomeric π-allylpalladium complexes reacts with the nucleophile in preference to the other, high enantioselection can be observed[205]. This route for asymmetric induction is not available to cyclic allylic systems because the required rotation about the single bond in the σ-allyl complex cannot occur.

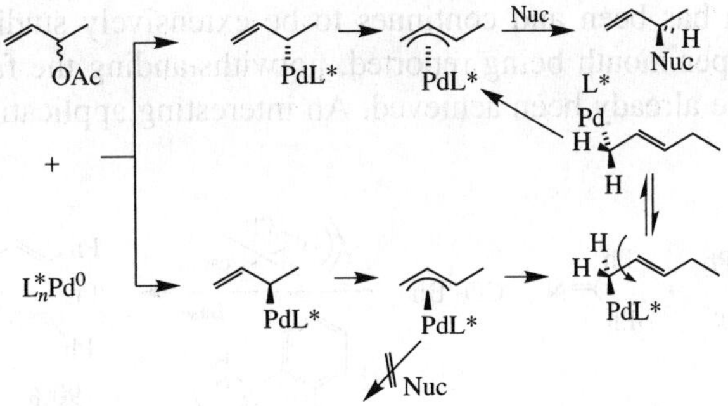

Scheme 14. Palladium(0)-catalyzed asymmetric induction.

Preparation of (4R,5R)-4-Nitro-5-vinyl-1,3,4,5-tetrahydrobenz[c,d]indole[205]

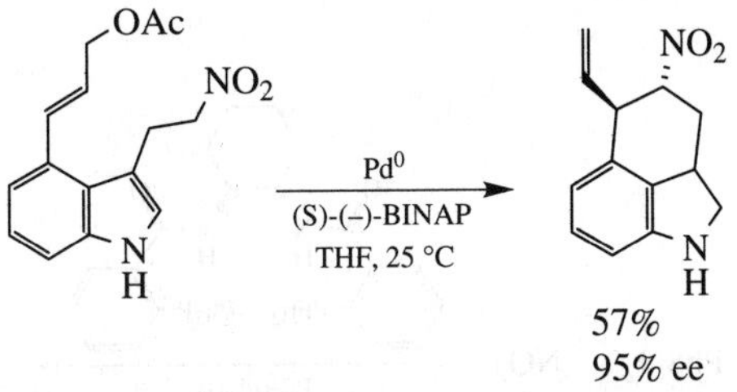

57%
95% ee

A mixture of K_2CO_3 (9.58 g, 69.4 mmol), $Pd(OAc)_2$ (0.187 g, 0.83 mmol) and (S)-(−)-BINAP (1.048 g, 1.66 mmol) in dry THF (30 mL) was stirred at 25 °C, under argon. The mixture turned from light orange to dark red. A solution of (E)-3-[(2-nitroethyl)-4-indolyl]propenyl acetate (8 g, 27.7 mmol) in dry THF (60 mL) was added drowise, then stirred for 6 h. The reaction mixture was filtered through silica gel, and the crude residue was washed with THF. The solvent was removed under reduced pressure. Flash chromatography (AcOEt/cyclohexane 1 : 9) yielded (4R,5R)-4-nitro-5-vinyl-1,3,4,5-tetrahydrobenz[c,d]indole (3.6 g, 57%) as a white solid (mp 210 °C (MeOH); TLC (AcOEt/cyclohexane 3 : 7) $R_f = 0.58$; $[\alpha]_D^{20} = -77$ ($c = 1$, CH_2Cl_2)).

A substantially less challenging case involves asymmetric induction in the alkylation of allyl systems which require only discrimination of the enantiotopic terminii of the π-allyl—that is, the π-allyl group is symmetrical[206]:

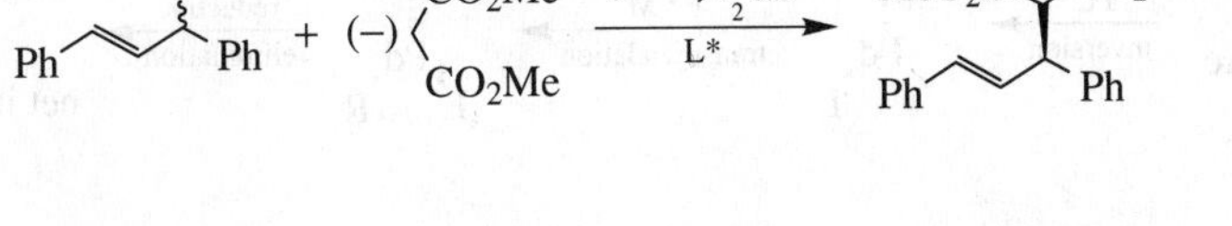

This exact system has been and continues to be extensively studied with one or two new ligand systems per month being reported, notwithstanding the fact the enantiomeric excess of >95% have already been achieved. An interesting application of this chemistry is the following[207]:

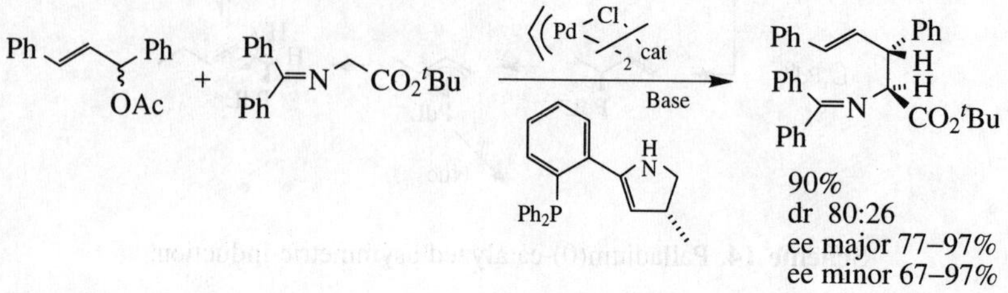

90%
dr 80:26
ee major 77–97%
ee minor 67–97%

Perhaps the most useful mode of asymmetric induction in π-allylpalladium chemistry involves the catalytic desymmetrization of meso substrates. Although most often used with heteronucleophiles (see below) a few useful examples involving carbanions have been reported[204]:

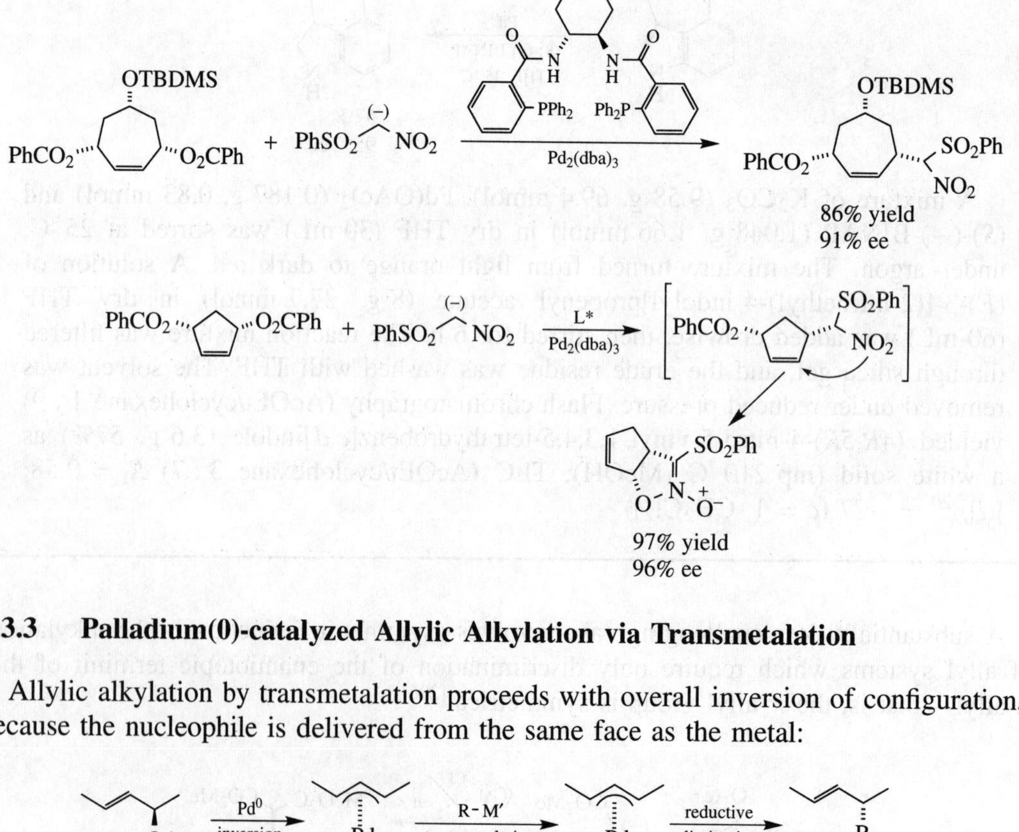

86% yield
91% ee

97% yield
96% ee

4.3.3 Palladium(0)-catalyzed Allylic Alkylation via Transmetalation

Allylic alkylation by transmetalation proceeds with overall inversion of configuration, because the nucleophile is delivered from the same face as the metal:

Transmetalation has been used in allylic systems much less frequently than with aryl or alkenyl systems, for several practical reasons. Allyl acetates are the most attractive class of allylic substrates, because of their ready availability from the corresponding alcohols, but they are substantially less reactive towards oxidative addition to palladium(0) than are allylic halides, and phosphine ligands are required to promote this process. Acetate coordinates quite strongly to palladium and this, along with the presence of excess phosphine, slows down the crucial transmetalation step, already the rate-limiting step. Finally, in contrast to dialkylpalladium(II) complexes, σ-alkyl-π-allylpalladium(II) complexes undergo reductive elimination only slowly, compromising this final step in the catalytic cycle. It is thus not surprising that alkylation of allylic acetates via transmetalation has been slow to develop.

However some progress has been made. A variety of allyl acetates were alkylated by aryl- and alkenyltin reagents under the special conditions of polar (DMF) solvent, added lithium chloride (3 equiv) to facilitate transmetalation, and 3% Pd(dba)$_2$ catalyst with no added phosphine; in fact, added phosphine inhibited the reaction[208]:

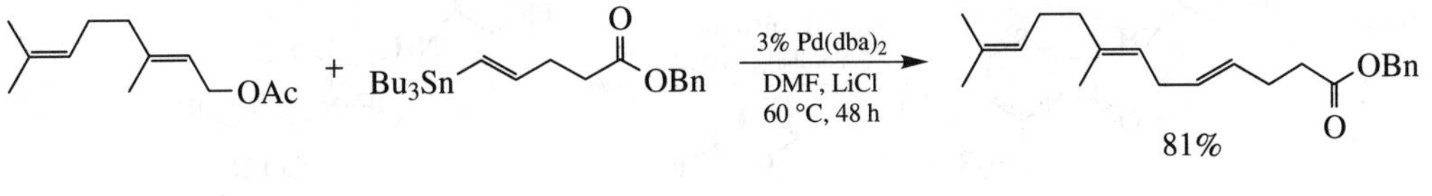

Coupling occurred exclusively at the primary position of the allyl group. The geometry of the alkene in both the allyl substrate and alkenyltin reagent was maintained, and modest functionality was tolerated in the tin reagent (CO$_2$R, OH, OSiR$_3$, OMe).

Vinyl epoxides are somewhat more tractable substrates and undergo clean allylic alkylation by a variety of aryl and vinylstannanes in good yield[209]. Predominately 1,4-addition is observed, and alkene geometry is maintained in the vinyltin component, but lost in the vinyl epoxide. Alkynic, benzylic and allylic tin reagents failed to transfer at all. Again, added ligands inhibited the reaction, and the best catalyst system was 3% PdCl$_2$(MeCN)$_2$ in DMF solvent:

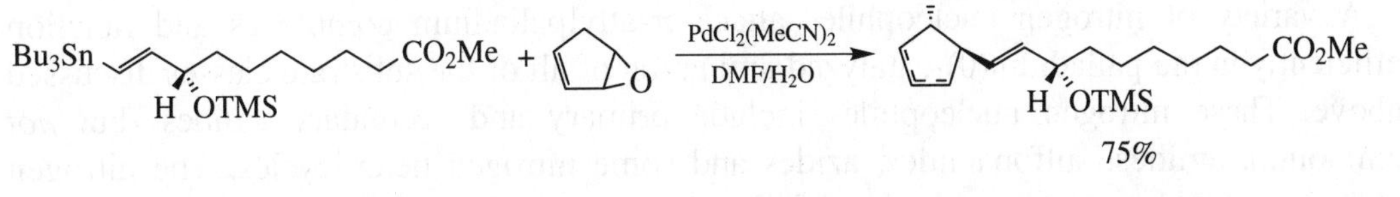

Related systems involving allylphosphonates[210] and allyl aryl ethers[211], as well as intramolecular versions involving transmetalation from zinc[212] are known:

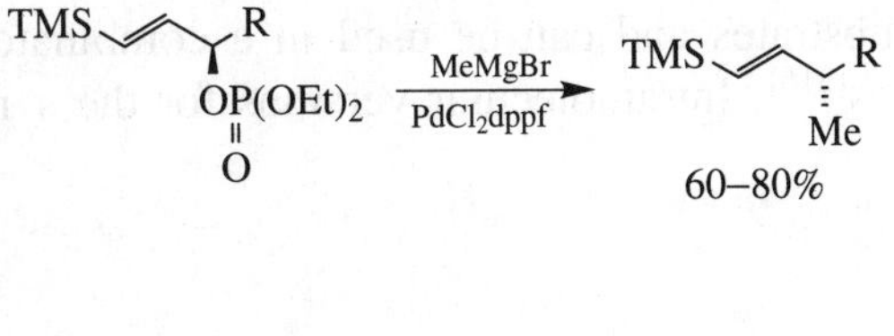

Allylic halides are much more amenable to transmetalation processes, because they both oxidatively add and transmetalate more readily than the corresponding acetates. This reaction proceeds under normal conditions (presence of phosphine ligands) and has been utilized in relatively complex systems[67,213]:

R' = pMeOPh, F$_2$C=CF

Carbonylative coupling between allylic halides and aryl, alkenyl[214] and allyltin reagents[215] to give ketones has also been achieved.

4.3.4   Palladium(0)-catalyzed Allylic Amination

A variety of nitrogen nucleophiles attack π-allylpalladium complexes and function efficiently in the palladium(0)-catalyzed amination of all of the substrate classes discussed above. These nitrogen nucleophiles include primary and secondary amines (but *not* ammonia), amides, sulfonamides, azides and some nitrogen heterocycles. The nitrogen almost invariably ends up at the less-hindered terminus of the allyl system, although that might not be the initial site of attack, as palladium(0) complexes catalyze rapid allylic rearrangement of allyl amines. As with alkylation, this is experimentally a straightforward and reliable reaction which proceeds under similar conditions. It is even efficient with polymer-bound allylic substrates and can be used in a combinatorial sense to generate a variety of allylic amines[216]. Intramolecular versions for the synthesis of macrocycles are efficient[217]:

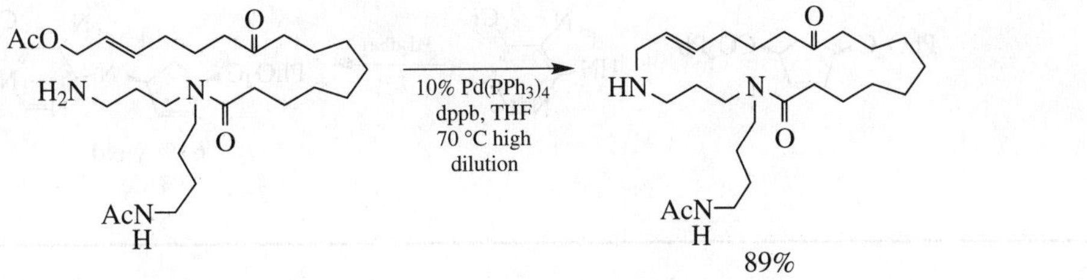

89%

Intramolecular allylic amination was beautifully combined with asymmetric allylic alkylation in a total synthesis of Lycorane[218]:

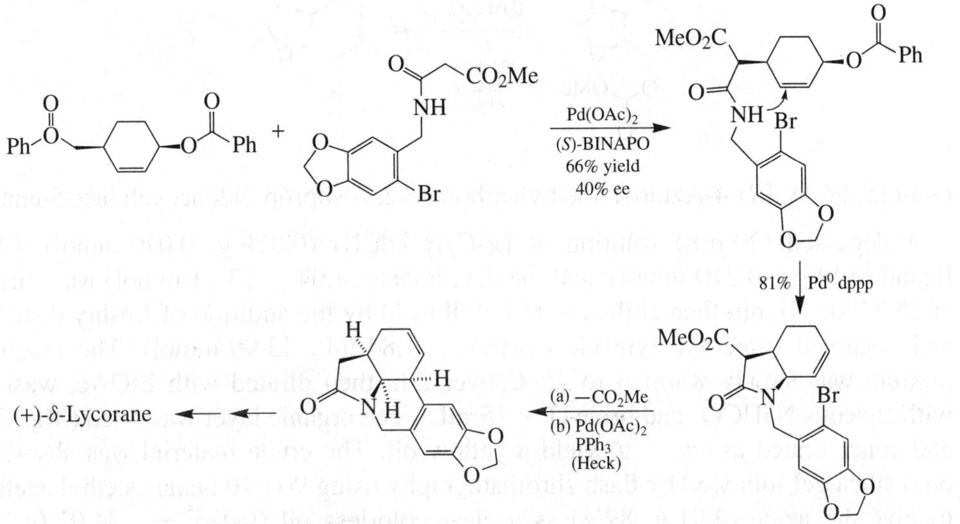

The development of asymmetric allylic amination increased the synthetic utility of the process substantially[219]. As with asymmetric alkylation, 3-acetoxy-1,3-diphenylpropene can be asymmetrically aminated in 99% ee[220], and a wide range of chiral ligands have been developed to achieve this. More useful is the enantioselective amination of meso allyl compounds[204,221]. Thus, (+)-Pancratistatin and Conduramine A′[222], carbocyclic nucleoside analogs[223], and (−)-Epibatidine[224] have all used asymmetric allylic amination as a key step:

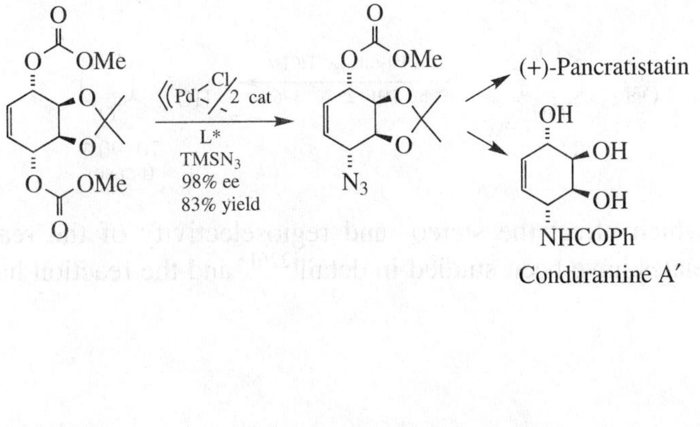

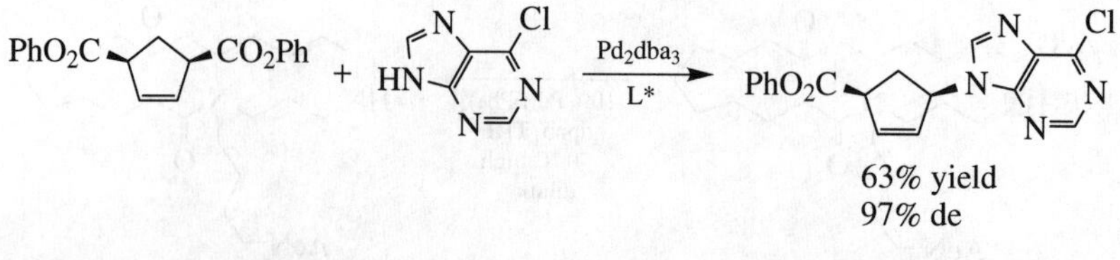

63% yield
97% de

Asymmetric Addition of Azide to meso-Allyl Dicarbonates

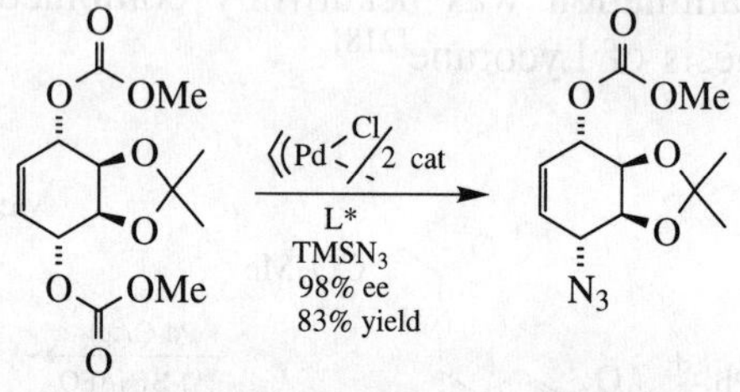

98% ee
83% yield

(−)-(1*S*,2*R*,3*S*,4*R*)-4-Azido-1-methylcarbonate-2,3-isopropylidenecyclohex-5-ene[222]

A degassed (20 mL) solution of [π-C$_3$H$_5$PdCl]$_2$ (0.026 g, 0.070 mmol), (+)-ligand (0.149 g, 0.210 mmol), and the dicarbonate (4.04 g, 13.24 mmol) was stirred at 25 °C for 10 min then chilled to 0 °C followed by the addition of freshly distilled and degassed trimethylsilylazide (TMSN$_3$) (1.84 mL, 13.90 mmol). The reaction mixture was slowly warmed to 25 °C over 2 h, then diluted with EtOAc, washed with aqueous NaHCO$_3$ and brine (1 × 15 mL). The organic layer was dried (MgSO$_4$) and concentrated *in vacuo* to yield a yellow oil. The crude material was absorbed onto silica gel followed by flash chromatography using 90 : 10 hexanes:ethyl acetate to give the azide (2.91 g, 82%) as a clear colorless oil ([α]$_D^{25}$ = −34.0° (*c* 2.1, CH$_2$Cl$_2$)).

4.3.5 Allylic Alkoxylation, Acetoxylation, Reduction, Elimination and Deprotection

Although nucleophiles other than stabilized carbanions and amines should be able to participate in palladium(0)-catalyzed reactions of allylic compounds, relatively few others have been studied. Alcohols have been used intramolecularly[225]:

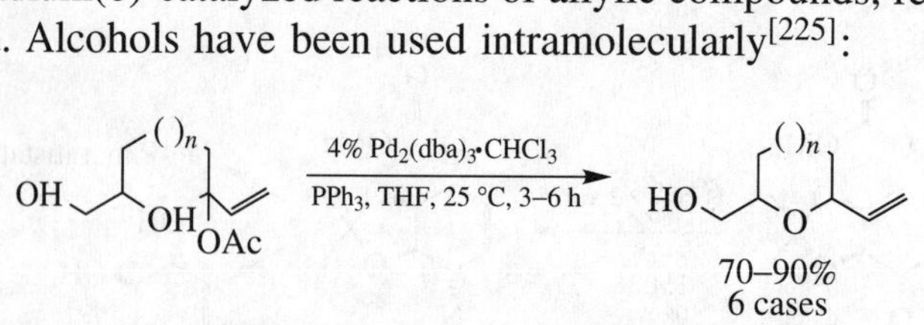

70–90%
6 cases

The factors which effect the stereo- and regioselectivity of the reaction of phenols with allyl carbonates have been studied in detail[226], and the reaction has been used with cyclic allyl epoxides[227]:

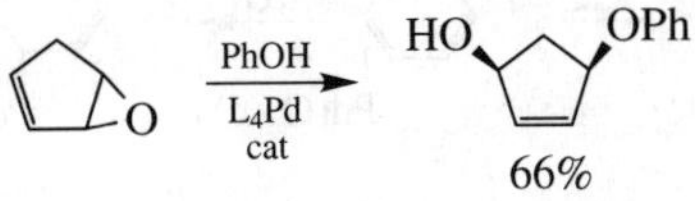

Glycal acetates were coupled to the anomeric OH of another carbohydrate[228] and alcohols were allylated[229] via palladium(0) catalyzed allylic alkoxylation:

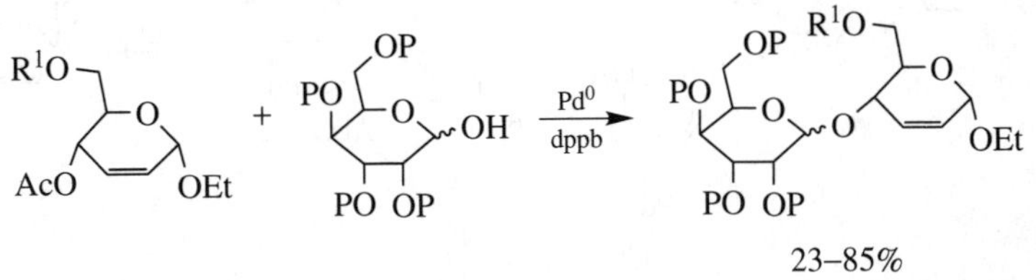

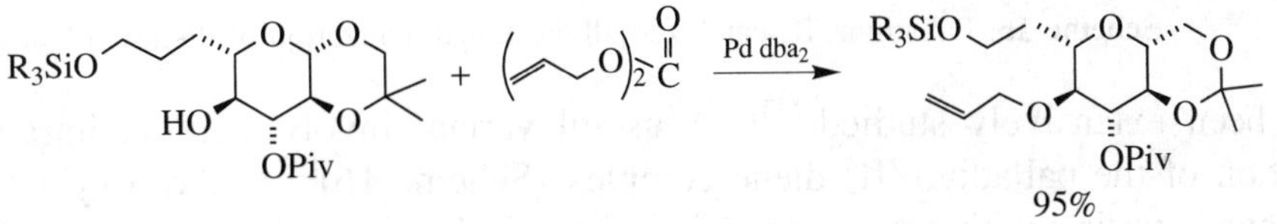

23–85%

95%

Triphenylsilanol was an efficient alkoxylation reagent for allyl epoxides[229,230]:

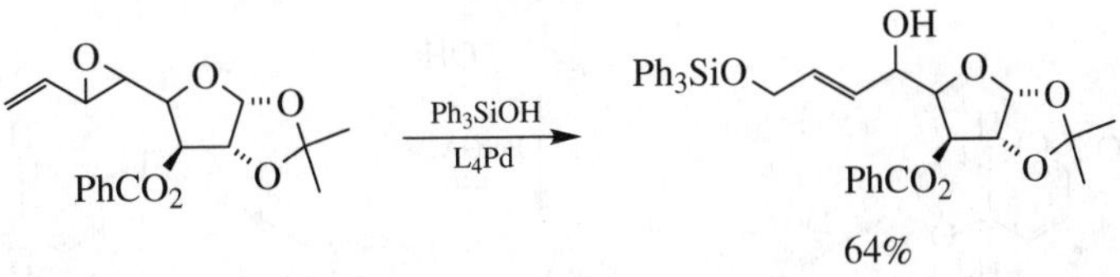

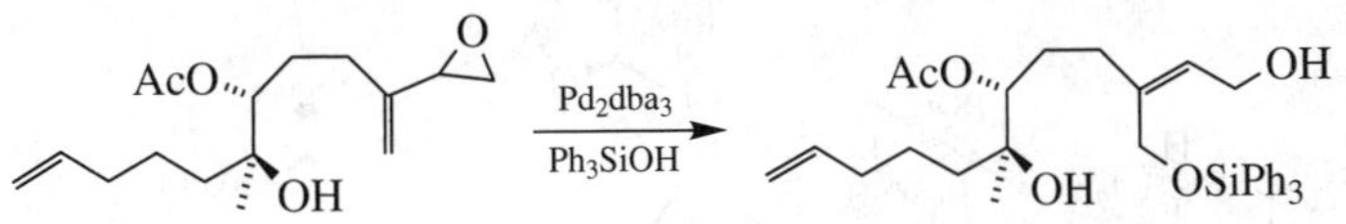

64%

Although acetate, by itself, is rarely used as a nucleophile in palladium(0)-catalyzed allylic displacement reactions (indeed, acetate is most often the leaving group!) it does play a major role in a very useful variant involving palladium(II)-catalyzed bis-acetoxylation or chloroacetoxylation of dienes (Scheme 15). The process begins with a palladium(II)-assisted acetoxylation of one of the two double bonds of the complexed diene, generating a π-allylpalladium(II) complex. In the absence of added chloride, the second acetate is delivered from the metal leading to the trans diacetate. Added chloride blocks this coordination site on the metal and leads to nucleophilic attack by uncomplexed acetate, from the face opposite the metal, giving the corresponding *cis*-diacetate and palladium(0)[231]. The reoxidation of palladium(0) to palladium(II) can be carried by benzoquinone/MnO_2 or more efficiently by O_2/catalytic hydroquinone/catalytic Co(salophen)[232]. Because the products of this reaction are allyl acetates, and because of the wide variety of palladium(0)-catalyzed reactions of allyl acetates this is a useful process

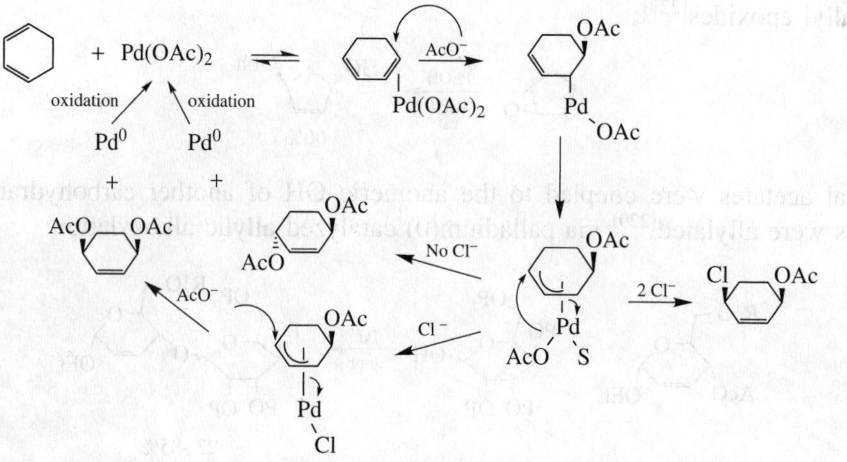

Scheme 15. Palladium(II)-catalyzed allylic functionalization of dienes.

that has been extensively studied[233]. A useful variant involves initial intramolecular alkoxylation of the palladium(II) diene complex (Scheme 16)[234]. Acetoxylactonization was achieved starting with a pendent carboxylic acid in place of the alcohol[235].

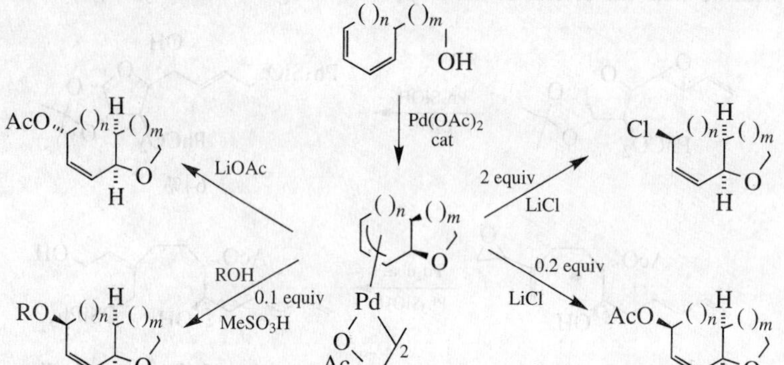

$$n = 1, 2 \qquad m = 1, 2 \qquad 60–90\%$$

Scheme 16. Palladium(II)-catalyzed alkoxy functionalization of dienes.

Preparation of trans-1,4-Diacetoxy-2-cyclohexene (Scheme 15)[231]

To a stirred solution of Pd(OAc)$_2$ (0.700 g, 3.1 mmol), LiOAc•2H$_2$O (6.8 g, 66.6 mmol) and *p*-benzoquinone (1.91 g, 17.6 mmol) in acetic acid (100 mL) was added MnO$_2$ (6.52 g, 74.9 mmol) followed by 1,3-cyclohexadiene (5.0 g, 62.5 mol) dissolved in pentane (200 mL). The reaction mixture, which separated into a pentane phase and an acetic acid phase, was moderately stirred at 25 °C for 8 h. The pentane phase was separated and collected and the remaining acetic acid phase was diluted with saturated NaCl (100 mL) and extracted with pentane (2 × 100 mL) and pentane/diethyl ether (1 : 1) (3 × 100 mL). The combined extracts were washed with saturated NaCl (3 × 40 mL), water (3 × 10 mL) and finally 2 M NaOH (3 × 30 mL). The organic phase was dried (MgSO$_4$) and evaporated to yield 11.51 g (93%) of crystalline product (>91% trans). Recrystallization from hexane gave 9.87 g (80%) of isomerically pure material (?99% trans; mp 49–50 °C).

Preparation of cis-1,4-Diacetoxy-2-cyclohexene (Scheme 15)[231]

Essentially the same procedure as for the preparation of the trans compound was used, but catalytic amounts of LiCl were added to a stirred solution of Pd(OAc)$_2$ (0.700 g, 3.0 mmol), LiCl (0.521 g, 12.3 mmol), LiOAc•2H$_2$O (21.5 g, 211 mmol) and *p*-benzoquinone (1.6 g, 14.8 mmol) in acetic acid (100 mL), followed by MnO$_2$ (6.8 g, 78 mmol) and then 1,3-cyclohexadiene (5 g, 62.5 mmol) in pentane (200 mL). After 22 h, the same work-up procedure as above gave 10.6 g (86%) of essentially pure *cis*-1,4-diacetoxy-2-cyclohexene (96% cis). Distillation afforded 9.8 g (79%).

Allylic sulfones can be prepared by palladium(0)-catalyzed reactions of allyl acetates with sulfinates[236]. In the presence of chiral ligands high enantiomeric excesses are obtained[237].

Allylic substrates are readily reduced by palladium(0) catalysts in the presence of a wide range of hydride sources[238]. Perhaps the mildest are alkyl ammonium formates, which reduced allylic substrates with clean inversion via the following route:

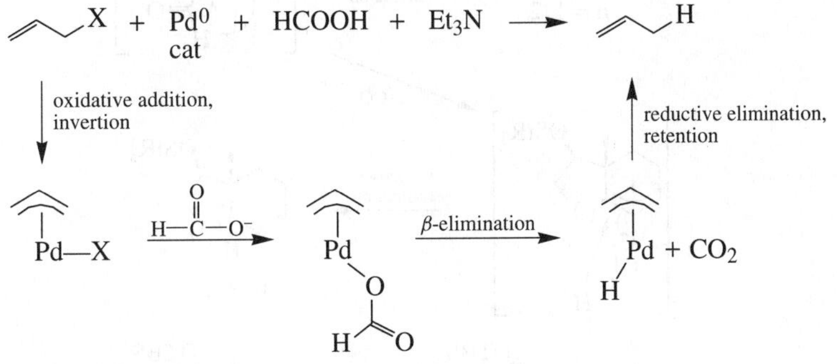

This is very useful in complex systems[239,240]:

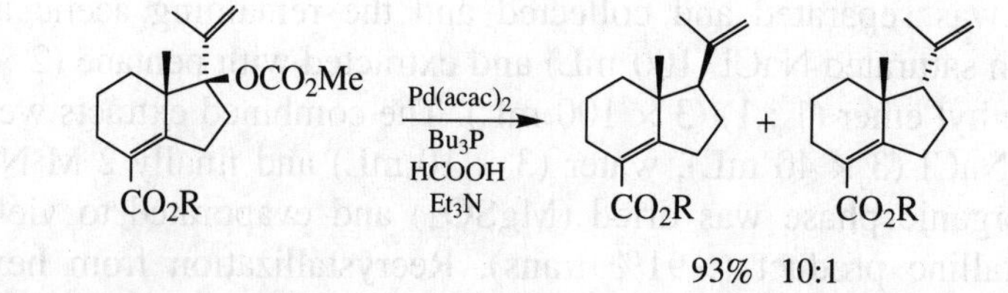

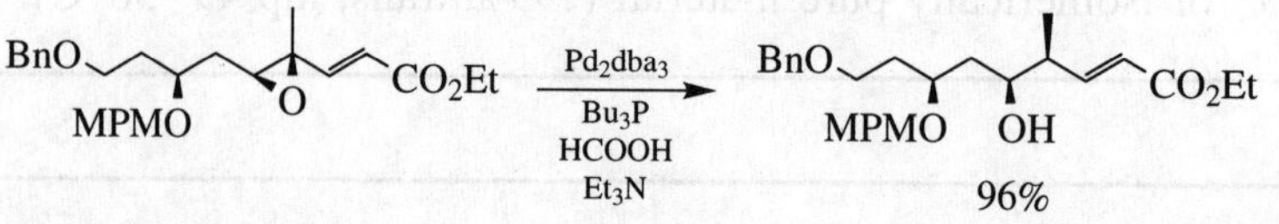

93% 10:1

96%

Intramolecular versions are quite stereoselective[241]. By using chiral ligands, asymmetry can be induced effectively[242]:

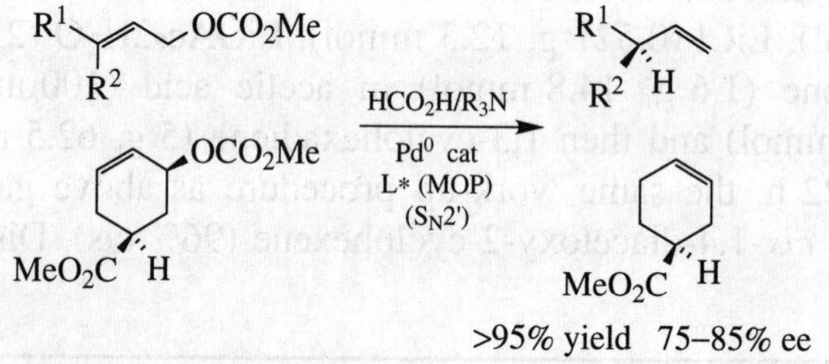

>95% yield   75–85% ee

MOP = monodentate optically active phosphine

Intramolecular Reduction of Allyl Formates[241 a]

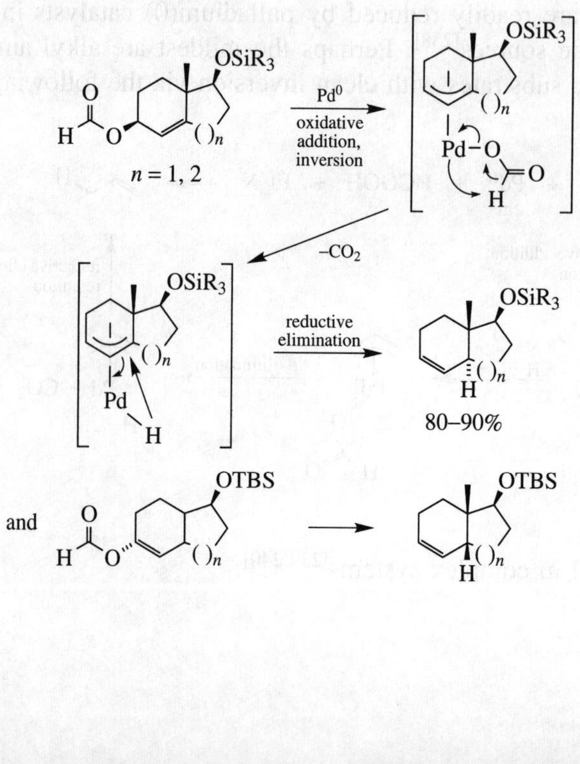

n = 1, 2

–CO₂

reductive elimination

80–90%

and

Preparation of the allylic formates

The formates used in this study were prepared by the treatment of the corresponding allylic alcohols with a mixture of formic acid, acetic anhydride and pyridine (1 : 1 : 1.5) at 25 °C.

Preparation of the palladium catalyst from Pd(acac)$_2$

To a solution of Pd(acac)$_2$ (0.046 g, 0.15 mmol) in THF (2 mL) was added Bu$_3$P (0.037 mL, 0.15 mmol) at room temperature under argon and the resulting pale yellow solution was stirred for 5 min. The hydrogenolyses were carried out as follows:

trans-Hydrindene ($n = 1$)

To a stirred solution of the β-formate (0.155 g, 0.5 mmol) in THF (0.5 mL) was added dropwise the stock solution of the catalyst (0.4 mL, 0.025 mol) at 25 °C. The reaction mixture was stirred for 30 min and passed through Florisil using ether (30 mL) as an eluent, and the combined solution was concentrated *in vacuo* to give a colorless oil (0.133 g). The crude product, contaminated with the heteroannular 3,4-diene (13%), was purified by flash column chromatography. The *trans*-hydrindene (0.109 g) was isolated in 82% yield.

Treatment of allyl carbonates with palladium(0) catalysts in the absence of nucleophiles results in elimination to give the diene. The most effective catalyst is generated by treating palladium acetate with one equivalent of tributylphosphine under argon. The resulting yellow solution has high catalytic activity[243]. The process occurs by oxidative addition with inversion, followed by syn elimination. In sterically biased systems high regioselectivity is observed[244]:

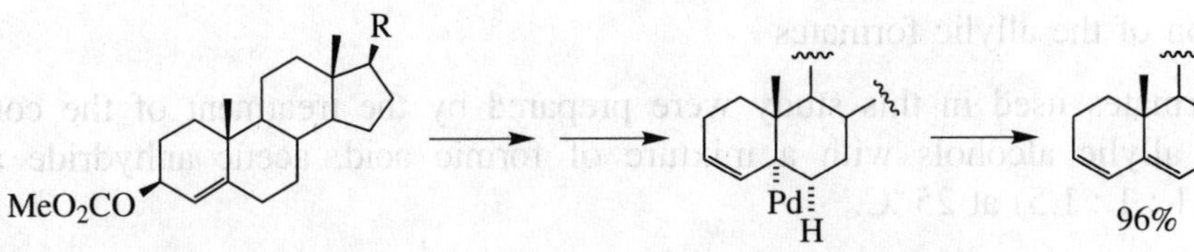

Decent asymmetric induction has been observed in appropriate symmetrical systems[245] (note that in the presence of base an anti elimination is possible[246]):

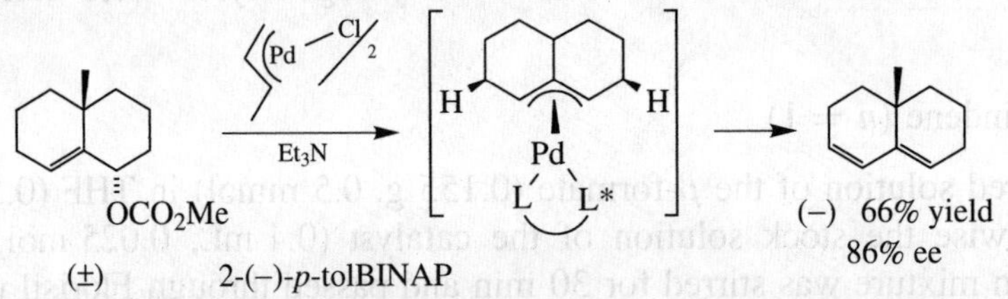

Finally, allyloxycarbonyl groups (alloc) have been used as protecting groups for alcohols, amines and carboxylic acid derivatives because they are easy to remove with palladium(0) catalysts via π-allyl chemistry. In these cases the allyl group is discarded with regeneration of the catalyst:

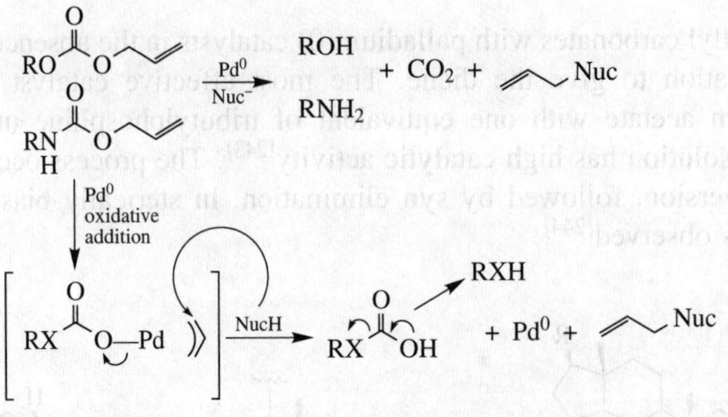

A variety of nucleophiles can be used including amines[247], stabilized carbanions[248], and nonnucleophilic allyl group acceptors such as phenylsilane and CF₃CON(TMS)(Me)[249]:

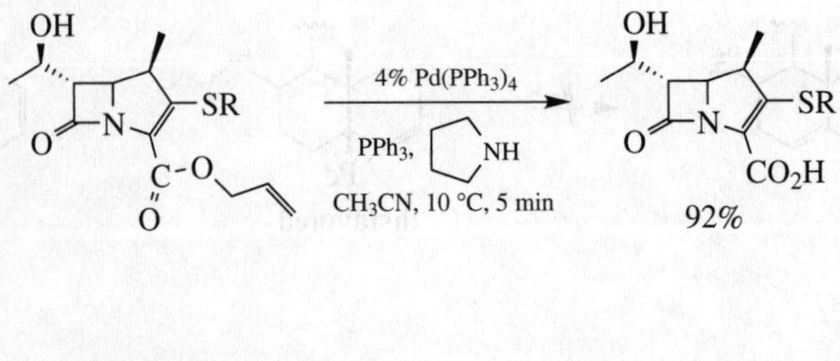

The most spectacular example of the utility of this protecting group strategy was the removal of 104 allylic protecting groups from NH_2 and phosphate moieties in a 60-mer oligonucleotide in almost 100% overall yield in a single palladium-catalyzed reaction[250].

Deprotection of Oligodeoxyribonucleotides: Allyl-AOC Method[250]

The controlled pore glass (CPG) supports binding the protected oligomer were washed with THF (1.0 mL) and dried *in vacuo*. To the resulting polymer were added $P(C_6H_5)_3$, $Pd_2(dba)_3$–$CHCl_3$, and a 1.2 M solution of $C_4H_9NH_2$–HCOOH (1 : 1) in THF. After being vigorously shaken with a mixer, the mixture was heated at 50 °C for 0.5–1 h. After the supernatant fluid was decanted, the resulting CPG supports were washed successively with THF (1 mL × 2), acetone (1 mL × 2), a 0.1 M sodium *N,N*-diethyl-dithiocarbamate (ddtc) aqueous solution (pH 9.7, 1 mL, 15 min), acetone (1 mL × 3), and water (1 mL). The ddtc washing was repeated once more. The cleaned CPG supports were treated with concentrated NH_4OH (1.5 mL) at 25 °C for 2 h to afford the fully deprotected oligo-DNA. Reaction of the protected 32-mer with $Pd_2(dba)_3$–$CHCl_3$ (0.140 g, 0.135 mmol), $P(C_6H_5)_3$ (0.360 g, 1.37 mmol), and $C_4H_9NH_2$–HCOOH in THF (6.7 mL, 8.04 mmol) gave 182 OD_{260} (optical density at 260 nm) of crude material. Reaction of the protected 43-mer with $Pd_2(dba)_3CHCl_3$ (0.190 g, 0.184 mmol), $P(C_6H_5)_3$ (0.490 g, 1.87 mol) and $C_4H_9NH_2$–HCOOH in THF (9.1 mL, 10.9 mmol) gave 256 OD_{260} of crude material. Reaction of the protected 60-mer (0.58 μmol) with $Pd_2(dba)_3$–$CHCl_3$ (156 mg, 0.151 mmol), $P(C_6H_5)_3$ (0.400 g, 1.53 mmol), and $C_4H_9NH_2$–HCOOH in THF (7.4 mL, 8.88 mmol) gave 209 OD_{259} of crude material.

4.3.6   Palladium(0)-catalyzed Allyl Complex Insertion Processes

Palladium(0) complexes catalyze a number of very useful cyclization reactions of allyl acetates having remote unsaturation, via oxidative addition/insertion processes (Scheme 17)[251]. These have been termed 'metalloene' reactions although they are probably not mechanistically related to the classic organic ene reaction. The reaction proceeds under very mild conditions and is highly stereoselective. Remarkably, acetic acid promotes the reaction in an unspecified manner. The mechanism is not known, and the one shown in Scheme 17 is speculative and based on known behavior of simpler systems. The requisite precursors for these cyclizations are almost invariably made quite directly using the palladium-catalyzed allylic alkylation reactions discussed in Section 4.3.2.

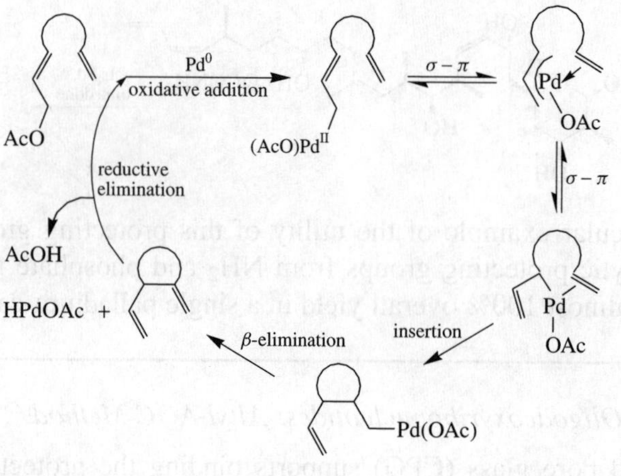

Scheme 17. Palladium(0)-catalyzed allyl insertion processes.

As initially disclosed, the process was terminated by β-hydride elimination[252] and resulted in the formation of a single ring and an exocyclic double bond. However, interception of the σ-alkylpalladium(II) intermediate prior to β-elimination has been utilized extensively. Interception by transmetalation from tin[253] or boron[254] results in simple alkylation of that position:

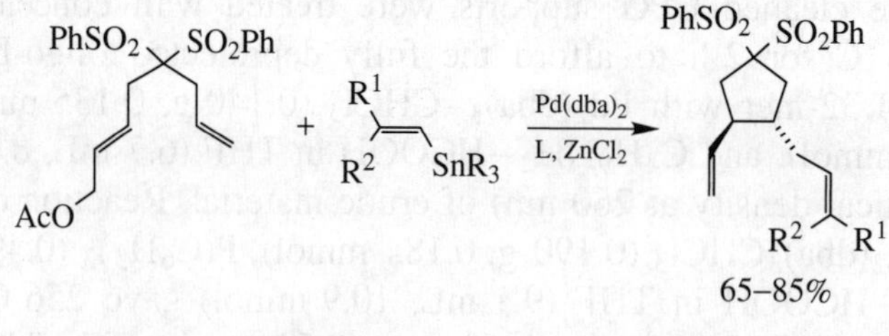

Preparation of 1,4-Diene Disulfone[252]

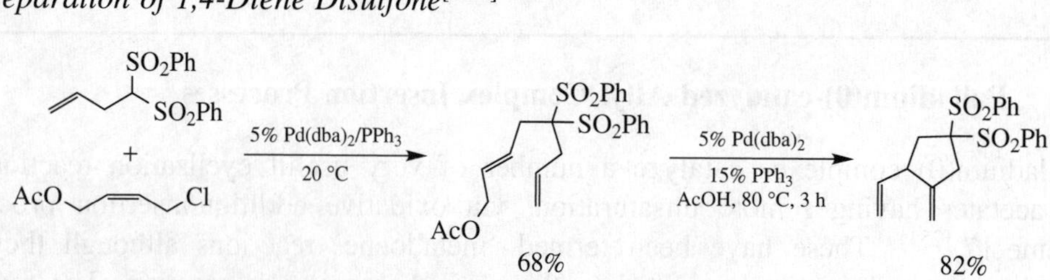

NaH (60% in mineral oil, 0.92 g, 22.9 mmol) was added in portions to a stirred solution of allyl disulfone (7.0 g, 20.8 mmol) in dry THF (90 mL) at 0 °C under argon. After stirring the mixture for 60 min at 25 °C, Pd(dba)$_2$ (0.60 g, 5 mol %) and triphenylphosphine (1.09 g, 20 mol %) were added and a solution of (Z)-1-acetoxy-4-chloro-2-butene (3.10 g, 20.9 mmol) in THF (20 mL) was rapidly dropped on to the reaction mixture. After stirring overnight (14 h) under argon, the reaction was quenched with water (200 mL) and the aqueous layer extracted with diethyl ether (4 × 100 mL). The combined organic phases were washed with water and brine (50 mL each) and dried (MgSO$_4$) and the solvent was evaporated *in vacuo*. The residue was purified by flash chromatography (silica gel) using hexane/ethyl acetate (3 : 1) as eluent. The product was obtained as yellow oil that slowly afforded white crystals on standing at 25 °C (mp 67–68 °C; yield 6.35 g, 68%).

To a solution of the substrate (0.500 g, 1.12 mmol) in 5 mL of HOAc (Fluka, puriss. p.a.) was added Pd(dba)$_2$ (0.032 g, 5 mol %) and triphenylphosphine (0.0439 g, 15 mol %) and the mixture was stirred at 80 °C under argon for 2 h. For work-up the reaction mixture was diluted with diethyl ether (100 mL) and extracted with water (40 mL), the organic layer was separated, washed with NaHCO$_3$ (10% in water, 30 mL), water and brine (30 mL each) and dried (MgSO$_4$) and the solvent was evaporated *in vacuo*. The residue was subjected to column chromatography (silica gel) using hexane/ethyl acetate (5 : 1) as eluent. The product (0.351 g, 81%) was obtained as a white crystalline solid (mp 113–114 °C; the melting point did not change on recrystallization of the product from ethanol).

With adjacent alkene groups, insertion followed by β-hydride elimination ensues forming two rings[255,256]:

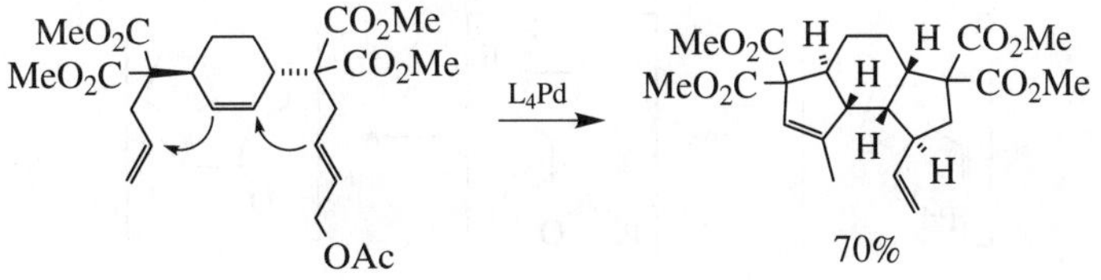

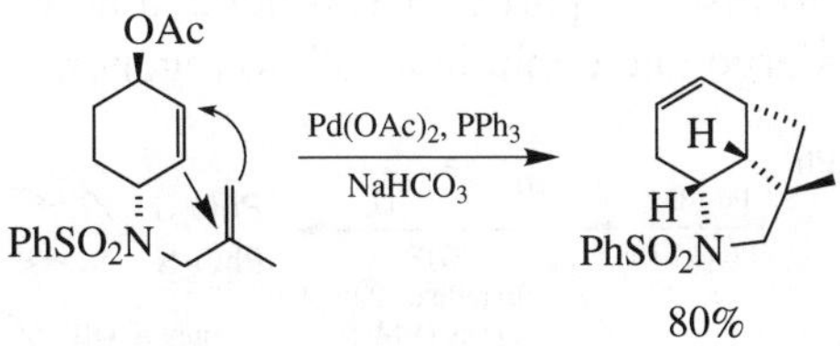

Even more facile is trapping with carbon monoxide, which inserts more readily than an alkene, and leads to terminal esters[257]:

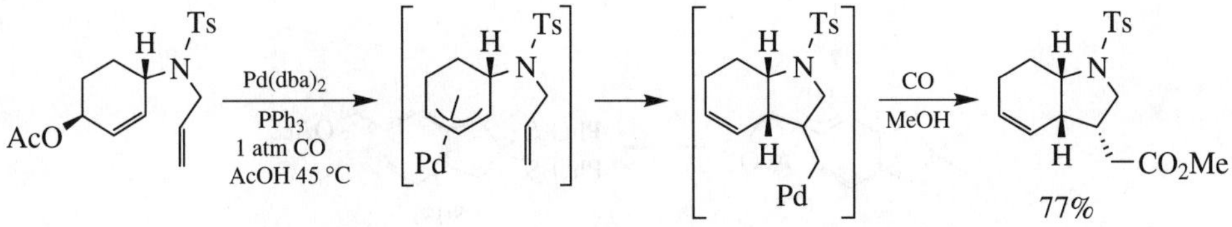

If the olefin group of the original allyl system is sterically accessible after cyclization/carbon monoxide insertion, it can also become involved[258]:

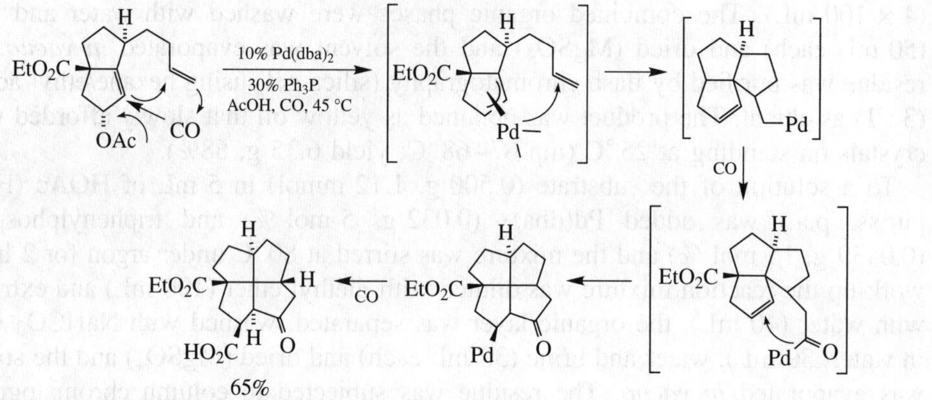

65%

Multiple insertions are similarly accessible by using alkynes as the initial insertion site[252]:

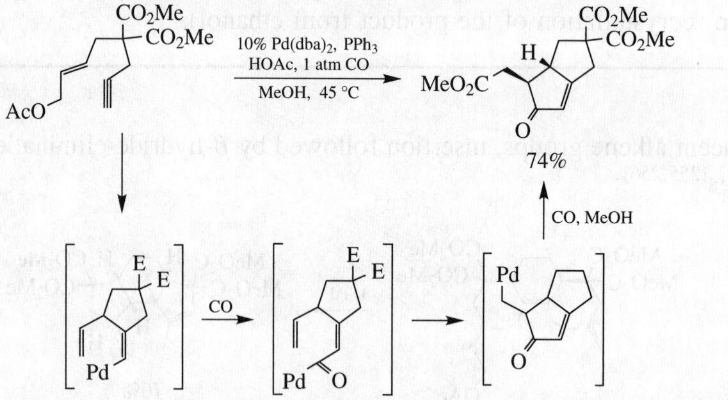

74%

When the initial alkene to insert is part of a 1,3-diene, a new π-allylpalladium complex is generated, which can undergo nucleophilic attack to regenerate an allylic acetate[259,260]:

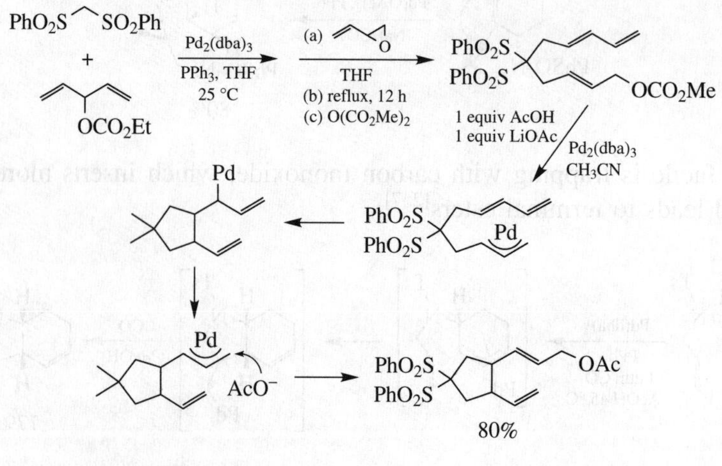

80%

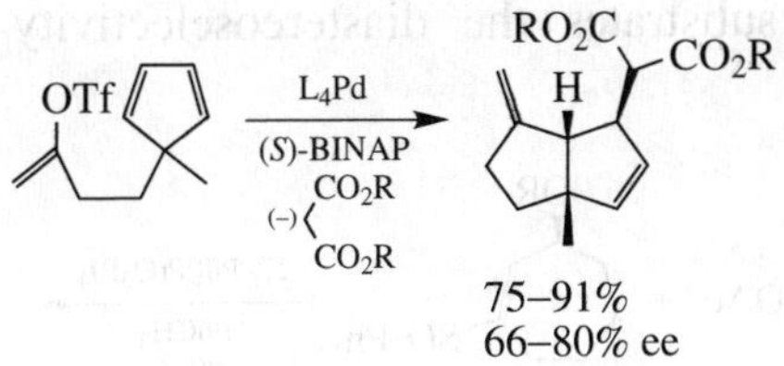

RO₂C CO₂R structure reaction:

OTf → L₄Pd, (S)-BINAP, (−)CO₂R CO₂R

75–91%
66–80% ee

4.3.7 Palladium(0)-catalyzed Cycloaddition Reactions via Trimethylene Methane Intermediates

Under appropriate conditions, π-allylpalladium complexes are electrophilic at the terminal positions, and undergo attack by a wide range of nucleophiles. By devising allyl substrates having a substituted methylene group that can generate, or at least stabilize, negative charge, a 1,3-dipolar system capable of a wide range of cycloaddition reactions is generated. The most extensively studied and exploited of these is derived from the palladium(0)-catalyzed reaction of bifunctional silyl acetate[261]:

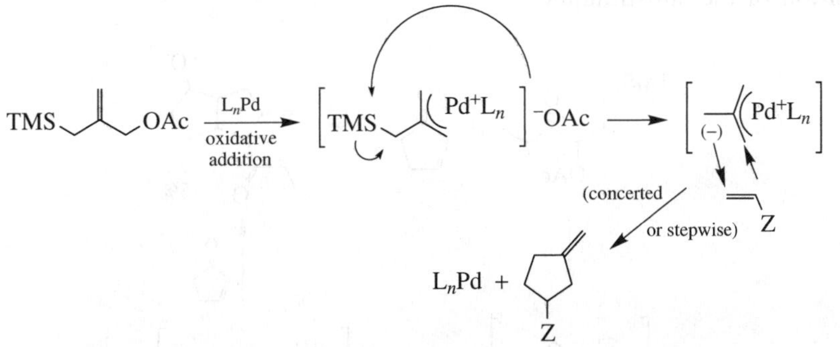

TMS—OAc →(LₙPd, oxidative addition) [TMS—(Pd⁺Lₙ)]⁻OAc → [(−)—(Pd⁺Lₙ)] →(concerted or stepwise, Z)

LₙPd + (methylenecyclopentane with Z)

The reaction is believed to proceed by oxidative addition of the allyl acetate to the palladium(0) catalyst forming a cationic π-allyl complex. Displacement of the adjacent silyl group by the acetate produces the zwitterionic intermediate, which undergoes reaction with (only) electron-deficient alkenes. The reaction is highly stereoselective with *trans*-alkenes, but some loss of alkene geometry occurs with *cis*-alkenes, suggesting a stepwise process[262]:

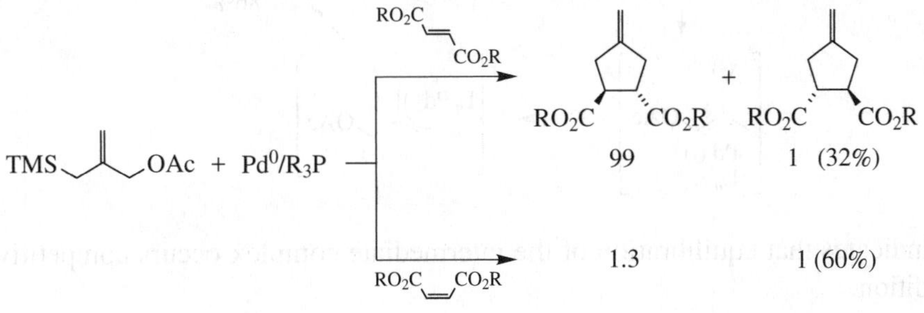

TMS—OAc + Pd⁰/R₃P →(RO₂C / CO₂R) products:

RO₂C''''CO₂R (99) + RO₂C''''CO₂R (1) (32%)

(RO₂C—CO₂R) → 1.3 : 1 (60%)

With optically active substrates the diastereoselectivity ranges from fair to excellent[263,264]:

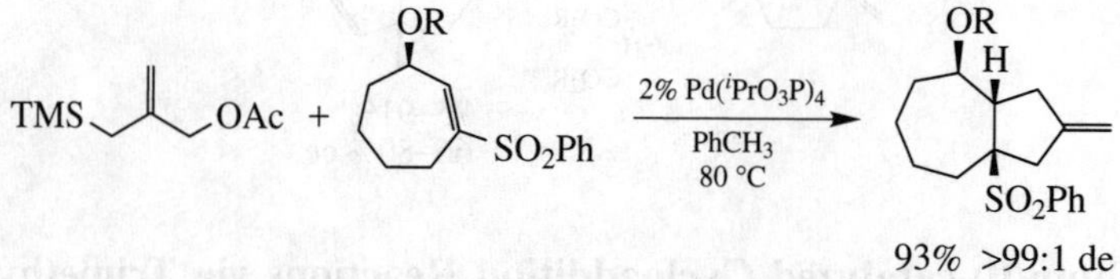

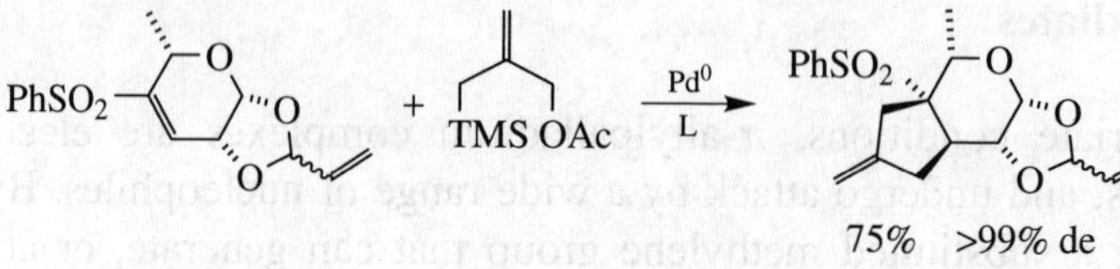

93% >99:1 de

75% >99% de

With unsymmetrical allylic acetate precursors, regioselective coupling occurs with both electron-withdrawing and electron-donating[265] substituents on the trimethylene methane fragment, the site of coupling always being the substituted terminus, regardless of the initial position of the substituents:

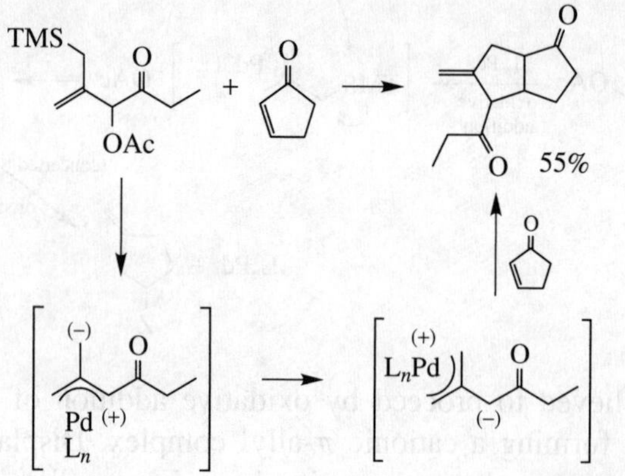

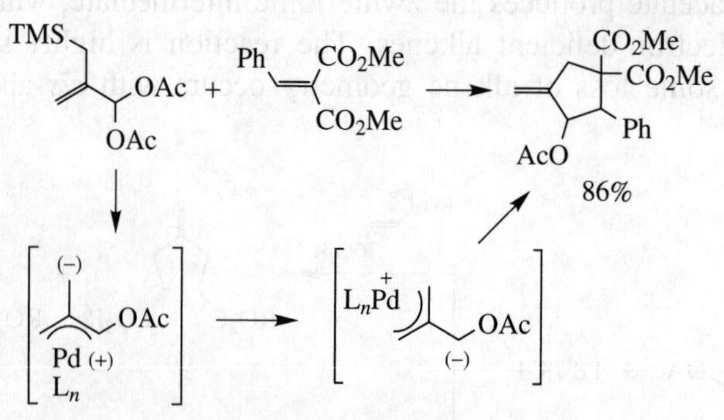

This indicates that equilibration of the intermediate complex occurs competitively with cycloaddition.

Preparation of Dimethyl-(±)-3a(R*),8(R*),8a(S*)-8-methyl-3-methylene-4-oxo-1,2,4,5,6,7,8,8a-octahydroazulene-1,1-dicarboxylate[266]

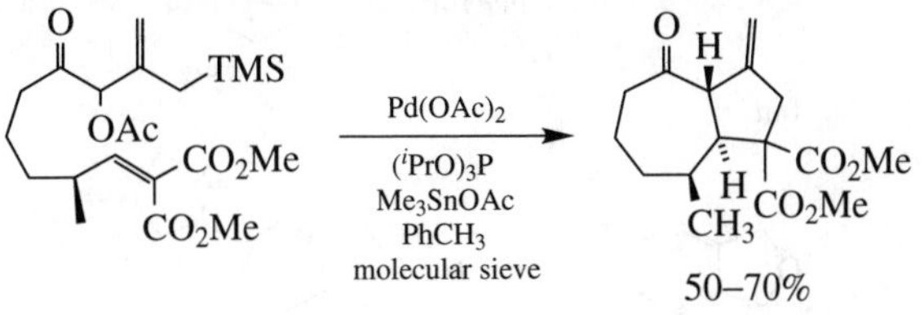

50–70%

To a mixture of the malonate (0.503 g, 0.118 mmol), trimethyltin acetate (0.0026 g, 0.012 mmol), and 3 Å powdered molecular sieves (0.025 g) in 0.7 mL of toluene was added 0.5 mL of a toluene solution containing palladium acetate (0.0013 g, 0.0059 mmol) and triisopropyl phosphite (0.0037 g, 0.018 mmol). The mixture was heated at reflux for 2 h and then passed through a plug of silica gel. Flash chromatography (15 : 4 : 1 hexanes/ether/benzene) afforded 0.0175 g (50%) of the cycloadduct as an oil (R_f 0.44 (1 : 1 ether/hexanes)).

With pyrones both $3 + 2$ and $3 + 4$ cycloaddition occurs, depending primarily on the structure of the acceptor:

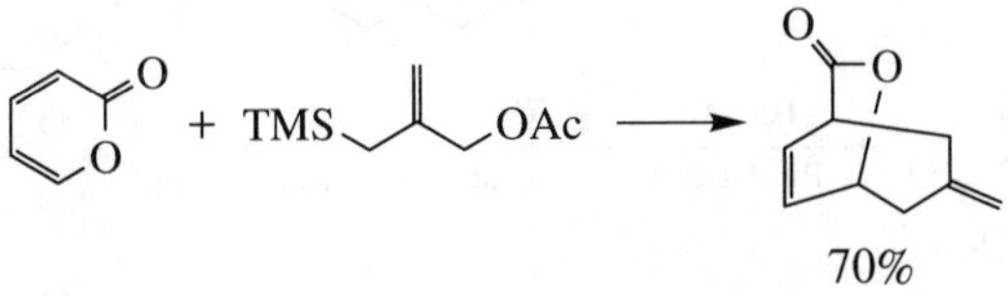

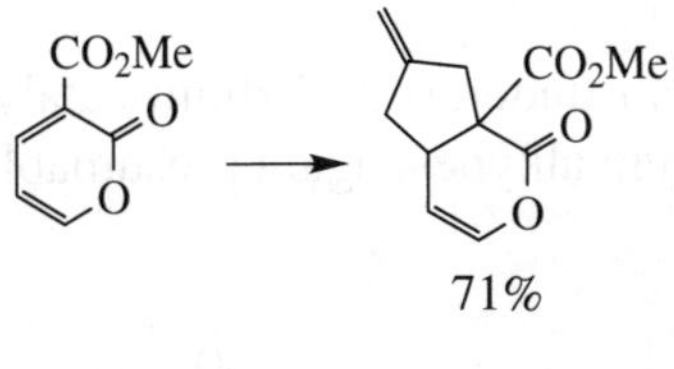

70%

71%

Tropone undergoes a $6 + 3$ cycloaddition in excellent yield[267]:

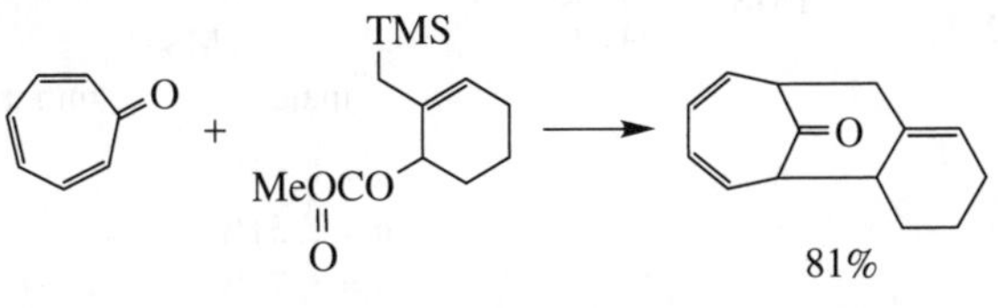

81%

Imines[268] and aldehydes[269] undergo this cycloaddition as well, in the presence of additives:

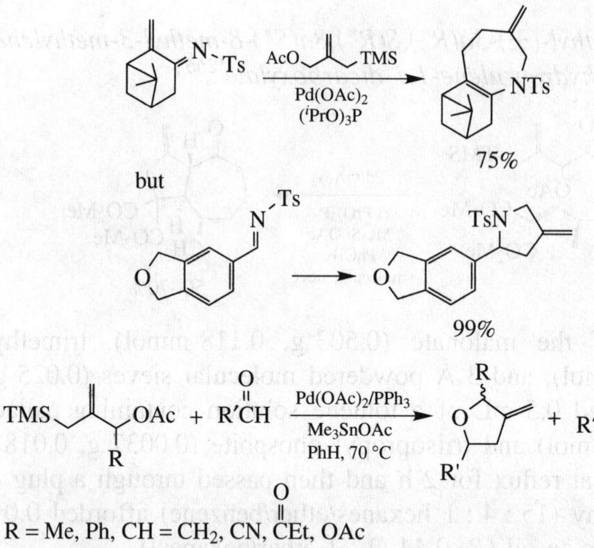

but

99%

R = Me, Ph, CH=CH₂, CN, CEt, OAc

The site of cycloaddition in the enone shown below can be controlled by altering reaction conditions[270]:

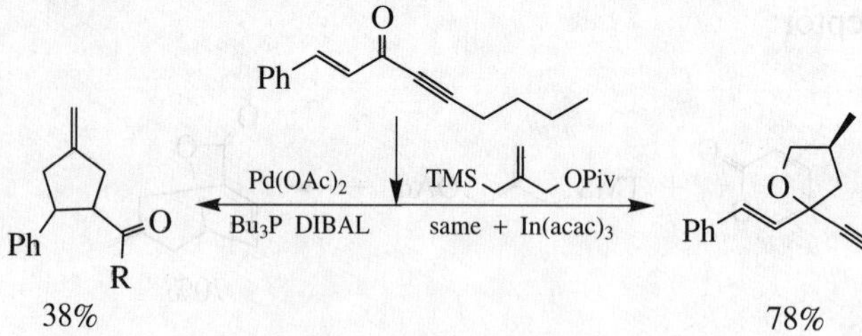

Methylenecyclopropanes also undergo palladium-catalyzed 3 + 2 cycloaddition reactions to electron-poor alkenes and alkynes, again presumably through trimethylenemethane complex intermediates[271]:

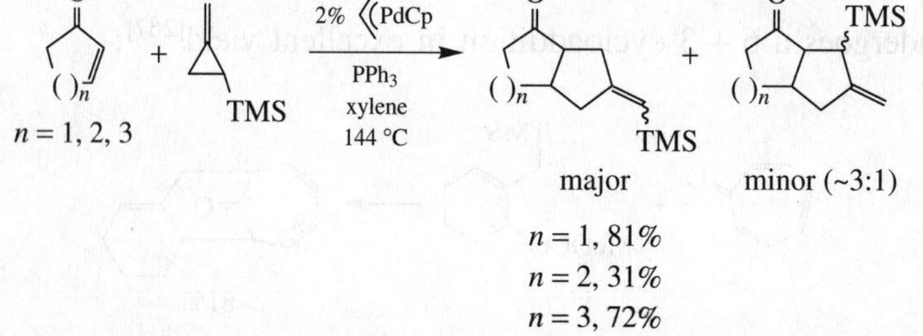

Intramolecular processes are particularly appealing because they are entropically favored and often occur with high regiochemical control[272,273]:

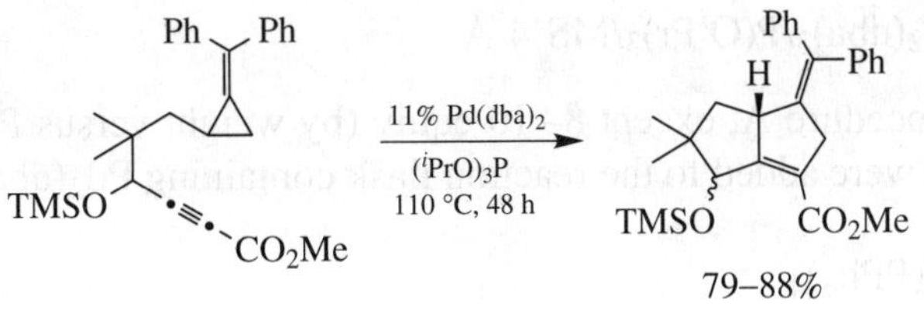

79–88%

With diastereoisomerically pure methylene cyclopropanes the reaction was stereospecific and proceeded with retention of configuration at the chiral carbon center at which reaction occured[274]:

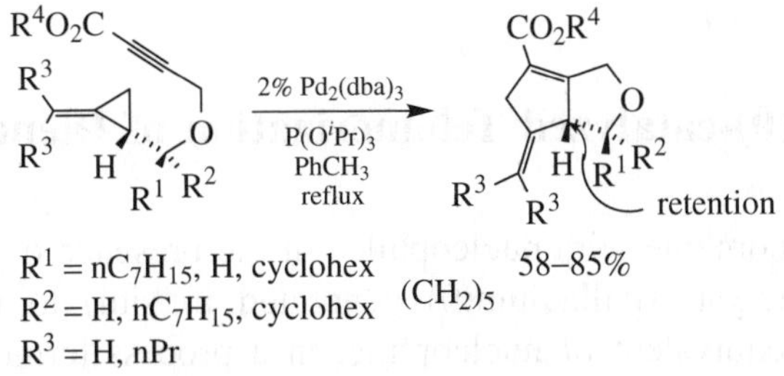

$R^1 = nC_7H_{15}$, H, cyclohex
$R^2 = H, nC_7H_{15}$, cyclohex (CH$_2$)$_5$
$R^3 = H, nPr$

58–85%

With alkene rather than alkyne acceptors, the reaction was stereospecific (retention) with respect to the preexisting cyclopropane stereocenter, and the configuration at the α-position determined the stereochemistry of the newly formed ring junction[275].

General Procedures for Palladium-catalyzed Methylene Cyclopropane Cycloadditions[275]

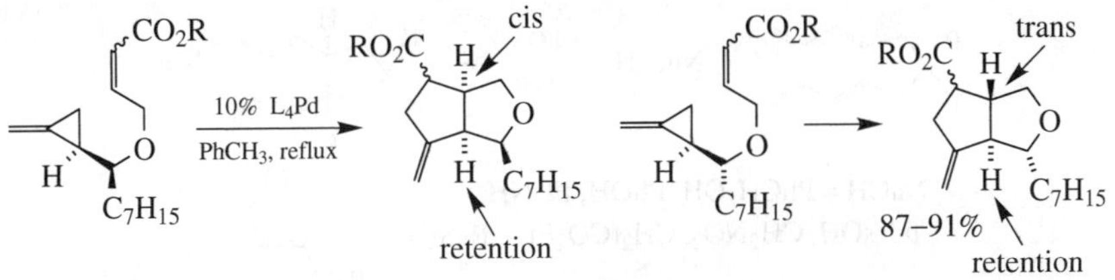

87–91%

Procedure A: Pd$_2$(dba)$_3$/P(O^{i}Pr)$_3$

A flame-dried round bottomed flask, equipped with a condenser was charged with Pd$_2$(dba)$_3$ and P(O^{i}Pr)$_3$ (P/Pd = 2 : 1 mol ratio). Toluene (desulfurized and freshly distilled over sodium) was introduced via a cannula followed by a solution of the substrate in toluene. The mixture was heated to 110 °C and stirred vigorously until the reaction was complete. Toluene was removed under vacuum and the crude mixture was purified by column chromatography. Elution of dibenzylideneacetone (dba) with toluene was followed by further elution using a mixture of ethyl acetate in hexanes to give the pure cycloadduct.

Procedure B: $Pd_2(dba)_3/P(O^iPr)_3/MS$ 4 Å

Identical to procedure A, except 8–10 equiv (by weight versus $Pd_2(dba)_3$) of 4 Å molecular sieves were added to the reaction flask containing $Pd_2(dba)_3$ and $P(O^iPr)_3$.

Procedure C: $Pd(PPh_3)_4$

Identical to procedure A, except $Pd(PPh_3)_4$ was added to the reaction flask instead of $Pd_2(dba)_3$ and $P(O^iPr)_3$.

4.4   Palladium(0)-catalyzed Telomerization of Dienes

Conjugated dienes combine with nucleophiles in the presence of palladium acetate/triphenyl phosphine catalysts (palladium(0) generated in situ) to produce dimers with incorporation of one equivalent of nucleophile, in a process termed telomerization[276]. Although the mechanism of this process has not been studied in detail, it is thought to involve reductive dimerization of two diene units, followed by addition of the nucleophile to the bis-π-allylpalladium complex thus formed (Scheme 18). Although this process has been long known and extensively studied, it has found little application in complex organic synthesis, except to provide very early starting materials. However, the recent development of an intramolecular version of this process promises to be of broader use[277–281]:

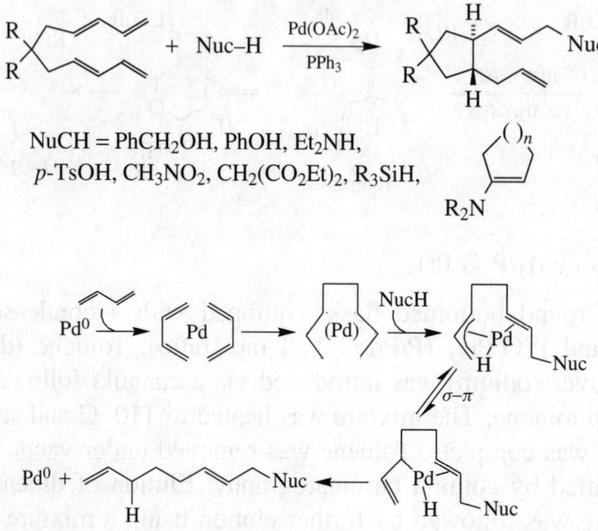

Scheme 18. Palladium(0)-catalyzed telomerization of dienes.

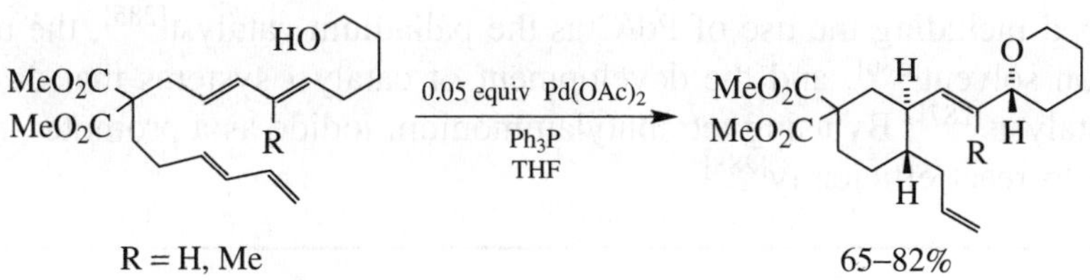

R = H, Me 65–82%

In the absence of nucleophiles, the bis-diene substrates undergo cycloisomerization[282]:

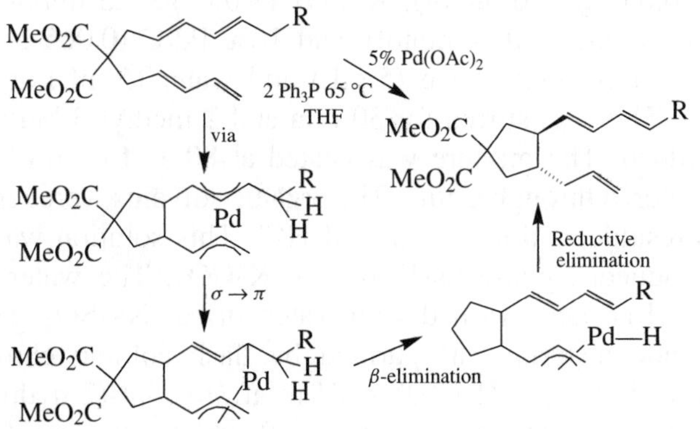

4.5 Palladium(0)/Copper(I)-catalyzed (Sonogashira) Coupling of Aryl and Alkenyl Halides with Terminal Alkynes

Terminal alkynes couple with aryl and alkenyl halides in the presence of palladium catalysts, copper(I) iodide, and a secondary or tertiary amine[283]. The reaction conditions are very mild and consist of simply stirring the alkyne and halide for a few hours at room temperature in the amine as solvent. The most common catalysts used are Pd(PPh$_3$)$_4$ or PdCl$_2$(PPh$_3$)$_2$, which is rapidly reduced to palladium(0) by the amine solvent. The mechanism is not known, nor is the role of the copper iodide understood. It is likely that copper acetylides are formed and transmetalate to the R–PdX formed by oxidative addition (Scheme 19).

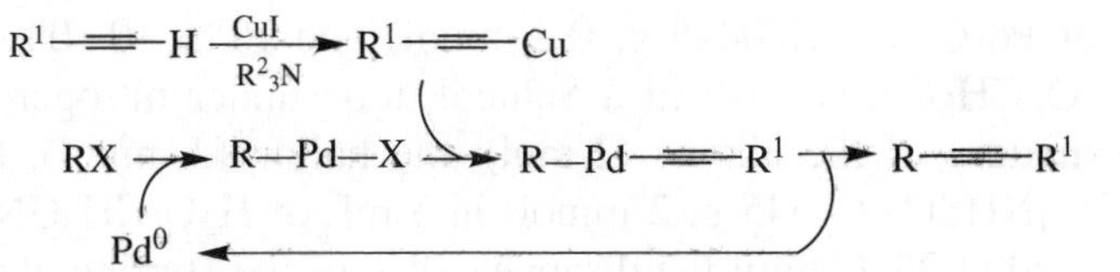

Scheme 19. Palladium/copper-catalyzed coupling of halides and alkynes.

As this process has become more important in the synthesis of highly functionalized organic compounds (e.g., enediyne antibiotics[284]), modified catalyst conditions have been developed including the use of Pd/C as the palladium catalyst[285], the use of water as the reaction solvent[286], and the development of catalyst systems that do not require copper co-catalysts[287]. By using tetrabutylammonium iodide as a promoter, aryl triflates can be made to react efficiently[288].

Aryl Halide–Alkyne Coupling Using Pd/C Catalysis[285]

3-Iodoaniline (0.657 g, 3.0 mmol), K_2CO_3 (1.035 g, 7.5 mmol), CuI (0.023 g, 0.12 mmol), PPh$_3$ (0.063 g, 0.24 mmol), and 10% Pd/C (0.065 g, 0.06 mmol Pd) were mixed in 1,2-dimethoxyethane (5 mL) and water (5 mL) at 25 °C under an argon atmosphere. This was stirred for 30 min and 2-methyl-3-butyn-2-ol (0.630 g, 7.5 mmol) was added. The mixture was heated at 80 °C for 16 h, cooled to room temperature and filtered through Celite. The organic solvents were removed *in vacuo*, and the aqueous residue acidified with 1 M HCl. This solution was extracted with toluene, and the aqueous phase basified with K_2CO_3. The water layer was then extracted with ethyl acetate, washed with water, dried (Na_2SO_4) and concentrated *in vacuo*. The crude material was purified by flash column chromatography on silica gel eluting with 95 : 5 CH_2Cl_2/MeOH to afford 3-[4-(2-hydroxy-2-methyl-3-butynyl)]aminobenzene (0.410 g, 78%) as an off-white solid (mp 119–120 °C).

Aryl Halide–Alkyne Coupling in Aqueous Solvent[286]

In a typical experiment the mixture of *p*-iodonitrobenzene (0.249 g, 1 mmol), phenyl acetylene (0.133 g, 1.3 mmol), $PdCl_2(PPh_3)_2$ (0.007 g, 0.01 mmol), CuI (0.0038 g, 0.02 mmol), Bu$_3$N (0.0185 g, 0.1 mmol) and 1 M K_2CO_3 (aq) (2 mL, 2 mmol) was stirred for 1 h at ambient temperature under argon. The product was extracted with ether and recrystallized from MeOH to give pure sample of 4-nitrotolan.

Aryl Halide–Alkyne Coupling Without Copper Co-catalyst[287]

A solution of $Pd(OAc)_2$ (0.0445 g, 0.2 mmol) and PPh$_3$ (0.1049 g, 0.4 mmol) in 5 mL of H_2O/CH_3CN (1 : 10) in a Schlenk tube under nitrogen was added to a well-stirred mixture of the alkyne (2 mol), the halide (4 mmol), Et$_3$N (700 µL, 5 mmol) and Bu$_4$NHSO$_4$ (0.645 g, 2 mmol) in 5 mL of H_2O/CH_3CN (1 : 10). Stirring was continued at 25 °C until the disappearance of the starting alkyne (observed by TLC). The mixture was then hydrolysed by 10 mL H_2O and extracted with 3 × 30 mL diethyl ether. Removing of solvent *in vacuo*, followed by column chromatography gave the expected product.

This coupling process has been used to synthesize a very wide range of complex molecules, from enediynes[289] to enyne-containing polyfurans[290] to phytochromes[291]:

Intramolecular versions are also efficient[292]:

This process has proved particularly useful for making oligomeric aryl alkynes[293] including dendrimers[294]:

$$C_{1134}H_{1146}$$
molecular weight 14776

It should find expanded use in the materials science area, based on these early successes.

References

[1] For an in depth treatment of the various processes catalyzed by palladium see: (a) Abel, E. W.; Stone, F. G. A.; Wilkinson, G.; (Eds), *Comprehensive Organometallic Chemistry II*, Vol. *12*, Pergamon, London, **1995**. (b) For a detailed (including procedures) but dated treatment of this topic see: Heck, R. F.; *Palladium Reagents in Organic Synthesis*, Academic Press, London, **1985**.

[2] Coulson, D. R; *Inorg. Synth.* **1972**, *13*, 121.

[3] Ukai, T.; Kawazawa, H.; Ishii, Y.; Bonnett, J. J.; Ibers, J. A.; *J. Organomet. Chem.* **1974**, *65*, 253.

[4] Amatore, C.; Jutland, A.; Khalil, F.; M'Barki, M. A.; Mottier, L.; *Organometallics* **1993**, *12*, 3168.

[5] Herrmann, W. A.; Brossmer, C.; Ofele, K.; Reisinger, C.-P.; Priermeier, T.; Beller, M.; Fischer, H.; *Angew. Chem. Int. Ed. Engl.* **1995**, *34*, 1844.

[6] Beller, M.; Riermeier, T. H.; *Tetrahedron Lett.* **1996**, *37*, 6535.

[7] Reetz, M. T.; Breinbauer, R.; Wanninger, K.; *Tetrahedron Lett.* **1996**, *37*, 4499.

[8] Beller, M.; Fischer, H.; Kühlein, K.; Reisinger, C.-P.; Herrmann, W. A.; *J. Organomet. Chem.* **1996**, *520*, 257.

[9] Tsuji, J.; *Synthesis* **1984**, 369.

[10] Celimene, C.; Dhimane, H.; Saboureau, A.; Lhornmet, G.; *Tetrahedron: Asymmetry* **1996**, *7*, 1585.

[11] Oikawa, M.; Ueno, T.; Oikawa, H.; Ichihara, A.; *J. Org. Chem.* **1995**, *60*, 5048.

[12] Pellissier, H.; Michellys, P.-Y.; Santelli, M.; *Tetrahedron Lett.* **1994**, *35*, 6481.

[13] Kang, S.-K.; Jung, K.-Y.; Chung, J.-N.; Namkoong, E.-Y.; Kim, T.-H.; *J. Org. Chem.* **1995**, *60*, 4678.

[14] For a review of palladium-catalyzed oxypalladation of alkenes see: Hosokawa, T.; Murahashi, S.-I.; *J. Synth. Org. Chem. Jpn.* **1995**, *53*, 1009, in English.

[15] Semmelhack, M. F.; Kim, C. R.; Dobler, W.; Meier, M.; *Tetrahedron Lett.* **1989**, *30*, 4925.

[16] Hosokawa, T.; Murahashi, S.-I.; Sonoda, A.; *J. Organomet. Chem.* **1989**, *370*, C13.

[17] van Benthem, R. A. T. M.; Hiemstra, H.; Michels, J. J.; Speckamp, W. N.; *J. Chem. Soc., Chem. Commun.* **1994**, 357.

[18] (a) Tenaglia, A.; Kammerer, F.; *Synlett* **1976**, 576; (b) Saito, S.; Hara, T.; Takahashi, N.; Hirai, M.; Moriwake, T.; *Synlett* **1992**, 237.

[19] Kraus, G. A.; Li, J.; Gordon, M. S.; Jensen, J. H.; *J. Org. Chem.* **1995**, *60*, 1154; based on Semmelhack, M. F.; Zask, A.; *J. Am. Chem. Soc.* **1983**, *105*, 2034.

[20] Gracza, T.; Jäger, V; *Synthesis* **1994**, 1359.

[21] Semmelhack, M. F.; Epa, W. R.; *Tetrahedron Lett.* **1993**, *34*, 7205.

[22] Larock, R. C.; Lee, N. H.; *J. Am. Chem. Soc.* **1991**, *113*, 7815.

[23] Korte, D. E.; Hegedus, L. S.; Wirth, R. K.; *J. Org. Chem.* **1977**, *42*, 1329.

[24] Minami, T.; Nishimoto, A.; Hanaoka, M.; *Tetrahedron Lett.* **1995**, *36*, 9505.

[25] Åkermark, B.; Bäckvall, J.-E.; Hegedus, L. S.; Zetterberg, K.; Siirala-Hansen, K.; Sjöberg, K.; *J. Organomet. Chem.* **1974**, *72*, 127.

[26] Hegedus, L. S.; McKearin, J. M.; *J. Am. Chem. Soc.* **1982**, *104*, 2444.

[27] (a) Hegedus, L. S.; Allen, G. F.; Bozell, J. J.; Waterman, E. L.; *J. Am. Chem. Soc.* **1978**, *100*, 5800; (b) for a review on the use of palladium in the synthesis and functionalization of indoles see: Hegedus, L. S.; *Angew. Chem., Int. Ed. Engl.* **1988**, *27*, 1113.

[28] Harrington, P. J.; Hegedus, L. S.; *J. Org. Chem.* **1984**, *49*, 2657.

[29] Irie, K.; Isaka, T.; Iwata, Y.; Yanai, Y.; Nakamura, Y.; Korzumi, F.; Ohigashi, H.; Wender, P. A.; Satomi, Y.; Nishmo, H.; *J. Am. Chem. Soc.* **1996**, *118*, 10733.

[30] Weider, P. R.; Hegedus, L. S.; Asada, H.; Andrea, S. V. D.; *J. Org. Chem.* **1985**, *50*, 4276.

[31] Larock, R. C.; Hightower, T. R.; Hasvold, L. A.; Peterson, K. P.; *J. Org. Chem.* **1996**, *61*, 3584.

[32] Hosokawa, T.; Takano, M.; Kuroki, Y.; Murahashi, S.-I.; *Tetrahedron Lett.* **1992**, *33*, 6643.

[33] (a) Kondo, Y.; Shiga, F.; Murata, N.; Sakumoto, T.; Yamanaki, H.; *Tetrahedron* **1994**, *50*, 11803. For related approaches to indoles from anilines and alkynes see: (b) Jeschke, T.; Wensbo, D.; Annby, U.; Gronowitz, S.; *Tetrahedron Lett.* **1993**, *34*, 6471; (c) Larock, R. C.; Yam, E. K.; *J. Am. Chem. Soc.* **1991**, *113*, 6689.

[34] Jacobi, P. A.; Brielmann, H. L.; Hauck, S. I.; *J. Org. Chem.* **1996**, *61*, 5013.

[35] (a) Hegedus, L. S.; Williams, R. E.; Hayashi, T.; *J. Am. Chem. Soc.* **1980**, *102*, 4973; (b) Hegedus, L. S.; Darlington, W. E.; *J. Am. Chem. Soc.* **1980**, *102*, 4980.

[36] (a) Hegedus, L. S.; Wieber, G.; Michalson, E.; Åkermark, B.; *J. Org. Chem.* **1989**, *54*, 4649; (b) Montgomery, J.; Wieber, G.; Hegedus, L. S.; *J. Am. Chem. Soc.* **1990**, *112*, 6255.

[37] Masters, J. J.; Hegedus, L. S.; Tamariz, J.; *J. Org. Chem.* **1991**, *56*, 5666.

[38] (a) Masters, J. J.; Hegedus, L. S.; *J. Org. Chem.* **1993**, *58*, 4547. For related intramolecular insertion of alkenes see: (b) Laidig, G. J.; Hegedus, L. S.; *Synthesis* **1995**, 527.

[39] (a) Toyota, M.; Nishikawa, Y.; Motoki, K.; Yoshida, N.; Fukumoto, K.; *Tetrahedron* **1993**, *49*, 11189; based on early work of Kende and co-workers: (b) Kende, A. S.; Roth, B.; Sanfilippo, P. J.; Blacklock, T. J.; *J. Am. Chem. Soc.* **1982**, *104*, 5808; (c) Kende, A. S.; Roth, B.; Sanfilippo, P. J.; *J. Am. Chem. Soc.* **1982**, *104*, 1784.

[40] Review: Overman, L. E.; *Angew. Chem., Int. Ed. Engl.* **1984**, *23*, 579.

[41] Clayden, J.; Warren, S.; *J. Chem. Soc., Perkin I* **1993**, 2913.

[42] (a) Panek, J. S.; Yang, K. M.; Solomon, J. S.; *J. Org. Chem.* **1993**, *58*, 1003. For earlier examples see: (b) Grieco, P. A.; Takegawa, T.; Bongers, S. L.; Tanaka, H.; *J. Am. Chem. Soc.* **1980**, *102*, 7587; (c) Martes, P.; Perfetti, P.; Zahra, J.-P.; Waegell, B.; *Tetrahedron Lett.* **1991**, *32*, 765.

[43] Nakazawa, M.; Sakomoto, Y.; Takahashi, T.; Tamooka, K.; Ishikawa, K.; Nakai, T.; *Tetrahedron Lett.* **1993**, *34*, 5923.

[44] Ikariya, T.; Ishikawa, Y.; Hirai, K.; Yoshikawa, S.; *Chem. Lett.* **1982**, 1815.

[45] (a) Tamru, Y.; Kagotani, M.; Yoshida, Z.; *J. Org. Chem.* **1980**, *45*, 5221; (b) Tamru, Y.; Kagotani, M.; Yoshida, Z.; *Tetrahedron Lett.* **1981**, 4245; (c) Garin, J.; Melendez, E.; Merchan, F. L.; Tejero, T.; Uriel, S.; Ayestaran, J.; *Synthesis* **1991**, 147.

[46] Auburn, P. R.; Whelan, J.; Bosnich, B.; *Organometallics* **1986**, *5*, 1533.

[47] Grigg, R.; Markandu, J.; *Tetrahedron Lett.* **1991**, *32*, 279.

[48] Overman, L. E.; Renaldo, A. F.; *J. Am. Chem. Soc.* **1990**, *112*, 3945.

[49] Bluthe, N.; Malacria, M.; Gore, J.; *Tetrahedron Lett.* **1983**, *24*, 1157.

[50] (a) Trost, B. M.; *Acc. Chem. Res.* **1990**, *23*, 34 and references cited therein; (b) Trost, B. M.; Hipskind, P. A.; Chung, J. Y. L.; Chan, C.; *Angew. Chem., Int. Ed. Engl.* **1989**, *28*, 1502.

[51] Trost, B. M.; Romero, D. L.; Rise, F.; *J. Am. Chem. Soc.* **1994**, *116*, 4268.

[52] Trost, B. M.; Tanourz, G. J.; Lautens, M.; Chan, C.; MacPherson, D. T.; *J. Am. Chem. Soc.* **1994**, *116*, 4255.

[53] Trost, B. M.; Krisake, M. J.; *J. Am. Chem. Soc.* **1996**, *118*, 233.

[54] Trost, B. M.; Li, Y.; *J. Am. Chem. Soc.* **1996**, *118*, 6625.

[55] Trost, B. M.; Fleitz, F. J.; Watkins, W. J.; *J. Am. Chem. Soc.* **1996**, *118*, 5146.

[56] Trost, B. M.; Shi, Y.; *J. Am. Chem. Soc.* **1991**, *113*, 701.

[57] Trost, B. M.; Shi, Y.; *J. Am. Chem. Soc.* **1992**, *114*, 791.

[58] Trost, B. M.; Yanai, M.; Hoogsteen, K.; *J. Am. Chem. Soc.* **1993**, *115*, 5294.

[59] Trost, B. M.; Hashmi, A.S.K; *Angew. Chem., Int. Ed. Engl.* **1993**, *32*, 1085.

[60] Trost, B. M.; Hashmi, A.S.K; *J. Am. Chem. Soc.* **1994**, *116*, 2183.

[61] For reviews on orthopalladation see: (a) Bruce, M. I.; *Angew. Chem., Int. Ed. Engl.* **1977**, *16*, 73; (b) Rehand, J.; Pfeffer, M.; *Coord. Chem. Rev.* **1976**, *18*, 327; (c) Omae, I.; *Chem. Rev.* **1979**, *79*, 287; (d) Omae, I.; *Coord. Chem. Rev.* **1980**, *32*, 235; (e) Ryabov, A. D.; *Synthesis* **1985**, 233.

[62] Brisdon, B. J.; Nair, P.; Dyke, S. F.; *Tetrahedron* **1981**, *37*, 173.

[63] Horino, H.; Inoue, N.; *J. Org. Chem.* **1981**, *46*, 4416.

[64] Review: Farina, V.; in *Comprehensive Organometallic Chemistry II*, Vol. *12*, (E. W. Abel, F. G. A. Stone, G. Wilkinson, Eds), Pergamon, Oxford, UK, **1995**, 161–240.

[65] Amatore, C.; Carre, E.; Jutland, A.; M'Barke, M.A.; *Organometallics* **1995**, *14*, 1818.

[66] Jutland, A.; Mosleh, A.; *Organometallics* **1995**, *14*, 1810.

[67] Tamao, K.; Sumitani, K.; Kiso, Y.; Zembayashi, M.; Fujioka, A.; Kodama S.-I.; Nakajima, I.; Minato, A.; Kumada, M.; *Bull. Chem. Soc. Jpn.* **1976**, *49*, 1958.

[68] Tamao, K.; Kodama, S.; Nakayima, I.; Kumada, M.; Minato, A.; Suzuki, K.; *Tetrahedron* **1982**, *38*, 3347.

[69] Dang, H. P.; Linstrumelle, G.; *Tetrahedron Lett.* **1978**, 191.

[70] Murahashi, S. I.; Yamamura, M.; Yanagesawa, K.; Mita, M.; Kondo, K.; *J. Org. Chem.* **1979**, *44*, 2408.

[71] (a) Hayashi, T.; Konishi, M.; Ito, H.; Kumada, M.; *J. Am. Chem. Soc.* **1982**, *104*, 4962. For a review see: (b) Sawamura, M.; Ito, Y.; *Chem. Rev.* **1992**, *92*, 857.

[72] Hayashi, T.; Niizuma, S.; Kamikawa, T.; Suzuki, N.; Hozumi, Y.; *J. Am. Chem. Soc.* **1995**, *117*, 9101.

[73] Kamikowa, T.; Nozumi, Y.; Hayashi, T.; *Tetrahedron Lett.* **1996**, *37*, 3161.

[74] Smith, A. B.; Qiu, Y.; Jones, D. R.; Kobayashi, K.; *J. Am. Chem. Soc.* **1995**, *117*, 12011.

[75] Takahashi, K.; Gunji, A.; Yanagi, K.; Miki, M.; *Tetrahedron Lett.* **1995**, *36*, 8055.

[76] Sakamoto, T.; Kondo, Y.; Murata, N.; Yamanaka, H.; *Tetrahedron* **1993**, *49*, 9713.

[77] Marquais, S.; Arlt, M.; *Tetrahedron Lett.* **1996**, *37*, 5491.

[78] Jabri, N.; Alexakis, A.; Normant, J. F.; *Tetrahedron Lett.* **1981**, *22*, 959.

[79] Dieter, R. K.; Li, S.; *Tetrahedron Lett.* **1995**, *36*, 3613.

[80] Schwartz, J.; Labinger, J.; *Angew. Chem., Int. Ed. Engl.* **1976**, *15*, 333.

[81] Okukado, N.; Van Horn, D. E.; Dlima, W.; Negishi, E.; *Tetrahedron Lett.* **1978**, 1027.

[82] Vincent, P.; Beaucourt, J. P.; Pichat, L.; *Tetrahedron Lett.* **1982**, *23*, 63.

[83] (a) Negishi, E.; Okukada, N.; King, A. O.; Van Horn, D. E.; Spiegel, B. I.; *J. Am. Chem. Soc.* **1978**, *100*, 2254; (b) Negishi, E.; *Acc. Chem. Res.* **1982**, *15*, 340.

[84] Baba, S.; Negishi, E.; *J. Am. Chem. Soc.* **1976**, *98*, 6729.

[85] Extensive review: Miyaura, N.; Suzuki, A.; *Chem. Rev.* **1995**, *95*, 2457.

[86] Narukawa, Y.; Nishi, K.; Onoue, H.; *Tetrahedron Lett.* **1996**, *37*, 2589.

[87] Humphrey, J. M.; Aggen, J. B.; Chamberlin, A. R.; *J. Am. Chem. Soc.* **1996**, *118*, 11759.

[88] Uenishi, J.-I.; Bean, J. M.; Armstrong, R. W.; Kishi, Y.; *J. Am. Chem. Soc.* **1987**, *109*, 4756.

[89] Wright, S. W.; Hageman, D. L.; McClure, L. D.; *J. Org. Chem.* **1994**, *59*, 6095.

[90] Jendralla, H.; Wagner, A.; Molrath, M.; Wunner, J.; *Liebigs Ann. Chem.* **1995**, 1253.

[91] Percec, V.; Chu, P.; Ungar, G.; Zhou, J.; *J. Am. Chem. Soc.* **1995**, *117*, 11441.

[92] Galda, P.; Rahalin, M.; *Synthesis* **1996**, 614.

[93] Keegstra, M. A.; De Feyter, S.; DeSchryver, F. C.; Müllen *Angew. Chem., Int. Ed. Engl.* **1996**, *35*, 774.

[94] Han, Y.; Walker, S. D.; Young, R. N.; *Tetrahedron Lett.* **1996**, *37*, 2703.

[95] Guiles, J. W.; Johnsen, S. G.; Murray, W. V.; *J. Org. Chem.* **1996**, *61*, 5169.

[96] Larhed, M.; Lindberg, G.; Hallberg, A.; *Tetrahedron Lett.* **1996**, *37*, 8219.

[97] Brown, S. D.; Armstrong, R. W.; *J. Am. Chem. Soc.* **1996**, *118*, 6331.

[98] Reviews: (a) Snieckus, V.; *Chem. Rev.* **1990**, *90*, 879; (b) Snieckus, V.; *Pure Appl. Chem.* **1994**, *66*, 2155.

[99] (a) Larsen, R. D.; King, A. O.; Chen, C. Y.; Corley, E. G.; Foster, B. S.; Roberts, F. E.; Yang, C.; Lieberman, D. R.; Reamer, R. A.; Tschaen, D. M.; Verhoeven, T. R.; Reider, P. J.; Lo, Y. S.; Rossano, L. T.; Brooks, A. S.; Meloni, D.; Moore, J. R.; Arnett, J. F.; *J. Org. Chem.* **1994**, *59*, 6391. For a mechanistic study of this process see: (b) Smith, G. B.; Dezeny, G. C.; Hughes, D. L.; King, A. O.; Verhoeven, T. R.; *J. Org. Chem.* **1994**, *59*, 8151.

[100] Muller, D.; Fleury, J. P.; *Tetrahedron Lett.* **1991**, *32*, 2229.

[101] Rocca, P.; Cochennee, C.; Marsais, F.; Thomas-dit-Dumont, L.; Mallett, M.; Godard, A.; Quéguiner, G.; *J. Org. Chem.* **1993**, *58*, 7832.

[102] Kelly, T. R.; Garcia, A.; Langa, F.; Walsh, J. J.; Bhaskar, K. V.; Boyd, M. R.; Götz, R.; Keller, P. A.; Walter, R.; Bringmann, G.; *Tetrahedron Lett.* **1994**, *35*, 7621.

[103] For very thorough review of the subject see: (a) Farina, V.; Krishnamurthy, V.; Scott, W. J.; *Org. React.* **1997**, *50*, 1; (b) Farina, V.; Roth, G. P.; in *Advances in Metal-Organic Chemistry*, Vol. 5, (L. S. Liebeskind, Ed), JAI Press, Greenwich, CT, **1995**, 1. For older reviews see: (c) Mitchell, T. N.; *Synthesis* **1992**, 803; (d) Stille, J. K.; *Angew. Chem. Int. Ed. Engl.* **1986**, *25*, 508.

[104] Curran, D. P.; Hoshino, M.; *J. Org. Chem.* **1996**, *61*, 6480.

[105] Farina, V.; Krishnan, B.; *J. Am. Chem. Soc.* **1991**, *113*, 9585.

[106] Liebeskind, L. S.; Fengl, R. W.; *J. Org. Chem.* **1990**, *55*, 5359.

[107] Farina, V.; Kapadia, S.; Krishnan, B.; Wang, C.; Liebeskind, L. S.; *J. Org. Chem.* **1994**, *59*, 5905.

[108] (a) Andrianome, M.; Delmond, B.; *J. Org. Chem.* **1988**, *53*, 542; (b) Gaare, K.; Repstad, T.; Bennecke, T.; Undheim, K.; *Acta Chem. Scand.* **1993**, *47*, 57.

[109] Labadie, J. W.; Tueting, D.; Stille, J. K.; *J. Org. Chem.* **1983**, *48*, 4634; (b) Labadie, J. W.; Stille, J. K.; *J. Am. Chem. Soc.* **1983**, *105*, 669.

[110] Four, P.; Guibe, F.; *J. Org. Chem.* **1981**, *46*, 4439.

[111] Stille, J. K.; Groh, B. L.; *J. Am. Chem. Soc.* **1987**, *109*, 813.

[112] Stille, J. K.; Simpson, J. H.; *J. Am. Chem. Soc.* **1987**, *109*, 2138.

[113] Kende, A. S.; Liu, K.; Kaldor, I.; Dorey, G.; Koch, K.; *J. Am. Chem. Soc.* **1995**, *117*, 8258.

[114] Smith, A. B.; Condon, S. M.; McCanley, J. A.; Leazer, J. L.; Leaky, J. W.; Maleczka, R. E.; *J. Am. Chem. Soc.* **1995**, *117*, 5402.

[115] Scott, W. J.; Stille, J. K.; *J. Am. Chem. Soc.* **1986**, *108*, 3033.

[116] Scott, W. J.; McMurray, J. E.; *Acct. Chem. Res.* **1988**, *21*, 47.

[117] Farina, V.; Krishnan, B.; Marshall, D. R.; Roth, G. P.; *J. Org. Chem.* **1993**, *59*, 5434.

[118] Paquette, L. A.; Wang, T.-Z.; Swik, M. R.; *J. Am. Chem. Soc.* **1994**, *116*, 11323.

[119] Nicolaou, K. C.; Sato, M.; Miller, N. D.; Gunzer, J. L.; Renaud, J.; Untersteller, E.; *Angew. Chem., Int. Ed. Engl.* **1996**, *39*, 887.

[120] (a) Shair, M. D.; Yoon, T.; Chou, T.-C.; Danishefsky, S. J.; *Angew. Chem., Int. Ed. Engl.* **1994**, *33*, 2477; (b) Shair, M. D.; Yoon, T.; Danishefsky, S. J.; *J. Org. Chem.* **1994**, *59*, 3755.

[121] (a) Kosugi, M.; Shimizu, Y.; Migita, T.; *Chem. Lett.* **1977**, 1423; (b) Kosugi, M.; Shimizu, Y.; Migita, T.; *J. Organomet. Chem.* **1977**, *129*, C23; (c) Kosugi, M.; Sasazawa, K.; Shimizu, Y.; Migita, T.; *Chem. Lett.* **1977**, 301.

[122] Roth, G. P.; Farina, V.; Liebeskind, L. S.; Peña-Cabrera, E.; *Tetrahedron Lett.* **1995**, *36*, 2191.

[123] Laborde, E.; Lesheski, L. E.; Kiely, J. S.; *Tetrahedron Lett.* **1990**, *31*, 1837.

[124] Hanan, G. S.; Lehn, J.-M.; Kyritsakes, N.; Fischer, J.; *J. Chem. Soc., Chem. Commun.* **1995**, 765.

[125] Wu, R.; Schumm, J.-S.; Pearson, D. L.; Tour, J. M.; *J. Org. Chem.* **1996**, *61*, 6906.

[126] van Mullekom, H. A. M.; Vekemens, J. A. J. M.; Meijer, E. W.; *J. Chem. Soc., Chem. Commun.* **1996**, 2163.

[127] Delnoye, D. A. P.; Sijkesina, R. P. S.; Vekemans, J. A. J. M.; Meijer, E. W.; *J. Am. Chem. Soc.* **1996**, *118*, 8717.

[128] Kosugi, M.; Kamezama, M.; Migita, T.; *Chem. Lett.* **1983**, 927.

[129] Guram, A. S.; Buchwald, S. L.; *J. Am. Chem. Soc.* **1994**, *116*, 7901.

[130] Paul, F.; Patt, J.; Hartwig, J. F.; *J. Am. Chem. Soc.* **1994**, *116*, 5969.

[131] (a) Wolfe, J. P.; Wagaw, S.; Buchwald, S. L.; *J. Am. Chem. Soc.* **1996**, *118*, 7215; (b) Wolfe, J. P.; Buchwald, S. L.; *J. Org. Chem.* **1996**, *61*, 1133; (c) For triflates see: Wolfe, J. P.; Buchwald, S. L.; *J. Org. Chem.* **1997**, *62*, 1264.

[132] (a) Driver, M. J.; Hartwig, J. F.; *J. Am. Chem. Soc.* **1996**, *118*, 7217. For triflates see: (b) Louis, J.; Driver, M. S.; Hamann, B. C.; Hartwig, J. F.; *J. Org. Chem.* **1997**, *62*, 1268; (c) Hartwig, J. F.; *Synlett* **1997**, 329.

[133] Zhao, S.-H.; Miller, A. K.; Berger, J.; Flippin, L.; *Tetrahedron Lett.* **1996**, *37*, 4463.

[134] Ward, Y. D.; Farina, V.; *Tetrahedron Lett.* **1996**, *37*, 6993.

[135] Wagaw, S.; Buchwald, S. L.; *J. Org. Chem.* **1996**, *61*, 7240.

[136] (a) Palucki, M.; Wolfe, J. P.; Buchwald, S. L.; *J. Am. Chem. Soc.* **1996**, *118*, 10333; (b) Mann, G.; Hartwig, J. F.; *J. Am. Chem. Soc.* **1996**, *118*, 13109.

[137] For an early, limited example see: Yoshida, J.; Tamao, K.; Takahashi, M.; Kumada, M.; *Tetrahedron Lett.* **1978**, 2161.

[138] (a) Hatanaka, Y.; Hiyama, T.; *J. Org. Chem.* **1988**, *53*, 970; (b) Hatanaka, Y.; Ebina, Y.; Hiyama, T.; *J. Am. Chem. Soc.* **1991**, *113*, 7075. For a review see: (c) Hatanaka, Y.; Hiyama, T.; *Synlett* **1992**, 845.

[139] (a) Gouda, K.-I.; Hagiwara, E.; Hatanaka, Y.; Hiyama, T.; *J. Org. Chem.* **1996**, *61*, 7232; (b) Hagiwara, E.; Gouda, K.-I.; Hatanaka, Y.; Hiyama, T.; *Tetrahedron Lett.* **1997**, *38*, 439.

[140] (a) Cacchi, S.; Ciattini, P. G.; Moreta, E.; Ortar, G.; *Tetrahedron Lett.* **1986**, *27*, 3931; (b) Dolle, R. E.; Schmidt, S. J.; Kruse, L. I.; *J. Chem. Soc., Chem. Commun.* **1987**, 904.

[141] Shishido, K.; Goto, K.; Miyoshi, S.; Takaishi, Y.; Shibuya, M.; *J. Org. Chem.* **1994**, *59*, 406.

[142] Edstrom, E. D.; Wei, Y.; *J. Org. Chem.* **1995**, *60*, 5069.

[143] McGuire, M. A.; Sorenson, E.; Owings, F. W.; Resnick, T. M.; Fox, M.; Baine, N. H.; *J. Org. Chem.* **1994**, *59*, 6683.

[144] (a) Cowell, A.; Stille, J. K.; *J. Am. Chem. Soc.* **1980**, *102*, 4193; (b) Martin, L. D.; Stille, J. K.; *J. Org. Chem.* **1982**, *47*, 3630.

[145] Mori, M.; Chiba, K.; Ban, Y.; *J. Org. Chem.* **1978**, *43*, 1684.

[146] (a) Mori, M.; Chiba, K.; Okita, M.; Ban, Y.; *J. Chem. Soc., Chem. Commun.* **1979**, 698; (b) Mori, M.; Chiba, K.; Okita, M.; Ban, Y.; *Tetrahedron* **1985**, *41*, 387.

[147] Tilley, J. W.; Coffen, D. L.; Shaer, B. H.; Lind, J.; *J. Org. Chem.* **1987**, *52*, 2469.

[148] Shimoyama, I.; Zang, Y.; Wu, G.; Negishi, E.-I.; *Tetrahedron Lett.* **1990**, *31*, 2841.

[149] Baillargeon, V. P.; Stille, J. K.; *J. Am. Chem. Soc.* **1986**, *108*, 452.

[150] Kotsuki, H.; Dalta, P. K.; Suenaga, H.; *Synthesis* **1996**, 470.

[151] Dewey, T. M.; Mundt, A. M.; Crouch, G. J.; Zyzniewski, M. C.; Eaton, B. E.; *J. Am. Chem. Soc.* **1995**, *117*, 8474.

[152] Knight, S. D.; Overman, L. E.; Parrandew, G.; *J. Am. Chem. Soc.* **1993**, *115*, 9293.

[153] Reviews: (a) Heck, R. F.; *Organic React.* **1982**, *27*, 345; (b) Heck, R. F.; *Palladium Reagents in Organic Synthesis*, Academic Press, London, UK, **1985**; (c) Söderberg, B. C.; in *Comprehensive Organometallic Chemistry II*, Vol. 12, (E. W. Abel, F. G. A. Stone, G. Wilkinson, Eds), Pergamon, Oxford, UK, **1995**, 241–297; (d) Overman, L. E.; *Pure Appl. Chem.* **1994**, *66*, 1423; (e) deMeijere, E.; Meyer, F. E.; *Angew. Chem., Int. Ed. Engl.* **1994**, *33*, 2379; (f) Cabri, W.; Candiani, I.; *Acc. Chem. Res.* **1995**, *28*, 2.

[154] (a) Daves, G. D.; Hallberg, A.; *Chem. Rev.* **1989**, *89*, 1433; (b) Daves, G. D.; *Acc. Chem. Res.* **1990**, *23*, 201.

[155] (a) Cabri, W.; Candiani, I.; Bedischi, A.; Penco, S.; Santi, R.; *J. Org. Chem.* **1992**, *57*, 1481; (b) Cabri, W.; Candiani, I.; Bedischi, A.; Santi, R.; *J. Org. Chem.* **1992**, *57*, 3558.

[156] McCrindle, R.; Ferguson, G.; Arsenault, G. J.; McAlees, A. J.; Stephanson, D. K.; *J. Chem. Res. Synop.* **1984**, 360.

[157] Andersson, C. M.; Karabelas, K.; Hallberg, A.; Andersson, C.; *J. Org. Chem.* **1985**, *50*, 3891.

[158] (a) Spenser, A.; *J. Organomet. Chem.* **1983**, *258*, 101; (b) Spenser, A.; **1984**, *265*, 323; (c) Spenser, A.; **1984**, *270*, 115.

[159] (a) Jeffery, T.; *Tetrahedron Lett.* **1985**, *26*, 2667; (b) Jeffery, T.; *J. Chem. Soc. Chem. Commun.* **1984**, 1287; (c) Jeffery, T.; *Synthesis* **1987**, 70.

[160] Larock, R. C.; Baker, B. E.; *Tetrahedron Lett.* **1988**, *29*, 905.

[161] Amatore, C.; Azzabi, M.; Jutland, A.; *J. Am. Chem. Soc.* **1991**, *113*, 8375.

[162] Jeffery, T.; *Tetrahedron* **1996**, *52*, 10113.

[163] (a) Casalnuovo, W. L.; Calabrese, J. C.; *J. Am. Chem. Soc.* **1990**, *112*, 4324; (b) Genet, J. P.; Blast, E.; Savignac, M.; *Synlett* **1992**, 1715; (c) Jeffery, T.; *Tetrahedron Lett.* **1994**, *35*, 3501; (d) Dibowski, H.; Schmidchen, F. P.; *Tetrahedron* **1995**, *51*, 2325.

[164] Kawataka, F.; Shimizu, J.; Yamamoto, A.; *Bull Chem. Soc. Jpn.* **1995**, *68*, 654.

[165] Andersson, C. M.; Karabelas, K.; Hallberg, A.; Andersson, C.; *J. Org. Chem.* **1985**, *50*, 3891.

[166] (a) Terasawa, M.; Kaneda, K.; Imanaka, T.; Teranishi, S.; *J. Organomet. Chem.* **1978**, *162*, 403; (b) Choudary, B. M.; Sarma, R. M.; Rao, K. K.; *Tetrahedron* **1992**, *48*, 719; (c) Zhwangyu, Z.; Yi, P.; Honwen, H.; Tsi-yu, K.; *Synthesis* **1991**, 539.

[167] Farr, R. N.; Outten, R. A.; Chen, J. C.-Y.; Daves, G. D.; *Organometallics* **1990**, *9*, 3151.

[168] Nishi, K.; Narukawa, Y.; Onoue, H.; *Tetrahedron Lett.* **1996**, *37*, 2987.

[169] Tietze, L. F.; T. Nöbel, Speacha, M.; *Angew. Chem., Int. Ed. Engl.* **1996**, *35*, 2259.

[170] Hiroshige, M.; Hauske, J. R.; Zho, P.; *Tetrahedron Lett.* **1995**, *36*, 4567.

[171] Reviews: (a) Overman, L. E.; *Pure Appl. Chem.* **1994**, *66*, 1423; (b) Grigg, R.; Sridharan, V.; Santhakumar, V.; Thornton-Pett, M.; Bridge, A. W.; *Tetrahedron* **1993**, *49*, 5177.

[172] (a) Sundberg, R. J.; Cherney, R. J.; *J. Org. Chem.* **1990**, *55*, 6028; (b) Okita, T.; Isobe, M.; *Tetrahedron* **1994**, *50*, 11143.

[173] Masters, J. J.; Link, J. T.; Snyder, L. B.; Young, W. B.; Danishefsky, S. J.; *Angew. Chem., Int. Ed. Engl.* **1995**, *34*, 1723; (b) Young, W. B.; Masters, J. J.; Danishefsky, S.; *J. Am. Chem. Soc.* **1995**, *117*, 5228.

[174] (a) Rigby, J. H.; Hughes, R. C.; Heeg, M. J.; *J. Am. Chem. Soc.* **1995**, *117*, 7834; (b) Bombrun, A.; Sageot, O.; *Tetrahedron Lett.* **1997**, *38*, 1057.

[175] Overman, L. E.; Abelman, M. M.; Kucera, D. J.; Tran, V. D.; Ricca, D. J.; *Pure Appl. Chem.* **1992**, *64*, 1813.

[176] Stocks, M. J.; Harrison, R. R.; Teauge, S. J.; *Tetrahedron Lett.* **1995**, *36*, 6553.

[177] Hiroshige, M.; Hauske, J. R.; Zhou, R.; *J. Am. Chem. Soc.* **1995**, *117*, 11590.

[178] Review: Shibasaki, M.; Sodeoka, M.; *J. Synth. Org. Chem. Jpn.* **1994**, *52*, 956.

[179] Tietze, L. F.; Raschile, T.; *Synlett* **1995**, 597.

[180] Ohrai, K.; Kondo, K.; Sodeoka, M.; Shibasaki, M.; *J. Am. Chem. Soc.* **1994**, *116*, 11737.

[181] (a) Loiseleur, O.; Meier, P.; Pfaltz, A.; *Angew. Chem., Int. Ed. Engl.* **1996**, *35*, 200; (b) Ozawa, F.; Kobatabe, Y.; Hayashi, T.; *Tetrahedron Lett.* **1993**, *34*, 2505 and references therein.

[182] Tang, X.-Q.; Harvey, R. G.; *Tetrahedron Lett.* **1995**, *36*, 6037.

[183] Burwood, M.; Davies, B.; Diaz, I.; Grigg, R.; Molina, P.; Sridharan, V.; Hughes, M.; *Tetrahedron Lett.* **1995**, *36*, 9053.

[184] Reviews: (a) Heumann, A.; Reglier, M.; *Tetrahedron* **1996**, *52*, 9289; (b) Grigg, R.; Sridharan, V.; in *Comprehensive Organometallic Chemistry II*, Vol. 12, (E. W. Abel, F. G. A. Stone, G. Wilkinson, Eds), Pergamon, Oxford, UK, **1995**, 299–321.

[185] Negishi, E.-I.; Sawada, H.; Tour, J. M.; Wei, Y.; *J. Org. Chem.* **1988**, *53*, 915.

[186] (a) Coperet, C.; Ma, S.; Negishi, E.-I.; *Angew. Chem. Int. Ed. Engl.* **1996**, *35*, 2125. For full papers on the subject see: (b) Coperet, C.; Ma, S.; Sugihara, T.; Negishi, E.-I.; *Tetrahedron* **1996**, *52*, 11529;

(c) Negishi, E.-I.; Coperet, C.; Ma, S.; Mita, T.; Sugihara, T.; Tour, J. M.; *J. Am. Chem. Soc.* **1996**, *118*, 5904; (d) Negishi, E.-I.; Ma, S.; Amanfu, J.; Coperet, C.; Miller, J. A.; Tour, J. M.; *J. Am. Chem. Soc.* **1996**, *118*, 5919.

[187] Maddaford, S. P.; Andersen, N. G.; Cristofoli, W. A.; Keays, B. A.; *J. Am. Chem. Soc.* **1996**, *118*, 10766.

[188] Reviews: (a) Malacria, M.; *Chem. Rev.* **1996**, *96*, 289; (b) Ojima, I.; Tzamarioudalsi, M.; Li, Z.; Donovan, R. J.; *Chem. Rev.* **1996**, *96*, 635; (c) Negishi, E.-I.; Copret, C.; Ma, S.; Liou, S.-Y.; Liu, F.; *Chem. Rev.* **1996**, *96*, 365.

[189] Zhang, Y.; Wu, G.-Z.; Agnel, G.; Negishi, E.-I.; *J. Am. Chem. Soc.* **1990**, *112*, 8590.

[190] Meyer, F.; Parsons, P. J.; deMeijere, A.; *J. Org. Chem.* **1991**, *56*, 6487.

[191] Grigg, R.; Loganathan, V.; Sridharan, V.; *Tetrahedron Lett.* **1996**, *37*, 3399.

[192] Oda, H.; Kobayashi, T.; Kosugi, M.; Migita, T.; *Tetrahedron* **1995**, *51*, 695.

[193] Kojima, A.; Takemoto, T.; Sodeoka, M.; Shibasaki, M.; *J. Org. Chem.* **1996**, *61*, 4876.

[194] Reviews: (a) Harrington, P. M.; in *Comprehensive Organometallic Chemistry II*, Vol. *12*, (E. W. Abel, F. G. A. Stone, G. Wilkinson, Eds), Pergamon, Oxford, UK, **1995**, 797–904; (b) Frost, C. G.; Howarth, J.; Williams, J. M. J.; *Tetrahedron Asymmetry* **1992**, *3*, 1089.

[195] Trost, B. M.; Klun, T. P.; *J. Am. Chem. Soc.* **1981**, *103*, 1864.

[196] Trost, B. M.; Kuo, G.-H.; Bennecki, T.; *J. Am. Chem. Soc.* **1988**, *110*, 621.

[197] (a) Boeckman, R. K.; Shair, M. D.; Vargas, J. R.; Stoltz, L. A.; *J. Org. Chem.* **1993**, *58*, 1295; (b) Michelet, V.; Besner, I.; Genet, J. P.; *Synlett* **1996**, 215.

[198] Naz, N.; Al-Tel, T. H.; Al-Abed, Y.; Voelter, W.; Fikes, R.; Hiller, W.; *J. Org. Chem.* **1996**, *61*, 3230.

[199] Review: Heumann, A.; Regher, M.; *Tetrahedron* **1995**, *51*, 975.

[200] Trost, B. M.; Ohmori, M.; Boyd, S. A.; Okawara, H.; Brickner, S. J.; *J. Am. Chem. Soc.* **1989**, *111*, 8281.

[201] Trost, B. M.; Hane, J. T.; Metz, P.; *Tetrahedron Lett.* **1986**, *27*, 5695.

[202] Roland, S.; Durand, J. O.; Savignac, M.; Genet, J. P.; *Tetrahedron Lett.* **1995**, *36*, 3007.

[203] (a) Schink, H. E.; Bäckvall, J.-E.; *J. Org. Chem.* **1992**, *57*, 1588; (b) Bäckvall, J.-E.; Gatti, R.; Schink, H. E.; *Synthesis* **1993**, 343.

[204] Reviews: (a) Trost, B. M.; *Acct. Chem. Res.* **1996**, *29*, 355; (b) Trost, B. M.; Van Vranken, D. L.; *Chem. Rev.* **1996**, *96*, 395.

[205] Kardos, N.; Genet, J.-P.; *Tetrahedron Asymmetry* **1994**, *5*, 1525.

[206] Short review: (a) Reiser, O.; *Angew. Chem., Int. Ed. Engl.* **1993**, *32*, 547. For mechanistic studies see: (b) Seebach, D.; Devaquet, E.; Ernst, A.; Hayakawa, M.; Kühnle, F. N. M.; Schweizer, W. B.; Weber, B.; *Helv. Chim. Acta* **1995**, *78*, 1636. For X-ray studies of the intermediate see: (c) von Matt, P.; Lloyd-Jones, G. C.; Minidis, A. B. E.; Pfaltz, A.; Macko, L.; Neubarger, M.; Zehinder, M.; Rüegger, H.; Pregosin, P.; *Helv. Chim. Acta* **1995**, *78*, 265. For effects of counterions see: (d) Burkhardt, U.; Baumann, M.; Togni, A.; *Tetrahedron Asymmetry* **1997**, *8*, 155. For studies dealing with Pd(0)-catalyzed racemization of allylmalonates see: (e) Bricot, H.; Carpenter, J.-F.; Mortreux, A.; *Tetrahedron Lett.* **1997**, *38*, 1053.

[207] Baldwin, I. C.; Williams, J. M. J.; Beckett, R. P.; *Tetrahedron Asymmetry* **1995**, *6*, 1515.

[208] Del Valle, L.; Stille, J. K.; Hegedus, L. S.; *J. Org. Chem.* **1990**, *55*, 3019.

[209] Tueting, D. R.; Echavarren, A. M.; Stille, J. K.; *Tetrahedron* **1989**, *45*, 979; Echavarren, A. M.; Tueting, D. R.; Stille, J. K.; *J. Am. Chem. Soc.* **1988**, *110*, 4039.

[210] Urabe, H.; Inami, H.; Sato, F.; *J. Chem. Soc., Chem. Commun.* **1993**, 1595.

[211] Moineau, C.; Bolitt, V.; Sinou, D.; *J. Chem. Soc., Chem. Commun.* **1995**, 1103.

[212] van der Louw, J.; Van der Baan, J. L.; de Kanter, F. J. J.; Bickelhaupt, F.; Klumpp, G. W.; *Tetrahedron* **1992**, *48*, 6087.

[213] Farina, V.; Baker, S. R.; Benigni, D. A.; Sapino, C.; *Tetrahedron Lett.* **1988**, *29*, 5739.

[214] Sheffy, F. K.; Godschalx, J. P.; Stille, J. K.; *J. Am. Chem. Soc.* **1984**, *106*, 4833.

[215] Merrifield, J. H.; Godschalx, J. P.; Stille, J. K.; *Organometallics* **1984**, *3*, 1108.

[216] Flegelova, Z.; Pátek, M.; *J. Org. Chem.* **1996**, *61*, 6735.

[217] Trost, B. M.; Casey, J.; *J. Am. Chem. Soc.* **1982**, *104*, 6881.

[218] Yoshigaki, H.; Satoh, H.; Sato, Y.; Nakai, S.; Shibasaki, M.; Mori, M.; *J. Org. Chem.* **1995**, *60*, 2016.

[219] Limited review: (a) Williams, J. M. J.; *Synlett* **1996**, 705. For a theoretical treatment see: (b) Blöchl, P.; Togni, A.; *Organometallics* **1996**, *15*, 4125.

[220] (a) Togni, A.; Burkhardt, U.; Gramlich, V.; Pregosin, P. S.; Salzmann, R.; *J. Am. Chem. Soc.* **1996**, *118*, 1031; (b) Jumnak, R.; Williams, A. C.; Williams, J. M. J.; *Synlett* **1995**, 821.

[221] Brief review: Trost, B. M.; *Pure Appl. Chem.* **1996**, *68*, 779.

[222] (a) Trost, B. M.; Pulley, S. R.; *J. Am. Chem. Soc.* **1995**, *117*, 10143; (b) Trost, B. M.; Pulley, S. R.; *Tetrahedron Lett.* **1995**, *36*, 8737.

[223] (a) Trost, B. M.; Madsen, R.; Guile, S. G.; Elia, A. E. H.; *Angew. Chem., Int. Ed. Engl.* **1996**, *35*, 1569; (b) Trost, B. M.; Shi, Z.; *J. Am. Chem. Soc.* **1996**, *118*, 3037.

[224] Trost, B. M.; Cook, G. R.; *Tetrahedron Lett.* **1996**, *37*, 7485.

[225] Trost, B. M.; Tenaglia, A.; *Tetrahedron Lett.* **1988**, *29*, 2927.

[226] Goux, C.; Massacret, M.; Lhosti, P.; Sinou, D.; *Organometallics* **1995**, *14*, 4585.

[227] Bradette, T.; Esher, J. L.; Johnson, C. R.; *Tetrahedron Asymmetry* **1996**, *7*, 2313.

[228] Sinou, D.; Frappa, I.; Lhoste, P.; Porwanski, S.; Krycza, B.; *Tetrahedron Lett.* **1995**, *36*, 6251.

[229] Shimizu, I.; Omura, T.; *Chem. Lett.* **1993**, 1759.

[230] (a) Trost, B. M.; Ito, N.; Greenspan, P. P.; *Tetrahedron Lett.* **1993**, *34*, 1421. (b) Trost, B. M.; Greenspan, P. D.; Geissler, H.; Kim, J. H.; Greeves, N.; *Angew. Chem., Int. Ed. Engl.* **1994**, *33*, 2182.

[231] (a) Bäckvall, J.-E.; Byström, S. E.; Nordberg, R. E.; *J. Org. Chem.* **1984**, *49*, 4619; (b) Nystrom, J.-E.; Rein, T.; Bäckvall, J.-E.; *Org. Synth.* **1989**, *67*, 105.

[232] (a) Bäckvall, J.-E.; Hopkins, R. B.; Grennberg, H.; Madder, M. M.; Awastä, A. K.; *J. Am. Chem. Soc.* **1990**, *112*, 5160; (b) Grennberg, H.; Gogoll, A.; Backväll, J.-E.; *J. Org. Chem.* **1991**, *56*, 5808.

[233] Bäckvall, J.-E.; *Advances in Metal-Organic Chemistry*, Vol. *1*, (L. S. Liebeskind, Ed.), JAI Press, **1989**, 135.

[234] Bäckvall, J.-E.; Andersson, P. G.; *J. Am. Chem. Soc.* **1992**, *114*, 6374.

[235] Bäckvall, J.-E.; Granberg, K. L.; Andersson, P. G.; Gatte, R.; Gogoll, A.; *J. Org. Chem.* **1993**, *58*, 5445.

[236] Trost, B. M.; Schmuff, N. R.; *J. Am. Chem. Soc.* **1985**, *107*, **396**, and references therein.

[237] (a) Trost, B. M.; Organ, M. G.; O'Doherty, G. A.; *J. Am. Chem. Soc.* **1995**, *117*, 9662. (b) Trost, B. M.; Kriscke, M. J.; Radinov, R.; Zanoni, G.; *J. Am. Chem. Soc.* **1996**, *118*, 6297.

[238] Review: Tsuji, J.; Mandi, T.; *Synthesis* **1996**, 1.

[239] (a) Mandai, T.; Kaihara, Y.; Tsuji, J.; *J. Org. Chem.* **1994**, *59*, 5847; (b) Lautens, M.; Delanghi, P. H. M.; *Angew. Chem., Int. Ed. Engl.* **1994**, *33*, 2448.

[240] Nagasawa, K.; Shimizu, I.; Nakata, T.; *Tetrahedron Lett.* **1996**, *37*, 6881.

[241] (a) Mandai, T.; Matsumoto, T.; Kawada, M.; Tsuji, J.; *Tetrahedron* **1993**, *49*, 5483; (b) Mandai, T.; Matsumoto, T.; Kawada, M.; Tsuji, J.; *J. Org. Chem.* **1992**, *57*, 1326 and 6090.

[242] Hayashi, T.; Iwaramura, I.; Naito, M.; Matsumoto, Y.; Nozumi, Y.; Miki, M.; Yanagai, K.; *J. Am. Chem. Soc.* **1994**, *116*, 775.

[243] Mandai, T.; Matsumoto, T.; Tsuji, J.; *Tetrahedron Lett.* **1993**, *34*, 2513.

[244] Mandai, T.; Matsumoto, M.; Nakao, Y.; Teramoto, A.; Kawada, M.; Tsuji, J.; *Tetrahedron Lett.* **1992**, *33*, 2549.

[245] Shimizu, I.; Matsumoto, Y.; Ono, T.; Satake, A.; Yamamoto, A.; *Tetrahedron Lett.* **1996**, *37*, 7115.

[246] Andersson, P. G.; Schab, S.; *Organometallics* **1995**, *14*, 1.

[247] Deziel, R.; *Tetrahedron Lett.* **1987**, *28*, 4371.

[248] Hang, D. T.; Nerenberg, J. B.; Schreiber, S. L.; *J. Am. Chem. Soc.* **1996**, *118*, 11054.

[249] Dessolin, M.; Guillerez, M.-G.; Thieret, N.; Guibe, F.; Loffet, A.; *Tetrahedron Lett.* **1995**, *36*, 5741.

[250] Hayakawa, Y.; Wakabayashi, S.; Kato, H.; Noyori, R.; *J. Am. Chem. Soc.* **1990**, *112*, 1691.

[251] Reviews: (a) Oppolzer, W.; in *Comprehensive Organometallic Chemistry II*, Vol. *12*, (E. W. Abel, F. G. A. Stone, G. Wilkinson, Eds), Pergamon, Oxford, UK, **1995**, 905–921; (b) Oppolzer, W.; *Angew. Chem., Int. Ed. Engl.* **1989**, *28*, 38; (c) Oppolzer, W.; *Pure Appl. Chem.* **1990**, *62*, 1941.

[252] Oppolzer, W.; Xu, J.-Z.; Stone, C.; *Helv. Chim. Acta* **1991**, *74*, 465.

[253] Oppolzer, W.; Ruiz-Montes, J.; *Helv. Chim. Acta* **1993**, *76*, 1266.

[254] Grigg, R.; Sukirthalingam, S.; Sridharan, V.; *Tetrahedron Lett.* **1991**, *32*, 2545.

[255] Oppolzer, W.; DeVita, R. J.; *J. Org. Chem.* **1991**, *56*, 6256.

[256] Grigg, R.; Sridharan, V.; Sukirthalingam, S.; *Tetrahedron Lett.* **1991**, *32*, 3855.

[257] (a) Oppolzer, W.; Gaudin, J. M.; Birkinshaw, T. N.; *Tetrahedron Lett.* **1988**, *29*, 4705; (b) Oppolzer, W.; Gaudin, J. M.; *Helv. Chim. Acta* **1987**, *70*, 1478.

[258] Keese, R.; Guidette-Grept, R.; Herzog, B.; *Tetrahedron Lett.* **1992**, *33*, 1207. For an earlier report see: (b) Oppolzer, W.; Bienayame, H.; Genevas-Borella, A.; *J. Am. Chem. Soc.* **1991**, *113*, 9660.

[259] Trost, B. M.; Luengo, J. I.; *J. Am. Chem. Soc.* **1988**, *110*, 8239.

[260] Ohshima, T.; Kagechika, K.; Adachi, M.; Sodeoka, M.; Shibasaki, M.; *J. Am. Chem. Soc.* **1996**, *118*, 7108.

[261] Reviews: (a) Trost, B. M.; *Angew. Chem., Int. Ed. Engl.* **1986**, *25*, 1; (b) Trost, B. M.; *Pure Appl. Chem.* **1988**, *60*, 1615; (c) Harrington, P. J.; in *Comprehensive Organometallic Chemistry II*, Vol. *12*, (E. W. Abel, F. G. A. Stone, G. Wilkinson, Eds), Pergamon, Oxford, UK, **1995**, 923–945; (d) Lautens, M.; Klute, W.; Tam, W.; *Chem. Rev.* **1996**, *96*, 49.

[262] Trost, B. M.; Chan, D. M. T.; *J. Am. Chem. Soc.* **1983**, *105*, 2315.

[263] (a) Trost, B. M.; Mignani, S. M.; *Tetrahedron Lett.* **1986**, *27*, 4137; (b) Trost, B. M.; Lynch, J.; Renant, P.; Steinman, D. H.; *J. Am. Chem. Soc.* **1986**, *108*, 284.

[264] Holzapfel, C. W.; van der Morwe, T.L.; *Tetrahedron Lett.* **1996**, *37*, 2303.

[265] Trost, B. M.; Nanninga, T. N.; Satoh, T.; *J. Am. Chem. Soc.* **1985**, *107*, 721.

[266] Trost, B. M.; Higuchi, R. L.; *J. Am. Chem. Soc.* **1996**, *118*, 10094 and references therein.

[267] (a) Trost, B. M.; Seoane, P. R.; *J. Am. Chem. Soc.* **1987**, *109*, 615; (b) Trost, B. M.; Matelich, M. C.; *J. Am. Chem. Soc.* **1991**, *113*, 9007.

[268] Trost, B. M.; Marrs, C. M.; *J. Am. Chem. Soc.* **1993**, *115*, 6636.

[269] Trost, B. M.; King, S. A.; *J. Am. Chem. Soc.* **1990**, *112*, 408.

[270] Trost, B. M.; Sharma, S.; Schmidt, T.; *Tetrahedron Lett.* **1993**, *34*, 7183.

[271] (a) Binger, P.; Büch, H. M.; *Top. Curr. Chem.* **1987**, *135*, 77. (b) For a stereochemical and mechanistic study see: Corley, H.; Motherwell, W. B.; Pennell, A. M. K.; Shipman, M.; Slawin, A. M. Z.; Williams, P. J.; Binger, P.; Stepp, M.; *Tetrahedron* **1996**, *52*, 4883.

[272] For use in the synthesis of [3.3.0]bicyclooctane and [4.3.0]bicyclononanes see: Corley, H.; Lewis, R. T.; Motherwell, W. B.; Shipman, M.; *Tetrahedron* **1995**, *51*, 3303.

[273] Lewis, R. T.; Motherwell, W. B.; Shipman, M.; Slawin, A. M. Z.; Williams, D. J.; *Tetrahedron* **1995**, *51*, 3285.

[274] Lautens, M.; Ren, Y.; *J. Am. Chem. Soc.* **1996**, *118*, 9597.

[275] Lautens, M.; Ren, Y.; *J. Am. Chem. Soc.* **1996**, *118*, 10668.

[276] Reviews: (a) Tsuji, J.; *Organic Synthesis with Palladium Complexes*; Springer-Verlag: Berlin, **1980**; (b) Tsuji, J.; *Top. Curr. Chem.* **1980**, *91*, 30; (c) Tsuji, J.; *Pure Appl. Chem.* **1981**, *53*, 2371; (d) Tsuji J.; *Pure Appl. Chem.* **1982**, *54*, 197; (e) Tsuji, J.; *Acc. Chem. Res.* **1973**, *6*, 8; (f) Tsuji, J.; *Adv. Organomet. Chem.* **1979**, *17*, 141; (g) Tsuji, J.; *Ann. N.Y. Acad. Sci.* **1980**, *333*, 250.

[277] Review: Takacs, J.; in *Comprehensive Organometallic Chemistry II*, Vol. *12*, (E. W. Abel, F. G. A. Stone, G. Wilkinson, Eds), Pergamon, Oxford, UK, **1995**, 785–796.

[278] Takacs, J. M.; Zu, J.; *J. Org. Chem.* **1989**, *54*, 5193.

[279] Takacs, J. M.; Chandramouli, S.; *Organometallics* **1990**, *9*, 2877.

[280] Takacs, J. M.; Zu, J.; *Tetrahedron Lett.* **1990**, *31*, 1117.

[281] (a) Takacs, J. M.; Chandramouli, S. V.; *Tetrahedron Lett.* **1994**, *35*, 9165; (b) Takacs, J. M.; Chandramouli, S. V.; *J. Org. Chem.* **1993**, *58*, 7315.

[282] Takacs, J. M.; Zhu, J.; Chandramouli, S.; *J. Am. Chem. Soc.* **1992**, *114*, 773.

[283] Review: Farina, V.; in *Comprehensive Organometallic Chemistry II*, Vol. *12*, (E. W. Abel, F. G. A. Stone, G. Wilkinson, Eds), Pergamon, Oxford, UK, **1995**, 222–225.

[284] For a review of enediyne procedures see: Maier, M. E.; *Synlett* **1995**, 13.

[285] (a) De la Rosa, M. A.; Velardi, E.; Grizman, A.; *Synth. Commun.* **1990**, *20*, 2059; (b) Bleicher, L.; Cosford, N. P. P.; *Synlett* **1995**, 1115.

[286] Bumagin, N.; Sukholminova, L. I.; Luzckova, E. V.; Tolstaya, T. P.; Beletskaya, I. P.; *Tetrahedron Lett.* **1996**, *37*, 897.

[287] Nguefack, J. F.; Bolitt, V.; Sinow, D.; *Tetrahedron Lett.* **1996**, *37*, 5527.

[288] Powell, N. A.; Rychnovsky, S. D.; *Tetrahedron Lett.* **1996**, *37*, 7901.

[289] (a) Wender, P. A.; Beckham, S.; O'Leary, J.-G.; *Synthesis* **1995**, 1279; (b) Nshikawa, T.; Yoshikai, M.; Kawai, T.; Unno, R.; Jomori, T.; Isobe, M.; *Tetrahedron* **1995**, *51*, 9339.

[290] Hoye, T. R.; Tan, L.; *Tetrahedron Lett.* **1995**, *36*, 1981.

[291] (a) Jacobi, P. A.; Guo, J.; Zhang, W.; *Tetrahedron Lett.* **1995**, *36*, 1197; (b) Jacobi, P. A.; Guo, J.; *Tetrahedron Lett.* **1995**, *36*, 2717.

[292] Heathcock, C. H.; Clasby, M.; Griffith, D. A.; Henke, B. R.; Sharp, M.; *Synlett* **1995**, 467.

[293] Bedard, T. C.; Moore, J. S.; *J. Am. Chem. Soc.* **1995**, *117*, 10662.

[294] (a) Moore, J.; *Angew. Chem., Int. Ed. Engl.* **1993**, *32*, 246; (b) Zeng, F.; Zimmerman, S. C.; *J. Am. Chem. Soc.* **1996**, *118*, 5326.

INDEX

Index compiled by Geoffrey Jones